TRANSACTIONS
OF THE
INTERNATIONAL
ASTRONOMICAL UNION

VOLUME XIIIB – PROCEEDINGS

INTERNATIONAL COUNCIL OF SCIENTIFIC UNION
INTERNATIONAL ASTRONOMICAL UNION
UNION ASTRONOMIQUE INTERNATIONALE

TRANSACTIONS
OF THE
INTERNATIONAL ASTRONOMICAL UNION
VOLUME XIIIB

PROCEEDINGS
OF THE THIRTEENTH
GENERAL ASSEMBLY
PRAGUE 1967

Edited by

L. PEREK

General Secretary of the Union

PUBLISHED FOR THE INTERNATIONAL ASTRONOMICAL UNION
WITH THE FINANCIAL ASSISTANCE OF UNESCO

SPRINGER-SCIENCE+BUSINESS MEDIA, B.V.
1968

Originally published by D. Reidel Publishing Company in 1967
Softcover reprint of the hardcover 1st edition 1967

ISBN 978-94-010-3478-4 ISBN 978-94-010-3476-0 (eBook)
DOI 10.1007/978-94-010-3476-0

PREFACE

The present part, Volume XIIIB, has the title Proceedings of the General Assembly. It contains the complete record of the Inaugural Ceremony; of the two sessions of the General Assembly; resolutions adopted by the General Assembly; Reports of 140 Meetings of Commissions; and a supplement to Astronomer's Handbook listing in principle only those parts which changed since the edition of IAU Transactions Vol. XIIC, (in particular the Membership of Commissions and Alphabetical List of Members of the Union).

Volume XIIIA, with the title Reports on Astronomy, has been published before the present volume. It contains the Report of the Executive Committee and the Draft Reports of Commissions as presented to the General Assembly and as subsequently corrected. Additions to the Draft Reports which would have required the repagination of the Draft Reports, appear in the present volume in the reports of the Commissions.

The Invited Discourses, the Joint Discussions, the proceedings of meetings of Commission 17 on the Moon Probes, and a report on the joint meeting of Commission 10 and 44 on 'Coordination of Solar Observations Made at Ground-based Observatories and with Space Vehicles' will appear in a separate volume entitled 'Highlights of Astronomy as presented at the XIIIth General Assembly of the IAU, Praha 1967'.

Other scientific results, in particular those presented at the meetings of Commissions, are reported in the present volume in a very condensed form, as it is assumed that the papers will be published in full in the current astronomical periodicals and publications. This step was considered essential by the Executive Committee in order to avoid duplication in publishing scientific results.

Three Symposia were held immediately after the General Assembly:

No. 33: Physics and Dynamics of Meteors

No. 34: Planetary Nebulae

No. 35: The Structure and Development of Solar Active Regions. The proceedings of these Symposia will appear as separate volumes in the series of IAU Symposia, published by the D. Reidel Publishing Company, Dordrecht, The Netherlands.

I wish to express my thanks to all those who contributed to the present volume: to the Presidents and Secretaries of Commissions who prepared the reports of Commissions; my special thanks are due to those who submitted their reports with the proverb "bis dat qui cito dat" in mind. The preparation of Parts 1 and 4 would have been impossible without the competency and skill of Miss G. Drouin and Mrs J. Daňková.

Our new printer and publishing house, D. Reidel Publishing Company, deserves the thanks of the Union for the high quality of presentation and for keeping within the short printing delays promised.

L. Perek
General Secretary

Astronomical Institute, Czechoslovak Academy of Sciences
Budečská 6
Praha 2, Czechoslovakia

21 November 1967

CONTENTS

Part 1

Part 2

Résolutions adoptées par l'Assemblée Générale
Resolutions adopted by the General Assembly

Part 3

Comptes rendus des séances des Commissions
Reports of Meetings of Commissions

Part 4

Astronomer's Handbook

POL SWINGS

Membre de l'Académie Royale de Belgique

Directeur de l'Institut d'Astrophysique, Sclessin

PRÉSIDENT DE L'UNION ASTRONOMIQUE INTERNATIONALE

1964–1967

PART 1

INAUGURAL CEREMONY

REPORT OF THE GENERAL ASSEMBLY

INAUGURAL CEREMONY

Tuesday, 22 August 1967 at $10^h 15^m$

The Inaugural Ceremony took place in the Congress Hall of Julius Fučík Park of Culture in Prague, in the presence of a distinguished gathering of representatives of the Government of Czechoslovakia, the Czechoslovak Academy of Sciences, the Czechoslovak National Astronomical Committee, and many local organizations.

The Union was honoured by the presence of His Excellency the Acting Prime Minister of the Czechoslovak Socialist Republic, F. Krajčír, and the President of the Czechoslovak Academy of Sciences, Academician F. Šorm.

Professor Pol Swings, President of the International Astronomical Union, acted as chairman for the ceremony.

After a few words of welcome to Members and guests by Dr B. Šternberk, President of the Czechoslovak National Astronomical Committee and President of the Local Organizing Committee, and the execution, by the Prague Chamber Orchestra, of the Bohemian Suite in D major by A. Dvořák, the President called upon Mr F. Krajčír, Acting Prime Minister of the Czechoslovak Socialist Republic, to address the Assembly.

ADDRESS BY THE ACTING PRIME MINISTER OF THE CZECHOSLOVAK SOCIALIST REPUBLIC MR F. KRAJČÍR

(English translation from Czech)

'Mr President, Ladies and Gentlemen, Dear Guests,

'As spokesman of the Government of the Czechoslovak Socialist Republic I warmly welcome your General Assembly to Prague. It is a great pleasure for me to be able to tell you how glad we are that this important scientific meeting is being held in the metropolis of our country.

'We recognize in you the most outstanding representatives of one of the oldest branches of science. Astronomy has helped mankind to unveil the secrets of the Universe and of natural laws, thus preparing the road for scientific and social progress. But truth and progress, as we know, often demanded heavy sacrifices also from astronomers, as they fought out-dated concepts and reactionary forces. Astronomy has always had a greatly stimulating effect upon the development of other scientific branches. Hence it is only natural that—especially with regard to the scientific revolution of our time—the best in science and technology serves the understanding of the world.

'The present-day achievements in space exploration have greatly stimulated the interest of the broad population in astronomy, too. This is borne out by numerous astronomical clubs at factories and schools as well as at public observatories—besides official astronomical institutions. The interest of Czechs and Slovaks in astronomy is traditional, and this tradition goes back to the oldest Central-European University in Prague with its astronomers of world-wide renown.

'During your stay in Czechoslovakia you will be certain to notice the large number of popular and scientific books, articles in newspapers and magazines, and radio and television programmes on astronomy. And now, the general public eagerly read and listen to the reports of your meetings. But you will also see, how much care and attention our Government devotes to the development of astronomy.

'The present rapid progress in science and engineering calls for close international co-operation. It is vitally important that the achievements of science serve progress and happiness, a better mutual understanding between nations, and that they are not abused for suppressing one nation by another or even threatening the very existence of mankind. Such international co-operation, however, is only possible in conditions of peace, goodwill and mutual confidence between nations. And here the International Astronomical Union plays a major role.

'Czechoslovak astronomers have always been doing their share in furthering international co-operation, and we feel deeply honoured that astronomers from all parts of the world have chosen Prague for their General Assembly.

'On behalf of the Government, permit me to express our firm belief that your deliberations here in Prague will contribute not only to progress in astronomy, but also towards a better understanding between nations and to their peaceful coexistence.

'The Government of the Czechoslovak Socialist Republic wishes your General Assembly lots of success and hopes that you will have a pleasant stay in our country.'

Professor Swings expressed his thanks to Mr Krajčír for his address, and invited Academician F. Šorm, President of the Czechoslovak Academy of Sciences, to address the gathering.

ADDRESS BY THE PRESIDENT OF THE CZECHOSLOVAK ACADEMY OF SCIENCES ACADEMICIAN F. ŠORM

'Mr President, Ladies and Gentlemen, distinguished Guests,

'It is a great honour and a pleasant task, indeed, to welcome all of you most cordially—Members, Invited Participants and Guests of the thirteenth General Assembly of the International Astronomical Union—the biggest international meeting in the almost fifty year existence of your Union and in all history of the world astronomy so far—on behalf of the Czechoslovak Academy of Sciences, of our whole scientific community and on my own behalf, in the Czechoslovak Socialist Republic and her historical capital.

'We consider your presence in this country as a high tribute paid to all our science and as a culmination of more than 600 years of development of astronomy in the territory of the present Czechoslovak Socialist Republic. Allow me to mention in this connection that from the very beginning of existence of Prague Charles University—the oldest university in Central and Eastern Europe—astronomy had been taught here and valuable original astronomical observations had been practised.

'At the outset of the 17th century Prague actually became one of the cradles of modern astronomy. Tycho Brahe lived here only for a short time at the end of his days, unfortunately, but his two years of activity in Prague were in a way a completion of his life-long labours. He invited Johann Kepler here and, perhaps for the first time in history, co-operation between an astronomer-observer and a theoretician took place. It is only thanks to the linking of Tycho's accurate observations with Kepler's tireless search for regularities that Kepler was able to discover the laws of planetary movements which finally confirmed Copernicus's heliocentric system and opened the way for Newton's law of gravitation. This linking of observation and theory is symbolic also for the present astronomers' meeting, and continues to be a live heritage of both great astronomers whose urgency is apparent particularly at a time when rapid development of science and technology compels all astronomers to specialize in an ever greater degree.

'It is a great pity that this important astronomical center in Prague ceased to exist after the Thirty Years' War. The world's harmony which Johann Kepler was seeking in Universe disappeared from his life and from all Central Europe. The Thirty Years' War brought to the Czech nation economic and political catastrophe and loss of independence. The consequences were so severe that even in the last century much effort of the Czech and Slovak nations was expended on the preservation of their national languages and cultures.

'No wonder that under those circumstances our nations could hardly compete with well-developed and more fortunate nations in the development of natural sciences. Nevertheless, there has always been a deep interest in science in this country, and those who were able to become scientists strove to contribute to the progress in science in spite of their modest circumstances.

'Restoration of national independence after World War I was the first condition of better development of science in Czechoslovakia. We can justly say, however, that the way to full development opened only after Socialist Czechoslovakia came into existence and after the Czechoslovak Academy of Sciences was founded.

'The Czechoslovak Academy of Sciences following the almost 200 years tradition of older scientific societies is not only the top scientific institution of the Czechoslovak Socialist Republic but also a big working scientific organism which, besides 200 Academy members—the most prominent Czech and Slovak scientists—comprises 115 scientific institutions with more than 12000 collaborators and an extensive net of working and managing scientific organs and societies. The Czechoslovak Academy of Sciences is active in all main fields of natural sciences, in basic theoretical branches of technical sciences, as well as social sciences.

'You will find that within the Czechoslovak Academy of Sciences astronomy has found unprecedently favourable conditions for its development. Two big astronomical institutes, at Ondřejov and at Skalnaté Pleso (Rocky Lake), were enlarged considerably and their scientific equipment is now incomparably better than even in recent past. It may suffice to mention only the big solar laboratory at Ondřejov, the coronograph on Lomnický Štít and the telescope of 79 inches (2 meters) in diameter being completed at Ondřejov. Before World War II there used to be two resident professional astronomers at Ondřejov, now there are almost thirty of them.

'Needless to say, we are pleased to hear compliments from our guests on the level of our astronomers' work. We appreciate the results of their research of the Sun, meteors, high atmosphere, comets, double stars and stellar systems.

'Scientific work in a small country has, however, always its specific problems. In no branch of science can we afford to carry out research in all directions like big powers. The Presidium of the Czechoslovak Academy of Sciences keeps urging astronomers—as well as scientists of other branches—to concentrate their efforts and means on a small number of fundamental, topical scientific problems. We believe that the situation in Czechoslovak astronomy in this respect is not bad; a few larger working groups with clear and systematic programs have been set up. However, the principle—not to disperse efforts—should be fought for all the time; there is always a certain strain between a rational directive towards concentrated collective work and natural centrifugal tendencies of individuals whose thinking, work enthusiasm and individual interest, in the long run, push scientific progress forward.

'From the standpoint of Czechoslovak astronomers' work, two types of foreign guests have come to this Congress. Many of you know our astronomers from joint work, from symposia and from publications. You work on the same similar problems; you will certainly have much experience, knowledge and suggestions to exchange with our scientists. The second category of astronomers, at least as numerous as the first one, works in branches which are not developed in this country, for one reason or another. Their contacts with Czechoslovak astronomers may be less direct and frequent but, believe us, we appreciate their presence just as much. They also bring much knowledge of great interest for every astronomer without regard to his specialized field of work.

'Many young Czechoslovak astronomers will have, for the first time, an opportunity to see or even to talk to personalities known to them hitherto only from literature. It will certainly be a stimulus for their further work. Not only astronomers themselves are looking forward to your Congress but also general Czechoslovak public. The Czechoslovak Astronomical Society attached to the Czechoslovak Academy of Sciences celebrates the 50th anniversary of its foundation this year.

'Society's conditions for admission of new members include active participation in a certain branch of astronomy, not only mere interest. In spite of that the Society has now about 800 members. Further thousands of other astronomy fans belong to scores of astronomical circles and clubs attached to plants, schools, and almost sixty public observatories.

'Three hundred sixty years ago Prague was a real capital of the astronomical world. Now it has become the capital again, at least for a few days. Allow me, then, to wish your session a complete success in the spirit of that tradition and to express my hope that this Assembly will contribute a great deal to the carrying out of the noble goals of the International Astronomical Union. Thank you.'

The Chairman thanked Professor Šorm and then invited Dr Šternberk, President of the Czechoslovak National Astronomical Committee and President of the Local Organizing Committee, to speak.

ADDRESS BY THE PRESIDENT OF THE CZECHOSLOVAK NATIONAL ASTRONOMICAL COMMITTEE, DR B. ŠTERNBERK

'Monsieur le Président, Mesdames et Messieurs, Chers Hôtes,

'Permettez-moi de vous saluer tout d'abord au nom du Comité Local d'Organisation de la treizième Assemblée Générale de l'Union Astronomique Internationale. Vous avez aimablement accepté notre invitation à y participer, ce qui nous fait un plaisir immense et nous vous souhaitons la bienvenue de tout notre coeur.

'Nous avions pour tâche et nous nous efforcions de préparer les conditions nécessaires à la réussite de ce Congrès. Toutefois, nous désirions également profiter de cette occasion pour vous familiariser avec les beautés de cette ville ancienne, de tout notre pays et avec la vie que nous y menons. Lorsque, il y a six ans, nous avons présenté pour la première fois au Comité Exécutif de l'UAI notre projet de choisir Prague pour la session de cette année, nous ne nous doutions pas que la face de notre capitale serait altérée par la construction du métro, juste en cette année. Heureusement ces travaux n'ont pas encore envahi la Vieille Ville qui vous intéressera sans doute le plus. C'est vrai qu'il est un peu difficile de se frayer le chemin depuis les hôtels de la place Venceslas jusqu'au siège de notre Congrès, mais nous espérons que les astronomes, qui connaissent bien les orbites des corps célestes, sauront également trouver leur chemin à travers les nombreux passages qui remplacent le carrefour de la place Venceslas.

'Le Comité Local d'Organisation vous a mis succinctement au courant de l'astronomie dans notre pays dans la brochure "La Tchécoslovaquie accueille les astronomes". En outre, il a distribué à tous les participants la brochure "L'astronomie en Tchécoslovaquie". Vous y trouverez des informations détaillées sur l'histoire et l'état actuel de notre astronomie, sur les observatoires scientifiques tchécoslovaques, sur leur travail et aussi sur l'ample vulgarisation de l'astronomie dans notre pays. Je ne les répéterai pas. Je vous demande toutefois de me permettre de vous saluer également au nom du Comité Astronomique National Tchécoslovaque et des instituts qu'il représente au sein de l'Union, à savoir les instituts de l'Académie tchécoslovaque des Sciences, de l'Académie slovaque des Sciences, de l'Enseignement Supérieur tchécoslovaque avec l'Université Charles à sa tête et les autres institutions.

'Je vous souhaite la bienvenue aussi au nom de la Société Astronomique Tchécoslovaque auprès de l'Académie tchécoslovaque des Sciences, qui groupe dans notre pays les travailleurs scientifiques et les spécialistes du domaine de l'astronomie. La Société est heureuse de pouvoir, en son année jubilaire, saluer à Prague les plus grands astronomes du monde entier.

'Je vous salue, enfin, au nom de la rédaction de notre revue scientifique "Bulletin of the astronomical Institutes of Czechoslovakia" qui a eu l'honneur de présenter sur ses pages les contributions des astronomes étrangers.

'Depuis des siècles déjà, notre pays partage l'idée qui constitue la raison d'être de notre Union, à savoir celle de la coopération internationale dans l'astronomie. L'astronome et médecin tchèque Tadeáš Hájek de Hájek a été le prédécesseur de notre collègue très estimé, le Président de la Commission 38 de l'UAI pour l'Echange des Astronomes, le Professeur Minnaert, car ce fut Hájek qui organisa le stage de Tycho Brahé à Prague, pour utiliser des termes actuels.

'Nous partageons l'idée de l'échange d'astronomes et de collaboration encore aujourd'hui, et, à l'heure actuelle, des astronomes de divers pays, par exemple soviétiques et américains, font des stages à nos observatoires. Permettez-moi de remercier à cette occasion l'Union, les savants et les observatoires étrangers d'avoir permis également à nos astronomes de travailler dans d'autres instituts.

'Pour terminer, voici une petite réminiscence. Il est bon de se rendre compte de l'accroissement spectaculaire des connaissances scientifiques au cours d'une seule génération. Lorsque, il y a 50 ans, je commençais ma carrière d'astronome, l'astrophysique était encore incapable de répondre à la question de savoir ce qu'étaient les nébuleuses spirales, si elles faisaient oui ou non partie de notre Voie Lactée. On commençait alors à se douter que l'astronomie se tenait au seuil d'une grande découverte qui devait par la suite extrêmement élargir les horizons de l'humanité. Je pense, et avec moi le pensent aussi d'autres collègues, que la situation est, peut-être, un peu analogue aujourd'hui et nous espérons que, pendant le Congrès, nous en serons amplement informés par les personnes les plus compétentes.

'Au nom des astronomes tchécoslovaques, je souhaite un plein succès aux discussions de la treizième Assemblée Générale de l'Union Astronomique Internationale.'

Professor Swings then expressed the thanks and the gratitude of the Union as follows:

'Messieurs les Représentants du Gouvernement de la République Socialiste Tchécoslovaque,

'Messieurs les Membres du Présidium de l'Académie Tchécoslovaque des Sciences,

'Monsieur le Président de l'Académie des Sciences, Professeur Šorm,

'C'est à vous, Messieurs, que l'Union Astronomique Internationale désire d'abord exprimer sa gratitude, car c'est vous qui avez rendu possible le treizième Congrès de notre Union. Tous les astronomes, venus à Prague, de tant de pays, parfois fort éloignés, vous sont aussi reconnaissants pour les paroles si aimables et si hospitalières que vous avez bien voulu prononcer. Votre pays est connu pour son hospitalité, autant que pour son niveau intellectuel et artistique élevé. Mais pour nous, astronomes, la Tchécoslovaquie est, avant tout, un centre très important de recherches astronomiques. Tous, nous savons que votre nation a un passé astronomique prestigieux et que le présent est, en tout point, digne de ce passé. Nous avons aussi la certitude que, grâce à l'aide du Gouvernement et de l'Académie, l'avenir de l'École Tchécoslovaque d'Astronomie sera très prospère. Au nom de tous mes collègues de l'Union, je vous remercie bien sincèrement.'

'Monsieur le Président du Comité d'Organisation et Cher Ami, Professeur Šternberk,

'Monsieur l'Ingénieur Rajský, cheville ouvrière du Comité, Messieurs les Membres du Comité,

'All of us here realize the tremendous amount of work which has been required for the organization of the present meeting. All the records of attendance have been beaten. All our expectations have been surpassed. This is partly due to the steady increase in the attractiveness of our Science. But I am also convinced that many of us have been attracted by the desire to get in closer contact with the renowned Astronomical School of Czechoslovakia. Let us be quite frank. Many of us also wanted to see your wonderful capital city, so full of artistic atmosphere and historical recollections, and to visit part of your beautiful country.

'To the President of the Organizing Committee, Professor Šternberk; to his principal associate, Engineer Rajský; to all the colleagues and their collaborators who took part in the preparation of our General Assembly we express our heart-felt thanks.'

Dr Šternberk then invited the President of the Union to address the Assembly.

ADDRESS BY THE PRESIDENT OF THE INTERNATIONAL ASTRONOMICAL UNION PROFESSOR POL SWINGS

'Lorsque, à la fin de la douzième Assemblée Générale de Hambourg, nous avons entendu le Professeur Guth inviter l'Union Astronomique Internationale, au nom de Monsieur le Président de l'Académie, le Professeur Šorm, à tenir notre treizième Assemblée à Prague, nous avons, tous, éprouvé une vive émotion et une grande joie. Le Professeur Guth nous a rappelé quelques souvenirs de savants éminents ayant trouvé à Prague l'hospitalité et la quiétude nécessaires à leurs travaux: Tycho Brahé, Johannes Kepler, Christian Doppler, Albert Einstein. Ces souvenirs illustraient l'hospitalité dont Prague fait preuve vis-à-vis des étrangers, ainsi que l'intérêt qui a toujours été accordé ici à l'Astronomie. En fait, cet intérêt a été manifesté depuis plus de 600 ans lors de la création de l'Université Charles en 1348. Nous savions que l'atmosphère intellectuelle et artistique bien connue de Prague serait propice au succès de notre treizième Assemblée Générale. D'ailleurs, depuis 45 ans, les astronomes tchécoslovaques ont joué un rôle très important au sein de l'UAI. Faut-il rappeler que le Professeur Nušl dont on fête cette année le centenaire de la naissance, fut Vice-Président de l'Union de 1928 à 1935 et premier directeur de l'Observatoire d'Ondřejov; que mon cher et honoré Collègue, le Professeur Šternberk qui, cette année, atteint comme l'Observatoire d'Ondřejov, son 70-ème anniversaire, remplit une tâche importante comme Vice-Président de 1958 à 1964; que de nombreux savants tchécoslovaques ont été présidents, vice-présidents ou membres de commissions depuis 1922; que mon ami, le Professeur Luboš Perek, oeuvre remarquablement, depuis trois ans, au sein du Comité Exécutif et qu'il va, sûrement, être amené à remplir la tâche la plus importante et la plus lourde au cours du prochain triennat; enfin, que la Société Astronomique de Tchécoslovaquie fête cette année son cinquantenaire?

'Nous verrons tous, dans peu de jours, le nouveau télescope de deux mètres grâce auquel nos collègues tchécoslovaques pourront s'attaquer aux importants problèmes d'actualité et nous sommes heureux que cette inauguration coïncide avec notre Assemblée Générale. Je profite de l'occasion aussi pour rendre hommage à la mémoire de Dr Bečvář, dont les Atlas rendent tant de services, notamment à nos élèves et à nos jeunes collègues entrant dans la carrière.

'L'histoire de l'Astronomie tchécoslovaque, il est inutile que je vous en parle en détail, puisque le Comité Organisateur a bien voulu remettre à chacun de nous sa magnifique petite brochure "La Tchécoslovaquie accueille les astronomes". Nous lui en sommes reconnaissants, comme pour toutes les autres marques d'hospitalité dont nous aurons l'occasion de reparler plus tard.

'I feel most unhappy not to have learned the Czech or the Slovak language. It was really my intention to do it, so that I could have expressed in a more proper and thankful way the gratitude of the Union. I acquired all the necessary books; but unfortunately I could really not find the time, and I apologize most sincerely for this.

'All the Members have received the *Astronomer's Handbook* prepared by our General Secretary. I hope that those of us who were not well acquainted with the history, Statutes, By-laws, etc. have read the corresponding sections. The article 1 of our Statutes reads:

"The objects of the International Astronomical Union are:

(*a*) to facilitate relations between astronomers of different countries where the organization of international co-operation is useful or necessary;

(*b*) to promote the study and development of Astronomy in all its departments."

'We are still as much in accord with these principles today as were our founders in Brussels in July 1919, when the IAU was established, together with three other sister Unions by the International Research Council. Only on one point have our views changed somewhat. In 1925 W.W. Campbell expressed the view that the purposes of our Union refer to future developments, rather than to a recital of the accomplishments of the past. At present, our IAU Symposia and Colloquia are after all more or less recitals, but they help us as a basis for our future programs.

'L'UAI, au moment de sa première Assemblée Générale à Rome en 1922, comprenait 19 Nations Adhérentes—dont la Tchécoslovaquie—, 32 Commissions et 207 Membres dont 83 étaient présents à Rome; le volume du Congrès avait 247 pages. Au dernier Congrès, avant la guerre, en 1938, l'UAI avait 554 Membres, dont 293 se trouvaient à Stockholm. Peu à peu, les Commissions consacraient plus de temps aux échanges scientifiques et proportionnellement moins aux questions administratives. Les travaux des 45 Commissions avaient été réunis en un volume de 518 pages. Lors de la dernière Assemblée, avant celle de Prague, il y avait 1630 Membres, dont 1160 se trouvaient à Hambourg; les 3 volumes des *Transactions* avaient un total de 1688 pages.

'Let me consider just one example:

'that of the commission which is closest to my heart, Commission 29 on Stellar Spectra, which had 9 members at the Rome Meeting in May 1922. When I became a member of Commission 29 in 1935 there were 19 members; three or four of us remain still active. Our draft report contained 10 pages, although two important matters were discussed in great detail: the spectral classification and the Wolf-Rayet Sequences. The summarized discussions required 1½ pages. There are now more than a hundred members; the draft reports to the twelfth General Assembly included 55 pages, and the summary of the discussions in Hamburg required 10 pages. Of course the draft reports still remain very useful. The fact that they are distributed before the Assembly is, I believe, unique among the Unions. They give us reviews prepared by qualified authors, on the progress in the course of the last three years. Possibly we shall have to demand shorter reviews, for financial reasons.

'This tremendous steady progression leads to awfully complex problems, one of the worst being of a financial nature. There will be much talk and work about finances at this Congress. I feel unhappy about it, but it is essential that this matter be studied with the greatest care. I know that our Finance Committee will help us efficiently; our General Secretary has investigated the problem in all its aspects; his endeavors will be a sound basis for the work of the Finance Committee.

'The difficult problems raised by the increase in membership have been recognized by all my predecessors of the postwar-era, and there is no doubt that changes will have to be made in the organization of our Union. At the Moscow Meeting, Professor Oort said: "The all-embracing congresses may to some extent have outlived themselves and... in a foreseeable future we may have to split up into smaller conferences". A few members of the present Executive Committee discussed the matter, and concluded that there was still a little time left before taking drastic measures. Doubtlessly many of us will regret the need for changes; the senior members already regret the atmosphere of the good old meetings in the thirties. The IAU is one of the rare Unions based on individual membership; we are proud of it, and most of us like to continue it that way. It may however not last for ever this way. Let us thus enjoy fully the present meeting which is still organized in the way decided in 1919, and in which all the subjects are still united.

'Une nouveauté de l'Astronomie, affectant fortement l'Union, est l'importance croissante de la coopération entre les astronomes et les spécialistes de disciplines voisines. Mes prédécesseurs ont déjà parlé de l'importance de la radioastronomie et de l'introduction des appareillages électroniques qui en résulte. Maintenant, il faut parler des recherches spatiales, des observations des rayons X ou gamma et, bien sûr, il en sera beaucoup question pendant le présent Congrès, d'autant plus que c'est aux points de contact de ces diverses branches que se font les découvertes les plus impressionnantes. En fait, comme presque tous les progrès sensationnels récents feront l'objet de discours sur invitation et de "joint discussions", je dois me dispenser d'en parler à la présente Cérémonie. Je me contenterai de parler *in fine* d'une comète brillante récente et laisserai aux orateurs invités, l'agréable tâche de parler de la Lune, des quasars et autres sujets de grande actualité.

'The IAU has always done its utmost to promote understanding and progress on an international scale. The astronomical co-operation has brought forth many fruitful results. Our Union is free of all political or governmental influences, and this is indeed the reason why the organization of various international projects has been so successful. All of us, astronomers, do all we can to contribute to international understanding, and thus to peaceful relations between all nations. In our field competition between nations must be replaced and indeed is more and more replaced by friendly collaboration; only in this way can our Science progress efficiently. Our Union is thus an instrument of good will, harmony, concord and peace among nations of different political background or ideology.

'En fait, la coopération mondiale et l'esprit international régnant au sein de notre Union existaient, parmi les astronomes, bien avant 1919 et c'est bien logique, car, de toutes les sciences, c'est, sans doute, la nôtre qui a le plus besoin de la collaboration internationale et celle-ci est, nous devons le reconnaitre, plus facile que dans les domaines économiques ou politiques.

'Un colloque d'Astrophysique sur la nature des comètes se tint à Paris en 1665, sous les auspices du roi Louis XIV et de sa Cour et fut, plus tard, imprimé par Pierre le Grand, à Amsterdam.

'La génération géniale des Cassini dirigea l'Observatoire de Paris de père, en fils, petit-fils et arrière-petit-fils aux 17e et 18e siècles. Dès 1761, le talentueux savant russe Lomonosov organisa des missions pour l'observation du passage de Vénus; des équipes étrangéres y participaient. Friedrich Wilhelm Struve quitta Hambourg en 1808 et, plus tard, fut le fondateur de l'Observatoire de Poulkovo où de nombreux étrangers, Argelander, Schiaparelli, Holden et d'autres travaillèrent avec leurs collègues russes. En 1824, Bessel tentait un programme international d'étude cartographique du ciel. L'Astronomische Gesellschaft, fondée en 1863, a, pendant plusieurs décennies, encouragé les entreprises de coopération internationale. Des colloques internationaux relatifs à des problèmes spécifiques divers se sont tenus à Washington en 1884 et à Paris en 1896 et 1911. Le programme de la Carte du Ciel a été établi il y a exactement 80 ans, en 1887. Le Programme des Selected Areas a été commencé par Kapteyn en 1904, la même année que l'Union Solaite Internationale. Et, comme je l'ai dit en débutant, Prague n'a-t-elle pas donné le plus bel exemple en invitant Tycho Brahé et Kepler? Actuellement, l'UAI, grâce à ses Commissions pour l'Echange des Astronomes et pour l'Enseignement de l'Astronomie, grâce aussi à ses symposia, colloquia et entreprises en collaboration peut servir de modèle pour toutes les autres Unions Internationales.

'At the Berkeley General Assembly Professor D.H. Menzel expressed our views forcefully: "Every gathering such as this is a significant breach in the walls that separate people of good will. Such gatherings make statesmen aware that, through science, nations can peacefully gain those material benefits they have fruitlessly sought to acquire through war."

'It is a source of extreme satisfaction to see so many young people in this audience. Their enthusiasm is gratifying. We may count on them to keep the friendly professional bonds between astronomers from North and South, East and West. To all our young colleagues the new techniques—radio, space, X-rays, etc. are taken for granted and this is how it should be. Yet it is only a little less than ten years ago, on October 4, 1957, that the first sputnik was launched and we are only on the threshold of the astronomical discoveries made by space instrumentation. We shall of course hear much on space astronomy at this Congress. We also hope that, as a result of the private contacts, especially between young astronomers, joint projects will be planned, combining the ideas and dreams of the astronomers who have or have not facilities for space experiments.

'It will again be fully realized that the progress in Astronomy is steady and rapid. Our science is privileged in that it never goes through periods lacking in excitement. At every General Assembly new exciting discoveries are announced and discussed.

'J'ai signalé tout à l'heure que l'état des finances de notre Union nous cause de grands soucis. A la deuxième séance de l'Assemblée Générale, nous pourrons avoir des informations précises et des suggestions du Comité des Finances. Peut-être souhaitez-vous, dès à présent, en vue d'éventuelles conversations, posséder quelques raisons de ces difficultés. Ce n'est certes pas le Secrétariat qui fait des dépenses extravagantes! Les dépenses de nature strictement administrative sont inférieures à 15% des ressources. Les principales dépenses concernent l'impression des publications, l'organisation des réunions scientifiques, les subsides destinés à favoriser les échanges d'astronomes (en particulier dans le cas des jeunes astronomes de pays en voie de développement), le financement de certaines opérations des commissions. Bien entendu, ces dépenses augmentent fortement avec le nombre de Membres. Vous aurez des indications détaillées et précises plus tard: je me contente de préparer le terrain, conscient du fait que ces difficultés financières se présentent pour la première fois.

'Dans de nombreux domaines, l'observateur ne peut pas savoir ce qu'il va trouver. Par exemple, de très nombreux spectres astronomiques révèlent des informations inattendues. C'est cela même qui, souvent, rend les observations astronomiques si excitantes. Les spectres de novae ou d'étoiles symbiotiques ne sont jamais identiques; ni ceux des comètes, d'ailleurs. Tous fournissent des renseignements physiques nouveaux.

'Nowadays there is a tendency toward teamwork. This is excellent in many cases. But let us beware: we shall always need the initiative of individualistic thinkers. Let us never discourage a scientist working in his ivory tower. We definitely need large and expensive telescopes, but let us still encourage astronomers belonging to modest institutions and using their brain. The training of young astronomers and the creation of enthusiasm are as honorable tasks as the elaborate research on cosmic bodies. The good administration of an institution is essential to the efficiency of all the workers; but let us avoid regimentation among the promising young scientists. The IAU has no power in itself; but it tries to help the young and old astronomers and to facilitate the progress of all the branches of Astronomy, whatever the glamour (or lack of it!) of the specific branch may be. In many cases we need observations at various longitudes; in other words Astronomy *must* be practised all over the world. In most countries the appropriations received by the astronomical institutions are only an extremely small fraction of the national product. When astronomical institutions are really too poor they should join their efforts.

'I would like to end on a purely astronomical note which will not be considered in any invited discourse or joint discussion, but which will obviously be discussed in Commission 15 (Physical Study of the Comets).

'En 1966, nous avons observé enfin une comète brillante passant très près du Soleil. Elle avait été découverte par deux amateurs japonais: Ikeya et Seki. Tous les astronomes intéressés à la physique cométaire attendaient un tel événement, semblable à l'apparition de 1882. Des spectres à haute dispersion ont pu être obtenus dans plusieurs observatoires. Au voisinage du Soleil, le spectre de la comète Ikeya-Seki a montré de nombreuses raies d'émission de Fe I, Ca I, Ca II, Na I, etc. Rappelons que, en 1882, observant visuellement en plein jour, deux astronomes irlandais, Lohse et Copeland, avaient vu des raies brillantes du fer. Pour des raisons semi-théoriques, cette observation a été souvent critiquée, parfois cruellement. Il y a une vingtaine d'années, j'ai pris la défense de ces excellents observateurs, Lohse et Copeland. Leur observation de 1882 a été pleinement confirmée par les spectres photographiés ou enregistrés photoélectriquement de la Comète Ikeya-Seki. L'étude détaillée des spectres de ce remarquable objet n'est pas encore terminée, mais elle se révèlera extrêmement fructueuse. C'est la revanche de Lohse et Copeland: une observation très soigneuse ne peut jamais être méprisée, même si elle semble en désaccord avec certaines considérations théoriques.

'Notre Union devra, sans doute, subir des modifications de ses Statuts et Règlements. Mais nous pouvons avoir confiance dans son avenir, car, depuis près d'un demi-siècle elle n'a cessé de prospérer et de rendre des services éminents à la Communauté astronomique toute entière. Et l'enthousiasme de nos jeunes Collègues nous prouve que nous pouvons compter sur eux pour continuer la grande oeuvre que représente notre Union Astronomique Internationale.'

Dr Šternberk thanked Professor Swings for his address.

After the execution, by the Prague Chamber Orchestra, of the Symphony in D major by J.V.A. Voříšek, the President formally declared open the thirteenth General Assembly of the International Astronomical Union.

GENERAL ASSEMBLY

AGENDA

1. Official opening of the General Assembly by the President.
2. Appointment of official interpreters.
3. Report of the Executive Committee:
   (*a*) Discussion of the printed Report (1964–66);
   (*b*) Report on the period January–August 1967;
   (*c*) Report of decisions taken at meetings in Prague, August 1967, in particular of the admission of new National Adhering Organizations.
4. Report by the General Secretary.
5. Report by the President of the proposals for membership of the Executive Committee.
6. Announcements of:
   (*a*) The names of representatives of Adhering Countries, empowered to vote on their behalf;
   (*b*) The names of representatives to serve on the Nominating Committee;
   (*c*) The names of Acting Presidents of Commissions.
7. The appointment of the Finance Committee.
8. The appointment of the Resolutions Committee.
9. Consideration of resolutions submitted by Adhering Organizations or National Committees of Astronomy (points 6 and 7 of the By-laws).
   Three such resolutions were formally submitted before the statutory deadline. They are shortly described, as follows:
   (*a*) (Germany–D.D.R.) to create a new IAU Commission for Relativistic Astrophysics and Cosmology;
   (*b*) (Belgium) to provide astronomical publications with abstracts of a standard type;
   (*c*) (Belgium) on the determination of short-period terms of nutation.
   The resolutions should be submitted for advice to the competent Commissions prior to the General Assembly.
10. Consideration of resolutions submitted by Commissions or by Inter-Union Commissions. Three such resolutions have been formally submitted. Their list follows. Other resolutions, not included in this list, have only been presented in a draft form, with the implication they will be discussed first by the Commissions.
   (*a*) (Commission 42) On a subvention to the Bibliography and Programme Notes for Eclipsing Binaries;
   (*b*) (Commission 42) On a grant for the third Colloquium on Variable Stars;
   (*c*) (Commission 42) On a grant for the publication of Eclipsing Binaries Circulars.
   These resolutions, having financial implications, should be submitted for advice to the Executive Committee and to the Finance Committee, and discussed under items 11 (*b*) or (*c*).
11. Report of the Finance Committee.
   (*a*) The accounts for 1964–66;
   (*b*) The additional budget for 1967;
   (*c*) The budget for the ensuing period.
12. Consideration of the unit of subscription for the ensuing period.
   The Executive Committee will propose to the Finance Committee that the unit of subscription be increased to its maximum value of 900 gold francs.

13. Consideration of resolutions submitted by the Executive Committee.
14. Consideration of resolutions submitted by the Commissions (subject to the recommendation of the Resolutions Committee).
15. The appointment of the Special Nominating Committee.
16. The nomination of new Members of the Union.
17. Commissions:
    (*a*) The dissolution and the creation of Commissions;
    (*b*) The election of Presidents and Vice-Presidents;
    (*c*) The membership of Organizing Committees.
18. The place and date of the fourteenth General Assembly.
19. The election of a President, six Vice-Presidents, a General Secretary, and an Assistant General Secretary.
20. Addresses by the retiring and newly-elected Presidents.
21. Closing Ceremonies.

At the first session of the General Assembly, held on 22 August 1967 from 14^h00^m to 16^h00^m, items 1 to 10 of the above agenda have been considered. The remaining items, and those which have been adjourned from the first session, have been considered at the second session, held on 31 August 1967 from 9^h00^m to 12^h00^m.

FIRST SESSION

Held on Tuesday 22 August at 14^h00^m, *in the Congress Hall in J. Fučík Park of Culture in Prague*

Professor Pol Swings, President, in the chair

1. *Official opening.* The President spoke a few words of welcome to the Members of the Union, invited participants and their guests, and formally declared open the first session of the thirteenth General Assembly of the Union.

2. *Official interpreters.* The General Secretary asked the General Assembly to approve of the appointment of the following four official interpreters:

from English to French: L. Houziaux and P.R. Demarque

from French to English: R. Kandel and H. Reeves

This was unanimously approved.

1. (*cont.*) The President then welcomed the official representatives of UNESCO and of the scientific Unions as follows:

UNESCO: Mr E.M. Fournier d'Albe

IUGS (International Union of Geological Sciences): Dr G. Fielder

IGU (International Geographical Union): Prof. Korčák

IUCr (International Union of Crystallography): Dr A. Línek

IUGG (International Union of Geodesy and Geophysics): Prof. P. Tardi

URSI (Union Radio Scientifique Internationale): Dr J.-L. Steinberg

The President then welcomed the representatives of various inter-union bodies and of other organizations represented in Prague, as well as representatives of several Academies of Sciences.

The President suggested that a message be sent to the President of the Socialist Republic of Czechoslovakia. This proposal was approved by acclamation.

The President then referred to the loss the Union had suffered during the past three years as a result of death; unable to pay individual tribute to all deceased Members, he however expressed the indebtedness of the Union to three of its past officers recently deceased, André Danjon, Bertil Lindblad, and Robert Methven Petrie.

The President then asked Members to stand while the General Secretary read the following list of Members who had died since the last meeting, or whose death was not known at the time of that meeting:

H.L. Alden, J.M. Aller, Sir Edward Appleton, F. Baldet, D. Barbier, A. Bečvář, D. Brouwer, L. Cadarso, L. Carnera, G. Cassinis, A. Danjon, A. Dermul, M.S. Eigenson, F. Fernandez de la Puente, E. Finlay-Freundlich, J. Génard, J. Gadomski, A. Gütler, C.B. Haselgrove, T. Hanaka, M.B.B. Heath, W.W. Heinrich, D. Hendrix, A.R. Hogg, G. Horn d'Arturo, I.G. Izsak, H.S. Jelstrup, F. Kepinski, E.A. Kreiken, D.K. Kulikov, G. Lemaître, B. Lindblad, J.C. Lindsay, D.B. McLaughlin, J.E. Mack, D.D. Maksutov, E.L. Martin, W.F. Meggers, S.K. Mitra, K. Moder, J.J. Nassau, N. Nechvile, B.A. Orlov, B.M. Peek, G. Pels, J.G. Perel, R.M. Petrie, V.F. Proskurin, G. Racah, F. Rigaux, N. Robinson, G.W. Rougoor, A.M. Rozis-Saulgeot, H. Schneller, E. Schoenberg, F.H. Seares, V.V. Sharonov, H. Siedentopf, E.C. Slipher, J. Stebbins, M.F. Subbotin, P.Y. Tsiang, Y. Tsukamoto, L. Vegard, J.H. Waddell, A.A. Weiss, F. Wünschmann.

At the request of the President, the General Secretary expressed the satisfaction of the Union to see in Prague so many astronomers, but also its regret of the absence, due to illness, of some eminent colleagues, in particular Professor Melnikov, President of Commission 9. As it was not possible to send individual messages, the sympathy of the Assembly to the absent colleagues was expressed through its unanimous applause.

The President then welcomed in particular those who have played an important role in the past life of the Union, Professor Ambarcumjan, Professor Oort, Professor Oosterhoff, Dr Sadler. He also welcomed Lady Spencer Jones and Mrs Lindblad, widows of the former Presidents of the Union, Sir Harold Spencer Jones and Professor Bertil Lindblad.

3. (*a*) *Report of the Executive Committee.* The General Secretary said that the Report of the Executive Committee, as printed in the *Agenda and Draft Reports*, was now open for discussion; and that he would be happy to answer any questions that might be asked. There were no questions, and the Report was approved.

(*b*) The General Secretary then read the following addendum to the Report of the Executive Committee, concerning the period January–August 1967.

RAPPORT COMPLÉMENTAIRE DU COMITÉ EXÉCUTIF SUR LA PÉRIODE JANVIER–AOÛT 1967

Introduction. Ce rapport complémentaire couvre la période qui s'étend du 1 janvier 1967 au 31 Juillet 1967. Il est rédigé d'un façon succincte, le volume qui sera publié à l'occasion de la quatorzième Assemblée Générale de l'Union Astronomique Internationale devant normalement contenir la période qui débute au 1 Juillet 1967.

Décisions prises par correspondance par le Comité Exécutif. Depuis le 1 janvier 1967, le Secrétaire Général a adressé aux membres du Comité Exécutif sept lettres circulaires, relatives à divers problèmes. Le Secrétaire Général Adjoint a également correspondu avec le Comité Exécutif sur des détails d'organisation des réunions et manifestations diverses liées à l'Assemblée Générale.

Les principales décisions concernent les points suivants:

(*a*) Préparation d'une liste de noms devant être proposés aux suffrages de l'Assemblée Générale comme Président, Vice-Présidents, Secrétaire Général, Secrétaire Général Adjoint de l'Union pour la période 1967–1970 (voir ci-après points 5 et 19 de l'Ordre du Jour).

(*b*) (Voir rapport financier, in *IAU Trans.*, **13A**, page xxxvii et page lii).

Le Comité Exécutif a approuvé la décision de demander à l'Assemblée Générale l'augmentation de 600 à 900 francs-or de l'unité annuelle de contribution des Organisations Adhérentes. Cette décision a été aussitôt portée à la connaissance des Comités Nationaux afin de leur permettre de prendre position à ce sujet avant l'Assemblée Générale où elle sera discutée sous le point 12 de l'Ordre du Jour. Elle correspond à l'augmentation maximum permise par les Statuts (article 11(*b*)), et il sera donc nécessaire de prévoir une nouvelle modification des Statuts, qui sera peut-être assez importante.

(*c*) Par ailleurs, le Comité Exécutif a approuvé une action conjointe du Président et du Secrétaire Général en vue de rassembler un fonds de réserve, grâce à des appels à des fondations publiques et privées. Un petit nombre d'organismes ont répondu; cette campagne est encore trop récente pour qu'elle ait pu porter tous ses fruits. Mais la décision du Comité Exécutif s'explique par la situation financière grave de l'UAI. Cette situation, après l'action en question, reste grave. Il en sera de nouveau question sous d'autres points (point 11) de l'Ordre du Jour.

(*d*) Le Comité Exécutif a accepté, par ailleurs,—avec gratitude—, le prêt sur 5 ans (sans intérêt) d'une somme de $10000 de la part du Conseil International des Unions Scientifiques.

(*e*) Un groupe de travail (Professeurs Swings, Président, Goldberg, et Gratton) a été constitué pour l'étude des effets, peut-être dangereux pour l'astronomie, du Projet 'Able', envisagé par des autorités spatiales et militaires des U.S.A.

(*f*) Enfin, le Comité Exécutif a décidé la modification du contrat passé avec nos éditeurs, ce qui impliquera sans doute le changement d'éditeur, et peut-être d'imprimeur.

Colloques et Symposiums. Depuis le 1 Janvier 1967, les réunions suivantes ont été organisées par l'UAI:

—Symposium no. 32, Continental Drift, the Motion of the Pole, and the Rotation of the Earth, Stresa, mars 1967.

—Colloquium on Spectroscopy, Bombay, janvier 1967.

—Colloquium on the Zodiacal Light and the Interplanetary Medium, Honolulu, janvier 1967.

Divers. Un grand nombre de discussions ont eu lieu, et des correspondances nombreuses ont été échangées notamment sur les importants problèmes suivants:

—Eventuelle création d'un 'solar-terrestrial service';

—Eventuelle création d'une Commission Interunion sur la Chimie Extra-terrestre;

—Organisation de la Commission Interunion sur la Physique des Relations Soleil-Terre.

Ces discussions sont encore en cours et aboutiront éventuellement au cours des mois à venir.

3. (*c*) *Report on decisions taken by the Executive Committee.*

(*i*) The General Secretary reported that the Executive Committee had admitted Columbia as a National Member, with the Academia Colombiana de Ciencias Exactas, Fisico Quimicas y Naturales as Adhering Organization. On behalf of the President, he welcomed Professor J. Arias de Greiff, national representative of Columbia, at this General Assembly.

(*ii*) On behalf of the Executive Committee the President then much regretted to inform of the resignation of the Lebanese Republic from adherence to the Union.

(*iii*) The President expressed the wishes of the Union to welcome in a near future new Adhering Countries, with many of which correspondence has been exchanged to this purpose, and from which astronomers have been able to attend this meeting.

4. *Report by the General Secretary.* The General Secretary stated that he did not wish to add anything to the published report of the Executive Committee, but that he would like to make some informal remarks, expressing his personal views on the present and future situation of the Union. He spoke as follows:

'Le rapport du Secrétaire Général qui figure au point 4 de l'Ordre du Jour est bien souvent tout d'abord l'expression de sa modestie, vraie ou fausse! Il se compare aux grands Anciens, tellement plus efficaces qu'il ne put jamais l'être; et il évoque avec complaisance la généreuse tradition de notre Union. Son caractère d'affectueuse intimité, hélas, diminue d'Assemblée Générale en Assemblée Générale, mais la cohérence en reste permanente. Dans une atmosphère certes difficile, elle est rendue toujours plus profonde par la solidarité désintéressée de tous ses Membres. Et la tradition se perpétue ainsi depuis un demi-siècle!

'Je ne faillirai certes pas à cette tradition. J'ai conscience d'arriver devant vous, en ce mois d'août 1967, avec le sentiment de n'avoir pas utilisé mon temps de Secrétariat Général, réduit maintenant,—et comme c'est justifié!—à trois années, aussi bien qu'il l'aurait fallu. Si le budget que je vous présente est équilibré, ce n'est pas comme celui de mes prédécesseurs, en raison d'une saine gestion, c'est par la vertu d'une arithmétique artificielle inspirée du budget célèbre d'un certain Procuste, fabricant de meubles dans la haute Antiquité.

'J'ai conscience aussi de devoir à l'Union plus qu'un budget équilibré. Je vous dois aujourd'hui une analyse brève, mais claire, de la situation actuelle; cette analyse, je l'ai présentée au Comité Exécutif en plus de détails; et le Comité Exécutif a eu la bonté d'approuver mon action, et d'approuver certaines de mes suggestions. Excusez-moi aujourd'hui si je me limite à l'essentiel.

'Les ressources de l'UAI, en augmentation limitée, sont essentiellement les cotisations des Etats membres; le plus cotisant des États verse actuellement pendant trois ans une somme qui serait insuffisante pour rembourser le voyage à Prague des Présidents de Commissions qui sont originaires de ce pays! En revanche, les dépenses de l'Union,—et je ne le regrette pas!—augmentent d'une façon exponentielle et rapide. Permettez-moi quelques exemples:

'Au temps des idylles, avant la Seconde Guerre Mondiale, le nombre des Membres de l'UAI était passé de 207 (en 1922) à 554 (en 1938) et le nombre de pages des *Transactions* de l'Union augmentait dans les mêmes proportions. Un budget assez large permettait une activité normale, un Secrétaire Général sans assistance aucune pouvait assurer normalement une administration simple.

'Après la Seconde Guerre Mondiale, le budget restait assez large, et les recettes suffisamment excédentaires, pour que des entreprises nouvelles soient favorisées, et notamment pour qu'un secrétariat plus étoffé puisse s'occuper entre les Assemblées Générales, et non plus seulement à l'occasion de celles-ci, des nouveaux problèmes de l'Union. De très nombreux projets étaient financés; mentionnons seulement certains d'entre eux: l'échange des astronomes, les Bulletins d'Information de l'Hémisphère Sud, les symposiums de toute nature (le premier symposium organisé en dehors d'une Assemblée Générale date, sauf erreur de ma part, de 1949; il s'agissait d'un colloque organisé en coopération avec l'Union Internationale de Mécanique Théorique et Appliquée; le suivant date de 1953) et bien sûr leur publication...

'Mais je vous l'ai dit, la courbe ascensionnelle lente des recettes était rattrapée, et je fixe ce rattrapage à la période qui vient de s'écouler, et très vite dépassée, par la courbe ascensionnelle rapide des dépenses.

'Ceci, c'est le premier problème: je vois en face de moi, les regards furieux de mes plus vieux amis, et j'en souffre comme eux: n'ai-je pas été contraint de demander au Professeur Minnaert de limiter l'aide à l'échange des astronomes, une des entreprises les plus utiles pourtant de l'UAI? N'aperçois-je pas le regard aigu de Dick Thomas qui me reproche le prix élevé de nos publications, et de n'en pas distribuer gratuitement un plus grand nombre? Le Dr Sahade ne s'attriste-t-il pas à juste titre de mes réticences devant les besoins financiers de l'excellent bulletin de liaison des astronomes de l'hemisphère austral? Les Présidents de Commissions,—et ils sont une quarantaine,—ne m'en veulent-ils pas un peu d'avoir été obligé de leur demander de trouver dans leur propre pays les ressources nécessaires à leur voyage à Prague, contrairement aux traditions les mieux établies?

'Il est donc clair que certaines traditions doivent disparaître; et je crains que la générosité excessive n'en soit une. Sur ma proposition, au cours de ces trois années, le Comité Exécutif a été obligé de prendre des décisions impopulaires à cet égard. Soyez assurés qu'il sera impossible au Comité Exécutif qui va entrer en fonctions de ne pas continuer une politique d'austérité, et je vous demande, instamment, de l'y aider, avec autant de continuité que de sincérité.

'Il est clair aussi que nos ressources doivent augmenter. Certes, les mesures que nous vous proposons sont de nature à résoudre momentanément le problème budgétaire; elles ne permettent pas de modifier le déséquilibre des croissances respectives des recettes et dépenses; mais seulement de retarder un peu—de l'espace entre deux Assemblées Générales,—le point critique. Les espoirs que nous avions de la vente des symposiums sont, je dois le dire, vains: si le nombre de Membres de l'UAI augmente, le nombre d'acheteurs n'augmente guère; il s'agit toujours des bibliothèques d'instituts; et malgré mes exhortations, publiées dans le *Bulletin d'Information* de l'UAI, et que je considère aujourd'hui comme du domaine du rêve éveillé, les Membres de l'UAI n'achètent pas les symposiums. Je ne vois pas clairement, compte tenu des efforts demandés aujourd'hui aux organisations nationales adhérentes, compte tenu de l'aide, certes estimable, mais encore très insuffisante, des organisations internationales, comment nous pourrions donner à nos ressources un taux de croissance égal à celui de nos dépenses.

'Le temps d'une grande mutation me semble donc venu. Je ne veux pas, à cette tribune, entrer dans le détail de la révolution qui peut aujourd'hui sembler nécessaire. Une discussion ouverte serait d'ailleurs inopportune avant un examen très approfondi de la situation par le Comité Exécutif.

'Mais plusieurs solutions paraissent évidentes. Tout d'abord une refonte complète de nos Statuts et de nos Règlements, donc de notre administration, devrait pouvoir nous rendre capables d'affronter l'explosion scientifique à laquelle nous participons, en 1967, en astronomie comme ailleurs. Les ressources devraient d'une façon ou d'une autre être rendues proportionnelles au nombre de Membres. Certaines libéralités d'un autre âge doivent cesser, par exemple la trop large distribution des *Draft Reports.* Les Assemblées Générales devraient, peut-être, se limiter à des réunions administratives de quelques centaines de personnes au plus,—tous les quatre ans peut-être, et non plus tous les trois ans. En revanche, de grands symposiums spécialisés, mais sans activité administrative, pourraient grouper, de temps à autre, plusieurs commissions. Plusieurs écueils sont à éviter, et je ne voudrais pas que ces suggestions soient prises de façon trop littérale. Il faut éviter l'éclatement de l'astronomie, happée par tant de forces centrifuges, et pour cela, il vaut mieux éviter la création de larges associations permanentes et choisir plutôt la formule de larges symposiums *ad hoc.* Il faut aussi, je crois, éviter la suppression des Membres individuels de l'Union: c'est grâce à cette originalité, me semble-t-il, qu'au lieu d'être un cénacle de notables, nous apparaissons vraiment comme la jeune astronomie vivante,—et l'âge moyen de cette assistance me donne clairement raison.

'De toutes façons, il faudra, ensemble, travailler à mieux coopérer au développement de l'astronomie, et au fonctionnement de notre Union.

'Le nouveau Président, mon successeur le nouveau Secrétaire Général, et le Comité Exécutif qui s'installera la semaine prochaine à cette tribune, auront, cela est certain, une tâche redoutable, et décisive. J'ai essayé, de mon mieux, de prévoir leurs difficultés et de ne pas leur laisser trop de problèmes mal résolus. Mais j'y ai incomplètement réussi, et mes efforts sont en vérité loin d'être suffisants. L'aide dont mon successeur aura besoin, c'est, avant tout, la vôtre. Pensez que le bureau administratif devra s'occuper de plus de 2000 Membres de l'UAI, de 38 Commissions, de 43 Organisations Adhérentes, et d'un réseau toujours plus compliqué d'organismes extérieurs, de commissions inter-unions, de problèmes délicats, de difficultés techniques graves. Aidez-le, d'abord en répondant à ses lettres, sans délai, et clairement; en l'informant sans délai de vos changements d'adresse, temporaires ou définitifs; en lisant avec soin le *Bulletin d'Information* de l'UAI... N'oubliez jamais certes que votre appartenance à l'UAI, si elle vous donne des droits, vous impose des devoirs. Nous appartenons à une collectivité solidaire, et les égoïsmes de chacun doivent s'effacer devant l'intérêt de tous. C'est l'astronomie toute entière que vous aiderez en aidant votre Secrétaire Général. Je vous en remercie à l'avance, comme je vous remercie de l'aide que depuis trois ans vous avez eu l'amitié de m'apporter'.

5. *Proposals for membership of the Executive Committee.* In accord with the provisions of the Statutes and By-laws, the President submitted for consideration to the General Assembly the names of those proposed by the present Executive Committee for the period 1967–1970. He reminded the Members that there was an agreed procedure by which a Special Nominating Committee, largely appointed by the Nominating Committee and approved by the General Assembly at its previous meeting, makes proposals to the Executive Committee. On this occasion the proposals by the Special Nominating Committee, to fill the positions of President, four Vice-Presidents, General Secretary and Assistant General Secretary had been unanimously adopted by the Executive Committee; and were those now before the General Assembly. The President accordingly submitted the name of Professor O. Heckmann for the office of President; the names of Dr V. Bappu, Professor L. Gratton, and Dr J. Sahade for the offices of Vice-Presidents. Professor W. Fricke, who was to serve till 1970, expressed the wish to retire after his first three-term as Vice-President; it is suggested that Professor W.N. Christiansen who was elected in Hamburg only to serve for three

years could replace Professor Fricke as Vice-President till 1970. The President then submitted the name of Dr Perek for the office of General Secretary; and the name of Professor C. de Jager for the office of Assistant General Secretary. The membership of the new Executive Committee would then be as follows:

PRESIDENT: O. Heckmann
VICE-PRESIDENTS: V. Bappu, W.N. Christiansen, L. Gratton, J. Sahade, M. Schwarzschild, A.B. Severny
GENERAL SECRETARY: L. Perek
ASSISTANT GENERAL SECRETARY: C. de Jager
ADVISORS: The departing President and General Secretary

According to the By-laws the formal election would take place at the final session of the General Assembly.

6. *Announcements.* At the request of the Chairman, the General Secretary made the following statement:

'Je dois tout d'abord rappeler que, conformément à l'article 6 du Règlement, l'Ordre du Jour doit être distribué aux Organisations Adhérentes et Comités Nationaux quatre mois au moins avant l'ouverture de la session. Il est clair que, dans la pratique, il a seulement été possible d'informer les Organisations Adhérentes et les Comités Nationaux des points les plus importants, ceux concernant les finances. Selon les Statuts, je suis donc obligé de demander l'approbation par l'Assemblée Générale de l'Ordre du Jour publié dans les *Agenda and Draft Reports.* De plus, je tiens à préciser que toute question additionnelle ne pourra être prise en considération que si cette prise en considération recueille l'approbation de la moité au moins des voix des pays représentés à l'Assemblée Générale.'

Then the General Secretary announced:

(*a*) the names of representatives of Adhering Countries, empowered to vote on their behalf;
(*b*) the names of representatives of Adhering Countries to serve on the Nominating Committee.

Country	(*a*) *Official Representative*	(*b*) *Representative on Nominating Committee*
Argentina	J. Sahade	C.M. Varsavsky
Australia	H.W. Wood	W.N. Christiansen
Austria	J. Hopmann	J. Hopmann
Belgium	H.L. Vanderlinden	S. Arend
Brazil	A. de Moraes	A. de Moraes
Bulgaria	N. Boneff	N. Boneff
Canada	K.O. Wright	D.A. MacRae
Chile	C. Anguita	J. Stock
Czechoslovakia	V. Guth	J.M. Mohr
Denmark	A. Reiz	K. Gyldenkerne
Finland	G. Järnefelt	G. Järnefelt
France	J.-F. Denisse	J. Delhaye
Germany (DDR)	H. Lambrecht	H. Lambrecht
Germany (DBR)	L. Biermann	R. Kippenhahn
Greece	J. Xanthakis	G. Contopoulos
Hungary	L. Detre	L. Detre
India	V. Bappu	V. Bappu
Ireland	P.A. Wayman	P.A. Wayman
Israel	R. Steinitz	G. Alter
Italy	F. Zagar	G. Righini
Japan	H. Hirose	Y. Fujita

Country	*(a) Official Representative*	*(b) Representative on Nominating Committee*
Korea (DPR)	no representative	no representative
Mexico	G. Haro	A. Poveda
Netherlands	C. de Jager	A. Blaauw
New Zealand	no representative	no representative
Norway	E. Jensen	P. Maltby
Poland	E. Rybka	J. Witkowski
Portugal	M.G.P. de Barros	M.G.P. de Barros
Rumania	C. Drâmbă	C. Drâmbă
South Africa	J. Hers	J. Hers
Spain	R. Carrasco	A. Orte Lledo
Sweden	Y. Öhman	T. Elvius
Switzerland	M. Waldmeier	M. Golay
Taiwan	M. Tsao	M. Tsao
Turkey	N. Gökdoğan	E. Balli
United Arab Republic	no representative	no representative
United Kingdom	H. Bondi	H. Bondi
U.S.A.	C.D. Shane	G.M. Clemence
U.S.S.R.	E.R. Mustel	A.G. Masevič
Vatican City State	P.J. Treanor S.J.	M.F. McCarthy S.J.
Venezuela	no representative	no representative
Yugoslavia	L. Randić	L. Randić

Note: In some cases, or in case of absolute necessity, National Committees have designated substitute delegates. In order to simplify the lists, the General Secretary has omitted the names of these substitute delegates from the above lists.

After the reading of these lists, the General Assembly, at the request of the General Secretary as earlier expressed, approved the Agenda through the votes of the National Representatives.

(c) *Acting Presidents.* The General Secretary announced that the Executive Committee had invited:

Dr F.L. Whipple, Vice-President of Commission 6, to take over the responsability of this Commission after the unfortunate death of Dr A.R. Hogg. He has also been designated as Acting President of this Commission during the meetings.

Dr D.S. Evans, Vice-President of Commission 30, to take over the responsibility of this Commission after the unfortunate death of Dr R.M. Petrie. He has also been designated as Acting President of this Commission during the meetings.

Dr J. Rösch to serve as Acting President of Commission 9, Professor O.A. Melnikov being unable to attend the General Assembly.

7. *Finance Committee.* In accord with the Statutes the following Finance Committee, consisting of one representative of each Adhering Country, was appointed by the General Assembly:

Argentina	J. Sahade	Finland	G. Järnefelt
Australia	B.Y. Mills	France	Ch. Fehrenbach
Austria	J. Hopmann	Germany (DDR)	J. Wempe
Belgium	A.G. Velghe	Germany (DBR)	H.H. Voigt
Brazil	A. de Moraes	Greece	L.N. Mavridis
Bulgaria	N. Boneff	Hungary	L. Dezsö
Canada	J. Halliday	India	G. Swarup
Chile	C. Anguita	Ireland	P.A. Wayman
Czechoslovakia	E. Buchar	Israel	R. Steinitz
Denmark	M. Rudkjøbing	Italy	M. Cimino

Japan	H. Hirose	Sweden	A. Wallenquist
Korea	no representative	Switzerland	J. Bonanomi
Mexico	P. Pishmish de Recillas	Taiwan	M. Tsao
Netherlands	P.T. Oosterhoff	Turkey	T. Gökmen
New Zealand	no representative	United Arab Republic	no representative
Norway	R. Brahde	United Kingdom	W.A. McCrea
Poland	W. Iwanowska	U.S.A.	J.S. Hall
Portugal	M.G.P. de Barros	U.S.S.R.	A.G. Haradze
Rumania	C. Drâmbă	Vatican City State	F.C. Bertiau
South Africa	J. Hers	Venezuela	no representative
Spain	J.O. Cardus	Yugoslavia	P.M. Djurković

Note: In some cases, or in case of absolute necessity, National Committees have designated substitute delegates. In order to simplify the list, the General Secretary has omitted the names of these substitute delegates from the above list.

8. *Resolutions Committee.* At the request of the President, the General Secretary explained that the duties of the Resolutions Committee were to advise the Executive Committee in respect of resolutions submitted by Commissions, with a view to recommending to the General Assembly which resolutions should be formally submitted for consideration by the General Assembly. The General Assembly appointed B.J. Bok, J.H. Oort, D.H. Sadler and A.B. Severny to serve on the Resolutions Committee. The General Secretary and the Assistant General Secretary will attend the Resolutions Committee in an advisory capacity.

9. *Consideration of resolutions submitted by Adhering Organizations or National Committees of Astronomy.* The General Secretary read the following three resolutions, as presented by the responsible National Committees.

(*a*) The National Committee of Astronomy of the German Democratic Republic at the German Academy of Sciences, Berlin,

recommends to the International Astronomical Union,

by reason of the facts that the theoretical and empirical (observing) researches on the structure of the Universe and on structural classification of astrophysical objects, for which the general relativity is a necessity to the theoretical interpretation, have been permanently growing in the last years, where the number of relevant publications and of international colloquia on relativistic astrophysics is a mirror of this growing,

as well as by reason of the fact that it seems in the highest degree desirable, especially nowadays, to coordinate all the astronomers and physicists working on theoretical and experimental researches on cosmology and relativistic astrophysics,

the foundation of a *Commission* of the International Astronomical Union on *Relativistic Astrophysics and Cosmology.*

(*b*) Le Comité National Belge d'Astronomie propose

que soit examinée la possibilité, pour les éditeurs des publications astronomiques, de joindre à celles-ci des fiches de format normalisé, donnant pour chaque article, en plus de la référence bibliographique, un très court résumé du sujet traité.

(*c*) Etant donné certains résultats obtenus en Belgique dans le domaine des marées terrestres, le Comité National Belge d'Astronomie propose l'examen du problème soulevé par la détermination des termes de nutation à courte période.

On behalf of the Executive Committee, the General Secretary suggested that resolution (*a*) be rejected as such and studied by the interested Commissions, notably by Commission 28. He then suggested to submit for advice resolution (*b*) to Commission 5 and resolution (*c*) to Commission 19.

According to the By-laws, the formal vote will have to take place at the final session of the General Assembly.

10. *Consideration of resolutions submitted by Commissions or by Inter-Union Commissions.* Three resolutions submitted by Commission 42 were read by the General Secretary. They have financial

and administrative implications and will therefore be submitted to the Executive Committee and to the Finance Committee.

The General Secretary then reminded the General Assembly of one resolution on the universal use of the metric system, already brought to knowledge of Members through the *IAU Information Bulletin*, and which will be submitted to the vote of the General Assembly at its last session.

The General Secretary then announced that all other items of agenda would be deferred until the next session, pending reports from the various Commissions and Committees.

The Chairman thanked the interpreters and formally adjourned the meeting at 16^h00^m.

FINAL SESSION

Held on Thursday 31 *August* 1967 *at* 9^h15^m, *in Lucerna Hall, in Prague*

Professor P. Swings, President, in the chair

11 and 12. *Report of the Finance Committee and consideration of the unit of subscription.* At the first session of the General Assembly a Finance Committee had been appointed to examine the accounts of the preceding three years and the budget for the coming three years, as submitted to it by the Executive Committee. The Chairman now called upon the President of the Finance Committee, Professor W. H. McCrea, to present his report to the General Assembly, which was read as follows:

REPORT OF THE FINANCE COMMITTEE TO THE GENERAL ASSEMBLY

1. The Finance Committee, as appointed at the first session of the General Assembly, appointed a Sub-Committee composed of Professor W. H. McCrea (Chairman), Professor E. Buchar, Professor Ch. Fehrenbach, Dr. J. S. Hall and Professor E.K. Haradze.

2. The Sub-Committee examined the accounts for the years 1964–1966 as given in the report of the Executive Committee in the light of the explanations given by the General Secretary, the Assistant General Secretary and the Assistant Secretary, and expressed its approval of them.

The Finance Committee took note of the fact that a very significant contribution to the administrative costs of the Union has been borne by the institutions and countries to which its administrative officers belong and considers that the General Assembly would wish to express its cordial thanks to all the authorities concerned.

The Finance Committee recommends to the General Assembly that the accounts for the years 1964–1966 be approved.

3. The Sub-Committee considered the supplementary budget for the calendar year 1967 as proposed by the Executive Committee. Clearly, the proposals adopted in 1964 were inadequate owing mainly to publication costs which amount to more than half the budget. The deficiency resulted mainly from difficulties and delays in printing and publishing and to a contract for this work that proved to be highly unfavorable in the resulting circumstances. The Sub-Committee is convinced that the revised budget is as accurate as possible and that the expenditure is the minimum required for commitments already incurred by the Executive Committee in carrying out the policy approved by the 1964 General Assembly. The Finance Committee, on the advice of the Sub-Committee, accordingly recommends that the General Assembly approve the supplementary budget for the year 1967 of $50000. In doing this, the Finance Committee implies its approval of a loan of $10000 from the International Council of Scientific Unions and of a diminution of the IAU funds in its savings accounts.

4. The Sub-Committee considered the budget for the three years 1968–1970 and prepared provisional figures for consideration by the full Committee. In doing so it recommended:

(*a*) that the unit of contribution should be raised to the figure of 900 gold francs for 1968, 1969, 1970;

(*b*) that the Executive Committee be authorized, as a matter of the utmost urgency, to prepare a revision of the Statutes to be adopted by the fourteenth General Assembly in such a way as to allow in the future the possibility of an adjustment of the contributions from Adhering Countries to the growth in the membership of the Union;

(*c*) that all measures compatible with the administrative and statutory possibilities be taken by the Executive Committee to limit the distribution of *Draft Reports* and to organize this distribution in the best possible way;

(*d*) that a new contract be entered into with a publisher for the publications of the Union implying the minimum financial risk for the Union, and that the acceptance of the present estimates be made conditional upon the conclusion of such a contract;

(*e*) that the General Secretary be urged to make payments by the IAU according only to the progress of the receipts and to the balance of the current accounts;

(*f*) that the Union refrain from subsidizing special projects resulting in free distribution of publications at a time when the IAU publications have to be subjected to severe restrictions;

(*g*) that in the term 1968–1970 the savings accounts be reimbursed as completely as possible, of the diminution they suffered in 1967, in order to build up a sufficient reserve for the further growth of the Union.

5. The Finance Committee as a whole considered the proposal of the Sub-Committee, accepted its recommendations and formally proposes the following budget for adoption by the General Assembly:

BUDGET OF INCOME AND EXPENDITURE, 1 JANUARY 1968–31 DECEMBER 1970

Estimated Income, 1968–1970

Heading	*Origin*	*U.S. dollars*
1.	National subscriptions	$157000
2.	Sales of publications	22300
3.	Subscriptions to IAU publications	0
4.	Interest on reserves in savings banks	4000
5.	Unesco subvention through the ICSU	42000
6.	Unesco contracts	*x*
	Total estimated income	$225300

Estimated Expenditure, 1968–1970

Heading	*Purpose*	*U.S. dollars*
1.	Administrative Office	$42000
2.	Subscription to ICSU	2000
3.	Expenses of Commissions	500
4.	Specific projects of Commissions	31500
5.	Expenses of the General Assembly	16300
6.	IAU publications	40000
7.	Meetings of the Executive Committee	10000
8.	Organization of Symposia	25000
9.	Publication of Symposia not by IAU	0
10.	Interunion Commissions }	7500
11.	Representation expenses }	
12.	Projects authorized by the Executive Committee	2000
13.	Unesco contracts	*x*
14.	Reimbursement of loan from ICSU	10000
15.	Reimbursement to savings accounts	38500
	Total estimated expenditure	$225300

6. Details of individual items under Item 4 (Specific projects of Commissions) are given below:

Commission No.	*Project*	*Sum requested (\$ U.S.)*	*Sum proposed by Finance Committee*
* 6	IAU Telegram Bureau	\$ 2000	\$ 1600
*10	Cartes héliographiques de la photosphère (3 × 1000 gold frs)	980	0
*10	Cartes synoptiques de la chromosphère (3 × 2700 gold frs)	1500	0
*16	Centre de documentation sur les planètes	2000	1400
*20	Minor Planets Center	1500	1300
*27	Catalogue of Variable Stars	6500	3500
*29	Information Bulletin for the Southern Hemisphere	3000	1500
*38	Exchange of Astronomers	24000	22000
*42	Bibliography and program notes for eclipsing binaries	225	200
		\$41705	\$31500

* Indicates items for which reductions are proposed.

Proposals for small sums and the grants for colloquia (\$100 for Commission 42 and \$500 for each colloquium accepted by the Executive Committee) can be provided for under other items of the budget.

W.H. McCrea
Chairman, Finance Committee
August 1967

Points 2 and 3 of this report were adopted unanimously. Point 4 was adopted at the unanimity minus one abstention.

The President then formally requested the General Assembly to give its approval to the whole report of the Finance Committee. One voting country abstained, all others voted in favour. The report was therefore adopted by the General Assembly.

The President expressed the thanks of the Union to Professor McCrea, to members of the Finance Committee, and to the General Secretary for their valuable work.

13. *Resolutions by the Executive Committee.* The General Secretary explained the procedure: before passing to the vote of the resolutions submitted to the General Assembly by the Executive Committee and by the National Committees, Professor Oort will present to the General Assembly the report of the Resolutions Committee concerning the various resolutions submitted by the Commissions, some of which being to be adopted by the Executive Committee itself.

Thus, the President called upon Professor Oort to read the report of the Resolutions Committee.

REPORT BY THE PRESIDENT OF THE RESOLUTIONS COMMITTEE

1. 39 resolutions have been received from the Commissions during the General Assembly, which make a total of 42 taking into account the 3 formally included in the Agenda, page v of the *Agenda and Draft Reports*; 3 resolutions also included in this Agenda came from National Committees or Adhering Organizations. The advice of a Commission has given place to an alternative resolution.

2. Out of these 42, 10 have been considered by the Finance Committee. In view of the difficult financial situation of the Union, stressed elsewhere, the Finance Committee has been obliged to recommend to the General Assembly a reduction, in some cases to zero, of the amount proposed for 8 of the specific projects of the Commissions so proposed. Two other requests have been recommended for inclusion in the budget under other items of the budget, and subject to the approval of the Executive Committee in the case of the organization of a colloquium.

The reasons for the reductions proposed by the Finance Committee are included in the report of this Committee.

3. The Resolutions Committee considered the remaining 32 resolutions with a view to recommending which should be formally submitted for consideration by the General Assembly.

As a result, the Resolutions Committee recommend to the General Assembly that the following 5 resolutions should be considered by the Executive Committee:

(1) (by Commission 41) relating to the conservation of old astronomical instruments and documents;
(2) (by Commission 41) relating to the preparation of an international history of astronomy;
(3) (by the Working Group on Photographic Materials) relating to the sizes of photographic plates;
(4) (by Commissions 4 and 31) relating to the definition of the second;
(5) (by Commissions 4 and 31) relating to the use of natural observational phenomena in determining the Ephemeris Time and the Universal Time.

4. On the 24 still remaining, the Resolutions Committee considers that 2 concerned administrative matters (arrangements of symposia, new structure of the Union...) which should be deferred to the Executive Committee for further consideration; 22 others concerned matters that were not of such a wide interest as to warrant individual consideration by the General Assembly. A resolution by Commission 37, of wide importance, was referred to the Executive Committee for incorporation within any statement or recommendation that it might wish to make.

A "blanket" resolution, to cover the resolutions not formally considered by the General Assembly, in the following terms was recommended to the Executive Committee:

'Considering the impracticability of giving individual attention to every resolution adopted by each of its 38 Commissions, and having full confidence in its Commissions, this General Assembly wishes to give its endorsement to the resolutions adopted by its individual Commissions, and recommends that astronomers give effect to these resolutions in so far as they are able.'

The President expressed the thanks of the Union to Professor Oort and to members of the Resolutions Committee for their valuable work.

13 (*a*). *Resolutions submitted by the Adhering Organizations and National Committees.* See point 9 above, page 21.

Resolution (*a*) proposed by the German Democratic Republic has been withdrawn by the National Committee of this country, and has not been therefore submitted to the vote of the General Assembly.

Resolution (*b*) (hereafter designated by A 2) has been discussed by Commission 5 and approved. Commission 5 will incorporate in its report remarks related to this resolution. The General Assembly then formally expressed its unanimous approval to this resolution.

Resolution (*c*) (hereafter designated by A 3) has been discussed by Commission 19, approved in its principle and was submitted to the vote of the General Assembly in a slightly modified version (see hereafter resolution A 3, page 38 and page 39). The General Assembly formally approved unanimously.

13 (*b*). *Resolutions submitted by the Executive Committee.* The Executive Committee has decided to propose two resolutions to the General Assembly; to adopt, as suggested by the Resolutions Committee, five resolutions of general interest, submitted by the Commissions themselves; and to adopt a sixth resolution submitted by Commission 17, which was also considered by the Executive Committee as of a general interest.

The texts of these eight resolutions, as adopted, are given, in French and in English, on pages 38 to 43.

All of them have been adopted unanimously, except resolution no. 7 on the photographic materials, which was adopted by all voting delegates except by the delegate of the Vatican City State, Dr Treanor, who expressed his opposition. Although adopted, the implementation of this resolution will be studied by the Working Group on Photographic Materials.

14. *Resolutions submitted by the Commissions.* The General Secretary reminded that the General Assembly had earlier expressed its approval on the report of the Resolutions Committee. According this report, a 'blanket' resolution, to cover the resolutions by Commissions not formally considered

by the General Assembly, was recommended and adopted unanimously by the General Assembly (see above page 25 and hereafter resolution no. 9, pages 42 and 43).

A resolution submitted by Commission 37 and deferred to the Executive Committee by the Resolutions Committee concerned the revision of the structure of the Union. The President expressed at this stage the intention of the Executive Committee to request from the incoming Executive Committee the careful study of the needs for changes and of the ways to meet these needs.

15. *Special Nominating Committee.* The President reminded the General Assembly of the agreed procedure by which a Special Nominating Committee—to consist of the President of the Union, the immediate past President, one retiring member of the Executive Committee (to be appointed by the Executive Committee) and four other members (to be appointed by the General Assembly)—is appointed within one year following each General Assembly to advise the Executive Committee on nominations for the new Executive Committee.

On this occasion there would be four members of the new Executive Committee who might retire in 1970, and the General Assembly had therefore to consider four alternative appointments of the four members.

The President then suggested, for the consideration of the General Assembly, the following alternative Committees, the final composition being dependent on which of the retiring members of the new Executive Committee (W.N. Christiansen, L. Perek, M. Schwarzschild, A.B. Severny) would be selected by the new Executive Committee:

PRESIDENT: O. Heckmann (German Federal Republic)

FORMER PRESIDENT: P. Swings (Belgium)

Members elected by the General Assembly, on the proposal of the Nominating Committee:

1. E.R. Mustel, L. Goldberg, M. Golay, D.H. Sadler, if W.N. Christiansen is appointed by the new Executive Committee.
2. L. Goldberg, B.Y. Mills, E.R. Mustel, D.H. Sadler, if L. Perek is appointed by the new Executive Committee.
3. H. Hirose, B.Y. Mills, E.R. Mustel, D.H. Sadler, if M. Schwarzschild is appointed by the new Executive Committee.
4. L. Goldberg, H. Hirose, B.Y. Mills, D.H. Sadler, if A.B. Severny is appointed by the new Executive Committee.

Note: During the General Assembly, through an error in the reading made by the General Secretary, a slightly different list has been approved by the General Assembly. A correspondence exchanged between the retiring General Secretary, Dr Pecker, and the delegates empowered to vote, led to an unanimous agreement on the above list, proposed by the Nominating Committee.

16. *New Members of the Union.* The General Secretary announced that, according to the Statutes, the Union is composed of Members approved, on the proposals of the Adhering Organizations or National Committees, by the Executive Committee on the basis of their achievement in some branch of Astronomy; and that, according to the By-laws, proposed changes in the list of Members are submitted by the Executive Committee to the Nominating Committee. He went on to say that the Executive Committee had acted according to these procedures and, on the advice of the Nominating Committee, had approved of the addition of 482 new Members and the deletion of 10 Members.

The names of the new Members had been prominently displayed and the General Secretary did not propose to read them. He informed the General Assembly that these names would be incorporated in the list of Members to be printed in Volume XIII B of the *Transactions.*

17. *Commissions.*

(*a*) The General Secretary announced that, as a result of proposal by Commission 43, the Executive Committee formally proposed to the General Assembly that:

Commission 43 be re-named 'Commission des Plasmas et de la Magnéto-Hydrodynamique en Astrophysique' (Commission on Astrophysical Plasmas and Magneto-Hydrodynamics).

This proposal was adopted by the General Assembly.

(*b*) The General Secretary, on behalf of the Executive Committee, submitted the following list of

Presidents and Vice-Presidents of Commissions for election by the General Assembly, subject to their willingness to serve:

No.	*President*	*Vice-President*	*No.*	*President*	*Vice-President*
4	G.A. Wilkins	J. Kovalevsky	26	P. Couteau	J. Dommanget
5	J.B. Sykes	K.F. Ogorodnikov	27	L. Detre	O.J. Eggen
6	F.L. Whipple	P. Simon	28	G.C. McVittie	E.M. Burbidge
7	W.J. Eckert	G.N. Dubošin	29	M.W. Feast	Y. Fujita
8	A.A. Nemiro	W. Fricke	30	D.S. Evans	H.A. Abt
9	J. Rösch	V.B. Nikonov	31	F. Zagar	G.M.R. Winkler
10	Z. Švestka	J.T. Jefferies	33	G. Contopoulos	S.W. McCuskey
12	M.N. Gnevyšev	R.G. Athay	34	D.E. Osterbrock	F.D. Kahn
14	M. Migeotte	A.H. Cook	35	A.G. Masevič	E. Schatzman
15	L. Biermann	V. Vanýsek	36	A.B. Underhill	R. Cayrel
16	J.S. Hall	B.J. Levin	37	M. Golay	G. Larsson-Leander
17	A. Dollfus	A.I. Lebedinskij †	38	M.G.J. Minnaert	A. Reiz
		G.H. Pettengill	40	J.P. Wild	D.S. Heeschen
19	P.J. Melchior	H.M. Smith	41	E. Rybka	O. Gingerich
20	G.A. Čebotarev	F.K. Edmondson	42	F.B. Wood	M. Plavec
21	M. Huruhata	H. Elsässer	43	L. Davis	R. Lüst
22	Z. Ceplecha	R.E. McCrosky	44	R. Wilson	V.K. Prokof'ev
23	W. Dieckvoss	P. Lacroute	45	Ch. Fehrenbach	B. Westerlund
24	W.J. Luyten	S. Vasilevskis	46	E.A. Müller	J. Kleczek
25	A.W.J. Cousins	D. Crawford			

This list was agreed by the General Assembly.

(*c*) The General Secretary explained that, according to the revised By-laws, the members of the Organizing Committees of the Commissions are appointed by the Commissions themselves, subject to the approval of the Executive Committee and the General Assembly. He accordingly submitted for general approval the list of Organizing Committees, which had been approved by the Executive Committee on the advice of the Nominating Committee. The Organizing Committees as approved by the General Assembly are given in the list of Commissions on pages 240–250.

18. *The place and date of the fourteenth General Assembly.* The Chairman called upon Professor H. Bondi to address the Assembly. Professor Bondi spoke as follows:

'I am speaking to you today as Chairman of the United Kingdom's National Committee for Astronomy. It is my pleasant duty to convey to you a joint invitation from the Royal Society and the University of Sussex to hold the fourteenth General Assembly of the Union in England at the University of Sussex. We would, indeed, be most happy if you were willing to accept the invitation to return to England after well over forty years. In these forty years there have been great advances in England in optical astronomy; the Isaac Newton telescope about to start work symbolizes our advances in optical astronomy; our work in radio astronomy from two great schools is well known to you all, and equally you know how much work in theoretical astronomy has been done in the United Kingdom. We therefore feel that we would be worthy hosts and it gives us particular pleasure to invite you to one of the most pleasant parts of the Country, to one of our new Universities, not far from the Royal Greenwich Observatory and within easy reach of London. I am sure you will find it a very good occasion and I can assure you of a very warm welcome, indeed. We shall all be delighted to see you in England.'

On the proposal of the President, the General Assembly unanimously agreed to accept this gracious invitation to meet in Sussex, near the city of Brighton, United Kingdom, in 1970, the precise dates being left for later decision by the Executive Committee. The President expressed the thanks of the Union to Professor Bondi.

19. *Election of the new Executive Committee.* The Chairman formally proposed that Professor Otto Heckmann be elected the new President of the Union. This was approved by acclamation.

The President reminded the Assembly of the names submitted for consideration at the first session, and formally proposed that Dr V. Bappu, Professor L. Gratton, and Dr J. Sahade be elected Vice-Presidents, in place of Professor W. Fricke, Professor Y. Hagihara, and Dr G. Haro, and that Professor W. N. Christiansen remain in office for a further term of three years. This was unanimously approved. He then formally proposed that Dr. L. Perek should be elected General Secretary in place of Professor J.-C. Pecker who now retired; this also was approved unanimously. He finally proposed that Professor C. de Jager be elected Assistant General Secretary in place of Dr L. Perek. The General Assembly expressed its approval of this appointment.

The Chairman then invited Professor Heckmann, Dr Bappu, Professor Gratton, Dr Sahade, and Professor de Jager to take seats on the platform.

20. *Addresses by the retiring and the newly-elected Presidents.* The retiring President, Professor P. Swings, addressed the Assembly as follows:

'Nous approchons de la fin de la partie administrative de cette treizième Assemblée Générale. Mais, avant d'inviter mon successeur, le Professeur Heckmann, à s'adresser à vous, je me permets de prendre une dernière fois la parole pour remplir quelques tâches agréables.

'L'entente au sein du Comité Exécutif que j'ai eu l'honneur et le privilège de présider a toujours été parfaite. Tous les Vice-Présidents et Conseillers ont rempli leur tâche avec un dévouement auquel je tiens à rendre hommage. Many valuable suggestions have been made by them in the interest of our Science. If our Union is one of the most active and flourishing it is due to a great extent to the harmony which has characterized our Executive Committee since 1920. I have belonged to three Executive Committees, for a total of about 20% of the life of the IAU, and I have always been impressed by the spirit of courtesy, devotion and concord which prevailed, even when we did not agree on specific points, which was very rare indeed. I have absolute confidence in the new Executive Committee, and especially in our newly elected President and General Secretary. Actually both have already been in close touch with the problems of the IAU. Professor Heckmann as Vice-President, Dr Perek as Assistant General Secretary. I also express the sincere thanks of the Union to the members of the Executive Committee who reach the end of their term: Professor Hagihara, Professor Haro and Professor Fricke.

'I also wish to express to the Special Nominating Committee my cordial thanks for their efficient help. We have faced together a few delicate problems, but have always found ways to solve them satisfactorily; at least we hope to have reached this goal.

'Dans un de ses discours, notre regretté ancien Président André Danjon avait défini le rôle du président comme suit: "Le rôle du président est celui du pilote vigilant qui cherche à éviter les écueils". J'ai essayé de remplir cette tâche. Mais si les problèmes journaliers de l'Union ont pu être résolus, nous le devons avant tout au Secrétaire Général, Dr J.-C. Pecker, au Secrétaire Général Adjoint, Dr L. Perek, et à leurs associés, particulièrement Mlle Dorothy Bell et Mr. Jappel à Nice, et Mlle Geneviève Drouin à Meudon. Vous avez sûrement tous conscience des lourdes tâches remplies par MM. Pecker et Perek. Mais je crois être autorisé à déclarer que c'est moi qui ai pu le mieux apprécier le travail énorme fourni par nos deux collègues, car ils ont bien voulu m'envoyer, régulièrement, copie de toute leur correspondance. Pour préciser les sommes de travail, je vous signale que la numérotation des lettres du Secrétaire Général a dépassé 3300, c'est-à-dire une moyenne de près de 100 par mois. De nombreuses lettres exigeaient beaucoup de réflexion, d'étude ou même de recherches. Beaucoup comprenaient plusieurs pages. Quant au Dr Perek, la numérotation de ses lettres a atteint le millier. Ce fut un grand privilège pour l'Union que ses affaires soient, pendant ce triennat, entre les mains de deux éminents astronomes, ayant, en même temps, un sens inné de l'organisation et de l'administration. Dr Pecker a été un parfait premier ministre et Dr Perek, un excellent vice-premier ministre de l'Union. En cela ils ont parfaitement suivi la tradition de tous leurs prédécesseurs, Fowler, Stratton, Oort, Strömgren, Oosterhoff et Sadler. Among these Mr. Sadler has continued to render us very great services as an experienced and devoted advisor, and I wish to thank him wholeheartedly. I ask all our colleagues in the audience to express their gratitude and friendship to Drs Pecker and Perek by their applause.

'This thirteenth General Assembly has been a tremendous success. The excellency of the scientific organization is essentially due to Drs Pecker and Perek. But the success of the practical arrangements is due to a considerable extent to our hosts, and especially to the Local Organizing Committee. Our thanks go especially to the President, our dear Colleague, Professor Šternberk, and to the Secretary our excellent friend, Dipl. Ing. Rajský. Drs Bumba, Kohoutek, Plavec and Vanýsek have collaborated greatly to the organizational work. But nobody will, I hope, criticize me if I express very special thanks to Mr. Rajský, who for three years has laboured strenuously for the success of the thirteenth General Assembly, neglecting no detail and accepting a lot of duties.

'I ask everybody in this audience to express their thanks by their applause.

'In a few minutes our colleague David S. Evans will give a comprehensive vote of thanks. I leave it to him to continue to express our gratitude to the numerous personalities who have contributed to the success of our Congress.

'Au moment de quitter la présidence de notre chère Union, j'ai plutôt le coeur gros. Je vous remercie une fois encore de l'honneur et de la confiance que vous m'avez témoignés. J'ai éprouvé de grandes satisfactions, en servant l'Union de mon mieux pendant ces trois années, au cours desquelles j'ai, d'ailleurs, appris maintes choses intéressantes. Je devrai m'habituer à ne plus recevoir, chaque mois, les copies de plus de cent lettres venant de Nice ou de Prague, dont une proportion considérable concernait des problèmes importants ou utiles.

'I wish to thank all my colleagues to have given me this opportunity to serve, for three years, the Science to which we are all so strongly bound.'

The newly-elected President, Professor Heckmann, then addressed the Assembly as follows:

'Dear Friends and Colleagues, Ladies and Gentlemen!

'When, some time ago, President Pol Swings told me that the pertinent committees of the IAU wanted to nominate me as his successor, I felt great hesitation to accept. But my reasons were deemed irrelevant.

'All preceding Presidents have expressed their deep gratitude to the Assembly for the great honour you bestowed on them. I want you to know that I have the same feeling of gratitude. I am glad, however, that my gratefulness may rest on two objective reasons: Pol Swings gave two arguments for the choice. First, he said, one wishes to express international esteem for the astronomical work done in my country. I am sure, this is meant and will be understood as a strong encouragement. The second reason Swings brought forward for the choice is my connection with the European Southern Observatory which in itself is an international enterprise, supported by France, Belgium, Holland, Denmark, Sweden and West Germany. I hope that the prestige of the Union will help to strengthen in the minds of our governments their appreciation of the scientific merit of this cooperative undertaking.

'Mon cher ami Pol Swings: Devenir votre successeur est un honneur que je ne pourrais pas exprimer de façon adéquate en quelques mots. Votre oeuvre scientifique, durant tant d'années, est en elle-même un grand monument aux fondations solides. La célèbre série de colloques d'astrophysique de Liège, dont vous êtes le père, est la preuve de votre intérêt universel et de vos efforts constants en faveur de la collaboration scientifique internationale. Vous m'avez promis votre aide et vos conseils, quand j'en aurai besoin; aussi, je me sens tout à fait rassuré contre les erreurs que pourrait commettre un président. Durant les années qui viennent, j'aurai toujours devant les yeux la façon magistrale dont vous avez dirigé l'orchestre de l'Union Astronomique Internationale.

'I feel happy and I feel even blessed having the support of two outstanding General Secretaries. The new Executive Committee will, I am sure, work in the same spirit of courtesy, devotion, and concord, which Pol Swings found in all of the Executive Committees to which he belonged.

'The IAU is growing incessantly. Naturally this growth gives concern to all of us. But such growth is spontaneous and therefore inescapable. Observatories are being enlarged, new instruments are under construction, new observatories are being founded, new techniques are being developed, and finally, all of us aim at opening up new fields of research year after year. Astronomy, therefore, is growing in all its dimensions—and our IAU grows with it. I shall not venture to delineate the asymptotic development of astronomy during centuries to come.

'It is my fervent hope, however, that in the foreseeable future the Union will not disintegrate; we must guard against fragmentation. It may be necessary to vest in the separate Commissions of the IAU an ever increasing responsibility to chart their own course. Nevertheless I refuse to believe, that a ship breaking up is the appropriate metaphor. Rather I would think of the Union as a host of sailing boats all flying the pennants of the same club. The club is their safeguard against the manifold dangers of isolation.

'This picture is intended to convey to you some of the thoughts evoked by the trust you have placed in me. With your support I can face with optimism our joint enterprise of insuring the growth and the health of the International Astronomical Union.'

21. *Closing ceremonies.* The Chairman called upon Dr D.S. Evans to propose a comprehensive vote of thanks, in the course of which he said:

'Monsieur le Président, Mesdames, Messieurs, Mes Chers Collègues,

'Partout dans le monde il y a des histoires, des faits, qui sont très bien connus, sauf dans les pays qu'ils concernent. Par exemple, tout le monde, sauf les Africains, sait bien que les autruches cachent la tête dans les sables en pensant qu'ainsi tout le corps est caché aux yeux de leurs ennemis. Tout le monde, sauf les Hollandais, connaît bien cette histoire du petit bonhomme, le gosse qui avait sauvé son pays d'une inondation en insérant le doigt dans la digue crevée.

'Et, de même, tout le monde, sauf sans doute les habitants de Tchécoslovaquie, connaît bien la vie de Bohème! Très mal dirigée à mon égard: Mimi dans sa mansarde, mourant de tuberculose des poumons, était quand même capable, elle, de remplir sans micro tous les vastes espaces d'une salle aussi énorme que celle-ci!... C'est une blague! et après deux semaines de la vraie vie de Bohème, nous la connaissons. Sans aucun doute, en fait, Mimi a été guérie, s'est mariée, et est partie en vacances dans la montagne avec grand-père, les enfants et les petits-enfants.

'Nous autres étrangers qui avons vécu la vraie vie de Bohème connaissons la vérité. Pendant deux semaines nous sommes reçus en princes. Chez vous, nous étions chez nous. Nous aussi sommes fiers de vos forêts, de vos montagnes, de vos lacs, de l'ancienneté de vos beaux monuments. Nous partageons avec vous votre satisfaction en regardant vos industries, vos universités, vos instituts de recherche, et, surtout, nous vous adressons nos félicitations pour les grands téléscopes nouveau-nés. Nous ressentons avec vous la même amitié non seulement envers vos savants, et académiciens, et grands administrateurs, mais également envers tous les habitants de ce beau pays que nous avons rencontrés—les chauffeurs de taxi, les conducteurs d'autobus, même le flic au coin de la rue et les jolies demoiselles en mini-jupe.

'Ma tâche est de vous remercier au nom de tous les délégués et hôtes. Il est impossible de rendre en détail cette dette énorme des individus aux individus et organisations, et de notre organisation aux individus et organisations tchèques. En quelques syllabes je dois canaliser une rivière de remerciements plus grande que la Vltava. Le rapport entre la réalité et ce que je peux exprimer par mes pauvres mots est plus grand que le rapport entre l'énergie d'un quasars et la petite tache qui marque sa position sur une plaque photographique.

'No congress as large as this one, where merely to greet everyone would take two full days' work, and as successful as this comes to pass without many hours, days and months of devoted and imaginative labour. Long before the first delegate thought of booking his ticket, our hosts were imagining what the General Assembly would look like—how many people to be housed and fed, how many meeting rooms would be needed; what to do about entertainment, postal services, medical services and a thousand other things. This has been a massive labour.

'I want you to know that we understand what this has implied and that we thank you most warmly and heartily. We thank especially the Government of the Socialist Republic of Czechoslovakia, the authorities of the Karlova Universita, the Czechoslovak Academy of Sciences, the Local Organizing Committee and all the ladies and gentlemen who have staffed the various bureaux.

'These are all important organizations; but organizations are necessarily composed of human beings who supply imagination, but who also can suffer from fatigue and headaches. We offer our sympathy and thanks to all the individuals in those organizations who must have suffered innumerable headaches in anticipating our every need.

'Let me mention especially the Acting Prime Minister, Mr Krajčír, who graciously spared time in the superhuman schedule of a highly placed statesman to bid us welcome at the Inaugural Ceremony.

'Our scientific hosts are the Czechoslovak Academy of Sciences and on your behalf I thank especially Academician Šorm, President of the Academy, and his staff. I hope the scientific level of our meeting has been sufficiently high to contribute yet a little more lustre to that august body.

'The president of the Local Organizing Committee is Dr Šternberk, already well-known to us for his many personal and scientific qualities. Now our thanks to him are a testimonial to his outstanding abilities as an organizer.

'A powerful influence for the smooth running of our meeting has been the effective personality of the secretary of that Committee, Mr Rajský, and his devoted staff. They all deserve our enthusiastic gratitude.

'If I do not mention by name any of the ladies who have answered our questions and solved our problems, this is because they are rather numerous, but every single one of them perfectly charming and most deserving of our thanks.

'I am here to thank our Czechoslovak hosts on your behalf and not to congratulate our own excellent officers and secretariat, but I make one exception. We are proud to have as our new General Secretary an eminent Czechoslovak astronomer, Dr Luboš Perek. It is an excellent augury that he can still smile after these gruelling days of work, following the tradition of his predecessor who smiled his way through everything. In his character as a citizen of our host nation I thank Dr Perek for his notable services. I could perhaps try to express our thanks in several of the native languages spoken by our delegates—but the excellent simultaneous translation service would spoil my fun and translate it all back again instantaneously.

'In the end what I have to say reduces to a single word spoken with all the sincerity and earnestness I can muster—and I hope the right pronunciation—Děkuji!'

On behalf of the ladies, Madame André Couder addressed the Assembly as follows:

'Monsieur le Président, Mesdames, Messieurs,

'C'est un grand honneur, mais aussi pour moi un grand plaisir, que de parler au nom des nombreuses femmes d'astronomes qui sont ici, et d'exprimer la gratitude que nous devons à nos hôtes.

'Ce n'est pas le fait que nous venions des quatre coins du monde qui rend la tâche difficile. Au contraire, je crois voir que les femmes des astronomes mènent, sous toutes les longitudes, des vies très analogues, et voici pourquoi, sans doute, nous avons tant de réactions communes, tant de goûts semblables. C'est donc un sentiment très unanime que nous avons éprouvé, et que je souhaite exprimer.

'Les observatoires les plus importants et les plus récents s'installent loin des lumières des villes en des lieux élevés, mal accessibles, de préférence désertiques, souvent enneigés. Bref, on y voit seulement le ciel.

'A l'opposé, vos Assemblées Générales se tiennent en de grandes villes, que la vie sociale, l'art et la science ont illustrées depuis des siècles. Ces prestiges terrestres ont désigné en premier lieu Rome. Nous venons de trouver l'éclat d'une antique culture associée aux formes les plus actives de la vie d'aujourd'hui. Nous avions rêvé de Prague, en feuilletant des images. La réalité est plus riche encore et elle nous a ravies.

'Dans cette ville magnifique nous étaient réservés par nos amies tchèques un accueil et une hospitalité pleine de chaleur. Le comité d'organisation avait préparé pour nous un programme remarquable. Nous avons vu les palais, les églises, les musées et leurs trésors artistiques. Des excursions nous ont fait découvrir au coeur des forêts les fiers châteaux, témoins d'un passé glorieux. Nous aimions déjà la musique si profondément originale de vos compositeurs; l'avoir entendue ici restera pour nous un souvenir inoubliable. Pas davantage nous n'oublierons les spectacles qui nous ont été présentés. Les tchécoslovaques ont acquis une réputation universelle pour ces fêtes de la vue, de la couleur, du mouvement qu'ils savent offrir, où s'expriment des sentiments à la fois populaires et raffinés. Vous n'avez rien négligé, pas même ce goût très ancien de la tentation, que les femmes vont satisfaire aujourd'hui, non dans un jardin célèbre, mais dans les magasins et les maisons de couture. Là, et dans les rues animées, nous nous sommes mêlées à la foule, ce qui est un grand plaisir des voyages.

'Notre gratitude et notre admiration vont à tous ceux et à toutes celles qui, par un travail long et difficile, ont organisé les réunions et les excursions si agréables et si intéressantes. Vraiment nous avons été comblées. Je ne puis exprimer nominativement notre gratitude à chacun et à chacune des membres du comité d'organisation; cependant je dois remercier particulièrement les personnalités qui en ont été les animateurs: Monsieur le Président du Comité, Dr Šternberk, le Secrétaire, Monsieur Rajský, Madame Perková, présidente du Comité des dames, et toutes les participantes de ce comité.

'Si les constructeurs de la tour de Babel avaient bénéficié de l'assistance d'interprètes aussi habiles et aussi aimables que celles que nous avons trouvées ici, leur oeuvre aurait connu un succès complet, comme ce congrès! Dans l'expression de notre reconnaissance, il faut leur joindre les dames secrétaires et les remercier de leur complaisance inlassable.

'Les femmes des astronomes ont parmi d'autres privilèges celui-ci: se connaître des amies sur tous les continents. Les amitiés anciennes retrouvées, les amitiés nouvelles que nous avons nouées ici ajoutent un prix infini au souvenir que nous emportons.

'Un court séjour à Prague ne permet pas de connaître les finesses de cette langue riche et complexe qu'est le tchèque. Nous en avons appris quelques mots seulement, par l'usage. Nous avons dû constamment employer celui qui signifie: merci.

'Je veux le dire ici, en tchèque, gauchement peut-être, mais de tout coeur: Děkuji!'

The Chairman then called upon the retiring General Secretary, Professor Pecker, who spoke as follows:

'A ces paroles de remerciements auxquelles rien n'est plus à ajouter, et auxquelles je souscris si complètement et si sincèrement, au milieu une fois encore de cette chaude amitié de nos collègues tchécoslovaques, je voudrais ajouter quelques mots, quelques remerciements, ceux d'un Secrétaire Général qui n'a pas toujours eu le temps de voir fonctionner la fascinante lanterne magique que fut cette Assemblée, mais qui a vu néanmoins derrière l'écran, à l'ombre des praticables, voire à travers l'excellent "Dissertatio cum Nuncio Sidereo", Seconde Série, de M.B. Bílek, tous ceux qui, au sein de l'Union Astronomique Internationale, ont eux aussi, fait de cette Assemblée le grand succès que nous savons.

'Comme Secrétaire Général, j'avais hérité de Donald H. Sadler la meilleure des secrétaires adjointes, Dorothy Bell. She has put all my files in order, and you could not imagine what accomplishment it was. Not only that, but her charm, her efficiency were without comparison. But it is difficult to be Miss IAU for life; and now, she has become Mrs Dorothy Brandt, who left the IAU fifteen months ago, but not so far as not to marry within the IAU! To my best congratulations (and note that I speak English not for her sake—her French slang is perfect!—but for that of her husband) to both Mr and Mrs Brandt, let me add my very deep thanks for the help I got from you, Dorothy, a few years ago,—and still quite recently during the meetings.

'Je me sens incapable de continuer en tchèque, ou en slovaque. Mais en français, seconde langue d'Arnošt Jappel, je peux lui dire que je me trompais tout à l'heure. Certes, il n'a pas apporté dans mon bureau le charme féminin de Miss IAU! Mais je peux aisément vous convaincre, je pense, que Monsieur UAI a mis à continuer à faire marcher nos affaires une efficacité complète. Sa maîtrise de toutes les langues, son expérience juridique et financière nous sont maintenant irremplaçables. En une période difficile pour l'UAI, cette transition s'est accomplie sans trop de heurts, grâce à la compréhension, à la compétence des uns et des autres, à leur sens de l'humour, indispensable qualité, grâce aussi à la continuité représentée dans le bureau de l'UAI par le sourire, la patience, l'activité de Christiane Caseneuve.

'Au bout du téléphone, en haut d'une tour, était aussi installé, pendant ces trois années, quelqu'un qui, silencieusement, modestement, sans désemparer, a édité plus de 6000 pages de textes et vérifié plusieurs dizaines de milliers de références bibliographiques. Elle a servi de plaque tournante entre les auteurs, les éditeurs, les imprimeurs. Vis-à-vis de Geneviève Drouin, l'Union a aussi une lourde dette, et je la remercie de toute mon amitié.

'Et je ne parle pas de toutes celles qui à Liège, à Paris, à Prague, à Meudon, à Nice, sans consacrer tout leur temps à l'UAI, ont néanmoins donné, chaque fois que c'était nécessaire, le coup de main indispensable.

'A tous mes adjoints, qui sont devenus mes amis, je vous demande maintenant d'exprimer, avec moi, toute notre gratitude.

'Laissez-moi encore ajouter un mot, très bref, puisque, pour la dernière fois, j'ai encore la parole. Ces trois années m'ont appris bien des choses, mais par-dessus tout, elles m'ont apporté des amitiés, dans l'UAI, plus encore peut-être au sein du Comité Exécutif, et tout particulièrement une amitié qui m'a été si précieuse, surtout aux heures de découragement, l'amitié de notre Président sortant. Qu'il me permette, très simplement, de l'en remercier, et d'associer Madame Swings à ces remerciements et à cette amitié.'

The Chairman then invited the new-elected General Secretary, Dr L. Perek, to pronounce a few words, as follows:

'As one to carry on in the fine art of conducting the business of the IAU I ask your permission to add a few words. I discovered behind my friend, Jean-Claude Pecker, a most considerate General Secretary whose decisions were taken with a view to the period to follow his term of office. For instance, when Dorothy Bell, now Mrs. Brandt, left, Jean-Claude Pecker offered the post of the Assistant Secretary to my countryman, Arnošt Jappel, who would be in this post also after the secretariat would have moved to Prague. Jean-Claude spared no effort to initiate me into the routine of the IAU office which certainly will add to the smooth transition from Nice to Prague.

'We had with Jean-Claude Pecker many extremely interesting discussions which raised some, as yet unsolved, theoretical problems. E.g. the problem of forty Commissions wishing to meet at the same time but not simultaneously. Or the problem of finding a system of coordinates where the parabola of the expenses would not catch up with the straight line of the receipts.

'I had a most agreeable time being in frequent contact, at least by correspondence, with so many distinguished Members of the Union, and I am looking forward to the future, fully confident that pleasant aspects will prevail and prepared to work hard towards this end.'

The President then closed the meeting around noon with the following words:

'Let us all plan to attend the fourteenth General Assembly in the United Kingdom.

'Je vous souhaite à tous un bon voyage de retour si vous rentrez immédiatement, une fructueuse participation aux symposiums si vous partez vers les merveilleuses montagnes des Tatry.

'I hereby declare the thirteenth General Assembly of the IAU closed.'

PART 2

RESOLUTIONS ADOPTEES
PAR L'ASSEMBLEE GENERALE

RESOLUTIONS ADOPTED
BY THE GENERAL ASSEMBLY

RESOLUTIONS ADOPTEES

PAR L'ASSEMBLEE GENERALE

SOMMAIRE DES DÉCISIONS CONTENUES DANS LE RAPPORT DE L'ASSEMBLÉE GÉNÉRALE

1. *Part Contributive Unitaire.* La part contributive unitaire a été fixée pour 1968, 1969, 1970, à 900 francs-or, conformément à l'article 11(*b*) des Statuts.

2. *Commissions.* La décision suivante a été prise (voir page 26):

La Commission 43 prend le nom de: "Plasmas et Magnéto-Hydrodynamique en Astrophysique".

3. *Résolutions proposées dans les délais statutaires par les Comités Nationaux d'Astronomie.* Trois telles résolutions figurent au point 9 de l'Ordre du Jour de l'Assemblée Générale sous une forme résumée. La résolution A 1, proposée par la République Démocratique Allemande, concernait la création d'une nouvelle Commission sur l'Astrophysique Relativiste et la Cosmologie. Après discussion par les Commissions 28 et 40, le Comité National de la République Démocratique Allemande a décidé le retrait pur et simple de cette résolution, désignée par A 1. Deux résolutions, A 2 et A 3, proposées par le Comité National d'Astronomie de Belgique ont été soumises pour avis aux Commissions 5 et 19, respectivement. Elles ont été adoptées par l'Assemblée Générale après avis de ces Commissions, sous une forme non modifiée en ce qui concerne la résolution A 2 (la Commission 5 incluant dans les comptes rendus de ses réunions des commentaires à cette résolution), et sous une forme modifiée (mais n'altérant pas le contenu scientifique de la résolution) en ce qui concerne la résolution A 3.

4. *Résolutions proposées par les Commissions.* Un certain nombre de Commissions ont présenté avant l'Assemblée Générale (point 10 de l'Ordre du Jour), ou ont voté, au cours de leurs réunions tenues pendant l'Assemblée Générale, des résolutions de caractère financier ou d'ordre général. Les résolutions financières ont été incluses dans le budget lorsqu'il a été jugé possible d'assurer ce financement. Les raisons des réductions apportées à ces demandes des Commissions (ou des suppressions de certaines entreprises spécifiques des Commissions) sont exprimées dans le rapport financier (ci-dessus, page 22). En acceptant le budget proposé par le Comité des Finances, et l'ensemble du rapport du Comité des Finances, l'Assemblée Générale a exprimé son opinion sur l'ensemble de ces résolutions.

De plus, le Comité des Résolutions a jugé certaines résolutions d'ordre assez général pour être considérées par l'Assemblée Générale. Elles sont incluses dans les résolutions 1 à 8 données ci-après.

Une résolution issue de la Commission 37 concernait l'organisation générale des travaux de l'Union. Elle a été déférée, pour étude, au nouveau Comité Exécutif.

Enfin les autres résolutions votées par les Commissions ont été appuyées globalement par la résolution no. 9 ci-après. Ces résolutions n'ont pas été reproduites séparément et se trouvent dans les rapports des Commissions, dans la troisième partie de ce volume, pp. 45–229.

RÉSOLUTION A 2

Proposée par le Comité National Belge d'Astronomie

Sur l'édition de fiches bibliographiques

L'Union Astronomique Internationale propose que soit examinée la possibilitée, pour les éditeurs

RESOLUTIONS ADOPTED

BY THE GENERAL ASSEMBLY

SUMMARY OF DECISIONS CONTAINED IN THE REPORT OF THE GENERAL ASSEMBLY

1. *Unit of Contribution.* The amount of the unit of contribution has been fixed, for 1968, 1969, 1970, at a value of 900 gold francs, according to item 11(*b*) of the Statutes.

2. *Commissions.* The following decision has been taken (see page 27):

To re-name Commission 43 "Astrophysical Plasmas and Magneto-Hydrodynamics".

3. *Resolutions submitted within the statutory delays by National Committees of Astronomy.* Three such resolutions are summarized under item 9 of the Agenda of the General Assembly. Resolution A1, submitted by the German Democratic Republic, proposed the creation of a new Commission on Relativistic Astrophysics and Cosmology. After discussion in Commissions 28 and 40, the National Committee of the German Democratic Republic decided to withdraw, completely and simply, this resolution designated as A1. Two resolutions, A2 and A3, proposed by the National Committee for Astronomy of Belgium, were submitted for advice to Commissions 5 and 19 respectively. Upon recommendation of the Commissions in question, the General Assembly adopted these resolutions, either unchanged, as in the case of resolution A2 (but with a comment in the report of the meetings of Commission 5), or slightly changed, without, however, affecting the scientific content, as in the case of resolution A3.

4. *Resolutions proposed by Commissions.* A certain number of Commissions proposed, before the General Assembly (point 10 of the Agenda), or in the course of their meetings held during the General Assembly, resolutions of a financial character or of general nature. The financial resolutions were included into the budget in so far as the project proposed was considered possible. The reasons for reducing the amounts proposed by Commissions (or for suppressing certain specific projects of Commissions) are given in the finance report (hereabove, page 22). Accepting the budget proposed by the Finance Committee and the report of this Committee, the General Assembly expressed its opinion as to these resolutions.

Moreover, the Resolutions Committee considered certain resolutions of a sufficiently general character to be dealt with by the General Assembly. These resolutions are included in resolutions 1 through 8, as hereinafter.

One resolution, proposed by Commission 37, concerned the over-all organization of the activities of the Union. This proposal has been deferred to the consideration of the new Executive Committee.

Finally, all the other resolutions adopted by Commissions are globally supported by resolution no. 9, as hereinafter. These resolutions have not been reproduced separately and are included in the reports of Commissions, given in the third part of the present volume, pp. 45–229.

RESOLUTION A 2

Proposed by the Belgian National Committee for Astronomy

On the publication of bibliographical cards

The International Astronomical Union proposes that the producers of astronomical publications des publications astronomiques, de joindre à celles-ci des fiches de format normalisé donnant pour chaque article, en plus de la référence bibliographique, un très court résumé du sujet traité.

RÉSOLUTION A 3

Proposée par le Comité National Belge d'Astronomie

Sur les marées terrestres

Etant donné certains résultats obtenus dans le domaine des marées terrestres, l'Union Astronomique Internationale recommande que les études des nutations à courte période soient poursuivies et développées.

RÉSOLUTION NO. 1

Proposée par le Comité Exécutif

Sur le système métrique

L'Union Astronomique Internationale recommande l'utilisation générale par les astronomes du monde entier, dans toutes leurs publications, des unités des systèmes métriques, soit exclusivement, soit concurremment à l'usage d'unités d'autres systèmes, les deux valeurs étant alors données pour chaque grandeur.

RÉSOLUTION NO. 2

Proposée par le Comité Exécutif

Sur les satellites porteurs de systèmes réflecteurs

L'Union Astronomique Internationale a noté avec une sérieuse inquiétude que des études sur les systèmes réflecteurs portés par des satellites étaient entreprises. L'Union note avec soulagement l'assurance donnée par les autorités compétentes que le Gouvernement des Etats-Unis d'Amérique n'est pas intéressé actuellement à la conception de systèmes réflecteurs orbitants et qu'aucune activité n'est envisagée dans cette direction.

RÉSOLUTION NO. 3

Proposée par la Commission 41

Sur les instruments et documents d'intérêt historique

L'Union Astronomique Internationale rappelle à tous les astronomes sa demande de protéger de tout dommage ou destruction les instruments astronomiques d'intérêt historique; ces instruments sont en effet des documents importants pour l'histoire des sciences. Quand il est impossible de protéger sur place ces instruments, les directeurs d'observatoires et autres autorités responsables sont priés de faire leur possible pour que les instruments soient conservés dans des musées.

L'Union regrette le fait que les papiers personnels de quelques astronomes aient été détruits par des personnes ignorant leur valeur, et de ce fait invite instamment les particuliers et les observatoires à protéger et conserver de tels manuscrits et lettres.

RÉSOLUTION NO. 4

Proposée par la Commission 41

Sur la préparation d'une histoire internationale de l'astronomie

L'Union Astronomique Internationale reconnaît qu'il est important et utile de préparer une histoire internationale de l'astronomie à partir de documents originaux, soutient dans ce domaine examine the possibility of including cards of standard format giving, for each article, a very short summary of the subject matter in addition to the bibliographical reference.

RESOLUTION A 3

Proposed by the Belgian National Committee for Astronomy

On Earth tides

Taking note of certain results obtained in the field of Earth tides, the International Astronomical Union recommends that studies of short-period terms of nutation be undertaken and followed up.

RESOLUTION NO. 1

Proposed by the Executive Committee

On the metric system

The International Astronomical Union recommends the general use by astronomers all the world over, and in all their publications, of units of the metric system, either exclusively, or together with units of other systems, both values being then given for each quantity.

RESOLUTION NO. 2

Proposed by the Executive Committee

On satellite-borne reflector systems

The International Astronomical Union has noted with serious concern the beginning of studies of satellite-borne reflector systems. It notes with relief the assurance given by the competent authority that the Government of the United States of America is not interested in the concept of orbiting reflector systems at the present time and that no activity in such a direction is visualized.

RESOLUTION NO. 3

Proposed by Commission 41

On instruments and documents of historical interest

The International Astronomical Union reminds all astronomers of the request to save from damage or destruction astronomical instruments of historical interest; these are considered to be important documents in the history of science. Where it is not possible to preserve such instruments *in situ*, directors of observatories and others are requested to do everything possible to ensure that they are preserved in museums.

It laments the fact that the personal papers of some astronomers have been destroyed by those unacquainted with their value, and therefore urges individuals and observatories to protect and preserve such manuscripts and letters.

RESOLUTION NO. 4

Proposed by Commission 41

On the preparation of an international history of astronomy

The International Astronomical Union recognizes the importance and usefulness of preparing an international history of astronomy based on original materials, supports the initiative in this endeavl'initiative de la Commission 41 sur l'Histoire de l'Astronomie et sollicite l'aide des Comités Nationaux d'Histoire de l'Astronomie et des spécialistes d'histoire des sciences, ainsi que des astronomes, pour mener cette tâche à bien.

RÉSOLUTION NO. 5

Proposée par les Commissions 4 *et* 31

Sur la définition de la seconde

(*a*) L'Union Astronomique Internationale note avec satisfaction que le Comité Consultatif pour la Définition de la Seconde a adopté le 13 juillet 1967 la Recommandation no. S-1 sur la définition de la seconde qui doit devenir l'unité de base du Système International des Unités et que la définition S-1 reconnaît ainsi l'existence de la seconde de temps des éphémérides. L'Union Astronomique Internationale exprime son accord avec cette définition proposée (S-1) de la seconde.

(*b*) Il est entendu que la Conférence Générale des Poids et Mesures peut adopter une définition légèrement différente de celle de la Recommandation S-1. En ce cas, l'Union Astronomique Internationale demande que, dans la portion du texte adopté qui précisera que la seconde de temps des éphémérides ne fait pas partie du Système International des Unités, soit incluse la phrase "la seconde de temps des éphémérides, qui est partie intégrante du Système des Constantes Astronomiques de l'Union Astronomique Internationale".

RÉSOLUTION NO. 6

Proposée par les Commissions 4 *et* 31

Sur l'utilisation de phénomènes observationnels naturels dans la détermination du Temps des Ephémérides et du Temps Universel

L'Union Astronomique Internationale désire insister sur le fait que, nonobstant

(*a*) la proposition faite à la Conférence Générale des Poids et Mesures d'adopter une définition de la seconde comme unité de base dans le Système International des Unités, qui soit établie à partir d'une transition atomique, et

(*b*) la possibilité, par voie de conséquence, de l'établissement d'une échelle intégrée de Temps Atomique obtenue par l'addition continue de multiples de cette unité,

les mesures de temps nécessaires à l'astronomie et aux sciences associées doivent continuer à être poursuivies à partir de phénomènes observationnels naturels, tels que ceux qui permettent la définition du Temps des Ephémérides (c'est-à-dire le mouvement orbital des objets du système solaire) et le Temps Universel (c'est-à-dire la rotation de la Terre).

RÉSOLUTION NO. 7

Proposée par le Comité Exécutif, sur avis du Groupe de Travail sur les Matériaux Photographiques

Considérant que la présente multiplicité des dimensions des plaques utilisées par les astronomes est extrêmement peu économique, et

Reconnaissant que les investissements dans des équipements difficiles à modifier sont trop importants pour que l'introduction d'une standardisation complète soit possible,

Mais désireuse d'éviter toute détérioration ultérieure de la situation actuelle,

L'Union Astronomique Internationale recommande

que, dans la mesure où cela est possible au point de vue de l'instrumentation, les astronomes et les ingénieurs chargés d'étudier les instruments futurs fassent des plans ou modifient les plans actuels de façon à utiliser seulement les dimensions de plaques dont la liste est donnée ci-après, ou des dimensions dérivant de celles-ci par un simple découpage.

our of Commission 41 on the History of Astronomy and requests the assistance of National Committees for the History of Astronomy and of specialists in the history of science, as well as astronomers, in carrying out this task.

RESOLUTION NO. 5

Proposed by Commissions 4 *and* 31

On the definition of the second

(*a*) The International Astronomical Union notes with satisfaction that the Consultative Committee for the Definition of the Second adopted on 13 July 1967 the Recommendation no. S-1 for the definition of the second which is to be the basic unit in the International System of Units, and that the definition S-1 recognizes the existence also of the second of ephemeris time. The International Astronomical Union concurs with this proposed definition S-1 of the second.

(*b*) It is understood that the General Conference of Weights and Measures may adopt a definition slightly different from that in Recommendation S-1. In this case the International Astronomical Union requests that, in the portion which states that the ephemeris second is not part of the International System of Units, there be included the phrase "the ephemeris second, which is part of the IAU System of Astronomical Constants".

RESOLUTION NO. 6

Proposed by Commissions 4 *and* 31

On the use of natural observational phenomena in determining the Ephemeris Time and the Universal Time

The International Astronomical Union wishes to emphasize that, notwithstanding

(*a*) the proposal before the General Conference of Weights and Measures to adopt the definition of the second as a basic unit in the International System of Units, in terms of an atomic transition, and

(*b*) the consequential possibility of setting up an integrated scale of Atomic Clock Time obtained by the continuous addition of multiples of this unit,

measures of time for the purpose of astronomy and associated sciences must continue to be based on natural observational phenomena such as give rise to Ephemeris Time (based on the orbital motions of bodies in the solar system) and Universal Time (based on the rotation of the Earth).

RESOLUTION NO. 7

of the Executive Committee, on the proposal of the Working Group on Photographic Materials

Whereas the present multiplicity of plate sizes used by astronomers is extremely uneconomic, and

Recognizing that the existing investment in equipment not easily modified is too large to introduce a thoroughgoing standardization,

Being desirous of preventing further deterioration of the situation in future,

The International Astronomical Union recommends,

that, in so far as may be instrumentally feasible, astronomers and instrument designers plan future instruments and modify existing ones so as to use only the plate dimensions listed below or dimensions derivable from these by cutting.

Dimensions en centimètres:

16 × 16, 18 × 13, 9 × 12, 24 × 24

Dimensions en inches:

$3\frac{1}{4} \times 4\frac{1}{4}$,	5 × 7,	8 × 10,	4 × 10
(8.3 × 10.8 cm)	(12.7 × 17.8 cm)	(20.3 × 25.4 cm)	(10.2 × 25.4 cm)

Il est reconnu que dans des circonstances spéciales, il sera nécessaire de commander des plaques de dimensions différentes, telles que celles qui sont utilisées par certains astrographes et télescopes de Schmidt, et telles que les dimensions inhabituelles des longues plaques utilisées en spectroscopie. La résolution implique seulement une restriction volontaire dans la variété des exigences relatives à des plaques de dimensions moyennes, et l'on espère ainsi obtenir des économies significatives.

RÉSOLUTION NO. 8

Proposée par la Commission 17

Sur la nomenclature lunaire

L'attribution de noms et de désignations permanents aux détails topographiques du côté de la Lune opposé à la Terre sera repoussée jusqu'à la quatorzième Assemblée Générale de l'UAI. A titre de mesure provisoire, un Groupe de Travail attribuera des numéros à environ 500 formations lunaires importantes.

RÉSOLUTION NO. 9

Sur les Résolutions adoptées par les Commissions

Prenant en considération l'impossibilité pratique d'accorder à chaque résolution adoptée par chacune de ses 38 Commissions une attention particulière, et affirmant la confiance complète qu'elle a en ses Commissions,

cette Assemblée Générale *désire exprimer son approbation* des résolutions adoptées par ses différentes Commissions, et *recommande* que les astronomes appliquent ces Résolutions dans toute la mesure du possible.

Centimetre sizes

16 × 16, 18 × 13, 9 × 12, 24 × 24

Inch sizes

$3\frac{1}{4}$ × $4\frac{1}{4}$,	5 × 7,	8 × 10,	4 × 10
(8.3 × 10.8 cm)	(12.7 × 17.8 cm)	(20.3 × 25.4 cm)	(10.2 × 25.4 cm)

It is recognized that in special circumstances there will be a need to order other plate sizes such as large ones for use in some astrographs and Schmidt telescopes and unusual sizes of long plates for spectroscopy. The resolutions implies only a voluntary restriction of demand variety in the medium size-range which it is hoped should lead to significant economies.

RESOLUTION NO. 8

Proposed by Commission 17

On the lunar nomenclature

The assignment of names and permanent designations to features on the far side of the Moon will be postponed until the fourteenth General Assembly. As an interim measure, a Working Group will assign numbers to about 500 major lunar formations.

RESOLUTION NO. 9

On the Resolutions adopted by the Commissions

Considering the impracticability of giving individual attention to every resolution adopted by each of its 38 Commissions, and having full confidence in its Commissions,

this General Assembly *wishes to give its endorsement* to the Resolutions adopted by its individual Commissions, and *recommends* that astronomers give effect to these Resolutions in so far as they are able.

PART 3

COMPTES RENDUS DES SEANCES DES COMMISSIONS

REPORTS OF MEETINGS OF COMMISSIONS

4. COMMISSION DES EPHEMERIDES

Compte-rendu de séance, 23 Août 1967

PRESIDENT: G. M. Clemence.
SECRETAIRE: J. Kovalevsky.

M. le Président ouvre la séance en rendant un dernier hommage aux membres de la commission décédés depuis le dernier congrès de l'U.A.I.

Prof. A. Danjon, ancien directeur de l'Observatoire de Paris, Membre du Bureau des Longitudes chargé de la Connaissance des Temps.

Dr. F. Fernandez de la Puente, Directeur du service des éphémérides de l'Instituto y Observatorio de Marina de San Fernando.

Dr. A. Tsukamoto, ancien directeur du service des Ephémérides du Japon (Hydrographic Division, Maritime Safety Board).

I. *Organisation de la Commission*

En plus des membres décédés, Dr. V. Planelles qui quitte le service des éphémérides espagnoles et Mme Gossner ne feront plus partie de la commission.

Par contre, les personnes suivantes sont proposées comme nouveaux membres: V. A. Abalakin, J. Arias de Greiff, R. Haupt, N. Owaki, M. Rodriguez, S. R. Trehan.

Enfin, Dr Freiesleben (Allemagne) est proposé comme membre consultant de la commission. En ce qui concerne la future organisation de la commission, il est proposé que le bureau soit constitué comme suit:

Président: G. A. Wilkins
Vice-président: J. Kovalevsky
Comité d'Organisation: G. A. Čebotarev, G. M. Clemence, R. L. Duncombe, W. Fricke.

La commission approuve ces propositions.

II. *Introduction of the IAU System of Astronomical Constants*

W. Fricke reported on the introduction of the new constants of aberration in the apparent places of fundamental stars starting from the volume for 1968.

D. H. Sadler supplemented the published report in respect of the content of the Astronomical Ephemeris (A.E.) for the years after 1968.

All appreciable changes will be introduced into the eclipses, the physical ephemerides of the Sun, Moon and planets and the ephemerides of the satellites, as from the *Astronomical Ephemeris* 1969. As from the A.E. 1970, it is proposed to incorporate the corrections to the lunar ephemerides in the tabulated values of ΔT.

The ephemerides of the Sun, Moon and planets will not be changed. Differential corrections to the lunar ephemeris together with the formulae from which differential corrections to the planetary ephemerides can be calculated will continue to be given in an appendix.

D. H. Sadler proposed that this procedure should be discontinued for the lunar ephemeris when there could be included W.J. Eckert's recalculation of solar perturbations. These would be therefore tabulated, without tables of corrections, in the A.E. 1972 and onwards.

However, since the corrections to the ephemerides of the Sun and planets from Mercury to Mars are smaller than the rounding errors, *D. H. Sadler* proposed that they continue to be given by means of formulae even after 1971. Not only are these corrections small, but they cannot, in any case, readily be computed for the change in the adopted value of the ratio of the masses of the Earth and Moon.

J. Kovalevsky supported this proposal and indicated that this problem arises even more acutely for Le Verrier's theories tabulated in the *Connaissance des Temps*, and asked that this publication be authorised to continue to publish the ephemerides in the present form with the eventual addition of correction formulae.

The Commission agreed with these two proposals. On the other hand, for the ephemerides of the five outer planets being based on a smooth and consistent numerical integration, it was agreed that it would be worth while recalculating the apparent ephemerides with the revised values of the constants, including the revised value of the mass-ratio Earth/Moon (the change in the value of the combined mass of the Earth and Moon has negligible effect).

III. *Further Improvement of the System of Astronomical Constants*

W. Fricke recalled the achievements of the working group on astronomical constants in 1963–64 that proposed the subsequently adopted IAU system of astronomical constants. The task of this working group is finished and the working group was dissolved. However, though the new constants are satisfactory, some others have not been changed (constant of precession and planetary masses). Consequently, *Fricke* suggested that a group should be formed in order to study how to resume the discussions and prepare a procedure for an eventual change in those constants.

G.M. Clemence noted that the system of planetary masses is not satisfactory. Some remarkable progress was recently made for the masses of the nearest planets from discussions of the motion of space probes. As for the outer planets, nothing new is to be expected in the immediate future. The best prospects are now in the study of the motions of some minor planets in commensurable orbits with that of Jupiter.

After some discussion, it appeared that the most adequate group would be the organizing committee of the Commission, provided that some members of other commissions might be consulted. It was decided that this group should be chaired by G. Wilkins.

IV. *Epoch of Ephemeris Time*

D.H. Sadler commented on the effects produced on the definition of the epoch of ephemeris time by the change in the adopted value of the constant of aberration χ. He proposed that in spite of this fact, no discontinuity should be introduced in the geometrical ephemerides of the Sun and planets and in the lunar ephemeris. This implies that the epoch of Ephemeris time should not be changed. A discontinuity in the apparent ephemerides of the Sun and planets will therefore arise but no effect will be introduced into the definition and actual determination of E.T. The actual effect on the theory of the motion of the Earth that would arise if the observations were reduced with $\chi = 20''.496$ instead of $20''.47$ is not straightforward and it is therefore better not to try to introduce this change.

D.H. Sadler also proposed that the observed Ephemeris Time as deduced from observations, should be distinguished by a reference number according to the particular ephemeris that is used to reduce the lunar observations. This would provide a natural extension to the already existing E.T.O. as defined at the IAU meeting of 1961 in Berkeley. Already two such reference numbers may be assigned in the near future.

He then read the proposed resolution (see below) which was supported by *G.M. Clemence* who stated that it was the result of a considerable correspondence and discussion between himself and some of the directors of the National Ephemerides. It was agreed that this resolution should be discussed informally before the next meeting of the commission.

Séance du 28 Août 1967

IV. *Epoch of Ephemeris Time* (*suite*)

No further comment was made on the solution proposed by *D.H. Sadler* and which further was supported by *R. Duncombe* and *J. Kovalevsky*.

The Commission unanimously approved this resolution (10 votes in favour, no abstention) which reads as follows:

Commission 4 recommends:

(1) That, consequent on the change, from 20″.47 to 20″.496, in the adopted value of the constant of aberration, and in spite of any effect that this change may have on the epoch from which Ephemeris Time is measured according to its adopted definition:

(*a*) no discontinuity be introduced into the lunar ephemeris;

(*b*) no discontinuity be introduced into the geometric ephemerides of the Sun and planets;

(*c*) a discontinuity be introduced into the apparent ephemerides of the Sun and planets, but the compilers of the international ephemerides be given discretion to decide whether these changes, arising solely from the changed aberration corrections, be formally incorporated into the ephemerides, or given as differential corrections.

(2) That (a) the time-arguments of the ephemerides of the Sun, Moon and planets be designated in the international ephemerides as Ephemeris Time (E.T.), without qualification;

(b) to each distinguishable lunar ephemeris (as recognised and designated by Commission 4) there be allocated a reference number j with the corresponding designations E.T.j and ΔT_j = E.T.j — U.T.;

(c) the first three such distinguishable lunar ephemerides be:

$j = 0$: the Improved Lunar Ephemeris (I.L.E.); (this ephemeris is that tabulated in the international ephemerides in the years 1960–1967);

$j = 1$: the I.L.E. corrected for

(i) the introduction of the IAU System of Astronomical Constants;

(ii) the error in the coefficient of term No. 182; (this ephemeris is that obtained by applying the corrections given in A.E. 1968 to the I.L.E. itself);

$j = 2$: the I.L.E. corrected for (i) and (ii), as for $j = 1$, for

(iii) the substitution of W. J. Eckert's (*Astr. J.*, **71**, 314, 1966) re-calculation of the solar perturbations instead of those used in I.L.E.

(d) the lunar ephemeris, specified above by $j = 2$, be used in the international ephemerides from 1972 onwards;

(e) the planetary ephemerides, as specified by resolutions 1 (b) and 1 (c), be used in the international ephemerides from 1968 onwards.

(3) That, pending a rediscussion of the observational data, the definition of the epoch of Ephemeris Time, as adopted by the tenth General Assembly (Moscow, 1958; *Trans. IAU*, **10**, 72, 1960) remain unchanged.

(4) That note be taken that:

(a) although the ephemerides of the Sun and Moon designated above are strictly in accord with the observational relationship between their mean longitudes as obtained by Spencer Jones (*Mon. Not. R. astr. Soc.*, **99**, 541, 1939), this relationship depends on the value (20″.47) of the constant of aberration used by Spencer Jones in his discussion;

(b) a discontinuity (new–old) of —0″.026 has been introduced into the apparent longitude of the Sun.

V. *The Introduction of a Modified Julian Day number*

G. Wilkins proposed a resolution against the official introduction of a modified Julian day in the form proposed during the joint meeting of commissions 4 and 31 (see p. 182). He argued that the chosen system is not a convenient one and that sufficient consideration had not been given to the proposition. *J. Kovalevsky* feared that the rejection of such a modified day count would prevent the IAU from improving the existing situation. The proposal was supported by *R.L. Duncombe*, *W. Fricke* and *D.H. Sadler*. It was agreed by 10 against 1.

VI. *Automatic Printing of the Ephemerides*

J. Kovalevsky described the tape converter which is used in the *Bureau des Longitudes* in order to convert 7 channel paper tape punched by the computer into 31 channel Monotype tape used to control a hot metal setting machine. The main advantages of the system are:

(1) a sizeable reduction of the cost of the setting,

(2) a considerable reduction of the number of misprints, though proofreading is still necessary,

(3) no change in the quality of the printing.

More than half of the tables published for 1968 are set by this system which will be used for 90% of the tables published in the *Connaissance des Temps* and in the *Ephémérides Nautiques* for 1969.

R. Haupt described some of the work done in the U.S. Naval Observatory with a magnetic tape to Linotype 15 channel paper-tape. Many tables have been published by this photo-composition method which will probably soon be used for the Air Almanac. A new system, in which rules can also be programmed was also described by *R. Haupt*. It would be possible to set the American Ephemeris in the present format.

W. Fricke recalled the method used for the *Apparent places of Fundamental stars* since 1960: the tables are printed with a card controlled typewriter and then photographed. No proof reading is carried out.

G. A. Wilkins stated that it was the intention to use a similar system to that described by J. Kovalevsky for the setting of the first part of the *Astronomical Ephemeris.* The editing and coding programs had been written in Fortran and a Monophoto filmsetter would be used.

A short discussion followed these presentations.

VII. *Working Group on Ephemerides for Space Research*

G. Wilkins presented the report of this working group, set up at the Hamburg meeting of the I.A.U. (see the annex, p. 51). It appears from the comments obtained by the working group that most space research organizations have larger computing possibilities than the Ephemeris Offices. The latter are ready to supply in machine readable form the best ephemerides that are in their possession.

A letter just received from the Jet Propulsion Laboratory showed the interest of this organization in the work of the group and it appeared that a collaboration could be fruitful for both parties. The commission therefore decided unanimously to maintain the group in existence. In addition to the present members, T. Lederle was nominated as member of the Working Group, which is also allowed to coopt other members outside Commission 4.

VIII. *Future Form of the Ephemerides*

D. H. Sadler introduced the question of the future form of tabulation of the fundamental ephemerides. He mentioned that the present procedure required that they be calculated at least five years in advance, which makes difficult the rapid introduction of a new theory on the new constants. Moreover, he questioned whether their presentation in annual printed volumes was the most useful as compared with, for example, availability in special publications for each body, as machine listings or in machine readable form. In addition, there was the problem of making these ephemerides available well in advance of publication for the preparation of navigational and other ephemerides for practical use. New methods could introduce difficulties.

The Commission agreed that the ephemerides distributed for the latter purpose need not be 'fundamental'.

There was also the question as whether the published ephemerides in their present form provided users with the pertinent data they request and *D. H. Sadler* suggested that this problem should be considered objectively, preferably in consultation with users.

R. L. Duncombe remarked that, in spite of a large distribution there are no complaints on the form of the present ephemerides.

G. M. Clemence suggested that *D. H. Sadler* initiate correspondence on these problems.

IX. *Ephemerides of Ceres, Pallas, Juno and Vesta*

R. L. Duncombe reported on a work by Jackson and himself; they had analysed 40 years of meridian observations of minor planets (1), (2), (3) and (4) and made a numerical theory of their motion until the year 2000 (*Astronomical papers of the American Ephemeris*, **XX**, part I). Compared with Herget's numerical theory, the new theory gives a considerable reduction in the mean residuals for these meridian observations and for 8 years of homogeneous photographic observations.

After some discussion on the fact that in the construction of this theory, older observations were not used, and despite the fact that Newcomb's value of $S/E + M$ was used, the Commission unanimously agreed that the national ephemerides be authorised to use this numerical theory in their publication.

X. *Jupiter Galilean Satellites*

J. L. Sagnier reported on the present and future status of the publication of the ephemerides and phenomena of the Galilean satellites of Jupiter. They will be computed directly from the analytic expressions instead of the tables by Sampson and consequently the method of Andoyer for the computation of the phenomena should be changed. The following schedule is foreseen:

Until 1969: ephemerides computed from Sampson tables by Andoyer's method. Phenomena deduced by Andoyer's method.

1970 and perhaps 1971: ephemerides computed from the analytic expressions in true coordinates. Phenomena deduced from these ephemerides by a modification of Andoyer's method.

As from 1971 or 1972: ephemerides in rectangular coordinates from the analytic expressions. Phenomena deduced from these ephemerides by purely geometrical methods.

The Commission unanimously authorised the national ephemerides to publish coordinates and phenomena of the Galilean satellites of Jupiter computed from the analytical expressions of Sampson.

ANNEX

REPORT OF THE WORKING GROUP ON EPHEMERIDES FOR SPACE RESEARCH

Introduction

The Group was set up by I.A.U. Commission 4 at Hamburg (*I.A.U. Trans.*, **12B** (1964), p. 103) with the principal function of finding out the current and likely future requirements for astronomical ephemerides for space research. The Group has not met between the two General Assemblies of the Union and so this report has been prepared by correspondence between the members.

We have assumed that the group should only be concerned with the ephemerides of natural astronomical bodies, i.e. that the ephemerides of artificial satellites and space probes are not within our terms of reference. Cospar Working Group I has not been able to gather any specific information about the requirements of space research organisations, but the general comments obtained have been included in this Report. In fact, most of the requirements of space research are not essentially different from those of dynamical, optical and radio astronomers. The principal exception to this statement is that in the case of launching of a space probe to another planet the relevant planetary ephemerides are required to the highest possible accuracy; they should therefore be based on the best estimates of the astronomical constants and other parameters of the system.

There is now, however, in astronomy as well as in space research a much greater emphasis in requiring the data to be in machine-readable form, rather than in annual printed volumes. We have therefore concerned ourselves with the more general problems of specifying what ephemerides already exist or need to be computed and of indicating appropriate ways of making the data more readily available. We hope that this Report will meet our terms of reference and be of value to Commission 4 in its considerations of future policy on the publications of ephemerides.

Review of Astronomical Ephemerides

(a) *The planets.* The first requirement is for heliocentric ephemerides of high precision (say, 10^{-8} a.u.); the most useful form is that of equatorial rectangular coordinates for a standard equinox, but such coordinates can be obtained from ecliptic longitude, latitude and radius vector by straightforward transformations. Ephemerides based on an analytic theory are most useful for theoretical purposes, since the theory can be used directly to determine the dependence of the coordinates on the fundamental parameters of the system. For many practical purposes, however, coordinates obtained by numerical integrations that have been fitted to existing, but less precise, theories are equally useful.

The second requirement is for geocentric ephemerides of the angular coordinates and of distance (range) and doppler velocity (range-rate) for radar astronomy. It is this latter requirement that demands a higher precision in distance than that previously thought necessary. At present the doppler velocities are calculated separately to meet special requirements, but for the convenience of radar astronomers consideration should be given to publishing them with the other coordinates. Consideration should be given in the construction of new analytical theories to the desirability of obtaining precise expressions for the derivatives. The precision of the geocentric ephemerides is, of course, dependent on the precision of the ephemeris of the Earth-Moon barycentre that is used; the further reduction to the centre of the Earth can be obtained directly to high precision from the lunar ephemeris.

Publication and Distribution of Ephemerides

It is clear that it is necessary, as well as desirable, that ephemerides should be made available in both printed form and in machine-readable-form, although the two forms need not be identical in content and arrangement. We consider the different requirements separately.

The Royal Greenwich Observatory is prepared to endeavour to meet requests for planetary and lunar ephemerides and for the smaller star catalogues. Its present computer is an ICT 1909, but it can handle a wide variety of codes, including many IBM codes. It is hoped that the R.G.O. will eventually have duplicate copies of the ephemerides in the U.S.N.O. list but at present the R.G.O.'s collection is incomplete.

The Astronomisches Rechen Institut will endeavour to meet requests for star catalogue data and the Bureau des Longitudes is prepared to act as a secondary data centre.

The only other known source of such data is NASA, but the extent to which the special ephemerides that have been produced in NASA establishments are available to other users is not known to us. Perhaps also other institutions, such as the European Space Data Centre, may be prepared to cooperate.

It should be made quite clear that organisations requesting data must be prepared to pay (directly or indirectly) for the costs of materials and postage. In some cases payment for computer time required for, say, copying magnetic tapes may be appropriate. Such matters are primarily for arrangement between the organisations concerned. The source of the data should, of course, be acknowledged in any report of work in which the data have been used.

Conclusions

We consider that there is no need for Commission 4 to initiate special action to provide ephemerides for space research, although we have drawn attention to the needs for further improvement for general astronomical purposes in the precision and scope of the published ephemerides and for a revised system of planetary masses.

We note that the organisations responsible for launching artificial earth satellites and space probes have much more powerful computing facilities than the almanac producing agencies, but we consider that the latter may be able to assist, either directly or by giving advice, when the need for a new ephemeris becomes apparent. The responsibility for such special ephemerides must, however, rest with the launching organisations.

We recommend that the distribution of ephemerides in machine-readable form be coordinated through a limited number of data centres in the manner that we have indicated.

G. A. ČEBOTAREV J. KOVALEVSKY
R. L. DUNCOMBE G. A. WILKINS

Note—The report on the Joint Meeting of Commissions 4 and 31 appears with the report of Commission 31.

5. COMMISSION DES ANALYSES DE TRAVAUX ET DE BIBLIOGRAPHIE

Report of Meeting, 23 August 1967

PRESIDENT: J. Kleczek.
VICE-PRESIDENT and SECRETARY: J. B. Sykes.

The Commission met at $8^h 45^m$ under the chairmanship of J. Kleczek.

The Draft Report and the Report of the Working Group were approved.

Ogorodnikov presented a new draft of several sections of part 52 (Astronomy) of the Universal Decimal Classification, prepared with the collaboration of several Soviet observatories and institutions. Copies are to be supplied to members of the Commission and Working Group for comment prior to submission to the Fédération Internationale de la Documentation.

The loss of two deceased members of the Commission (Baldet and Dermul) was noted with regret. The following proposals for membership were accepted, subject to approval by the Executive Committee: Aoki, Chandrasekhar, Dewhirst, Heintz, Wempe, and the following as Consulting Members: Dunn, Feuillebois, Kemp, Lavrova, Radlova, Ščerbina-Samojlova. The proposal that Sykes should be the next President and Ogorodnikov the next Vice-President was accepted.

Sykes proposed that the title of the Commission should be changed to the more contemporary one of 'Commission for Documentation' (in French, Commission de Documentation), and it was agreed that such a request should be made to the Executive Committee.

Ogorodnikov described a proposal to make the Astronomy section of the Russian abstract journal *Referativnyj žurnal* available in English at a cost of the order of $1 per month in a total time of as little as six months following the publication of the original articles abstracted. This proposal will be circulated to Members of the Union and members of the Working Group after further discussions with VINITI (the U.S.S.R. Scientific & Technical Information Institute).

Two letters were read, one from *F. G. Smith* concerning the appearance of a new journal which some astronomers considered redundant, and suggesting the listing of major astronomical journals and their principles of operation; the other from *McNally*, suggesting (1) a current-awareness journal for astronomy, (2) a survey of astronomical publications and the use made of them, and the possibility of publishing only abstracts of some papers, (3) a consideration of the feasibility of automatic data retrieval in astronomy.

These suggestions led to considerable discussion at this and the second meeting (see below). *Steinberg* commented that the *Annales d'Astrophysique* is already publishing abbreviated papers with full versions available on demand. *Sykes* stated that a study of the use made of astronomical publications was already in progress at the University of Leicester (U.K.). The preparation of a list of primary publications dealing with astronomy is already under consideration by the ICSU Abstracting Board (Draft Reports, p. lxxv). The existence of the lengthy list by *Feuillebois* in *Transactions IAU*, **12C**, 76–95 was also mentioned.

Wempe stated that the late Dr. Schneller's *Geschichte und Literatur des Lichtwechsels der veränderlichen Sterne* was to be published (3rd edition) and continued at the Institute for Stellar Physics (Potsdam and Sonneberg) but that it would not be possible to send copies on exchange as hitherto. A recommendation that the work should be available on subscription would be sought by resolution from Commissions 5 and 27 jointly.

Wempe put forward the view that the rejection of the abbreviation of 'magnitude' as 'mag' when representing a difference in values (*Transactions IAU*, **12C**, 109) was not justified, and called attention to the analogy with the abbreviations for units of time (e.g. 12^h, 12 hours) and temperature (e.g. 10°, 10 deg).

Second meeting, 23 August 1967

The Commission and the Working Group met at 10^h30^m under the chairmanship of J.B. Sykes.

Velghe reported on the poor response (50 acceptances out of 650 notifications) to an offer by the Observatoire Royal de Belgique to publish the *Bibliography of Astronomy* 1881–1898 [Bibliographie générale de l'astronomie] in 9 volumes at $ 125 the set, for orders received by 1 August. It was agreed that he should send one further reminder to those who had not replied, extending the closing date to 1 December. If this did not produce the necessary 250 to 300 subscriptions, *Steinberg* suggested that the slips (numbering 52166) might be microfilmed.

Velghe described the possibilities concerning the re-publication of *Les observatoires astronomiques et les astronomes.* It was noted that Professor T.L. Page is now editing an annual *Observatories of the World* listing main instruments and projects, and that this (with perhaps an extended listing of instruments) would take care of these topics, while *Les observatoires* could give locations, addresses, and names of staff, on loose sheets (one per institution) revised as appropriate, with occasionally revised name indexes. *McNally* asked for a standard form of inquiry to facilitate replies. *Velghe* undertook to investigate collaboration with Page on the compilation of these works.

Feuillebois described the Working Group of the Commission and its relation to the astronomical section of the International Federation of Librarians' Associations. She commented on the points proposed for discussion (*Draft Reports*, p. lxxvi–lxxviii), and a discussion followed on problems of (1) transliteration of the Cyrillic alphabet, (2) abbreviations of periodical titles, (3) observatory publication policy, (4) multiplication of journals, (5) the desirability of asking publishers of astronomical journals to include copies of the abstract of each article on cards (as proposed by the Belgian National Committee for Astronomy), (6) the prospects for a current-awareness journal in astronomy on the lines of *Current Papers in Physics*, and (7) the possibilities of mechanising the abstracts journals in astronomy.

The conclusions arrived at were as follows:

(1) The Union should adhere to the recommendation (already given in *Transactions IAU*, **12C** 68–69/113–115) for the use of the ISO system of transliteration in international matters, but should not seek to impose it for use within a country where a system easier to type and pronounce is in favour. In this and other matters pertaining to the IAU *Style Book* the Commission should be consulted at the drafting stage.

(2) The ISO revised recommendation on the abbreviation of periodical titles should be awaited. The recent appearance of a British Standard (4148:1967) on the subject was noted. The matter is also under consideration by the ICSU Abstracting Board.

(3) Observatories should be encouraged to reduce the number of publication series, but it was recognised that a variety of forms is often unavoidable. The survey being conducted by Mme van de Poll de Lange (Leiden Observatory) is still in progress, and a notice will be placed in the *Information Bulletin* asking those who have not yet responded to do so.

(4) It was accepted that the increasing volume of astronomical publications must result in an increased charge on libraries and that only a trend towards publication in abbreviated form could reduce this.

(5) Some librarians were in favour of ready-made abstract cards, but the possible effect of providing these on prices of journals was also pointed out.

(6) Little enthusiasm was shown for the production of a current-awareness journal in astronomy or the feasibility of doing so as a large-scale undertaking (but see the report of the fourth meeting, below).

(7) It was felt that the abstracts journals, and in particular the *Astronomischer Jahresbericht*, were fully aware of the possibilities of mechanisation and were still being held back by lack of abstracting staff and some delays in receipt of publications.

Third meeting, 23 August 1967

The Commission met (in parallel with the second meeting) at 10^h30^m under the chairmanship of J. Kleczek.

The use of Esperanto in astronomical publications was discussed. The following proposal was made by the Commission: "The editors of astronomical journals are asked to afford authors the opportunity of including with their articles summaries in Esperanto".

Fourth meeting, 24 August 1967

The Working Group held an informal meeting at $10^h 30^m$.

Topics (5), (6) and (7) were further discussed.

It was agreed that the possibility of a rapid supply of cards bearing bibliographical details of published articles and observatory publications (current-awareness service) should be investigated both with VINITI and with Feuillebois' Paris project. Should this project not be feasible, journal editors should be approached to ascertain their willingness to include abstract cards, in accordance with the (in itself excellent) proposal by the Belgian National Committee.

It was resolved that the Commission should give every encouragement to abstracts journals in whatever progress these could make in automating their work.

Kulikovskij answered questions concerning the occasional omission of papers from the *Referativnyj žurnal* and the varying estimates of delay between publication of an original paper and the appearance of an abstract in the *Referativnyj žurnal. Heintz* urged that the Commission should once more reiterate the resolution that authors should send one copy of every publication to each of the abstracting services (*Transactions IAU*, **11B**, 91).

The question of a change of name for the Working Group consequent upon the proposed change in the title of the Commission was discussed, and it was decided that the name Working Group of Astronomical Librarians (Groupe de travail des bibliothécaires astronomiques) should be adopted. This does not imply that, for example, representative editors of astronomical journals and abstracts journals would be excluded from the group.

6. COMMISSION DES TELEGRAMMES ASTRONOMIQUES

Report of Meeting, 23 August 1967

ACTING PRESIDENT: F.L. Whipple.
SECRETARY: O.J. Gingerich.

The Acting President regretted that his unanticipated role as chairman had been brought about by the untimely death of the President, Dr A.R. Hogg, on 31 March 1966.

There being no corrections to the Draft Report, it was accepted without change.

Dr *Gingerich* formally announced his resignation from the Directorship of the Central Bureau and also from the Commission. The meeting expressed its appreciation to him for his services to the Bureau and his efficient handling of the move from Copenhagen to Cambridge. Dr Marsden will be the new Director, and Dr Gingerich agreed to act as an Associate Director.

The Acting President suggested that in view of the small size of the Commission, an Organizing Committee was no longer necessary. The Commission concurred.

The Commission discussed briefly the advantages of collecting combined fees for the telegrams and circulars on a yearly basis, and decided to leave the matter in the hands of Dr Marsden.

The Commission resolved as follows:

Because of the current financial situation of the IAU, and especially in view of the impassioned plea of the General Secretary for all possible economies, Commission 6 has voted to request only the usual $ 667 per annum for the support of the Central Bureau for Astronomical Telegrams and not the higher sum of $1000 per annum mentioned in the Draft Reports.

On the other hand, the Smithsonian Astrophysical Observatory is at present subsidizing at least half of the actual cost of operating the Central Bureau, excluding the actual telegram expenses. To ease this load partially, Commission 6 recommends that the subscription rates to the IAU Circulars be increased by 20%.

(The support was reduced to $533 per annum: see the report of the Finance Committee.)

7. COMMISSION DE LA MECANIQUE CELESTE

Report of Meetings

PRESIDENT: Y. Hagihara.
SECRETARY: W. J. Eckert.

Most of the first session was devoted to the business meeting. The Colloquium on The Use of Electronic Computers for Analytical (not numerical) Developments in Celestial Mechanics constituted sessions 2, 3, 4 and part of 5. A brief review of the Paris Colloquium was presented at session 5. Technical communications were presented at sessions 1, 5 and 6. Dr Morando acted as interpreter at all sessions.

Business Meeting

The President read the names of members deceased since the last General Assembly: D. Brouwer, W.W. Heinrich, I.G. Izsak, G. Lemaître, E.L. Martin, M.F. Subbotin.

The President's recommendations for the composition of the Commission for the next three years were read and approved:

President: W. J. Eckert

Vice President: G. N. Dubošin

Organizing Committee: G.A. Čebotarev, Y. Hagihara, J. Kovalevsky, Y. Kozai, P. J. Message, K. Stumpff, V.G. Szebehely, F. Zagar.

New Members: G. Bozis, A. Deprit, I.V. Galibina, G. Giacaglia, P. Goldreich, C.L. Goudas, J.D. Hadjidemetriou, W.M. Kaula, B. Marsden, J. Meffroy, B. Morando, Sh.G. Šaraf, R. Sibahara, E. Stiefel, J.P. Vinti.

Consulting Members: V.I. Arnold, S.P. Diliberto, D.G. King-Hele, J. Moser, I.D. Žongolovič.

The resignation of V.V. Michkovitch from the Commission was reported.

The President reviewed the financial and clerical problems of the Union caused by the rapid growth of astronomy and the necessity for brevity in the Proceedings.

Minor corrections to the Draft Report were submitted to the President; corrections will be made in volume XIIIA of the IAU Transactions.

PARIS COLLOQUIUM

The colloquium, The Gravitational Problem of *n* Bodies, sponsored by Commission 7 and held in Paris August 16–18 was described briefly by Drs Contopoulos and Szebehely. The material presented at the colloquium is scheduled for publication in March 1968 in the Bulletin Astronomique.

PRAGUE COLLOQUIUM

The program for the colloquium on the Use of Electronic Computers for Analytical Developments in Celestial Mechanics was as follows:

(1) M.S. Davis: Programming systems available for analytical developments.

(2) J. Kovalevsky: Revue de quelques méthodes de programmation de calculs littéraux de mécanique céleste.

(3) A. Deprit: Lindstedt's series on a computer.

(4) J. Chapron and L. Ghertzman: Problèmes pratiques posés par la programmation d'une théorie littérale du mouvement de la lune.

(5) P. Sconzo: The Formac language and its application to celestial mechanics.

(6) G.B. Ephimov: The construction of the limit solution in the low-thrust optimal acceleration problem.

The editor of the Astronomical Journal has agreed to publish papers from this colloquium in a single issue of the Journal during the first half of 1968.

TECHNICAL SESSIONS

The following communications were presented at the sessions for papers:

G. Giacaglia: Secular variation in the motion of Trojan asteroids.

A. Deprit: Brown's conjecture concerning the long period Trojan orbits.

Y. Kozai: Stationary solutions for Hestia-type asteroids in 3-dimensional restricted problem of three bodies.

C. Oesterwinter: Planetary elements for 300000 years.

D. Eckhardt: The use of a computer to develop solutions for the physical libration of the Moon.

W. J. Klepczynski: The mass of Jupiter as derived from the motion of 52 Europa.

D. A. O'Handley: The determination of the mass of Jupiter from the motion of 65 Cybele.

H. G. Hertz: The mass of Vesta.

H. G. Walter: Conversion of osculating elements into mean elements.

E. Rabe: Periodic Trojan librations of the 12:1 commensurability type.

8. COMMISSION DE L'ASTRONOMIE DE POSITION

Report of Meetings, 23, 24 and 25 August 1967

PRESIDENT: R.H. Stoy.
SECRETARIES: W. Gliese,
C.A. Murray,
G. van Herk.

The greater part of the meetings of Commission 8 at Prague was devoted to scientific matters. Formal business was limited to the selection of a new organizing committee, the adoption of the Draft Report and the consideration of a number of resolutions.

The new organizing committee as accepted by the Executive of the Union is as follows:

President: A.A. Nemiro

Vice-President: W. Fricke

Organizing Committee: A.N. Adams, J.E.B. von der Heide, P. Lacroute, R.H. Stoy, R.H. Tucker, H. Yasuda, M.S. Zverev.

The Draft Report was adopted without amendment. Relevant additions to it have been incorporated at appropriate places in the present report. The four resolutions which were passed are:

(1) Commission 8 strongly recommends that, for the time being, all meridian observations continue to be reduced to the equator and equinox of 1950·0.

(2) Commission 8 reaffirms its recommendation that observers include time and latitude stars in their observing programmes.

(3) Commission 8 stresses the great importance of photographing the southern heavens in their entirety with good astrometric cameras during the period of the S.R.S. programme.

(4) Commission 8 notes with great appreciation the progress already made with the S.R.S. programme and renews its encouragement to other southern observatories to take part in this work.

The scientific reports fell into three groups: those relating to work in hand, those reviewing the accuracy of the fundamental system, and those surveying new instrumental methods and techniques. An attempt was made to avoid unnecessary duplication of material both with the Joint Discussion on Modern Problems of Fundamental Astrometry and with meetings of cognate commissions.

WORK IN PROGRESS

Scott reported on the work of the committees of which he was Chairman. There was nothing to add to the Draft Report of the Double Star Committee. Very satisfactory progress had been made in the various Reference Star programmes. A catalogue of final positions of 21499 AGK3R stars, reduced to the FK4 system, had been formed. In summarizing the accuracy of the Catalogue positions, *Scott* pointed out that while the stars which were common to the USNO list and Zverev's list had each received the full 10 observations, most of those which were in Zverev's list alone had received only 8 observations; this resulted in a somewhat larger mean error for the latter. The average mean errors for all stars were $\pm 0^s.0050 \sec\delta$ and $\pm 0''.116$ in R.A. and Dec. respectively; 95% of all stars had mean errors less than $\pm 0^s.0095 \sec\delta$ and $\pm$ 0·214 in the two co-ordinates. So far, there had only been time to compare AGK3R with a catalogue of northern PZT stars observed with the USNO 7-inch transit circle by Dr. J.A. Hughes. The actual mean errors of the differences were in satisfactory agreement with those expected from the known accidental errors of the two catalogues. Further comparisons with the Yale Zone Catalogues, the G.C. and the S.A.O. Catalogue would be carried out.

Good progress had been made with the observations of the SRS programme. The Bergedorf meridian circle arrived in Perth in July 1967 and was expected to be operational later in the year.

The USNO 7-inch transit circle arrived at El Leoncito during January 1967, and observations were begun in July.

The President remarked that the successful completion of the AGK3R programme, and the excellent progress with the SRS programme owed much to the hard work and enthusiasm of Scott and his colleagues at the Washington Naval Observatory. We were all most grateful to them and to the astronomers throughout the world who have cooperated so wholeheartedly in these projects.

Dieckvoss gave an account of the progress of the AGK3 photographic programme. Now that the final AGK3R had been received from Scott, it was hoped that the first of the five volumes of AGK3 (+ 90° to + 55°) would be ready for the printer by the end of 1967. The AGK3 data, including proper motions and spectral types, would be available in machine readable form to interested institutions. Proper motions of individual stars, specified by AGK2 number, could be supplied on request. The mean square error of the new epoch appears to be not larger than $\pm 0\overset{''}{.}18$ in both coordinates while the mean errors of the proper motion components is $\pm 0\overset{''}{.}008$ in both coordinates.

In the discussion on these reports, *van Herk* asked whether the 11 observational catalogues, which together formed AGK3R, showed any systematic differences depending on magnitude, to which *Scott* replied that it was difficult to separate errors depending on magnitude from those arising from different combinations of observatories, since the fainter Zverev stars had received fewer observations than the majority of stars in the catalogue.

In reply to a further question by *van Herk, Dieckvoss* said that AGK3 results could be supplied for stars of a specified spectral type.

Hoffleit then described the current state of the Yale Zone Catalogue work. She displayed a preliminary copy of the catalogue for − 30° to − 35°. The positions were rather less accurate than those in most of the Yale Catalogues, but they had been of great value in indicating large errors in the early Cordoba catalogues, (many entries in which depend on one meridian observation only), with consequential corrections to proper motions in CPC. All high proper motion stars had been especially checked and no new ones, which were not already in the Luyten Surveys, were found. A feature of the catalogue was the series of charts showing the magnitude and direction of proper motion of all stars whose motions were appreciable. This would be of value in searching for common proper motion stars, and members of moving clusters. The catalogue for − 35° to − 40° would probably be published in 1968. The zones between − 40° and − 50° and between − 60° and − 70° had been measured but not yet reduced. The plates south of − 70° overlap appreciably in both right ascension and declination. A large number of faint Melbourne stars selected by Eichhorn are being measured on these South Polar Cap plates in addition to the CPC stars which constitute the programme stars measured in the other zones.

Stoy said that it was most desirable that the southern hemisphere should be re-photographed, preferably by three or more separate cameras, during the SRS programme. The Cape had undertaken to provide one such series of plates with generous overlaps between them. Commission 8 should consider a resolution on this matter.

Eichhorn requested that one or two stars per square degree, reaching the magnitude limit of the Astrographic Catalogue, should be included in the observing lists for the photographic zone catalogues. These would provide material for determination of the magnitude equation in the Astrographic Catalogue when their proper motions became available.

Vasilevskis reported on the second-epoch work of the Lick Proper Motion Programme which has been started. The first phase will be a pilot programme which will not exceed two years; proper motions of approximately 20,000 stars are expected to be obtained. The second phase will cover completely the sky photographed during the first epoch. It will be possible to obtain proper motions in step with taking the second epoch photographs due to automatic measurements of plates and to high speed reduction of measurements. The Automatic Measuring Engine has been completed and installed at Lick Observatory by the Gaertner Scientific Corporation.

Wesselink spoke on the Yale-Columbia Proper Motion Programme which is the southern counterpart to the Lick Programme. He described buildings, telescopes, and the plates used in El Leoncito. The programme has 600 centres covering the sky between − 25° and − 90° declination. The plate centres are 5° apart in declination. So far first epoch plates have been obtained for 180 centres; the completion of the observing programme for the first epoch plates is expected within 3 more years. The taking of the second epoch plates will begin 20 years from now. It is expected that the resulting proper motions will have an accuracy of $\pm 0''.005$ per year.

Vasilevskis suggested an overlap of 3 zones of the two programmes.

THE ACCURACY OF THE FUNDAMENTAL SYSTEM

Fricke asked that any observations relevant to the improvement of the FK4 should be sent to the Astronomisches Rechen-Institut at Heidelberg where investigations of possible improvements are carried out continuously. He drew attention to the paper "Desiderata for FK5" to be presented at the Joint Discussion.

Adams discussed the progress being made by F.S. Danse at USNO toward the construction of an instrumental catalogue of positions and proper motions of fundamental stars. Six absolute catalogues of positions, observed with the six-inch transit circle between 1910 and 1962, were now available.

Afanasjeva described work by Pavlov, Staritsyn and herself on the Reference Catalogue (RC) of right ascensions formed from the latest three photo-electric catalogues of the Pulkovo time service. Comparison with FK4 showed small systematic variations with declination which were confirmed by N30 and several recent astrolabe catalogues. There was also a systematic difference between RC and FK4 depending on magnitude which was confirmed by observations made at Moscow and Paris.

Guinot reviewed briefly the progress of the various astrolabe observing programmes, which were designed to improve the fundamental system. Five catalogues had already appeared, two were in press, while observations were continuing at seven observatories between latitudes + 49° and − 34°.

Vassiliev gave an account of astrolabe observations made at Pulkova and *Anguita* described the results of observations made with the astrolabe at Santiago, which is operated jointly by the Universidad de Chile and the E.S.O. The differences of the form $\Delta\alpha_\delta$ between the astrolabe observations and FK4 agreed well with those obtained from the meridian observations presented at the Joint Discussion. The difference appeared to increase numerically with increasing southern declination, reaching about $0^s.03$ at $\delta = -60°$. This result was confirmed in a later paper by Thomas, presented by *Murray*, on results derived from observations of FK4 stars and also GC stars reduced to the FK4 system, made with the Cape astrolabe.

van Herk drew attention to the systematic differences in μ_α of the FK4 stars as given in the FK4 and in N30. The differences show a marked dependence on magnitude amounting to $-0''.160 \pm 0''.026$ per 100 years per magnitude in the region $-28° > \delta > -80°$.

A smaller, but nevertheless significant, magnitude dependence was found in the region round the Hyades. van Herk's investigation was restricted to stars brighter than the seventh magnitude as the FK4 contains no fainter stars. It would appear dangerous, therefore, to reduce any catalogue of faint star proper motions to the system of either the FK4 or N30 as one might apply corrections which are erroneous to the extent of the average values of proper motions for stars of the ninth magnitude.

Fricke remarked that the differences between the systems of FK3, N30 and FK4 are smaller in μ_δ than in μ_α and that there was a high probability that the μ_δ systems are much more reliable than the μ_α systems because of the controlling effect of the Sun and planet observations. The μ_δ systems of N30 and FK4 are free of noticeable errors due to the neglect of latitude variations prior to 1900.

The next two papers dealt with the improvement of the fundamental co-ordinate system by observations of minor planets. *Aoki* gave an account of Yasuda's work on the four bright minor planets, which is published in *Publ. astr. Soc. Japan*, **18**, 297, 1966. *Polojentsev* read a paper by Orelskaya, reviewing the present state of the programme for using ten minor planets, which had been elaborated by Samoilova-Yakhontova of the Leningrad Institute of Theoretical Astronomy. Sixteen observatories had so far contributed over 7000 photographic observations, but further progress could not be made until the systematic errors in the Yale catalogues could be determined from comparison with the results from the current reference star programmes.

DEVELOPMENTS IN INSTRUMENTATION AND TECHNIQUES

Lacroute discussed the possibilities of making angular measures from satellites. Such measures being made from outside the atmosphere would be completely free from the complications and systematic errors introduced by refraction. Lacroute described equipment weighing about 150 kilograms which could be used on a satellite that was stabilized to about one minute of arc. A system of mirrors superimposes on the same objective two fields at a predetermined angle of the order of 90° apart. The focal plane of the objective is scanned by two photon counters. Angles between stars brighter than magnitude 7 could be measured within one second of time to an accuracy of 0\".01. This apparatus would have immediate and important applications. It is clear that the moment has come when astrometrists must take an interest in measures made in space.

Teleki discussed the influence of refraction in fundamental astrometry. (See also *Trudy* 17 *Astrometric Conference SSSR,* 1966). He pointed out that it was unjustifiable to use a conventional refraction table at all observatories since anomalous refraction of a systematic character, and even normal refraction, was known to depend on the local topography. Refraction tables in current use were based on theories devised in the nineteenth century and ignored much new information on the structure of the atmosphere that was now available. He urged that the whole refraction problem should be studied intensively and that as much care should be given to the siting of astrometric instruments as is now given to the location of large reflectors.

Høg described the Bergedorf 19-cm Repsold meridian instrument which has been equipped with photoelectric micrometers employing fast data logging for observing the star and reading the declination circle. Results obtained during test measurements in Bergedorf gave internal mean errors for a photoelectric observation of $\pm$ 0.ˢ012 and $\pm$ 0\".25 as compared with $\pm$ 0.ˢ016 and $\pm$ 0\".43 for a visual observation made during the AGK3R programme. The 7200 corrections of the 3′ divisions of the two gold circles were determined with a mean error of $\pm$ 0\".14 by the general symmetrical method in a very short time.

Requième reported on the first results obtained with the photoelectric micrometer of the Bordeaux transit circle which was described to Commission 8 at Hamburg in 1964. Observations of FK4 stars between − 20° and + 75° gave a mean square error for a single observation of $\pm$ 0.ˢ010 sec δ. SRS stars in the equatorial zone as faint as magnitude 9.0 gave a mean error of $\pm$ 0.ˢ0126 as compared with $\pm$ 0.ˢ0183 for visual observations made with the same instrument before 1964. The size of this mean error seems to be practically independent of magnitude and spectral type. There is, however, a systematic difference between photoelectric and visual observations of the same stars which seems to depend not on magnitude but on spectral type. It is practically zero for K stars but reaches 8.5 millisecond for A stars and may be caused by the chromatic properties of the transit circle objective.

Laustsen outlined instrumental developments and results obtained with the Copenhagen transit circle at Brorfelde of which a full account has been published in Nr. 190 of the *Publikationer og mindre Meddelelser fra Københavns Observatorium.* The chief points of interest are the photographic micrometer and the photoelectric recording of the circle. The observed internal mean errors of a single observation of a star brighter than magnitude 11 are $\pm$ 0.ˢ015 sec δ and $\pm$ 0\".22. With the method of photoelectric scanning, which has been in use since December 1965, a circle position can be determined with the six photoelectric micrometers in 16 seconds with a mean error of less than $\pm$ 0\".03. Division errors have been determined for all 2160 diameters with a mean error for the correction to a single diameter of $\pm$ 0\".026. The measures for this investigation took only 15 hours and the reductions 20 hours. *Laustsen* remarked that systematic errors of the order of 0\".1 to 0\".2 and extending over several degrees might well exist in division error corrections determined by the older methods and that these might give rise to catalogue errors of the type $\Delta\delta_\delta$.

Høg remarked that the accuracy of the reading of the declination circle claimed by Laustsen was unbelievably high. He wanted to stress, however, that he did in fact believe it. Such high accuracy is produced by using a glass circle illuminated from behind, by the scanning data logging technique proposed by himself in 1960, and by the accurate alignment of the microscopes perpendicular to the circle. They had had trouble at Hamburg because a gold circle reflects like a mirror.

Plans for a new transit circle under consideration for the USNO were discussed by *Klock*. It is intended that the instrument should incorporate the most advanced optical and mechanical techniques available. Readings from the photoelectric micrometers and digitized circle will be fed straight to a computer. The conventional long refracting telescope, which is particularly susceptible to thermal and flexural errors, will be replaced by a compact folded reflector system.

Høg said that the use of an on line computer with a meridian instrument as planned by Klock is a solution for the very near future. The rapid development of cheap process computers like the PDP-8 should make the simple data logging systems a short intermezzo in meridian instrumentation techniques.

De Barros reported that the new axis for the Porto mirror transit circle which was brought into use in April 1967 showed no appreciable flexure and that the various micrometers are now digitized.

Woolsey made a brief report on the progress that has been made with bringing the Ottawa mirror transit into service. An examination of the instrumental constants has revealed an instability in the central section which produces a large change in azimuth during a night's work. There is also a scatter in the registration of both co-ordinates which is attributable to thermal effects between the two collimators.

Harin gave some results of the application of his modified Courvoisier method for measuring the flexure of a tube (*Trudy* 17 *Astrometric Conference SSSR,* 1966.) Preliminary observations with the Wanschaff vertical circle showed that the ends of the tube are not directed exactly along a vertical line when the tube is pointed at zero zenith distance. The ends deviated to north or south according to the direction of the rotation of the tube when setting it to the zenith. The mean deformation for one end was about 4 microns as compared with the horizontal deformation of 170 microns. It seems that this '*Biegungsrückstand*' (residue of flexure) is due to peculiarities in the construction of the tube.

Levy spoke briefly on the search for a suitable site for the French National Astrometric Observatory. The 19 cm meridian instrument of the Liège Observatory has been installed on an isolated plateau about 1700 m above sea level near Font-Romeu in the Eastern Pyrenees and has been working there since April, 1967. If the site proves satisfactory, two meridian instruments with photoelectric registration, one for right ascension and the other for declination, will be installed and also a large astrograph.

Some considerations on the optimum design and location for transit instruments were discussed by Pavlov in a paper read by *Afanasjeva.*

Debarbat showed the results of observations of Mars (1963), Vesta (1963), Uranus (1964), and Jupiter (1964 and 1965/66) which were made with the Danjon astrolabe. Only observations made in two passages (east and west) were used to compute the deviations $\Delta\alpha$ and $\Delta\delta$ from the elements published in the American Ephemeris. In most cases comparisons with meridian observations made at the same epoch are very satisfactory. The discussion of the observations of Mars manifested that the accuracy was of the same order as that reached in the observations of stars.

9. COMMISSION DES INSTRUMENTS ASTRONOMIQUES

Compte Rendu des réunions

PRESIDENT: J. Rösch.
SECRETAIRE: G. Wlérick.

La commission a tenu quatre séances, deux le 24 Août, les autres les 29 et 30 Août 1967; la première et la troisième ont été consacrées aux questions administratives et aux problèmes d'optique, tandis que la seconde et la dernière traitaient des récepteurs photoélectriques d'images. En l'absence de O.A. Melnikov, le président par intérim, J. Rösch, a ouvert la première réunion en demandant à la commission de rendre hommage à la mémoire de trois de ses membres disparus: A. Danjon, D. Maksutov, N. Kutcherov.

Administration (24 Août 1967)

(*a*) *Présidence et vice Présidence*

J. Rösch signale que la première tâche consiste à revoir la composition du comité d'organisation.
En principe, le Vice-Président est appelé à succéder au Président; le Comité Exécutif a demandé à J. Rösch d'assumer cette charge; celui-ci a accepté et cette nomination est ratifiée par la Commission.
Il faut nommer un vice-président; pour cette fonction, le Comité Exécutif a proposé V.B. Nikonov, et J. Rösch indique qu'il est heureux de cette proposition qui est ratifiée par les membres de la Commission.

(*b*) *Comité d'organisation*

La Commission est entièrement libre du choix du Comité d'organisation. Le Comité actuel s'est réuni et a proposé que le nouveau Comité comprenne un nombre plus important de membres car la Commission 9 doit informer les astronomes de l'ensemble des progrès techniques qui se produisent dans des directions très diverses, en des lieux très variés.
Les noms suggérés pour le nouveau Comité sont: Babcock, Baum, Giovanelli, Mc Gee, Meinel, Melnikov, Valníček.
Ils sont accepté à l'unanimité, moins une abstention.

(*c*) *Nouveaux membres de la Commission*

Diverses propositions, émanant de Melnikov, Rösch, Walker, Wlérick, ..., sont ratifiées à l'unanimité; les noms des nouveaux membres sont les suivants: Bhattacharyya, Chincarini, D.L. Crawford, Delbouille, Delsemme, Einasto, Gratton, Ioanisiani, Kozyrev, Kurt, Mihelson, Moroz, Odgers, P.V. Ščeglov, Walenquist, Zacharov.

(*d*) *Membres consultants*

Le Président rappelle la possibilité de proposer au Comité Exécutif des membres consultants. A son avis, cette procédure, qui a été largement utilisée par la Commission de l'Histoire de l'Astronomie, peut renforcer la représentativité et l'efficacité de la Commission 9. Sur proposition de A. Couder, Mc Gee, Kohler, Rösch, les noms suivants seront transmis au Comité Exécutif: A. Bayle, Flory, Schwesinger, R. Wilson, C. Wynne.

(*e*) *Nature des groupes de travail*

Le Président indique qu'ils doivent être renouvelés à chaque Assemblée Générale.

Il pense qu'il faut maintenir le groupe 'Récepteurs photoélectriques d'Images' qui est très actif; il est moins certain qu'il faille conserver le groupe 'Qualité des Images et Choix des Sites' et propose la nomination d'un nouveau Président de ce groupe qui jugera s'il faut le maintenir.

D'autre part, il semble logique de rattacher à la Commission 9, le groupe 'Matériaux photographiques' qui est actuellement autonome et ne comprend pas de membres de la Commission.

L. Gratton indique qu'on pourrait considérer d'autres groupes de travail, par exemple un groupe 'Instrumentation sur la Lune' et un groupe 'Automation'.

J. Rösch suggère de ne pas aller plus avant, actuellement, dans la création des groupes de travail, mais indique qu'il faudra, à brève échéance, considérer une réorganisation générale de la Commission.

D'autre part, *O.A. Melnikov* a fait parvenir à J. Rösch une proposition de création d'un groupe de travail sur 'l'utilisation planifiée des grands téléscopes' dans divers domaines, en particulier les vitesses radiales, les quasars, les classifications spectrales. *Melnikov* pense que les mesures seraient beaucoup plus homogènes si les divers Observatoires utilisaient le même équipement spécialisé. *J. Rösch* a consulté le Comité d'organisation, qui pense que cette proposition est actuellement difficilement réalisable, mais que par contre, l'échange des plans des instruments serait très utile. *Mihelson,* qui a discuté avec *Melnikov,* indique que, dans la pensée de ce dernier, c'est surtout dans le domaine des vitesses radiales qu'un équipement homogène serait souhaitable. La Commission approuve finalement l'opinion de J. Rösch qui indique qu'il vaudrait mieux que ce soit la Commission des vitesses radiales qui prenne des mesures dans ce domaine.

(f) Présidence et appartenance aux groupes de travail

Il n'est pas opportun de fournir au Comité exécutif une liste trop longue des membres des groupes de travail. Mais les Présidents sont libres de faire appel à des membres supplémentaires.

J.D. Mc Gee a demandé à ne plus présider le groupe 'Récepteurs Photoélectriques d'Images' et J. Rösch, prenant la présidence de la Commission, doit abandonner celle du groupe 'Qualité des Images'.

Les propositions suivantes sont acceptées:

Récepteurs Photoélectriques d'Images: Président: G. Wlérick. Membres: W. Baum, K. Ford, G. Kron, Mc Gee, P.V. Ščheglov, M. Walker.

Qualité des Images: Président: H.W. Babcock. Membres: Dommanget, Kotchinski, J. Rösch, Tatarski, M. Walker.

Matériaux Photographiques: Président: D.S. Evans. Membres: Mme Y. Andrillat, W.C. Miller, Stoy.

(g) Appartenance à la Commission 9

A propos du groupe de travail 'Matériaux Photographiques', *D.S. Evans* signale qu'il voudrait bien devenir membre de la Commission 9 mais qu'il est déjà membre de trois autres commissions; cela lui est donc impossible, compte tenu des règles en usage. J. Rösch signale que le secrétariat général lui a demandé de signaler au Comité exécutif les cas particuliers de cette nature. Il devrait être possible d'appartenir à la Commission 9 en plus de trois autres commissions, en raison de son caractère technique.

Optique (29 Août 1967)

Neuf communications ont traité de problèmes d'optique:

(*a*) *H. Kohler* et *R. Wilson*: Correcteurs pour téléscopes pseudo-Cassegrain.

(*b*) *A. Delsemme*: Optimisation de l'ensemble Téléscope-Spectrographe.

(*c*) *A. Baranne et M. Duchesne*: Spectrographie électronique à grande dispersion.

(*d*) O.A. Melnikov, présenté par J. Rösch: plaidoyer en faveur de la mesure de la structure des images stellaires données par les téléscopes moyens et grands et corrélation avec les éléments météorologiques.

J. Rösch rappelle que, en 1964, à Hambourg, il y a déjà eu accord sur l'importance de ce travail.

Par contre, il signale qu'on n'a pas eu connaissance de mesures, à part celles qui sont effectuées au Pic du Midi.

(*e*) *Boyce*: Photomètre photoélectrique à balayage

(*f*) *Hall*: Polarimètre photoélectrique à deux faisceaux

(*g*) *Treanor*: Dispositif pour déceler dans un champ stellaire, les étoiles dont la lumière est polarisée.

(*h*) *Vandekerkhove*: Dispositif permettant la rotation d'un prisme objectif.

(*i*) *Rösch*: Le programme français de recherche de sites.

Récepteurs photoélectriques d'Images.

Récepteurs photoélectriques d'Images (24 et 30 Août 1967)

Les 15 communications présentées se divisent en deux groupes: celles qui traitent de l'électronographie d'une part, celles qui concernent les amplificateurs à cascade et les tubes de télévision d'autre part.

(*a*) *A. Lallemand*: Photométrie des étoiles très faibles avec la caméra électronique.

(*b*) *G. Wlérick*: Photométrie des objets étendus avec la caméra électronique.

(*c*) *Fehlenbok* et *M. Combes*: Construction d'une caméra électronique embarquable, en versions électrostatique et magnétique.

(*d*) *G. Kron* et *Papiashvili*: Linéarité de l'électronographie. Optimisation de l'emploi des récepteurs électronographiques.

(*e*) et (*f*) *J.D. Mc Gee et al*:

1ère communication: Propriétés et utilisation du Spectracon.

2ème communication: Propriétés des tubes à cascades.

(*g*) *M. Walker*: Comparaison de la camera électronique et du Spectracon.

(*h*) *J. Fredrick*: Etude spectroscopique de divers types d'étoiles avec un convertisseur d'images infrarouge.

(*i*) *Hynek*: Emploi de la télévision pour la surveillance de la Lune et la détection des supernovae.

(*j*) *Smyth*: Evaluation des paramètres principaux d'un récepteur photoélectrique d'images.

(*k*) *P. V. Ščeglov*: Recherche des domaines où l'utilisation des récepteurs photoélectriques d'images est la plus féconde.

(*l*) *K. Ford*: Propriétés des 24 tubes 'Carnegie' actuellement en service.

(*m*) *Hewitt*: Propriétés des tubes à cascade. Application à la spectroscopie des étoiles variables.

(*n*) *W. Baum*: Application d'un tube 'Carnegie' à la stabilisation des images téléscopiques.

(*o*) *W. Wilcock*: Mesure précise du rendement quantique.

WORKING GROUP ON PHOTOGRAPHIC MATERIALS

Meetings at Prague, 24, 25 and 29 August 1967

CHAIRMAN: R.H. Stoy.
SECRETARY: D.S. Evans.

In opening the proceedings, the Chairman gave a special welcome to Messrs. William C. Miller of the Mount Wilson and Palomar Observatories, Harry Mohr of ORWO and William Swann of Eastman Kodak.

The draft report was adopted and Dr *Evans* gave an account of the information that had been gathered from two questionnaires, the first of which had been sent to all known manufacturers, the second to all observatories. There had been a good response from the manufacturers but only 27 observatories had replied. Although this sample is small, some of the results obtained are interesting. Over a three year period, these 27 observatories ordered 9752 dozen plates of which 7276 were manufactured by Eastman Kodak. Altogether, 35 different sizes were demanded with 'medium' sizes (from 50 × 100 mm to 180 × 127 mm) accounting for 6677 dozen. Such diversity of sizes is obviously uneconomic. The two sizes most used were 160 × 160 mm (1839 dozen) and 3¼ × 4¼ inches or 83 × 108 mm (1986 dozen). Eighty per cent of the Kodak plates were of eight emulsion types, fifty per cent being of type O. The astronomical profession clearly owes an immense debt to Eastman Kodak Company for its work on the development of new emulsions.

Mr *Miller* said that the limitation in the variety of sizes ordered for the Mount Wilson and Palomar Observatories to 4 × 5, 4 × 10, 5 × 7, 8 × 10, 10 × 10, and 14 × 14 inches, that is to 102 × 127, 102 × 254, 127 × 178, 203 × 254, 254 × 254 and 356 × 356 mm, had effected considerable economy. Even so, only about thirty per cent of the plates ordered finally arrived in the files with useful exposures on them. Salvage of silver from time-expired unused plates produced a significant sum and was of special importance now that the price of silver had risen sharply. Mr *Miller* also described the essentials of plate cutting and stressed the superiority of a diamond cutter which did not produce a ragged edge with incipient cracks in it leading to unwanted fractures when plates were bent. He suggested that an attempt should be made to organise the manufacture of the Mount Wilson and Palomar design of plate cutter, though this would have to be for supply to observatories only. Mr *Miller* recommended cutting at least ½ inch, i.e. 13 mm, from plate edges which were edges during the original coating. This was to avoid the non-uniformity of photometric properties near the edges of an emulsion coating.

Mr *Swann* took up the points that had been raised at the Naples (N.Y.) meeting of July 1966 which had been attended by representatives of the I.A.U. Working Group, of a similar A.A.S. Committee, and of Eastman Kodak. Topics discussed included packaging, labelling, sealing of boxes (not now done in plastic because the heat required fogs the plates), and abrasion of plate edges during delivery. The baking of plates by the manufacturer is not practicable. Baking should be done in dry air with no packaging around so that there is complete freedom for moisture to escape from the emulsion. Packaging plates in plastic boxes or in aluminium foil envelopes would be expensive though this might avoid the fogging by unsaturated hydrocarbons from cardboard box materials that had been reported by Mr Miller. There was a demand from the meeting for the labelling on box ends to be in characters large enough to be easily legible in dim light.

Mr *Swann* asked for guidance from astronomers as to the requirements for an F-sensitisation. A faster N-plate with normal response in the green was probably desirable. Kodak had produced film with a 'rail-edge' which kept successive turns apart when the film was rolled. This had been used as a base for almost gelatine-free emulsions for use in space work in the far ultraviolet (<2200 Å).

Mr *Mohr* described the products and policy of ORWO and distributed appropriate literature.

There followed a protracted discussion on the possibility of standardising plate sizes which crystallised in the resolution submitted to the General Assembly recommending the use, as far as is practical, of the following sizes of plates:

160 × 160	180 × 130	90 × 120	240 × 240	mm
$3\frac{1}{4}$ × $4\frac{1}{4}$	5 × 7	4 × 10	8 × 10	inches
(83 × 108	127 × 178	102 × 254	203 × 254	mm)

The Working Group on Photographic Materials is to continue its work under the Chairmanship of Dr. Evans but will be regarded as a working group of Commission 9 (Instruments). It will concern itself, inter alia, with

(1) the organisation of the manufacture and distribution of the Mount Wilson and Palomar plate cutter.
(2) procuring and handling photographic material.
(3) causes of apparent plate defects.

As before, the results of its investigations will be made available to all manufacturers and other interested parties.

10. COMMISSION DE L'ACTIVITE SOLAIRE

Report of Meetings

PRESIDENT: Z. Švestka.
SECRETARY: A.D. Fokker.

Business Meeting, 23 August 1967

The President, in his opening address, thanks for the cooperation he has experienced in preparing the draft report.

I. CO-OPERATIVE STUDY OF SOLAR ACTIVE REGIONS (CSSAR)

R. Michard gave the following report.

The programme of CSSAR, accepted by IAU Commission 10 and the IQSY Committee called for:

(1) Selection of some interesting active regions for study;

(2) Compilation and publication of a catalogue of data available in all participating stations;

(3) Compilation of the material by a number of 'collectors', with final compilation in Meudon;

(4) Publication of the material in the form of microfilms.

Later it was agreed that interested scientists would be given the opportunity to study the collected material in Meudon, to insure an immediate and efficient use of the material.

A number of 10 suitable active regions were selected. It turned out that point 2 of the above programme would involve too great an amount of work for its expected value.

Point 3 of the programme has been achieved with reasonable success. The cooperativeness of all collectors should be gratefully acknowledged. Between February and June 1967 the CSSAR material at Meudon was studied by G. Godoli, V. Bumba, R. Hedeman, T. Fortini, M. Torelli, T. Tsap, M.J. Martres, I. Iscovici, J. Leroy and M. Trellis. Some papers dealing with the CSSAR will be presented at the IAU Symposium No. 35 in Budapest.

The CSSAR project seems to have given already most of its expected results and the scientific information has been largely extracted from the collected data. Therefore point 4 of the initial programme can be cancelled.

The following is a short enumeration of the aspects studied.

For the first time it was possible to describe without gaps the birth and development of an active region with a resolving power in time of about 1 hour in white light. Sunspot evolution was studied by Waldmeier in Zürich and Mrs Martres and Iscovici in Meudon. The chromospheric development, from K spectroheliograms and filtergrams, was studied by Godoli and Fortini. Interesting comparisons between spots, K and Hα structures could be made in Meudon.

Unfortunately the magnetic field data were more limited. They consisted of whole disk magnetograms of Mt Wilson, detailed maps of longitudinal fields in active regions from Crimea and Meudon and data on sunspot polarities and peak values from a few other observatories. Comparisons of magnetic field structure, morphology of the active regions, flare production, etc. were made by the visiting scientists in Meudon. Synoptic charts for the coronal lines were discussed by Leroy in connection with white light coronal observations of Haleakala and Pic du Midi, with Stanford heliographic maps and with chromospheric data.

As to the organization of such a project as CSSAR one has the choice between a rather loose planning which leaves much initiative and freedom to the data collectors (as was the case with CSSAR) or a more rigid type of organization, with detailed scheduling. A balance between detailed planning and free initiative must be found which is proper to each particular project.

It should be avoided to have different international projects going on at the same time. This puts too many obligations on the participating institutes and individuals.

The CSSAR, in the reporter's opinion, was a reasonable success in achieving world wide compilation and leading to substantial scientific results. The cooperation of all those who contributed, collected and studied the material is gratefully acknowledged.

Consequently to the report of R. Michard, *M. Waldmeier* presented some pictures illustrating the development of sunspots in various active regions. Most of the spots selected for study in the CSSAR belonged to the new cycle.

II. THE PROTON FLARE PROJECT (PFP)

P. Simon gave the following report.

The idea of organizing a comprehensive study of all the solar-geophysical effects associated with a proton flare was initiated by IAU Commission 10 and subsequently the IQSY Committee decided to have organized a special Proton Flare Project as part of the IQSY program. The organizational framework was established by Z. Švestka on the recommendations of several members of Commission 10.

The scope of the project was the following:

(1) to stimulate the observations of a particular proton flare and of its geophysical effects;

(2) to prepare a series of papers on this event for publication in the IQSY Annals.

It was decided that, prior to publication, the results of the cooperative undertaking would be presented during the IQSY/COSPAR meeting in London and at the IAU Symposium in Budapest.

Early in 1966 I accepted the invitation to act as coordinator of the PFP. In the period of between 1 May and 1 October 1966 one or more alerts for imminent proton flare events were to be issued from Meudon Observatory. The two alerts which were given to the cooperating observers were both successful and the proton flare events of July 7, Aug. 28 and September 2 were assigned for a common study. Especially on the occasion of the July 7 event good optical observations were secured, while information was available from several satellites at many different places in space.

The cooperative study was subdivided in the following sections:

(1) the magnetic and the velocity field before the formation of the active region;

(2) the birth and development of this region (magnetic, optical and radio aspects);

(3) the activity of the region (optical, radio and X-ray aspects);

(4) the flare event proper, in its various aspects;

(5) the associated high energy particles (satellite, balloon, neutron monitor and polar absorption data);

(6) the low energy particle effects (satellite observations in the magnetosphere, geomagnetic storms etc.).

Almost 60 reports on the various aspects of the event were received; for the papers belonging to each of the above sections a summarizing paper was prepared.

This joint study has been a success in providing the fullest possible description of all the aspects—optical, radio, X-ray and corpuscular—of the event and the centre of activity to which it was associated. Although no spectacular discoveries were made, a very comprehensive picture was obtained of all the processes which lead to the occurrence of a proton flare event; very valuable observations of protons both outside and inside the magnetosphere were obtained.

The cooperation of many scientists all over the world was very fine indeed.

III. THE COOPERATIVE STUDY OF FAST VARIATIONS OF LOCAL MAGNETIC FIELDS

V.A. Krat gave the following report.

In this cooperative study seven observatories in the USSR, and the observatories in Ondřejov, Bucuresti, Debrecen, Potsdam and Berlin (HHI) participated by obtaining directed photographs in white light, Hα spectroheliograms, radio noise records and maps of magnetic fields near sunspots. The study of a large body of observational material obtained during the summer of 1966 is now in progress. A preliminary report is given in '*Solnečnije dannije*', 1967 nr. 3. Large fluctuations of the sunspot magnetic fields, of the order of several hundred gauss, were established. A fieldstrength variation of 10 gauss per minute is common. From this study it appears that sporadic measurements of fieldstrengths in sunspot regions cannot be considered as representative for periods of the order of several hours. Also the finestructure of magnetic field configurations is subject to appreciable fluctuations.

IV. PROSPECTIVE INTERNATIONAL PROJECTS

The President gives the word to *E. R. Mustel* for presenting a recommendation of the Soviet Committee on Solar-Terrestrial Physics.

The Soviet Committee on STP proposes to launch an international enterprise, 'International Years of the Active Sun', devoted to a study of solar phenomena during the years 1968–1970, aiming at a better understanding of solar activity and its influences on radiocommunications and corpuscular radiation. This enterprise would be different from the IGY and IQSY programs by

(1) emphasis on special projects (like the PFP);

(2) use of advanced methods of space research;

(3) emphasis on uniform flare patrols, for instance to be made in polar regions.

In the statement of the Soviet Committee a number of specific research items are recommended.

In the discussion following this recommendation, *C. de Jager* appreciates the idea, but he wonders whether the start of the Active Sun Years can be postponed until 1969 or even to the declining phase of the solar cycle. *K. O. Kiepenheuer* reminds that IGY and IQSY were largely geophysical undertakings; solar people need not feel themselves responsible for launching such enterprises. *R. Michard* warns that solar observers should not bear the burden of too many projects. *R. Howard* thinks the initiative should come from the Inter-Union Commission on Solar-Terrestrial Physics (IUCSTP); Commission 10 could then endorse the plans submitted by IUCSTP. *H. W. Dodson-Prince* and *R. Michard* express willingness for rendering good services, but these should not hamper too much the pure solar research.

The President closes the discussion, promising to prepare a suitable resolution of Commission 10, taking into account these comments, and to be approved later.

V. SOLAR-TERRESTRIAL SERVICE

A. H. Shapley gives information on a newly proposed Solar-Terrestrial Service.

In recent meetings of URSI and COSPAR the need of a solar-terrestrial service has been discussed; this item will be on the agenda of the UGGI Assembly in Switzerland (Sept. 1967), while also the IUCSTP Commission has paid attention to this matter.

Apart from the already existing services, like IUWDS, Quarterly Bulletin on Solar Activity, there is need of a central body which cares for such matters as are important for workers in different fields. This service should look into questions like

uniformity of reporting;

publication of solar-terrestrial data;

improvement of the communication network;

forecasts of solar activity.

In this connection *A. H. Shapley* calls attention to the monthly bulletin 'Solar-Geophysical Data', published by the U.S. Environmental Science Service Administration (ESSA). He suggests that Commission 10 makes a statement, stressing the importance which this bulletin has for both solar- and geo-physicists. The following statement is approved. 'Commission 10 notes that a prompt publication of solar activity data from world sources is highly useful to solar astronomers as well as to solar-terrestrial scientists; it considers the bulletin "Solar-Geophysical Data", published by the U.S. Environmental Science Service Administration, as an outstanding specimen of such a short-term publication and therefore urges the authorities concerned to maintain this valuable service and recommends the solar astronomers to cooperate in providing their data to ESSA whenever practical'.

VI. THE COSPAR PANEL ON FLARES

A. B. Severny gives some information on the activities of the COSPAR Panel on Flares.

The Panel on Solar Flares of COSPAR was created in 1965 as one of the sections of Working Group II, which is concerned with the physical processes in space and in the environment of the Earth. The main purpose of the Panel is to make the fullest possible use of space experiments for understanding the physical nature of the flare phenomenon and the disturbances produced by flares in space and in the upper atmospheric layers of the Earth. Data obtained from space experiments should be combined with ground-based observations of flares and their effects. The Panel is also concerned with the possible methods of flare forecasting, in view of the radiation hazards which flares may produce. During the COSPAR meeting in London (1967) a resolution, drafted by the Flare Panel, was adopted which urges the appropriate organizations to increase their efforts in the field of rapid interchange of solar and geophysical data. Working Group II also adopted the proposal to organize a symposium on solar flares at the next COSPAR meeting in Japan, to be held in May 1968.

R. Michard suggests that Commission 10 endorses the resolution of COSPAR Working Group II. This suggestion is approved.

VII. PHOTOGRAPHIC JOURNAL OF THE SUN

M. Cimino introduces a new publication, called *Photographic Journal of the Sun.*

During the last year a photographic supplement with daily Hα and K-line photographs of the Sun has been inserted in the Monthly Bulletin of the Monte Mario Observatory (Rome). It has now been decided to start a more complete photographic journal as from January 1968. This will be made possible through the cooperation of the Solar observatories at Manila, Kodaikanal and Teheran. It is intended also to include photographs in white light. Some specimen photographs are presented to the meeting.

On the proposal of *R. Michard*, Commission 10 will give moral support to the new journal by an appropriate resolution.

VIII. SPECTRAL DIAGRAMS OF TYPE IV SOLAR RADIO BURSTS

A.D. Fokker makes a recommendation on the preparation and the publication of spectral diagrams of type IV bursts.

A spectral diagram gives a representation in the time-frequency plane, by equal intensity contours, of the overall course of development of a radio event. Such a representation may be derived if a sufficient number of single frequency records, all over the spectrum, is available.

A number of such diagrams for type IV events have been published by different authors. Since type IV events deserve great interest on account of their association with high-energy corpuscular radiation from flares, it seems desirable to prepare their spectral diagrams more systematically and in a standardized fashion. The whole set of diagrams, to be obtained in a cooperative effort, should be published towards the end of the present cycle.

C. de Jager suggests to prepare a resolution, asking observers to contribute to a better coverage in frequency. The President will look into this.

IX. FORECASTS OF SOLAR ACTIVITY

A.H. Shapley gives some details on the forecasts of solar activity and the experience obtained. He discusses the various needs of forecasts and the possibilities for giving different kinds of forecasts. There is still insufficient agreement about the evaluation of different aspects of solar activity; for instance, there is no unanimity on what solar radio outbursts should be rated as a type IV event. The President remarks that a simple classification system of radio bursts is hard to establish, in view of the enormous complexity of many events. *V. A. Krat* points out that the Sun is such a capricious object as to seriously limit the feasibility of giving detailed forecasts. In any forecasting the structure of magnetic fields is the most important piece of information.

X. THE INTER UNION COMMISSION ON SOLAR TERRESTRIAL PHYSICS (IUCSTP)

The President gives some information on the activities of IUCSTP, in which he himself represents the I.A.U.

This Commission (President: H. Friedman) was formed by ICSU in 1966; it is meant to coordinate activities in the field of solar terrestrial physics, as a substitute for the IQSY organizational framework. IUCSTP is now preparing a list of research programmes to be recommended for the years of solar maximum. The commission is assisted by representatives from different disciplines. Representatives for the domain of solar activity are C. de Jager, E.R. Mustel and M. Pick-Gutmann.

XI. ORGANIZING COMMITTEE

The President announces the proposed composition of the organizing committee of Commission 10:
Švestka (President), Jefferies (Vice President), Fokker (Secretary),
Kiepenheuer, Michard, Nagasawa, Newkirk, Rösch, Severny, Waldmeier.

Joint Meeting of Commissions 10 and 44 on the Cooperation of Solar Observations made at Ground-based Observatories and made with Space Vehicles, 24 August 1967

The following papers were presented to this meeting.

Z. Švestka:	Introduction
* *K.O. Kiepenheuer*:	The need of ground correlated high resolution space observations in the visible.
* *A.B. Severny*:	Space observations needed for improving our knowledge of solar flares.
* *Y. Öhman*:	Space observations of flare sprays and related surge phenomena.
* *E.M. Reeves*:	The needs and requirements from the standpoint of the ultraviolet solar observations.
J.R. Winckler:	Hard solar X-rays in the deka-keV range (read by de Jager).
* *C. de Jager*:	Hard solar X-rays in the deci-MeV range.
* *G.G. Fazio*:	Solar gamma rays and their correlation with ground based observations.
* *C. de Jager*:	Concluding remarks.

Papers marked by asterisks will be published in the volume Highlights of Astronomy Presented at the XIIIth General Assembly of IAU, Prague 1967. The contribution by Winckler appears in Proceedings of the IAU Symposium No 35.

Business Meeting, 26 August 1967

I. RESOLUTIONS

The President reads the following resolutions, drafted for approval by Commission 10.

1. The Working Group of Commission 10 on Coronal Intensity Standardization,

Recognizing the importance of patrol measurements for a better knowledge of the solar corona and of its time evolution, as well as for an early warning of the arrival of active centers at the East limb,

Stressing that these measurements should be expressed in a homogeneous scale and be taken at a sufficiently large number of stations well distributed in longitude,

Aware of the difficulty of time-sharing problems on instruments during the limited observing periods available,

Recommends:

— That the authorities responsible for the scheduling of the instruments already in operation, or soon to be operative, stress equally the following three types of work:
  — patrol observations;
  — special purpose observations of the corona;
  — observations needed for technical improvements;
— That, in all stations which presently assure (or have in the past assured) patrol observations, the time allocated to these observations be at least maintained, or reestablished, to the minimum level required for a meaningful coverage; this applies to the K-corona as well as to the emission corona for those stations which are now equipped (or would soon be equipped) with a K-coronameter;
— That patrol observations be organized at the stations of Huancayo and the Hawaian Islands, which are particularly important on account of their longitudes and favourable meteorological conditions;
— That patrol observations be organized also at the stations of Irkutsk and Abastumani;
— That patrol observations be made with the Kodaikanal coronagraph;
— That steps leading to the temporary exchange of observers between different stations be encouraged;
— That, because of the important technical work (to be described somewhere else) which remains to be done in order to reach a homogeneous system of measurement, and which has been divided among its members, the Working Group continues its activities past the XIII General Assembly.

2. Recognizing the usefulness of well-specified international cooperative projects, particularly for better understanding of the problems of solar-terrestrial physics, but also for the study of solar activity itself, and bearing in mind the great significance of close international cooperation for present-day science,

Commission 10 of the I.A.U. recommends that IUCSTP promotes and organizes short-term cooperative projects during the coming years of the active Sun, provided that these projects are well prepared, will not overlap in time and will not place such a burden on the solar observers that their own research work would be influenced in an unfavourable way.

3. Commission 10, being informed about the resolution adopted by the COSPAR Working Group II during the London meeting (1967), in which the continuations and the rapid exchange of information on solar activity through regional warning centres is recommended,

wishes to endorse this resolution and recommends that solar observers continue their efforts to keep the Sun under permanent observation and to make data on current solar activity available for rapid world-wide dissemination.

4. Commission 10

— notes that the *Cartes Synoptiques de la Chromosphere Solaire* and associated catalogues give in a concise form an exhaustive description of solar activity including the location, size and progressive evolution of active regions and filaments;
— considers that, although similar information has recently been made available on a day to day basis, the synoptic presentation is more directly useful for the study of long term changes;
— notes that the *Cartes Synoptiques* have been used during recent years in a number of investigations of 27 days, yearly and cyclic variations of solar activity and associated terrestrial effects;
— considers that their value in such work is due to their availability for nearly 5 eleven-year cycles of solar activity and increases with time;
— therefore recommends the continuation of this publication under partial financial support from I.A.U.;

being aware of severe limitations in the resources of the Union and of the necessity to help new projects,

— proposes to reduce the subvention to this publication to 1500 U.S. dollars for the period until the XIV General Assembly.

5. Commission 10 recommends to continue the annual subvention of gold francs 1000,— for the *Heliographic maps of the photosphere* for the coming 3 years.

The *Heliographic maps* contain the sunspot groups, the plages and the evolution tables. They are compiled from observations of the Zurich Observatory and its branch stations in Locarno and Arosa as well as from many collaborating observatories. They are published on an international basis and since 1966 printed in English. The *Heliographic maps* are distributed in 500 copies to all astronomical observatories and to a large number of magnetic, ionospheric and geophysical observatories that are interested in solar activity. This series covers the last 30 years and was widely used in many geophysical researches, in statistics on sunspot activity and in recent years for the study of local magnetic fields on the sun. The maps are a necessary addition to the heliographic maps of the chromosphere published by the Meudon Observatory and to the heliographic maps of the corona published in the Quarterly Bulletin on Solar Activity.

The annual subvention granted by the I.A.U. covers about half of the printing costs in a year of low solar activity and about one third in a year of solar maximum. If the subvention granted in the last 20 years should be stopped, the Zurich Observatory would be faced with severe financial difficulties. The special character of the heliographic maps does not allow to publish them in an astronomical journal. Therefore, the continuation of the subvention would be highly appreciated.

6. Commission 10, having been presented a sample of the series of photographs of the Sun prepared by the Monte Mario Observatory in Rome and being informed about the plan to publish these as a Photographic Journal of the Sun, starting January 1968.

wishes to express its appreciation for this initiative and welcome this Journal as a valuable contribution to making basic information on the development of solar activity available to the scientific community.

All these resolutions were accepted.

II. DISCUSSION ON THE QUARTERLY BULLETIN ON SOLAR ACTIVITY

(*a*) *History and present status*

M. Waldmeier gives some information on the history and the present status of the *Quarterly Bulletin on Solar Activity.* In 1928 this bulletin started as a publication of character figures for solar phenomena. Since 1938 it contains a list of flares. Since 1938 the name *Quarterly Bulletin* is used. In 1947 tables for coronal observations and for radio emission appeared in the *Q.B.* Since 1956 the *Q.B.* is financed by the Federation of Astronomical and Geophysical Services (FAGS). In 1961 the Inter Union Commission for Solar Terrestrial Relationships was assigned as an advising committee to the *Q.B.* This function was transferred in 1966 to the IUCSTP. The *Q.B.* consists at present of four parts: Sunspots, flares, coronal observations and radio emission.

The *Q.B.* has always been meant to be a final publication. Certain aspects of solar activity are not represented in the *Q.B.*: filaments, magnetic fields and X-rays. If one should like to have these data included in the *Q.B.* as well, a recommendation to this effect should be presented to FAGS and a centre should be found for preparing the data.

(*b*) *Publication of magnetic maps*

R. Howard gives an account of the publication of magnetic maps.

Daily magnetograms are now being published in the Solar-Geophysical Data of ESSA (Boulder). A publication of synoptic maps, covering the period 1959–1966, is in preparation, but this series will not be continued.

A.B. Severny shows some examples of comparisons of magnetic field measurements made by the Crimean observatory with magnetograms from Irkutsk, Mt. Wilson and Meudon. In order to give regular standardized information in the form of magnetic maps of separate active regions, one should know how to reduce the different systems of measurement to each other. At present five observatories make measurements of the magnetic field in active regions on a regular basis. It seems possible, by further cooperative study and intercomparison of the maps, to arrive at an adequate system for the publication of magnetic data.

V. A. Krat, on account of the rapid variation of magnetic fields, doubts whether a combination of measurements by different observatories is possible.

Severny however thinks that the general features vary sufficiently slowly so as to permit the combination of the results from different observatories into one presentation.

H. W. Dodson-Prince recommends to give some educational aid to those who want to study the magnetic data.

E. Tandberg-Hansen stresses the need of having a magnetic map for each day.

(*c*) *Flare data*

R. Michard gives the following report.

A new system of classification for the Hα flares was adopted at the Hamburg meeting (1964). This system became effective 1 Jan. 1965. Practically all patrol stations shifted from the old to the new system of assigning importance at the proper time. At present the World Data Centers receive more than 1000 flare reports each month.

Although the observatories generally do their best to comply with the IAU recommendations, there are still large inhomogeneities in the reported flare data. One method to get rid of the inconsistencies between reports from different stations is to derive correction factors for reducing the data from individual observatories to a common scale. This procedure has been introduced in the *Solar-Geophysical Data*, published by ESSA (Boulder). But even after such reductions the number of reported flares per hour is still unevenly distributed over the 0^h–24^h U.T. interval. Another method is to urge the observers to adhere to the definition of a flare as a sudden and strong brightening of part of the Hα plage and to the recommended procedures for assigning the importance. It is obvious that a few observatories report as flares certain brightness fluctuations which by most stations are not rated as a flare.

At the Hamburg Assembly it was recommended that WDC-C at Meudon should look into the trends of each station and should bring the results to the attention of the observers. This was done in a preliminary way for the flare reports of 1966. The usual method of defining a 'Mean Observatory' was followed. The average number of flares per time interval reported by a given station i was compared with the corresponding quantity for the Mean Observatory; the ratio is to be called K_i. For flares of importance 1 we note the following: The range of K_i is enormous, ranging from 0.39 to 5.59. For cinematographic stations the dispersion of K_i is much smaller. For flares of importance 2 the dispersion is still larger, even for cinematographic stations.

In many cases it is possible to trace the cause of anomalies for certain stations.

First, visual observers often see more details on the sun than does the usual cine-patrol with small image; they also have difficulties in telling apart the more gradual brightness fluctuations from the sudden ones. So, Hα variations may be reported as a flare of importance S or 1 which are not flares according to the definition. Statistical methods of correction are powerless against this kind of error.

Secondly, a number of stations easily shift subflares into class 1. This error is to some extent corrected by averaging the different reports relating to the same flare.

In order to improve the homogeneity of flare data the most important problem is to eliminate the spurious flares. For instance, if someone sees a flare of importance 2, but three other stations do not report any flare whatsoever, one is justified to eliminate this flare. It is tried to adapt the computer programme for treating flare data to make such checks. The effort towards improvement of the supply of data by all colleagues participating in the flare patrol is gratefully acknowledged.

In the discussion to Michard's report *M. Rigutti* gives some details on photometric analysis of flare photographs made at Arcetri. It appeared that the area of a flare is strongly dependent on the threshold brightness level. There is a tendency to overestimate the area of flares near the limb. From isodensitography a more meaningful flare classification could be derived.

C. Popovici confirms the fact that many observatories do not adhere to the conventions set for flare classification. He makes some comments on the feasibility of adopting the "mean observation" as a reference standard.

H. W. Dodson-Prince notes that flare importance rating can also be influenced by the sort of instrument used. Photometry, according to her, is difficult to standardize.

(*d*) *Solar radio data*

A.D. Fokker gives the following report.

The part solar radio emission now comprises about one half of the Quarterly Bulletin on Solar Activity. The biggest and most important of the tabulations in the part Solar Radio Emission is the one which contains single-frequency data of the so-called distinctive events. This table has greatly expanded during recent years, not only due to the increase of solar activity, but also because the number of contributing observatories has increased. For a given event, more frequencies can now be listed than a few years ago. The data are so arranged that one can appreciate the approximate extent in frequency and the overall character of each radio event at a glance. Particularly at frequencies near 3000 and 200 Mc/s there is a good deal of redundancy of information. From the several reports that are available for each of these, or other frequencies, a representative one is selected for insertion in the table.

At the higher frequencies, about 500 Mc/s and up, the world-wide patrol coverage is quite satisfactory. One can be sure that practically all solar microwave radio bursts at frequencies above 2600 Mc/s do figure in the table. At lower frequencies the situation is less satisfactory. Especially on the American continent there is a certain neglect of single frequency observations at frequencies below 500 Mc/s. As a consequence, the low frequency representation of distinctive events that occur between, say, 18·00 and 26·00 UT, is often insufficient.

The second largest tabulation is the one of spectral classification. Recently a new system of tabulation has been introduced, in which separate entries are provided for metric and decametric events. The radio spectrographic patrol coverage was maintained essentially by stations in USA and Australia, leaving a gap in European a.m. hours. Recently, spectrographs have been put in operation that will probably help to bridge this gap.

The Harvard spectral data figured both in the *QB* and in the ESSA *Solar-Geophysical Data.* At present the duplication of spectral information in these two periodicals has become complete, since now also the Sydney data are contained in the SGD, while the Boulder spectral data are being entered in the QB. The QB listings have the advantage, though, that the three series of observations are combined on one table.

There is also duplication with regard to the synoptic charts of Nançay interferometer observations at 169 and 408 Mc/s.

Radio-heliographic maps are lacking in the *QB*. At present 9.1 cm heliograms from Stanford and 21 cm heliograms from Fleurs are both published in the *SGD*. Since it does not seem to have sense to enter more duplicated data in the *QB* I would not exactly advise to enter the heliograms in the *QB*. Yet it is a pity that the *QB* does not comprehensively contain all the data relating to solar radio noise activity.

It is very fortunate that more and more observatories include copies of selected radio events in their reports. It would be somewhat troublesome to enter all these in the *QB* and moreover, such a collection of selected events would have a somewhat haphazard character.

From the figures of daily flux density at several frequencies one can see at a glance that there are still discrepancies among the different stations resulting from imperfect intensity calibration. Very few data are available at frequencies below 200 Mc/s and little is known about their reliability. A working group of URSI Commission 5 pays attention to the question of absolute calibration.

The typescript of solar radio data generally is finished within 6 months after the end of the period it covers.

Business Meeting, 30 August 1967

DISCUSSION ON THE QUARTERLY BULLETIN ON SOLAR ACTIVITY (CONTINUED)

(*e*) *Solar index for the central part of the disk*

G. Newkirk, on behalf of S.I. Akasofu, S. Chapman and W.O. Roberts, presents a recommendation for the renewal of the publication of a solar index (based on sunspots) for the central part of the solar disk in the Quarterly Bulletin. This recommendation is argumented by the fact that solar activity in the central part of the disk has been found to correlate appreciably better with various geophysical effects than the activity of the sun as a whole.

M. Waldmeier remarks that, consequently to an advice of Commission 10, no central disk index for sunspots has been entered in the Quarterly Bulletin since 1939. He asks whether, on resumption of a central disk index, it should be tried to fill the gap which then will exist for the period 1939 until now.

C.W. Allen comments that central disk data, in his opinion, are not particularly useful; studies can better be concentrated on individual active areas.

H.W. Dodson-Prince recommends to reconstitute the central values in order to be able to compare the solar cycles 18 and 19.

K.O. Kiepenheuer notes that the central index is obviously of interest to geophysicists only; interested people might look themselves more closely into the solar data.

S. Chapman finds it reasonable that the solar people provide the data; he calls also for the publication of a flare index. *H.W. Dodson-Prince* remarks that the derivation of a suitable flare index is still a research problem.

The President concludes the discussion by requesting M. Waldmeier to consider the possibility of resuming the publication of a central disk index and of filling in the gap. Waldmeier declares himself willing to look into this matter.

(*e*) *Further discussions*

H. Nussbaumer stresses the desirability to publish in the Quarterly Bulletin data on the Sun's ultraviolet and X-ray emission. The data, now scattered over various sources, should be collected in one publication as soon as possible.

A.B. Severny notes that Commission 44 should be consulted on this matter. The President promises to discuss this with the President of Commission 44; perhaps a working group can be formed.

P.J. Cardus points to the duplication of data that are published both in the ESSA Solar-Geophysical Data and in the Quarterly Bulletin. He wonders whether such duplication should not be avoided.

The President thinks it is reasonable to have a fast appearing bulletin of a preliminary nature and a more retarded one giving definitive data.

H.W. Dodson-Prince notes that the Solar-Geophysical Data is a service bulletin whereas the Quarterly Bulletin is a contribution to prosperity.

M. Waldmeier notes that the flare tables in the two bulletins are different; coronal data figure in the *Q.B.* only. The *Q.B.* should contain such data as can be contributed for a long time on the basis of a regular and uniform system. In the *Q.B.* publication of magnetic data and of ultraviolet and X-ray data should only be started once one can be sure of a continuous and standardized supply.

J.T. Jefferies proposes that a working group will study the philosophy for the *Q.B.* and will look into the possibilities of entering new data in it. It is agreed that the following scientists will be asked to become members of a Working Group on the Presentation of Solar Data:

Fokker, Howard, J. Virg. Lincoln, Mandelshtam, Sawyer, Severny, H.J. Smith, Waldmeier.

III. RECOMMENDATION PRESENTED BY THE REPRESENTATIVES OF THE SOLAR PHYSICS DISCIPLINE IN THE IUCSTP

The representatives of the solar physics discipline in IUCSTP (de Jager, Mustel, Pick) have been asked by IUCSTP to present recommendations for research in the coming years of solar maximum.

Therefore, during this IAU-meeting the representatives discussed some aspects of international cooperation in the field of solar physics and solar terrestrial relations and formulated the following recommendation, which they have requested to be submitted for the approval of Commission 10 of the IAU, at its meeting on August 30th, 1967:

Apart from special projects which have already been recommended by Commission 10, the main emphasis during the coming years of the active Sun should be placed on the following observations:

(1) Observation and analysis of changes of local magnetic fields on the Sun, using the highest possible spatial and time resolution. The behaviour of magnetic fields during chromospheric flares is an especially important matter for study – as is the question of the origin of solar magnetic fields.

(2) Non-stationary and non-thermal processess inside the solar corona. This study should include a detailed analysis of the following:

(*a*) Decametric and metric wave-length observations during chromospheric flares and at times of other non-stationary phenomena. Intensity measurements with radio-spectrographic, and polarimetric observations are recommended using the highest possible angular and time resolution.

(*b*) Radar studies of the Sun outside and during active periods.

(*c*) Coronal cinematographic observations of active events in different coronal lines.

(*d*) Direct observations of active regions using geophysical rockets.

(*e*) Satellite observations of the evolution of short wave-length spectral lines during periods of solar activity. Intensity and polarimetric measurements should be made.

(*f*) Photography of the Sun in those ultra-violet and X-ray regions to which the solar corona is the most important contributor.

(*g*) Satellite observations of the continuum radiation arising from active coronal regions in the short wave-length range. Simultaneous polarimetric observations are highly recommended.

It is very important that attempts be made to carry out these studies simultaneously.

This recommendation has been approved.

IV. REPORT OF THE WORKING GROUP ON THE STANDARDIZATION OF INTENSITY MEASUREMENTS OF THE SOLAR CORONA

J. Rösch presents the report of this Working Group.

The Working Group, after having treated the subject by correspondence, had a meeting in Bagnères de Bigorre from 7 to 11 August 1967. The results of the discussions will be published in *Solar Physics*. On the basis of the general conclusions, a resolution has been drafted for adoption by Commission 10 (see page 74). In the following, a summary of the technical aspects, relating to the emission corona only, is given.

The Working Group realizes that it is premature and difficult in practice to prescribe a unique method of observation to all the stations. Nevertheless, considerable progress towards uniformization can be made if each individual observatory makes frequent comparisons of its measurements, made by its normal methods, with measures made according to a uniform method.

To this end the Working Group recommends to photograph the spectrum at position angles 5° apart, with the slit of the spectrograph perpendicular to the solar limb. Such spectra can be obtained, by practically all stations with their available equipment, without drastic modifications.

The Working Group stresses the fundamental importance of knowing the altitude above the limb to which the measurements pertain. The difficulty of establishing the height is no doubt one of the main causes of discrepancies between the measurements made at different stations. The following possibilities are suggested for improving the evaluation of height:

(1) The use of an occultation disk of glass or metalized silicum with a transmission of 10^{-4} through which the solar limb may be seen.

(2) The use of a metallic disk, which is centered with the solar image in such a way as to obtain the most regular distribution of chromatic brilliance around it. The height at the base of the spectra is given by the difference between the disk's radius and the photospheric radius.

(3) If, in order to reduce scattered light, an over-sized disk is used, this disk should be first centered along with a disk of minimum size as referred to in (2).

The Working Group prefers to await the results of the above mentioned observations before making a definitive choice of the standard height at which the patrol measurements will have to be made. In the meantime, it recommends that preparative studies be made in a range of heights containing the distance of 1′ to the limb. The standard height, to be fixed ultimately, will probably not exceed this level.

The amount of information provided by observations with radial slits and at many different position angles is very great indeed. As long as there will be no means of treating the material automatically, one may restrict oneself to derive only the information which is needed to serve as a reference standard for the patrol measurements. The Working Group recommends that such methods of analysis will be applied as allow the evaluation of possible causes of systematic discordances.

The Working Group recommends that, preliminary to achieving full uniformization of the observational methods, the measurements be reduced to a common scale, according to methods which aim at eliminating the following causes of error:

(*a*) the small weight of certain observations;

(*b*) the fact that measurements are not made simultaneously;

(*c*) the errors in position angle;

(*d*) the effect introduced by averaging over a certain length of the slit;

(*e*) the discrepancies among the heights above the limb to which the various measurements refer.

With regard to (*a*) and (*b*), the stations are requested to derive 6-monthly averages for position angles 5° apart, giving appropriate weights to the individual measurements.

In view of (*c*) it is recommended to investigate the causes of systematic differences in position angle. To this end the 6-monthly averages should be plotted in polar coordinates. Intercomparisons can be made by the superposition of two polar plots, rotating them until the best coincidence between maxima and minima is obtained.

It should be investigated how effect (*d*) depends on the gradient of coronal intensity. For the present, the procedure of measurements should be kept as constant as possible.

In view of (*e*) it is very desirable to have the intensities derived for a height which, at a given observatory, is kept constant throughout the series of measurements.

Utilizing the 6-monthly polar diagrams, each station *Si* can be compared with an arbitrary station *So* by plotting against each other the measured intensities at corresponding position angles. A mean curve drawn through the points defines a relation between the values of *Si* and those of *So* and this curve can be used to transform the polar diagram of *Si* in a polar diagram corresponding to the scale of *So*. Taking the average of all these transformed diagrams (including the diagram of *So*), one obtains a representation of a 6-monthly mean corona in an arbitrary scale. Coronal data should then be reduced to this common scale before entering them in the Quarterly Bulletin on Solar Activity.

The Working Group estimates that a 6-months period suffices for the series of green coronal line measurements. Over a longer period certain stations might not keep their scale sufficiently constant. For the red line the average should be taken over at least one year since the number of available measurements is smaller.

As a matter of fact, the described comparisons can be applied already to the observations made during the past several years, notably with regard to the effects (*a*), (*b*) and (*c*).

Scientific Meeting, 30 August 1967

This meeting was devoted to a discussion of dynamic solar radio events. The following communications were presented.

1. *Optical observations.*

E. Tandberg Hansen:	The interaction of dark active filaments and flares.
D.H. Menzel:	Solar flares and shock waves.
L. Křivský:	The evolution structure of proton flares near the limb and associated coronal effects.

2. *Radio observations.*

A. Boischot:	A source of synchrotron radiation at a large distance from the Sun.
F. Fürstenberg and A. Krüger:	On the acceleration of particles in the post-maximum phase of solar microwave bursts.
A. Krüger:	On homologous bursts of continuous solar radio emission.
T. de Groot:	Remarks on type I bursts.
A.D. Fokker:	On the connection between solar electron events and noise storms.
M. Vinokur:	The 169 Mc/s multilobe interferometer at Nançay.
J.M. Malville:	Starting frequencies of type III bursts.

12. COMMISSION DE LA RADIATION ET DE LA STRUCTURE DE L'ATMOSPHERE SOLAIRE

PRESIDENT: R. Michard.
SECRETARY: L. Delbouille.

First Meeting, 24 August 1967

The president opened the session by announcing that Professor E. Müller had asked to be replaced as secretary of the Commission 12. After having thanked Dr. Müller for her excellent work, the president announced that Dr. Delbouille has accepted the secretaryship of the Commission.

1. WORKING GROUP ON SOLAR ECLIPSES

The chairman of this working group, *Dr J. Houtgast*, presented the following administrative report of its activities.

The resolution no. 5 (see Draft Report p. 227) to the effect that interference by vapour trail from airplanes be avoided, was brought to the attention of the authorities of the countries concerned with regard to the eclipses of May 20 and November 12, 1966. It appeared that additional measures have to be taken to avoid also incidental private flights over the sites of eclipse observations during an eclipse.

The organizational work with regard to the expeditions for the observation of the 12 November eclipse, passing over South America, done by Dr R. Fleischer of the National Science Foundation, should be mentioned here. The need is felt for a similar organization for the total solar eclipse on March 7, 1970, visible near Mexico and in Florida.

There is some interest for the eclipse of September 22, 1968, being total in the U.S.S.R. Circular nr. 116 of the U.S. Naval Observatory gives information of the eclipse-path.

Some changes have to be made in the composition of the Working Group. Professor Mihailov and Dr Parijskij wish to retire because they are no more active as eclipse observers. Professor Mihailov suggests as new members M.N. Gnevyšev and G.B. Gelfreich.

In the course of the past three years M. Cimino, R. Haupt, F. Q. Orrall and M. Rigutti were proposed as new members.

All proposals are approved, so that now the Working Group has the following members:

J. Houtgast (chairman), Athay, Cimino, R.L. Duncombe, Gelfreich, Gnevyšev, Haupt, Jefferies, von Klüber, Kristenson, Orrall, G. Righini, Rigutti, Suemoto, Waldmeier.

2. WORKING GROUP ON CENTRAL INTENSITIES OF FRAUNHOFER LINES

A.K. Pierce, chairman of this working group, presented the following report of the work done since 1964.

The precision determination of line profiles and hence central intensities of Fraunhofer lines required high dispersion, a narrow well-defined instrumental profile, freedom from scattered light, and a linear system of scanning and recording. The three-year period since the Hamburg meeting has been one of implementation and application of new technological advances, the building of new equipment or the modernization of old. Lasers have helped to make the determination of the instrumental profile much easier though there are still unanswered questions: polarization and coherence. When observing the sun high speed repetitive scanning reduced the effect of low frequence observational noise. This has only been possible using on-line computers.

The members of the Working Group have suggested for standards a list of about 40 lines distributed in the wavelength interval $\lambda\lambda$ 3000 to 7700. These lines have been selected for a number of reasons. Some of them because they are close to laser lines, others because they have been observed many times, often by very different techniques, such as the strontium line λ 4607 or the sodium D lines. The list includes lines of widely different half-widths, different central intensities and different excitation conditions. They should be suitable for the determination of spectrographic parameters, and if observed over a period of time will give us a measure of any change within the sunspot cycle.

The reduction of the observations has received much study. The filtering and deconvolution problems have been solved by the Fourier transform method. At a meeting of the Working Group held in Prague, the deconvolution problem was discussed by Delbouille and Brault. Professor Krat presented the work in Pulkovo, Minnaert commented about polarization.

The actual composition of the Working Group is:

A.K. Pierce (chairman), M.K. Bappu, J. Brault, A.H. Cook, L.D. Delbouille, G.H. Elste, J. Houtgast, C. de Jager, V. Karpinsky, V.A. Krat, M.G. Minnaert, E.A. Müller, L. Neven, G. Righini, F.R. Roddier, G.F. Roland, H.H. Voigt, O.R. White, G.G. Wlérick, C. Pecker-Wimel.

3. SCIENTIFIC REPORTS

(a) *C.E. Moore-Sitterly—The solar spectrum* 2935–8770 *Å*, *NBS Monograph* **61**, 1966

Presentation of the new solar tables, published in collaboration with M.G. Minnaert and J. Houtgast. The volume contains 24000 lines of which 73% are wholly or partially identified.

(b) *L. Delbouille—Recent progress in observations of the solar spectrum at the Jungfraujoch*

Information about this work, by Delbouille and Roland, is under print in the *Phil. Trans. R. Soc. Lond.* A.

(c) *A.B. Severny—The general magnetic field of the quiet sun*

By integrating the high resolution measurements, representations of the general (polar) magnetic field of the sun have been obtained for the period 1963–1967.

(d) *G. Newkirk—Comments on the general magnetic field of the sun*

Spherical harmonic expansion of the solar surface field in June 1965 and November 1966 have been computed by Dr. M. Altschuler of High Altitude Observatory (Boulder).

Second Meeting, 24 August 1967

MODELS OF THE PHOTOSPHERE AND LOW CHROMOSPHERE

Scientific reports

(a) *C. de Jager—Review of results of Utrecht Study Week*

Presentation of the 'Bilderberg continuum atmosphere', a reference model of the solar photosphere and low chromosphere, established during an international study week held in Holland in April 1967.

(b) *O. Gingerich* gave a more complete description of the Bilderberg model.

(c) *G. Elste—Critique of the Bilderberg continuum atmosphere model*

Between 5000 and 16000 Å, the B.C.A. does not provide a satisfactory explanation of limb darkening measurements.

(d) *D. Labs* and *H. Neckel—Critical comparison of available observations of the photospheric continuum*

Extensive survey of all the observing material available today about solar absolute intensities, solar irradiance and the solar constant.

(e) *C. Rouse—Balmer decrement in the solar photosphere and chromosphere*

The decrease in intensity of hydrogen lines in the solar photosphere and chromosphere can be considered as due to the screened coulomb interaction of the bound electron with the nucleus.

(f) *P.R. Wilson*

Discussion of two problems. Model of the mean photosphere cannot be computed without taking into account the horizontal fluctuations in temperature and opacity. Calculations have been made of the source function for a spectral line formed by a two levels atom in an atmosphere with properties varying in the horizontal plane.

Third and Fourth Meetings, 29 August 1967

The president invited Dr J. Houtgast to be the chairman of these two sessions devoted to eclipse studies, chromosphere and corona.

SCIENTIFIC REPORTS

(a) *R.F. Haupt* made an announcement: the Nautical Almanac Office, U.S. Naval Observatory, Washington DC 20390, publishes circulars giving predicted data on solar eclipses. Upon written request, local circumstances are computed for specific locations.

Observer reports about the times of contacts will be greatly appreciated.

(b) *Dr Vinogradov* commented about the ratio Fe/N in the corona.

(c) *E. Tandberg-Hansen—The excitation of* Ba II *in the chromosphere*

(d) *G. Newkirk, M.D. Altschuler* and *J. Harvey—Influence of magnetic fields on the structure of the solar corona*

A current-free approximation of the coronal magnetic field has been calculated from Mount Wilson photospheric magnetic observations traced by computer and compared to a white-light photograph of the 12 November 1966 eclipse.

(e) *J.M. Malville—Some results of coronal studies*

A ratio Fe over H in the corona of 1·1 to $1{\cdot}9 \times 10^{-5}$ has been derived from eclipse observations made in Bolivia.

(f) *G. Henderson—Interferometric studies of the corona*

Line profiles of the λ 5303 emission have been obtained, using a Fabry-Perot spectrometer, during the eclipse of November 12, 1966, observed from the NASA aircraft 'Galileo'.

(g) *M.N. Gnevyšev, G.B. Gelfreich* and *E.K. Makarova—Recent observations of eclipses*

The solar eclipses of 30.V.65, 20.V.66 and 9.V.1967 were used by the soviet astronomers mostly for radio-astronomical observations (brightness distribution on the solar disk, fine structure of local radioemission sources connected with sunspots groups).

(h) *O. Jefferies* announced that the results of the observations of 1962 eclipse, made from New-Guinea jointly by the Harvard Astr. Observatory - Sacramento Peak Observatory and the National Bureau of Standards will be published soon.

(i) *J. Houtgast* reported about work made in collaboration with *O. Namba* on the self reversals of chromospheric lines in the wavelength region 3438–4101 Å and on the ratio of line strengths in the chromosphere and in the photosphere.

(j) *J. Houtgast, O. Namba, R.J. Rutten—Preliminary Report of the Dutch Eclipse Expeditions in* 1966

Two expeditions have been successful: one in Greece and one in Brazil. A specially designed spectrograph was used to record the spectral region λ 4450 to λ6100.

(k) *J. Blamont, A. Maligne* (presented by *R. Bonnet*): *Assombrissement au bord solaire aux longueurs d'onde* 2000–1200 *Å, observé à l'éclipse du* 12 *novembre* 1966

Observations made from a rocket launched, in South America, during the eclipse.

(l) *R. Falciani, A. Righini, M. Rigutti—On the spectrum of the Corona at the total eclipse of November* 12, 1966

Search for the existence of coronal emission of Ca II H- and K-lines.

(m) *R. Falciani, A. Righini, M. Rigutti—A refined isodensitometric technique*

The results of this research have been published in the *Mem. Soc. astr. ital.*, **38**, 1967.

(n) *M. Cimino* and *de Biase—Results of the eclipse expeditions of Rome Observatory in* 1966

Accurate solar-limb darkening photoelectric observations have been made during the total eclipse of November 12, 1966 at Bagi, Rio Grande do Sul, Brazil.

(o) *J.T. Jefferies* and *F.Q. Orrall—Eclipse experiments in May* 1965 *and November* 1966.
Spectroscopic observations between 2950 and 9200 Å have been made in order to measure coronal lines and derive coronal abundances.

(p) *M. Waldmeier—Structure of polar rays and coronal streamers*
Temperature and electronic densities seem, in general, greater in the northern part of the corona.

Fifth Meeting

FINE STRUCTURE OF THE PHOTOSPHERE

Scientific Reports

(a) *J.M. Beckers* presented high resolution observations of the photospheric velocity fields obtained in observing the Fe line at 6569·23 Å.

(b) *J. Rösch* showed excellent quality motion pictures of photospheric granules. On some sequences, taken during an eclipse, the dark edge of the moon was recorded to show the high obtained contrast.

(c) *E. Frazier* discussed the results of Fourier analysis of photospheric velocities obtained in observing the steepest parts of line profiles.

(d) *F. Roddier* observed periodical oscillations (of the order of 300 sec) in the position of Sr I and Ca I photospheric lines.

(e) *R. Howard* and *J.M. Wilcox* have recorded the 5250 Fe I line and observed velocity oscillations, with a period of about 300 sec, in fine details of the photosphere.

(f) *M. Waldmeier* presented a motion picture film taken at the occasion of the November 12, 1966 eclipse.

Correction to the draft report

p. 199—*Atlas et tables du spectre de Fraunhofer—Identifications* paragraph 2—to read:
La préparation du nouvel atlas photométrique du spectre solaire, commencée par Delbouille, Roland et Neven à l'aide du spectromètre à double passage installé au ....

14. COMMISSION DES DONNEES SPECTROSCOPIQUES FONDAMENTALES

Report of Meetings, 23 and 25 August 1967

PRESIDENT: Charlotte Moore-Sitterly.
SECRETARY: J.G. Phillips.

Silent tribute was paid to the memory of three members lost by death since the 1964 meeting: J.E. Mack, W.F. Meggers, G. Racah; and of one consultant, G.H. Dieke.

Corrections to the Draft Report were submitted in writing to the President. They were not extensive.

Unfortunately, no offprints of the Report will be available for general distribution, but the President hopes to have a number of copies made and distributed to interested workers.

It was decided that the present subcommittees 2a and 2b be combined to form one committee on 'Atomic Collision Cross Sections and Line Broadening'. An additional committee on 'Structure of Atomic Spectra' was, also, recommended.

R.W. Nicholls, Director, Centre for Research in Experimental Space Science, York University, Toronto, Canada, has been given the support of Commission 14 as a sponsor for an International Symposium on Laboratory Astrophysics, to be held at York University in 1969.

A.G.W. Cameron brought to the attention of Commission 14 the question of the possible formation of a Joint Commission on Extraterrestrial Chemistry.

CODATA held its second annual meeting in Moscow on May 31, June 1, 2, 1967. It was suggested that smaller countries be given membership in CODATA and on National Committees, with the Academy of the country as a link. The Executive Staff has begun to assemble a directory containing information on continuing data projects all over the world.

Relevant to this, a report by *Nils Åslund* has been submitted to the President 'On the Problem of Sorting and Presenting Results of Molecular Spectroscopic Studies'. The problem is one of communicating numerical information from one experimentalist to another or from experimentalists to theoreticians. The idea of an information bank from which information could be distributed, was discussed at a seminar on molecular physics held in Blaricum, Holland, March 29–31 under the leadership of C.M. Moser. This report has been submitted by the President to G. Waddington to be considered by CODATA.

The Secretary General of IUPAC, *R. Morf*, submitted a list of suggestions on Nomenclature, Symbols and Usage in Analytical Atomic Spectroscopy. It was pointed out that this general question should be handled by the newly-formed Triple Commission for Spectroscopy.

The President made the following interim suggestions regarding the list:

The use of small caps to denote a spectrum, i.e. Mg I rather than Mg I.

The use of *EP* rather than V_e for excitation potential.

The use of *IP* rather than V_i for ionization potential.

Scientific Meeting 23 August, 1967

B. Edlén, Chairman of the Committee on Standards of Wavelength, proposed for discussion the following topics:

(1) Stabilization of laser wavelengths.
(2) The primary standard:
  (a) The wavelength of the Kr^{86} line in different light sources
  (b) Origin of the observed asymmetry in the profile of the Kr^{86} line

(3) Rare-gas standards.
(4) Wavelength standards in the vacuum ultraviolet.
(5) The refractive index of air.

With regard to topic 5, reference was made to a paper by J. C. Owens (6) confirming the validity of the Edlén equation (1) within the usual ranges of temperature and pressure.

The discussion in the Draft Report concerning vacuum-ultraviolet standards was briefly reviewed.

The question of the actual wavelength of the Kr^{86} primary standard line in light sources other than the Engelhard hot-cathode lamp (topic 2a) had arisen in connection with Norlén's measurements referred to in the Draft Report in which he produced the primary standard by means of a microwave-excited electrodeless lamp obtained from Dr. Engelhard.

The suggestion that the wavelength of the primary standard line emitted by this electrodeless lamp should be calibrated against that of the adopted standard lamp met with a gratifying response from members of the committee.

Terrien pointed out that the primary standard is defined in terms of conditions which are impossible to attain in practice. However, it has been shown that there is a $1:10^8$ agreement between theory and practical lamps. The same meter steel gauge measured at different laboratories using different sources have agreed within 10^8. The electrodeless lamp mentioned by Edlén is not a standard, so one does not know the actual wavelength, but it could be very easily checked against a standard lamp. Terrien reported that agreement between Hg^{198} lamps was 0·0012 Å.

Cook asked what was the width of the line in the electrodeless Engelhard lamp. *Edlén* replied that the profile had not been measured. *Cook* commented that the broadening is probably mainly Doppler; that he could undertake to make wavelength and width checks. *Cook* agreed with Terrien regarding the measurement of the meter, finding a few parts in 10^8 agreement.

Baird reported finding a reproducibility of 1 to $2:10^9$ in the wavelength of lamps producing the standard source, and he offered to make wavelength measurements for Edlén at NRC.

To the question of the observed asymmetry of the primary standard line and other Kr^{86} lines Engelhard reported that the purity of the Kr^{86} samples was better than 99.5% as measured by mass spectroscopy. This would appear to exclude the possibility that the asymmetry could be caused by a satellite due to one of the lighter isotopes. Cook confirmed that the observed profiles of the standard line could hardly be explained in terms of isotopic contamination. The cause of the asymmetry remains unknown.

Cook reported that solar spectroscopy requires the establishment of several standard line profiles of Kr^{86} and Hg^{198}. Of value are the yellow and violet lines of Hg^{198} and its green line, although the temperature must be controlled very carefully.

Terrien reported the publication (7) of profiles of lines of Kr^{86}. Baird called attention to an agreement with the Australian workers that the a symmetry of the Kr^{86} standard line is 0·02 m^{-1}.

The chairman called on Dr. Engelhard to report on the stability of la ser wavelengths. *Engelhard* stated that they have had considerable experience at PTB using the He–Ne laser in metrology and gave a detailed presentation of their studies concerning the dependence of the laser wavelength on various experimental conditions. With all variables taken into account the wavelength of the $He–Ne^{20}$ laser had been found to range from 6329·9136 to 6329·9146 Å. This accuracy is sufficient for practical measuring purposes, and for such purposes the $He–Ne^{20}$ laser is considered as an excellent working standard.

Baird plans to correlate emission from lasers with the primary standard as a control when using them in metrology. He wonders whether one should ever consider a laser source as more than a poor secondary standard. He is planning to use high power tunable lasers and will seek to find a ground state line in the visible. Such a line, he believes, will have the most freedom from the troubles indicated above.

Cook agreed with the comments of Baird. He asked whether a rare gas might not be used to make a stable maser. He intends to try to stabilize a molecular maser. He emphasized, (1) that the width of the line was important in the He–Ne laser, and (2) that the electronic properties of the servo synchronization system are very important for the stability of the wavelengths to be measured.

Terrien agreed with both Baird and Cook. He plans to use the He–Ne laser only when the wavelength is not important. However, he thinks that eventually standard lines will be found. For purpose of metrology, it would be an advantage to find a longer wavelength line.

Kessler reported that at NBS they are using absorption techniques for stabilizing laser wavelengths.

Second Meeting, 23 August 1967

C. J. Humphreys reported on improved interferometric wavelength measurements of Xe^{136} and Kr^{86}.

The pair-coupling detected in the *f–g* transitions of the rare gases in the 4 μ region is described in detail in a recently published paper (3).

To the above report by Humphreys on rare-gas standards the chairman commented that G. Norlén at Lund had recently measured many of the same Kr^{86} lines (5). The agreement with the values now reported by Humphreys is very good except for a small systematic difference in part of the region.

Transition Probabilities

R. H. Garstang, Chairman of the Committee on Transition Probabilities, reviewed recent work on this subject and on cross sections. He drew attention to experimental work on forbidden transitions in Pb I and Pb II (2) and I I (4) which materially increase our confidence in the correctness of the transition probabilities of forbidden lines. For allowed transitions attempts to include configuration interaction in the theory are very important.

The President closed the session with some remarks on Future Trends in Spectroscopy.

Standards of Wavelength

Sustained effort must be made to extend the work on standards to shorter wavelengths, since optical stellar spectra now overlap the soft X-ray region. It is, also, conceivable that standards of wavelength for X-ray spectra may well come under the domain of this Commission, particularly in the range where optical and soft X-ray spectra overlap. The work of Commission 44 points to future needs that must be recognized.

At the time when plans were being made to initiate the large program on 'Atomic Energy Levels' at the National Bureau of Standards, there was a request that the program include X-ray, as well as, optical spectra. This is already becoming a problem of astrophysical interest.

Transition Probabilities

The demand for experimental data of high quality is so great that work in this field will continue to expand.

Solar and Other Stellar Spectra

Present solar spectra extend from near 6 Å to well beyond 25000 Å. In the 'accessible' and near infrared region the new solar atlases in course of preparation at Liège, based on observations at the Jungfraujoch, as well as those from Michigan and elsewhere, all indicate the presence of many more faint solar lines than are at present recorded. At least 25% of the known solar lines over this range are still unidentified. It is hoped that the real solar spectrum in the near infrared can be observed above the earth's atmosphere. Most of this region is still unexplored.

To shorter waves Tousey and his associates report more than 7000 solar lines between 2000 Å and 3000 Å, of which only about 52% are identified. Other stellar spectra will, also, require interpretation in the ultraviolet as observing techniques improve. Ultraviolet stellar spectroscopy is still in its infancy.

Many lines to longer waves in high dispersion stellar spectra still defy identification. It is shocking to realize that at least 88 atomic lines in the sun spot spectrum remain unidentified, to say nothing of the thousands of molecular spot lines that have never been measured. Here is a serious challenge to the spectroscopists, both atomic and molecular.

Laboratory Spectra

A sudden revival of interest in highly-ionized atomic spectra has resulted from the observations of the ultraviolet solar spectrum. A wide gap exists between the demand and the supply for analyses of spectra of abundant elements. An illustration will serve to describe the general picture. The total number of silicon spectra to be expected is 14. Lines from each of the 14 spectra are present in the solar spectrum. The span of wavelengths covered is 6 Å to 25129 Å, and the range of ionization potentials is 8 to 2673 eV.

In the Draft Report, selected spectra of a number of elements are mentioned, which are being studied at Lund, Culham, and other laboratories. The present remarks are limited mostly to those spectra which would well repay further study, but are not on scheduled programs. As is the case with Si, more emphasis should be given to the spectra of intermediate stages of ionization (IV and greater) from K thru Ni.

Some third spectra have been seriously neglected, notably the spectra of the inert gas type, but this is now being recognized. Cr III is scheduled, but it is still in an unfinished state so far as analysis is concerned. It may well be of interest to stellar workers. The spectra of phosphorus above the second stage of ionization should be completely reobserved and analyzed.

A number of "well-known" spectra should be reobserved with modern sources and equipment. Analyses could be extended for Sc I and Sc II. Notable additions could be made to the first and second spectra of the familiar elements Fe, Co, Ni. Of these, Fe II is being investigated by L. C. Green; and Ni II has been reobserved by A. G. Shenstone, who is finding hundreds of new lines that as yet have defied interpretation. Monographs are needed also, on Sr I and Ba I.

As revisions of individual spectra are completed, the writer hopes to include the data in the new series of publications which will contain revised Tables of 'Atomic Energy Levels' and of 'Multiplets'.

Rare Earth Spectra

These spectra provide both an interesting and a difficult challenge. At long last Ce I is being analyzed. For years this very complex spectrum defied analysis in the hands of talented spectroscopists. Now Ce I, as observed with improved sources, is known to have some 22000 lines of which about 17000 have been classified, according to W. C. Martin. The progress with other lanthanon spectra is summarized on pp. 253–254 of the Draft Report.

Special mention should be made of Yb II. The late W. F. Meggers worked on this spectrum for over 30 years and collaborated with the late G. Racah for more than 15 years on the interpretation. The final Monograph has just gone to press. Unfortunately, Meggers did not live to see this 'magnum opus' thru publication. It was his desire that Yb II be the model to inspire and guide others in their studies of rare earth spectra. Meggers' painstaking care in observing and interpreting this complex spectrum and the Zeeman patterns has resulted in the classification of more than 4000 lines, in the range 2018 Å to 24377 Å, from 141 even levels and 174 odd levels.

Enough illustrations have been given to indicate that with suitable modern light sources such as the electrodeless lamp, sliding spark, etc. and good equipment, improved line lists of correctly separated spectra of atoms and ions can be obtained. Many spectra formerly classed as 'well-known' can be studied further to great advantage.

Summary

Many other topics might be mentioned: autoionization, forbidden lines, laser research, etc. Critically compiled spectroscopic data will be in greater demand as time goes on, and international cooperation will become more essential as the demand increases. It is the responsibility of Commission 14 to be alert to the needs of the astrophysicist and to collaborate to the utmost extent in fulfilling these needs. Not only is this a responsibility; it is a rewarding duty in the interesting era ahead.

Third Meeting, 25 August 1967

CHAIRMAN: M. Migeotte.

The last two scientific sessions dealt with molecular spectra. Recent work on diatomic molecules at the National Research Council of Canada was reported by *G. Herzberg*. New bands of B_2, BH, Si_2 and SiH were observed by the method of flash photolysis. In CH a Rydberg Series was observed, giving the ionization potential 10·64 eV.

Absorption spectra of molecular ions were observed by means of flash discharges. Two new states of C_2 were found but they are not easy to explain. SiC_2 has also been observed at low temperature in absorption at Ottawa.

G. Herbig gave a general talk on 'Interstellar Molecules' in which he described possible processes of formation of diatomic molecules in space; suggesting that two-body radiative recombination may be operating. From observations of CH (the 0·0 band at 3143) an f-value $= 3 \times 10^{-3}$ has been derived.

A search for accessible triatomic molecules in the spectrum of ζ Oph has been unfruitful.

He stressed the need for more laboratory observations and analyses, particularly of the ground states of the hydrides; for absolute f-values (of CH^+, for example); and for photoionization and photodissociation cross sections.

Fourth Meeting, 25 August 1967

CHAIRMAN: G. Herzberg.

In the last session, *A. Lagerqvist* reported on recent laboratory work at Stockholm on 'Diatomic Spectra'. Many diatomic molecules of hydrides and oxides have been observed and analyzed. The papers from this laboratory will be in more and more demand by astrophysicists. *P. Swings* made a special request that FeH be put on the program at Stockholm.

The final talk was by *J.G. Phillips*, Chairman of the Committee on Molecular Spectra, who described the Berkeley Program. The first Monograph of the Berkeley Series, CN, was published in 1963. In the second Monograph of this series, 35 bands of C_2 are tabulated and analyzed. Successive publications will give similar data on TiO and ZrO.

The program is designed to help the astrophysicist. For example, the C_2 analysis is being used to study some sixteen carbon stars. The low ratio of C^{12}/C^{13}, 2 to 1, in Y Can Ven is not permitted by Fowler's present theory. Much more study of this problem is needed. The C_2 work is being used, also, to revise the identifications in the new solar Monograph. Other illustrations could be given, but these serve to show the present need for such programs.

REFERENCES

1. Edlén, B. 1966, *Metrologia*, **2,** 71.
2. Hults M.E. 1966, *J. opt. Soc. Am.*, **56,** 1298.
3. Humphreys, C.J., Paul Jr., E., Cowan, R.D., Andrew, K.L. 1967, *J. opt. Soc. Am.*, **57,** 855.
4. Husain, D., Wiesenfeld, J.R. 1966, *Nature*, **213,** 1227.
5. Norlén, G. 1967, *Ark. Fys.*, in press.
6. Owens, J.C. 1967, *Appl. Opt.*, **6,** 51.
7. Terrien, J. 1967, *J. Phys.* (Paris) **28,** suppl. p. C 2–3.

15. COMMISSION POUR L'ETUDE PHYSIQUE DES COMETES

Report of Meetings

PRESIDENT: F.L. Whipple.
SECRETARY: V. Vanýsek.

Commission 15 met in two announced sessions on 24 and 29 August, and in one additional session in the afternoon on 29 August.

PART 1

Administrative Business, Meeting 24 August 1967

The President introduced the general plans of the agenda as well as the methodology of correcting Draft Reports. The members were asked to inform the Secretary only about errors, misinterpretations and missing papers in the bibliography, which would then be corrected. The duly corrected Draft Report was finally approved by the President and by the Secretary. Some proposed changes of whole paragraphs were refused to reduce printing costs.

Administrative Business, Meeting 29 August 1967

The second session was devoted to the discussions about the progress of the proposition which was recommended during the previous Hamburg General Assembly (see *Trans. IAU*, **XIIB**, 186ff., 1964) and to the new propositions and suggestions made by the members.

Progress of 'The Atlas of Characteristic Comet Photographs'
(Committee: Donn, Dossin, Haser, Vanýsek and Wurm)

Dr Wurm briefly summarized the work done by him, Dr Donn and Dr Rahe in the last year on the Atlas of Cometary Forms on film prints. The technique proposed by Dr Wurm at the last IAU meeting is used, and Dr Rahe continues preparing the final redaction of the Atlas. It was concluded that a Commission recommendation might be of help in the support of this useful work.

Because the new plan of Dr Richter to publish the Isophotometric Atlas of Cometary Forms seems to be valuable, too, the joint final text of the Recommendation included Richter's as well as Wurm's plans on Atlases (see Recommendations). Dr Richter intends to publish an isophotometric morphological atlas of comets on the base of about 100–150 isophotometric pictures which describe the photometric structure and development of these comets by means of equidensities.

New Proposals and Suggestions Made by Members

The President received before the Prague General Assembly the following propositions, which are presented in summarized form to the Commission:

(1) A Coordinated International Observational Programme for Comets (Donn and Wurm);
(2) Telegraphic Notices on Special Comet Characteristics (Miller);
(3) Cometary Position Plates with Filters for CN, C_2 or Continuum (Vanýsek);
(4) Establishing of a Small Standing Committee for Coordination (Donn);
(5) Cometography; a compilation kept current, of positional and physical observations of comets (Marsden);

During the Assembly an additional suggestion was presented by Dr. Vsekhsvyatsky:

(6) A Comet Searching Service on Sky Photographs.

The President outlined the goals of proposals and stated that the propositions made by Wurm and Donn practically included suggestions made by Miller and Vanýsek and might be discussed simultaneously.

In the discussion among Drs Dossin, Levin, Swings, Vanýsek, Whipple and Wurm the principles of Recommendation II were agreed upon.

Dr Swings pointed out that the above-mentioned suggestions of detailed observational goals had been presented by other astronomers in other formulations previously. The fact that similar ideas reappear confirms their necessity but also the difficulties in the realization of proposed plans.

For the international cooperation in the physical observation of comets, the following working group was established: Biermann (chairman), Bappu, Dossin, Hirose, Konopleva, Miller, Mrkos, Roemer.

Dr Marsden's plan was found to be a very useful one, but the number of the data to be published seemed extremely high.

Everyone agreed with Dr Vsekhsvyatsky that cometary studies are becoming increasingly important and that it would be desirable to search all sky photographs immediately after exposure for possible new comets. But Dr *Roemer* and others pointed out that such an effort would be very time-consuming and too expensive for practical implementation.

The Commission tabled the recommendation.

Actions of Commission 15

Recommendation 1: Noting the extreme value of central repositories for and dissemination of cometary photographs and spectrograms, Commission 15 recommends the publication of an isophotometric atlas of comets on the basis of equidensities by the method of Richter and Hoegner, the publication of an Atlas of Characteristic Comet Photographs prepared by Donn, Rahe and Wurm, and the cooperation of all observatories in making their cometary material available for these purposes.

Recommendation 2: In view of the expanding potentials for cometary observations by new ground-based and space techniques, Commission 15 appoints a Coordination Committee for Observation of Comets and recommends the full cooperation of all observatories and the Central Bureau for Astronomical Telegrams, Cambridge, Mass., in planning observing programs, in reporting unusual cometary activity and in making special observations of comets.

The following communications were presented at the Scientific Meetings:

W. Schlosse: Statistical Properties of Rays in Comet Morehouse 1908 III.

K. Pflug: Spectrophotometric Measurements in the Head of Comet Arend-Roland (1957 III).

R.F. Probstein: A Theory of Dust Comets (joint paper with M.L. Finson).

J.C. Brandt: The Distribution of Aberration Angles of Type-I Comet Tails.

E. Roemer: Frequency Functions of Nuclear Dimensions in Comets.

R. Lüst: Artificial Ion Cloud Experiments.

L. Biermann: The Emission Lines of Atomic Hydrogen in Comets.

C. Arpigny: On the Interpretation of Some Features Observed in Cometary Spectra.

A.H. Delsemme: Progress Report on the Theory of Evaporation of the Cometary Nucleus.

O.V. Dobrovolsky: Statistical Dependence Between the Brightness and Coma Diameter.

L.M. Šulman: Gas Dynamics of the Nuclear Regions of Comets.

V. Vanýsek: Lifetimes of CN and C_2 Molecules.

D.O. Mohnač: Photodissociation and Lifetimes of the C_2 Molecules in the Cometary Atmospheres and Related Problems.

S.K. Vsekhsvyatsky: New Facts About the Origin of Comets.

J. Witkowski: Statistical Investigations on the Distribution of Perihelia of Cometary Orbits.

Z. Sekanina: Non-gravitational Effects in Comets and Cometary Splits.

16. COMMISSION POUR L'ETUDE PHYSIQUE DES PLANETES ET DES SATELLITES

PRESIDENT : R. Wildt.
SECRETARY: J.H. Focas.

Commission 16 met at Prague during the following Sessions.
Administrative Session I: General business, resolutions, invited papers.
Administrative Session II: Activities of the Committee 'International Cooperation for The Observation of Planets'.
Administrative Session III: Martian nomenclature.
Scientific Session I: Communications on Venus, Mars, Jupiter.
Scientific Session II: Communications on Venus, Mars, Jupiter.

Administrative Session I

The President announces the death of the following members of the Commission:

Dr B. M. Peek, he was one among the most distinguished visual observers of Jupiter. His extensive work is summarized in his classical book *Jupiter* published in 1958.

Professor V. V. Sharonov; he was a great pioneer in the physical study of the planets, by photometry, colorimetry and comparative laboratory investigations on terrestrial substances.

The Draft report was accepted with no modifications.

Resolutions: At the request of *Dr G. de Vaucouleurs*, the following Resolution was accepted by the Commission:

Resolution 1—'Commission 16 recommends that the revised Physical Ephemeris of Mars by V. Meiller in U.S. Naval Observatory Circular N° 98, 1964, the basis of which was explained by G. de Vaucouleurs in *Icarus*, **3**, 236, 1964, be recommended for adoption in the National Ephemerides and that it should be used henceforth by all concerned with Martian Mapping and exploration'.

Résolution 1—'La Commission 16 recommande que les 'Physical Ephemeris' de Mars révisées par V. Meiller dans la Circulaire N° 98, 1964 du Naval Observatory, dont la base a été expliquée par G. de Vaucouleurs dans *Icarus*, **3**, 236, 1964, soient recommandées pour adoption dans les "National Ephemerides" et utilisées dorénavant par toute personne concernée en cartographie Martienne et exploration'.

At the request of *Dr A. Dollfus*, President of the Committee 'International Cooperation for the Observation of Planets', the following Resolution was accepted by the Commission:

Résolution 2—'La Commission 16 demande le renouvellement de la subvention de $2000 pour 3 ans, destinée à faciliter le fonctionnement des centres de Données sur les Planètes et l'étude par des astronomes de visite, des clichés qui y sont groupés'.

Resolution 2—'Commission 16 asks for the renewal of the subvention of $2000 for 3 years, destined to facilitate the functioning of the Planetary Data Centers and the study, by visiting astronomers, of the photographic documents collected'.

INVITED DISCUSSIONS

Report on Mariner-Mars Television Experiment
R. B. Leighton

The pictorial data from the Mariner IV television experiment, corrected for (1) the intensity and spectral transfer characteristics of the camera system, (2) the geometric distortions due to non-linear TV line-scan, (3) the modulation transfer function of the camera, and (4) scattered light, are presented.

The resulting pictures show more than 300 craters and an additional 300 possible craters, as compared with less than 100 identified in the preliminary pictures previously published. The principal processes that modify martian surface relief are suggested to be meteoroidal pelting and rubble mantling, covering by wind-blown dust, and thermal creep of loose surface debris. Crater wall slopes up to 12° were found. Large scale (10 km) slopes and albedo fluctuations, measured statistically, are found to be 3° rms and 2% rms, respectively.

Spectres infrarouges planétaires et stellaires par spectroscopie de Fourier
J. et P. Connes

Le principe de la spectroscopie par transformation de Fourier est dû à Michelson mais ses avantages actuels du point de vue du rapport signal sur bruit n'ont été établis que récemment par Felgett et Jacquinot.

Les principales difficultés techniques ont été résolués au cours d'un programme poursuivi en collaboration avec le Jet Propulsion Laboratory, de Pasadena (Californie).

Des spectres de Vénus, Mars, Jupiter et quelques étoiles froides ont été obtenus entre 1 et 2,5 μ. Le gain de résolution par rapport aux meilleurs spectres existant préalablement est égal ou supérieur à 100. Les résultats déjà acquis comprennent la détection de CO sur Mars et Vénus et celle de HCl et HF sur Vénus; ils sont exposés dans 3 Articles déjà parus: *JOSA*, **56,** 896, 1966; *Astrophys. J.* **147,** N° 3, 1230, 1967; et *J. Phys.* **28,** C2, 120, 1967. Leur extension nécessite—et justifie—la réalisation de grands collecteurs de lumière spécialisés pour ce travail. La construction d'un prototype a été entreprise à Bellevue.

Administrative Session II

Dr A. Dollfus, President of the Committee 'International Cooperation for the Observation of Planets' points out that this Committee has two purposes: (1) to organize programs of planetary observations on an international scale, (2) to coordinate the work of the two Planetary Data Centers. The general outline of the programs developed as above and the most part of the results obtained, are given in Part A. of these Proceedings (Comm. 16) Supplementary data having reached the Committee and the Data Centers after release of the Draft Report, are given in the following.

The President invites Dr W. Baum, Director of the Lowell Data Center, and Dr J. H. Focas, Director of the Meudon Data Center, to make a restatement on the development and activities of the two centers since 1964.

W. A. Baum: In accordance with a resolution passed by the IAU Executive Committee at Berkeley in 1961, a center for the collection of Planetary photographs and planetary data has been established at Lowell Observatory. The construction and operation of this new Center has been generously supported by the National Aeronautics and Space Administration. Astronomers wishing to use these facilities should communicate with the director, Dr William A. Baum, Planetary Research Center, Lowell Observatory, Flagstaff, Arizona 86001, USA.

Equipment at the Flagstaff Center includes modern darkrooms with automatic control of water temperature, an automatic film-processing machine, an IBM 1130 electronic computer, a laboratory for the design and construction of electronic instruments, and various instruments for the assessment of planetary observations.

The first undertaking of the Planetary Research Center was to organize and catalogue the large collection of planetary photographs obtained at Lowell Observatory during the past 60 years. This collection includes about 10000 plates and films with more than 300000 planet images. Historical collections from several other observatories, including Lick Observatory, have also been copied and catalogued. Duplicates of everything are being supplied to the Meudon Center. As of August 1967, about 6700 film copies (two from each original plate) have been sent from Flagstaff to Meudon. About 1270 of these were made from composites, while the rest were contact copies.

The Flagstaff Center has also received and catalogued the following planetary film copies (two from each original) produced by other NASA-supported observatories and by the Meudon Center:

Meudon Center Composites	776
New Mexico State University	1974
Table Mountain Observatory	762

The photographic observation of the planets at Lowell Observatory is being continued by the staff of the Planetary Research Center, using a new electronically controlled 35-mm film camera that takes 30-cm-long filmstrips in place of the array of images formerly distributed on a photographic plate. More than 500 filmstrips of Mars were obtained in various colors on 85 nights in 1967 with the 24-inch Lowell refractor. Films were calibrated with a new sector-disk sensitometer designed at the Center. Work is in progress on a more sophisticated planet camera that utilizes a two-stage image-converter tube with magnetic deflection coils to compensate rapidly for the turbulent motions of a planet image due to atmospheric disturbances.

Another portion of the work of the Flagstaff Center is the systematic study of the Lowell plate collection. This work currently includes a statistical analysis of the distribution of cloudiness on Mars, and it is supported in part by the Jet Propulsion Laboratory. A special optical projector was designed and built for such studies in order that planet images can be superimposed optically with a coordinate reticle. Areas of Mars where surface features are obscured can thus be identified, coded, and recorded on IBM cards. Computer programs have been written for mapping cloudiness indices and for sorting out seasonal and diurnal effects.

Much of the effort of the Flagstaff Center is devoted to the design and construction of optical and electronic instruments for making physical observations of the planets. One of the recently completed instruments is a two-channel photoelectric polarimeter for simultaneously reading two planes of polarization while the entrance aperture is scanned repeatedly across a planet image. Another instrument is a combination photoelectric area scanner and spectrum scanner with an output signal that is accumulated in a multichannel analyzer, where each channel corresponds to a point along the line of scan.

J.H. Focas: A total of approximately 10000 documents are now available for consultation at the Meudon Data Center, under the form of original negatives, counter-copies of original negatives, negative and positive composites, reproduction on film or paper, etc.... To these, planetary maps and drawings as well as many thousands of polarimetric determinations on the major planets, should be added.

The Meudon Center is attached administratively to the Service 'Physique du Système Solaire', the director of which is Dr A. Dollfus. The permanent staff of the Center consists of a Scientific Director, (Dr J.H. Focas), a documentalist, two photographers and a computer.

Routine work, as a rule, comprises: (*a*) evaluation of documents for compositing, photometry, positional work, i.e., triangulation, rotations, flattening, etc...; (*b*) computing the geometrical data of the planets; (*c*) listing by planets and apparitions; (*d*) preparing data for computer cards; (*e*) preparing composite images; (*f*) mounting composites for consultation, labelling, coding, filing and boxing them by planets and apparitions. Two photographic dark-rooms are available.

Research work on documents ready for consultation is carried out at the Center. Visiting collaborators participate in the research. Laboratory and observational work is made through instruments kindly placed at the Center's disposal by the 'Physique du Système Solaire' Service. The Milano-Merate Astronomical Observatory, under the direction of Professor F. Zagar, kindly assumed the publication of the photographic maps of Mars, prepared at the Meudon Center, covering the period 1905–1967. Number of such maps appeared in the *Pubblicazioni del l'Osservatorio Astronomico di Milano-Merate—Nuove Series.*

Results concerning the research made at the Meudon Center or obtained through the International Cooperation for the observation of the Planets, not included in Part A of the Proceedings, are the following:

MERCURY

Rotation et cartographie de la planète Mercure
A. Dollfus

L'analyse des photographies et dessins recueillis au Pic-du-Midi depuis 1942 par H. Camichel et A. Dollfus donne une période de rotation de $58{\cdot}67 \pm 0{\cdot}03$ jours qui sont exactement les 2/3 de la période de révolution. Un planisphère complet des taches de la surface est dressé (publié dans *C. R. Acad. Sci. Paris*, **264**, 1765, 1967. A paraître dans *Icarus*.

VENUS

Ultra-violet photographic survey of clouds in Venus' Atmosphere
A. Dollfus

Report about the results of the International Cooperation organized in 1959 and 1962 by the present Committee. Eight observatories obtained altogether several hundreds of U.V. pictures of Venus, achieving several sequences of uninterrupted observations during several tenths of hours. Venusian atmosphere can evolve quickly. Clouds are not likely composed of solid dust, but of particles capable of condensation and evaporation. Motions of the clouds indicate a slight tendency for an overall drift similar to a rotation of the whole atmosphere in 3 to 5 days. Report in press in *Proceedings of the Conference on Atmospheres of Mars and Venus, Tucson* 28–30 *Feb.* 1967.

Motions of Ultraviolet clouds on Venus
B.A. Smith

Measured drift rates of Ultraviolet clouds infer a retrograde rotation period of 113 ± 5 hours for the planet's atmosphere, in sharp contrast to the 244 day (radar) rotation period for the solid globe.

La rotation rétrograde en 4 *jours de l'atmosphère nuageuse de Vénus*
P. Guérin (in cooperation with H. Camichel and Ch. Boyer)

Toutes les séries de photographies ultraviolettes de Vénus obtenues au Pic-du-Midi à l'élongation du matin (1966) et celle du soir (1967) pendant 4 à 10 heures d'affilée, montrent une rotation d'ensemble rétrograde des taches nuageuses (un tour en 4 jours environ). Il ne s'agit pas d'une reconstitution sur place, ni de vents symétriques par rapport au méridien subsolaire. On suspecte des fluctuations de la vitesse instantanée.

MARS

G. de Mottoni

Martian photographic maps prepared till now allow a detailed study of the seasonal variations of the dark areas of the planet. They moreover show the evolution of secular variations which occurred during the period 1907–1954 in the areas of Cyclopia, Margaritifer Sinus, Oxia Palus, Thymiamata S., Nilosyrtis and Moeris Lacus, Nepenthes-Thoth-Casius.

J.H. Focas
Photovisual map of Mars 1967 *based on observations made through the* 83 *cm Refractor of the Meudon Observatory*

The map contains items up to $0''\!.2$ size. The projection used allows the study of the phenomena occurred in the north circumpolar and polar areas and their fine structure.

A. Dollfus

Presentation of a selection of high quality martian photographs obtained in different observatories in the scheme of the International Cooperation in 1965 and 1967, and available at the IAU Data Center.

JUPITER

Evidence of vorticity in the great Red Spot on Jupiter
B.A. Smith

Observations of the interaction of small dark features with the periphery of the Red Spot show the latter to have a counter-clockwise vortical motion with a *nine-day* rotation period.

International Cooperation for the Photographic Observation of Jupiter
J.H. Focas

This campaign was organized by the Committee 16-C since 1964, on the purpose to collect observational data allowing hydrodynamical investigations allied with the zonal and the meridian circulation, to which the cataclysmic epigenic activity in the atmosphere of Jupiter should be added. Such observational data refer in principle to: (*a*) the revision of the rotation period of the planet determined by single rotations in one and the same night, to avoid the effect of the proper motions of the atmospheric markings; (*b*) the evolution and repartition of bright and dark matter; (*c*) the study of the proper motions of the markings and mixing conditions of the atmospheric currents allied with vertical and vortical motions of bright and dark matter; (*d*) The periodic drift of the belts in latitude related with the variation of their intensity considered to be the criteria of meridian circulation in the atmospheric cells; (*e*) the cycles of atmospheric activity.

SATURN

V. Bobrov (communicated by A. Dollfus)

The present Committee organized an international cooperation for the observation of the Saturn ring during the crossing of its plane by the Earth in 1966. Several hundreds of plates and measurements were sent or reported to the Committee at Meudon Observatory. Measurements of the plates are undertaken at the original observatories, or at the Meudon IAU Data Center. Preliminary results give a thickness of the order of 1 km for the Saturn ring.

Thickness of the Ring
Kiladze

Photometric measurements of the plates of Saturn taken at Abastumani, Georgia S.S.R., gave the precise time of the two Earth transits in the plane of the ring: 1966, Oct. 28, 23, h2 ± 1h. U.T. and 1966, Dec. 18, 05 h 3 ± 3h, 3 U.T. The thickness of Saturn's ring is 0·92 km.

Découverte d'un nouveau satellite de Saturne
A. Dollfus

Un nouveau satellite de Saturne a été découvert sur les clichés obtenus au Pic-du-Midi les 15, 16 et 17 Décembre 1966. Il a été retrouvé sur un cliché du 18 Décembre obtenu par Walker au Naval Research Observatory U.S.A., et sur quatre clichés obtenus par Texereau à McDonald Observatory U.S.A., le 29 Octobre 1966. L'orbite est pratiquement circulaire et dans le plan equatorial, son rayon vaut 2·65 fois le rayon équatorial de Saturne, la période vaut 17 h, 975. Une plus grande élongation à l'Est s'est produite le 15 Décembre 1966 à 18 h 20 m. La magnitude est voisine de 14. Le nom proposé est 'Janus'. (Cf. *C.R. Acad. Sci. Paris*, **264**, 822, 1967).

Administrative Session III

MARTIAN NOMENCLATURE

A Working Group comprising Professor G. P. Kuiper, Professor J. S. Hall, and Dr A. Dollfus was appointed by Comm. 16 to work out the martian nomenclature, after the Mariner IV Experiment.

Professor G.P. Kuiper, President of the Group, summarized the developments occurred in this line before the Space Era and the necessity of paving the way for securing a nomenclature of martian features, satisfying the requirements created after the first photographs of craters and finer details of the surface of the planet were taken by Mariner IV. Such requirements are expected to grow with increasing resolution of photographs to be taken in the future. In this spirit the Working Group proposed the following resolution which was accepted by Commission 16.

Resolution III
Guide Lines for Naming Martian Craters

(1) The 1958 I.A.U. Martian Nomenclature (Trans. IAU, X, 259–263, 1958) is hereby reaffirmed with the understanding that 'features' refers to telescopically observed spots, dark and bright, not to craters. The approx. 120 names of dark and bright areas are regarded as 'provinces' whose boundaries may be defined more precisely as a result of increased image resolution.

(2) Within these 'provinces', bright and dark, craters and other tectonic structures occur. The most prominent of these may be named after deceased scientists and members of other suitable professions, as is customary for lunar craters.

(3) The number of names should be held reasonably small.

(4) Among the appropriate professional categories are: astronomers, mariners, geologists and geophysicists, physicists, mathematicians, biologists, chemists, astronauts, engineers, philosophers, artists and historians. Since only 10–20 categories will be introduced versus some 120 provinces, groups of adjacent provinces may be combined as larger districts in the assignments of crater names. An effort may be made to relate in an interesting manner professional groups to major surface features.

(5) Craters not named may be designated by coordinates or—as on the Moon—by reference to a nearby major crater, with an appropriate letter attached.

(6) The large area of Mare Sirenum (Areographic Longitude 163°, Latitude −33°) allows one category names only. Five craters are named (Figs. 1–4):

(a) Mariner IV picture 8: Ejriksson (50 km dia.).
(b) Mariner IV picture 10: Columbus (110 km dia.).
(c) Mariner IV picture 10: Magelhaens (110 km dia.).
(d) Mariner IV picture 11: Mariner (160 km dia.). Symbolic name.
(e) Mariner IV picture 16: Nansen (90 km dia.).

Résolution III
Directives pour la nomination des cratères de Mars

(1) La Nomenclature Martienne établie en 1958 par l'U.A.I. (*Trans. I.A.U.*, **X**, 259–263, 1958) est réaffirmée ici, entendu que les 'accidents' se réfèrent à des taches sombres et claires observées téléscopiquement, et pas à des cratères. Les 120 noms environ, de régions sombres et claires sont considérés comme des 'provinces' dont les limites pourraient se définir avec une plus grande précision avec l'augmentation de la résolution des images.

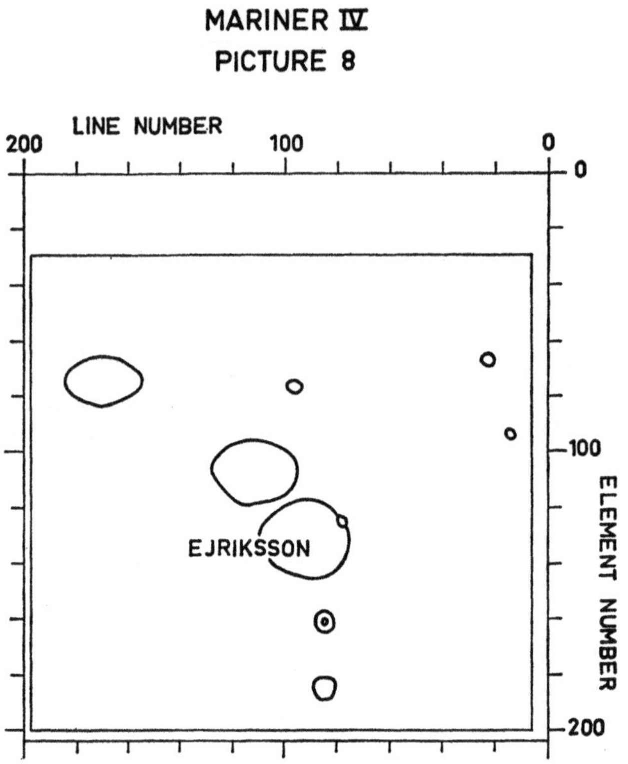

FIG. 1. Crater Ejriksson.

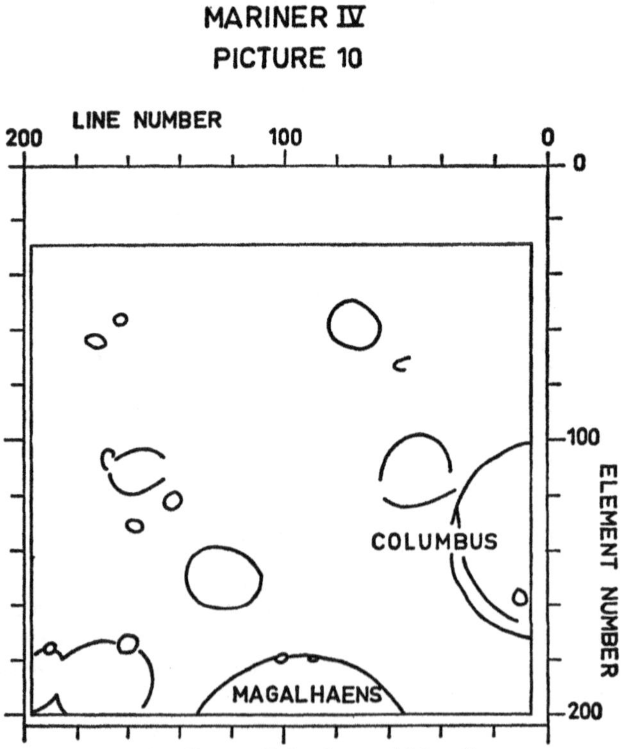

FIG. 2. Craters Columbus and Magelhaens.

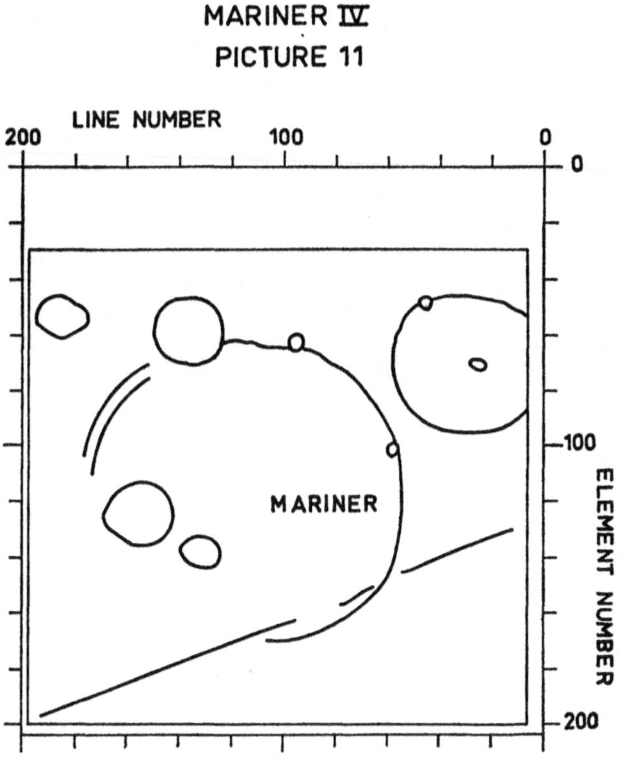

FIG. 3. Crater Mariner.

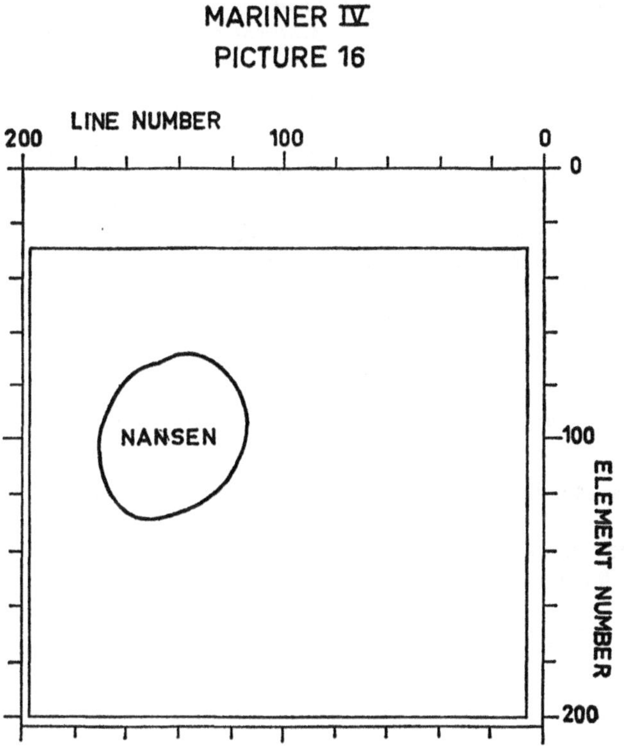

FIG. 4. Crater Nansen.

(2) Parmi ces 'provinces', claires et sombres, il y a des cratères et d'autres structures tectoniques. Les plus saillants d'entre eux peuvent être nommés d'après des scientifiques décédés et des personnes d'autres professions appropriées, comme c'est l'habitude pour les cratères lunaires.

(3) Le nombre des noms doit être raisonnablement restreint.

(4) Parmi les catégories des professions appropriées se trouvent celles des: Astronomes, Navigateurs, Géologues et Géophysiciens, Physiciens, Mathématiciens, Biologues, Chimistes, Astronautes, Ingénieurs, Philosophes, Artistes et Historiens. Puisque 10–20 catégories seront introduites contre 120 'provinces', des groupes de provinces adjacentes peuvent être combinées en districts plus vastes dans la désignation de noms de cratères. Un effort pourrait être fait pour rapporter d'une manière intéressante des groupes professionnels à des accidents majeurs de la surface.

(5) Des cratères non nommés peuvent être désignés par des coordonnées ou—comme sur la Lune—par référence à un plus grand cratère voisin avec une lettre appropriée attachée.

(6) La large région de Mare Sirenum (Longit. Aréographique 163°—Latitude − 33°) ne permet qu'une catégorie de noms seulement. Cinq cratères ont été nommés (figs. 1–4):

(a) Mariner IV Photo 8: Ejriksson (50 km. diam.)
(b) Mariner IV Photo 10: Columbus (110 km. diam.)
(c) Mariner IV Photo 10: Magelhaens (110 km. diam.)
(d) Mariner IV Photo 11: Mariner (160 km. diam.). Nom symbolique.
(e) Mariner IV Photo 16: Nansen (90 km. diam.).

Commission 16 decides the continuation of the life of the Working Group 'Martian Nomenclature' under the following form: G. P. Kuiper—Chairman; A. Dollfus, J. S. Hall, R. B. Leighton—Members.

Scientific Sessions I and II

The following scientific papers were presented:

F. Link—Sur la parenté des Circulations Atmosphériques Terrestre et Jovienne.

C. Sagan et J. Pollack—Elevation differences on Mars.

J. Pollack and C. Sagan—A wind-blown dust model of the surface and seasonal changes of Mars.

W. M. Irvine, T. H. Simon and D. H. Menzel—Multicolor Photoelectric Photometry of Venus, Mars and Jupiter.

L. A. Bugajenko, I. K. Koval, A. V. Moroženko (communicated by D. Y. Martynov—*Photoelectric observation of Mars in* 1967.

V. V. Avramchuk (communicated by D. Y. Martynov)—*On the distribution of methane absorption in the band* 6190 Å *over the disc of Jupiter.*

O. I. Bugajenko, I. K. Koval, E. G. Janovitsky (communicated by D. Y. Martynov)—*The effect of turbulent image motion on photometric observation of planets.*

M. Combes—Etude de haute atmosphère de Jupiter.

P. B. Boyce—Wavelength Dependance of the Brightness of the Individual Martian Features determined by Area Scanning Spectrophotometry.

T. Gehrels—Polarimetry of Planets.

Correction to Draft Report, Commission 16—*Page* 300—*Line* 8

Instead of:
'that the value of the abundance of molecular hydrogen ranges between 75 and 150 km/atm.—(66)',
to read:
'that the value of the abundance of molecular hydrogen ranges between 25 and 45 km/atm. and that of helium 75 and 150 km/atm.'

17. COMMISSION DE LA LUNE

PRESIDENT: Dr D. H. Menzel.
SECRETARY: B. M. Middlehurst.

First Meeting, 23 August 1967

The President opened the meeting with a few preliminary announcements. The President proposed that Miss B. Middlehurst act as secretary during the XIIIth General Assembly. Ten new members have been proposed for Commission 17. Dr O. Günther retires. Two sessions would be held for consideration of lunar nomenclature for the far side by a working group under the chairmanship of Dr Z. Kopal. On August 26, invited papers and reports on the results of the Surveyors I, II, and III, Orbiters I through IV, and Zond 3 would be given in a meeting open to all Congress participants. A joint session with Commissions 16 and 40 was scheduled for Monday, August 28th under the chairmanship of G. Pettengill to report on radar and radiometric observations of the Moon and planets. The sessions of August 29 and 30 were to be devoted to papers; titles are given below. An additional meeting devoted to lunar transient events was called for August 25. The President then stepped down and the chair was taken by Dr Kopal for the remainder of the meeting for discussion of lunar nomenclature.

SESSION 1 ON LUNAR NOMENCLATURE

CHAIRMAN: Dr Z. Kopal.
MEMBERS OF WORKING GROUP: A. Dollfus, F. J. Heyden, Z. Kopal, K. Koziel, G. Kuiper, D. Martynov, D. Menzel, A. Mikhailov, M. Minnaert, E. Shoemaker.

Details of the present status of lunar data were surveyed by the Chairman and members of the Working Group, and new ACIC and Russian maps of the far side of the moon were displayed. The coverage, including the recently completed series of photographs by Orbiter V, is now complete. It was stressed that, above all, we need a control system to establish beyond doubt the identity of named or numbered features; also that the reality of the features themselves must be recognized unambiguously. Suggestions made as to the method to be adopted were: (1) that the same method as that used on the front side should be used for the far side (*Mikhailov*); (2) that we should divide the far side into geometrical regions, perhaps defined by a 10° × 10° grid, with some flexibility for adjustment to accommodate the borders of topographical features (*Kuiper* and others), and that key features should be designated for reference within these areas.

The relative merits of the use of names, letters, and numbers as designations for the lunar features were discussed, also the suitability of classes of names (astronomers, allied scientists, and a wider group to include names associated with the present space efforts, particularly those of cosmonauts, astronauts, and others who have lost their lives in space vehicles). The need to avoid undue haste in the adoption of names was stressed (*Urey*) and suggestions for a waiting period of from five to one-hundred years were made (*Minnaert, Urey,* and others).

Second Meeting, 24 August 1967

SESSION 2 ON LUNAR NOMENCLATURE

CHAIRMAN: Dr Z. Kopal.

The second part of the *Atlas of the Far Side of the Moon* [*Атлас оьратной стороны луны частьп,* Издателъство 'Наука', Москва, 1967], was distributed to members of the Working Group on lunar nomenclature at the beginning of the meeting.

The first speaker, Dr *Schingareva*, referred to the new USSR maps of the far side of the Moon and to the Russian nomenclature already in use. She welcomed the opportunity of comparing the USSR results with the ACIC charts and stated that after reception of the data of Zond 3 in 1965, it became necessary to have some named objects for reference. The principles followed in naming the features were (1) that the same density of named objects as on the front side was desirable; (2) it was not necessary to limit the size of the named objects, but efforts were made to keep the diameter greater than 25 km; the least is 18 km. With regard to the names and referring to Dr Kuiper's statement in the first session that it was necessary to include names prominent in connection with a variety of sciences, the USSR agrees and their proposal has included the names of Soviet rocket engineers, etc. Examples were given. This tribute is being paid to workers engaged in promoting space research. The USSR is grateful it they can have better data from ACIC so that mistakes can be avoided.

The Chairman then opened a discussion as to the relative merits of classes of names, on the necessity for unambiguous identification, and a description of the coverage by Orbiters I through V. Dr *Kuiper* pointed out necessary revisions in the light of new material following examination of the names adopted by the IAU at the 1964 meeting for the back side of the moon. He pointed out that many of these features had been identified from full moon photographs and stressed the necessity for good oblique lighting conditions; but at that time only the Zond data (full moon) had been available. In the following discussion, Dollfus, Kopal, Kuiper, Martynov, Menzel, Mikhailov, Minnaert, Shane, Shoemaker, Urey, and others took part. Dr. *Shane* made the following comments on the American position. The US National Committee on Lunar Mapping and Nomenclature have felt that (1) we should be very conservative in assigning names; (2) because the names are to be permanent, we should try to use names of permanent renown. There should be no hurry, but on the contrary, the scheme should be worked out as fully and carefully as possible. In addition to names, a system of numbers and/or letters would allow for later additions by future generations. A start towards assigning the designations could be made at a later meeting of the IAU.

Dr *Minnaert* then proposed that a start should be made now and this motion was seconded by Kopal. Dr *Menzel* stated that he agreed essentially with Dr Minnaert, but that there was a practical problem as to procedure in the time before the next meeting. There is one possibility that does not commit us to any system of names, as the ACIC have already placed about 450 numbers on the charts of the far side as temporary identifications. It is possibly easier to work by number. We should try to keep the naming as simple as possible and avoid duplication, and for the time being, we can do no better than to adopt these numbers to which we can add as better material becomes available. Dr *Mikhailov* stated that he was in agreement with Dr Minnaert and that we should let the future generations take care of themselves; Dr Menzel's proposal was a very reasonable one. To assign names within a very short time would be perhaps premature. We could appoint a small committee to assign names to be approved by the next IAU meeting. Dr *Mikhailov* added that it is clear that it is better not to have to revise names as happened in the first excitement of the original probes in 1964. This is not likely to happen again, but it is still wiser to proceed cautiously. A motion that an international committee be set up for such a purpose was proposed by Dr *Menzel.* Amendments suggested by Dr *Kopal* were as follows: (1) we should not plan to adopt a greater number of names than on the front side of the Moon; (2) for formations below the minimum size for which names are adopted, designations should be by number. Both amendments were carried unanimously. Dr Menzel's motion was then passed.

Details of nomenclature were then considered. The use of the term *thalassoids* was debated, but it was generally agreed that for the time being the name 'thalassoides' should not be adopted officially as description of a class of features. Dr *Kuiper* pointed out that though the far side has such features, the near side has them also and on the near side they are called 'basins.'

The distribution of the ACIC map was then considered. Dr Brunk, Chief of Planetary Astronomy, NASA, USA, distributed 30 copies of the ACIC Lunar Far Side Charts (LFC-2, scale 1:10000000, 1st edition, August 1967) with designated numbers overlaid in red to the members of the Working

Group. He stated that the chart without overlay is generally available and additional copies can be had by providing name and address. The special edition would be available through Dr Brunk's office (address: Lunar and Planetary Program, Office of Space Science and Applications, National Aeronautics and Space Administration, Washington, D.C., 20546, USA). Fifteen-hundred copies of the map without overlay would be available for distribution to Congress participants after the third session.

Dr *Dollfus* stressed the importance of continuity in the procedures used for lunar nomenclature and pointed out that the rules previously used by astronomers were reformulated and improved by Commission 16 (Physical Study of Planets) in 1961. The IAU General Assembly in Berkeley in 1961 had adopted a resolution referring to procedures for lunar nomenclature (*Trans. IAU.*, **XIB**, 235). The text of this resolution was then read by Dr *Kopal* and reaffirmed by the present committee.

A resolution was then put forward by Dr *Menzel* and seconded by Dr *Mikhailov* to be passed to the General Assembly for consideration: that the assignment of names for lunar features should be postponed until the next meeting of the IAU. The resolution was carried unanimously and its formal expression is given in the section 'Resolutions' under No. 8 on page 43.

Third Meeting, 26 August 1967

PRESENTATION OF LUNAR PROBE DATA

This session was open to all participants of the Congress

CHAIRMAN: D. Menzel.

The following invited papers were presented:

Paper	*Author*
The Surveyor spacecraft and its scientific instruments (Presented by E. Shoemaker)	H. Haglund, Surveyor Project Manager, Jet Propulsion Laboratory, USA
Details of Orbiter spacecraft and methods used for photography	I. Taback, Lunar Orbiter Spacecraft Manager, NASA, USA
The lunar Orbiter photographs, including those of the limbs and near and far sides	W. Brunk, NASA, USA
Scientific results regarding the lunar surface from pictures relayed from Surveyors I and III	E. Shoemaker, USGS, USA
UV and IR results from Zond 3	A. I. Lebedinsky, Moscow University, USSR
Zond 3 Maps	V. V. Ševčenko, GAIŠ, Moscow, USSR
Orbiter IV photographs of sinuous rilles: evidence for the presence of water on the Moon	H. Urey, La Jolla, USA

At this meeting, 1500 copies of the ACIC map, LFC-2 (1st ed. Aug. 1967) were distributed.

Fourth Meeting, 28 August 1967

RADAR AND RADIOMETRIC OBSERVATIONS OF THE MOON AND PLANETS

CHAIRMAN: G. Pettengill.

This session is reported by Commission 40.

Fifth Meeting, 29 August 1967

PHYSICAL STUDIES OF THE MOON

CHAIRMAN: D. Menzel.

The following papers were presented:

Paper	*Author*
Report on the *Consolidated Lunar Atlas*	G.P. Kuiper
Early results from the magnetic field experiment on lunar Explorer 35 (Presented by J. Brant)	N.F. Ness, K.W. Behannon, C.S. Scearce, S.C. Cantarano
Infrared measurements of the lunar surface including Surveyor I and III results	J. Saari
Data processing of lunar IR results	H. Ingrao, *et al.*
Thermal response of lunar surface at landing site (Presented by H. Ingrao)	J.T. Holland, H. Ingrao
Lunar surface models including temperature-dependent thermal properties	J.L. Linsky
Comments on the nature of the lunar surface	D. MacRae
Opposition effect and the lunar luminescence during eclipses	F. Link
Orbiter photographs showing morphological correlations at lunar transient event sites	B. Middlehurst
Lunar transient event reports: the need for truly independent confirmation	P. Moore
Lunar volcanism	G. Fielder
Origin of lunar craters	N. Boneff

Sixth Meeting, 29 August 1967

SELENODESY AND RELATED QUESTIONS

CHAIRMAN: Dr K. Koziel.

The following papers were presented:

Paper	*Author*
Absolute and relative heights and shapes on the Moon	J. Hopmann
On libration	A.A. Nefediev
A new secondary selenodetic net (Presented by E.A. Whitaker)	D.W.G. Arthur
Theory of physical librations	D. Eckhart
Shape, size and hypsometric peculiarities	I.V. Gavrilov
Determination of parameters of the lunar rotation by a method independent of its profile	A.A. Gorynya
The annular solar eclipse of May 20, 1966	I.M. Demenko, A.S. Duma, D.P. Duma
Cracow libration papers	K. Koziel
Lunar libration	N. Moutsoulas

Seventh Meeting, 30 August 1967

INSTRUMENTS, PHYSICAL STUDIES, AND OTHER PAPERS

CHAIRMAN: Professor A.I. Lebedinsky.

The following papers were presented:

Paper	*Author*
A polarizer-discriminator	A.K. Korol
Role of secondary craters	B. Levin (1)
Role of slopes and albedo	B. Levin, L.N. Radlova (2)
Lunar photometric function	G. Fielder
Evidence for the presence of water on the Moon	H. Urey
Laboratory investigation of the proton-excited luminescence of material of types which might exist on the lunar surface	J.E. Geake
The lunar eclipse of April 24, 1967; earth- and lunar-based photographs suggest that diffraction in the lower levels of the earth's atmosphere may appreciably increase the brightness of the eclipse during totality	B.M. Middlehurst

An additional session was called on August 25 to discuss a proposal by Miss Middlehurst to coordinate efforts of many observatories and other groups of lunar workers in order to promote the greatest efficiency in observing lunar transient events. It is hoped by such a cooperative effort to increase the present small number (16) (**3, 4**) of permanent records and also to obtain records of good quality. The Smithsonian Observatory has offered the use of its satellite communications network, once the communications project is properly funded, to facilitate rapid communication between participating groups. *Dr Urey* spoke strongly in support of the research and said that volcanic activity must have occurred on the Moon. What is still uncertain is the frequency of lunar events now. *Patrick Moore* described the work of the British lunar observers who are participating in the watch for events and stressed the need for confirmed observations. *Dr Hopmann* discussed some of the earlier reports which he said could not possibly be dismissed as erroneous. *Dr T. W. Rackham* and *Mr E.A. Whitaker* described filter techniques. Summing up, *Miss Middlehurst* said that, in view of the widespread support and interest, the research organization and communications links could be improved in the near future. It is to be hoped that, as in the case of cometary and variable star research, observers with small telescopes in many parts of the world could materially assist the research by providing a constant watch and that, by making use of a system of alerts, institutions with auxiliary apparatus such as spectrographs, photoelectric photometers and other apparatus could more easily obtain good permanent records.

REFERENCES

1. Levin, B. 1967, *Astr. Zu.*, **44**, 828.
2. Levin, B., Radlova, L.N. 1967, *Astr. Zu.*, **44**, 1296.
3. Middlehurst, B.M. 1967, *Icarus*, **6**, 141.
4. Middlehurst, B.M. 1967, *Rev. Geophys.*, **5**, 185.

19. COMMISSION DE LA ROTATION DE LA TERRE

PRÉSIDENT: B. Guinot.
SECRÉTAIRE: R. Vicente.

Comptes-rendus des séances

La commission a tenu 4 séances, les 23, 25, 26 et 28 août 1967.

I. BIH

Il est annoncé que tous les problèmes touchant le BIH seront discutés lors des réunions de la commission 31.

II. SYMPOSIUM UAI 32 ET RÉSOLUTIONS DU COMITÉ NATIONAL DE GÉODÉSIE D'URSS

W. Markowitz présente les résolutions adoptées par le symposium 32 tenu à Stresa en mars 1967 dont le sujet était: mouvement séculaire du pôle, dérive des continents, rotation de la Terre. Le texte de ces résolutions apparaît dans les comptes-rendus de ce symposium. Une des résolutions proposait le choix du 'nouveau pôle de 1900–05' de Cecchini comme pôle-origine, une autre la constitution de nouvelles chaînes d'instruments destinées à éliminer les erreurs introduites par les catalogues d'étoiles.

A.A. Mihailov présente ensuite un rapport du Comité National de Géodésie d'URSS relatif également à la constitution de chaînes d'instruments sur des parallèles communs: chaîne de PZT sur le parallèle international + 39°, chaînes de PZT et d'astrolabes à installer dans l'hémisphère austral.

Comme le montre le rapport du groupe de travail constitué à Stresa, la synthèse des points de vue sur les chaînes d'instruments fut aisément réalisée (recommandation 1).

A propos de l'origine du pôle, *E.P. Fedorov* s'oppose à la résolution de Stresa car, selon lui, on ne peut pas affirmer qu'il existe réellement une dérive du pôle géographique et parce que, pour certains travaux de géophysique, le pôle 'fixe' est, en fait, le pôle défini par le petit axe d'inertie de la Terre. *P. Melchior* préférerait l'appellation 'pôle origine' au lieu de pôle fixe. *E.P. Fedorov*, cependant, reconnait que pour des raisons de commodité, conformes aux traditions astronomiques, on peut exprimer les coordonnées du pôle instantané dans un système défini par les latitudes conventionnelles des 5 stations du parallèle + 39°, sans mentionner que ce système définit un pôle à une date quelconque. Ce point de vue est adopté unanimement par la commission (résolution 1).

III. RÉSOLUTION PROPOSÉE PAR LE COMITÉ NATIONAL BELGE SUR LES NUTATIONS À COURTE PÉRIODE ET LES MARÉES TERRESTRES

Un exposé de synthèse présenté par *P. Melchior* montre comment les perturbations extérieures exercées sur la Terre donnent lieu simultanément à la précession et aux nutations de l'axe de rotation dans l'espace, aux nutations de l'axe de rotation par rapport à la Terre et aux marées terrestres et insiste sur l'intérêt de ne pas dissocier l'étude de ces divers phénomènes.

La commission 19, pleinement consciente de la valeur des travaux de P. Melchior et de ses collaborateurs a unanimement recommandé que ces travaux soient poursuivis et étendus; elle a approuvé dans son principe une résolution soumise par le Comité National Belge à l'Assemblée Générale qui proposait l'examen de ces phénomènes, mais elle en a légèrement modifié la forme à la demande de P. Melchior et W. Markowitz (résolution A2).

IV. RAPPORTS DU SIL ET DU SIMP

T. Nicolini a définitivement calculé les coordonnées du pôle pour la période 1941–1949. Le volume

X des résultats du SIL est sous presse. Les coordonnées du pôle publiées dans le vol. XIIIA des Transactions de l'UAI ont pu être améliorées par l'usage de séries d'observations discontinues de Carloforte et Kitab. Voir Annexe 1.

S. Yumi présente le rapport sur les travaux du SIMP. Les résultats préliminaires paraissent dans les rapports annuels auxquels on voudra bien se reporter. Les coordonnées du pôle de 1962·00 à 1966·35 ont été publiées dans le volume XIIIA; on donne (Annexe II) les coordonnées de 1966·00 à 1967·20. Les coordonnées de 1966·00 à 1966·35 ont été très légèrement modifiées.

La commission exprime unanimement ses remerciements à *T. Nicolini* et *S. Yumi.*

Nous n'avons pas reçu d'informations plus récentes que celles publiées dans le vol. XIIIA sur la publication des résultats de la période 1949–1962.

V. PRÉSENTATION DE TRAVAUX SCIENTIFIQUES

E.P. Fedorov rappelle les traits caractéristiques des séries de latitude:

(1) il est impossible de caractériser la précision des observations par un seul nombre, la précision étant fonction de la fréquence;

(2) les erreurs accidentelles peuvent simuler des variations périodiques.

Il présente ensuite les travaux de ses collaborateurs:

(*a*) I.I. Glagoleva et Ya.S. Yatskiv ont développé pratiquement une méthode d'analyse qui exclut tout arbitraire;

(*b*) S.P. Major a réduit à un système commun les mesures de latitude de Mizusawa, Carloforte et Ukiah, de 1900 à 1934 et les a analysées, il donne une liste des termes périodiques trouvés;

(*c*) V.K. Tarady a tenté de déduire les termes principaux de la nutation de l'analyse des observations du SIL, l'accord entre les valeurs observées et théoriques n'est pas satisfaisant.

S. Débarbat a analysé les mesures de temps et de latitude faites à Paris depuis 1956·5 jusqu'à 1963·0 avec l'astrolabe Danjon; elle en a déduit les variations lunaires de la verticale et a mis en évidence le terme de nutation diurne prévu par Molodensky et déjà trouvé par d'autres auteurs.

S. Yumi a présenté les travaux des astronomes de l'observatoire de Mizusawa: (*a*) T. Okuda, d'après l'étude des termes Z, est conduit à penser qu'il pourrait exister des variations globales du potentiel de la Terre. (*b*) Y. Wako suppose que le terme Z est dû à des termes erronés de la nutation; (*c*) M. Kirita montre que le mouvement annuel du pôle est dû principalement à la circulation atmosphérique.

N.S. Sidorenkov estime que la circulation atmosphérique rend compte des irrégularités à longue période; il attire l'attention sur la nécessité d'avoir des données météorologiques plus précises et plus complètes.

M. Torao présente l'étude de S. Iijima et S. Okazaki sur la composante biannuelle de la vitesse de rotation de la Terre. Cet effet, d'amplitude totale 18 ms, est attribué aux vents stratosphériques.

N. Stoyko discute d'abord de la précision des coordonnées du pôle instantané pour la période 1941–1949. Il étudie ensuite les diverses méthodes de représentation de la rotation de la Terre: ces méthodes ne permettent pas une extrapolation. Cependant la corrélation entre la vitesse de la rotation et la longitude des noeuds de la lune, de Mme A. Stoyko semble se confirmer.

A. Stoyko a trouvé de fortes corrélations entre l'activité du Soleil, la nutation libre, la vitesse de rotation de la Terre, la circulation générale de l'atmosphère et l'énergie des séismes profonds.

H. Enslin étudie les méthodes pour calculer les corrections de groupes. Ces diverses méthodes donnent des résultats équivalents pour un programme donné. Mais il y a intérêt à choisir le programme qui correspond le mieux au but recherché, il y a en particulier, intérêt à réduire le nombre des groupes d'un programme pour éliminer les erreurs en fonction de l'ascension droite.

H.M. Smith présente le projet d'astrolabe à réflexion de D. Thomas. La réflexion sur un prisme de CER-VIT permettrait d'éliminer les effets thermiques sur la distance zénithale, constatés sur les astrolabes Danjon et de choisir une distance zénithale instrumentale quelconque. Les dimensions de l'instrument pourraient être accrues. Un montage expérimental sur un astrolabe Danjon est en cours.

Annexe I. Coordonnées du pôle de 1941 à 1949 (valeurs révisées), rapportées à l'origine conventionnelle internationale
unité 0″.001

Fraction d'année	1941		1942		1943		1944	
	x	y	x	y	x	y	x	y
060	+ 66	+ 65	+ 156	+ 62	+ 89	+ 207	− 11	+ 184
143	+ 45	+ 37	+ 98	+ 74	+ 86	+ 195	+ 3	+ 229
226	+ 22	+ 90	+ 75	+ 45	+ 132	+ 177	+ 25	+ 222
310	− 15	+ 101	− 3	+ 113	+ 108	+ 134	+ 87	+ 228
393	+ 8	+ 134	+ 35	+ 100	+ 153	+ 140	+ 194	+ 203
476	+ 29	+ 177	+ 9	+ 110	+ 135	+ 82	+ 285	+ 171
560	+ 69	+ 196	+ 23	+ 113	+ 122	+ 46	+ 302	+ 59
643	+ 172	+ 209	+ 15	+ 112	+ 118	+ 43	+ 292	− 8
726	+ 137	+ 192	− 5	+ 171	+ 40	+ 26	+ 161	− 83
810	+ 129	+ 154	− 19	+ 170	− 30	+ 50	+ 2	− 88
893	+ 148	+ 97	+ 33	+ 172	− 85	+ 81	− 59	− 56
976	+ 144	+ 83	+ 25	+ 225	− 28	+ 91	− 133	+ 12

Fraction d'année	1945		1946		1947		1948	
	x	y	x	y	x	y	x	y
060	− 162	+ 98	− 121	− 57	+ 103	− 75	+ 264	+ 108
143	− 158	+ 131	− 189	+ 2	+ 7	− 77	+ 237	+ 7
226	− 113	+ 238	− 184	+ 86	− 76	− 28	+ 191	− 19
310	− 53	+ 301	− 145	+ 203	− 150	+ 91	+ 106	− 62
393	+ 94	+ 310	− 74	+ 285	− 154	+ 160	+ 17	− 23
476	+ 200	+ 311	+ 62	+ 330	− 84	+ 246	− 32	+ 27
560	+ 266	+ 172	+ 174	+ 346	+ 10	+ 285	− 29	+ 67
643	+ 286	+ 105	+ 219	+ 324	+ 119	+ 324	− 38	+ 139
726	+ 329	+ 10	+ 254	+ 268	+ 189	+ 311	− 78	+ 184
810	+ 252	− 95	+ 309	+ 122	+ 263	+ 272	− 32	+ 253
893	+ 114	− 137	+ 323	− 30	+ 286	+ 227	+ 26	+ 305
976	+ 64	− 104	+ 233	− 40	+ 278	+ 148	+ 129	+ 324

Annexe II. Coordonnées du pôle de 1966·00 à 1967·20 rapportées à l'origine conventionnelle internationale
unité 0″.001

Date de Bessel	x	y	Date de Bessel	x	y	Date de Bessel	x	y
1966·00	+ 75	+ 114	1966·40	− 119	+ 215	1966·85	+ 125	+ 291
05	+ 33	+ 103	45	− 115	+ 255	90	+ 123	+ 273
10	0	+ 98	50	− 104	+ 298	95	+ 115	+ 253
15	− 29	+ 100	55	− 86	+ 330	1967·00	+ 99	+ 233
20	− 58	+ 108	60	− 57	+ 344	05	+ 79	+ 213
25	− 86	+ 124	65	− 10	+ 345	10	+ 56	+ 194
30	− 105	+ 149	70	+ 52	+ 337	15	+ 31	+ 177
35	− 116	+ 181	75	+ 96	+ 324	20	+ 12	+ 165
			80	+ 117	+ 308			

Annexe III. Complément au rapport du vol. XIIIA

INFORMATION ON THE LATITUDE OBSERVATIONS OF THE PULKOVO OBSERVATORY

Latitude observations were renewed at Pulkovo in 1948 when the old zenith-telescope of 135 mm aperture was put in operation after the rebuilding of the observatory during the war. The program of observation compiled for a whole period of nutation was completed in 1967. During this time 47.4 thousand determinations of latitude were made. Beginning from 1955 the results were communicated to the SIL and SIR. In 1955–61 the observations were made on an extended program lasting through the whole night (in winter up to 18 hours), in order to determine latitude variations of short period. These observations are reduced with the new value of the constant of aberration and taking into account Oppolzer's terms and the tidal deflection of the vertical. Much attention was paid to the determination of the value of the screw.

In order to determine the second coordinate of the pole's motion a new latitude station was organized at Blagoveshchensk in the Far East, 97° east of Pulkovo. Regular observations with a zenith-telescope of 180 mm aperture were begun in 1959. From 1959·17 to 1967·42, 19528 latitude determinations were obtained. From the beginning of 1965 the results are sent to the international services at Mizusawa and Paris.

BIBLIOGRAPHY

Kostina, L. 1966, On the increase of precision of latitude calculations. Analysis of latitude observations, *Trudy Tashkent astr. Obs.*, **12,** 85–104.

Résolutions proposées par la commission no. 19

Résolution

(1) Les coordonnées du pôle instantané doivent être rapportées à une origine définie par les coordonnées initiales suivantes:

Mizusawa	+ 39°8′ 3″.602
Kitab	1″.850
Carloforte	8″.941
Gaithersburg	13″.202
Ukiah	12″.096

Cette origine est appelée Origine Conventionnelle Internationale (OCI).

Recommandations

(1) Considérant que le problème de la dérive continentale est d'une grande importance et que les vitesses et les directions des dérives relatives des continents pourraient être confirmées et mesurées par des observations astronomiques, l'UAI recommande aux autorités concernées:

(*a*) que les stations existantes qui emploient soit des lunettes photographiques zénithales, soit des astrolabes Danjon et qui sont situées sur des parallèles suffisamment proches fassent des observations conjointes du temps et de la latitude. Cela implique que les mêmes étoiles appartiennent aux programmes d'observation de chacune des stations de la chaîne,

(*b*) que toutes les stations internationales des latitudes en fonction sur le parallèle nord de 39°8′ soient équipées de lunettes photographiques zénithales et qu'elles fassent avec ces instruments des observations régulières et conjointes du temps et de la latitude sans cesser les observations de la latitude avec les instruments visuels,

(*c*) qu'une chaîne similaire de lunettes photographiques zénithales soit établie dans l'hémisphère austral à une latitude voisine de − 34°, en Australie, en Afrique et en Amérique du Sud (en Argentine et au Chili), (*d*) que les observatoires du Mont Stromlo, du Cap, de La Plata et de Santiago forment une chaîne d'astrolabes et fassent des observations conjointes du temps et de la latitude,

(*e*) que, lorsque de nouvelles stations sont établies, elles soient placées en des lieux tels qu'elles puissent faire partie d'une chaîne existante ou en former une nouvelle,

(*f*) qu'une distribution d'astrolabes régulière en latitude soit conservée afin d'étudier les dérives entre stations de latitudes différentes,

(*g*) que soit étudiée la possibilité d'utiliser des satellites artificiels et des cataphotes sur la lune, afin de mesurer la dérive des continents avec grande précision.

Remerciements

(1) La commission 19 de l'UAI, appréciant la qualité des travaux du Service International du Mouvement Polaire et la rapidité de la publication de leurs résultats, prie le Dr. S. Yumi, directeur de ce service, d'accepter l'expression de sa reconnaissance.

(2) La commission 19 de l'UAI, ayant pris connaissance des valeurs définitives des coordonnées du pôle pour la période critique 1941–1949 prie le Professeur T. Nicolini qui s'est chargé du travail difficile de leur calcul, d'accepter l'expression de ses remerciements.

20. COMMISSION DES POSITIONS ET DES MOUVEMENTS DES PETITES PLANETES, DES COMETES ET DES SATELLITES

PRESIDENT: P. Herget.
SECRETARY: G.A. Wilkins.
INTERPRETERS: J. Kovalevsky and E. Rabe.

First Meeting, 24 August 1967

The meeting was opened at $14^h 00^m$ by the President with a word of welcome, the introduction of the Secretary and Interpreters and a recognition of each of the 21 members present. The names of new members of the Commission were read: Debehogne, Galibrina, Marsden, Milet, Sekanina and Sitarski. The following members have retired: Jeffers, Kahrstedt, König and Reinmuth.

During a standing tribute to the memories of the members who had passed away during the last triennium the President read the following short obituaries:

Dirk Brouwer (1902–1966) was educated at Leiden and spent his entire astronomical career at Yale. He was an ardent worker, he published prolifically, and he became the dean of celestial mechanicians in America. He succeeded Schlesinger as director of the Yale Observatory in 1941, he carried out the completion of the Yale Zone Catalogues, and he has been a respected and admired leader in the current large astrometric programs. The advent of artificial satellites directed his attention to the peculiar problems of their general perturbations, and he set a high standard for all workers in this field. He also served as editor of the Astr. Journal since 1941. He was elected a member of the U.S. National Academy of Sciences in 1951, and he served as President of Commission 20 during the critical years from 1948 to 1955.

W. W. Heinrich (1884–1965) received his Doctors' degree from the Charles University in Prague in 1908, having published his first paper on the libration of Patroclus (the second known Trojan asteroid) in 1907. He spent short periods at Strasbourg, Göttingen and Königstuhl before returning in 1912 to Prague, where he became Professor in 1926 and remained until his retirement in 1957. He published about fifty papers on the three-body problem and was an active member of Commissions 7 and 20; at the age of eighty, he presented a paper at IAU Symposium No. 25 in Greece and afterwards attended the General Assembly at Hamburg.

F. Kepinski (1885–1966) studied at Warsaw and Göttingen. He was an assistant in Berlin-Babelsberg and then at Warsaw Observatory. In 1927 he became professor and head of the Observatory of the Warsaw College of Technology. In 1926 he started his investigations of the motion of the Comet Kopff. His research in this field was crowned by the rediscovery of that comet in 1958 after its close approach to Jupiter in 1954. He published two textbooks on spherical and practical astronomy and 88 scientific papers on problems of geodetic astronomy and on motion of comets and asteroids.

Gerrit Pels (1893–1966) from Woerden, The Netherlands, attended secondary school at Utrecht and in 1919 was appointed as a computer at Leiden. He continued his studies of mathematics, and he was a highly valued assistant to the senior staff. He worked with De Sitter on the satellites of Jupiter, with J. Woltjer on the satellites of Saturn, with Oort in stellar dynamics, and he became very skilled in computing the orbits of comets and minor planets, and in measuring and reducing the plates taken by Van Gent in South Africa. He used the photographic refractor to observe the selected minor planets of Brouwer's program, and selected proper motion stars from the lists of Platt and Blaauw. He searched two large regions of the sky for new members of the Hyades and Pleiades clusters. In 1954 he was appointed to the senior staff at Leiden in recognition of his outstanding contributions, and he continued work until his final illness.

M.F. Subbotin was born in Ostrolenka (Poland) in 1893. He studied in Warsaw University. In 1922 he was appointed Director of the Tashkent Observatory, from 1930–1960 he was professor of

Celestial Mechanics at the Leningrad University. In 1942–1964 he was Director of the Institute of Theoretical Astronomy. His contributions in Celestial Mechanics concerned the fundamental problem of improving the convergence of the expansions of the disturbing function and the determination of comet orbits. His well known three-volume manual of Celestial Mechanics and his activity in guiding astronomical institutions inspired many young people to work in this branch of astronomy. He worked also in pure mathematics. He was stricken with a heart attack on 26 December 1966 in Leningrad.

The appointment by the Executive Committee of the officers of the Commission for the coming triennium were announced: Professor G. A. Chebotarev, as President; and Professor F. K. Edmondson, as Vice-President. The President expressed his regrets that Dr S. G. Makover, the present Vice-President was not in good health and would not be able to serve as President.

The agenda for the sessions were announced. Z. Sekanina invited the members of the Commission to visit the Center for Numerical Mathematics in the University of Prague during the afternoon of 28 August.

The volume of Ephemerides of Minor Planets for 1968 had just been published by the Institute of Theoretical Astronomy, Leningrad, and copies were available for the members of the Commission.

DRAFT REPORT

The President expressed his thanks to those members who had contributed to the Draft Report, which he considered to be a very important activity of the Commission. The following contributions were, however, received too late for inclusion in the Report.

The Institute of Theoretical Astronomy, Leningrad, reports that the Crimean Astrophysical Observatory has obtained 908 positions of 135 minor planets in 1966 and 450 positions of 53 planets in the first half of 1967. The rate may be expected to remain at about 1000 positions per year.

Miss J. Gill (NASA) and Mrs B. Gault (Yale) have determined a new position of the pole of Neptune's equator using more than 3000 visual and photographic observations of Triton's position angle and separation. These observations encompassed 85 seasons from 1887 to 1958. The reciprocal mass of Neptune was found to be $1/19296 \pm 9$, which is 3 per cent smaller than the value 1/18730 published by Van Biesbroeck in 1957.

R. B. Hunter (*Mon. Not. R. astr. Soc.*, **136**, 245, 267, 1967), using hypothetical starting conditions, has investigated the evolution of the orbits of possible satellites of Jupiter and close minor planets, by means of numerical integration. His conclusions regarding stability are in agreement with the known satellites, and he is led to suspect the possibility of a ring of minor planets between Jupiter and Saturn.

PRESIDENTIAL ADDRESS

The President briefly reviewed the present status and problems of minor planet work. The Naval Ordnance Research Calculator (NORC) is being withdrawn from service, but it has completed the computation of some 33000 ephemerides at opposition for 450 numbered minor planets up to the year 2000 and for many unnumbered objects. The Minor Planet Center at Cincinnati has completed much of its set task of improving the orbits of numbered minor planets, and those based on recent identifications. The comparison with observations gives residuals that on average are less than $2\overset{''}{.}5$; the orbits of some of these minor planets were previously not secure and one 'lost planet' was recovered. The Institute of Theoretical Astronomy at Leningrad continues to produce the annual volume of ephemerides and accepts responsibility for three-quarters of the numbered minor planets; about 300 orbits have been improved recently.

The completion of the Leiden survey of faint minor planets would raise the question as to whether it is worthwhile continuing to study minor planets individually once the properties of the minor planet population are known statistically. Certainly it is desirable to re-examine the aims and procedures of minor planet work. It is imperative to keep up the work on selected minor planets and it is desirable that observations down to the 17th magnitude should be made. On the other hand the annual volume should mark those minor planets for which observations are no longer required. The Minor Planet Circulars give ephemerides for minor planets for which further observations are urgently needed; observations should be made over an extended arc to ensure correct identification. No progress had been made on the proposal (see *Trans. IAU*, **12B**, 236) to resurvey the sky since it has not been possible to find a suitable instrument or site.

GENERAL DISCUSSION AND REPORTS

The President suggested three topics for general discussion: (*a*) the accuracy with which positions should be given, (*b*) cooperation in reductions and (*c*) reference star catalogues.

(*a*) *Accuracy of Positions.* In the case of short-focus telescopes it seems desirable that the maximum precision should be obtained at the first attempt but otherwise it is sufficient to publish approximate positions and leave the accurate measurement and reduction until a request is made by a computer. *H. Hertz* asked whether it would be feasible for observers with much material to list the dates and centers of unmeasured plates. *Herget* commented that such a list is difficult to use, and that the limiting magnitude would also be significant. The situation should not exist, but perhaps the computer requiring the observations should circulate search ephemerides and let the observers look on any relevant plates.

(*b*) *Cooperative reductions.* The need for cooperative reductions is becoming less acute as electronic computers become more widespread. *E. Roemer* pointed out, however, that the problem will continue to exist for long-focus plates of faint minor planets and comets since ordinary reference stars cannot be used.

(*c*) *Reference star catalogues.* The situation would be considerably improved once the Bergedorf AGK2/3 program was completed. For the time being consideration might be given to the use of the Smithsonian Astrophysical Observatory catalogue which exists in printed form and on magnetic tape.

Leiden Survey. C.J. van Houten gave a brief report on the Leiden Survey of faint minor planets. The survey is based on plates taken with the 48-inch Palomar-Schmidt telescope to a limiting magnitude of 20·5. Reliable orbits had been determined for 85 per cent of the 2100 objects for which at least 3 positions had been obtained. The material was being used to determine the number/-magnitude relationships in more detail, to look for fine structure in the number/mean-distance relationship near the Kirkwood gaps, and to find new families. In the ensuing discussion, he stated that 15 new Trojans had been found.

Absolute Magnitudes. T. Gehrels reported on his recent work on the photo-electric light curves of individual minor planets and on his new tabulation (see pp. 121–132) of the photographic magnitudes of the numbered minor planets. A full report of the work will appear in the *Astronomical Journal.*

First Session on Comets, 24 August 1967

CHAIRMAN: Dr E. Roemer.

Use of large telescopes. In the discussion of the draft of Resolution 4 (see below) *the Chairman* pointed out that faint comets are presently being observed only at Tucson and Tokyo, although she congratulated the observers at Skalnaté Pleso on the skilful work done with a smaller instrument. *Hirose* said that the 74-inch reflector could be used for observations of comets only 3 weeks a year but that the 36-inch Dodaira reflector is available for 2 weeks each month. Positions for faint reference stars pose a difficulty, for the Tokyo Observatory lacks, and has been unable to obtain, many volumes of the Astrographic Catalogue. *Dobrovol'skij* asked whether image converters would be useful, but it was pointed out that the small field of view and changing geometrical distortions render such devices presently suitable only for physical observations. *Roemer* drew attention to the need for accurate predictions (to 5′) for observations with large reflecting telescopes and stated that it was still useful to have two independent predictions. It was agreed that the resolution should be put forward.

Repository of positions. Herget stated that the Cincinnati Observatory can now offer more flexible facilities for the repository of positions. The data, to be supplied in a standard format specified by the Working Group, would be kept on a magnetic tape.

Current Comet Catalogue. The Chairman congratulated *Marsden* on the production of the Supplement to the BAA Catalogue of Orbits. After discussion it was agreed that bibliographic material should be published in the *Transactions IAU*, but that elements should continue to be given in the progress reports in *Quarterly Journal RAS*, preferably in the format of the Catalogue of Orbits.

Magnitude predictions. There was a lengthy discussion on the bases and forms of magnitude predictions, different systems being appropriate to different types of comet (according to the degree of central condensation) and to different instruments. There was also the further difficulty that the appearance of a comet, especially new comets, might change unpredictably. A review article giving practical guidance to observers and computers would be useful.

Cometography. Marsden suggested that consideration be given to the compilation of a new cometography. He pointed out that there is no recent cometography that has the completeness of Pingré's work, now almost two centuries old, even though the compendia by Holetschek, and more recently by Vsehsvjatskij, are very useful. An important feature of the new cometography would be an ephemeris for each comet giving both positions and geocentric and heliocentric distances. Such ephemerides would be helpful in studies of statistics of cometary discoveries and of nuclear sizes. The distances are often omitted from the first ephemerides of a new comet, or, if they are included, they may be substantially in error. Prediscovery ephemerides are obviously useful for identifying possible prediscovery observations. Information should also be included concerning comets for which orbits have *not* been calculated. At present, the most complete sources of data for such comets are the lists by Baldet in the *Annuaire* of the Bureau of Longitudes, but the data there are very brief and no references to the original literature are given. Since so many comets are involved (almost 900 for which orbits have been calculated and an equal number for which they have not), it would be desirable to incorporate only the more reliable observational reports in some form of code. It is also important to consider how the cometography might be kept up to date. *Roemer* noted that a very considerable effort would be involved and that costs of formal publication probably would be very high for a limited distribution. She suggested that it might be possible to issue information in loose-leaf form, starting currently and working backwards. *Candy* drew attention to the *Provisional List of Uncertain Comets* published by the Yamamoto Observatory, Kyoto, in 1956.

Catalogue of Original and Future Comet Orbits. Sekanina reported that he had completed, in 1966, a comprehensive catalogue of comets having nearly parabolic orbits, together with original and future orbits. He now has 15 additional orbits and requests that other computers send him their results, including sufficient detail for assessment of the accuracy, so that the catalogue may be kept current.

Evolution of orbits. Kazimirčak-Polonskaja described her recent numerical studies of the evolution during 400 years of the orbits of some 45 short-period comets under perturbations by the major planets. The work will be published in full elsewhere.

Second Session on Comets, 28 August 1967

CHAIRMAN: Dr E. Roemer.

ACTING SECRETARY: M. P. Candy.

Magnitude predictions. Since no concensus regarding procedure had been arrived at, the *Chairman* proposed that the Working Group study the problem further. Suggestions and comments, to be addressed to the chairman of the Working Group, were solicited.

Cometography. The feasibility of preparing a new cometography will be considered by a committee chosen from among the members of Commissions 15 and 20: Marsden (Chm.), Candy, Dossin, Kresák, Roemer, Sitarski, Vsehsvjatskij.

Non-gravitational forces. The motion of a number of periodic comets has now been studied with sufficient care, so that non-gravitational effects can be recognized with some confidence. There is, however, a divergence of opinion as to whether the events that produce the anomalous accelerations are continuous or discontinuous and as to the extent to which the activity may be concentrated near perihelion. A number of short reports intended to be representative of recent work and current thinking were presented.

Marsden described a new study of the orbits of six comets at their recent returns. The excellent n-body integration program of Schubart and Stumpff was used, the attractions by all nine planets being considered. The coefficients for the differential corrections, which include the full effect of the planetary perturbations, were obtained by a completely numerical procedure.

Marsden then reported on behalf of *J.L. Brady* regarding P/Halley: In contrast to *Zadunaisky*, who found it necessary to utilize two orbits to represent the observations of P/Halley at the 1910 return, Brady finds that one orbit will suffice. The principal difference between the two investigations is that Zadunaisky made use of normal places while Brady used the individual observations. Brady remarks: 'One can force selected groups of residuals below a realistic threshold by using two orbits, but such a treatment is certainly not justified until an exhaustive search for a single gravitational model has failed.' He concludes that non-gravitational forces are not detectable in the observations of any one apparition, and possibly not in two apparitions. Evidence for significant non-gravitational effects arises, however, in trying to link three or more successive returns: On tracing back the orbit determined by linking the returns of 1910 and 1835, Brady has found that the calculated time of perihelion in 1759 was 4·2 days too early; in 1682 it was 8·2 days too early. A very similar result has been obtained by H.F. Michielsen, who, in tracing the orbit back to 1378, found it necessary to adjust the period by 4 days at each perihelion passage.

In discussion, *Marsden* commented regarding the hitherto normal supposition that a careful discussion of observations at two apparitions would lead to a reliable prediction at the third. His prediction at the third apparition of P/Honda-Mrkos-Pajdušáková had been almost 30′ in error, almost entirely due to non-gravitational effects. Taking note of Roemer's remark that recovery of comets with large reflectors is troublesome if the predicted positions are as much as 5′ in error, *Marsden* suggested that non-gravitational forces would have to be allowed for in some way if recoveries of affected comets at fourth and subsequent apparitions are to be made without undue waste of time. 'Perhaps one can take them into account by simply including a secular term in the mean motion, as has often been done in the past. My own preference is to include a velocity term in the original equations of motion. I am not convinced that a specific impulse has ever manifested itself in the positional observations of a comet. Even if impulses do occur, it seems to me that, from the point of view of representing the observations and of making satisfactory predictions for future returns, one might just as well assume that the non-gravitational forces act continuously.' *Roemer* pointed out that the largest non-gravitational effects seemed to occur in the comets of more diffuse appearance—the effects were less apparent in the well-condensed comets. *Candy* asked whether errors in the planetary masses could account for the discrepancies. *Marsden* said that this had been considered and was found not to be the case. *Hertz* asked whether there was any evidence of secular effects in the eccentricities. *Marsden* replied that the effects seemed to be mainly in the motion.

Herget, on the other hand, considers the non-gravitational effects to be discontinuous. He reported that the experience with P/Pons-Brooks (1883–1884) showed a distinct deviation of the residuals after perihelion. Comet Schwassmann-Wachmann 1, which moves in a nearly circular orbit and therefore is observable for several months of each year, was studied initially at Cincinnati on the basis of observations 1934–1944 for the purpose of obtaining an accurate ephemeris for observers. No anomalies in the motion were found. A recent, more comprehensive study (*Astr. J.* **66**, 266, 1961), in which a sample of all observations since discovery were represented, revealed several rather sharp discontinuities in the run of residuals. The largest, which occurred in 1927, shortly after discovery, corresponds to a change of velocity of one meter per second if it is represented as a single impulse. P/Schwassmann-Wachmann 1 is an especially valuable example for investigation of the nature of non-gravitational effects because the time continuity of observations permits the recognition of discontinuous events that could be smoothed over in other comets. What is needed to test the hypothesis of a discontinuous trajectory is to base a solution on all observations *after* perihelion at one apparition combined with those *before* perihelion at the next apparition. Then representation of the unused observations, those before the first perihelion passage and after the second, would show the importance of non-gravitational effects near perihelion. We need to find the magnitude of the "equivalent impulse" in as many cases as possible. In P/Reinmuth 2 *Rabe* found no evidence of non-gravitational forces.

Sekanina believes that the splitting of cometary nuclei is the strongest evidence supporting the existence of observable non-gravitational effects. He reported on his detailed study of three 'splits': Comet Ikeya-Seki in 1965, P/Biela, and Comet Wirtanen in 1957.

Sekanina suggested that (1) special attention be paid to astrometric observations of all the components of split comets, (2) independent orbits be calculated in such a way as to identify discontinuities for both short-periodic and near-parabolic comets, and (3) investigation be made of possible correlation of changes in physical characteristics with dynamical discontinuities.

In the general discussion that ensued, *Whipple* remarked that it was hard to see, from the physical point of view, how such large impulses could take place. The events in P/Schwassmann–Wachmann 1 must be mild ones. *Sitarski* commented that the form of the gravitational equation that Marsden suggested would preclude studies of evolutionary changes in orbital elements. *Sekanina* reported, in answer to a question, that he had tried to investigate correlations of impulse events with brightness changes, but specific associations could not be made because of insufficient observational data. He noted further that generally decreasing importance of non-gravitational effects in individual comets seemed reasonably associated with decreasing physical activity as an age effect. *Whipple* stated his view that splitting occurs most commonly as a result of increased rotation. It would appear, then, that the material of the comet nucleus must possess appreciable strength. A density of the order of 1·3 gm cm^{-3} would be associated with dirty-ice composition.

Second Meeting, 28 August 1967

The President opened the meeting at $10^h\ 30^m$ by outlining the substance of the letter that he was sending to the President of the Union in response to his request for suggestions for limiting the rapid growth in the attendance at General Assemblies.

PROPOSALS

The following resolutions were then passed without objection.

(1) 'Commission 20 recommends that the Minor Planet Center at Cincinnati continue to issue the *Minor Planet Circulars*, and that a sum of \$ 500·00 per annum be made available to the Minor Planet Center for defraying the necessary expenditures.'

(2) 'Commission 20 recommends that the new improved list of values of g (magnitude at unit distances) prepared by Dr T. Gehrels be adopted for the magnitudes of these minor planets, and that the list be printed in the *Transactions IAU.*'

(3) 'Commission 20 recommends to all observers who have the appropriate facilities the urgent need for making early plans to observe (1566) Icarus by optical ($12 <$ magn. < 20) and electronic means during the coming apparition and the close approach (0·04 AU) of 1968 June 14–15, and notes that a reliable ephemeris based on the observations of 1966 is now available.'

(4) 'Commission 20 recognizes that the use of large telescopes for astrometric observations is essential to the study of the origin and physical evolution of comets and minor planets and therefore:

calls the attention of observers to the values of this area of work;

urges sympathetic consideration by the appropriate authorities of requests for observing time on suitable instruments;

solicits the cooperation of those observatories that have facilities for measurement and reduction of plates, including collections of star catalogues;

recommends the development of additional facilities, automatic insofar as practicable, for the measurement of plates up to 25 × 25 cm; and

commends the efforts of Commission 23 toward calculation of improved values of plate constants for the Astrographic Catalogue.'

The Commission then agreed to the recommendation of the President on the compositions of the Organizing Committee (namely: Arend, Herget, Hirose, Wood) and the Working Group on Comets (namely: Roemer (Chairman), Candy, Kresak, Makover, Marsden, Sitarski).

GENERAL DISCUSSION AND REPORTS

Misidentifications. Herget drew the attention of observers to the desirability of re-examining the plate when a *Minor Planet Circular* indicated a large residual after an orbit improvement had been made; this might be due to a misidentification and the correct object might be on the same plate at a position indicated by the residuals.

Icarus. Herrick drew attention to his new 4-day ephemeris of (1566) Icarus for 1968. The minor planet would not be visible until about 6 June, that is only a few days before the close approach on 15 June. The residuals should not exceed a few seconds of arc. He hoped that Icarus would be observed by radar and he would prepare special ephemerides if the requirement were made known to him. Observations should be reported as early as possible. *Herget* expressed his thanks to the observers at Pretoria who had made special efforts to obtain observations in 1966.

Eros. Herget stated that a new ephemeris of Eros by Dr Rabe was now available from the Minor Planet Center.

Observations. Hertz said that he had been encouraged by the response of observers to his appeal for more positions of the minor planet (197) Arete. There have been 69 oppositions since discovery in 1879 but only 71 accurate positions in 27 oppositions have been published in the open literature. He hoped that accurate positions for a further 13 oppositions would be made available, since he knew of at least 55 approximate positions. He suggested that one way in which additional observations might be found for a particular minor planet of interest was to look for occasions when it was close to other minor planets at times when they had been observed. Another approach calls for the preparation of lists of plates by observatories which have sizeable collections, giving date, plate center, field size, and magnitude limit. From such data, a computer could determine on which plates a particular asteroid could be found, and *Hertz* would like to find out whether any efforts in that direction could be undertaken.

Satellites. D. Pascu reported that he had taken a series of astrometric plates of the satellites of Mars with the 61-inch reflector of the U.S. Naval Observatory. A special filter was used to diminish the light from Mars in order to obtain an image of the planet to which the positions of the satellites could be referred. About 60 of the 88 exposures had measurable images for both satellites. At least five stellar images are present on each plate, and their positions will be determined from 25 astrographic plates taken with the Lowell and Naval Observatory astrographs. Photographs and further details are given in *Sky and Telescope*, **34**, 22, July 1967. *P. V. Sudbury* reported that he had obtained in 1967 a series of observations of Jupiter V at the Cassegrain focus of the 74-inch reflector of the Kottamia Observatory. A small neutral density filter was used to attenuate the light from the planet. Measures were made on a Ferranti-Zeiss digitized measuring engine, and computerized reduction procedures were developed. Although the coordinates of the satellites were well determined with respect to the background stars, the determination of the center of the planet has not so far been satisfactory. *Gehrels* said that a neutral density filter had been similarly used by Dollfus in his recent discovery of a new satellite of Saturn. He himself had used a small 'bucket' for such work.

Final Remarks. President *Čebotarev* then took the chair, and introduced Dr *Edmondson*, who gave a brief appreciation of the work of the retiring President and in particular of his responsibility for increasing the amount of cooperation in the reduction of observations and the calculation of orbits. In reply, *Herget* said that the cooperation in astronomy, in particular that between the Minor

Planet Center and the Institute of Theoretical Astronomy, w as a fine example, and that he was proud to have served as President after such men as Brouwer and Leuschner.

Addendum

PHOTOGRAPHIC MAGNITUDES OF THE MINOR PLANETS

Prepared by *T. Gehrels*

The listing below replaces the one published in *Trans. IAU* **10**, 305, 1958, as about 1500 magnitude determinations were added from the Indiana Asteroid Program, 14 May 1958 – 31 December 1965. The results of the Indiana Asteroid Program are published regularly in the *Minor Planet Circulars*. A detailed description of this combination of asteroid magnitudes is in *Astronomical Journal* 72, (Dec. 1967;, in press).

The table contains the asteroid number and the absolute magnitude g, which is the magnitude at a distance of one astronomical unit from the Sun and from the Earth. The mean opposition magnitude, in the second column, is computed with $p_0 = g + 5 \log a(a-1)$, and a is the semimajor axis of the orbit. The last column contains the weight; unit weight corresponds to a probable error of ± 0.13 mag. Magnitudes that have relatively large weight and yet a colon (:) in the last column are uncertain, perhaps because of appreciable brightness variation due to large obliquity of the pole. The magnitudes are on the International Photographic System.

The cooperation of Dr F. K. Edmondson and his staff on the Indiana Asteroid Program, and the support by the National Aeronautics and Space Administration are gratefully acknowledged.

No.	p_0	g	wt.	No.	p_0	g	wt.	No.	p_0	g	wt.
1	7·45	4·00	4·0	51	11·11	8·56	5·6	101	12·30	9·24	1·8
2	8·52	5·06	6·0	52	11·58	7·52	6·0:	102	13·56	10·34	3·7:
3	9·57	6·33	6·0	53	12·84	9·71	3·9	103	11·76	8·45	2·3
4	6·74	4·20	8·0	54	12·05	8·72	2·2	104	13·56	9·42	8·0
5	10·95	7·90	2·9	55	12·38	8·95	3·3	105	12·23	9·66	1·4
6	9·30	6·60	3·5	56	12·57	9·48	5·2	106	12·84	8·66	7·0:
7	9·34	6·74	8·0	57	12·42	8·25	2·4	107	12·80	8·11	4·8
8	9·49	7·38	6·0	58	13·13	9·82	5·7:	108	13·47	9·21	2·2
9	9·77	7·17	5·5	59	11·97	8·63	2·6	109	13·33	10·03	2·0
10	10·62	6·46	4·5	60	12·57	9·95	2·1	110	11·83	8·45	2·7
11	10·44	7·68	8·3	61	12·54	8·67	5·0	111	12·05	8·97	4·9
12	11·17	8·71	2·2	62	13·85	9·72	5·3:	112	13·31	10·60	3·1
13	10·91	7·87	1·5	63	10·72	8·10	3·0:	113	12·15	9·58	2·6
14	10·38	7·31	2·0	64	12·02	8·75	1·8	114	12·62	9·36	4·3
15	9·38	6·19	6·0	65	12·34	7·77	4·5	115	11·60	9·02	1·5
16	10·53	6·78	6·1	66	13·65	10·45	6·9	116	12·12	8·68	3·1
17	11·39	8·59	4·0	67	12·47	9·79	2·5	117	12·95	9·07	2·4
18	10·06	7·69	6·1	68	11·67	8·19	2·2	118	12·59	9·87	3·9
19	10·98	8·25	3·3	69	12·03	8·18	1·8	119	12·38	9·33	4·8
20	10·03	7·38	2·0	70	12·18	9·05	1·8	120	12·92	8·82	5·0
21	11·30	8·58	2·3	71	11·79	8·37	4·0	121	12·87	8·23	2·7
22	11·09	7·37	4·1	72	12·52	10·23	2·5	122	13·23	8·97	7·7
23	11·39	8·24	3·9	73	13·43	10·19	3·5	123	13·34	10·04	3·2
24	12·21	8·08	2·8	74	13·49	10·02	3·6	124	12·15	8·99	5·1
25	11·61	8·98	4·0	75	13·14	9·89	2·7	125	12·91	9·51	2·1
26	11·76	8·54	4·4	76	13·39	8·88	1·9	126	13·04	10·31	4·6
27	10·96	8·46	2·2	77	12·79	9·55	5·2	127	12·86	9·44	5·8
28	11·52	8·05	3·5	78	12·17	9·03	5·7	128	12·10	8·68	7·8
29	10·15	7·16	4·4:	79	11·92	9·18	2·1	129	11·36	7·70	2·2
30	11·22	8·68	4·8	80	11·56	9·19	2·6	130	11·96	7·85	1·8
31	11·92	7·73	1·8	81	13·26	9·64	2·7	131	13·65	10·94	2·6
32	11·71	8·64	2·9	82	12·76	9·32	4·5	132	13·33	10·21	1·8
33	13·38	9·75	3·8	83	12·37	9·66	2·6:	133	12·92	8·92	3·9
34	12·76	9·48	4·4:	84	12·71	10·17	4·1	134	12·67	9·65	4·7
35	13·58	9·68	5·5	85	12·02	8·81	4·1	135	11·81	9·11	6·3
36	13·33	9·92	1·0	86	13·72	9·65	7·4	136	13·12	10·78	3·9
37	11·58	8·39	6·6	87	12·84	8·16	6·0	137	12·99	8·89	1·1
38	12·97	9·58	4·4	88	11·57	8·12	2·6	138	13·35	10·60	2·0
39	10·76	7·31	12·6	89	11·08	8·09	2·3	139	12·55	9·07	2·0
40	10·64	8·35	2·4	90	13·37	9·24	4·4	140	12·75	9·37	5·3:
41	11·65	8·21	4·3:	91	12·71	9·64	2·1	141	12·79	9·55	2·0
42	11·47	8·74	2·7	92	12·18	7·94	2·7	142	14·02	11·34	2·2
43	11·20	9·08	2·1	93	12·07	8·65	4·8	143	13·88	10·45	2·4
44	10·60	7·91	3·9	94	12·88	8·72	1·8	144	12·07	8·86	5·4
45	11·77	8·42	3·5	95	12·80	8·79	1·8	145	12·08	8·83	4·4
46	12·13	9·20	3·7	96	12·97	8·98	2·2	146	12·49	9·14	3·2
47	12·74	9·08	2·4	97	11·81	8·57	2·1	147	13·85	9·72	7·8
48	12·07	7·98	4·4	98	13·53	10·25	2·1	148	12·11	8·66	2·2:
49	12·63	8·56	5·6:	99	14·64	11·41	1·5	149	14·09	12·05	3·5
50	13·50	10·30	3·0	100	13·07	9·02	4·8	150	13·01	9·15	2·0

No.	p_0	g	wt.	No.	p_0	g	wt.	No.	p_0	g	wt.
151	13·41	10·33	4·2	201	12·64	9·38	2·6:	251	15·22	11·16	5·2
152	13·71	9·58	2·0	202	12·84	8·82	4·7	252	14·70	10·54	2·0
153	14·11	8·75	4·6	203	13·49	10·10	4·6	253	14·60	11·40	3·3
154	12·58	8·37	2·3	204	13·48	10·23	2·3	254	15·24	13·15	1·8
155	15·18	11·45	0·6	205	13·83	10·36	2·1	255	14·76	11·36	3·4
156	12·87	9·50	3·4	206	13·39	9·90	4·2	256	14·85	10·96	2·6
157	15·39	12·34	1·2	207	13·27	10·93	5·5	257	14·24	10·13	2·1
158	14·21	10·57	3·8:	208	14·05	10·36	5·1	258	12·58	9·45	3·4
159	13·37	9·29	2·3	209	13·16	8·99	5·7	259	13·20	9·07	5·7
160	13·39	10·02	6·4	210	13·82	10·46	2·2	260	14·97	10·33	4·5
161	12·74	10·16	2·1	211	12·90	8·94	3·4:	261	12·98	10·52	7·5
162	13·92	9·98	4·5	212	13·47	9·37	3·5	262	15·73	12·73	1·0
163	12·92	10·37	4·1	213	13·33	9·91	4·0	263	15·28	11·60	5·9:
164	12·82	9·65	1·0	214	13·46	10·34	2·5	264	13·43	9·92	2·2:
165	12·80	8·68	5·1	215	14·26	10·81	4·9	265	15·37	12·69	1·3
166	13·94	10·66	1·8	216	11·52	8·02	3·4	266	12·99	9·47	2·6
167	14·18	10·57	6·2	217	14·62	10·97	2·6	267	15·47	12·01	2·0
168	13·59	9·07	3·1	218	13·04	9·80	2·6	268	13·50	9·44	6·1
169	13·08	10·55	4·1	219	12·87	10·35	3·6	269	14·35	11·22	2·1
170	13·61	10·62	1·2:	220	14·83	12·32	3·1	270	12·14	10·04	5·0
171	13·79	9·66	1·7	221	12·84	8·92	3·1	271	14·77	10·86	3·1
172	12·12	9·54	3·9	222	14·42	10·27	4·4	272	15·15	11·68	1·7
173	12·30	8·89	1·9	223	15·10	11·04	2·6	273	14·09	11·47	3·2:
174	13·09	9·47	4·9	224	12·99	9·80	5·4	274	15·12	11·16	3·9
175	14·18	9·90	7·2	225	14·02	9·54	2·7	275	13·38	9·92	6·5
176	13·60	9·42	1·3	226	14·31	10·97	2·8	276	13·41	9·31	2·1
177	13·96	10·50	2·2	227	14·18	10·05	3·4	277	14·83	11·15	3·1
178	13·34	10·56	7·4:	228	15·99	13·88	2·6	278	14·02	10·60	6·7
179	13·42	9·58	4·4:	229	15·02	10·44	6·0	279	15·42	9·68	4·6
180	14·77	11·41	4·0	230	11·10	8·51	1·9:	280	15·64	11·86	2·9
181	13·05	8·95	2·0	231	14·24	10·49	5·1	281	14·88	12·81	1·8
182	12·62	9·95	5·0	232	14·52	11·53	3·9	282	14·34	11·86	2·5
183	14·37	10·87	2·5	233	12·75	9·52	3·9	283	13·40	9·43	2·4
184	13·63	9·43	6·2	234	12·87	10·27	4·1	284	13·98	11·45	3·8
185	11·82	8·43	3·1	235	13·51	9·84	2·0	285	15·92	11·78	2·0
186	12·82	10·28	2·4	236	12·99	9·48	3·1	286	14·49	10·26	3·3
187	12·83	9·45	2·3	237	14·17	10·73	2·8	287	11·90	9·38	1·8
188	13·79	10·35	3·0:	238	12·96	9·24	2·6	288	14·36	10·93	3·7
189	13·10	10·35	3·3	239	15·52	11·69	1·9	289	14·37	10·72	2·0
190	13·82	8·48	3·1	240	12·92	9·68	3·9	290	15·63	13·15	1·2
191	13·59	9·90	3·2	241	12·66	8·68	2·5	291	14·87	12·70	2·0
192	10·94	8·30	1·9	242	14·10	10·46	4·1	292	13·88	10·94	1·6
193	14·00	10·91	3·9:	243	14·72	11·09	2·2	293	14·53	10·90	3·3
194	11·96	8·83	5·8	244	15·39	13·35	4·0	294	15·16	10·99	1·8
195	13·83	10·16	2·3	245	13·67	9·60	3·4	295	14·96	11·45	1·6
196	11·67	7·58	2·1	246	13·14	9·84	2·1	296	15·65	13·46	2·8
197	14·17	10·78	3·5	247	12·52	9·13	2·5	297	14·54	10·35	5·7
198	12·34	9·56	1·8	248	14·22	11·42	3·2	298	14·65	12·37	3·5
199	14·05	9·86	2·6	249	15·02	12·44	2·2	299	15·68	12·97	6·4
200	12·73	9·34	2·1	250	12·80	8·66	2·0	300	14·79	10·54	8·2

No.	p_0	g	wt.	No.	p_0	g	wt.	No.	p_0	g	wt.
301	14·66	11·30	4·5	351	13·64	10·20	2·8	401	14·52	10·06	4·7
302	14·82	12·17	2·8:	352	13·57	11·48	2·6	402	12·89	9·89	2·1
303	14·01	9·90	3·9	353	15·58	12·20	4·8	403	13·72	10·19	2·5
304	13·58	10·94	4·8	354	10·97	7·47	3·8	404	12·89	9·81	1·1
305	14·26	10·20	5·7	355	14·60	11·64	2·0	405	12·68	9·62	3·7
306	12·59	10·06	2·9:	356	12·44	9·01	3·2	406	15·05	11·31	3·6
307	14·61	10·88	2·3	357	13·56	9·42	2·7	407	13·35	10·20	2·6
308	12·40	8·99	5·8:	358	13·90	10·23	8·5	408	14·88	10·72	7·4
309	14·63	11·39	2·6	359	13·78	10·41	3·6	409	11·57	8·53	4·8
310	15·04	11·60	3·8	360	13·38	9·49	2·3	410	12·98	9·62	1·8
311	14·83	11·13	2·5	361	14·86	9·55	3·3	411	13·91	10·14	2·5
312	13·68	10·21	3·2	362	12·86	9·81	2·3	412	13·63	10·19	4·6
313	12·16	9·59	4·3	363	13·38	9·97	3·1	413	14·05	10·99	0·9
314	15·25	11·10	2·0	364	13·16	10·99	3·9	414	15·31	10·59	3·1
315	15·99	13·77	3·6	365	13·84	10·32	2·2	415	13·77	10·28	3·6
316	15·05	10·87	6·2	366	13·87	9·73	5·8	416	12·75	9·26	2·5
317	13·48	11·14	3·9	367	14·13	11·97	3·3	417	14·09	10·58	5·3
318	14·53	10·28	2·5	368	15·06	11·03	3·4	418	13·78	10·70	2·6
319	15·79	11·25	1·2	369	12·72	9·52	1·8	419	12·53	9·45	4·3
320	15·55	11·63	2·4	370	14·07	11·63	3·6	420	13·91	9·32	3·8
321	14·96	11·28	6·4	371	13·23	9·86	3·1	421	15·89	12·93	1·0
322	13·80	10·33	2·3	372	12·60	8·43	3·0	422	14·19	12·00	2·6
323	13·89	11·30	1·3	373	14·33	10·22	3·6	423	12·43	8·42	2·6
324	11·31	8·04	4·5	374	13·40	9·93	8·8:	424	14·15	10·69	3·3
325	14·22	10·00	2·0	375	12·41	8·28	3·6	425	14·40	10·72	4·9
326	12·45	10·03	1·2	376	12·80	10·45	2·8	426	13·18	9·49	1·8
327	14·75	11·29	5·7	377	13·06	9·77	6·7	427	14·41	10·56	3·3
328	14·00	9·91	1·7	378	14·44	10·98	4·8	428	15·22	12·82	1·9
329	13·45	10·64	1·5	379	14·12	9·97	4·3	429	13·93	10·82	2·2
330	15·19	13·40	0·6	380	13·76	10·50	5·9	430	15·26	11·66	2·4
331	14·25	10·31	2·0	381	13·72	9·48	2·3	431	14·10	9·98	5·8
332	14·18	10·72	4·1:	382	13·79	9·67	3·5	432	12·44	9·88	3·7
333	14·48	10·36	5·2	383	14·86	10·75	1·3	433	11·43	12·31	2·1
334	13·73	8·48	3·4	384	13·82	10·61	2·5	434	13·34	12·02	1·2
335	12·71	9·90	3·7	385	12·40	8·80	3·9	435	13·99	11·24	3·5
336	13·13	10·88	3·6:	386	12·00	8·30	0·8	436	15·26	11·03	3·9
337	12·54	9·95	2·8	387	12·29	8·90	1·9	437	14·13	11·53	5·4
338	13·43	9·70	3·3	388	13·15	9·25	3·1	438	13·49	10·50	2·1
339	14·28	10·37	3·6	389	12·29	9·18	1·3	439	14·72	10·60	1·5
340	14·45	11·04	1·7	390	14·59	11·38	2·0	440	14·93	12·79	2·9
341	14·66	12·55	4·8	391	14·67	12·24	1·7	441	13·20	9·67	4·7:
342	14·16	11·14	2·6	392	14·20	10·52	1·6	442	13·56	11·06	2·1
343	15·23	12·57	2·5	393	12·58	9·11	3·5	443	13·65	11·50	6·8
344	12·30	9·21	3·3	394	14·24	10·80	7·0	444	12·79	9·34	4·2
345	12·47	10·03	5·1	395	14·86	11·37	4·4	445	14·28	10·08	2·4
346	12·52	9·02	2·5	396	14·55	11·15	5·5	446	13·39	9·90	2·4
347	13·25	10·13	2·1	397	13·38	10·21	2·7	447	14·15	10·29	3·1
348	14·50	10·67	2·7	398	15·24	11·86	0·8	448	15·08	10·95	3·1
349	10·95	7·20	4·7	399	14·47	10·47	2·0	449	13·63	10·64	3·0:
350	14·00	9·92	2·2	400	15·43	11·31	3·7	450	15·16	11·24	1·6

No.	p_0	g	wt.	No.	p_0	g	wt.	No.	p_0	g	wt.
451	12·16	8·16	1·9	501	14·26	10·10	3·5	551	14·28	10·44	2·6
452	16·99	13·35	0·6	502	14·81	12·22	2·3:	552	14·49	10·33	4·2:
453	13·94	11·88	5·7	503	13·57	10·21	4·4	553	15·56	13·37	2·0
454	13·14	9·99	1·9	504	14·39	11·04	3·4	554	11·96	9·39	4·3
455	13·13	9·91	3·1	505	13·32	10·04	2·8	555	15·80	11·63	2·0
456	14·35	10·87	3·5	506	13·67	9·70	3·3	556	13·26	10·47	3·1
457	16·99	12·94	0·6	507	14·45	10·30	4·8	557	15·70	12·97	4·2
458	14·36	10·49	2·6	508	13·59	9·42	2·4	558	13·70	9·98	4·0
459	14·95	11·81	2·5	509	13·68	9·67	2·0	559	13·87	10·53	1·8
460	15·26	11·91	4·6	510	14·04	10·93	6·9	560	15·11	11·70	1·8
461	15·62	11·52	2·0	511	11·24	7·02	6·5:	561	16·04	11·87	0·8
462	14·39	10·74	6·3	512	14·13	12·05	2·5	562	14·72	10·80	3·1
463	15·43	12·80	1·9	513	14·38	10·47	4·9	563	12·80	9·47	3·5
464	13·80	10·28	2·6	514	14·16	10·18	2·2	564	15·10	11·70	1·5
465	15·01	10·95	2·3	515	15·91	11·80	5·9	565	14·74	12·01	1·4
466	13·62	9·11	3·5	516	12·68	9·42	1·6	566	13·59	9·05	3·0
467	15·79	12·01	6·2	517	14·51	10·37	5·3	567	14·53	10·37	6·0
468	14·70	10·54	4·9	518	15·16	12·20	5·6	568	14·01	10·33	2·1
469	14·05	9·89	2·6	519	13·68	10·18	2·1	569	14·04	10·82	3·8
470	14·00	11·36	4·0	520	15·74	11·83	3·1:	570	14·62	10·02	4·9
471	11·44	7·76	2·2	521	13·25	9·86	2·1	571	15·47	12·81	3·1
472	13·46	10·49	2·6	522	14·83	9·93	8·2	572	14·41	11·78	5·8
473	14·95	11·10	0·6	523	14·45	10·63	3·4	573	14·57	10·66	4·2
474	14·58	11·82	2·5	524	13·95	10·78	3·4	574	15·99	13·74	3·2
475	15·41	12·32	1·2					575	15·25	12·26	2·6
476	12·86	9·66	2·4	526	14·95	10·83	3·3	576	14·48	10·61	3·6
477	14·06	11·39	6·2	527	14·64	11·28	4·4	577	14·78	10·69	5·4
478	12·71	8·79	2·7:	528	14·40	9·85	3·5	578	13·84	10·43	1·7
479	14·29	10·94	2·6	529	14·97	11·04	4·9	579	13·18	9·26	4·4
480	13·05	9·86	2·1	530	14·21	9·95	3·7	580	15·15	10·87	3·6
481	13·14	9·74	3·6	531	15·74	12·22	0·6	581	15·12	10·85	3·8
482	13·87	9·99	2·8	532	11·34	7·88	3·4:	582	13·50	10·38	3·6:
483	14·17	9·57	1·7	533	14·92	11·06	1·1	583	14·47	10·25	2·6
484	14·62	11·37	2·3	534	14·53	10·85	3·1	584	12·52	9·95	3·1
485	12·97	9·56	2·9	535	13·46	10·43	2·3	585	14·11	11·40	5·4
486	14·69	12·18	3·6	536	13·97	9·25	3·3	586	14·26	10·30	5·8
487	12·77	9·52	3·4	537	14·01	10·00	3·4	587	15·95	13·48	0·8
488	13·09	8·94	2·6	538	14·67	10·49	4·1	588	16·05	9·34	9·1
489	13·78	9·63	2·5	539	14·41	11·02	2·6	589	14·10	9·97	2·7
490	13·63	9·44	4·4	540	14·29	12·13	7·3	590	14·96	11·07	2·0
491	14·14	9·91	2·5	541	14·78	11·24	5·4	591	15·09	11·83	2·2
492	14·96	10·88	8·2:	542	13·92	10·20	2·6	592	14·45	10·52	2·3
493	15·84	11·75	2·8	543	14·52	10·52	5·3	593	13·71	10·40	3·2
494	13·85	9·98	7·6	544	14·27	11·19	4·6	594	16·87	13·71	2·4
495	14·29	11·45	1·8	545	13·61	9·42	7·4	595	13·41	9·16	2·0
496	15·03	12·92	3·8	546	13·78	10·69	3·3	596	13·51	9·75	2·5
497	14·22	10·61	2·5	547	14·31	10·84	3·6	597	13·94	10·69	1·6
498	13·09	9·88	4·6	548	14·80	12·47	2·3	598	13·28	9·84	3·1
499	15·47	10·12	2·6	549	15·00	11·73	1·9	599	13·42	9·96	3·3
500	13·57	10·45	6·1	550	13·37	10·30	3·4	600	14·57	11·35	3·6

No.	p_0	g	wt.	No.	p_0	g	wt.	No.	p_0	g	wt.
601	14·46	10·33	3·7	651	15·09	11·15	2·6	701	14·41	10·50	3·2
602	13·64	9·59	1·8	652	15·29	12·29	2·0	702	12·85	8·62	2·2:
603	16·44	13·46	1·3	653	14·45	10·53	3·3	703	15·63	13·59	1·9
604	14·40	10·22	2·4	654	12·15	9·78	2·0	704	11·50	7·50	1·1
605	14·40	10·52	2·3	655	14·40	10·53	3·6	705	13·47	9·72	2·0
606	14·56	11·50	3·8	656	15·29	11·11	5·6	706	15·37	12·00	1·3
607	14·37	10·76	5·4	657	15·04	11·92	4·3	707	15·41	13·36	1·7
608	15·72	11·78	2·6	658	15·24	11·62	5·1	708	14·99	11·74	2·0
609	15·21	11·17	6·0	659	16·31	9·59	6·6	709	13·50	9·77	2·0
610	17·23	13·20	1·3	660	12·89	9·94	4·2:	710	16·29	12·15	3·2
611	14·35	10·50	2·1	661	14·52	10·61	2·9	711	14·75	12·54	1·7
612	16·43	12·29	0·6	662	14·76	11·77	7·9	712	12·51	9·47	2·3
613	14·66	10·92	4·1	663	14·45	10·45	1·5	713	14·43	9·86	3·5
614	15·29	11·99	2·6	664	15·46	11·28	3·3	714	13·14	10·19	3·6
615	14·38	11·22	7·1	665	13·85	9·69	3·3	715	14·44	10·99	6·5
616	14·60	11·61	3·5	666	14·99	11·91	6·5	716	15·32	11·79	4·1
617	15·76	9·06	10·7:	667	14·73	10·50	3·5	717	16·01	11·88	4·1
618	13·71	9·50	3·3	668	16·76	13·30	2·1	718	14·71	10·73	2·6
619	14·04	11·12	2·6	669	15·24	11·33	3·6	719	19·83	16·77	0·6
620	15·12	12·40	1·8	670	14·60	11·08	3·8	720	14·58	10·90	2·6
621	15·79	11·68	3·9	671	15·10	11·04	2·0	721	15·19	10·40	4·0
622	14·33	11·67	2·3	672	15·28	12·28	1·9	722	15·08	13·05	2·0
623	14·21	11·44	1·5	673	14·65	11·11	3·9:	723	15·19	11·31	1·8
624	15·18	8·56	10·8	674	12·13	8·39	1·3	724	17·45	14·74	0·6
625	14·17	10·98	2·3	675	12·61	9·16	4·1:	725	15·30	12·27	1·5
626	13·02	9·99	1·2	676	14·47	10·47	3·9	726	15·06	12·04	2·6
627	14·73	11·02	3·3	677	14·68	10·87	6·1	727	14·20	11·18	3·0
628	13·41	10·36	2·1	678	13·59	10·56	2·7:	728	16·08	13·78	0·6
629	14·90	10·80	2·6	679	12·20	9·13	1·0	729	14·38	10·95	0·8
630	15·53	12·38	2·0	680	14·90	10·80	3·1	730	16·89	14·66	1·4
631	13·55	10·06	2·1	681	15·97	11·89	2·1	731	14·38	10·52	3·1
632	16·41	13·18	1·8	682	16·66	13·47	0·6	732	14·65	11·88	1·1
633	14·93	11·01	2·6	683	13·58	9·48	1·8	733	14·63	10·09	0·8
634	14·98	11·01	1·8	684	14·76	12·05	3·3	734	15·07	10·91	5·1
635	14·10	9·97	3·0	685	15·07	12·86	2·2	735	14·23	10·86	2·0
636	14·33	10·61	2·6	686	14·50	11·43	2·1:	736	14·24	12·13	3·3
637	16·02	11·86	2·5	687	16·30	12·95	1·6	737	12·74	9·67	2·3
638	14·35	10·97	7·0:	688	14·95	11·64	4·4	738	14·96	11·01	5·8
639	13·21	9·30	3·1	689	15·68	13·26	3·1	739	13·39	10·00	1·9
640	14·48	10·30	1·7	690	12·96	8·79	1·6	740	14·28	10·30	2·2
641	15·90	13·74	1·3:	691	14·34	10·43	2·6	741	14·61	11·25	2·0
642	14·91	10·72	2·5	692	14·60	10·10	1·0	742	14·29	10·38	5·0
643	14·93	10·47	3·1:	693	14·16	10·37	3·2	743	14·70	11·20	7·2
644	14·80	11·71	3·1	694	13·34	10·09	5·4	744	15·34	11·15	3·7
645	15·22	11·01	2·0	695	12·68	9·72	2·3	745	15·30	10·99	2·1
646	16·64	14·20	1·0	696	14·47	10·24	5·4	746	14·60	10·51	2·6
647	15·32	12·58	1·8	697	13·96	10·29	1·0	747	12·61	8·73	3·1
648	14·89	10·71	2·6	698	15·45	11·80	1·5	748	15·06	9·75	4·7
649	17·08	14·10	1·0	699	16·23	13·10	2·1	749	15·08	12·85	1·0
650	16·24	13·47	3·7	700	14·55	12·36	2·2	750	15·67	12·94	2·5

No.	p_0	g	wt.	No.	p_0	g	wt.	No.	p_0	g	wt.
751	12·95	9·96	1·6	801	15·51	12·40	2·0	851	15·05	12·86	2·6
752	14·19	11·41	2·0	802	15·61	13·51	2·0	852	13·85	11·31	1·8
753	14·26	11·81	2·6	803	14·82	10·57	3·1	853	14·94	12·53	7·7
754	14·24	10·38	1·9	804	12·74	9·15	2·7	854	15·90	13·35	4·3
755	14·78	10·60	4·5	805	14·77	10·50	2·4	855	15·35	12·81	4·2
756	15·45	11·16	0·8	806	15·33	11·09	1·0	856	14·63	11·91	6·6
757	14·00	11·43	3·1	807	15·73	11·81	4·4:	857	14·75	12·67	2·5
758	13·48	9·23	2·4	808	14·30	10·90	2·3	858	14·92	11·39	3·5
759	15·09	11·96	2·0	809	15·41	13·08	2·6	859	15·11	10·86	3·6:
760	13·61	9·46	4·2	810	16·18	14·13	2·4	860	14·25	10·75	2·3
761	15·39	11·76	3·8	811	15·45	11·75	5·7	861	14·91	10·76	4·9
762	13·33	9·18	2·0	812	15·63	12·41	0·6	862	14·75	11·23	1·4
763	16·03	13·81	3·9	813	15·24	13·07	2·5	863	14·45	10·22	2·3
764	14·83	10·62	4·6	814	14·05	9·85	2·4				
765	16·99	14·01	0·6	815	15·15	11·93	3·0	865	15·81	13·14	3·6
766	14·78	10·84	3·9	816	15·21	11·32	3·6	866	14·46	10·35	3·2
767	15·18	11·09	6·8	817	14·95	11·88	4·3	867	16·10	12·09	2·1
768	15·38	11·26	1·0	818	14·52	10·32	2·3	868	14·45	11·13	4·6
769	14·36	10·14	3·8	819	15·25	13·15	5·7	869	16·51	13·22	2·8
770	14·26	12·09	2·0	820	15·26	11·14	1·8	870	15·34	12·91	1·1
771	14·77	11·56	2·0	821	15·84	12·38	2·8	871	15·86	13·69	2·6
772	13·54	9·65	1·0	822	14·81	12·55	5·2:	872	14·51	11·14	4·7
773	14·15	10·53	2·3	823	14·85	12·68	2·6	873	15·46	12·31	3·6
774	13·86	9·90	3·4	824	14·87	11·37	3·1	874	15·05	10·88	2·9
775	15·20	11·29	3·1	825	15·21	13·03	4·6	875	15·72	12·73	1·6
776	12·57	8·80	3·0	826	15·72	12·38	1·8:	876	15·78	11·86	2·0
777	15·49	11·22	2·7	827	16·15	13·84	2·6	877	14·58	11·74	10·1
778	15·69	11·50	1·8	828	15·27	11·06	5·7	878	19·04	16·50	0·6
779	12·89	9·65	0·8	829	14·97	11·92	4·0	879	15·63	12·67	0·6
780	14·23	10·14	2·6	830	14·74	10·50	2·0	880	16·87	12·96	0·8
781	14·75	10·45	2·0	831	15·63	13·48	0·6	881	16·62	13·50	1·1
782	14·59	12·54	2·3	832	15·67	12·03	3·9	882	15·66	11·55	2·6
783	14·57	12·08	3·3	833	16·14	12·23	1·2	883	15·78	13·57	1·5
784	14·34	10·27	4·7	834	15·53	10·37	4·1	884	16·48	9·77	4·3:
785	13·41	10·37	2·7	835	16·16	11·93	2·8	885	15·73	11·67	4·0
786	14·21	10·02	3·1	836	16·35	14·27	0·8	886	14·32	10·17	1·8:
787	14·28	11·32	1·3	837	15·32	12·95	1·9	887	19·15	16·23	0·9
788	13·57	9·46	2·1:	838	14·88	11·18	3·5	888	14·18	10·85	2·2
789	15·54	12·27	2·4	839	14·79	11·66	2·0	889	14·93	12·19	2·2
790	13·56	9·02	3·1	840	14·61	10·50	0·6	890	15·21	11·27	1·5
791	14·75	10·62	2·1	841	15·60	13·34	1·6:	891	14·83	11·20	4·6
792	14·21	11·07	1·3	842	15·97	11·71	1·0	892	14·92	10·62	3·1
793	14·36	10·86	2·2	843	16·54	14·22	0·6	893	14·72	10·74	3·4
794	16·45	12·29	2·4	844	14·83	10·61	5·7	894	14·99	10·89	2·5
795	14·12	10·71	3·2	845	14·89	11·11	1·8	895	13·99	9·74	2·5
796	13·42	10·25	3·5	846	15·51	11·37	4·4	896	15·29	12·95	3·3
797	14·56	11·61	5·1	847	14·81	11·33	2·2	897	15·11	12·14	1·5
798	14·49	10·58	2·6	848	15·93	11·84	2·0	898	16·74	13·37	1·3
799	14·37	11·41	6·2	849	13·06	8·90	1·5	899	15·10	11·38	1·7
800	14·65	12·56	3·1	850	14·51	10·62	2·6	900	15·68	12·87	3·0

No.	p_0	g	wt.	No.	p_0	g	wt.	No.	p_0	g	wt.
901	14·93	12·75	3·5:	951	15·14	13·01	2·3	1001	14·72	10·50	3·1
902	16·28	13·54	1·1	952	14·15	10·29	4·9	1002	15·53	12·04	1·2
903	15·06	10·75	0·6	953	14·98	11·49	2·6	1003	15·39	11·23	2·9
904	15·22	11·34	1·3	954	15·82	11·70	1·0	1004	15·24	10·70	2·6
905	14·39	12·24	1·8	955	15·71	12·63	1·0	1005	15·08	10·91	5·3
906	14·33	10·63	3·2	956	15·87	13·50	3·3	1006	16·92	12·79	2·6
907	14·15	10·64	2·9	957	14·70	10·96	3·5	1007	15·87	12·54	2·9
908	14·83	12·02	3·0	958	16·35	11·03	1·6	1008	15·91	11·86	1·6
909	14·31	9·54	2·0	959	16·02	11·80	0·8	1009	19·95	16·82	0·6
910	14·94	11·17	6·6	960	16·33	14·09	3·5	1010	15·42	11·66	1·8
911	15·45	8·82	3·2	961	15·68	12·38	2·6	1011	16·25	13·63	2·0
912	13·38	9·27	2·4	962	16·31	12·60	4·5	1012	15·96	13·13	2·0
913	15·79	13·69	1·0	963	15·75	13·51	2·2	1013	13·85	10·58	1·9
914	13·15	10·39	1·6	964	15·92	11·94	1·4	1014	16·48	12·95	2·3
915	15·22	13·04	3·6	965	15·88	11·70	2·7:	1015	14·47	10·23	2·6
916	15·14	12·60	2·6	966	14·43	11·08	1·4	1016	15·35	13·19	3·8
917	15·08	12·49	4·3	967	15·42	13·24	2·0	1017	15·28	12·17	1·1
918	15·60	11·96	6·0	968	14·93	11·29	4·6	1018	14·59	11·63	3·5
919	15·76	12·30	2·4	969	16·11	13·33	1·3	1019	15·06	13·85	0·6
920	15·07	11·93	1·7	970	16·45	13·43	1·6	1020	15·59	12·11	1·1
921	15·36	11·15	2·3	971	14·36	11·18	3·4:	1021	13·27	9·89	3·1
922	16·29	13·01	2·6	972	14·68	10·68	2·5	1022	14·76	11·24	0·8
923	15·78	12·65	1·0	973	15·20	10·91	4·9	1023	15·15	10·97	4·1
924	14·14	10·37	4·1	974	14·61	11·66	2·2	1024	15·53	11·88	3·2
925	11·95	8·64	2·4	975	14·89	11·31	1·8	1025	15·48	14·04	0·8
926	15·46	11·61	1·5	976	14·65	10·44	6·9	1026	16·76	14·52	0·8
927	14·30	10·05	2·3	977	14·81	10·71	2·5	1027	16·10	11·92	4·2
928	14·87	10·72	3·1	978	14·97	10·74	2·7	1028	14·90	10·33	2·6
929	15·82	13·61	2·1	979	15·07	10·92	3·9	1029	15·59	11·90	5·7
930	15·20	12·49	1·1	980	12·58	9·19	1·6	1030	15·64	11·54	1·8
931	14·61	10·44	2·9	981	16·13	12·06	3·8	1031	14·67	10·69	1·2
932	13·35	10·68	2·3	982	15·28	11·27	2·6	1032	15·06	10·94	2·7
933	15·97	13·41	2·7	983	14·85	10·66	2·1	1033	16·02	12·13	3·6
934	14·84	11·43	1·5:	984	14·18	10·66	5·4	1034	16·02	13·66	2·3
935	16·35	14·19	3·0	985	16·53	14·15	1·9	1035	15·83	11·69	2·0
936	15·22	11·08	4·6	986	14·83	10·70	3·4	1036	13·98	10·76	1·8
937	15·29	13·10	3·1	987	14·77	10·64	6·5	1037	17·12	15·05	0·8
938	16·52	12·33	3·1	988	16·55	12·39	0·8	1038	16·88	11·59	0·6
939	15·53	13·29	2·0	989	16·54	13·32	0·6	1039	15·44	12·17	3·8
940	14·96	10·40	2·0	990	16·15	12·91	3·4	1040	15·64	11·53	1·5
941	15·47	12·00	3·2:	991	15·96	11·83	1·0	1041	14·96	10·93	1·8
942	15·66	11·48	2·4	992	16·08	12·15	1·6	1042	15·20	10·95	2·3
943	14·97	10·88	2·5	993	16·95	13·31	1·1	1043	14·98	10·92	5·8
944	19·18	11·95	0·9	994	14·35	11·41	2·7	1044	15·19	12·15	6·4
945	14·49	11·32	2·5	995	14·58	14·45	3·6:	1045	17·13	14·60	1·3
946	15·41	11·29	2·5	996	15·68	11·63	1·8	1046	15·34	11·47	3·9
947	14·26	10·85	1·8	997	16·23	12·99	2·1	1047	15·62	13·40	2·8
948	16·22	12·27	2·6	998	16·20	12·11	0·6	1048	13·98	10·60	2·6
949	14·78	10·90	6·7	999	15·46	12·34	2·5	1049	15·81	11·75	0·8
950	14·97	12·41	1·0	1000	15·54	11·29	0·6	1050	16·98	13·83	0·9

No.	p_0	g	wt.	No.	p_0	g	wt.	No.	p_0	g	wt.
1051	15·46	11·19	2·2	1101	16·32	11·98	1·3	1151	17·45	14·81	0·6
1052	15·25	13·04	2·9:	1102	14·86	10·85	3·2	1152	14·93	12·23	3·6
1053	16·49	13·36	3·1	1103	14·79	13·51	1·2	1153	15·42	13·32	2·6
1054	15·32	11·57	4·9	1104	16·66	13·50	1·5	1154	15·92	11·35	2·4
1055	14·73	12·63	2·4	1105	15·07	11·16	3·8	1155	15·78	13·00	2·5
1056	15·00	12·81	2·3	1106	15·96	12·87	1·0	1156	16·16	13·89	1·1
1057	15·57	11·88	2·6	1107	14·41	10·18	5·7:	1157	15·42	11·17	3·3
1058	15·23	13·13	4·1	1108	15·04	12·34	1·6:	1158	15·19	12·17	5·1
1059	15·46	12·27	2·4:	1109	15·18	10·94	3·6	1159	15·50	12·92	2·9
1060	16·48	14·27	2·1	1110	15·41	13·25	3·3	1160	15·78	12·77	1·5
1061	16·09	11·98	2·9	1111	15·31	11·43	5·2	1161	16·88	12·70	0·6
1062	15·07	11·17	6·5	1112	14·80	10·88	2·6	1162	15·58	10·24	2·8
1063	14·83	12·41	2·9	1113	14·70	10·60	6·2	1163	15·83	11·58	3·1
1064	15·23	12·25	2·0	1114	14·73	10·67	2·6	1164	16·42	14·03	1·3
1065	16·31	13·78	2·1:	1115	14·62	10·54	2·6	1165	15·63	11·49	1·8
1066	16·74	14·10	1·2	1116	14·56	10·81	2·0	1166	15·54	12·58	1·5
1067	15·72	12·06	2·9	1117	15·35	13·11	2·5	1167	15·52	10·93	2·6
1068	15·76	12·03	4·1:	1118	15·22	10·98	2·0	1168	16·00	13·01	1·9
1069	14·84	10·72	2·6	1119	15·46	12·32	5·0	1169	16·72	14·29	1·9
1070	16·34	12·08	1·6	1120	15·48	13·33	1·2	1170	15·55	13·11	2·0
1071	14·83	11·32	4·4	1121	15·50	12·52	0·8	1171	14·91	10·75	2·2
1072	15·90	11·68	1·3	1122	15·85	12·74	1·5	1172	16·01	9·32	2·7
1073	16·66	12·47	2·0	1123	15·06	12·88	5·9	1173	16·63	10·03	4·6
1074	15·41	11·25	3·6	1124	15·81	12·05	1·6	1174	16·77	12·84	0·6
1075	15·40	11·48	1·8	1125	18·36	14·20	0·8	1175	15·86	11·59	1·6
1076	15·80	12·98	7·1	1126	16·01	13·70	1·5	1176	15·48	12·18	1·5
1077	16·53	13·92	1·6	1127	14·72	11·64	1·1	1177	14·93	10·44	2·4
1078	15·06	12·76	4·6	1128	15·23	11·74	3·1	1178	16·16	12·89	0·7
1079	15·69	12·03	3·0	1129	14·96	11·03	2·6	1179	18·13	15·00	0·6
1080	16·23	13·55	1·6	1130	15·61	13·42	2·5	1180	15·51	10·13	4·7
1081	16·23	12·16	4·4	1131	17·51	15·32	1·0	1181	15·86	12·63	0·6
1082	15·69	11·57	4·2	1132	15·10	11·82	1·7	1182	14·85	12·58	2·3
1083	16·36	13·91	1·0	1133	15·17	13·10	2·5	1183	15·58	12·99	3·9
1084	15·00	11·72	2·6	1134	18·56	15·29	1·6:	1184	15·38	12·14	4·1:
1085	14·91	10·71	4·4	1135	14·91	11·67	2·5	1185	15·47	13·26	3·7
1086	14·80	10·62	2·6	1136	15·24	12·22	3·6	1186	14·65	10·73	1·7
1087	14·81	10·89	2·3	1137	14·67	11·98	2·6	1187	16·00	12·82	2·6
1088	14·79	12·68	2·6	1138	16·35	12·20	1·7	1188	15·11	13·03	4·2
1089	14·97	12·82	4·4	1139	15·58	14·25	1·1	1189	14·87	11·11	3·1
1090	16·50	13·97	1·4	1140	14·65	11·19	2·0	1190	15·88	13·17	1·1
1091	16·45	11·85	0·9	1141	16·81	14·51	0·8	1191	15·33	11·64	1·9
1092	15·45	11·74	3·5	1142	15·72	11·53	3·0	1192	16·12	13·58	1·0
1093	14·06	9·91	2·1	1143	16·04	9·34	5·5	1193	16·42	13·22	1·9
1094	15·91	12·93	1·8	1144	15·97	10·90	2·1	1194	15·25	11·52	2·3
1095	14·54	12·56	1·1	1145	14·91	12·22	2·6	1195	16·78	14·51	0·6
1096	14·40	11·30	2·1	1146	14·87	10·89	1·1	1196	14·68	11·47	4·0
1097	16·26	13·08	1·8	1147	15·75	13·45	2·0	1197	14·80	11·12	1·8
1098	15·16	11·88	1·6	1148	15·15	11·23	1·7	1198	18·93	16·69	0·6
1099	15·83	11·67	1·3	1149	15·13	11·42	3·6	1199	15·23	11·31	3·8
1100	15·95	12·25	6·9	1150	16·66	14·58	2·6	1200	15·72	11·72	3·9

No.	p_0	g	wt.	No.	p_0	g	wt.	No.	p_0	g	wt.
1201	15·87	12·56	3·9	1251	15·14	11·79	3·1	1301	15·22	11·78	1·4
1202	16·61	11·31	1·8	1252	14·96	11·66	2·6	1302	15·98	11·88	2·9
1203	16·80	13·12	3·6	1253	17·38	13·22	0·8	1303	14·68	10·43	1·5
1204	15·76	13·48	4·3:	1254	15·67	11·54	3·3	1304	14·64	10·41	2·9
1205	18·13	15·19	0·6	1255	15·75	11·59	4·4	1305	15·39	11·48	3·1
1206	15·03	11·38	1·6:	1256	16·07	10·80	1·8	1306	14·93	10·78	3·4
1207	16·11	12·19	3·2	1257	15·74	12·85	2·2	1307	15·42	13·17	1·1
1208	16·35	9·69	2·4	1258	15·94	11·73	4·1	1308	15·61	11·89	3·4
1209	15·69	11·48	2·1	1259	15·81	11·74	3·6	1309	15·50	11·23	2·1
1210	15·17	11·26	1·8	1260	16·00	12·87	2·3	1310	15·30	12·69	1·3
1211	15·81	12·05	3·8:	1261	15·96	11·83	1·0	1311	16·47	13·77	1·6
1212	16·16	10·80	1·9	1262	15·19	11·29	1·5	1312	16·80	12·74	1·6:
1213	16·27	12·14	1·7	1263	14·66	11·42	1·3	1313	16·16	12·94	1·1
1214	15·37	12·04	2·8:	1264	14·48	10·84	1·3	1314	16·42	14·05	0·9
1215	14·86	11·81	1·9	1265	14·95	11·03	0·6	1315	15·31	11·06	4·9
1216	15·52	13·32	3·9	1266	14·85	10·34	5·4	1316	17·45	14·79	0·6
1217	16·99	14·48	0·6	1267	16·18	13·39	2·0	1317	13·83	9·65	0·8
1218	16·62	14·34	1·6	1268	15·19	9·89	4·7	1318	15·53	13·13	0·6
1219	15·35	13·21	4·9	1269	14·98	9·65	3·6	1319	15·57	11·71	1·1
1220	16·14	12·24	2·1	1270	16·25	14·05	2·0	1320	15·75	11·89	0·8
1221	20·30	19·06	0·6	1271	15·85	11·72	3·6	1321	14·83	11·05	3·0
1222	16·66	13·16	0·6	1272	16·99	13·51	0·6	1322	16·77	14·08	1·0
1223	15·28	11·63	3·9	1273	16·60	13·99	3·4	1323	15·36	11·15	3·5:
1224	15·16	12·77	2·1	1274	15·34	13·15	3·8	1324	15·63	13·56	0·6
1225	15·71	13·51	2·6	1275	15·05	11·78	4·1	1325	16·17	12·31	1·2
1226	16·26	13·21	1·1	1276	16·01	11·83	1·0	1326	15·14	11·90	0·7
1227	15·75	11·52	4·1	1277	15·73	12·42	3·5	1327	16·33	12·89	2·0
1228	16·16	12·71	1·0	1278	15·04	12·40	3·3	1328	15·98	11·26	1·0
1229	17·14	12·91	0·8	1279	16·31	13·75	2·2	1329	14·50	11·37	2·2
1230	17·67	14·64	0·6	1280	15·59	11·02	4·3	1330	15·74	11·55	0·6
1231	15·92	12·68	2·2	1281	15·46	12·46	2·3	1331	15·61	11·54	4·9
1232	15·47	11·28	2·6	1282	15·53	11·42	2·8	1332	15·22	11·21	4·0
1233	15·32	12·32	1·2	1283	16·20	11·92	0·6	1333	16·00	12·83	2·8
1234	15·91	12·00	2·9:	1284	14·59	11·40	3·6	1334	15·09	11·36	2·6
1235	16·50	15·30	0·9	1285	15·22	11·34	3·0	1335	17·11	14·89	0·6
1236	15·52	12·81	2·6	1286	15·46	11·53	2·1	1336	15·63	12·02	2·0
1237	15·08	11·96	2·4	1287	15·94	12·03	3·6	1337	15·95	12·22	1·8
1238	16·25	13·01	1·0	1288	16·36	12·69	1·7:	1338	16·30	14·02	2·0
1239	16·77	13·54	3·1	1289	15·12	11·49	1·7	1339	15·42	11·49	5·0
1240	14·64	10·99	5·6	1290	16·20	13·65	0·8	1340	16·66	12·47	1·6
1241	14·61	10·38	2·6	1291	15·28	11·36	5·7	1341	15·37	11·98	1·1
1242	14·62	11·23	7·1	1292	15·52	12·55	6·0	1342	15·70	13·35	0·8
1243	15·24	11·18	5·6:	1293	17·32	15·13	2·1	1343	15·52	12·49	2·9
1244	15·12	12·63	2·3	1294	15·21	11·92	4·6	1344	16·26	14·02	3·0
1245	14·61	10·92	4·1	1295	16·17	11·64	3·8	1345	16·07	10·71	1·3
1246	15·86	12·72	1·1:	1296	15·30	12·62	2·5	1346	15·53	12·37	0·9
1247	15·71	11·58	4·8:	1297	16·30	12·38	1·6	1347	15·17	12·13	4·6
1248	14·29	10·94	4·2	1298	15·61	11·48	3·6	1348	15·59	12·09	2·6
1249	15·03	12·85	2·9	1299	16·41	12·89	2·4	1349	15·47	11·55	2·8
1250	17·14	14·15	1·3	1300	15·77	12·29	3·6	1350	15·74	12·11	2·5

No.	p_0	g	wt.	No.	p_0	g	wt.	No.	p_0	g	wt.
1351	15·20	10·97	2·8	1401	14·97	12·79	1·9	1451	15·88	13·76	1·0
1352	15·78	12·31	3·6	1402	17·79	14·52	0·6	1452	17·09	13·00	1·0
1353	15·00	11·08	2·4	1403	16·88	13·56	0·6	1453	14·85	13·70	1·0
1354	16·21	12·09	3·1	1404	16·83	10·17	0·9	1454	16·78	14·24	1·3
1355	14·79	13·80	1·2	1405	16·56	14·31	0·5	1455	16·65	14·41	0·6
1356	15·30	11·26	2·0	1406	15·77	12·47	3·1	1456	16·00	11·78	1·0
1357	14·84	10·61	0·6	1407	15·72	12·28	1·9	1457	15·66	12·36	0·8
1358	15·65	12·84	2·0	1408	16·08	12·00	2·2	1458	15·58	12·43	3·5
1359	14·96	10·86	1·0	1409	15·01	11·75	5·7	1459	16·03	11·86	3·1
1360	15·59	12·42	1·1	1410	16·29	12·36	3·1	1460	16·70	13·74	2·0
1361	15·98	11·93	0·8	1411	15·82	11·93	2·6	1461	14·83	10·72	3·8
1362	15·75	11·38	1·3	1412	15·74	13·59	2·9	1462	16·25	12·09	0·5
1363	16·26	12·55	2·6	1413	16·34	12·41	4·5	1463	16·15	12·02	1·6
1364	15·78	11·87	4·0	1414	17·22	13·74	0·6	1464	16·19	12·30	3·0
1365	15·48	13·24	3·3	1415	15·55	13·38	5·0	1465	16·07	12·14	0·6
1366	14·83	11·18	2·5	1416	15·65	11·73	1·9	1466	16·60	14·02	0·8
1367	16·69	14·20	1·0	1417	16·15	12·30	3·0	1467	14·33	9·79	1·8
1368	14·80	11·88	1·5	1418	15·24	13·02	4·1	1468	16·75	14·54	3·0
1369	15·47	11·38	1·7	1419	15·01	12·65	2·9	1469	14·98	10·87	4·6
1370	17·11	14·86	0·6	1420	16·23	12·82	1·3	1470	16·36	12·19	1·6
1371	16·51	12·26	1·2	1421	15·49	11·44	1·9	1471	15·69	12·35	4·9
1372	16·14	12·70	4·0	1422	15·93	13·69	1·9	1472	16·08	13·88	1·7
1373	18·81	14·23	0·6	1423	16·17	12·54	2·1	1473	16·56	13·52	1·8
1374	16·95	14·70	0·8	1424	14·91	10·70	2·4	1474	16·30	12·92	0·8
1375	15·54	12·79	3·4	1425	15·88	12·76	1·8	1475	16·60	14·09	0·8
1376	15·94	13·75	3·5	1426	15·10	12·05	2·2	1476	17·12	14·79	0·8
1377	16·45	14·18	2·0	1427	15·29	11·88	6·3	1477	16·47	12·27	0·7
1378	15·73	13·16	2·0	1428	15·05	11·52	2·2	1478	16·28	13·49	1·0
1379	15·04	12·10	2·6	1429	16·32	13·33	2·8	1479	15·70	12·44	2·5
1380	17·22	13·08	0·6	1430	16·16	13·16	1·2	1480	16·39	14·28	1·7
1381	15·78	12·93	2·6	1431	15·64	12·51	1·0	1481	15·75	11·83	2·1
1382	15·62	13·46	2·0	1432	15·86	13·27	2·5	1482	15·70	12·05	5·4
1383	16·87	12·84	1·6	1433	16·25	12·76	1·2	1483	15·94	12·59	3·0
1384	16·06	12·80	3·8	1434	15·35	11·43	2·9	1484	15·61	12·22	0·8
1385	15·33	11·94	2·6	1435	17·90	14·70	0·6	1485	16·40	12·47	1·1
1386	17·22	14·68	0·6	1436	15·52	11·37	2·0	1486	16·59	14·49	5·2
1387	16·61	14·33	0·9	1437	15·76	9·12	6·1	1487	15·88	11·75	4·2
1388	15·81	11·89	1·9	1438	16·92	12·70	1·0	1488	15·93	11·97	1·2
1389	16·16	12·52	2·6	1439	16·27	10·93	0·6	1489	17·21	13·02	0·6
1390	14·63	10·03	2·1	1440	16·98	12·80	2·1	1490	15·17	12·66	1·7
1391	15·87	12·89	1·0	1441	17·30	14·13	1·3	1491	16·80	12·57	2·3
1392	16·01	12·89	0·9	1442	16·19	12·53	2·8	1492	16·36	14·33	2·7:
1393	15·84	13·13	3·5:	1443	16·09	12·31	1·5	1493	15·28	12·58	2·3:
1394	15·51	12·78	3·6	1444	16·32	12·14	0·6	1494	15·96	13·88	2·7
1395	16·91	12·67	0·8	1445	15·96	11·87	2·6	1495	16·60	13·42	0·9
1396	15·27	13·03	4·1	1446	16·10	13·87	2·8	1496	15·89	13·77	2·3
1397	15·99	12·72	2·6	1447	15·78	12·83	2·1	1497	16·56	12·86	3·5
1398	15·54	11·37	1·7	1448	16·85	14·29	0·5	1498	17·06	13·00	0·7
1399	17·33	15·17	0·6	1449	15·90	13·73	2·2	1499	15·89	12·64	1·4
1400	16·99	12·88	0·6	1450	15·67	12·55	1·5	1500	16·62	14·40	1·9

No.	p_0	g	wt.	No.	p_0	g	wt.	No.	p_0	g	wt.
1501	16·62	13·64	2·5	1551	16·22	13·60	5·5	1601	15·78	13·58	3·2
1502	16·30	12·92	1·9	1552	16·54	12·63	1·1	1602	15·37	13·14	1·9
1503	14·99	11·84	2·5	1553	16·42	12·70	4·2	1603	15·46	12·04	2·4
1504	15·56	12·93	2·2	1554	15·85	12·71	2·0	1604	15·62	11·68	5·6
1505	15·58	12·36	1·8	1555	15·82	12·53	1·3	1605	15·23	11·31	4·2
1506	16·27	13·24	1·8	1556	15·92	11·33	1·7	1606	15·95	12·66	1·5
1507	17·05	14·59	1·0	1557	16·11	12·20	3·4	1607	15·77	12·79	5·6
1508	16·60	13·15	1·5	1558	15·74	11·47	4·0	1608	15·74	13·59	1·6
1509	15·04	14·00	1·8	1559	15·67	13·06	2·2	1609	14·99	11·93	2·6
1510	15·77	12·52	2·4	1560	16·06	12·79	2·1	1610	16·81	14·70	4·2
1511	16·60	14·07	1·0	1561	16·17	11·98	0·8	1611	16·06	11·82	2·4
1512	15·71	10·40	6·8	1562	15·55	13·37	1·8	1612	16·17	12·10	1·3
1513	16·45	14·36	1·3	1563	15·84	13·76	1·0	1613	16·31	12·92	2·6
1514	15·76	13·54	1·0	1564	16·25	12·09	2·1	1614	15·61	11·73	4·9
1515	16·88	13·85	0·6	1565	16·30	13·68	1·3	1615	15·47	11·37	3·7
1516	16·06	12·91	2·0	1566	12·35	17·74	0·5	1616	16·08	12·35	3·6
1517	15·47	12·12	3·7	1567	15·17	10·90	1·4	1617	16·25	12·03	1·2
1518	15·84	13·66	1·9	1568	15·63	13·12	1·7	1618	16·22	12·58	4·2
1519	16·95	12·84	0·5	1569	16·61	12·43	2·2:	1619	14·99	12·77	2·6:
1520	15·70	11·62	2·4	1570	16·13	12·53	2·2	1620	13·28	15·87	1·4
1521	16·80	13·19	1·0	1571	17·26	13·13	1·2	1621	15·02	12·83	1·9
1522	16·19	13·64	1·4	1572	15·40	11·32	3·5	1622	15·58	13·38	2·6
1523	15·56	13·34	1·6	1573	16·46	13·90	1·1	1623	15·91	11·78	1·8
1524	15·94	11·86	5·8	1574	16·26	11·50	2·9	1624			
1525	16·87	13·57	2·2:	1575	16·47	13·90	0·8	1625			
1526	17·15	14·73	0·5	1576	15·88	11·75	3·3	1626			
1527	15·76	13·58	2·9	1577	17·43	15·24	1·8	1627	15·17	14·13	1·3:
1528	16·19	13·52	1·6	1578	17·08	11·74	2·5	1628	15·48	11·56	1·6
1529	16·58	11·19	1·1	1579	15·63	11·02	1·1	1629	16·21	14·00	0·6
1530	16·70	14·46	1·0	1580	17·65	15·56	0·5	1630			
1531	16·18	13·02	0·8	1581	15·46	11·27	2·2	1631	15·73	13·52	1·4
1532	15·75	11·85	4·3	1582	17·18	12·99	1·1	1632	15·82	12·61	0·8
1533	15·70	11·80	3·2	1583	16·48	9·71	3·4	1633	15·73	11·54	0·7
1534	16·33	12·95	1·5	1584	14·82	12·25	1·3	1634	16·77	14·54	1·5
1535	16·93	12·80	0·9	1585	15·45	11·69	1·9	1635	16·34	12·72	0·5
1536	16·61	14·49	0·7	1586	16·08	13·38	2·4	1636	15·58	13·38	0·8
1537	17·03	13·06	2·4:	1587	15·78	12·81	1·4	1637	15·28	11·26	3·1:
1538	18·01	15·48	0·6	1588	16·01	12·07	2·2	1638	16·17	12·76	1·1
1539	16·13	11·99	5·3	1589	15·88	13·21	4·9	1639	15·04	12·00	0·3
1540	15·43	11·82	3·6	1590	15·23	13·04	2·6	1640			
1541	15·91	12·46	2·5	1591	15·79	13·18	0·9	1641			
1542	15·73	11·68	3·8	1592	16·31	12·86	1·4	1642			
1543	16·70	13·54	1·8:	1593	16·70	14·52	1·8	1643	16·57	13·72	2·1
1544	15·43	12·86	1·8	1594	15·68	13·38	1·2	1644	15·08	12·11	0·5
1545	16·26	12·80	2·1	1595	15·67	12·48	1·4	1645	16·57	12·58	0·5
1546	15·83	11·65	1·2	1596	15·45	11·76	3·7	1646			
1547	15·92	12·73	1·0:	1597	16·88	13·27	1·6:	1647			
1548	16·14	12·65	3·7	1598	16·78	14·32	2·0	1648			
1549	15·79	13·60	1·3	1599	16·32	12·19	2·6	1649			
1550	16·46	13·48	2·1	1600	15·15	14·17	0·5	1650			

No.	p_0	g	wt.
1651			
1652			
1653	15·59	12·47	0·5
1654	15·96	12·04	2·7
1655			
1656			
1657			
1658			
1659	14·85	11·36	0·3
1660	16·75	14·13	0·9
1661	16·06	14·00	0·3
1662	16·37	12·97	0·3
1663			
1664			
1665			
1666			
1667			
1668			
1669			
1670			
1671			
1672	17·24	13·05	2·1
1673			
1674			
1675			

No.	p_0	g	wt.
1676			
1677	16·28	13·34	0·3
1678	15·69	11·50	0·3
1679	16·45	12·35	0·3
1680			
1681			
1682			
1683			
1684			
1685			
1686			
1687			
1688			
1689			
1690			
1691			
1692			
1693			
1694			
1695			

No.	p_0	g	wt.
1696			
1697			
1698			
1699		14·32	0·2
1700			
1701			
1702			
1703			
1704	15·99	13·82	0·3
1705			
1706			
1707			
1708			
1709			
1710	16·98	14·54	0·5
1711			
1712			
1713			
1714	15·76	12·74	0·5
1715			

21. COMMISSION DE LA LUMINESCENCE DU CIEL

PRESIDENT: F. E. Roach.

Scientific Session, 23 August 1967

The following presentations were made to the Commission, all on August 23 except for that of Kordylewski which was presented informally to a small group on August 29.

Speaker	*Topic*	*Reference*
Ščeglov	H alpha observations	*Astr. J. USSR* **41**, 331, 1964 *Astr. Circ. USSR* **44**, 414, 1967 *Astr. Circ. USSR* **44**, 411, 1967
Huruhata	Rocket Observations of 6300, 5577, 5300	*Space Research* **VIII** (in press)
Dachs	Airglow observations at Toumeb (S.W. Africa)	
Roach	(*a*) Summary of IQSY Observations (*b*) Galactic Light (*c*) Cosmic Light (*d*) New photometric map of the Milky Way	(*a*) *Annals of the IQSY*, 1967 (in press) (*b*) *Modern Astrophysics*, 1967 (in press) (*c*) *The Geophysical Jour*, 1968 (in press)
Weill	(*a*) Predawn enhancement of 6300 (*b*) Observations of NI 5200	(*a*) *C. r. Acad. Sci.*, **264**, 1685, 1967 (*b*) *C. r. Acad. Sci.*, **264**, 1286, 1967
Lebedinsky	Airglow observations from Cosmos 92	*Space Research* VII (1967)
Weinberg	Zodiacal Light Conference, Jan–Feb, 1967	Summary Report, Hawaii Institute of Geophysics HIG 67-13
Hoffmeister	Visual Observations of the Zodiacal Light	
Simpson	Observations of the Earth-Moon Libration Clouds	*Physics Today*, **20**, 39, 1967
Kordylewski	Observations of the Earth-Moon Libration Clouds	
Courtés	Ultraviolet Observations of the Zodiacal Light	*L'Astronomie*, Juin–Juillet, 1967

The following resolution was adopted by the Commission 21:

It is recommended that, during the next triennium, there should be a concentration of airglow observations at chains of stations along selected longitude meridians. New stations are particularly needed in the Southern Hemisphere.

Pairs of magnetically conjugate stations should be established wherever possible, and the interrelationships between the airglow and the ionosphere should be studied by coordinated experiments.

22. COMMISSION DES METEORES ET METEORITES

PRESIDENT: P.M. Millman.
SECRETARY: A.F. Cook.

First Session, Thursday, 24 August 1967, 10:30 hours

The business meeting of Commission 22 was called to order by *P.M. Millman*, who was in the chair. Approximately 50 members and guests were present. The Chairman noted with regret the loss by death of one of the members of the Commission, Monsieur F. Baldet from France.

There was general discussion of the Draft Report, and it was noted that a number of 1964–1966 references had been omitted. Since only corrections of errors in the text of the Draft Report were allowed by the Executive Committee, it was agreed that a supplementary list of 1964–1966 references, including titles, should be appended to these proceedings. The Chairman pointed out the duty of all members of the Commission to supply to the President complete lists of references from their respective countries.

The appointments by the Executive Committee of Z. Ceplecha as the new President of Commission 22 and of R.E. McCrosky as the new Vice-President, were announced. New members of the Commission, approved by the Executive Committee, were noted as follows:

Members: E. Anders, J. Davis, W.G. Elford, C. Ellyett, C.S.L. Keay, T.R. Kaiser, M. Kresáková, B.A. McIntosh, T.N. Nazarova, H. Oleak, Z. Plavcová, J. Rajchl, R.B. Southworth, I.F. Verniani.

The following have been asked to serve for the coming term as consulting members:

Consulting Members: W.M. Alexander, E.P. Henderson, M.H. Hey, B.H. Mason, U.B. Marvin, K.I. Sztroky, F. Wickman, J. Zähringer.

It was agreed that the current Organizing Committee of the Commission, together with the President, Vice-President and Vice-President Elect, should act as a Nominating Committee to select a slate of five members to be nominated as the Organizing Committee to serve during the 1967–1970 term.

C.L. Hemenway announced that a new international panel under the auspices of COSPAR had been formed to attempt to resolve some of the confusion that exists in the observational data on interplanetary dust, obtained by observations from rockets and satellites, and from collection techniques on the earth's surface.

Second Session, Friday, 25 August 1967, 10:30 hours

The Nominating Committee presented the following slate for the 1967–70 term.

Organizing Committee: P.B. Babadžanov, W.G. Elford, C.L. Hemenway, P.M. Millman, A.A. Yavnel.

Meteorite Committee: Chairman: A.A. Yavnel. Vice-Chairman: E. Anders.

There were no further nominations and the above slate was elected unanimously.

The items appearing on pages cviii–cix of the I.A.U. Agenda were now taken up.

Item 1—It was unanimously agreed that the Executive of Commission 22 produce an up-to-date mailing list of astronomers active in research in meteoric astronomy. *Z. Ceplecha* undertook to have this revised each year and mailed to all on the list.

Item 2—*L. Kresák* noted that arrangements had been made for the early publication (6 to 8 months) of the proceedings of I.A.U. Symposium No. 33, the Physics and Dynamics of Meteors.

Item 3—Members of the Commission were urged to circulate freely their published papers, making use of the mailing list mentioned under Item 1.

Item 4—The policy of the Executive Committee of the I.A.U., that funds to assist young astronomers in attending General Assemblies and Symposia should not be cut, was duly noted and approval for this policy expressed.

Item 5—It was agreed unanimously that the current name of Commission 22 remains suitable and should not be changed at the present time.

Item 6—Note was taken of the importance of recording bright fireballs, particularly in connection with the falls of meteorites.

Item 7—This item was withdrawn by the originator.

Item 8—No new suggestions for further symposia or colloquia were made at this time. All those interested in meteorite research were requested to send comments to Dr. G. Harbottle, International Atomic Energy Agency, Vienna, re possible subjects for inclusion in the program of the international meteorite symposium to be held in Vienna, August 7–13, 1968.

In view of the IAU Symposium No. 33, to be held at Tatranská Lomnica, September 4–9, 1967, no papers were presented at the Commission 22 sessions in Prague.

SUPPLEMENTARY LIST OF 1964–1966 REFERENCES

Astapovič, I.S., Terentjeva, A.K. 1966, The Leonid Meteor Shower, *Komety Meteory*, **14**, 24.

Astavin-Razumin, D.L. 1966, A Method for Determining the Stellar Magnitudes of Meteors, *Meteornaja materija v atmosfere Zemli*, Nauka, Moscow, 93.

Bronšten, V.A. 1966, Theory of the Spectroscopic Investigation of Meteorite-Dropping Bolides, *Meteoritika*, **27**, 97.

Bronšten, V.A., Liubarsky, K.A. 1966, Radiation from Meteors and Bolides, *Meteornaja materija v atmosfere Zemli*, Nauka, Moscow, 3.

Chebotarev, R.P. 1966, On Simple Measurements of the Range of Meteors, *Bjull. Inst. Astrofiz. Dušanbe-Stalinabad*, no. 47, 3.

Cvetkov, V.I., Bojarkina, A.P. 1966, Results of the Investigation of New Evidence on the Fall of the Tunguska Meteorite in 1908, *Meteornaja materija v atmosfere Zemli*, Nauka, Moscow, 81.

Delov, I.A. 1966, Equipment and Methods of Investigation of Irregular Movements in the Meteor Zone, *Resultaty issledovanij po MGP, issl. meteorov*, no. 1, 133.

Delov, I.A. 1966, Turbulent Motions in the Atmosphere at Heights of 90–110 km and their Connection with other Phenomena, *Meteory*, Kiev Univ., 126.

Doğan, N. 1966, On the Diurnal and Seasonal Variations in the Number of Micrometeoritic Particles, and their Origin, *Commun. Fac. Sci. Ankara Univ.*, **15A**, 25.

Fesenkov, V.G. 1964, On the Orbit of the Tunguska Meteorite, *Meteoritika*, **25**, 163.

Fialko, E.I. 1966, On the Evaluation of the De-ionization Rate in Meteor Trails, *Komety Meteory*, **13**, 22.

Fialko, E.I. 1966, Rates of Meteor Echoes on $\lambda = 10$ m During the IGY-IGC Period, *Komety Meteory*, **13**, 41.

Fialko, E.I. 1966, On the Comparison of Meteor Radars According to their Sensitivity, *Komety Meteory*, **14**, 38.

Fialko, E.I., Chumak, Iu.V. 1966, On the Use of Intermediate Type Echoes for the Measurement of the Diffusion Coefficient and Electron Line Density, *Komety Meteory*, **14**, 16.

Fialko, E.I., Moisya, R.I., Kolomiets, G.I., Melnik, V.I., Chumak, Iu.V. 1966, The Experimental Study of the Energy Distribution of Meteor Bodies, *Meteory*, Kiev Univ., 144.

Fialko, E.I., Moisya, R.I., Melnik, V.I., Kolomiets, G.I. 1966, Radar Observations of Geminids in 1964, *Komety Meteory*, **14**, 19.

Glöde, P. 1966, Results of Radar Meteor Observations of the Quadrantid Shower, *Kleinheubacher Berichte*, **11**, 157.

Ivanov, K.G. 1966, Reply to an Article by A.V. Zolotov Entitled 'Thermal Ionization of the Ionosphere by the Shock Wave of the Tunguska Explosion of 1908', *Geomagn. Aeronom.*, **6**, 777.

Kaščeev, B.L., Nečitailenko, V.A., Suverov, Iu.I. 1966, The Drift of Trains, *Meteory*, Kiev Univ., 119.

Katasev, L.A. 1966, *Issledovanije meteorov v atmosfere Zemli fotograficheskim metodom*, Leningrad.

Kirova, O.A., Zaslavskaja, N.I. 1966, Some Results on the Dispersion of Material from the Fall Area of the Tunguska Meteorite, *Meteoritika*, **27**, 119.

Kovšun, I.N. 1966, On the Number of Particles of Fragmenting Meteors, *Meteory*, Kiev Univ., 54.

Kovšun, I.N. 1966, On the Density of Meteor Bodies, *Meteory*, Kiev Univ., 63.

Kramer, E.N. 1966, The Orbit, Luminosity and Spectrum of the Bright Meteor of 5 July, 1962, *Meteory* Kiev Univ., 75.

Kručinenko, V.G. 1966, On the Drag of Meteor Bodies, *Meteory*, Kiev Univ., 47.

LaPaz, L. 1965, Catalog of the Collections of The Institute of Meteoritics, The University of New Mexico, *Univ. New Mexico Pub. in Meteoritics*, no. 4, Univ. New Mexico Press, Albuquerque.
Liubarsky, K.A. 1966, Variations in Telescopic Meteor Heights, *Resultaty issledovanij po MGP, issl. meteorov*, no. 1, 160.
Liubarsky, K.A. 1966, More on the Determination of the Luminosity Function of Meteors, *Meteornaja materija v atmosfere Zemli*, Nauka, Moscow, 97.
Liubarsky, K.A., Cvetkov, V.I. 1966, Luminosity Function of Meteors Observed 1962–1963, *Meteornaja materija v atmosfere Zemli*, Nauka, Moscow, 103.
Martynenko, V.V. 1966, Amateur Meteor Astronomy in the Crimea, *Meteory*, Kiev Univ., 147.
Nekrasov, V.I., Emelianov, Iu.M. 1966, Pre-fall Characteristics of the Forest in the Area of the Tunguska Meteorite, *Meteoritika*, **27**, 128.
Nemirova, E.K. 1966, Results of Observations of the Polarization Effect for Sporadic Meteors, *Komety Meteory*, **13**, 29.
Pokrovsky, G.I. 1966, On the Explosion of Meteor Bodies Moving in the Atmosphere, *Meteoritika*, **27**, 103.
Romig, M., Lamar, D. 1966, Anomalous Sounds and Electromagnetic Phenomena Associated with the Flight of Bolides, *Meteornaja materija v atmosfere Zemli*, Nauka, Moscow, 38.
Rubcov, L.N. 1966, On the Limiting Magnitude of Meteors Registered at a Wavelength of 17 m, *Bjull. Inst. Astrofiz. Dušanbe-Stalinabad*, no. 45, 12.
Sidorova, V.V. (editor) 1966, *Radio-wave Propagation by Meteors, Parts* 3–4, Kazan Univ., Kazan.
Smirnov, V.A. 1966, On Methods of Obtaining and Studying Spectrograms of Meteors, *Meteory*, Kiev Univ., 90.
Smirnov, V.A. 1966, On Radiation and Methods of the Spectrophotometry of Meteors, *Meteory*, Kiev Univ., 104.
Šodiev, U. 1966, Investigation of the Train of an Orionid Meteor Observed on the 22nd of October, 1962, *Resultaty issledovanij po MGP, issl. meteorov*, no. 1, 156.
Terentjeva, A.K. 1966, A Possible Meteor Stream of the Family from Pons-Winnecke Periodic Comet, *Meteory*, Kiev Univ., 40.
Vinogradov, A.P., Devertz, A.L., Dobrina, E.I. 1966, Concentration of C^{14} in the Atmosphere at the Time of the Tunguska Fall, and a Discussion of Antimatter, *Dokl. Akad. Nauk SSSR*, **168**, 900.
Watkins, C.D. 1966, High Sensitivity Radar Investigations of Magnetic Field-Aligned Ionization in the E Region, *Electron Density Profiles in Ionosphere and Exosphere*, edited by E.V. Frihagen, North Holland Publ. Co., 208.
Watkins, C.D., Sutcliffe, H.K. 1965, Radar Observations of Weak Field-Aligned Ionization at a Frequency of 300 Mc/s, *J. atmos. terr. Phys.* **27**, 309.
Zotkin, I.T. 1966, The Trajectory and Orbit of the Tunguska Meteorite, *Meteoritika*, **27**, 109.
Zotkin, I.T. 1966, A Bolide over the Urals, *Meteornaja materija v atmosfere Zemli*, Nauka, Moscow, 69.

23. COMMISSION DE LA CARTE DU CIEL

Compte rendu des séances du 23 août et du 24 août 1967

Président: P. Sémirot.
Sécrétaire: W. J. Luyten.

Sémirot fait part du décès de deux membres de la Commission 23, MM. Danjon et Fernandez de la Puente.

Il présente ensuite un compte rendu financier. En accord avec le Secrétaire Adjoint Jappel, les comptes sont arrêtés en recettes à une somme égale à 1422,15 dollars.

Le calcul des constantes pour les zones de déclinaison supérieure à 32° a été confié à l'Observatoire de Hambourg. *Hans Kox* présente un bref rapport sur le travail qu'il a exécuté. Les étoiles repères ont été choisies dans le catalogue AGK_2 et dans le catalogue préliminaire de l'AGK_3. Dans le but de tester les méthodes de calcul la zone de Catane a été choisie. Une première réduction a été faite en tenant compte des termes linéaires et les résidus ont fait l'objet d'une étude très détaillée. Une réduction exécutée en tenant compte des termes du deuxième ordre a montré que ces termes ne sont pas négligeables et sont de la forme:

$$px^2 + qxy \quad \text{en } x$$
$$pxy + qy^2 \quad \text{en } y$$

En outre, les résidus de la première réduction ont été étudiés suivant les formules

$$urx + vr^2x + wr^3 \quad \text{en } x$$
$$ury + vr^2y + wr^3 \quad \text{en } y$$
$$(\text{avec} \quad r^2 = x^2 + y^2)$$

En tenant compte de ces termes les résidus deviennent plus petits. En réponse aux questions de Deutsch, Dieckvoss, Eichhorn, Herget et Lacroute, il est précisé par Kox que les constantes u, v, w sont significatives.

Lacroute fait un exposé sur l'amélioration des constantes de clichés en utilisant leurs recouvrements. Les positions des étoiles repères du catalogue photographique de la Carte du Ciel sont déduites des positions de l'AGK_2 et de l'AGK_3. La méthode de réduction par recouvrement permet de multiplier le poids par quatre. *Deutsch* fait remarquer que les clichés de la Carte n'ont pas été pris à la même époque et que, par conséquent, cette méthode de recouvrement peut être en défaut à cause des mouvements propres.

Une discussion s'engage sur la forme à donner à la publication des résultats. Prennent part à cette discussion Deutsch, Dieckvoss, Eichhorn, Herget, Lacroute, Murray, Sémirot et Wood. Il est décidé de publier les ascensions droites et les déclinaisons de toutes les étoiles et pour chacun des clichés de façon à tenir compte de ce qu'ils n'ont pas été pris à la même époque. Faut-il choisir une publication sur bandes magnétiques ou sur cartes perforées? La durée de vie d'une bande magnétique est inconnue mais si ce procédé est choisi il est proposé de publier les α, δ et les coordonnées standards *X*, *Y*. *Murray* souhaite que lors de la publication de l'AGK_3 les mouvements propres soient donnés non seulement sous la forme μ^s_α, μ''_δ mais aussi sous la forme $(\mu_\alpha \cos\delta)$.

Lors de la deuxième réunion de la Commission 23 *Sémirot* pose le problème de la détermination des coordonnées équatoriales des étoiles repères pour l'époque de la prise des clichés et, par conséquent, celui de la détermination des mouvements propres.

Faut-il utiliser seulement les positions de l'AGK_2 et de l'AGK_3 ou tous les catalogues qui sont à la disposition des astronomes?

Lacroute n'utilisera que l'AGK_2 et l'AGK_3. Herget formera un 'catalogue général' de toutes les étoiles repères du catalogue photographique de Bordeaux (12000) en utilisant les catalogues AGK_1,

AGK_2, AGK_3, Yale, Bordeaux 1953 et les positions du catalogue photographique de Bordeaux. La plupart des données y compris les α, δ, les coordonnées mesurées du catalogue x, y ont été perforées. Il faut attendre l'AGK_3. Le plan de travail d'Herget est le suivant: il calculera par approximations successives

(*a*) les positions et les mouvements propres à partir du 'catalogue général',

(*b*) les positions des étoiles repères du catalogue astrographique par comparaison au "catalogue général",

(*c*) les constantes des plaques à partir des positions (b),

(*d*) les positions des étoiles du catalogue photographique à partir de (c),

(*e*) puis recommencera les calculs (a), (b), (c).

A l'unanimité on convient de choisir pour tous les travaux astrographiques les positions de l'AGK_3; par exemple: positions de petites planètes, comètes, etc.

Pour la zone Sud on convient d'attendre la fin de la réobservation des étoiles repères et la publication d'un AGK_3 pour l'hémisphère Sud.

Eichhorn expose son travail sur les constantes de la zone d'Hyderabad.

L'époque de la prise des clichés est sensiblement l'époque des observations de l'AGK_2. Il n'y a donc pas de problème de mouvements propres. Les coefficients des termes quadratiques et ceux qui dépendent de la couleur sont négligeables. Seuls des termes linéaires en x, y apparaissent et dépendent du diamètre de l'image et de la coma. En tenant compte de ces termes l'erreur moyenne passe de $0''38$ à $0''28$ et Eichhorn fait remarquer que le terme dépendant de la coma varie avec chaque plaque.

Eichhorn expose sa méthode de calcul des constantes par recouvrement des clichés.

Le problème de la répétition de la Carte du Ciel sera repris ultérieurement. Des Progrès importants ont été accomplis dans la construction des instruments et il faut en tenir compte.

Il apparaît nécessaire à tous les membres de la Commission de disposer d'une machine à mesurer automatique. *Dieckvoss* et *Lacroute* traitent de la machine de Hambourg et de celle de Lick (Vasilevsky), *Herget* décrit celle de Strand de l'U.S. Naval de Washington et *Luyten* donne les principes de celle qui est en cours de construction pour l'Observatiore de l'Université de Minnesota.

Les discussions qui ont eu lieu au cours des réunions de la Commission 23 mettent surtout en évidence l'évolution des méthodes de calcul des constantes des clichés photographiques en liaison étroite avec les progrès techniques accomplis dans le domaine des machines à calculer. On a d'abord réduit plaque par plaque, puis par zone de 2° allant de 0^h à 24^h d'ascension droite et enfin par des méthodes de recouvrement intéressant une grande partie du ciel, voir les travaux de Eichhorn, Lacroute et de Vegt.

La nouvelle commission de la Carte du Ciel sera ainsi composée:

PRÉSIDENT: W. Dieckvoss

VICE-PRÉSIDENT: P. Lacroute.

COMITÉ D'ORGANISATION: P. Couderc, A.N. Deutsch, Herget, H. Kox, W.J. Luyten, P. Sémirot, H.W. Wood.

MEMBRES: Abhyankar, Bouigue, Eichhorn, Fracastoro, Gallouet, Gunther, Harris, Landi Dessy, Lourens, Murray, O'Connel.

24. COMMISSION DES PARALLAXES STELLAIRES ET DES MOUVEMENTS PROPRES

Report of Meetings, 23 and 25 August 1967

PRESIDENT: A. N. Dejč.
SECRETARY: S. Vasilevskis.

In opening the President pointed out that several members of the Commission have not responded to his requests for reports and proposals. Three members have not shown any sign of activity during the last six years, and the President raised the question of possible deletion of these three persons from the list of membership of Commission 24. After some discussions the Commission decided that although the particular persons have retired from astronomical activity, they should be retained as *de facto* honorary members because of their past contributions.

The following changes in membership were proposed and approved: (1) Delete V. Osvalds on his request. (2) Add A. Blaauw, W. Dieckvoss, N. V. Fatčihin, L. W. Fredrick, A. N. Goyal, W. F. van Altena.

The President announced that on the basis of the membership response to his inquiry, the following officers of the Commission have been nominated for approval by the General Assembly:

President: W. J. Luyten.

Vice-President: S. Vasilevskis.

The *Organizing Committee* was nominated and approved as follows: N. A. Dejč, V. V. Lavdovskij, J. v. B. Lourens, W.W. Morgan, C. A. Murray, K. Aa. Strand, P. van de Kamp.

The following additions to and changes in the *Draft Reports* were proposed: *Murray* of Royal Greenwich Observatory: Symms has determined the parallax and the masses of the components of Polaris by combining visual double star observations with measurements of Greenwich and Allegheny plates.

Meurers of Vienna Observatory: About 2000 proper motions have been determined in the region of h and χ Per; they are being analyzed for internal motions. The large refractor has been prepared for work on proper motions.

Rodriguez of San Fernando Observatory: 132 plates have been taken of 48 fields with galaxies, on the Pulkovo astrometric program.

Upgren of Van Vleck Observatory: The reference to work on M37 should be changed to *A.J.* **71**, 736, 1966.

The amendments above were approved for inclusion in this report. The President then proposed an agenda for reports and discussions. The agenda was approved and the scientific session proceeded as follows.

Eichhorn reported on reduction of parallax photographs. By using plate constants instead of dependencies one may introduce proper motions of reference stars as unknowns under a condition that the sum of these proper motions is zero. This method yields parallaxes with a higher weight than the conventional procedure of assuming the proper motion of reference stars to be zero.

Lacroute indicated that the method above may be advantageous for large epoch differences but probably there is no advantage if the reference stars have not moved appreciably. Several other members expressed skepticism regarding the merit of the inclusion of proper motions of reference stars as unknown into the plate solution.

Hoffleit reported that photography on the Yale parallax program has been discontinued after leaving the 26-inch refractor in Australia. There are, however, almost 7000 unmeasured photographs in various groups, as stars with only one or no parallax determination, sets of insufficient number of plates for parallax determination, long-series observations of binary stars for orbital motion, and plates taken for proper motions. The present small Yale staff connected with the program is not able to measure and reduce the plates in a reasonable time, and the material can be made available to those interested.

In discussions the Commission expressed its concern regarding the Yale program and also regarding the future of the 26-inch refractor. The Commission encourages interested persons to cooperate with Yale.

Fredrick reported that the procedure of parallax determination and the format of publication will be changed at McCormick. He also expressed the need for standard fields of stars for instrumental calibrations. *Dieckvoss* pointed out that Selected Areas at + 45° have been suggested earlier (*A.J.* **65**, 174, 1960) for the purpose of calibration.

Upgren reported that the Van Vleck parallax work has been reactivated with an emphasis on faint *M*-dwarfs. Plate constants in x and y will be derived from about 10 reference stars.

Gliese informed that the second revised edition of the Catalogue of Stars within 20 pc. will be published. Only approximately 40 stars have been added since 1957. This number, however, would need a drastic revision upward if the proposals by Schilt and Strand were adopted, i.e. not to reduce the results of various observatories to a common system, in spite of some large systematic differences. In that case all the Yale and Cape parallaxes should be increased by $0\overset{''}{.}005$ and $0\overset{''}{.}006$ respectively. As long as the systematic differences are not explained, there seems to be no reason for preferring a particular observatory. On the other hand, the explanation of these differences is urgent for a reliable system of parallaxes.

Dejč reported that Kiselev at Pulkovo has derived the value of $+ 0\overset{''}{.}075 \pm 0\overset{''}{.}012$ (m.e.) for the parallax of LP 9-231 which has been called a blue pigmy star. The Pulkovo result indicates that it is a white dwarf.

Morgan reported on the present status of spectroscopic parallaxes. Reliable stellar luminosities can be derived spectroscopically for a limited number of types, not for all the spectral classes. It is not advisable therefore to include spectroscopic parallaxes into a parallax catalogue at present.

After discussions the following resolution was passed:

Commission 24 requests W.W. Morgan to organize and to chair a working group on spectroscopic parallaxes. Members of the working group to be coopted by the Chairman in cooperation with the President of the Commission. It is recommended that the group be formed and a report on spectroscopic parallaxes be prepared before the next General Assembly of the IAU.

Vasilevskis reported that the equipment for automatic measurement of plates has been completed and installed at Lick Observatory by the Gaertner Scientific Corporation. Extensive tests have proved that its performance is in accordance with the original plans (*A.J.* **65**, 208, 1960) and specifications. The equipment consists of the Survey Machine for inspection and selection of objects for measurement, and the Automatic Engine for completely automatic measurement of positions and magnitudes, in a specially conditioned separate room without the presence of an operator. The range of measurement is 17 inches (430 mm), and the mean error of a measurement over this range is less than one micron; the repeatability is of the order of 0.3 microns.

Strand reported on the new measuring machine for plates up to 10×10 in. The positioning, centering and readout is done automatically; each of the three steps is initiated by an operator at the machine. The presence of the operator has no appreciable effect because the main structural components, including the carriages, are made of granite. Air bearings are used for support of moving parts. Centering is performed by a rotating sector with 24 slits 400 microns wide. Ferranty moiré fringe system is used for positioning and readout, the latter to one micron. Images from 70 to 270 microns can be measured, with an accuracy of 0.8 microns. Speed is about 5 times and accuracy about 24% higher than with the conventional machines.

Luyten discussed briefly his new derivation of the Luminosity function based on proper motion data obtained from a survey made on 117 plates of the Palomar Schmidt Survey. More than 50000 moving objects were found brighter than 21.5 pg and with motions larger than $0\overset{''}{.}04$ annually. Using solar apex determinations to calculate mean parallaxes a luminosity function was derived which is found to have a maximum at $M = + 15.7$ pg. The total number of stars nearer than 10 parsecs was found to be between 550 and 600 and the total mass of luminous objects at 0.064 $\odot$ per cubic parsec.

Fatčihin reported on preliminary results from stellar proper motions determined with reference to galaxies. The absolute motions of 14000 stars, including 1283 AGK3 stars were related to 271 galaxies in 82 areas photographed with the Pulkovo normal astrograph, the average epoch difference being 22.4 years. The values of $A = 302\overset{\circ}{.}5 \pm 4\overset{\circ}{.}5$, $D = 32\overset{\circ}{.}6 \pm 2\overset{\circ}{.}7$ were found for the solar apex from stars of approximate photographic magnitude 15. The mean secular parallax of these stars increases with the galactic latitude, from $0\overset{''}{.}0047$ to $0\overset{''}{.}00112$, with the probable errors of the order of $\pm 0\overset{''}{.}001$ for each of three zones investigated. A comparison with the AGK3 yielded the correction to the Newcomb precession: $\Delta m = -0\overset{''}{.}0053 \pm 0\overset{''}{.}0011$ and $\Delta n = 0\overset{''}{.}0044 \pm 0\overset{''}{.}0011$. Systematic differences with the AGK3 were also studied.

Dieckvoss reported that AGK3 proper motions of 163811 stars were statistically investigated for solar motion, secular parallax, galactic rotation and dispersion. The solutions gave the following values for correction to centennial precession and for Oort's constants in km/sec/kpc, with their mean errors: $\Delta m = -0\overset{''}{.}32 \pm 0\overset{''}{.}01$, $\Delta n = +0\overset{''}{.}51 \pm 0\overset{''}{.}01$, $A = +14.8 \pm 0.7$, $B = 11.3 \pm 0.6$.

Giclas reported on the Lowell proper motion program. More than 12000 stars with annual motions larger than $0\overset{''}{.}26$ have been listed in 12 Lowell Bulletins, and an additional number of approximately 5000 stars is expected to be published in the same way. A question was raised whether eventually a catalog of all the stars arranged in RA should be published.

After discussions Commission 24 recommended:

It would be very useful to publish a catalogue of Lowell Survey stars, after the survey is completed. All the duplicate measurements should be combined, and the stars listed in the sequence of their right ascension, with references to the original finding charts.

Dejč reported on the progress of the Pulkovo program of proper motions with reference to galaxies in the zone − 25° to − 90°. The Observatories at Santiago, Perth, Cape and Cordoba are participating with their normal astrographs; it is hoped that La Plata will also join. In addition, a new Maksutov type telescope has been installed 100 km north of Santiago, with the focal length, meniscus diameter and field of 2 m, 70 cm and 5° respectively. A list of 153 fundamental stars of FKSZ in the same zone has been prepared for observing with normal astrographs, or preferably with longer telescopes of focal length 5 to 6 m.

After discussions Commission 24 concluded:

That it is important to observe urgently the 153 fundamental stars with telescopes mentioned above. The Commission recommends that southern observatories having such telescopes participate in the program.

25. COMMISSION DE PHOTOMETRIE STELLAIRE

PRESIDENT: V.B. Nikonov.
VICE-PRESIDENT: A.W.J. Cousins.

Scientific Meeting, 23 August 1967

SCIENTIFIC REPORTS

A. W. J. Cousins and R.H. Stoy—The general problem of the establishment of photometric standards for stellar photometry. Dr Stoy presented a compilation of UBV-data near the equator obtained from various lists and proposed that these data be accepted as a new definition of the UBV-system.

J. Stock—The atmospheric extinction in the UBV-system. New methods for determining and eliminating the atmospheric extinction effects are presented and it is proposed that the UBV-system be defined by strictly extinction-free data.

It is proposed that a working group for a new definition of the UBV-system be formed.

Business and Scientific Meeting, 25 August 1967

BUSINESS MEETING

The following organisation of Commission 25 was agreed upon: President, A.W.J. Cousins; Vice-President, D.L. Grawford: Organizing Committee: A.N. Argue, M. Golay, R.H. Hardie, G.E. Kron, V.B. Nikonov, E. Rybka, J. Stock.

The working group for the new definition of the UBV-system is composed of: President, A.W.J. Cousins; members: H.L. Johnson, J. Stock, V. Straižys.

The Draft Report was adopted with the following additions.

Reduction methods

K. Serkowski reports that 'The difficulties in the transformation of $U–B$ colours to the standard system can be avoided if the B and U magnitudes are transformed separately. For each of these spectral regions the difference between the directly observed and standard magnitudes is represented as linear function of the mean gradient of the energy distribution in this spectral region; using the spectrophotometric data these gradients can be tabulated as a function of spectral type, luminosity class, and reddening. If such gradients are used as coefficients in the transformation (and atmospheric extinction) equations for U and B magnitudes good transformation is achieved even if the instrumental system differs appreciably from the standard UBV system.'

Photometric Systems

M. Golay reports: 'Le système photometrique que j'ai créé à l'observatoire de Genève est intermédiaire entre un système à large bande et un système à bande intermédiaire (**127a**), (**128**). Il se compose de filtres (U), (B), (V) proches des UBV de Johnson et de 4 autres filtres B_1, B_2, V_1, G, qui partagent les zones couvertes par les filtres (B) et (V). Les bandes U, B_1, B_2, V_1, G, servent à établir divers paramètres. Ceux-ci sont: $B_2–V_1$, $V_1–G$, $B_1–B_2$, $U–B_2$, d, Δ, m_2, g.

Les 4 derniers sont définis par les relations:

$$\Delta = (U-B_2) - 1{,}055\ (B_2-G)$$
$$g = (B_1-B_2) - 1{,}520\ (V_1-G)$$
$$d = (U-B_1) - 1{,}6\ (B_1-B_2)$$
$$m_2 = (B_1-B_2) - 0{,}69\ (B_2-V_1)$$

L'indice de couleur B_2–V_1 (Golay et Goy (**136**)) est intéressant car il est peu dépendant de la composition chimique (ou plutôt de l'effet de blanketing attribué à la composition chimique). L'établissement de divers diagrammes, par exemple B_1–B_2 versus Δ, permettent de bien séparer les étoiles 'weak' des 'strong', les sous-naines (**129**) et (**138**) et les métalliques (**129**) et (**139**a). Le paramètre d'est particulièrement sensible à la magnitude absolue pour les types spectraux *A* à *G* (Hauck (**139**)). Les mesures sont publiées dans divers catalogues (**214**), (**140**a), (**140**b). Actuellement près de 1200 étoiles sont publiées et environ 1000 sont en cours de dépouillement. Le programme comprend:

(*a*) les étoiles A_m des catalogues suivants.

(*b*) les sous-naines de Eggen et Sandage MN 119.

(*c*) les étoiles A, F, G des catalogues de Chalonge et de Johnson et Morgan, *Ap. J.* **117**.

(*d*) les étoiles de Wallerstein de *Ap. J.* suppl. **6**, N61.

(*e*) les étoiles géantes K des catalogues de Wallerstein et Helfer, *A.J.* **71** N5, de R. et G. Cayrel, de *Annual Review of Astronomy and Astrophysics*, **4**, de la thèse de Spite (*Ann. Astrophys.*).

(*f*) les étoiles 'weak' et 'strong' des listes de N. Roman, *Ap. J.* **112** et **116**.

(*g*) les étoiles magnétiques de Babcock.

La paramètre m_2 est une mesure du blanketing. Il permet de mesurer le degré de métallicité et séparer les 'weak', 'strong', métalliques, sous-naines.

Le diagramme *d* versus B_2–V_1 est l'équivalent d'un diagramme Hertzsprung-Russel.

Les paramètres *d*, *g*, Δ, m_2 dépandent peu de l'extinction interstellaire et peuvent être corrigés dans les cas d'extinction très élevées.

Ces divers paramètres ont été particulièrement appliqués à l'étude des étoiles des amas galactiques, (bibliographie dans le rapport à la commission 37). Le problème des étoiles métalliques a été exposé et plusieurs étoiles métalliques ont été trouvées dans les Pleiades (Golay (**135a**)).

Photometric Standard

M. Golay reports: 'MM. F. Rufener, B. Hauck, G. Goy, E. Peytremann, M. Golay, (**214**), (**140a**), (**140b**), ont particulièrement étudié l'établissement d'un groupe d'étoiles standard. Ces étoiles couvrent tous les types spectraux de *O* à *G*, rougies et non rougies, brillantes et faibles (jusqu'à la 8ème magnitude), dans des régions très diversées du ciel. Ces étoiles sont mesurées dans mon système photometrique (*U*), (*B*), (*V*), B_1, B_2, V_1, *G*, en 7 couleurs (**127**a) et (**128**). Les résultats de ces mesures sont dans les catalogues publiés ou en cours de publication. Les standards sont celles qui ont le plus grand nombre de mesures. La précision des couleurs atteint 0·005.'

Extraterrestrial Stellar Photometry

M. Golay reports: '1° Messieurs A. Gaide et D. Huguenin (**271a**), (**271b**) et (**271c**), ont développé une nacelle semi-stabilisée destinée à transporter par ballon un téléscope de Maksutov. Les étoiles s'inscrivent sous forme de traces sur le film placé dans le plan focal du téléscope. Des filtres limitent les mesures à l'ultima-violet. Environ 400 étoiles differentes ont été enregistrées au cours de plusieurs vols atteignant des altitudes comprises entre 33 et 37 km. La nacelle pèse environ 150 kg et les ballons avaient un volume de 50000 m^3. La détermination des indices ultra-violet est en cours.

2° Un groupe d'étude constitué d'équipes francaise, belge et de l'Observatoire de Genève, a étudié un projet de satellite astronomique pour l'Organisation ESRO (**271d**) (Organisation européenne de recherches spatiales), (**271d**).

Dans une étude préliminaire (**271a**), j'ai étudié les diagrammes en 7 couleurs qu'il serait intéressant d'établir on ultra-violet.

Scientific Reports

V. Straižys and K. Zdanavičius—'Photometric system for two-dimensional classification of Population I stars'. From absolute energy curves an optimum seven-color system for the classification of Population I star was calculated.

V. B. Nikonov and K. Zdanavičius—'Practical test of Straižys' system'. A test of Straižys' system using interference filtres shows satisfactory agreement between theoretical and empirical data.

M. J. Smith—*'The performance of the Edinburgh twin telescope'*.

Scientific Meeting, 29 August 1967

SCIENTIFIC REPORTS

M. Walker—*'Determination of stellar magnitudes by electronography'*. It was shown that satisfactory stellar magnitude sequences can be obtained from electronographic images alone without the need of an external sequence.

I. King—*'A method for deriving standard magnitudes directly from in-focus star images'*. Magnitude sequences can be derived directly from in-focus stellar images obtained with telescopes of a long focal length, if the plate is properly calibrated with a sensitometer.

J. Efimov—*'On the application of TV techniques for stellar photometry'*. Satisfactory photometric precision can be reached on photographs obtained by television techniques.

A. G. Velghe—*'Infrared magnitude sequences.'* The need for more infrared magnitude sequences was demonstrated.

J. Hall—*'The scanning polarimeter.'* Results of the application of a scanning photometer to polarization measurements in planets and galaxies were presented.

M. Jerzykiewicz—'Further evidence of absorption by the interplanetary medium'. Evidence for the existence of an absorbing cloud in the position of the Gegenschein is presented.

BIBLIOGRAPHY

127a. Golay, M. 1963, Essai de classification stellaire bi-dimensionnelle à l'aide d'une photométrie à large bande, Publ. Obs. Genève, Fasc. 64.

135a. Golay, M. Seven-colour photometry at the Geneva Observatory, dans: Spectral classification and multicolour photometry, I.A.U. Symposium, N 24.

139a. Hauck, B. 1967, Classification tri-dimensionnelle effectuée avec le système photométrique de l'Observatoire de Genève. Cas particulier des étoiles A_m. Publ. Obs. Genève (sous presse).

140a. Rufener, F. Hauck, B. Goy, G. Peytremann, E. Maeder, A. 1966, Catalogue et edition augmentée, J. Observateurs (sous presse).

140b. *ibid.* 1967, Catalogue, suite. J. Observateurs (en préparation).

271a. Gaide, A., Huguenin, D. 1965, Mesures astronomiques par ballons stratosphériques, *Orion* N 90 et 91.

271b. Gaide, A. 1967, Détermination d'indices de couleur ultraviolets, Publ. Obs. Genève (sous presse).

271c. Gaide, A., Huguenin, D. 1966, Recherche sur l'ultra-violet stellaire au moyen de ballons stratosphériques, *Communication à la Société de Physique de Genéve.*

271d. Groupe franco-belgo-suisse 1965, *Projet d'un satellite astronomique*, ESRO, Paris.

271e. Golay, M. 1964, Photométrie stellaire en ultra-violet, Rapport N1 d'ESRO, Paris.

26. COMMISSION DES ETOILES DOUBLES

Report of Meeting, 25 August 1967

PRESIDENT: K. Aa. Strand.
SECRETARY: J. Dommanget.

In opening the meeting, *the President* announced with regret the death of Dr E.L. Martin and Dr R.M. Petrie, both of whom had been active members of the Commission. Dr Petrie had also been a member of the Organizing Committee.

The appointment of Dr P. Couteau, as President, and Dr J. Dommanget, as Vice-President, as proposed by the Executive Committee, was approved. An Organizing Committee, consisting of Eggen, van de Kamp, Kulikovsky and Strand was approved.

The Draft Report was adopted with the correction of a few misprints.

The President reported the following additions to the Draft Report, which were received after its preparation:

(1) Dick had provided the information that Horst Pauscher and Karl-Heinz Hintze had carried out observations, with the 65-cm Babelsberg refractor, which were published in Veröff. Sternw. Babelsberg, **XV**, Heft 4.

(2) Wieth-Knudsen had constructed a new double-image micrometer for telescopes with a focal ratio of 1/7·5.

(3) Wierzbiński had prepared a catalog of Δm's for 2535 pairs concerning 2379 binary systems. He also reported that Bem had started systematic observations with a 20-cm refractor at the Wroclaw Astronomical Observatory.

(4) Hopmann had expressed his regrets that a number of refractors suitable for double star observations were not being utilized.

Dommanget reported that the 45-cm refractor at Uccle had been provided with a new objective from Zeiss, and that a new observing program had been established.

Couteau announced the replacement of the 38-cm refractor at Nice with a 50-cm refractor, furnished by the Paris Observatory.

Djurkovic reported on his activities at the Observatory at Belgrade. The observations made between 1961 and 1964 would be published in *Bull. Obs. astr. Beograd*, **26**, No. 1 (1967).

The President reported that proceedings of the colloquium 'On the Evolution of Double Stars', held at Uccle in September 1966, had been published (*Commun. Obs. r. Belgique*, Ser. B, No. 17, 1967). He expressed his thanks to Dommanget, as the editor, for the expeditious publication of the proceedings.

On the basis of a proposal by *Couteau*, the Commission approved a resolution to hold a colloquium at Nice in the fall of 1969 on the subject of different methods of observing double stars.

On the proposal by *Couteau* for the establishment of a working group to periodically select and publish a list of orbits considered definitive or well-determined, it was felt that the information in the orbit catalogues, usually giving the quality of the orbit, would provide the necessary information.

It was also suggested that consideration be given to reporting such orbits in the Information Circular published by Muller, but the Commission decided not to take action on this proposition.

Muller reported that he intended to continue publishing his catalog of ephemerides, as he had done in the past.

Couteau called attention to the variation in the delays in publication of observations. Certain observers grouped long series of ten or more years into one publication, while others published their measures at frequent intervals, sometimes annually. In the case of lengthy delays in publication, *Couteau* would like the authors to provide the observations ahead of publication to certain members of the Commission, to whom others could address themselves.

Strand and *Worley* felt that it would not pay to include unpublished observations in the General Catalogue because of the extra work involved, and especially because of the inevitable mistakes which would occur when provisional means were replaced with final ones.

It was unanimously agreed that it was necessary to accelerate the frequency of publications, preferably after each series of measures have been completed.

In regard to his proposal of new surveys, *Worley* remarked that the great discovery surveys of double stars were carried out many decades ago, and, in particular, much of the sky north of + 60° declination was searched with only a 12-inch refractor. Many double stars could be discovered today by a repeat of at least part of the sky with a telescope of reasonable size, and this would also provide some measure of completeness of the original survey.

In addition, the importance of surveys of special kinds of objects, such as variable stars or nearby stars, was stressed.

Couteau reported that during a three-year period at Nice he had examined 8400 stars between + 18° and + 24° of declination at the rate of 80 stars per hour, and had discovered 200 new pairs. He had examined half of the stars in the AGK2 Catalog between + 20° and + 23° declination and planned to complete soon the zone between + 18° and + 25° declination. He expected the 76-cm refractor at Nice to be back in operation by the end of 1968, and planned to examine all the stars in the BD between declinations + 30° and + 42° with this instrument.

Couteau mentioned that among the pairs discovered in this manner, 4 to 5 percent already showed a perceptible motion after an interval of one year.

Luyten reported on the search for double stars from common proper motion among nearby stars. From the examination of a series of plates with the 48-inch Palomar Schmidt telescope, he had discovered about 1000 pairs where the components had common proper motions, with magnitudes brighter than 21. He expected to find 20000 such pairs from the complete survey.

Strand remarked that the parallax program at the U. S. Naval Observatory was indirectly a survey. 30 to 40 binaries with separations less than 1″5 had been discovered in the program.

Worley remarked that in making his proposal of new surveys he had not thought of general surveys, but rather of systematic surveys of special objects.

The President said that it would be difficult for the Commission to determine without further deliberation what surveys should be undertaken. Moreover, they would depend almost entirely upon the choice of the individual observer.

As to the problem regarding the use of computers for large scale orbit computations, this problem had already been taken care of at the Hamburg General Assembly (*Trans. IAU.*, **XIIA**, 357, 1965), and no new case had occurred since.

The President, calling attention to the proposal by Arend and Dommanget in regard to modification of the definition of the orbital element as adopted by the Commission at the 1935 meeting, invited Dommanget to explain the problem. *Dommanget* said that the 1935 definition of the orbital element was in full accord with the one adopted in celestial mechanics, and also agreed with the usage in regard to spectroscopic orbits of always giving the semi-amplitude of the radial velocity curve with a positive sign.

The modification which was proposed applied only to the cases where it is not possible to determine which of the nodes is the ascending one, and where, in that case, the definition states that the ascending node should be chosen between 0° and 180°, with the double sign ± assigned to the inclination. This had led to the wrong impression by too many orbit computers, that it is permissible to use a negative value for the inclination.

In response to a request by the majority of members present to make a minimum amount of modifications to the 1935 definitions, *Dommanget* suggested that only that part dealing with the ambiguity of the position of the ascending node should be changed.

After rejection of a proposal to establish a working group to study the question further, the Commission agreed to modify the 1935 definition of the elements, as follows:

i Inclination at the ascending node. For direct motion in the first quadrant, for retrograde motion in the second.

Ω Position angle of the ascending node for equinox 1900.

ω Longitude of periastron. Angle in plane of true orbit, always counted in the direction of motion, from the node, as given, to periastron.

When data referring to the third coordinate are not available, one of the nodes is provisionally chosen as the ascending one. In each case computers should state in their publications whether the given node is the ascending node.

By publishing at the same time, as proposed by Arend and Dommanget, and agreed to by other members of the Commission, the Campbell and the Thiele-Innes elements, no doubt would exist in regard to the definitive or provisional character of the ascending node. The presence of the double sign for the elements C, H, pL, pN, showing the ambiguity of the position of the ascending node, the upper sign should correspond logically to the value published for Ω and ω, and the lower sign to the values obtained by the addition of 180° to each.

Muller requested that in the case where a publication gives both Campbell and Thiele-Innes elements, the author should indicate which series of elements were used in the computations.

A special meeting was held in regard to the working group proposed by Eggen of Commissions 26 and 42. It was concluded that a liaison between Commissions 26 and 30 would be more desirable.

Upon the conclusion of the business meeting, brief summaries were given by Batten and Dommanget on recent statistical investigations made by them.

Batten reported that he and Ovenden had investigated the distribution of the values of the longitude of periastron of orbits of spectroscopic binaries with reference to the well-known Barr Effect. They had used the visual binaries listed in Worley's Orbit Catalog as a control group and found, to their surprise, that the orbits in this catalog also appeared to have a non-random distribution of the longitudes of the periastron, in that for the primary orbits they seem to be more frequently between 0° and 180° than between 180° and 360°. The authors did not believe that the ambiguity in the node would produce such an effect, neither did they think that it was a selection effect.

Dommanget reported that at the completion of the catalog of radial velocity ephemerides for visual binaries with known orbits (*Obs. r. Belgique*, Série B, No. 15) a list of 70 binaries was established for which the position of the true orbital plan was well defined in space.

A statistical investigation in regard to the distribution of the poles of the planes of the orbits brought out two important results.

The first was the absence of poles in a region around the galactic coordinates $l^{II} = 35°$, $b^{II} = +60°$.

Having considered this point as the pole of a new galactic reference system, the Chi square test applied to the values of the new latitudes of the various orbital poles gave a probability of 0·3 percent, a value very significant of a non-random distribution.

A second result concerned the nine binaries nearest to the solar system. If one excludes α Centauri, the poles of their orbits group themselves quite close to the direction $l^{II} = 100°$ and $b^{II} = -15°$, while their space motion is more or less parallel to the direction $l^{II} = 5°$, $b^{II} = -10°$. It is realized that the two directions are nearly perpendicular to each other or that the average direction of the space motion is within 5° of the mean orbital plane.

27. COMMISSION DES ETOILES VARIABLES

Report of Meetings

PRESIDING AT BOTH SESSIONS: G. H. Herbig.
SECRETARY: J. D. Fernie.

During the Prague General Assembly, the Commission conducted two administrative sessions, as well as one general scientific meeting and two specialized discussions devoted to flare stars and to symbiotic objects.

First business session, 23 August 1967

Following the introduction of the Secretary and the interpreters (M. Bretz for French and German, V. Cessevič for Russian), the chairman reported on the activities of Commission 27 during the past 3 years; some of these are described in the main Commission report and will not be repeated here. It was announced that the next Variable Star Colloquium in the triennial Bamberg-Budapest series will be held in Budapest in early September 1968. Commissions 27 and 42 have requested of the Union a grant of $ 500 in support of the Colloquium. Furthermore, the Liège Astrophysical Symposium in 1969 will be devoted to pre-main sequence stellar evolution, a subject which partially falls in the domain of Commission 27.

Attention was called to the depository for unpublished photoelectric observations of variable stars, now established in the Library of the Royal Astronomical Society in London. The establishment of a second such center in Odessa, to serve the astronomers of Eastern Europe, is under consideration.

Mention was also made of the fact that, through the courtesy of the National Geographic Society and the Mt. Wilson and Palomar Observatories, the rules for publication of Palomar Atlas copies had been eased, in cases where the reproductions are used for purposes of variable star identification. The details of this arrangement were described in a circular sent to members of the Commission in March 1965.

The President then turned to questions having to do with the *General Catalogue of Variable Stars.* He read extracts from a letter from Dr B. Kukarkin, which described this work and some of the problems of the Moscow bureau. The work there began in 1940, and was taken up again in 1946 following the interruption caused by the war and in response to a proposal from the Executive Committee of the Union that it accept the task of naming new variables, and publishing the necessary catalogues. The first edition of the *GCVS* (listing 10912 variables) appeared in 1948, and in the following ten years the bureau issued 9 *Supplements*, published (in Leiden) 5 lists of new designations, and in 1950 issued the *Catalogue of Stars Suspected of Variability.* The second edition, in two volumes and containing even more complete information than the first, appeared in 1958. It lists 14708 stars. Two *Supplements* have been published, with designations for 4083 new variables and more precise data on 1647 others. (A copy of the *Second Supplement* was presented by Cessevič during the meeting). A table of galactic coordinates of variable stars, compiled by L. Plaut, and a *Second Catalogue of Stars Suspected of Variability* with information on 3907 stars have been published.

The third edition of the *GCVS* is now nearing completion. It will contain information on $18791 + N$ variables, where N is several hundred, representing the last group of new variables that will be assigned names in the fall of 1967.

The bureau's work of collecting and analyzing variable star data has been progressively improved, some processes having been automatized. Despite this, the volume of work has greatly increased. The staff of the bureau must still read and analyse the literature and letters; this cannot be done by machines. On account of the increasing complexity of variable star work, highly qualified personnel are essential for this task. The result is that the astronomical staff have become slaves to this flood of information, and have little time left for their own scientific work. It is evident that this small group, which has not been enlarged since 1946, will soon be unable to cope with the task, despite the unselfish help given them by their colleagues of the Sternberg Astronomical Institute.

Therefore Commission 27 has been asked by Kukarkin to support the formal request made to the Union by the Astronomical Council, USSR Academy of Sciences, for financial support. A grant of $5000 was requested for extraordinary assistance toward the preparation of the third edition of the *GCVS*, and an annual subvention of $500 for continuing technical support. (Kukarkin mentioned, incidentally, that the bureau has to spend much time on checking and correcting star coordinates that have been determined carelessly.)

In addition to these activities, the Variable Star Commission of the Astronomical Council is responsible for publishing the bulletin *Variable Stars* (*Peremennye Zvezdy*) and the *Astronomical Circulars*, which impose an additional financial burden. *Kukarkin* asked for a statement by Commission 27 whether it would favor the transformation of *Variable Stars* into a subscription journal. This would make it possible, if enough subscriptions were obtained, not only to release the staff of the bureau from this technical labor but to issue the journal more regularly, to expand its format, and to publish papers in languages other than Russian.

Finally, *Kukarkin* expressed on behalf of himself and his staff the deepest gratitude to the world-wide community of variable star investigators for their friendly help.

The Commission formally endorsed the principle that *Variable Stars* be placed on a subscription basis. The President commented that the financial position of the Union was such that it was very unlikely that support in these amounts could be expected. The President made the following interim suggestion: the Organizing Committee of Commission 27 wishes to recommend, and asks approval of the Commission, that in the future not only *Variable Stars* be placed on a subscription basis, but that the *GCVS* and its *Supplements* be sold instead of being distributed without charge, as is the practice at present. If this were done, presumably the amount of the grant requested for assistance to the Moscow bureau could be decreased. The Commission could consider the matter again at its second business meeting, by which time the reaction of the Soviet delegation to this counter proposal would be known. The Commission unanimously endorsed this suggestion.

A proposal by *W.J. Miller* was presented, to the effect that the Moscow bureau modify its present procedure of assigning variable star designations. In essence, present practice is that new variables are not named until just before a new *Supplement* is completed, at which time the designations are assigned in order of right ascension. Miller's proposals were (1) that the assignment of names be performed more frequently, almost as soon as the reality of the variation is accepted, without waiting for a final ordering of stars according to right ascension; and (2) that these new designations be immediately available on inquiry. In the discussion which followed, the possibility of publishing frequent *Benennungsliste* was raised. Cessevič indicated that it would be impossible to publish full details on a short time scale, but that publication of simple lists of new names might be feasible. It was mentioned that use might be made of the *Information Bulletin of Commission* 27 for this purpose. The President finally proposed that the details be left to the consideration of Kukarkin and his associates, and that the Commission simply convey its opinion to them that the assignment of designations should be made more frequently, and that the possibility of reasonably rapid announcement of these designations be examined. This proposal carried unanimously.

In 1961, a Working Group of Commission 27 under the chairmanship of Dr A. Wesselink was appointed to recommend a system of nomenclature for variable stars in extragalactic systems. Since the most urgent problem exists in the case of the Magellanic Clouds, the Group has given its main attention to that question. Wesselink presented its final recommendations to the Commission: (1) An area surrounding each cloud be defined by the following corner points (for the equinox 1875·0, in order to be compatible with the official constellation boundaries), connected by meridians or small circles:

Small Cloud:	*Large Cloud:*
$0^h\,00^m$, $-69°00'$	$4^h\,35^m$, $-64°00'$
1 40 , − 69 00	6 35 , − 64 00
1 40 , − 77 00	6 35 , − 74 00
0 00 , − 77 00	4 35 , − 74 00

(2) Within these two areas, all variable stars will be assigned designations not according to constellation but on the new system, regardless of whether they are physical members of the Clouds or not. Non-variable stars in these areas will not be affected in any way. (3) A new system of designations will be devised which will contain an indication whether the star falls in the LMC or SMC area.

In other words, all stars in these two areas will be removed from future *GCVS*, and will appear only in a special catalogue of Magellanic Cloud variables. Some special indication may be given to stars that are known to be foreground objects. This mixing of Cloud members and non-members is an annoyance, but it is conventionally accepted in galactic and globular clusters and in some extragalactic systems. After some discussion, the President proposed (a) that the Commission accept the recommendations of the Working Group; (b) that the criteria for the acceptance of a new variable into the new system be fully as stringent as those of the *GCVS*; (c) that technical details, such as the designations themselves, be left to a new Working Group which the Organizing Committee would appoint for the task of actually compiling the new Catalogue. He would report on some administrative details in the second business session. The Commission endorsed these proposals unanimously.

Dr Cessevič exhibited a copy of a limited photographic edition of finding charts for some 5000 variable stars that had been prepared by the Odessa Observatory. Copies of this Atlas had been deposited at 20 observatories active in variable star work.

In conclusion, Dr *Oosterhoff* complimented Drs Herbig, Feast, and Sawyer-Hogg on the Report of Commission 27.

Second business session, 29 August 1967

The President announced that a small bureau will be established at the Mt. Stromlo Observatory, under the direction of Dr S. Gascoigne, for the preparation of a Catalogue of Magellanic Cloud variables following the recommendations of the Wesselink Group (which had now been discharged, with thanks for their labors). To assist and advise Dr. Gascoigne, a new Working Group had been established under his chairmanship, composed of representatives of the major centers of work on Cloud Variables: S. Gaposchkin (Harvard), B. Kukarkin (Moscow), J. Landi Dessy (Córdoba), P. Th. Oosterhoff (Leiden), A. Thackeray (Pretoria), R. Woolley (Herstmonceux).

It was reported that as a result of the Commission's proposal at the earlier session, the Soviet request for financial support for the work of the Moscow bureau had been modified. The President explained that it would be desirable to identify for the Executive Committee the precise terms on which the Commission might agree to support this request, and so he moved the following resolution:

> Commission 27 recommends that, in view of the important services of cataloguing and systematizing variable star data that are performed by the Variable Star bureau of the Astronomical Council, USSR Academy of Sciences, the IAU give partial financial support to this work in the amount of a single grant of $2000, plus an annual subvention in the amount of $500. We recommend these funds specifically to support (1) the reorganization of the work of the bureau, and (2) continued free distribution of publications to members of Commissions 27 and 42, with the publications being sold to all other individuals and institutions.

The Commission unanimously endorsed this resolution.

Dr *Detre* discussed the need for a concentrated observational effort to determine secondary periods for RR Lyrae stars in a few selected globular clusters, particularly M3, M5, M15. About 3000 plates should be obtained of these three clusters over a period of a few months, which implies an observatory located in an excellent climate. He pointed out also the importance for the same purpose of old plate material, especially that in the files of the Harvard and Mt. Wilson Observatories. After some discussion of the accessibility of those plates to outside investigators, the Commission endorsed a resolution supporting and encouraging such a intensive observational program on globular cluster RR Lyrae stars.

The President described the identification atlas of old novae that was being prepared by Dr Ch. Bertaud. Several sample charts from this atlas were displayed. This enterprise is especially important because of great interest being shown in these stars at the present time. The Commission voted a resolution expressing strong support and encouragement for Dr Bertaud's work. Attention was also called to a list prepared by Dr Bertaud of possibly recurrent novae which deserve continuing attention. Copies were available at the meeting for those interested, and a condensed version appears as an appendix to this report.

The subject of possible new activities for amateur associations was introduced by the President. He pointed out that Commission 27 has no authority or intention of interfering with the activities of volunteer observers. For amateurs, variable star work is a hobby like collecting stamps, and professional astronomers cannot tell amateurs what to do with their spare time. The only reason for bringing the subject up is this: the major activity of the associations is following Mira variables. These programs were initiated over a half-century ago, and met a real need of the astronomy of that era. Without minimizing the usefulness of such observations even today, one must admit that astronomy has since become a much broader subject. Are there some other programs suitable for amateur observers that might be equally profitable in the light of today's science? Almost every astronomer could suggest a few possibilities; the following is a somewhat biased list:

(1) A more thorough patrol of recurrent novae, both known and suspected (see the Bertaud list for candidates).

(2) A thorough and systematic study of the brighter flare stars, to detect both individual flares and prolonged periods of high activity.

(3) Accurate timing of minima of short-period binaries of special interest: for example, U Gem, V Sge, RW Tri.

(4) Systematic observation of T Tauri stars that are known to have occasional nova-like outbursts: VY Tau, UZ Tau, EX Lup.

(5) A number of interesting irregular variables are apparently not being followed at all: WW Vul, BF Ori, XY Per A + B, BN Ori.

Feast stressed the point that continuing work by amateurs on long-period variables was very important and useful, but that an improvement in photometric sequences that they used was also very necessary. Mrs. Mayall agreed, and said that the amateur groups would be glad to add other stars to their programs, if only good sequences could be provided. She indicated that perhaps those amateurs who have photoelectric photometers might be able to supply such data.

Mrs *Mayall* also reported that it was planned eventually to record all 2·5 million AAVSO observations on punched cards, and a beginning has been made. She welcomed advice on the best format for publication (10-day means, plots, or individual observations), and whether it was desirable to combine these results with those of other amateur organizations. The President suggested that interested persons discuss these questions directly with Mrs Mayall.

Hoffmeister presented a short report on the comprehensive program carried out at Sonneberg to investigate the statistical properties of variable stars, and their light curves. The significance of such work was emphasized by the Commission in a resolution:

Commission 27 stresses the great importance of statistical studies of the types and distribution of variable stars for investigations of the structure and evolution of the galaxy.

The President noted that the late Dr H. Schneller had left unfinished in Potsdam the manuscript of a new edition of the *Geschichte und Literatur*. Such bibliographical monographs were extremely useful to variable star investigators, and everyone would surely agree on the desirability of completing and publishing this new edition. It was unanimously resolved that:

Commission 27 strongly supports and encourages the preparation of a new edition of the *Geschichte und Literatur der Veränderlichen Sterne*.

The Commission was informed of the discussion of flare star problems that had taken place a few days previously in a small meeting of specialists. *L. Geršberg* stressed the importance of continuous photoelectric observations of flare stars. A continuous 24-hour coverage from observatories all around the earth is desirable on some special occasions, as had been organized in the past to support radio observations of flare stars. Although the organization of such cooperative efforts is in the province of the IAU, Commission 27 is too large a body to direct such a program. It was also pointed out that there was a real need for more systematic exchange of data, preprints, information on programs, etc. For these reasons, the President appointed a Working Group on Flare Stars to discuss these problems in detail, and organize cooperative programs as they saw fit. The following astronomers were designated as the nucleus of the Working Group: A.D. Andrews, P.F. Čugainov, L. Geršberg, V. Oskanian. They were empowered to select their own chairman (Dr Čugainov was subsequently elected) and to co-opt any other members that could assist in their task, with special attention being given to radio astronomers. The Commission supported this work through the following resolution:

Commission 27 emphasizes the importance of cooperative studies of flare stars, especially those based on simultaneous photoelectric, spectroscopic and radio observations during periods of 24 hours and longer.

Miss *Hoffleit* reported to the Commission an inquiry she had received (from scientists outside the variable star field) regarding the possibility of compiling a catalogue of all existing astronomical direct photographs. She felt that this would be an impractically large task, but asked for opinion whether a list of observatories which have such material might be compiled, together with a general description of the nature and time coverage of the photographs. Following some discussion, the question was deferred until a more detailed and specific proposal could be put before the Commission.

Hill discussed briefly the significance of the hot helium stars $+10°2179$, HD 124448, HD 168476,... which seem to have chemical compositions much like the R CrB variables. He pointed out the importance of a careful photometric study of the hot He stars to find if any are indeed variable.

P. Kulikovsky called attention to a list of double T Tauri stars published in *Information Circular of Commission* 26, Nos. 31 and 32. He urged special attention be given to these stars.

In conclusion, the President presented the slate of new officers of Commission 27 that was proposed by the present Organizing Committee, in consultation with the General Secretary:

President: L. Detre.

Vice-President: O. J. Eggen.

Organizing Committee: A. Bojarčuk, R. Kippenhahn, K. Kwee, G. Odgers, G. Herbig.

The Commission indicated its approval of this new panel of officers. The meeting then adjourned.

Appendix

POSTNOVAE TO BE EXAMINED FOR RECURRENCE

Ch. Bertaud

All the known recurrent novae have amplitudes less than 9 magnitudes (T Pyx: $\Delta m = 6^m_.6$; RS Oph: $7^m_.2$; T CrB: $8^m_.0$; WZ Sge: $8^m_.1$) except possibly VY Aqr, for which the amplitude is not known. The old novae in the following list conform to this $\Delta m < 9^m$ limit, and thus deserve particular attention.

HR Lyr 1919 ($\Delta m = 8^m_.5$). This amplitude corresponds to a minimum magnitude of 15·0, but the star is photographically very active. Between 1947 and 1952, it varied between $14^m_.25$ and $15^m_.30$ (L. Rosino and A. Betti, *Pubbl. Bologna* **5**, No. 21, 1952). One Meudon plate of 1962 Aug. 24 gave $15^m_.0$.

IM Nor 1920 ($7^m_.5$:). The identification of the prenova at $16^m_.5$ is uncertain (S.I. Bailey, *Harv. Bull.* 734, 1920). There are only a few observations of this star during the month-long maximum at $9^m_.0$.

V1016 Sgr 1899 ($6^{m}.5$). The maximum may not have been observed; an extrapolated maximum of $7^{m}.0$ corresponds to an amplitude of $7^{m}.9$.

HS Sgr 1900 ($6^{m}.5$:). The probable magnitude of the prenova according to Miss Woods is $16^{m}.5$, and the extrapolated maximum according to McLaughlin is $10^{m}.0$.

V999 Sgr 1910 ($8^{m}.4$). The magnitude 26 years after the outburst was $16^{m}.6$ (H. H. Swope, *Harv. Bull.* 913, 11, 1942); the maximum magnitude is assumed to be 8·2.

V1017 Sgr 1919 ($7^{m}.0$). The star now is slowly variable near mag. 14 (M. W. Mayall, *J. R. astr. Soc. Can.*, **57**, 127, 1963). The small amplitude in 1919, and the existence of a minor outburst in 1901 (D. B. McLaughlin, *Publ. astr. Soc. Pacific*, **58**, 46, 1946), justifies continued surveillance.

FN Sgr 1925 ($5^{m}.0$). The spectrum resembles that of an old nova (G. H. Herbig, *Publ. astr. Soc. Pacific*, **62**, 215, 1950).

FM Sgr 1926 ($8^{m}.5$:). The extrapolated maximum magnitude is 8·0, while the prenova seemed to be of mag. 16 or 17. The amplitude is therefore uncertain.

V441 Sgr 1930 ($7^{m}.3$). The amplitude given here is obtained from a maximum magnitude of 8.7, but it is probable that the real maximum magnitude was about 8.0; the amplitude was on this basis about 8 mag.

EU Sct 1949 ($8^{m}.4$). The minimum magnitude of 16.8 follows from observations of the prenova between 1918 and 1949 (M. Harwood, *Leiden Ann.*, **21**, 387, 1962).

X Ser 1903 ($6^{m}.0$). The maximum of $9^{m}.0$ was attained in 1903. At minimum the star oscillates around 15^{m} with a range of 2 mag., in a fashion resembling Z And. There is a resemblance to the quasiperiodic oscillations of RR Tel before its outburst (C. Payne-Gaposchkin, *The Galactic Novae*, p. 192, 1957).

28. COMMISSION DES GALAXIES

Reports of Meetings

PRESIDENT: R. Minkowski.
SECRETARY: E. Holmberg.

The Commission met on 23, 24, and twice on 29 August. The successive sessions dealt with (1) business matters, (2) stellar composition of galaxies, (3) presentation and discussion of a series of papers on extragalactic research, and (4) reports of Working Groups and presentation of a second series of papers. The second session was a joint meeting with Commission 45.

A Joint Discussion on X-Ray Astronomy was held under the sponsorship of Commissions 28, 33, 34 and 44, a Joint Discussion on Extragalactic Radio Sources under the sponsorship of Commissions 28, 40 and 45.

Business Meeting, 23 August 1967

The Draft Report was adopted by the Commission subject to minor editorial corrections.

On page 571, *line* 6,

for $Q \propto M^{5/3}$ read $Q \propto M^{7/4}$

for $Q \propto M^{5/3}$ read $Q \propto M^{3/2}$.

The President proposed, and the Commission approved, that the following list of names should be sent to the Executive Committee of the Union as those of persons who should form the Organizing Committee of the Commission: D.S. Evans, D.S. Heeschen, E. Holmberg, B.E. Markarjan, M. Schmidt. At the final General Assembly on 31 August, this list was approved, with the addition of R. Minkowski as past President, and the following appointments were made: President: G.C. McVittie. Vice-President: E.M. Burbidge. The new President appointed E. Holmberg to continue as Secretary of the Commission.

The Commission discussed and rejected by vote the proposal submitted to the Executive Committee by the National Committee of the DDR to form a new IAU Commission for Relativistic Astrophysics and Cosmology and a proposal, submitted by the President following a suggestion by the Secretary General, to create a Working Group on Cosmology. A proposal by the President to change the name of the Commission to 'Commission for Galaxies and Cosmology' was rejected by a vote.

A recommendation, adopted at the IAU Symposium no. 29 at Burakan and approved by the Executive Committee (see IAU Information Bulletin, no. 17), was discussed and rejected by the Commission. Instead, the following resolution was submitted by the President and approved by a vote of the Commission: 'Commission 28 recommends that the services of the Central Bureau of Astronomical Telegrams be used to announce without delay significant events and changes in quasistellar sources and objects in order to achieve effective coordination of observations in all frequency ranges'.

A resolution concerning the definition of radio magnitudes at different frequencies had been submitted by J. Heidmann to Commissions 28 and 40. It was decided that Commission 28 should refrain from discussing this matter, and that the Commission would concur with the resolution adopted by Commission 40. The resolution subsequently was withdrawn.

Following the wishes of a number of members of the Commission, a proposal had been sent by the President to the Assistant Secretary General to hold a Symposium on Extragalactic Research in 1970 immediately before or after the XIVth General Assembly. The Commission endorsed this proposal by vote, and authorized and requested the new President to take all necessary steps for the arrangement of the Symposium.

1st Scientific Meeting on the Stellar Composition of Galaxies, 24 August 1967

In this joint meeting with Commission 45, R. Minkowski acted as chairman, and V. Reddish as secretary. Two invited papers with the common title 'On the Stellar Composition of Galaxies' were given by *W. W. Morgan* and by *H. Spinrad.* A number of commission members, among them W. Baum, G. de Vaucouleurs, R. Minkowski, and T. Page, participated in the following discussion.

2nd Scientific Meeting: Presentation of Papers, 29 August 1967

The following 24 papers had been submitted by members of the Commission for presentation and discussion at the Commission meetings:

1. *Abell, G.O.*, Luminosity Functions and Structures of Several Clusters of Galaxies.
2. *Bertola, F.*, Variations of the Mass to Light Ratio within the Galaxies.
3. *Bottinelli, L., Gouguenheim, L., Heidmann, J., and Heidmann, N.*, 21 cm Observations of Small Galaxies.
4. *Burbidge, E.M., and Burbidge, G.R.*, Velocity Fields in NGC 1808 and an Anonymous 'Integral Sign' Galaxy.
5. *Carranza, G., Courtès, G., Georgelin, Y., and Monnet, G.*, Interferometric Studies of Ionized Hydrogen in M33.
6. *Demoulin, M.H.*, NGC2685.
7. *Gallouët, L., and Heidmann, N.*, Optical Positions of Brightest Galaxies.
8. *Heidmann, J., and Heidmann, N.*, The Mean Optical Emissivity of Galaxies.
9. *Hodge, P.W.*, The Distribution of H II Regions in Galaxies.
10. *Iskudarjian, S.G., Kalloghljan, A.T., Sahakjan, C.A., and Tovmassjan, H.M.*, The Burakan Classification of the Central Parts of Galaxies.
11. *Kalinkov, M.*, Autocorrelation Functions in the Surface Distribution of Galaxies (by the Palomar Atlas).
12. *Kwast, A., Zonn, W., Neyman, J., and Scott, E.L.*, Characteristics of Galaxies Forming a Pair.
13. *Markarjan, B.E.*, A Survey of the Spectra of Faint Galaxies.
14. *Page, T.*, Line Intensities in Galaxy Spectra with an Image Tube.
15. *Polikarov, A., and Kalinkov, M.*, A Possible Classification of Quasi Hypotheses.
16. *Richter, N.*, The Variability of Blue Objects as a Test of Their Galactic or Extragalactic Nature.
17. *Rubin, V.C., and Ford, W.K. Jr.*, Image Tube Spectra of Extragalactic Objects.
18. *Sĕrsic, J.L.*, The Formation of Galaxies by Fragmentation.
19. *Souffrin, S.*, Remark about a Possible Variation in the Spectrum of the Nucleus of the Seyfert Galaxy NGC3516.
20. *Veleva, B., and Kalinkov, M.*, Results of Counting of Galaxies on the Palomar Atlas.
21. *Whitrow, G.J., and Thompson, I.H.*, On Radial Gravitational Motions, Including Collapse, of Spherically Symmetrical Bodies of Perfect Fluid.
22. *Wild, P.*, 'Superexplosions' and Supernovae.
23. *Zelmanov, A.*, On the Differential Criterion of Homogeneity in Cosmology.
24. *Zwicky, F.*, Classification of Clusters of Galaxies.

Since the limited time available did not permit the presentation of all papers at the sessions of the Commission, the papers to be presented were selected by the Organizing Committee. The papers no. 1; 3; 5; 9; 14; 17; and 20 were read at the 2nd Scientific Meeting. In the short discussion periods following each paper remarks and comments were made by G.O. Abell, V.A. Ambarcumjan, E.M. Burbidge, T. Kiang, W.W. Morgan, T. Page, M.S. Roberts, E.L. Scott, P. Wild, F. Zwicky.

3rd Scientific Meeting: Reports of Working Groups and Presentation of Papers, 29 August 1967

A report on the activities of the Working Group for Research on Supernovae was read by F. Zwicky. The membership of the Group has been reconstituted and now consists of the following:

V.A. Ambarcumjan, Ch. Bertaud, A.A. Bojarčuk, L. Detre, E.A. Dibaj, G. Haro, B.V. Kukarkin, H. Lambrecht, L. Rosino, K. Rudnicki, J.L. Sěrsic, A.D. Thackeray, B.E. Westerlund, P. Wild, F. Zwicky (chairman).

The Working Group on Supernovae expressed its support to a request by K. Rudnicki for endorsement of his project to assemble a catalog of essential characteristics of supernovae. The Commission gave its approval to this project.

B.E. Westerlund summarized the report of the Working Group on the Magellanic Clouds. The reconstituted list of membership includes the following: J. Landi Dessy, Ch. Fehrenbach, S.C.B. Gascoigne, E.M. Lindsay, B.Y. Mills, C.H. Payne-Gaposchkin, A.D. Thackeray (chairman), P.A. Wayman, B.E. Westerlund.

The Working Group on Galaxy Photometry submitted a brief report. The reconstituted list of membership includes the following: H.D. Ables, F. Bertola, J. Bigay, W. Bronkalla, G. de Vaucouleurs (chairman), S. D'Odorico, C. Fraser, N. Heidmann, I.R. King, D. Koelbloed, G.E. Kron, B.E. Markarjan, P. Pishmish de Recillas, N. Richter, H. Rood, C.D. Shane, R.R. Shobbrook, S. Simkin, D.W.N. Stibbs, E. Vandekerkhove, S. van den Bergh, B.A. Voroncov-Vel'jaminov, F. Zwicky.

V.A. Ambarcumjan reported on work at the Burakan Observatory, including the papers no. 10 and 13. The papers no. 2; 16; 19; 22 and 23 also were presented in this session. In the discussion periods remarks and comments were made by V.A. Ambarcumjan, A. Elvius, E. Holmberg, I.R. King, T. Page, and V.C. Rubin.

APPENDIX. SUMMARY OF WORK DONE IN USSR ON COSMOLOGY AND RELATED TOPICS OF GENERAL RELATIVITY

(Prepared by A.L. Zelmanov)

I. GENERAL PROBLEMS

I.M. Halatnikov (1) found a class of solutions of gravitation equations with a time singularity in energy density; the singularity is due to scale contraction for one of the space coordinates. These solutions are not contained in that, found earlier by E.M. Lifšic and I.M. Halatnikov, which describes an anisotropic collapse and contains a physical singularity both in presence and in absence of matter. The behaviour of electromagnetic fields under anisotropic collapse is also discussed.

V.A. Belinskij and I.M. Halatnikov (2) proposed a solution for empty world with a fictitious singularity in a synchronous (semi-geodetic) coordinate system, attained in all points of the space simultaneously, the solution depending on the largest number (four) of physically arbitrary functions of three variables. L.P. Griščuk (3) showed, that for the space, filled with dust-like matter, the solution of gravitation equations depending on the largest number (eight) of physically arbitrary functions of three variables had a true (physical) time singularity.

According to H. Keres (4) there exist two kinds of solutions of Einstein equations, Newtonian and non-Newtonian. Solutions of the first kind in non-relativistic approximation do satisfy the Newtonian gravitational theory (invortical fields), while those of the second kind do not satisfy it (vortical fields). The last ones are characterized not only by the Newtonian scalar potential, satisfying the Poisson equation, but also by a vector potential, satisfying Laplace equation. All solutions of Einstein equations for empty space were found which satisfy in a suitable chosen coordinate system also the equations of the Newtonian theory (5). A conclusion is also made according to which Petrov's classification of gravitational fields has no analogue in the Newtonian theory.

Proceeding from the fundamental character of the symmetry properties, G.I. Naan (6) assumed that the Universe was completely symmetrical and consisted of the world and anti-world. The Naan anti-world differs from the world not only by replacing the particles by anti-particles, but also by the space reflection and by the time inversion. Mass, energy and other integral quantities in the world and respectively in the anti-world can vary, but each of these quantities remain equal to zero in summary, i.e. in the whole Universe composed of the world and anti-world. The interaction (in its usual sense) between the world and anti-world is absent. G. I. Naan suggested that in connection with the possibility of the appearance of the mass and energy in each of the two parts of the Universe, all phenomena which are characterized by an enormous energy-output (cosmic rays, supernovae, quasars, nuclei of some galaxies, radio-galaxies), as well as the beginning of the expanding of Metagalaxy were of interest from the point of view of the proposed hypothesis.

According to Y.P. Terleckij (7) macroscopic violations of thermodynamics do not contradict Gibbs statistics. In particular, the existence of a thermodynamic machine of the second kind is possible with heat sources possessing negative energy and negative temperature. The causality principle can become invalid, and the existence of macroscopic systems with negative time direction can be possible. Particles can exist not only with positive, but also with negative and even imaginary masses. These general ideas were applied by Y.P. Terleckij also to the problem of quasars, radio-galaxies and other phenomena. The essence of these applications is an appeal to such processes, the possibility of which follows from these ideas. In particular, it was indicated that the idea of the existence of a new type of energy sources (such as the thermodynamic machine of the second kind) can be used for explaining the energy emission by radiogalaxies and quasars.

E.B. Gliner (8) suggesting a physical interpretation of some algebraic structures of the energy-momentum tensor, assumed that a kind of matter is possible, for which this tensor is equal to the metrical one, multiplied by a constant (μ). This matter called 'μ-vacuum', macroscopically possesses the properties of vacuum. The supposition was considered according to which the real vacuum was a μ-vacuum. The homogeneous μ-vacuum world possesses the de Sitter metrics. Further (9) E.G. Gliner proposed a generalized gravitation law, expressed by a system of 20 equations (in partial derivatives of the second order) in accordance with the fact that the space-time curvature is described by 20 quantities. Einstein gravitation law, expressed by a system of 10 equations, is a consequence of the generalized law.

B.L. Altšuler (10) suggested a method for presenting the Einstein equations in an integral form with the aid of a covariant Green function. Under a definite choice of the equation for these function the integral form appears to be a covariant expression of the Mach principle. This principle becomes equivalent to the requirement that the Einstein equations should be valid in the integral rather than in the differential form. It is shown in particular that the homogeneous isotropic cosmological models are incompatible with the proposed form of the Mach principle.

K.P. Stanjukovič (11) assumed the existence of stable particles—planckeons—with modern values of mass, radius and density ($1{\cdot}1 \times 10^{-5}$ g, $1{\cdot}3 \times 10^{-33}$ cm, 10^{95} g cm^{-3} respectively). Connecting this hypothesis with his version of the theory of the universal constants variability, K.P. Stanjukovič supposed the existence of closed particles also with other definite values of mass, radius and density. K.P. Stanjukovič's ideas (on the radiation by elementary particles and others) caused the polemics between him on the one hand and Y.B. Zeldovič and Y.A. Smorodinskij on the other (12), (13).

M.A. Markov (14) considered characteristic masses $m_0 = (\hbar c/\gamma)^{1/2}$ and $m_1 = e/\gamma^{1/2}$ as possible largest values of elementary particles masses (maximons). These masses are of the order of 10^{-5} g and 10^{-6} g respectively.

The ratio of masses of electron and neutron is approximately equal to that of electromagnetic and strong interactions constants. Proceeding from this empiric correlation, D.F. Kurdgelaidze (15) introduced the universal mass (equal to $1{\cdot}25 \times 10^{-25}$ g) and the mass, connected with the gravitational interaction (equal to 4×10^{-66} g). Using them, he considered the known empiric correlations between microphysical constants and metagalactic parameters. The connection between symmetry properties of elementary particles and cosmology was considered by D. Ivanenko and D. Kurdgelaidze (16).

G.M. Idlis (17) developed the idea of analogy between the Metagalaxy and a certain particle, which is elementary from the outer side. He considered also the problem of the causal determinacy of the three-dimensional character of space and of the universal nature of the gravitation law.

II. HOMOGENEOUS ISOTROPIC UNIVERSE AND ITS PHYSICAL EVOLUTION. ORIGIN OF GALAXIES

A. D. Černin (**18**) obtained exact solutions of the gravitational equations (without the cosmological constant) for homogeneous isotropic cosmological models, filled with a non-interacting mixture of dust-like matter and radiation, and found for these models expressions for the red shift, for amount of matter till the given red shift, for apparent brightness and for apparent angular dimension.

K. P. Stanjukovič (**19**) considered geodetics in homogeneous isotropic models. (See also (**69**)).

The investigation of the problem on angular dimensions of distant objects in the Universe with discrete distribution of the mass (i.e. with mass, concentrated in the galaxies), executed earlier by Y. B. Zeldovič and by V. M. Dashevsky and Y. B. Zeldovič was generalized by V. M. Daševskij and V. I. Sliš (**20**) for the case, in which, besides the matter concentrated in galaxies, the matter distributed continuously (radiation, intergalactic gas) does exist also. The authors came to the conclusion according to which in this case the curve of the dependence of angular diameter on distance must always have a minimum.

A. D. Černin (**21**) considered homogeneous isotropic cosmological models with a chaotic magnetic field. He derived exact solutions of the gravitation equations for such models and investigated the gravitational stability of these models.

Considering the possibility for determination of the metagalactic mean matter density from quasars observations V. M. Daševskij and Y. B. Zeldovič (**22**) analyzed the shortcomings of usual methods, based on observations of distributions by the apparent magnitude or by the red shift, and proposed a new method. It consists of the calculation of the number of objects with a given red shift and with apparent magnitude, brighter (not greater) than some given one. It was pointed out that the conclusions depend so much on the proposed evolution law of the observed objects, that the observational data can be conformed with any cosmological model by accepting a suitable supposition.

A. V. Zasov (**23**) examined for homogeneous isotropic models general correlations, connecting the metrical and luminosity distances, red shifts and aberrational time.

Y. B. Zeldovič, L. B. Okun' and S. B. Pikelner (**24**) examined the fate of quarks in the hot and the cold models of the Universe and in stars. They suggested, that at present the concentration of quarks in relation to nucleons in the hot and the cold models must be of the order of respectively 10^{-10}–10^{-13} and 10^{-18}.

E. M. Lifšic's relativistic theory of gravitational instability was completed in the work by E.M. Lifšic and I. M. Halatnikov (**25**). Examining the possibility of the formation of stars and galaxies as a result of the gravitational instability of the homogeneous Universe I. D. Novikov (**26**) obtained a positive answer by the supposition, that small density fluctuations, accompanied by small metrics fluctuations, arose at a sufficiently early stage of the expansion of the homogeneous Universe, i.e. under densities much greater that the nuclear one. According to this author, it is possible in this way to explain the formation of cosmical objects of various sizes including galaxy clusters with dimensions of the order of hundreds of megaparsecs. A. D. Saharov (**27**) assumed (in the limits of the cold model), that the initial inhomogeneities arose as a result of the quantum fluctuations of cold baryon-lepton matter at densities of the order of 10^{98} baryons (i.e. 10^{74} g) per 1 cm^3. It is supposed also that at such densities gravitational effects are of decisive importance in the equation of state and as a result of this fact at a certain baryon density the energy density vanishes. A theoretical estimation results in such initial non-homogeneities, which can explain the origin of clusters with masses of the order of 10^5—10^6 solar masses, that is close to the masses of globular clusters. (Hence the dimensions of initial non-homogeneities are deduced from some initial positions, and are not selected as usually in order to bring the results in accordance with the empirical data). According to A. D. Saharov's hypothesis the gravitational collapse of these primary stellar clusters led to the appearance of gas, possessing non-homogeneities in the motion and distribution. An increase of these non-homogeneities led to the formation of galaxies. According to the proposed hypothesis the spherical component of galaxies consists mainly of clusters of primary stars, captured by the gravitational field of the rotating gas cloud, while the stars of the flat component are formed from the gas.

V. A. Ambarcumjan (**28**) continued to develop his cosmogonic ideas, based on the analysis of the observational data, concerning processes occurring in galaxies and systems of galaxies and on the interpretation of these data as an evidence of non-stationarity of the systems of galaxies. According to these ideas the cosmical objects arise directly from the superdense matter, the processes of their formations (including the formation of galaxies and their systems) continuing at present. The community between the non-stationarity of the system of galaxies on one hand and the expansion of Metagalaxy on the other is supposed.

I. D. Karačencev (**29**), counting the galaxies up to different angular diameters, came to the conclusion, that the result of this counting gives an independent confirmation to the idea of the continuous formation of new galaxies. I. D. Karačencev (**30**) found also, that the degree of non-stationarity (i.e. the virial mass-luminosity ratio) of different systems of galaxies (pairs, triple systems, groups and clusters of galaxies) is the higher, the greater is the luminosity of the system, and that this correlation includes the Metagalaxy also. The last result was interpreted as a confirmation of V. A. Ambarcumjan's idea on the connection between the non-stationarity of systems of galaxies and the expansion of the Metagalaxy. A hypothesis conforming to V. A. Ambarcumjan's idea on the formation of cosmical objects immediately from superdense matter with the relativistic cosmological theory, was proposed by I. D. Novikov (**31**). According to this hypothesis, quasars are core-parts of the expanding homogeneous isotropic Universe, delayed in their expansion. The primary linear dimensions of such cores were smaller than their gravitational radius. In the proper time of the cores the delay in the expansion is absent. For an external observer the appearance of the cores can occur at different times. The author of the hypothesis derived an exact solution of the gravitational equations, describing a cosmological model with delayed cores. The possibility of formation of such cores on the contraction stage which could precede the expansion stage was considered. See also (**35**).

In the work devoted to computation of the spectrum and of the intensity of metagalactic radiation produced by stars and by radio-sources in a homogeneous isotropic universe (**32**) A. G. Doroškevič and I. D. Novikov suggested, that observations in the region 0·06–30 cm could give a possibility to check the hypothesis of the 'hot' Universe.

Accepting the interpretation of the black-body radiation with $T=3\,°\mathrm{K}$ (background) as a confirmation of the hot Universe hypothesis, Y. B. Zeldovič and I. D. Novikov (**33**) showed that another explanation of the origin of the detected black-body cosmic radiation, according to which this radiation resulted finally from the energy release in cosmic bodies in a late stage of the cosmological expansion, is very incredible because it requires a rapid release of a considerable part of the nuclear energy—in the supposition of its further re-emission in the form of equilibrium radiation. To confirm the correctness of the interpretation of the detected radiation by means of the 'hot' universe hypothesis Y. B. Zeldovič (**34**) suggested observations of neutrino: relict neutrinos should reach us from the early expansion stage with the matter density of 35 orders higher than the modern one. But for their detection the experiment precision must be million times higher than the present one. Proceeding from the hot Universe hypothesis Y. B. Zeldovič and I. D. Novikov (**35**) presented critical comments on the hypothesis on cores delayed in the expansion (see (**31**)). According to these comments the existence of bodies with dimensions smaller than their gravitational radius at the early stages of expansion of the Universe led to a strong accretion of radiation by these bodies. If further calculations confirm that the accretion is catastrophically high, the hypothesis of cores delayed in expansion would contradict the observational data.

L. M. Ozernoj (**36**) examined the development of perturbations in the hot cosmological model. It is obvious that for explaining the formation of such condensations as galaxy clusters resulting from the gravitational instability of a gaseous medium, it is necessary to assume it to be of a very high temperature. This condition is valid for the earlier stages of expansion of the 'hot' Universe. But then the gas was almost fully ionized, its density was high, and it strongly interacted with the radiation, which was of a still higher density. Hence, adiabatic perturbations under which the matter contracted together with the radiation, did not increase because of the great elasticity of the mixture of the matter and radiation, due to the radiative pressure which was greater than that of the matter (the critical mass was too great). Even isothermic perturbations, under which only the matter contracts (the critical mass for them was much smaller) increased slowly. The interaction of the metagalactic matter with the radiation became insignificant, when the ionized gas recombined. A theory of the thermal balance of expanding metagalactic gas for different variants of heating and cooling is also given. The idea is proposed of the heating of the recombined gas due to the dissipation of turbulent motions and to explosions of rapidly evolving objects of the quasars type (see also (**41**)). The role of initial conditions in the formation of galaxies of different morphologic types is considered.

A.G. Doroškevič, Y.B. Zeldovič and I.D. Novikov (**37**) studied the formation of clusters and galaxies due to gravitational instability after the recombination of the metagalactic gas. This gas was cold and the critical mass was of the order of 10^6 solar masses. According to the authors' hypothesis, cosmic objects with such masses—'parent stars'—were formed, and they heated the gas so, that the critical mass became of the order of 10^9-10^{10} solar masses. As a result metagalactic objects of the second generation arose. These objects lead to the formation of non-uniformities of the order of $10^{13}-10^{14}$ solar masses. These non-uniformities lead to the isolation of clouds, from which galaxy clusters formed.

The problem on parent stars explosions was considered by G.S. Bisnovaty-Kogan (**38**). The necessity of the epoch of neutral hydrogen in the evolution of the Universe was studied by R.A. Syunjajev (**39**). The distribution function of the appearing cosmic objects by the moment and by the mass was examined by A.G. Doroškevič (**40**). Assuming the intergalactic matter to be composed of gas (mainly of hydrogen) V.L. Ginzburg and L.M. Ozernoj (**41**) analyzed its thermal regime. The authors took into account both cooling of the gas due to emission and cosmological expansion and heating of it due to the dissipation of energy of the plasma waves originated by cosmic rays, formed under the explosions of galaxies and radiogalaxies. At present the main mechanism of cooling is the adiabatic expansion. According to the authors conclusions the gas is almost fully ionized. This fact explains why the greatest part of the intergalactic matter yet avoids observations.

An attempt of detecting the ultraviolet radiation of the metagalactic ionized hydrogen and helium with the aid of the apparatus mounted at Soviet interplanetary automatic stations in 1964–1966, was made by V.G. Kurt (**42**). According to V.G. Kurt and R.A. Syunjajev (**43**) it can be deduced from the negative result of this attempt, that metagalactic density, in any case (if special assumptions concerning the gas temperature are not made) is less than the treble critical one.

M.S. Kardašov and G.B. Šolomickij (**44**) estimating the limiting distance and the corresponding value of the red shift (z), up to which observations of extragalactic objects are possible, came to the conclusion, that this value can be determined by Thomson dispersion of radiation in the intergalactic medium. This value, depending on the medium density and on the type of the model, lies between 3·5 and 7.

Y.B. Zeldovič, I.D. Novikov and R.A. Syunjajev (**45**) proposed methods for the investigation of the intergalactic helium and suggested the cosmological significance of the possible results, because the present helium content in the intergalactic medium, firstly, depends on the character of the expansion at its earlier stages and, secondly, influences strongly the thermal balance and the ionization state of the gas. Hence, the investigation of the intergalactic helium will allow to determine firstly, the degree of isotropy of the expansion at the earlier stages, i.e. at the mean density of the order of 1 g cm^{-3} (when the formation of hydrogen and helium occurs), and secondly, the present density and state of the intergalactic gas. R.A. Syunjajev (**46**) considered possible sources and methods of observation of helium and deuterium (its formation is an intermediate stage in the process of transforming hydrogen into helium). A.G. Doroškevič and R.A. Syunjajev (**47**) calculated the thermal balance of the intergalactic medium and pointed out the important role of helium in cooling the medium.

Assuming the existence of both matter and antimatter in the observed part of the Universe N.A. Vlasov (**48**) proposed an optical method for detecting the antimatter.

According to Y.B. Zeldovič and I.D. Novikov (**49**) the charge asymmetry of the Universe at the present epoch is due to a similar asymmetry at the contraction epoch (preceding to the present expansion) when under the density of the order of 10^{-30} g cm^{-3} and as result of common nuclear reactions (for instance, in cosmic bodies) an amount of energy of the order of 10^{16} erg g^{-1} was emitted. When approaching the singularity almost complete symmetry due to the creation of particles and anti-particles was reached. When the expansion began and hence the energy density declined the former antisymmetry of particles and anti-particles reappeared.

For explaining the observed charge asymmetry, A. D. Saharov (**50**) supposed that in the superdense state of the hot universe in particular baryon and muon conservation laws were broken (instead of them the conservation law is supposed to be valid for a combined baryon-muon charge).

III. ANISOTROPIC MODELS

The work by I.R. Pijr (**51**) is devoted to the static axial-symmetrical solution of the problem of many particles in general relativity and to clarifying the physical meaning of these solutions.

A.S. Kompaneec and A.S. Černov (**52**) derived solutions of the Einstein equations for a non-stationary homogeneous axial-symmetric model of the Universe in two limited cases: for the dust-like matter and for the ultrarelativistic gas. The solutions possess a singularity at a certain moment of time. A solution for a non-relativistic gas with a constant isentropy index is also derived.

Generalizing the ideas by S.B. Pikelner (**53**), J.H. Piddington *et al.* on existing of intergalactic magnetic fields, Y.B. Zeldovič (**54**) assumed the existence of a primary homogeneous magnetic field in the Universe. Solutions for a homogeneous, but anisotropic axially symmetrical Universe were considered, the anisotropy being due to the presence of a magnetic field. This magnetic field turns out to be frozen into the medium. Its strength at present is, suppositionally, less than 3×10^{-9} gauss. The vortical electric field, created by the variation of the magnetic field, is absent for an observer co-moving with the matter.

A.G. Doroškevič (**55**) considered a series of exact solutions of the gravitation equations for anisotropic homogeneous cosmological models with the magnetic field (to be more precise, homogeneous axial-symmetric models with magnetic field, oriented along the symmetry axis) and without it. The properties of the solutions near the singularity and in the late stages of the expansion were studied. It was found that the magnetic field can strongly influence the behaviour of the model near the singularity (i.e. in the initial stages) under any equations of state, but in the later stages of the expansion this influence becomes inappreciable. It is pointed out that the well known objection by F. Hoyle (according to which the ratio of the magnetic field energy to the full matter energy increases infinitely when approaching the singularity) against including the magnetic field in cosmological solutions is wrong.

Homogeneous axial-symmetrical models with a magnetic field, directed along the symmetry axis, were considered also by I. S. Šikin (**56**).

The gravitational instability of anisotropic homogeneous cosmological models was considered briefly by A.G. Doroškevič (**57**). It is shown that in such models the amplification factor of the initial inhomogeneities is much greater than that of the homogeneous isotropic models. The influence of the magnetic field is discussed also. It is shown, that the account for the possible anisotropy of the expansion can turn out to be important for the theory of cosmic objects formation not only in the cold universe, but in the hot one also.

A.G. Doroškevič, Y.B. Zeldovič and I.D. Novikov (**58**) showed that in an anisotropically expanding Universe the weakly interacting particles (neutrinos, gravitons) moving in different directions, change their momentum in different ways and this, in its turn, strongly influences the picture of the anisotropic expansion. In particular, a higher (than in case of isotropic Universe) neutrino density turns out to be possible.

I. M. Halatnikov (**59**) considers a stationary cylindric-symmetric model, filled with matter obeying the equation of state $p = n\rho c^2$, and with a magnetic field. In connection with this model the question was discussed concerning the possibility of considering quasars as magnetogravitational formations.

IV. SPHERICAL SYMMETRY

R.I. Hrapko (**60**) considered various co-ordinate systems applicable for describing the semi-closed world, in connection with the Lichnerowich continuity conditions. He (**61**) also solved analytically the problem of the sight of the emitting surface of a gravitating sphere, expanding with parabolic velocity from inside of its Schwarzschild sphere. The same author (**62**) investigated the structure of the world of a special type with spatial spherical symmetry, including the Schwarzschild, de Sitter, Kottler and Nordström metrics, as particular cases.

V.S. Brežnev (**63**), V.S. Brežnev, D.D. Ivanenko and B.N. Frolov (**64**) considered solutions of Einstein equations, describing spherically symmetrical fields of condensations in the expanding universe and possessing Friedmanian asymptotics at infinity. Spherically symmetrical motions of the matter were considered in general relativity also by K.P. Stanjukovič and S.M. Kolesnikov (motions of the dust-like matter (**65**) and adiabatic motions of non-dust matter (**66**)), by K.P. Stanjukovič, O. Šaršekejev and V.Z. Gurovič (automodel motions of relativistic gas (**67**)), by K.P. Stanjukovič and V.Z. Gurovič (ultra-relativistic gas scattering (**68**)).

K.P. Stanjukovič and O. Šaršekejev (**69**) transformed Schwarzschild and Friedman line elements to a form, in which the metrics of spatial sections, non-orthogonal to the time lines, is euclidean.

V. COLLAPSE, ROTATION, QUASARS

A. G. Doroškevič, Y.B. Zeldovič and I.D. Novikov (**70**) considered the gravitational collapse of non-symmetric and rotating masses. It was shown that, as P.A. Bergman had supposed, the characteristic picture of gravitational selfclosing, true for the spherical case, is the same for the general case. In the case of collapse of a non-rotating body the quadrupole and higher field moments (which are due to asymmetry), decrease for an external observer inversely to the time elapsed. But in the case of collapse of a rotating body the deviations in metrics connected with the rotation when damping do not approach zero. Static non-spherical solutions of the Einstein equations were also considered, and in particular, the properties of the surface, analogous to that of Schwarzschild.

Y.B. Zeldovič and M.A. Podurec (**71**) considered the evolution of the system of heavy point particles. As effects, leading to the evolution, the following ones were taken into account: evaporation of particles, and gravitational radiation and capture. According to the authors' conclusion the system, evolving slowly, passes the sequence of quasi-equilibrium states, the evolution being inevitably terminated in a collapse. The applicability of the results to astronomical objects was also discussed.

Considering the behaviour of an electrically charged dust-like sphere, I.D. Novikov (**72**) came to the conclusion, that its contraction inside the Schwarzschild surface is followed by the expansion, both the preceding contraction stage, forecoming before the sphere is plunged inside the surface, and the following stage of expansion occuring after the sphere appears outside this surface, taking place in similar, but not coinciding space-time regions, disconnected in time, and euclidean on spatial infinity. The maximal mean density is of the order of $c^6 m^4/\varepsilon^6$, where m is the mass, ε the charge. The infinite density is reached in the center only. It was supposed that the increase of perturbations or quantum effects must lead to a similar behaviour of a non-charged material sphere.

In contradiction to the abovementioned views, M.E. Gercenštein (**73**) considered the gravitational collapse as a stage in an oscillatory process with passage through the center, the maximal radius of the central body exceeding the gravitational one. Such an oscillatory process is regarded as a behaviour of a quasar. According to this author, oscillatory collapse, in principle, is visible for an outer observer and does not exclude the emission of large amounts of energy. M.E. Gercenštein and Yu.M. Ajvazjan (**74**) considered the picture of the oscillatory collapse from the point of view of an external observer, and came to the conclusion that in this picture the symmetry of the contraction and expansion is not disturbed.

Criticizing F. Michel's idea concerning the role of neutrino emission in the phenomenon of quasars, Y.B. Zeldovič (75) pointed out that the gravitational collapse considerably extinguished not only the energy output through electromagnetic emission, but also that through the neutrino one. So establishing the impossibility of a sufficient output of the energy under a spherically symmetric collapse, Y.B. Zeldovič simultaneously pointed out the significance of non-symmetric motions of the matter in the process of the transformation of the gravitational energy of a collapsing star into the energy of radiation and cosmic rays, and in other processes accompanying the collapse.

In connection with the collapse phenomena I.S. Šklovskij and N.S. Kardašov (**76**), Y.B. Zeldovič and I.D. Novikov (**77**) considered energy emission in the form of gravitational waves, arising as a by-product in the case of non-symmetric motions of matter. L.M. Ozernoj (**78**) considered gravitational emission due to non-symmetric motions of the matter under explosion phenomena.

Y.B. Zeldovič (**79**) pointed out the process of liberation of gravitational energy in non-symmetric accretion through the glow of gas in a shock wave, arising under its falling in the gravitation field of a collapsed star. Observations of this phenomenon should particularly be a method for detecting collapsed stars. Another method of their detection (among spectral binaries with invisible companions) was suggested by O.H. Gusejnov and Y.B. Zeldovič (**80**).

Y.B. Zeldovič and I.D. Novikov (**81**) assuming that in the quasar atmosphere the radiative pressure is balanced by the gravitation, found the mass of the quasar core to be of the order of 10^8 solar masses. A disturbance of the equilibrium in outer layers leads to their ejection. The accretion of the matter from the atmosphere (about three solar masses per year) to the near vicinity of the collapsed core is suggested as a possible mechanism of energy generation by the quasar.

Y.B. Zeldovič (**82**) showed that, because of the dependence of the gravitational field of a body on its rotation, a gravitational analogue of Zeeman-effect in the field of the rotating star must exist. As it is shown by A.G. Doroškevič (**83**) the dependence of the gravitational field of a body on its rotation acts also in the process of accretion: the moment of momentum of a rotating body decreases during the accretion of the matter, free of the moment of momentum. The decrease of the moment is found to be most rapid for collapsed stars.

V.L. Ginzburg (**84**) proposed a hypothesis, according to which quasars are radiation belts around large collapsing magnetic protostars. V.L. Ginzburg and L.M. Ozernoj (**85**) investigated the gravitational collapse of a spherically symmetrical mass of gas with zero pressure, possessing a magnetic moment. According to the drawn conclusions the magnetic moment of a star decreases for an exterior observer as a power function, and the increasing magnetic field draws to the Schwarzschild surface.

I.D. Novikov (**86**) considered the variation of the magnetic field during the collapse of a magnetic star as a possible mechanism of electromagnetic waves emission by quasars. N.S. Kardašov (**87**) considered the strengthening of the magnetic field of a rotating plasma cloud under the gravitational contraction as the cause of the observed features of powerful sources of the non-thermal emission.

L.M. Ozernoj (**88**) proposed the theory of the magnetoid – a massive plasma quasi-stationary configuration, its equilibrium being supported by the ordered circular streaming of plasma, and came to the conclusion that it is possible to explain the properties of a quasar within the limits of the magnetoid model.

Massive configurations stabilized by the rotation under the slow loss of the matter from the equator, were considered by A.S. Bisnovaty-Kogan, Y.B. Zeldovič and I.D. Novikov (**89**).

BIBLIOGRAPHY

1. Halatnikov, I.M. 1965, *Zu. eksp. teor. Fiz.*, **48**, iss. 1, 261.
2. Belinskij, V.A., Halatnikov, I.M. 1965, *Zu. eksp. teor. Fiz.*, **49**, iss. 3 (9), 1000.
3. Griščuk, L.P. 1966, *Zu. eksp. teor. Fiz.*, **51**, iss. 2 (8), 475.
4. Keres, H. 1965, *Zu. eksp. teor. Fiz.*, **48**, iss. 5, 1319.
5. Keres, H. 1966, *Zu. eksp. teor. Fiz.*, **50**, iss. 2, 493.
6. Naan, G.I. 1964, *Publik. Tartu. astr. Obs.*, **34**, N 6, 423.
7. Terleckij, Y.P. 1965, *Izv. vyših učebnyh zavedenij, Fizika*, N 6, 122, 1964; in Robinson *et al.*, *Quasistellar Sources and Gravitational Collapse*, Chicago, p. 466.
8. Gliner, E.B. 1965, *Zu. eksp. teor. Fiz.*, **49**, iss. 2 (8), 542.

9. Gliner, E.B. 1965, *Zu. eksp. teor. Fiz.*, Pis'ma v red., **2,** iss. 2, 53.

10. Altšuler, B.L. 1966, *Zu. eksp. teor. Fiz.*, **51,** iss. 4, (10), 1143.

11. Stanjukovič, K.P. 1966, *Dokl. Akad. Nauk SSSR* **168,** N 4, 781.

12. Stanjukovič, K.P. 1966, *Usp. Fiz. Nauk*, **89,** iss. 4, 731; *Gravitacionnoje pole i elementarnyje časticy.* Nauka, Moskva, 1965.

13. Zeldovič, Y.B., Smorodinskij, Y.A. 1966, *Usp. Fiz. Nauk* **89,** iss. 4, 734.

14. Markov, M.A. 1966, *Zu. eksp. teor. Fiz.* **51,** iss. 3 (9), 878.

15. Kurdgelaidze, D.F. 1964, *Zu. eksp. teor. Fiz.* **47,** iss. 6, (12), 2313.

16. Ivanenko, D., Kurdgelaidze, D. 1965, Preprint, *Užgorodsk. Gos. Universitet, Kaf. teor. Fiz. i labor, teorii elem. častic*, p. 8.

17. Idlis, G.M. 1965, *Vest. Akad. Nauk Kaz. SSR*, N 5; *Trudy Astrofiz. inst. Akad. Nauk Kaz. SSR*, **5,** 105, 1965; in: *Filosofskije problemy relativistkoj kosmologii i teorii tiagotenija Einsteina*, Kiev, 1965, p. 302.

18. Černin, A.D. 1965, *Astr. Zu.* **42,** 1124.

19. Stanjukovič, K.P. 1966, *Problemy teorii gravitacii i elementarnyh čhastic*, Atomizdat, Moskva, p. 140.

20. Daševskij, V.M., Sliš, V.I. 1965, *Astr. Zu.* **42,** 863.

21. Černin, A.D. 1966, *Astr. Zu.* **43,** 797.

22. Daševskij, V.M., Zeldovič, Y.B. 1967, *Astr. Zu.* **44,** 67.

23. Zasov, A.V. 1967, *Soobšč. Trudy gos. astr. Inst. Sternberga*, N 147, 26.

24. Zeldovič, Y.B., Okun', L.B., Pikelner, S.B. 1965, *Usp. Fiz. Nauk* **87,** issl. 1, 113; *Phys. Letters*, **17,** 164, 1965.

25. Lifšic, E.M., Halatnikov, I.M. 1963, *Usp. Fiz. Nauk* **80,** iss. 3, 391.

26. Novikov, I.D. 1964, *Zu. eksp. teor. Fiz.* **46,** iss. 2, 686.

27. Saharov, A.D. 1965, *Zu. eksp. teor. Fiz.* **49,** iss. 1 (7), 345.

28. Ambarcumjan, V.A. 1964, *13e Conseil de Physique Solvey*, Bruxelles, p. 1.

29. Karačencev, I.D. 1964, *Dokl. Akad. Nauk SSSR*, **158,** N 4, 815.

30. Karačencev, I.D. 1966, *Astrofizika* **2,** 81.

31. Novikov, I.D. 1964, *Astr. Zu.* **41,** 1075.

32. Doroškevič, A.G., Novikov, I.D. 1964, *Dokl. Akad. Nauk SSSR*, **154,** 809.

33. Zeldovič, Y.B., Novikov, I.D. 1967, *Astr. Zu.*, **44,** 663.

34. Zeldovič, Y.B. 1966, *Usp. Fiz. Nauk*, **89,** iss, 4, 647.

35. Zeldovič, Y.B., Novikov, I.D. 1966, *Astr. Zu.*, **43,** 758.

36. Ozernoj, L.M. 1967, Transactions of the Symposium 'Peremennye zviozdy i zviozdnaja evolucija' (Moscow, Nov. 24–27, 1964), Nauka.

37. Doroškevič, A.G., Zeldovič, Y.B., Novikov, I.D. 1967, *Astr. Zu.* **44,** 295.

38. Bisnovaty-Kogan, G.S. 1967, *Astr. Zu.* **44** (in press).

39. Syunjajev, R.A. 1967, *Dokl. Akad. Nauk SSSR*, (in press).

40. Doroškevič, A.G. 1967, *Astrofizika* **3,** iss. 2 (in press).

41. Ginzburg, V.L., Ozernoj, L.M. 1965, *Astr. Zu.* **42,** 943.

42. Kurt, V.G. 1965, in: *Issledovanija kosmič prostranstva*, Nauka, p. 576.

43. Kurt, V.G., Syunjajev, R.A. 1967, *Zu. eksp. teor. Fiz.*, Pis'ma v red., **5,** iss. 9, 299; *Kosm. Issled.*, **5,** iss. 4, 611, 1967.

44. Kardašov, N.S., Šolomickij, G.B. 1965, *Astr. Cirk.* N 336, 3.

45. Zeldovič, Y.B., Novikov, I.D., Syunjajev, R.A. 1966, *Astr. Cirk.*, N 371, 1.

46. Syunjajev, R.A. 1966, *Astr. Zu.* **43,** iss. 6, 1237.

47. Doroškevič, A.G., Syunjajev, R.A. 1968, *Astr. Zu.* (in press).

48. Vlasov, N.A. 1964, *Astr. Zu.* **41,** 893.

49. Zeldovič, Y.B., Novikov, I.D. 1966, *Zu. eksp. teor. Fiz.*, Pis'ma v red. **4,** iss. 3, 117.

50. Saharov, A.D. 1967, *Zu. eksp. teor. Fiz.*, Pis'ma v red., **5,** iss. 1, 32; Corrections see, iss. 4, 136.

51. Pijr, I.R. 1964, *Trudy Inst. Fiz. i Astr. Akad. Nauk Est. SSR*, N 25, 36 and 75.

52. Kompaneec, A.S., Černov, A.S. 1964, *Zu. eksp. teor. Fiz.*, **47,** iss. 5 (11), 1939.

53. Pikelner, S.B. 1965, *Astr. Zu.* **42,** 3, 515.

54. Zeldovič, Y.B. 1965, *Zu. eksp. teor. Fiz.*, **48,** iss. 3, 986.

55. Doroškevič, A.G. 1965, *Astrofizika*, **1,** 255.

56. Šikin, I.S. 1966, *Dokl. Akad. Nauk SSSR*, **171,** N 1, 73.

57. Doroškevič, A.G. 1966, *Astrofizika*, **2,** 37.

58. Doroškevič, A.G., Zeldovič, Y.B., Novikov, I.D. 1967, *Zu. eksp. teor. Fiz.*, Pis'ma v red., **5,** iss. 4, 119.

59. Halatnikov, I.M. 1965, *Zu. eksp. teor. Fiz.*, Pis'ma v red. **5**, iss. 6, 195.
60. Hrapko, R.I. 1965, *Zu. eksp. teor. Fiz.*, **49**, iss. 6 (12), 1884.
61. Hrapko, R.I. 1966, *Zu. eksp. teor. Fiz.*, **50**, iss. 4, 971.
62. Hrapko, R.I. 1966, *Zu. eksp. teor. Fiz.*, **50**, iss. 6, 1936.
63. Brežnev, V.S. 1966, *Problemy teorii gravitacii i elementarnyh čhastic*, Atomizdat, Moskva, p. 158.
64. Brežnev, V.S., Ivanenko, D.D., Frolov, E.N. 1968, *Trudy vtoroj sovetskoj gravitacionnoj konferencii, Tbilisi* (in press).
65. Stanjukovič, K.P., Kolesnikov, S.M. 1965, *Zu. prikladnoj Matematiki i Mekhaniki*, **29**, iss. 4, 717.
66. Kolesnikov, S.M., Stanjukovič, K.P. 1966, *Problemy teorii gravitacii i elementarnykh častic*, Atomizdat, Moskva, p. 135.
67. Stanjukovič, K.P., Šaršekejev, O., Gurovič, V.Z. 1965, *Dokl. Akad. Nauk SSSR*, 165, N 3, 510.
68. Stanjukovič, K.P., Gurovič, V.Z. 1965, *Dokl. Akad. Nauk SSSR*, **165**, N 4, 806.
69. Stanjukovič, K.P., Šaršekejev, O. 1966, *Problemy teorii gravitacii i elementarnykh častic*, Atomizdat Moskva, p. 206 and 241.
70. Doroškevič, A.G., Zeldovič, Y.B., Novikov, I.D. 1965, *Zu. eksp. teor. Fiz.*, **49**, iss. 1 (7), 171.
71. Zeldovič, Y.B., Podurec, M.A. 1965, *Astron. Zu.*, **42**, 963.
72. Novikov, I.D. 1966, *Zu. eksp. teor. Fiz.*, Pis'ma v red., **3**, iss. 5, 223; *Astr. Zu.* **43**, iss. 5, 911, 1966.
73. Gercenštein, M.E. 1966, *Zu. eksp. teor. Fiz.*, **51**, iss. 1 (7), 129, iss. 4 (10), 1127.
74. Gercenštein, M.E., Ajvazjan, Yu.M. 1966, *Zu. eksp. teor. Fiz.*, **51**, iss. 5 (11), 1405.
75. Zeldovič, Y.B. 1963, *Astr. Cirk.* N 250, 1.
76. Šklovskij, I.S., Kardašov, N.S. 1964, *Dokl. Akad. Nauk SSSR* **155**, N 5.
77. Zeldovič, Y.B., Novikov, I.D. 1964, *Dokl. Akad. Nauk SSSR*, **155**, N 5, 1033.
78. Ozernoj, L.M. 1965, *Zu. eksp. teor. Fiz.*, Pis'ma v red., **2**, 83.
79. Zeldovič, Y.B. 1964, *Dokl. Akad. Nauk SSSR*, **155**, N 1, 67.
80. Gusejnov, O.H., Zeldovič, Y.B. 1966, *Astr. Zu.* **43**, iss. 2, 313.
81. Zeldovič, Y.B., Novikov, I.D. 1964, *Dokl. Akad. Nauk SSSR*, **158**, N 4, 811.
82. Zeldovič, Y.B. 1965, *Zu. eksp. teor. Fiz.*, Pis'ma v red., **1**, iss. 3, 40.
83. Doroškevič, A.G. 1966, *Astron. Zu.* **43**, iss. 1, 105.
84. Ginzburg, V.L. 1964, *Dokl. Akad. Nauk SSSR*, **156**, N 1, 43.
85. Ginzburg, V.L., Ozernoj, L.M. 1964, Preprint, Fiz. Inst. AN A-30.
86. Novikov, I.D. 1964, *Astr. Cirk.*, N 290, 1.
87. Kardašov, N.S. 1964, *Astr. Zu.* **41**, 807.
88. Ozernoj, L.M. 1967, Transactions of the Symposium 'Peremennyje zviozdy i zviezdnaja evolucija' (Moscow, Nov. 24–27, 1964), Nauka, *Astr. Zu.* **43**, 300, 1966.
89. Bisnovaty-Kogan, G.S., Zeldovič, Y.B., Novikov, I.D. 1967, *Astr. Zu.* **44**, 525.

29. COMMISSION DES SPECTRES STELLAIRES

Report of Meetings held on 24, 28 and 29 August 1967

PRESIDENT: J. Sahade.
SECRETARY: M. K. V. Bappu.

Business Meeting

(*a*) The President asked the audience to stand in memory of the Commission members Drs Mc Laughlin, Nassau and Rozis-Saulgeot, who passed away since the Hamburg meeting.

(*b*) The following astronomers were proposed as members of the new Organizing Committee: V. Bappu, Herbig, Houziaux, Kopylov, Przybylski and Sahade.

(*c*) The Draft Report as presented, was accepted with minor corrections.

(*d*) *Miss A. B. Underhill* reported on the conclusions of the Working Group appointed to examine the need of an Atlas of high dispersion tracings. She considered it a useful aid for teaching purposes. The Atlas could either be a compilation of tracings or be available in digitized form. At least 55 stars would be required for such a purpose and the provision of tracings for each would be very expensive. The Working Group prefers the use of digitization from the standpoint of low cost of production of such an atlas. Since such a publication would be most useful for teaching purposes, Miss Underhill was requested to continue her efforts on the feasibility of such an enterprise in close collaboration with Commission 46.

(*e*) *J. B. Oke* presented the report of the Working Group on spectrophotometry. A considerable amount of absolute spectrophotometry of stars has been carried out since the Hamburg meeting, the references to which are given in the Draft Report, Appendix I. A list of stars suitable for absolute standards was published by Oke in 1964. These are all near the celestial equator, but not all of them have been measured from 6000 Å to 11000 Å. These standards are too bright for use with large telescopes and pulse counting techniques. Fainter standards near the celestial equator are now being measured at Palomar.

Further work on the comparison of standard stars with laboratory sources is continuing. Work on this aspect is currently in progress at the Lick Observatory by Hayes and at the Mount Wilson and Palomar Observatories by Schikd and Oke.

During the IAU meetings at Praha a meeting was held by all interested in the problems of absolute spectrophotometry. *Oke*'s report of this meeting follows: '*Willstrop* outlined his recent work using a 10 channel narrow band system which covered the range 3300 Å to 10000 Å. Reductions are not yet completed. *Kienle* reported that Labs has compared the energy distribution of the Sun and of stars of the same spectral class. The agreement is extremely good. *Stecher* outlined the calibration methods used for far ultra-violet work from rockets. At 1216 Å the ionization effects in gases are known and can be used for calibration. At 2537 Å a calibrated mercury lamp is used. *Oke* outlined briefly the methods being used at Palomar to compare stars and laboratory sources. *Davis* discussed the work which has been done on radii and effective temperatures of stars using the Hanbury Brown stellar interferometer in Australia. They have completed work on 15 B, A and F type stars. *Kienle* strongly urged that absolute fluxes of stars be presented in the form log f_ν versus $1/\lambda$ (μ). *Divan* indicated that ε Persei is an excellent absolute standard star since the Balmer jump is small'.

(*f*) *Mme Cayrel* briefly outlined the report of the Committee on Line Intensity Standards. She reported that only the coudé intensity calibration systems at Kitt Peak, Haute-Provence, Mount Wilson and Palomar, and Mount Stromlo have been calibrated against the standard Kienle step filter. Many other high dispersion facilities await similar calibration check with the standard. The measured equivalent widths are in better agreement when step calibrations are used instead of a continuous calibration system.

It was decided by the Commission that the committee, as presently constituted, would continue to function until the next General Assembly.

(*g*) It was also decided by the Commission that the Working Group on Be stars be incorporated as an official Working Group of Commission 29.

Scientific Meetings

The following papers were presented:

P. Swings: Remarks and Recollections of a Stellar Spectroscopist.

A.B. Underhill: Some Problems Concerning the He I Lines in B Type Spectra.

L. Houziaux: Quelques problèmes relatifs aux étoiles des premiers types à raies d'émission.

P.S. Conti, A.J. Deutsch: Colour Anomalies and Metal Deficiencies in Solar Type Disk-Population Stars.

J.P. Swings: The Ultra-violet Spectrum of CH Cygni.

Y. Fujita: Some Problems in Connection with High Dispersion Spectra of Carbon Stars.

T.P. Stecher: Stellar Spectrophotometry from a Pointed Rocket.

D.C. Morton: Stellar Spectra in the Far Ultra-violet.

K.G. Henize: Ultra-violet Objective Prism Spectra from Gemini XI and XII Manned Space Flights.

J.B. Oke, A. Vaughan: High Resolution Scanning.

C. Fehrenbach: The Spectrum of Nova Delphini 1967.

J.D. McGee: Image Tubes for Stellar Spectroscopy.

A. Vaughan, H. Zirin: The He I 10830 Line in Late Type Stars.

A.V. Hewitt: Spectrometry with a Cascade Intensifier.

W. Kent Fort, V. Rubin, P.B. Boyce: Image Tube Spectroscopy.

M. Walker: Electron Spectroscopy with the Lallemand Electronic Camera and McGee Spectracon.

G.E. Kron: The Electronic Camera and Spectrum Scanning.

R. Herman: Profiles de raies à haute résolution obtenues avec la camera électronique.

W. Livingston: High Dispersion Stellar Spectroscopy and the Role of the Image Tube at the McMath Solar Telescope (read by *A.K. Pierce*).

G. Münch, G. Neugebauer: The Spectra of Cool Stars in the PbS Infra-red.

P. Connes: High Resolution Infra-red Stellar Spectra.

30. COMMISSION DES VITESSES RADIALES

Report of meeting, 25 August 1967

PRESIDENT: D. S. Evans.
ACTING SECRETARY: R. F. Griffin.

Dr D. S. Evans, the Acting President, paid tribute to the late President, Dr R. M. Petrie. The text of the draft report with typographical corrections was accepted, and the composition of the Commission discussed.

Dr R. F. Griffin—Radial Velocities by the photoelectric method. A description of the radial velocity spectrometer used at the coudé focus of the Cambridge 36-inch telescope has been published.

In reply to questions by Fehrenbach, MacRae and Fletcher, *Griffin* expressed preference for the use of reference stars over an iron arc. He stated that the observation rate was 7 ± 1 stars per hour at all magnitudes and that the radial velocity diaphragm had a proportional transparency of 11·3 %.

Dr G. de Vaucouleurs—Scanning methods. At the Toronto symposium in June 1966, an experimental two-channel pulse counting spectrum scanner using a tuneable interference filter to measure radial velocities of extended sources was described (McDonald Obs. Preprint No. 2). Further development of the system and examples of applications to emission regions in nearby galaxies (M 33, M 101) were presented.

Mrs V. Rubin—Applications of Image Tubes. An image tube spectrograph designed at the Department of Terrestrial Magnetism has been used by myself and W. Kent Ford Jr. to obtain plates for determination of accurate radial velocities.

de Vaucouleurs. There is image-tube work in progress at Flagstaff and at McDonald, so it should soon be possible to compare image-tube velocities. We have observed NGC 4038 with the McDonald scanner.

Mrs Penston—A0 stars in the Galactic Caps. The aim of this Herstmonceux investigation is to determine whether the velocity dispersion increases with height above the galactic plane. If the dispersions are in the ratios of those found by Alexander the value of σ_w is found to increase with height from about 9 ± 1 km/s below 200 parsecs to about 18 ± 3 at about 500 parsecs.

Eelsalu. You selected a group of stars of type A0 from the Henry Draper Catalogue and then classified them on the MK system?

Mrs Penston. Yes. Of the 64 stars we observed, we found 15 Am stars, 20 A0 V and the rest scattered between B8 and A5.

Eelsalu. The HD types have systematic errors which are dependent on magnitude, so the sample of HD type A0 stars will also be magnitude-dependent.

MacRae. Are these the actual velocity dispersions of the stars or do they include the errors of observation?

Mrs Penston. They do not include observational errors.

MacRae. How great were those?

Mrs Penston. About 5 km/s for the mean of three plates.

Fehrenbach. Observations at Haute Provence tend to support the suggestion that the velocity dispersion increases with height above the galactic plane.

Dr F. K. Edmondson—Solar Motion from 11th magnitude A-stars. The result for the solar motion indicates that the 11th magnitude A-stars are moving toward the central region of the galaxy ($l_0^{II} = 335°$) with a velocity of 14 ± 3 km/s relative to the naked-eye A-stars.

An analysis of *UBV* photoelectric measures of 50 stars by Young leads to a mean distance of 640 parsecs for this sample. The galactic rotation effect is small and not well determined ($rA = 3{\cdot}5 \pm 2$ km/sec).

Feast. Thackeray has been working on southern B8 stars up to about 600 pc distance and has got indications of galactic rotation. Are the corrections you apply from your standards 'night corrections' or do they apply to the whole programme?

Edmondson. They apply to individual observing runs.

Dr D. A. MacRae—Proposal by Heard for 9th magnitude standard stars. At IAU Symposium No. 30 on Radial Velocities it was agreed that a set of standard-velocity stars, of magnitude 9 or thereabouts, was desirable. It was proposed that observatories engaged in radial-velocity determinations with slit spectrographs should select candidates for such a list and, if necessary, observe them further to establish their velocities with greater accuracy.

At the David Dunlap Observatory we have chosen 24 stars of spectral types F, G and K in the declination zone + 25° to + 30° which appear to have fairly well-established constant velocities as determined by us from four or more 66 Å/mm and 33 Å/mm prism spectrograms (*Publ. D.D.O.* **2**, No. 4, 107, 1956). New observations with the 40 Å/mm grating spectrograph will be made to give improved mean velocities from eight or more spectrograms for those stars which appear still to have constant velocities. Collaboration by other observatories is welcome and we are willing to observe standards proposed elsewhere.

Table 1

HD (or AG)	α(1950)	δ(1950)	Ptg mag	Spectral type	No. of plates	Weighted mean
4388	00ʰ43ᵐ·8	30°40′	8·80	K3 III	4	− 24·8 ± 0·2
12029	01 55·8	29 08	8·96	K2 III	5	+ 40·0 ± 0·3
14969	02 22·6	29 39	8·96	K3 III	4	− 27·8 ± 0·3
23169	03 40·9	25 34	9·39	G2 V	4	+ 17·4 ± 0·3
32963	05 04·8	26 16	8·36	G2 V	4	− 62·0 ± 0·4
42397	06 08·5	25 01	8·68	G0 IV	4	+ 39·8 ± 0·2
(AG4044)	07 31·4	28 51	9·29	G0 IV	4	+ 0·6 ± 0·5
65934	07 59·1	26 47	8·87	G8 III	4	+ 34·4 ± 0.5
75935	08 50·9	27 06	9·35	G8 V	4	− 18·7 ± 0·7
86801	09 58·7	28 48	9·48	G0 V	4	− 4·6 ± 0·5
90861	10 27·1	28 50	8·36	K2 III	5	+ 39·8 ± 0.4
102494	11 45·3	27 37	8·26	G8 IV	4	− 21·1 ± 0.7
112299	12 53·0	26 01	9·19	F8 V	4	+ 3·6 ± 0·4
122693	14 00·5	24 48	8·74	F8 V	4	+ 1·0 ± 0·4
132737	14 57·7	27 21	9·03	K0 III	4	− 20·7 ± 0·6
140913	15 43·1	28 37	8·81	G0 V	4	− 14·5 ± 0·2
149803	16 33·9	29 51	8·90	F7 V	4	+ 1·2 ± 0·7
160952	17 39·7	29 37	9·04	G8 III	4	+ 30·6 ± 0·3
171232	18 30·6	25 27	8·66	G8 III	4	− 30·6 ± 0·9
(AG10133)	19 33·0	28 59	9·55	F7 V	4	− 37·0 ± 0·4
194071	20 20·5	28 05	9·06	G8 III	4	− 12·8 ± 0·3
204934	21 29·0	28 09	9·64	K1 III	4	+ 2·5 ± 0·3
213947	22 32·3	26 20	8·93	K4 III	4	+ 20·0 ± 0·4
223094	23 43·9	28 26	8·97	K5 III	4	+ 20·8 ± 0·5

Reply to *Popper*. The velocities are on the system of the David Dunlap Observatory, which contributed about 25% of the velocities in the Wilson Catalogue, so that their relationship to any other system should be well defined.

Feast. Thackeray pleads that standards should be within 15° of the equator. Kitt Peak people should also be interested in this since they are going to observe from both northern and southern hemispheres. Thackeray points out that you can get different velocity systems by the use of different emulsions even in the same spectrograph.

Dr D.S. Evans—Southern Standards. We find that most reliable velocities of ninth magnitude southern stars have been observed only in the Cape programmes at the Radcliffe Observatory, that they are mainly of late type and of rather high southern declinations. Later type standards are of value in that they simultaneously check spectrographic stability and the system of wavelength standards. A choice of stars including more in low declinations and some earlier types, observed from several different places, would be desirable. For the present, the following list of 14 solar-type stars which have given good results, may be of interest to other observers.

Table 2

FDS No		HD	1950 R.A.	1950 Dec.	*V*	*B–V*	$(U–B)_c$	Type	RV & PE	No. Plates
V	819	6655	$01^h03^m.7$	– 72°49	8·04	+ 0·55	1·66	G3 IV–V	+ 15·5 ± 0·5	(4)
I	52	24331	03 48 · 9	– 42 43	8·59	+ 0·91		K2 V	+ 22·4 ± 0·5	(4)
IV	668	39194	05 45 · 1	– 70 11	8·09	+ 0·76	1·83	K0 V	+ 14·2 ± 0·4	(4)
V	880	– 43°2527	06 30 · 7	– 43 29	8·61	+ 1·15	2·18	K1 III	+ 13·1 ± 0·5	(4)
V	883	48381	06 39 · 9	– 33 25	8·48	+ 1·05	2·08	K0 IV	+ 39·5 ± 0·5	(4)
II	373	83443	09 35 · 3	– 43 03	8·21	+ 0·81		K0 V	+ 27·6 ± 0·5	(9)
V	904	83516	09 35 · 9	– 34 51	8·63	+ 0·97	1·98	G8 IV	+ 42·0 ± 0·5	(4)
II	386	101266	11 36 · 4	– 45 05	9·29	+ 0·65		G5 IV	+ 20·6 ± 0·5	(5)
II	396	111417	12 46 · 7	– 45 33	8·33	+ 1·41		K3 IV	– 16·0 ± 0·5	(5)
IV	723	120223	13 46 · 1	– 43 29	8·97	+ 0·97	1·96	G8 IV–V	– 24·1 ± 0·6	(5)
IV	759	176047	18 56 · 4	– 34 32	8·10	+ 0·96	2·04	K1 III	– 40·7 ± 0·5	(4)
IV	768	193231	20 17 · 8	– 54 58	8·39	+ 0·73	1·78	G8 V	– 29·1 ± 0·6	(5)
V	1053	196983	20 38 · 7	– 34 04	9·08	+ 1·18	2·23	K2 III	– 8·0 ± 0·6	(4)
IV	794	219509	23 14 · 2	– 67 11	8·69	+ 1·06	2·20	K5 V	+ 62·3 ± 0·5	(4)
						(Nf component of wide binary)				

Dr J. Dommanget—Orientation of the Orbits of Binaries. Although no significantly non-random distribution is found in either heliocentric or galactic latitude, there is a region centred near $l^{II} = 35°$, $b^{II} = + 60°$, towards which no orbital poles are directed.

Of nine binary systems within 10 pc of the Sun, eight (including the solar system) have orbital poles near $l^{II} = 100°$, $b^{II} = - 15°$, if the sense of rotation is disregarded; the exception is α Cen.

Dr D.S. Evans—Revision of the General Catalogue of Radial Velocities. I show a part of the manuscript of the revision of the Radial Velocity Catalogue, and also the layout of the notes and bibliography. Further parts will be produced over the next few months and submitted in the form of proof copies to a small number of astronomers who have undertaken to read them critically.

RAPPORT DE LA SESSION MIXTE 30 ET 42

Monday, 28 August 1967 — 8:45 a.m. to 10:15 a.m.

CHAIRMAN: A. H. Batten.
SECRETARY: J. M. Fletcher.

Ninety-six persons attended this session which was devoted to a discussion of the 'Determination of Orbital Elements and Stellar Masses'.

The Chairman introduced the session by explaining that the 30-42 Joint Working Group had felt that the time was ripe for a discussion of how far the derived orbital elements of close binary systems represented the true physical nature of the systems. Not all the speakers invited had agreed that this was an important topic, but he hoped this topic might emerge in discussion. He discussed the distribution of the longitudes of periastron in close binaries, as found in the recently completed *Sixth Catalogue of the Orbital Elements of Spectroscopic Binary Systems.* This distribution clearly shows non-random features in addition to the well-known Barr Effect, and clearly indicates that the longitude of periastron is often a mathematical parameter of the velocity-curve rather than a true measure of the orientation of the orbit. He briefly described the new *Catalogue.*

D. Ya. Martynov expressed his concern that many observatories in the United States had apparently not received copies of his *Bibliography of Spectroscopic Binaries* and asked all participants to check this matter on their return home and to notify him if they had not received their copies. He discussed the work of Dr Cherepaschuk on the problem of eclipsing binaries with extended atmospheres, with particular reference to the Wolf-Rayet system V444 Cygni. It had been possible to obtain consistent solutions from four-colour photoelectric observations. The electron density in the atmosphere appeared to follow an exponential relation not in agreement with that derived by Chandrasekhar.

R. H. Koch reported on the co-operative programme of observations of SX Cassiopeiae sponsored by Commission 42 during 1964–67. Narrow-band and UBV photometry have shown the period of the system to have been constant since its discovery. The light-curve is intrinsically variable, but the width of primary eclipse, at a given time, is independent of the wavelength of observation. Analysis of the light-curve will probably prove to be very difficult. Over one hundred old spectrograms obtained by Joy have been discovered in the Mount Wilson Observatory files. The velocity curves compiled from these do not agree with Struve's published results. The star has a faint companion of possibly variable radial velocity. Astrometric observations, however, indicate that this companion is optical.

K. O. Wright discussed recent Victoria observations of 32 Cygni. Both photoelectric and spectroscopic observations are desired during the interval August 1 to November 1, 1968 to cover the eclipse of September 1968. Variations in the satellites of the K-line have been found, and it is desirable to establish the time-scale of these. The chromospheric lines in the spectra of 31 Cygni, 32 Cygni and VV Cephei seem to show large changes of intensity from eclipse to eclipse. It is not clear whether these are real changes or of instrumental origin.

M. Plavec pleaded for more observations of bright eclipsing binaries. The theory of close binary systems has made great strides in the last few years and the need for new and better observational data is urgent. Of 191 systems brighter than $8^{m}_{\cdot}5$ at maximum light, it is impossible to say of 46 whether they are Algol-type or main-sequence systems. For a further 59 the available information permits only a tentative classification. He intended to circulate a list of systems for which observations are particularly desirable. *Martynov* questioned the value of such lists, in his experience they had produced little or no response. *Plavec* felt that a short list, with notes on each system might serve to stimulate interest among observers. The question of the availability of suitable comparison stars for bright systems was raised. *Plavec* was of the opinion that this would be no problem in the vast majority of cases.

D. M. Popper emphasized the importance of using sufficiently high spectrographic dispersions for the determination of stellar masses. A dispersion of at least 20 Å/mm, and often of 10 Å/mm, is necessary for this purpose. He reiterated Plavec's remarks on the need for new photometric observations. He felt that such observations should be on a well-standardized colour system—such as *UBV*—and usually in at least two colours. He commented on the existence of an apparent group of systems in which both components are subgiants of spectral types G or early K and which show H and K emission in one or both spectra. The system AR Lacertae is the best known example. In some of these systems, SS Bootis being a case in point, both evolved components may be undermassive for their luminosity with masses less than one solar mass. The so-called R Canis Majoris systems are binaries with both components reportedly having abnormally low masses for their main-sequence spectral types. For one of these systems, TU Monocerotis, the K-line of the secondary spectrum blends with an interstellar line on plates of inadequate dispersion. Use of adequate dispersion removes the mass anomaly. Koch commented that it is difficult to standardize one's photometric system each night. Popper replied this was not necessary so long as the observations were on a standard system.

M. W. Ovenden presented a discussion by himself and A. H. Batten on the mass data in the *Sixth Catalogue*. Masses (m) have been computed for those systems in the Catalogue for which inclinations can be found from visual or eclipsing orbits, and for which absolute magnitudes can be found from visual orbits, trigonometric parallaxes or Hγ equivalent widths. For absolute visual magnitudes $-6 < M_v < +2$, the (M_v, log$m$) relation is linear, with no systematic differences between the magnitudes derived from Hγ and those derived from dynamical and trigonometric parallaxes. For $M_v > +4$ there is slight evidence for a dual (M_v, log m) relation. The observed (M_{bol}, log m) relation is in good agreement with a series of main-sequence models with (X, Y) = (0·68, 0·30).

Mrs A. E. Ringuelet-Kaswalder presented work by herself and J. Sahade on the determination of masses of Algol-type systems. Two methods are commonly used. Either one assumes that the primary component is a main-sequence star, or one assumes that the secondary component fills its equipotential lobe. In the second method, the mass-ratio derived is very sensitive to the assumed radius of the lobe, and therefore quite small photometric errors in the determination of the radius of the secondary component can lead to serious errors in the derived masses. Uniform use of the first method might eliminate the apparent anomaly of the R Canis Majoris systems. Plavec commented that the important point was to obtain good observations. For 19 out of 25 systems with well-determined elements, the two possible assumptions led to closely similar masses. Sahade replied that such agreement was to be expected in many cases, but the second method could fail badly; and, furthermore, it should be remembered that the equipotential surfaces refer to ideal stars and not to real ones.

A later communication was received from *V.P. Cesevič* concerning recent work on eclipsing binaries in the U.S.S.R. V. Tabačnik (Odessa) has worked on the computation of elements by means of an electronic computer, and has applied his method to S Cancri. M. Kiperman and A. Šulberg (Odessa) have investigated the possibility of determining the non-linear coefficients of limb darkening, as discussed by Grygar. Cesevič himself has investigated the analysis of atmospheric eclipses with particular reference to the systems RW Trianguli and U Geminorum. He found much difficulty in the use of Linnell's tables and questions whether they can represent an observed light-curve at all. Poostylnik (Tartu) has also considered the problems of atmospheric eclipses, and computed tables based on theoretical values of the coefficient of absorption.

31. COMMISSION DE L'HEURE

Report of Meetings

PRESIDENT: H. M. Smith.
SECRETARY: R. G. Hall.
INTERPRETER: J. Bonanomi.

First session, 23 August 1967

The loss was noted of two distinguished members of the Commission; A. Danjon and F. de la Puente.

The President reported that the following people were proposed for membership in the Commission: G.M.R. Winkler, E. Proverbio, and G. Hemmleb. All were approved by the members.

The President reported that the following two names were proposed as consulting members of the Commission: U. Adelsberger and J. Terrien. The transfer of L. Essen from member to consulting member was under discussion. This was approved.

The President asked for comments on the *Draft Report*. There being none, the *Report* was approved.

The Director of the B.I.H. reported on the activities of the B.I.H. which are given in the *Draft Report*. *H. M. Smith*, President of the Directing Board of the B.I.H., reported that it is not a function of the B.I.H. to report or note legal times of the various Governments. There was general agreement that this was the case. The report of the Director of the B.I.H. was approved with thanks to B. Guinot.

Resolutions that had been submitted to the President were then considered.

A resolution from the French Commission of Geodesy and Geophysics concerning the form of the emission of time signals from HBG, Switzerland, was taken note of by the Commission. The resolution reads:

> Le Comité National Français de Géodésie et Géophysique a pris connaissance avec grand intérêt du programme d'émissions de signaux horaires continus diffusés par l'émetteur suisse HBG sous contrôle de l'Observatoire Cantonal de Neuchâtel. La poursuite de ces émissions serait de la plus grande importance tant pour les stations séismologiques fixes que pour les travaux de séismologie expérimentale. Le Comité souhaite que des marques d'identification de l'heure soient ajoutées au programme d'émission.

Resolutions on the sign convention of time and frequency differences were submitted by *G. M. R. Winkler* and *B. Guinot*. After some discussion a working group of B. Guinot, G. M. R. Winkler, W. Markowitz, and L. Essen was set up to draft a single resolution incorporating the points raised.

A resolution on the need for carrying atomic clocks, in aircraft, to synchronize clocks world-wide was submitted by *N. Stoyko*. Discussions on this resolution were postponed, and the subject was dealt with at a Joint Meeting of Commissions 4 and 31. (Resolution 4.)

Second Session, 28 August 1967

At an earlier joint meeting (26 August 1967) of Commissions 4 and 31, it was proposed that the following officers of Commission 31 be nominated: President, F. Zagar; Vice President, G. M.R. Winkler; Organizing Committee, D. Belocerkovskij, J. Bonanomi, H. Enslin, W. Markowitz, H. M. Smith, and M. Torao. All votes were in favour of this.

After extended discussion a proposal by *G. M. R. Winkler* and *R. G. Hall* that the Commission should recognize Modified Julian Day Numbers was rejected. It was proposed that this be studied further for future meetings.

Resolution 1 on the Julian Day Number to be used for reporting 5 and 10-day values was submitted by *B. Guinot* and approved unanimously.

L. Essen reported on the uncertainties of Atomic Standards and reported the accuracies, the advantages, and disadvantages of the various types. He suggested direct tests of relativity effects on atomic standards.

J. Bonanomi reported on comparisons of national standards. He referred to flying clock experiments which facilitate time intercomparisons between standards, and frequency comparisons between different types of atomic standards located at the various laboratories. He finds all laboratory models of atomic standards to agree to at least 3×10^{-12}. He reported various determinations of frequency of Hydrogen Masers.

Discussion on the above two papers was postponed as the meeting was adjourned.

Third Session, 29 August 1967

The meeting was opened by the President who read Resolution 2 on the sign conventions in expressing time and frequency differences, which had been drafted by the working group. After some discussion the resolution was passed unanimously.

R. G. Hall presented the current results of the U.S. Naval Observatory Moon-position program. Preliminary results of ET0-UT2, corrected for irregularities of the limb of the Moon, were presented as determined from over 5000 plates taken at 17 observatories located internationally. It was pointed out by *W. Markowitz* that the use of artificial satellites has probably supplanted the use of the dual-rate Moon-position camera for geodetic positions.

W. Markowitz presented the progress during the last three years in synchronizing clocks by means of VLF, Omega, Loran-C, artificial satellites, and by flying clocks, both by landing and by automatic means. Certain specialized methods of synchronization using the Moon and meteor bursts as reflectors were mentioned.

S. Puskhin presented data on the use of atomic standards in the USSR and gave the calculated frequency differences between the USSR standard (Hydrogen Maser) and standards in France, the United Kingdom, and the USA. These were 1 or 2 parts in 10^{11}. The comparison was made via the VLF stations GBR and NBA.

In a general discussion of caesium beam atomic standards it was pointed out by *H.M. Smith* that the short beam standards of the Hewlett Packard Company are independently made and adjusted and hence can be averaged to obtain a mean frequency. *W. Markowitz* and *J. Bonanomi* pointed out that these modern short beam tubes were as accurate as the old long beam tubes. *L. Essen* suggested that further work is needed on new long beam tubes to increase the accuracy of caesium standards.

The meeting adjourned.

Fourth Session, 29 August 1967

The President opened the session by introducing *M. Cohen* from Commission 40 (Radio Astronomy) who discussed the use of long base line interferometer measures of radio sources in time-keeping. Present systems in use allow relative time measurements to 1 microsecond. It is hoped that future systems would give a better determination of the rotation of the Earth than can be obtained with optical equipment. In the discussion it was pointed out that this equipment is very expensive and that different air-mass paths of the different stations would introduce systematic effects, perhaps larger than for optical equipment.

H. Torao presented results of many careful experiments on the propagation of radio waves. He pointed out that it would be desirable for all to use the same velocities of propagation for the determination of the travel time of time signals.

F. Zagar reported on measurement techniques of radio time signal comparisons. It is necessary to calibrate very carefully the delay introduced by the receiving equipment.

After thanks to the interpreter and secretary, the President closed the meeting.

Resolutions adopted by Commission 31:

(1) La Commission 31 de l'Union Astronomique Internationale recommande que les données à l'usage des services horaires ayant le temps universel comme argument

(*a*) soient rapportées, dans la mesure du possible, à 0^hTU,

(*b*) si elles sont tabulées de 5 en 5 jours, soient rapportées aux dates juliennes se terminant par 4,5 et 9,5,

(*c*) si elles sont tabulées de 10 en 10 jours, soient rapportées aux dates juliennes se terminant par 9,5.

(2) 1. To avoid any possible error in the sign of a difference in time between clocks, or in frequencies between oscillators, algebraic quantities should be given.

2. The following definitions and conventions may be used in conjunction with the algebraic expressions:

2.1. Let A denote either a clock or its reading; similarly for B. At some instant let

$$A - B = D \tag{1}$$

Then D is positive if reading A is greater than reading B, and D is negative if reading A is smaller than B.

2.2 Let the frequency of an oscillator a be f_a; similarly for oscillator b and frequency f_b. Let

$$f_a - f_b = \Delta f. \tag{2}$$

Then a is *higher in frequency* than b if Δf is positive. The frequency of a is *lower* if Δf is negative.

2.3 The *deviation from nominal* of an oscillator a is defined as

$$d = f_a - f_n \tag{3}$$

where f_n is the nominal frequency. The frequencies f_a and f_n are expressed in terms of the second equal to 9 192 631 770 periods of caesium radiation.

2.4. The *fractional difference in frequency* between two oscillators H and K, with respective frequencies f_H and f_K, which have the same nominal frequency, f_n, is computed from

$$S = (f_H - f_K)/f_n \tag{4}$$

and is conventionally designated

$$H - K = S, \tag{5}$$

The same sign conventions apply as in 2.2.

2.5. A time comparison between a clock and a time signal should be expressed as in section 2.1.

2.6. A frequency comparison between an oscillator and a radio frequency emission should be expressed as in sections 2.2, 2.3, and 2.4.

BUREAU INTERNATIONAL DE L'HEURE. SUPPLEMENTARY REPORT

DIRECT REDUCTION OF INDIVIDUAL MEASURES OF TIME AND LATITUDE

Principles

As it could be easily foreseen from the new transmission forms of data to the BIH and the IPMS, our aim was to compute directly the coordinates of the pole x, y and UT1 from the individual values (for one group or one night) of the latitude ϕ and UT0.

This computation is only feasible if it is possible to take into account the systematic errors attached to each instrument. By applying the method described in the 'Draft Reports' we reduced the data for every 0·05 year for 1964, 1965 and 1966. It appeared that the systematic errors are stable enough to allow a one-year extrapolation.

Methods of Reduction

We are using the following procedure:

(I) Reduction for every 0·05 year by use of conventional system of initial latitudes, longitudes and weights which realise the 'mean observatory' (as explained in the 'Draft Report'). Every instrument is treated independently (80 instruments; among them 21 are measuring simultaneously time and latitude).

(II) Analysis of the residuals r of ϕ and UT0. These residuals are represented by the formula.

$$r = a + b \sin 2\pi t + c \cos 2\pi t + d \sin 4\pi t + e \cos 4\pi t \quad (1)$$

(t in Besselian years)

(III) Reduction for every 5 day interval, starting from the individual values of ϕ and UT0, corrected according to formula (1). As the corrected values are statistically equivalent, their repartition among the various observatories does not matter.

Reduction I maintains the 'mean observatory'. It is regularly made as soon as *all* the data are received.

Analysis II is to be made periodically (at least twice a year) to keep fresh values of the systematic corrections.

Reduction III must be used for the publication of the 'definitive time' (heure definitive) with very short delay. It allows, if necessary, not to wait the results of any observatory which could have been delayed.

Obtained Results

This method was started in January 1967, with the systematic corrections obtained for 1966. It was used actually for the publication of the 'definitive time', starting from June 1967 (circular D10).

Table 1 gives the unsmoothed values of the results for every five-day interval, with their standard errors. Every set of values for one five-day interval is independently obtained from the others.

It seems that the external agreement of the values of x, y and UT1 is not so good as it could be hoped from the standard errors (1ms in UT1, 0".015 in x and y). That could be due to the existence of short-period irregularities that we are studying.

Future computations of the BIH

For the present time, coordinates of the pole are computed from latitude measurements only and published earlier than UT1. Therefore we must correct the results of the above computation to bring them to the system of polar coordinates already published.

Starting from January 1968, UT1 and coordinates of the pole simultaneously obtained from reduction III will be published every month with a delay of the order of one month after the last observations (for instance, results of January will be published at the beginning of March.).

Smoothed values will be published as the current results. However, computed values of five-day intervals will be published later in an annual report.

B. GUINOT

Tableau 1

Calcul des coordonnées du pôle (x, y) et de TU1-TUC par usage simultané des mesures individuelles de temps et de latitude

Date (0^h TU)	x	y	TU1-TUC en $0.^{s}0001$	écarts-type sur x	écarts-type sur y	écarts-type sur TU-TUC
1967 jan. 4	+ 0″.124	− 0″.044	0152	0″.016	0″.014	$0.^{s}0011$
9	99	72	0172	20	17	15
14	124	48	0158	17	16	12
19	139	76	0204	13	14	10
24	77	153	0216	19	19	15
29	102	101	0171	17	15	12
fév. 3	127	109	0182	15	16	12
8	109	67	0218	16	16	12
13	72	95	0255	16	15	12
18	81	55	0257	15	16	12
23	85	91	0222	14	14	11
28	64	45	0189	18	16	12
mars 5	67	88	0227	16	15	14
10	53	64	0212	14	13	10
15	71	50	0177	14	15	10
20	48	78	0211	17	17	12
25	19	68	0184	13	13	10
30	41	48	0191	17	16	13
avr. 4	21	60	0148	17	15	12
9	55	38	0159	14	14	11
14	36	68	0166	12	12	9
19	19	51	0142	12	13	10
24	39	37	0094	17	15	12
29	+ 28	25	0097	13	13	10
mai 4	− 15	47	0142	15	14	10
9	+ 13	− 9	0065	13	15	12
14	− 2	+ 28	0045	16	14	11
19	+ 9	− 10	0095	14	14	11
24	+ 28	− 42	0078	14	13	10
29	+ 58	− 54	0095	18	16	13
juin 3	+ 13	− 34	0089	14	13	10
8	18	− 27	0115	17	13	13
13	22	− 17	0220	13	13	11
18	2	− 50	0221	14	13	13
23	60	+ 14	0211	14	13	11
28	35	− 9	0273	14	13	11
juil. 3	29	− 25	0323	14	13	11
8	25	− 32	0409	16	14	12
13	46	+ 14	0425	16	14	12

SESSIONS MIXTES 31 ET 4

Held on 24, 25 and 26 August 1967

CHAIRMAN: H. M. Smith.
SECRETARIES: R. G. Hall and J. Kovalevsky.
INTERPRETER: J. Bonanomi.

I. DEFINITION OF THE SECOND

H. Barrell presented the recommendation made by the *Comité Consultatif pour la Définition de la Seconde* (CCDS) for an atomic definition of the second. The recommendation reads as follows:

Recommandation S-1

Le Comité Consultatif pour la Définition de la Seconde

Récommande

que la seconde, unité de base du Système International d'Unités, soit définie dans les termes suivants:

La seconde est la durée de 9192631770 périodes de la radiation correspondant à la transition entre les deux niveaux hyperfins de l'état fondamental de l'atome de césium 133,

que la seconde telle qu'elle fut définie par décision du Comité International des Poids et Mesures à sa session de 1956 soit désignée sous le nom de 'seconde des éphémérides'.

H. Barrell pointed out that two other recommendations call for further investigations of atomic frequency standards and for joint meetings of representatives of the various international scientific or technical Unions concerned, to study the application of this new definition under the coordination of the *Comité International des Poids et Mesures.*

He drew attention to the declaration of the CCDS which develops these points insisting, in particular, on the necessity of maintaining Ephemeris Time.

W. Markowitz and *N. Stoyko*, representatives of the IAU on the CCDS, expressed their accord with this new definition.

J. Terrien described the procedure that would be followed before the adoption of the final text of this recommendation by the General Conference of Weights and Measures. The form of the text may be modified to exclude the Ephemeris Second from the International System, but the Conference will not interfere with the astronomical use of Ephemeris Time. The International System is a consistent and limited system and does not exclude other units. Since there will not be a formal abrogation of the ephemeris second, the astronomers should be free to use Ephemeris Time.

W. Markowitz expressed the wish that, in the case of a statement of the General Conference that the ephemeris second is not part of the International System of Units, it should be formally recognized that it remains part of the IAU system of astronomical constants. This suggestion was discussed and it was generally agreed that such a statement should be transmitted to the General Conference.

W. Markowitz drafted a resolution concurring with the proposed definition of the second, but pointing out that the ephemeris second is part of the IAU system of astronomical constants. After discussion this resolution was modified and adopted by the commissions. It was later approved by the General Assembly in the following form:

Resolution 1

Commissions 4 and 31 recommend that the following views and information be transmitted to the General Conference of Weights and Measures. (See Resolution No 5, p. 182.).

There was more discussion on the third recommendation of the CCDS on the future activity of the *Comité International des Poids et Mesures* in the application of the new definition. *J. Terrien* assured the members of the commissions that there would be no modification of the third recommendation of the CCDS, which reads as follows:

Recommandation S-3

Le Comité Consultatif pour la Définition de la Seconde

Récommande

que le Comité International des Poids et Mesures suscite une réunion comprenant des représentants de diverses organisations telles que Bureau International de l'Heure, Union Astronomique Internationale, Union Géodésique et Géophysique Internationale, Union Radioscientifique Internationale, Union Internationale des Télécommunications (Comité Consultatif International des Radiocommunications), pour étudier les problèmes soulevés par l'application des décisions prises concernant la nouvelle définition de l'unité de temps.

II. EPHEMERIS TIME

D.H. Sadler made the following remarks on the various time-scales now available.

The atomic time-scale, that he proposes to call *atomic clock time* (temps atomique intégré), will be a valuable means of measuring intervals of time and will replace E.T. as the reference in the study of the irregularities of the rotation of the Earth.

The measure of Ephemeris time (and hence A.T.–E.T.) is subject to four kinds of limitations.

a) *Limitation of principle*: it can be obtained practically only through observations of the Moon.

b) *Limitation of physical knowledge*: the Sun-Earth-Moon system is not conservative.

c) *Limitation of theory*: errors in the theory of the motion of the Moon and in the astronomical constants.

d) *Limitation of observations*: arising in the practice of observing the Moon with respect to the stars, or on the meridian.

L. Essen was not in favour of the expression 'atomic clock time' and doubted whether it will be possible to measure significantly A.T.–E.T. *T.C. van Flandern* pointed out that this would allow the identification of the tidal friction.

A short spirited discussion of the inherent accuracies of the various time scales followed. *G.M.R. Winkler* and *J. Terrien* stressed the opinion that both atomic and ephemeris times are physical uniform times. Only the practical precision with which they can be measured is different.

In a later meeting, *D.H. Sadler* proposed a resolution emphasizing that the new definition of the second does not suppress the necessity of observing U.T. and E.T. This was supported by *G.M. Clemence*, and after some discussion the resolution was approved by the commission and was subsequently adopted by the IAU General Assembly.

Resolution 2

(See Resolution No. 5, p. 182).

J. Terrien indicated that the International Committee of Weights and Measures will probably approve such a resolution.

III. RELATIVISTIC EFFECTS

G.M. Clemence, in his introductory remarks, stressed the fact that, in the last 10 years, experiments have shown the reality of the slowing of clocks in motion with respect to others (1) as well as in a stronger gravity field (2). However, some attacks on Einstein's theory of general relativity have recently been made, for example by R. H. Dicke, who found a minor flattening of the Sun that would suffice to explain part of Mercury's motion of the perihelion. On this point, *C. G. McVittie* criticized the theory of the Sun's interior used by Dicke and suggested that the observation of the solar disc should be continued during two solar cycles.

G.M. Clemence reported on work with *V. Szebehely* in which they computed the difference between the parameters t and s of the basic formula.

$$\frac{dt}{ds} = 1 + \left(\frac{1}{r} - \frac{1}{2a}\right) \times \frac{2GM}{c^2}$$

where a is the semi-major axis of the Earth's orbit, r the radius vector, GM is the heliocentric constant of gravitation, and c the speed of light.

They provisionally identified s with the proper time of an orbiting clock and t with Ephemeris Time. They deduced that the time indicated by the clock departs periodically from its mean value by 0·0017 second.

G.C. McVittie commented on this identification which he had also made using the Schrodinger approximation (3).

N. Stoyko stressed the importance of measuring A.T.–E.T. regardless of their theoretical significance. He also pointed out that O. Costa de Beauregard gave formulae analogous to Clemence in 1957 while Fokker found the same numerical results by numerical integration.

W.H. McCrea commented on the formula giving the relativistic red shift and pointed out that one cannot observe the gravitational potential, but only its gradient. One has:

$$h\nu = h\nu_0\left(1 - \frac{M}{r}\right)$$

ν_0 is observed by an observer linked with the emitter, while ν is the frequency at zero potential.

L. Essen recommended an experimental approach to the problem and stressed the difficulty introduced by the fact that the reference frames linked with various clocks are different.

G. Becker commented upon Aoki's and Clemence's formula and wished to modify it, taking into account the presence of the clock on the Earth. He believed that the Moon revolving around the Earth is comparable to an atomic clock and that, therefore, E.T. observed with the Moon should be identical to A.T. and will not present the annual term given by those authors.

S. Aoki agreed with this point, but pointed out that Ephemeris Time is being defined with the Sun and, consequently this term is real in the definition. He believes that the theory of the motion of the Sun should include relativistic effects in order to define a truly uniform coordinate time E.T.

W. Markowitz pointed out that from a practical point of view it is unlikely we will soon be able to check any annual variation of atomic time. The problem may become important when a precision of 10^{-14} is attained in atomic clocks.

M. Missana remarked that in the computation of the effect, it should be noted that in relativity theory, the distances must not be identified with the radius-vectors as in classical mechanics.

In a written contribution which arrived too late for presentation at the meeting, *R.H. Dicke* deduced a variation of the rate of an atomic clock of $6{\cdot}6 \times 10^{-10}$ (total increase in clock rate from winter to summer).

References

1. Frische, D.H., Smith, J.H. 1963, *Am. J. Phys.* **31,** 322.
2. Pound, R.V., Rybka, G.A. 1960, *Phys. Rev. Lett.* **4,** 337.
3. McVittie, G.C. 1932, *Mon. Not. R. astr. Soc* **92,** 868.

IV. COORDINATED UNIVERSAL TIME

G.M.R. Winkler put forward a proposal to increase the tolerance of the representation of UT2 by UTC to 300 ms and to authorize the Director of the Bureau International de l'Heure (BIH) to change the frequency off-sets at the beginning of any month.

H.M. Smith recalled that two systems of coordinated time exist: the one of the BIH and the one in use in the USSR. They are similar in principle.

D. Belocerkovskij confirmed that the coordination with the BIH in frequency will continue, but that the maximum tolerance in UT2–UTC is limited to 50 milliseconds.

G.M.R. Winkler stated that he would withdraw his proposal if the present systems of the BIH were adopted universally.

A general discussion on the proposal by Winkler took place. The following points were raised:

The navigators would accept a tolerance of 200 ms (instead of the present 100 ms) but not larger (*D.H. Sadler*).

No major change in the principles of UTC should be introduced since it has proved useful (*H.M. Smith, W. Markowitz, L. Essen*).

An increase of the tolerance, as well as the possibility of changing the frequency other than once a year, would facilitate the work of the BIH and permit less frequent changes (*G.M.R. Winkler, B. Guinot, N. Stoyko*).

Some improvements in the UTC system appear necessary to satisfy the needs of the highest-precision users (*G.M.R. Winkler*).

Such changes should be accepted by all users and hence should be referred to UIGG, URSI and CCIR (*W. Markowitz*).

B. Guinot asked for statements by users on whether they prefer offsets in frequency or steps in time.

The discussion was continued in a subsequent session. After some informal discussions *D. Belocerkovskij* indicated that, if possible, the tolerance of the USSR system will be raised to 100 ms. As a consequence it was decided to stress the desirability of a single uniform system and it was agreed that it was inopportune to propose an increase in the tolerance. *B. Guinot* proposed the following resolution which was adopted unanimously.

Resolution 3

L'Union Astronomique Internationale

reconnaissant

(a) la nécessité de la coordination des émission de signaux horaires

(b) la nécessité de réduire le nombre des décalages de fréquence et d'ajustements par sauts

recommande

(1) que la coordination des émissions de signaux horaires dans le système appelé 'temps coordonné' (TUC) défini par le CCIR à sa réunion plenière de 1966 (Recommendation 374-1, CCIR Documents of the XIth Plenary Assembly, Oslo 1966, Volume III, p. 281, 282), soit universellement étendue,

(2) que la possibilité d'accroître la tolérance pour la différence TU2–TUC soit étudiée par les organisations concernées.

Two other resolutions, one presented by *N.* and *A. Stoyko* and the other by *G.M.R. Winkler* were discussed. Their aim was to define and to establish a unique international atomic time scale under the responsibility of the BIH and to encourage national time services to make a broader use of atomic frequency standards in order to build local time scales.

B. Guinot indicated that the BIH had experience in constructing an atomic time scale (A.3) by integration and that this technique gives rise to errors due to the instability of the standards. It is therefore necessary to compare as many time scales as possible. These time scales should be based on counts and their origin should not be linked to a given phenomenon such as an epoch in U.T.

After discussion the final wording of the resolutions was presented and adopted.

Resolution 4

Commissions 4 *et* 31 *recommandent*

1. l'établissement d'une échelle du temps atomique en prenant pour fréquence de l'étalon à césium, 9192631770 Hz et qui soit d'accord approximativement avec le TU2 à 0 Heure 1er Janvier 1958, compte tenu du changement des longitudes conventionnelles,

2. que le BIH coordonne la publication des résultats de la comparaison des horloges atomiques de différents établissements en utilisant le transport des horloges atomiques par avion.

Resolution 5

Commissions 4 *and* 31 *of the International Astronomical Union recommend:*

(i) That the BIH compute an International Atomic Time Scale comprised of independent time scales of the major national time services.

(ii) That this time scale, based on the experiences gained with the experimental scale A3, be computed similarly and published in the form of corrections to the contributing time scales with respect to the international scale.

(iii) That the frequency averaging and integrating as well as time averaging should be used in the computation, but greater weight should be given to time averaging wherever possible.

(iv) That the national time services be encouraged in turn to maintain their time scales independently, based on a broader use of atomic frequency standard resources, and that they also study in cooperation with the BIH the conditions under which different time scales in different countries can be brought into coincidence with that of the BIH.

This resolution was adopted unanimously.

B. Guinot wished to change the name of A.3 to 'International Atomic Time Scale'.

V. PUBLICATION OF E.T.–U.T.

B. Guinot proposed that values of ΔT should be published as soon as they are available. The BIH could centralize and disseminate this information.

R. Duncombe indicated that this is done in the Astronomical Ephemeris but believed that it might be done faster by other channels. As far as investigations are concerned, the table of AT–UT2 that is now given in these ephemerides should suffice.

J. Kovalevsky stressed the necessity of the centralization and dissemination of partial results obtained in different places by various techniques and *B. Guinot* expressed the belief that such a centralization would encourage new observations, accelerate their transmission and encourage new studies.

D.H. Sadler and *R. Duncombe* thought it would be difficult for one agency to publish observations as the work would be considerable.

It was suggested by *H.M. Smith* that correspondence between the director of the BIH and other interested organizations on this matter would be useful.

VI. MODIFIED JULIAN DAY

R.G. Hall and G.M.R. Winkler noted that there is a confusion in the use of Julian Day Number due to a widespread use of various continuous day counts. They suggested a Modified Julian Day number, already in informal use, that equals the Julian day minus 2400000·5, in addition to the Julian Day Numbers.

Several other propositions were made in order to have a more distinctive name (CDC: continuous day count). *R. Duncombe* objected because of the possible confusion between the two systems. He did not wish to give official notice to this new system. *G. Wilkins* feared confusion for people using other systems.

No definite opinion of a majority appeared and, after further consideration at subsequent meetings of both commissions, the proposal was rejected.

33. COMMISSION DE LA STRUCTURE ET DE LA DYNAMIQUE DU SYSTEME GALACTIQUE

Report of Meeting, 23 August 1967

PRESIDENT: B.J. Bok.
SECRETARY: Th. Schmidt-Kaler.

The Draft Report was discussed and adopted with minor revisions. The proposed Membership of the Organizing Committee was discussed. Nominations were announced for 27 new members of Commission 33.

The Proposal for a separate Commission on Dynamics of the Galaxy (Nahon) was brought forward. Contopoulos reported that the participants of the Colloquium on New Methods in Stellar Dynamics (Besançon, Sept. 21–24, 1966) had discussed the matter and that they had agreed that a single unified Commission on the Structure and Dynamics of the Galaxy should be maintained. A vote was taken and there appeared to be unanimous agreement in favor of a single Commission.

The question of the superscripts for galactic coordinates, l^{II} and b^{II}, was discussed at length. *van Herk* recommended that these should be retained indefinitely, in the same manner as Catalogue designations. He commented that, eventually, there might well be a third system of galactic coordinates. Others disagreed. No firm action is to be taken at this stage and the continued use of the superscripts is recommended for another three year period. The subject is to be brought up again in three years, when, hopefully, a firm decision can be made by Commission 33.

A proposal to urge the preparation of new southern hemisphere charts was presented by *Blaauw*, and there was general agreement that this be done. *Varsavsky* announced that charts for selected regions near southern radio sources (to $V = 18$) can be provided by the Instituto Argentina de Radioastronomia. *M. Schmidt* said that a limited number of sets of the Whiteoak Extension of the Palomar Sky Survey (red prints only, $\delta = -33°$ to $-45°$) will be available for sale in 1968. Announcements in the leading astronomical journals will be forthcoming. *Haffner* announced that a Photographic Atlas of the Southern Milky Way, obtained from plates made with the Boyden Observatory's 10-inch Metcalf Refractor, complete to $B = 16{\cdot}5$, will be offered for sale by ESO in 1968. Mention was also made of Westerlund's Photo-Visual Atlas of the Southern Milky Way (to $V = 16.5$) made from photographs with the Uppsala Schmidt Telescope at Mount Stromlo Observatory (see, Minutes of the Fourth Meeting). The meeting recommended that Commission 33 should favor the completion of the Lick Observatory Sky Atlas to declinations south of $-45°$. The need for charts of many varieties for the southern hemisphere is obviously becoming an urgent one.

BRIEF PAPERS

U. Haug: Luminous Stars in the Northern Milky Way.
C. Hunter and A. Toomré: Dynamics of the Bending of the Galaxy.

Second Meeting, 24 August 1967

OBSERVATIONAL ASPECTS OF THE SPIRAL STRUCTURE OF OUR OWN AND OTHER GALAXIES

M.S. Roberts: Observed Features of some External Galaxies.
A. Blaauw: Observational Data on the Spiral Structure of the Galaxy.

J. Isserstedt and Th. Schmidt-Kaler: Galactic Structure as Shown by a New Type of Stellar Aggregate.
J.S. Miller: Radial Velocities and Kinematics of Galactic H II Regions.
M. E. Dixon: An Observational Approach to Interstellar Gas Dynamics.

Third Meeting, 24 August 1967
(Dr Contopoulos presiding)

THEORETICAL APPROACHES TO SPIRAL STRUCTURE

C.C. Lin: The Density Wave Theory of Galactic Spirals.
S.B. Pikelner: Magnetic Approaches to Spiral Structure.
M. Fujimoto: Self-Gravitating Spiral Models of the Galaxy. (presented by *G. Contopoulos*)

Fourth Meeting, 28 August 1967

SHORT PAPERS

I.R. King and G. Contopoulos: A Report on the IAU Colloquium on the Gravitational *n*-Body Problem (Paris, August 16 to 18, 1967).
F.K. Edmondson and A. Young: The Solar Motion for 11th Magnitude Stars.
D. Crampton: Be Stars and Galactic Rotation.
F.C. Bertiau, M.F. McCarthy and P.J. Treanor: A General Catalogue of B-Emission Stars.
B.E. Westerlund: A Photo-Visual Atlas of the Southern Milky Way (V=16.5).
H. Haffner: A Photographic Atlas of the Southern Milky Way.
A. Ardeberg and C. Roslund: The Lund Galactic Program in Scorpius.
A.P. Henderson: A 21-cm Study of the Large-Scale Deformations of the Galactic Plane (presented by *G. Westerhout*).
J. Rickard: Radio and Optical Studies of the Perseus Arm (presented by *G. Westerhout*).
W.J. Luyten: On the Faint End of the Luminosity Function.

Note: At the conclusion of the presentation by Dr Luyten, the President of Commission 33 congratulated the speaker heartily on these latest results of a life-time devoted to research on intrinsically faint stars.

K.A. Innanen: The Groombridge 1830 Moving Group.
C.M. Varsavsky: Atomic Hydrogen and Interstellar Dust.
L. Marochnik: On the Influence of Gravitating Condensations of Interstellar Gas on the Distribution of Peculiar Velocities of the Stars.
H. Elsässer: Programs for Southern OB Stars.
J. Dommanget: La Cinematique des Couples Stellaires Visuels dans le Voisinage du Soleil.

ADDITIONS TO THE DRAFT REPORTS

P. 688, add at the end of the second paragraph in the section b):
Apriamašvili has also classified the spectra of about 3900 stars in a large region near $l^{II}=40°$.

P. 689, line 17 from the bottom, add:
Similar work has been done by Bartaja and Haradze (**36**) who classified 376 OB stars in the region of the association III, IV, VII Cas, and 1240 stars of all classes in the region of I Vul.

P. 689, line 2 from the bottom:
The work contains now three catalogues of about 7600 stars for one region at the anticentre and two regions in Taurus.

P. 692, add references:

21a. Apriamašvili, S.P. 1966, *Abastumani Astroph. Obs. Bull.* **35**, 3.

36a. Bartaja, R.A., Haradze, E.K. 1966, *Abastumani Astroph. Obs. Bull.* **35**, 105, 109.

39. Also 1966, *Abastumani Astroph. Obs. Bull.* **35**, 53, 72.

P. 694, end of first paragraph, add:

But Slavenas, and Bogdanovič, (**9**a) have found only a three-arm structure with a pitch angle of 81°.

P. 697, add the reference:

9a. Slavenas P., Bogdanovič A. 1966, *Vilnius Astr. Obs. Bull.* **17**, 24.

P. 715, line 21, read:

Kuzmin (**165**) has considered also the application of the virial theorem in its generalized form to the nonhomogeneous ellipsoids, the conditions for the existence of symmetry planes in stellar systems and the formation of the Schwarzschild velocity distribution under the influence of the irregular gravitational forces. The theoretical explanation for the observed ratio of the velocity dispersions in the z and R directions has been found.

P. 715, at the end of 'Miscellaneous Problems':

Antonov (**169**) has shown that for a system with n degrees of freedom there exist n invariants of a continuum element in the phase-space.

P. 717, additions to references:

59. (Ogorodnikov) 1966, *Dokl. Akad. Nauk SSSR*, **170**, 297.

98. (Maročnik) 1966, *Astroph. Inst. Bull. (Dušanbe)* **48**, 3.

P. 719, additions to references:

169. Antonov, V.A. 1965, *Vestnik Leningrad Univ.* **13**, 136.

COMMITTEE OF 'SELECTED AREAS'

Report of Meeting, 25 August 1967

CHAIRMAN: T. Elvius.
SECRETARY: K. Lodén.

Draft Report

The Draft report was accepted without corrections.

Committee of 'Selected Areas'

It was decided that the present Selected Area Committee of Commission 33 (*T. Elvius*, chairman, *Haradze*, *McCuskey* and *Plaut*) should continue its work for another period of three years and that *Fehrenbach* should be included in the Committee. This was proposed by *B.J. Bok* after consultations with the Committee and the new President of Commission 33. Bok also pointed out that Working Groups of this kind are by definition of a temporary nature and the question of continuation should be reconsidered every three years.

Reference Lists and Maps for 'Selected Areas'

These questions were the main subjects discussed at the meeting. *Terzan* presented a detailed proposal for publication of reference lists and maps in Selected Areas. In view of the importance for various investigations that these fields have, it would be of value to have a catalogue in which chronological reference lists are given. For each reference the field area, the number of stars investigated and a description of the published data (magnitudes, spectra, proper motions etc.) should be listed. There is also need for identification of the individual stars from various catalogues within every Selected Area. Maps of Selected Areas with numbers from a suitable catalogue covering a relatively large area around the central star would also be important. Terzan and the staff at the Lyon observatory are prepared to start the work necessary for preparing lists and charts. The cost for the project, and also the time needed, are difficult to estimate.

The question of maps was first discussed. *Elvius* suggested that reproductions from the Palomar Sky Survey maps could be used. Somewhat different opinions (*Blaauw*, *Roman*) about limiting magnitudes caused Elvius and Bok to ask Terzan to start preparing sample charts for a few centers at different galactic latitudes and to distribute them for comments to those interested in this project. Thereafter a largescale project could be planned. *Bouigue* supported Terzan's project and emphasized its importance. Elvius also mentioned that Uppsala has plates which could possibly be used for the southern sky. Terzan accepted the task of preparing trial maps for some areas.

With regard to the *lists*, it was generally agreed that Terzan should start the work. *Haradze*, who, together with Bartaja, has been working on spectral classification in Selected Areas in order to reclassify HD stars, stressed the importance of Terzan's and Bouigue's proposal. The Selected Area Committee members should be considered as advisors. Plaut and the Kapteyn Astronomical Laboratory can assist in a variety of ways.

Photometry

Rybka asked for opinion concerning the usefulness of an extension of his programme on photoelectric standards in Selected Areas to fainter stars. Magnitudes can be measured near $7^{m}.5$, $9^{m}.0$ or $10^{m}.5$ for two stars in each field, one of spectral type A, the other of type K. It was agreed that the inclusion of fainter standard stars would be advantageous, but it would be preferable to add stars of F (or G) type in order to avoid colour effects, as was pointed out by *W. Becker* and *T. Elvius.*

Blaauw talked about research started at Groningen on narrow-band photometry in 1 square degree fields in some Selected Area Zones (0°+15°, +30° in declination).

The quality of the photometry is now bring investigated. There will be a separation of giants and dwarfs. The limiting magnitude is not yet fixed. This work is similar to that commenced by Oja and Häggkvist at the Uppsala Observatory in areas of the northern galactic polar cap.

Astrometry

Murray talked briefly about his programme of remeasuring proper motions in the Radcliffe Catalogue. Good photometry is needed for these stars. For details on this and other astrometric programmes in the Selected Areas, see the report of the specially-arranged Joint Meeting of Commissions 23 and 24 together with the Selected Area Committee, (p. 137 of the present volume).

Cleveland and Parenago fields

It was agreed that tables and maps should be prepared giving the coordinates and boundaries of these fields with indication of positions for the Kapteyn Selected Areas in or near these fields.

33. GROUPE DU TRAVAIL SUR REGIONS CHOISIES, 23 ET 24 SESSION MIXTE

Report of Meeting, 29 August 1967

PRESIDENT: B.J. Bok.
SECRETARY: K. Loden.

At the request of Vasilevskis and Murray, the Presidents of Commissions 23 and 24 planned a Joint Meeting of the two Commissions, together with the Working Group on Selected Areas, for the purpose of discussing the problems that arise in the planning for future large-scale astrometric programs of proper motions.

In his Introductory Remarks, *Murray* reminded the Meeting that modern techniques of automatic measurement of photographic plates for stellar positions, coupled with new and efficient methods of data processing (reductions and analysis) are making it possible to measure very large numbers of stars on many plates with early epochs. Such measurements can be combined, for the determinations of mass proper motions, with results from recent plates. The Kapteyn Selected Areas hold special promise for work on faint stars, since for these centers large numbers of excellent early-epoch plates are available. It is essential that astronomers and observatories interested in the work should coordinate their efforts. *Murray* spoke specifically about the present plan of Royal Greenwich Observatory to remeasure the proper motions for the stars in the Radcliffe Catalogue of Selected Areas. The 26-inch Refractor at Herstmonceux seems ideally suited for obtaining the second-epoch plates.

Vasilevskis gave a survey of the Lick proper motion program, in which the proper motions are referred to galaxies. This program consists of 1246 fields covering the sky from $\delta = +90°$ to $\delta = -23°$. The instrument is the 20″ Carnegie Astrograph. The fields are of size 6° × 6°, on 17″ × 17″ plates, which gives a scale of $55\overset{''}{.}14$ per mm. The first epoch plates have been taken with a blue (photographic) lens in 1947–1954. The second epoch plates will be taken in 1967 and onwards in two colors, blue and yellow. The final catalogue will contain proper motions, magnitudes and colors. The limiting magnitude will be approximately $18\overset{m}{.}5$. The brightest limit is $7\overset{m}{.}5$. All plates will be measured automatically. There will be two initial phases of the project: First, about 150 fields will be treated in order to develop the best routines; Second, the measurements for all fields will be repeated after 1–2 years.

During the discussion that followed the presentations, the southern hemisphere needs were stressed. The new Yale-Columbia Program in Argentina will ultimately extend the Lick Program to far southern declinations, but it will be many years before second-epoch plates with sufficiently long base lines in time can be obtained. The possibility should be explored of obtaining second-epoch plates for the parallax centers of the original Johannesburg, Yale-Columbia, Program, initiated by F. Schlesinger. The 26-inch Yale-Columbia Refractor is now the property of Mount Stromlo Observatory and the second-epoch plates might well be obtainable with the original instrument. It was recommended that any future Working Group should examine how and where these second-epoch plates might be photographed and measured.

It was further noted that many good early collections of northern and southern Selected Area photographs are available at the Harvard, Yale, Mount Wilson, Pulkovo and other Observatories—and that there is also a fine collection of early Selected Area plates at the Kapteyn Astronomical Laboratory. These can be important sources for first-epoch plates for the measurement of proper motions of faint stars.

Plaut summarized the Yale-Columbia program of southern Selected Areas. There are first-epoch plates available for all southern areas and second-epoch plates for the declination zone $-15°$. This latter zone has been measured differentially at Groningen but the reductions are not yet finished.

The second-epoch plates were taken in the early 1930's. If a new series of plates were taken in the near future, then the weight of the results would be increased four-fold. It is obviously important that this be done.

Reference was also made to the northern Selected Area needs and prospects. *Vasilevskis* mentioned that Lick Observatory has a special series of first-epoch plates, made with the Carnegie Astrograph, for all Selected Areas north of − 15°. There are no immediate plans for second-epoch plates, nor for the measurement of the Selected Area material, but, if desired, preference might be given to some of the Selected Area centers. Murray stressed the importance of having available at the earliest possible date absolute (Lick) proper motions for 30 to 40 stars, which would then serve as standards for the reduction of measurements from other Selected Area photographs. It would then become possible to reduce all Selected Area relative proper motions to absolute values on the Lick system referred to faint galaxies. Luyten stressed the importance of star-like quasars as reference objects for the measurement of stellar proper motions.

Luyten spoke briefly about the Minnesota Program for the detection of stars with proper motions in excess of $0\overset{\prime\prime}{.}1$ per year and about the measurements of the proper motions for the faint stars of this sort. 48-inch Palomar-Schmidt plates, covering 40 square degrees each, are searched, and large proper motion stars are found at the average rate of 12 per square degree. An automatic measuring engine is under construction and all reductions will be by modern data-processing techniques.

The difference between the epochs will be 14–17 years. The magnitude interval is 4^m–21^m and the mean error is $\pm 0\overset{\prime\prime}{.}020$. The accuracy for the faint stars is better than for the bright ones. A reduction to absolute proper motions by means of quasars will be performed; it is also intended to have normal galaxies as a reference system. *Luyten* mentioned that Schmidt telescopes are good for astrometric work, but that the measurements should preferably be limited to stars relatively near the center of the plate.

Blaauw spoke briefly about the new McCormick Observatory proper motion survey. Present plans call for a repetition of the Vyssotsky-Williams program under a joint program involving the McCormick Observatory and the Kapteyn Astronomical Laboratory (*Blaauw*). The original plates are all centered on bright (parallax) stars, dimmed to $V = 10{\cdot}5$ or fainter with the aid of a rotating sector. The first-epoch plates go back to the period 1915 to 1930, and these are to be repeated once again, thus providing for a very long base-line in time. Since the central stars (700 fields) all have known proper motions on the FK4 system, absolute proper motions of high precision and on the FK4 system can be found for the 30000 faint stars. In addition to the proper motion measurements an effort will be made to obtain for these stars precise photometric data and, possibly, also objective-prism radial velocities. *Blaauw* urges improved meridian circle observations for the central stars in the 700 McCormick fields.

In conclusion, the Meeting (with 43 in attendance) went on record as favoring the creation of a *Working Group on Proper Motions of Faint Stars*, with Plaut as Chairman, and as members: A.N. Deutsch, Dieckvoss, T. Elvius, Luyten, McCuskey, Murray and Vasilevskis.

34. COMMISSION DE LA MATIERE INTERSTELLAIRE ET DES NEBULEUSES PLANETAIRES

Report of Meeting, 26 and 29 August 1967

PRESIDENT: S.B. Pikelner.
SECRETARY: T.K. Menon.

Business Meeting

At the first meeting on 26 August the Draft Report was approved subject to minor corrections or additions.

The President announced that the present Vice-President D.E. Osterbrock will be the next President and the next Vice-President will be F.D. Kahn. The following were suggested as the members of the Organizing Committee: L.H. Aller, J.M. Greenberg, L. Rosino and S.B. Pikelner. S.A. Colgate was suggested as a Consulting Member.

Scientific Meetings

INTERGALACTIC GAS

R.A. Syunyaev reviewed the properties of intergalactic gas from the theoretical and the observational points of view. He suggested that the intergalactic gas is most probably in the form of ionized hydrogen and helium. He pointed out that the observations so far have only suggested an upper limit of $P_{gas} < 4 \times 10^{-29}$ gms/cm^3 which is still too high compared to the critical matter density in the universe of about 2×10^{-29} gm/cm^3.

Y.B. Zel'dovič reported on attempts to measure intergalactic radiation in the bands λ1050–1340 Å and λ1225–1340 Å using two detectors on the Interplanetary Probe Venus II. The first band includes unshifted Ly α and the second band observes gas at a great distance such that λ1216 $(1 + Z)$ should go into that region. Theory predicts intensity as a function of P_{gas} and temperature. The upper limit from observations $I < 10^{-7}$ ergs/cm^2/sec/Steradian (Kurt and Synyaev).

G.B. Field stressed that the information on neutral intergalactic gas depends on the cosmological hypothesis. His calculations on the abundance of H_2 molecules indicated that it is low but not negligible. He emphasized the need to repeat Kohler's observations of the 21-cm absorption in the Virgo Cluster.

The Nancay group reported on their repetition of Kohler's observations with, so far, negative results.

J. Baldwin discussed the attempts at Cambridge to detect the step in the spectrum at about 1420 Mc/s expected from the 21-cm radiation of intergalactic neutral atomic hydrogen. The upper limit for H *I* is $\leqslant 8 \times 10^{-30}$ gm/cm^3.

Field pointed out that at the low densities involved the spin temperature of atomic hydrogen is in equilibrium with the cosmic background and hence there is no reason to expect significant radiation from them.

C. Varsavsky reported on unsuccessful attempts by W.G. Popped to detect 21-cm radiation from Hoffmeister's extragalactic absorbing cloud believed to be at a distance of 10^5 to 10^6 parsecs.

Syunyaev remarked that, since the U.V. spectrum of quasars is relatively flat, the absorption by H *I* is more critical and hence the upper limit can be better determined in the case of such observations.

MAGNETIC FIELD IN THE GALAXY

R.D. Davies presented a review of the present status in our knowledge of the magnetic field in the spiral arms.

Recent measurements of the polarized background radio emission of the Galaxy and of the Faraday rotation of the linearly polarized radiation from extragalactic radio sources on passing through the Galaxy, have made a significant contribution to our understanding of the magnetic field in the Galaxy. These data can be combined with information on the polarization of starlight, on the synchrotron emission from the galactic disk and on the Zeeman effect in interstellar clouds to provide a new picture of the magnetic field structure in the local spiral arm.

The field in the local arm appears to be directed towards $l^{II} = 70°$ and there is evidence that the field is converging in this direction. In addition the field at intermediate latitudes above and below the plane is directed in opposite senses. The field below the plane is directed towards $l^{II} = 70°$ approximately. At present it is not clear whether this field is associated with the spiral arm as a whole or is a more local phenomenon. Irregular magnetic field structures are found in the spiral arm. The largest are the 'spurs' of synchrotron emission which contain fields aligned parallel to the outer edge of the 'spurs'. In addition a fine structure seems to be intrinsic to the field. Studies of the optical polarization of starlight in galactic clusters and of the depolarization of radio emission from the extragalactic sources at low latitudes indicate a structure of about 0·3 pc.

An estimate of the mean strength of the magnetic field along the spiral arm can be obtained from the Faraday rotation data of extragalactic sources. The value which is about 5×10^{-6} gauss is dependent on the assumed clumpiness of the electron density in the spiral arm. This is much lower than 3×10^{-5} gauss normally estimated for the field strength necessary to produce the observed synchrotron emission from the galactic disk. The difference between these values again is consistent with a tangled field on the small scale which nevertheless has a net mean value.

Seymour discussed his spherical harmonic analysis of the data on Faraday notation, optical polarization and background polarization. His results are reported in M.N.

H.C. Van de Hulst pointed out that one can make estimates of magnetic field based on cosmic ray data alone.

S.B. Pikelner remarked that a tangled magnetic field tends to increase the velocity dispersion which is not observed. Hence the reversals of magnetic field will have to be on a large scale.

J.R. Shakeshaft suggested that, if the curvature of the radio-frequency spectrum of the galactic background is related to the steepening of the cosmic ray electron energy spectrum beyond 3 GeV, then we get a value of the magnetic field to be 5×10^{-6} G.

DYNAMICS OF GAS CLOUDS

G.B. Field reviewed the results of the Field-Saslaw model of interstellar cloud mass spectrum. For $\sigma\varepsilon$ independent of mass, the mass spectrum is predicted to be a power law $m^{-\beta}$, with $\beta = 3/2$. Recent work by Field and Hutchins shows that if $\sigma\varepsilon$ depends on mass as expected, if all clouds have the same density and if there is equi-partition of energy, $\beta \simeq 3/2$, contrary to initial expectation.

These predictions are compared with observations. 21-cm observations by Heiles disclose some 800 'cloudlets' ($1 < m < 30 M_{\odot}$) and 12 'concentrations' ($300 < m < \infty$). There are no objects in the range 30–300 $M_{\odot}$ contrary to prediction. Moreover, the cloudlets follow a power law closely, with $\beta = 1{\cdot}65$. The relative numbers of cloudlets and concentrations are characterized by $\beta = 1{\cdot}25$.

Observations of dust and Ca^{+} absorption also show that a range of masses is present. According to Scheffler statistics of color excesses of 4700 stars indicate $\beta = 2 \pm 0{\cdot}2$. Therefore there is some degree of agreement between the model ($\beta = 3/2$) and the observations ($\beta = 1{\cdot}65$–2). Nevertheless, there are considerable problems. On the observational side, there is the question of the clouds in the mass range 30–300 $M_{\odot}$. On the theoretical side, there is the question of fragmentation of clouds during collision (ignored in the model) and residual elasticity of cloud collisions due to magnetic field (also ignored in the model). Various mechanisms for cloud formation, including thermal and hydromagnetic instability were discussed.

P. V. Ščeglov reported that his studies using a narrow band Fabry-Perot etalon, with a bandwidth of less than 3 Km/sec at Hα, indicated existence of high velocities in the range of − 120 to + 150 Km/sec in a number of emission nebulae.

M. Smith reported on his measurements of the line profiles of Hα and [N II] lines in a number of emission nebulae using a Fabry-Perot technique.

H. Zimmermann discussed the processes occurring during the collision and cooling of interstellar clouds. He considered in turn the initial compression period, then the subsequent expansion and the interval between collisions. His calculations suggest a mean number density of H_2 molecules of at most 5×10^{-4} times the number density of hydrogen atoms. The details of his theory are published in *Astronomische Nachrichten*.

S. B. Pikelner stressed the importance of magnetic fields in such considerations. Motion of the clouds along the magnetic field is possible but perpendicular to the field lines only small amplitude oscillations are possible. He suggested that the collisions perpendicular to the galactic plane are elastic but along the lines collisions can make clouds grow. Heavy clouds bend magnetic lines down and collect clouds, until new stars are formed.

INTERSTELLAR GRAINS

T.P. Stecher reviewed observations pertaining to interstellar grains.

'The wavelength dependence of interstellar extinction has been extended from the far infrared to the rocket ultraviolet with the result that no proposed particle will satisfactorily reproduce the observations with the existing theoretical treatments. Regional extinction curves have been extensively studied and show a large variation in the R value. There is some suggestion, however, that this may be a property of very young objects and for the general interstellar medium R may be more nearly a constant.

'Excellent observations of both color and polarization angle have been made on several reflection nebula which should reveal the angular scattering dependence of the particles when given a thorough theoretical treatment. In the special case of backscattering the theoretical interpretation is greatly simplified. The diffuse galactic light has been measured by several observers and the results seem to indicate particles that are isotropic scatterers with a high albedo. This appears contrary to the theoretical properties of previously proposed particles.

'The wavelength dependence of interstellar polarization shows a considerable scatter from star to star but the maximum polarization usually occurs in the yellow. A wavelength dependence on the direction of the magnetic field appears to be established. The dependence of position angle on color is not clear at this time. The correlation of polarization with color excess shows scatter and indicates there is a maximum degree of polarization.

'The infrared ice band at $3{\cdot}05\,\mu$ was not observed although it should have been if ice is a major component of the interstellar grains. The Mira variables have been found to be intrinsically polarized lending support to the hypothesis that grains are formed in late type stars. Some infrared observations have been interpreted as thermal emission of grains heated by young stars embedded in the clouds. The temperature of these grains is over 10^3 degK.

'The diffuse interstellar lines now number 28 and remain unidentified. Accurate profiles of λ4430 have been obtained. This line was found not to correlate physically with the change of slope at 4300 in the extinction curve. λ4430 does, however, show a correlation with extinction with a different line for high velocity clouds.

'The work of the last three years tends to reject ice and support some form of carbonaceous material. The observation pertaining to the interstellar grains as interpreted by existing theoretical models are, however, inconsistent and we must therefore conclude that the composition of the grains is unknown. Whether the difficulty lies in the theory and its approximations and measured physical parameters or the observations or both remains to be seen. We can, however, be hopeful that the next three years could provide the solution or at a minimum considerable additional progress.'

J. Mayo Greenberg reviewed theories of interstellar grains.

'During the past few years the observation of extinction, polarization and scattering by interstellar grains has been very considerably extended and this faces us with the requirement of more definitive theories.

'The theoretical problems may be put into two categories: (1) direct confrontation with observation. (2) physical and chemical interactions of grain with stars and with the interstellar medium. Although the latter are often only indirectly related to observations they are the ultimate foundation on which all theories must be based. An apparently satisfactory grain model—from an optical point of view—may be entirely inconsistent with physical or chemical criteria. (As an absurd example consider the motion that the grains are made of uranium hexofluoride particles.) Problems of direct confrontation include interstellar extinction and polarization, light from reflection nebulae, diffuse galactic light, grain emission in the infrared. Physical and chemical problems include theories of grain growth, interactions with stellar radiation, interactions with the interstellar gas atoms and ions, interaction with low energy cosmic rays (superthermal particles), interactions with magnetic fields. Finally there are problems which are sufficiently uncertain that they are difficult to categorize. In particular, attention is directed to the diffuse interstellar lines which are almost certainly intimately connected with the grains.

'No attempt will be made here to discuss all of the recent theoretical work. The Draft Report of Commission 34 has a rather complete survey of the literature in this field. We limit ourselves here to only a few points.

'Very extensive calculations have been carried out on model interstellar grains consisting of graphite cores with dielectric mantles or purely dielectric elongated (infinite cylinder) grains. Detailed considerations have been given to the optical properties of graphite and dirty ice throughout the spectral range of astronomical observations. It now seems that both the graphite core plus dielectric mantle and the elongated dielectric grains can give a reasonable representation of the interstellar extinction from the far infrared to about $\lambda^{-1} = 5\mu^{-1}$. Beyond this region (toward the far U.V.) neither model seems to predict a continued rise in the extinction as has been reported by Stecher. The reason for this lies in the fact that the dielectric materials which one can conceive of as either existing by themselves or as growing on graphite cores have a high absorptivity in the far U.V. In this wavelength region the grains are relatively large compared with the wavelength and consequently give rise to a reduced extinction relative to similar sized particles which are non-absorbing. In the far infrared (below $\lambda^{-1} = 0{\cdot}5\mu^{-1}$) the very anomalous types of extinction which have been suggested can not be produced by normal sized grains nor even by moderate modifications in the sizes (or size distributions) of ordinary grains as might be produced by interaction of the grains with stellar fluxes in H II regions.

'The observed maximum polarization relative to extinction can be matched by graphite core-dielectric mantle particles as well as by dielectric cylinders. The latter type of particles have been shown to give a moderately good representation of the wavelength dependence of polarization but with two difficulties which are not yet completely resolved; namely, the predicted position of the maximum polarization is a bit off and the sharp polarization drop in the infrared is not reproduced. Explanations of these discrepancies may possibly be found in the variations of degree of orientation with size. The flat graphite by itself can not produce the wavelength dependence of polarization. This has been shown by calculations for the small flakes and for larger flakes by use of microwave scattering results (unpublished). For graphite core ice mantle particles approximate calculations indicate that the polarization dependence on wavelength may be reasonably given. However microwave experiments on such particles indicate that the theory may not yet be reliable and this result is therefore questionable.

'A very significant advance in our understanding of the Davis-Greenstein magnetic orientation mechanism has been made by Jones and Spitzer. On the basis of application of this theory to the orientation of dirty ice grains and to the orientation of graphite core-ice mantle grains it may now be stated that: (1) For dielectric grains whose magnetic susceptibility is as estimated by Davis and Greenstein the magnetic field required to produce the highest observed polarization is $B = 2 \times 10^{-5}$ gauss; (2) for graphite core-ice mantle grains the minimum magnetic field requirement is higher by perhaps a factor of three. Using "super paramagnetism" the magnetic field requirement for dielectric grains may be reduced by orders of magnitude.

'Among the unsolved problems the explanation of the diffuse interstellar lines is outstanding and we can hope that when we know their origin we will have better defined the interstellar grains.

'Further observations of interstellar extinction and polarization in the ultraviolet beyond $\lambda^{-1} = 3\mu^{-1}$ are clearly indicated. Metallic and graphite particles will produce significantly less polarization in this region than dielectric ones. Beyond $\lambda^{-1} = 5\mu^{-1}$ the situation does not appear to be so distinctively different. Observations and theoretical calculations of the scattering and extinction by grains in filamentary structures of reflection nebulae would be extremely desirable because here one needs to explain simultaneously the principal optical properties of the interstellar grains. The answer to the question of orientation of grains by possible magnetic fields defining these filaments is of considerable interest.

'Finally and without discussion we list some additional problems for future work where present solutions are either non-existent or inadequate: interaction of superthermal particles with the grains, effects of surface or general irregularities on the optical characteristics of small particles, the chemistry and physics of the origin and growth of interstellar particles, detailed studies of grains in H II regions.'

Appendix

ADDENDA AND CORRIGENDA TO THE REPORT OF COMMISSION 34 PUBLISHED IN TRANSACTIONS IAU, VOLUME XIII A

On page 730, paragraph 2, line 5, after: '...on the central star (**4**)', add: 'Kohoutek (**4a**) identified 109 new faint planetaries, and positions of 322 planetaries were measured (**4b**)'.

On page 731, at the end of paragraph 5 beginning with: 'A central star of WN type...', add 'Variability of nuclei of nebulae was studied by Kohoutek (**66a, b**)'.

On page 740, at the end of paragraph 1 after: '...in Cygnus Loop.', add: 'Poveda (**241a**) argued that filaments are thin sheets. The total energy in the beginning evaluated as high as $5 \cdot 10^{49}$ ergs.'

On page 744, at the end of paragraph 2, after: '...and no nucleus forms.', add: 'Collapse of isothermal gas sphere was studied and used to formations of stars clusters, quasars and galaxies (**327a**).'

On page 745, in the *Bibliography*, the following references should be added:

4a. Kohoutek, L. 1965, *Bull. astr. Inst. Csl.*, **16**, 221.

4b. Kohoutek, L., Pěkný, Z., Perek, L. 1965, *Bull. astr. Inst. Csl.*, **16**, 189.

13a. Kohoutek, L. 1964, *Publ. astr. Soc. Pacif.*, **76**, 435.

66a. Kohoutek, L. 1964, *Bull. astr. Inst. Csl.*, **15**, 161.

66b. Kohoutek, L. 1966, *Bull. astr. Inst. Csl.*, **17**, 318.

106a. Griem, H. R. 1967, *Astrophys. J.*, **148**, 547.

241a. Poveda, A. 1964, *Ann. Astrophys.*, **27**, 522.

327a. Penston, M. V. 1966, *R. Obs. Bull.*, no. 117, E 300.

35. COMMISSION DE LA CONSTITUTION INTERNE DES ETOILES

Reports of Meetings, 23 and 26 August 1967

PRESIDENT: P. Ledoux.
VICE-PRESIDENT: C. Hayashi.
SECRETARIES: A. Masevič, M.H. Wrubel.

A. Administration Matters

Four meetings were held, part of the first meeting being devoted to administrative matters.

a. The Draft Report was approved enthusiastically.

b. The members indicated that they would be unhappy if the formation of a new commission on Cosmology and General Relativity would affect the discussion of stellar models based upon general relativity within Commission 35.

c. The Commission expressed its regret that Professor Hayashi, Vice President of the Commission, could not attend the meeting because of health.

B. Scientific Sessions

1. FIRST SCIENTIFIC SESSION

Demarque discussed the determination of ages of globular clusters by using the 'knee' of the curve which does not depend upon the uncertainties of convection theory. He also noted that one needs a large value of Z to produce the observed gap in NGC188. *Masevič* showed how different initial chemical compositions affect the evolution of stars of from four to thirty solar masses. *Kippenhahn* emphasized the uncertainties in the treatment of semi-convection. *Faulkner* noted that to get agreement with the horizontal branches of globular clusters one must use small masses. *Baker* discussed the determination of the parameters of RR Lyrae stars from their periods, colors, and the mass-luminosity relation.

2. SECOND SCIENTIFIC SESSION

Schwarzschild commented on the production of neutrons when hydrogen is brought into the carbon-rich convection zone during the pulses caused by thermal instability. *Weigert* described thermal pulses during the evolution of a close binary.

J. Cox described non-linear computations of radial oscillations of models studied previously in linearized form by Baker and Kippenhahn.

Christy interpreted the observed progression of features in the velocity curves of Cepheids with periods of seven to fifteen days, in terms of a compressional wave that moves inward, is reflected, and returns to the surface where it is observed. *A. N. Cox* compared small amplitude oscillations in non-linear computations with the results of linear theory.

3. THIRD SCIENTIFIC SESSION

Roxburgh described calculations for stars in uniform rapid rotation. *Hardorp* noted that calculations of line profiles based upon improved models give equatorial velocities for a given half-width that differ from previous results and therefore that quoted values of $v \sin i$ are uncertain for rapidly rotating stars. *Ostriker* gave the results of calculations for differentially rotating white dwarfs.

Fricke discussed the secular stability of rotation.

4. FOURTH SCIENTIFIC SESSION

Mestel discussed the general interaction between magnetic fields and internal circulations and noted that large-scale circulation can prevent interior magnetic fields from leaking to the surface. *Skumanich* described non-radial oscillations of rotating polytropes. *Bisnovsky-Kogan* described mass loss under the action of high opacity in a stellar atmosphere. *Wrubel* gave a report of an informal conference held in October 1966 on the evolution of stars of low mass. *Paczinski* discussed the dynamical stability of stellar envelopes near the limiting Hayashi track. *Cameron* described a model of an X-ray source based upon accretion.

5.

An additional informal session was devoted to the problem of semi-convection in massive stars.

6. ADDITION TO THE DRAFT REPORT, SECTION III

It is now possible to report on the work of Carson *et al.* on new opacity calculations. Two separate and independent attacks—one using the Thomas-Fermi statistical model of the atom and numerically computed wave-functions (1), the other retaining the general approach of Mayer subjected to a critical analysis (2)—lead to the same conclusion namely that while there is broad agreement, at the upper end of the temperature-pressure scale, with previous calculations, the latter are consistently low at the lower end (by a factor of about 3 for $T \simeq 10^6$ °K).

References

1. Carson, T.R., Mayers, D.F., Stibbs, D.W.N. 1967, *The Calculation of Stellar Radiative Opacity* (in press).
2. Carson, T.R., and Hollingsworth, H.M. *A critique of the Hydrogenic Approximation* (in press).

36. COMMISSION DE LA THEORIE DES ATMOSPHERES STELLAIRES

Report of Meetings

PRESIDENT: K.H. Böhm.
SECRETARY: E. Böhm-Vitense.

Business Meeting, 23 August 1967

The Draft Report was unanimously accepted. The members of the Commission were informed that the Organizing Committee had agreed to hold one business and only one scientific session, in which Dr. Nagirner (Leningrad University) should present a summary of recent work on the theory of line formation done in the USSR. The commission members agreed to this proposal.

The suggestions for a new president and a new vice-president were announced by the president. The following astronomers were elected by the commission as members of the new organizing committee: K.H. Böhm, J.T. Jefferies, V.V. Sobolev, R.N. Thomas, S. Ueno.

K.H. Böhm suggested to have a general discussion about plans for further colloquia to be organized by Commission 36 and about the next Harvard-Smithsonian Conference. *O. Gingerich* announced the plans for the next Harvard-Smithsonian Conference on stellar atmospheres which will be held in Cambridge (Massachusetts) in April 1968. The duration of the meeting will be three days. The suggested topic is: Theory and observation of normal stellar atmospheres. *O. Gingerich* and *C.A. Whitney* explained that the Harvard-Smithsonian group was especially interested in a discussion of the relation between spectral classification and the predictions from theoretical model atmospheres. *Gingerich* suggested to prepare a grid of model atmospheres in advance. There followed a discussion between Underhill, Thomas, Gingerich, and Whitney about the possible definition of normal stars. *Miss Underhill* suggested to exclude explicitly the discussion of atmospheres of intrinsic variable stars from this conference. On the other hand it was agreed to include supergiant atmospheres. *Thomas* stated that one of the aims of this meeting should be a clear definition of normal stellar atmospheres. The commission members agreed in principle to the proposed topic for the next Harvard-Smithsonian Conference.

Underhill and *Böhm* suggested to have the next colloquium of Commission 36 in 1969 somewhere in Europe. Tentatively they proposed 'Chromospheres and Coronas' as a possible subject of this meeting. *Mrs Böhm-Vitense* and *Miss Underhill* suggested to treat not only the usual radiative transfer and spectroscopic problems but to include also the fundamental hydrodynamic problems of stellar chromospheres. There followed a detailed discussion about the problem whether early type as well as late type stars should be the subject of this conference. There were strong arguments by some members of the commission to include both types of stars.

Thomas proposed to have no short contributed papers during this meeting but to limit the agenda to detailed summaries and discussion from the floor. *Miss Underhill* suggested to hold the meeting in Utrecht.

Scientific Meeting, 29 August 1967

As a supplement to the communications made at the business meeting the president announced the names of the suggested new members of Commission 36.

Dr D.I. Nagirner gave his review on recent work in 'Resonance Radiation Transfer Theory' done in the USSR.

He paid special attention to the recent work by V.V. Sobolev and his collaborators at the Leningrad University. Especially those problems for which an analytical solution can be found were discussed. The basic physical assumption used in most papers by the Leningrad astrophysicists are as follows:

(1) The gas consists only of one kind of two-level atoms with no continua.

(2) There is complete frequency redistribution.

(3) The frequency dependence of the absorption coefficient as well as the probability of quantum survival are space independent.

(4) The atmosphere is plane-parallel and either infinite, semi-infinite or finite.

Many of the arising problems for infinite atmospheres can be solved using the generalized H-function originally introduced by V. V. Sobolev. The cases in which the line absorption coefficient has either a Doppler-, Lorentz- of a Voigt-profile has been considered. Special attention has been paid to the asymptotic behavior (for large optical depth) of the solutions. In the case of finite atmospheres the solutions could be expressed in terms of certain types of *X*- and *Y*-functions. A comparison of the analytical solutions found by the Leningrad group with the purely numerical solutions by Avrett and Hummer has been carried out and excellent agreement has been found.

In conclusion *Dr Nagirner* emphasized that certain multi-level problems have also been considered by Sobolev. Exact analytical solutions for some non-stationary problems have been found by V.V. Ivanov and Nagirner.

In the following discussion *Dr Nagirner* restated (answering a question by Dr Houtgast) that the solutions given are really analytical solutions and that no iteration like in some numerical procedures is required. *Miss Busbridge* emphasized that, though a number of simplifying assumptions go into these computations, the mathematics is really very complex. Answering a question by Cayrel, *Nagirner* stated that no analytical solution would be possible if the space-dependence of the absorption coefficient were taken into account. In answer to a question by Böhm, *Nagirner* suggested that an analytical solution should be possible if the continuous absorption were taken into account. Kalkofen drew attention to the fact that analytical solutions can be found for interlocking problems with a common lower level if the stimulated emission can be neglected.

Y. Sobouti communicated briefly some of his own results on non-coherent scattering in finite atmospheres. He has applied principles of invariance. He was able to consider more general cases of the redistribution function.

Miss Busbridge drew attention to the large number of papers which have been written on analytical solution of non-coherent scattering problems and which had not explicitly been mentioned in this session. *J. T. Jefferies* emphasized the importance of the analytical work, but also drew attention to the fact there is not much hope to treat line transfer problems for more realistic model atmospheres by these methods. He strongly emphasized the importance of the numerical work on line transfer problems which has been done during recent years. He gave a very brief summary of this work.

SESSION MIXTE 12 ET 36

Report of Meeting, 28 August 1967

CHAIRMAN: E. Böhm-Vitense.

A joint session of commissions 12 and 36 was held on the following subject: 'Observations and Theory of the Continuous Spectrum of the Solar Photosphere and Chromosphere Below 4000 Å'.

The program included the presentation and discussion of the following brief papers:

1) *J. Houtgast:* 'A Survey of Observations of the Continuous Spectrum of the Sun between 3000 and 4000 Å'.

2) *O. Gingerich* and *J. Rich:* 'The Far Ultraviolet Spectrum of the Sun and the Identification of the Continuum of the First Excited Level of Si'. (Dr. Gingerich also discussed in some detail the predictions of the ultraviolet spectrum from the 'Bilderberg' model.)

3) *J. Blamont, R. Bonnet:* 'Données expérimentales sur la transition photosphère-chromosphère et leur discussion'.

4) *A. Sauval* (presented by *L. Neven*): 'The Solar Ultraviolet Continuum'.

5) *K. Nishi* (presented by *F. Moriyama*) 'The Energy Distribution and Center-to-Limb Variation of the Solar Ultraviolet Continuum'.

6) *O. Namba* and *J. Houtgast*: 'Continuum Location and Total Line Absorption in the Solar Spectrum between 3000 and 4100 Å'.

7) There was also a very brief communication by *H. Neckel,* in which he compared Houtgast's results to those of Labs and Neckel and drew attention to the excellent agreement.

37. COMMISSION DES AMAS STELLAIRES ET DES ASSOCIATIONS

Report of Meetings, 25, 28, and 29 August 1967

PRESIDENT: M.F. Walker.
SECRETARIES: J. Ruprecht, H.H. Swope.

A. Administrative Sessions

During the meetings, the following administrative matters were discussed:

(1) The Draft Report was approved, subject to certain corrections proposed by members of the Commission.

(2) The following new officers of the Commission were approved by the Executive Committee of the Union: President, Golay; Vice-President, Larsson-Leander.

(3) The new Organizing Committee of the Commission was elected by the members of the Commission and consists of: Barhatova, Gascoigne, I.R. King, Rosino, Ruprecht, Swope, and Walker. It is expected that the membership of the Organizing Committee will normally not exceed six. However, for the coming three-year period, Walker will serve as an additional member of the Organizing Committee since both the President and Vice-President are new.

(4) The following new members of the Commission were approved by the Organizing Committee: Aarseth, Artjuhina, Bouvier, Feinstein, Kadla, Lodén, Lynden-Bell, Morgan, Šarov, van den Bergh, G. A. H. Walker, and Woolley.

(5) Following a discussion of the problems currently faced by the Union, the Commission adopted the following resolution for transmission to the Executive Committee: 'Commission 37 views with great concern the problems arising from the continued growth of the Union. The Commission therefore urges the Executive Committee to set up a special Committee which shall, before the next General Assembly, make a thorough study of all aspects of these problems and prepare recommendations regarding the future structure of the Union for adoption at that time'.

(6) The Commission considered a letter from H. Wilkens proposing the calculation of special tables for the derivation of more accurate galactic coordinates of clusters for the year 2000. The Commission concluded that extremely accurate galactic coordinates of clusters are unnecessary.

(7) The question was raised (by Golay) as to when Trumpler's spectroscopic and photometric observations of stars in open clusters, being prepared by Weaver, would be published. Since Weaver was not present at the session, it was agreed that I. R. King should convey to him the feeling of the Commission that early publication of this important material is mandatory.

(8) Blaauw called attention to the fact that not all authors are following the recommendations regarding the nomenclature of Associations adopted by the Commission at Hamburg, and urged that this nomenclature should be used in all future publications.

B. Scientific Sessions

The following is a list of the authors and titles of the scientific papers presented during the meetings of the Commission, in order of their presentation:

I. R. King: 'Report on the I.A.U. Paris Colloquium on the Gravitational N-Body Problem'.

G. E. Kron: 'Surface Photometry of Globular Clusters with an Electronic Camera'.

M. F. Walker: 'Determination of Stellar Magnitudes by Electronography'.

R. D. Cannon and C. A. Murray: 'Recent Cluster Proper Motion Work at the Royal Greenwich Observatory'.

I. R. King: 'Star Counts in Globular Clusters'.

T. Schmidt-Kaler: 'Integrated Colors and Magnitudes of Globular and Open Clusters'.

G.A. Sundman: 'A Search for Southern Stellar Associations at the Stockholm Observatory'.
A.J. Wesselink: 'The Color-Magnitude Diagrams of the Open Clusters NGC6067 and NGC4103'.
A. Feinstein: 'The Open Cluster NGC6231'.
U. Lindoff: 'A Very Red Star in NGC2533'.
M.F. McCarthy: 'Image-tube Observations of Faint Stars in the Pleiades'.
G.A. Walker: 'Hγ Observations of Stars in Associations'.

38. COMMISSION POUR L'ECHANGE DES ASTRONOMES

Report of Meeting, 24 August 1967

PRESIDENT: M. Minnaert.
SECRETARY: F. B. Wood.

Professor P. Swings, President of the IAU., expressed the gratitude of the Executive Committee to commission 38 and promised as much financial help as was possible despite the present financial difficulties of the Union.

Professor Minnaert expressed thanks and stated that any monies granted the commission would be very carefully expended. He then presented a financial report of the activities of the commission in the past three years. During this interval, a total of 53 requests had been granted; in past years, out of approximately 100 grants, only two could have been called unsuccesful. Some of the administrative details were described; one of the great advantages was in being able to make decisions and payments rapidly without undue formalities.

A discussion of the draft report then followed. Professor B. J. Bok raised the question of giving names of recipients of aid in the draft and final reports on the grounds that these were honors similar to fellowships. This was seconded by Professor F. B. Wood and a motion by Bok to give names of recipients in the future passed without dissenting vote. A list of those who received grants in the past will be mimeographed and sent to the members of the commission.

Dr M.V. Migeotte then gave the report of the sub-commission created at the 12th General Assembly to aid travel of students desiring to specialize in astronomy but finding no facilities in their own countries. Details are given in the Draft Report (pp. 1xxxv–1xxxvi). Dr Migeotte felt the sub-commission should be kept in existence even if it handled only one or two cases each year. The report of the sub-committee was accepted with thanks.

The next item was the composition of Commission 38 for the next three years. The Executive Committee of the union had nominated Professor Minnaert as president. As vice-president, in place of Dr Migeotte who resigned in order to take the presidency of another commission, it had named Professor A. Reiz.

After some discussion, D. Ya. Martynov, P. M. Routly, J. Sahade, J. P. Wild, were recommended to the Executive Committee for membership of the new organizing committee. Some changes in the membership of the commission wil also be proposed.

Professor Minnaert then presented the following resolution for presentation to the executive committee:

'The members of Commission 38 and other members of the Union present at the August 24 session of this commission stress the importance of the program of Commission 38 and recommend that great efforts should be made in order to ensure the continuation of its activities and to avoid interruption of of its work.

During the next three years a minimum subsidy of $8000 will be necessary for this purpose. Smooth development of these activities requires a certain reserve, which should be transferred each year to the next year.

New efforts should be made to obtain an increased additional subsidy (from UNESCO).'

This motion was carried.

A prolonged discussion followed on the possibility of sending either distinguished senior astronomers or younger but well-trained astronomers on visits to countries less advanced in astronomical training. *Dr Bappu* noted two situations exist. One was countries with no astronomical facilities where matters must be handled first at a local level before such visits would be profitable. The second was countries where observatories with specialized facilities already exist, which could profit greatly by a program of foreign visitors. After many suggestions, the committee agreed that while the details could not be settled, an international program of visiting professors would be of considerable importance and requested that the president prepare a letter to be sent to possible sources of funding.

Finally, *Bok* requested that minutes be prepared, mimeographed, and distributed to members of the Executive Committee and presidents and vice-presidents of all commissions.

40. COMMISSION DE LA RADIOASTRONOMIE

Report of Meetings

PRESIDENT: M. Ryle.
SECRETARY: J. R. Shakeshaft.

Business Meetings, 23 and 29 August 1967

After welcoming both members and non-members of the Commission the President drew attention to the death of Dr Lloyd Berkner, who was deeply involved in the foundation of the U.S. National Radio Astronomy Observatory. The following matters were then discussed:

1. THE DRAFT REPORT

Members were asked to inform the secretary of any minor amendments required: the report was then accepted *nem con.*

The President asked for members' views on the future form of the Draft Report. Ideally this should be a comprehensive, critical, review of the entire field but such a review would necessarily be longer than was feasible. Alternatives were a report with a complete bibliography but little comment as in the present case or else a more personal view of selected aspects of the subject.

F. G. Smith and *Christiansen* spoke in favour of the present form, valuing the bibliography highly, though Christiansen noted that the procedure was very expensive. *Denisse* then questioned whether Draft Reports need be circulated other than to members of the particular Commission concerned. *Minkowski* pointed out that the Executive Committee were in a position to influence the form by their instructions to Presidents. In the absence of further opinions the sense of the meeting seemed to be that no change was desirable.

2. NEW MEMBERS

Nomination of the following new members was approved: Abrami, Dieter, Hughes.

The President mentioned that there were many proposed new members of the Union who were likely to want to join Commission 40 and that consideration might have to be given to some limitation of numbers.

3. IUCAF

F. G. Smith reported on the activities of the Inter-Union Commission on Frequency Allocations for Radio Astronomy and Space Science. Although IUCAF were continuing to press for protection of the 322–329 MHz deuterium band there was no hope of it becoming available in the foreseeable future. It was therefore particularly important for all radio astronomers to request from their national administrations better protection in the 406–410 MHz band. He enquired as to whether any radio astronomers had observed in one of the low frequency 'guard bands'. Apparently no one had and the President felt that they were too full of transmitters to be usable.

4. CALIBRATION OF 21-CM LINE RECEIVERS

Van Woerden, the convenor of the Working Group set up at the previous General Assembly to study the standardisation of 21 cm surveys, had written to the President to say that the report on the subject had unfortunately not yet been prepared. It was agreed that a report was still desirable and a reconstituted Working Group was set up, consisting of Van Woerden (convenor), J. A. Roberts, Varsavsky, and Westerhout with co-opted members L. Bottinelli and D. R. W. Williams (not members of the IAU).

5. LUNAR OCCULTATIONS OF THE GALACTIC CENTRE REGION

Kerr reminded members of the opportunities (*Ap. J.*, **148**, 923, 1967) for observations during the next four years. Cooperation was desirable at frequencies of about 250 MHz, at 1420 MHz and at 1667 MHz, both in the continuum and lines. An informal meeting of those interested was arranged.

6. PROPOSED RESOLUTION ON RADIO MAGNITUDE SCALES

Heidmann put forward a recommendation concerning the definition of radio magnitudes, i.e. that the radio monochromatic magnitude m_f at frequency f be defined by $m_f = -53{\cdot}45 - 2{\cdot}5 \log_{10} S_f$ where S_f is the flux density in M.K.S. units, with a corresponding expression for absolute radio magnitude.

This resolution was supported by *Laffineur* but opposed by *Scheuer* and *Burke*. *Scheuer* then moved an amendment, seconded by *M. Schmidt* and supported by *Minkowski* 'That the use of radio magnitudes be discouraged'. After further discussion both the original resolution and the amendment were withdrawn without any vote being taken.

7. VARIABLE SOURCE OBSERVATIONS

Minkowski spoke briefly on his reasons for believing that an 'International Year for Variable Source Observations', as suggested in the IAU Bulletin No. 17, was not appropriate. The President then put the resolution that 'Commission 40 recommends that the services of the Central Bureau of Astronomical Telegrams be used to announce without delay significant events and changes in quasi-stellar sources and objects in order to achieve effective communication of observations in all frequency ranges'. This was passed *nem con.*

A meeting of those interested was subsequently held at which *K. I. Kellermann* volunteered to act as a clearing centre for radio observations of variable sources.

8. PROPOSED COMMISSION ON 'RELATIVISTIC ASTROPHYSICS AND COSMOLOGY'

The President suggested that Commission 40 was one of those particularly concerned with this proposal (item 9(*a*) of agenda for the General Assembly) and should consider the matter. The IAU Executive Committee inclined rather towards a working group than a new Commission.

Minkowski spoke strongly against either possibility since Commission 28, of which he was President, had been looking after Cosmology for many years. He was in fact recommending a change of its title to 'Galaxies and Cosmology'.

Christiansen said that the IAU Executive Committee had thought of a joint working group of Commissions 28, 40 and perhaps one or two more. The President felt personally that such a working group was the best solution. There were no other comments.

9. REORGANIZATION OF THE COMMISSION

The President reminded members of the discussion in Hamburg concerning a possible reorganization of the Commission. This had resulted in a decisive rejection of any split-up but a recommendation for the arrangement of joint sessions and discussions with other Commissions. He had tried, with some success, to do this for the present meeting.

10. INTERNATIONAL RADIO ASTRONOMY OBSERVATORY

The President said that there had been discussions at the 1966 URSI meeting at Munich concerning an International Observatory but no formal approach had been made to the IAU. He felt that discussions could most usefully remain informal for the time being.

11. ORGANIZING COMMITTEE

To assist the new President J.P. Wild and Vice-President D.S. Heeschen the following Organizing Committee was nominated: Kardashev, Kundu, Lequeux, Smerd, Westerhout.

Scientific Meetings

In addition to being one of the Commissions sponsoring the Joint Discussions on 'New Techniques in Space Astronomy' and 'Extra-galactic Radio Sources' of which accounts are given elsewhere in this volume Commission 40 had the following scientific meetings:

1. A joint session with Commission 10 on 'Dynamic Solar Radio Events' (30 August). For details see the report of Commission 10.

2. Two sessions on 'Planetary and Lunar Radio and Radar Astronomy' (28 August), jointly with Commissions 16 and 17. These were organized by A.H. Barrett, F.D. Drake and G.H. Pettengill. The programme was as follows:

Radar Astronomy

V.A. Kotelnikov: 'Joint USSR-Jodrell Bank Observations of Venus in 1966'.
G. Pettengill: 'Short Pulse 3.8 cm Radar Observations of Mars'.
B.S. Yaplee (read by *R.W. Hobbs*): 'Comparison of Lunar Topography from Optical and Radar Measurements'.
J.H. Thomson: 'Supersynthesis as applied to Lunar Mapping by Radar'.
J.H. Thomson: 'Three-dimensional Holographic Imaging Techniques in Lunar Radar Astronomy'.
G. Pettengill: 'Lunar Mapping by Radar at 70, 23 and 3·8 cm'.

Radio Astronomy

A.E. Salomonovitch: 'Phase Effect of Mercury at $\lambda = 8$ mm'.
E.E. Epstein: 'Phase Effect of Mercury and Venus at $\lambda = 3$ mm'.
C. Sagan: 'The Microwave Phase Effect of Mercury'.
J.K. Alexander: 'Decametre-Wave Observations of Jupiter'.
N.J.B.A. Branson: 'High Resolution Observations of Jupiter at 1407 MHz'.
A.G. Smith: 'Recent Studies of Jovian Millisecond Pulses'.
J.A. Roberts: 'Linear Polarization of Jupiter at 2650 MHz'.
J.A. Roberts: 'Circular Polarization of Jupiter at 530 MHz'.

3. Three sessions on 'Radio Studies of H II Regions' (24 and 28 August), jointly with Commission 34. These were organized by F. J. Kerr. The programme was as follows:

Recombination Lines

B.F. Burke: 'The large-scale distribution of H II Regions in our Galaxy'.
N.H. Dieter: 'Survey of recombination line sources'.
F.F. Gardner: 'CSIRO work on recombination lines'.
D.R.W. Williams: 'Observations of higher-order recombination lines'.
L. Goldberg: 'Departures from thermodynamic equilibrium'.
Y. Terzian: 'Electron temperatures of H II regions derived from radio observations'.
D.H. Menzel: '*f*-values for high-level hydrogen lines'.
H.C. van Woerden: 'Some observations of the recombination line at 1424 MHz'.

Continuum

B.Y. Mills: 'High-resolution observations of the Orion nebula at 408 MHz'.

R.D. Davies: 'Interferometry of H II regions associated with OH sources'.
D.S. Heeschen: 'High-density H II regions associated with OH emission clouds'.
V.A. Hughes: 'The radio source W 49 and anomalous emission'.

OH *Lines*

A.H. Barrett: 'Review of MIT work on OH lines'.
H.F. Weaver: 'Review of Berkeley work on OH lines'.
F.F. Gardner: 'CSIRO work on OH lines'.
R.D. Davies: 'Long-baseline interferometric studies of OH emission sources'.
H.F. Weaver: 'OH in the galactic center region'.
I.S. Šklovskij: 'Theory of OH emission'.
T.K. Menon: 'Time variation in OH sources'.

4. One session (29 August) on miscellaneous topics, organized by Y.N. Parijskij. The programme was as follows:
T.W. Cole: 'Interplanetary Scintillations at Cambridge'.
V.V. Vitkevich and T.D. Antonova: 'Interplanetary Scintillation at metre wavelengths'.
V.V. Vitkevich and V.I. Vlasov: 'The Motion of Interplanetary Irregularities'.
M.H. Cohen: 'Interplanetary Scintillation at centimetre wavelengths'.
A. Braccesi: 'Scintillation of Weak Sources at 408 MHz'.
A.E. Salomonovich: 'Cosmic Background Radiation at 8.2 mm'.
B.F. Burke: 'Cosmic Background Radiation at 9.2 mm'.
J.R. Shakeshaft: 'Cosmic Background Observations at Cambridge'.
S.Ya. Braude: 'Low Frequency Spectra of Radio Sources'.
V.A. Sanamian: 'Observations of Radio Sources at Low Frequencies at Serpukhov'.
M. Ceccarelli: 'Clustering of Very Weak Sources'.
I. Kazès: 'Study of Normal Galaxies with High Resolution'.
L. Weliachew: '21-cm Line Observations of Nearby Galaxies'.

41. COMMISSION D'HISTOIRE DE L'ASTRONOMIE

Report of First Meeting, 25 August 1967

PRESIDENT: E. Rybka.
SECRETARY: Z. Horský.

Before the beginning of this session the President distributed among the members and guests copies of the Information Circular concerning the General History of Astronomy, vol. III, and Prof. Kulikovsky distributed copies of the Bibliography of books and papers, published in 1966, on the history of astronomy.

Present were more than 35 members and guests.

Members and guests standing, the President announced that Commission 41 suffered a loss by the death of the members of our Commission Yu. Perel and E.F. Dijksterhuis and consulting member L. Thorndike.

The Agenda of this meeting was:

1. ADOPTION OF THE DRAFT REPORT AND DISCUSSION OF THE POLICY OF COMMISSION 41

The Draft Report was adopted without changements. Professor Kulikovsky remarked that the Bibliography published in the Draft Report is only a supplementary one, the complete Bibliography being published by him and distributed among the members of the Commission.

2. DISCUSSION OF THE NEED OF ANNUAL BIBLIOGRAPHY

A lengthy discussion showed that all members felt the need of an Annual Bibliography of books and papers on the history of astronomy. Mrs Whitrow (Editor, Isis Bibliography Project) asked whether there existed any co-operation between the editors of the Isis Critical Bibliography and the Bulletin Signalétique on the one hand and the editor of the Bibliography on the History of Astronomy published by Commission 41 of the IAU, on the other hand.

3. DISCUSSION OF THE NEED OF INFORMATION CIRCULAR

In the discussion, the members found the Information Circular issued by the President of the Commission very helpful and suggested that it should be published.

4. DISCUSSION OF THE GENERAL HISTORY OF ASTRONOMY

The President of Commission 41, *Professor E. Rybka* outlined to the members of the Commission the tentative plan of Volume III of the General History of Astronomy. This volume shall, in principle, be devoted to the astronomy of the XVIIIth and XIXth centuries. This tentative plan should also be discussed during the International Congress of the History of Science in Paris, 1968. Many members of Commission 41 requested the co-operation of the national committees of history of astronomy in countries, where they exist; such co-operation is desirable. The meeting proposed a resolution on this subject for adoption.

5. DISCUSSION ON INSTRUMENTS OF HISTORICAL INTEREST

It was unanimously agreed that instruments and manuscripts of historical interest require protection from damage. After a discussion, a resolution concerning this subject was prepared for adoption.

6. MICROFILMING OF DOCUMENTS ON THE HISTORY OF ASTRONOMY

In the discussion, *Dr Forbes* drew attention to the importance of securing permission and providing the means for copying the Royal Greenwich Observatory's records of the Astronomers Royal of the past three centuries mentioned in the Draft Report of this Commission, and asked the members to discuss the following questions:

(*a*) whether it was agreed, in principle, that a formal request should be made to the Astronomer Royal for permission to microfilm these valuable and unique documents and

(*b*) if so, to discuss in the meanwhile which steps might also be taken to secure support for such a project.

Professor Kulikovsky strongly supported this suggestion and proposed that a sub-committee from Commission 41 should be set up to investigate this particular matter and also the general question of making more manuscript material readily accessible to scientists all over the world. *Dr Sykes* said that Commission 5 was interested in this project because of the possibility that the collection of some 50000 slips relating to the Bibliography of Astronomy 1881–1898, now held at the *Observatoire Royal de Belgique,* would be made available on microfilms. He asked that Commission 5 should be kept informed of any progress by Commission 41 in this matter.

Mrs Helen Wright spoke of the programme of the American Philosophical Society in Philadelphia to have some of the papers at the Royal Society in London microfilmed. With the sponsorship of an organisation like the American Philosophical Society, she suggested that it might be possible to obtain funds to carry out this programme. The papers would thus become available to historians of science not only in England, but also in the United States, and in other countries the world over. The Institute for History and Philosophy of Science of the American Institute of Physics in New York worked on a joint programme with the American Philosophical Society for the recording of the history of quantum physics; it is now starting a similar programme for the recording of the history of astronomy. The programme includes the tape-recording of leading figures in the history of contemporary astronomy and the organisation and deposition of their papers in suitable quarters. Thus, in the summer of 1966, several tape-recording sessions were held with Dr Harlow Shapley, and afterwards the tapes were transcribed and indexed, and are now available to qualified historians at the American Institute of Physics. It is hoped that this programme can be extended not only in the United States, but also in many other countries. Tape-recording of the leading figures in astronomy is an invaluable record to future historians, and every effort should be made to encourage the members of the Commission to aid this programme. *Professor Rybka* read a letter from Professor Zagar. Professor Zagar writes:

L'Observatoire de Brera à Milan a accompli la rédaction du second volume de la 'Correspondance sur Mars' de G.V. Schiaparelli; il s'agit d'un volume qui contient à peu près 300 lettres échangées de 1890 à 1910 entre Schiaparelli et les plus importants aérographes contemporains. Le troisième volume, qui sera aussi le dernier, sera publié ensuite.

Report of Second Meeting, 25 August 1967

PRESIDENT: E. Rybka.
SECRETARY: Z. Horský.

Present were more than 30 members and guests.

7. COMMUNICATIONS ON THE DEVELOPMENT OF STUDIES ON THE HISTORY OF ASTRONOMY

Professor František Link read a paper concerning the astronomical observations recorded in the twelfth-century chronicle written by a Czech anonymous author, known as the Canon of Vyšehrad. This author recorded many observations of planets and an observation of a sunspot in 1139. The existence of this sunspot is confirmed by Chinese observations. This observation of a sunspot is the first made in Bohemia.

Dr Tadeusz Przypkowski described, in his paper, a very original pendulum clock dating from 1654 and built by Gohier in Paris for the King of Poland Jan Casimir (The clock being signed: Gohier à Paris 1654 Pour le roi de Pologne Jean Casemir). This clock, still in very good repair, is now in the collections of the Museum Przypkowski in Jędrzejów (Poland). It is undoubtedly the oldest pendulum clock surviving in perfect state of preservation.

Mrs. E.S. Parsamjan read a paper written jointly with K.A. Mkztchjan concerning a possible ancient 'observatory' on the territory of Armenia.

Dr Winfried Petri informed the meeting of his new studies concerning the history of Tibetan astronomy and the history of astronomy in Asian countries.

Dr Carlos Juan Lavagnino discussed in his paper "The qualitative development of astronomy and the scientific place of cosmonautics" the problems concerning the periodisation of the history of astronomy and the influence of cosmonautics on the development of astronomy. He supposes that the periods of the evolution of human society may be linked in parallel to the periods of development of astronomy.

8. RESOLUTIONS

The resolutions adopted are as follows:

1. *For the General Assembly*

(See resolution No. 4, p. 38–9.)

2. *For the General Assembly*

(See resolution No. 3, p. 38–9.)

3.

Commission 41 considère comme très utile la composition et l'édition de la Bibliographie annuelle sur l'histoire de l'astronomie qui mérite tout le secours.

Commission 41 voulant assurer la totalité de la Bibliographie demande à leurs membres et membres-consultants de bien vouloir prendre la tâche et la responsibilité d'envoyer aux compilateurs de la Bibliographie ses résultats de depouillement des éditions du caractère générale et celles des académies et des universités de leurs pays.

Commission 41 s'adresse aux Comités Nationaux de l'UAI des pays qui ne sont pas présentés à la Commission 41 de choisir une personne qui pourra être responsable pour le même sujet.

Commission 41 prie le Président de la Commission 41 de s'adresser aux institutions astronomiques des pays qui ne sont pas présentés à l'UAI, en les demandant le même secours.

Commission 41 rapelle à tous auteurs la nécessité d'envoyer aux compilateurs de la Bibliographie les tirages-à-part ou les données bibliographiques précises.

9. MEMBERSHIP OF THE COMMISSION

Prof. E. Rybka was asked by the Executive Committee to continue as President of Commission 41 during the coming three years, with O.J. Gingerich as Vice-President.

The following Organizing Committee was elected: V. L. Čenakal, J.-O. Fleckenstein, C. D. Hellman, Z. Horský, P.G. Kulikovsky, W. Petri.

New members of Commission 41: J.G. Argyrakos, J. Dobrzycki, D. Kotsakis, J.R. Lévy.

Died: Yu. Perel, E. F. Dijksterhuis.

Consulting members of Commission 41: Aaboe, Burke-Gaffney, Daumas, Fischer, Forbes, Freiesleben, Hartner, Horský, Hoskin, Kennedy, Labat, Maddison, Millas-Vallicrosa, Musto, Needham, Nevskaja, Pedersen, Price, Przypkowski, Ronan, Ronchi, Santillana, Schove, Taton, Thoren, Tumanjan, Veselovsky, van den Waerden, Wattenberg, Woolf, Wright.

Working group for the description and microfilming of astronomical manuscripts: President: O.J. Gingerich. Members: V.L. Čenakal, E. Forbes, O. Pedersen.

42. COMMISSION DES ETOILES DOUBLES PHOTOMETRIQUES

Report of Meetings

PRESIDENT: J.E. Merrill.
SECRETARY: R.H. Koch.

Business Sessions

The business meetings of Commission 42 at the XIII General Assembly of the IAU were held August 23 and 29, 1967, Dr. Merrill presiding. This summary has been compiled from the minutes of the separate meetings.

A resolution of condolence and remembrance was offered by Mrs Kron:

'Commission 42 has lost one of its finest as well as one of its relatively rare Southern Hemisphere members in the death of Arthur R. Hogg.

In 1927 he joined the staff of the then Commonwealth Observatory at Mt. Stromlo, Australia. As a talented painstaking observational astronomer he made many contributions of permanent value. These include atlas photographs of the southern skies and photoelectric photometric work on open clusters that he had discovered, light curves of eclipsing binaries and at least one photoelectric standard magnitude sequence. The last task, a fundamental but often thankless one, was typical of his unselfish considerate nature.

The last few years of his life were mainly devoted to carrying on the duties of deputy director of Mt. Stromlo. A heart attack a few months after the Hamburg meeting of the IAU meant the too-early end of a fully active career. Yet in April 1966, he had just finished delivering a speech the evening before he passed away quietly at dawn.

Gone now is the chance to enjoy his brilliant metaphors in both verbal and 'private' communications. We shall miss them and the colleague who made them—Arthur R. Hogg.'

A resolution of condolence and remembrance was offered by Dr Wright:

'Commission 42 lost one of its founding members and a great observer of double and variable stars in the death of Dean B. McLaughlin. For forty years he combined visual magnitude observations with spectrographic observations and was one of the very few who were able to compare visual magnitude and radial velocity curves obtained at the same time. This commission has repeatedly urged this practice and should continue to do so.

Dean McLaughlin worked at Michigan most of his life with a small telescope by modern standards, but he accomplished more than most of us with his studies of spectroscopic binaries, Be-stars, the ratio V/B in the hydrogen lines, and novae. From the point of view of our Commission, probably the most interesting of his achievements was the discovery of the zeta Aur-type binaries—VV Cep, 31 Cyg and 32 Cyg. He demonstrated clearly the possibility of determining dimensions of these systems combining early and late-type components.

McLaughlin was an almost perfect researcher and inspired many of his students to follow his path. We shall miss him.'

A resolution of condolence and remembrance was offered by Dr Popper:

'In suddenly losing R.M. Petrie the astronomical profession has lost one of its most respected members at the height of an exceptionally fruitful career. While he made lasting contributions in numerous fields of stellar astronomy, his impact on the interpretation of the spectra of eclipsing binaries was unique and will continue to have profound influence.

Consciously or subconsciously, each of us in writing his scientific papers must be looking to someone whose insight and judgement serve as prods and guides in our own work. For me, R.M. Petrie has been such a person since I heard his first paper nearly thirty years ago. He never wrote a trivial paper. His advice, sought in the highest councils of our profession, was always thoughtful and wise. We shall miss him.'

A resolution of condolence and remembrance was offered by Dr O'Connell:

'The unexpected death of Professor Herbert Schneller represents a grievous loss for all variable star astronomers and, in particular, for those working in the field of eclipsing binaries. By his long-continued observations and original studies he contributed greatly to the development of this branch of astronomy. His greatest contribution, however, was his bibliography of variable stars. The series of splendid volumes of the *Geschichte und Literatur der veränderlichen Sterne* which he published must always be an indispensable work of reference for variable star astronomers as long as active work is being done in this field. I cannot imagine a time when they will not be needed so that Schneller has indeed, together with his predecessors—Müller, Hartwig and Prager—a lasting monument in the *Geschichte und Literatur*.'

These four resolutions were adopted unanimously and are to be sent to the respective widows.

Consideration of the 'Bibliography and Program Notes for Eclipsing Binaries' (hereafter BPN) resulted in the creation of a permanent standing committee on the BPN, with Dr Larsson-Leander as chairman. Some minor changes in format and in mailing procedures will be made.

Dr Wood summarized the past and continuing cooperative observing campaigns on W Ser, SX Cas, VV Cep, epsilon Aur, zeta Aur, 31 Cyg, 32 Cyg, U CrB and HD 128220. For next triennium AR Cas and GL Car were chosen as the newly added systems with the programs to be coordinated by Drs Kitamura and Serkowski, if possible. A new Committee on Cooperative Programs was named with Mrs Kron as chairman, Drs Gyldenkerne, Kitamura, Magalishvili, Sahade, Wood as additional members.

It was voted to disband the Joint (26, 42) Working Group. The Joint (30, 42) Working Group is to be continued and will co-opt one astronomer from Commission 26. From Commission 42 representation is Drs Batten (chairman), Plavec, Popper, Sahade. The Committee on Times of Minima was re-formed under Dr Herczeg, chairman, and will attempt to obtain wider cooperation in observation of stars on its selected list. Additional committee members are Drs Binnendijk, Cester, Knipe, Plavec, Szafraniec. The following members represent Commission 42 on the Organizing Committee for the Fourth Variable Star Colloquium: Drs Merrill, O'Connell and Wood.

The Commission voted to approve the following requests for IAU subventions toward partial support for: (1) mailing expenses for the BPN, (2) administrative expenses for the Fourth Variable Star Colloquium, and (3) publication of the Cracow Eclipsing Binaries Circulars. The Commission also adopted resolutions identical to those passed by Commission 27 on the desirability of continuing the publication of the *Geschichte und Literatur der veränderlichen Sterne* and of an IAU subvention toward streamlining and continuing the work of the Variable Star Bureau of the Astronomical Council of the USSR Academy of Sciences.

It was voted to accept the Draft Report, with such minor changes as had been submitted.

With Dr F. B. Wood and Dr M. Plavec named as President and Vice-President, respectively, for 1967–1970, it was voted that Drs Batten, Herczeg, Kitamura, Larsson-Leander, Merrill and Česevič constitute the Organizing Committee for the triennium.

The meeting concluded with resolutions of thanks to the several persons who contributed materially to the work of the Commission during the sessions and during the preceding three years.

43. COMMISSION DE PLASMAS ET MAGNETO-HYDRODYNAMIQUE EN ASTROPHYSIQUE

PRESIDENT: T. G. Cowling.
SECRETARY: R. J. Tayler.

Report of Meetings, 23 and 28 August 1967

Business Meeting

The Draft Report of the Commission was approved.

The President reported on discussions which he had had about the scope and name of the Commission and the organising committee was asked to look further into this matter, and so bring proposals to the IAU Executive Committee.

On the proposal of L. Davis, Jr., the Commission agreed to express in a resolution the desirability of prompt and regular publication of solar wind data in a similar way to geomagnetic indices. The resolution (drafted by T. G. Cowling, L. Davis, Jr. and R. J. Tayler) was as follows:

If all experimenters studying the solar wind were to pool their data on wind velocities, it would now be possible to produce an almost continuous record, similar, for example, to the record of indices of geomagnetic activity. This would be of great value to a wide range of astronomical and geophysical research.

We therefore ask the International Astronomical Union to express its full support for the prompt and regular publication of solar wind data in one of the journals already publishing data on solar or geomagnetic activity.

P. A. Sturrock reported that a conference, sponsored by the American Astronomical Society and (probably) the American Physical Society, would be held in the United States in the Autumn of 1968 on 'Plasma Instabilities in Astrophysics'. Welcoming such a conference, the Commission agreed to the following motion:

Commission 43 warmly welcomes the proposal to hold a Conference on Plasma Instabilities in Astrophysics under the sponsorship of the American Astronomical Society and the American Physical Society in the early Autumn of 1968. The Commission regards such a conference as very timely, and would wish to associate itself fully with all steps needed to ensure its success as a truly international conference.

It was proposed by the President and the organising committee that L. Davis, Jr. and R. Lüst be nominated as future President and Vice-President, respectively, of Commission 43, and that the organising committee consist of T. G. Cowling, S. Kaplan and one further member yet to be nominated to succeed H. Alfvén. This was approved.

Scientific Meetings

23 August 1967

The following short papers were presented:

L. Davis, Jr.: Comments on the various theories of the azimuthal components of the motion of the solar wind.

L. Mestel: Magnetic braking and the solar wind.

P. A. Sturrock: A model of the solar wind.

Professor Alfvén also spoke about the role of electric fields in astrophysics.

28 August 1967

The following short papers were presented:

A. B. Severny: 'Comparison of flares and electric currents'.

P. A. Sturrock: 'Theory of solar flares'.

E. Chvojková: 'Some features of particles spiralling along magnetic fields in large gravity fields'. 'Comments on Babcock's theory of the solar magnetic field'.

M. Kopecký: 'New possibility of explaining the sunspot butterfly diagram'.

J. V. Tuominen: 'Comments on solar magnetic field model and possible application to magnetic stars'.

F. G. E. Krause: 'A solution of the dynamo problem by means of a linear theory of M.H.D. turbulence'.

Because of shortage of time the following two papers were unable to be presented:

E. Dobov: 'Similarity relations in magnetohydrodynamics'.

V. V. Vitkevič: 'Some data on an experimental investigation (by methods of radioastronomy) of the velocity and direction of the interplanetary plasma'.

44. COMMISSION DES OBSERVATIONS ASTRONOMIQUES AU-DEHORS DE L'ATMOSPHERE TERRESTRE

Reports of Meetings

PRESIDENT: L. Goldberg.

Scientific Meeting, 23 August 1967

In opening the meeting, the President stated that in view of a number of recent highly successful space missions it had been decided to devote the entire meeting of the Commission to the reporting of new results. The solar results were to be given at this meeting, which was being held jointly with Commissions 10 and 12, while the results of other space investigations would be reported at an extra session scheduled for Saturday, 26 August.

W. M. Neupert reported preliminary results obtained from observation of the solar flare X-ray emission line spectrum of iron from 1·3 to 20 Å with two crystal spectrometers carried on board the third Orbiting Solar Observatory (OSO-III). Intensity variations with time were monitored for a series of new emission lines tentatively identified as transitions in Fe XXVI through Fe XX.

H. E. Hinteregger next described observations with a second scanning spectrometer carried on OSO-III and recording ultraviolet radiation from the whole sun in the spectral region 250–1300 Å.

R. G. Teske gave an account of his monitoring measurements of the soft solar X-ray flux from the entire solar disc using a standard ion chamber photometer with a pass band in the wavelength range 8–12 Å. This instrument is carried in the wheel section of OSO-III.

W. M. Glencross reported results obtained by the University College-Leicester group on the spectral distribution of soft X-ray emission as a function of position on the solar disc in the spectral region from 6 to 20 Å.

L. E. Peterson reported the analysis of the first two weeks of observations of solar and cosmic X-rays at energies greater than about 8 kilovolts in an experiment carried in the wheel section of OSO-III.

K. A. Pounds discussed possible models of the solar corona that would be in accord with the Leicester measurements of absolute intensities of solar X-ray spectra below 20 Å in the absence of solar flares.

R. Tousey then summarized the work of the Naval Research Laboratory in photographing the sun with rocket coronagraphs. In the most recent flight, very interesting observations were made of the outer solar corona, the zodiacal light and the earth-lit new moon as seen projected against the corona.

Preliminary results from a radio astronomy experiment on the OGO-III spacecraft were presented in a paper by F. T. Haddock. The objective of the experiment was to determine the kinds of dynamic spectra that the sun displays at frequencies in the range from 4 MHz to 2 MHz. Many interesting solar events have been recorded.

A. Vinogradov concluded the meeting with an account of two investigations in the USSR, the first on theoretical calculations on the spectrum of the solar corona in the spectral region 5–30 Å and the second on observations of the spectra of laboratory plasmas in the spectral region 120–200 Å, with special reference to the spectra of Ca X–XVI.

Scientific Meeting, 26 August 1967

The session consisted of a series of contributed papers concerned with observations of radiation from non-solar sources made with rockets, satellites and space probes.

D. C. Morton gave an account of Princeton University rocket observations of ultraviolet spectra obtained on five flights over a period of two years beginning in June, 1965. The spectral resolution is 1–3 Å and the wavelength range 1130–3670 Å. The objects observed include eleven early type stars as well as the planets Venus and Jupiter.

T.P. Stecher next presented the results of similar observations on rocket stellar spectra obtained with a photoelectric recording scanning spectrophotometer.

H. Friedman reported results obtained by G. Carruthers at the Naval Research Laboratory in which a spectrograph was used in conjunction with an electronic image converter to obtain stellar spectra in the region 1000–1300 Å. The results were consistent with those reported earlier by Dr. Stecher and Dr. Morton.

V.G. Kurt described measurements of ultraviolet radiation from the Milky Way measured with photon counters on board the spaceships Venus II and Venus III. Measurements were made in two narrow spectral regions, one including Lyman alpha and the other just outside.

G. Courtès then gave a paper on the rocket photography of emission nebulosities in the ultraviolet. The zodiacal light was also recorded with good density at a distance of about 8° from the sun.

G.W. Clark presented preliminary results of an experiment aboard the OSO-III Satellite directed at the measurement of cosmic gamma rays above 70 MeV.

The next two papers dealt with measurements of low frequency cosmic noise. F.T. Haddock described results obtained by D. Walsh from a rocket launched on June 30, 1965 containing three receivers operating at 0·75 MHz, 1·5 MHz, and 2 MHz. The apogee of the rocket orbit was 1701 kilometers. J.K. Alexander described results obtained on board the Advanced Technology Satellite (ATS) launched on 6 April 1967 into an elliptical orbit with an apogee altitude of about 11000 kilometers and a perigee altitude of about 180 kilometers. The experiment operates at 7 discrete frequencies from 450 kHz to 4 MHz.

The session was concluded with a brief report by I. Tindo on measurements of soft X-ray radiation from the moon-orbiting satellite Luna-12.

The complete list of papers and titles follows:

Wednesday, 23 August 1967

W.M. Neupert, W. Gates, M. Swartz, R. Young: 'Observation of the Solar Flare X-ray Emission Line Spectrum of Iron From 1·3 Å to 20 Å.'

H.E. Hinteregger: 'Solar EUV Spectra'.

R.G. Teske: 'Solar X-rays'.

P.J. Bowen, R.L.F. Boyd, W.M. Glencross, C.R. Negus, K.A. Pounds, A.P. Willmore: 'Spectral Distribution of Soft X-ray Emission as a Function of Position on the Solar Disc'.

H.S. Hudson, D.A. Schwartz, L.E. Peterson: 'Observations of Solar and Cosmic X-rays on the OSO-III'.

K.A. Pounds, P.C. Russell: 'Some Recent Solar X-ray Photographs'.

R. Tousey: 'Photography with a Rocket Coronagraph'.

F.T. Haddock: 'Solar Bursts 2000–4000 kHz'.

N. Basov, V. Boyko, J. Voinov, A. Kononov, S. Mandel'štam, J. Sklizkov, A. Vinogradov: 'Laboratory Investigations of Ca x–xvi Spectra'.

J. Bergman, L. Vainštein, A. Vinogradov: 'Spectrum of Solar Corona in Region 5–30 Å'.

K.A. Pounds, K. Evans: 'The X-ray Emission Spectrum of a Solar Active Region'.

Saturday, 26 August 1967

D.C. Morton: 'Rocket Stellar Spectra'.

T.P. Stecher: 'Rocket Stellar Spectra with a Scanning Spectrophotometer'.

H. Friedman: 'Stellar Spectra 1000–1300 Å'.

V.G. Kurt: 'The UV Observations of the Milky Way'.

G. Courtès: 'Ultraviolet Emission Nebulosities'.

G. Clark: 'Gamma Rays'.

D. Walsh: 'Low Frequency Cosmic Noise'.

J.K. Alexander: 'Advanced Technology Satellite-Radio Astronomy Results'.
S. Mandel'štam, I. Tindo, G. Čeremuhin, L. Sorokin, A. Dmitriev: 'Soft X-ray Radiation of the Moon and Cosmic Rays Background—as observed by Luna-12'.

SERVICE D'AÉRONOMIE DU CNRS

SOLAR OBSERVATIONS IN THE NEAR ULTRA VIOLET

A study of the solar emission between 2000 Å and 3000 Å has been actively pursued by a group of the Service d'Aéronomie du CNRS under J.E. Blamont.

Monochromatic pictures of the sun at 2200 Å, 2700 Å, and 2900 Å were obtained from a Veronique flight in November 1964 (**1, 2**). During this flight it was demonstrated that even at 40 km of altitude, data could be obtained. The main effort of the group was then shifted towards balloon observations. The following results have been obtained.

(1) first observations of the granulation at 2000 Å (resolution of one to two seconds of arc) by J.E. Blamont and G. Carpentier, with the discovery of bright elements of small size (a few seconds) and long lifetime (more than 30 minutes) (**3**). These inhomogeneities should be taken into account when computing radiative transfer in this spectral range.

(2) limb darkening observations in the continuum and in the lines between 1800 Å and 2900 Å by R. M. Bonnet and J.E. Blamont (**4, 5**); the continuous opacities of the photosphere layers were deduced from these observations and are not in agreement with usual models. From the source functions obtained for the shortest wavelengths, a value for the temperature minimum was proposed for inclusion in the Bilderberg model. (**5, 6**). The depth of formation of faculae seems to be situated deep in the photosphere, since faculae can be observed close to the center of the disk from 2200 Å downwards, where the brightness temperature is 5500 °K.

(3) metallic photoionization discontinuities were observed at least for Mg I but the most conspicuous feature revealed was the discontinuity around 2080 Å, tentatively attributed to the photoionization of Aluminum. This discontinuity is much larger at the center than at the limb where it practically vanishes. It is responsible for the change in the aspect of the sun between 2100 Å and 2050 Å (disappearance of limb darkening, appearance of chromospheric features and faculae) (**7**) and it explains also the absence of the discontinuities of Si(I) (^3P, ^1D, ^1S states) in the Tousey spectra. Optical depth unity at 2000 Å seems to appear at the chromospheric level.

(4) stigmatic spectra of the Mg II doublet were obtained from a balloon borne spectrograph, providing the profile of the Mg II lines with a resolution of 0.04 Å and limb center variations of the profile of the lines with a space resolution of one minute of arc. (**8**).

Date	*Vehicle*	*Experiment*
November, 8, 1964	Veronique	limb darkening by pictures at 2200 Å, 2700 Å and 2900 Å
May 2, 1966	Balloon	granulation at 2000 Å
October 3, 1966	Balloon	limb darkening by pictures at 1976, 2235 Å.
January 13, 1967	Veronique	stigmatic spectrum from 1800 to 2800 Å.
March 22, 1967	Balloon	discontinuity at 2080 Å.
June 13, 1967	Balloon	stigmatic spectra of Mg II doublet

References

1. Blamont, J.E., Bonnet, R.M. 1965, Solar Limb Darkening in Ultraviolet Continuum (2000 Å–3000 Å), *Space Research*, **VI**, 895.
2. Blamont, J.E., Bonnet, R.M. 1966, Mesure de l'assombrissement du centre au bord du disque solaire entre 2000 et 3000 Å. *C. r. Acad. Sci. Paris*, **262**, 152.

3. Blamont, J.E., Carpentier, G. 1967, Mise en évidence de la granulation solaire à 2000 Å, *Solar Physics*, **1**, 180.
4. Bonnet, R.M., Blamont, J.E., Gildwarg, Ph. 1967, Limb Darkening Observation from 1800 Å to 2800 Å, *Ap. J. Letters*, **148**, L 115.
5. Bonnet, R.M., Blamont, J.E. 1967, Limb Darkening Observations between 1800 Å and 2900 Å. *Solar Physics*, **2**, 3–33.
6. Bonnet, R.M., Lago, B. 1966, Inversion de l'intégrale de Laplace. Application aux courbes d'assombrissement centre bord obtenues dans l'ultraviolet, *C. r. Acad. Sci. Paris*, **262**, 1636.
7. Blamont, J.E., Bonnet, R.M. 1967, Mesure de l'assombrissement centre-bord du soleil entre 2000 et 3000 Å, *C.r. Acad. Sci. Paris*, **264**, 1158–1161.
8. Lemaire, Ph., Blamont, J.E., Stigmatic Balloon spectra of the Solar MgII Doublet (to be published, *Ap. J. Letters*)

Appendix

ADDENDA TO THE REPORT OF COMMISSION 44 PUBLISHED IN TRANSACTIONS IAU, VOLUME XIII A

Addenda to Section I.1., page 993

K.G. Henize signale que son groupe vient de terminer la construction d'un télescope *f*/3 de 15 cm d'ouverture muni d'un prisme objectif. Ce dernier sera utilisé par les astronautes du programme d'Applications Apollo pour obtenir des spectres d'étoiles dans la région 1400–3000 Å. La dispersion sera de 90 Å/mm à 1500 Å.

G.R. Carruthers rapporte que son groupe au US. Naval Research Laboratory projette deux vols supplémentaires de spectrographes munis de convertisseurs d'images. Le premier sera du type 'Schmidt' et couvrira la région de 1230 à 1800 Å. Le second système sera composé uniquement de surfaces réfléchissantes et couvrira le domaine de 950 à 1400 Å.

Miss Roman apporte les précisions suivantes relatives au satellite OAO-B auquel il est fait allusion p. 993, pénultième paragraphe.

'Le but de cette expérience, dont le responsable est A. Boggess, est de déterminer la composition et les propriétés des étoiles A et B, d'étoiles variables et des étoiles B particulières en utilisant des données spectrophotométriques absolues dans le domaine de 1000 à 4000 Å avec une résolution de 2 Å. Un télescope Cassegrain de 36 pouces suivi d'un spectromètre très ouvert en constitue le système optique. Six photomultiplicateurs utilisés en compteurs d'impulsions seront utilisés pour l'enregistrement des données'.

Addendum to bibliography II.1., page 1000.

Carruthers, G.R. 1967, An upper limit on the concentration of molecular hydrogen in interstellar space, *Astrophys. J.*, **148**, L141.
Henize, K.G., Wackerling, L.R., O'Callaghan, F.G. 1967, Ultraviolet stellar spectroscopy on Gemini 11, *Science*, **155**, 1407.
Jenkins, E.B. 1967, An observation of the interstellar Lyman-α absorption profile for several Orion stars. *Astr. J.*, **72**, 304.
Peytreman, E., Janin, G. 1966, Problems concerning the scientific program of an astronomical satellite. *C.r. Séanc. Soc. Phys. Hist. nat. Genève*, **1**, no. 2, 57.
Roman, N.G. 1967, Observatory reports. NASA, Washington, D.C. *Astr. J.*, **72**, 280.
Garton, W.R.S. 1966, Spectroscopy in the vacuum ultraviolet, in *Advances in Atomic and Molecular Physics*, **2**, D.R. Bates and I. Estermann, eds. (New York: Academic Press), p. 93.
Stone, M.E., Morton, D.C. 1967, Far ultraviolet interstellar absorption lines in Delta and Pi Scorpii, *Astrophys. J.*, **149**, 29.
Underhill, A.B. 1967, Near ultraviolet spectrum from a B star. *J. R. astr. Soc. Can.*, **61**, 9.

APPENDIX V. MEASUREMENTS OF RADIATION FROM THE EARTH

Dr A. I. Lebedinskij calls attention to the important series of observations of radiation from the Earth, which have been carried out from Soviet Earth satellites, and are described in the publications referenced below.

References

1. Lebedinskij, A.I., Iozenas, V.A. Krasnopolskij, V.A., Kuznetcov, A.P. 1967, The Earth's Ultraviolet Spectrum According to the Measurements from the 'Cosmos-65' Satellite, *Space Research*, **7**, North-Holland Publ. Co., Amsterdam.
2. Lebedinskij, A.I., Boldyrev, V.G., Tulupov, V.I., Kudinova, G.N., Levčenko, A.D., Švidkovskaja, T.E. 1967, The Spectrum of the Earth's Heat Radiation According to the Observation from the 'Cosmos-45', 'Cosmos-65', and 'Cosmos-92' Satellites, *Space Research*, **7**, North-Holland Publ. Co., Amsterdam.
3. Iozenas, V.A., Krasnopolskij, V.A., Kuznetcov, A.P., Lebedinskij, A.I. The features of the Ozone Planetary Distribution According to Observations from the Artificial Earth Satellites, *Space Research*, **8**, North-Holland Publ. Co., Amsterdam (in press).
4. Lebedinskij, A.I., Andrianov, Yu.G., Barkova, G.N., Boldyrev, V.G., Karavaev, I.I., Leljakina, T.A., Poljakova, T.G., Safronov, Yu.P., Tulupov, V.I. *Space Research*, **8**, North-Holland Publ. Co., Amsterdam (in press).
5. Lebedinskij, A.I., Krasnopolskij, V.A., Kuznetcov, A.P., Iozenas, V.A. 1965, Study of the Geomagnetospheric Radiation in the Visible and Ultraviolet Ranges, *Issledovanija kosmičeskogo prostranstva* (Space exploration), ed. 'Nauka', Moscow, p. 77.
6. Lebedinskij, A.I., Glovatskij, D.N., Tulupov, V.I., Hlopov, B.V., Fomičev, A.A., Šuster, G.I. 1965, Infrared Spectrophotometry of the Earth's Heat Radiation, *Issledovanija kosmičeskogo prostranstva* (Space exploration), ed. 'Nauka', Moscow, p. 65.
7. Krasnopolskij, V.A., Kuznetcov, A.P., Lebedinskij, A.I. 1966, The Earth's Ultraviolet Spectrum According to the Measurements from 'Cosmos-65' Satellite, *Geomagnetism and Aeronomy*, **6**, no. 2, 185.
8. Lebedinskij, A.I., Tulupov, V.I., Safronov, Yu.P., Andrianov, Yu.G., Karavaev, I.I. 1967, Statistical Characteristics of the Earth's Radiation to the Space in the 7–26 micron Wavelength Range, *Geomagnetism and Aeronomy*, **7**, no. 3, 423.

45. COMMISSION DES CLASSIFICATIONS SPECTRALES ET INDICES DE COULEUR A PLUSIEURS BANDES

Reports of Meetings

PRESIDENT: W.P. Bidelman.
SECRETARY: V.C. Reddish.

Business Meeting, 23 August 1967

DRAFT REPORT

The Draft Report was approved as submitted, subject to the addition of several supplementary reports which appear as an Appendix to the present document.

OFFICERS

The President reported that the following recommendations were being made for the Commission's officers for the next triennium:
President: Ch. Fehrenbach.
Vice-President: B.E. Westerlund.
Organizing Committee: W.P. Bidelman, E.K. Haradze, C.O.R. Jaschek, H.L. Johnson, W.W. Morgan.

CATALOGUES

M.F. McCarthy reported that the Vatican Observatory was setting up a card index of references to emission-line stars contained in various published catalogues. C.O.R. Jaschek and K.G. Henize reported similar reference catalogues in preparation. A working group consisting of these three investigators was appointed to coordinate these efforts and to arrange for eventual publication.

W.P. Bidelman expressed the hope that positions for all new discoveries would be given using the equinox of 1900. M.F. McCarthy stated that at least some coordinates should always be given, even if only X, Y positions on the Palomar Sky Survey charts. W.P. Bidelman spoke of the need for a general reference catalogue giving full bibliographic data for individual stars. About 10^6 stars would be involved, and the resources of a large institution such as NASA would certainly be required to provide such a centralized information service. Several members of the Commission expressed interest in, and support for, this proposal.

SPECTRAL ATLASES

W.W. Morgan reported that the new Yerkes spectral atlas will be ready about July 1968, and also that the Arizona-KPNO spectral atlas, in which he is also somewhat involved, should be ready at approximately the same time. Ch. Fehrenbach's objective-prism spectral atlas, with a dispersion of 80 Å/mm, is nearly completed. Work is also continuing on the Córdoba spectral atlas at a dispersion of 42 Å/mm. It was stated by W.C. Seitter that her Bonn objective-prism spectral atlas was very near completion, and that it would soon be available. The dispersions involved are 240 Å/mm, 645 Å/mm, and 1280 Å/mm. The President drew attention to the high quality of the spectra reproduced in this work.

STANDARDS

The need for further faint spectral standards was discussed at some length, and it was pointed out that the extensive work on faint selected-area stars already carried out by N.G. Roman could well provide the basis for the selection of a substantial number of standard regions. This problem was referred to the officers of the Commission for further action.

COLLOQUIA AND SYMPOSIA

It was suggested that the Commission consider the advisability of sponsoring meetings on the subjects of 'Stellar Rotation' and 'Automated Spectral Classification' within the next few years.

Scientific Meetings, 24, 25, 29 August 1967

The Commission participated in three scientific sessions, the first of which was a joint meeting with Commission 28. In view of the fact that the papers given at these sessions will shortly be published, and also since it is very difficult to adequately summarize the lively discussions that took place at these sessions, it is thought best to simply acknowledge here the presentation of the following papers:

W. W. Morgan: 'Stellar compositions of giant galaxies'.
H. Spinrad: 'Spectrophotometry and stellar content of the nuclei of galaxies'.
J.G. Davies: 'Stellar diameters determined by use of the Brown-Twiss interferometer'.
C.J. Durrant: 'Cambridge narrow-band studies of silicon stars'.
A.G.D. Philip: 'Photoelectric photometry of field horizontal-branch stars'.
J.A. Graham: 'Photoelectric photometry of globular cluster horizontal-branch stars'.
D.L. Crawford: 'Narrow-band photometry of peculiar B stars'.
H.J. Wood: 'Photoelectric studies of spectrum variability in peculiar A stars'.
W.P. Bidelman: 'The detection and photometry of weak-metal stars'.
E.K. Haradze: 'Spectral classification work at the Abastumani Observatory'.
P. Pesch: 'Studies of M dwarfs in the Hyades'.
P.J. Treanor: 'Studies of M dwarfs in the Pleiades'.
W.P. Bidelman: 'The work of the Michigan Curtis Schmidt telescope at Cerro Tololo, Chile'.
K. Gyldenkerne: 'Narrow-band photometry of late-type stars'.
J. Rousseau: 'Six-color observations of late-type main sequence stars'.
P.B. Boyce: 'Spectral scans and photometry of late-type dwarfs'.

Appendix

MATERIAL SUPPLEMENTARY TO DRAFT REPORT

A. Beer (Cambridge): Work has been carried out at Cambridge to assemble the material for a reduction to a unified system of all measurements of indicators of equivalent width of Balmer lines in several thousand early-type stars, essentially O7-B9—dealt with in about sixty observational series—that have so far been listed in the existing literature, covering some three decades, or have been supplied by colleagues from their unpublished observations. The dependence of these Balmer-line intensities on the different instruments and observational methods, as well as on apparent brightness, spectrum, luminosity class, the system of assigned weights, on the errors due to photographic, spectroscopic and random effects, etc.—altogether nine independent parameters—is being analysed in joint work with Professor Th. Schmidt-Kaler in Bonn. A complete computer analysis of the intercorrelation between these series, and their necessary adjustment, provides corrected standard values for all of these stars. The corresponding distance scale is calibrated with the aid of clusters. It is hoped that the final outcome of this homogenization will be improved insight into the structural features of the Milky Way in our neighborhood, spiral arms or otherwise, out to about 4 kpc and beyond.

J.S. Neff (State U. of Iowa): An intermediate-bandpass photometric system of broad scope has been established under the direction of Dr W.W. Morgan; the fundamental work was done at the McDonald Observatory in the years 1963–64. Laboratory calibrations were carried out using the facilities of the Washburn Observatory and the spectroscopy department of the University of Wisconsin. Details of this work are described by Neff and Travis (*Astr. J.* **72**, 48, 1967), who also describe a new reduction method that yields a substantial increase in accuracy. Some applications of this system, which is called the CHI system, have been made by Morgan and Neff (*Astr. J.* **69**, 145, 1964) and Neff (*Astr. J.* **71**, 202, 1966). More detailed descriptions of these applications are in preparation. At the University of Iowa Observatory the CHI system is used with a 24-inch Cassegrain reflector and an integrating digital photometer. Tests have indicated that observations can be made of stars as faint as $m_{0.55} = 14$.

A method has been developed to derive T_e, g, blocking by lines in the various bandpasses, and reddening for RR Lyrae stars under the assumption that the variable is in a quasi-stationary state near minimum light. This method can also be applied to non-variable stars of spectral types earlier than G8. For stars of later type empirical methods must be used.

V. Straižis (Vilnius Observatory): For the past several years I have been attempting to establish a new multicolor system that will permit a purely photometric two-dimensional classification of stars and a determination of interstellar absorption. Preliminary results have been given in *Trans. IAU.* **12B**, 261, 1966. The first part of the work consisted of a theoretical analysis of stellar energy curves of stars of different spectral types and luminosity classes, as determined by various investigators. The properties of different magnitude systems with intermediate bandwidths were examined in the wavelength range λλ3000–6500. As a result a seven-color system with central wavelengths λ3500 (U), λ3750 (P), λ4050 (X), λ4600 (Y), λ5100 (Z), λ6200 (T), and λ6400 (S) and with half-widths of the order of 200 Å was chosen. The system seems to render possible the purely photometric determination of spectral and luminosity classes for B to M stars, the estimation of their metallicity, and the unambiguous evaluation of interstellar absorption for each individual star. Only five magnitudes, U, P, X, Y, Z, are needed to classify the O, B, A, and F stars, while five others, X, Y, Z, T, S, are adequate to classify the K and M stars. For the classification of the G-type stars a combination of both five-color systems may be necessary. The classification is accomplished by different versions of the Q method as used by the Walravens, Borgman, Strömgren and Golay. The system can be used with either interference or glass filters, which makes it simple to adapt it to photographic methods. Observational tests of the new system are being carried out both in Vilnius and in the Crimea. An indirect confirmation of the system's efficiency has been obtained by comparison with other multicolor systems of intermediate bandwidth.

The system is expected to be useful in enabling two- or three-dimensional classification of faint stars in densely crowded fields of the Milky Way where it is difficult to obtain objective-prism spectra.

46. COMMISSION POUR L'ENSEIGNEMENT DE L'ASTRONOMIE

Report of Meetings

PRESIDENT: E. L. Schatzman.
VICE-PRESIDENT: Miss E. A. Müller.
TEMPORARY SECRETARY: T. L. Page.

First Meeting, 23 August 1967

The President in chair.

Concerning the Draft Report M. Minnaert objected to the statement on p. lxxxix which said that Commission 46 should not be concerned with secondary school education. E. Schatzman explained that this was the result of a questionnaire sent to members of the IAU, but agreed that this paragraph in the Draft Report should be altered. A discussion followed which ended with the general agreement that (1) astronomy should be taught well in secondary schools, (2) professional astronomers had neglected this in the past, (3) secondary school teachers need help in preparing courses (like ESCP in the United States), (4) the mathematics teachers might be encouraged to include astronomy in their courses, (5) the Commission 46 must promote the teaching of astronomy in schools, integrating astronomy with other sciences such as in USA ('Earth Sciences') and Australia ('Integration of Science').

The Commission's President proposed that the organizing committee should remain small, but that each member nation of the IAU designates one member interested in the teaching of astronomy to join the Commission as a member in order to enable the Commission to keep in touch with the educational programs in all countries. It was agreed that each member nation of the IAU be invited to designate one member of Commission 46 who is to be selected according to his capabilities and his interest in promoting the teaching of astronomy in his own country. The President will inform the Executive Committee that some of the member nations may not have a suitable person and that Commission 46 may ask for 2 or 3 persons from some other nations when that seems desirable. The total membership of Commission 46 should be kept below 45.

The President announced his decision to retire as president of Commission 46, and proposed that the Commission's Vice-president, Miss Professor E. A. Müller, should succeed him and that Dr J. Kleczek should become Vice-president. The President's propositions were accepted by the Commission. J. Kleczek had been acting as Secretary of the 'Summer Schools for Young Astronomers' sponsored by UNESCO, and he will continue doing so during the period he will be holding the Vice-presidency of Commission 46. It is planned to appoint an Assistant Secretary to the Secretary of the UNESCO Summer Schools who will in turn become secretary of these schools and Vice-president of Commission 46 when the Vice-president takes over the presidency upon retirement of the current president.

After a lengthy discussion the following members were proposed to form the organizing Committee of Commission 46: G. O. Abell, B. J. Bok, E. K. Haradze, V. Kourganoff, M. G. J. Minnaert, T. L. Page, E. L. Schatzman.

The President discussed the report of new mathematics teaching in Belgium and the pamphlet 'Observatories of the World' published by the Smithsonian Astrophysical Observatory, Cambridge, Mass., U.S.A., which should be kept up to date for the use of astronomers wishing to study or observe elsewhere.

J. Kleczek reported on the 6½-week 'Summer School in Astronomy for Young Astronomers' at Manchester University, England. He explained that students of poor preparation gained more at the Summer School but that they could not reach the level of the better prepared students. Miss E. A. Müller pointed out the advantages of having a few well prepared students join these Summer Schools for the benefit of the other students.

With a vote of appreciation to E. Schatzman for his services as President of the Commission the meeting adjourned until August 28.

Second Meeting, 28 August 1967

The Vice-president in chair.

Because the retiring President was unable to be present, the Vice-president took over the chair for the second and third meeting of the Commission. She pointed out that the following two main topics should be discussed during the two planned sessions of the Commission: I. The teaching of integrated sciences and of astronomy at all levels [i.e. (a) elementary school, (b) secondary school, (c) university—undergraduate level, (d) university—graduate level, (e) general public], and II. the aims and goals and the future programs of Commission 46. In order to start and stimulate these discussions, she read the following address of the retiring President, *E. Schatzman*, to the Commission and interested participants:

'I would like to express my views on the general problem of the teaching of astronomy and astronomical education. After three years of activity as President of Commission 46 on the Teaching of Astronomy, I have reached a certain number of conclusions.

(1) The problems concerning the teaching of astronomy differ very much when we consider the elementary school, the high school, the university (undergraduate), and the formation of research workers.

I shall insist several times in this speech on the following point: at the elementary level, from the elementary school to undergraduate students, astronomy has not to be taught as a separate subject; astronomy should be integrated with the other sciences, and I shall later give several examples, in order to explain more clearly what I mean by integration.

On the other hand, at the graduate level the essential problem is to teach basic methods and to suggest ideas in order to promote what will be original work five years from now. It may appear a fantastic ambition to teach today what is yet unknown. However, I see no other way than the paradox to express what I think is the basic nature of graduate education.

(2) Let us consider first the elementary level, when knowledge is essentially qualitative and descriptive.

The sciences of the earth, geography, geology, are taught at a very elementary level. To place the earth in the solar system, and the solar system in the universe, to explain gravitational attraction, and let the child understand that people of Prague are not upside down with respect to the people of Mount Stromlo, are fundamental aspects of our knowledge, and one part of the education in geography. I shall leave to the representatives of the IAU at the meeting of the Inter-Union Committee on Scientific Education, in Varna in 1968, the task of writing in great detail the possible program of an integration of astronomy, geography and geology into a unique program of earth sciences for the junior high school.

I would like to draw attention to the work of the mathematicians. For more than fifty years the union of mathematics has studied with great care the problems of the teaching of mathematics in the secondary school. As it is well known, this care comes purely to a research carried into the field of the logics of mathematics, and to the idea that the young child is very logical and does not understand what is not logical. Mathematicians think that the difficulty of many children in understanding mathematics is not due to the difficulty of the concepts which are introduced, but to the fact that many of them are introduced in a quite irrational manner, and at a pace which does not follow the development of the mind.

Physicists are just beginning to worry about the pedagogy of physics, but we have the good luck that astronomy, in fact, begins much before physics. Are we unable to imitate the efforts of the physicists, and shall we ignore the longlasting work of the mathematicians? The difficulty is not so much a difficulty of choosing a program, but a difficulty of choosing a method for teaching the basic concepts of space, time, motion, matter. Measuring the classroom with a rod, and mapping it, is a simple and exciting task at the age of 8 or 9. Shall we immediately conclude that it is just as easy to understand how the shape and size of the earth have been determined, and how the distance to the sun has been measured?

(3) A second level begins with the introduction of the quantitative sciences, physics and chemistry.

The USA, with the American Institute of Physics, and the Australians have published the results of a very interesting work. The USA now has a textbook for the teaching of the earth sciences; the Australians have a textbook for the teaching of sciences.

These attempts are quite interesting for the astronomer and for the scientist. The astronomer finds for the first time books in which astronomy has a very large place, about one-third of the space in the American book on the earth sciences, about one-fifth of the space in the Australian book. From the point of view of the scientist, this is a remarkable effort to present the unity of the world to children.

This, however, cannot have a world-wide utilization. Even at the level of the last three years of high school, these books avoid almost entirely the use of mathematics, and remain qualitative and descriptive. As far as the IAU is concerned, I believe that a very interesting task is ahead of us, with the discussions which shall begin with the representatives of the other unions.

However, this does not avoid the task that has to be undertaken in each country, if not in each university; in the present situation, it seems impossible to incorporate in the curriculum 60 hours of astronomy, as Minnaert suggested several years ago; but it seems possible to incorporate astronomy in physics, mechanics, geography, natural history (especially geology). However, I shall repeat that the task is not simply to determine which subjects have to be taught, but to find the proper pedagogical methods.

Physicists have begun to worry about the way in which physical concepts are understood and acquired, and I would like to stress, in this connection, the importance of the work of Bachelard in '*La formation de l'esprit scientifique* (4ème Edition, Paris, Vrin, 1960). The book of Bachelard is known among philosophers rather than among scientists. It is interesting to see how the basic concepts of force and mass are related to the experiments that people have made with their bodies since the time they were babies and children. Deeply rooted in the unconscious part of the mind is the emotional knowledge of weight and inertia.

The problem of the teaching of science is not simply to have the most gifted to understand mass and force, but to have all children understand these concepts. I believe that it is only by the connection of rational concepts with unconscious knowledge that it will be possible to overcome the difficulty.

Similarly, astronomers may wonder how to teach the basic concepts of space, time, matter and evolution, and how to include these questions in education. The immensity of space, the unity of space and time are yet difficult to grasp, even for people having a good education. The teaching of physics is traditionally related to instruments and machinery and not turned towards the basic laws of physics. The program of integration naturally emphasizes the basic laws of physics.

(4) It is probably in the university, at the undergraduate level, that the most interesting task can be undertaken. The Feynman lectures in physics, in which a certain amount of astronomy and astrophysics is included, are a good example of what can be done. However, it may not be an easy task to discuss with a physicist whether the Rutherford formula is more or less important than the laws of the Keplerian motion. The best solution would be to have astronomers or astrophysicists undertaking the task of teaching an elementary course in physics and including in such a course the basic knowledge of the universe.

(5) I shall not develop the question of the teaching of astronomy at the graduate level. This is already a research problem and each astronomer, each institution, should have the freedom of deciding its own research program and of recruiting with its own goals.

The Vice-president then opened the discussion on the topics of the agenda.

T. Page spoke on integrated science courses used in the United States for secondary school and college students not specializing in science. He described the Earth Sciences Curriculum Project, which cost about $5000000.—, and which brought over 50 scientists and teachers together to write a new text book called "Investigating the Earth". This book has been tested in more than 100 US high schools for an eight-month course, of which 20% is astronomy and the rest is geology, meteorology and oceanography. The astronomy writers were C.M. Huffer, J.A. Hynek, T.L. Page and A. Slettebak. Although the high school students were very interested in astronomy (including solar physics, stellar interiors, origin of the solar system, galactic structure and cosmology), the teachers did not like to teach it, because they knew very little about the subject. Integrated science courses for university students at Chicago, Wesleyan and Yale Universities also used astronomy to capture student interest. T. Page described one he gave at Yale for 220 students which developed three broad topics: (1) *Theories of motion*, a historical treatment from Ptolemy through Kepler, Galileo, Newton and planetary orbits to relativity theory; (2) *Matter and energy*, leading to the theory of stellar evolution through thermodynamics, nuclear physics, and stellar interiors; (3) *The origin of life*, including geology, the formation of the solar system, early conditions on earth and the formation of organic compounds. A student poll showed that about 75% found this course very interesting, but 5% found it confusing and dull.

In answer to questions from Reeves, Abell and Minnaert, Page said that these courses are so called 'problem oriented', intended to show students the current research problems in astronomy and other sciences, and to show how much previous work led up to these problems. The laboratory work required of the students was concentrated in two-week projects. Each student spent two weeks following up some problem of his choice (such as Flying Saucers, Pairs of Galaxies, the Law of Redshifts, Sizes of Lunar Craters etc.).

Miss V. Reade commented that, in general, students do not like the historical approach because they consider it requires them to learn 'out-dated, wrong facts'. Minnaert, Page and others replied that this misconception should be eliminated; the historical development of such concepts as gravitation or the galaxy or stellar structure is characteristic of science, and provides basic understanding of advanced astronomical problems.

M. Minnaert spoke on laboratory exercises used in first-year Dutch university courses. He stressed the fact that secondaryschool preparation in astronomy varies widely, so that experience in elementary, fundamental observations must be given in the university. Most of the exercises are carried out by pairs of students working together for one or two nights, all students working on the same exercise at the same time. Calculations are performed with slide rules. About 20 small telescopes (4-inch, *f*/12) with hour-angle and declination circles were built in the University Observatory shop. About 15 small microdensitometers were also built for student use; the purchase price of parts totalled about $ 100.– each. These student instruments were used in exercises on variable stars, spectral line profiles, count of stars in globular clusters etc.

Valuable laboratory manuals mentioned by M. Minnaert include those by Stetson, Dussheimer, Jaschek, the Utrecht manual (soon to be published in English) and Atanasijevic (for advanced courses). It was agreed by the commission that a list of these manuals and other teaching aids should be compiled by a working group.

E. Kononovič stressed the need for astronomy textbooks that proceed rapidly to the modern research problems, and described the new Soviet textbook *General Astronomy* by Bakulin, Moroz and Kononovič (Moscow, 1967). The first few chapters cover coordinates and positional astronomy, and the remainder is devoted to the sun, planets, stars, galaxies and cosmogony. The Vorontsov-Velyaminov exercise book can be used with this text.

The Vice-president in chair then adjourned the second session and proposed the discussions to be continued at the scheduled hour of the following day.

Third Meeting, 29 August 1967

The Vice-president in chair.

The Vice-president reminded the members of the Organizing Committee who were able to be present, and the audience consisting of about 80 people, that during this last session of Commission 46 the future programs of the Commission were to be discussed.

Referring to the teaching methods *M. Lavagnino* spoke in favor of the historical method of integrated science courses. He recommended Pannekoek's textbook which he found most useful for his courses in La Plata, Argentina. *Miss B. Middlehurst* commented that introductory courses should always start with observational facts familiar to the students.

P. Routly spoke about the American Astronomical Society's Committee on Education and its efforts to improve the general public understanding of astronomy and astronomical research. He described two color motion pictures with sound made by the American Astronomical Society. The first shows how a radio astronomer (M.S. Roberts) works. This film, called 'Radio View of the Universe' was shown at the Twelfth IAU. General Assembly in Hamburg in 1964. The second film, 'Exploring the Milky Way', shows how an optical astrophysicist (G.W. Preston) works at Lick Observatory. Both films are now ready for distribution through Modern Learning Aide, New York City. Teachers may contact P. Routly at Princeton University Observatory, Princeton, New Jersey, U.S.A. for further information.

In this connection the Vice-president mentioned several excellent French films on various astronomical subjects. Other films were mentioned such as the Canadian 'Universe' by D. McRea, and the Australian on Radio Astronomy by T. Gold. It was agreed that such educational films should also be listed by the Commission.

P. Moore spoke about popular books on astronomy written for the general public with serious interest in astronomical developments. Many young people are attracted to professional astronomy by reading these books. Several comments were made agreeing that such books are needed so that the general public will become more familiar with current astronomical research goals. *G. Hawkins* mentioned the need for simpler books for very young readers, several of which have recently been published in the United States.

The Vice-president then proposed three continuing programs for Commission 46. She described them in the following way and asked for comments:

(1) Summer School for Young Astronomers, supported by UNESCO, to be continued by J. Kleczek and his Working Group, as decided in the first meeting of the Commission.

(2) Cooperation with Commission 38 on Exchange of Astronomers which should include the possibility of

(a) sending some of the good students of the UNESCO Summer Schools to work at some larger observatory or institution; and

(b) arranging for visiting professors to smaller observatories or institutes who will give lecture courses and advise the students on research programs with the available telescopes. The Vice-president stressed the importance of these visiting professorships and how greatly the smaller observatories and institutes may benefit from them.

(3) Improvement of the teaching of astronomy at all levels by

(a) compiling lists of educational material (books, motion pictures, photographic plates, lab exercises etc.);

(b) exchanging data on courses, exam questions at various levels, and student performances;

(c) exchanging data on teacher preparation.

Many comments and suggestions followed the presentation of the proposed programs. In particular, the formation of working groups was discussed, and the availability of educational material at various observatories was commented. It was agreed that the detailed planning of the future programs of Commission 46 was to be made by the Organizing Committee, and that it was up to the Organizing Committee to form working groups once the detailed programs have been established.

Asking the audience, the Vice-president learned that 12 countries were represented at the third meeting of Commission 46. One member of each of these countries agreed to help the Commission by establishing a list of educational material etc. in his own country.

Thanking the audience for the many comments and suggestions and calling the members of the Organizing Committee for a special session the Vice-president adjourned the meeting.

Meeting of the Organizing Committee, 30 August 1967

The following members of the Organizing Committee attended the special meeting: Miss E. A. Müller, Vice-president, G. O. Abell, E. V. Kononovič (replacing E. K. Haradze who was unable to attend), M. G. J. Minnaert and T. L. Page. They decided to collect data for the list of Educational Materials by a letter (drafted by T. L. Page) to the 12 persons who had offered their collaboration during the third meeting of the Commission, and to other persons interested in the teaching of astronomy. A draft list will be collected by Miss E. A. Müller, the new president, and sent in mimeographed form to the Commission members, to all national representatives, and to the persons who submit items for the list. This will be done by about March 15, 1968. The national representatives will be asked to make corrections and additions, so that a final list can be published by the I.A.U. in September 1968.

After this list is prepared, the Organizing Committee plans to establish 2 or 3 working groups to arrange further exchange of data under point (3) of Miss E. A. Müller's programs for Commission 46.

ADDENDUM TO THE REPORT OF COMMISSION 46

(1) The following Institutions have answered to the questionnaire after completion of the report:
Observatory and Astronomical Institute, Jena
Commission pour l'Enseignement de l'Astronomie (Comité National Belge d'Astronomie).

Le Comité Belge s'est particulièrement intéressé aux incidences de la réforme de l'enseignement des mathématiques dans l'enseignement moyen. Ce problème présente un intérêt suffisant pour donner ici un bref rapport sur les travaux de la Commission Belge.

Réforme de l'Enseignement des Mathématiques dans le secondaire en Belgique

Parallèlement à son travail d'enquête générale concernant l'enseignement de l'Astronomie en Belgique à tous les dégrés, la Commission ad-hoc créée par le Comité National Belge d'Astronomie s'est vivement préoccupée du projet de réforme de l'enseignement des mathématiques dans le secondaire. Elle a émis deux voeux; insistant particulièrement sur *l'applicabilité* des notions enseignées pour les sciences d'observation et a entretenu une correspondance suivie avec les Ministres de l'Education Nationale ayant la réforme de l'enseignement des mathématiques dans leurs attributions. Ces voeux ont été transmis pour information, aux commissions universitaires chargées d'élaborer la liste des désidérata des facultés des sciences et des sciences appliquées qui a été portée devant la Commission Ministérielle pour l'Enseignement des Mathématiques.

Le projet de réforme repose sur les bases suivantes:

(*a*) Donner aux grandes structures mathématiques toute l'importance qu'elles méritent. Ceci suppose un enseignement basé sur des notions modernes coordonnées.

(*b*) Eviter à tout prix un enseignement qui ne serait pas basé sur des réalités naturelles. En particulier, insister sur la nécessité d'une maîtrise complète des techniques du calcul (calcul algébrique, matriciel, différentiel, intégral). Réduire aussi la brêche séparant la géométrie de l'algèbre. Donner aussi la possibilité de mathématiser une situation.

Il est hors de propos de décrire ici le programme proposé, celui-ci devant être approuvé et amendé encore par la Commission du Programme. Un accord a été obtenu dès à présent au sein des Universités belges sur la base du document fourni par les Facultés des Sciences et des Sciences Appliquées de l'Université de Bruxelles.

La Commission Belge a dressé un tableau complet des enseignements d'Astronomie dans les universités belges mais sans indication de recrutement.

Princeton University Observatory has given the titles of the courses, the number of students attending the courses and the number of B.A. and Ph.D. awarded every year: 4 and 2.

(2) A list of observatories where training of astronomers could be undertaken has been obtained by Dr Page. A limited distribution to the members of the organizing committee has been made.

(3) *Ecoles de jeunes astronomes.* 12 étudiants ont suivi les cours de l'Ecole de Manchester qui ont duré 6 semaines 1/2 (Inde: 3, Egypte: 2, Portugal: 1, USA: 1, Roumanie: 1, Tchécoslovaquie: 1, Pologne: 2, Hollande: 1).

Un rapport a été établi, indiquant les résultats atteints par chacun des étudiants. Il semble que les étudiants des pays en voie de développement aient grandement bénéficié du contact avec des étudiants de pays développés. Les étudiants des pays développés ont en général aisément atteint le niveau du Ph.D. mais les étudiants des pays en voie de développement ont fait des progrès considérables au cours de ces 6 semaines et les meilleurs d'entre eux peuvent être valablement recommandés pour un stage de recherche en Observatoire.

Ceci correspond tout à fait à ma propre expérience. Il faut d'abord arracher les étudiants à de très mauvaises habitudes intellectuelles et les obliger à apprendre une foule de choses qu'ils ignorent avant de les amener à un niveau comparable au Ph.D.

L'expérience peut-elle être continuée?

Les propositions suivantes peuvent être examinées:

Michigan. Participation des étudiants au trimestre printemps-été (spring-summer term May 1 to August 20). La subvention devrait couvrir les frais de voyage et de séjour des jeunes astronomes.

Arcetri. Professor Righini offers to organize a course for 10 students in June-July 1968, in solar astronomy.

We can also plan to have courses organized later at the following places:

JILA, 1969, Solar Physics

Edinburgh 1970, spectroscopy (solar, stellar), photometry, polarimetry.

On the other hand, the following institutions have given positive answers concerning the accommodation of students recommended by the IAU schools:

Kitt Peak	a few graduate students	research only
Caltech	a few good students	courses, research
Cambridge	a few good postgraduates	informal astronomical training
Lick	visitors	research
Arcetri	2	research
Max Planck München	well prepared students	research
Abastumani	1 to 3	astronomical training
Harvard	1 to 2	entire Ph. D. program

I believe that quite a few other institutions might accommodate for 1 or several years one or two young astronomers under recommendation of the IAU.

I suggest therefore the following:

The IAU school continues for 3 more years. I suggest that the executive committee appoints an assistant secretary to Dr Kleczek. When Dr Kleczek will become President of the Commission 46, the assistant secretary will become secretary.

Proposal should be made to the executive committee within 1 year.

(4) *CIES.* E. Schatzman a représenté l'UAI à une réunion de la CIES le 9 Mars 1967. L'opération la plus intéressante de la CIES est sans doute la réunion de Varna en 1968 sur l'intégration des Enseignements scientifiques. Les principes de la réunion de Varna seront exposés dans la réunion de discussion de la commission 46 sur la pédagogie. L'UAI pourrait désigner deux délégués à Varna.

PART 4

ASTRONOMER'S HANDBOOK

ASTRONOMER'S HANDBOOK

The Astronomer's Handbook (IAU Transactions Vol. XIIC, edited by J.-C. Pecker in 1966) contains many documents of permanent value. Only those parts which changed after the XIIIth General Assembly will be given here in the form of supplements. The numbering of paragraphs in XIIC will be adhered to in order to facilitate the reference from one volume to the other.

The reader is referred to the IAU Transactions Vol. XIIC in particular in all matters relating to the Style Book (p. 55 in French, p. 101 in English), to the Organization of Symposia and Colloquia (p. 129), and to the Statutes and By-laws (p. 138).

PART 1 OF ASTRONOMER'S HANDBOOK

I. SHORT HISTORY OF THE INTERNATIONAL ASTRONOMICAL UNION

A, B, C no change against Trans. IAU, XIIC, 3.

D. Table 2 should be supplemented by the following data:

XIIIth General Assembly, 1967, Prague (Czechoslovakia)	
Members of the Union (after the G. A.)	2009
Participants (788 Members)	1835
Number of Commissions	38
Volumes of Transactions A	1156 pages
B	319 pages
Highlights 1967	558 pages
President: O. Heckmann	1967–1970
General Secretary: L. Perek	1967–1970
Assistant General Secretary: C. de Jager	1967–1970
Vice-Presidents: W. Fricke	1964–1967
W. N. Christiansen	1964–1970
M. K. Vainu Bappu	1967–1973
L. Gratton	1967–1973
J. Sahade	1967–1973

II. ADMINISTRATION AND FINANCES OF THE UNION

A. ADMINISTRATION

The address of the Secretariat is

International Astronomical Union
Czechoslovak Academy of Sciences
Budečská 6
PRAHA 2
Czechoslovakia.

The Secretariat is headed by the General Secretary, L. Perek. The staff of the office consists of the Assistant Secretary, A. Jappel, and a secretarial employee, Mrs J. Daňková. All problems related to Symposia and Colloquia are dealt with by the Assistant General Secretary,

Prof. C. De Jager
Astronomical Observatory "Sonnenborgh"
Zonnenburg 2
UTRECHT
The Netherlands.

He also maintains the contact with the publisher.

B. FINANCES

(*a*) The XIIIth General Assembly raised the unit of contribution to 900 gold francs, that is to 293.94 U.S. dollars, for the years 1968, 1969, 1970.

III. THE INTERNATIONAL ASTRONOMICAL UNION AND THE OTHER INTERNATIONAL SCIENTIFIC ORGANIZATIONS

A. THE INTERNATIONAL COUNCIL OF SCIENTIFIC UNIONS

The composition of other international bodies changes at other intervals than does that of the IAU. The reader is therefore referred to the Yearbook of the ICSU for the current year. Only IAU representatives are mentioned in the following paragraphs.

The IAU is represented by the General Secretary, L. Perek, in the Executive Committee of the ICSU.

B. IAU REPRESENTATIVES IN SCIENTIFIC AND SPECIAL COMMITTEES AND COMMISSIONS OF ICSU

1. *Committee on Space Research (COSPAR)*: L. Gratton (till 1968).
2. *Special Committee for the International Year of the Quiet Sun (IQSY).*
   The Committee was discontinued as from 1 January 1968.
3. *International Abstracting Board (IAB)*: J.B. Sykes.
4. *Federation of Astronomical and Geophysical Sciences (FAGS)*: D.H. Sadler *(Vice-President)*, B. Guinot.
   IAU representatives in FAGS services:
   Bureau International de l'Heure (BIH): H.M. Smith, F. Zagar.
   International Polar Motion Service: Organizing Committee of IAU Commission 19.
   Quarterly Bulletin on Solar Activity: Organizing Committee of IAU Commission 10.
   International Ursigrams and World Day Service (IUWDS): R. Michard.
   Solar Particles and Radiation Monitoring Organization (SPARMO): C. De Jager.

C. IAU REPRESENTATIVES IN INTER-UNION COMMITTEES AND COMMISSIONS OF ICSU

1. *Comité International de Geophysique (CIG)*: G. Righini (Vice-President), R. Michard.
2. *Inter-Union Committee on Frequency Allocations for Radio Astronomy and Space Science (IUCAF)*: J.F. Denisse, V.A. Sanamjan, F.G. Smith (Chairman), O. Hachenberg.
3. *Inter-Union Committee on the Ionosphere (IUCI)*: discontinued.
4. *Inter-Union Commission on Solar Terrestrial Physics (IUCSTP)* created in 1966: Z. Švestka.
5. *Inter-Union Commission on Science Teaching (IUCST)*: Miss Edith Müller.
6. *Committee for Data on Science and Technology (CODATA)*: Mrs. Ch. Moore-Sitterly.
7. *Inter-Union Commission on Spectroscopy*: A.H. Cook, B. Edlén, J.G. Phillips, M.J. Seaton.

D. IAU REPRESENTATIVES IN OTHER ORGANIZATIONS

1. *La Fondation Internationale du Pic du Midi*: A. Lallemand.
2. *Le Comité Consultatif pour la Définition de la Seconde (CCDS) du Bureau International des Poids et Mesures*: B. Guinot, H.M. Smith.
3. *Le Comité Consultatif pour la Définition du Mètre du Bureau International des Poids et Mesures*: A.H. Cook.
4. *Le Comité Consultatif International des Radiocommunications*: H.M. Smith.

IV. SERVICES AND FUNCTIONS OF THE IAU

A, B, C. No change of principle against Trans. IAU XIIC, 17.

D. THE CENTRAL BUREAU FOR ASTRONOMICAL TELEGRAMS

Smithsonian Astrophysical Observatory, Cambridge, Massachusetts 02138, *U.S.A.*
Director of the Bureau: *Dr B.G. Marsden*
Associate Directors: *Dr O.J. Gingerich, Dr R.B. Southworth*

For nearly a century astronomers have maintained a telegraph and cable network for rapid communication of such celestial news as the discovery of comets and novae. When the International Astronomical Union was established the Central Bureau for Astronomical Telegrams was placed under its jurisdiction (through Commission 6 of the Union), initially at the Brussels Observatory, and then for a great many years at the Copenhagen Observatory. Since 1965 the Bureau has been at the Smithsonian Astrophysical Observatory.

Telegrams or cables can be received any time of day or night at the Smithsonian Observatory. The cable address is SATELLITES, NEWYORK, the telex number is 92-1428, and the Western Union address for astronomers in North America is RAPID SATELLITE, CAMBMASS. In North America, telegrams are sent out collect via Western Union to any astronomers agreeing to pay for the messages. The larger share of other addresses are reached through the Regional Warning Centers of the International Ursigram and World Days Service (the AGIWARN network). The financial arrangements depend on the particular Regional Warning Centers, whose addresses appear below; observatories served through the Meudon center are expected to place a deposit with the Central Bureau.

France, British Isles, Spain, Portugal, Italy, Greece, Middle East, North Africa, Germany, Austria, Belgium, Denmark:
Dr P. Simon, Observatoire de Paris, Section d'Astrophysique, 92-Meudon, France.

Sweden, Norway, Finland:
Mr Per Akerlind, Board of Swedish Telecommunications, Stockholm 16, Sweden.

The Netherlands:
Mr H. Van Lohuizen, Radio Receiving Station NERA, Nederhorst den Berg, The Netherlands.

Czechoslovakia, Poland, Hungary, Rumania:
Mr P. Tříska, Czechoslovak Academy of Sciences, Geophysical Institute, Ionosphere Department, Praha, Spořilov, Czechoslovakia.

U.S.S.R., People's Republic of China, Bulgaria:
Mrs L.N. Liakhova, Institute of Terrestrial Magnetism, Ionosphere and Radio Propagation (IZMIRAN), Leninskii r-p, P.O. Vatutenki, Moscow, U.S.S.R.

Japan:
Dr H. Uyeda, Radio Research Laboratories, Ministry of Posts and Telecommunications, Kokubunji P.O., Koganei-shi, Tokyo, Japan.

Australia, New Zealand:

Mr F.E. Cook, Ionospheric Prediction Service, Commonwealth Center, Elizabeth Street, Sydney, N.S.W., Australia.

It is not necessary for observatories to receive all the "astrograms"; instead, subscribers can specify only the desired standard categories (the identifying telegram code word appears in the second column):

A.	ALPHA	Comet discoveries, brighter than magnitude 12.
B.	BETA	Comet discoveries, fainter than magnitude 12.
C.	CHARLY	Comet observations and brief finding ephemerides (primarily for observers).
D.	DELTA	Precise positions, orbits and ephemerides (primarily for computers).
E.	ECHO	Novae or supernovae, brighter than magnitude 12.
F.	FRANCE	Novae or supernovae, fainter than magnitude 12.
G.	GAMMA	All others, including planetary or lunar phenomena, spectroscopic binaries, etc.

Almost all the messages are transmitted according to a telegraphic cipher code. The code is described in *Trans. IAU*, **XIIC**, pp. 34–38, to which the corrections in *Trans. IAU*, **XIIIA**, pp. lxxxii–lxxxiii should be noted. Corrected reprints of the code may be obtained from the Central Bureau. Some messages cannot be cast in the standard coded form, but observers are nevertheless urged to include sufficient redundancy to prevent errors.

In addition to the astrograms, the Central Bureau maintains an even wider coverage by means of its postcard *Circulars*, which are issued irregularly about 50 times a year. These contain confirmations of the telegrams, predicted ephemerides for comets and unusual asteroids as well as observed positions, observations of novae and supernovae, and other astronomical information that requires rapid dissemination. About 600 observatories and amateurs subscribe to the *Circulars*.

The *Circulars* are available by advance subscription only, for 50 or 100 issues (not corresponding to a calendar year). These financial arrangements are entirely separate from the telegrams. The subscription rates below are those as of 1 December 1967 and depend on the destination and mode of transport:

	North America	Other
100 consecutive issues, airmail	\$15.50	\$24.50
100 consecutive issues, surface mail	13.50	18.50
50 consecutive issues, airmail	9.00	13.50
50 consecutive issues, surface mail	8.00	10.50

There will be a reduction of \$1.00 for subscribers who do not require invoices. Checks, bank drafts or money orders should be made payable to the "Central Bureau for Astronomical Telegrams".

V. THE PUBLICATIONS OF THE INTERNATIONAL ASTRONOMICAL UNION

A. TRANSACTIONS OF THE IAU

XIIIA—Prague 1967; pp. 1151, Reidel Publishing Company, Dordrecht, 1968
XIIIB—Prague 1967; pp. 319, Reidel Publishing Company, Dordrecht, 1968

B. PROCEEDINGS OF SYMPOSIA

24. Spectral Classification and Multicolour Photometry; ed. by K. Lodén, U. Sinnerstad; pp. 383; Academic Press (London), 1966; 120s.
25. The Theory of Orbits in the Solar System and in Stellar Systems; ed. by G. Contopoulos; pp. 380; Academic Press (London), 1966; 120s.
26. Abundance Determinations in Stellar Spectra; ed. by H. Hubenet; Academic Press (London), 1966; pp. 374; 120s.
27. The Construction of Large Telescopes; ed. by D.L. Crawford; pp. 234; Academic Press (London), 1966; 70s.

28. Fifth Symposium on Cosmical Gas Dynamics: Aerodynamic Phenomena in Stellar Atmospheres; ed. by R. N. Thomas; pp. 483; Academic Press (London), 1967; 140s.
29. Instability Phenomena in Galaxies; ed. by M. Arakeljan (in Russian). Orders to the Meždunarodnaja Kniga, Moskva, USSR, or agencies.
30. Determination of Radial Velocities and their Applications; ed. by A. H. Batten and J. F. Heard; pp. 262; Academic Press (London), 1967; 84s.
31. Radio Astronomy and the Galactic System; ed. by H. van Woerden; pp. 501; Academic Press (London), 1967; 140s.
32. Continental Drift, Secular Motion of the Pole, and Rotation of the Earth
33. Physics and Dynamics of Meteors
34. Planetary Nebulae
35. The Structure and Development of Solar Active Regions

Proceedings of Symposia Nos. 24 through 28, 30 and 31 are distributed by Academic Press, Inc. (London) Limited, Berkeley Square House, Berkeley Square, London W.1. In the Western Hemisphere by Academic Press, Inc., 111 Fifth Avenue, New York, 10003, N.Y. Orders should be sent to this address.

Proceedings of Symposia from No. 32 onwards are published and distributed by D. Reidel Publishing Company, Dordrecht, The Netherlands. Orders should be sent to this address.

C. OTHER PUBLICATIONS PUBLISHED AND SPONSORED BY THE IAU

1. *Circulars of Commissions of the IAU.* In addition to Commissions 26, 27, and 42 (see IAU Transactions XIIC, p. 44), Commission 41 (History of Astronomy) publishes the current bibliography on the history of Astronomy (editor P. G. Kulikovskij) and Information Circular. In 1964–1967, Nos. 7–14 appeared, edited by E. Rybka.

28. *Proceedings of IAU Colloquia*

(*a*) *Blanketing Effect*, edited by K. H. Böhm, *Journal of Quantitative Spectroscopy and Radiative Transfer*, vol. 6, no. 5, 1966

(*b*) *Late-type-Stars*, edited by Margherita Hack, Osservatorio Astronomico di Trieste, 1966

(*c*) *The Fine Structure of the Solar Atmosphere*, edited by K. O. Kiepenheuer, *Forschungsberichte* 12, Deutsche Forschungsgemeinschaft, Franz Steiner Verlag, Wiesbaden 1966.

(*d*) *Evolution of Double Stars*, edited by J. Dommanget, *Observatoire Royal de Belgique, Communications*, Série B, no. 17, 1967.

(*e*) *Les nouvelles méthodes de la dynamique stellaire*, edited by F. Nahon and M. Hénon, Editions du Centre National de la Recherche Scientifique, Paris, 1967.

(*f*) *International Conference on Spectroscopy*, 1967. Mimeographed proceedings in two volumes. Requests for copies should be addressed to: Scientific Information Officer, Bhabha Atomic Research Centre, Modular Laboratories, Trombay, Bombay-74, India.

VI. THE SYMPOSIA OF THE INTERNATIONAL ASTRONOMICAL UNION

Table 1

No.	Title	Date and place	Editor and Publishing House	Organized in co-operation with
29.	Instability Phenomena in Galaxies	May, 1966 Bjurakan, USSR	M. Arakeljan, Meždunarodnaja kniga (in Russian)	
30.	Determination of Radial Velocities and their Applications	June, 1966 Toronto, Canada	A. H. Batten and J. F. Heard IAU–Academic Press	

31. Radio Astronomy and the Galactic System	August, 1966 Noordwijk, The Netherlands	H. van Woerden IAU–Academic Press	URSI
32. Continental Drift, Secular Motion of the Pole, and Rotation of the Earth	March, 1967 Stresa, Italy	Wm. Markowitz and B. Guinot IAU–Reidel	IUGG
33. Physics and Dynamics of Meteors	September, 1967 Tatranská Lomnica, Czechoslovakiá	L. Kresák, P.M. Millman IAU–Reidel	
34. Planetary Nebulae	September, 1967 Tatranská Lomnica, Czechoslovakia	D.E. Osterbrock and R.C. O'Dell IAU–Reidel	
35. The Structure and Development of Solar Active Regions	September, 1967 Budapest, Hungary	K.O. Kiepenheuer, IAU–Reidel	IQSY

PART 2 OF ASTRONOMER'S HANDBOOK

STYLE BOOK

Authors are urgently requested to adhere to the Style Book as published in the IAU Transactions Vol. XIIC (French version, p. 56 to 100; English version, p. 101 to 127), especially to the part concerning

Transliteration
References
Abbreviations.

There is only one change against Vol. XIIC: Manuscripts will be sub-edited by the publisher and not by the IAU secretariat.

PART 3 OF ASTRONOMER'S HANDBOOK

EXECUTIVE COMMITTEE

MEMBERSHIP OF COMMISSIONS

ALPHABETICAL LIST OF MEMBERS

EXECUTIVE COMMITTEE — COMITÉ EXÉCUTIF

Appointed August 1967 — Élu Août 1967

PRESIDENT — PRÉSIDENT

Professor O. Heckmann, Director of the European Southern Observatory, Bergedorfer Straße 131, 205 Hamburg 80, G.F.R.

VICE-PRESIDENTS — VICE-PRÉSIDENTS

Professor M.K. Vainu Bappu, Director of the Astrophysical Observatory, Kodaikanal, India.
Professor W.N. Christiansen, School of Electrical Engineering, University of Sydney, Sydney, N.S.W., Australia.
Professor L. Gratton, Laboratorio di Astrofisica, Casella Postale 67, Frascati, Rome, Italy.
Professor J. Sahade, Observatorio Astronomico, La Plata, Argentina.
Professor M. Schwarzschild, Princeton University Observatory, Peyton Hall, Princeton, New Jersey 08540, U.S.A.
Professor A.B. Severnyj, Director of the Crimean Astrophysical Observatory, USSR Academy of Sciences, P/O Naučnyj, Crimea, U.S.S.R.

GENERAL SECRETARY — SECRÉTAIRE GÉNÉRAL

Dr L. Perek, Astronomical Institute of the Czechoslovak Academy of Sciences, Budečská 6, Praha 2, Czechoslovakia

ASSISTANT GENERAL SECRETARY — SECRÉTAIRE GÉNÉRAL ADJOINT

Professor C. de Jager, Astronomical Observatory "Sonnenborgh", Zonnenburg 2, Utrecht, The Netherlands.

COMPOSITION DES COMMISSIONS

MEMBERSHIP OF COMMISSIONS

† Members deceased since September 1967.

4. Commission des Ephémérides. (Ephemerides)

PRÉSIDENT: G. A. Wilkins.

VICE-PRÉSIDENT: J. Kovalevsky.

COMITÉ D'ORGANISATION: G. M. Clemence, G. A. Čebotarev, R. L. Duncombe, W. Fricke.

MEMBRES: Abalakin, Cunningham, de Greiff, Fayet, Gondolatsch, R. Haupt, Herrick, Lahiri, Lederle, Owaki, Rodriguez, Sadler (F. M. McBain), Sato, Sinzi, Trehan, Woolard.

MEMBRES CONSULTANTS: H. C. Freiesleben.

5. Commission des Analyses de Travaux et de Bibliographie. (Abstracts and Bibliography)

(Committee of the Executive Committee)

PRÉSIDENT: J. B. Sykes.

VICE-PRÉSIDENT: K. F. Ogorodnikov.

MEMBRES: Alter, Aoki, Beer, Belorizky, Bourgeois, Bouška, Chandrasekhar, Collinder, Dewhirst, Felber, Fricke, Harwood, Heinemann, Heintz, Henn, Hirose, Kleczek, Luplau-Janssen, Martynov, Pecker (J.-C.), Sticker, Wempe, Witkowski, Zinner.

MEMBRES CONSULTANTS: A. Dunn, G. Feuillebois, A. Kemp, N. B. Lavrova, L. N. Radlova, I. S. Ščerbina-Samojlova.

6. Commission des Télégrammes Astronomiques. (Astronomical Telegrams)

(Committee of the Executive Committee)

PRÉSIDENT: F. L. Whipple.

VICE-PRÉSIDENT: P. Simon.

MEMBRES: Buchar, Candy, Cunningham, Hers, Marsden, Martynov, Roemer.

7. Commission de la Mécanique Céleste. (Celestial Mechanics)

PRÉSIDENT: W. J. Eckert.

VICE-PRÉSIDENT: G. N. Dubošin.

COMITÉ D'ORGANISATION: G. A. Čebotarev, Y. Hagihara, J. Kovalevsky, Y. Kozai, P. J. Message, K. Stumpff, V. G. Szebehely, F. Zagar.

MEMBRES: Aksenov, Barlier, Belorizky, Böhme, Bozis, Brumberg, Cesco, Clemence, Contopoulos, Cook (A. H.), Cox (J. F.), Cunningham, Danby, Deprit, Dubošin, Duncombe, Fabre, Ferraz-Mello, Galibina, Garfinkel, Giacaglia, Goldreich, Goudas, Hadjidemetriou, Hamid, Herget, Herrick, Hertz, Hori, Jeffreys, Katsis, Kaula, Kozai, Kustaanheimo, Marsden, Meffroy, Merman (G. A.), Morando, Musen, Popović, Rabe, Rjabov, Roy, Sadler (D. H.), Schubart, Sconzo, Sehnal, Sibahara, Siegel, Slavenas, Stiefel, Šaraf, Thiry, Vinti, Wilkens, Wilkins.

MEMBRES CONSULTANTS: V. I. Arnold, S. P. Diliberto, D. G. King-Hele, J. Moser, I. D. Žongolovič.

8. Commission de l'Astronomie de Position. (Positional Astronomy)

PRÉSIDENT: A. A. Nemiro.

VICE-PRÉSIDENT: W. Fricke.

COMITÉ D'ORGANISATION: A. N. Adams, P. Lacroute, R. H. Stoy, R. H. Tucker, J. E. B. Von der Heide, H. Yasuda, M. S. Zverev.

MEMBRES: Adams, Anguita, Argyrakos, Atkinson, Barros, Bohrmann, Dambara, Débarbat, Demetrescu, De Moraes, Dick, Dieckvoss, Dubois-Chevalier, Duncombe, Eichhorn, Fayet, Gliese, Gordon, Guinot, Haas, Hansson, Heintz, Hoffleit, Kosin, Larink, Laustsen, Lévy, Maître, Marcus, Melchior, Murray, Nefed'eva, Novopašennyj, Okuda, Petrov, Pilowski, Podobed, Položencev, Postoiev, Proverbio, Quijano, Reiz, Requième, Rybka (P.), Sandig, Scheepmaker, Schmeidler, Scott (F. P.), Sémirot, Slaucitajs, Symms, Ševarlić, Tavastšerna, Teleki, Thomas (D. V.), Tsubokawa, Tucker, Tuzi, Van Herk, Verbaandert, Wood (H. W.), Woolsey, Zimmermann.

9. Commission des Instruments Astronomiques. (Astronomical Instruments)

PRÉSIDENT: J. Rösch.

VICE-PRÉSIDENT: V. B. Nikonov.

COMITÉ D'ORGANISATIONS: H.W. Babcock, W. A. Baum, R. G. Giovanelli, J. D. McGee, A. B. Meinel, O. A. Mel'nikov, B. Valníček.

MEMBRES: Aly, Atkinson, Bakos, Baranne, Barros, Beck, Behr, Bhattacharyya, Blitzstein, Brahde, Brealey, Brückner, Bowen (I. S.), Chincarini, Code, Couder, Cousins, Crawford, Delbouille, Delsemme, Dewhirst, Dimitroff, Dobronravin, Dollfus, Duchesne, Dunham, Ejnasto, Evans (J. W.), Fellgett, Ford, Gottlieb, Hall (J. S.), Høg, Hooghoudt, Ingrao, Ionnisiani, Jensch, Karandikar, Köhler, Kozyrev, Krasovskij, Kron (G. E.), Kurt, Lallemand, Linfoot, Linnik, Livingston, Meinel, Merman (N. V.) Mihel'son, Mohler, Monin, Moroz, Nikonov, Odgers, Petford, Pierce, Platzeck, Ramberg, Richardson (E. H.), Ring, Samaha, Scheffler, Seddon, Sisson, Smyth, Strand, Suharev, Ščeglov (V. P.), Väisälä, Wallenquist, Walker (M. F.), Walraven, Wayman, Wellgate, Wilcock, Wlérick, Zacharov.

MEMBRES CONSULTANTS: A. Bayle, L. E. Flory, Schwesinger, R. Wilson, C. J. Wynne.

10. Commission de l'Activité Solaire. (Solar Activity)

PRÉSIDENT: Z. Švestka.

VICE-PRÉSIDENT: J. T. Jefferies.

COMITÉ D'ORGANISATION: K. O. Kiepenheuer, R. Michard, S. Nagasawa, G. Newkirk, J. Rösch, A. B. Severnyj, M. Waldmeier.

SECRÉTAIRE: A. D. Fokker.

MEMBRES: Allen, Aly, Athay, Babcock (H. W.), Balli, Banin, Bell (B.), Belorizky, Bhatnagar (A.), Billings, Bonov, Bray, Brückner, Bruzek, Bumba, Cardus, Cimino, Coutrez, De Feiter, De Groot, De Jager, Dezső, Dizer, Dodson-Prince, Dollfus, Dubov, Dunn, Elste, Ferraro, Fracastoro, Friedman, Fritzová-Švestková, Gaizauskas, Giovanelli, Gleissberg, Gnevyšev, Gnevyševa, Godoli, Goldberg, Gopasjuk, Gullón, Gurtovenko, Hanasz, Hedeman, Hirayama, Howard (R. F.), Hyder, Iošpa, Jäger, Jakimiec, Jakovkin (N. A.), Jensen, Kleczek, Kopecký, Krat, Křivský, Krüger (A.), Kuklin, Kundu, Künzel, Lehnert, Leighton, Leroy, Letfus, Loughhead, McKenna, Macris, Makita, Martres, Mathias, Mattig, Maxwell, Menzel, Mergentaler, Mitra, Mogilevskij, Mohler, Moiseev, Moreton, Mustel, Nishi, Notuki, Obridko, Öhman, Orrall, Pick-Gutmann, Popovici, Razmadze (T. S.), Richardson (R. S.), Ringnes, Roberts (W. O.), Rompolt, Saito (K.), Sakurai, Sarabhai, Schlüter, Schmidt (H. U.), Servajean, Shapley (A. H.), Sheeley, Simon (P.), Sitnik, Slonim, Smerd, Smith (E. v. P.), Smith (Henri J.), Stepanov, Stešenko, Sturrock, Suemoto, Swarup, Šapošnikova, Šklovskij, Tandberg-Hanssen, Trellis, Tuominen, Valníček, Van 't Veer (F.), Vitinskij, Von Klüber, Warwick (C. S.), Wild (J. P.), Wlérick, Xanthakis, Zirin, Zwaan.

12. Commission de la Radiation et de la Structure de l'Atmosphère Solaire. (Radiation and Structure of the Solar Atmosphere)

PRÉSIDENT: M.N. Gnevyšev.

VICE-PRÉSIDENT: R.G. Athay.

COMITÉ D'ORGANISATION: L. Delbouille, R. Michard, M. Minnaert, Miss Edith A. Müller, A.K. Pierce, Z. Suemoto.

MEMBRES: Adam, Allen, Bappu, Billings, Blackwell, Blaha, Blamont, Böhm, Böhm-Vitense, Bray, Brück (H.A.), Bruzek, Christiansen, De Jager, Dubov, Edlén, Edmonds, Elste, Evans (J.W.), Gingerich, Giovanelli, Godoli, Gökdogan, Goldberg, Haupt (R.F.), Hotinli, Houtgast, Hubenet, Jäger, Jefferies, Kanno, Kato, Kawaguchi, Kiepenheuer, Konovovič, Kopecký, Laborde, Labs, Leighton, Leroy, Locke, Loughhead, Lüst (R.), Mathias, Matsushima, Mattig, Mergentaler, Migeotte, Moore-Sitterly, Mugglestone, Namba, Neven, Newkirk, Nicolet, Nishi, Orrall, Pande, Pajdušáková, Pecker (J.-C.), Pecker-Wimel, Peyturaux, Redman, Righini, Rigutti, Roland, Schröter, Seaton, Severnyj, Sitnik, Sobolev (V.M.), Swensson, Švestka, Tandberg-Hanssen, Thomas (R.N.), Tousey, Uchida, Unsöld, Voigt, Von Klüber, Waldmeier, Warwick (J.W.), Wilson (P.R.), Wlérick, Zirin, Zirker.

MEMBRES CONSULTANTS: Mrs J.S. Duncombe.

14. Commission des Données Spectroscopiques Fondamentales. (Fundamental Spectroscopic Data)

PRÉSIDENT: M. Migeotte.

VICE-PRÉSIDENT: A.H. Cook.

COMITÉ D'ORGANISATION: R.H. Garstang, G. Herzberg, W. Lochte-Holtgreven, Mrs Ch. Moore-Sitterly, J.G. Phillips, R. Tousey.

MEMBRES: Allen, Andrew, Baird, Bates, Blaha, Branscomb, Burgess, Carver, Douglas (A.E.), Dobronravin, Dressler, Dufay, Edlén, Engelhard, Essen, Felenbok, Garton, Green (L.C.), Herman, Hindmarsh, Humphreys, Jacquinot, Jordan (H.L.), Junkes, Kessler, Kiess †, King (R.B.), Lawrence, Layzer, Littlefield, Lochte-Holtgreven, Martin Jr. (W.C.), Melnikov, Milford, Minnaert, Mohler, Monfils, Naqvi, Nevin, Nicholls, Obi, Peach, Prokof'ev (V.K.), Rigutti, Rosen, Salpeter (E.W.), Schadee, Seaton, Smit, Somerville, Swings, Takayanagi, Terrien, Traving, Trefftz, Van Bueren, Van Regemorter, Varsavsky, Wares, Wiese, Wilson (R.), Zirin.

MEMBRES CONSULTANTS: W.S. Benedict, R.D. Cowan, H.M. Crosswhite, A. Dalgarno, H. Griem, V. Kaufman, A. Lagerquist, N.P. Penkin, I. Sobelman.

15. Commission pour l'Étude Physique des Comètes. (Physical Study of Comets)

PRÉSIDENT: L. Biermann.

VICE-PRÉSIDENT: V. Vanýsek.

COMITÉ D'ORGANISATION: G. Herzberg, B.J. Levin, E. Roemer, F.L. Whipple, K. Wurm.

MEMBRES: Arpigny, Bertaud, Beyer, Bobrovnikoff, Bouška, Brown (P. Lancaster), Dobrovolskij, Dossin, Dufay (J.) †, Guiguay, Haser, Hunaerts, Liller (W.), Lüst (Rhea), Lyttleton, Martel-Chossat, Miller (F.D.), Pflug, Poloskov, Probstein, Rémy-Battiau, Richter (N.), Rijves, Sivaraman, Swings, Vsehsvjatskij, Waterfield.

16. Commission pour l'Étude Physique des Planètes et des Satellites. (Physical Study of Planets and Satellites)

PRÉSIDENT: J.S. Hall.

VICE-PRÉSIDENT: B.J. Levin.

COMITÉ D'ORGANISATION: A. Dollfus, G. Herzberg, G.P. Kuiper, C.H. Mayer, C.E. Sagan, R. Wildt.

MEMBRES: Barabašov, Barreto, Baum, Bobrov, Brunk, Bullen, Camichel, Chamberlain, Collinson, Connes (J.), de Marcus, de Mottoni, de Vaucouleurs, Drake, Focas, Fox,

Gehrels, Giclas, Goody, Guérin, Günther, Hunten, Ingrao, Irvine, Jeffreys, Jelley, Kaplan (L.), Kellog, Kiess †, Kopal, Koval, Levin, Liddell, Link, Lipskij, Luplau-Janssen, Martynov, Menzel, Middlehurst, Millman, Miyamoto, Narayana, O'Keefe, Öpik, Pettengill, Rösch, Runcorn, Safronov, Shoemaker, Sinton, Smith (B.A.), Smith (Harlan J.), Spinrad, Strong, Tombaugh, Whitaker, Wilson (A.G.).

MEMBRES CONSULTANTS: R. Hide, G. Kennedy, G. McDonald, J. Mintz, J.B. Pollack.

17. Commission de la Lune. (The Moon)

PRÉSIDENT: A. Dollfus.

VICE-PRÉSIDENT: A.I. Lebedinskij †, G.H. Pettengill.

SECRÉTAIRE: B.M. Middlehurst.

COMITÉ D'ORGANISATION: G.P. Kuiper, D.H. Menzel, M.G.J. Minnaert.

MEMBRES: Arthur, Ashbrook, Barabašov, Bell (B.), Boneff, Botelheiro, Cameron (W.), Drofa, Džapiašvili, Eckert, Elston, Ezerskij, Fielder, Focas, Gold, Green (J.), Guth, Habibullin, Hall (R.G.), Haupt (R.F.), Heyden, Hopmann, Ingrao, Jakovkin (A.A.), Jeffreys, Karandikar, Kopal, Levin, Liddell, Markowitz, Martynov, Matsushima, Millman, Moore (P.), Nefed'ev, Nicholson, O'Keefe, Potter, Rösch, Sadler (F.M. McBain), Sagan, Sato, Schrutka-Rechtenstamm, Shoemaker, Sinton, Strom (R.), Sytinskaja, Troickij, Ueta, Urey, Watts, Weimer, Whitaker.

19. Commission de la Rotation de la Terre. (Rotation of the Earth)

PRÉSIDENT: P.J. Melchior.

VICE-PRÉSIDENT: H.M. Smith.

COMITÉ D'ORGANISATION: E.P. Fedorov, B. Guinot, W. Markowitz, S. Yumi.

MEMBRES: Abraham, Aksent'eva, Billaud, Bonanomi, Brkić, Buchar, Cecchini, Čudovičeva, Demetrescu, Dramba, Enslin, Essen, Fichera, Fleckenstein, Gama, Hall (R.G.), Hers, Iijima, Jeffreys (H.), Kalmykov, Koebcke, Kulikov, Lagrula, Lederle, Levallois, Meinig, Mihajlov, Miyadi, Narayana, Nicolini, Okuda, Opalski, Orte, Parijskij (N.N.), Pavlov, Popov, Postoiev, Proverbio, Randić, Rice, Romanskaja, Saharov, Scheepmaker, Sekiguchi, Stoyko (A.), Stoyko (N.), Sugawa, Ševarlić, Takagi, Tanner, Tardi, Teleki, Thomas (D.), Torao, Tsao, Vicente, Witkowski, Young (A.), Yumi.

20. Commission des Positions et des Mouvements des Petites Planètes, des Comètes et des Satellites. (Positions and Motions of Minor Planets, Comets and Satellites)

PRÉSIDENT: G.A. Čebotarev.

VICE-PRÉSIDENT: F.K. Edmondson.

COMITÉ D'ORGANISATION: S. Arend, P. Herget, H.H. Hirose, H.W. Wood.

MEMBRES: Ahmed, Böhme, Boyer, Bruwer, Candy, Clemence, Cunningham, Debehogne, Evdokimov, Galibina, Gehrels, Giclas, Gutiérrez-Moreno, Herrick, Hertz, Itzigsohn, Jahontova, Kamieński, Kazimirčak-Polonskaja, Kovalevsky, Kresák, Kuiper, Makover, Marsden, Michkovitch, Milet, Missana, Orlov, Oterma, Popović, Porter, Protitch, Rabe, Rasmusen, Robertson, Roemer, Schmitt, Schrutka-Rechtenstamm, Schubart, Sekanina, Sitarski, Strobel, Torroja, Väisälä, Van Biesbroeck, Van Houten, Van Houten-Groeneveld, Wilkins, Zadunaisky.

21. Commission de la Luminescence du Ciel. (The Light of the Night Sky)

PRÉSIDENT: M. Huruhata.

VICE-PRÉSIDENT: H. Elsässer.

COMITÉ D'ORGANISATION: J. Dufay †, M.F. Ingham, A. Lebedinskij †, F.E. Roach, G. Weill, J.L. Weinberg.

MEMBRES: Asaad, Barber, Bates, Blackwell, Chamberlain, Chapman, Cook (A.F.), Dauvillier, Dufay (M.), Dumont, Elvey, Fesenkov, Fesenkova, Galperin, Gauzit, Grandmontagne,

Harang, Haug, Hoffmeister†, Hunten, Jarrett, Kaplan (J.), Karandikar, Karjagina, Kastler, Krasovskij, Megrelišvili, Neužil, Nicolet, Pearse, Pettit-Knaflich, Robley, Sinvhal, Steiger, Tanabe, Vassy, Wallace, Weill, Weinberg.

MEMBRES CONSULTANTS: Nguyen Doan.

22. Commission des Météores et Météorites. (Meteors and Meteorites)

PRÉSIDENT: Z. Ceplecha.

VICE-PRÉSIDENT: R.E. McCrosky.

COMITÉ D'ORGANISATION: P.B. Babadžanov, W.G. Elford, C.L. Hemenway, A.A. Javnel', P.M. Millman.

MEMBRES: Abbott, Anders, Astapovič, Cook (A.F.), Davies, Davis (J.), Ellyett, Fedinskij, Fesenkov, Fireman, Gauzit, Guigay, Guth, Halliday, Hawkins, Hey, Hirose, Hodge, Hoffmeister†, Hoppe, Jacchia, Kaiser, Katasev, Keay, Kizilirmak, Kostylev, Kresák, Kresáková, Krinov, Levin, Lindblad (B.A.), Lovell, McIntosh, Nazarova, Nielsen, O'Keefe, Oleak, Olivier, Öpik, Plavcová, Rajchl, Southworth, Verniani, Whipple.

MEMBRES CONSULTANTS: Alexander, Henderson, Hey, Marvin, Mason, Sztrokay, Wickmann, Zahringer.

23. Commission de la Carte du Ciel. ('Carte du Ciel')

PRÉSIDENT: W. Dieckvoss.

VICE-PRÉSIDENT: P. Lacroute.

COMITÉ D'ORGANISATION: P. Couderc, A.N. Dejč, P. Herget, H. Kox, W.J. Luyten, P. Sémirot, H.W. Wood.

MEMBRES: Abhyankar, Bouigue, Dessy, Eichhorn, Fracastoro, Gallouët, Günther, Harris, Lourens, Murray, O'Connell.

24. Commission des Parallaxes Stellaires et des Mouvements Propres. (Stellar Parallaxes and Proper Motions)

PRÉSIDENT: W.J. Luyten.

VICE-PRÉSIDENT: S. Vasilevskis.

COMITÉ D'ORGANISATION: A.N. Dejč, V.V. Lavdovskij, J.v.B. Lourens, W.W. Morgan, C.A. Murray, K.Aa. Strand, P. van de Kamp.

MEMBRES: Blaauw, Cecchini, Delhaye, Dieckvoss, Eichhorn, Fatčihin, Franz, Fredrick, Giclas, Gliese, Goyal, Haas, Harris, Hoffleit, Kohlschütter, König, Lippincott, Meurers, Missana, Moreno, Paloque, Schilt, Smart, Stearns, Vyssotsky, Wagman, Wilkens.

MEMBRES CONSULTANTS: W.F. Van Altena.

25. Commission de Photométrie Stellaire. (Stellar Photometry)

PRÉSIDENT: A.W.J. Cousins.

VICE-PRÉSIDENT: D.L. Crawford.

COMITÉ D'ORGANISATION: A.N. Argue, M. Golay, R.H. Hardie, G.E. Kron, V.B. Nikonov, E. Rybka, J. Stock.

MEMBRES: Bahng, Becker (W.), Behr, Bigay, Bok (B.J.), Bok (P.F.), Borgman, Brück (H.A.), Chis, Dufay (J.) †, Eelsalu, Elvius (T.), Feinstein, Fellgett, Fitch, Gallouët, Graham, Gutiérrez-Moreno, Haffner, Hertzsprung †, Hiltner, Hoag, Holmberg, Irwin, Jackisch, Johnson (H.L.), McCarthy, Masani, Mayer (P.), Mianes, Moreno, Muller (A.B.), Mumford, Notni, Osawa, Philip, Quijano, Schmidt (H.), Smyth, Steinlin, Strajžic, Strohmeier, Velghe, Wallenquist, Walraven, Wesselink, Willstrop, Young (A.T.).

26. Commission des Étoiles Doubles. (Double Stars)

PRÉSIDENT: P.C. Couteau.

VICE-PRÉSIDENT: J. Dommanget.

COMITÉ D'ORGANISATION: O. J. Eggen, P. Kulikovskij, P. Muller, K. Aa. Strand, P. van de Kamp.

MEMBRES: Arend, Baize, Batten, Cester, Dejč (A. N.), Dhonomidjojo, Dick, Djurković, Finsen, Franz, Güntzel-Lingner, Hadjidemetriou, Heintz, Hertzsprung †, Holden, Hopman, Jeffers, Johnson (M. C.), Jonckheere, Kuiper, Kumar, Lippincott, Luplau-Janssen, Luyten, Mourao, Van Albada, Van Biesbroeck, Van den Bos, Wierzbinski, Wieth-Knudsen, Worley, Zagar.

27. Commission des Étoiles Variables. (Variable Stars)

PRÉSIDENT: L. Detre.

VICE-PRÉSIDENT: O. J. Eggen.

COMITÉ D'ORGANISATION: A. A. Bojarčuk, G. H. Herbig, R. Kippenhahn, K. Kwee, G. Odgers.

MEMBRES: Ahnert, Alanija, Arp, Ashbrook, Balázs-Detre, Banerji, Bateson, Bertaud, Bhatnagar (P. L.), Cesevič, Chavira, Chis, Christy, Daube-Kurzemniece, de Kock, Deutsch (A. J.), Efremov, Eskioglu, Feast, Fernie, Fitch, Gaposchkin, Geyer, Haro, Harwood, Hertzsprung †, Heyden, Hoffleit, Hoffmeister †, Holopov, Huffer, Huruhata, Ikaunieks, Iščenko, Jarzebowski, Jones, Kopylov, Kordylewski, Kraft, Kukarkin, Kumsišvili, Lacchini, Ledoux, Lortet-Zuckermann, McNamara, Mannino, Masani, Mavridis, Mayall (M. W.), Miller (W. J.), Nielsen, O'Connell, Oosterhoff, Opolski, Oskanjan, Payne-Gaposchkin, Peltier, Piotrowski, Plaut, Plavec, Rakos, Richter (G.), Rodgers, Romano, Rosino, Sawyer-Hogg, Shapley (H.), Slavenas, Smak, Smith (Harlan J.), Strohmeier, Swope, Tempesti, Vandekerkhove, Van Hoof, Wachmann, Walker (M. F.), Walraven, Wenzel, Wesselink, Wright (F. W.), Xanthakis, Zinner, Zwicky.

28. Commission des Galaxies. (Galaxies)

PRÉSIDENT: G. C. McVittie.

VICE-PRÉSIDENT: E. M. Burbidge (Mrs).

SECRÉTAIRE: E. Holmberg

COMITÉ D'ORGANISATION: D. S. Evans, D. Heeschen, E. Holmberg, B. E. Markarjan, R. L. Minkowski, M. Schmidt.

MEMBRES: Abell, Alfvén, Ambarcumjan, Arp, Baum, Bertola, Bigay, Boischot, Bok, Bondi, Bonnor, Burbidge (G. R.), Contopoulos, Courtès, Dessy, de Vaucouleurs, Elvius (A. M.), Field, Gascoigne, Haro, Heckmann, Heidmann, Hindman, Hodge, Hoyle, Humason, Johnson (H. M.), Kalinkov, King (I. R.), Kundu, Kustaanheimo, Layzer, Lindblad (P. O.), McCrea, Madwar, Mayall (N. U.), Mills, Morgan, Narlikar, Neyman, Oort, Omer, Osterbrock, Pachner, Page, Pishmish de Recillas, Poveda, Prendergast, Reaves, Reddish, Richter (N. B.), Rindler, Roberts (M. S.), Robinson (I.), Rosino, Rubin, Rudnicki, Sandage, Schmidt (M.), Schücking, Schürer, Sciama, Scott (E. L.), Sěrsic, Shakeshaft, Shane (C. D.), Shapley (H.), Spinrad, Stibbs, Thackeray, Treder, Vandekerkhove, Van den Bergh, Voroncov-Vel'jaminov, Westerlund, Whitford, Whitrow, Wild (P.), Wilson (A. G.), Zel'dovič, Zwicky.

29. Commission des Spectres Stellaires. (Stellar Spectra)

PRÉSIDENT: M. W. Feast.

VICE-PRÉSIDENT: Y. Fujita.

COMITÉ D'ORGANISATION: M.K.V. Bappu, G. Herbig, L. Houziaux, I. M. Kopylov, A. Przybylski, J. Sahade.

MEMBRES: Abhyankar, Abt, Aller, Andrillat (H.), Andrillat (Y.), Asaad, Babcock (H. W.), Berger, Bertaud, Bidelman, Bloch, Boggess, Bojarčuk, Bouigue, Bretz, Brück (H. A.), Buscombe, Cayrel-de-Strobel (G.), Cayrel (R.), Chalonge, Code, Conti, de Groot, Deutsch (A. J.), Divan, Dobronravin, Dolidze, Dunham, Edmonds Jr., Evans (D. S.), Fringant, Gollnow, Gorbackij, Gratton, Greenstein, Griffin (R. F.), Groth, Guthrie, Hack,

Haro, Heard, Heintze, Henize, Herman, Huang, Iwanowska, Jaschek (C.O.R.), Jaschek (M.), Jugaku, Keenab, Kienle, King (R.B.), Kraft, Larsson-Leander, McCuskey, McNamara, Mannino, Mel'nikov, Milligan, Morton (D.C.), Mustel, Nicholls, Nikitin, Oetken, Oke, Osawa, Payne-Gaposchkin, Pedoussaut, Ringuelet-Kaswalder, Rosen, Sargent, Searle, Slettebak, Smak, Stawikowski, Stecher, Svolopoulos, Swensson, Swings, Taffara, Thackeray, Thompson (G.I.), Underhill, Vandekerkhove, Van 't Veer-Menneret, Voigt, Wallerstein, Wehlau, Wellmann, Wempe, Willstrop, Wilson (O.C.), Wilson (R.), Wright (K.O.), Wyller, Yamashita.

30. Commission de Vitesses Radiales. (Radial Velocities)

PRÉSIDENT: D.S. Evans.

VICE-PRÉSIDENT: H.A. Abt

SECRÉTAIRE: W. Buscombe.

COMITÉ D'ORGANISATION: A.H. Batten, A. Blaauw, R. Bouigue, A.D. Thackeray.

MEMBRES: Abt, Beardsley, Bertiau, Boulon, de Vaucouleurs, Edmondson, Fehrenbach, Gollnow, Gratton, Griffin (R.F.), Harding, Heard, Kraft, Northcott, Oetken, Pagel, Preston, Rubin, Sahade, Underhill, Wayman.

31. Commission de l'Heure. (Time)

PRÉSIDENT: F. Zagar,

VICE-PRÉSIDENT: G.M.R. Winkler

COMITÉ D'ORGANISATION: D.J. Belocerkovskij, J. Bonanomi, H. Enslin, W. Markowitz, H.M. Smith, M. Torao.

MEMBRES: Abraham, Arbey, Bakulin, Billaud, Blaser, Brkić, Decaux, Demetrescu, Dingle, Dramba, Dubois-Chevallier, Fuchs, Gama, Gökmen, Gougenheim, Guinot, Hall (R.G.), Hemmleb, Hers, Iijima, Koebcke, Lacombe, Lederle, Lorón, Madwar, Melchior, Mihailov, Miyadi, Opalski, Orte Lledo, Pavlov, Postoiev, Proverbio, Randić, Sandig, Stoyko (A.), Stoyko (N.), Ščeglov, Širjaev, Šternberk, Takagi, Tardi, Thomson (M.M.), Tsao, Verbaandert, Von der Heide, Winkler.

MEMBRES CONSULTANTS: U. Adelsberger, L. Essen, J. Terrien.

33. Commission de la Structure et de la Dynamique du Système Galactique. (The Structure and Dynamics of the Galactic System)

PRÉSIDENT: G. Contopoulos.

VICE-PRÉSIDENT: S.W. McCuskey.

COMITÉ D'ORGANISATION: A. Blaauw, B.J. Bok, E.K. Haradze, F.J. Kerr, G.G. Kuzmin, C.C. Lin, M. Schmidt.

MEMBRES: Aarseth, Agekjan, Ambarcumjan, Baldwin, Bidelman, Bigay, Barbanis, Becker (W.), Beer, Blanco, Boulon, Burke, Crawford, Delhaye, Dieter, Dzigvašvili, Edmondson, Ejnasto, Elsässer, Elvius (A.M.), Elvius (T.), Fehrenbach, Fenkart, Fricke, Goldreich, Gyldenkerne, Haug, Heckmann, Hénon, Holopov, Hunter (C.), Idlis, Ikaunieks, Irwin, Iwaniszewska-Lubienska, Jaschek (C.O.R.), Johnson (H.M.), Kaburaki, King (I.R.), Kinman, Kukarkin, Kurth, Lecar, Lindblad (P.O.), Lunel, Luyten, Lynden-Bell, McCarthy, McRae, Malmquist, Mavridis, Mirzojan, Mohr, Morgan, Münch (G.), Murray, Nahon, Ogorodnikov, Ollongren, Oort, Ovenden, Perek, Perry, Pesch, Philip, Pilowski, Priester, Ramberg, Roman, Ruben, Rudnicki, Schalén, Schilt, Schmidt (H.), Schmidt (K.H.), Schmidt-Kaler, Šarov, Shimizu, Slettebak, Steinlin, Stibbs, Szebehely, The (Pik Sin), Toomré, Upgren, Vanderlinden, Vandervoort, Van Hoof, Varsavsky, Velghe, Vetešnik, Wayman, Weaver, Westerhout, Westerlund, Woltjer, Woolley.

34. Commission de la Matière Interstellaire et des Nébuleuses Planétaires. (Interstellar Matter and Planetary Nebulae)

PRÉSIDENT: D.E. Osterbrock.

VICE-PRÉSIDENT: F. D. Kahn.

COMITÉ D'ORGANISATION: L. H. Aller, M. J. Greenberg, S. B. Pikel'ner, L. Rosino.

MEMBRES: Andrillat (H.), Andrillat (Y.), Baldwyn, Behr, Bok (B. J.), Chopinet, Chvojková, Courtès, Davies (R. D.), Dieter, Dibaj, Divan, Dunham, Friedemann, Elvius (A. M.), Gehrels, Goldsworthy, Grzedzielski, Gurzadjan, Hall (J. S.), Haradze, Havtasi, Henize, Hiltner, Herbig, Johnson (H. M.), Kaplan (S. A.), Kerr, Kipper, Kogure, Kohoutek, Lambrecht, Liller, Loden, McCrea, McNally, Martel-Chossat, Menon, Minin, Minkowski, Münch (G.), Nandy, Osaki, Ozernoj, Pottasch, Razmadze (N. A.), Rožkovskij, Safronov, Savedoff, Schalén, Schatzman, Schmidt (K. H.), Schmidt-Kaler, Seaton, Seddon, Serkowski, Sharpless, Šklovskij, Sobolev, Spitzer, Strömgren, Syrovatskij, Takakubo, Thackeray, Thompson, Treanor, Van de Hulst, Van Woerden, Vanýsek, Viswanathan, von Hoerner, Voroncov-Vel'jaminov, Walker (G. A. H.), Westerhout, Whitford, Wickramasinghe, Zimmermann (H.).

MEMBRES CONSULTANTS: S. A. Colgate

35. Commission de la Constitution des Étoiles. (Stellar Constitution)

PRÉSIDENT: A. G. Masevič (Mrs).

VICE-PRÉSIDENT: E. Schatzman.

MEMBRES: Bhatnagar (L.), Biermann, Bondi, Boury, Brownlee, Burbidge (G. R.), Cameron (A. G. W.), Carson, Chandrasekhar, Chiu, Christy, Cowling, Cox (A. N.), Cox (J. P.), Demarque, Dingens, Epstein (J.), Ezer, Ferraro, Fowler, Gamow, Gjellestad, Hayashi, Henyey, Hitotuyanagi, Hoyle, Kaplan (S. A.), Kippenhahn, Kothari, Krook, Kumar, Kushwaha, Lebedinskij †, Lebovitz, Ledoux, McCrea, Marx, Masani, Mestel, Reeves (H.), Reiz, Rosseland, Roxburgh, Ruben, Salpeter (E. E.), Savedoff, Schwarzschild, Sears, Spiegel, Strömgren, Suda, Sweet, Tayler, Temesváry, Tuominen, Van der Borght, Vardya, Weigert, Wrubel, Ževakin.

36. Commission de la Théorie des Atmosphères Stellaires. (The Theory of Stellar Atmospheres)

PRÉSIDENT: A. B. Underhill (Miss).

VICE-PRÉSIDENT: R. Cayrel.

COMITÉ D'ORGANISATION: K. H. Böhm, V. V. Ivanov, J. T. Jefferies, R. N. Thomas, S. Ueno.

MEMBRES: Abhyankar, Aller (L. H.), Athay, Avrett, Baschek, Biermann, Böhm-Vitense, Busbridge, Carson, Cayrel (G.), de Feiter, Edmonds, Elste, Gebbie, Gingerich, Godoli, Gökdogan, Grant, Greenstein, Groth, Gussmann, Hack, Hardorp, Hearn, Henyey, Hitotuyanagi, Hotinli, Houtgast, Houziaux, Huang, Hummer, Hunger, Johnson (H. R.), Kalkofen, Kumar, Krishna Swamy, Kushwaha, McDonald (J. K. Petrie), Matsushima, Menzel, Mihalas, Mirzojan, Miyamoto, Müller (E. A.), Münch (G.), Mustel, Neven, Orrall, Pagel, Pecker (J.-C.), Pecker-Wimel, Petrie (J. K. McD.), Phillips, Pottasch, Rudkjøbing, Saito (S.), Schatzman, Sitnik, Sobolev, Sobouti, Spiegel, Stibbs, Strom (S.), Swihart, Traving, Uesugi, Unno, Unsöld, Van Regemorter, Van 't Veer (F.), Vardya, Weidemann, Wellmann, Whitney (C. A.), Wright (K. O.), Wrubel, Wyller, Zirker, Zwaan.

37. Commission des Amas Stellaires et des Associations. (Star Clusters and Associations)

PRÉSIDENT: M. Golay.

VICE-PRÉSIDENT: G. Larsson-Leander.

COMITÉ D'ORGANISATION: Mrs K. A. Barhatova, S. B. C. Gascoigne, I. R. King, L. Rosino, J. Ruprecht, Miss H. H. Swope, M. F. Walker.

MEMBRES: Aarseth, Alter, Artjuhina, Balázs, Becker (W.), Blaauw, Bouvier, Cuffey, Eggen, Feast, Feinstein, Golay, Haffner, Hénon, Holopov, Hoag, Johnson (H. L.), Kadla, Lavdovskij, Lodén (K.), Lohmann, Lynden-Bell, Mavridis, Menon, Meurers, Morgan, Pishmish de Recillas, Reddish, Šarov, Sandage, Sawyer-Hogg, Serkowski, Shapley (H.), Smyth, Takase, The, Van den Bergh, Vanderlinden, Van Schewik, von Hoerner, Walker (G. A. H.), Wallenquist, Weaver, Woolley.

38. Commission pour l'Échange des Astronomes. (Exchange of Astronomers)

(Committee of the Executive Committee)

PRÉSIDENT: M.G.J. Minnaert

VICE-PRÉSIDENT: A. Reiz

COMITÉ D'ORGANISATION: D.Ja. Martynov, P.M. Routly, J. Sahade, J.P. Wild.

MEMBRES: Abetti, Babcock (H.D.)†, Bappu, Bok, Bowen, Delhaye, Keller, Kienle, Kourganoff, Mihajlov, Miyamoto, Mohr, Page, Righini, Rosseland, Stoy, Swings, Witkowski, Wood (F.B.), Wyatt.

MEMBRES CONSULTANTS: E.M. Fournier d'Albe

40. Commission de la Radioastronomie. (Radio Astronomy)

PRÉSIDENT: J.P. Wild.

VICE-PRÉSIDENT: D.S. Heeschen.

COMITÉ D'ORGANISATION: N. Kardašev, M.R. Kundu, J. Lequeux, M. Ryle, S.F. Smerd, G. Westerhout.

MEMBRES: Abrami, Akabane, Atanasijević, Avignon, Baldwin, Baltá-Elias, Barret, Blum, Boischot, Bolton, Bracewell, Brown (R. Hanbury), Burbidge (G.R.), Cardus, Carr, Christiansen, Chvojková, Cohen, Costain, Coutrez, Daene, Davies (J.G.), Davies (R.D.), de Groot, Denisse, Dewhirst, Dieter, Dodson-Prince, Drake, Elwert, Eriksen, Field, Findlay, Firor, Fokker, Friedman, Galt, Ginzburg, Gold, Gorgolowski, Grahl, Hachenberg, Haddock, Hagen, Hajkin, Harang, Harrower, Heidmann, Hewish, Hey, Hill (E.R.), Howard (W.E.), Hughes (J.A.), Ihsanova, Kahn, Kakinuma, Kazès, Kerr, Kraus, Laffineur, Le Squeren, Letfus, Lilley, Little (A.G.), Locke, Lovell, McClain, McGee (R.X.), McKinley, MacRae, Macris, McVittie, Mannino, Maxwell, Mayer (C.H.), Menon, Mills, Minkowski, Moriyama, Muller (C.A.), Oort, Osterbrock, Parijskij (Ju.N.), Pick-Gutmann, Priester, Rabben, Righini, Roberts (J.A.), Rydbeck, Sanamjan, Scheuer, Schmidt (M.), Scott (P.F.), Seeger, Shakeshaft, Šklovskij, Simon (P.), Slee, Smith (A.G.), Smith (F.G.), Smith (Harlan J.), Soboleva, Stahr-Carpenter, Steinberg, Swarup, Takakura, Thompson (A.R.), Tlamicha, Townes, Troickij, Van Bueren, Van de Hulst, Van Woerden, Varsavsky, Vitkevič, Wade, Walsh, Weaver, Yen, Železnjakov.

41. Commission de l'Histoire de l'Astronomie. (History of Astronomy)

PRÉSIDENT: E. Rybka.

VICE-PRÉSIDENT: O. Gingerich.

COMITÉ D'ORGANISATION: J.O. Fleckenstein, C.D. Hellman, Z. Horský, P.G. Kulikovskij, W. Petri, V.L. Čenakal.

MEMBRES: Abetti, Argyrakos, Ashbrook, Baehr, Beer, Birkenmajer †, Collinder, Dick, Dingle, Dobrzycki, Douglas (A.V.), Erpylev, Evans (D.S.), Felber, Ferrari d'Ochieppo, Filliozat, Hirose, Kamienski, King (H.C.), Kotsakis, Lévy, Link, Michel, Michkovitch, Nielsen, Norlind, Nørlund, Omer, Ovenden, Pelseneer, Pogo, Samaha, Slavenas, Sticker, Whitrow, Yabuuti, Zagar, Zinner.

MEMBRES CONSULTANTS: Aaboe, Burke-Gaffney, Daumas, Fischer, Forbes, Freiesleben, Hartner, Horský, Hoskin, Kennedy, Labat, Maddison, Millas-Vallicrosa, Musto, Needham, Nevskaja, Pedersen, Price, Przypkowski, Ronan, Ronchi, Santillana, Schove, Taton, Thoren, Tumanjan, Veselovsky, van den Waerden, Wattenberg, Woolf, Wright.

42. Commission des Étoiles Doubles Photométriques. (Photometric Double Stars)

PRÉSIDENT: F.B. Wood.

VICE-PRÉSIDENT: M. Plavec.

COMITÉ D'ORGANISATION: A.H. Batten, T. Herczeg, M. Kitamura, G. Larsson-Leander, J.E. Merrill, V.P. Cesevič.

MEMBRES: Abhyankar, Bappu, Beer, Binnendijk, Blitzstein, Broglia, Cester, Chis, Cillié,

Cousins, Dadaev, de Kort, Ebbighausen, Ferrari d'Occhieppo, Fracastoro, Fredrick, Fresa, Gaposchkin (S.), Grygar, Günther, Gyldenkerne, Hardie, Heard, Huang, Huffer, Irwin, Kizilirmak, Koch, Kopal, Knipe, Krat, Kristenson, Kron (K.G.), Kruszewski, Kwee, Landolt, Lavrov, Linnell, Magalašvili, Martynov, Mumford, Nelson, O'Connell, Ovenden, Piotrowski, Plaut, Popovici, Popper, Rakos, Sahade, Schmidt (H.), Serkowski, Shapley (M.B), Shulberg, Sinvhal, Strohmeier, Szafraniec, Van Hoof, Walter, Wellmann, Wesselink, Whitney (B.S.), Widorn, Wright (K.O.).

43. Commission des Plasmas et Magnéto-Hydrodynamique en Astrophysique. (Astrophysical Plasmas and Magnetohydrodynamics)

PRÉSIDENT: L. Davis Jr.

VICE-PRÉSIDENT: R. Lüst.

COMITÉ D'ORGANISATION: T.G. Cowling, S.A. Kaplan, B. Lehner.

MEMBRES: Alfvén, Axford, Banerji, Bel, Bhatnagar (P.L.), Bierman, Bumba, Chandrasekhar, Csada, Ferraro, Ginzburg, Jensen, Kopecký, Krause, Kushwaha, Ledoux, Mestel, Parker, Piddington, Pikel'ner, Schlüter, Severnyj, Simon (R.), Smerd, Spitzer, Sturrock, Sweet, Šklovskij, Talwar, Tayler, Trehan, Van Bueren, Wentzel.

MEMBRES CONSULTANTS: C. P. Sonett, S. N. Vernov.

44. Commission des Observations Astronomiques au-dehors de l'Atmosphère Terrestre. (Astronomical Observations from outside the Terrestrial Atmosphere)

PRÉSIDENT: R. Wilson.

VICE-PRÉSIDENT: V.K. Prokof'ev.

COMITÉ D'ORGANISATION: J.E. Blamont, C. de Jager, Goldberg, R. Lüst, F.G. Smith, R. Tousey.

MEMBRES: Alexander (J.K.), Bogess, Boneff, Bosman-Crespin, Boyd, Brandt (J.C.), Butler, Carver, Chamberlain, Chubb, Code, Coutrez, Danielson, Deutsch (A.J.), Fazio, Fisher (P.C.), Fredga, Friedman, Giacconi, Ginzburg, Gold, Gratton, Haddock, Hallam, Hartz, Heddle, Henize, Herzberg (G.), Hinteregger, Houziaux, Huguenin, Ivanov-Holodnuj, Kraushaar, Kuiper, Kupperian, Lindblad (B.A.), Lines, Link, Lovell, McGee (J.D.), Mandel'štam, Masevič, Mehltretter, Meinel, Migeotte, Mihailov, Milford, Monfils, Morton (D.C.), Müller (Edith A.), Neupert, Newkirk, Nicolet, Nikolsky, Öhman, Parkinson, Pecker (J.-C.), Pounds, Priester, Reeves (E.M.), Rense, Righini, Rogerson, Roman, Rosen, Sarabhai, Schwarzschild, Smith (Henry J.), Steinberg, Underwood, Van de Hulst, Walsh, Whipple, Wilson (A.G.), Zwicky, Žongolovič.

MEMBRES CONSULTANTS: M. Florkin, H. Glasser, E.A. Lauter, G. Occhialini, B.F. de Veubeke.

45. Commission des Classifications Spectrales et Indices de Couleur à Plusieurs Bandes. (Spectral Classifications and Multi-band Colour Indices)

PRÉSIDENT: Ch. Fehrenbach.

VICE-PRÉSIDENT: B. Westerlund.

COMITÉ D'ORGANISATION: W.P. Bidelman, E.K. Haradze, C.O.R. Jaschek, H.L. Johnson, W.W. Morgan.

MEMBRES: Bahng, Barbier-Brossat, Bartaja, Beer, Blanco, Borgman, Boulon, Boyce, Buscombe, Cameron (R.C.), Canavaggia, Cester, Chalonge, Cowley, Dessy, Divan, Duflot (M.), Elvius (T.), Evans (D.S.), Feast, Golay, Griffin (R.F.), Gyldenkerne, Hack, Hallam, Henize, Herman, Hiltner, Hoag, Iwanowska, Jaschek (M.), Keenan, Kron (G.E.), Lodén (L.O.), Lodén (K.), Low, McCarthy, McCuskey, McNamara, Malmquist, Meinel, Mendoza, Morgan, Nikonov, Notni, Oosterhoff, Osawa, Perry, Philip, Preston, Reddish, Redman, Roman, Sanwal, Schmidt-Kaler, Seitter, Sharpless, Sinnerstad, Sinvhal, Slettebak, Steinlin, Stephenson, Strajžis, Strömgren, Thackeray, The, Treanor, Van den Bergh, Velghe, Walker (G.A.H.), Warner, Williams (J.A.), Wilson (O.C.), Yoss.

46. Commission pour l'Enseignement de l'Astronomie. (The Teaching of Astronomy)
Président: E. A. Müller (Miss).
Vice-Président: J. Kleczek.
Comité d'Organisation: G. O. Abell, E. K. Haradze, E. V. Kononovič, V. Kourganoff, M. G. J. Minnaert, T. L. Page, E. Schatzman, T. L. Swihart.
Membres: Dinulescu, Lambrecht, Rodgers.

LISTE ALPHABETIQUE DES MEMBRES

ALPHABETICAL LIST OF MEMBERS

Names are given under the first letter of the first word. For example the name Van de Hulst will be found under V.

The names of new members, elected at the XIII General Assembly, are prefixed by an asterisk.

The names of the countries are given in the English form.

Names of astronomers using Cyrillic alphabet in their own country (U.S.S.R.) are given with the new transliteration rules adopted by the International Organization for Standardization (I.S.O.) (see Volume XIIC, page 113). In all cases, the previously used transliteration follows the new one, in brackets; appropriate cross-referencing has been attempted. Example: YAKHONTOVA [see JAHONTOVA] is found under Y, while JAHONTOVA (and the address) is found under J. Note that Č, Š, Ž follow C, S, Z in the alphabetical list.

Please inform the Assistant Secretary of the I.A.U., Budečská 6, Praha 2, Czechoslovakia, of any errors you may find in this list.

Les noms figurent à la première lettre du premier mot; par exemple, le nom Van de Hulst se trouve à la lettre V.

Les noms des nouveaux membres élus par la XIIIe Assemblée générale sont annotés d'un astérisque.

Les noms de pays sont données sous leur forme anglaise.

Les noms des astronomes utilisant l'alphabet cyrillique dans leur pays (U.R.S.S.) sont donnés selon les nouvelles règles de translitération adoptées par l'I.S.O. (International Organization for Standardization) (voir dans le volume XIIC, page 68). Dans tous les cas, les translitérations usuelles de ces noms suivent, entre parenthèses, les nouvelles formes. Quand ces formes usuelles sont placées, dans l'ordre alphabétique, différemment des nouvelles, le nom est également cité à sa place usuelle; par exemple, YAKHONTOVA [see JAHONTOVA] se trouve à la lettre Y, tandis que la nouvelle forme JAHONTOVA (et l'adresse) se trouve à la lettre J. A noter que les initiales Č, Š, Ž suivent les initiales C, S, Z dans la liste alphabétique.

Les Membres de l'Union sont priés de bien vouloir informer le Secrétaire Adjoint de U.A.I., Budečská 6, Praha 2, Czechoslovakia, de toutes les erreurs qu'ils pourraient relever dans la présente liste.

* AARSETH (Dr S.), Institute of Theoretical Astronomy, Madingley Road, Cambridge, U.K.
* ABALAKIN (Dr V.K.), Institute of Theoretical Astronomy, USSR Academy of Sciences, Leningrad, U.S.S.R.

ABBOT (Dr C.G.), Astrophysical Observatory of the Smithsonian Institution, Washington, D.C., U.S.A.

ABBOTT (Prof. Dr W.N.), 42 Leophoros A. Michalacopoulou, Ilisia, Athens 612, Greece

ABDALA (Dr J.), Edificio "Pio XII" primer piso No. 1, San Francisquito a Puente Ayacucho, Caracas, Venezuela

ABELL (Dr G.O.), Astronomy Department, University of California, Los Angeles 24, California, U.S.A.

ABETTI (Prof. G.), Instituto Naz. Ottica-Arcetri, Largo Enrico Fermi 7, 50125 Florence, Italy

ABHYANKAR (Dr K.D.), Director, Observatory of Hyderabad, Hyderabad, Deccan, India

ABRAHAM (H.J.), Mount Stromlo Observatory, Canberra, A.C.T., Australia

ABRAMI (Prof. A.), Osservatorio Astronomico, via G.B. Tiepolo No. 11, Trieste (409), Italy

ABREU (Dr J.C. de Brito e), Observatorio Astronomico de Lisboa, Tapada, Lisbon, Portugal

ABT (Dr H.A.), Kitt Peak National Observatory, P.O. Box 4130, Tucson, Arizona 85717, U.S.A.

ADAM (Miss Dr M.G.), University Observatory, South Parks Road, Oxford, England

ADAMOPOULOS (Dr G.), rue Lamias No. 8, Athens (6), Greece

ADAMS (A.N.), U.S. Naval Observatory, Washington 25, D.C., U.S.A.

ADEL (Prof. A.), Atmospheric Research Observatory, Arizona State College, Flagstaff, Arizona, U.S.A.

ADOLFSSON (Dr T.), Lund Observatory, Lund, Sweden

AGEKJAN (Prof. Dr J.A.), [Agekian], University of Leningrad, Leningrad, U.S.S.R.

AHMED (Iman Ibrahim), Helwan Observatory, Helwan, near Cairo, U.A.R.

AHNERT (Dr P.), Sternwarte, Sonneberg, Thüringen, G.D.R.

AKABANE (Dr K.), Tokyo Astronomical Observatory, Mitaka, Tokyo, Japan

AKSENOV (Dr E. P.), Sternberg Astronomical Institute, Moscow, U.S.S.R.

AKSENT'EVA (Mrs Dr Z. N.), [Aksentjeva], Director, Gravimetrical Observatory of the Ukrainian Academy of Sciences, Poltava, U.S.S.R.

ALANIJA (Dr I. F.), [Alania], Abastumani Astrophysical Observatory, Abastumani, Georgian SSR, U.S.S.R.

* ALEXANDER (Dr A.F. O'D.), 1 Athelstan Road, Dorchester, Dorset, U.K.

ALEXANDER (Dr J. K.), Radio Astronomy Section, Laboratory for Space Science, Goddard Space Flight Center, Greenbelt, Maryland 20771, U.S.A.

ALFVÉN (Prof. H.), Royal Institute of Technology, Stockholm, Sweden

ALKSNIS (Dr A.K.), Astrophysical Laboratory, Latvian Academy of Sciences, Riga, U.S.S.R.

* ALLADIN (S. M.), Department of Astronomy, Osmania University, Hyderabad, India

ALLEN (Prof C. W.), University of London Observatory, Mill Hill Park, London, N.W.7, U.K.

ALLER (Prof. L. H.), Astronomy Department, University of California, Los Angeles, California, 90024, U.S.A.

ALMÁR (Dr I.), Konkoly Observatory, Szabadsághegy, Budapest XII, Hungary

ALTER (Dr G. A.), Bet Yizhaq, near Natanya, Israel

ALY (Dr M.K. M.), Deputy Director, Helwan Observatory, Helwan, near Cairo, Egypt

AMBARCUMJAN (Prof. Dr V.A.), [Ambartsumian], President of the Academy of Sciences of Armenian SSR, Erevan, Armenia, U.S.S.R.

* ANDERS (Dr E.), Enrico-Fermi Institute for Nuclear Studies, University of Chicago, Chicago, Illinois 60637, U.S.A.

* ANDREW (K. L.)

* ANDREWS (D. H.), Dominion Astrophysical Observatory, Victoria, B.C., Canada

ANDRILLAT (Prof. Dr H.), Laboratoire d'Astronomie de la Faculté des Sciences de Montpellier, Hérault, France

ANDRILLAT (Mme Dr Y.), Laboratoire d'Astronomie de la Faculté des Sciences de Montpellier, Hérault, France

ANGUITA (Dr C. A.), Universidad de Chile, Observatório Astronómico Nacional, Casilla 36-D, Santiago, Chile

* AOKI (Dr Shinko), Tokyo Astronomical Observatory, University of Tokyo, Mitaka, Tokyo, Japan

* ARAKELJAN (Dr M. A.), [Arakelyan], Byurakan Astrophysical Observatory, Byurakan, Armenia, U.S.S.R.

ARBEY (Prof. L. M. J.), Directeur de l'Observatoire, 25 Besançon, Doubs, France

AREND (Dr S.J. V.), Observatoire Royal de Belgique, 3 avenue Circulaire, Uccle-Brussels 18, Belgium

ARGUE (A. N.), The Observatories, Madingley Road, Cambridge, England

ARGYLE (Dr P. E.), Dominion Radio Astrophysical Observatory, P. O.Box 248, Penticton, B.C., Canada

ARGYRACOS (Prof. Dr J. G.), 193 Patission Street, Athens (816), Greece

ARP (Dr H. C.), Mount Wilson and Palomar Observatories, 813 Santa Barbara Street, Pasadena, California, U.S.A.

* ARPIGNY (Dr C.), Institut d'Astrophysique, Université de Liège, Cointe-Sclessin, Belgium

ARROYO (Dr M. L.), Observatorio Astronómico, Alfonso XII, 3, Madrid 7, Spain

ARSAC (Dr J.), Observatoire de Paris, 61 avenue de l'Observatoire, Paris-14e, France

ARTHUR (D.W.G.), Lunar and Planetary Laboratory, University of Arizona, Tucson, Arizona, U.S.A.

ARTJUHINA (Mrs Dr N.M.), [Artiukhina], Sternberg Astronomical Institute, Moscow, U.S.S.R.

ASAAD (Dr A.S.), Helwan Observatory, Helwan, near Cairo, Egypt

ASHBROOK (Dr J.), 16 Summer Street, Weston 93, Massachusetts, U.S.A.

ASTAPOVIČ (Dr I.S.), [Astapovich], Nikolsko-Botaničeskaja, 15–17a log. 45, Kiev-33, U.S.S.R.

ATANASIJEVIĆ (Dr. I.), Department of Astronomy, Faculty of Sciences, Driehuizerweg 200, Nijmegen, The Netherlands

ATHAY (Dr R.G.), High Altitude Observatory of the University of Colorado, Boulder, Colorado, U.S.A.

ATKINSON (R. d'E.), Astronomy Department, Indiana University, Swain Hall West, Bloomington, Indiana 47405, U.S.A.

* AUMAN (Dr J.R.), Princeton University Observatory, 265 FitzRandolph Road, Princeton, New Jersey 08540, U.S.A.

AVIGNON (Miss Dr Y.), Observatoire de Paris, Section d'Astrophysique, 92-Meudon, France

* AVRETT (Dr E.H.), Smithsonian Astrophysical Observatory, 60 Garden Street, Cambridge, Massachusetts 02138, U.S.A.

* AXFORD (W.I.), Department of Astronomy, Cornell University, Ithaca, New York, U.S.A.

BABADŽANOV (Dr P.B.), [Babadjanov], Director, Astrophysical Institute, Dušanbe, U.S.S.R.

BABCOCK (Dr H.D.)†, Mount Wilson and Palomar Observatories, 813 Santa Barbara Street, Pasadena, California, U.S.A.

BABCOCK (Dr H.W.), Mount Wilson and Palomar Observatories, 813, Santa Barbara Street, Pasadena, California, U.S.A.

BACCHUS (Prof. P.), 1, impasse de l'Observatoire, 59 Lille (Nord), France

BAEHR (Dr U.), Astronomisches Rechen-Institut, Mönchhofstrasse 12–14, 69 Heidelberg, G.F.R.

* BAHCALL (Dr J.N.), Kellogg Radiation Laboratory, California Institute of Technology, Pasadena, California 91109, U.S.A.

BAHNER (Dr K.), Landessternwarte, Königstuhl-Heidelberg, G.F.R.

BAHNG (Prof. J.D.R.), Dearborn Observatory, Northwestern University, Evanston, Illinois, U.S.A.

BAILLAUD (Prof. R.), 22, rue des Fontenottes, Besançon (Doubs), France

BAIRD (Dr K.M.), Division of Applied Physics, National Research Council, Sussex Drive, Ottawa, Ontario, Canada

BAIZE (Dr P.), 26, rue Daubigny, Paris-17e, France

BAKER (Dr J.G.), 60 Garden Street, Cambridge, Massachusetts 02138, U.S.A.

* BAKER (Dr N.), Astronomy Department, Columbia University, New York, N.Y. 10027, U.S.A.

BAKER (Dr R.H.), 179, E. Villanova Drive, Claremont, California, California, U.S.A.

BAKOS (Prof. G.A.), Department of Physics, University of Waterloo, Waterloo, Ontario, Canada

BAKULIN (Dr P.I.), Sternberg Astronomical Institute, Moscow, U.S.S.R.

BALÁZS (Dr B.), Konkoly Observatory, P.O. 114, Box 67, Budapest XII, Hungary

BALÁZS-DETRE (Mrs Dr J.), Konkoly Observatory, Szabadsághegy, P.O. 114, Box 67, Budapest XII, Hungary

BALDWIN (Dr J.E.), Cavendish Laboratory, Free School Lane, Cambridge, England

* BALKLAV (Dr A.E.), Astrophysical Laboratory, Latvian Academy of Sciences, Riga, U.S.S.R.

BALLARIO (Prof. M.C.), Osservatorio Astrofisico di Arcetri, Via S. Leonardo 75, Florence, Italy

BALLI (Prof. Dr E.), University Observatory, Beyazit, Istanbul, Turkey

BALTÁ-ELIAS (Prof. J.), Universidad de Madrid, Madrid, Spain

BANERJI (Dr A.C.), Muir Central College, Allahabad University, Allahabad, India

* BANIN (Dr V.G.), Post Office Meget, Sibizmiran, Irkutsk, U.S.S.R.

BAPPU (Dr M.K. Vainu), Director, Astrophysical Observatory, Kodaikanal, India, Vice-President of the Union.

BARABAŠOV (Prof. Dr N.P.), [Barabashov], Director, Astronomical Observatory, Harkov, U.S.S.R.

BARANNE (A.), Observatoire de Marseille, 2, place Le Verrier, 13-Marseille-4e, France

BARBANIS (Dr B.), Astronomical Department, University of Thessaloniki, Thessaloniki, Greece

BARBER (D. R.), "Craigwell", Coulsdon Road, Sidmouth, Devon, U.K.

* BARBIER-BROSSAT (Mme M.), Chargée de Recherche, Observatoire de Marseille, 2, place Le Verrier, 13-Marseille-4e, France

BARHATOVA (Mrs Dr K. A.), [Barkhatova], State University, Sverdlovsk, U.S.S.R.

BARLIER (F.), Observatoire de Paris, Section d'Astrophysique, 92-Meudon, France

BARNEY (Miss Dr I.), Yale University Observatory, Box 2023, Yale Station, New Haven, Connecticut, U.S.A.

BARRETO (Prof. L.M.), Observatorio Nacional, rua General Bruce, 586-São Cristóvão, Rio de Janeiro, Brazil

BARRETT (Prof. A.H.), Research Laboratory of Electronics, Massachusetts Institute of Technology, Cambridge, Massachusetts 02139, U.S.A.

BARROS (Prof. Dr M.G.P. de), Observatorio Astronomico, Universidade do Porto, Monte da Virgem, Vila Nova de Gaia, Portugal

BARTAJA (Mrs Dr R.A.), [Bartaya], Astrophysical Observatory, Abastumani, Georgia, U.S.S.R,

BASCHEK (Dr B.), Institut für Theoretische Physik, Ohlshausenstrasse, Neue Universität, Bau C 4, 23 Kiel, G.F.R.

BATES (Prof. D.R.), Department of Applied Mathematics, Queen's University, Belfast, U.K.

BATESON (F.M.), Astronomer-in-Charge, Mount Joh University Observatory, P.O.Box 32, Lake Tekapo, New Zealand

BATRAKOV (Dr Ju.V.), Institute of Theoretical Astronomy, Leningrad, U.S.S.R.

BATTEN (Dr A. H.), Dominion Astrophysical Observatory, Victoria, B.C., Canada

BAUER (C.A.), Pennsylvania State University, Department of Physics, State College, Pennsylvania, U.S.A.

BAUM (Dr W.A.), Planetary Research Center, Lowell Observatory, Flagstaff, Arizona 86001, U.S.A.

BAUSTIAN (W. W.), Kitt Peak National Observatory, 950 North Cherry Avenue, Tucson, Arizona 85717, U.S.A.

BEALS (Dr C.S.), Manotick, Ontario, Canada

* BEARDSLEY (W. R.), Allegheny Observatory, Riverview Park, Pittsburgh 14, Pennsylvania, U.S.A.

* BEAVERS (Dr W.), Iowa State University, Ames, Iowa, U.S.A.

BECK (H.-G.), Carl Zeiss Observatory, Jena, G.D.R.

BECKER (Prof. Dr F.), Klingsorstrasse 3/96, 8 München 61, G.F.R.

* BECKER (Dr R.A.), Aerospace Corporation, P.O.Box 95085, Los Angeles, California, U.S.A.

BECKER (Prof. Dr W.), Astronomisch-Meteorologische Anstalt, Venusstrasse 7, Binningen (BL), Switzerland

* BECKERS (Dr J. M.), Sacramento Peak Observatory, Sunspot, New Mexico 83349, U.S.A.

BEER (Dr A.), The Observatories, Madingley Road, Cambridge, U.K.

* BEGGS (D. W.), The Observatories, Madingley Road, Cambridge, U.K.

BEHR (Dr A.), Universitäts-Sternwarte, Geismarlandstrasse 11, 34 Göttingen, G.F.R.

BEL (Miss Dr N.), Institut d'Astrophysique, 98 bis, boulevard Arago, Paris-14e, France

BELL (Miss Dr Barbara), Harvard College Observatory, Cambridge, Massachusetts 02138, U.S.A.

* BELL (Dr R. A.), Department of Physics and Astronomy, University of Maryland, College Park, Maryland, U.S.A.

BELOCERKOVSKIJ (Dr D. J.), [Belotserkovsky], Institute of Physico-Technical Measurements, Moscow, U.S.S.R.

BELORIZKY (Dr D.), Observatoire de Marseille, 2, place Le Verrier, 13-Marseille-4e, France

* BELTON (Dr M. S.), Kitt Peak National Observatory, 950 North Cherry Avenue, Tucson, Arizona 85717, U.S.A.

* Bely (O.), Observatoire de Nice, Le Mont-Gros, 06-Nice, France
* Benevides (P.), Observatoire, São Paulo, Brazil
Berger (Dr J.), Institut d'Astrophysique, 98 bis, boulevard Arago, Paris-14e, France
Berkner (Dr L. V.), 3632 N.E. 24th Ave, Fort Lauderdale, Florida, U.S.A.
Bertaud (Dr Ch.), Observatoire de Paris, Section d'Astrophysique, 92-Meudon, France
Bertiau (Dr F. C.), Specola Vaticana, Castel Gandolfo, Vatican City State
Bertola (Dr F.), Osservatorio astrofisico di Asiago (Vicenza), Italy
Beyer (Dr M.), Justus-Brinckmann-Strasse 101, Hamburg-Bergedorf, G.F.R.
* Bhatnagar (A.), Astrophysical Observatory, Kodaikanal 3, India
Bhatnagar (Dr P. L.), Department of Applied Mathematics, Indian Institute of Science, Bangalore 3, South India
* Bhattacharyya (J.C.), Astrophysical Observatory, Kodaikanal 3, India
* Bhonsle (R. V.), Physical Research Laboratory, Ahmedabad, India
Bidelman (Dr W. P.), The Observatory, University of Michigan, Ann Arbor, Michigan, U.S.A.
Bielicki (Dr M.), Astronomical Observatory of Warsaw University, Al. Ujazdowskie 4, Warsaw, Poland
Biermann (Prof. Dr L.), Max-Planck Institut für Physik und Astrophysik, Institut für Astrophysik, Föhringer Ring 6, München 23, G.F.R.
Bigay (Prof. J.H.), Directeur de l'Observatoire de Lyon, 69 Saint-Genis-Laval, France
Billaud (G.), Observatoire de Paris, 61, avenue de l'Observatoire, Paris-14e, France
Billings (Dr D.E.), High Altitude Observatory of the University of Colorado, Boulder, Colorado, U.S.A.
Binnendijk (Prof. Dr L.), Flower and Cook Observatory, University of Pennsylvania, Philadelphia 4, Pennsylvania, U.S.A.
Blaauw (Prof. Dr A.), Director, Kapteyn Astronomical Laboratory, Broerstraat 7, Groningen, The Netherlands
Blackett (Prof. P.M.S.), Imperial College of Science and Technology, South Kensington, London, S.W.7, U.K.
Blackwell (Prof. D. E.), University Observatory, Oxford, U.K.
Blaha (Dr M.), Astronomical Institute, Czechoslovak Academy of Sciences, Observatory Ondřejov, Czechoslovakia
Blamont (Prof. J. E.), Service d'Aéronomie, Réduit de Verrières, 91-Verrières-le-Buisson, France
Blanco (Dr V. M.), Observatorio Inter-Americano de Corro Tololo, Casilla 63-D, La Serena, Chile
Blaser (Prof. J.P.), Swiss Institute of Technology, Gloriastrasse 5, Zürich, Switzerland
* Bless (Dr R. C.), Washburn Observatory, University of Wisconsin, Madison, Wisconsin 53705, U.S.A.
Blitzstein (Dr W.), Flower and Cook Observatory, University of Pennsylvania, Philadelphia 4, Pennsylvania, U.S.A.
Bloch (Mlle Dr M.), Observatoire de Lyon, Saint-Genis-Laval (Rhône), France
Blum (Dr E. J.), Observatoire de Paris, Section d'Astrophysique, 92-Meudon, France
Bobrov (Dr M. S.), Astronomical Council, U.S.S.R. Academy of Sciences, Vavilova Street 20, Moscow V-312, U.S.S.R.
Bobrovnikoff (Dr N. T.), The Ohio State University, Columbus, Ohio, U.S.A.
Boggess (Dr A. III), Goddard Space Flight Center, Washington 25, D.C., U.S.A.
Bogorodskij (Dr A. F.), [Bogorodsky], Astronomical Observatory of the Kiev University, Kiev, U.S.S.R.
Böhm (Prof. Dr. K.-H.), Astronomy Department, University of Washington, Seattle, Washington 98105, U.S.A.
Böhm-Vitense (Mrs Dr E.), Astronomy Department, University of Washington, Seattle, Washington 98105, U.S.A.
Böhme (Dr S.), Astronomisches Rechen-Institut, Mönchhofstrasse 12–14, 69 Heidelberg, G.F.R.

BOHRMANN (Prof. Dr A.), Landessternwarte, Königstuhl-Heidelberg, G.F.R.
BOISCHOT (Dr A.), Observatoire de Paris, Section d'Astrophysique, 92-Meudon, France
BOJARČUK (Dr. A. A.), Crimean Astrophysical Observatory, USSR. Academy of Sciences, P/O Naučnyj, Crimea, U.S.S.R.
BOK (Prof. B. J.), Steward Observatory, University of Arizona, Tucson, Arizona 85721, U.S.A.
BOK (Lie Rak), Director of Phyongyang Astronomical Observatory, Phyongyang, Korea
BOK (Mrs Dr P. F.), Steward Observatory, University of Arizona, Tucson, Arizona 85721, U.S.A.
* BOLDT (Dr E. A.), NASA, Goddard Space Flight Center, Greenbelt, Maryland, U.S.A.
BOLTON (J. G.), Radiophysics Laboratory, C.S.I.R.O., University Grounds, Sydney, N.S.W., Australia
BONANOMI (Dr J.), Directeur de l'Observatoire de Neuchâtel, Neuchâtel, Switzerland
BONDI (Dr H.), King's College, Strand, London W.C.2, U.K.
BONEFF (Prof. Dr N.), Section d'Astronomie, Académie des Sciences Bulgare, Sofia, Bulgaria
BONEV (B.), Observatoire Populaire, Stara-Zagora, Bulgaria
* BONNEAU (M.), Faculté des Sciences, 25-Besançon, France
BONNOR (Prof. W. B.), Queen Elizabeth College, Campden Hill Road, London W.8, U.K.
BONOV (A.D.), Université de Sofia, Sofia, Bulgaria
BONSACK (Prof. W. K.), Hawaii Institute of Geophysics, Rm 436 Astrophysics, 2525 Correa Rd, Honolulu, Hawaii 96822, U.S.A.
BORGMAN (Dr J.), "Kapteyn" Sterrewacht, Mensingenweg 20, Roden (Drente), The Netherlands
BOSMAN-CRESPIN (Mme D.), Institut d'Astrophysique, Université de Liège, Cointe-Sclessin (Liège), Belgium
BOTELHEIRO (Dr A.), Observatorio Astronomico, Lisbon, Portugal
BOUIGUE (Prof. R.), Directeur de l'Observatoire de Toulouse, 1, avenue Camille-Flammarion, 31-Toulouse (Haute-Garonne), France
BOULON (Dr J.), Observatoire de Marseille, 2, place Le Verrier, 13-Marseille-4e, France
BOURGEOIS (Prof. P.), 31, rue Paul Hankar, Uccle-Brussels 18, Belgium
* BOURY (Dr A.), Institut d'Astrophysique de l'Université de Liège, Cointe-Sclessin, Belgium
BOUŠKA (Dr J.), Astronomical Institute of the Charles University, Švédská 8, Prague 5-Smíchov, Czechoslovakia
BOUVIER (Prof. P. B.), Observatoire de Genève, 1290 Sauverny (GE), Switzerland
BOWEN (Dr E. G.), C.S.I.R.O., Division of Radiophysics, University Grounds, Chippendale, N.S.W., Australia
BOWEN (Dr I. S.), Mount Wilson and Palomar Observatories, 813 Santa Barbara Street, Pasadena 4, California, U.S.A.
BOYARČHUK (Dr A. A.), [see Bojarčuk]
* BOYCE (Dr P. B.), P.O.Box 1269, Flagstaff, Arizona 86001, U.S.A.
BOYD (Prof. R.L. F.), University College, Gower Street, London W.C.1, England
BOYER (Dr L.), Observatoire de Nice, Le Mont-Gros, Nice (A.-M.), France
* BOZIS (Dr G.), Astronomical Department, University of Thessaloniki, Thessaloniki, Greece
* BRACCESI (A.), Laboratorio di Radioastronomia, Instituto di Fisica, Bologna, Italy
BRACEWELL (Prof. R. N.), Radio Astronomy Institute, Stanford University, Stanford, California, U.S.A.
BRAES (L.L. E.), University Observatory, Sterrewacht Leiden, The Netherlands
BRAHDE (R.), Institute of Theoretical Astrophysics, University of Oslo, Blindern, Oslo 3, Norway
BRANDT (Dr J. C.), Code 614, Goddard Space Flight Center, Greenbelt, Maryland 20771, U.S.A.
* BRANDT (Dr V. E.), Sternberg Astronomical Institute, Moscow, U.S.S.R.
BRANSCOMB (Dr L. M.), National Bureau of Standards, Washington 25, D.C., U.S.A.
BRAY (De R. J.), National Standards Laboratory, C.S.I.R.O., University Grounds, Sydney, N.S.W., Australia
BREALEY (G. A.), Dominion Observatory, Ottawa, Ontario, Canada
BRETZ (Miss Dr M.), c/o Observatoire de Marseille, 2, place Le Verrier, 13-Marseille-4e, France

BRICARD (Dr J.), Maître de Conférences à la Faculté des Sciences de Paris, Institut de Physique du Globe, 191, rue St. Jacques, Paris-5e, France

BRKIĆ (Z.M.), Astronomska Observatorija, Veliki Vracar, Beograd, Yugoslavia

BROGLIA (Dr P.), Osservatorio Astronomico, Merate (Como), Italy

* BROSTERHUS (Dr E.B.F.), Dominion Observatory, Ottawa, Ontario, Canada

* BROTEN (N.W.), Radio and Electrical Engineering Division, National Research Council, Ottawa, Ontario, Canada

BROWN (Dr Harrison), Division of Geological Sciences, California Institute of Technology, Pasadena 4, California, U.S.A.

* BROWN (Dr P. Lancaster), Karlsvik, 30 Enghams Wood Road, Beaconsfield, Bucks., U.K.

BROWN (Dr R. Hanbury), Chatterton Astronomy Department, School of Physics, University of Sydney, Sydney, N.S.W., Australia

BROWNLEE (Dr R.R.), P.O.Box 1663, Los Alamos, New Mexico, U.S.A.

BRÜCK (Prof. H.A.), Astronomer Royal for Scotland, Royal Observatory, Blackford Hill, Edinburgh 9, U.K.

BRÜCK (Mrs Dr M.T.), Royal Observatory, Blackford Hill, Edinburgh 9, U.K.

BRÜCKNER (Dr G.), University of Maryland, Department of Physics and Astronomy, College Park, Maryland, U.S.A.

BRUMBERG (Dr V.A.), Institute of Theoretical Astronomy, USSR. Academy of Sciences, Leningrad, U.S.S.R.

* BRUNK (Dr W.), Lunar and Planetary Programs, NASA Headquarters, Washington, D.C. 20546, U.S.A.

BRUWER (J.A.), Republic Observatory, Johannesburg, South Africa

BRUZEK (Dr A.), Fraunhofer Institut, Schöneckstrasse 6, Freiburg i. Br., G.F.R.

BUCHAR (Prof. Dr E.), Astronomical Institute, Technical University, Karlovo nám. 13, Prague 2, Czechoslovakia

BULLARD (Sir Edward C.), University of Cambridge, Department of Geodesy and Geophysics, Madingley Rise, Madingley Road, Cambridge, England

BULLEN (Prof. K.E.), Department of Applied Mathematics, University of Sydney, Sydney, Australia

BUMBA (Dr V.), Astronomical Institute, Czechoslovak Academy of Sciences, Observatory Ondřejov, Czechoslovakia

BURBIDGE (Mrs Dr E.M.), University of California, La Jolla, California, U.S.A.

BURBIDGE (Dr G.R.), University of California, La Jolla, California, U.S.A.

BURCH (Dr C.R.), H.H. Royal Fort Wills Physics Laboratory, Bristol 8, U.K.

BURGESS (Dr A.), University of Cambridge, Department of Applied Mathematics and Theoretical Physics, Silver Street, Cambridge, U.K.

BURKE (Dr B.F.), Department of Terrestrial Magnetism, Carnegie Institution of Washington, Washington, D.C., U.S.A.

BUSBRIDGE (Miss Dr I.W.), St. Hugh's College, Oxford, U.K.

BUSCOMBE (Dr W.), Mount Stromlo Observatory, Canberra, A.C.T., Australia

BUTLER (Dr H.E.), Royal Observatory, Blackford Hill, Edinburgh 9, U.K.

CALAMAI (Prof. G.), Osservatorio astrofisico di Arcetri, Florence, Italy

CAMERON (Dr A.G.W.), Belfer Graduate School of Science, Yeshiva University, Amsterdam Ave & 186th St., New York, N.Y. 10033, U.S.A.

* CAMERON (Dr R.C.), NASA, Goddard Space Flight Center, Greenbelt, Maryland 20771, U.S.A.

* CAMERON (Mrs W.), NASA, Goddard Space Flight Center, Greenbelt Maryland 20771, U.S.A.

CAMICHEL (Dr H.), Observatoire de Toulouse, 31-Toulouse (Haute-Garonne), France

CAMM (Dr G.L.), Department of Mathematics, The University, Manchester 13, U.K.

CANAVAGGIA (Mlle Dr R.), Observatoire de Paris, 61 avenue de l'Observatoire, Paris-14e, France

CANDY (M.P.), 97 Hawkswood Drive, Hailsham, Sussex, U.K.

CAPRIOLI (Dr G.), Osservatorio astronomico di Monte Mario, Via Trionfale 204, Rome, Italy
* CAPRIOTTI (Dr E. R.), Astronomy Department, Ohio State University, Columbus, Ohio 43210, U.S.A.
CARDUS (Rvd°. P. J.), Astronome del Observatorio del Ebro, Tortosa, Spain
* CARPENTER (Dr L. H.), NASA, Goddard Space Flight Center, Greenbelt, Maryland 20771, U.S.A.
CARR (Dr T. D.), Department of Physics, University of Florida, Gainesville, Florida, U.S.A.
CARRASCO (Dr R.), Director, Observatorio Astronómico de Madrid, Alfonso XII, 3, Madrid (7), Spain
CARROLL (Sir John A.), K.B.E., 14, Belvedere Grove, Wimbledon, London, S.W.19, U.K.
CARSON (Dr T. R.), University Observatory, Buchanan Gardens, St. Andrews, Fife, U.K.
* CARVER (Prof. J. H.), Department of Physics, University of Adelaide, Adelaide, Australia
CAYREL (Dr R.), Observatoire de Paris, Section d'Astrophysique, 92-Meudon, France
CAYREL-DE-STROBEL (Mme Dr G.), Institut d'Astrophysique, 98bis boulevard Arago, Paris-14e, France
CECCARELLI (Prof. M.), Istituto di Fisica generale, Università di Bologna, Italy
CECCHINI (Prof. G.), Direttore, Osservatorio Astronomico di Torino in Pino Torinese, Italy
CEPLECHA (Dr Z.), Astronomical Institute, Czechoslovak Academy of Sciences, Observatory Ondřejov, Czechoslovakia
CESCO (Dr C. U.), Observatorio Félix Aguilar, Facultad de Ingeniería, Universidad Nacional de Cuyo, San Juan, Argentina
CESCO (Dr R. P.), Observatorio Astronomico, La Plata, Argentina
CESEVIČ (Prof. Dr V. P.), [Tsesevich], Director, Astronomical Observatory, Odessa, U.S.S.R.
CESTER (Prof. Dr B.), Osservatorio Astronomico, Via G.B. Tiepolo, 15, Trieste (409), Italy
CHALONGE (Dr D.), Institut d'Astrophysique, 98 bis boulevard Arago, Paris-14e, France
CHAMBERLAIN (Prof. J. W.), Kitt Peak National Observatory, P.O.Box 4130, Tucson, Arizona 85717, U.S.A.
* CHANDRA (S.), General Electric Co., Pasadena, California, U.S.A.
CHANDRASEKHAR (Prof. S.), Laboratory for Astrophysics and Space Research, 933 East 56th Street, Chicago, Illinois 60637, U.S.A.
CHAPMAN (Prof. S.), High Altitude Observatory of the University of Colorado, Boulder, Colorado 80302, U.S.A.
* CHARVIN (P.), Observatoire de Meudon, 5 place Janssen, 92-Meudon, France
CHAVIRA (E.), Observatorios de Tonantzintla y Tacubaya Apartado Postal 25264, Ciudad Universitaria, México 20, D.F., Mexico
CHEBOTAREV (Prof. G. A.), [see ČEBOTAREV]
* CHINCARINI (G.), Osservatorio Astrofisico, Asiago (Vicenza), Italy
CHIS (Prof. G.), Observatoire Astronomique, rue de la République 109, Cluj, Roumania
* CHIU (Dr H. Y.), Institute for Space Studies, 475 Riverside Drive, New York, N.Y. 10027, U.S.A.
* CHIUDERI (Dr F. D.), Osservatorio Astrofisico di Arcetri Largo E. Fermi, 5, Firenze, Italy
CHOPINET (Mlle M.), Observatoire de l'Université de Bordeaux, 33-Floirac, France
CHRISTIANSEN (Prof. W. N.), School of Electrical Engineering, University of Sydney, Sydney, N.S.W., Australia, Vice-President of the Union.
CHRISTY (Prof. R. F.), Physics Department, California Institute of Technology, Pasadena, California, U.S.A.
CHUBB (Dr T. A.), 5023 N 38th Street, Arlington 7, Virginia, U.S.A.
CHUDOVICHEVA (Mrs Dr N. A.), [see ČUDOVIČEVA]
CHUVAEV (Dr K. K.), [see ČUVAEV]
CHVOJKOVÁ (Dr E.), Astronomical Institute, Czechoslovak Academy of Sciences, Budečská 6, Prague 2, Czechoslovakia
CICHOWICZ (Dr L.), Polytechnic School, Koszykowa 75, Warsaw, Poland
CILLIÉ (Prof. G. G.), Department of Mathematics, University of Stellenbosch, Cape Province, South Africa

CIMINO (Prof. Dr M.), Direttore, Osservatorio Astronomico, su Monte Mario, Rome, Italy
CIMMERMAN (Prof. Dr G. K.), [Zimmerman], Nikolaev Observatory, Nikolaev, U.S.S.R.
* CLARK (Dr B.C.), National Radio Astronomy Observatory, P.O.Box 2, Green Bank, West Virginia 24944, U.S.A.
* CLARK (Dr G.), Physics Department, Massachusetts Institute of Technology, Cambridge, Massachusetts 02139, U.S.A.
* CLAYTON (Dr D. D.), Space Science Department, Rice University, Houston, Texas, U.S.A.
CLEMENCE (Dr G. M.), Yale University Observatory, Box 2023, Yale Station, New Haven, Connecticut, U.S.A.
* CLIMENHAGA (Dr J. L.), University of Victoria, Victoria B.C., Canada
CODE (A. D.), 6524 Sterling Hall, University of Wisconsin, Madison, Wisconsin, U.S.A.
COHEN (Prof. Dr M.H.), Department of Applied Electrophysics, University of California, San Diego, La Jolla, California 92037, U.S.A.
* COLGATE (Dr S.), New Mexico Institute of Mining and Technology, Socorro, New Mexico, U.S.A.
COLLINDER (Dr P.), Odensgatan 5 B, IV, Uppsala, Sweden
* COLLINS II (Dr G.W.), Department of Astronomy, Ohio State University, Columbus, Ohio 43210, U.S.A.
COLLINSON (E. H.), Culpho End, Culpho, Ipswich, Suffolk, U.K.
* CONDON (Dr E. U.), Joint Institute for Laboratory Astrophysics, University of Colorado, Boulder, Colorado 80310, U.S.A.
CONNES (Mme J.), Service de Calcul Numérique, Observatoire de Meudon, 5, place Janssen, 92-Meudon, France
* CONNES (P.), Laboratoire du C.N.R.S., Bellevue, France
* CONTI (Dr P. S.), Lick Observatory, University of California, Santa Cruz, California 95060, U.S.A.
CONTOPOULOS (Prof. Dr G.), Astronomical Department, University of Thessaloniki, Thessaloniki, Greece
CONWAY (Dr R.G.), Nuffield Radio Astronomy Laboratories, Jodrell Bank, Macclesfield, Cheshire, U.K.
COOK (Dr A.F.), Smithsonian Astrophysical Observatory, Cambridge, Massachusetts 02138, U.S.A.
COOK (Dr A. H.), National Physical Laboratory, Teddington, Middlesex, U.K.
* CORLISS (Dr C. H.), National Bureau of Standards, Washington D.C. 20234, U.S.A.
COSTAIN (Dr C. H.), Dominion Radio Astrophysical Observatory, Penticton B.C., Canada
COUDER (Dr A.), Observatoire de Paris, 61 avenue de l'Observatoire, Paris-14e, France
COUDERC (Dr P.), Observatoire de Paris, 61 avenue de l'Observatoire, Paris-14e, France
COURTÈS (Dr G.), Observatoire de Marseille, 2, place Le Verrier, Marseille-IV, France
COUSINS (Dr A. W. J.), Royal Observatory, Observatory, Cape Province, South Africa
COUTEAU (Dr P. C.), Observatoire de Nice, Le Mont-Gros, 06-Nice, France
COUTREZ (Prof. Dr R.A. J.), Directeur, Institut d'Astronomie et d'Astrophysique, Université de Bruxelles, 50 avenue F. D. Roosevelt, Brussels, Belgium
COVINGTON (A. E.), National Research Council, Ottawa, Ontario, Canada
* COWLEY (Mrs A. P.),
COWLING (Prof. T. G.), Department of Mathematics, The University, Leeds 2, U.K.
COX (Dr A. N.), Box 1663, Los Alamos, New Mexico, U.S.A.
COX (Prof. J.F.), Recteur Honoraire, Professeur Honoraire, 351, avenue Louise, Brussels, Belgium
COX (Dr J.P.), Joint Institute for Laboratory Astrophysics, 1511 University Avenue, Boulder, Colorado, U.S.A.
CRAWFORD (Dr D.L.), Kitt Peak National Observatory, 950 North Cherry Avenue, Tucson, Arizona 85717, U.S.A.
CRISTALDI (Dr S.), Piazza Vaccarini 11, Catania, Italy
* CRISTESCU (Dr C.), Observatoire Astronomique, rue Cutitul de Argint 5, Bucarest 28, Roumanie

CSADA (Dr I. K.), Konkoly Observatory, Konkoly Thege U. 13–17, Budapest XII, Hungary

* CUDABACK (Dr D. D.), Radio Astronomy Laboratory, 633 Campbell Hall, University of California, Berkeley, California 94720, U.S.A.

CUFFEY (J.), Goethe Link Observatory, Indiana University, Bloomington, Indiana, U.S.A.

CUNNINGHAM (Dr L. E.), Department of Astronomy, University of California, Berkeley 4, California, U.S.A.

CUREA (Prof. Dr I.), Recteur de l'Université de Timisoara, Timisoara, Roumania

CZYZAK (Prof. S. J.), Astronomy Department, The Ohio State University, Columbus, Ohio 43210, U.S.A.

ČEBOTAREV (Prof. G. A.), [Cheboratev], Institute of Theoretical Astronomy, Leningrad B-164, U.S.S.R.

ČEKIRDA (Dr A. T.), [Tchekirda], Harkov Observatory, Harkov, U.S.S.R.

ČENAKAL (Dr V. L.), [Tchenakal], Lomonosov Museum of the USSR Academy of Sciences, Leningrad, U.S.S.R.

* ČERNEGA (Dr N. A.), [Tchernega], Astronomical Observatory, Kiev University, Kiev, U.S.S.R.

ČUDOVIČEVA (Mrs Dr N. A.), [Chudovicheva], Engelhardt Observatory, Kazan, U.S.S.R.

* ČUGAJNOV (Dr P. F.), [Tchugainov], Crimean Astrophysical Observatory, USSR Academy of Sciences, Crimea, U.S.S.R.

ČUVAEV (Dr K. K.), [Chuvaev], Crimean Astrophysical Observatory, USSR Academy of Sciences, P/O Naučnyj, Crimea, U.S.S.R.

DADAEV (Dr A. N.), Pulkovo Observatory, Leningrad, U.S.S.R.

* DADIĆ (Dr Ž.), Institut za povijest nauka Jazu, Demetrova 18, Zagreb 3, Yugoslavia

DAENE (Dr H.), Astrophysikalisches Observatorium Potsdam, Abtlg, Radioastronomie, Tremsdorf (Potsdam-Land), G.D.R.

DALLAPORTA (Prof. N.), Istituto di Fisica teorica, Università di Padova, Italy

* DAMBARA (Dr T.), Geographical Survey Institute, 7-1000 Kamimeguro, Meguro-ku, Tokyo, Japan

DANBY (Dr J.M.A.), Yale University Observatory, Box 2023, Yale Station, New Haven, Connecticut, U.S.A.

DANIELSON (Dr R. E.), Princeton University Observatory, Princeton, N.J., U.S.A.

* DANZIGER (Dr I. J.), Harvard College Observatory, 60 Garden Street, Cambridge, Massachusetts, U.S.A.

DA SILVA (Dr A.V.C.S.), Observatorio Astronomico de Universidade Santa Clara, Coimbra, Portugal

DAUBE-KURZEMNIECE (Mrs Dr I. A.), Astrophysical Laboratory of the Academy of Sciences of the Latvian S.S.R., Riga, U.S.S.R.

DAUVILLIER (Prof. A.), Observatoire du Pic-du-Midi, 65-Bagnères-de-Bigorre, France

DAVIES (Dr J. G.), Nuffield Radio Astronomy Laboratories, Jodrell Bank, Macclesfield, Cheshire, U.K.

DAVIES (Dr R. D.), Nuffield Radio Astronomy Laboratories, Jodrell Bank, Macclesfield, Cheshire, U.K.

DAVIS (Dr J.), Department of Physics, University of Sydney, Sydney, N.S.W., Australia

DAVIS Jr, (Prof. L.), Physics Department, California Institute of Technology, Pasadena, California 91109, U.S.A.

DAVIS (Dr M. S.), Triangle Universities Computation Center, P.O.Box 12175, Research Triangle Park, North California 27709, U.S.A.

D'AZAMBUJA (Dr L.), Laclotte, Quartier Arribourdès, Salies-de-Béarn, (Basses-Pyrénées), France

D'AZAMBUJA (Mme M. L.), Laclotte, Quartier Arribourdès, Salies-de-Béarn, (Basses-Pyrénées), France

DÉBARBAT (Miss S.), Observatoire de Paris, 61 avenue de l'Observatoire, Paris-14e, France

* DEBEHOGNE (Dr H.), Observatoire Royal de Belgique, 3 avenue Circulaire, Brussels, Belgium

DECAUX (B.), Bureau International de l'Heure, Observatoire de Paris, 61, avenue de l'Observatoire, Paris-14e, France

* DEEMING (Dr T. J.), Astronomy Department, 404 Physics Building, University of Texas, Austin, Texas 78712, U.S.A.

DE FEITER (L. D.), Astronomical Observatory, "Sonnenborgh", Zonnenburg 2, Utrecht, The Netherlands

* DE GRAAFF (Dr W.), Laboratorium voor Ruimte-Onderzoek, Huizingalaan 121, Utrecht, The Netherlands

* DE GREIFF (Dr J. A.), Observatorio Astronomico Nacional, Apartado Postal 2584, Bogota, Colombia

DE GROOT (Dr T.), Sterrewacht "Sonnenborgh", Servaas Bolwerk 13, Utrecht, The Netherlands

* DEINZER (Dr W.), Institut für Theoretische Astrophysik der Universität Heidelberg, Mönchhofstrasse 12–14, 69 Heidelberg, G.F.R.

DE JAGER (Prof. Dr C.), Astronomical Observatory "Sonnenborgh", Zonnenburg 2, Utrecht, The Netherlands. Assistant General Secretary of the Union.

DEJČ (Prof. Dr A.N.), [Deutsch], Pulkovo Observatory, Leningrad M-140, U.S.S.R.

DE KOCK (R.P.), Royal Observatory, Observatory, Cape Province, South Africa

DE KORT, S.J. (Rev. Dr J.), Department of Astronomy, Faculty of Sciences, Catholic University, Driehuizerweg 200, Nijmegen, The Netherlands

* DELANNOY (J.), Observatoire de Meudon, 5, place Janssen, 92-Meudon, France

DELBOUILLE (L.), Institut d'Astrophysique, Université de Liège, Cointe-Sclessin (Liège), Belgium

DELHAYE (Prof. J.), Observatoire de Paris, 61, avenue de l'Observatoire, Paris-14e, France

DELSEMME (Prof. A. H.), Department of Physics and Astronomy, The University of Toledo, Toledo, Ohio 43606, U.S.A.

DE MARCUS (Prof. W. C.), Department of Physics, University of Kentucky, Lexington, Kentucky, U.S.A.

DEMARQUE (Dr P. R.), Ryerson Physical Laboratory, 1100-14 East 58th Street, Chicago, Illinois 60637, U.S.A.

DEMETRESCU (Dr G.), Directeur, Observatoire de Bucarest, 5, rue Cutitul de Argint, Bucarest, Roumania

DE MORAES (Dr A.), Caixa Postal 30627, São Paulo, SP, Brazil

DE MOTTONI (Dr G.), Via Fratelli Rosselli 15/23, Genova, Italy

DENISSE (Dr J. F.), Directeur, Observatoire de Paris, 61 avenue de l'Observatoire, Paris-14e, France

DENNISON (Dr E. W.), Mount Wilson and Palomar Observatories, 813, Santa Barbara Street, Pasadena, California, U.S.A.

* DENT (Dr W. A.), Astronomy Department, Physics-Astronomy Building, University of Michigan, Ann Arbor, Michigan 48104, U.S.A.

* DEPRIT (Dr A.), Professor at the University of Louvain; on leave at Boeing Scientific Research Laboratories, P.O.Box 3981, Seattle, Washington 98124, U.S.A.

DESSY (Dr J. Landi), Observatorio Nacional Argentino, Cordoba, Argentina

DETRE (Dr L.), Director, Konkoly Observatory, Szabadsághegy, Budapest XII, Hungary

DEUTSCH (Dr A. J.), Mount Wilson and Palomar Observatories, 813 Santa Barbara Street, Pasadena 4, California 91106, U.S.A.

DEUTSCH (Prof. Dr A. N.), [see DEJČ]

DE VAUCOULEURS (Prof. G.), Department of Astronomy, University of Texas, Austin 12, Texas, U.S.A.

DEWHIRST (Dr D. W.), The Observatories, Madingley Road, Cambridge, U.K.

DE WITT (Dr J. H.), Arthur J. Dyer Observatory, Vanderbilt University, Nashville 5, Tennessee, U.S.A.

DEZSŐ (Prof. L.), Heliophysical Observatory of the Hungarian Academy of Sciences, Debrecen 10, Hungary

DIBAJ (Dr E. A.), [Dibay], Sternberg Astronomical Institute, Moscow, U.S.S.R.

DICK (Prof. Dr J.), Rote Kreuz-Strasse 7, Potsdam-Babelsberg 2, G.D.R.
* DICKEL (Dr J. R.), University of Illinois Observatory, Urbana, Illinois 61801, U.S.A.
DIECKVOSS (Dr W.), Hamburger Sternwarte, Hamburg-Bergedorf, G.F.R.
DIETER (Mrs Dr N. H.), Radio Astronomy Laboratory, University of California, Berkeley, California 94720, U.S.A.
DIJKSTERHUIS (Dr E. J.), Wagnerlaan 27, Bilthoven, The Netherlands
DIMITROFF (Prof. G. Z.), Hartland, Vermont 05048, U.S.A.
* DIMOV (Dr N. A.), Crimean Astrophysical Observatory, USSR Academy of Sciences, Crimea, U.S.S.R.
DINGENS (Prof. P.S. A.), Sterrenkundig Instituut, J. Plateaustraat 22, Gent, Belgium
DINGLE (Prof. H.), 104, Downs Court Road, Purley, Surrey, U.K.
DINULESCU (Prof. N.), Directeur-adjoint, Observatoire de Bucarest, 5 rue Cutitul de Argint, Bucarest, Roumania
DIVAN (Mlle Dr L.), Institut d'Astrophysique, 98 bis boulevard Arago, Paris-14e, France
* DIVARI (Dr N. B.), Odessa Politechnical Institute, Odessa, U.S.S.R.
* DIXON (Dr M. E.), Department of Astronomy, University of Edinburgh, Edinburgh, U.K.
DIZER (Prof. Dr M.), Astronomical Observatory, Kandilli, Çengelköy, Istanbul, Turkey
DJURKOVIĆ (P. M.), Astronomical Observatory, Volgina 7, Belgrade, Yugoslavia
* DOAZAN (Mme V.), Observatoire de Paris, 61, avenue de l'Observatoire, 75-Paris-14e, France
DOBRONRAVIN (Dr P. P.), Crimean Astrophysical Observatory of the USSR Academy of Sciences, P/O Naučnyj, Crimea, U.S.S.R.
DOBROVOL'SKIJ (Dr O. V.), [Dobrovolsky], Institute of Astrophysics, Dušanbe, U.S.S.R.
* DOBRZYCKI (Dr J.), Institute of the History of Science and Technics, Nowy Swiat 72, Warsaw, Poland
DODSON-PRINCE (Mrs Dr H. W.), McMath Hulbert Observatory, University of Michigan, Pontiac 4, Michigan, U.S.A.
* DOHERTY (Dr L. R.), Astronomy Department, Sterling Hall, University of Wisconsin, Madison, Wisconsin 53706, U.S.A.
DOLIDZE (Dr M. V.), Abastumani Observatory, Abastumani, Georgia, U.S.S.R.
DOLLFUS (Dr A.), Observatoire de Paris, Section d'Astrophysique, 92-Meudon, France
DOMBROVSKIJ (Dr V. A.), [Dombrovsky], Leningrad University Observatory, Leningrad, U.S.S.R.
DOMINKO (Prof. Dr F.), Saranovićeva 11, Ljubljana, Yugoslavia
DOMMANGET (Dr J.), Observatoire Royal de Belgique, 3 avenue Circulaire, Uccle-Brussels 18, Belgium
DONITCH (N.), Address unknown
* DONN (Dr B. D.), Code 613, Goddard Space Flight Center, Greenbelt, Maryland, U.S.A.
DOS REIS (Prof. M.), Director, Observatory, Coimbra, Portugal
DOSSIN (Dr F.), European Southern Observatory, Am Bahnhof 21, 205 Hamburg 80, G.F.R.
DOUGLAS (Dr A. E.), Division of Physics, National Research Council, Ottawa, Ontario, Canada
DOUGLAS (Miss Dr A. V.), 127 King St. W., Kingston, Ontario, Canada
DOUGLAS (Prof. J. N.), Department of Astronomy, The University of Texas, Austin, Texas, U.S.A.
DRAKE (Prof. F. D.), Department of Astronomy, Cornell University, Ithaca, New York 14850, U.S.A.
DRÂMBĂ (Prof. Dr C.), Observatoire de Bucarest, 5, rue Cutitul de Argint, Bucarest 28, Roumania
* DRESSLER (Dr K.), Department of Astrophysical Sciences, 265 FitzRandolph Road, Princeton, New Jersey 08540, U.S.A.
DROFA (Dr V. K.), Astronomical Observatory, Kiev University, Kiev, U.S.S.R.
DUBOIS-CHEVALLIER (Mme R.), Observatoire de Paris, 61, avenue de l'Observatoire, Paris-14e, France
DUBOŠIN (Prof. Dr G. N.), [Duboshin], Sternberg Astronomical Institute, Moscow, U.S.S.R.
DUBOV (Dr E. E.), Crimean Astrophysical Observatory, USSR Academy of Sciences, P/O Naučnyj, Crimea, U.S.S.R.

DUCHESNE (Dr M.), Observatoire de Paris, 61 avenue de l'Observatoire, Paris-14e, France
DUERKSEN (J. A.), Chief, Section of Gravity and Astronomy, U.S. Coast and Geodetic Survey, Washington 25, D.C., U.S.A.
DUFAY (M.), Institut de Physique Générale de la Faculté des Sciences de Lyon, 18 quai Claude Bernard, Lyon, France
DUFLOT (Mme Dr M.), Observatoire de Marseille, 2, place Le Verrier, Marseille-4e, France
* DUFLOT (R.), Chargée de Recherches CNRS, Observatoire de Marseille, 2, place Le Verrier, 13-Marseille-4e, France
* DUMONT (R.), Observatoire de Bordeaux, 33-Floirac, France
DUNCOMBE (Dr R. L.), Director, Nautical Almanac Office, U.S. Naval Observatory, Washington, D.C., 20390, U.S.A.
DUNHAM, Jr. (Dr T.), Department of Physics, University of Tasmania, Hobart, Tasmania, Australia
DUNN (Dr R. B.), Sacramento Peak Observatory, Sunspot, New Mexico, U.S.A.
DYER (Dr E. R.), National Academy of Sciences, 2101 Constitution Avenue, N.W., Washington 25, D.C., U.S.A.
DŽAPIAŠVILI (Dr V. P.), [Dzapiashvili], Abastumani Observatory, Abastumani, Georgia, U.S.S.R.
DZIGVAŠVILI (Dr R. M.), [Dzigvashvili], Abastumani Astrophysical Observatory, Georgian Academy of Sciences, Abastumani, U.S.S.R.

EBBIGHAUSEN (Prof. E. G.), Department of Physics, University of Oregon, Eugene, Oregon, U.S.A.
ECKERT (Dr W. J.), Watson Scientific Computing Bureau, 612 West 115th Street, New York, N.Y. 10025, U.S.A.
* EDDY (Dr J.), High Altitude Observatory, Boulder, Colorado 80302, U.S.A.
EDLÉN (Prof. B.), Fysiska Institutionen, Lund, Sweden
EDMONDS Jr. (Dr F. N.), Department of Astronomy, The University of Texas, Austin, Texas 78712, U.S.A.
EDMONDSON (Dr F. K.), Director, Goethe Link Observatory, Indiana University, Bloomington, Indiana 47401, U.S.A.
EELSALU (Dr H. T.), W. Struve Tartu Astrophysical Observatory, Tartu, U.S.S.R.
EFREMOV (Dr Ju. I.), Astronomical Council, U.S.S.R. Academy of Sciences, Vavilova Str. 20, Moscow V-312, U.S.S.R.
EGGEN (Dr O. J.), Director, Mount Stromlo Observatory, Canberra, A.C.T., Australia
* EGYED (Prof. L.), The Eötvös Lorand University, Budapest, Hungary
EICHHORN (Prof. H.K.v.W.), Department of Astronomy, University of South Florida, Tampa, Florida 33549, U.S.A.
EJNASTO (Dr J. E.), [Einasto], Institute of Physics and Astronomy of the Estonian Academy of Sciences, Tartu, U.S.S.R.
* ELFORD (Dr W. G.), Department of Physics, University of Adelaide, Adelaide, Australia
ELGARÖY (Dr Ö.), Institute of Theoretical Astrophysics, University of Oslo, Blindern, Oslo 3, Norway
* ELLIOT (Dr I.), Sacramento Peak Observatory, Sunspot, New Mexico, U.S.A. and Dunsink Observatory, Castleknock, Co. Dublin, Ireland
ELLIS (Prof. G. R. A.), University of Tasmania, Hobart, Tasmania, Australia
* ELLYETT (Prof. C. D.), Department of Physics, University of Newcastle, Newcastle, Australia
ELSÄSSER (Prof. Dr H.), Director, Landessternwarte, Heidelberg-Königstuhl, G.F.R.
* ELSMORE (B.), Mullard Radio Observatory, Cavendish Laboratory, Free School Lane, Cambridge, U.K.
ELSTE (Dr G.), Department of Astronomy, University of Michigan, Ann Arbor, Michigan 48104, U.S.A.
ELVEY (Dr C. T.), 5359 E. Hawthorne Street, Tucson, Arizona 85711, U.S.A.
ELVIUS (Mrs Dr A. M.), Stockholm Observatory, Saltsjöbaden, Sweden

ELVIUS (Dr T.), Astronomiska Observatoriet, Uppsala, Sweden
ELWERT (Prof. Dr. G.), Astronomisches Institut der Universität, Gmelinstrasse 6, Tübingen, G.F.R.
* EMINZADE (Dr T.A.), Šemaha Astrophysical Observatory, Azerbaidžan Academy of Sciences, Baku, U.S.S.R.
ENGELHARD (Dr E.), Physikalisch-Technische Bundesanstalt (PTB), Bundesallee 100, 33 Braunschweig, G.F.R.
ENSLIN (H.), Deutsches Hydrographisches Institut, 2 Hamburg 4, G.F.R.
* EPSTEIN (Dr E.E.), Aerospace Corporation, P.O.Box 95085, Los Angeles, California 90045, U.S.A.
EPSTEIN (Dr I.), Rutheford Observatory, Columbia University, New York 27, N.Y., U.S.A.
ERICKSON (Dr W.C.), University of Maryland, Department of Physics and Astronomy, College Park, Maryland, U.S.A.
ERIKSEN (G.), Institute of Theoretical Astrophysics, University of Oslo, Blindern, Oslo 3, Norway
ERPYLEV (Dr N.P.), Astronomical Council, USSR Academy of Sciences, Vavilova Street 20, Moscow V-312, U.S.S.R.
ESHLEMAN (Prof. V.R.), Radioscience Laboratory, Stanford University, Stanford, California U.S.A.
ESKIOĞLU (Dr A.N.), Fen Fakültesi Astronomi, Ankara, Turkey
ESSEN (Dr L.), National Physical Laboratory, Teddington, Middlesex, U.K.
EVANS (Dr D.S.), Royal Observatory, Observatory, Cape Province, South Africa
EVANS (Dr J.W.), Director, Sacramento Peak Observatory, Sunspot, New Mexico, U.S.A.
EVDOKIMOV (Dr Ju.V.), [Dr Yu.V.], Astronomical Observatory, Kazan University, Kazan, U.S.S.R.
EWEN (Dr H.I.), Ewen Knight Corporation, Oak and Pine Streets, East Natick, Massachusetts, U.S.A.
EZER (Mrs Dr D.), Research Associate, Institute for Space Studies, 475 Riverside Drive, New York 27, N.Y., U.S.A.
EZERSKIJ (Dr V.I.), [Ezersky], Astronomical Observatory, Harkov University, Harkov, U.S.S.R.

FABRE (Dr H.), Observatoire de Nice, Le Mont-Gros, 06 Nice, France
FATČIHIN (Dr N.V.), [Fatchikhin], Pulkovo Observatory, Leningrad, U.S.S.R.
* FAULKNER (Dr J.), Peterhouse, Cambridge, U.K.
* FAULKNER (Dr J.), Kellogg Radiation Laboratory, California Institute of Technology, Pasadena, California, U.S.A.
FAYET (Dr G.), 3 villa Hallé, 36 rue Hallé, Paris-14e, France
FAZIO (Dr G.G.), Smithsonian Astrophysical Observatory, 60 Garden Street, Cambridge, Massachusetts 02138, U.S.A.
FEAST (Dr M.W.), Radcliffe Observatory, P.O.Box 373, Pretoria, South Africa
FEBRER (Dr J.), Observatorio Fabra, Barcelona, Spain
FEDOROV (Prof. E.P.), Director, Main Astronomical Observatory, Academy of Sciences, Ukrainian S.S.R., Kiev 41, U.S.S.R.
FEDINSKIJ (Prof. Dr V.V.), [Fedynsky], Astronomical Council, USSR Academy of Sciences, Vavilova Street 20, Moscow V-312, U.S.S.R.
FEHRENBACH (Prof. Ch.), Directeur, Observatoire de Marseille, 2, place Le Verrier, 13-Marseille-4e, France
* FEINSTEIN (Dr A.), Observatorio Astronómico, La Plata, Argentina
FELBER (Dr H.-J.) Academy Observatory, Babelsberg, G.D.R.
FELENBOK (P.), Observatoire de Paris, Section d'Astrophysique, 92-Meudon, France
FELLGETT (Prof. P.B.), Department of Applied Physical Sciences, University of Reading, Whiteknights Park, Reading, Berkshire, U.K.

* FELTEN (Dr J.E.), Department of Physics, University of California-San Diego, La Jolla, California 92038, U.S.A.

FENKART (Dr R.P.), Astronomisch-Meteorologische Anstalt der Universität Basel, Binningen, Switzerland

FERNIE (Dr J.D.), David Dunlap Observatory, Richmond Hill, Ontario, Canada

FERRARI D'OCCHIEPPO (Prof. Dr K.), Universitäts-Sternwarte, Türkenschanzstrasse 17, A-1180 Vienna, Austria

FERRARO (Prof. V.C.A.), Queen Mary College, University of London, Mile End Road, London, E.1., U.K.

FERRAZ-MELLO (Dr S.), Instituto Tecnológico de Aeronáutica, Observatório, Sao José dos Campos, São Paulo, Brazil

FESENKOV (Prof. Dr V.G.), Meteorite Committee, USSR Academy of Sciences, Stroitelej Street 3, Building 1, Doorway 2, Moscow V-313, U.S.S.R.

FESENKOVA (Mrs Prof. Dr E.V.), Astronomical Council, Vavilova Street 20, USSR Academy of Sciences, Moscow V-312, U.S.S.R.

FICHERA (Prof. E.), Osservatorio Astronomico di Capodimonte, Napoli, Italy

* FICHTEL (Dr C.E.), NASA, Goddard Space Flight Center, Greenbelt, Maryland 20771, U.S.A.

FIELD (Dr G.B.), Department of Astronomy, University of California, Berkeley, California 94720, U.S.A.

FIELDER (Dr G.), University of London Observatory, Mill Hill Park, London, N.W.7., U.K. (Until 20 September 1968: Lunar and Planetary Laboratory, University of Arizona, Tucson, Arizona 85721, U.S.A.)

FILLIOZAT (Dr J.), Professeur au Collège de France, 11, place Marcelin Berthelot, 75-Paris-5e, France

FINDLAY (Dr J.W.), National Radio Astronomy Observatory, Edgemont Dairy Road, Charlottesville, Virginia, U.S.A.

FINSEN (Dr W.S.), Director, Republic Observatory, Johannesburg, South Africa

FIREMAN (Dr E.L.), Smithsonian Astrophysical Observatory, 60 Garden Street, Cambridge, Massachusetts 02138, U.S.A.

FIROR (Dr J.W.), High Altitude Observatory, University of Colorado, Boulder, Colorado, U.S.A.

* FISCHEL (Dr D.), Code 613, NASA, Goddard Space Flight Center, Greenbelt, Maryland 20771, U.S.A.

* FISHER (P.C.), Lockheed Research Laboratory, 3251 Hanover Street, Palo Alto, California 93404, U.S.A.

FITCH (Dr W.S.), Steward Observatory, University of Arizona, Tucson, Arizona, U.S.A.

FLECKENSTEIN (Prof. Dr J.O.), Directeur de l'Institut d'Histoire des Sciences et des Techniques de l'École Polytechnique de Munich, Munich, G.F.R.

FLEISCHER (Dr R.), 2121 P Street, N.W., Washington, D.C. 20037, U.S.A.

FOCAS (Dr J.H.), Centre de Documentation sur les Planètes de l'Union Astronomique Internationale, Observatoire de Meudon, 92-Meudon, France

FOKKER (Dr A.D.), Sterrewacht "Sonnenborgh", Servaas Bolwerk 13, Utrecht, The Netherlands

FORD Jr. (Dr W.K.), Department of Terrestrial Magnetism, Carnegie Institution of Washington, Washington, D.C., U.S.A.

FORTINI (Prof. T.), Osservatorio Astronomico, Rome, Italy

FOWLER (Prof. W.A.), California Institute of Technology, 1201 East California Street, Pasadena, California, U.S.A.

FOX (W.E.), 40, Windsor Road, Newark, Nottinghamshire, U.K.

FRACASTORO (Prof. M.G.), Direttore, Osservatorio astronomico di Torino, Pino Torinese, Italy

FRANK-KAMENECKIJ (Prof. D.A.), [Frank-Kamenetsky], Institute for Atomic Energy of the USSR Academy of Sciences, Moscow, U.S.S.R.

* FRANKLIN (Dr F.A.), Smithsonian Institution Astrophysical Observatory, 60 Garden Street, Cambridge, Massachusetts 02138, U.S.A.

FRANZ (Dr O. G.), Lowell Observatory, Flagstaff, Arizona 86002, U.S.A.
* FREDGA (Miss Dr K.), Institute of Plasma Physics, Royal Technical University, Stockholm 70, Sweden
FREDRICK (Prof. L. W.), Leander McCormick Observatory, University of Virginia, Charlottesville, Virginia, 22903, U.S.A.
FRESA (Prof. A.), Osservatorio Astronomico di Capodimonte, Napoli, Italy (Retired residence: Via Porto 23, Salerno, Italy).
FRICKE (Prof. Dr W.), Direktor, Astronomisches Rechen-Institut, Mönchhofstrasse 12–14, 69 Heidelberg, G.F.R.
* FRIEDEMANN (Dr C.), Universitätssternwarte und Astrophysikalisches Institut, der Friedrich-Schiller Universität, Jena, G.D.R.
FRIEDMAN (Dr H.), Code 7320, U.S. Naval Research Laboratory, Washington, D.C. 20390, U.S.A.
FRINGANT (Miss Dr A. M.), Institut d'Astrophysique, 98bis boulevard Arago, Paris-14e, France
* FRITZOVÁ-ŠVESTKOVÁ (Mrs Dr L.), Astronomical Institute, Czechoslovak Academy of Sciences, Observatory Ondřejov, Czechoslovakia
FUCHS (Prof. Dr J.), Universitäts-Sternwarte, Universitätsstrasse 4, Innsbruck, Austria
FUJITA (Dr Y.), Department of Astronomy, Faculty of Science, University of Tokyo, Bunkyo-ku, Tokyo, Japan

* GABRIEL (Dr M.), Institut d'Astrophysique de l'Université de Liège, Cointe-Sclessin, Belgium
* GAIDE (A.), Observatoire Cantonal de Genève, Genève, Sauverny, 1290, Switzerland
GAIZAUSKAS (Dr V.), Dominion Observatory, Ottawa, Ontario, Canada
* GALIBINA (Mrs Dr I. V.), Institute of Theoretical Astronomy, USSR Academy of Sciences, Leningrad, U.S.S.R.
GALLET (R. M.), National Bureau of Standards, Boulder, Colorado, U.S.A.
GALLOUËT (Dr L.), Observatoire de Paris, 61, avenue de l'Observatoire, 75-Paris-14e, France
GALPERIN (Dr G. I.), Institute of Physics and of the Atmosphere, B. Gruzinskaja 10, Moscow, U.S.S.R.
GALT (Dr J. A.), Dominion Radio Astrophysical Observatory, Box 248, Penticton, B.C., Canada
GAMA (Dr L. I.), Observatorio Nacional, R. General Bruce 586, S. Cristovâo, Rio de Janeiro, Brazil
GAMOW (Prof. G.), Professor of Physics, University of Colorado, Boulder, Colorado, U.S.A.
GAPOSCHKIN (Dr S.), Harvard College Observatory, Cambridge, Massachusetts 02138, U.S.A.
GARDNER (Dr F. F.), C.S.I.R.O. Division of Radiophysics, University Grounds, Chippendale, N.S.W., Australia
GARFINKEL (Dr B.), Ballistic Research Laboratories, Aberdeen Proving Ground, Maryland, U.S.A.
GARSTANG (Dr R. H.), Joint Institute for Laboratory Astrophysics, University of Colorado, Boulder, Colorado 80304, U.S.A.
GARTLEIN (Dr C. W.), Department of Physics, Cornell University, Ithaca, N.Y., U.S.A.
GASCOIGNE (Dr S. C. B.), Mount Stromlo Observatory, Canberra, A.C.T., Australia
GAUZIT (Dr J.), Observatoire de Lyon, Saint-Genis-Laval (Rhône), France
GAVIOLA (Dr E.), Instituto Fisico Balseiro, Bariloche, Rio Negro, Argentina
GEAKE (Dr J. E.), Faculty of Technology, University of Manchester, Manchester 1, U.K.
* GEBBIE (Dr K.), Department of Physics, University College, Gower Street, London, W.C.1, U.K.
GEHRELS (Dr T.), Space Sciences Building, University of Arizona, Tucson, Arizona 85721, U.S.A.
* GEL'FREJH (Dr G. B.), [Gelfreikh], Main Astronomical Observatory USSR Academy of Sciences, Pulkovo, U.S.S.R.
* GENT (H.), Royal Radar Establishment, St Andrews Road, Great Malvern, Worcestershire, U.K.
GENTILI DE GIUSEPPE (M.), Observatoire du Pic-du-Midi, 65-Bagnères-de-Bigorre (Hautes-Pyrénées), France

* Geršberg (Dr R. E.), [Gershberg], Crimean Astrophysical Observatory, USSR Academy of Sciences, Crimea, U.S.S.R.

Geyer (Dr E. H.), Observatorium Hoher List, (Universität Bonn), 5568 Daun/Eifel, G.F.R.

* Ghabrus (Dr Roushdy Azer), Helwan Observatory, Helwan, near Cairo, U.A.R.

* Giacaglia (Dr G.), Ecole Polytechnique, Université de São Paulo, São Paulo, Brazil

* Giacconi (Dr R.), American Science and Engineering Inc., 11 Carleton Street, Cambridge, Massachusetts 02142, U.S.A.

* Giannone (P.), Osservatorio Astronomico, Rome, Italy

Giannuzzi (Prof. M. A.), Corso Matteotti 190, Albano, Rome, Italy

Giclas (H. L.), Lowell Observatory, Flagstaff, Arizona, U.S.A.

* Giese (Dr R. M.), Institut für extraterrestrische Physik des Max-Planck-Instituts für Physik und Astrophysik, 8046 Garching b. München, G.F.R.

Gingerich (Dr O. J.), Smithsonian Astrophysical Observatory, 60 Garden Street, Cambridge, Massachusetts 02138, U.S.A.

Ginzburg (Prof. V. L.), Corresponding Member of the USSR Academy of Sciences, Physical Institute, Leninskij Prospekt 53, Moscow, U.S.S.R.

Giovanelli (Dr R. G.), National Standards Laboratory, C.S.I.R.O. University Grounds, Sydney, N.S.W., Australia

Gjellestad (Miss G.), Universitetet i Bergen, Geofysisk Instituut, Avd. C., Bergen, Norway

Gleissberg (Prof. Dr W.), Buchenweg 12, 6375 Oberstedten (Taunus), G.F.R.

Gliese (Dr W.), Astronomisches Rechen-Institut, Mönchhofstrasse 12–14, 69 Heidelberg, G.F.R.

Gnevyšev (Dr M. N.), [Gnevyshev], Astronomical Observatory, Leningrad M–140, U.S.S.R.

Gnevyševa (Mrs Dr R. S.), [Gnevysheva], Main Astronomical Observatory, USSR Academy of Sciences, Pulkovo, U.S.S.R.

Godart (Prof. O.), Université de Louvain, 10, rue Père Damien, Brussels 14, Belgium

Godoli (Prof. Dr G.), Direttore, Osservatorio Astrofisico, Catania, Italy

Gökdoğan (Mrs Prof. Dr. N.), Director, University Observatory, Beyazit, Istanbul, Turkey

Gökmen (Dr T.), Kandilli Observatory, Çengelköy, Istanbul, Turkey

Golay (Prof. Dr M.), Directeur de l'Observatoire de Genève, 1290 Sauverny (GE), Switzerland

Gold (Prof. T.), Space Science Bldg., Cornell University, Ithaca, N.Y. 14850, U.S.A.

Goldberg (Prof. L.), Harvard College Observatory, Cambridge, Massachusetts 02138, U.S.A.

* Goldreich (Dr P.), California Institute of Technology, Pasadena, California 91109, U.S.A.

* Goldstein (Dr S. J.), Leander McCormick Observatory, University of Virginia, Charlottesville, Virginia 22903, U.S.A.

Goldsworthy (Prof. F. A.), Department of Mathematics, University of Leeds, Leeds 2, Yorkshire, U.K.

Gollnow (Dr H.), Mount Stromlo Observatory, Canberra, A.C.T., Australia

Gomes (Dr A. M.), University of Brazil, Rua Ipiranga, 25–40 Andar, Rio de Janeiro, ZC–01, Brazil

* Gompertz (Dr G.), School of Physics, University of Newcastle upon Tyne, Newcastle-upon-Tyne, 1, U.K.

Gondolatsch (Prof. Dr F.), Astronomisches Recheninstitut, Mönchhofstrasse 12–14, 69 Heidelberg, G.F.R.

González (G.), Observatorio Astronomico, Universidad Nacional de Mexico, Tonantzintla, Mexico

* Goody (Dr E. M)., Director, Blue Hill Observatory, Harvard University, Cambridge, Massachusetts 02138, U.S.A.

* Gopasjuk (Dr S. I.), [Gopasyuk], Crimean Astrophysical Observatory, USSR Academy of Sciences, Crimea, U.S.S.R.

Gorbackij (Dr V. G.), [Gorbatzky], Leningrad University Observatory, Leningrad, U.S.S.R.

Gordon (Dr J. E.), Director, Nikolaev Observatory, Nikolaev, U.S.S.R.

Gorgolewski (Dr S.), Astronomical Observatory, N. Copernicus University, Sienkiewicza 30, Torun, Poland

GORŠKOV (Prof. Dr P. M.), [Gorshkov], State University, Leningrad, U.S.S.R.
GORYNJA (Dr A. A.), [Gorynya], Kiev University Observatory, Kiev, U.S.S.R.
GOSSNER (Mrs S. D.), U.S. Naval Observatory, Washington 25, D.C., U.S.A.
GOTTLIEB (K.), Mount Stromlo Observatory, Canberra, A.C.T., Australia
* GOUDAS (Dr C. L.), University of Patras, Greece
GOUGENHEIM (Prof. A.), Directeur du Service Hydrographique de la Marine, 30, boulevard Flandrin, 75-Paris-16e, France
* GOULD (Dr R.), Physics Department, University of California-San Diego, La Jolla, California, U.S.A.
* GOY (G.), Observatoire Cantonal de Genève, Genève, 1290 Sauverny, Switzerland
* GOYAL (A. N.), Department of Mathematics, University of Rajasthan, India
* GRAHAM (Dr J. A.), Kitt Peak National Observatory, P.O.Box 4130, Tucson, Arizona 85717, U.S.A.
GRAHL (Dr B. H.), Universitäts-Sternwarte Bonn/Rhein, Poppelsdorfer Allee 49, 53 Bonn, G.F.R.
* GRAINGER (Dr J.), University of Manchester, Institute of Science and Technology, Sackville Street, Manchester 1, U.K.
GRANDMONTAGNE (Dr R.), Faculté des Sciences, 18 quai Claude Bernard, Lyon (Rhône), France
GRANDON (Prof. R.), Carlos Edwards 1270, Santiago de Chile, Chile
* GRANT (Dr I. P.), Pembroke College, Oxford, U.K.
GRATTON (Prof. L.), Laboratorio di Astrofisica, Casella Postale 67, Frascati, Rome, Italy. Vice-President of the Union.
GREDZIELSKI (Dr S.), Astronomical Observatory of Warsaw University, Al. Ujazdowskie 4, Warsaw, Poland
* GREEN (Dr J.), Research Scientist, Advanced Research Laboratory, Douglas Aircraft Company, Inc., 5251 Bolsa Avenue, Huntington Beach, California 92646, U.S.A.
GREEN (Prof. L. C.), Strawbridge Observatory, Haverford College, Haverford, Pennsylvania 19041, U.S.A.
* GREEN (Dr R. M.), Department of Astronomy, The University, Glasgow, W.2., Scotland, U.K.
GREENBERG (Dr J. M.), Department of Physics, Rensselaer Polytechnic Institute, Troy, N.Y., U.S.A.
GREENSTEIN (Prof. J. L.), Mount Wilson and Palomar Observatories, 1201 East California Street, Pasadena, California, U.S.A.
GRIFFIN (Dr R. F.), University Observatories, Madingley Road, Cambridge, U.K.
* GRIFFIN (Dr R. M.), The Observatories, Madingley Road, Cambridge, U.K.
* GRIGOREVSKIJ (Dr V. M.), [Grigorevsky], Kišinev State University, Kišinev, U.S.S.R.
GROTH (Dr H. G.), Universitäts-Sternwarte München, Sternwartstrasse 23, 8 Munich 27, G.F.R.
* GRUŠINSKIJ (Prof. Dr N. P.), [Grushinsky], Sternberg Astronomical Institute, Moscow, U.S.S.R.
* GRYGAR (Dr J.), Astronomical Institute, Czechoslovak Academy of Sciences, Observatory Ondřejov, Czechoslovakia
GUÉRIN (Dr P.), Institut d'Astrophysique, 98bis boulevard Arago, Paris-14e, France
GUIGAY (Dr G.), Observatoire de Marseille, 2, place Le Verrier, 13-Marseille-4e, France
GUINOT (Dr B.), Observatoire de Paris, 61, avenue de l'Observatoire, Paris-14e, France
GULLÓN (Dr E.), Observatorio Astronomico de Madrid, Spain
GÜNTHER (Dr O.), Astrophysical Observatory, Potsdam, G.D.R.
GÜNTZEL-LINGNER (Dr U.), Astronomisches Rechen-Institut, Mönchhofstrasse 12–14, 69 Heidelberg, G.F.R.
GURTOVENKO (Dr E. A.), Main Astronomical Observatory, Ukrainian Academy of Sciences, Kiev, U.S.S.R.
GURZADJAN (Dr G. A.), [Gurzadian], Bjurakan Astrophysical Observatory, Armenian S.S.R., Erevan, Armenia, U.S.S.R.
* GUSSMANN (Dr E.-A.), Astrophysikalisches Observatorium Potsdam der Deutschen Akademie der Wissenschaften, Berlin, G.D.R.

GUTH (Prof. V.), Astronomical Institute, Czechoslovak Academy of Sciences, Observatory Ondřejov, Czechoslovakia

GUTHRIE (B.N.G.), The Royal Observatory, Edinburgh 9, U.K.

GUTIERREZ-MORENO (Mrs Dr A.), Observatorio Astronómico de la Universidad de Chile, Casilla 36-D, Santiago de Chile, Chile

GYLDENKERNE (K.), Universitets Observatoriet, Broerfelde, pr. Tølløse, Denmark

HAAS (Prof. Dr J.), Lipschitz-Strasse 1, 53 Bonn, G.F.R.

HABIBULLIN (Prof. Dr Š. T.), [Habibulin], Director, Astronomical Observatory of State University, Lenina 18, Kazan, U.S.S.R.

HACHENBERG (Prof. Dr O.), Radiosternwarte Bonn, Poppelsdorfer Allee 49, 53 Bonn, G.F.R.

HACK (Mrs. Prof.), Director, Astronomical Observatory, Via Tiepolo 11, Trieste (34131), Italy

* HAČIKJAN (Dr E.E.), [Khachikyan], Bjurakan Astrophysical Observatory, Armenian Academy of Sciences, Bjurakan, U.S.S.R.

HADDOCK (Dr F.), The Observatory, University of Michigan, Ann Arbor, Michigan, U.S.A.

* HADJIDEMETRIOU (Dr J.D.), Astronomical Department, University of Thessaloniki, Thessaloniki, Greece

* HAERENDEL (Dr G.), Institut für extraterrestrische Physik des Max-Planck-Instituts für Physik und Astrophysik, 8046 Garching b. München, G.F.R.

HAFFNER (Prof. H.), Astronomisches Institut der Universität Büttnerstrasse 72, 87 Würzburg, G.F.R.

HAGEN (Dr J.P.), Professor of Radio Astronomy, The Pennsylvania State University, University Park, Pennsylvania, U.S.A.

HAGIHARA (Prof. Y.), Department of Astronomy, Faculty of Science, University of Tokyo, Bunkyo-ku, Tokyo, Japan

HAJKIN (Prof. Dr S.E.), [Khaikin], Pulkovo Observatory, Leningrad, U.S.S.R.

HALL (Dr J.S.), Director, Lowell Observatory, Flagstaff, Arizona, U.S.A.

HALL (Dr R.G.), U.S. Naval Observatory, Washington, D.C. 20390, U.S.A.

HALLAM (Dr K.L.), Solar Physics Branch, NASA, Goddard Space Flight Center, Greenbelt, Maryland, U.S.A.

HALLIDAY (Dr I.), Dominion Observatory, Ottawa, Ontario, Canada

* HÄMEEN-ANTTILA (Prof. Dr A.), The Astronomical Institute of the Oulu University, Oulu, Finland

HAMID (Dr S. El-Din), Department of Astronomy, Faculty of Science, University Fouad, Giza, Cairo, U.A.R.

* HANASZ (Dr J.), Astronomical Observatory, Toruń-Piwnice, Poland

HANSSON (N.), Lund Observatory, Lund, Sweden

HARADZE (Prof. Dr E.K.), [Kharadze], Director, Astrophysical Observatory, Abastumani, Georgia, U.S.S.R.

HARANG (Prof. L.), Institute of Geophysics, University of Oslo, Blindern, Oslo 3, Norway

HARDIE (Dr R.H.), Dyer Observatory, Vanderbilt University, Nashville 5, Tennessee, U.S.A.

* HARDING (G.A.), Royal Greenwich Observatory, Herstmonceux Castle, Hailsham, Sussex, U.K.

HARDORP (Dr J.), Hamburger Sternwarte, 205 Hamburg-Bergedorf, G.F.R.

HARGREAVES (F.J.), Mirastelle, Woodland Way, Kingswood, Surrey, U.K.

HARO (Dr G.), Observatorio Astronómico Nacional, Apartado Postal 70–264, Ciudad Universitaria, Mexico 20, D.F. Mexico

HARRIS (B.J.), Perth Observatory, Bickley, Western Australia

* HARRISON (Dr E.R.), Goddard Space Flight Center, Greenbelt, Maryland 20910, U.S.A.

HARROWER (Prof. G.A.), Department of Physics, Queen's University, Kingston, Ontario, Canada

* HARTE (Dr E.A.), Department of Applied Mathematics, University College of Swansea, Singleton Park, Swansea, Wales, U.K.

HARTZ (Dr T.R.), Telecommunications Division, Defence Research Board, Ottawa, Ontario, Canada

* HARWIT (Dr M.), Cornell University, Center for Radiophysics and Space Research, Clark Hall, Ithaca, New York 14850, U.S.A.

HARWOOD (Miss M.), "The Hutch", Nantucket, Massachusetts 02554, U.S.A.

HASER (Dr L.), Max-Planck-Institut für Physik und Astrophysik, Föhringer Ring 6, 8 Munich, G.F.R.

HATTORI (Dr A.), Kwasan Observatory, University of Kyoto, Yamashina, Kyoto, Japan

* HAUCK (B.), Observatoire Cantonal de Genève, 1290 Sauverny (GE), Switzerland

HAUG (Dr U.), Astronomisches Institut der Universität, Waldhäuserstrasse 64, 74 Tübingen, G.F.R.

HAUPT (Dr H.F.), Universitäts-Sternwarte Wien, Türkenschanzstrasse 17, A-1180 Vienna, Austria

* HAUPT (R.), Assistant Director, American Ephemeris and Nautical Almanac, U.S. Naval Observatory, Washington, D.C. U.S.A.

HAUPT (R.F.), U.S. Naval Observatory, Washington, D.C. 20390, U.S.A.

HAVTASI (Dr J.Š.), [Khavtasi, Dr J.Sh.], Institute Cybernetics, Academy of Sciences, Chitadze Street 6, Tbilisi, Georgia, U.S.S.R.

HAWKINS (Dr G.S.), Harvard College Observatory, Cambridge, Massachusetts 02138, U.S.A.

HAYASHI (Prof. C.), Department of Nuclear Science, Faculty of Science, Kyoto University, Kyoto, Japan

* HAZLEHURST (Dr J.), Department of Astronomy, The University, Manchester 13, U.K.

HEARD (Dr J.F.), David Dunlap Observatory, Richmond Hill, Ontario, Canada

* HEARN (Dr A.G.), The Culham Laboratories, Abingdon, Berkshire, U.K.

HECKMANN (Prof. Dr O.), Director, The European Southern Observatory, Bergedorfer Straße 131, 205 Hamburg 80, G.F.R. President of the Union.

HEDDLE (Dr D.W.O.), Department of Physics, University of York, Heslington, York, U.K.

HEDEMAN (Miss E.R.), McMath-Hulbert Observatory, Lake Angelus Road, Pontiac, Michigan, U.S.A.

HEESCHEN (Dr D.S.), National Radio Astronomy Observatory, Edgemont Dairy Road, Charlottesville, Virginia 22901, U.S.A.

HEIDMANN (Dr J.), Observatoire de Paris, Section d'Astrophysique, 92-Meudon, France

HEINEMANN (Prof. Dr K.), Bergstrasse 77, 69 Heidelberg, G.F.R.

HEINTZ (Dr W.D.), University Observatory, Sternwartstrasse 23, Munich 27, G.F.R. (Until August 1968: Sproul Observatory, Swarthmore, Pennsylvania 19081, U.S.A.)

* HEINTZE (Dr J.R.W.), Sterrewacht "Sonnenborgh", Servaas Bolwerk 13, Utrecht, The Netherlands

HEISKANEN (Prof. V.A.), Geodetic Institute, Helsinki, Finland

HELFER (Dr H.L.), Department of Physics and Astronomy, University of Rochester, Rochester, N.Y., U.S.A.

HELLMAN (Prof. C.D.), 1065 Lexington Avenue, New York, N.Y. 10021, U.S.A.

HEMENWAY (Dr C.L.), 140 S. Lake Avenue, Albany, New York, U.S.A.

* HEMMLEB (Ing G.), Geodätisches Institut Potsdam der Deutschen Akademie der Wissenschaften, Berlin, G.D.R.

HENIZE (Dr K.G.), Astronaut Office, NASA, Manned Spacecraft Center, 2101 Webster-Seabrook Road, Houston, Texas 77058, U.S.A.

HENN (Miss F.), Astronomisches Rechen-Institut, Mönchhofstrasse 12–14, 69 Heidelberg, G.F.R.

HÉNON (M.), Institut d'Astrophysique, 98bis boulevard Arago, Paris-14e, France

HENYEY (Prof. L.G.), Astronomical Department, University of California, Berkeley, California 94720, U.S.A.

HERBIG (Dr G.H.), Lick Observatory, University of California, Santa Cruz, California 95060, U.S.A.

HERCZEG (Dr T.), Hamburger Sternwarte, 205 Hamburg-Bergedorf, G.F.R.

HERGET (Dr P.), Cincinnati Observatory, Observatory Place, Cincinnati, Ohio 45208, U.S.A.

HERMAN (Mme Dr R.), Observatoire de Paris, Section d'Astrophysique, 92-Meudon, France
HERRICK (Dr S.), University of California, Los Angeles 24, California, U.S.A.
HERS (J.), Republic Observatory, Johannesburg, South Africa
HERTZ (Dr H. G.), 2112 Florida Avenue, N.W., Washington., 8 D.C., U.S.A.
HERZBERG (Dr G.), Director, Division of Pure Physics, National Research Council, Sussex Drive, Ottawa, Ontario, Canada
* HERZBERG (Mrs Dr L. H.), Defence Research Board, Ottawa, Ontario, Canada
HEWISH (Dr A.), Cavendish Laboratory, Free School Lane, Cambridge, U.K.
HEY (Dr J. S.), Falconhurst, 12 Lodge Drive, Malvern, Worcestershire, U.K.
HEYDEN S. J. (Dr F. J.), Georgetown College Observatory, Washington 7, D.C., U.S.A.
HIBBS (A. R.), Jet Propulsion Laboratory, California Institute of Technology, Pasadena, California 91109, U.S.A.
* HIEI (Dr E.), Tokyo Astronomical Observatory, University of Tokyo, Mitaka, Tokyo, Japan
* HIGGS (Dr L. A.), Radio and Electrical Engineering Division, National Research Council, Ottawa, Ontario, Canada
HILL (Dr E. R.), Division of Radiophysics, C.S.I.R.O., University Grounds, Sydney, N.S.W., Australia
* HILL (Dr P. W.), University Observatory, St Andrews, Fife, U.K.
HILTNER (Dr W. A.), Yerkes Observatory, Williams Bay, Wisconsin, U.S.A.
HINDMAN (J. V.), Sinding Spring Observatory, Coonabarabran, N.S.W., Australia
HINDMARSCH (Prof. W. R.), Physics Department, The University, Newcastle upon Tyne 1, U.K.
HINTEREGGER (Dr H. E.), Hq. AFCRL (CRUU) Stop 30, L.G.Hanscom Field, Bedford, Massachusetts 01731, U.S.A.
* HIRAYAMA (Dr T.), Tokyo Astronomical Observatory, University of Tokyo, Mitaka, Tokyo, Japan
HIROSE (Dr H.), Tokyo Astronomical Observatory, Mitaka, Tokyo, Japan
HIRST (P. W.), Water's Edge, Greenbanks Road, Rondebosch, Cape Province, South Africa
HITOTUYANAGI (Prof. Dr Z.), Department of Astronomy, Faculty of Sciences, Tohoku University, Sendai, Japan
HOAG (Dr. A. A.), Kitt Peak National Observatory, P.O.Box 4130, Tucson, Arizona 85717, U.S.A.
* HOBBS (Dr R. W.), Code 7133e, U.S. Naval Research Laboratory, Washington, D.C. 20390, U.S.A.
HODGE (Prof. P. W.), Astronomy Department, University of Washington, Seattle, Washington 98105, U.S.A.
HOFFLEIT (Miss Dr E. D.), Yale University Observatory, Box 2023 Yale Station, New Haven, Connecticut, U.S.A.
HØG (E.), Hamburger Sternwarte, 205 Hamburg 80, G.F.R.
HÖGBOM (Dr. J. A.), University Observatory, Sterrewacht, Leiden, The Netherlands
* HOGG (Dr D. E.), National Radio Astronomy Observatory, Edgemont Dairy Road, Charlottesville, Virginia 22901, U.S.A.
* HOGLUND (Dr B.), Institute of Electron Physics, Chalmers Technical University, Gribraltargatan 5 G, Göteborg S, Sweden
HOHLOVA (Mrs Dr V. L.), [Khokhlova], Astronomical Council, USSR Academy of Sciences, Vavilova Str. 20, Moscow, U.S.S.R.
* HOLDEN (F.), Astronomer-in-charge, Lamont-Hussey Observatory, Bloemfontein, P.O.Box 875, Bloemfontein, South Africa
HOLMBERG (Prof. E. B.), Director of the Observatory, Uppsala, Sweden
HOLOPOV (Dr P. N.), [Kholopov], Astronomical Council of the USSR Academy of Sciences, Vavilova Street 20, Moscow V-312, U.S.S.R.
HOOGHOUDT (B. G.), University Observatory, Sterrewacht, Leiden, The Netherlands

Hopmann (Prof. Dr J.), Universitäts-Sternwarte, Türkenschanzstrasse 17, A-1180 Vienna, Austria

Hoppe (Prof. Dr J. A.), Sonnenbergstrasse 12, 69 Jena, G.D.R.

Hori (Dr G.), Yale University Observatory, Box 2023, Yale Station, New Haven, Connecticut, U.S.A.

* Hosokawa (Dr Y.), Yamagata University, Koshirakawa-cho, Yamagata-shi, Yamagata, Japan

Hotinli (Dr M.), University Observatory, Beyazit, Istanbul, Turkey

* Houck (Dr T. E.), Washburg Observatory, University of Wisconsin, Madison, Wisconsin 53706, U.S.A.

* House (Dr L. L.), High Altitude Observatory, Boulder, Colorado 80302, U.S.A.

Houtgast (Dr J.), Astronomical Observatory "Sonnenborgh", Zonnenburg 2, Utrecht, The Netherlands

Houziaux (Dr L.), Institut d'Astrophysique, Université de Liège, Cointe-Sclessin (Liège), Belgium

Howard (Dr R. F.), Mount Wilson and Palomar Observatories, 813 Santa Barbara Street, Pasadena, California 91106, U.S.A.

Howard (Dr W. E. III), National Radio Astronomy Observatory, Edgemont Road, Charlottesville, Virginia 22901, U.S.A.

Hoyle (Prof. F.), St. John's College, Cambridge, U.K.

Huang (Dr Su-Shu), Dearborn Observatory, Northwestern University, Evanston, Illinois 60201, U.S.A.

Hubenet (Dr H.), Sterrewacht "Sonnenborgh", Servaas Bolwerk 13, Utrecht, The Netherlands

Huffer (Dr C. M.), San Diego State College, Department of Astronomy, San Diego 15, California, U.S.A.

* Hughes (Dr J. A.), Observatorio Astronomico Austral, Yale-Columbia el Leoncito, Barreal, Prov. San Juan, Argentina

Hughes (Dr V.A.), Department of Physics and Astronomy, Queen's University, Kingston, Ontario, Canada

* Huguenin (Dr G. R.), Harvard College Observatory, 60 Garden Street, Cambridge, Massachusetts 02138, U.S.A.

Humason (Dr M. L.), P.O.Box 165, Mendocino, California, U.S.A.

* Hummer (Dr D.), Joint Institute for Laboratory Astrophysics, University of Colorado, Boulder, Colorado, U.S.A.

Humphreys (C. J.), Head, Research Department, U.S. Naval Ordnance Laboratory, Corona, California 91720, U.S.A.

Hunaerts (Dr J. J.), Observatoire Royal de Belgique, 3 avenue Circulaire, Uccle-Brussels 18, Belgium

Hunger (Prof. Dr K.), Institut für Theoretische Physik, Technische Hochschule, Hannover, G.F.R.

Hunten (Dr D. M.), Kitt Peak National Observatory, P.O.Box 4130, Tucson, Arizona 85717, U.S.A.

Hunter (Dr A.), Royal Greenwich Observatory, Herstmonceux Castle, Hailsham, Sussex, U.K.

* Hunter (Dr C.), Department of Mathematics, Massachusetts Institute of Technology, Cambridge, Massachusetts 02139, U.S.A.

* Hunter (Dr J. H.), Yale University Observatory, Box 2023, Yale Station, New Haven, Connecticut 06520, U.S.A.

Huruhata (Prof. M.), Tokyo Astronomical Observatory, Mitaka, Tokyo, Japan

* Hyder (Dr C. L), Sacramento Peak Observatory, Sunspot, New Mexico 88349, U.S.A.

Hynek (Dr J. A.), Dearborn Observatory, Northwestern University, Evanston, Illinois 60201, U.S.A.

* Ianini (Dr G.), Observatorio Astronómico, Cordoba, Argentina

* IBEN Jr. (Dr I.), Kellog Radiation Laboratory, California Institute of Technology, 1201 East California Street, Pasadena, California 91109, U.S.A.
* IBEN (Prof. Icko), Department of Physics, Massachusetts Institute of Technology, Cambridge, Massachusetss 02139, U.S.A.

IDLIS (Dr G. M.), Astrophysical Institute, Alma Ata, U.S.S.R.

IHSANOVA (Mrs Dr V. N.), [Ikhsanova], Main Astronomical Observatory, USSR Academy of Sciences, Pulkovo, U.S.S.R.

IIJIMA (Dr S.), Tokyo Astronomical Observatory, Mitaka, Tokyo, Japan

IKAUNIEKS (Dr J.), Director, Astrophysical Laboratory, Academy of Sciences, Turgeneva 19, Riga, Latvija, U.S.S.R.

IKEDA (Dr T.), International Latitude Observatory, Mizusawa-shi, Iwate-Ken, Japan

IKHSANOVA (Mrs Dr V. N.), [see IHSANOVA]

INGHAM (Dr M. F.), Department of Astrophysics, University Observatory, South Parks Road, Oxford, U.K.

INGRAO (H. C.), Harvard College Observatory, Cambridge, Massachusetts 02138, U.S.A.

IONNISIANI (Dr B. K.), Main Astronomical Observatory, Academy of Sciences, Pulkovo, U.S.S.R.

* IOŠPA (Dr B. A.), [Ioshpa], Institute of Terrestrial Magnetism, Ionosphere and Propagation of Radiowaves, USSR Academy of Sciences, Moscow, U.S.S.R.
* IRELAND (Dr J. G.), Royal Observatory, Edinburgh 9, U.K.

IRIARTE (B.), Tonantzintla and Tacubya Observatories, Mexico

* IRVINE (Dr W. M.), Department of Physics and Astronomy, University of Massachusetts, Amherst, Massachusetts 01002, U.S.A.

IRWIN (Prof. Dr J. B.), Steward Observatory, University of Arizona, Tucson, Arizona 85721, U.S.A.

IŠČENKO (Dr I.M.), [Istchenko], Astronomical Observatory, Taškent, U.S.S.R.

ITZIGSOHN (M.), Observatorio Astronómico, La Plata, Argentina

IVANOV (Dr V. V.), Astronomical Observatory, Leningrad University, Leningrad, U.S.S.R.

IVANOV-HOLODNYJ (Dr G. S.), [Ivanov-Kholodny], Institute of Applied Geophysics, Academy of Sciences of the USSR, Moscow, U.S.S.R.

IWANISZEWSKA-LUBIENSKA (Mrs Dr C.), Astronomical Observatory, N. Copernicus University, Sienkiewicza 30, Torun, Poland

IWANOWSKA (Miss Prof. W.), Astronomical Observatory, N. Copernicus University, Sienkiewicza 30, Torun, Poland

JACCHIA (Dr L.), Smithsonian Astrophysical Observatory, 60 Garden Street, Cambridge, Massachusetts 02138, U.S.A.

JACKISCH (Dr G.), Academy Observatory, Sonneberg, Thüringen, G.D.R.

JACKSON (C.), Yale-Columbia Station, Mount Stromlo Observatory, Canberra, A.C.T., Australia

JACKSON (Dr P.), Universitäts-Sternwarte Wien, Türkenschanzstrasse 17, A-1180 Vienna, Austria

JACOBSEN (Prof. T. S.), University of Washington, Seattle 5, Washington 98115, U.S.A.

JACQUINOT (P.), Laboratories de Bellevue, 1, place Aristide-Briand, Bellevue, 92-Meudon, France

JÄGER (Prof. Dr F. W.), Astrophysikalisches Observatorium, Telegraphenberg, Potsdam, G.D.R.

JAHONTOVA (Mrs Dr N. Samojlova), [Yakhontova], Institute of Theoretical Astronomy, Leningrad, U.S.S.R.

JAKIMIEC (Dr J.), Wroclaw University Astronomical Institute, Kopernika 11, Wroclaw 9, Poland

JAKOVKIN (Prof. Dr A. A.), [Yakovkin)], Main Astronomical Observatory of the Academy of Sciences of the Ukrainian SSR, Kiev 41, U.S.S.R.

JAKOVKIN (Dr N. A.), [Yakovkin], Kiev University Observatory, Kiev, U.S.S.R.

* JAMES (Dr R. A.), Department of Astronomy, The University, Manchester 13, U.K.

JÁNOSSY (Prof. L.), Director, Central Physical Institute of Budapest, Budapest, Hungary

JÄRNEFELT (Prof. G. J.), Director of the Observatory, Helsinki, Finland

JARRETT (Dr A. H.), Department of Applied Mathematics, Queen's University, Belfast, Northern Ireland, U.K.

* JARZEBOWSKI (Dr T.), Astronomical Observatory, Kopernika 11, Wroclaw, Poland

JASCHEK (Dr C.O. R.), Observatorio Astronómico, La Plata, Argentina

JASCHEK (Mrs Dr M.), Observatorio Astronómico, La Plata, Argentina

JASTROW (Dr R.), NASA, Institute for Space Studies, 2880 Broadway, New York, N.Y. 10025

JAVET (Prof. Dr P.), Chantemerle 19, 1010 Lausanne, Switzerland

JAVNEL' (Dr A. A.), [Yavnel], Meteorite Committee, USSR Academy of Sciences, Moscow, U.S.S.R.

JEFFERIES (Dr J. T.), Hawaii Institute of Geophysics, 2525 Correa Road, University of Hawaii, Honolulu, Hawaii 96822, U.S.A.

JEFFERS (Dr H. M.), Lick Observatory, Mount Hamilton, California, U.S.A.

JEFFREYS (Lady), Girton College, Cambridge, U.K.

JEFFREYS (Sir Harold), 160 Huntingdon Road, Cambridge, U.K.

JELLEY (Dr J. V.), Nuclear Physics Division, Building 8, Atomic Energy Research Establishment, Harwell, Didcot, Berkshire, U.K.

JENKINS (Miss L. F.), Yale University Observatory, Box 2023, Yale Station, New Haven, Connecticut, U.S.A.

* JENNISON (Prof. R. C.), The Electronics Laboratory, University of Kent, Canterbury, Kent, U.K.

JENSCH (A.), Carl Zeiss Observatory, Jena, G.D.R.

JENSEN (Prof. E.), Director, Institute of Theoretical Astrophysics, University of Oslo, Blindern, Oslo 3, Norway

JOHNSON (Dr H. L.), Lunar and Planetary Laboratory, University of Arizona, Tucson, Arizona, U.S.A.

JOHNSON (Dr H. M.), Dept. 52–10, Bldg. 202, Lockheed Missiles and Space Co., 3251 Hanover Street, Palo Alto, California 94304, U.S.A.

JOHNSON (Prof. H. R.), Department of Astronomy, Swain Hall West, University of Indiana, Bloomington, Indiana, U.S.A.

JOHNSON (Dr M. C.), Department of Physics, University, Edgbaston, Birmingham 15, U.K.

JONCKHEERE (R.), 34, boulevard National, 13-Marseille-1er, France

JONES (A. F.), 14 Main Road, Tahunanui, Nelson, New Zealand

* JORDAN (Dr C.), The Culham Laboratories, Abingdon, Berkshire, U.K.

* JORDAN (H. L.), Director, Instituto Europeo per le Ricerce Spaziali, Via Sulpicio Galba, Casella Postale 64, Frascati, Rome, Italy

* JOSHI (M. N.), Tata Institute of Fundamental Research, Bombay, India

JOY (Dr A. H.), Mount Wilson and Palomar Observatories, 813 Santa Barbara Street, Pasadena, California, U.S.A.

JUGAKU (Dr J.), Tokyo Astronomical Observatory, Mitaka, Tokyo, Japan

JUNKES S.J. (Dr J.), Specola Vaticana, Castel Gandolfo, Vatican City State

* JURKEVICH (Dr I.), Department of Astronomy, David Rittenhouse Laboratory, University of Pennsylvania, Philadelphia, Pennsylvania 19104, U.S.A.

KABURAKI (Prof. M.), Department of Astronomy, Faculty of Science, University of Tokyo, Yayoicho, Bunkyo-ku, Tokyo, Japan

KADLA (Mrs Dr Z.), Pulkovo Observatory, Leningrad M-140, U.S.S.R.

KAHN (Dr F. D.), Department of Astronomy, The University, Manchester 13, U.K.

KAHRSTEDT (Prof. Dr A.), Potsdamer Strasse 43, 1 Berlin 45, Germany

* KAI (Dr K.), Tokyo Astronomical Observatory, University of Tokyo, Mitaka, Tokyo, Japan

KAIDANOVSKIJ (Dr N. L.), [Kaidanovsky], Main Astronomical Observatory, Academy of Sciences, Pulkovo, U.S.S.R.

* KAISER (Prof. T. R.), Department of Physics, The University, Sheffield 10, U.K.

KAKINUMA (Dr T.), Research Institute of Atmospherics, Nagoya University, Toyokawa, Aichi, Japan

KALANDADZE (Mrs Dr N. B.), Astrophysical Observatory, Abastumani, Georgia, U.S.S.R.

* KALER (Dr J. B.), University of Illinois Observatory, Urbana, Illinois 61801, U.S.A.

KALINJAK (Dr A. A.), [Kaliniak], Main Astronomical Observatory, Academy of Sciences, Pulkovo, U.S.S.R.

KALINKOV (M.), Section d'Astronomie de l'Académie Bulgare des Sciences, rue 7 Novembre No 1, Sofia, Bulgaria

* KALKOFEN (Dr W.), Smithsonian Astrophysical Observatory, 60 Garden Street, Cambridge, Massachusetts 02138, U.S.A.

KALMYKOV (Dr A. M.), Taškent Astronomical Observatory, Uzbek Academy of Sciences, Taškent, U.S.S.R.

KAMIEŃSKI (Prof. Dr M.), Kopernik's Str. N 8/18, Apart. 2, Warsaw, Poland

* KAMIJO (Dr F.), Department of Astronomy, University of Tokyo, Yayoi, Bunkyo-ku, Tokyo, Japan

* KANNO (Dr M.), Kwasan Astronomical Observatory, University of Kyoto, Yamashina, Kyoto, Japan

KAO (Dr Ping-tse), 12, Lane 65, Ling I Str, Taipei, Taiwan

KAPLAN (Dr J.), Department of Physics, University of California at Los Angeles, Los Angeles, California, U.S.A.

KAPLAN (L.), Jet Propulsion Laboratory, 4800 Oak Grove Drive, Pasadena, California, U.S.A.

KAPLAN (Prof. S. A.), Radio Physical Research Institute of Gorkij University, Gorkij, U.S.S.R.

KARANDIKAR (Dr R. V.), Director, Nizamiah Observatory, Begumpet, Hyderabad-16, India

KARDAŠEV (Dr N. S.), [Kardashev], Sternberg Astronomical Institute, Moscow, U.S.S.R.

KARIMOV (Dr M. G.), Astrophysical Institute, Alma Ata, U.S.S.R.

KARJAGINA (Miss Dr Z. V.), [Karyagina], Astrophysical Institute, Alma Ata, U.S.S.R.

KASTLER (Prof. A.), Ecole Normale Supérieure, 45 rue d'Ulm, Paris-14e, France

KATASEV (Dr L. A.), Commission on Comets and Meteors, Astronomical Council, USSR Academy of Sciences, Vavilova Street 20, Moscow V-312, U.S.S.R.

* KATO (Dr S.), Department of Astronomy, University of Tokyo, Yayoi, Bunkyo-ku, Tokyo, Japan

KATSIS (Dr D. N.), 12, rue Varnis, Nea Smyrne, Athens, Greece

* KAUFMANN (Prof. P.), Directeur de l'Observatoire de Radio-Astronomie, Umuarama, São Paulo, Brazil

* KAULA (Dr W. M.), Institute of Geophysics and Planetary Physics, University of California, Los Angeles, California 90024, U.S.A.

* KAWABATA (Dr Kin-aki), Tokyo Astronomical Observatory, University of Tokyo, Mitaka, Tokyo, Japan

KAWAGUCHI (Dr I.), Kwasan Observatory, University of Kyoto, Yamashina, Kyoto, Japan

KAZES (Dr I.), Observatoire de Paris, Section d'Astrophysique, 92-Meudon, France

KAZIMIRČAK-POLONSKAJA (Mrs Dr E. I.), [Kazimirchak-Polonskaya], Institute of Theoretical Astronomy, USSR Academy of Sciences, Leningrad, U.S.S.R.

* KEAY (Dr C. S. L.), University of Newcastle, New South Wales, Australia

KEENAN (Dr P. C.), Perkins Observatory, Delaware, Ohio 43015, U.S.A.

KELLER (Dr G.), Program Director for Astronomy, National Science Foundation, Washington 25, D.C., U.S.A.

* KELLERMAN (Dr K. I.), National Radio Astronomical Observatory, Box 2, Green Bank, West Virginia 24944, U.S.A.

KELLOG (Dr W. W.), National Center for Atmospheric Research, Boulder, Colorado, U.S.A.

* KENDERDINE (Dr S.), Mullard Radio Astronomy Observatory, Cavendish Laboratory, Free School Lane, Cambridge, U.K.

KERES (Prof. H. P.), University of Tartu, L. Koidula 6, Tartu, Estonia, U.S.S.R.

KERR (Dr F.J.), University of Maryland, College Park, Maryland 20742, U.S.A.
KESSLER (Dr K.G.), Atomic Physics Division, National Bureau of Standards 13.00, Washington D.C., 20234, U.S.A.
KHACHIKYAN (Dr E.E.), [see HAČIKJAN
KHAIKIN (Prof. Dr S.E.), [see HAJKIN]
KHARADZE (Prof. Dr E.K.), [see HARADZE]
KHAVTASI (Dr J.Sh.), [see HAVTASI]
KHOKHLOVA (Mrs Dr V.L.), [see HOHLOVA]
KHOLOPOV (Dr P.N.), [see HOLOPOV]
KIANG (Dr T.), Dunsink Observatory, Castleknock, Co. Dublin, Ireland
KIENLE (Prof. Dr H.), Ziegelhäuser Landstrasse 31, 6900 Heidelberg, G.F.R.
KIEPENHEUER (Prof. Dr K.O.), Direktor des Fraunhofer Instituts, Schöneckstrasse 6, Freiburg i. Breisgau, G.F.R.
* KILADZE (Dr R.I.), Abastumani Astrophysical Observatory, Georgian Academy of Sciences, Abastumani, U.S.S.R.
KING (Dr H.C.), McLaughlin Planetarium, Royal Ontario Museum, University of Toronto, 100 Queen's Park, Toronto 5, Canada
KING (Dr I.R.), Berkeley Astronomy Department, University of California, Berkeley, California, U.S.A.
KING (Dr R.B.), Department of Physics, California Institute of Technology, Pasadena, California, U.S.A.
KINMAN (Dr T.D.), Lick Observatory, University of California, Santa Cruz, California 95060, U.S.A.
KIPPENHAHN (Prof. R.), Universitäts Sternwarte Göttingen, Geismarlandstrasse 11, 34 Göttingen, G.F.R.
KIPPER (Prof. Dr A.J.), Director, Institute for Physics and Astronomy of the Academy of Sciences of the Estonian SSR, Tartu, U.S.S.R.
KIRAL (Dr A.), University Observatory, Beyazit, Istanbul, Turkey
KISELEV (Dr A.A.), Main Astronomical Observatory, USSR Academy of Sciences, Pulkovo, U.S.S.R.
* KISLJAKOV (Dr A.G.), [Kisliakov], Radiophysical Research Institute, Gorkij, U.S.S.R.
* KISSEL (Dr K.E.), Director, Aerospace Research Laboratories, Wright-Patterson Air Force Base, Ohio, U.S.A.
KITAMURA (Dr M.), Tokyo Astronomical Observatory, Mitaka, Tokyo, Japan
KIZILIRMAK (Prof. A.), Ege University, Observatory, P.K. 21, Bornova, Izmir, Turkey
KLECZEK (Dr J.), Astronomical Institute, Czechoslovak Academy of Sciences, Observatory Ondřejov, Czechoslovakia
* KLEMOLA (Dr A.R.), Lick Observatory, University of California, Santa Cruz, California 95060, U.S.A.
KLIMIŠIN (Dr I.A.), [Klimishin], Astronomical Observatory, Lvov University, Lvov, U.S.S.R.
* KLOCK (Dr B.L.), U.S. Naval Observatory, Department of Navy, Washington, D.C. 20390, U.S.A.
* KNIPE (G.F.G.), Republic Observatory, Gill Street, Johannesburg, South Africa
KNOX-SHAW (Dr H.), Fresh Woods, Elgin, Cape Province, South Africa
* KO (Dr Hsien-Ching), Department of Electrical Engineering, Ohio State University, Columbus, Ohio 43210, U.S.A.
KOCH (Dr R.H.), Flower and Cook Observatory, University of Pennsylvania, Philadelphia, Pennsylvania 19104, U.S.A.
KOEBCKE (Dr F.), Astronomical Observatory, Sloneczna 36, Poznan, Poland
KOELBLOED (Dr D.), Astronomical Institute, Roeterstraat 1a, Amsterdam, The Netherlands
* KOGURE (Dr T.), 2 Somoncho, Kinugasa, Kitagu, Kyoto, Japan
KÖHLER (Prof. Dr H.), Sauerbruchstrasse 6, Heidenheim a.d. Brenz, G.F.R.

KOHLSCHÜTTER (Prof. Dr A.), Königstrasse 16, Bonn, G.F.R.

KOHOUTEK (Dr L.), Astronomical Institute, Czechoslovak Academy of Sciences, Budečská 6, Prague 2, Czechoslovakia

* KOLČINSKIJ (Dr I. G.), [Kolchinsky)], Main Astronomical Observatory, Ukrainian Academy of Sciences, Kiev, U.S.S.R.

* KOLESOV (Dr A. K.), Astronomical Observatory, Leningrad, U.S.S.R.

KÖNIG (Dr A.), Mühldamm 9, 6904 Ziegelhausen, G.F.R.

KONOPLEVA (Dr V. P.), Main Astronomical Observatory, Academy of Sciences, Ukrainian SSR, Kiev 127, U.S.S.R.

KONOVOVIČ (Dr E. V.), [Konovovich], Sternberg Astronomical Institute, Moscow V-234, U.S.S.R.

KOPAL (Prof. Z.), Department of Astronomy, The University, Manchester 13, U.K.

KOPECKÝ (Dr M.), Astronomical Institute, Czechoslovak Academy of Sciences, Observatory Ondřejov, Czechoslovakia

KOPYLOV (Dr I. M.), Observatory, Sojuza Pechatnikov 25-a, Leningrad F-121, U.S.S.R.

KORDYLEWSKI (Dr K.), Astronomical Observatory, Kopernika 27, Cracow, Poland

* KOROL' (Dr A. K.), Main Astronomical Observatory, Ukrainian Academy of Sciences, Kiev, U.S.S.R.

KOROL'KOV (Dr D. V.), Main Astronomical Observatory, USSR Academy of Sciences, Pulkovo, U.S.S.R.

* KOSIN (Dr G. S.), [Kossin], Main Astronomical Observatory, USSR Academy of Sciences, Pulkovo, U.S.S.R.

KOSTYLEV (Dr K. V.), Engelhardt Observatory, Kazan, U.S.S.R.

* KOTEL'NIKOV (Acad. V.A.), Commission for Radioastronomy, USSR Academy of Sciences, Moscow, U.S.S.R.

KOTHARI (Dr D. S.), Department of Physics, University of Delhi, New Delhi, India

KOTSAKIS (Dr D.), 189 Hippocrates Street, Athens (708), Greece

KOURGANOFF (Prof. V.), Faculté des Sciences, Service d'Astronomie, 91-Orsay, France

KOVAL' (Dr I. K.), Main Astronomical Observatory, Ukrainian Academy of Sciences, Kiev, U.S.S.R.

KOVALEVSKY (Dr J.), Bureau des Longitudes, 3, rue Mazarine, Paris-6e, France

KOWATCHEW (B. J.), Section d'Astronomie de l'Académie Bulgare dés Sciences, rue 7 Novembre No 1, Sofia, Bulgaria

KOX (Dr H.), Hamburger Sternwarte, 205 Hamburg-Bergedorf, G.F.R.

KOZAI (Dr Y.), Tokyo Astronomical Observatory, Mitaka, Tokyo, Japan

KOZIEŁ (Prof. Dr K.), Professor of Astronomy, Jagellonian University, Kopernika 27/3, Cracow, Poland

KOZYREV (Dr N. A.), Pulkovo Observatory, Leningrad, U.S.S.R.

KRAFT (Dr R. P.), Mount Wilson and Palomar Observatories, 813 Santa Barbara Street, Pasadena, California, U.S.A.

* KRAMER (Dr K. N.), Astronomical Observatory, Odessa University, Odessa, U.S.S.R.

KRANJC (Dr A.), Osservatorio Astronomico di Bologna, Italy

KRASOVSKIJ (Prof. Dr I.), [Krassovsky], Box 1269, Moscow K 9, U.S.S.R.

KRAT (Prof. V. A.), Pulkovo Observatory, Leningrad M-140, U.S.S.R.

KRAUS (Dr J. D.), Radio Observatory, Ohio State University, Columbus, Ohio, U.S.A.

* KRAUSE (Dr F.), Institut für Magnetohydrodynamik der Deutschen Akademie der Wissenschaften zu Berlin, Jena, G.D.R.

KRAUSHAAR (Prof. W. L.), 20 Holden Wood Road, Concord, Massachusetts, U.S.A.

KRESÁK (Dr L.), Astronomical Institute, Slovak Academy of Sciences, Dúbravská cesta A/11 Bratislava, Czechoslovakia

* KRESÁKOVÁ (Mrs Dr M.), Astronomical Institute, Slovak Academy of Sciences, Dúbravská cesta A/11, Bratislava, Czechoslovakia

KRINOV (Dr E.L.), Committee on Meteorites of the Academy of Sciences of the USSR, ul. M. Uljanovoj 3, korpus 1, Moscow V-313, U.S.S.R.

* KRISHNAN (Dr T.), Stanford Radioastronomy Institute, Stanford University, Stanford, California, U.S.A.

* KRISHNASWAMY (K.S.), Yerkes Observatory, Williams Bay, Wisconsin, U.S.A.

KRISTENSON (Dr H.), Ole Rømer Observatory, Aarhus C., Denmark

* KRISTIEN (Dr J.), Washburn Observatory, University of Wisconsin, Madison, Wisconsin 53706, U.S.A.

KŘIVSKÝ (Dr L.), Astronomical Institute, Czechoslovak Academy of Sciences, Observatory Ondřejov, Czechoslovakia

KROGDAHL (Prof. W.S.), Department of Astronomy, University of Kentucky, Lexington, Kentucky, U.S.A.

KRON (Dr G.E.), U.S. Naval Observatory, Flagstaff Station, Flagstaff, Arizona 86001, U.S.A.

KRON (Mrs K.G.), P.O.Box 217, Flagstaff, Arizona 86001, U.S.A.

KROOK (Dr M.), Harvard College Observatory, Cambridge, Massachusetts 02138, U.S.A.

* KRÜGER (Dr A.), Heinrich-Hertz-Institut für Schwingungsforschung der Deutschen Akademie der Wissenschaften, Berlin, G.D.R.

KRÜGER (Prof. E.), Via Mauro Macchi 65, Milano, Italy

KRUSZEWSKI (Dr A.), Astronomical Observatory of Warsaw University, Al. Ujazdowskie 4, Warsaw, Poland

* KRZEMINSKI (Dr W.), Zakład Astronomii P.A.N., Aleje Ujazdowskie 4, Warszawa, Poland

KUBIKOWSKI (Dr J.J.), Astronomical Observatory, Kopernika 11, Wroclaw 9, Poland

KUČEROV (Dr N.I.), [Kutcherov], Main Astronomical Observatory, Academy of Sciences, Pulkovo, U.S.S.R.

* KUHI (Dr L.V.), Berkeley Astronomy Department, University of California, Berkeley, California 94720, U.S.A.

KUIPER (Dr G.P.), Lunar and Planetary Laboratory, University of Arizona, Tucson, Arizona, U.S.A.

KUKARKIN (Prof. Dr B.V.), Sternberg Astronomical Institute, Moscow-V-234, U.S.S.R.

* KUKLIN (Dr G.V.), Post Office Meget, Sibizmiran, Irkutsk, U.S.S.R.

KULIKOV (Prof. Dr K.A.), Sternberg Astronomical Institute, Moscow, U.S.S.R.

KULIKOVSKIJ (Dr P.G.), [Kulikovsky], Sternberg Astronomical Institute, Moscow V-234, U.S.S.R.

KUMAR (Dr S.S.), Leander McCormick Observatory, University of Virginia, Charlottesville, Virginia, U.S.A.

KUMSIŠVILI (Dr J.I.), [Kumsishvili], Abastumani Astrophysical Observatory, Georgian Academy of Sciences, Abastumani, U.S.S.R.

KUNDU (Dr M.R.), Tata Institute of Fundamental Research, Colaba, Bombay 5, India

KÜNZEL (H.), Heinrich-Hertz-Institut für Solar-terrestrische Physik, Sonnenobservatorium Einsteinturm, Potsdam, G.D.R.

KUPERUS (M.), Astronomical Observatory, "Sonnenborgh", Servaas Bolwerk 13, Utrecht, The Netherlands

KUPPERIAN Jr. (Dr J.E.), Astrophysics Branch, NASA, Goddard Space Flight Center, Greenbelt, Maryland, U.S.A.

KUPREVIČ (Dr N.F.), [Kuprevich], Pulkovo Observatory, Leningrad, U.S.S.R.

* KURT (Dr V.G.), Sternberg Astronomical Institute, Moscow, U.S.S.R.

KURTH (Dr R.), Professor of Mathematics, Georgia Institute of Technology, Atlanta, Georgia, U.S.A.

KUSHWAHA (Prof. R.S.), Department of Mathematics, University of Jodhpur, Jodhpur (Raj), India

KUSTAANHEIMO (Prof. P.E.), Astronomical Observatory, Helsinki, Finland

KUZ'MIN (Dr G.G.), Observatory of the Academy of Sciences of the Estonian SSR, Tartu, U.S.S.R.

KWEE (Dr K.K.), University Observatory, Leiden, The Netherlands

LABORDE (Dr G.), Observatoire de Paris, Section d'Astrophysique, 92-Meudon, France
LABS (Dr D.), Landessternwarte, Königstuhl, 69 Heidelberg, G.F.R.
LACCHINI (Prof. G.B.), Borgo Durbecco, via A. Cicognani No 10, Faenza (Ravenna), Italy
LACOMBE (Dr C.G.), Observatorio Nacional, Rua General Bruce, 586, Sao Cristovao (ZC-08), Rio de Janeiro, Gb., Brazil
LACROUTE (Prof. P.), Directeur, Observatoire de Strasbourg, 11, rue de l'Université, 67-Strasbourg (Bas-Rhin), France
LAFFINEUR (Dr M.), Institut d'Astrophysique, 98bis boulevard Arago, Paris-14e, France
* LAGERQVIST (Prof. A.), Institute of Physics, Stockholm University, Vanadisvägen 9, Stockholm Va, Sweden
LAGRULA (Prof. J.), Directeur de l'Institut de Météorologie, et de Physique du Globe de l'Algérie, B.P. 1296, Alger R.P., Algeria
LAHIRI (Sir N.C.), Officer-in-charge, Nautical Almanac Division, Meteorological Department, Alipore, Calcutta, India
LALLEMAND (Dr A.), Observatoire de Paris, 61, avenue de l'Observatoire, Paris-14e, France
* LAMBERT (Dr D.), Department of Astrophysics, University Observatory, South Parks Rd, Oxford, U.K.
LAMBRECHT (Prof. Dr H.), Direktor der Universitäts-Sternwarte, Schillergässchen 2, 69 Jena, G.D.R.
LAMLA (Dr E.), Hamburger Sternwarte, 205 Hamburg-Bergedorf, G.F.R.
* LANDINI (M.), Osservatorio Astrofisico di Arcetri, Florence, Italy
* LANDOLT (Dr A.), Department of Astronomy and Physics, Louisiana State University, Baton Rouge, Louisiana 70803, U.S.A.
LA PAZ (Dr L.), Director, Institute of Meteoritics, University of New Mexico, Albuquerque, New Mexico, U.S.A.
LA POINTE (Dr S.M.), Department of Physics, University of Montreal, Montreal, Quebec, Canada
* LARGE (Dr M.I.), Chatterton Astronomy Department, University of Sydney, Sydney, Australia
LARINK (Prof. Dr J.), Hamburger Sternwarte, 205 Hamburg-Bergedorf, G.F.R.
LARSSON-LEANDER (Dr G.), Lund Observatory, Lund, Sweden
LAUGIER (Mme M.), 77 boulevard Gambetta, 06-Nice, France
LAUSTSEN (Mr S.), Universitets, Observatoriet, Brorfelde, pr. Tølløse, Denmark
LAUTMAN (Dr D.A.), Smithsonian Astrophysical Observatory, 60 Garden Street, Cambridge, Massachusetts 02138, U.S.A.
LAVDOVSKIJ (Dr V.V.), [Lavdovski]), Pulkovo Observatory, Leningrad, U.S.S.R.
LAVROV (Dr M.I.), Engelhardt Astronomical Observatory, Kazan University, Kazan, U.S.S.R.
* LAWRENCE (Dr G.), Department of Astrophysical Sciences, 265 FitzRandolph Road, Princeton, New Jersey 08540, U.S.A.
LAYZER (Prof. Dr D.), Harvard College Observatory, 60 Garden Street, Cambridge, Massachusetts 02138, U.S.A.
* LEBOVITZ (Dr N.), Department of Mathematics, Eckhart Hall, Room 309, University of Chicago, 1118–32 East 58th Street, Chicago, Illinois 60637, U.S.A.
* LECAR (Dr M.), Smithsonian Astrophysical Observatory, 60 Garden Street, Cambridge, Massachusetts 02138, U.S.A.
LEDERLE (Dr T.), Astronomisches Rechen-Institut, Mönchhofstrasse 12–14, 69 Heidelberg, G.F.R.
LEDOUX (Prof. P.), Institut d'Astrophysique, Université de Liège, Cointe-Sclessin, Liège, Belgium
LEHNERT (Dr B.P.), Royal Institute of Technology, Department of Electronics, Stockholm 70, Sweden
LEIGHTON (Prof. R.B.), California Institute of Technology, Pasadena, California 91109, U.S.A.
LEJKIN (Dr G.A.), [Leikin], Astronomical Council of the USSR Academy of Sciences, Vavilova Street 20, Moscow V-312, U.S.S.R.
* LENGAUER (Dr G.G.), Main Astronomical Observatory, USSR Academy of Sciences, Pulkovo, U.S.S.R.

LEQUEUX (J.), Observatoire de Paris, Section d'Astrophysique, 92-Meudon, France
LE ROUX (Dr E.), Observatoire de Paris, Section d'Astrophysique, 92-Meudon, France
LEROY (Dr J.-L.), Observatoire du Pic-du-Midi, 65 Bagnères-de-Bigorre, Hautes-Pyrénées, France
LE SQUEREN (Mme A. M.), Observatoire de Paris, Section d'Astrophysique, 92-Meudon, France
LETFUS (Dr V.), Astronomical Institute, Czechoslovak Academy of Sciences, Observatory Ondřejov, Czechoslovakia
LEVALLOIS (J. J.), Institut Geographique National, 2 avenue Pasteur, 94 Saint-Mandé, France
LEVIN (Dr B. J.), O. Schmidt Institute of Physics of the Earth, USSR Academy of Sciences, B. Gruzinskaja 10, Moscow, U.S.S.R.
LÉVY (Dr J. R.), Observatoire de Paris, 61, avenue de l'Observatoire, Paris-14e, France
* LIDDELL (U.), National Aeronautics and Space Administration, Lunar and Planetary Program, Office of Space Science and Applications, Washington, D.C. 20546, U.S.A.
* LILLER (Dr M.), Harvard College Observatory, 60 Garden Street, Cambridge, Massachusetts 02138, U.S.A.
LILLER (Prof. W.), Harvard College Observatory, 60 Garden Street, Cambridge, Massachusetts 02138, U.S.A.
LILLEY (Prof. A. E.), Harvard College Observatory, Cambridge, Massachusetts 02138, U.S.A.
LIMBER (Dr D. N.), Yerkes Observatory, Williams Bay, Wisconsin, U.S.A.
* LIN (Dr C. C.), Department of Mathematics, Massachusetts Institute of Technology, Cambridge, Massachusetts, U.S.A.
LIN (Rong-an), 64, Kung Yuen Road, Taipei, Taiwan
LINDBLAD (Dr B.A.), Observatory, Lund, Sweden
LINDBLAD (Dr P. O.), Stockholm Observatory, Saltsjöbaden, Sweden
LINDLEY (W. M.), Pentonwarra, Trevone, Padstow, Cornwall, U.K.
LINDSAY (Dr E.M.), Director, Armagh Observatory, Armagh, Northern Ireland, U.K.
LINES (Dr A. W.), Royal Aircraft Establishment, Farnborough, Hampshire, U.K.
LINFOOT (Dr E. H.), The Observatories, Madingley Road, Cambridge, U.K.
LING (Dr Chih-bing), Institute of Mathematics, Academia Sinica, P.O.Box No. 143, Taipei, Taiwan
LINK (Dr F.), Astronomical Institute, Czechoslovak Academy of Sciences, Budečská 6, Prague 2, Czechoslovakia
LINNELL (Dr A. P.), Department of Astronomy, Michigan State University, East Lansing, Michigan 48823, U.S.A.
LINNIK (Prof. Dr V. P)., Pulkovo Observatory, Leningrad, U.S.S.R.
LIPPINCOTT (S. L.), Sproul Observatory, Swarthmore College, Swarthmore, Pennsylvania, U.S.A.
LIPSKIJ (Dr Y. N.), [Lipsky], Sternberg Astronomical Institute, Moscow, U.S.S.R.
* LISZKA (Dr L.), Geophysical Observatory, Kiruna C, Sweden
LITTLE (A. G.), Chatterton Astronomy Department, School of Physics, University of Sydney, Sydney, N.S.W., Australia
LITTLE (C. Gordon), Chief, Division 87, National Bureau of Standards, Room 3001, Radio Bldg. Boulder, Colorado, U.S.A.
LITTLEFIELD (Dr T. A.), Department of Physics, King's College, Newcastle upon Tyne, U.K.
LIVINGSTON (Dr W. C.), Kitt Peak National Observatory, P.O.Box 4130, Tucson, Arizona 85717, U.S.A.
LOCHTE-HOLTGREVEN (Prof. Dr W.), Institut für Experimentalphysik der Universität, Kiel, G.F.R.
LOCKE (Dr J. L.), Radio Astronomy Section, Radio and Electrical Engineering Division, National Research Council, Ottawa 2, Ontario, Canada
* LODÉN (Dr K.), Stockholm Observatory, Saltsjöbaden, Sweden
LODÉN (Dr L. O.), Stockholm Observatory, Saltsjöbaden, Sweden
LOHMANN (Prof. Dr W.), Astronomisches Rechen-Institut, Mönchhofstrasse 12–14, 69 Heidelberg, G.F.R.

LORÓN (Dr M.M.), Observatorio Astronomico, Alfonso XII, No 3, Madrid, Spain
LORTET-ZUCKERMANN (Mme M.-C.), Institut d'Astrophysique, 98bis, boulevard Arago, Paris-14e, France
LOUGHHEAD (R.E.), National Standards Laboratory, C.S.I.R.O., University Grounds, Sydney, N.S.W., Australia
LOURENS (J.v.B.), Royal Observatory, Observatory, Cape Province, South Africa
LOVELL (Sir Bernard), Nuffield Radio Astronomy Laboratories, Jodrell Bank, Macclesfield, Cheshire, U.K.
* LOW (Dr F.J.), Rice University, Department of Space Science, Houston, Texas 77001, U.S.A.
LOZINSKIJ (Dr A.M.), [Lozinsky], Astronomical Council of the USSR Academy of Sciences, Vavilova Street 20, Moscow V-312, U.S.S.R.
LÜBECK (K.), Hamburger Sternwarte, 205 Hamburg-Bergedorf, G.F.R.
LUNDQUIST (Dr C.A.), 218 Parker Street, Newton, Massachusetts, U.S.A.
LUNEL (Mme M.), Maître-Assistant d'Astronomie, Faculté des Sciences de l'Université de Lyon, Lyon (Rhône), France
LUPLAU-JANSSEN (Dr C.E.), Urania Observatory, Dr Olgas Vej 25, Copenhagen F, Denmark
* LÜST (Dr Rhea), Max-Planck-Institut für Physik und Astrophysik, Föhringer Ring 6, 8 München 23, G.F.R.
LÜST (Prof. Dr R.), Max-Planck-Institut für extraterrestrische Physik, 8046 Garching b. München, G.F.R.
LUYTEN (Prof. W.J.), The Observatory, University of Minnesota, Minneapolis 14, Minnesota, U.S.A.
LYNDEN-BELL (Dr D.), Royal Greenwich Observatory, Herstmonceux Castle, Hailsham, Sussex, U.K.
* LYNDS (Dr B.T.), Steward Observatory, University of Arizona, Tucson, Arizona 85721, U.S.A.
LYNDS (Dr C.R.), Kitt Peak National Observatory, 950 North Cherry Avenue, P.O.Box 4130, Tucson, Arizona 85717, U.S.A.
* LYNGA (Dr G.), Uppsala Southern Station, Mount Stromlo Observatory, Canberra A.C.T., Australia
LYTTKENS (Dr E.), Observatory, Uppsala, Sweden
LYTTLETON (Dr R.A.), St John's College, Cambridge, U.K.

MCCARTHY, S.J. (Dr M.F.), Specola Vaticana, Castel Gandolfo, Vatican City State
MCCLAIN (E.F.), Head, Radio Astronomy Branch, Naval Research Laboratory, Washington 25, D.C., U.S.A.
MCCREA (Prof. W.H), The University of Sussex, School of Mathematical and Physical Sciences, Falmer, Sussex, U.K.
MCCROSKY (Dr R.E.), Smithsonian Astrophysical Observatory, 60 Garden Street, Cambridge, Massachusetts 02138, U.S.A.
MCCUSKEY (Dr S.W.), Warner and Swasey Observatory, Case Institute of Technology, East Cleveland, Ohio 44112, U.S.A.
* MCDONALD (Dr F.B.), NASA, Goddard Space Flight Center, Greenbelt, Maryland 20771, U.S.A.
MCDONALD (Mrs Dr J.K. Petrie), Dominion Astrophysical Observatory, Victoria, B.C., Canada
MCDONALD (Mrs S. Levy), 2818 Webster Street, Berkeley 5, California, U.S.A.
* MCELROY (Dr M.B.), Kitt Peak National Observatory, 950 North Cherry Avenue, Tucson, Arizona, U.S.A.
MCGEE (Prof. J.D.), Physics Department, Imperial College of Science and Technology, Prince Consort Road, London, S.W. 7, U.K.
MCGEE (R.X.), Division of Radiophysics, C.S.I.R.O., University Grounds, Sydney, N.S.W., Australia
* MCINTOSH (Dr B.A.), National Research Council, Ottawa, Ontario, Canada

McKenna (Miss S.M.P.), Dunsink Observatory, Co. Dublin, Ireland
McKinley (Dr D.W.R.), c/o National Research Council, Ottawa, Ontario, Canada
McNally (Dr D.), University of London Observatory, Mill Hill Park, London N.W.7, U.K.
McNamara (Dr D.H.), Department of Physics, Brigham Young University, Provo, Utah, U.S.A.
* McNarry (L.R.), National Research Council, Ottawa, Ontario, Canada
MacRae (Dr D.A.), Director, David Dunlap Observatory, Richmond Hill, Ontario, Canada
McVittie (Prof. G.C), University of Illinois, Observatory, Urbana, Illinois, U.S.A.
Machado (J.M.A. Braz), Serviços Radioelectricos dos C.T.T., Lisbon, Portugal
Macris (Prof. Dr C.J), National Observatory, Athens (306), Greece
Madeira (Dr J.A.), Observatorio Astronomico, Tapada, Lisbon, Portugal
Madwar (Prof. M.R), 15 rue Pahlewi, Dokki, Cairo, U.A.R.
Maffei (Dr P)., Laboratorio di Astrofisica, Casella Postale 67-Frascati (Rome), Italy
Magalašvili (Dr N.L.), [Magalashvili], Abastumani Observatory, Abastumani, Georgia, U.S.S.R.
Maître (Dr V.), Observatoire, 25-Besançon, France
* Makita (Dr M)., Tokyo Astronomical Observatory, University of Tokyo, Mitaka, Tokyo, Japan
Makover (Dr S.G.), Institute for Theoretical Astronomy, Mendelejevskaja linia 1, Leningrad 164, U.S.S.R.
* Malacara (Dr D.), Observatorio Astronómico Nacional, Apdo. Postal 70–264, Ciudad Universitaria, México 20, D.F., México
* Malitson (Dr H.H.), NASA, Goddard Space Flight Center, Greenbelt, Maryland 20771, U.S.A.
Malmquist (Prof. K.G.), S:t Olofsgatan 10 A, Uppsala, Sweden
Maltby (Dr P.), Institute of Theoretical Astrophysics, University of Oslo, Blindern, Oslo 3, Norway
* Malville (Dr J.M.), High Altitude Observatory, Boulder, Colorado, U.S.A.
* Mamedbejli (Dr G.D.), [Mamedbeili], Šemaha Astrophysical Observatory, Azerbaidžan Academy of Sciences, Baku, U.S.S.R.
* Mammano (A.), Osservatorio Astrofisico, Asaigo (Vicenza), Italy
Mandel'štam (Dr S.L)., [Mandelshtam)], Laboratory of Spectroscopy, Lebedev Institute, Academy of Sciences, of the USSR Moscow, U.S.S.R.
* Mangeney (A.), Observatoire de Meudon, 5, place Janssen, 92-Meudon, France
Mannino (Prof. G.), Osservatorio Astronomico Universitario, Via Zamboni 33, Bologna, Italy
* Maran (Dr S.P.), Kitt Peak National Observatory, P.O.Box 4130, Tucson, Arizona 85717, U.S.A.
Marcus (Prof. E.), Observatoire de Bucarest, 5 rue Cutitul de Argint, Bucarest, Roumania
* Margoni (R.), Osservatorio Astrofisico, Asiago (Vicenza), Italy
Markarjan (Dr B.E.), [Markarian], Bjurakan Observatory, Erevan, Armenia, U.S.S.R.
Markov (Dr A.V.), Main Astronomical Observatory, Academy of Sciences, Pulkovo, U.S.S.R.
Markowitz (Prof. Wm.), Department of Physics, Marquette University, Milwaukee, Wisconsin 53233, U.S.A.
* Marsden (Dr B.), Smithsonian Astrophysical Observatory, 60 Garden Street, Cambridge, Massachusetts 02138, U.S.A.
Martel-Chossat (Mme Dr M.T.), Observatoire de Lyon, Saint-Genis-Laval (Rhône), France
* Martin (N.), Astronome Adjoint, Observatoire de Marseille, 2, place Le Verrier, 13-Marseille-4e, France
* Martin, Jr. (W.C.), Chief, Spectroscopy Section, National Bureau of Standards, Physics Building A-165, Washington D.C. 20234, U.S.A.
* Martinet (L), Observatoire Cantonal de Genève, Genève, Sauverny 1290, Switzerland
* Martres (Mme M.J.), Observatoire de Meudon, place Jules Janssen, 92-Meudon, France
Martynov (Prof. Dr D.Ja.), Director of the Sternberg Astronomical Institute, Moscow University, Leninskije Gory, Moscow V-234, U.S.S.R.

* Marvin (Dr U. B.), Smithsonian Astrophysical Observatory, Cambridge, Massachusetts, U.S.A.

Marx (Prof. G.), Roland Eötvös University, Puskin u. 5–7, Budapest VIII, Hungary

Masani (Prof. A.), Osservatorio Astronomico di Brera, Via Brera 28, Milano, Italy

Masevič (Mrs Dr A.G.), [Massevitch], Astronomical Council of the USSR Academy of Sciences, Vavilova Street 20, Moscow V-312, U.S.S.R.

* Mathews (Dr W. G.), Physics Department, University of California-San Diego, La Jolla, California 92038, U.S.A.

Mathewson (Dr D. S.), C.S.I.R.O., Division of Radiophysics, University Grounds, Chippendale, N.S.W., Australia

Mathias (Prof. Dr O.F. W.), Sonnenstrasse 12/II, Graz, Austria

* Mathis (Dr J. S)., Astronomy Department, University of Wisconsin, Madison, Wisconsin 53706, U.S.A.

Matsushima (Prof. Dr S.), Department of Astronomy, The Pennsylvania State University, University Park, Pennsylvania 16802, U.S.A.

Matthews (Dr T. A.), Department of Radio Astronomy, California Institute of Technology, Pasadena 4, California, U.S.A.

Mattig (Dr W.), Fraunhofer Institut, Schöneckstrasse 6, 78 Freiburg i. Breisgau, G.F.R.

Mavridis (Prof. Dr L. N.), Department of Geodetic Astronomy, University of Thessaloniki, Thessaloniki, Greece

Maxwell (Dr A.), Harvard Radio Astronomy Station, Fort Davis, Texas, U.S.A.

Mayall (Mrs M. W.), 187 Concord Avenue, Cambridge, Massachusetts 02138, U.S.A.

Mayall (Dr N. U.), Kitt Peak National Observatory, 950 North Cherry Avenue, Tucson, Arizona, U.S.A.

Mayer (C. H.), Code 7131, U.S. Naval Research Laboratory, Washington, 25 D.C., U.S.A.

* Mayer (Dr P.), Astronomical Institute of the Charles University, Švédská 8, Prague 5-Smíchov, Czechoslovakia

* Mayer (Dr U.), Astronomisches Institut der Universität, Waldhäuser Strasse 64, 74 Tübingen, G.F.R.

* Meaburn (Dr J.), Department of Astronomy, The University, Manchester 13, U.K.

* Medd (W. J.), Radio and Electrical Engineering Division, National Research Council, Ottawa, Ontario, Canada

* Meeks (Dr M. L.), Millstone Hill Field Station, MIT-Lincoln Laboratory, Lexington, Massachusetts, U.S.A.

* Meffroy (J.), Chef de Travaux, Faculté des Sciences, 34-Montpellier, France

Megrelišvili (Mrs Dr T. G.), [Megrelishvili], Abastumani Astrophysical Observatory, Georgian Academy of Sciences, Abastumani, U.S.S.R.

Mehltretter (Dr J. P.), Fraunhofer Institut, Schöneckstrasse 6, 78 Freiburg i. Breisgau, G.F.R.

* Mein (P.), Observatoire de Meudon, 5, place Janssen, 92-Meudon, France

Meinel (Dr A. B.), Kitt Peak National Observatory, 950 North Cherry Avenue, Tucson, Arizona, U.S.A.

* Meinig (Dr M.), Geodätisches Institut Potsdam der Deutschen Akademie der Wissenschaften, Berlin, G.D.R.

Melchior (Dr P. J.), Observatoire Royal de Belgique, 3 avenue Circulaire, Uccle-Brussels 18, Belgium

Mel'nikov (Prof. Dr. O. A.), Pulkovo Observatory, Leningrad, U.S.S.R.

* Mendez (Dr M.), Observatorio Astronómico Nacional, Apdo. Postal 70–264, Ciudad Universitaria, México 20, D.F., México

Mendoza V., (Dr E.), Observatorio Astronomico Nacional, 20 Piso Torre de Ciencias, Ciudad Universitaria, Mexico 20, D.F., Mexico

Menon (Dr T. K.), Institute for Astronomy, University of Hawaii, 2525 Correa Road, Honolulu, Hawaii 96822, U.S.A.

MENZEL (Dr D.H.), Harvard College Observatory, 60 Garden Street, Cambridge, Massachusetts 02138, U.S.A.

MERGENTALER (Prof. Dr J.), Astronomical Observatory, Kopernika 11, Wroclaw, Poland

MERMAN (Dr G.A.), Institute of Theoretical Astronomy, Leningrad, U.S.S.R.

MERMAN (Mrs Dr N.V.), Main Astronomical Observatory, USSR Academy of Sciences, Pulkovo, U.S.S.R.

MERRILL (Dr J.E.), Morrison Observatory, Central Methodist College, Fayette, Missouri 65248, U.S.A.

MERTON (Dr G.), 17 Holywell, Oxford, U.K.

MESSAGE (Dr P.J.), Mathematics Department, University of Liverpool, Liverpool 3, U.K.

MESTEL (Dr L.), Department of Applied Mathematics and Theoretical Physics, Free School Lane, Cambridge, U.K.

MEURERS (Prof. Dr J.), Direktor der Universitätssternwarte, Türkenschanzstrasse 17, A-1180, Vienna, Austria

* MEYER (Dr F.), Max-Planck-Institut für Physik und Astrophysik, Föhringer Ring 6, 8 Munich 23, G.F.R.

MEZGER (Dr P.), Universitäts-Sternwarte, Poppelsdorfer Allee 49, 53 Bonn, G.F.R.

MIANES (P.), Observatoire de Bordeaux, 33-Floirac, France

MICHARD (Dr R.), Observatoire de Paris, Section d'Astrophysique, 92-Meudon, France

MICHEL (H.), 54, rue Ten Bosch, Brussels 5, Belgium

MICHIE (Dr R.W.), Kitt Peak National Observatory, 950 North Cherry Avenue, Tucson, Arizona, U.S.A.

MICHKOVITCH (Prof. V.V.), Serbian Academy of Sciences and Arts, Knez Mihailova 35, Belgrade, Yugoslavia

MICZAIKA (Dr G.R.), Aeronutronic, Ford Motor Company, Ford Road, Newport Beach, California, U.S.A.

MIDDLEHURST (Miss B.M.), Lunar and Planetary Laboratory, University of Arizona, Tucson, Arizona 85719, U.S.A.

MIGEOTTE (Dr M.V.), Institut d'Astrophysique de l'Université de Liège, Cointe-Sclessin (Liège), Belgium

MIHAJLOV (Prof. Dr A.A.), [Mikhailov], Pulkovo Observatory, Leningrad M 140, U.S.S.R.

* MIHALAS (Dr D.), Princeton University Observatory, 265 FitzRandolph Road, Princeton, New Jersey 08540, U.S.A.

MIHEL'SON (Dr N.N.), [Mikhelson], Main Astronomical Observatory, Academy of Sciences, Pulkovo, U.S.S.R.

* MIKESELL (Dr A.H.), U.S. Naval Observatory, Washington D.C. 20390, U.S.A.

* MILET (B.), Aide Astronome, Observatoire de Nice, Le Mont-Gros, 06-Nice, France

MILFORD (Dr S.N.), Astrophysics and Geophysics Research, Grumman Aircraft Engineering Corporation, Bethpage, Long Island, New York, U.S.A.

MILLER (Dr F.D)., Department of Astronomy, The University of Michigan, Physics-Astronomy Building, Ann Arbor, Michigan 48104, U.S.A.

MILLER (Dr R.H.), Institute for Computer Research, University of Chicago, Chicago, Illinois, U.S.A.

MILLER, S.J. (Rev. W.J.), Director, Fordham University Astronomical Laboratory, Box 88, Fordham University, New York, N.Y. 10458, U.S.A.

MILLIGAN (J.E.), Astrophysics Branch, Goddard Space Flight Center, Greenbelt, Maryland, U.S.A.

MILLMAN (Dr P.M)., National Research Council, Sussex Drive, Ottawa, Ontario, Canada

MILLS (Dr B.Y.), School of Physics, University of Sydney, Sydney, N.S.W., Australia

MININ (Dr I.N.), Leningrad University Observatory, Leningrad, U.S.S.R.

MINKOWSKI (Dr R.L.), Radio Astronomy Laboratory, University of California, Berkeley, California 94720, U.S.A.

MINNAERT (Prof. M. G. J.), Director, Observatory "Sonnenborgh", Zonnenburg 2, Utrecht, The Netherlands

MIRZABEKJAN (Dr E. G.), [Mirzabekian], Astrophysical Observatory, Bjurakan, Armenia, U.S.S.R.

MIRZOJAN (Dr L. V.), [Mirzoyan], Bjurakan Astrophysical Observatory, Erevan, Armenia, U.S.S.R.

MISSANA (Prof. N.), Osservatorio Astronomico di Pino Torinese, (Torino), Italy

MITCHELL Jr., (Dr W. E.), Perkins Observatory, Delaware, Ohio, U.S.A.

MITIĆ (L. A.), Astronomical Observatory, Volgina 7, Belgrade, Yugoslavia

MITRA (Dr A. P.), Assistant Director, National Physical Laboratory, New Delhi, India

MITROFANOVA (Miss Dr L. A.), Main Astronomical Observatory, USSR Academy of Sciences, Pulkovo, U.S.S.R.

MIYADI (Dr M.), Tokyo Astronomical Observatory, Mitaka-shi, Tokyo, Japan

MIYAMOTO (Prof. Dr S.), Director, Kwasan Observatory, Faculty of Science, University of Kyoto, Yamashina, Kyoto, Japan

* MOERDIJK (Dr W.), Université de Gand, Gand, Belgium

* MOFFET (Dr A. T.), Radio Astronomy Laboratory, California Institute of Technology, Pasadena, California 91109, U.S.A.

* MOGILEVSKIJ (Prof. E. I.), [Mogilevsky], Institute of Terrestrial Magnetism, Ionosphere and Propagation of Radiowaves, USSR Academy of Sciences, Moscow, U.S.S.R.

MOHLER (Prof. O. C.), McMath-Hulbert Observatory, 895 Lake Angelus Road North, Pontiac 4, Michigan, U.S.A.

MOHR (Prof. Dr J. M.), Astronomical Institute of the Charles University, Švédská 8, Prague 5-Smíchov, Czechoslovakia

MOISEEV (Dr I. G.), Crimean Astrophysical Observatory, USSR Academy of Sciences, P/O Naučnyj, Crimea, U.S.S.R.

MOLČANOV (Prof. A. P.), [Molchanov], Radio Astronomical Laboratory, Institute of Physics, Leningrad State University, V-164, Leningrad, U.S.S.R.

* MØLLER (O.), Ole Rømer Observatory, Aarhus C, Denmark

MONFILS (A.), Institut d'Astrophysique, Université de Liège, Cointe-Sclessin (Liège), Belgium

MONIN (Dr G. A.), Crimean Astrophysical Observatory of the Academy of Sciences of the USSR, P/O Naučnyj, Crimea, U.S.S.R.

* MOORE (P.), Director, Armagh Planetarium, Armagh, U.K.

MOORE-SITTERLY (Mrs Dr Ch.), National Bureau of Standards 222.00, Washington D.C. 20234, U.S.A.

* MORANDO (B.), Bureau des Longitudes, 3, rue Mazarine, 75-Paris-6e, France

MOREAU (Dr F.), 115, avenue George Bergmann, Brussels, Belgium

MORENO (Prof. H.), Observatório Astronómico Nacional, Casilla 36-D, Santiago, Chile

MORETON (G. E.), Division of Physics, C.S.I.R.O., University Grounds, Chippendale, N.S.W., Australia

MORGAN (Dr W. W.), Yerkes Observatory, Williams Bay, Wisconsin 53191, U.S.A.

* MORIMOTO (Dr M.), Tokyo Astronomical Observatory, University of Tokyo, Mitaka, Tokyo, Japan

MORIYAMA (Dr F.), Tokyo Astronomical Observatory, Mitaka, Tokyo, Japan

MOROZ (Dr V. I.), Sternberg Astronomical Institute, Moscow, U.S.S.R.

* MORRIS (Dr S. C.), Dominion Astrophysical Observatory, Victoria, B.C., Canada

* MORRISON (Dr P.), Department of Physics, Massachusetts Institute of Technology, Cambridge, Massachusetts 02139, U.S.A.

MORTON (Dr D. C.), Department of Astrophysical Sciences, Princeton University Observatory, 265 FitzRandolph Road, Princeton, New Jersey, U.S.A.

MORTON (Dr G. A.), R.C.A. Laboratories, Princeton, New Jersey, U.S.A.

* MOTZ (Dr L.), Rutherfure Observatory, Columbia University, Box 57 Pupin Hall, New York, N.Y. 10027, U.S.A.

* MOURADIAN (Z.), Observatoire de Meudon, 5, place Janssen, 92 Meudon, France
MOURÃO (R.R. de Freitas), Observatorio Nacional, Rua General Bruce, 586-Sao Cristóvao, Rio de Janeiro, Brazil
MUGGLESTONE (Dr D.), University of Queensland, Brisbane, Queensland, Australia
MULDERS (Dr G.F.W.), c/o National Science Foundation, Washington 25 D.C., U.S.A.
* MULLALY (Dr R.F.), School of Electrical Engineering, University of Sydney, Sydney, Australia
MÜLLER (Miss Prof. E.A.), Observatoire de Genève, 1290 Sauverny (GE), Switzerland
MÜLLER (Prof. Dr R.), Postfach 8, 8204 Degerndorf/Inn, G.F.R.
MULLER (Dr A.B.), Kapteyn Astronomical Laboratory, Broerstraat 7, Groningen, The Netherlands
MULLER (Prof. Ir. C.A.), Radio Observatory, Bosrand 25, Dwingeloo, The Netherlands
MULLER (Dr P.), Observatoire de Paris, Section d'Astrophysique, 92-Meudon, France
* MUMFORD (G.S.), Department of Mathematics, Tufts University, Medford, Massachusetts 02155, U.S.A.
MÜNCH (Prof. G.), California Institute of Technology, 1201 E. California Street, Pasadena, California, U.S.A.
MURRAY (C.A.), Royal Greenwich Observatory, Herstmonceux Castle, Hailsham, Sussex, U.K.
MUSEN (Dr P.), National Aeronautics and Space Administration, Goddard Space Flight Center, Greenbelt, Maryland, U.S.A.
MUSTEL (Prof. Dr E.R.), Astronomical Council, USSR Academy of Sciences, Vavilova Street 20, V 312, U.S.S.R.
* MUTSCHLECNER (Dr P.), Department of Astronomy, Indiana University, Bloomington, Indiana 47401, U.S.A.

NADOLSKI (Prof. Dr V.), Université de Iasi, rue M. Sadoveanu 5, Iasi, Roumania
NAGASAWA (Dr S.), Tokyo Astronomical Observatory, Mitaka, Tokyo, Japan
NAHON (Prof. F.), Faculté des Sciences, 35 rue des Docks, 25-Besançon, France
NAKANO (S.), 47, Myogadani, Bunkyo-ky, Tokyo, Japan
NAMBA (Dr O.), Astronomical Observatory "Sonnenborgh", Zonnenburg 2, Utrecht, The Netherlands
* NANDY (Dr K.), Royal Observatory, Edinburgh 9, U.K.
NAQVI (Dr A.M.), Office of Research, Aerospace Corporation, San Bernardino, California 92402, U.S.A.
NARAYAN (Dr A.L.), Maharaja's College, Vizianagram, India
NARAYANA (J.V.), Meteorologist, Astrophysical Observatory, Kodaikanal 3, India
* NARLIKAR (Dr J.V.), King's College, Cambridge, U.K.
NAUR (Dr P)., Danish Institute for Computing Machinery, Gl. Carlsbergvej 2, Valby, Denmark
* NAZAROVA (Dr T.N.), Institute of Geochemistry and Analytic Chemistry, USSR Academy of Sciences, Moscow, U.S.S.R.
NECKEL (Dr H.), Hamburger Sternwarte, 205 Hamburg-Bergedorf, G.F.R.
NEFED'EV (Dr A.A.), [Nefediev], Director, Engelhardt Observatory, Kazan, U.S.S.R.
NEFED'EVA (Dr A.I.), [Nefedieva], Engelhardt Observatory, Kazan, U.S.S.R.
NELSON (Prof. B.), Department of Astronomy, San Diego State College, San Diego 15, California, U.S.A.
NEMIRO (Dr A.A.), Pulkovo Observatory, Leningrad, U.S.S.R.
* NESS (Dr N.F.), NASA, Goddard Space Flight Center, Greenbelt, Maryland 20771, U.S.A.
* NEUGEBAUER (Dr G.), Physics Department, California Institute of Technology, 1201 East California Street, Pasadena, California 91109, U.S.A.
NEUPERT (Dr W.M.), Solar Physics Branch, Goddard Space Flight Center, Greenbelt, Maryland, U.S.A.
* NEUŽIL (Dr L.), Astronomical Institute, Czechoslovak Academy of Sciences, Observatory Ondřejov, Czechoslovakia

NEVEN (Dr L.), Observatoire Royal de Belgique, 3 avenue Circulaire, Uccle-Brussels 18, Belgium
NEVIN (Prof. T. E.), University College, Dublin, Ireland
NEWKIRK Jr., (Dr G. A.), High Altitude Observatory of the University of Colorado, Boulder, Colorado, U.S.A.
NEWTON (H. W.), "Alandale", Dean Road, Seaford, Sussex, U.K.
* NEWTON (Dr R. R.), Applied Physics Laboratory, Johns Hopkins University, 8621 Georgia Avenue, Silver Spring, Maryland, U.S.A.
NEYMAN (Prof. J.), Statistical Laboratory, University of California, Berkeley 4, California, U.S.A.
NICHOLLS (Prof. R. W.), Chairman, Department of Physics, York University, Toronto, Ontario, Canada
NICHOLSON (W.), Royal Greenwich Observatory, Herstmonceux Castle, Hailsham, Sussex, U.K.
NICOLET (Prof. M.), Institut d'Aéronomie Spatiale, 3 avenue Circulaire, Uccle-Brussels 18, Belgium
NICOLINI (Prof. T.), Direttore, Osservatorio Astronomico di Capodimonte, Napoli, Italy
* NICOLOV (N. S.), Université de Sofia, Faculté Physique, Sofia, Bulgaria
* NICOLSON (G. D.), c/o National Institute for Telecommunications Research, P.O.Box 3718, Johannesburg, South Africa
NIELSEN (A. V.), Ole Rømer Observatorium, Aarhus C., Denmark
NIKITIN (Dr A. A.), Observatory, State University, Leningrad, U.S.S.R.
NIKOLSKY (Dr G. M.), Institute of Terrestrial Magnetism, Ionosphere and Radio Wave Propagation, Academy of Sciences of the USSR., Moscow, U.S.S.R.
NIKONOV (Dr V. B.), Crimean Astrophysical Observatory of the Academy of Sciences of the USSR., P/O Naučnyj, Crimea, U.S.S.R.
* NILSSON (Dr C.), Smithsonian Astrophysics Observatory, 60 Garden Street, Cambridge, Massachusetts 02138, U.S.A.
NINGER-KOSIBOWA (Dr. S.), Astronomical Observatory, Kopernika 11, Wroclaw, Poland
NININGER (Dr H. H.), Sedona's Meteorite Museum, P.O.Box 146, Sedona, Arizona, U.S.A.
* NISHI (Dr K.), Tokyo Astronomical Observatory, University of Tokyo, Mitaka, Tokyo, Japan
NISSEN (Prof. J. J.), Observatorio Astronomico, San Juan, Argentina
NOCI (Dr G.), Osservatorio astrofisico di Arcetri, Florence, Italy
NORDSTRÖM (Dr H.), Gjörwellsgatan 15, Stockholm, Sweden
NORLIND (Dr W.), Svanegatan 16, Lund, Sweden
NØRLUND (Prof. N. E.), Director, Geodetic Institute, Malmögade 6, Copenhagen, Denmark
NORTHCOTT (Miss R. J.), David Dunlap Observatory, Richmond Hill, Ontario, Canada
* NOTNI (Dr P.), Sternwarte Babelsberg, Institut für relativistische und extragalaktische Forschung der Deutschen Akademie der Wissenschaften, Berlin, G.D.R.
NOTUKI (Dr M.), Tokyo Astronomical Observatory, Mitaka, Tokyo, Japan
* NOVIKOV (Dr I. D.), Mathematical Institute of the USSR Academy of Sciences, Moscow, U.S.S.R.
NOVOPAŠENNYJ (Dr B. V.), [Novopashenny], Astronomical Observatory, Odessa University, Park Ševčenko, Odessa, U.S.S.R.
* NOVOSELOV (Prof. Dr V. S.), Astronomical Observatory, Leningrad University, Leningrad, U.S.S.R.
NOWACKI (Dr H.), Astronomisches Rechen-Institut, Mönchhofstrasse 12–14, 69 Heidelberg, G.F.R.
* NOYES (Dr R. W.), Smithsonian Institution, 60 Garden Street, Cambridge, Massachusetts 02138, U.S.A.

O (Choe Zai), Kim Il Sen University, Phyongyang, Democratic People's Republic of Korea
OBI (Dr S.), Department of Earth Science, College of General Education, University of Tokyo, Komaba, Shibuya, Tokyo, Japan

* OBRIDKO (Dr V. N.), Institute of Terrestrial Magnetism, Ionossphere, and Propagation of Radiowaves, USSR Academy of Sciences, Moscow, U.S.S.R.
* OCCHIONERO (F.), Laboratorio di Astrofisica, Frascati (Roma), Italy
O'CONNELL S.J. (Dr D. J. K.), Direttore della Specola Vaticana, Castel Gandolfo, Vatican City State
* O'DELL (Dr R. C.), Yerkes Observatory, Williams Bay, Wisconsin 53191, U.S.A.
ODGERS (Dr G. J.), Dominion Astrophysical Observatory, Victoria B.C., Canada
OETKEN (Dr L.), Astrophysical Observatory, Potsdam, G.D.R.
OGORODNIKOV (Prof. Dr K. F.), University, Leningrad, U.S.S.R.
ÖHMAN (Prof. Y.), Stockholms Observatorium, Saltsjöbaden, Sweden
* OJA (Dr T.), Astronomical Observatory, Uppsala, Sweden
OKE (Dr J. B.), Mount Wilson and Palomar Observatories, 1201 East California Street, Pasadena 4, California, U.S.A.
O'KEEFE (J. A.), 3712 Thornapple Street, Chevy Chase 15, Maryland, U.S.A.
OKUDA (Dr T.), Director, International Latitude Observatory of Mizusawa, Mizusawa-shi, Iwate-ken, Japan
ÖLANDER (Prof. V. R.), Geodetic Institute, Helsinki, Finland
* OLEAK (Dr H.), Sternwarte Babelsberg, Institut für relativistische und extragalaktische Forschung der Deutschen Akademie der Wissenschaften, Berlin, G.D.R.
OLIVIER (Dr C. P.), American Meteor Society, 521 N. Wynnewood Avenue, Narberth, Pennsylvania, U.S.A.
OLLONGREN (Dr A.), Centraal Rekeninstituut, Stationsweg 46, Leiden, The Netherlands
* OLSEN (Dr K. H.), Los Alamos Scientific Laboratory, Box 1663 Group J-15, Los Alamos, New Mexico, U.S.A.
* OLSON (Dr E. C.), University of Illinois Observatory, Urbana, Illinois 61801, U.S.A.
* OMAROV (Dr T. B.), Astrophysical Institute of the Kazah Academy of Sciences, Alma-Ata, U.S.S.R.
OMER, Jr. (Dr G. C.), Department of Physical Sciences, University of Florida, Gainesville, Florida, U.S.A.
ONDERLIČKA (Dr B.), Department of Astronomy, Purkyně University, Brno, Czechoslovakia
* ŌNO (Dr Y.), Department of Physics, Hokkaido University, Sapporo, Hokkaido, Japan
OORT (Prof. Dr J. H.), Director, University Observatory, Leiden, The Netherlands
OOSTERHOFF (Prof. Dr P. T.), University Observatory, Leiden, The Netherlands
OPALSKI (Dr W.), Institut Astronomique de l'École Polytechnique, Koszykowa 75, Warsaw, Poland
ÖPIK (Dr E. J.), Armagh Observatory, Armagh, U.K.
OPOLSKI (Dr A.), Astronomical Observatory, Kopernika 11, Wroclaw, Poland
ORLIN (Dr H.), Gravity and Astronomy Branch; Geodesy Division U.S. Coast and Geodetic Survey, Washington Science Center, Rockville, Maryland 20852, U.S.A.
ORLOV (Dr A. A.), Sternberg Astronomical Institute, Moscow, U.S.S.R.
ORLOVA (Mrs Dr N. S.), The Pulkovo Observatory, Leningrad, M-140, U.S.S.R.
ORRALL (Prof. F. Q.), Hawaii Institute of Geophysics, 2525 Correa Road, Honolulu, Hawaii 96822, U.S.A.
ORTE LLEDO (D. A.), Observatorio de Marina, San Fernando, Cádiz, Spain
* ORUS (J. J.), Université, Barcelona, Spain
* ŌSAKI (Dr T.), Ryukoku University, Shimo-gyo-ku, Kyoto, Japan
OSAWA (Dr K.), Tokyo Astronomical Observatory, Mitaka, Tokyo, Japan
OSKANJAN (V.), Astrophysical Observatory, Bjurakan, Armenian SSR, U.S.S.R.
OSTER (Prof. L.), Joint Institute for Laboratory Astrophysics, University of Colorado, Boulder, Colorado, U.S.A.
OSTERBROCK (Prof. D. E.), Washburn Observatory, University of Wisconsin, Madison, Wisconsin 53706, U.S.A.

* Ostriker (Dr J.P.), Princeton University Observatory, Peyton Hall, Princeton, New Jersey 08540, U.S.A.

Oterma (Dr L.), Sirkkalank 31, Turku 6, Finland

* Ottelet (Dr I.), Institut d'Astrophysique de l'Université de Liège, Cointe-Sclessin, Belgium

* Otzen-Petersen (J.), Copenhagen Observatory, Øster Voldgade 3, Copenhagen K, Denmark

Ovenden (Prof. Dr M.W.), Professor of Astronomy, c/o Department of Geophysics, The University of British Columbia, Vancouver 8, B.C., Canada

* Owaki (Dr N.), Marine Research Laboratory, Hydrographic Office, Tsukili, Chuo-ku, Tokyo, Japan

* Owen (Dr T.), IIT Research Institute (ASC), 10 West 35th Street, Chicago, Illinois 60616, U.S.A.

Owren (Prof. L.), Radiophysics Laboratory, Dartmouth College, Hanover, New Hampshire 03755, U.S.A.

Özemre (Dr K.), University Observatory, Beyazit, Istanbul, Turkey

* Ozernoj (L.M.), Lebedev Physical Institute, Moscow, U.S.S.R.

Ozsvath (Prof. I.), Earth and Planetary Sciences Laboratory, Graduate Research Center, P.O. Box 30365, Dallas, Texas, 75230, U.S.A.

* Pachner (Dr J.), Astronomical Institute, Czechoslovak Academy of Sciences, Budečská 6, Prague 2, Czechoslovakia

Pacholczyk (Prof. Dr A.G.), Steward Observatory, University of Arizona, Tucson, Arizona, U.S.A.

* Paczynski (Dr B.), Astronomical Observatory, Al. Ujazdowskie 4, Warsaw, Poland

Page (Prof. T.L.), Director, Van Vleck Observatory, Wesleyan University, Middletown, Connecticut, 06457, U.S.A.

Pagel (Dr B.E.J.), Royal Greenwich Observatory, Herstmonceux Castle, Hailsham, Sussex, U.K.

* Pajdušáková (Mrs Dr L.), Director, Skalnaté Pleso Observatory, Slovak Academy of Sciences, Tatranská Lomnica, Czechoslovakia

* Pal (Dr A.), Université, Cluj, Roumanie

Palmer (Dr H.P.), Nuffield Radio Astronomy Laboratories, Jodrell Bank, Macclesfield, Cheshire, U.K.

Paloque (Dr E.), Directeur Honoraire de l'Observatoire de Toulouse, 3 ter rue de Belfort, Toulouse (Haute-Garonne), France

Panajotov (Dr L.A.), [Panayotov], Pulkovo Observatory, Leningrad, U.S.S.R.

* Pande (M.C.), U.P. State Observatory, Nainital, India

Parijskij (Dr Ju.N.), [Parijsky, Dr Yu.N.], Main Astronomical Observatory, USSR Academy of Sciences, Pulkovo, U.S.S.R.

Parijskij (Dr N.N.), [Parijsky], O. Schmidt Institute of Physics of the Earth, USSR Academy of Sciences, B. Gruzinskaja 10, Moscow, U.S.S.R.

Parker (Dr E.N.), Institute for Nuclear Studies, University of Chicago, Chicago, Illinois 60637, U.S.A.

* Parker (Dr R.A.R.), Washburn Observatory, University of Wisconsin, Madison, Wisconsin 53705, U.S.A.

Parkinson (Dr W.H.), Harvard College Observatory, 60 Garden Street, Cambridge, Massachusetts 02138, U.S.A.

* Pasinetti (L.), Osservatorio Astronomico di Merate (Como), Italy

Paton (J.), Department of Natural Philosophy, The University, Edinburgh 8, U.K.

Pavlov (Prof. Dr N.N.), Pulkovo Observatory, Leningrad, U.S.S.R.

Payne-Gaposchkin (Mrs Dr C.H.), Harvard College Observatory, Cambridge, Massachusetts 02138, U.S.A.

* Peach (Dr G.), Department of Physics, University College, Gower Street, London W.C.1, U.K.

* Peale (Dr S.J.), Astronomy Department, University of California, Los Angeles, California 90024, U.S.A.

PEARSE (Prof. R.W.B.), Imperial College of Science and Technology, Prince Consort Road, London, S.W. 7, U.K.

* PEAT (Dr D.W.), The Observatories, Madingley Road, Cambridge, U.K.

PECKER (Prof. Dr J.-C.), Observatoire de Paris, Section d'Astrophysique, 92-Meudon, France; Observatoire de Nice, Le Mont-Gros, 06-Nice, France

PECKER-WIMEL (Mme Prof. C.), Institut d'Astrophysique, 98bis boulevard Arago, Paris-14e, France

PEDOUSSAUT (Dr A.), Observatoire de Toulouse, 31-Toulouse (Haute-Garonne), France

* PEERY (Dr B.F.), Department of Astronomy, Indiana University, Bloomington, Indiana 47405, U.S.A.

* PEKERIS (Prof. C.L.), The Weitzmann Institute of Science, Department of Mathematics, Rehovoth, Israel

PELS-KLUYVER (Mrs H.A.), University Observatory, Leiden, The Netherlands

PELSENEER (Dr J.), 76, avenue des Grenadiers, Ixelles, Brussels, Belgium

PELTIER (L.C.), 327 S. Bredeick Street, Delphos, Ohio, U.S.A.

* PENSADO (J.), Observatoire, Alfonso XII, 3, Madrid, Spain

PEREK (Dr L.), Astronomical Institute, Czechoslovak Academy of Sciences, Budečská 6, Prague 2, Czechoslovakia. General Secretary of the Union.

* PERRY (C.L.), Louisiana State University

PESCH (Prof. P.), Warner and Swasey Observatory, Case Institute of Technology, Taylor and Brunswick Roads, East Cleveland, Ohio, 44112, U.S.A.

* PETERSON (Dr L.E.), Physics Department, Revelle College, University of California-San Diego, La Jolla, California 92038, U.S.A.

PETFORD (Dr A.D.), University Observatory, South Park Road, Oxford, U.K.

PETRI (Dr W.), Unterleiten 2, Post-Box 106, 8162 Schliersee (OBB), G.F.R.

* PETROV (Dr G.M.), Nikolaiev Department of the Main Astronomical Observatory, USSR Academy of Sciences, Nikolaev, Ukrainian SSR, U.S.S.R.

PETTENGILL (Dr G.H.), Arecibo Ionospheric Observatory, Puerto Rico, U.S.A.

PETTIT (Dr E.), Mount Wilson and Palomar Observatories, 813 Santa Barbara Street, Pasadena 4, California, U.S.A.

PETTIT (Mrs H.B. Knaflich), 4131–1st Avenue, N.W. Seattle, Washington 98107, U.S.A.

PEYTURAUX (Dr R.), Institut d'Astrophysique, 98bis boulevard Arago, Paris-14e, France

* PFLUG (Dr K.), Astrophysikalisches Observatorium Potsdam der Deutschen Akademie der Wissenschaften, Berlin, G.D.R.

* PHILIP (Dr A.G. Davis), Dudley Observatory, 140 South Lake Avenue, Albany, New York 12208, U.S.A.

PHILLIPS (Prof. J.G.), Department of Astronomy, University of California, Berkeley, California, U.S.A.

PIASKOVSKY (Prof. D.V.), [see PJASKOVSKIJ]

PICK-GUTMANN (Mme M.), Observatoire de Paris, Section d'Astrophysique, 92-Meudon, France

PIDDINGTON (Dr J.H.), National Standards Laboratory, Division of Physics, University Grounds, Chippendale, N.S.W., Australia 2008.

PIERCE (Dr A.K.), Kitt Peak National Observatory, 950 North Cherry Avenue, P.O. Box 4130, Tucson, Arizona 85717, U.S.A.

PIKEL'NER (Prof. S.B.), Sternberg Astronomical Institute, Moscow V-234, U.S.S.R.

PILOWSKI (Prof. Dr K.), Geodätisches Institut der Technischen Hochschule, Nienburger Strasse 1, Hannover, G.F.R.

PINTO (Dr G.), Osservatorio Astronomico, Padova, Italy

PIOTROWSKI (Dr S.), Astronomical Observatory, Ujazdowskie 4, Warsaw, Poland

PISHMISH DE RECILLAS (Mrs Dr P.), Observatorio Astronómico Nacional, Universidad Nacional de Mexico, Ciudad Universitaria, Apartado Postal 70264, Mexico 20, D.F., Mexico

PJASKOVSKIJ (Prof. D.V.), [Piaskovsky], University Astronomical Observatory, Kiev, U.S.S.R.

PLAKIDIS (Prof. Dr S.), 4B Eridanou Street, Athens (612), Greece
PLASKETT (Prof. H.H.), 48 Blenheim Drive, Oxford, U.K.
PLASSARD (Dr J.), Director, Ksara Observatory, Ksara, Lebanon
PLATZECK (Dr R.), Comision Nacional de la Energia Atomica, San Carlos de Bariloche, Argentina
PLAUT (Dr L.), Kapteyn Astronomical Laboratory, Broerstraat 7, Groningen, The Netherlands
* PLAVCOVÁ (Mrs Dr Z.), Astronomical Institute, Czechoslovak Academy of Sciences, Observatory Ondřejov, Czechoslovakia
PLAVEC (Dr M.), Astronomical Institute, Czechoslovak Academy of Sciences, Observatory Ondřejov, Czechoslovakia
PODOBED (Dr V.V.), Sternberg Astronomical Institute, Moscow, U.S.S.R.
POGO (Dr A.), Mount Wilson and Palomar Observatories, 813 Santa Barbara Street, Pasadena 4, California, U.S.A.
POLOSKOV (Dr S.M.), Publishing House of Foreign Literature, Novo-Aleksejevskaja 52, Moscow, U.S.S.R.
POLOŽENCEV (Dr D.D.), [Polozhentsev], Pulkovo Observatory, Leningrad M-140, U.S.S.R.
* POLUPAN (Dr P.N.), Astronomical Observatory, Kiev University, Kiev, U.S.S.R.
POPOV (Dr N.A.), Gravimetrical Observatory, Ukrainian Academy of Sciences, Poltava, U.S.S.R.
POPOVA (Mrs Dr M.D.), Section d'Astronomie de l'Académie Bulgare des Sciences, rue 7 Novembre No 1, Sofia, Bulgaria
POPOVIĆ (Prof. Dr B.), Teknika Fakultato, NIŜ, Yugoslavia
POPOVICI (Prof. Dr C.), Observatoire de Bucarest, 5 rue Cutitul de Argint, Bucarest, Roumania
POPPER (Prof. D.M.), Department of Astronomy, University of California, Los Angeles 24, California, U.S.A.
PORTER (Dr J.G.), Whitestones, Hempstead Lane, Hailsham, Sussex, U.K.
POSTOIEV (Dr A.), Instituto Astronômico e Geofísico, Universidade de São Paulo, Caixa Postal 8137, São Paulo, Brazil
POTTASCH (Dr S.R.), Sterrenkundig Laboratorium Kapteyn, Broerstraat 7, Groningen, The Netherlands
POTTER (Dr H.I.), Main Astronomical Observatory, Academy of Sciences, Pulkovo, U.S.S.R.
POUNDS (Dr K.A.), University of Leicester, Leicester, U.K.
POVEDA (Dr A.), Observatorio Astronómico, Universidad Nacional de Mexico, Apartado Postal 70264, Mexico 20, D.F., Mexico
PRADHAN (Dr), Nizamiah Observatory, Hyderabad, India
PRENDERGAST (Dr K.H.), 1402 Pupin, Columbia University, New York, N.Y., U.S.A.
PRENTICE (J.P.M.), Star Ridge, Battisford, Needham Market, Suffolk, U.K.
PRESTON (Dr G.W.III), Lick Observatory, University of California, Santa Cruz, California 95060, U.S.A.
* PRICE (Dr M.J.), Kitt Peak National Observatory, 950 North Cherry Avenue, P.O.Box 4130, Tucson, Arizona 85717, U.S.A.
PRIESTER (Prof. W.), Universitäts-Sternwarte, Poppelsdorfer Allee 49, 53 Bonn, G.F.R.
* PROBSTEIN (Dr R.F.), Department of Mechanical Engineering, Massachusetts Institute of Technology, Cambridge, Massachusetts 02139, U.S.A.
PROKOF'EV (Prof. V.K.), [Prokofiev], Crimean Astrophysical Observatory, USSR Academy of Sciences, P/O Naučnyj, Crimea, U.S.S.R.
* PROKOF'EV (Dr V.V.), [Prokofiev], Main Astronomical Observatory Ukrainian Academy of Sciences, Kiev, U.S.S.R.
PROKOF'EVA (Mrs Dr I.A.), [Prokofieva], Main Astronomical Observatory, Academy of Sciences, Pulkovo, U.S.S.R.
* PRONIK (Dr V.I.), Crimean Astrophysical Observatory, USSR Academy of Sciences, P/O Naučnyj, Crimea, U.S.S.R.
PROTHEROE (Dr W.M.), Department of Astronomy, 174 W. Eighteenth Street, Ohio State University, Columbus, Ohio 43210, U.S.A.

PROTITCH (M.B.), Astronomical Observatory, Veliki Vračar, Belgrade, Yugoslavia
PROVERBIO (Dr E.), Osservatorio astronomico di Brera, Milano, Italy
PRZYBYLSKI (Dr A.), Mount Stromlo Observatory, Canberra, A.C.T., Australia
PURGATHOFER (Dr A.Th.), Universitäts-Sternwarte Wien, Türkenschanzstrasse 17, A-1180 Vienna, Austria

* QUIJANO (L.), Observatoire, San Fernando, Spain
QVIST (Dr B.), Åbo Akademie, Åbo, Finland

RABBEN (Dr H.H.), Max-Planck-Institut für Physik und Astrophysik, Abt. Extraterrestrische Physik, Garching b. München, G.F.R.
RABE (Prof. Dr E.), Cincinnati Observatory, Observatory Place, Cincinnati 8, Ohio, U.S.A.
RADHAKRISHNAN (V.), California Institute of Technology, Radio Astronomy Department, 1201 East California Street, Pasadena, California, U.S.A.
* RAGHAVAN (N.), Department of Physics, Indian Institute of Technology, Kanpur, India
* RAHIM (Dr M.H.A.), Helwan Observatory, Helwan, U.A.R.
RAIMOND (Dr E.), University Observatory, Sterrewacht, Leiden, The Netherlands
* RAJCHL (Dr J.), Astronomical Institute, Czechoslovak Academy of Sciences, Observatory Ondřejov, Czechoslovakia
* RAJU (P.K.), Tata Institute of Fundamental Research, Bombay, India
* RAKAVY (Prof. G.), Einstein Institute of Physics, Dept. of Theoretical, The Hebrew University of Jerusalem, Jerusalem, Israel
RAKOS (Dr K.D.), University of Graz, Universitäts-Sternwarte, Universitätsplatz 5, Graz, Austria
RAKSHIT (Prof. H.), Bengal Engineering College, Sibpore, Hewrah, India
RAMBERG (Prof. Dr J.M.), European Southern Observatory, Am Bahnhof 21, 205 Hamburg 80, G.F.R.
RAMSEY (Dr W.H.), Department of Physics, University of Witwatersrand, Milner Park, Johannesburg, South Africa
RANDIĆ (Dr L.), Tehnićki fakultet, Zagreb, p.p. 89, Yugoslavia
RASMUSEN (H.Q.), Vaerslevgaarden, pr. Vaerslev, Denmark
* RAYROLE (J.), Observatoire de Meudon, 5, place Janssen, 92-Meudon, France
RAZMADZE (Dr N.A.), Abastumani Observatory, Abastumani, Georgia, U.S.S.R.
RAZMADZE (Dr T.S.), Institute Geophysics, Academy of Sciences, Cosmic Ray Station, Tbilisi, U.S.S.R.
* REAVES (Dr G.), Astronomy Department, University of Southern California, Los Angeles, California 90007, U.S.A.
REBER (Dr G.), Research Corporation, 405 Lexington Avenue, New York 17, N.Y., U.S.A.
REDDISH (Dr V.C.), Royal Observatory, Edinburgh 9, U.K.
REDMAN (Prof. R.O.), Director of the Observatories, Madingley Road, Cambridge, U.K.
REEVES (Dr E.M.), Harvard College Observatory, 60 Garden Street, Cambridge, Massachusetts 02138, U.S.A.
REEVES (Dr H.), Faculté des Sciences, Bâtiment Sidonie, 91-Orsay, France
REID (Dr J.H.), Lockheed Electronics Company, Houston Aerospace Systems, 16811 El Camino Road, Houston, Texas 77058, U.S.A.
REINMUTH (Dr K.), Häusserstrasse 61, Heidelberg, G.F.R.
REIZ (Prof. A.), University Observatory, Østervoldgade 3, Copenhagen-K, Denmark
* REMY-BATTIAU (Mme Dr L.), Institut d'Astrophysique, Université de Liège, Cointe-Sclessin, Belgium
* RENSE (W.A.), Physics Department, University of Colorado, Boulder, Colorado, U.S.A.
* REQUIÈME (Y.), Observatoire de Bordeaux, 33-Floirac, France
* REUNING (Dr E.), Department of Physics and Astronomy, University of Georgia, Athens, Georgia 30601, U.S.A.

* REYNOLDS (Dr J.H.), Department of Physics, University of California, Berkeley, California, U.S.A.

RICE (D.A.), Department of Commerce, U.S. Coast and Geodetic Survey, Washington 25, D.C., U.S.A.

RICHARDSON (Dr E.H.), Dominion Astrophysical Observatory, Victoria, B.C., Canada

RICHARDSON (R.S.), Griffith Observatory, P.O.Box 27787, Los Felix Station, Los Angeles 27, California, U.S.A.

* RICHTER (Dr G.), Sternwarte Sonneberg (Thüringen) der Deutschen Akademie der Wissenschaften, Berlin, G.D.R.

RICHTER (Dr N.), Karl Schwarzschild Observatorium Tautenburg, Kreis Jena, G.D.R.

RIGHINI (Prof. G.), Direttore, Osservatorio Astrofisico di Arcetri, Largo S. Fermi 5, Florence, Italy

RIGUTTI (Dr M.), Osservatorio Astrofisico di Arcetri, Largo E. Fermi 5, Florence, Italy

* RIIHIMAA (Dr J.J.), c/o Prof. Tiuri, Technical University, Department of Electrotechnics, Albert Street 40–42, Helsinki, Finland

RIJVES (Dr W.G.), Institute of Physics, Mathematics and Mechanics, Academy of Sciences of the Estonian S.S.R., Tartu, U.S.S.R.

RINDLER (Prof. W.), Graduate Research Center, P.O.Box 30365, Dallas, Texas 75230, U.S.A.

RINEHART (Dr J.S.), Senior Research Fellow, Institutes for Environmental Research, ESSA, Boulder, Colorado 80302, U.S.A.

RING (Prof. J.), Department of Applied Physics, The University, Hull, U.K.

RINGNES (Dr T.S.), Institute of Theoretical Astrophysics, University of Oslo, Blindern, Oslo 3, Norway

* RINGUELET-KASWALDER (Dr A.), Observatorio Astronomico, Universidad Nacional de la Plata, La Plata, Argentina

RJABOV (Dr Ju.A.), [Ryabov Yu.A.], Sternberg Astronomical Institute, Moscow, U.S.S.R.

ROACH (Dr F.E.), National Bureau of Standards, Boulder, Colorado 80301, U.S.A.

ROBERTS (Dr J.A.), Department of Astronomy, University of Toronto, Toronto 5, Ontario, Canada

ROBERTS (Dr M.S.), National Radio Astronomy Observatory, Edgemont Road, Charlottesville, Virginia, U.S.A.

ROBERTS (Dr W.O.), Director, National Center for Atmospheric Research, Boulder, Colorado, U.S.A.

ROBERTSON (W.H.), Sydney Observatory, Sydney, N.S.W., Australia

* ROBINSON (Dr B.J.), Principal Research Scientist, C.S.I.R.O., Radiophysics Laboratory, University Grounds, Chippendale, Australia

ROBINSON (Prof. I.), Graduate Research Center, P.O.Box 30365, Dallas, Texas 75230, U.S.A.

ROBLEY (Dr R.), Observatoire du Pic-du-Midi, Bagnères-de-Bigorre (Hautes-Pyrénées), France

* RODDIER (F.), Faculté des Sciences de Nice, avenue Valrose, 06-Nice, France

RODGERS (Dr A.W.), Mount Stromlo Observatory, Canberra, A.C.T., Australia

* RODRIGUEZ (M.), Director, Observatorio San Fernando, Spain

* ROEDER (Dr R.C.), David Dunlap Observatory, Richmond Hill, Ontario, Canada

ROEMER (Miss Dr E.), Lunar and Planetary Laboratory, University of Arizona, Tucson, Arizona 85721, U.S.A.

ROGERSON, Jr (Prof. J.B.), Princeton University Observatory, 265 FitzRandolph Road, Princeton, New Jersey, U.S.A.

* ROLAND (Dr G.), Institut d'Astrophysique de l'Université de Liège, Cointe-Sclessin, Belgium

ROMAN (Miss Dr N.G.), NASA Headquarters, Washington D.C. 20546, U.S.A.

ROMANÁ PUJO S.J. (Dr A.), Direcctor, Observatorio del Ebro, Apartado 9, Tortosa, Spain

ROMANO (Dr G.), Viale S. Francesco 7, Treviso, Italy

ROMANSKAJA (Mrs Dr S.V.), [Romanskaya], Pulkovo Observatory, Leningrad, U.S.S.R.

* ROMPOLT (Dr B.), Astronomical Institute of Polish Academy of Sciences, Kopernika 11, Wroclaw, Poland

RÖSCH (Prof. J.), Directeur de l'Observatoire du Pic-du-Midi, Bagnères-de-Bigorre (Hautes Pyrénées), France

* ROSE (Dr W.K), Astrophysical Science Department, Princeton University, 265 FitzRandolph Road, Princeton, New Jersey 08540, U.S.A.

ROSEN (Dr B.), Institut d'Astrophysique, Université de Liège, Cointe-Sclessin (Liège), Belgium

ROSINO (Prof. L.), Director of the Astrophysical Observatory, Asiago (Vicenza), Italy

* ROSLUND (Dr C.), Astronomical Observatory, Lund, Sweden

ROSSELAND (Prof. S.), Institute of Theoretical Astrophysics, University of Oslo, Blindern, Oslo 3, Norway

* ROSSI (Dr B.), Department of Physics, Massachusetts Institute of Technology, Cambridge, Massachusetts 02139, U.S.A.

ROTTENBERG (Dr J. A.), Agr. Physics, MacDonald College, Quebec, Canada

ROUTLY (Dr P.M.), Princeton University Observatory, 211 Fitz-Randolph Road, Princeton, New Jersey 08540, U.S.A.

ROXBURGH (Dr I.), The University of Sussex, School of Mathematical and Physical Sciences, Falmer, Brighton, Sussex, U.K.

ROY (Dr A.E.), Department of Astronomy, University of Glasgow, Glasgow, W.2., U.K.

ROŽKOVSKIJ (Dr D. A.), [Rozhkovski], Astrophysical Institute, Alma Ata, U.S.S.R.

RUBAŠEV (Dr B. M.), [Rubashev] Pulkovo Observatory, Leningrad, U.S.S.R.

RUBEN (Dr G.), Astrophysical Observatory, Potsdam, G.D.R.

RUBIN (Mrs Dr V.C.), Department of Terrestrial Magnetism, Carnegie Institution of Washington, 5241 Broad Branch Road N.W., Washington D.C., U.S.A.

RUDKJØBING (Prof. M.), Director, Ole Rømer Observatory, Aarhus C, Denmark

RUDNICKI (Dr K.), Obserwatorium Astronomiczne U.W., Aleje Ujazdowskie 4, Warsaw, Poland

RUFENER (Dr F.-G.) Observatoire de Genève, 1290 Sauverny (GE), Switzerland

RULE (B.R.), Mount Wilson and Palomar Observatories, 813 Santa Barbara Street, Pasadena, California, U.S.A.

* RUNCORN (Prof. S. K.), Physics Department, The University, Newcastle-upon-Tyne, U.K.

RUPRECHT (Dr J.), Astronomical Institute, Czechoslovak Academy of Sciences, Budečská 6, Prague 2, Czechoslovakia

* RUSKOL (Mrs Dr E. L.), Institute of Physics of the Earth, USSR Academy of Sciences, Moscow, U.S.S.R.

* RUSSELL (Dr J. A.), Astronomy Department, University of Southern California, University Park, Los Angeles, California 90007 U.S.A.

RUTLLANT (Prof. F.), Director, Observatorio Astronomico de la Universidad de Chile, Casilla 36-D, Santiago de Chile, Chile

RYABOV (Dr Yu. A.), [see RJABOV]

RYBKA (Prof. Dr E.), Director, Cracow Astronomical Observatory, Kopernika 27, Cracow, Poland

RYBKA (Dr P.), Astronomical Institute of the Wroclaw University, Dembowskiego 19, Wroclaw 9, Poland

RYDBECK (Dr O.E.H.), Research Laboratory of Electronics, Chalmers University of Technology, Gibraltargata 5G, Götheborg, Sweden

RYLE (Prof. Sir M.), Cavendish Laboratory, Free School Lane, Cambridge, U.K.

* SACK (Dr N.), Einstein Institute of Physics, Dept. of Theoretical Physics, the Hebrew University of Jerusalem, Jerusalem, Israel

SADLER (D.H.), Superintendent, H.M. Nautical Almanac Office Royal Greenwich Observatory, Herstmonceux Castle, Hailsham, Sussex, U.K.

SADLER (Mrs F.M. McBain), H.M. Nautical Almanac Office, Royal Greenwich Observatory, Herstmonceux Castle, Hailsham, Sussex, U.K.

SAFRONOV (Dr V.S.), O.J. Schmidt Institute of the Physics of the Earth, USSR Academy of Sciences, B. Gruzinskaja 10, Moscow 242, U.S.S.R.

SAGAN (Prof. C.), Harvard College Observatory and Smithsonian Astrophysical Observatory, 60 Garden Street, Cambridge, Massachusetts 02138, U.S.A.

* SAGITOV (Dr M. U.), Sternberg Astronomical Institute, Moscow, U.S.S.R.

SAHADE (Dr J.), Observatorio Astronomico, La Plata, Argentina, Vice-President of the Union

SAHAROV (Dr V. I.), [Sakharov], Main Astronomical Observatory, USSR Academy of Sciences, Pulkovo, U.S.S.R.

SAITO (Dr K.), Tokyo Astronomical Observatory, Mitaka, Tokyo, Japan

SAITO (Dr S.), Kwasan Observatory, University of Kyoto, Yamashina, Kyoto, Japan

* SAKURAI (Dr K.), Ionosphere Research Laboratory, Kyoto University, Kyoto, Japan

SALPETER (Dr E. E.), Laboratory of Nuclear Studies, Cornell University, Ithaca, New York, U.S.A.

SALPETER, S. J. (Dr E. W.), Specola Vaticana, Castel Gandolfo, Vatican City State

* SALUKVADZE (Dr G. N.), Abastumani Astrophysical Observatory, Georgian Academy of Sciences, Abastumani, U.S.S.R.

SAMAHA (Prof. A. H. M.), Director, Helwan Observatory, Helwan, near Cairo, U.A.R.

SANAMJAN (Dr V. A.), [Sanamian], Bjurakan Astrophysical Observatory, Erevan, Armenia, U.S.S.R.

SANDAGE (Dr A.), Mount Wilson and Palomar Observatories, 813 Santa Barbara Street, Pasadena 4, California, U.S.A.

SANDAKOVA (Dr E. V.), Kiev University Observatory, Kiev, U.S.S.R.

SANDERS (P.), Observatoire Royal de Belgique, 3 avenue Circulaire, Uccle Brussels, 18, Belgium

SANDIG (Prof. H.-U.), Observatory of the Technical University, Lohrmann Institut, Dresden 8027, G.D.R.

SANTOS (Dr A. J. Baptista dos), Observatorio Astronomico, Tapada, Lisbon, Portugal

* SANWAL (N. B.), Department of Astronomy, Osmania University, Hyderabad–7 (A. P.), India

* SAPAR (Dr A. A.), Institute of Physics and Astronomy of the Estonian Academy of Sciences, Tartu, U.S.S.R.

SARABHAI (V. A.), Physical Research Laboratory, Navrangpura, Ahmedabad, India

* SARGENT (Dr W.L.), California Institute of Technology, Pasadena, California 91109, U.S.A.

* SARMA (M.B. K.), Department of Astronomy, Osmania University, Hyderabad, India

* SASTRY (C. V.), Astrophysical Observatory, Kodaikanal -3, India

SATO (Dr Y.), Tokyo Astronomical Observatory, Mitaka, Tokyo, Japan

SAVEDOFF (Prof. M. P.), University of Rochester, River Campus, Rochester 20, New York, 14627, U.S.A.

SAWYER HOGG (Mrs Prof. H. B.), David Dunlap Observatory, Richmond Hill, Ontario, Canada

SCHADEE (Dr A.), Astronomical Observatory "Sonnenborgh", Servaas Bolwerk 13, Utrecht, The Netherlands

SCHAIFERS (Dr K.), Landessternwarte, 69 Heidelberg-Königstuhl, G.F.R.

SCHALÉN (Prof. C.), Director of the Observatory, Lund, Sweden

SCHATZMAN (Prof. E. L.), Institut d'Astrophysique, 98bis, boulevard Arago, Paris-14e, France

SCHEEPMAKER (Dr A. C.), c/o Randweg 201, Bussum, The Netherlands

SCHEFFLER (Dr H.), Landessternwarte, 69 Heidelberg-Königstuhl, G.F.R.

SCHEUER (Dr P.A. G.), Mullard Radio Astronomy Observatory, Cavendish Laboratory, Free School Lane, Cambridge, U.K.

SCHILLER (Prof. Dr K.), Leiter der Sternwarte Frankfurt, Senckenberg-Anlage 23, Frankfurt (Main), G.F.R.

SCHILT (Prof. J.), 481 Greenville Avenue, Centerdale 11, Rhode Island, U.S.A.

SCHLÜTER (Prof. Dr A.), Max-Planck-Institut für Physik und Astrophysik, Institut für Astrophysik, Aumeisterstrasse 6, Munich 23, G.F.R.

* SCHALBERGER (Dr D. S.), Physics and Astronomy Department, University of Rochester, River Campus Station, Rochester, New York 14267, U.S.A.

SCHMEIDLER (Prof. Dr F.), Universitäts-Sternwarte, Sternwartstrasse 23, Munich 27, G.F.R.

SCHMIDT (Prof. Dr H.), Direktor der Universitäts-Sternwarte, Poppelsdorfer Allee 49, 53 Bonn, G.F.R.

SCHMIDT (Dr H.-U.), Max-Planck-Institut für Physik und Astrophysik, Föhringer Ring 6, 8 Munich 23, G.F.R.

SCHMIDT (Dr K.-H.), University Observatory, Jena, G.D.R.

SCHMIDT (Prof. M.), Mount Wilson and Palomar Observatories, 1201 East California Street, Pasadena, California 91109, U.S.A.

* SCHMIDT (Dr Th.), Landessternwarte Heidelberg-Königstuhl, 69 Heidelberg, G.F.R.

SCHMIDT-KALER (Dr Th.), Universitäts-Sternwarte Bonn, Poppelsdorfer Allee 49, 53 Bonn, G.F.R.

SCHMITT (A.), Observatoire de Strasbourg (Bas-Rhin), France

SCHRÖTER (Dr E. H.), Universitäts-Sternwarte, 34 Göttingen, G.F.R.

SCHRUTKA-RECHTENSTAMM (Prof. Dr G.), Universitäts-Sternwarte Wien, Türkenschanzstrasse 17, A 1180 Vienna, Austria

SCHUBART (Dr J.), Astronomisches Rechen-Institut, Mönchhofstrasse 12–14, 69 Heidelberg, G.F.R.

SCHÜCKING (Dr E.), Astronomy Department, University of Texas, Austin, Texas, U.S.A.

SCHULTE (Dr D. H.), Itek Corporation, 10 Maguire Road, Lexington, Massachusetts, U.S.A.

SCHÜRER (Prof. Dr M.), Astronomisches Institut der Universität Bern, Sidlerstrasse 5, Bern, Switzerland

SCHWARZ (Dr U. J.), Kapteyn Astronomical Laboratory, Broerstraat 7, Groningen, The Netherlands

SCHWARZSCHILD (Prof. M.), Princeton University Observatory, Peyton Hall, Princeton, New Jersey 08540, U.S.A. Vice-President of the Union

SCIAMA (Prof. D. W.), University of Cambridge, Department of Applied Mathematics and Theoretical Physics, Free School Lane, Cambridge, U.K.

SCONZO (Dr P.), Federal Systems Division I.B.M., 1730, Cambridge Street, Cambridge, Massachusetts 02138, U.S.A.

SCOTT (Prof. E. L.), Statistical Laboratory, University of California, Berkeley 4, California, U.S.A.

SCOTT (F. P.), U.S. Naval Observatory, Washington, D.C. 20390, U.S.A.

SCOTT (Dr P. F.), Mullard Radio Astronomy Observatory, Cavendish Laboratory, Free School Lane, Cambridge, U.K.

SEARLE (Dr L.), Mount Stromlo Observatory, Mount Stromlo, Canberra, A.C.T., Australia

SEARS (Dr R. L.), Astronomy Department, Physics-Astronomy Building, University of Michigan, Ann Arbor, Michigan 48104, U.S.A.

SEATON (Prof. M. J.), Department of Physics, University College London, Gower Street, London, W.C.1, U.K.

SEDDON (H.), The Royal Observatory, Edinburgh 9, U.K.

SEEGER (Dr C. L.), Department of Astronomy, 403 Physics Building, The University of Texas, Austin, Texas 78712, U.S.A.

SEHNAL (L.), Astronomical Institute, Czechoslovak Academy of Sciences, Observatory Ondřejov, Czechoslovakia

* SEIELSTAD (Dr G. E.), California Institute of Technology, Pasadena, California 91109, U.S.A.

SEITTER (Dr W. C.), Universitäts-Sternwarte Bonn, Observatorium Hoher List, 5568 Daun/Eifel, G.F.R.

* SEKANINA (Dr Z.), Computing Center, Faculty of Mathematics and Physics, Charles University, Malostranské nám. 25, Prague 1, Czechoslovakia

* SEKIGUCHI (Dr N.), Tokyo Astronomical Observatory, University of Tokyo, Mitaka, Tokyo, Japan

SEMENOV (Prof. Dr L. I.), Nikolaev Branch of the Pulkovo Observatory, Leningrad, U.S.S.R.

SÉMIROT (Prof. Dr P.), Directeur de l'Observatoire de Bordeaux, 33-Floirac, France

SEN (Dr H. K.), High Temperature Section, Thermal Radiation Laboratory, Air Force Cambridge Research Center, Bedford, Massachusetts, U.S.A.
SERKOWSKI (Dr K.), Mount Stromlo Observatory, Canberra, A.C.T., Australia
SĔRSIC (Dr J. L.), Observatorio Astronómico, Cordóba, Argentina
SERVAJEAN (Dr R.), Observatoire de Paris, Section d'Astrophysique, 92-Meudon, France
* SETTI (G.), Laboratorio di Radioastronomia, Istituto di Fisica, Bologna, Italy
SEVARLIĆ (Dr B. M.), [see ŠEVARLIĆ]
SEVERNYJ (Prof. Dr A. B.), [Severny], Director, Crimean Astrophysical Observatory, USSR Academy of Sciences, P/O Naučnyj, Crimea, U.S.S.R., Vice-President of the Union.
SHAKESHAFT (Dr J. R.), St Catharine's College, Cambridge, U.K.
SHANE (Dr C. D.), P.O.Box 582, Santa Cruz, California, U.S.A.
SHANE (W.W.), University Observatory, Sterrewacht, Leiden, The Netherlands
SHAO (Cheng-yuan), Harvard College Observatory, 60 Garden Street, Cambridge, Massachusetts 02138, U.S.A.
* SHAPIRO (Dr I. I.), Massachusetts Institute of Technology, Lincoln Laboratory, P.O.Box 73, Lexington 73, Massachusetts, U.S.A.
SHAPLEY (A. H.), C.R.P.L., National Bureau of Standards, Boulder, Colorado, U.S.A.
SHAPLEY (Prof. Dr H.), Sharon Cross Road, Peterboro, New Hampshire, U.S.A.
SHAPLEY (Mrs M. B.), Sharon Cross Road, Peterboro, New Hampshire, U.S.A.
SHAPOSHNIKOVA (Mrs Dr E. F.), [see ŠAPOŠNIKOVA]
* SHARAF (Mrs Dr S. G.), [see ŠARAF]
SHAROV (Dr A. S.), [see ŠAROV]
SHARPLESS (Dr S. L.), Department of Physics and Astronomy, The University of Rochester, Rochester, New York 14627, U.S.A.
SHAW (Prof. J. H.), Department of Physics and Astronomy, Ohio State University, 174 W. 18th Avenue, Columbus, Ohio, U.S.A.
SHAW (Prof. R. W.), Clark Hall of Science, Cornell University, Ithaca, New York, U.S.A.
* SHCHEGLOV (Dr P. V.), [see ŠČEGLOV]
SHCHEGLOV (Prof. Dr V. P.), see [ŠČEGLOV]
SHCHEGOLEV (Dr D. E.), [see ŠČEGOLEV]
SHCHIGOLEV (Prof. B.M.), [see ŠČIGOLEV]
* SHEELEY Jr (Dr N. R.), Kitt Peak National Observatory, 950 North Cherry Avenue, Tucson, Arizona 85717, U.S.A.
* SHEN (Dr B. S. P.), Flower and Cook Observatory, University of Pennsylvania, Philadelphia, Pennsylvania 19104, U.S.A.
SHEN (Dr Zee), Department of Mathematics, National Taiwan University, Taipei, Taiwan
SHERIDAN (K. V.), C.S.I.R.O. Division of Radiophysics, University Grounds, Chippendale, N.S.W., Australia
SHIMIZU (Dr T.), Institute of Astrophysics, College of Science, University of Kyoto, Kyoto, Japan
SHIRJAEV (Dr A. V.), [see ŠIRJAEV]
SHKLOVSKY (Dr I. S.), [see ŠKLOVSKIJ]
SHOEMAKER (E.), U.S. Geological Survey, Menlo, Park California, U.S.A.
* SHORE (Dr B. W.), Harvard College Observatory, 60 Garden Street, Cambridge, Massachusetts 02138, U.S.A.
SHTEINS (Dr K. A.), [see ŠTEJNS]
SHULBERG (Dr A. R.), [see ŠUL'BERG]
* SIBAHARA (Dr R.), n° 30, Kawashima-Sakurazonocho, Ukyo-ku, Kyoto, Japan
SIEGEL (C. L.), Göttingen University, Bunsenstrasse 3–5, Göttingen, G.F.R.
SILVA (Prof. E. C.), Professor da Escola Naval, Alfeite, Lisbon, Portugal
SILVA (Prof. Dr I. F. da), Faculdade de Ciencias, Lisbon, Portugal
* SIMON (Dr G. W.), Sacramento Peak Observatory, Sunspot, New Mexico, U.S.A.
SIMON (Dr P.), Observatoire de Paris, Section d'Astrophysique, 92-Meudon, France

SIMON (R.), Institut d'Astrophysique, Université de Liège, Cointe-Sclessin (Liège), Belgium
* SIMOVLJEVIĆ (Dr J.), Prirodno matem. fakultet, katedra za astronomiju Beograd, Studentski trg 16, Belgrade, Yugoslavia
SINNERSTAD (Dr. U.), Stockholm Observatory, Saltsjöbaden, Sweden
SINTON (Dr W.M.), Hawaii Institute of Geophysics, University of Hawaii, 2525 Correa Road, Honolulu, Hawaii 96822, U.S.A.
SINVHAL (Dr S.D.), Director, Uttar Pradesh State Observatory, Naini Tal, India
SINZI (Dr A.M.), Hydrographic Office, 5 chome, Tsukiji, Chuo-ku, Tokyo, Japan
SISSON (G.M.), Planetrees, Wall, Hexham, Northumberland, U.K.
* SITARSKI (Dr G.), Astronomical Institute of Polish Academy of Sciences, Al. Ujazdowskie 4, Warsaw, Poland
SITNIK (Dr G.F.), Sternberg Astronomical Institute, Moscow, U.S.S.R.
SIVARAMAN (K.R.), Astrophysical Observatory, Kodaikanal-3, India
* SKUMANICH (Dr A.), High Altitude Observatory, Boulder, Colorado 80301, U.S.A.
SLAUCITAJS (Prof. Dr S.), Observatorio Astronómico, La Plata, Argentina
SLAVENAS (Prof. Dr P.V.), Kestucio gatve 17-3, Vilnius 4, Lithuanian S.S.R., U.S.S.R.
* SLEBARSKI (T.B.), University Observatory, St Andrews, U.K.
SLEE (O.B.), Division of Radiophysics, C.S.I.R.O., University Grounds, Sydney, N.S.W., Australia
SLETTEBAK (Prof. Dr A.), Perkins Observatory, Delaware, Ohio, U.S.A.
SLONIM (Mrs Dr E.M.), Astronomical Observatory, Taškent, U.S.S.R.
SMAK (Dr J.), Astronomical Observatory of Warsaw University, Al. Ujazdowskie 4, Warsaw, Poland
SMART (Prof. W.M.), 8 Wellington Road, Edgbaston, Birmingham 15, England
SMERD (S.F.), Radiophysics Laboratory, C.S.I.R.O., University Grounds, Sydney, N.S.W., Australia
* SMEYERS (Abbé Dr P.), Institut Astronomique, rue de Namur, 61, Louvain, Belgium
SMILEY (Dr C.H.), Ladd Observatory, Brown University, Providence, Rhode Island, U.S.A.
SMIT (Prof. J.A.), Physical Laboratory of the University, Bijlhouwerstraat 6, Utrecht, The Netherlands
SMITH (Dr A.G.), Department of Physics, University of Florida, Gainesville, Florida, U.S.A.
SMITH (B.A.), New Mexico State University Observatory, Box Y, University Park, New Mexico 88070, U.S.A.
* SMITH (Dr G.), Department of Astrophysics, University Observatory, South Parks Road, Oxford, U.K.
SMITH (Mrs Dr E.v.P.), Department of Physics and Astronomy, University of Maryland, College Park, Maryland 20742, U.S.A.
SMITH (Prof. F. Graham), Nuffield Radio Astronomy Laboratories, Jodrell Bank, Macclesfield, Cheshire, U.K.
SMITH (Prof. Harlan J.), Director, McDonald Observatory, Astronomy Department, Room 407 Physics Building, The University of Texas, Austin, Texas 78712, U.S.A.
SMITH (Dr Henry J.), National Aeronautics and Space Administration, Code SG, Washington 25, D.C., U.S.A.
SMITH (H.M.), Royal Greenwich Observatory, Herstmonceux Castle, Hailsham, Sussex, U.K.
SMYTH (Dr M.J.), Royal Observatory Blackford Hill, Edinburgh 9, U.K.
* SOBIESKI (Dr S.), Astrophysics Branch, Code 613, NASA, Goddard Space Flight Center, Greenbelt, Maryland 20771, U.S.A.
SOBOLEV (Dr V.M.), Main Astronomical Observatory, USSR Academy of Sciences, Pulkovo, U.S.S.R.
SOBOLEV (Prof. Dr V.V.), The University, Leningrad, U.S.S.R.
SOBOLEVA (Mrs Dr N.S.), Main Astronomical Observatory, USSR Academy of Sciences, Pulkovo, U.S.S.R.

SOCHER (Dr H.), Universitäts-Sternwarte Wien, Türkenschanzstrasse 17, A-1180 Vienna, Austria
SOFRONICKIJ (Dr A.V.), [Sofronitski], Moscow Aviation Institute, Moscow D 80, U.S.S.R.
* SOLOMON (Dr P.M.), Astronomy Department, Columbia University, Box 110 Pupin Hall, New York, New York 10027, U.S.A.
SOMERVILLE (Dr W.B.), University College, Gower Street, London W.C.1, England
* SOUFFRIN (P.), Institut d'Astrophysique de Paris, 98 bis boulevard Arago, 75-Paris-14e, France
* SOUTHWORTH (Dr R.B.), Harvard College Observatory, 60 Garden Street, Cambridge, Massachusetts 02138, U.S.A.
* SPIEGEL (Dr E.), Courant Institute of Mathematical Sciences, 251 Mercer Street, New York, New York 10003, U.S.A.
SPINRAD (Dr H.), Berkeley Astronomical Department, University of California, Berkeley, California, U.S.A.
* SPITE (F.), Observatoire de Meudon, 5, place Janssen, 92-Meudon, France
SPITZER (Dr L.), Director, Princeton University Observatory, Princeton, New Jersey, U.S.A.
STAHR-CARPENTER (Mrs Dr M.), Leander McCormick Observatory Box 3818, University Station, Charlottesville, Virginia 22903, U.S.A.
* STANILA (Dr G.), Observatoire Astronomique, rue Cutitul de Argint 5, Bucarest 28, Roumanie
STANLEY (G.J.), California Institute of Technology, Pasadena 4, California, U.S.A.
STAWIKOWSKI (Dr A.), Astronomical Observatory of Toruń University, Sienkiewicza 30, Torun, Poland
STEARNS (Dr C.L.), Van Vleck Observatory, Middletown, Connecticut, U.S.A.
STECHER (T.P.), 10 812 Margate Road, Silver Spring, Maryland, U.S.A.
STEIGER (Prof. W.R.), Hawaii Institute of Geophysics, University of Hawaii, 2525 Correa Road, Honolulu, Hawaii 96822, U.S.A.
* STEIN (Dr W.), Princeton University Observatory, 265 Fitz-Randolph Road, Princeton, New Jersey 08540, U.S.A.
STEINBERG (Dr J.L.), Service de Radioastronomie, Observatoire de Paris, Section d'Astrophysique, 92-Meudon, France
STEINITZ (Dr R.), Institute of Planetary and Space Physics, Tel-Aviv University, Ramat-Aviv, Israel
STEINLIN (Dr U.W.), Astronomische Anstalt der Basel Universität, Binningen, Switzerland
STEPANOV (Dr W.E.), Post Office Meget, Sibizmiran, Irkutsk, U.S.S.R.
STEPHENSON (Prof. C.B.), Warner and Swasey Observatory, Case Institute of Technology, Taylor and Brunswick Roads, East Cleveland, Ohio 44112, U.S.A.
STERNBERK (Dr B.), [see ŠTERNBERK]
STERNE (Dr T.E.), Smithsonian Astrophysical Observatory, 60 Garden Street, Cambridge, Massachusetts 02138, U.S.A.
STEŠENKO (Dr N.V.), [Steshenko], Crimean Astrophysical Observatory, USSR Academy of Sciences, P/O Naučnyj, Crimea, U.S.S.R.
STEWART (Dr J.Q.), Princeton University Observatory, Princeton, New Jersey, U.S.A.
* STEYAERT (Dr H.), Université de Gand, Gand, Belgium
STIBBS (Prof. D.W.N.), University Observatory, Buchanan Gardens, St Andrews, Fife, U.K.
STICKER (Prof. Dr B.), Direktor des Instituts für Geschichte der Naturwissenschaften, Hartungstrasse 5, Hamburg 13, G.F.R.
* STIEFEL (E.), Mathematische Abteilung, Eidgenösische Technische Hochschule, Zurich, Switzerland
STOCK (Dr J.), Warner and Swasey Observatory, East Cleveland 12, Ohio, U.S.A.
STODOŁKIEWICZ (Dr J.S.), Astronomical Observatory of Warsaw University, Al. Ujazdowskie 4, Warsaw, Poland
* STONE (Dr R.G.), Sacramento Peak Observatory, Sunspot, New Mexico, U.S.A.
STOY (Dr R.H.), H.M. Astronomer, Royal Observatory, Observatory, Cape Province, South Africa

STOYKO (Mme A.), Observatoire de Paris, 61 avenue de l'Observatoire, Paris-14e, France
STOYKO (Dr N.), Observatoire de Paris, 61 avenue de l'Observatoire Paris-14e, France
* STRAJŽIS (Dr V. L.), [Straizhis], Astronomijos observatorija, Vilnius 31, Čiurlionio 29, Lithuania, U.S.S.R.
STRAND (Dr K. Aa.), U.S. Naval Observatory, Washington, D.C. 20390, U.S.A.
STRASSL (Prof. Dr H.), Direktor des Astronomischen Instituts der Universität, Steinfurterstrasse 107, Münster (Westfalen), G.F.R.
* STRITTMATTER (P. A.), Peterhouse, Cambridge, U.K.
STROBEL (Dr W.), Astronomisches Rechen-Institut, Mönchhofstrasse 12–14, 69 Heidelberg-Königstuhl, G.F.R.
STROHMEIER (Dr W.), Direktor der Remeis Sternwarte, Sternwartstrasse 7, Bamberg, G.F.R.
STRÖMGREN (Prof. B.), Observatoriet, Østervoldgade 3, Copenhagen K, Denmark
* STROM (Dr R. G.), Lunar and Planetary Laboratory, University of Arizona, Tucson, Arizona 75721, U.S.A.
* STROM (Dr S. E.), Smithsonian Astrophysical Observatory, 60 Garden Street, Cambridge, Massachusetts 02138, U.S.A.
STRONG (Prof. J. D.), Laboratory of Astrophysics and Physical Meteorology, Johns Hopkins University, Baltimore 18, Maryland, U.S.A.
STUMPFF (Prof. Dr K.), Am Sölenborn 12, 34 Göttingen, G.F.R.
STUMPFF (Dr P.), National Radio Astronomy Observatory, Edgemont Dairy Road, Charlottesville, Virginia, U.S.A.
* STURROCK (Dr P. A.), Institute for Plasma Research, Stanford University, Stanford, California, U.S.A.
SUDA (Dr K.), Astronomical Institute, Tohoku University, Sendai, Japan
SUEMOTO (Dr Z.), Tokyo Astronomical Observatory, Mitaka, Tokyo, Japan
SUGAWA (Dr C.), International Latitude of Mizusawa, Mizusawa-shi, Iwate-ken, Japan
SUHAREV (Dr L. A.), [Sukharev], Pulkovo Observatory, Leningrad, U.S.S.R.
* SULTANOV (Dr G. F.), Šemaha Astrophysical Observatory, Azerbaidžan Academy of Sciences, Baku, U.S.S.R.
SVESTKA (Dr Z.), [see Švestka]
* SVOLOPOULOS (Prof. Dr S. N.), University of Ioannina, Ioannina, Greece
* SWAMY (Dr Krishna K. S.), Astrophysics Branch Code 613, Laboratory for Space Studies, NASA, Goddard Space Flight Center, Greenbelt, Maryland, U.S.A.
SWARUP (Dr G.), Tata Institute of Fundamental Research, Bombay, India
SWEET (Prof. P. A.), The University Observatory, Glasgow, W.2., U.K.
SWENSSON (Dr J. W.), Fysiska Institutionen, Lund, Sweden
SWIHART (Dr T. L.), Steward Observatory, University of Arizona, Tucson, Arizona, U.S.A.
SWINGS (Prof. P.), Directeur de l'Institut d'Astrophysique, Université de Liège, Cointe-Sclessin (Liège), Belgium
SWOPE (Miss H. H.), Mount Wilson and Palomar Observatories, 813 Santa Barbara Street, Pasadena, California 91106, U.S.A.
SYKES (Dr J. B.), 20 Milton Lane, Steventon, Abingdon, Berkshire, U.K.
SYKES-HART (Mrs A. B.), University Observatory, Oxford, U.K.
SYMMS (L.S.T.), Royal Greenwich Observatory, Herstmonceux Castle, Hailsham, Sussex, U.K.
* SYROVACKIJ (S. I.), [Sirovatski], Lebedev Physical Institute, Moscow, U.S.S.R.
SYTINSKAJA (Mrs Dr N. N.), [Sytinskaya], University Observatory, Leningrad, U.S.S.R.
SZAFRANIEC (Miss Dr R.), Astronomical Observatory, Kopernika 27, Cracow, Poland
SZEBEHELY (Prof. V. G.), Yale University Observatory, Box 2023 Yale Station, New Haven, Connecticut 06520, U.S.A.
SZEIDL (Dr B.), Konkoly Observatory, P.O.Box 114, Box 67, Budapest XII, Hungary
ŠAPOŠNIKOVA (Mrs Dr E. F.), [Shaposhnikova], Crimean Astrophysical Observatory, USSR Academy of Sciences, P/O Naučnyj, Crimea, U.S.S.R.

* ŠARAF (Mrs Dr Sh. G.), [Sharaf], Institute of Theoretical Astronomy, USSR Academy of Sciences, Leningrad, U.S.S.R.
ŠAROV (Dr A. S.), [Sharov], Sternberg Astronomical Institute, Moscow, U.S.S.R.
* ŠČEGLOV (Dr P. V.), [Shcheglov], Sternberg Astronomical Institute, Moscow, U.S.S.R.
ŠČEGLOV (Prof. Dr V. P.), [Shcheglov], Director of the Astronomical Observatory, Taškent, U.S.S.R.
ŠČEGOLEV (Dr D. E.), [Shchegolev], Pulkovo Observatory, Leningrad, U.S.S.R.
ŠČIGOLEV (Prof. B. M.), [Shchigolev], Sternberg Astronomical Institute, Moscow, U.S.S.R.
ŠEVARLIĆ (Dr B. M.), Assistant Professor, Astronomska Observatorija, Veliky Vracar, Belgrade, Yugoslavia
ŠIRJAEV (Dr A. V.), [Shirjaev], Astronomical Observatory, Leningrad University, Leningrad, U.S.S.R.
ŠKLOVSKIJ (Dr I. S.), [Shklovsky], Sternberg Astronomical Institute, Moscow, U.S.S.R.
ŠTEJNS (Dr K. A.), [Shteins], Latvian University, Riga, U.S.S.R.
ŠTERNBERK (Dr B.), Director, Astronomical Institute, Czechoslovak Academy of Sciences, Budečská 6, Prague 2, Czechoslovakia
ŠUL'BERG (Dr A. R.), [Schulberg], Astronomical Observatory, Odessa University, Odessa, U.S.S.R.
ŠVESTKA (Dr Z.), Astronomical Institute, Czechoslovak Academy of Sciences, Observatory Ondřejov, Czechoslovakia

TAFFARA (Prof. S.), Osservatorio Astronomico, Padova, Italy
* TAI (Dr Yuin-Kwei), Director, Institute of Geophysics, Central University, Taipei, Taiwan
TAGAKI (Dr S.), International Latitude Observatory of Mizusawa, Mizusawa-shi, Iwate-ken, Japan
TAKAKUBO (Dr K.), Astronomical Institute, Faculty of Science, Tohoku University, Sendai, Japan
TAKAKURA (Dr T.), Tokyo Astronomical Observatory, Mitaka, Tokyo, Japan
TAKASE (Dr B.), Department of Astronomy, Faculty of Science, University of Tokyo, Bunkyo, Tokyo, Japan
* TAKAYANAGI (K.), Institute of Space and Aeronautical Science, University of Tokyo, Komaba, Meguro-ku, Tokyo, Japan
TAKENOUCHI (Dr T.), Tokyo Astronomical Observatory, Mitaka, Tokyo, Japan
* TALWAR (S. P.), Department of Physics, University of Delhi, Delhi, India
* TAMMAN (Dr G. A.), Astronomisch-Meteorologische Anstalt der Universität Basel, Venusstrasse 7, Binningen, Switzerland
* TANABE (Dr H.), Tokyo Astronomical Observatory, University of Tokyo, Mitaka, Tokyo, Japan
* TANAKA (Dr H.), Research Institute of Atmospherics, Nagoya University, Ichida-machi, Tokyo-wa-shi, Aichi, Japan
* TANAKA (Dr Y.), Werkgroep Kosmische Straling, Kamerlingh Onnes Laboratorium, Nieuwsteeg 18, Leiden, The Netherlands
TANDBERG-HANSSEN (Dr E.), High Altitude Observatory, University of Colorado, Boulder, Colorado, U.S.A.
TANNAHILL (Dr T. R.), University Observatory, Glasgow, W.2, U.K.
TANNER (R. W.), Dominion Observatory, 3 Observatory Crescent, Ottawa, Ontario, Canada
TARDI (Prof. P.), 4 villa de Ségur, Paris-7e, France
* TASSOUL (Dr J.), University of Chicago, Department of Astronomy, Chicago, U.S.A.
TAVARES (Dr J. C. T. L.), University of Coimbra, Coimbra, Portugal
TAVASTŠERNA (Dr K. N.), [Tavastsherna], Main Astronomical Observatory, USSR Academy of Sciences, Pulkovo, Leningrad M-140, U.S.S.R.
TAYLER (Dr R. J.), Astronomy Centre, University of Sussex, School of Mathematical and Physical Sciences, Falmer, Sussex, U.K.
TCHEKIRDA (Dr A. T.), [see ČEKIRDA]

TCHENAKAL (Dr V. L.), [see ČENAKAL]
TCHERNEGA (Dr N. A.), [see ČERNEGA]
TCHUGAINOV (Dr P. F.), [see ČUGAJNOV]
TÉHÉRANY (D.), 9 avenue des Scarabées, Brussels 1, Belgium
* TEJFEL' (Dr V. G.), [Teifel], Astrophysical Institute of the Kazah Academy of Sciences, Alma Ata, U.S.S.R.
TELEKI (Dr G.), Astronomical Observatory, Volgina 7, Belgrade, Yugoslavia
TEMESVÁRY (Dr S.), Max-Planck-Institut für Physik und Astrophysik, Institut für Astrophysik, Aumeisterstrasse 6, Munich 23, G.F.R.
TEMPESTI (Prof. P.), Osservatorio Astronomico di Collurania, Teramo, Italy
TER HAAR (Dr D.), Magdalen College, Oxford, U.K.
TERRAZAS (L. R.), Tonantzintla and Tacubaya Observatories, Mexico
TERRIEN (Dr J.), Directeur, Bureau International des Poids et Mesures, Pavillon de Breteuil, 92-Sèvres, France
* TERZAN (A.), Observatoire de Lyon, 69-St-Genis-Laval, France
* TERZIAN (Dr Y.), Center for Radiophysics and Space Research, Space Sciences Building, Cornell University, Ithaca, New York 14850, U.S.A.
* TESKE (Dr R. G.), Astronomy Department, University of Michigan, Ann Arbor, Michigan, U.S.A.
* TEXEREAU (J.), Observatoire de Paris, 61, avenue de l'Observatoire, 75-Paris, 14e, France
THACKERAY (Dr A. D.), Director, Radcliffe Observatory, P.O.Box 373, Pretoria, South Africa
THE (Dr Pik Sin), Bosscha Observatory, Lembang, Java, Indonesia
THERNÖE (K. A.), University Observatory, Østervoldgade 3, Copenhagen-K, Denmark
THIRY (Prof. Y.), Professeur à la Faculté des Sciences de l'Université de Paris, 7–9 quai St-Bernard, Paris-5e, France
THOMAS (Dr D. V.), Royal Greenwich Observatory, Herstmonceux Castle, Hailsham, Sussex, U.K.
THOMAS (Dr R. N.), Joint Institute for Laboratory Astrophysics, University of Colorado, Boulder, Colorado 80304, U.S.A.
THOMPSON (Dr A. R.), Radio Astronomy Institute, Stanford University, Stanford, California 94305, U.S.A.
THOMPSON (Dr G. I.), The Royal Observatory, Edinburgh 9, U.K.
THOMSON (Dr J. H.), Nuffield Radio Astronomy Observatory, Jodrell Bank, Macclesfield, Cheshire, U.K.
THOMSON (M. M.), Dominion Observatory, 3 Observatory Crescent, Ottawa, Ontario, Canada
* THORNE (Dr K.), Physics Department, California Institute of Technology, Pasadena, California 91109, U.S.A.
TIFFT (Dr W. G.), Lowell Observatory, Flagstaff, Arizona, U.S.A.
* TIFREA (Dr E.), Observatoire Astronomique, rue Cutitul de Argint 5, Bucarest 28, Roumania
TILLES (Dr D.), Oceanography Department, Oregon State University, Cornvallis, Oregon, 97331, U.S.A.
* TING (Yeou-Tswen), Chief of the Astronomy Section, Taipei, Taiwan
* TIURI (Prof. Dr M.), Technical University, Department of Electrotechnics, Albert Street 40–42, Helsinki, Finland
TLAMICHA (Ing A.), Astronomical Institute, Czechoslovak Academy of Sciences, Observatory Ondřejov, Czechoslovakia
* TODORAN (Dr I.), Observatoire, Université, Cluj, Roumania
* TOLBERT (Dr C. R.), Leander McCormick Observatory, University of Virginia, Charlottesville, Virginia 22903, U.S.A.
TOMBAUGH (C. W.), New Mexico State University, Research Center, University Park, New Mexico, U.S.A.
* TOOMRE (Dr A.), Department of Mathematics, Massachusetts Institute of Technology, Cambridge, Massachusetts 02139, U.S.A.
TORAO (M.), Tokyo Astronomical Observatory, Mitaka, Tokyo, Japan

TORGARD (Miss Dr I. H. M.), Observatory, Lund, Sweden

TORONDŽADZE (Dr A.F.), [Torondzhadze], Abastumani Observatory, Abastumani, Georgia, U.S.S.R.

TORROJA (Dr J. M.), University of Madrid, Madrid 3, Spain

TOUSEY (Dr R.), U.S. Naval Research Laboratory, Washington 25, D.C., U.S.A.

* TOVMASJAN (Dr G. M.), [Tovmasyan], Bjurakan Astrophysical Observatory, Armenian Academy of Sciences, Burakan, U.S.S.R.

TOWNES (Dr C. H.), Department of Physics, Columbia University, New York, N.Y., U.S.A.

TRAVING (Dr G.), Institut für Theoretische Physik, Olshausenstrasse, Neue Universität, Bau 13/I, Kiel, G.F.R.

TREANOR S.J. (Dr P. J.), Specola Vaticana, Castel Gandolfo, Vatican City State

* TREDER (Prof. Dr H.-J.), Deutsche Akademie der Wissenschaften, Berlin, G.D.R.

TREFFTZ (Miss Dr E.), Max-Planck-Institut für Physik und Astrophysik, Aumeisterstrasse 6, Munich 23, G.F.R.

TREHAN (Dr S. K.), Department of Physics and Astrophysics, University of Delhi, Delhi, India

TRELLIS (M.), Observatoire de Nice, Le Mont-Gros, 06 Nice, France

* TREMKO (Dr J.), Skalnaté Pleso Observatory, Slovak Academy of Sciences, Tatranská Lomnica, Czechoslovakia

TREXLER (J. H.), U.S. Naval Research Laboratory, Washington 25, D.C., U.S.A.

TROICKIJ (Dr V. S.), [Troitsky], Scientific Radiophysical Institute, Gorkij, U.S.S.R.

TS'AI (Chang-hsien), Taipei Observatory, Yuan Shan, Taipei, Taiwan

TSAO (Prof. Mo), 47, 3rd Section, Hsin-I Road, Taipei, Taiwan

TSESEVICH (Prof. Dr V. P.), [see CESEVIČ]

TSUBOKAWA (Dr I.), Geographical Survey Institute of Japan, 7-1000 Kamimeguro, Meguro-ku, Tokyo, Japan

* TSUCHIYA (A.), Tokyo Astronomical Observatory, University of Tokyo, Mitaka, Tokyo, Japan

TUCKER (R. H.), Royal Greenwich Observatory, Herstmonceux Castle, Hailsham, Sussex, U.K.

TUOMINEN (Prof. J. V.), Pihlajatie 49 B 20, Helsinki-Töölö, Finland

* TURŁO (Dr Z.), Astronomical Observatory, Torun-Piwnice, Poland

TUVE (Dr M. A.), Carnegie Institution of Washington, Department of Terrestrial Magnetism, 5241 Broad Branch Road, N.W. Washington 15, D.C., U.S.A.

TUZI (Dr K.), Tokyo Astronomical Observatory, Mitaka, Tokyo, Japan

TWISS (R. Q.), National Physical Laboratory, Teddington, Middlesex, U.K.

* UCHIDA (Dr Y.), Tokyo Astronomical Observatory, University of Tokyo, Mitaka, Tokyo, Japan

UENO (Dr S.), Institute of Astrophysics, College of Science, University of Kyoto, Kyoto, Japan

* UESUGI (Dr A.), Yamagami-cho 2–8, Otsu, Shiga, Japan

UETA (Dr J.), Department of Astronomy, Faculty of Science, Kyoto University, Kyoto, Japan

UNDERHILL (Prof. Dr A. B.), Astronomical Observatory "Sonnenborgh", Servaas Bolwerk 13, Utrecht, The Netherlands

* UNDERWOOD (J. H.), Goddard Space Flight Center, Greenbelt, Maryland 20771, U.S.A.

UNNO (Dr W.), Department of Astronomy, Faculty of Science, University of Tokyo, Bunkyo, Tokyo, Japan

UNSÖLD (Prof. Dr A.), Direktor des Instituts für Theoretische Physik, Olshausenstrasse, Neue Universität, Bau 13/I Kiel, G.F.R.

* UPGREN (Dr A.), Department of Astronomy, Wesleyan University, Middletown, Connecticut 06457, U.S.A.

* UPTON (Dr E. K. L.), Astronomy Department, University of California, Los Angeles, California 90024, U.S.A.

* URBARZ (Dr H.), Astronomisches Institut der Universität, Waldhäuser Strasse 64, 74 Tübingen, G.F.R.

UREY (Prof. H. C.), University of California, La Jolla, California, U.S.A.

VÄISÄLÄ (Prof. Y.), Puolalanpuisto 1, Turku, Finland
VALNÍČEK (Dr B.), Astronomical Institute, Czechoslovak Academy of Sciences, Observatory Ondřejov, Czechoslovakia
VAN ALBADA (Prof. G. B.), Astronomical Institute, Roeterstraat 1a, Amsterdam-C, The Netherlands
VAN BIESBROECK (Prof. G.), Lunar and Planetary Laboratory, University of Arizona, Tucson, Arizona, U.S.A.
VAN BUEREN (Prof. H. G.), Astronomical Observatory, "Sonnenborgh", Zonnenburg 2, Utrecht, The Netherlands
VAN DE HULST (Prof. Dr H. C.), University Observatory, Leiden, The Netherlands
VAN DE KAMP (Prof. Dr P.), Director, Sproul Observatory, Swarthmore College, Swarthmore, Pennsylvania 19081, U.S.A.
VANDEKERKHOVE (Dr E.), 7 avenue de Sumatra, Brussels 18, Belgium
VAN DEN BERGH (Prof. S.), David Dunlap Observatory, Richmond Hill, Ontario, Canada
VAN DEN BOS (Dr W. H.), c/o Republic Observatory, Johannesburg, South Africa
* VAN DER BORGHT (Prof. R.), Department of Mathematics, Monash University, Melbourne, Australia
VAN DER LAAN (Dr H.), Department of Astronomy, University of Western Ontario, London, Ontario, Canada
VANDERLINDEN (Prof. H. L.), Hertogsdreef 15, Bosvoorde, Brussels 17, Belgium
* VANDERVOORT (Prof. P. O.), The Leiden Observatory, Leiden, The Netherlands
VAN DIGGELEN (Dr J.), Sterrewacht "Sonnenborgh", Zonnenburg 2, Utrecht, The Netherlands
VAN HERK (Dr G.), Sterrewacht 6, Kaiserstraat, Leiden, The Netherlands
VAN HOOF (Prof. A.), Sterrenkundig Instituut van de Universiteit, 61 Naamsche Straat, Leuven, Belgium
VAN HOUTEN (Dr C. J.), University Observatory, Leiden, The Netherlands
VAN HOUTEN-GROENEVELD (Mrs I.), University Observatory, Leiden, The Netherlands
VAN REGEMORTER (Dr H.), Observatoire de Paris, Section d'Astrophysique, 92-Meudon, France
VAN SCHEWICK (Dr H.), Universitäts-Sternwarte Bonn, Poppelsdorfer Allee 49, Bonn, G.F.R.
VAN 'T VEER (Mme Dr C.), Institut d'Astrophysique, 98bis boulevard Arago, Paris-14e, France
VAN 'T VEER (Dr F.), Institut d'Astrophysique, 98bis, boulevard Arago, Paris-14e, France
VAN WIJK (U.)†, Princeton University Observatory, Princeton, New Jersey 08540, U.S.A.
VAN WOERDEN (Dr H.), Kapteyn Astronomical Laboratory, Broerstraat 7, Groningen, The Netherlands
VANÝSEK (Dr V.), Astronomical Institute of the Charles University, Švédská 8, Prague 5, Czechoslovakia
VARDYA (Dr M. S.), Astrophysics Division, Tata Institute of Fundamental Research, Colaba, Bombay 5, India
VARSAVSKY (Dr C. M.), Las Heras 1975 (6° A), Buenos Aires, Argentina
VASILEVSKIS (Dr S.), Lick Observatory, University of California, Santa Cruz, California 95060, U.S.A.
* VASIL'EV (Dr V.M.), Main Astronomical Observatory, USSR Academy of Sciences, Pulkovo, U.S.S.R.
VASSY (Prof. E.), Professeur à la Faculté des Sciences, Physique de l'Atmosphère, 1 quai Branly, Paris-7e, France
* VAUGHAN Jr. (Dr A.H.), Mount Wilson and Palomar Observatory, 813 Santa Barbara Street, Pasadena, California 91106, U.S.A.
VAUQUOIS (B.), Maître de Conférences à la Faculté des Sciences, Grenoble, Isère, France
VELGHE (Prof. A. G.), Directeur de l'Observatoire Royal de Belgique, 3 avenue Circulaire, Uccle-Brussels 18, Belgium
VENTER (S. C.)†, P.O.Box 1416, Pretoria, Transvaal, South Africa
VERBAANDERT (Dr J.), Observatoire Royal de Belgique, 3 avenue Circulaire, Uccle-Brussels 18, Belgium

VERGNANO (Prof. A.), Osservatorio Astronomico, Pino Torinese, Italy
* VERNIANI (F.F.), Harvard College Observatory, University of California, Berkeley, California 94720, U.S.A.
* VETEŠNÍK (Dr M.), Astronomical Institute, Purkyně University, Kotlářská 2, Brno, Czechoslovakia
VICENTE (Prof. Dr R.O.), R. Mestre Aviz, 30, R/c, Algés, Portugal
VIDAL (Dr J.M.C.), Avenida José Antonio 677, Barcelona, Spain
* VINTI (J.P.), Department of Mathematics, North Carolina Observatory at Raleigh, North Carolina 27607, U.S.A.
* VIRGOPIA (N.), Osservatorio Astronomico, Rome, Italy
* VISWANATHAN (N.), Mount Wilson and Palomar Observatories, 813 Santa Barbara Street, Pasadena, California, U.S.A.
* VITINSKIJ (Dr Ju.I.), [Vitinsky, Dr Yu.I.], Main Astronomical Observatory, USSR Academy of Sciences, Pulkovo, U.S.S.R.
VITKEVIČ (Dr V.V.), [Vitkevich], Chief, Department of Radio-Astronomy, Physical Institute, Moscow, U.S.S.R.
VOIGT (Prof. Dr H.H.), Direktor der Universitäts-Sternwarte, Geismarlandstrasse 11, 34 Göttingen, G.F.R.
* VOLLAND (Dr H.), Astronomisches Institut der Universität, Poppelsdorfer Allee 49, 53 Bonn, G.F.R.
VON DER HEIDE (Dr J.E.B.), Hamburger Sternwarte, Hamburg-Bergedorf, G.F.R.
VON HOERNER (Dr S.), National Radio Astronomy Observatory, P.O.Box 2, Green Bank, West Virginia, U.S.A.
VON KLÜBER (Dr H.), The Observatories, Madingley Road, Cambridge, U.K.
VON WEIZSÄCKER (Prof. Dr C.F.), Schwarzbuchenweg 40, Hamburg-Wellingsbüttel, G.F.R.
VORONCOV-VEL'JAMINOV (Prof. Dr B.A.), [Vorontsov-Velyaminov] Sternberg Astronomical Institute, Moscow, U.S.S.R.
VSEHSVJATSKIJ (Prof. Dr S.K.), [Vsekhsvyatsky], Astronomical Observatory, Kiev 53, U.S.S.R.
VYSSOTSKY (Prof. A.N.), 905 Garden Drive, Winter Park, Florida, U.S.A.

WACHMANN (Prof. Dr A.A.), Hamburger Sternwarte, Hamburg-Bergedorf, G.F.R.
* WACKERNAGEL (Dr H.B.), University of Colorado, Colorado Springs, Colorado, U.S.A.
WADE (Dr C.M.), National Radio Astronomy Observatory, P.O.Box 2, Greem Bank, West Virginia, U.S.A.
WAGMAN (Dr N.E.), Allegheny Observatory, Riverview Park, Pittsburgh, Pennsylvania 15214, U.S.A.
WALDMEIER (Prof. Dr M.), Director, Swiss Federal Observatory, Schmelzbergstrasse 25, 8006 Zürich, Switzerland
WALKER (Dr G.A.H.), Dominion Astrophysical Observatory, Victoria, B.C., Canada
WALKER (Dr M.F.), Lick Observatory, University of California, Santa Cruz, California 95060, U.S.A.
WALLACE (Dr L.V.), Kitt Peak National Observatory, 950 North Cherry Avenue, Tucson, Arizona, U.S.A.
WALLENQUIST (Prof. Dr A.A.E.), Uppsala Observatory, Kvistaberg Station, Bro, Sweden
WALLERSTEIN (Dr G.N.), Department of Astronomy, University of California, Berkeley 4, California, U.S.A.
WALRAVEN (Dr T.), Leiden Observatory, Sterrewacht, Leiden, The Netherlands
WALSH (Prof. D.), Astronomy Department, University of Michigan, Ann Arbor, Michigan, U.S.A.
WALTER (Prof. Dr K.), Astronomisches Institut der Universität, Waldhäuserstrasse 64, 74 Tübingen, G.F.R.
* WAMPLER (Dr E.J.), Lick Observatory, Mount Hamilton, California 95140, U.S.A.

WAPSTRA (Prof. A.), Voltaplein 19, Amsterdam-O, The Netherlands

* WARES (Dr G. W.), CFR Astrophysics, Space Physics Laboratory, Air Force Cambridge Research Laboratory, Laurence G. Hanscom Field, Bedford, Massachusetts, 01730, U.S.A.

* WARNER (Dr B.), Department of Astrophysics, University Observatory, South Parks Rd, Oxford, U.K.

WARWICK (Mrs Dr C. S.), Sun-Earth Relationships Section, Ionosphere Research and Propagation Division, National Bureau of Standards, Boulder Laboratories, Boulder, Colorado, U.S.A.

WARWICK (Dr J. W.), High Altitude Observatory of the University of Colorado, Boulder, Colorado, U.S.A.

WARZÉE (Dr J.), Observatoire Royal de Belgique, 3 avenue Circulaire, Uccle-Brussels 18, Belgium

WATERFIELD (Dr R. L.), Swinley Road, Ascot, Berkshire, U.K.

WATTS (Dr C. B.), U.S. Naval Observatory, Washington 25, D.C., U.S.A.

WAYMAN (Dr P. A.), Director, Dunsink Observatory, Castleknock, Co. Dublin, Ireland

WEAVER (Dr H. F.), Radio Astronomy Laboratory, University of California, Berkeley 4, California, U.S.A.

* WEHLAU (Dr A.), Department of Astronomy, University of Western Ontario, London, Ontario, Canada

WEHLAU (Prof. Dr W. H.), Department of Mathematics and Astronomy, University of Western Ontario, London, Ontario, Canada

WEIDEMANN (Prof. Dr V.), Institut für Theoretische Physik und Sternwarte der Universität, 23 Kiel, Olshausenstrasse, Neue Universität, Haus C 4/1, G.F.R.

WEIGERT (Dr A.), Universitäts-Sternwarte, Geismarlandstrasse 11, 34 Göttingen, G.F.R.

* WEILL (G.), Institut d'Astrophysique, 98 bis, boulevard Arago, 75-Paris-14e, France

WEIMER (Dr T.), Professeur à la Faculté des Sciences, Directeur de l'Observatoire d'Alger, Bouzaréa, Algeria

* WEINBERG (Dr J. L.), Hawaii Institute of Geophysics, Haleakala Observatory, P.O.Box 135, Kula Maui, Hawaii 96790, U.S.A.

WELLGATE (G. B.), Royal Greenwich Observatory, Herstmonceux Castle, Hailsham, Sussex, U.K.

WELLMANN (Prof. P.), Universitäts-Sternwarte, Sternwartstrasse 23, Munich 27, G.F.R.

WEMPE (Prof. Dr J.), Direktor des Astrophysikalischen Observatoriums, Telegraphenberg, Potsdam, G.D.R.

WENTZEL (Dr D. G.), Astronomy Program, University of Maryland, College Park, Maryland 20740, U.S.A.

WENZEL (Dr W.), Academy Observatory, Sonneberg, Thüringen, G.D.R.

WESSELINK (Dr A.J.), Yale University Observatory, Box 2023, Yale Station, New Haven, Connecticut 06520, U.S.A.

WESTERHOUT (Dr G.), Astronomy Program, University of Maryland, College Park, Maryland 20740, U.S.A.

WESTERLUND (Dr B. E.), Steward Observatory, University of Arizona, Tucson, Arizona 85721, U.S.A.

WEYMANN (Prof. R.J.), Steward Observatory, University of Arizona, Tucson, Arizona, U.S.A.

WHIPPLE (Dr F. L.), Director, Smithsonian Astrophysical Observatory, 60 Garden Street, Cambridge, Massachusetts 02138, U.S.A.

WHITAKER (E. A.), Lunar and Planetary Laboratory, University of Arizona, Tucson, Arizona, U.S.A.

* WHITE (Dr O. R.), Sacramento Peak Observatory, Sunspot, New Mexico 88349, U.S.A.

* WHITEOAK (Dr J. B.), C.S.I.R.O. Radiophysics Laboratory, University Grounds, Chippendale, Australia

WHITFORD (Dr A. E.), Director of the Lick Observatory, University of California, Santa Cruz, California 95060, U.S.A.

WHITNEY (Prof. B. S.), University of Oklahoma, Observatory, Norman, Oklahoma, U.S.A.

WHITNEY (Dr C. A.), Smithsonian Astrophysical Observatory, 60 Garden Street, Cambridge, Massachusetts 02138, U.S.A.

WHITROW (Dr C. J.), Department of Mathematics, Imperial College of Science and Technology, Prince Consort Road, London, S.W.7, U.K.

* WICKRAMASINGHE (Dr N. C.), Jesus College, Cambridge, U.K.

* WIDING (Dr K. D.), Code 7144, U.S. Naval Research Laboratory, Washington, D.C. 20390, U.S.A.

WIDORN (Dr T.), Universitäts-Sternwarte Wien, Türkenschanzstrasse 17, A-1180, Vienna, Austria

* WIEDLING (Dr T.), Skjutsargatan 24, Nyköping, Sweden

WIERZIBINSKI (Prof. Dr S.), Astronomical Observatory, Kopernika 11, Wroclaw, Poland

* WIESE (W. L.), Chief, Plasma Spectroscopy Section, National Bureau of Standards, Physical Building A-149, Washington, D.C. 20234, U.S.A.

WIETH-KNUDSEN (Dr N. P.), Svend Trostvej 12^{III}, Copenhagen-V, Denmark

WILCOCK (Dr W. L.), Department of Physics, University College of North Wales, Bangor, Caernarvonshire, U.K.

* WILCOX (Dr J. M.), Space Sciences Laboratory, University of California, Berkeley, California, U.S.A.

WILD (Dr J. P.), Radiophysics Laboratory, C.S.I.R.O. University Grounds, Chippendale, N.S.W., Australia

WILD (P.), Astronomisches Institut der Universität Bern, Sidlerstrasse 5, Bern, Switzerland

WILDT (Prof. R.), Yale University Observatory, Box 2023, Yale Station, New Haven, Connecticut 06520, U.S.A.

WILKENS (Prof. Dr A.), Oberföhringerstrasse 10, Munich 27, G.F.R.

WILKINS (Dr G. A.), H.M. Nautical Almanac Office, Royal Greenwich Observatory, Herstmonceux Castle, Hailsham, Sussex, U.K.

* WILLIAMS (Dr D.), Radio-Astronomy Laboratory, University of California, Berkeley, California, U.S.A.

* WILLIAMS (Dr J. A.), Physics-Astronomy Building, University of Michigan, Ann Arbor, Michigan, U.S.A.

WILLSTROP (Dr R. V.), University Observatories, Madingley Road, Cambridge, U.K.

WILSON (Dr A. G.), P.O.Box 113, Topanga, California, U.S.A.

WILSON (Dr O. C.), Mount Wilson and Palomar Observatories, 813, Santa Barbara Street, Pasadena 4, California, U.S.A.

* WILSON (Dr P. R.), Department of Applied Mathematics, University of Sydney, Sydney, Australia

WILSON (Dr R.), Spectroscopy Division, Culham Laboratory, Abingdon, Berkshire, U.K.

* WINKLER (Dr G. M. R.), U.S. Naval Observatory, Department of the Navy, Washington, D.C., U.S.A.

WITKOWSKI (Prof. Dr J.), Director, Observatory of Poznan, Ul. Sloneczna 36, Poznan, Poland

WLÉRICK (Dr G.), Observatoire de Paris, Section d'Astrophysique, 92-Meudon, France

* WOLSTENCROFT (Dr R. D.), Royal Observatory, Edinburgh 9, U.K.

WOLTJER (Prof. L.), Department of Astronomy, Columbia University, New York 27, N.Y., U.S.A.

WOOD (Prof. F. B.), Flower and Cook Observatory, University of Pennsylvania, Philadelphia 4, Pennsylvania, U.S.A.

WOOD (Dr H. W.), Government Astronomer, The Observatory, Sydney, N.S.W., Australia

* WOOD (Dr J. A.), Smithsonian Astrophysical Observatory, 60 Garden Street, Cambridge, Massachusetts 02138, U.S.A.

WOOLARD (E. W.), Director, Nautical Almanac, U.S. Naval Observatory, Washington 25, D.C., U.S.A.

WOOLF (Dr N. J.), Princeton University Observatory, Department of Astrophysical Sciences, 265 FitzRandolph Road, Princeton, New Jersey, U.S.A.

Woolley (Sir Richard), Astronomer Royal, Royal Greenwich Observatory, Herstmonceux Castle, Hailsham, Sussex, U.K.
Woolsey (E. G.), Dominion Observatory, Ottawa, Ontario, Canada
Worley (C. E.), U.S. Naval Observatory, Washington D.C., 20390, U.S.A.
* Worrall (Dr G.), The Observatories, Madingley Road, Cambridge, U.K.
Woszcyk (Dr A.), Astronomical Observatory of Toruń University, Sienkiewicza 30, Torun, Poland
Wright (Miss Dr F. W.), Harvard College Observatory, Cambridge, Massachusetts 02138, U.S.A.
Wright (Miss H.), 175 West 12th Street, New York, N.Y. 10011, U.S.A.
* Wright (Dr J. P.), Smithsonian Astrophysical Observatory, 60 Garden Street, Cambridge, Massachusetts 02138, U.S.A.
Wright (Dr K. O.), Director, Dominion Astrophysical Observatory, R.R.7, Victoria, B.C., Canada
Wrubel (Dr M. H.), Goethe Link Observatory, Indiana University, Bloomington, Indiana, U.S.A.
Wurm (Prof. Dr K.), Hamburger Sternwarte, Hamburg-Bergedorf, G.F.R.
Wyatt Jr., (Dr S. P.), University of Illinois Observatory, Urbana, Illinois, U.S.A.
Wyller (Dr A. A.), Bartol Research Foundations, Swarthmore, Pennsylvania, U.S.A.

Xanthakis (Prof. J.), Research and Computing Center, Academy of Sciences of Athens, Athens, Greece

Yabuuti (Dr K.), Research Institute of Humanistic Science, Kyoto University, Kyoto, Japan
Yakhontova (Mrs Dr N. Samoilova), [see Jahontova]
Yakovkin (Prof. Dr A. A.), [see Jakovkin]
Yakovkin (Dr N. A.), [see Jakovkin]
* Yamashita (Dr Y.), Department of Astronomy, University of Tokyo, Yayoi, Bunkyo-ku, Tokyo, Japan
Yaplee (B.S.), Code 7134, U.S. Naval Research Laboratory, Washington, D.C. 20390, U.S.A.
Yasuda (Dr H.), Tokyo Astronomical Observatory, Mitaka, Tokyo, Japan
Yavnel (Dr A. A.), [see Javnel']
Yen (Prof. J. L.), Department of Electrical Engineering, University of Toronto, Toronto, Ontario, Canada
* Yilmaz (Dr F.), Observatoire de l'Université, Istanbul, Turkey
Yoss (Dr K. M.), University of Illinois Observatory, Urbana, Illinois, U.S.A.
Young (Dr A.), Mathematics Department, University of Liverpool, Liverpool 3, U.K.
* Young (Dr A. T.), Astronomy Department, University of Texas, Austin, Texas 78712, U.S.A.
Yü (Dr Ching-Sung), Williams Observatory, Hood College, Frederick, Maryland, U.S.A.
Yul (K.), Phyongyang Astronomical Observatory, Phyongyang, Korea
Yumi (Dr S.), International Latitude Observatory of Mizusawa, Mizusawa-shi, Iwate-ken, Japan

Zabriskie (Prof. Dr F. R.), Department of Astronomy, 102 Whitmore Laboratory, Pennsylvania State University, University Park, Pennsylvania 16802, U.S.A.
* Zacharov (Dr I.), Astronomical Institute, Czechoslovak Academy of Sciences, Observatory Ondřejov, Czechoslovakia
Zadunaisky (Prof. P. E.), Instituto "Torcuato di Tella", Florida 936, Buenos Aires, Argentina
Zagar (Prof. F.), Direttore dell'Osservatorio Astronomico, Via Brera 28, Milano, Italy
Zahn (J.-P.), Observatoire de Nice, Le Mont-Gros, 06-Nice, France
Zanstra (Prof. H.), Zomerzorgerlaan 9, Bloemendaal, The Netherlands

* ZEL'DOVIČ (Acad. Ja. B.), [Zeldovich, Acad Ya. B.), Mathematical Institute of the USSR Academy of Sciences, Moscow, U.S.S.R.

ZEL'MANOV (Dr A. L.), Sternberg Astronomical Institute, Moscow, U.S.S.R.

ZHELEZNIAKOV (Dr V. V.), [see ŽELEZNJAKOV]

ZHEVAKIN (Dr S. A.), [see ŽEVAKIN]

ZHONGOLOVICH (Prof. Dr I. D.), [see ŽONGOLOVIČ]

ZIMMERMANN (Prof. Dr G. K.), [see CIMMERMAN]

ZIMMERMANN (Dr H.), University Observatory, Schillergässchen 2, Jena, G.D.R.

ZINNER (Prof. Dr E.), Remeis Sternwarte, Sternwartstrasse 7, Bamberg, G.F.R.

ZIRIN (Prof. H.), Mount Wilson and Palomar Observatories, 1201 East California Street, Pasadena, California, U.S.A.

ZIRKER (Dr J. B.), Hawaii Institute of Geophysics, 2525 Correa Road, University of Hawaii, Honolulu, Hawaii 96822, U.S.A.

ZONN (Prof. W.), Astronomical Observatory, Ujazdowskie 4, Warsaw, Poland

ZVEREV (Prof. Dr M. S.), Pulkovo Observatory, Leningrad M-140, U.S.S.R.

ZWAAN (Dr C.), Astronomical Observatory, "Sonnenborgh", Zonnenburg 2, Utrecht, The Netherlands

ZWICKY (Prof. Dr F.), California Institute of Technology, Pasadena, California 91109, U.S.A.

ŽELEZNJAKOV (Dr V. V.), [Zhelezniakov], Scientific Radiophysical Institute, Gorkij, U.S.S.R.

ŽEVAKIN (Dr S. A.), [Zhevakin], Scientific Radiophysical Institute, Gorkij, U.S.S.R.

ŽONGOLOVIČ (Prof. Dr I. D.), [Zhongolovich], Institute of Theoretical Astronomy, Leningrad, U.S.S.R.

GPSR Compliance

The European Union's (EU) General Product Safety Regulation (GPSR) is a set of rules that requires consumer products to be safe and our obligations to ensure this.

If you have any concerns about our products, you can contact us on ProductSafety@springernature.com

In case Publisher is established outside the EU, the EU authorized representative is:

Springer Nature Customer Service Center GmbH
Europaplatz 3
69115 Heidelberg, Germany

Batch number: 10762209

Printed by Printforce, the Netherlands

TRANSACTIONS
OF THE
INTERNATIONAL
ASTRONOMICAL UNION

VOLUME XIIIA – REPORTS

INTERNATIONAL COUNCIL OF SCIENTIFIC UNIONS
INTERNATIONAL ASTRONOMICAL UNION
UNION ASTRONOMIQUE INTERNATIONALE

TRANSACTIONS
OF THE
INTERNATIONAL ASTRONOMICAL UNION
VOLUME XIIIA - REPORTS

REPORTS
ON
ASTRONOMY

Edited by
Luboš PEREK
General Secretary of the Union

PUBLISHED FOR THE INTERNATIONAL COUNCIL OF SCIENTIFIC UNIONS
WITH THE FINANCIAL ASSISTANCE OF UNESCO

SPRINGER-SCIENCE+BUSINESS MEDIA, B.V.
1967

Originally published by D. Reidel Publishing Company in 1967
Softcover reprint of the hardcover 1st edition 1967

Unesco subvention – 1967 – AVS/414/25

ISBN 978-94-010-3478-4 *ISBN 978-94-010-3476-0 (eBook)*
DOI 10.1007/978-94-010-3476-0

PREFACE

The present volume, XIII A, contains the corrected Reports of IAU Commissions, as presented to the XIII General Assembly of the Union, held in Prague, Czechoslovakia, from 22 to 31 August 1967. The volume, entitled "Reports on Astronomy", is an authoritative, detailed, and unique survey of the accomplishments in all branches of astronomy from 1964 to 1967. For technical reasons, the volume also contains the Report of the Executive Committee 1964–1967, which originally was to appear in Volume B.

The page proofs of the present volume, assembled into the Agenda and Draft Reports, had been distributed to the Members of the Union before the General Assembly. Only errors and misprints in the Draft Reports have been corrected in the present volume: additions had to be deferred to volume XIII B in order to speed up publication.

Volume XIII B, subtitled "Proceedings of the Thirteenth General Assembly", will contain: the report of the Inaugural Ceremony and of the sessions of the General Assembly; the resolutions adopted by the General Assembly and by Commissions; the proceedings of Commission meetings, and those parts of the Astronomer's Handbook which underwent changes (i.e. representation of the IAU in international scientific bodies; composition of the Executive Committee; Symposia, Colloquia, and Publications of the Union; list of Commissions and of their membership; list of Members of the Union).

A separate volume, not numbered in the series of IAU Transactions, and entitled "Highlights of Astronomy as presented at the Thirteenth General Assembly of the Union 1967" will contain the Invited Discourses, the proceedings of the Joint Discussions, and those of one or two other meetings.

The proceedings of Symposia No. 33—Physics and Dynamics of Meteors, No. 34—Planetary Nebulae, and No. 35—Development of Solar Active Regions, held immediately after the General Assembly, will be published in the Symposium series of the Union.

The present volume is a collective work of forty Presidents of Commissions. The personal style was preserved in preference to an entirely homogenous presentation. The reports were edited by the former General Secretary, Professor Jean-Claude Pecker, during his term of office. The extremely difficult task of subediting the volume was entrusted to the experience, patience, and competence of Miss Geneviève Drouin. All those who assisted in preparing this volume, as well as the printer, Willmer Brothers Limited, and the publisher, D. Reidel Publishing Company, deserve the gratitude and appreciation of the Union and of the readers as well.

Praha
4 October 1967

L. Perek
General Secretary

CONTENTS

Report of the Executive Committee 1964–1967

Reports of Commissions

RAPPORT DU COMITE EXECUTIF 1964-1967*

ACTIVITÉ DE L'UNION ASTRONOMIQUE INTERNATIONALE

(See English version on page xvii)

Introduction

Ce rapport couvre la période qui s'étend du 1 janvier 1964 au 31 décembre 1966. Par souci de continuité, il a été composé selon des principes identiques à ceux qui ont guidé la rédaction des rapports précédents et notamment celui relatif à la période 1961–63, publié dans le Volume XII B des *Transactions*, et approuvé formellement par la douzième Assemblée Générale.

La période couverte comprend le début de l'année 1964 (jusqu'au 25 août 1964) pendant laquelle le Comité Exécutif actuel n'était pas encore en fonction. Symétriquement, elle exclut le début de l'année 1967, qui pourtant dépend du Comité Exécutif actuel: un rapport complémentaire provisoire, portant sur cette période (1 janvier–21 août 1967) sera oralement présenté devant la treizième Assemblée Générale, et, en cas d'approbation, publié dans le Volume XIII B des *Transactions*.

Pendant la période couverte par le présent rapport, un certain nombre de décisions ont été prises par le Comité Exécutif. Les unes, liées à la douzième Assemblée Générale, ou prises lors de cette Assemblée, ont été incluses dans le Volume XII B des *Transactions*, et ne seront pas à nouveau mentionnées ci-après. Les autres ont été publiées dans le *Bulletin d'Information* de l'UAI et sont à nouveau mentionnées ci-après dans les paragraphes appropriés. Le *Bulletin d'Information* de l'UAI a publié un grand nombre d'informations en provenance des Commissions de l'UAI, des Commissions Inter-Unions, des Services Astronomiques, etc. Ces informations sont essentiellement contenues dans les *Draft Reports* des Commissions appropriées de l'UAI, dans la mesure ou elles restent valides actuellement.

Les problèmes les plus graves rencontrés par l'Union concernent les Publications et les Finances: à chacune de ces deux questions est consacré un appendice au présent rapport. Un troisième appendice est consacré à quelques données relatives aux Organisations Adhérentes et aux Membres de l'Union. De plus, un certain nombre de Commissions Administratives (5, 6, 38, 46) sont considérées comme des comités du Comité Exécutif: leurs rapports complètent le présent rapport, auquel s'ajoutent également le rapport final du Président de l'IUCSTR (Inter-Union Commission on Solar and Terrestrial Relationships), l'UAI ayant joué le rôle d'union-mère de cet organisme inter-union, et le rapport du Président du Groupe de Travail sur les Émulsions Photographiques.

Structure et Administration de l'UAI

L'Union Astronomique Internationale fonctionne depuis la douzième Assemblée Générale dans le cadre des Statuts nouveaux adoptés lors de cette Assemblée.

Administration

Le fait dominant qui a commandé depuis lors la politique du Comité Exécutif est l'augmentation rapide des activités de l'Union: augmentation du nombre de ses Membres, de ses Commissions, des Pays Adhérents, complexité croissante des relations avec les autres organisations.

**Note*. Le texte du Rapport et de ses deux premiers Appendices sont publiés ci-après en deux langues. Seul le texte français fera autorité, en cas de besoin, le texte anglais n'étant qu'une traduction du texte français.

Depuis l'été 1964, afin de faire face à cette situation difficile, une sténo-dactylographe a été adjointe au Secrétaire Adjoint. De plus le Secrétaire Général et le Secrétaire Général Adjoint ont bénéficié des services de plusieurs personnes, à Meudon et à Prague, non salariées par l'UAI, grâce à la compréhension des autorités nationales françaises et tchécoslovaques, ce qui prouve que, même augmenté d'une personne, le personnel administratif de l'UAI est, en nombre, très insuffisant.

Il est apparu au Comité Exécutif qu'il était, pour des raisons financières claires, impossible de faire mieux, dans l'avenir immédiat. C'est pourquoi des solutions ont été recherchées en vue de simplifier l'administration, parfois très lourde, de certains problèmes: c'est en partie pour cette raison que la Liste de Distribution des publications de l'UAI a été considérablement réduite, et que l'édition des Symposiums de l'UAI postérieurs au no. 31 sera confiée, si possible, à une firme privée (voir ci-après page lx).

Réunions du Comité Exécutif

Le Comité Exécutif sortant a tenu sa vingt-cinquième réunion à Hambourg, pendant la douzième Assemblée Générale (24 août-3 septembre 1964), sous la présidence du Professeur V. A. Ambarcumjan. Tous les membres et conseillers étaient présents. Le Comité Exécutif actuel a tenu sa vingt-sixième réunion à Hambourg, le 3 septembre 1964, sous la présidence du Professeur P. Swings, Président. Tous les membres et conseillers étaient présents. Le Comité Exécutif a tenu, sous la présidence du Professeur P. Swings, les deux réunions suivantes: La vingt–septième session a eu lieu à l'Observatoire de Nice, France, du 15 au 19 septembre 1965; tous les membres et conseillers étaient présents sauf les Professeurs G. Haro et V. A. Ambarcumjan. La vingt-huitième session a eu lieu à Prague, Tchécoslovaquie, du 16 an 20 septembre 1966; tous les membres et conseillers étaient présents, sauf les Professeurs Y. Hagihara et V. A. Ambarcumjan, dont l'état de santé avait empêché la participation.

Entre les réunions, les affaires de l'Union ont été conduites par correspondance. Vingt-quatre lettres circulaires ont été envoyées par le Secrétaire Général et le Secrétaire Général Adjoint aux membres du Comité Exécutif.

Membres de l'UAI et Organisations Adhérentes

La Nouvelle–Zélande a adhéré à l'UAI en 1964. Aucune autre modification depuis la onzième Assemblée Générale n'est intervenue, malgré des contacts préliminaires pris avec des représentants qualifiés des pays suivants: Algérie, Cuba, Indonésie, Iran, Niger, Pakistan. La Colombie a demandé son adhésion qui sera proposée à la treizième Assemblée Générale.

L'Appendice III (page lxxi) rappelle la liste des Pays Adhérents et des Organisations Adhérentes correspondantes.

Il est à noter que la création de 'Pays Adhérents Associés' (dont la cotisation aurait été nulle, et qui n'auraient pas eu droit de vote sur les questions ayant des implications financières) a été longuement envisagée. Le Comité Exécutif a créé en son sein en 1965 un groupe de travail chargé d'étudier cette question. Mais après discussion, le Comité Exécutif, en 1966, a décidé de ne demander à l'Assemblée Générale aucune modification des Statuts à cet effet.

Au 1 janvier 1964, le nombre des Membres de l'UAI était de 1276. Depuis cette date, la douzième Assemblée Générale a élu 352 Membres nouveaux, cependant que nous étions informés du décès de 56 Membres; au 31 décembre 1964, le nombre de Membres de l'Union est donc de 1572.

Une liste des Membres du décès desquels nous avons été informés depuis le 1 janvier 1964 figure dans l'Appendice III. Cette liste comprend malheureusement tant d'astronomes distingués qu'il est seulement possible de rendre ici hommage à ceux qui ont contribué à l'administration de l'Union.

BERTIL LINDBLAD, Président de l'Union Astronomique Internationale de 1948 à 1952, a été Directeur de l'Observatoire de Stockholm, à Saltsjöbaden, depuis 1927. Ses travaux dans le domaine de la structure galactique (au sens le plus large) lui ont valu, entre autres distinctions, la médaille d'or de la Royal Astronomical Society, et la Bruce Gold Medal de la Société Astronomique du Pacifique. Il a été Président de l'Académie Royale Suédoise des Sciences. Il laissera le souvenir d'un ami bienveillant et humain, autant que celui d'un astronome distingué (1895–1965).

ROBERT METHVEN PETRIE, Vice-Président de l'Union Astronomique Internationale de 1958 à 1964, était Directeur du Dominion Astrophysical Observatory, à Victoria, Canada. Ses travaux de spectroscopie stellaire, appliquée notamment aux problèmes des parallaxes spectroscopiques, lui ont valu de nombreuses distinctions. Il était honoré du titre de 'Dominion Astronomer', et était Président de la Royal Astronomical Society of Canada. Ses collègues garderont de lui le souvenir d'un savant éminent, mais aussi d'un ami plein d'humour et de sagesse (1906–1966).

ANDRÉ DANJON, Président de l'Union Astronomique Internationale de 1958 à 1931, était, avant sa retraite, Directeur de l'Observatoire de Paris et Professeur à la Sorbonne depuis 1945. Ses travaux d'optique et de photométrie, puis ses recherches d'astronomie fondamentale, et particulièrement la conception et l'utilisation de l'astrolabe impersonnel à prisme qui porte son nom, lui ont valu de très nombreuses distinctions, notamment la médaille d'or de la Royal Astronomical Society et la médaille d'or du C.N.R.S. Il était Membre de l'Académie des Sciences de Paris, associé étranger de plusieurs institutions académiques, et, en France, il a présidé de nombreux comités. Ses collègues et ses élèves n'oublieront jamais ce qu'ils doivent à son énergie, et en garderont un très ému souvenir (1890–1967).

Commissions de l'UAI

Ce volume des *Agenda and Draft Reports* contient des rapports excellents et détaillés, préparés par les Présidents de chacune des 38 Commissions de l'Union, assistés par leur Vice-Président, leur Comité d'Organisation, les responsables des divers groupes de travail ou comités ayant été créés au sein de ces Commissions par les Commissions elles-mêmes.

Depuis la douzième Assemblée Générale deux nouvelles commissions ont été créées: la Commission 45 (Classifications Spectrales et Indices de Couleur à Plusieurs Bandes) et la Commission 46 (Enseignement de l'Astronomie); cette dernière fonctionne comme comité du Comité Exécutif.

En raison du décès du Professeur R. M. Petrie, le Professeur D. S. Evans a été nommé Président par interim de la Commission 30 (Les Vitesses Radiales), cependant que le Professeur F. L. Whipple était nommé Président par interim de la Commission 6 (Télégrammes Astronomiques) en raison du décès du Dr A. G. Hogg.

Entre les Assemblées Générales, les Commissions ont coopté parmi les Membres de l'UAI un certain nombre de membres. De plus, des Présidents ont souhaité s'adjoindre les services d'astronomes non-membres de l'UAI. En vue de simplifier la procédure, il a été demandé aux Présidents de les considérer pour tous les problèmes pratiques comme des Membres de l'UAI. Des propositions relatives aux nouvelles compositions des Commissions, comprenant ces membres cooptés et les nouveaux Membres élus par l'Assemblée Générale seront faites par les Présidents, les listes publiées dans le Volume XII C des *Transactions* servant de point de départ à ces propositions.

Entre les Assemblées Générales, les Commissions ont pris l'initiative d'un certain nombre de Colloques et de projets financés par l'UAI.

Colloques des Commissions de l'UAI

(tenus entre le 1 janvier 1964 et le 31 décembre 1966).

IAU Colloquium on *The Position of Variable Stars in the Hertzsprung-Russell Diagram*, Bamberg, 10–15 August 1965 (Commissions 27 and 42).

IAU Colloquium on *Interstellar Grains*, Troy, New York, 24–26 August 1965 (Commission 34)

IAU Colloquium on *The Blanketing Effect*, Heidelberg, 17–19 March 1966 (Commission 36)

IAU Colloquium on *Late-Type Stars*, Trieste, 13–16 June 1966 (Commissions 14, 29, 35, 36).

IAU Colloquium on *Atomic Collisions*, Boulder, 11–16 July 1966 (Commission 14).

IAU Colloquium on the *Evolution of Double Stars*, Uccle, 29 August–2 September 1966 (Commission 26).

Entreprises spécifiques des Commissions, approuvées par la douzième Assemblée Générale

La liste de ces entreprises, et des dépenses correspondantes, se trouvent page xxx, dans l'Annexe au présent rapport traitant des Finances de l'Union. Pour la plupart, ces entreprises constituent le prolongement d'entreprises précédentes, et n'appellent aucun commentaire.

Nous noterons cependant que les crédits attribués aux Cartes d'Étoiles Variables de l'Hémisphère Sud, et ceux attribués au Catalogue sur Cartes des Amas et Associations n'ont pas été utilisés. Cette dernière entreprise a dû être retardée, et le Comité Exécutif a donné son accord pour la réalisation, à partir de 1967, de ce projet par les éditions de l'Académie des Sciences de Budapest. La somme attribuée au Dr Bateson pour les Cartes d'Étoiles Variables de l'Hémisphère Sud devra être versée en 1967.

Entreprises spécifiques des Commissions, approuvées, entre les douzième et treizième Assemblées Générales, par le Comité Exécutif

Il s'agit:

(*a*) de la réédition de l'*Atlas of Representative Cometary Spectra*, par l'Institut d'Astrophysique de Liège, demandée par de nombreux chercheurs;

(*b*) de l'édition, par le Dr Sahade, de l'*Information Bulletin for the Southern Hemisphere*, publication très lue dans l'Hémisphère Sud, et particulièrement bien préparée et présentée;

(*c*) de la réédition, sous forme de feuilles mobiles, du livre ancien, et maintenant dépassé, de Rigaux, *Les Observatoires Astronomiques et les Astronomes*, par l'Observatoire Royal de Belgique à Uccle. Cette nouvelle forme de publication permettra l'actualisation permanente de l'ouvrage.

Symposiums

Symposiums de l'UAI

Pendant la période couverte par le présent rapport, les Symposiums suivants ont été organisés par l'UAI:

no.	Sujet	Lieu	Date
23	Astronomical Observations from Space Vehicles	Liège	Août 1964
24	Spectral Classification of Multi-Colour Photometry	Saltsjöbaden	Août 1964
25	The Theory of Orbits in the Solar System and in Stellar Systems	Thessaloniki	Août 1964
26	Abundance Determination in Stellar Spectra	Utrecht	Août 1964
27	The Construction of Large Telescopes	Tucson, Pasadena	Avril 1965

28	Cosmical Gas Dynamics. V. Aerodynamical Phenomena in Stellar Atmospheres	Nice	Septembre 1965
29	Instability Phenomena in Galaxies	Burakan	Mai 1966
30	Determination of Radial Velocities and their Applications	Toronto	Juin 1966
31	Radio Astronomy and the Galactic System	Noordwijk	Septembre 1966

Les Symposiums nos. 23, 24, 25 and 26 ont été organisés peu de temps avant la douzième Assemblée Générale de l'UAI. Le Symposium no. 28 a été organisé en coopération avec l'Union Internationale de Mécanique Théorique et Appliquée (IUTAM), le Symposium no. 31 avec l'Union Radio Scientifique Internationale (URSI).

Le Symposium no. 32, organisé conjointement avec l'Union Internationale de Géodésie et de Géophysique (IUGG), aura lieu à Stresa, Italie, du 21 au 25 mars 1967 et portera sur 'Continental Drift, Secular Motion of the Pole and Rotation of the Earth'.

Les Symposiums nos. 33, 34, 35, portant respectivement sur 'Physics and Dynamics of Meteors', 'Planetary Nebulae', 'The Structure and Development of Solar Active Regions', auront lieu dans la semaine qui suivra la treizième Assemblée Générale de l'UAI à Prague, et se tiendront respectivement à Tatranska Lomnica, Tchécoslovaquie (nos. 33 et 34) et à Budapest, Hongrie (no. 35).

Symposiums auxquels l'UAI a participé

The History of Astronomy (IUPHS-UAI)	Hambourg	Août 1964
The Trajectories of Artificial Celestial Bodies as Determined from Observations (IUTAM-COSPAR-UAI)	Paris	Avril 1965
Planetary Atmospheres and Surfaces as Deduced from Radio Observations (URSI-UAI)	Dorado Beach	Mai 1965
COSPAR Sixth International Space Science Symposium (sessions on Galactic and Extragalactic Phenomena and Space Research) (COSPAR-UAI)	Buenos Ayres	Mai 1966
Solar Terrestrial Physics (URSI, IUGG-IAGA, UAI)	Belgrade	Août 1966

L'UAI était de plus représentée à de nombreux symposiums, et aux assemblées de nombreux organismes de l'ICSU; ces représentations ont été annoncées dans les *Bulletins d'Information* successifs de l'UAI.

Relations avec les autres Organisations

Représentation de l'UAI

1. *Conseil International des Unions Scientifiques.* Le Secrétaire Général (D. H. Sadler) a représenté l'Union à la seconde réunion du Comité Exécutif de l'ICSU à Londres (juin 1964). Le Secrétaire Général a représenté l'Union à la 3ème réunion du Comité Exécutif de l'ICSU à Munich (avril 1965). Le Président (P. Swings) et le Secrétaire Général (J.-C. Pecker) ont représenté l'Union à la onzième Assemblée Générale de l'ICSU à Bombay (janvier 1966). Le Secrétaire Général a représenté l'Union aux 4ème et 5ème réunions du Comité Exécutif de l'ICSU à Bombay (janvier 1966). Le Secrétaire Général Adjoint (L. Perek) a également assisté à la onzième Assemblée Générale de l'ICSU. Le Secrétaire Général a représenté l'Union à la 6ème réunion du Comité Exécutif de l'ICSU à Monaco (septembre 1966).

Le Secrétaire Général est actuellement le représentant désigné de l'Union au Comité Exécutif de l'ICSU.

2. *Organismes de l'ICSU.* L'Union participe aux travaux d'un certain nombre de comités spéciaux et scientifiques, de comités et de commissions interunions fondés sous les auspices de l'ICSU. On trouvera dans le Volume XII C des *Transactions*, p. 13 et sq., des données sur ces organismes.

La liste suivante donne seulement l'état actuel de la représentation de l'UAI au sein de ces organismes.

(*a*) Committee on Space Research (COSPAR): L. Gratton.

(*b*) Special Committee for the International Years of the Quiet Sun (IQSY): G. Righini (Vice-Président); R. Michard (reporter for 'Solar Activity').

(*c*) Comité International de Géophysique (CIG): G. Righini (Vice-Président); R. Michard.

(*d*) Federation of Astronomical and Geophysical Services (FAGS): D. H. Sadler (Vice-President); B. Guinot.

(*e*) Inter-Union Commission on Frequency Allocations for Radioastronomy and Space Science (IUCAF): J. F. Denisse; V. A. Sanamian; F. G. Smith (Chairman); A. Unsöld.

(*f*) Inter-Union Commission on Solar-Terrestrial Physics (IUCSTP): Z. Svestka.

(*g*) Inter-Union Commission on Spectroscopy: B. Edlén; G. Herzberg; J. G. Phillips; M. J. Seaton.

(*h*) Commission Inter-Union sur l'Enseignement des Sciences (CIES-IUCST): E. Schatzman.

(*i*) ICSU Abstracting Board (IAB): J. Kleczek.

(*j*) Committee for Data on Science and Technology (CODATA): Ch. Moore-Sitterly.

3. *Autres organisations.* L'Union est représentée dans:

(*a*) La Fondation Internationale du Pic-du-Midi, par A. Lallemand.

(*b*) Le Comité Consultatif pour la Définition de la Seconde (CCDS), du Bureau International des Poids et Mesures, par W. Markowitz et N. Stoyko.

(*c*) Le Comité Consultatif pour la Définition du Mètre, du Bureau International des Poids et Mesures, par Ch. Moore-Sitterly.

(*d*) Les différents services de FAGS:

Bureau International de l'Heure (BIH): H. M. Smith; N. N. Pavlov.
International Polar Motion Service: (Organizing Committee of IAU Commission 19).
Quarterly Bulletin of Solar Activity: (Organizing Committee of IAU Commission 12).
International Ursigrams and World Days Service (IUWDS): R. Michard.
Solar Particles and Radiations Monitoring Organizations (SPARMO): C. de Jager.

Remarque: Les Commissions Inter-Unions sur les Relations Soleil-Terre (IUCSTR) et sur l'Ionosphère (IUCI) ont achevé leurs travaux. L'Union Astronomique Internationale y était représentée, respectivement par C. W. Allen (Président), R. Giovanelli, E. Mustel, M. Waldmeier pour la première, et par C. W. Allen (Secretary), R. N. Bracewell, V. G. Fesenkov, K. O. Kiepenheuer, pour la seconde.

Relations directes avec l'UNESCO

L'Union a bénéficié de contrats directs avec l'UNESCO, dans des buts variés; ils figurent, ci-après, dans les bilans financiers.

Il s'est agi d'abord d'aider la réunion de deux Symposiums de l'UAI (nos. 23 et 25) sur des sujets d'un intérêt général suffisant pour légitimer l'aide de l'UNESCO.

Ce fut ensuite une aide régulière à la Commission 38, par le moyen d'un contrat forfaitaire, destiné à couvrir une partie des dépenses de la Commission, spécialement celles entraînées par les échanges d'astronomes intéressant des pays en voie de développement.

Enfin, le principe d'une 'Ecole Annuelle de Jeunes Astronomes' a été retenu, à la suite d'une réunion d'un groupe d'experts, travaillant sous l'égide de la Commission 46 de l'UAI. Un crédit prévu à cet effet pour 1966 n'a pu être utilisé en raison du faible délai imparti à la préparation de l'école; un contrat de remplacement a permis l'attribution à un étudiant d'un pays en voie de développement d'une bourse d'études dans un observatoire bien équipé. En 1967, la première école d'été de l'UNESCO-UAI aura lieu à Manchester, sous la direction du Professeur Z. Kopal. Le Secrétaire Général en est le Dr J. Kleczek.

Il s'agit là d'une expérience, qui a été longuement débattue au sein du Comité Exécutif. Celui-ci aurait préféré sans doute que l'UNESCO aidât plutôt l'UAI à assumer ses charges normales, devenues trop lourdes. D'autres collègues pensent qu'une école d'été n'est efficace que si elle joue le rôle de pré-sélection, les meilleurs élèves qu'elle permet de détecter effectuant ensuite de longs séjours dans des observatoires équipés.

Quoiqu'il en soit, les résultats de l'expérience seront examinés avec attention, en vue des actions à venir.

L'UAI est désireuse de maintenir avec l'UNESCO de tels contrats: cette possibilité s'est révélée fructueuse et l'UAI tient à en exprimer à l'UNESCO sa reconnaissance.

Relations avec l'ICSU et ses organismes

L'ICSU est l'organisme dont dépend l'Union le plus directement: il verse à l'UAI la subvention de l'UNESCO (non compris les contrats directs) et coordonne les actions interunions.

Cette aide est précieuse, et nécessaire à l'UAI.

Malheureusement le nombre d'Unions membres de l'ICSU augmente, et la subvention de l'UNESCO à l'ICSU reste sans changement: il est donc difficile que la subvention reçue par l'UAI puisse croître.

De plus la création de nouveaux organismes interunions, en forçant l'UAI à rembourser les frais de ses représentants, qui doivent participer à des réunions assez fréquentes et souvent lointaines, impose une charge à l'Union.

C'est pourquoi, tout en poussant au développement de la coopération interdisciplinaire, l'UAI a toujours eu pour politique de favoriser les efforts de regroupements, et d'oeuvrer pour une grande efficacité. Ainsi a-t-elle joué un rôle important dans la réorganisation des comités intéressés à la physique des relations Soleil-Terre: suppression de l'IUCSTR, de l'IUCI, fin prochaine de CIG et de l'IQSY, création de l'IUCSTP qui devrait avoir avant tout un rôle de coordination: autant de faits récents qui contribueront, sans doute, à l'amélioration de la coopération interdisciplinaire dans un domaine essentiel.

Le Comité Exécutif de l'UAI a aussi souhaité renforcer le rôle des Unions, et par conséquent a toujours manifesté son opposition au développement de comités nationaux indépendants consacrés à des questions interdisciplinaires, les structures nationales devant conserver le caractère interdisciplinaire des structures internationales.

L'UAI a participé, avec de nombreuses Unions (IUTAM, UGGI, URSI) et Comités (COSPAR, IQSY) à l'organisation de symposiums (voir ci-dessus p. xiii).

CONCLUSIONS GÉNÉRALES

La période de trois ans qui vient de s'écouler a été réellement difficile.

Certes, les Membres de l'Union, et parmi eux, principalement les Présidents des Commissions, les Vice-Présidents, les membres des Comités d'Organisation des Commissions, les responsables des symposiums et colloques, les responsables de divers services, les représentants de l'UAI au sein de divers organismes, ont accompli un travail admirable.

Symétriquement, le Bureau Administratif de l'UAI, avec le souci permanent de ne pas excéder les possibilités financières de l'UAI, a fait de son mieux pour les y aider, et pour contribuer à l'essor de l'astronomie mondiale. On a notamment tenté, dans ce but, d'améliorer et de simplifier certaines méthodes administratives, cependant qu'un grand effort était fait pour maintenir les publications de l'UAI à un niveau élevé.

Pourtant tout ne va pas pour le mieux dans le meilleur des mondes: la nouvelle politique des publications de l'UAI, qui trouve son origine dans le désir légitime de diffuser mieux des publications d'une qualité améliorée, est coûteuse. De plus, sa mise en application, qui implique des relations étroites entre le Secrétaire Général, en France, l'imprimeur et l'éditeur, au Royaume-Uni, rencontre des difficultés pratiques, accrues encore par des difficultés internes (d'ailleurs momentanées) à la firme chargée de l'impression, et par le changement, en 1966, du Secrétaire Adjoint de l'UAI. Les publications sont en vérité d'une qualité satisfaisante, mais la production est trop lente, et des difficultés de distribution se présentent. Ces difficultés gênent bien évidemment les relations entre le Bureau Administratif et les Membres de l'UAI. De plus la vente reste insuffisante.

Ces relations, à vrai dire, sont fondées sur une tradition ambiguë. L'UAI est une des rares Unions à être constituée de membres individuels. Ceux-ci ne payent aucune cotisation, aucun droit; mais, de plus en plus nombreux, ils reçoivent gratuitement des publications de plus en plus épaisses, ils utilisent les services d'organismes financés par l'Union, ils bénéficient des contrats scientifiques que l'Union favorise largement.

De plus la contribution des pays est modeste, eu égard aux budgets sans cesse croissants de l'Astronomie dans ces pays, surtout à l'ère du développement rapide de la Radioastronomie et de la Recherche Spatiale.

La contribution annuelle des pays dont la cotisation est la plus élevée est égale à cinq fois le montant du voyage aller et retour entre l'Amérique du Nord et l'Europe; ou au coût de l'impression et de la distribution de deux *Bulletins d'Information*; ou encore à la contribution de l'UAI à l'organisation d'un seul symposium!

Il n'y a que deux solutions possibles.

Ou bien une certaine justice financière est rétablie, et la tradition généreuse de l'Union tempérée: les Membres devraient payer une cotisation; ses publications seraient toutes vendues et non plus partiellement distribuées gratuitement (cette mesure est déjà en cours d'application); les observatoires accepteraient de prendre le relais de l'Union dans le financement des entreprises spécifiques des Commissions.

Ou bien, une certaine libéralité est maintenue, notamment en ce qui concerne l'adhésion des Membres et les entreprises spécifiques: mais les organisations adhérentes devraient accepter une augmentation de leur cotisation (voir ci-dessous page xxv).

Le Comité Exécutif est aujourd'hui convaincu qu'une attitude différente est impossible, et qu'un certain archaïsme trop généreux des budgets de l'UAI n'est plus aujourd'hui de mise. De plus, il est certain que le 'gigantisme' croissant de l'Union devra dans peu de temps faire place à de nouvelles structures. Préparer cette mutation sera le rôle du prochain Comité Exécutif.

REPORT OF THE EXECUTIVE COMMITTEE 1964–1967*

ACTIVITY OF THE INTERNATIONAL ASTRONOMICAL UNION

(voir texte français, page ix)

INTRODUCTION

The present report covers the period from 1 January 1964 to 31 December 1966. For the sake of continuity, it has been prepared in accord with the principles governing previous reports, especially that of the period 1961–1963, formally approved by the twelfth General Assembly, and published in Volume XII B of the *Transactions*.

The period under report includes the first half of the year 1964 (until 25 August 1964) which fell under the responsibility of the former Executive Committee. On the other hand, it does not include the first half of the year 1967, still within the responsibility of the present Executive Committee. An oral complementary report to cover this period (1 January to 21 August 1967) will be presented to the thirteenth General Assembly and, if approved, published in Volume XIII B of the *Transactions*.

The Executive Committee has reached a number of decisions during this period. Those in connection with the twelfth General Assembly, or approved by it, were published in Volume XII B of the *Transactions* and will not be repeated here. Those taken at later dates have been published in the *IAU Information Bulletin*, and will be referred to in the appropriate paragraphs that follow. Reports on the activity of IAU Commissions, Inter-Union Commissions, Astronomical Services, etc., published in the *Information Bulletin*, have been incorporated into the Draft Reports of Commissions, in so far as they still apply.

The most serious problems of the Union are those concerning publication and finance; they are dealt with in two separate appendices.

A third appendix is devoted to factual data concerning Adhering Countries and IAU Members. Four administrative Commissions (5, 6, 38 and 46) are now treated as committees of the Executive Committee; their reports follow immediately after this Report, on page lxxv. The final reports of the President of IUCSTR (Inter-Union Commission on Solar and Terrestrial Relationships), of which the IAU was the parent Union, and of the President of the Working Group on Photographic Materials, are also added to the present Report, on page xcix.

Structure and Administration of the IAU

Since the twelfth General Assembly, the International Astronomical Union has operated in accord with the provisions of the new Statutes, as adopted at that General Assembly.

Administration

The Executive Committee has had to cope with a great intensification in the activity of the Union: the number of its Members increased, as did that of its Commissions and Adhering Organizations; relations with other organizations became more involved.

This difficult situation was met, in the summer of 1964, by adding a shorthand-typist to the existing staff. Moreover, the General Secretary and the Assistant General Secretary had the benefit of the services of several assistants, in both Meudon and Prague, generously

*Note. The text of the report and of its Appendices I and II are hereafter published in two languages. The French text being the original, from which the English text has been translated, is the only one to be authoritative, in case of need.

provided by the French and Czechoslovak authorities. Evidently, though re-inforced by one additional member, the staff of the Administrative Office of the Union is still largely insufficient.

The Executive Committee realizes that, the financial situation of the Union being as it is, no improvement is possible in the near future. The only remedy seems to be in a simplification of IAU administration, rather cumbersome the last years: partly for this reason, it has been decided to curtail the IAU Distribution List and to arrange for the Symposium volumes to be published by a commercial firm from Symposium no. 32 onwards (see page lxviii below).

Meetings of the Executive Committee

The former Executive Committee held its twenty-fifth meeting in Hamburg during the twelfth General Assembly (from 24 August to 3 September 1964) under the presidency of Professor V. A. Ambarcumjan. All members and councillors were present. The present Executive Committee held its twenty-sixth meeting in Hamburg, on 3 September 1964, under the presidency of Professor P. Swings. All members and councillors were present. The Executive Committee, presided over by Professor P. Swings, met as follows during the period under report: The twenty-seventh meeting was held at the Nice Observatory, France, from 15 to 19 September 1965; all members and councillors were present except Professor G. Haro and Professor V. A. Ambarcumjan. The twenty-eighth meeting was held in Prague, Czechoslovakia, from 16 to 20 September 1966; all members and councillors were present, except Professor Y. Hagihara and Professor V. A. Ambarcumjan, who was prevented from attending by illness.

Between meetings, the business of the Union was conducted by correspondence. Twenty-four circular letters were sent from the General Secretary and the Assistant General Secretary to the members of the Executive Committee.

Members of the IAU and Adhering Countries

New Zealand has adhered to the Union since 1964. No other modification has occurred since the twelfth General Assembly, though preliminary consultations have been held with representatives of the following countries: Algeria, Cuba, Indonesia, Iran, Nigeria and Pakistan. Columbia has applied for adhesion, and its application will be submitted to the thirteenth General Assembly.

The list of Adhering Countries and corresponding Adhering Organizations is given in Appendix III (page lxxi).

It should be noted that much thought was given to the creation of a new category of membership, that of 'Associated Adhering Countries'. Countries adhering in this category would have paid no contribution, but would have had no right to vote on questions with financial implications. The Executive Committee formed a working group from among its members to investigate this problem; but after a lengthy discussion, the Executive Committee decided, in 1966, not to ask the General Assembly to modify the Statutes to this effect.

The number of IAU Members was 1276 on 1 January 1964. Since then, the twelfth General Assembly has elected a further 352 new Members. On the other hand, the General Secretary has been informed of the decease of 56 Members, so that the Union had 1572 Members on 31 December 1964.

The list of the Members whose decease has been notified since 1 January 1964 is given in Appendix III. Unfortunately this list includes so large a number of eminent scientists that it is possible only to pay hereafter separate tribute to the memory of those who have contributed greatly to the administration of the Union:

Bertil Lindblad, President of the International Astronomical Union (1948–1952), was Director of the Stockholm Observatory, Saltsjöbaden, from 1927. His achievements in the domain of the Galactic Structure (in the broadest sense) were rewarded by the award of the Gold Medal of the Royal Astronomical Society, and the Bruce Gold Medal of the Astronomical Society of the Pacific, beside other distinctions. He was President of the Royal Swedish Academy of Sciences. He will be remembered as a good friend, full of understanding and humanity, as well as a distinguished astronomer (1895–1965).

Robert Methven Petrie, Vice-President of the International Astronomical Union (1958–1964), was Director of the Dominion Astrophysical Observatory, Victoria, Canada. His work in the domain of stellar spectrography, applied especially to the problem of spectroscopic parallaxes, was rewarded by numerous distinctions. He was honored with the title of 'Dominion Astronomer' and was President of the Royal Astronomical Society of Canada. His colleagues will remember him as an outstanding scholar, but also as a friend full of humour and wisdom (1906–1966).

André Danjon, President of the International Astronomical Union from 1958 to 1961, was, before he retired, Director of the Observatoire de Paris and Professor at the Sorbonne from 1945. His achievements in the domain of photometry, his research into fundamental astronomy and, in particular, his conception and application of the impersonal prismatic astrolabe, which was given his name, brought him many distinctions, including the Gold Medal of the Royal Astronomical Society and the Gold Medal of the C.N.R.S. He was member of the Academie des Sciences de Paris, associate member of numerous foreign academic institutions and president of a number of committees in France. His colleagues and pupils will never forget how much they owe to him, and they will always treasure his memory (1890–1967).

Commissions of the IAU

The volume of *Agenda and Draft Reports* includes as usual the excellent reports prepared by the Presidents of each of the 38 Commissions of the Union, assisted by their Vice-Presidents, Organizing Committees, those responsible for the various working groups or committees formed within the Commissions, and by the Commissions themselves.

At the twelfth General Assembly, two new commissions were established: Commission 45 (Spectral Classification and Multi-Band Colour Indices), and Commission 46 (Teaching of Astronomy); the latter is regarded as a committee of the Executive Committee.

Professor R. M. Petrie has been succeeded by Professor D. S. Evans, nominated as Acting President of Commission 30 (Radial Velocities), and Professor F. L. Whipple has been nominated Acting President of Commission 6 (Astronomical Telegrams) to replace the late Dr A. G. Hogg.

Between the General Assemblies, Commissions have co-opted a number of new members from among the Members of the Union. Moreover, some Presidents wished to obtain assistance from astronomers who are non-members of the Union; they have been asked to regard them as Members for all practical purposes. The Presidents will prepare proposals for the new composition of Commissions including co-opted members and new Members to be elected by the General Assembly. The list published in Volume XII C of the *Transactions* will serve as a starting point for these proposals.

Between the General Assemblies, Commissions organized several Colloquia, and other projects financed by the IAU.

Colloquia of IAU Commissions
(held between 1 January 1964 and 31 December 1966)

IAU Colloquium on *The Position of Variable Stars in the Hertzsprung-Russell Diagram*, Bamberg, 10–15 August 1965 (Commissions 27 and 42).

IAU Colloquium on *Interstellar Grains*, Troy, New York, U.S.A., 24–26 August 1965 (Commission 34).

IAU Colloquium on *The Blanketing Effect*, Heidelberg, 17–19 March 1966 (Commission 36).

IAU Colloquium on *Late-Type Stars*, Trieste, Italy, 13–16 June 1966 (Commissions 14, 29, 35, 36).

IAU Colloquium on *Atomic Collisions*, Boulder, Colorado, U.S.A., 11–16 July 1966 (Commission 14).

IAU Colloquium on the *Evolution of Double Stars*, Uccle, 29 August–2 September 1966 (Commission 26).

Specific Projects of Commissions, approved by the twelfth General Assembly

The list of such projects, with the corresponding expenditure, is given on page xlvi in Appendix I to the present report, dealing with the finances of the Union. These projects are, as a rule, continuations of previous projects and thus need no comment.

It will be seen, however, that the grants allocated to *Southern Variable Stars Charts*, and to the *Catalogue of Star Clusters* have not been used. The latter project has been delayed, and the Executive Committee has now agreed to a proposal that it be published by the Publishing Office of the Academy of Sciences, Budapest, from 1967 onwards. The grant made to Dr Bateson for the *Southern Variable Stars Charts* will have to be paid in 1967.

Specific Projects of Commissions approved, between the twelfth and thirteenth General Assemblies, by the Executive Committee

The projects are as follows:

(*a*) A new edition of the *Atlas of Representative Cometary Spectra* by the Institut d'Astrophysique de Liège, to meet the needs of many scientists.

(*b*) The publication, by Dr Sahade, of the *Information Bulletin for the Southern Hemisphere*. This publication, in great demand in the Southern Hemisphere, is particularly well prepared and presented.

(*c*) A new edition, in the form of a loose-leaf catalogue, of the now outdated catalogue by Rigaux, *Les Observatoires Astronomiques et les Astronomes*; this is a project of the Observatoire Royal de Belgique in Uccle. The revised form of publication will allow the catalogue to be kept up to date.

Symposia

IAU Symposia

During the period covered by the present report, the following symposia were organized by the Union:

no.	Subject	Place	Date
23	Astronomical Observations from Space Vehicles	Liège	August 1964
24	Spectral Classification and Multi-Colour Photometry	Saltsjöbaden	August 1964
25	The Theory of Orbits in the Solar System and in Stellar Systems	Thessaloniki	August 1964
26	Abundance Determination in Stellar Spectra	Utrecht	August 1964
27	The Construction of Large Telescopes	Tucson, Pasadena	April 1965
28	Cosmical Gas Dynamics. V. Aerodynamical Phenomena in Stellar Atmospheres	Nice	September 1965

no.	Subject	Place	Date
29	Instability Phenomena in Galaxies	Bjurakan	May 1966
30	Determinations of Radial Velocities and their Applications	Toronto	June 1966
31	Radio Astronomy and the Galactic System	Noordwijk	September 1966

Symposia nos. 23, 24, 25, and 26 had been held shortly before the twelfth General Assembly of the IAU. Symposium no. 28 was organized in co-operation with the International Union of Theoretical and Applied Mechanics (IUTAM), Symposium no. 31 in co-operation with the International Union of Scientific Radio (URSI).

Symposium no. 32, organized jointly with IUGG, will be held in Stresa, Italy, from 21 to 25 March 1967, and is to deal with 'Continental Drift, Secular Motion of the Pole, and Rotation of the Earth'.

Symposia nos. 33, 34, 35, on 'Physics and Dynamics of Meteors', 'Planetary Nebulae' and 'The Structure and Development of Solar Active Regions' respectively, will be held in Tatranska Lomnica, Czechoslovakia (nos. 33 and 34), and in Budapest, Hungary (no. 35).

Symposia with IAU participation

The History of Astronomy (IUPHS-IAU), Hamburg, August 1964.

The Trajectories of Artificial Celestial Bodies as Determined from Observations (IUTAM-COSPAR-IAU), Paris, April 1965.

Planetary Atmospheres and Surfaces as Deduced from Radio Observations (URSI-IAU), Dorado Beach, Puerto-Rico, U.S.A., May 1965.

COSPAR Sixth International Space Symposium (sessions on Galactic and Extragalactic Phenomena and Space Research) (COSPAR-IAU), Buenos-Aires, May 1966.

Solar-Terrestrial Physics (URSI-IUGG-IAGA-IAU), Belgrade, August 1966.

Moreover, the IAU was represented at numerous Symposia and at Assemblies of many ICSU Organizations; such representations were announced in the *IAU Information Bulletins*.

Relations to Other Organizations

IAU Representation

1. *International Council of Scientific Unions.* The General Secretary (then D. H. Sadler) represented the Union at the second meeting of the ICSU Executive Committee in London (June 1964). The General Secretary represented the Union at the third meeting of the ICSU Executive Committee in Munich, April 1965. The President and the General Secretary represented the Union at the eleventh General Assembly of ICSU in Bombay (January 1966). The General Secretary represented the Union at the fourth and fifth meetings of the ICSU Executive Committee (Bombay, January 1966). The Assistant General Secretary also attended the eleventh General Assembly of ICSU. The General Secretary represented the Union at the sixth meeting of the ICSU Executive Committee, in Monaco, September 1966.

The General Secretary is the present representative of the Union on the Executive Committee of ICSU.

2. *ICSU Organizations.* The Union participates in the activity of a number of Special and Scientific Committees, Inter-Union Commissions and Commissions established and sponsored by ICSU. Particulars of these organizations are given on pages 13 onwards of Volume XII C of the *Transactions*.

The following list gives only the present state of IAU representation in these organizations:

(*a*) Committee on Space Research (COSPAR): L. Gratton.

(*b*) Special Committee for the International Years of the Quiet Sun (IQSY): G. Righini (Vice-President); R. Michard (reporter for 'Solar Activity').

(*c*) Comité International de Géophysique (CIG): G. Righini (Vice-President); R. Michard.

(*d*) Federation of Astronomical and Geophysical Services (FAGS): D. H. Sadler (Vice-President); B. Guinot.

(*e*) Inter-Union Commission on Frequency Allocations for Radioastronomy and Space Science (IUCAF): J.-F. Denisse; V. A. Sanamian; F. G. Smith (Chairman); A. Unsöld.

(*f*) Inter-Union Commission on Solar-Terrestrial Physics (IUCSTP): Z. Svestka.

(*g*) Inter-Union Commission on Spectroscopy: B. Edlén; G. Herzberg; J. G. Phillips; M. J. Seaton.

(*h*) Inter-Union Commission on Science Teaching (CIES-IUCST): E. Schatzman.

(*i*) ICSU Abstracting Board (IAB): J. Kleczek.

(*j*) Committee for Data on Science and Technology (CODATA): Ch. Moore-Sitterly.

3. *Other Organizations.* The Union is represented on the following Organizations:

(*a*) La Fondation Internationale du Pic-du-Midi, by A. Lallemand.

(*b*) Le Comité Consultatif pour la Définition de la Seconde (CCDS) du Bureau International des Poids et Mesures, by W. Markowitz and N. Stoyko.

(*c*) Le Comité Consultatif pour la Définition du Mètre du Bureau International des Poids et Mesures, by Ch. Moore-Sitterly.

(*d*) Various Services of FAGS, that is on the directing boards of:

Bureau International de l'Heure (BIH), by H. M. Smith, N. N. Pavlov.

International Polar Motion Service, by the Organizing Committee of IAU Commission 19.

Quarterly Bulletin of Solar Activity, by the Organizing Committee of IAU Commission 12.

International Ursigrams and World Days Service (IUWDS), by R. Michard.

Solar Particles and Radiations Monitoring Organization (SPARMO), by C. de Jager.

Note: The Inter-Union Commission on Solar and Terrestrial Relationships (IUCSTR), and that on the Ionosphere (IUCI), have terminated their activities. The IAU was represented by C. W. Allen (President), R. Giovanelli, E. Mustel and M. Waldmeier on the former, and by C. W. Allen (Secretary), R. N. Bracewell, V. G. Fesenkov, K. O. Kiepenheuer on the latter.

Direct Relations with UNESCO

The Union had the benefit of various direct contracts with UNESCO, as indicated in the financial report.

The first contract provided for a grant towards two IAU Symposia (nos. 23 and 25), whose subjects were sufficiently broad to justify UNESCO assistance.

The second contract provided for regular financial assistance to Commission 38, partly to cover the Commission's expenditure especially in connection with the exchange of astronomers from or to developing countries.

Finally, an *ad hoc* working group of experts organized by Commission 46 of the IAU, agreed on the organization of an annual 'School of Young Astronomers'. A UNESCO grant provided for this purpose in 1966 could not be fully used owing to the short time available for the preparation of the curriculum. A substitute contract allowed the Union to allocate a scholarship to a student from a developing country towards his studies at a well equipped Observatory. The first UNESCO–IAU Summer School will open in Manchester, in 1967, under the direction of Professor Z. Kopal. The General Secretary of the school is Dr J. Kleczek.

This proposal was studied in great detail by the Executive Committee. The Committee would certainly have preferred that UNESCO, if it had been possible, should have assisted the normal activities of the Union, which have become rather burdensome. Other colleagues believe that a summer school is only justified as a pre-selection agent which would single out the most promising students for a longer stay at well-equipped observatories.

Be it as it may, the results of the experiment will be carefully examined with regard to future policy.

The IAU desires to maintain such contracts with UNESCO, to their mutual benefit. The present possibilities are likely to be fruitful, and the Union wishes to express its thanks and appreciation to UNESCO.

Relations with ICSU and its organisms

The ICSU is the organization on which the Union is most directly dependent: it transmits to the Union the UNESCO subvention (direct contracts not included) and co-ordinates the activity of Inter-Union Organizations.

This help is valuable and necessary to the Union.

Unfortunately the number of Unions within the ICSU increases, while the UNESCO subvention to ICSU remains essentially unchanged: thus the portion of the UNESCO subvention to the IAU is not likely to increase.

The Union is responsible for travel grants for its representatives on the increasing number of Inter-Union bodies whose meetings are very frequent and held at places often very remote.

While continuing to support the development of inter-disciplinary co-operation, the Union has therefore advocated efforts directed towards regrouping and greater efficiency. This was the Union's attitude in the re-organization of the Committees concerned with the physics of solar-terrestrial relationships; leading to the discontinuation of the IUCSTR and IUCI, the imminent termination of the activities of CIG and IQSY, and the formation of IUCSTP, whose principal objective will be the co-ordination of various activities. This recent action of the IAU is certain to contribute to a better inter-disciplinary co-operation in a domain of fundamental importance.

The Executive Committee of the IAU, in its effort to strengthen the position of the Unions, opposed the development of independent national committees to deal with inter-disciplinary problems, as it is desired that national structures should conform to the inter-disciplinary character of international structures.

The IAU participated with numerous Unions (IUTAM, IUGG, URSI) and Committees (COSPAR, IQSY) in the organization of Symposia (see above, page xxi).

GENERAL CONCLUSIONS

The three years covered by the present report were difficult in many respects.

It is certain, on the one hand, that the Members of the Union, and among them especially the Presidents of Commissions, Vice-Presidents, members of Organizing Committees of Commissions, members responsible for the organization of Symposia and Colloquia, and of numerous Special Services, representatives of the Union in different organizations, etc. have achieved admirable results.

The Administrative Office of the Union, always anxious to keep within the narrow limits of the IAU budget, did its utmost to help Members in these efforts and thus to contribute to the promotion of the interests of Astronomy all the world over. In this respect attempts were made to simplify certain administrative procedures, whilst maintaining the high standard of the IAU publications.

On the other hand, however, there is nothing perfect in this best of all worlds: the new IAU publication policy, adopted with the legitimate desire to improve the quality and distribution of IAU publications, is expensive. Moreover, its application, which calls for a narrow contact between the General Secretary in France, and the printers and publishers in the United Kingdom, met with difficulties in practice which were aggravated by a minor temporary crisis in the organization at the printers and by the change in the person of the IAU Assistant Secretary in 1966; the publications have been slow to come out of the press and their distribution has met with difficulties. This, naturally, puts a strain on the relations between Administration and Members. Moreover, the sales have, so far, been unsatisfactory.

In this connection, it should be emphasized that these relations with Members are based on a rather biased tradition. The IAU is one of the few international organizations having individual members. Members of the Union pay no contribution and no fees; yet, steadily increasing in number, they obtain, free of charge, publications ever ampler in volume; and they have the benefit of the numerous services organized and financed by the Union, and of the scientific contacts the Union facilitates.

Moreover, the contributions of Adhering Organizations are modest compared with the sums their countries spend on Astronomy in their own countries, especially in this age of the rapidly-developing Radio-Astronomical and Space Sciences.

The highest single annual national contribution is equal to only five times the return fare between North America and Europe, and corresponds to the cost of printing and distributing two issues of the *Information Bulletin*, which is also equal to the IAU contribution to the costs of a single Symposium!

There are only two possible solutions.

An attempt may be made to introduce more financial justice by relaxing the generous tradition of the Union and having Members pay a contribution; IAU publications being sold to Members and not distributed free of charge (this measure is already being put into operation); observatories assisting the Union by financing part of the Specific Projects of the Commissions.

Or, if a certain liberality is to be maintained, especially as regards membership and Specific Projects, the Adhering Organizations must agree to accept an increase in the contributions that they pay (see below, Appendix I, page xli).

The Executive Committee is convinced that there is no other alternative, and that the traditional generosity of the IAU budgets is obsolete. Moreover, it is certain that the growing 'gigantism' of the Union will soon have to give way to a new structure. It will be the task of the future Executive Committee to work towards this change.

APPENDICE I. RAPPORT FINANCIER

(See English version on page xli)

A. INTRODUCTION

Conformément à la politique financière introduite en 1961, le Comité Exécutif présente aux Membres de l'UAI un seul 'Compte Résumé des Recettes et Dépenses' couvrant la période du 1 janvier 1964 au 31 décembre 1966. Toutes les sommes ont été converties en dollars U.S.; le taux de change adopté, pour suivre les règles de l'Unesco, a été le suivant:

£ sterling: \$2·80   \$ U.S.: 4·90 francs français

Certaines opérations (virements d'un compte à un autre, remboursements de sommes excédant les besoins) ont été faites au change du jour par les banques: il s'ensuit des erreurs de change.

Le compte résumé a été obtenu à partir des neuf comptes (un pour chaque devise, et un par an) établis à partir des cahiers. Ceux-ci seront disponibles pour examen par le Comité des Finances. Ils ont été examinés et certifiés corrects par les experts comptables ayant travaillé pour l'Union. Le présent rapport se termine (ci-dessous page xxxix) par la déclaration d'expertise pour l'ensemble des comptes des trois années en considération, suivi par des suggestions relatives au budget des années suivantes.

Le rapport ci-après comporte, en commentaire du compte résumé, des notes nécessaires à sa compréhension complète point par point. Les sommes dues à la fin de l'exercice 1966, ou les recettes correspondant à des paiements arriérés pour la période 1961–1964, sont notamment indiquées dans chaque cas. Contrairement à ce qui s'est passé dans la période 1961–1964, ces sommes ne sont pas négligeables, en raison de la modification importante apportée au système des publications de l'Union; nous avons consacré un appendice détaillé à cette question.

Le rapport enfin est conclu par une mise en garde sérieuse, la situation financière étant en effet *particulièrement difficile*, et par des propositions relative au budget de 1967 et des années 1968, 1969, 1970.

B. COMPTE RÉSUMÉ DES RECETTES ET DÉPENSES

Recettes

1. *Cotisations des pays adhérents.* La XIe Assemblée Générale de l'UAI, à Berkeley, a décidé de fixer l'unité de contribution à 600 francs-or pour 1962 et les années suivantes. On rappelle qu'un franc-or est égal à 0,2903225 gramme d'or fin, et à 0,3266 dollar. L'unité de cotisation est donc de \$195·96.

A la suite de la modification des statuts adoptée par la XIIe Assemblée Générale, il a été demandé aux Pays Adhérents de choisir une nouvelle catégorie d'adhésion, conformément au nouvel article 4 des Statuts, tel qu'il figure dans le Volume XII C des *Transactions*. Le nombre total d'unités de contribution a donc augmenté nettement, d'une façon qui n'a toutefois pas correspondu entièrement aux espoirs du Comité Exécutif de la période 1961–1964, qui avait suggéré la modification des Statuts, notamment dans ce but.

Il est à noter qu'en 1966 un effort particulier a été fait, avec succès, pour le recouvrement des cotisations dues, et pour un choix meilleur, dans certains cas particuliers, de la catégorie d'adhésion.

Au 31 décembre 1966, *six* pays sont en dette vis-à-vis de l'Union (soit au total 20 unités de contribution dues). *Un* pays en revanche s'est déjà acquitté du montant (6 unités) de la cotisation 1967. Le compte d'un autre pays est créditeur de \$140·40.

COMPTE RÉSUMÉ DES RECETTES ET DÉPENSES

(Exprimé en U.S. Dollars)

Pour la période 1964–1966

Recettes

			U.S. Dollars	U.S. Dollars
1. Cotisation des Pays Adhérents	*Nombre d'unités*			
	Dues	*Reçues*		
1963 et antérieures	16	13		
1964	126	119		
1965	152	146		
1966	157	150		
(1967)	0	7		85 341·91
2. Vente des Publications				11 882·78
3. Abonnements aux Publications de l'UAI				4 469·79
4. Agios créditeurs bancaires				
Comptes courants			319·54	
Comptes bancaires			8 513·35	
				8 832·89
5. Subventions de l'UNESCO, accordées par l'intermédiaire de l'ICSU				
Imputation: Symposiums, publications				
1964			14 000·00	
1965			14 000·00	
1966			14 000·00	
				42 000·00
6. Contrats avec l'UNESCO				
Participation à l'organisation du Symposium no. 23			1 154·86	
Participation à l'organisation du Symposium no. 25			4 499·46	
Échange des Astronomes (Commission 38)			2 000·00	
Ecole d'Eté (organisation) (Commission 46)			1 306·00	
Bourse d'étude – 1966 – (Commission 46)			3 000·00	
				11 960·32
7. Divers				
Subvention de l'ICSU pour l'IUCSTR			3 564·40	
Participation de l'URSI à l'organisation du Symposium no. 20			200·00	
Remboursements divers				
– à la Commission 38			3 072·04	
– de la subvention à l'IUCI			180·00	
– du fonds de réserve du Bureau des Télégrammes			2 267·04	
				9 283·48
Total des recettes effectives				$173 771·17
8. Excès des Dépenses sur les Recettes				19 700·84
				$193 472·01

Etat des Comptes Bancaires au 1 janvier 1964		
– en banque (épargne)	62 730·84	
– en banque (comptes courants)	46 798·81	
– en petite caisse	0·37	
		$109 530·02

COMPTE RÉSUMÉ DES RECETTES ET DÉPENSES
(Exprimé en U.S. Dollars)
Pour la période 1964–1966

Dépenses

		U.S. Dollars	U.S. Dollars
1.	Bureau Administratif		30 231·60
2.	Cotisation à l'ICSU		1 490·74
3.	Dépenses de Fonctionnement des Commissions		452·29
4.	Entreprises spécifiques des Commissions approuvées par l'Assemblée Générale		36 514·62
5.	Assemblées Générales – Hambourg, août 1964		
	(*a*) Membres du Comité Exécutif (frais de voyages)	1 714·80	
	(*b*) Présidents des Commissions (frais de voyages)	7 219·41	
	(*c*) jeunes astronomes (frais de voyages)	5 044·71	
	(*d*) divers	1 582·54	
	– Prague, août 1967		
	(*d*) divers (voyage à Prague du Secrétaire Général)	133·26	
			15 694·72
6.	Frais d'impression des Publications éditées par l'UAI		56 712·00
7.	Réunions du Comité Exécutif		
	(*a*) Nice, septembre 1965	3 092·36	
	(*b*) Prague, septembre 1966	2 130·49	
			5 222·85
8.	Organisation des Symposiums et Colloques de l'UAI et Participation à d'autres réunions scientifiques		20 014·30
9.	Publications des Symposiums (lorsque l'éditeur n'est pas l'UAI)		13 088·36
10.	Commissions Interunions – Comités de l'ICSU, etc		7 473·11
11.	Autres frais de représentation		858·00
12.	Entreprises Spécifiques approuvées par le Comité Exécutif		350·00
13.	Programme sous contrat direct avec l'UNESCO		
	(*a*) Commission 38, Echange d'Astronomes	2 000·00	
	(*b*) Commission 46, Organisation des Ecoles d'été	1 306·24	
	(*c*) Commission 46, Bourse d'étude	2 000·00	
			5 306·24
	Total des dépenses effectives		193 408·83
14.	Différence de change dans les virements internes		63·18
		Total	$193 472·01
	Excès des Dépenses sur les Recettes effectives		19 700·84
	Soldes au 31 *décembre* 1966		
	(*a*) existant en banques (épargne)	71 244·19	
	(*b*) existant en banques (comptes courants)	18 518·92	
	(*c*) existant en petite caisse	66·07	
			89 829·18
			$109 530·02

2. *Ventes des publications.* En raison de la complexité des comptes, il est difficile de donner un bilan par ouvrage vendu. Mais nous avons tenté, en un appendice séparé, de présenter un bilan relatif aux opérations (recettes et dépenses) liées aux publications de l'UAI, que l'UAI ait été ou non l'éditeur.

3. *Abonnements aux publications de l'UAI.* Une expérience a été tentée, à partir de l'année 1963: il s'agissait de distribuer les publications de l'UAI contre le versement d'une souscription forfaitaire, à toutes les institutions de la 'liste de distribution'. La difficulté du recouvrement, et la complexité administrative de ces opérations ont conduit à supprimer ces abonnements dès 1966. Voir appendice relatif aux Publications, page lv.

4. *Agios créditeurs bancaires.* Les intérêts sur les comptes d'épargne ont été laissés à ces comptes, et incorporés aux capitaux déposés.

De plus, le compte courant, en francs français, a rapporté des intérêts, comptabilisés séparément.

5. *Subvention de l'UNESCO.* Cette subvention est versée par l'intermédiaire du Conseil International des Unions Scientifiques. Les comptes détaillés relatifs à cette subvention sont adressés chaque année à l'UNESCO. La ventilation des dépenses est laissée à la discrétion de l'UAI pourvu qu'elles restent imputées à des paragraphes budgétaires acceptables par l'UNESCO.

6. *Autres subventions de l'UNESCO.* L'Union a négocié avec l'UNESCO plusieurs contrats séparés, définis par une convention écrite et détaillée. Ces contrats ont été les suivants:

(*a*) Participation à l'organisation du Symposium no. 23, 'Astronomical Observations from Space Vehicles'. (UNESCO/NS/1966/64; 135-5674)
... 'The funds shall be used for meeting all or part of the cost of travel of participants'.

Attribution	\$1500
Sommes dépensées (figurant au budget)	\$1154·86

(le trop perçu a été remboursé à l'UNESCO).

(*b*) Participation à l'organisation du Symposium no. 25, 'The Theory of Orbits in the Solar System and in Stellar Systems'. (UNESCO/NS/1997/64; 135-5675).

Attribution	\$4500
Sommes dépensées (figurant au budget)	\$4499·46 (Somme réellement encaissée par l'UAI)
Somme remboursée à l'UNESCO à l'expiration du contrat	0

(*c*) Participation aux échanges d'astronomes (Commission 38) (UNESCO/NS/2477/65; ED.SS.0085)

Attribution (1965-1966)	\$2000

(Sommes dépensées en totalité. Voir paragraphes 4 et 13 des Dépenses.)

(*d*) Organisation d'une réunion d'experts chargés d'élaborer un programme à long terme d'Ecoles d'Eté Internationales d'Astronomie (Nice, 1965). (Commission 46) (UNESCO/NS/2468/65; EC.SC.0085)

Attribution	\$2000
Dépense réelle (figurant au budget)	\$1306

(L'UNESCO n'a pas déboursé la différence)

(*e*) Organisation de stages d'études individuels (Commission 46) (UNESCO/NS/3034/66; EC.DOC.5555)

Attribution	\$4000
Versé en 1966 (figurant au budget)	\$3000
Dépensé en 1966 (figurant au budget)	\$2000

Le compte sera clos en 1967, conformément au contrat.

7. *Divers.* Dans ce paragraphe on a groupé:

(*a*) les subventions d'organismes divers.

(*i*) Il s'agit de la subvention de l'ICSU pour le fonctionnement de l'IUCSTR. La somme figurant dans les comptes représente le bilan d'un versement de $3500·00, et d'un reversement de $64·40, conformément au compte détaillé ci-après:

Recettes		*Dépenses*	
Subvention	3500·00	Frais de voyages	1360·63
Reversement par le Professeur Allen	64·40	Publication du Xe Rapport de l'IUCSTR	1194·20
		Frais administratifs et de secrétariat	562·55
	$3564·40		$3117·38

La somme de $64·40, reversée à l'UAI à la fin de l'exercice 1966, sera reversée à l'ICSU en 1967, de même que les sommes non dépensées: le total s'en monte à $447·02, montant du trop perçu à rembourser par l'UAI à l'ICSU.

L'IUCSTR ayant été supprimée en 1966, cette somme ne peut en effet être utilisée dans l'avenir.

(*ii*) Il s'agit aussi de la subvention de l'URSI pour l'organisation du Symposium no. 20. L'URSI a également contribué à l'organisation du Symposium UAI-URSI no. 31, mais cette subvention a été directement versée aux organisateurs. Il en est de même pour la subvention de l'IUTAM au Symposium UAI-IUTAM no. 28.

(*b*) Au moment du déplacement du Bureau des Télégrammes de Copenhague à Cambridge (Mass.), il a été remboursé à l'UAI les sommes non dépensées de roulement, constitué antérieurement à 1961.

La subvention de l'UAI à l'IUCI, commission interunion, supprimée en 1966, a été en partie reversée à l'UAI.

Enfin certains bénéficiaires des bourses de voyage de la Commission 38 ont remboursé une partie des fonds reçus, en raison de modifications intérieures dans leurs projets après le versement des sommes attribuées.

Il est clair que, sur les $9283·48 figurant à ce paragraphe, seulement $3700 sont des recettes effectives; le reste correspond à des reversements et devrait en fait être soustrait des dépenses effectives, et des recettes effectives. Il s'agit d'un trop faible pourcentage (2% environ) de ces sommes pour que nous ayons cru devoir compliquer les comptes en les présentant de cette façon, malgré le caractère plus véridique qu'aurait eu cette présentation.

8. *Excès des dépenses sur les recettes.* Cet excès n'a jamais été aussi important, malgré l'accroissement notable des recettes. Nous y reviendrons ci-après, au paragraphe D, page xxxv.

Dépenses

Les numéros de chaque paragraphe correspondent à ceux du budget adopté par l'Assemblée Générale, à Hambourg, en août 1964.

1. *Dépenses du bureau administratif.* Ces dépenses sont constituées uniquement par les salaires du Secrétaire Adjoint et d'une sténodactylo, par les charges sociales diverses y afférentes, par les frais de poste et télégraphe, ainsi que par l'achat de papeterie.

Bien que très réduites (grâce à l'aide fournie par l'Observatoire Royal de Greenwich, puis par les Observatoires de Nice et de Paris, et par l'Académie des Sciences de Tchécoslovaquie: mobilier, téléphone, personnel, locaux, mis sans contrepartie à la disposition de l'UAI) ces dépenses excèdent très légèrement les prévisions budgétaires, mais de façon non significative.

Entreprises Spécifiques des Commissions

Commission no.	Entreprises	Solde 1964	Attribué 1965–1967	Payé en $ 1964	1965	1966	Total $	Solde 1967 $ (*d*)
5	Bibliographie de l'Astronomie		250·00	—	—	—	—	250·00
6	Bureau des Télégrammes		2 000·00	666·66	666·66	666·66	1 999·98	0·02
10	Cartes Synoptiques		2 646·00	882·00	882·00	882·00	2 646·00	—
10	Cartes Héliographiques		980·00	326·66	326·66	326·66	979·98	0·02
16	Centre de Documentation sur les Planètes (Meudon)		2 000·00	—	665·31	665·31	1 330·62	669·38
20	Minor Planet Center		1 500·00	—	—	1 000·00	1 000·00	500·00
23	Carte du Ciel (*a*)	4 365·91	2 000·00	1 476·54	—	3 467·22	4 943·76	1 422·15
27	English Version of Remarks to the Second Supplement of J.C.V.S. (*b*)		700·00	—	—	—	—	—
27	English Version of Remarks to the Second Catalogue of Stars suspected of variability (*b*)		700·00					
27	Southern Variable Stars		1 000·00	—	—	—	—	1 000·00
37	Card Catalogue of Clusters and Associations		1500·00	—	—	—	—	1500·00
38	Exchange of Astronomers (*c*)	8 006·47	—					
	IAU grant 1964	7 500·00	—	—	—	—	—	—
	Reimbursed grants (1964–1966)	3 072·04	—	—	—	—	—	—
		18 578·51	20 000·00 (+2 000·00 contrat UNESCO)	7 200·58	5 200·67	13 213·03	23 614·28 (+2 000·00 contrat UNESCO)	14 964·23
				Ces sommes comprennent les versements effectués à l'aide du contrat UNESCO de $2 000.				
	Total	22 944·42	35 276·00	10 552·44	6 891·30 (+ 850·00 contrat UNESCO)	19 070·88 (+1 150·00 contrat UNESCO)	36 514·62 (+2 000·00 contrat UNESCO)	20 305·80

2. *Cotisation* à *l'ICSU*. L'Union verse annuellement à l'ICSU, pour contribuer à ses dépenses administratives, une somme de 2% des cotisations nationales versées à l'UAI l'année précédente.

3. *Dépenses des Commissions*. Cette somme reste minime. Seules les Commissions 42 et 46 en ont fait usage pendant la période 1961–1964; les pièces justificatives sont à la disposition du Comité des Finances.

4. *Entreprises spécifiques des Commissions approuvées par l'Assemblée Générale*. Les détails relatifs aux projets en question figurent dans le rapport du Comité des Finances à la XIIe Assemblée Générale (*Trans. UAI*, XI B, pour 1964, *Trans. UAI*, XII B, 1966, pour 1965 et 1966).

L'état des dépenses pour ces différentes entreprises est donné dans la liste ci-après (voir tableau page xxx).

Cette liste appelle quelques commentaires:

(*a*) La Carte du Ciel fonctionne avec des fonds de roulement, dont un compte détaillé est tenu par le Président de la Commission 23.

(*b*) Les sommes en question ($1400) attribuées par l'UAI ne seront pas en fait dépensées: elles seront dépensées par l'Académie des Sciences de l'U.R.S.S., en contrepartie d'une portion de dettes diverses de l'U.R.S.S. vis-à-vis de l'UAI.

(*c*) Les sommes relatives à l'Echange des Astronomes proviennent, pour $2000, d'un contrat avec l'UNESCO. Nous avons donc retranché cette somme de la colonne 'total dépensé', correspondant à ce paragraphe du budget.

(*d*) Le solde au 1 janvier 1967, représente les sommes dues aux différentes entreprises; il n'est pas certain que toutes ces sommes seront dépensées; il est clair que certaines d'entre elles correspondent à des erreurs de change; d'autres à des entreprises dont la réalisation a dû prendre un certain retard.

5. *Assemblée Générale, Hambourg, août* 1964. En 1964, les dépenses ont été plus faibles qu'en 1961; il y a à cela une raison principale, à savoir la position géographique de Hambourg, plus centrale que celle de Berkeley. De plus la générosité de nos hôtes, comme dans le passé, a permis de limiter les dépenses administratives sur place. Les sommes indiquées en(*d*), Divers, correspondent principalement aux frais de voyage du personnel du secrétariat de l'UAI ayant assisté le Secrétaire Général et le Secrétaire Général Adjoint.

6. *Frais d'impression des publications éditées par l'UAI*. Il s'agit du paragraphe le plus lourd des dépenses, celui où les dépassements budgétaires sont les plus grands. Un chapitre spécial de ce rapport est consacré au problème des publications. On s'y reportera pour plus de détails.

7. *Réunions du Comité Exécutif*. Les dépenses de la réunion de 1964, à Hambourg, sont comprises dans celles de l'Assemblée Générale, paragraphe 5*a*, ci-dessus.

La somme totale dépensée est faible, grâce au fait que les pays hôtes et les institutions de ces pays ont reçu généreusement le Comité Exécutif, mais aussi aux efforts des membres du Comité Exécutif pour combiner la réunion du Comité Exécutif avec quelque autre réunion scientifique, financée par ailleurs.

8. *Organisation des symposiums et colloques*. La majeure partie des dépenses est consacrée aux frais de voyage de certains participants, choisis par le Comité d'Organisation de chaque symposium, et dans une mesure beaucoup plus faible, à des dépenses administratives. Il est souvent arrivé que les sommes attribuées aux différentes réunions par le Comité Exécutif n'aient pu être dépensées dans leur totalité, certaines subventions ayant souvent été reversées en un moment où leur utilisation n'était plus possible. Le tableau ci-dessous donne les détails des sommes dépensées.

Symposium no.	Sujet, Lieu, Date	Paiements effectués
23	Astronomical Observations from Space Vehicles, Liège, août 1964	$ 1 155·05
24	Spectral Classification of Multi-colour Photometry, Saltsjöbaden, août 1964	2 403·89
25	The Theory of Orbits in the Solar System and in Stellar Systems, Thessaloniki, août 1964	5 459·57
26	Abundance Determination in Stellar Spectra, Utrecht, août 1964	300·00
27	The Construction of Large Telescopes, Tucson (Pasadena), avril 1965	00·00
28	Cosmical Gas Dynamics, Aerodynamical Phenomena in Stellar Atmospheres, Nice, septembre 1965	1 450·00
29	Instability Phenomena in Galaxies, Burakan, mai 1966	1 100·01
30	Determination of Radial Velocities, Toronto, juin 1966	1 500·00
31	Radio Astronomy and the Galactic System, Noordwijk, septembre 1966	1 972·25
		$15 340·77

Le Symposium 27 a pu être organisé grâce à la générosité des organisations invitantes, sans qu'aucune subvention de l'UAI ait été nécessaire. Les Symposiums 23 et 25 ont été financés en presque totalité par des contrats spéciaux avec l'UNESCO.

De plus, les sommes suivantes ont été dépensées pour d'autres réunions:

Colloque UAI sur les Étoiles Variables (Bombay 1965)	$ 500·00
Colloque UAI sur les Atmosphères Stellaires (août 1962)	150·00
Colloque UAI sur les Étoiles de Types Avancés (Trieste)	373·53
Colloque UAI sur le 'blanketing effect' (Heidelberg)	500·00
Colloque UAI sur les Étoiles Doubles (Uccle)	500·00
Colloque UAI sur les Collisions Atomiques (Boulder)	500·00
Symposium IUPHS (Histoire de l'Astronomie)	400·00
Symposium IUTAM-COSPAR-UAI (Trajectories of Artificial Celestial Bodies)	1 000·00
Symposium COSPAR-UAI-URSI-UGGI (Solar Terrestrial Physics, Belgrade, 1966)	750·00
	$ 4 673·53

9. *Publication des symposiums, lorsque l'éditeur n'est pas l'UAI.* L'UAI a décidé d'éditer tous les symposiums à partir du Symposium 24. On se reportera à leur sujet à l'appendice consacré aux Publications de l'Union, page lv.

Ce paragraphe concerne donc uniquement les volumes antérieurs:

Symposium no.	Titre	Editeur	Sommes payées
9	Radio Astronomy	Stanford U. Press	$ 80·28
14	The Moon	Academic Press	12·09
19	Site Testing	*Bulletin Astronomique*	1 648·12

Symposium	Titre	Editeur	Sommes payées
20	The Galaxy and the Magellanic Clouds	Australian Academy of Sciences	2 240·00
21	The System of Astronomical Constants	*Bulletin Astronomique*	1 005·83
22	Stellar and Solar Magnetic Fields	North Holland Pub. Co.	3 000·00
23	Astronomical Observations from Space Vehicles	*Annales d'Astrophysique*	5 102.04
			$13 088·36

En ce qui concerne le Symposium 9, il s'agit du rachat par l'UAI du stock, et, pour le Symposium 14, d'un seul volume.

Pour les Symposiums 19, 21, 23, publiés sans frais dans des revues scientifiques, il s'agit de l'achat des exemplaires que l'UAI distribuait gratuitement (participants et auteurs aux symposiums, Liste de Distribution) avant la décision relative à la limitation de cette distribution. Voir pour plus de détails l'appendice consacré aux Publications de l'UAI (page lv). Il n'est pas à espérer de recettes notables issues des ventes de ces volumes, dans les années à venir.

10. *Commissions interunions, comités de l'ICSU, etc.* Les sommes dépensées peuvent l'être soit sous forme d'une contribution forfaitaire au fonctionnement de l'organisme interunion, soit sous forme de remboursement de frais de voyage aux représentants de l'UAI. Le détail des dépenses a été le suivant:

Dépenses diverses de fonctionnement général pour IUCSTR (voyages des représentants de l'UAI, frais administratifs, impression des rapports IUCSTR) (1964, 1965, 1966)	$3 117·38
Frais de représentation (voyage des représentants de l'UAI) à COSPAR (1966)	1 070·00
Frais de représentation à CIG (pour les années antérieures à 1964)	286·14
Cotisations (forfaitaires) à CIES (1964, 1965, 1966)	599·59
à IUCAF (1964, 1965, 1966)	1 200·00
à IUCI (1964, 1965)	200·00
à IQSY (1963, 1964)	1 000·00
	$7 473·11

Plusieurs remarques doivent être faites:

Le cas de l'IUCSTR est spécial, puisque l'UAI y a joué le rôle d'union-mère. La subvention de l'ICSU a été administrée par l'UAI, et les comptes détaillés en sont les suivants, pour les trois années sur lesquelles portent le rapport.

Subvention accordée		Dépenses	
$3000	pour 1964	1964	$1 500·53
500	pour 1966	1965	772·80
remboursement du trop perçu: $64·40	pour 1966	1966	844·05
			$3 117·38

L'IUCI n'a eu qu'une existence éphémère. Sur les $200 versés par l'UAI, $180 nous ont été remboursés: voir Recettes, paragraphe 7.

Des dettes vis-à-vis du CIG (voyages de délégués) correspondant à l'exercice 1963 ont dû être réglées en 1964. Il en est de même pour la cotisation à l'IQSY de 1963, réglée en 1964; nous noterons qu'elle n'a pas été réclamée en 1965 et 1966, et nous la considérons comme due (soit $1500 à régler en 1967, nos engagements vis-à-vis de l'IQSY étant d'un maximum de cinq années).

11. *Autres frais de représentation.* La somme indiquée a permis au Président d'assister à l'Assemblée Générale du Conseil International des Unions Scientifiques à Bombay, en janvier 1966.

Il est à noter que les paragraphes 10 et 11, qui correspondent à ceux du budget, sont facilement confondus; il est proposé dans l'avenir de limiter le paragraphe 10 aux cotisations forfaitaires, et de faire entrer les frais de déplacements des représentants de l'UAI aux réunions des Commissions interunions dans le paragraphe 11.

12. *Entreprises spécifiques approuvées par le Comité Exécutif.* Un petit nombre d'entreprises ont été approuvées par le Comité Exécutif après l'Assemblée Générale de Hambourg.

Commission 15	*Atlas of Representative Cometary Spectra* (Institut d'Astrophysique de Liège) Cette somme n'a pas encoré été utilisée	$500
Commission 29	*Information Bulletin for the Southern Hemisphere* (Dr Sahade) Cette somme correspond à $700 pour 1965, $700 pour 1966; le Comité Exécutif a accepté la demande de Monsieur Sahade de reporter sur 1966–1967 l'utilisation de cette somme. Toutefois les chèques adressés en 1966 à Monsieur Sahade n'ont pas encore été encaissés par lui au 1 décembre 1966. Seule une somme de $350 a été utilisée.	$1400
Commission 5	*Les Observatoires Astronomiques et les Astronomes* (Observatoire Royal de Belgique) Il s'agit de la réédition, sous forme de feuilles mobiles, du livre de Rigaux. Cette somme n'a pas encore été utilisée.	$500

En résumé, $2400 ont été attribués, mais $2050 restent encore à être versés, au cours de l'année 1967. S'il n'en n'était pas ainsi, faute d'un avancement suffisant des projets, l'Assemblée Générale devrait se prononcer sur la reconduction des attributions proposées par le Comité Exécutif.

13. *Programmes sous contrat direct avec l'UNESCO.* L'Unesco a financé, par l'intermédiaire de contrats directs, 5 opérations (voir Recettes, paragraphe 6). Les dépenses relatives aux Symposiums 23 et 25 figurent au paragraphe 8. Le détail des dépenses relatives aux autres contrats figure dans les rapports adressés à l'Unesco et sont à la disposition du Comité des Finances.

14. *Différences de change dans les virements internes.* Cette somme correspond essentiellement à l'opération de change défavorable qui a accompagné, en 1966, le virement d'une importante partie du compte en livres sterling aux deux autres comptes, en francs et en dollars; ce virement a été effectué en un moment où circulaient en Grande-Bretagne des informations faisant état d'une dévaluation possible de la livre sterling.

Nous remarquerons que, par la suite, les factures en livres sterling ont été payées par l'intermédiaire de notre compte français, à un taux de change lui aussi différent du taux de l'Unesco. Si nous avions conservé le compte en livres, le déficit de change serait donc légèrement différent de celui porté dans les comptes. A ces différences de change, ont été incorporés (sans que la séparation soit possible) les changes bancaires des virements internes.

Solde au 31 *décembre* 1966. L'examen des avoirs bancaires fait apparaître un lourd excès des dépenses de $19 700.84 et un accroissement de $8 513.35 des comptes d'épargne.

Le déficit est considérable, malgré une politique d'économie stricte, particulièrement en ce qui concerne le bureau administratif. Il est essentiellement dû (voir ci-après xxxvii) à l'importance des frais d'impression et d'édition des publications de l'UAI.

C. DÉCLARATION D'EXPERTISE RELATIVE AUX COMPTES 1964–1966

Messieurs,

Conformément à la mission que vous avez bien voulu nous confier, nous vous présentons le résultat de notre examen des comptes de l'Union Astronomique Internationale pour la période du 1.1.64 au 31.12.66.

L'examen a porté, notamment, sur les documents suivants:

- comptes des recettes et dépenses pour l'année 1964 vérifiés par Monsieur G. Chambraud, Expert-Judiciaire à Nice (Alpes-Maritimes);
- comptes des recettes et dépenses pour l'année 1965 vérifiés par Monsieur W. E. Copus, F. C. A.-Chartered Accountant, à Mougins (Alpes-Maritimes);
- comptes des recettes et dépenses pour l'année 1966;
- comptes des recettes et dépenses et situation des comptes bancaires pour la période de trois ans: 1964, 1965 et 1966.

Nous avons également examiné les six comptes séparés des dépenses insérées dans le rapport financier du Comité Exécutif.

Les comptes ont été comparés par Messieurs Chambraud (année 1964) et Copus (année 1965) et par nous-mêmes (année 1966) avec les cahiers de comptes, reçus et pièces comptables diverses.

Nous certifions que ces comptes sont en accord complet avec les informations et explications que nous avons demandées pour nous permettre d'effectuer notre examen.

Le compte, ci-inclus, des recettes et dépenses, résumant les comptes des années 1964, 1965 et 1966, est la représentation exacte des dépenses et recettes de l'Union Astronomique Internationale pour la période du premier janvier 1964 au trente et un décembre 1966.

A Paris, le 8 Mars 1967

ROGER BACLE
H. E. C.—Expert-Comptable
Inscrit au Tableau de l'Ordre
de la Région Parisienne

D. CONSIDÉRATIONS RELATIVES AU BUDGET DE 1967

Afin de préparer le budget des années 1968–1970 et un budget additionnel pour 1967 les tableaux ci-dessous rappellent (sans répéter la définition des différents paragraphes) les prévisions budgétaires antérieures et les différences relevées avec les comptes réels.

Remarques

Les prévisions 1967 ont été faites, compte tenu de la situation difficile des finances de l'Union, en astreignant le budget à des économies sévères sur certains chapitres, comme suit.

Révision des prévisions pour la période 1965-1967

Nouvelles prévisions relatives au Budget 1967 (établies en mars 1967)		Nouvelles prévisions relatives au Budget 1965-1967 (établies en mars 1967)			
Recettes	Dépenses	Recettes	$P_2 - P_1$	Dépenses	$P_2 - P_1$
1. 35 000	1. 8 000	1. 93 309	− 691	1. 30 112	+ 112
2. 12 000	2. 500	2. 20 011	−21 989	2. 1 510	− 490
3. 0	3. 0	3. 1 496	− 8 504	3. 277	− 1 722
4. 3 000	4. 10 000	4. 9 146	+ 1 146	4. 35 962	+ 686
5. 14 000	5. 14 000	5. 42 000	0	5. 14 134	+ 134
6. (pour mémoire)	6. 63 600	6. (6 306 + x)	(+ 6 306 + x)	6. 109 153	+79 153
7. (divers)	7. 0	7. (6 083)	(+ 6 083)	7. 5 223	+ 223
	8. 4 500	8. 10 000	+10 000	8. 14 216	− 2 784
	9. 0			9. 10 768	−39 232
	10. 2 000			10. 6 472	− 38
	11. 0			11. 858	− 142
	12. 2 000			12. 2 350	− 2 650
	13. (pour mémoire; égal au §6 des Dépenses, + 1000)			13. (6 306 + x)	(+6 306 + x)
				14. 56	+ 56
64 000	105 600	190 351 + x		237 397 + x	
8. 10 000					
74 000					

P_1 = Prévisions faites en mars 1967. P_2 = Prévisions faites en août 1964.
(Sommes en dollars U.S.)

Le licenciement d'une personne, en janvier 1967 et jusqu'au transfert du Bureau Administratif à Prague, licenciement rendu possible par l'aide apportée par le Personnel de l'Observatoire de Nice, permet de réduire à $8000 les besoins du Bureau Administratif à la charge de l'UAI. Mais les dépenses réelles sont plus élevées; si l'on voulait assurer l'indépendance du Bureau Administratif, il faudrait prévoir $20 000 à 25 000 par an. Si l'on désirait, de plus, améliorer les conditions de travail du Bureau Administratif, il faudrait 2 ou 3 personnes supplémentaires, et un budget de $35 000 à 40 000 par an.

Les soldes des subventions aux entreprises spécifiques se montent à $20 306. Il est suggéré d'économiser sur ces soldes et de ne dépenser que $4500 pour la Commission 38, à quoi s'ajoutera le montant du contrat de l'UNESCO.

Un emprunt fait à l'ICSU, et remboursable d'ici 1972, de $10 000, a de plus été contracté (Recettes, § 8).

Les estimations relatives aux publications (Recettes, §2 et 3; Dépenses, §6 et 9) résultent des calculs décrits ci-après, dans le chapitre consacré à la politique des publications de l'UAI, pp. liii-lx. Ces chiffres peuvent apparaître catastrophiques. Ils sont fort différents de ceux qu'impliquait le Budget établi au mois d'août 1964: les recettes sont insuffisantes et les dépenses excessives. Les ventes n'ont pas correspondu aux espoirs. L'accroissement du volume de littérature imprimée et le coût de la distribution gratuite des *Draft Reports* et des *Transactions* ont pesé largement sur les dépenses.

Les nouvelles estimations pour 1967 permettent de réévaluer les estimations pour la période de trois ans 1965–1967 et de les comparer avec le Budget. Mis à part les subventions de l'UNESCO, non incluses dans le Budget autorisé, nous noterons que les estimations budgétaires de 1964 relatives aux dépenses du Bureau Administratif, de l'Assemblée Générale, du Comité Exécutif, et aux entreprises spécifiques des Commissions, correspondent presque exactement aux estimations actuelles. Des économies importantes ont été réalisées sur l'organisation des Symposiums, grâce à l'aide des organisations nationales, – et sur les entreprises approuvées par le Comité Exécutif. Mais les publications (§6 + 9) montrent un excès de dépenses de $40 000 par rapport aux prévisions de 1964.

Les estimations budgétaires de 1964 relatives aux recettes correspondent presque exactement aux estimations pour les cotisations. Le surplus de $10 000 au §8 correspond au prêt remboursable de l'ICSU. On notera (§2 + 3) un manque à gagner de $30 000 sur les ventes des publications par rapport aux prévisions. Les recettes diverses (§7) correspondent surtout à des reversements de sommes trop perçues au titre des §4 et 8 des dépenses, qui sont donc en fait bénéficiaires d'une économie supplémentaire correspondant à cette somme.

Le tableau ci-dessous montre alors les éléments du calcul du Budget additionnel demandé en 1967.

Calcul du Budget Additionnel 1967

		Recettes	Dépenses	
1.	Montant réel (1962–63)	$96 184	$87 109	
2.	Montant réel 1964	57 420	61 675	
3.	Total réel 1962–64 (1 + 2)	153 604	148 784	
4.	Disponible fin 1964 (Recettes moins dépenses)			$4 820
5.	Budget estimé 1962–64 (estimation février 1964)	143 384	150 219	
6.	Supplément budgétaire autorisé en août 1964 par l'Assemblée Générale (calculé en comparant recettes et dépenses du Budget estimé)			6 500
7.	Budget 1965–1967 approuvé par la XIIe Assemblée Générale	198 000	197 776	
8.	Report de l'avance budgétaire (4 + 6)	11 320		
9.	Budget autorisé 1965–1967 (non compris contrats UNESCO) (8 + 7)	209 320	197 776	
10.	Total subventions UNESCO sous contrat 1965–1966	6 306	5 306	
11.	Montant réel 1965–1966 (y compris contrats UNESCO)	116 351	131 797	
12.	Estimation 1967, sans les contrats UNESCO (estimation mars 1967)	74 000	105 600	
13.	Estimation 1965–1967 (11 + 12) (y compris contrats UNESCO 1965–1966)	190 351	237 397	
14.	Supplément budgétaire demandé à la XIIIe Assemblée Générale (calculé en comparant recettes et dépenses du Budget estimé)			50 000

C'est donc la somme considérable de $50 000 qui sera proposée au Comité des Finances, comme budget additionnel de l'UAI pour 1967. Bien que cette somme soit légitimée, en importante partie, par le fait que des publications payées en 1967 ne produisent de recettes qu'en 1968, il est clair que nous devons proposer une méthode qui permette de corriger cette ponction (aux détriments des sommes figurant dans les comptes d'épargne) dans la période 1968–1970, faute de quoi la situation deviendra réellement catastrophique.

E. CONCLUSIONS ET BUDGET 1968–1970

Il est donc clair qu'en raison principalement de l'accroissement considérable des charges dues aux publications, des sources nouvelles de revenus doivent être trouvées.

Sans entrer ici dans le détail, nous remarquerons que les recettes sont proportionnelles, à peine, au nombre de Membres (proportionnalité acceptable par les Organisations Adhérentes), les dépenses dues à la distribution gratuite, à tous les Membres, de diverses publications (*Draft Reports*, *Transactions*, volumes B et C, *Bulletins d'Information*) sont approximativement proportionnelles au *carré* du nombre de Membres: le volume de la publication augmente, comme le nombre de destinataires, proportionnellement au nombre de Membres.

Le déséquilibre est donc fatal, et s'accroît nécessairement. Il est dans la logique de la croissance de l'Union. On peut sans doute reprocher au Secrétaire Général actuel, responsable des comptes depuis l'été 1964, de n'avoir pas pris conscience assez tôt de cet état de fait et de n'avoir pas suffisamment éclairé le Comité Exécutif à ce sujet. Mais il était inévitable, un jour ou l'autre, de faire face à un tel déséquilibre.

Une première solution est dans une réduction de la générosité de l'Union. Les Membres de l'UAI, sans verser la moindre cotisation, bénéficient d'avantages nombreux, reçoivent des publications gratuites, utilisent les services des Commissions de l'UAI. On pourrait appliquer strictement les Statuts, exiger une rentabilité des entreprises spécifiques des Commissions, réduire l'aide apportée à l'organisation des Assemblées Générales, etc.

Mais il serait souhaitable de réduire le plus possible les effets de l'austérité: priver d'aide les entreprises spécifiques des Commissions, réduire l'aide apportée aux jeunes astronomes, sont des opérations contraires à l'esprit de l'UAI.

Le Secrétaire Général, au nom du Comité Exécutif, proposera donc de tempérer l'austérité grâce à des ressources nouvelles. Conscient de l'effort qu'il demande aux Organisations adhérentes, mais certain que la vie même de l'UAI étant en cause, ces organisations auront à coeur de comprendre la nécessité de cette action, le Comité Exécutif propose l'augmentation de l'unité de contribution jusqu'à sa valeur maximum tolérée par les Statuts (soit de 600 à 900 Francs-or).

Si cette proposition est adoptée par l'Assemblée Générale, alors un budget raisonnable pourra être maintenu, et le budget additionnel de 1967 compensé en très large partie d'ici 1970.

Si cette proposition est rejetée, alors il faudra que l'UAI en mesure les conséquences. La suppression des publications de l'UAI, celle des Membres individuels, peuvent être nécessaires, dès après le Volume XIII des *Transactions* et la XIIIe Assemblée Générale.

Des budgets précis seront proposés, pour chaque éventualité, au Comité des Finances. Comme base de discussion, on peut partir du Budget suivant pour 1968–1970, établi, en ce qui concerne les recettes, à partir du Budget 1967, où le budget des publications correspond à celui du Rapport II ci-dessous sur la politique des publications de l'UAI (page lv), et où des compressions sur les dépenses ont été opérées (notamment au § 4).

	Recettes		Dépenses
1.	\$105 000 + x ?	1.	\$30 000
2.	73 400	2.	1 500
3.	0	3.	500
4.	8 000	4.	20 000
5.	42 000	5.	14 000
6.	(x)	6.	99 000
7.	?	7.	5 000
		8.	20 000
	\$228 400 + x ?	9.	0
		10.	7 500
		11.	0
		12.	2 000
		13.	(x)
		14.	10 000 (Remboursement à l'ICSU)
		15.	50 000 (Remboursement aux comptes épargnes)
			\$259 500

Il est nécessaire d'épargner $30 000, ou de susciter une recette de cet ordre de grandeur.

Solutions possibles:

(*a*) Augmentation de 600 à 900 Francs-or, de la valeur de l'unité de cotisation, sans changement de catégorie d'adhésion des Organisations Adhérentes.

Ressource supplémentaire: $52 000

(*b*) Cotisation des Membres individuels (mesure à introduire) de $10 par an.

Ressource supplémentaire: $48 000 environ

(*c*) Suppression du *Bulletin d'Information*; remplacement par un bulletin ronéotypé.

Économie, environ: $8000

(*d*) Suppression des publications (y compris les *Transactions*, mais à l'exception du *Bulletin d'Information*, des *Draft Reports* 1967, et et du Symposium 32, antérieurs à l'Assemblée Générale de Prague) (voir ci dessous, p. lv.)

Économie, environ: $30 000

D'autres économies devront être trouvées, si les mesures adoptées par l'Assemblée Générale s'avèrent insuffisantes. Il est suggéré à l'Assemblée Générale d'autoriser le Comité Exécutif à étudier et décider ces nouvelles économies, pour un montant de $50 000, si aucune solution n'est définitivement adoptée à l'issue de l'Assemblée Générale.

APPENDIX I. REPORT ON IAU FINANCES

(Voir texte français, page xxv)

A. INTRODUCTION

In accord with the policy adopted in 1961, the Executive Committee presents to the Members of the Union a single 'Summarized Account of Receipts and Expenditure' covering the period from 1 January 1964 to 31 December 1966. All amounts are given in U.S. dollars converted from other currencies at the following UNESCO conversion rates, treated as exact:

£ sterling: \$2·80   U.S. \$: 4·90 French francs

Certain bank operations, such as transfers from one account to another, reimbursements of excess funds, were effected at current rates of exchange and thus resulted in exchange differences.

The summarized account has been derived from nine (one for each currency and each year) individual accounts. These accounts, which will be available for examination by the Finance Committee, have been audited and certified as correct by the Union's professional accountants. At the end of the present report (page liv) there is printed an auditor's certificate covering the account for the three years under consideration.

The present report includes, appended to the summarized account, explanatory notes to each point concerned; the amounts due at the end of the financial year 1966, or the receipts corresponding to outstanding payments for the period 1961–1964, are given separately for each case. In contrast to 1961–1964, these amounts are not negligible, owing to the important modification in the IAU publication system; this problem is dealt with in more detail in a separate appendix (see page lxiii).

The report ends with a serious warning, the financial situation of the Union being particularly difficult, and with proposals as to the budget for 1967 and the years 1968, 1969, 1970.

B. SUMMARIZED ACCOUNT OF RECEIPTS AND PAYMENTS

(see appended table)

Receipts

1. *Subscriptions of Adhering Countries.* The eleventh General Assembly of the IAU, held in Berkeley, resolved to fix the amount of the unit of contribution at 600 gold francs, the gold franc being equivalent to 0·3266 U.S. dollar. Thus, the unit of contribution is 195·96 U.S. dollars.

Arising from the modifications in the Statutes adopted by the twelfth General Assembly, Adhering Organizations were approached with the request to choose a new category of adherence, in accord with the new article 4 of the Statutes, as published in Volume XII C of the *Transactions*. This appeal led to an increase in the total number of units of contribution which, however, failed fully to live up to the hopes of the Executive Committee in office between 1961 and 1964, the one to propose the modifications adopted.

It should be noted that a special and successful effort was made in 1966 to collect the outstanding contributions and to have the category of adherence raised in certain individual cases.

As on 31 December 1966, six countries still owed contributions to the Union (the units unpaid amounting to 20). On the other hand, one country had already paid its contribution (6 units) for 1967, and the account of another country shows a balance of \$140·40 to its credit.

SUMMARIZED ACCOUNT OF RECEIPTS AND PAYMENTS
(in U.S. Dollars)

for the period 1964–1966

Receipts

			U.S. dollars	U.S. dollars
1. Subscriptions from Adhering Countries				
	Number of units			
	due (outstanding)	*received*		
(1963 and before) ..	16	13		
1964	126	119		
1965	152	146		
1966	157	150		
(1967)	0	7		
				85 341·91
2. Sales of publications (for details see page lxviii)				11 882·78
3. Subscriptions to IAU publications (for details see page xliv)				4 469·79
4. Interest on savings and other accounts				
Current accounts			319·54	
Saving accounts			8 513·35	
				8 832·89
5. Subvention from UNESCO, through ICSU				
1964			14 000·00	
1965			14 000·00	
1966			14 000·00	
				42 000·00
6. Contracts with UNESCO				
Grant towards the organization of Symposium no. 23 ..			1 154·86	
Grant towards the organization of Symposium no. 25 ..			4 499·46	
Exchange of Astronomers (Commission 38)			2 000·00	
Summer School (Organization) (Commission 46)			1 306·00	
Scholarship 1966 (Commission 46)			3 000·00	
				11 960·32
7. Other receipts				
ICSU grant for IUCSTR			3 564·40	
URSI participation in the organization of Symposium no. 20			200·00	
Various refunds				
–to Commission 38			3 072·04	
–of the IUCI grant			180·00	
–of the Telegram Bureau reserve fund			2 267·04	
				9 283·48
Total of effective receipts				$ 173 771·17
8. Excess of Payments over Receipts				19 700·84
				$ 193 472·01
Bank accounts as on 1 January 1964				
in bank (savings)			62 730·84	
in bank (current accounts)			46 798·81	
in petty cash			0·37	
				$ 109 530·02

SUMMARIZED ACCOUNT OF RECEIPTS AND PAYMENTS

(in U.S. Dollars)

for the period 1964–1966

Payments

	U.S. dollars	U.S. dollars
1. Administrative Office		30 231·60
2. Subscription to the ICSU		1 490·74
3. Expenses of Commissions		452·29
4. Specific projects of Commissions approved by the General Assembly (for details see page xlvii)		36 514·62
5. General Assemblies – Hamburg, August 1964		
(*a*) Members of Executive Committee (travel expenses)	1 714·80	
(*b*) Presidents of Commissions (travel expenses)	7 219·41	
(*c*) Young astronomers (travel expenses)	5 044·71	
(*d*) Miscellaneous – Prague, August 1967	1 582·54	
(*d*) Other expenses (Travel expenses of General Secretary to Prague)	133·26	
		15 694·72
6. Printing of publications by the IAU (for details see page xlvii)		56 712·00
7. Meetings of Executive Committee		
(*a*) Nice, September 1965	3 092·36	
(*b*) Prague, September 1966	2 130·49	
		5 222·85
8. Organization of IAU Symposia and Colloquia, and participation in other scientific meetings (for details see page xlvii)		20 014·30
9. Publication of Symposia (if the IAU is not the publisher) (for details see page xlviii)		13 088·36
10. Inter-Union Commissions, Committees on the ICSU, etc. (see separate list on page xlix)		7 473·11
11. Other expenses of representation		858·00
12. Specific projects approved by the Executive Committee (for details see page l)		350·00
13. Programmes under direct UNESCO contract		
(*a*) Commission 38, Exchange of Astronomers	2 000·00	
(*b*) Commission 46, Organization of Summer Schools	1 306·24	
(*c*) Commission 46, Scholarship	2 000·00	
		5 306·24
Total of effective expenses		$ 193 408·83
14. Exchange losses due to internal transfers		63·18
Total		$ 193 472·01
Excess of payments over receipts, carried down		19 700·84
Balances as on 31 December 1966		
(*a*) in banks (savings)	71 244·19	
(*b*) in banks (current accounts)	18 518·92	
(*c*) in petty cash	66·07	
		89 829·18
		$ 109 530·02

2. *Sales of publications.* Owing to the complexity of the accounts it is difficult to give separate statements for each volume sold. However, an attempt has been made in Appendix II to present a balance-sheet of all operations (receipts and payments) in connection with IAU publications, whether or not the Union was financially responsible.

3. *Subscriptions to IAU publications.* A new procedure was introduced in 1963: it consisted in requesting annual subscriptions from all institutions on the Distribution List. The difficulties connected with collecting these subscriptions and the administrative complexity of the operations led to the termination of this system in 1966. See Appendix II on Publications, page lxiii.

4. *Interest on savings accounts.* The interest on the savings accounts was retained in these accounts and has been incorporated into their respective balances.

In addition, the current account in French francs attracted interest, as shown in a separate account.

5. *Subvention from UNESCO.* This subvention is paid through the International Council of Scientific Unions. Detailed accounts of the expenditure from this subvention are submitted to UNESCO each year. The distribution of the subvention is left to the discretion of the Union, provided the individual allocations are within the categories approved by UNESCO.

6. *Contracts with UNESCO.* The Union negotiated a number of separate contracts with UNESCO, of which the details are as follows:

(*a*) Participation in the organization of Symposium no. 23, 'Astronomical Observations from Space Vehicles' (UNESCO/NS/1966/64; 135–5674)... 'The funds shall be used for meeting all or part of the cost of travel of participants'.

Grant	$1500
Expenditure (as appearing in the accounts)	$1154·86

(the balance was reimbursed to UNESCO).

(*b*) Participation in the organization of Symposium no. 25, 'Theory of Orbits in the Solar System and in Stellar Systems' (UNESCO/NS/1997/64; 135–5675).

Grant	$4500
Expenditure (as appearing in the accounts)	$4499·46 (amount actually received by IAU)
Reimbursed to UNESCO	$ 0

(*c*) Contribution to Exchange of Astronomers (Commission 38) (UNESCO/NS/2477/65; ED.SS. 0085)

Grant (1965–1966)	$2000

(Total grant spent. See headings 4 and 13 of Payments).

(*d*) Organization of meeting of experts to work out long-term programme of International Astronomical Summer Schools (Nice, 1965) (Commission 46) (UNESCO/NS/2468/65; EC.SC. 0085)

Grant	$2000
Expenditure	$1306

(UNESCO did not pay the balance).

(*e*) Organization of individual scholarships (Commission 46) (UNESCO/NC/3034/66; EC.DOC. 5555)

Grant	$4000	
Received in 1966	$3000	as appearing in the accounts
Spent in 1966	$2000	

The account will be closed in 1967, according to contract.

7. *Miscellaneous.* This heading includes:

(*a*) Subventions from various organizations.

(*i*) A subvention from ICSU towards the expenses of IUCSTR. The figure in the accounts is the balance of a receipt of $3500·00 and of a reimbursement of $64·40, as follows:

Receipts		*Payments*	
Grant	$3500·00	Travel expenses	$1360·63
Reimbursement by Professor Allen	64·40	Publication of the Xth IUCSTR Report	1194·20
		Administrative and secretariat expenses	562·55
	$3564·40		$3117·38

The amount of $64·40 reimbursed to the IAU, will be returned to ICSU in the course of 1967, in the same way as other unspent sums: the total balance to refund to ICSU is $447·02. IUCSTR having terminated in 1966, this amount cannot be used in the future.

(*ii*) Subvention from URSI towards the organization of Symposium no. 20.

URSI also contributed to the organization of IAU–URSI Symposium no. 31, but this contribution was paid direct to the organizers. The same applies to the IUTAM contribution to IAU–IUTAM Symposium no. 28.

(*b*) When the Telegram Bureau was transferred from Copenhagen to Cambridge (Massachusetts, U.S.A.), the active balance of the funds attributed to it in 1961 was reimbursed to the Union. The IAU subvention to IUCI, an Inter-Union Commission terminated in 1966, was partly reimbursed to the IAU. Some recipients of travel grants made by Commission 38 later returned part of the funds received, as their personal programmes had changed after receipt of the grants.

It should be emphasized that out of the $9283·48 under this heading only $3700 are actual receipts: the remainder corresponds to reimbursements and should be subtracted from both the actual receipts and payments. This amount represents a very small percentage of the budget (about 2%) and it is preferred not to complicate the accounts by presenting them in this way, though this would add to their accuracy.

8. *Excess of Expenditure over Receipts.* The excess has never yet been so high, in spite of the marked increase in receipts. This problem will be dealt with in more detail in paragraph D, page li.

Payments

The headings have been numbered to correspond to those of the Budget approved by the General Assembly in Hamburg in August 1964.

1. *Expenses of the Administrative Office.* These expenses include only the salaries of the Assistant Secretary and of one typist; social security payments; postage and telegraph charges; and purchase of stationery.

Although much smaller than they might otherwise be, thanks to the material aid of the Royal Greenwich Observatory, the Observatoire de Nice and the Czechoslovak Academy of Sciences (in the form of furniture, telephone, personnel, accommodation), the expenses exceed slightly the budgetary estimates, but not significantly to affect the budget.

2. *Subscription to the ICSU.* The Union makes an annual contribution towards the administrative expenses of ICSU of an amount equal to 2% of the income from subscriptions from Adhering Countries for the previous year.

Specific Projects of Commissions

Commission no.	Projects	Balance 1964	Allocated 1965–1967	Paid in $ 1964	Paid in $ 1965	Paid in $ 1966	Total $	Balance 1967 $ (*d*)
5	Bibliography of Astronomy		250·00	—	—	—	—	250·00
6	IAU Telegram Bureau		2 000·00	666·66	666·66	666·66	1 999·98	0·02
10	Cartes Synoptiques		2 646·00	882·00	882·00	882·00	2 646·00	—
10	Cartes Héliographiques		980·00	326·66	326·66	326·66	979·98	0·02
16	Centre de Documentation sur les Planètes (Meudon)		2 000·00	—	665·31	665·31	1 330·62	669·38
20	Minor Planet Center		1 500·00	—	—	1 000·00	1 000·00	500·00
23	Carte du Ciel (*a*)	4 365·91	2 000·00	1 476·54	—	3 467·22	4 943·76	1 422·15
27	English Version of Remarks to the Second Supplement of J.C.V.S. (*b*)		700·00	—	—	—	—	—
27	English Version of Remarks to the Second Catalogue of Stars suspected of variability (*b*)		700·00					
27	Southern Variable Stars		1 000·00	—	—	—	—	1 000·00
37	Card Catalogue of Clusters and Associations		1 500·00	—	—	—	—	1 500·00
38	Exchange of Astronomers (*c*)	8 006·47	—	—	—	—	—	—
	IAU grant 1964	7 500·00	—	—	—	—	—	—
	Reimbursed grants (1964–1966)	3 072·04	—	—	—	—	—	—
		18 578·51	20 000·00 (+2 000·00 UNESCO contract)	7 200·58	5 200·67	13 213·03	23 614·28 (+2 000·00 UNESCO contract)	14 964·23
				these sums include payments from the $2000 UNESCO contract				
	Total	22 944·42	35 276·00	10 552·44	6 891·30 (+ 850·00 UNESCO contract)	19 070·88 (+1 150·00 UNESCO contract)	36 514·62 (+2 000·00 UNESCO contract)	20 305·80

3. *Expenses of Commissions.* This amount is minimum. Grants were only made to Commissions 42 and 46 over the period 1964–1966. The respective documents will be available to the Finance Committee for examination.

4. *Specific Projects approved by the General Assembly.* The details of these projects are given in the Report of the Finance Committee to the twelfth General Assembly (*Transactions IAU*, XI B, for 1964; *Transactions IAU*, XII B, for 1965 and 1966).

Details of actual payments on the different projects are given in the following list (see table appended).

This list requires the following comments:

(*a*) The publication of the Carte du Ciel is subsidized by funds administrated by the President of Commission 23.

(*b*) The IAU allocations concerned ($1400) will actually not be spent: they are partly compensated by sums due to the Union from the Academy of Sciences of the U.S.S.R.

(*c*) $2000 is a grant attributed by UNESCO contract towards the Exchange of Astronomers. This amount has been subtracted from the 'total of payments' corresponding to this heading of the budget.

(*d*) The balance as on 31 December 1966 is made up of credits in favour of different projects. It is not certain whether all these credits will be spent. Some represent exchange differences, others cover projects not yet realized.

5. *General Assembly, Hamburg, August 1964.* Expenditure was less in 1964 than in 1961 owing to the favourable position of Hamburg, more central than that of Berkeley. Moreover, the generosity of the host country permitted us, as in the past, to limit incidental expenses to a minimum. The items under (*d*), Miscellaneous, correspond, in the main, to travel expenses of the Secretarial staff present at the General Assembly and to those of the Assistant General Secretary.

6. *Printing of publications, published at the expense of the IAU.* This is the weightiest heading of the budget, the one in which over-expenditure is most serious. The problem of publication is dealt with in more detail in a special Appendix to the present report (Appendix II).

7. *Meetings of the Executive Committee.* The expenses arising from the 1964 meeting in Hamburg are included in those for the General Assembly, heading 5, as above. Total expenditure is small owing to the generosity of the host countries and their institutions, and to the efforts of the members of the Executive Committee to combine these with other scientific meetings subsidized from other than IAU sources.

8. *Organization of Symposia and Colloquia.* The bulk of the payments consists of travel grants to participants selected by the Organizing Committee of each Symposium; organizational expenses were much smaller. Allocations were frequently not fully spent: individual grants were often returned to the Union too late to be re-allocated. The following table gives the payments in detail.

Symposium no.	Subject, Place Date	Payments made $
23	*Astronomical Observations from Space Vehicles*, Liège, August 1964	1155·05
24	*Spectral Classification of Multi-Colour Photometry*, Saltsjöbaden, August 1964	2403·89
25	*The Theory of Orbits in the Solar System and in Stellar Systems*, Thessaloniki, August 1964	5459·57
26	*Abundance Determination in Stellar Spectra*, Utrecht, August 1964	300·00

continued on next page

Symposium no.	Subject, Place, Date	Payments made $
27	*The Construction of Large Telescopes*, Tucson (Pasadena), April 1965	0
28	*Cosmical Gas Dynamics, Aerodynamical Phenomena in Stellar Atmospheres*, Nice, September 1965	1450·00
29	*Instability Phenomena in Galaxies*, Burakan, May 1966	1100·01
30	*Determination of Radial Velocities*, Toronto, June 1966	1500·00
31	*Radio Astronomy and the Galactic System*, Noordwijk, September 1966	1972·25
		$15 340·77

Owing to the generosity of the host organizations, the Union incurred no expenses in respect of the organization of Symposium no. 27. Symposia nos. 23 and 25 were financed almost completely from separate UNESCO contracts. In addition, the following amounts were spent on other meetings:

IAU Colloquium on *Variable Stars* (Bamberg, 1965)	$500·00
IAU Colloquium on *Late-Type Stars* (Trieste, 1966)	373·53
IAU Colloquium on *The Blanketing Effect* (Heidelberg, 1966)	500·00
IAU Colloquium on *Double Stars* (Uccle, 1966)	500·00
IUPHS Symposium on *The History of Astronomy* (Hamburg, 1964)	400·00
IUTAM–COSPAR–IAU Symposium on *Trajectories of Artificial Celestial Bodies* (Paris, 1965)	1000·00
COSPAR–IAU–URSI–UGGI Symposium on *Solar Terrestrial Physics* (Belgrade, 1966)	750·00
	$4 673·53

9. *Publication of the proceedings of Symposia not published at the expense of the IAU.* The IAU has decided to publish, at its own expense, all symposia from Symposium no. 24 onwards; for details see Appendix II on the Publications of the Union, page lxiii. Thus, the present paragraph deals only with earlier volumes:

Symposium no.	Title	Published by	Payment in $
9	Radio Astronomy	Stanford U. Press	80·28
14	The Moon	Academic Press	12·09
19	Site Testing	*Bulletin Astronomique*	1648·12
20	The Galaxy and the Magellanic Clouds	Australian Academy of Sciences	2240·00
21	The System of Astronomical Constants	*Bulletin Astronomique*	1005·83
22	Stellar and Solar Magnetic Fields	North Holland Pub. Co.	3000·00
23	Astronomical Observations from Space Vehicles	*Annales d'Astrophysique*	5102·04
			$13 088·36

As to Volume no. 9, the IAU has repurchased the stock; in the case of Volume no. 14 a single copy was concerned.

Symposia nos. 19, 21 and 23 were published free of cost to the Union in scientific journals; the Union purchased the copies it required for free distribution to participants, authors and institutions on the Distribution List, prior to the decision to limit this distribution. For more details see Appendix on Publications, on page lxiii. No appreciable revenue may be expected from future sales of these volumes.

10. *Inter-Union Commissions, ICSU Committees, etc.* The payments were made either as contributions to the expenses of the inter-union bodies concerned, or as refunds of travel expenses to IAU representatives. Expenditure was as follows:

Various contributions to the general expenses of IUCSTR (travel of IAU representatives, organizational expenses, printing of IUCSTR Reports 1964, 1965, 1966)	$3117·38
Representational expenses (travel expenses of IAU representatives) to COSPAR (1966)	1070·00
Representational expenses to CIG (for years preceding 1964)	286·14
Contributions to CIES (1964, 1965, 1966)	599·59
Contributions to IUCAF (1964, 1965, 1966)	1200·00
Contributions to IUCI (1964, 1965)	200·00
Contributions to IQSY (1963, 1964)	1000·00
	$7473·11

The following notes should be added:

IUCSTR is a special case, the IAU having been the parent Union. The ICSU subvention was administrated by the IAU, and the respective accounts are as follows for the three years covered by the present report:

Receipts			Payments	
ICSU grant	$3000	for year 1964	1964	$1500·53
	500	for year 1966	1965	772·80
Refund:	64·40	for year 1966	1966	844·05
				$3117·38

The existence of IUCI was short-lived. From the $200 paid by the IAU, $180 was repaid: see Receipts, paragraph 7.

Our debits to CIG (travel grants to delegates) for the year 1963 had to be paid in 1964. The same applies to the contribution to IQSY for 1963, paid in 1964; it should be noted that this contribution has not been claimed for 1965 and 1966, but it is still as a debit (that is $1500 to be paid in 1967, our engagement to IQSY being for a maximum of five years).

11. *Other representational expenses.* The budget provision permitted the President to participate in the General Assembly of the International Council of Scientific Unions in Bombay, in January 1966.

It should be noted that headings 10 and 11, which correspond to those of the budget, are easily confused: it is proposed henceforth to limit heading 10 to contributions, and to include travel expenses of IAU representatives to meetings of Inter-Union Commissions under heading 11.

12. *Specific projects approved by the Executive Committee.* A small number of projects were authorized by the Executive Committee after the General Assembly in Hamburg, as follows:

Commission 15. *Atlas of Representative Cometary Spectra* (Institut d'Astrophysique de Liège) $500

This sum has not yet been used.

Commission 29. *Information Bulletin of the Southern Hemisphere* (Dr Sahade) $1400

This grant is made up of $700 for 1965 and $700 for 1966; the Executive Committee approved Dr Sahade's proposal to carry over this credit to the years 1966–1967. However, the cheques sent to Dr Sahade had not been passed through the bank as on 31 December 1966. The amount actually paid was $350.

Commission 5. *Les Observatoires Astronomiques et les Astronomes* (Observatoire Royal de Belgique) $500

This is a grant towards a new edition, on cards, of the book by Mr Rigaux. The credit is still unspent.

To sum up, $2400 was granted, but $2050 is still to be paid in the course of 1967. If the actual progress of the projects requires a further deferment of payments, the General Assembly will have to reconsider the allocations made by the Executive Committee.

13. *Programmes under direct UNESCO contracts.* The UNESCO financed five projects by direct contracts (see Receipts, heading 6). Expenditure on Symposia 23 and 25 is included under heading 8. Details of payments in respect of the contracts are given in the reports submitted to UNESCO. They are available to the Finance Committee for examination.

14. *Exchange differences in internal transfers.* The item refers to a currency operation in 1966 when a considerable sum was transferred from the account in pounds sterling to two other accounts, that in French francs and that in U.S. dollars. The transfer was made subsequent to rumours in Great Britain indicating an imminent devaluation of the pound.

It should be noted that invoices in pounds had thus to be paid from our French account at a rate of exchange different from that of UNESCO. If the account in pounds had been maintained, the currency losses would have differed slightly from those actually incurred. Incorporated into the exchange differences are also the bank expenses incurred with the internal transfers (their separation was impossible).

Balance as on 31 December 1966. An examination of the bank balances reveals an excess of $19 700·84 in payments over receipts, and an increase of $8513·35 in the savings accounts.

The deficit is appreciable, in spite of a policy of rigid economies, especially as regards the Administrative Office. It is due, in the main (see below), to the high costs of printing and publishing IAU publications.

C. AUDIT OF THE ACCOUNTS 1964–1966

In compliance with the mission with which you kindly entrusted us, we present the result of our audit of the accounts of the International Astronomical Union for the period from 1 January 1964 to 31 December 1966.

We have examined, in particular, the following documents:

- the accounts of receipts and payments for the year 1964, as audited by Mr G. Chambraud, Chartered Accountant, Nice, Alpes-Maritimes;
- the accounts of receipts and payments for the year 1965, as audited by Mr W. E. Copus, F.C.A., Chartered Accountant, Mougins, Alpes-Maritimes;
- the account of receipts and payments for the year 1966;
- the accounts of receipts and payments and the state of the bank accounts for the period of three years 1964, 1965 and 1966.

We have also examined the six separate accounts of payments included in the financial report of the Executive Committee.

The accounts were compared by Messrs Chambraud (year 1964) and Copus (year 1965) and by ourselves (year 1966) with the account books, receipts and other documents produced.

We certify that the accounts are in full accord with the information and explanations we have required for the purpose of our audit.

The enclosed account of receipts and payments, summarizing the accounts for the years 1964, 1965 and 1966, is an accurate representation of the payments and receipts of the International Astronomical Union for the period from 1 January 1964 to 31 December 1966.

ROGER BACLE

Paris,
8 March 1967

H.E.C.—*Chartered Accountant Inscribed in the Tableau de l'Ordre of the Region of Paris*

(*Note*: The original statement by the Chartered Accountant is written in French, and signed)

D. CONSIDERATIONS AS TO THE BUDGET FOR 1967

In order to prepare the budget for the years 1968–1970, and an additional budget for 1967, the tables below summarize (without repeating the titles of the individual headings) the previous budgetary estimates and the differences as compared with the actual accounts.

Survey of the estimates for the period from 1965 *to* 1967

New estimates as to the budget for 1967 (as established in March 1967)		New estimates as to the budget for 1965–1967 (as established in March 1967)			
Income	Expenditure	Income	$E_1 - E_2$	Expenditure	$E_1 - E_2$
1. 35 000	1. 8 000	1. 95 309	− 691	1. 30 112	+ 112
2. 12 000	2. 500	2. 20 011	− 21 989	2. 1 510	− 490
3. 0	3. 0	3. 1 496	− 8 504	3. 277	− 1 722
4. 3 000	4. 10 000	4. 9 146	+ 1 146	4. 35 962	+ 686
5. 14 000	5. 14 000	5. 42 000	+ 0	5. 14 134	+ 134
6. (pro memoria)	6. 63 600	6. (6 306 +x)	(+ 6 306 +x)	6. 109 153	+79 153
7. (miscellaneous)	7. 0	7. (6 083)	(+6 083)	7. 5 223	+ 223
	8. 4 500	8. 10 000	+ 10 000	8. 14 216	− 2 784
	9. 0			9. 10 768	− 39 232
	10. 2 000			10. 6 472	− 38
	11. 0			11. 858	− 142
	12. 2 000			12. 2 350	− 2 650
	13. (pro memoria, equal to heading 6 of income + 1000)			13. (6 306 +x)	(+ 6 309 +x)
				14. 56	+ 56
64 000	105 600	190 351 +x		237 397 +x	
8. 10 000					
74 000					

(All sums in U.S. dollars)

E_1: Estimates made in March 1967
E_2: Estimates made in August 1964

Notes

The estimates for 1967 have been made with due regard to the difficult financial situation of the Union, and impose on the budget rigid economies under certain headings. One employee was discharged in January 1967 and will not be replaced until the Administrative Office moves to Prague. This step could only be taken as a result of the help given by the staff of the Observatoire de Nice. It permitted a reduction of the estimated IAU expenditure on the Administrative Office to \$8000. However, this expenditure would rise to about \$20 000 to \$25 000 a year if the Administrative Office were to be independent of extraneous financial assistance. And if it were desired to improve the working conditions of the Administrative Office, its staff would have to be increased by two or three members, and the budget raised to \$35 000–\$40 000.

The balance in favour of the Specific Projects is \$20 306. It is suggested to economize on this balance by not spending for Commission 38 more than \$4500, in addition to the grant from the UNESCO contract.

ICSU has granted the Union a loan of 10 000 U.S. dollars to be repaid by 1972.

The estimates as to publications (Income, headings 2 and 3, Expenditure, headings 6 and 9) result from calculations given in more detail in the Appendix II, on the IAU publication policy, pages lxiii-lxx.

No matter how well justified by the fact that the publications paid for in 1967 will only sell in 1968, these figures may seem disastrous. They differ widely from those estimated in the budget set up in August 1964: receipts are low, expenditure is high. Sales of IAU publications did not live up to expectation. Expenses are enhanced by the increase in the volume of IAU literature and the costs of the free distribution of the *Transactions*.

The new estimates for 1967 permit a re-evaluation of the estimates for the three-year period 1965–1967 and a comparison with the budget. Disregarding the UNESCO grants, not included in the approved budget, the following should be noted:

The budgetary estimates of 1964 as regards the expenses of the Administrative Office, the General Assembly, the Executive Committee and for the Specific Projects of Commissions, correspond almost exactly to the present estimates. Significant economies were achieved in the organization of Symposia (owing to help of national organizations), and on the projects approved by the Executive Committee. On the other hand, publications (headings 6 and 9) show an excess of \$40 000 in expenditure over the estimates from 1964.

The budgetary estimates established in 1964 of income correspond almost exactly to the estimates of the contribution by Adhering Organizations. The surplus of \$10 000 under heading 8 corresponds to the loan obtained from ICSU. On the other hand, it should be noted (headings 2 and 3) that, in comparison with the estimates, the Union failed to obtain \$30 000 through sales of IAU publications. The miscellaneous receipts (heading 7) correspond mainly to reimbursed grants, made under headings 4 and 8 of Expenditure; thus, these headings include additional savings corresponding to this amount.

The details of the additional budget proposed for 1967 are given in the following table:

Additional Budget for 1967

	Income	Expenditure	
1. Actual amount (1962–63)	\$96 184	\$87 109	
2. Actual amount (1964)	57 420	61 675	
3. Total actual amount 1962–64 (1 + 2)	153 604	148 784	
4. Liquid assets by end 1964 (income minus expenditure)			\$4 820
5. Budget estimate 1962–64 (estimate of February 1964)	143 384	150 219	

	Income	Expenditure	
6. Budgetary supplement authorized by General Assembly in August 1964 (calculated from comparison with income and expenditure of budget estimate)			6 500
7. Budget 1965–1967, as approved by the XIIth General Assembly	198 000	197 776	
8. Surplus, from 4 and 6, not used in 1964	11 320		
9. Budget authorized for 1965–1967 (UNESCO contracts excluded) (8 + 7)	209 320	197 776	
10. Total subventions from UNESCO under contract 1965–66	6 306	5 306	
11. Actual amount 1965–66 (UNESCO contracts included)	116 351	131 797	
12. Estimation 1967, without UNESCO contracts (estimation of March 1967)	74 000	105 600	
13. Estimation 1965–67 (11 + 12) (UNESCO contracts 1965–66 included)	190 351	237 397	
14. Budgetary supplement proposed to the XIIIth General Assembly (calculated from comparison of income and expenditure of the budget estimate)			50 000

Thus, the amount of $50 000 will be proposed to the Finance Committee, as an additional IAU budget for the financial year 1967. It need not be emphasized that proposals will have to be made to remedy this situation (to the detriment of the balance on the savings accounts) in the period 1968–1970, if the situation is not to become catastrophic.

E. CONCLUSION AND BUDGET 1968–1970

The estimated expenses, especially high due to the marked increase in the costs of publication, call for new sources of income.

Without entering into more detail it should be noted that the receipts are proportional, or almost so, to the number of Members (such a proportionality should be acceptable to the Adhering Organizations in computing a raise of their annual contribution to the Union) while the expenses increased by the free distribution to all Members of IAU publications (*Draft Reports*, *Transactions* Volumes XII B and C, *Information Bulletins*), are approximately proportional to the square of the number of Members. The amount published increases (like the number of recipients) in proportion to the number of Members.

Thus, an imbalance is inevitable, and it will become even more serious in the future. This, however, is implied in the logic of the Union's growth. One might reproach the present General Secretary, responsible for the accounts since the summer of 1964, with the fact that he did not realize this state of things early enough and thus could not inform the Executive Committee accordingly.

A first solution is to reduce the generosity of the Union. Members of the IAU, without paying any fee, have numerous advantages, in that they receive free of charge IAU publications, and use the services of IAU Commissions. It would be possible to apply the Statutes rigorously, to require that the specific projects of Commissions be profitable, to reduce the contribution to the organization of General Assemblies, etc.

However, it is desirable to minimize the effects of austerity: it is contrary to the spirit of the IAU not to assist the Special Projects of Commissions or to reduce the help given to young astronomers, etc.

Thus, the General Secretary will propose, on behalf of the Executive Committee, to moderate austerity by finding new sources of financial support. Fully aware of the sacrifices asked for from the Adhering Organizations, but certain that the very existence of the IAU is endangered, and that these organizations will not fail to understand the necessity of the measures to be taken, the Executive Committee proposes an increase in the unit of contribution to the maximum value allowed by the Statutes, that is from 600 to 900 gold francs.

If this proposal is approved by the General Assembly, it will be possible to formulate a reasonable budget and to redress the effect of the additional budget for 1967 by 1970.

If this proposal is rejected, it will be necessary for the IAU to face the consequences. It might become necessary to suppress IAU publications, to suppress the individual membership, etc.

Budgets for both alternatives will be proposed to the Finance Committee. Discussions may be based on the following budget proposed for 1968–1970, in which income has been estimated on the same bases as for 1967 and expenditure on publications has been taken from the report on the IAU publication policy (Appendix II); reductions on certain expenses (§ 4 for example) have been taken into account.

	Income		Expenditure	
1.	\$105 000 + x	1.	\$30 000	
2.	73 400	2.	1 500	
3.	0	3.	500	
4.	8 000	4.	20 000	
5.	42 000	5.	14 000	
6.	(x)	6.	99 000	
7.	?	7.	5 000	
		8.	20 000	
	\$228 400 + x	9.	0	
		10.	7,500	
		11.	0	
		12.	2 000	
		13.	(x)	
		14.	10 000	Re-imbursement to ICSU
		15.	50 000	Re-imbursement to savings accounts
			\$259 500	

It is imperative either to save 30 000 U.S. dollars or to raise additional funds of this order of magnitude.

Possible solutions:

(*a*) Increase the unit of contribution by the Adhering Organizations from 600 to 900 gold francs, without change in the category of adherence.
Additional income for three years about: \$52 000.

(*b*) Contribution of \$10 a year by individual Members (to be introduced).
Additional income for three years about: \$48 000.

(*c*) Suppression of the *Information Bulletin* and its replacement by a mimeographed Bulletin.
Economies of about: \$8 000.

(*d*) Suppression of publications (*Transactions* included, with the exception, however, of the *Information Bulletin, Draft Reports* 1967, and Symposium no. 32, which are to be prepared previous to the General Assembly in Prague).
Savings of about: \$30 000.

Other possibilities of achieving economy will have to be sought if the measures adopted by the General Assembly prove insufficient. It is accordingly proposed to the General Assembly that it authorize the Executive Committee to study possibilities of finding new ways of saving 50 000 U.S. dollars, and to introduce the necessary measures, if no definitive solution is otherwise adopted by the General Assembly itself.

APPENDICE II. LA POLITIQUE DES PUBLICATIONS DE L'UNION ASTRONOMIQUE INTERNATIONALE

(See English version on page lxiii)

1. TRANSACTIONS, SYMPOSIUMS, BULLETINS D'INFORMATION

Toute publication doit être éditée, imprimée, distribuée, et vendue. Ces quatre fonctions essentielles sont accompagnées de fonctions secondaires, par exemple la publicité liée à la distribution.

Les publications de l'UAI (voir Appendice III.C, page lxxii) sont de trois sortes. On exclura de cette analyse celles que l'UAI subventionne à fonds perdus, sans d'ailleurs s'occuper en rien des problèmes techniques de leur édition. Elles figurent le plus souvent parmi les 'Entreprises Spécifiques des Commissions' (au § 4 ou au § 12 du Budget, Dépenses, voir Appendice I, page xxv), ou encore au § 8 du Budget (Dépenses).

(*a*) *Transactions de l'Union*

(*i*) L'UAI a la responsabilité de l'édition, l'édition scientifique proprement dite étant du ressort des Présidents de Commissions, des responsables des Discussions, etc.

Le manuscrit préparé par l'UAI est imprimé par la firme britannique Willmer Brothers, dont les Membres de l'UAI ont pu apprécier la compétence technique remarquable. Les relations entre l'imprimeur et les auteurs (épreuves, commandes de tirages à part) sont entièrement assurées par les services de l'UAI responsables des Publications, et situés à l'Observatoire de Meudon.

La reliure, les couvertures, la publicité, la vente sont du ressort d'Academic Press, avec qui un contrat définit les relations de l'UAI.

Les *Agenda and Draft Reports*, publiés sans reliure, jouent le rôle d'épreuves pour le Volume A des *Transactions*.

Pour la onzième Assemblée Générale, les *Transactions* ont été publiées en 2 volumes, distribués gratuitement.

Pour la douzième Assemblée Générale, les *Transactions* ont été publiées en 3 volumes; seuls les Volumes B et C ont été distribués gratuitement aux Membres de l'Union, le Volume A ayant été constitué essentiellement par la réimpression des *Draft Reports*, et par les Conférences Solennelles; sans reliure, ces derniers volumes ont été distribués gratuitement aux Membres de l'UAI.

Il faut noter que le Volume B a été distribué gratuitement aux Membres anciens (ceux qui faisaient partie de l'Union lors de la douzième Assemblée Générale) cependant que le Volume C a été distribué gratuitement à tous les Membres de l'UAI, y compris ceux élus lors de la douzième Assemblée Générale.

Il est clair qu'au point de vue financier, les *Transactions* sont une lourde charge pour l'UAI, les ventes ne pouvant compenser que de façon infime les dépenses liées à cette considérable distribution gratuite.

(*ii*) Il est donc envisagé de réduire les charges imposées par la publication des *Transactions*, en supprimant la distribution aux Membres des volumes autre que le Volume B, qui seul, d'après les Statuts, doit leur être distribué gratuitement.

Cependant, le volume total imprimé augmente régulièrement proportionnellement au nombre de Membres,—et même plus vite. De plus le coût de la production a augmenté sensiblement. Il faut donc faire les bilans suivants:

Transactions	Pages	Dépenses	Ventes
Volumes XI A, B (y compris les *Draft Reports*)	1112	$21 930·17 (1961–64)	env. $6000* (1961–66)
Volumes XII A, B, C (y compris les *Draft Reports*)	1690	38 378·52 (1964–66) 4 599·63 (1967)	5 533·47* (1966) +4 000 ? (1967)
Volumes XIII A, B, C (y compris les *Draft Reports*)	2000?	15 000·00 (1967) 50 000·00 (1968–69)	0 (1967) 30 000 ? (1968–70)

* à quoi il faut ajouter environ $2500 en tout, provenant de la contribution forfaitaire versée par les Institutions de la Liste de Distribution (1964–66).

Le bilan, au 31 décembre 1966, dans le cadre de la période 1964–66, se traduit par un excès de dépenses de $28 000 environ.

(*b*) *Symposiums de l'Union*

(*i*) Symposiums jusqu'au No. 23 inclus.

L'édition scientifique est toujours du ressort des responsables de l'organisation du symposium, ou plus précisément de l'un d'entre eux, choisi avec l'accord du Comité Exécutif, comme éditeur scientifique. C'est, depuis 1961, le Secrétaire Général Adjoint qui a la responsabilité, pour l'UAI, de mener à bien les négociations préliminaires à ces désignations.

Jusqu'au Symposium No. 23 inclus, l'édition a été la responsabilité entière de diverses maisons d'éditions privées, ou de journaux scientifiques. Des contrats divers, tous différents, ont lié l'UAI à ces éditeurs.

Sans entrer dans les détails, disons que ce système a conduit à des irrégularités graves dans la date de distribution des symposiums, à des présentations et des formats hétérogènes, parfois même à des volumes n'indiquant aucunement la responsabilité de l'UAI dans la publication. Il s'en est suivi des réclamations nombreuses de la part des bibliothécaires de bibliothèques scientifiques, et une forte limitation des ventes, du fait qu'il ne s'agissait pas d'une véritable collection.

Au point de vue financier, le tableau ci-après montre que cette solution, quoique économique, conduisait néanmoins à une dépense notable pour chaque volume.

Symposiums No.	Contribution de l'UAI
12	$2750 (1961)
13	600 (1961)
14	4300 (1962)
15	100 (1962)
16	2600 (1963)
17	0 (1962)
18	1500 (1963)
19	1650 (1965)
20	2250 (1964)
21	1000 (1965)
22	3000 (1965)
23	5100 (1966)

Les ventes n'ont pratiquement rien rapporté. Dans la plupart des cas, l'UAI s'est bornée à acheter les exemplaires nécessaires aux distributions gratuites, sans plus. Dans d'autres cas, un petit bénéfice a été enregistré. Globalement, la vente des Volumes 12 à 23 (bien qu'il soit difficile de donner des chiffres détaillés) a rapporté à l'Union, pour la période 1961–1966, une somme d'environ $8000.

Il est clair que les abonnements forfaitaires souscrits (§ 3 des Recettes) par les institutions figurant sur la Liste de Distribution de l'UAI ne compensent pas ces dépenses.

Abonnements (pendant la période 1961–1966)

1963	$3271·61
1964, 1965	6367·70
1966	30·11
	$9669·42

Ceci correspond approximativement à 215 institutions cotisant en fait, alors que 264 institutions ont reçu les publications; la difficulté du recouvrement des versements forfaitaires est clairement exprimée par ces chiffres.

De plus, il reste des stocks de certains de ces symposiums, dont la vente peut encore rapporter quelques revenus à l'UAI. En toute objectivité, ces revenus doivent être considérés comme négligeables.

(*ii*) Pour ces diverses raisons, ce système a été abandonné, à partir du Symposium No. 24 inclus. Le Comité Exécutif espérait, d'une part, une régularité plus franche des publications, d'autre part, une vente plus efficace.

A partir du Symposium No. 24, le système adopté est devenu identique à celui adopté pour les *Transactions*. L'éditeur est devenu le Service des Publications de l'UAI, sis à Meudon. L'imprimeur a été la firme Willmer Brothers. La firme Academic Press a eu la responsabilité de la reliure, des couvertures, de la publicité, de la distribution, de la vente.

Certains des espoirs fondés sur ce nouveau système ont été en partie déçus. La lenteur de la publication n'a pas été améliorée (ni d'ailleurs aggravée!): seuls sont sortis des presses à ce jour les Symposiums No. 24, 25, 26.

Les ventes ont été meilleures qu'auparavant; mais le progrès n'est guère sensible: il est vrai qu'un stock important reste à vendre, et que les revenus à en attendre sont sans doute, cette fois, non négligeables, et ont été estimés compte tenu de la publication des volumes au cours de l'été 1966 et des stocks existants.

Enfin la qualité de la publication a été unanimement considérée comme très satisfaisante.

Pour ces symposiums, le détail des comptes est comme suit:

Symposium	Impression	Reliure, distribution	Vente en 1966	Produit des ventes futures
24	$5990·25	$738·55	$2411	env. $1000
25	7800·94	728·34	2434	env. 1000
26	6102·81	716·00	2336	env. 1000

Dépenses:	$22 076·89	Recettes:	$10 400
dont	14 112·92 en 1966	dont	8000 (env.) en 1967
Reste dû env.:	8000 à payer en 1967	et	2400 (env.) en 1968
			(y compris la vente des tirages à part).

Il faut noter que $2000 environ viennent s'ajouter aux recettes, provenant de la contribution forfaitaire des Institutions de la Liste de Distribution.

Le bilan (arrondi) au 31 décembre 1966 est, en excès de dépenses: $14 100; mais en 1967 et 1968, il y aura un excès de recettes de $2400 environ. Le bilan total reste nettement déficitaire.

Les sommes indiquées pour les ventes en 1966 sont des valeurs exactes. Mais ces ventes affecteront seulement les comptes de 1967. Il n'est pas facile de comparer ces comptes aux comptes réels; dans certains cas en effet, les sommes provenant des ventes ont été déduites des factures de la firme Academic Press; les nombres figurant aux recettes (2) et aux dépenses (6) sont donc, en réalité l'un et l'autre sous-estimés, mais le bilan en est correct.

Pour les Symposiums 27 à 31, imprimés en 1966 et 1967, nous avons estimé à $7000 en moyenne le coût du symposium, et à $6000 environ les recettes par symposium.

(*iii*) Pour les Symposiums 32 et suivants, une politique différente a été envisagée par le Comité Exécutif. Il a semblé en effet, au vu des résultats de l'expérience commencée avec le Symposium 24, que la faible vente (se traduisant par une lourde insécurité pour l'UAI) était due à deux raisons: la lenteur de la publication, l'excès des distributions gratuites, souvent injustifiées. La suppression de la Liste de Distribution doit aider, en partie seulement, à résoudre ce problème.

La lenteur de la publication des Volumes 24, 25, 26 a été due en partie à un manque de coordination entre l'imprimeur et la maison Academic Press. A la suite de longues discussions, il a été envisagé de passer avec cette firme un contrat valable pour une durée de temps limité, et selon lequel elle prendrait à sa charge la responsabilité complète de l'édition, le service des publications de l'UAI n'intervenant plus que pour assurer les relations des auteurs avec Academic Press. L'UAI recevrait un pourcentage des ventes et ne débourserait en contrepartie que le prix des volumes distribués gratuitement. Les Membres de l'UAI continueraient à pouvoir se procurer à prix réduit les publications de l'Union Astronomique Internationale. Il est prévu, dans des conditions, que la publication de chaque volume de symposium pourrait ne coûter qu'une somme assez faible, de l'ordre de $1000, d'ailleurs compensée, l'année suivant la publication, par des recettes assez faibles, mais du même ordre de grandeur.

En 1967, ces dépenses seront nulles. En 1968, pour les Symposiums 32 à 35, elles seront de l'ordre de $4000. Des dépenses du même ordre doivent être prévues pour les années à venir ($4000 en 1969).

Il est important ici de noter que les publications astronomiques se vendent mal, par rapport, par exemple, aux publications de chimie, ou même de mathématiques. Un éditeur professionnel arrive à une compensation entre différentes sciences; l'UAI, en tant qu'éditeur, est au contraire nécessairement perdante.

Précisons enfin que le présent rapport, et tout particulièrement le présent paragraphe, ne sont que provisoires. Des discussions sont menées actuellement avec Academic Press (et d'autres éditeurs), et un rapport plus actuel sera présenté devant l'Assemblée Générale, au mois d'août 1967.

(*c*) *Bulletin d'Information de l'UAI*

Cette publication, qui compte deux livraisons chaque année, (sans que ce soit une règle!), est éditée et distribuée par le Bureau Administratif; elle est imprimée à Nice par la maison Pierotti depuis 1965, et distribuée gratuitement. Il semble souhaitable de maintenir le système actuel. Le bilan est le suivant depuis 1965 (antérieurement à cette date, les dépenses correspondantes ont été incluses dans le § 1).

	1965	$2146·89	(No. 13 et 14)
	1966	$1226·72	(No. 15 et 16)
Prévisions:			
	1967	Paiement du No. 17 publié en 1966, plus deux livraisons environ	$3000
	1968–1969	4 livraisons environ	$6000

La somme relative à 1966 est faible, car à partir du No. 15, l'imprimeur n'a plus été chargé de l'expédition qui a été faite par le Bureau Administratif; les frais de poste ont donc été, en 1966, contrairement à ce qui s'est passé en 1965, inclus dans le § 1 des Dépenses.

Les prévisions tiennent compte de l'augmentation du nombre des Membres de l'UAI, mais supposent que le prix des travaux d'impression effectués en Tchécoslovaquie restera le même que celui des travaux effectués en France.

2. LES TIRAGES À PART

Il s'agit essentiellement de la fabrication des tirages à part compensés partiellement par leur vente (un certain nombre de tirages à part étaient édités pour les besoin du Bureau Administratif).

Les sommes dépensées dans ce but sont les suivantes:

		Dépenses			Récupération		
Transactions	XII A	£147	10	0	£0	0	0
	XII B	1042	3	2	805	17	7
	XII C	64	18	0	0	0	0
Symposium	24	519	11	6	627	10	3
	25	327	18	3	453	9	3
	26	453	4	9	592	11	9
Struve Requiem		9	5	0	0	0	0
	Total	£2564	10	8	£2479	8	10

Les sommes se compensent à peu près. On notera que les sommes payées par les auteurs sont légèrement, d'après les tarifs utilisés (tarifs moyens), supérieures au prix coûtant.

Nous devons noter que le système de commande des tirages à part a été grandement compliqué par une distinction entre 'offprints' et 'reprints'. Malheureusement, cette distinction, de nature technique, ne permet pas toujours facilement de prévoir si tel article pourra être tiré à part comme 'offprint' ou comme 'reprint'; et les prix en sont fort différents! Le secrétariat de l'UAI a essayé d'établir deux échelles différentes de tarifs; ces tarifs n'ont pu être appliqués de façon stricte; d'où les différences entre les débours et les récupérations.

Par ailleurs, il est souvent clair que des distributions excessives de tirés à part, si elles permettent d'aider les bibliothèques de petits établissements, ou des chercheurs isolés, sont une lourde charge pour l'UAI, en limitant la vente des volumes, et en imposant un travail considérable au Bureau Administratif.

Pour cette raison, les principes suivants seront appliqués, par décision du Comité Exécutif, à partir du Symposium 27 et du Volume XIII des *Transactions*.

(*a*) Un seul type de tirages à part est utilisé.

Il s'agira de documents sans couverture, sans repagination, et sans surcharges imprimées (ce qu'en anglais on désigne par 'offprints' par opposition avec les 'reprints' qui sont repaginés et même recomposés parfois en partie.)

(*b*) Chaque Président de Commission recevra gratuitement 25 exemplaires du Rapport de sa Commission (Volume A) et du compte-rendu des séances (Volume B). Cette quantité devrait suffire, les Membres de l'UAI recevant tous ces textes par ailleurs.

(*c*) Les Présidents des Discussions Communes recevront gratuitement 50 exemplaires de l'ensemble de la Discussion qu'ils auront présidée. Tous les Membres de l'UAI reçoivent ces textes dans le Volume B des *Transactions*.

(*d*) Les auteurs des Conférences Solennelles recevront gratuitement 25 exemplaires de leur Conférence; tous les Membres de l'UAI recevront ces textes en accompagnement du Volume B (ou C dans le cas du Volume XII) des *Transactions*.

(*e*) Les auteurs de communications aux symposiums recevront autant d'exemplaires qu'ils le désireront de leurs communications; les prix leur seront communiqués avec le bulletin de commande qui leur sera adressé au moment de la correction des épreuves. Les discussions ayant eu lieu pendant les symposiums ne feront pas l'objet de tirages à part.

(*f*) Aucune autre des publications de l'UAI (et notamment pas le Volume XII C) ne sera l'objet d'une édition de tirages à part.

3. DISTRIBUTION ET VENTE DES PUBLICATIONS DE L'UAI

La politique traditionnelle de l'UAI était une politique de générosité. Quoi en effet de plus généreux que la distribution gratuite des *Draft Reports* et des *Transactions*, à tous les Membres de l'UAI, – sans aucune contrepartie?

De plus, toutes les publications de l'UAI sont vendues à ses Membres à prix réduits depuis des années.

Le *Bulletin d'Information*, depuis sa création, est aussi distribué gratuitement.

En 1962, il a paru souhaitable de conserver le bénéfice de cette générosité aux institutions dont la liste constitue la Liste de Distribution, mais en la tempérant du paiement forfaitaire d'un abonnement très avantageux aux souscripteurs. Cette Liste de Distribution, en 1966, comprenait 264 institutions; environ 20% de ces institutions ne versaient pas leurs contributions, malgré des rappels fréquents. De plus, il était clair que la liste elle-même n'était pas réellement à jour, ni représentative de l'état mondial des bibliothèques astronomiques. Enfin, le Bureau Administratif devait faire face à une administration très lourde, en raison même des carences de nombreuses institutions de la liste, et de son caractère toujours provisoire.

Enfin, les charges croissantes de l'Union (en conclusion du rapport principal, on trouvera des remarques générales sur ce thème) ont impressionné le Comité Exécutif. Malgré un évident serrement de coeur à l'idée de renoncer, même partiellement à une tradition de générosité, le Comité Exécutif a donc décidé une diminution massive des dépenses liées à la distribution, devenue trop libérale, des publications de l'Union, et ce de trois façons.

(*a*) *Liste de distribution*

Une première mesure est la suppression des avantages des institutions de la Liste de Distribution. Les institutions intéressées, prévenues par circulaire de cette décision regrettable mais nécessaire, se procureront les publications de l'UAI par les moyens commerciaux normaux, à partir du Symposium 27 et du Volume XIII des *Transactions*. Elles continueront à recevoir les *Bulletins d'Information* de l'UAI, mais ne souscriront aucun abonnement.

Cependant 21 institutions situées dans des pays en voie de développement ont été exclues de cette mesure. Une Liste de Distribution réduite a été établie. Les institutions de cette liste reçoivent désormais (sans avoir besoin de souscrire un abonnement), toutes les publications de l'UAI: il a semblé en effet au Comité Exécutif que les ressources des différents pays varient beaucoup d'un pays à l'autre, et que celles de cette liste réduite avaient des difficultés particulières.

Enfin, certaines institutions, appartenant à des pays où des contrôles stricts sont appliqués aux changes, ont des difficultés à souscrire des abonnements, comme à utiliser les moyens commerciaux internationaux normaux, sans pour autant manquer de ressources. Pour ces pays (et d'ailleurs tous les autres), le Comité Exécutif a proposé aux Comités Nationaux d'Astronomie respectifs le paiement d'une souscription forfaitaire (s'ajoutant à la cotisation à l'UAI) pour toutes les institutions de la liste de distribution établie par les comités nationaux intéressés. Ces institutions sont ajoutées à la Liste de Distribution réduite. Elles ne sont actuellement encore qu'au nombre de 7. Mais tous les pays intéressés par ce nouveau système n'ont pas encore pris position à cet égard.

Enfin la Liste de Distribution comprend les membres du Comité Exécutif, et les conseillers (11 personnes), l'UNESCO (10 exemplaires) et certaines institutions internationales comme l'ICSU. La Liste de Distribution contient ainsi environ 100 institutions.

Il est bon de garder en mémoire que le système de publication proposé, à partir du Symposium 32, qui consiste à faire appel aux services d'un éditeur professionnel, aura pour conséquence de contraindre l'UAI à acheter les volumes distribués aux institutions de la Liste de Distribution réduite.

Il a semblé au Comité Exécutif que la nouvelle politique de distribution, nettement plus économique que l'ancienne, d'une administration plus simple, était également plus juste que l'ancienne, dans la mesure où elle favorise (plus qu'auparavant d'ailleurs) les seules institutions ayant réellement des difficultés. Certaines injustices subsistent nécessairement (certaines institutions de pays développés peuvent n'avoir que de très faibles ressources), mais le Comité Exécutif a le sentiment que la nouvelle politique de distribution, d'une impérieuse nécessité, était aussi la moins mauvaise solution possible.

(*b*) *Membres de l'UAI*

L'article 24 du Règlement de l'UAI prévoit que:

'Le volume des *Transactions* de l'Union contenant les comptes rendus de l'Assemblée Générale est fourni gratuitement:

(*a*) à tous ceux qui, à l'époque de la réunion de l'Assemblée Générale, étaient Membres de l'Union, y compris les Membres cooptés au titre de l'article 2 du présent Règlement;

(*b*) aux institutions dont la liste est arrêtée par le Comité Exécutif;

(*c*) aux Organismes ou Comités Nationaux adhérents.'

Ceci veut dire que seul le Volume B des *Transactions*, sous leur forme actuelle, doit être distribué gratuitement aux Membres. Aussi le Comité Exécutif a-t-il dû, à regret, prendre la décision de limiter effectivement cette distribution. Le seul Volume XIII B sera certainement distribué gratuitement; rappelons que déjà le Volume XII A avait été vendu et exclu de la distribution gratuite aux Membres de l'Union Astronomique Internationale. Evidemment, les *Draft Reports* (qui essentiellement constituent les épreuves du Volume A) continueront à être distribués gratuitement; ceci est essentiel à la préparation de l'Assemblée Générale de l'UAI.

(*c*) *Participants aux Symposiums de l'UAI*

Enfin le Comité Exécutif a décidé de ne plus distribuer les volumes des symposiums aux participants, souvent très nombreux, des symposiums de l'UAI, mais seulement à un nombre très réduit d'entre eux, une dizaine environ. Cette décision est évidemment très regrettable: mais là encore, il s'agit d'une nécessité qui a paru impérieuse au Comité Exécutif.

4. PRÉVISIONS FINANCIÈRES RELATIVES AUX PUBLICATIONS

En rassemblant les chiffres ci-dessus, il faut prévoir pour les années 1967 et suivantes, les Dépenses et Recettes suivantes relatives aux publications.

		Budget des publications		Excès de
Année		Recettes	Dépenses	Dépenses
1967	Transactions XII A, B, C	$4 000	$4 600	
	Draft Reports and Transactions XIII A	0	15 000	
	Symposiums 24, 25, 26	8 000	8 000	
	Symposiums 27 à 31	0	33 000	
	Bulletins d'Information	0	3 000	
		12 000	63 600	+$51 600
1968	Transactions XIII A, B, C	26 000	50 000	
	Symposiums 24, 25, 26	2 400	0	
	Symposiums 27 à 31	30 000	2 000	
	Symposiums 32 à 36	0	4 000	
	Bulletins d'Information	0	3 000	+ 600
1969	Transactions XIII A, B, C	2 000	0	
	4 Symposiums	4 000	4 000	
	Bulletins d'Information	0	3 000	+1 000
1970	Transactions XIII A, B, C	2 000	0	
	Draft Reports and Transactions XIV A	3 000	25 000	
	4 Symposiums	4 000	4 000	
	Bulletins d'Information	0	4 000	+24 000
		$73 400	$99 000	+$25 600

Il est clair que ce type d'estimations est hautement imprécis, et qu'elles peuvent être aussi bien trop optimistes que trop pessimistes.

Ces estimations ont servi de base au Budget qui sera proposé au Comité des Finances, lors de la treizième Assemblée Générale, et à l'Assemblée Générale elle-mème (voir, ci-dessus pages xxxv et xxxviii).

NOTE AJOUTÉE AUX ÉPREUVES

Une réestimation de certaines des valeurs ci-dessus a été faite par le Secrétaire Général et le Secrétaire Général Adjoint avant la réunion de Prague et a servi de base aux discussions du Comité des Finances. Les nouvelles estimations sont incorporées dans le rapport du Comité des Finances, volume XIII B.

APPENDIX II. PUBLICATION POLICY

(Voir texte français, page lv)

I. TRANSACTIONS, SYMPOSIA, INFORMATION BULLETINS

All IAU publications have to be edited, printed, distributed, and sold. These four principal operations are accompanied by secondary ones, such as publicity.

The IAU publications may be divided into three groups, excluding those subsidized by the Union on a non-profit basis along with the technical problems to which they lead. Payments in respect of the latter are normally included among the 'Specific Projects of Commissions' (headings 4 and 12 of the Budget and Accounts), or in heading 8.

(a) *Transactions of the IAU*

(*i*) The IAU is responsible for the editing of its publications, the scientific editing in the proper sense of the word being the province of the Presidents of Commissions, organizers of Discussions, etc.

The manuscript, sub-edited by the IAU, is printed by the British printers Willmer Brothers, whose outstanding achievements are generally appreciated by Members. The contact between printer and authors (proofs, orders for reprints) is maintained through the IAU Publication Service in Meudon.

Binding, covers, publicity and sales are looked after by Academic Press, with whom the Union has a contract.

The *Agenda and Draft Reports*, published with paper covers, are simultaneously the page proofs of Volume A of the *Transactions*.

The *Transactions* of the eleventh General Assembly were published in two volumes, distributed free of charge to Members of the Union.

The *Transactions* of the twelfth General Assembly were published in three volumes, of which Volumes B and C were distributed free of charge to Members of the Union. Volume A included the re-imposed Draft Reports and the Invited Discourses of which reprints, unbound and without covers, were also distributed to Members free of charge.

It should be noted that Volume B was distributed free only to those Members who belonged to the Union at the time of the twelfth General Assembly, while Volume C was distributed also to new Members, that is those admitted at that General Assembly.

From the economic standpoint, the *Transactions* are obviously a heavy burden on IAU finances, sales making up for only an insignificant part of the large-scale courtesy distribution.

(*ii*) Thus, steps have already been taken to reduce the expenses of the publication of the *Transactions* by discontinuing the free distribution of volumes other than Volume B, the only one required by the Statutes for such distribution.

It should be emphasized that the amount of IAU printed matter increases in proportion to the number of Members, if not faster, and that the rise in production costs is considerable. This gives the following balance:

Transactions	Pages	Cost (in $)	Sales (in $)
Volumes XI A, B (*Draft Reports* included)	1112	21 930·17 (1961–64)	Approx. 6 000* (1961–66)
Volumes XII A, B, C (*Draft Reports* included)	1690	38 378·52 (1964–66) 4 599·63 (1967)	5 533·47*(1966) +4 000 (1967) ?
Volumes XIII A, B, C (*Draft Reports* included)	2000?	15 000·00 (1967) ? 50 000·00 (1968–69) ?	0 30 000 (1968–70) ?

* to which one has to add about $2500, coming from the forfaitary contribution paid by the Institutions on the Distribution List (1964–66).

The balance on 31 December 1966 for the period from 1964 to 1966 shows an excess of expenditures of about $28 000.

(*b*) *IAU Symposia*

(*i*) Symposia up to and including no. 23.

Scientific editing is the responsibility of the organizers of the particular symposium or, more precisely, of one of them, selected with the approval of the Executive Committee as Scientific Editor. Since 1961, the arrangements for the designation of Scientific Editors have been made by the Assistant General Secretary.

Up to and including Symposium no. 23, publication was the exclusive responsibility of different private publishers or scientific journals. The contracts of the Union with such publishers varied widely in scope.

Without going into more details, it should be said that this system led to serious trouble: the distribution dates of symposium volumes were irregular, their presentation and format were non-uniform, IAU prerogatives were not mentioned on them, etc. Complaints by librarians and scientific libraries were numerous, and sales diminished as the symposium volumes formed no uniform series in the proper sense of the word.

The following table shows that, no matter how economical, this system resulted in considerable expenses for each volume.

Symposium no.	IAU contribution
12	$2750 (1961)
13	600 (1961)
14	4300 (1962)
15	100 (1962)
16	2600 (1963)
17	0 (1962)
18	1500 (1963)
19	1650 (1965)
20	2250 (1964)
21	1000 (1965)
22	3000 (1965)
23	5100 (1966)

Sales brought virtually no financial return to the IAU. In most cases they were limited to the purchase made by the Union for free distribution, benefits going to the publisher. In other cases, the records show some very small benefit. Altogether, the income from the sales (although difficult to estimate in detail) brought to the IAU, for the period 1961–66, about $8000.

Subscriptions (heading 3 of Receipts) by institutions on the Distribution List were obviously insufficient to balance expenditure.

Subscriptions from 1961 to 1966

1963	$3271·61
1964, 1965	6367·70
1966	30·11
	$9669·42

This corresponds to about 215 subscribers, while publications were distributed to 264 institutions. The figures are illustrative of the difficulty of collecting outstanding subscriptions.

A number of symposium volumes still in stock may represent some future income to the Union, but probably quite insignificant.

(*ii*) The above considerations called for suppression of the existing system from and including Symposium no. 24. The Executive Committee had hoped for more regularity in production and more efficient sales.

The system adopted for and from Symposium no. 24 onwards is the same as for the *Transactions*. Editing has become an IAU service, and is done at Meudon. The volumes have been printed by Willmer Brothers. Academic Press has been responsible for binding, covers, publicity, distribution and sales.

Many a hope based on this new system was frustrated. Publication was not speeded up (nor slackened, either): so far only Symposia nos. 24, 25 and 26 have come from the press.

Sales were better than previously, but the improvement is not very marked. On the other hand, there still remains a stock of volumes to be sold, and the income may be appreciable. It has been estimated from the sales effected during the summer of 1966.

The quality of the publications has unanimously been considered very satisfactory.

For these Symposia, the following details are given:

Symposium no.	Printing $	Binding, Distribution $	Sales in 1966 $	Future sales $
24	5990·25	738·55	2411	About 1000
25	7800·94	728·34	2434	About 1000
26	6102·81	716·00	2336	About 1000

Payments:	22 076·89	Receipts:	10 400
out of which	14 112·92 paid in 1966	out of which about	8 000 in 1967
and about	8 000·00 to be paid in 1967	and	2 400 in 1968
			(sales of reprints included)

It should be noted that about $2000 must be added to the product of sales (1964–65) coming from the forfaitary contribution of the Institutions on the Distribution List.

The balance (rounded off) as on 31 December 1966 shows an excess of $14 100 over revenue. However, it is expected that the years 1967 and 1968 will bring an excess of $2400 over expenditure. Nevertheless, the overall balance shows a large deficit.

The figures given for the sales in 1966 are exact. But they will only be credited to the accounts in 1967.

It is not easy to compare these estimates with the actual accounts, the amounts resulting from sales having, in certain cases, been subtracted from the invoices of Academic Press; thus the figures under heading 2 of Receipts and under heading 8 of Payments are actually underestimates, but the balance is correct.

The estimates for Symposiums volume 27 to 31, printed in 1966 or 1967, are about $7000 for production and $6000 as revenue, as an average for each volume.

(*iii*) As to Symposium volumes from and including no. 32 onwards, the Executive Committee is examining the possibility of introducing a new policy.

Experience with Symposium no. 24 showed that sales are insufficient (adding to the insecurity of the IAU) owing to two facts: the long publication delay and the too liberal courtesy distribution, often unjustified. The abolition of the Distribution List may be a partial solution of the problem.

The long delay in the publication of Symposia nos. 24, 25, and 26 was due, in part, to insufficient coordination between printers and publishers. After lengthy discussions it has been considered to negotiate a contract of limited duration with Academic Press by which it would be fully responsible for publication, so that the Union's only responsibility would consist in arranging contact between authors and publishers. The Union would receive part of the revenue from sales, but would pay for the volumes distributed free. Members would be entitled to purchase IAU publications at reduced prices. It is expected that publication of each Symposium volume would not cost the IAU more than $1000 in such conditions, compensated, the year following publication, by an income of the same order of magnitude, though not very large.

In 1967, this expenditure will be zero. In 1968, for Symposia nos. 32 to 35, it will be of the order of $4000. Payments of the same order have to be expected for the years to come ($4000 in 1969).

In this connection it should be noted that astronomical publications have a thin market in comparison with those of other branches, such as chemistry or even mathematics. A professional publisher may balance his losses in one branch by gains in another, but the Union may only lose as its own publisher.

The present report, and especially the above paragraph, are only tentative. Discussions now under way with Academic Press and other publishers will permit the Executive Committee to prepare a more definitive report before the General Assembly in August 1967.

(*c*) *IAU Information Bulletin*

This publication, which appears twice a year (not, however, compulsorily), is edited and distributed by the Administrative Office. It has been printed by Messrs. Pierotti, Nice, since 1965. It is distributed free of charge. The present system is satisfactory and should be maintained. The balance for 1965 and 1966 is as follows (prior to 1965, the corresponding expenditure was included under heading 1):

1965	$2146·89 (nos. 13 and 14)
1966	1226·72 (nos. 15 and 16)

Estimates:

1967	no. 17 printed in 1966 and about two further editions	$3000
1968–1969	about 4 editions	$6000

The total paid in 1966 is reasonably low, the distribution having been made by the Administrative Office from no. 15 onwards, not by the printer. However, postage for 1966 has been included under heading 1 of Payments, not heading 8 as in 1965.

The estimates have taken into account the anticipated increase in the number of IAU Members, and assume that the costs of printing in Czechoslovakia will be the same as in France.

2. OFFPRINTS AND REPRINTS

The production of offprints and reprints is partly compensated by their sale (a number of offprints is ordered for the use of the Administrative Office).

The balance in this respect is as follows:

	Payments			Receipts		
Transactions XII A	£147	10	0	£0	0	0
XII B	1042	3	2	805	17	7
XII C	64	18	0	0	0	0
Symposium no. 24	519	11	6	627	10	3
25	327	18	3	453	9	3
26	453	4	9	592	11	9
Struve Requiem	9	5	0	0	0	0
Total	£2564	10	8	£2479	8	10

Payments and receipts are virtually in balance. It is seen that the prices, as given in the price lists and paid by the authors, are, on the average, slightly above production costs.

The system of ordering and selling prints has been much complicated by the distinction between 'offprints' and 'reprints'. Unfortunately, this distinction, of a purely technical character, does not always indicate whether a certain paper is better printed as an 'offprint' or a 'reprint'; and the price difference is appreciable. The Secretariat undertook to establish two different price lists; these lists, however, could not be applied strictly, and hence the difference between outlay and revenue.

It is certain that excessive distribution of offprints, though welcomed by minor libraries and isolated scientists, is a heavy burden to the Union. Moreover, it reduces the volume of sales and gives much trouble to the Administrative Office.

This prompted the Executive Committee to adopt the following principles for and from Symposium 27 and *Transactions* XIII onwards:

(*a*) Offprints are standardized.
Henceforward only offprints will be used, with no cover, without re-pagination and without added material (in contradistinction to reprints, which are re-paginated and sometimes partly re-composed).

(*b*) Each Commission President will obtain, without charge, 25 copies of the Report of his Commission (Volume A) and of the proceedings of its meetings (Volume B). This number may be sufficient, the Members of the Union receiving all this material, either in advance or in Volume B.

(*c*) The Presidents of Joint Discussions will obtain free of charge 50 copies of the complete Discussion over which they presided. All IAU Members will receive these texts in Volume B of the *Transactions*.

(*d*) The authors of the Invited Discourses will obtain free of charge 25 copies of their discourses. All IAU Members will receive these texts simultaneously with Volume B (or C in the case of *Transactions* XII) of the *Transactions*.

(*e*) The authors of contributions to symposia may obtain any desired number of copies of their contributions: the respective prices will be marked on the order sheet they will receive with the proofs. No offprints will be made of the discussions held during symposia.

(*f*) IAU publications other than those mentioned above, will not be available as offprints (especially not Volume C of *Transactions*).

3. DISTRIBUTION AND SALES OF IAU PUBLICATIONS

The traditional policy of the IAU has been that of generosity. For, actually, what is more generous than to distribute the *Draft Reports* and *Transactions* to all Members without reciprocity?

Moreover, all IAU publications have been sold to Members at reduced prices for years. Ever since it started, the *Information Bulletin* has also been distributed free of charge.

In 1962 it appeared reasonable to maintain the advantage of this generosity as regards the institutions which make up the IAU Distribution List, but to moderate it by introducing a subscription system, very favourable for the subscribers. The Distribution List included 264 subscribers in 1966, of which about 20% failed to pay their fees, in spite of numerous reminders. In addition, it became evident that the list was not up to date and did not represent that state of astronomical libraries the world over. And, eventually, the Administrative Office was faced with difficult administrative problems, for many institutions on the list did not pay, and the list itself was changing continuously.

The ever-increasing responsibilities of the Union (for general notes, see end of the main report) led the Executive Committee to reconsider the position. Although completely reluctant to renounce the tradition of generosity, the Executive Committee resolved to reduce the expenditure incurred through publication in three ways.

(*a*) *Distribution List*

The first step consisted in abolishing the advantages of the institutions on the Distribution List. The institutions concerned, advised of this regrettable but necessary step by circular letters, will have to purchase IAU publications through normal commercial channels, as from Symposium 27 and *Transactions* XIII onwards; they will continue receiving the *IAU Information Bulletin* free of charge.

However, 21 institutions in developing countries were exempt from this measure. A new, curtailed Distribution List was prepared. The institutions of this list will obtain all IAU publications without any subscription fee: the Executive Committee is fully aware that resources vary widely between countries, and that the institutions on the curtailed list have especially great difficulties.

Other institutions, that is those in countries with strict currencies, have difficulties in subscribing to controlled international publications, though they do not lack funds. For such countries (not excluding others), the Executive Committee has proposed that the respective National Committees of Astronomy should contribute to the distribution of IAU publications to their institutions by adhering to the Union in a slightly increased category (*b*). So far only seven such institutions have responded favourably, the other countries having not yet replied and notified their final decision.

The curtailed Distribution List includes also the members and advisors of the Executive Committee (11), UNESCO (10 copies) and certain international organizations, such as ICSU. The curtailed Distribution List thus contains about a hundred addresses.

It should be realized that the publication system proposed as from Symposium 32 onwards (i.e. to use the services of a professional publisher) implies that the Union will have to purchase the volumes distributed to the institutions on the curtailed Distribution List.

The Executive Committee felt that the new distribution policy, much more economical and simpler from the administrative standpoint, is also much fairer than the previous system, as it only favours those institutions facing real difficulties. A certain injustice cannot be avoided (some institutions of developed countries may be very low in funds), but the Executive Committee feels that the new distribution policy, imperative indeed, is also the best solution.

(*b*) *IAU Members*

Article 24 of the By-laws reads as follows:

'The volume of the *Transactions* of the Union containing the Proceedings of the General Assembly is distributed free of cost to:

(*a*) all those who, at the time of the meeting of the General Assembly, were Members of the Union including those co-opted by virtue of article 2 of the present By-laws;

(*b*) institutions on a list approved by the Executive Committee;

(*c*) Adhering Organizations or National Committees.'

Hence it follows that only Volume B of the *Transactions*, in their present form, has to be distributed free to Members. The Executive Committee has unfortunately to agree to a reduction of free distribution as much as possible. Only Volume XIII B will be distributed free of charge; it should be remembered that the first volume distributed through normal commercial channels, and thus exempt from free distribution, was Volume XII A. The *Draft Reports*, which actually are the page proofs of Volume A will continue to be distributed free of charge; the preparation for the General Assembly so requires it.

(*c*) *Participants in IAU Symposia*

The Executive Committee has also agreed to discontinue free distribution of Symposium volumes to participants, often very numerous, in IAU Symposia, and to present courtesy copies only to a limited number (about 10) of them. This decision is very regrettable, but it is a step the Executive Committee felt imperative.

4. FINANCIAL ESTIMATES IN RESPECT OF PUBLICATIONS

An analysis of the figures given above permits the following estimates to be made of receipts from and payments for IAU publications for 1967 and the following years:

Continued overleaf.

		Receipts $	Payments $	Balance $
1967	Transactions XII A, B, C	4 000	4 600	
	Draft Reports and Transactions XIII A	0	15 000	
	Symposia 24, 25, 26	8 000	8 000	
	Symposia 27 to 31	0	33 000	
	Information Bulletins	0	3 000	
		12 000	63 600	+51 600
1968	Transactions XIII A, B, C	26 000	50 000	
	Symposia 24, 25, 26	2 400	0	
	Symposia 27 to 31	30 000	2 000	
	Symposia 32 to 36	0	4 000	
	Information Bulletins	0	3 000	+ 600
1969	Transactions XIII A, B, C	2 000	0	
	4 Symposia	4 000	4 000	
	Information Bulletins	0	3 000	+1 000
1970	Transactions XIII A, B, C	2 000	0	
	Draft Reports and Transactions XIV A	3 000	25 000	
	4 Symposia	4 000	4 000	
	Information Bulletins	0	4 000	+24 000
		$73 400	$99 000	+$25 600

The figures are commented upon in more detail in paragraph (*d*) of the Financial Report, Appendix I. Naturally, such estimates are rather inaccurate and may be either over-optimistic or too pessimistic. They have been taken as basis for the Budget which will be submitted to the Finance Committee during the thirteenth General Assembly and to the General Assembly itself (see above, pages li and liii).

NOTE ADDED IN PROOF

A re-estimation of some of the above values has been made by the General Secretary and the Assistant General Secretary before the meeting in Prague and has been used as a basis for the discussions of the Finance Committee. These new estimations are incorporated in the report of the Finance Committee, volume XIII B.

APPENDIX III

A. LIST OF COUNTRIES ADHERING TO THE UNION

The following is a list of the 43 countries that adhere to the Union in December 1966, giving also the year of adherence, the approximate number of Members, and the Adhering Organizations.

Country	Year	Members	Adhering Organizations
Argentina	1927	14	Comité Nacional de Astronomía, Buenos Aires.
Australia	1939	39	Australian Academy of Sciences, Canberra.
Austria	1955	14	Österreichische Akademie der Wissenschaften, Wien.
Belgium	1920	34	Académie Royale de Belgique, Bruxelles.
Brazil	1961	8	Conselho Nacional de Pesquisas, Rio de Janeiro.
Bulgaria	1957	6	Académie des Sciences Bulgare, Sofia.
Canada	1920	40	Department of Energy, Mines and Resources, Ottawa.
Chile	1947	6	Universidad de Chile, Santiago de Chile.
Czechoslovakia ..	1922	25	Academy of Sciences, Praha.
Denmark	1922	13	Det Kongelige Danske Videnskabernes Selskab, København.
Finland	1948	8	Académie des Sciences et des Lettres, Helsinki.
France	1920	123	Académie des Sciences, Paris.
Germany (GDR)	1951	21	Deutsche Akademie der Wissenschaftenzu Berlin.
Germany (GFR)	1951	100	Rat Westdeutscher Sternwartenzu München.
Greece	1920	11	Academy of Athens, Athens.
Hungary	1947	8	Hungarian Academy of Sciences, Budapest.
India	1946	17	Ministry of Education, Government of India, New Delhi.
Ireland	1947	3	Dublin Institute for Advanced Studies, Dublin.
Israel	1954	2	The Israel Academy of Sciences and Humanities, Jerusalem.
Italy	1921	43	Consiglio Nazionale delle Ricerche, Roma.
Japan	1920	47	Science Council of Japan, Tokyo.
Korea (PDR) ..	1961	3	Academy of Sciences, Phyongyang.
Lebanese (Republic)	1954	1	National Lebanese Committee for Astronomy, Ksara.
Mexico	1921	9	Universidad Nacional de Mexico, Mexico, D.F.
Netherlands ..	1922	47	Koninklijke Nederlandse Academie van Wetenschappen, Amsterdam.
New Zealand ..	1964	2	Royal Society of New Zealand, Wellington.
Norway	1922	10	Det Norske Videnskaps-Akademi, Oslo.
Poland	1922	33	Polska Akademia Nauk, Warszawa.
Portugal	1924	12	Secção Portuguesa das Uniões Internacionais, Astronomica, Geodésica e Geofísica, Lisboa.
Roumania ..	1928	8	Conseil Scientifique de l'Observatoire de Bucarest, Bucarest.
South Africa ..	1938	16	Council for Scientific and Industrial Research, Pretoria.
Spain	1922	14	Comision Nacional de Astronomia, Madrid.
Sweden	1925	26	Kungl. Vetenskaps-Akademien, Stockholm.
Switzerland ..	1923	14	Schweizerische Naturforschende Gesellschaft, Basel.
Taiwan (Rep. of China)	1959	7	Academia Sinica, Nankang, Taipei.
Turkey	1961	10	Türk Astronomi Dernegi, Istanbul.
United Arab Republic	1925	6	Astronomical Center at Helwan, Egyptian Region.
United Kingdom	1920	133	The Royal Society, London.
U.S.A.	1920	396	The National Academy of Sciences, Washington, D.C.

Country	Year	Members	Adhering Organizations
U.S.S.R.	1920	207	Academy of Sciences of the U.S.S.R., Moscow.
Vatican City State	1932	6	Pontificia Accademia delle Science, Città del Vaticano.
Venezuela ..	1953	1	Academia de Ciencias Fisicas, Matematicas y Naturales, Caracas.
Yugoslavia ..	1935	12	Conseil Fédéral pour la Coordination de la Recherche Scientifique, Belgrade.

B. LIST OF DECEASED MEMBERS

The Executive Committee regrets to record the loss by death of the following 58 Members of the Union (at the date of 1967, January 1st).

H. L. Alden	3 February 1964
Sir Edward Appleton	21 April 1965
F. Baldet	8 November 1964
D. Barbier	1 April 1965
A. Bečvař	10 January 1965
D. Brouwer	31 January 1966
L. Carnera	30 July 1962
G. Cassinis	13 January 1964
M. S. Eigenson	15 August 1962
F. Fernandez de la Puente	4 January 1965
E. Finlay-Freundlich	24 July 1964
J. Génard	6 March 1963
J. Gadomski	2 January 1966
A. Gütler	19 March 1964
C. B. Haselgrove	27 May 1964
T. Hanala	10 November 1963
M. B. B. Heath	21 November 1963
W. W. Heinrich	30 May 1965
D. Hendrix	26 December 1961
A. R. Hogg	31 March 1966
I. G. Izsak	21 April 1965
H. S. Jelstrup	6 April 1964
F. Kepinski	8 April 1966
E. A. Kreiken	16 August 1964
D. K. Kulikov	11 June 1964
G. Lemaître	19 June 1966
B. Lindblad	25 June 1965
J. C. Lindsay	26 September 1965
D. B. McLaughlin	8 December 1965
J. E. Mack	15 April 1966
D. D. Maksutov	12 August 1964
E. L. Martin	9 August 1966
W. P. Meggers	19 November 1966
S. K. Mitra	27 February 1967
K. Moder	13 November 1965
J. J. Nassau	11 May 1965
N. Nechvile	5 July 1964
B. A. Orlov	16 September 1963
B. M. Peek	5 May 1965
G. Pels	2 November 1966
J. G. Perel	28 November 1964
R. M. Petrie	8 April 1966
V. F. Proskurin	21 May 1964
J. Racah	28 August 1965
F. Rigaux	21 September 1962
N. Robinson	7 September 1964
A.-M. Rozis-Saulgeot	1965
E. Schoenberg	23 January 1965
F. H. Seares	20 July 1964
V. V. Sharonov	27 November 1964
H. Siedentopf	28 November 1963
E. C. Slipher	7 August 1964
J. Stebbins	16 March 1966
Y. Tsukamoto	20 December 1965
L. Vegard	21 December 1963
J. H. Waddell	1966
A. A. Weiss	30 December 1964
F. Wünschmann	17 January 1966

C. LIST OF IAU PUBLICATIONS

Transactions of the IAU, Volume XII A, pp. 809, price 168s. (Members 104s.)
Editor: J.-C. Pecker
Published by Academic Press, Inc. (London) Ltd.

Transactions of the IAU, Volume XII B, pp. 672, price 140s. (Members 100s.)
Edited and published as Volume XII A.

Transactions of the IAU, Volume XII C, pp. 207, price 60s. (Members 40s.)
Edited and published as Volume XII A.

IAU Symposium Volumes

No. 19. *Le Choix des Sites d'Observatoires* (Site Testing), pp. 386, price 40 Frs (Members 24 Frs) — Editors: J. Rösch, G. Courtès, J. Dommanget. Reprinted from *Bulletin Astronomique* (Gauthier-Villars, Paris).

No. 20. Joint IAU-URSI Symposium on *The Galaxy and the Magellanic Clouds*, pp. 404, price 105s. (Members 65s.) — Editors: F. J. Kerr and A. W. Rodgers. Published by the Australian Academy of Sciences.

No. 21. *Le Système des Constantes Astronomiques* (The System of Astronomical Constants), pp. 330, price 22 Frs (Members 14 Frs) — Editor: J. Kovalevsky. Reprinted from *Bulletin Astronomique* (Gauthier-Villars, Paris)

No. 22. *Stellar and Solar Magnetic Fields*, pp. 460, price \$15·75 (Members \$10·35) — Editor: R. Lüst. Published by North-Holland Publishing Company, Amsterdam.

No. 23. Joint IAU-COSPAR Symposium on *Astronomical Observations from Space Vehicles*, pp. 370, price 50 Frs (Members 30 Frs) — Editor: J.-L. Steinberg. Reprinted from *Annales d'Astrophysique*.

No. 24. *Spectral Classification and Multicolour Photometry*, pp. 383, price 120s. (Members 80s.) — Editors: K. Lodén, L. O. Lodén, U. Sinnerstad. Published by Academic Press, Inc. (London) Ltd.

No. 25. *The Theory of Orbits in the Solar System and in Stellar Systems*, pp. 380, price 120s. (Members 80s.) — Editor: G. Contopoulos. Published by Academic Press, Inc. (London) Ltd.

No. 26. *Abundance Determination in Stellar Spectra*, pp. 373, price 120s. (Members 80s.) — Editor: H. Hubenet. Published by Academic Press, Inc. (London) Ltd.

IAU Colloquium Volumes

Commission 36:
Proceedings of the Colloquium on *The Theory of Stellar Atmospheres* — Editor: C. de Jager. Reprinted from *Journal of Quantitative Spectroscopy and Radiative Transfer* (Pergamon Press).

Commission 37:
Proceedings of the Colloquium on *Clusters and Stellar Evolutions* — Editors: O. J. Eggen, C. H. Herbig. Published as no. 82 of the *Royal Observatory Bulletin*.

Commission 27 and 42:
Proceedings of the Colloquium on *The Position of Variable Stars in the Hertzsprung-Russell Diagram* — Published by the Remeis-Sternwarte Bamberg. *Kleine Veröff. Remeis-Sternw.*, **4**, no. 40.

Publications by, or on behalf of, IAU Commissions and by Astronomical Services

Commission 6:
IAU (Telegram Bureau) *Circulars* nos. 1874–2000 — Issued by the IAU Central Bureau for Astronomical Telegrams.

Commission 10: *Heliographic Charts of the Photosphere* (Volume XII, nos. 3 and 4)	Published by the Eidg. Sternwarte in Zürich, under the direction of M. Waldmeier.
Commission 10: *Cartes Synoptiques de la Chromosphère Solaire* (Volume IV, fascicules 3 and 4)	Prepared by Observatoire de Paris at Meudon, under the direction of R. Michard.
Commission 20: *Minor Planet Circulars*	Mimeographed Circulars issued by the Minor Planet Center, Cincinnati.
Commission 23: *Astrographic Catalogues*. Information regarding the different sections (Hyderabad, Perth-Edinburgh, Oxford-Potsdam, Melbourne, Catania, Tacubaya, Uccle-Paris, etc.) may be obtained from the respective Observatories or from the President of Commission 23	P. Sémirot, President of Commission 23, Observatoire de Bordeaux, 33-Floirac, France.
Commission 26: *Circulaires d'Information* no. 34 through no. 41	Prepared by P. Müller, Observatoire de Meudon, France.
Commission 27: *Information Bulletin on Variable Stars*, up to no. 165	Prepared by L. Detre, Konkoly Observatory, Budapest.
The Solar Spectrum 2935 Å *to* 8779 Å	Ch. E. Sitterly, M. G. J. Minnaert, J. Houtgast *National Bureau of Standards Monograph* 61.
Monthly Notes of the IPMS, no. 7 (1964) through no. 10 (1966)	Published by International Polar Motion Service, Mizusawa-shi.
IAU Information Bulletin, nos. 12–17	Prepared by the General Secretary.

5. COMMISSION DES ANALYSES DE TRAVAUX ET DE BIBLIOGRAPHIE

(Comité du Comité Exécutif)

PRÉSIDENT: Dr J. Kleczek, Astronomical Institute, Czechoslovak Academy of Sciences, Observatory Ondřejov, Czechoslovakia.

VICE-PRÉSIDENT: Dr J. B. Sykes, 20 Milton Lane, Steventon, Abingdon, Berkshire, England.

PAS DE COMITÉ D'ORGANISATION.

MEMBRES: Alter, Baldet†, Beer, Belorizky, Bourgeois, Bouska, Collinder, Dermul, Felber, Fricke, Harwood, Heinemann, Henn, Hirose, Luplau-Janssen, Martynov, Ogorodnikov, Pecker (J.-C.), Sticker, Witkowski, Zinner.

FORM OF PAPERS

Not only the content, but also the presentation of a scientific paper is important for the reader. A Style Book for astronomers has been published (*Trans. IAU*, **12 C,** 55 and 101, 1966). It contains:

A. How to prepare manuscripts for printing.

B. Spelling and transliteration.

C. Notations, formulae, units.

D. Abbreviations of titles of scientific periodicals (*Trans. IAU*, **12 C,** 76).

E. How to correct proofs.

For the comments of librarians of astronomical observatories see section C of the Report of the Working Group, which follows. Useful hints for authors may be found also in a booklet published recently by the Royal Society (1965).

It has been remarked that English abstracts in foreign journals are very useful, even if the English is sometimes not very good. Such abstracts should be encouraged. Our Commission recommends also that Russian abstracts be published with astronomical papers.

ECONOMY OF PUBLICATION

In some cases, there is a tendency for the same work to be published at least four times: as research material in a scientific journal; repeated in a slightly different form at one or more symposia, either before or after the previous publication; summarized, or expanded, in one or more progress reports on astronomy; and, finally, incorporated in a textbook. In the opinion of some astronomers, it is the second stage, namely publication in the proceedings of symposia, that is confusing and wasteful. In this connection it would be interesting to extrapolate the gloomy comments of Barnabe Rich from 1613: 'One of the diseases of this age is the multiplicity of books; they doth so overcharge the world that it is not able to digest the abundance of idle matter that is everyday hatched and brought forth into the world.' (Swanson 1966, p. 12).

The average readership of a scientific article is surprisingly low in chemistry and biosciences (Swanson 1966) and this seems also to be true for astronomy. It has been therefore suggested to publish papers only in the form of lengthy (about one page) abstracts, with the full text available on request. The Russian journal *Atomnaya Energiya* regularly publishes some papers in this way. This new possibility of economic publication has of course also its drawbacks.

BIBLIOGRAPHY OF ASTRONOMY 1881–98

All the material concerning this bibliography collected with great patience and diligence by Mr Dermul in Anvers has been transferred to the Observatoire Royal de Belgique. The present state and future plans for publication follow from a quotation by the Director of the Observatory, Professor Velghe (1966): 'L'Observatoire Royal de Belgique envisage la publication de la Bibliographie de l'Astronomie pour la période 1881–98, lorsqu'il sera en possession de toutes les fiches y afférentes. Le Directeur de l'Observatoire Royal fait un pressant appel à tous ceux qui sont engagés à dépouiller la littérature astronomique de leur pays, pour que les fiches lui parviennent avant le ler juin 1967, date à laquelle le fichier sera considéré comme complet. Le projet de publication sera communiqué aux membres de la Commission 5, au cours de la prochaine Assemblée Générale de l'UAI, à Prague.'

ABSTRACTING JOURNALS

Recorded knowledge about the Universe accumulates at an increasing rate while the rate at which it can be read remains constant. In the astronomer's endeavour to keep abreast of the increasing flow of recorded knowledge, the abstracting journals have the most important role. The four abstracting journals for astronomy have continued to abstract world literature on astronomy on a routine basis. They have all become member-services of the ICSU Abstracting Board.

ABSTRACTING BOARD OF THE INTERNATIONAL COUNCIL OF SCIENTIFIC UNIONS (IAB)

The IAU is a member of the IAB. The aim of the IAB is to initiate and further develop the exchange and publication of primary and secondary scientific information. By secondary publications are meant abstracts and indexes drawn up from the already published material.

Since 1964 an annual list of publications issued by ICSU and its members is prepared by IAB. The IAB acts as an advisory body for ICSU and its members on problems of scientific information. It represents the interests of ICSU and its members in other international institutions dealing with scientific information.

There are two categories of members of the IAB:

(1) representatives of international unions and

(2) representatives of the main secondary publications (abstracting and reference services —the so-called member-services). At present there are four member-services for astronomy: *Astronomischer Jahresbericht*, *Bulletin signalétique du Centre National de la Recherche Scientifique*: Section 2, Astronomie, Astrophysique, Physique du Globe, *Physics Abstracts* and *Referativnyi Zhurnal.*

At the July 1966 full board meeting in Paris the IAU was represented by the President of Commission 5. His proposals were all accepted and it was resolved (Resolution 8, cooperation with IAU) that:

(1) The abstracting journal *Astronomischer Jahresbericht* was unanimously elected member-service of the Board for the German language in astronomy.

Bulletin signalétique du CNRS and *Referativnyi Zhurnal* are the two other member-services for this field, for the French and Russian languages respectively.

There is no abstracting journal in the English language covering astronomy as such.

(2) The IAU will keep the IAB informed of collective works and of publications resulting from scientific meetings in the field of astronomy.

The Secretariat will communicate this information to member-services in astronomy with a view to improving the coverage of non-periodical publications in this field by member-services of the Board.

The IAU recommends to member-services for astronomy to envisage the possibility of announcing these publications not only under the names of scientific editors, but also under the names of the individual authors themselves.

(3) With the help of the IAU the Board will try to establish a list of all primary periodicals dealing with astronomy.

The action on hand in the other fields covered by the Board to improve primary publications will be extended to these publications in astronomy.

A Working Group on Documentation in Astronomy was formed at the Hamburg meeting. This Working Group forms a link between IAU Commission 5 and the 'Fédération Internationale des Associations de Bibliothécaires'. On the basis of correspondence with all the astronomical libraries of the world Mme Feuillebois has carefully prepared the hereafter appended information, suggestions and comments for the Draft Report of Commission 5.

J. KLECZEK
President of the Commission

REFERENCES

The Royal Society 1965, *General Notes on the Preparation of Scientific Papers*, 2nd edition, London, 32 pages.
Swanson, D. R. 1966, *Bulletin of the Atomic Scientists*, **22** (2), Feb. 1966, pp. 8–12.
Velghe, A. G. 1966, A note on the astronomical bibliography 1881–98, to be published in the *IAU Information Bulletin*.

APPENDICE. GROUPE DE TRAVAIL SUR LA DOCUMENTATION ASTRONOMIQUE

A. Informations

1. La Fédération Internationale des Associations de Bibliothécaires (FIAB) a créé une sous-section des bibliothèques d'observatoires et de sociétés astronomiques, qui travaillera en étroit rapport avec le groupe de travail, et qui a tenu ses premières réunions pendant la 32ème session de la FIAB à La Haye, les 13 et 15 Septembre 1966.

2. Son statut prévoit, dans le cadre des sessions de la FIAB:

(*a*) une réunion annuelle;

(*b*) une assemblée générale tous les 3 ans (l'année précédant l'Assemblée Générale de l'UAI), pour préparer les travaux du groupe de travail.

3. Le programme des séances tenues à La Haye était le suivant:

(*a*) Présentation de la nouvelle sous-section et établissement des statuts.

(*b*) Communications:

– Emploi de la Flexowriter pour le catalogage en astronomie (Mr Kemp, Edinburgh).

– Etat actuel et problèmes de la bibliographie astronomique (Mme Lavrova, Institut Sternberg, Moscou).

– 'The U.S. Naval Observatory Library at Washington, D.C.' (Mrs L. S. Multhauf, U.S. Naval Observatory).

– Présentation du catalogue par matières et de la liste des 'vedettes-matières' d'astronomie de l'Observatoire de Paris (Mme Feuillebois).

– Classification de la 'Library of Congress' et les vedettes-matières pour l'astronomie. Son emploi dans les bibliothèques américaines (Mr Weigel, University of Michigan).

(*c*) Séance de travail:

– Analyse et critique du 'style book' à paraître dans le volume XII C des *Transactions of the IAU.*

– Publications et distribution des publications d'observatoires.

– Etude des possibilités de limitation des revues commerciales.

– Etablissement d'un Répertoire International des Bibliothèques d'Observatoires et de Sociétés Astronomiques. Un important matériel est déjà réuni. Pour la publication préliminaire de ce répertoire, sur feuillets mobiles, une demande est faite au Conseil Exécutif de la F.I.A.B. d'une subvention de 5000 francs suisses.

B. Suggestions pour la réunion de la Commission 5 à Prague

Au cours de l'Assemblée Générale de l'UAI à Prague, le groupe de travail souhaite que soit prévue une séance de travail qui lui permettra de mettre au point le rapport de ses travaux.

C. Commentaires

– Le groupe de travail note avec plaisir la publication d'un 'Style Book' dans le volume XII C des *Transactions of the IAU.*

– Elle suggère les commentaires suivants en vue de l'élaboration du volume XIII C: (*N.B.:* Les paragraphes et pages indiqués ci-dessous sont ceux du volume XII C).

1. *Présentation des articles* (*Form of papers*)

– 'Résumés' (paragr. *b*, p. 57; English version, p. 103): Le groupe de travail suggère 'que les articles . . . soient toujours accompagnés d'un sommaire dans la langue employée par l'auteur, ainsi que d'un sommaire plus court, soit en anglais, soit en français (ou au moins dans l'une des quatre langues suivantes: français, anglais, allemand, russe). L'idéal consisterait d'ailleurs, de la part des éditeurs de revues, à insérer dans chaque fascicule des fiches détachables 125 × 75 mm pour chaque article publié (solution déjà adopté par le *Bulletin Astronomique*, 3ème série, 1966.

– 'Bibliographie', paragr. *e*, p. 59; English version, p. 105): Le groupe de travail trouve très souhaitable les recommandations de l'UAI. Toutefois, à propos du paragraphe 'Présentation des références' elle demande à l'UAI:

– d'étudier avec précision la question de la date de la référence: s'agit-il de la date de la soumission du manuscrit ou de celle de l'édition? Cette précision est particulièrement importante lorsqu'il s'agit d'un livre à auteurs multiples ou d'un livre réédité (*iii*, p. 60; English version, p. 106).

– de distinguer (δ, p. 61; English version, p. 107) les mentions 'in preparation' et 'in press', la seconde ne devant être employée que lorsque l'article est effectivement en cours d'impression, c'est-à-dire après correction des épreuves, ce qui permet d'ailleurs d'apporter certaines précisions à cette référence.

– l'indication 'in preparation' ne permettant pas de se reporter à une référence bibliographique utilisable, il serait même préférable de ne la mentionner qu'au cours de l'article et non à la fin.

2. *Translittération*

– Le groupe de travail souhaite que l'Union Astronomique Internationale, conformément à sa vocation d'association internationale, adopte la translittération ISO/R9, 1954.

– Si cette translittération est favorablement accueillie à Prague, elle suggère que le document ISO/R9 soit intégralment reproduit dans le volume XIII C (de préférence au tableau donné p. 69; English version, p. 115).

3. *Abréviations des titres de périodiques scientifiques* (*pp.* 74 *et suiv.*)

Lors de la préparation du volume XII C il n'existait pas de liste internationale d'abréviations des titres de périodiques scientifiques. C'est pourquoi la *World List of Scientific Periodicals* pouvait être recommandée. Mais une liste internationale est actuellement préparé par l'International Standard Organization, tandis que la World List semble en défaveur.

Dans ces conditions, le groupe de travail pense qu'il serait prudent d'attendre la publication de la liste ISO et de surseoir à toute décision jusqu'à cette date.

4. *Revues de références astronomiques* (*Astronomical abstracting journals*)

La création d'un nouvel 'Abstracting Journal' de langue anglaise ne paraît pas souhaitable (Les *Physics Abstracts* analysent d'ailleurs les grandes revues d'astronomie).

Par contre, *l'Astronomischer Jahresbericht*, qui est actuellement la bibliographie la plus complète et la plus utilisée, pourrait être améliorée: (*a*) en abrégeant le délai de publication; (*b*) en translittérant les titres russes au lieu de les traduire en allemand; (*c*) en publiant des 'abstracts' en langue française ou anglaise, ce qui rendrait la consultation beaucoup plus aisée, en particulier pour les jeunes astronomes rarement familiarisés avec la langue allemande; (*d*) en élargissant le domaine des publications analysées: congrès, livres à auteurs multiples; publications d'observatoires (chaque observatoire devrait vérifier que ses publications figurent au paragraphe 4 de *l'Astronomischer Jahresbericht*); (*e*) enfin en étudiant les moyens d'automatiser la bibliographie actuellement existante. Etudier l'automatisation de la bibliographie internationale courante devrait en effet être une préoccupation de la Commission 5 dans un proche avenir. L'idéal serait évidemment de n'avoir qu'une seule bibliographie internationale courante d'astronomie.

5. *Publications*

La question de la distribution des publications d'observatoires, et de la parution de nouvelles revues commerciales, a été longuement discutée à La Haye. Une enquête préalable avait d'ailleurs été faite et nous reproduisons ci-dessous l'essentiel des réponses.

(*a*) *Distribution des publications d'observatoires.* Bien que la multiplicité des publications d'observatoires, généralement constituées de 'reprints', soit un lourd handicap pour les bibliothèques, les bibliothécaires souhaitent en général leur maintien. Ces publications permettent en effet: (*i*) d'avoir une idée d'ensemble du travail effectué dans chaque observatoire; (*ii*) de constituer une monnaie d'échange pour les observatoires qui n'ont pas assez de crédits pour acquérir les revues dont sont extraits les articles reproduits sous forme de 'reprints' (en particulier lorsqu'il s'agit de publications d'Académies ou de revues non astronomiques); (*iii*) de multiplier les possibilités de prêt dans les observatoires où les lecteurs sont très nombreux. Toutefois le groupe de travail suggère que soit réduit dans la mesure du possible le nombre de titres publiés par chaque observatoire. De plus, afin de faciliter la distribution et l'échange des publications d'observatoires, la sous-section des bibliothèques d'observatoires et de sociétés astronomiques de la FIAB prépare pour 1967 une liste mise à jour de toutes les publications non commerciales d'observatoires (y compris les catalogues d'étoiles)*. Cette

*Mme Van de Poll de Lange, bibliothécaire de l'Observatoire de Leyde, a bien voulu se charger de ce travail.

liste précisera, pour chaque publication l'adresse exacte de l'organisme distributeur, ainsi que les numéros disponibles et les numéros épuisés. Cette liste sera présentée à Prague par le groupe de travail.

(*b*) *Etude des possibilités de limitation des revues commerciales.* Le groupe de travail s'inquiète de la parution, au cours de ces dernières années, de nouvelles revues astronomiques prises en charge par des éditeurs commerciaux et qui viennent s'ajouter aux revues déjà existantes. Mais les bibliothécaires sont obligés de constater que les astronomes eux-mêmes souhaitent l'acquisition de ces nouvelles revues. De plus, les revues existantes (comme *l'Astrophysical Journal*) sont de plus en plus volumineuses, et l'on conçoit mal comment elles pourraient s'accroître encore.

Toutefois ce problème doit, semble-t-il, être pris en considération par la Commission 5 de l'UAI: c'est en effet aux astronomes eux-mêmes qu'il appartient de prendre la décision de limiter de telles revues.

G. FEUILLEBOIS

6. COMMISSION DES TELEGRAMMES ASTRONOMIQUES

(Comité du Comité Exécutif)

PRÉSIDENT: Dr A. R. Hogg†.

VICE-PRÉSIDENT: Dr F. L. Whipple, Director of the Smithsonian Observatory, 60 Garden Street, Cambridge, Massachusetts 02138, U.S.A.

DIRECTEUR DU BUREAU: Dr O. Gingerich, Smithsonian Astrophysical Observatory, Cambridge, Massachusetts 02138, U.S.A.

DIRECTEURS ADJOINTS DU BUREAU: Dr R. B. Southworth and Dr B. G. Marsden, Smithsonian Astrophysical Observatory, Cambridge, Massachusetts 02138, U.S.A.

COMITÉ D'ORGANISATION: A. G. P. Velghe, K. A. Thernöe, O. J. Gingerich.

MEMBRES: Buchar, Candy, Cunningham, Hers, Martynov, Roemer.

Le Comité Exécutif de l'Union Astronomique Internationale a appris avec peine le décès du Dr A. R. Hogg, Président de la Commission 6 depuis 1961.

Le Professeur F. L. Whipple a bien voulu accepter de préparer ce rapport et d'assumer le rôle de Président par interim de la Commission. Il a préparé le présent rapport, assisté par le Dr O. Gingerich, Directeur du Bureau des Télégrammes Astronomiques.

INTRODUCTION

At the Twelfth General Assembly of the IAU in Hamburg, arrangements were made to move the Central Bureau from its traditional home in Copenhagen to Cambridge. The gradual increase in the number of subscribers, coupled with the erratic nature of the Bureau's work—periods of intense activity when spectacular comets appear alternating with many weeks of inactivity—made it desirable to locate the Bureau at a larger observatory. In its new headquarters at the Smithsonian Astrophysical Observatory, the Central Bureau can call upon the services of the round-the-clock communication center, the network of astrophysical observing stations (with their Baker-Nunn cameras, which are ideal for confirming bright comets), the computations center, and the publications division.

Dr Gingerich, whose unusually able services brought about the smooth and efficient transfer of the Bureau to its new location, has indicated his intention to resign as the Bureau's Director, since his own interests lie primarily elsewhere. Dr Marsden, who has played an increasingly active role as an Associate Director and who is a specialist in the computation of comet orbits, would appear to be the logical choice for the new Director. Dr Gingerich and Dr Southworth would then serve as Associate Directors. The Commission is truly indebted to Dr Gingerich for his skill and perseverance in effecting the transfer of the Central Bureau.

F. L. WHIPPLE
Acting President of the Commission

REPORT OF THE CENTRAL BUREAU FOR ASTRONOMICAL TELEGRAMS

In its first year in Cambridge, the Central Bureau distributed 64 numbered *Circulars* on 53 occasions. These carried information about 16 supernovae and ten comets discovered during 1965. The most extensive activity resulted from the spectacular sun-grazing comet, Ikeya-Seki, for which 11 *Circulars* were issued in a period of six weeks. In its second year, the Central Bureau issued 39 numbered *Circulars* on 30 occasions. These carried information about five supernovae and six comets discovered during 1966. Since 17 comets are predicted to come to perihelion during 1967, we can anticipate considerable activity in the Bureau.

When the Central Bureau moved to Cambridge, the *Harvard Announcement Cards* were discontinued, and their subscribers began to receive the *Circulars*. This more than doubled the number of subscribers, so that at present over 600 *Circulars* are mailed to 50 countries.

The telegram activity of the Bureau is somewhat harder to measure; 43 'telegram books' were distributed to a large block of domestic subscribers as well as to the AGIWARN system during 1965 and 1966. It is difficult to count the actual number of recipients because the telegrams are relayed to partial lists of subscribers according to various categories. In addition, on special occasions messages were sent to only a few selected observatories.

The international distribution of telegrams has worked out magnificiently through the AGIWARN system of the International Ursigram and World Days Service. In particular, the arrangements made by Dr R. Michard and the effort of Dr P. Simon at Meudon have facilitated the redistribution of most of the European telegrams. Telegrams to South Africa, Japan, and some points of South America have been relayed through the telegram-link connecting the Smithsonian Observatory with its astrophysical observing stations.

The task of collecting advance telegram deposits for the subscribers served from the Meudon Center has proved rather formidable, and several European subscribers are still delinquent in sending in their deposits. Similarly, many observatories have been lax in responding to our expiration notices. Since the *Circulars* are addressed by a computer program, the failure to renew promptly results automatically in an interruption of service.

At the Hamburg Meeting of the IAU, suggestions for an improved telegram code were discussed, and fortunately it was possible to introduce the revisions at the same time that the transfer of the Central Bureau took place. The success of the new code is perhaps most vividly shown by the awkward situation that almost inevitably arises when an occasional astronomer transmits a message in the old form: it is then generally necessary to telegraph back asking for the equinox of the observations. In addition to the explicit inclusion of the equinox in the revised code, the time of observation is given in decimals of day, there is a continuous ordering of data in the accurate position code, and a more precise description of magnitudes is provided. An explanation of the new code appears in the IAU Handbook (*Transactions*, XII C); a list of errata are attached to this report. Corrected reprints of the code are available through the Central Bureau.

During 1965 and 1966, $7328.00 were received for subscriptions to the *Circulars*, which covered only the printing and postage. The continuing operation of the Bureau requires more than the equivalent of one full-time worker, which is supplied by the Smithsonian Astrophysical Observatory. The subvention of $666.00 per year from the IAU pays for very little of the actual operation of the Bureau, but it provides considerable moral support and gives the operation a desirable flexibility that it would not otherwise have. It seems entirely reasonable to ask that the IAU subvention be raised to at least $1000 per year.

OWEN GINGERICH
Director of the Bureau

APPENDIX. ERRATA ET CORRIGENDA FOR IAU TRANSACTIONS XIIC.

p. 33 Strike out category H and for category G read:
G. GAMMA All others, including planetary or lunar phenomena, spectroscopic binaries, etc.

p. 34 The Air Mail rates in the column 'Other' are now $8.50 and $16.50. *Under Middle groups for a position*, the fifth approximate group should read: XNMMP.

p. 35 The fifth line should read:
8. X = X, the character for a nonsignificant digit.

p. 36 The fourth line should read:
K. KKKK = q in A.U.
The last line of the first example should read:
R.A. − $1^{m}.2$, in Dec. + 12′.

p. 37 The first line should read:
3. CANDY COMET CANDY 19503 10208 58335 13625 17656 15092 10640 45059 CANDY.

38. COMMISSION POUR L'ECHANGE DES ASTRONOMES

(Comité du Comité Exécutif)

PRÉSIDENT: Professor M. G. J. Minnaert, Sterrewacht 'Sonnenborgh', Zonnenburg 2, Utrecht, The Netherlands.

VICE-PRÉSIDENT: Professor M. Migeotte, Institut d'Astrophysique, Université de Liège, Cointe-Sclessin, Belgium.

COMITÉ D'ORGANISATION: B. J. Bok, D. Ya. Martynov, T. L. Page.

MEMBRES: Abetti, Babcock (H. D.), Bappu, Bowen, Danjon, Keller, Kienle, Kourganoff, Mikhailov, Miyamoto, Mohr, Righini, Rosseland, Sahade, Stoy, Swings, Thackeray, Witkowski, Wood (F. B.), Wyatt.

MEMBRE CONSULTANT: Fournier d'Albe (E. M.).

In the last report of this Commission, written at the end of 1963, we mentioned with some surprise that the number of travel grants awarded has shown but little variations in the preceding ten years. Since then the situation has drastically changed. In the last three years a total of 52 awards has been made, *twice as much* as in each of the previous periods. It is clear that the wonderful possibilities available have gradually become known and that more and more institutes are taking advantage of them. In 16 other cases the request could not be granted or was sent only for provisional information and did not result in an allocation.

In the period 1964–66 the following grants were made:

125. A British astronomer to Austin.
6. A Czechoslovakian astronomer to Herstmonceux.
7. A Belgian astronomer to Tucson.
8. A British astronomer to Paris.
9. An Italian astronomer to Mt Hamilton.
130. A Japanese astronomer, returning from Heidelberg.
1. A Japanese astronomer to Santa Monica.
2. An English astronomer to Groningen.
3. A Polish astronomer to Sydney.
4. A Belgian astronomer to Mt Stromlo.
5. An Argentine astronomer to Hamburg and Praha.
6. A French astronomer to Mt Palomar.
7. A Czechoslovakian astronomer to Liège.
8. A Japanese astronomer to Pic-du-Midi.
9. A Japanese astronomer to Utrecht.
140. An Indian astronomer to Tucson.
1. A Czechoslovakian astronomer to Victoria.
2. A French astronomer to Cairo.
3. A Dutch astronomer to Nice.
4. An Italian astronomer to Pic-du-Midi.
5. An Italian astronomer to Utrecht.
6. A Belgian astronomer to Edinburgh.
7. An Australian astronomer to McDonald Observatory.
8. An Indian astronomer to Utrecht.
9. An Italian astronomer to Leiden.
150. A Czechoslovakian astronomer to Asiago.

1. A Japanese astronomer to Los Angeles.
2. A Dutch astronomer to Praha.
3. A British astronomer to Mt Hamilton.
4. An Egyptian astronomer to several European observatories.
5. A Finnish astronomer to Meudon.
6. A Greek astronomer to Meudon.
7. An Israeli astronomer, returning from Tucson.
8. An Italian astronomer to Leiden.
9. A French astronomer to Berkeley.
160. A Jugoslavian astronomer to Nijmegen.
1. Two Argentine astronomers to U.S.A. National Radio Observatory.
2. An Indonesian astronomer to Mt Stromlo.
3. Two German astronomers to Berkeley.
4. A Belgian astronomer to Nice.
5. A Greek astronomer to Meudon.
6. A Japanese astronomer to Pasadena.
7. An Indian astronomer to Toronto.
8. An Egyptian astronomer to Herstmonceux.
9. A French astronomer to Tucson.
170. An Indian astronomer, returning from Utrecht.
1. A French astronomer to La Jolla.
2. A Japanese astronomer to Groningen.
3. An Australian astronomer to the U.S.S.R.
4. A British astronomer to Kingston.
5. An Israeli astronomer to Utrecht.
6. An Italian astronomer to the U.S.A.

In surveying this list, we notice that a considerable fraction of the grants, actually 67%, have been allocated for *intercontinental* travel. It seems that gradually more and more opportunities for nearby travels become available, while the IAU is almost the only body from which help on a larger scale can be expected. In the same way we notice that there were no requests concerning travels from one of the U.S.A. states to the other, nor from one of the U.S.S.R. republics to the other. From the total number of grants, 33% concerned travels to the U.S.A.

The purposes for which the subsidies were granted may be summarized as follows:

To get acquainted with the instruments or methods of an institute, specialized in a certain field; or to discuss theoretical problems with the members of such an institute.

To establish contacts between an observer and theorists.

To learn a special technique, in view of an instrument to be erected at the home observatory.

To obtain high dispersion spectra or other observations with big instruments.

To make special observations in the Southern Hemisphere or at a station with an exceptionally clear sky.

To combine a special, self-built apparatus with a big telescope.

To inspect the plate collections.

To make use of a big computer.

To prepare a book, written together with a foreign astronomer. To give final shape to a monograph.

During the three years 1964–66, the grants totalled $22 542·24, including those (see hereafter) payed with the UNESCO subsidy, and taking into account partial reimbursements. Some of the grants listed above where not yet cashed on 31 December, 1966, and are therefore still included in the balance at hand on 31 December, 1966, which is of $14 964·23.

During the years 1965–66 the committee received from UNESCO an important subsidy of 2000 U.S. dollars, to be granted to astronomers of developing countries. (Cf. *IAU Inf. Bull.* no. 16). This subsidy was used to pay the allocations nos. 140, 151, 162 in the above list and the allocation made by the Sub-Commission.

In the course of the year 1966, the Commission has for the first time exceeded to a rather considerable extent its yearly subsidy. This is considered to be a healthy development, since former years have left a certain accumulated balance, which should be gradually used.

As yet, the action of the Commission is still restricted in two respects: (1) only costs of travel are paid, no costs of living at the host observatory; (2) no grants are given for participation in symposia, unless if combined with scientific work of longer duration. Very roughly speaking, the subsidies granted do seldom exceed \$300 $\sqrt{t}$, where t is the duration of the stay, expressed in months. On the whole these restrictions are unavoidable; one might, however, consider the possibility of exceptions in very special cases.

Information concerning the activities of the Commission were given on request to cultural administrations in Belgium and in Peru.

The list of 'Facilities for Visiting Astronomers', prepared for the 1964 meeting as an appendix to the report, has been published only at the end of 1966 in volume XII A of the *Transactions*, p. 82. In order to avoid a similar delay, the next list will be published directly after the Prague meeting in the *IAU Information Bulletin*.

Since the last General Assembly the status of Commission 38 was modified: it became a committee of administrative character, working directly under the direction of the Executive Committee. The Executive Committee will have in future a certain responsibility in determining the general policy of the Commission. The collaboration with the General Secretary proved excellent.

On this same occasion the decision was taken to create a Sub-Commission, intended to help graduate students who want to specialize in astronomy, but who have not yet reached the doctor's degree. (Cf. *IAU Inf. Bull.*, no. 14.)

A report of this Sub-Commission is added below. The coordination of the activity of the Sub-Commission with the work of Commission 46 and its summer-schools will have to be carefully planned.

M. G. J. MINNAERT
President of the Commission

RAPPORT DE LA SOUS-COMMISSION

PRÉSIDENT: Professeur M. Migeotte, Institut d'Astrophysique, Université de Liège, Cointe-Sclessin, Belgique.

MEMBRES: H. Kienle, Mlle E. A. Müller, P. M. Routly.

La Sous-Commission a été créée à l'occasion du congrès de Hambourg 1964, pour faciliter le séjour à l'étranger de *licenciés en sciences*, qui désirent se spécialiser en astronomie et qui n'ont pas la possibilité de faire ces études dans leur pays. On a songé en particulier aux pays en voie de développement.

Deux demandes de subsides ont été reçues par le sous-comité.

(1) Un étudiant du Nigéria, ayant le grade de B.Sc., se proposait d'aller étudier à Berkeley, où il pouvait bénéficier d'un 'research assistantship'. Il a demandé l'intervention de l'UAI pour couvrir ses frais de voyage. Le crédit demandé a été accordé en se

basant sur le curriculum vitae et sur des recommandations très élogieuses. Cependant, par suite de circonstances locales et personnelles, le candidat en question a dû décider d'ajourner d'un an son séjour à l'étranger.

(2) Un étudiant d'Israël a demandé l'intervention du sous-comité pour poursuivre ses études à l'Université de Manitoba (Canada), où il peut obtenir une assistance financière. Il étudiera l'astronomie, en particulier la cosmologie et la relativité générale.—Après avoir examiné le curriculum vitae et vu les recommandations de quatre de ses professeurs, le sous-comité a décidé de mettre à la disposition de ce candidat une somme couvrant les frais de voyage.

Il est clair que la création du sous-comité n'a pas encore suffisamment attiré l'attention, quoiqu'elle ait été annoncée dans le *Bulletin d'Information* no. 14 (juin 1965). D'autre part il est difficile, pour la Sous-Commission, de payer les frais de séjour d'étudiants qui désirent poursuivre leurs études à l'étranger pendant un ou deux ans, vu l'importance des crédits qui devraient être accordés.

M. MIGEOTTE
Président de la Sous-Commission

46. COMMISSION POUR L'ENSEIGNEMENT DE L'ASTRONOMIE

(Comité du Comité Exécutif)

PRÉSIDENT: Professeur E. L. Schatzman, Institut d'Astrophysique, 98bis, Boulevard Arago, Paris 14e, France.

VICE-PRÉSIDENT: Professor Miss E. A. Müller, Observatoire Cantonal, Geneva, Switzerland.

COMITÉ D'ORGANISATION: L. H. Aller, B. J. Bok, E. K. Kharadze, V. Kourganoff, M. G. J. Minnaert, T. L. Page, H. C. Van de Hulst.

MEMBRES: La Commission est en cours d'organisation.

MEMBRES CONSULTANTS: E. P. Levitan, G. Walusinsky.

A la suite de la discussion de Hambourg sur l'Enseignement de l'Astronomie a été créée la Commission 46. La discussion de Hambourg proposait à la Commission des tâches résultant des obligations internationales de l'UAI.

Les problèmes qui se sont posés à la Commission 46 se sont montrés extrêmement difficiles. En effet, la situation de l'Astronomie diffère considérablement d'un pays à l'autre, et on voit jouer l'influence du niveau général de développement, de l'organisation de l'enseignement, des moyens de recherche. Les traditions ont un rôle considérable et, à l'intérieur d'un même pays, la situation peut varier d'une Université à l'autre.

Les problèmes d'un pays en voie de développement ne sont pas ceux des Etats-Unis ou de l'U.R.S.S. Ainsi en est-il par exemple pour l'Inde, où le Président de la Commission a entrepris une enquête spéciale. Sur ces pays, on consultera avec fruit la référence (1).

ENQUÊTE SUR LA SITUATION DE L'ENSEIGNEMENT DE L'ASTRONOMIE

La Commission 46 a fait une enquête sur la situation de l'Enseignement de l'Astronomie au moyen d'un questionnaire paru dans le *Bulletin d'Information* no. 14 de l'UAI. Nous reproduisons ci-dessous le questionnaire:

Questionnaire

1. Name?

2. Institution?

3. Teaching programme
- (*a*) *Place (observatory, university, institute)?*
- (*b*) *Subject(s)?*
- (*c*) *Level (graduate, undergraduate, etc.)?*
- (*d*) *Schedule (number of hours annually)?*
- (*e*) *Number of students?*

4. Do you have at your disposal course notes, books, duplicated notes, or notes that could be photographically reproduced?

5. Enrollment:
- (*a*) *Present (1963–65)?*
  - *M.Sc.: (about) . . .*
  - *Ph.D.: (about) . . .*
- (*b*) *Expected (1966)?*
  - *M.Sc.: (about) . . .*
  - *Ph.D.: (about) . . .*
- (*c*) *Desired?*

6. Present schedules at high school or secondary school level?

Les réponses reçues appellent les observations suivantes:

(1) Quarante-deux institutions seulement ont répondu; un grand nombre de centres importants n'ont pas donné de réponse. La liste des institutions ayant répondu est donnée en Annexe I.

(2) Les réponses ont été faites du point de vue du professeur et non pas du point de vue des étudiants. On sait en général le nombre d'heures d'enseignement qui sont données, mais on ne sait pas le nombre d'heures d'enseignement que doivent suivre les étudiants.

Exceptionnellement, le rapport d'Utrecht donne les précisions suivantes:

Undergraduate: 4^h lecture + 3^h practical work, during 2 years.

Graduate: 4^h per week annually; 60^h of observations (solar and stellar) annually; 1^h per week seminar and colloquia; 5 afternoons per week of practical and theoretical work during 3 years.

(3) Quelques institutions disposent de notices détaillées sur l'Enseignement de l'Astronomie (Harvard College Observatory, Faculté des Sciences de Paris), mais il n'est pas possible, la plupart du temps, de savoir exactement quelle est l'organisation de l'Enseignement.

(4) Le project d'échange de notes de cours se heurte à de grandes difficultés:

– la plupart des cours de 3ème cycle ne sont pas rédigés, et les notes de cours ne sont pas disponibles.

– les notes de cours, quand elles existent, sont toujours rédigées dans la langue où est fait l'enseignement, c'est-à-dire en Hollandais, Italien, Polonais, Russe, etc.

(5) Les statistiques sur le recrutement et les perspectives de recrutement ne permettent pas de se faire une idée de la situation mondiale, pour les raisons suivantes:

– les renseignements sont incomplets,

– les différences de niveau sont très grandes d'une institution à l'autre.

Cependant, sur les 42 Institutions ayant répondu au questionnaire, seulement 15 donnent le nombre d'étudiants pour le M.Sc.et le Ph.D. pour les périodes 1963–64, 1964–65 et les perspectives. Les données cumulatives déduites des 15 institutions sont les suivantes:

	1963–64	1964–65	Perspectives
M.Sc.	75	89	124
Ph.D.	148	165	241

Le nombre beaucoup plus élevé de Ph.D. que de M.Sc. provient du fait que dans quelques grandes Universités des Etats-Unis, le degré de 'Master of Science' n'existe pas, la fin des études conduisant à l'obtention d'un doctorat en sciences (par exemple: Harvard, Caltech, Yerkes).

En d'autres termes, dans la plupart des cas, les collègues qui ont répondu au questionnaire l'on fait comme s'ils s'adressaient à un collègue connaissant déjà parfaitement la situation locale. Il est donc extrêmement difficile de rendre compte de l'enquête d'une façon utilisable par les collègues.

L'ASTRONOMIE DANS L'ENSEIGNEMENT SECONDAIRE

La discussion de 1964, à Hambourg, avait montré un très grand intérêt pour les questions relatives à l'Enseignement secondaire. Il sera à nouveau question de ce problème à propos de la CIES (Commission Internationale pour l'Enseignement des Sciences). De même que l'Union Internationale de Mathématiques prend une part considérable dans l'organisation et le renouvellement de l'enseignement mathématique secondaire, de même l'Union Astronomique Internationale a pour devoir de s'occuper des problèmes analogues pour l'astronomie.

Dans beaucoup de pays le Ministère de l'Education Nationale consulte les universités pour l'établissement des programmes pour écoles secondaires. Souvent les comités nationaux d'astronomie ont pris des initiatives pour obtenir des modernisations. De plus, les futurs enseignants des écoles secondaires sont en général formés dans les universités et nous sommes responsables de leur formation scientifique et didactique.

Les problèmes de l'Enseignement Secondaire exigent la collaboration d'astronomes professionnels et d'enseignants, au courant de la pédagogie et au niveau des écoles. Les avis de nos Membres Consultants seront de grande valeur pour une telle collaboration.

D'autre part il est d'un grand intérêt de coopérer pour ces questions avec des représentants d'autres sciences. Nous rejoignons ici le problème de l'Enseignement intégré des sciences dont il a été question à la CIES.

CRÉATION D'UNE ÉCOLE ANNUELLE DE FORMATION

La constitution de la Commission 46 a permis d'entreprendre la réalisation d'un projet auquel s'intéressait particulièrement l'UNESCO: la création d'une Ecole annuelle de formation de jeunes astronomes. Un contrat avec l'UNESCO, pour une somme de $2000, en 1965, a permis la réunion d'une commission d'experts à Nice, les 8 et 9 juillet 1965. Participaient à cette réunion: Mlle Edith Müller, MM. Kharadze, Kourganoff, Liller, Pecker (hôte), Van de Hulst, Woolley, Fournier d'Albe (UNESCO) et Schatzman (président).

A la suite de cette réunion, une résolution a été proposée au Comité Exécutif de l'UAI en vue de la création de ces Ecoles (Annexe II). Après cette résolution une enquête a été faite auprès des diverses institutions astronomiques. On trouvera en Annexe III le rapport résumant les résultats de cette enquête. Une Ecole d'un type un peu différent a été organisée par l'U.R.S.S. à Tashkent en 1966. On trouvera en Annexe IV le rapport sur l'Ecole pour l'observation des satellites artificiels.

Actuellement, le Dr J. Kleczeck a été chargé du secrétariat scientifique de l'Ecole, et les négociations sont en cours avec Manchester pour son organisation en 1967.

Il semble être dans l'esprit de la résolution de Nice que la Commission 46 n'assume plus de responsabilités après la nomination du secrétaire de l'Ecole de jeunes astronomes. Eventuellement, un groupe d'experts désignés par l'UAI pourrait assurer le rôle de conseiller scientifique auprès du Dr Kleczeck.

Le budget 1965–66 de l'UNESCO prévoyait le démarrage des Ecoles d'été dès 1966. Comme cela n'a pas été possible, un contrat a été passé entre l'UNESCO et l'UAI, la Commission 46 étant chargée de l'application de ce contrat, pour permettre l'attribution de bourses à des étudiants de pays en voie de développement; cette solution permettait l'emploi de sommes disponibles, et répondait à un besoin analogue à celui qui avait entraîné la formation des écoles d'été. L'existence de ces bourses a été annoncée par le truchement du *Bulletin d'Information* de l'UAI, mais les demandes d'allocation ont été peu nombreuses; de plus la somme attribuée par l'UNESCO ne permettait d'attribuer qu'une bourse: l'examen des dossiers a permis au Secrétaire Général de l'UAI, après consultation avec le Président de la Commission, d'attribuer cette bourse à Monsieur de la Reza, étudiant bolivien désirant étudier l'astronomie à l'Université de Genève, pendant l'année scolaire 1966–67.

Un contrat avec l'UNESCO, de $8000 par année, est en préparation pour les années 1967 et 1968, pour permettre l'organisation des Ecoles d'été.

COMMISSION INTERNATIONALE POUR L'ENSEIGNEMENT DES SCIENCES (CIES)

La représentation de l'UAI à l'ICSU entraînait immédiatement une demande d'admission de l'UAI à la CIES, émanation de l'ICSU.

La CIES a tenu depuis 1964, trois réunions: Dakar, janvier 1965; Paris, septembre 1965; Paris, septembre 1966, réunions auxquelles l'UAI était représentée par MM. Minnaert et Schatzman (Dakar), par M. Schatzman (Paris, 1965) et par M. Kourganoff (Paris, 1966), représentant le Président de la Commission.

Le colloque et la réunion de Dakar ont fait l'objet d'un compte-rendu publié dans le *Bulletin d'Information* no. 14 de l'UAI et reproduit ici.

Rapport sur la réunion de Dakar

Du lundi 18 janvier au vendredi 22 janvier 1965, a eu lieu à Dakar (Sénégal) le congrès de la CIES. Dans cette Commission collaborent les sections d'enseignement de 10 Unions scientifiques internationales, parmi lesquelles l'Union Astronomique Internationale. La CIES est directement responsable envers le Conseil International des Unions Scientifiques (ICSU).

Le but spécifique de la CIES est l'étude des relations mutuelles entre l'enseignement des mathématiques et des différentes sciences naturelles. Le congrès de Dakar avait encore un but plus étendu, choisi en rapport avec les problèmes locaux: 'L'Enseignement des Sciences et le Progrès Economique'.

Le délégué de l'UAI était le Professeur E. Schatzman, président de la section d'enseignement nouvellement créée; le Professeur M. Minnaert, qui avait déjà collaboré avec la CIES, a pu également assister à ce congrès' Ces deux astonomes étaient les seuls représentants de notre Science. En vue du congrès, des rapports préliminaires avaient été préparés sur l'enseignement de chaque science en particulier. Un tel rapport avait aussi été rédigé pour l'astronomie par M. Minnaert.

Les jours du 14 au 18 étaient consacrés aux séances de groupes d'études, auxquelles prirent part environ 40 personnes. Chaque participant était membre de deux des six groupes. M. Schatzman assista: (1) au groupe C, étudiant les rapports entre les mathématiques et la physique; (2) au groupe E, examinant la question de la formation des techniciens. M. Minnaert prit part: (1) au même groupe C; (2) au groupe D concernant l'enseignement aux niveaux élémentaire et secondaire. Un programme particulièrement détaillé fut élaboré en vue de l'introduction des mathématiques modernes dans l'enseignement secondaire, spécialement pour la section scientifique.

La partie officielle du congrès fut ouverte solennellement en présence du Ministre de l'Education Nationale, de membres de l'Université de Dakar et d'autres personnalités officielles. Environ 75 participants étaient inscrits, parmi lesquels un assez grand nombre de scientifiques et d'enseignants, appartenant à des pays en voie de développement. Pendant le congrès, les rapporteurs des six groupes d'étude firent part au congrès en son entier de leurs conclusions, auxquelles de nombreux compléments et modifications furent présentés par les participants.

Le congrès était bien d'accord sur la nécessité d'enseigner l'astronomie dans toutes les écoles secondaires, pendant un minimum d'une heure par semaine durant une année, soit dans la dernière ou dans l'avant-dernière classe; de cette manière on peut faire usage des connaissances déjà obtenues en physique et en mathématiques. On y ajouta le souhait d'introduire des leçons occasionnelles d'astronomie dans les écoles élémentaires et au commencement des études secondaires.

Lorsque les géologues et les météorologistes proposèrent à leur tour d'introduire des notions de leurs sciences dans l'enseignement secondaire, on considéra la possibilité d'un cours intégré des 'sciences de la Terre' comprenant l'astronomie, la géologie, la géophysique, la météorologie;

chacune de ces sciences étant enseignée par un professeur spécial, mais dans l'ensemble étant strictement coordonné.

Le congrès de Dakar donna également l'occasion de se réunir au comité restreint de la CIES, qui vérifia les comptes et qui décida de proposer à l'ICSU les trois actions suivantes: (1) rédiger les rapports complets des séances du congrès, rapports qui seront présentés à l'UNESCO pour publication; (2) étudier plus en détail les questions concernant la formation des enseignants et des techniciens; (3) préparer un 'Source Book', contenant un grand nombre d'exemples d'applications des mathématiques dans l'enseignement secondaire et dans les deux premières années universitaires pour la Physique, la Mécanique, l'Astronomie et la Cristallographie. Ce livre sera également édité par l'UNESCO.

Une journée spéciale, très intéressante, fut consacrée aux contributions que l'enseignement des sciences peut apporter au développement économique des pays nouveaux.

Questions posées aux réunions de Paris

A la réunion de Paris, 1965, la question principale qui s'est trouvé posée est celle d'un 'Source Book' dans lequel des sujets de problèmes de physique, cristallographie, astronomie, et mécanique seraient proposés et résolus en illustration du cours de mathématiques du niveau 'fin des études secondaires, début des études supérieures'. Le professeur Kristansen, de Copenhague, a été chargé de l'unification de l'ouvrage du point de vue des énoncés et du traitement mathématique de la solution.

En même temps, les représentants de l'UNESCO ont soumis à la CIES un projet d'étude de l'Enseignement intégré des Sciences dans l'Enseignement secondaire.

Ce projet paraît présenter le plus grand intérêt, et il faut signaler à ce sujet le remarquable travail accompli aux Etats-Unis pour donner dans les Ecoles secondaires un enseignement intégré des Sciences de la Terre.

Il paraît difficile que la CIES joue le rôle de rédacteur d'ouvrages d'enseignement secondaire. Il paraît important par contre que la CIES joue le rôle de conseiller scientifique auprès de l'UNESCO. Le problème de l'Enseignement intégré des Sciences dépasse largement par son importance les problèmes de l'Enseignement secondaire. Il se pose comme un problème philosophique; des recherches pédagogiques doivent être entreprises à son sujet, il est extrêmement actuel et fait partie du progrès même des sciences. Il mérite d'être étudié au niveau universitaire, aussi bien du point de vue des débutants que du point de vue de la formation des chercheurs.

L'étude de l'enseignement intégré des sciences pose aussi le conflit entre l'unité de la Science et la spécificité des sciences; d'un côté la profonde unité des processus naturels, de l'autre la spécificité des méthodes associées aux concepts de champ et de force, d'onde, de particule, de processus élémentaire et de processus collectif, qu'il s'agisse des méthodes de mesure, ou des méthodes de traitement théorique de ces concepts.

L'étude de ces questions, à la fois scientifiques et pédagogiques, doit être entreprise avec le but de donner, en 1967, l'enseignement qui permettra aux étudiants de faire des recherches d'avant-garde en 1977.

Un résumé de cette tendance peut être trouvé en particulier dans le programme de travail proposé à la Commission.

TERMES DE RÉFÉRENCE DE LA COMMISSION 46

La Commission 46 n'administre pas les bourses, n'organise pas l'échange des astronomes, n'établit pas la liste des situations disponibles.

La Commission 46 coopère avec la Commission 38 pour élaborer la liste des Institutions, des chercheurs qui y travaillent et des directions de recherches dans lesquelles de jeunes astronomes pourraient être guidés.

La Commission 46 a pour tâche principale l'étude de la pédagogie de l'astronomie en vue de la préparation à la recherche. Elle examine les résultats obtenus au moyen d'une formation physique et mathématique convenable, et le rôle de la formation astronomique au cours des études universitaires.

Elle s'efforcera de promouvoir la publication d'un recueil de questions d'examen, d'exercices de travaux pratiques et de problèmes, afin de faciliter la tâche de formation des astronomes.

Ceci pose le problème de l'enseignement intégré des sciences qui devrait être étudié par la CIES. Toutefois les questions de l'Enseignement secondaire (enseignement moyen, ou secondary level) ne sont pas du ressort de la Commission 46. Celle-ci pourrait tout au plus agir comme conseiller scientifique en ces matières.

On trouvera maintenant quelques propositions.

(1) Le recrutement des astronomes serait facilité par la distribution aux étudiants d'une brochure sur l'astronomie, les matières enseignées, les directions de recherches existantes, les carrières possibles. Cela peut être une affaire locale ou nationale; la Commission 46 lance un pressant appel pour que de telles brochures soient rédigées.

Une telle brochure devrait comporter des conseils aux étudiants sur la meilleure formation.

(2) Le recrutement de bons étudiants et de futurs chercheurs de bonne qualité repose beaucoup plus sur l'organisation d'un enseignement de haut niveau que sur l'existence d'un enseignement élémentaire et de caractère semi-populaire.

(3) L'étude des problèmes pédagogiques peut s'accompagner d'une aide pédagogique aux petites institutions. L'UAI pourrait charger un astronome d'éditer un ouvrage réunissant des questions d'examen, des énoncés d'exercices pratiques et de problèmes, éventuellement accompagnés de leur solution.

A cet égard, il convient de signaler les problèmes posés par Irwin (**2**); la formation des astronomes doit comprendre un entraînement aux techniques d'observations. Cet entraînement ne peut se faire sans instruments modernes. S'il s'agit d'instruments optiques ils doivent aussi opérer dans un ciel clair. On peut signaler à cet égard le succès de l'Observatoire de Genève dans la formation de ses étudiants au moyen de l'instrument de recherche de 1 mètre situé à 360 kilomètres de Genève, à l'Observatoire de Haute-Provence, en France.

E. SCHATZMAN
Président de la Commission

BIBLIOGRAPHIE

1. Salam, A. *Phys. Today*, March 1965.
2. Irwin, J. B. 1966, *Science*, **152**, 1597.
3. McNally, O. 1965, The Education of Astronomers in the United Kingdom, *Q.J.R. astr. Soc.*, **7**, 22.
4. 1965, The Earth Science Curriculum Project Investigating the Earth, The American Society of Physics.

ANNEXE I. LISTE DES INSTITUTIONS AYANT RÉPONDU AU QUESTIONNAIRE

Afrique du Sud: Univ. Obs., Cape Town: Univ. South Africa, Pretoria.
Allemagne (DBR): Univ. Kiel.
Allemagne (DDR): Techn. Univ. Dresden.
Australie: Monash Univ. Clayton; Perth Obs.; Sydney Obs.
Bulgarie: Sofia Univ.

Canada: David Dunlap Obs., Univ. Toronto; Univ. Western Ontario, London.
Espagne: Inst. Marina San Fernando, Cadiz.
Etats-Unis: Boston Univ., Boston, Mass.; Calif. Inst. Techn., Pasadena, Calif.; Yerkes Obs., Univ. Chicago, Illinois; Univ. Colorado, Boulder, Col.; Harvard Univ., Cambridge, Mass.; Indiana Univ., Bloomington, Ind.; University Mich., Ann Arbor, Mich.; Pomona College, Claremont, Calif.; Swarthmore College Sproul Obs., Swarthmore, Penna.
Finland: Astron. Obs., Helsinki.
France: Univ. Aix-Marseille; Univ. Montpellier; Univ. Paris.
Grande-Bretagne: Royal Obs., Edinburgh; Royal Greenwich Obs., Herstmonceux; Univ. London Obs.
Italie: Oss. Astrof. Arcetri, Firenze; Oss. Astron., Trieste.
Japon: Univ. Tohoku, Sendai; Univ. Tokyo; Univ. Kyoto.
Mexique: Univ. Mexico.
Pays-Bas: Kapteyn Astron. Lab., Groningen; Sonnenborgh Obs. Univ., Utrecht.
Pologne: Astron. Obs. Univ. Copernicus, Torun.
Portugal: Astron. Obs. Univ. Lisbonne; Astron. Obs. Univ., Porto.
Suède: Univ. Obs. Stockholm.
Suisse: Univ. Obs. Basel; Obs. Genève.
Turquie: Univ. Ankara.

Received after completion of the report

U.S.A.: Princeton Univ.; Southern Missionary College, Collegedale, Tennessee; UCLA.
U.R.S.S.: Conseil astronomique.

ANNEXE II. RÉUNION DE DÉLÉGUÉS DE LA COMMISSION 46 ET D'EXPERTS

(Nice, 8–9 juillet 1965)

Propositions relatives à la formation des jeunes astronomes

I. La formation de jeunes astronomes exige un entraînement pratique et théorique de longue durée.

Organisée sur le plan international, notamment à l'intention de pays en voie de développement, cette formation pourrait se faire au moyen d'une Ecole de formation pratique d'Astronomie de l'UAI. Cela comporterait une formation essentiellement pratique 'A', reçue pendant 3 mois. Dans cette période, les étudiants seraient astreints à un contrôle strict de leur travail.

Il serait souhaitable que des étudiants reçoivent une formation 'B', en étant envoyés pour une année dans une institution où ils trouveraient une formation pratique et une formation théorique plus spécialisées.

II. L'organisation de cette Ecole Internationale pourrait se faire de la façon suivante:

(1) Trouver, dans chaque institution astronomique, le nombre et le niveau des jeunes astronomes qui pourraient faire l'objet d'un encadrement scientifique sérieux pendant un an.

(2) Choisir, chaque année, le pays et l'institution qui pourraient donner la formation intense de 3 mois. Si cela est à la fois souhaitable et possible, l'institution pourrait inviter des assistants (instructeurs) d'autres institutions.

(3) Adresser aux différents pays et institutions, en vue du recrutement des étudiants, un programme de l'Ecole, indiquant notamment les exigences du niveau (graduate level en mathématiques ou en physique, et au moins une connaissance de base de l'astronomie), les conditions de travail pendant la période d'entraînement intense, les conditions d'admission dans les institutions pour les deux types d'entraînement, une demande séparée devant être faite pour les 2 stages.

III. Le fonctionnement de cette Ecole Internationale demande les conditions suivantes:

(1) Désignation par le Comité Exécutif d'un secrétaire général pour 3 ans, prenant en charge l'organisation des Ecoles de formation de l'UAI.

(2) Désignation, dans l'Institution qui prend en charge la formation de 3 mois, d'un directeur.

(3) Préparation d'un budget de l'Ecole. Il conviendra de déterminer la contribution de l'institution recevant l'Ecole de 3 mois et la contribution des institutions recevant les stagiaires d'un an.

(4) Préparation d'un contrat UAI-UNESCO, permettant notamment d'aider au recrutement des étudiants pour le stage de formation intense. On rappelle qu'il existe déjà des ressources appréciables pour les stages de longue durée.

ANNEXE III. RAPPORT SUR L'ENQUÊTE CONCERNANT LA FORMATION DE JEUNES ASTRONOMES

A la date du 19 novembre 1965, j'avais reçu vingt et une réponses à l'enquête commencée en septembre.

Vingt réponses approuvent en termes souvent chaleureux, l'intention de la Commission 46. Toutes les Institutions qui estiment pouvoir contribuer à l'initiative de la Commission 46 indiquent ce qu'elles peuvent faire.

Le classement suivant résume les propositions reçues:

(1) *Ecole de trois mois*

L'Observatoire de Bucarest demande à l'Academie des Sciences de Roumanie d'approuver l'organisation d'une Ecole pratique d'Astrométrie en 1967.

Le Dublin Institute for Advanced Studies (Professeur Wayman) envisage d'accueillir une telle Ecole dans quelques années.

L'Observatoire d'Edinbourg (Professeur Bruck) envisage de participer à l'organisation de ces Ecoles après 1967.

L'Observatoire de Harvard (Professeur Menzel) envisage de tenir une telle Ecole en 1968, moyennant quelques arrangements de durée. La lettre du Professeur Menzel pose clairement le problème du budget de l'Ecole.

Le Joint Institute for Laboratory Astrophysics, Boulder (Professeur Garstang) pourrait tenir une Ecole de Physique solaire après 1966.

Le Department of Astronomy de l'Université de Manchester (Professeur Kopal) pourrait s'engager à tenir une Ecole de trois mois en 1967.

Le Comité National d'Astronomie de Prague (Professeur Sternbeck) envisage de participer à ces Ecoles après 1968.

L'Observatoire du Mont Stromlo (réponse indicative du Professeur Bok) tient chaque année, à plus petite échelle (8 étudiants), une telle Ecole. L'extension paraît possible au cours d'une des prochaines années.

L'Observatoire de Wesleyan (Professeur Page) pourrait prendre une fois cinq étudiants (astrométrie, astrophysique, galaxies).

L'Observatoire de Yerkes (Professeur Morgan) pourrait participer aux Ecoles dans quelques années.

(2) *Entraînement d'un an*

Les Institutions suivantes peuvent contribuer à l'entraînement d'un an:

Royal Greenwich Observatory (Professeur Woolley); jusqu'à trois étudiants par an.

Observatoire de Liège (Professeur Swings): peut accueillir des étudiants à condition que soit résolu le problème financier.

National Radio Observatory (Green Bank, Professeur Heeschen). Un certain nombre de postes de 'Research associate' sont disponibles, pour des étudiants ayant déjà reçu un entraînement. Il est probable que seuls les meilleurs étudiants des Ecoles de l'UAI pourraient bénéficier des bourses du NRAO.

(3) *Autres réponses*

Les institutions suivantes: Berkeley, Johannesburg, Kitt Peak, Royal Society (au nom des Institutions Britanniques), Saltsjöbaden, Torun, pour des raisons variées, ne peuvent prendre part à l'organisation des Ecoles elles-mêmes.

Les réponses laissent penser que le stage de formation de un an serait possible, mais l'absence de précision à ce sujet rend nécessaire une nouvelle enquête.

L'Académie des Sciences de Paris (au nom des Institutions Françaises) fait une réponse évasive; le Professeur Rienäcker, au nom de l'Académie des Sciences de Berlin, annonce une réponse précise avant le 31 décembre 1965 concernant une Ecole en 1967.

En résumé, on peut faire les prévisions suivantes:

1967: Manchester ou Bucarest
1968: Harvard
1969: JILA
1970: Edinburgh ou Mont Stromlo

Il paraît vraisemblable que dès maintenant des plans peuvent être faits pour six Ecoles consécutives.

Il avait été convenu de chercher un secrétaire pour l'Ecole. Après consultation, le Comité Exécutif de l'UAI a désigné le Dr J. Kleczek.

ANNEXE IV. A SUMMER SCHOOL ON SATELLITE TRACKING FOR YOUNG OBSERVERS FROM THE COUNTRIES OF EASTERN EUROPE AND ASIA

(Tashkent, 7–15 September 1965)

In agreement with the plans of multilateral co-operation between the Academies of Sciences of East European and Asian countries on the problem 'Scientific investigations based on satellite tracking data' the Astronomical Council of the U.S.S.R. Academy of Sciences organized in Tashkent jointly with the Tashkent Astronomical Observatory during 7–13 September 1965 a summer school on satellite tracking for young observers. Tashkent was chosen since a good tracking station functions at the Tashkent Astronomical Observatory and because fine weather is typical there for this season, which is very important for training in tracking of satellites. Young astronomers from Bulgaria, Hungary, the DDR, Mongolia, Poland, Romania, the Soviet Union and Czechoslovakia (altogether 40 students) participated at the school.

The program included 14 lectures on general topics related to: methods of satellite tracking with the aid of photographic cameras; reduction methods of data obtained; and possibilities for using the data for scientific purposes. The detailed program was the following.

On 7 September three lectures were delivered on the utilization of precise photographic satellite tracking data for cosmic geodesy and for the study of the figure of the Earth and the Earth atmosphere.

On 8 September the NAPA camera used in the Soviet Union for precise satellite tracking was demonstrated and a review of methods for taking photographs of artificial satellites presented. Besides, the students were informed in a special lecture about possibilities to insure precise timing.

The third and the fourth days (9–10 September) were dedicated to two important items: predictions of satellite passages and methods of the reduction of photographic data. These problems were considered also in two lectures delivered on 13 September.

Then separate groups of students (5–6 persons in a group) were trained daily in handling the cameras, plate reduction, reception of precise time signals and calculation of ephemeris. Each student was given the opportunity to take several photographs of the balloon-satellite Echo-2 and to reduce the obtained plates. The lectures were delivered and the training carried out by the scientific staff of the Astronomical Council of the U.S.S.R. Academy of Sciences of the Institute of Theoretical Astronomy and of the Pulkovo and Tashkent Astronomical Observatories.

As a result, young observers, who as a rule had not been enough experienced in photographic tracking had a good training under the guidance of experienced observers and were given theoretical knowledge necessary for the work in this domain.

On 13 September a closing lecture was given by Professor V. P. Shcheglov, the Director of the Tashkent Observatory on the history of the astronomy of middle Asia. The lecture was followed by a two-day tour to the ancient Observatory of Ulughbok near Samarkand. The students had also the opportunity to see the beautiful architectural monuments of Samarkand.

Sight-seeing tour of the city of Tashkent and a visit to the Institute of Ancient Oriental Manuscripts were also arranged for the participants. Several documentary films about Middle Asia and a new picture about Ulughbok—the ancient founder of the Uzbok astronomy—were shown.

On the closing day all the participants were given special diplomes.

N. P. SLOVOKHOTOVA,
Scientific collaborator
of the Astronomical Council,
U.S.S.R. Academy of Sciences

INTER-UNION COMMISSION ON SOLAR AND TERRESTRIAL RELATIONSHIPS

The Commission held its final meetings in the House of Youth, Belgrade, on 25–26 August 1966, prior to the Inter-Union Symposium on Solar-Terrestrial Physics.

The meetings to be reported are:

(*i*) Business meeting of the Commission.

(*ii*) Open discussion on 'Suitability of Solar Data for Terrestrial Correlations'.

(*iii*) Individual contributions on solar-terrestrial relations.

At the time of the meetings, the membership was as follows:

Members

IAU	IUGG	URSI
C. W. Allen (U.K.), *President*	Mrs N. P. Benkova (U.S.S.R.)	G. M. Allcock (N.Z.)
R. Giovanelli (Australia)	T. Obayashi (Japan)	D. K. Bailey (U.S.A.), *Secretary*
E. R. Mustel (U.S.S.R.)	W. O. Roberts (U.S.A.)	R. Coutrez (Belgium)
M. Waldmeier (Switzerland)	E. J. Vassy (France)	A. H. Shapley (U.S.A.)

Corresponding Members

E. K. Bigg (Australia)
Mrs. H. Dodson-Prince (U.S.A.)
F. Link (Czechoslovakia)
R. Michard (France)
M. Nicolet (Belgium)

There were 155 names on the Roster of Active Workers.

Business Meeting of the Commission (IUCSTR)

After some informal discussion the meeting began at $10^h\ 20^m$, 25 August 1966. It was attended by the following Members or Corresponding Members:

C. W. Allen, *President* (in the Chair), D. K. Bailey, *Secretary*, E. R. Mustel (represented by N. W. Pushkov), T. Obayashi, W. O. Roberts, A. H. Shapley, M. Waldmeier.

The President of the newly formed Inter-Union Commission on Solar-Terrestrial Physics (IUCSTP), Dr H. Friedman, attended by invitation.

It was accepted that since the IUCSTR is to terminate in 1966 by ICSU resolution the present meetings would be the last held by the present Commission.

The report of the previous meeting at Hamburg, 24 August, 1964 (Trans. IAU, **12B**, 651, 1966) was adopted.

The President reported briefly on the use of available funds for travel and subsistence allowances. The funds are from ICSU sources and accounting is in the hands of the IAU Secretariat.

Future organization of solar terrestial relations

The main concern of the meeting was to consider the situation occasioned by various efforts to coordinate the science of solar-terrestrial relations (STR) with the wider science now using the title solar-terrestrial physics (STP). These activities were briefly reviewed and will be recorded. The Working Group on Solar-Terrestrial Physics, mentioned in the report on the Hamburg Meeting (p. 652), had met in September 1964 and recommended the formation of a Symposium Committee on STP to coordinate symposia in this field. This was considered by the IUCSTR through the medium of circulars in November 1964 and February 1965. The IUCSTR approved the Working Groups recommendation but considered it was *not* necessary to form a *new* Symposium Committee. In February 1965 the IUCSTR reported its willingness to modify itself with the view to accepting the responsibility of coordinating STP. This proposal was considered by ICSU at its meeting in Munich, April 1965 and resulted in Resolution 5 (ICSU Executive Committee 3rd Meeting, Summary Record, p. ii, 1965) which stated:

> The Executive Committee resolves that the Inter-Union Commission on Solar and Terrestrial Relationships (IUCSTR) be urged to make appropriate additions to its membership, and modify its terms of reference so as to provide representation for all interested Unions, Commissions of ICSU; and that the Unions and Commissions concerned be likewise urged to coordinate the arranging of symposia through this Commission.

In accordance with this resolution the IUCSTR submitted in July 1965 Draft Proposals for its own modification in order to undertake coordination of STP symposia. Several discussions with interested parties were held and in November 1965 the 'Proposals for an International Solar-Terrestrial Commission (ISTC)' was submitted to ICSU in time for consideration at the ICSU Bombay Meeting in January 1966. However, these proposals were not actually used as basis for the discussions in Bombay. The future of STP research and STP symposia was conjoined with the future of IQSY and led to the Resolution VII (ICSU Eleventh General Assembly, Summary Record, p. 37, 1966) which might be stated in a reduced form as follows:

(1) That the IUCI and IUCSTR be terminated in 1966,

(2) That an Inter-Union Commission on Solar-Terrestrial Physics (IUCSTP) be set up in 1966:

  (*i*) to promote, organize and coordinate international research in solar-controlled disciplines of geophysics,

  (*ii*) to coordinate all ICSU symposia in the field of solar-terrestrial physics, and

  (*iii*) to provide certain advisory services on operational activities arising out of IQSY.

The ICSU resolution was discussed and was considered to be a satisfactory basis for the future promotion of STR. It was noted that no Parent Union had been proposed and satisfaction was expressed that the IUCSTP would be directly responsible to ICSU.

The Commission then discussed what recommendations, arising out of the work of the IUCSTR, should be sent forward to the IUCSTP. The following recommendations were made:

(*a*) That the IUCSTP be asked to accept advisory responsibilities with respect to the *Quarterly Bulletin of Solar Activity* and other relevant Services.

(*b*) That the IUCSTP might at times be expected to promote operational planning, associated working group activities, and timely assessments of particular subjects such as problems arising from overlapping areas of interest in various scientific Unions under ICSU.

(*c*) That the IUCSTP give attention to maintenance of continuity in STP programmes and data.

(*d*) That the Roster of Active Workers maintained by the IUCSTR (containing 155 names) be sent to the IUCSTP Executive. This represents a useful list of scientists actively interested in STR.

(*e*) That, although the Reports **1** to **10** on Solar-Terrestrial Relations constitute an important part of the IUCSTR (and previous JCSTR) output, it is *not* recommended that the IUCSTP necessarily continue such reports.

(*f*) That archive material relating to the IUCSTR (and previous JCSTP) be retained for the present by C. W. Allen at the University of London Observatory. The following spares of the Reports on STR still exist 8 of **2**, 55 of **4**, 15 of **5**, 14 of **6**, 34 of **7**, 29 of **8**, 10 of **9**, and 13 of **10**.

It was recommended that the Report of the Belgrade meetings of IUCSTR should be prepared for presentation to the ICSU Executive Committee meeting in Monte Carlo (October 7–8, 1966), and that the points relevant to the IUCSTP should be sent to its President (Dr H. Friedman) as soon as possible in order that the points may be considered in the planning of future activities.

As a conclusion to the Meeting the Commission noted with satisfaction the creation of the new IUCSTP and commended the President (IUCSTR) for his leadership in the Commission's activities and for his participation in the arrangements for specification of terms of reference for the new Commission which should serve as an excellent basis for future international activities not only in the field of solar-terrestrial physics but also for the continuation and extension of activities of the character that have in the past occupied our Commission.

The activities and responsibilities of the IUCSTR *will close* when the various reports and recommendations have been prepared, circulated and submitted. A statement is to be circulated to the Roster of Active Workers. The President is to report to ICSU in October. He will also arrange with the IAU to close the accounts after the ICSU meeting in October.

Open Discussion on Suitability of Solar Data for Terrestrial Correlations

Meeting held at 2 p.m. 25 August 1966.

Attended by Allen, Bailey, Bernard, Buzchanov, Campbell, Cardús, Davies, Dieminger, Dinulescu, Djurkovič, Ehmert, Fortini, Friedman, Jacchia, Karabin, Knuth, Kundu, Landstreet, Lauter, Lincoln, Lüst, Matsushita, Milogradov, Minnis, Mustel, Obayashi, Parepeanu, Piccardi, Pick, Popovici, Przeuijervic, Pushkov, Ramanathan, Righini, Roberts, Roederer, Sawyer, Senatre, Shapley, Svestka, Tandberg-Hanssen, Valniček, van Zandt, Waldmeier, White, Zhulin.

The President introduced the discussion by saying that the solar data to be discussed related to the extended routine data intended for use rather than exploration or speculation. The purposes of such data could be summarized: (*i*) for studying effects on the Earth and surrounding Earth-space, (*ii*) for coordinating specialized observations and researches, and (*iii*) for studying the Sun itself. The discussion would take this order of priorities.

Points that might be considered in discussing the data were, (*i*) the measurement often was no more than an *indication* of the physical quality required, (*ii*) measurements were frequently subjective and required standardization and comparability, (*iii*) data were often applied through correlations and secondary correlations may require elimination, (*iv*) it was often necessary to segregate a total solar effect into individual activity contributions, (*v*) there is need for numerical indices for descriptively complex phenomena.

The discussion was subdivided into (*i*) generalities, such as the points mentioned above, (*ii*) slowly varying phenomena, active areas, (*iii*) rapidly varying phenomena, flares, and (*iv*) phenomena not associated with active areas including the quiet Sun.

Discussion points

Shapley: Difficulty of dealing with very extensive and almost continuous data. There may be need for more analysis before publishing.

Waldmeier: Filament should be reported in a numerical form suitable for correlation with *M* regions.

Mustel: More whole-disk magnetic field measurements should be undertaken (recommendation A made). Early magnetic storm data should be sorted into recurrent and sporadic.

Lincoln: The CRPL are preparing to publish the Mount Wilson magnetic data.

Pushkov: Introduced the question of solar terrestrial services (recommendation C made).

Valniček: Introduced the question of computer processing of data.

Jacchia: Suggested that steps be taken to segregate the active and quiet Sun components of 10·7 cm radiation (recommendation B made).

Friedman: Referred to the variations in 2 to 10 Å solar emissions and their relations to ionosphere responses. These variations challenge solar observers to detect analogous variations in spectroheliographic material.

Kundu: Suggested 3 cm solar-radio recordings might give indication of similar variations.

Godoli (Submitted): The peculiarities of long-term correlations can be ascribed to the behaviour of solar indices.

IUCSTR Recommendations arising out of Discussion on Solar Data

A. *Solar Disk Magnetic Observations*

In view of the importance of relating the whole picture of the Sun's magnetic field to the magnetic field in the neighbourhood of the Earth and in interplanetary space

It is *recommended:*

That observatories with facilities for high spectroscopic resolution consider the possibility of making magnetic observations of the whole solar disk on a regular basis.

B. *Quiet and Active Radio Observations*

In view of the great use made of solar radio observations at various wavelengths for correlation with solar XUV flux measurements at all stages of the sunspot cycle

It is *recommended:*

That radio observatories offering suitable data be asked to endeavour to express the results quantitatively in such a way that the two components, quiet and active, can be segregated.

C. *International Solar-Terrestial Services*

Considering:

(1) That the collection and publishing of solar-terrestrial data is prepared by many different organizations in several countries,

(2) That this work has received some coordination in the past, but that this coordination needs to be maintained and improved with particular reference to such questions as standards and definitions,

(3) That it is by no means certain that all data thus published are of sufficient use to the scientific and operating communities to justify their publication,

(4) That it is possible that additional observations could usefully be made,

(5) That in the field of meteorology the WMO provides services for all user purposes,

The IUCSTR *recommended:*

(1) That a clear need now exists for a more closely coordinated *international solar-terrestrial services* such that the needs of both the scientific and operational communities requiring solar-terrestrial information will be effectively served.

(2) That the IUCSTP be asked to consider how the need for such services can be met.

Individual Contributions on Solar-Terrestrial Relations

Meeting held at 9.30 a.m. 26 August 1966.

It was an open meeting attended by about 45 persons. The papers submitted were reported very briefly since it may be expected that many of them will be published elsewhere.

T. Obayashi dealt with the identification of *M*-regions from their influence on the diurnal variations of cosmic rays. The phasing of the various disturbances was illustrated on a 'floral' diagram.

S. Matsushita gave an analysis showing the detailed effects on the geomagnetic S_q and L variations that could be associated with solar activity.

M. Pick described a study of the slowly varying component of solar radio emission in relation to the structure of associated centres of solar activity. The distribution in the radio spectrum is found to be related to a factor measuring certain geometric properties of sunspots and their magnetic fields.

E. Tandberg-Hanssen gave results on the magnetic fields of prominences obtained from the High Altitude Observatory magnetograph at Climax.

B. Valniček illustrated the features of an active flare prominence event observed with high spectroscopic resolution in Hα.

G. Piccardi questioned the causality of correlating phenomena. Previously his F-test had been related to flares and he now finds a good relation between the P-tests and proton flares.

P. Bernard discussed a 2-year period in geomagnetic variations but could not ascribe it to a solar activity effect; he find the magnetic field of ionospheric winds quantitatively sufficient to account for it.

Constance Sawyer described the preference for certain heliographic longitudes of 45 proton-flare regions of cycle 19. With 15 famous flares of the last century they define a rotation period of 27·2135 days. The rigid rotation and long lifetime of this pattern suggest a deep-seated magnetic field of large scale.

T. E. van Zandt had studied the variations in the upper atmospheric ratios $O^+/He^+/H^+$ throughout the minimum of the solar cycle. The interpretation of these variations led to the suggestion that the coronal variations lagged a year behind those of the chromosphere. Support for this suggestion was found in the E and F layer critical frequency variations.

C. Popovici, A. Parepeanu and V. Dinulescu considered the major flares during 1958–61 which did *not* give rise to geomagnetic storms. Certain aspects and morphology regularities were detected.

K. Davies studied the sudden ionosphere frequency deviations resulting from flares and showed that the characteristic electron increase occurred mostly above 100 km. Time variation studies suggest that the D and E–F effects originate from separate parts of the solar ionizing spectrum. The relaxation time of the E region is found to be about one minute.

M. L. White described how the mechanism of rotation and drift in the solar photosphere should give rise to a 3-leaf Rossby wave pattern such as occurs in 'dishpan' experiments. Many solar phenomena, including the equatorial drift, can be explained on this basis.

D. H. Menzel submitted (in his absence) a paper on the nature of sunspots and solar activity.

Invited Address

An invited address by Dr E. G. Bowen, Chief of Division of Radiophysics, CSIRO, Sydney, entitled 'Solar and Planetary Interactions' had been planned for the afternoon of 26 August.

Unfortunately this address had to be cancelled as Dr Bowen was unable to make the trip from Australia.

Closure of IUCSTR Meetings

The Belgrade final Meetings of the IUCSTR closed at the end of the Individual Contributions meeting on the morning of 26 August 1966.

REPORT OF THE WORKING GROUP ON PHOTOGRAPHIC MATERIALS

1. The formation of this group dates from a meeting convened during the Hamburg General Assembly, 31 August 1964, by R. H. Stoy, supported by I. S. Bowen and H. Haffner. There was a large and enthusiastic attendance, which passed with acclamation the following resolution: 'In view of the great importance to astronomy of the supply of the most suitable photographic material, this meeting of astronomers from many countries requests the Executive Committee of the International Astronomical Union to set up a working group responsible either to itself or to some nominated Commission to investigate, (*a*) the present supply of photographic materials and the best methods of procuring and handling them, (*b*) the potential consumption of specialized photographic material, with a view to interesting the various manufacturers in its production.'

2. By letter of 15 September 1964, the General Secretary conveyed the approval of the Executive Committee of the IAU to the establishment of the Working Group, with the following membership:

President: R. H. Stoy
Secretary: D. S. Evans
Members: Aly, Mme Andrillat, Bertola, Gollnow, Griffin, Haffner, Mendoza, Mikhailov, Savedoff, Sharpless, Harlan J. Smith, Steinlin, Valniček, Velghe, Wempe.

3. The operations of the Working Group so far undertaken, demonstrate the great extent of the problems involved. Some of these are capable of being solved almost immediately by the action of astronomers themselves. Others require consultation with manufacturers. Yet others, which are concerned with problems of currency allocation, the possibility that some materials may be designated as of strategic importance by manufacturing countries, and so forth, may require discussion at governmental level. It is clear that there is a continued necessity for the work of the Group, and that numbers of problems will require a long time for their evaluation even before solutions can be sought. The investigations undertaken so far have been admirably supported by the manufacturers of photographic materials, and by a considerable number of observatories. The number of the latter is, however, not as great as it should be. It must be emphasized that, in order to get meaningful statistics describing the supply position, it is just as important for observatories not now facing difficulties, to report their situation, as it is for those who have problems in need of solution.

We take this opportunity of thanking those manufacturers and observatories who have collaborated in the work of the Group, and urge those who have not yet sent in reports to do so.

The foregoing remarks make it clear that the present report cannot be comprehensive. Its aim is to draw attention to a few of the problems which have revealed themselves, in the hope that these can be tackled first, leaving others for the continuing work of the Group.

4. A circular letter was sent to all manufacturers known to be active in the production of photographic materials suitable for astronomical purposes. A report incorporating their replies was produced in January 1966 and circulated with a questionnaire to a large proportion of the world's observatories. Copies of these documents and of an analysis of the replies may be had on request from the Cape Observatory.

5. The American Astronomical Society has set up a Committee of its own in the same general field. It should be clear that because of the relatively close proximity of the U.S. and Canadian observatories to the Kodak works at Rochester, N.Y., and because of the large numbers of observatories and astronomical departments of universities in these two countries, the problems encountered are somewhat different from those in the world at large. Nevertheless there are

problems, many of which have been discussed at an expert level in the admirable memoranda prepared for private circulation by Mr William C. Miller of Mount Wilson and Palomar Observatories. It is evident that the U.S. and Canada constitute a natural area for the establishment of a system of cooperative ordering and supply, and the practical problem is one in which the IAU Working Group as such can have very little constructive role, except to offer encouragement.

It is understood that a cooperative ordering scheme is being operated in West Germany and one may be begun in Australia.

A Joint Meeting of the AAS and IAU Groups was held in 1966 July 29, together with representatives of Eastman Kodak, Rochester, at the Mees Observatory, Naples, N.Y. Members of the Working Group who attended were Evans, Sharpless, Smith and Savedoff. The discussion ranged over the whole field of supply and storage. Items from the conference summary which are of particular interest are incorporated in the report referred to in paragraph 4. It seems desirable for the Group to organize discussion leading to definite recommendations on particular topics.

The obvious first choices would seem to be:

(*a*) Standardization of plate sizes for astronomical purposes.

(*b*) Canalization of demand for emulsion varieties to reduce unnecessary diversity.

(*c*) Definition of the next tasks to be undertaken by the Group.

6. *Publications.*

It is understood that Mr William C. Miller may be contemplating the production of a book. At the moment his vast experience and knowledge of photographic emulsions as applied in astronomy is scattered through various published papers and memoranda produced for private circulation. The Working Group considers that a book by Mr Miller would be invaluable, and wishes him success in his labours.

The following is a list of some publications likely to be of interest in the field:

F. Bertola, Sull'aumento di rapidità delle lastre fotographiche per uso astronomico, *Contr. Oss. astrofis. Univ. Padova Asiago*, No. 154, *Mem. Soc. astr. ital.*, **35**, no. 2, 1964.

I. S. Bowen and L. T. Clark, Hypersensitisation and Reciprocity Failure of Photographic Plates, *J. opt. Soc. Am.*, **30**, 508, 1940.

J. D. Fernie, A New Photographic Developer for Astronomical Use, *Publ. astr. Soc. Pacif.*, **74** 238, 1962.

J. D. Fernie and Sandra Holm, Studies in Astronomical Photography. I. On the Use of baked IIa-O Spectroscopic Plates, *J. R. astr. Soc. Can.*, **58**, 13, 1964.

II. On the Use of Diafine Developer, *Ibid.*, **58**, 201, 1964.

A. A. Hoag, Cooled-Emulsion Experiments, *Publ. astr. Soc. Pacif.*, **73**, 301, 1961.

William C. Miller, The Application of Pre-Exposure to Astronomical Photography, *Publ. astr. Soc. Pacif.*, **76**, 328, 1964.

: A Pre-Flasher for Photographic Plates, *Ibid.*, **76**, 433, 1964.

: Memorandum: The Mount-Wilson Palomar Photographic Plate Speed System.

: Preliminary Report on the Reduction of Reciprocity Failure in Astronomical Plates by Controlled Baking, Photographic Laboratory, Mount Wilson and Palomar Observatories, 1956: Appendix, 1966.

N. Richter and W. Hogner, Eine neue Plattensorte für Astroaufnahmen im roten Spektralbereich, *Monatsbericte der Deutschen Akademie der Wissenschaften*, **6**, 809, 1964.

R. Shaw, The Equivalent Quantum Efficiency of Aerial Films, *J. photogr. Sci.*, **13**, 308, 1965.

W. F. Swann, Summary of AAS, IAU, Kodak meeting at Naples, N.Y., 1966.

Report of Meeting on Photographic Plate Testing, Mount Wilson and Palomar Observatories, 1966.

Kodak Plates and Films for Science and Industry.

R. H. STOY
President of the Working Group

4. COMMISSION DES EPHEMERIDES

PRÉSIDENT: Professor G. M. Clemence, Yale University Observatory, Box 2023, Yale Station, New Haven, Connecticut, U.S.A.

VICE-PRÉSIDENT: Dr G. A. Wilkins, H.M. Nautical Almanac Office, Royal Greenwich Observatory, Herstmonceux Castle, Hailsham, Sussex, England.

COMITÉ D'ORGANISATION: G. A. Čebotarev, A. Danjon†, R. L. Duncombe, F. Fernandez de la Puente†, W. Fricke, J. Kovalevsky, D. H. Sadler.

MEMBRES: Cunningham, Fayet, Gondolatsch, Gossner, Herrick, Lahiri, Lederle, Planelles, Sadler (F. M. McBain), Sato, Sinzi, Tsukamoto†, Woolard.

The General Assembly, at its meeting in 1964, adopted the following Resolution No. 4, which had been proposed by a joint discussion of Commissions 4, 7, 8, 19, 20, and 31: 'The International Astronomical Union endorses the final list of constants proposed by the Working Group on the System of Astronomical Constants and recommends that it be used in the national and international astronomical ephemerides at the earliest practicable date.'

The report of the Working Group mentioned and the proceedings of the joint discussion have been published in *Transactions of the International Astronomical Union*, Volume XIIB, pages 591–625.

The appended reports of the various offices preparing ephemerides show the extent to which it has so far been practicable to introduce the new system of constants. While excellent progress has been made, a few troublesome questions remain, most of them relating to the basic ephemerides of the Sun and Moon and the definition of Ephemeris Time.

The principal accomplishments so far effected may be summarized as follows:

The revised apparent places of fundamental stars are introduced as from 1968 as are also the day numbers for the reduction of stars from mean place to apparent place. For the ephemerides of the Sun, Moon, and planets, differential corrections (or simple formulae from which they may be calculated) are provided in those cases where the corrections are appreciable, also as from 1968. So far it has not been practicable to modify the planetary ephemerides so as to include the effect of the new mass-ratio of the Sun to the Earth and Moon. Furthermore, the new value of the constant of aberration has not been introduced into the ephemerides of the Sun, Moon, and planets, nor into the measure of Ephemeris Time. Questions relating to these matters will be discussed at the meeting of the Commission.

G. M. CLEMENCE
President of the Commission

APPENDIX: REPORTS OF THE DIRECTORS OF THE NATIONAL EPHEMERIDES

Instituto y Observatorio de Marina, San Fernando (Cadiz), Spain

The work in this Ephemeris Office has been to continue the publication of the 'Efemerides Astronómicas', the 'Almanaque Nautico' and the 'Almanaque Aeronautico', with little modifications not being worth mentioning.

VICENTE PLANELLES

Astronomisches Rechen-Institut, Heidelberg, Germany

Fundamental Catalogue. Reports on the formation of FK4 were given by Gliese (**1**) and Kopff, Nowacki, and Strobel (**2**). Formulae and tables for the reduction of small proper motions and co-ordinate differences, which had been used in the construction of FK4, were published by Kopff, Nowacki, Strobel and Mürle (**3**). For the reduction of positions and proper motions of GC and N30 it is suggested to use the tables of systematic differences FK4-GC and FK4-N30 computed by Brosche, Nowacki and Strobel (**4**). Heinemann (**5**) compiled a list of star catalogues from 1900 to 1962 which is a continuation and supplement to previous lists.

In preparation of further improvements of the fundamental catalogue and of a revision of the GC an analytical method for deriving systematic differences in star positions between different catalogues was developed by Brosche (**6**). This method is applicable in machine computation. In continuation of previous work, and in co-operation with the U.S. Naval, the Yale, and the Potsdam-Babelsberg Observatories many star catalogues have been placed on punch cards.

Further information on the fundamental catalogue and related subjects is given in the Report of Commission 8.

APFS and exchange of ephemerides. Within the period under review four volumes of 'Apparent Places of Fundamental Stars' were published (1965–68). Within the year 1967, the volumes for 1969 and 1970 will appear. In view of the early appearance of the volumes it is recommended again that the data may be photomechanically reproduced by Ephemeris Offices wishing to include apparent places of stars in their almanacs.

As recommended by IAU Commission 4 (*Trans. IAU*, **12B,** 105), the new value of the constant of aberration adopted in the IAU System of Astronomical Constants was introduced in the APFS from the volume 1968 onwards. Since in the IAU System of Astronomical Constants the constants of precession and nutation remained unchanged, the calculation of apparent star places is affected by the revision of the constant of aberration only. For the reduction of observations made before *and* after 1968, observers are advised to use the formulae given in APFS 1968, page v, for reducing the star places of the previous volumes of APFS to the new value of the aberration.

The apparent places of all stars included in FK4 Sup (**7**) have been computed at Heidelberg from the year 1965 onwards. They will *not* be published, but are available to observers on request. They are definitely not recommended for inclusion in national ephemerides.

The communication of apparent and mean places to Ephemeris Offices has been continued. The number of recipients has increased from 14 to 19 by the addition of the following institutions: Observatorio Astronomico Nacional (Bogotá), Observatorio Astronomico (Coimbra), Direktorat Hidrografi Angkatan Laut (Djakarta), Hidrografski Institut, Jugoslavenski Ratne Mornarice (Split), Institut für Meereskunde (Warnemünde).

Bibliography

1. Gliese, W. The Right Ascension System of the Fourth Fundamental Catalogue (FK4). *Veröff. astr. Rechen-Inst. Heidelberg*, no. 12, 1963.

2. Kopff, A., Nowacki, H., Strobel, W. Individual corrections to FK3 and the declination system of the Fourth Fundamental Catalogue (FK4). *Veröff. astr. Rechen-Inst. Heidelberg*, no. 14, 1964.

3. Kopff, A., Nowacki, H., Strobel, W., Mürle, A. Formeln und Tafeln zur differentiellen Präzession und zur Übertragung von Eigenbewegungen. *Veröff. astr. Rechen-Inst. Heidelberg*, no. 13, 1964.

4. Brosche, P., Nowacki, H., Strobel, W. Systematic differences FK4-GC and FK4-N30 for 1950.0. *Veröff. astr. Rechen-Inst. Heidelberg*, no. 15, 1964.

5. Heinemann, K. Verzeichnis von Sternkatalogen 1900–1962. *Veröff. astr. Rechen-Inst. Heidelberg*, no. 16, 1964.
6. Brosche, P. Representation of systematic differences in positions and proper motions of stars by spherical harmonics. *Veröff. astr. Rechen-Inst. Heidelberg*, no. 17, 1966.
7. *Preliminary Supplement to the FK4 (FK4 Sup). Veröff. astr. Rechen-Inst. Heidelberg*, no. 11, 1963.

W. FRICKE

Nautical Almanac Office, U.S. Naval Observatory, Washington, D.C., U.S.A.

The Nautical Almanac Office is responsible for the publication of four annual ephemerides, the American Ephemeris—for astronomers, the Nautical Almanac—for surface navigation, the Air Almanac—for air navigation, and the Ephemeris—for land surveyors, as well as the pamphlet Astronomical Phenomena.

The American Ephemeris and Nautical Almanac, unified with the Astronomical Ephemeris issued by H.M. Nautical Almanac Office, Royal Greenwich Observatory commencing with the volume for 1960, has continued to be published.

Beginning with the year 1968 the IAU System of Astronomical Constants (*Trans. IAU*, **12B**, 105, 1966), was introduced wherever practicable. An explanation of these constants with references and formulae was printed in the *Supplement to the A.E. 1968*, 'The Introduction of the IAU System of Astronomical Constants'. This supplement was included in the 1968 edition and also published separately. A section in this supplement is devoted to the changes in, and corrections to, the Explanatory Supplement to the Astronomical Ephemeris and the American Ephemeris and Nautical Almanac, due to the introduction of the IAU System of Astronomical Constants. Also in the 1968 edition, the new position of the north pole of Mars was adopted in the computation for the Physical Ephemeris (**1**).

In cooperation with the U.S. Naval Oceanographic Office and H.M. Nautical Almanac Office a new set of *Sight Reduction Tables for Marine Navigation* is being designed and produced (*Navigation*, **13**, 97, 1966). Volumes **4**, **5** and **6** are now being photo-composed. The remaining volumes are being computed. Extension of the computer-controlled photo-composition method to the Air Almanac and portions of the American Ephemeris is planned for the near future.

The navigational almanacs have continued to be published with minor changes. Beginning with 1965 the precision of the Sun and stars in the Air Almanac was increased to $0\overset{'}{.}1$.

The preparation of advance predictions of solar eclipses in accordance with the recommendation of the International Astronomical Union has been continued (**2, 3**). Circumstances of solar eclipses with argument in U.T. have also been prepared (**4, 5, 6**).

Progress of a cooperative project with Yale University Observatory to put star catalogs in machine readable form has been continued (**7, 8**).

The special Ephemeris of the Radio Longitude of the Central Meridian of Jupiter, System III (1957·0) is being continued for the years 1968–71.

Additional astronomical information has been prepared and published to meet public demand (**9, 10, 11**).

A new electronic computer, the IBM 360/40 with four magnetic tape units was installed, in August 1966, in the new G. W. Hill Scientific Computing Center.

The research project on the motions of the principal planets conducted jointly by Yale University Observatory, the Watson Scientific Computing Laboratory and the U.S. Naval Observatory has been continued (**12, 13, 14, 15, 16**).

Bibliography

1. Meiller, V. Physical Ephemeris of Mars, 1877–1967, *U.S. nav. Obs. Circ.*, 98, 1964.
2. Duncombe, J. S. Solar Eclipses, 1971–75, *U.S. nav. Obs. Circ.*, 101, 1964.

3. Duncombe, J. S. Solar Eclipses, 1976–80, *U.S. nav. Obs. Circ.*, 113, 15 December 1966.
4. Duncombe, J. S., Morrison, B. L. Total Solar Eclipse of 30 May 1965, *U.S. nav. Obs. Circ.*, 102, 1964.
5. Duncombe, J. S. Annular Solar Eclipse of 20 May 1966, *U.S. nav. Obs. Circ.*, 109, 1965.
6. Duncombe, J. S. Total Solar Eclipse of 12 November 1966, *U.S. nav. Obs. Circ.*, 110, 1965.
7. Meiller, V. Durchmusterung and Henry Draper numbers of Albany General Catalog Stars, *U.S. nav. Obs. Circ.*, 104, 1964.
8. Elvove, S. Astronomical Data in Machine Readable Form, *U.S. nav. Obs. Circ.*, 111, 1966.
9. Tables of Polaris and the Equation of Time 1965–66, *U.S. nav. Obs. Circ.*, 97, 1964.
10. Weston, L. B. Sunlight, Moonlight, and Twilight for Antarctica 1966–68, *U.S. nav. Obs. Circ.*, 107, 1965.
11. Morrison, B. L. Phases of the Moon 1800–1959, *U.S. nav. Obs. Circ.*, 112, 1966.
12. Franz, O. G., Mintz, B. F. Tables of X and Y, Elliptic Rectangular Coordinates. *Astr. Papers, Am. Eph.*, **19**, pt. I, 1964.
13. Duncombe, R. L., Klepczynski, W. J., Van Flandern, T. Geocentric Distance and Velocity of Venus 1961–70, *U.S. nav. Obs. Circ.*, 96, 1964.
14. Jarnagin, M. P. Expansions in Elliptic Motion. *Astr. Papers, Am. Eph.*, **18**, 1965.
15. Duncombe, R. L., Tufekcioglu, Z., Larson, G., Rectangular Coordinates of Mercury 1800–2000, *U.S. nav. Obs. Circ.*, 106, 1965.
16. Eckert, W. J. Lunar Theory. *Astr. Papers, Am. Eph.*, **19**, pt. II (in preparation).

R. L. DUNCOMBE

H.M. Nautical Almanac Office, Royal Greenwich Observatory,
Herstmonceux, England

The Office has continued, without substantial change, the production of the four ephemerides:

The Astronomical Ephemeris—for astronomers,
The Nautical Almanac—for surface navigation,
The Air Almanac—for air navigation,
The Star Almanac—for land surveyors.

The first three ephemerides are completely unified with the corresponding U.S. editions, which are printed separately from identical reproducible material; this material can be made available, at a small charge, to the official almanac-producing agency in any other country that desires to publish similar ephemerides in this form.

Advanced proofs of the first part of *The Astronomical Ephemeris* have been circulated to ensure that the tabulated data are available, in ample time, to all agencies responsible for the preparation of navigational and other almanacs.

As recommended by Commission 4, at its meeting in Hamburg on 28 August 1964 (*Trans. IAU*, **12B**, 105, 1966), the IAU System of Astronomical Constants was introduced into *The Astronomical Ephemeris* for 1968, in so far as this was practicable. A thorough study of the effects of the introduction of the new system on the ephemerides has been included in the *Supplement to the A.E. 1968*; copies of the Supplement are included in the A.E. 1968, and are also available separately on application. The full effects of the changes have been incorporated only in the tabulations of the Day Numbers, for the Observatories and for Table VII (the factors S and C), and formally (since the changes are not significant to the tabular precision) in the ephemeris for the physical observation of Mars, for satellite V of Jupiter and for the satellites of Uranus and Neptune. For all other ephemerides, for which the corrections are appreciable, differential corrections, or simple formulae from which such corrections can be calculated, are included in an appendix (*Formulae and Corrections for Conversion to the IAU*

System of Astronomical Constants), so that the printed ephemerides can be readily converted to the new system.

As envisaged at the meeting in Hamburg, it has proved to be impracticable to incorporate the changes in the ephemerides of the planets due to the changed value of the mass ratio $S/(E + M)$. Further, pending international discussion of, and agreement on, the possible redefinition of the epoch of Ephemeris Time, it was decided not to introduce the effect of the change in the adopted value of the constant of aberration either into the measure of E.T. or into the ephemerides.

As from the edition of 1969, it has been found possible to include the full effects of the changes in all ephemerides other than the fundamental ephemerides of the Sun, Moon and planets; these are already in print and it has been considered adequate to give differential corrections, or formulae for them, in an appendix. Differential corrections for the ephemeris of the Moon have been evaluated for the years 1968 to 1971, and it is possible that a corrected ephemeris for the Moon will be introduced in *The Astronomical Ephemeris* for 1972.

Considerable thought has been given to the possible redesign of *The Astronomical Ephemeris* to accord more with current requirements; but present indications are that the only resulting changes will be some additions, particularly in respect to geocentric distances, and some consequential rearrangements. It is hoped that, by 1972, the fundamental ephemerides will be printed from computer-controlled photo-set copy.

A new electronic computer, the ICT 1909, with punched-card and punched paper-tape input and output, and with four magnetic tape units, was installed in May 1966; this will, in due time, make practicable the systematic calculation and storage, in machine-readable form, of the fundamental ephemerides of the Sun, Moon and planets in accord with the theories and constants as adopted by the Commission.

As from the editions for 1965, an increased precision (of $0\overset{'}{.}1$) has been included in the ephemerides of the Sun and the stars in *The Air Almanac;* some other additions, changes and consequential rearrangements were also made. No changes have been made in the other publications. But the Office has cooperated with the U.S. Naval Oceanographic Office and the Nautical Almanac Office at the U.S. Naval Observatory in the design and production of a new set of *Sight Reduction Tables for Marine Navigation* (*J. Inst. Navig.*, **19**, 281, 1966).

The prediction of occultations by the Moon and planets has been somewhat extended, particularly in respect of the radio sources. A comprehensive programme for the prediction of lunar occultations is now available on the new ICT 1909 computer.

As part of the British contribution to the tracking and orbital analysis of artificial satellites, the Office is providing a comprehensive service of 'look-data' predictions for selected satellites and stations; as a result of this service, the number of useful observations has been more than doubled. Special ephemerides have also been supplied for the use of radio astronomers.

The Office is currently celebrating the two-hundredth anniversary of the first publication of *The Nautical Almanac and Astronomical Ephemeris*; this was prepared and published by the fifth Astronomer Royal, Nevil Maskelyne, in 1766 for the year 1767. It made practicable the method of lunar distances for the determination of longitude at sea. To mark this event, special supplements were included in both *The Astronomical Ephemeris* and *The Nautical Almanac* for 1967.

D. H. SADLER

Bureau des Longitudes, Paris, France

Le travail de rédaction des publications du *Bureau des Longitudes* a été poursuivi au cours des trois dernières années selon un rythme analogue à celui des périodes précédentes. Le contenu de la *Connaissance des Temps* n'a pas subi de modifications autres que celles qui ont été annoncées dans le dernier rapport et la présentation, depuis 1966, du système UAI de constantes astronomiques.

Par suite l'automatisation des calculs et des procédés d'impression, les intervalles de tabulation des données concernant les planètes seron iniformisés à partir de la *Connaissance des Temps* pour 1968: de jour en jour pour l'ascension droite et la déclinaison de toutes les planètes, de 2 jours en 2 jours pour les coordonnées écliptiques héliocentriques et géocentriques (sauf Mercure qui conserve la périodicité de 1 jour).

A partir de 1968, les constantes pour la réduction des étoiles seront publiées avec la valeur de la constante de l'aberration du système UAI. Les coordonnées des observatoires et les données relatives à l'ellipsoïde terrestre seront aussi ramenées à ce système. Par contre, aucun changement dans les valeurs numériques servant à calculer les éclipses et les éphémérides du Soleil, de la Lune et des planètes ne sera introduit *au moins jusqu'en 1971* et les éphémérides publiées ne présenteront donc aucune discontinuité avant cette date. Des tables donnant les corrections aux éphémérides de la Lune pour les rendre compatibles avec le système UAI seront publiées à partir de 1968.

Depuis septembre 1965, le *Bureau des Longitudes* est équipé d'un calculateur électronique Bull Gamma 30 S à disques magnétiques. Tous les calculs des tables publiées par la *Connaissance des Temps* et les *Ephémérides Nautiques* sont calculés sur cette machine. La publication des résultats des calculs ainsi programmés sera effective dans sa totalité à partir de la *Connaissance des Temps* pour 1968. Un convertisseur de bandes SEA, permettant de convertir des bandes à 7 canaux perforés par l'ordinateur en bandes Monotype commandant la fonte des caractères, a été mis en fonction en avril 1966. Il assurera dans l'avenir l'impression automatique de tous les tableaux de nos publications. Déjà, 164 pages de la *Connaissance des Temps* pour 1967 ont été composées de cette manière et ce nombre s'accroîtra d'année en année. La moitié des *Ephémérides Nautiques* pour 1968 seront composés par le même procédé.

Parallèlement, au Bureau des Longitudes il a été effectué des recherches concernant les tables du Soleil de Le Verrier et le temps employé dans ces tables (1). Un grand effort est aussi déployé vers l'étude du problème des satellites de Jupiter (2) avec l'espoir d'obtenir des tables provisoires qui pourraient améliorer celles de Sampson en attendant qu'un nombre d'observations modernes suffisant permette de construire une théorie complète entièrement nouvelle.

Aucun changement n'est intervenu dans la présentation des Ephémérides Nautiques et les Ephémérides Aéronautiques.

Bibliographie

1. Sagnier, J. L. Comparaison numérique d'expressions issues de deux théories du Soleil, *Bull. astr.*, 3e série, **1**, fasc. 2, 9.
2. Ferraz-Mello, S. Recherches sur le mouvement des satellites galiléens de Jupiter, *Bull. astr.*, 3e série, **1**, fasc. 4, 287.

Pour A. DANJON,
J. KOVALEVSKY

Ephemeris Department of the Institute of Theoretical Astronomy, Leningrad, U.S.S.R.

During the period under review four issues of the 'Astronomical Almanac of the U.S.S.R.' for the years 1967, 1968, 1969, 1970 have been published. The positions and proper motions of stars are given in the FK4 system as decided by Commission 4 (Ephemerides) at the XIth IAU Meeting (Berkeley, 1961).

In accordance with the recommendations of the IAU adopted at the XIIth Meeting (Hamburg, 1964) some ephemerides have been based on the IAU System of Astronomical Constants.

As the above mentioned resolutions of the IAU envisaged the introduction of the IAU System of Astronomical Constants beginning with ephemerides for 1968 the corresponding differential corrections as well as recomputed ephemerides have been published in the Appendices to the Astronomical Almanacs of the U.S.S.R. for the years 1969 and 1970.

An ephemeris of the lunar crater Mösting A has been regularly published in these issues of the Astronomical Almanac of the U.S.S.R. being based from 1969 onwards on the corrected formulae due to F. Hayn and K. Kozieł.

Ephemerides of the Sun, Moon and planets have been printed in full conformity with the first part of the Astronomical Ephemeris circulated in advance by the British Nautical Almanac Office, the corrections for the transform to the IAU System of Astronomical Constants being listed in the Appendix to the Astronomical Almanac of the U.S.S.R. for the year 1970.

The navigational almanacs have been regularly published as before. The Nautical Almanac has been issued for the years 1966, 1967, 1968, 1969, and the Air Almanac for 1965, 1966, 1967, 1968.

Computations of right ascensions of 664 stars for the Soviet Time Service programme for the epoch 1970·0 have been completed and published (**1**).

Investigations concerning the methods of computations of the ephemerides of the Moon and planets as well as ones dealing with estimation of the accuracy of planetary ephemerides have been continued (**2–5**).

A series of papers coupled with elaboration of analytical theories of planetary motion as well as with numerical methods of Celestial Mechanics should also be mentioned as forming the basis for use in computation of ephemerides (**6–13**).

Several papers were devoted to research on a new theory of the physical libration of the Moon (**14, 15**) and to the use of high-speed computers for prediction of solar and lunar eclipses (**16**).

Bibliography

1. Zheleznyak, M. B., Mal'kova, A. G., Rumyantseva, L. I. Positions and Star Constants for the Stars of the Time Service Program for the Epoch 1970·0, App. to the *Bjull. Inst. teoret. Astr.*, **10,** no. 2, 1965.

2. Domanskaya, N. S. The Methods of Computation of Ephemerides of the Inner Planets and an Estimation of their Accuracy, *Bjull. Inst. teoret. Astr.*, **10,** no. 2, 1965.

3. Foursenko, M.A. Methods of the Ephemeris Calculation of the Moon, *Bjull. Inst. teoret. Astr.*, **10,** no. 4, 1965.

4. Kulikov, D. K. Ephemerides of Major Planets and Astrometry, *Bjull. Inst. teoret. Astr.*, **9,** no. 10, 1964.

5. Kulikov, D. K. A Preliminary Estimation of the Accuracy of the Inner Planets' Coordinates, *Bull. astr.*, **25,** fasc. 1–3, 1965.

6. Foursenko, M. A. The Formulae for Transition from the Center of Gravity of the Earth-Moon System to that of the Earth, *Bjull. Inst. teoret. Astr.*, **11,** no. 2, 1966.

7. Izvekov, V. A. On the Computation of the Lunar Inequality in the Sun's Motion, *Bjull. Inst. teoret. Astr.*, **11,** no. 2, 1966.

8. Izvekov, V. A. Program for Solution of a System of Equations of Observation with Formal Analysis of Data, *Bjull. Inst. teoret. Astr.*, **9,** no. 8, 1964.

9. Miatchine, V. Th. Sur une appréciation rigoureuse de l'erreur de la méthode de M. Stoermer. 1ère et 2ème partie, *Bjull. Inst. teoret. Astr.*, **9,** no. 10; **10,** no. 3, 1965.

10. Kaverin, A. A. Calculation of Solar Eclipses for the Moon, *Bjull. Inst. teoret. Astr.*, **9,** no. 10, 1965.

11. Sharaf, Sh. G., Budnikova, N. A. Théorie du mouvement de Pluton, *Trudy Inst. teoret. Astr.*, **10,** 1964; **11,** 1966.

12. Brumberg, V. A. Representation of the Coordinates of the Planets by the Trigonometric Series, *Trudy Inst. teoret. Astr.*, **11,** 1966.

13. Glebova, N. I. The Determination of the Second-Order Perturbations by Hill's Method on High-Speed Computers, *Bjull. Inst. teoret. Astr.*, **11**, no. 3, 1966.
14. Polanouere, M. D. La théorie de la libration physique de la Lune, *Bjull. Inst. teoret. Astr.*, **9**, no. 8, 1954.
15. Polanouere, M. D. Sur la détermination des constantes de la libration physique de la Lune, *Bjull. Inst. teoret. Astr.*, **10**, no. 3, 1965.
16. Rumjanzewa, L. I. Verwendung der elektronischen Rechenmaschinen zur Vorausberechnung der Sonnenfinsternisse, *Bjull. Inst. teoret. Astr.*, **11**, no. 2, 1966.

V. K. ABALAKIN

Hydrographic Office, Tokyo, Japan

The Japanese Ephemeris has continued to be published for the years 1966 to 1968. The ephemerides of the Sun, Moon and planets have been reproduced from the advanced proofs of the Astronomical Ephemeris provided by the H.M. Nautical Almanac Office, and the mean places of 1535 fundamental stars have been provided by the Astronomisches Rechen-Institut, Heidelberg; the remaining parts have been computed in the Hydrographic Office.

From the volume 1968 the new IAU System of Astronomical Constants is adopted for computation of day numbers and apparent places of circumpolar stars. Correction tables or formulae for reduction to the new system are shown at the end of the volume.

The Nautical Almanac, the *Abridged Nautical Almanac* and the *Polaris Almanac for Surveying* for the years 1965 to 1967 have continued to be published without substantial change.

In view of the reopening of the antarctic expedition *The Altitude-Azimuth Almanac* has been compiled for 1965 to 1967; it shows the apparent altitude and azimuth of the Sun at 69°00′22″S, 39°35′24″E. At the request of the National Defence Agency the *Abstract from the Japanese Ephemeris* has been compiled annually, showing apparent places of bright stars and some data for the Sun.

A program for enlarging the computer HIPAC 103, whose main parts were installed in 1963, will be completed by the end of March 1967, and international and domestic exchanges of computational data will be possible afterwards.

The value of ΔT for use in predictions of eclipses and the compilation of almanacs has been settled by arrangement with the Tokyo Astronomical Observatory from results of occultation observations made at both institutes (**1, 2, 3**).

Constant difficulty exists in keeping the list of Standard Times in the almanacs up to date because of lack of an international channel of information.

In May 1966, it was resolved by the National Committee of Astronomy, Science Council of Japan, that computations of apparent places of stars for various investigations in positional astronomy in Japan should be carried out at a specified institute using theoretically rigorous formulae. And the Hydrographic Office agreed officially to take over this responsibility. The work is to begin by the end of March 1967. Concerning this work Inoue (**4**) treated the effects of coordinate-transformations on the E-terms of day-numbers C and D.

Bibliography

1. Yamazaki, A., Sugimoto, K., Mori, T., Nisimoto, T. 1964, *Hydrogr. Bull.*, Tokyo, no. 76, p. 34 (for 1958–59).
2. Yamazaki, A., Sugimoto, K., Mori, T. 1965, *Hydrogr. Bull.*, Tokyo, no. 79, p. 43 (for 1960–61).
3. Yamazaki, A., Sugimoto, K., Haruyama, T. 1966, *Data Report of Hydrographic Observations*, Series of Astronomy and Geodesy, no. 1, p. 1 (for 1962–64).
4. Inoue, K. 1964, *Proc. Intern. Lat. Obs. Mizusawa*, no. 4, p. 101.

A. M. SINZI

Nautical Almanac Unit, Regional Meteorological Centre,
India Meteorological Department, Calcutta, India

During the period under review, two volumes of the Indian Ephemeris and Nautical Almanac, viz., for 1966 and 1967, have been published. The issue for 1968 is expected to be published by January 1967.

No material change in respect of the contents of the publication has been introduced during these years except that calendars of the Jews and Indian Parsis have been added in Part V from the 1967 issue.

As regards the IAU System of Astronomical Constants recommended at the XIIth General Assembly, the revised value of the flattening factor for Earth, viz. 1/298·25, has been introduced in the volume for 1968 and the relevant tables have been revised. The prediction of local circumstances of eclipses and occultations have also been done on this basis. The ephemerides of the Sun, Moon and planets are given in conformity with the first part of the Astronomical Ephemeris. In tabulating the solar parallax and solar aberration revised values of $8''.794$ and $20''.496$ respectively have been used, but the effect of the latter has not been included in the ephemerides of the apparent place of the Sun. The values of the Day Numbers C and D as revised on the basis of the new aberration constant have been used from this issue.

Detailed local predictions of eclipses with maps for the Annular Solar Eclipse of 23 November 1965, and the Annual Solar Eclipse of 20 May 1966 as visible in India were made.

In addition to the above-mentioned main publication the following two other publications continue to be issued annually.

1. Tables of Sunrise, Sunset and Moonrise, Moonset—a reprint from the Indian Ephemeris and Nautical Almanac. The issue for 1967 was published in 1966.
2. Rashtriya Panchangs in twelve languages, viz. in English, Sanskrit, Hindi and nine regional languages of India giving details of the Indian Calendar. The issues for 1880 Saka era (1966–67 A.D.) were published in 1966.

A new item of work undertaken is the editing and publication of an English translation of Dixit's Bharatiya Jyetish-Sastra, a notable book in Marathi language on the history of Indian astronomy.

N. C. LAHIRI

7. COMMISSION DE LA MECANIQUE CELESTE

PRÉSIDENT: Professor Y. Hagihara, Department of Astronomy, Faculty of Science, University of Tokyo, Bunkyo-ku, Tokyo, Japan.

VICE-PRÉSIDENT: Dr W. J. Eckert, Watson Scientific Computing Bureau, 612 West 115th Street, New York 25, N.Y., U.S.A.

COMITÉ D'ORGANISATION: G. A. Chebotarev, J. Kovalevsky, G. Lemaître†, P. J. Message, K. Stumpff, F. Zagar.

MEMBRES: Aksenov, Barlier, Belorizky, Böhme, Brouwer†, Brumberg, Cesco, Clemence, Contopoulos, Cook (A. H.), Cox (J. F.), Cunningham, Danby, Duboshin, Duncombe, Fabre, Ferraz Mello, Garfinkel, Hamid, Heinrich†, Herget, Herrick, Hertz, Hori, Izsak†, Jeffreys (H.), Katsis, Kozai, Kustaanheimo, Martin†, Merman (G. A.), Michkovitch, Musen, Popovič, Rabe, Rjabov, Roy, Sadler (D. H.), Schubart, Sconzo, Sehnal, Siegel, Slavenas, Subbotin, Szebehely, Thiry, Wilkens, Wilkins.

MEMBRES CONSULTANTS: Arnold, Diliberto, Kaula, King-Hele, Moser, Vinti.

INTRODUCTION

The number of works on celestial mechanics has been growing rapidly in these three years owing to the development of powerful computers and the modern progress in mathematics, being stimulated by the launching of space vehicles.

The revision of Brown's lunar theory by Eckert and his colleagues is unprecedentedly important because of the increase of the accuracy of observations in recent years, notably in radio technique, not only for the theory itself but also for unveiling the past history of the Moon.

The planetary theory on the motion of major planets is being continued at the Naval Observatory by Duncombe and his colleagues by its tradition. It is noted that Musen's modification of Hansen's method of general perturbation allows to compute the perturbation by iteration to any desired high orders. Several papers were published on the motion of the satellites of Mars. Maxwell's theory of Saturn's rings has been reviewed. The discovery of Mercury's rotation period by means of radio-echo observations invoked a new dynamical problem of resonance. It is a matter of joy that the theory of Jupiter's Galilean satellites is being attacked for completing the unfinished work of de Sitter.

Since the radio-echo observations of artificial Earth satellites are now available, the determination of the geopotential with higher tesseral harmonics has become possible by means of satellite observations, both photographic and radio-echo. Izsak, Kozai, King-Hele and Cook, Kaula, and Guier and Newton determined the tesseral harmonic coefficients by discussing satellite observations, as well as the corrections to the station coordinates. The theory of motion of satellites under the influence of such complicated geopotential is very interesting. At the same time the clarification of the nature of orbits with high eccentricities and inclinations is requested. The study of motion in interplanetary probes is now opening a new era in the history of celestial mechanics. New problems of optimization have been raised for a space vehicle with the minimum time of transit and with the minimum fuel or thrust for transfer or rendez-vous between two types of space trajectories.

By the use of electronic computers the complete survey of periodic and non-periodic orbits in the restricted three-body problem is in its culmination, not only for clearing up Brown's conjecture on the horse-shoe orbits by Rabe, Deprit and Goodrich, and for testing the hypothesis of capture or escape of an asteroid from Jupiter, but also for the whole domains of the plane of motion by Bartlett and Hénon. Hénon applied Birkhoff's idea of surface transformations for studying the whole evolution of periodic, asymptotic and ergodic orbits and greatly enlarged the scope of study by Strömgren on the restricted three-body problem.

Marvellously enough algebraic calculations, addition, multiplication, differentiation, and integration, in a completely literal fashion can now be programmed and the manipulation of multiple Fourier series was made possible. The operations of Delaunay can be carried out on computers.

Also modern electronic computers facilitate the numerical solution of the n-body problem even for n sufficiently large and a new tendency is opened of studying the global characteristics of stellar assemblies by tracing each individual trajectories and of comparing with the theory of statistical stellar dynamics. The nature of the third integral of motion, originally found in some specialized potential fields for galactic study, is now being cleared up by numerical works.

The regularizing transformations of Thiele, Levi-Civita, Sundman and Birkhoff are generalized by Lemaître, Arenstorf and are applied for computing close encounters of space vehicles.

The recent progress in the theories of non-linear differential equations, non-linear integral equations, almost-periodic functions in analysis, as well as in the theories of invariant points for surface transformations, measurability, ergodicity in topology is enabling us to attack some simplified models of the three-body problem. Krylov-Bogoliubov's averaging method in non-linear analysis and Diliberto's periodic surface theory are now being applied to the motion of celestial bodies. Wintner's and Lichtenstein-Hölder's theories are based on non-linear integral equations. Siegel-Moser's theory on the behaviour of motion around equilibrium points are founded on Birkhoff's surface transformations and Poincaré's invariant point theorem. Merman's revision of Chazy's work, Kolmogorov-Arnold's theorems on Hamiltonian systems and the generalization of Liapounov's idea on stability are worth mentioning.

Publications in book-form appeared during this tri-annual period

J. Kovalevsky, *Introduction à la Mécanique Céleste*, éd. Armand Colin, Paris, 1963.
D. G. King-Hele, *Theory of Satellite Orbits in an Atmosphere*, Butterworths, London, 1964.
R. Deutsch, *Orbital Dynamics of Space Vehicles*, Prentice-Hall, Engelwood Cliffs, N.J., 1963.
K. Stumpff, *Himmelsmechanik II*, Deutscher Verlag d. Wiss., Berlin, 1964.
P. R. Escobal, *Methods of Orbit Determination*, John Wiley, New York, 1965.
I. I. Muller, *Introduction to Satellite Geodesy*, F. Ungar, 1964.
G. V. Groves, ed., *Dynamics of Rockets and Satellites*, North-Holland Pub. Co., Amsterdam, 1965.
W. M. Kaula, *Theory of Satellite Geodesy*, Blaisdell, 1966.
V. G. Szebehely, *Theory of Orbits in the Restricted Problem of Three Bodies*, Acad. Press, New York, 1967.
V. G. Szebehely, ed., *Celestial Mechanics and Astrodynamics*, AIAA Astrodynamics Specialist Conference held at Yale University in 1963, Acad. Press, New York, 1964.
R. L. Duncombe and V. G. Szebehely, ed., *Methods in Astrodynamics and Celestial Mechanics*, AIAA Astrodynamics Specialist Conference held at Montery, Calif. in 1965, Acad. Press, New York, 1966.
E. Stiefel, ed., *Mathematische Methoden der Himmelsmechanik und Astronautik*, Bericht einer Tagung des Math. Forschungsinstitut Oberwolfach, 1964, Bibliographisches Inst. AG, Mannheim, 1966.
J. Kovalevsky, ed., IUTAM-IAU-COSPAR Symposium on the *Trajectories of Artificial Celestial Bodies as determined by Observations*, held in Paris, 1965, Springer, Berlin, 1966.

Z. Kopal, ed., Proc. *International Conference on Selenodesy*, Reidel, Utrecht, 1966.
W. N. Hess, ed., *Introduction to Space Science*, Gordon and Breach, New York, 1965.
G. Veis, ed., Proc. IAG Symposium on the *Use of Artificial Satellites for Geodesy*, Athens, 1966.
M. Roy, ed., Proc. IUTAM Symposium on *Dynamics of Satellites* at Paris, Springer, Berlin, 1963.
G. Contopoulos, ed., *The Theory of Orbits in the Solar System and in Stellar Systems*, Proc. IAU Symposium No. 25, Thessaloniki, 1964, Acad. Press, 1966.
Y. Hagihara, *Celestial Mechanics*, I. Transformation Theory, MIT Press, 1967; II. Perturbation Theory; III. Form of Integrals and Convergency; IV. Periodic Solutions; V. Topological Study.
Lectures in Applied Mathematics, V, VI, VII, *Space Mathematics*, I, II, III, Amer. Math. Soc., 1966, Rhode Is.

The following monographs have been published in Soviet Union, as reported by Chebotarev:

M. B. Bank, *Elements of Dynamics of a Space Flight*, Moscow, 1965.
V. V. Beletski, *Motion of an Artificial Satellite with respect to the Centre of Mass*, Moscow, 1965.
E. A. Grebenikov and V. G. Demin, *Interplanetary Flights*, Moscow, 1965.
G. N. Duboshin, *Celestial Mechanics, Analytical and Qualitative Methods*, Moscow, 1964.
V. A. Egorov, *Non-Plane Problem of reaching the Moon*, Moscow, 1965.
G. A. Chebotarev, *Analytical and Numerical Methods of Celestial Mechanics*, Moscow-Leningrad, 1965.
P. E. Eliasberg, *Introduction to the Theory of the Flight of Artificial Earth Satellites*, Moscow, 1965.

The National Commission on Celestial Mechanics, U.S.S.R., organized a conference on the theory of the motion of artificial Earth satellites in Riga on May 19–23, 1964. About 20 papers were read. The Commission sponsored a conference on the general problems in celestial mechanics held in Kiev on January 25–27, and 16 papers were read and discussed. A Plenum of the Commission on Celestial Mechanics was held in Tbilisi in October, 1965. The program included the scientific topics and the organizational part. The following were elected at the Plenum as new staffs of the Bureau of the Commission on Celestial Mechanics: G. W. Duboshin (chairman), V. A. Brumberg (vice-chairman), E. P. Aksenov (scientific secretary), D. E. Okhozimski, G. A. Chebotarev, I. D. Zhongolovich, E. A. Grebenikov.

TWO-BODY PROBLEM

New methods of orbit determination have been proposed by Sconzo (1) and Kranjc (2). Sconzo referred to the expressions for p and q of Charlier and applied Lambert's theorem to elliptic motions. Fredrich (3) published a new approximation process for solving Kepler's equation. Dommanget (4) studied the evolution of a binary system by considering the two-body problem of variable mass. Hadjidemetriou (5) obtained analytic solutions of the two-body problem with variable mass and Huang (30) those with mass ejection, and both applied the idea to binary systems.

Goodyear (6) obtained the solution in a completely general closed form for coordinates and their derivatives of the two-body problem. The solutions are valid for circular, elliptic, parabolic, hyperbolic and rectilinear orbits of the attracting force, and for the hyperbolic and rectilinear orbits of the repulsive force. A digital computer programme capable of generating and manipulating symbolic mathematical expressions has been used by Sconzo and his colleagues (7) to derive explicit expressions for the coefficients of the f- and g- series of Keplerian motions. Terms up to the twenty-seventh order have been obtained as polynomials in the local invariants of Stumpff (8). Recurrent formulae for computing the coefficients of f- and g- series were given by Bond (26). Herrick (9) reviewed and generalized the universal variables to be used with elliptic, hyperbolic and parabolic orbits, including circular and rectilinear orbits, and showed their relationship to Stumpff's formulae. Herget (10) has devised a programme for an IBM 1620 electronic computer for computing preliminary orbits by the variation of geocentric distances on the basis of as many observations as may be available. Stumpff (11) obtained unified expressions for elliptic, hyperbolic and parabolic motions by taking a new variable in place of the time and by developing the rectangular coordinates in Taylor series.

Jarnagin (12) published tables of the expansion of functions for the elliptic motion by revising the tables by Cayley.

Barlier (13) published a method for determining the osculating elements of an artificial satellite by using short geocentric arcs of the orbit of about ten degrees with the assumed value of the semi-major axis. The same problem by using the whole revolution is solved for determining the semi-major axis. He showed that a refined analysis of observations, passage by passage, permits the determination of zonal harmonics with a small number of observations and the detection of short-period perturbations. Muller, Barlier, Chassaing (14) studied theoretically the positions of the instantaneous pole and the values of the velocity on an arc along the apparent orbit of an artificial satellite at transit. Mrs Morael-Courtois (15) has given an automatic process and the computing programme for reconstructing a trajectory from visual, photographic and radio observations, and applied to the first French satellite D–1. She studied the use of interferometric observations and gave an useful procedure for treating numerically the atmospheric drag, while Barlier (16) gave a synthetic discussion on satellite orbits by means of a small number of observations. On the other hand, Zhongolovich (17) published a method for determining the position of an artificial satellite from simultaneous observations of its topocentric directions from known stations on the Earth's surface. Sejnalov (18) gave a schema for circular orbit computation of an artificial satellite by means of an electronic computer and Mamedov (19) that for a nearly parabolic orbit from three observations of a comet.

Popovič published papers on orbit determination method in a Yugoslavian journal.

Sconzo is working on the Fourier series expansion in terms of the mean anomaly of a product of the form $r^m \cdot \dot{r}$ by his symbols, in order to use in a guidance problem for low thrust powered trajectories.

The Smithsonian Astrophysical Observatory, the Royal Aircraft Establishment, the Leningrad Institute of Theoretical Astronomy and the Centre National d'Etude Spatiale have determined the orbits of a number of artificial satellites by means of a new and improved programme (20) and carried out the orbit corrections. Kranjc (21) proposed three new methods for orbit determination. Hertz (22) used short arcs in orbit determination.

Sochilina and Makarova (23) discussed the problem of accuracy of the determination of the elements of an artificial Earth satellite from optical observations. Miss Sochilina (24) obtained analytical formulae for errors of the determined elements, which depend on the distribution of observations along an arc of the orbit. On the basis of these formulae a criterion for the exclusion of some elements from the improvement was derived for an ill-conditioned normal system. She also considered the problem of choosing a time interval in which the systematic errors due to the imperfection of the theory are small.

Batrakov (25) derived a unique system of formulae for the improvement of an orbit by using optical observations as well as range and range velocity data, which are suitable both in the case of small and large eccentricities. Herrick (27) substituted, in place of the difference of two single integrals for a double integral in the equations for the variation of parameters, other combinations of integrals which are more effective with the parameters associated with the universal variables. Böhme (28) published tables for computing the initial values in the special perturbation computation by Kulikov's method (29) which is based on Cowell's.

BIBLIOGRAPHY

1. Sconzo, P. *Astr. Nachr.*, **287**, 149, 1963.
2. Kranjc, A. *Pubbl. Oss. astr. Univ. Bologna*, **8**, no. 18, 1963.
3. Fredrich, J. *Astr. Nachr.*, **288**, 141, 1964.
4. Dommanget, J. *Ann. Obs. R. Belgique*, (3) **9**, no. 5, 1963.
5. Hadjidemetriou, J. *Icarus*, **5**, 34, 1966; *Z. Astrophys.*, **63**, 116, 1966; Proc. *IAU Symposium No.* 25, 129, Thessaloniki, 1964.
6. Goodyear, W. H. *Astr. J.*, **70**, 189, 1965.
7. Sconzo, P., LeShack, A. R., Tobey, R. *Astr. J.*, **70**, 269, 1965.
8. Stumpff, K. *Himmelsmechanik*, Bd. I. Deutscher Verlag d. Wiss., Berlin, 1959; *NASA Techn. Note*, D–1415, 1962.
9. Herrick, S. *Astr. J.*, **70**, 309, 1965.
10. Herget, P. *Astr. J.*, **70**, 1, 1965.
11. Stumpff, K. *Mathematische Methoden der Himmelsmechanik und Astronautik*, 1, 1966.
12. Jarnagin, M. P. *Astr. Pap. Amer. Eph.*, **18**, 1965.
13. Barlier, F. *Bull. astr.*, **25**, 325, 1965.
14. Muller, P., Barlier, F., Chassaing, J. P. *Bull. astr.*, (3) **1**, 37, 1966.
15. Morael-Courtois, N. Centre Nat. Etude Spatiale, Note Techn. 5, 1966.
16. Barlier, F. *Trajectories of Artificial Celestial Bodies as determined by Observations*, 36, 1966.
17. Zhongolovich, I. D. *Bjull. Inst. Teor. Astr.*, **10**, no. 8 (121), 509, 1966.
18. Sejnalov, R. A. *Bjull. Inst. Teor. Astr.*, **10**, no. 8 (121), 537, 1966.
19. Mamedov, M. A. *Bjull. Inst. Teor. Astr.*, **10**, no. 8 (121), 549, 1966.
20. For example:
    Gooding, R. H. *Planet. Space Sci.*, **12**, 37, 1964; **14**, 1173, 1966.
    Gooding, R. H., Taylor, R. J. *R. Aircraft Estab. Techn. Rep.*, 66190, 1966.
    Hiller, H. *R. Aircraft Estab. Techn. Rep.*, 66020, 1966.
    Merson, B. H. *Dynamics of Satellites*, 83, Springer, Berlin, 1963.
    Merson, B. H. *R. Aircraft Estab. Techn. Rep.*, 66255, 1966; Proc. IUTAM–IAU–COSPAR Symposium in Paris, on *Trajectories of Artificial Celestial Bodies as determined by Observations*, 1966.
    Gaposchkin, E., in the *Smithsonian Astrophys. Obs., Special Reports.*
21. Kranjc, A. *Pubbl. Oss. astr. Univ. Bologna*, **8**, no. 18, 1963.
22. Hertz, H. G. *NASA Techn. Note*, D–3172, 1966.
23. Sochilina, A. S., Makarova, E. N. Beobachtungen künstlicher Erdsatelliten, no. 3, Berlin, 1964.
24. Sochilina, A. S. *Bjull. Inst. Teor. Astr.*, 1966 (in press).
25. Batrakov, Yu. V. Proc. XVI. Intern. Astronaut. Congress, 1965.
26. Bond, V. R. *Astr. J.*, **71**, 8, 1966.
27. Herrick, S. Proc. *IAU Symposium No.* 25, 274, Thessaloniki, 1964.
28. Böhme, S. *Astr. Nachr.*, **287**, 241, 1963.
29. Kulikov, D. K. *Bjull. Inst. Teor. Astr.*, **7**, no. 10 (93), 770, 1960.
30. Huang, S.-S. *Astrophys. J.*, **138**, 471, 1963.

PERTURBATION THEORY

Bohan (**1**) applied Wilkens' numerical-analytic method (**2**) of expanding the disturbing function. Martynenko (**3**) published tables for Newcomb's operators in the expansion of the disturbing function in powers of eccentricities in the elliptic problem of three bodies. Liakh (**4**) derived the general expressions for Newcomb's operators.

Goodyear (**5**) has applied the method of variation of parameters for computing the perturbations. The initial coordinates of an osculating two-body trajectory were taken as parameters. Perlin (**6**) published tables for computing the perturbation of elements with eccentric anomaly as the independent variable.

Danby (7) used his matrix method to derive formulae for the calculation of planetary theory in rectangular coordinates. Now he (8) developed a theory in a set of polar coordinates which, he says, is simpler and more convenient in practice, after practical evaluation of the constants.

Musen (9) expanded the perturbation of position vectors of planets in purely periodic series in powers of planetary masses. The effect of lower order perturbations to higher order perturbations is expressed as the corresponding terms in the multiple potential in Maxwell's expansion. He reduced the differential equations in the form which can be integrated by mere quadratures. Further Musen (10) has taken Cartan's integral invariants as foundation of the theory of variation of elements and obtained the differential equations for the general perturbation as the first system of Pfaffian equations associated with the linear differential form appearing in the integral invariant. The equations for general perturbations of the Gibbsian unit vector, of the Gibbsian rotation vector and of Euler's parameters are defined. The utilization of the Gibbsian position vectors represents an extension of B. Strömgren's and Musen's ideas on special perturbations to the problems of general perturbations. Euler's parameters find their application in Hansen's lunar theory. Bailie and Fisher (11) obtained an analytical representation of Musen's theory of artificial satellites in terms of the orbital true longitude, including terms with small divisors derived from the third and fourth harmonics of the geopotential. Musen (12) included the long-period terms caused by near-commensurability of mean motions in his theory of the long-range effect based on Halphen's method, at first by removing all short-period terms by numerical process, because no convenient expansion of the disturbing function is available for large values of eccentricity and inclination and the ratio of semi-major axes.

The theory of general perturbations in rectangular coordinates is the most direct of all methods of expansion of the perturbations into series, because it is intimately associated with the computation of ephemerides and has not the singularity of the zero eccentricity. Brouwer's theory of general perturbations in rectangular coordinates makes use of the variation of elements in the canonical form. But this is not of any advantage if the perturbations are expanded in trigonometric series with purely numerical coefficients. Davis (13) rewrote Brouwer's formulae (14) in terms of the standard elliptic elements but the formulae contain two terms of degree -1 in the eccentricity. Musen (15) suggests to use Eckert-Brouwer's formulae (16) for the orbit correction as a foundation of the planetary theory. Musen and Carpenter (17) referred to a vectorial expression for perturbation, which is free from those disadvantages and is convenient for numerical computation, and suggested to use the method of iteration to compute the effects of higher orders. Such effects are important not only in the planetary case but also in the case of artificial celestial bodies moving in orbits in cis-lunar space far away from the Earth.

Musen (18) developed a numerical lunar theory which can be used to obtain the rectangular coordinates of a satellite moving in a highly inclined orbital plane. The arguments of the theory are of the Laplacian type, that is, the linear functions of the true orbital longitudes of the satellite and of the Sun. He made use of Hansen's device to perform the integration and introduced a fictitious satellite whose true orbital longitude is considered as a constant until the integration is completed. The perturbations in the orbital plane are obtained by means of a W-function analogous to the one of the classical Hansen theory. Musen has shown that the combination of the ideas of Laplace, Hansen and Hill represents a convenient way to obtain a numerical lunar theory, at the same time as this work represents a further development and a simplification of the results given by Musen in his previous article (19). Musen's modification (20) of Hansen's solution of the lunar problem has been programmed and verified by duplicating some of Hansen's series. Charnow (21) developed methods for manipulating trigonometric series with numerical coefficients and literal arguments. The programme described will be used to calculate perturbations and ephemerides of both natural and artificial satellites.

Hori (22) applied a theorem by Lie in canonical transformations to the transformation theory of perturbed dynamical systems and worked out a theory of general perturbations with unspecified canonical variables. All formulae are in the form of canonical invariance, the osculating variables are given explicitly in terms of the mean variables and the theory is applicable to the case in which the unperturbed portion of the Hamiltonian depends on angular variables as well as the momentum variables.

Grebenikov (23) discussed the averaging method in celestial mechanics originally devised by Krylov-Bogoliubov, in particular by considering the case of near-commensurability of mean motions. He has given a mathematical proof for the justification of Delaunay-Hill's method of averaging. They proved the stability of the Lagrangian triangular solution in the elliptic restricted three-body problem for small eccentricities of the disturbing body for an infinite time interval, provided that the triangles formed by the undisturbed three bodies and by the disturbed three bodies differ by an infinitesimal amount and that the masses are subject to certain restrictions. Musen (24) extended the method of Krylov-Bogoliubov to higher orders and compared with Poincaré's method. The determination of the elements affected by the long-period and secular perturbations, as well as the elimination of the short-period effects, is reduced to the solution of a set of partial differential equations step by step.

Morrison (25) applied the generalized method of averaging to a perturbed vector system of differential equations by assuming no resonance. Von Zeipel's procedure is a particular case of the generalized method of averaging corresponding to an appropriate choice of the arbitrary functions arising in the averaged equations.

Gormally (26) comments van der Corput's theory of asymptotic series as a consistent basis for the perturbation method.

Ritz's variational method has been applied by Galerkin (27) for solving non-linear periodic systems of differential equations. Cesari (28) and Urabe (29) worked out the method for obtaining the solution in trigonometric polynominals with the estimation of errors.

Brumberg (30) discussed the theory of planetary motion represented in purely trigonometric form. Two methods by using the rectangular coordinates are suggested for the practical elaboration of such theory with the aid of digital computers. The first method is based on the use of an intermediary orbit according to Hill's method in his lunar theory. The second method consists in determining the formal trigonometric series for the coordinates and substituting in the differential equations of the planetary motion, thus it is reduced to solving an infinite non-linear algebraic system. The disturbing function is expanded by Tisserand-Lebeuf polynomials, which are simpler and more suitable for close lunar satellites than by an ordinary method (31).

Kovalevsky (32) has worked on the problem of the motion of a satellite with large inclination and eccentricity. He showed (33) how one can eliminate the terms of the first and second orders in the ratio of the mean motions in a closed form in inclination and eccentricity by von Zeipel's method. He also discussed the character of the motion of the pericentre whether there is revolution or libration as a function of the eccentricity and inclination. The theory is now being applied to Nereid.

Meffroy (34, 35) eliminated short-period terms of a planetary theory of the first order by means of von Zeipel's method. Newcomb's operators are used in the expansion of the disturbing function with the Delaunay variables. He carried out the computation by keeping the third degree terms in eccentricity and inclination in the disturbing function.

Ferraz Mello (36, 37) analysed von Zeipel's method and showed that it is equivalent to the method of Lindstedt-Poincaré in which only the reduction of the degree of freedom is intended. The Hamiltonian systems are classified according to the possibility of total reduction or a partial reduction. In the first case of lunar type the disturbing function does not contain long-period terms of the first order. This is the case of non-canonical equations including the effect of solar radiation pressure on the motion of an Earth satellite. In the second case of planetary type the disturbing function contains long-period terms of the first order with moderate eccentricity and inclination.

Miatchine (**38**) estimated the error of numerical integration of a differential equation by Störmer's method. He (**39**) also discussed a criterion for changing the integration steps by Runge-Kutta's method. Subbotin (**40**) studied the estimate of accuracy in the methods of computing the ephemeris of the inner planets. Foursenko (**41**) discussed the computation of the Moon's ephemeris on Brown's expansions and estimated the accuracy of the Moon's coordinates.

Gröbner (**42**) proposed a method for numerical treatment of ordinary differential equations by means of Lie's series. The method has been applied by Knapp (**43**) to the *n*-body problem. Filippi (**44**) discussed the accuracy of the method of Runge-Kutta-Fehlberg in the solution of the *n*-body problem.

Seidelmann of Naval Observatory is developing an iterative solution of Hansen's method.

BIBLIOGRAPHY

1. Bohan, N. A. *Bjull. Inst. Teor. Astr.*, **9**, no. 6 (109), 448, 1964.
2. Wilkens, A. *Astr. Nachr.*, **190**, 289, 1911.
3. Martynenko, B. K. *Bjull. Inst. Teor. Astr.*, **10**, no. 7 (120), 457, 1965.
4. Liakh, R. A. *Publ. Astr. Obs. Univ. Leningrad*, **20**, 1964.
5. Goodyear, W. H. *Astr. J.*, **70**, 524, 1965.
6. Perlin, F. H. *Bjull. Inst. Teor. Astr.*, **9**, no. 9 (112), 607, 1964.
7. Danby, J. M. A. *Astr. J.*, **67**, 287, 1962.
8. Danby, J. M. A. *Astr. J.*, **70**, 155, 1965.
9. Musen, P. *J. Observateurs*, **48**, 11, 1965.
10. Musen, P. *NASA Techn. Note*, D–2301, 1964.
11. Bailie, A. E., Fisher, D. *NASA Techn. Note*, D–1468, 1963.
12. Musen, P. *NASA Techn. Note*, D–2152, 1964.
13. Davis, M. *Astr. J.*, **56**, 188, 1952.
14. Brouwer, D. *Astr. J.*, **51**, 37, 1944.
15. Musen, P. *J. geophys. Res.*, **71**, 1966 = NASA X–641–66–119, 1966.
16. Eckert, W. J., Brouwer, D. *Astr. J.*, **46**, 125, 1937.
17. Musen, P., Carpenter, L. *J. geophys. Res.*, **68**, 2727, 1963.
18. Musen, P. *J. Observateurs*, **49**, 54, 1966.
19. Musen, P. *J. geophys. Res.*, **66**, 403, 1961.
20. Musen, P. *J. geophys. Res.*, **68**, 1439, 1963.
21. Charnow, M. L. *NASA Techn. Note*, D–3378, 1966.
22. Hori, G. *Publ. astr. Soc. Japan*, 18, no. 4, 1966.
23. Grebenikov, E. A. *Astr. Zu.*, **41**, no. 3, 567, 1964; **42**, no. 1, 190, 1965; *Publ. Sternberg Inst.*, **35**, 1966.
24. Musen, P. *J. Astronaut. Sci.*, **12**, 129, 1965.
25. Morrison, J. A. *Methods in Celestial Mechanics and Astrodynamics*, 117, 1966; *SIAM J.*, **13**, 96, 1965. cf., Kyner, W. T. *SIAM J.*, **13**, 136, 1965.
26. Gormally, J. M. *Methods in Celestial Mechanics and Astrodynamics*, 199, 1966.
van der Corput, J. G. *J. Analyse Math.*, **4**, 341, 1956; **5**, 315, 1957.
27. Kantorowitsch, L. W., Krylov, W. I. *Näherungsmethoden der höheren Analysis*, Deutsche Verlag d. Wiss., Berlin, 1956.
28. Cesari, L. *Contr. to Diff. Eqs.*, **1**, 149, 1963.
29. Urabe, M. *Archive Rat. Mech. Analysis*, **20**, 120, 1965.
30. Brumberg, V. A. *Trudy Inst. Teor. Astr.*, **9**, 3, 1966.
31. Brumberg, V. A. *Bjull. Inst. Teor. Astr.*, **11**, no. 2 (125), 1966.
32. Kovalevsky, J. *C. R. Acad. Sci.*, Paris, **258**, 4435, 1964.

33. Kovalevsky, J. Proc. *IAU Symposium No.* 25, 326, Thessaloniki, 1964.
34. Meffroy, J. *C. R. Acad. Sci.*, Paris, **260,** 1571, 1965.
35. Meffroy, J. *C. R. Acad. Sci.*, Paris, **260,** 1863, 1965.
36. Ferraz Mello, S. *Mem. Soc. astr. Italiana*, **37,** no. 2, 1966.
37. Ferraz Mello, S. *C. R. Acad. Sci.*, Paris, **258,** 463, 1964.
38. Miatchine, V. F. *Bjull. Inst. Teor. Astr.*, **9,** no. 10 (113), 668, 1964; **10,** no. 3 (116), 1965.
39. Miatchine, V. F. *Bjull. Inst. Teor. Astr.*, **7,** no. 4 (87), 257; **8,** no. 8 (101), 134, 1961.
40. Subbotin, N. S. *Bjull. Inst. Teor. Astr.*, **10,** no. 2 (115), 143, 1965.
41. Foursenko, M. A. *Bjull. Inst. Teor. Astr.*, **10,** no. 4 (117), 272, 1965.
42. Gröbner, W. *Die Lie-Reihen und ihre Anwendungen*, Berlin, 1960.
43. Knapp, H. *Monatshefte f. Math.*, **68,** 33, 1964.
44. Filippi, S. *Mathematische Methoden der Himmelsmechanik und Astronautik*, 307, Math. Forschungsinstitut, Oberwolfach, 1966.

PLANETARY AND SATELLITE THEORY

Cohen and Hubbard (**1**) computed the orbits of five outer planets by means of the method of special perturbation over 120 000 years. A remarkable libration of the close approaches of Pluto to Neptune was noticed such that the distance between these planets is never less than 18 A.U. It is concluded that the orbit of Pluto is safe from any very close approach to Neptune and no particular instability resulting from the fact that the perihelion of Pluto is inside the orbit of Neptune. The period of libration is about 19 670 years and the amplitude is about 76°. In addition there is a 2° or 3° modulation with a 4300 year period which corresponds to the period of the great inequality of Neptune and Uranus. A detailed analytical theory of the motion of Pluto has been published by Sharaf and Budnikova (**2**). The improved values of Pluto's elements based on the observations for the period 1914–18 are found to be in good accord with observations. Hori and Giacaglia (**3**) studied the secular perturbation of Pluto. After short-period terms are eliminated the equations of motion for the secular perturbation are of two degrees of freedom. They eliminated the critical argument of period 20 000 years by a method analogous to that of Hori (**3a**) applied to the motion of an artificial satellite with critical inclination. The first order theory agrees with the results of numerical integration by Cohen and Hubbard (**1**). In particular the theory shows the revolution of the argument of perihelion in a period of 15 000 000 years, the change of the eccentricity $0{\cdot}243 < e < 0{\cdot}266$ and that of the inclination $14\overset{\circ}{.}1 < I < 15\overset{\circ}{.}4$.

Choudhry (**4**) studied the motion of asteroids of the Hecuba type by means of von Zeipel's method. Kovalevsky (**5**) investigated the behaviour of the long-period terms in the motion of a satellite as disturbed by the Sun, and found two types of motion: the one type is the class of revolutionary motions with rotating pericentres and the orbit is oscillatory in which the pericentre oscillates around the maximum latitude. Orlov (**6**) applied Delaunay-von Zeipel's method to the differential equations in Hill's lunar theory and obtained an approximate analytic representation of the coordinates. These two authors did not refer to the expansion in powers of the eccentricities and inclinations, similarly to Hori's method (**7**) on the motion of the Moon.

Herget's programme (**8**) for the method of variation of arbitrary vectorial constants for the IBM 1410 includes the perturbations by any or all of the planets from Mercury to Pluto or Venus to Pluto and has been applied with complete success to more than a dozen asteroids of small eccentricity and to about 40 other asteroids in all. He also formed a differential correction programme associated with this programme.

Morando (**9**) developed a numerical method for a general theory of an asteroid disturbed by Jupiter, in which secular terms appear only in the mean motions and the time is not present in the coefficients of periodic terms. The preliminary orbit is not Keplerian, but includes the motions of the node and perihelion. The first order long-period terms are then obtained. Finally the short-period terms as well as higher order long-period terms are computed. The method has been successfully applied to Vesta, for which the maximum amplitude of long-period terms in eccentricity and inclination does not exceed their mean values and for which there is no resonance. The results are comparable with those by Perrotin in 1880 computed by means of Le Verrier's method as far as short-period terms are concerned. The method can be easily generalized to more complete planetary problems.

Wilkens is continuing his work on the commensurabilities in the system of asteroids through an investigation of the spatial motion of the Hecuba type asteroids, after having discussed in former papers (**10**) the plane motion of asteroids near the Hestia and Hecuba gaps. The study is extended to the spatial motion of the Hestia type asteroids. Schubart has generalized his theoretical work on the characteristic asteroids by including the influence of the orbital eccentricity of Jupiter, mainly for the Hilda group by neglecting the effect of short-period terms. He has seen no indication of instability in the motion of asteroids of this group.

Kotsakis (**11**) compared the computed orbit in the restricted three-body problem and the corresponding Kepler orbit and found that the orbits will fill rings round the Sun. In the case of an exploding planet at a certain distance from the centre of mass of the Sun and Jupiter the difference is small, but at certain distances the area filled by the fragments is larger in the restricted three-body problem.

R. B. Hunter, according to a letter from Roy, has carried out a preliminary computation dealing with the application of the three-dimensional elliptical restricted three-body problem to the motions of asteroids and Jovian satellites. From the numerical integration of a large number of orbits an attempt has been made to map out the regions around Jupiter of stable and unstable orbits, also the region of the asteroid distribution highly unstable due to the Jovian perturbation. The Glasgow group, R. B. Hunter, M. W. Ovendon, A. E. Roy, has obtained two results of interest: (*i*) orbits have been found which take asteroids from the asteroid region to become temporary satellites of Jupiter before being sent into a narrow belt of quasi-stable orbits about 7 A.U. from the Sun, (*ii*) the group of real asteroids at the mean motion commensurability 3/2 separates an inner group of fictitious asteroids which become direct satellites of Jupiter from an outer group of fictitious asteroids which become retrograde satellites of Jupiter.

Galibina (**12**) studied the original and future orbits of comets with eccentricities near unity. The result of computation of 26 original and 13 future orbits of long-period comets with the true anomaly as the independent variable shows that 21 of the original orbits have been elliptic and 5 hyperbolic, and that 8 of the future orbits are elliptic and 5 hyperbolic. Kastel (**13**) computed the close approach of Comet Brooks II with Jupiter in 1886 by taking into account the perturbation due to five planets from Venus to Saturn and the oblateness of Jupiter, and found that the capture to a major planet of a short-period comet is a more favourable hypothesis. Brady (**14**) studied the motion of 30 nearly parabolic cometary orbits and showed that 75% of the original orbits were elliptic before the comet entered the region of the planet and picked up energy by increasing their eccentricities and might attain hyperbolic velocity and be ejected. He suggested to study the planetary perturbations after the perihelion passage.

Makover (**15**) points out that there exists a stellar encounter causing considerable changes in cometary orbits owing to the presence of the comet cloud of Oort (**16**) and thus shows that there is no difficulty against the capture hypothesis of comets in the solar system, the ejection hypothesis as the origin of comets being inconsistent. Chebotarev (**17**) studied the motion of an infinitesimal mass in the outer region of the solar system under the action of the galactic centre and saw considerable perturbation caused on the orbital elements. The same problem has been attacked by Chebotarev (**18**) by assuming the disturbing body to be the galactic nucleus. It is shown that the stable motion of a comet with $e_0 = 0{\cdot}6$ is possible at a distance of 80 000 A.U. from the Sun and that the boundaries of the cloud of comets are approximately 60 000 to 100 000 A.U.

Ferraz Mello (**19**) has studied the planar motion of the four Galilean satellites of Jupiter in rectangular coordinates. He has shown that the use of Hill's method leads to difficulties and that higher order harmonics of Jupiter's gravitational potential as well as the relativity corrections are negligible. He thus proposed a new method based on a normalized form of the equations by using a functional relation which leads to integro-differential equations as suggested by the Laplace integral for the motion of the perijoves. The method takes into account from the very first approximation the effects of the resonances between the satellites I and II as well as between II and III. His numerical application gave perturbations of I by II in good agreement with the classical results. The second order theory is now being worked out by Ferraz Mello, while Sagnier is engaged in the generalization of the method for orbits with non-vanishing inclinations.

Lanzano (**20**) computed the third order theory for the equilibrium configuration of a rotating planet and its effect on the secular variation of the orbital elements.

Marsden (**21**) studied the motion of the Galilean satellites of Jupiter. Von Zeipel's method was applied for the elimination of the short-period terms and the number of degrees of freedom was reduced from fourteen to twelve. The long-period and critical terms were handled essentially by the *ad hoc* procedure first utilized by Laplace. He is now engaged in the programming, the addition, multiplication, differentiation and integration of Fourier series for a computer and planning to extend to a higher degree of approximation in order to determine new constants by comparing the theory with observations.

Miss A. Bec is working on the determination of the mass of Jupiter by means of the motion of the satellites VI to XII, while Kovalevsky computed the planetary perturbation of Jupiter VIII. An important program of algebraic operations on literal trigonometric series of Delaunay type with four angles and five small parameters in the coefficients is now being carried out in the Bureau des Longitudes by means of an electronic computer for future theoretical works on the lunar and satellite theories.

Wilkins (**58**) is continuing his analysis of the observations of the satellites of Mars.

Fish (**22**), Redmond and Fish (**23**) discussed whether the observed secular acceleration of Phobos is caused by the bodily tidal friction in Mars. They think the hypothesis not inconsistent that Phobos has been captured or formed in an orbit just inside a synchronous orbit $4{\cdot}5 \times 10^9$ years ago. Kotsakis (**24**) discussed the dispersion of the fragments of an exploding planet. Alfvén (**25**) by basing on the Yerkes' photometric observations of asteroids pointed out that the rotational period was not systematically related with the size of asteroids and hence that the current hypothesis on the fragmentation of a planet into asteroids could not be reconciled. He thinks that there must have been some process by which all planets including asteroids obtain this period of rotation when they are formed by condensation of interplanetary material. Alfvén suggested a stepwise condensation process. Jaschek and Jaschek (**26**) found relations among the absolute magnitudes, frequency and the age of asteroidal families. Anders (**27**) found the absolute magnitude distribution of asteroids to be similar to the Gaussian for brighter members and grading into a logarithmic curve for the fainter. If the asteroids belonging to Hirayama families were re-assembled into their parent asteroids, then the Gaussian portion of the frequency curve would be enhanced at the expense of the logarithmic portion. Anders concluded that the present state of the asteroidal belt is not in a highly fragmented one and estimated the life of Hirayama families.

Sconzo (**28**) integrated the equations of motion of a secondary body in the equatorial plane of a rotational symmetric central body rigorously by using the Weierstrassian elliptic functions. The results obtained are used to evaluate the shift of the apsidal line of the secondary body due to the oblateness of the central body. Applications are made to the two Mars satellites as well as to the fifth satellite of Jupiter. The numerical results are in good agreement with observation. The computed perihelion shift of Mercury due to the Sun's oblateness with the zonal harmonics coefficient $J_2 \sim 0.001$ amounts to 34% of the observed value.

Vinogradova and Radzievskii (29) tried to explain the discrepancy of the motion of Phobos and Deimos by the solar light pressure due to the non-homogeneity of the figure of Mars. The light pressure accelerates or decelerates the motion according as the orientation of Mars's figure.

An interesting paper was published by Goldreich (30) on the secular change of the eccentricity of satellite orbits in the solar system. One cause will be the tidal effect on a planet by its satellite which has been the subject of past frequent discussions, and the other will be the tidal effect on the satellite by the planet. Urey, Elsässer and Rochester (31) proposed this latter mechanism for facing the satellite's same phase towards the planet and decreasing the eccentricity of the satellite's orbit. Goldreich has computed the dissipation due to the tidal effect on the satellite and seen the importance of the second effect, except for Phobos, Deimos, the Moon and Jupiter V.

McCord (53) studied the Neptunian system. By assuming the tidal friction as the dominant mechanism he has seen that Triton's orbit was nearly parabolic in the past.

Antonakopoulos (54) studied the restricted three-body problem in a resisting medium and showed that all three-dimensional orbits developed into planar circular forms, by thinking it to be the cause of the present configuration of the solar system.

In the absence of satellites an oblate spinning planet will precess about the normal to its orbital plane. Goldreich (32) considered how a planet might keep satellite orbit in its equatorial plane as it precesses. He discarded the tidal effect for the cause. The major effect produced by the oblateness of a planet on its satellite orbits is a secular motion of their pericentres and nodes. If the satellite's nodal period is much shorter than the planet's precessional period, then the inclination of the satellite orbit relative to the planet's equator will not vary as the planet precesses. A satellite formed in an equatorial orbit or brought into the equatorial plane by the tide will continue to move in the equatorial plane as its planet precesses. Goldreich showed why satellites are formed in equatorial orbits, thus explained the reason why Phobos and Deimos are always on the equatorial plane of Mars.

A. F. Cook and Franklin (33) discussed in detail Maxwell's theory on the stability of Saturn's rings. Maxwell's two uniformly dense models with collisions neglected on one hand and incompressibility assumed on the other hand are supplemented by three additional models, based on a Maxwellian velocity distribution of the constituent particles. The first model neglects collisions, the second assumes adiabatic compression, and the third isothermal compression. They criticized Maxwell's assumptions that the tangential force resulting from tangential displacements was more important in determining the stability than the radial force resulting from radial displacements, and that the variation of the angular velocity across the ring did not significantly alter any stability criterion. They showed that the shearing effect of differential rotation must not be so underestimated and that radial oscillations which are unaffected by shear are the ones that govern the ring stability. They made new photometry of Saturn's rings and interpreted the result in the most general way. They preferred their first model for which the radius of frozen transparent droplets on particles is 7 μ.

The stability of infinitely thin self-gravitating galaxies, in analogy to Saturn's ring, has been considered by Toomre (34), Hunter (35) and Yabushita (36). Toomre applied Fourier-Bessel analysis to obtain equilibrium figures for galaxies with infinite extension and replaced the galaxies by a finite number of concentric rings for stability analysis. Hunter referred to a spheroidal coordinate system and employed Legendre polynomials, and obtained the frequencies of free oscillations as functions of an infinite matrix. Yabushita considered the stability of a ring with differential rotation and that of disk galaxies by making use of a cylindrical coordinate system and appropriate Bessel functions. Yabushita showed that the frequencies of free oscillations are given by the eigen-values of a certain infinite matrix and solved the eigen-value problem numerically. The upper limit of the mass of the ring which is stable against axisymmetric perturbations was given for several model rings. The numerical values differed greatly from Maxwell's due to the consideration of the effect of both edges of the ring, which was entirely neglected by Maxwell.

Chebotarev and Bozhkova (37) have obtained trigonometric formulae giving precessional data for long time intervals. The perturbed values of the eccentricity of the Earth are given up to 3×10^7 years backward. Volkov (38) obtained the equilibrium surface of a liquid mass slightly different from an ellipsoid in the gravitational field of a remote material point (Roche satellite problem) by means of Lichtenstein-Liapounov's method.

Carpenter (39) programmed the computation of first order planetary perturbations using Musen's method (loc. cit. PERTURBATION THEORY), and is now working on higher order effects. He obtained accurate integrations of the equations of motions of the five outer planets from 1800 through 2000 using the rapidly convergent Chebyshev series. These series were also used in the method of variation of elements for studies of asteroidal motions. Kolenkiewics applied the method to periodic orbits obtaining a harmonic solution of the very restricted four-body problem.

O'Keefe and Liu (41), Colombo (42), Colombo and Shapiro (43) suggested that Mercury's rotation rate, which had been considered to be 3/2 times faster than its orbital mean motion, required for stabilizing this resonant spin a sufficient deviation from axial symmetry. Pettengill and Dyce (45), Peale and Gold (46) determined the sidereal rotation period of Mercury to be 59 days in contrast to its revolution period of 88 days by means of radar-echo observation. Liu (40) studied theoretically the libration of Mercury. Goldreich and Peale (44) showed that tidal friction would bring an axially symmetric planet to an asymptotic rotation rate which is somewhat faster than its orbital mean motion. In an asymptotic spin state tidal torque averaged over an orbital period vanishes, and the maximum torque occurs at the perihelion. The precise value of the final spin is determined by the amplitude and frequency dependence of the planet's dissipation function. The existence of resonant spin states at rotation rates of any half-integer, negative or positive, multiple of the orbital mean motion has been shown by Goldreich and Peale (44). They derived $(B - A)/C \approx 10^{-8}$ for stabilizing the resonance against the disruptive influence of the solar tidal torque, and thought that there may exist stable spin states which are both faster and slower than the observed, since it is known that for the Moon there exist stable resonant spin states 0·5, 1·5, 2, 2·5 times the orbital mean motion, and hence that Mercury and the Moon have bypassed some of these stable resonances before attaining the present spin. It is remarked that Iapetus rotates synchronously because of the high orbital eccentricity.

Goldreich (48) discussed the spin of a planet or a satellite, which is losing its angular momentum through tidal friction and may approach one of its possible final states, and derived a criterion whether the final state attained is one of synchronous rotation.

Goldreich and Peale (47) studied in more detail the spin-orbit coupling and the capture probability by assuming the type of tidal torques with which MacDonald (49) re-discussed the tidal effect on the orbital elements of the Moon studied formerly by Darwin (51) and Jeffreys (52). They noticed two types of resonant spin rate for planets and satellites: the first occurs in eccentric orbits at rotation rate of each half-integer multiple of the orbital mean motion, the simplest being the synchronous rotation; and the second involves the presence of another planet or satellite with a resonant spin, in which case the planet or satellite always aligns the same axis toward the second planet or satellite at each conjunction. They derived the averaged equations of motion for the complete revolution period and formulated the stability criterion for both types of resonance. Probabilities of capturing a planet or satellite into one of the commensurable rotation states as it is being despun by tidal friction are calculated. Application to Mercury reveals the very small value of $(B - A)/C \approx 10^{-8}$ to be sufficient to stabilize Mercury's rotation period at two-thirds of its orbital period. The probability that Mercury would be captured at this resonance is calculated for several assumed forms of tidal torques. They say that Venus may be in a resonant spin state of the second type, and that the sidereal rotation period of 243·16 days retrograde, determined by Carpenter (**50**) and others, would be commensurable with its synodic motion, requiring $(B - A)/C \geqslant 10^{-4}$ for stabilizing this rotation, so that the capture probability at this resonance appears to be very small.

According to a letter from Miss Roman of NASA, D. O. Muhleman of the Jet Propulsion Laboratory has constructed a least-square computer programme that allows for the computation of the astronomical unit and corrections to eight orbital elements of the Sun and Venus relative to a provisional theory. JPL radar observations of Venus taken during the period from 1961 to 1964 yield almost 55 000 normal equations resulting in an estimate for the A.U. of 149 598 388 ± 50 km. The most significant orbital element correction is to the longitude of Venus relative to that of the Earth, which amounts to 0″96 ± 0″02 nearly twice the value derived by Duncombe. A second solution included the optical observations of Venus from 1943 to 1949. The resulting A.U. estimate was 149 598 439 ± 50 km. These estimates are based on the value 299 792·5 km s^{-1} for the speed of light. Future progress on this problem awaits an improvement in the provisional theory and the inclusion of modern optical observations.

Hertz wrote me that he was now engaged in an attempt to determine the mass of Vesta from perturbations on the asteroid (197) Arete which has approached Vesta to within a few hundredths of A.U. five times since its discovery in 1879. A very preliminary result for the mass of Vesta is $(1{\cdot}17 \pm 0{\cdot}10) \times 10^{-10}$ solar mass.

According to a letter from P. O. Lindblad, Thomas Giuli at the Royal Institute of Technology in Stockholm is investigating by means of numerical computations the effect of particle capture on the rotation of a planet. The rotational angular momentum acquired by a planet is calculated as it gravitationally attracts particles which are initially on heliocentric orbits in the ecliptic. The calculations have been completed for the case where the particles are initially on circular heliocentric orbits. The case for particles initially on eccentric heliocentric orbits is now under investigation.

Duncombe wrote me about the activity of the Naval Observatory. Duncombe himself has completed nearly the fitting of Clemence's new theory of the motion of Mars to observations extending from 1750 to 1960. Jackson has derived improved elements of Ceres, Pallas, Juno and Vesta from a discussion of over three thousand meridian transit observations extending from 1920 to 1958. Generation of precise ephemerides and comparison with observations since discovery, for six asteroids with nearly commensurable mean motions in the ratio 2:1 with Jupiter, has been continued by Klepczynski, O'Handley, Fiala and Duncombe. It is hoped to derive corrections to the mass of Jupiter from these analysis. Pascu is determining new positions of the satellites of Mars from photographs with the 61-inch astrometric reflector.

Message (**55**) and Schubart (**56**) discussed analytically the motion in the restricted three-body problem in which the mean motions of the infinitesimal mass and the smaller mass around the larger mass of the two finite masses are nearly in the ratio $(p + 1)/p$. There exist periodic solutions of the second sort of Poincaré when the commensurability is very close and of the first sort when it is less close. Message studied numerically the equations for the long-period variations of elements for the Hecuba type asteroids, while Schubart isolated the secular and critical terms by a numerical averaging process. Message has seen that the effect of the large amplitude perturbations near commensurability on a distribution of asteroids, which was originally uniform over mean motions, shows a draining off from the vicinity of exact commensurability of a magnitude large enough to account for the observed gap in the mean motion distribution. On the other hand, Goldreich (**57**) showed that special cases of near-commensurate mean motions are stable under tidal forces.

Lyttleton and Yabushita (**59**) studied the effect of stellar encounters on planetary motions by the method of variation of elements. In order to estimate the cumulative effects for a long series of encounters they assumed a Gaussian distribution of star velocities and derived the standard deviations of the changes of the orbital elements in terms of the elements, the average stellar velocity, the mean stellar density, the age of the solar system and the lower bound of the encounter parameter. Yabushita (**60**) examined the changes in orbital elements of a binary star by a distant stellar encounter and their cumulative effects.

It is added that an interesting paper was published by Urey (**61**) on the chemical evidence relative to the origin of the solar system.

BIBLIOGRAPHY

1. Cohen, C. J., Hubbard, E. C. *Astr. J.*, **70**, 10, 1965.
2. Sharaf, S. G., Budnikova, N. A. *Trudy Inst. Teor. Astr.*, **10**, 1964; **11**, 89, 1966.
3. Hori, G., Giacaglia, G. E. O. *Astr. J.*, (recent issue).
3a. Hori, G. *Astr. J.*, **65**, 291, 1960.
4. Choudhry, R. K. *J. Indian Math. Soc.*, new series, **28**, 99, 1964.
5. Kovalevsky, J. *C. R. Acad. Sci.*, Paris, **258**, 4435, 1964.
6. Orlov, A. A. *Bjull. Inst. Teor. Astr.*, **10**, no. 5 (118), 360, 1965.
7. Hori, G. *Astr. J.*, **68**, 125, 1963.
8. Herget, P. *Astr. J.*, **67**, 16, 1962.
9. Morando, B. *Bull. astr.*, (3) **1**, fasc. 4, 1966.
10. Wilkens, A. *Sitzungsber. Bay. Akad. Wiss., Math-Phys. Kl.*, 1963.
11. Kotsakis, D. *Z. Astrophys.*, **60**, 273, 1965; *Bull. Soc. Math. Grèce*, **5**, fasc. 1, 1964.
12. Galibina, I. V. *Bjull. Inst. Teor. Astr.*, **9**, no. 1 (104), 46, 1963; **9**, no. 7 (110), 465, 1964.
13. Kastel, G. R. *Bjull. Inst. Teor. Astr.*, **10**, no. 2 (115), 118, 1965.
14. Brady, J. L. *Astr. J.*, **70**, 279, 1965.
15. Makover, S. G. *Bjull. Inst. Teor. Astr.*, **9**, no. 8 (111), 525, 1964.
16. Oort, J. H. *Bull. Astr. Inst. Netherlands*, **11**, 91 and 259, 1950.
17. Chebotarev, G. A. *Astr. Zu.*, **40**, no. 5, 1963; **41**, no. 5, 983, 1964.
18. Chebotarev, G. A. *Astr. Zu.*, **43**, 435, 1966.
19. Ferraz Mello, S. *Bull. astr.*, (3) **1**, fasc. 4, 1966.
20. Lanzano, P. *Icarus*, **1**, 121, 1962.
21. Marsden, B. G. The Motion of the Galilean satellites of Jupiter, Dissertation, Yale Univ., 1966.
22. Fish, F. F. *Icarus*, **3**, 87, 1964.
23. Redmond, J. C., Fish, F. F. *Icarus*, **4**, 442, 1965.
24. Kotsakis, D. D. *Bull. Soc. Math. Grèce*, **5**, 3, 1964; *Z. Astrophys.*, **60**, 273, 1965.
25. Alfvén, H. *Icarus*, **3**, 52 and 57, 1964.
26. Jaschek, C., Jaschek, M. *Astr. J.*, **68**, 108, 1963.
27. Anders, E. *Icarus*, **4**, 399, 1965.
28. Sconzo, P. IBM Pub. Independent Res. and Develop. Program, Washington System Center, Task No. 0232–G, 1965.
29. Vinogradova, V. P., Radzievskii, V. V. *Astr. Zu.*, **42**, no. 2, 424, 1965.
30. Goldreich, P. *Mon. Not. R. astr. Soc.*, **126**, 257, 1963.
31. Urey, H. C., Elsässer, W. M., Rochester, M. G. *Astrophys. J.*, **129**, 842, 1959.
32. Goldreich, P. *Astr. J.*, **70**, 5, 1965.
33. Cook, A. F., Franklin, F. A. *Astr. J.*, **69**, 173, 1964; **70**, 704, 1965.
34. Toomre, A. *Astrophys. J.*, **138**, 385, 1963; **139**, 1217, 1964.
cf., Milder, D. M. *Astrophys. J.*, **145**, 109, 1966.

35. Hunter, C. *Mon. Not. R. astr. Soc.*, **126,** 299, 1963; **129,** 321, 1965.
36. Yabushita, S. *Mon. Not. R. astr. Soc.*, **133,** 247, 1966.
37. Chebotarev, G. A., Bozhkova, A. I. *Bjull. Inst. Teor. Astr.*, **9,** no. 3 (106), 169, 1963.
38. Volkov, M. S. *Bjull. Inst. Teor. Astr.*, **11,** no. 3 (126), 1967.
39. Carpenter, L. *NASA Techn. Note*, D–3168, 1964.
40. Liu, H. S. *J. geophys. Res.*, **71,** 3099, 1966.
41. O'Keefe, J. A., Liu, L. S. *Science*, **150,** 1717, 1965.
42. Colombo, G. *Nature*, **208,** 575, 1965.
43. Colombo, G., Shapiro, I. I. *Smithson. astrophys. Obs., Spec. Rep.*, no. 1881, 1965; *Astrophys. J.*, **145,** 296, 1966.
44. Goldreich, P., Peale, S. J. *Nature*, **209,** 1078 and 1117, 1966.
45. Pettengill, G. H., Dyce, R. B. *Nature*, **206,** 1240, 1965.
46. Peale, S. J., Gold, T. *Nature*, **206,** 1241, 1965.
47. Goldreich, P., Peale, S. J. *Astr. J.*, **71,** 425, 1966.
48. Goldreich, P. *Astr. J.*, **71,** 1, 1966.
49. MacDonald, G. J. F. *Revs. of Geophys.*, **2,** 467, 1964.
50. Carpenter, R. L. *Astr. J.*, **69,** 2, 1964; **71,** 142, 1966.
51. Darwin, G. H. *Phil. Trans.*, **170,** 447, 1879; **171,** 713, 1880 = Sci. Pap., II, 36 and 208, Cambridge Univ. Press, 1908.
52. Jeffreys, H. *Mon. Not. R. astr. Soc.*, **122,** 339, 1961.
Ruskol, E. L. *Astr. Zu.*, **43,** 829, 1966.
53. McCord, T. B. *Astr. J.*, **71,** 585, 1966.
54. Antonakopoulos, G. Thesis, Univ. of Athens, 1966.
55. Message, P. J. Proc. *IAU Symposium No.* 25, 197, Thessaloniki, 1964.
56. Schubart, J. *ibid.*, 187.
57. Goldreich, P. *ibid.*, 268; *Mon. Not. R. astr. Soc.*, **130,** 159, 1965.
58. Wilkins, G. A. See Report of Commission 20, p. 399.
59. Lyttleton, R. A., Yabushita, S. *Mon. Not. R. astr. Soc.*, **129,** 105, 1965.
60. Yabushita, S. *Mon. Not. R. astr. Soc.*, **133,** 133, 1966.
61. Urey, H. C. *Mon. Not. R. astr. Soc.*, **131,** 199, 1966.

LUNAR THEORY

A series of important works in the lunar theory is being continued by Eckert and his colleagues (**1**, **2**, **3**, **4**).

Because of the high quality and long acceptance as the standard of comparison Brown's lunar theory will, for years to come, play a key role in the discussion of the observed lunar motion and in the critical examination of new and more precise theoretical developments. In the modification of Brown's basic solution to facilitate the comparison with observation the full precision of the solution was not preserved since this was not at first considered necessary. Some of this loss of precision was regained by the 'Improved Lunar Ephemeris' in 1952. Eckert, Walker, Eckert (**1**) have made the full accuracy of Brown's solution available for the comparison with observation and to increase the precision of the relations between the computed coordinates and the parameters on which they are based. The precision of the solar terms in sine of the parallax is improved by more than one order of magnitude.

Eckert (**2**) discovered a large concentration of mass immediately inside the lunar surface by comparing his new determination of the motion of the Moon's perigee and node with observations. Eckert based his important discovery on the new solution of the lunar theory by Eckert and Smith (**3**) and on the dynamical ellipticity of the Earth determined by artificial satellites. The parameter giving the radial distribution of the lunar mass is known to be g', which is 0·6 for a homogeneous body and 1·00 for a hollow spherical shell. Brown has taken $g' = 0{\cdot}50$, the same value as for the Earth, but the value of the ellipticity of the Earth came out a little larger than the one usually adopted. Spencer Jones determined $g' = 0{\cdot}87$ from the adopted value of the ellipticity. From Eckert's values $d\pi = -3''.1$, $d\Omega = -27''.9$, he obtained $g' = 0.965$.

Eckhardt (5) solved the rotationary motion of the Moon by numerical integration and studied the forced physical libration.

Klock and Scott (6) determined the coefficients of periodic term 182 in Brown's lunar theory from the 6-inch transit observations during 1952–64. The result is in good accord with the theory of Eckert and Smith (3).

Curott (7) analysed 32 ancient solar eclipse reports to determine the secular decrease in angular velocity of the Earth. The computation is based on Ephemeris Time and recently adopted astronomical constants. Secular trends in the Earth's acceleration are noted although ambiguous. He obtained $\dot{\Omega}/\Omega = -(1.13 \pm 0.20) \times 10^{-10}$.

Musen's attempt in the lunar theory has been already described in the section of PERTURBATION THEORY, page 16.

Petrovskaya (8) continued her study on the convergence of Hill's series and estimated the magnitude of neglected terms in the power series in powers of $\mathfrak{m}$ when the expansion is truncated at the term of $\mathfrak{m}^6$. Schubart (9) solved Hill's problem and obtained plane and space periodic solutions. He referred to the method of variation of elements in terms of the components of the disturbing force. He also discussed the convergence of the solution. Hochstadt (10) discussed the discriminant $\Delta(\lambda)$ of Hill's equation as regards the stability of the solution. The λ-axis is separated to an infinite sequence of the stability and instability intervals. Hochstadt obtained the representation of $\Delta(\lambda)$ in terms of λ, based on the fact that $\Delta(\lambda)$ is an entire function of order 1/2. Blumen (11) studied the eigen-values and Levy and Keller (12) the instability intervals.

It is remarked that Brillouin (13) proposed new methods for solving Hill's equation, one is based on recurrence formulae and the other on the Brillouin-Wentzel-Kramers' method often employed in quantum mechanics. De Vogelaere (14) proposed again a new method in connection with the Störmer problem of the motion of charged particles in a magnetic field.

Szebehely (15) cleared up with negation the question whether zero-velocity curves are orbits, and gave the restriction to the initial condition and the field in order to obtain the affirmative answer, that is, the restriction is that the absolute value of the gradient of the potential should be proportional to the curvature of the equi-potential line.

It is remarkable that Barton (16) of Cambridge has written a scheme of programmes for performing the manipulation of multiple Fourier series in an entirely literal fashion on a computer. The programmes have been used to derive the lunar disturbing function in terms of the elliptic elements and the mean anomalies to the tenth order of small quantities. He has duplicated the first Delaunay operations completely and reproduced a large part of the second operation, and thus reproduced Delaunay's lunar theory by means of these programmes. He was successful in producing Hill's variational curve on the computer. Stumpff (17) discussed again his third order differential equation for the radius vector in Hill's lunar problem.

Polanuer (18) worked out a new theory on the physical libration of the Moon, which is free from resonance terms, on the basis on Lindstedt's method.

Lyttleton (19) supplemented the effect of meteoritic accretion on the lunar orbit, causing the contraction of the orbital size, in addition to the tidal dissipative effect. Lyttleton (20) further discussed the reduction of the lunar distance through meteroritic impart. He estimated that in order to reduce the lunar distance by 1/3 an amount of matter of about 1/6 of the lunar mass would be necessary to interact with the Moon. Smalley (23) concluded the maximum capture distance of the Moon was 1·3 times the present distance. Kopal and Lyttleton (21) argued against the permanent binary hypothesis as the remote history of the Earth-Moon system that the closed nature of the zero-velocity curve did not guarantee the permanent binary hypothesis.

Belorizky (**22**) compared the acceleration of the Moon in its non-perturbed motion, and the gravity acceleration at the Earth's equator by basing on the new values of the astronomical constants.

Van Flandern is making a combined solution of grazing occultations and selected central occultations by the Moon for the period 1956–66.

BIBLIOGRAPHY

1. Eckert, W. J., Walker, M. J., Eckert, D. *Astr. J.*, **71**, 314, 1966.
2. Eckert, W. J. *Astr. J.*, **70**, 787, 1965.
3. Eckert, W. J., Smith, H. F. *Astr. Pap. Amer. Ephem.*, **19**, Part 2, 1966; Proc. *IAU Symposium No.* 25, 242, Thessaloniki, 1964.
4. Eckert, W. J. *Trans. IAU*, **12 B**, 113, 1964.
5. Eckhardt, D. H. *Astr. J.*, **70**, 466, 1965.
6. Klock, B. L., Scott, D. K. *Astr. J.*, **70**, 335, 1965.
7. Curott, D. R. *Astr. J.*, **71**, 264, 1966.
8. Petrovskaya, M. S. *Bjull. Inst. Teor. Astr.*, **7**, no. 6 (89), 441, 1959; **9**, no. 4 (107), 257, 1963.
9. Schubart, J. *Math. Annalen*, **149**, 131, 1963.
10. Hochstadt, H. *Math. Zeitschr.*, **82**, 237, 1963; *Archive Rat. Mech. Analysis*, **19**, 353, 1965; *Trans. N. Y. Acad. Sci.*, (2) **26**, 887, 1964; *Comm. Pure Appl. Math.*, **17**, 25, 1964.
11. Blumen, L. E. *Comm. Pure Appl. Math.*, **16**, 261, 1963.
12. Levy, D. M., Keller, J. B. *Comm. Pure Appl. Math.*, **16**, 469, 1963.
13. Brillouin, L. *Quart. Appl. Math.*, **6**, 167, 1948.
14. de Vogelaere, R. *Can. J. Math.*, **2**, 440, 1950.
15. Szebehely, V. G. *Amer. Rocket Soc. J.*, **32**, 1938, 1962.
16. Barton, D. *Astr. J.*, **71**, 438, 1966.
17. Stumpff, K. *Mathematische Methoden der Himmelsmechanik und Astronautik*, 51, 1960; *NASA Tech. Note* D–1540, 1541, 1963.
18. Polanuer, D. *Bjull. Inst. Teor. Astr.*, **9**, no. 8 (111), 550, 1964; **10**, no. 3 (116), 230, 1965.
19. Lyttleton, R. A. *Icarus*, **1**, 137, 1962.
20. Lyttleton, R. A. *Icarus*, **5**, 162, 1966.
21. Kopal, Z., Lyttleton, R. A. *Icarus*, **1**, 455, 1963.
22. Belorizky, D. *J. Observateurs*, **49**, 46, 1966.
23. Smalley, V. G. *Icarus*, **5**, 491, 1966.

RELATIVITY EFFECT

Aoki (**1**) derived the difference of the coordinate time and the proper time for the planetary motion on the basis of Schwarzschild's line element of the general relativity theory, and identified the coordinate time as the Ephemeris time. Geisler (**2**) obtained the orbit of a particle with the coordinates expressed parametrically in terms of the coordinate time, and Geisler and McVittie (**3**) derived the orbital period in Schwarzschild's field of gravitation, in order to meet the observational test of the relativity theory proposed by Kustaanheimo and Lehti (**10**), that is, the difference between the sidereal period and the anomalistic period for an orbit of high eccentricity.

Podurets (**4**) derived Einstein's equations for a spherically symmetrical motion of a continuous medium, and Zeldovich (**5**) considered the motion of matter under the mutual gravitational force in uniform density and velocity distribution. The Newtonian and Einsteinian solutions of the problem are compared. Atkinson (**6**) derived two general integrals for Einstein's field equations in static spherically symmetric case, and computed light tracks near a very massive star.

Dicke (7) proposed the regression of the node of Icarus as sensitive measure of the oblateness of the Sun. Francis (8) derived the observable difference of 50 km between the Newtonian and Einsteinian theories at a close approach of Icarus to the Earth in 1968. Kustaanheimo (9), Kustaanheimo and Lehti (10) pointed out the route dependence of the gravitational red shift.

Florides and Synge (11) computed by successive approximation, in a modified procedure from that of Das, Florides and Synge (12) the stationary gravitational field due to a fluid mass, not necessarily of constant density, rotating steadily and slowly about an axis of symmetry. It was found (12) unexpectedly that the internal structure affects the metric outside the mass.

Geisler and McVittie (13) have found a coordinate transformation which transforms the Schwarzschild metric for the field of a spherical body into a special case of the general axi-symmetric metric originally derived by Levi-Civita (14) and by Weyl (15). The transformed Schwarzschild metric is perturbed in such a way as to obtain this special form and the field equations are solved approximately for the perturbing terms. Krause (16) obtained the relativistic potential of an oblate spheroid by referring to the energy tensor in the interior. Geisler and McVittie (13) transformed so as not to include the energy tensor in the interior. With values of the arbitrary constants appearing in the solution, which is formed by a reduction to the Newtonian theory, the metric obtained is taken to be that of the oblate Earth's field. The equatorial orbits of a free particle are discussed and the advance of perigee is calculated.

Roxburgh (17) raised a hypothesis that the inner layers of the Sun rotate more rapidly than the outer layers which, he says, is not inconsistent with the current cosmogonical hypothesis and tried to explain the perihelion advance of Mercury as gravitational effect. Similarly Winer (18) computed the advance of Mercury's perihelion due to the polytropic internal structure of the Sun.

BIBLIOGRAPHY

1. Aoki, S. *Astr. J.*, **69**, 221, 1964.
2. Geisler, P. A. *Astr. J.*, **68**, 715, 1963.
3. Geisler, P. A., McVittie, G. C. *Astr. J.*, **70**, 14, 1965.
4. Podurets, M. A. *Astr. Zu.*, **41**, no. 1, 28, 1964.
5. Zeldovich, Y. B. *Astr. Zu.*, **41**, no. 5, 873, 1964.
6. Atkinson, R. d'E. *Astr. J.*, **70**, 513 and 517, 1965.
7. Dicke, R. H. *Astr. J.*, **70**, 395, 1965.
8. Francis, M. P. *Astr. J.*, **70**, 449, 1965.
9. Kustaanheimo, P. Information, Dept. Appl. Math., Univ. Helsinki, No. 1, 1965.
10. Kustaanheimo, P., Lehti, R. *Astr. J.*, **68**, 392, 1963.
11. Florides, P. S., Synge, J. L. *Proc. R. Soc.*, **A 270**, 467, 1962.
12. Das, A., Florides, P. S., Synge, J. L. *Proc. R. Soc.*, **A 263**, 451, 1961.
13. Geisler, P. A., McVittie, G. C. *Mon. Not. R. astr. Soc.*, **131**, 483, 1966.
14. Levi-Civita, T. *Atti Accad. Lincei, Rendiconti*, **26**, 307, 1917; **27**, 3, 183, 220, 240, 283, 343, 1918; **28**, 3, 101, 1919.
15. Weyl, H. *Annalen d. Physik*, **54**, 117, 1917; **59**, 101, 1919.
16. Krause, H. G. L. *NASA Techn. Note*, R–188, 1964.
17. Roxburgh, I. W. *Icarus*, **3**, 92, 1964; **4**, 220, 1965.
18. Winer, I. M. *Mon. Not. R. astr. Soc.*, **132**, 401, 1966.

ARTIFICIAL SATELLITES

Vinti (1) computed his intermediary orbit (2) under the special geopotential, for which the separation of variables in Hamilton-Jacobi's method is capable, to the third order harmonic terms (3). By the inclusion of the third zonal harmonics the potential corresponds to a potential fitted exactly through the third zonal harmonics and to about two-thirds of the fourth, and gains in accuracy due to the absence of small denominators in the eccentricity and inclination. Vinti (4) studied the invariant properties of the spheroidal potential of an oblate planet by means of a metric-preserving transformation of the associated Cartesian system. This presumes separability when the coefficients J_2 and J_3 are taken into account. The inclusion of J_3 depends on translating the origin of the spheroidal coordinates by a distance $\frac{1}{2} J_2^{-1} | J_3 |$ equatorial radii, which amounts to 7 km.

Vinti (5) discussed by von Zeipel's method the effects of a constant force on a Keplerian orbit, similarly to the case of Keplerian motion of an electron around a charged nucleus in a constant electrostatic field. He referred to the parabolic coordinates in analogy to Epstein's theory of Stark effect in quantum-mechanics.

Differential corrections to Vinti's and Izsak's (6) solution were derived by Allen and Knolle (7), and the Izsak intermediary orbit specified by only six parameters was shown to be accurate enough compared with the rotating and precessing ellipse which requires eight parameters. Borchers (8) has, by extending the result of Izsak on the second order solution of Vinti's problem, presented a computational procedure for obtaining the coordinates and velocity of a satellite from the initial values. Borchers also derived the exact expression for the velocity in Vinti's intermediary orbit by using Izsak's orbital elements, for comparing with other methods of perturbation by means of the so-called Newton-Raphson iteration method. One of the drawbacks of the ordinary perturbation method is that they necessitate frequent multiplications of Fourier series which are inevitably accompanied with the question of convergence or divergence of the expansions.

Aksenov, Grebenikov, Demin (9) discussed the two fixed centre problem in connection with Vinti's and Kislik's (10) potentials. Aksenov (11) developed an analytical theory of the motion of an artificial satellite in the Earth's gravitational field, with the disturbing action of the Moon and the Sun taken into account. New formulae are derived for providing with the perturbations of the elements of the intermediary motion, which is described by the equations of the two fixed centre problem. The properties of the disturbed motion are studied. They (12) and Yarov-Yarovoy (13) solved the problem and, as well as Vinti (14), showed that the physical significance of Vinti's potential is the translation of the origin of the spheroidal coordinates by a distance $-\frac{1}{2} J_3/J_2$ times the equatorial radii, which is larger than any change produced by J_3. According to Vinti (14) it is related to the long-period terms of perturbation theory and turns out to be equal to the displacement by J_3 of the plane of symmetry of those exactly elliptical polar orbits which are possible solutions with the spheroidal model. Recently Alekseev (15) and Deprit (16) discussed in detail the motion of a particle attracted by two fixed mass centres and classified various kinds of possible real motions in three dimensions. Degtjarov and Ewdokimowa (17) analysed the stability of circular orbits in the two-fixed centre problem and plotted the regions of stability and instability. Marchel (18, 19) studied the effect of perturbations due to other bodies in this problem. If the masses of those centres are different, then the odd harmonics J_3 can be accounted for. Perturbations of this solution due to other causes are developed.

Garfinkel (20) published an improved theory of his former method of 1959. He revised the potential for his intermediary orbit so as to incorporate the known secular variations up to the second order furnished by the perturbation theory. The assumed form of the geopotential preserved the gross features of the actual geopotential such as axial and equatorial symmetry, singularity at the origin and the vanishing at infinity like the reciprocal of the distance, at the same time as it satisfies the Stäckel conditions for Hamilton-Jacobi separability by leading to a solution in terms of elliptic integrals. The revised algorithm provides a more accurate final orbits, incorporating a substantial fraction of the Keplerian secular variation of the third order. An internal check of the theory is carried out by comparison of the natural frequencies of the orbit with those of a perturbed ellipse. Garfinkel and McAllister (21) extended the known solution of the main problem of the artificial satellite theory to include the effects of all higher order zonal harmonics of the geopotential. By supposing that J_m, $m > 2$, are at least of order J_2^2 Garfinkel and McAllister used the von Zeipel method to derive general expressions for all the secular and long-period variations of orders J_m and J_m/J_2 respectively for any m. With the aid of the addition theorem due to Groves (22), Garfinkel (23) and Izsak (24) and of Legendre's integral form for the Legendre polynomials the result is exhibited in terms of the associated Legendre polynomials in $(1 - e^2)^{-1/2}$ and in cos I.

Then Garfinkel (25) extended the research to tesseral harmonic perturbation of an artificial satellite. For a close satellite the method of von Zeipel is used for calculating the long-period variations of order $n J_{m,\lambda}/\omega$, where n is the mean motion and ω the angular velocity of the Earth's rotation. With the aid of the addition theorem (23) for spherical harmonics the results are expressed in a most compact form. Further Garfinkel (26) expanded the spherical harmonics of the type $r^{-m-1}\ Y_{m,\lambda}\ (\Theta, \varphi)$ in Fourier polynomials of argument $\Theta = \alpha v + \mu g + \lambda h$, where v is the true anomaly, g and h are the Delaunay variables, and then of argument $\Theta = \alpha l + \mu g + \lambda h$ with Newcomb's polynomials in e as coefficients. He also formulated the long-period part for an Earth satellite in resonance with the Earth's rotation.

Aksnes (27) is attempting to include some of the main secular variations in an intermediary orbit and to solve in finite form with the separation of variables.

Kovalevsky (28) studied various applications of separability of the equations of motion in the solution of the artificial satellite problem. A. H. Cook (29) discussed the condition under which exact solutions may be obtained to the problem of orbits in free space. He considered the problem of finding potentials, which satisfy Laplace's equation and enable the Hamilton-Jacobi equation to be separated.

Grebenikov (30) applied Hill's method of rectangular coordinates to the motion of artificial satellites. Giacaglia (31) extended Brouwer's theory (32) to higher order terms. Von Zeipel's method is shown to be very powerful still. The secular and long-periodic variations in Delaunay's elements of an artificial satellite are obtained for zonal harmonics of any order. Fisher (33) compared the von Zeipel and the modified methods as applied to artificial satellites. Formulae were derived for osculating elements when the modified Hansen theory of Musen (34) is expressed in terms of orbital true longitude, instead of eccentric anomaly as in Bailie and Bryant's (35) work, to any prescribed order of J_2, in which high precision is aimed at.

Orlov (36) computed the secular and long-period perturbations which agreed with Brouwer's (32) and Kozai's (37), and developed an analytical theory of the influence of the Sun's attraction on the motion of satellites. On the basis of the circular and elliptical restricted three-body problem the formulae are derived for defining the perturbations up to the third order. Kholshevnikov (38, 39), by studying the secular perturbations in the disturbed motion of celestial bodies, discussed the stability of the orbital motion in the field of an oblate planet. Allan and Cook (40), Cook and Scott (41) and Gooding (42) discussed the long-periodic motion of the plane of a distant circular orbit by taking the oblateness and the luni-solar gravitational force into account, where the two perturbing effects become comparable, in analogy to the motion of Iapetus. A general solution is obtained for simultaneous precession of any number of fixed axes around which the precessions occur. Cook (43) has shown that for nearly circular orbits with $e < 0{\cdot}005$ the rate of rotation of the major axes can be markedly non-linear or even oscillatory while the variation in eccentricity is no longer sinusoidal, in contrast to the case of moderately eccentric orbits. Especially Cook discussed the effect of general odd harmonics on the motion in orbits of small eccentricity.

Kaula (44) discussed the gravitational and other perturbations, including radiation pressure but excluding drags either mechanical or electromagnetic, of a satellite orbit and formulated analytically the partial derivatives of the orbital rectangular components of position and velocity.

The usual expressions, which are suitable for not too small values of eccentricity, for perturbation of the Keplerian elements are extended by Batrakov (**45**) for the computation of coordinates without loss of accuracy to the case of small eccentricity.

Miss Shute (**52**), for studying Moon-to-Earth trajectories, employed a reduced form, obtained by considering the selenocentric velocity asymptotes, of the patch conic method for determining the initial elements of a particle launched or ejected from the Moon's surface with any arbitrary starting conditions, and derived explicit analytic functions for geocentric energy, the Jacobi constant, the angular momentum, standard orbital elements and the conditions for Moon-to-Earth trajectories. Percents of randomly ejected material which initially strike the Earth are seen to be in retrograde motion or to go into heliocentric orbits.

Miss Shute and Chiville (**46**) estimated the lunar-solar effect on the orbital life-times of artificial Earth satellites with highly eccentric orbits.

Gaska (**47**) determined periodic orbits of artificial satellites by considering the harmonics up to the sixth order in the geopotential and obtained the radius of convergence of the periodic series.

Sconzo and Benedetto (**48**) obtained for the motion of an equatorial satellite a simplification of expressing the radius vector in a form similar to that for a revolving conic, and determined the apsidal motion which agreed not only with the orbits of artificial Earth satellites but also with those of Martian satellites. They say that the apsidal motion of Mercury is partly due to the oblateness of the Sun.

Finally it is remarked that King-Hele and Quinn (**49**) published tables of the launched artificial Earth satellites. Herzberg (**50**) found unpredicted period, and Hiller (**51**) discrepancies in the eccentricity and argument of perigee, in the orbital motion of the Alouette I artificial Earth satellite.

Recently Kyner and Bennett (**53**) modified the classical Encke method for integrating the equations of motion of a near-Earth satellite so that the intermediary, which they call nominal, and the instantaneous orbits are forced to remain close for many revolutions by rotating the nominal ellipse slowly in a plane which is rotating slowly in space.

BIBLIOGRAPHY

1. Vinti, J. P. *J. Res. nat. Bur. Stand.*, **67 B,** 191, 1963; *Mathematische Methoden der Himmelsmechanik und Astronautik*, 97, 1966.

2. Vinti, J. P. *Trans. IAU*, **12 A,** 22, 1965.

3. Vinti, J. P. *J. Res. nat. Bur. Stand.*, **70 B,** 17, 1966.

4. Vinti, J. P. *J. Res. nat. Bur. Stand.*, **70 B,** 1, 1966.

5. Vinti. J. P. Proc. *IAU Symposium No.* 25, 355, Thessaloniki, 1964.

6. Izsak, I. G. *Smithson. Astrophys. Obs., Spec. Rep.*, no. 52, 1960.

7. Allen, W. A., Knolle, W. E. *Astr. J.*, **69,** 393, 1964; **70,** 152, 1965.

8. Borchers, R. V. *NASA Techn. Note*, D–1539, 1963.

9. Aksenov, E. P., Grebenikov, E. A., Demin, V. G. *Astr. Zu.*, **40,** no. 2, 363, 1963.

10. Kislik, M. D. *Iskusstvennye Sputniki Zemli*, **4,** 3, 1960; **13,** 23, 1961.

11. Aksenov, E. P. *Space Research USSR* (Kosm. Issledovanie), **1,** 1963; *Vestnik Sternberg Inst.*, no. 137, 1965; *Publ. Sternberg Inst.*, **35,** 1966.

12. Aksenov, E. P., Grebenikov, E. A., Demin, V. G. *Iskusstvennye Sputniki Zemli*, **16,** 163 and 173, 1963.

13. Yarov-Yarovoy, M. S. *Bjull. Inst. Teor. Astr.*, **8,** no. 9 (102), 647, 1962.

14. Vinti, J. P. *J. Res. nat. Bur. Stand.*, **70 B,** 1, 1966.

15. Alekseev, V. M. *Bjull. Inst. Teor. Astr.*, **10,** no. 4 (117), 241, 1965.

16. Deprit, A. *Bull. Soc. Math. Belg.*, **14,** 12, 1962.

17. Degtjarov, W. G., Ewdokimowa, L. S. *Bjull. Inst. Teor. Astr.*, **10,** no. 8 (121), 516, 1966.

18. Marchel, Ch. Actes du 15e Congrès Intern. d'Astronautique, **1**, 159, 1964.
19. Marchel, Ch. *Bull. astr.*, (3) **1**, fasc. 3, 1966.
20. Garfinkel, B. *Astr. J.*, **69**, 223, 1964.
21. Garfinkel, B., McAllister, G. T. *Astr. J.*, **69**, 453, 1964.
22. Groves, G. V. *Proc. R. Soc.*, **A 254**, 48, 1960.
23. Garfinkel, B. *Astr. J.*, **69**, 567, 1964.
24. Izsak, J. G. *J. geophys. Res.*, **69**, 2623, 1964.
25. Garfinkel, B. *Astr. J.*, **70**, 784, 1965.
26. Garfinkel, B. *Astr. J.*, **70**, 699, 1965.
27. Aksnes, K. *Astrophys. Norv.*, **10**, no. 4, 1965.
28. Kovalevsky, J. *Rendiconti Sem. Mat. e Fisico di Milano*, **35**, 1965.
29. Cook, A. H. *Mon. Not. R. astr. Soc.*, **134**, 253, 1966.
30. Grebenikov, E. A. *Bjull. Inst. Teor. Astr.*, **7**, 811, 1960.
31. Giacaglia, G. E. O. *Astr. J.*, **69**, 303, 1964; Correction, *Astr. J.*, **71**, 228, 1966.
32. Brouwer, D. *Astr. J.*, **64**, 378, 1959.
33. Fisher, D. *Astr. J.*, **68**, 718, 1963.
34. Musen, P. *J. geophys. Res.*, **64**, 2271, 1959; **66**, 403, 1961.
35. Bailie, A., Bryant, R. *Astr. J.*, **65**, 451, 1960.
36. Orlov, A. A. *Bjull. Inst. Teor. Astr.*, **10**, no. 1 (114), 6, 1965; **10**, no. 5 (118), 360, 1965; Proc. XV. Intern. Astronaut. Congr., **1**, 1965.
37. Kozai, Y. *Astr. J.*, **67**, 440, 1962.
38. Kholshevnikov, K. V. *Bjull. Inst. Teor. Astr.*, **10**, no. 5 (118), 332, 1965.
39. Kholshevnikov, K. V. *Publ. Astr. Obs. Univ. Leningrad*, no. 20, 1964; no. 21, 1964; *Vestnik Univ. Leningrad*, no. 13, 1965.
40. Allan, R. R., Cook, G. E. *Proc. R. Soc.*, **A 280**, 97, 1964.
41. Cook, G. E., Scott, D. W. *R. Aircraft Estab. Techn. Rep.*, 65292, 1965.
42. Gooding, R. H. *R. Aircraft Estab. Techn. Rep.*, 66018, 1966.
43. Cook, G. E. *Planet. Space Sci.*, **14**, 433, 1966.
44. Kaula, W. M. *Dynamics of Rockets and Satellites*, ed. by Groves, G. V., 178, 1965; *Aerospace Corp. Spec. Proj. Techn. Rep.*, SP-TR-65-4, 1965.
45. Batrakov, Yu. V. Symposium on the *Trajectories of Artificial Celestial Bodies as determined by Observations*, Paris, 1965 (in press).
46. Shute, B. E., Chiville, J. *Planet. Space Sci.*, **14**, 361, 1966.
47. Gaska, S. *Bull. Polish Observ. Artif. Earth Satell.*, Warsaw, no. 10, 26, 1964; *Vestnik Sternberg Inst.*, Moskow, no. 138, 321, 1966.
48. Sconzo, P., Benedetto, J. Federal System Div., Intern. Business Machine Corp., Task No. 0232-G, 1965.
49. King-Hele, D. G., Quinn, E. *Planet. Space Sci.*, **12**, 681, 1964; **13**, 707, 1965; **14**, 817, 1966.
50. Herzberg, L. *Planet. Space Sci.*, **14**, 451, 1966.
51. Hiller, H. *R. Aircraft Estab. Techn. Rep.*, 66020, 1966.
52. Shute, B. E. *Astr. J.*, **71**, 602, 1966.
53. Kyner, W. T., Bennett, M. M. *Astr. J.*, **71**, 579, 1966.
Kyner, W. T. *Astr. J.*, **71**, 583, 1966.

Critical inclination

Chapront (**1**) has shown that the libratory motion in the orbits of the critical inclination is produced around the value of the argument of latitude of the perigee equal to $\pm 90°$ if the coefficient J_2 is only considered, but it occurs around different points if higher harmonics are taken into account, especially the harmonics J_6 and J_8 play a fundamental role, the character of the libration of the perigee around the node can be reversed, and the odd harmonics shift the libration centre by a large amount. Chapront (**2**) then used the energy integral and saw how great is the influence either of the variation of the value of the even harmonics or of the introduction of odd harmonics.

BIBLIOGRAPHY

1. Chapront, J. *C. R. Acad. Sci.*, Paris, **260**, 2131, 1965.
2. Chapront, J. *Bull. astr.*, (3) **1**, 23, 1966.

Effect of tesseral harmonics

The asymmetric character of the Earth's equator has been discovered by Kozai, Izsak, Kaula, King-Hele and others (**1**) by the observable effects on the orbits of Earth satellites. In the first approximation the motion is studied by taking the tesseral harmonic expansion of the geopotential. If J_{22} is only considered, then there are two equilibrium points in the orbit of a satellite along the major and the minor axes of the equatorial section of the geopotential level surface, the former being unstable and the latter stable. The motion of a 24-hour satellite is of libration around the stable equilibrium point and is of revolution around the unstable one. Allan (**2**) discussed the libration of period about two years by a differential equation in the form of the one for a spherical pendulum by taking the regression of the orbital plane due to the luni-solar gravitational force into account. Stumpff (**3**) discussed the libration at the equilibrium points of a 24-hour satellite in analogy to the libration around the equilibrium points of the restricted three-body problem.

Cook (**4**) computed the long-period effect of all the coefficients of the tesseral harmonic expansion. Blitzer (**5**, **6**), on the other hand, computed the perturbation of the polar coordinates. He takes the first approximate positions of the equilibrium points as those corresponding to J_{22}. The latitudinal deviation of the first approximate equilibrium points occurs from odd zonal harmonics and tesseral harmonics with $n - m$ odd in J_{nm}, and the longitudinal deviation from tesseral harmonics with $n - m$ even. There are $2m$ equilibrium longitudes corresponding to J_{nm} in the form

$$\lambda_0 = \lambda_{22} + \frac{s\pi}{2} \qquad (s = 0, 1, 2, 3)$$

for $m = n = 2$, for example. The equations of motion for the deviation from the equilibrium positions are solved.

Recently Doppler shifts in radio observation of satellites revealed the coefficients of harmonic terms of the geopotential. The experiment and method of observation have been described by Newton (**7**) and Guier (**8**). Guier (**9**) determined the non-zonal tesseral harmonics from Doppler data. Guier and Newton (**10**) gave the coefficients of the tesseral harmonic terms. They noticed the resonance effects of satellite orbits with the geopotential. Anderle (**11**) analysed the Doppler data and gave values for the gravity coefficients for $(n, m) = (15, 13)$, $(13, 13)$ and $(15, 14)$. Yionoulis (**12**) studied more in detail and gave the equations which enable to determine which harmonics will contain near-resonant contributions for a given satellite orbit. The coefficients associated with the harmonics of degree and order $(n, m) = (13, 13)$, $(14, 14)$ and $(15, 14)$ are found. The resonance effects provide with a means of obtaining additional harmonic coefficients whose contributions might otherwise have been too small to be detected.

In order to explain the resonance effects of satellite orbits with the geopotential Blitzer (**13**) discussed the motion of a satellite under the influence of the longitude-dependent terms of the geopotential in a frame of reference rotating with the mean rotation of the satellite by using the differential equations of Kaula (**14**) and the coordinate transformation formulae of B. Jeffreys (**15**). The orbits are assumed to be nearly circular but unrestricted with respect to inclination. In general, the effect of the tesseral harmonics is to induce short-period perturbations of small amplitude. However, when the satellite's mean motion is commensurable with the Earth's rotation, Blitzer has seen two distinct types of resonance to set up. Dynamical resonances occur when the impressed frequency due to some J_{nm} term is equal to the natural frequency of the satellite's orbital motion, and the amplitude of the induced oscillation builds up with time, the eccentricity increases and the orbit is unstable in the sense that it is driven out of the resonance state. For the other commensurabilities librational resonances occur when the satellite is in a localized potential well in the moving frame and the motion is analogous to a pendulum motion, and is a long-period libration in longitude of the ground track of the satellite. For nearly equatorial motion only the 24-hour orbit is librational and orbits in dynamical resonance are confined to the range from 12 to 48 hours, clustering around but not including the 24-hour orbit.

Bennett (**16**), Steg and De Vries (**17**) studied the effect of the Sun on the Earth-Moon libration point.

Garfinkel (**18**) treated the theory of resonance in dynamics in a formal solution in the problem of small divisors. There the Bohlin-von Zeipel technique was used to solve the ideal resonance problem to $O(k^{1/2})$, where k is the small parameter, and to express the results in terms of the elliptic functions sn, cn, dn, and the Jacobi Zeta function. The general results have been applied to the problem of a 24 hour satellite and the critical inclination. The results confirm the work of Kevorkian *et al.* (**19**) who need the methods of non-linear mechanics to match the resonance solution in the vicinity of the critical inclination with the classical solution with a singularity.

BIBLIOGRAPHY

1. *Trans. IAU*, **12 A**, 23, 1965.
2. Allan, R. R. *Planet. Space Sci.*, **11**, 1325, 1963; **12**, 283, 1964; *R. Aircraft Estab. Techn. Rep.*, 64078, 1965; *Proc. R. Soc.*, **A 288**, 60, 1965; Proc. 16th IAF Conference, Athens, 1965, **6**, 119, 1966; *Planet. Space Sci.*, 1966 (to appear).
3. Stumpff, K. *Mathematische Methoden der Himmelsmechanik und Astronautik*, 113, 1966.
4. Cook, G. E. *Planet. Space Sci.*, **11**, 797, 1963.
5. Blitzer, L. *J. geophys. Res.*, **70**, 3987, 1965.
6. Blitzer, L. *Amer. Rocket Soc. J.*, **32**, 714, 1962.
7. Newton, R. R. *Appl. Phys. Lab. Rep.*, TG–571, 1963; TG–627, 1964; *J. geophys. Res.*, **69**, 3619, 1964.
8. Guier, W. H. *Appl. Phys. Lab. Rep.*, TG–503, 1963.
9. Guier, W. H. *Nature*, **200**, 124, 1963.
10. Guier, W. H., Newton, R. R. *J. geophys. Res.*, **70**, 4613, 1965.
11. Anderle, R. J. *J. geophys. Res.*, **70**, 2453, 1965.
12. Yionoulis, S. M. *J. geophys. Res.*, **70**, 5991, 1965; **71**, 1289, 1966; *Appl. Phys. Lab. Rep.*, TG–633, 1964.
13. Blitzer, L. *J. geophys. Res.*, **71**, 3557, 1966.
14. Kaula, W. M. *Geophys. J.*, **5**, 104, 1961.
15. Jeffreys, B. *Geophys. J.*, **10**, 141, 1965.
16. Bennett, A. G. Boeing Sci. Res. Lab., D2–23307–1964.
17. Steg, L., De Vries, J. P. *Space Sci. Rev.*, **5**, 210, 1966.
18. Garfinkel, B. *Astr. J.*, **71**, 657, 1966.
19. Eckstein, M., Shi, Y., Kevorkian, J. Proc. *IAU Symp. No.* 25, 291, Thessaloniki, 1964.

Light-pressure effect

The study of the effect of the solar-light pressure on the motion of an Earth satellite has been continued by Poljakchova (**1**) by the method of variation of elements with special reference to the resonance case in which the motion is affected by long-periodic variations with large amplitude. Sehnal (**2**) discussed the Poynting-Robertson effect (**3**). He (**4**) also computed the perturbations of the orbital elements of artificial satellites caused by the radiation pressure reflected from the Earth. Cunningham (**5**) defined and computed the eclipse factor for the solar radiation effect. His eclipse factor is the time interval between the immersion in and the emersion from the shadow of the Earth divided by the orbital period. Escobal (**6**) computed the orbital entrance and emergence from the shadow.

Eremenko (**7**) described a method of solving the equations for obtaining the point of intersection of an orbit of a satellite with the Earth's shadow. These points of intersection are necessary for the calculation of the perturbation due to the solar radiation pressure. The same problem was discussed by Batrakov (**8**) for the case of small eccentricities.

BIBLIOGRAPHY

1. Poljakchova, E. N. *Bjull. Inst. Teor. Astr.*, **9**, 15, 1963.
2. Sehnal, L. *Bull. astr. Inst. Csl.*, **14**, 133, 1963.
3. Robertson, H. P. *Mon. Not. R. astr. Soc.*, **97**, 423, 1937.
4. Sehnal, L. Proc. COSPAR-IUTAM-IAU Symposium on the *Trajectories of Artificial Satellites determined from Observations*, Paris, 1965.
5. Cunningham, F. G. *Amer. Rocket Soc. J.*, **32**, 1399, 1962.
6. Escobal, P. R. *Amer. Rocket Soc. J.*, **32**, 1939, 1962.
7. Eremenko, R. P. *Bjull. Inst. Teor. Astr.*, **10**, no. 6 (119), 1965.
8. Batrakov, Yu. V. *Bjull. Inst. Teor. Astr.*, 1966 (in press).

Effect of air drag

King-Hele (**1, 2, 3**) with his colleagues continued a detailed study of the effect of the Earth's atmosphere on the motion of satellites. Cook, King-Hele and Walker (**4**) and King-Hele (**5**) at first supposed that the air density at a given height in a spherically symmetric atmosphere varies sinusoidally with the geocentric angular distance from the maximum density direction which occurs almost overhead of the Sun's altitude. They showed how the perigee distance and the orbital period vary with the eccentricity through the satellite's life and how the eccentricity varies with time, and expressed the life-time and the air density at the perigee in terms of the ratio of change of the orbital period. King-Hele and Quinn (**6**) estimated the life of a satellite of nearly circular orbit. Then Cook, King-Hele, Walker (**7**) and Cook and King-Hele (**8**) considered the atmospheric oblateness. Together with Groves (**9**), Parkyn (**10**), Lee (**11**) and Po (**12**), King-Hele (**13**) derived formulae for determining the density distribution in a rotating oblate atmosphere. The evidence for the oblateness has been shown by Nigam (**14**) and Anderson (**14a**). Allan and Cook (**15**) showed the symmetry of a geocentric dust belt and the zodiacal light, and Cook (**16**) studied the effect of aerodynamic lift on satellite orbits. A more generalized theory, in which the effect of luni-solar perturbations and the Earth's oblateness can be expressed in similar form, has been completed.

Fominov (**17**) considered the problem of determining the parameters of the atmospheric model proposed by himself. He obtained analytical expressions for the perturbation of the elements of a satellite caused by the atmospheric drag for the revolutionary period of the satellite in the case of large orbital eccentricity and also when the oblateness of the atmosphere depends on the altitude. The formulae obtained are applicable for the case when the atmospheric oblateness is considerable, as well as for the case when it is small.

Cook, King-Hele (**18**) and King-Hele (**3a**) discussed the day-to-night variation of the air density. Cook and Plimmer (**19**), Cook (**20, 21**) and King-Hele (**22**) considered the oblate rotating atmosphere, King-Hele (**23**) treated further the effect of a meridional wind on satellite orbits.

Fominov (**24**) considered the effect of the solar activity, in addition to the diurnal effect on the oblate Earth atmosphere, by distinguishing the pole of the diurnal effect and the pole of the Earth's rotation. He also discussed the variation of density distribution in the atmosphere by the tidal effect of the Moon and the Sun. The orbital evolution due to the tidal dissipation by solid friction of the Earth's mantle has been considered by Kaula (**25**).

Cook (26), Smith (27), Jacchia and Slowey (28) and Jacchia (28a) discussed the drag effect and the life-time of a satellite. Katsis (29) followed Brouwer-Hori's method (30) for computing the drag effect. Sehnal and Mills (31) computed the short-period drag perturbations of the orbits of artificial satellites.

Sconzo, Rossoni, Greenfield, Champion (32) determined the atmospheric density from satellite observations. The result shows that the method is sensitive enough to reveal small oscillations of the atmospheric density which are in phase with the decimetric solar radio flux.

Cook and King-Hele (18) observed that dissipative effect, either tidal or due to Poynting-Robertson drag for a micro-satellite, gives a mechanism for permanent or semi-permanent capture leading to secular changes in orbital inclination. They say that this appears to give a partial explanation of the formation and structure of Saturn's rings.

King-Hele (33) with his collaborators determined the upper atmosphere density and its time variation between the recent sunspot maximum and minimum. Cook and Scott (34) derived the exospheric densities near solar minimum, and King-Hele (3a) the semi-annual variation.

BIBLIOGRAPHY

1. King-Hele, D. G. *Planet. Space Sci.*, **12**, 4835, 1964.
2. King-Hele, D. G. *Theory of Satellite Orbits in an Atmosphere*, Butterworths, 1964, London.
3. King-Hele, D. G. *R. Aircraft Estab. Techn. Rep.*, 66010, 1966.
3a. King-Hele, D. G. *ibid.*, 66076, 1966; *Planet. Space Sci.*, **14**, 863, 1966.
Anderson, A. D. *Planet. Space Sci.*, **14**, 849, 1966.
4. Cook, G. E., King-Hele, D. G., Walker, D. M. C. *Proc. R. Soc.*, **A 257**, 224, 1960.
5. King-Hele, D. G. *Proc. R. Soc.*, **A 267**, 541, 1962; **275**, 357, 1963.
6. King-Hele, D. G., Quinn, E. *Planet. Space Sci.*, **13**, 707, 1965; **14**, 107, 1966.
7. Cook, G. E., King-Hele, D. G., Walker, D. M. C. *Proc. R. Soc.*, **A 264**, 88, 1961.
8. Cook, G. E., King-Hele, D. G. *Proc. R. Soc.*, **A 275**, 357, 1965.
9. Groves, G. V. *Proc. R. Soc.*, **A 254**, 48, 1959; **252**, 16, 1959; *Planet. Space Sci.*, **5**, 314, 1961; *Mathematische Methoden der Himmelsmechanik und Astronautik*, 147, 1966.
10. Parkyn, D. G. *J. geophys. Res.*, **65**, 1, 1960.
11. Lee, V. A. *Amer. Rocket Soc. J.*, **32**, 102, 1962.
12. Po, Nam Tum *Bjull. Inst. Teor. Astr.*, **10**, no. 5 (118), 317 and 324, 1965.
13. King-Hele, D. G. *Planet. Space Sci.*, **11**, 261, 1963; *Ann. Géophys.*, **22**, 40, 1966.
14. Nigam, R. C. *J. geophys. Res.*, **69**, 1361, 1964.
14a. Anderson, A. D. *Nature*, **209**, 656, 1966.
15. Allan, R. R., Cook, G. E. *J. geophys. Res.*, (in press), 1966.
16. Cook, G. E. *Planet. Space Sci.*, **12**, 1009, 1964.
17. Fominov, A. M. *Bjull. Inst. Teor. Astr.*, **9**, no. 3 (106), 185, 1963; *ibid.*, **9**, no. 7 (110), 499, 1964; *ibid.*, **9**, 1966 (in press).
18. Cook, G. E., King-Hele, D. G. *Phil. Trans. R. Soc.*, **A 259**, 33, 1965.
19. Cook, G. E., Plimmer, R. N. A. *Proc. R. Soc.*, **A 258**, 516, 1960.
20. Cook, G. E. *Proc. R. Soc.*, **A 261**, 246, 1961.
21. Cook, G. E. *Planet. Space Sci.*, **13**, 929, 1965.
22. King-Hele, D. G. *Planet. Space Sci.*, **12**, 835, 1964.
23. King-Hele, D. G. *R. Aircraft Estab. Techn. Rep.*, 66010, 1966; *Proc. R. Soc.*, **A 294**, 261, 1966.
24. Fominov, A. M. *Bjull. Inst. Teor. Astr.*, **9**, no. 3 (106), 185; no. 7 (110), 499, 1963.
25. Kaula, W. M. *J. geophys. Res.*, **68**, 4959 and 4967, 1963.
26. Cook, G. E. *Planet. Space Sci.*, **13**, 929, 1965; **15**, 627, 1967.
27. Smith, D. E. *Planet. Space Sci.*, **13**, 1283, 1965; **14**, 455, 1966.
28. Jacchia, L. G., Slowey, J. *Smithson. astrophys. Obs., Spec. Rep.*, no. 136, 1963; no. 242, 1967.
28a. Jacchia, L. G. *ibid.*, no. 184, 1965.
29. Katsis, D. N. *Epimegetoy Ergasterioy Astronomias Panepistemioy Athenon*, 1965.

30. Brouwer, D., Hori, G. *Astr. J.*, **66**, 193 and 264, 1961.
31. Sehnal, L., Mills, S. B. *Smithson. astrophys. Obs., Spec. Rep.*, 1966.
32. Sconzo, P., Rossoni, J. P., Greenfield, R. J., Champion, K. S. *Progress Astronaut. Aeronaut.*, **14**, 157, 1964.
33. King-Hele, D. G., Rees, J. M. *J. atmosph. terr. Phys.*, **25**, 495, 1963.
King-Hele, D. G., Quinn, E. *J. atmosph. terr. Phys.*, **27**, 197, 1965; *Planet. Space Sci.*, **13**, 693, 1965; **14**, 107, 1966.
34. Cook, G. E., Scott, D. E. *R. Aircraft Estab., Techn. Rep.* (to be issued).

Rotation of a satellite

Beletski (1), Beletski and Zonov (2), and Zonov (3) discussed in detail the rotational motion of an Earth satellite along its orbit and classified various kinds of rotational motion. Hagihara (4) studied the rotational motion of an Earth satellite during its flight along its orbit by taking into account of the geomagnetic field, the field of the solar wind of charged particles and the effect of electrostatic and magnetic induction to the body of the satellite. The effect of Earth's oblateness to the rotational motion was studied by Sarychev (5). Morgan (6) studied the effect of librating motion on the orbit of a dumb-bell-shaped satellite. Volkov (7) discussed the translational-rotational motion of a satellite by Hamilton-Jacobi's method and deduced the series representation of the periodic motion in a gravitational field of a sphere, then constructed a general approximate solution in the vicinity of such a periodic translational and rotational motion, which corresponds to a periodic solution of the first sort of Poincaré.

Morrison (8) discussed the damping due to a roll-vee, gyrostabilized system, wherein the pitch axis of the satellite remains perpendicular to the orbital plane. It is shown that, no matter how large the initial local angular velocity of the satellite is, this velocity reaches any given smaller value in a finite time, and how it is possible to obtain a bound on their damping time and how the satellite settles in one of two possible Earth-pointing positions.

Demin (9) studied the satellite problem by means of qualitative methods. In particular the stability in Liapounov's sense and in Kolmogorov-Arnold's sense has been discussed for the case of translational-rotational motion.

Kondurar (10) is continuing his work on the theory of the translational-rotational motion of artificial satellites in the gravitational field of a planet.

BIBLIOGRAPHY

1. Beletski, V. V. *Iskusstvennye Sputniki Zemli*, **3**, 13, 1959; **6**, 11, 1961; **13**, 53, 1962.
2. Beletski, V. V., Zonov, V. *ibid.*, **7**, 32, 1961.
3. Zonov, V. *ibid.*, **3**, 118, 1959.
4. Hagihara, Y. *Smithson. Contr. Astrophys.*, **5**, 113, 1961.
5. Sarychev, V. A. *Iskusstvennye Sputniki Zemli*, **6**, 3, 1961.
6. Morgan, J. P. *Amer. Rocket Soc. J.*, **31**, 1089, 1961.
7. Velkov, M. S. *Bjull. Inst. Teor. Astr.*, **8**, no. 5 (98), 343, 1962; **9**, no. 9 (105), 120, 144; no. 4 (108), 74, 283, 1963.
8. Morrison, J. A. *Quart. Appl. Math.*, **22**, 148, 1964.
9. Demin, V. G. *Space Research USSR* (Kosm. Issledovanie), **2**, 1964; *Publ. Lumumba Univ.*, Ser. Mechanics, 1966.
10. Kondurar, V. T. *Bjull. Inst. Teor. Astr.*, **10**, no. 9 (122), 582, 1966; *Publ. Pedagogical Inst. Ivanovo*, 1966.

Lunar orbiters and artificial satellites of planets

Kaula (1) computed the life-time of a lunar orbiter by means of the Runge-Kutta method by taking into account the important part of the disturbing function which is of long-periodic Earth effect with respect to both the lunar satellite and the Earth-Moon orbit. He noticed the importance of the J_3 term of the Moon's gravity field. Chebotarev and Kirpichnikov (2) studied by numerical integration the motion of an artificial lunar satellite and its stability. Forga is carrying out his work on the application of Hill's method on the motion of a lunar satellite. Roy is continuing his attempt to produce an analytical theory of a close artificial lunar satellite taking into account the disturbing effect due to the second harmonics in the Moon's gravitational field, together with the Earth and the solar effects. The secular part and the main periodic part of the second order theory have been derived. The theory is now being programmed for computation.

A particular attention should be paid to the problem of the motion of the lunar satellites, as was pointed out by Kovalevsky (3), for an orbit with small inclination, due to the fact that there is no dominant second harmonics in the lunar gravitational field as in the Earth's field and that the disturbing force produced by other harmonics is comparable in magnitude with the attraction due to the Earth. For moderate inclinations and eccentricities the problem has been tackled by the technique already used for large inclinations and eccentricities by Kovalevsky (4). By discussing the motion of the peri-selenium he showed that the perturbation can be extremely large so that the life-time of the orbit may become short.

Goudas (5) discussed the moments of inertia, the gravity field and the non-homogeneity of the Moon. Poljakchova (6) studied the effect of solar radiation pressure on the motion of an artificial lunar satellite.

Milder (7) estimated the probability of the lunar impact of a lunar orbiter or the escape of a satellite from its primary on the basis of the ensemble averages of statistical mechanics. The restricted three-body configuration is used as a model in calculating the mean orbit life-times between impacts of a lunar satellite in bounded orbits of various energies.

Szebehely and Pierce (8) studied a group of Earth-to-Moon trajectories with consecutive collisions, that is, those which pass through the centres of both the Moon and the Earth, by means of regularization of the equations of motion for a model restricted three-body problem.

Mrs Stellmacher-Amilhat (9) studied the motion of an asteroid during close approach to a planet. She compared various analytical solutions obtained by considering planeto-centric coordinates within the spheres of different radii. Numerical comparison was made also for the orbit of Mariner II. This method permits to define more accurately the idea of the sphere of influence of a planet. A similar problem was discussed by Stiefel (10) who put emphasis on problems of regularization.

Breakwell and Ralph (11), and Wolaver (12) extended the restricted three-body problem for the Earth-Moon system to include the effect of the Sun by means of von Zeipel's method. Breakwell and Perko (13) matched heliocentric ellipses, corrected for the influence of planetary attraction, in the vicinity of a planet with local hyperbolas. It is found that the attraction due to the destination planet alone causes both a displacement of the arrival asymptote and a time-of-arrival correction which consists of two parts: a gross time bias and a local time bias. The latter depending only on the eccentricity of the arrival hyperbola. The computation of fly-by interplanetary trajectories, such as a trajectory from the Earth to Mars via Venus, would be greatly simplified if the planetary attractions could be neglected and the trajectory be idealized as two heliocentric Keplerian orbits joined as one massless planet point.

Dusek (14) studied the motion in the vicinity of the equilibrium points of a generalized restricted three-body problem by including accelerations caused by low-thrust forces in such a way that the conservative character of the problem is preserved, and discussed the stability of the motion by Liapounov's method.

Chebotarev and Bozkova (15) studied the motion of artificial satellites of Mars, Venus and Mercury in the sphere of activity of the respective planet, and in particular the motion of polar satellites.

Kaula (16) recently determined the tesseral harmonic coefficients of the lunar gravitational field by analysing lunar satellite orbit perturbations. Chebotarev and Kirpichnikov (2) computed numerically the orbit of a lunar orbiter. Yarov-Yarovoi (17) invented an analytical theory of the motion of a space vehicle to the Moon. A method was given for finding the polynomials of the lowest power in an auxiliary variable, which represent with a given accuracy the rectangular coordinates, the velocity components and the time in which the orbits pass close to the Earth and the Moon.

Magnaradze (18) studied the motion of a space vehicle with a variable mass during the flight to Venus in the gravitational field of the Sun, Earth, Venus, and Jupiter. The power series in time have been built up and their convergence has been proved for a sufficiently small time interval. To define the coefficients of the series, recurrent relations are derived which may be used for calculations on modern computers. The errors due to the truncating at the first several terms are estimated.

Giacaglia (19) and Oesterwinter (20) studied the motion of a lunar orbiter characterized by a relatively large value of J_{22}, the former by taking the lunar equator as reference plane and the latter the ecliptic.

Miss Shute (21) studied the initial conditions for Moon-to-Earth trajectories.

BIBLIOGRAPHY

1. Kaula, W. M. *Trajectories of Artificial Celestial Bodies as Determined from Observations*, La Lune à l'Ère 1966; Proc. Intern. Conf. on Selenodesy, 1966.
2. Chebotarev, G. A., Kirpichnikov, S. N. *Bjull. Inst. Teor. Astr.*, **8**, no. 6 (99), 402, 1962; **10**, no. 2 (115), 109, 1965.
3. Kovalevsky, J. *La Lune à l'Ère Spatiale*, Presses Universitaires de France, p. 140, 1966.
4. Kovalevsky, J. Proc. *IAU Symposium No.* 25, 326, Thessaloniki, 1964.
5. Goudas, C. L. *Icarus*, **3**, 375, 1964; **4**, 218, 1965.
6. Poljakchova, E. N. *Bjull. Inst. Teor. Astr.*, **9**, no. 6 (109), 440, 1964.
7. Milder, D. M. *Astr. J.*, **68**, 724, 1963.
8. Szebehely, V., Pierce, D. A. *Progress Astronaut. Aeronaut.*, **14**, 35, 1964.
9. Stellmacher-Amilhat, I. *Bull. astr.*, (3) **1**, fasc. 3, 1966.
10. Stiefel, E. *Dynamics of Rockets and Satellites*, North Holland Pub. Co., Chap. 4, 1965.
11. Breakwell, J. V., Ralph, P. *Methods in Celestial Mechanics and Astrodynamics*, p. 55, 1966.
12. Wolaver, L. E. *ibid.*, p. 75.
13. Breakwell, J. V., Perko, L. M. *ibid.*, p. 159.
14. Dusek, H. M. *ibid.*, p. 37.
15. Chebotarev, G. A., Bozkova, A. *Bjull. Inst. Teor. Astr.*, **9**, no. 3 (106), 169; no. 6 (109), 388, 1963.
16. Kaula, W. M. Publ. no. 502, Inst. Geophys. Planet. Phys., Univ. Calif., 1966.
17. Yarov-Yarovoi, M. S. *Publ. Sternberg Inst.*, **36**, 1967; *Differential Equations* (USSR), **1**, no. 7, 1965; **1**, no. 9, 1965.
18. Magnaradze, N. S. *Bjull. astr. Obs. Abastumani*, **30**, 1964.
19. Giacaglia, G. E. O. *NASA Techn. Rep.*, X-547-65-218, 1965.
20. Oesterwinter, C. Dissertation, Yale, 1966.
21. Shute, B. E. *Astr. J.*, **71**, 602, 1966. See (**52**) of section on ARTIFICIAL SATELLITES, page 32.

Optimization problem

The problem of moving in a minimum time between two points in a central gravitational field located at set distances from the centre of attraction has been solved by Lebedev and Rumyantsev (1) by assuming that the space vehicle's motor develops an acceleration which is constant in magnitude. The Lagrange method is used leading to the solution of the boundary value problem for a set of differential equations. Kelley (2) used an adjoint technique for obtaining the values of the Green function at the terminal points. Jurovics (3, 4) discussed the optimum steering problem for the entry of a multi-stage vehicle into a circular orbit. Kholshevnikov (5) studied the astro-correction of the encounter of two vehicles at a fixed time by means of one-impulse transfer. Novoselov (6) studied the optimal two-impulse transfer between orbits of small inclinations and eccentricities. Novoselov (7) considered the problem of applying the methods of analytical dynamics to the theory of optimal transfers. The problem of the determination of non-planar optimal transfer to a given circular orbit around a central body with spherical symmetry is studied (8). Kirpichnikov (9) is dealing with some problems of the construction of optimal impulse trajectories in the central field with complicated boundary values. Tarushkin (10) proposed a method of successive optimization for solving non-linear problems.

Hiller (11) discussed the so-called Hohmann transfer between non-coplanar circular orbits and also dealt with the transfer between a circular and a non-coplanar elliptic orbits. Recently Hiller (12) considered the transfer between one or two semi-elliptic paths which needs at most three impulses. He assumed the transfer to occur only at the apsides of the ellipses and the whole plane change to take place at coincident apocentres of these elliptic paths, the assumption which had been justified by Barrar (13). Hiller optimized the total characteristic velocity for impulsive transfer between non-coplanar elliptic orbits having a common centre of attraction and collinear major axes in the same sense. He has seen that for three-impulse transfers the optimum mode is to transfer between pericentres of the initial and final elliptic orbits and for two-impulse transfer the optimum mode is to transfer from the pericentre of the inner ellipse to the apocentre of the outer ellipse. Krasinsky (14) studied the optimum transfer between close coplanar Keplerian orbits. Kirpichnikov (15) discussed the cosmic trajectories of minimum time and gave expenditure of mass. Sconzo (16) solved the one impulse rendez-vous problem by using Lambert's theorem and the transition matrix. The 36 elements of the matrix have been computed explicitly.

Hiller (17) studied the optimum impulsive transfers between non-coplanar elliptic and circular orbits and between non-coplanar elliptic orbits with collinear major axes. King-Hele (18) considered the enlargement of elliptic satellite orbits by continuous micro-thrust.

Mission analysis requires the selection of initial conditions and control parameters giving a trajectory that satisfies a set of objectives subject to numerous constraints. In addition to the non-linearity of the dependence of the objectives and constraints on the inputs, their respective numbers may not agree, and the optimization is necessary. Campbell, Moore and Wolf (19) have given a general formulation of the problem which reduces to finding the minimum length of a vector. The method of solution is with iteration. Van Dine, Fimple and Edelbaum (20) gave a new numerical approach to the solution of the non-linear two-point boundary value problem with application to optimum low-thrust space trajectories, consisting of a finite-difference Newton-Raphson algorithm. Munick (21) discussed the problem of finding the thrust programme so that a specified payload is delivered in vertical flight to a desired altitude with the least fuel expenditure. He considered this Goddard problem for a wide class of drags of physical interest. Ehricke (22) studied interplanetary manoeuvers in manned helio-nautical missions. The manoeuvers considered are perihelion brake, off-perihelion acceleration and retro-manoeuvers, and heliocentric planet approach manoeuvers. Goldstein and Seidman (23) considered fuel optimal controls for a ferry vehicle attempting rendez-vous with an orbital satellite under the additional constraint.

BIBLIOGRAPHY

1. Lebedev, V. N., Rumyantsev, B. N. *Iskusstvennye Sputniki Zemli*, no. 16, 252, 1963.
2. Kelley, J. H. *Amer. Rocket Soc. J.*, **30**, 947, 1960.
3. Jurovics, S. A. *Amer. Rocket Soc. J.*, **31**, 518, 1961.

4. Jurovics, S. A., McIntyre, J. E. *Amer. Rocket Soc. J.*, **32**, 1354, 1962.
5. Kholshevnikov, K. W. *Bjull. Inst. Teor. Astr.*, **10**, no. 5 (118), 103, 1965.
6. Novoselov, V. S. *Bjull. Inst. Teor. Astr.*, **9**, no. 5 (108), 295, 1963.
7. Novoselov, V. S. *Vestnik Univ. Leningrad*, **13**, 1964.
8. Novoselov, V. S. *Publ. Astr. Obs. Univ. Leningrad*, **20**, 1965; **22**, 1965.
9. Kirpichnikov, S. N. *Vestnik Univ. Leningrad*, no. 1, 1964; no. 7, 1964.
10. Tarushkin, V. T. *Vestnik Univ. Leningrad*, no. 13, 1965.
11. Hiller, H. *Planet. Space Sci.*, **13**, 147, 1233, 1965.
12. Hiller, H. *Planet. Space Sci.*, **14**, 773, 1966.
13. Barrar, R. B. *Astronaut. Acta*, **9**, 1, 1963.
14. Krasinsky, G. A. *Bjull. Inst. Teor. Astr.*, **9**, no. 6 (109), 425, 1964.
15. Kirpichnikov, S. N. *Bjull. Inst. Teor. Astr.*, **10**, no. 1 (114), 27, 1965.
16. Sconzo, P. *Astronaut. Acta*, **9**, 367, 1963.
17. Hiller, H. *Planet. Space Sci.*, **13**, 1233, 1965; **14**, 773, 1966.
18. King-Hele, D. G. Proc. 14th Int. Astronaut. Congr., **4**, 155, Springer, Berlin, 1965.
19. Campbell, J. H., Moore, W. E., Wolf, H. *Methods in Astrodynamics and Celestial Mechanics*, 355, 1966.
20. Van Dine, C. P., Fimple, W. R., Edelbaum, T. N. *ibid.*, 377.
21. Munick, H. *ibid.*, 401.
22. Ehricke, K. A. *ibid.*, 325.
23. Goldstein, A. A., Seidman, T. I. *Mathematische Methoden der Himmelsmechanik und Astronautik*, 257, 1966.

Numerical values of the coefficients of geopotential

New determinations of the harmonic coefficients of the geopotential were made by Kozai (**1**), Smith (**2**), King-Hele (**3**), King-Hele and Cook (**4**), and their collaborators (**5, 6, 7, 8**). Kozai (**1**) used Baker-Nunn camera observations of nine satellites. For odd harmonics King-Hele, Cook and Scott (**6**) analysed the change in the eccentricities of six satellites with orbital inclinations spaced as uniformly as possible between 28° and 96° and found the most satisfactory representation of the potential to be in terms of four coefficients $10^6 J_3 = -2{\cdot}56$, $10^6 J_5 = -0{\cdot}15$, $10^6 J_7 = -0{\cdot}44$, $10^6 J_9 = 0{\cdot}12$. For even zonal harmonics (**4, 7**) the authors took the inclinations to be higher than 28°. The numerical values are given in Table 1.

Table 1

	Kozai (**1**)	Smith (**2**)	King-Hele *et al.* (**3, 4, 5, 6, 7, 8**)
$10^6 J_2$	1082·645	1082·64	1082·64
$10^6 J_4$	−1·649	−1·70	−1·52
$10^6 J_6$	0·646	0·73	0·57
$10^6 J_8$	−0·270	−0·46	0·44
$10^6 J_{10}$	−0·054	−0·17	—
$10^6 J_{12}$	−0·357	−0·22	—
$10^6 J_{14}$	0·179	0·19	—
$10^6 J_3$	−2·546	—	−2·56
$10^6 J_5$	−0·210	—	−0·15
$10^6 J_7$	−0·333	—	−0·44
$10^6 J_9$	−0·053	—	0·12
$10^6 J_{11}$	0·302	—	—
$10^6 J_{13}$	−0·114	—	—

The coefficients of tesseral harmonic expansion of the geopotential are determined by means of photographic observations by Izsak (9) with the notation

$$C_{nm} = N_{nm}\ \bar{C}_{nm}, \qquad S_{nm} = N_{nm}\ \bar{S}_{nm},$$

$$N_{n0} = (2n + 1)^{1/2}, \qquad N_{nm} = \left[\frac{2(2n+1)\cdot(n-m)!}{(n+m)!}\right]^{1/2}, (m \neq 0),$$

The result is shown in Table 2.

Table 2

Izsak (9)

$10^6. C$	$10^6. S$
$\bar{C}_{22} = 1{\cdot}17$	$\bar{S}_{22} = -0{\cdot}95$
$\bar{C}_{31} = 0{\cdot}81$	$\bar{S}_{31} = -0{\cdot}25$
$\bar{C}_{32} = 0{\cdot}24$	$\bar{S}_{32} = -0{\cdot}25$
$\bar{C}_{33} = -0{\cdot}50$	$\bar{S}_{33} = 0{\cdot}93$
$\bar{C}_{41} = -0{\cdot}18$	$\bar{S}_{41} = -0{\cdot}25$
$\bar{C}_{42} = -0{\cdot}11$	$\bar{S}_{42} = 0{\cdot}23$
$\bar{C}_{43} = 0{\cdot}28$	$\bar{S}_{43} = -0{\cdot}08$
$\bar{C}_{44} = -0{\cdot}08$	$\bar{S}_{44} = 0{\cdot}29$
$\bar{C}_{51} = -0{\cdot}09$	$\bar{S}_{51} = 0{\cdot}19$
$\bar{C}_{52} = 0{\cdot}31$	$\bar{S}_{52} = -0{\cdot}50$
$\bar{C}_{53} = -0{\cdot}72$	$\bar{S}_{53} = 0{\cdot}11$
$\bar{C}_{54} = -0{\cdot}18$	$\bar{S}_{54} = 0{\cdot}51$
$\bar{C}_{55} = 0{\cdot}18$	$\bar{S}_{55} = -0{\cdot}42$
$\bar{C}_{61} = -0{\cdot}01$	$\bar{S}_{61} = 0{\cdot}13$
$\bar{C}_{62} = 0{\cdot}16$	$\bar{S}_{62} = -0{\cdot}37$
$\bar{C}_{63} = 0{\cdot}14$	$\bar{S}_{63} = -0{\cdot}17$
$\bar{C}_{64} = -0{\cdot}20$	$\bar{S}_{64} = -0{\cdot}41$
$\bar{C}_{65} = -0{\cdot}40$	$\bar{S}_{65} = -0{\cdot}28$
$\bar{C}_{66} = -0{\cdot}53$	$\bar{S}_{66} = -0{\cdot}41$

(six degree solution)

Table 3

Guier and Newton (11)

$10^6. A$	$10^6. B$
$A_2^2 = 7{\cdot}53$	$B_2^2 = -3{\cdot}79$
$A_3^1 = 6{\cdot}90$	$B_3^1 = 0{\cdot}80$
$A_3^2 = 4{\cdot}56$	$B_3^2 = -2{\cdot}54$
$A_3^3 = 2{\cdot}47$	$B_3^3 = 3{\cdot}66$
$A_4^1 = -2{\cdot}39$	$B_4^1 = -1{\cdot}87$
$A_4^2 = 1{\cdot}77$	$B_4^2 = 1{\cdot}88$
$A_4^3 = 3{\cdot}59$	$B_4^3 = 0{\cdot}03$
$A_4^4 = -0{\cdot}89$	$B_4^4 = 0{\cdot}81$
$A_5^1 = 0{\cdot}64$	$B_5^1 = -0{\cdot}78$
$A_5^2 = 1{\cdot}26$	$B_5^2 = -1{\cdot}58$
$A_5^3 = 0{\cdot}43$	$B_5^3 = 0{\cdot}49$
$A_5^4 = -2{\cdot}29$	$B_5^4 = -1{\cdot}22$
$A_5^5 = -0{\cdot}16$	$B_5^5 = -3{\cdot}14$
$A_6^1 = 0{\cdot}00$	$B_6^1 = 0{\cdot}51$
$A_6^2 = -0{\cdot}82$	$B_6^2 = -0{\cdot}79$
$A_6^3 = 2{\cdot}70$	$B_6^3 = 0{\cdot}26$
$A_6^4 = -1{\cdot}56$	$B_6^4 = -2{\cdot}59$
$A_6^5 = -0{\cdot}90$	$B_6^5 = -2{\cdot}60$
$A_6^6 = 0{\cdot}07$	$B_6^6 = -1{\cdot}18$

(eight degree solution)

On the other hand, Guier (10), Guier and Newton (11) deduced from the Doppler tracking of five satellites the coefficients of zonal harmonics of odd orders through the ninth, the non-zonal harmonics of orders from the second through the eighth and the sectorial harmonics to the thirteenth order. By combining these values with King-Hele's (3) they computed a shape of the geoid and studied the distribution of the magnitude of the harmonics. They found no apparent relation with the harmonics of the topography and saw that the harmonic coefficients of the geoid are consistent with random density variations with negligible spatial correlation beyond about 0·01 of the Earth's radius, which begin near the top of the mantle and continue to an undetermined depth. Guier and Newton expanded the geopotential in the form

$$\sum_{n,\,m} \left(\frac{R}{r}\right)^{n+1} \left[\frac{(n-m)!}{(n+m)!}\right]^{1/2} (1 - x^2)^{m/2}\, \frac{\mathrm{d}^m P_n\,(\cos\theta)}{\mathrm{d}\,x^m}\, [A_n^m \cos m\lambda + B_n^m \sin m\lambda]$$

and gave the coefficients A_n^m and B_n^m as in Table 3. Yionoulis (12) computed the coefficients $(n, m) = (13, 12), (13, 13), (15, 13)$.

Kozai (13) discussed all recent determinations of the harmonic coefficients by means of artificial satellites and the future programmes for improving the results and detecting the effects due to Earth tide and any other time variations of the potential to the satellite motion.

A. H. Cook (14) published an account of a comparison between various determinations of the even zonal harmonics of the geopotential from satellite observations by paying particular attention to the problem of estimation that arises for the fact that there are more parameters to determine than the independent data. He (15) discussed also these questions in connection with the motion of the Moon.

Simultaneous observations of artificial satellites facilitate the determination of the correction to station coordinates at the same time as the geopotential. Zhongolovich (16), Kaula (17), and Izsak (19) discussed the problem. Kaula (18) also discussed the Earth mantle, and the tidal dissipation of solid friction, including thermal effect and even a similar effect on the Moon (20).

Zhongolovich (21) studied the general theory of using simultaneous observations of artificial Earth satellites in geodesy and geophysics. Zhongolovich (16) derived the general equations and then the methods of space triangulation (21). The deviation of the geodetic system of coordinates (16) and the satellite positions in geodetic system (22, 23, 24) are determined. The project of the unique world space triangulation is proposed (21). He also dealt with the methods of discussion and the use of INTEROBS observations for determining the atmospheric fluctuations of short periods (22, 25). Batrakov (26) considered the problem of the possibility of using resonant satellites for the determination of the coefficients of the geopotential. He (26a) also developed a general theory of defining the mutual positions of observational stations by simultaneous or nearly simultaneous optical and radio observations of a satellite at several stations.

Kaula (27) reviewed the tesseral harmonic determination from the dynamics of satellite orbits and compared and combined the satellite determination of the Earth's gravity field with other results for geodetic parameters, such as the gravimetric determination. Kaula (28) published his new determination of the 44 tesseral harmonic coefficients, together with 36 station coordinates and 511 orbital elements, from 7234 Baker-Nunn camera determinations of five satellites. Supplementary observations on the accelerations of 24-hour satellites and directions between tracking stations for simultaneous observations are also included in the reduction.

It is noted that the factor GM of the expression for the geopotential, among other constants, has been determined from the radio tracking of Mariner (Venus) and Ranger III–VII lunar missions to be $3{\cdot}986009 \times 10^{14}$ m^3 s^{-2} by Sjogren and Trask (29).

BIBLIOGRAPHY

1. Kozai, Y. *Publ. astr. Soc. Japan*, **16**, 263, 1964.
2. Smith, D. E. *Planet. Space Sci.*, **13**, 1151, 1965.
3. King-Hele, D. G. *Planet. Space Sci.*, **12**, 835, 1964; *Nature*, **207**, 575, 1965.
4. King-Hele, D. G., Cook, G. E. *Geophys. J.*, **10**, 17, 1965.
5. King-Hele, D. G., Cook, G. E., Rees, J. M. *Geophys. J.*, **8**, 119, 1963.
6. King-Hele, D. G., Cook, G. E., Scott, D. W. *Planet. Space Sci.*, **13**, 1213, 1965.
7. King-Hele, D. G., Cook, G. E., Scott, D. W. *Planet. Space Sci.*, **14**, 49, 1966.
8. King-Hele, D. G., Cook, G. E., Watson, H. M. *Nature*, **202**, 996, 1964.
9. Izsak, I. G. *J. geophys. Res.*, **69**, 2621, 1964.
10. Guier, W. H. *Nature*, **200**, 124, 1963.
11. Guier, W. H., Newton, R. R. *J. geophys. Res.*, **70**, 4613, 1965.
12. Yionoulis, S. M. *J. geophys. Res.*, **71**, 1768 and 4064, 1966.
13. Kozai, Y. *Space Sci. Rev.*, **5**, 818, 1966.

14. Cook, A. H. *Geophys. J.*, **10**, 181, 1965.
15. Cook, A. H. *Bull. astr.*, **25**, 33, 1965.
16. Zhongolovich, I. D. *Astr. Zu.*, **41**, no. 1, 156, 1964.
17. Kaula, W. M. Proc. IUTAM Symposium on *Dynamics of Satellites*, at Paris, ed. by M. Roy, Springer, Berlin, 1963; *J. geophys. Res.*, **68**, 473, 1963; *Geophys. J.*, **5**, 104, 1961; *Advance Geophys.*, **9**, 191, 1962; *Dynamics of Rockets and Satellites*, 179, 1965; *Introduction to Space Science*, ed. by W. N. Hess, 297, 1965; Proc. IAG Symposium, *Use of Artificial Satellites for Geodesy*, Athens, 1966; *J. geophys. Res.*, **71**, 4377, 1966.
18. Kaula, W. M. *J. geophys. Res.*, **68**, 4959 and 5183, 1963.
19. Izsak, Y. G. *J. geophys. Res.*, **68**, 4967, 1963; **69**, 2621, 1964; *Trajectories of Artificial Celestial Bodies as Determined by Observations*, 1966.
20. Kaula, W. M. *The Earth-Moon System*, ed. by B. G. Marsden and A. G. W. Cameron, Plenum, New York, 1966; *Revs. Geophys.*, **2**, 661, 1964.
21. Zhongolovich, I. D. *Studia Geoph. et Geodet.*, **9**, Paris, 1965.
22. Zhongolovich, I. D. *Beobachtungen Künstlicher Erdsatelliten*, no. 3, Berlin, 1965; no. 4, Prague, 1966.
23. Zhongolovich, I. D. *Bull. Stations Optical Observ. AES* (USSR), no. 46, 1966.
24. Zhongolovich, I. D. *Bjull. Inst. Teor. Astr.*, **10**, no. 8 (121), 509, 1966.
25. Zhongolovich, I. D. COSPAR Symp., Paris in 1965, Berlin in 1966.
26. Batrakov, Yu. V. Proc. XIV. Intern. Astronaut. Congr., **4**, Paris, 1964.
26a. Batrakov, Yu. V. *Astr. Zu.*, **42**, no. 1, 195, 1965.
27. Kaula, W. M. Proc. IAG Symposium, *Use of Artificial Satellites for Geodesy*, Athens, 1966; *J. geophys. Res.*, **71**, 5303, 1966; Publ. no. 508, Inst. Geophys. Planet. Phys., Univ. Calif., 1966; Proc. *IAU Symposium No.* 21, *Astronomical Constants, Bull. astr.*, **25**, 21, 1965.
28. Kaula, W. M. Publ. no. 487, Inst. Geophys. Planet. Phys., Univ. Calif., 1966.
29. Sjogren, W. L., Trask, D. W. *J. Spacecraft*, **2**, 689, 1965.

THREE-BODY PROBLEM: REGULARIZATION

Lemaître's transformation (1) for the regularization of the three-body problem has been worked out by Deprit and Delie (2), Deprit and Roels (3). Deprit and Broucke (4) applied Lemaître's regularizing transformation to the restricted problem of three bodies in the form convenient for numerical work on electronic computers. Broucke (5) generalized the regularizing transformation in the form

$$Z = \tfrac{1}{4}\,(\zeta^n + \zeta^{-n}), \qquad Z = \tfrac{1}{2}\cos n\zeta,$$

where n is a finite non-zero real number and ζ, Z are complex variables. The first form reduces to Birkhoff's for $n = 1$ and to Lemaître's or to Arenstorf's (6) for $n = 2$. The second form reduces to Thiele's for $n = 2$. Broucke deduced the formulae through conformal mapping in the general case and formulated the canonical equations of motion.

Sconzo (7) carried out the formal series expansion up to the eighteenth order terms of the solution of the three-body problem when the Sundman type variable is used. In particular, Levi-Civita's variable u defined by $dt = V\,du$, where V is the total potential, is used throughout the expansion. This expansion will be obtained by means of iterative procedure similar to that in the method of Peano-Picard for constructing the solution of a system of differential equations.

Szebehely and Giacaglia (8, 9) presented the equations of motion of the restricted three-body problem by allowing non-zero eccentricity of Jupiter's orbit with the true anomaly of the primaries as independent variable and of dimensionless pulsating coordinates as dependent variables, first introduced by Scheibner in 1866 and later by Nechvile. The short- and long-period effects of the eccentricity are discussed in connection with the generalized Jacobi integral. The similarity between the equations for the circular and for the elliptic cases permits the introduction of regularizing transformations by following the methods applicable for the circular case. The major effect of the eccentricity is that the regularized equations are in the form of integro-differential equations. Szebehely (**10**) then expanded the terms appearing in the regularized equations according to the powers of a dimensionless parameter associated with the mass-ratio of the primaries.

Szebehely (**11**) attempted to give the power series solution of the restricted three-body problem by using the Thiele-Burrau regularization transformation. Since real singularities are eliminated in this way, the only remaining singularities are on the imaginary regularized time axis. They may be eliminated by Poincaré's time transformation and the general solution of the restricted problem may be represented in convergent Taylor series.

According to Sundman it is known that, if the total angular momentum of the three bodies is not zero, the coordinates and the time are expressed by analytic functions of the regularizing variable τ in the strip of the complex plane τ defined by $|\mathfrak{Re}\,\tau| < \infty$, $|\mathfrak{Im}\,\tau| < \Omega$, where Ω depends on the masses and the initial condition, that the coordinates and the time are expanded in convergent Taylor series about any real value of τ within a circle of radius larger than or equal to Ω, and that the series can be continued analytically.

Cesco (**12**, **13**) has shown by an example that Borel's integral method of summability of divergent series, conveniently adapted for numerical computation, can be used for this purpose even outside the circle of convergence of the Taylor series solution.

Brumberg (**14**) has numerically applied Mittag-Leffler's theorem on the expansion of an analytic function defined by a Taylor series in series of polynomials. The infinite strip of Sundman is contained in Mittag-Leffler's star region so that the series of polynomials are convergent for any real value of τ and accordingly of t.

Lóper García (**15**) improved and extended the classical ephemeris computed by Zumkley (**16**) of a planar problem with equal masses. Cesco (**17**) applied this method to find the quasi-asymptotic solution of a hyperbolic-elliptic type of the three-body problem. Cesco (**18**) derived the differential equations for the Keplerian elements in the n-body problem and solved without referring to the method of variation of elements. Cesco says that this straightforward derivation simplifies and completes the results obtained by Chazy (**19**).

Kustaanheimo (**20**), Kustaanheimo and Stiefel (**21**) transformed by conformal mapping the Keplerian motion in the three-dimensional space to that in the four-dimensional manifold in which the equations of motion reduce to linear differential equations with constant coefficients completely regular at the attracting centres. The theory is based on spinor analysis developed by Kustaanheimo himself (**22**) and Kustaanheimo and Nuoto (**23**) in real matrix form. They tried to generalize to higher dimensions by applying Hurwitz-Radón theorem for quadratic forms.

Rössler (**24**) applied this Kustaanheimo-Stiefel transformation for computing the perturbation by combining with the three-dimensionally generalized form of Levi-Civita's regularization. Numerical experiment has been made and it is shown that the regularized computation is in many cases, particularly for orbits of high eccentricity, superior to classical methods. The disturbing function is expanded in terms of eccentric anomalies after Hansen, and the theory is applied to Vesta and compared with Leveau's classical work (**25**).

Further Stiefel and Waldvogel (**26**) generalized Birkhoff's regularizing transformation to three dimensions along the line of Kustaanheimo and Stiefel. A Kustaanheimo and Stiefel transformation is combined with two inversions. A mapping can be constructed which makes possible a simultaneous regularization of both singularities in the three-dimensional restricted three-body problem. An extension to the three-dimensional elliptic restricted three-body problem is suggested. Stiefel (**27**) discussed some methods for coordinate perturbation with regularized coordinates. Waldvogel (**28**) showed the connection between regularization and the theory of functions of complex variables.

A four-dimensional spinor regularization was given by Kustaanheimo (**29**) by basing on the Lorenz-invariant spinor square root of a four-dimensional vector in the Minkowski space time. He (**30**) also gave the spinor form of the energetic identities which he treated in vector form in 1963.

Belorizky (**31**) proved once that the regularizing variable ω of Sundman can not in the whole ω-plane make the coordinates of the three bodies holomorphic functions of ω in the triangular equilibrium solution of Lagrange. Now he (**32**) extended it to the general three-body problem.

Recently Pierce (**33**) obtained a solution in the form of trigonometric series of Levi-Civita's regularized equations by the method of general perturbations. The solution shows some of the characteristics of typical lunar or planetary theory. Then he computed four different orbits both circular and collision orbits without restriction on the eccentricity. He has seen the improvements of long-term accuracy by replacing the osculating reference orbit with a mean reference orbit.

BIBLIOGRAPHY

1. Lemaître, G. *Bull. Cl. Sci. Acad. Belg.*, **38**, 582, 1218, 1952; **40**, 759, 1954; *Vistas in Astronomy*, **1**, 207, 1955.
2. Deprit, A., Delie, A. *Ann. Soc. Sci. Bruxelles*, **75**, 5, 1961; *Bull. Cl. Sci. Acad. Belg.*, **48**, 652, 1962; *Archive Rat. Mech. Analysis*, **12**, 325, 1963.
3. Deprit, A., Roels, J. *Bull. Cl. Sci. Acad. Belg.*, **48**, 432, 1962.
4. Deprit, A., Broucke, R. *Icarus*, **2**, 207, 1963.
5. Broucke, R. *Icarus*, **4**, 8, 1965.
6. Arenstorf, R. F. *Astr. J.*, **68**, 548, 1963.
7. Sconzo, P. unpublished yet.
8. Szebehely, V., Giacaglia, G. E. O. *Astr. J.*, **69**, 230, 1964.
9. Szebehely, V. *Mathematische Methoden der Himmelsmechanik und Astronautik*, 21, 1966.
10. Szebehely, V. *Astr. J.*, **69**, 309, 1964.
11. Szebehely, V., Pierce, D. A. *Astr. J.*, **71**, 886, 1966.
12. Cesco, R. P. *Obs. astr. La Plata, ser. Astr.*, **25**, no. 4, 1963.
13. Cesco, R. P. *Obs. astr. La Plata, ser. Astr.*, **31**, no. 14, 1965.
14. Brumberg, V. A. *Bjull. Inst. Teor. Astr.*, **9**, no. 4 (107), 234, 1963.
15. Lóper García, F. *Commission Sci. Res. Province of Buenos Aires*, La Plata, no. 21, 1966.
16. Zumkley, J. *Astr. Nachr.*, **272**, 66, 1941.
17. Cesco, R. P. *Obs. astr. La Plata, ser. Astr.*, (in press).
18. Cesco, R. P. *Obs. astr. La Plata, ser. Astr.*, (in press).
19. Chazy, J. *Mécanique Céleste*, Presses Universitaires, Paris, 221, 1953.
20. Kustaanheimo, P. *Ann. Univ. Turkuens*, A. I, 73, 1964 = *Publ. astr. Obs. Helsinki*, no. 102, 1964.
21. Kustaanheimo, P., Stiefel, E. *J. reine angew. Math.*, **218**, 204, 1965 = *Publ. Astr. Obs. Helsinki*, no. 110, 1965.
22. Kustaanheimo, P. Lectures on Celestial Mechanics I, *Publ. Astr. Obs. Helsinki*, no. 117, 1966.
23. Kustaanheimo, P., Nuoto, V. S. *Publ. Astr. Obs. Helsinki*, no. 109, 1965.
24. Rössler, M. *Z. angew. Math. Phys.*, **16**, 395, 1965.
25. Rössler, M. Dissertation, Techn. Hochsch. Zürich, 1966.
26. Stiefel, E., Waldvogel, J. *C. R. Acad. Sci.*, Paris, **260**, 805, 1965.
27. Stiefel, E. *Mathematische Methoden der Himmelsmechanik und Astronautik*, 87, 1966.
28. Waldvogel, J. Dissertation, Techn. Hochsch. Zürich, 1966.
    Waldvogel, J. *Mathematische Methoden der Himmelsmechanik und Astronautik*, 33, 1966.
29. Kustaanheimo, P. *Comm. Soc. Sci. Fenn., Phys.-Math.*, **30–9**, 1965.
30. Kustaanheimo, P. *Mathematische Methoden der Himmelsmechanik und Astronautik*, 333, 1966.
31. Belorizky, D. *Bull. astr.*, **16**, 327, 1952.
32. Belorizky, D. *C. R. Acad. Sci.*, Paris, **262**, 1133, 1966.
33. Pierce, D. A. *Astr. J.*, **71**, 545 and 562, 1966.

THREE-BODY PROBLEM: NUMERICAL SURVEY

Extensive numerical works have been carried out in search for periodic orbits not only around the equilibrium points and the masses but also remote from these singularities in the restricted three-body problem due to the recent development of electronic computers which facilitate the numerical treatment in a great deal.

Szebehely and Williams (**1**) published tables showing pertinent information regarding the collinear equilibrium points of the restricted three-body problem. General applications to problems of cosmogony, stellar dynamics, space science are pointed out by Szebehely (**2**). He (**3**) also reviewed the linear and non-linear aspects of the problem of motion around the equilibrium points with special reference to the restricted problem.

Danby (**4**) discussed the variational stability of the triangular equilibrium points in the elliptic restricted problem by taking the mass ratio μ and the eccentricity e of the primary's orbit as two parameters, and found the region in the (μ, e)-plane in which there is stability. For the general problem of three bodies with masses m_1, m_2, m_3 it is known that this configuration is stable if $(m_2m_3 + m_3m_1 + m_1m_2)/(m_1 + m_2 + m_3)^2$ is less than some quantity depending only on e. Grebenikov (**5**) discussed the same problem on the stability in the sense of Liapounov. Bennett (**6**) computed the characteristic exponents of the same problem, and refined and extended Danby's transition curve for the stability and instability regions by the use of the symmetry character of the Hamiltonian systems. In the unstable region not considered by Danby it is found that the instabilities are due to three different types of the characteristic roots. The curves separating these three regions are extensions of the curves for transition between the stable and unstable regions. For the collinear points there is no value of μ and e for which variational stability exists. Bernstein and Ellis (**7**), by a method similar to the one by Danby and Bennett, gave a criterion for the boundedness of all solutions of a linearized non-homogeneous equations with periodic coefficients for the motion of an infinitesimal mass in the neighbourhood of the triangular equilibrium points. Szebehely (**8**) discussed the effect of the Coriolis term on the stability characteristics of the triangular equilibrium points. It is known that the stability of these points is due to the Coriolis term, which when neglected gives unstable solutions. The relation between the critical mass-ratio at which instability sets in and the magnitude of the Coriolis force is discussed.

Rabe (**9**) is continuing his research on the restricted Trojan problem. He takes the rather accurately determined long-period solutions of the restricted problem as reference or intermediate orbits in his work for the existence of additional classes of periodic librations about the triangular equilibrium points (**9**) in the fundamental planes of the circular and the elliptic problems (**10, 11**) as well as in the three-dimensional space. For the plane restricted problem a third order stability analysis of the most general non-periodic Trojan librations incorporating long- and short-period terms has been undertaken also on the basis of some selected reference solutions of long period. Certain stability limits were found to exist in the form of an upper e_{max} for the heliocentric eccentricity e of such Trojans. This limit e_{max} decreases with the increasing size of the basic libration of long period. The approximate results are $e_{max} = 0{\cdot}19$ for every small basic amplitude, $e_{max} = 0{\cdot}16$ for a reference libration extending over almost 21° in longitude, and $e_{max} = 0{\cdot}08$ for a basic libration of 43° total amplitude. Since e_{max} seems to decrease roughly in proportion to the second power of the increasing libration amplitude, these findings tend to explain rather well the actually observed maximum eccentricities and librations of the Trojan asteroids. As a product of this stability study the existence of several families of commensurability-related periodic solutions incorporating short-period oscillations has been also studied, from the occurrence of certain integral values of the characteristic exponents as a function of the amplitude of these superposed short-period terms (**12**).

Another series of works have been published by Deprit and his colleagues. Deprit and Price (13) started the integration of the differential equations by referring to the method of Steffensen (14), in which the coordinates are expanded in powers of time, and computed the characteristic exponents by the variational equations. Deprit (15), Deprit and Henrard (16) computed symmetric orbits which are doubly asymptotic to one of the three collinear equilibrium points. Numerical interpolation gave thirty-five special values of the mass-ratio which give rise to doubly asymptotic orbits: three for the L_1-point, two for the L_2-point, and thirty for the L_3-point. Two of these values are of particular interest since they are near the mass-ratio of the Earth-Moon system. Then Deprit (17) and Deprit and Delie (18) studied numerically the motion in the vicinity of the triangular equilibrium points. It has been seen that short- and long-period elliptic oscillations of the first order theory can be extended to two families of periodic motions except for a denumerable set of critical mass-ratios for which the long-period librations are in resonance with the short-period oscillations. Thus Brown's conjecture is justified after Rabe on the horse-shoe orbits which go through orbits doubly asymptotic to the L_1, L_2, L_3-points successively.

Deprit (19) has shown further that Routh's criterion on the existence of periodic solutions around L_4 is valid only in the first order, by expanding the analytical relation between the mass-ratio and the lower bound of the orbital parameter in power series up to the fourteenth order, and that the limiting orbits are of the stable type for mass-ratios as high as 0·044, the orbits evolving from infinitesimal ellipses into large-sized asymmetrical ovals. Deprit and Palmore (20) studied that analytic continuation of the family of short-period orbits around L_4 and saw that the characteristic exponents are of the stable type, in contradiction to Goodrich's conclusion (21).

Roels has studied the resonance effect in the vicinity of the triangular equilibrium points (22), in particular, those periodic orbits with periods twice that of Jupiter (23). An expansion of the solution in Fourier series is suggested, which is not of d'Alembert characteristic and which can be obtained by leaving some parameters at first arbitrary and then adjusted at the next approximation. The behaviour of the families of orbits is given in the neighbourhood of the critical mass-ratio. Delie (24) has given the d'Alembert series for the long-period solutions around those equilibrium points. Roels (25) gave a new method which enables to discover families of periodic orbits for all mass-ratios even for those discrete values for which the classical series expansions do not exist. A general solution in the vicinity of the collinear equilibrium points is discussed by Henrard (26). Henrard and Rom are applying the same technique for normalizing the problem at L_4.

Deprit, Henrard, Rom (27) carried out Birkhoff's (28) normalization about a stable equilibrium for a conservative Hamiltonian system with two degrees of freedom by building up explicitly the necessary canonical transformation by the method of indeterminate coefficients. The normalizing canonical transformation expresses the Cartesian phase variables in a form of double Fourier series in two angular coordinates whose coefficients are power series in the square roots of two action momenta. In the normalized part of the transformed Hamiltonian the angular coordinates are ignorable. They implemented such a Birkhoff's normalization on an electronic computer. The characteristic exponents along the singular families of long and short period orbits around L_4 are developed as power series of the orbital parameter. Then the procedure was applied to the equilateral triangular centre of libration for the Earth-Moon system up to order 13. Breakwell and Pringle (29) also had computed the same problem up to order four with only partial normalization. The question on resonance has been treated by Gustavson (30).

Goodrich (31) studied particularly the short-period periodic orbits around the triangular equilibrium points, in contrast to Rabe's work for long-period periodic orbits. Power series expansions are used and the coefficients are determined by recurrence formulae. Goodrich found two types of short-period orbits. The Jacobi constant increases as the orbital size increases. The Jacobi constant is smaller and the initial velocity is larger for a short-period orbit compared with a long-period orbit. For small values of the orbital parameter representing the deviation from the equilibrium point the orbit begins with elliptical (type I), and then for a certain value of the orbital parameter there occur two branches, type I and type II, of which the latter type orbits are asymmetrical and cut the JS-axis. Later for a larger value of the orbital parameter the type I orbits can cross the JS-axis.

These years are flourishing in the survey of families of periodic orbits by means of powerful electronic computers. Bartlett (32) computed periodic solutions with Thiele's variable and by means of a modified Runge-Kutta method, in particular asymmetric periodic orbits in the restricted problem with two equal masses for which Strömgren and his colleagues have made very extensive study. Bartlett found seven new classes different from Strömgren's. Hénon (34), after the manner in which Hénon and Heiles (33) studied the third integral in the dynamical problem of the galaxy, obtained periodic orbits which cut twice the straight line JS joining the two finite masses. Hénon's procedure is to study the behaviour of the successive intersection of a trajectory with the surface of section. Thus Hénon found twenty-one classes, of which six are new. He discussed the stability character of the invariant points, which represent periodic orbits, for the transformation of a point on the surface of section to its consecutive intersection of the trajectory with the surface. G. Zech of the Astronomisches Recheninstitut is studying several families of periodic orbits of the restricted three-body problem by numerical integration.

Goudas (35) computed numerically nine doubly-symmetric periodic orbits in three dimensions and examined their stability through the eigen-values of the Jacobian. Goudas called the least unstable orbits to be quasi-stable, for which all eigen-values are unity. He showed that the mass-ratio μ has only a quantitative effect on the problem such that it does not produce any new type of periodic orbits, that almost all periodic orbits which are highly inclined to the plane of motion of the two primaries are very unstable, and that all periodic orbits going round the straight line equilibrium points are very unstable whatever their inclinations are and in general that the families of periodic orbits change in a more or less regular way as μ varies. The discussion is greatly facilitated by the assumption of the doubly-symmetric character.

Then Hénon (36) started an exhaustive and systematic explorative study of the periodic and non-periodic orbits of the problem for two equal masses. He found three types of orbits, quasi-periodic, semi-ergodic and escape orbits. For high values of Jacobi's constant C all orbits in the vicinity of the main bodies are quasi-periodic. As C decreases, the region occupied by the semi-ergodic orbits appears and extends. When $C < 3{\cdot}4568 \ldots$, Jacobi's limit opens and the semi-ergodic orbits become escape orbits, at the same time the region occupied by quasi-periodic orbits shrinks and eventually vanishes. The behaviour of small island sub-regions is quite interesting. This elegant work of Hénon is promising an extensive theory on the behaviour of various types of orbits as C varies and as the period passes through rational and irrational numbers. A work of Wintner discussing the genealogy of various classes of Strömgren's periodic orbits is reminded.

Hénon and Heiles (33) studied by numerical experiment the problem of the existence of the third integral and found that the third integral exists only for a limited range of initial conditions. A trajectory is integrated to the next point at which the trajectory cuts a fixed surface. An invariant point for such a transformation to the next intersection corresponds to a periodic trajectory. The transformed points are seen to cluster in some limited regions, which they call islands. The set is dense everywhere but the islands do not cover the whole area but leave a so-called sea between islands in which the ergodic trajectory is dense everywhere. Barbanis (37) studied the behaviour of the hodographs of galactic orbits by means of a formal third integral. Contopoulos (48) gave tables of this third integral. Contopoulos (39) classified the integrals of motion into isolating, quasi-isolating and ergodic, and proved by means of von Zeipel's method that near any given Hamiltonian in the form of a series there is a separable Hamiltonian.

Contopoulos (40), Contopoulos and Montsoulas (41) discussed the general case of resonance in an axially symmetric potential field in which the unperturbed frequencies in two directions are in a rational ratio p/q. It has been seen that the general form of the third integral is not valid but a new integral is found which can be used as a third integral. For $p + q < 4$ the case shows quite a peculiar character. In particular Contopoulos discussed the case $p = q$ and the role of small divisors in a third integral. Contopoulos and Woltjer (42) discussed the third integral in the case of non-smooth potentials. Contopoulos (43) studied by numerical process the periodic orbits. Periodic orbits in the vicinity of a periodic orbit, which oscillates p times along the x-axis and q times along the y-axis, are called tube orbits, which can be studied by the third integral. He discussed the transition from a tube orbit to an ordinary box orbit. Barbanis (44) studied the transition from the case of the isolating third integral to that of the non-isolating third integral by numerical experiment with a potential similar to Contopoulos's (40) or to Hénon-Heiles's (33). He has seen that the third integral ceases to be isolating for a large range of initial conditions with increasing total energy, but that even for energies higher than the energy of escape there is a set of orbits which do not escape to infinity, indicating the existence of a very nearly isolating integral. Anderle (45) discussed a special form of the third integral.

Contopoulos (38) applied the idea to the restricted three-body problem and also in the elliptic problem in the form of series. Numerical integration by Bozis (46) indicates the existence of a new non-isolating integral besides the isolated Jacobi integral in the restricted three-body problem. Bozis transformed the equations of motion by Birkhoff's regularizing transformation in order to include collision and near-collision orbits. With various values of the initial conditions he computed various orbits. The boundaries and the invariant curves are given analytically in the regularized system for the two-body problem where the two integrals are known to exist. He found the boundaries of the orbit by using a series-type integral of Contopoulos.

Barbanis (44) studied the behaviour and topology of the family of periodic orbits of the problem. Hénon's papers (34, 36) on the numerical exploration of the restricted three-body problem have been discussed by Bozis and Szebehely (47).

BIBLIOGRAPHY

1. Szebehely, V., Williams, C. A. *Astr. J.*, **69**, 460, 1964.
2. Szebehely, V. *Space Sci. Rev.*, **2**, 219, 1963.
3. Szebehely, V. *Methods in Astrodynamics and Celestial Mechanics*, 3, 1966.
4. Danby, J. M. A. *Astr. J.*, **69**, 165 and 294, 1964.
5. Grebenikov, E. A. *Astr. Zu.*, **41**, no. 3, 567, 1964.
6. Bennett, A. *Icarus*, **4**, 177, 1965.
7. Bernstein, I. S., Ellis, J. *Icarus*, **5**, 299, 1966.
8. Szebehely, V. Intern. Congress of Mathematicians, Moscow, 1966.
9. Rabe, E. Lectures in Applied Mathematics, Vol. 6, 79, Amer. Math. Soc., 1966.
10. Rabe, E. *ibid.*, 102.
11. Rabe, E. Proc. *IAU Symposium No.* 25, 176, Thessaloniki, 1964.
12. Rabe, E. *Astr. J.*, **70**, 329 and 687, 1965.
13. Deprit, A., Price, J. F. *Astr. J.*, **70**, 836, 1965.
14. Steffensen, J. F. *K. Danske Videnskab. Selskab., Math. Fys. Medd.*, **30**, no. 18, 1956.
15. Deprit, A. *Icarus*, **4**, 273, 1965.
16. Deprit, A., Henrard, J. *Astr. J.*, **70**, 271, 1965.

17. Deprit, A. Boeing Sci. Res. Lab., D1-82-0335, 1963.
18. Deprit, A., Delie, A. *Icarus*, **4**, 242, 1965.
19. Deprit, A. *Astr. J.*, **71**, 77, 1966.
20. Deprit, A., Palmore, J. *Astr. J.*, **71**, 94, 1966.
21. Goodrich, E. F. *NASA Techn. Note*, X-643-65-420, 1965.
22. Roels, J. Dissertation, Univ. Louvain, 1965.
23. Roels, J. *Bull. astr.*, (3) **1**, fasc. 3, 1966.
24. Delie, A. Dissertation, Univ. Louvain, 1966.
25. Roels, J. *Bull. astr.*, (3) **1**, fasc. 2, 67, 1966.
26. Henrard, J. Dissertation, Univ. Louvain, 1966.
27. Deprit, A., Henrard, J., Rom, A. R. M. Boeing Sci. Res. Lab., D1-82-0556, 1966; Boeing Sci. Res. Lab., D1-82-0564, 1966; *Icarus*, **5**, 1966 (in press).
28. Birkhoff, G. D. *Dynamical Systems*, Amer. Math. Soc., 1927.
29. Breakwell, J. V., Pringle, R. *Methods in Celestial Mechanics and Astrodynamics*, 55, 1966.
30. Gustavson, F. G. *Astr. J.*, **71**, 670, 1966.
31. Goodrich, E. F. *Astr. J.*, **71**, 88, 1966.
32. Bartlett, J. H. *Publ. Medd. Københavns Obs.*, no. 179, 1964.
33. Hénon, M., Heiles, C. *Astr. J.*, **69**, 73, 1964.
34. Hénon, M. *Ann. Astrophys.*, **28**, 499 and 992, 1965.
35. Goudas, C. L. *Icarus*, **2**, 1, 1963.
36. Hénon, M. *Bull. astr.*, (3) **1**, fasc. 1, 57, 1966; fasc. 2, 49, 1966.
37. Barbanis, B. *Astr. J.*, **70**, 285, 1965.
38. Contopoulos, G. *Astrophys. J.*, **142**, 802, 1965.
39. Contopoulos, G. *Astrophys. J.*, **138**, 1297, 1963.
40. Contopoulos, G. *Astr. J.*, **68**, 763, 1963; Proc. *IAU Symposium No.* 25, 3, Thessaloniki, 1964.
41. Contopoulos, G., Montsoulas, M. *Astr. J.*, **70**, 817, 1965; **71**, 687, 1966.
42. Contopoulos, G., Woltjer, L. *Astrophys. J.*, **140**, 1106, 1964.
43. Contopoulos, G. *Astr. J.*, **70**, 526, 1965.
44. Barbanis, B. *Astr. J.*, **71**, 415, 1966; Proc. *IAU Symposium No.* 25, 19, Thessaloniki, 1964.
45. Anderle, P. *Bull. astr. Inst. Csl.*, **17**, no. 4, 166 and 169.
46. Bozis, G. *Astr. J.*, **71**, 404, 1966.
47. Bozis, G., Szebehely, V. Proc. *IAU Symposium No.* 25, 163, Thessaloniki, 1964.
48. Contopoulos, G. *Astrophys. J.*, **145**, 109, 1966; *Astrophys. J. Suppl.*, **13**, no. 122, 1966.

THREE-BODY PROBLEM: ANALYTICAL THEORY

Nahon (**1**) derived a simplified form of the differential equations for the planar three-body problem when the position variable is the virial of Sundman defined as the moment of inertia of the three masses with respect to their centre of mass. D'Ambly (**41**) discussed the collinear central figure of the *n*-body problem.

Pius (**2**) obtained the bounds of the variation of the semi-major axes in the restricted three-body problem for small values of the disturbing mass.

Jefferys (**3**) proved the existence of doubly-symmetric periodic orbits in the three-dimensional restricted problem by the method of analytic continuation for sufficiently small mass-ratios. Krasinsky (**4**) studied the closed path of double collision with the larger mass in the plane circular restricted problem and obtained the existence of some types of paths both symmetric and asymmetric with respect to the *x*-axis by the method of a small parameter.

Richards (**5**) showed the asymptotic instability of the triangular equilibrium points, not by means of the linearized system of variational equations of motion but by means of the Jacobi integral only, by discussing the variation of the potential and the behaviour of the zero-velocity curves. Lanzano (**6**) studied the periodic motions around one of the primaries in the restricted problem, which can be generated from circular Keplerian orbits and are valid for any mass-ratio, by extending the procedure of Siegel for the lunar theory. The Fourier series representation is obtained, in which the coefficients are expanded in powers of the parameter related to the orbital period, and is shown to be convergent for small values of the period and for any value of the mass-ratio μ of the primaries. On the other hand, Arenstorf (7) found periodic solutions representing the analytic continuation of a Keplerian elliptic motion by a more general periodicity condition than Poincaré's, and proved the existence of such periodic solutions as holomorphic functions of the mass-ratio μ and the parameter ϵ which represents the deviation of the complex radius vector from the generating orbit. Danby (8) discussed the inclusion of extra factors, such as the variation of the law of attraction, the non-sphericity of the bodies, and the addition of a fourth distant body. The possible existence of such solutions is considered and where they exist, their stability is studied. Omarov (9) solved the restricted problem with variable mass of the primaries.

Choudhry (10) proved the existence of analytic continuation and its parametric representation for $\mu \neq 0$ corresponding to the circular orbits for $\mu = 0$ in the restricted problem in a three-dimensional coordinate system by referring to Birkhoff's work on the three-body problem. Barrar (11) proved the existence of periodic solutions of the second sort of Poincaré in the restricted problem.

Jefferys (12) has shown the existence of a class of inclined periodic solutions of the restricted problem, which form a continuous sequence with eccentricity as parameter and a discrete sequence with mean motion as parameter. They exist only for special values of the inclination and if the ratio of the mean motions to that of Jupiter is sufficiently small or sufficiently great. These are Poincaré's periodic solutions of the third sort.

Grémillard (13) published in a series of papers the existence of periodic solutions of the third sort of Poincaré by revising the former work of von Zeipel. He proved the existence of periodic solutions in which two eccentricities are zero at the initial epoch but are expanded in powers of a parameter ρ for $p - q$ even, where $n/n' = p/q$ is the ratio of mutually prime integers, and the existence of periodic solutions in which the initial eccentricities are not zero but inclinations are small for $p - q$ odd and greater than unity. He also proved von Zeipel's result that there does not exist any solution other than those for which $\bar{\lambda} = r(\pi/2)$ for $p - q$ odd and $\bar{\lambda} = r\pi$ for $p - q$ even ($r = 0, \pm 1, \pm 2, \ldots$), where $\bar{\lambda} = p\lambda_0' - q\lambda_0$ with the initial mean longitudes λ_0 and λ_0'. He is now constructing numerical examples for the case $p - q = 2$. Choudhry (14) showed the existence of periodic solutions of the third sort in the elliptic restricted problem and proved the stability of the generating solution.

Hénon is now studying periodic orbits of the second species of Poincaré which exist in the restricted problem when the second mass vanishes and which are characterized by a series of encounters between the second and the third masses.

Alekseev (15) estimated by Merman's method (16) the perturbation of a hyperbolic motion in the three-body problem basing on Merman's theorems in the theory of perturbed motions.

Pius (17) applied the theory of symmetric periodic orbits of the Schwarzschild type as a first approximation to the motion of Hecuba. Barrar (18) gave a new proof for the existence of periodic orbits which are closed after several revolutions after the fashion of Moser (19). Moser referred to the invariant curves of area-preserving mappings of a torus according to Birkhoff.

Petrovskaya (20) constructed a periodic solution of the first sort in the plane restricted problem in the form of power series in μ and $\beta^{1/2}$ where μ denotes the mass-ratio and β the ratio of the radii of the undisturbed orbit of the infinitesimal mass and of the relative orbits of the primaries. She found that the series are convergent for $0 \leqslant |\mu| \leqslant 0{\cdot}001$, $0 \leqslant |\beta| \leqslant 0{\cdot}4480$ for the inferior motion of the infinitesimal mass and $0{\cdot}26 \leqslant |\beta| \leqslant 0{\cdot}42785$ for the exterior motion.

Eckstein, Shi and Kevorkian (21) have studied the motion of a close satellite to the smaller of the two finite masses by taking as the independent variable, instead of the time, the angle between the radius vector of the satellite and the instantaneous node of its orbit around the primary. The equations of motion first given by Struble (23) are solved by a generalized asymptotic expansion procedure involving three variables to the second order as regards the smaller of the two finite masses. The three variables differ each other by one order of magnitude in the small parameter, and give the measure of the periods of the node and the pericentre. It is shown that rather large amplitude oscillations in the eccentricity, inclination and pericentre occur over the longest of these three periods for a moderate value of the inclination. They (22) demonstrated that the energy integral or any known exact integral of motion can be used to determine certain higher-order terms in the time history of the satellite motion after the geometry of the orbit is obtained. They showed how the loss of accuracy due to the presence of secular and long-period terms in such cases is to be overcome.

Szebehely, Pierce, Standish (24) computed by using Birkhoff's regularization a family of orbits with the property that they are non-periodic and collide with both primaries. Szebehely (25) discussed the problem of capture and of so-called 'swing-by' orbits and outlined the application of Tisserand's criterion for identifying comets to a new problem of space research.

Szebehely (26) reviewed recent results on the restricted problem as an irreducible dynamical system with two degrees of freedom.

Shibahara and Yoshida (27) sharpened the theorem of Birkhoff and Merman (28) in the general three-body problem. Alekseev (29) discussed with examples the interchange of a hyperbolic motion and an elliptic motion, and he (30) derived a criterion for hyperbolic and hyperbolic-elliptic motions according to the idea of Chazy and Merman.

Arenstorf (31) expanded the solution of the differential equations in the perturbation theory as functions of the initial values from which useful estimates and a sufficiently detailed description can be obtained by comparing with the unperturbed solutions so as to enable solution of the boundary-value problem for the perturbed system, and applied to the periodic solutions in the restricted three-body problem.

Conley (32) obtained a denumerably infinite number of long-periodic solutions of the planar restricted problem on the basis of Poincaré's fixed point theorem for torus mappings.

Schubart and Stumpff (33) developed an n-body programme of high accuracy for the calculation of disturbed ephemerides of planets and comets, and tested it successively by the high-speed computers at Heidelberg and Darmstadt. Schubart (34) succeeded in rediscovering the lost comet Tempel-Tuttle 1866 I by identifying it with 1965 *i* according to a calculation of the perturbations of 1866 I on the basis of this n-body problem. By means of another programme which allows generally the calculation of the orbits of a system of numerous mass points Wielen (35) studied the motion in a fictitious open cluster consisting of 100 stars. He considered three models differing in the distributions of the initial velocities and of masses. The analytical and statistical properties of stellar clusters known from stellar dynamics, such as time of relaxation and rate of escape, have been verified.

The n-body problem when n is sufficiently great is now being attacked by means of computers in view of application to stellar dynamics for unveiling the global characteristics by examining the behaviour of each individual constituent. Indeed, Miller (36) has shown that stellar dynamical systems possess the macroscopic irreversibility characteristic of statistical mechanical systems by means of a computer calculation of an n-body system. Aarseth (37) studied numerically the evolution of an isolated n-body system with $n = 100$, in order to determine the dynamical age of the system. The equipartition effect leads to pronounced mass segregation with the formation of a dense nucleus of heavy members and an extended halo of mainly light bodies. Von Hoerner (38), by taking $n = 16$ and $n = 25$ for globular clusters, found that within two to four times of relaxation all clusters with different starting conditions approach an isothermal polytrope and a Maxwellian velocity distribution, and that in much longer times the deviations from such a configuration become important. The virial theorem and Chandrasekhar's time (39) of relaxation are approximately verified. Miller (40) took $n = 4, 8, 12, 16, 24, 32$ and observed that two similar systems evolve simultaneously and the separation of their representative points in phase space grows exponentially with time because of encounter effects. The time constant of this exponential growth is short, about four times the mean time between binary collisions of one particle divided by the number n. The n-body systems with inverse-square law forces are seen to behave as tightly coupled systems.

Aarseth (42) further studied the advanced stages of dynamical evolution. The overall relaxation time is in good agreement with that predicted by Chandrasekhar's theory for centrally concentrated clusters but the mean relaxation time for each mass group is well correlated with a pronounced mass segregation, the heavy members forming a dense nucleus. Aarseth does not find any approach to the equipartition of kinetic energy for the clusters as a whole. The total velocity distribution develops an excess at high and low velocities as compared with the Maxwellian. He found some evidence for selective escape among light members in isolated clusters and the total mass loss to be greater than in the case of an equal mass distribution. The dispersion of the binding energy increases with time and may be used to show the dynamical age of simulated clusters.

BIBLIOGRAPHY

1. Nahon, F. *C. R. Acad. Sci.*, Paris, **258**, 815, 1964.
2. Pius, L. Yu. *Bjull. Inst. Teor. Astr.*, **11**, no. 2 (125), 1967 (in press).
3. Jefferys, W. H. *Astr. J.*, **70**, 393, 1965.
4. Krasinsky, G. A. *Bjull. Inst. Teor. Astr.*, **9**, no. 2 (105), 154, 1963.
5. Richards, P. B. *Icarus*, **1**, 334, 1963.
6. Lanzano, P. *Icarus*, **2**, 364, 1963.
7. Arenstorf, R. F. *Amer. J. Math.*, **85**, 27, 1963.
8. Danby, J. M. A. *Astr. J.*, **70**, 181, 1965.
9. Omarov, T. B. *Astr. Zu.*, **41**, no. 1, 170, 1964.
10. Choudhry, R. K. *J. Indian Math. Soc.*, **28**, 119, 1964.
11. Barrar, R. B. *Astr. J.*, **70**, 3, 1965.
12. Jefferys, W. H. *Astr. J.*, **71**, 99 and 566, 1966.
13. Grémillard, J. *C. R. Acad. Sci.*, Paris, **234**, 2339, 1952; **236**, 49, 1953; **239**, 153, 1954; **244**, 1011, 1957; **247**, 1307, 1958; **251**, 2121, 1960; **252**, 1424, 1961; *Bull. astr.*, **24**, 483, 1965.
14. Choudhry, R. K. *Bjull. Inst. Teor. Astr.*, **10**, no. 8 (121), 523, 1966.
15. Alekseev, V. M. *Astr. Zu.*, **38**, no. 2, 325; no. 4, 726; no. 6, 1099, 1961; **39**, no. 4, 724, 1962.
16. Merman, G. A. *Bjull. Inst. Teor. Astr.*, **6**, no. 2 (75), 73, 1955.
17. Pius, L. Yu. *Bjull. Inst. Teor. Astr.*, **8**, no. 1 (94), 11, 1961.
18. Barrar, R. B. *Math. Annalen*, **160**, 363, 1965.
19. Moser, J. *Math. Annalen*, **126**, 325, 1953; *Nachr. Akad. Wiss. Göttingen, Math.-Phys. Kl.*, **1962**, 1.
20. Petrovskaya, M. S. *Bjull. Inst. Teor. Astr.*, **9**, no. 2 (105), 85, 108, 1963; **10**, no. 6 (119), 1965.
21. Eckstein, M. C., Shi, Y. Y., Kevorkian, J. *Astr. J.*, **71**, 248, 1966; *Douglas Report*, no. 3078, 1964; *Methods in Astrodynamics and Celestial Mechanics*, 183, 1966.
22. Eckstein, M. C., Shi, Y. Y., Kevorkian, J. *Astr. J.*, **71**, 301, 1966; Proc. *IAU Symposium No.* 25, 291, Thessaloniki, 1964.
    Kevorkian, J. *Astr. J.*, **71**, 878, 1966.
23. Struble, R. A. *J. Math. Analysis Appl.*, **1**, 1960.
24. Szebehely, V., Pierce, D., Standish, M. *Celestial Mechanics and Astrodynamics*, 35, 1964.

25. Szebehely, V. Conference on the Exploration of Mars and Venus, NASA–VPI, VIII 1, 1965.
26. Szebehely, V. *Lectures in Applied Mathematics*, **5**, 150, 1966, Amer. Math. Soc.
27. Shibahara, R., Yoshida, J. *Publ. astr. Soc. Japan*, **15**, 87, 1963.
28. Merman, G. A. *Astr. Zu.*, **30**, no. 3, 332, 1953; *Bjull. Inst. Teor. Astr.*, **6**, no. 2 (75), 69, 73, 1955.
cf., Khilmi, G. F., *Qualitative Methods in the Many-Body Problem*, Gordon-Breach, New York, 1961.
29. Alekseev, V. M. *Astr. Zu.*, **38**, no. 2, 325; no. 4, 726; no. 6, 1099, 1961; **39**, no. 4, 724; no. 6, 1102, 1962.
30. Alekseev, V. M. *Vestnik Moskov. Univ.*, no. 1, 67, 1961 = *Amer. Rocket Soc. J.*, **32**, 1488, 1962.
31. Arenstorf, R. F. *Amer. J. Math.*, **85**, 27, 1963; *Mathematische Methoden der Himmelsmechanik und Astronautik*, 75, 1966.
32. Conley, C. C. *Comm. Pure Appl. Math.*, **16**, 449, 1963.
33. Schubart, J., Stumpff, P. *Astr. Rechen-Inst. Veröff.*, **18**, 1966.
34. Schubart, J. *Sterne und Weltraum*, **4**, 277, 1965.
35. Wielen, R. Dissertation, Heidelberg, 1965.
36. Miller, R. H. Proc. *IAU Symposium No.* 25, 137, Thessaloniki, 1964.
37. Aarseth, S. J. *ibid.*, 141; *Mon. Not. R. astr. Soc.*, **126**, 223, 1963.
38. von Hoerner, S. *ibid.*, 146; *Z. Astrophys.*, **50**, 184, 1960; **57**, 47, 1963.
39. Chandrasekhar, S. *Principles of Stellar Dynamics*, Dover Pub., New York, 1960.
40. Miller, R. H. *Astrophys. J.*, **140**, 250, 1964.
41. D'Ambly, C.-G. *Astr. Nachr.*, **288**, 81, 1965.
42. Aarseth, S. J. *Mon. Not. R. astr. Soc.*, **132**, 35, 1966.

PERIODICITY, ERGODICITY, STABILITY

Nahon (**1**) studied by referring to the virial a class of potential which is a generalization of radial potential and which admits trajectories on the equi-potential surfaces with constant velocity. In some cases (**2**) the Lagrangian function shows that all trajectories of a given energy have damping oscillations around a unique iso-kinetic trajectory of the same energy. Nahon gave a procedure of constructing all fields which admit equi-potential orbits. Further Nahon (**3**), by limiting to conservative systems of two degrees of freedom, defined a field of directions and gave the necessary and sufficient condition for an initial field to be permanent. He (**4**) then, starting from particular solutions of the Hamilton-Jacobi equation, deduced their asymptotic property and showed the invariance with respect to Birkhoff's conformal transformation. The study is intended to draw conclusions on stellar dynamics. Mrs Losco found by this method force fields in which level lines are spirals (**4a**). In such cases a quantitative study can be made as to the asymptotic behaviour of trajectories for $t \to +\infty$, independently of the knowledge of the first integral, either isolating or non-isolating. Miss Ghertzman (**5**) studied the motion of a particle in a force field with axial symmetry as a function of a parameter which represents, if it is not small, the J_2 term of the potential of an oblate planet. She followed numerically the solution for various values of the parameter and found its critical value which limits the two regions of different characteristics of motion, i.e. revolution and libration, of the mean anomaly.

Moser (**6**) discussed the combination tones for Duffing's problem with forcing terms by referring to Kolmogorov and Arnold's theory, and Struble and Yionouli (**7**) the general perturbation of the harmonically forced Duffing's equation, and Struble (**8**) the oscillations of a pendulum under parametric excitation. Hale (**9**) discussed the successive approximations of the solution of differential equations containing a small parameter. Harvey (**10**) and Villari (**11**) showed the existence of periodic solutions of the same type of differential equations with forced terms, and Ezeilo the existence of almost-periodic solutions of a certain type of differential equations with a dissipative term. Silnikov (12) studied the case of the existence of a denumerable number of periodic solutions. Knobloch (13) presented after Wazewsky a new method of approximation for periodic solutions of a non-linear differential equation of the second order.

Brumberg (14) discussed the solution in the form of series of polynomials in the three-body problem according to the idea of Mittag-Leffler (15), Painlevé (16), Goursat (17), Belorizky (18). Merman (19) studied the asymptotic solutions of a canonical system of one degree of freedom in the case in which the characteristic exponents are zero, by referring to the indices of a singular point. He (20) proved the existence of unstable regions of periodic solutions of canonical systems in the near-resonance cases.

Moser is working on the convergence proof of the expansions for quasi-periodic motions. Moser (21) showed that an expansion of the motion in trigonometric series is convergent provided that the frequencies are rationally independent, and accordingly succeeded in the rigorous existence proof of quasi-periodic solutions and justified the series expansions commonly used in celestial mechanics if properly interpreted in his own sense. Jefferys and Moser (22) applied these results to the three-body problem and established the existence of inclined quasi-periodic motions with three frequencies (two mean motions and a nodal precession) in the unstable case where the inclination I exceeds I_0 such that $\sin^2 I_0 = 3/5$. Also they found inclined quasi-periodic motions with four independent frequencies for smaller inclination. These results hold for small masses or in the lunar case.

Krasinski (23) proved the existence of quasi-periodic solutions close to circular orbits in the plane n-body problem.

The stability of a Hamiltonian system of differential equations, especially a system of linear canonical differential equations with periodic coefficients, has been studied recently by Moser (24), Gelfand and Lidskii (25), and Segal (26) with the variational equations. Glimm (27), by extending the idea of Kolmogorov, Moser and Arnold (28), discussed the formal stability of a Hamiltonian system. Salvadori (29) studied the stability of an equilibrium in the critical case according to Liapounov. Pliss (30) showed that the stability problem of the trivial solution of the system

$$\frac{dx}{dt} = Ax + p(x, y), \qquad \frac{dy}{dt} = By + q(x, y)$$

in n-dimensional vector notation is reduced to that for the existence and construction of an n-dimensional vector function $p^*(x)$ such that the above stability problem is equivalent to the stability problem of the trivial solution of a system

$$\frac{dx}{dt} = Ax + p^*(x).$$

The construction of Liapounov's functionals (31, 32, 33), by means of which the stability statement is obtained, has been described by Brayton and Moser (34) with application to electric circuit net works. By extending this idea to partial differential equations for a non-linear mixed initial boundary value problem Brayton and Miranker (35) constructed Liapounov's functionals and showed that complete stability follows from the existence of a Liapounov functional, and gave a number of examples. Yoshizawa (36) constructed Liapounov's functional, and showed the existence of a bounded solution and discussed the asymptotic behaviour of the solution of a system of differential equations. He also obtained theorems on the extreme stability and almost-periodic solutions of functional-differential equations after Hale (37) and La Salle (38). Halkin (39) extended a theorem of Liapounov on the closeness and connectivity of a set. Dearman and Le May (40) made a survey of various techniques for generating Liapounov functionals.

Fomin (41) studied the perturbation method in the theory of dynamical stability of systems with distributed parameters by reducing the stability problem to the study of linear Hamiltonian systems in separable Hilbert space. Almkvist (42) discussed the stability of linear differential equations with periodic coefficients in Hilbert space by means of the perturbation method. Ura and Kimura (43) studied the stability in the topological dynamics of Gottschalk and Hedlund (44) on the basis of Zubov's condition for the stability of a closed invariant set. Ura (45) studied the stability, relative stability and saddle points on the flow, that is, the trajectories outside a closed invariant set. Schwartz (46) considered a compact analytical mapping of Banach space and Conley (47) applied the idea to an area-preserving disk mapping into itself associated with the satellite problem in an axially symmetric field. Minty (48), Browder (49) and Shinprot (50) proved the fixed point theorem in Hilbert space.

Moser (51) and Arnold (52) studied the area-preserving mapping of a torus onto itself by developing the idea of Birkhoff. The results assert the existence of curves near the circle $r =$ constant, which are invariant under the mapping in polar coordinates

$$r' = r + \mu F(r, \theta), \qquad \theta' = \theta + \beta + \mu\,[\alpha(r) + G(r, \theta)],$$

$$a \leqslant r \leqslant b, \qquad b - a \geqslant 1, \qquad 0 < \mu \leqslant 1, \qquad \beta = \text{constant},$$

where F and G are $p \geqslant 4$ times continuously differentiable. The invariant curves exist not for every rotation number (53) ω, but only for those ω which lie in the interval $\alpha(a) + \epsilon < (\omega - \beta) / \mu < \alpha(b) - \omega$ and which satisfy the infinitely many inequalities $| n\omega - 2\pi m | \geqslant \mu \epsilon n^{-(1+\sigma)}$ for some $\sigma > 0$ and for every non-zero integers m and n. The theorem is applied by Jefferys (54) with numerical computation of periodic orbits in the restricted three-body problem. By judging from the behaviour of the trajectories in the neighbourhood of an invariant point Jefferys studied the stability and the ergodicity of the motion and the transition between stability and instability.

A motion near a periodic solution is characterized by the eigen-values of the linear terms of the differential equations in local coordinates. Diliberto (55) has shown that when these local coordinates have purely imaginary characteristic roots the possibility of stability exists and that when these roots are commensurable with the frequency of the periodic solution the system is in general unstable. It was believed that there were an infinite set of algebraic conditions necessary for formal stability. Diliberto (56) reduced these conditions to two for a Hamiltonian system with two degrees of freedom.

Kyner (70) studied the relationships among the classical Delaunay theory, the Diliberto periodic surface theory and the Krylov-Bogoliubov method of averaging, and saw that the first two methods produce the same second order approximations and that there exists a formal expansion of a family of periodic two-surfaces if a monotonicity condition is satisfied.

Diliberto's theory (57, 58) of periodic surfaces has been applied to the satellite orbit problem. It was shown by Haseltine (59) in the case of satellite motion that the process of iteration did not break down at higher approximations and the dynamical system was formally stable. In the case of incommensurable periods there always exist formal expansions and no non-trivial example of convergent expansions. For the case of commensurable periods there is an algebraic difficulty. Diliberto (55) recently solved this difficulty in two cases and he (56) proved that the doubly-periodic expansions in the case of commensurable periods are possible for satellite motion around an oblate Earth with even zonal harmonics.

The ergodic theorem (65) has been discussed recently by several authors, such as Dowker (60), Brunk (61), Blum and Hanson (62), Chacon and Ornstein (63). Furstenberg (64) studied the strict ergodicity and transformations in almost-periodic processes.

Let $\mathfrak{A}$ be a diffeomorphism (52) of a plane ring which conserves the area, and γ be a simple curve in the ring which is not homologous to zero. Then Poincaré's theorem tells that the curves γ and $\mathfrak{A}\gamma$ have at least two common points. Arnold (**66**) calls the operation $\mathfrak{A}: \Omega \to \Omega$ is globaly canonical if it is homotopic to the identity and

$$\oint_{\gamma} p\mathrm{d}q = \oint_{\mathfrak{A}\gamma} p\mathrm{d}q. \qquad (p\mathrm{d}q = p_1\mathrm{d}q_1 + \ldots + p_n\mathrm{d}q_n),$$

for each closed curve $\gamma \subset \Omega$. Let $P(x) = p(\mathfrak{A}x)$, $Q(x) = q(\mathfrak{A}x)$. The operation $\mathfrak{A}$ is globaly canonical if and only if

$$\mathfrak{A}(x) = \int_{x_0}^{x} (Q - q)\,\mathrm{d}P + (p - P)\,\mathrm{d}q$$

defines a monovalent function $\mathfrak{A}(x)$. Let T be a torus $p = 0$ in Ω and $\mathfrak{A}T$ be the image of T by a globaly canonical operation $\mathfrak{A}$. The tori T and $\mathfrak{A}T$ have at least 2^n common points provided that $\mathfrak{A}T$ satisfies $p = p(q)$, $|\mathrm{d}p / \mathrm{d}q| < \infty$. Arnold proved that, if $\mathfrak{A}$ is a globaly canonical operation sufficiently near $\mathfrak{A}_0 : p,\ q \to p,\ q + \omega(p)$, such that for $\det \left|\frac{\partial \omega}{\partial p}\right| \neq 0$ there exists a point p_0 for which $\omega_i(p_0)$ are commensurable, i.e., $\omega_i(p_0) = 2\pi m_i/N$, with integers N and m_i, $i = 1, 2, \ldots, n$, then the operation $\mathfrak{A}^N$ has at least 2^n fixed points in the neighbourhood of the torus $p = p_0$.

Arnold (**67**), by extending his former theory (**68**), considered the k-dimensional vector differential equations containing a small parameter ϵ and with an analytic $\omega(I)$

$$\dot{\varphi} = \omega(I) + \epsilon f(I, \varphi),$$
$$\dot{I} = \epsilon F(I, \varphi),$$

where the components of ω are non-commensurable and f, F are analytic and periodic with period 2π in the angular variable φ, such that $|\mathfrak{Im}\,\varphi| < \varrho$, in comparison with

$$\dot{J} = \epsilon \bar{F}(J), \qquad \bar{F}(J) = (2\pi)^{-k} \int\int F(J, \varphi)\,\mathrm{d}\varphi.$$

He proved for $k = 2$, but not yet for $k > 2$, that with positive constants C and C' we have

$$C\sqrt{\epsilon} < |I(t) - J(t)| < C'\sqrt{\epsilon}\,(\log 1/\epsilon)^2$$

for $0 \leqslant t \leqslant 1/\epsilon$.

Arnold (**69**) discussed stationary motions and their stability on the modern version of differential geometry of Lie groups of infinite dimensions, in contrast to the groups of diffeomorphism, and applied it to hydrodynamics of perfect fluid.

BIBLIOGRAPHY

1. Nahon, F. *C. R. Acad. Sci.*, Paris, **257A**, 2403, 1963; **259A**, 65, 1964.
2. Nahon, F. Proc. *IAU Symposium No.* 25, 43, Thessaloniki, 1964.
3. Nahon, F. *C. R. Acad. Sci.*, Paris, **262A**, 669, 1966.
4. Nahon, F. *ibid.*, **262A**, 1428, 1966.

4a. cf., Szebehely, V. Proc. *IAU Symposium No.* 25, 34, 1964.

5. Ghertzman, L. *Bull. astr.*, (3) **1**, fasc. 3, 1966.
6. Moser, J. *Comm. Pure Appl. Math.*, **18**, 167, 1965; Intern. Symposium on Nonlinear Diff. Eqs. and Nonlinear Mech., Acad. Press, New York, 71, 1963.
7. Struble, R. A., Yionouli, S. M. *Archive Rat. Mech. Analysis*, **9**, 422, 1962.
8. Struble, R. A. *Quart. Appl. Math.*, **21**, 121, 1963.
9. Hale, J. K. *Contr. Diff. Eqs.*, **1**, 215, 1963.
10. Harvey, C. A. *Contr. Diff. Eqs.*, **1**, 425, 1963.
11. Villari, G. *Annali Mat. Pura Appl.*, **65**, 153, 389, 1964; **69**, 171, 191, 1965.

12. Silnikov, L. P. *Doklady Mat.*, **6**, 163, 1965.
13. Knobloch, H. W. *Math. Zeitschr.*, **82**, 177, 1963; *Michigan Math. J.*, **9**, 303, 1962.
14. Brumberg, V. A. *Bjull. Inst. Teor. Astr.*, **9**, no. 4 (107), 234, 1963.
15. Mittag-Leffler, G. *Acta Math.*, **23**, 43, 1900; **24**, 183, 205, 1901; **26**, 353, 1902; **29**, 101, 1905; **42**, 285, 1920.
16. Painlevé, P. *Leçons sur la Théorie Analytique des Equations Différentielles*, Hermann, Paris, 1895; *Bull. astr.*, **15**, 81, 1898.
17. Goursat, E. *Bull. Sci. Math.*, (2) **27**, 226, 1903.
18. Belorizky, D. *J. Observateurs*, **16**, 109, 149, 189, 1933; *Bull. Sci. Math.*, (2) **62**, 271, 1938.
19. Merman, G. A. *Bjull. Inst. Teor. Astr.*, **9**, no. 6 (109), 394, 1964.
20. Merman, G. A. *ibid.*, **11**, no. 4 (127), 1967 (in press).
21. Moser, J. *Math. Annalen*, 1967 (in press); *SIAM Rev.*, **8**, 145, 1966.
22. Jefferys, W. H., Moser, J. *Astr. J.*, **71**, 568, 1966.
23. Krasinski, G. A. *Bjull. Inst. Teor. Astr.*, **11**, no. 2 (125), 1967. (in press).
24. Moser, J. *Comm. Pure Appl. Math.*, **9**, 673, 1956; **11**, 81, 257, 1958; *Nachr. Akad. Wiss. Göttingen*, Math-Phys. Kl., **1955**, 87; **1962**, 1.
25. Gelfand, I. M., Lidskii, V. B. *Uspekhi Mat. Nauk*, **10**, no. 1 (63), 3, 1955 = *Amer. Math. Soc. Transl.*, (2), **8**, 143, 1958.
26. Segal, M. *Comm. Pure Appl. Math.*, **17**, 401, 1964.
27. Glimm, J. *Comm. Pure Appl. Math.*, **17**, 509, 1964.
28. Arnold, V. I. *Uspekhi Mat. Nauk*, **18**, 13, 91, 1963; = *Soviet Math. Survey*, **18**, no. 5, 9; no. 6, 85, 1963; *Doklady Mat.*, Akad. Nauk, **156**, 9, 1964.
cf. Janet, M. Proc. *IAU Symposium No.* 25, 33, Thessaloniki, 1964.
29. Salvadori, L. *Annali Math. Pura Appl.*, **69**, 1, 1965.
30. Pliss, V. A. *Doklady Mat.*, Akad. Nauk, **154**, 1044, 1964.
31. La Salle, J., Lefschetz, S. *Stability of Liapounov's Direct Method with Applications*, Acad. Press, New York, 1961.
32. Hahn, W. *Theorie und Anwendung der direkten Methode von Liapunov*, Springer, Berlin, 1959.
33. Cesari, L. *Asymptotic Behavior and Stability Problems in Ordinary Differential Equations*, Springer, Berlin, 1959.
34. Brayton, R. K., Moser, J. *Quart. J. Appl. Math.*, **22**, 9, 81, 1964.
35. Brayton, R. K., Miranker, W. L. *Archive Rat. Mech. Analysis*, **17**, 358, 1964.
36. Yoshizawa, T. *Mem. Coll. Sci.*, Univ. Kyoto, **33**, 301, 1960; *Contr. Diff. Eqs.*, **1**, 371, 1963; *Archive Rat. Mech. Analysis*, **6**, 409, 1960; **17**, 148, 1964.
37. Hale, J. K. *Contr. Diff. Eqs.*, **1**, 401, 1963; *Archive Rat. Mech. Analysis*, **15**, 289, 1964.
38. La Salle, J. P. *Annals of Math.*, **65**, 571, 1957.
39. Halkin, H. *Archive Rat. Mech. Analysis*, **17**, 272, 1964; *J. Math. Analysis Appl.*, **7**, 272, 1964; **10**, 325, 1965; *J. Analyse Math.*, **12**, 1, 1964.
40. Dearman, C. C., Le May, A. R. *Mathematische Methoden der Himmelsmechanik und Astronautik*, 213, 1966.
41. Fomin, V. M. *Doklady Mat.*, Akad. Nauk, **163**, 830, 1965.
42. Almkvist, G. *Pacific J. Math.*, **16**, 383, 1966.
43. Ura, T., Kimura, I. *Proc. Japan Acad.*, **40**, 703, 1964.
44. Gottschalk, W., Hedlund, G. Topological Dynamics, Amer. Math. Soc., 1955.
45. Ura, T. *Contr. Diff. Eqs.*, **3**, 249, 1965.
46. Schwartz, J. *Comm. Pure Appl. Math.*, **16**, 25, 1963.
47. Conley, C. C. *Comm. Pure Appl. Math.*, **17**, 237, 1964.
48. Minty, G. J. *Duke Math.* J., **29**, 341, 1962.
49. Browder, F. E. *Duke Math. J.*, **30**, 557, 1963.
50. Shinprot, M. *Archive Rat. Mech. Analysis*, **17**, 225, 1964.
51. Moser, J. *Nachr. Akad. Wiss. Göttingen*, Math. Phys. Kl., **1962**, no. 1. Cf., (**24**).
52. Arnold, V. I. *Uspekhi Mat.*, Akad. Nauk, **18**, no. 6, 91, 1963.
53. Birkhoff, G. D. *Acta Math.*, **43**, 1, 1920; *Annali Scuola Norm. Sup. Pisa*, (2) **4**, 267, 1935; **5**, 1, 1936 = *Collec. Math. Pap.*, II, 111, 466 and 668, Amer. Math. Soc., 1950.

54. Jefferys, W. H. *Astr. J.*, **71**, 306, 1966.
55. Diliberto, S. P. *Techn. Rep.*, Contract Nour 222 (88) and NR 041–255, 1966.
56. Diliberto, S. P. *Astr. J.*, **71**, 246, 1966.
57. Diliberto, S. P., Kyner, W. T., Freund, R. B. *Astr. J.*, **66**, 118, 1961.
58. Diliberto, S. P. *Naval Ordnance Lab. Rep.*, 6445, 1959.
59. Haseltine, W. R. *Astr. J.*, **68**, 522, 1963.
60. Dowker, Y. N. *Duke Math. J.*, **14**, 105, 1947.
61. Brunk, H. D. *Trans. Amer. Math. Soc.*, **78**, 482, 1955.
62. Blum, J. R., Hanson, D. L. *Proc. Amer. Math. Soc.*, **16**, 413, 1965.
63. Chacon, R. V., Ornstein, D. S. *Illinois J. Math.*, **4**, 153, 1960.
64. Furstenberg, H. *Amer. J. Math.*, **83**, 573, 1961.
65. Yosida, K. *Functional Analysis*, Springer, Berlin, 1966.
66. Arnold, V. I. *C. R. Acad. Sci.*, Paris, **261**, 3719, 1965.
67. Arnold, V. I. *Doklady Akad. Nauk*, **161**, no. 1, 9, 1965.
68. Arnold, V. I. *Uspekhi Mat. Nauk*, **18**, no. 5 (113), 13; no. 6 (114), 91, 1963. Cf., (28).
69. Arnold, V. I. *C. R. Acad. Sci.*, Paris, **260**, 5668, 1965; **261**, 17, 1965; *Ann. Inst. Fourier*, Univ. Grenoble, **16**, 319, 1966.
70. Kyner, W. T. Proc. *IAU Symposium No.* 25, 26, Thessaloniki, 1964.

Y. HAGIHARA
President of the Commission

8. COMMISSION DE L'ASTRONOMIE DE POSITION

PRÉSIDENT: Dr R. H. Stoy, H. M. Astronomer, Royal Observatory, Observatory, Cape Province, South Africa.

VICE-PRÉSIDENT: Dr A. A. Nemiro, Pulkovo Observatory, Leningrad, U.S.S.R.

COMITÉ D'ORGANISATION: W. Fricke, J. Lévy, F. P. Scott, S. Slaucitajs, E. G. Woolsey, M. S. Zverev.

MEMBRES: Adams, Anguita, Atkinson, Barros, Bohrmann, Brouwer†, Danjon†, Débarbat, Demetrescu, de Moraes, Dick, Dubois-Chevallier, Duncombe, Fayet, Gliese, Gordon, Guinot, Haas, Lacroute, Larink, Laustsen, Maître, Marcus, Melchior, Murray, Nefedieva, Novopashenny, Okuda, Pilowski, Planelles, Podobed, Polozhentsev, Postoiev, Reiz, Rybka (P.), Sandig, Scheepmaker, Schmeidler, Sémirot, Šervarlic, Symms, Tavastjerna, Teleki, Thomas (D. V.), Tsubokawa, Tucker, Tuzi, Van Herk, Verbaandert, von der Heide, Wood (H. W.), Wünschmann†, Yasuda, Zimmerman.

Groupes de Travail:

1. Photographic Catalogues of Stars to the 9th Magnitude.
   Chairman: D. Brouwer† (W. Dieckvoss)
   Members: Barney, Deutsch (A. N.), Dieckvoss, Eckert, Fatchikhin, Kox, Lourens, Scott, Vasilevskis, von der Heide, Wood (H. W.)
2. AGK3R Committee.
   Chairman: F. P. Scott
   Members: Brouwer†, Fricke, Heckmann, Nemiro, Symms, Zverev
3. SRS Committee.
   Chairman: F. P. Scott
   Members: Brouwer†, Fricke, Sémirot, Stoy, von der Heide, Zverev
4. Double Star Committee.
   Chairman: F. P. Scott
   Members: Dieckvoss, Lourens, Stoy, Zverev

It is with deep regret that we record the deaths of Professor Dr F. Wünschmann on 17 January 1966 and of Professor D. Brouwer on 31 January 1966. Dr W. Dieckvoss has kindly consented to take Professor Brouwer's place as Chairman of the Working Group on Photographic Catalogues of Stars to the Ninth Magnitude until a new Chairman can be appointed at Prague.

This report is based mainly on replies to a circular letter sent to all members of the Commission in September 1966 and should refer to the position about that time. I would like to record my thanks to those many members who responded so promptly and so fully and especially to Professor Nemiro for his comprehensive report on Positional Astronomy in the U.S.S.R., and to Dr Dieckvoss and Mr Scott for the reports of the Working Groups over which they preside. The whole of Commission 8 should be grateful to Mr Scott and his colleagues not only for the magnificent work they are doing at the U.S. Naval Observatory but also for the unselfish and unstinted help they have freely given to many of their fellow workers in less endowed observatories and for the way in which they are co-ordinating the work on the AGK3R and SRS projects. These projects still form the principal activity for many of the observatories engaged on astrometric work.

With the AGK3R nearing completion, attention in the northern hemisphere seems to be turning to the improvement of instrumental equipment and to a more critical examination of the fundamental system. The publication of the FK4 just over three years ago has provided the starting point and stimulus for much of this latter work. A comprehensive and independent observational check of the FK4 system is being provided by the gradual realization of the programme envisaged by Professor Danjon soon after he introduced the prismatic astrolabe to astronomy.

CATALOGUES

The past three years have seen the completion by the Babelsberg Observatory of both the Geschichte des Fixsternhimmels, Abteilung II (Southern Sky) (**1**) and the Index der Sternörter 1925–60, Index II (**2**). They have also seen the first publication of an extremely useful comprehensive star catalogue by the Astrophysical Observatory of the Smithsonian Institution (**3**). This catalogue gives the positions and proper motions on the FK4 system of 258 997 stars between declinations +80° and −80° for the epoch and equinox of 1950·0. It is based on a combination of the FK4, FK3, GC, AGK2, AGK1, Greenwich AC, Yale, CPC50, CAZ, Me3 and Me4 catalogues. Unfortunately it did not also incorporate the section of the CPC50 between − 64° and − 80° (**69**) which would have greatly strengthened what is, at present, its weakest part. Much of the work of preparation of the SAOSC was done on an electronic computer as it was intended to provide the fullest possible list of reliable star places in computer-accessible form to speed up the derivation of accurate positions of artificial satellites from the thousands of photographs being taken by the Smithsonian Baker-Nunn cameras all over the world. The catalogue is available on magnetic tape compatible with IBM 729 tape units and also in book form.

At the Astronomisches Rechen-Institut, Heidelberg, work on problems relating to the preparation of fundamental catalogues has gone on steadily. A report on the formation of the declination system of FK4 and on the derivation of individual corrections to positions and proper motions from differential observations was given by Kopff, Nowacki and Strobel (**42**). The same authors, together with Mürle (**41**), published formulae and tables for the reduction of small proper motions and co-ordinate differences which had been used in the construction of FK4. While the systematic differences FK4−FK3 were given in FK4, tabulated differences FK4−GC and FK4−N30 were published by Brosche, Nowacki and Strobel (**12**). In continuation of previous lists of star catalogues Heinemann (**32**) compiled a list of catalogues from 1900 to 1962.

An investigation of the stability of systems of positions was carried out by Gliese (**30**) in making use of the α_8-systems of fundamental observations with transit instruments from 1840 to 1958.

The apparent places of the stars included in FK4 Supp are not being given in the annual international volumes 'Apparent Places of Fundamental Stars'. They have, however, been computed for all 1987 FK4 Supp stars at Heidelberg from 1965 onwards. They are available to observers on request.

A method for deriving systematic differences in star positions between different catalogues was developed by Brosche (**13**). The differences are expanded in spherical harmonics, and the highest order of the expansion is determined by applying the χ^2−or F-test. The method, slightly modified, will be used in the machine computation of the systematic differences between FK4 and important observational catalogues. The work is part of a current project to revise the Albany General Catalogue and to prepare future improvements of FK4. For the same purpose the Astronomisches Rechen-Institut, in co-operation with the U.S. Naval, the Yale and Potsdam-Babelsberg Observatories, has continued to place a large number of star catalogues on punch cards. Lists of other star catalogues and astronomical data available in machine readable form are given in U.S. Naval Observatory Circulars Nos. 99 and 111.

Fricke (**27, 29**) has carried out comparisons of the distance scales which can be established from secular parallaxes on the basis of the proper motion systems of GC, N30, FK3 and FK4.

Work on the determination of the correction to lunisolar precession and of the galactic constants A and B on the basis of the various proper motion systems is in progress. For further details see report of Commission 33.

At Uccle, Professor Melchior has prepared a catalogue of all the meridian observations of latitude stars and has deduced the best positions and proper motions for 1950·0 on the FK4 system. The calculations were carried out on an IBM 1620 and can easily be revised when new observations of the stars become available. A catalogue of the positions of 939 PZT stars as observed with the Washington seven-inch transit circle has been completed by J. A. Hughes. It appears in University Microfilms, Chicago, Illinois, 1966 and a discussion of it will be given in the Publications of the U.S. Naval Observatory.

A catalogue of the positions of 240 stars in the Mizusawa, Richmond, Tokyo and Washington PZT lists has been published in Tokyo by H. Yasuda and H. Hara (**88**). The Mizusawa PZT star positions have been revised by utilizing the results of observations made during the period from 1958 to 1961. The revised declination system seems to be closer to the GC than to FK3, while the revised right ascension system is consistent with that of Washington.

By analysing the meridian observations of the four brightest asteroids made at the Cape, Tokyo and Washington during the period 1950–62, H. Yasuda has shown that it is possible to derive not only improved equinox and equator points but also improved elements of the Earth's orbit (**87**). At Washington the results of a discussion of the observations of these four brightest asteroids made between 1920 and 1960 are being prepared for publication in the Astronomical Journal.

Professor Melchior has made a theoretical investigation of the close relationship between the phenomenon of precession and nutation and the terrestrial tides (**51**). He found that corrections to the amplitudes of the short period nutations are necessary and that these can be obtained from observations made with horizontal pendulums.

OBSERVATORIES — PROGRAMMES AND INSTRUMENTS

The observations for the AGK3R with the Babelsberg meridian circle came to an end with the death of Professor Wünschmann. It has been decided not to complete the reductions of the observations already made partly because they can not be ready in time for inclusion in the final catalogue and partly because they seem to be affected by an instrumental fault that was not detected until after many of the observations had been made. The meridian circle has now been thoroughly examined and after overhauling is to be transferred to the Potsdam Geodetic Institute where its first work will be the observation of reference stars for the PZT. This latter instrument (250/3780 mm) was built in the workshops of the Babelsberg Observatory but after testing is being transferred with the transit circle. Professor Dick has considered some of the problems of a horizontal transit circle (**18**).

At Belgrade the Askania meridian circle (190/2578 mm) is being used for differential observations to obtain the declinations of 3245 stars which form the latitude programmes of 19 observatories. The Askania transit circle (190/2578 mm) is being investigated in preparation for the determination of absolute right ascensions while absolute declinations will be found with the Askania vertical circle (190/2578 mm) as soon as some technical difficulties with the instrument have been sorted out. The systematic determination of the declinations of 36 zenith stars according to the programme of Ševarlic and Teleki (**66**) is being continued with the Askania zenith telescope (110/1285 mm). This programme was begun in 1960 and is expected to run to 1980.

Observations for the third Besançon meridian catalogue were completed in January 1964 and the catalogue itself published in 1965 (**50**). It contains 1090 stars of which 326 are included in the FK3, 429 of the others in the GC and 806 in Blaauw's list. A second publication, now in preparation, will give the proper motions of 764 of the stars.

At Bordeaux, M. Requième has built a micrometer incorporating a Lallemand photomultiplier cell that automatically registers the times of star transits across the meridian. With it one can observe stars down to magnitude 8·9 with equal accuracy for all declinations between $-20°$ and $+75°$. M. Requième's work is described in a thesis which is to be submitted in October 1966.

The Brorfelde transit circle has been equipped with a new photographic micrometer, by means of which about 10 000 observations have been made since 1964. High accuracy in the tracking of stars during transit has been obtained through an electronic velocity-generating device which makes the plateholder-carriage to move uniformly and with a rate matched to that of any star within 0·1%. A new and accurate method of photoelectric recording of the circle in machine-readable form has been in use since December 1965. Recently all 2160 diameter corrections were determined by this method with an accuracy of $0\overset{''}{.}03$ (m.e.); the diameter correction measurements were carried out in about 15 hours and the complete reduction in 20 hours, and it is therefore intended to repeat the determination with short intervals in order to investigate the stability of the glass circle. Night-to-night comparisons of differential observations give the internal mean errors $\epsilon_\alpha \times \cos\delta = 0\overset{s}{.}013$ and $\epsilon_\delta = 0\overset{''}{.}20$ for a single observation of a star brighter than the visual magnitude 8·5 and somewhat larger errors for fainter stars.

The Cooke transit circle at Herstmonceux is engaged on a programme of stars selected from the General Catalogue of Radial Velocities as having well determined radial velocities but poor proper motions. The graduation errors of the 'primary divisions' (0°, 20°, 40° etc.) of the circle have been determined photographically and the results found to be in good agreement with the values adopted from the visual measurements made in 1936–40. A new determination of the pivot errors made with an auto-collimator agrees very well with that made in 1948. A definitive investigation of the instrumental system during the years 1957–61 based on an analysis of over 21 700 observations of FK4 stars made during the AGK3R programme is in progress.

Revised proper motions on the FK4 system have been determined for 182 semi-regular and RV Tauri-type variable stars. Of these, 44 were selected as being in particular need of modern positions. These were obtained differentially from observations made with the Cooke transit circle of each variable and six adjacent FK4 stars. The results will be published in a forthcoming Royal Observatory Bulletin.

Meridian observations at La Plata have been temporarily interrupted because the transit circle has been transferred to the Southern Astrometric Station at La Leona where it has been used since 1965 for the observations of the absolute declinations of southern FK4 stars and also of 'Kustner Series'. The transit circle lent by the Lick Observatory is being reconditioned prior to installation at La Plata. The reduction of observations made in former years has continued and the sytematic differences between the 'instrumental' and 'fundamental' declination systems have been evaluated for the FKSZ stars.

The Munich vertical circle continues to be used for the observations of the absolute declinations of FK4 stars, the major planets and the four brightest asteroids.

The Ottawa mirror transit was brought into operation in February 1965. Results indicate a probable error in right ascension of slightly less than $0\overset{''}{.}25$; the probable error in declination was not known at the time of reporting as the automatic engine for measuring the circle films was being modified. A new PZT is to be established a few miles due west of the present instrument. When the new PZT is in operation, the old one will be moved to a site near Calgary, Alberta where it will be on the same latitude as Herstmonceux.

Trials on the mirror transit circle of the University of Porto Observatory continue. A study of the flexure of the original axis indicated that it was dependent on the reaction of the pivots.

A new axis is therefore being built in the hope of eliminating this flexure. A grant from NATO has provided apparatus for the digital read out of the micrometers.

At Strasbourg, the observations of the double stars on Zverev's list and of the bright stars between + 30° and + 72° declination are nearly finished and it is hoped that the reductions will be completed within a year. P. Lacroute and P. Bacchus (**44, 45, 8, 46**) have shown that it is possible to improve the results of differential meridian observations significantly by reducing all the series of observations together and using a method they call 'synthèse'. This method has been applied to the Strasbourg AGK3R observations and the resulting gain in the weight is being investigated. P. Bacchus (**7**) has considered the problem of determining the systematic differences between catalogues by a statistical method. P. Lacroute (**47, 48**) has continued his studies of the use of overlapping plates in the formation of a photographic catalogue and has applied his method to plates of the AGK2 and AGK3.

Mr Harley Wood reports from Sydney that the Grubb-Parsons two co-ordinate measuring machine has now been received and will be used in the production of catalogues especially in the Sydney astrographic zone (− 51° to − 65°). The measuring agent in each co-ordinate is a moire scale with digital read out so that all measures can be recorded on punched cards.

The positions of the Moon, exterior planets and the four brightest asteroids as observed at Tokyo during 1960 and 1962 have been published (**85, 86**). Observations of the Moon, inner planets, outer planets and the four brightest minor planets referred to the FK4 system are being continued with the Gautier meridian circle, as also are the observations for the SRS project begun in March 1963. This programme includes not only the SRS and 'Bright' stars between − 10° and − 30° but also fundamental stars between 0° and − 35°, near the zenith and in the northern sky. The readings of the circle films with the automatic photoelectric measuring machine are now registered in computer-readable form.

The Askania meridian circle at Uccle has been returned to the makers for various modifications including digitization. These modifications are now complete and the instrument is being remounted.

At Washington the programme of fundamental observations of the Sun, Moon, planets, four brightest asteroids, FK4 and FK4 Supp stars and 'Bright' stars north of − 30° declination, started with the six-inch transit circle in 1963, has continued. In conjunction with this programme, two observations are planned for the SRS stars north of − 20°. Provisional positions of the Sun, Moon, planets and asteroids are being published in the U.S. Naval Observatory Circulars a few months after the completion of the observations. The results of observations made with the six-inch transit circle between 1949 and 1956 have been published (**5**). A catalogue of the positions of the Sun, Moon, planets and the four brightest asteroids, FK4 and FK4 Supp stars as observed with the six-inch transit circle between 1956 and 1962 is nearly completed. The six-inch transit circle observations of the Moon made between 1925 and 1966 have been corrected for limb irregularities using Watt's charts and reduced to the FK4 system. B. L. Klock and D. K. Scott (**39**) have analysed the six-inch transit circle observations of the Moon made between 1952 and 1964 for a correction to the coefficient of periodic term 182 of Brown's Tables of the Moon. The correction found was in good agreement with the discordance found by Eckert and Smith (**19**) in their solution of the main problem of lunar theory.

The Washington seven-inch transit circle was taken out of service on 29 June 1965 to equip it for a programme of observations of fundamental and SRS stars at El Leoncito, Argentina, the site of the Yale-Columbia Southern Observatory. An excellent clock system and an improved photographic device for recording the readings of the micrometers have been acquired and the electrical connections pre-wired for rapid installation. It is expected that the instrument and associated equipment will be shipped from Washington during November 1966. A catalogue of the Parenago stars and close double stars as observed with the seven-inch transit circle is being compiled by E. S. Jackson.

The pivot errors of the six-inch transit circle have been measured with a digitized auto-collimator designed automatically to monitor reflections from a mirror attached to one end of the axis. A full set of measures in one co-ordinate taken at 10° intervals can be obtained in less than ten minutes. A sidereal clock system planned by A. N. Adams and B. L. Klock has been installed for timing star transits. It has been possible to hold the epoch of the clock to within ± 0·5 millisecond for periods of several months.

A programme for the development of a new transit circle is in progress under the supervision of Adams and Klock. An electronic circle and automatic star and planet tracker are now being constructed at the Farrand Optical Company, Inc. A modified Cassegrain configuration is being considered for the instrument itself and special attention is being given to reducing the flexural and thermal errors known to exist in present transit circles.

ASTROLABES

Several Danjon Prismatic Astrolabes are now in use throughout the world. Since their principal task is the precise determination of latitude and time, their work is normally reported to Commissions 19 and 31. A by-product of these routine observations is, however, of great interest to Commission 8 since an analysis of them over an extended period can yield relative star positions within a band of declination 60° wide. These positions have high individual accuracy and their zonal errors appear to be small. They are, therefore, eminently suitable for checking the systematic accuracy of star positions obtained from transit circle observations.

The first catalogue of astrolabe star positions was compiled at the Paris Observatory in 1961 and clearly demonstrated how valuable as a check such catalogues would be, especially if they could be compiled from a series of astrolabes well spread in latitude. Several observatories are now actively engaged on this work and have greatly extended their star lists so as to include as many stars suitable for making comparisons with the standard catalogues as possible.

From Paris, B. Guinot reports that independent catalogues are being compiled from observations made with the astrolabes at Paris, Haute-Provence and Besançon. Catalogues are also being compiled from the astrolabe observations made during the IGY at Tananarive and at Alger. They will also be compiled from astrolabe observations still being made (in co-operation with the Paris Observatory) at Alger, Quito and Santiago (Chile).

A collection of all such material is being made at Paris. At present it includes results from Paris, Neuchâtel, Curaçao, Tananarive and Potsdam. It is still too early to try to combine these data but a preliminary study of them has shown good agreement between them both as regards individual and systematic errors. It appears that the standard deviation of the angular errors (combination of errors in both right ascension and declination) in FK4 for 1962 was fairly homogeneous and about 0ʺ10 between declinations + 80° and − 25°. South of − 25° it increases rapidly and becomes about 0ʺ25 at declination − 60°.

An analysis of the results of the astrolabe observations made at Herstmonceux during 1959–63 is being prepared for publication. The catalogue contains 230 right ascensions and 141 declinations for a mean epoch of approximately 1962·5. Comparisons with other catalogues show that the systematic accuracy of the Herstmonceux results is comparable with those obtained from the Paris astrolabe and the Pulkova photoelectric transit instrument. The probable errors of the individual stars, including an allowance for systematic errors, compare favourably with the probable errors of FK4 for the epoch of observation.

Individual and systematic corrections to FK4 star places derived from observations made with the astrolabe at Potsdam have been published (**34**, **38**). The removal of this instrument from Potsdam, where it has been working in the time and latitude service since 1957, to an equatorial site is under consideration.

Also under consideration is the possibility of moving the Besançon astrolabe to a station at Rio Grande in the Argentine in latitude 53° 48′ south and longitude 67° 47′ west. This astrolabe has been used intensively for the past three years at Besançon.

At Quito, Dr Scheepmaker has continued without interruption the series of astrolabe observations he began in October 1963. His apparatus has been improved however by the installation on 1 May 1964 of new time equipment provided by the American Science Foundation of the United States and on 1 September 1964 a new Belin chronograph registering to 1/1000 second. Three of the 11 fundamental groups, each consisting of 28 FK4 stars, are observed per night as well as special 'catalogue groups' of FK4 and FK4 Supp stars. The observations are reduced in Paris on an IBM 7040 but the interpretation is done in Quito. The results of an analysis of the observations made between October 1963 and April 1967 will be published in Boletin Astronomico del Observatorio Astronomico de Quito, No. 1. This publication will give the individual corrections $\Delta\alpha$ of 187 FK4 and 40 FK4 Supp stars, $\Delta\delta$ of 113 FK4 and 6 FK4 Supp stars and systematic corrections to FK4, all between declinations $-30°$ and $+30°$. It is hoped that the number of stars for which individual corrections can be found will rise to 600 within the next four years.

A Danjon astrolable was installed at the International Latitude Observatory of Mizusawa in 1964 and the observing programme, begun early in 1966, is made up of FK4 and FK4 Supp stars, 131 of which are in common with Alger.

Mr R. D. Adams reports from New Zealand that observations with the Danjon astrolabe have been made regularly since 1964. Preliminary reductions for the 1964 observations have been carried out on an IBM 650 computer using a programme devised by B. Guinot. Reductions of the 1965 and 1966 observations will be made when the appropriate programme for the new Elliott 503 computer has been written. Mr Adams remarks:

'Our original plan of work with the astrolabe was to complete observations of basic star groups, then to assist in the compilation of a catalogue of southern stars, and eventually to pursue a study of stellar proper motion, but unless we can obtain the services of a qualified astronomer, it may be necessary to curtail this programme and even stop observations. In view of the contribution we can make to positional astronomy, the equipment we have available, and the large amount of work already done, I would be very sorry to see this done. If you know of any suitable astronomer who might be interested in undertaking this work for us, either on a permanent basis or for a period of a year or two, I would be very grateful if you could put us in touch with him.'

The astrolable from Herstmonceux was transferred to the Cape in 1965 (**71, 72, 73**). The observing programme contains eight standard groups of 28 stars each and more than 1000 'catalogue' stars representing all suitable stars brighter than $5^{m}.6$ in the FK4 and the 'Bright Star' list associated with the SRS project. The aim is for each star to be observed at least twice by each of five observers. Present indications are that the programme will be substantially completed by the end of 1968. The relatively small number of standard groups is to ensure the determination of reliable group corrections (**74**). Attempts to eliminate variation of prism angle by thermostatic control of the astrolabe building during the day have proved partially successful. The closing error in prism angle has been virtually eliminated, but individual nights still show significant prism angle variations. This is due to the difficulty of predicting night-time temperatures, and to variations of ambient temperature during the night.

The quality of the results obtained so far has been rather disappointing, due to the generally poor seeing. Allowing for the errors of the adopted star places, the average probable error of the observation of one star is about 20% larger than it was at Herstmonceux for the same observers. A large seeing disk presents the astrolabe with what is essentially an extended object. The large effective f-ratio ($f/90$) means that there is frequently insufficient contrast to enable faint stars to be observed. Observations are sometimes impossible even in clear conditions, either because of oscillation or blurring-out of the images.

This dependence of the quality of astrolabe results on the seeing has been utilized by French astronomers for testing possible observatory sites since the installation of an astrolabe on an expeditionary basis is a relatively cheap and simple operation.

Partly because of the scarcity of trained observers and partly to ensure the best possible conditions, several observatories have decided to co-operate in the establishment of a French National Station for Fundamental Astrometry. The quality of the seeing at three sites is being examined by a study of the internal agreement of observations made with astrolabes that have been temporarily placed there. These sites are at Briançon (Alpes), Font-Romeu (Pyrénées-Orientales) and Nice. The quality at all these sites is very variable with time so that it is difficult to arrive at a definite conclusion. Nevertheless, the results from these expeditions added to those from astrolabes in service elsewhere indicate that uneven country is an unfavourable factor, whilst nearness to the sea is favourable. These experimental observations are being continued and, in addition, the Gautier transit circle borrowed from Liège is being installed at Font-Romeu alongside the astrolabe so that the results from the two instruments can be critically compared. The transit circle will be used by observers from the Bordeaux Observatory for making observations in connection with the SRS project.

At Paris, Mlle Débarbat has investigated the possibilities of observing the planets with a Danjon astrolabe (**16, 17**). The results are very encouraging and seem to be of the same precision as for stellar observations. The positions found for Mars in 1963 are in good agreement with the Washington meridian observations. Several observatories have agreed to take part in a special series of planetary observations, particularly of Jupiter.

At the Cape, Dr Thomas is making a design study of an astrolabe in which the prism is replaced by a pair of plane mirrors in the hopes that

1. Such an instrument would give greater stability of the effective altitude of observation when the ambient temperature is changing;
2. The mirror angle can be chosen to make the instrumental almucantar any convenient altitude, e.g. 50°;
3. With the limitation on the size of the prism removed, there is the possibility of constructing a larger instrument capable of reaching fainter magnitudes.

The aperture of the instrument for which the design study is being made is 25 cm which should enable it to reach two magnitudes fainter than the standard Danjon astrolabe.

SUGGESTED TOPICS FOR DISCUSSION

1. The results of the AGK3 and AGK3R projects.
2. Progress with the SRS project.
3. The desirability of obtaining a homogeneous photographic cover of the southern hemisphere to supplement the SRS programme. (W. Dieckvoss).
4. The desirability of moving further transit circles to the southern hemisphere. (J. Dick).
5. Methods of using astrolabe results for the correction of fundamental catalogues and, in particular, the systematic correction to the FK4 system implied by the preliminary discussions of astrolabe results. (D. V. Thomas).
6. Other evidence on the accuracy of the FK4 system.
7. The desirability of some sort of uniformity about the presentation of the discussion given

in the introduction of meridian catalogues. (V. Maître as a result of a conversation with W. Fricke).

8. The desirability of meridian circle observers publishing the average apparent magnitude of stars as they appear during observation in order that magnitude effects can be properly evaluated. (G. van Herk).

9. Desiderata for future meridian circle observations.

10. New instrumental methods and techniques. (P. Sémirot).

11. The possibilities of making accurate astrometric measures from artificial satellites. (P. Lacroute who points out that the angle between two stars can be measured with an accuracy of $0\overset{''}{.}01$, or less, provided that the angle in question is within 15′ of a given reference angle and that the satellite is not oscillating more quickly than once in ten seconds.)

12. The design of new wide angle objectives for astrometric purposes and their performance in use. (W. Dieckvoss).

13. A critical survey of the methods of reduction of co-ordinates measured on plates, with particular attention to those methods using field overlaps to reduce large regions simultaneously. (W. Dieckvoss).

14. The desirability of setting up one or more centres for the automatic measurement of large numbers of plates. (W. Dieckvoss).

15. The desirability of the compilation of a 'General Catalogue of Stars to the Ninth Magnitude' as suggested by Brouwer and as partially accomplished by the Smithsonian Astrophysical Observatory (**3**). (W. Dieckvoss).

R. H. STOY

President of the Commission

REPORTS OF THE WORKING GROUPS

Photograph Catalogues of Stars to the Ninth Magnitude

At Bergedorf the AGK2 has been reduced afresh in the FK4 system after the application of appropriate systematic corrections to the original rectangular co-ordinates measured at Bergedorf and Bonn. Also completed there is a preliminary version of the AGK3 based on a preliminary AGK3R provided by the Washington U.S. Naval Observatory. The Bergedorf Observatory is willing on request to provide transcripts from the magnetic tapes for the positions and proper motions of selected group of stars.

Part of the AGK2 and AGK3 data was sent to P. Lacroute who applied his method of reducing overlapping fields to the polar cap north of + 40° declination and derived plate constants for both the 1930 and the 1958 epochs.

The fifth volume of the Cape Photographic Catalogue for 1950·0, which covers the zone between − 64° and − 80°, has been published (**69**) while the sixth, and final, volume, which covers the sky south of − 80°, is ready for publication.

The Yale Observatory work on the zones between − 30° and − 90° continues. H. K. Eichhorn has concentrated on the zone − 70° to − 90° using measures of plates taken by H. Wood in Sydney. Together with C. F. Lukac of the U.S. Army Map Service, he has worked out procedures for the simultaneous reduction of large areas using overlaps of neighbouring plates.

W. DIECKVOSS

International Reference Star Programs

AGK3R Program. By late 1964 all observatories except Babelsberg, which did not complete its commitment, had sent the results of their observations to the U.S. Naval Observatory for the compilation of a catalog of final positions of the reference stars. With the exceptions of Heidelberg and Strasbourg, which sent mean results for each star, all other observatories sent individual results as recommended by Resolution No. 5, adopted by Commission 8 during the Berkeley Meeting of the IAU, 1961.

The intent of Resolution No. 5 was to permit a night-by-night reduction of the observations to the FK4 and to enable other analyses of the observations to be carried out in detail. The wisdom of this Resolution, as it worked out in practice, is now questioned. The writer is firmly convinced that the final catalog would have been essentially as accurate and much simpler to compile if each observer had completely reduced his own observations before forwarding them to Washington. Too much time had to be spent in transferring the voluminous data to punch cards and in correcting thousands of errors by correspondence.

All observations have been reduced to the FK4 on a night-by-night basis whenever possible. Clamp differences have been derived and applied where they seemed to be necessary.

In the course of deriving night corrections and in studying their effect on the AGK3R stars, the mean error of a single observation was estimated in one or more ways for each observatory that had sent individual results. Corresponding mean errors for Heidelberg and Strasbourg were estimated by other methods. The resulting mean errors are shown in Table 1. The mean errors from the FK4 stars reflect the fact that the parameters for the nightly reductions were based on them. With the exception of the Pulkovo declinations, the mean errors from different data for an observatory are in fairly satisfactory agreement.

Systematic differences and the relative weights of results from different observatories are now being derived.

Table 1. AGK3R Program: Internal Mean Errors of a Single Observation

	AGK3R STARS				FK4 STARS	
Observatory	with 2 obs'ns		with 4 obs'ns		many obs'ns	
	$\xi_\alpha \cos\delta$	ξ_δ	$\xi_\alpha \cos\delta$	ξ_δ	$\xi_\alpha \cos\delta$	ξ_δ
Bergedorf	$\pm0^{s}.016$	$\pm0''.42$	$\pm0^{s}.$—	$\pm$ $''$—	$\pm0^{s}.017$	$\pm0''.38$
Bordeaux	.018	.34	.018	.33	.016	.29
Greenwich	.019	.45	.—	.—	.019	.43
Heidelberg	.017	.40	.—	.—	.—	.—
Nicolaiev	.020	.31	.—	.—	.016	.30
Ottawa	.025	.44	.—	.—	.023	.43
Paris	.016	.37	.016	.34	.014	.31
Pulkovo	.016	.61	.—	.—	.012	.44
Strasbourg	.014	.33	.014	.33	.—	.—
USNO 6-inch	.017	.32	.—	.—	.014	.27
USNO 7-inch	.013	.24	.—	.—	.012	.21

SRS Program. Table 2 shows the status of the SRS program as of 1 July 1966.

Table 2. Status of the SRS Program—1 July 1966

Observatory	Zone	Commitment		Date Started		Completed	
		stars	*obs'ns*	*mo.*	*yr.*	*obs'ns*	*red'ns*
Abbadia	$+5°$ to $-15°$	1 560	4	4	62	84%	90%
Bordeaux	$+5$ to -15	1 560	4	6	62	72	100
Bucharest	$+5$ to -10	1 176	4	7	62	96	29
Nicolaiev	0 to -20	5 984	2	3	64	100	25
San F'ndo	-10 to -30	3 709	4	3	63	33	53
Tokyo	-10 to -30	3 560	4	3	63	73	70
Cape	-30 to -90	10 082	4	4	61	35(1)	90
Santiago-Pulkovo	-25 to -90	11 496	4	1	63	81(2)	20

(1) All observations of the reference stars in the zone $-40°$ to $-52°$ have been completed and a preliminary catalog of the results has been prepared. Observations in the zone $-30°$ to $-40°$ commenced during March 1966.

(2) The zone $-47°$ to $-90°$ will be completed by the end of the year.

It is expected that the Hamburg-Bergedorf meridian circle will be shipped to Perth, Australia during March or April 1967. The U.S. Naval Observatory's seven-inch transit circle is expected to be installed at El Leoncito, Argentina, the site of the Yale-Columbia Southern Observatory, during the latter part of 1966 or early 1967. Both of these instruments will be engaged in programs of fundamental work and in the observation of the southern reference stars.

The U.S. Naval Observatory has computed over 134 000 apparent places, 60 000 refractions and 57 000 m, n and c reductions for observations made in connection with the SRS program.

Double Star Program. A revised list of 2292 double stars that might be troublesome to measure on photographic plates was compiled at the U.S. Naval Observatory from the Index Catalogue of Visual Double Stars, 1961·0, by Hamilton M. Jeffers and Willem H. van den Bos with Frances M. Greeby, *Publ. Lick Obs.*, **31**, Parts I and II.

The following criteria, based on recommendations of the Double Star Committee appointed by Commission 8 during the Hamburg Meeting of the IAU, were used in the selection of double stars.

(*a*) The magnitude of the brighter component shall lie within the limits, $6^{m}.0 \leqslant m_B < 9^{m}.0$.

(*b*) The magnitude difference between the components shall lie within the limits,

$$0^{m}.0 \leqslant (m_F - m_B) \leqslant 4^{m}.0.$$

(*c*) The separation, s, of the components shall lie within the limits, $2''.0 \leqslant s < d$ where, $d = 29{\cdot}0 - (m_B + m_F)$.

(*d*) The brighter component must have a Durchmusterung number.

Interested observers may obtain copies from the U.S. Naval Observatory if requested before 1970.0.

F. P. SCOTT

BIBLIOGRAPHY

1. *Geschichte des Fixsternhimmels*, Abteilung II, Vols. 18, 19.
2. *Index der Sternörter* 1925–60, *Index II.* Band III, IV, V, VI, VII, VIII, IX.
3. *Star Catalogue. Positions and Proper Motions of 258 997 stars for the epoch and equinox of 1950·0* (four volumes). Smithsonian Institution Washington, D.C. 1966.
4. Adams, A. N., Scott, D. K. 1964–65, Observations of the Sun, Moon and planets – six-inch transit circle results. *U.S. nav. Obs. Circ.* no. 103, 105, and 108.

5. Adams, A. N., *et al.* 1964, Results of observations made with the six-inch transit circle 1949–56. *U.S. nav. Obs. Publ.* **19**, Part 1.

6. Atkinson, R. d'E., Symms, L. S. T., Blackwell, K. C. 1964, Pivot Errors and Axis Flexure in the seven-inch Cooke Transit Circle. *R. Obs. Bull.*, no. 109.

7. Bacchus, P. 1964, Sur la comparaison des catalogues. *Ann. Obs. Strasbourg*, **6**, 75.

8. Bacchus, P., Lacroute, P. 1964, Réduction des observations méridiennes avec 'synthèse'. *Ann. Obs. Strasbourg*, **6**, 49.

9. Böhme, S., Fricke, W. 1965, Astronomical constants. A survey of determined values. *Bull. astr.*, **25**, no. 3, 269.

10. Boshniakovich, N. A. 1965, The precessional constant, parameters of solar motion and galactic rotation in the FK4 system. *Bull. astr.*, **25**, no. 2, 119.

11. Brealey, G. A. 1964, The Digital Velocity-Generating Computer for the Dominion Observatory Mirror Transit. *Publ. Dom. Obs.*, **25**, 6.

12. Brosche, P., Nowacki, H., Strobel, W. 1964, Systematic differences FK4 − GC and FK4 − N30 for 1950·0. *Veröff. astr. Rechen Inst. Heidelberg*, no. 15.

13. Brosche, P. 1966, Representation of systematic differences in positions and proper motions of stars by spherical harmonics. *Veröff. astr. Rechen Inst. Heidelberg*, no. 17.

14. Brouwer, D. 1965, Relations among some important astronomical constants. *Bull. Astr.*, **25**, no. 3, 241.

15. Clemence, G. M. 1965, The System of Astronomical Constants. *A. Rev. Astr. Astrophys.*, **3**, 93.

16. Débarbat, S., Kovalevsky, J. 1963, Discussion des observations de Mars faites à l'astrolabe impersonnel A. Danjon. *Bull. astr.*, **24**, fasc. 1.

17. Débarbat, S. 1966, Discussion des observations de Mars effectuées en 1963 à l'astrolabe A. Danjon. *Bull. astr.*, **1**, (3ème série), fasc. 4.

18. Dick, J. 1966, Gedanken zur Frage eines horizontalen Meridiankreises. *Sterne* (in press).

19. Eckert, W., Smith, H. F. 1966, The Equations of Variation in a Numerical Lunar Theory, IAU Symposium no. 25, 242.

20. Eichhorn, H. 1963, The relationship between standard coordinates of stars and the measured coordinates of their images. *Appl. Opt.*, **2**, 17.

21. Eichhorn, H., Jefferys, W. H. III. 1963, Algorithm for the rigorous determination of differential astrometric data. *Astr. J.*, **68**, 71, Ref. AAS.

22. Eichhorn, H., Williams, C. A. 1963, Systematic accuracy of photographic star positions and proper motions. *Astr. J.*, **68**, 72, Ref. AAS.

23. Eichhorn, H. 1963, On the use of panoramic cylindrical projection in photographic astrometry. *Astr. J.*, **68**, 210.

24. Eichhorn, H., Williams, C. A. 1963, On the systematic accuracy of photographic astrometric data. *Astr. J.*, **68**, 221.

25. Engelbrecht, J. 1966, Das Photo-Zenit-Telescop (PZT) der Sternwarte Babelsberg (Konstruktion und Funktion). *Feingerätetechnik* 15.1, S. 8–12.

26. Federov, E. P. 1965, Some recent studies of nutation in the U.S.S.R. *Bull. astr.*, **25**, no. 2, 125.

27. Fricke, W. 1965, The influence of proper motion systems on the determination of Statistical Parallaxes. *Vistas in Astronomy*, **8**, 205.

28. Fricke, W. 1965, Astronomical constants as an object of recent research. Inaugural address delivered at the IAU Symposium no. 21. *Bull. Astr.*, **25**, no. 1, 5.

29. Fricke, W. 1965, Ein Vergleich der Entfernungsskalen verschiedener Eigenbewegungssysteme. *Z. Astrophys.*, **61**, 20.

30. Gliese, W. 1965, Stability of systems of positions obtained with transit instruments, derived from fluctuations in the α_δ–systems. *Astr. J.*, **70**, 162.

31. Gliese, W. 1966, Cape Catalogue of Circumpolar Stars for the Equinox 1950·0. *R. Obs. Bull.*, no. 106.

32. Heinemann, K. 1964, Verzeichnis von Sternkatalogen 1900–62. *Veröff. astr. Rechen Inst. Heidelberg*, no. 16.

33. Hemenway, P. D. 1966, Washington six-inch transit circle. *Sky Telesc.*, **31**, 72.

34. Höpfner, J. 1966, Sternkoordinatenkorrektionen aus dem Beobachtungsmaterial am Astrolab Danjon. Geod. Inst. Potsdam, Arbeiten Nr. 12.
35. Hurukawa, K. On the precise calculation of the apparent places of fixed stars using the rectangular coordinates. *Proc. Intern. Lat. Obs. Mizusawa,* Nr. 3., S. 69–92 (Japan. mit. Engl. Ref.).
36. Hurukawa, K. On the precise calculation of the apparent places of fixed stars using the rectangular coordinates (II). *Proc. Intern. Lat. Obs. Mizusawa,* Nr. 4, S 73–80 (Japan. mit. Engl. Ref.).
37. Jefferys, W. H. III. 1963, On computational technique for photographic astrometry with overlapping plates. *Astr. J.,* **68,** 111.
38. Kirschner, H. Katalogkorrektionen der Form $\Delta\alpha_\alpha$ des FK4 (in press).
39. Klock, B. L., Scott, D. K. 1965, An investigation of the coefficients to periodic term 182 of Brown's lunar theory. *Astr. J.,* **70,** 335.
40. Klock, B. L. Instrumentation program for an improved meridian circle at the U.S. Naval Observatory (Abstract). *Astr. J.,* **71** (in press).
41. Kopff, A., Nowacki, H., Strobel, W., Mürle, A. 1964, Formeln und Tafeln zur differentiellen Präzession und zur Übertragung von Eigenbewegungen. *Veröff. astr. Rechen Inst. Heidelberg,* no. 13.
42. Kopff, A., Nowacki, H., Strobel, W. 1964, Individual corrections to FK3 and the declination system of the Fourth Fundamental Catalogue (FK4). *Veröff. astr. Rechen Inst. Heidelberg,* no. 14.
43. Labrecque, J. J. 1960, A Schmidt Camera built at the Dominion Observatory. *J. R. astr. Soc. Can.,* **54,** 2.
44. Lacroute, P. 1964, Etude des erreurs systématiques à craindre dans l'établissement d'un catalogue méridien relatif. *Ann. Obs. Strasbourg,* **6,** 33.
45. Lacroute, P. 1964, Amélioration dans les réductions d'un catalogue méridien, méthode de 'synthèse'. *Ann. Obs. Strasbourg,* **6,** 39.
46. Lacroute, P. 1964, Etude externe des observations méridiennes de Strasbourg. *Ann. Obs. Strasbourg,* **6,** 85.
47. Lacroute, P. 1964, Amélioration des constantes de clichés en utilisant leur recouvrement. *Ann. Obs. Strasbourg,* **6,** 97.
48. Lacroute, P. 1964, Etude sur l'établissement d'un catalogue photographique. *Ann. Obs. Strasbourg,* **6,** 113.
49. Loeser, H. G. 1964, Eine neue Tabelle der Normalrefraktion und die Berechnung der Refraktion nach Sondenaufstiegen der Wetterämter. Deutsche Geod. Kommission, Reihe D, Nr. 6.
50. Maître, V. 1965, Troisième Catalogue de l'Observatoire de Besançon. *Ann. Obs. Besançon,* **6,** fasc. 6.
51. Melchior, P. 1965, Déduction du Phénomène de Précession-Nutation de la Marée Terrestre. *Commun. Obs. r. Belgique,* no. 240.
52. Melchior, P. 1966, Détermination expérimentale des effets dynamiques au noyau liquide de la Terre dans les marées terrestres diurnes. *Commun. Obs. r. Belgique,* Série B, no. 4.
53. Menzies, A. 1966, Astronomy of Position. *Mon. Notes astr. Soc. Sth. Afr.,* **25,** 59.
54. Pilowski, K., Soltau, G. 1963, Zur Fixierung der Kippachse eines transportablen Passageinstrumentes. *Mitt. astr. Ges.,* p. 116.
55. Pilowski, K. 1963, Zur Verwendung transportabler Meridiankreise für fundamentale Positionbestimmungen. *Mitt. astr. Ges.,* p. 117.
56. Pjaskowskij, D. W. 1964, Lehrbuch der sphärischen Astronomie für Astronomiestudenten der Ukrainischen Universitäten. Kiew Univ., Kiew. 136S. Preis 26 Kop. (Russ.)
57. Podobed, V. V. 1965, *Fundamental astrometry,* Translated from the Russian by Scripta Technica, Inc. English translation edited by A. N. Vyssotsky. Univ. Chicago Press, Chicago.
58. Poloshenzew, D. Astrometrie – Wissenschaft der Genauigkeit. Vasiona 12, 49–52 (Serb.-Kroat.).
59. Proverbio, E. Riduzione degli errori sistematici nelle osservazioni meridiane di tempo e di longitudine. *Mem. Soc. astr. ital.,* **35,** 183 = *Contr. Oss. astr. Milano-Merate* (NS) Nr. 229.

60. Requième, M. 1965, (automatic registration of star transits). *C. r. Acad. Sci. Paris*, **260**. 2421.

61. Scott, F. P. 1963, Estimates of the accuracy of positions taken from photographic star catalogues, present and future. The use of Artificial Satellites for Geodesy. Symposium, Washington D.C., 26–28 April 1962. North Holland Publ. Co., Amsterdam.

62. Scott, F. P. 1963, The system of fundamental proper motions. In *Stars and Stellar Systems*, III, *Basic Astronomical Data*, Ch. 2, Univ. Chicago Press.

63. Scott, F. P. 1964, Limitations imposed on celestial navigation due to inaccuracies of star positions. *Navigation*, **2**, 20.

64. Scott, F. P., Hughes, J. A. 1964, Computation of apparent places for the southern reference star program. *Astr. J.*, **69**, 368.

65. Scott, F. P. 1964, A method of evaluation of the elliptic E terms of aberration. *Astr. J.*, **69**, 372.

66. Ševarlic, B., Teleki, G. 1959, Le programme de détermination de déclinaisons des 36 étoiles à la lunette zénithale. *Bull. astr. Obs. Belgrade*, **24**, no. 3–4, 27.

67. Slaucitajs, S. J. 1963, Astronomical Refractión Constant, Refraction Index and Expansion Coefficient and Optical Density of the Air. *Inf. Bull. Southern Hemisphere*, no. 4.

68. Slaucitajs, S. J. 1963, Sobre la Constante de Refracción Astronómica, el Indice de Refracción, el Coeficiente de Dilatación y la Densidad Optica del Aire. *Asoc. argent. Astr. Bol.*, no. 6, 1964.

69. Stoy, R. H. 1966, Cape Photographic Catalogue for 1950.0. Zones – 64° to – 80°. *Cape Annals*, **21**.

70. Teleki, G. 1966, Declination System of Latitude Stars. *Astr. Zu.*, **43**, no. 3, 678.

71. Thomas, D. V. 1964, Results obtained with a Danjon Astrolabe at Herstmonceux, I. *R. Obs. Bull.*, no. 92.

72. Thomas, D. V. 1965, The Astrolabe for the Cape Observatory. *Mon. Notes. astr. Soc. Sth Afr.*, **24**, 36.

73. Thomas, D. V. 1966, Results of Observations made with the Danjon Astrolabe O.P.L. no. 11 at the Cape. *Mon. Notes astr. Soc. Sth. Afr.*, **25**, 115; 146; 172.

74. Thomas, D. V. Group Corrections and the Optimization of an Observing Programme. *R. Obs. Bull.* (in press).

75. Thomson, M. M. 1960, Positional Astronomy, Dominion Observatory. *J. R. astr. Soc. Can.*, **54**, 2.

76 van Herk, G. Hoe is de stand in de astrometrie? *Hemel en Dampkring*, **61**, 101.

77 Wayman, P. A.. Symms, L. S. T., Blackwell, K. C. 1963, Proper Motions and Radial Velocities of Hyades Stars. *R. Obs. Bull.*, no. 98.

78. Wilkins, G. A. 1965, Review of the system of astronomical constants. *Bull. astr.*, **25**, no. 3, 251.

79. Weller, W. J. 1964, Rapid Computer Search Technique for Astronomical Catalogues. *Publ. astr. Soc. Pacific*, **76**, 152.

80. Wood, H. 1963, Positional astronomy. The Donovan astronomical lecture for 1963. *J. Proc. R. Soc. New South Wales*, **97**, 135 = *Sydney Obs. Papers* Nr. 47.

81. Woolard, W. W., Clemence, G. M. 1966, *Spherical Astronomy*. Academic Press, New York.

82. Woolsey, E. G. 1964, Catalogue Contributions of the Ottawa Meridian Circle. *J. R. astr. Soc. Can.*, **58**, 2.

83. Woolsey, E. G. 1966, Results of Observations made with the Reversible Meridian Circle 1956–61. Catalogue of 3753 AGK3 Reference Stars. Corrections to 930 FK3 Stars. *Publ. Dom. Obs.*, **25**, no. 8.

84. Woolsey, E. G. 1966, Results of Observations made with the Reversible Meridian Circle 1954–62. Catalogue of 2665 Stars. *Publ. Dom. Obs.*, **25**, 9.

85. Yasuda, H. *et al.* 1963, Meridian observations of the positions of the Moon, major planets, and some minor planets during the year 1962. *Tokyo astr. Bull.*, no. 161.

86. Yasuda, H., Hara, H. 1964, Meridian Observations of the positions of the Moon, major planets, and some bright minor planets during the year 1960. *Tokyo astr. Bull.*, no. 162.

87. Yasuda, H. 1964, A method for determining the systematic corrections to a fundamental catalogue from the differential observations of planets. *Publ. astr. Soc. Japan*, **16**, 45.

88. Yasuda, H., Hara, H. 1964, The catalogue of 240 stars in the PZT programs of Mizusawa, Richmond, Tokyo and Washington. *Ann. Tokyo astr. Obs.*, (II), **8**, 162.

89. *Fragen der Astrometric.* Herausgegeben vom Astr. Hauptobs. Akad. Wiss. Ukrain SSr. Kiew, 'Naukowa Dumka', 1964. 96S. Preis 38 Kop. (Russ.)

90. Anonym. 1965, *Geodätische Astronomie.* II. Internationales Kolloquium. *Wiss. Z. techn. Univ. Dresden*, **14**, 635 = *Mitt. de. Lohrmann-Institutes*, Nr. 11.

91. Preliminary Supplement to the FK4 (FK4 Sup.) 1963, *Veröff. astr. Rechen Inst. Heidelberg*, no. 11.

92. Proceedings of Conference on *Construction and Use of Star Catalogs*, held at the University of Maryland, College Park, Md. 3–5 October 1966, to be published in a forthcoming issue of the *Astr. J.*.

APPENDIX. REPORT ON THE PROBLEMS OF POSITIONAL ASTRONOMY STUDIED IN THE U.S.S.R (JULY 1963–AUGUST 1966)

1. *Studies in accordance with recommendations of the IAU and other international organizations*

(*a*) *The southern reference stars* (*SRS*). The observations of the SRS in the declination zone − 5° to − 20° have been completed at the Nikolayev Observatory by means of the Repsold Meridian Circle (observers: V. V. Konin and E. V. Khrutskaja). The reduction of observation is being carried on intensively and nearing completion. A semi-automatic electronic measuring machine is used for measurements of the films with the images of circle readings (**38**).

The observations of 13 489 stars (SRS, BS and DS) in the declination zone − 25° to − 90° are being continued at the Cerro-Calan Observatory by the Chilean and Soviet astronomers. More than 60 000 observations were obtained by the end of 1965 (observers: C. Anguita, V. S. Bedin, M. S. Zverev, G. Carrasco, P. Loyola, A. A. Naumova, D. D. Polojentsev and T. A. Polojentseva).

(*b*) *Bright stars* (*BS*). The observations of bright stars are being made in Chile (see point (*a*)) and also in the Kiev University Observatory (observers: A. G. Gregul, V. K. Drofa and N. A. Chernega) in the declination zone − 10° to + 90°. Altogether 7500 observations have been already obtained. In the Tashkent Observatory the bright stars are observed in the right ascensions only ($-20° < \delta < 90°$). The observations in the declination zones − 20° to − 10° and 0° to + 10° were completed and published (**124, 125**). In total the work will be completed in 1967.

(*c*) *The stars of the PZT and Latitude programs.* The observations of the latitude stars and stars in the PZT programs are being made with the vertical circle at Pulkovo (observers: G. S. Kossin and B. K. Bagildinsky) and with the meridian circle at the Engelhardt Observatory. Determinations of the right ascensions and declinations of the PZT stars and also the declinations of the latitude stars were commenced at the Sternberg Institute with the new meridian circle. The determination of the declinations of the Pulkovo zenith stars in the program of observations with the zenith-telescope ZTL–180 are being made at the Odessa Observatory (observer: L. A. Klepikova).

(*d*) *The KSZ program.* The reduction of observations of the KSZ and FKSZ stars are being continued. The catalogues of the Kiev (**107**), Engelhardt (**114**), Tashkent (**119**) and Odessa (**98**) Observatories have been published. The following editions have been prepared for publication: (1) L. F. Gorel, 'The Catalogue of 2600 Stars (KSZ) in the Declination Zone − 5° to − 20°, compiled from Observations with the Meridian Circle at Nikolayev'; (2) B. V. Novopashenny, 'Catalogue of Right Ascensions of 2963 KSZ Stars in the Declination Zone − 5° to − 25°, compiled from Observations at Odessa'.

A re-reduction of the catalogues of the KSZ and FKSZ stars in the FK4 system are being made at the Pulkovo, Kiev and Engelhardt observatories (zone $-10°$ to $+90°$). The compiling of the general KSZ catalogue in the declination zone $+90°$ to $-30°$ was commenced at Pulkovo, all the obtained observations being used.

The observations of the selected minor planets according to the KSZ program were made at Pulkovo, Moscow (Sternberg Institute), Golosseyevo, Tashkent and Nikolayev. A part of these observations is published (**18, 20, 21, 127–133**). The observations of minor planets, made at the Cape Observatory during 1956–63 were reduced and published at Pulkovo (**19, 128**).

The photographic observations of the first epoch of 205 areas with the galaxies and 240 areas with the fundamental KSZ stars in the declination zone $+90°$ to $-25°$ have been completed. These observations were made at Pulkovo, Moscow (Sternberg Institute), Golosseyevo and Tashkent.

II. *Absolute determinations. Observations of the solar system bodies*

At Pulkovo the compiling of the absolute catalogues of right ascensions of 507 bright and 522 faint stars (large transit instrument) and declinations of 542 bright and 531 faint stars (vertical circle) have been completed (**33, 34, 61**).

The catalogue of right ascensions as regards to $\Delta\alpha_\delta$ is close to N30. A comparison of declinations of the new Pulkovo catalogue with the fundamental systems shows that it is also in good agreement with N30.

The mean error of one observation is represented by the formula

$$\sigma(z)^2 = (0\overset{''}{.}34)^2 + (0\overset{''}{.}17 \text{ tg } z)^2$$

The absolute declinations of 1181 bright stars (in the declination zone $+90°$ to $-30°$) and 611 faint FKSZ stars (in the declination zone $+90°$ to $-25°$) were determined at Golosseyevo with the Vanschaff vertical circle. The catalogue is being prepared for print.

The absolute determinations of the declinations of 1443 bright stars of the southern sky ($-90°$ to 30° of declination) made by the Chilean and Soviet astronomers at the Cerro-Calan Observatory are nearing completion. The observations are being made by means of the photographic vertical circle (PVC), designed by M. S. Zverev. The mean error of one observation is $\pm 0\overset{''}{.}25$.

The construction of the large transit instrument for absolute determinations of the right ascensions of the southern sky stars will be completed at Pulkovo in the near future. By the end of 1966 this instrument will be forwarded to Chile.

The regular meridian observations of the Sun, Moon and major planets were made at Pulkovo, Nikolayev, Moscow (Sternberg Institute), Tashkent, Kharkov and Engelhardt Observatory (**12, 13, 35, 43, 44, 52, 53, 62–67, 86**).

The photographic observations (**17, 51**) of the major planets (except Mercury) were made at Nikolayev.

III. *Separate series of observations and their reduction*

The compiling of three catalogues of the absolute right ascensions of about 800 stars (**55, 57, 59**) has been completed at Pulkovo. They are based on the observations made during 1957–63 with three photoelectric transit instruments of the Time Service. The compiling of the general catalogue of right ascensions of the stars in the U.S.S.R. Time Service program is nearing completion. This catalogue is based on more than 150 000 observations.

The systematic observations with the Danjon astrolabe are in progress at the Pulkovo Observatory. The reduction of two annual series (1963–64 and 1964–65) has been completed.

A reduction of the absolute observations obtained by M. A. Grachev during 1909–17 has been completed at the Engelhardt Observatory. The Catalogue of 2500 stars has been compiled.

The compiling of the catalogue of right ascensions of the stars in the declination zone + 90° to + 70° has been completed at the Kiev Observatory (**112**). For 79 stars of this catalogue the proper motions were determined (**113**). Each 0°5 division of the meridian circle of the Kiev Observatory was investigated using the photographic device (**111**).

Three catalogues of right ascensions of about 200 stars in the U.S.S.R. Time Service program were compiled from the observations with two instruments of the Time Service in Tashkent. One of them was published (**120**); the others are in print.

IV. *Instruments*

An original device for photographing the circle was constructed by A. M. Stafeev at the Odessa Observatory. A detailed investigation of the errors of the circle readings through 10′ were completed.

A transistor photoelectric device for registering star transits, permitting to observe the faint stars, was designed at Nikolayev by M. I. Ilkiv.

The investigation of the L. A. Sukharev horizontal meridian circle is being made at Pulkovo.

Preliminary observations of the stars in α have been started.

A double-meniscus astrograph (D = 70 cm; f = 2·2 m), made in the U.S.S.R., for the photographic observations of galaxies and fundamental stars is mounted in Chile.

V. *Conferences*

The 17th Astrometrical Conference was held at Pulkovo during 31 May–3 June 1966. Altogether more than 100 astronomers, representative of 26 astronomical institutions of the U.S.S.R., German Democratic Republic, Mongolia, Romania and Yugoslavia participated. Two main problems were discussed: (1) on the compiling of the general catalogue KSZ and (2) on the constructing of new astrometrical instruments. 'Transactions' of the conference are being prepared for print.

A. A. NEMIRO
Vice-President of the Commission

BIBLIOGRAPHY

In *Izv. glav. astr. Obs. Pulkovo (Leningrad)*, **23**, *no.* 174, *Part* 4, 1964 (*In Russian with English abstracts*).

1. Baturina, G. D., Bedin, V. S., Varina, V. A., Gnevyscheva, K. G., Zverev, M. S., Izvekova, A. A., Murri, S. A., Naumova, A. A., Polozhentsev, D. D. Observations of AGK3R Stars with the Toepfer Meridian Circle at Pulkovo.
2. Zverev, M. S., Murri, S. A. On the System of Right Ascensions of the Toepfer Meridian Circle.
3. Baturina, G. D. The System of Declinations of the Toepfer Meridian Circle.
4. Baturina, G. D., Varina, V. A., Gnevyscheva, K. G., Naumova, A. A., Polozhentsev, D. D. The Method of Reduction of Differential Observations of Declinations by Means of Punch-Card Machines.
5. Varin, M. P. The Application of Electronic Computers for Computations of the Apparent Places of Stars.
6. Petrov, G. M. On the Determination of the Absolute Azimuth and Inclination of the Transit Instrument.
7. Kim Gun-der A Determination of the Right Ascensions of 88 Circumpolar Stars.
8. Petrov, G. M. On the Determination of the Collimation Error of a Transit Instrument.

9. Bagildinsky, B. K., Kossin, G. S., Medvedeva, L. I. An Investigation of the Flexure of the Struve-Ertel Vertical Circle at Pulkovo.
10. Markina, O. T. Results of Utilisation of a Mirror Filter for Observations of the Sun.
11. Medvedeva, L. I. Preliminary Results of Observations of Polarissima BD + 89°3 with the Pulkovo Vertical Circle.
12. Brovenko, V. Y., Kalinina, O. F., Markina, O. T., Petrov, G. M., Fedorova, R. T. The Right Ascensions of Bodies of the Solar System Determined from Observations with the Freiberg-Kondratyev Transit Instrument at the Nikolayev Observatory during 1961.
13. Gordon, Y. E., Gorel, L. F., Dzuba, I. P. Observations of the Sun, Moon and Major Planets with the Meridian Circle at the Nicolayev Observatory.
14. Ilkiv, M. I. An Improved Photoelectric Device for the Registration of Star Transits.
15. Ivakin, V. M., Ilkiv, M. I. An Electronic-Mechanical Device for the Determination of the Mean Moments of Star Transits.
16. Zimmermann, H. K. Über einen Fall der Bestimmung von systematischen Differenzen.
17. Brovenko, V. Y. Results of Photographic Observations of Uranus and Neptune at the Nicolayev Observatory during 1961–62.
18. Bronnikova, N. M., Koroleva, L. S., Strugatskaya, A. A., Chudivicheva, O. N. Precise Positions of Minor Planets from Photographic Observations at Pulkovo.
19. Kiseleva, T. P., Sokolova, L. A., Koroleva, L. S. Precise Positions of Minor Planets from Photographic Observations at the Cape Observatory.
20. Kalihevich, Th. Th. Results of Photographic Observations of Minor Planets at the Nikolayev Observatory.
21. Chudovicheva, O. N. Results of Photographic Observations of the Minor Planet Iris at the Nikolayev Observatory.
22. Sokolova, V. A. The Program for Computations of Precise Positions of Minor Planets by Means of an Electronic Computer.

In *Izv. glav. astr. Obs. Pulkovo* (*Leningrad*), **22**, *no.* 176, *Part* 6, 1965 (*In Russian with English abstracts*).

23. Nemiro, A. A. Astrometry at Pulkovo (1945–63).
24. Gordon, Y. E. The Nikolayev Branch of the Pulkovo Observatory (Fifty Years since its Foundation).
25. Semenov, L. I. Observations Made with the Freiberg Transit Instrument during the Past 50 Years.
26. Zimmerman, G. K. The Determination of Declinations with the Vertical Circle of the Nikolayev Branch of the Pulkovo Observatory during Fifty Years.
27. Gordon, Y. E., Gorel, L. F. Observations of AGK3R Stars with the Repsold Meridian Circle at the Nikolayev Observatory.
28. Konin, V. V. The Practical Application of Zimmerman's Method for the Investigation of the System of an Instrument.
29. Polojentsev, D. D. Computations of Refraction from the Pulkovo Tables by Means of Automatic Computers.
30. Petrov, G. M. On one Systematic Error of an Eyepiece Micrometer.
31. Ilkiv, M. I. A Balance Scheme for Increasing the Stability of a Photoelectric Device for Registration of Star Transit.
32. Kiseleva, T. P. Photographic Positions of Mars obtained with Pulkovo Short-Focus Astrograph during 1960–61.

In *Izv. glav. astr. Obs. Pulkovo* (*Leningrad*), **24**, *no.* 179, *Part* 3, 1966 (*In Russian with English abstracts*).

33. Pljugina, A. I. The System of Right Ascensions of the Equatorial Stars of the New Polkovo Absolute Catalogue.
34. Kossin, G. S. A Comparison of the Catalogue of Declinations $Pu_{50}F$ and $Pu_{50}Z$ with the Fundamental Systems.

35. Kossin, G. S. The Observations of the Sun and Planets made with the Vertical Circle of the Pulkovo Observatory.
36. Medvedeva, L. I. The Investigations of the Divisions of the Struve-Ertel Vertical Circle of the Pulkovo Observatory made in Connection with Observations of Polarissima BD + 89°3.
37. Godisov, N. P. The Program of Observations made with the Danjon Prismatic Astrolabe at the Pulkovo Observatory.
38. Ivakin, V. M., Ilkiv, M. I., Nechaev, V. M., Tilk, E. M. An Electronic Mounting for the Measurement of the Photographs of a Limb.
39. Bedin, V. S. The Determination of the Aberrational Constant from Latitude Observations at Pulkovo during 1920–28.

In *Izv. glav. Obs. Pulkovo (Leningrad)*, **24**, *no.* 181, *Part* 5, 1966 (*In Russian with English abstracts*).

40. Zverev, M. S., Naumova, A. A., Naumov, V. A., Torres, K. The Preliminary Results of Observations with the Pulkovo Photographic Vertical Circle.
41. Naumov, V. A. Instrumental Errors of the Photographic Vertical Circle.
42. Gnevysheva, K. G. The Operation of the Toepfer Meridian Circle during 50 years.
43. Baturina, G. D., Gnevysheva, K. G., Chubey, M. S. The Right Ascensions of the Sun, Mercury, Venus and Mars obtained from Observations with the Toepfer Meridian Circle of the Pulkovo Observatory during 1961.
44. Kossin, G. S. The Results of Observations of the Sun and Planets with the Vertical Circle at Pulkovo.
45. Bozhko, I. I. The Day-Time Observations with the Vertical Circle of the Nikolayev Observatory.
46. Ilkiv, M. I. Preliminary Results of Observations of the Right Ascensions of Major Planets by Photoelectric Method.
47. Kossin, G. S. The Results of the Reduction of Observations of the Sun made with the Vertical Circle of the Pulkovo Observatory during 1956–63.
48. Isvekova, A. A. The Most Advantageous Distribution of Reference Stars for the Meridian Observation of Right Ascensions.
49. Konin, V. V. The Reduction of Meridian Observations using the Results of the Stars under Determination.
50. Varin, M. P. The Computation of the Bessel Reduction Values with Electronic Computers.
51. Brovenko, V. G., Kalikhevich, F. F. The Photographic Observations of Major Planets of the Nicolayev Observatory during 1961–62.

In *Trudy glav. astr. Obs. Pulkovo (Leningrad)*, *Ser. II*, **74**, 1966 (*In Russian with English abstracts*).

52. Semenov, L. I. Right Ascensions of the Sun, Moon and Large Planets as derived from Observation at the Nikolayev Observatory during 1939–59.
53. Petrov, G. M., Dzjuba, I. P. Right Ascensions of the Solar System Bodies from Observations with the Freiberg-Kondratiev Transit Instrument at the Nikolayev Observatory during 1954–59.
54. Parshin, S. N. The Catalogue of Absolute Right Ascensions of the 135 Circumpolar Stars compiled from Observations made by S. S. Lebedev and A. P. Sokolov during 1892–95.

In *Trudy glav. astr. Obs. Pulkovo (Leningrad)*, *Ser. II*, **75**, 1966 (*In Russian with English abstracts*).

55. Afanasjeva, P. M., Pavlov, N. N., Staritsyn, G. V. A Catalogue of Right Ascensions of the 508 Stars obtained from Observations with the F-3 Photoelectric Transit Instrument at the Pulkovo Observatory during IGY and IGC.

56. Afanasjeva, P. M., Pavlov, N. N., Staritsyn, G. V. The Reference Catalogue of Right Ascensions of 513 Stars for the General Catalogue of the U.S.S.R. Time Services.

57. Chelombitko, A. P. The Catalogue of Right Ascensions of 431 Stars Obtained with the New Transit Instrument П П И–1 at the Pulkovo Observatory during 1959–63.

58. Karetnikova, E. N. The Reduction of the Zenith Stars of the Catalogue by the Chain Method.

59. Vassiljev, V. M., Karetnikova, E. N. The Catalogue of Right Ascensions of 501 Stars, obtained from Observations with the Photoelectric Zeiss Transit Instrument N 14 561 during the IGY at Pulkovo.

60. Gordon, Y. E., Gorel, L. F., Dzjuba, I. P., Kalinina, O. F. The Catalogue of Right Ascensions and Declination of the AGK3R Stars Compiled at Nikolayev from Observations with the Meridian Circle.

In *Trudy glav. astr. Obs. Pulkovo* (*Leningrad*). *Ser. II*, **76**, 1966 (*In Russian with English abstracts*).

61. Kossin, G. S., Bagildinsky, B. K. The Results of Observations of Bright and Faint Fundamental Stars with the Vertical Circle of the Pulkovo Observatory.

62. Kalinina, O. F., Markina, O. T., Petrov, G. M., Fedorova, R. T. The Right Ascensions of the Sun, Moon, Mösting A and Major Planets observed with the Transit Instrument at Nikolayev during 1962.

63. Kalinina, O. F., Markina, O. T., Petrov, G. M., Fedorova, R. T. The Right Ascensions of the Sun, Moon, Mösting A and Major Planets observed with the Transit Instrument at Nikolayev during 1963.

64. Markina, O. T., Petrov, G. M. The Right Ascensions of the Sun, Mercury, Venus and Mars observed with the Transit Instrument at Nikolayev during 1964.

65. Gordon, Y. E., Gorel, G. K., Gorel, L. F. The Results of Observations of the Sun, Moon and Major Planets made with Meridian Circle at Nikolayev during 1962–63.

66. Bozjko, I. I., Zimmerman, G. K. The Declinations of the Sun, Moon and Major Planets in the FK4 System, obtained with the Vertical Circle of the Nikolayev Observatory during 1961–62.

67. Bozhko, I. I., Zimmerman, G. K. The Declinations of the Sun, Moon and Major Planets in the FK4 System obtained with the Vertical Circle of the Nikolayev Observatory during 1963–64.

In *Trudy* 16. *astrometr. Konf. SSSR* (*Kazan*, 1963), *issued in* 1965 (*In Russian with English abstracts*).

68. Nemiro, A. A. The Determination of Positions of Stars of Special Programs at the U.S.S.R. Observatories.

69. Tavastsherna, K. M. Problems of Meridian Astrometry discussed at the IAU XIth General Assembly and Organization of Observations of Reference Stars of the Southern Sky.

70. Zverev, M. S. The Work of the Pulkovo Astronomers in Chile.

71. Izvekov, V. A., Izvekova, A. A. The Determination of Parameters of the Meridian Circle with the Б Э С М–2.

72. Nenakhova, E. M. The Program of Observations of PZT Stars.

73. Angita, C., Zverev, M. S. Preliminary Results of Observations with the Repsold Meridian Circle of the Cerro-Calan Observatory (Chile).

74. Petrov, G. M. Meridian Observations of the Sun and Major Planets.

75. Kossin, G. S. Day-Time Observations of Stars.

76. Deutsch, A. N. Photographic Determinations of Positions of Major Planets.

77. Strugatskaja, A. A. Photographic Observations of Venus made at the High-Altitude Solar Station of the Pulkovo Observatory.

78. Onegina, A. B., Sereda, E. M. A Determination of Positions of Mars and Deimos.

79. Brovenko, B. Y. Photographic Observations of Major Planets with the Nicolayev Zonal Astrograph.

80. Kabaeva, N. N. Experimental Photography of Jupiter with the Wide-Angle Astrograph of the Sternberg Astronomical Institute.
81. Zysjakova, R. F. An Application of the Local Desensibilization Method for Positional Observations of Planets.
82. Sharonov, V. V. Photographic Irradiation as a Source of Errors of Astrometric Observations of the Moon and Planets.
83. Jakovkin, A. A., Demenko, I. M., Miz, L. M. The Formulas and Methods of Practical Astrometry on the Moon.

In *Soobšč. gos. astr. Inst. P. K. Sternberga, no.* 145, 1966.

84. Oborneva, A. G. Investigation of Local Parts of the Limb of the Repsold Meridian Circle using Podobed's Method.
85. Hommik, L. M. On the Determination of the Inclinations of the Horizontal Threads.
86. Kosina, O. A., Kabaeva, N. N., Guljaev, A. P., Podobed, V. V. Observations of the Right Ascensions of the Sun and Major Planets with the New Meridian Circle.
87. Guljaev, A. P. Notes on the Weights of the Catalogues with regard to Accidental Errors.

In *Astr. Zu.*, **40,** (1963), **41,** (1964), **42** (1965) (*In Russian with English abstracts*).

88. Mikhailov, A. A. 1963, On the Constant of Aberration and Solar Parallax. **40,** no. 6.
89. Efimov, A. A. 1964, A Photoelectric Spiral Micrometer for Limb Readings. **41,** no. 3.
90. Blinov, N. S. 1964, A Possible System of the General Catalogue of U.S.S.R. Time Service. **41,** no. 5.
91. Zverev, M. S. 1964, A Photographic Method for the Determination of the Corrections of all the Diameters of a Divided Circle. **41,** no. 6.
92. Kiselev, A. A. 1965, A Vector Interpretation of the Main Methods of Photographic Astrometry. **42,** no. 2.
93. Kulikov, K. A. 1965, The System of Astronomical Constants. **42,** no. 3.
94. Zverev, M. S. 1965, The Determination of Systematic Errors of a Fundamental Catalogue from Meridian Observations of Series of Fundamental Stars. **42,** no. 4.
95. Podobed, V. V., Nesterov, V. V. 1965, Astrometry. **42,** no. 5.

In *Voprocy Astrometrii*, '*Naukova Dumka*', *Kiev*, 1966 (*In Russian with English abstracts*).

96. Kharin, A. S. On the Improvements of Differential Catalogue in Systematic and Accidental Respects.
97. Korol, A. K., Nenakhova, E. M. Coordinates of the Polarissima BD + 89°3 from Observations with the Vertical Circle at Goloseyevo.
98. Nechayev, V. M. Catalogue of Declinations of KSZ_2 Stars in the Zone − 12° to − 20°.
99. Cherniev, L. F. Proper Motions in Declination of Circumpolar Stars.
100. Zoteyev, E. S. Investigation of the Graduation Errors of the Meredian Circle of the Odessa Astronomical Observatory in Degree Diameters.
101. Kharin, A. S. A Method for Simultaneous Determination of a Level Constant and the Periodic Errors of the Screw of a Level Trier.
102. Volianskaya, M. U. Tables of Annual Variation of the Star Constants '*a*' and '*a*''.

In *Figura i dvizhenie Luny*. '*Naukova Dumka*', *Kiev* 1965 (*In Russian with English abstracts*).

103. Duma, D. P. On the Determination of Equinox Corrections from Lunar Observations by Newcomb's Method.
104. Duma, D. P. On the Effect of Errors of the Elements of the Moon's Orbit on the Determination of Zero Points of Fundamental Catalogues.
105. Duma, D. P. Determination of the Equinox Correction from Observations of the Crater Moesting A.

106. Kharin, A. S. *A Catalogue of Declinations of Zenith-telescopes' Program Stars in the FK4 System for the Epoch of Observations and Equinox* 1950·0 (In Russian). Published by the Glav. Astr. Obs. Akad. Nauk Ukrainskoy SSR, 1963.
107. Drofa, V. K., Tchernaga, N. A. *Catalogue of Positions of Faint Stars in the PFKSZ System (Zone* + 10° to + 30°). (In Russian). Kiev University Press, 1964.

In *Visnik Kijvsk. Univ., no.* 6, *Ser. Astr.* (*In Ukrainian with Russian abstracts*), *Kiev*, 1964.

108. Drofa, V. K., Tchernega, N. A. Corrections to the Right Ascensions and Declinations of the 242 Stars of PFKSZ. p. 56.
109. Gregul, A. J. Investigations of the Pivots of the Meridian Circle of the Kiev Astronomical Observatory. p. 62.
110. Gregul, J. On the Methods of Reductions of Circumpolar Stars Observations. p. 68.

In *Visnik Kijvsk. Univ., no.* 7, *Ser. Astr.* (*In Ukrainian with Russian abstracts*), *Kiev*, 1966.

111. Tchernega, M. Y. The Determination of the Diameter Corrections of the Meridian Circle of the Kiev Astronomical Observatory. p. 87.
112. Gregul, A. J. Catalogue of the Right Ascensions of 197 Polar Stars. p. 97.
113. Gregul, A. J. The Determination of the Proper Motions in Right Ascensions of the 79 Polar Stars. p. 111.

114. Nefedyeva, A. J. 1966, Differential Catalogue of 2288 KSZ Stars. Kazan. ***Izv. astr.*** *Obs. V. P. Engel'gardta*, no. 33.

In *Uchenye Zapiski. Lat. Gos. Univ. im Petra Stucki*, **68** (*In Russian with English abstracts*), *Riga*, 1964.

115. Cirulis, K. A. A photoelectric Device for Registration of Star Transit.
116. Steins, K. A., Rose, L. A. A photoelectric Device with two Independent Photocascades for Registration of Star Transits.
117. Abele, M. K. The Project of a Photoelectric Zenith Tube.

118. Bykov, M. F. 1961, *Catalogue of the Absolute Right Ascensions of the* 623 *Faint Stars in the FK3 System*. Published by the Acad. of Uzbek. SSR.
119. Kim Gun-der, Borodicki, I. M. 1964, Catalogue of Right Ascensions of 814 KSZ_2 Stars in Zone − 10° to − 15° of Declination. *Cirk. Tashk. astr. Obs.*, no. 325.

In *Trudy Tachk. astr. Obs., Ser. II*, **10**, (*Issledovania po Astrometrii*) (*In Russian*), *Taschkent*, 1964.

120. Jacevich, B. V. Catalogue of the Right Ascensions of 190 Stars of the Time Service List, Observed at the Transit Instrument A П M–10 in Taschkent. p. 7.
121. Rachimov, A. G. Proper Motions of Stars in 23 Selected Areas. p. 31.
122. Borodicky, I. M. Investigation with the Interferometer of the Pivots of the Taschkent Meridian Circle. p. 87.

123. Tursunov, O. 1966, The Investigation of the Temperature Coefficient of Collimation of the Taschkent Meridian Circle. *Cirk. Tashk. astr. Obs.*, no. 335 (in Russian).
124. Zaugolnikova, I. G. 1966, A Differential Catalogue of the Right Ascensions of 331 Bright Stars in Zone − 10° to − 20° of Declination. *Cirk. Tashk. astr. Obs.*, no. 336 (in Russian).
125. Chodak, G. G. 1966, A Differential Catalogue of Right Ascensions of 337 Bright Stars in Zone 0° to + 10° of Declination. *Cirk. Tashk. astr. Obs.*, no. 338 (in Russian).
126. Podobed, V. V. 1966, On the Observations of the Stars in the Latitude Programs with the Meridian Circle and Zenith Telescope. In *Diskyssia Resultatov Shirotnyh nabljudenii*, Taschkent.

Varia.

127. Perezhogina, T. P. 1964, Observations of the Minor Planets by wide-angle Astrograph in the Moscow Observatory. *Bjull. Inst. teor. Astr.*, **9**, no. 9, 617.

128. Kisseleva, T. P., Koroleva, L. S., Sokolova, V. A. 1965, Accurate Photographic Positions of Minor Planets observed at Cape Observatory and measured at Pulkovo. *Bjull. Inst. teor. Astr.*, **10**, no. 1, 76.

129. Bronnikova, N. M., Kiseleva, T. P., Strugatskaya, A. A., Chudivicheva, O. N. 1965, Accurate Photographic Positions of Minor Planets observed at Pulkovo. *Bjull. Inst. teor. Astr.*, **10**, no. 1, 81.

130. Ivakina, T. Y., Kalikhevich, F. F., Fedorova, R. T. 1965, Results of Photographic Observations of Minor Planets at the Nikolayev Observatory. I. *Bjull. Inst. teor. Astr.*, **10**, no. 2, 164.

131. Perezhogina, T. P. 1965, Photographic Observations of the Minor Planets by Wide-angle Astrograph in the Moscow Observatory. *Bjull. Inst. teor. Astr.*, **10**, no. 5, 380.

132. Alieva, N., Ischmuhamedov, H., Primkulov, S. 1966, Accurate Positions of Minor Planets according to Photo Observations with Astrographic Telescope in Taschkent in 1956–61. *Bjull. Inst. teor. Astr.*, **10**, no. 8, 552.

133. Alieva, N. 1966, Accurate Positions of Minor Planets According to Photo Observations with Astrographic Telescope in Taschkent in 1962. *Bjull. Inst. teor. Astr.*, **10**, no. 8, 566.

Books.

134. Zagrebin, D. V. 1966, *Vvedenie v astrometriju* (The Introduction to Astrometry) (In Russian), 'Nauka'.

9. COMMISSION DES INSTRUMENTS ASTRONOMIQUES

PRÉSIDENT: Professor Dr O. A. Mel'nikov, Pulkovo Observatory, Leningrad, U.S.S.R.

VICE-PRÉSIDENT: Professeur J. Rösch, Directeur de l'Observatoire du Pic-du-Midi, 65-Bagnères-de-Bigorre, France.

COMITÉ D'ORGANISATION: W. A. Baum, I. S. Bowen, J. S. Hall, G. E. Kron.

MEMBRES: Aly, Atkinson, Babcock (H. W.), Bakos, Baranne, Barros, Beck, Behr, Blitzstein, Brahde, Brealey, Bruckner, Code, Couder, Cousins, Danjon†, Dewhirst, Dimitroff, Dobronravin, Dollfus, Duchesne, Duham, Evans (J. W.), Fellgett, Ford, Giovanelli, Gottlieb, Hiltner, Høg, Hooghoudt, Ingrao, Jensch, Karandikar, Köhler, Krassovsky, Lallemand, Linfoot, Linnik, Livingston, McGee (J. D.), Maksutov†, Meinel, Merman (N. V.), Mohler, Monin, Nikonov, Petford, Pierce, Platzeck, Ramberg, Richardson (E. H.), Ring, Samaha, Scheffler, Seddon, Shcheglov, Sisson, Smyth, Strand, Sukharev, Väisälä, Valniček, Walker (M. F.), Walraven, Wayman, Wellgate, Wilcock, Wlérick.

INTRODUCTION

The years under report 1964–67 are noted by a rapid development of the theory of optics in general and astrometrical instruments in particular. This period was characterized by the application of a new method to the study of the quality of optical systems on the basis of frequency-contrast characteristics, etc. The method of control and adjustment of optical instruments improved considerably (**1, 2**).

The design and construction of instruments also improved. Optical systems of new large instruments with diameters of 1 to 1·5, 2·5 to 3·0, 3·3 to 3·5, 5 to 6 m are being developed and there is a tendency to design 7·5 to 15 m telescopes (**3**) for the Kitt Peak Observatory!

Some of the above instruments are under construction and will be mounted in the near future. Much attention is being devoted to the selection of suitable sites as regards seeing, meteorological data and the astroclimate in general. Much progress was attained in the development of the theory of star visibility in telescopes, and especially in the problem of limiting exposure and magnitude.

It was found that the focal length is of primary importance in this respect (**4**).

This was the reason for changing the primary focus $f/3{\cdot}7$ (with a correction lens) of 200-inch (508 cm) Mt Palomar telescope to the Cassegrainian focus $f/9$ (with a correction lens) for the field of 15′. Experimental and observational tests showed that this gave a gain of about one magnitude (**4**).

There was also progress in the development of detectors of radiation (for wide range of the spectrum), in particular photographic plates. The new sort of Kodak plates IIIa-J (087), specially baked, which have a sensitivity less by a factor of two than Kodak 103a-O (at identical detecting quantum efficiency), permit longer exposures and a gain in limiting magnitude from $23^m_{\cdot}8$ to $24^m_{\cdot}5$ or even $25^m_{\cdot}0$ (200-inch telescope, $f/3{\cdot}7$) (**5, 6, 7**).

Special attention is given to automatic control of temperature during the development of plates, which is of great importance (**8**).

A new method for the reception of information, the holographic (**9**) imaging method, has found application. The possibilities of auxiliary apparatus have increased greatly: photoelectric (new photomultipliers with multilayer cathodes and new infrared detectors for 8–14μ), spectroscopic (Fourier transformation, SISAM, diffraction grating with a surface of 320 × 280 mm, échelle, etc.), interferometric (wide and narrow band filters), and other techniques. Significant progress has been made in automatic driving and guiding of large telescopes (**10**), possible compensation of the instability of star images, etc. Automatic telescopes and spectral apparatus on balloons, rockets and artificial satellites for outside the atmosphere observations also attained a high technical level.

At present, it is possible to devote serious attention to the development of astronomical instruments and auxiliary apparatus for observations on the Moon. In connection with this it seems desirable to discuss with members of Commission 9 the possibility of organizing a working group, which might be named 'Astronomical Instruments for Observations on the Moon'.

REPORTS FROM OBSERVATORIES OUTSIDE U.S.S.R.

Yerkes Observatory

Professor W. A. Hiltner reported:

1. A 24-inch (61 cm) Cassegrain reflecting telescope has been installed in the northeast dome of the Observatory. This telescope is unique in that it is rotatable on its optical axis for the elimination of systemic errors in polarization measurements. This instrument has been entirely successful for any residual systematic error is less than 0·0001 magnitude.

2. A small laboratory-observatory has recently been completed at Yerkes Observatory. The building consists of two domes with a laboratory in between. An equatorial mounting is available in either dome. At the present time a seven-inch *f*/4 Schmidt telescope is in one dome and large-aperture (6-inch/15 cm) spectrometer with a 10-inch (25 cm) telescope in the other. The Schmidt is used for the identification of faint OB stars with an ultra low-dispersion objective prism, and the spectrometer is used for spectrophotometric observations of comets and nebulae.

3. A 40-inch (102 cm) Cassegrain-coudé reflector is now under construction for the Observatory. It will be installed in the southeast dome, originally occupied by the Ritchey 24-inch (61 cm) reflector. (The Ritchey instrument has been donated to the Smithsonian Institute, Washington, D.C.)

Stockholm's Observatory

Professor Dr Y. Öhman reported:

1. Zeiss in Oberkochen is considering manufacturing our flare photometers if this would prove of value to solar research.

2. S. E. Nilson has tried a double image line shifter (see also (**11**)) for the purpose of determining Doppler velocities of dark and bright surges, etc. (**12**).

3. Kerstin Fredga who has been research associate at Goddard Space Flight Center in 1964, 1965 and 1966 has made in 1965 several recordings of the solar disk in the Mg II line at 2802·7 using a Solc-type birefringent filter of about 4Å (**13**).

4. Y. Öhman has designed a spectrograph for recording from an unstabilized rocket the spectrum of various parts of the solar surface. The instrument uses a great number of small lenses placed on a spherical surface. The focus of all these lenses is situated in the centre of curvature. Here a very small (0·5 mm or so) reflecting sphere is used, functioning as slit for a spectrograph. By using a continuously moving film tracings are obtained when the various lenses project, by chance, the solar image on the small sphere. The instrument is mainly intended for the study of the Mg II lines at 2800 in flares.

5. J. O. Stenflo has started the design of a spectrograph which he would like to use in space research for studying solar magnetic fields in small structure elements of the solar surface.

6. The coronograph of the Stockholm station in Anacapri has been used recently by B. Rydgren (**14**) for photometric measurement of the Earth shine on the Moon. Rydgren has found

a very good agreement with Danjon's previous measurements and has been able to extend the curve between intensity and phase angle.

7. Y. Öhman and B. Rydgren published a brief paper in 1963 on the method of measuring radiation by means of the thermomagnetic effect (**15**).

8. U. Kusoffsky is following up a suggestion made by Y. Öhman in (**16**) for reducing scattered light in sunspot photometry by using two diaphragms, one with a size equal to the Airy disk of the telescope and placed in its focal plane, the second one of greater size and placed in front of the phototube. Observations show that the influence of diffracted light from the circumference of the objective is reduced to a minimum with these simple devices.

9. Finally work has been done by Y. Öhman on shearing interferometry of the diffraction image (**17**) in collaboration with U. Kusoffsky (**18**). This work aims at increasing resolution (and may have particular applications in spectroscopy and in space research) but seems to open up some new ways for studying atmospheric seeing also.

Stockholm's Observatory

L. O. Lodén reported:

A new type of combined starlight photometer and polarimeter is under construction by myself at the Stockholm Observatory. The instrument will secure automatic sky background compensation like the previous one used by the author (see (**19**)), but it has only one multiplier. A special mechanism will switch rapidly between star and sky and the subtraction of the sky deflection from the star deflection is made automatically in the amplifier.

The Lick-Stromlo-U.S. Naval Observatory Electronic Camera

(U.S. Naval Observatory, Flagstaff, Arizona)

Gerald E. Kron and I. I. Papiashvili (personal report)

(*a*) *General.* This electronic camera is identical to the Lallemand camera in principle, but quite different in operation, as the photocathode can be protected by closing a special valve during plate change, thus making possible the use of the same photocathode for many plates. The equipment was moved from the Lick Observatory to the U.S. Naval Observatory at Flagstaff in October 1965, where development and use of the camera is continuing.

(*b*) *Photocathodes and tube background.* Formerly we were able to prepare tubes with low electronic background only by starving the photocathode of caesium during its preparation, thus limiting our photocathode efficiencies to a maximum of 6% at 4200Å. During the last 18 months the camera has been modified by rebuilding the electronic lenses with larger radial clearances and with larger radii on the ends of the electrode terminal surfaces both for the purpose of reducing the electrostatic field strength in the lens gaps. In addition to this, we have modified our cathode preparation technique so as to give us some control over unwanted caesium deposition on tube walls and electrodes. As a result of these changes, we can now prepare tubes with fully caesiated Cs-Sb photocathode whose electronic background, at an operating potential of 30 kV, is determined by the thermal emission of the photocathode. Photocathodes can now be made with efficiencies of from 6 to 12% at 4200Å, and they maintain 90% of their initial sensitivity for from 2 to 12 months. There is a tendency for the more efficient cathodes to have the longer lives.

(*c*) *Photometric possibilities.* Further laboratory studies of the growth of density with exposure for Ilford K.5, and L.4 nuclear track emulsions strengthen the belief held by us and others that there is a very nearly linear relationship between density and exposure up to densities of between 3 and 4. The electronic camera therefore offers the possibility of accurate photoelectric photometry on star and nebula fields, and the equivalent of photoelectric scanning of spectra. The electronic camera has the obvious advantage over the photomultiplier of recording a whole star field or an entire spectrum with one exposure. However, experimental application of our electronic camera to surface photometry of E type galaxies has brought to our attention a new difficulty. The light from the field is, of course, focused on the semi-transparent photocathode, where its work is done. Unfortunately, the light then continues into the interior of the electronic section of the camera, where it suffers a number of reflections from very highly polished steel electrodes; some of the light then returns to the inside surface of the photocathode, where it reacts a second time to produce a vague, faint but troublesome pattern. The intensity and shape of this optical background pattern depend upon the characteristics of the original subject; it can be said that this spurious pattern can reach a density of as high as 0·1 on L.4 emulsion after a two-hour exposure with the 61-inch (155 cm) astrometric telescope of the U.S. Naval Observatory. The presence of this background pattern interferes with accurate photometry, and we are now experimenting with methods for reducing or eliminating the effect. We have built an experimental camera with a cathode support surface that has been ground with 500 mesh carborundom, instead of being polished as usual. The definition of this tube is only slightly inferior to that of a tube with a polished cathode surface, but the incoming light is widely scattered before encountering the reflective electrode surfaces. We also plan to experiment with elimination of the polish on certain critical areas of the inside of the electrodes, with the hope that return of light to the photocathode can be decreased without an increase in electronic background resulting from the lack of polish.

(*d*) *Some performance results.* We have obtained a number of one hour exposures of E type galaxies on L.4 and K.5 emulsion employing the 61-inch (155 cm) *f*/10 astrometric reflector. The camera, which had an 11% cathode, was operated at 30 kV. The plates were developed for six minutes in D-19 developer at 68°F. Equal exposures were made by photography with Eastman 103a–O plates for comparison. Photographs and electrographs were both scanned with a microdensitometer for determining both signal and noise levels. We found that the L.4 electrographs were somewhat more dense than the photographs, whereas the photographs had a noise level about 3·6 times larger than the electrographs. At the sky level of brightness (density 0·1 on the photographs) the information gain of the electronic camera over the 103a–O emulsion was about 40/1. At the brightness level at the centre of the E galaxy (density 2·3 on the photograph, 3·7 on the electrograph) the gain on the light from the galaxy in favour of the electronic camera was about 16/1. We were able to follow the light in the outer portion of the E galaxy down to a level of about magnitude 28 per square second of arc with the electronic camera, as compared to about magnitude 26·5 per square second of arc generally reported for photography. When K.5 emulsion is employed in the electronic camera the noise level and density growth rate both increase, so that the information gain is the same as for L.4 emulsion, but the K.5 quite naturally saturates with less exposure, owing to its fewer, coarser grains.

When one makes a visual comparison between a photograph and a fine-grained electrograph, one notes less contrast in the electrograph, and the almost complete absence of granularity; otherwise the two pictures look much alike. The superiority of the electrograph as a repository of scientific data can be fully appreciated only by comparing scans made with a microdensitometer.

(*e*) *The future.* We plan to finish our experiments on the reduction of optical background, and then build a background-free functional tube, which we plan to use on various difficult photometric problems, employing the 61-inch (155 cm) astrographic reflector. We also plan to continue development of the tube by trying to produce more sensitive cathodes especially of the multi-alkali types. In addition, we hope to improve the definition at the edge of the 34 mm diameter photocathode, now only about 35 l.p./mm as compared with about 80 l.p./mm for the central region of the cathode.

Uppsala Observatory

Professor Dr A. Wallenquist reported:

1. The 100/135/300 cm Schmidt telescope at the Uppsala Observatory Kvistaberg station is in operation since 1963 and several hundred plates of good quality have been secured for various investigations. The spherical mirror is of a low expansion borosilicate glass from Pilkington Brothers Ltd, England. The final polishing was carried out by Professor Y. Väisälä, Turku, Finland. He also figured the 100 cm correcting plate of borosilicate crownglass of type BK 7 from Schott and Gen. Jena.

The telescope is also furnished with an objective prism with a diameter of 80 cm and a refracting angle of 7°, figured by Carl Zeiss Jena from UBK 7-glass. The dispersion on the plate at $H\gamma$ is 262 Å/mm.

2. A detailed description of the instrument will be published in the beginning of 1967, a preliminary description is given in (**20**).

Mt Wilson and Palomar Observatories

Professor I. S. Bowen, Dr H. W. Babcock, J. Boke, E. W. Dennison, B. H. Rule reported:

1. The construction of a 60-inch (152 cm) photometric telescope to be located on Palomar Mountain has been started by the Mt Wilson and Palomar Observatories. The mirrors of fused quartz will be figured as a Ritchey-Chrétien system with a Cassegrain focal ratio of 8·75 and a coudé focal ratio of 30. A fork mounting and a conical steel telescope tube will be used. A small primary focal length of 3·8 meters permits the use of a very short tube and allows the Cassegrain focus to be located only 92 centimeters from the intersection of the declination and polar axes. These design features will permit very rapid shifts from one object to another. The telescope will be equipped for wide-field photography at the Cassegrain focus and with the latest multichannel photoelectric system for pulse counting. A coudé spectrograph will also be provided.

2. During the last three years efforts have been made to provide much more versatile and powerful data-acquisition system for use with various photoelectric and photo-conductive detectors. These systems have been designed and built under the direction of Dennison and are now installed on the 200-inch (508 cm), 100-inch (254 cm), and 60-inch (152 cm) telescopes. From λ 3200 to λ 11 000, the impressive success of pulse-counting technique indicates that almost all future photoelectric measurements are likely to be made in this way. All systems are being built with either reversible counters or dual counters. By using two apertures, one for sky, the other for star plus sky, and by switching rapidly between them, it is possible by commuting the counters to obtain the difference signal only. The continuous compensation for the sky leads to much higher stability and accuracy than was previously possible.

Although pulse-counting techniques have proven very successful, one limitation has persisted for work on brighter standard stars: the pulse amplifiers and discriminators have been limited to count rate of at most 10^5 counts per second. Recent experiments with commercial and other amplifiers have shown that count rates of 4×10^6 counts per second can now be readily achieved. This gain in counting speed now makes existing standard stars much easier to measure with large telescopes.

3. A large Cassegrain observing facility has been mounted below the 200-inch (508 cm) mirror cell. The system includes a universal chair for the observer, precision off-set guiders and direct-viewing optics. A rigid mounting base has been incorporated which will readily accept large and heavy new instruments.

4. An $f/9$ conversion lens system has been built and is in use at the new Cassegrain station.

A new Cassegrain nebular image-tube spectrograph employing a short focal length concentric mirror system has been designed by Bowen. Both the input camera optics for the image-tube and the re-imaging camera optics are provided with thermal-expansion compensation. A wide variety of image-tubes can be used. The instrument will be completed by the end of 1966.

5. A 33-channel photoelectric spectrometer has been designed by Oke, Rule, and Dennison; construction is well under way. The instrument will cover the spectral range from λ 3200 to λ 11 500 with band passes of 2 Å or larger. The instrument will employ only pulse-counting and will feed into a versatile and automatic data system. Spectrophotometry will be possible down to 22nd magnitude.

6. The 100-inch (254 cm) coudé scanner has been completely modernized to use pulse-counting techniques and complete digital controls. It can be used either in the conventional fashion or with Fabry-Pérot interferometers. A two-channel scanner is being built for the 200-inch coudé spectrograph. When used with a Fabry-Pérot étalon a resolving power of 150 000 should be achieved.

7. A new equatorial solar telescope designed primarily by Rule is under construction. When finished the telescope will include a 15-inch (38 cm) off-axis reflection system for spectroscopy, a 9-inch (23 cm) reflecting coronagraph, and two 10-inch (25 cm) refracting systems for cinematography.

Sacramento Peak Observatory

Dr J. W. Evans reported:

1. Construction has been started on a solar tower telescope designed by R. B. Dunn with an all reflecting or quartz optical system of 75 cm aperture and 64 m focal length. All reflecting surfaces are enclosed in an evacuated tube with quartz windows to keep the surfaces clean for minimizing scattered light, and to avoid image degradation due to convection inside the instrument. The contract completion date is September 1968.

2. A punched card digital recorder has been added to our microphotometer, the recording Doppler comparator, and the photoelectric scanning spectrograph. These recorders have been very useful, but are now being replaced by digital magnetic tape recorders because of their greater convenience.

3. J. Beckers has prepared a computer programme by means of which the state of polarization of light as a function of wavelength can be traced through any system of partial (or complete) polarizers and birefringent elements. This programme has been useful in the design of special birefringent filters, and is available to anyone wishing to use it.

4. J. Beckers has built a 22 cm aperture photoelectric coronameter for scanning the emission lines in polar coordinates. The instrument has a threshold considerably less than 10^{-6} Å equivalent width through a sky of brightness 10^{-4} that of the Sun.

5. R. B. Dunn designed an eclipse unit consisting of a 25 cm reflecting telescope and $f/5$ grating spectrograph with an average dispersion of 35 Å/mm. Its purpose is to obtain coronal spectrograms showing the continuum as well as emission lines. Two samples were built and successfully used at the 1965 eclipse. One of these was operated by Dunn from the NASA aircraft. It was guided to about 10″ accuracy by a gyro actuated servo system designed by Dunn. Both instruments are to be used again at the November eclipse in South America.

6. H. Mauter built a spectroheliographic attachment for our 13 m spectrograph to photograph a small area of the solar surface in Fraunhofer lines of small width. Mauter has successfully used bandwidths down to 0·03 Å.

Washburn Observatory at the University of Wisconsin

Dr. A. D. Code reported:

1. *Scanning Spectrometer.* A new scanning spectrometer has been placed into operation at the Pine Bluff Observatory. This instrument is of the type described by Schröder (1965), having only two reflections, one from a concave mirror, the other from a plane grating. This particular instrument operates at $f/12$ with a reciprocal dispersion at the exit slit of approximately 70 Å/mm in the first order. The spectrum is scanned rapidly by rotating the plane grating by means of a stepping motor, in four angströms steps. The output from a photomultiplier mounted behind the exit slit is amplified and stored in a core memory multichannel digital analyser. Each four-angström interval is stored in a separate location in the core memory and the output of successive scans are added to the accumulative spectrum. A single scan of 4000 angströms is executed in about 0·5 seconds.

In operation, three distinct advantages have been found over the method of single slow scans. It was found that it was possible to obtain spectra of fainter objects for a given accuracy in the same total exposure time. More impressive, however, was the ability to record high quality data through moderate cloud cover, and essentially doubling the number of useful observing nights. Finally, it was noted that the rapid comparison of different spectral regions decreased the scintillation noise on bright objects, resulting in higher precision than that obtained with single scans. The technique is also applicable to high resolution line profile measurements without requiring seeing compensation. In short, the system has the integrating properties of the photographic plate and the inherent accuracy of the photoelectric technique. Any night on which photographic spectroscopy can be done is suitable for photoelectric spectrophotometry by this technique.

2. *Automatic telescope.* An automatic telescope has been installed at the Pine Bluff Observatory. This instrument automatically sets on programmed stars and makes multicolour photoelectric measurements. The instrument is used to continually observe a number of standard stars throughout the night, to provide data on atmospheric extinction, and therefore reduce the amount of time required for this purpose by the larger instruments. The automatic telescope is an 8-inch (20 cm) reflector under the control of a PDP–8 general purpose computer. The stars to be observed are read in on punched paper tape. The telescope sets on the coordinates, and then centres the star. If approximate colours and magnitudes are inputed to the computor, the telescope is capable of identifying and centring on the particular star in the neighbourhood of brighter objects. When the object is acquired, a set of measurements is obtained with three filters, a Čerenkov standard source, and a dark measurement. This output is recorded both digitally and in analogue fashion; and the data is processed by the control computer.

3. *Wide field high resolution spectrometer.* A crossed échelle spectrometer has been constructed and tested in the laboratory by Schröder. This instrument was designed to give a very high ratio angular field to spectral purity. By operating in very high orders, correspondingly large areas of the sky may be observed for a given spectral resolution. The instrument may be used as a photoelectric scanner or operated photographically. It is ideally suited for use with image tubes, since the crossed dispersion puts a large spectral region within the physical dimensions of the photocathode.

4. *Standard radiation sources.* We have recently completed an extensive calibration of tungsten ribbon filament lamps for use as stellar radiation sources. The absolute spectral distribution of these lamps was directly compared with a fundamental platinum black body. The internal accuracy of these measurements is less than the fundamental uncertainties that exist in the determination of the radiation constant in Planck's law. The agreement between our calibrations and those of the National Bureau of Standards are excellent. The reason for this note, however, is to point out that we found the pyrometric technique for determining the

spectral distribution of tungsten ribbon lamps to be unreliable. Apparently the emissivity of tungsten varies among filaments of different manufacture. A description of this work by R. C. Bless, A. D. Code and D. J. Schröder is now in press.

Observatoire de Paris-Meudon—Instruments solaires

1. G. Wlérick signale que le laboratoire solaire horizontal de l'Observatoire de Meudon, décrit dans le rapport précédent, est entré en service en Février 1965. Il est composé de quatre parties: coelostat, télescope, monochromateur, spectrographe. Un soin particulier a été apporté pour supprimer l'astigmatisme. Les barillets des miroirs du coelostat comportent les dispositifs antiflexion. Dans chacune des autres parties, l'astigmatisme est corrigé, grâce à la déformation d'une pièce optique, suivant le dispositif imaginé par A. Couder. L'ensemble télescope-monochromateur-spectrographe est isotherme pour minimiser les turbulences instrumentales. La qualité optique de l'instrument est révélée par le fait qu'il a permis d'obtenir en plusieurs occasions de très bons spectres montrant la structure fine des raies photosphériques.

2. G. Wlérick signale aussi que le bâtiment de la Tour Solaire est achevé. Sa hauteur est de 34 mètres et il est constitué par deux chemises entièrement séparées. Le télescope vertical, qui doit entrer en service en 1967, a une pupille de 60 cm de diamètre et une longueur focale de 45 mètres. L'objectif est un miroir sphérique qui travaille en-dehors de l'axe; le faisceau est replié grâce à un miroir plan, légèrement déformé pour corriger l'astigmatisme de l'objectif. Le télescope est fermé à ses deux extrémités par des glaces pour réduire la turbulence. Le spectrographe associé doit entrer en service en 1968.

3. P. Charvin a construit une nouvelle version du coronomètre Lyot-Charvin décrit dans le rapport précédent. L'instrument est muni d'un objectif de qualité coronographique; il travaille 'dans l'axe' et mesure l'intensité et la polarisation de la raie coronale λ 5303 Å. Sa mise en service, à l'Observatoire de Nice, est prévue en 1967.

4. R. Michard fournit les renseignements suivants:

(*a*) Nouveaux héliographes monochromatiques. L'héliographe Hα de Lyot, utilisé à Meudon depuis 1954 à la surveillance cinématographique des éruptions solaires, a été remplacé en 1964 par un instrument plus moderne, permettant d'obtenir simultanément une image correctement posée du disque solaire et des protubérances.

D'autre part, un 'héliographe à longueur d'onde variable' a été mis en service en 1965; il permet d'enregistrer en succession rapide trois images de la chromosphère en Hα, Hα + 0·7 Å, et Hα − 0·7 Å à l'échelle de 42 mm pour le Soleil entier et avec une bande passante de 0·7 Å. Cet appareil s'est révélé très utile pour l'étude des phénomènes à grande vitesse radiale normalement invisibles au centre de la raie Hα.

(*b*) Les laboratoires de spectroscopie solaire de Meudon (spectrohéliographe de 7 mètres) et du Pic du Midi ont été améliorés; remplacement de tous les anciens miroirs en verre par des miroirs de silice, guidages photoélectriques, automatisation des observations, nouveau coelostat pour le Pic du Midi.

(*c*) Pour mesurer les petits déplacements locaux des raies sur un spectrogramme (effet Doppler ou effet Zeeman en lumière polarisée), on a construit deux 'lambdamètres' photo-électroniques qui ont trouvé de nombreuses applications en physique solaire. Ils permettent aussi de mesurer les champs magnétiques stellaires pour des étoiles à raies larges.

Observatoire Royal de Belgique

Professor E. Vandekerkhove reported on the following items:

1. Un déviateur (to be published in the November-December issue of the Revue *Ciel et Terre*).
2. Etude du grand prisme de l'Observatoire d'Uccle (just finished for publication).
3. Le grand télescope de l'Observatoire d'Uccle (in preparation).

U.S. Naval Observatory

Professor K.Aa. Strand reported:

1. An extensive programme in the development of new transit circle instrumentation is in progress at the U.S. Naval Observatory under the supervision of A. N. Adams and B. L. Klock. An electronic circle and automatic star and planet tracking micrometer are under construction. This instrumentation should have a repeatability of 0·05 seconds of arc. The photoelectric micrometer will be able to observe, in both coordinates, night and day objects, including the Sun. The new system will be connected to an electronic control and data acquisition for initial pointing of the telescope and for storage of the observational data. The equipment is to be initially installed and tested on the 6-inch (15 cm) Transit Circle.

2. Designs for an automatic transit circle are in progress. The new instrument will incorporate the electronic circle and photoelectric tracker mentioned above. Special attention is being given to reducing the thermal and flexural errors known to exist in the present transit circles. The new transit circle will be installed at the U.S. Naval Observatory, Flagstaff Station, in Arizona.

3. A two-coordinate measuring machine, capable of high speed precise measurements on photographic plates over a 10 × 10-inch area, has been acquired by the U.S. Naval Observatory. Based upon design specifications by K. Aa. Strand, the machine incorporates the following new features: granite structural members, granite-air bearing stages, granite guideways, automatic acquisitioning and centring, and moiré fringe position recorders. Output data are obtained on punched cards, and repeatability of centring is within one micron. So far, this machine has been used principally in the parallax and photographic double star programmes.

Royal Greenwich Observatory, Herstmonceux

Professor R. Woolley reported:

Construction of the Isaac Newton telescope is complete. The fully-assembled instrument has passed its works tests satisfactorily, and erection on the site at Herstmonceux should be completed in 1967, April. The building housing should be finished later in the summer. The optical train employed is standard: the 98-inch (250 cm) mirror is an $f/3{\cdot}0$ paraboloid and observing will be possible at the prime focus (from a removable cage carrying the observer), at a Cassegrain focus ($f/14{\cdot}2$) and at a coudé focus ($f/32$). The main innovations are in the mechanical design. The polar axis consists of a squat cylinder fabricated from mild steel plate (the 'polar disk') 6·7 m in diameter and 0·9 m deep, rotating on pressurized oil pads. The telescope tube is hung between stout tines springing from this disk. The mirror is supported axially in its cell by an air-bag, 6 mm thick, inflated to a pressure varying automatically with the elevation of the telescope tube. In use, the control system ensures that the axial component of the weight of the mirror (4000 kg) is supported to ± 0·5 kg by the air-bag.

Cassegrain and coudé spectrographs are being designed and built in the Engineering Department of the Observatory. Work on the first is well advanced and it should be ready when the telescope itself is commissioned. Its optical axis will be at 90° to that of the telescope and two Newtonian collimators, both working at $f/14{\cdot}2$, are being provided. One of 11 cm aperture will work with either of two field-flattened Schmidt cameras of equivalent focal length 14 and 33 cm; the other of 6·5 cm aperture will work with a semi-solid Schmidt camera of e.f.l. 4·4 cm. Several gratings will be available, giving a range of linear dispersions from 12Å/mm to 36Å/mm in the blue and 25 to 60Å/mm in the red.

2. The 30-inch (76 cm) Thompson reflector has been modified to illuminate a vertical coudé spectrograph by the addition of two flat mirrors, one moving on the declination axis and one fixed on a tripod support to the north of the polar axis. The focal ratio of the telescope, with this arrangement, is $f/47$. The collimator of the spectrograph is off axis with aperture of 12 cm and focal length 570 cm. The grating of 600 lines/mm carries the twice-through corrector plate

of the off-axis Schmidt camera of 81 cm focal length. The 18 cm × 2·5 cm plates are bent to the required radius, and 1500 Å of spectrum at 10 Å/mm in the 2nd order (blue) are recorded. Observations of IAU standard velocity stars show an s.e. per plate of ± 0·3 km s^{-1} for this instrument.

3. Experimental work in co-operation with Professor McGee's department at Imperial College, London, has been carried out on this telescope with a Lenard-window image intensifier. Stellar spectra were obtained on Ilford XM emulsion with a resolution of 40 l.p. per mm and exposure gain of x18, compared with baked IIa-O photographic emulsion exposed at the same focus. G5 emulsion with an improved resolution of 80 l.p. per mm gave a gain of x6. The image quality was reduced for long exposures by the changing component of the ambient magnetic field due to the movement of dome and telescope.

4. A 5 cm lens grating spectrograph of low dispersion, for use with a magnetically shielded 'Spectracon' image tube, is being installed on the 30-inch reflector. An alternative lens camera using the full collimator aperture of 13 cm will provide further focus experiments with image intensifier tubes at higher dispersions of up to 3·5 Å/mm in the blue.

5. A linear measuring engine fitted with digitization and printed read-out of position has been modified to give oscillograph display of the line profiles on stellar spectrograms. Two separate photocells are used for star and comparison display.

Kitt Peak National Observatory

Dr L. Crawford reported:

1. Now in operation at the Kitt Peak National Observatory are two 16-inch (41 cm) reflectors. One, called No. 3, is a Cassegrain *f*/7·6 equipped with a standard, single channel photoelectric photometer. The telescope is housed in the 16-foot (488 cm) diameter rotating dome, adjacent to the 84-inch (213 cm) telescope building. The other 16-inch (41 cm) telescope, called No. 4, is an *f*/18 Cassegrain, equipped with a standard, single channel photoelectric photometer. This instrument is equipped with a digital read-out system, which gives not only the intensity data on a Brown strip recorder chart, but also gives the star number, sidereal time, hour angle and declination, comments, and the output of the integrators with their gain, on punch paper tape suitable for reduction on the digital computer in Tucson. This 16-inch telescope is located in a 16-foot (488 cm) diameter rotating dome midway between the two 36-inch (91 cm) telescopes.

2. The No. 1 and 2 16-inch telescopes are both *f*/13·5 and were used during the site survey and first years at Kitt Peak. They are now located at the Cerro Tololo Inter-American Observatory in Chile. They are equipped with single channel photometers identical to those on the No. 4 16-inch telescope at Kitt Peak. They do not have provision for punch paper tape output, but the electronics consists of a precision timer and programmer for single channel or double channel work, to control one or two integrators in parallel. The circuits are digital and solid state Integrators or DC amplifiers are used.

3. The No. 1 36-inch (91 cm) telescope has been in operation since 1960 and is equipped with a single channel photoelectric photometer and with a digital read-out system identical to that on the No. 4 16-inch (41 cm) telescope. The telescope tube is equipped with a flip top that gives the choice of a *f*/16·6 Cassegrain or an *f*/7·6 Cassegrain. A Cassegrain spectrograph is available for the *f*/13·6. A direct camera with guider is now under design and construction and will be available for the *f*/7·6 focus. Instruments used on the 84-inch (213 cm) can be used at this *f*/7·6 focus. The No. 2 36-inch telescope is about ready to go into limited operation. It is also equipped with a flip top, and gives the choice of a *f*/13·6 Cassegrain or *f*/36 coudé focus. No instruments are presently planned for the coudé. A single and double channel photometer

will be available for the Cassegrain focus, and the telescope will be equipped with digital read-out similar to that of the No. 4 16-inch telescope.

4. The 84-inch telescope is now in full operation and was described in the previous Commission report. Additional instruments now available, not described in that report, are a single channel Cassegrain photometer, a two-channel (star-sky) photometer, and a Cassegrain spectrograph.

5. Reduction programmes are available for computer reduction of the photometry done at Kitt Peak. Programmes are available for Hα, Strömgren four-colour, and *UBV* type photometry. Input can be from punched paper tape produced at the telescope, punched cards, or punched paper tape produced in Tucson after the Brown recorder tracings have been read. Output is in any of several formats, depending on the wishes of the observer.

6. The design work for the 150-inch (381 cm) telescope is now well under way. Preliminary design was described in (**22**). The telescope will incorporate a 150-inch diameter fused quartz mirror, which will give *f*/2·8 at the prime focus through any of several corrector systems for wide photography. A *f*/14 Cassegrain focus will be available, with the observer riding in the telescope in an observing cage, and an *f*/30 coudé will be available. It uses a permanent five-mirror system with the beam in a horizontal plane. A coudé optical laboratory will be available, in addition to the spectrograph. It should be possible to change any of the foci to another in about ten minutes. The *f*/9 Cassegrain will be a Ritchey-Chrétien type system for wide field limit photography. Most photoelectric photometry and spectroscopy will be done at this focus. The telescope will be located in a dome approximately 100 feet (30 m) in diameter, and with the declination axis approximately 125 feet (38 m) above the ground. The detailed design of the telescope mounting is now under way by Westinghouse Electric Corp., Sunnyvale, California, and should be finished about March 1967. Detailed design of the rotating dome and building are under way by Skidmore, Owings and Merrill, Chicago, Illinois, and should also be finished about March 1967. The construction of both the mounting and the dome and building should therefore start by the summer or fall of 1967. The fused quartz mirror is being fabricated by General Electric Corp. in Cleveland and delivery is expected in March 1967. The optical work on the large primary mirror will be done in the Tucson headquarters. The optical shop facility is now finished and awaiting delivery of the mirror. The grinding machine for the large mirror will be fabricated by L. and F. Machine Co., Huntington Park, California.

Steward Observatory, the University of Arizona, Tucson

Dr D. J. Taylor reported:

1. A 90-inch (229 cm) telescope is being constructed for the Steward Observatory by Boller and Chivens Company. It will have an *f*/9 Ritchey-Chrétien focus and an *f*/30 coudé focus. The fork mounting will be 65 feet (20 m) above ground in a simplified dome consisting of cylindrical sections. The drive is a digitally controlled torque motor. The telescope is scheduled to be in operation at the Kitt Peak Station in 1968.

2. Development of auxiliary instrumentation for the 90-inch telescope is already under way. A 3½-inch (8·9 cm) beam Cassegrain spectrograph with folded Schmidt and image tube cameras is currently being constructed. A horizontal 8-inch (20 cm) beam coudé spectrograph with 12, 30, 75, and 188-inch (30, 76, 191, 240 cm) focal length cameras has been designed. The three shortest cameras are arranged in a vertical stack and are changed by means of an elevator. Image tubes are being tested and evaluated, and an automatic-tracking offset guider utilizing an image dissector tube being developed.

3. Two photoelectric spectrum scanners have recently been completed for use with Steward Observatory's existing 36-inch (91 cm) telescope. One is located at the coudé focus and is designed for high resolution line profile spectrophotometry. The other may be used either at

the Cassegrain or Newtonian focus and is intended for low resolution energy distribution measurements.

Lunar and Planetary Laboratory

Dr Tom Gehrels reported:

The new 154-cm reflector of the Lunar and Planetary Laboratory is being used at $f/13$ and $f/45$ for polarimetry when it is not in use for the primary programme of planetary photography. The mirrors are nearly free from instrumental polarization thanks to careful aluminization. The internal probable error of the polarization measurements is $\pm$ 0·0006 magnitude for objects brighter than 7th magnitude and observations lasting about half an hour per filter. The filters are at 0·33, 0·36, 0·43, 0·52, 0·64, 0·83 and 0·95μ. A 71-cm reflector is being used with high-altitude balloons for polarimetry at 0·22 and 0·28μ.

Royal Observatory, Edinburgh

Professor H. A. Brück reported:

1. The twin 16-inch (41 cm) programmatic telescope of the Royal Observatory, Edinburgh, has now been commissioned and the effectiveness of the system has been demonstrated by a series of measurements in Cygnus by Dr V. C. Reddish, who obtained measures on different nights under poor atmospheric conditions.

2. Improvements are being made in the light of operational experience such as an improvement of the accuracy of the digital position measurements and of the form of pulse counting photometer.

3. A second Becker iris photometer, modified for digital punch-out, and a servo-controlled iris is being commissioned and will be the first machine in Edinburgh to be 'on-line' to the 4100 Elliott computer now being installed. The photometer will then be fully automatic after initial measurements have been taken.

4. The prime purpose of the 4100 computer installation is for the full 'on-line' automation of measuring machines and telescopes. It includes a 16 384 word magnetic core store, three magnetic tape units, paper tape input/output equipment, and a control typewriter.

5. Priority is being given to the control of measuring machines by computer as the number of Schmidt plates to be measured is likely to increase when the new Outstation of the Royal Observatory at Monte Porzio, Italy, becomes operational in 1967. This station will contain a 16/24-inch (41/61 cm) Schmidt telescope with a focal length of 60-inches (152 cm) made by Messrs. Grubb Parsons.

6. The first phase of automation of the Edinburgh 16/24-inch Schmidt telescope is complete. This provides for analogue control of right ascension, declination and shutter operation, while design studies have been initiated for autoguiding and the automatic loading of plates. These are essential first steps to fully automatic 'on-line' control.

7. There has been some delay in the construction of the automatic luminosity and XY co-ordinates measuring machine 'Galaxy' as previously reported. The major mechanical assembly, weighing some 3·6 tons is complete and the electronics are in an advanced state of construction. Delivery of the machine to the Observatory is now scheduled for May, 1967, when it will be installed in an environmentally controlled room which is now being built.

La Plata Astronomical Observatory, Argentina

Dr J. Sahade reported:

In regard to the La Plata Observatory 85-inch (216 cm) reflecting telescope project, the construction of the mounting and the control consoles has been constructed by Boller and Chivens, a division of the Perkin-Elmer Corporation, in the United States. It is planned that the project become sponsored by all the Argentine National Universities.

Torun Astronomical Observatory of N. Copernicus University

Professor W. Iwanowska reported:

A photoelectric photometer of standard type has been built for the use with the 90 cm telescope as well as with some smaller instruments.

Astronomical Institute of the Czechoslovak Academy of Sciences

Dr. B. Sternberk reported:

A two-meter reflector made by VEB Carl Zeiss Jena, equipped with the primary, Cassegrain and coudé foci is under construction at the Ondřejov Observatory, Czechoslovakia. It will be formally inaugurated at the XIIIth General Assembly of the IAU.

Tokyo Astronomical Observatory

Dr H. Hirose reported:

In 1964, a Schmidt camera of the ordinary type was completed and erected at the Dodaira Station of the Tokyo Astronomical Observatory. Principal features of the camera are as follows:

Aperture of the correcting plate	50 cm
Diameter of the main mirror	65 cm
Focal length	100 cm
Diameter of the field	10·5 cm (+ 5·3)

Flat plate can be used.

Manufacturer: Nippon Kögaki K.K. (Japan Optical Co.), Tokyo, Japan.

The International Latitude Observatory of Mizusawa, Japan

Dr H. Hirose reported:

The Observatory has completed an automatic measuring machine for the photographic plates obtained with the floating zenith telescope in order to make measurements free from personal errors. The aimed precision is $\pm 1\mu$ and this precision is realized by combination of two photomultipliers with a decatron counter (**23**).

Besides, following improvements have been applied to our instruments, respectively:

(1) The Talcott level of the visual zenith telescope (V.Z.T.) has been replaced by the level of electromagnetic type in order to set the whole telescope in fine equilibrium position with the use of servo-mechanism. The micrometer reading of the V.Z.T. is counted electrically with the use of a digitizer.

(2) Another new unit produced is for Danjon's astrolabe. This unit counts interval of the time-contacts electrically and is connected with the input-unit of an electronic computer (**24**).

The Hydrographic Office of Marine Safety Agency, Japan

Dr H. Hirose reported:

A new timing device for the observation of artificial satellites has been developed by Mr F. Ono (**25**) as given below:

A satellite trail is chopped by a narrow slit travelling in front of the photographic plate of the satellite tracking camera, while the serial position marks of the very slit are photographed on the same photographic plate by the aid of a multiple discharging flash lamp. With this device, we can obtain a dot-like image of a satellite corresponding to the time when the position of satellite's image on the photographic plate coincides with that of the moving slit. The actual apparatus has three slits and observations are made with an equatorially mounted camera having this device. The time of the flash of the discharge lamp are recorded with an oscillograph with the standard time signal. Accuracy of timing is expected to be 0·1 ms.

Mr F. Ono (**26**) developed a measuring device for personal equation by applying a principle that terminal voltage of a condenser of C–R integrated circuit is proportional to duration of charged time. By marking a time-scale on a meter of moving-coil-type, small time interval can be directly measured. This device has been employed regularly to determine personal equation for visual observation of occultations at the Hydrographic Office of Japan.

The Geographical Survey Institute, Japan

Dr H. Hirose reported:

A new satellite tracking camera has been completed in the Geodetic Observatory at Kanozan since 1963. This camera is equipped with a photoelectric timing device designed by Professor Tsubokawa and mounted equatorially.

It takes the photograph of the trail of a satellite and background stars simultaneously. Several knives are placed parallel with each other in front of the photographic plate with their axes oriented almost perpendicular to the direction of the satellite motion, and thus make sharp breaks of the trail of the satellite.

The photoelectric timing device is consisted of the knives and two photomultiplier tubes. A knife, 0·2 mm in width, has two optically flat surfaces, the intersection of which makes the knife-edge. The image of the satellite projected on the surface of one side is reflected and the total light comes in one photo-tube. Then, according to the movement of the satellite, the image moves to the other side, and the light comes in the other phototube. The effective sensitivity of the two tubes is made equal, and the difference of the outputs of them is amplified. Then the instant of the change of the current direction corresponds to the transit time of the satellite over the knife-edge. An oscillograph (visigraph) is used for recording the output current (**27**).

Thus, the time and the corresponding position of the satellite referred to the background stars are obtained.

Milano-Merate Observatory, Italy

Dr J. H. Focas, Secretary of Commission 16, reported:

Etude et Construction d'un nouveau type de miroir métallique pour l'Observatoire de Merate (Milan, Italie) d'un diamètre de 75 centimètres, ouvert à 1 : 3. Travail de construction d'un autre miroir du même type, d'un diamètre de 137 centimètres, ouvert à 1 : 4·5 pour le même observatoire. Construction d'une monture équatoriale d'un type nouveau pour ce miroir.

Astrophysical Institute of the Liège University, Belgium

Dr M. Migeotte reported:

A 76 cm telescope has been installed in 1966 at the Sphinx Observatory, a close dependance of the Jungfraujoch International Scientific Station, situated at an altitude of 3580 m in the Swiss Alps.

This instrument has been built by Grubb-Parsons, Newcastle-upon-Tyne (England), to be used at the Cassegrain and coudé focus, the relative apertures being respectively *f*/15 and *f*/40.

Acquired through Belgian and Swiss funds, the installation of this new instrument is the result of a fruitful collaboration between the Institut d'Astrophysique of the University of Liège (Belgium) and the Observatory of Geneva (Switzerland).

Boyden Observatory (internationally sponsored), Bloemfontein in South Africa

Dr J. Dommanget (Observatoire Royal de Belgique) reported:

The 32/36-inch (81/91 cm), *f*/3·75 telescope (system Baker-Schmidt) jointly established by the Armagh, Dunsink and Harvard Observatories (ADH telescope). The telescope has been installed with an objective prism and tested (**28, 29**) for a wide angle and flat field.

Recent programmes include surveys of star clusters and Hα emission lines objects in the Magellanic Clouds, searches for faint variables in the wing of the Small Magellanic Cloud, photographic photometry of galactic clusters, studies for the spectral classification of stars, etc. The limiting magnitude, for Kodak 103a–O (blue) plates, is 20 m for an exposure of 30 minutes. Magnitude correction for distance from the centre of the field of 2·3 degrees is only + 0·34 (the edge of a circular plate of 10·5-inch diameter). With the special aperture shopper the vignetting is reducing these errors to zero. The plates 16 × 16 cm are used (field 2·9 × 2·9 degrees); the field error at full aperture is only 0·1 magnitude (at the edge of a plate). A. D. Andrews (Armagh Obs.) and J. Dommanget give (**29**) critical and practical remarks on the possible and existing methods of optical alignment and adjustment of the telescope in general and the interesting ADH telescope in particular.

Dominion Astrophysical Observatory, Victoria

Professor Dr K. O. Wright reported:

1. At the 72-inch (183 cm) telescope a new all-mirror Cassegrain spectrograph is being tested which will replace the two prism arrangement.

2. At the 48-inch (122 cm) coudé spectrograph a large mosaic four-grating composite is being prepared which will have a 17-inch (43 cm) collimator which will greatly increase the efficiency of that spectrograph.

3. The most interesting development may be a 'Superpositioning Image Slicer' which is being prepared by Dr E. H. Richardson, who describes it as follows:

'Rather than stacking slices of the stellar image along the length of the slit, this device produces elongated slices and superimposes them on the slit. This eliminates between-slice interruptions in the illumination of the slit, making it practical to use the device without the necessity of placing a cylindrical lens near the focus of the spectrograph camera.

4. 'A multiple-reflection mirror system, fabricated from four identical mirrors, is used to perform the slicing operation. Thus, it is not necessary to aid one or more optical elements to produce each additional slide.

5. 'The device is being tested at the coudé spectrograph of the 48-inch telescope with the *f*/5 spectrograph camera. For stellar spectrograms 0·4 mm in width (i.e. spectral line length) the gain in speed is equivalent to that which would be achieved by replacing the grating with one having from four to nine times the ruled area: that is, exposure times are reduced by a factor of two or three.'

6. Additionally, Dr G. I. Odgers reported: 'In September 1964 Dr R. M. Petrie and I began design studies for a 150-inch new Queen Elizabeth II reflecting telescope. Last April 1966 this has become G. I. Odgers responsibility (the well known astronomer Dr R. M. Petrie, the famous specialist in different sections of astrophysics and astronomical instruments, passed away on 8 April 1966) in consultation with Professor I. S. Bowen and Mr. Rule who have provided very useful advice. The design study is given in form of a report by Dr G. I. Odgers (the report covers the design studies for the period up to 30 January 1966). The primary quartz fused silica mirror *f*/2.8 (due to arrive from Corning Glass Co.) has been built into a stubby stiff structure which will permit good observing at both the prime focus and Cassegrain. The optical system of the telescope for the prime focus is a Ritchey-Chrétien. The parameters

of the optical path are the following: prime focus $f/2{\cdot}8$, Cassegrain $f/8$ and $f/15$, coudé $f/30$. An horizontal coudé room of 60 ft between slit and collimation will contain a five mirror system. The automation and safety systems have been incorporated; all the preliminary engineering studies for this telescope has been finished. However the final design will be only in April 1967.

A design team composed of three men from Dilworth, Second, Meager and Associates Ltd., Toronto, one from A. B. Sanderson and Company Ltd., B.C., and the astronomers of the Dominion Astrophysical Observatory, was established in Victoria, British Columbia, at the end of January 1966. The telescope will be located at Mt Kobou (latitude 49°51, altitude 1905 m).'

National Aeronautics and Space Administration, Washington, U.S.A.

N. G. Roman reported:

1. The Manned Spacecraft Center is setting up solar and fixed frequency radio telescopes at the Apollo tracking and communication sites to provide 24-hour monitoring of solar activity. The data will be transmitted to the Apollo control centre to provide warning of solar proton events.

2. Goddard Space Flight Center. GSFC has ordered a standard 36-inch (91 cm) telescope from Boller and Chivens. Modification for radio astronomy of the 85-ft (26 m) antenna at the Rosman Data Acquisition facility has been completed. A programme of lunar occultation observations has been initiated.

3. Jet Propulsion Laboratory. JPL has ordered a 24-inch (61 cm) reflector with both Cassegrain and coudé foci for Table Mountain Observatory. Improvements in the dome and optics of the 16-inch (41 cm) telescope are underway. A millimeter radio astronomy antenna, 18 ft (5·4 m) in diameter and capable of operating at 4 mm is also being added; one of the radiometers to be used with the facility is an 8 mm, 300°K parametric amplifier. Observations of the Moon and nearer planets are planned.

4. Other work supported by NASA:

P. Baumeister (University of Rochester) has produced, essentially at will, narrow-band multilayer ultraviolet filters suitable for flight experiments to meet particular design specifications.

E. Eberhardt (ITT, Ft Wayne) is investigating the pulse counting capabilities of their tubes to answer such questions as: 'What is the source of dark pulses?' 'What role does ageing play?' 'Is a based tube superior to an unbased tube?'

5. Ames Research Center. G. Augason has constructed a Michelson-type infrared interferometer operated in liquid helium for wavelength between 10 and 40μ. Performance tests are being conducted. G. Augason and F. West have designed and are now having constructed a filter photometer for the near infrared. A. Margossi and F. West have conducted a series of tests on the performance of an infrared interferometer (Michelson-type) spectrometer capable of 5 cm resolution in the region out about 3μ.

Development of Astronomical Optics and Instruments in France

1. Dr F. Roddier (he is working mostly on solar physics) has quite recently developed (**30**) a new kind of spectrograph using atomic beam techniques. The profiles of the Fraunhofer solar line can be obtained by measuring the intensity of the fluorescence of an atomic beam excited by the light to be analysed. The source frequency is shifted by the Zeeman effect. The experiment uses calcium (4226·7 Å), strontium (4607·3 Å) and barium (5535·5 Å) atomic beams, in order to observe the solar spectrum in the neighbourhood of these wavelengths. The resolving

power is about 10^7. Using the resonance in zero field as a reference point, the absolute value of the shift of the solar line can be obtained; this method was used in 1953 by J. E. Blamont (**31**) for the measurement of the atmospheric Na lines. For the solar lines the method was used (Sr I 4607·3 Å) by J. E. Blamont and F. Roddier (**32, 33**). The result is: there is real limb-effect (in general accordance with M. G. Adam's interferometric and other results), there is a clear increase of the microturbulent velocities with height in the solar atmosphere, etc.

2. The same authors have made also some experimental studies on photomultipliers (**34**) for example the statistical study of the impulse in photomultipliers (experimental and mathematical distribution of the impulse).

3. Report by G. Courtès, C. Fehrenbach, E. Hughes and J. Romand (the total report is published in (**35**)). The authors give list of literature and non-exhaustive review of development and progress in certain fields of optics, concerning itself mainly with the conception and realization of instruments. The previous review (and also literature) of optical work in France was also published.

(*a*) ***Réalisations optiques.***

(*i*) La Société Angénieux vient d'ajouter à la gamme des nombreuses focales variables qu'elle produit, deux objectifs de caractéristiques remarquables par l'extension importante du rapport des focales extrêmes ('range') qui atteint maintenant la valeur 20 (pour l'Orticon, dont la focale varie de 35 à 350 mm, le champ est exceptionnel avec 64° d'ouverture, $f/3{\cdot}8$, etc.). Les photographies de la Lune par Ranger ont été prises avec des objectifs Angénieux 25 mm/0·95. Les satellites Tiros et Nimbus sont équipés d'objectifs Kinoptik (TGA) $f = 5{\cdot}7/1{\cdot}8$ et l'une des caméras utilisées pour filmer la rencontre spatiale des capsules Gemini était équipée d'un objectif Kinoptik $f = 18/1{\cdot}8$.

(*ii*) La Société Kinoptik, spécialisée dans la réalisation d'objectifs pour le cinéma et la télévision, vient d'ajouter à sa gamme les objectifs suivants: Apochromat 9 mm, $f/1{\cdot}5$, champ total 80°; Tegea 9·8 mm, $f/1{\cdot}8$; Lynxar 60 mm, $f/0{\cdot}7$, etc.

(*iii*) Parmi les récentes réalisations de la Société Sopelem, nous citerons des objectifs à focale variable pour le cinéma et la télévision.

(*b*) ***Instruments optiques.***

(*i*) L'Observatoire de Haute-Provence du CNRS a mis en service en 1958 un télescope de 1·93 m ($f/5$) de diamètre (Cassegrain $f/15$; coudé $f/20$) dont les miroirs ont été taillés au Laboratoire d'Optique de l'Observatoire de Paris (**37**). Un grand spectrographe coudé est disposé dans un grand laboratoire annexe (**38**). Le spectrographe est équipé de deux réseaux de Bausch and Lomb, comprenant l'un 774, l'autre 1200 traits au millimètre. La hauteur des traits est de 150 mm. Le changement de réseau et le passage de l'une à l'autre des diverses chambres se font automatiquement et ne demandent que quelques secondes. Les dispersions de 83 Å/mm à 3 Å/mm peuvent être réalisées. Tout le domaine spectral de 3100 Å à 17 000 Å est couvert par les diverses combinations. Ce télescope comporte de nombreux appareils auxiliaires (caméra électronique de Lallemand, etc.). Un instrument analogue est en voie de réalisation pour l'Observatoire Européen Austral (ESO). Il sera installé au Chili au cours de l'année 1967.

(*ii*) Parmi les autres instruments installés à l'Observatoire de Haute-Provence et en d'autres observatoires, signalons un prisme-objectif à champ normal de Fehrenbach de 40 cm de diamètre pour la mesure des vitesses radiales stellaires (**39**). Un instrument identique a été installé en Afrique du Sud, dans le cadre de l'Observatoire Européen Austral. Un télescope de Schmidt de 60 cm de lame (optique de l'Observatoire de Paris) est installé à l'Observatoire de Haute-Provence par l'Institut d'Astrophysique de Liège. Un autre télescope de Schmidt de même diamètre a été installé à Paris-Meudon. Un télescope de 1·07 m a été installé à l'Observatoire du Pic du Midi. Il est surtout destiné à des études du système solaire. Les deux grandes lunettes des Observatoires de Paris-Meudon et de Nice ont été remises en état et la première est de nouveau en service régulier; la seconde le sera bientôt.

Une grande tour solaire, comprenant un coelostat à deux miroirs de 80 cm et 70 cm de diamètre, est installée à Paris-Meudon. Elle comprend un télescope bouché de 69 cm de diamètre et de 45 m de distance focale qui permettra la photographie et l'étude directe des images solaires. Un grand spectrographe avec un collimateur de 14 m de distance focale comprenant des réseaux de 15 cm × 30 cm, permettra la spectrographie solaire.

(*iii*) Les opticiens et les astronomes francais ont été engagés avec leurs collègues européens dans la réalisation de grands systèmes optiques.

Dans une étude faite à Marseille par A. Baranne du système Ritchey-Chrétien (**40, 41**), cet auteur a pu montrer que cette combinaison n'est pas plus sensible aux flexions qu'un système parabolique classique de même rapport d'ouverture. On a calculé l'optique du télescope Ritchey-Chrétien ouvert à *f*/3, de 3·50 m de diamètre (*f*/3 primaire, *f*/8 Cassegrain), destiné à l'Observatoire Européen du Chili. M. Baranne a calculé pour le foyer direct un système de correction à trois verres dont un seul légèrement asphérisé. Il donne un champ de 1° de diamètre, avec des images de 0″5. Ce système correcteur a 25 cm de diamètre, et n'est situé qu'à 33 cm du foyer primaire.

(*iv*) D'un autre côté, A. Baranne, après avoir réalisé un spectrographe intéressant à très haute dispersion (**42**), a indiqué récemment un montage des spectrographes astronomiques à pupille conjuguée. Dans ce montage, un miroir auxiliaire sphérique reforme une image non dispersée du collimateur sur l'objectif de chambre. Ceci permet de donner à cet objectif une ouverture qui correspond à l'ouverture efficace. Le réseau est monté en Littrow, ce qui permet d'employer les réseaux échelles, et d'utiliser de très hautes dispersions angulaires.

(*c*) *Optique complémentaire des instruments astronomiques au sol et dans l'espace*:

(*i*) Réducteur focal (**43**) du téléscope de 193 cm.

(*ii*) Système interférentiel à systèmes d'anneaux multiples (études extragalactiques en lumière monochromatique).

(*iii*) Filtre à bandes multiples.

(*iv*) Caméras à grand champ (catadioptrique, 66° × 120°) pour les observations astronomiques spatiales).

(*v*) Utilisation spectrographique du réducteur focal.

(*vi*) Spectromètre interférentiel pour l'étude des nébuleuses.

(*d*) Divers travaux ont été effectués dans le domaine de l'appareillage et des méthodes, et publiés pour la plupart dans les *Publications de l'Observatoire de Haute-Provence.*

VEB Carl Zeiss

This firm finished many new interesting instruments, in particular 800/1900/2400 mm Schmidt telescope field 5°15 × 5°15 (24 × 24 cm) with 4° objective prism for the Observatory of the Latvian Academy of Sciences, S.S.R., near Riga.

O. A. MELNIKOV
President of the Commission

BIBLIOGRAPHY

1. Vany, A. S. de 1966, *Appl. Opt.*, **5**, 867.
2. King, J. R. 1966, *Publ. astr. Soc. Pacif.*, **78**, 35.
3. Cox, R. E. 1966, *Rev. pop. Astr.*, **60**, 4.
4. Bowen, I. S. 1964, *Astr. J.*, **69**, 816.

5. Marchant, I. C., Millikan, A. G. 1965, *J. opt. Soc. Am.*, **55**, 907.
6. Report of Meeting on 'Photographic Plate Testing', Mt Wilson and Palomar Obs., 17–18 January 1966.
7. Sandage, A., Miller, W. C. 1966, *Astrophys. J.*, **144**, 1238.
8. Miller, W. C. 1966, *Publ. astr. Soc. Pacif.*, **78**, 128.
9. Stroke, G. W. 1965, *Astr. J.*, **70**, 693.
10. 1965, *New Technics in Astronomy*, Moscow. Ed. N. N. Michelson.
11. Öhman, Y. 1964, *Atti del Convegno sui campi magnetici solari*, Roma, in press.
12. Nilsson, S. E. 1966, *Stockholm Obs. Medd.*, no. 153.
13. Fredga, K. 1966, *Astrophys. J.*, **144**, 854.
14. Rydgren, B. 1966, *Stockholm Obs. Medd.*, no. 154.
15. Öhman, Y., Rydgren, B. 1963, *Stockholm Obs. Medd.*, no. 142.
16. Kusoffsky, U. 1966, *Stockholm Obs. Medd.*, no. 156.
17. Öhman, Y. 1965, *Stockholm Obs. Medd.*, nos. 146, 150, 151.
18. Öhman, Y., Kusoffsky, U. 1966, *Stockholm Obs. Medd.*, no. 157 = *Revue d'Optique*, **2**, 49.
19. Lodén, L. O. 1964, *Stockholm Obs. Ann.*, **22**, 7.
20. Wallenquist, A. 1965, *Sky Telesc.*, **29**, 136.
21. Lallemand, A. *et al.* 1966, *C. r. Acad. Sci. Paris*, **262**, 838.
22. Crawford, D. L. 1965, *Sky Telesc.*, **29**, 268.
23. Sugawa, Ch., Hurukawa, K., Okawa, H. 1966, On the Automatic Measuring Machine for the Photographic Plates of the Floating Zenith Telescope at Mizusawa. *Proc. Int. Latit. Obs. Mizusawa*, no. 6, 117 (in Japanese).
24. Sugawa, Ch., Hurukawa, K. 1966, Data Processing System for the Time Recording in the Astrolabe Observations. *Ibid.*, no. 6, 76 (in Japanese).
25. Ono, F. A new timing device of satellite tracking camera for geodetic purpose, *Rep. hydrogr. Res.*, no. 1, 63.
26. Ono, F. A simple device to determine personal equation, *Ibid.*, no. 2 (in press).
27. Tsubokawa, I. 1965, A precise satellite tracking camera with a photoelectric timing device, *Space Res.*, **5**, 916.
28. Lindsay, E. M. 1953, *Irish astr. J.*, **2**, 140 = *Armagh Obs. Leaflet*, no. 17.
29. Andrews, A. D., Dommanget, J. 1965, *Irish Astr. J.*, **7**, 73.
30. Roddier, F. 1965, *Ann. Astrophys.*, **28**, 463.
31. Blamont, J. E. 1953, *C. r. Acad. Sci. Paris*, **237**, 1320.
32. Blamont, J. E., Roddier, F. 1961, *Phys. Rev. Letters*, **7**, 437.
33. Blamont, J. E., Roddier, F. 1964, *C. r. Acad. Sci. Paris*, **258**, 449.
34. Roddier, F., Roussin, A. 1964, *C. r. Acad. Sci. Paris*, **259**, 4593.
35. Courtès, G., Fehrenbach, C., Hughes, E., Romand, J. 1966, *Appl. Opt.*, **5**, 1349.
36. Vany, A. S. de 1963, *Appl. Opt.*, **2**, 200.
37. Couder, A. 1960, *Ann. Astrophys.*, **23**, 311 = *Publ. Obs. Hte-Provence*, **5**, no. 21.
38. Fehrenbach, C. 1959, *Publ. Obs. Hte-Provence*, **5**, no. 1.
39. Fehrenbach, C., Rebeirot, E. 1963, *Publ. Obs. Hte-Provence*, **5**, no. 43.
40. Baranne, A. 1963, *C. r. Acad. Sci. Paris*, **257**, 3825 = *Publ. Obs. Hte-Provence*, **7**, no. 1.
41. Baranne, A. 1965, Thèse, Université de Marseille.
42. Baranne, A. 1965, *C. r. Acad. Sci. Paris*, **260**, 3283 = *Publ. Obs. Hte-Provence*, **7**, no. 43.
43. Courtès, G. 1960, *Ann. Astrophys.*, **23**, 115.

REPORTS FROM OBSERVATORIES OF THE U.S.S.R.

(prepared by N. N. Michelson)

I. Instruments

1. *Pulkovo Observatory of the U.S.S.R. Academy of Sciences*

(*a*) The new large stellar interferometer (with a base of 6 m) is in general completed. It is to be used for double star work.

(*b*) The second horizontal solar telescope with a diffraction spectrograph (diameter of mirror: 0·44 m) is completed in general.

(*c*) At the expedition in Chile the vertical circle (system of M. S. Zverev) is in operation and the 0·7 m double meniscus astrometric astrograph (system of D. D. Maksutov) is being mounted.

(*d*) The 0·45 m double channel television telescope is completed. It has an aperture of *f*/1·34 and also aperture from *f*/20 to *f*/270. The instrument is used for observations of the Moon and the study of seeing.

(*e*) At the Pulkovo Station near Kislovodsk, the 0·23 m astrometrical astrograph with *f*/10 is completed and will be used for observations of the Moon and planets.

(*f*) The astrolabe (system of A. Danjon) is completed.

(*g*) The 0·2 m refractor, with *f*/15, is completed.

2. *Crimean Astrophysical Observatory of the U.S.S.R. Academy of Sciences.*

The system of the Nasmith and coudé foci for spectroscopic observations with the 2·6 m reflector (memorial of J. A. Shayn) have been assembled.

3. *Sternberg Astronomical Institute of the Moscow University.*

(*a*) At the Southern Station in the Crimea, the 1·25 m reflector with numerous special apparatus (photometric, photographic and spectroscopic) was mounted in 1960 and is used for observations with apertures *f*/4 and *f*/17. In the Cassegrainian focus, there is a shortening system with *f*/2. Contact image tubes, spectrographs, etc. are widely used for different purposes.

(*b*) The 0·3 m Zeiss refractor was mounted in Moscow.

4. *Tartu Observatory.*

The new Tartu Observatory (memorial of W. Struve) of the Estonian Academy of Sciences was inaugurated in 1964. The main building, mechanical shop, two observing pavilions and several smaller ones are completed.

The following instruments have been mounted or are being mounted:

(*a*) the 0·7 m reflector with *f*/4, *f*/16 and *f*/40 (completed);

(*b*) three 0·48 m reflectors with *f*/4·5 and *f*/16, one of which is mounted in Tallin, are used for photoelectric observations with distance control;

(*c*) the 0·35 m reflector with *f*/21 (completed);

(*d*) the 0·2 m reflector with *f*/15 (completed);

(*e*) the 1·5 m reflector is being designed;

(*f*) an experimental plastic dome has been constructed (**4, 5, 6, 7**).

5. *The Astrophysical Institute of the Kazakhstan Academy of Sciences.*

The following instruments have been completed and mounted:

(*a*) the 0·8 m reflector;

(*b*) the 0·17 m Schmidt camera with *f*/1;

(*c*) the 0·34 m reflector with *f*/5·5 and *f*/3·4 with an optical magnifier (camera) for observations of planets. The instrument is mounted on the mounting of the 0·2 m reflector.

6. *Odessa Astronomical Observatory.*

(*a*) the 0·425 m reflector with the P. P. Argunov correcting system (*f*/10) is completed (**8**);

(*b*) the 0·5 m reflecting telescope (*f*/7) with an electrophotometer is completed;

(*c*) one sky patrol for observations of bolides and three for observations of meteors have been mounted.

7. *The Main Astronomical Observatory of the Ukrainian Academy of Sciences* (Goloseyevo near Kiev).

The following instruments have been mounted:

(*a*) the horizontal solar telescope with a 0·44 m mirror;

(*b*) the horizontal lunar telescope (*f*/60).

8. *The Abastumani Astrophysical Observatory of the Georgian Academy of Sciences.*

The following instruments have been mounted:

(*a*) the horizontal solar telescope with a 0·44 m mirror;

(*b*) the 0·12 m non-eclipse coronograph (design of I. A. Prokofyeva).

9. *Astronomical Station at Haute Sayannes.*

The 0·2 m coronograph (system of B. Lyot) has been mounted.

10. *The Astrophysical Laboratory of the Latvian Academy of Sciences.*

(*a*) The 0·55 m reflector (*f*/13·5) has been mounted;

(*b*) The large Schmidt telescope 0·8/1·2 m is being completed by the firm VEB, Carl Zeiss, Jena.

11. *Southern Station of the Leningrad University at Burakhan.*

The 0·48 m and 0·50 m reflecting telescopes with photoelectric photometers and other apparatus have been mounted (**9**).

12. *Astrophysical Observatory of the Azerbaijan Academy of Sciences at Shemaha (near Baku).*

The 2 m reflecting telescope made by the firm VEB, Carl Zeiss, Jena, has been mounted. The instrument is to be used together with spectroscopic, photoelectric, photographic and other apparatus for observations of nonstable stars, galaxies, old novae, variable stars, quasars (QSS, QSG, QSO) and other peculiar and normal objects.

13. *Astronomical Observatory of the Tadjikistan Academy of Sciences at Dushanbe.*

The 0·4 m astrograph (*f*/5) made by the firm Carl Zeiss, Jena, with a field of $8^\circ\!.5 \times 8^\circ\!.5$ has been mounted.

14. The series of *solar horizontal telescopes* with a 0·44 m mirror (partly quoted above) have been mounted at Pulkovo, Sverdlovsk, Shemaha, the Far East, Sayannes, Alma-Ata and Abastumani.

15. The series of 0·48 m *reflecting telescopes* (partly quoted above) have been mounted at the observatories in Lvov, Shemaha, Tartu, Kiev and the Southern Stations of the Leningrad and Moscow Universities.

16. The small series of 0·7 m *reflecting telescopes* (partly quoted above) are being used very successfully at the Simeis (Crimea), Dushanbe, Tartu and Kharkov (not mounted as yet) Observatories.

17. The 0·35 m *meniscus camera* have been mounted at the Engelhardt (Kazan) and Shemaha Observatories.

II. Scientific Work

1. *Pulkovo Observatory.*

(*a*) Numerical calculations of the 2·6 m (*f*/3·67) wide angle ($2^\circ\!.6$) telescope with a hyperbolic prime mirror have been completed and a scheme for testing the latter elaborated (**10, 11**). A three-component afocal corrector is used in the convergent beam.

(*b*) Numerical calculations of the 0·3 m objective with an aperture of $f/2$ for the scheme of the 'concentric meniscus Cassegrain' with a 6° field have been completed. The aberration of the field is less than 3·5 microns at an asphericity of only 5 microns (**12**).

(*c*) The deformations of the large mirrors of telescopes are being investigated theoretically (by methods of the theory of elasticity, in particular by the photoelastic method) and experimentally (by the method of a dynamic scale model of epoxides resine and also by the stereometric method—igdontin) (**13-18**).

(*d*) The model of an automatic coordinate measuring machine with photoelectric setting on the star image and two-step reading of the coordinates has been constructed on the basis of a universal standard microscope (**19**).

(*e*) The coarse diffraction grating for the 0·65 m telescope to be used for diminishing the light of bright stars has been completed in the form of a rotating magnitude screen or 'jalousie' (**20**).

(*f*) An apparatus for measurements of the negatives of circle readings of the photographic vertical circle at the expedition in Chile has been constructed.

2. *The Crimea Astrophysical Observatory.*

(*a*) A method for air operated unloading of prime mirrors of telescopes has been developed (G. A. Monin).

(*b*) A series of various high-speed meniscus systems has been developed and computed numerically: an 80 mm system with $f/0{\cdot}6$ for a camera spectrograph with a field of 36° and resolving power of 30 microns (for the 1·22 m reflecting telescope), a 100 mm system with $f/5$ and a field of 20°, a 100 mm system with $f/0{\cdot}4$, etc. (**21**).

3. *Image converters, image intensifiers and other tubes* have had very significant and wide application at the Sternberg Astronomical Institute, the Crimea, Pulkovo and other Observatories (**27**).

4. At the *Tartu Observatory* great attention was devoted to the automatization of the observational process and the data processing of observations.

(*a*) Universal standard drive mechanisms have been assembled for small telescopes with motors for setting correction and the possibility of application of a photo-guiding system (**23**).

(*b*) An electrophotometer with a digit-printing system and a d.c. amplifier has been constructed for the 0·48 m reflector. It is possible to record the data by means of a self-recording (on paper) electric potentiometer or a digit printing accessory (**24**).

5. At the *Engelhardt Observatory* an iris photometer has been constructed on the basis of the MΦ–2 standard-type microphotometer (**25, 26**).

6. At the *Astrophysical Observatory near Alma-Ata* the system of Cassegrain has been introduced for the 0·5 m meniscus telescope, providing for $f/6$ and 10° field (**27**).

7. *Siberian Institute of Terrestrial Magnetism, Ionosphere and Propagation of Radio Wave.*

(*a*) A solar magnetograph has been constructed for recording the three vector components of the magnetic field strength (or intensity) and also radial velocities (**28**).

(*b*) An automatic scanner has been assembled for the horizontal solar telescope, and provides for the scanning of the solar disk on the slit of the spectrograph.

III. Additional Reports

1. *Crimean Astrophysical Observatory, U.S.S.R Academy of Sciences* (compiled by I. M. Kopylov):

A stellar spectrograph with an échelle grating was completed and mounted at the coudé focus of the 2·6 m reflector (memorial of G. A. Shajn) in 1963. The échelle operates in the 74–37 orders (λ λ 3450–7000 Å). The camera provides reciprocal dispersions of 1, 1–2, 2 Å/mm and resolution of 0·66 Å. The pre-dispersion is obtained by a 22°5 light crown prism (**30**). The large coudé spectrograph for the 2·6 m reflector was put into operation in the summer of 1966. The diffraction grating (with a 320 × 280 mm surface and 600 lines per mm) is used in the second and third orders. Three Schmidt cameras of focal length 27 (69 cm), 52 (132 cm), and 131 (333 cm) inches are now being used. Reciprocal dispersion are respectively 12–8, 7·5–5 and 2·4–1·6 Å/mm.

2. *Institute of Terrestrial Magnetism, Ionosphere and Propagation of Radio-Wave* (*near Moscow*), U.S.S.R. Academy of Sciences (compiled by Professor N. P. Benkova):

The tower solar telescope 0·38/17/27 m is now in operation. The focal length of the collimator – camera (I) – camera (II) is 10 m, the diffractional grating 15 × 15 cm, 600 lines per mm (**31**). There is a photoelectric magnetograph: (*a*) One measures the total vector of photospheric magnetic field (all the Stokes parameters are measured simultaneously) (**32**) and the transversal component only (**33**). The radial velocity and the intensity of the selected point of the line profile is measured simultaneously. (*b*) Measurements of photospheric and chromospheric magnetic field are achieved simultaneously by two independent magnetographs (**34**) and two selected lines. During the years 1963–65, the series of observations of the magnetic field were obtained with the complex magnetograph for the collection of original observational data and the solution of several methodological problems (**35**). The design and construction of B. Lyot coronagraph with a stationary spectrograph is completed. The single-lens object glass has a diameter of 0·53 m and 8 m focal length and reciprocal dispersion of 1 Å/mm (**35**). An identical instrument has been completed for the Pulkovo Observatory and will be soon completed for several other observatories.

IV. General Remarks

1. The work of U.S.S.R. astronomers, during the three-year period since the last report of the IAU, in the fields of construction, engineering, assembling, mounting, etc. of telescopes was successful.

2. We regret that, during the period elapsed, two well-known astronomers working in the field covered by our Commission passed away: D. D. Maksutov (12 August 1964) – the famous specialist in astronomical optics, and N. I. Kutcherov (17 October 1964) – a specialist in the comparatively new and now very important problem of site testing and astroclimate.

BIBLIOGRAPHY

1. Boyarchuk, A. A., Esipov, V. F., Moroz, I. V. 1966, *Astr. Zu.*, **43**, 421.
2. Dibai, E. A., Esipov, V. F. 1965, *Astr. Zu.*, **42**, 281.
3. Dibai, E. A. 1966, *Astr. Zu.*, **43**, 52.
4. Vilman, Ch., Zhelnin, G., Einasto, R., Ergenson, R. 1964, Tartu Astronomical Observatory, Tallin.
5. Eelsaln, H. T., Kutusov, S. A. 1965, *Vestnik Akad. Nauk SSSR*, no. 9, 35, 68.
6. Einasto, J., Steinbach, M. 1965, *Sterne*, **41**, no. 7–8.
7. Kubar, T. E. 1965, *New Technics in Astronomy*, Moscow, no. 2, 61.
8. Argunov, P. P. 1965, *New Technics in Astronomy*, Moscow, no. 2, 8.
9. Dombrowsky, V. A., Gagen-Torn, V. A., Gutkevich, S. M., Polakova, T. A., Svechnikov, M. A., Shulov, O. S. 1965, *Uchen. Zap. Leningrad Univ.*, no. 328, 83.
10. Belorosova, T. S., Maksutov, D. D., Mermann, N. V., Sosnina, M. A. 1964, *Izv. glav. astr. Obs. Pulkove*, no. 175.
11. Sosina, M. A. 1966, *Izv. glav. astr. Obs. Pulkove*, no. 184.
12. Merman, N. V., Sosnina, M. A. 1965, *Izv. glav. astr. Obs. Pulkove*, no. 178.

13. Grosswald, E. G., Tavastsherna, K. S. 1964, *Izv. glav. astr. Obs. Pulkove*, no. 177.
14. Grosswald, E. G., Tavastsherna, K. S. 1964, *Izv. glav. astr. Obs. Pulkove*, no. 180.
15. Grosswald, E. G. 1965, *New Technics in Astronomy*, Moscow, no. 2, 31.
16. Chernina, V. S. 1964, *Izv. glav. astr. Obs. Pulkove*, no. 177.
17. Fridman, V. M. 1964, *Izv. glav. astr. Obs. Pulkove*, no. 177.
18. Budnikova, T. V., Fridman, V. M. 1964, *Izv. glav. astr. Obs. Pulkove*, no. 177.
19. Zaciorsky, L. M., Nicolaeva, L. V., Bubnov, G. A. Patent no. 420508.
20. Kanaev, I. I. 1964, *Izv. glav. astr. Obs. Pulkove*, no. 174.
21. Popov, G. M. 1965, *New Technics in Astronomy*, Moscow, no. 2, 5.
22. Scheglov, P. V. 1965, *New Technics in Astronomy*, Moscow, no. 2, 42.
23. Veisman, U. K., Kubar, T. E. 1965, *New Technics in Astronomy*, Moscow, no. 2, 48.
24. Veisman, U. K. 1965, *New Technics in Astronomy*, Moscow, no. 2, 40.
25. Urasin, U. N. 1965, *New Technics in Astronomy*, Moscow, no. 2, 48.
26. Urasin, U. N. 1964, *Bjull. astr. Obs. V. P. Engel'gardta*, no. 38.
27. Rozhkowsky, D. A., Gluschkov, U. N., Kurchakov, A. V. 1963, *Izv. Akad. Nauk KazSSR*, Ser. Fiz.-Matem., **16**, 52.
28. Stepanov, E. V., Knulin, G. V., Kuznecov, D. A. 1965, *Solar Activity*, Moscow, no. 2.
29. Alexandrovich, D. A., Kuznetcov, D. A., Maksutov, D. D. 1965, *Solar Activity*, Moscow, no. 2.
30. Kopylov, I. M., Steschenko, N. V. 1965, *Izv. Krym. astr. Obs.*, **33**, 208.
31. Mogilevsky, E. I., Zhulin, I. A., Yoshpa, B. A. 1965, *Solar Activity* (sbornik), no. 2, ed. by 'Science', Moscow; B. A. Yoshpa, I. A. Zhulin, E. I. Mogilevsky, Proc. Symp. Solar magn. fields, Rome, Sept. 1964.
32. Yoshpa, B. A., Obridko, V. N. *Astr. Zu.*, **41**, no. 6, 1964; *Solar Activity* (sbornik), no. 2, 1965, ed. by 'Science', Moscow; Proc. Symp. Solar magn. field, Rome, Sept. 1964.
33. Yoshpa, B. A., Mogilevsky, E. I. 1965, *Solar Activity* (sbornik), no. 2, ed. by 'Science', Moscow.
34. Zhulin, I. A., Mogilevsky, E. I. 1965, *Solar Activity* (sbornik), no. 2, ed. by 'Science', Moscow.
35. Yoshpa, B. A., Obridko, V. N. *Solar Data*, no. 5, 1965; Obridko, V. N., Shilova, N. S., *Solar Activity*, no. 3, 1966 (in press).
36. Nicolsky, G. M., Sasonov, A. A. 1966, *Astr. Zu.*, **43** (in press). Sasonov, A. A., 1966, *Nature*, U.S.S.R. Acad. of Sci., no. 9.

APPENDIX I. WORKING GROUP ON IMAGE TUBES FOR ASTRONOMY

(prepared by J. D. McGee, Chairman)

In the following Draft Report the subject matter is presented under headings indicating the place where the work was done rather than the method used.

This is considered more appropriate in view of the volume of work being done and reported and the inherent difficulty of extracting and summarizing the individual reports to give a balanced account of the work. It is considered better to let the individual authors describe their own work.

Every effort has been made to obtain a report from known workers in this field. If any report has been overlooked, or not sent in in time, it is hoped that it will be submitted at the IAU General Assembly in Prague, in August 1967.

A. Lallemand, Observatoire de Paris, Paris

La photographie électronique permet de réaliser un récepteur d'images plus sensible que la photographie classique. Dans le domaine des grandes longueurs d'onde, le gain peut être considérable et il est possible d'obtenir des images d'intérêt astronomique pour des longueurs d'onde inaccessibles à la photographie classique, par exemple vers 14 000Å.

Nous avons, au cours de ces dernières années, développé des procédés de fabrication des couches photosensibles destinées à être utilisées dans la caméra électronique et sensibles au proche infrarouge. Elles permettent d'enregistrer des spectres avec plusieurs heures de pose et avec un rapport signal-bruit satisfaisant au-delà de 14 000 Å.

Un autre aspect de la photographie électronique, que nous considérons aujourd'hui comme peut-être le plus important, a été étudié; c'est la possibilité d'obtenir des mesures photométriques de sources extrêmement faibles et ponctuelles (étoiles, quasars, etc.) sans avoir besoin d'utiliser des standards dans la gamme des magnitudes considérées (1).

La caméra électronique permet, en utilisant convenablement la méthode de Schilt dans le cas d'objets de magnitudes élevées, une détermination facile des magnitudes les plus élevées enregistrées par la plaque à électrons. Il suffira de connaître la magnitude de quelques étoiles, les plus brillantes du cliché, pour étalonner le récepteur. D'autre part, la caméra électronique reproduisant l'image des objets qu'on mesure, les erreurs d'identification sont beaucoup plus faciles à éviter que dans le cas des mesures faites avec des photomultiplicateurs, la correction du fond du ciel nocturne est beaucoup plus aisée, sans risque d'introduire une autre étoile dans le champ de l'appareil de mesures. On obtient dans le même temps de pose toutes les étoiles ou objets enregistrés sur le cliché.

Les photocathodes utilisées dans la caméra électronique étant de même nature que celles utilisées dans les photomultiplicateurs, le rattachement aux standards offre le maximum de facilités, en particulier dans le système *UBV*.

M. Duchesne et ses collaborateurs, Observatoire de Paris, Paris

Le gain en sensibilité de la photographie électronique, par rapport à la photographie classique, a été déterminé dans le domaine visible, pour différentes bandes spectrales, par M. Duchesne et A. Cosson. Ils ont montré que les gains maximaux étaient obtenus pour les faibles densités photographiques et que les poses de longues durées favorisaient la photographie électronique. Des gains de l'ordre de 200 ont pu être réalisés vers 5400 Å. Ce travail a été poursuivi dans le domaine infra-rouge, entre 11 000 et 12 000 Å; des gains très importants, de l'ordre de 200 à 400, ont été obtenus par rapport à la plaque 1Z Kodak sensibilisée.

Les photocathodes utilisées dans la caméra électronique sont fabriquées avec un excès de césium; cet excès de césium produit une instabilité des couches photoémissives qui peuvent perdre une partie importante de leur sensibilité pendant le stockage. M. Duchesne (2) a montré qu'une illumination intense permet de restituer la sensibilité initiale. Cette sensibilisation s'explique, en partie, par la sublimation du césium superficiel; il y a équilibre, par l'intermédiaire de la tension superficielle du césium, entre le césium libre présent à la surface de la couche photoémissive et celui qui recouvre la paroi interne de l'ampoule. Dans la caméra électronique, les photocathodes sensibilisées ont un comportement identique à celui des cathodes stables.

M. Duchesne, M. Feissel et B. Guinot (3) ont terminé l'étude systématique de la distorsion en coussinet introduite par l'optique electronique de la camera électronique. Cette étude a montré que les mesures astrométriques pouvaient être faites avec une précision de l'ordre de 1 μ, précision nécessaire à la réduction des clichés de Mercure en cours d'exploitation pour la détermination de la constante de l'aberration.

Dans le but de remplacer par des photocathodes planes les photocathodes sphériques actuellement utilisées, C. Hézard (4) a commencé l'étude d'un objectif à immersion électrostatique à symétrie cylindrique. Dans le cas particulier d'un objectif à fentes croisées, il a déterminé expérimentalement la valeur des aberrations de champ.

M. Duchesne et B. Guinot ont poursuivi l'enregistrement de spectres de Mercure à l'aide du télescope de 193 cm de l'Observatoire de Haute-Provence, du spectrographe spécialement conçu pour ce travail et de la caméra électronique. Plusieurs missions seront encore nécessaires pour la détermination de la constante de l'aberration avec la précision excellente que laissent espérer les clichés déjà exploités.

M. Duchesne, en collaboration avec A. Baranne et J. M. Lecontel, a procédé aux réglages nécessaires à l'adaptation de la caméra électronique à un spectrographe à très grande résolution placé au foyer coudé du télescope de 193 cm. Un dispositif spécialement étudié permet de mettre aisément et avec sécurité la caméra électronique en position de travail. En 1966, des recherches portant sur deux programmes ont pu être entreprises:

1. Recherches des composantes interstellaires des raies D_1D_2 du sodium et de la raie Ca II 3933. Les spectres obtenus avec plusieurs étoiles ont permis de tester les possibilités offertes par l'association des deux appareils.

2. Etude de la raie $H\beta$ d'étoiles Be. Ce programme permet de préciser des observations effectuées sur des spectres pris en photographie classique avec les chambres IV et V du spectrographe placé au foyer coudé du télescope de 193 cm. Ces études seront poursuivies et étendues à l'observation de certaines étoiles variables du type β Canis Majoris.

G. Wlérick et ses collaborateurs, Observatoire de Paris, Meudon, France

En liaison avec le Professeur A. Lallemand et son laboratoire, les recherches suivantes ont été entreprises:

(*a*) Continuation à Meudon du programme de photométrie précise avec la caméra électronique (G. Wlérick, M. Combes, L. Vapillon). La source de lumière uniforme qui sert pour déterminer la sensibilité locale de la photocathode a été contrôlée. Son défaut d'uniformité est inférieur à 0·5%. Cette source a été utilisée régulièrement dans les divers programmes de photographie électronique. La précision avec laquelle il faut faire correspondre les passages au microphotomètre pour un cliché astronomique et pour un cliché de lumière uniforme a été déterminée: elle est de l'ordre de 100 microns si on désire une précision supérieure à 1%.

(*b*) Continuation à l'Observatoire du Pic du Midi du programme de photographie électronique des planètes Jupiter et Saturne dans les 3 couleurs *U*, *B*, *V* (G. Wlérick et M. Combes).

L'exploitation des clichés de Jupiter a servi de base à la thèse de 3ème cycle de Combes sur les propriétés physiques de l'atmosphère de cette planète (5).

(*c*) Adaptation d'une caméra électronique à un spectrographe stellaire ouvert à *f*/2 et destiné au foyer coudé du télescope de 1·93 mètre de l'Observatoire de Haute-Provence (J. P. Picat, F. Gex, G. Wlérick). Une mise au point satisfaisante a été obtenue pour l'ensemble de l'intervalle spectral utilisé (3900–5100 Å). Les premiers spectres ont été obtenus en Octobre 1966.

(*d*) Préparation d'un dispositif permettant de prendre des photographies électroniques au foyer Newton du télescope de 193 cm de l'Observatoire de Haute-Provence (G. Wlérick, F. Gex).

Le but des observations projetées est de mesurer des astres très faibles, de magnitude supérieure à 22. La mise en service est prévue en 1967.

M. Walker, Lick Observatory, California, U.S.A.

Since the last Draft Report, work has continued with the Lallemand electronic camera. This work has comprised both laboratory tests and improvements, together with related optical developments, and observations with the tube at the focus of the coudé spectrograph of the 120-inch (305 cm) reflector.

I. Improvements in Technique

A. The tube has been fitted experimentally with a metal valve, to eliminate the grease in the glass stopcock used previously. That grease remains at room temperature during operation of the tube and probably sets the limit to the vacuum attainable. Preliminary experiments with a modified Veeco valve having a teflon seat indicate that indeed a substantially lower pressure may be achieved in the tube when the valve is used.

B. A new preparation stand has been constructed for the electronic camera, employing absorption pump and large ion pump instead of the usual diffusion pump and cold trap. The new system is virtually oil-free and eliminates the danger of back streaming from the diffusion pump or cold trap. Tests indicate that the pumping speed is somewhat lower than with the regular system, but that the ultimate vacuum obtained is substantially better.

C. Experiments have been made replacing the activated charcoal used by Lallemand *et al.*, as a getter in the tube with zeolite ('molecular sieve'). This material has the advantage of being a dielectric so that small particles falling into the insulating ('triple-seal') section of the tube do not induce electrical discharges as is the case with the charcoal. In addition, there is evidence that the ultimate vacuum obtained with the zeolite getter is substantially better than with the activated charcoal.

D. The preparation procedure has been still further elaborated since the last report to improve the ultimate vacuum and hence cathode life. At present, a preliminary 5-hour bake at 350°C is given upon sealing the new plates and cathode ampoule in the tube. Then, the tube is cooled and tested for electrical discharges under a total voltage of 46 kV (operating potential being 26 kV). Then, before use a second 5-hour bake at 350° is given, followed by liquid air cooling, cathode placement, and the 24-hour ageing process before beginning observations at the telescope. Cathode life with Sb–Cs cathodes (S–11) is now very long. Usually, no change is found at the end of a two-night run, so that in principle the tube could probably be used for one or two weeks if kept constantly refrigerated with liquid air.

II. Instrumentation

A. Image tube. Since the last report, Lallemand has succeeded in preparing S–1 photocathodes for use in the electronic camera. These cathodes are chemically more active than the S–11 and hence harder to preserve in the tube. However, by means of the techniques described in this and the preceding reports, it is possible to preserve them for a maximum of about two days. This is important as tests have shown that the same ageing process used with the S–11 to reduce background is required with the S–1. Several runs have been made at the telescope using these cathodes and the results are outlined below.

B. Optical. As indicated in the last report, the fact that the information gain, at least in the blue region of the spectrum, is only about 15 instead of the much higher values quoted earlier, means that every effort must be made to optimize the efficiency of the optical system ahead of the tube. In general, this means going to a very fast spectrograph camera and this is difficult to do using ordinary Schmidt optics because of the size of existing image tubes. A new type of camera has now been designed by Dr I. Bowen to fill this need. It employs two concentric spherical mirrors and operates as an 'inverse Cassegrain', with the (collimated) beam first striking the smaller convex mirror, then diverging to fill the large concave mirror, and coming to a focus beyond the convex. The advantage of this system is that the focal plane is completely accessible. Also, it is completely achromatic, as no corrector is used. By choosing the ratio of the radii of the two mirrors properly, the 3rd order aberrations are completely eliminated, while by re-introducing some 3rd order aberrations, these may be used to balance off 5th and 7th order terms, so that very good definition may be obtained at a focal ratio of $f/1$. Disadvantages are the small field (not too important for low-dispersion applications) and the fact that the field is curved, concave toward the incoming beam.

Such a camera for a beam diameter of 6½ inches (16·5 cm) and having a speed of $f/1{\cdot}1$, has now been constructed at Lick. Preliminary optical tests were carried so far to the point of obtaining an axial resolution of 20 microns, and it is expected that with final adjustment the resolution will be appreciably closer to the calculated value of 12 microns.

III. Scientific Programme.

A. Using Sb–Cs cathodes. 1. The programme of measurement of line intensities in planetary nebulae mentioned in the last report has been continued in collaboration with Dr. Aller. Following the completion of the first programme in the blue region, a new programme was undertaken covering the wavelength region from 4600 Å to 5900 Å. The first programme served to indicate the possibilities of the tube in spectrophotometric work since a number of the nebulae had independent determinations of line intensities in the blue region. The present programme takes advantage of the large information gain of the tube in the green-yellow region (in the range of 50 to 100 over Kodak 103aD at 5500 Å) to permit us to observe faint lines not otherwise observable in these objects. As before, advantage is taken of the linear relationship between intensity and plate density and calibration was obtained from observations of a very hot star of known energy distribution.

2. High time-resolution spectroscopy of variable stars. Using a single trail technique, it has proved possible to obtain continuous coverage of the spectroscopic variations in a rapid variable during one or two nights. The time resolution obtainable depends upon the brightness of the star and the quality of the seeing. In good seeing, a time resolution of about five minutes is possible for photographic magnitude 12·5, at a dispersion of 48 Å/mm on the photocathode. Using this technique, new observations of the rapidly varying, short-period spectroscopic binary AE Aqr have been obtained which show a wealth of detail. Observations of this type have also been obtained of SS Cyg. McRae + 43°1, T CrB, RW Tri, and UX UMa.

3. *Nuclei of galaxies.* Observations have been obtained of the nuclear regions of a number of galaxies, with interesting results. Here, advantage is taken of the speed-gain of the tube to permit use of the large scale of the coudé to study the spectroscopic details in the nuclei of rather faint galaxies. Comparison with spectra of the same objects taken with the conventional prime-focus spectrograph shows that the increased scale at the coudé does bring out detail otherwise not detectable. The linear response of the tube has been used to derive energy curves for the nuclei of several galaxies, comparing them to a hot star of known energy distribution.

4. *High dispersion spectroscopy.* Some tests have been made with the tube mounted at the focus of the 160-inch (406 cm) camera and a dispersion of 2 Å/mm on the cathode. Observations were made of the K line of several stars in which structure of the interstellar K line was known to exist. The known structure plus some new features were detected, and the application shows promise.

B. Using S–1 cathodes. Several runs have now been made using the tube at the focus of the 20-inch (51 cm) Schmidt and a grating giving a dispersion of 32 Å/mm on the cathode. Observations have been made of Jupiter, Saturn, and Venus, of a number of Mira stars at minimum light and several of the so-called 'dark brown stars' found by Leighton *et al.*, at Mt Wilson. In Jupiter, very good observations were obtained of the methane band at 10 900 Å which should lead to a temperature determination. Venus has been observed out to 13 000 Å which shows well resolved $C^{12}O_2^{16}$ and $C^{13}O_2{}^{6}$ bands in the region of 12 200 Å. One experiment has also been made of photographing Venus at 10 500 Å at the 160-inch camera with a dispersion of 4 Å/mm. Here, a 3-hour exposure shows the band system of CO_2, with enough dispersion that line widths of individual rotation-vibration levels can be determined, leading to a determination of atmospheric pressure. The speed-gain of the S–1 cathode over Kodak 1Z (ammoniated) was measured to be 100–300 at 9400 Å at Lick, depending upon cathode sensitivity as described elsewhere (**6**).

C. Application of Electronography to Stellar Photometry. Dr Kron and I have used this tube at the 61-inch astrometric reflector of the U.S. Naval Observatory, Flagstaff, to verify Lallemand's result that electronographic plates of star fields can be used for photometry and give a linear relation between star intensity and plate density as measured in a Schilt-type photometer. We find that if the plate (Ilford G5, developed five minutes in D–19) is traced in a microphotometer operated as a densitometer and one integrates the areas under the star-image intensity-profiles, a linear relationship exists between − 5/2 log (area) and magnitude over a range from the plate limit to at least five magnitudes over the plate limit. The importance of this result for stellar photometry is obvious. Using electronography, it is not necessary to have a photoelectric sequence to calibrate photographic photometry, as is usually done. All that is required is to set up in the field a few bright stars by photoelectric transfer. Then, the magnitudes of all fainter stars can be determined directly from the electronographic plates. The speed of the electronographic process means that very faint stars—heretofore accessible only to the largest telescopes—can be reached with moderate sized instruments of long focus, for example a 60-inch (150 cm) Cassegrain. Also, a 'photoelectric sky' is not required except during the setting up of the bright standards, a real advantage to many observers located in less than the best climatic conditions.

Finally, under the best conditions, one may expect by this technique to obtain higher accuracy and to observe to a fainter limit than is possible photographically owing to the finer 'grain' (track size) of the electronographic plates.

C. R. Lynds and W. C. Livingston, Kitt Peak National Observatory, Tucson, U.S.A.

Activity of the past two years have been primarily in the use of the TSE and Cascade type tubes for actual astronomical observation and with a minimum of laboratory experimentation and tube evaluation work.

As reported in London (7) Lynds is regularly (three scheduled nights a month) using a TSE tube made by English Electric Valve Co., England, in connection with the 84-inch (213 cm) telescope for the spectroscopic observation of quasi-stellar objects. This programme continues (**8**). A significant improvement in the system was made by Lynds and Villere with the development of an $f/2$ semi-solid fused quartz Schmidt spectrograph camera. Approximately $4\frac{1}{2}$ inches (11·4 cm) in diameter this camera fits into the tube focusing solenoid. Superior imagery, now limited mainly by the transfer lens, plus response to 3200 Å is achieved.

On the very high dispersion spectrograph of the McMath Solar Telescope, guest investigators have used an RCA-Carnegie cascade tube for interstellar and emission line problems (**9**).

In the laboratory Livingston studied the granularity as a function of sample size of photographic materials and available image tube recordings (**10**). The results show that, while modern phosphor output tubes have a substantially finer 'grain' than competing photographic materials, the effective granularity of the image tube record increases over the photographic emulsion at low spatial frequencies. This large scale granularity function must be taken into account when predicting the S/N of an observational record.

Plans for the future include special image tube Cassegrain spectrographs for the 84- and 150-inch (213 cm and 381 cm) telescopes and an alternate horizontal spectrograph at the solar telescope. The latter should prove of special value for tube evaluation. Being a fixed installation, laboratory type tests can be conducted on actual stellar objects. In the past, conflict with daytime work has seriously limited image tube experiments. Now an image tube system can remain intact for the prolonged intervals required for the development of equipment.

J. D. McGee and colleagues, Applied Physics Department, Imperial College, London

Electronographic tube—The Spectracon

The development of the Spectracon—a mica-window, single-stage, electronographic image tube for recording spectra has continued (**11**). The following features have been investigated:

1. Introduction of pre-prepared photocathodes has been well established of S.9, S.11 and S.1 photocathode types but the S.20 cathode tends to deteriorate after this operation. Detailed study of the reasons for this exceptional behaviour of the S.20 cathodes have been carried on without any conclusive findings.

2. Improvements in the electrostatic electron accelerating electrode system in the tube and of the solenoid producing the focusing magnetic field have resulted in improved image resolution and geometry.

3. A pneumatic device for pressing the emulsion-coated Melinex film against the mica window has been investigated and after lengthy tests a compromise between the rubber-coated roller and the pneumatic system appears the most satisfactory.

4. Tubes with S.1 cathodes were tested and compared with Kodak Z-type emulsions. For exposure times of about 10 minutes, the background due to tube thermal emission was acceptable while the speed was from 500 to 5000 times greater than the Z-type emulsion as the wave length varied from 8000 to 12 000 Å.

Other tests showed that the Spectracon tube can be cooled at the cathode end to − 50°C or lower quite safely and under these circumstances the background from S.1 cathodes should be greatly reduced. These tests are continuing.

5. Screening to exclude the effects of stray magnetic fields on the image resolution by transverse deflection has shown that it is quite feasible to reduce this effect sufficiently that a variation of about 0·5 gauss in the worst direction produces a negligible deterioration of the image resolution.

6. Telescope trials with Spectracon tubes have been carried out in co-operation with the staff of the Royal Greenwich Observatory, Herstmonceux, using a spectrograph at the coudé focus of the 30-inch (76 cm) telescope. These have confirmed the earlier tests at Mt Wilson and in the laboratory that the gain in speed over direct photography using baked IIa–O is about 10 times when Ilford G5 emulsion is used.

7. The transmission of electrons through mica sheets (**12**) comparable to Spectracon windows has been measured by yet another method and the results confirm those already reported. In summary it may be stated that the Spectracon operating at 40 kV with a 4μm window is as 'fast' as a direct electronographic recording system operating at 25 kV and that the signal/noise ratios of images recorded under these two conditions are indistinguishable.

8. As it became clear that, in practice, much of the inherent gain in speed of the image tube system was lost because it had to be used with inferior optical systems it was decided that a special optical system designed to operate at $f/1$ with the Spectracon, should be designed and built. This is in the early design stages which is a joint project between the staff of the Royal Greenwich Observatory, the Applied Optics Section (Drs C. G. Wynne and W. T. Welford) and The Applied Physics Section of Imperial College.

9. A shortened tube, 15 cm instead of 28 cm long, has been made successfully and preliminary assessment of its characteristics are promising. With such a tube the complete image tube and focus coil unit can be about 10 cm in diameter and 25 cm long and weigh about 7 kg.

Cascade Image Intensifier

The development of the cascade image intensifier (**13**) has continued and tubes of very satisfactory performance can now be made. The performance may be summarized as follows:

three-stage tube with 38 mm diameter useful field. Gain $\sim 10^6$ (blue light), S.9 or S.11 cathodes, (50 to 100μA/l m). Resolution 50 lines per mm at centre and remaining > 30 lines per mm over field of 30 mm diameter. Background so low that exposures of about one hour are practicable with transfer optics which give a speed gain of about 100 times.

Probably the outstanding defect of these tubes is the presence of bright scintillations, as appears to be the case in all tubes of this type. These spurious signals produce a background which would be objectionable in high quality images. Investigations continue into the causes and possible means of removal of these signals.

Similar three-stage cascade tubes are being made with S.1 or S.20 primary photocathodes.

T. Dunham, Jr., Mt Stromlo Observatory

A five-stage secondary-emission image intensifying tube has been mounted in the coudé spectrograph of the 74-inch (188 cm) telescope, at the focus of the 120-inch (305 cm) camera. The optical axis of the camera is 26 inches (66 cm) above the heavy steel frame of the instrument, and measurements show that distortion of the field of the 400-gauss permanent magnet in which the tube is mounted is insignificant. The tube was carefully selected by 20th Century Electronics to have a very low dark count (less than 50 scintillations cm^{-2} s^{-1}). The resolving power, measured with a Baum microprojector is 25–30 line-pairs per mm. The display at the output was photographed with a transfer lens on stationary 35 mm film, and also on moving film to record individual scintillations (**14**). Due to the wide range in brightness of the scintillations, granularity is severe unless the aperture ratio of the transfer system is held to *f*/8 to *f*/11. Under these conditions the image tube showed only a moderate gain over direct photography. The moving film scintillation counting technique, however, allows use of the full aperture (*f*/1·9) of the transfer system. Performance is being compared with that for the stationary film technique. A Fabry-Pérot étalon in the collimated beam of the spectrograph has been provided with mechanical scanning, synchronized with a precision grating drive, so that approximately 50 fringes can be recorded simultaneously with the image tube (**15**). The spectra of stars to 4·0 magnitudes have been recorded at 1·8 Å/mm in 20 minutes.

M. J. Smyth and P. W. J. L. Brand, Royal Observatory, Edinburgh, Scotland

A Lenard window image tube (Spectracon) constructed at Imperial College is being evaluated as a spectrophotometric detector, in the Cassegrain spectrograph of the 36-inch (91 cm) reflector (**16**).

A suitable environment for the image tube on the telescope, free from electrical discharge due to damp air, has been achieved by the use of airtight compartments for the high voltage components, and a thermo-electric cold trap between the spectrograph camera and the tube cathode; and the image shift caused by motion of the tube with respect to the geomagnetic field had been investigated. The equivalent quantum efficiency of the tube is being determined, as a function of absolute spectral sensitivity, spatial frequency response, and granularity noise. Granularity is derived from computer analysis of density measurements made with a digitized microphotometer. A calibrated tungsten lamp is used for sensitivity determination, and the modulation transfer function is also being studied. The methods being developed will be immediately applicable to other types of image tube, and to conventional photography.

The tube is being used in a stellar spectrophotometric programme that is also being carried out by direct photography, thus enabling a comparison of photometric effectiveness to be made.

D. J. Bradley, Physics Dept., Queen's University, Belfast

1. The performance of image tubes for recording high resolution spectrograms has been evaluated (**17**). Tubes tested include a five-stage T.S.E.M. tube (**18**), single and three-stage

phosphor-photocathode tubes (19) and a Lenard window tube (20). For spectroscopy in general the Lenard window tube has outstanding signal-to-noise performance with, in addition, the great advantage of a linear density-exposure relationship. Signal induced background, at an overall voltage of 40 kV, was found to be negligible and absorption lines were faithfully recorded.

Operating with a voltage of only 26 kV, the signal-induced background and noise of the T.S.E.M. tube, and the wide spread in the multiplication factor for individual photoelectrons, resulted in such a decrease in the effective signal-to-noise ratio in the recorded image as to make it unsuitable for quantitative spectroscopy.

The signal-to-noise performance of the cascaded phosphor-photocathode tube was much superior to that of the T.S.E.M. tube.

2. A single stage magnetically focused image converter with a 'Spectrosil' window, S–9 photocathode and a phosphor screen output has been employed to record Fabry-Pérot interferograms of λ 1849Å of Hg I for the first time (21). The performance of the image tube at this wavelength was entirely satisfactory. Cs-I photocathodes have been tested in photocells before employing them in solar blind tubes.

3. A development channelled image intensifier (22) has also been tested for spectral recording. With channels 0·5 mm diameter and an effective resolving limit of 1–2 line pairs per mm, a spectral resolving power of 10^4 was obtained (23) for a camera focal length of 30 cm, employing the second ring of a Fabry-Pérot étalon with an effective finesse of 10. Such a combination of interferometer-channelled image intensifier would be particularly useful for rocket and satellite spectroscopy of extended sources.

W. A. Hiltner, Yerkes Observatory, University of Chicago, U.S.A.

Experimental work on image tubes has been discontinued here, principally because of lack of time, certainly not because of any lack of faith in the electronographic process. When opportunity presents itself, I want to apply a system to astronomical research.

Something over one year ago Drs Kron and Kent Ford and I intercompared several methods of image intensification. Our conclusion at that time was that the gain of electronography over classical photography in the blue is set at near one times the quantum efficiency of the photocathode in percent, in agreement with measurements of the quantum efficiency of photographic emulsions, and probably slightly less for the cascaded image tube although any loss is essentially compensated for by easier and more efficient operation of the cascaded tube.

During the past year we built a spectrograph for the cascaded image tube. It has a *f*/2 Schmidt camera with field flattener. We used it on the McDonald 82-inch reflector for the observation of quasi-stellar and Haro-Luyten objects. Spectrograms of 200 Å/mm were obtained of 18·4 magnitude objects (about the limit of visual setting) in about two hours when unbaked Kodak IIa-O emulsion was used (24). The spectrograms were quite satisfactory for radial velocity measurements. We will use the spectrograph again at the next opportunity.

Dr W. Kent Ford, Carnegie Institution of Washington, U.S.A.

The Carnegie Image Tube Committee, with support from the National Science Foundation, has sponsored the development of a two-stage, cascaded image intensifier at RCA. The tubes have multialkali photocathodes, 38 mm diameter useful areas, and provide information gains of 10 over conventional photography. Several dozen tubes have been fabricated by RCA for the Carnegie Committee. These are being distributed with the necessary auxiliary equipment to various observatories. Thus far, sets have been installed at Yerkes, Kitt Peak, Mt Wilson, Lick, Lowell, University of Texas, Naval Observatory, Dominion Astrophysical, Palomar, University of Arizona, and Mt Stromlo.

Dr R. E. Danielson and colleagues, Princeton University Observatory, U.S.A.

RCA engineers and members of the Princeton Observatory are currently completing tests of an image orthicon camera system for use on the Stratoscope II Program (**25, 26**). Uncooled high resistivity glass target tubes with 0·001 mesh spacing have yielded integration periods in excess of ten minutes. Sensitivity comparison, using stellar sources, indicates at least a factor of ten improvement in speed over 103a–G film used in Stratoscope II. Tests now in progress, operating the image orthicon in the 'isocon' mode, indicates further improvement. Analysis of the data also indicates good linearity of exposure versus the volume of the video pulse.

Within the next year the Princeton Observatory expects to operate a SEC Vidicon, built by Westinghouse, comparable in image and target size to the 3-inch (7·6 cm) image orthicon, under the similar test conditions. The SEC Vidicon tests will also involve cooled operation and spectral readout.

M. F. Ingham, the Department of Astrophysics, Oxford

Mr J. E. Beckman investigated the usefulness of image intensifier tubes to astronomy (**27**). Two types of tube were evaluated: an English Electric Valve Co. transmission secondary emission tube (the so-called Imaging Photomultiplier) and one of the Lenard window type. For both tubes the gain, spectral sensitivity, resolution and dark emission were measured. The distortion introduced by the secondary-emission tube was also studied. Finally, the English Electric tube was used in conjunction with the solar spectrograph to obtain high dispersion (4·2 mm/Å) spectra of Venus and Jupiter. With these spectra it was possible to place an upper limit on the abundance of oxygen in the atmosphere of Venus and to measure the abundance of hydrogen in the atmosphere of Jupiter.

At the Atomic Energy Research Establishment at Harwell, Mr A. V. Hewitt of this Department, and Dr W. N. Charman have been conducting similar experiments with an E.M.I. cascade tube (**28**) which they will shortly use with the 30-inch (76 cm) telescope at the Royal Greenwich Observatory Herstmonceux for experiments in stellar spectroscopy.

Dr R. J. Davis, Project Celescope, Astrophysical Observatory, Smithsonian Institution, Cambridge, U.S.A.

A special-purpose television camera tube (the Uvicon) and associated optics and digital television system (the Celescope) have been developed to meet the Smithsonian Astrophysical Observatory's requirement for measuring the ultraviolet brightnesses and the positions of a very large number of objects throughout the sky. The wavelength region of interest, from 1000 to 3000 Å, is observable only from above the Earth's atmosphere, thus precluding our utilization in this wavelength region of the well-developed photometric techniques used in ground-based astronomy.

The unique characteristic of the Uvicon tube is its combination of accurate photometric measurement ability, ultraviolet sensitivity and ruggedness of construction. In order to achieve the necessary photometric accuracy, range and sensitivity the Uvicon incorporates a secondary electron conduction (SEC) target to amplify and store the photoelectric image from the photocathode. The tube is available with two photocathode materials: caesium iodide, which is insensitive to wavelengths longer than about 2000 Å, and caesium telluride, insensitive to wavelengths longer than about 3000 Å. The short wavelength cut-off is that of the lithium-fluoride face-plate (1050 Å at room temperature). The basic design of the tube allows considerable flexibility in regard to its geometrical and electrical configuration; in the case of the Uvicon, the face-plate is incorporated as a refractive element in the optical design of the telescope system, and both the imaging section and the electron scanning gun utilize only electrostatic focusing and scanning techniques. The Uvicon was developed and manufactured by Westinghouse Electric Corporation under contract to the Smithsonian Astrophysical Observatory. Another Westinghouse tube, an LEM camera tube developed for the Lunar Excursion Module, if modified to provide ultraviolet sensitivity, would provide improved performance over the existing Uvicon configuration.

In order to achieve full photometric accuracy, all operating parameters of the tube must be very carefully controlled – including the timing of the voltage variations required for operating the tube. In this way, we have been able to achieve a photometric precision of $\pm$ 5% in the measurement of laboratory light sources during calibration of the Celescope experiment. Our investigation has also indicated that this accuracy can be improved, by a significant but unknown amount, by more careful control of operating voltages and sequences. In order to achieve full photometric accuracy, the Uvicon must be calibrated under conditions as near as possible to those under which it is to be used. The transfer function is far from linear, and varies from one part of the raster to another. Variations with temperature are less severe, but must also be determined by individual calibration if the tube is to be used outside of the range + 30° to + 10°C. There is, fortunately, no reciprocity failure, total exposure being exactly proportional to the product of intensity and exposure time between one and 150 seconds.

Typical operating characteristics of the Celescope system, utilizing the Uvicon, are as follows:

	A-Type	D-Type
Photocathode Quantum Efficiency	3% at 2000 Å	8% at 1400 Å
Spectral passband	1050 to 3200 Å	1050 to 1900 Å
Photometric Range	6×10^{-16}	to 6×10^{-12} ampere
Photometric accuracy	$\pm$ 5%	
Field of view*	1·1 inch (2·8 cm) diameter circle at photocathode	
Positional measurement accuracy*	0·01 inch (0·25 mm) at photocathode	
Limiting resolution*	0·01 inch (0·25 mm) at photocathode	

*These characteristics were not critical to the Celescope experiment, and can fairly easily be improved if required for other experiments.

In order to take full advantage of the speed and accuracy with which a television system such as this can generate data, the signal must be analysed by a high-speed digital computer. We attain our highest accuracy by using digital techniques to control the Uvicon and digitize its output signal before transmission. By generating our video signal in digital form, it can be recorded on magnetic tape in a format acceptable as input to a digital computer. The equipment and techniques required for operating the Uvicon as a stellar photometer have been developed by Electro-Mechanical Research Inc., under contract to the Smithsonian Astrophysical Observatory.

This research has been supported in part by Contract NAS 5–1535 from the National Aeronautics and Space Administration.

L. W. Frederick, Leander McCormick Observatory, University of Virginia, Charlottesville, U.S.A.

The mica window S–1 tube has been completely remounted so that it can be operated with the potential on. The system is in excellent operating order and is being used to continue the late-type stellar studies previously reported. Some 380 additional plates have been taken on the new McCormick reflector primarily by George Lockwood, a graduate student here.

A cascade S–1 image intensifier converter is currently being mounted and will use a circulatory slush cooling system. This tube appears to be an excellent tube.

An eight-foot non-magnetic optical bench has been fabricated to assist in the aligning and testing of our equipment. A Baum type pattern projector has been purchased from Hobbs, Ltd. to assist in the evaluation of tubes.

Our primary objective is to apply the Carnegie tubes to astronomical problems. Hence, we do very little experimental and development work.

BIBLIOGRAPHY

1. Lallemand, A., *et al.* 1966, *C. r. Acad. Sci. Paris*, **262**, 838.
2. Duchesne, M. 1965, *J. Phys.*, **26**, 117.
3. Duchesne, M., Feissel, M., Guinot, B. 1965, *Notes Inf. Publ. Obs. Paris*, fasc. 26.
4. Hézard, C. 1965, Thèse de 3ème cycle, Paris.
5. Combes, M. 1966, Contribution à l'étude de l'atmosphère de Jupiter, Thèse de 3ème cycle, Paris.
6. 1966, *Adv. Electronics Electron Phys.*, Academic Press, **22A**, 761.
7. Livingston, W. C., Lynds, C. R., Doe, L. A. 1966, Recent Astronomical Research Utilizing a High Gain Image Intensifier Tube. *Adv. Electronics Electron Phys.*, ed. L. Marton (New York and London: Academic Press), **22B**, 705.
8. Lynds, C. R., Stockton, A. N., Livingston, W. C. 1965, New Spectroscopic Observations of Quasi-Stellar Sources, *Astrophys. J.*, **142**, 1667.
   Lynds, C. R., Stockton, A. N. 1966, The Large Redshift of the Quasi-Stellar Source 1116 + 12, *Astrophys. J.*, **144**, 446.
   Lynds, C. R., Hill, S. J., Heere, K., Stockton, A. N. 1966, New Spectroscopic Observations of Fourteen Quasi-Stellar Sources, *Astrophys. J.*, **144**, 1244.
9. Thaddeus, P., Clauser, J. F. 1966, Cosmic Microwave Radiation at 2·63 mm from Observations of Interstellar CN, *Phys. Rev. Letters*, **16**, 819.
10. Livingston, W. C. *Adv. Electronics Electron Phys.*, ed. L. Marton (New York and London: Academic Press), **23**, in press.
11. McGee, J. D., Khogali, A., Ganson, A., Baum, W. A. 1966, *Adv. Electronics Electron Phys.*, Academic Press, London, **22A**, 11.
12. McGee, J. D., Khogali, A., Ganson, A. 1966, *Adv. Electronics Electron Phys.*, Academic Press, London, **22A**, 31.
    Jeffers, S., McGee, J. D. 1966, *Adv. Electronics Electron Phys.*, Academic Press, London, **22A**, 41.
13. McGee, J. D., Airey, R. W., Aslam, M., Powell, J. R., Catchpole, C. E. 1966, *Adv. Electronics Electron Phys.*, Academic Press, London, **22A**, 113.
14. Dunham, T. 1960, *J. opt. Soc. Amer.*, **50**, 1129.
15. Dunham, T. 1962, *Astr. J.*, **67**, 575.
16. Brand, P. W. J. L., Smyth, M. J. 1966, Use of a Lenard-window image tube for astronomical spectrophotometry, *Adv. Electronics Electron Phys.*, **22B**, 741.
17. Bradley, D. J., Bates, B., Juulman, C. O. L., Majumdar, S. 1964, *Appl. Opt.*, **3**, 1461.
18. Wilcock, W. L., Emberson, D. L. Weekley, B. 1960, *Nature*, **185**, 370.
19. McGee, J. D., Catchpole, C. E. 1962, *British IEE Conf. Rept.*, Series 5, p. 182.
20. McGee, J. D., Wheeler, B. E. 1962, *Adv. Electronics Electron Phys.*, Academic Press, London, **16**, 47.
21. Bradley, D. J., Bates, B., Juulman, C. O. L., Majumdar, S. 1964, *Nature*, **202**, 579.
22. McGee, J. D. 1961, *Rep. Progr. Phys.*, **24**, 167.
23. Bradley, D. J. 1966, *Sciences et Industries Spatiales*, no. 7/8.
24. Hiltner, W. A., Cowley, A. P., Schild, R. E. 1966, *Publ. astr. Soc. Pacif.*, **78**, 464.
25. Cope, A. D., Luedicke, E. 1965, *Adv. Electronics Electron Phys.*, Academic Press, London, **22A**, 175.

26. Flory, L. E., Pike, W. S., Morgan, J. M., Boyer, L. A. 1966, *Adv. Electronics Electron Phys.*, Academic Press, London, **22B**, 885.
27. 1966, *Adv. Electronics Electron Phys.*, Academic Press, London, **22A**, 369.
28. 1966, *Adv. Electronics Electron Phys.*, Academic Press, London, **22A**, 101.

APPENDICE 2. GROUPE DE TRAVAIL SUR LA QUALITÉ DES IMAGES ET LE CHOIX DES SITES

(préparé par J. Rösch, président)

La dénomination de ce groupe pourrait, à la rigueur, ne mentionner que le Choix des Sites, car le rôle de la qualité des images dans presque tous les genres d'observation est suffisamment reconnu maintenant pour que toute opération de prospection en tienne compte, parmi un certain nombre d'autres facteurs; et par contre, toute étude sur la qualité des images est destinée, directement ou indirectement, à rendre plus efficace la recherche des sites les meilleurs.

Les méthodes appliquées à la sélection des sites ne se codifient que lentement, sans doute parce qu'en général, jusqu'ici, le temps disponible pour chaque opération a été limité par la hâte de passer à l'implantation de l'observatoire projeté. Cet état de choses a aussi pour conséquence que faute de conclusions de portée générale certaine, la littérature est assez pauvre en publications dans ce domaine, et les résultats des prospections restent souvent consignés dans des documents peu diffusés en dehors des organismes responsables des opérations. Le présent rapport sera donc limité aux travaux parvenus à la connaissance de l'auteur et ne saurait prétendre être exhaustif.

La délimitation préliminaire des régions propres à l'implantation d'un observatoire de programme déterminé s'oriente progressivement vers une utilisation plus approfondie des données météorologiques et des quelques principes dont la validité se précise, notamment l'intérêt des altitudes d'au moins 2000 mètres pour la photométrie, l'importance significative de la faiblesse des écarts des températures diurne et nocturne, l'avantage des sommets isolés pour la stabilité de la température nocturne, le rôle néfaste des inhomogénéités dans les couches voisines du sol.

La publication d'études de ce genre sur des régions limitées, telles que celle de Hansen *et al.* pour les Iles Hawaï (1), celle qui paraîtra prochaînement sur les campagnes de l'ESO en Afrique du Sud (2) ou celles en cours en France et en Espagne, et au Brésil, constituera, par confrontation avec les résultats obtenus dans les observatoires futurs, une source précieuse pour la compréhension des phénomènes.

L'un des problèmes les plus irritants dans les débuts d'une prospection reste l'évaluation de la nébulosité nocturne. J. Saissac a tenté, pour le Sud de la France, de dresser une carte du nombre d'heures utilisables en partant du nombre de jours de pluie relevé dans les stations seulement pluviométriques dont le réseau est dense, et des corrélations établies entre ce nombre et le nombre d'heures utilisables pour les stations météorologiques complètes qui sont beaucoup moins nombreuses (3). Pour un travail plus direct, R. de Freitas-Mourao a réalisé, selon les suggestions faites au Symposium no. 19 de l'UAI, une chambre 'tout-ciel' susceptible d'une autonomie de plusieurs jours et permettant, sur des images exposées environ 1 minute, de mesurer l'absorption par les magnitudes limites atteintes (4).

Après le nombre d'heures photométriquement utilisables dans l'année, l'élément essentiel (à part le point particulier de la diffusion atmosphérique pour les observations de la couronne solaire) est la qualité des images. Devant la difficulté pratique (surtout en campagne) d'obtenir un spectre de Fourier à deux dimensions des déformations de la surface d'onde, on se contente d'un simple nombre, dont on cherche cependant une valeur objective attachée à une interprétation bien définie. C'est le but de l'ASM (Automatic Seeing Monitor) de Babcock (5) ou de la lunette polaire de Walker (6). Dans le premier cas, on obtient directement une mesure de l'agitation, donc de la fluctuation de la direction de la normale à la surface d'onde. L'ASM a été largement utilisé au Chili (Cerros Tololo, Morado, Pachon, La Silla), où Irwin a notamment étudié la variation de la quantité mesurée avec la distance zénithale (**7**), ainsi qu'au Mt Palomar et à Siding Spring (Australie) (**8, 9**). Dans le cas de la lunette polaire, on pourrait obtenir la même donnée, mais moyennant un procédé d'exploitation quantitative des traînées photographiques ou une traduction photoélectrique directe de l'agitation. Un travail dans ce sens est en cours à l'Observatoire du Pic du Midi.

Une telle estimation de la valeur d'un site, quant à la qualité des images, repose sur le postulat que les spectres de Fourier des déformations des surfaces d'onde réelles appartiennent à une famille à un seul paramètre. Le moyen le plus simple de s'assurer de la légitimité de la prévision de la qualité des images dans un grand instrument à partir de mesures d'agitation est d'établir empiriquement une corrélation entre les deux phénomènes; c'est ce qui a été entrepris à l'Observatoire du Pic du Midi par des mesures quasi-simultanées d'une part, au moyen de *l'imagemètre*, du diamètre équivalent de l'image d'une étoile donnée par un réflecteur de 105 cm, d'autre part, au moyen du chercheur du même instrument, de l'agitation fournie par des traînées photographiques.

Une étude observationnelle des corrélations s'impose encore plus si l'on veut prévoir une qualité d'image d'après des mesures de fluctuations ou de gradients de température dans les premières dizaines de mètres de l'atmosphère au-dessus du sol. Un travail de ce genre fait à l'Observatoire de Haute-Provence (**10**) montre bien une corrélation limite entre les fluctuations à une hauteur égale à celle de l'ouverture du télescope de 193 cm et le diamètre estimé de l'image observée dans ce même instrument, du moins jusqu'à une certaine valeur de ce diamètre; mais on trouve aussi des diamètres de l'image *plus grands* que ceux qui suivent la corrélation, ce qui révèle des détériorations dans des couches plus élevées que la prise de température. Un tel résultat ne peut que confirmer que la solution la plus sûre consisterait à faire un prélèvement optique à un niveau élevé aussi proche que possible de celui auquel se trouvera, en moyenne, l'ouverture du futur instrument.

RÉFÉRENCES

1. Hansen, R. T., Hansen, S. F., Price, S. 1966, *Publ. astr. Soc. Pacif.*, **78**, 14.
2. *Publ. Obs. austr. Européen* (à paraître).
3. Saissac, J. 1966, Documents du Comité National Français d'Astronomie.
4. Freitas Mourao, R. R. de Thèse, Université de Paris (à paraître).
5. Babcock, H. W. 1963, Mt Wilson and Palomar Observatories.
6. Harlan, E. A., Walker, M. F. 1965, *Publ. astr. Soc. Pacif.*, **77**, 246.
7. Irwin, J. B. 1966, *Astr. J.*, **71**, 28.
8. *A. Rep. Dir. Mt Wilson Palomar Obs.*, 1963–64, et 1964–65.
9. 1966, *CARSO Rep.*, Carnegie Institution of Washington.
10. Bourlon, Ph. 1966, Documents du Comité National Français d'Astronomie.
11. Bateson, F. M. 1964, *Publ. Univ. Pennsylvania*, **10**, 1–139.
12. Matsushima, S. 1964, *Publ. astr. Soc. Pacif.*, **76**, 224.
13. Hogg, A. R. 1965, *Mt Stromlo Publ.*, AST. 65/169.

10. COMMISSION DE L'ACTIVITE SOLAIRE

PRÉSIDENT: Dr Z. Švestka, Astronomical Institute of the Czechoslovak Academy of Sciences, Ondřejov, Czechoslovakia.

VICE-PRÉSIDENT: Dr J. T. Jefferies, Hawaii Institute of Geophysics, 2525 Correa Road, University of Hawaii, Honolulu, Hawaii 96822, U.S.A.

SECRÉTAIRE: Dr A. D. Fokker, University Observatory 'Sonnenborgh', Servaas Bolwerk 13, Utrecht, the Netherlands.

COMITÉ D'ORGANISATION: R. G. Giovanelli, K. O. Kiepenheuer, W. O. Roberts, J. Rösch, A. B. Severny, Henry J. Smith, M. Waldmeier.

MEMBRES: Allen, Aly, Athay, d'Azambuja (L.), Babcock (H. W.), Balli, Bell, Belorizky, Billings, Bogorodsky, Bonov, Bray, Brückner, Bruzek, Bumba, Cardus, Cimino, Coutrez, Dezsö, Dizer, Dodson-Prince, Dollfus, Dubov, Dunn, Elste, Ferraro, Fracastoro, Gaizauskas, Gleissberg, Gnevyshev, Gnevysheva, Godoli, Goldberg, Gullon, Gurtovenko, Howard (R. F.), Jensen, Karimov, Kleczek, von Klüber, Kopecký, Krat, Kristenson, Krook, Künzel, Lehnert, Leighton, Letfus, Loughhead, Macris, Mathias, Mattig, Maxwell, McKenna, Menzel, Mergentaler, Michard, Mitra (A. P.), Miyamoto, Mohler, Moiseev, Moreton, Müller (H.), Mustel, Nagasawa, Nedungadi, Newkirk, Notuki, Öhman, Orrall, Pecker (Ch. W.), Pick, Popovici, Prokofiev, Ramanathan, Razmadze (T. S.), Richardson (R. S.), Ringnes, Saito (K.), Schlüter, Schröter, Servajean, Shapley (A. H.), Shaposhnikova, Shklovsky, Simon (P.), Sitnik, Slonim, Smerd, Smith (E. v. P.), Stepanov, Steshenko, Suemoto, Tandberg-Hanssen, Trellis, Tuominen, Valníček, Wild (J. P.), Wlérick, Xanthakis, Zirin, Zwaan.

The research activity in the field of Commission 10 has been enormously increasing as one can easily observe when comparing the present list of references with the Draft Reports of the preceding three-year periods. Therefore, some limitation of the contents of the Draft Report is necessary. We have not included any paper presenting only observations without any discussion of them and as the solar-terrestrial physics is concerned, only those papers are included, which contribute to our knowledge of the Sun and its wave and particle emission. That means that we have not mentioned here the numerous papers which treat the influence of the solar activity on the terrestrial magnetosphere, ionosphere and troposphere, unless some information on the problems of solar physics or solar cyclic variations can be deduced from them. We have included all the remaining papers concerning solar activity published since the beginning of 1964 and very few papers which appeared in 1963 and clearly could not be included into the last Professor Severny's Report.

Most of the members of the Organizing Committee and many members of the Commission participated in preparing the Draft Report. Particularly: Section 6 is based on the report on sunspots prepared by W. Mattig, section 8 on the report on prominences prepared by J. T. Jefferies, section 9 on the report on the active corona prepared by J. Rösch, and section 10 on the report on active radio phenomena prepared by A. D. Fokker. E. v. P. Smith substantially contributed to sections 11, 12, and 13. A. B. Severny's report on magnetic field measurements has been used extensively when preparing sections 4, 6, and 7.2. A. B. Severny and N. V. Steshenko also provided me with a list of references and abstracts of papers published in the U.S.S.R. and S. Nagasawa informed me in a similar way about papers published in Japan. A detailed report on terrestrial effects of solar activity prepared by C. Sawyer unfortunately arrived too late to be included in this Report. My thanks are due to all these contributors as well as to all other solar physicists who have provided me with useful information on the recent progress of their work.

1. GENERAL PROGRESS REPORT

1.1. *International projects*

During the reported period four international projects were organized, closely connected with problems of solar activity:

International Years of the Quiet Sun (IQSY) started on 1 January, 1964 and lasted for two years. From the point of view of solar physicists, the main aims of IQSY were (1) to maintain a complete patrol service during the period of the solar minimum, and (2) to profit by observations of fairly isolated active processes on the Sun both as the study of solar active regions and the solar-terrestrial physics are concerned.

The study of isolated active regions was the main purpose of the Cooperative Study of Solar Active Regions (CSSAR), organized by R. Michard for half a year during the IQSY period, from 1 April to 30 September 1965. Its main aim was to study the development of active regions without any disturbing effects of adjacent solar activity, with particular emphasis on the final decay of local magnetic fields.

Commission 10 also organized the Proton Flare Project (PFP) from 1 May to 30 September, 1966, with P. Simon as the Chief Coordinator. Very successful proton flare forecasts enabled to get many data on proton flares of 7 July and 2 September, 1966, and on the associated active regions.

The project on 'Rapid Variations of Solar Magnetic Fields' organized by V. A. Krat, had many features similar to CSSAR and PFP.

Final results of these projects had not yet been known when this Draft Report was submitted for press. It is supposed that the scientists responsible for these projects give general information on them during the Prague Assembly and some of the results will be reported at the IAU Symposium on the Structure and Development of Solar Active Regions in Budapest, in September 1967.

1.2. *Books and Symposia*

Three important books, of general interest for those working in the field of solar activity, appeared during the reported period: A monograph about sunspots by Bray and Loughhead (1964) gives a description of a wide range of sunspot phenomena and their interpretation. Kundu (1965*a*) published 'Solar Radio Astronomy' and gave here a very detailed account of our present knowledge of the radio emission from the Sun. And the book 'Solar Activity' written by Tandberg-Hanssen (1967) has tried 'to give a treatment of all the different manifestations of activity on the Sun in a unique framework: as the inevitable results of the interaction between solar magnetic fields and the solar plasma'. In all these books and most extensively in the last one, readers can find a very complete account of the results obtained in the last decade in the study of solar activity, with extensive lists of references.

Papers presented at six symposia, conferences and summer schools organized in 1963 or earlier, were published during the investigated period: The Symposium on Stellar and Solar Magnetic Fields = IAU Symposium No. 22 (Lüst 1965), the AAS–NASA Symposium on the Physics of Solar Flares (Hess 1964), the Plasma Space Science Symposium (Chang and Huang 1965), the Summer School 'Introduction to Solar Terrestrial Relations' (Ortner and Maseland 1965), the Utrecht Symposium 'The Solar Spectrum' (de Jager 1965*a*), and the 1st Consultation on Heliophysics at Tatranská Lomnica (Mergentaler 1964). The following meetings were held during the reported period: The Symposium on Astronomical Observations from Space Vehicles = IAU Symposium No. 23, 17–20 August 1964 (published in *Ann. Astrophys.* **27** and **28**); the Meeting on Sunspots in Florence, 9–12 September, 1964 (Righini 1966); the Meeting on Solar Magnetic Fields and High Resolution Spectroscopy in Rome,

14–16 September, 1964 (edited by M. Cimino; not yet published and consequently, some papers presented there could not be included in this Report); the 3rd Consultation on Solar Physics at Tatranská Lomnica, 13–16 October 1964 (Kopecky 1965*a*); Solar Physics NATO Summer Course at Lagonissi, 12–26 September, 1965 (Xanthakis 1966*d*); Colloquium on 'The Fine Structure of the Solar Atmosphere', Anacapri, 6–8 June, 1966 (Kiepenheuer 1966*b*); the Inter-Union Symposium on Solar-Terrestrial Physics in Belgrade, 29 August–2 September, 1966 (not yet published); and the 4th Conference on Solar Physics held in Sopot and organized by the Wroclaw Observatory in September 1966 (not yet published).

1.3. *Scientific progress report*

(For details, see sections 2–14).

Remarkable progress has been made in our understanding of the general development of active regions. It seems to be clear enough that all solar magnetic fields are basically bipolar units and that the supergranular pattern of the solar atmosphere plays a fundamental role in the development of active regions. The rapidly improving techniques of magnetographic measurements has permitted to resolve relatively homogeneous magnetic fields in complicated multipolar structures and to detect the very fine structure of solar magnetic fields with height in the solar atmosphere. Strong fields in small limited regions in the photosphere as well as isolated chromospheric fields are of particular interest. Chromospheric pictures obtained with the domeless coudé refractor at Anacapri (Kiepenheuer 1966*b*) give an example of the progress achieved in the photographic study of the fine structure of the solar chromosphere.

When studying sunspots, one still meets with the most serious difficulty of the stray light, which contributes more than 50% in the umbra light. Both the experiments carried out with balloon-borne telescopes as well as new methods of theoretical analysis proposed by several authors, have tried to solve this problem. Nevertheless, some of the results of theoretical analyses are still in contradiction, particularly as to the existence or non-existence of sub-hydrostatic pressures in sunspots. Magnetographic measurements indicate the existence of twisted fields in sunspots. Generally, the variation of the magnetic field strength with distance seems to be similar to that of a dipole submerged beneath the photosphere, at least as the intensity distribution is concerned. Great attention has been paid to the fine structure of the Evershed effect, but the authors did not arrive at a unique solution of the problem. Progress has been made in the difficult transfer problem for lines formed in the magnetic field, where solutions have been obtained for greatly generalized conditions.

Attention has been paid to variations of the magnetic field associated with flares as well as to the location of flares with respect to the magnetic field. While our knowledge of the first problem has not significantly improved since the last Draft Report, several authors now agree that flare brightenings are distributed along both sides of a neutral line of the field and usually extend parallel to the neutral line, in their later development. A remarkable progress has been achieved in our knowledge of the characteristic features of proton flares and of the active centres capable of producing proton flares. Successful forecasts of proton flare events during the period of the Proton Flare Project give good evidence of this.

The many years' standing problem of the broadening mechanism of Balmer lines in flares has been definitely settled in favour of the Stark effect, at flares projected on the disk. The filamentary structure of flares, first suspected by Suemoto and Hiei, has been proved by several authors. The great interest of solar physicists in flares is well demonstrated by the high number of newly proposed or modified flare theories and many of them have already taken the filamentary structure of flares into account. One has to admire many ingenious ideas contained in these theories but, on the other hand, one cannot escape the feeling that we are still far from being able to understand what the flare phenomenon actually looks like.

Observations of longitudinal magnetic fields in prominences have progressed in the past three years. They indicate that field strengths in quiescent prominences range from about 2 to 10 G, while active prominences and loops are associated with substantially stronger fields. Great attention has been paid to loop systems which imply a conveyance of field into the corona and are a manifestation of the storage in the corona of fast, non-thermal particles. Another problem which has been extensively studied, is the filament activation by active regions. Newly detected characteristic oscillations of 'winking' filaments, sometimes starting tens of minutes before a flare appearance, might be very significant for our understanding the active processes on the Sun.

Analyses of monochromatic solar corona indicate a filamentary structure of coronal condensations, matter in the condensation being largely concentrated in loop systems. A combination of Coronascope II, solar eclipse, and nearly synoptic K-coronameter observations permitted to get a unique series of coronal data covering several solar rotations. In this way, a streamer co-rotating with the chromospheric features could be followed for about two months. June 14, 1966 was the historical date, when on the first time, the Sun was photographed from the surface of the Moon.

The role of gyro-resonance absorption and emission by electrons has become better understood, both with regard to the microwave slowly varying component and to the emission of type IV bursts at cm and dm wavelengths. Progress has also been achieved towards an understanding of the harmonic structure of type II bursts, an agreed-upon mechanism, however, has not yet been accepted. On the other hand, most of the properties of type I noise storms are still unexplained and the question, how to account for elliptical polarization of type III bursts, is not yet satisfactorily answered. Thanks to the great number of single-frequency records, made by many observatories on a regular scheme, it is now possible to derive spectral diagrams for most of the major radio events. On the other hand, there is still a great want of more data on source positions and polarizations of type IV emission.

Direct X-ray spectroheliograms at a variety of wavelengths have verified close correlation of increased X-emission with plage activity; at short wavelengths, beneath ~ 15 Å, the emission arises entirely from centres of activity and seems to be concentrated in small, hot elements embedded in a larger lower temperature region of the coronal condensation. The degree of concentration to active regions as observed by slitless spectra in the XUV wavelengths, also depends markedly on the ionization equilibrium temperature of the emitting ion and seems to be a function of the stage of development of the spot group. It has been agreed that the X-ray emission from active regions as well as many X-ray enhancements associated with flares are of thermal origin. Contradictory conclusions appear, however, as to the height of these emissions in the solar corona, as well as to the mechanism producing possibly non-thermal X-ray bursts. The observation of the solar eclipse in 1966 by means of the Explorer 30 satellite by the Arcetri group has demonstrated the wide applicability of the RTT monitoring system.

The number of papers dealing with observations of solar particles from space vehicles has increased many-fold since the last report. Very important studies were made with the Pioneer 6 space probe; the data indicate the existence of large numbers of filamentary tubes, originating at the Sun but intertwined in interplanetary space. In spite of twisting and intertwining of these filaments, the spiral configuration remains. Cosmic rays are envisioned as constrained to the flux tubes. A number of proton increases was observed aboard several space probes, only partly associated with flares. Some events were even recorded aboard two or more Soviet or American space probes and deep-space satellites. The important role of α-particles in streams of high-energy particles, recognized during the past years, throws some doubt on the terminology of the 'proton' flare events. Nuclei with $Z \geqslant 3$ have been observed after major flares on at least four occasions and possibly in some other events, as well. It is not yet clear, however, to what extent the composition of the nuclei reflects the photospheric abundance.

Evidence for solar-flare electrons was also obtained from IMP and Electron satellites and from the Mariner 4 space probe.

A comparison of IMP-1 magnetic field data with solar magnetograph observations has shown that the interplanetary magnetic field is directly tied to and co-rotates with the solar photospheric magnetic fields. During the minimum period of solar activity, since 1962, the interplanetary magnetic field was divided into 2, and since late 1963, into 4 sectors, alternating in direction toward and away from the Sun, and this structure had correspondence to grossly averaged magnetic features in the photosphere. This low-activity structure of the interplanetary space becomes less stable as the solar activity increases.

Summarizing these discoveries, one observes that, generally, two basic features characterize the past period in the research of solar activity: a progressive tendency to discover fine structure in all the studied phenomena and a gradual transformation of our old conception of the interplanetary space into the new one of an extensive and expanding solar atmosphere, directly studied with deep space probes.

2. EQUIPMENT AND METHODS

2.1. *Optical observations*

A new solar research station of the Cambridge Observatory was completed at Rabat (Malta). It is equipped with a large horizontal solar telescope of about 40 meter length with a large Babcock grating and all facilities for photographic, photometric and magnetographic work. At the Observatory of the University of Hawaii four new solar instruments have been installed: A 10-inch aperture (25 cm) Lyot coronagraph which feeds a coudé spectrograph with dispersion 1 or 6Å/mm; a dual coronagraph which is to obtain simultaneously photographs through 2Å interference filters of the corona in the green line and of active prominences in Hα; a moderately large-scale Hα camera which is used to study individual active regions under conditions of good seeing; and a photographic polarimeter for determining Stokes parameters of the radiation in broad spectral bands. The Manned Spacecraft Center installed in 1966 three Hα solar telescopes employing a Lyot-Öhman filter, one on a seventy foot tower at Houston (Texas) and the other two in Carnarvon (Australia) and in the Canary Islands. New solar observatory of the Deutsche Forschungsgemeinschaft near Locarno, equipped for observations of active solar phenomena, was described by Brückner (1964).

The new domeless coudé refractor for solar work on Capri was described by Kiepenheuer (1964*b*). Inner tube contains a 35 cm objective of 15 m focal length and an outer tube protects this first one against wind and rain. Through the polar axis the solar light is guided to a spectrograph. A new solar telescope and spectrograph with 70 cm coelostat and 50 cm objective was also installed at the Kwasan Observatory (Nakai and Kubota 1964). The McMath solar telescope at Kitt Peak was described by Pierce (1964), and the description of an evacuated tower telescope planned for Sacramento Peak Observatory was given by Dunn (1964).

A triple heliograph, which allows to take simultaneously Hα and K line filtergrams together with a photospheric picture was installed at the National Observatory in Athens. At this Observatory also a new automatic solar telescope equipped with Halle Hα filter and TV technique was built. At the Monte Mario Observatory in Rome, regular patrol service has been started in 1964–65 both in the Hα and K lines. A new Halle Hα filter at Irkutsk allows to get automatically Hα filtergrams at 0, $\pm$ 0·25, and $\pm$ 0·50Å from the line centre (Skomorovsky 1966). The composite filtergrams taken in three wavelengths (0 and $\pm$ 0·50Å) in the Hα line at Lockheed were described by Larmore and Ramsey (1964). New equipment, which allow to photograph simultaneously solar disk and solar limb, were described by Michard (1965) and by Carroll (1965). A new space-velocity coronagraph, which allows to get simultaneous photographs of the prominence image and of the Hα line in its spectrum was installed at Ondřejov (Valníček and Kleczek 1964, Valníček *et al.* 1965). The new spectroheliograph of the Catania Observatory, applicable to Hα, Hδ, K and λ4247 lines was described by Fracastoro and Cristaldi (1964). A new electrophotometer for fast recording of the Hα line profiles of limb phenomena was built at the Kiev Observatory of the Academy of Sciences (Gurtovenko and Skorik 1965).

A new Zeiss (Jena) coronagraph was installed at Lomnický Štít in High Tatra Mountains (Lexa 1963). The new coronagraph and spectrograph at Climax was described by Rush and Schnable (1964). Modifications of the Lyot coronagraph, leading to an increase of speed of measurement of the absolute intensities of coronal lines in various coronal regions have been realized at Pic-du-Midi (Demarcq *et al.* 1965). The Astrophysical Observatory Arosa has been equipped in 1965 with a new Zeiss (Oberkochen) coronagraph. It is combined with a powerful grating spectrograph; using a dispersion of 5 Å/mm and a slit width of 0·03 mm, the exposure time for the λ5303 line is one second only. The K-coronameter of the Norikura Corona Station and the method of observation and reduction of the observed results were described by Nagasawa *et al.* (1965). A description of the optical scheme and the main characteristics of a large Lyot-type coronagraph which is being made in U.S.S.R., was given by Nikolsky and Sazanov (1966).

A combined narrow-band filter consisting of 3 Å multilayer filter and the three thickest elements of Lyot-Öhman filter was described by Zirin (1966). Ramsay (1966) gave a description of a new monochromatic filter based on a series of plane-parallel Fabry-Pérot interferometers aligned in a parallel mode. Lyot-Öhman and Šolc filters produced in Czechoslovakia were described by Valníček (1965*a*). Hamana and Suzuki (1966) studied a method of reducing the effect of finite filter pass-band on the intensity measurements of flares. A double image line shifter proposed by Öhman (1966*a*) has been used by Nilsson (1966) for the purpose of determining Doppler velocities of dark and bright surges. Öhman (1966*b*) also made a suggestion for reducing scattered light in sunspot photometry by a special system of two diaphragms and tests have shown that the influence of diffracted light from the circumference of the objective is reduced to a minimum with this device (Kussovsky 1966). Öhman (1966*c*) also proposed other methods for the same purpose. An electronic device for discriminative measurements of sunspot areas was described by Boev *et al.* (1965). Giovanelli (1966*b*) discussed methods and techniques for the observation of sunspots in white light, including discussions of seeing, scattered light, thermal deformation of mirrors and photographic techniques.

2.2. *Magnetic field measurement*

Since the beginning of 1964 considerable effort has gone into the construction of new instrumentation at the 150-foot solar tower telescope at Mt Wilson. A guider and scanning system have been installed which now enable to scan the solar image very accurately in any desired raster pattern. The solar magnetograph has been completely rebuilt, and two spectrum lines can now be used simultaneously. Data are recorded on a magnetic tape recorder for later analysis with a digital computer (Howard 1966). A double magnetograph recording simultaneously magnetic field in two different lines (λ5250 and λ6103, or Hα) also started to work in 1965 at the Crimean Observatory (Severny 1966*a*) and a similar device has also been developed at Nizmir (Moscow), for simultaneous observations in λ5250 and Hβ lines (Zhulin and Mogilevsky 1965*b*, Mogilevsky *et al.* 1965). An earlier description of the electronic scheme of the Crimean magnetograph was published by Nikulin (1964) and further details concerning the magnetograph at Nizmir were described by Iospha and Mogilevsky (1965) and by Ioshpa and Obridko (1965*d*).

A new magnetograph recording all components of the solar magnetic field has been designed at Irkutsk (Kuznetsov *et al.* 1966). The domeless coudé refractor at the German station at

Capri (Kiepenheuer 1964*b*) is combined with a longitudinal and transverse magnetograph with automatic mapping of the field distribution (Deubner 1966). The McMath-Hulbert Observatory has improved its measurements of solar magnetic fields through the addition of a Glan Thompson prism and mica quarter-wave-plate to the optical train of the vacuum spectrograph. A magnetometer for the study of weak longitudinal magnetic fields has been developed by Bhattacharyya at Kodaikanal. A Babcock-type magnetograph has been established in the solar tower on Monte Mario in Rome. At this Observatory also the measurement of sunspot magnetic fields has been modernized and is now carried out by means of a photomultiplier and a study of the Zeeman effect by atomic beam without using high dispersing spectrographs is in progress.

A detailed summary of methods of measurements of solar magnetic fields was presented by Severny (1966*c*). Bruns *et al.* (1965) described a method used at Crimea for measurements of transverse magnetic field. The method of observation of the total vector magnetic field with the Pulkovo Observatory magnetograph was described by Kotljar (1964). The method of simultaneous measurements of solar magnetic fields on two levels in the solar atmosphere was explained by Zhulin (1965) and the selection of proper lines for this purpose was discussed by Dubov (1966*a*) and Severny (1966*a*). Kuklin (1966*b*) described the electro-optical modulator of the solar magnetograph at Irkutsk. The degree of reliability of radial velocities obtained by means of a magnetograph was discussed by Ioshpa and Obridko (1965*b*) and by Ikhsanov (1966) and the influence of seeing on measurements of radial velocities with a magnetograph was investigated by Gopasyuk and Severny (1964). A method for deriving the complete magnetic field from magnetographic measurements of the longitudinal Zeeman effect was proposed by Schmidt (1964, 1965) and modified by Semel (1967). Possible effects of steep field gradients between close spots on results obtained with a scanning magnetograph were discussed by Teske *et al.* (1964). The method of magnetic field measurements at Meudon has been analysed by Semel (1967).

A method of measuring solar magnetic fields by using a birefringent filter with twin transmission bands was proposed by Öhman (1965) and a very similar method actually has been applied in Meudon. A method of the determination of three components of the magnetic field of sunspots by means of a 'lambdameter' was described by Rayrole (1964). The same author (Rayrole 1966) has designed and proven a method for deriving the full magnetic field vector from photographic measurements of the Zeeman displacements of lines observed in various polarization states. The method uses simultaneously two lines from the same multiplet but with different Zeeman patterns, which gives the possibility to determine also the scattered light in spots.

The magnetograph of the High Altitude Observatory at Climax, used for measurements of magnetic fields in prominences using the Hα line was described by Lee *et al.* (1965). Sensitivity of the equipment equals two gauss. Smolkov (1966*a*) has shown that, in principle, it is possible to get information on the coronal magnetic field when applying a sensitive photoelectric method to the λ5303 line.

2.3. *Solar radio astronomy*

New radiotelescopes have been installed at several observatories: 8 mm wavelength at the Crimean Observatory (Efanov and Moiseyev 1965), 2700 MHz at the Dominion Radio Astronomical Observatory in Penticton (Covington and Locke 1965), 2000 MHz at New Delhi (Sarma and Joshi 1965), 930 MHz at Bordeaux (Bernyer *et al.* 1966), 700 MHz at Fleurs, 408 MHz at Nançay, 87·4 MHz in Catania, and 74 MHz solar radio polarimeter in Ottawa (Bhonsle and McNarry 1964*b*). Eight-foot disk antennas for continuous solar observation at 1420, 2700, and 5000 MHz have been installed in 1966 by the Manned Spacecraft Center at Houston, Carnarvon and in the Canary Islands. A method permitting records of very weak microwave bursts was described by Durasova and Yudin (1966).

At Fleurs Observatory, in 1964, additions have been made to the east-west arm of the 21 cm crossed grating interferometer to enable simultaneous operation at 43 cm wavelength. A 16 element 408 MHz array giving a resolving power of 1·6′ in the E–W direction with separation between lobes of 25′ was installed at Nançay (Clavelier 1966). The large radiotelescope completed at this observatory has been used for solar observations at 5000 MHz with a resolving power of 1·2′ × 6·5′. At Toyokawa, the 16 + 2 element compound interferometer for use at 9400 MHz was extended to 32 + 2 elements in 1966, and the maximum resolving power was increased from 0·7′ to 0·35′. A new 32 + 2 element compound interferometer for use at 3750 MHz and a two-element interferometer at 1000 MHz were completed in 1967 (H. Tanaka 1966). A multi-element interferometer for 2700–2900 MHz has been built at Algonquin Radio Observatory (Covington *et al.* 1965). An interferometric method which can easily measure the position, motions and intensity of an isolated active region on the disk was described by Turlo (1965).

The Dapto radio spectrograph, which worked in the frequency range from 5 to 210 MHz has been extended to 2000 MHz (Suzuki *et al.* 1964). A radio spectrograph working in the frequency range from 50 to 210 MHz was completed in 1966 at Ondřejov (Tlamicha 1966). A 20 channel spectrum radiometer operating between 20–120 MHz at a sampling rate of 100 per second with a fibre-optic real time direct recording system has been installed in Ottawa in 1966. The design of a receiver with 60 channels intended for the frequency band 160–320 MHz as a radio spectrograph with high sensitivity and high time resolution, was described by Van Nieuwkoop (1964). A description of the radio spectrograph equipment at Boulder was given by Lee and Warwick (1964).

2.4. *Observations from space vehicles*

Apart from many launchings of balloons carrying equipment for recording high-energy particles, two balloon-borne solar telescopes were launched during the investigated period. The first of them, Coronascope II, was described by Newkirk and Bohlin (1964). In the balloon-borne coronograph the scattered light was suppressed to about 10^{-9} of the mean solar disk brightness. Two successful flights of Coronascope II took place on 3 June and 1 July, 1965. Solar corona was also observed from the Moon on 14 June, 1966, with the Surveyor I TV camera (Newkirk 1967). The second balloon-borne telescope, constructed at the Fraunhofer Institute, was described by Kiepenheuer and Mehltretter (1964). It contains an 0·5 Å Hα Lyot-Öhman filter and a grating spectrograph. Stenflo has started the design of a spectrograph to be used in space research for studying solar magnetic fields in small structure elements of the solar surface. Quite a number of rocket flights were made in U.S.A., U.S.S.R., Great Britain and France for the purpose of obtaining EUV and X-ray spectra and photographs of the Sun. The techniques used for the measurement of X-radiation were reviewed by Boyd (1965).

Only one solar observatory was functioning in orbit during the first year of IQSY. This was the SOLRAD 1964–01 D satellite, launched in January 1964. It was equipped with ionization chambers to cover the ranges 1–8 Å, 8–14 Å, and 44–60 Å, in the RTT system (Kreplin 1965). In 1965, three more solar satellites were launched: OSO–2 (1965–7A), launched on 3 February, which provided a month of continuous observation of the solar X-ray emission and also yielded spectroheliograms and white corona measurements; SOLRAD 1965–16 D, launched on 9 March, which carried the same measurements as the 1964–01 D satellite, but the 8–14 Å detectors were considerably more sensitive; and SOLRAD 1965–93A (Explorer 30), launched on 19 November, which was the most sophisticated satellite in the SOLRAD series. Its digital memory system worked for about one month, but the RTT system provided very useful data on solar X-ray emission within the ranges 0·5–3, 1–8, 1–20, and 8–16 Å until August

1966, when the satellite began a precessional motion. Solar X-ray flux within 2–10 and 8–18 Å was also measured aboard the Electron 2 and 4 satellites (Vasilev 1966).

Many other satellites have provided data on interplanetary magnetic fields, solar wind disturbances and solar high-energy particles. Particularly, these were IMP–1 (launched on 26 November 1963), IMP–3 (29 May 1965), IMP–4 (1 July 1966), OGO–1 (5 September 1964) and Vela 5 and 6 (20 July 1965). The most important information on solar wind disturbances and energetic proton bursts, however, were obtained with deep space probes. Mariner–4, launched on 28 November 1964, provided data on solar particle events for 10 months prior to the Mars fly-by. Soviet space probes Zond–3 (launched on 18 July 1965), Venus–2 (12 November 1965) and Venus–3 (16 November 1965) carried detectors recording 1–5 MeV protons and a gas discharge counter giving information on protons with energy $\gtrsim$ 40 MeV (Vernov *et al.* 1966). Pioneer–6, launched on 16 December 1965, has given valuable data on the anisotropy of solar protons events. Due to this 'high density' of space probes and highly excentric satellites in 1965, some solar wind disturbances and proton events could be recorded aboard several space vehicles in different distances from the Earth.

The network of ground-base RTT monitoring stations has been substantially increased and the observations intensified. As an example, about 700 transits of the Explorer–30 satellite were telemetered in April–August 1966 by the network of the Centre National d'Etudes Spatiales. A solar eclipse was observed by means of the RTT monitoring by the Arcetri group (Landini *et al.* 1966*a*).

3. CYCLIC VARIATIONS OF SOLAR ACTIVITY

3.1. *General contributions*

A monograph, particularly discussing the cyclic variations of the solar activity and the structure and dynamics of the subphotospheric levels was written by Rubashev (1964*a*). Kopecký (1967) summarized our present knowledge on the periodicity of sunspot groups.

Rotational means of the Wolf numbers for 'geophysical' rotation periods 0–1797 (1832–1964) were computed by Schnur and Wagner (1965). Godoli and Allen (1964) summarized and discussed the indices used for describing various characteristics of the solar activity phenomena and the correlation between different indices of solar activity was investigated by Zabza (1964). Autocorrelation of several indices of solar activity was made by Altschuler and Sastry (1965), who compared the years 1937–38, 1947–48, and 1957–58. A power spectrum analysis of time series of several solar indices taking into account only phenomena within $\pm$ 1 day distance from the central meridian was made by Shapiro (1965*a*). The result showed a pronounced peak around 27 days and appreciable peaks at the first few harmonics. The mechanism producing the harmonics was also discussed (Shapiro 1965*b*). Eigenson's recurrence indices have been revised by Ringnes (1964*b*).

The east-west asymmetry of sunspots was studied by Dezsö (1965) and by Dezsö and Gerlei (1964*b*). Developing groups prevail on the eastern and declining spots on the western solar hemisphere. On the contrary, Pajdušáková (1964) could not find any east-west asymmetry of sunspots. Xanthakis (1965) discussed the asymmetry in the distribution of the sunspot magnetic field strengths on the northern and southern solar hemispheres and Gleissberg (1964) compared the north-south asymmetry in solar wind (deduced from the geomagnetic activity) and in solar spottedness. Fredga (1965) found evidence of a coupling between the sunspot activity of the two hemispheres in consecutive cycles (12–19) when the latitude distribution was taken into account. She considers this result as a support of Alfvén's magnetohydrodynamic sunspot theory. Godoli and Tagliaferri (1965) pointed out that the increase of maximum sunspot numbers during the past four cycles was not reflected in other phenomena of solar activity and some of them even decreased.

The fact discovered by Trellis (1963) that the axis of sunspot belts is not identical with the axis of the solar rotation was independently confirmed by Kuleshova (1964). The axis deviation amounts to 0°53 according to Trellis and to 0°80 according to Kuleshova.

3.2. *The 11-year solar cycle*

Gnevyshev (1967) has found that each 11-year cycle of solar activity consists of two processes with different physical properties. The variety of shapes of the 11-year curves depends on the way these processes overlap. The existence of such a double 11-year cycle has been shown for coronal observations (Gnevyshev 1965*a*, *b*), sunspots (Antalová and Gnevyshev 1965, Gnevyshev and Antalová 1965), solar radio flux (Gnevyshev 1965*c*) and proton flares (Gnevyshev and Křivský 1966, Křivský and Krüger 1966). Secondary maxima in the spot formation activity connected with the activity development in low-latitude zones were detected by Schegoleva (1965*a*). Vitinsky (1965*c*) has shown that the latitude factor is more important for the indices of solar activity than the factor of the phase of the 11-year cycle. Chistiakov (1965*d*) found five breaks in the 11-year cycle sunspot curve.

Several authors discussed the variations of various aspects of solar activity with the 11-year cycle. Thus Ringnes (1964*c*) studied the life-time and Arroyo (1965) the area distribution of sunspots, Dezsö and Gerlei (1964*a*, 1965) the ratio of penumbral-umbral areas, Xanthakis (1966*c*) different indices of solar activity, Babin *et al.* (1965) the solar supercorona, Tousey *et al.* (1964) the Lyman-alpha profile, Bolshakova (1965) the solar wind velocity, Le Squeren-Malinge (1964*b*) the coronal electron density, Švestka (1966) the proton flare occurrence, and Parmenter (1966) the density of spicules. Giovanelli (1964) has concluded that the frequency of appearance of new spot groups is the prime variable during the cycle. The mutual relations between the differential rotation and the solar cycle were discussed by N.O. Weiss (1965*b*).

The Waldmeier relation between the maximum Wolf number and the time of rise of the cycle were modified and partly applied to the 20th solar cycle by Rubashev (1964*b*), King-Hele (1966), Alexiou and Poulakos (1966) and Xanthakis (1966*b*, *c*). King-Hele has predicted $R_{max} = 140$ in 1968 and 110 in 1978. According to Xanthakis (1966*b*) the next maximum will not appear before 1968·5. Xanthakis (1966*a*) also discussed the unusually high maxima in the 9th and 19th solar cycles.

Differences between the latitude drift of sunspots on the northern and southern hemispheres were pointed out by Schegoleva (1965*b*) and Gudzenko *et al.* (1965). On the basis of a detailed study of the latitude and longitude drift of sunspots Ramanathan and Natarajan (1965) have concluded that sunspots originate in a layer, which takes part in the differential rotation and consequently must be very close to the photosphere. A modified Babcock theory was used by Kopecký (1966*b*) to an interpretation of the butterfly diagram. The fine structure of this diagram was discussed by the same author (1965*b*, 1966*a*). Tuominen (1965*c*) modified the Babcock theory of the Sun's magnetic field as to the latitude field strength distribution. On the basis of Kopecký's and Tuominen's modifications of the Babcock theory the butterfly diagram was discussed by Godoli (1964*b*, 1965, 1966*b*). Chvojková (1965*a*, 1966*b*) also contributed to the Babcock theory having extended the investigation beyond the latitude of 30°. The drift of sunspots beyond 30° to higher latitudes could be qualitatively explained.

Two zones of coronal activity discovered by Waldmeier in 1940 have been confirmed in 1963 after the beginning of the new spot cycle (Waldmeier 1964). The main zone and the polar zone were well separated from each other by a belt of solar prominences. Hyder (1965*a*) showed that in 1957–58 the reversal of Sun's polar fields was synchronized with the poleward drift of high-latitude filaments. A coincidence of this reversal with the maximum Ca plage activity for each hemisphere was found by Godoli (1964*a*). Although bright surges at the limb should be typical phenomena of the activity centres, observations of Godoli and Mazzucconi

(1967) showed that during 1957–64, 6% of them were polar ($\varphi \geqslant 50°$). They seem to follow the polar faculae activity. Leighton (1964) tried to clarify some aspects in the Babcock's theory, particularly the expansion and migration of unipolar and bipolar magnetic regions and discussed the relation of this process to the 11-year solar cycle. The behaviour of the polar fields over several solar cycles was discussed by Sheeley (1966).

Sunspot minima were investigated by Giovanelli (1964, 1966*a*). On separating spot groups of the old and new cycles, it becomes apparent that the main distinction between different solar minima are the time displacements between the decay- and rise curves. The solar activity during the IQSY period was described in detail by Dodson and Hedeman (1964*d*, 1966*b*) and by Dodson *et al.* (1964, 1965). The last minimum was also discussed by Vitinsky (1965*b*), Chistiakov (1965*a*) and Dinulescu (1965). According to most of these authors the minimum occurred in July 1964 and it was unique by the lowest number of spotless days since the limit of available data in 1818 (Dodson and Hedeman 1966*b*). Bell and Wolbach (1965) have shown that Danjon's equation of the brightness variation of lunar eclipses with the phase of the solar cycle predicts dates of sunspot minima with a high degree of accuracy. The next minimum has been predicted by this method for 1974·9.

3.3. *Other cyclic variations*

A critical summary on the statistics of sunspots, with particular reference to the long-term solar cycle and secular variations of the solar activity was presented by Waldmeier (1966). He concludes that the long-term solar cycle seems to be produced by variations of the relative abundance of spot groups of different life times (in agreement with Ringnes, see below) and areas. According to Dezsö and Kovácz (1965) the ratio of umbra and penumbra area in sunspots shows a long-term variation, the period of which seems to equal eight solar 11-year cycles. Kopecký (1964*a*) proved the existence of the 80-year cycle in the average importance of sunspot groups and discussed some hydromagnetic hypotheses on its interpretation (1964*b*). Schove's auroral numbers were used by Gleissberg (1966) for studying some properties of the 80-year cycles. 38 and 41 years, respectively, have been found as the mean lengths of the period of ascent and descent. Secular variations, distinct from the 11-year period, were found by Ringnes (1963, 1964*b*, 1965) for sunspot groups of various life-times. Groubé (1965) has called attention to the fact that properties of solar cycles before and after 1856 differ in several aspects.

Longer cycles of solar activity were discussed by Golubtsov (1965) on the basis of the data given by Schove. A cycle of 650 years consisting of two 325-year cycles found earlier by Rubashev has been verified. A long-term relation between the solar activity and the Earth's rotation was discussed by Link (1964*a*). A very detailed analysis of indirect records of solar activity in the past 2600 years was made by the same author (Link 1964*b*). According to Link's results, three cyclic variations are clearly recognizable: 11, 80 and 400 years.

3.4. *Physical and statistical interpretations*

General characteristics of the solar activity were tentatively explained by Das (1964), who assumed the existence of a rapidly rotating core of the Sun and outbursts of matter from this core to the surface layers. Gudzenko and Chertoprud (1965) proposed an explanation of solar cyclic activity, according to which a cellular belt originates and gradually dissipates in the convective layer. Separate cells, which detach from the belt and ascend, appear in the form of spot groups. This model was later specified by a detailed analysis of the dissipation of such a belt into cells (Chertoprud 1966). A mechanism of oscillatory instability with a period of the order of ten years has been proposed as another interpretation of the cycle of solar activity by Iroshnikov (1965). The oscillations are due to an interaction of the differential rotation and the magnetic forces. A similar mechanism, with a 22-year period of oscillations, was considered by Krause (1965). A more simplified dynamo theory of the solar magnetic cycle was proposed by Csada (1965). A magneto-convective mechanism for the solar cycles has been suggested by M. L. White (1965). Proshnikov (1965) investigated a possible oscillatory instability of the gas in the region surrounding the surface of the maximum of unperturbed entropy. An oscillatory unstable regime appears in the case when negative dissipation in the convective layer predominates over the positive dissipation in nonconvective layers. The period of the resulting oscillations, however, does not seem to be of interest from the point of view of the known solar cyclicity.

Jakimcowa (1967) has shown that the solar activity fluctuations may be represented by means of a stationary stochastic process. This method makes possible prognoses of solar activity. Chen Biao and In Chun Lin (1965), as well as Paveliev and Pavelieva (1965) tried to find periodic terms in the cyclic variation of the solar activity, which determine the variable heights of the 11-year solar cycles. Some phenomenological properties of the cyclic activity of the Sun were investigated by Gudzenko and Chertoprud (1964) using the cybernetic 'black box' scheme method. Chertoprud and Kotov (1965), Kotov *et al.* (1965) and Gudzenko and Chertoprud (1966) interpreted statistically the variations of solar activity as due to an autonomous generator with one degree of freedom synchronized with a weak periodic force.

Trellis (1966*a*) deduced a formula which gives the height of tide on the Sun caused by planets. The greatest effect is due to Venus and Jupiter and Earth follows as to the importance. A correlation was obtained between the height of tide on the Sun calculated according to this formula and sunspot areas or the numbers of newly born spot groups, respectively (Trellis 1966*b*, *c*). Jose (1965) has pointed out that the variation in the motion of the Sun around the centre of mass of the solar system has a periodicity of 178·7 years, which seems to be related to the cyclic variations of the solar activity. R. M. Wood and K. D. Wood (1965) also tried to explain the 11-year cycle as due to solar motion around the centre of mass of the solar system and found the theoretical period of one solar cycle equal to 11·08 years. Planetary influences on the unstable layers of the Sun have also been considered by Romanchuk (1965*b*) who made some forecasts of solar activity based on the constellations of planets.

4. ACTIVE CENTRES

(The radio, X-ray and corpuscular emission from active regions are reported in Sections 10.2, 11, and 13, respectively.)

4.1. *The structure of active regions*

Morphology of an active region on the Sun was discussed by Kiepenheuer (1966*d*). The chromosphere in active regions seems to be filled by threads of the length of 10^4–10^5 km and thinner than 1″. Sometimes these threads form a loop-like structure (Kiepenheuer 1966*c*). Loop features in all three layers of the solar atmosphere were discussed by Kleczek (1964*d*). Tsap (1965*a*) confirmed that the orientation of fine chromospheric Hα structure is in good agreement with the transverse magnetic fields, which however, is not the case for the orientation of the penumbra filamentary structure observed in integrated light. According to Howard and Harvey (1964), dark fibrils seem to lie perpendicular to isogauss lines of the longitudinal field measured in the photosphere, while the filaments in general lie parallel to these isogauss lines and over the null line of the field. An excellent correlation between the Hα fine structure in an active centre and the longitudinal component of weak and moderately strong magnetic fields was found by Smith and Ramsey (1966), generally in agreement with an earlier Leighton's (1965) result for chromospheric network outside the active regions.

Bumba and Howard (1965*a*, 1966*a*) have found that the supergranular pattern of the solar atmosphere plays a very fundamental role in the development of active regions. New regions are formed in or immediately adjacent to expanding old magnetic fields. The first appearance of the emission of a new active region takes place in the 'intersupergranular' space. This is in agreement with Leighton's (1965) suggestion that the supergranulation currents concentrate the magnetic fields into the chromospheric network.

After reducing solar magnetograms covering a period of seven solar rotations, Bumba *et al.* (1966*b*) have shown that the Sun would appear as a magnetic variable star if viewed near the plane of its equator. Measurements of the integrated solar light in the K_2 line lead to the conclusion that also in the K-line light the Sun can be observed as a variable star (Bumba and Růžičková-Topolová 1967). H and K line profiles in active regions were studied by Paciorek (1965) and Engvold (1966). According to Korobova (1965) the area of spots and the area of plages on the phase of growth of an active region are connected by a linear relation. Similar study for the phase of decline was made by Yilmaz (1964*a*).

Kuznetsov and Stepanov (1966*a*) found motion velocities in local magnetic fields increasing from -50 m s^{-1} at the photospheric level to -500 m s^{-1} at $h \simeq 2000$ km (Hβ line) and to -2 km s^{-1} at $h \simeq 4500$ (K_3 line) and identified them as velocities of the sporadic solar wind. Motions in an old centre of activity at the limb were studied by Zirker (1965). He found larger line-of-sight velocities than for spicules in quiet regions. Kaplan and Petrukhin (1964) tried to interpret peculiar motions of gases observed far from any sunspot group in a region of increased magnetic field.

Parker (1964*e*) has discussed the problem whether the presence of magnetic fields enhances the generation of sound waves. He has shown how the magnetic fields are directly responsible for the enhancement of the chromosphere and corona above plages. Pikelner (1965, 1966) has summarized older ideas of himself and other authors about the appearance of active regions (plages) caused by a 10 to 20% increase in the convection velocity, this increase being due to the fact that a weak magnetic field suppresses small-scale motions and decreases turbulent viscosity. Pikelner and Livshitz (1964) concluded that above active regions the energy is transmitted to the chromosphere by decelerated Alfvén waves, which are transformed in the corona to magneto-sound waves and dissipate. The manner in which the dissipation can occur was investigated by Gold (1964, 1965). Hydromagnetic oscillations in an isothermal atmosphere with a vertical magnetic field were studied by Stepién (1966). The energy flow through an active region was discussed by Kiepenheuer (1966*a*).

The nature of M-regions was studied by Piddington (1964) and Mustel (1966). While Piddington supposes that these are regions on the Sun above which the magnetic field is unusually weak or lies perpendicularly to the geomagnetic equatorial plane, the analysis made by Mustel has shown that M-regions seem to be final, very prolonged 'magnetic-tail' stages of active regions.

4.2. *Magnetic field measurements*

Review papers on solar magnetic fields were presented by Severny (1964*b*, 1965*d*), Howard (1965), Schatzman (1965*c*), Zirker (1965) and von Klüber (1966).

Fine scan magnetographic observations made by several authors (Severny 1964*a*, *b*, 1965*c*, Bumba and Howard 1965*d*, Stenflo 1966) show that what appears with small resolution to be a relatively homogeneous structure (e.g. unipolar regions) breaks down to a complicated multipolar structure at high resolution, with a complicated demarcation line between polarities and rapid change in orientation of the field vector. The maximum field strength and the number of magnetic elements resolved increases with increasing resolution. The disbalance between N and S fluxes characteristic for weak magnetic regions is also found to be increasing with increasing resolution.

Using the new double magnetograph at Crimea, Severny (1966*a*) observed many facts which are evidence of the very fine structure of solar magnetic fields with the heights in solar atmosphere. He has found appreciable differences in the magnetic field structure of active regions in the photosphere and chromosphere (the Hα line) and even in the photosphere at the levels of λ5250 and λ6103 lines. Particularly the appearance of isolated fields of one sign only at the intermediate level (λ6103), with no fields of opposite sign in their vicinity is difficult to understand. Much work in this way has also been done at Izmiran in Moscow. Hβ measurements of local magnetic fields made it possible to detect isolated fields in the chromosphere, of considerable strength (~ 150 G), without connection with the photospheric fields (Ioshpa *et al.* 1964, Mogilevsky and Schelting 1966), which strongly indicates a force-free structure of local magnetic fields in the chromosphere. A scheme of permanent generation in active regions of corpuscular streams consisting of discrete plasmoids with quasi force-free magnetic field has been offered (Kalinin and Mogilevsky 1965, Mogilevsky 1966*b*). This hypothesis of force-free fields was criticized by Wallis (1965*b*). Severny (1966*a*) and Ioshpa and Obridko (1966) raised some objections against the use of the Hβ line for measuring the chromospheric field.

Zhulin and Mogilevsky (1965*a*) investigated magnetic field variations in active regions and have concluded that the magnetic fields in plages and sunspots are mutually independent. Bumba and Howard (1965*a*, 1966*a*) have found that the increase of the magnetic flux takes place only during the first few days of the development of an active region. Semel (1967) has studied the 'balance' of magnetic fluxes for various active regions. On the basis of an investigation of the trajectories of individual knots in an active prominence, Rompolt, (1965*a*, *b*) constructed a picture of the magnetic lines of force above the active region.

In fields of intermediate field strength, Bumba and Howard (1965*a*) have found that the magnetic flux of the predominant polarity is concentrated around the boundaries of the supergranules, and the opposite polarity is seen weakly in the centre of each cell. Using the photographic technique of Leighton, Sheeley (1966, 1967) has found that during the development of a bipolar magnetic region the flux density is distributed in successively smaller fragments. Field strength as high as 700 G was detected in some of them. The semiregular pattern of weak magnetic fields in the photosphere has been explained by Bumba and Howard (1965*b*) and by Bumba *et al.* (1966*a*) as the result of the expansion, weakening and stretching by differential rotation of magnetic fields of old active regions and of continuing development of new regions within the pattern. The largest of the complexes of activity, consisting of many active regions, result in formation of very complex features of unipolar magnetic regions. Leighton (1964) interpreted the dispersal and migration of unipolar and bipolar regions as a random-walk, diffusion-like process caused by supergranulation convection currents. All these measurements are generally consistent with the hypothesis that bipolar magnetic regions are the sources of all the flux on the solar surface (Sheeley 1966).

4.3. *Flare activity*

A monograph on the morphology of solar activity was published by Vitinsky (1966). A general review of the different phenomena occurring in solar active regions was presented by de Jager (1965*b*).

Some general characteristics of solar active centres were studied by Vitinsky (1965*a*), who modified the Švestka's earlier classification of active regions, particularly based on their flare activity. A magnetic classification of solar active regions, which characterizes well the production of flares, has been proposed by Martres *et al.* (1966*a*). The probability of flare occurrence in relation to the area of spot groups was discussed by Garczyńska *et al.* (1967).

Characteristics of active regions capable of producing proton flares were described by Avignon *et al.* (1964*a*, 1965, 1966) and by Levitzky (1964, 1965, 1966). Kopecký and Křivský

(1966) have found indications that the occurrence of small sunspot groups in the vicinity of large E and F groups may be the stimulating factor, which gives rise to proton flares in the large groups. The 'repetition centres', which give rise to more than one type IV burst, were studied by Lapointe (1964) and Caroubalos (1964*b*). Strong enhancements of 3·3 mm emission were observed by M. Simon (1965) in regions that later flared. Tanaka and Kakinuma (1964) have found that the 3·2 to 7·5 cm flux ratio of the slowly varying component is greater than unity in most active regions in which proton flares appear. Elliott and Reid (1965) suppose that the class 3 flare on 26 September 1963 did not produce any GLE due to the late phase of development of the associated active region. C. S. Warwick (1965*a*) has found that proton flare regions cluster in certain preferred heliographic longitudes.

Bruzek (1965) studied the evolution of one of the most active regions of the last solar cycle, which appeared in July 1961 and an analysis of flares, magnetic fields and activity in the sunspot group of September 1963 was presented by Zirin and Werner (1967) and by Moreton and Severny (1967). This active centre was also studied by Anastassiadis *et al.* (1964).

5. FACULAE AND PLAGES

Tokuya and Nagasawa (1964) drew attention to the fact that there are systematic differences between the areas of Ca plages as given by various observatories. The foreshortening correction for Ca plages was discussed by Godoli and Monsignori Fossi (1967). The same authors (1966) also performed a research on Ca plage evolution. Maps of Ca plages for the first year of IQSY were published by Godoli (1966*a*). Preferential longitudes in the occurrence of Ca plages were found by Godoli *et al.* (1966).

V. A. Krat (1965*a*) studied the H and K line profiles in the calcium chromospheric faculae. He concludes that calcium faculae cannot be regarded as the tops of photospheric faculae and finds that chromospheric faculae are localized in the lower chromosphere below 1000 km; H_3 and K_3 originate in higher levels in the coldest chromospheric elements with $T_e < 7500°K$. Conditions for Ca II emission in chromospheric formations were discussed by the same author in an earlier paper (V. A. Krat 1964). On the basis of hydrogen and calcium chromospheric lines, cold and hot elements in faculae in various parts of active regions were discussed by Kuznetsov and Stepanov (1966*b*). Transversal and longitudinal magnetic fields of $\sim$ 200 G were found in a bright plage by Stepanov and Grigorev (1966). Regions with a purely longitudinal field usually are very small. The authors suppose that the increase of such an area is the necessary condition for a spot occurrence in the plage. Tsap (1965*b*) has found that the brightness of flocculi in the K-line increases with the strength of all components of the magnetic field. The brightness does not increase after some maximum brightness is reached. The corresponding limiting field strength is different for various plages and even for different places of one plage. Livshitz (1965) has shown that the increase of the emission flux in the faculae can be well explained by an increase of the velocity of convection in the sub-photospheric layers by 30%.

The depth distribution of the temperature in faculae was investigated by several authors. Kuzminykh (1964*a*, *b*) and Kozhevnikov and Kuzminykh (1964) worked out a facula model using data on the continuous spectrum of the faculae. Polonsky (1965, 1966) studied infrared lines of the molecule CO and deduced the rotation temperature in one facula, very close to the temperature of the neighbouring photosphere in the uppermost layers. Curves of growth were constructed for four faculae by Voyhanskaya (1966). An inhomogeneous facula model made up of hot facular granules embedded in regions with physical conditions identical to those in the normal undisturbed photosphere has been suggested by Withbroe (1967). Bachmann (1967) discussed Hα, Hβ, H and K-line profiles in plages and evaluated the excitation temperature in them.

The He λ10 830 line profiles in plages were studied by Gulyaev (1964) and by Namba (1965*a*). The non-thermal velocity in plages was found by Namba lower than in the quiet chromosphere. This fact has not been confirmed by V. A. Krat (1965*a*), who found the non-thermal velocity in calcium plages equal to 15 km s^{-1}. Voyhanskaya (1965) determined the non-thermal velocity gradient above calcium plages, 7 m s^{-1} km^{-1}. A broadening of the cores of D Na and b Mg lines in spectra of faculae in comparison with the photosphere was observed by Teplitzkaya and Effendyeva (1966). They interpret it as due to an increase of the non-thermal velocity in faculae, by the factor of about 2. Orrall (1966) investigated vertical velocity fluctuations in weak plages as inferred from observed wavelength displacements of the K line core. Much longer periods of these fluctuations were found in bright calcium flocculi and particularly in one weak plage than in the undisturbed chromosphere. On the other hand, a similar study made in the photospheric λ3931 Fe I line did not show any difference (Orrall 1965*a*).

Observation of the hydrogen λ21 655Å and λ12 818Å lines in a facula were described by Mitropolskaya (1966). An upper limit of the ratio $N(He^3)/(N(He^4)$ in plages equal to 10^{-2} was determined by Namba (1965*b*).

A negative correlation between the mean annual numbers of polar faculae and the sunspot relative numbers was found by Y. Tanaka (1964) and by Sheeley (1964, 1966). Sheeley concludes that the number of polar faculae can be considered for a measure of the magnetic field in the polar regions.

6. SUNSPOTS

6.1. *General problems*

A general discussion of the sunspot phenomenon was presented by Kiepenheuer (1966*a*) and review papers on the structure of sunspots and solar magnetodynamic theories were given by de Jager (1966) and Ferraro (1966). Reviews of statistics and evolution of sunspots were presented by Waldmeier (1966) and Xanthakis (1966*e*).

An extension of the Waldmeier classification of sunspot groups has been proposed by Künzel (1964) in order to describe the trend of the group development. This new classification takes into account development indices, the frequency of flares, plage areas and magnetic classification. According to Martres (1964, 1966), γ and $\beta\gamma$ regions are formed as products of an accidental superposition of normal sunspot groups. Dezsö (1964) classified evolutionary phases of sunspots by variations in the umbral and penumbral areas of single spots and spot groups. According to Dezsö and Gerlei (1964*a*), the spot and group development can be predicted from the ratio of penumbral-umbral areas. The lifetimes and some other basic characteristics of sunspot groups were discussed by Ringnes (1963, 1964*a*, *b*, 1965) and by Kopecký and Suda (1965). A large sunspot group of 8–18 June 1963, was described by Belorizky (1964).

Kozhevnikov (1964*a*) has shown that the area of a spot group increases in the same manner as the area of a section of a sphere rising in a liquid. The ascent velocity is proportional to the magnetic field strength of the group. From this point of view, Kozhevnikov (1964*b*, 1965*a*) also discussed some geometrical characteristics of sunspot groups. Some of them indicate that the magnetic field is concentrated in the lower parts of the convective zone and is carried upward by convective currents in the initial phase of the solar cycle (Kozhevnikov 1965*b*). Vitinsky and Ihsanov (1964*a*, Vitinsky 1965*a*) have shown that the alteration of the area of a sunspot group before the maximum of development cannot be described by one law, and three types of sunspot groups have been distinguished according to the rate of the area increase.

The proper motion of large bipolar sunspot groups was investigated by Yilmaz (1964*b*) with the result that the follower spot had a larger eastward proper motion than the leader spot. On the other hand, this divergent motion depends on the state of evolution (Dezsö *et al.* 1964), the average velocity of developing spots being higher than that of declining ones. Vitinsky and Ihsanov (1964*b*) have shown that the change of distance between the main spots can be explained as the raising of a magnetic tube to the solar surface, with an average velocity of ~ 115 m s^{-1}. It has further been shown by Ward (1966) that the average longitudinal motion of a sunspot is an inverse function of the spot size and that the motion decreases in the proportion in which the group develops in area (and on this basis, the solar-rotation rate for sunspots has been corrected). Bumba (1964*b*) found two types of motion of small sunspots in the group relatively to the main spot: shifts in spirals around the main spot with velocities approaching 50 m s^{-1} and motions from East to West usually in the following part of the group with velocities of 100–200 m s^{-1}. Both types of motion take place practically parallel to isogausslines of the magnetic field. Gopasyuk and Moreton (1967) have found that the direction of relative motions of sunspots in an active group tends to coincide with that of transversal field.

Jakimiec (1964*b*) has deduced from micrometric measurements that single spots are markedly elongated in the direction parallel to the equator and the same tendency is indicated in spots of bipolar groups, but only during the growth period. Menzel and Moreton (1963) compared stereoscopically photographs taken in 6-hour intervals and clearly observed depressions in sunspots and in their surroundings.

High definition photographs obtained by Stratoscope I have shown many tiny bright dots in the umbra (Danielson 1964, 1966). The diameter is 300 km or less and the lifetime is of the order of 0·5 hour. Bumba (1965*b*) described the process of penumbra formation and disintegration as well as the decrease of umbra on the basis of high resolution photographs. He also studied the development of light bridges from photospheric granulation. Korobova (1966) has shown that one may distinguish three types of photospheric bridges in sunspots. The fine structure of sunspots has also been discussed by Rösch (1966) including a detailed study of umbral granuli. Aly and Galal (1967) studied details in the sunspot structure profitting by the suppressed scattered photospheric light during a solar eclipse.

The relationship of regular sunspots to supergranular cells was demonstrated statistically and on good quality photographs by Bumba (1965*a*). These photographs show light bridges as prolongation of facular network in sunspots. The sudden formation of penumbra appears as the interaction between the spot magnetic field and the field facular network cell. New small sunspots arise always from the peripheries of supergranular cells (Bumba and Howard 1966*a*). Two ways, in which sunspots can be formed directly as the result of the circulation in the supergranulation, have been suggested by G. W. Simon and Leighton (1964). The same authors conclude that the gradual disintegration of sunspots may be due to the 'erosion' of the penumbral boundaries by the large cell currents. According to Leighton (1964), the relative compactness of *p* spots is due to the fact that the lines of force of *p* spots are more tightly twisted and hence offer more resistance toward being 'sliced away' by the supergranulation currents.

The influence of false light in sunspots produced by image motions was discussed by Birkle and Mattig (1965), who calculated intensity profiles in dependence of the diffusion constant. The same influence was studied using photoelectric and photographic observations. A proposal has been made to approximate the scattering function by a sum of two Gaussian functions. Some further information on the atmospheric turbulence were given by Makita and Morimoto (1964). The influence of scattered light on sunspot observations was discussed by Zwaan (1965*a*). Using observations of photospheric Fe II lines fictitiously observed in sunspots he finds that the spot spectrum contains 60% photospheric and 40% umbral light. He also discussed the influence of stray light on the curves of growth.

Bright rings around sunspots were discussed by Chistiakov (1964) and by Kasinsky (1964*a*). Photometric measurements show that the increase of brightness in the ring is due to the increase of the area covered by the granules (Chistiakov 1965*c*) and that the maximum brightness occurs around spots with 30 000 km diameter (Kasinsky 1964*b*). This suggests that sunspots might be identified with supergranuli with strong magnetic fields in their centres. Kasinsky (1964*c*, 1965) and Surkov (1966*a*) also discussed heating mechanisms of bright rings and Kasinsky (1966) found a correlation between the magnetic field variations and the brightness intensity in the rings.

6.2. *Sunspot magnetic fields*

Review papers on the measurements and the structure of the sunspot magnetic fields were given by Severny (1964*b*), von Klüber (1966), Evans (1966) and Jäger (1966).

The magnetic field of unipolar sunspots has been studied extensively by Severny (1965*b*) with the magnetograph of the Crimean Astrophysical Observatory. The appearance of strong transverse fields inside spots is noted again and the run of the field strength and the inclination of the line of force with distance from the spot centre has been analysed. In the penumbra and outside the spot the field is mainly transversal. The variation of the magnetic field strength with distance is similar to that of a dipole submerged beneath the photosphere as has also been found by Ioshpa and Obridko (1965*a*). Severny (1965*b*) paid special attention to the investigation of the system of vertical electric currents in unipolar spots. Polarization measurements with the magnetograph are inconsistent with the transversal field measurements by a polarization detector in a wide spectral region as done by Leroy (Severny 1964*c*). The rapid rotation of the transverse field-strength components leads to the presence of strong vertical currents. From measurements in different lines, Severny (1965*a*, *c*) has found that the rotation might be 80° over a depth difference of 100 km and the existence of twisted field is possible. On the other hand, the appearance of small regions with transverse fields inside sunspots is associated with the occurrence of a characteristic fine structure emission. Examination of high resolution spectra of polarization of sunspots, made by Steshenko (1966) showed that magnetic field of small ($\leqslant 2''$) pores never falls below 1000 G. In one case the magnetic field of $\sim$ 5000 G has been found in a small ($\sim 2''$) area inside a sunspot. Beckers and Schröter (1966) also found isolated points ($< 2''$) outside the spots with strong magnetic fields, $\sim$ 1200 G. They conclude that these magnetic knots fall in dark intergranular space but are clearly distinct from ordinary pores. Sheeley (1966) has found that the source flux of sunspot groups increases linearly with the group size. During the decline phase the flux density is distributed in successively smaller fragments and field strength as high as 700 G was detected in some of them.

First observations of sunspots with the new Capri vector-magnetograph show that in the penumbra the inclination of the magnetic field vector is of the order of 45° and that the central component of a Zeeman triplet is hardly linearly polarized but mainly circular (Deubner 1967). Rayrole (1966) and Semel (1967) investigated the colinearity between the field vector and the gas motion. The motion tends in all cases to be more horizontal than the field. Gopasyuk's (1966*b*) comparison of radial velocities and magnetic fields for bipolar groups has shown that during the stage of development of the group, gases are flowing down inside sunspot at both chromospheric and photospheric levels. According to Gopasyuk (1965*a*), irregular rotation of spots leading to the winding of the magnetic field in a direction opposite to that of the spot rotation is sometimes observed. It leads to a sausage instability of the plasma and in consequence of it to the increase of the frequency in the appearance of surges.

Photographic magnetic field measurements were composed to isogausslines in one sunspot group by Teske *et al.* (1964). The usual picture was obtained but the scanning by photographic procedure with 210 points was too rough. Supplementary data to Nishi's earlier measurements of the orientation of the sunspot magnetic fields were presented by Nishi and Unno (1965). Variations of the sunspot magnetic field as high as 900 G during one day and even as high as 800 G in the course of several hours were observed by Künzel (1967) in some sunspots. Kuklin (1964*b*) found that sometimes magnetic fields of sunspots were rapidly changing although the other active processes were absent.

Magnetic field distribution in three different levels of the solar atmosphere for geophysically active and inactive sunspot groups was compared and discussed by Bumba (1964*a*). From an analysis of the photospheric magnetic field structure and surge trajectories Gopasyuk (1965*b*) concludes that in the corona the spot magnetic field has an azimuthal component which does not depend on the rotation of the spot at photospheric level and the magnetic field decreases more slowly in the corona than in the case of a dipole field. The spot fields are closing in the corona, even in the case when there are large spots between them. From measurements of the chromospheric magnetic field in the Hα line, V. A. Krat (1965*b*) has found that the sunspot magnetic field in the chromosphere is usually greatly displaced from the position of the sunspot in the photosphere; the values of the displacement range up to 17″. It seems, therefore, that the magnetic field of sunspots penetrates into the chromosphere in curved streams of matter. The mean gradient of the magnetic field in the low chromosphere above sunspots was found as 1·2 G km^{-1} (Vyalshin and Krat 1965). On the basis of type IV burst observations, Sakurai (1964) has concluded that sunspot magnetic fields consist of bipolar fields tilted westward.

Godovnikov and Smirnova (1965) made an attempt to compute the sunspot magnetic field by means of elongated magnetic tubes. Kuklin (1964*a*) has shown that the magnetic field of some sunspots can be approximated by an axial force-free field. Mogilevsky (1966*b*) discussed the general properties of force-free magnetic fields and proved the general theorem that isolated magnetic plasma acquires always a force-free magnetic field. Molodensky (1966) described a method of computing the magnetic field in the chromosphere above an axial-symmetrical sunspot on the basis of magnetometer measurements. Zhugzhda (1966) has proposed models of potential fields fur sunspots, by which some peculiarities of spot magnetic fields can be explained. Schatzman (1965*a*) discussed some consequencies of the model of a force-free field in sunspots for the character of fibres seen in Hα spectroheliograms.

6.3. *Motions inside sunspots*

From the study of weak Fraunhofer lines in 62 different sunspots the velocity field has been derived by Maltby (1964). An upper limit of 0·15 km s^{-1} is found for the vertical velocity component and a comparison between observed and calculated line profiles shows a strong velocity change with optical depth. An angle less than 6° with the horizontal plane in the outer part of the penumbra may satisfy the observations as well as the equation of continuity. At the outer limit of the penumbra the velocity falls abruptly to zero (Brekke and Maltby, 1963). At the Kodaikanal Observatory the Evershed effect was studied with Zeeman insensitive lines (Bhatnagar 1964, 1966). The mean velocity gradient was found to be 4 m s^{-1} km^{-1} in depth. Besides the well-known radial velocity a small vertical downward component (0·3 km s^{-1}) in the penumbra was found. Darker penumbral regions show larger equivalent widths in the lines than bright regions. Highly resolved spectra with brightness and sight-line velocity variations and the comparison of lines with different temperature sensitivity lead Beckers (1966) to the picture that the bright penumbral filaments are stationary and the dark regions are moving. This, however, is in contradiction with the results obtained by Schröter (1965*a*, 1966*a*), who studied the centre-to-limb variation of the Evershed flow in one sunspot in weak and strong lines and determined the velocity and angles in dependence of the position in the spot. He concludes that the Evershed effect is produced by streaming in the penumbral fine structure: in the bright structure we have the outstreaming and in the dark channels the in-streaming material. The 'resolved' velocities are about 6 km s^{-1}. The downward motion of the order of 0·3 km s^{-1} is not real, due to a relativistic shift of the photospheric comparison lines. Schröter (1965*b*) estimated the temperature difference between the bright and dark features to 700°. The net flux of kinetic energy is roughly equal to the radiation flux deficit in the spot.

The fine structure of radial velocity distribution in line discontinuities in the Evershed effect was studied by Bumba (1965c). Observations at the Kitt Peak Observatory (Bumba 1966) show the absence of the normal Evershed effect in small and young sunspots. Downward motions with velocities up to 6 km s^{-1} in weak lines and up to 14 km s^{-1} in Na D lines have been found. Berdichevskaya (1964) considered the Evershed effect as an outflow of matter in response to a horizontal pressure gradient established between the spot and the surrounding photosphere. The depth-dependence of the velocity corresponds to St John's picture. The rates of outflow from subphotospheric layers are estimated and lead to the low temperature of sunspots. Gopasyuk (1966a) has found that the magnetic axis of the spot and the normal to the plane of velocity vectors of the spot coincide and are inclined by 7°–8° to the east of the normal to the solar surface. The difficulties in explaining the peculiarities of the Evershed effect can be overcome by assuming that the magnetic field rises to the surface of the Sun during all the time of existence of the spot.

6.4. *Physical state of sunspots*

Spectroscopic observations of sunspots were reviewed and physical conditions in a standard sunspot were summarized by Van't Veer (1966a). Rödberg (1966) measured the intensity ratio between spot and photosphere and its centre-to-limb variation in three wavelength regions with a pinhole camera and found a very good agreement with earlier Makita's and Morimoto's observations. Similar measurements were also made by Surkov (1966b). The intensity ratio between spot and photosphere in the Hα line, the equivalent width of some Fraunhofer lines and the colour temperature in a spot were determined by Bergsjö (1963).

Zemanek and Stefanov (1964a) constructed a curve of growth for Fe I lines in the spectrum of a large spot, taking into account magnetic strengthening. The results were compared with the curve of growth constructed for the same spot from Ti I lines (Zemanek and Stefanov (1964b) and for another spot from Fe I lines (Zemanek and Stefanov 1966a). Results from three spots lead to the mean excitation temperature of 3900°K and the velocity of turbulence 3·0 km s^{-1} (Zemanek and Stefanov 1966b). Olijnyk's (1965, 1966) calculation of equivalent widths shows that magnetic strengthening of the different Fe I lines is from 3% to 10%. A photometry of the continuous spectrum made by Stefanov and Zemanek (1965, 1966) shows that the development of a spot leads to an increase of the brightness in the short-wave region by $\sim$ 1·5 times. Stepanyan (1966) has found that during the development of a spot the difference of brightness temperature and excitation temperature at λ6250Å decreases and, at the moment of the maximum spot area and minimum temperature, this difference is almost vanishing.

A report on the existence of molecular lines in the spot spectra has been given by Moore-Sitterly (1966). Lines of CaH, SiH, TiO and ZrO are present only in the spot spectra, not on the solar disk. Intensity measurements of TiO lines in the spectral region $\lambda > 7054$Å on Kitt Peak and Mt Wilson plates have been carried out by Hitchcock (1965). The slope of the straight line between intensity and rotational quantum number corresponds to a sunspot temperature lower than 1000°K. Some arguments against this very low spot temperature were brought into discussion by Schadee (1966) who showed that a temperature of about 3600 °K was also in agreement with the observations.

The possibility of utilizing sunspot spectra for the determination of elements on the Sun has been discussed by Dubov and Khromova (1964), the main problem being the scattered light. The Li line λ6103·6Å was identified, with 4·6 + 2·5 mÅ equivalent width. The lithium abundance in sunspots was found 7·5 times higher than in the quiet photosphere (Dubov 1964). Schmahl and Schröter (1965) redetermined the lithium abundance and the isotope ratio Li^6/Li^7 using the λ6708Å contour of four sunspot spectra obtained by Stumpff. They found $1{\cdot}0 \times 10^{11} < H/Li < 2{\cdot}3 \times 10^{11}$ and $Li^6/Li^7 \simeq 0{\cdot}05$.

The contours of the central part of the H and K lines at different points in one sunspot and its surroundings were determined by Paciorek (1965). The profiles on the two sides of the spot were quite different and the maximum single peak of emission did not lie above the umbra. The comparison between the observed and the theoretical profiles led to a turbulent velocity of 7·9 km s^{-1} in the spot chromosphere. From halfwidth measurements of the K_3 peak in spot spectra turbulent velocities of about 15 km s^{-1} were obtained by Mattig (1966*b*). The absorption D_3 line in the penumbra of one spot was discussed by T. V. Krat (1965).

Equivalent widths of 148 neutral and ionized lines and profiles of the Balmer lines $H\alpha$–$H\delta$ have been determined from two concave spectra of one spot by Fricke and Elsässer (1965). The amount of scattered light was derived from the wings of Balmer lines. The variation of temperature and pressure with optical and geometrical depth was deduced; in the spot umbra a remarkably lower gas pressure was found than in the photosphere at the same geometrical depth. Therefore, deviations from hydrostatic equilibrium must exist in the sunspot. Zwaan (1965*b*) discussed the determination of a sunspot model and determined equivalent widths of neutral and ionized lines. A new method suggested by him yields the amount of stray light directly from the spectroscopic measurement. A model was constructed. In contradiction to Fricke and Elsässer, however, Zwaan (1965*b*, 1966) has found that the spot is in hydrostatic equilibrium; models which deviate from hydrostatic equilibrium cannot explain the observed equivalent widths. Zwaan suspects that sub-hydrostatic pressures obtained by other authors, have been due to unproperly corrected stray light. According to Zwaan (1965*b*) there is no dependence of the physical conditions in the umbra on its area, perhaps with the exception of pores and very small spots. On the other hand, Van't Veer (1965) suspects that the gas pressure is lower for larger spots. Furthermore, Zwaan (1965*b*) discussed the sunspot as a stable structure in the solar 'mantle' taking magnetic forces and the Wilson effect into account. A new determination of the gas pressure in the spot has been carried out by Van't Veer-Menneret and Van't Veer (1965), with the result that the gas pressure is about the same or slightly lower than in the photosphere at the same level. The same problem was discussed once more by Van't Veer (1966*b*) with a new statistical method which makes it possible not only to find the best model but also to estimate the validity of the method. The obtained gas pressure is similar to the results of the earlier paper. This method was also applied by Van't Veer (1966*c*) to the Mt Wilson sunspot No. 11 730 and the results were compared with some other recent models.

On the basis of spot photographs showing the Wilson effect, Wilson (1965, 1966*a*) has derived that the absorption coefficient in the umbra is between one or two orders of magnitude less than that in the photosphere. A ratio of one order of magnitude is not inconsistent with the assumption of radiative equilibrium in the umbra. Jensen and Maltby (1965) calculated the optical radius of a sunspot as a function of the vertical optical depth. It is pointed out that the vertical temperature distribution cannot be deduced without taking into account the finite size of the spot. The results indicate that the umbra is about as opaque as the photosphere. Mattig (1967) investigated the same problem of the transparency or the geometrical height-scale from sunspot spectra near the limb. From the geometrical height difference between the origin of the line centre of different lines and the continuum he has found that the spot height scale is of the same order as that of the photosphere. Buslavsky (1966) solved the transfer problem for a cylindrical sunspot taking into account the flux of radiation through the side walls. The temperature distribution in this case depends on the radius of the spot and on the ratio of the absorption coefficient in the sunspot centre to that in the photosphere. Makita (1966) discussed the non-existence of a homogeneous sunspot model and proposed again an umbral model consisting of hot and cold elements. A tentative analysis gives the hot components the flux fraction of 0·5 and space fraction of 0·1. Stankiewicz (1966) studied the relation between the magnetic field strength and the temperature in the sunspot. The temperature difference was found proportional to H^2.

6.5. *Theory*

Hydromagnetic stability calculations which are applicable to sunspots and some recent theoretical sunspot models were summarized by Danielson (1965). Deinzer (1965, 1966) has constructed theoretical sunspot models in magnetohydrostatic equilibrium by assuming that a magnetic field in the hydrogen convection zone inhibits the convective energy transport. The system of differential equations was solved numerically. The results agree with observations to within 50%, the maximum possible field strength at the surface is about 5000 G. Some remarks on this theory have been presented by Wilson (1966*b*), who points out that it is necessary to take into account better values of the sunspot structure in the visible layers. Inhibition of the convective energy transport was also discussed by Chitre (1963). He shows that it is impossible to obtain a reasonable sunspot model assuming a complete suppression of convection. The magnetohydrostatic equilibrium has also been discussed by Jakimiec (1965) and by Jakimiec and Zabza (1965, 1966). With the assumption of a short scale spot model and a depth depression of about 1000 km an equilibrium is possible but not without difficulties; the non-twisted magnetic field is able to balance the pressure difference of about 10^6 dyn/cm^2. In an earlier paper, Jakimiec (1964*a*) presented a sunspot model for Michard's 'typical spot'. The spot was found to be a depression in the photosphere and a medium with the density and transparency nearly the same as in the photosphere.

Oster (1964) proposed a sunspot magnetic field model, in which a single spot corresponds to the end-field of a solenoid approximated by the field of a single current loop. De Jager (1964*b*) assumes that the magnetic field in a sunspot is constant with depth and compares this field with the equipartition magnetic field, which has an energy density equal to the convective energy density in the undisturbed Sun. He finds that below 10^4 km, the tension in the lines of force in a 3000 G sunspot is no longer sufficient to prevent them from being turned over by the convective buoyancy forces. Thus below 10^4 km, the magnetic lines of force are very twisted and disordered in this model. Makarov (1964) made an attempt to connect in one system the magnetic field strength in umbra of a sunspot, the area of umbra and the depth of umbra.

Some observational and theoretical aspects of the dynamics of the evolution in the fine structure of sunspots were discussed by Bumba *et al.* (1966*c*) in relation to the magnetic field structure. It has been shown that neither a pure magnetic field diffusion nor a pure transport of lines-of-force can explain the sunspot development. A scheme of interrelation of turbulent elements of the convection zone and the magnetic tube of the spot was suggested by Chistiakov (1965*b*) as an explanation of the filamentary structure of the penumbra. Tuominen (1964, 1965*a*, 1966) has shown that large horizontal whirls connected with the differential rotation may be created in the photospheric and subphotospheric layers of the Sun. These whirls explain the high magnetic fields as well as the observed movements of individual spots. The relationship of the observed local magnetic fields to convection in the Sun was considered by N. O. Weiss (1964, 1965*a*). Arguments in support of the outward transport of magnetic fields of spots and bipolar regions into the corona were presented by Romanchuk (1965*a*).

A possible mode of nonradiative transport, that of Alfvén waves, was examined by Musman (1965). He has concluded that it is not likely that these waves contribute substantially to the energy transport in sunspots. Hydromagnetic waves in a nonisentropic medium in the presence of a gravitational field were discussed by Getling (1965). Such conditions can occur in the convective zone beneath sunspots. The generation of magnetohydrodynamic waves in sunspots has also been discussed by Marik (1966). The Alfvén wave flux is found higher above the spot than the flux outside spots in active regions, by one order of magnitude.

The perturbing effect of static magnetic fields on the normal distribution of pressure, temperature and density in the solar atmosphere was discussed by Menzel and Shore (1966). The theory was applied to sunspots taking velocity fields into account. Menzel (1964) has also shown that magnetohydrodynamic flow in a force-free medium leads to a form of Bernouilli's theorem: a hot gas flowing at slow speeds along magnetic lines of force, is greatly accelerated and undergoes rapid cooling as it encounters a strong magnetic field. The possible application of this phenomenon to sunspots is evident. Clark (1965) presented some magnetohydrodynamic equations with exact solutions, which might be applied to conditions to be expected in sunspots.

A new discussion of the electric conductivity leads to the conclusion that in the visible layer of sunspots the conductivity is only of the order 10^8–10^9 (Schröter 1966*b*). This value is more than two orders of magnitude lower than assumed before. The consequences are an anisotropic conductivity and a short decay time comparable with the observed sunspot lifetime. A new very careful calculation of the conductivity with different sunspot models and different magnetic field configurations in dependence of the optical depth, carried out by Kopecký and Kuklin (1966*a*), however, leads again to a decay time longer than that obtained by Schröter. A simplified method of computing the electric conductivity in sunspots has been proposed by Kopecký (1966*c*). Tables of electric conductivity in various models of sunspots are presented by Kuklin (1966*a*). Kopecký and Kuklin (1966*b*) also computed the electrostatic field strength for different depths and phenomena in the solar atmosphere. In sunspots, they get 10^{-12} to 10^{-10} cgse.

Obridko (1965*b*) solved the transfer equation for a spectral line with a split upper level in the magnetic field. The first approximation was compared with the Unno solution and quite large discrepancy at the line centre was found. The derived resonance scattering matrix in a magnetic field differs from the absorption matrix (Obridko 1965*a*). Results of numerical computations of the line formation were given by Rachkovsky (1965). For the lines with a non-split upper level true absorption and radiative scattering were taken into account. Very large discordances were noted between the Unno theory and that of the author. Later on, Rachkovsky (1966) has solved the most general transfer problem in case of arbitrary splitting of both quantum levels producing the spectral line in a homogeneous magnetic field. Kats' and Stepanov's (1966) numerical calculations of the intensity profiles for different solutions of the equation of transfer show that especially for magnetic field strength larger than 1500 G the exact solutions of Rachkovsky must be used. A review paper about the line formation in a magnetic field was presented by Mattig (1966*a*). Here, a solution of the equation of transfer with true absorption was given for the special case that the absorption coefficient was composed of a depth- and wavelength-dependent factor.

7. SOLAR FLARES

(Radio, X-ray and particle emission from solar flares are reported in sections 10, 11, and 12, respectively.)

7.1. *General problems*

Review papers on optical observations of solar flares were presented by Byrne *et al.* (1964), Ellison and Reid (1964), H. J. Smith (1964), de Jager (1965*e*), Zirin (1965*b*) and Švestka (1966*a*). The kinematics of flares, with a particular emphasis on explosive-type flares, was reviewed by Moreton (1965*c*). The energy and mass problem of flares was discussed by Bruzek (1966*b*).

The influence of the intermittent pattern of periods of observation on the number of discovered flares of different duration was studied by C. S. Warwick (1965*b*). Some observational and physical aspects of the effective Hα line-width measured with spectrohelioscope were discussed by Fritzová-Švestková (1964*a*, *b*). C. S. Warwick (1964) tried to obtain values of flare importance that would be free from systematic errors, applying approximate corrections to different observatories. A law of foreshortening of the apparent flare area has been deduced by Miyazawa *et al.* (1966). Zirin (1965*c*) and Akinyan (1965) tried to divide solar flares in several morphological classes. Isophotes of flares have been studied by Dizer (1967). A photometric and morphological study of 132 flares observed during IGY was presented by Michard *et al.* (1964). Besides other results, the authors have verified that flares covering umbrae are very active in producing microwave radiation and SID's. Dodson and Hedeman (1964*b*) reviewed the problem of differentiation of flares with respect to SID's and pointed out the influence of inhomogeneities in the set of observed data on any investigation of problems of this kind. They have concluded that the relatively important flares without any associated SID tend to consist of many small parts, primarily far from major spots and are not associated with any significant microwave radiation. Mitra (1964) described the effect of a sudden increase in long-wave field intensity and considered it for the best indirect method of observation of flares. This conclusion was opposed by J. H. Reid (1965). Short periodic regularities in the occurrence of solar flares have been suspected by Paghis (1964) and by Tifrea (1965). Obashev (1966) has found that the area of a flare is proportional to its lifetime t, and the total energy of a flare in the optical region is proportional to $t^{5/2}$.

The series of large flares which occurred in July 1961 was discussed by Bruzek (1965), who also constructed a typical 'complete' large flare event. Detailed analysis of flares in the sunspot group of September 1963 was presented by Zirin and Werner (1967). They have found that most flares are homologous and some are triggered by disturbances elsewhere in the region. Observations of two active events on the Sun in the morning of 22 June, 1962 were described by Ballario and Tagliaferri (1964). Dodson and Hedeman (1965, 1966*a*) have suggested that some flares which appear in one active region within two hours, may be actually two phases of one complex flare phenomenon, while the late phase of major flares may be a development of phenomena initiated during the early phase of the complex event. Major homologous flares can repeat in less than five hours. McKenna (1965*c*) pointed out that a number of flares which appeared in the large group of July 1959 consisted of brightenings of apparently the same portions of the calcium plage. The relative structure of the chromospheric network remained fairly unchanged in spite of the great variations in the underlying sunspot pattern. According to Bumba and Howard (1965*c*), the flare development in each of its stages is in a close relation to the pre-existing calcium network.

Hallam (1964) reported on observations of flares in the hydrogen Lyman-alpha line aboard OSO–1. The Lyman-alpha contrast of the flare to the surrounding background is estimated at 5–150.

7.2. *Association of flares with magnetic fields and sunspots*

Magnetic observations relating to solar flares were reviewed by Howard (1964) and by Severny (1964*a*). Generally the occurrence of a flare results in the decrease of the magnetic gradients in the vicinity of the neutral points and this effect is most pronounced for proton flares. Godovnikov *et al.* (1964) have found indications of this effect in smaller flares as well. A simplification of the magnetic pattern following a class 2 + flare appearance has also been reported by Deubner (1965). Schmidt (1964), who investigated the great flare of 16 July 1959, found large and rapid changes in the total magnetic flux, but less variations in the geometrical flux distribution. On the other hand, no significant changes in the sunspot field strength were found by Teske (1965) during a class 2 flare. The stability of solar magnetic fields in the regions of flares was briefly discussed by J. W. Warwick (1964).

From simultaneous magnetographic records of magnetic fields in the Hα and λ5250 lines and central intensities of these lines Severny (1966*a*, *b*) concluded that the class 1 + flare of 4 October 1965 appeared at the place of opposite directions of the photospheric and chromospheric magnetic fields. Moreton and Severny (1966, 1967) carried out a careful examination of the position of the first brightenings of flares on maps of longitudinal and transversal magnetic fields for the active group of 17–24 September 1963. Their analysis shows: (1) That the first brightenings occur simultaneously in widely separated regions being separated by neutral line of the magnetic field; (2) At least one of these bright knots lies immediately adjacent to the neutral line in the region of the greatest field gradient; (3) All flares tended to occur in two centres: the first one where the transverse field vectors were crossing and another one where strong oppositely directed currents appeared. In a similar study, Martres *et al.* (1966*b*) have shown that if a flare consists of two or more bright points, these are distributed along both sides of a line of change of magnetic polarities. Ioshpa and Obridko (1965*c*) found bright flare filaments extended along the neutral line and parallel to it. Azimuths of the transverse field were found perpendicular each to the other in the two flare filaments. Most of these results were also confirmed by S. F. Smith and Ramsey (1966). The association of a class 2 flare with the longitudinal magnetic field of the active region was also discussed by Banos (1967). McKenna (1965*b*) called attention to the fact that a part of the class 3 flare of 10 November 1960 originated above a region containing a strong transverse magnetic field.

The macroscopic and microscopic structures of flares and their relations to spots, magnetic fields and filaments, were reviewed by Kiepenheuer (1964*a*, 1965). Gopasyuk (1964) has found that the velocities of descending and ascending gases in sunspots increase during the flares. A model of the magnetic field in major flares, particularly applicable to proton flares of the A type, has been proposed by Banin (1966*b*).

The appearance of strong flares is often preceded by the appearance of new spots or by an intensification of spots, previously existing near the seats of flares (Shaposhnikova and Ogir 1966). Antalová (1965) found an interdependence between the proper motion of sunspots and the occurence of flares in their close vicinity as well as changes in the area of sunspots close to ribbon-shaped flares. In disagreement with this conclusion, no striking dependence between motions in complex sunspot groups and flare occurrence could be found by Kuklin and Syklen (1966). Howard (1963) studied flares associated with GLE. In each case the nearby spot group decreased in area from the day before to the day after the flare. The association of solar flares with sunspots has also been discussed by McIntosh (1965).

7.3. *Proton flares and forecasts*

Lists of proton flares were published by Malitson (1963), Švestka and Olmr (1966) and Dvoryashin (1966). General characteristics of proton flares were summarized by Dvoryashin (1964*a*, 1965), Malville (1964) and Pick (1966). Thirty-two proton flares before 1955 have been detected on the basis of vertical incidence ionospheric soundings (Švestka 1966*d*). Proton flares during the sunspot cycles 12 to 18 were identified by Bhargava and Subrahmanyan (1966) from an analysis of the amplitudes of geomagnetic sudden commencements.

Avignon *et al.* (1964*a*, *b*, 1965) have described typical configurations of proton flares. The configuration A, which is most efficient in producing PCA's, appears over spots belonging to a centre consisting of two sets of spots of opposite polarities separated by a small distance. Bruzek (1964*a*, *b*) has proved a close relationship between loop prominence systems and proton flares. Vinogradov (1964) could not find any difference in SID's produced by proton and non-proton flares. On the other hand, some associations between X-ray emission and proton flares were found by Dvoryashin (1964*b*, 1965).

Predictions of solar flares were discussed by Firor and Lilliequist (1965). In the U.S.A., ESSA and SFCRL are each engaged in forecasting solar disturbances on an experimental basis (U.S.A. Report 1966). On the basis of PCA observations during three solar cycles, the proton flare occurrence in 1966–68 has been predicted by Švestka (1966*b*, *d*). Weddell (1964) suggested a statistical method of the prediction of flares about one month in advance. Dodson and Hedeman (1964*a*) found an unexplained variation of solar proton events in a period of 29·5 days.

7.4. *Flare spectra*

Review papers on spectrographic observations of flares were presented by Zirin (1964*a*) and Severny (1965*e*). Švestka (1965*a*) reviewed the methods of the spectral analysis of flares and critically summarized the results obtained until 1964.

Systematic analysis of the Balmer spectrum of solar flares observed in projection against the solar disk were presented by Švestka (1965*a*) and by de Feiter (1966). Evidence of Stark broadening has been given and the electron density was computed by the halfwidth method for several flares. Švestka also proposed several methods for determining the optical thickness of flares and estimated the electron temperature. Deviations from thermal equilibrium (b_n factors) were determined for several flares (Švestka 1964*c*, 1965*a*) and a filamentary structure of flares appears to be the only possible explanation of the found results. Another evidence of the filamentary flare structure is the non-existence of Balmer continuum emission in disk flares. The observational evidence of the filamentary structure was summarized (Švestka 1965*b*) and a schematic flare model with filamentary structure was presented (Švestka 1965*a*). From the point of view of the filamentary structure, unavailable and available methods of the spectral analysis of flares have been distinguished. Ciurla and Rompolt (1965) computed the influence of the filamentary structure on the line profiles of prominences and flares.

The halfwidth method of determining n_e in flares was corrected for electron damping by de Feiter (1964, 1966). Time variation of n_e in a proton flare was determined (de Feiter and Švestka 1964) and n_e values for a number of flares deduced (Fritzová-Švestková and Švestka 1966*b*). They vary from 4×10^{13} cm^{-3} in the flash state of a class 3+ proton flare to 5×10^{12} cm^{-3} in limb flares. The state of excitation in a flare, as it can be derived from the observed Balmer decrement, has been compared by de Feiter (1966) with theoretical computations of the population of the energy levels of hydrogen for various physical conditions outside local thermodynamic equilibrium, taking 20 discrete levels and the continuum into account. As one of the results, the filamentary structure of flares is found again.

Hydrogen Balmer spectra of individual flares on the disk were studied by Kurochka (1964*a*), Guseynov (1965*b*) and Banin (1965*c*). While Banin's results ($n_e = 3 \times 10^{13}$ cm^{-3}) are in agreement with those mentioned above, Kurochka and Guseynov found Doppler broadening of Balmer lines. Highest velocities are found for hydrogen and lowest velocities for metals. Banin has also discussed profiles which appear if both Doppler and Stark broadening are present.

Many authors investigated hydrogen flare spectra on the limb. Kurochka (1965, 1966) found all lines broadened by Doppler effect with nonthermal velocities of several tens of kilometers per second and electron temperature lower than 10 000 °K. He gives $N_2(\mathrm{H}) \simeq 2 \times 10^{14}$ cm^{-2} and $N_1(\mathrm{Ca\ II}) \simeq 5 \times 10^{13}$ cm^{-2}. Doppler broadening was also found by Polupan (1964), who discussed several hundreds of photometric profiles for the flare of 27 July 1961. Polupan and Yakovkin (1965) solved stationarity equations for the first six levels of the hydrogen atom and compared these data with observations of this flare. As the result, they found $n_e = 2 \times 10^{12}$ cm^{-3} and $T_e = 7500$ °K for $h = 10''$ above the limb and confirmed the filamentary structure of flares. In a bright pulsating phenomenon on the limb, Gurtovenko and Kostik (1964) found $n_e = 2 \times 10^{11}$ cm^{-3} and $T_e = 11\,000$ °K. Shilova (1966) estimated the Lyman-alpha radiation flux in a limb flare as 30–100 times higher than in the quiet chromosphere. Differences between disk and limb flares were discussed by Švestka (1964*a*, 1965*a*). He also computed the influence of seeing on the line profiles, which might change in some cases the true Stark profiles in Doppler-like contours (1965*a*).

Helium lines in flare spectra were studied by Banin (1965*c*), Goldberg-Rogozinskaya (1965) and Kubeš (1965). In the helium regions of a disk flare, Banin found $T_e = 17\,000$ °K and nonthermal velocity equal to 5 km s^{-1}. The optical thickness of the D_3 line in spectra of 8 flares was found by Kubeš within the range of 1 to 5 and nonthermal velocity resulted to 5–20 km s^{-1} assuming T_e = 15 000 °K. Quite different results were deduced by Goldberg-Rogozinskaya for a limb flare: $n_e = 10^{10}$ cm^{-3}, T_e = 30 000 °K for He I and 50 000 °K for He II regions. Zirin's (1964*b*) study of three He II lines in the limb proton flare of 20 November 1960, showed that the spectrum of He II was due to pure recombination. From continuum radiation, Zirin deduced $n_e = 10^{11}$ cm^{-3}. Most of the material in the observed part of the flare was at coronal temperature. Bray (1964) found a different time-development of two flares in the hydrogen and helium (D_3) lines and suggested that this difference was due to differences in the mode of excitation of these lines. Banin and Prokofieva (1966) also observed that helium and metal emission in the flare of 26 September 1963, were delayed in comparison with the hydrogen lines.

A method of determination of the flare optical thickness in the H and K lines has been proposed by Kurochka (1964*b*). A method of constructing and solving the emission curve of growth for a flare on the disk was developed by Letfus (1964). An application of this method to the flare of 20 July 1958 led to the following results: $n_e = 6 \times 10^{13}$ cm^{-3}, T_{exc} = 4690 °K for Fe I and 5350 °K for Fe II, T_{ion} = 6850 °K, nonthermal velocity equal to 5·8 km s^{-1}. When analysing Ca II, Al I, and Mg I emission lines in flares, Alikaeva (1965) found n_e = 1 to 5×10^{13} cm^{-3} and T_e = 9000–13 000 °K, varying during the flare development. Lower electron temperature and higher electron density were deduced from Na I lines (Alikaeva 1966). Geometrical thickness of calcium threads in a class 2 flare was found equal to 10^5–10^6 cm (Alikaeva 1964). According to Banin (1964) the emission of metals in coronal flare regions is similar to that observed in sunspot prominences and differs from that in chromospheric flare regions. Banin (1965*a*) has also found that metal lines with the upper excitation potential of $\sim$ 5·5 eV brighten most rapidly during the flare increase and those with $\sim$ 3·3 eV somewhat more slowly. The brightening of other lines is considerably slower.

The line asymmetry in flare spectra has been explained by Fritzová-Švestková (1964*c*) as due to an asymmetrical turbulent motion and by Ballario (1964, 1965) as due to the loop-type motions in the flare. According to Banin (1965*a*, *b*) the asymmetry is caused by a strengthening of the red wing of the line as a result of superposition of the additional emission of matter inflowing into the flare. Guseynov (1966*b*), too, explains the asymmetry as due to an additional emission. A flare model consisting of two loops has been proposed by Banin (1965*b*, 1966*a*).

A review paper on the continuous emission of flares in the visible spectral region was presented by Švestka (1966*a*). Korchak (1965) has shown that an interpretation of the visible flare continuum by synchrotron radiation of relativistic electrons meets with serious difficulties, if compared with the X-ray and radiowave spectral regions. Švestka (1966*c*) has called attention to the fact that the continuous emission of flares in the visual region could be explained by the H^- emission if it appeared in low parts of flares, in which the Lyman-alpha radiation temperature is lower than 7500 °K and the density of hydrogen atoms exceeds 10^{15} cm^{-3}.

7.5. *Flare theories*

Flare theories were reviewed by Wentzel (1964*a*) and by Sweet (1965) and brief general comments on theories of flares were made by Parker (1964*a*) and by Goldberg (1966).

The manner in which the dissipation of magnetic energy can occur in the sudden events of flares was discussed by Gold (1964, 1965). Sweet (1964) pointed out the possibility of a large-scale resistive instability which could account for the sudden onset of flares. A plasma instability, which is able to produce dissipation of the magnetic field energy within an extended region and a sufficiently short time, has been suggested by Jaggi (1964, 1965). It appears if two dipole magnetic fields in the solar atmosphere approach each other and—as the result—a finite conductivity instability occurs. Dungey (1965) has discussed hyperbolic magnetic null points in a plasma, in which a thin sheet of intense currents can be formed. When the current density has grown, various kinds of instability may occur.

A very significant contribution to the Sweet's mechanism was presented by Petschek (1964) who took into account standing magneto-hydrodynamic waves as a possible mechanism for converting magnetic energy to plasma energy. The found rapid rate at which energy can be released removes the criticism that the release of magnetically stored energy is too slow. A simple two-dimensional model of Sweet's mechanism was discussed by Green (1965) to determine the rate of interpenetration of two anti-parallel magnetic fields. A dynamic dissipation mechanism of the magnetic field, leading to a direct transition of magnetic energy to that of fast particles, has been proposed by Syrovatsky (1966). The process of the ejection of condensations observed after flares was discussed by Livshitz and Pikelner (1964). The condensations are connected with the formation and separation of loops on lines of force of the sunspot field.

Sturrock and Coppi (1964, 1966) have proposed a new model of solar flares, in which the preflare state consists of a mass of gas supported by a sheared magnetic field. The flash phase of a solar flare is identified with a gravitational resistive instability of this plasma-field configuration. The instability tends to break up the plasma into filaments with dimensions of a kilometer or less. In another model, proposed by Wentzel (1964*b*), the flare is caused by the collapse of an unstable chromospheric region under the action of magnetic fields. The collapse is stopped by forces which are due to the compression of the magnetic field and as soon as the velocity of collapse becomes comparable to Alfvén speeds, turbulence is likely to occur. The optical emission is localized in thin surfaces at the eddy boundaries. Both these theories are able to explain the filamentary structure of flares as well as many flare-associated phenomena.

A theory supported by laboratory experiments has been proposed by Jacobsen and Carlqvist (1964) and by Alfvén and Carlqvist (1967): An energy comparable with the total energy of a solar flare, is stored in the magnetic fields connected with filamentary electric currents, which flow through the solar atmosphere. An interruption of the current-carrying plasma filaments leads to a rapid release of the stored magnetic energy at the point of interruption. Dubov (1966*b*) published some comments on the laboratory modelling of solar events. A theory of a powerful point explosion as the source of a flare was suggested by Guseynov (1964*a*, 1965*a*). The temperature in the centre of the explosion was estimated to $5 \times 10^6 - 8 \times 10^7$ °K depending of the flare importance. The brightening of flocculi in the surroundings of a flare (Guseynov 1964*b*) and turbulence in flares (Guseynov 1966*a*) were also discussed from the point of view of this theory. An explanation of flares by point explosions in a non-homogeneous atmosphere was also discussed by Obashev (1965*b*). Kiepenheuer (1964*a*, 1965), on the other hand, does not believe that the flare phenomenon is a macroscopic explosion or collapse. He suspects that subtelescopic distorsions and twisting of the associated magnetic field are the basic processes leading to a flare. Zwaan (1965*b*) finds the origin of flares in spasmodic changes in sunspots, by means of which matter is ejected into the chromosphere. Oster (1966) and Oster and Sofia (1966) suspect that an origin of the flare in the photosphere or even below the photosphere cannot be excluded. This conclusion is based on the results of computation of radiative cooling under conditions to be expected in flares. Also Blizard (1965) presented a hypothesis on the origin of flares in subphotospheric levels.

On the other hand, many authors have suspected than an acceleration of fast particles is the primary process in flares (Dubov 1963, Carmichael 1964, H. Elliot 1964, Sakurai 1965*a*, 1966*a*, and de Feiter 1966). Carmichael suggests that the primary energy involved in flares is the mechanical energy of the solar wind coming from the active region. When a collapse occurs, the energy of the magnetic field is first utilized in the acceleration of particles in the corona and the optical flare results from the descent of these high energy particles into the chromosphere. According to Sakurai, a collapse of the magnetic field occurs in either the upper chromosphere or the lower corona and the released energy is consumed in the acceleration of trapped particles. In association with a penetration of the stream of accelerated electrons into the lower chromosphere, the Hα flare originates. Accelerated electron streams also produce X-ray bursts by bremstrahlung, excite plasma oscillations in the corona and generate type III radio bursts. De Feiter suggests that the optical emission of flares is brought about by a downward propagating ionizing agent which originates in the high energy region whose existence is inferred from X-ray and radio bursts. The inhomogeneous structure of the Balmer flare is due to irregularities of this ionizing stream, which propagate along lines of force. According to Elliot the energetic protons are permanently present in the active region as trapped particles in the sunspot magnetic fields and the flare is the direct consequence of a catastrophic loss of these particles in the chromosphere. Kawabata (1966*a*) has suggested that ejection of a plasma cloud producing a geomagnetic storm is of primary importance with respect to the origin of the flare phenomena. A magnetically isolated cloud, which appears due to a reconnection of magnetic lines of force, is accelerated in consequence of inhomogeneities in the magnetic pressure and squeezed out from the sunspot region. The consequent collapse of the coronal condensation gives rise to the observed flare phenomenon. Assuming a sudden heating of the corona above flare, Wentzel and Solinger (1966) consider a shock moving downward from the heated region. Its effect may relate to the flare nimbus and to the disappearance of striation patterns.

From the high tritium and helium-3 content in Discoverer–17 material, Fireman (1964) concluded that nuclear reactions had occurred on the solar surface during the 12 November 1960 flare. The production of neutrons and gamma rays by nuclear reactions on the Sun was estimated by Chubb (1964). Wentzel (1965) has demonstrated that Fermi acceleration in a limited region near solar flares involves many attractive features so that it should remain under consideration.

7.6. *Optical flare—associated phenomena*

Observations of coronal and prominence phenomena associated with flares were reviewed by Zirin (1965*a*). A general discussion of transient flare-associated phenomena in the solar atmosphere was presented by G. F. Anderson (1966). Surges accompanying flares were studied by Zhang He-qi and Cao Tian-jun (1964), by Morgante and Torrisi (1965) and by Obashev (1965*a*). The Chinese authors pointed out the close relation of radio events to these ejections. Obashev estimated the kinetic energy of a limb surge. High-velocity surges associated with a limb flare were described by Valníček (1964*a*). An intrinsically close connection between surges and the small circular flares on their bases was emphasized by Bruzek (1966*a*), who found some similarity between the fine structure of the surge flare and the structure inside the associated surge.

When discussing the series of large flares which occurred in July 1961, Bruzek (1965) paid particular attention to associated loop prominences; the moving material accompanying the flare of 16 July 1959 was discussed by Dodson and Hedeman (1964*a*). For other studies of the association of loop prominences with flares see section 8. Slonim (1964, 1965) has concluded that in flares associated with loop prominence systems, the photospheric, chromospheric and coronal gases are mixed up in the flare region.

Observations of individual explosive-type flares and associated filament activations were described by Moreton (1964*b*) and by Morgante and Torrisi (1965). Moreton (1965*b*) also summarized the observed filament changes closely linked with the flare process. Other papers dealing with the filament activation by flares are mentioned in section 8.

Motions connected with moustaches were investigated by Koval (1964). In absorption lines moustaches are accompanied by shifts to the shortwave side and by jets to the longwave side, and the velocity increases with the height. The brightness asymmetry of the emission lines in spectra of moustaches is explained as due to a weakening of the moustache emission by absorption ejections (Koval 1965*a*). The position of moustaches in a spot group relative to the magnetic field was studied by Howard and Harvey (1964) and by Koval (1965*b*). According to Koval, moustaches appear mainly near the line dividing the magnetic field polarities $H_{\parallel} = 0$ and at the border of the penumbra of large spots in the regions of interspersion of transversal fields.

Öhman (1966*d*) has found several cases where the Hα spectrum of a flare shows the two emission peaks of the low chromosphere at the limb suggesting central absorption due to mottles situated above the flare.

8. PROMINENCES

Prominence classifications have been extended through sunspot cycle No. 19 (Menzel and Jones 1963). Also of general interest is the production of a catalogue of eruptive prominences for the years 1938 to 1961 (Kleczek 1964*c*). Comparative studies in this catalogue have shown that the most active eruptive prominences occur during solar maxima. The indices employed for describing solar filament activity were analysed by Forti *et al.* (1965).

Extensive observational data on prominence line spectra have been given by Tandberg-Hanssen (1964*a*). The line lists given here seem comprehensive enough to form a catalogue of solar prominence lines. A list of equivalent widths of 194 lines in the spectrum of an active prominence observed at the solar eclipse on 15 February 1961, was given by Sotirovski (1965). Of particular interest is the observational material presented by Tandberg-Hanssen (1967) on the relative strengths of a number of He I prominence lines. It is established that the intensity ratio of corresponding singlet and triplet lines is strongly dependent on the activity of the prominence. Tandberg-Hanssen (1964*b*) also discussed the emission of Ba II lines observed in dense quiescent prominences, which pointed to conditions far removed from LTE.

Monochromatic studies of prominence forms have led to further observations of dark structures in the Hα and D_3 images (Valníček 1965*b*, Gahm 1966).

Spectra of quiescent prominences were studied by Melnikov *et al.* (1964) and by Zeldina and Sergeeva (1966). Morozhenko (1964, 1965, 1966) analysed hydrogen and metallic filaments in quiescent prominences, Zeldina (1964) applied the curve of growth, Piunov (1964*a*) proposed a method of determining Doppler halfwidths of Balmer lines taking selfabsorption into account, and Gurtovenko and Rakhubovsky (1965) discussed the Balmer decrement in prominences. Theoretical studies of a quiescent prominence spectrum obtained at the 1958 eclipse (Hirayama 1964) indicate that hydrogen ionization in prominences is due to Lyman continuum radiation which penetrates into the gas. According to Kawaguchi's (1964, 1965) discussion, this result implies a finely threated structure for a quiescent prominence. Yakovkin and Zeldina (1964*a*, *b*, 1965) have concluded that the prominence emission in the low Balmer lines are due to resonance scattering of photospheric radiation, while in higher lines, the emission is due to recombination. According to Stoyanova (1964), the observed relative populations correspond to excitation by photospheric radiation up to $n = 9$. The influence of systematic motions on the Hα profile was studied by Rompolt (1967). Kostik and Orlova (1964) studied profiles of Balmer lines in the spectrum of a dark filament and the same authors (1966) explained the bright hem of dark filaments in the Hα line as due to an additional excitation of the chromosphere caused by the photospheric radiation reflected from the filament.

Spectra of an active prominence were studied by Piunov (1964*b*, 1965), who found electron density $\sim 8 \times 10^{11}$ cm^{-3} and neutral hydrogen concentration $\sim 4 \times 10^{12}$ cm^{-3}. Teryaeva-Geichenko (1963, 1964) studied spectra of surges both on the disk and above the limb. The mechanism of excitation of the 3rd to 6th hydrogen levels is the photospheric radiation and radiative recombinations (Geichenko 1966). Yakovkin and Zeldina (1964*c*) have concluded that also the Hα radiation of eruptive prominences can be fully explained by scattered photospheric light. Švestka (1964*d*) pointed out the different behaviour of halfwidths of high Balmer lines in spectra of active prominences and flares. The effect on line equivalent widths of a filamentary prominence model was discussed by Rompolt and Ciurla (1967).

Recent progress in our knowledge of the condensation process of solar prominences has been surveyed by Kleczek (1965). Some general considerations on the energy balance as a factor in the evolution and structure of prominences have been set down by Menzel and Doherty (1963), Doherty and Menzel (1965) and Menzel (1965). Progress to an understanding of the generation and development of prominences from coronal material is, however, still held up pending a quantitative understanding of the mechanical heat input into the corona—especially its dependence on density and temperature. The question is considered further by Kuperus (1965) in a study of the initial phases of prominence formation and of the inbalance between mechanical heating by hydromagnetic waves and radiative cooling. The character and rate of heating of a solar prominence by the surrounding corona was considered by Ioshpa (1965). According to his considerations, not accounting for the radiative heat loss, a prominence without any internal magnetic field would be heated within less than one hour.

Observations of longitudinal magnetic fields in prominences have progressed in the past three years. Reports presented by Ioshpa (1963), Hyder (1964*a*), Lee *et al.* (1965), and Rust (1966) indicate that field strengths in quiescent prominences range from about 2 to 10 G, while active prominences and loops are associated with substantially stronger fields. In active prominences associated with flares, the magnetic field can reach to some hundred of gauss (Ioshpa *et al.* 1964). Ioshpa (1966) found a two-layer structure of the magnetic vector in the region of a quiescent filament. The supposition that the magnetic fields in prominences change their polarity in subsequent cycles, seems to have been confirmed by Hyder (1964*b*). The association of magnetic fields specifically with loop prominence configurations was discussed by Hyder (1964*a*) and by Kleczek (1964*a*). Kleczek (1963*b*) has shown that, since the motion of the prominence plasma is controlled by magnetic fields, rising loop systems imply a conveyance of field into the corona. There remains some question, however, as to whether the observed upward growth of loop systems is to be attributed to growth of individual loops or to the formation, at successively greater elevations, of new prominence loops. Rompolt (1965*a*) constructed a scheme of the magnetic field above an active region on the basis of recurrent trajectories followed by the active prominence matter.

A theoretical study of the emission line polarization produced by resonance scattering is given by J. W. Warwick and Hyder (1965). A comparison of the predictions with observational data shows that this cannot be accounted for without incorporating the influences on the degree and direction of polarization of magnetic fields and electron collisions. Hyder's (1965*b*) polariscope observations of quiescent prominences give promise of useful data if extended over a solar cycle. Active prominences show a larger polarization degree than the quiescent ones (Brückner 1965). Observations of strong polarization in the continuous spectrum of a flare spray are reported by Öhman *et al.* (1967). The (linear) polarization is consistent with an origin in Thomson scattering by free electrons.

Studies of the association of loop prominence formation with large proton flares and type IV radio bursts indicate that loops are a manifestation of the storage in the corona of fast, non-thermal, particles (Jefferies and Orrall 1964, Bruzek 1964*a*, *b*). This view gains credibility from observations of extremely broad, non-gaussian Hα wings in loop systems (Jefferies and Orrall 1965*a*, Orrall 1965*b*). Further support is found from comparison of the mass in a coronal condensation and that which is drained down the loops during the lifetime of the system (Kleczek 1964*a*, Jefferies and Orrall 1965*b*). A possible mode of expansion of loop prominences was discussed by Fokker (1964).

Filament activation by active regions has also been extensively studied. Vertical Doppler shifts giving rise to the appearance of 'winking' filaments have been discussed by Dodson and Hedeman (1964*a*), S. F. Smith and Ramsey (1964), Prokopieva (1964) and Ramsey and S. F. Smith (1965, 1966). Observations in the red and blue wings of Hα show that the material seems to exhibit up and down motion of an oscillatory nature. Each filament appears to have its own characteristic frequency of oscillation, repeating this frequency with each subsequent disturbance. Some typical changes in the filaments appear already several minutes or tens of minutes before the flare appearance: Hyder's (1966) theoretical study has aimed at interpreting this oscillatory phenomenon as a damped harmonic oscillation of conducting material resting in a magnetic field whose strength varies with height. It is suggested that winking filaments and 'disparitions brusques' may be different manifestations of a common phenomenon, the one being a stable, the other an unstable oscillation. The nature of the driving force generated by a flare and initiating the oscillation has been considered by G. F. Anderson (1966) and by Meyer (1966). Valníček (1964*b*) compared the velocity changes of matter emanating from explosive-type flares and blow-off prominences and found a difference in the nature of these two processes. A different type of activation leading to filament breakup by developing sunspots is reported by Bumba and Howard (1965*a*).

Yet another connection between filaments and active regions is found by S. F. Smith (1965) in the observations of 'sympathetic' flares formed along (or under) filaments which themselves had been flare activated within the preceding 30 minutes or so. As pointed out by Tandberg-Hanssen (1967), this observation, if confirmed, can have highly important implications to the flare mechanism.

Prominence activity stimulated by or associated with adjacent surge and flare activity has also been discussed by Liszka (1963, 1964). The circumstances surrounding the occurrence of a giant eruptive prominence, and the subsequent history of the gas are set down by Hedeman and Dodson (1966).

An interesting study of the suppression (by about 0·3 km s^{-1}) of the chromospheric oscillatory velocity field near a quiescent prominence has been made by Kleczek (1964*b*). The lifetime by quiescent prominences was studied by Forti and Godoli (1965) and found equal to 1·5 rotations, on the average.

Sturrock and Woodbury (1965) suggest that quiescent filaments result from slippage of the photosphere, which occurs preferentially along a neutral line, and a model filament has been calculated by them. McLean (1964) has considered equilibrium configurations of the magnetic field which may apply to quiescent arch prominences. Chvojková (1965*b*, 1966*a*) and Chvojková and Klepešta (1965) have shown that some phenomena observed in eruptive prominences can be explained if the effect of gravity is taken into account when considering plasma particles frozen in the magnetic field.

9. CORONAL CONDENSATIONS AND ACTIVITY

In 1963, two Symposia were held, which contained many papers concerning the problem of this section: *The Solar Corona* (Evans 1963) and *The Solar Spectrum* (de Jager 1965*a*). A review paper on the interpretation of coronal observations in X-ray, UV, visible and radio wavelengths was presented by Seaton (1964) and contemporary ideas on physical properties of the solar corona in connection with chromospheric activity and radio bursts were reviewed by J. W. Warwick (1965*b*).

In the solar corona, the distinguishing between active and 'quiet' conditions is particularly difficult. Therefore, at least four basic papers which perhaps do not seem to be directly connected with the active corona, have to be mentioned here, because, in fact, they are very important for studies of the active coronal regions as well: Burgess and Seaton (1964) and Burgess (1964) pointed out the role of dielectronic recombination in the process of formation of the emission coronal lines. And Pottasch (1963, 1964) presented very important papers on the element abundances in the corona.

Perturbing agents in the chromosphere and corona causing magnetic storms with and without ssc were studied by Garczynska (1965). The same author also investigated the velocities of shock waves in solar corona at the time of flares (Garczynska 1967). Parker (1964*c*) discussed the response of the coronal medium to shock waves produced by flares.

Many papers have been published which deduce physical properties of the solar corona from radiowave and X-ray observations. These are mentioned in sections 10 and 11.

Papers discussing monochromatic solar corona are more frequent than those concerning the K-corona, which is obviously due to difficulties arising from observations of the K-corona outside the eclipse. Nevertheless, many papers on the monochromatic corona are also based on eclipse observations. Suzuki and Hirayama (1964) and Letfus (1965) have presented models of coronal condensations, which are in accordance with the temperature computations based on the abovementioned Burgess' dielectronic recombination and Pottasch's abundances. Suzuki and Hirayama have assumed an inhomogeneous temperature distribution in the studied coronal condensation observed at the eclipse of 15 February, 1952, and find the abundances of iron and nickel 10 times higher than in the photosphere. Letfus has discussed the Fe XIV $\lambda 5303$ and Fe X $\lambda 6374$ emission line intensities in two condensations observed in 1952 and 1962; for a cylindrical model he finds central temperature 4 to 5×10^6 °K from the green line and about half of this value from the red line. These temperature differences are explained by a filamentary structure of the coronal condensation. Spectra of coronal condensations containing loop prominences were studied by Orrall (1965*c*). Electron temperature 2×10^6 °K and electron density 2×10^9 cm^{-3} were found in the condensation, the density being 50 times lower than that for the loop prominence in the Hα emitting region. Assuming rough pressure equilibrium between the prominence and the rest of the condensation, a model for the loop structure of a sporadic condensation has been proposed. The possibility that matter in the coronal condensation is largely concentrated in the loop system has also been considered by Tsubaki (1966). Saito and Billings (1964) presented a model of the condensation based on polarimetric observations of the condensation on the western limb during the February 5, 1962 eclipse.

Coronagraphic observations have been used, particularly, in studies of associations between various forms of the solar activity. Kleczek (1963*a*) investigated relations between green ($\lambda 5303$) loops and arches with the underlying photospheric magnetic fields. The same problem was also discussed by Bumba *et al.* (1965). The authors have found that the brightest coronal features tend to be associated with the highest gradients in the photospheric magnetic fields. Křivský and Makarov (1966) have studied the deformation of the green corona after nine selected proton flares which appeared close to the limb and conclude that the condensations are deformed in non-radial directions. Nagasawa and Nakagomi (1965) studied the intensity of the green coronal line in connection with the birth and decay of a calcium plage. They have shown that the coronal activity has a tendency to appear before the birth and to disappear only after the death of a calcium plage. An attempt has been made by Shpitalnaya (1964) to analyse events preceding the formation of bright knots in the green coronal line. She observed a stream collision at the boundary separating the coronal condensation from the undisturbed corona resulting in 'dashes' developed in the He D_3 line at the heights from 30″ to 60″ above the limb. Zirin's (1964*b*) observations of the limb proton flare of 20 November 1960, have led to the conclusion that a substantial fraction of the coronal condensation may be at temperatures exceeding 4×10^6 °K. Vsechsvyatsky *et al.* (1966) used pictures of the corona taken during the eclipse on 30 May 1965, to study correlations between the corona and the chromospheric and photospheric activity. Kernoa *et al.* (1965) compared intensities of 12 coronal lines photographed at the eclipse of 15 February 1961, in 'hot' and 'cold' points of the corona. Schmieder *et al.* (1967) have shown that the intensity ratio of Ni XV λλ6702 and 8024 lines is more sensitive to the electron density in coronal condensations than the ratio of Fe XIII lines discussed earlier by Perche and Dumont (1964).

Some papers based on coronagraphic observations and studying correlations of coronal characteristics with the 11-year cycle are mentioned in section 3. Apart from them, Gentili *et al.* (1966) compared variations of the λλ5303 and 6374 coronal line intensities during the 18th and 19th solar cycles, on the basis of Pic-du-Midi observations. Significant differences could only be found in the λ6374 line behaviour near the solar minimum.

Almost all studies of the K-corona have been based on eclipse observations. At the eclipse of 1961, Kawaguchi (1967) studied a coronal dome formation overarching a prominence. The condensation photographed by Saito during the eclipse of 1962, was discussed by Saito and Billings (1964). At the eclipse of 1963, Stoddard *et al.* (1966) observed polar rays and J. W. Harvey (1965) made an attempt to associate them with the polar magnetic field of the Sun on the basis of K_3 spectroheliograms. And several papers are based on observations of the eclipse of 1965: Leroy and Servajean (1966) investigated the brightness of the K-corona in the immediate neighbourhood (3000 km) of a prominence. No appreciable brightness variation larger than 5% could be found. Saito (1964) photographed the polar rays and having compared these photographs with the earlier ones he deduced an 11-year variation of the length of the hypothetic dipole whose lines of force might be identical with these rays. Finally, Bohlin *et al.* (1966) used solar eclipse photographs of 30 May 1965, photographs from the balloon-borne coronagraph flights (Coronascope II, Newkirk and Bohlin 1965) of 3 June and 1 July 1965 and nearly synoptic K-coronameter scans from 1 May to 31 July 1965 and achieved in this way a unique combination of coronal data covering several solar rotations. This has allowed basic phenomenological features of one major streamer to be established in the region $1 < \rho < 5$. The streamer lifetime was at least two months, its rotation was in agreement with that of surface features at the streamer latitude and the streamer exhibited a spiraling of the classical Achrimedian form due to the rotation of the Sun. Surveyor I made the first observation of the solar corona from the lunar surface on 14 June 1966 (Newkirk 1967).

Regular observations of the K-corona are not yet frequent enough to make any study on the cyclic variations possible. Nevertheless, Nishi and Nagasawa (1964) have succeeded to find a correlation between coronal streamers and the quiescent prominences. Nakagomi and Nishi (1965) have made a statistical study comparing the solar electron corona and the green coronal line intensity with various photospheric and chromospheric phenomena. They conclude that regions of the maximum polarized component of the K-corona appear over active regions where they correspond to coronal condensations as well as outside active regions where they indicate the coronal streamers.

Kuperus (1965) made an attempt to explain the origin of permanent coronal condensations in terms of the excess heating of these parts of the atmosphere, due to an enhancement of the mechanical energy flux. He has shown that fluctuations of the mechanical energy flux give rise to coronal inhomogeneities. Bird's (1965) theoretical considerations lead to the conclusion that the maximum coronal temperature does not vary significantly with the initial heat imput within very broad limits. If however, the heat imput exceeds some critical limit, as may be the case with solar flares, the maximum coronal temperature begins to increase rapidly. The energy balance in a coronal condensation, considered as a magnetic tube of force, has been investigated by Obashev (1964). His study of the loss of energy of the electronic component of coronal condensations shows that the cooling caused by radiation and thermal conductivity takes a very short time (about 10 minutes).

10. RADIO ASPECTS OF SOLAR ACTIVITY

10.1. *Reviews and general papers*

General review papers on solar radio emission were written by Maxwell (1965) and Smerd (1964*a*). Relationships between solar radio emission and X-emission were reviewed by Kundu (1963, 1964), and theoretical aspects of particle and radio emission were discussed in a critical review paper by Schatzman (1965*b*). General papers on flare-associated radio phenomena are those of Ellison and Reid (1964), Wild (1964), de Jager (1965*c*), and McLean (1965). A review paper on the observation and physical interpretation of solar radio bursts was presented by Wild *et al.* (1963) and by Smerd (1964*b*).

The origin of type I and II bursts and of reverse-drift pairs connected with type III emission was discussed by Zheleznyakov (1965*b*). Frequencies of electromagnetic waves originating when fast-moving particles pass through a magnetized plasma were computed by Ginzburg (1964). The possibility of amplifying radio waves in a plasma was discussed by Smerd (1965). Gelfreikh (1964) studied the depolarization and transfer of radiation in plasma in the presence of an arbitrary magnetic field. The magneto-ionic theory in relation to solar radio emission was critically discussed by Wallis (1965*a*). Fung and Yip (1966) have shown that in the solar corona, electrons moving in the magnetic field of a sunspot may generate cyclotron radiation, which will be observable mainly in the ordinary mode in the vicinity of the second harmonic of the cyclotron frequency.

Kawabata (1966*a*) tried to explain the flare-associated bursts from the point of view of his new flare theory (cf. section 7.5). Takakura (1964) made an attempt to derive information on the magnetic field intensity in the corona from known characteristics of type I, II, III, and IV bursts and on the basis of various assumptions as to the mechanisms involved. The same author (Takakura 1966) discussed the general implications of solar radio bursts for the study of the corona.

10.2. *The slowly varying component*

Review papers on slowly varying component at microwaves were written by Swarup (1964), Pick (1965) and Hachenberg (1965).

Caroubalos (1964*a*, 1965), Das Gupta and Basu (1964), Krüger *et al.* (1964) and H. Tanaka (1964) dealt with the variation of the slowly varying component through the solar cycle. Two maxima of the solar radio emission flux during the 19th cycle were found by Gnevyshev (1965), in agreement with other manifestations of the solar activity (cf. section 3). Attention has been drawn to the fact that the degree of correlation between the 10·7 cm flux and the EUV solar radiation varies with solar cycle (A. D. Anderson 1964, 1965). A phase-shift of the 27-day variation for 10·7 cm and 21 cm flux was discussed by Sastry (1966).

Combining the results of observations made at Toyokawa and at Nançay, H. Tanaka and Steinberg (1964) derived the structure of the polarized microwave source above a bipolar sunspot. Khangildin (1964) studied the Sun's radio brightness distribution at 8 mm and obtained evidence for the existence of small sources right above sunspots with brightness temperatures of 7 to 10×10^4 °K. A reduced radio brightness was found to be connected with certain filaments. The same author (Khangildin 1965) also measured the polarization of 8 mm radiation from selected active regions. On the basis of observations carried out with a 17 GHz grating interferometer, Tsuchiya and Nagane (1967) found the sources of the slowly varying component optically thin at this frequency. The spectral distribution of the slowly varying component was discussed by Molchanov (1964*a*), Krüger and Michel (1965), and Abbasov *et al.* (1965). Krüger and Michel found evidence for the existence of a secondary maximum near 1500 MHz, apart from the well-known maximum near 6000 MHz.

Solar eclipses were used to measurements of discrete sources by Higgs and Broten (1964)

Gosachinsky *et al.* (1964) and Hett (1966). The Soviet authors detected an aborption region, which could be identified with a limb prominence. High-resolution observations of a partial solar eclipse at 9·1 cm were described by Asper *et al.* (1964). Characteristic properties of solar radio sources associated with plages without sunspots were discussed by Abbasov and Yasnov (1966).

M. Simon (1965) found regions of enhanced 3·2 mm emission to correspond with Hα active regions, particularly so with those that are prone to flare production. Active regions that later flared were considerably enhanced for a day or more prior to the flare. Avignon *et al.* (1966) established a relationship between certain properties of the slowly varying component and the optical structure of the associated centre of activity. On the contrary, Soboleva (1965) did not find any dependence of the centimeter and dekameter radio emission on the type of the sunspot group. H. Tanaka and Kakinuma (1964) noted a relation between the spectrum of the slowly varying component and the occurrence of proton flare events. Most of the proton flares occurred in active regions which had the 3·2 to 7·5 cm flux ratio greater than unity. On the other hand, Eliseyev and Moiseyev (1965) found that regions with and without proton flares do not differ with regard to the intensity of the slowly varying component of radio emission at 9 and 21 cm. The centres of 21 cm condensations are shifted to the east relative to the spot with the largest magnetic field. Ikhsanova (1964) who carried out simultaneous measurements at 3·15 and 8·7 cm, found a shift of coronal condensations towards the equator with regard to the sunspots. An increase of the radio emission associated with rapid revival of a sunspot group was studied by Borovik *et al.* (1965).

Magneto-bremsstrahlung has been considered as the mechanism producing the $2 \text{ cm} < \lambda < 20 \text{ cm}$ slowly varying component by Molchanov (1964*b*) and by Grebinsky and Molchanov (1964).

10.3. *Noise storms*

A review paper on noise storms was presented by Fokker (1965). Work on noise storms done with the 169 MHz interferometer at Nançay was summarized by Le Squeren (1964*a*). Elgaröy (1965) summarized narrow-band studies of solar bursts made at Oslo. De Groot (1966) gave an extensive account of the narrow-band spectra and other properties of weak stormbursts as observed with a 7-channel receiver. In addition to many examples of observed bursts he presented a discussion on possibilities of interpretation. Some properties of the storm-centres at 408 MHz were deduced by Clavelier (1967). Noise storms at 200 MHz were studied by Yurovskaya (1964). A new classification of solar radio storms in the meter range has been proposed by Tlamicha *et al.* (1964).

Wild and Tlamicha (1964, 1965) called attention to the occurrence of chains of stormbursts which often show a drift in frequency. This phenomenon was also discussed by Hanasz (1966). Eckhoff (1966) introduced the name 'flash burst' for bursts with a duration of the order of 0·1 second which occur isolated during ordinary noise storms or, very rarely in groups. These flash bursts seem to be identical to the 'spike bursts' which were noted previously by de Groot (1962, 1966).

Trachtengertz (1966) criticized the Takakura's (1963) theory of type I solar radio bursts and suggested that interaction of accelerated coronal electrons with magnetohydrodynamic impulses could be considered for a possible mechanism of the generation of type I radio bursts. Mogilevsky (1966*a*) has presented a hypothesis that narrow-band type I bursts are due to forbidden transitions between split Zeeman sub-levels of the ground state in hydrogen atoms.

10.4. *Bursts of type II, III, and V*

J. W. Warwick (1965*a*) summarized the results of sweep-frequency observations in the range of 7·6 to 41 MHz, made at the High Altitude Observatory, Boulder. Observational material of type II bursts was studied by the late A. A. Weiss (1965) to derive the velocity of the sources as a function of height. He discussed the results in terms of a magneto-hydrodynamic shock front. The question how radio emission can be generated in shock fronts was discussed by Zaitsev (1965) and Tidman (1965). Both theories account for the existence of a first and a second harmonic. Zaitsev, in addition, ascribes band splitting to electron drift in the shock front. Takakura (1964), A. A. Weiss (1965), Formichev and Chertok (1965) and Krishnamurthi *et al.* (1965) made estimates of the strength of the magnetic field on the basis of characteristics of type II bursts and various hypotheses about their mechanism. Takakura estimates a field strength of about 15 G at a height of $1{\cdot}5 \times 10^5$ km; for the same height Fomichev and Chertok derive a value of 5 to 8 G. According to these authors the field strength at a height of 4×10^5 should be 1·2 to 2 G, while Krishnamurthi *et al.* found 3 to 8 G. The estimate of Weiss lies between 2 and 20 G for a height of $3{\cdot}5 \times 10^5$ km. Stewart (1966) found that the fast drifting spectral features known as 'herring bones' that diverge from a type II bursts, sometimes exhibit strong circular polarization, in contrast to the type II bursts themselves. Drago and Tagliaferri (1966) found that the association percentage between subflares and type II bursts was increasing towards the minimum of solar activity, which is interpreted as a consequence of coronal electron density variation with the solar cycle.

Observations of type III bursts were extended to as low a frequency as 1·5 MHz by means of sweep-frequency recordings on board of the Alouette satellite (Hartz 1964). Frequency drift curves indicate velocities in the range of 0·1 *c* to 0·15 *c*. De Groot (1966) dealt with high-speed records of weak type III bursts. With the help of observations made with the panoramic spectrograph at Boulder, some characteristics of type III bursts were studied by Boischot (1966). Solar type III bursts were also observed with a swept frequency receiver in the 200 MHz range by Elgaröy and Rödberg (1964). Evidence of an increase in the source size on the way out through the corona was presented.

The polarization of type III bursts was studied by several authors. Bhonsle and McNarry (1964*a*) determined polarization ellipses at 74 MHz. Daene and Voigt (1964) reported on both linear and circular polarization of type III bursts at 23 MHz. Gopala Rao (1965), from sweep-frequency observations from 40 to 60 MHz, noted a double structure of the type III burst with differently polarized components. The sporadic circular polarization of type III bursts at 200 MHz was studied by Enome (1964).

Velocities in the range $c/6$ to $c/2$ were derived by Stewart (1965) for the outward motion of sources of type III bursts. He concluded that this should be the actual velocity of the electron cloud. Malitson and Erickson (1966) determined positions of type III bursts at 26·3 MHz with a multi-element interferometer. From these observations, the coronal electron density above the active region was found 10 times higher than in the Van de Hulst model of the equatorial maximum corona. Likewise Morimoto (1964) found a density 3 times as great as the density in a streamer, according to the Newkirk model.

Tlamicha and Takakura (1963, 1964) found a good coincidence in position between active prominences and filaments and groups of type III bursts at 200 MHz. In the sunspot group of September 1963, Zirin and Werner (1967) found for almost every type III burst a corresponding Hα brightening in the active region, but not all brightenings produced bursts. In another active region McKenna (1966) could discover a specific brightening within $\pm$ 1 minute of the start of 96% of all the radio events studied. Records of intense type III bursts on the night hemisphere of the Earth were reported by Fortini *et al.* (1966).

The properties of type V bursts were studied by A. A. Weiss and Stewart (1965). Heights and sizes are similar to those of type III bursts. The authors suggest that type V emission stems from coherent Čerenkov plasma waves excited by fast ($c/3$) electrons that oscillate between magnetic mirror points.

A. A. Weiss and Wild (1964) computed trajectories of charged particles in the vicinity of magnetic neutral planes. They were able to account for observed features of type III bursts assuming an exponential decrease of magnetic field strength with height. Sturrock (1964) discussed several properties of type III bursts on the supposition that type III emission is related, by coupling, to plasma oscillations excited by electron streams. A similar study was made by Oster and Altschuler (1964).

In March 1966, many narrow-band width bursts, apparently of a type not so far reported, were observed by Ellis and McCulloch (1966) in the frequency range 20–50 MHz.

10.5. *Type IV bursts*

A compilation of flux density, polarization and positions for type IV bursts was published by Kai (1965*a*). A list of 174 type IV events recorded from 1956 to 1963 was prepared by Švestka and Olmr (1966). A collection of spectral diagrams, based on single frequency records, was published by Fokker (1966). Some aspects of type IV continuum radiation on decimeter wavelengths were discussed by Kundu (1965*b*).

Sakurai (1964, 1965*b*) found an east-west asymmetry in the differences in onset time of the microwave and metric continuum components, which he explains by the assumption of a westward-inclined magnetic field structure. Malitson and Erickson (1965, 1966) reported on a moving type IV burst on 11 August 1963 and on the location of solar continuum emission at 26·3 MHz during the period from 18 to 25 September 1963. A fast moving type IV event was also reported by Philip (1964) according to measurements at frequencies from 13 to 23 MHz made with a multichannel interferometer. During a class 3 flare time variations in the radio emission were found by McKenna (1965*a*) closely associated with fluctuations of the Hα intensity of an optical flare region over the umbra of a major spot. In the course of the type IV burst of 5 February 1965 observed by Aller *et al.* (1966) at 230 MHz, four short-lived absorption events were recorded, each with the duration of about 1 second.

In two papers, Kai (1965*b*, *c*) first gave a thorough account of observed polarization characteristics of type IV bursts. He then deals with the possibilities of interpreting the polarizations: (*a*) on the basis of gyroresonance radiation; (*b*) on the basis of radiation from coherent plasma waves. In an appendix the emissivities and absorption coefficients for gyrating electrons are derived on the basis of expressions worked out by Kawabata (1964). Other contributions to the theory of type IV bursts were given by Böhme (1964*a*) and McLean (1964). On account of the influence which the ambient plasma exerts on synchrotron emission, Zheleznyakov and Trakhtengertz (1965) specified a condition which is imposed on the generation of synchrotron radiation.

Křivský (1964, 1965*a*) made a supposition that the delayed type IV burst on 16 September 1963 was produced by the cloud of ejected particles during its passage through the interplanetary space. This supposition was modified by J. Elliott and I. H. Reid (1966).

For the correlation of type IV bursts with proton flares cf. section 7.3. The radio evidence for solar corpuscular emission was summarized by Maxwell *et al.* (1964). The occurrence of type IV bursts in the course of the 11-year solar cycle was studied by Křivský and Krüger (1966); two maxima of occurrence have been found.

Radio spectra of selected type IV bursts were described by Tsimakhovich (1964), Suzuki *et al.* (1964), Yurovskaya (1965), Hagen and Barney (1966), and Castelli and Michael (1967). Machalski and Zieba (1965) investigated the correlation between 810 MHz radio bursts and solar flares.

10.6. *Microwave bursts*

The continuous spectrum of solar microwave bursts was studied by Hachenberg and Krüger (1964). They found flux densities increasing monotonically towards higher frequencies, generated by thermal radiation, as well as broad emission bands with maxima at various frequencies, produced by synchrotron radiation and in most cases a combination of both these two mechanisms occurred. Similar results were obtained by Dravskikh (1964). Takakura and Kai (1966) presented inferences on the energy distribution of electrons that produce impulsive microwave bursts and X-ray bursts. From the study of the microwave emission of flares, which appeared beyond the limb, Erushev (1966) has concluded that upper borders of sources of microwave bursts are localized at the height of some tens of thousands kilometers above the photosphere.

McKenna (1964) commented on the 'late' 10 cm maximum in particle producing flares. According to Eliseyev and Moiseyev (1965), in active regions with proton flares the number of flares accompanied by microwave bursts exceeds that in regions without proton flares by a factor of 3 to 4. G. A. Harvey (1965) has concluded that proton flares are accompanied by very characteristic broad profile bursts at 2800 MHz, which might enable an easy recognizing of proton events. This observation is in good agreement with the Fletcher's (1964) conclusion that the most important frequency interval for the correlation of type IV bursts with proton flares is 1000–3750 MHz. Haurwitz (1964) found that particularly flares which activated dark filaments were highly productive of 2800 MHz radio bursts.

Erushev and Eliseyeva (1965) compared the microwave bursts at 3·2 cm with SID's and direct X-ray records. A similar comparison was made by G. A. Harvey (1964) for microwave bursts at 10·7 cm. Bursts unaccompanied by SWF seem to originate in lower layers than the SWF producing bursts. The different centre-limb variation in the numbers of flares producing SID's and microwave bursts, has been tentatively explained by Mergentaler (1965).

Berger and Malville (1966) have proved that the hydrogen 3 cm emission line cannot be observed in the flare radio spectrum.

II. SOLAR EUV RADIATION, X- AND GAMMA-RAYS

General review papers on this subject were presented by Friedman (1963), Tousey (1964), de Jager (1964*a*), Winckler (1964), and Mandelshtam (1965*a*). A very detailed review paper was prepared by Mandelshtam (1965*b*) and a critical one by de Jager (1965*f*) who also suggested a classification for flare-associated X-ray bursts. Relationship between solar radio and X-emission were reviewed by Kundu (1963) and de Jager (1965*d*). Allen (1965) summarized variations of the XUV solar spectrum due to solar activity. The review paper written by Tousey *et al.* (1965) deals with EUV spectroheliograms and with results of the research carried out with rocket-borne grating spectrographs. Results summarized by Lindsay (1964) are particularly based on the OSO–1 measurements. Naval Research Laboratory results were summarized by Friedman (1964*a*) and results achieved in the United Kingdom were reviewed by Pounds (1965).

Direct spectroheliograms at a variety of wavelengths show close correlation with plage activity. Monochromatic pictures of the Sun in Mg II λ2803 line were obtained during two rocket flights, using a Šolc birefringent filter with ~ 4Å passband (Fredga 1966). The older regions were found more intense than the younger ones, which is opposite to the effect observed in the Hα line. A solar photograph obtained from a rocket flight in the spectral region within 170 to 400Å was discussed by Zhitnik *et al.* (1964). Blake *et al.* (1964, 1965) found that X-ray sources observed at ~ 50Å were spread over areas of the order of 6′ in diameter and larger, comparable to the underlying plages, whereas strong ~ 10Å sources were always smaller than 2′. These results imply that a coronal condensation is a complex structure containing small, hot centres embedded in a large lower temperature region. Essentially all the radiation at ~ 10Å originated in active regions only, while about one half of ~ 50Å emission was distributed over the disk. An estimate for the size of the X-ray emitting region was also given by Reidy (1966).

Four flights of Skylark rockets carrying equipments for X-ray solar photographs, were launched from August to October 1965. Results achieved with the first flight (passband 44–70 Å and < 25 Å) were described by Russell (1965*a*) and by Black *et al.* (1965). The second flight has shown that bright X-ray plages are observed even one day before the appearance or one day after disappearance of optical plages (Russell 1965*b*). During the third flight the Sun was photographed in He II λ304 and Fe IX λ171 lines (Burton and Wilson 1965). With the fourth flight, the < 50 Å source dimensions were found smaller than 1′ for three active regions (Russell and Pounds 1966). For the first time, dark regions were clearly visible on the disk, probably marking regions of lower coronal density. Photographs of the Sun within 3 to 75 Å region obtained with glancing incidence telescopes aboard an Aerobee rocket, were described by Underwood and Muney (1967). Earlier results deduced from X-ray photographs of the Sun were summarized by Mandelshtam (1965*c*).

Observations of EUV line emission carried out aboard OSO–1 were discussed by Neupert (1964, 1965, 1967). It has been estimated that the Fe XV and Fe XVI emission in plages is approximately 150 to 200 times that of an equivalent area of the quiet Sun, while for the He II λ304 line this ratio is about 15. The intensity ratio Fe XVI/Fe XV depends on the phase of development of active regions and increases during flares. The observations are inconsistent with an increase of electron temperature, but they are consistent with a model for active regions, which postulates both an increase of electron density together with an increase in the vertical range over which the Fe XV and Fe XVI radiation is emitted. Electron temperature of $1{\cdot}75 \times 10^6$ °K best fits the observed data. This can be compared with Widing's (1966) estimate of $T_e = 3$ to 5×10^6 °K in an active region emitting Fe XV and Fe XVI. Blake *et al.* (1965) have also concluded that the X-ray emission of plages is dominated by X-ray line radiation from iron of ionization states XV and above. Fe XVII was found to come from the active plage only. This has also been confirmed by flux variations in different lines with solar activity (Neupert *et al.* 1964, Hinteregger *et al.* 1964, Hall *et al.* 1965*a*, *b*, Austin *et al.* 1966). The flux ratio of Fe XV λ284 and He II λ304 lines decreased from 0·25 to ~0·04 during the period August 1960–March 1964. Between 170 and 300 Å, a continuum emission has been recorded from intense centres of activity (Tousey *et al.* 1965). Strong lines in the 170–180 Å region were discussed by Elston *et al.* (1964) and the range 72–105 Å was photographed on the first time in February 1966 (Austin *et al.* 1966).

Suemoto and Moriyama (1964) have pointed out that EUV line intensities are too strong if compared with the radio intensities. This discrepancy disappears if one assumes that the solar disk is bright only in patches covering about 0·15 of its total area. Stockhausen (1965) has shown that an existence of inhomogeneities in the corona with densities of $\sim 10^{10}$ cm^{-3} can be inferred from the observed correlation of the Fe XIV line strengths with solar activity.

Variations of the solar X-ray spectrum below 20 Å under non-flare and flare-active conditions were studied by Culhane *et al.* (1964). Chubb *et al.* (1964) compared the spectral energy of solar X-rays from 2 to 20 keV under quiet conditions and during subflare activity. In the second case, an emission peak was found near 10 to 12 keV, which might be evidence of an intensification of emission associated with Fe XXV and Fe XXVI ions. Typical non-flare and flare-enhanced X-ray spectra were obtained with the satellite Ariel in the range 4 to 14 Å (Bowen *et al.* 1964). The non-flare spectra are supposed to arise from discrete hot regions in the corona, with $T_e = 5 \times 10^6$ °K. In flares, as far as their emission is of thermal origin, T_e corresponds to $\sim 1{\cdot}2 \times 10^7$ °K. In case the X-ray emitting region has the extent of the visible flare, $n_e = 4 \times 10^{10}$ cm^{-3} is found. Tindo and Shurygin (1965) measured X-ray radiation between 2 and 18 Å during two rocket flights and compared the results with the distribution and brightness temperature of active regions present on the disk.

An analysis of X-ray data (< 10 Å) from the OSO–1 satellite has shown that the slowly varying component of X-rays and the non-impulsive X-ray bursts can be well related to their microwave counterparts (W. A. White 1964). Correlation of X-ray flares with 2800 MHz events is better than with Hα observation. Lindsay (1965) has also concluded that bursts of X-ray radiation associated with flares coincide extremely well with 3750 MHz data even as far as common secondary maxima are concerned. On the other hand, the correlation between X-ray and Hα flares is much worse (Van Allen *et al.* 1965, Conner *et al.* 1965), and particularly bad as to the flare importances. There are many Hα flares without any X-ray bursts. X-ray measurements made aboard Electron–2 have been described by Tindo (1965) and they confirm many of the conclusions mentioned above. Very good correlation is found with 10·7 cm radio flux again.

X-ray measurements carried out with the SOLRAD satellites were discussed by Kreplin and Gregory (1965) and by Thomas *et al.* (1965). According to the X-ray flux the solar minimum occurred in July 1964. Friedman (1964*b*) gives that during the solar cycle the X-ray emission varies by the factor 7 for $\lambda < 100$ Å, 70 for $\lambda < 12$ Å, and 300 for $\lambda < 8$ Å. Landini *et al.* (1966*a*, *b*) used the 1965–93 A satellite for an observation of the partial solar eclipse in X-rays. Sudden and deep changes of X-ray intensity were observed when the Moon's limb covered or uncovered active regions. No changes of this type were observed in the UV (1225–1350 Å) region.

Five X-ray flare bursts observed aboard OSO–1 were discussed by Frost (1964, 1965). He has found, in agreement with Kundu's (1963) analysis that an impulsive microwave burst was associated with every X-ray burst, which was not the case for type III bursts and other meter wave events. This fact has also been confirmed by Chubb *et al.* (1966) who have concluded from it that the X-ray emission region is located in the chromosphere rather than in the corona. Korchak (1965*b*) also places the X-ray source in the upper chromosphere. On the contrary, Kawabata (1966*b*) shows that the X-ray source must be located in the coronal region, since an X-ray event on 4 September 1960, was associated with a flare behind the solar limb. A similar conclusion has been drawn by Tindo (1965) from the fact that X-regions appear earlier on the eastern limb and disappear later on the west than optical active regions. Zirin (1964*b*) estimates that the X-ray emission comes from 30 000 km in the corona. Moreton (1964*a*, 1965*a*) drew attention to the fact that all high-energy (> 20 keV) X-ray events were connected to flares of the explosive type. The extraordinary enhancements of X-ray emission associated with the flares of 6 and 7 August 1960, were explained by Švestka (1964*b*), on the basis of flare spectra, as due to an increased density in the active region in which the flares appeared.

Chubb *et al.* (1966) presented an improved analysis of X-ray measurements in the $h\nu > 20$ keV region, carried out from rockets during three class 2+ flares in 1959. Two of these flares showed spectra equivalent to that expected from a thermal plasma with $T_e \simeq 10^8$ °K. Other events can be explained by thermal radiation, too, if the balloon measurements are properly corrected. The authors stress that there currently exists no definitive observational evidence for non-thermal X-ray emission.

SID effects were used to a study of solar X-ray bursts by several authors: Fortini (1963) has found similar trend of SEA and SCNA events if produced by flares recurring in the same area of an active region and she ascribes it to the influence of persistent magnetic fields that control the particle motion; a thermal model does not seem capable of explaining these recurrent characteristics. On the basis of ionospheric measurements, Odincova (1964) has made an estimate of the intensity distribution below 1000 Å in the spectrum of the flare of 28 August 1958. SEA data were analysed by Erushev (1965) who estimated that the maximum energy X-ray flux below 2 Å varied within 10^{-7} to 10^{-3} erg cm^{-2} s^{-1} in various flares. Neshpor (1965) deduced a time variation of the X-ray spectrum during the flare of 23 November 1957, from SID records. Eliseyeva (1966) has confirmed that X-ray flares do not correspond to type III bursts.

On the basis of an analysis of simultaneous observations of X-ray emission of the Fe XV and Fe XVI lines and 10·7 cm radio emission, Shklovsky (1964) has shown that a coronal temperature of $\sim 1{\cdot}5 \times 10^6$ °K is sufficient to explain the X-ray emission on the quiet Sun. The X-ray emission of flares is explained by the inverse Compton effect on relativistic electrons with energies $\sim 10^7$ to 10^8 eV. Zheleznyakov (1965*a*), too, presented the same explanation. On the other hand, Acton (1965*a*) concludes that the X-ray emission of flares cannot be explained by the inverse Compton effect and shows that the solar flare radiation can be explained by thermal processes if $T_e = 4 - 5 \times 10^6$ °K (Acton 1965*b*). Acton (1965*c*) also suggested that X-rays in flares might be produced by electron transitions among inner *K*–, *L*–, and *M*– spheres, in consequence of an ionization by energetic electrons. Korchak (1965*b*) concludes that the X-ray emission is due to bremsstrahlung of non-relativistic and sub-relativistic electrons. The X-ray burst accompanying the flare of 28 September 1961, could be explained by Korchak (1965*a*) by bremsstrahlung if the particle density in the breaking levels was two orders higher than in the undisturbed atmosphere. If the electron energy spectrum is extrapolated to relativistic energies, the observed centimeter burst can be explained as well. Elwert (1964), too, tried to combine X-ray, radio and particle observations of a flare in one homogeneous system. According to the later Korchak's (1965*b*) paper, however, it seems unlikely that a single mechanism will provide a smooth fit from X-rays to radio bursts.

Possible mechanisms of γ-ray production in solar flares were discussed by Dolan (1964) and by Dolan and Fazio (1965). Until now, however, all trials to detect γ-rays from the Sun, have given negative results (Fazio 1964, Kraushaar *et al.* 1965, Frye and Reines 1966).

12. HIGH ENERGY PARTICLES

A detailed general summary on solar proton events was presented in the NASA Solar Proton Manual edited by McDonald (1963). Review papers on high-density particles emitted from the Sun were published by Obayashi (1964*a*, *b*), K. A. Anderson (1964*b*), Roederer (1964*a*, *b*), de Jager (1965*e*) and Ogilvie (1965). A very illustrative critical review paper on particle and radio emission from the Sun was presented by Schatzman (1965*b*). Malitson and Webber (1963) and Webber (1964) described general characteristics of solar cosmic ray events and Fichtel *et al.* (1963) summarized details of individual solar particle events. Lists of PCA events were given by Bailey (1964), Basler and Owren (1964), Cummings (1965), Fritzová-Švestková and Švestka (1966*a*) and, for the period before 1956, by Švestka (1966*d*). Solar proton events during the IQSY period were summarized by Goedecke *et al.* (1967).

Křivský (1965*c*, 1966) studied the flight time of solar protons to the Earth and discussed a space model of the clouds of high-energy protons ejected from flares (1965*b*). Van Allen *et al.* (1964), Hakura (1965), Leinbach *et al.* (1965), Chivers and Burrows (1966), and Adams and Masley (1966) investigated relations of PCA records to the energy spectrum of high-energy particles and the progressive softening of the spectrum during the PCA development. Yoshida and Akasofu (1965) have shown that solar flares associated with cosmic-ray increases are the most energetic throughout the entire range of the energy spectrum of solar particles. Shea and Smart (1965) proposed an analytical method to analyse flare-associated cosmic ray increases, effectively including as many stations as possible.

Statistical comparison of flares and flare-associated phenomena with ground-base measurements of cosmic rays were made by Vladimirsky and Pankratov (1964), Dorman *et al.* (1964), Yoshida (1965) and Vladimirsky *et al.* (1966). In the last two papers, a definite statistical increase of solar cosmic rays ($\sim 0{\cdot}5\%$) has been found for meter type IV bursts and X-ray flares. Cosmic-ray increases not associated with solar flares have been suspected by Dorman *et al.* (1965). Chirkov *et al.* (1965) and Kuzmin *et al.* (1965) studied cosmic-ray variations associated with active regions passing over the solar disk in December 1957 and in July 1961, respectively. Vladimirsky (1965*a*) expressed an idea that all solar flares produce low-energy cosmic rays.

The number of papers dealing with observations of high-energy particles from space vehicles has increased many-fold since the last report. In this section we discuss mostly the flare-associated events; for other observations see section 13.

K. A. Anderson (1964*a*) re-examined the solar proton flare events of late August 1957. Arnoldy *et al.* (1964) made a comparison of the cosmic radiation measurements from the Earth, balloon ion chambers and Pioneer-5 during the March-April 1960 events. Four solar proton events observed by Explorers XII and XIV were discussed by Bryant *et al.* (1964, 1965*a*). They have found that the rate of propagation of protons is linearly related to particle velocity, such that the time to reach maximum intensity is inversely proportional to velocity and conclude that the propagation involves a degree of scattering that is independent of the energy. Charakhchan A. N. and T. V. (1965*a*) have shown that the differential energy spectrum of solar flare particles has the same exponent (~ 3) in the several hundred MeV range for various flares. In the higher energy range the spectrum exhibits a discontinuity and the exponent increases to 5-6. The location of the discontinuity is different in different flares. Softening also occurs after the onset of a magnetic storm (Vakulov *et al.* 1964, Charakhchan A. N. and T. V. 1965*b*). Spectra of high-energy solar protons were also studied by Ogilvie and Bryant (1964), Bennett (1964), Dorman and Miroshnichenko (1965, 1966), Laxutin (1965), Paulikas *et al.* (1966) and Charakhchan *et al.* (1966). The last authors found an indication that the diffusion coefficient of solar protons in the interplanetary space increased at about three-times from 1960 to 1963. Increases of lower-energy (~ 1 MeV) protons were recorded with several space probes and analysed by Krimigis and Van Allen (1966) and by Vernov *et al.* (1966*a*, *b*). Only some of these increases can be directly associated with solar flares.

The anisotropy of solar cosmic rays was studied by McCracken (1963) and by Kuzmin *et al.* (1965), but the most important anisotropy studies were made with the Pioneer 6 space probe, particularly at the proton event of 30 December 1965 (Fan *et al.* 1966, Bartley *et al.* 1966, McCracken and Ness 1966). It has been found that the cosmic ray anisotropy is well aligned with the interplanetary magnetic field and the direction of anisotropy is shown to exhibit marked and abrupt changes, well correlated with the changes in the direction of the magnetic field. The data from Pioneer 6 indicate the existence of large numbers of filamentary tubes, originating at the Sun but intertwined in interplanetary space. Cosmic rays are envisioned as constrained to the flux tubes.

The important role of α-particles in producing PCA's was shown by Weir and Brown (1964), and by Hakura (1965). The high-altitude balloon measurements of the relativistic solar α-particles in the late phase of the 12-13 November 1960 event were discussed by Yates (1964), who found a power law instead of the exponential rigidity spectrum. This result was criticized by Waddington and Freier (1965). The increased participation of α-particles after the onset of a sc geomagnetic storm was discussed by Sakurai (1965*d*).

Review papers on the composition of solar cosmic rays were presented by Fichtel (1964) and by Biswas and Fichtel (1965). Reviews of the results obtained from the study of cosmic rays with Soviet satellites and space probes, with a particular emphasis to the detection of nuclei heavier than protons were given by Ginzburg *et al.* (1963, 1964). The participation of heavy nuclei in solar cosmic rays was also discussed by Vladimirsky (1964). Cosmic-ray nuclei with $Z > 2$ have been observed after major flares on at least four occasions (Biswas *et al.* 1963, 1966, Biswas and Fichtel 1964). During the 18 July 1961 event the increase in the flux of medium nuclei ($6 \leqslant Z \leqslant 9$) was a factor 40 over the normal cosmic ray flux. Biswas and Fichtel (1964, 1965) concluded that the composition of the nuclei appeared to reflect the photospheric abundances. Quite an opposite conclusion, however, was made by Charakhchan A. N. and T. N. (1965). Increases of the flux of nuclei with $Z \geqslant 5$ and $Z \geqslant 15$ were also recorded aboard the Electron-2 satellite (Bloch *et al.* 1965, Kurnosova *et al.* 1966) and with the second Soviet cosmic rocket (Vladimirsky 1965*b*). Most of these increases, however, do not seem to be connected with solar flares and Vladimirsky suspects that one might meet there with a new effect associated with some solar events different from the flare phenomenon.

Solar neutrons may have been detected by means of balloon observations after an importance 3 flare. A flux of 148 ± 60 neutrons $m^{-2}\,s^{-1}$ in the 20–160 MeV range was determined (Krishna Apparao *et al.* 1966). On the other hand, Vela satellites operating since October 1963 have not detected any neutron bursts in excess of $1{\cdot}5 \times 10^3$ neutrons/cm^2 integrated intensity even after importance 3 flares accompanied by type IV radio bursts (Blame and Asbridge 1966). Lingenfelter *et al.* (1965) calculated the intensity and energy spectrum of neutrons above 1 MeV produced by flare-accelerated protons and α-particles in the photosphere. They have found that bursts of solar neutrons following major flares should be observable at high altitudes. Roelof (1966) has shown that due to diffusion, neutron-decay protons are insensitive indicators of the presence of neutrons in solar flares and it is unlikely that neutrons from a flare can be detected by means of proton observations.

Evidence for solar-flare electrons was obtained from the IMP satellites (Cline *et al.* 1964, Cline 1966, K. A. Anderson and Lin 1966), Electron–2 satellite (Tindo 1965) and Mariner 4 space probe (Van Allen and Krimigis 1965). While the solar origin of the electron flux reported by Cline *et al.* is doubtful, the other authors found direct relations of the recorded bursts of electrons to bursts of radio noise and X-rays associated with flares. The western longitudes of all the flares suggest a restriction of the electrons to field lines with a half angle of 15–20°.

The acceleration of protons by hydromagnetic shocks induced in the corona by heating due to flares was analysed by Weddell (1965). Solution of the relativistic equations of motion of protons subject to the moving magnetic field accompanying a shock leads to results, which agree in several aspects with the observed characteristics of solar proton events, nevertheless the resulting proton flux seems to be much too small. Wentzel (1965) and Sakurai (1965*d*) consider the Fermi acceleration mechanism as the predominating process in accelerating solar cosmic rays from flares. Korff (1964) has concluded that the acceleration must take place at levels of about 100 000 km above the photosphere. The initial phase of a solar proton event was investigated by G. C. Reid (1964) and an attempt was made to treat the diffusion through the solar atmosphere in a quantitative fashion. The possibility of a storage of energetic particles in the corona was discussed by Malville (1964). Willis (1966) examined the motion of an isolated energetic charged particle in a force-free magnetic field with rectilinear lines of force. He has found that under certain conditions a particle travels predominantly across the magnetic lines of force and it is suggested that some energetic charged particles may be able to escape from force-free magnetic fields by this process. The propagation and diffusion of solar flare protons and the modulation of the energy spectrum in the interplanetary space were discussed by Parker (1964*d*), Ivanov and Kolomeets (1965), Fibich and Abraham (1965) and Shishov (1966). Krimigis (1965) suggested a unified interplanetary diffusion model and used it for an interpretation of the time history of the intensity of several cosmic ray events. Wibberenz (1966) pointed out the significance of drift motions for solar energetic particle propagation in interplanetary space.

13. SOLAR WIND DISTURBANCES

Review papers on solar corpuscular radiation including enhanced solar corpuscular emission were written by Krimigis (1963), Parker (1964*b*), Mustel (1964*a*, *b*, *c*) and by Colburn and Sonett (1966). Scientific findings of Mariner 2 were reviewed by Sonett (1963), and by Snyder and Neugebauer (1964, 1965). Some papers concerning solar wind disturbances also appeared in 'The Solar Wind' symposium edited by Mockin and Neugebauer (1966). Parker (1966) summarized his earlier ideas in a synthetic picture: Disturbed conditions in the interplanetary space result when there is more turbulence in the wind than usual and/or when there is a sudden increase of coronal temperature in association with a solar flare. The 27-day recurring disturbances are associated with narrow regions of turbulence between broad regions of wind. Sudden heating of the corona at the time of a solar flare results in a sudden increase in the rate of expansion of the corona giving a blast wave which propagates outward through the wind. Nevertheless, some authors still prefer the magnetic bottle model in its original Golds form, which has some advantage in explaining the flare-associated corpuscular emission in an illustrative way. Some common features of the Parker's and Gold's models were discussed by Santarelli (1965).

Using measurements aboard Mariner 2, Vela 2, IMP 1 and Explorer 14, new determinations of solar wind velocity and temperature were made by Wolfe and Silva (1965), Wolfe *et al.* (1966), Neugebauer and Snyder (1966) and Strong *et al.* (1966). From quiet to disturbed conditions, proton's velocity increases from slightly more than 300 km s^{-1} up to 800 km s^{-1}, temperature from the order of 10^4 °K up to 9×10^5 °K and proton density from ~ 5 to $80/cm^3$. The maximum density usually appears at the leading (western) edge of each stream. The magnetic field temporarily increases well above 10 gamma in association with disturbances from the Sun (Ness *et al.* 1966*a*).

The interplanetary magnetic field is directly tied to and co-rotates with the solar photospheric magnetic fields. This has been shown dramatically from a comparison of IMP–1 magnetic field data with solar magnetograph observations. The magnetic lines of force extend in Archimedes spirals in accordance with Parker's theory of the solar wind. During the three-month interval of IMP–1 observations (1963 November 27 to 1964 February 17) the interplanetary field divided into 4 sectors, alternating in direction toward and away from the Sun. These sectors had correspondence to grossly averaged magnetic features in the photosphere (Ness and Wilcox 1964, Wilcox and Ness 1965, Ness 1966*b*, Ness and Wilcox 1966*a*, Smolkov 1966*b*, Sakurai 1966*b*, Ness *et al.* 1966*b*). Co-rotation of the systems of corpuscular streams also appears from the stability of the 27-day recurrence of geomagnetic disturbances (Vsechsvyatsky 1964, Lapointe 1964), cosmic ray intensity (Mori *et al.* 1964, Balasubrahmanyan *et al.* 1965), and proton increases (Gregory and Newdick 1964, Bryant *et al.* 1965*b*), and it has also been supported by measurements of Pioneer 6 and IMP–3 (Ness 1966*a*). Davis (1964, 1965, 1966), Davis *et al.* (1964, 1966), Coleman *et al.* (1966), Lazarus *et al.* (1966) and E. J. Smith *et al.* (1966) discussed observations made aboard Mariner 2 and 4, IMP–1 and OGO–1. It appears that the polarity pattern of the interplanetary field changed considerably between August 1962, where two reversals occurred per solar rotation, and November 1963, when the number of reversals was doubled. In late 1964, the Mariner 4 magnetometer recorded a 4-sector polarity pattern similar to that observed with IMP–1. Starting early in 1965, however, the Mariner-4 data show a considerable evolution of the patterns, which become less definite than they were earlier. This suggests that the pattern becomes less stable as the Sun becomes more active. However, at least one feature of the patterns may have persisted for the whole period of more than two years. Some observations of recurrent proton increases suggest a trapping of protons over certain regions (Bryant *et al.* 1963, 1965, McDonald and Desai 1966). Interplanetary magnetic fields co-rotating with the Sun apparently confine the protons from flares and 27-day recurring regions to given sectors and thus influence the appearance or non-appearance of protons at the Earth (O'Gallagher and Simpson 1966). Helium of solar origin in these recurrent particle streams was detected by Gloeckler (1965).

Snyder *et al.* (1963) have concluded that M–regions are the emitters of high-velocity plasma and Snyder (1964) tried to associate the solar wind velocity variations observed by Mariner 2 with observed plages on the Sun. Mustel (1965*b*) examined the two theories of the formation of quasistationary corpuscular streams from the Sun: (1) the streams originate in active regions, and (2) particles generated in undisturbed regions are deflected by neighbouring active regions. According to Mustel, all geomagnetic, satellite and space probe data confirm the first hypothesis. Nevertheless, many other authors consider the second alternative, i.e. the hypothesis of 'cone of avoidance' as the more favourable one, particularly Allen (1964), Saito (1964), Sinno (1964) and Nojima *et al.* (1964). The outflow of particles from M-regions was discussed by Piddington (1964) and Mustel (1966) and the structure that would be expected to develop when a fast stream in the solar wind is embedded in a slower ambient wind was considered by Hirshberg (1965). McCracken *et al.* (1967) report the observation, made with Pioneer 6, of a number of recurrent phenomena having all the characteristics of Forbush decreases apart from any flare-association. They were, however, ultimately correlated with recurrent M-regions.

The north-south asymmetry in the solar wind as observed with IMP-1 and 3 was discussed by Ness and Wilcox (1966*b*). Křivský and Letfus (1965) suppose that this asymmetry is due to an inclined interplanetary magnetic field caused by the galactic wind. The existence of a solar proton belt in a distance of 1·5–2 A.U. has been suggested by Vernov *et al.* (1966*c*) on the basis of proton flux measurements carried out aboard Zond 3 and Venus 2 space probes.

By comparing PCA events and Forbush-type decreases, Gosling (1964) proved a storage of high-energy protons in the cloud ejected from a flare and Haurwitz *et al.* (1966) and Švestka (1966*e*, 1967) deduced an asymmetrical magnetic field distribution in the cloud, with a magnetically strong western boundary. Geometrical dimensions and kinematic characteristics of solar corpuscular streams have been estimated by Mustel and Maisuradze (1966). A possibility of connection of two magnetic bottles in the interplanetary space was pointed out by Gopasyuk and Křivský (1967). From the interaction point plasma moves towards the Sun, as has actually been observed in some cases by Vitkevitch and Vlasov (1966) from the radio sources scintillation. A 'cleaning' of the interplanetary space by solar particle emissions was discussed by Lebeau *et al.* (1964) for the case of flares of 3 July 1957. The influence of the interplanetary magnetic fields on the propagation of solar protons was examined by Sakurai (1965*c*). Production of pairs of shock waves by the flare-associated enhanced solar wind was suggested by Sonett and Colburn (1965) and discussed by Sturrock and Spreiter (1965) and by M. Simon and Axford (1966). A scheme of permanent generation in active regions of corpuscular streams consisting of discrete plasmoids with own quasi force-free magnetic field has been offered by Mogilevsky (1964, 1965, 1966, Kalinin and Mogilevsky 1965). His analysis of the observations of magnetic fields in the interplanetary space confirms that the corpuscular streams with own magnetic field have a discrete structure.

Akasofu (1964) has tried to show that the solar plasma ejected from an active region can contain an appreciable amount of neutral hydrogen atoms. This supposition has been criticized both from the solar and geophysical points of view by Brandt and Hunten (1966) and by Cloutier (1966). Cloutier shows that the flux of neutral hydrogen is less than 10^{-5} of the total solar wind flux at 1 A.U.

14. SOLAR-TERRESTRIAL RELATIONS

Quite a lot of papers relating to the problems of solar-terrestrial physics, have already been mentioned in the preceding sections. Particularly, reader's attention is called to sections 7.3, 12, and 13.

Jonah *et al.* (1965) have prepared a five-volume catalogue which brings together the major optical, radio frequency, ionospheric and geomagnetic phenomena and events for the 19th solar cycle. Pick (1966) summarized the main optical and radioelectric properties of flares associated with SID, PCA and ssc magnetic storms. A synthetic study of severe solar-terrestrial disturbances of 9–12 February 1958, was presented by Hakura and Nagai (1964) and the same events were also described by Rivera and Gonzáles (1964).

Connections between the fine structure of type IV bursts and the production of PCA'S with long and short delay times were investigated by Böhme (1966). Bell (1963) has pointed out that type IV bursts which cover the full frequency range from microwaves to meter wavelength are strikingly more successful in the production of PCA events (as well as geomagnetic storms) than are type IV bursts limited to two or three octaves of the spectrum. Fritzová-Švestková and Švestka (1966*a*) investigated in detail the association between type IV burst flares and PCA events. They have concluded that strong PCA's are caused predominantly by flares between 10° E and 90° W on the solar disk and give the probability of PCA production by type IV burst flares as a function of the flare position on the disk. There seems, however, to be a general tendency of some active regions to emit or not to emit high-energy protons from type IV burst flares to the Earth.

Statistical association of flares with Forbush decreases during two solar cycles was studied by Krymsky and Shafer (1965). The series of Forbush effects of September 1963 was studied by Fischer and Křivský (1965). A close correlation between the integral flux of high-energy solar protons and the amplitude of the associated Forbush-type decrease has been found by Pankratov (1965).

Associations between flares, type IV bursts, and geomagnetic disturbances were studied by Akinjan and Dolginova (1965). Böhme (1964*b*) has found that intense type IV bursts produce, on the average, more intense geomagnetic storms and the corresponding delay time is shorter. Sudden commencement magnetic storms recorded at Zô-se during 1936 to 1962 were studied in association with solar flares and type IV bursts by Fan Chen (1965). Fritzová-Švestková and Hřebík (1964) have confirmed that there exist very important flares without any magnetic storm and major storms without any flare. They have also confirmed, however, that all flares associated with type IV bursts are sources of corpuscular streams. Fritzová-Švestková (1965) also studied the delay times of major and weak storms. Caroubalos (1964) established a statistical relationship between the delay time of geomagnetic storms and the radio energy emitted by the associated type IV burst.

Mustel and Egorova (1964, 1965) and Mustel (1965*a*) have shown that the recurrent geomagnetic disturbances are connected with the passage of active regions through the central solar meridian. Geoactive chromospheric flares are uniformly distributed over the solar disk, but the energy of the geomagnetic disturbance decreases with the increasing distance of the flare from the central meridian. According to Bell (1963), 80% of the major storms arise from flares located within 20° of the central meridian. Fritzová-Švestková (1966) deduced from the longitude distribution of flares producing major geomagnetic storms some conclusions on the density distribution and the trail curvature of the corpuscular streams ejected from flares. Billings and Roberts (1964) discussed the origin and some properties of the 27-day variations of geomagnetic activity from the point of view of Parker's model of the solar wind. The close correlation of central-meridian passages of unipolar magnetic regions with recurrent geomagnetic storms during the last solar minimum was discussed by Bumba and Howard (1966*b*). Ballario (1966) has examined the solar and geomagnetic activity during the IQSY Retrospective World Intervals for 1964 and 1965. She concludes that the observed K_p maxima are correlated with central meridian passages of regions in which plages, mostly associated with spot groups, show remarkable development changes.

Šťastná (1964) and Bateman and Oster (1965) verified the earlier Becker's result: If two sunspot groups are located at the same longitude but on the opposite hemispheres, their passage through the central meridian causes a decrease of the K_p index within approximately two days. If, however, the longitude difference increases to about 10°, the A_p index strongly increases. Šťastná suggests that this is connected in some way with the fact that in the first case opposite polarities, while in the second case identical polarities are located at the same longitude. Halenka (1967) makes an attempt to explain yearly variations of geomagnetic activity by an existence of two opposite active intervals of heliographic latitudes on the Sun.

Bednářová-Nováková (1964) and Halenka (1964) have tried to show that flares are not sources but only indicators of some events of the ejections of corpuscular streams from the Sun. They believe that the actual source of the corpuscular emission are filaments and the most geoactive phenomena on the Sun are filaments associated with unstable sunspot groups (Bednářová-Nováková 1964*a*, 1966*a*, Bednářová-Nováková *et al.* 1964, Halenka 1966*b*). Halenka (1966*a*) even tries to prove that geomagnetic storms erroneously ascribed to proton flares were caused, in fact, by central meridian passage of unstable filaments and Bednářová-Nováková (1967) opposes the opinion that sudden disappearance of filaments is caused by flares. Bednářová-Nováková (1966*b*) and Bednářová-Nováková *et al.* (1964) have also pointed out that when forecasting geomagnetic activity it is necessary to know the coronal formation extending over the solar central meridian at the given moment. This is in agreement with the conclusion made by Gnevyshev and Ol (1965) that the mean annual values of geomagnetic disturbances are connected with the intensity of the λ 5303 Å coronal line more closely than with any other index of solar activity.

15. ADMINISTRATIVE REPORT

Two working groups of Commission 10 were active during the reported period:

(*a*) The working group on CSSAR was established at the Hamburg meeting in 1964, in the following composition:

Michard (chairman), Dodson-Prince, Fokker, Giovanelli, Jefferies, Kiepenheuer, Righini, Roberts, Rösch, Severny, H. J. Smith, Švestka, Waldmeier.

(*b*) The working group on coronal intensity standardization was established in 1966 sponsored both by the IAU Commissions 10 and 12, in the following composition:

Rösch (chairman), Gnevyshev (secretary), Nagasawa, Newkirk, Waldmeier.

In fact, this is a re-establishment of an earlier Working Committee, which treated some problems of coronal observations in the period 1958–61. Many observers of the solar corona have been of the opinion that some more work on the standardization of coronal data is urgently needed. This working group has planned to hold a meeting at Bagnères-de-Bigorre before the Prague General Assembly, in August 1967.

Since important parts of the activity of these working groups are envisaged for 1967, i.e. after submitting the present Report for press, we have preferred not to include any partial reports here, and the chairmen of both these working groups will present verbal reports at the meetings of the Commission 10 in Prague.

The following recommendations are to be considered:

(1) A. H. Shapley recommends to discuss, from the point of view of Commission 10, the proposal to establish a solar-terrestrial service and its proposed status. This proposal is based on a resolution taken at the URSI XV Assembly, Munich 1966, calling on IUCSTP to make the study and potentially arrange for the establishment of such a service. Commission 10 could recommend for or against the establishment of such a service and make comments on its needs and terms of reference.

(2) A. H. Shapley recommends to discuss the experience with the new IAU classification of flare importances and to have a discussion whether other international standards may be needed as regards description of other forms of solar activity in addition to Hα flares. The possible need for such additional work on standardization of descriptions could benefit from a general discussion at Prague and eventually lead to establishment of an appropriate working group if there appears a need for it.

(3) A. H. Shapley would also be interested in having a discussion on the success of the IQSY solar activity forecasts and also whether the post-IQSY forecasts made and distributed

by IUWDS and potentially modified to serve non-solar-minimum conditions are useful or expressed in terms appropriate to the present understanding of the development of solar activity.

(4) A. D. Fokker recommends to devote a discussion to the *Quarterly Bulletin on Solar Activity*, with a particular emphasize on Resolution 3 of the last Hamburg meeting. This resolution concerns the possibility of contributing solar magnetograms to the Quarterly Bulletin.

(5) W. O. Roberts recommends that the *Quarterly Bulletin on Solar Activity* resume publication of solar data for the central region, discontinued in December 1938, and that the central region data for the period 1938 to the present be compiled and published.

(6) A. D. Fokker recommends to publish spectral diagrams of type IV bursts on a regular basis.

(7) A. D. Fokker has pointed out some highly desirable research activities to be carried out in the study of solar radio events: (*a*) Observations should aim at a more precise determination of the spectral profile of individual type I bursts. (*b*) Two-dimensional position determinations of individual stormbursts and determinations of their angular sizes will be most rewarding. (*c*) It is desirable that more detailed information on source sizes be obtained for all types of solar radio events and at several frequencies. (*d*) It is desirable to extend determinations of polarization ellipses for type III bursts to such an extent that, for one and the same burst, polarimetric observations be available at more than two different frequencies. (*d*) A knowledge of the actual form of the intensity curve of impulsive microwave bursts in its rising part may well be important for an understanding of the explosive phase of the flare phenomenon. (*f*) There is still a great want of more data on source positions and polarizations of type IV emission.

BIBLIOGRAPHY

Abbasov, A. R., Yasnov, L. V. 1966, *Soln. Dann. Bjull.*, no. 1, 58.
Abbasov, A. R., Grebinsky, A. S., Durasova, M. S. 1965, *Vest. Leningrad. gos. Univ.*, no. 1, 102.
Acton, L. W. 1965*a*, *Nature*, **204**, 64.
Acton, L. W. 1965*b*, *Trans. am. geophys. Un.*, **46**, 129.
Acton, L. W. 1965*c*, *Nature*, **207**, 737.
Adams, G. W., Masley, A. J. 1966, *Planet. Space Sci.*, **14**, 277.
Akasofu, S. I. 1964, *Planet. Space Sci.*, **12**, 905.
Akinyan, S. T. 1965, *Soln. Dann. Bjull.*, no. 10, 58.
Akinyan, S. T., Dolginova, Y. N. 1965, *Soln. Aktivnost*, **2**, 183.
Alexiou, P. G., Poulakos, C. P. 1966, *Praktika Acad. Athènes*, in press.
Alfvén, H., Carlqvist, P. 1967, *Solar Phys.*, **1**, in press.
Alikaeva, K. V. 1964, *Spektrofot. Issl. aktivn. Obraz. na Solnce*, Kiev, p. 73.
Alikaeva, K. V. 1965, *Soln. Dann. Bjull.*, no. 10, 50.
Alikaeva, K. V. 1966, *Soln. Dann. Bjull.*, no. 7, 63.
Allen, C. W. 1964, *Planet. Space Sci.*, **12**, 487.
Allen, C. W. 1965, *Space Sci. Rev.*, **4**, 91.
Aller, H. D., Jensen, C. J., Malville, J. M. 1966, *Nature*, **209**, 1014.
Altschuler, M. D., Sastry, Ch. V. 1965, *Nature*, **206**, 1035.
Aly, M. K., Galal, A. A. 1967, in preparation.
Anastassiadis, M., Ilias, D., Caroubalos, C., Macris, C., Elias, D. P. 1964, *Nature*, **201**, 357.
Anderson, A. D. 1964, *J. atmos. Sci.*, **21**, 1.
Anderson, A. D. 1965, *J. geophys. Res.*, **70**, 3231.
Anderson, G. F. 1966, Dissertation, Univ. of Colo., Techn. Rep., Contract no. Nonr 1147 (13).
Anderson, K. A. 1964*a*, *J. geophys. Res.*, **69**, 1743.

Anderson, K. A. 1964*b*, in *Space Physics*, John Wiley and Sons, New York, p. 611.
Anderson, K. A., Lin, R. P. 1966, *Phys. Rev. Lett.*, **16**, 1121.
Antalová, A. 1965, *Bull. astr. Inst. Csl.*, **16**, 32 (also in *Publ. astr. Inst. Czech. Acad. Sci.*, **51**, 98).
Antalová, A., Gnevyshev, M. N. 1965, *Astr. Zu.*, **42**, 253.
Arnoldy, R. L., Winckler, J. R., Hoffman, R. A. 1964, *J. geophys. Res.*, **69**, 1679.
Arroyo, M. L. 1965, *Bull. astr. Inst. Csl.*, **16**, 244.
Asper, H. K., Bracewell, R. N., Denter, J., Krishnan, T., Picken, J., Rose, S. 1964, *J. geophys. Res.*, **69**, 1805.
Austin, W. E., Purcell, J. D., Snider, C. B., Tousey, R., Widing, K. G. 1966, *Space Res.*, **7**, in press.
Avignon, Y., Caroubalos, C., Martres, M. J., Pick, M. 1964*a*, in AAS–NASA Symp. on *Physics of Solar Flares*, p. 25.
Avignon, Y., Martres, M. J., Pick, M. 1964*b*, *Ann. Astrophys.*, **27**, 23.
Avignon, Y., Caroubalos, C., Martres, M. J., Pick, M. 1965, in *IAU Symp. no.* 22, p. 373. North-Holland Publ. Co., Amsterdam.
Avignon, Y., Martres, M. J., Pick, M. 1966, *Ann. Astrophys.*, **29**, 33.
Babin, V. I., Vitkevich, V. V., Vlasov, V. I., Gorelova, M. V., Sukhovey, A. G. 1965, *Astr. Zu.*, **42**, 107.
Bachmann, G. 1967, Proc. of the 4th Consultation on Heliophysics, Wroclaw, in press.
Bailey, D. K. 1964, *Planet. Space Sci.*, **12**, 495.
Balasubrahmanyan, V. K., Ludwig, G. H., McDonald, F. B., Palmeira, R. A. R. 1965, *J. geophys. Res.*, **70**, 2005.
Ballario, M. C. 1964, in AAS–NASA Symp. on *Physics of Solar Flares*, p. 65.
Ballario, M. C. 1965, *Mem. Soc. astr. ital.*, **36**, 41.
Ballario, M. C. 1966, *IQSY Notes* no. 17, 2, and no. 18, 4.
Ballario, M. C., Tagliaferri, G. L. 1964, *Mem. Soc. astr. ital.*, **35**, 363.
Banin, V. G. 1964, *Izv. Krym. astrofiz. Obs.*, **32**, 38.
Banin, V. G. 1965*a*, *Izv. Krym. astrofiz. Obs.*, **33**, 118.
Banin, V. G. 1965*b*, *Soln. Dann. Bjull.*, no. 6, 47.
Banin, V. G. 1965*c*, *Izv. Krym. astrofiz. Obs.*, **34**, 252.
Banin, V. G. 1966*a*, *Izv. Krym. astrofiz. Obs.*, **35**, 190.
Banin, V. G. 1966*b*, *Trudy SibIZMIR*, in press.
Banin, V. G., Prokopjeva, L. V. 1966, *Soln. Dann. Bjull.*, no. 6, 52.
Banos, G. 1967, *Solar Phys.*, **1**, in press.
Bartley, W. C., Bukata, R. P., McCracken, K. G., Rao, U. R. 1966, *J. geophys. Res.*, **71**, 3297.
Basler, R. P., Owren, L. 1964, *Sci. Rep. geoph. Inst. Univ. Alasca*, UAG–R, 152.
Bateman, G., Oster, L. 1965, *Astr. J.*, **70**, 132.
Beckers, J. M. 1966, in Proc. of the Meeting on *Sunspots*, Florence, p. 186.
Beckers, J. M., Schröter, E. H. 1966, AAS Meeting in Boulder, in preparation.
Bednářová-Nováková, B. 1964*a*, *Geofys. Sb. Praha* 1963, 383.
Bednářová-Nováková, B. 1964*b*, *Geofys. Sb. Praha* 1963, 399 (also in *Studia geophys. et geodet.*, **8**, 63).
Bednářová-Nováková, B. 1966*a*, *Geofys. Sb. Praha* 1965, in press.
Bednářová-Nováková, B. 1966*b*, *Studia geophys. et geodet.*, **10**, 437.
Bednářová-Nováková, B. 1967, *Geofys. Sb. Praha* 1966, in press.
Bednářová-Nováková, B., Bucha, V., Halenka, J., Konečný, M. 1964, *Geofys. Sb. Praha* 1963, 407.
Bell, B. 1963, *Smithson. Contr. Astrophys.*, **8**, 119.
Bell, B., Wolbach, J. G. 1965, *Icarus*, **4**, 409.
Belorizky, D. 1964, *Astronomie*, **78**, 408.
Bennett, S. M. 1964, AVCO TR 64–12.
Berdichevskaya, V. S. 1964, *Astr. Zu.*, **41**, 80.
Berger, R. A., Malville, J. M. 1966, *Publ. astr. Soc. Pacific*, **78**, 343.
Bergsjö, N. J. 1963, *Ark. Astr.*, **3**, 203.
Bernyer, J., Poumeyrol, F., Rousseau, J. M. 1966, *Ann. Astrophys.*, **29**, 307.

Bhargava, B. N., Subrahmanyan, R. V. 1966, *Planet. Space Sci.*, **14**, 871.
Bhatnagar, A. 1964, Thesis, Kodaikanal.
Bhatnagar, A. 1966, *Z. Astrophys.*, **64**, 460.
Bhonsle, R. V., McNarry, L. R. 1964*a*, *Astrophys. J.*, **139**, 1312.
Bhonsle, R. V., McNarry, L. R. 1964*b*, *Can. J. Phys.*, **42**, 292.
Billings, D. E., Roberts, W. O. 1964, *Astrophys. norw.*, **9**, 147.
Bird, G. A. 1965, *Astrophys. J.*, **141**, 1455.
Birkle, K., Mattig, W. 1965, *Z. Astrophys.*, **60**, 264.
Biswas, S., Fichtel, C. E. 1964, *Astrophys. J.*, **139**, 941.
Biswas, S., Fichtel, C. E. 1965, *Space Sci. Rev.*, **4**, 709.
Biswas, S., Fichtel, C. E., Guss, D. E., Waddington, C. J. 1963, *J. geophys. Res.*, **68**, 3109.
Biswas, S., Fichtel, C. E., Guss, D. E. 1966, *J. geophys. Res.*, **71**, 4071.
Black, W. S., Booker, D., Burton, W. M., Jones, B. B., Shenton, D. B., Wilson, R. 1965, *Nature*, **206**, 654.
Blake, R. L., Chubb, T. A., Friedman, H., Unzicker, A. E. 1964, *Space Res.*, **4**, 785.
Blake, R. L., Chubb, T. A., Friedman, H., Unzicker, A. E. 1965, *Astrophys. J.*, **142**, 1 (also in *Ann. Astrophys.*, **28**, 583).
Blame, S. J., Asbridge, J. R. 1966, *J. geophys. Res.*, **71**, 4605.
Blizard, J. B. 1965, *Astr. J.*, **70**, 667.
Bloch, Y. L., Dorman, L. I., Kurnosova, L. V., Logachev, V. I., Plamonov, G. F., Razorenov, L. A., Sinitzina, V. G., Suslov, A. A., Fradkin, M. I. 1965, *Issled. kosm. Prostranstva*, Nauka Moskva, p. 514.
Boev, K., Mitrani, L., Ormandzhev, S. 1965, *Astr. Zu.*, **42**, 861.
Bohlin, J. D., Hansen, R. T., Newkirk, G. A. 1966, AAS Special Meeting on Solar Astronomy, Oct. 1966, in preparation.
Böhme, A. 1964*a*, *Mitt. astrophys. Obs. Potsdam*, no. 108.
Böhme, A. 1964*b*, *Monatsber. Dtsch. Akad. Wiss.*, Berlin, **6**, 1.
Böhme, A. 1966, presented at the Inter-Union Symp. on Solar Terr. Phys., Belgrade, in preparation.
Boischot, A. 1966, *Ann. Astrophys.*, **29**, in press.
Bolshakova, O. V. 1965, *Astr. Zu.*, **42**, 859.
Borovik, V. N., Korzhavin, A. N., Peterova, N. G. 1965, *Soln. Dann. Bjull.*, no. 10, 67.
Bowen, P. J., Norman, K., Pounds, K. A., Sanford, P. W., Willmore, A. P. 1964, *Proc. R. Soc. London* Ser. A, 281, 538.
Boyd, R. L. F. 1965, *Space Sci. Rev.*, **4**, 35.
Brandt, J. C., Hunten, D. M. 1966, *Planet. Space Sci.*, **14**, 95.
Bray, R. J. 1964, *Z. Astrophys.*, **60**, 207.
Bray, R. J., Loughhead, R. E. 1964, *Sunspots*, Chapman and Hall, London.
Brekke, K., Maltby, P. 1963, *Ann. Astrophys.*, **26**, 383.
Brückner, G. 1964, *Sterne und Weltraum*, **3**, 196.
Brückner, G. 1965, in *IAU Symp. no.* 22, p. 189. North-Holland Publ. Co., Amsterdam.
Bruns, A. V., Nikulin, N. S., Severny, A. B. 1965, *Izv. Krym. astrofiz. Obs.*, **33**, 80.
Bruzek, A. 1964*a*, *J. geophys. Res.*, **69**, 2386.
Bruzek, A. 1964*b*, *Astrophys. J.*, **140**, 746.
Bruzek, A. 1965, in AAS–NASA Symp. on *Physics of Solar Flares*, p. 301.
Bruzek, A. 1966*a*, in *The Fine Structure of the Solar Atmosphere*, Colloquium Anacapri, Franz Steiner Verlag, p. 126.
Bruzek, A. 1966*b*, in *Solar Physics*, Lagonissi Summer Course, John Wiley and Sons, New York, in press.
Bryant, D. A., Cline, T. L., Desai, U. D., McDonald, F. B. 1963, *Phys. Rev. Lett.*, **11**, 144.
Bryant, D. A., Cline, T. L., Desai, U. D., McDonald, F. B. 1964, in AAS–NASA Symp. on *Physics of Solar Flares*, p. 289.
Bryant, D. A., Cline, T. L., Desai, U. D., McDonald, F. B. 1965*a*, *Astrophys. J.*, **141**, 478.
Bryant, D. A., Cline, T. L., Desai, U. D., McDonald, F. B. 1965*b*, *Phys. Rev. Lett.*, **14**, 481.
Bumba, V. 1964*a*, *Contr. Wroclaw astr. Obs.*, **14**, 11.
Bumba, V. 1964*b*, *Contr. Wroclaw astr. Obs.*, **14**, 31.

Bumba, V. 1965*a*, in *IAU Symp. no.* 22, p. 192. North-Holland Publ. Co., Amsterdam.
Bumba, V. 1965*b*, in *IAU Symp. no.* 22, p. 305. North-Holland Publ. Co., Amsterdam.
Bumba, V. 1965*c*, in *The Solar Spectrum* (ed. C. de Jager), p. 266.
Bumba, V. 1966, in *The Fine Structure of the Solar Atmosphere*, Colloquium Anacapri, Franz Steiner Verlag, p. 114.
Bumba, V., Howard, R. 1965*a*, *Astrophys. J.*, **141**, 1492.
Bumba, V., Howard, R. 1965*b*, *Astrophys. J.*, **141**, 1502.
Bumba, V., Howard, R. 1965*c*, *Astrophys. J.*, **142**, 796.
Bumba, V., Howard, R. 1965*d*, *Publ. astr. Inst. Czech. Acad. Sci.*, **51**, 24.
Bumba, V., Howard, R. 1966*a*, in Proc. of Meeting on *Sunspots*, Florence, p. 220.
Bumba, V., Howard, R. 1966*b*, *Astrophys. J.*, **143**, 592.
Bumba, V., Růžičková-Topolová, B. 1967, *Solar Phys.*, **1**, no. 2, in press.
Bumba, V., Howard, R., Kleczek, J. 1965, *Publ. astr. Soc. Pacific*, **77**, 55.
Bumba, V., Howard, R., Smith, S. F. 1966*a*, In Proc. of the Meeting on *Solar Magnetic Fields*, Rome, in press.
Bumba, V., Howard, R., Smith, S. F. 1966*b*, Proc. of the Goddard Magnetic Star Symp., in press.
Bumba, V., Kopecký, M., Kuklin, G. V. 1966*c*, *Bull. astr. Inst. Csl.*, **17**, 57.
Burgess, A. 1964, *Astrophys. J.*, **139**, 776.
Burgess, A., Seaton, M. J. 1964, *Mon. Not. R. astr. Soc.*, **127**, 355.
Burton, W. M., Wilson, R. 1965, *Nature*, **207**, 61.
Buslavsky, V. 1966, *Izv. Krym. astrofiz. Obs.*, **35**, 179.
Byrne, F. N., Ellison, M. A., Reid, J. H. 1964, *Space Sci. Rev.*, **3**, 319.
Carmichael, H. 1964, in AAS–NASA Symp. on *Physics of Solar Flares*, p. 451.
Caroubalos, C. 1964*a*, in AAS–NASA Symp. on *Physics of Solar Flares*, p. 203.
Caroubalos, C. 1964*b*, *Ann. Astrophys.*, **27**, 333.
Caroubalos, C. 1965, in *Solar System Radio Astronomy* (ed. J. Aarons), Plenum Press, New York, p. 109.
Carroll, G. 1965, *Publ. astr. Soc. Pacific*, **77**, 431.
Castelli, J. P., Michael, G. A. 1967, *Solar Phys.*, **1**, 125.
Chang, C. C., Huang, S. S. 1965, editors, Proc. of the *Plasma Space Symposium* in Washington, D. Reidel Publ. Comp., Dordrecht, Holland.
Charakhchan, A. N., Charakhchan, T. N. 1965*a*, *Geomagn. Aeronom.*, **5**, 220.
Charakhchan, A. N., Charakhchan, T. N. 1965*b*, *Issled. kosm. Prostranstva*, Nauka Moskva, 533.
Charakhchan, A. N., Kvashnin, A. N., Charakhchan, T. N. 1966, *Geomagn. Aeronom.*, **6**, 126.
Chen Biao, In Chun Lin 1965, *Acta astr. Sin.*, **13**, 89.
Chertoprud, V. E. 1966, *Astr. Zu.*, **43**, 390.
Chertoprud, V. E., Kotov, V. A. 1965, *Astr. Cirk.*, no. 318, 1.
Chirkov, N. P., Kuzmin, A. I., Krymsky, G. F. 1965, in *Kosmich. Luchi i Probl. Kosmofiz.*, Novosibirsk, p. 192.
Chistiakov, V. F. 1964, *Soln. Dann. Bjull.*, no. 2, 56, and no. 3, 56.
Chistiakov, V. F. 1965*a*, *Astr. Cirk.*, no. 331.
Chistiakov, V. F. 1965*b*, *Soln. Aktivnost*, **2**, 157.
Chistiakov, V. F. 1965*c*, *Soln. Aktivnost*, **2**, 168.
Chistiakov, V. F. 1965*d*, *Izv. glav. astr. Obs. Pulkovo*, **24**, no. 178, 60.
Chitre, S. M. 1963, *Mon. Not. R. astr. Soc.*, **126**, 431.
Chivers, H. J. A., Burrows, J. R. 1966, *Planet. Space Sci.*, **14**, 131.
Chubb, E. L. 1964, in AAS–NASA Symp. on *Physics of Solar Flares*, p. 445.
Chubb, E. L., Friedman, H., Kreplin, R. W. 1964, *Space Res.*, **4**, 759.
Chubb, T. A., Kreplin, R. W., Friedman, H. 1966, *J. geophys. Res.*, **71**, 3611.
Chvojková, E. 1965*a*, *Bull. astr. Inst. Csl.*, **16**, 57.
Chvojková, E. 1965*b*, *Bull. astr. Inst. Csl.*, **16**, 63.
Chvojková, E. 1966*a*, in Proc. of the Meeting on *Sunspots*, Florence, p. 209.
Chvojková, E. 1966*b*, in Proc. of the Meeting on *Sunspots*, Florence, p. 211.

Chvojková, E., Klepešta, J. 1965, *Bull. astr. Inst. Csl.*, **16**, 70.
Ciurla, T., Rompolt, B. 1965, *Publ. astr. Inst. Czech. Acad. Sci.*, **51**, 100.
Clark, A. 1965, *Physics Fluids*, **8**, 644.
Clavelier, B. 1966, *C. r. Acad. Sci.*, Paris, **262**, 225.
Clavelier, B. 1967, *Ann. Astrophys.*, **30**, in press.
Cline, T. L. 1966, in *The Solar Wind*, in press.
Cline, T. L., Ludwig, G. H., McDonald, F. B. 1964, *Phys. Rev. Lett.*, **13**, 786.
Cloutier, P. A. 1966, *Planet. Space Sci.*, **14**, 809.
Colburn, D. S., Sonett, C. P. 1966, *Space Sci. Rev.*, **5**, 439.
Coleman, P. J., Davis, L., Smith, E. J., Jones, D. E. 1966, *J. geophys. Res.*, **71**, 2831.
Conner, J. P., Evans, W. D., Montgomery, M. D., Singer, S., Stogsdill, E. E. 1965, *Space Res.*, **5**, 546.
Covington, A. E., Locke, J. L. 1965, *J. R. astr. Soc. Can.*, **59**, 101.
Covington, A. E., Legg, T. H., Kennedy, W. A. G. 1965, *Inf. Bull. solar radio Obs. Europe*, **19**, 7.
Csada, I. K. 1965, in *IAU Symp. no.* 22, p. 434. North-Holland Publ. Co., Amsterdam.
Culhane, J. L., Willmore, A. P., Pounds, K. A., Sanford, P. W. 1964, *Space Res.*, **4**, 741.
Cummings, P. 1965, *AFCRL*–66–140, *Scient. Rep.* no. 7, Fort Davis, Texas.
Daene, H., Voigt, W. 1964, *Inf. Bull. solar radio Obs. Europe*, no. 17.
Danielson, R. E. 1964, *Astrophys. J.*, **139**, 45.
Danielson, R. E. 1965, in *IAU Symp. no.* 22, p. 314. North-Holland Publ. Co., Amsterdam.
Danielson, R. E. 1966, in Proc. of the Meeting on *Sunspots*, Florence, p. 120.
Das, A. K. 1964, *Proc. nat. Inst. Sci. India*, **30**, 283.
Das Gupta, M. K., Basu, D. 1964, *Nature*, **203**, 626.
Davies, K., Donelly, R. F. 1966, *J. geophys. Res.*, **71**, 2843.
Davis, L. 1964, in AAS–NASA Symp. on *Physics of Solar Flares*, p. 443 (also in 1965, *IAU Symp. no.* 22, p. 202).
Davis, L. 1966, in *The Solar Wind*, in press.
Davis, L., Smith, E. J., Coleman, P. J., Sonett, C. P. 1964, *Trans. am. geophys. Un.*, **45**, 79.
Davis, L., Smith, E. J., Coleman, P. J., Jones, D. E., Siscoe, G. L. 1966, presented at the Inter-Union Symp. on Solar Terr. Phys., Belgrade, in preparation.
de Feiter, L. D. 1964, in AAS–NASA Symp. on *Physics of Solar Flares*, p. 81.
de Feiter, L. D. 1966, *Rech. astr. Obs. Utrecht*, **18**, 79.
de Feiter, L. D., Švestka, Z. 1964, *Bull. astr. Inst. Csl.*, **15**, 117.
de Groot, T. 1962, *Inf. Bull. solar radio Obs. Europe*, no. 9.
de Groot, T. 1966, *Rech. astr. Obs. Utrecht*, **18** (1).
Deinzer, W. 1965, *Astrophys. J.*, **141**, 548.
Deinzer, W. 1966, in Proc. of the Meeting on *Sunspots*, Florence, p. 236.
de Jager, C. 1964*a*, *Research in Geophysics*, MIT Press, **1**, 1.
de Jager, C. 1964*b*, *Bull. astr. Inst. Netherl.*, **17**, 253.
de Jager, C. 1965*a*, editor, *The Solar Spectrum*, D. Reidel Publ. Comp., Dordrecht, Holland.
de Jager, C. 1965*b*, in *Introduction to Solar-Terr. Relations* (ed. by Ortner and Maseland), p. 55.
de Jager, C. 1965*c*, ibid., p. 78.
de Jager, C. 1965*d*, ibid., p. 86.
de Jager, C. 1965*e*, ibid., p. 96.
de Jager, C. 1965*f*, *Ann. Astrophys.*, **28**, 125.
de Jager, C. 1966, in Proc. of the Meeting on *Sunspots*, Florence, p. 242.
Demarcq, J., Hugon, M., Rösch, J., Trellis, M. 1965, *C. r. Acad. Sci.*, Paris, **261**, 4629.
Deubner, F. L. 1965, in *IAU Symp. no.* 22, p. 369. North-Holland Publ. Co., Amsterdam.
Deubner, F. L. 1966, in *The Fine Structure of the Solar Atmosphere*, Colloquium Anacapri, Franz Steiner Verlag, p. 26.
Deubner, F. L. 1967, in preparation.
Dezsö, L. 1964, *Publ. Debrecen Obs.*, **1**, 1.
Dezsö, L. 1965, *Publ. astr. Inst. Czech. Acad. Sci.*, **51**, 49.
Dezsö, L., Gerlei, O. 1964*a*, *Publ. Debrecen Obs.*, **2**, 3.

Dezsö, L., Gerlei, O. 1964*b*, *Publ. Debrecen Obs.*, **2**, 35.
Dezsö, L., Gerlei, O. 1965, *Publ. astr. Inst. Czech. Acad. Sci.*, **51**, 41.
Dezsö, L., Kovácz, Á. 1965, *Bull. astr. Inst. Csl.*, **16**, 76.
Dezsö, L., Gerlei, O., Sípos, V. 1964, *Publ. Debrecen Obs.*, **2**, 27.
Dinulescu, V. 1965, *Publ. astr. Inst. Czech. Acad. Sci.*, **51**, 58.
Dizer, 1967.
Dodson, H. W., Hedeman, E. R. 1964*a*, in AAS–NASA Symp. on *Physics of Solar Flares*, p. 15.
Dodson, H. W., Hedeman, E. R. 1964*b*, *Planet. Space Sci.*, **12**, 393.
Dodson, H. W., Hedeman, E. R. 1964*c*, *J. geophys. Res.*, **69**, 3965.
Dodson, H. W., Hedeman, E. R. 1964*d*, *Science*, **143**, 237.
Dodson, H. W., Hedeman, E. R. 1965, *Astr. J.*, **70**, 673.
Dodson, H. W., Hedeman, E. R. 1966*a*, *Astrophys. J.*, **145**, 224.
Dodson, H. W., Hedeman, E. R. 1966*b*, *Astr. J.* (abstract), in press.
Dodson, H. W., Hedeman, E. R., Stewart, F. L. 1964, *Science*, **145**, 1050.
Dodson, H. W., Hedeman, E. R., Stewart, F. L. 1965, *IQSY Notes*, no. 13, 63 (also in *Science*, **148**, 1328).
Doherty, L. R., Menzel, D. H. 1965, *Astrophys. J.*, **141**, 251.
Dolan, J. F. 1964, *Astr. J.*, **69**, 137.
Dolan, J. F., Fazio, G. G. 1965, *Rev. Geophys.*, **3**, 319.
Dorman, L. I., Miroshnichenko, L. I. 1965, *Geomagn. Aeronom.*, **5**, 377.
Dorman, L. I., Miroshnichenko, L. I. 1966, *Geomagn. Aeronom.*, **6**, 215.
Dorman, L. I., Ivanov, V. I., Kolomieets, E. V. 1964, *Geomagn. Aeronom.*, **4**, 399 and 575.
Dorman, L. I., Koridze, V. G., Shatashvili, L. H. 1965, *Geomagn. Aeronom.*, **5**, 159.
Drago, F. G., Tagliaferri, G. L. 1966, *Z. Astrophys.*, **63**, 202.
Dravskikh, A. F. 1964, *Izv. glav. astr. Obs. Pulkovo*, **23**, no. 172, 40.
Dubov, E. E. 1963, *Izv. Krym. astrofiz. Obs.*, **29**, 86.
Dubov, E. E. 1964, *Izv. Krym. astrofiz. Obs.*, **32**, 26.
Dubov, E. E. 1966*a*, *Soln. Dann. Bjull.*, no. 2, 53.
Dubov, E. E. 1966*b*, *Izv. Krym. astrofiz. Obs.*, **37**, in press.
Dubov, E. E., Khromova, T. P. 1964, *Izv. Krym. astrofiz. Obs.*, **31**, 247.
Dungey, J. W. 1965, in Proc. *Plasma Space Sci. Symp.*, D. Reidel Publ. Comp., Dordrecht, Holland, p. 160.
Dunn, R. B. 1964, *Appl. Opt.*, **3**, 1353.
Durasova, M. S., Yudin, O. I. 1966, *Soln. Dann. Bjull.*, no. 7, 72.
Dvoryashin, A. S. 1964*a*, *Space Res.*, **4**, 780.
Dvoryashin, A. S. 1964*b*, *Geomagn. Aeronom.*, **4**, 626.
Dvoryashin, A. S. 1965, *Soln. Aktivnost*, **2**, 92.
Dvoryashin, A. S. 1966, *Izv. Krym. astrofiz. Obs.*, **36**, in press.
Eckhoff, H. K. 1966, *Inst. theor. Astroph. Blindern Report*, no. 18.
Efanov, V. A., Moiseyev, I. G. 1965, *Izv. Krym. astrofiz. Obs.*, **34**, 53.
Elgaröy, Ö. 1965, in *Solar System Radio Astronomy* (ed. J. Aarons), Plenum Press, New York, p. 201.
Elgaröy, Ö., Rödberg, H. 1964, *Astrophys. norw.*, **8**, 271.
Eliseyev, G. F., Moiseyev, I. G. 1965, *Izv. Krym. astrofiz. Obs.*, **34**, 3.
Eliseyeva, L. A. 1966, *Izv. Krym. astrofiz. Obs.*, **35**, 262.
Elliot, H. 1964, *Planet. Space Sci.*, **12**, 657.
Elliott, I., Reid, J. H. 1965, *Planet. Space Sci.*, **13**, 163.
Elliott, I., Reid, J. H. 1966, *Observatory*, **86**, 63.
Ellis, G. R. A., McCulloch, P. M. 1966, *Nature*, **211**, 1070.
Ellison, M. A., Reid, J. H. 1964, *Research in Geophysics*, MIT Press, **1**, 43.
Elston, R. C., Kolb, A. C., Austin, W. E., Tousey, R., Widing, K. G. 1964, *Astrophys. J.*, **140**, 436.
Elwert, G. 1964, in AAS–NASA Symp. on *Physics of Solar Flares*, p. 365.
Engvold, O. 1966, *Inst. theor. Astrophys. Blindern Rep.*, no. 17.
Enome, S. 1964, *Publ. astr. Soc. Japan*, **16**, 135.

Erushev, N. N. 1965, *Izv. Krym. astrofiz. Obs.*, **33**, 164.
Erushev, N. N. 1966, *Izv. Krym. astrofiz. Obs.*, **36**, in press.
Erushev, N. N., Eliseyeva, L. A. 1965, *Izv. Krym. astrofiz. Obs.*, **34**, 42.
Evans, J. W. 1963, editor, *The Solar Corona.* IAU Symposium no. 16. Academic Press, New York and London.
Evans, J. W. 1966, in Proc. of the Meeting on *Solar Magnetic Fields*, Rome, in press.
Fan, C. Y., Lamport, J. E., Simpson, J. A., Smith, D. R. 1966, *J. geophys. Res.*, **71**, 3289.
Fan Chen 1965, *Acta Sci. natur. Univ. Nanking*, **9**, 32.
Fazio, G. G. 1964, in AAS–NASA Symp. on *Physics of Solar Flares*, p. 145.
Ferraro, V. C. A. 1966, in Proc. of the Meeting on *Solar Magnetic Fields*, Rome, in press.
Fibich, M., Abraham, P. B. 1965, *J. geophys. Res.*, **70**, 2475.
Fichtel, C. E. 1964, in AAS–NASA Symp. on *Physics of Solar Flares*, p. 263.
Fichtel, C. E., Guss, D. E., Ogilvie, K. W. 1963, *Solar Proton Manual*, NASA TR–169, p. 19.
Fireman, E. L. 1964, in AAS–NASA Symp. on *Physics of Solar Flares*, p. 279.
Firor, J. W., Lilliequist, C. 1965, Proc. of the Third Intern. Symp. on *Bioastronautics and the Exploration of Space*, Bedwell and Strughold eds., Chapter III.
Fischer, S., Křivský, L. 1965, *Bull. astr. Inst. Csl.*, **16**, 316.
Fletcher, J. D. 1964, *AIAA J.*, **2**, 2193.
Fokker, A. D. 1964, *Bull. astr. Inst. Csl.*, **15**, 119.
Fokker, A. D. 1965, in *Solar System Radio Astronomy* (ed. J. Aarons), Plenum Press, New York, p. 171.
Fokker, A. D. 1966, *Bull. astr. Inst. Netherl. Suppl.*, **1**, 309.
Fomichev, V. V., Chertok, I. M. 1965, *Astr. Zu.*, **42**, 1256.
Forti, G., Godoli, G. 1965, *Mem. Soc. astr. ital.*, **36**, 399.
Forti, G., Godoli, G., Noci, G. 1965, *Ann. Geofis.*, Roma, **13**, 45.
Fortini, T. 1963, *Astrophys. J.*, **138**, 278.
Fortini, T., Garcia, Ch., Gay, M., Glover, F. 1966, *J. geophys. Res.*, **71**, 1938.
Fracastoro, M. G., Cristaldi, S. 1964, *Atti Fondaz.* 'Giorgio Ronchi', **19**, 210.
Fredga, K. 1965, in *IAU Symp. no.* 22, p. 310. North-Holland Publ. Co., Amsterdam.
Fredga, K. 1966, *Astrophys. J.*, **144**, 854.
Fricke, K., Elsässer, H. 1965, *Z. Astrophys.*, **63**, 35.
Friedman, H. 1963, *A. Rev. Astr. Astrophys.*, **1**, 59.
Friedman, H. 1964*a*, in AAS–NASA Symp. on *Physics of Solar Flares*, p. 147.
Friedman, H. 1964*b*, *Research in Geophysics*, MIT Press, **1**, 197.
Fritzová-Švestková, L. 1964*a*, *Bull. astr. Inst. Csl.*, **15**, 34.
Fritzová-Švestková, L. 1964*b*, *Bull. astr. Inst. Csl.*, **15**, 163.
Fritzová-Švestková, L. 1964*c*, *Contr. Wroclaw astr. Obs.*, **14**, 47.
Fritzová-Švestková, L. 1965, *Publ. astr. Inst. Czech. Acad. Sci.*, **51**, 171.
Fritzová-Švestková, L. 1966, *Bull. astr. Inst. Csl.*, **17**, 64.
Fritzová-Švestková, L., Hřebík, F. 1964, *Bull. astr. Inst. Csl.*, **15**, 222.
Fritzová-Švestková, L., Švestka, Z. 1966*a* *Bull. astr. Inst. Csl.*, **17**, 249.
Fritzová-Švestková, L., Švestka, Z. 1966*b*, in *The Fine Structure of the Solar Atmosphere*, Colloquium Anacapri, Franz Steiner Verlag, p. 124.
Frost, K. J. 1964, in AAS-NASA Symp. on *Physics of Solar Flares*, p. 139.
Frost, K. J. 1965, *Space Res.*, **5**, 513.
Frye, G. M., Reines, F. 1966, *J. geophys. Res.*, **71**, 3119.
Fung, P. C. W., Yip, W. K. 1966, *Planet. Space Sci.*, in press (also in *Austr. J. Phys.*, in press).
Gahm, G. 1966, *Stockholm Obs. Medd.*, in press.
Garczynska, I. 1965, *Publ. Astr. Inst. Czech. Acad. Sci.*, **51**, 155.
Garczynska, I. 1967, Proc. of the 4th Consultation on Heliophysics, Wroclaw, in press.
Garczynska, I., Jakimiec, M., Kordolewski, Z., Mergentaler, J., Rompolt, B. 1967, Proc. of the 4th Consultation on Heliophysics, Wroclaw, in press.
Geichenko, M. S. 1966, *Vop. astrofiz.*, Kiev, 48.
Gelfreikh, G. B. 1964, *Izv. glav. astr. Obs. Pulkovo*, **23**, no. 175, 59.

Gentili, M., Hugon, M., Leroy, J. L., Rösch, J., Trellis, M. 1966, *Ann. Astrophys.*, **29**, 43.
Getling, A. V. 1965, *Astr. Zu.*, **42**, 568.
Ginzburg, M. A. 1964, *Geomagn. Issled.*, no. 6, 5.
Ginzburg, V. L., Kurnosova, L. V., Razorenov, L. A., Fradkin, M. I. 1964, *Usp. fiz. Nauk*, **82**, 585 (also in 1963, *Space Sci. Rev.*, **2**, 778).
Giovanelli, R. G. 1964, *Observatory*, **84**, 57.
Giovanelli, R. G. 1966*a*, in Proc. of the Meeting on *Sunspots*, Florence, p. 57.
Giovanelli, R. G. 1966*b*, ibid., p. 68.
Gleissberg, W. 1964, *Naturwissenschaften*, **51**, 309 (also in *Veröff. astr. Inst. Univ. Frankfurt*, no. 8).
Gleissberg, W. 1966, *J. Br. astr. Ass.*, **76**, 265.
Gloeckler, G. 1965, *J. geophys. Res.*, **70**, 5333.
Gnevyshev, M. N. 1965*a*, *Vest. Akad. Nauk SSSR*, no. 5, 67.
Gnevyshev, M. N. 1965*b*, *Soln. Aktivnost*, **2**, 99.
Gnevyshev, M. N. 1965*c*, *Astr. Zu.*, **42**, 488.
Gnevyshev, M. N. 1967, *Solar Phys.*, **1**, 107.
Gnevyshev, M. N., Antalová, A. 1965, *Publ. astr. Inst. Czech. Acad. Sci.*, **51**, 47.
Gnevyshev, M. N., Křivský, L. 1966, *Astr. Zu.*, **43**, 385.
Gnevyshev, M. N., Ol, A. I. 1965, *Astr. Zu.*, **42**, 992.
Godoli, G. 1964*a*, *Atti Accad. naz. Lincei Rc.*, **36**, 51.
Godoli, G. 1964*b*, *Atti Accad. naz. Lincei Rc.*, **37**, 155.
Godoli, G. 1965, *Atti Accad. naz. Lincei Rc.*, **37**, 197.
Godoli, G. 1966*a*, *Osserv. Mem. Oss. astrofis. Arcetri*, 84.
Godoli, G. 1966*b*, *Observatory*, in press (also in Proc. of the Meeting on *Solar Magnetic Fields*, Rome, in press).
Godoli, G., Allen, C. W. 1964, *Planet. Space Sci.*, **12**, 149.
Godoli, G., Mazzucconi, F. 1967, *Astrophys. J.*, in press.
Godoli, G., Monsignori Fossi, B. 1966, *Osserv. Mem. Oss. astrofis. Arcetri*, 86.
Godoli, G., Monsignori Fossi, B. 1967, *Solar Phys.*, **1**, 148.
Godoli, G., Tagliaferri, G. 1965, *Nature*, **207**, 1283.
Godoli, G., Mazzucconi, F., Monsignori Fossi, B. 1966, *Ann. Geofis.*, Roma, in press.
Godovnikov, N. V., Smirnova, E. P. 1965, *Izv. Krym. astrofiz. Obs.*, **33**, 86.
Godovnikov, N. V., Ogir, M. B., Shaposhnikova, E. F. 1964, *Izv. Krym. astrofiz. Obs.*, **31**, 216.
Goedeke, A. D., Masley, A. J., Adams, G. W. 1967, *Solar Phys.*, **1**, in press.
Gold, T. 1964, in AAS-NASA Symp. on *Physics of Solar Flares*, p. 389.
Gold, T. 1965, in *IAU Symp. no.* 22, p. 390. North-Holland Publ. Co., Amsterdam.
Goldberg, L. 1966, *Adv. Earth Sci.*, MIT Press, **3**.
Goldberg-Rogozinskaja, N. M. 1965, *Izv. glav. astr. Obs. Pulkovo*, **24**, no. 178, 35.
Golubstov, V. V. 1965, *Soln. Dann. Bjull.*, no. 6, 70.
Gopala Rao, U. V. 1965, *Austr. J. Phys.*, **18**, 283.
Gopasyuk, S. I. 1964, *Izv. Krym. astrofiz. Obs.*, **32**, 14.
Gopasyuk, S. I. 1965*a*, *Izv. Krym. astrofiz. Obs.*, **33**, 100.
Gopasyuk, S. I. 1965*b*, *Izv. Krym. astrofiz. Obs.*, **34**, 288.
Gopasyuk, S. I. 1966*a*, *Izv. Krym. astrofiz. Obs.*, **35**, 139.
Gopasyuk, S. I. 1966*b*, *Izv. Krym. astrofiz. Obs.*, **37**, in press.
Gopasyuk, S. I., Křivský, L. 1967, *Solar Phys.*, **1**, 145.
Gopasyuk, S. I., Moreton, G. E. 1967, in preparation.
Gopasyuk, S. I., Severny, A. B. 1964, *Soln. Dann. Bjull.*, no. 12, 43.
Gosling, J. T. 1964, *J. geophys. Res.*, **69**, 1233.
Gossachinsky, I. V., Egorova, T. M., Ryzhkov, N. F. 1964, *Izv. glav. astr. Obs. Pulkovo*, **23**, no. 172, 79.
Grebinsky, A. S., Molchanov, A. P. 1964, *Geomagn. Aeronom.*, **4**, 635.
Green, R. M. 1965, in *IAU Symp. no.* 22, 398. North-Holland Publ. Co., Amsterdam.
Gregory, J. B., Newdick, R. E. 1964, *J. geophys. Res.*, **69**, 2383.
Groubé, W. 1965, *Astronomie*, **79**, 87.

Gudzenko, L. I., Chertoprud, V. E. 1964, *Astr. Zu.*, **41**, 697.
Gudzenko, L. I., Chertoprud, V. E. 1965, *Astr. Zu.*, **42**, 267.
Gudzenko, L. I., Chertoprud, V. E. 1966, *Astr. Zu.*, **43**, 113.
Gudzenko, L. I., Medvedeva, N. A., Chertoprud, V. E. 1965, *Astr. Cirk.* no. 342, 1.
Gulyaev, R. A. 1964, *Astr. Zu.*, **41**, 313.
Gurtovenko, E. A., Kostik, R. I. 1964, *Spektrofot. Issled.*, Kiev, **33**.
Gurtovenko, E. A., Rakhubovsky, A. S. 1965, *Soln. Dann. Bjull.*, no. 9, 49.
Gurtovenko, E. A., Skorik, K. E. 1965, *Soln. Dann. Bjull.*, no. 12, 56.
Guseynov, R. E. 1964*a*, *Soln. Dann. Bjull.*, no. 7, 54 (also in *Izv. Akad. Nauk Azerb. SSR*, Ser. fiz.-techn., no. 3, 121).
Guseynov, R. E. 1964*b*, *Trudy Šemach. (Shemakh.) astrofiz. Obs.*, **3**, 5.
Guseynov, R. E. 1965*a*, *Soobšč. (Soobshch.) Šemach. (Shemakh.) astrofiz. Obs.*, **4**, 121.
Guseynov, R. E. 1965*b*, *Soln. Dann. Bjull.*, no. 3, 45, and no. 5, 54.
Guseynov, R. E. 1966*a*, *Soln. Dann. Bjull.*, no. 5, 56, and no. 6, 48.
Guseynov, R. E. 1966*b*, *Izv. Akad. Nauk Azerb. SSR*, ser. fiz.-techn., no. 2–4, in press.
Hachenberg, O. 1965, in *Solar System Radio Astronomy* (ed. J. Aarons), Plenum Press, New York, p. 95.
Hachenberg, O., Krüger, A. 1964, *Z. Astrophys.*, **59**, 261.
Hagen, J. P., Barney, W. M. 1966, *Astr. J.*, **71**, 386.
Hakura, Y. 1965, *J. Radio Res. Lab.*, **12**, 231.
Hakura, Y., Nagai, M. 1964, *J. Radio Res. Lab.*, **11**, 197.
Halenka, J. 1964, *Studia geophys. geodet.*, **8**, 255.
Halenka, J. 1966*a*, presented at the Inter-Union Symp. on Solar Terr. Phys., Belgrade, in preparation.
Halenka, J. 1966*b*, *Geofys. Sb. Praha* 1965, in press.
Halenka, J. 1967, *Geofys. Sb. Praha* 1966, in press.
Hall, L. A., Schweizer, W., Hinteregger, H. E. 1965*a*, *J. geophys. Res.*, **70**, 2241.
Hall, L. A., Schweizer, W., Heroux, L., Hinteregger, H. E. 1965*b*, *Astrophys. J.*, **142**, 13.
Hallam, K. L. 1964, in AAS–NASA Symp. on *Physics of Solar Flares*, p. 63.
Hamana, S., Suzuki, T. 1966, *Tokyo astr. Bull.*, no. 174.
Hanasz, J. 1966, *Austr. J. Phys.*, **19**, no. 5, in press.
Hartz, T. R. 1964, *Ann. Astrophys.*, **27**, 831.
Harvey, G. A. 1964, *Astrophys. J.*, **139**, 16.
Harvey, G. A. 1965, *J. geophys. Res.*, **70**, 2961.
Harvey, J. W. 1965, *Astrophys. J.*, **141**, 832.
Haurwitz, M. W. 1964, *Astrophys. J.*, **140**, 1236.
Haurwitz, M. W., Yoshida, S., Akasofu, S. I. 1966, *J. geophys. Res.*, **70**, 2977.
Hedeman, E. R., Dodson, H. W. 1966, *Astr. J.*, in press.
Hess, W. N. 1964, editor, AAS–NASA Symp. on the *Physics of Solar Flares*, Washington.
Hett, J. H. 1966, *Astr. J.*, **71**, 387.
Higgs, L. A., Broten, N. W. 1964, *Astr. J.*, **69**, 543.
Hinteregger, H. E., Hall, L. A., Schweizer, W. 1964, *Astrophys. J.*, **140**, 319.
Hirayama, T. 1964, *Publ. astr. Soc. Japan*, **16**, 104.
Hirschberg, J. 1965, *J. geophys. Res.*, **70**, 5353.
Hitchcock, J. L. 1965, *Publ. astr. Soc. Pacific*, **77**, 290.
Howard, R. 1963, *Astrophys. J.*, **138**, 1312.
Howard, R. 1964, in AAS–NASA Symp. on *Physics of Solar Flares*, p. 89.
Howard, R. 1965, in *IAU Symp. no.* 22, 129. North-Holland Publ. Co., Amsterdam.
Howard, R. 1966, in *The Fine Structure of the Solar Atmosphere*, Colloquium Anacapri, Franz Steiner Verlag, p. 107.
Howard, R., Harvey, J. W. 1964, *Astrophys. J.*, **139**, 1328.
Hyder, C. L. 1964*a*, *Astrophys. J.*, **140**, 817.
Hyder, C. L. 1964*b*, *Astr. J.*, **69**, 545.
Hyder, C. L. 1965*a*, *Astrophys. J.*, **141**, 272.
Hyder, C. L. 1965*b*, *Astrophys. J.*, **141**, 1374.
Hyder, C. L. 1966, *Z. Astrophys.*, **63**, 78.

Ikhsanov, R. N. 1966, *Izv. Krym. astrofiz. Obs.*, **35**, 150.
Ikhsanova, V. N. 1964, *Izv. glav. astr. Obs. Pulkovo*, **23**, no. 172, 31.
Ioshpa, B. A. 1963, *Geomagn. Aeronom.*, **3**, 903.
Ioshpa, B. A. 1965, *Astr. Zu.*, **42**, 754.
Ioshpa, B. A. 1966, *Soln. Aktivnost*, **3**, in press.
Ioshpa, B. A., Mogilevsky, E. I. 1965, *Soln. Aktivnost*, **2**, 118.
Ioshpa, B. A., Obridko, V. N. 1965*a*, *Soln. Dann. Bjull.*, no. 2, 62, and no. 3, 54.
Ioshpa, B. A., Obridko, V. N. 1965*b*, *Soln. Dann. Bjull.*, no. 5, 62.
Ioshpa, B. A., Obridko, V. N. 1965*c*, *Soln. Dann. Bjull.*, no. 11, 46.
Ioshpa, B. A., Obridko, V. N. 1965*d*, *Soln. Aktivnost*, **2**, 131.
Ioshpa, B. A., Obridko, V. N. 1966, *Soln. Dann. Bjull.*, no. 5, 68.
Ioshpa, B. A., Mogilevsky, E. I., Obridko, V. N. 1964, *Space Res.*, **4**, 789.
Iroshnikov, R. S. 1965, *Astr. Zu.*, **42**, 494.
Ivanov, V. I., Kolomeets, E. V. 1965, *Izv. Akad. Nauk SSSR*, **29**, 1807.
Jacobsen, C., Carlqvist, P. 1964, *Icarus*, **3**, 270.
Jäger, F. W. 1966, in the Meeting on *Solar Magnetic Fields*, Rome, in press.
Jaggi, R. K. 1964, in AAS–NASA Symp. on *Physics of Solar Flares*, p. 419.
Jaggi, R. K. 1965, in *IAU Symp. no.* 22, p. 384. North-Holland Publ. Co., Amsterdam.
Jakimcova, M. 1967, Proc. of the 4th Consultation on Heliophysics, Wroclaw, in press.
Jakimiec, J. 1964*a*, *Contr. Wroclaw astr. Obs.*, **14**, 17.
Jakimiec, J. 1964*b*, *Acta astr.*, **14**, 107.
Jakimiec, J. 1965, *Acta astr.*, **15**, 145.
Jakimiec, J., Zabza, M. 1965, *Publ. astr. Inst. Czech. Acad. Sci.*, **51**, 39.
Jakimiec, J., Zabza, M. 1966, *Acta astr.*, **16**, 73.
Jefferies, J. T., Orrall, F. Q. 1964, in AAS–NASA Symp. on *Physics of Solar Flares*, p. 71.
Jefferies, J. T., Orrall, F. Q. 1965*a*, *Astrophys. J.*, **141**, 505.
Jefferies, J. T., Orrall, F. Q. 1965*b*, *Astrophys. J.*, **141**, 519.
Jensen, E., Maltby, P. 1965, *Astrophys. norw.*, **10**, 17.
Jonah, F. C., Dodson, H. W., Hedeman, E. R. 1965, *Solar Activity Catalogue* 1954–1963, 5 volumes, NASA (Houston).
Jose, P. D. 1965, *Astr. J.*, **70**, 193.
Kai, K. 1965*a*, *Ann. Tokyo astr. Obs.*, **9**, 195.
Kai, K. 1965*b*, *Publ. astr. Soc. Japan*, **17**, 294.
Kai, K. 1965*c*, *Publ. astr. Soc. Japan*, **17**, 309.
Kalinin, Y. D., Mogilevsky, E. I. 1965, *Issled. kosm. Prostranstva*, Nauka, Moskva, p. 368.
Kaplan, S. A., Petrukhin, N. S. 1964, *Soln. Dann. Bjull.*, no. 10, 63.
Kasinsky, V. V. 1964*a*, *Soln. Dann. Bjull.*, no. 9, 53.
Kasinsky, V. V. 1964*b*, *Soln. Dann. Bjull.*, no. 11, 42.
Kasinsky, V. V. 1964*c*, *Soln. Dann. Bjull.*, no. 12, 54.
Kasinsky, V. V. 1965, *Soln. Dann. Bjull.*, no. 5, 58.
Kasinsky, V. V. 1966, *Soln. Dann. Bjull.*, no. 6, 56.
Kats, J. M., Stepanov, V. E. 1966, Proc. of the Meeting on *Sunspots*, Florence, p. 213.
Kawabata, K. 1964, *Publ. astr. Soc. Japan*, **16**, 30.
Kawabata, K. 1966*a*, *Rep. Ionosph. Space Res. Japan*, **20**, 107.
Kawabata, K. 1966*b*, *Rep. Ionosph. Space Res. Japan*, **20**, 118.
Kawaguchi, I. 1964, *Publ. astr. Soc. Japan*, **16**, 86.
Kawaguchi, I. 1965, *Publ. astr. Soc. Japan*, **17**, 367.
Kawaguchi, I. 1967, *Solar Phys.*, **1**, in press.
Kernoa, A., Michard, R., Servajean, R. 1965, *Ann. Astrophys.*, **28**, 716.
Khangildin, U. V. 1964, *Astr. Zu.*, **41**, 302.
Khangildin, U. V. 1965, *Soln. Dann. Bjull.*, no. 12, 49.
Kiepenheuer, K. O. 1964*a*, in AAS–NASA Symp. on *Physics of Solar Flares*, p. 323.
Kiepenheuer, K. O. 1964*b*, *Appl. Opt.*, **3**, 1363.
Kiepenheuer, K. O. 1965, in *The Solar Spectrum* (ed. C. de Jager), p. 240.
Kiepenheuer, K. O. 1966*a*, in Proc. of the Meeting on *Sunspots*, Florence, p. 26.

Kiepenheuer, K. O. 1966*b*, editor, *The Fine Structure of the Solar Atmosphere*, Colloquium Anacapri, Franz Steiner Verlag, Wiesbaden.
Kiepenheuer, K. O. 1966*c*, in Kiepenheuer 1966*b*, p. 11.
Kiepenheuer, K. O. 1966*d*, in *Solar Physics*, Lagonissi Summer Course, John Wiley and Sons, New York, in press.
Kiepenheuer, K. O., Mehltretter, J. P. 1964, *Appl. Opt.*, **3**, 1359.
King-Hele, D. G. 1966, *Nature*, **209**, 285.
Kleczek, J. 1963*a*, *Publ. astr. Soc. Pacific*, **75**, 9.
Kleczek, J. 1963*b*, *Bull. astr. Inst. Csl.*, **14**, 167.
Kleczek, J. 1964*a*, in AAS–NASA Symp. on *Physics of Solar Flares*, p. 77.
Kleczek, J. 1964*b*, *Astrophys. J.*, **140**, 742.
Kleczek, J. 1964*c*, *Bull. astr. Inst. Csl.*, **15**, 42.
Kleczek, J. 1964*d*, *Bull. astr. Inst. Csl.*, **15**, 123.
Kleczek, J. 1965, *Vistas in Astronomy*, **6**, 137.
Kopecký, M. 1964*a*, *Bull. astr. Inst. Csl.*, **15**, 44.
Kopecký, M. 1964*b*, *Bull. astr. Inst. Csl.*, **15**, 178.
Kopecký, M. 1965*a*, editor, The 3rd Consultation on *Solar Physics and Hydrodynamics*, Tatranská Lomnica, *Publ. astr. Inst. Czech. Acad. Sci.*, **51**.
Kopecký, M. 1965*b*, *Publ. astr. Inst. Czech. Acad. Sci.*, **51**, 46.
Kopecký, M. 1966*a*, in Proc. of the Meeting on *Sunspots*, Florence, p. 61.
Kopecký, M. 1966*b*, in Proc. of the Meeting on *Solar Magnetic Fields*, Rome, in press.
Kopecký, M. 1966*c*, *Bull. astr. Inst. Csl.*, **17**, 270.
Kopecký, M. 1967, *Adv. Astr. Astrophys.*, **5**, in press.
Kopecký, M., Křivský, L. 1966, *Bull. astr. Inst. Csl.*, **17**, 360.
Kopecký, M., Kuklin, G. V. 1966*a*, *Bull. astr. Inst. Csl.*, **17**, 45.
Kopecký, M., Kuklin, G. V. 1966*b*, *Bull. astr. Inst. Csl.*, **18**, in press.
Kopecký, M., Suda, J. 1965, *Bull. astr. Inst. Csl.*, **16**, 78.
Korchak, A. A. 1965*a*, *Geomagn. Aeronom.*, **5**, 32.
Korchak, A. A. 1965*b*, *Geomagn. Aeronom.*, **5**, 601 (also in *Izv. Akad. Nauk SSSR*, **29**, 1813).
Korff, S. A. 1964, *Astr. J.*, **69**, 548.
Korobova, Z. B. 1965, *Soln. Dann. Bjull.*, no. 2, 42.
Korobova, Z. B. 1966, *Astr. Zu.*, **43**, 480.
Kostik, R. I., Orlova, T. V. 1964, *Soln. Dann. Bjull.*, no. 3, 51.
Kostik, R. I., Orlova, T. V. 1966, in *Vop. Astrofiz.*, Kiev, p. 54.
Kotljar, L. M. 1964, *Soln. Dann. Bjull.*, no. 10, 66.
Kotov, V. A., Gudzenko, L. I., Chertoprud, V. E. 1965, *Astr. Cirk.*, no. 333.
Koval, A. N. 1964, *Izv. Krym. astrofiz. Obs.*, **32**, 32.
Koval, A. N. 1965*a*, *Izv. Krym. astrofiz. Obs.*, **33**, 138.
Koval, A. N. 1965*b*, *Izv. Krym. astrofiz. Obs.*, **34**, 278.
Kozhevnikov, N. I. 1964*a*, *Astr. Zu.*, **41**, 934.
Kozhevnikov, N. I. 1964*b*, *Soln. Dann. Bjull.*, no. 3, 61, and no. 4, 65.
Kozhevnikov, N. I. 1965*a*, *Soln. Dann. Bjull.*, no. 1, 58.
Kozhevnikov, N. I. 1965*b*, *Astr. Zu.*, **42**, 78.
Kozhevnikov, N. I., Kuzminykh, V. D. 1964, *Astr. Zu.*, **41**, no. 2.
Krat, V. A. 1964, *Izv. glav. astr. Obs. Pulkovo*, **24**, no. 177, 57.
Krat, V. A. 1965*a*, *Izv. glav. astr. Obs. Pulkovo*, **24**, no. 178, 19.
Krat, V. A. 1965*b*, *Publ. astr. Inst. Czech. Acad. Sci.*, **51**, 22 (also in *Publ. glav. astr. Obs. Pulkovo*, no. 76).
Krat, T. V. 1965, *Soobšč.* (*Soobshch.*) *Šemach.* (*Shemakh.*) *astrofiz. Obs.*, **4**, 15.
Krause, F. 1965, in *IAU Symp. no.* 22, p. 426. North-Holland Publ. Co., Amsterdam.
Kraushaar, W., Clark, G. W., Garmire, G., Helmken, H., Higbie, P., Agogino, M. 1965, *Astrophys. J.*, **141**, 845.
Kreplin, R. W. 1965, *Space Res.*, **5**, 951.
Kreplin, R. W., Gregory, B. N. 1965, *U.S. nav. Res. Lab. Rep.*, NRL Progr., Aug. 1.
Krimigis, S. M. 1963, *Proc. Iowa Acad. Sci.*, **70**, 393.
Krimigis, S. M. 1965, *J. geophys. Res.*, **70**, 2943.

Krimigis, S. M., Van Allen, J. A. 1966, *Phys. Rev. Lett.*, **16**, 419.
Krishna Apparao, M. V., Daniel, R. R., Vijayalakshmi, B., Bhatt, V. L. 1966, *J. geophys. Res.*, **71**, 1781.
Krishnamurthi, M., Sastry Sivarama, G., Rao Seshagiri, T. 1965, *Austr. J. Phys.*, **18**, 473.
Křivský, L. 1964, *Bull. astr. Inst. Csl.*, **15**, 115.
Křivský, L. 1965*a*, *Space Res.*, **5**, 1081.
Křivský, L. 1965*b* *Publ. astr. Inst. Czech. Acad. Sci.*, **51**, 176.
Křivský, L. 1965*c*, *Bull. astr. Inst. Csl.*, **16**, 27.
Křivský, L. 1966, *Bull. astr. Inst. Csl.*, **17**, 141.
Křivský, L., Krüger, A. 1966, *Bull. astr. Inst. Csl.*, **17**, 243.
Křivský, L., Letfus, V. 1965, *J. atmos. terr. Phys.*, **27**, 1307 (also in *Bull. astr. Inst. Csl.*, **16**, 319).
Křivský, L., Makarov, V. I. 1966, *Bull. astr. Inst. Csl.*, **17**, 234.
Krüger, A., Michel, H. 1965, *Nature*, **206**, 601.
Krüger, A., Krüger, W., Wallis, G. 1964, *Z. Astrophys.*, **59**, 37.
Krymsky, G. F., Shafer, G. V. 1965, *Izv. Akad. Nauk SSSR*, **29**, 1891.
Kubeš, P. 1965, *Publ. astr. Inst. Czech. Acad. Sci.*, **51**, 89.
Kuklin, G. V. 1964*a*, *Soln. Dann. Bjull.*, no. 7, 45.
Kuklin, G. V. 1964*b*, *Soln. Dann. Bjull.*, no. 8, 45.
Kuklin, G. V. 1966*a*, *Rez. Nabljud. i Issled. v Period MGSS*, Moskva, **4**, in press.
Kuklin, G. V. 1966*b*, *Rez. Nabljud. i Issled. v Period MGSS*, Moskva, **4**, in press.
Kuklin, G. V., Syklen, A. E. 1966, *Rez. Nabljud. i Issled. v Period MGSS*, Moskva, **4**, in press.
Kuleshova, K. F. 1964, *Soln. Dann. Bjull.*, no. 4, 70, no. 5, 72, and no. 6, 53.
Kundu, M. R. 1963, *Space Sci. Rev.*, **2**, 438.
Kundu, M. R. 1964, in AAS–NASA Symp. on *Physics of Solar Flares*, p. 335.
Kundu, M. R. 1965*a*, *Solar Radio Astronomy*, Interscience Publishers, a division of John Wiley and Sons, New York.
Kundu, M. R. 1965*b*, in *The Solar Spectrum* (ed. C. de Jager), p. 408.
Künzel, H. 1964, *Astr. Nachr.*, **288**, 177.
Künzel, H. 1967, *Astr. Nachr.*, in press.
Kuperus, M. 1965, *Rech. astr. Obs. Utrecht*, **17** (1).
Kurnosova, L. V., Mandelshtam, S. L., Razorenov, L. A., Tindo, I. P., Fradkin, M. I. 1966, *Kosm. Issled.*, **4**, 170.
Kurochka, L. N. 1964*a*, *Spektrofot. Issled. aktivn. Obrazov. na Solnce*, Kiev, 81.
Kurochka, L. N. 1964*b*, *Inf. Bjull. MGG*, no. 6, 62.
Kurochka, L. N. 1965, *Soln. Dann. Bjull.*, no. 2, 49, and no. 7, 45.
Kurochka, L. N. 1966, *Soln. Dann. Bjull.*, no. 3, 70.
Kusoffsky, U. 1966, *Stockholm Obs. Medd.*, no. 156.
Kuzmin, A. I., Krymsky, G. F., Skripin, G. V. 1965, *Kosmich. Luchi i Probl. Kosmofiz.*, Novosibirsk, p. 153.
Kuzminykh, V. D. 1964*a*, *Soobšč.* (*Soobshch.*) *gos. astr. Inst. P.K. Sternberga*, **133**, 10.
Kuzminykh, V. D. 1964*b*, *Astr. Zu.*, **41**, 692.
Kuznetsov, L. A., Stepanov, V. E. 1966*a*, *Rez. Nabljud. i Issled. v Period MGSS*, Moskva, **4**, in press.
Kuznetsov, L. A., Stepanov, V. E. 1966*b*, *Rez. Nabljud. i Issled. v Period MGSS*, Moskva, **4**, in press.
Kuznetsov, L. A., Kuklin, G. V., Stepanov, V. E. 1966, *Rez. Nabljud. i Issled. v Period MGSS*, Moskva, **4**. in press.
Landini, M., Noci, G., Russo, D., Tagliaferri, G. L. 1966*a* *Osserv. Mem. Oss. astrofis. Arcetri*, **85**, 85.
Landini, M., Russo, D., Tagliaferri, G. L. 1966*b*, *Nature*, **211**, 393.
Lapointe, S. M. 1964, *Ann. Géophys.*, **20**, 454.
Larmore, L., Ramsey, H. 1964, *Appl. Opt.*, **3**, 1379.
Laxutin, L. L. 1965, *Geomagn. Aeronom.*, **5**, 432.
Lazarus, A. J., Bridge, H. S., Davis, J., Snyder, C. W. 1966, *Space Res.*, **7**, in press.

Lebeau, A., Cartron, S., Olivieri, G., Pick, M., Schlich, R., Weill, G. 1964, *Ann. Géophys.*, **20**, 309.
Lee, R. H., Warwick, J. W. 1964, *Radio Sci. J. Res. NBS/USCN–URSI*, **68D**, 807.
Lee, R. H., Rust, D. M., Zirin, H. 1965, *Appl. Opt.*, **4**, 1081.
Leighton, R. B. 1964, *Astrophys. J.*, **140**, 1547.
Leighton, R. B. 1965, in *IAU Symp. no.* 22, p. 158. North-Holland Publ. Co., Amsterdam.
Leinbach, H., Venkatesan, D., Parthasarathy, R. 1965, *Planet. Space Sci.*, **13**, 1075.
Leroy, J. L., Servajean, R. 1966, *Ann. Astrophys.*, **29**, 263.
Le Squeren, A. M. 1964*a*, in AAS–NASA Symp. on *Physics of Solar Flares*, p. 199.
Le Squeren-Malinge, A. M. 1964*b*, *Ann. Astrophys.*, **27**, 183.
Letfus, V. 1964, *Bull. astr. Inst. Csl.*, **15**, 211.
Letfus, V. 1965, *Bull. astr. Inst. Csl.*, **16**, 231.
Levitsky, L. S. 1964, *Izv. Krym. astrofiz. Obs.*, **32**, 84.
Levitsky, L. S. 1965, *Izv. Krym. astrofiz. Obs.*, **34**, 16.
Levitsky, L. S. 1966, *Izv. Krym. astrofiz. Obs.*, **35**, 253.
Lexa, J. 1963, *Bull. astr. Inst. Csl.*, **14**, 107.
Lindsay, J. C. 1964, *Planet. Space Sci.*, **12**, 379.
Lindsay, J. C. 1965, *Ann. Astrophys.*, **28**, 586.
Lingenfelter, R. E., Flamm, E. J., Canfield, E. H., Kellman, S. 1965, *J. geophys. Res.*, **70**, 4077 and 4087.
Link, F. 1964*a*, *Bull. astr. Inst. Csl.*, **15**, 187.
Link F. 1964*b*, *Planet. Space Sci.*, **12**, 333.
Liszka, L. 1963, *Stockholm Obs. Medd.*, no. 143.
Liszka, L. 1964, *Stockholm Obs. Medd.*, no. 144.
Livshitz, M. A. 1965, *Soln. Aktivnost*, **2**, 103.
Livshitz, M. A., Pikelner, S. B. 1964, *Astr. Zu.*, **41**, 464.
Lüst, R. 1965, editor, *Stellar and Solar Magnetic Fields*, IAU Symp. no. 22, North-Holland Publ. Co., Amsterdam.
Machalski, J., Zieba, S. 1965, *Publ. astr. Inst. Czech. Acad. Sci.*, **51**, 162.
Makarov, V. I. 1964, *Izv. glav. astr. Obs. Pulkovo*, **23**, no. 175, 40.
Makita, M. 1966, in Proc. of the Meeting on *Sunspots*, Florence, p. 88.
Makita, M., Morimoto, M. 1964, *Publ. astr. Soc. Japan*, **16**, 1.
Malitson, H. H. 1963, in *Solar Proton Manual*, NASA TR R–169, p. 89 and 109.
Malitson, H. H., Erickson, W. C. 1965, *Inf. Bull. solar radio Obs. Europe*, no. 19.
Malitson, H. H., Erickson, W. C. 1966, *Astrophys. J.*, **144**, 337.
Malitson, H. H., Webber, W. R. 1963, in *Solar Proton Manual*, NASA TR R–169, p. 1.
Maltby, P. 1964, *Astrophys. norw.*, **8**, 205.
Malville, J. M. 1964, in AAS–NASA Symp. on *Physics of Solar Flares*, p. 257.
Mandelshtam, S. L. 1965*a*, *Issled. kosm. Prostranstva*. Nauka Moskva, p. 531.
Mandelshtam, S. L. 1965*b*, *Space Sci. Rev.*, **4**, 587.
Mandelshtam, S. L. 1965*c*, *Ann. Astrophys.*, **28**, 614.
Marik, M. 1966, *Astr. Zu.*, **43**, 400.
Martres, M. J. 1964, *C. r. Acad. Sci.*, Paris, **259**, 3445.
Martres, M. J. 1966, in press.
Martres, M. J., Michard, R., Soru-Iscovici, I. 1966*a*, *Ann. Astrophys.*, **29**, 245.
Martres, M. J., Michard, R., Soru-Iscovici, I. 1966*b*, *Ann. Astrophys.*, **29**, 249.
Mattig, W. 1966*a*, in Proc. of the Meeting on *Sunspots*, Florence, p. 194.
Mattig, W. 1966*b*, *Mitt. astr. Ges.*, **20**, 34.
Mattig, W. 1967, in preparation.
Maxwell, A. 1965, in *The Solar Spectrum* (ed. C. de Jager), p. 342.
Maxwell, A., Defouw, R. J., Cummings, P. 1964, *Planet. Space Sci.*, **12**, 435.
McCracken, K. G. 1963, in *Solar Proton Manual*, NASA TR R–169, p. 57.
McCracken, K. G., Ness, N. F. 1966, *J. geophys. Res.*, **71**, 3315.
McCracken, K. G., Rao, U. R., Bukata, R. P. 1967, in press.
McDonald, F. B. 1963, editor, *Solar Proton Manual*, NASA TR R–169.

McDonald, F. B., Desai, U. D. 1966, presented at the Inter-Union Symp. on solar-terr. Phys., Belgrade, in preparation.
McIntosh, P. S. 1965, *Astr. J.*, **70**, 683.
McKenna, S. 1964, *Inf. Bull. solar radio Obs. Europe*, no. 17, 7.
McKenna, S. 1965*a*, *Mon. Not. R. astr. Soc.*, **129**, 437.
McKenna, S. 1965*b*, *Observatory*, **85**, 121.
McKenna, S. 1965*c*, *Astr. J.*, **70**, 684.
McKenna, S. 1966, *Observatory*, in press.
McLean, D. J. 1964, *Ann. Astrophys.*, **27**, 195.
McLean, D. J. 1965, in *Solar System Radio Astronomy*, (ed. J. Aarons), Plenum Press, New York, p. 117.
Melnikov, O. A., Zhuravlev, S. S., Aslanov, I. A. 1964, *Trudy Šemach. (Shemakh.) astrofiz. Obs.*, **3**, 40.
Menzel, D. H. 1964, *Astrophys. norw.*, **9**, 225.
Menzel, D. H. 1965, in *The Solar Spectrum* (ed. C. de Jager), p. 254.
Menzel, D. H., Doherty, L. R. 1963, in *The Solar Corona* (ed. J. W. Evans), Academic Press, New York, p. 159.
Menzel, D. H., Jones, F. S. 1963, Sci. Rep. under Contract AF 19(604)–4962 and AF 19(628)–3322.
Menzel, D. H., Moreton, G. E. 1963, in *The Solar Corona* (ed. J. W. Evans), Academic Press, New York, p. 315.
Menzel, D. H., Shore, B. W. 1966, in Proc. of the Meeting on *Sunspots*, Florence, p. 226.
Mergentaler, J. 1964, editor, First Consultation of Polish and Czechoslovak Astronomers on Heliophysics and Magnetohydrodynamics, Contr. *Wroclaw astr. Obs.*, no. 14.
Mergentaler, J. 1965, *Publ. astr. Inst. Czech. Acad. Sci.*, **51**, 169.
Meyer, F. 1966, *Mitt. astr. Ges.*, in press.
Michard, R. 1965, *Astronomie*, **79**, 131.
Michard, R., Mouradian, Z., Olivieri, G., Soru-Iscovici, I. 1964, *Notes Inf. Publ. Obs. Paris*, 20.
Mitra, S. N. 1964, *J. Inst. Telecommun. Engrs.*, **10**, 135.
Mitropolskaya, O. N. 1966, *Astr. Zu.*, **43**, 963.
Miyazawa, M., Higashi, K., Oe, T., Yamaguchi, K., Nagasawa, S. 1966, *Tokyo astr. Obs. Bull.*, 2nd Ser., in press.
Mockin, R. J., Neugebauer, M. 1966, editors, *The Solar Wind*, J.P.L. Symp.
Mogilevsky, E. I. 1964, *Geomagn. Aeronom.*, **4**, no. 2.
Mogilevsky, E. I. 1965, *Space Res.*, **5**, 544 (also in *Izv. Akad. Nauk SSSR*, **29**, 1881).
Mogilevsky, E. I. 1966*a*, *Geomagn. Aeronom.*, **6**, 809.
Mogilevsky, E. I. 1966*b*, *Soln. Aktivnost*, **3**, in press.
Mogilevsky, E. I., Schelting, B. D. 1966, in Proc. of the Meeting on *Solar Magnetic Fields*, Rome, in press.
Mogilevsky, E. I., Zhulin, I. A., Ioshpa, B. A. 1965, *Soln. Aktivnost*, **2**, 108.
Molchanov, A. P. 1964*a*, *Izv. glav. astr. Obs. Pulkovo*, **24**, no. 177, 38.
Molchanov, A. P. 1964*b*, *Učen. (Uchen.) Zap. Leningrad gos. Univ.*, no. 323, 107.
Molodensky, M. M. 1966, *Astr. Zu.*, **43**, 727.
Moore-Sitterly, Ch. E. 1966, in Proc. of the Meeting on *Sunspots*, Florence, p. 181.
Moreton, G. E. 1964*a*, in AAS–NASA Symp. on *Physics of Solar Flares*, p. 209.
Moreton, G. E. 1964*b*, *Astr. J.*, **69**, 145.
Moreton, G. E. 1965*a*, in *The Solar Spectrum* (ed. C. de Jager), p. 252.
Moreton, G. E. 1965*b*, in *IAU Symp. no.* 22, p. 371. North-Holland Publ. Co., Amsterdam.
Moreton, G. E. 1965*c*, in Proc. *Plasma Space Sci. Symp.* (ed. C. C. Chang and S. S. Huang), p. 24.
Moreton, G. E., Severny, A. B. 1966, *Astr. J.*, **71**, 172.
Moreton, G. E., Severny, A. B. 1967, *Solar Phys.*, **1**, in press.
Morgante, O., Torrisi, S. 1965, *Pubbl. Oss. astrofis. Catania*, no. 70.
Mori, S., Ueno, H., Nagashima, K., Sagisaka, S. 1964, *Rep. Ionosph. Space Res. Japan*, **18**, 275.

Morimoto, M. 1964, *Publ. astr. Soc. Japan*, **16**, 163.
Morozhenko, N. N. 1964, *Soln. Dann. Bjull.*, no. 4, 50, no. 5, 63, and no. 10, 45.
Morozhenko, N. N. 1965, *Soln. Dann. Bjull.*, no. 12, 45.
Morozhenko, N. N. 1966, *Soln. Dann. Bjull.*, no. 7, 58.
Musman, S. 1965, *Astr. J.*, **70**, 145.
Mustel, E. R. 1964*a*, *Space Res.*, **4**, 77.
Mustel, E. R. 1964*b*, *Space Sci. Rev.*, **3**, 139.
Mustel, E. R. 1964*c*, *Astr. Zu.*, **41**, 777.
Mustel, E. R. 1965*a*, *Astr. Zu.*, **42**, 276.
Mustel, E. R. 1965*b*, *Astr. Zu.*, **42**, 473.
Mustel, E. R. 1966, presented at the Inter-Union Symp. on solar terr. Phys., Belgrade, in preparation.
Mustel, E. R., Egorova, N. B. 1964, in AAS–NASA Symp. on *Physics of Solar Flares*, p. 285.
Mustel, E. R., Egorova, N. B. 1965, *Soln. Aktivnost*, **2**, 5.
Mustel, E. R., Maisuradze, P. A. 1966, *Geomagn. Aeronom.*, **6**, 815.
Nagasawa, S., Nakagomi, Y. 1965, *Rep. Ionosph. Space Res. Japan*, **19**, 339.
Nagasawa, S., Nishi, K., Ohki, Y., Shimizu, I., Yajima, S., Shimizu, Y. 1965, *Tokyo astr. Obs. Rep.*, **13**, 641.
Nakagomi, Y., Nishi, K. 1965, *Tokyo astr. Obs. Rep.*, **13**, 559.
Nakai, Y., Kubota, J. 1964, *Mem. Coll. Sci. Kyoto Univ.* Ser. A, 30, 323.
Namba, O. 1965*a*, in *The Solar Spectrum* (ed. C. de Jager), p. 256.
Namba, O. 1965*b*, *Astrophys. J.*, **141**, 827.
Neshpor, Y. I. 1965, *Izv. Krym. astrofiz. Obs.*, **34**, 313.
Ness, N. F. 1966*a*, *J. geophys. Res.*, **71**, 3319.
Ness. N. F. 1966*b*, in *The Solar Wind* (ed. R. J. Mockin, M. Neugebauer), in press.
Ness, N. F., Wilcox, J. M. 1964, *Phys. Rev. Lett.*, **13**, 461.
Ness, N. F., Wilcox, J. M. 1966*a* *Astrophys. J.*, **143**, 23.
Ness, N. F., Wilcox, J. M. 1966*b*, presented at the Inter-Union Symp. on solar terr. Phys., Belgrade, in preparation.
Ness, N. F., Scearce, C. S., Cantarano, S. 1966*a*, *J. geophys. Res.*, **71**.
Ness, N. F., Scearce, C. S., Seek, J. B., Wilcox, J. M. 1966*b*, *Space Res.*, **6**, in press.
Neugebauer, M., Snyder, C. W. 1966, *J. geophys. Res.*, **71**, 4469.
Neupert, W. M. 1964, in AAS–NASA Symp. on *Physics of Solar Flares*, p. 49 (also in *Space Res.*, **4**, 730).
Neupert, W. M. 1965, *Ann. Astrophys.*, **28**, 446.
Neupert, W. M. 1967, *Solar Phys.*, **1**, in press.
Neupert, W. M., Behring, W. E., Lindsay, J. C. 1964, *Space Res.*, **4**, 719.
Newkirk, G. 1967, in preparation.
Newkirk, G., Bohlin, J. D. 1964, *Appl. Opt.*, **3**, 543 (also in *Sky Telesc.*, **28**, 1.).
Newkirk, G., Bohlin, J. D. 1965, *Ann. Astrophys.*, **28**, 234.
Nikolsky, G. M., Sazanov, S. A. 1966, *Astr. Zu.*, **43**, 868.
Nikulin, N. S. 1964, *Izv. Krym. astrofiz. Obs.*, **31**, 209.
Nilsson, S. E. 1966, *Stockholm Obs. Medd.*, no. 153.
Nishi, K., Nagasawa, S. 1964, *Publ. astr. Soc. Japan*, **16**, 285.
Nishi, K., Unno, W. 1965, in *IAU Symp. no.* 22, p. 273. North-Holland Publ. Co., Amsterdam.
Nojima, Y., Nakagomi, Y., Nagasawa, S. 1964, *Rep. Ionosph. Space Res. Japan*, **18**, 319.
Obashev, S. O. 1964, *Izv. Akad. Nauk. Kazach. SSR*, Ser. fiz.-mat., **3**, 91.
Obashev, S. O. 1965*a*, *Dokl. Akad. Nauk SSSR*, **163**, 599.
Obashev, S. O. 1965*b*, *Astr. Zu.*, **42**, 1017.
Obashev, S. O. 1966, *Soln. Aktivnost*, **3**, in press.
Obayashi, T. 1964*a*, *Space Sci. Rev.*, **3**, 79.
Obayashi, T. 1964*b*, *Planet. Space Sci.*, **12**, 463.
Obridko, V. N. 1965*a*, *Astr. Zu.*, **42**, 102.
Obridko, V. N. 1965*b*, *Astr. Zu.*, **42**, 502.

Odincova, I. N. 1964, *Geomagn. Aeronom.*, **4**, 1035.
O'Gallagher, J. J., Simpson, J. A. 1966, *Phys. Rev. Lett.*, **16**, 1212.
Ogilvie, K. W. 1965, in Proc. *Plasma Space Sci. Symp.* (ed. C. C. Chang and S. S. Huang), p. 115.
Ogilvie, K. W., Bryant, D. A. 1964, *J. geophys. Res.*, **69**, 393.
Öhman, Y. 1965, in *IAU Symp. no.* 22, 173. North-Holland Publ. Co., Amsterdam.
Öhman, Y. 1966*a*, in Proc. of the Meeting on *Solar Magnetic Fields*, Rome, in press.
Öhman, Y. 1966*b*, *Stockholm Obs. Medd.*, no. 156.
Öhman, Y. 1966*c*, in Proc. of the Meeting on *Sunspots*, Florence, p. 147.
Öhman, Y. 1966*d*, *Stockholm Obs. Medd.*, in press.
Öhman, Y., Stiber, G., Kusoffsky, U. 1967, *Solar Phys.*, **1**, 60.
Olijnyk, P. A. 1965, *Soln. Dann. Bjull.*, no. 11, 54.
Olijnyk, P. A. 1966, *Soln. Dann. Bjull.*, no. 2, 42.
Orrall, F. Q. 1965*a*, *Astrophys. J.*, **141**, 1131.
Orrall, F. Q. 1965*b*, in *The Solar Spectrum* (ed. C. de Jager), p. 254.
Orrall, F. Q. 1965*c*, ibid., p. 308.
Orrall, F. Q. 1966, *Astrophys. J.*, **143**, 917.
Ortner, J., Maseland, H. 1965, editors, *Introduction to Solar Terrestrial Relations*, D. Reidel Publ. Comp., Dordrecht, Holland.
Oster, L. 1964, *Ann. Astrophys.*, **27**, 14.
Oster, L. 1966, *Astrophys. J.*, **143**, 928.
Oster, L., Altschuler, M. 1964, in AAS–NASA Symp. on *Physics of Solar Flares*, p. 377.
Oster, L., Sofia, S. 1966, *Astrophys. J.*, **143**, 944.
Paciorek, J. 1965, *Publ. astr. Inst. Czech. Acad. Sci.*, **51**, 69.
Paghis, I. 1964, *Nature*, **202**, 992.
Pajdušáková, L. 1964, *Contr. Wroclaw astr. Obs.*, **14**, 75.
Pankratov, A. K. 1965, *Izv. Krym. astrofiz. Obs.*, **33**, 156.
Parker, E. N. 1964*a*, in AAS–NASA Symp. on *Physics of Solar Flares*, p. 417.
Parker, E. N. 1964*b*, *Research in Geophysics*, MIT Press, **1**, 99.
Parker, E. N. 1964*c*, *Astrophys. J.*, **139**, 690.
Parker, E. N. 1964*d*, *J. geophys. Res.*, **69**, 1755.
Parker, E. N. 1964*e*, *Astrophys. J.*, **140**, 1170.
Parker, E. N. 1966, presented at the Inter-Union Symp. on solar terr. Phys., Belgrade, in preparation.
Parmenter, B. C. 1966, *Publ. astr. Soc. Pacific*, **78**, 250.
Paulikas, G. A., Freden, S. C., Blake, J. B. 1966, *J. geophys. Res.*, **71**, 1795.
Paveliev, S. .V., Pavelieva, Z. S. 1965, *Trudy glav. geofiz. Obs.*, **181**, 92.
Perche, J. C, Dumont, J. P. 1964, *Mém. Soc. R. Sci. Liège*, V-ème Sér., **9**, 186.
Petschek, H. E. 1964, in AAS–NASA Symp. on *Physics of Solar Flares*, p. 425.
Philip, K. W. 1964, *Astrophys. J.*, **139**, 723.
Pick, M. 1965, in the *Solar System Radio Astronomy* (ed. J. Aarons), Plenum Press, New York, p. 81.
Pick, M. 1966, *Ann. Géophys.*, **22**, 310.
Piddington, J. H. 1964, *Planet. Space Sci.*, **12**, 113.
Pierce, A. K. 1964, *Appl. Opt.*, **3**, 1337.
Pikelner, S. B. 1965, *Vistas in Astronomy*, **6**, 131.
Pikelner, S. B. 1966, *Usp. fiz. Nauk*, **88**, 505.
Pikelner, S. B., Livshitz, M. A. 1964, *Astr. Zu.*, **41**, 1007.
Piunov, I. D. 1964*a*, *Izv. glav. astr. Obs. Pulkovo*, **23**, no. 175, 49.
Piunov, I. D. 1964*b*, *Soln. Dann. Bjull.*, no. 2, 43.
Piunov, I. D. 1965, *Soln. Dann. Bjull.*, no. 4, 43.
Polonsky, V. V. 1965, *Astr. Zu.*, **42**, 857.
Polonsky, V. V. 1966, *Astr. Zu.*, **43**, 235.
Polupan, P. N. 1964, *Inf. Bjull. MGG*, no. 6, 55 (also in *Spektrofot. Nablj. akt. Proces. na Solnce*, Kiev, p. 61).
Polupan, P. N., Yakovkin, N. A. 1965, *Astr. Zu.*, **42**, 764.

Pottasch, S. R. 1963, *Mon. Not. R. astr. Soc.*, **125**, 543.
Pottasch, S. R. 1964, *Mon. Not. R. astr. Soc.*, **128**, 73.
Pounds, K. A. 1965, *Ann. Astrophys.*, **28**, 132.
Prokopieva, L. V. 1964, *Soln. Dann. Bjull.*, no. 4, 57.
Proshnikov, R. S. 1965, *Astr. Zu.*, **42**, 259.
Rachkovsky, D. N. 1965, *Izv. Krym. astrofiz. Obs.*, **33**, 111.
Rachkovsky, D. N. 1966, *Izv. Krym. astrofiz. Obs.*, **37**, in press.
Ramanathan, A. S., Natarajan, V. 1965, *Observatory*, **85**, 188.
Ramsay, J. V. 1966, in *The Fine Structure of the Solar Atmosphere*, Colloquium Anacapri, Franz Steiner Verlag, p. 39.
Ramsey, H., Smith, S. F. 1965, *Astr. J.*, **70**, 688.
Ramsey, H., Smith, S. F. 1966, *Astr. J.*, **71**, 197.
Rayrole, J. 1964, *C. r. Acad. Sci.*, Paris, **258**, 1161.
Rayrole, J. 1966, in *The Fine Structure of the Solar Atmosphere*, Colloquium Anacapri, Franz, Steiner Verlag, p. 104.
Reid, G. C. 1964, *J. geophys. Res.*, **69**, 2659.
Reid, J. H. 1965, *J. atm. terr. Phys.*, **27**, 127.
Reidy, W. P. 1966, *Space Res.*, **7**, in press.
Righini, G. 1966, editor, Proceedings of the Meeting on *Sunspots* (Atti del convegno sulle macchie solari), G. Barbèra, Firenze.
Ringnes, T. S. 1963, *Astrophys. norw.*, **8**, 127.
Ringnes, T. S. 1964*a*, *Astrophys. norw.*, **8**, 161.
Ringnes, T. S. 1964*b*, *Astrophys. norw.*, **8**, 303.
Ringnes, T. S. 1964*c*, *Astrophys. norw.*, **9**, 95.
Ringnes, T. S. 1965, *Astrophys. norw.*, **10**, 27.
Rivera, T. L., Gonzáles, C. G. 1964, *Bol. Obs. Tonantzintla Tacubaya*, **3**, 325.
Rödberg, H. 1966, *Nature*, **211**, 394.
Roederer, J. G. 1964*a*, *Research in Geophysics*, **1**, 115.
Roederer, J. G. 1964*b*, *Space Sci. Rev.*, **3**, 487.
Roelof, E. C. 1966, *J. geophys. Res.*, **71**, 1305.
Romanchuk, P. R. 1965*a*, *Astr. Zu.*, **42**, 775.
Romanchuk, P. R. 1965*b*, *Soln. Dann. Bjull.*, no. 5, 65, no. 7, 65, and no. 8, 74.
Rompolt, B. 1965*a*, *Publ. astr. Inst. Czech. Acad. Sci.*, **51**, 120.
Rompolt, B. 1965*b*, *Bull. astr. Inst. Csl.*, **16**, 227.
Rompolt, B. 1967, Proc. of the 4th Consultation on Heliophysics, Wroclaw, in press.
Rompolt, B., Ciurla, T. 1967, Proc. of the 4th Consultation on Heliophysics, Wroclaw, in press.
Rösch, J. 1966, in Proc. of the Meeting on *Sunspots*, Florence, p. 88.
Rubashev, B. M. 1964*a*, *Problemy solnechnoj Aktivnosti*, Izdat. Akad. Nauk SSSR.
Rubashev, B. M. 1964*b*, NASA TT F-244, 13.
Rush, J. H., Schnable, G. K. 1964, *Appl. Opt.*, **3**, 1347.
Rust, D. 1966, *Astr. J.*, **71**, 178 (also in Dissertation, Univ. of Colorado).
Russell, P. C. 1965*a*, *Nature*, **205**, 684.
Russell, P. C. 1965*b*, *Nature*, **206**, 281.
Russell, P. C., Pounds, K. A. 1966, *Nature*, **209**, 490.
Saito, K., Billings, D. E. 1964, *Astrophys. J.*, **140**, 760.
Saito, T. 1964, *Rep. Ionosph. Space Res. Japan*, **18**, 260.
Sakurai, K. 1964, *Rep. Ionosph. Space Res. Japan*, **18**, 366.
Sakurai, K. 1965*a*, *Rep. Ionosph. Space Res. Japan*, **19**, 408 (also in *Publ. astr. Soc. Japan*, **17**, 403).
Sakurai, K. 1965*b*, *J. geophys. Res.*, **70**, 3235.
Sakurai, K. 1965*c*, *Planet. Space Sci.*, **13**, 745.
Sakurai, K. 1965*d*, *Planet. Space Sci.*, **13**, 867.
Sakurai, K. 1966*a*, *Publ. astr. Soc. Japan*, **18**, 77.
Sakurai, K. 1966*b*, *Rep. Ionosph. Space Res. Japan*, **20**, 33.
Santarelli, A. 1965, *Pubbl. Oss. astrofis. Catania*, no. 74.

Sarma, N. V. G., Joshi, M. N. 1965, *Radio Propag. Unit Sci. Rep.*, no. 21.
Sastry, Ch. V. 1966, *J. atmos. terr. Phys.*, **28**, 117.
Schadee, A. 1966, *Astrophys. J.*, **145**, 348.
Schatzman, E. 1965*a*, in *IAU Symp. no.* 22, p. 337. North-Holland Publ. Co., Amsterdam.
Schatzman, E. 1965*b*, in *The Solar Spectrum* (ed. C. de Jager), p. 313.
Schatzman, E. 1965*c*, in the Introduction to *Solar Terrestrial Relations* (ed. J. Ortner and H. Maseland), p. 133.
Schegoleva, G. P. 1965*a*, *Soln. Dann. Bjull.*, no. 4, 46.
Schegoleva, G. P. 1965*b*, *Soln. Dann. Bjull.*, no. 8, 70.
Schmahl, G., Schröter, E. H. 1965, *Z. Astrophys.*, **62**, 143.
Schmidt, H. U. 1964, in AAS–NASA Symp. on *Physics of Solar Flares*, p. 107.
Schmidt, H. U. 1965, *Mitt. astr. Ges.*, p. 89.
Schmidt, H. U. 1966, in Proc. of the Meeting on *Solar Magnetic Fields*, Rome, in press.
Schmieder, B., Perche, J. C., Wlérick, G. 1967, *Ann. Astrophys.*, in press.
Schnur, G., Wagner, G. 1965, *Astr. Nachr.*, **288**, 219.
Schröter, E. H. 1965*a*, *Z. Astrophys.*, **62**, 228.
Schröter, E. H. 1965*b*, *Z. Astrophys.*, **62**, 256.
Schröter, E. H. 1966*a*, in Proc. of the Meeting on *Sunspots*, Florence, p. 190.
Schröter, E. H. 1966*b*, ibid., p. 222.
Seaton, M. J. 1964, *Planet. Space Sci.*, **12**, 55.
Semel, M. 1967, *Ann. Astrophys.*, in press.
Severny, A. B. 1964*a*, in AAS–NASA Symp. on *Physics of Solar Flares*, p. 95.
Severny, A. B. 1964*b*, *Space Sci. Rev.*, **3**, 451.
Severny, A. B. 1964*c*, *Izv. Krym. astrofiz. Obs.*, **31**, 126.
Severny, A. B. 1965*a*, *Izv. Krym. astrofiz. Obs.*, **33**, 3.
Severny, A. B. 1965*b*, *Izv. Krym. astrofiz. Obs.*, **33**, 34.
Severny, A. B. 1965*c*, *Astr. Zu.*, **42**, 217.
Severny, A. B. 1965*d*, *Trans. IAU*, **12A**, 755.
Severny, A. B. 1965*e*, in *The Solar Spectrum* (ed. C. de Jager), p. 221.
Severny, A. B. 1966*a*, *Astr. Zu.*, **43**, 465.
Severny, A. B. 1966*b*, in *The Fine Structure of the Solar Atmosphere*, Colloquium Anacapri, Franz Steiner Verlag, p. 109.
Severny, A. B. 1966*c*, *Usp. fiz. Nauk*, **88**, 3.
Shapiro, R. 1965*a*, *J. geophys. Res.*, **70**, 245.
Shapiro, R. 1965*b*, *J. geophys. Res.*, **70**, 3581.
Shaposhnikova, E., Ogir, M. 1966, *Izv. Krym. astrofiz. Obs.*, **34**, 272.
Shea, M. A., Smart, D. F. 1965, *Am. geophys. Un. Trans.*, **46**, 128.
Sheeley, N. R. 1964, *Astrophys. J.*, **140**, 731.
Sheeley, N. R. 1966, *Astrophys. J.*, **144**, 723.
Sheeley, N. R. 1967, *Solar Phys.*, **1**, in press.
Shilova, N. S. 1966, *Astr. Zu.*, **43**, 108.
Shishov, V. I. 1966, *Geomagn. Aeronom.*, **6**, 223.
Shklovsky, I. C. 1964, *Astr. Zu.*, **41**, 676.
Shpitalnaya, A. A. 1964, *Izv. glav. astr. Obs. Pulkovo*, **24**, no. 177, 60.
Simon, G. W., Leighton, R. B. 1964, *Astrophys. J.*, **140**, 1120.
Simon, M. 1965, *Astrophys. J.*, **141**, 1513.
Simon, M., Axford, W. I. 1966, *Planet. Space Sci.*, **14**, 901.
Sinno, K. 1964, *Rep. Ionosph. Space Res. Japan*, **18**, 314.
Skomorovsky, V. I. 1966, *Res. Nabljud. i Issled. v Period MGSS*, Moskva, **4**, in press.
Slonim, Y. M. 1964, *Soln. Dann. Bjull.*, no. 12, 48.
Slonim, Y. M. 1965, *Soln. Dann. Bjull.*, no. 1, 51.
Smerd, S. F. 1964*a*, *Research in Geophysics*, MIT Press, 1, 65.
Smerd, S. F. 1964*b*, in AAS–NASA Symp. on *Physics of Solar Flares*, p. 343.
Smerd, S. F. 1965, in *The Solar Spectrum* (ed. C. de Jager), p. 398.
Smith, E. J., Holzer, R. E., Coleman, P. J., Davis, L., Siscoe, G. S. 1966, presented at the Inter-Union Symp. on solar terr. Phys., Belgrade, in preparation.

Smith, H. J. 1964, in AAS–NASA Symp. on *Physics of Solar Flares*, p. 1.
Smith, S. F. 1965, *Lockheed Rep.* no. 19038.
Smith, S. F., Ramsey, H. E. 1964, *Z. Astrophys.*, **60**, 1.
Smith, S. F., Ramsey, H. E. 1966, *Astr. J.*, **71**, 399.
Smolkov, G. Y. 1966*a*, *Issled. Geomagnet. Aeronom.*, Nauka Moskva, 212.
Smolkov, G. Y. 1966*b*, presented at the Inter-Union Symp. on solar terr. Phys., Belgrade, in preparation.
Snyder, C. W. 1964, in AAS–NASA Symp. on *Physics of Solar Flares*, p. 273.
Snyder, C. W., Neugebauer, M. 1964, *Space Res.*, **4**, 89.
Snyder, C. W., Neugebauer, M. 1965, in Proc. *Plasma Space Sci. Symp.* (ed. C. C. Chang and S. S. Huang), p. 38.
Snyder, C. W., Neugebauer, M., Rao, U. R. 1963, *J. geophys. Res.*, **68**, 6361.
Soboleva, N. S. 1965, *Izv. glav. astr. Obs. Pulkovo*, **24**, no. 177, 73.
Sonett, Ch. P. 1963, *Space Sci. Rev.*, **2**, 751.
Sonett, Ch. P., Colburn, D. S. 1965, *Planet. Space Sci.*, **13**, 675.
Sotirovski, P. 1965, *Ann. Astrophys.*, **28**, 106.
Stankiewicz, A. 1966, *Postepy Astronomii*, **14**, 102 (also in *Acta astr.*, in press).
Šťastná, A. 1964, *Studia geophys. geodet.*, **8**, 174.
Stefanov, A. P., Zemanek, E. N. 1965, *Vest. Kiev. Univ.*, Ser. astr., no. 6, 41.
Stefanov, A. P., Zemanek, E. N. 1966, *Vest. Kiev. Univ.*, Ser. astr., no 8, in press.
Stenflo, J. O. 1966, *Observatory*, **86**, 73.
Stepanov, V. E., Grigorev, V. M. 1966, *Trudy SibIZMIR*, in press.
Stepanyan, N. N. 1966, *Izv. Krym. astrofiz. Obs.*, **35**, 166, and **37**, in press.
Stepień, K. 1966, *Acta astr.*, in press.
Steshenko, N. 1966, *Izv. Krym. astrofiz. Obs.*, **37**, in press.
Stewart, R. T. 1965, *Austr. J. Phys.*, **18**, 67.
Stewart, R. T. 1966, *Austr. J. Phys.*, **19**, 209.
Stockhausen, R. 1965, *Astrophys. J.*, **141**, 277.
Stoddard, L. G., Carson, D. G., Saito, K. 1966, *Astrophys. J.*, **145**, 796.
Stoyanova, M. N. 1964, *Soln. Dann. Bjull.*, no. 4, 45.
Strong, I. B., Asbridge, J. R., Bame, S. J., Heckman, H. H., Hundhausen, A. J. 1966, *Phys. Rev. Lett.*, **16**, 631.
Sturrock, P. A. 1964, in AAS–NASA Symp. on *Physics of Solar Flares*, p. 357.
Sturrock, P. A., Coppi, B. 1964, *Nature*, **204**, 61.
Sturrock, P. A., Coppi, B. 1966, *Astrophys. J.*, **143**, 3.
Sturrock, P. A., Spreiter, J. R. 1965, *J. geophys. Res.*, **70**, 5345.
Sturrock, P. A., Woodbury, E. A. 1965, *Astr. J.*, **70**, 694.
Suemoto, Z., Moriyama, F. 1964, *Ann. Astrophys.*, **27**, 775.
Surkov, E. P. 1966*a*, *Soln. Dann. Bjull.*, no. 1, 63.
Surkov, E. P. 1966*b*, *Soln. Dann. Bjull.*, no. 5, 76.
Suzuki, T., Hirayama, T. 1964, *Publ. astr. Soc. Japan*, **16**, 58.
Suzuki, T., Attwood, C. F., Sheridan, K. V. 1964, *Observatory*, **84**, 55.
Švestka, Z. 1964*a*, *Contr. Wroclaw astr. Obs.*, **14**, 41.
Švestka, Z. 1964*b*, *Space Res.*, **4**, 768.
Švestka, Z. 1964*c*, *Bull. astr. Inst. Csl.*, **15**, 38.
Švestka, Z. 1964*d*, *Bull. astr. Inst. Csl.*, **15**, 162.
Švestka, Z. 1965*a*, *Adv. Astr. Astrophys.*, **3**, 119.
Švestka, Z. 1965*b*, *Publ. astr. Inst. Czech. Acad. Sci.*, **51**, 95.
Švestka, Z. 1966*a*, *Space Sci. Rev.*, **5**, 388.
Švestka, Z. 1966*b*, *Bull. astr. Inst. Csl.*, **17**, 95.
Švestka, Z. 1966*c*, *Bull. astr. Inst. Csl.*, **17**, 137 (also in *The Fine Structure of the Solar Atmosphere*, Colloquium Anacapri, p. 121).
Švestka, Z. 1966*d*, *Bull. astr. Inst. Csl.*, **17**, 262.
Švestka, Z. 1966*e*, presented at the Inter-Union Symp. on solar terr. Phys., Belgrade (also in *Bull. astr. Inst. Csl.*, **18**, in press).
Švestka, Z., Olmr, J. 1966, *Bull. astr. Inst. Csl.*, **17**, 4.

Swarup, G. 1964, in AAS–NASA Symp. on *Physics of Solar Flares*, p. 179.
Sweet, P. A. 1964, in AAS–NASA Symp. on *Physics of Solar Flares*, p. 409.
Sweet, P. A. 1965, in *IAU Symp. no.* 22, p. 377. North-Holland Publ. Co., Amsterdam.
Syrovatsky, S. I. 1966, *Astr. Zu.*, **43**, 340.
Takakura, T. 1963, *Publ. astr. Soc. Japan*, **15**, 462.
Takakura, T. 1964, *Publ. astr. Soc. Japan*, **16**, 230.
Takakura, T. 1966, *Space Sci. Rev.*, **5**, 80.
Takakura, T., Kai, K. 1966, *Publ. astr. Soc. Japan*, **18**, 57.
Takakura, T., Tlamicha, A. 1964, in AAS–NASA Symp. on *Physics of Solar Flares*, p. 207.
Tanaka, H. 1964, *Proc. Res. Inst. Atmos. Nagoya Univ.*, **11**, 41.
Tanaka, H. 1966, *Proc. Res. Inst. Atmos. Nagoya Univ.*, **13**, 69.
Tanaka, H., Kakinuma, T. 1964, *Rep. Ionosph. Space Res. Japan*, **18**, 32.
Tanaka, H., Steinberg, J. L. 1964, *Ann. Astrophys.*, **27**, 29.
Tanaka, Y. 1964, *Publ. astr. Soc. Japan*, **16**, 336.
Tandberg-Hanssen, E. 1964*a*, *Astrophys. norw.*, **9**, 13.
Tandberg-Hanssen, E. 1964*b*, *Astrophys. J. Suppl.*, **9**, 107.
Tandberg-Hanssen, E. 1967, *Solar Activity*, Blaisdell Publishing Comp., in press.
Teplitzkaya, R. B., Effendyeva, S. A. 1966, *Res. Nabljud. i Issled. v Period MGSS*, Moskva, **4**, in press.
Teryaeva, M. S. 1963, *Soln. Dann. Bjull.*, no. 7, 58.
Teryaeva-Geichenko, M. S. 1964, *Soln. Dann. Bjull.*, no. 3, 47.
Teske, R. G. 1965, *Astr. J.*, **70**, 695.
Teske, R. G., Dodson, H. W., Hedeman, E. R. 1964, *Astrophys. J.*, **139**, 1336.
Thomas, L., Venables, F. H., Williams, K. M. 1965, *Planet. Space Sci.*, **13**, 807.
Tidman, D. A. 1965, *Planet. Space Sci.*, **13**, 781.
Tifrea, E. 1965, *Publ. astr. Inst. Czech. Acad. Sci.*, **51**, 92.
Tindo, I. P. 1965, *Issled. kosm. Prostranstva*, Nauka Moskva, p. 533.
Tindo, I. P., Shurygin, A. I. 1965, *Kosm. Issled.*, **3**, 262.
Tlamicha, A. 1966, *Inf. Bull. solar radio Obs. Europe*, no. 21, 1.
Tlamicha, A., Takakura, T. 1963, *Nature*, **200**, 999.
Tlamicha, A., Křivský, L., Olmr, J. 1964, *Bull. astr. Inst. Csl.*, **15**, 49.
Tokuya, A., Nagasawa, S. 1964, *Rep. Ionosph. Space Res. Japan*, **18**, 122.
Tousey, R. 1964, *Q.J.R. astr. Soc.*, **5**, 123.
Tousey, R., Purcell, J. D., Austin, W. E., Garrett, D. L., Widing, K. G. 1964, *Space Res.*, **4**, 703.
Tousey, R., Austin, W. E., Purcell, J. D., Widing, K. G. 1965, *Ann. Astrophys.*, **28**, 755.
Trachtengertz, V. Y. 1966, *Astr. Zu.*, **43**, 356.
Trellis, M. 1963, *C. r. Acad. Sci.*, Paris, **256**, 2300.
Trellis, M. 1966*a*, *C. r. Acad. Sci.*, Paris, **262**, 221.
Trellis, M. 1966*b*, *C. r. Acad. Sci.*, Paris, **262**, 312.
Trellis, M. 1966*c*, *C. r. Acad. Sci.*, Paris, **262**, 372.
Tsap, T. T. 1965*a*, *Izv. Krym. astrofiz. Obs.*, **33**, 92.
Tsap, T. T. 1965*b*, *Izv. Krym. astrofiz. Obs.*, **34**, 296.
Tsimakhovich, N. 1964, *Soln. Dann. Bjull.*, no. 5, 54.
Tsubaki, T. 1966, *Publ. astr. Soc. Japan*, **18**, 1.
Tsuchiya, A., Nagane, K. 1967, *Solar Phys.*, **1**, 121.
Tuominen, J. 1964, *Astrophys. norw.*, **9**, 71.
Tuominen, J. 1965*a*, in *IAU Symp. no.* 22, p. 236. North-Holland Publ. Co., Amsterdam.
Tuominen, J. 1965*b*, *Z. Astrophys.*, **60**, 231.
Tuominen, J. 1965*c*, *Observatory*, **85**, 82.
Tuominen, J. 1966, in Proc. of the Meeting on *Sunspots*, Florence, p. 59.
Turlo, Z. 1965, *Publ. astr. Inst. Czech. Acad. Sci.*, **51**, 156.
Underwood, J. H., Muney, W. S. 1967, *Solar Phys.*, **1**, 129.
Vakulov, P. V., Vernov, S. N., Gorchakov, E. B., Logachev, Y. I., Charakhchan, A. N., Charakhchan, T. N., Chudakov, A. E. 1964, *Space Res.*, **4**, 26.
Valníček, B. 1964*a*, *Contr. Wroclaw astr. Obs.*, **14**, 55.

Valníček, B. 1964*b*, *Bull. astr. Inst. Csl.*, **15**, 207.
Valníček, B. 1965*a*, *Publ. astr. Inst. Czech. Acad. Sci.*, **51**, 62.
Valníček, B. 1965*b*, *Bull. astr. Inst. Csl.*, **16**, 258.
Valníček, B., Kleczek, J. 1964, *Appl. Opt.*, **3**, 1383.
Valníček, B., Kleczek, J., Topol, F. 1965, *Bull. astr. Inst. Csl.*, **16**, 44.
Van Allen, J. A., Krimigis, S. M. 1965, *J. geophys. Res.*, **70**, 5737.
Van Allen, J. A., Lin, W. C., Leinbach, H. 1964, *J. geophys. Res.*, **69**, 4481.
Van Allen, J. A., Frank, L. A., Maehlum, B., Acton, L. W. 1965, *J. geophys. Res.*, **70**, 1639.
Van Nieuwkoop, J. 1964, *Nature*, **201**, 380.
Van't Veer, F. 1965, in *The Solar Spectrum* (ed. C. de Jager), p. 257.
Van't Veer, F. 1966*a* in Proc. of the Meeting on *Sunspots*, Florence, p. 150.
Van't Veer, F. 1966*b*, *Ann. Astrophys.*, **29**, 7 and 119.
Van't Veer, F. 1966*c*, *Ann. Astrophys.*, **29**, 223.
Van't Veer-Menneret, C., Van't Veer, F. 1965, *Ann. Astrophys.*, **28**, 1026.
Vasilev, B. N. 1966, *Kosm. Issled.*, **4**, 748.
Vernov, S. N., Logachev, Y. I., Lyubimov, G. P., Nikolaev, A. G., Gorchakov, E. V., Pereslegina, N. V., Vakulov, P. V. 1966*a*, *Space Res.*, **7**, in press.
Vernov, S. N., Chudakov, A. E., Vakulov, P. V., Logachev, Y. I., Lyubimov, G. P., Nikolaev, A. G., Pereslegina, N. V. 1966*b*, presented at the Inter-Union Symp. on solar terr. Phys., Belgrade, in preparation.
Vernov, S. N., Chudakov, A. E., Vakulov, P. V., Logachev, Y. I., Lyubimov, G. P., Pereslegina, N. V. 1966*c*, presented at the Inter-Union Symp. on solar terr. Phys., Belgrade, in preparation.
Vinogradov, Y. I. 1964, *Izv. Krym. astrofiz. Obs.*, **32**, 67.
Vitinsky, Y. I. 1965*a*, *Izv. glav. astr. Obs. Pulkovo*, **24**, no. 178, 49.
Vitinsky, Y. I. 1965*b*, *Geofiz. Bjull. mezhduved. geofiz. Kom. Akad. Nauk SSSR*, no. 16, 60.
Vitinsky, Y. I. 1965*c*, *Soln. Dann. Bjull.*, no. 11, 62, and no. 12, 53.
Vitinsky, Y. I. 1966, *Morfologia soln. Aktivnosti*, Nauka Moskva.
Vitinsky, Y. I., Ihsanov, R. N. 1964*a*, *Soln. Dann. Bjull.*, no. 7, 66.
Vitinsky, Y. I., Ihsanov, R. N. 1964*b*, *Soln. Dann. Bjull.*, no. 10, 57, and no. 12, 63.
Vitkevich, V. V., Vlasov, V. I. 1966, presented at the Inter-Union Symp. on solar terr. Phys., Belgrade, in preparation.
Vladimirsky, B. M. 1964, *Izv. Krym. astrofiz. Obs.*, **31**, 271.
Vladimirsky, B. M. 1965*a*, *Izv. Krym. astrofiz. Obs.*, **33**, 151.
Vladimirsky, B. M. 1965*b*, *Izv. Krym. astrofiz. Obs.*, **34**, 305.
Vladimirsky, B. M., Pankratov, A. K. 1964, *Izv. Krym. astrofiz. Obs.*, **32**, 46 (also in *Izv. Akad. Nauk SSSR*, **28**, 2019).
Vladimirsky, B. M., Pankratov, A. K., Stepanyan, A. A. 1966, *Izv. Krym. astrofiz. Obs.*, **35**, 222.
von Klüber, H. 1966, in *Solar Physics*, Lagonissi Summer Course, John Wiley and Sons, New York, in press.
Voyhanskaya, N. F. 1965, *Soln. Dann. Bjull.*, no. 9, 57.
Voyhanskaya, N. F. 1966, *Astr. Zu.*, **43**, 413.
Vsechsvyatsky, S. K. 1964, *Geomagn. Aeronom.*, **4**, 328.
Vsechsvyatsky, S. K., Gurtovenko, E. A., Dziubenko, N. I., Ivanchuk, V. I. 1966, *Soln. Dann. Bjull.*, in press.
Vyalshin, G. F., Krat, V. A. 1965, *Izv. glav. astr. Obs. Pulkovo*, **24**, no. 178, 26.
Waddington, C. J., Freier, P. S. 1965, *J. geophys. R* ., **70**, 230.
Waldmeier, M. 1964, *Z. Astrophys.*, **59**, 205.
Waldmeier, M. 1966, in Proc. of the Meeting on *Sunspots*, Florence, p. 41.
Wallis, G. 1965*a*, *Publ. astr. Inst. Czech. Acad. Sci.*, **51**, 145.
Wallis, G. 1965*b*, *Publ. astr. Inst. Czech. Acad. Sci.*, **51**, 184.
Ward, F. 1966, *Astrophys. J.*, **145**, 416.
Warwick, C. S. 1964, in AAS–NASA Symp. on *Physics of Solar Flares*, p. 27.
Warwick, C. S. 1965*a*, *Astrophys. J.*, **141**, 500.
Warwick, C. S. 1965*b*, *Astrophys. J.*, **142**, 767.

Warwick, J. W. 1964, in AAS–NASA Symp. on *Physics of Solar Flares*, p. 441.
Warwick, J. W. 1965*a*, in *Solar System Radio Astronomy* (ed. J. Aarons), Plenum Press, New York, Chapter 8.
Warwick, J. W. 1965*b*, *NCAR Techn. Note*, no. 8, 215.
Warwick, J. W., Hyder, C. L. 1965, *Astrophys. J.*, **141**, 1362.
Webber, W. R. 1964, in AAS–NASA Symp. on *Physics of Solar Flares*, p. 215.
Weddell, J. B. 1964, in AAS–NASA Symp. on *Physics of Solar Flares*, p. 103 (also in *Astronaut. Acta*, **10**, 339).
Weddell, J. B. 1965, *Planet. Space Sci.*, **13**, 789.
Weir, R. A., Brown, R. R. 1964, *J. geophys. Res.*, **69**, 2193.
Weiss, A. A. 1965, *Austr. J. Phys.*, **18**, 167.
Weiss, A. A., Stewart, R. T. 1965, *Austr. J. Phys.*, **18**, 143.
Weiss, A. A., Wild, J. P. 1964, *Austr. J. Phys.*, **17**, 282.
Weiss, N. O. 1964, *Mon. Not. R. astr. Soc.*, **128**, 225.
Weiss, N. O. 1965*a*, in *IAU Symp. no.* 22, p. 330. North-Holland Publ. Co., Amsterdam.
Weiss, N. O. 1965*b*, *Observatory*, **85**, 37.
Wentzel, D. G. 1964*a*, in AAS–NASA Symp. on *Physics of Solar Flares*, p. 397.
Wentzel, D. G. 1964*b*, *Astrophys. J.*, **140**, 1563.
Wentzel, D. G. 1965, *J. geophys. Res.*, **70**, 2716.
Wentzel, D. G., Solinger, A. B. 1966, *Astr. J.*, **71**, 402.
White, M. L. 1965, *Astr. J.*, **70**, 333.
White, W. A. 1964, in AAS–NASA Symp. on *Physics of Solar Flares*, p. 131 (also in *Space Res.*, **4**, 771).
Wibberenz, G. 1966, presented at the Inter-Union Symp. on solar terr. Phys., Belgrade, in preparation.
Widing, K. G. 1966, *Astrophys. J.*, **145**, 380.
Wilcox, J. M., Ness, N. F. 1965, *J. geophys. Res.*, **70**, 5793.
Wild, J. P. 1964, in AAS–NASA Symp. on *Physics of Solar Flares*, p. 161.
Wild, J. P., Tlamicha, A. 1964, *Nature*, **203**, 1128.
Wild, J. P., Tlamicha, A. 1965, *Bull. astr. Inst. Csl.*, **16**, 73.
Wild, J. P., Smerd, S., Weiss, A. A. 1963, *A. Rev. Astr. Astrophys.*, **1**, 291 (also in 1964, *Usp. fiz. Nauk*, **84**, 99).
Willis, D. M. 1966, *Planet. Space Sci.*, **14**, 483.
Wilson, P. R. 1965, *Astrophys. J.*, **142**, 773.
Wilson, P. R. 1966*a*, in Proc. of the Meeting on *Sunspots*, Florence, p. 146.
Wilson, P. R. 1966*b*, *Astrophys. J.*, **143**, 993.
Winckler, J. R. 1964, in AAS–NASA Symp. on *Physics of Solar Flares*, p. 117.
Withbroe, G. L. 1967, in preparation (presented at the Special Meeting on *Solar Astronomy* of AAS, October, 1966).
Wolfe, J. H., Silva, R. W. 1965, *J. geophys. Res.*, **70**, 3575.
Wolfe, J. H., Silva, R. W., Myers, M. A. 1966, *J. geophys. Res.*, **71**, 1319.
Wood, R. M., Wood, K. D. 1965, *Nature*, **208**, 129.
Xanthakis, J. 1965, *Mem. Soc. astr. ital.*, **36**, 25.
Xanthakis, J. 1966*a*, in Proc. of the Meeting on *Sunspots*, Florence, p. 63.
Xanthakis, J. 1966*b*, *Nature*, **210**, 1242.
Xanthakis, J. 1966*c*, *Bull. astr. Inst. Csl.*, **17**, 215.
Xanthakis, J. 1966*d*, editor, *Solar Physics*, Lagonissi Summer Course, John Wiley and Sons, New York, in press.
Xanthakis, J. 1966*e*, in Xanthakis (1966*d*), in press.
Yakovkin, N. A., Zeldina, M. Y. 1964*a*, *Astr. Zu.*, **41**, 336.
Yakovkin, N. A., Zeldina, M. Y. 1964*b*, *Astr. Zu.*, **41**, 914.
Yakovkin, N. A., Zeldina, M. Y. 1964*c*, *Soln. Dann. Bjull.*, no. 9, 62.
Yakovkin, N. A., Zeldina, M. Y. 1965, *Soln. Dann. Bjull.*, no. 5, 50.
Yates, G. K. 1964, *J. geophys. Res.*, **69**, 3077.
Yilmaz, F. 1964*a*, *Publ. Univ. Obs. Istanbul*, no. 78.
Yilmaz, F. 1964*b*, *Publ. Univ. Obs. Istanbul*, no. 79.

Yoshida, S. 1965, *J. geophys. Res.*, **70**, 2065.
Yoshida, S., Akasofu, S. I. 1965, *Planet. Space Sci.*, **13**, 435.
Yurovskaya, L. I. 1964, *Izv. Krym. astrofiz. Obs.*, **32**, 76.
Yurovskaya, L. I. 1965, *Izv. Krym. astrofiz. Obs.*, **34**, 9.
Zabza, M. 1964, *Contr. Wroclaw astr. Obs.*, **14**, 57.
Zaitsev, V. V. 1965, *Astr. Zu.*, **42**, 740.
Zeldina, M. Y. 1964, *Spektrofotom. Issled. aktivn. Obraz. na Solnce*, Kiev, p. 19.
Zeldina, M. Y., Sergeeva, A. N. 1966, *Voprosy Astrofiziky*, Kiev, p. 36.
Zemanek, E. H., Stefanov, A. P. 1964*a*, *Spektrofotom. Issled. aktivn. Obraz. na Solnce*, Kiev, p. 94.
Zemanek, E. N., Stefanov, A. P. 1964*b*, *Inf. Bjull. MGG*, no. 6, p. 103.
Zemanek, E. N., Stefanov, A. P. 1966*a*, *Voprosy Astrofiziki*, Kiev, p. 72.
Zemanek, E. N., Stefanov, A. P. 1966*b*, in Proc. of the Meeting on *Sunspots*, Florence, p. 216.
Zhang He-qi, Cao Tian-jun 1964, *Acta astr. Sinica*, **12**, 138.
Zheleznyakov, V. V. 1965*a*, *Astr. Zu.*, **42**, 96.
Zheleznyakov, V. V. 1965*b*, *Astr. Zu.*, **42**, 244.
Zheleznyakov, V. V., Trakhtengertz, V. Y. 1965, *Astr. Zu.*, **42**, 1005.
Zhitnik, I. A., Krumov, V. V., Malyavkin, L. P., Mandelshtam, C. L. 1964, *Kosm. Issled.*, **2**, 920.
Zhugzhda, Y. D. 1966, *Astr. Zu.*, **43**, 967.
Zhulin, I. A. 1965, *Soln. Dann. Bjull.*, no. 11, 48.
Zhulin, I. A., Mogilevsky, E. I. 1965*a*, *Geomagn. Aeronom.*, **5**, 1092.
Zhulin, I. A., Mogilevsky, E. I. 1965*b*, *Soln. Aktivnost*, **2**, 149.
Zirin, H. 1964*a*, in AAS–NASA Symp. on *Physics of Solar Flares*, p. 33.
Zirin, H. 1964*b*, *Astrophys. J.*, **140**, 1216.
Zirin, H. 1965*a*, Proc. of the *Plasma Space Sci. Symp.* (ed. C. C. Chang and S. S. Huang), p. 38.
Zirin, H. 1965*b*, in *IAU Symp. no.* 22, p. 339. North-Holland Publ. Co., Amsterdam.
Zirin, H. 1965*c*, *Am. geophys. Un. Trans.*, **46**, 190.
Zirin, H. 1966, *Appl. Opt.*, **5**, 474.
Zirin, H., Werner, S. 1967, *Solar Phys.*, **1**, 66.
Zirker, J. B. 1965, in *IAU Symp. no.* 22, p. 208. North-Holland Publ. Co., Amsterdam.
Zwaan, C. 1965*a*, in *IAU Symp. no.* 22, p. 277. North-Holland Publ. Co., Amsterdam (also in *The Solar Spectrum* – ed. C. de Jager – p. 262).
Zwaan, C. 1965*b*, *Rech. astr. Obs. Utrecht*, **17**, part 4.
Zwaan, C. 1966, in Proc. of the Meeting on *Sunspots*, Florence, p. 169.

Z. ŠVESTKA,
President of the Commission

12. COMMISSION DE LA RADIATION ET DE LA STRUCTURE DE L'ATMOSPHERE SOLAIRE

PRÉSIDENT: Dr R. Michard, Observatoire de Paris, Section d'Astrophysique, 92-Meudon, France.

VICE-PRÉSIDENT: Dr M. N. Gnevyshev, Astronomical Observatory, Leningrad M-140, U.S.S.R.

COMITÉ D'ORGANISATION: R. G. Athay, M. G. J. Minnaert, E. A. Müller, M. J. Seaton, Z. Suemoto.

MEMBRES: Allen, Babcock, Billings, Blackwell, Blaha, Blamont, Böhm, Böhm-Vitense, Bray, Brück (H. A.), Bruzek, Chamberlain, Christiansen, de Jager, Delbouille, Edlén, Edmonds, Elste, Evans (J. W.), Friedman, Giovanelli, Godoli, Gökdoğan, Goldberg, Hotinli, Houtgast, Hubenet, Jäger, Jefferies, Kawaguchi, Kiepenheuer, Kononovich, Kopecký, Laborde, Labs, Leighton, Locke, Loughhead, Lüst, Mathias, Matsushima, Mattig, Mergentaler, Migeotte, Mugglestone, Namba, Neven, Newkirk, Nicolet, Pecker (J.-C.), Pecker-Wimel, Peyturaux, Pierce, Redman, Righini, Rigutti, Schröter, Severny, Sitnik, Sitterly, Sobolev (V. M.), Švestka, Swensson, Tandberg-Hanssen, Thomas (R. N.), Tousey, Unsöld, Voigt, von Klüber, Waldmeier, Warwick (J. W.), Zirin, Zirker.

INTRODUCTION

Ce rapport et la liste bibliographique associée décrivent les progrès réalisés dans l'étude de l'atmosphère solaire 'normale' de Janvier 1964 à Octobre 1966. Quelques travaux publiés dès 1963, mais malheureusement omis dans notre précédent rapport, ont été inclus.

Le lourd travail nécessaire pour passer en revue un domaine scientifique aussi vaste que celui de notre Commission a été partagé entre C. W. Allen (couronne), R. G. Athay (chromosphère) et R. Michard. Ce dernier adresse ses remerciements aux deux collègues qui l'ont aidé si généreusement.

Nous sommes particulièrement reconnaissants à tous les membres de la Commission qui ont bien voulu nous adresser un compte rendu de leurs travaux ou de ceux de leurs collègues afin de rendre notre revue plus actuelle et plus vivante.

RÉFÉRENCES GÉNÉRALES

Notons les divers symposiums organisés récemment sur des sujets intéressant la Commission:

- *The Solar Spectrum*, Symposium d'Utrecht, 1963 (**1**).
- *Stellar and Solar Magnetic Fields*, Symposium no. 22 de l'UAI, Rottach-Egern, 1963 (**2**).
- *Astronomical Observations for Space Vehicles*, Symposium no. 23 de l'UAI, Liège, 1964 (**3**).
- *Abundance Determinations in Stellar Spectra*, Symposium no. 26 de l'UAI, Utrecht, 1964 (**4**).
- *Campi magnetici solari e la spettroscopia ad alta risoluzione*, Rome, 1964 (**9**), l'une des nombreuses manifestations organisées pour le 4ème Centenaire de la naissance de Galilée.
- *Cosmical Gas Dynamics*, Symposium no. 28 de l'UAI, Nice, 1965 (**5**).
- First Harvard-Smithsonian Conference on Stellar Atmospheres, Cambridge, 1964 (**7**).
- Second Harvard-Smithsonian Conference on Stellar Atmospheres, Cambridge, 1965 (**8**).

– *Solar Physics*, an Advanced Study Institute sponsored by NATO, Lagonissi, 1965 (**6**).

– *The Fine Structure of the Solar Atmosphere*, Anacapri, 1966 (**10**), organisé pour l'inauguration du télescope solaire de la station d'Anacapri du Fraunhofer Institut.

Les références ultérieures à ces symposiums et colloques seront données sous la forme (RG **n**) **n** étant le numéro d'ordre de la publication dans la liste de Références Générales.

Un traité intitulé 'Astrophysique et Astronomie Stellaire' est publié par l'Académie des Sciences de l'U.R.S.S. Le tome III contient une description de la physique solaire par cinq spécialistes éminents. Une monographie sur la couronne solaire et le rayonnement corpusculaire dans l'espace interplanétaire a été éditée sous la direction de Vsekhviatsky (**11**). Un excellent livre sur la couronne a été récemment publié par Billings (**12**).

LE SPECTRE CONTINU SOLAIRE

Les efforts pour parvenir à une courbe d'énergie précise du continuum solaire au centre du disque se poursuivent activement. Labs annonce une publication (**1**) récapitulant l'ensemble des résultats de son programme. Une calibration en unités absolues du continuum adoptée dans les Atlas photométriques du spectre solaire de l'Observatoire d'Utrecht et de l'Observatoire de Göttingen sera établie.

Peyturaux a terminé en 1965 ses comparaisons de Soleil au corps noir pour le domaine spectral visible, mais seuls des résultats préliminaires ont été publiés (**2**). L'extension de ces mesures absolues jusqu'à 15 μ est prévue.

En U.R.S.S., l'Institut Sternberg est particulièrement actif dans ce domaine. Murashova et Sitnik (**3**) donnent les résultats d'une série d'observations obtenues en 1957-59 et les comparent à ceux de 1952-55. Ultérieurement Sitnik combine ses résultats en une courbe unique pour $\lambda\lambda$ 3280–12 500 Å (**4**) et présente des mesures dans l'IR (2·7 to 5 μ) (**5**). Il discute de la meilleure méthode pour contrôler la stabilité de la transparence du ciel (**6**). Enfin de nouvelles observations pour $\lambda\lambda$ 3100–6600 Å sont effectuées par Makarova (**7**). Le même auteur a discuté les données disponibles, lesquelles présentent des écarts systématiques très importants (**8**). Lambert et Willstrop (**9**) ont cherché à établir la courbe d'énergie solaire en combinant leurs mesures absolues des étoiles avec la magnitude V et la couleur $B-V$ du Soleil.

Houtgast (**10**) a publié les résultats de ses mesures absolues portant sur 32 'fenêtres' du domaine 3000–4000 Å. Il trouve une discontinuité de Balmer Δ log $I \simeq 0{\cdot}03$ en contradiction avec la valeur 'classique' d'environ 0·12.

Un grand intérêt s'attache à l'extension de la courbe d'énergie solaire à l'IR lointain, ce qui nécessite des observations en ballon. Les résultats déjà obtenus par Murcray *et al.* (**11**) et par Beer (**12**) indiquent que la température de brillance moyenne du disque solaire décroît de 5400° à 4270° entre $\lambda = 4\,\mu$ et $\lambda = 100\,\mu$.

L'assombrissement centre-bord dans l'IR constitue un test très sensible du modèle thermique des couches supérieures de la photosphère. Des prédictions ont été faites pour divers modèles par Léna (**13**) et par Noyes *et al.* (**14**). Pour l'instant les mesures disponibles ne s'étendent que jusqu'à 10·2 μ (Pierce) mais imposent déjà des limitations strictes aux modèles.

Le même problème peut être étudié par l'analyse du continuum UV mais la théorie du coefficient d'absorption est alors plus difficile (cf. Matsushima, RG 7). Blamont et Bonnet (**15**) ont tracé les courbes d'assombrissement à 2880, 2660 et 2190 Å grâce à des images en lumière filtrée obtenues en fusée. Des travaux plus récents du même groupe montrent que le domaine spectral 3000–2000 Å est accessible aux ballons sondes ce qui permet l'emploi d'instruments plus importants et mieux stabilisés.

La dépression du continuum solaire à 2085 Å a été discutée quantitativement par Kodaira (**16**) qui l'attribue à la discontinuité d'ionisation de Al I à partir du niveau $3p^2P_0$.

De nouvelles mesures d'assombrissement dans le spectre visible et le proche UV ont été obtenues au Pic du Midi par Mouradian (**17**), elles s'étendent jusqu'à $\mu = 0{\cdot}10$. Le même auteur (*loc. cit.*) a appliqué la méthode d'Unsöld lors de l'éclipse partielle du 15 Février 1961 pour déterminer le profil du bord lui-même. Les résultats demeurent assez incertains mais la méthode pourrait conduire à de bons résultats si les observations étaient obtenues par 'très bonnes images', comme ce fut le cas au Pic du Midi le 20 Mai 1966 (Rösch, RG **10**).

Diverses méthodes de détermination du profil du bord par photométrie d'une portion de croissant au voisinage des contacts intérieurs d'une éclipse totale ont été appliquées par Kristenson (**18**) et par Heintze (**19**) (ainsi que par plusieurs expérimentateurs à l'éclipse du 20 Mai 1966). Ce dernier auteur trouve un anneau brillant assez prononcé, ce qui le conduit à un modèle pour lequel le minimum de température se situe à une profondeur optique assez grande, $\tau_{5000} = -0.07$. Ce modèle est incompatible avec l'assombrissement centre-bord à 10 μ.

La question d'éventuelles différences thermiques entre les pôles et l'équateur du Soleil continue d'être à l'ordre du jour, les observateurs employant d'ailleurs aussi bien des mesures des largeurs équivalentes et intensités centrales des raies de Fraunhofer que des mesures du continuum, comme critère de température. Selon Mulders et Slaughter (**20**), l'écart thermique éventuel est insignifiant. Il en est de même pour Appenzeller et Schröter (**21**) qui donnent une limite supérieure de 6°. Un programme de comparaison pôle-équateur a été entrepris par Houtgast (non publié).

Les remarques de K. H. Böhm (cf. Rapport précédent) sur le contenu d'information des lois de variations centre-bord de l'intensité émergente ont donné lieu à quelque controverse (**22, 23**). Par ailleurs une élégante méthode d'inversion de la relation $I(\mu) \leftrightarrows S(\tau)$ a été proposée par Delache (**24**).

Notons enfin que les travaux récents d'interprétation du continuum solaire ont bénéficié des nouveaux calculs détaillés des coefficients d'absorption continue effectués par Bode (**25**).

ATLAS ET TABLES DU SPECTRE DE FRAUNHOFER. IDENTIFICATIONS

Mrs Moore-Sitterly annonce que Minnaert, Houtgast et elle-même . . . 'have now concluded our publication of the Second Revision of Rowland's preliminary table of solar spectrum wavelengths. This compendium is in press and should be in print before 1967. The title is "The Solar Spectrum 2935 Å to 8770 Å". It contains approximately 24 000 lines, and includes a number of new lines. Measured equivalent widths from the records of the Utrecht Photometric Atlas replace the Rowland estimated line intensities. Reduced widths derived from these directly measured equivalent widths are listed. Other columns indicate the spot behavior of the atomic lines, and give revised identifications as to chemical origin of the lines. For classified atomic lines the lower excitation potential and multiplet number are listed. For molecular lines the rotation branch and quantum number, and the vibration band are indicated. Complete band designations are entered in notes. About 73% of the lines are wholly or partially identified. This volume will appear as Monograph 61 of the National Bureau of Standards (349 pp).'

La préparation du nouvel atlas photométrique du spectre solaire, commencée par Delbouille, Neven et Roland à l'aide du spectromètre à double passage installé au Jungfraujoch par l'équipe de l'Institut d'Astrophysique de Liège, est très avancée. Un rapport spécial sera présenté à la Commission par les auteurs.

A Kitt Peak, on a entrepris une nouvelle détermination de l'échelle des longueurs d'onde du Catalogue de Rowland révisé et de son extension à l'IR.

Pierce annonce en outre la publication d'un catalogue de plusieurs milliers de raies observées en émission à l'extrême bord du Soleil, raies dont la plupart sont dues aux terres rares.

Il est clair que l'exploration du spectre solaire ne sera jamais achevée, car toute amélioration technique (résolution, rapport signal/bruit du récepteur photographique ou photoélectrique) fait apparaître des raies nouvelles. Sur un spectre obtenu à Sacramento Peak et couvrant seulement 5 Å (5047·2–5052·9) l'auteur de ce rapport a compté 22 raies qui ne figuraient pas dans la Première Révision du Catalogue de Rowland! Un travail systématique de recherche des raies faibles est dû à Ahmad Kiasatpoor qui a publié pour le domaine $\lambda\lambda$ 6000–7000 un catalogue contenant de très nombreuses raies nouvelles (**1**). Blackwell nous annonce l'obtention à Oxford de spectres solaires où le rapport signal/bruit est poussé jusqu'à la limite imposée par le nombre de photons.

On assiste à un regain d'intérêt pour le problème de l'identification des raies du spectre de Fraunhofer. Grâce à de nouvelles données de laboratoire, Swensson (**2**) a révisé l'identification du spectre de Mg I et attribué à cet atome 91 raies solaires jusqu'alors non identifiées. Le même auteur trouve 26 raies de P I dans le proche IR (**3**) et étudie les raies de l'azote dans le même domaine spectral (**3 bis**). Grevesse (**4**) discute de l'identification du bismuth.

La présence de raies interdites dans le spectre de Fraunhofer a été abondamment étudiée. Kodaira (**5**) considère le cas de [Fe II]. Les raies prédites de cet élément ont été systématiquement recherchées par J. P. Swings sur des enregistrements photoélectriques à haute résolution obtenus au Jungfraujoch: une vingtaine de raies très faibles peuvent être attribuées à [Fe II] et deux autres à [Ni II] (**6, 7**).

Les observations modernes mettent en évidence des particularités intéressantes du spectre de Fraunhofer. Ainsi des dépressions diffuses à 6362, 6344 et 6319 Å sont attribuées par Mitchell et Mohler à des multiplets de Ca I en présence d'auto-ionisation à partir du niveau supérieur (**8**). Selon Swensson (**9**), les profils larges et complexes de certaines raies de Mn I vers 8700 Å peuvent être expliqués par la structure hyperfine. Aller indique que les profils des raies de C I dans le visible sont anormalement larges, ce qui pourrait être dû à la formation de quasi-molécules CH.

L'étude spectroscopique du Soleil dans l'UV et l'EUV se poursuit activement et de grands progrès ont été faits dans l'identification de ce domaine spectral. Une revue récente est due à Tousey *et al.* (RG **3**). Comme le spectre d'émission de l'EUV est dû à la couronne ou à la couche de transition, on trouvera de nombreuses références sur ce domaine spectral dans les parties de ce rapport consacrées aux couches extérieures de l'atmosphère solaire (cf. ci-dessous). Pottasch (**10**) a préparé une revue très détaillée des problèmes de l'interprétation du spectre EUV. Goldberg *et al.* (**11**) discutent la question de l'identification de CO dans l'UV.

Il est particulièrement satisfaisant que de nouvelles équipes scientifiques de divers pays soient maintenant en mesure de contribuer à la spectroscopie 'spatiale' du Soleil. De bons spectres du domaine 2500–1000 Å ont été obtenus par Black *et al.* (**12**). Par la suite la même équipe de Culham Laboratory a réussi à photographier le spectre de la chromosphère-couronne de 850 à 2950 Å, la fente étant maintenue à 10″ du limbe avec une précision de 2″. Le spectre montre un continuum chromosphérique et de très nombreuses raies d'émission dont l'identification est en cours; des résultats préliminaires ont été publiés par Burton et Wilson (**13**).

En Grande-Bretagne également, une équipe de l'Université de Leicester a obtenu de remarquables résultats dans la spectrographie des rayons X solaires (région 25–11 Å). L'Observatoire d'Utrecht a developpé un spectromètre de résolution élevée pour le domaine 44–64 Å. Enfin en France, Bonnet (Service d'Aéronomie du CNRS) a obtenu le spectre du limbe solaire dans le domaine 3000–1900 Å à l'aide d'un coronographe embarqué sur fusée-sonde.

LONGUEURS D'ONDE. ROTATION. CIRCULATION.

La question des longueurs d'onde des raies de Fraunhofer sous l'effet de la rotation de l'astre, du déplacement relativiste et des diverses autres causes de décalages systématiques, a fait l'objet de nombreux travaux en U.R.S.S. De nouvelles observations spectroscopiques de la rotation solaire ont été effectuées par Aslanov (**1, 2**). Le même auteur a préparé un 'Catalogue des déplacements et des profondeurs optiques des raies de Fraunhofer dans l'atmosphère solaire' (**3**) en vue de l'étude du déplacement relativiste (**4, 5**).

La discussion de Melnikov (**6**) montre que les variations systématiques du décalage des raies avec leur intensité, le potential d'excitation et la longueur d'onde peuvent se ramener à un effet de hauteur dans l'atmosphère solaire, lié aux variations du champ de vitesses. Après correction de cet effet on obtient un parfait accord avec la théorie de la relativité.

Dans un travail déjà cité plus haut, Appenzeller et Schröter ont obtenu la variation centre-bord des longueurs d'onde de plusieurs raies. Roddier (cf. ci-dessous) a mesuré le profil des raies de résonance de Sr I, Ba I et Ca I par la méthode du jet atomique dans une échelle absolue de longueur d'onde. Il serait sans doute très utile d'étendre des mesures de ce type à un grand nombre de raies, car les problèmes des 'longueurs d'onde', des profils asymétriques et de la structure hétérogène de l'atmosphère solaire sont indissolublement liés.

Il est intéressant de signaler ici les travaux relatifs à la circulation générale des couches superficielles du Soleil, bien que ces mouvements semblent difficiles à mettre en évidence par la spectroscopie et ne soient révélés que par l'analyse du mouvement des taches.

Une telle analyse a été effectuée par Ward (**7, 8, 9**) qui montre la présence d'une corrélation marquée entre les mouvements propres en latitude et longitude des taches (données de Greenwich) et présente un modèle de circulation sans symétrie axiale constitué d'un ensemble d' 'ondes' qui transportent un certain moment angulaire de l'équateur vers le pôle. Les aspects énergétiques de ce modèle ont été discutés par Starr et Gilman (**10, 11**).

La théorie de la rotation différentielle, basée sur l'hypothèse de Biermann qui l'attribue à la viscosité non isotrope des éléments turbulents dans la zone convective, a été développée par Sakurai (**12**); il critique le traitement antérieur de Kippenhahn et aboutit à un bon accord avec la loi de rotation observée. La théorie de la rotation dans les couches internes du Soleil en équilibre radiatif a été traitée par Rubashev (**13**), par intégration numérique des équations exprimant le bilan du moment angulaire (équations de Jeans).

LA PHOTOSPHÈRE

(*a*) *Granulation. Structures cinématiques et magnétiques*

Le nombre des recherches expérimentales nouvelles sur la granulation est relativement faible, sans doute parce que tous les spécialistes sont conscients de l'extrême difficulté d'obtenir des documents suffisamment bons pour mériter une étude quantitative! Le point des connaissances actuelles sur la granulation a été fait dans une monographie de Bray et Loughhead (**1**).

A Poulkovo, Vassiljeva et Yudina (**2**) ont mesuré photoélectriquement les fluctuations de brillance de la surface solaire simultanément pour deux longueurs d'onde λ_1 et λ_2: la corrélation entre les fluctuations observées décroît fortement quand l'écart $\lambda_2 - \lambda_1$ augmente. Ce résultat est confirmé par Krat et Petrova qui utilisent une technique photographique (**3**). Ces auteurs esquissent une explication basée sur la propagation de perturbations de longueur d'onde 10^3 km à partir des éléments convectifs. Cependant, les deux méthodes ayant une constante de temps très courte, il n'est pas impensable que des effets de *distorsion aléatoire* de l'image due à l'atmosphère terrestre affectent les résultats.

Le spectre de puissance à deux dimensions (fréquences spatiale et temporelle) a été déterminé par Lévy (**4**) à l'aide d'un excellent film de granulation obtenu au Pic du Midi. Dans ce travail, les effets de distorsion atmosphérique évoqués ci-dessus se sont révélés très gênants et limitent la signification des résultats aux hautes fréquences spatiales. L'auteur confirme la présence d'une faible composante oscillatoire dans les fluctuations de brillance de la photosphère, composante dont l'existence probable a été annoncée par Edmonds et McCullough (**5**).

Rösch (RG **10**) a présenté de remarquables images de la granulation obtenues au Pic du Midi lors de l'éclipse partielle du 20 Mai 1966. On pourra obtenir le 'profil instrumental' de la combinaison atmosphère-télescope grâce à l'étude du bord lunaire et établir ainsi une photométrie plus rigoureuse des granules.

A Sacramento Peak, Beckers (RG **10**) a étudié les fluctuations de brillance photosphérique d'une échelle plus grande que la granulation. Il obtient un spectre de puissance aux basses fréquences spatiales qui ne montre aucune fréquence privilégiée, contredisant ainsi les résultats anciens de Stuart et Rush, et suggérant que la supergranulation n'a pas d'effet sur le spectre continu photosphérique.

En revanche le phénomène cinématique de supergranulation a pu être corrélé avec d'autres structures. Simon et Leighton (**6**) ont montré que les détails brillants du réseau des spectro-héliogrammes K_{232}, ainsi que les champs magnétiques associés, tendent à apparaître aux frontières des cellules de la supergranulation. Celle-ci serait le résultat d'une convection non-stationnaire dont les éléments refouleraient le champ magnétique à leur périphérie selon le mécanisme proposé par Parker. Simon (RG **5**) a mis en évidence les vitesses verticales faibles dans la photosphère aux frontières des supergranules et a trouvé (RG **10**) que les granules sont entraînés dans le mouvement général de la photosphère associé à la supergranulation. Quelques observations de la supergranulation ont été obtenues à Meudon (**7**).

L'étude du champ de vitesses verticales dans l'atmosphère solaire a été étendu à la photosphère profonde par Edmonds, Michard et Servajean (**8**) qui ont étudié sur un film pris par Evans à Sacramento Peak les fluctuations de vitesse, intensité centrale et largeur équivalente pour la raie 5052 C I, en corrélation avec d'autres raies de niveaux plus superficiels. Outre les oscillations, une importante composante 'convective' est présente dans les couches profondes. Les mesures spatio-temporelles de vitesse radiale effectuées sur les films de Sac'Peak ont été analysées à nouveau par P. Mein (**9**) qui a obtenu des spectres de puissance à deux dimensions directement comparables aux diagrammes diagnostiques de la théorie des ondes. Cette analyse confirme la séparation en plusieurs composantes du champ de vitesses; elle montre l'absence de propagation horizontale des oscillations, question que l'auteur discute par une méthode nouvelle.

Edmonds étudie à nouveau sur plusieurs séries de mesures le problème des corrélations entre vitesses verticales et granulation (**10**). D'autre part il examine les écarts à une loi gaussienne des fonctions de distribution des fluctuations de brillance et de vitesse de la photosphère. Ces écarts sont faibles mais significatifs (**11**). Selon cet auteur, les variations des propriétés statistiques des structures fines solaires selon l'échantillon considéré excèdent les limites permises par divers critères de signification, et l'on doit envisager des variations locales ou temporelles de ces propriétés.

Dans ce sens, Orrall (**12**) a montré que les propriétés des oscillations photosphériques sont pratiquement les mêmes dans les régions brillantes du réseau K_{232} (autre que les plages proprement dites) que dans les régions normales. Ultérieurement (**13**), il a étudié les vitesses radiales dans la raie K_3 elle-même, obtenant leur spectre de puissance qu'il compare avec les données relatives aux raies de profondeur de formation supérieure.

Les structures chromosphériques dans les raies de Balmer et H et K sont aussi discutées par Sivaraman et Vainu Bappu (non publié) et par Dodson-Prince et Mohler (RG **10**).

L'étude de la structure fine des champs magnétiques dans l'atmosphère solaire *hors des régions actives* a été poursuivie par plusieurs auteurs. Vasiljeva (**14**) étudie ces structures du point de vue statistique: fluctuations de vitesses et fluctuations magnétiques ne sont pas corrélées. Krat et Vjalshin comparent les champs et les vitesses en H β (**15**). A Kitt Peak, Livingston a mesuré simultanément la structure du champ en H α et 6569 Fe I; pour le domaine $5 < H < 100$ gauss les champs photosphérique et chromosphérique sont parfaitement corrélés et égaux. Des observations à très haute résolution de Beckers et Schröter confirment que, loin des régions actives, le champ fluctuant est concentré en points isolés de 3 à 5″ ayant un champ de plusieurs dizaines de gauss, non corrélé aux vitesses radiales ou à la granulation photosphérique (**16**).

Une attention particulière a été portée aux champs polaires. Severny décrit ainsi ses résultats (**17, 18, 19**): 'The examination of general magnetic field of the Sun carried out with high resolution (2″ × 5″) showed that there exists no coherent field. General m.f. of the Sun consists of a number of small (5″ of arc and less) elements of different polarity, field strengths and size. The net (mean) magnetic field on S–pole were absent since the summer 1963 up to September 1964, when weak m.f. of N–polarity appeared (+ 1, 1 gauss). On the northern polar cap and in northern hemisphere the field of S–polarity predominated for this period with the mean strength 4–5 gauss. The net magnetic flux at N–pole and in N–Hemisphere was negative due to the larger sizes of S–elements in N–Hemisphere, so that the ratio of these fluxes were:

$$F_{-}(\mathrm{N}) : F_{+}(\mathrm{N}) : F_{-}(\mathrm{S}) : F_{+}(\mathrm{S}) = 2 : 0{,}5 : 1 : 1.$$

'In 1965 this ratio was variable and the net flux for the whole disk was sometimes negative or positive. The observed distribution of general m.f. with latitude cannot be described by a dipole field or by the field of uniformly magnetized sphere.'

Cette description est pleinement en accord avec les résultats de Livingston (**20**), Grigorjev et Stepanov (**21**) et Stenflo (**22**).

Bumba *et al.* (**26**) ont utilisé les magnétogrammes du Mount Wilson pour calculer comment le Soleil apparaîtrait en tant qu'étoile variable magnétique vue dans son plan équatorial.

Des revues de la question des structures fines de l'atmosphère solaire ont été publiées par Michard (**23**) (RG **6**) et par Noyes (RG **5**). Dans la référence (RG **5**) on trouvera aussi d'importantes mises au point théoriques par Spiegel, Böhm, Moore, Lighthill, Rybicki. On notera aussi la revue de Pikelner (**24**) et, sur la question des champs magnétiques, celle de Severny (**25**).

Dans le présent rapport, les travaux théoriques sont plus spécialement traités dans un paragraphe ultérieur.

(*b*) *Profils et largeurs équivalentes des raies de Fraunhofer*

Les problèmes évoqués dans les paragraphes (*b*) (*c*) (*d*) de ce rapport sont évidemment étroitement liés. Nous n'avons pu éviter un certain arbitraire dans la répartition des travaux cités entre ces trois rubriques.

De Jager et Neven (**1**) ont terminé leur programme d'enregistrement au spectromètre du Jungfraujoch des profils de 50 raies environ du proche IR, observées pour 6 positions sur le disque, de $\mu = 0{\cdot}30$ à $\mu = 1$. Les mêmes auteurs et Namba ont étudié le profil de 10 830 He I sur la photosphère normale (**2**). Les résultats sont interprétés à l'aide d'un modèle hétérogène de la chromosphère. Le délicat problème de la correction de l'effet du profil instrumental est discuté par De Jager et Neven (**3**).

Les données sur les profils du multiplet 10 700 C I de $\mu = 1$ à $\mu = 0{\cdot}3$ publiées par ces mêmes auteurs (**4**) leur ont servi d'exemple pour exposer une méthode d'analyse 'empirique' des profils qui permet en principe de déterminer un paramètre de macroturbulence (modèle à 2 colonnes), l'abondance, l'amortissement, la fonction-source et la microturbulence (**5**). La macroturbulence est tirée de l'asymétrie des profils par une procédure un peu sommaire dont la validité a été discutée par Kulander et Jefferies (**6**). Les longueurs d'onde précises de ce même multiplet de C I ont été déterminées à Göttingen par Brückner.

Olson (**20**) a discuté les asymétries observées des raies de Fe I, O I et C I (données de Jager et Neven). Il montre que les oscillations photosphériques ne jouent qu'un rôle négligeable dans l'asymétrie. Un modèle de vitesses convectives variables avec la profondeur rend compte de l'ensemble des observations.

Les profils des raies de Balmer $H\alpha$ à $H\delta$ ont été mesurés, au centre du disque et au bord ($\sin \theta = 0{\cdot}99$), et discutés par Kuli-Zade (**7**) qui obtient un bon accord entre observation et théorie. Le même auteur (**8**) a effectué des observations similaires pour diverses raies très fortes du spectre de Fraunhofer: raies D de Na I, b de Mg I, 4227 Ca I, 3944 et 3961 Al I, H et K (voir aussi les références (**9**) et (**10**). Khetsuriani (**11**) a publié de nouvelles mesures photoélectriques des profils de 7 raies de Paschen ainsi que de la raie Brackett γ pour plusieurs positions sur le disque.

A Göttingen, Bubke a mesuré photoélectriquement en fonction de $\sin \theta$ les intensités centrales des raies du multiplet $a^3F - y^3F_0$ de Fe I (**12**). Il en déduit la variation de la température d'excitation avec la profondeur optique, température qui passe par un minimum de 3800 à 3900° vers $\log \tau_0 = -5{\cdot}7$. Voigt a aussi mesuré les intensités centrales de raies de Ba I et II, Sr I et II, Eu I et II et des résultats préliminaires ont été publiés (**13**).

Une nouvelle courbe de croissance pour le centre du disque solaire a été déterminée par Cowley et Cowley (**14**) qui emploient les forces d'oscillateur de Corliss-Bozman et les largeurs équivalentes de 612 raies déterminées photoélectriquement à l'Observatoire MacMath.

E. A. Müller, Baschek et Holweger signalent qu'un programme d'observations centre-bord et d'analyse des profils de raies permises et interdites de l'oxygène est en voie d'achèvement: les auteurs cherchent à déterminer le modèle thermique et le modèle de turbulence qui rendent compte au mieux des observations. Une comparaison entre modèles et observations des largeurs équivalentes de quelques raies de Ti a été effectuée par Nissen (**15**).

L'étude des bandes moléculaires du spectre de Fraunhofer a été l'objet de bien des travaux au cours des trois dernières années. L'article de Schadee (**16**) sera très utile car il rassemble sous une forme claire et complète la plupart des notions physiques et des méthodes astrophysiques nécessaires à la discussion des spectres moléculaires dans le Soleil. En outre il passe en revue les identifications de nombreux composés dans le spectre de la photosphère *normale* en tenant compte de critères basés sur les intensités probables des raies correspondantes.

Les largeurs équivalentes de raies de CH au centre du disque ont été mesurées par Ruzickova-Topolova (**17**) qui détermine une température de rotation. Withbroe (**18**) a effectué, toujours pour la bande G de CH, un travail plus étendu comportant la mesure photoélectrique des largeurs équivalentes, mais aussi des profils de raies bien isolées, en fonction de la position sur le disque. Les observations peuvent être interprétées dans l'hypothèse de l'ETL à l'aide d'un modèle thermique dû à Elste et d'un modèle de turbulence.

Poletto et Rigutti ont étudié (**19**) deux bandes du système rouge de CN: ils trouvent une température de rotation et tentent une estimation de la force d'oscillateur de la transition électronique correspondante. A Kodaikanal, Nirupama Raghavan a observé photoélectriquement pour divers $\sin \theta$ les profils de raies choisies de CN et C_2: il analyse les largeurs en termes de microturbulence et trouve une vitesse presque isotrope de 3 km s^{-1} (non publié).

(*c*) *Théorie du spectre de Fraunhofer*

Une grande partie des travaux dans ce domaine continuent d'être centrés sur le problème des écarts à l'équilibre thermodynamique local (ETL), qui peut être considéré de deux points de vue différents:

– recherche empirique de contradictions auxquelles pourrait conduire l'approximation de l'ETL, et influence de cette approximation sur la détermination de diverses caractéristiques de l'atmosphère (en particulier les abondances);

– analyse théorique directe grâce à la solution simultanée des équations de l'équilibre statistique et des équations de transfert; la comparaison ultérieure aux observations permet de tester les diverses hypothèses de la théorie et le choix du modèle d'atmosphère.

Cette dernière approche est utilisée dans le cas des raies de Na I par Johnson (**1**) qui montre que l'écart à l'ETL est très faible dans la photosphère: les températures d'excitation trouvées empiriquement par le même auteur ne sont pas explicables par la théorie. Le problème a été repris par Chamaraux (**2**) qui obtient un bon accord entre profils calculés et profils observés pour les raies D. Selon Mugglestone (RG **8**) le centre de ces raies est très influencé par les écarts à l'ETL.

L'analyse empirique des profils des mêmes raies D permet à Curtis (RG **8**) de déterminer les fonctions-sources dans la raie et le continuum qui se révèlent égales pour $\tau\,(5000) > 0{\cdot}01$. Par ailleurs Curtis détermine un paramètre d'élargissement Doppler très faible, explicable par le seul élargissement thermique en l'absence de toute turbulence appréciable.

Une analyse similaire a été faite pour le multiplet *b* de Mg I par Hénoux (**3**) d'après les profils observés en fonction de la position sur le disque par Waddell. La fonction-source est bien déterminée, mais le coefficient d'élargissement Doppler n'est pas en accord avec celui de Curtis. D'autre part Hénoux a mesuré les profils dans la basse chromosphère ($h < 1500$ km) et cherche à les interpréter avec la fonction-source obtenue sur le disque: les contradictions rencontrées ne pourraient être levées que par un modèle hétérogène de la basse chromosphère.

A propos de ces études empiriques des multiplets, notons qu'une controverse théorique sur la question de l'égalité des fonctions-sources dans les raies d'un multiplet a opposé Waddell et Athay (cf. les références dans le chapitre sur la Chromosphère).

Pour en revenir au traitement théorique, notons l'analyse des raies de résonance de Ca II et Mg II par Dumont (**4**), qui d'autre part perfectionne le traitement numérique du transfert hors ETL pour tenir compte de l'amortissement (**5**). Le cas de l'hydrogène est traité par Cuny (**6**) ainsi que par plusieurs autres auteurs (RG **7**) dans le cadre d'une analyse théorique de l'atome à 3 niveaux.

La discussion empirique de l'approximation de l'ETL pour les raies métalliques a été considérée par de nombreux auteurs dans le cadre des recherches sur les abondances (cf. ci-dessous). Ici nous noterons plus spécialement l'établissement par Holweger (**7**) d'un modèle empirique de la photosphère et de la basse chromosphère basé sur les intensités centrales des raies de Fraunhofer: le modèle obtenu est à température constamment décroissante jusqu'à $\tau\,(5000) \simeq 10^{-6}$ ($T = 3900°$). L'emploi d'observations et de valeurs *f* récentes conduit à des résultats compatibles avec l'approximation de l'ETL. L'étude des largeurs équivalentes conduit aussi Teplitskaya (**8**) et Pagel (RG **8**) à des températures d'excitation très voisines des températures électroniques adoptées.

Mitropolskaya (**9, 10, 11**) applique la théorie de l'élargissement par effet Stark et effet de pression à ses observations des profils des raies de l'hydrogène Hα et Hβ, Paschen β et Brackett γ. Un accord raisonnable est obtenu avec le modèle classique de de Jager; le cas des raies de Balmer dans les facules est aussi traité.

Pande et Sitnik (**12**) et Pande (**13, 14**) ont étudié les raies des bandes de CO dans la photosphère et insistent sur le fait que les molécules CO sont concentrées au voisinage du minimum de température. Les observations nouvelles (**14**) conduisent à une température de rotation pour CO.

Rachkovsky a poursuivi l'étude du transfert dans les raies en présence d'un champ magnétique (15). Le calcul numérique (16) conduit à des écarts significatifs entre les résultats de la théorie de Unno et de celle de l'auteur.

La discussion des profils des raies de Fraunhofer peut être améliorée ou simplifiée de diverses manières. On notera une discussion de la représentation du profil des ailes des raies fortes par Gussman (17) qui critique la formule à deux paramètres de Mattig et Schröter et lui préfère une formule à un seul paramètre. Le même auteur (18) présente une modification de la méthode des fonctions de poids permettant de simplifier son emploi.

Enfin, bien qu'il porte sur le cas stellaire, nous citerons ici un travail de Traving (19) qui étudie l'effet sur la courbe de croissance d'une épaisseur optique $\Delta\tau$ *finie* des éléments turbulents de l'atmosphère: il s'agit là d'une tentative bienvenue pour dépasser les concepts classiques de micro- et macroturbulence où $\Delta\tau$ est supposée nulle ou infinie.

(*d*) *Abondances des éléments*

Dans la mesure où un effort plus grand est fait pour la mesure des forces d'oscillateurs, la détermination de valeurs plus précises des abondances des éléments dans le Soleil devient possible et est activement poursuivie.

Ainsi les valeurs *f* de Corliss, Bozman et Warner ont été employées par Goldberg *et al.* (1) à une nouvelle estimation de l'abondance de Fe I. Leftus (2, 3) étudie en détail le cas du titane. Moyennant l'emploi des données les plus récentes sur les forces d'oscillateurs, les spectres de Ti I et Ti II conduisent aux mêmes abondances, et l'approximation de l'ETL semble tout à fait acceptable.

Nous avons reçu d'Aller un rapport sur un programme important de détermination des abondances dans le Soleil et certains types d'étoiles, programme auquel participent plusieurs astronomes. Les donnécs solaires sont des profils mesurés, avec double passage sur le réseau, par Mitchell et Mohler (4) aux Observatoires MacMath et du Mount Wilson. Le programme comporte une étude approfondie des causes d'erreurs d'ordre physique (valeurs *f*) et astrophysique: cf. Aller *et al.* (5) et la revue détaillée de Aller (6). Mugglestone et O'Mara (7) ont montré que l'élargissement par amortissement (effet Stark et effets de pression) peut jouer un rôle notable dans l'estimation des abondances, ce qui conduit à rejeter la méthode des fonctions de poids. Les auteurs préfèrent le calcul complet des profils à l'aide de deux modèles thermiques extrêmes avec variation de l'abondance jusqu'à ce que le profil calculé s'accorde au profil observé. Pour le cas des raies faibles où l'on rencontre souvent le problème des 'blends' la méthode comporte également le calcul d'un spectre 'synthétique' couvrant l'ensemble du 'blend'. Aller note que 'the effects of uncertainties in the *f*-values mask the errors which may arise from the uncertainties in the model or theoretical treatment'. Ceci conduit à un programme de calculs quantiques des probabilités de transition, calculs très difficiles mais possibles pour certains métaux comme Al et Mg: cf. Chapman *et al.* (8). Aller communique aussi des valeurs préliminaires d'abondances pour Ru, In, Au.

Les abondances des terres rares ont été étudiées par A. Righini et Rigutti (9) ainsi que par Wallerstein (10). De nouvelles mesures de largeurs équivalentes conduisent Grevesse (11) à réviser légèrement l'abondance admise pour le strontium.

Le lithium, représenté seulement par sa faible raie de résonance fortement 'blendée' à 6708 Å, pose un problème intéressant discuté par Lynds (12). Cependant Dubov (13) a remarqué que la raie 6708 est très renforcée dans les taches solaires, ce qui permet une détermination meilleure de l'abondance. La question a été reprise par Schmahl et Schröter (14); une analyse soignée montre que l'abondance de Li est supérieure à la valeur admise antérieurement; le rapport isotopique Li^6/Li^7 est de l'ordre de 5%.

E. A. Müller nous annonce les premiers résultats d'un programme étendu de recherches sur les abondances poursuivi en collaboration avec Mutschlecner. Le cas du nickel a été traité par l'analyse des largeurs équivalentes observées en plusieurs positions sur le disque: une même abondance représente correctement les spectres obtenus du centre au bord et les raies de divers potentiels d'excitation. D'autre part le même auteur passe en revue les problèmes des abondances solaires dans deux publications récentes (RG **4**, RG **6**).

Sur la base d'observations et de données de laboratoire nouvelles, L. Herzberg, Delbouille et Roland montrent que les raies de $C^{13}N$ sont indécelables dans le spectre de Fraunhofer et obtiennent une limite inférieure du rapport C^{12}/C^{13} (**16**).

Gökdoğan et Pecker (**15**) ont signalé que deux multiplets de Fe I conduisent à des courbes de croissance distinctes, ce qui jetterait quelques doutes sur la détermination des abondances par la méthode classique. L'amélioration des observations et des forces d'oscillateurs disponibles tend à réduire les discordances que Pecker et ses collaborateurs attribuaient aux écarts à l'ETL: mais la mise en garde demeure salutaire!

(*e*) *Modèles et théories de la photosphère et de la zone convective*

Il n'existe pas actuellement de 'modèle' de la photosphère susceptible de rendre compte de l'ensemble des faits d'observation. A fortiori n'existe-t-il pas de modèle qui aurait *en outre* l'avantage d'être basé sur une interprétation physique complète et cohérente. Il est vrai que si un tel modèle existait la physique de l'atmosphère solaire normale serait 'achevée'!

Heintze *et al* (**1**) ont proposé un nouveau modèle empirique, le 'Utrecht Reference Model of the solar photosphere and low chromosphere' en vue de faciliter la comparaison et la discussion de résultats obtenus par l'analyse des divers types d'observations. Il est certain que l'emploi d'un tel modèle de référence présente des avantages considérables à condition que ce modèle soit fréquemment révisé. Une première révision sera préparée à l'issue de discussions prévues pour 1967 d'une part à Utrecht (International Study Week on the Quiet Photosphere), d'autre part au sein de la Commission 12.

L'équilibre radiatif dans la photosphère a été discuté par Waddell sur la base des observations du continuum solaire, en considérant le flux d'énergie et aussi le flux de photons (**2**). Gingerich a calculé un nouveau modèle en équilibre radiatif de la photosphère en tenant compte soigneusement de l'effet de blanketing (**14**): des écarts notables et inexpliqués subsistent avec les observations de la courbe d'énergie solaire. Frisch (**15**) étudie l'influence des fortes raies de Fraunhofer sur l'équilibre radiatif dans les couches supérieures où il n'y a plus équilibre thermodynamique local.

Cependant les contributions les plus importantes aux problèmes de transfert dans la photosphère portent sur l'interprétation des fluctuations locales du rayonnement émergent. Wilson (**3**, **4**) traite cette question en vue de l'interprétation des données sur la variation centre-bord des écarts quadratiques moyens de brillance dans la granulation en termes de fluctuations de température. L'analyse du même problème par Edmonds (**16**) conduit à des fluctuations thermiques qui croissent régulièrement avec la profondeur et atteignent une valeur quadratique moyenne de 450° (à τ constant) vers $\tau = 1{\cdot}3$ ou 740° à profondeur géométrique constante. Dans un article plus récent (**5**) Wilson montre que les *valeurs moyennes* de la fonction-source S dans la photosphère déduites des valeurs moyennes de l'intensité émergente en appliquant l'équation de transfert de l'atmosphère homogène (négligeant la granulation) ne sont pas correctes, parce que les relations entre S et T, et surtout entre τ et T, ne sont pas linéaires. Une analyse entièrement cohérente du spectre continu solaire compte tenu de ses fluctuations locales reste à faire.

Des méthodes mathématiques pour l'analyse du transfert de rayonnement dans un milieu fluctuant ont été développées par Rybicki (**6**). Il établit les équations qui lient les fluctuations d'opacité et d'énergie locale (décrites par leurs fonctions de corrélation) aux fluctuations du champ de rayonnement, en particulier du rayonnement émergent. La méthode s'applique donc au cas où une partie de l'énergie n'est pas sous forme radiative. La théorie suggère que certaines fluctuations spatiales à grande échelle du rayonnement émergent sont liées à l'épaisseur de la couche convective plutôt qu'à l'échelle adoptée pour la convection.

De nouveaux efforts sont déployés pour obtenir une théorie hydrodynamique meilleure de la zone convective et on consultera avant tout dans ce domaine les mises au point de Spiegel et Böhm (RG **5**). Dans un travail non encore publié Böhm a développé une méthode pour analyser les modes convectifs compte tenu de l'influence de la viscosité et de la conductivité thermique liées à la turbulence, ainsi que des échanges radiatifs; la méthode permet une évaluation du spectre d'énergie des modes convectifs.

De nombreux travaux ont été entrepris en vue d'expliquer les observations des oscillations photosphériques et d'analyser la propagation et la dissipation dans l'atmosphère solaire des divers modes possibles.

Kato (**7**) a calculé le spectre du champ acoustique produit dans une zone de convection turbulente et modifié par son transfert dans l'atmosphère stable. Le spectre obtenu a un pic au voisinage de la fréquence critique. Par ailleurs le même auteur a analysé la réponse d'une atmosphère stable à une perturbation ayant une source ponctuelle, perturbation harmonique (**8**) ou impulsive (**9**).

Uchida (**10**) a étudié la propagation des divers modes d'ondes gravitationnelles et aérodynamiques dans une atmosphère non isotherme et trouve que certains modes de période 300 s sont piégés dans la couche voisine du minimum de température. Il décrit d'autre part (**11**) l'excitation des oscillations par les mouvements des granules. Enfin Uchida étudie (**12**) la propagation dans la couronne de perturbations magnéto-hydrodynamiques: l'intersection du front de l'onde rapide avec la surface solaire parcourt cette dernière à une vitesse de 10^3 km s^{-1} et peut être identifiée aux perturbations se propageant à partir des éruptions solaires (films de Moreton, Lockheed Solar Observatory).

Souffrin (**13**) a étudié le 'filtrage' par une atmosphère convectivement stable du champ de vitesse aléatoire qui est imposé à sa base par une couche convective sous-jacente. L'atmosphère se comporte comme un filtre coupant les basses fréquences et le spectre des oscillations photosphériques peut être expliqué comme conséquence de ce filtrage. La dissipation radiative s'oppose à l'existence d'ondes de gravité dans la photosphère.

Enfin Meyer et Schmidt (RG **10**) ont analysé la réponse de l'atmosphère stable à l'excitation par un 'granule', considéré comme un soulèvement local de la couche limite inférieure. Pour un choix convenable des paramètres décrivant ce granule, il y a excitation des modes d'oscillations propres de l'atmosphère; la solution numérique permet d'interpréter les oscillations observées.

INSTRUMENTS

On notera avec satisfaction la multiplication des grands instruments solaires modernes.

Dans l'île de Capri le Fraunhofer Institut a installé un télescope d'une conception révolutionnaire dont on trouvera la description par Kiepenheuer et Mattig dans la référence (RG **10**). Un magnétographe photoélectrique permettant de mesurer le vecteur champ magnétique lui est associé (Deubner, RG **10**).

A l'Observatoire Astrophysique de Crimée le magnétographe a été adapté à la mesure des champs transverses (Bruns *et al.* (**1**)) ainsi qu'à la mesure des composantes longitudinales pour deux raies. Les mêmes facilités existent à l'IZMIRAN (Joschpa, Obridko (**2**)). D'autre part le spectromètre à double passage pour la mesure des profils des raies de Fraunhofer à Poulkovo est décrit par Karpinsky (**3**).

Livingston a établi un remarquable magnétographe en association avec le télescope McMath de l'Observatoire de Kitt Peak. Deux raies peuvent être mesurées simultanément; avec 5250 Fe I le bruit est seulement de 0·5 gauss pour un trou d'exploration de 1″ × 1″ et une constante de temps de 1 s. L'appareil peut être aussi adapté à la mesure des trois composantes du vecteur champ.

A Kodaikanal également un magnétographe photoélectrique a été construit par Battacharya.

A Meudon un laboratoire de spectroscopie solaire à télescope horizontal (D = 40 cm, F = 13 m) construit par G. Wlérick est entré en service en 1965. Le même auteur dirige la construction de la nouvelle tour solaire (D = 60 cm, F = 45 m) qui pourra être utilisée en 1967–68. Un nouveau coronomètre photoélectrique amélioré, construit par P. Charvin, doit être installé à l'Observatoire de Nice.

Enfin à Sacramento Peak la nouvelle tour solaire sera la première à comporter un télescope dans le vide, ainsi que nombre de particularités révolutionnaires. Sa construction est activement poursuivie sous la direction de Dunn et Evans.

Nous avons aussi été informé de nouvelles réalisations instrumentales pour l'étude du Soleil à Edinburgh (Brück), à Kanzelhöhe (Mathias) et à Arosa où un nouveau coronographe a été installé (Waldmeier).

BIBLIOGRAPHIE

Références Générales

1. *The solar spectrum*, Ed. C. de Jager, D. Reidel Publ. Co., Dordrecht, 1965.
2. *Stellar and solar magnetic fields*, Ed. R. Lüst, IAU Symposium no. 22, North Holland Publ. Co., Amsterdam, 1965.
3. *Astronomical observations from Space vehicles*, Ed. J. L. Steinberg, IAU Symposium no. 23, in *Ann. Astrophys.*, **27, 28**; also in book from Publ. CNRS, Paris, 1965.
4. *Abundance determinations in stellar spectra*, Ed. H. Hubenet, IAU Symposium no. 25, Acad. Press, London and New York, 1966.
5. *Cosmical gas dynamics*, Ed. R. N. Thomas, IAU Symposium no. 28, Acad. Press, London and New York, 1967.
6. *Solar physics*, Ed. J. Xanthakis, John Wiley and Sons Ltd., London, in press.
7. First Harvard-Smithsonian Conference on stellar atmospheres, Cambridge, Mass., 1964, *Smithson. Inst. astrophys. Obs.*, spec. Rep., no. 167, 1964.
8. Second Harvard-Smithsonian Conference on stellar atmospheres, Cambridge, Mass., 1965, *Smithson. Inst. astrophys. Obs.*, spec. Rep., no. 174, 1965.
9. *Campi magnetici solari e la spettroscopia ad alta risoluzione*, Ed. M. Cimino, Barberi, Firenze, *sous presse*.
10. *The fine structure of the solar atmosphere*, Ed. K. O. Kiepenheuer, Forschungsberichte Deuts. Forschungsgemeins., F. Steiner Verl. Wiesbaden, 1966.
11. *Solar corona and corpuscular radiation in interplanetary space*, S. K. Vsekhviatsky *et al.*, Kiev University, 1965.
12. Billings, D. E. *A guide to the solar corona*, Acad. Press, London, New York, 1966.

Le spectre continu solaire

1. Labs, D. *Z. Astrophys.*, sous presse.
2. Peyturaux, R. 1964, *C.R. Acad. Sci.*, Paris, **258**, 1159.
3. Murashova, M. S., Sitnik, G. F. 1963, *Astr. Zu.*, **40**, 819.
4. Sitnik, G. F. 1965, *Astr. Zu.*, **42**, 59.
5. Sitnik, G. F. 1965, *Astr. Cirk.*, no. 344, 1.
6. Sitnik, G. F. 1965, *Astr. Zu.*, **42**, 996.

7. Makarova, E. A. 1965, *Astr. Zu.*, **42**, 681.
8. Makarova, E. A. 1964, *Astr. Zu.*, **41**, 288.
9. Lambert, D. L., Willstrop, R. V. 1965, *Observatory*, **85**, 124.
10. Houtgast, J. 1965, *Koninkl. Nederl. Akad. Wetensch.*, Amsterdam Proceed., Series B, **68**, no. 5.
11. Murcray, F. H., Murcray, D. G., Williams, W. J. 1964, *Appl. Opt.*, **3**, 1373.
12. Beer, R. 1966, *Nature*, **209**, no. 5029, 1226.
13. Léna, P. 1966, *Ann. Astrophys.*, **29**, 361.
14. Noyes, R., Gingerich, O., Goldberg, L. 1966, *Astrophys. J.*, **145**, 344.
15. Blamont, S., Bonnet, R. 1966, *C.R. Acad. Sci.*, Paris, **262**, 3152.
16. Kodaira, K. 1965, *Z. Astrophys.*, **60**, 240.
17. Mouradian, Z. 1965, *Ann. Astrophys.*, **28**, 805.
18. Kristenson, H. 1964, *Mon. Not. R. astr. Soc.*, **128**, 13.
19. Heintze, J. R. W. 1965, *Rech. astr. Obs. Utrecht*, **17**.
20. Mulders, G. F. W., Slaughter, C. D. 1965, *Publ. astr. Soc. Pacif.*, **77**, 295.
21. Appenzeller, I., Schröter, E. H. *Astrophys. J.*, sous presse.
22. Cowley, C. 1965, *Astr. J.*, **70**, 671.
23. Delache, Ph. *Astrophys. J.*, (soumis à).
24. Delache, Ph. 1966, *Ann. Astrophys.*, **29**, 109.
25. Bode, G. 1965, *Veröff. Inst. Theoret. Phys. Sternw.*, Kiel.

Atlas et Tables du Spectre de Fraunhofer. Identifications

1. Ahmad Kiasatpoor 1962, *Georgetown Obs. Monograph*, no. 19.
2. Swensson, J. W., Risberg, G. 1966, *Ark. Fys.*, **31**, no. 16.
3. Swensson, J. W. 1966, *Z. Astrophys.*, **64**, 11.
3 **bis.** Swensson, J. W. *Bull. Acad. Roy. Belg*, (sous presse.)
4. Grevesse, N. 1966, *Ann. Astrophys.*, **29**, 365.
5. Kodaira, K. 1964, *Z. Astrophys.*, **60**, 24.
6. Swings, J. P. 1965, *Ann. Astrophys.*, **28**, 703.
7. Swings, J. P. 1966, *Ann. Astrophys.*, **29**, 371.
8. Mitchell, W. E., Jr., Mohler, O. C. 1965, *Astrophys. J.*, **141**, 1126.
9. Swensson, J. W. 1960, *Ark. Fys.*, **32**, no. 25, 463.
10. Pottasch, S. R. 1964, *Space Sci. Rev.*, **3**, 816.
11. Goldberg, L., Parkinson, W. H., Reeves, E. M. 1965, *Astrophys. J.*, **141**, 1293.
12. Black, W. S., Booker, D., Burton, W. M., Jones, B. B., Shenton, D. B., Wilson, R. 1965, *Nature*, **206**, no. 4985, 654.
13. Burton, W. M., Wilson, R. 1965, *Nature*, **207**, no. 4992, 61.

Longueurs d'onde. Rotation. Circulation

1. Aslanov, I. A. 1963, *Astr. Zu.*, **40**, 1036.
2. Aslanov, I. A. 1964, *Uchen. Zap. Leningrad. gos. Univ.*, Ser. mat. Nauk, **21**, no. 38, 53.
3. Aslanov, I. A. 1964, *Acad. Sci. Az. SSR*, Bakou.
4. Melnikov, O. A., Zhuravlev, S. S., Aslanov, I. A., Kuliev, D. M., Salmanzade, R. Kh. 1964, *Trans. Shamakha astr. Obs.*, **3**, 63.
5. Melnikov, O. A., Zhuravlev, S. S., Aslanov, I. A., Kuliev, D. M., Salmanzade, R. Kh. 1964, *Uchen. Zap. Leningrad. gos. Univ.*, Ser. mat. Nauk, **21**, no. 38, 27.
6. Melnikov, O. A. 1964, *Izv. glav. astr. Obs. Pulkove*, **23**, no. 175.
7. Ward, F. 1964, *Astr. J.*, **69**, 562.
8. Ward, F. 1964, *Pure appl. Geophys.*, **58**, 157.
9. Ward, F. 1965, *Astrophys. J.*, **141**, 534.
10. Starr, V. F., Gilman, P. A. 1965, *Astrophys. J.*, **141**, 1119.
11. Starr, V. P., Gilman, P. A. 1965, *Tellus*, **17**, 334.
12. Sakurai, T. 1966, *Publ. astr. Soc. Japan*, **18**, 174.
13. Rubashev, B. M. 1965, *Izv. glav. astr. Obs. Pulkove*, **24**, no. 178, 41.

La Photosphère

(a) Granulation. Structures cinématiques et magnétiques

1. Bray, R. J., Loughhead, R. E. *The solar granulation*, International Astrophysics Series, Chapman and Hall, London (in press).
2. Vassiljeva, G. Y., Yudina, I. V. 1965, *Soln. Dann.*, no. 3, 58.
3. Krat, V. A., Petrova, N. N. 1965, *Soln. Dann.*, no. 10, 48.
4. Lévy, M. *Ann. Astrophys.* (soumis aux).
5. Edmonds, F. N., Jr., McCullough, J. R. 1966, *Astrophys. J.*, **144**, 754.
6. Simon, G. W., Leighton, R. B. 1964, *Astrophys. J.*, **140**, 1120.
7. Lévy, M. 1965, *C. R. Acad. Sci.*, Paris, **260**, 806.
8. Edmonds, F. N., Jr., Michard, R., Servajean, R. 1965, *Ann. Astrophys.*, **28**, 534.
9. Mein, P. 1966, *Ann. Astrophys.*, **29**, 153.
10. Edmonds, F. N., Jr. 1966, *Astrophys. J.*, **144**, 733.
11. Edmonds, F. N., Jr. 1967, *Solar Physics*, **1**, 5.
12. Orrall, F. Q. 1965, *Astrophys. J.*, **141**, 1131.
13. Orrall, F. Q. 1966, *Astrophys. J.*, **143**, 917.
14. Vasiljeva, G. Y. 1964, *Izv. glavn. astr. Obs. Pulkove*, **23**, no. 175, 28.
15. Krat, V. A., Vjalshin, G. F. 1965, *Izv. glavn. astr. Obs. Pulkove*, **24**, no. 178, 26.
16. Beckers, J. M., Schröter, E. H. Boulder AAS Meeting, October, 1966.
17. Severny, A. 1965, *Astr. Zu.*, **42**, 217.
18. Severny, A. 1966, *Izv. Krym. astrofiz. Obs.*, **35**.
19. Severny, A. 1966, *Izv. Krym. astrofiz. Obs.*, **37** (in press).
20. Livingston, W. C. 1966, *Scientific American*, **215**, no. 5, 54.
21. Grigorjev, V. M., Stepanov, V. E. 1966, Reports on Solar Physics, *Sib. IZMIR. Acad. Sci. USSR*, **32**.
22. Stenflo, J. O. 1966, *Observatory*, **86**, 73.
23. Michard, R. 1964, *Trans. IAU*, **12B**, 532.
24. Pikelner, S. B. 1966, *Usp. fiz. Nauk*, **88**, 505.
25. Severny, A. 1966, *Usp. fiz. Nauk*, **88**, 3.
26. Bumba, V., Howard, R., Smith, S. F. Proc. Goddard Magnetic Stars Symposium, 1965, in press.

(b) Profils et largeurs équivalentes des raies de Fraunhofer

1. De Jager, C., Neven, L. *Bull. astr. Inst. Netherl.* (sous presse).
2. De Jager, C., Namba, O., Neven, L. 1966, *Bull. astr. Inst. Netherl.*, **18**, 128.
3. De Jager, C., Neven, L. 1966, *Bull. astr. Inst. Netherl.*, **18**, 306.
4. De Jager, C., Neven, L. 1964, *Mém. Soc. R. Sci. Liège*, **9**, 151.
5. De Jager, C., Neven, L. 1967, *Solar Physics.*, **1**, 27.
6. Kulander, J. L., Jefferies, J. T. 1966, *Astrophys. J.*, **146**, 194.
7. Kuli-Zade, D. M. 1964, *Astr. Zu.*, **41**, 920.
8. Kuli-Zade, D. M. 1965, *Astr. Zu.*, **42**, 1022.
9. Kuli-Zade, D. M. 1964, *Vestnik Leningrad Univ.*, no. 19.
10. Kuli-Zade, D. M. 1964, *Soln. Dann.*, no. 11, 56.
11. Khetsuriani, T. S. 1965, Acad. Sci. Georg. SSR, **37**, no. 3, 563.
12. Bubke, O. 1965, *Z. Astrophys.*, **62**, 83.
13. Föppl, H. *et al.* 1964, Max Planck Inst. Phys. Astrophys. München, MPI–PA–23/64.
14. Cowley, C. R., Cowley, A. P. 1964, *Astrophys. J.*, **140**, 713.
15. Nissen, P. E. 1965, *Ann. Astrophys.*, **28**, 556.
16. Schadee, A. 1964, *Bull. astr. Inst. Netherl.*, **17**, 311.
17. Ruzickova-Topolova, B. 1965, *Bull. astr. Inst. Csl.*, **16**, 23.
18. Withbroe, G. L. *Astrophys. J.* (sous presse).
19. Poletto, G., Rigutti, M. 1964, *Z. Astrophys.*, **60**, 199.
20. Olson, E. C. 1966, *Astrophys. J.*, **143**, 904.

(*c*) *Théorie du spectre de Fraunhofer*

1. Johnson, H. R. 1965, *Ann. Astrophys.*, **27**, 695.
2. Chamaraux, P. *Ann. Astrophys.*, 1966 (sous presse).
3. Hénoux, J.-C. 1966, *Ann. Astrophys.*, **29**, 271.
4. Dumont, S. 1966, *C.R. Acad. Sci.*, Paris, **263**, 85.
5. Dumont, S. 1966, *C.R. Acad. Sci.*, Paris, **262**, 740.
6. Cuny, Y. 1964, *C.R. Acad. Sci.*, Paris, **258**, 3292.
7. Holweger, H. *Z. Astrophys.* (sous presse).
8. Teplitskaya, R. B. 1964, *Astr. Zu.*, **41**, 907.
9. Mitropolskaya, O. N. 1963, *Astr. Zu.*, **40**, 427.
10. Mitropolskaya, O. N. 1964, *Astr. Zu.*, **41**, 930.
11. Mitropolskaya, O. N. 1966, *Astr. Zu.*, **43**, 694.
12. Pande, M. Ch., Sitnik, G. F. 1965, *Astr. Zu.*, **42**, 1250.
13. Pande, M. Ch. 1966, *Astr. Zu.*, **43**, 407.
14. Pande, M. Ch. 1966, *Astr. Zu.*, **43**, 708.
15. Rachkovsky, D. N. 1963, *Izv. Krym. astrofiz. Obs.*, **30**, 267.
16. Rackhovsky, D. N. 1965, *Izv. Krym. astrofiz. Obs.*, **33**, 111.
17. Gussman, E. A. 1964, *Z. Astrophys.*, **59**, 66.
18. Gussman, E. A. *Z. Astrophys.* (sous presse).
19. Traving, G. 1964, *Z. Astrophys.*, **60**, 167.

(*d*) *Abondances des éléments*

1. Goldberg, L., Kopp, R. A., Dupree, A. K. 1964, *Astrophys. J.*, **140**, 707.
2. Leftus, V. 1965, *Bull. astr. Inst. Csl.*, **16**, 311.
3. Leftus, V. 1965, *Publ. astr. Inst. Csl.*, no. 51, 17.
4. Mitchell, W. E., Mohler, O. C. 1964, *Appl. Optics*, **3**, 467.
5. Aller, L. H., O'Mara, B. J., Little, S. 1964, *Proc. nat. Acad. Sci.*, **51**, 1238.
6. Aller, L. H. 1965, *Adv. Astr. Astrophys.*, **3**, 1.
7. Mugglestone, D., O'Mara, B. J. 1966, *Mon. Not. R. astr. Soc.*, **132**, 87.
8. Chapman, R. D., Clarke, W. H., Aller, L. H. 1966, *Astrophys. J.*, **144**, 376.
9. Righini, A., Rigutti, M. 1966, *Ann. Astrophys.*, **29**, 379.
10. Wallerstein, G. 1966, *Icarus*, **5**, 75.
11. Grevesse, N. 1966, *Ann. Astrophys.*, **29**, 287.
12. Lynds, C. R. 1965, *Astrophys. J.*, **142**, 396.
13. Dubov, E. E. 1964, *Izv. Krym. astrofiz. Obs.*, **32**, 26.
14. Schmahl, G., Schröter, E. H. 1965, *Z. Astrophys.*, **62**, 143.
15. Gökdoğan, N., Pecker, J.-C. 1964, *Ann. Astrophys.*, **27**, 417.
16. Herzberg, L., Delbouille, L., Roland, G. *Astrophys. J.* (sous presse).

(*e*) *Modèles et théories de la photosphère et de la zone convective*

1. Heintze, J. R. W., Hubenet, H., De Jager, C. 1964, *Bull. astr. Inst. Netherl.*, **17**, 442.
2. Waddell, J. 1965, *Astrophys. J.*, **142**, 326.
3. Wilson, P. R. 1964, *Astrophys. J.*, **139**, 929.
4. Wilson, P. R. 1964, *Astrophys. J.*, **140**, 1148.
5. Wilson, P. R. 1965, *Astrophys. J.*, **142**, 1195.
6. Rybicki, G. B. 1965, *Smithson. Inst. astrophys. Obs., spec. Rep.*, no. 180.
7. Kato, S. 1966, *Astrophys. J.*, **143**, 372.
8. Kato, S. 1966, *Astrophys. J.*, **143**, 893.
9. Kato, S. 1966, *Astrophys. J.*, **144**, 326.
10. Uchida, Y. 1966, *Astrophys. J.*, **142**, 335.
11. Uchida, Y. 1967, *Astrophys. J.*, **147**, (sous presse).
12. Uchida, Y. (sous presse, revue non précisée).
13. Souffrin, P. 1966, *Ann. Astrophys.*, **29**, 55.
14. Gingerich, O. 1966, *J. quant. Spect. rad. Transfer*, **6**, 609.
15. Frisch, H. 1966, *J. quant. Spect. rad. Transfer*, **6**, 629.
16. Edmonds, F. N. Jr. 1966, *Astrophys. J.*, **144**, 733.

Instruments

1. Bruns, A. V., Nikulin, N. S., Severny, A. B. 1965, *Izv. Krym. astrofiz. Obs.*, **23**, 80.
2. Joschpa, B. A., Obridko, V. N. 1965, in *Regul. issled. Progr. mejd. geofiz. God.*, *Soln. Aktiv.*, p. 131, Ed. Izdatel. Nauka, Moscou.
3. Karpinsky, V. N. 1965, *Izv. glav. astr. Obs. Pulkove*, **24**, no. 178, 84.

THE CHROMOSPHERE

Report prepared by R. G. Athay

(*a*) *Trends in Research and Reviews*

The period since the previous Draft Report seems to be characterized by somewhat more emphasis on disk observations of chromospheric structure and chromospheric heating and somewhat less emphasis on eclipse data. However, new spectrophotometric data from eclipses still plays a vital role in chromospheric research. Much important data from earlier eclipses still remain unpublished. Eclipse observers are urged to make data available to other astronomers on a shorter time scale than has been characteristic of the past.

Review articles on chromospheric phenomena by J. T. Jefferies, R. G. Athay, H. E. Hinteregger and R. O. Redman are contained in the proceedings of the Utrecht Conference on The Solar Spectrum (1) held in 1963. Other important reviews include a review of the extreme ultra-violet spectrum of the chromosphere by Tousey (2).

(*b*) *Spectrophotometry at Eclipse*

Optical continuum intensities near the solar limb recorded at the 1954 total eclipse have been reported at four wavelengths, $\lambda\lambda$ 3860, 4155, 4760 and 6190, by Kristenson (1) and at five wavelengths, $\lambda\lambda$ 4605, 4910, 5200, 5700 and 6430, by Heintze (2). Narrow band continuum intensities in the blue and red regions of the spectrum were measured at the 1961 eclipse by Mouradian (3). All of these data are in fair quantitative agreement with each other and with earlier continuum data. The continuum data now provide a rather good basis for realistic models of the lower chromosphere. However, improved photometric precision is still needed.

Radio continuum brightness temperatures at 3 cm were observed at the 1961 eclipse by Hackenberg, Popowa and Prinzler (4) and by Drago, Noci and Piatelli (5). The latter authors report limb brightening along equatorial diameters and limb darkening along polar diameters, a result that may be due partially to coronal contributions but may also have a strong chromospheric component.

Makita (6) has reported intensities and profiles of H α and the H and K lines observed at the 1958 eclipse and intensities of Hβ, Hγ, Hδ and H and K observed at the 1961 eclipse have been reported by Stolov (7). Takayanagi (8) has reported intensities in the CN λ 3800 band observed at the 1958 eclipse.

Successful observations at the May 1965 eclipse were made by Houtgast (9) and several groups have reported successful observations at the 1966 May 20 and November eclipses. Data for these more recent eclipses are not yet available.

(*c*) *Spectrophotometry of the Disk and at the Solar Limb Outside of Eclipse*

Observers at the Kitt Peak Observatory under the supervision of Pierce (1) are preparing a catalogue of approximately 8000 emission lines observed in the low chromosphere at the solar limb with the McMath tower telescope. We look forward to the publication of this important catalogue in the near future.

Mouradian (cf. ref. 3 §*b*) and Guljaev (**10**) have observed the profiles of the Paschen-gamma and λ 10 830 lines for the mean (spicules unresolved) chromosphere. Profiles and equivalent widths are given as a function of height. White and Wilson (**2**) report similar data for Hα and conclude that unit opacity at line center is found at about 1500 km above the solar limb.

Considerable discussion of the phenomena of a 'double-limb' observed in Hα filtergrams occurred during this period. Both Bhavilai, Norton and Giovanelli (**3**) and Simon and White (**4**) report that the 'double-limb' effect disappears when proper allowances and corrections are made for scattered light in the optical system.

Disk profiles of the λ 10 830 line were observed by de Jager, Namba and Neven (**5**). They conclude that this line is formed predominately in spicules covering about 10% of the solar disk, that the optical thickness of spicules at line center is 1·3 to 2·6 and that the mean broadening velocity is 9–11 km s^{-1}. Fisher (**6**) finds for the same line $\tau_0 = 2$ to 4 and a broadening velocity of 15 km s^{-1} has published a profile of the mean Hα line on the disk with a central intensity of 15·5%. Zirin and Howard (**7**) show from spectroheliograms made at λ 10 830 that 10 830 absorption is limited to the edges of the chromospheric network. They interpret this to mean that the network boundaries have relatively high temperature and density.

Suemoto (**8**) has studied variations in the H and K profiles observed with high resolution from point to point on the solar disk. He interprets the profile as a superposition of emission and absorption lines.

White (**9**) has published data on the mean Hα profile on the disk showing a central intensity of 15·5%.

Measurements of chromospheric polarization in Hα by Brückner (RG **9**) give information on polar magnetic fields.

(*d*) *Spectrophotometry in the XUV*

Tousey's (ref. **2**, §*a*) review article includes profiles of the Mg II resonance lines, showing the central emission features well resolved but with insufficient quantitative information. Profiles of Lyman-α and Lyman-β lines observed at different positions on the disk are given by Tousey, Purcell, Austin, Garrett and Widing (**1**). The wavelength interval from λλ 310–55 Å containing the Lyman-like lines of He II has been observed by Hinteregger, Hall and Schweizer (**2**) and by Hall, Schweizer, Heroux and Hinteregger (**3**). Other references on the XUV spectra are given in other parts of this Draft Report.

(*e*) *Spicules and Chromospheric Structure*

Improved spatial and spectral resolution in disk filter- and spectroheliograms have led to serious attempts to identify spicules on the solar disk. Beckers (**1**) and Zirin (**2**) have concluded that spicules occur predominately in the dark features outlining the chromospheric network. The spicules occur in clumps with several spicules found within a given dark mottle. Statistical properties of these dark features according to Beckers, agree well with the expected statistics for spicules. Bhavilai (**3**), however, concludes from similar observational data that the spicules are bright against the disk and are to be identified with small bright mottles found within the dark mottles, which are generally larger. Thus, the different observers agree as to the general location of the spicules but differ in the detailed identifications.

Mouradian (ref. **3**, §*b*) studied time sequences of Doppler shifts of spicules observed spectroscopically in Hα at the limb. He found only ascending motions. Beckers, Noyes and Pasachoff (**4**) report from similar studies, however, that they find both ascending and descending motions and that a given spicule may reverse its direction of motion, i.e., first rise then fall. Nikolskii and Sazanov (**5**) also find that the sign of the Doppler shift may reverse for a given spicule.

Widths of spectral lines emitted by spicules has received considerable attention. Athay and Bessey (**6**) indicate that the wide profiles reported earlier for the K line are not satisfactorily explained as a collection of narrower profiles of unresolved features. Giovanelli, Michard and Mouradian (**7**) dispute this claim as do also Gulyaev and Livshits (**8**). On the other hand, Beckers, Noyes and Pasachoff (**4**) agree with the conclusions of Athay and Bessey. More recently Livshits (**8 bis**) places emphasis on velocities inside spicules.

Nikolskii (**9**) finds narrower profiles for λ 5876 than reported by other observers. He attributes this to lack of resolution in previous observations. Nikolskii and Sazanov (**5**) note a similar effect in Hα profiles arising from closely spaced spicules moving relative to one another. Spicules spectra are also being studied in Kodaikanal by Sivaraman.

Contradictory results on the nature of the interspicule medium also persist. Beckers, Noyes and Pasachoff (**4**) agree with an earlier suggestion by Michard that the interspicular medium has a non-negligible opacity and produces observable absorption in the spicule Hα profiles. This is at variance with most models of the chromosphere which portray the interspicule medium as coronal-like, i.e., relatively hot and tenuous.

Oscillatory motions in the chromosphere have been studied in the K line by Jensen and Orrall (**10**) and by Orrall (**11**). They find that the oscillatory period for the K_3 component is only 180 seconds as contrasted with a period of 300 seconds for lines formed deeper in the atmosphere. Orrall identifies this period with a high-frequency tail on the spectrum of periodicities observed in the photosphere. For lines formed progressively higher in the atmosphere, this high-frequency tail becomes progressively more important and dominates in the chromosphere.

Quasi-periodic intensity fluctuations in K_2 and K_3 were also found by Jensen and Orrall (**10**). Simon and Leighton (**12**) have found that the borders of the chromospheric network coincide with local regions of enhanced magnetic field strength and descending matter as inferred from Doppler shifts in Hα and Hβ. Field strengths are of the order of 20 gauss and descending velocities are 1–2 km s^{-1}. They find a lifetime for the network cells of about 20 hours.

Godoli, Mazzucconi, Fossi and Torrisi (**13**) have studied the size of Hα chromospheric mottles utilizing the progressive occultation of the mottles by the moon during the eclipse of 1966 May.

(*f*) *Theoretical Spectroscopy and Models*

Models of the low chromosphere derived from the continuous spectrum by Hiei (**1**) and Heintze (**2b**) both show an outwards increase of temperature in the lowest layers of the chromosphere. Heintze places the temperature minimum at $\tau_c \approx 3 \times 10^{-2}$. De Jager (**2**) derives the location of the temperature minimum from the transition from absorption lines to emission lines at λ 1800. He places the minimum at $\tau_c = 2 \times 10^{-3}$.

The Utrecht reference model photosphere extends through the temperature minimum to $\tau_c = 10^{-4}$. It shows a temperature minimum of 4250° at $\tau_c = 2 \times 10^{-2}$ and a temperature of 5100° at $\tau_c = 10^{-4}$. The concept of a single reference model seems desirable and the Utrecht reference model serves as a useful first step. It could profitably be extended higher in the chromosphere and critically appraised by those deriving chromospheric models.

Takayanagi (**8b**) derives electron temperatures in the low chromosphere from CN data. He finds $\tau_c \approx 4000°$ at the base of the chromosphere and $T_e \approx 5000°$ near 500 km. Kodaira (**3**) identifies the depression in the continuum at λ 2085 with the ionization limit of Al I $3P^2P^0$.

Zheleznyakov (**4**) finds electron temperatures from radio data at mm wavelengths of about 5700° between heights of 2000 km and 2800 km and rising to about 8000° at 3000 km. He points out, however, that these temperatures, as well as an indicated drop in temperature above 3000 km, may reflect uncertainties due to non-uniform structure in the chromosphere.

Dietz and House (5) have discussed the Lyman continuum of hydrogen formed in the chromosphere but they do not arrive at a definite temperature. Johnson (6) discusses the Na D lines and concludes that on the basis of present models and theory he is unable to reconcile the excitation temperatures inferred from the line profiles. Athay (7) discusses the XUV data of hydrogen, He I and He II in relation to the chromospheric model. He finds a hot, tenuous interspicule regime that exhibits a rapid transition to coronal conditions just above the Lyman-α emitting region which ends near $T_e \approx 50\,000°$. The He II λ 304 line is formed in the sharp chromosphere-corona transition region. Similar conclusions are reached by Athay (8) from a study of the XUV lines of heavier elements.

Suemoto and Moriyama (9) conclude from XUV and radio data that the Sun emits these radiations from only a small fraction of the surface, which is an alternative to the rapid transition region. Nikolskii (ref. 9, §*e*) criticizes the hard-shell model of Dietz and Zirin (10) which eliminates the chromospheric component to the interspicule medium at lower heights.

Miyamoto (11) concludes from a discussion of the thermal stability of the chromosphere-corona transition region that the region is dynamically unstable and must give rise to phenomena resembling spicules. Problems of thermal stability are also discussed by Hunter (26).

Excitation of hydrogen in the chromosphere has been discussed by Kingston (12) and Athay (13). Helium I excitation has been discussed by Athay (14) and Nikolskaya (15) and helium II excitation by Athay (13). Controversy still exists as to the required electron temperatures for He I excitation resulting mainly from difference in the photoionization rates adopted. Nikolskaya (15) and several earlier authors conclude that relatively low electron temperatures (6000–10 000°) are sufficient whereas others (Athay, 14) have concluded that temperatures of 30 000° or higher are required.

The subject of source function equality and inequality has stimulated considerable discussion. Waddell (16) demonstrated empirically that the source functions for different multiplet members in the Na D lines and the Mg *b* lines were equal at a given point in the atmosphere. White (17) concluded from his analysis of Balmer line profiles that within the Balmer series the source functions were not equal. This conclusion was challenged by Pierce and Waddell (18) and defended by White (19). On the basis of the empirical results for Na D and Mg *b* and from numerical calculations, Waddell (20) concluded that the inferred source function equality required that collision rates between excited levels must exceed spontaneous transition rates to the ground state. Athay (21) challenged this conclusion, which was again defended by Waddell (22). Numerical calculations by Avrett (23) demonstrated that the required equality could be achieved with collision rates considerably lower than those inferred by Waddell.

A possible mechanism to explain the K_2 and H_2 emission peaks was advanced by Goldberg (24). He suggested that dielectronic recombinations may produce a relatively wide emission feature shifted slightly to the violet of the normal absorption line. Tandberg-Hanssen (25) has investigated the H and K lines of Ba II in chromospheric spectra from the 1952 and 1962 eclipses. He finds an overpopulation of the $J = 3/2$ level relative to the $J = 1/2$ level which he suggests may possibly arise from interlocking transitions to the 6^2P levels and by spin change collisions with hydrogen atoms.

(*g*) *Theory of Chromospheric Heating and Structure*

The period since the last Draft Report has been marked by an interest in mechanical energy transport mechanisms in the chromosphere; however, many authors still propose coronal heating mechanisms that traverse the chromosphere without dissipation.

Dubov (1) pointed out that Osterbrock's previous estimate of chromospheric energy requirements was very likely overestimated by a large factor. A similar result was found by

Athay (**2**) who also found that the rate of energy per proton (or electron) was independent of height in the chromosphere.

Heating of the chromosphere by hydrodynamic and hydromagnetic shock waves has been discussed by Kuperus (**3**), Bird (**4, 5**), Saito (**6**), Lüst and Scholer (**7**), Livshits (**8**), Pikel'ner and Livshits (**9**). Saito suggests that shock waves may produce spicules in the upper chromosphere and Lüst and Scholer find a radial steeping of wave fronts in the presence of magnetic fields and a strong channeling of the waves along lines of force. Pikel'ner and Livshits suggest that difficulties associated with the strong refraction of magnetosonic shock waves in the middle chromosphere may be eliminated by using an appropriate chromospheric model. Interference of wave fronts produces a two-dimensional pattern whose scale is comparable with the wavelength of the wave disturbance. Energy is transferred to the chromosphere mainly by the Alfvén mode. Recent work by Biermann (RG **10**) suggests that more energy is dissipated near the bottom than in higher layers of the chromosphere.

Pikel'ner (**10**) has suggested a model for the supergranulation and chromospheric network which couples the material motions with the magnetic field structure. Magnetic lines of force are concentrated at the cellular boundaries by the outwards flow of gases.

Various modes of standing waves in the chromosphere have been discussed. Bahng and Schwarzschild (**11**) have attributed the chromospheric oscillations to standing acoustic waves and Uchida (**12**) has pointed out that an alternative possibility is a standing compressional gravity wave. Jensen and Orrall (ref. **10,** §*e*) note that the periods of oscillation are consistent with standing sonic waves or isobaric oscillations.

REFERENCES

(*a*) *Chromosphere—Review*

1. Jefferies, J. T., Athay, R. G., Hinteregger, H. E., Redman, R. O. 1965, *The Solar Spectrum*, ed. C. de Jager, Reidel, Dordrecht, Holland.
2. Tousey, R. 1963, *Space Sci. Rev.*, **2,** 3.

(*b*) *Chromosphere—Eclipse*

1. Kristenson, H. 1964, *Mon. Not. R. astr. Soc.*, **128,** 113.
2. Heintze, J. R. W. 1965, Thesis, Utrecht Observatory.
3. Mouradian, Z. 1965, *Ann. Astrophys.*, **28,** 805.
4. Hackenberg, O., Popowa, M., Prinzler, H. 1964, *Z. Astrophys.*, **58,** 36.
5. Drago, F., Noci, G., Piatelli, M. 1964, *Ann. Astrophys.*, **27,** 708.
6. Makita, M. *Publ. astr. Soc. Japan*, in press.
7. Stolov, A. L. 1963, *Astr. Zu.*, **40,** 532.
8. Takayanagi, Y. *Publ. astr. Soc. Japan*, in press.
9. Houtgast, J. 1967, private communication.

(*c*) *Chromosphere—Disk and Limb*

1. Pierce, A. K. 1967, private communication.
2. White, O. R., Wilson, P. 1966, *Astrophys. J.*, **146,** 241.
3. Bhavilai, R., Norton, D. G., Giovanelli, R. G. 1965, *Astrophys. J.*, **141,** 274.
4. Simon, G. W., White, O. R. 1966, *Astrophys. J.*, **143,** 38.
5. Jager, C. de, Namba, O., Neven, L. 1966, *Bull. astr. Inst. Netherl.*, **18,** 128.
6. Fisher, R. R. 1964, *Astrophys. J.*, **140,** 1326.
7. Zirin, H., Howard, R. 1966, *Astrophys. J.*, **146,** 367.
8. Suemoto, Z. 1966, *Proc. Japan Acad.*, in press.
   Suemoto, Z. 1967, *Solar Physics*, submitted for publication.
9. White, O. R. 1964, *Astrophys. J.*, **139,** 1340.
10. Guljaev, R. A. 1964, *Astr. Zu.*, **41,** 313.

(*d*) *Chromosphere—XUV*

1. Tousey, R., Purcell, J. D., Austin, W. E., Garrett, D. L., Widing, K. G. 1963, *Space Res.*, **4**, 703.
2. Hinteregger, H. E., Hall, L. A., Schweizer, W. 1964, *Astrophys. J.*, **140**, 319.
3. Hall, L. A., Schweizer, W., Heroux, L., Hinteregger, H. E. 1965, *Astrophys. J.*, **142**, 13.

(*e*) *Chromosphere—Spicules and Structure*

1. Beckers, J. M. 1963, *Astrophys. J.*, **138**, 648.
   Beckers, J. M. 1964, Thesis, Utrecht Observatory.
2. Zirin, H. 1964, private communication.
3. Bhavilai, R. 1965, *Mon. Not. R. astr. Soc.*, **130**, 411.
4. Beckers, J. M., Noyes, R. W., Pasachoff, J. M. 1966, *Astr. J.*, **71**, 155.
5. Nikolskii, G. M., Sazanov, A. A. 1966, *Astr. Zu.*, **43**, 928.
6. Athay, R. G., Bessey, R. J. 1964, *Astrophys. J.*, **140**, 1174.
7. Giovanelli, R. G., Michard, R., Mouradian, Z. 1965, *Ann. Astrophys.*, **28**, 871.
8. Gulyaev, R. A., Livshits, M. A. 1965, *Astr. Zu.*, **42**, 661.

8 bis. Livshits, M. A. 1966, *Astr. Zu.*, **43**, 718.

9. Nikolskii, G. M. 1965, *Astr. Zu.*, **42**, 64.
10. Jensen, E., Orrall, F. Q. 1963, *Astrophys. J.*, **138**, 252.
11. Orrall, F. Q. 1966, *Astrophys. J.*, **143**, 917.
12. Simon, G. W., Leighton, R. B. 1964, *Astrophys. J.*, **140**, 1120.
13. Godoli, G., Mazzucconi, F., Monsignori Fossi, B. C., Torrisi, S. 1966, *Mem. Oss. Arcetri*, **85**, 73.

(*f*) *Chromosphere—Theoretical Spectroscopy and Models*

1. Hiei, E. 1967, in press.
2. Jager, C. de. 1963, *Bull. astr. Inst. Netherl.*, **17**.

2b. Heintze,

3. Kodaira, K. 1965, *Z. Astrophys.*, **60**, 240.
4. Zheleznyakov, V. V. 1965, *Astr. Zu.*, **41**, 1021.
5. Dietz, R., House, L. L. 1965, *Astrophys. J.*, **141**, 1393.
6. Johnson, H. R. 1964, *Ann. Astrophys.*, **27**, 695.
7. Athay, R. G. 1965, *Astrophys. J.*, **142**, 755.
8. Athay, R. G. 1966, *Astrophys. J.*, **145**, 784.

8b. Takayanagi,

9. Suemoto, Z., Moriyama, F. 1964, *Ann. Astrophys.*, **27**, 775.
10. Dietz, R., Zirin, H. 1963, *Astrophys. J.*, **138**, 664.
11. Miyamoto, S. 1965, *Publ. astr. Soc. Japan*, **17**, 127.
12. Kingston, A. E. 1964, *Astrophys. J.*, **140**, 736.
13. Athay, R. G. 1965, *Astrophys. J.*, **142**, 732.
14. Athay, R. G. 1963, *Astrophys. J.*, **137**, 931.
15. Nikolskaya, K. I. 1966, *Astr. Zu.*, **43**, 936.
16. Waddell, J., III. 1963, *Astrophys. J.*, **137**, 1210.
17. White, O. R. 1963, *Astrophys. J.*, **137**, 1217.
18. Pierce, A. K., Waddell, J., III. 1964, *Astrophys. J.*, **140**, 1160.
19. White, O. R. 1964, *Astrophys. J.*, **140**, 1164.
20. Waddell, J., III. 1963, *Astrophys. J.*, **138**, 1147.
21. Athay, R. G. 1964, *Astrophys. J.*, **140**, 1579.
22. Waddell, J., III. 1964, *Astrophys. J.*, **140**, 1586.
23. Avrett, E. H. 1966, *Astrophys. J.*, **144**, 59.
24. Goldberg, L. 1964, *Astrophys. J.*, **140**, 384.
25. Tandberg-Hanssen, E. 1966, private communication.
26. Hunter, J. H. 1966, *Icarus*, **5**, 321.

(g) *Chromosphere—Heating and Structure*

1. Dubov, E. E. 1965, *Astr. Zu.*, **42,** 1014.
2. Athay, R. G. 1966, *Astrophys. J.*, **146,** 223.
3. Kuperus, M. 1965, Thesis, Utrecht Observatory.
4. Bird, G. A. 1964, *Astrophys. J.*, **137,** 684.
5. Bird, G. A. 1964, *Astrophys. J.*, **137,** 340.
6. Saito, M. 1964, *Publ. astr. Soc. Japan*, **16,** 179.
7. Lüst, R., Scholer, M. 1966, *Z. Naturforschung*, **21,** 1098.
8. Livshits, M. A. 1965, *Astr. Zu.*, **41,** 473.
9. Pikel'ner, S. B., Livshits, M. A. 1965, *Astr. Zu.*, **41,** 1007.
10. Pikel'ner, S. B. 1963, *Astr. Zu.*, **39,** 973.
11. Bahng, J., Schwarzschild, M. 1963, *Astrophys. J.*, **137,** 601.
12. Uchida, Y. 1965, *Astrophys. J.*, **142,** 335.

THE CORONA

Report prepared by C. W. Allen

The corona is an object of outstanding interest both in its own right and as a source of particle and XUV emissions that have significant effects in interplanetary space and in the ionosphere. Although the interests are wide it is necessary to restrict the scope of the corona for the purpose of this review. My intended limits for coronal space are between the transition layer (separating corona and chromosphere) and the upper level beyond which no direct visible or radio radiations are detected. Problems of interplanetary space are considered only if they refer back to this more directly detectable corona. Scattering of radio sources is omitted. Furthermore it is only the normal and persistent features of the corona that are to be included. The permanent condensations are included but the sporadic condensations (**39**) are omitted as being more suited to the consideration of Commission 10.

The period covered by this report represents literature from about August 1963 until October 1966; that is, from those given in *Trans. IAU*, **12A,** p. 120, up to the time of writing October 1966. During this period I. S. Shklovski's book 'Physics of the Solar Corona' has been published in English, but it refers mainly to work completed by 1961. The fine 'Guide to the Solar Corona' by D. Billings should also be mentioned (RG **12**).

(a) *Continuous Spectrum*

Since most of the coronal light comes from the continuous spectrum representing electron scatter the photometry of heterochromatic images is used mainly for the purpose of indicating electron density distribution. The best opportunities for the study of such images are at total solar eclipses. The types of observations and analyses made can be summarized as, (*i*) image details and delineation, (*ii*) photometry, radial distribution, particular regions, complete isophotes, (*iii*) polarimetry, (*iv*) spectral distribution, colour, (*v*) electron density distribution, (*vi*) magnetic field distribution.

Results from recent eclipses are as follows:

1961 *February* 15
Polarization and electron densities: Fracassini and Hack (**1**).
Colour, polarization of low corona: Chalonge and Andrillat (**2**).
Image, isophotes: Löchel and Högner (**3**).
Isophotes, polarization: Michard and Sotirovsky (**4**).
Photometry: Cimino and Croce (**5**).

1962 *February* 4–5

Isophotes, polarization: densities of coronal condensation: Saito and Billings (**6**).
General polarization, photometry, densities: Saito and Hata (**7**).
Polar plumes, polarization, magnetic field, density variations: Saito (**8**).
Image, polar plumes, photometry: Tsubaki *et al.* (**9**).
Density and loops in condensation: Tsubaki (**10**).
Polar plumes and magnetic field: Waldmeier (**11**).

1963 *July* 20

Image, isophotes: Waldmeier (**12**).
Spectral distribution: Falciani and Rigutti (**13**).
F–K separation and density at great distance: Blackwell and Petford (**14**).
Polar rays and magnetic field details: Harvey (**15**).
Polar rays and magnetic field details: Stoddard *et al.* (**16**).

1965 *May* 30

Image, isophotes: Waldmeier (**17, 18**).
Polar plumes: Saito (**19**).

There have been results from the earlier eclipses of 1936 by Falchi *et al.* (**20**), and of 1952 by Falciani and Rigutti (**21**). A review of accumulated eclipse data has been made by Hata and Saito (**22**).

In addition to eclipse observations the electron corona has been studied for many years at Climax, Colorado, and the results made available in graphical form up to November 1960 in the Data Centre A reports (**23**). Data since that time although lodged at the Data Centre does not appear to be published, except for the days near the 1963 eclipse (**24**). The K-coronameter has since been installed at Mount Haleakala, Hawaii, and the data for the days near the 1965 eclipse have been reported from this station (**25**). Data from Pic-du-Midi have been assembled since about 1963 and sent to data centres. Studies of the Climax routine data have been made, for example, by Newkirk (**26**) who found the mean coronal brightness to be about twice that given by the van de Hulst model (**27**). This was probably associated with the very high solar activity during the period of analysis 1956–58.

From 1963 white light coronal observations are being reported from Japan. Nishi and Nagasawa (**28**) find the intensity to be in fair agreement with the van de Hulst model (**27**), and the streamers to coincide with quiescent prominences.

The increasing evidence and importance of the streaming of material outward from the Sun has led to heightened interest in the white light coronal features at large distances from the Sun. Recently Bohlin *et al.* (**29**) obtained the 3-dimensional structure of a large streamer by combining eclipse and coronameter data. Moreover efforts have been made to observe from great heights by using balloons, rockets or satellites.

The work of Coronascope II, observing white light corona from the stratosphere at 30 km, is described by Newkirk and Bohlin (**30**). An electron streamer extending to 5 solar radii appeared to be associated with a recurrent geomagnetic storm. The use of balloons during an eclipse in 1963 enabled Gillett, Stein and Ney (**31**) to trace the corona to a distance of 8°.

The observations of a coronagraph flown in a rocket by Tousey (**32**) give white light intensities in polar and equatorial directions to a distance of 9°. This experiment did not appear to reveal any obvious streamers.

Proposals to fly a coronagraph in OSO satellites have been made since 1961 but whereas there has been some partial success this has not yet led to useful results. Later OSO flights are awaited to obtain the full benefit of satellite coronagraph possibilities. In the meanwhile the corona has been photographed from the Moon with the TV camera on board Surveyor I while the Sun was setting beyond the lunar horizon (personal communication from G. Newkirk).

The main intentions of coronal continuum photometry are to convert the measurements into electron and gas densities. The integral equations for this conversion have been reconsidered by Fracassini (**33**). The density estimates from the papers quoted do not lead to any significant change from earlier models but the results are becoming more reliable and they extend good density measurements to about 20 solar radii (**14**). Blackwell (**34**) shows that we are near the limiting extension that can be measured and that improvement will depend on the use of satellites. Density measurements are used mainly for coronal physics and solar wind problems. Chamberlain (**35**) and Parker (**36**) have considered a hypothetical problem concerning a slow wind (breeze) emission. Brandt *et al.* (**37**) have shown that coronal temperature derived from density gradient is similar to that derived spectroscopically using dielectronic recombination.

Eclipse photometry can be applied also to the measurement of the density and size of coronal features. An outstandingly clear condensation seen on the west limb at the eclipse of 1962 has been studied by Tsubaki (**10**), Saito and Billings (**6**), and Waldmeier (**38**). The central density has been found to be about 10 times that of the normal corona and the size about 10° of solar circumference. Tsubaki (**10**) shows that the main structure is composed of a loop system. Such loops appear normal to stable condensations as well as 'sporadic condensations' (**39**).

The thermal radio emission from the Sun comes from the electrons of the Sun's atmosphere and, in the metre waves, from the corona. However, the fact, that the radiation is emitted from a wide range of depth, with many irregularities making it difficult to trace the paths of the rays, means that radio observations have not in the past provided any very satisfactory estimates of coronal temperature nor density. Recently Barkov (**40**) solved the transfer equation for a spherically symmetrical corona at $\lambda = 9{\cdot}1$ cm and found a fairly satisfactory agreement with observations. For many years the metre wave solar emissions have led to coronal temperature estimates of about 800 000 °K but a new attack on this problem by Oster and Sofia (**41**) using their new theory of radiative transfer in a dispersive medium led to about 2×10^6 °K in agreement with recent spectroscopic estimates. The problem has now been reconsidered by Fokker (**42**) who contends that the Oster dispersion theory is incorrect and arrives at 640 000 °K for the coronal temperature. This is to be compared with spectroscopic evaluation in the next section. The Oster theory has also been criticized by Cronyn (**43**). The burst emission has led Morimoto (**44**) to an estimate of coronal streamer (condensation) density that is much higher than those derived from electron scatter or line emission.

In recent years it has been possible to observe the electron corona by radar reflection. Chisholm and James (**45**) observing at 38·26 MHz have found the effective target size of the Sun, 15 to 20 $\times 10^{17}$ m^2; the outward streaming velocity, 16 km s^{-1}; and random mass motions, $\pm$ 35 km s^{-1}. Mass motion of this order could account for the coronal emission line widths and would imply that line measurements gave no evidence on thermal broadening. Billings (**46**) points out that if the mass motions are radial the line broadening effect will be much smaller.

(*b*) *Spectral Emission*

Many coronal structures observable in the continuum can also be seen in emission lines and the two types of observation can give a complementary interpretation. The continuum is proportional to electron density and the line emission is approximately proportional to the square of the density. The density can also be derived from the intensity ratios of certain emission lines, e.g. those of Fe XIII, and spacial distributions have been determined from such ratios by Dumont and Perche (**50**). This work has been extended to Ni XV by Schmiedler, Perche and Wlérick (in preparation).

Line emission observations are well suited to the study of coronal condensations. Kleczek (**51**) showed the 5303 line to be composed mainly of loops and arches in agreement with

Tsubaki's (**10**) white light condensations. However, Waldmeier (**52**) prefers the line 5694 of Ca XV for giving a representation of the size and shape of condensations. Hori (**53**) has shown that the 5694 coronal line is seen with greater frequency for the more active spots (Class F). Although the 5694 line statistics look much the same as those of slowly changing condensations, dm wave emission areas, and XUV ionizing emission areas, Hori is inclined to associate this line physically with the flare mechanism. Nishi and Nakagomi (**54**) and also Suzuki and Hirayama (**55**) find the form of a condensation varies considerably with the emission line used, being largest for the 5303 line. Models have been derived. Bumba *et al.* (**56**) found 5303 emission closely associated with photospheric magnetic fields.

Systematic coronal line observations have been maintained and considered by Gentili, *et al.* (**57**), Rösch (**58**), Gnevyshev (**59**), Kalinkov and Raikova (**60**), and Lexa (**60a**). Waldmeier has studied in (**61**) the variation of the polar zone through the sunspot cycle, and in (**62**) the 5303 radial intensity distribution.

The 1961 eclipse visible coronal lines were observed for distribution with heliographic latitude by Kernoa *et al.* (**63**), and a difference between 'hot' and 'cold' regions was detected.

A spectroscopic measurement of the wavelengths of the yellow coronal lines has been made by Boardman and Billings (**64**) giving the values 5694·44 and 5646·44Å. Billings (**65**) has continued his studies of coronal line profiles in an attempt to discriminate between mass motions of (*a*) an oscillatory, or (*b*) a radial type. There is evidence of (*a*) in the sharp features and (*b*) in the more diffuse features. Billings (**66**) does not consider that the random Doppler velocities found in the Chisholm and James (**45**) radar studies would necessarily have a significant line of sight component. Line profiles of 5303 and 6374 with the Fraunhofer line effect eliminated have been made by Nikolsky and Sassanov (**67**).

Hyder (**68**) has considered earlier measurements of coronal line polarization together with his own. There are disagreements amongst the observations and moreover there are theoretical complications in that the polarization might be caused by both radiatively and collisionally excited atoms. A full theoretical discussion of 5303 polarization has been given by Charvin (**69**). Data are insufficient for clear conclusions but there is some agreement with expectations on the magnetic field and on collision excitation. Such calculations have been extended to Fe XIII lines by Perche (**69a**).

A study of the identification of coronal lines in the ground based observable region has been made by Pryce (**70**) and Collins (**71**) who identify 5926 as Ar XV and 7144 as Ni IX. This suggests a total of 39 identified lines. Ten other possible identifications with Ne VIII, Na IX, Mg X, Al XI and Si XII have been proposed by Ch. Pecker (**72**). Lines of Si X, Al IX, Mg VIII and Si IX that should be detectable in the infra-red have been proposed by Münch (**73**).

However it is in the far ultra-violet and X-ray regions (the XUV) that the greatest efforts at identification have been concentrated. Progressively improved and extended spectra have become available from rocket and satellite spectroscopy (fairly recent review by Tousey (**74**) and Lindsay (**75**)). Line spectra have been obtained by Hinteregger *et al.* (**76**), Blake *et al.* (**77**), Hall *et al.* (**78**), Tousey *et al.* (**79**), Austin *et al.* (**80**). Standardizations are mainly due to Hinteregger (**81**). For comparison with these measurements, wavelengths have been predicted mainly by extrapolating spectroscopic data along isoelectronic sequences. Extrapolations and resulting identification lists have been prepared by Pecker and Rohrlich (**82**), Rohrlich and Pecker (**83**), Jordan (**84**), Nikitin (**85**), Zirin (**86**), Feldman *et al.* (**87**).

The greatest resistance to identification was provided by the strong lines in the region 170–220Å. The first step towards recognition came from the discovery of a similar set of lines in the spectrum of the Zeta torus by Fawcett *et al.* (**88**) who suspected the iron of the torus walls. This was soon accepted, Elton *et al.* (**89**). However, before the degree of ionization and spectroscopic transitions could be determined it was necessary to establish further laboratory techniques to produce the same lines under better controlled conditions. The theta-pinch was used by House (**90**) and, together with various spark discharges, by Fawcett and Gabriel (**91**) to produce not only the solar lines but also the spectra of many iron group ions. This made it possible to sort out the lines of the complex ion mixture producing the spectra. In the meantime Neupert (**92**) effected a separation of ions by their intensity changes with solar activity. As a result of these analyses a classification of most of the solar lines in the 170–220Å region has been made; Cowan and Peacock (**93**), Gabriel, Fawcett and Jordan (**94**), and Gabriel and Fawcett (**95**), Alexander *et al.* (**96, 97**). Probably all the identifications of (**94**) can be accepted.

(*c*) *Coronal Physics*

The new attack on coronal physics of recent years has been stimulated by the increase of XUV spectroscopic data. Physical evaluations depend heavily on the ionic excitation, ionization, and recombination coefficients which were reviewed by Seaton (**100**). Seaton (**101**) also determined the excitation associated with proton collisions. Since Seaton's review there has not been much change proposed for the effective ionizing rate except that Goldberg *et al.* (**102**) and van Rensbergen (**103**) have determined the increase associated with autoionization. On the other hand the recombination coefficients have been profoundly modified by the introduction of dielectronic recombination by Burgess (**104, 105, 106**).

The degree of ionization in relation to the temperature is affected by the Burgess results and consequently the calculations of House (**107**) are rendered out of date. Ionization temperatures using dielectronic recombination are found by Burgess and Seaton (**108**) to be about double the previous values. There is no longer any serious difference between the ionization temperature and the Doppler or equilibrium temperature of the corona. There is no need to postulate that ion temperature is greater than electron temperature although House and Billings (**109**) consider this to be a partial explanation. Ionization temperatures in coronal condensations by Letfus (**110**) have been found higher than in the quiet corona. Shklovsky (**111**) finds $1{\cdot}5 \times 10^6$ °K is high enough to account for coronal X-ray emission. Hughes (**111a**) suggests measuring temperature from $L\alpha$ scattering.

XUV line intensities have been used extensively to determine coronal ion numbers or densities. For this the relative populations of upper and ground levels are needed and can in some cases be determined by Jordan's (**112**) method.

Renewed efforts have been made to determine the coronal chemical composition. From the visible and XUV lines, Pottasch (**113**) and Widing and Porter (**114**) determine the relative abundances of certain heavy elements and compare results with the photosphere and meteorites. There has been much evidence by Letfus (**110**), Jefferies and Orrall (**115**) and Jordan (**116**) confirming Pottasch's earlier conclusion (**117**) that Fe and other heavy atoms were more abundant in the corona than the photosphere. On the other hand Dupree and Goldberg (**118**) find agreement. The possibility that the difference is real has led to some search for an explanation. The infall of meteoritic material has been considered (**100**), and also the thermal diffusion process (**100, 119**) whereby the abundance of heavy multiple ions can be increased in the hot regions of a plasma. In a number of papers Brandt *et al.* (**120, 121, 122, 123**) have discussed analogous problems connecting composition, temperature, solar wind, heliocentric latitude, etc. Neupert (**124**) and Delache (**125, 125a**) also relates electron temperature to coronal expansion.

Each XUV or visible line intensity measurement may be used to derive the amount of coronal material within a certain temperature range and this information may then be used to determine a temperature, density, height model of the corona (**126**). Various attempts by Athay (**127**), Dupree and Goldberg (**118**), Jordan (**128**) and Kanno (**129**) have led consistently to the conclusion that there is a very thin transition layer separating the corona from the chromosphere. For such computations it is usual to assume approximate constancy of pressure. On the other hand Livshits and Nikolskii (130) find a $N_e^{2.5}\,T = const.$ law for the transition layer. The model so produced has been used by Livshits (131) to balance ultraviolet emission against dissipation of shock wave energy. In this work self absorption of lines is usually neglected but on Stockhausen's evidence (131a) some correction may be necessary.

The very existence of the corona demands a heating mechanism and it is now widely agreed that acoustical energy from the convection zone is transmitted through the photospheric layer and dissipated in the corona. This process has been reviewed by Schatzman (132). Kuperus (133) has made a quantitative study of the various propagation and dissipative processes and derives a coronal model that agrees well with spectroscopic results (128). Birds' analyses (134, 135, 136) give a fairly natural explanation of the rather consistent temperature of the corona as a whole and the rapid drop at the transition zone. Uchida (137) has investigated the effect of facular magnetic fields and finds a result that is rather similar to a coronal condensation. Whitaker (138) finds that energy could be transmitted to the corona by gravity waves.

The physical reality of the Parker hydrodynamic theory of the outward wind through the corona has been supported by an alternative analysis by Caravillano and King (139). New numerical solutions relating to the viscous effects and energy input have been made by Scarf and Noble (140). Jensen (141) has applied Coulomb collision cross-sections to coronal evaporation and finds a mass loss comparable with space probe requirements.

INSTRUMENTS

The recent instrumental developments have been mainly concerned with applications for observations with balloons, rockets, or artificial satellites. Coronagraphs have necessarily been provided with occulting disks which are apodized either by the saw-tooth method or by several nearly equal disks on the axis: (150), Newkirk and Bohlin (151), Zanoni and Hill (152). Zirin and Newkirk (153) have shown that a reflecting coronagraph is feasible. Noxon (154) has developed a coronagraph which can make regular measurements of coronal lines in poor skies.

There have been several developments in XUV reflecting telescopes and spectrographs but, in general, these were not designed specifically for the corona. The increased accuracy of rocket pointing control has, however, enabled Burton, Ridgeley and Wilson (155) to set the slit of their spectrograph 10″ outside the solar limb and obtain spectra that are coronal in the region 950–2950Å.

BIBLIOGRAPHY

1. Fracassini, M., Hack, M. 1963, *Mem. Soc. astr. ital.*, **34**, 277.
2. Chalonge, D., Andrillat, Y. and H. 1963, *Ann. Astrophys.*, **26**, 439.
3. Löchel, K., Högner, W. 1965, *Z. Astrophys.*, **62**, 121.
4. Michard, R., Sotirovsky, P. 1965, *Ann. Astrophys.*, **28**, 96.
5. Cimino, M., Croce, V. 1965, *Mem. Soc. astr. ital.*, **36**, 3.
6. Saito, K., Billings, D. E. 1964, *Astrophys. J.*, **140**, 760.
7. Saito, K., Hata, S. 1964, *Publ. astr. Soc. Japan*, **16**, 240.
8. Saito, K. 1965, *Publ. astr. Soc. Japan*, **17**, 1.
9. Tsubaki, T., *et al.* 1964, *Publ. astr. Soc. Japan*, **16**, 13.
10. Tsubaki, T. 1966, *Publ. astr. Soc. Japan*, **18**, 1.
11. Waldmeier, M. 1965, *Z. Astrophys.*, **61**, 186.
12. Waldmeier, M. 1964, *Z. Astrophys.*, **60**, 28.
13. Falciani, R., Rigutti, M. 1966, *Mem. Soc. astr. ital.*, **37**, 189.
14. Blackwell, D. E., Petford, A. D. 1966, *Mon. Not. R. astr. Soc.*, **131**, 383 and 399.
15. Harvey, J. W. 1965, *Astrophys. J.*, **141**, 832.

16. Stoddard, L. G., *et al.* 1966, *Astrophys. J.*, **145**, 796.
17. Waldmeier, M. 1966, *Z. Astrophys.*, **63**, 242.
18. Waldmeier, M. 1966, *Astr. Mitt., Zurich*, no. 268.
19. Saito, K. 1965, *Publ. astr. Soc. Japan*, **17**, 421.
20. Falchi, A., *et al.* 1966, *Mem. astr. Soc. ital.*, in print.
21. Falciani, R., Rigutti, M. 1966, *Z. Astrophys.*, **63**, 85.
22. Hata, S., Saito, K. 1966, *Ann. Tokyo astr. Obs.*, **10**, 16.
23. Data Centre A. 1962, *Solar Activity Rep.*, Series 16.
24. 1963, *High Alt. Obs. Rep.* no. 59.
25. 1965, *High Alt. Obs. Rep.* no. 62.
26. Newkirk, G. 1961, *Astrophys. J.*, **133**, 983.
27. van de Hulst, H. C. 1950, *Bull. astr. Inst. Netherl.*, **11**, 135.
28. Nishi, K., Nagasawa, S. 1964, *Publ. astr. Soc. Japan*, **16**, 285.
29. Bohlin, J. D., Hansen, R. T., Newkirk, G. A. 1966, Communic. AAS Special Meeting on Solar Astronomy, October 1966.
30. Newkirk, G., Bohlin, J. D. 1965, *Ann. Astrophys.*, **28**, 234.
31. Gillett, F. C., *et al.* 1964, *Astrophys. J.*, **140**, 292.
32. Tousey, R. 1965, *Ann. Astrophys.*, **28**, 600.
33. Fracassini, M. 1963, *Mem. Soc. astr. ital.*, **34**, 15.
34. Blackwell, D. E. 1964, *Space Sci. Rev.*, **1**, 612.
35. Chamberlain, J. W. 1965, *Astrophys. J.*, **141**, 320.
36. Parker, E. N. 1965, *Astrophys. J.*, **141**, 322.
37. Brandt, J. C., *et al.* 1965, *Astrophys. J.*, **141**, 809.
38. Waldmeier, M. 1963, *Z. Astrophys.*, **56**, 291.
39. Jefferies, J. T., Orrall, F. Q. 1965, *Astrophys. J.*, **141**, 505 and 519.
40. Barkov, V. I. 1965, *Astr. Zu.*, **42**, 749.
41. Oster, L., Sofia, S. 1965, *Astrophys. J.*, **141**, 1139.
42. Fokker, A. D. 1966, *Bull. astr. Inst. Netherl.*, **18**, 359.
43. Cronyn, W. M. 1966, *Astrophys. J.*, **144**, 834.
44. Morimoto, M. 1963, *Ann. Tokyo astr. Obs.*, **8**, 125.
45. Chisholm, J. H., James, J. C. 1964, *Astrophys. J.*, **140**, 377.
46. Billings, D. E. 1965, *Astrophys. J.*, **141**, 325.
50. Dumont, J. P., Perche, J. C. 1964, *Mém. Soc. R. Sci. Liège*, **9**, 186.
51. Kleczek, J. 1963, *Publ. astr. Soc. Pacif.*, **75**, 9.
52. Waldmeier, M. 1963, *Z. Astrophys.*, **58**, 57.
53. Hori, M. 1963, *Z. Astrophys.*, **58**, 117.
54. Nishi, K., Nakagomi. 1963, *Publ. astr. Soc. Japan*, **15**, 56.
55. Suzuki, T., Hirayama, T. 1964, *Publ. astr. Soc. Japan*, **16**, 58.
56. Bumba, V., *et al.* 1965, *Publ. astr. Soc. Pacif.*, **77**, 55.
57. Gentili, M., *et al.* 1966, *Ann Astrophys.*, **29**, 43.
58. Rösch, J. 1963, *C. r. Acad. Sci.*, Paris, **256**, 1221.
59. Gnevyshev, M. N. 1963, *Astr. Zu.*, **40**, 401.
59a. Rybansky, M. *Bull. astr. Inst. Csl.*, in press.
60. Kalinkov, M. P., Raikova, D. V. 1964, *C. r. Acad. Sci.*, Paris, **259**, 2181.
60a. Lexa, J. 1965, *Publ. astr. Inst. Csl.*, no. 51, 125.
61. Waldmeier, M. 1964, *Z. Astrophys.*, **59**, 205.
62. Waldmeier, M. 1965, *Z. Astrophys.*, **61**, 144.
63. Kernoa, A., *et al.* 1965, *Ann. Astrophys.*, **28**, 716.
64. Boardman, W. J., Billings, D. E. 1965, *Astrophys. J.*, **141**, 1289.
65. Billings, D. E. 1964, *Astrophys. J.*, **139**, 710.
66. Billings, D. E. 1965, *Astrophys. J.*, **141**, 325.
67. Nikolsky, G. M., Sassanov, A. A. 1964, *Solar Data* 1963 Nov. 12, p. 57.
68. Hyder, C. L. 1965, *Astrophys. J.*, **141**, 1382.
69. Charvin, P. 1965, *Ann. Astrophys.*, **28**, 877.
69a. Perche, J. C. 1965, *C. r. Acad. Sci.*, Paris, **260**, 6037, and **261**, 5319.
70. Pryce, M. H. L. 1964, *Astrophys. J.*, **140**, 1192.

71. Collins, P. D. B. 1964, *Astrophys. J.*, **140,** 1206.
72. Pecker, Ch. 1963, *Contr. Inst. Astrophys.*, Paris, Sér. A, no. 302.
73. Münch, G. 1966, *Astrophys. J.*, **145,** 237.
74. Tousey, R. 1964, *Q.J.R. astr. Soc.*, **5,** 123.
75. Lindsay, J. C. 1964, *Planet. Space Sci.*, **12,** 379.
76. Hinteregger, H. E., *et al.* 1964, *Astrophys. J.*, **140,** 319.
77. Blake, R. L., *et al.* 1965, *Astrophys. J.*, **142,** 1.
78. Hall, L. A., *et al.* 1965, *Astrophys. J.*, **142,** 13.
79. Tousey, *et al.* 1965, *Ann. Astrophys.*, **28,** 755.
80. Austin, W. E., *et al.* 1966, *Astrophys. J.*, **145,** 373.
81. Hinteregger, H. E. 1965, *Space Sci. Rev.*, **4,** 461.
82. Pecker, Ch., Rohrlich, F. 1963, *Astrophys. J. Suppl.*, **8,** no. 79, 227.
83. Rohrlich, F., Pecker, Ch. 1963, *Astrophys. J.*, **138,** 1246.
84. Jordan, C. 1965, *Comm. Univ. London Obs.*, no. 68.
85. Nikitin, A. A. 1964, *Astr. Zu.*, **41,** 482.
86. Zirin, H. 1964, *Astrophys. J.*, **140,** 1332.
87. Feldman, V., Fraenkel, B. S., Hoory, S. 1965, *Astrophys. J.*, **142,** 719.
88. Fawcett, B. C., *et al.* 1963, *Nature*, **200,** 1303.
89. Elton, R. C., *et al.* 1964, *Astrophys. J.*, **140,** 390.
90. House, L. L., *et al.* 1964, *Astrophys. J.*, **140,** 814.
91. Fawcett, B. C., Gabriel, A. H. 1965, *Astrophys. J.*, **141,** 343.
92. Neupert, W. M. 1965, *Ann. Astrophys.*, **28,** 446.
93. Cowan, R. D., Peacock, N. J. 1965, *Astrophys. J.*, **142,** 390.
94. Gabriel, A. H., *et al.* 1965, *Nature*, **206,** 390.
95. Gabriel, A. H., Fawcett, B. C. 1965, *Nature*, **206,** 808.
96. Alexander, E., *et al.* 1965, *Phys. Letters*, **14,** 40.
97. Alexander, E., *et al.* 1965, *Nature*, **206,** 176.
100. Seaton, M. J. 1965, *The Solar Spectrum*, ed. de Jager, p. 273, Reidel.
101. Seaton, M. J. 1964, *Mon. Not. R. astr. Soc.*, **127,** 191.
102. Goldberg, L., *et al.* 1965, *Ann. Astrophys.*, **28,** 589.
103. van Rensbergen, W. 1967, submitted to *Solar Physics*.
104. Burgess, A. 1964, *Astrophys. J.*, **139,** 776.
105. Burgess, A. 1965, *Ann. Astrophys.*, **28,** 774.
106. Burgess, A. 1965, *Astrophys. J.*, **141,** 1588.
107. House, L. L. 1964, *Astrophys. J. Suppl.*, **8,** no. 81, 307, 1964.
108. Burgess, A., Seaton, M. J. 1964, *Mon. Not. R. astr. Soc.*, **127,** 355.
109. House, L. L., Billings, D. F. 1964, *Astrophys. J.*, **140,** 1182.
110. Letfus, V. 1965, *Bull. astr. Inst. Csl.*, **16,** 231.
111. Shklovsky, I. S. 1964, *Astr. Zu.*, **41,** 676.
111a. Hughes, C. J. 1965, *Astrophys. J.*, **142,** 321.
112. Jordan, C. 1965, *Phys. Letters*, **18,** 259.
113. Pottasch, S. R. 1964, *Mon. Not. R. astr. Soc.*, **128,** 73.
114. Widing, K. G., Porter, J. R. 1965, *Ann. Astrophys.*, **28,** 779.
115. Jefferies, J. T., Orrall, F. R. 1966, *Astrophys. J.*, **145,** 231.
116. Jordan, C. 1966, *Mon. Not. R. astr. Soc.*, **132,** 463.
117. Pottasch, S. R. 1964, *Space Sci. Rev.*, **3,** 816.
118. Dupree, A. K., Goldberg, L. 1966, submitted to *Solar Physics*.
119. Chapman, S. 1958, *Proc. Phys. Soc.*, **72,** 353.
120. Brandt, J. C., *et al.* 1965, *Astrophys. J.*, **141,** 809.
121. Brandt, J. C., *et al.* 1965, *Icarus*, **4,** 19.
122. Brandt, J. C. 1966, *Observatory*, **86,** 138.
123. Brandt, J. C. 1966, *Astrophys. J.*, **143,** 266.
124. Neupert, W. M. 1964, *Astrophys. J.*, **139,** 935.
125. Delache, Ph. 1965, *C. r. Acad. Sci.*, Paris, **261,** 643.
125a. Delache, Ph. 1966, in Symp. on Interdisciplinary Aspects of Radiative Transfer, Philadelphia, February, 1966.

126. Allen, C. W. 1965, *Space Sci. Rev.*, **4**, 91.
127. Athay, R. G. 1966, *Astrophys. J.*, **145**, 784.
128. Jordan, C. 1965, Thesis.
129. Kanno, M. 1966, *Publ. astr. Soc. Japan*, **18**, 103.
130. Livshits, M. A., Nikolskii, G. M. 1964, *Astr. Zu.*, **41**, 75.
131. Livshits, M. A. 1964, *Astr. Zu.*, **41**, 473.
131a. Stockhausen, R. 1965, *Astrophys. J.*, **141**, 277.
132. Schatzman, E. 1965, *Introduction to Solar Terrestrial Relations*, Eds. J. Ortner and H. Masseland, p. 112, Reidel.
133. Kuperus, M. 1965, *Rech. astr. Obs. Utrecht*, **17**, no. 1.
134. Bird, G. A. 1964, *Astrophys. J.*, **139**, 684.
135. Bird, G. A. 1964, *Astrophys. J.*, **140**, 288.
136. Bird, G. A. 1964, *Astrophys. J.*, **141**, 1455.
137. Uchida, Y. 1963, *Publ. astr. Soc. Japan*, **15**, 376.
138. Whitaker, W. A. 1963, *Astrophys. J.*, **137**, 914.
139. Carovillano, R. L., King, J. H. 1965, *Astrophys. J.*, **141**, 526.
140. Scarf, F. L., Noble, L. M. 1965, *Astrophys. J.*, **141**, 1479.
141. Jensen, E. 1963, *Astrophys. Norv.*, **8**, 99.
150. *Sky Telesc.*, 1962, **24**, 197, 315, 327; 1964, **28**, 16.
151. Newkirk, G., Bohlin, J. D. 1964, *Appl. Optics*, **3**, 543.
152. Zanoni, C. A., Hill, H. A. 1965, *J. opt. Soc. Am.*, **55**, 1608.
153. Zirin, H., Newkirk, G. 1963, *Appl. Optics*, **2**, 977. *High Alt. Obs. Repr.*, no. 298.
154. Noxon, J. F. 1966, *Astrophys. J.*, **145**, 400.
155. Burton, W. M., *et al.* 1966–67, *Mon. Not. R. astr. Soc.*, in press.

WORKING GROUP ON SOLAR ECLIPSES

CHAIRMAN: J. Houtgast

The total eclipses of 1965 May 30, and 1966, November 12, the annular-total eclipse of 1966, May 30, were actively observed from the ground, from aircrafts and with sounding rockets.

In connection with the total eclipses, steps were taken by the General Secretary of IAU to implement the Resolution no. 5 passed by the IAU General Assembly in Hamburg, in order to avoid interference caused to ground based observations by vapour trails from airplanes.

A comprehensive report of the work performed during the 1965–66 eclipses will be prepared after the next IAU General Assembly.

WORKING GROUP ON CENTRAL LINE INTENSITIES

CHAIRMAN: A. K. Pierce

The Working Group actively pursued its goals during the last three years and intermediate progress reports were circulated to members and interested colleagues.

A comprehensive report will be prepared after the coming IAU General Assembly.

14. COMMISSION DES DONNEES SPECTROSCOPIQUES FONDAMENTALES

PRÉSIDENT: Mrs Dr Ch. Moore Sitterly, National Bureau of Standards, Washington, D.C. 20234, U.S.A.

VICE-PRÉSIDENT: Dr G. Herzberg, Director, Division of Pure Physics, National Research Council, Sussex Drive, Ottawa 2, Ontario, Canada.

COMITÉ D'ORGANISATION: A. H. Cook, B. Edlén, R. H. Garstang, T. A. Littlefield, J. G. Phillips.

MEMBRES: Adam, Allen, Baird, Barrell, Bates, Branscomb, Burgess, Dobronravin, Engelhard, Essen, Felenbok, Garton, Green, Herman, Hindmarsh, Humphreys, Jacquinot, Junkes, Kessler, Kiess, King (R. B.), Layzer, Lochte-Holtgreven, Mack†, Meggers†, Melnikov, Migeotte, Milford, Minnaert, Mohler, Monfils, Naqvi, Nevin, Nicholls, Obi, Prokofiev, Racah†, Rigutti, Rosen, Salpeter (E. W.), Seaton, Smit, Swings, Terrien, Tousey, Traving, Trefftz, Van Bueren, Van Regemorter, Varsavsky, Wilson (R.)*, Zirin.

MEMBRES CONSULTANTS: Andrew (K. L.), Bovey, Dalgarno, Dieke†, Griem, Kuhn, Lagerqvist, Penkin.

1. Committee on Standards of Wavelength.
   Chairman: B. Edlén.
   Members: Baird, Barrell, Cook (A. H.), Engelhard, Herzberg, Humphreys, Kessler, Littlefield, Terrien.

2. Committee on Transition Probabilities.
   Chairman: R. H. Garstang.
   Members: Allen, Bates, Branscomb, Green, King (R. B.), Layzer, Lochte-Holtgreven, Melnikov, Milford, Minnaert, Penkin, Smit, Prokofiev, Zirin.

2a. Sub-Committee on Broadening Parameters of Spectral Lines.
   Chairman: G. Traving.
   Members: Griem, Trefftz.

2b. Sub-Committee on Cross Sections.
   Chairman: M. Seaton.
   Members: Branscomb, Burgess, Dalgarno, Hindmarsh, Van Regemorter.

3. Committee on Molecular Spectra.
   Chairman: J. G. Phillips.
   Members: Herzberg, Lagerqvist, Nicholls, Rosen, Swings.

*AERE, Harwell, U.K.

GENERAL COMMENTS

The responsibility of Commission 14 is increasing so rapidly that an appraisal of the present activity and future trends may be in order. The Commission is unique in its basic purpose, namely, to provide a strong liaison between astronomers and laboratory spectroscopists. In the United States, at least four data centers exist primarily to provide critical data of astrophysical interest: (1) Atomic Energy Levels and (2) Transition Probabilities (both at the National Bureau of Standards, Washington, D.C.), (3) Cross Sections (Joint Institute for Laboratory Astrophysics, Boulder, Colorado), (4) Diatomic Molecules (University of California, Berkeley, California).

Other problems are coming into the foreground, such as the use of lasers in precision spectroscopy—a topic suggested by a number of our members. The coupling of astrophysical needs with laboratory programs is one of our prime obligations.

ACTIVITIES OF THE COMMISSION

Since the 1964 General Assembly the Commission has acted as sponsor for three symposia:

1. In Trieste, 13–15 June, 1966, the IAU was represented by President Swings, and the Commission by Mrs Moore-Sitterly as co-sponsor of a colloquium on Late-Type Stars. This was organized by Mrs Hack, Director of the Trieste Observatory.
2. In Boulder, Colorado, 11–15 July, 1966, the Commission sponsored a Colloquium on Atomic Collision Processes, which was well represented by a number of our members under the leadership of Seaton (see below).
3. In Bombay, India, 9–18 January, 1967, an International Conference on Spectroscopy was organized by Asundi and attended by Herzberg and Mrs Moore-Sitterly as representatives of Commission 14.

Two International Commissions require representation from Commission 14.

1. The Triple Commission for Spectroscopy has served to coordinate the needs and efforts of the three International Unions, Physics, Chemistry, and Astronomy, in the field of spectroscopy. In 1964, this Commission convened in Hamburg immediately after the Twelfth General Assembly of the IAU. At this meeting progress reports on atomic and molecular spectra were presented with detailed bibliographies (**388**)*.

At the next meeting, held in Copenhagen, 19 August, 1965, it was decided to reorganize the TCS. Since two of the three International Unions concerned have Spectroscopy Commissions (IUPAC and IAU) it was recommended that IUPAP should, similarly, form its own Spectroscopic Commission in order to have three 'parallel' Commissions, one for each Union. It was proposed that each Union name four members to serve on the TCS. For the IAU the representatives will be: Edlén, Mrs Moore-Sitterly, Phillips, and Seaton. The normal term of service will be six years. In order to preserve continuity it was suggested that each group of four select two to serve only three years.

The function of the TCS as reconstituted will be to decide on matters of concern to the three International Unions and to coordinate the work of the three separate Commissions.

2. At Unesco House in Paris, 16–17 June, 1966, an ICSU Committee on Data for Science and Technology was established, with Harrison Brown as Chairman *pro tem*. This committee is made up of both National and Union Representatives. The President of IAU Commission 14 is the IAU Representative. Rossini was chosen as President, Klemm and Vodar as Vice-Presidents, and Sutherland, Secretary-Treasurer. Further information can be obtained from the office of the Executive Director, Guy Waddington, 2101 Constitution Ave., N.W., Washington, D.C., 20418, (U.S.A.). The Reports of Commission 14 furnish much useful data and reference material for this ICSU Committee.

STANDARDS OF WAVELENGTH

(Committee 1)

B. Edlén

The metrological properties of the He–Ne laser radiation, with the approximate vacuum wavelength 6329·914Å, are being investigated at NBS (**315a**), NRC, NPL (**404a**) and PTB (**135a**). The results indicate that this radiation is likely to become an exceedingly useful secondary standard for length measurements.

*Reference numbers in parentheses refer to the general bibliography which follows the report of the Commission.

Work on the determination of line profiles of selected Hg 198 and Kr 86 lines, advocated at the 1964 meeting of this Committee, is reported by Terrien (BIPM) and Rowley (NPL). The results from BIPM and NPL agree on the existence of a small but detectable asymmetry in all the investigated Kr 86 lines, similar to the one previously found in the Kr 86 primary standard line. Terrien found (**452a**) that the asymmetry was independent of all changes made in the experimental arrangements, and that it could be described in terms of a faint satellite about 0·007 cm^{-1} to the longwave side of the main line. No explanation for the asymmetry has been offered. We wish to point out, however, that approximately this kind of asymmetry would be expected from a small admixture of Kr 84.

Engelhard further reports that, after having studied the original laboratory records of Kösters and Lampe, he has revised their formula for the refractive index of air, whereby the difference with the formula of Barrell and Sears is reduced from 9 to 6 units in $(n - 1)\ 10^8$. A comprehensive review of our present knowledge of the refractive index of air, published by Edlén (**134**), includes an improved dispersion formula for standard air, as well as formulas for the dependence of refractivity on temperature and pressure and on relative contents of CO_2 and H_2O. Since the differences between the new dispersion formula and that of 1953 are very small, it was suggested that a replacement of the 1953 formula as standard in spectroscopic work would not be necessary for the time being. Regarding the absolute values of the refractive index it was concluded that there are as yet no definite indications that the values adopted in 1953 should be changed, but new experiments aiming at reducing the present uncertainty of $\pm\ 5 \times 10^{-8}$ would be desirable.

Rare Gas Standards

The results of measurements on 37 Kr 86 lines from 6458 to 4264 Å, made at the NPL and previously reported to this Commission, have been recalculated by Cook (**96**) to give wavelengths as observed through the anode end of an Engelhard lamp. These values are compared with those previously obtained by Kaufman and by Phelps, and it is estimated that means of the three sets of data may be relied upon to 0·0001 Å.

Humphreys and Paul (**225**) have communicated for this Report the results of photographic interferometer measurements on selected lines, mainly in the infrared, in the spectra of Kr 86 and Xe 136 (Tables 1 and 3). Provisional values for the 1*s* and 2*p* terms which they have derived from these measurements are shown in Tables 2 and 4. Microwave-excited electrodeless tubes were used as light sources, and the Hg 198 lines at 4047·7147 and 4359·5625 Å, supplemented by some argon wavelengths, served as standards. When comparing the new Kr 86 wavelengths with those reported in 1964 to this Commission by Littlefield and Sharp one finds a systematic difference, the new values being about 0·003 Å lower.

Interferometric measurements on 84 lines of Ar I and 69 lines of Ar II, in the region 5000–7000 Å, have been completed by Norlén (**343a**) at Lund. The light source was a water-cooled hollow cathode, the Fabry-Pérot etalons ranged up to 49 mm, and as reference line served the Kr 86 line at 6057 Å, produced by a microwave-excited electrodeless lamp obtained from the PTB. The error limits are estimated to ± 0·0004 Å. When comparing the results with the two previous sets of interferometric measurements on Ar I lines in this region (both from 1953) one finds no significant systematic difference with Littlefield and Turnbull, while the values of Burns and Adams are on the average 0.002 Å larger.

Table 1. ^{86}Kr I vacuum wavelengths

(1966 interferometric measurements by Humphreys and Paul)

4275·1716	7696·6568	8115·1306	8300·3880
7589·5005	7856·9828	8192·3055	8511·2072
7603·6375	8061·7200	8265·5111	8779·1576
7687·3590	8106·5929	8283·3254	8931·1424

Table 2. Energy levels of ^{86}Kr I

(Humphreys and Paul, 1966)

$1s_5$	0·0000	$2p_{10}$	11 196·7759*	$2p_4$	17 624·1762
$1s_4$	945·0243	$2p_9$	12 322·6605*	$2p_3$	17 947·4074
$1s_3$	5219·8755	$2p_8$	12 335·6386	$2p_2$	17 973·4277
$1s_2$	5874·9615	$2p_7$	12 992·6540	$2p_1$	18 883·3301*
		$2p_6$	13 151·6002		
		$2p_5$	14 121·1219*		

*Only one interferometrically observed combination available for evaluation.

Table 3. ^{136}Xe I vacuum wavelengths

(1966 interferometric measurements by Humphreys and Paul)

4502·2424	4735·4775	7969·5326	8282·3956	8821·8359
4525·9510	4808·3635	8063·5534	8349·1198	8864·7497
4625·5719	4917·8799	8208·5947	8411·5043	8954·7131
4672·5344	7644·1300	8233·9003	8650·8817	9047·9337
4698·3365	7889·5649	8268·7944	8741·7743	9165·1711

Table 4. Energy levels of ^{136}Xe I

(Humphreys and Paul, 1966)

$1s_5$	0·0000	$2p_9$	11 052·2472	$2p_4$	21 311·5744
$1s_4$	977·6092	$2p_8$	11 335·5089	$2p_3$	22 094·8038
$1s_3$	9129·2210	$2p_7$	11 888·4800	$2p_2$	22 211·1544
$1s_2$	10 117·4932	$2p_6$	12 144·9127	$2p_1$	22 792·4634
		$2p_5$	13 051·4106		

Thorium Standards

The measurements by Littlefield and Wood, previously reported to this Commission, have now been published (**287a**). Giacchetti, at the Argonne Laboratory, has furnished a typewritten compilation and comparison of all the five sets of precision measurements on thorium lines that have been published up to this date. This compilation, which contains four-decimal wavelengths of 701 lines from 2566 to 9050 Å, will make it possible to extract a list of wavelengths measured with consistent results by two or more observers and thus being acceptable as standards.

Vacuum-ultraviolet Standards

Substantial progress has recently been achieved regarding wavelength standards in the vacuum ultraviolet. This is largely due to the efficient use of the NBS 10·7 m vacuum spectrograph in the hands of Kaufman and Ward. In the case of N I, these authors (**246**) have derived calculated wavelengths, accurate to 0·001 Å or better, for 78 lines from 908 to 1745 Å, based on their measurements of the multiplets at 1492 and 1742 Å combined with Eriksson's (**136b**) measurements of the forbidden transition at 3466 Å. The same authors (**245**) have checked the suitability of calculated Cu II standards and have revised the value of the ground level a^1S_0 which permits a considerable extension of available Cu II standards toward shorter wavelengths. In the same paper they report precision measurements of some key multiplets which lead to revised values of the calculated wavelengths for Si II (Table 5) and Ge II (Table 6) and confirm the C I wavelengths calculated by L. Johansson (**231a**).

A paper on Si I by Radziemski, Andrew, Kaufman and Litzén (**391**) gives the final values of low-pressure-source levels as well as recalculated wavelengths of 100 lines and measured wavelengths of 41 lines, thus covering the region from 1991 to 1560 Å with standards accurate to 0·0016 Å or better.

Wavelengths of the Lyman lines, 1s − np, have been calculated for the spectra from H I to Si XIV by Edlén and Svensson (**132**) in connection with their redetermination of the X-unit. The values for Be IV through Si XIV were later revised (**133**) in view of the information given by Garcia and Mack in an exhaustive treatise on levels and wavelengths of the spectra from H I to Ca XX (**168**). This important work by Garcia and Mack represents an all out effort to push the accuracy of these data to the limit set by our present knowledge of the atomic constants and by the present state of the theory of one-electron spectra.

Table 5. Si II vacuum wavelengths (calculated)

(Kaufman and Ward)

820·9210	889·7228	1023·7002	1260·4212	1526·7073
822·8613	899·4063	1190·4157	1264·7374	1533·4319
843·7192	901·7359	1193·2894	1265·0010	1808·0125
848·0700	989·8730	1194·5001	1304·3716	1816·9283
850·1409	1020·6988	1197·3936	1309·2772	1817·4512

Table 6. Ge II vacuum wavelengths

(Kaufman and Ward)

Calculated	Calculated	Calculated	Calculated	Measured
843·7165	905·9771	1017·0600	1576·8547	1938·0077
856·4880	920·5537	1055·0261	1581·0698	1938·8906
862·2339	920·7195	1075·0720	1602·4863	1979·2736
872·3075	926·4736	1237·0589	1649·1942	2007·6883
875·4927	941·8962	1261·9053		
875·5766	999·1011	1264·7096		
885·9663	1016·6377	1538·0907		

TRANSITION PROBABILITIES

(Committee 2)

Work in this field is so active that no attempt will be made to cover it completely. The existing data centers maintain current reference files from which information can be obtained.

Atomic transition probabilities. Wiese and his staff have published a critical compilation of atomic transition probabilities for about 4000 spectral lines of the first ten elements, H thru Ne, based on all available literature sources (**485**). Glennon and Wiese issued a current bibliography on this subject (**182**) in April 1966, which supersedes their Monograph 50, 1962.

Two survey articles may be mentioned. Foster (**159a**) has given a comprehensive account of the experimental methods for the determination of oscillator strengths, and Garstang (**172a**) has surveyed the period 1960–64.

1. *Experimental Determinations of Transition Probabilities*

R. B. King reports a continuation of work on oscillator strengths derived from absorption spectra. Relative values for 300 lines of Fe I have been published (**254**). The work by Moise on Transition Rates of Atomic Transitions, reported in 1964, has since been published (**319**).

Similarly, the earlier report on the absolute oscillator strengths of lines in the spectra of ten elements is now in print (**283**). The results by Link on lifetimes of upper states obtained by using the 'level crossing' technique has stimulated the effort to improve the earlier determinations of absolute *f*-values. Ashenfelter is using the atomic beam apparatus for this purpose.

In collaboration with Whaling, who has constructed an ion source capable of producing a beam of metallic ions, an attempt is being made to observe the rate of decay of Fe ions emitted in the second and third stages of ionization, as observed with a Van de Graff accelerator and carbon foil target. This work is in its initial stage, according to King.

Kuhn reports studies at Oxford of self broadening and pressure shift of lines in the visible and near infrared spectra of He (**464**), Ne, Ar (**434**), and Kr, observed by means of direct recording interferometers. Measurements of resonance broadening constants can be used to obtain *f*-values of strong resonance lines.

Peach has determined relative *f*-values of certain pairs of Fe I lines, with accuracies estimated from 2 to 15%. All lines have low excitation potentials, below 3·5 eV (**370**).

Garstang has stressed the importance of recent work on lifetime determinations. The method of phase shift has been used in some of the more important investigations of the last few years. Intensity modulated light is used to excite an atomic state, and the delay in the phase of the emission fluorescence is related to the lifetime of the state. Link (**287b**) has determined the lifetimes of the upper states of the resonance lines of Na I, K I, Rb I and Cs I, with results in good agreement with earlier work. Lawrence and Savage (**283a**) have worked on B I, B II, C I, C II, N I and N II. Cunningham and Link (unpublished) have measured several states in Na I, Tl I, In I, Cu I, Ga I, Ag I, Pb I and Bi I, and Savage and Lawrence (unpublished) have studied Si I, Si II, P I, P II, S I, S II, O I, Ne II and Ar II. The method has been applied also to molecules. A delayed coincidence technique was used by Klose in work on Ne I and he has done further work on Ar I.

Prokofiev has reported as follows: The lifetime of an excited state of Cs 7 $^2P_{3/2}$ has been determined by the double-resonance method (**9**). Lifetimes of some states of Hg and Cd have been determined by the delayed-coincidence method (**471**). Oscillator strengths in the spectra of Fe I (**476**), Al I and Ga I (**376**) and Cu I and Ag I (**425**) have been determined by the anomalous-dispersion method. Relative values of oscillator strengths in the spectra of Fe II (**326**), Ti I, V I, Cr I, Fe I, Co I and Ni I (**325**) have been determined by the radiation method. The transition probability of the Ne I line designated $2s_2 - 2p_4$ (**118**), oscillator strengths of Cu I (**364**), and resonance peaks of Xe I (**82**) have been determined by the absorption method. Use of the combined measurement of full and linear absorption in a layer of vapours formed in the heated graphite tray has made it possible to determine the absolute values of oscillator strengths of a number of resonance peaks of Bi I, Sb I and Te I (**288**). Lifetimes of the states H I with $n = 3$, 4 and 5 (**11**) have been measured by the method of excitation of a beam of high-velocity hydrogen atoms in helium.

Corliss has prepared a Supplement to NBS Monograph 32 on Spectral Line Intensities, which contains a calibration of some 1400 lines short of 2450 Å (**102**). In Ni I he has reduced relative intensities and oscillator strengths of 888 lines between 2800 and 9900 Å to absolute values (**100**). He and Warner have carried out similar work on 2000 lines of Fe I between 2080 and 4150 Å (**101**).

II. *Theoretical Investigations of Oscillator Strengths*

Czyzak, Krueger and others (**105, 106**) have done extensive theoretical work on the calculation of atomic wave functions, transition probabilities (allowed and forbidden), and cross sections which are primarily of astrophysical interest. Most of the wave functions have been used for determining transition probabilities for lines in selected spectra of P, S, Cl,

Ar, K, Ca, Cr, Mn, Fe and Ni in various stages of ionization extending as high as Ar xv, K xiv, Ca xv, Mn xiii, Fe xiii, Ni xvii.

In collaboration with the above authors, Aller and his associates have calculated oscillator strengths for some 50 transitions of astrophysical interest in Na I, Mg II, Al III and Si IV, by using Hartree-Fock self-consistent field wave functions with exchange (**81**). A. B. Underhill has carried out similar calculations of *gf*-values for 25 selected lines of Fe II, Fe III and Ni II by using both the Hartree-Fock and Bates-Damgaard methods (**459**).*

Trefftz reports work on wave functions and oscillator strengths of Be-like ions, C III, N IV, O V (**381**) and similar work taking into account configuration mixing for some Fe- and Be-like ions (**437**).

Very extensive calculations of screened hydrogenic wave functions, based upon Layzer's theory, have been completed by Naqvi and co-workers (**333, 334**). The screening parameters were calculated for the ground state and a large number of excited states along isoelectronic sequences with 2 (He) through 55 (Cs) electrons. The calculations for each isoelectronic sequence were done for twenty members, starting at the negative ion, in some cases. The screening parameters for still higher members can be determined by interpolation. Thus, calculations have been completed for a large number of atoms and ions of the periodic table, including moderately and highly ionized atoms.

These hydrogenic wave functions are currently being used to calculate allowed and forbidden transition probabilities. The results for some allowed transitions of the lithium, beryllium, boron and carbon isoelectronic sequences have been published (**335**) and others are in preparation. The calculations are done in intermediate coupling wherever appropriate. Work is also in progress on calculations of transition probabilities by using Hartree-Fock wave functions including configuration interaction.

Forbidden transition probabilities for ions of the nitrogen isoelectronic sequence (N I to Fe XX) have been calculated by Naqvi (**336**) in intermediate coupling by using Hartree-Fock wave functions with configuration interaction and also screened hydrogenic wave functions. There is good agreement between these two sets of calculations for multiply charged ions and this agreement improves as the ionic charge increases. The approximate screened hydrogenic wave functions are quite good for calculations of forbidden transition probabilities for ions with an ionic charge of four or more.

Prokofiev has also reported progress on theoretical work. A criterion of quantum defect, is suggested as useful for the application of a Coulombian approximation in calculating transition probabilities (**197, 344**): it has been used for calculating the oscillator strengths in the sharp and diffuse series of Al I, Ga I, In I and Tl I, as well as for resonance peaks in the isoelectronic series from He I to Ar I (**343**).

Transition probabilities have been calculated for the magnetic-dipole and other forbidden transitions in isoelectronic series, particularly in B I and Al I (**343**).

Garstang has discussed forbidden transitions in excitation by electron impact (**170**), with special emphasis on interesting transitions in He. Because of the astrophysical importance of Ba II, he and Hill have published all *gf* values for a number of Ba II lines (**171**). He has also calculated energy levels, oscillator strengths and transition probabilities for selected configurations and lines in the Fe XVII spectrum. This contribution is particularly useful to those who are interpreting the far ultraviolet solar spectrum (**172**).

Levinson and Nikitin (**287**) have published a convenient *Handbook for Theoretical Computation of Line Intensities in Atomic Spectra.*

For molecular spectra, a review of work on intensities and transition probabilities is given by Phillips (see §11 p. 250).

*See Addendum on page 266.

SPECTRAL LINE BROADENING

(Committee 2a)

W. R. Hindmarsh

Of the causes of the broadening of spectral lines only Doppler broadening and the various types of 'pressure' broadening are important in astrophysical conditions. Doppler broadening is readily calculable if the velocity distribution of the emitting atoms is known; this report is concerned only with pressure broadening.

Considerable progress has been made in recent years in understanding the mechanism of broadening of spectral lines by collision between the radiating (or absorbing) atom and its neighbours. Weisskopf, as long ago as 1933, clearly appreciated the nature of the impact approximation, in which it is assumed that only the integrated effect over a collision on the phase and amplitude of the radiation matters; and of the quasistatic approximation, in which the perturbation of the frequency of the radiating atom by the fields due to neighbouring atoms is treated as though it were changing very slowly. The regions of validity of these approximations were early discussed by a number of authors, including Unsöld. The adiabatic theory of Lindholm and Foley was for some time the only available description which allowed numerical calculations in the impact approximation. Since then, the quantum mechanical formalism has been developed, especially by Baranger and Kolb and Griem; almost all numerical calculations have been based on the classical path approximation.

There are two distinct aspects of this problem – broadening due to interaction with charged particles (electrons and ions) on the one hand, and broadening due to interaction with neutral particles (atoms) on the other hand.

The theory of broadening due to charged particles has been greatly developed in recent years, especially by Griem and his collaborators, who were the first to include correctly the effect of collision-induced transitions. Earlier review articles by Böhm, Traving, Margenau and Lewis, Mazing, and Baranger, and more recent ones by Breene (**50**), van Regemorter (**399**), Wiese (**484**) and Griem (**195**) treat the subject in some detail. The last four discuss the influence of non-adiabatic collisions, and the last article includes extensive tables for calculation of the profiles (so far as they are determined by collisions with charged particles) of a large number of hydrogenic and non-hydrogenic ('isolated') lines of astrophysical importance. Comparisons between theory and experiment for broadening by charged particles have been reviewed by Baranger, Griem (**195**), Mazing and Wiese (**484**).

For hydrogenic lines a quasi-static theory, corrected for shielding effects (Ecker, Baranger and Mozer) is used for the broadening by ions. The quantum-mechanical non-adiabatic impact theory, including a correction due to Lewis, is used for the broadening by electrons. Approximate analytical expressions for the profiles of hydrogen lines have been given by Griem, although doubts have been expressed as to their validity for the electron densities and temperatures encountered in stellar atmospheres (Pfennig *et al.* (**383**)). The same difficulty seems also to be present in the work of Nguyen *et al.* (**338**). Griem's most recent wing formula (**196a**) is claimed to be free of these defects. For the lines $L\alpha$, $L\beta$, Hα, Hβ, and Hγ, experiments carried out at rather high electron densities ($\geqslant 10^{16}$ cm^{-3}) and temperatures ($\geqslant 10^4$ °K) give satisfactory agreement with theory (**195, 484**), although Boldt and Cooper (**44**) did not find agreement with Griem's wing formula for $L\alpha$. Careful comparisons of theoretical and experimental profiles for high Balmer and Paschen lines have been made by Ferguson and Schlüter (**152**) and by Vidal (**472**) at electron densities of 'particular astrophysical interest (about 10^{13} cm^{-3}), and an electron temperature of about 2000°K. More recent work by Vidal (**473**), Pfennig *et al.* (**383**), Pfennig and Trefftz (**382**), Schülter *et al.* (**414**), Schlüter and Avila (**415**) and Edmonds *et al.* (**135**) reaffirms that the use of a quasi-static theory for electrons as well as ions gives good results whenever it is expected to be applicable as estimated by Unsöld.

These studies also include investigations deep in the impact regime for electron broadening (**414**), where the electron contribution is observed to be weaker than the quasi-static one. The most recent impact calculations by Griem show the same tendency, although the electron impact contribution observed at low electron densities seems to be definitely smaller than calculated (**196a**).

For non-hydrogenic lines the most important contribution is due to impact broadening by electrons; inelastic collisions are taken into account (Vainshtein and Sobel'man), (Griem, Kolb, Baranger and Oertel), (Bréchot and van Regemorter (**49**)). Experimental work has indicated that measured widths do not agree well with the earlier theories for ion lines.* This is no doubt because, as theoretical calculations indicate (Griem and Shen), (Bréchot and van Regemorter (**49**)), straight-line trajectories should be replaced by hyperbolic trajectories in the case of lines from ions broadened by interactions with charged particles. Griem (**196**) has recently discussed the available experimental results, and shown that most of the disagreement with theory disappears if Coulomb effects and the broadening of the lower state are taken into account. Sobel'man (**428**) has examined the conditions in which spectral line-broadening can give information on the inelastic scattering of electrons by excited atoms.

The broadening due to charged particles of hydrogen lines in the radio-frequency region, corresponding to transitions $n \rightarrow n + 1$, has recently been treated by Griem (**196b**). Here most of the broadening is due to inelastic collisions with electrons, since ion and elastic collision electron broadening are reduced because of cancellation between the upper and lower states of the transition. The calculations are consistent with observations of such lines from H II regions.

Broadening due to interactions between the radiating atom and neutral atoms is much less well understood in detail than that due to charged particles. If the interaction between the radiating and perturbing atoms can be described by an inverse power potential $V(r) = C_p/r^p$ then, for a given $p > 3$, the Lindholm-Foley theory predicts a definite value for the ratio of broadening to shift for the spectral lines, provided that the impact approximation is valid. It is often assumed that the forces between different atoms are dispersion forces, $p = 6$, but a casual glance at experimental results (see, for example, Ch'en, and Takeo) shows that the ratio, broadening : shift, is far from constant. This is partly because many of the experiments have been carried out at pressures so high that the impact approximation is unlikely to give a good description of the line profile. It is emphasized that only results obtained at sufficiently low densities of the perturbing atom can be used for comparison purposes. The actual condition is that the profile should be emcompassed in a wave-number range $\Delta\nu \ll \frac{1}{2\pi c} (v^p/C_p)^{1/p-1}$ where v is the mean relative velocity of radiating and perturbing atoms; this means $N \lesssim 10^{19}$ cm^{-3} in most laboratory experiments. Most astrophysical conditions are such that the impact approximation for neutral atom broadening is valid. Even when only low-pressure results are considered, the ratio of broadening to shift does not have an even approximately constant value. The most probable explanation is that the interaction between neutral atoms must include a short-range repulsive term, so that the whole interaction is described by, possibly, the Lennard-Jones potential $V(r) = C_{12}/r^{12} - C_6/r^6$. The profile is always Lorentzian in the impact approximation, but the measured values of broadening and shift can be used to derive values of C_{12} and C_6. (Hindmarsh (**217**), Behmenberg (**30**), G. Smith, Hindmarsh, Petford and Smith.) The values of C_6 obtained in this way agree reasonably well (to within a factor of 2 in most cases) with calculated values, and the values of C_{12} show regularities which may also make possible the development of simple theoretical calculations for them. Most of these measurements have been made with rare gases used as the broadening agent, while in stellar atmospheres atomic hydrogen is the most important neutral broadener.

*See Addendum on page 266.

It is impossible at present to deduce within any certainty the short-range interaction for hydrogen from measures of the interaction for rare gases. A crucial experiment, not yet carried out, would be the laboratory measurement of the collision broadening and shift of a spectral line of astrophysical importance, by atomic hydrogen at low enough densities to be certain of the validity of the impact approximation. The measurements of Kusch, who used a water-stabilized arc, do not come into this category.

Much attention has recently been devoted to the 'resonance' broadening caused by collisions between atoms of the same kind. In conditions where the impact approximation is valid, theory and experiment now seem to agree to within a few per cent (Vaughan). A useful theoretical expression, which should be accurate to 10%, has been developed by Ali and Griem (**5**).

It is emphasized that all experiments for the determination of line profiles are very difficult; they must always be critically examined before the results are used for the interpretation of astrophysical data.

ATOMIC COLLISION CROSS SECTIONS
(Committee 2b)

The Sub-Committee on Cross Sections organized a Colloquium held at JILA in Boulder, Colorado, on 11–15 July, 1966. Here, four main topics were discussed by experts in each field: The Physics of Atomic Collision Processes; Line Broadening and Photon Scattering; Ionization and Recombination, and Data Centers; Astrophysical Problems Requiring Atomic Collision Data.

Plans for publishing the proceedings were discussed. Some review articles may be found in the *Reviews of Modern Physics* and others in JILA Reports. A review on atomic processes in stellar atmospheres prepared by B. E. J. Pagel (**365**) is of special interest to astrophysicists.

ATOMIC COLLISION CROSS SECTIONS
M. J. Seaton

The present report has been prepared by a group whose assistance is gratefully acknowledged: Branscomb, Burgess, Dalgarno, Hindmarsh, Van Regemorter, Bely and Geltman. The subject is a large and rapidly growing one. Only the more recent references are listed. A number of earlier ones can be found in general review papers and in the 1964 IAU Report for Commission 14.

I. *General Review*

Recently published books include a third much enlarged edition of the well-known text book by Mott and Massey on *The Theory of Atomic Collisions* (**331**); a book on the physics of atomic collisions in which the emphasis is mainly experimental (**207**); an introduction to atomic collision phenomena in ionized gases (**306**); and a book on the theory of electron-atom collisions (**127**). Review articles and contributed papers are published in two conference reports (**194, 307**), and a new review journal has been started (**21**). A number of reviews are being published by the JILA Data Center; these are concerned with experimental (**252**) and theoretical (**407**) work on electron impact ionization (not discussed further in the present report), and work on electron impact excitation (**320**). Two reviews have been published on resonance phenomena in electron collisions (**65, 427**). The Riga group have published collected papers on collision theory (**465, 467**) and a review of recent work on electron-atom collisions by Ya Veldre and Peterkop (**21**).

II. *Data Centers*

A service of systematic collection and assessment of atomic collision data is provided by two Data Centers:

1. The Information Center of the Joint Institute for Laboratory Astrophysics maintains a complete collection of papers reporting measurements or calculations of low energy ($\leqslant$ 10 keV) electron collision cross sections for atoms and astrophysically interesting molecules. Several reports have been published (**250, 251, 252, 320, 407, 465**). A similar programme is under way for photo-ionization. Further information about the Center can be obtained from Dr L. J. Kieffer, Information Center, Joint Institute for Laboratory Astrophysics, University of Colorado, Boulder, Colorado 80302, U.S.A.

2. The Atomic and Molecular Processes Information Center, Oak Ridge National Laboratory, provides a service of collecting, storing, evaluating and disseminating information, with particular reference to work concerned with heavy particle collisions, particle interactions with electric and magnetic fields, and particle penetration into matter. Several reports have been published (**346, 347, 348**) and further information may be obtained from Dr C. F. Barnett, Director, Atomic and Molecular Processes Information Center, Oak Ridge National Laboratory, Post Office Box Y, Oak Ridge, Tennessee 37831, U.S.A.

III. *Electron Collisions*

(*a*) *Collisions with neutral atoms.* Although the energy of the He atom can be calculated to 7 significant figures, and phases for $e - \mathrm{H}$ elastic scattering can be calculated to 3 or 4 figures, very considerable effort is required in order to calculate cross sections for near threshold electron impact excitation of neutral atoms to an accuracy much better than a factor of two (**67**). It is clear that the development of new methods of making such calculations is much to be desired. On the experimental side, developments in atomic beam and electron energy selection techniques have resulted in advances in the study of low energy inelastic scattering processes, and of resonances in elastic scattering, but much more could be done in obtaining reliable absolute cross section data.

There has been a great deal of interest in experimental and theoretical studies of elastic scattering resonances (see the review articles (**65, 427**)). Since the resonances have widths $< 0{\cdot}1$ eV, they will generally not be of importance for electron scattering phenomena in astrophysics, for which the electron energy spread is much greater than 0·1 eV. The associated radiative processes of photon absorption in diffuse auto-ionization lines, and of dielectronic recombination, are of much greater astrophysical interest.

Calculations of elastic scattering cross sections have been made by using the method of polarized orbitals (**169, 401, 439**) and calculations for the alkalis have been made by Karule and others by the method of eigenfunction expansions (**465**). An elaborate study of elastic scattering in C, N and O and of excitation of the ground configuration terms has been made by Smith, Henry and Burke.

It is predicted by theory (**167**) that, due to the degeneracy of all hydrogen states with the same principal quantum number n, $e - \mathrm{H}$ inelastic cross sections will have an oscillatory behavior in the near threshold region and will remain finite at threshold. This has been confirmed experimentally for $1s$–$2s$ and $1s$–$2p$ transitions in H (**80, 194**).

Systematic experimental studies have been made of excitation functions of observable lines in Na and K (**490, 491**) and a number of excitation functions of He have been measured (**219, 309**). Further experiments have been performed on the polarization of electron impact radiation (**194, 200**): for the alkalis there is good agreement with theory (**194**) but in a number of other cases the polarization is found to vary rapidly in the near threshold region, and further theoretical clarification is required.

(*b*) *Approximate formulae.* The Born approximation gives useful results for transitions in which exchange is unimportant, and is exact in the limit of high energies. A number of new calculations have been made for H (**359, 468**) and He (**161, 276**).

By using measured electron angular distributions at high energies, it is possible to deduce Born approximation cross sections for all values of the energy and, on extrapolating the data, to obtain optical oscillator strengths (**277, 281, 423**).

The Born-Oppenheimer approximation gives poor results for exchange processes, and attempts have been made to obtain alternative simple approximations for exchange scattering (**3**). One approach is to impose certain orthogonality conditions on the wave functions; this has been applied to various transitions in H and He (**31, 32, 33**). An approximation first proposed by Ochkur (**350**) consists essentially of retaining only the leading term in the expansion of the exchange amplitude in inverse powers of the energy (**351, 405**). Extensive calculations for excitation of H (**406**) and for excitation of He from the ground state (**352**) and from the 2 ^{3}S state (**353**) show good agreement with experiment at high and moderate energies. Another simple alternative to the Born and Oppenheimer approximations is to use a wave function in which allowance is made for the electron-electron interaction (**460, 461**). The method gives good agreement with experiment for excitation of H (**104, 360**).

Extensive developments have been made in the use of classical theory for the calculation of electron impact excitation and ionization cross sections. In most of this work exact classical expressions are used for binary electron-electron collisions, and various assumptions are made concerning the relative velocities of the impinging and atomic electrons (**61, 255, 256, 349, 433, 474**). In addition to this work on approximate classical theories, some work has been done by Percival on the exact solution of the classical problem corresponding to electron impact ionization of H. It is important to realize that classical theories can be used for electron exchange processes (**61, 349**). Extensive tabulations have been made for classical approximations to cross sections in N, O, Ne, Ar, Xe and Kr (**466**).

In semi-classical impact parameter theories, similar to the Bethe approximation, classical orbits are used for the colliding electron, and the probability of an atomic transition is calculated by using quantum perturbation theory. Such theories are superior to classical binary encounter theory at high energies (**257, 416**). Burgess (**61**) used a combination of binary encounter theory and semi-classical impact parameter theory.

Approximate formulae for allowed transitions, in which the cross section is taken to be proportional to the oscillator strength multiplied by an empirical factor $\bar{g}$, can give useful estimates but may be in error by factors of 2 or 3. Some improvement should result from using a form of the Bethe approximation involving a cut-off in the impact parameter (**61**). When formulated in such a way that conservation conditions are satisfied, this method should give reasonable cross sections for transitions with strong coupling, such as $n \rightarrow n \pm 1$ transitions in H (**258, 412**).

(*c*) *Excitation of positive ions.* The $\bar{g}$ formula may be used for approximate estimates but may give cross sections in error by as much as a factor of 2 or 3 (**66**). Modified classical and semi-classical impact parameter-methods may also be used for positive ion excitation (**61**). Quantum mechanical calculations are generally more accurate for positive ions than for neutrals. The following methods may be used:

(*i*) The Coulomb-Born (CB) approximation uses Coulomb waves in place of plane waves.

(*ii*) The Close Coupling (CC) approximation requires the solution of coupled integro-differential equations.

(*iii*) The Coulomb-Exchange (CE) approximation (**35**) is similar to the approximation of Ochkur (**350, 405**) for exchange excitation of neutrals.

(*iv*) A number of improvements have been made in the Distorted Wave (DW) approximation (**42**).

(*v*) The Exact Resonance (ER) approximation is used for transitions between states of the same configuration (**42**).

(vi) In the Quantum Defect (QD) method, information obtained from observed energy levels is extrapolated to positive energies, and used to calculate cross sections (**421**).

Table 9 gives a summary of calculations for electron impact excitation of positive ions. Further references are given in a previous report (**420**).

Table 9. Calculations for electron impact excitation of positive ions

(see also (**420**))

Individual Ions

Ion	States	Approximation	Reference
He^+	$1s, 2s, 2p$	CC	**64, 305**
Li^+	$1\,^1S, 2\,^1S$	CB	**445**
N^+	$2s, 2p, 3s, 3p, 3d$	CB, CC	**66**
O^{5+}	$2s, 2p, 3p, 4p$		**463**
Fe^{15+}	$3s, 3p$	CB	**269**
Fe^{7+}	$3d, 4s, 4p, 4f, 5s, 5f, 6s, 6f, 7f$	CB	**270**
Fe^{14+}	$3s^2\,^1S, 3s\,3p\,^1P, 3s\,3p\,^3P, 3s\,3d\,^1D, 3s\,3d\,^3D$	CB	**38**
Fe^{13+}	$3s^2\,3p\,^2P$ and $3s\,3p^2\,^2P, ^2D, ^2S$	CB	**380**
C^+	$2p\,^2P_{1/2}, ^2P_{3/2}$	QD	
Si^+	$3p\,^2P_{1/2}, ^2P_{3/2}$	QD	
N^{2+}, O^{3+}	$2p\,^2P_{1/2}, ^2P_{3/2}$	QD	**363**
Ne^+	$2p\,^2P_{1/2}, ^2P_{3/2}$	QD	**363**

Isoelectronic Sequences

Sequence	Transitions	Approximation	Reference
Li	$2s–ns, 2s–np, 2s–nd$	CB	**36, 37**
B	$2p$ } $^2P_{1/2} - ^2P_{3/2}$	QD and DW	**42**
F	$2p^5$ } $^2P_{1/2} - ^2P_{3/2}$	QD and DW	**42**
Al	$3p$ } $^2P_{1/2} - ^2P_{3/2}$	QD and DW	**42**
Cl	$3p^5$ } $^2P_{1/2} - ^2P_{3/2}$	QD and DW	**42**
C	$2p^2$ } $^3P - ^1D$	CE	**413**
Si	$3p^2$ } $^3P - ^1D$	CE	**413**
C	$2p^2\,^3P - ^1D, ^3P - ^1S, ^1D - ^1S$	DW, ER	**107**
P	$3p^2\,^3P - ^1D, ^3P - ^1S, ^1D - ^1S$	DW	

(*d*) *Electron collisions with molecules.* Calculations for electron collisions with molecules are obviously more difficult than calculations for collisions with atoms. The Born approximation has been used to calculate the cross sections for dissociating transitions in H_2^+ and H_2 (**249, 372, 492**), and measurements have been made for dissociation of H_2^+ by electron impact (**129**). Classical methods have been used to calculate excitation cross sections in heavier molecules (**26**). Elastic scattering by H_2 has been considered in connection with the problem of calculating the free-free absorption coefficient of H_2^- (**431**).

Inelastic differential cross sections have been measured for O_2, CO and N_2 (**278, 279, 280**): oscillator strengths can be obtained from extrapolation of the data. Beam measurements have been made of total cross sections in H_2, O_2 and N_2 (**2, 185**) in the energy range 0·5 to 15 eV. Swarm measurements in molecular gases (**136**) are of importance for studies of energy loss mechanisms. Some calculations have been made of scattering lengths (**286**).

Resonant structure has been observed in the scattering of electrons by H_2 (**184, 274**) and N_2 (**417**), and the theory has been discussed in terms of compound negative ion states (**84**).

Much computational effort has been devoted to the determination of cross sections for vibrational (**53, 446**) and rotational (**85, 109, 357, 410, 447**) excitation of H_2, N_2 and O_2. Experimental cross sections have been obtained (**396, 418**) for the dissociative attachment process $e + H_2 \rightarrow H + H^-$. In this work the target H_2 is mainly in the lowest vibrational state. For the inverse process, $H + H^- \rightarrow e + H_2$, which is of astrophysical interest (**365**), many final vibrational states must be considered.

IV. *Recombination*

A recent general survey by Bates (**17**) may not be readily available; material cited there will, therefore, be included here.

1. *Radiative recombination* coefficients for separate hydrogenic *nl* states have been tabulated by Burgess (**60**) (for $n \leqslant 20$ and electron temperature 0 to ∞, by Boardman (**43**) (for $n \leqslant 10$ and $10^3 \leqslant T \leqslant 10^6$) and by Glasco and Zirin (**181**) (for $n \leqslant 15$ and $2{\cdot}5 \times 10^3 \leqslant T \leqslant 2{\cdot}56 \times 10^6$).

2. The importance of *dielectronic recombination* has been discussed by Burgess (**59, 62**) and a simple general formula for the coefficient in low density plasmas given (**63**). A simple treatment given by Tucker and Gould (**458**) involves assumptions which are not entirely correct.

3. Electron-molecular ion *dissociative recombination* has been partially reviewed by Biondi (**307**). The basic mechanism suggested by Bates appears to have been confirmed by the experiments of Biondi and associates (**40, 95**) on Ar_2^+ and Ne_2^+. Their experiments on He_2^+ (**402**) also support Bate's hypothesis, but these have been criticised by Ferguson *et al.* (**153**).

Experimentally determined recombination coefficients have been reported for: He_2^+ (**362**); N_2^+, N_4^+ (**199, 242**) (in good agreement); NO^+ (**198**); O_2^+ (**41**); Ne_2^+ (**95, 215, 362**) (in good agreement); Ar_2^+, Kr_2^+, and Xe_2^+ (**362**); Cs_2^+ (**203, 206**) (not at comparable temperatures). Approximate temperature variations of the coefficients are given in (**198, 199, 215**). Farhat (**140**) claims that the temperature variation for neon may be deduced from experiments of Taylor and Herkowitz but this is disputed (**452**).

The situation with respect to He_2^+ is rather complicated and is partially reviewed in (**153**) (see also comments in (**95**)); the abnormally small rates observed (**83, 362**), which may be due to the absence of any potential curve 'crossing' in the low vibrational states, means that dissociative recombination may not be the dominant process—Ferguson *et al.* (**153**) suggest collisional-radiative recombination instead, while Collins (**94**) suggests *collisional-dissociative recombination* in which a crossing is not necessary, the transitions to a repulsive state of the neutral molecule taking place due to collision with a third body.

A perturbation theory calculation for H_2^+ has been made by Wilkins (**487**). His conclusion that the calculation is in excellent agreement with experiment is unjustified—the experiments provide only an upper bound and may concern H_3^+ predominantly.

4. *Collisional-radiative recombination* has been reviewed briefly by Bates (**307**). Calculations which take into account the plasma boundary conditions to eliminate the electron temperature as an independent variable have been carried out for decaying $He^+ - e - H$ and $He^+ - e - He$ plasmas by Bates and Kingston (**20**). Simple approximate theories which agree well with the more detailed calculations have been given by several authors (**10, 18, 295**).

Several dense plasma experiments in fair to good agreement with theory have been reported (**330**) on He, (**4** and **206**) on Cs, (**99**) on H, but the lower density Cs plasma measurements of Hammer and Aubrey (**203**) disagree with theory by a factor ~ 20.

5. *Three-body electron-ion recombination* in which the third body is a neutral atom has been treated in an elegant energy diffusion approximation by Pitaevskii (**384**). Detailed statistical calculations have been made by Bates and Khare (**22**).

The case in which one has homopolar molecules instead of atoms has been treated by Dalidchik and Sayasov (**113**) by using Pitaevskii's method.

6. *Ionic recombination.* The simple classical theory of Thomson (corrected by Loeb) for three-body recombination at low densities has been refined to some extent by Natanson (who also treats high densities), Bruckner (**56**) and Mahan and Person (**293**). Contrary to Bruckner's own estimates, more detailed calculations (**150**) following Bruckner's model agree quite well with Thomson. Keck and Carrier (**247**) have carried out calculations based on ideas similar to those used by Pitaevskii (**384**). Detailed statistical calculations (**22, 307**), in principle exact at low densities, have been carried out by Bates and Moffett (**23, 24**). Their results agree very closely with those of Thomson.

Experimental data on three-body ion-ion recombination, and on two-body recombination by electron transfer, have been reported by Mahan and Person (**292**) and Carlton and Mahan (**76**) for a wide variety of molecules. Two-body recombination in iodine has been measured by Greaves (**192**). Recombination in oxygen has been measured by Jütting *et al.* (**240**).

v. *Heavy Particle Collisions*

Atom-atom collision processes are of astrophysical interest at thermal energies and at very high energies. The thermal energy phenomena require special consideration, but at high energies the Born approximation may be used. Born cross sections for atomic excitation and ionization by proton impact are comparable to, but larger than, the corresponding cross sections for electron impact at the same velocity. Excitation and ionization by heavy particles other than bare nuclei is complicated by the possibility of excitation and ionization of the incident particle. Collisions in which both incident and target particles undergo transitions may be more probable, at high energies, than transitions in which only one particle undergoes a transition. High energy neutrals are of little astrophysical interest since they are readily ionized and the reverse processes of charge transfer or electron capture are improbable (**47, 300**). Heavy particles other than bare nuclei can induce intercombination transitions. At high energies the valence electron of the incident heavy particle may be regarded as a free electron: calculations have been made by Bates and Crothers (**25**) for $H + He\,(1\,^1S) \rightarrow H + He\,(2\,^3S)$.

Reviews of heavy particle collision phenomena have been given by Bates and by Wilets and Gallaher (**307**).

(*a*) *High Energy Collisions.*

H^+ and H incident on H and He are of particular astrophysical interest. The Born approximation is generally used, but more refined calculations are of interest in assessing its accuracy and range of validity (**19, 86, 93, 163, 226, 289, 486**).

Excitation of H *by* H^+. Born calculations have been made for $H^+ + H\,(1s) \rightarrow H^+ + H\,(2s, 2p, 3s, 3p, 3d)$ and for $H^+ + H \rightarrow H^+ + H^+ + e$ (see (**369**)). The calculations for ionization are in agreement with experiment (**179**) above 40 keV but at lower energies give cross sections greater than those determined experimentally. The Born cross section for $2p$ excitation is about twice the measured cross section (**436**) between 5 and 30 keV, but is smaller than the measured cross section below 3 keV.

Cross sections for other transitions in H produced by H^+ impact may be deduced from electron impact data (**74**). A simple formula has been obtained (**303**) for excitation from the ground state to any level with large principal quantum number n. The total cross section for excitation of all discrete states has also been obtained (**71**). Cross sections have been calculated for $H^+ + H\,(n) \rightarrow H^+ + H\,(n')$, $n = 2, 3, 4, 5$ and 10 and $n - n' = 1$ or 2: as n increases proton impact for such transitions becomes more important than electron impact (**74**).

Excitation and ionization of H *by* H. Cross sections have been calculated by Bates and others for the excitation processes H (1*s*) + H (1*s*) → H (n_1l_1) + H (n_2l_2) and for the ionization processes H (1*s*) + H (1*s*) → H (n_1l_1) + H^+ + *e* or H^+ + *e* + H^+ + *e* (**303**) and useful formula have been obtained for high energy cross sections summed over various final states.

Calculations have been made of velocity distribution of electrons produced in ionization of target atoms and projectile atoms. Studies have been made of collisions for which the target atom is in an excited state (**45**) and for which both projectile and target atoms are in excited states (**387**). The charge transfer process, H + H → H^+ + H^-, has also been investigated (**299**).

Charge transfer of H^+ *in* H. Cross sections for electron capture, H^+ + H (1*s*) → H (*nl*) + H^+, have been computed by Bates and Dalgarno by using the Brinkman-Kramers form of the Born approximation for states up to 4*f*, and this work has been extended to higher final states and to capture from excited atoms (**70, 218, 302, 304**). The absolute accuracy is open to question but relative values for capture into different excited states appear to be satisfactory (**218**). Numerous comparisons of different first-order approximations have been made.

Stopping power of H^+ *in* H *and of* H^+ *in an* H^+, *e plasma.* The efficiency with which a beam of protons is slowed down in a gas of ground state hydrogen atoms has been examined in detail and the mean energy expended in producing an ion pair has been calculated by Dalgarno and others.

The energy loss of fast protons in an H^+, *e* plasma has also been calculated; the presence of free electrons produces a marked increase in the loss rate.

Excitation and ionization of He *by* H^+ *and* H. Calculations for excitation and ionization are much less extensive for He than for H. The following processes taken mostly from earlier work have been studied:

$H^+ + He\,(1\,^1S) \rightarrow H^+ + He\,(2\,^1P)$
$H^+ + He\,(1\,^1S) \rightarrow H^+ + He\,(3\,^1P)$
$H^+ + He\,(1\,^1S) \rightarrow H^+ + He^+\,(1s) + e$
$H^+ + He\,(1\,^1S) \rightarrow H^+ + He^+\,(nl) + e$
$H + He\,(1\,^1S) \rightarrow H\,(\Sigma) + He\,(2\,^1P)$, $H + He\,(2\,^3S) \rightarrow$
$H\,(\Sigma) + He\,(2\,^3P)$, $H + He\,(2\,^3S) \rightarrow H\,(\Sigma) + He\,(3\,^3P)$
$H + He\,(1\,^1S) \rightarrow H + e + He\,(\Sigma)$.

Σ here denotes a sum over all final states.

The calculated ionization cross sections agree with measured cross sections at energies above 400 keV, but are larger than measured cross sections at lower energies. These results are prior to 1963 and individual references are not included.

Charge transfer of H^+ *in* He. The Born approximation has been used (**298**) to calculate cross sections for $H^+ + He\,(1\,^1S) \rightarrow H\,(nl) + He^+\,(n'l')$. The calculated total charge transfer cross section agrees with experiment above 40 keV, and the calculated cross section for capture into the 3*s* state of H agrees with experiment (**224**) above 100 keV but is greater than the experimental cross section at lower energies. Experimental evidence (**224**) suggests that the cross section for capture into the *n*th state of H varies as n^{-3} for impact energies above 20 keV. At energies below 20 keV, the calculated cross sections for capture into H (2*s*) and H (2*p*) are considerably larger than measured values (see (**227, 307**, p. 618)).

The capture process for the case of H and He^+ in their ground states has been studied by using higher order approximations (**46, 48, 193**).

(*b*) *Low Energy Collisions.*

Excitation by proton impact. Cross sections for excitation by electron impact and by proton impact are broadly similar as functions of the velocity of the electron or the proton. For the electron case, cross sections rise rapidly from threshold to a maximum which occurs at a velocity corresponding to about twice the threshold energy, ΔE; the cross sections then decrease asymptotically as $E^{-1} \ln E$ for optically allowed transitions or E^{-1} for optically forbidden transitions. It follows that, in thermal conditions, proton impact will be more important than electron impact only if $kT \gg \Delta E$. A process for which this condition is satisfied is $H^+ + H(2s\frac{1}{2}) \rightarrow H^+ + H(2p\frac{1}{2}, 2p\frac{3}{2})$. The corresponding charge transfer process, $H^+ + H(2s\frac{1}{2}) \rightarrow H(2p\frac{1}{2}, 2p\frac{3}{2}) + H^+$, has a much smaller cross section. Proton impact can also be of importance in redistributing angular momentum, $H^+ + H(nl) \rightarrow H^+ + H(nl \pm 1)$, for higher excited states (**374**). For proton collisions with positive ions the repulsive Coulomb field reduces the cross sections, but under coronal conditions the reaction rate for $H^+ + Fe^{13+}(3p\frac{1}{2}) \rightarrow H^+ + Fe^{13+}(3p\frac{3}{2})$ is still comparable with the rate for direct excitation by electron impact (**419**).

The condition $kT \gg \Delta E$ can also be satisfied for transitions between molecular rotational states. Calculations have been made (**453**) for $H^+ + CN(J = 0) \rightarrow H^+ + CN(J = 1)$, which may be of importance in connection with the cosmic microwave radiation at $\lambda = 2{\cdot}63$ mm (**158**). It may be expected that the cross sections for excitation of higher rotational levels will be of a similar order of magnitude.

Charge transfer. The process of symmetrical resonance charge transfer, $H^+ + H \rightarrow H + H^+$, does not involve an electronic transition and its cross section can be predicted reliably (**373**). The process is of importance in considering diffusion of H^+ in H.

Asymetric charge transfer, $H^+ + X \rightarrow H + X^+$, involves an electronic transition and will, in general, proceed very slowly at thermal energies. Exceptions occur and one such is $H^+ + O(^3P) \rightarrow H + O^+(^4S)$, which is accidentally resonant. Cross section estimates for the reaction at thermal energies, which are of interest for astrophysical and geophysical problems, have been obtained from an analysis of experimental data at higher energies and from an analysis of upper atmosphere data.

Radiative charge transfer. Radiative charge transfer may proceed more rapidly than ordinary charge transfer at thermal energies. Calculations have been made by using a simple theory. Reactions investigated are $He^{2+} + H \rightarrow He^+ + H^+ + h\nu$ and $He^{2+} + He \rightarrow He^+ + He^+ + h\nu$ (**6**).

Radiative association. Calculations have been made for $H^+ + H \rightarrow H_2^+ + h\nu$, $C + H \rightarrow CH + h\nu$, $C^+ + H \rightarrow CH^+ + h\nu$ and for the free-free transition $H^+ + H \rightarrow H^+ + H + h\nu$. The rate of formation of HeH^+ has not been calculated though the molecule undoubtedly exists (**205, 315**).

The process $H + H + h\nu \rightarrow H_2'$, free-bound absorption, has been investigated by Soloman and others (**429**) for the transition $1s\sigma\, 2p\sigma\, {}^3\Sigma_u^+ \rightarrow 1s\sigma\, 2s\sigma\, {}^3\Sigma_g^+$, and does not appear to be an astrophysically important source of opacity.

Radiative association of two hydrogen atoms, $H + H \rightarrow H_2 + h\nu$, has a very small rate coefficient (**297**).

Collision-induced radiative deactivation. Because of the large polarizability of $He(2\,^1S)$, the reaction $H^+ + He(2\,^1S) \rightarrow H^+ + He(1\,^1S) + h\nu$ proceeds quite rapidly. A similar study has been made for $He(1\,^1S) + He(2\,^1S) \rightarrow He(1\,^1S) + He(1\,^1S) + h\nu$ (**7**).

Associative detachment. Reactions of the type $X + Y^- \rightarrow XY + e$ provide efficient mechanisms for the removal of negative ions. A simple model for $H + H^- \rightarrow H_2 + e$ suggests a rate coefficient of 10^{-10} cm^3 s^{-1}. Developments in the theory of resonating states should yield a more precise estimate. The process is of importance in connection with the H^- abundance in stellar atmospheres (**365**) and in that it provides a mechanism for the formation of H_2.

Mutual neutralization. Reactions of the type $X^+ + Y^- \rightarrow X + Y$ usually proceed rapidly. The case of $H^+ + H^- \rightarrow H + H$ has been studied in detail.

Penning ionization. Processes of the type $X^* + Y \rightarrow X + Y^+ + e$ can also occur with high efficiency. Rate coefficients for He (2 ^{1}S) + H → He (1 ^{1}S) + H^+ + e and He (2 ^{3}S) + H → He (1 ^{1}S) + H^+ + e may be of the order 10^{-11} cm^3 s^{-1}. Much larger values may occur, especially when a pair of excited atoms is involved.

Excitation of rotation of H_2 *by* H *impact.* Calculations have been made of the rate coefficients for H + H_2 (J = 0) → H + H_2 (J = 2), which is of importance for the cooling of the interstellar gas. Further refinements (**8, 112**) give reaction rates which are in good general agreement with the earlier work but somewhat smaller at the higher temperatures considered. The main uncertainty is in the H – H_2 interaction potential. Rate coefficients have also been obtained for other rotational transitions in H_2 and D_2 (**112**), and for 0 → 1 rotational transitions in CH, CN and OH produced by H atoms.

Calculations have been made for H_2 (J = 0) + H_2 (J = 0) → H_2 (J) + H_2 (J') (**115**).

Transitions between fine and hyperfine structure levels by H *atom impact.* Calculations have been made (**108**), in varying degrees of refinement, of the reaction rate for H + H (F = 0) → H + H (F = 1), which is of importance for 21 cm studies.

The processes H + O (^3P_J) → H + O ($^3P_{J'}$) and H + C^+ (^2P_J) → H + C^+ ($^2P_{J'}$) and H + Si^+ (^2P_J) → H + Si^+ ($^2P_{J'}$) contribute to the cooling mechanism in the interstellar gas. Recent work (**426**) suggests that the H + O reaction may be the most important, but the theory used was one developed for the study of collision-induced changes in hyperfine structure (**111**) and may not be adequate for changes in fine structure (**73, 342**).

Quenching collisions. Reactions of the type X + Na ($^2P_{\frac{1}{2}}$) → X + Na ($^2S_{\frac{1}{2}}$) have been studied theoretically (**341**) but the lack of adequate wave functions for the quasi-molecule XNa has prevented quantitative predictions. Experimental results are available for the case in which X is an inert gas (**435**). Further study of the case for which X is atomic hydrogen may be of importance for work on the formation of strong lines (**365**).

Chemical reactions. Exothermic reactions which produce H_2 and CH, OH, NH, CN, CO and N_2 have been discussed. It is conventional practice to express the rate coefficient in Arrhenius form.

Transport properties. The thermal conductivity and viscosity of atomic hydrogen have been calculated to high accuracy at high temperatures and low temperatures (**57**). Formulae are available for the transport coefficients of fully ionized plasmas (**116**).

Ion-molecule reactions. New laboratory techniques have been developed for the study of reactions of ground state atomic and molecular ions (positive and negative) with neutral species, including unstable species such as O, N and O_3 (**148**). Measurements are made at 300°K. Results have been obtained for a large number of reaction rates which are of interest for studies of the upper atmospheres of the Earth (**154, 155**) and of Mars (**345**) and of processes in comet tails (**148**). It is found that exothermic charge transfer to molecules is generally fast. In the case of negative ions, many fast associative detachment reactions are found to occur (**149**).

A comprehensive survey and review is being prepared (**78**).

MOLECULAR SPECTRA

(Committee 3)

J. G. Phillips

For the sake of brevity, this report will be limited to progress in the study of those molecules that are of immediate interest to astrophysicists. This means that the main emphasis will be on certain diatomic molecules, though a few of the simpler polyatomic molecules will have to be included. Even with this limitation a reasonably complete bibliography of papers that have appeared since the Hamburg meeting would include over 220 titles. Not all of these are listed here; instead, the emphasis will be on papers summarizing information on the various molecules.

Before turning to the individual studies, it might be worth mentioning that a complete card file bibliography is maintained at Berkeley, and inquiries are welcome. In addition, a bi-monthly Newsletter is distributed from Berkeley in collaboration with S. P. Davis of the Physics Department. This Newsletter includes information on work currently in progress at various laboratories.

1. *Analyses and Molecular Constants*

N_2. Several investigations of the nitrogen spectrum in the vacuum ultraviolet have been carried out. Tilford, Vanderslice, Wilkinson and others at NRL have observed the absorption spectrum between 1060 and 1520Å (**455**). In addition, the vacuum ultraviolet nitrogen afterglow has been described by Tanaka *et al.* (**448**), and Codling has investigated the structure in the photo-ionization continuum of N_2 near 500Å (**90**). Miller has described some high-resolution Vegard-Kaplan bands observed in emission (**317**). Isotope shifts have been observed by Mahan-Smith and Carroll in some weaker systems (**294**), while Ogawa *et al.* have studied isotope shifts in the vacuum ultraviolet (**354**). As far as theoretical calculations are concerned, Rydberg levels have been calculated by Lefebvre-Brion and Moser (**285**). Sahni and De Lorenzo have carried out self-consistent field calculations of the ground, ionized, and excited states of both N_2 and O_2 (**409**).

CN. The rotational and vibrational constants of the Red System are being re-evaluated by Rigutti and Poletto at Arcetri; Fishburne and others at Ohio State University have extended the observation of the Red System beyond 1 μ. The results of the latter are to be published in the *J. molec. Spectrosc.* Published studies include one by Radford on the hyperfine structure of the B $^2\Sigma^+$ state (**390**), and a report by Evenson *et al.* on the optical detection of microwave transitions between excited electronic states and the identification of the transitions involved (**138**). By using high resolution solar spectra obtained at the Jungfraujoch by Delbouille and Roland, Benedict has identified about 2860 lines between 7498 and 12 016Å as being due entirely or partly to solar CN. With the aid of these observations, it has been possible to complete the laboratory work by Phillips concerning this molecule.

H_2, HD, D_2. An important series on the absorption spectra of these molecules is being published by Monfils (**321**). General studies of electronic excited states of H_2 have been carried out by Browne (**54**), Rothenberg and Davidson (**404**), and Wakefield and Davidson (**475**). The absorption spectrum in the vacuum ultraviolet is being investigated by Namioka (**332**); the infrared spectra of hydrogen and helium have been described by Gloerson and Kieke (**183**). Calculations of fine structure constants of metastable H_2 in the C $^3\Pi_u$ state have been carried out by Chiu (**87**); Ross and Phillipson have calculated force constants for both H_2 and H_2^+ (**403**). A molecular orbital description of the lowest excited singlet state of H_2 has been published by Kato *et al.* (**243**). Wave-functions for H_2 and H_2^+ have been calculated by Goodisman (**186**).

He_2. Of primary importance is a series of articles by Ginter on the spectrum and structure of this molecule (**180**). In addition, Browne has studied some excited states of the helium molecule (**55**).

C_2. (See also p. 250). The electronic structure of C_2 has been described by Fougere and Nesbet (**160**). Some new bands in the Swan system of C_2 have been reported by Bugrim and his co-workers (**58**). Solomon and Stein have studied the infrared absorption of C_2, as well as CO and CN (**430**). The analysis of 35 bands of the Swan system of C_2 has been completed at Berkeley, and is being published by the University of California Press.

Hydrides and Deuterides. Milligan and Jacox have reported on the infrared spectrum of NH as produced in the photolysis of NH_3 in inert and reactive matrices (**318**). A rotational extension of the ($c\,^1\Pi - a\,^1\Delta$) system of NH and ND has been published by Shimauchi (**422**). Asundi writes that the Spectroscopy Division, Atomic Energy Establishment, Trombay, has submitted for publication an analysis of the (0 – 1) and (1 – 1) bands of the ($d\,^1\Sigma - c\,^1\Pi$) system of NH, and has completed the rotational analysis of the (0 – 0) band of the (A $^3\Pi$ – X $^3\Sigma$) system of ND.

The OH radical has been studied in the infrared and vacuum ultraviolet by Ogilvie (**356**) and Felenbok and Czarny (**151**), respectively. The latter identified the (C $^2\Sigma$ – X $^2\Pi$) system of OH and OD. Finally, the observation of a sequence in the (A $^2\Sigma$ – X $^2\Pi$) system of OH has been reported by Stoebner *et al.* (**438**).

The recent interest in discrete lines in the radio spectrum of interstellar matter has resulted in laboratory and theoretical investigations of the radio frequency lines of CH by Douglas and Elliott (**123**) and Goss (**188**).

The analysis of a CaH band at 2883 Å and CaD bands at 2720 Å, 2789 Å and 2877 Å has been published by Khan (**248**).

Oxides. The past two years have seen the appearance of many studies of the spectrum of NO. Of fundamental importance is a series by Miescher and various collaborators (**126, 237, 275, 316**). Horn and Dickey have reported on the near-infrared emission spectrum of NO (**220**); observations in the same region have led to the calculation of molecular constants of NO by Olman *et al.* (**358**). Rydberg series in the NO spectrum have been investigated by Lefebvre-Brion and Moser (**284**) and by Huber (**221**). Representative of papers on the rotation-vibration spectrum of NO are reports by Meyer and Haeusler (**313**), Aubel and Hause (**13**) and Meyer *et al.* (**314**).

Krupenie of NBS has published a review entitled *The Band Spectrum of Carbon Monoxide* (**272**). It includes a description of CO, CO^+, and CO^{++}. He is, also, preparing a review on the spectrum of molecular oxygen. Also of importance are molecular constants derived by Rank *et al.* for CO by using a heated absorption tube in the infrared (**394**). Other studies of the CO infrared spectrum include work by Weinberg *et al.* on a hot band at 4·7 μ (**478**), and by Solomon and Stein (**430**). At the extreme end of the spectrum, we have a report on the radio-frequency spectrum of the $a\,^3\Pi$ state of CO by Freund and Klemperer (**162**). The high resolution vacuum ultraviolet spectrum has been investigated by Tilford *et al.* (**454**). The rotational structure of the $d\,^3\Delta$ state of CO has been examined by Kovacs (**267**).

Most of the work that has appeared on the O_2 molecule has involved the ultraviolet. The absorption spectrum in the extreme ultraviolet has been described by de Reclhac and Damany-Astoin (**398**). Codling and Madden have discovered a new Rydberg series near 500 Å (**89**). Self-consistent field calculations of O_2 have already been mentioned above in connection with N_2 (**409**).

Several metallic oxides have been studied during the past two years. Becart and Mahieu have discovered new heads in the visible system of AlO (**29**). Asundi reports the discovery of a new band system of AlO in the region 2300–2800 Å, by the group at Trombay. Weltner and McLeod have identified the ground state of ZrO from neon matrix investigations at 4°K (**479**). The ultraviolet spectrum of MgO has been studied by Pesic and Kliska (**379**). Kasai and Weltner have investigated the ground states and hyperfine-structure separations of ScO, YO, and LaO (**241**). Dunn (Michigan) and Klemperer (Harvard) have reanalyzed the LaO and ScO spectra in order to resolve the question of the doublet or quartet structure of the ground state. Their results show that the ground state is $^2\Sigma$. In the case of TiO, Kovacs has carried out a thorough investigation of the triplet terms (**268**), and Toros has studied the anomalous multiplet splitting of these triplet terms (**457**). Asundi writes that the group at Trombay has carried out vibrational and rotational analyses of five bands of the orange system of FeO.

Molecular Ions. Stuart and Matsen have calculated a one-center wave function for the ground state of the HeH^+ molecular ion (**440**). H_2^+ force constants have been calculated by Ross and Phillipson (**403**), while Peek has calculated eigen parameters for certain orbitals of H_2^+ (**371**), and Wind has investigated vibrational states of that same ion (**488**). A study of the electronic structure of CH^+ has been made by Morre *et al.* (**327**). For CO^+, publications include that of Krupenie and Weissman on potential energy curves (**271**) and of Ogawa describing a Rydberg series in CO converging to the $B\ ^2\Sigma^+$ state of CO^+ (**61**). Two papers have appeared on the C–X system of N_2^+ as the result of work by Joshi (**235**). Finally, Asundi writes that a reinvestigation of the rotational structure of the bands of the second negative system of O_2^+ in the 2100–6100Å region is in progress at Trombay.

Polyatomic Molecules. A description of the near infrared spectrum of water and aqueous solutions has been published by Yamatera *et al.* (**489**). Bayly *et al.* have carried out a study of the absorption spectra of liquid phase H_2O, HDO and D_2O from 0·7 to 10 μ (**28**). At the other end of the spectrum, Bell has investigated the spectra of H_2O and D_2O in the vacuum ultraviolet (**34**). An important analysis of the 4050Å group of the C_3 molecule has been made by Gausset *et al.* (**177**), see p. 250. Several publications have appeared on various aspects of the infrared spectrum of CO_2; examples are a description of the 2·8 μ bands by Gordon and McCubbin (**187**), of the 4·3 μ bands by Rao and Oberly (**395**), and a new measure of the $3\ \nu_3$ band by Fayt *et al.* (**147**). Dixon has shown that the CO flame bands represent transitions from the lowest excited state of CO_2, $^1B_2$ to the ground state (**117a**).

Herzberg and his colleagues at the National Research Council of Canada are actively engaged in studying the spectra and structures of diatomic molecules, simple polyatomic molecules and free radicals. A brief summary of work on selected molecular spectra under investigation, prepared by Herzberg, is presented below.

C_2. It has been established by Ballik and Ramsay (**15**) that the ground state of the C_2 molecule is not the lower state of the Swan bands ($^3\Pi_u$) but the lower state of the Mulliken bands ($^1\Sigma_g^+$), and the energy difference between these two states is 610 cm^{-1}. Several other new electronic states have been found and their constants have been determined. Herzberg has recently found a simple new band system which represents a $\Sigma_u^+ - \Sigma^+$ transition of C_2 or C_2^+ or C_2^- (**214**). Neither the upper nor lower state has been observed previously.

Absorption Spectra, Diatomic Molecules: Si_2, P_2, BH, SiH, NF, NCl. The spectrum of Si_2 has been studied by Verma and Warsop (**469**) and that of P_2 by Creutzberg (**103**). In both cases a number of new electronic states have been found.

Several investigations on diatomic hydrides have been made in the near and far ultraviolet regions. Bauer *et al.* (**27**) have made a detailed study of the absorption spectrum of BH, and additional results on this molecule have been obtained by Johns *et al.* (**233**). Two new systems of SiH and SiD have been found and analyzed by Verma (**470**). The spectrum of HS has been studied by Morrow (**328**).

Until recently no spectroscopic information about the molecules NF and NCl has been available. Douglas *et al.* (**125**) have now observed very simple and interesting spectra of these molecules which are analogues of the visible and infrared absorption bands of the O_2 molecule.

Polyatomic Molecules and Radicals: CH_2, BH_2, PH_2 AlH_2, SiH_2, HCP, HCO, HCF, HCCl, HSiCl, HSiBr, C_3, NO_2, CS_2, CF_2, NH_3, CH_4.

The investigation of the red bands of CH_2 has been completed (**213**) and a similar spectrum has been found for the BH_2 radical, which is new (**213**). Dixon *et al.* (**117**) have succeeded in analyzing one of the bands of PH_2 originally discovered by Ramsay (**392**), and they have established the structure of the molecule in the upper and lower state. Spectra of AlH_2 and SiH_2 have also been obtained, but the analyses are not yet complete.

The HCP molecule (an analogue of HCN) has been studied by Johns *et al.* (**234**) and the absorption spectrum of HCO has been further studied by Johns *et al.* (**232**).

Phillips has referred to the extensive analyses of the 4050 Å group of the C_3 radical, based on new absorption spectra obtained by the flash photolysis technique (**177**). The electronic transition has been definitely identified as $^1\Pi_u - {}^1\Sigma_g^+$, and improved molecular constants have been determined. Extensive spectra of the two isomeric molecules CCN and CNC have been obtained and studied by Merer and Travis (**311**), and fairly complete analyses of the spectra of NCN and N_3 have been completed by Herzberg and Travis (**211**) and Douglas and Jones (**124**).

A new study of NO_2 in the visible region has been made by Douglas and Huber (**122**). One progression of simple bands has been found, but the remainder of the spectrum is extremely difficult to analyze because of the presence of innumerable perturbations. This spectrum is of particular astrophysical interest because of the suggested possibility of its presence in Mars (**253**).

The near ultraviolet spectrum of CS_2 has been reinvestigated by Kleman (**260**), Douglas and Milton (**120**), and Douglas and Zanon (**121**).

Douglas has continued work on the spectra of NH_3 in the vacuum ultraviolet (**119**). The infrared and Raman spectra of CH_4 have been studied by Herranz and others (**209, 210**).

Other molecular spectra are also on the Ottawa program but they are less likely to be of astrophysical interest than those mentioned above.

Mme R. Herman and her collaborators at Meudon report the following:

Mg_2 – Bands observed at high temperature near 2852 Å (**296, 480**).
Al_2, AlH – Diffuse bands observed near 2700 Å (**481**).
CaH – Studies of satellites in red, and pressure effects (**482**).
C_2 – Absorption spectra observed in an electric discharge source reveal short lifetimes (**208**).

At the Dominion Astrophysical Observatory, Wright reports some work on diatomic molecules of astrophysical interest. Tatum and Nichols are preparing papers on two subjects: one on 'Partition Functions and Association Equilibria' (**450**) and one on 'Intensity Data' (**449**).

II. *Intensities and Transition Probabilities*

J. G. Phillips

N_2, N_2^+. The following are representative of the numerous investigations that either have been or are in the process of being carried out on the nitrogen spectra: (*a*) An extensive series of studies at the University of Maryland and NRL by Benesch *et al.* (**39**); (*b*) the use of an electron beam excitation by Jeunehomme to study the oscillator strengths of the first negative and second positive systems (**230**); (*c*) the observation of the intensity distribution of the Lyman-Birge-Hopfield band system by McEwen and Nicholls (**308**), and the calculation by Nicholls of Franck-Condon factors for the Gaydon-Green band system (**339**); (*d*) an excellent calculation of Franck-Condon factors by Zare *et al.* (**493**); (*e*) a recent determination of vibrational transition probabilities and *r* centroids for the C–X system by Joshi (**236**).

CN. The oscillator strength of the CN red system has been investigated by Jeunehomme (**229**). Other studies of the red system include the observation of the radiative lifetime of the $^2\Pi$ state by Wentink *et al.* (**483**) and an interesting derivation of *f*-values for the red system from the solar spectrum by Poletto and Rigutti (**386**). Reis has investigated the oscillator strength of the violet system (**400**), while Kudryavtsev *et al.* have determined the matrix element of the dipole moment for the electronic transition of the violet system (**273**).

H_2, D_2, T_2. A general description of the properties of the hydrogen molecule has been published by Dalgarno and Williams (**110**). Rank, Wiggins and their collaborators have observed the intensities of hydrogen quadrupole bands, as well as intensities of features in the spectrum of H_2O and CO_2 (**159, 393**). A general description of transition probabilities in the electronic and vibrational spectra of the hydrogen molecule has been produced by Geiger and collaborators (**178**). For the ultraviolet, calculations have been made of vibrational overlap integrals by Patch (**367**), an experimental study of the continuum emission by Hamberger and Johnson (**202**), and of total absorption cross sections of H_2, N_2 and O_2 in the 550–200 Å region by Samson and Cairns (**411**).

C_2. The Swan system has been the subject of four investigations. Absolute band strengths for the Swan system have been derived by Mentall and Nicholls (**310**). Jain has published transition probability parameters of the Swan and the Fox-Herzberg band systems (**228**). In addition, shock-tubes have been used by Fairbairn (**139**) and by Harrington *et al.* (**204**), to obtain experimentally the oscillator strengths of the Swan bands. On the theoretical side, Franck-Condon factors have been calculated for the C_2 band systems by Ortenberg, as well as for band systems of NO and CO (**361**).

O_2. The photo-ionization and absorption cross sections of O_2 and N_2 in the 600 to 1000 Å region have been observed by Cook and Metzger (**97**). In the same spectral region the absorption cross section of oxygen has been studied by Kosinskaya and Startsev (**265**), and Nicholls reports that he and Degen have almost completed intensity measurements on bands of the Herzberg system. Halmann and Laulicht have studied the vibrational transition probabilities in the Schumann-Runge absorption bands of $^{16}O_2$ and $^{18}O_2$ (**201**). In the infrared, Badger *et al.* have observed the absolute intensities of the discrete and continuous absorption bands of oxygen at 1·26 and 1·065 μ, and the radiative lifetime of the $^1\Delta_g$ state (**14**). Finally, two recent studies of the negative band systems of O_2^+ have been published by Nicholls (**340**) and Jeunehomme (**231**).

Oxides. The absolute transition probabilities in the ultraviolet spectrum of CO have been observed by Hesser and Dressler (**216**), and by Cook *et al.* who observed the photo-ionization and absorption coefficients in the 600 to 1000 Å region (**98**). Kovacs and Toros have discussed the intensity distribution in the triplet bands of CO (**266**), while two recent works discuss the $A\,^1\Pi - X\,^1\Sigma^+$ system; the first is by Skerbele *et al.* on vibrational intensities (**424**), and the second is by Tilford and Simmons on the electric quadrupole component (**456**). In the infrared, Breeze and Ferriso have studied the integrated intensities of the fundamental and first overtone bands of CO between 2500° and 5000°K (**52**). Finally, for the ion CO^+ there is an investigation of lifetimes and transition probabilities by Lawrence (**282**).

The NO spectrum has been the subject of several investigations. Of general interest are (*a*) the experimental study of transition probabilities in the β and γ band systems by Antropov *et al.* (**12**), and (*b*) a somewhat more recent parallel study of the δ, γ and β band systems of NO by Erkovich and Ageshin (**137**). The effect of temperature and pressure or path length on band strengths of NO has been investigated by Carpenter and Franzosa (**77**) and by Churchill and Meyerott (**88**); the latter, in discussing the spectral absorption of heated air, also provides data on N_2 and O_2 absorption. The rotation-vibration NO spectrum has been investigated by Breeze and Ferriso, who measured the integrated intensities of the 5·3 μ fundamental and the 2·7 μ overtone bands at temperatures between 1400°K and 2400°K (**51**). In somewhat more detail, the widths and strengths of vibration-rotation lines in the fundamental band of NO have been measured by Abels and Shaw (**1**).

Relatively little has been done recently on intensities in the spectra of metallic oxides. Tawde and Korwar have studied the effective vibrational temperature as a variant of the electronic transition moment in AlO (**451**). Somewhat earlier there appeared a calculation of wave functions and binding energies of the TiO molecule by Carlson and Nesbet (**75**).

Polyatomic Molecules. Reference has already been made to the work of Rank and his co-workers on CO_2 (**393**). Additional publications of a general nature include the report by

Penner and Varanasi on approximate band absorption and total emissivity calculations for CO_2 (**377**), and the work of Gray and Selvidge on relative intensity calculations (**189**). Finally, several studies involve measures of intensities of specific CO_2 features: (*a*) of the 4·3 μ band, by Gray (**190**), and Ferriso *et al.* (**157**); (*b*) of the 9·4 μ and 10·4 μ bands, by Rasool (**397**) and Gray (**191**); (*c*) of the 15 μ band, by Varanasi and Lauer (**462**).

Many investigations have been carried out during the past two years on various aspects of the emission or absorption spectrum of H_2O. It was one of the molecules investigated by Metzger and Cook on continuous absorption, photo-ionizations and fluorescences in the 600 to 1000Å region (**312**). In addition to H_2O, they studied NH_3, CH_4, C_2H_2, C_2H_4, and C_2H_6. The absorption and photo-ionization cross sections of H_2O were measured by Watanabe and Jursa (**477**), while the emissivity of water vapor at temperatures to 1000°K has been observed by Podkladenko (**385**). Penner and Varanasi have made approximate band absorption and total emissivity calculations for H_2O (**378**). Studies of specific bands of H_2O include one by Ferriso and Ludwig on spectral emissivities and integrated intensities of the 1·87 μ, 1·38 μ and 1·14 μ bands between 1000°K and 2200°K (**156**). Patch has carried out measurements of the absolute intensity of the 2·7 μ band of water vapor in a shock tube (**368**), and this same band has been investigated by Maclay in an observation of integrated absorptances (**291**). At still longer wavelengths there are observations of spectral emissivities and integrated intensities of the 6·3 μ fundamental band by Ludwig *et al.* (**290**), and a study of the line breadth of the 1·64 mm absorption in water vapor by Rusk (**408**).

III. *Atlases*

A very useful series of identification atlases of molecular spectra is being produced by Nicholls and his co-workers. To the present time, three atlases have been published: (*a*) the AlO $A\,^2\Sigma - X\,^2\Sigma$ blue-green system; (*b*) the N_2 second positive system; (*c*) the N_2^+ first negative system.

Volume III of *Molecular Spectra and Molecular Structure*, by Herzberg, (**212**) has been completed. A revision of the Table of Molecular Constants in his book entitled Spectra of Diatomic Molecules is being prepared by Huber and Herzberg.

VACUUM-ULTRAVIOLET ATOMIC SPECTRA

Observations of the XUV solar spectrum and of stellar spectra (**329**) have stimulated renewed interest in laboratory spectra of highly-ionized atoms. R. Wilson reports on the work at Culham laboratory, as follows:

High Temperature Plasmas. These sources include a range of thetatron plasmas and the pinch discharge Zeta, supplemented by a high voltage spark. Most of the intense solar lines reported earlier in the range 170 to 220Å have been identified by Gabriel and Fawcett as due to Fe VIII to Fe XII, and Fe XIV (**144, 145, 164, 166**). Isoelectronic sequences from Ca through Ni have resulted in the identification of highly-ionized Ni in the Sun to the short-wave limit 60Å (**165**). Recent work on Fe XII, XIII, XV, XVI is leading to further solar identifications.—In the soft X-ray region from 15 to 100Å, newly observed spectra of Ne VII, Ne VIII, Ar IX to Ar XII, Kr IX, Kr X, Xe IX, Sc XII, Sc XIII, Ti XIII, Ti XIV, and V XIV, V XV are being investigated (**141, 142, 143**).—*Laser-Produced Plasmas.* The production of hot dense plasmas by focussing a high-powered laser beam on a solid target in vacuum is proving to be a very powerful source of excitation that permits a good isolation of ionization stages. Results for Fe XV and Fe XVI have been published (**146**), and the spectra Cr XV, Cr XVI; Mn XVI, Mn XVII; Fe XVII, Fe XVIII; Co XVIII and Co XIX are being analyzed.

The spectra of the iron-group elements in high ionization stages are being studied also at the Hebrew University, Jerusalem, by Alexander, Feldman, Fraenkel and Hoory. Their latest results concern Co VIII, Ni IX, Cu X (**4a**) and Cr VI, Mn VII, Fe VIII (**150a**).

At Lund University, L. Å. Svensson and J. O. Ekberg are using a 5 m grazing incidence spectrograph for a systematic investigation of the vacuum-spark spectra below 400 Å of elements in the iron group. The work on titanium has been completed, the result being a list of some 500 wavelengths accurate to about 0·003 Å and comprising ionization stages from Ti v to Ti xii.

At Uppsala, Bockasten and his group have developed a small theta pinch into a highly efficient light source for intermediate stages of ionization. They have applied it to the study of N iii – N vi and O iii – O vi. Some preliminary results are given in (**43a**), and the definitive description of N iv and N v has been published by Hallin (**200a, 200b**). This work covers the entire region from the shortest wavelengths to the near infrared.

At NRL in Washington, Tilford reports the observation of highly-ionized spectra of N and O in the region below 400 Å, with a resolution 0·04 Å. Asymmetrical auto-ionized emission lines have been observed above the first ionization limit in N iii. Multiplets of Ne iv have also been observed.

In the Far Ultraviolet Physics Section of the National Bureau of Standards, Madden, Codling and Ederer have continued their observations of atomic spectra made with synchrotron light as a background source. In the case of Ne i two basic types of transition have been found in the range 150 to 275 Å: the excitation of a subshell $2s$-electron to outer p-orbits and the simultaneous excitation of two outer $2p$-electrons (**92**). A 3-meter grazing incidence monochromator has been constructed for use with the synchroton in the region 50 to 300 Å (**131**). It has been used to determine the absolute absorption cross section near selected broader resonance profiles in the photo-ionization continua of He, Ne (**91**) and Ar (**130**).

Garton reports active research in his laboratory at Imperial College. J. M. Wilson has accumulated vacuum ultraviolet absorption spectra of Si, Ge, Sn, Pb. The work on Pb i has been completed (**176**). Connerade has greatly extended the observations in two spectra which were originally studied by Beutler, Hg i and Cs i. He plans to reobserve Tl i and Rb i, similarly. The range is 600 to 900 Å. He is using six times the dispersion of the early work, and thus can greatly extend the series. In this region, Goldstein has found interesting series in Pb i converging on the limits $sp^2\ {}^{2,4}L$. Learner and Morris have found new autoionizing levels in Hg i from emission spectra.

In the grazing incidence region, Burgess, Jenkins, and Mansfield have used a large capacitor bank known as 'Maggi I' as a source to produce the spectra C iv and N v. Line broadening and shifts are being studied.

In collaboration with colleagues at Argonne and Harvard the principal series of Ra i have been extended to $n = 50$; and Ba i series to $n = 80$ according to Garton. Some corrections in Ca i (**173, 337**) have been made, and new auto-ionizing resonances in Tl i (**174**) and In i (**366**) have been observed. Important new series in La i have been published (**175**).

GENERAL WORK ON ATOMIC SPECTRA

At the National Bureau of Standards emphasis is being given to first and second rare-earth spectra of the lanthanon group in order to provide data for Volume IV of *Atomic Energy Levels*. A summary of this work is in print and need not be repeated here (**324**). Suffice to say that the first and second spectra of Ce, Pr, Pm, Ho, Tm and Yb are on the Bureau Program.

In Amsterdam, Klinkenberg and his associates are working on Nd i, Nd ii, Tb i (**261**) and Tb ii. Er i (**301, 389, 432**) and Er ii are being studied by a number of workers.

Third and fourth spectra of this group are being investigated at The Johns Hopkins University. This program was started by the late G. H. Dieke in collaboration with Crosswhite. Third spectra may prove to be of astrophysical importance, since they have strong infrared lines; fourth spectra may, also, be of interest. Recent analyses of La III (**444**), Ce III (**442**), Pr III (**441**), Pr IV (**443**) and Gd III (**72**) have been published.

Mention should be made of the program in progress at the Radiation Laboratory in Berkeley. Here, Conway and his associates have extensive observations of Dy I and Dy II. They are also progressing with the observation and analysis of spectra in the actinon group: Cm I, Cm II, Bk and Cf.

At the Laboratoires de Bellevue, Blaise has continued his work on Pu I and Pu II in collaboration with Argonne. A detailed report of all spectra on their program has been published by Chabbal and Jacquinot (**79**).

Prokofiev reports that series regularities in spectra have been studied experimentally; perturbations, being reflected by both transition probabilities and quantum defect, have been found in the main series of Ca I, Sr I and Ba I; in the diffuse series of Al I, and the sharp series of Tl I (**375**). Long absorption series of Al I, Ca I, In I and Tl I in the region $\geqslant$ 2000 Å have been observed. A great number of new absorption lines have been classified (**375**). Lines of multiply-ionized krypton, Kr XIII, have been obtained in the region 65–110 Å (**262**). The sequence of the atomic shell-filling and multiplicity of the main term has been studied theoretically (**259**). Hyperfine and isotopic structures for 275 lines of Pu have been analyzed. New data on the classification of the Pu spectrum have been obtained (**264**).

Of recent results of astrophysical interest obtained at Lund there should be mentioned the comprehensive description of the C I spectrum by L. Johansson (**231a**), the precision measurements by K. B. S. Eriksson of the red, green and ultraviolet forbidden lines of O I (**136a**) as well as the ultraviolet forbidden line of N I (**136b**), and work on S I by L. Jakobson (**228a**). A complete reinvestigation of Mg I by G. Risberg (**400a**) has enabled Swensson and Risberg (**445a**) to identify numerous new Mg I lines in the solar spectrum. The analysis of Fe IV by Edlén has yielded, so far, a practically complete set of levels of the ground configuration $3d^5$, which has provided identification of a number of forbidden transitions in RR Telescopii.

The urgent problem of providing the astrophysicist with data on analyses involves a revision of the published material in Volumes I, II, and III of *Atomic Energy Levels* and of the *Multiplet Tables*. This has been started in a new Series of pamphlets prepared for individual spectra for which analyses are essentially complete. Section 1 (**322**) deals with Si II, Si III and Si IV and is in print. Section 2 contains similar data on Si I; it is in press. Forthcoming Sections will include the spectra in Table 10.

Table 10. Spectra scheduled for Revised Compilation

H I sequence (4)					Na I			
Li I					Mg I	Mg II		
Be I	Be II				Al I		Al III	
C I	C II	C III	C IV		Si I			
N I	N II		N IV	N V	P I	P II		
O I					Ca I	Ca II		
	F II						Ti III	Ti IV

Revised analyses of most of these spectra have come from Lund. For Si I the analysis is chiefly from Purdue University, and for P I and P II it has been done at Princeton.

New Multiplet Tables are partially completed, also, for several more complex spectra listed in Table 11.

Table 11. Spectra – Further Analysis Unlikely

Mn I	Mn II		Nb I	Nb II
		Fe III		Mo II
		Co III	Ru I	Ru II
		Ni III	I I	I II
Cu I	Cu II			Hf II

Kiess has directed work on a new description and analysis of Ti I (**16**) and is completing monographs on Mo I and Ta I, respectively. A new analysis of Hf I by Meggers was nearly finished at the time of his death.

At the National Bureau of Standards the reference catalogue on atomic spectra is being maintained as completely as possible.

ATLASES

Two important Atlases have been published at Castel Gandolfo by Fr. Junkes and his staff. Part One of *Atomic Spectra in the Vacuum Ultraviolet from* 2250–1100 Å (**239**) contains nine photographic charts (30 × 40 cm) of hollow cathode spectra of seven elements: Al, C, Cu, Fe, Ge, Hg, Si and the Lyman bands of molecular hydrogen from 1650–1260 Å as excited by a hydrogen lamp.

The second Atlas consists of sixteen photographic charts (24 × 30 cm) on which the Spectrum of Thorium from 9400–2000 Å is recorded, juxtaposed to the spectrum of the iron arc. The regions 9000–4000 Å and 2400–2000 Å are taken in the first order with a reciprocal dispersion 5 Å/mm; the region 4300–2400 Å is taken in the second order (**238**).

THE SOLAR SPECTRUM

The atlas of the Ultraviolet Solar Spectrum that is being prepared by Tousey and his staff at the Naval Research Laboratory is nearing completion. Some 7000 lines are listed between 3000 and 2090 Å. They have been observed on rocket spectrograms taken in flights in 1961 and 1964. About half of the lines have been identified, which indicates the urgent need for further laboratory work on atomic spectra.

A comparison of the solar spectrum in the range 170 to 220 Å with spectra of the Zeta plasma (Harwell) and with spectra of an Fe-loaded θ-pinch (NRL) shows that many of the strong solar lines must be due to Fe. Similar work on these identifications has been carried on independently by other groups, see for example p. 252, and papers by Gabriel and others at Culham (**166**). Spectroheliograms obtained in the light of these lines at NRL showed that they were emitted by the whole solar disk, indicating that intermediate ionization stages of Fe were mainly responsible. It was finally proposed by a number of workers that the solar lines could be produced by $3p$-$3d$ transitions in Fe IX to Fe XIII.

Workers at Culham report that on 9 April, 1965 a spectrum of the chromosphere and corona in the range 950–2950 Å was obtained from the flight of a stabilized Skylark rocket (**68**). About 300 emission lines were observed, of which some 75% have been identified. New identifications include intersystem combinations in C III, N IV, O V and forbidden transitions in Fe XI and Fe XII (**69**).

It is a pleasure to report that the current revision of the 1928 edition of Rowland's Table of Solar Spectrum Wavelengths has at long last been completed. In 1948 the IAU requested Minnaert at Utrecht and Mrs Moore-Sitterly in Washington to carry out this project on a cooperative basis. Since then the work has gone forward slowly but steadily. Minnaert and Houtgast have furnished the equivalent widths for more than 20 000 lines from the Utrecht Atlas, while Moore has revised and extended the identifications with the aid of the data collected for the 'Atomic Energy Levels' Program. Copies of the Monograph are available for distribution (**323**).

CHARLOTTE MOORE-SITTERLY
President of the Commission

BIBLIOGRAPHY

1. Abels, L. L., Shaw, J. H. 1966, *J. molec. Spectrosc.*, **20,** 11.
2. Aberth, W., Sunshine, G., Bederson, B. 1964, see ref. **307,** p. 53.
3. Abiodun, R. A., Seaton, M. J. 1966, *Proc. phys. Soc.*, London, **87,** 145.
4. Aleskovskii, Yu. M. 1963, *Soviet Phys. JETP*, **17,** 570.
4a. Alexander, E., Feldman, U., Fraenkel, B. S., Hoory, S. 1966, *J. opt. Soc. Am.*, **56,** 651.
5. Ali, A. W., Griem, H. R. 1965, *Phys. Rev.*, **140A,** 1144; 1966, corrected in *Phys. Rev.*, **144,** 366.
6. Allison, D. C., Dalgarno, A. 1965, *Proc. phys. Soc.*, London, **85,** 845.
7. Allison, D. C., Browns, J. C., Dalgarno, A. 1966, *Proc. phys. Soc.*, London, in press.
8. Allison, D. C., Dalgarno, A. 1967, *Proc. phys. Soc.*, London, **89,** 41.
9. Alman, E. L., Chaika, M. P. 1965, *Optika Spektrosk.*, **19,** 968.
10. d'Angelo, N. 1965, *Phys. Rev.*, **140A,** 1488.
11. Ankudinov, V. A., Bodashev, S. V., Andreev, E. P. 1965, *Soviet Phys. JETP*, **48,** 40.
12. Antropov, E. T., Dronov, A. P., Sobolev, N. N. 1964, *Optics Spectrosc.*, **17,** 355.
13. Aubel, J. L., Hause, C. D. 1966, *J. chem. Phys.*, **44,** 2659.
14. Badger, R. M., Wright, A. C., Whitlock, R. F. 1965, *J. chem. Phys.*, **43,** 4345.
15. Ballik, E. A., Ramsay, D. A. 1963, *Astrophys. J.*, **137,** 61, 84.
16. Banks, H. W., Bozman, W. R., Wilson, C. M. 1966, *Georgetown Obs. Monograph* no. 20, 176 pp.
17. Bates, D. R. 1964, Proc. Symposium on *Atomic Collision Processes in Plasmas*, Culham Lab.
18. Bates, D. R., Kingston, A. E. 1964, *Proc. phys. Soc.*, London, **83,** 43.
19. Bates, D. R., Williams, D. A. 1964, *Proc. phys. Soc.*, London, **83,** 425.
20. Bates, D. R., Kingston, A. E. 1964, *Proc. R. Soc. London* Ser. A, **279,** 10, 32.
21. Bates, D. R., Esterman, I. ed. 1965, *Advances in Atomic and Molecular Physics*, **1,** Academic Press, New York and London; 1966, ibid., **2.**
22. Bates, D. R., Khare, S. P. 1965, *Proc. phys. Soc.*, London, **85,** 231.
23. Bates, D. R., Moffett, R. J. 1965, *Nature*, **205,** 272.
24. Bates, D. R., Moffett, R. J. 1966, *Proc. R. Soc. London* Ser. A, **291,** 1.
25. Bates, D. R., Crothers, D. S. F. 1966, in preparation.
26. Bauer, E. B., Bartky, C. 1966, *J. chem. Phys.*, **43,** 2466.
27. Bauer, S. H., Herzberg, G., Johns, J. W. C. 1964, *J. molec. Spectrosc.*, **13,** 256.
28. Bayly, J. G., Kartha, V. B., Stevens, W. H. 1963, *Infrared Phys.*, **3,** 211.
29. Becart, M., Mahieu, J. 1964, *J. Phys.*, Paris, **25,** 873.
30. Behmenberg, W. 1964, *J. quantit. Spectrosc. radiat. Transfer*, **4,** 177.
31. Bell, K. L., Moiseiwitsch, B. L. 1963, *Proc. R. Soc. London* Ser. A, **276,** 346.
32. Bell, K. L. 1965, *Proc. phys. Soc.*, London, **86,** 246.
33. Bell, K. L., Eissa, H., Moiseiwitsch, B. L. 1966, *Proc. phys. Soc.*, London, **88,** 57.
34. Bell, S. 1965, *J. molec. Spectrosc.*, **16,** 205.
35. Bely, O. 1966, *Proc. phys. Soc.*, London, **87,** 1010.
36. Bely, O. 1966, *Proc. phys. Soc.*, London, **88,** 587.
37. Bely, O. 1966, *Ann. Astrophys.*, **29,** 131.
38. Bely, O., Blaha, M. 1966, in preparation.

39. Benesch, W., Vanderslice, J. T., Tilford, S. G., Wilkinson, P. G. 1965, *Astrophys. J.*, **142**, 1203; 1966, ibid., **143**, 236; 1966, ibid., **144**, 408.
40. Biondi, M. A. 1963, *Phys. Rev.*, **129**, 1181.
41. Biondi, M. A. 1964, *Ann. Geophys.*, **20**, 5.
42. Blaha, M. 1964, *Bull. astr. Inst. Csl.*, **15**, 33.
43. Boardman, W. J. 1964, *Astrophys. J. Suppl.*, **9**, 185.
43a. Bockasten, K., Hallin, R., Johansson, K. B., Tsui, P. 1964, *Phys. Letters*, **8**, 181.
44. Boldt, G., Cooper, W. Z. 1964, *Z. Naturforsch.*, **19a**, 968.
45. Bouthilette, D. B., Healey, J. A., Milford, S. N. 1964, in *Atomic Collision Processes*, North-Holland Publ. Co., Amsterdam, p. 1081.
46. Bransden, B. H., Cheshire, I. 1963, *Proc. phys. Soc.*, London, **81**, 820.
47. Bransden, B. H. 1966, *Adv. atom. molec. Phys.*, **1**, 85.
48. Bransden, B. H., Sin Fie Lam, L. T. 1966, *Proc. phys. Soc.*, London, **87**, 653.
49. Bréchot, S., van Regemorter, H. 1964, *Ann. Astrophys.*, **27**, 432, 739.
50. Breene, R. G. 1964, see *Hand. Physik* **27**, Springer, Verlag, Berlin.
51. Breeze, J. C., Ferriso, C. C. 1964, *J. chem. Phys.*, **41**, 3420.
52. Breeze, J. C., Ferriso, C. C. 1965, *J. chem. Phys.*, **43**, 3253.
53. Breig, E. L., Lin, C. C. 1965, *J. chem. Phys.*, **43**, 3839.
54. Browne, J. C. 1964, *J. chem. Phys.*, **41**, 1583.
55. Browne, J. C. 1965, *J. chem. Phys.*, **42**, 2826.
56. Bruckner, K. A. 1964, *J. chem. Phys.*, **40**, 439.
57. Buckingham, R. A., Fox, J. W., Gal, E. 1965, *Proc. R. Soc. London* Ser. A, **284**, 237.
58. Bugrim, E. D., *et al.* 1965, *Optics Spectrosc.*, **19**, 292.
59. Burgess, A. 1964, *Astrophys. J.*, **139**, 776.
60. Burgess, A. 1964, *Mem. R. astr. Soc.*, **69**, 1.
61. Burgess, A. 1964, Proc. Symposium on *Atomic Collision Processes in Plasmas*, Culham Lab.; 1964, see also ref. **307**, p. 237.
62. Burgess, A. 1965, *Ann. Astrophys.*, **28**, 774.
63. Burgess, A. 1965, *Astrophys. J.*, **141**, 1588.
64. Burke, P. G., McVicar, D. D., Smith, K. 1964, *Proc. phys. Soc.*, London, **83**, 397.
65. Burke, P. G. 1965, *Adv. Phys.*, **14**, 521.
66. Burke, P. G., Tait, J. H., Lewis, B. A. 1965, *Proc. phys. Soc.*, London, **87**, 209.
67. Burke, P. G., Taylor, A. J. 1966, *Proc. phys. Soc.*, London, **88**, 549.
68. Burton, W. M., Wilson, R. 1965, *Nature*, **207**, 61.
69. Burton, W. M., Ridgeley, A., Wilson, R. 1967, *Mon. Not. R. astr. Soc.*, in press.
70. Butler, S. T., Johnston, I. D. S. 1964, *Nucl. Fusion*, **4**, 196.
71. Butler, S. T., Parcell, L. 1965, *Phys. Lett.*, **14**, 110.
72. Callahan, W. R. 1963, *J. opt. Soc. Am.*, **53**, 695.
73. Callaway, J., Bauer, E. 1965, *Phys. Rev.*, **140A**, 1072.
74. Carew, J., Milford, S. N. 1963, *Astrophys. J.*, **138**, 772.
75. Carlson, K. D., Nesbet, R. K. 1964, *J. chem. Phys.*, **41**, 1051.
76. Carlton, T. S., Mahan, B. H. 1964, *J. chem. Phys.*, **40**, 3683.
77. Carpenter, R. O'B., Franzosa, M. A. 1965, *J. quantit. Spectrosc. radiat. Transfer*, **5**, 465.
78. Cermak, V., Dalgarno, A., Ferguson, E. E., Friedman, L., McDaniel, E. W. 1966, *Ion-Molecule Reactions*. Review being prepared under the auspices of the Atomic and Molecular Processes Information Center, Oak Ridge National Laboratory.
79. Chabbal, R., Jacquinot, P. 1964, Rapp. d'Act. Lab. Aimé Cotton, 18 pp.
80. Chamberlain, G. E., Smith, S. J., Heddle, D. W. O. 1964, *Phys. Rev. Lett.*, **12**, 647.
81. Chapman, R. D., Clarke, W. H., Aller, L. H. 1966, *Astrophys. J.*, **144**, 376.
82. Chashchina, G. I., Shreider, E. Ja. 1966, *Optika Spectrosk.*, **20**, 511.
83. Chen, C. L., Leiby, C. C., Goldstein, L. 1961, *Phys. Rev.*, **121**, 1391.
84. Chen, J. C. Y. 1964, *J. chem. Phys.*, **40**, 350, 3513.
85. Chen, J. C. Y. 1966, *Phys. Rev.*, **146**, 61.
86. Cheshire, I. 1965, *Phys. Rev.*, **138A**, 992.
87. Chiu, Leu-Yung Chow 1965, *Phys. Rev.*, **137A**, 384.

88. Churchill, D. R., Meyerott, R. E. 1965, *J. quantit. Spectrosc. radiat. Transfer*, **5**, 69.
89. Codling, K., Madden, R. P. 1965, *J. chem. Phys.*, **42**, 3935.
90. Codling, K. 1966, *Astrophys. J.*, **143**, 552.
91. Codling, K., Madden, R. P. 1966, *J. opt. Soc. Am.*, **56**, 552; Abstr. ThG 13.
92. Codling, K., Madden, R. P., Ederer, D. L. 1967, *Phys. Rev.*, **155**, 26.
93. Coleman, J. P., McDowell, M. R. C. 1965, *Proc. phys. Soc.*, London, **85**, 1097; 1966, ibid., **87**, 879.
94. Collins, C. B. 1965, *Phys. Rev.*, **140A**, 1850.
95. Connor, T. R., Biondi, M. A. 1965, *Phys. Rev.*, **140A**, 778.
96. Cook, A. H. 1965, *J. opt. Soc. Am.*, **55**, 780.
97. Cook, G. R., Metzger, P. H. 1964, *J. chem. Phys.*, **41**, 321.
98. Cook, G. R., Metzger, P. H., Ogawa, M. 1965, *Can. J. Phys.*, **43**, 706.
99. Cooper, W. S. III, Kunkel, W. B. 1965, *Phys. Rev.*, **138A**, 1022.
100. Corliss, C. H. 1965, *J. Res' nat. Bur. Stand.*, **69A**, 87.
101. Corliss, C. H., Warner, B. 1966, *J. Res. nat. Bur. Stand.*, **70A**, 325.
102. Corliss, C. H. 1967, *Monograph nat. Bur. Stand.*, 32, Suppl., 33 pp.
103. Creutzberg, F. 1966, *Can. J. Phys.*, **44**, 1583.
104. Crothers, D., McCarroll, R. 1965, *Proc. phys. Soc.*, London, **86**, 753.
105. Czyzak, S. J., Krueger, T. K. 1965, *Mem. R. astr. Soc.*, **69**, Part 4.
106. Czyzak, S. J., Krueger, T. K., McDavid, W. L. 1965, Tech. Rep. ARL 65–10.
107. Czyzak, S. J., Krueger, T. 1966, *Proc. phys. Soc.*, London, in press.
108. Dalgarno, A., Henry, R. J. W. 1964, *Proc. phys. Soc.*, London, **83**, 157.
109. Dalgarno, A., Henry, R. J. W. 1965, *Proc. phys. Soc.*, London, **85**, 679.
110. Dalgarno, A., Williams, D. A. 1965, *Proc. phys. Soc.*, London, **85**, 685.
111. Dalgarno, A., Rudger, M. R. H. 1965, *Proc. R. Soc. London* Ser. A, **286**, 519.
112. Dalgarno, A., Henry R. J. W., Roberts, C. S. 1966, *Proc. phys. Soc.*, London, **88**, 611.
113. Dalidchik, F. I., Sayasov, Yu. S. 1966, *Soviet Phys. JETP*, **22**, 212.
115. Davison, W. D. 1964, *Proc. R. Soc. London* Ser A, **280**, 227.
116. Devoto, R. S. 1966, *Physics Fluids*, **9**, 1230.
117. Dixon, R. N., Duxbury, G., Ramsay, D. A. 1967, *Proc. R. Soc. London*, **296**, 137.
117a. Dixon, R. N., 1963, *Proc. R. Soc. London* Ser A, **275**, 431.
118. Donin, V. I., Chebotarev, V. P. 1966, *Optika Spektrosk.*, **20**, 740.
119. Douglas, A. E. 1963, *Disc. Faraday Soc.*, **35**, 158.
120. Douglas, A. E., Milton, E. R. V. 1964, *J. chem. Phys.*, **41**, 357.
121. Douglas, A. E., Zanon, I. 1964, *Can. J. Phys.*, **42**, 627.
122. Douglas, A. E., Huber, K. P. 1965, *Can. J. Phys.*, **43**, 74; and in preparation.
123. Douglas, A. E., Elliott, G. A. 1965, *Can. J. Phys.*, **43**, 496.
124. Douglas, A. E., Jones, W. J. 1965, *Can. J. Phys.*, **43**, 2216.
125. Douglas, A. E., Jones, W. E. 1966, *Can. J. Phys.*, **44**, 2251; and in preparation.
126. Dressler, K., Miescher, E. 1965, *Astrophys. J.*, **141**, 1266.
127. Drukarev, G. F. 1965, *The Theory of Electron-Atom Collisions*, Academic Press, London (translated from the Russian edition of 1962).
129. Dunn, G. H., Van Zyl, B. 1966, *JILA Rep.* no. 84; 1966, *Phys. Rev.*, **154**, 40.
130. Ederer, D. L., Codling, K., Madden, R. P. 1966, *Bull. am. phys. Soc.*, **11**, 456; Abstr. BB 8.
131. Ederer, D. L., Madden, R. P. 1966, *J. opt. Soc. Am.*, **56**, 552; Abstr. ThG 14.
132. Edlén, B., Svensson, L. Å. 1965, *Ark. Fys.*, **28**, 427.
133. Edlén, B. 1966, *Ark. Fys.*, **31**, 509.
134. Edlén, B. 1966, *Metrologia*, **2**, 71.
135. Edmonds, F. N., Schlüter, H., Wells, D. C. 1967, *Mem. R. astr. Soc.*, in press.
135a. Engelhard, E. 1966, *Zs. angew. Phys.*, **20**, 404.
136. Engelhardt, A. G., Phelps, A. V., Risk, C. G. 1964, *Phys. Rev.*, **135A**, 1566.
136a. Eriksson, K. B. S. 1965, *Ark. Fys.*, **30**, 199.
136b. Eriksson, K. B. S. 1967, *Ark. Fys.*, in press.
137. Erkovich, S. P., Ageshin, F. S. 1965, *Optics Spektrosc.*, **18**, 551.
138. Evenson, K. M., Dunn, J. L., Broida, H. P. 1964, *Phys. Rev.*, **136A**, 1566.

139. Fairbairn, A. R. 1966, *J. quantit. Spectrosc. radiat. Transfer*, **6**, 325.
140. Farhat, N. H. 1963, *Proc. Inst. electr. electron. Eng.*, **51**, 1063.
141. Fawcett, B. C., Gabriel, A. H., Jones, B. B., Peacock, N. J. 1964, *Proc. phys. Soc.*, London, **84**, 257.
142. Fawcett, B. C., Gabriel, A. H. 1964, *Proc. phys. Soc.*, London, **84**, 1038.
143. Fawcett, B. C. 1965, *Proc. phys. Soc.*, London, **86**, 1087.
144. Fawcett, B. C., Gabriel, A. H. 1965, *Astrophys. J.*, **141**, 343.
145. Fawcett, B. C., Gabriel, A. H. 1966, *Proc. phys. Soc.*, London, **88**, 262.
146. Fawcett, B. C., Gabriel, A. H., Irons, F. E., Peacock, N. J., Saunders, P. A. H. 1966, *Proc. phys. Soc.*, London, **88**, 1051.
147. Fayt, A., Courtoy, C. P., de Hemptinne, M. 1965, *Ann. Soc. scient. Bruxelles*, **79**, III, 233.
148. Fehsenfeld, F. C., Ferguson, E. E., Schmeltekopf, A. L. 1966, *J. chem. Phys.*, **44**, 3022.
149. Fehsenfeld, F. C., Ferguson, E. E., Schmeltekopf, A. L. 1966, *J. chem. Phys.*, **45**, 1844.
150. Feibelman, P. J. 1965, *J. chem. Phys.*, **42**, 2462.
150a. Feldman, U., Fraenkel, B. S. 1966, *Astrophys. J.*, **145**, 959.
151. Felenbok, P., Czarny, J. 1964, *Ann. Astrophys.*, **27**, 244.
152. Ferguson, E., Schlüter, H. 1963, *Ann. Phys.*, New York, **22**, 351.
153. Ferguson, E. E., Fehsenfeld, F. C., Schmeltekopf, A. L. 1965, *Phys. Rev.*, **138A**, 381.
154. Ferguson, E. E., Fehsenfeld, F. C., Goldan, P. D., Schmeltekopf, A. L. *J. geophys. Res.*, **70**, 4323.
155. Ferguson, E. E., Fehsenfeld, F. C., Goldan, P. D., Schmeltekopf, A. L., Schiff, H. I. 1965, *Planet. Space Sci.*, **13**, 823.
156. Ferriso, C. C., Ludwig, C. B. 1964, *J. chem. Phys.*, **41**, 1668.
157. Ferriso, C. C., Ludwig, C. B., Acton, L. 1966, *J. opt. Soc. Am.*, **56**, 171.
158. Field, G. B., Hitchcock, J. L. 1966, *Phys. Rev. Lett.*, **16**, 817.
159. Fink, U., Wiggins, T. A., Rank, D. H. 1965, *J. molec. Spectrosc.*, **18**, 384.
159a. Foster, E. W. 1964, *Rep. Progr. Phys.*, **27**, 469.
160. Fougere, P. F., Nesbet, R. K. 1966, *J. chem. Phys.*, **44**, 285.
161. Fox, M. A. 1965, *Proc. phys. Soc.*, London, **86**, 789; 1966, ibid., **88**, 65.
162. Freund, R. S., Klemperer, W. 1965, *J. chem. Phys.*, **43**, 2422.
163. Fulton, M. J., Mittleman, M. H. 1965, *Ann. Phys.*, New York, **33**, 65.
164. Gabriel, A. H., Fawcett, B. C., Jordan, C. 1965, *Nature*, **206**, 390.
165. Gabriel, A. H., Fawcett, B. C. 1965, *Nature*, **206**, 808.
166. Gabriel, A. H., Fawcett, B. C., Jordan, C. 1966, *Proc. phys. Soc.*, London, **87**, 825.
167. Gailitis, M. K., Damburg, R. 1963, *Proc. phys. Soc.*, London, **82**, 192.
168. Garcia, J. D., Mack, J. E. 1965, *J. opt. Soc. Am.*, **55**, 654.
169. Garrett, W. R., Mann, R. A. 1964, *Phys. Rev.*, **135A**, 580.
170. Garstang, R. H. 1966, *J. chem. Phys.*, **44**, 1308.
171. Garstang, R. H., Hill, S. J. 1966, *Publ. astr. Soc. Pacif.*, **78**, 70.
172. Garstang, R. H. 1966, *Publ. astr. Soc. Pacif.*, **78**, 399.
172a. Garstang, R. H. 1966, *Abundance Determinations in Stellar Spectra*, p. 57. Hubenet, H. ed., IAU Symposium no. 26, Academic Press, New York and London.
173. Garton, W. R. S., Codling, K. 1965, *Proc. phys. Soc.*, London, **86**, 1067.
174. Garton, W. R. S. 1965, *Proc. phys. Soc.*, London, **86**, 1077.
175. Garton, W. R. S., Wilson, M. 1966, *Astrophys. J.*, **145**, 333.
176. Garton, W. R. S., Wilson, M. 1966, *Proc. phys. Soc.*, London, **89**, 841.
177. Gausset, L., Herzberg, G., Lagerqvist, A., Rosen, B. 1965, *Astrophys. J.*, **142**, 45.
178. Geiger, J., Topschowsky, M. 1966, *Z. Naturforsch.*, **21a**, 626.
179. Gilbody, H. B., Ireland, J. V. 1964, *Proc. R. Soc. London* Ser. A, **277**, 137.
180. Ginter, M. L. 1966, *J. chem. Phys.*, **45**, 248.
181. Glasco, H. P., Zirin, H. 1964, *Astrophys. J. Suppl.*, **9**, 193.
182. Glennon, B. M., Wiese, W. L. 1966, *Misc. Publ. nat. Bur. Stand.*, 278, 92 pp.
183. Gloerson, P., Kieke, G. H. 1965, *J. molec. Spectrosc.*, **16**, 191.
184. Golden, D. E., Bandel, H. W. 1965, *Phys. Rev. Lett.*, **14**, 1010.
185. Golden, D. E., Bandel, H. W., Salemo, J. A. 1966, *Phys. Rev.*, **146**, 40.

186. Goodisman, J. 1964, *J. chem. Phys.*, **41**, 3889.
187. Gordon, H. R., McCubbin, Jr., T. K. 1966, *J. molec. Spectrosc.*, **19**, 137.
188. Goss, W. M. 1966, *Astrophys. J.*, **145**, 707.
189. Gray, L. D., Selvidge, J. E. 1965, *J. quantit. Spectrosc. radiat. Transfer*, **5**, 291.
190. Gray, L. D. 1965, *J. quantit. Spectrosc. radiat. Transfer*, **5**, 569.
191. Gray, L. D. 1966, *J. opt. Soc. Am.*, **56**, 555.
192.* Greaves, C. 1964, *J. Electronics and Control*, **17**, 171.
193. Green, T. A., Stanley, H. E., Chang, Y. C. 1965, *Helv. phys. Acta.*, **38**, 109.
194. Grenoble Conference Report, 1966. See Flower, D. R., Seaton, M. J., 1967, *Proc. Phys. Soc.*, **91**, 59.
195. Griem, H. R. 1964, *Plasma Spectroscopy*, Ch. 4, McGraw Hill, New York.
196.* Griem, H. R. 1966, *Phys. Rev. Lett.*, **17**, 509.
197. Gruzdev, P. F. 1966, *Optika Spektrosk.*, **20**, 377.
198. Gunton, R. C., Shaw, T. M. 1965, *Phys. Rev.*, **140A**, 756.
199. Hackam, R. 1965, *Planet. Space Sci.*, **13**, 667.
200. Hafner, H., Kleinpoppen, H., Krüger, H. 1965, *Phys. Lett.*, **18**, 270.
200a. Hallin, R. 1966, *Ark. Fys.*, **31**, 511.
200b. Hallin, R. 1966, *Ark. Fys.*, **32**, 201.
201. Halmann, M., Laulicht, I. 1965, *J. chem. Phys.*, **42**, 137; 1965, ibid., **43**, 438.
202. Hamberger, S. M., Johnson, A. W. 1965, *J. quantit. Spectrosc. radiat. Transfer*, **5**, 683.
203. Hammer, J. M., Aubrey, B. B. 1966, *Phys. Rev.*, **141**, 146.
204. Harrington, J. A., Modica, A. P., Libby, D. R. 1966, *J. chem. Phys.*, **44**, 3380.
205. Harris, F. E. 1966, *J. chem. Phys.*, **44**, 3636.
206. Harris, L. P. 1965, *J. appl. Phys.*, **36**, 1543.
207. Hasted, J. B. 1964, *Physics of Atomic Collisions*, Butterworths, London.
208. Herman, L., *et al.* 1965, *J. quantit. Spectrosc. radiat. Transfer*, **5**, 559.
209. Herranz, J., Stoicheff, B. P. 1963, *J. molec. Spectrosc.*, **10**, 448.
210. Herranz, J., Morcillo, J., Gómez, A. 1966, *J. molec. Spectrosc.*, **19**, 266.
211. Herzberg, G., Travis, D. N. 1964, *Can. J. Phys.*, **42**, 1658.
212. Herzberg, G. 1966, *Molecular Spectra and Molecular Structure*, **3**, Van Nostrand, Princeton, N.J.
213. Herzberg, G., Johns, J. W. C. 1966, *Proc. R. Soc. London*, **295**, 107.
214. Herzberg, G. 1966, in preparation.
215. Hess, W. 1965, *Z. Naturforsch.*, **20a**, 451.
216. Hesser, J. E., Dressler, K. 1965, *Astrophys. J.*, **142**, 389.
217. Hindmarsh, W. R. 1963, *Phys. Lett.*, **7**, 115.
218. Hiskes, J. R. 1965, *Phys. Rev.*, **137A**, 361.
219. Holt, H., Krotkov, R. 1966, *Phys. Rev.*, **144**, 82.
220. Horn, E. F., Dickey, F. P. 1964, *J. chem. Phys.*, **41**, 1614.
221. Huber, M. 1964, *Helv. phys. Acta*, **37**, 329.
224. Hughes, R. H., Dawson, H. R., Doughty, B. M., Kay, D. B., Stigers, C. A. 1966, *Phys. Rev.*, **146**, 53.
225. Humphreys, C. J., Paul, Jr., E. 1966, Naval Ord. Lab. Corona, NOLC Rep., **688**, 7.
226. Ingber, L. 1965, *Phys. Rev.*, **139A**, 35.
227. Jaecks, D., Van Zyl, B., Geballe, R. 1965, *Phys. Rev.*, **137A**, 340.
228.* Jain, D. C. 1964, *J. quantit. Spectrosc. radiat. Transfer*, **4**, 427.
229. Jeunehomme, M. 1965, *J. chem. Phys.*, **42**, 4086.
230. Jeunehomme, M. 1966, *J. chem. Phys.*, **44**, 2672.
231. Jeunehomme, M. 1966, *J. chem. Phys.*, **44**, 4253.
231a. Johansson, L. 1966, *Ark. Fys.*, **31**, 201.
232. Johns, J. W. C., Priddle, S. H., Ramsay, D. A. 1963, *Disc. Faraday Soc.*, **35**, 90.
233. Johns, J. W. C., Grimm, F. A., Porter, R. F. 1966, in preparation.
234. Johns, J. W. C., Tyler, J. K., Shurvell, H. F. 1966, in preparation.
235. Joshi, K. C. 1966, *Proc. phys. Soc.*, London, **87**, 285, 561.

*See additional references on page 266.

236. Joshi, K. C. 1966, *Proc. phys. Soc.*, London, **87,** 989.
237. Jungen, C., Miescher, E. 1965, *Astrophys. J.*, **142,** 1660.
238. Junkes, J., Salpeter, E. W. 1964, Specola Vaticana, Vatican City.
239. Junkes, J., Salpeter, E. W., Milazzo, G. 1965, Specola Vaticana, Vatican City.
240. Jütting, H., Koepp, R., Booz, J., Ebert, H. G. 1965, *Z. Naturforsch.*, **20a,** 213.
241. Kasai, P. H., Weltner, Jr., W. 1965, *J. chem. Phys.*, **43,** 2553.
242. Kasner, W. A., Biondi, M. A. 1965, *Phys. Rev.*, **137A,** 317
243. Kato, Y., Hayes, E. F., Duncan, A. B. F. 1964, *J. chem. Phys.*, **41,** 986.
244. Kaufman, V., Radziemski, Jr., L. J., Andrew, K. L. 1966, *J. opt. Soc. Am.*, **56,** 911.
245. Kaufman, V., Ward, J. F. 1966, *J. opt. Soc. Am.*, **56,** 1591.
246. Kaufman, V., Ward, J. F. 1967, *Appl. Opt.*, **6,** 43.
247. Keck, J., Carrier, G. 1965, *J. chem. Phys.*, **43,** 2284.
248. Khan, M. A. 1966, *Proc. phys. Soc.*, London, **87,** 569.
249. Khare, S. P., Moiseiwitsch, B. L. 1966, *Proc. phys. Soc.*, London, **88,** 605.
250. Kieffer, L. J. 1966, A Bibliography of Low Energy Electron Collision Cross Section Data, *JILA Inf. Center Rep.* no. 2.
251. Kieffer, L. J. 1966, Addendum to *JILA Rep.* no. 30, 27 pp.
252. Kieffer, L. J., Dunn, G. H. 1966, *Rev. mod. Phys.*, **38,** 1.
253. Kiess, C. C., Corliss, C. H., Kiess, H. K. 1962, *Astr. J.*, **67,** 579.
254. King, R. B., Olsen, K. H., Corliss, C. H. 1965, *Astrophys. J.*, **141,** 354.
255. Kingston, A. E. 1964, *Phys. Rev.*, **135A,** 1529, 1527.
256. Kingston, A. E. 1965, *Proc. phys. Soc.*, London, **86,** 467, 1279; 1965, ibid., **87,** 193.
257. Kingston, A. E., Schram, B. L., de Heer, F. J. 1965, *Proc. phys. Soc.*, London, **86,** 1374.
258. Kingston, A. E., Laver, J. E. 1966, *Proc. phys. Soc.*, London, **87,** 399.
259. Klechkovsky, V. M. 1965, *Optika Spektrosk.*, **19,** 441.
260. Kleman, B. 1963, *Can. J. Phys.*, **41,** 2034.
261. Klinkenberg, P. F. A. 1966, *Physica*, **32,** 1113.
262. Kokonov, E. Ja., Mandelshtam, S. L. 1965, *Optika Spektrosk.*, **19,** 145.
264. Korostyleva, L. A., Striganov, A. R. 1966, *Optika Spektrosk.*, **20,** 545.
265. Kosinskaya, I. V., Startsev, G. P. 1965, *Optics Spektrosc.*, **18,** 416.
266. Kovacs, I., Toros, R. 1965, *Acta Phys.*, **18,** 101.
267. Kovacs, I. 1965, *Acta Phys.*, **18,** 107.
268. Kovacs, I. 1965, *J. molec. Spectrosc.*, **18,** 229.
269. Krueger, T. K., Czyzak, S. J. 1965, *Mem. R. astr. Soc.*, **69,** 145.
270. Krueger, T. K., Czyzak, S. J. 1966, *Astrophys. J.*, **144,** 1194.
271. Krupenie, P. H., Weissman, S. 1965, *J. chem. Phys.*, **43,** 1529.
272. Krupenie, P. H. 1966, Nat. Stand. Ref. Data System – NBS, 5, 87 pp.
273. Kudryavtsev, E. M., Gippius, E. F., Pechenov, A. N., Sobolev, N. N. 1963, *High Temperature*, **1,** 60, 196.
274. Kuyatt, C. E., Simpson, J. A., Mielczarek, S. R. 1965, Quebec Conf. Proc., p. 113.
275. Lagerqvist, A., Miescher, E. 1966, *Can. J. Phys.*, **44,** 1525.
276. Lassetre, E. H., Jones, E. A. 1964, *J. chem. Phys.*, **40,** 1218.
277. Lassetre, E. H., Krasnow, M. E., Silverman, S. M. 1964, *J. chem. Phys.*, **40,** 1242.
278. Lassetre, E. H., Krasnow, M. E. 1964, *J. chem. Phys.*, **40,** 1248.
279. Lassetre, E. H., Silverman, S. M. 1964, *J. chem. Phys.*, **40,** 1256.
280. Lassetre, E. H., Silverman, S. M., Krasnow, M. E. 1964, *J. chem. Phys.*, **40,** 1261.
281. Lassetre, E. H. 1965, *J. chem. Phys.*, **43,** 4479.
282. Lawrence, G. M. 1965, *J. quant. Spectrosc. radiat. Transfer*, **5,** 359.
283. Lawrence, G. M., Link, J. K., King, R. B. 1965, *Astrophys. J.*, **141,** 293.
283a. Lawrence, G. M., Savage, B. D. 1966, *Phys. Rev.*, **141,** 67.
284. Lefebvre-Brion, H., Moser, C. M. 1965, *J. molec. Spectrosc.*, **15,** 211.
285. Lefebvre-Brion, H., Moser, C. M. 1965, *J. chem. Phys.*, **43,** 1394.
286. Lenander, C. J. 1966, *Phys. Rev.*, **142,** 1.
287. Levinson, I. B., Nikitin, A. A. Handbook for *Theoretical Computation of Line Intensities in Atomic Spectra*, 242 pp. Israel Program for Scientific Translations; Daniel Davey and Co., Inc., 257 Park Ave. South, New York, N.Y.

287a. Littlefield, T. A., Wood, A. 1965, *J. opt. Soc. Am.*, **55**, 1509.
287b. Link, J. K. 1966, *J. opt. Soc. Am.*, **56**, 1195.
288. Ljivov, B. V. 1965, *Optika Spektrosk.*, **19**, 507.
289. Lovell, S. E., McElroy, M. B. 1965, *Proc. R. Soc. London* Ser. A, **283**, 100.
290. Ludwig, C. B., Ferriso, C. C., Abeyta, C. N. 1965, *J. quantit. Spectrosc. radiat. Transfer*, **5**, 281.
291. Maclay, G. G. 1965, *J. chem. Phys.*, **43**, 185.
292. Mahan, B. H., Person, J. C. 1964, *J. chem. Phys.*, **40**, 392.
293. Mahan, B. H., Person, J. C. 1964, *J. chem. Phys.*, **40**, 2851.
294. Mahan-Smith, D., Carroll, P. K. 1964, *J. chem. Phys.*, **41**, 1377.
295. Makin, B., Keck, J. C. See ref. **307.**
296. Mallard, Mlle. 1966, Diplôme d'Etudes Supérieures de Physique, Paris.
297. Malville, J. M. 1964, *Astrophys. J.*, **139**, 198.
298. Mapleton, R. A. 1963, *Phys. Rev.*, **130**, 1839.
299. Mapleton, R. A. 1965, *Proc. phys. Soc.*, London, **85**, 841.
300. Mapleton, R. A. 1966, *Phys. Rev.*, **145**, 25.
301. Marquet, L. C., Davis, S. P. 1965, *J. opt. Soc. Am.*, **55**, 471.
302. May, R. M. 1964, *Nucl. Fusion*, **4**, 207.
303. May, R. M. 1965, *Phys. Lett.*, **14**, 98, 198.
304. May, R. M., Lodge, J. G. 1965, *Phys. Rev.*, **137A**, 699.
305. McCarroll, R. 1964, *Proc. phys. Soc.*, London, **83**, 409.
306. McDaniel, E. W. 1964, *Collision Phenomena in Ionized Gases*, Wiley, New York.
307. McDowell, M. R. C. ed. 1964, *Atomic Collision Processes*, Proc. 3rd Int. Conf. on Physics of Electronic and Atomic Collisions, North Holland Publ. Co., Amsterdam.
308. McEwen, D. J., Nicholls, R. W. 1966, *Nature*, **209**, 902.
309. McGowen, W., Clarke, E. M., Curley, E. K. 1965, *Phys. Rev. Lett.*, **15**, 917.
310. Mentall, J. E., Nicholls, R. W. 1965, *Proc. phys. Soc.*, London, **86**, 873.
311. Merer, A. J., Travis, D. N. 1965, *Can. J. Phys.*, **43**, 1795; 1966, ibid., **44**, 353.
312. Metzger, P. H., Cook, G. R. 1964, *J. chem. Phys.*, **41**, 642.
313. Meyer, C., Haeusler, C. 1964, *C. r. Acad. Sci.*, Paris, **259**, 748; 1965, ibid., **260**, 4182.
314. Meyer, C., Haeusler, C., Barchewitz, P. 1965, *J. Phys.*, Paris, **26**, 799.
315. Michels, H. H. 1966, *J. chem. Phys.*, **44**, 3834.
315a. Mielenz, K. D., Stephens, R. B., Gillilland, K. E., Nefflen, K. F. 1966, *J. opt. Soc. Am.*, **56**, 156.
316. Miescher, E. 1966, *J. molec. Spectrosc.*, **20**, 130.
317. Miller, R. E. 1965, *J. chem. Phys.*, **43**, 1695.
318. Milligan, D. E., Jacox, M. E. 1964, *J. chem. Phys.*, **41**, 2838.
319. Moise, N. L. 1966, *Astrophys. J.*, **144**, 763, 774, 782.
320. Moiseiwitsch, B. L., Smith, S. J. 1966, in preparation.
321. Monfils, A. 1965, *J. molec. Spectrosc.*, **15**, 265.
322. Moore, C. E. 1965, Nat. Stand. Ref. Data System – NBS, 1965, 3, Sec. 1.
323. Moore, C. E., Minnaert, M. G. J., Houtgast, J. 1966, *The Solar Spectrum* 2935Å *to* 8770Å, Monograph nat. Bur. Stand. 61, 349 pp.
324. Moore, C. E. 1966, *Astr. J.*, **71**, 796.
325. Morozova, N. G., Startsev, G. P. 1964, *Optika Spektrosk.*, **17**, 327.
326. Morozova, N. G., Startsev, G. P. 1965, *Optika Spektrosk.*, **18**, 899.
327. Morre, P. L., Browne, J. C., Matsen, F. A. 1965, *J. chem. Phys.*, **43**, 903.
328. Morrow, B. A. 1966, *Can. J. Phys.*, **44**, 2447.
329. Morton, D. C., Spitzer, Jr., L. 1966, *Astrophys. J.*, **144**, 1.
330. Motley, R. W., Kuckes, A. F. 1962, Proc. 5th int. Conf. on *Ionization Phenomena in Gases*, Munich, Ed. H. Maecker, North-Holland Publ. Co., Amsterdam.
331. Mott, N. F., Massey, H. S. W. 1965, *Theory of Atomic Collisions*, 3rd ed., Clarendon Press, Oxford.
332. Namioka, T. 1965, *J. chem. Phys.*, **43**, 1636.
333. Naqvi, A. M., Victor, G. A. 1963, *GCA Tech. Rep.* 63–15–A, Vols. I and II, Geophys. Corp. America, Bedford, Mass.

334. Naqvi, A. M. 1964, *J. quantit. Spectrosc. radiat. Transfer*, **4**, 597.
335. Naqvi, A. M. 1965, Proc. 7th int. Conf. on *Ionization Phenomena in Gases*, Belgrade.
336. Naqvi, A. M. 1966, Conf. on UV and X-ray Spectrosc. of Lab. and Astrophys. Plasma, Culham Laboratory, England.
337. Newsom, G. H. 1966, *Proc. phys. Soc.*, London, **87**, 975.
338. Nguyen, H., Drawin, H. W., Herman, L. 1966, Rep. no. CEA–R2912, C.E.N., France.
339. Nicholls, R. W. 1965, *J. chem. Phys.*, **42**, 804.
340. Nicholls, R. W. 1965, *Can. J. Phys.*, **43**, 1390.
341. Nikitin, E. E., Bykhovskii, V. K. 1964, *Optics Spectrosc.*, **17**, 444.
342. Nikitin, E. E. 1965, *J. chem. Phys.*, **43**, 744.
343. Nikitin, A. A., Jakubovsky, O. A. 1966, *Vest. Leningrad. gos. Univ.*, no. 7, 142, 146.
343a. Norlén, G. 1967, *Ark. Fys.*, in press.
344. Norman, G. E. 1965, *Optika Spektrosk.*, **19**, 657.
345. Norton, R. B., Ferguson, E. E., Fehsenfeld, F. C., Schmeltekopf, L. A. 1966, *Planet. Space Sci.*, **14**, 969.
346. Oak Ridge National Laboratory 1964, *Atomic and Molecular Collision Cross Sections of Interest in Controlled Thermonuclear Research*, ORNL–3113 Rev.
347. Oak Ridge National Laboratory. 1963, *Bibliography of Atomic and Molecular Processes*, ORNL–AMPIC–1; 1964, ORNL–AMPIC–3.
348. Oak Ridge National Laboratory. 1965, *Directory of International Workers in the Field of Atomic and Molecular Collisions*, ORNL–AMPIC–2.
349. Ochkur, V. I., Petrunkin, A. M. 1963, *Optics Spectrosc.*, **14**, 245.
350. Ochkur, V. I. 1964, *Soviet Phys. JETP*, **18**, 503.
351. Ochkur, V. I. 1965, *Optics Spectrosc.*, **19**, 258.
352. Ochkur, V. I., Brattsev, V. F. 1965, *Optics Spectrosc.*, **19**, 274.
353. Ochkur, V. I., Brattsev, V. F. 1966, *Astr. Zu.*, **42**, 1034.
354. Ogawa, M., Tanaka, Y., Jursa, A. S. 1964, *Can. J. Phys.*, **42**, 1716.
355. Ogawa, M. 1965, *J. chem. Phys.*, **43**, 2142.
356. Ogilvie, J. F. 1964, *Nature*, **204**, 572.
357. Oksyuk, Ya. D. 1966, *Soviet Phys. JETP*, **22**, 873.
358. Olman, M., McNelis, M., Hause, C. 1964, *J. molec. Spectrosc.*, **14**, 62.
359. Omidvar, K. 1965, *Phys. Rev.*, **140A**, 26, 38.
360. Omidvar, K. 1966, *NASA Rep.* X–641–66–245.
361. Ortenberg, F. S. 1964, *Optics Spektrosc.*, **16**, 398.
362. Oskam, H. J., Mittelstadt, V. R. 1963, *Phys. Rev.*, **132**, 1445.
363. Osterbrock, D. E. 1965, *Astrophys. J.*, **142**, 1423.
364. Ostroumov, P. P., Possikhin, V. S. 1965, *Optika Spektrosk.*, **19**, 653.
365. Pagel, B. E. J. 1967, to be published.
366. Parkinson, W. H., Reeves, E. M., Garton, W. R. S. 1966, *Can. J. Phys.*, **44**, 1745.
367. Patch, R. W. 1964, *J. chem. Phys.*, **41**, 1881.
368. Patch, R. W. 1965, *J. quantit. Spectrosc. radiat. Transfer*, **5**, 137.
369. Peach, G. 1965, *Proc. phys. Soc.*, London, **85**, 709.
370. Peach, J. V. 1966, Thesis, Oxford.
371. Peek, J. M. 1965, *J. chem. Phys.*, **43**, 3004.
372. Peek, J. M. 1965, *Phys. Rev.*, **140A**, 11.
373. Peek, J. M. 1966, *Phys. Rev.*, **143**, 33.
374. Pengelly, R. M., Seaton, M. J. 1964, *Mon. Not. R. astr. Soc.*, **127**, 165.
375. Penkin, N. P., Shabanova, L. N. 1965, *Optika Spektrosk.*, **18**, 749, 941; *Optics and Spectrosc.*, **18**, 425, 535.
376. Penkin, N. P., Shabanova, L. N. 1965, *Optika Spektrosk.*, **18**, 896; *Optics and Spectrosc.*, **18**, 504.
377. Penner, S. S., Varanasi, P. 1964, *J. quantit. Spectrosc. radiat. Transfer*, **4**, 799.
378. Penner, S. S., Varanasi, P. 1965, *J. quantit. Spectrosc. radiat. Transfer*, **5**, 391.
379. Pesic, D., Kliska, M. 1963, *Glasnik Hem. Drustva*, **28**, 347.
380. Petrini, D. 1966, in preparation.
381. Pfennig, H., Steele, R., Trefftz, E. 1965, *J. quantit. Spectrosc. radiat. Transfer*, **5**, 335.

382. Pfennig, H., Trefftz, E. 1966, *J. quant. Spectrosc. radiat. Transfer*, **6**, 549; 1966, *Z. Naturforsch.*, **21a**, 697, 1648; 1966, *Z. Phys.*, **190**, 253.
383. Pfennig, H., Trefftz, E., Vidal, C. R. 1966, *J. quantit. Spectrosc. radiat. Transfer*, **6**, 557.
384. Pitaevskii, L. P. 1962, *Soviet Phys. JETP*, **15**, 919.
385. Podkladenko, M. V. 1964, *Optics Spectrosc.*, **17**, 359.
386. Poletto, G., Rigutti, M. 1964, *Z. Astrophys.*, **60**, 199.
387. Pomilla, F. R., Milford, S. N. 1966, *Astrophys. J.*, **144**, 1174.
388. Price, W. C., Moore, C. E., Phillips, J. G. 1965, *J. opt. Soc. Am.*, **55**, 744.
389. Racah, G., Goldschmidt, Z. B., Toaff, S. 1966, *J. opt. Soc. Am.*, **56**, 407.
390. Radford, H. E. 1964, *Phys. Rev.*, **136A**, 1571.
391. Radziemski, Jr., L. J., Andrew, K. L., Kaufman, V., Litzén, U. 1967, *J. opt. Soc. Am.*, **57**, 336.
392. Ramsay, D. A. 1956, *Nature*, **178**, 374.
393. Rank, D. H., Fink, U., Foltz, J. V., Wiggins, T. A. 1964, *Astrophys. J.*, **140**, 366.
394. Rank, D. H., St. Pierre, A. G., Wiggins, T. A. 1965, *J. molec. Spectrosc.*, **18**, 418.
395. Rao, K. N., Oberly, R. 1966, *J. opt. Soc. Am.*, **56**, 555.
396. Rapp, D., Sharp, T. E., Briglia, D. D. 1965, *Phys. Rev. Lett.*, **14**, 533.
397. Rasool, S. I. 1964, *Mém. Soc. R. Sci.*, Liège, **9**, 55.
398. de Reclhac, L., Damany-Astoin, N. 1964, *C. r. Acad. Sci.*, Paris, **258**, 519.
399. van Regemorter, H. 1965, *A. Rev. Astr. Astrophys.*, **3**.
400. Reis, V. H. 1965, *J. quantit. Spectrosc. radiat. Transfer*, **5**, 585.
400a. Risberg, G. 1965, *Ark. Fys.*, **28**, 381.
401. Robinson, E. J., Geltman, S. 1966, *JILA Rep.* no. 65.
402. Rogers, W. A., Biondi, M. A. 1964, *Phys. Rev.*, **134A**, 1215.
403. Ross, W. R., Phillipson, P. 1966, *J. chem. Phys.*, **44**, 844.
404. Rothenberg, S., Davidson, E. R. 1966, *J. chem. Phys.*, **44**, 730.
404a. Rowley, W. R. C., Wilson, D. C. 1966, *J. opt. Soc. Am.*, **56**, 259.
405. Rudge, M. R. H. 1965, *Proc. phys. Soc.*, London, **85**, 607.
406. Rudge, M. R. H. 1965, *Proc. phys. Soc.*, London, **86**, 763.
407. Rudge, M. R. H. 1966, in preparation.
408. Rusk, J. R. 1965, *J. chem. Phys.*, **42**, 493.
409. Sahni, R. C., De Lorenzo, E. J. 1965, *J. chem. Phys.*, **42**, 3612.
410. Sampson, D. H., Mjolsness, R. C. 1965, *Phys. Rev.*, **140A**, 1466; 1966, ibid., **144**, 116.
411. Samson, J. A. R., Cairns, R. B. 1965, *J. opt. Soc. Am.*, **55**, 1035.
412. Saraph, H. E. 1964, *Proc. phys. Soc.*, London, **83**, 763.
413. Saraph, H. E., Seaton, M. J., Shemming, J. 1966, *Proc. phys. Soc.*, London, **89**, 27.
414. Schlüter, H., Avila, C., Durham, J. 1965, Proc. 7th int. Conf. on *Ionization Phenomena in Gases*, Belgrade.
415. Schlüter, H., Avila, C. 1966, *Astrophys. J.*, **144**, 785.
416. Schram, B. L., Vriens, L. 1965, *Physica*, **31**, 1431.
417. Schulz, G. J. 1964, *Phys. Rev.*, **135A**, 988.
418. Schulz, G. J., Asundi, R. K. 1965, *Phys. Rev. Lett.*, **15**, 946.
419. Seaton, M. J. 1964, *Mon. Not. R. astr. Soc.*, **127**, 191.
420. Seaton, M. J. 1965, *Trans. int. astr. Un.*, **12A**, 152.
421. Seaton, M. J. 1966, *Proc. phys. Soc.*, London, **88**, 801, 815.
422. Shimauchi, M. 1964, *Sci. Light*, **13**, 53.
423. Silverman, S. M., Lassetre, E. H. 1964, *J. chem. Phys.*, **40**, 1265.
424. Skerbele, A., Meyer, V. D., Lassettre, E. N. 1966, *J. chem. Phys.*, **44**, 4069.
425. Slavenas, I. Ju. Ju. 1966, *Optika Spektrosk.*, **20**, 485.
426. Smith, F. J. 1966, *Planet. Space Sci.*, **14**, 929, 937.
427. Smith, K. 1966, *Rep. Prog. Phys.*, **29**, Part II.
428. Sobel'man, I. I. 1965, *Soviet Phys. JETP*, **21**, 642.
429. Solomon, P. M. 1964, *Astrophys. J.*, **139**, 999.
430. Solomon, P. M., Stein, W. 1966, *Astrophys. J.*, **144**, 825.
431. Somerville, W. B. 1964, *Astrophys. J.*, **139**, 192.
432. Spector, N. 1966, *J. opt. Soc. Am.*, **65**, 341.

433. Stabler, R. C. 1964, *Phys. Rev.*, **133A**, 1268.
434. Stacey, D. N., Vaughan, J. M. 1964, *Phys. Lett.*, **11**, 105.
435. Starr, W. L., Shaw, T. M. 1966, *J. chem. Phys.*, **44**, 4181.
436. Stebbings, R. F., Young, R. A., Oxley, C. L., Ehrhardt, H. 1965, in preparation.
437. Steele, R., Trefftz, E. 1966, *J. quantit. Spectrosc. radiat. Transfer*, **6**, 833.
438. Stoebner, A., Delbourgo, R., Laffitte, P. 1965, *C. r. Acad. Sci.*, Paris, **261**, 5044.
439. Stone, P. M. 1966, *Phys. Rev.*, **141**, 137.
440. Stuart, J. D., Matsen, F. A. 1964, *J. chem. Phys.*, **41**, 1646.
441. Sugar, J. 1963, *J. opt. Soc. Am.*, **53**, 831.
442. Sugar, J. 1965, *J. opt. Soc. Am.*, **55**, 33.
443. Sugar, J. 1965, *J. opt. Soc. Am.*, **55**, 1058.
444. Sugar, J., Kaufman, V. 1965, *J. opt. Soc. Am.*, **55**, 1283.
445. Sural, D. P., Sil, N. C. 1966, *Proc. phys. Soc.*, London, **87**, 201.
445a. Swensson, J. W., Risberg, G. 1966, *Ark. Fys.*, **31**, 237.
446. Takayanagi, K. 1965, *J. phys. Soc. Japan*, **20**, 562.
447. Takayanagi, K., Geltman, S. 1965, *Phys. Rev.*, **138A**, 1003; 1966, ibid., **143**, 25.
448. Tanaka, Y., Innes, F. R., Tursa, A. S., Nakamura, M. 1965, *J. chem. Phys.*, **42**, 1, 83.
449. Tatum, J. B., Nichols, R. W. 1967, *Astrophys. J. Suppl.*, **14**, 21.
450. Tatum, J. B., Nichols, R. W. 1966, *Publ. Dom. astrophys. Obs.*, **13**, no. 1, in press.
451. Tawde, N. R., Korwar, V. M. 1965, *Ind. J. pure appl. Phys.*, **3**, 198.
452. Taylor, R. L., Herkowitz, S. B. 1965, *Proc. Inst. electr. electron. Eng.*, **53**, 657.
452a. Terrien, J. Proceedings of Symposium 'Méthodes nouvelles de spectroscopie instrumentale (Paris et Orsay 1966)', to be published.
453. Thaddeus, P., Clauser, J. F. 1966, *Phys. Rev. Lett.*, **16**, 819.
454. Tilford, S. G., Vanderslice, J. T., Wilkinson, P. G. 1965, *Can. J. Phys.*, **43**, 450.
455. Tilford, S. G., *et al.* 1966, *Astrophys. J. Suppl.*, **13**, no. 115.
456. Tilford, S. G., Simmons, J. D. 1966, *J. chem. Phys.*, **44**, 4145.
457. Toros, R. 1966, *Acta Phys. Hungar.*, no. 1–2, 91.
458. Tucker, W. H., Gould, R. J. 1966, *Astrophys. J.*, **144**, 243.
459. Underhill, A. B. 1966, private communication, August.
460. Vainshtein, L. A., Presnyakov, L., Sobelman, I. 1964, *Soviet Phys. JETP*, **18**, 1383.
461. Vainshtein, L. A., Presnyakov, L., Opykhtin, V. 1965, *Soviet Phys. JETP*, **20**, 1542.
462. Varanasi, P., Lauer, J. L. 1966, *J. quantit. Spectrosc. radiat. Transfer*, **6**, 127.
463. Varsavsky, C. M. 1963, *Planet. Space Sci.*, **11**, 1001.
464. Vaughan, J. M. 1966, *Proc. R. Soc. London*, **295**, 164.
465. Veldre, V. Ya., Damburg, R. Ya., Peterkop, R. K. 1963, *Tr. Inst. Fiz. Akad., Nauk Latv. SSR*, **13**; (1966, English transl. *Atomic Collisions*, Butterworths; see also, ref. **467**).
466. Veldre, V. Ya., Lyash, A. V., Rabik, L. L., Fridken, L. A. 1965, *Latv. PSR Zinat. Akad. Vestis Fiz.*, Tehn Ser., USSR, **4**, 3.
467. Veldre, V. Ya. 1966, *JILA Inf. Center Rep.* no. 3, 153 pp.
468. Veldre, V. Ya., Rabik, L. L. 1966, *Optics Spectrosc.*, **19**, 265.
469. Verma, R. D., Warsop, P. A. 1963, *Can. J. Phys.*, **41**, 152.
470. Verma, R. D. 1965, *Can. J. Phys.*, **43**, 2136.
471. Verolainen, Ja. F., Osherovitch, A. L. 1966, *Optika Spektrosk.*, **20**, 929; 1965, *Dokl. Akad. Nauk. SSSR*, **164**, 1022.
472. Vidal, C. R. 1964, *Z. Naturforsch.*, **19a**, 947.
473. Vidal, C. R. 1965, Proc. 7th int. Conf. on *Ionization Phenomena in Gases*, Belgrade.
474. Vriens, L. 1965, *Physica*, **31**, 385, 1081.
475. Wakefield, C. B., Davidson, E. R. 1965, *J. chem. Phys.*, **43**, 834.
476. Walters, A. K., Startsev, G. P. 1964, *Optika Spektrosk.*, **17**, 483.
477. Watanabe, K., Jursa, A. S. 1964, *J. chem. Phys.*, **41**, 1650.
478. Weinberg, J. M., Fishburne, E. S., Rao, K. N. 1965, *J. molec. Spectrosc.*, **18**, 428.
479. Weltner, Jr., W., McLeod, Jr., D. 1965, *Nature*, **205**, 87.
480. Weniger, S. 1964, *J. Phys.*, Paris, **25**, 946.
481. Weniger, S. 1965, *Ann. Astrophys.*, **28**, 117.
482. Weniger, S. 1966, *Colloque de Trieste sur les étoiles froides.*

483. Wentink, Jr., T., Isaacson, L., Morreal, J. 1964, *J. chem. Phys.*, **41**, 278.
484. Wiese, W. L. 1965, *Plasma Diagnostic Techniques*, Ch. 6, Eds. Huddlestone, R. H., Leonard, S. L., Academic Press, New York and London.
485. Wiese, W. L., Smith, M. W., Glennon, B. M. 1966, Nat. Stand. Ref. Data System–NBS, 4, Vol. I, 153 pp.
486. Wilets, L., Gallaher, D. R. 1966, *Phys. Rev.*, **147**, 13.
487. Wilkins, R. L. 1966, *J. chem. Phys.*, **44**, 1884.
488. Wind, H. 1965, *J. chem. Phys.*, **43**, 2956.
489. Yamatera, H., Fitzpatrick, B., Gordon, G. 1964, *J. molec. Spectrosc.*, **14**, 268.
490. Zapesochnyi, I. P., Shimon, L. L., Soshnikov, A. K. 1965, *Optics Spectrosc.*, **19**, 480.
491. Zapesochnyi, I. P., Shimon, L. L. 1966, *Optics Spectrosc.*, **19**, 268.
492. Zare, R. N. 1965, *JILA Rep.* no. 31.
493. Zare, R. N., Larsson, E. O., Berg, R. A. 1965, *J. molec. Spectrosc.*, **15**, 117.

ADDENDA

On page 235, after the first alinea: 'In collaboration with . . . ", add the following alinea:

Green, Johnson, and Kolchin (**192a**) have computed oscillator strengths in He I for all possible electric dipole transitions between the terms $n^{1,3}\ S$, $n^{1,3}\ P^0$, $n^{1,3}\ D$, where $n = 1$ to 9 for S terms, 2 to 8 for P^0 terms and 3 to 8 for D terms. Central-field wave functions with exchange and configuration interaction, were used. Oscillator strengths were computed by the length and velocity expressions.

On page 237, line 9, after: " . . . with the earlier theories for ion lines", add: "while typically consistent to $\pm$ 20% with calculations for the case of arc lines."

Additional References

192a. Green, L. C., Johnson, N. C., Kolchin, E. K. 1966, *Astrophys. J.*, **144**, 369.
196a. Griem, H. R. 1967, *Astrophys. J.*, **147**, 1092.
196b. Griem, H. R. 1967, *Astrophys. J.*, **148**, 547.
228a. Jakobson, L. 1967, *Ark. Fys.*, **34**, 19.

15. COMMISSION POUR L'ETUDE PHYSIQUE DES COMETES

PRÉSIDENT: Dr F. L. Whipple, Director, Smithsonian Astrophysical Observatory, 60 Garden Street, Cambridge, Massachusetts, 02138, U.S.A.

VICE-PRÉSIDENT: Professor Dr L. Biermann, Max-Planck-Institut für Physik und Astrophysik, Direktor des Instituts für Astrophysik, Fohringer Ring 6, 8 Munich 23, Germany.

COMITÉ D'ORGANISATION: G. Herzberg, B. J. Levin, E. Roemer, V. Vanýsek, K. Wurm.

MEMBRES: Baldet†, Bertaud, Beyer, Bobrovnikoff, Bouska, Dobrovol'skij, Dossin, Dufay (J.), Guigay, Haser, Houziaux, Hunaerts, Liller, Lyttleton, Martel-Chossat, Miller (F.D.), Poloskov, Richter, Rijves, Swings, Vsehsvjatskij, Waterfield.

I. NARRATIVE REPORT

A. Introduction

The form of this Draft Report follows the practice of the 1964 Report. The first is a Narrative, briefly presenting progress as evaluated by the writer. Second is a short Administrative Report including results and plans for cooperative activities. Third is a comprehensive Bibliography, with extremely short abstracts. The writer is greatly indebted to the members of the Commission who have provided him with important assistance in the preparation of this report, particularly C. Arpigny (spectroscopy), G. Herzberg (laboratory work), and V. Vanýsek and B. J. Levin (Slavic languages). The writer, however, takes complete responsibility for the accuracy and relevancy of the contents.

A major event was the 1965 sun-grazing comet (Ikeya-Seki, 1965*f*), whose perihelion distance was only 0·008 A.U. As the eighth member of the sun-grazing comet group that includes the great comet of 1882, it aroused worldwide interest and contributed valuable knowledge concerning the stability of cometary nuclei and concerning activities in and about the coma.

Another event of great cometary importance was the Thirteenth Liège Symposium devoted to the nature and origin of comets. This international symposium organized by the current President of the IAU took place in July 1965 and brought together an international group of researchers in this field.

Its proceedings and other important books and monographic articles are listed at the beginning of the Bibliography under the heading: A. Books and Monographs. The references are alphabetized under each subsection separately, and the subsections identified by capital letter, A., B., C., etc., for cross-referencing.

B. Photometry

The rapidly increasing pace of cometary photometry and its improved quality obtained by photoelectric and other methods of measurement are highly gratifying.

Here are listed contributions to magnitude measures, surface photometry, special observations, studies of brightness changes and correlations, polarization and measures in the radio and infrared regions.

At the Haute-Provence Observatory, high-resolution spectrograms have been used (Malaise 1966) to determine photometric radial profiles of CN, C_2, C_3 and CH, and a special photometer for this use is being designed.

N. Richter and W. Högner (1964, 1966) use contrast printing of positives and negatives to produce isophotes from comet photographs. An atlas is being prepared to include all bright comets.

V. G. Rijves (1964, 1965) has also developed new photometric parameters and techniques for calibrating cometary photographs and presents photometric observations. D. A. Rožkovskij (1965) describes a new procedure for photographic photometry and processes photographs of Comets Ikeya, Humason, Alcock and Kopff.

For the IQSY, S. K. Vsehsvjatskij (1964*a*, 1964*b*) makes recommendations about programs and methods of comet observation. In the same program R. B. Southworth (1966) describes the use of the 12-station satellite-tracking system of the Smithsonian Astrophysical Observatory for measuring integrated magnitudes and colors of comets and gives some results for Comet Ikeya (1964*f*).

E. Roemer continues her valuable regular reports on cometary brightness and appearances, and (1965) with R. E. Lloyd (1966) and M. Thomas (Roemer *et al.* 1966) has presented nearly 1450 observations of 75 comets. She is planning an atlas of direct photographs of comets and will analyze comet photographs of the U.S. Naval Observatory for head and tail structure.

S. Vsehsvjatskij (1964*a*, 1964*b*, 1964*c*) has published two extensions (to 1965) of his catalogue of photometric parameters of comets. For 167 comets in the catalogue, A. A. Demenko (1966) has classified their tails according to Bredikhin's types. V. Rijves (1964, 1965) presents photometric measures for several comets.

Isophotes of the heads (*h*) and/or tails (*t*) of comets have been made by the following investigations: A. A. Demenko (1964) and D. O. Mokhnach (*t*, Seki-Lines, 1962*c*); O. V. Dobrovol'skij and H. Ibadinov; and V. P. Džapiašvili and G. N. Salukvadze (1965) (*h* and *t*, Arend-Roland, 1957 III); G. N. Salukvadze and V. P. Džapiašvili (1966) (*h* and *t*, Mrkos, 1957 V); G. K. Nazarchuk (1965) (*h* and *t*, Arend-Roland with theory for stream formation); L. M. Genkina (1964, 1965*a*, 1965*b*, 1965*c*) (*h*, 1961*e*, 1963*i*, 1964*h*, and P/Kopff); T. A. Polyakova and G. S. Romashkin (1965) (*h*, 1965*b*, 1965*c*, 1964*h*, 1964*g*); and E. E. Khachikian and R. A. Epremian (1966) (*h*, 1963*e*). S. Grudzinska (1965) has investigated photometrically 1963*a* and 1963*b*.

Photoelectric photometry has been carried out by J. Bouska and P. Mayer (1966) for 1964*h* (Everhart), in the UBV system with a special filter, and by Mrkos, Tremko and Vanýsek (1966) for 1966*b* (Kilston) and 1966*c* (Barbon). On 1964*f* (Ikeya) N. A. Kovar and R. P. Kovar (1965*a*, 1965*b*) have made photoelectric scans for CN, C_3, C_2 and CH.

The remarkable photographs by T. Hirayama and F. Moriyama (1965) of Comet Ikeya-Seki (1965*f*) with a coronograph show the tail highly curved as the comet swings through the corona. Their observations of the splitting of the nucleus are of great value. The coronograph observation by J. M. Malville and C. Evans (1966) show striking changes with time in the ratio of the D_1 and D_2 lines of Na, suggesting changes in the optical depth. V. Vanýsek (1966) made photoelectric measures in the infrared, and Z. Ceplecha and B. Valniček (1966) in the near red. The infrared measures at 1·65, 2·2, 3·4 and 10μ for 19 days by E. E. Becklin and J. A. Westphal (1966) show a calculated blackbody temperature that is nearly 50% too high for expected dust temperatures. They suggest iron or iron-like particles as the cause.

Other observations of Comet Ikeya-Seki have been reported by M. Bader and L. C. Haughney (1966) from a jet aircraft at 13 km, photographed several times; P. B. Boyce and W. M. Sinton (1966) before and after perihelion with spectra scanner; J. L. Weinberg (1966), polarization measures and Stokes parameters at λ 5300 Å; and A. D. Thackeray (1965) on the split nucleus.

M. Beyer continues his long series of comet observations with the 20-cm refractor of the Hamburg Observatory. These observations of cometary brightnesses and Beyer's determination of brightness-variation parameters are particularly valuable because of their uniformity, prolonged over so many years. Results (1964, in press) for several comets show two very interesting outbursts of Comet Alcock (1965*h*). Beyer plans an index list for his measures of about 100 comets. Bertaud (1965) studied 'flares' of the Comet Schwassmann-Wachmann I.

For Comet Humason (1961*e*) C. E. Kearns and R. Rudnicki (1966) have published 28 fine photographs with the 48-inch (121 cm) Palomar Schmidt (five in yellow). They show the extraordinary changes occurring in the head and tail of this distant comet.

Correlations between cometary brightnesses and solar activity are reported by Vsehsvjatskij (1965*a*, 1965*b*) and (1964*d*, 1964*e*, 1965*c*, 1965*d*, 1966) for several comets in the last 40 years, including a correlation for P. Schwassmann-Wachmann I with solar rotation. P. Notni photoelectrically (1961) and S. Grudzińska (in press) also find correlations of cometary brightness with solar activity. Correlations with the brightness of the zodiacal light are suggested by M. Fracassini and L. E. Pasinetti (in press), and with meteorological conditions in Europe and North America by Sekanina (1964*a*, 1964*b*).

The secular fading in the absolute brightnesses of P/Encke is indicated not to exceed 1 magnitude per century by L. Krésak (1965) from maximum brightness observed at each apparition, whereas D. H. Douglas-Hamilton and the writer (Whipple and Douglas-Hamilton 1966) derive a much more rapid fading. The latter find no solar-correlated variations for P/Encke nor firm choice between a $1/\Delta^2$- and the $1/\Delta$-law suggested by Öpik (1963). J. Bouska (1965) expects P/Oterma to become unobservable after its recent orbital changes.

V. G. Rijves (1964, 1965) finds a surprisingly low value of the parameter n, $(1/r^n)$ for comets.

T. Gehrels (1966) and M. T. Martel and M. Saute (in press) made polarization measures of comets. T. Takakura *et al.* (1965) could detect no radio emission from Comet Ikeya-Seki (1965*f*) at 160 hertz.

F. Dossin (1966) found a probable comet near the Sun during the eclipse of 20 July 1963.

Questions as to the true secular rates of fading for periodic comets and possible correlations of nuclear or head luminosity with solar activity continue to remain without definitive answers, at least to the writer. Improved photometric methods and their extensive application still appear necessary.

C. Spectroscopy

The George Darwin Lecture, which was delivered by the President of our Union in 1964, was devoted to cometary spectra (Swings, A 1965). In the first part an historical resumé is given of the spectroscopic observations of comets during the hundred years that have elapsed from the first such observation of a comet in 1864 until the recent progress brought about by the use of some of the largest telescopes for cometary observations. Then a number of important desiderata are expressed and plans for the future spectroscopic work on comets are described. These concern observations from the ground as well as from balloons or aircraft, from rockets or satellites.

In two important survey papers C. Arpigny (A 1965) and (1965) has studied the Swings mechanism (resonance fluorescence excitation, including the effects of the Fraunhofer lines) in detail on the basis of the theory of the steady-state conditions. His predictions agree very satisfactorily with observed cometary spectra. He has also discussed the Greenstein effect (different relative intensities of some rotational lines in different regions of a comet), and presents (1965) a critical revision of abundance determinations in comets, with new results as well. Furthermore, he reviews the question of the forbidden oxygen lines present in the spectra of a number of comets and questions the identification of cometary [O I] lines in the cases when the green line is stronger than the red doublet ($I_G > I_R$). In this connection, a series of spectra of Comet Bester (1948 I) are being studied photometrically in Liège with a view to establishing the origin of the strong line observed at λ 5577 on these spectra. This is the only case with $I_G > I_R$ where there remains any strong doubt as to the origin of the oxygen lines. Here we should also point out an error that has slipped into the Bibliography appearing in the 1964 Draft Report: The comment 'green [O I] line present' given for reference 57 (p. 178 of the Report), which concerns Comet Seki-Lines (1962*c*) (not *e*), is wrong; only the red line λ 6300 has been identified in the spectrum of this comet.

The advantage of cometary observations with high-resolution equipment can be no better illustrated than by the result of A. S. Stawikowski and J. L. Greenstein (1964), who measured the λ 4745 line in Comet Ikeya (1963*a*) for the C^{12}/C^{13} isotopic abundance ratio. They found $C^{12}/C^{13} = 70 \pm 15$, comparable to the values for the Earth and meteorites. This result is perhaps the only direct evidence pointing (still not definitively) to a solar-system origin for comets.

Comet Ikeya-Seki, 1965*f*, the sun-grazing comet, produced some startling spectroscopic results. V. Livingston *et al.* (1966) with the McMath Solar Spectrograph at 5 mm/Å (!), identified Si, Ca, Na, Cr, Mn as well as many Fe lines in the spectrum near perihelion. The Na D lines in emission and Fraunhofer absorption show a huge displacement arising from the high velocity of the comet. Na was being ejected some 4 km s^{-1} radially from the nucleus and reached a maximum velocity of 18 km s^{-1} along the tail. J. Dufay, P. Swings and Ch. Fehrenbach (1965, 1966) at the Haute-Provence Observatory find from a preliminary study that about 90 very sharp metallic emission lines appear on the 4Å/mm spectra of 21 October. Most of these are Fe I lines, while the rest belong to Na I, Ca I, Ca II, Mn I, Ni I, and probably Al I. The Fe I lines are short, their extension being about 8 seconds of arc on each side of the central nuclear condensation. On the contrary, the Na I D lines and the Ca II H and K lines are rather long (about 25 and 60 seconds of arc, respectively). A detailed study of the excitation of these emissions has been undertaken in Liège, and it will be possible to compare the results of the post-perihelion observations with pre-perihelion spectra lent by Dr A. D. Thackeray from the Radcliffe Observatory (Thackeray *et al.* 1966). Spectroscopic material has also been lent from the Sacramento Peak and the Kitt Peak Observatories.

Low-dispersion spectra showed a weak continuum both on 13 October and 21 October 1965. A. D. Thackeray *et al.* (1966) also found that near perihelion the CN bands were weaker, compared to greater solar distance, and that the presence of C_2 and C_3 was uncertain. On 20 and 21 October 1965 G. W. Curtis and Sacramento Peak Observatory Staff (1966) observed four spectra at 2·5Å/mm. They observed both the comet and the solar corona, and made polarization measures of the comet and photoelectric scans in Na D. These measures are being analyzed. J. M. Malville and C. Evans (B 1966) measured the Na D_2/D_1 ratio and the Na/continuum ratio near perihelion, finding rapid changes in the gas/dust ratio.

Comet Candy (1960*n*). Let us point out a note (Wyller 1962) concerning the identification of some near infrared emissions. This was published in 1962, but was not mentioned in the Draft Report of the 1964 General Assembly of the IAU. Flynn (1961) had observed strong emissions in the wavelength regions λλ 7765–7850 and λλ 8013–8070, and he had attributed these emissions to the red system of CN in spite of the large difference between these wavelengths and the expected positions of the (2, 0) and (3, 1) CN red bands. Wyller (1962) gave some arguments in favor of the identification of the observed emissions with the (3, 0) and (4, 1) bands of the C_2 Phillips system. Actually, Wyller's main argument should be revised because it used a value of 2500°K for the rotational 'temperature' of the C_2 Swan bands, whereas a recent determination of this parameter by C. Arpigny (1966*a*) indicates a value of about 5000°K. Thus, the value of J (total angular momentum quantum number) at which the intensity maxima should occur in the various branches of the Phillips bands is a factor of about $\sqrt{2}$ larger than indicated by Wyller, with the result that the wavelength coincidences between these maxima and the observed emission features is now less convincing. However, the intensity distributions are certainly very broad and only a detailed examination could settle the question of a possible contribution of the C_2 Phillips bands to the observed infrared emissions. In connection with the problem of the formation of C_2, it is important to establish whether these bands are present or not in the spectra of comets. L. J. Stief and V. J. DeCarlo (1965) suggest that the Swan bands arise from the photodisassociation of acetylene or acetylene-like molecules.

Comet Humason (1961*e*). C. Arpigny (1964) has made some corrections to Warner and Harding's (1963) discussion of their spectroscopic observations. It has been shown that an anomaly mentioned by these authors concerning the CO^+ comet-tail bands (absence of the *R*-branches) disappears when the correct line strengths are used for the various branches and when the effects of the Fraunhofer lines are taken into account in the fluorescence-excitation mechanism. At 5 A.U., F. V. Dossin (1966) finds a faint continuum at the center and emission of CO^+ N_2^+ and CN (faint).

Comet Seki (1961*f*). Y. Andrillat (1966), spectrum 3500–5000 Å, 29 Å/mm, 15 and 18 October 1961.

Comet Ikeya (1963*a*). F. D. Miller (1964), two objective-prism spectra. C. Arpigny (1966*b*) shows that the λ 4300 band of CH was excited by fluorescence. The electric dipole moment of CH appears to be about 1 Debye. He finds (1966*a*) from high-resolution spectra also of Seki-Lines (1962 III) and Burnham (1960 II) that the 'temperatures' of the rotational and vibrational distribution of cometary C_2 Swan bands are higher (5–6000°K) than earlier studies indicate and virtually independent of heliocentric distance or comet. See also Malaise (B 1966) and Grudzińska (B 1965).

Comet Alcock (1963*b*). D. Chalonge and M. Bloch (1966) find from low-dispersion spectra that scattered continuum is slightly redder than solar radiation.

Comet P/Kearns-Kwee (1963*d*). F. Dossin (1964) describes the spectrum. The theoretical paper on the C_2 Swan bands in comets by R. E. Stockhausen and D. E. Osterbrock (1965) is criticized by C. Arpigny (1966*c*) on both observational and theoretical grounds.

G. Righini (1966) discusses early observations of cometary spectra.

In his contribution to this report C. Arpigny stresses the following desiderata in cometary spectroscopy:

(*a*) Monochromatic radial profiles of the various atoms or radicals observed in comets. Dependence of these profiles and of the 'radii' of the coma in monochromatic light upon the heliocentric distance of the comet would yield invaluable information concerning the production of these atoms or radicals (Arpigny 1965, Malaise B 1966).

(*b*) Observational data on the spectral energy distribution of the scattered continuum and on the polarization of this continuum. When compared with the results of the theoretical investigations, this would shed light upon the problem of the nature of the scattering particles. See e.g., L. Remy-Battiau (1964, 1966).

(*c*) Radial profiles of the forbidden oxygen lines and intensities of these lines relative to those of the other emissions. This would be useful for the study of the excitation mechanism responsible for the emission of these cometary [O I] lines as well as for the determination of abundances in cometary atmospheres.

The writer believes that an amplification of the compiled cometary spectral and other cometary characteristics, so well begun by L. Remy-Battiau and P. Swings (1966), would greatly advance our understanding of the nature of comets.

D. Structure and origin of the nucleus

The writer is delighted with the increasing attention, both observational and theoretical, being given to physical studies of cometary nuclei. A. H. Delsemme (1966*a*, 1966*b*) has made the most comprehensive effort to date in developing a specific icy model. He explores the vital role of hydrates in the chemistry of the parent molecules, and limits his basic model to H_2O, CH_4, NH_3 and earthy dust. He calculates rather successfully the surface conditions at various solar distances to predict the onset of activity for an incoming comet. Much depends upon the abundances, particularly whether there is an excess of CH_4. He suggests that the albedos may be appreciably higher than for the Moon.

W. F. Hübner (1965*a*, 1965*b*, 1966) and with A. Weigert (1966) have laid particular stress on the role of ice grains in the immediate neighborhood of the nucleus as producing opacity to reduce solar radiation and to induce observable effects near the nucleus. They develop a theory for the rate of gas production in comets, for brightness outbursts of P/Schwassmann-Wachmann I at great solar distances, and for the strength of [O I]. Dobrovol'skij (1966) also demonstrates the great possible opacity of dust to solar heating of the nucleus. He also (1964*a*, 1964*b*) discusses the action of solar corpuscular radiation on ices of the nucleus and a possible mechanism for solar-flare reactions involving free-radical formation on the nucleus. V. Vanýsek (1966) discusses the nature of cometary dust and its marked variation in observability; e.g., Comets Arend-Roland (1957 III) and Mrkos (1957 V) show 100 times the dust/gas ratio of P/Encke and others.

B. J. Levin (1966) raises the important problem of the transfer of volatiles from beneath the surface to adsorption on the meteoritic surface, particularly for periodic comets. He concludes that the fall in absolute magnitude ($\sim 1^m$) repeatedly noted for Comet P/Encke shortly after perihelion is caused by the nearly complete desorption of gases from the surface and exhaustion of available volatiles just below. He stresses that this phenomenon occurs frequently in comets. In 1964 V. Vanýsek (1965) noted that for P/Encke this brightness loss was extraordinary ($\sim 4^m$) and suggested that it arose from a complete cessation of nuclear activity due to complete exhaustion of the active layer, the rate of luminosity loss being about that expected for molecules of lifetimes some 10^5 s. The writer here points out a possible alternative, viz., that periodic comet nuclei are active at any time only in a finite number of areas, and that momentarily all were inactive for Comet Encke. Observed jets from nuclei tend possibly to support this alternative. The answer to this problem is important.

W. Liller (1966) finds that desorption energies calculated by the Levin formula appear to apply well for Comet Arend-Roland (1957 III). J. Remy-Battiau (1964) concludes that cometary bursts are not due to solar particles but arise from some phenomenon in the nuclei, possibly involving dust.

E. Roemer (1966*a*, 1966*b*) made a significant advance in our knowledge of the dimensions of cometary nuclei by measuring limiting magnitudes for 18 periodic and 11 nearly parabolic comets. Depending on the comet and the assumed albedo, the radii lie in the range 0·1 to 65 km.

From P. E. Zadunaisky's (1966) definitive orbits of Halley's comet (1910 II) before and after perihelion, Z. Sekanina (1967*a*), using his general theory (1967*b*), shows that the orbital changes could have arisen by selective loss of 0·5% of the mass of the nucleus. He discusses in detail the splitting of Comet Ikeya-Seki (1965*f*) (1966*a*). B. Marsden (1966) is investigating the orbits of the sun-grazing comets and finds tentatively no evidence for physical accelerations.

From a study of 13 split comets, R. B. Stefanik (1966) shows that splitting of cometary nuclei without apparent physical cause is frequent among 'new' comets in any observable part of their orbits. Velocities of separation are of the order of 10–30 m s^{-1}. He and the writer (Whipple and Stefanik 1966) investigated the possibility that icy nuclei were originally heated by radioactivity to drive the most volatile materials out to the surface layers where they froze. Splitting could then result by heat shock when they first approach the Sun.

B. Donn (1964, 1965) discusses the growth of icy comet nuclei in space and the physics of compaction under gravity. N. Richter (1966) summarizes the general problems related to the origin of comets. D. McNally (1966) compares various common problems of the interstellar medium and the origin of comets.

In a series of important articles, E. J. Öpik (1966*a*, 1966*b*, 1966*c*) develops in more detail the orbital consequences of comets crossing the orbits of planets and his estimate (1966*a*) of 10^4 to 10^5 old nuclei trapped within the orbit of Jupiter. He shows that lunar ejecta cannot account for the meteorites and concludes that they come from the old comets (1966*b*). He expands theories of cometary accretion (1966*c*) within a Laplace-type nebula and almost disproves the possibility of the gravitational lens as a formative mechanism. He shows that accretion of comets in the present Oort cloud is impossible and shows how planetary perturbations could disperse comets from the planetary plane to this diffuse cloud. There is much food for thought, calculation and observation in Öpik's writings. The question of cometary and/or asteroidal contributions to the Apollo group of Earth-crossing bodies is vital to resolve.

The most recent general qualitative summary of theories of the origin and formation of comets is that by P. Lancaster-Brown (1964). The writer believes, however, that his hope for comet regeneration by any accretion process within or near the solar distance of the planets must be abandoned in view of the space measures of the solar wind. All isolated small bodies (possibly including the Moon) appear to be dissipating their mass.

J. L. Greenstein (1966) finds that possible nuclear processes appear at present to be unobservable in comets and hence of little current interest.

A number of contributions on the motions of comets and their source but not directly related to their physics are of interest to this Commission. They are presented by reference only, but starred (*) with slightly longer than usual abstracts (Arnold 1965, Brady 1965, Dufay in press, Harwit 1966*a*, 1966*b*, Krésak 1965, 1966, Lyttleton and Hammersley 1964, Makover 1964, Piotrowski 1965, Sekanina 1966*b*, 1966*c*, 1966*d*, Sitarski 1964, Šteins 1964, Šteins and Kronkalne 1964*a*, 1964*b*, Vsehsvjatskij 1965, 1966*a*, 1966*b*, 1966*c*). It is encouraging to see so much attention given to these basic problems of cometary orbits and perturbations.

E. Coma, excitation and ionization

K. Wurm (1966*a*, 1966*b*) with A. Mammano (in preparation) and with J. Rahe (in preparation) and the latter (Rahe 1965) for various comets, particularly Comet Donati, 1858 VI, find major excitation and ionization within 500 km of the nucleus with time scales of only 100 s. See also Öpik (D 1966*a*, D 1966*b*, D 1966*c*). Nearly symmetrical streaming occurs for neutral particles; for CN it is $1/2$ km s^{-1} corresponding to $T \sim 500°K$! Expulsion to $\sim 150\,000$ km requires the turning of ions to the tail axis. These processes show no clear dependence on solar distance. The ions stream only towards Sun, so that type-I tails originate in the nuclear region. Wurm and his collaborators still find no explanation for such rapid and extensive excitation, ionization and motions in conventional processes, including the solar wind, and suggest intrinsic processes such possibly as effects of electrical fields.

In an important contribution on the mechanism of ionization and excitation in comet heads, L. Biermann and E. Trefftz (1964) find that the excitation of [O I] lines must predominantly take place in photodissociation to produce the atom from parent molecules. They require copious production of parent molecules from the nucleus, more than is apparently required to account for the C_2, CN and CO^+ bands. In bright comets they find as do others that gas densities must be high enough to permit exothermic chemical reactions up to distance of 10^4 km or more; in this way they form CO^+. They find a long time scale (>10 d) for some primary ionization. They extrapolate their theory to the ultraviolet, predicting radiation flux, and discuss detailed processes of ionic dissociation of several parent molecules such as H_2O^+, H_2^+, C_2H_2, CO_2, and C_3H_4. V. Čeredničenko (1965, 1966), for a comet at $r = 1$ A.U. analyzes all possible transformation processes for C_2N_2, $C_2N_2^+$, CN, CN^+, N_2, N_2^+, C_3H_6, $C_3H_6^+$, C_2H_2, $C_2H_2^+$, C_3, C_2, CH_2, CH_2^+, CH, H_2, H_2O, H_2O^+, OH and OH^+ in the field of solar photon and corpuscular radiation (using old data on solar radiation). W. M. Jackson and B. Donn (1966*a*) carried out Monte-Carlo calculations on molecular collisions near a comet nucleus and discuss possible parent molecules for C_3. They calculate (1966*b*) the photochemical yield near the nucleus for parent molecules. They expect CO^+ near the nucleus.

J. Bouska (1965*a*, 1965*b*, 1966) and E. B. Kostyakova (1965) determine the intensity distribution in the head for Comet Arend-Roland (1957 III). Ju. N. Gnedin and A. Z. Dolginov (1966*a*, 1966*b*) and D. O. Mokhnach (1966*a*, 1966*b*) derive theoretical formulae for the distribution of parent and luminous molecules (mostly C_2 and CN near the nucleus) and compare them with photometric observations. They calculate numbers of molecules produced and lost, lifetimes, velocities and general distribution of molecules in the coma. Much of this work follows L. Haser's model discussed by this Commission in 1964. He has continued and expanded his 'fountain model' in the meantime (1966).

F. D. Miller (1965*a*, 1965*b*) discusses the forms of the coma and velocities for particles and gas, showing how a bright leading edge may be produced by a suitable repulsive force (Comets Mrkos, 1957 V, and Seki-Lines, 1962 III). V. Rijves (1960*a*, 1960*b*, 1963) develops the problem of the high rate of expansion of gases and the deviation of ions into the tail from the heads of comets while N. G. Ptitsina calculates the geometry of cometary envelopes on the basis of hydrodynamical processes. J. Dufay (in press) involves molecular excitation by electrons in the outer corona for sun-grazing comets. For Comet Mrkos (1957 III) M. F. A'Hearn (1964) analyzes C_2 densities.

For the intensity ratios of the [O I] lines in a comet, N. S. Kovar and R. P. Kovar (1966) develop a theory in terms of the electron density and temperature within the coma. They too require molecular processes. They determine (B 1965*a*, B 1965*b*) the ratios of C_3/CN and CH/CN in Comet Ikeya, 1964*f*.

V. Vanýsek (1965) finds that systematic changes in observed color indices for several comets depend upon the diameter of the photoelectric photometer used and ascribes the effect to different lifetimes of CN and C_2 molecules coupled with effects due to ice grains ejected from the nucleus. A similar study by M. E. Dewey and F. D. Miller (1966) for Comet Seki, 1961 VIII, from isophotes gives valuable information on lifetimes of C_2 and its parent molecule.

For Comet Morehouse, 1908 III, W. Schlosser (1966) has measured 210 Greenwich plates to determine motions in and about the head. He finds a velocity expansion of 2·4 km s^{-1} coming to zero within about a day after an outburst. He has many measures on condensations and development of tail rays, including sinusoidal motion. Studies of forms and motions in the heads of comets have been carried out by O. V. Dobrovol'skij (1966) for Comet Honda, 1955 V, and P. Egibekov (1964) for Comet Ikeya, 1963*a*. Dobrovol'skij utilizes observed solar UV radiation (300–1700 Å) and finds that the lifetimes of CO and CO_2 molecules before ionization are too long ($\sim 10^6$ s) for them to produce CO^+ and CO_2^+ by solar photoionization.

J. Rahe (1966) not only finds parabolic structures about the nucleus but points out again certain extraordinary curved arcs radiating from the nucleus of Comets 1861 II, 1862 III and Halley. Special observations suppressing CN at λ 3883 would expedite observations of complicated head features.

The contributions of this section generally avoid magnetohydrodynamic processes of ionization, dissociation or excitation by solar wind. Among many overlapping papers discussed in other sections of this Report, those by Bouska and Mayer (B 1966), Hübner (D 1965*a*, D 1965*b*, D 1966), Hübner and Weigert (D 1966), Ioffe (F 1965, F 1966*a*, F 1966*b*), Nazarchuk (B 1965), Malaise (B 1966) and Vanýsek (D 1966) should perhaps be mentioned. Much of the theoretical material of this section received thorough treatment in the 1964 Report of this Commission. Since that Report we find considerable progress in measures and in the treatment of molecular lifetimes. A great spurt of progress should occur when radiations from the parent molecules can be observed directly from space stations. We still observe too few of the many processes occurring in the heads of comets for a truly satisfactory understanding of them.

F. Type-I tails

Opinion is still divided as to possible correlations between cometary plasma tails and general solar activity. D. Antrack, L. Biermann and Rh. Lüst (1964) in an extensive statistical treatment find no such correlation and a probably smaller decrease with heliocentric latitude than Stumpff had found. But for Comet Tomita-Gerber-Honda (1964 VI) they (Biermann and Lüst 1966) and M. F. A'Hearn (1965) suggest strongly that a tail knot and curvature were related to the solar event of 4 July 1964. R. Pflug (1966) from the orientation of 323 observations of 11 comet tails found no correlations with the phase of sunspot activity or the heliocentric distance. The scatter in direction, however, was greater at sunspot maximum. He suggests that more active regions (not flares) produce lower solar-wind velocity and greater comet brightness.

Regardless of uncertainties in detailed relations between solar activity and type-I tails, few investigators question that the solar wind controls the motions of the ionic tails. The analogy with the Earth's magnetosphere (e.g., Ness 1966) is compelling. The massive effort by J. C. Brandt and M. J. S. Belton (A 1966) in measuring carefully the orientation of 1607 plasma tails, in addition to the work of Pflug (1966), has led (Brandt 1966*a*, 1966*b*, and 1966*c* with M. J. S. Belton and M. W. Stephens 1966) to striking conclusions. The measures confirm the assumption that statistically a plasma tail is directed nearly in the plane of the orbit in a direction measured by the combined velocity vectors of the solar wind and cometary motion, the aberration angle. There is a sharp velocity minimum to the solar wind of some $150 \pm 50\ \mathrm{km\ s^{-1}}$ (Brandt 1966*c*, Brandt *et al.* 1966) and a mean velocity of $500\ \mathrm{km\ s^{-1}}$ with only a $50\ \mathrm{km\ s^{-1}}$ probable spread in means between solar maxima and minima. These values agree broadly with the directly measured velocities from space probes by, e.g., C. W. Snyder, C. W. Neugebauer and U. R. Rao (1963) and N. F. Ness and J. M. Wilcox (1964).

Brandt (1966*c*) finds that the comet tails, Mariner II and storm data are consistent with the following expressions for the solar-wind velocity:

$$w = 330 + 9\ \Sigma K_{\mathrm{p}} \qquad (\mathrm{km\ s^{-1}})$$

or

$$w = 316\ A_{\mathrm{p}}^{0.20} \qquad (\mathrm{km\ s^{-1}}),$$

where ΣK_{p} is the sum of the geomagnetic index K_{p} for the day in question and A_{p} is the surface magnetic-disturbance index. A residual tangential component (Brandt 1966*a*, 1966*b*, 1966*c*, Brandt *et al.* 1966) of about $10\ \mathrm{km\ s^{-1}}$ in the effective solar wind is directed forward in the sense of the solar rotation. Brandt (1966*c*) doubts that a latitude dependence of the plasma velocity can be determined at present and suggests that Pflug's (1966) results of a minimum velocity at the latitude of the sunspot zone may be due to the non-uniform distribution of cometary data.

G. Guigay (1966*a*) confirms similarly a wind velocity of some $400\ \mathrm{km\ s^{-1}}$ for the tail of Comet Ikeya-Seki (1965*f*) but notes (1966*b*) that for distant Comet Humason (1961*e*) the tail is sometimes well in advance of the radius vector to the Sun as does E. J. Öpik (1956, 1964) in his report on motions in the tail of Halley's comet. Similarly, K. Wurm and A. Mammano (1964) find that for Comets 1908 III and 1911 V the tails varied irregularly in direction while for 1907 IV the tail axis lay in the plane of the orbit. They find, however, a systematic lag (25°).

The theory of type-I tails has developed conceptually from solar light pressure (inadequate), through L. Biermann's (1951) direct interaction of solar and cometary ions (wind flux inadequate) to H. Alfvén's (1957) magnetohydrodynamic (MHD) coupling. He visualized the supersonic collisionless MHD shock wave developing about the ions in a comet head, with dimensions comparable to the observed head. Our knowledge of the solar wind and the ultraviolet solar radiation have become precise through the measures made from space vehicles (e.g. Gringauz *et al.* 1960, Synder *et al.* 1963 and Ness and Wilcox 1964) while improved physical theory has shown that solar radiation is inadequate to produce the cometary ions copiously enough to fit the improved observational data of comets. We now know that the type-I tails are essentially ionic and that the solar wind carries enough momentum to carry along the observed ion clouds if the MHD net can selectively couple with the cometary ions independently of the neutral molecules over the large areas involved. The next step by M. Harwit and F. Hoyle (1962*a*, 1962*b*) attempted to solve these problems by charge exchange from the protons. Theoreticians now appear to agree that electrons not protons must and can produce this ionization, and that electrons also play a major role in the MHD coupling, chiefly because of their enormous mobility compared to protons or heavier ions. Other problems concern the complex ray structures and their motions.

An excellent qualitative view of plasma theory can be obtained from the discussions of L. Biermann (1966) and N. F. Ness and B. D. Donn (1966). The latter (and Ptitsina, below) show how the complexity of the ray structure is qualitatively explained by the MHD neutral tubes and sheaths as pointed out by W. I. Axford, H. E. Petschek and G. L. Sicoe (1965) while the enormous role of electrons, including their ability to excite [O I] lines, was stressed by Axford (1964). Major contributions to this theory have been made by L. Maročnik (1963, 1964) in studying the disturbed shock region for both pressure and ionization effects; N. N. Kazantseva (1966) who considers an oblique collision between the corpuscular stream and the cometary plasma and N. G. Ptitsina (1964, 1965) who develops an MHD approach to the study of moving envelopes and ray structures in plasma tails, comparing them with observations of Comet Morehouse, 1908 II; Z. M. Ioffe (1965, 1966*a*, 1966*b*), who calculated the shape of the coma and plasma tails from a theory of supersonic plasma flux about a cometary head for a three-dimensional compressible medium; D. B. Beard (1966), who stressed the role of electrons and charge separation in tail structure formation; L. Biermann, B. Brosowski and H. U. Schmidt (1966), who developed in detail the shock front or 'contact' surface concept some 10^6 km ahead of the cometary head; and N. S. Kovar and J. W. Kern (1966) who have developed an MHD model for comets.

A most detailed theory has been developed by M. K. Wallis (1967) showing how transverse plasma instabilities couple the motions of the solar wind to the comet plasma. He finds viscous heating important in maintaining the ray structures some 100 times their widths against diffusion *without* a magnetic field. This theory shows promise of clearing up some of the major problems of tail ray structure, while almost all of the theories, in analogy with the Earth's magnetosphere, show how extended tail streamers can converge back on the major axis of flow as so frequently observed. Great progress is being made!

Other theoretical papers concerning the interaction of photons, corpuscular radiation and molecules magnetic-field induced in comet heads are by V. I. Čeredničenko (1965*a*, 1965*b*, 1965*c*), L. S. Maročnik (1963, 1964) V. M. Južakov (1963). Other observational and theoretical papers besides Belton and Brandt (A 1966), Nazarchuk (B 1965) include E. Fajziev (1964), solar distance effects; Z. Sekanina (1966), diameters of coma of many comets and action of solar wind; Bernasconi (1965), method for calculation of true cone of cometary tail; J. Bouska (1964), velocities and acceleration in the tail of Comet Arend-Roland 1957 III and Nazarchuk (B 1965), on the theory of corpuscular radiation on the tail of the same comet; O. V. Dobrovol'skij (1964*a*, 1964*b*), accounting for cometary bursts by sudden increases in the solar wind; E. Öpik (1965), on theoretical failure of solar radiation or exchange mechanisms with the solar wind to explain the abundance of CO^+ ions in type-I tails; C. E. Kearns and K. Rudnicki (1965), velocities in tail of Comet Humason 1962 VIII; Ch. Cailliatte (1966), method for determining the true divergence of gas and dust tails; and Rh. Lüst (in press), streaming of CO^+ features fairly parallel to tail axis of Comet Morehouse 1908 III, but *no* evidence that CO^+ formed close to the nucleus.

G. Type-II and -III tails

Ejection, motions and properties of dust in comet tails were studied by a few investigators: O. Dobrovol'skij (in preparation), from an analysis of all published data on simultaneous determinations of ejection velocities and accelerations of dust particles, found a uniform numerical relationship in the head and tails of comets; for Comet Arend-Roland (1957 III) E. Fajziev (1964*a*, 1964*b*) determined graphically the initial velocity of particles ejected into the tail; M. L. Finson and R. F. Probstein (1966) studied the motions of dust carried out of a cometary nucleus by the expanding gas and pressed backward by light pressure.

With the exception of Finson and Probstein, most investigators are dissatisfied with the application of the Bessel-Bredikhin mechanical theory to type-II comet tails, particularly to the formation of synchronic bands, e.g., A. A. Demenko (1966) from a study of three photographs of Comet Seki-Lines, 1962 III. He further develops the theory for analysis of these bands. B. J. Levin (1964, 1966), M. J. S. Belton (1965, 1966) and P. Notni (1964, 1966) agree that the classical theory is not satisfactory and that the evidence points almost conclusively to forces arising from the solar wind. They do not, however, agree on the detailed physical mechanism at work.

Levin (1964) notes that Na is highly accelerated in the tail of some comets but that the lifetime of the neutral atom is too short. He argues (1966) for the recognition of Bredikhin's type-III tails and suggests strongly that molecules are also present and active in the so-called dust tails. Belton (1965, 1966) shows that type-II tails lie in the orbital plane but are directed away from the Sun much forward ($h \sim 1$) to the direction expected on the basis of light pressure alone. For ten comets the effects are similar and independent of solar distance. A postulated interplanetary drag is unsatisfactory. Ejection velocities to explain synchronic bands are too high for dust. He suggests that electrons may interact with the dust and solar plasma. Notni (1964, 1966) independently notes the theoretical difficulties of the classical theory and suggests that dust particles in a plasma may encounter high electron density (10^7–10^8/cm^3) and attain surprisingly high negative charges (up to $-$ 100 eV) so that the MHD plasma can interact to produce strong repulsive forces radially from the Sun. CO^+ ions may interact.

Thus it appears that some uncertainties still remain in dust-tail theory. Also there remain serious theoretical doubts as to the major processes for dissociation and ionization and even observational questions as to the location and time scales of these processes.

H. Laboratory work

In addition to the monographic papers on cometary spectra already mentioned, by P. Swings (A 1965) and C. Arpigny (A 1965), we point out J. G. Phillips' (1966) valuable summary of related laboratory spectroscopy, including a bibliography of recent spectroscopic data and physical properties of a number of possible parent molecules in comets. From laboratory studies he presents dissociation energies and ionization potentials for C_2, CH, CH^+, CN, CN^+, CO, CO^+, NH, NH^+, N_2, N_2^+, OH, OH^+, and oscillator strengths for certain systems of CO, CO^+, CO_2^+, and N_2^+. He found that he could derive OH emission by bombarding methanol (CH_3OH) so that H_2O is not necessarily the parent molecule. L. J. Stief and V. J. DeCarlo (1965) can obtain C_2 bands by dissociating CH_4 and C_2H_2 but not from C_2H_4 and C_2H_6. W. M. Vaidya (1966) compares spectra of comets and hydrocarbon flames to find certain cometary bands of CH and OH in a Bunsen flame. In an ethylene flame he finds a new spectrum, of HCO; possibly in comets?

In the Spectroscopic Laboratory of the National Research Council of Canada the study of the spectra and structures of diatomic molecules and simple polyatomic molecules and free radicals has been continued. Recently G. Herzberg (1966) has found a new simple band system that represents a $\Sigma_u^+-\Sigma_g^+$ transition of either C_2 or C_2^+ or C_2^-. The 0–0 band of this system is at 5416Å. It would be clearly of interest to see whether this spectrum occurs in comets.

Several studies of diatomic hydrides have been carried out during the last three years. The work by Herzberg and Johns on CH, mentioned in the earlier report, has not yet been published, but the value of the ionization potential (10·64 eV) given earlier must be considered as reliable. Clear evidence of predissociation in the $B^2\Sigma^-$ state has been obtained in absorption. Furthermore, it was established that in fluorescence the $C^2\Sigma^+-\mathrm{K}^2\Pi$ transition of CH does not occur or is very weak, while the corresponding transition in CD is strong. This observation suggests very strongly that the $C^2\Sigma^+$ state is subject to strong predissociation in CH but much less in CD. Two new absorption systems of SiH (and SiD) have been found and analyzed by Verma (1965). Both upper states are $^2\Sigma^+$. The first system is at 3250Å and might be of interest for cometary spectra. In addition, considerable work has been done on the spectra of BH (Bauer *et al.* 1964) and SH (Morrow in press). Most of these spectra are in the vacuum ultraviolet.

New and very interesting spectra have been found of NF by Douglas and Johns (in press) and Johns, and NCl (Colin and Johns in preparation). These spectra are analogues of the atmospheric oxygen bands. A brief investigation of the spectrum of CCl has shown that the excited $A^2\Delta$ state is a regular state.

In an unpublished investigation by Herzberg, the spectra of N_2^+, CO^+, and CO_2^+ have been observed for the first time in absorption.

With regard to triatomic molecules, the following spectra may be of interest. The investigation of the red band of CH_2 found several years ago has now been completed and is about to appear (Herzberg and Johns in press). The red bands correspond to the singlet form of CH_2. A similar spectrum of BH_2 involving the ground state of this free radical and occurring in the red and near infrared regions has been observed and studied by Herzberg and Johns (in press).

An extensive study of the 4050Å group of the C_3 radical on the basis of absorption spectra obtained in the laboratory has been published by Gausset *et al.* (1965, 1966). This spectrum is now rather well understood. The paper mentioned contains also a spectrogram of the fluorescence of the flash photolysis of diazomethane which shows a striking resemblance to a cometary spectrum since it contains not only very strongly the spectrum of C_3, but also those of C_2 and CN.

Other triatomic molecules recently studied at Ottawa and of possible interest to cometary physics are: CCN (Merer and Travis 1965), CNC (Merer and Travis 1966), NCN (Herzberg and Travis 1964), N_3 (Douglas and Jones 1965) and CO_2^+ (Johns 1964). All these spectra are quite extensive with characteristic bands that should be readily identifiable in cometary spectra if they occur. E. Stokes Fishburne and K. N. Rao (1966) show reproductions of high-dispersion spectra of CO^+, 0·2Å/mm.

Other related spectroscopic work: Ch. Jungen and E. Miescher (1966) on emission NO in the UV; H. Papazian (1966), an unknown absorption band, 210–230 mμ, from photolysis of solid HN_3 and, in progress, a study of radicals from photochemical reactions; A Monfils (1965*a*, 1965*b*) on absorption spectra of H_2, HD and D_2; W. Jackson, S. Glicker and B. Donn, in progress, photochemistry of benzene and other aromatic compounds as potential cometary parent molecules.

Laboratory studies of the bombardment of cometary molecules by ions and electrons are of particular importance in interpreting cometary spectra in terms of physical processes.

M. Dufay, Marie-Claude Poulizac (1966) and with J. Desesquelles (Poulizac *et al.* in press) have studied the spectra at λλ 2500–8900 Å of CO and CO_2 excited by H^+ and H_2^+ at 30–600 keV and pressures of 10^{-5} to 10^{-1} mm Hg. They find interesting line and band structures for C I, C II, O I, O II, CO^+, CO and CO_2^+. The vibration levels correspond to high temperatures. Measurements are made of cross sections for ionization and charge exchange for H^+ ions.

Other molecules currently under study in this fashion are CH_4, C_2H_4, C_2H_6, C_2H_2 and CH_4. A. G. Koval', V. T. Koppe and Ja. M. Fogel' (1966*a*, 1966*b*) have studied the emission spectra of N_2^+, CO^+, CO_2, CO_2^+ and others excited by 13 keV electrons.

R. Herman (1966) is conducting an interesting experiment in the polymerization and growth of carbon particles from hydrocarbon molecules in an electrical discharge. Results are somewhat like those from photolysis. She finds growth, C_2, C_3, etc. and also direct catalytic dissociation of hydrocarbon molecules on the carbon surface.

I. Space experiments

P. T. Gronstal and R. P. Bukata (1966) looked for 0·1 to 1·0 MeV protons from Comet Ikeya-Seki (1965*f*) by balloon-supported equipment on 20 October 1965. No cometary effects could be detected.

H. Föppl *et al.* (1965, 1967) and G. Haerendel *et al.* (in press) have carried out seven artificial plasma experiments as proposed in 1961. Reacting compounds, $Ba(NO_3)_2 + 4$ Al or 4 La, $Sr(NO_3)_2 + 6$ Mg or $Ba(NO_3)_2$ or $3\,BaO_2 + n$Al or nMg were sent by rocket as high as 2000 km. Approximately 100 g of Ba II could be observed at such heights, even when distributed over 10^5 km². The diameter of the clouds increased faster than by purely molecular diffusion. Wind shears could be measured. Ionization of 5 to 7% could be reached. The cloud of neutral Sr was observed over the range of optically thick to thin. The time constant for removal of the radiating material by the atmosphere was $\sim 10^3$ s. The ionized clouds tended to diffuse along the lines of the geomagnetic field. Attempts will be made to reach heights $> 10^5$ km.

H. Bredohl (1966) reports on Professor Rosen's equipment that released liquid ammonia at an altitude of 200 km from an ESRO rocket, July 1964. Various spectrophotometric and photometric instruments detected feeble emission of NH_2, and measured the expansion and polarization of the cloud. The experiment is encouraging for future such research and points up certain needed ground-based experiments and observations.

A. Potter (1966) has produced a cloud of CN radicals at 150 km in a rocket experiment and has investigated the origin of free radicals in comet heads.

Although E. J. Öpik (1965) has shown sound reasons for doubting the observability and therefore the value of artificial comet experiments in deep space, especially those involving icy masses, the liquid experiments in the high atmosphere have proved to be of great interest and value. As he points out, however, many ground-based experiments are sorely needed to clarify actual physical processes in comets, regarding both observed phenomena and mode of formation.

In planning for possible space missions, 'fly-bys' at 10^3 to 10^4 km from a cometary nucleus, R. L. Roberts, F. Narin and P. M. Pierce (1966) have calculated missions for 110 periodic comets during the interval to Halley's in 1986. They discuss possible scientific objectives, instrumentation and specifications. About one such mission a year is theoretically possible. See also P. Swings (A 1965).

II. ADMINISTRATIVE REPORT

It is very encouraging to note the degree to which past recommendations to Commission 15 have been applied by observers. Narrow-band and monochromatic photometry have been used effectively for several comets. A broad use of the photographic archives of comets has been a source of very real contribution during the past years. A number of observatories have been most generous in allowing visiting astronomers to utilize their photographic collections.

Professor Wurm's distribution of copies of cometary photographs has been a start toward a wider distribution of this sort. More discussion is needed concerning the bringing together of the proposed Atlas of Characteristic Cometary Forms.

In spite of the existence of astronomical observatories over most of the globe they are still far from successful in even the simple photography of cometary forms on a continuous 24-hour basis. Also we are still far from such a continuous photoelectric photometric program as was suggested by Dr Vsehsvjatskij. These matters should receive more consideration to ascertain whether important bright comets cannot be followed more completely than in the past.

It is not clear whether further Commission recommendations are essential to the production and study of 'artificial comets' by various national or international space programs. Further discussion of this matter appears to be needed, perhaps in consultation with representatives of Commission 44.

F. L. WHIPPLE
President of the Commission

III. BIBLIOGRAPHY

Note that short abstracts of the quoted papers are frequently included in parentheses after the reference; these are generally not the exact titles of the paper. Because of the extreme difficulty of locating a reference in the text when a purely alphabetical bibliography is used the writer is experimenting with an alternative compromise. Each section of the report is accompanied by its separate alphabetical bibliography. Cross-indexing is indicated by the appropriate section letter heading.

A. Books and monographs

Arpigny, C. 1965, in *Annual Reviews of Astronomy and Astrophysics*, Ed. L. Goldberg, Annual Reviews, Inc., Palo Alto, Calif., Vol. 3, p. 351. (Spectra of comets and interpretation.)

Belton, M. J. S., Brandt, J. C. 1966, *Astrophys. J. Suppl.*, **13**, 125. (Catalogue of comet-tail orientations.)

Dobrovol'skij, O. *Comets*, 'Nauka', Moscow. (On the physical nature of comets.)

Dufay, J. 1966, *Les Comètes*, Presses Univ. de France, Paris. (Monograph on comets.)

Guigay, G. 1966, Les Processus non stationnaires dans les comètes et l'activité solaire. (Transl. from Russian of O. Dobrovol'skij's, *Trudy Inst. Astrofiz. Stalinabad*, **8**, 1961.)

Marsden, B. G., Porter, J. G. 1966, *Mem. Br. astr. Ass.*, **40**, no. 2. (Comet orbits through 1965.)

Swings, P. (*a*) 1965, in *Advances in Space Science and Technology*, Vol. VII, Ed. F. I. Ordway, III, N. Y. Academic Press, New York, p. 393. (Astronautical investigations of comets, a review and projection.)

(*b*) 1965, *Q. J. R. astr. Soc.*, **6**, 28. (George Darwin lecture on cometary spectra.)

(*c*) 1966, Ed., *Nature et origine des comètes*, Soc. roy. des Sciences de Liège. Mémoires. Collection in – 8°, 5ème Sér., Vol. 12, fasc. unique; also in Univ. Liège Inst. Astrophys. Coll. 8°, Vol. 21, no. 510.

U.S.S.R. (*a*) 1964, *Investigations of comets according to the program of the International Year of Quiet Sun*, Naukova Dumka, Kiev, in Russian. (Collection of papers.)

(*b*) 1965, *Fizika Komet i Meteorov*, Naukova Dumka, Kiev, in Russian. (Collection of papers.)

(*c*) 1966, *Physics of comets*, Naukova Dunka, Kiev, in Russian. (Collection of papers.)

B. Photometry

Bader, M., Haughney, L. C. 1966, *Astr. J.*, **71**, 844. (Photographs 1965*f* from aircraft.)
Becklin, E. E., Westphal, J. A. 1966, *Astrophys. J.*, **145**, 445. (Infrared photometry at 1·65, 2·2, 3·4 and 10μ, 1965*f*.)
Bertaud, Ch. 1965, *L'Astronomie*, **79**, 130.
Beyer, M. 1964, *Astr. Nachr.*, **288**, 113. (Physical observations of 1963*a*, 1963*b*, 1963*d*, Schwassmann-Wachmann I.)
Bouska, J. (*a*) 1965, *Bull. astr. Inst. Csl.*, **16**, 358. (Visibility of P/Oterma.)
(*b*) 1966, *Publ. astr. Inst. Charles Univ.*, Prague, no. 49. (Observations, 1965*f*.)
Bouska, J., Mayer, P. 1966, *Bull. astr. Inst. Csl.*, **17**, 84. (Photoelectric photometry, 1964*h*.)
Boyce, P. B., Sinton, W. M. 1966, *Astr. J.*, **71**, 847 (abstr.). (Spectral scans of 1965*f*.)
Ceplecha, Z., Valniček, B. 1966, *Bull. astr. Inst. Csl.*, **17**, 211. (Photoelectric red photometry, 1965*f*.)
Demenko, A. A. (*a*) 1964, *Investigations of comets by IQSY programme*, 128. (Tail structure isophotes, 1962*c*.)
(*b*) 1966, Types of cometary tails, in *Physics of the Moon and planets*, Jerusalem, p. 94, translated from the Russian by Israel Program for Scientific Translations (*Fizika luny i planet*, Ed. I. K. Koval, Naukova Dumka, Kiev, 1964). (167 comets, tail classifications.)
Dobrovol'skij, O. V., Ibadinov, H. *Bjull. Inst. Astrofiz., Dušanbe-Stalinabad.* (Photometry of Arend-Roland.)
Dossin, F. 1966, Univ. Liège Inst. Astrophys. Coll. 8°, Vol. 21, no. 510, 189. (Probable comet during solar eclipse.)
Džapiašvili, V. P., Salukvadze, G. N. 1965, *Abastumansk. astrofiz. Obs. Gora Kanobili Bjull.*, no. 32, 155. (Photometry of Arend-Roland.)
Fracassini, M., Pasinetti, L. E. 1966, in press. (Correlation of comet brightnesses with zodiacal light.)
Gehrels, T. 1966, Univ. Liège Inst. Astrophys. Coll. 8°, Vol. 21, no. 510, 193. (Wavelength dependence of polarization observed in comets.)
Genkina, L. M. 1964, *Kometnyi Cirkulyar*, no. 14. (1965*a*, *ibid.*, no. 23; 1965*b*, *ibid.*, no. 25; 1965*c*, *ibid.*, no. 27.) (Photometry, 1961*e*, 1963*i*, 1964*h*, and P/Kopff.)
Grudzińska, S. (*a*) 1965, *Bull. astr. Obs. Univ. N. Copernicus*, no. 34. (Photometry and some objective prism spectra, 1963*a*.)
(*b*) *ibid.*, in press. (Outburst of 1963*b*, coincides with a solar flare.)
Hirayama, T., Moriyama, F. 1965, *Publ. astr. Soc. Pacif.*, **17**, 433. (Coronograph photographs of 1965*f* near Sun.)
Kearns, C. E., Rudnicki, K. 1966, *Z. Astrophys.*, **64**, 337. (28 photographs of Comet Humason in 1962.)
Khachikian, E. E., Epremian, R. A. 1966, *Soobšč. Bjurak. Obs.*, no. 37, 48. (Photometry 1965*e*.)
Konopleva, V. 1966, *Physics of comets*, Naukova Dumka, Kiev. (Magnitude 1955 IV.)
Kovar, N. A., Kovar, R. P. 1965*a*, *Astrophys. J.*, **142**, 1191 (1965*b*, *Astr. J.*, **70**, 142). (Photoelectric photometry 1964*f*.)
Krésak, L'. 1965, *Bull. astr. Inst. Csl.*, **16**, 348. (Secular change in brightness for Comet P/Encke.)
Malaise, D. 1966, Univ. Liège Inst. Astrophys. Coll. 8°, Vol. 21, no. 510, 199. (Profils photométriques des têtes de comètes en lumière monochromatique, Comets 1959*k* and 1963*a*.)
Malville, J. M., Evans, C. 1966, *Astr. J.*, **71**, 169 (abstr.). (D_1 and D_2 of Na, coronograph for 1965*f*.)
Martel, M. T., Saute, M. In press. (Polarization 1965*f*.)
Miller, F. D. 1966, *Publ. astr. Soc. Pacif.*, **78**, 57. (Observations 1964*c*.)
Mokhnach, D. O. *Kometnyi Cirkulyar*, no. 34. (Photographic photometry 1962 III.)
Mrkos, Tremko, Vanýsek, V. 1966, unpublished. (Photoelectric photometry, 1966*b*, 1966*c*.)
Nazarchuk, G. K. 1965, in *Fizika komet i meteorov*, Naukova Dumka, Kiev, p. 35. (Photometry, Arend-Roland.)

Notni, P. 1961, *Astr. Nachr.*, **286**, 169. (Photoelectric photometry Arend-Roland correlation with sunspots.)
Öpik, E. J. 1963, *Irish astr. J.*, **6**, 93. (On photometry and nature of comets.)
Polyakova, T. A., Romashkin, G. S. 1965, *Kometnyi Cirkulyar*, no. 28. (Photometry, 1965*b*, 1965*c*, 1964*h*, 1964*g*.)
Richter, N., Högner, W. 1964, *Sterne*, **40**, heft 1–2 (1966, Univ. Liège Inst. Astrophys., Vol. 21, no. 510, 243). (Isophotes by contrast printing.)
Rijves, V. G. (*a*) 1964, *Bjull. Kom. Komet. Meteor.*, no. 9, 23. See also Rijves, 1965. (New methods of cometary photometry.)
(*b*) 1965, *ibid.*, no. 11, 3. See also Rijves, 1964. (Low value of parameter *n* for comets.)
Roemer, E. (*a*) 1965, *Astr. J.*, **70**, 397.
(*b*) Regularly, *Publ. astr. Soc. Pacif.*, (Comet reports.)
Roemer, E., Thomas, M., Lloyd, R. E. 1966, *Astr. J.*, **71**, 591.
Roemer, E., Lloyd, R. E. 1966, *ibid.*, **71**, 443.
Rožkovskij, D. A. 1965, in *Fizika komet i meteorov*, Naukova Dumka, Kiev, p. 22. (A new procedure of comet photometry.)
Salukvadze, G. N., Džapiašvili, V. P. 1966, *Abastumansk. astrofiz. Obs. Gora Kanobili Bjull.*, no. 32, 161. (Photometry, 1957 V.)
Sekanina, Z. 1964*a*, *Publ. astr. Inst. Charles Univ.*, Prague, no. 41. (1964*b*, *ibid.*, no. 42). (Comets and meteorology.)
Southworth, R. B. 1966, Univ. Liège Inst. Astrophys. Coll. 8°, Vol. 21, no. 510, 247. (Defocuser for satellite-tracking cameras.)
Takakura, T., Tsuchiya, A., Morimoto, M. 1965, *Publ. astr. Soc. Japan*, **17**, 437. (Negative radio observation of 1965*f* at 160 H.)
Thackeray, A. D. 1965, *Mon. Notes astr. Soc. Sth. Afr.*, **24**, 159. (Split nucleus of 1965*f*.)
Vanýsek, V. 1966, *Bull. astr. Inst. Csl.*, **17**, 212. (Photoelectric IR measures of 1965*f*.)
Vsehsvjatskij, S. K. (*a*) 1964*a*. *IQSY Instruction Manual*, no. 8, Comets, London.
(*b*) 1964*b*, *Acad. Sci. Ukrainian SSR, Information Bull.*, no. 6, 3.
(*c*) 1964*c*, *Astr. Zu.*, **41**, 539. (Comet brightnesses.)
(*d*) 1964*d*, *Kometnyi Cirkulyar*, no. 17, 12 IX (1964*e*, *Geophys. Bull. Moscow*, no. 15, 61; 1965*c*, *Kometnyi Cirkulyar*, no. 20, 7 I; 1965*d*, *ibid.*, no. 25, 27 IY; 1966, Univ. Liège Inst. Astrophys., Vol. 21, no. 510, 53).
(*e*) 1965*a*, in *Fizika komet i meteorov*, Naukova Dumka, Kiev, p. 52 (1965*b*, *ibid.*, p. 56). (Magnitude of comets 1963*a* and 1963*b*.)
Weinberg, J. L. 1966, *Astr. J.*, **71**, 875 (abstr.). (Polarization, 1965*f* at 5300Å.)
Whipple, F. L., Douglas-Hamilton, D. H. 1966, Univ. Liège Inst. Astrophys. Coll. 8°, Vol. 21, no. 510, 469. (Secular and other brightness variations in P/Encke and P/Faye.)

C. Spectroscopy

Andrillat, Y. 1966, *J. Observateurs*, Marseille, **49**, 325. (Blue spectrum 39Å/mm, 15 and 18 October 1961.)
Arpigny, C. (*a*) 1964, *Observatory*, **84**, 118. (On R-branches of CO^+ comet-tail bands.)
Arpigny, C. (*b*) 1965, *Acad. r. Belg. Cl. Sci. Mem.* 8°, 35, fasc. 5. (Important survey of cometary line and band production.)
(*c*) 1966*a*, Univ. Liège Inst. Astrophys. Coll. 8°, Vol. 21, no. 510, 165. (The rotational and vibrational 'temperatures' of C_2 in comets.)
(*d*) 1966*b*, *ibid.*, 155. (Theory of λ 4300 band of CH in 1963*a*.)
(*e*) 1966*c*, *Astrophys. J.*, **144**, 424. (Criticism of Stockhausen and Osterbrock, 1965.)
Chalonge, D., Bloch, M. 1966, Univ. Liège Inst. Astrophys. Coll. 8°, Vol. 21, no. 510, 281. (Low dispersion spectra 1963*b* for continuum, slightly redder than Sun.)
Curtis, G. W., Sacramento Peak Obs. Staff 1966, *Astr. J.*, **71**, 194. (20–21 October spectra 25Å/mm 1965*f* and other measures.)
Dossin, F. (*a*) 1964, *C. r. Acad. Sci. Paris*, **258**, 3965. (Strong continuum and CN violet, weak Swan bands in 1963*d*.)
(*b*) 1966, *Astr. J.*, **71**, 853. (Spectrum of 1961*e* at 5 A.U.)

Dufay, J., Swings, P., Fehrenbach, Ch. (*a*) 1965, *C. r. Acad. Sci. Paris*, **261**, 3971. (Spectra, 1965*f*, 13 October, low dispersion, and 21 October 1965, high and low dispersion.)
(*b*) 1966, *Astrophys. J.*, **142**, 1698.

Flynn, F. H. 1961, *Observatory*, **81**, 247. (Infrared emissions from 1960*n*.)

Livingston, V., Roddier, F., Spinrad, H., Slaughter, C., Chapman, D. 1966, *Sky Telesc.*, **31**, 24. (Fe, Si, Ca, Na, Cr, Mn lines identified in spectrum, 1965*f* and high-velocity displacements for Na I near perihelion.)

Miller, F. D. 1964, *Astrophys. J.*, **139**, 766. (Two objective prism spectra.)

Remy-Battiau, L. 1964, *Acad. r. Belg. Bull. Cl. Sci.*, 5ème Sér., **50**, 74. (From continuous spectra derive properties of dust particles in comets.); 1966, *Acad. R. Belg. Bull. Cl. Sci.* 5e série, **52**, no. 11.

Remy-Battiau, L., Swings, P. 1966, Univ. Liège Inst. Astrophys. Coll. 8°, Vol. 21, no. 510, 517. (Valuable compilation of spectra and other data for 10 comets.)

Righini, G. 1966, Univ. Liège Inst. Astrophys. Coll. 8°, Vol. 21, no. 510, 129. (On first cometary spectra.)

Stawikowski, A., Greenstein, J. L. 1964, *Astrophys. J.*, **140**, 1280. (C^{12}/C^{13} ratio = 70 in 1963*a*.)

Stief, J., DeCarlo, V. J. 1965, *Nature*, **205**, 1197. (On the production of C_2 Swan bands in cometary emission.)

Stockhausen, R. E., Osterbrock, D. E. 1965, *Astrophys. J.*, **141**, 287. (Theory of C_2 Swan bands in comets.)

Thackeray, A. D., Feast, M. W., Warner, B. 1966, *ibid.*, **143**, 276. (Spectra of 1965*f*.)

Warner, B., Harding, G. A. 1963, *Observatory*, **83**, 219. (Spectroscopic observation of 1961*e*.)

Wyller, A. A. 1962, *Observatory*, **82**, 73. (Infrared spectrum of 1960*n*.)

D. Structure and origin of the nucleus

*Arnold, J. R. 1965, *Astrophys. J.*, **141**, 1536. (Monte-Carlo calculations of earth-crossing orbits, including lunar ejecta.)

*Brady, J. L. 1965, *Astr. J.*, **70**, 279. (30 long-period comets calculated entirely through planetary system. 23 attain negative 1/*a*.)

Delsemme, A. H. 1966*a*, Univ. Liège Inst. Astrophys. Coll. 8°, Vol. 21, no. 510, 69 (1966*b*, *ibid.*, 77). (Important detailed theory of comet nucleus stressing the vital role of hydrates.)

Dobrovol'skij, O. V. 1964*a*, *Bjull. Kom. Komet. Meteor.*, no. 9, 3 (1964*b*, *ibid.*, no. 9, 9; 1966, *Bjull. Inst. Astrofiz. Dušanbe-Stalinabad*, no. 41–42, 47). (On the theory of comet nuclei and heads including effects of dust and mechanism of solar flare interaction.)

Donn, B. 1964, *Astr. J.*, **69**, 137 (1965, *ibid.*, **70**, 320, abstr.). (Cometary and crystal growth in a primordial solar nebula.)

*Dufay, J. *C. r. Acad. Sci. Paris*, **263**, in press. (Statistics elements periodic comet orbits as function of period.)

Greenstein, J. L. 1966, Univ. Liège Inst. Astrophys. Coll. 8°, Vol. 21, no. 510, 483. (Possible nuclear processes in comets.)

*Harwit, M. (*a*) 1966*a*, *Astr. J.*, **71**, 857 (abstr.). (Orbits of sun-grazing comets.)
(*b*) 1966*b*, Univ. Liège Inst. Astrophys. Coll. 8°, Vol. 21, no. 510, 489. (On statistics of short- and long-period comet orbits and concentration of long-period near the ecliptic.)

Hübner, W. F. 1965*a*, *Bayerische Akad. Wiss.* (*Math. und Naturwiss.*), **12**, 147 (1965*b*, *Z. Astrophys.*, **63**, 22; 1966, *Astr. J.*, **71**, 858). (Theory of icy comet nucleus and processes therein.)

Hübner, W. F., Weigert, A. 1966, *Z. Astrophys.*, **64**, 185. (Role of ice grains in physics of cometary nuclei.)

*Krésak, L'. (*a*) 1965, *Bull. astr. Inst. Csl.*, **16**, 292. (On commensurability of short-period comet orbits, rediscovery statistics and conclusion that nuclei must be very small.)
(*b*) 1966, Univ. Liège Inst. Astrophys. Coll. 8°, Vol. 21, no. 510, 459. (Distribution of orbits of short-period comets with various interesting conclusions about their changes.)

Lancaster-Brown, P. 1964, *J. Br. astr. Ass.*, **75**, 17. (A general review of theories of comet origin.)

Levin, B. J. 1966, Univ. Liège Inst. Astrophys. Coll. 8°, Vol. 21, no. 510, 65. (Sublimation and adsorption near surfaces of cometary nuclei.)

Liller, W. 1966, *ibid.*, 195. (Desorption in nucleus of 1957 III.)

*Lyttleton, R. A., Hammersley, J. M. 1964, *Mon. Not. R. astr. Soc.*, **127**, 257. (Stochastic approach to a distribution in comet orbits.)

*Makover, S. G. 1964, *Bjull. Inst. teor. Astr.*, **9**, 525. (On Oort cloud of comets, perturbations by stars and capture theory for short-period comets.)

Marsden, B. 1966, Private communication.

McNally, D. 1966, Univ. Liège Inst. Astrophys. Coll. 8°, Vol. 21, no. 510, 581. (Problems of the interstellar medium and comets.)

Öpik, E. J. (*a*) 1966*a*, The stray bodies in the solar system, Part II, The cometary origin of meteorites, in *Advances in astronomy and astrophysics*, Ed. Z. Kopal, Academic Press, Inc., New York, Vol. 4, p. 301. (On cometary nuclei, orbital perturbations, meteorites and problems of origin.)

(*b*) 1966*b*, Univ. Liège Inst. Astrophys. Coll. 8°, Vol. 21, no 510, 523. (Detailed article on the origin of comets.)

(*c*) 1966*c*, *ibid.*, 575. (Identifying Apollo asteroids with old comet nuclei.)

*Piotrowski, S. L. 1965, *Acta astr.*, **15**, 281. (Accretion of comets by dust surrounding the solar system and greater chance of discovery of retrograde than direct long-period comets.)

Remy-Battiau, L. 1964, Aug. 20, Tech. Note No. 6. (Dust and bursts in cometary nuclei.)

Richter, N. 1966, Univ. Liège Inst. Astrophys. Coll. 8°, Vol. 21, no. 510, 439. (Report on the origin of comets.)

Roemer, E. 1966*a*, *ibid.*, 15 (1966*b*, *ibid.*, 23). (Measures of diameters of 29 comets by limiting magnitudes.)

Sekanina, Z. (*a*) 1966*a*, *Bull. astr. Inst. Csl.*, **17**, 207. (Splitting of 1965*f*.)

*(*b*) 1966*b*, *ibid.*, **17**, 67 (1966*c*, *Publ. astr. Inst. Charles Univ.*, Prague, no. 48). (Calculates new original and future comet orbits.)

*(*c*) 1966*d*, Univ. Liège Inst. Astrophys. Coll. 8°, Vol. 21, no. 510, 443. (On Oort cloud of comets and possibility of an 'interstellar swarm'.)

(*d*) 1967*a*, *Z. Astrophys.*, in press. (Calculation of non-gravitational effects on nucleus of P/Halley.)

(*e*) 1967*b*, *Bull. astr. Inst. Csl.*, in press. (Theory of physical forces on cometary nuclei.)

*Sitarski, G. 1964, *Acta astr.*, **14**, 323. (Orbit P/Grigg-Skjellerup, a secular acceleration seems necessary from 1947 to 1961.)

Stefanik, R. P. 1966, Univ. Liège Inst. Astrophys. Coll. 8°, Vol. 21, no. 510, 29. (Study of 13 split comets.)

*Šteins, K. A. 1964, *Latviiskii Gos. im. P. stuchki Uchenye Zapiski*, **68**, 39. (Review of his previous results on the evolution of cometary orbits and the origin of comets from Oort's cloud.)

*Šteins, K. A., Kronkalne, S. (*a*) 1964*a*, *Acta astr.*, **14**, 4. (Calculations of perturbations by Jupiter of $1/a$ and q and theory of diffusion of comets.)

*(*b*) 1964*b*, *ibid.*, 311. (Calculations of 20 000 fictitious comets completely through planetary system, follow Štein's diffusion laws.)

Vanýsek, V. (*a*) 1965, *Bull. astr. Inst. Csl.*, **16**, 355. (Stoppage of activity in nucleus of P/Encke.)

(*b*) 1966, Univ. Liège Inst. Astrophys. Coll. 8°, Vol. 21, no. 510, 255. (Dust grains and gas in cometary nuclei.)

*Vsehsvjatskij, S. K. 1965, *Kometnyi Cirkulyar*, no. 27, 30 YI (1966*a*, *ibid.*, no. 40, 1 YI; 1966*b*, Univ. Liège Inst. Astrophys. Coll. 8°, Vol. 21, no. 510, 53; 1966*c*, *ibid.*, 495). (Influence of nearby stars on Oort cloud of comets. Conclusion negative to theory. Holds to ejection theory although accepting icy comet model.)

Whipple, F. L., Stefanik, R. P. 1966, Univ. Liège Inst. Astrophys. Coll. 8°, Vol. 21, no. 510, 33. (Splitting of cometary nuclei possibly caused by radioactive heating of icy nuclei.)

Zadunaisky, P. E. 1966, *Astr. J.*, **71**, 20. (Orbits of P/Halley, before and after perihelion.)

E. Coma, excitation and ionization

A'Hearn, M. F. 1964, *Publ. astr. Soc. Pacif.*, **76**, 106. (Analysis of C_2 densities.)

Biermann, L., Trefftz, E. 1964, *Z. Astrophys.*, **59**, 1. (On the ionization and excitation mechanism in comet heads, especially [O I], CO^+ and the UV.)

Bouska, J. (*a*) 1965*a*, *Acta Univ. Carolinae–Math. et Physica*, no. 1, 1; also in *Publ. astr. Inst. Charles Univ.*, Prague, no. 43 (1965*b*, *Bull. astr. Inst. Csl.*, **16**, 358). (Measures and theory of C_2 and CN in 1957 III.)
(*b*) 1965*b*, *Bull. astr. Inst. Csl.*, **16**, 358 (1966, Univ. Liège Inst. Astrophys. Coll. 8°, Vol. 21, no. 510, 185). (Structure of tail rays.)

Čeredničenko, V. (*a*) 1965, in *Fizika komet i meteorov*, Naukova Dumka, Kiev, p. 37 (1966, *ibid.*, p. 51). (Transformation processes for 21 molecules and ions by solar photon corpuscular radiation.)

Dewey, M. E., Miller, F. D. 1966, *Astrophys. J.*, **144**, 1170. (Isophotes of C_2 in 1961 VIII, and lifetimes of C_2 and its parent.)

Dobrovol'skij, O. V. (*a*) 1966, *Bjull. Inst. Astrofiz. Dušanbe-Stalinabad*, no. 41, 37. (Growth of the head of 1955 V.)
(*b*) *ibid.* (Lifetimes of CO and CO_2 under solar radiation.)

Dufay, J. *Atomes*, in press. (Excitation for molecules in sun-grazing comets.)

Egibekov, P. 1964, in *Investigations of comets according to the program of the IQSY*, Naukova Dumka, Kiev. (1963*a*, absolute energy distribution in the spectrum.)

Gnedin, Ju. N., Dolginov, A. Z. 1966*a*, *Astr. Zu.*, **43**, 181 (1966*b*, *Icarus*, **5**, 64). (Theoretical distribution of molecules, or atoms, in cometary heads.)

Haser, L. 1966, Univ. Liège, Inst. Astrophys. Coll. 8°, Vol. 21, no. 510, 233. (Extension of 'fountain model' theory of cometary heads.)

Jackson, W. M., Donn, B. (*a*) 1966*a*, Univ. Liège Inst. Astrophys. Coll. 8°, Vol. 21, no. 510, 133. (Collisions near the nucleus and parent molecules for C_3.)
(*b*) 1966*b*, *Astr. J.*, **71**, 859 (abstr.). (Photochemistry near the nucleus.)

Kovar, N. S., Kovar, R. P. 1966, *Astr. J.*, **71**, 166. (On the theoretical intensity ratios of the [O I] lines in comets.)

Kostyakova, E. B. 1965, *Soobšč. gos. astr. Inst. Sternberga*, no. 139, 24. (Computed flux for continuum and bands 1957 III.)

Miller, F. D. 1965*a*, *Astr. J.*, **70**, 283 (1965*b*, *ibid.*, 685). (Forms of isophotes and velocities in comets for gas and particles.)

Mokhnach, D. O. 1966*a*, *Kometnyi Cirkulyar*, no. 35, 2 (1966*b*, *ibid.*, 3). (Lifetime of C_2 and initial velocity in Comet Burnham, 1960 II.)

Rahe, J. (*a*) 1965, Thesis, Hamburger Sternwarte. (Analysis of remarkable activity < 1000 km from nucleus of 1858 VI.)
(*b*) 1966, Univ. Liège Inst. Astrophys. Coll. 8°, Vol. 21, no. 510, 141. (Visual and photographic structures in the coma region of comets.)

Rijves, V. 1960*a*, *Tartu Publ.*, **33**, no. 4, 282 (1960*b*, *ibid.*, 289; 1963, *Bjull. Kom. Komet. Meteor.*, no. 8, 3). (Motions in cometary heads and tail formation.)

Schlosser, W. 1966, Univ. Liège Inst. Astrophys. Coll. 8°, Vol. 21, no. 510, 335. (Motions of structures in 1908 III.)

Vanýsek, V. 1965, *Acta Univ. Carolinae–Math. et Physica*, no. 1, 23; also in *Publ. astr. Inst. Charles Univ.*, Prague, no. 44. (Lifetime of C_2 and CN molecules and solid grains in cometary atmospheres.)

Wurm, K. (*a*) 1966*a*, Univ. Liège Inst. Astrophys. Coll. 8°, Vol. 21, no. 510, 119. (Report on general problems of the cometary head.)
(*b*) 1966*b*, *Icarus*, **5**, 219. (Further study of observations strengthens the argument that ionization in comets is an intrinsic property.)

Wurm, K., Mammano, A. In preparation. (Find narrow source for particles excitation and ionization < 100 s near nucleus < 1000 km.)

Wurm, K., with Rahe and Mammano In preparation. (In 1858 VI find very important results concerning head.)

F. Type-I tails

A'Hearn, M. F. 1965, *Publ. astr. Soc. Pacif.*, **77**, 257. (A solar event and a plasma tail knot.)
Alfvén, H. 1957, *Tellus*, **9**, 92. (MHD approach to the theory of plasma tails.)
Antrack, D., Biermann, L., Lüst, Rh. 1964, in *Annual reviews of astronomy and astrophysics*, Ed. L. Goldberg, Annual Reviews, Inc., Palo Alto, California, Vol. 2, p. 327. (Statistics of plasma tails, no correlation with solar activity.)
Axford, W. I. 1964, *Planet. Space Sci.*, **12**, 719. (Role of electrons in plasma tails.)
Axford, W. I., Petschek, H. E., Sicoe, G. L. 1965, *J. geophys. Res.*, **70**, 1231. (The magnetosphere and plasma tails.)
Beard, D. B. 1966, *Planet. Space Sci.*, **14**, 303.
Bernasconi, A. 1965, *Icarus*, **4**, 214. (Method of analysis of a comet's tail.)
Biermann, L. (*a*) 1951, *Z. Astrophys.*, **29**, 274. (Solar corpuscular interaction with tail ions.) (*b*) 1966, Univ. Liège Inst. Astrophys. Coll. 8°, Vol. 21, no. 510, 297. (Survey of plasma tails and the solar wind.)
Biermann, L., Brosowski, B., Schmidt, H. U. 1966, *Ber. Bayer. Akad. Wiss.*, July. ('Contact' surface in MHG plasma tail formation.)
Biermann, L., Lüst, Rh. 1966, Univ. Liège Inst. Astrophys. Coll. 8°, Vol. 21, no. 510, 329. (A solar event and a plasma tail knot.)
Bouska, J. 1964, *Bull. astr. Inst. Csl.*, **15**, 113. (Velocity and acceleration tail in Arend-Roland.)
Brandt, J. C. (*a*) 1966*a*, Univ. Liège Inst. Astrophys. Coll. 8°, Vol. 21, no. 510, 309. (First results on solar wind and the orientation of plasma tails.)
(*b*) 1966*b*, *Astrophys. J.*, **144**, 1221. (Torque on the Sun of solar wind by measured comets.)
(*c*) 1966*c*, *G.S.F.C.* (Gross plasma velocities for the orientation of ionic comet tails.)
Brandt, J. C., Belton, M. J. S., Stephens, M. W. 1966, *Astr. J.*, **71**, 157 (abstr.). (Velocities of solar wind plasma from the orientations of ionic comet tails.)
Cailliatte, Ch. 1966, *Bull. astr.*, Paris, Sér. 3, fasc. 2, 37. (Method in analysis of gas and dust tails.)
Čeredničenko, V. I. 1965*a*, in *Fizika komet i meteorov*, Naukova Dumka, Kiev, p. 45 (1965*b*, *ibid.*, p. 50; 1965*c*, *Kometnyi Cirkulyar*, no. 26). (C_3H_6, $C_2H_2C_2$ and CO molecules in the photon and corpuscular radiation field of the Sun.)
Dobrovol'skij, O. V. 1964*a*, *Bjull. Kom. Komet. Meteor.*, no. 9, 3 (1964*b*, *ibid.*, 9). (Corpuscular solar wind and cometary outbursts.)
Fajziev, E. 1964, *Bjull. Inst. Astrofiz. Dušanbe-Stalinabad*, no. 37, 3 (1964*b*, *ibid.*, no. 38, 32). (Tail velocities of Arend-Roland decrease with distance from the Sun.)
Gringauz, K. I., Kurt, V. G., Moroz, V. I., Šklovskij, I. S. 1960, *Astr. Zu.*, **37**, 716. (Measures of solar wind from space.)
Guigay, G. (*a*) 1966*a*, *C. r. Acad. Sci. Paris*, Sér. B, **262**, 1574. (Solar wind and direction of tail, 1965*f*.)
(*b*) 1966*b*, Univ. Liège Inst. Astrophys. Coll. 8°, Vol. 21, no. 510, 369. (On the form of the tail of 1961*e*.)
Harwit, M., Hoyle, F. 1962*a*, *Astrophys. J.*, **135**, 867 (1962*b*, *ibid.*, 875). (Theory of type-I tails and solar wind.)
Ioffe, Z. M. 1965, *Bjull. Inst. Astrofiz. Dušanbe-Stalinabad*, no. 39, 84 (1966*a*, *Astr. Zu.*, **43**, 175; 1966*b*, *ibid.*, 655). (Supersonic MHD flow about comet plasma.)
Južakov, V. M. 1963, *Astr. Zu.*, **40**, 779. (Calculates field of 5×10^{-6} G from CO^+ ions in head 1957*h*.)
Kazantseva, N. N. 1966, *Bjull. Astrofiz. Inst. Dušanbe-Stalinabad*, no. 41–42, 51. (Shock front in cometary plasma.)
Kearns, C. E., Rudnicki, K. 1965, *Acta astr.*, **15**, 1. (Velocity components for 35 features in the tail of 1962 VIII.)
Kovar, N. S., Kern, J. W. 1966, *Astr. J.*, **71**, 861 (abstr.). (Proposed hydromagnetic model for comets.)
Lüst, Rh. 1967, Letter, in press. (CO^+ features in 1908 III.)
Maročnik, L. S. 1963, *Astr. Zu.*, **40**, 714 (1964, *Bjull. Kom. Komet. Meteor.*, no. 9, 19). (Interaction of the solar wind with the plasma head of a comet.)

Ness, N. F. 1966, *Science*, **151**, 1041. (Earth's magnetosphere and similarity with comet tails.)
Ness, N. F., Donn, B. D. 1966, Univ. Liège Inst. Astrophys. Coll. 8°, Vol. 21, no. 510, 343. (Excellent description of plasma tail theory.)
Ness, N. F., Wilcox, J. M. 1964, *Phys. Rev. Letters*, **13**, 461. (Observations of the solar wind.)
Öpik, E. J. (*a*) 1956, *Irish astr. J.*, **4**, 49; also in *Armagh Obs. Leaflet* no. 44 (1964, *Z. Astrophys.*, **58**, 192). (Motions of condensations in the tail of P/Halley.)
(*b*) 1965, *Irish astr. J.*, **7**, 38. (On the production of CO^+ ions in type-I tails.)
Pflug, R. P. 1966, Univ. Liège Inst. Astrophys. Coll. 8°, Vol. 21, no. 510, 363. (Statistics of plasma tail directions and solar activity.)
Ptitsina, N. G. 1964, *Bjull. Kom. Komet. Meteor.*, no. 9, 12 (1965, *ibid.*, no. 11, 11). (MHG approach to moving envelopes, and rays in type-I tails.)
Sekanina, Z. 1966, Univ. Liège Inst. Astrophys. Coll. 8°, Vol. 21, no. 510, 145. (Evidence for coma diameters affected by solar wind.)
Snyder, C. W., Neugebauer, C. W., Rao, U. R. 1963, *J. geophys. Res.*, **68**, 6361. (Observations of the solar wind.)
Wallis, M. K. 1967, *Planet. Space Sci.*, **15**, 137. (Plasma instability and viscosity in plasma-tail ray production.)
Wurm, K., Mammano, A. 1964, *Icarus*, **3**, 1. (Axes of type-I comet tails.)

G. Type-II and -III tails

Belton, M. J. S. 1965, *Astr. J.*, **70**, 451 (1966, Univ. Liège Inst. Astrophys. Coll. 8°, Vol. 21, no. 510, 317). (Evidence that direction and velocities in dust tails indicate an effect of solar wind.)
Demenko, A. A. 1966, *Vestnik Kiev Univ.*, no. 8, Ser. Astron. (Theory and measures of synchronic formations in comet tails.)
Dobrovol'skij, O. In preparation. (Analysis of all published data on accelerations of dust particles.)
Fajziev, E. 1964*a*, *Bjull. Inst. Astrofiz. Dušanbe-Stalinabad*, no. 37, 3 (1964*b*, *ibid.*, no. 38, 32). (Initial velocity of particles ejected into the tail of 1957 III.)
Finson, M. L., Probstein, R. F. 1966, *Amer. Inst. Aeron. Astronaut. Paper*, no. 66, New York. (Motions of gas from cometary nucleus and light pressure.)
Levin, B. J. (*a*) 1964, *Icarus*, **3**, 497. (Lifetime of Na too short to account for motion in tails of comets.)
(*b*) 1966, Univ. Liège Inst. Astrophys. Coll. 8°, Vol. 21, no. 510, 323. (Arguments for molecules in type-II tails.)
Notni, P. 1964, *Veröff. Sternw. Babelsberg*, **15**, heft 1 (1966, Univ. Liège Astrophys. Coll. 8°, Vol. 21, no. 510, 379).

H. Laboratory work

Bauer, S. H., Herzberg, G., Johns, J. W. C. 1964, *J. molec. Spectrosc.*, **13**, 256. (Spectrum of BH.)
Colin, John, J. W. C. In preparation. (Spectrum of NCl.)
Douglas, A. E., Johns, J. W. C. *Can. J. Phys.*, in press. (Spectrum of NF.)
Douglas, A. E., Jones, W. J. 1965, *ibid.*, **43**, 2216. (Spectrum of N_3.)
Dufay, M., Poulizac, M.-C. 1966, Univ. Liège Inst. Astrophys. Coll. 8°, Vol. 21, no. 510, 427. (Spectra of CO and CO_2 excited by accelerated H^+ and H_2^+.)
Fishburne, E., Narahari-Rao, K. 1966, *ibid.*, 409. (High-dispersion spectra of CO^+.)
Gausset, L. 1966, Univ. Liège Inst. Astrophys. Coll. 8°, Vol. 21, no. 510, 413. (See also Gausset *et al.*, 1965.)
Gausset, L., Herzberg, G., Lagerquist, L., Rosen, B. 1965, *Astrophys. J.*, **142**, 45. (Intension study of the spectrum of C_3 λ 4050 group.)

Herman, R. 1966, *ibid.*, 411 (abstr.). (Polymerization and growth of carbon from an electric discharge.)
Herzberg, G. 1966, Letter. (New band system of C, C_2^+ or C_2.)
Herzberg, G., Johns, J. W. C. *Proc. R. Soc. London*, in press. (Red bands of CH_2.)
Herzberg, G., Travis, D. N. 1964, *Can. J. Phys.*, **42**, 1658. (Spectrum of NCN.)
Johns, J. W. C. 1964, *ibid.*, 1004. (Spectrum of CO_2^+.)
Jungen, Ch., Miescher, E. 1966, *Astrophys. J.*, **142**, 1660. (Emission spectrum of NO, 1400–1600Å.)
Koval', A. G., Koppe, V. T., Fogel', Ja. M. 1966*a Kosm. Issled.*, **4**, no. 1 (1966*b*, *Astr. Zu.*, **43**, 209). (Emission spectra of several molecules including excited by 13 keV electrons.)
Merer, A. J., Travis, D. N. (*a*) 1965, *Can. J. Phys.*, **43**, 1795. (Spectrum of CCN.) (*b*) 1966, *ibid.*, **44**, 353. (Spectrum of CNC.)
Monfils, A. 1965*a J. molec. Spectrosc.*, **15**, 265 (1965*b*, *Tech. Note*, no. 7, Liège, AFCRL–65–684, 15 July). (H_2, HD, D_2 in absorption.)
Morrow, *Can. J. Phys.*, in press. (Spectrum of SH.)
Papazian, H. 1966, Letter from N. Roman. (Study of radicals in photochemical reactions.)
Phillips, J. G. 1966, Univ. Liège Inst. Astrophys. Coll. 8°, Vol. 21, no. 510, 393. (Important summary of spectroscopic data and physical properties of many cometary molecules and ions.)
Poulizac, M.-C., Desesquelles, J., Dufay, M. *C. r. Acad. Sci. Paris*, **263**, in press. (See also Dufay and Poulizac, 1966.)
Stief, L. J., DeCarlo, V. J. 1965, *Nature*, **205**, 1197. (Laboratory work on CH_4, C_2H_2, C_2H_4 and C_2H_6.)
Vaidya, W. M. 1966, Univ. Liège Inst. Astrophys. Coll. 8°, Vol. 21, no. 510, 433. (Cometary CH and OH bands in hydrocarbon flame; HCO in comets?)
Verma, R. D. 1965, *Can. J. Phys.*, **43**, 2136. (Absorption system of SiH and SiD.)

I. Space experiment

Bredohl, H. 1966, Univ. Liège Inst. Astrophys. Coll. 8°, Vol. 21, no. 510, 417. (NH_3 artificial comet experiment to 200 km.)
Föppl, H., Haerendel, G., Loidl, J., Lüst, Rh., Melzner, F., Meyer, B., Neuss, H., Rieger, E. 1965, *Planet. Space Sci.*, **13**, 95 (*ibid.*, **15**, 357). (Report on observations of artificial plasma sent as high as 2000 km by rockets.)
Gronstal, P. T., Bukata, R. P. 1966, *Nature*, **211**, 176. (Balloon search for low-energy protons from Comet Ikeya-Seki, none identified.)
Haerendel, G., Lüst, Rh., Rieger, E. *Planet. Space Sci.*, in press. (See Föppl *et al.*, 1965.)
Öpik, E. J. 1965, *Irish astr. J.*, **7**, 31. (On cometary physics and related experimentation.)
Potter, A. 1966, Letter from N. Roman. (A cloud of CN radicals was produced at 150 km by a chemical reaction.)
Roberts, D. L., Narin, F., Pierce, P. M. 1966, Univ. Liège Inst. Astrophys. Coll. 8°, Vol. 21, no. 510, 261. (Planning for cometary intercept space missions.)

16. COMMISSION POUR L'ETUDE PHYSIQUE DES PLANETES ET DES SATELLITES

PRESIDENT: Professor R. Wildt, Yale University Observatory, Box 2023 Yale Station, New Haven, Connecticut 06520, U.S.A.

VICE-PRESIDENT: Dr G. Herzberg, Director, Division of Pure Physics, National Research Council, Sussex Drive, Ottawa, Ontario, Canada.

COMITÉ D'ORGANISATION: A. Dollfus, G. P. Kuiper, C. H. Mayer, C. E. Sagan, V. V. Sharonov†.

MEMBRES: Ashbrook, Barabashov, Barreto, Bobrov, Bullen, Camichel, Chamberlain, Collinson, DeMarcus, De Mottoni, de Vaucouleurs, Drake, Focas, Fox, Gehrels, Giclas, Gold, Guérin, Günther, Hall (J. S.), Hunten, Ingrao, Jeffreys, Jelley, Kaplan (L), Kellog, Kiess, Kopal, Koval, Levin, Link, Lipsky, Luplau-Janssen, Martynov, Menzel, Middlehurst, Miyamoto, Murray, O'Keefe, Öpik, Peek†, Pettengill, Rösch, Safronov, Shoemaker, Sinton, Smith (B. A.), Smith (Harlan J.), Spinrad, Strong, Tombaugh, Urey, Whitaker, Wilson (A. G.).

INTRODUCTION

Owing to the formation of Commission 17, which has, since 1964, cognizance of all lunar studies, this report deals only with research on other satellites and the planets published during the period from 1964 until 1966. Unforeseen circumstances have forced the President of Commission 16 to distribute the work of compiling this report among several members, to whom he is greatly obliged for assuming this responsibility. Their respective contributions are:

- Photometry, Polarimetry of the Planets and Observations of Planetary Surfaces, by J. H. Focas.
- Spectroscopic Studies of the Planets, by T. Owen.
- Radio Studies of the Planets, by C. H. Mayer and M. Spangler.
- Internal Constitution of the Planets, by W. C. DeMarcus.
- Exobiology, by C. Sagan.

An appendix has been added, prepared by D. Ja. Martynov, on the Physics of the Planets in the U.S.S.R. (1963–66), although some of the above sections take Soviet publications into full account.

AD HOC COMMITTEE ON MARTIAN NOMENCLATURE

Professor Robert B. Leighton of the California Institute of Technology, who is the principal investigator of the TV experiment conducted during the Mariner IV flight, has sought the advice of Commission 16 in regard to the problem of devising a nomenclature suitable for the Martian surface features discovered on the prints developed from the Mariner IV data.

Accordingly, the President of Commission 16 appointed an Ad Hoc Committee to advise Professor Leighton, consisting of the following members of Commission 16: G. P. Kuiper (chairman), A. Dollfus, J. S. Hall. This committee met in Pasadena in September 1965 and thereafter continued deliberations by correspondence. It is hoped that report can be submitted to Commission 16 at the next General Assembly.

PHOTOMETRY, POLARIMETRY OF THE PLANETS AND OBSERVATIONS OF PLANETARY SURFACES

(prepared by J. H. Focas)

A. Planetary Data Centers and International Cooperation for the Observation of the Planets

An extensive work was developed by the two Data Centers established at the Lowell and the Meudon Observatories since 1961 by the IAU. Dr J. H. Focas is in charge of the Meudon Planetary Data Center, and Dr. W. A. Baum in charge of the Lowell Observatory Planetary Data Center. For the time being, a total of more than 15 000 documents are available at the Lowell Center and approximately 10 000 documents at the Meudon Center. Such documents are listed and filed according to the geometrical data of the individual planets. They are consultable under the form of original negatives, or counter-copies of original negatives, original composite images, negatives of the composite images or reproductions on paper or/and films.

The observatories which have contributed by exchanging documents with the Centers and/or performing observations, in the line of the International Cooperation for the observation of the planets, are the following:

Africa: Brazzaville, Johannesburg.
Asia: Kwassan, Tokyo (Japan).
Europe: Athens (Greece); Genova, Rome (Italy); Juvisy, Meudon, Nice, Pic-du-Midi (France).
U.S.A.: Harvard (Mass.); Lick, Mt Wilson, Table Mountain (Calif.); Lowell, Tucson (Arizona); MacDonald (Texas); New-Mexico (New-Mexico).
U.R.S.S.: Kiev, Volgograd, Alma-Ata.

'The Brighter Planets' published in 1964 by the Lowell Observatory is the last work of the late Dr E. C. Slipher. It contains a great number of photographs of Venus, Mars, Jupiter and Saturn taken by him in the last 60 years. Such photographs are among the best ever taken and are accompanied by an extensive explanatory text reflecting the most salient characters of the planetary phenomena.

This Commission, through Committee 16*c* 'International Cooperation for the Observation of the Planets' has organized the following observational campaigns:

Venus: Photography in short wavelengths for studying evident periodicities of atmospheric phenomena.

Mars: Apparition 1964–65. Photography and polarization in different wavelengths for studying the nature of the soil, clouds and their motion; high resolution mapping; coordinated ground based observations with the flight of Mariner IV to Mars.

Jupiter: Photographic patrol of the planet at one minute intervals for many hours per night, for the revision of the rotation period as a function of the zenographical latitude, study of proper motions of atmospheric formations as well as the short or long term atmospheric activity of the planet.

Saturn: Photographic, visual and photometric observations during the transits of the Sun and the Earth from the plane of the Rings in 1966–67. Professor Bobrov and Dr P. Muller mostly contributed to the preparation of the program of this international campaign.

B. Specific Results Concerning Individual Planets

The results obtained during the period 1964–66 for each one of the planets, including those obtained by the International Cooperation for the Observation of the Planets and research carried out by the two Planetary Data Centers are summarized in the following.

The Planet Mercury

Observations: G. P. Kuiper and T. Gehrels using a 154 cm reflector at $F/13$ are performing a program of photometric and polarimetric measurements on the planet with seven filters between 0·3 and 1·0 microns and with experiments at shorter and longer wavelengths.

A. Hämeen-Anttila using photographs of the planet taken at Pic-du-Midi Observatory in 1942–44 by B. Lyot and H. Camichel, carried out at the Meudon Observatory a photographic study on the purpose to compare the contexture of the soil of Mercury to that of the Moon.

B. Smith and cooperators have been successful in photographing the planet in red light (77 plates) at the New-Mexico Observatory. Surface markings which may be seen on many of the plates, are being studied as a possible source for an optically derived rotation period. To-date, no conclusive results have been obtained.

Dr Menzel and W. M. Irvine report that multicolor photoelectric photometry of the planet was made by the Harvard College Observatory at the Stations of Boyden (South-Africa) and the Houga (Southern France), using ten narrow band filters between 0·315 and 1·06μ plus *UBV* on the purpose to obtain monochromatic phase curves and albedos.

A. Dollfus carried out in 1966 systematic visual observations at the Pic-du-Midi Observatory for the study of the rotation period of the planet.

Theoretical investigations: C. Sagan studying the photometric properties of Mercury, suggests a common cause for the similarities of the surfaces of this planet and the Moon. The solar proton wind cannot penetrate the atmosphere of Mercury implied by spectrometric and polarimetric observations, and contemporary solar cosmic ray protons have sufficient intensity to account for the observed properties with the age of the solar system. A conceivable alternative source of proton irradiated powders is accretion of interplanetary dust slowly falling through the thin Mercurian atmosphere (**1**).

The Planet Venus

Observations: Ch. Boyer, H. Camichel and P. Guérin carried out at the Pic-du-Midi Observatory (106 cm reflector) an extensive photographic patrol of the planet in violet and ultra-violet light during the evening elongation in 1965 and the morning elongation in 1966. High quality series of photographs taken uninterruptedly during one or two, up to seven hours, allowed stereo-photogrammetric measurements, tracing of isophote maps and composite images of the planet, for the study of the motion of bright or dusky areas through the disk. The motion of such atmospheric formations is in general retrograde; the recurrence of a Y-shaped formation normal to the line joining the cusps follows a cycle of approximately four days. Some connection seems to exist between the position of the sub-solar point on the disk and the aperture of the Y-shaped formation, which is variable (**2, 3**).

G. P. Kuiper and T. Gehrels using a 154 cm reflector at $F/13$, are performing a photo-polarimetric program on the planet with seven filters between 0·3 and 1·0 micron and with experiments at shorter and longer wavelengths.

R. Younkin has conducted spectrophotometric measurements on the planet at JPL from 0·32 to 1·1μ.

B. Smith and cooperators at the New Mexico Observatory photographed the planet (1081 plates) in ultra-violet, blue and red light during the last three years. Measures of the height of the scattering layer in the atmosphere and studies of motions of atmospheric clouds have been continued (**4**).

D. Menzel and W. M. Irvine report that multicolor photoelectric photometry of the planet was made by the Harvard College Observatory at the Stations of Boyden (South Africa) and the Houga (Southern France) using ten narrow band filters between 0·315 and 1·06μ plus *UBV* on the purpose to obtain monochromatic phase curves and albedos.

A. Dollfus and E. Maurice carried out a photometric study of photographs of Venus taken by him in 1964 at the Pic-du-Midi Observatory during the inferior conjunction of the planet. The photometry of the aureola lengthening the horns shows that the Venusian atmosphere overlying the clouds layer is charged with particles of approximately $1{\cdot}5\,\mu$ diameter, the number of which decreases by a factor 2 over 2·8 km approximately; their diffusion factor per cubic centimeter opposite to the Sun amounts to $3{\cdot}10^{-8}$ stilb/phot at the level of the clouds layer (**5**).

A. Dollfus and collaborators carried out during the period 1964–66 systematic polarimetric observations of the planet in the spectral ranges $\lambda = 0{\cdot}47, 0{\cdot}55, 0{\cdot}65, 0{\cdot}84, 0{\cdot}95, 1{\cdot}05\,\mu$ at the Pic-du-Midi and the Meudon Observatories for the study of the variation of the polarization with the wavelength. This variation is connected with the size of the particles diffusing the light in the atmosphere of the planet. The non-uniform distribution of the polarization over the disk was studied; the deviation of the direction of the polarization at the limb can be explained by multiple scattering in the atmosphere overlying the clouds layer.

C. Capen at the Table Mountain Observatory carried out systematic photographic observations of the planet during the last four years.

H. Prinz and H. Obendorfer carried out respectively photographic (*UV*) and visual observations of the planet, at the Volkssternwarte in Munich in 1964.

I. L. Thomsen at the Carter Observatory, Wellington, New Zealand, conducted visual observations of the bright cusp-caps and terminator peculiarities of the planet.

Theoretical investigations: According to J. B. Pollack and Carl Sagan, an intercomparison of passive microwave observations at a variety of phase angles and frequencies, coupled with the radar reflectivities, can potentially provide information on the chemistry and structure of the Cytherean surface material. A variety of silicates, carbonates and oxides and one variety of organic molecule-polycyclic aromatic hydrocarbons—can be postulated as primary constituents of the Cytherean surface. The thermal, magnetic and electrical properties of these materials have been analyzed as a function of frequency, temperature, porosity, granularity and impurity content. The comparison of Venus and laboratory data shows that fused quartz and a wide range of powdered oxides, carbonates and silicate may, to within observational uncertainty be primary Cytherean surface materials. Magnetic materials, granite and the hydrocarbons are excluded. The surface is likely to be a desert. The low radar reflectivity at 3·6 cm cannot be due to a general 3·6 cm absorption by the atmosphere and clouds. The result can be attributed to a localized high absorption above a surface cold spot in the first radar Fresnel zone, or to variation of porosity of surface material and therefore of the dielectric constant with depth (**6**).

C. Sagan and J. Pollack established surface temperature distributions on Venus needed to match the microwave phase and polarization observations which are consistent with observations made by Mariner II (**7, 8**).

According to C. Sagan and J. Pollack, balloon observations of near infrared reflectivity of the clouds of Venus and ground based infrared spectrometric observations show the presence of condensed water in the clouds and water in the atmosphere above the clouds. The amount of water vapor in the underlying atmosphere of Venus required to maintain the clouds is of the order of $10\ \mathrm{g\ cm^{-2}}$. The opacity of this amount of water vapor, combined with that of carbon dioxide and of the clouds themselves, provides an efficient greenhouse effect, quantitatively capable of maintaining surface temperatures $\sim 700°$ Kelvin (**9**).

E. R. Lipincott, R. Eck, M. O. Dayhoff and C. Sagan, by studies of the thermodynamic equilibrium chemistry of the planet, indicated the absence of significant quantities of organic materials, particularly polycyclic aromatic hydrocarbons on the surface of Venus (**10**).

N. P. Barabashov (Kharkov Astron. Observatory), performed a theoretical analysis of data on brightness distribution across the disk of Venus which made it possible to estimate the value of the optical thickness of the layer situated above the cloud cover, $\tau \leqslant 0{\cdot}07$ (**11**).

The Planet Mars

Martian cartography: Continuing the program of Martian cartography established by the Meudon Planetary Data Center, G. de Mottoni prepared, on the basis of photographic documents taken at the Lowell Observatory, three photographic maps of Mars covering the apparitions of 1907, 1909 and 1911. This work is being published by the Milan Observatory in cooperation with the Meudon Planetary Data Center. The photographic maps of Mars prepared till now by G. de Mottoni from documents existing at the Meudon Planetary Data Center cover the apparitions of 1907, 1909, 1911, 1941, 1943, 1946, 1948, 1950, 1952, 1954, 1956 (two), 1958. The details contained in these maps are of the order of < 0·5 second of arc. They show the evolution of the seasonal phenomena of the dark areas of the planet and accidental variations of the surface.

The maps covering the apparitions of 1914, 1916 and 1918 are in preparation.

A high resolution (up to 0·25 second of arc) photovisual map of Mars was established at the Meudon Planetary Data Center on the basis of photographic documents and visual observations taken by H. Camichel, A. Dollfus and J. H. Focas, at the Pic-du-Midi Observatory (107 cm reflector) and Meudon Observatory (83 cm refractor) during the 1964–65 apparition of the planet. This map, as well as ground based polarimetric observations, taken by A. Dollfus and J. H. Focas at the Meudon Observatory, were designed to help in the interpretation of the photographs taken by the spacecraft Mariner IV.

An Atlas containing rectified photographs of Mars taken by Mariner IV was prepared by G. de Mottoni; the same studied the altitude of a cloud appearing on Mariner's IV photographs at the Amazonis region.

During the 1964–65 opposition, Ch. Capen *et al.*, at the Table Mountain Observatory carried out a systematic photographic and visual patrol of the planet and especially of the Amazonis desert—Trivium Charontis—Elysium Regions and related atmospheric phenomena. A photovisual map of Mars of the Mariner IV scan longitudes has been prepared by the same as well as a pictorial map of Mars Atlas 1964–65 (358 photographs). The former established also the regression curve of the north polar cap (**12, 13**).

Observations: R. G. Tull through spectral scan of Mars made at the 82-inch (208 cm) telescope in the range of $0{\cdot}5$ to $1{\cdot}2\mu$ obtained a spectral reflectivity curve of this planet extrapolated with the aid of published data to $1{\cdot}52\mu$. He finds the Russell-Bond albedo rising to a maximum of 39% at $0{\cdot}84\mu$. A shallow minimum occurs near 1μ followed by a rise to 49% at $1{\cdot}52\mu$. The radiometric albedo is found to be 0·287. Earlier conclusions that the surface layer of Mars consists largely of limonite are consistent with these observations (**14**).

G. P. Kuiper and T. Gehrels using a 154 cm reflector at $F/13$, are performing a photopolarimetric program with seven filters between 0·3 and 1·0 micron and with experiments at shorter and longer wavelengths.

R. Younkin has conducted at the JPL spectrophotometric measurements of the planet from $0{\cdot}32$ to $1{\cdot}1\mu$. The monochromatic reflectance of Mars has been measured as a function of phase angle and the specific intensity of the maria compared to the continents.

H. Camichel, A. Dollfus and J. H. Focas carried out systematic photographic observations of the planet during the apparition of 1964–65 at the Pic-du-Midi Observatory through the 106 cm reflector in the spectral ranges $0{\cdot}32$ to $0{\cdot}85\mu$. High resolution visual observations of the planet were made through the 106 cm reflector of the Pic-du-Midi Observatory and the 83 cm refractor of the Meudon Observatory to complement the photographic observations.

A. Dollfus and J. H. Focas conducted systematic polarimetric observations of the planet during the 1964–65 apparition of the planet in the spectral ranges 0·47 to 1·05 μ at the Pic-du-Midi and the Meudon Observatories.

A. Dollfus and E. Bowell carried out laboratory work by studying simultaneously the polarimetric and photometric properties of samples for the purpose of matching them with analogous properties of the soil of Mars. They conclude that the surface of the soil of the ochre colored desertic bright areas of Mars is similar to a mixture of fine grains of any size, wrapping each other and forming a sufficiently areated deposit of a low roughness. The grains consist of limonite, or a pulverized absorbent and opaque material with the surface of the grains coated by a deposit of hydrated iron oxide of the limonite type.

As regards the dark areas, new polarimetric data confirm previous results as per which the seasonal darkening of these areas is associated with the change of their microscopic contexture.

A. Dollfus studied the scattering coefficient of the atmosphere of Mars in 0·61 μ. He infers the depth for the Martian atmosphere, under normal conditions to be $7 \cdot 10^4$ cm and an atmospheric pressure on the ground of approximately 30 millibars (**15**).

A. Dollfus and J. H. Focas studied the transparency of the Martian atmosphere in the blue and concluded that the luminance of the atmosphere amounts, for the phase angle zero, to 0·07 times the luminance of the soil (**16**), proving a fairly good agreement with the Rayleigh scattering law λ^{-4} for the atmosphere of Mars.

J. H. Focas studied the visibility of the surface markings of Mars in blue and ultra-violet light, on photographs taken mainly at the Lowell Observatory during the last 40 years. He concludes that: semi-tone markings show an increasing intensity in the short wavelengths and provoke deformation of the usual aspect of the planet; the surface of the planet remains for the most part of the time visible irrespectively of the wavelength; there are tenuous veils of variable opacity seen in short wavelengths which seem to follow a seasonal cycle (**17**).

J. H. Focas using a polarizing micro-photometer at the Meudon Observatory measured the contrast between bright and dark areas of Mars on photographs taken at Pic-du-Midi Observatory in the spectral area 0·32 to 1·08 μ during the apparition of the planet in 1963. This contrast decreases with decreasing wavelength, a limb darkening is evident in the absence of bright areas at the limb.

C. Banos using a microdensitometer at the Meudon Observatory obtained tracings from photographs of the planet taken at the Pic-du-Midi Observatory in 1963 and 1965, confirming the existence of limb darkening as above.

B. Smith and cooperators at the New Mexico Observatory took photographs of the planet (946 plates) in UV, blue, green, red and infrared light. Studies of photographs have included measures of the aerographic coordinates of features recorded during the 1965 apparition and a more general investigation relating to the morphology of the Martian surface. A correlation between reported radar reflectivity and optical features has been found. Comparison between the dark areas recorded in red light and the so-called 'atmospheric' features recorded in blue light are being thoroughly investigated. Photoelectric-photometric observations made during 1965 apparition tend to support the opinion of J. C. Robinson and B. A. Smith that many, perhaps most, phenomena observed in blue and ultraviolet light are related to the Martian surface and not to its atmosphere (**18, 19**).

Photoelectric observations of the planet were taken at the New Mexico Observatory (145 *UBV* sets).

At the Main Astronomical Observatory of the Ukrainian Academy of Sciences, U.R.S.S., I. K. Koval and A. V. Morozhenko obtained during the period 1963–65 series of polarimetric observations of the planet in eight spectral intervals (355–600 mμ) through an automatic photoelectric polarimeter attached to the 28-inch (71 cm) reflector.

It has been found from the phase curves of polarization that the angle of inversion shifts from 27° ($\lambda = 600$ mμ) to 15° ($\lambda = 355$ mμ). Analogous laboratory investigations on 16 terrestrial samples having the same photometric properties as Mars show that the angle of inversion is independent from wavelength. They explained the displacement of the angle of inversion as due to the Martian atmosphere and determined the atmospheric pressure at the surface (19 ± 8 mb). The comparison of the polarization properties of Mars and terrestrial samples shows the best resemblance of Fe_2O_3, $2H_2O$ to the Martian surface (**20, 21, 22**).

Mrs L. A. Bugaenko, O. I. Bugaenko, I. K. Koval, A. V. Morozhenko, during the period 1964–65 carried out photoelectric scans of the planet. These observations were paralleled by scanning turbulent disks of stars at approximately the same zenith distances. A conclusion is that photometric investigations of planets (contrasts, limb darkening) have to be accompanied by scanning stars as their image profiles undergo essential time-variations due to variations in the state of the Earth's atmosphere. Preliminary calculations show that the results based on photometric measurements of contrast and limb-darkening are to be revised critically for the planets of small angular size (Mars for example). A mathematical method has been developed taking into account the distorting effect of the Earth's atmosphere in the photometric observations of Mars.

E. G. Yanovitsky reports that 'the problem of diffuse reflection and transmission of light in planetary atmospheres for an arbitrary but not very elongated indicatrix is solved approximately.' Calculations indicate that the discrepancy between the pressure of the Martian atmosphere obtained by photometric and spectroscopic methods may be explained as due to a great amount of aerosol particles with radii $r \leqslant 10^{-6}$ cm in the Martian atmosphere. The full amount of aerosol particles of radius 10^{-6} cm in the column of the atmosphere is estimated as 10^{15} cm^{-2} (**23**).

D. Menzel and W. M. Irvine report that multicolor photoelectric photometry of the planet was made by the Harvard College Observatory at the Stations of Boyden (South Africa) and the Houga (Southern France) using ten narrow band filters between 0·315 and 1·06μ plus *UBV* for the purpose of obtaining monochromatic phase curves and albedos.

F. A. Gifford Jr. suggests that the Martian canals are chains of desert sand dunes. The necessary physical conditions for dune formation probably exist on Mars. Sand driving winds are a factor of seven higher and so Martian dune chains should be longer than on the Earth (**24**).

Tombaugh, Reese and Robinson (New Mexico State University Observatory) carried out systematic visual observations of the planet along with photographic ones. Temporary observations were noted in the Trivium Charontis area and whitishness over certain desert areas diminishing around Martian noon and reformed in later afternoon. At the time Mariner IV was sweeping over Mars, a good appraisal of Martian atmospheric conditions and surface details was possible owing to superb seeing. Tombaugh was able to identify on Mariner's pictures features recorded telescopically.

G. de Mottoni carried out at the Collurania-Teramo Observatory, Italy visual observations and color estimates of Mars, during the 1965 apparition (**25**).

R. Prinz and M. Oberndorfer carried out, respectively, photographic and visual observations of the planet during the 1964–65 apparition at the Volkessternwarte of Munich.

Theoretical investigations: Accurate aerographic coordinates of several hundred stations in the revised physical ephemeris system were determined by G. de Vaucouleurs and his collaborators at the University of Texas (**26, 27, 28**).

The geometric and photographic parameters of the terrestrial planets were discussed by G. de Vaucouleurs and best values derived, including the radiometric albedo of Mars $A = 0{\cdot}29$ (**29**).

C. Sagan, J. P. Phaneuf and M. Ihnat compared ultra-violet, visible and near infrared spectra of the Martian bright areas with the corresponding laboratory reflectivities of a variety of minerals containing ferric oxides and silicates, as solids and in pulverized form. The results obtained show that pulverized limonite, a ferric oxide polyhydrate, matches the shape and amplitude of the Martian Russell-Band albedo (**30**). The large amounts of limonite suggested by photometric polarimetric and spectrometric observations of Mars are not inconsistent with geophysical expectation. Iron can be considered as a principal constituent of the Martian surface (**31**).

A comprehensive theoretical analysis of the photometry and polarimetry of the Martian surface was made by J. Pollack and C. Sagan according to which: the material of the surface of the bright areas of Mars is a powder rather than porous rock; the spectra of the bright and dark areas are controlled by the Fe_2O_3 moiety in the wavelength region between 3000 and 11 000Å, as is indicated by the low, nearly constant reflectivity between 3000 and 4500Å and the good fit to the laboratory spectral data between 4500Å and 11 000Å; a large drop in the reflectivity between 2·4 and 3·1 microns in both the dark and the bright areas indicates the presence of sizeable hydration in both regions. Neither Martian ice clouds nor carbonates are capable of accounting for this reflectivity decline; comparison of theoretical spectra between 3·0 and 3·8 microns that allow for thermal emission with observed spectra and the index of refraction implied by the polarization phase angle curve, for the bright and dark areas show that there is one water of hydration molecule attached to Fe_2O_3 although others may be present, attached to other mineral; the maximum in the Martian reflectivity near 1·4 microns also implies the presence of water of hydration; goethite is a principal component of the bright and dark areas, their absolute reflectivities giving a first approximation of the actual particle sizes; the bright areas are characterized by particle sizes of about 50 microns, while the dark areas have an average particle diameter of 200 microns outside the seasonal wave of darkening and about 400 microns during the darkening (**32**).

C. Sagan, J. Pollack and R. Goldstein studied elevation differences between bright and dark areas for the opposition of 1963 and 1965; there is a strong correlation between high reflectivity at radar frequencies and dark areas at optical frequencies; there are occasional displacements between the reflectivity maximum and the centers of dark areas which can be understood in terms of a systematic elevation difference between bright and dark areas; some regions undergoing marked secular changes have very shallow slopes 1 to 2 degrees and elevations of the order of 6 km; canals have steeper slopes larger than 4° and elevations of the same order while large dark areas such as Syrtis M and Moeris Lacus have elevations between 10 and 20 km (**33**). They attribute the discrepancy between the dynamical and optical oblateness of Mars to the concentration of dark highlands near the Martian equator; the results may also bear on the discrepancy between ground based infrared spectrometric and polarimetric values, on the one hand, and Mariner IV occultation values on the other hand, of Martian surface pressures. The infrared pressures $\sim$ 12 mb refer to an average of the bright and dark areas. The occultation pressures $\sim$ 6 mb are necessary biased towards elevations; both ingress and egress occurred in or very near dark areas (**34–38**).

According to C. Sagan and J. Pollack, secular changes observed near dark areas characterized by shallow slopes are due to the drifting wind-blown dust alternatively covering and uncovering dark areas at slightly higher altitudes (**35**).

The progressive springtime darkening of the dark areas of Mars was discussed by J. Pollack, E. Greenberg and C. Sagan in terms of two models, one in which the darkening is due to biological activity in response to the increased temperature and humidity, the other in which fine dust is wind-blown off dark Martian highlands in spring. The observational data of Focas on darkening waves of Mars were subjected to a statistical significance analysis which shows very significant correlation between latitude and time of maximum darkening. On the other hand, correlations were found on the dust model. The obtained data do not permit a choice between the biological and the wind-blown dust models (**36**). The particle size modulation required in the dark areas to explain the seasonal photometric and polarimetric changes can be accounted for by Martian winds; according to the same model, the intrinsic contrast between bright and dark areas becomes very small at 4500 Å; the blue haze is not an atmospheric phenomenon but an intrinsic loss of surface contrast (**37**).

Using the photographic, polarimetric and visual observations of the planet collected in ten observatories throughout the world in the scheme of the International Cooperation of IAU, during the apparitions of 1954, 1956 and 1958, A. Dollfus prepared a synthesis giving the following results: diameters of the planet $d_e = 6790$ km; $d_p = 6710$ km; volume: $1620 \cdot 10^{26}$ cm^3 $\pm$ 1%; mean density: 4·09; ellipticity of the globe: 0·0117. The distribution in altitude of the opaque yellow veils can not be proportional to the atmospheric pressure; such particles should remain in the lower atmosphere layers. The 'blue' clouds are in general very tenuous. A light and transparent haze persists in the polar areas by the end of the Martian spring, discernible only polarimetrically. The yellow clouds appeared in 1956, as in the past, above some privileged bright temperate areas: they were, at start associated with white ice crystal clouds, the size of the constituing particles ranging up to 10 or 15 microns; their reflecting power, when in suspension in the atmosphere is approximately the same as that of the bright areas of the soil. The atmospheric circulation during the Martian spring and summer is of the laminar type, due to the topography of the planet and not to turbulence (**48**).

A wide program of investigation of the Martian photographs taken in short wavelengths by E. C. Slipher at the Lowell Observatory during the last 40 years is in progress at that observatory.

A. Palm and B. Basu investigated the extent of blue clearings as a function of the number of the meteor showers intersecting Mars' orbit at 10° intervals of heliometric longitude. A rank correlation method has been employed. The analysis indicates a small but statistically significant negative correlation between the extent of blue clearing and meteor shower activity. This result, combined with the optical properties of the Martian atmosphere and the close resemblance between the characteristics of the blue haze and those of the terrestrial noctilucent clouds suggests that variable amounts of interplanetary dust are suspended in the Martian atmosphere and that the occasional clearings of the blue haze are caused by a diminishing influx of these dust particles (**39**).

B. T. O'Leary and D. G. Rea computing the radiation balance on Mars and the observed duration of the Mountains of Michell suggest that they are CO_2 condensations in depressions with depths of $\sim$ 6–9 km. The appearance of bright spots on Hellas and other areas in the southern hemisphere can also be explained by depressions where the minimum night time temperature is low enough for CO_2 to condense. The possibility of H_2O condensation in depressions is also discussed. Such results support the hypothesis that the Martian deserts are lower than their surroundings (**40**).

D. G. Rea presents evidence from polarization and brightness measurements to improve the model which requires transport of dust on and off certain areas to explain the seasonal darkening of Mars. He suggests that variations in brightness for all times and areas may be due not to variations in chemical properties but to variations in particle size. He discusses the consequences of the possibility of the Martian surface being composed of finely divided goethite or limonite (**41**).

J. A. Roth describes a method of making colored globes of Mars from drawings, photographs and observations (**42**).

R. A. Wells discussed observations of Martian cloud occurrences during the past century; yellow clouds show a maximum occurrence several weeks after perihelion and constant occurrence for the remainder of the orbit: white clouds show two groups of maximum occurrence;

insolation occurrency at perihelion produces the strongest winds and yellow clouds; a cyclic relationship appears between white clouds frequency and waxing and waning of polar caps; aerographic positions of clouds reveal interesting topographic implications; yellow clouds show definite relations to the dark maria. Select occurrences of white clouds may reveal the possible presence of volcanoes and mountain ranges. The 'W' shaped clouds in Tharsis may be explained by the presence of mountain ridges. An examination of certain white clouds implies that at least some of the dark maria are elevations (**43, 44**).

R. A. Wells suggests white clouds formations to be physical evidence of the presence of mountain ranges of Mars (**45**). Re-evaluating Wright's plates for the 1924 and 1926 oppositions of Mars, he derives a diameter of the planet of 6682 km (**46**).

The Planet Jupiter

Observations: G. P. Kuiper and T. Gehrels using a 154 cm reflector at $F/13$ are performing a polarimetric and photometric program with seven filters between 0·3 and 1·0 micron and with experiments at shorter and longer wavelengths, covering the poles of the planet.

R. Younkin has conducted at the JPL spectrophotometric measurements of the planet from 0·32 to 1·1 μ.

D. Menzel and W. M. Irvine report that multicolor photoelectric photometry of the planet was made by the Harvard College Observatory at the Stations of Boyden (South Africa) and the Houga (Southern France) using ten narrow band filters between 0·315 and 1·06μ plus *UBV* for the purpose of obtaining monochromatic phase curves and albedos.

At the Lowell Observatory, S. Jones and colleagues carried out systematic photographic work of the planet in the course of the International Cooperation. High quality photographs, and a cinematographic film showing 1/4 of rotation of the planet, were taken.

C. Banos (Athens) working on photographs taken at the Athens Observatory in 1964–66 through the 25-inch (63 cm) refractor obtained variations of the coefficient of short term atmospheric activity of the order of three months.

J. B. Orr (New Zealand) followed the phenomena connected with regression of veils covering the equatorial area of the planet since 1960.

R. A. McIntosh (New Zealand) analyzed his own observations of the peculiar disturbance phenomena occurred in the southern hemisphere of Jupiter in 1962 (**49**).

P. Guérin obtained in November 1963 at the Pic-du-Midi Observatory high resolution photographs of the planet and the Galilean satellites. Such photographs show the fine structure of filamentary strips, the Red Spot, etc. (**50**).

A. Dollfus, J. H. Focas and M. Marin obtained, at the Pic-du-Midi Observatory, photographs of the planet in different spectral areas during the apparitions of the planet of 1964–65 and 1965–66.

A. Dollfus and J. H. Focas using the 83-cm refractor of the Meudon Observatory took series of photographs of the planet during the 1965–66 observational campaign prepared by the IAU.

J. H. Focas using the 16-inch (38 cm) refractor of the Nice Observatory developed a program of photographic observations of the planet during the 1965–66 observational campaign.

S. Ferraz-Mello and P. Sudbury started at the Pic-du-Midi Observatory a program of photographic observations of Jupiter and its satellites for the determination of the positions of the Galilean satellites and particularly the Vth.

T. Coupinot working at the Pic-du-Midi Observatory developed a general method of photometric scanning of the planet.

Photoelectric observations of the planet were taken at the New Mexico Observatory (177 *UBVO* sets.)

B. Smith and cooperators at the New Mexico Observatory photographed the planet (3409 plates) in *UV*, blue, green, red and infrared light. Patrol photographs have been measured to obtain zenographic coordinates and corresponding proper motions of the Red Spot, the three bright oval spots in the South Temperate Zone and other features of special interest. Special photographic observations followed a rapidly moving current perhaps analogous to the 'Jet Stream' moving along the South edge of the South component of the North Temperate Belt. A spectral photographic program also documented the interaction between a small spot in the 'circulation current' of the South Tropical Zone and the Red Spot. The photographs show conclusively that there exists a counter-clockwise vortical motion of the Red Spot (**51, 52, 53**).

At the Astrophysical Institute of the Academy of Sciences of Kazakstan (U.R.S.S.): V. G. Teifel studied the photometric and spectral properties of the Red Spot for different distances to the disk's limb. He concludes that the Red Spot is a gaso-aerosol formation which, optically, resembles the surrounding clouds (**54, 55**). V. G. Teifel, A. N. Aksenov, Z. N. Grigor'eva, and Z. G. Romanenko carried out a photographic-photometric work on the short-term atmospheric activity of the planet. They conclude for the period 1965–66 a periodic variation of the coefficient of activity of the order of three months. Such periodicity was found in the green, orange and red. This is not the case for the blue and the violet (**56, 57**).

A. Binder and D. P. Cruikshank report: Photometric observations of eclipse reappearances of Jupiter I and II were made in 1962–64 to search for a possible anomalous brightening of the satellite after eclipse. A brightening, if present, would suggest a frost or snow deposit or a haze layer caused by a surface temperature drop during eclipse. In each of four cases of J I eclipse reappearances, a brightness anomaly was indeed observed, having an average value of 0·09 stellar magnitudes. The anomaly decayed in about 15 minutes. A single observation of J II showed no anomaly (**58**).

A. B. Binder and D. P. Cruikshank, report: An observation of an eclipse reappearance of J II with the Kitt-Peak 36 inch (91 cm) telescope showed a brightness anomaly of 0·03 ± 0·01 stellar magnitudes. The anomaly decayed in 10 minutes. Two high quality observations of J III showed no brightness anomaly greater than 0·01 stellar magnitudes (**59**).

S. K. Vsekhsviatsky (Kiev State University) noted an unusual activity of Jupiter in 1962–65 and explained it as a display of volcanic activities (**60, 61, 62, 63**).

O. R. Bolkvadze at the Abastumani Astrophysical Observatory carried out electro-polarimetric investigations of the planet in the focus of the 40 cm refractor which showed: 0·1% to 0·5% of polarization for the planet's integral light with no dependance on the phase angle. Polarization in the central and polar regions is comparatively high depending on the phase angle; a stronger polarization (9·4%) was recorded at the southern region of the planet (**64**).

Tombaugh and Reese (New Mexico State University Observatory) carried out systematic visual observations of the planet. Reese made approximately 5000 central meridian transits of Jovian markings. Evolution of colors of Jovian markings was particularly studied as well as the structure of the belts. Two dark spots have been observed whirling around the Red Spot counter-clockwise. The Red Spot faded markedly in 1966 and its North perimeter became deformed.

A. P. Lehnan (B.A.A.) studied the rotation period and location of the Red Spot and concludes that there is a periodicity in the rotation period and a relationship between the variation in rotation periods and latitude fluctuations (**65**).

R. Prinz and H. Obendorfer at the Volkssternwarte Münich carried out, respectively, photographic and visual observations of the planet in the 1964–66 period.

E. H. Collinson, W. E. Fox, A. W. Heath, N. V. Jones, J. B. Murray, members of the B.A.A. carried out systematic visual observations of the planet in 1964–66.

C. Capen at the Table Mountain Observatory, conducted systematic multicolor photographic observations of the planet in 1964–66.

T. Akabane at Kwasan Observatory (Japan) using the 24-inch (61 cm) reflector, carried out systematic multicolor photography of the planet in 1966.

Theoretical investigations: M. Combes using the Lallemand electronic camera mounted on the 40-inch reflector at the Pic-du-Midi Observatory obtained high quality photographs of Jupiter allowing an accurate photometry for establishing the center-limb darkening curve. He obtained the following values of the center-limb darkening coefficient: $q\ (U) = 0{\cdot}65 \pm 0{\cdot}05$; $q\ (B) = 0{\cdot}75 \pm 0{\cdot}10$; $q\ (V) = 1{\cdot}10 \pm 0{\cdot}07$. He concludes: that the optical depth in the *UV* is comprised between 0·50 and 0·80; that the value of the abundance of molecular hydrogen ranges between 75 and 150 km atm **(66).**

The Planet Saturn

Observations: G. P. Kuiper and T. Gehrels using a 154 cm reflector at *F*/13, are performing a photoelectric program with seven filters between 0·3 and 1 micron and with experiments at shorter and longer wavelengths, covering the poles of the planet.

R. Younkin has conducted spectrophotometric measurements of the planet at the JPL from 0·32 to 1·1 μ.

G. Wlérick and cooperators, obtained in December 1965 at the Pic-du-Midi Observatory through the electronic camera series of photographs of the planet in different spectral regions (Vapillon, **67**).

B. A. Smith and cooperators photographed the planet at the New Mexico Observatory in blue, green, red and UV light (225 plates). Successful photographic observations were made of the rings of Saturn during the Earth's passage through the ring plane on 29 October 1966. A continuous sequence beginning several days before 29 October and lasting more than a week afterwards, will permit a determination of the exact time of passage through the plane and a lower limit for the thickness of the rings. Photoelectric observations of the planet, made during the latter part of 1966 (16 sets *WUBVRI*) will establish the phase coefficient for the globe of Saturn in six colors, without interference from the rings.

D. Menzel and W. M. Irvine report that multicolor photometry of the planet was made by the Harvard College Observatory at the stations of Boyden (South Africa) and the Houga (Southern France) using ten narrow band filters between 0·315 and 1·06μ plus *UBV* on the purpose to obtain monochromatic phase curves and albedos.

O. R. Bolkvadze at the Abastumani Astrophysical Observatory carried out electro-polarimetric investigations of the planet in the focus of the 40 cm refractor which showed 0·1% to 0·5% of polarization for the planet's integral light with no dependence on the phase angle; polarization in the central and polar regions is comparatively high depending on the phase angle; a stronger polarization (11·7%) was recorded at the southern region of the planet. Especially high polarization (up to 14–15%) is displayed by the east and west sides of the outer ring of the planet, and in this respect the west side always dominates over the east side **(68).**

Transit of the Earth and Sun through the plane of the Rings in 1966. The Sun crossed the plane of the Rings on 15 June and the Earth on 29 October and 17 December. According to information received till now (December 1966) by the Committee of Commission 16 'International Cooperation for the Observation of the Planets', the following observers collected observations of the above said phenomena:

C. Banos (Athens, Greece)	63 cm refr.	Photographic Yellow	Tricolor
C. Capen (Table Mountain, U.S.A.)	40 cm refr.	Photog. and Vis.	Multicolor
A. Dollfus (Pic-du-Midi	106 cm refl.	Photog. and Vis.	Yellow blue
and Meudon, France;	60 cm refl.	Photog. and Vis.	Yellow blue
U.R.S.S.)	70 cm refl.	Photog. and Vis.	Yellow blue
J. Focas (Meudon, France)	60 cm refl.	Photog. and Vis.	Vis.

D. Read (New Zealand)	20 cm refr.	Visual	Yellow
McIntosh (New Zealand)	30 cm refr.	Visual	Yellow
I. Thomson (New Zealand)	20 cm refr.	Visual	Yellow
I. Kiladze (Abastumani, U.R.S.S.)	70 cm refl.	Photographic	Yellow
M. Martin (Pic-du-Midi, France)	106 cm refl.	Photographic	Yellow
R. Prinz (Munich, W. Germany)	30 cm refl.	Photographic	Yellow blue
H. W. Oberndorfer (Munich, W. Ger.)	20 cm refl.	Visual	Yellow
J. Texereau (McDonald, U.S.A.)	200 cm refl.	Photographic	Yellow
C. Tombaugh (New Mexico, U.S.A.)	40 cm refl.	Visual	Yellow
E. Reese (New Mexico, U.S.A.)	40 cm refl.	Visual	Yellow

Theoretical investigations: M. S. Bobrov, analyzing photometric data, showed that Saturn's rings are a system of a certain thickness, composed of many particles. Taking into account on the inter-darkening theory the dispersion of particle sizes, he obtains a good agreement between theory and observation (**69, 70**).

F. A. Franklin and A. F. Cook using photometric data of the Rings A and B of Saturn taken through the 60-inch (152 cm) reflector of the Boyden Observatory (S. Africa) studied the optical thickness of five representative portions of the ring and the phase variation and albedo of the ring particles. The non-linear and color-dependent phase curves of the ring near opposition allow two possible interpretations. The more likely model gives the fraction of the ring volume occupied by particles (Ring A = $1{\cdot}0 \times 1{\cdot}0^{-3}$, Ring B = $1{\cdot}3 \times 10^{-3}$) and a characteristic size of 7μ. The other model gives values of the former quantity (Ring A = 6×10^{-3}, Ring B = 4×10^{-3}), and in addition particle radii of about 300μ and a ring thickness of about 10 cm.

I. N. Minin presented an approximate method of solving the problem of polarized light dispersion in the planet's atmosphere taking into account aerosol component and light reflection from the surface (**71**).

The Planet Uranus

Observations: R. Younkin has conducted at the JPL spectrophotometric measurements of the planets from 0·32 to $1{\cdot}1\mu$.

G. Coupinot obtained at the Pic-du-Midi Observatory through rapid photoelectric scanning records allowing the determination of the diameter and the limb darkening of the planet (**72**).

B. Smith and cooperators obtained at the New Mexico Observatory photographs (19 plates) of the planet in blue light.

BIBLIOGRAPHY

1. Sagan, C. 1966, The photometric properties of Mercury. *Astrophys. J.*, **144**, 1218.
2. Boyer, Ch., Camichel, H. 1965, Etude photographique de la rotation de Vénus. *C. r. Acad. Sci. Paris*, **260**, 809.
3. Boyer, Ch., Guérin, P. 1966, Mise en évidence directe par la photographie d'une rotation rétrograde de Vénus en 4 jours. *C. r. Acad. Sci. Paris*, **263**, 253.
4. Smith, B. A. 1964, A Photographic Optical Semidiameter of Venus. *Astr. J.*, **69**, 150.
5. Dollfus, A., Maurice, E. 1965, *C. r. Acad. Sci. Paris*, **260**, 427.
6. Pollack, J. B., Sagan, C. 1965, The Microwave Phase Effect of Venus. *Icarus*, **4**, 62.
7. Pollack, J. B., Sagan, C. 1965, Polarization of Thermal Emission from Venus. *Astrophys. J.*, **141**, 1161.
8. Sagan, C., Pollack, J. B. 1967, An Analysis of the Mariner II Microwave Observations of Venus. To be published.

9. Sagan, C., Pollack, J. B. 1966, On the Nature of the Clouds and the Origin of the Surface Temperature of Venus. *Astr. J.*, **71**, 178.
10. Lippincott, E. R., Eck, R., Dayhoff, M. O., Sagan, C. Thermodynamic Equilibria in Planetary Atmospheres. *Astrophys. J.*, **147**, 753.
11. Barabashov, N. P. 1965, Sur les particularités photométriques de la lumière réfléchie venant de la surface visible de Vénus. *Vestnik Karkovsk. gos. Univ.*, no. 4, *Ser. Astr.*, Fasc. 1, 13.
12. Capen, C. F. Physical appearance of the planet Mars during the period of Mariner IV encounter. *JPL Lab. Report.*
13. Capen, C. F. 1965, Mars Patrol 1964. *JPL, SPS*, no. 37.
14. Tull, R. G. 1966, *Icarus*, **5** (in press).
15. Dollfus, A. 1966, *C. r. Acad. Sci. Paris*, **262**, 519.
16. Dollfus, A., Focas, J. H. *C. r. Acad. Sci. Paris*, **262**, 1024.
17. Focas, J. H. 1966, Transparency of the Martian atmosphere and visibility of the surface details in the blue and the ultra-violet. COSPAR Meeting.
18. Robinson, J. C. 1963, Photographic observations of Mars at New Mexico State University in 1960–61. *Icarus*, **2**, 403.
19. Robinson, J. C. 1966, Ground based Photography of the Mariner IV Region of Mars. *Icarus*, **5**, 245.
20. Morozhenko, A. V. 1964, Résultats des observations polarimétriques de Mars en 1962–63. Collection 'Physique de la Lune et des Planètes', Edition Naukova Dounka, Kiev, p. 58.
21. Morozhenko, A. V. 1960, Propriétés de la polarisation de l'atmosphère et de la surface de Mars. Collection 'Physique de la Lune et des Planètes', Edition Naukova Dounka, Kiev.
22. Morozhenko, A. V. 1966, Recherche des propriétés de polarisation de la lumière réfléchie par des formes variées des couches terrestres. Collection 'Physique de la Lune et des Planètes'. Edition Naukova Dounka, Kiev.
23. Yanovitsky, E. G. 1965, Sur l'aérosol composant l'atmosphère de Mars. Colloque d'Astrophysique, Edition Naukova Dounka, Kiev.
24. Gifford, Jr., F. A. 1964, The Martian canals according to a purely Aeolian hypothesis. *Icarus*, **3**, 130.
25. Mottoni G. de 1966, Osservazioni visuali di Marte durante l'opposizione afelica del 1965, **2**, Fasc. 2.
26. Vaucouleurs, G. de 1964, *Icarus*, **3**, 236.
27. Meiller, V. 1964, *U.S. nav. Obs. Circ.*, no. 98.
28. Vaucouleurs, G. de 1965, *Sky Telesc.*, **30**, 196.
29. Vaucouleurs, G. de 1964, *Icarus*, **3**, 187
30. Sagan, C., Phaneuf, J. P., Ihnat, M. 1965, Total Reflection Spectrophotometry and Thermogravimetric Analysis of Simulated Martian Surface Materials. *Icarus*, **4**, 43.
31. Sagan, C. 1966, Mariner IV Observations and the Possibility of Iron Oxides on the Martian Surface. *Icarus*, **5**, 102.
32. Pollack, J. B., Sagan, C. 1967, Photometric and Polarimetric Properties of Mars. To be published.
33. Sagan, C., Pollack, J. B., Goldstein, R. 1967, Radar Doppler Spectroscopy of Mars. I. Elevation Differences between Bright and Dark Areas. *Smithson. Inst. astrophys. Obs. Res. Space. Sci. Spec. Rep.*, no. 221, 1966. *Astr. J.*, In press.
34. Sagan, C., Pollack, J. B. 1966, Elevation Differences on Mars. *Smithson. Inst. astrophys. Obs. Spec. Rep.*, no. 224.
35. Pollack, J. B., Sagan, C. 1967, Secular Changes and Dark Area Regeneration on Mars. *Icarus* (in press).
36. Pollack, J. B., Greenberg, E., Sagan, C. 1966, A Statistical Analysis of the Martian Wave of Darkening and Related Phenomena. Submitted to *Planet. Space Sci.*
37. Sagan, C., Pollack, J. B. 1966, An inorganic model of Martian phenomena. *Astr. J.*, **71**, 178.
38. Sagan, C., Pollack, J. B. 1966, On the nature of the canals of Mars. *Nature*, **212**, 117.

39. Palm, A., Basu, B. 1965, The Blue Haze of Mars. *Icarus*, **4**, 111.
40. O'Leary, B. T., Rea, D. G. The influence of topography on the formation of temporary bright patches on Mars. Space Sci. Lab., University of California Berkeley.
41. Rea, D. G. 1964, The darkening wave on Mars. *Nature*, **201**, 1014.
42. Roth, J. A. 1964, Precision globes of Mars. *Sky Telesc.*, **27**, 19.
43. Wells, R. A. 1966, An analysis of Martian clouds and their topographical relationships. *ESRO scient. Notes*, no. 54.
44. Wells, R. A. 1966, Some aspects of Martian clouds and their relation to the topography. Vienna COSPAR Proceedings, 10–19 May.
45. Wells, R. A. 1965, Evidence that the dark areas on Mars are elevated Mountain Ranges. *Nature*, **207**, 735.
46. Wells, R. A. 1965, A re-evaluation of W. H. Wright's plates of the 1924 and 1926 oppositions of Mars. *Planet. Space Sci.*, **13**, 261.
47. Dollfus, A., Focas, J. 1966, Polarimetric Study of the Planet Mars. AFCRL Final Report, Contract AF 61(052)508.
48. Dollfus, A. 1965, Etude de la Planète Mars de 1954 à 1958. *Ann. Astrophys.*, **28**, 722.
49. McIntosh, R. A. 1965, The 1962 eruption on Jupiter. *Sth Stars*, **21**, 20.
50. Guérin, P. 1965, Photographies de Jupiter à très haute résolution. *Ann. Astrophys.*, **28**, 395.
51. Reese, E. J., Smith, B. A. 1966, A rapidly moving spot on Jupiter's north temperate belt. *Icarus*, **5**, 248.
52. Reese, E. J., Solberg, H. G. 1966, Recent measures of the latitude and longitude of Jupiter's Red Spot. *Icarus*, **5**, 266.
53. Smith, B. A. 1966, Rotation Numbers for Jupiter. Apparitions of 1965–66, 1966–67. *Contr. Obs. New Mexico State Univ. Spec. Series*, no. 2 and 3.
54. Teifel, V. G. 1964, Sur les propriétés photométriques de la Tache Rouge sur Jupiter. *Astr. Zu.*, **41**, no. 3.
55. Teifel, V. G. 1967, Spectrophotométrie de la Tache Rouge de Jupiter. Fizika planet. AN Kaz. SSR. Nauka (in press).
56. Aksenov, A. N. 1966, Résultats préliminaires de l'étude photométrique de l'activité de l'atmosphère de Jupiter. Issledovanija planety Jupiter, AN-SSSR.
57. Aksenov, A. N., Grigor'eva, Z. N., Priboeva, N. V., Romanenko, Z. G., Tejfel, V. G. 1967, Activité périodique de Jupiter d'après les données d'observations photométriques. Astronomiceskij Vestnik.
58. Binder, A. B., Cruikshank, D. P. 1964, Evidence for an atmosphere on Io. *Icarus*, **3**, 299.
59. Binder, A. B., Cruikshank, D. P. 1966, Photometric search for atmospheres on Europe and Ganymede. *Icarus*, **5**, 7.
60. Vsetsviatsky, S. K. 1965, *Astr. Cirk.*, no. 317, 2.
61. Vsetsviatsky, S. K. 1965, Sur la nature des changements sur la surface de Jupiter. *Astr. Zu.*, **42**, 639.
62. Vsetsviatsky, S. K. 1965, Changements sur Jupiter. *Geofiz. Astr. Inf. Bjull.*, no. 8, 3.
63. Vsetsviatsky, S. K. Résultats des observations de Jupiter en 1962–65. Travaux A.F.I. (in press).
64. Bolkvadze, O. R. Recherches sur les polarisations particulières de Jupiter et de Saturne.
65. Lehnan, A. P. The motion of the Red Spot on Jupiter. *J. Br. astr. Ass.*, **74**, 203
66. Combes, M. 1966, Contribution à l'étude de la haute atmosphère de Jupiter. Thèse, Paris.
67. Vapillon, L. Thesis, in preparation.
68. Bolkvadze, O. R. Recherches sur les polarisations particulières de Jupiter et de Saturne.
69. Bobrov, M. S. Situation actuelle de la question de la structure et de l'épaisseur des anneaux de Saturne. Travaux A.F.J. (in press).
70. Bobrov, M. S. Sur les observations des anneaux de Saturne, etc... *Cirk. VAGO* (in press).
71. Minin, I. N. 1965, Sur la diffusion de la lumière dans les atmosphères des Planètes. *Trudy astr. Obs. Leningrad. gos. Univ.*, **22**, no. 328, 39.
72. Coupinot, G. Détermination du diamètre d'un petit disque par balayage photoélectrique. *C. r. Acad. Sci. Paris* (in press).

A REVIEW OF SPECTROSCOPIC STUDIES OF THE PLANETS, 1964-67

(prepared by T. Owen, IIT Research Institute)

Explanatory Remarks

This review covers the period from 1 January 1964 to 1 December 1966, with the exception of those papers (published in early 1964) already reviewed in the previous Commission 16 report. The review is based on a survey of current periodicals supplemented by communications with astronomers working in this area. Where preprints were available, papers accepted for publication have been included in order to extend the currency of the review. The dedicated assistance of Mr Ron Wojcik in carrying out the overwhelming majority of the survey work is gratefully acknowledged.

The review is organized to follow the usual plan of considering each planet in turn, proceeding outward from the Sun. Each section is independent, having its own page numbers and list of references. In several cases, papers have been published which discuss observations of more than one planet. In such instances some redundancy has been permitted in order that each topic may be complete. Because of the relative scarceness of published material, however, the major planets have been considered in two groups: Jupiter-Saturn and Uranus-Neptune.

A. Mercury

Several attempts were made to determine whether or not Mercury possesses an atmosphere and, if so, what its composition might be. N. A. Kozyrev (**1**) presented spectra of Mercury indicating emission components in the cores of Fraunhofer hydrogen lines. He concluded that this was evidence for an accreted hydrogen atmosphere. H. Spinrad and P. W. Hodge (**2**) showed that this effect could be produced by the superposition of the spectrum of the twilight sky on the Mercury spectrum, when the latter exhibits a pronounced Doppler displacement. Under these conditions, each absorption line appears as an unresolved doublet, while the small space between the cores of each pair of lines imitates an emission line. Kozyrev (**3**) has rejected this interpretation since he states that he monitored the twilight spectrum very carefully and rejected those plates on which the contribution from twilight was appreciable.

V. I. Moroz (**4**) observed the spectrum of Mercury from 1·0 to 3·9μ. His most significant finding was an apparent enhancement of the telluric CO_2 bands near 1·6μ which he has interpreted as evidence for the presence of a CO_2 atmosphere about Mercury. The amount of CO_2 remains uncertain because the strength of these bands makes their intensity pressure-dependent. Moroz estimates 0·3–7 g cm^{-2}. H. Spinrad, G. B. Field, and P. W. Hodge (**5**) have looked for the weak CO_2 band at 8689Å that would give the abundance directly, but found no trace, setting an upper limit of 57 m-atm (11·2 g cm^{-2}), consistent with Moroz' observations. Reviewing available evidence, Spinrad *et al.* concluded that Mercury probably does have a tenuous atmosphere, although its mass remains uncertain.

REFERENCES

1. Kozyrev, N. A. 1964, *Sky Telesc.*, **27**, 339.
2. Spinrad, H., Hodge, P. W. 1965, *Icarus*, **4**, 105.
3. Kozyrev, N. A. 1965, *Sky Telesc.*, **30**, 360.
4. Moroz, V. I. 1964, *Astr. Zu.*, **41**, 1108.
5. Spinrad, H., Field, G. B., Hodge, P. W. 1965, *Astrophys. J.*, **141**, 1155.

B. Venus

The ultraviolet spectrum of Venus below 3000Å was observed for the first time by D. Evans (**1**). He found a broad absorption-like feature extending from 3300Å to the short-wave limit of his observation (2300 Å) with a maximum at 2500 Å. Evans points out that this feature can be interpreted as either the result of true absorption by ozone (1/3 to 1/10 the amount present in the Earth's atmosphere) or as the combined effect of a Rayleigh scattering atmosphere above a cloud layer having a reflectivity which decreases toward shorter wavelengths. It is not yet possible to determine which is the correct interpretation. I. N. Glushneva (**2**) reported observations indicating that the albedo of Venus diminishes slightly with decreasing wavelength in the region 4500–3200 Å.

H. Spinrad and E. H. Richardson (**3**) determined an upper limit of 57 cm-atm of O_2 in the atmosphere of Venus using high resolution spectra in the region of the A band (7600 Å). This result is in disagreement with the tentative identification of oxygen previously reported by Prokofiev and Petrova (**4**) and suggests a relative abundance of $< 8 \times 10^{-5}$ for O_2 in the planet's atmosphere. H. Spinrad (**5**) reported the discovery of a new hot band of CO_2 in the region near 8750 Å from which he deduced a rotational temperature of ~ 400°K.

M. Bottema, W. Plummer, and J. Strong (**6, 7, 8**) used a balloon-borne telescope to obtain spectra of Venus in the region in the strong water vapor band near 1·13μ. They detected water vapor in excess of the residual amount present above the balloon in the Earth's atmosphere and concluded that the atmosphere of Venus above the clouds contains between $5{\cdot}2 \times 10^{-3}$ and $22{\cdot}2 \times 10^{-3}$ g cm^{-2}, depending on the pressure at the cloud tops. A. Dollfus (**9**) observed the spectrum of Venus in the region of the 1·4μ water vapor band, using interference filters to compare the intensity in this region with the neighboring continuum. He found evidence for absorption due to water vapor in the atmosphere of Venus and derived an abundance of 7×10^{-3} g cm^{-2}. M. Bottema, W. Plummer, J. Strong, and R. Zander (**10**) obtained a low resolution spectrum of Venus in the 1·7–3·4μ region which they have shown to be closely matched by the laboratory spectrum of an ice cloud. They conclude that the Venus clouds consist of ice crystals.

M. J. S. Belton and D. M. Hunten (**11**) announced the discovery of Doppler shifted components in the wings of telluric water vapor lines near 8200 Å. They attributed these components to absorption by water vapor in the atmosphere of Venus. This result was confirmed by H. Spinrad and S. J. Shawl (**12**) and appears to lend strong support to the identifications at longer wavelengths.

J. A. Westphal, R. L. Wildey, and B. C. Murray (**13**) were able to produce detailed maps of isotherms on the planet's disk by means of observations in the 8–14μ region. From an examination of these maps, the authors found that rather large, complex variations occur in the atmosphere of Venus at the levels producing the 8–14μ emission. J. A. Westphal (**14**) presented reduced data from diametric scans of Venus carried out with 8-14μ radiation. These data indicate an unexpected brightening near the planet's limb, suggesting a more complex atmospheric structure than that postulated by previous models.

V. I. Moroz (**15**) reobserved the 1·2–3·8μ spectral region with improved resolution and presented a table of wavelengths and equivalent widths of the planetary CO_2 absorption bands. He deduced an abundance of 25 g cm^{-2} for the CO_2 above the cloud layer and a total pressure at the cloud tops of 0·3 atm. An earlier tentative identification of CO absorption at 2·35μ was confirmed; the derived abundance was 1·5 cm-atm. Strong absorption for $\lambda > 3\mu$ was observed leading to an albedo at 3·3μ of $\sim 1\%$. The identity of the absorber remains unknown but Moroz suggests that it may contribute to a strong greenhouse effect in the planet's atmosphere.

B. Guinot (**16**) used a specially designed spectrograph employing a Fabry-Pérot etalon to examine different parts of the disk of Venus in the 5500-5700 Å region. He found evidence for an equatorial velocity of 110 ± 8 m s^{-1} corresponding to a period of retrograde rotation of $4{\cdot}1 \pm 0{\cdot}7$ days.

REFERENCES

1. Evans, D. C. *Astr. J.*, **70**, 321, 1965 and *NASA-Goddard Space Flight Center Report*, X-613-66-172, May 1966.
2. Glushneva, I. N. 1966, *Astr. Zu.*, **41**, 720.
3. Spinrad, H., Richardson, E. H. 1965, *Astrophys. J.*, **141**, 282.
4. Prokofiev, V. K., Petrova, N. N. 1962, *Mém. Soc. R. Sci. Liège*, 5ème *ser.*, **7**, 311.
5. Spinrad, H. 1966, *Astrophys. J.*, **145**, 943.
6. Bottema, M., Plummer, W., Strong, J. 1964, *Astrophys. J.*, **139**, 1021.
7. Bottema, M., Plummer, W., Strong, J. 1965, *Ann. Astrophys.*, **28**, 225.
8. Strong, J. 1966, *Proceedings of the Cal Tech-JPL Lunar and Planetary Conference*. p. 147 June 15.
9. Dollfus, A. *Ibid.*, p. 187.
10. Bottema, M., Plummer, W., Strong, J., Zander, R. 1964, *Astrophys. J.*, **140**, 1640.
11. Belton, M. J. S., Hunten, D. M. 1966, *Astrophys. J.*, **146**, 307.
12. Spinrad, H., Shawl, S. J., 1966, *Astrophys. J.*, **146**, 328.
13. Westphal, J. A., Wildey, R. L., Murray, B. C. 1965, *Astrophys. J.*, **142**, 799.
14. Westphal, J. A. 1966, *J. geophys. Res.*, **71**, 2693.
15. Moroz, V. I. 1964, *Astr. Zu.*, **41**, 711.
16. Guinot, B. 1965, *C. r. Acad. Sci.*, Paris, **260**, 431.

C. Mars

The ultraviolet spectrum of Mars was observed by D. Evans (**1**) by means of a rocket-borne spectrograph. The region covered was 2400 to 3500 Å with a resolution of about 50 Å. The ultraviolet reflectivity derived from the observations was interpreted in terms of a Rayleigh scattering atmosphere composed of some mixture of N_2, CO_2, and Ar, with a surface pressure in the range of 5–20 mb (a value of 10 mb was favored). No evidence of absorption was detected; Evans concluded that solar UV radiation in the 2000–3000 Å region will reach the planet's surface.

B. T. O'Leary (**2**) reinterpreted Martian spectra (obtained by Kiess, Karrer, and Kiess) with the help of laboratory spectra of NO_2 and derived an upper limit of 0·1 mm-atm for the abundance of NO_2 in the Martian atmosphere. J. V. Marshall (**3**), using spectra of Mars obtained at the Mt Wilson Observatory and his own laboratory spectra, set an upper limit of 0·008 mm-atm on the NO_2 abundance. T. Owen (**4**) obtained a new Martian spectrum showing no trace of the NO_2 absorption feature studied by O'Leary and Marshall in the laboratory, thereby confirming their conclusions that detectable concentrations of the gas are not present in the Martian atmosphere.

The determination of the composition and surface pressure of the Martian atmosphere by means of curve-of-growth studies of the CO_2 absorptions has continued. T. Owen (**4**) obtained new observations of the 8689 and 10 486 Å bands of CO_2 from which he derived a CO_2 abundance of 65 ± 20 m-atm. Using this value of the abundance and a curve-of-growth determined in the laboratory to interpret observations of the 1·6μ CO_2 band obtained by Kuiper, Owen derived an atmospheric surface pressure of 9^{+11}_{-4} mb. Belton and Hunten (**5**) observed the 10 486 Å band in detail and fitted a synthetic spectrum to the observations. In this way they derived a CO_2 abundance of 68 ± 26 m-atm and a rotational temperature of 194°K. Relying on an analysis of the strong 2·0μ CO_2 bands in the Martian spectrum carried out by L. D. Gray (**6**), Belton and Hunten derived a surface pressure in the range of 5–13 mb, with 6 mb as a preferred value. H. Spinrad, R. A. Schorn, R. Moore, L. P. Giver and H. J. Smith (**7**) carried out an extensive series of observations of the 8689 Å band. They determined a CO_2 abundance of 90 ± 27 m-atm which they used in conjunction with previous analyses of the 2·0μ and 1·6μ bands to deduce a surface pressure of 10^{+10}_{-5} mb. Comparing their work with the results obtained from the Mariner IV occultation (5 or 6 mb), the authors concluded that Mars must have an almost pure CO_2 atmosphere. L. D. Gray (**6**) has used a random Elsässer band model for the interpretation of the $2{\cdot}0\mu$ bands. She was able to derive an expression for the absorption in terms of the product of the amount of gas (m) and the pressure (p), viz. $mp = 500 \pm 100$ m-atm mb. For m in the range of 60 to 85 m-atm, Gray derived a surface pressure of $7{\cdot}1 \pm 2{\cdot}2$ mb.

R. A. Schorn, H. Spinrad, R. Moore, H. J. Smith, and L. P. Giver (**8**) reported new observations confirming the presence of water vapor in the Martian atmosphere. They were able to detect Doppler-shifted Martian components in the water vapor band at 8200 Å and derived an abundance of 10 to 20μ precipitable water. The concentration of water vapor appears to vary with time and location on the planet, and to depend on the size of the Martian polar caps. F. J. Heyden, C. C. Kiess, and W. R. Willauer (**9**) have cautioned that care must be taken in interpreting planetary vapour absorptions near 8200 Å because of the presence of many faint telluric features in this region. A. Dollfus (**10, 11**), using the same technique employed in studying Venus, deduced an amount of 45μ precipitable water in the Martian atmosphere from observations of the $1{\cdot}4\mu$ absorption band.

R. L. Younkin (**12**) made spectrophotometric measurements of Mars from $0{\cdot}5$ to $1{\cdot}1\mu$ in a search for reflectance features typical of limonite as observed in the laboratory. He found no evidence of such features in the Martian spectrum, placing an upper limit of 2% on their intensity. R. G. Tull (**13**) also observed this region of the spectrum and used published data to extend the observed spectrum to $1{\cdot}52\mu$. His results essentially confirm Younkin's work but add a distinct minimum at 1μ. Tull concludes that his observations are not inconsistent with the presence of limonite on the Martian surface. A. B. Binder and D. P. Cruikshank (**14, 15**) compared the spectrum of Mars in the $0{\cdot}8$ to $2{\cdot}4\mu$ region with naturally occurring limonite found on desert rocks. They found that the limonite provided a good match for the Martian bright areas, and that basalts and basaltic volcanic ash did not have a reflection spectrum comparable with that of the Martian dark areas. These results are consistent with the findings of A. L. Draper, J. A. Adamcik and E. K. Gibson (**16**) who used 1 to 2μ reflection spectra of mixtures of goethite and hematite as comparisons for the Martian spectrum.

V. I. Moroz (**17**) reported an extensive series of observations of the Martian spectrum in the $1{\cdot}1$ to $4{\cdot}1\mu$ region. From his own measures of the $1{\cdot}6\mu$ and $2{\cdot}0\mu$ CO_2 bands and the CO_2 abundance determined by Kaplan, Münch, and Spinrad, Moroz derived a surface pressure of 15^{+10}_{-5} mb. The relative concentration of CO_2 was found to be given by $\alpha = 40/p^2$, such that a pure CO_2 atmosphere would exert a pressure of $6{\cdot}3$ mb. Moroz confirmed Kuiper's discovery of ice absorption bands in spectra of the Martian polar caps. He also found evidence of the bands near $3{\cdot}5\mu$ originally discovered by Sinton. Moroz considered his determinations of the planet's infrared reflectivity to be in satisfactory agreement with the spectrum of limonite.

J. S. Shirk, W. A. Haseltine, and G. C. Pimentel (**18**) suggested that the bands observed by Sinton near $3{\cdot}5\mu$ in the Martian spectrum might be due to deuterated water vapor in the Martian atmosphere. D. G. Rea, B. T. O'Leary, and W. M. Sinton (**19**) re-examined the observational evidence and found a correlation between the intensities of the $3{\cdot}5\mu$ bands and the amount of telluric water vapor in the optical path. They concluded that these features were probably due to telluric water vapor and should not be attributed to the Martian atmosphere or surface.

L. P. Giver (**20**) pointed out the need for additional observations of the CO_2 hot band at $10{\cdot}4\mu$ in the spectrum of Mars, since presently available results do not appear to be consistent with the latest values for the surface pressure and CO_2 abundance.

J. Connes, P. Connes, and L. D. Kaplan (**21**) announced the discovery of new absorption bands in the near infrared spectrum of Mars which they tentatively attributed to reduced gases in the Martian atmosphere.

REFERENCES

1. Evans, D. C. 1965, *Science*, **149**, 969.
2. O'Leary, B. T. 1965, *Publ. astr. Soc. Pacif.*, **77**, 168.
3. Marshall, J. V. 1964, *Comm. Lunar Planet. Lab.*, **2**, 167.
4. Owen, T. 1966, *Astrophys. J.*, **146**, 257.
5. Belton, M. J. S., Hunten, D. M. 1966, *Astrophys. J.*, **145**, 454.
6. Gray, L. D. 1966, *Icarus*, **5**, 390.
7. Spinrad, H., Schorn, R. A., Moore, R., Giver, L. P., Smith, H. J. 1966, *Astrophys. J.*, **146**, 331.
8. Schorn, R. A., Spinrad, H., Moore, R., Smith, H. J., Giver, L. P. *Astrophys. J.* (in press).
9. Heyden, F. J., Kiess, C. C., Willauer, W. R. 1966, *Astrophys. J.*, **143**, 595.
10. Dollfus, A. 1965, *C. r. Acad. Sci.*, Paris, **261**, 1603.
11. Dollfus, A. 1966, *Proceedings of the CALTECH–JPL Lunar and Planetary Conference*, p. 288 (15 June).
12. Younkin, R. L. 1966, *Astrophys. J.*, **144**, 809.
13. Tull, R. G. 1966, *Icarus*, **5**, 505.
14. Binder, A. B., Cruikshank, D. P. 1964, *Comm. Lunar Planet. Lab.*, **2**, 193.
15. Binder, A. B., Cruikshank, D. P. 1966, *Icarus*, **5**, 521.
16. Draper, A. L., Adamcik, J. A., Gibson, E. K. 1964, *Icarus*, **3**, 63.
17. Moroz, V. I. 1964, *Astr. Zu.*, **41**, 350.
18. Shirk, J. S., Haseltine, W. A., Pimentel, G. C. 1965, *Science*, **147**, 48.
19. Rea, D. G., O'Leary, B. T., Sinton, W. M. 1965, *Science*, **147**, 1286.
20. Giver, L. P. 1965, *Publ. astr. Soc. Pacif.*, **77**, 128.
21. Connes, J., Connes, P., Kaplan, L. D. 1966, *Science*, **153**, 739.

D. Jupiter and Saturn

The ultraviolet spectrum of Jupiter was recorded by T. Stecher (**1, 2**) from 2100 to 3000 Å, and verified by D. Evans (**3**) to 2300 Å. Stecher found that the continuum observed in this region could be explained by a Rayleigh scattering atmosphere consisting of 10 km atm of H_2. However, the constituents responsible for an absorption feature at 2600 Å and a sharp drop in intensity below 2100 Å remain to be identified. I. N. Glushneva (**4**) reported measurements of the albedo of Jupiter from 3200 to 4500 Å. The albedo falls off sharply at 4100 Å reaching a minimum at 3500 Å with a recovery toward shorter wavelengths.

D. J. Taylor (**5**) determined a bolometric Bond albedo of 0·45 for Jupiter from a series of spectrophotometric measurements in the 3400 to 10 000 Å region. Using this albedo, Taylor derived an equilibrium temperature of 105°K, considerably lower than measured infrared temperatures.

D. Clarke and J. F. Grainger (**6**) observed the profile of Hβ in the spectrum of Jupiter and found no differences between solar and Jovian profiles to within 1%. This negative result is in agreement with a search for aurorae carried out by G. A. Dulk and J. A. Eddy (**7**) who looked for evidence of emission components in the profile of Hα.

V. G. Teifel' (**8**) studied the behavior of several methane absorption bands in the 0·7 to 1·0μ region of the spectrum of Jupiter. Teifel' found that the absorption declines slightly toward the planet's limb, increases slightly at intermediate latitudes and declines toward the poles. A two layer model was suggested as necessary to account for the absorption pattern. T. Owen, E. H. Richardson, and H. Spinrad (**9**) found a new ammonia band in the spectrum of Jupiter near 7600 Å.

T. Owen (**10, 11**) compared laboratory absorption spectra of several gases with the spectrum of Jupiter in the regions 9700 to 11 200 Å, and 7750 to 8800 Å. All non-solar absorptions were identified as due to NH_3 or CH_4 and upper limits were set on the abundances of the other gases studied as well as on the value of D/H in the Jovian atmosphere. The $3\nu_3$ band of CH_4

at 1·1 μ was analyzed to obtain a rotational temperature of 200 ± 25°K. A methane band which overlaps the 7900 Å ammonia band was considered to be responsible for the misidentification of ammonia in the spectrum of Saturn. L. P. Giver (**12**) announced marginal detection of ammonia in the spectrum of Saturn at 6470 Å.

L. P. Giver (**12**) reported no anomalous inclination of ammonia or methane lines in spectra of Jupiter and Saturn obtained in 1964–65. H. Spinrad and L. P. Giver (**13**) confirmed this result in a later report.

R. E. Danielson (**14**) reported observations of Jupiter's spectrum from 0·8 to 3·1 μ obtained from a balloon at an altitude of 84 000 feet (26 km). Broad absorption features observed at 0·85, 0·99, 1·15, 1·37, and 1·7 μ were attributed to CH_4, while NH_3 was considered to be responsible for an absorption at 3 μ. A feature observed at 2·25 μ was ascribed to the (1–0) band of 40 km-atm of H_2.

B. C. Murray, R. L. Wildey, and J. A. Westphal (**15**) mapped the 8 to 14 μ brightness temperature distribution across the disk of Jupiter. They found no latitudinal variation of brightness temperature greater than 1°K nor variations greater than 0·5°K across the band structure although an enhancement of as much as a factor 30 was found associated with the position on the disk of a satellite shadow. Subsequent studies of satellite shadows by Wildey (**16, 17**) failed to show the enhancement originally discovered. On the other hand, improved resolution permitted Wildey, Murray and Westphal (**18**) to detect variations in brightness temperature across the disk of Jupiter. The light bands were found to be about 0·5° cooler than the dark bands and the Red Spot appeared to be 1·5 to 2·0° cooler than the surrounding disk.

REFERENCES

1. Stecher, T. 1965, *Astrophys. J.*, **142**, 1186.
2. Stecher, T. 1965, *Ann. Astrophys.*, **28**, 788.
3. Evans, D. C. *NASA-Goddard Space Flight Center Report* X–613–66–172, May 1966.
4. Glushneva, I. N. 1964, *Astr. Zu.*, **41**, 720.
5. Taylor, D. J. 1965, *Icarus*, **4**, 362.
6. Clarke, D., Grainger, J. F. 1965, *Planet. Space Sci.*, **13**, 1164.
7. Dulk, G. A., Eddy, J. A. 1966, *Astr. J.*, **71**, 160.
8. Teifel', V. G. 1966, *Astr. Zu.*, **43**, 154.
9. Owen, T., Richardson, E. H., Spinrad, H. 1964, *Astrophys. J.*, **139**, 1374.
10. Owen, T. 1965, *Astrophys. J.*, **141**, 444.
11. Owen, T. 1965, *Astrophys. J.*, **142**, 782.
12. Giver, L. P. 1965, *Publ. astr. Soc. Pacif.*, **77**, 128.
13. Spinrad, H., Giver, L. P. 1966, *Publ. astr. Soc. Pacif.*, **78**, 175.
14. Danielson, R. E. 1966, *Astrophys. J.*, **143**, 949.
15. Murray, B. C., Wildey, R. L., Westphal, J. A. 1964, *Astrophys. J.*, **139**, 986.
16. Wildey, R. L. 1965, *Science*, **47**, 1035.
17. Wildey, R. L. 1966, *Astrophys. J.*, **144**, 1241.
18. Wildey, R. L., Murray, B. C., Westphal, J. A. 1965, *J. geophys. Res.*, **70**, 3711.

E. The Galilean Satellites of Jupiter

A. B. Binder and D. P. Cruikshank (**1**) observed an anomalous brightening of Io (in the UV) after it emerged from Jupiter's shadow. They interpreted this result as possible evidence for the presence of a tenuous atmosphere. B. C. Murray, R. L. Wildey, and J. A. Westphal (**2**) found that the observed integral brightness temperatures of Ganymede and Callisto exceeded the values expected from simple equilibrium with solar radiation. T. Owen (**3**) found no evidence of the strong 8873 Å absorption of methane in spectra of Io, Ganymede, and Callisto indicating an upper limit of 100 cm atm for the abundance of methane in possible satellite atmospheres.

V. I. Moroz (4) examined the spectrum of all four Galilean satellites in the region 0·8 to 2·5 μ. He found no evidence of atmospheric absorptions but pointed out that the reflection spectra of Io and Callisto resembled that of Mars, while the spectra of Europa and Ganymede were similar to spectra of Saturn's ring and the Martian polar caps, i.e., indicative of snow or ice covering.

A. A. Kalinyak (5) presented observations of Io, Europa, and Ganymede in the region 5800–6600 Å in which he found several new lines in addition to the solar Fraunhofer lines. However, A. B. Binder and D. P. Cruikshank (6), obtained spectra having superior resolution which did not show the lines in question.

REFERENCES

1. Binder, A. B., Cruikshank, D. P. 1964, *Icarus*, **3**, 299.
2. Murray, B. C., Wildey, R. L., Westphal, J. A. 1964, *Astrophys. J.*, **139**, 986.
3. Owen, T. 1965, *Science*, **149**, 974.
4. Moroz, V. I. 1965, *Astr. Zu.*, **42**, 1287.
5. Kalinyak, A. A. 1965, *Astr. Zu.*, **42**, 1067
6. Binder, A. B., Cruikshank, D. P. 1966, *Astrophys. J.*, **144**, 1240.

F. Uranus and Neptune

F. Low (1) measured the brightness temperature of Uranus at 20 μ and found a value of 55 ± 3°K. T. Owen (2) derived a temperature of 60 ± 15°K from an analysis of the fine structure of a methane band at 6800 Å. T. Owen (3) has identified the bands discovered by Kuiper near 7500 Å in spectra of Uranus and Neptune; they are due to methane in the atmospheres of these planets.

REFERENCES

1. Low, F. 1966, *Astrophys. J.*, **146**, 326.
2. Owen, T. 1966, *Astrophys. J.*, **146**, 611.
3. Owen, T. *Icarus* (in press).

RADIO STUDIES OF THE PLANETS

(prepared by C. H. Mayer and M. Spangler)

Radio studies of the planets in the past three years have been particularly fruitful, with such outstanding developments as the discovery from radar observations that the rotation period of Mercury is 59 days rather than the long accepted value of 88 days (1); the observational identification of the 10 cm radio emission of Venus with the solid surface (2); the very successful demonstration of a new method of radio investigation of planets by the observation of the occultation of the radio signals from the Mariner IV spacecraft by Mars (3); the discovery that the occurrence of the decameter bursts from Jupiter is dependent on the orbital position of the satellite Io (4); and the observation for the first time of radio emission of the planets Uranus (5, 19) and Neptune (6).

The papers (or summaries) presented at the Symposium on Planetary Atmospheres and Surfaces, Dorado, Puerto Rico, 24 to 27 May 1965, sponsored by the International Scientific Radio Union and the International Astronomical Union were published in a special issue of Radio Science (*NBS Journal of Research*, Section D), and references to those papers not concerned exclusively with the Moon are listed together (7–38) in the bibliography.

A number of reviews of planetary investigations by their radio emission and radar reflection properties have been published in the report period (7, 11, 17, 27, 29, 207, 208, 39–49).

Mercury

Radar observations at JPL in 1963 at 12·5 cm wavelength indicated a radar cross-section for Mercury of 5%, a surface rougher than that of Venus or Mars, and spectra consistent with an 88-day rotation period (**50, 35**). Radar observations at MIT at 23 cm wavelength in 1964 indicated a radar cross-section of 10% and a surface roughness greater than that of Venus (**51**).

In 1965, Pettengill and Dyce made radar observations at 430 MHz at Arecibo (**1, 32**) which showed that the rotation period of Mercury was not synchronous as had long been believed, but instead was 59 days direct. The shorter than synchronous rotation period for Mercury was shown to be a likely consequence of solar tidal friction with an expected range of period between 56 and 88 days (**52, 53**). It was demonstrated that the new radar rotation period could be reconciled with the reported visual observations (**54**), and further theoretical investigation indicated a stable, locked-in rotation at 2/3 the orbital period of Mercury, 58·65 days, very close to the radar period, as the most likely rotation period (**55, 56, 57, 58, 59**).

The discovery that the rotation of Mercury is not synchronous and that all parts of the planet receive direct solar radiation at some time provides an explanation for the lack of sizeable phase effect in the radio emission observed at 11 cm wavelength (**60, 19, 5**) whereas a larger phase effect is indicated by observations at 8 mm wavelength (**20, 61**). Reported observations at 3·4 mm wavelength appear to be at variance as they suggest a smaller phase effect at this wavelength (**62, 63**).

The implications of the non-synchronous rotation of Mercury with regard to the characteristics of the atmosphere are considered by Rasool *et al.* (**64**).

A test of general relativity by the effect of solar gravity on the delay times for radar echoes from the planets Mercury and Venus has been proposed (**65, 66**).

Venus

A large number of observations have again been made of the radio emission of the dark side of Venus confirming the general features of the spectrum including the small decrease of brightness temperature at wavelengths longer than 21 cm (**5, 18, 19, 67–82**). The region of the spectrum in the vicinity of the 1·35-cm water vapor resonance was examined in detail, but the results were conflicting and no definite evidence for the line was found (**17, 21, 22, 23, 83**). A number of studies were made to investigate absorption in the Venus atmosphere and clouds as the explanation of the decrease in brightness temperature at short centimeter and millimeter wavelengths and the greenhouse effect (**25, 26, 39, 84–94**).

Measurements of the brightness temperature of the illuminated hemisphere of Venus confirm an appreciable phase effect at short wavelength, but indicate a very small amplitude of phase dependence at 10 cm and 20 cm (**5, 71, 75, 95, 96, 97**). A theoretical derivation of the microwave phase effect has been made to relate the observations to planetary characteristics (**98**), and the analytical treatment of the lunar phase effect has been applied to Venus (**99**).

The polarization of the thermal radiation from a solid planetary surface has been analyzed theoretically (**100, 101, 28, 102**), and observations of the integrated radiation over the disk of Venus show a very small polarization (**2, 103, 104**).

A crucial experiment was performed by Clark and Kuz'min (**2**) at the 1964 inferior conjunction using the high-resolution of the Cal Tech interferometer at 10 cm wavelength to measure the distribution of polarization over Venus. They observed differential polarization over Venus as predicted for thermal radiation from a solid surface with a dielectric constant of 2·5, providing the first direct observational link between the radio emission of Venus and the solid surface. They also found a slightly higher brightness temperature at the limb than at the center of the un-illuminated disk, a lower brightness temperature at the pole, and a diameter for the radio disk slightly smaller than the diameter of the visible disk.

The 1·9 cm radiometer carried by the Mariner-2 spacecraft to the vicinity of Venus provided three scans across Venus with a resolution of 1/6 the diameter of the disk which indicated limb darkening (**105**).

Several attempts to reconcile the ionospheric hypothesis for the origin of the radio emission of Venus with the experimental observations have been published (**106, 107, 108, 109**). Walker and Sagan (**110**) conclude that the difficulties of explaining the required ionization are too great and that the ionosphere of Venus cannot have an important effect on the microwave spectrum.

Two investigators have reported possible correlations of the radio emission of Venus with solar activity (**69, 111**).

Alternative radiation sources to supplement the thermal radiation of the surface and atmosphere of Venus have been investigated by Vakhnin and Lebedinskiy (**112**) who speculate that a general planetary glow discharge in the atmosphere of Venus may generate radio waves, and by Plummer and Strong (**113**) who speculate that electrical discharge radiation may be observable at radio wavelengths in the atmospheres of both Venus and the Earth.

A very small radar cross-section of 1% is measured for Venus at 3·6 cm wavelength (**114**) which is only 1/10 that measured at 12·5 cm wavelength indicating the possibility of some absorption in the atnosphere of Venus. The surface of Venus appears uniformly rough at 3·6 cm, like the surface of the Moon at optical wavelength.

Radar observations of Venus at 12·5 cm wavelength made in 1962 gave a rotation period of 230 days retrograde from the day to day motion of a feature across the spectrum and 266 days retrograde from the bandwidth of the spectrum, a radar cross-section of 10% and a surface dielectric constant of 3·75, and scattering characteristics of a surface smoother than the Moon (**115, 116, 117, 118**). The Faraday rotation of the Venus echo was measured and found consistent with rotation in the terrestrial ionosphere with no evidence for extraterrestrial rotation (**119**).

Radar observations of Venus at 12·5 cm in 1964 gave a rotation period of 250 days retrograde from spectral width of the echoes (**30, 120**), but a rotation period from comparison of surface features which appeared on the spectra in both 1962 and 1964 of 244 days retrograde (**120**). The reflectivity was measured as 0·114 corresponding to a surface dielectric constant of 3·75 (**120**), and scattering theory applied to the echo spectrum shows that Venus deviates only slightly from isotropic scattering (**35**).

Radar observations in 1964 at 23 cm wavelength (**51**) gave distance measurements accurate to 1·5 km, a radar cross-section of 15% and scattering properties indicating a surface smoother than the Moon with mean slope of 8°.

Radar observations at 700 MHz in 1962 (**121**) and 1964 (**37, 122**) found a value for the astronomical unit of 149 598 000 ± 400 km, a radar cross-section of 19%, and a rotation period from the spectral width of 230 ± 25 days.

Radar measurements of Venus at 410·25 MHz in 1962 (**123**) and 1964 (**124**) gave a value of the astronomical unit of 149 596 600 ± 900 km and showed that the rotation period is between 100 and 300 days.

Radar observations at 70 cm wavelength gave a rotation period for Venus of 247 ± 5 days and indicate surface slopes about one-half the average inclination of the Moon (**29**).

A variable radar cross-section was measured at 50 MHz (**125**) with an average value of 20%, and at 38 MHz (**126**) where the average value was about 15%.

It has been suggested that the rotation period of Venus is locked at 243·61 days retrograde so that the same face of Venus is presented to Earth at successive inferior conjunctions (**127**).

The analysis of radar observations to obtain surface roughness (**128, 38**) and rotation information (**36**) has been discussed.

Mars

In 1963, radar observations at 700 MHz in Russia (**129, 205**) showed narrow band echoes suggesting smooth reflecting areas, and different radar reflectivity at different Martian longitudes with the average 7%.

Arecibo radar observations in 1965 at 430 MHz (**33**) provided measures of the radar reflectivity as a function of Martian longitude with values which varied between 3% and 13%, a tendency for the strongest echoes when dark regions of Mars were near the sub-radar point, and an especially strong reflection at Martian longitude 200 degrees.

Radar observations at JPL at 12·5 cm wavelength (**31, 130**) showed a strong narrow band echo from Martian longitude 200–210°, and a somewhat wider band echo from the region 240–250°. Analysis of these observations (**131**) indicates that the correlation of high reflectivity near dark regions may correspond to the dark regions being elevated areas like continental blocks and the bright areas being lower elevations similar by analogy to dry ocean basins.

The radio emission of Mars was observed over the wavelength range from 3·2 mm to 21·3 cm and unsuccessful measurement attempts made to 154 cm, with the following observed disk brightness temperature or upper limits of intensity: 3·2 mm, 240°K (**132**); 3·4 mm, 190°K (**63**); 8 mm, 225°K (**20, 133**); 8·6 mm, 230°K (**132**); 3·75 cm, 190°K (**134**); 6 cm, 190°K (**75**); 6 cm, 192°K–11 cm, 162°K–21·3 cm, 190°K (**5, 19, 135**); 12·5 cm, 225°K (**24**); 21·2 cm, 271°K (**71**); 70 cm, $S < 0{\cdot}024 \times 10^{-26}$ w m^{-2} Hz^{-1}–154 cm, $S < 0{\cdot}05 \times 10^{-26}$ w m^{-2} Hz^{-1} (**18, 74**). These observations describe a spectrum consistent with thermal radiation from the surface of Mars with no evidence for non-thermal radiation or for ionospheric or atmospheric effects.

The atmosphere and ionosphere of Mars have been investigated (**3**) using a new technique of observing the occultation of the radio signals of a spacecraft by a planet (**136, 137**), and a number of interpretations of the results have been presented (**138–142**). A recent analysis indicates a tenuous atmosphere with a surface molecular number density only 0·8% that of the Earth, carbon dioxide at the principal atmospheric constituent with principally atomic oxygen above 90 km, the principal ion in the main daytime ionosphere is O^+, the temperature is low about 180°K near the surface and about 80°K at ionospheric heights, and the atmosphere is confined near the planet (**143**).

Jupiter

Radar contact with Jupiter in 1963 was reported by the Russian group (**144**) at 700 MHz who found a radar reflectivity of 10%, and by JPL (**145**) who found an echo from a zone centered on Jovian longitude 32° corresponding to a reflectivity of 60%, but no echo from other longitude zones. However, JPL did not find any echoes from Jupiter in 1964 using the same apparatus (**146**), and no echoes were observed at 430 MHz in 1964 using the Arecibo radar (**33**).

It was discovered by Bigg (**4**) using the observations of the High Altitude Observatory (**147, 148**) that the probability of occurrence of the decameter radiation depends on the position of the satellite Io in its orbit. This correlation has been confirmed by other observations (**8, 149–153**), and it has been pointed out (**150, 151**) that the decameter activity of Jupiter is most likely when the longitude of the central meridian in System III is either 110° or 250°, and at the same time Io is opposite longitude 200° on Jupiter which is in the plane of the magnetic axis. An investigation of the radiation from Jupiter at 3·75 cm shows no dependence on Io (**154**). One study (**153**) suggests weaker effects on the decameter radiation by positions of the satellites Europa and Ganymede, but another (**152**) shows no dependence. Evidence is presented that the decameter emission may depend on the relative positions of the Sun, the Earth, and Jupiter, and may be related to a magnetospheric tail (**155**). No pronounced correlation with solar activity was found (**156, 157**).

Catalogs have been published of the observations of Jupiter with the swept-frequency interferometer at the High Altitude Observatory for 1960–63 (**147**), and 1964–65 (**148**). Results of the high resolution spectral observations of the decameter radiation made at the University of Helsinki including the effects of Io are available in (**158–161**). Observations of the decameter radiation at six frequencies at the University of Tasmania are presented and analyzed (**162**), and the recent decameter observations of the University of Florida are given in (**8, 163, 164**). The polarization of the decameter radiation has been investigated (**165–168**). The very short time and narrow band decameter bursts from Jupiter have been the subject of a number of investigations (**10, 169, 170**). The apparent sizes of the decameter sources on Jupiter have been measured (**9, 171**), and effects of the Earth's ionosphere in producing scintillations (**172**) and Faraday rotation (**173**) have been studied. Observations of weak decameter radiation from Jupiter (**174, 175**) suggest that there may be a steady low-level component.

A study of the variable period of rotation found from the decameter sources (**176**) indicates that the apparent rotation period varies cyclically about a constant mean value in 11·9 years, Jupiter's orbital period, and the variation may be associated with beaming of the radiation.

Mechanisms of origin of the decameter radiation are discussed in (**7, 177, 178**), and the propagation of the radiation in the source is considered in (**179**).

Observations at shorter wavelengths have given the intensity of radiation from Jupiter (or disk brightness temperature) at 3·2 mm, 4·3 mm, and 8·6 mm (**132**); 8·35 mm (**180**); 2·07 cm (**67**); 3·02 cm (**181**); 6 cm (**75**); 6·5 cm (**182**); 13 and 21 cm (**68**); 43 cm (**183, 206**); 70 cm (**184**); and 70·16 cm (**185, 186**). The spectrum between 49 and 170 cm is discussed in (**16, 187**). Anomolously intense radiation at 8·6 mm was reported (**14**).

The intensity and polarization of the radiation has been measured at 21·2 cm (**188**), and at 70 and 155 cm (**15**). The intensity, polarization and beaming has been measured at 21·5 cm (**189**). Observations of the intensity, polarization, beaming, and rotation period have been made at 10 cm (**190**) and 21·2 cm (**71**). These parameters and in addition the angular diameter of the radio source have been determined at 49 cm (**191**), and a comprehensive study of the above parameters has been made at wavelengths of 6, 10, 11, 21, 74, and 100 cm and the characteristics of the radiation belts derived (**192**).

Further information about Jupiter's radiation belts is given by the measurements of the brightness distribution of the decimeter radiation (**12**), and other observations (**193–196**). The theory of synchrotron radiation in a dipolar field has been developed (**13, 197**).

Related theoretical work is available on the thermal emission of Jupiter (**198**), the atmosphere of Jupiter (**199**), and the magnetic field (**200**).

Saturn

Measurements of the brightness temperature of Saturn confirm the upward trend with increasing wavelength: 3·2 mm, 97°K–4·3 mm, 103°K – 8·6 mm, 116°K (**132**); 8 mm, 132°K (**20, 201**); 6 cm, 190°K (**75**); 11·3 cm, 182°K (**202**); 6 cm, 179°K–11·3 cm, 196°K–21·3 cm, 303°K (**5, 19**); and 21·2 cm, 286°K (**71**).

Attempts to measure linear polarization of the radiation at 11·3 cm (**5, 19, 202**) gave negative results with an upper limit of about 6%.

Uranus and Neptune

Uranus and Neptune have been added to the list of planets studied by their radio emission. An attempt to detect Uranus at 11·3 cm (**203**) gave an upper limit to the disk brightness temperature of 320°K. Later observations provided measurements of the disk temperature of 130°K at 11·3 cm (**5, 19**), 159°K at 3·75 cm (**204**), and 220°K at 1·9 cm (**6**). Observations of Neptune at 1·9 cm gave a disk brightness temperature of 180°K.

BIBLIOGRAPHY

1. Pettengill, G. H., Dyce, R. B. 1965, 'A Radar Determination of the Rotation of the Planet Mercury', *Nature*, Lond., **206**, 1240.
2. Clark, B. G., Kuz'min, A. D. 'The Measurement of the Polarization and Brightness Distribution of Venus at 10·6-cm Wavelength', *Astrophys. J.*, **142**, 23, 1965; *Astr. Zu.*, **42**, 595, 1965; *Dokl., SSSR*, **161**, 551, 1965; *Soviet Phys. Dokl.*, **10**, 180, 1965.
3. Kliore, A., Cain, D. L., Levy, G. S., Eshleman, R. Von, Fjeldbo, G., Drake, F. D. 1965, 'Occultation Experiment: Results of the First Direct Measurement of Mars' Atmosphere and Ionosphere', *Science*, **149**, 1243.
4. Bigg, E. K. 1964, 'Influence of the Satellite Io on Jupiter's Decametric Emission', *Nature*, Lond., **203**, 1008.
5. Kellermann, K. I. 1966, 'The Thermal Radio Emission from Mercury, Venus, Mars, Saturn, and Uranus', *Icarus*, **5**, 478.
6. Kellermann, K. I., Pauliny-Toth, I. I. K. 1966, 'Observations of the Radio Emission of Uranus, Neptune, and Other Planets at 1·9 cm'. *Astrophys. J.*, **145**, 954.

Symposium on Planetary Atmospheres and Surfaces, Dorado, Puerto Rico, Special Issue, Radio Science, Vol. 69D, December, 1965.

7. Ellis, G. R. A. 'The Decametric Radio Emission of Jupiter', p. 1513.
8. Carr, T. D., Gulkis, S., Smith, A. G., May, J., Lebo, G. R., Kennedy, D. J., Bollhagen, H. 'Results of Recent Investigations of Jupiter's Decametric Radiation', p. 1530.
9. Slee, O. B., Higgins, C. S. 'Results from CSIRO, Sydney, Australia', p. 1536.
10. Warwick, J. W., Gordon, M. A. 'Frequency and Polarization Structure of Jupiter's Decametric Emission on a 10-millisecond Scale', p. 1537.
11. Roberts, J. A. 'Jupiter, as Observed at Short Radio Wavelengths', p. 1543.
12. Berge, G. L. 'An Interferometric Study of Jupiter at 10 and 21 cm', p. 1552.
13. Thorne, K. S. 'Dependence of Jupiter's Decimeter Radiation on the Electron Distribution in its Van Allen Belts', p. 1557.
14. Gibson, J. E. 'Observations of Jupiter at 8·6 mm', p. 1560.
15. Kazès, I. 'Simultaneous Observations of Jupiter on Three Frequencies', p. 1561.
16. Barber, D., Gower, J. F. R. 'A Report of Measurements', p. 1563.
17. Barrett, A. H. 'Passive Radio Observation of Mercury, Venus, Mars, Saturn, and Uranus', p. 1565.
18. Hardebeck, H. E. 'Mars and Venus at 70-cm Wavelength', p. 1573.
19. Kellermann, K. I. 'Radio Observations of Mercury, Venus, Mars, Saturn, and Uranus', p. 1574.
20. Salomonovich, A. E. 'The Observations of Radio Emission from the Planets Mercury, Mars, and Saturn at Wavelength of 8 mm', p. 1576.
21. Drake, F. D. 'A Search for the 1·36-cm Water-Vapor Line in Venus', p. 1577.
22. Gibson, J. E., Corbett, H. H. 'Radiation of Venus at the 13·5-mm Water Line', p. 1577.
23. Welch, W. J. 'Observations of the 1·35-cm Water-Vapor Line in Venus', p. 1580.
24. Muhleman, D. O., Sato, T. 'Observations of Mars at 12·5-cm Wavelength, p. 1580.
25. Basharinov, A. E., Kutuza, B. G. 'On the Nature of the Cloud Layer of Venus (From Radiometric Observations at Microwave)', p. 1580.
26. Sagan, C., Pollack, J. B. 'An Analysis of Microwave Observations of Venus', p. 1583.
27. Troitsky, V. S. 'Investigation of the Surfaces of the Moon and Planets by the Thermal Radiation', p. 1585.
28. Hagfors, T., Moriello J. 'The Effect of Roughness on the Polarization of Thermal Emission from a Surface', p. 1614.
29. Pettengill, G. H. 'A Review of Radar Studies of Planetary Surfaces', p. 1617.
30. Goldstein, R. M. 'Preliminary Venus Radar Results', p. 1623.
31. Goldstein, R. M. 'Preliminary Mars Radar Results', p. 1625.
32. Pettengill, G. H. 'Recent Arecibo Observations of Mercury', p. 1627.
33. Dyce, R. B. 'Recent Arecibo Observations of Mars and Jupiter', p. 1628.
34. Sagan, C., Pollack, J. B. 'Radio Evidence on the Structure and Composition of the Martian Surface', p. 1629.

35. Muhleman, D. O. 'Radar Scattering from Venus and Mercury at 12·5 cm', p. 1630.
36. Shapiro, I. I. 'Application of Planetary Measurements to Planetary Radius and Rotation Rate Determinations', p. 1632.
37. Kotelnikov, V. A. 'Radar Observations of Venus in the Soviet Union in 1964', p. 1634.
38. Beckmann, P., Klemperer, W. K. 'Interpretation of the Angular Dependence of Backscattering from the Moon and Venus', p. 1669.

39. Barrett, A. H., Staelin, D. H. 1964, 'Radio Observations of Venus and the Interpretations' *Space Sci. Rev.*, **3**, p. 109, D. Reidel Publ. Co., Dordrecht-Holland.
40. Douglas, J. N. 'Decametric Radiation from Jupiter', *IEEE Trans.*, **MIL-8**, 173, 1964; *IEEE Trans.*, **AP-12**, 839, 1964.
41. Goldstein, R. M. 'Radar Investigations of the Planets', *IEEE Trans.*, **MIL-8**, 199, 1964; *IEEE Trans.* **AP-12**, 865, 1964.
42. Haddock, F. T. 1966, 'Radar and Radio Studies of the Moon and the Planets: Recent Progress (1960–63)', *Progress in Radio Science* 1960–63, Vol. V, *Radio Astronomy*, p. 86 E. Herbays *et al.*, editors, Elsevier Publ. Co., Amsterdam.
43. Mayer, C. H. 'Thermal Radio Radiation from the Moon and Planets', *IEEE Trans.*, **MIL-8**, 236, 1964; *IEEE Trans.*, **AP-12**, 902, 1964.
44. Mayer, C. H. 1966, 'Radioastronomy Studies of Venus and Mars', *Astronaut. Aeronaut.*, **4**, 13.
45. Pettengill, G. 1964, 'Spectral Measurement Techniques in Planetary Radar', *Radio Science*, **68D**, 1025.
46. Pettengill, G. H., Shapiro, I. I. 1965, 'Radar Astronomy', *A. Rev. Astr. Astrophys.*, **3**, 377.
47. Smith, A. G., Carr, T. D. 1964, *Radio Exploration of the Planetary System*, D. Van Nostrand, Princeton.
48. Thomson, H. H. 1963, 'Planetary Radar', *Q. J. R. astr. Soc.*, **4**, 347.
49. Warwick, J. W. 1964, 'Radio Emission from Jupiter', *A. Rev. Astr. Astrophys.*, **2**, 1.
50. Carpenter, R. L., Goldstein, R. M. 1963, 'Radar Observations of Mercury', *Science*, **142**, 381.
51. Evans, J. V., Brockelman, R. A., Henry, J. C., Hyde, G. M., Kraft, L. G., Reid, W. A., Smith, W. W. 1965, 'Radio Echo Observations of Venus and Mercury at 23 cm Wavelength', *Astr. J.*, **70**, 486.
52. Peale, S. J., Gold, T. 1965, 'Rotation of the Planet Mercury', *Nature*, Lond., **206**, 1240.
53. Goldreich, P. 1965, 'Tidal De-Spin of Planets and Satellites', *Nature*, Lond., **208**, 375.
54. McGovern, W. E., Gross, S. H., Rasool, S. I. 1965, 'Rotation Period of the Planet Mercury', *Nature*, Lond., **208**, 375.
55. Colombo, G. 1965, 'Rotational Period of the Planet Mercury', *Nature*, Lond., **208**, 575.
56. Liu, H.-S., O'Keefe, J. A. 1965, 'Theory of Rotation for the Planet Mercury', *Science*, **150**, 1717.
57. Laslett, L. J., Sessler, A. M. 1966, 'Rotation of Mercury: Theoretical Analysis of the Dynamics of a Rigid Ellipsoidal Planet', *Science*, **151**, 1384.
58. Jefferys, W. H. 1966, 'Rotation of the Planet Mercury', *Science*, **152**, 201.
59. Colombo, G., Shapiro, I. I. 1966, 'The Rotation of the Planet Mercury', *Astrophys. J.*, **145**, 296.
60. Kellermann, K. I. 1965, '11-cm Observations of the Temperature of Mercury', *Nature*, Lond., **205**, 1091.
61. Kutuza, B. G., Lossovsky, B. Ya., Salomonovich, A. E. 1965, *Astr. Cirk.*, no. 327.
62. Epstein, E. E. 1966, 'Mercury: Anomalous Absence from the 3·4 Millimeter Radio Emission of Variation with Phase', *Science*, **151**, 445.
63. Epstein, E. E. 1966, 'Disk Temperature of Mercury and Mars', *Astrophys. J.*, **143**, 597.
64. Rasool, S. I., Gross, S. H., McGovern, W. E. 1966, 'The Atmosphere of Mercury', *Space Sci. Rev.*, **5**, 565.
65. Shapiro, I. I. 1964, 'Fourth Test of General Relativity', *Phys. Rev. Letters*, **13**, 789.
66. Shapiro, I. I. 1966, 'Testing General Relativity with Radar', *Phys. Rev.*, **141**, 1219.

67. Baars, J. W. M., Mezger, P. G., Wendker, H. 1965, 'The Flux Density of the Strongest Thermal Radio Sources at the Frequency of 14·5 GHz', *Z. Astrophys.*, **61**, 134.

68. Boischot, A., Ginat, M., Kazès, I. 1963, 'Observations du Rayonnement des Planètes Vénus et Jupiter sur 13 cm et 21 cm de Longueur d'Onde', *Ann. Astrophys.*, **26**, 385.

69. Copeland, J. 1966, 'Observations of Venus at 8·6 mm Wavelength Near Inferior Conjunction', *Astrophys. J.*, **143**, 996.

70. Copeland, J., Tyler, W. C. 1964, 'Preliminary Results from Measurements of 8·6 mm Radiation from Venus', *Astrophys. J.*, **139**, 409.

71. Davies, R. D., Williams, D. 1966, 'Observations of the Continuum Emission from Venus, Mars, Jupiter, and Saturn at 21·2 cm Wavelength', *Planet. Space Sci.*, **14**, 15.

72. Drake, F. D. 1964, 'Microwave Observations of Venus, 1962–63', *Astr. J.*, **69**, 62.

73. Epstein, Eugene E. 1965, '3·3-mm Observations of Venus in 1964', *Astr. J.*, **70**, 721.

74. Hardebeck, H. E. 1965, 'Radiometric Observations of Venus and Mars at 430 Mc/s', *Astrophys. J.*, **142**, 1696.

75. Hughes, M. P. 1966, 'Planetary Observations at Wavelength of 6 cm', *Planet. Space Sci.*, **14**, 1017.

76. Lynn, V. L., Meeks, M. L., Sohigian, M. D. 1964, 'Observations of Venus, the Region of Taurus A, and the Moon at 8·5-Millimeter Wavelength', *Astr. J.*, **69**, 65.

77. McCullough, T. P., Boland, J. W. 1964, 'Observations of Venus at 2·07 cm', *Astr. J.*, **69**, 68.

78. Staelin, D. H., Barrett, A. H., Kusse, B. R. 1964, 'Observations of Venus, the Sun, Moon, and Tau A at 1·18-cm Wavelength', *Astr. J.*, **69**, 69.

79. Thornton, D. D., Welch, W. J. 1964, 'Radio Emission from Venus at 8·35 mm', *Astr. J.*, **69**, 71.

80. Tolbert, C. W., Straiton, A. W. 1964, '35-Gc/s, 70-Gc/s and 94-Gc/s Cytherean Radiation', *Nature*, Lond., **204**, 1242.

81. Tolbert, C. W., Straiton, A. W. 1964, 'An Investigation of 35 Gc, 70 Gc, and 94 Gc Cytherean Radiation', *Univ. Texas Elec. Engr. Res. Lab. Rpt.* no. 1 – 02.

82. Vetukhnovskaya, Yu. N., Kuz'min, A. D., Kutuza, B. G., Losovskiy, B. Ya., Salomonovich, A. E. 1963, 'Measurements of the Spectrum of Radio Emission of the Night Side of Venus in the Centimeter Band', *Izv. Vyssih Učeb. Zaved., Radiofiz.*, **6**, 1054.

83. Staelin, D. H., Barrett, A. H. 1966, 'Spectral Observations of Venus near 1-Centimeter Wavelength', *Astrophys. J.*, **144**, 352.

84. Basharinov, A. E., Kutuza, B. G. 'The Nature of the Cloud Layer on Venus from Microwave Observations', *Astr. Zu.*, **43**, 149, 1966; *Soviet Astr.*, **10**, 117, 1966.

85. Deirmendjian D. 1964, 'A Water Cloud Interpretation of Venus' Microwave Continuum', *Icarus*, **3**, 109.

86. Ho, W., Kaufman, I. A., Thaddeus, P. 1966, 'Laboratory Measurement of Microwave Absorption in Models of the Atmosphere of Venus', *J. geophys. Res.*, **71**, 5091.

87. Kuz'min, A. D. 1965, 'On the Theory of Venus Radio Emission', *Izv. Vyssih Učeb. Zaved., Radiofiz.*, **8**, 7.

88. Kuz'min, A. D. 1964, 'The Venus Model with a "Cold" Absorbing Atmosphere', *Izv. Vyssih Učeb. Zaved., Radiofiz.*, **7**, 1021.

89. Ohring, G., Mariano, J. 1964, 'The Effect of Cloudiness on a Greenhouse Model of the Venus Atmosphere', *J. geophys. Res.*, **69**, 165.

90. Rasool, S. I., Jastrow, R. 1964, 'The Atmospheres of Mars, Venus and Jupiter', *Life Sciences and Space Research II*, M. Florkin and A. Dollfus, editors, p. 3, North Holland Publ. Co., Amsterdam.

91. Salomonovich, A. E. 1964, 'Concerning the Detection of Water in the Atmosphere of Venus', *Izv. Vyssih Učeb. Zaved., Radiofiz.*, **7**, 51.

92. Strelkov, G. M. 1965, 'Effect of the Venusian Cloud Cover on its Radio Emission in the Centimeter and Millimeter Wave Bands', *Issledovaniye Atmosfer Venery i Marsa*, p. 21, Izdatel'stvo Naukova Dumka, Kiev.

93. Thaddeus, P. 1964, 'Recent Space Probe and Earth-Based Studies of Mars and Venus', *J. British Interplanet. Soc.*, 1963–64, **19**, 419.

94. Thaddeus, P. 1965, 'The Atmosphere of Venus', *Unmanned Exploration of the Solar System, Advances in the Astronautical Sciences,* Vol. 19, G. W. Morgenthaler and R. G. Morra, editors, p. 201, American Astronautical Soc. Publ.
95. Basharinov, A. E., Vetukhnovskaya, Yu. N., Kuz'min, A. D., Kutuza, B. G., Salomonovich, A. E. 'Measurements of the Brightness Temperature of Venus at 8 mm', *Astr. Zu.*, **41**, 707, 1964; *Soviet Astr.*, **8**, 563, 1965.
96. Dickel, J. R. 1966, 'Measurement of the Temperature of Venus at a Wavelength of 3·75 cm for a Full Cycle of Planetary Phase Angles', *Icarus*, **5**, 305.
97. Kuz'min, A. D. 'Measurements of the Brightness Temperature of the Illuminated Side of Venus at 10·6 cm', *Astr. Zu.*, **42**, 1281, 1965; *Soviet Astr.*, **9**, 995, 1966.
98. Pollack, J. B., Sagan, C. 1965, 'The Microwave Phase Effect of Venus', *Icarus*, **4**, 62.
99. Troitskiy, V. S. 1964, 'The Theory of Radio Emission of Venus and Mars', *Izv. Vyssih Učeb. Zaved., Radiofiz.*, **7**, 208.
100. Soboleva, N. S., Parijsky, Yu. N. 'The Possibility of Observing the Polarization of Thermal Radio Emission of Planets', *Astr. Zu.*, **41**, 363, 1964; *Soviet Astr.*, **8**, 282, 1964.
101. Heiles, C. E., Drake, F. D. 1963, 'The Polarization and Intensity of Thermal Radiation from a Planetary Surface', *Icarus*, **2**, 281.
102. Pollack, J. B., Sagan, C. 1965, 'Polarization of Thermal Emission from Venus', *Astrophys. J.*, **141**, 1161.
103. Seielstad, G. A., Morris, D., Radhakrishan, V. 1963, 'Linear Polarization Measurements of the 10·6-cm Wavelength Radiation from Small-Diameter Radio Sources', *Astrophys J.*, **138**, 602.
104. Kuz'min, A. D., Dent, W. 1966, 'Measurements of the Brightness Temperature and Polarization of the Radio Emission of Venus at 3·75 cm', *Astr. Zu.*, **43**, 692.
105. Barath, F. T., Barrett, A. H., Copeland, J., Jones, D. E., Lilley, A. E. 1964, 'Mariner 2 Microwave Radiometer Experiment and Results', *Astr. J.*, **69**, 49.
106. Danilov, A. D. 1964, 'Radioastronomical Investigations and Modern Concepts Concerning the Venusian Atmosphere', *Kosmichiskie Issledovaniya*, **2**, 121.
107. Danilov, A. D. 'Radioastronomical Studies and the Ionosphere of Venus', *Dokl., SSSR*, **162**, 774, 1965; *Soviet Phys. Dokl.*, **10**, 483, 1965.
108. Danilov, A. D., Yatsenko, S. P. 'The Ionospheric Interpretation of the Results of Radar Observations of Venus – Parts I and II', *Geomagn. i Aeronom.*, **3**, 585, 1963; *Geomagn. and Aeronomy*, **3**, 475, 1963.
109. Kuz'min, A. D. 1963, 'Concerning the Ionospheric Model of Venus', *Izv. Vyssih Učeb. Zaved., Radiofiz.*, **6**, 1090.
110. Walker, R. G., Sagan, C. 1966, 'The Ionospheric Model of the Venus Microwave Emission: An Obituary', *Icarus*, **5**, 105.
111. Letfus, V. 1966, 'Radio Emission of Venus at 3 cm and the Solar Activity', *Nature*, Lond., **211**, 176.
112. Vakhnin, V. M., Lebedinskiy, A. I. 1965, 'The Nature of Radio Noise Emission from the Surface of Venus', *Kosmischeskiye Issledovaniya, Izdatel'stov Nauka*, **3**, 917.
113. Plummer, W., Strong, J. 1966, 'A New Estimate of the Surface Temperature of Venus', *Astrophys. J.*, **144**, 422.
114. Karp, D., Morrow, W. E., Jr., Smith, W. B. 1964, 'Radar Observations of Venus at 3·6 Centimeters', *Icarus*, **3**, 473.
115. Carpenter, R. L. 1964, 'Study of Venus by CW Radar', *Astr. J.*, **69**, 2.
116. Goldstein, R. M. 1964, 'Venus Characteristics by Earth-Based Radar', *Astr. J.*, **69**, 12.
117. Levy, G. S., Schuster, D. 1964, 'Further Venus Radar Depolarization Experiments', *Astr. J.*, **69**, 29.
118. Muhleman, D. O. 1964, 'Radar Scattering from Venus and the Moon', *Astr. J.*, **69**, **34**.
119. Schuster, D., Levy, G. S. 1964, 'Faraday Rotation of Venus Radar Echoes', *Astr. J.*, **69**, 42.
120. Carpenter, R. L. 1966, 'Study of Venus by c w Radar – 1964 Results', *Astr. J.*, **71**, 142.
121. Kotel'nikov, B. A., Dubrovin, V. M., Dubinskii, V. A., Kislik, M. D., Kusnetsov, B. I., Lishin, I. V., Morosov, V. A., Petrov, G. M., Rzhiga, O. N., Sytsko, G. A., Shakhovskoi,

A. M. 'Radar Observations of Venus in the Soviet Union in 1962', *Dokl., SSSR,* **151,** 532, 1963; *Soviet Phys. Dokl.,* **8,** 642, 1964.

122. Kotel'nikov, V. A., Aleksandrov, Yu. N., Apraksin, L. V., Dubrovin, V. M., Kislik, M. D., Kuznetsov, B. I., Petrov, G. M., Rzhiga, O. N., Frantsesson, A. V., Shakhovskoi, A. M. 'Radar Observations of Venus in the Soviet Union in 1964', *Dokl., SSSR,* **163,** 50, 1965; *Soviet Phys. Dokl.,* **10,** 578, 1966.

123. Ponsonby, J. E. B., Thomson, J. H., Imrie, K. S. 1964, 'Radar Observations of Venus and a Determination of the Astronomical Unit, 1962', *Mon. Not. R. astr. Soc.,* **128,** 1.

124. Ponsonby, J. E. B., Thomson, J. H., Imrie, K. S. 1964, 'Rotation Rate of Venus Measured by Radar Observations, 1964', *Nature,* Lond., **204,** 63.

125. Klemperer, W. K., Ochs, G. R., Bowles, K. L. 1964, 'Radar Echoes from Venus at 50 Mc/sec', *Astr. J.,* **69,** 22.

126. James, J. C., Ingalis, R. P. 1964, 'Radar Observations of Venus at 38 Mc/sec', *Astr. J.,* **69,** 19.

127. Goldreich, P., Peale, S. J. 1966, 'Resonant Rotation for Venus?', *Nature,* Lond., **209,** 1117.

128. Hagfors, T. 1966, 'Relationship of Geometric Optics and Autocorrelation Approaches to the Analysis of Lunar and Planetary Radar', *J. geophys. Res.,* **71,** 379.

129. Kotel'nikov, V. A., Dubrovin, V. M., Dubinskii, B. A., Kislik, M. D., Kuznetsov, B. I., Petrov, G. M., Rabotyagov, A. P., Rzhiga, O. N., Shakhovskoi, A. M. 1964, 'Radar Studies of the Planet Mars in the Soviet Union', *Dokl., SSSR,* **151,** 811, 1963; *Soviet Phys. Dokl.,* **8,** 760.

130. Goldstein, R. M. 1965, 'Mars: Radar Observations', *Science,* **150,** 1715.

131. Sagan, C., Pollack, J. B., Goldstein, R. M. 1966, 'Radar Doppler Spectroscopy of Mars, I. Elevation Differences Between Bright and Dark Areas', *Smithson. astrophys. Obs. Special Rpt.* no. 221.

132. Tolbert, C. W. 1966, 'Observed Millimeter Wavelength Brightness Temperatures of Mars, Jupiter, and Saturn', *Astr. J.,* **71,** 30.

133. Kutuza, B. G., Losovskii, B. Ya., Salomonovich, A. E. 'Observations of the Radio Emission of Mars at 8 mm', *Astr. Zu.,* **43,** 236, 1966; *Soviet Astr.,* **10,** 190, 1966.

134. Dent, W. A., Klein, M. J., Aller, H. D. 1965, 'Measurements of Mars at λ 3·75 CM from February to June, 1965', *Astrophys J.,* **142,** 1685.

135. Kellermann, K. I. 1965, 'Radio Observations of Mars', *Nature,* Lond., **206,** 1034.

136. Fjeldbo, G., Eshleman, V. R. 1965, 'The Bistatic Radar-Occultation Method for the Study of Planetary Atmospheres', *J. geophys. Res.,* **70,** 3217.

137. Fjeldbo, G., Eshleman, V. R., Garriott, O. K., Smith, F. L. III. 1965, 'The Two-Frequency Bistatic Radar-Occultation Method for the Study of the Planetary Ionospheres', *J. geophys. Res.,* **70,** 3701.

138. Atchley, D. W. 1966, 'Atmosphere of Mars: Mariner IV Models Compared', *Science,* **153,** 1518.

139. Chamberlain, J. W., McElroy, M. B. 1966, 'Martian Atmosphere: The Mariner Occultation Experiment', *Science,* **152,** 21.

140. Donahue, T. M. 1966, 'Upper Atmosphere and Ionosphere of Mars', *Science,* **152,** 763.

141. Gross, S. H., McGovern,. W. E., Rasool, S. I. 1966, 'Mars: Upper Atmosphere', *Science,* **151,** 1216.

142. Johnson, F. S. 1965, 'Atmosphere of Mars', *Science,* **150,** 1445.

143. Fjeldbo, G., Fjeldbo, W. C., Eshleman, Von R. 1966, 'Models for the Atmosphere of Mars Based on the Mariner 4 Occultation Experiment', *J. geophys. Res.,* **71,** 2307.

144. Kotel'nikov, V. A., Apraksin, L. V., Dubrpvin, V. M., Kislik, M. D., Kuznetsov, B. I., Petrov, G. M., Rzhiga, O. N., Frantsesson, A. V., Shakhovskoi, A. M. 'Radar Observations of the Planet Jupiter', *Dokl., SSSR,* **155,** 1037, 1964; *Soviet Phys. Dokl.,* **9,** 250, 1964.

145. Goldstein, R. M. 1964, 'Radar Observations of Jupiter', *Science,* **144,** 842.

146. Goldstein, R. M. 1965, 'Discussion', *Radio Science,* **69D,** 1629.

147. Warwick, J. W., Kreiss, W. T. 1964, 'Observations of Jupiter's Sporadic Radio Emission in the Range 7·6–41 Mc/s for the Years 1960–63', *High Alt. Obs. IGY Solar Activity Rpt.* no. 28.

148. Warwick, J. W., Dulk, G. A. 1965, 'Observation of Jupiter's Sporadic Radio Emission in the Range 7·6–41 Mc/s January 1964 through May 1965', *High Alt. Obs. IGY Solar Activity Rpt.* no. 32.

149. Alexander, J. K., Stone, R. G. 1965, 'Influence of Satellite Io on the 26·3-Mc/s Emission from Jupiter', *Astrophys. J.*, **141**, 1587.

150. Dulk, G. A. 1965, 'Io-Related Radio Emission from Jupiter', *Science*, **148**, 1585.

151. Duncan, R. A. 1965, 'Modulation of Jupiter Decametric Emission by the Satellite Io', *Planet. Space Sci.*, **13**, 997.

152. Duncan, R. A. 1966, 'Comments on the Modulation of Jovian Decametric Emission by Jupiter's Satellites', *Planet. Space Sci.*, **14**, 173.

153. Lebo, G. R., Smith, A. G., Carr, T. D. 1965, 'Jupiter's Decametric Emission Correlated with the Longitudes of the First Three Galilean Satellites', *Science*, **148**, 1724.

154. Dickel, J. R. 1965, 'Lack of Effect of Io on Jovian 3·75-cm Emission', *Nature*, Lond., **206**, 1241.

155. Gruber, G. M. 1965, 'Possible Contribution of Jupiter's Magnetrospheric Tail to the Radio Emissions of the Planet in the Decametric Region', *Nature*, Lond., **208**, 1271.

156. Barrow, C. H., Resch, G. M., Hyde, F. W., Gruber, G. M., Bosch, M. C. 1964, 'Spaced-Site Observations of Jupiter in 1963', *Nature*, Lond., **204**, 637.

157. Gledhill, J. A., Gruber, G. M., Bosch, M. C. 1963, 'Radio Observations of Jupiter during 1962', *Nature*, Lond., **197**, 474.

158. Riihimaa, J. J. 1964, 'High-Resolution Spectral Observations of Jupiter's Decametric Radio Emission', *Nature*, Lond., **202**, 476.

159. Riihimaa, J. J. 1964, 'Observations of Fine Structure in Jupiter's Decametric Radio Emission', *Ann. Acad. Sci. Fennicae, Series A, VI. Physica*, **156**, 1.

160. Riihimaa, J. J. 1966, 'High-Resolution Spectra of Decametric Radio Bursts from Jupiter', *Nature*, Lond., **209**, 387.

161. Riihimaa, J. J. 1966, 'High-Resolution Spectral Observations of Decametric Radio Bursts from Jupiter', *Ann. Acad. Sci. Fennicae, Series A, VI. Physica*, **206**, 1.

162. McCulloch, P. M., Ellis, G. R. A. 1966, 'Observations of Jupiter's Decametric Radio Emissions', *Planet. Space Sci.*, **14**, 347.

163. Carr, T. D., Brown, G. W., Smith, A. G., Higgins, C. S., Bollhagen, H., May, J., Levy, J. 1964, 'Spectral Distribution of the Decametric Radiation from Jupiter in 1961'. *Astrophys. J.*, **140**, 778.

164. Smith, A. G., Lebo, G. R., Six, N. F., Jr., Carr, T. D., Bollhagen, H., May, J., Levy, J. 1965, 'Decameter-Wavelength Observations of Jupiter—The Apparitions of 1961 and 1962', *Astrophys. J.*, **141**, 457.

165. Barrow, C. H. 1964, 'Polarization Observations of Jupiter at Decameter Wavelengths', *Icarus*, **3**, 66.

166. Barrow, C. H. 1964, 'Polarization of 16-Mc/s Radiation from Jupiter', *Nature*, Lond., **201**, 171.

167. Sherrill, W. M. 1965, 'Polarization of Jovian Emission at Decameter Wavelengths', *Nature*, Lond., **205**, 270.

168. Sherrill, W. M. 1965, 'Polarization Measurements of Decameter Emission from Jupiter', *Astrophys. J.*, **142**, 1171.

169. Baart, E. E., Barrow, C. H., Lee, R. T. 1966, 'Millisecond Radio Pulses from Jupiter', *Nature*, Lond., **211**, 808.

170. Olsson, C. N., Smith, A. G. 1966, 'Decametric Radio Pulses from Jupiter: Characteristics', *Science*, **153**, 289.

171. Slee, O. B., Higgins, C. S. 1966, 'The Apparent Sizes of the Jovian Decametric Radio Sources', *Austr. J. Phys.*, **19**, 167.

172. Smith, A. G., Block, W. F., Morton, W. A., Lebo, G. R., Carr, T. D., Olsson, C. N. 1966, 'Influence of the Terrestrial Environment on the Temporal and Statistical Characteristics of Jovian Decametric Radiation', *Radio Science*, **1**, 1167.

173. Warwick, J. W., Dulk, G. A. 1964, 'Faraday Rotation of Decametric Radio Emissions from Jupiter', *Science*, **145**, 380.

174. Stone, R. G., Alexander, J. K., Erickson, W. C. 1964, 'Low-Level Decameter Emissions from Jupiter', *Astrophys. J.*, **140**, 374.
175. Dulk, G. A., Clark, T. A. 1966, 'Almost-Continuous Radio Emission from Jupiter at 8·9 and 10 MHz', *Astrophys. J.*, **145**, 954.
176. Gulkis, A., Carr, T. D. 1966, 'Radio Rotation Period of Jupiter', *Science*, **154**, 257.
177. Chang, D. B. 1963, 'Amplified Whistlers as the Source of Jupiter's Decameter Radiation', *Astrophys. J.*, **138**, 1231.
178. Zheleznyakov, V. V. 'The Origin of Jovian Radio Emission', *Astr. Zu.*, **42**, 798, 1965; *Soviet Astr.*, **9**, 617, 1966.
179. Gulkis, S., Carr, T. D. 1966, 'Asymmetrical Stop Zones in Jupiter's Exosphere', *Nature*, Lond., **210**, 1104.
180. Thornton, D. D., Welch, W. J. 1963, '8·35 mm Radio Emission from Jupiter', *Icarus*, **2**, 228.
181. Korolkov, D. V., Parijsky, Yu. N., Timofeeva, G. M. 1964, 'Radioastronomical Observations of Jupiter with High Resolution', *Astr. Cirk.*, no. 283.
182. Gol'nev, V. Ya., Lipovka, N. M., Pariiskii, Yu. N. 'Observation of Radio Emission from Jupiter at a Wavelength of 6·5 cm at Pulkovo', *Dokl. SSSR*, **157**, 554, 1964; *Soviet Phys. Dokl.*, **9**, 512, 1965.
183. Rzhiga, O. N., Trunova, Z. G. 'Measurement of Jupiter's Intrinsic Decimeter-Wavelength Radiation', *Astr. Zu.*, **42**, 121, 1965; *Soviet Astr.*, **9**, 93, 1965.
184. Hardebeck, H. E. 1965, 'Radiometric Observations of Jupiter at 430 Mc/s', *Astrophys J.*, **141**, 837.
185. Krotikov, V. D., Troitskii, V. S., Tseitlin, N. M. 'Radio Emission Temperature of the Moon and Jupiter at 70·16 cm', *Astr. Zu.*, **41**, 951, 1964; *Soviet Astr.*, **8**, 761, 1965.
186. Zakharov, A. V., Krotikov, V. D., Troitskiy, V. S., Tseytlin, N. M. 1964, 'Measurement Results of the Radioemission Intensity of a Series of Discrete Sources, the Moon, and Jupiter at a Wavelength of 70·16 cm', *Izv. Vyssih Učeb. Zaved., Radiofiz.*, **7**, 553.
187. Barber, D., Gower, J. F. R. 1965, 'The Spectral Index of the Radiation from Jupiter between 178 and 610 Megacycles/Second', *Planet. Space Sci.*, **13**, 889.
188. Miller, A. C., Griffin, J. 1966, '1414 Mc/sec Jupiter Observations', *Astr. J.*, **71**, 755.
189. Gary, B. 1963, 'An Investigation of Jupiter's 1400 Mc/sec Radiation', *Astr. J.*, **68**, 568.
190. Bash, F. N., Drake, F. D., Gundermann, E., Heiles, C. E. 1964, '10-cm Observations of Jupiter, 1961–63', *Astrophys. J.*, **139**, 975.
191. Barber, D. 1966, 'The Polarization, Periodicity and Angular Diameter of the Radiation from Jupiter at 610 Mc/s', *Mon. Not. R. astr. Soc.*, **133**, 285.
192. Roberts, J. A., Komesaroff, M. M. 1965, 'Observations of Jupiter's Radio Spectrum and Polarization in the Range from 6 cm to 100 cm', *Icarus*, **4**, 127.
193. Berge, G. L., Morris, D. 1964, 'Decimeter Measurements Relating to the Possible Displacement of Jupiter's Magnetic Dipole', *Astrophys. J.*, **140**, 1330.
194. Roberts, J. A., Ekers, R. D. 1966, 'The Position of Jupiter's Van Allen Belt', *Icarus*, **5**, 149.
195. Roberts, J. A., Komesaroff, M. M. 1964, 'Evidence for Asymmetry of Jupiter's Van Allen Belt', *Nature*, Lond., **203**, 827.
196. Berge, G. L. 1965, 'Circular Polarization of Jupiter's Decimeter Radiation', *Astrophys. J.*, **142**, 1688.
197. Ortwein, N. R., Chang, D. B., Davis, L., Jr. 1966, 'Synchrotron Radiation from a Dipole Field', *Astrophys. J.*, *Suppl.*, **12**, 323.
198. Naumov, A. P., Khizhnyakov, I. P. 'Thermal Emission of Jupiter', *Astr. Zu.*, **42**, 629, 1965; *Soviet Astr.*, **9**, 480, 1965.
199. Gross, S. H., Rasool, S. I. 1964, 'The Upper Atmosphere of Jupiter', *Icarus*, **3**, 311.
200. Hide, R. 1966, 'Planetary Magnetic Fields', *Planet. Space Sci.*, **14**, 579.
201. Kutuza, B. G., Lossovsky, B. Ya., Salomonovich, A. E. 1965, '8 mm Radio Wave Emission by Saturn', *Dokl. SSSR*, **161**, 1301.
202. Davies, R. D., Beard, M., Cooper, B. F. C. 1964, 'Observation of Saturn at 11·3 Centimeters', *Phys. Rev. Letters*, **13**, 325.

203. Slee, O. B. 1964, 'A Search for Radio Emission from Uranus at 11·3 cm', *Astrophys. J.*, **140**, 823.
204. Klein, M. J., Seling, T. V. 1966, 'Radio Emission from Uranus at 8 Gc/s', *Astrophys. J.*, **146**, 599.
205. Alexandrov, Y. N., Rzhiga, O. N. 1966, 'A Comparison of the Reflection Characteristics of Mars at 40 and 12·5 cm from Radar Observations at Opposition in 1963', *Astr. Zu.*, **43**, 813.
206. Rzhiga, O. N., Slobodenyuk, G. I., Titov, V. N., Trunova, Z. G. 1965, 'A Microwave Radiometer and Measurement of the Natural Radiation of Jupiter', *Radio Engr. and Elect. Phys.*, **10**, 306.
207. Kotelnikov, V., Doubrovine, V., Kouznetsov, B., Petrov, G., Rjiga, O., Chakhovski, A. 1965, 'Etude des Planètes à l'Aide du Radar', *Onde Electrique*, **45**, 161.
208. Koz'min, A. D. 1966, 'Results of Radio Observations of Mercury, Venus, and Mars', *Uspekhi Fizich. Nauk*, **90**, 303.

PLANETARY INTERIORS

(prepared by W. C. DeMarcus)

A set of new models for Jupiter and Saturn has been published by P. J. E. Peebles (**1**). In his work the effects of temperature were included in more detail than earlier works. The hydrogen abundances derived by Peebles for these two planets are in very good agreement with previous models published by Ramsey (**2**), Fessenkov and Massevitch (**3**), DeMarcus (**4**) and Öpik (**5**). However the thermal effects, although significant only in the outermost layers, do alleviate a discrepancy in the ratio of the two accepted non-spherical terms in the external potential of Saturn. R. T. Reynolds and A. L. Summers (**6**) have published a set of models for Uranus and Neptune. In agreement with previous work, (**7, 8**) they find that hydrogen atoms (in whatever chemical form) are present only to the extent of 1/5 and 1/6 the mass of the total planet.

For the terrestrial planets, new models of the Earth have been calculated by F. Birch (**9**), R. G. McQueen, J. N. Fritz and S. P. Marsh (**10**) and S. P. Clark and A. E. Ringwood (**11**). In general these models incorporate new data from high pressure experiments and use the much more precise value of the coefficient of the second zonal harmonic derived from artificial satellite observations. Although the change in this coefficient is only about 1/3%, the density of the Earth's core is increased by roughly 5% which is in the direction and almost of the required amount to allow the assumption that the core is 'pure iron'. Detailed models of the other terrestrial planets have been published by S. Plagemann (**12**) (Mercury), D. L. Lamar (**13**) (Mars), R. A. Lyttleton (**14**) (Mars) and K. E. Bullen (**15**) discussed the interior of Mars.

Review articles on the Interiors of the Terrestrial Planets have been published by G. J. F. MacDonald (**16**) and R. L. Kovach and D. L. Anderson (**17**).

Some mention of recent observational results, presumably significant for the theory of planetary interiors in ways as yet not completely understood, may be made. Of necessity such citations can only be made in a fragmentary way. B. C. Murray and R. L. Wildey (**18**) measured the 8–14μ flux from Jupiter at the bottom of the Earth's atmosphere and reported that a 128°K black body at Jupiter's position would give the same results. F. E. Low (**19**) reported a set of monochromatic fluxes in the same region and deduced that a 132°K black-body would produce the same total 8–14μ flux but found the monochromatic fluxes gave a decidedly non-black spectral distribution. Low also measured the 8–14μ flux of Saturn and found it to agree with that of a 103°K black-body. W. J. Welch and D. D. Thornton (**20**) found an 'equivalent disk temperature' of 144°K $\pm$ 23°K at 8·35 mm wavelength. D. J. Taylor (**21**) studied the bolometric geometric albedo of Jupiter using data of essentially constant epoch. The time homogeneity of such data is rendered important by Harris (**22**) substantiation of

W. Becker's conclusion that the opposition visual magnitude of Jupiter varies with an amplitude of perhaps 0·5 magnitude. He found $p_B = 0{\cdot}28$ ($\pm$ 10%). Assuming a conventional phase function this would imply an equilibrium temperature for Jupiter, if there were no internal heat sources, of 150°K. Spectroscopic estimates of 'temperatures' are summarized and discussed by T. Owen (**23**), T. Owen and T. E. Walsh (**24**) and more recently by R. E. Danielson (**25**) who reported some new measurements at the same time. In general the spectroscopic 'temperatures' are much larger than the bolometric ones. T. E. Stecher (**26**) utilized rocket measurements of Jupiter's reflectivity in the ultraviolet and found that they could be interpreted as implying about 11 km atm of H_2 above Jupiter's Cloud Layer. By way of contrast Rank, Fink and Wiggins (**27**), by other methods, estimate 270 km atm of H_2 above Jupiter's cloud level.

The physics of planetary interiors relies heavily on the physics of phase equilibria, (e.g. fusion, polymorphism, etc.). It would be impossible to incorporate a bibliography for these fields in a report of this size. Consequently a very few works are reported either because their results were surprising or because they may serve as key references for entry to more extensive bibliographical information.

K. F. Sterrett, W. Klement and G. C. Kennedy (**28**) redetermined the melting curve of iron and warned of the risks involved in extra-polation to pressures in the few megabar range. Also E. A. Kraut and G. C. Kennedy (**29**) have formulated a new empirical fusion law which, if extrapolated, implies a finite fusion temperature at infinite pressure. This result contrasts markedly with the Simon empirical fusion curve in current use which extrapolates to infinite melting temperature at infinite pressure. As 'key' references to a vast array of other physico-chemical work one can list two papers by A. E. Ringwood (**30, 31**).

The known existence of equatorial currents on Jupiter and Saturn pose vexing problems in fluid dynamics indeed. A paper by G. Carrier (**32**) indicates some of the perhaps surprising phenomena which occur in rotating fluids with density stratification and geometrical constraints. The possibility of even more complex hydrodynamic behavior for multi-component compressed gases is mentioned by W. C. DeMarcus and R. Wildt (**33**) in the last paragraph of a speculation on the nature of Jupiter's Red Spot.

REFERENCES

1. Peebles, P. J. E. 1964, *Astrophys. J.*, **140**, 328.
2. Ramsey, W. H., Miles, B. 1952, *Mon. Not. R. astr. Soc.*, **112**, 234.
3. Fessenkov, V. G., Massevitch, A. G. 1951, *Astr. Zu.*, **28**, 317.
4. DeMarcus, W. C. 1954, *Astr. J.*, **63**, 2.
5. Öpik, E. J. 1962, *Icarus*, **1**, 200.
6. Reynolds, R. T., Summers, A. L. 1965, *J. geophys. Res.*, **70**, 199.
7. Ramsey, W. H. 1963, *Mon. Not. R. astr. Soc.*, **125**, 469.
8. DeMarcus, W. C., Reynolds, R. T. 1962, Proc. 11th Int. Astrophysical Colloquium at Liege.
9. Birch, F. 1965, *J. geophys. Res.*, **69**, 4377.
10. McQueen, R. G., Fritz, J. N., Marsh, S. J. 1964, *J. geophys. Res.*, **69**, 2947.
11. Clark, S. P., Ringwood, A. E., 1964, *Rev. Geophys.*, **2**, 35.
12. Plagemann, S. 1965, *J. geophys. Res.*, **70**, 985.
13. Lamar, D. L. 1962–63, *Icarus*, **1**, 258.
14. Lyttleton, R. A. 1965, *Mon. Not. R. astr. Soc.*, **129**, 21.
15. Bullen, K. E. 1966, *Mon. Not. R. astr. Soc.*, **133**, 229.
16. MacDonald, G. J. F. 1963, *Space Sci. Rev.*, **2**, 473.
17. Kovach, R. L., Anderson, D. L. 1965, *J. geophys. Res.*, **70**, 2873.
18. Murray, B. C., Wildey, R. L. 1963, *Astrophys. J.*, **137**, 692.
19. Low, F. E. 1965, *Lowell Obs. Bull.*, no. 128, 184.

20. Welch, W. J., Thornton, D. D. 1965, *Astr. J.*, **70**, 149.
21. Taylor, D. J. 1965, *Icarus*, **4**, 362.
22. Harris, D. L. 1961, *The Solar System*, III: *Planets and Satellites*, Univ. of Chicago Press, Ch. 8.
23. Owen, T. 1965, *Astrophys. J.*, **142**, 782.
24. Owen, T., Walsh, T. E. 1965, *Nature*, **208**, 476.
25. Danielson, R. E. 1966, *Astrophys. J.*, **143**, 949.
26. Stecher, T. E. 1965, *Astrophys. J.*, **142**, 1186.
27. Rank, D. H., Fink, U., Wiggins, T. A. 1966, *Astrophys. J.*, **143**, 980.
28. Sterrett, K. F., Klement, W., Kennedy, G. C. 1965, *J. geophys. Res.*, **70**, 1979.
29. Kraut, E. A., Kennedy, G. C. 1966, *Phys. Rev. Letters*, **16**, 608.
30. Ringwood, A. E. 1966, *Advances in Earth Science*, P. M. Hurley (Editor), M.I.T. Press, p. 357–399.
31. Ringwood, A. E. 1966, *Advances in Earth Science*, P. M. Hurley (Editor), M.I.T. Press, p. 287–355.
32. Carrier, G. F. 1965, *J. Fluid Mech.*, **23**, 145.
33. DeMarcus, W. C., Wildt, R. 1966, *Nature*, **209**, 62.

EXOBIOLOGY

(prepared by C. Sagan)

Four recent books provide general discussions of many aspects of exobiology. A lively symposium on prebiological organic chemistry and the origin of life, with attention to areas of common interest with astronomy is *The Origins of Prebiological Systems*, edited by S. W. Fox (Academic Press, 1965). A comprehensive discussion of exobiology in general, and the possibility of life on Mars in particular, written by 66 scientists of various disciplines over two years, is *Biology and the Exploration of Mars*, edited by C. S. Pittendrigh, W. Vishniac and J. P. T. Pearman (U.S. National Academy of Sciences, Washington, D.C., 1966). A companion volume containing a selection of representative papers, including a discussion of the Mariner IV results, and an extensive bibliography is *Extraterrestrial Life: An Anthology and Bibliography*, compiled by E. A. Shneour and E. A. Ottesen (U.S. National Academy of Sciences, Washington, D.C., 1966). See also *Intelligent Life in the Universe* by I. S. Shklovskii and C. Sagan (Holden-Day, 1966).

The prevailing belief—by no means rigorously proved—is that wherever we have a non-diffuse atmosphere of approximately cosmic composition at moderate temperatures and supplied with energy sources, organic molecules are produced in such profusion that self-replicating molecular systems capable of evolution by natural selection invariably arise (see for example, Pittendrigh *et al.*, 1966). Since the primitive atmospheres of planets in the solar system are, in many theories of origin, thought to contain reduced gases, and to have moderate temperature regions and energy sources at some atmospheric level, it is possible that the origin of life and life-related organic molecules is a pervasive accompaniment to the early evolution of planetary atmospheres. To determine whether the contemporary planets support indigenous life, three general procedures can be used:

(1) Determine the physical characteristics of the planetary environment. Some conditions, for example, very high temperatures, will be so severe as to exclude the possibility of life.

(2) Simulate the environment and test the capabilities of terrestrial microorganisms to survive there. A positive result will indicate that there are no insuperable physical or chemical constraints on indigenous life, but will certainly not prove the presence of indigenous life. It will, however, underscore the need for sterilization of space vehicles intended for such planets.

(3) Search for direct evidence for indigenous extraterrestrial life.

Progress in these three topics since 1963 are treated below:

(1) *Planetary Environments*

(*a*) *Mercury*

Studies of the disk-integrated microwave emission at wavelengths less than 2 cm indicate brightness temperatures varying with phase between 210° and 360°K (Kellermann, 1967, Epstein, 1967). These results combined with the radar reflectivities and the solution of the one-dimensional equation of heat conduction, indicate that some decimeters below the surface of Mercury on both the bright and dark sides biologically moderate temperatures prevail.

(*b*) *Venus*

Identification of water vapour above the clouds of Venus has been accomplished (Bottema, Plummer and Strong, 1964; Dollfus, 1963; Belton and Hunten, 1966; Spinrad and Shawl, 1966). The upper levels of the clouds have been shown to consist of ice (Bottema, Plummer, Strong and Zander, 1965; Sagan and Pollack, 1967*a*). Temperatures at greater depths in the clouds appear to be high enough to permit the formation of supercooled water droplets (Pollack and Sagan, 1965*a*; Sagan and Pollack, 1967*a*), which in turn seem capable of explaining the microwave spectrum (Basharinov and Kutuza, 1965). This combination of water, moderate temperatures and the known carbon dioxide content makes the clouds a possible biological habitat, since the major requirements for photosynthesis are provided, the principal uncertainty is the amount of inorganic minerals supplied by convection from the surface (see Sagan, 1961). Thermodynamic equilibrium studies have suggested that no substantial quality of hydrocarbons or other organic materials are present either in the clouds or on the surface (Mueller, 1964; Lippincott *et al.*, 1967). This result is consistent with an analysis of radar and passive microwave observations (Pollack and Sagan, 1965*b*).

The possible biological interest in the surface of Venus depends on the origin of the centimeter and decimeter microwave emission. The principal alternatives to a hot surface are the ionospheric model (Jones, 1961); the glow discharge model (Vakhnin and Lebedinsky, 1966) and the cloud electrical discharge model (Tolbert and Straiton, 1962; Plummer and Strong, 1966). The ionospheric model is inconsistent with the 19 mm limb darkening found by Mariner II (Barath *et al.*, 1964); in addition it encounters very serious difficulties because of the high ionospheric electron densities required (Walker and Sagan, 1966). The glow discharge conjecture assumes charge separation between the bright and dark sides of the planet, implying a maximum in the brightness temperature at dichotomy for the disk-integrated phase observations, and at the terminator for observations which resolve the disk. These predictions are inconsistent with the observations (Pollack and Sagan, 1965*c*; Clark and Kuz'min, 1965; Pollack and Sagan, 1965*b*). The electrical discharge model in all reasonable variants predicts limb brightening at 19 mm (Pollack and Sagan, 1967*a*); this is inconsistent with the Mariner II microwave observations. Hot surface models are in agreement with these various observables. If the microwave emission does arise from the surface of Venus, then there appears to be little likelihood of indigenous life on the planet, except for a bare possibility at high elevations in the polar region.

(*c*) *Mars*

The presence of water vapor in the Martian atmosphere has been confirmed spectroscopically, and has been shown to increase in the vicinity of the vaporizing polar cap in the Martian spring. (Schorn *et al.*, 1967). Here again, water, CO_2, and light are present, as well as a range of minerals at the surface and suspended in the atmosphere. Lower pressures than had previously been estimated have been found on Mars by ground-based spectroscopy and by the Mariner IV occultation experiment (see, for example, H. Brown, *et al.*, editors, 1965). Computer studies of insolation, heat conduction and carbon dioxide phase changes on Mars suggest that the polar cap is composed largely of condensed carbon dioxide (Leighton and Murray, 1966; Leovy, 1966). Both of these studies have no direct bearing on the possibility of life. The triple point pressure of water is just under 6 mb; therefore, where the total atmospheric pressure is 6 mb or less, liquid water at any temperature is forbidden. Variations of atmospheric pressure with topography (Sagan and Pollack, 1966) and with season (Leighton and Murray, 1966) have been suggested. The lower pressures imply that solar and galactic cosmic rays arrive at the Martian surface with relatively undiminished intensity (O'Gallagher and Simpson, 1965). However, the resulting increase in mutation rate is not significant compared to the spontaneous rate in terrestrial organisms (Horowitz, 1966). The Mariner IV results seem to have established a substantial topographical heterogeneity of the Martian surface. Therefore, it can be argued that even if the average conditions are inclement for indigenous biology, there may exist microenvironments (Lederberg and Sagan, 1962) which are more favorable.

The initial results of the Mariner IV photographic mission were converted into cratering statistics and compared with similar statistics for the Moon; it was concluded that the mean age of the fraction of the Martian surface viewed was comparable to the age of the planet (Leighton, Murray, Sharp, Allen and Sloan, 1965). Since no striking examples of water erosion features were found in these photographs, it was concluded that no liquid water had been present during the history of the planet. Since liquid water would seem an essential prerequisite for the origin of life on Mars, these results seemed to imply that life never arose on Mars. However, re-evaluations of the cratering statistics (Anders and Arnold, 1965; Baldwin, 1965; Witting, Narin and Stone, 1965) have reduced the age to several hundred million years, thus permitting several billions of years for extensive liquid water on early Mars (see also Hartmann, 1966). The radar implication that the Martian bright areas are lowlands analogous to terrestrial ocean basins (Sagan, Pollack and Goldstein, 1967) does not necessarily imply the existence of early oceans on Mars. If the Martian bright areas are composed—as the photometric and polarimetric data suggest—of limonite or goethite—hydrated ferric oxides—there is a very significant amount of water weakly chemically bound to the Martian surface; it is not out of the question that this source of water could be tapped by Martian organisms. This and related exercises in hypothetical Martian biochemical ecology are treated by Vishniac *et al.* (1966). The Martian exospheric temperature implied by the Mariner IV occultation experiment (Kliore, Cain, Levy, Eshleman, Fjeldbo and Drake, 1965) seems too low to permit the escape of any significant quantities of atomic oxygen during the age of the planet. The origin of limonite or goethite seems to imply the presence of substantial amounts of liquid water on primitive Mars (Sagan, 1966), now bound to the crust. Rocket observations (Evans, 1965) have indicated that solar ultraviolet light in the 2000–3000 Å region is falling unimpeded on the Martian surface. While the solar flux at this wavelength is lethal to many varieties of terrestrial microorganisms, possible protection mechanisms—some of them involving limonite or goethite—exist, and this observation cannot be considered to exclude life on Mars (see, for example, Vishniac *et al.*, 1966).

(d) The Jovian Planets

Computer exercises on quenched thermodynamic equilibrium (Lippincott *et al.*, 1967) as well as some laboratory experiments (Ponnamperuma, 1967) indicate that high temperatures, electric discharges or ultraviolet light applied to the presumed constituents of the atmospheres of the Jovian planets produce substituted benzenes and substituted polycyclic aromatic hydrocarbons in significant yield. Some of these compounds are brightly colored, and may contribute to the observed variable coloration of the clouds of the Jovian planets. Despite the low temperatures recorded in the infrared, the 2 cm brightness temperatures of the Jovian planets appear approximately constant from Jupiter to Neptune (Kellermann and Pauliny-Toth, 1966); a circumstance explicable in terms of the microwave absorption properties and the vapor pressure curve of ammonia. These and related observations seem to imply that long wavelength thermal emission is arising from deeper and warmer layers in the atmospheres of the Jovian planets, and that biologically more clement conditions exist deep in the atmospheres of these planets, as had been envisioned some years earlier (Sagan, 1964).

(2) *Simulation Experiments*

The wide range of habitats populated by terrestrial organisms – especially microorganisms – implies that very severe conditions are required to exclude the possibility of life. In experiments simulating Martian pressures, atmospheric composition, moisture content, diurnal temperature variation and ultraviolet irradiation, every sample of terrestrial soil tested contained some varieties of microorganisms capable of surviving the simulated Martian conditions indefinitely (Packer, Scher and Sagan, 1963). When the available amount of water is increased, growth and replication of microorganisms take place (Young *et al.*, 1964). However, at pressures below 6 mb, growth of terrestrial microorganisms does not take place, as anticipated (Hawrylewicz, 1967). There seems little doubt that at least in favored microenvironments, where the water content is higher than the Martian average, terrestrial microorganisms introduced to Mars may replicate. Planet-wide dissemination of contaminants by winds in a period of the order of a few years seems possible. To avoid interference with subsequent life detection experiments, and for other reasons, sterilization of Mars-bound spacecraft is mandatory (see Chapters 25–29 in Pittendrigh *et al.*, 1966). Growth of bacteria and fungi in ammonia-methane and ammonia-hydrogen atmospheres have been demonstrated by Siegel and Giumarro (1965). These conditions are not dissimilar from conditions expected in the Jovian atmosphere below the clouds. It seems likely that terrestrial microorganisms can grow in the warmer lower regions of the atmospheres of all the Jovian planets.

(3) *Direct Evidence*

The only samples of extraterrestrial materials currently accessible for laboratory investigation for organic matter and indigenous life are the meteorites. In the carbonaceous chondrites there is unambiguous indigenous organic matter of high molecular weight. While some mass spectrometric and other similarities have been reported with biogenic organic matter, the bulk of recent evidence suggests that these molecules can be produced at thermodynamic equilibria at temperatures of a few hundred degrees C in the solar nebula. Highly structured, so-called organized elements have been found embedded in the carbonaceous chondrites. Some of these have later proved to be terrestrial contaminants; others remain of unknown origin. There are no reliable cases of growth of viable organisms which are indigenous to the carbonaceous chondrites. For a discussion of these topics, see Anders and Fitch (1962); Fitch and Anders (1963); Urey and Arnold (1966); Staplin (1965); Studier, Hyatsu and Anders (1965).

A variety of previously supposed lines of evidence for life on Mars have now been shown to have alternative explanations. The Sinton bands at 3·45, 3·58 and 3·69μ attributed by Sinton (1957, 1959) and Colthup (1961) to methyl, methylene and aldehyde groups in the Martian dark areas now appear to have other explanations. The two longer wavelength features match the spectrum of HDO (Shirk, Haseltine and Pimentel, 1965) and similar features are found in spectra of the Sun taken through the Earth's atmosphere (Rea, O'Leary and Sinton, 1965). The 3·45μ features, if real, can be explained in terms of inorganic carbonates (Rea, Belsky and Calvin, 1963). The Martian seasonal changes previously attributed to biological activity appear to be quantitatively explicable in terms of windblown dust and elevation differences on Mars (Sagan and Pollack, 1966*b*, 1966*c*) following earlier suggestions of Sharanov (1957) and Rea (1964). Alternative a biological models of the seasonal changes include frost heaving of microhills (Otterman and Bronner, 1966) and presumably reversible ultraviolet interaction with crystals in the Martian surface (Smoluchowsky, 1965). Martian secular changes can also be understood in terms of drifting and windblown dust on regions of low slope (Pollack and Sagan, 1967*b*).

An inability to detect Martian life unambiguously over interplanetary distances does not of course exclude life on Mars. Studies of the Earth by meteorological satellites have shown it is extremely difficult to detect life on our planet even with resolution substantially better than the best ground-based or space-borne photography of Mars (Kilston, Drummond and Sagan, 1966). Substantial improvements in resolution are necessary for the remote photographic detection of life on Earth (Sagan *et al.*, 1966).

Despite a fair amount of speculation, a very small amount of actual observational work has been performed in attempts to detect intelligible radio transmission from planets of other stars. Despite an occasional provocative result — such as the apparent sinusoidal variation of the emission of radio source CTA 102 at 32·5 cm, but at no other wavelengths—nothing remotely approaching an unambiguous detection of intelligible radio signals has been performed. These questions are further discussed in Cameron (1963) and by Shklovskii and Sagan (1966).

BIBLIOGRAPHY

Anders, E., Arnold, J. R. 1965, *Science*, **149**, 1494.
Anders, E., Fitch, F. W. 1962, *Science*, **138**, 1392.
Baldwin, R. B. 1965, *Science*, **149**, 1498.
Barath, F. T., Barrett, A. H., Copeland, J., Jones, D. E., Lilley, A. E. 1964, *Astr. J.*, **69**, 49.
Basharinov, A. E., Kutuza, B. G. 1965, *Radio Science*, **69D**, 1580.
Belton, M. J. S., Hunten, D. M. 1966, *Astrophys. J.*, **146**, 307.
Bottema, M., Plummer, W., Strong, J. 1964, *Ann. Astrophys.*, **28**, 225.
Bottema, M., Plummer, W., Strong, J., Zander, R. 1965, *J. geophys. Res.*, **70**, 4401.
Brown, H., *et al.*, eds. 1966, *Proc. JPL Lunar and Plan. Conf.*, JPL, Pasadena.
Cameron, A. G. W. 1963, *Interstellar Communication*, W. A. Benjamin, Inc.
Clark, B. G., Kuz'min, A. D. 1965, *Astrophys. J.*, **142**, 23.
Colthup, N. B. 1961, *Science*, **132**, 529.
Dollfus, A. 1963, *C. r. Acad. Sci.*, Paris, **256**, 3250.
Epstein, E. 1967, to be published.
Evans, D. C. 1965, *Science*, **149**, 969.
Fitch, F. W., Anders, E. 1963, *Ann. New York Acad. Sci.*, **108**, Art. 2, 495.
Hartmann, W. K. 1966, *Icarus*, **5**, 565.
Hawrylewicz, E. 1967, to be published.
Horowitz, N. H. 1966, *Science*, **151**, 789.
Jones, D. E. 1961, *Planet. Space Sci.*, **5**, 166.
Kellermann, K. 1967, to be published.
Kellermann, K., Pauliny-Toth, I. K. 1966, *Astrophys. J.*, **145**, 954.
Kilston, S. D., Drummond, R. R., Sagan, C. 1966, *Icarus*, **5**, 79.
Kliore, A., Cain, D. L., Levy, G. S., Eshleman, V. R., Fjeldbo, G., Drake, F. D. 1965, *Science*, **149**, 1243.
Lederberg, J., Sagan, C. 1962, *Proc. nat. Acad. Sci. U.S.A.*, **48**, 1473.
Leighton, R. B., Murray, B. C. 1966, *Science*, **153**, 136.
Leighton, R. B., Murray, B. C., Sharp, R. P., Allen, J. B., Sloan, R. K. 1965, *Science*, **149**, 627.
Leovy, C. 1966, *Science*, **154**, 1178.
Lippincott, E. R., Eck, R. V., Dayhoff, M. O., Sagan, C. 1967, *Astrophys. J.*, **147**, 753.
Mueller, R. R. 1964, *Icarus*, **3**, 285.
O'Gallagher, J. J., Simpson, J. A. 1965, *Science*, **149**, 1233.
Otterman, J., Bronner, F. E. 1966, *Science*, in press.
Packer, E., Scher, S., Sagan, C. 1963, *Icarus*, **2**, 293.
Plummer, W., Strong, J. 1966, *Astrophys. J.*, **144**, 422.

Pollack, J. B., Sagan, C. 1965*a*, *J. geophys. Res.*, **70**, 4403.
Pollack, J. B., Sagan, C. 1965*b*, *Icarus*, **4**, 62.
Pollack, J. B., Sagan, C. 1965*c*, *Astrophys. J.*, **141**, 1161.
Pollack, J. B., Sagan, C. 1967*a*, to be published.
Pollack, J. B., Sagan, C. 1967*b*, *Icarus*, in press.
Ponnamperuma, C. 1967, to be published.
Rea, D. G. 1964, *Nature*, **201**, 1014.
Rea, D. G., Belsky, T., Calvin, M. 1963, *Science*, **141**, 923.
Rea, D. G., O'Leary, B. T., Sinton, W. M. 1965, *Science*, **147**, 1286.
Sagan, C. 1961, *Science*, **133**, 849.
Sagan, C. 1964, *Life Sciences and Space Research II*, Florkin, M. and Dollfus, A., eds., North Holland Publishing Co., Amsterdam, p. 35.
Sagan, C. 1966*a*, *Icarus*, **5**, 102.
Sagan, C., Pollack, J. B. 1966*b*, *Smithson. astrophys. Obs. Spec. Rep.*, no. 224.
Sagan, C., Pollack, J. B. 1966*c*, *Astr. J.*, **71**, 178.
Sagan, C., Pollack, J. B. 1967, *J. geophys. Res.*, 15 January issue, in press.
Sagan, C., Pollack, J. B., Goldstein, R. M. 1967, *Astr. J.*, **72**, 20.
Sagan, C. *et al.* 1966, Chapter 9 in *Biology and the Exploration of Mars*, eds., Pittendrigh, C. S., Vishniac, W. and Pearman, J. P. T., U.S. National Academy of Sciences, Washington, D.C.
Schorn, R., *et al.* 1967, *Astrophys. J.*, in press.
Sharanov, V. V. 1957, *Priroda Planet*, Moscow.
Shirk, J. S., Haseltine, W. A., Pimentel, G. C. 1965, *Science*, **147**, 48.
Siegel, S. M., Giumarro, G. 1965, *Icarus*, **4**, 37.
Sinton, W. M. 1957, *Astrophys. J.*, **126**, 321.
Sinton, W. M. 1959, *Science*, **130**, 1234.
Smoliuchowsky, R. 1965, *Science*, **148**, 946.
Spinrad, H., Shawl, S. J. 1966, *Astrophys. J.*, **146**, 328.
Staplin, F. L. 1965, Chapter 2, *Current Aspects of Exobiology*, Mamikinian, G. and Briggs, M. A. eds., Pergamon Press, London.
Studier, M. H., Hyatsu, R., Anders, E. 1965, *Science*, **149**, 1455.
Tolbert, C. W., Straiton, A. W. 1962, *J. geophys. Res.*, **67**, 1741.
Urey, H. C., Arnold, J. R. 1966, Chapter 4, *Biology and the Exploration of Mars*, eds. Pittendrigh, C. S., Vishniac, W. and Pearman, J. P. T., U.S. National Academy of Sciences, Washington, D.C.
Vakhnin, V. M., Lebedinsky, A. I. 1966, *Zemlya i Vslennaya*, no. 1, 79.
Vishniac, W., *et al.* 1966, Chapter 11, *Biology and the Exploration of Mars*, eds. Pittendrigh, C. S., Vishniac, W. and Pearman, J. P. T. U.S. National Academy of Sciences, Washington, D.C.
Walker, R. G., Sagan, C. 1966, *Icarus*, **5**, 105.
Witting, J., Narin, F., Stone, C. A. 1965, *Science*, **149**, 1496.
Young, R. S., Deal, P. H., Bell, J., Allen, J. L. 1964, *Life Sciences and Space Research II*, Florkin, M. and Dollfus, A., eds., North Holland Publishing Co., Amsterdam, p. 105.

APPENDIX. STUDIES ON THE PHYSICS OF PLANETS IN U.S.S.R.

(prepared by D. Ya. Martynov)

This is a brief summary of the most important works published in U.S.S.R. within the period of 1963–66.

Mercury

V. I. Moroz investigated Mercury's spectrum between 1 and 13 μ. It was found that CO_2 bands at 1·6 μ are somewhat stronger than telluric bands. Possible contents of CO_2 are within limits from 0·3 up to 7 g cm^{-2}. Mercury's spectra were obtained close to perihelion, when its temperature was highest. Already in the 2–2·5 μ region thermal radiation is more intense than reflected radiation. Brightness temperature near the $2{\cdot}2\mu$ region is $650° \pm 20°$K (reduced to the sub-solar point), colour temperature in the interval of $2{\cdot}2$–$3{\cdot}4\mu$ is 670°K. Integral albedo makes up about 0·15 (1, 2, 3).

Brightness temperature of $530 \pm 130°$K on 8-mm wave, averaged over Mercury's disk and reduced to the standard distance, was obtained with the 22 meter radio-telescope RT-22 at the P. N. Lebedev Physical Institute in 1964–65.

The corresponding temperature of the sub-solar point of the planet turned out to be $T_0 = 660 + 120°$K, the temperature lapse to the terminator following the law $T = T_0 \cos^{1/4} \varphi$. On uniformly brightly illuminated part of the disk T equals $540 \pm 85°$K. The received data are in agreement with the measurements in the infra-red range and shows the possibility of interpreting the millimetre radiation as caused by heating only at the expense of solar radiation (4).

Venus

T. A. Polojentseva (Pulkovo) on the basis of photographic spectrophotometry confirmed the absorption of violet terminal of solar spectrum by Venus and drew attention to the variability of this absorption (5). N. P. Barabashov, I. L. Belkina came to the same results (8, 9).

N. A. Kozyrev (Pulkovo) received spectra of night sky of Venus.

Luminescence of atmosphere low layers was preliminarily identified with formadehyde luminescence. On 28 May 1964 he fixed a bright flash on the night part of Venus. It had a continuous spectrum which corresponded to much lower temperature than spectral energy distribution in the crescent of Venus. There are also emission lines and bands, especially in the region Hp. The full flash energy was estimated to be equal to 10^{22} erg (6).

N. P. Barabashov (Kharkov Astronomical Observatory) performed a theoretical analysis of data on brightness distribution across the disk of Venus which made it possible to estimate the value of the optical thickness of the layer situated above the cloud cover, $\tau \leqslant 0{\cdot}07$ (7).

V. I. Jezersky and V. N. Jevsjukov (Kharkov Astronomical Observatory) considered the problem of interpreting Regulus brightness attenuation curve for the polytropic model of upper atmosphere during an occulation by Venus (10).

In 1963–66 V. I. Moroz obtained and processed a large number of records of infra-red spectra of Venus. In the region of 3–4μ albedo of the planet does not exceed a few per cent. Probably in this interval a strong real absorption takes place which remains present also in the longer wavelength region. It is doubtful whether the cloud layer is composed of ice particles because ice bands are not found at 1·5 and 2μ region and they are clearly noticed in the terrestrial cirrus spectrum (11, 12, 13, 14, 15). Intensity of CO_2 bands in the region of 1–$2{\cdot}5\mu$ is considerably reduced near the lower conjunction. Changes in the bands with phase make us inevitably to reject the 'simple reflection' model in which the cloud layer is considered to be a sharp optically equivalent to the solid surface above which an absorption layer CO_2 is situated. Observations are well presented by the model supposing that CO_2 absorption bands are formed inside the cloud layer possessing the properties of homogeneous dispersing medium. Optical thickness of the atmosphere of Venus in the visible and neighbouring infra-red spectral region (up to 2μ) makes up apparently 10–30, parameter $\frac{2\pi_2}{\lambda} \approx 20$, dispersion factor is about 10^{-5} cm^{-1}, relative concentration of CO_2 is about 2% (17).

Radio-emission spectrum of Venus in the wavelength range from 4 mm to 10 cm was measured with radio-telescope RT-22 of the Physical Institute. It was proved for the first time that brightness temperature of Venus is considerably lower in the millimetre range than in the centimetre one.

To explain the received spectrum, models of Venus are worked out in FIAN and NIRFI

(**18, 19, 20, 21, 22**). They include the model with 'cold' atmosphere, according to which high-temperature radiation of the centimetre region is emitted from the planet's hot surface, but the lowering of the brightness temperature in the millimetre region, is caused by absorption and re-radiation by the planet's colder atmosphere.

Joint Soviet-American measurements made by the Physical Institute of the Academy of Sciences and the California Technological Institute discovered differencial polarization and thus proved that radio-emission from Venus in 10 cm wavelength region is really caused by the hot surface of the planet; temperature of the equatorial part at Venus' midnight accounts for 650 ± 70°K (**23**). The day-time temperature on the surface of Venus slightly differs from the night temperature. Near the planet's poles there were discovered some regions, the temperature of which is approximately by 150°K lower than the temperature of the disk's equatorial part.

It was determined that dielectric constant of the planet's surface is $\epsilon = 2{\cdot}5$ and the planet's surface radius is 6060 ± 60 km.

In the Institute of Radio-Engineering and Electronics radio-emission and absorption in water cloud formations and in rain-drops were investigated, which made it possible to determine their electric parameters (**25, 26**). These investigations gave ground for making a conclusion on the origin of the cloud layer of Venus as consisting of overcooled water drops (**27**).

In the period of 11–30 June 1964 in the Soviet Union radar observations of Venus on frequency of about 700 MHz were continued.

Distance measurement accuracy as compared with 1962 was increased by eight times and as to speed—by two times, they made up accordingly 2 km and 2·5 cm s^{-1}.

Comparison of the results of different years showed that Venus was reflecting as more smooth sphere in the observation period in 1964 than in 1962.

Reflection factor of the surface of Venus determined by full energy of the received signal averages 19%.

Measurements on determining the rotation period of Venus carried out in 1964, confirm the conclusion on reverse rotation, and the value of this period, as follows from the results of 1962 and 1964, is equal to 230 days $\pm$ 25 days (**28**).

Judged by pictures of Venus in the region 7000Å obtained in 1964 in the Crimean Astrophysical Observatory, the presence of oxygen in the atmosphere of Venus (upper-cloud layer) was confirmed; equivalent width of the most intensive lines of the band amounts to 5 mÅ. Oxygen content is estimated as not more than 150 cm-atm (**29**).

Mars

Polarimetric observation series of Mars were obtained at the time of oppositions in 1963 and 1965 (Kiev, Golosejevo). The series were received on the 28-inch (71 cm) reflector with the help of an automatic electric polarimeter in eight regions of spectrum which are cut out by light filters in spectrum interval of 355–600 mμ. Phase polarization curves were formed, they showed clearly the decrease of polarization inversion angle from 27° ($\lambda = 600$ mμ) up to 15° ($\lambda_{\text{eff}} = 355$ mμ). Analogous laboratory measurements of 16 samples of ground bed-rocks, which are close to the Martian ones by their photometric properties, showed that the inversion angle value for them does not depend on wavelength. Explaining the change of the inversion angle by the influence of the Martian atmosphere, an attempt was made to determine the optical parameters of the Martian atmosphere. The atmospheric pressure near the surface of Mars proved to be 19 ± 8 mb. Xantoderith-material of $Fe_2O_3 \cdot 2H_2O$ compound—comes closer of all to the Martian surface by its polarization properties (J. K. Koval, A. V. Morejenko) (**30, 31, 32, 33**).

Calculations were made which showed that the difference between total atmosphere pressure on Mars obtained from photographic investigations, on the one hand, and from spectroscopic ones, on the other hand, could be explained by the presence in the Martian atmosphere of a large quantity of aerosol particles with radius $r \leqslant 10^{-6}$ cm. The total quantity of aerosol particles with radius of 10^{-6} cm in the atmosphere column of 10^{15} cm^{-2} was estimated by E. G. Janovitskij (**34**).

In 1963 and 1965, V. I. Moroz measured the equivalent width of CO_2 bands in the interval of 1·1–2·5μ (**11, 35, 36, 37, 38**). Equivalent width of CO_2 bands near 1·6μ were used to estimate total pressure near the Martian surface. If we assume that the content of CO_2 in the Martian atmosphere is equal to 60 m-atm, then full pressure will be $p \approx 20 \pm 10$ mb (**37, 38**). Contents ratio $O^{18} : O^{16}$ in the Martian atmosphere in the limits of error does not differ from that of the Earth. Dependence of the Martian albedo on the wavelength is close to limonite reflecting power (albedo). Spectrum of polar capes contains details resembling snow reflecting spectrum (**37**).

Photographic comparison of spectra made at Pulkovo made it possible to establish dependence of the reflecting power of the Martian atmosphere on the wavelength with maximum at $\lambda = 5200$ Å. Comparison of snow and ice parameters with the results of laboratory researches made it possible to conclude that particles which disperse light in the Martian atmosphere are snow (**39**).

In the FIAN of the Academy of Sciences of the U.S.S.R. brightness temperature of the illuminated part of Mars was determined. The temperature turned out to be 225°K ± 10°K, which shows that the temperature of the layer responsible for millimeter emission is superior to planet's average temperature (**40**).

In the Institute of Radio-Engineering and Electronics, in February 1963 radio-location of the planet Mars was carried out on the frequency of 700 MHz. Average reflection factor of Mars' surface is 7% (**41**).

Jupiter and Saturn

At the Astrophysical Institute of the Kaz. SSR Academy of Sciences (Alma-Ata) researches were carried out concerning latitudinal distribution of methane absorption across Jupiter's and Saturn's disk. The absence of an increase in absorption towards the edges of Jupiter's disk was again confirmed and, contrary to Hess' data, a decrease in absorption towards Saturn's poles is found similar to Jupiter.

Theoretical calculations on absorption distribution were carried out with allowance for the change of absorption facts with regard to depth in the Jupiter's cloud layer. It was found that the Jovian outer atmosphere should be by approximately four times more rarefied than it is usually assumed, hence pressure on the cloud layer surface amounts to about 0·5 atm (**42, 43, 44, 45, 46**).

Photometric and spectrum properties of the Red Spot, changes in contrast with the Spot approaching the edge of the disk were studied in different wavelengths. The Red Spot by its optical properties is similar to the surrounding cloud layer and is apparently of gas-aerosol formation (**47, 48**).

On the basis of radiometric observation data on Jupiter, obtained by Murray, Wildey and others, V. G. Tejfel made an attempt to determine the value of adiabatic temperature gradient in the Jovian atmosphere. The deduced value $4{\cdot}3 \times 10^{-5}$ degree/cm is in good agreement with theoretically calculated values (**49**).

N. P. Pribojeva measured intensity distribution in spectra of three Galilean satellites of Jupiter. Colour-indices were calculated (**50**).

According to data, obtained by V. I. Moroz, strong absorption bands of Jupiter's spectrum situated in the region of 1–2·5μ are identifiable with NH_4 and CH_4, but at 2μ and 1·23μ contribution of induced H_2 absorption is possible. Analysis of intensities conducted in the framework of the model of simple reflection gives total pressure near the upper limit of the cloud layer $P = 2$ atm, the amount of $CH_4 \approx 75$ m-atm, $H_2 \leqslant 50$ km-atm (**11, 3, 51**).

Spectra of Jupiter's satellites II and III possess peculiarities which can be explained by the presence of ice or snow on their surfaces (decrease of albedo in the region of $1 > 1{\cdot}4\mu$). In the spectra of satellites I and IV these peculiarities are absent (**3, 52, 53**).

U. V. Alexandrov (Kharkov Astronomical Observatory) analysed theoretically data received by the Kharkov Astronomical Observatory in 1933–58 concerning brightness distribution across the Jovian disk for one-layer and two-layer models of atmosphere and determined the values of the optic parameters (**54**).

At the Main Astronomical Observatory (Pulkovo), inner radiation belts of Jupiter were discovered (**55**) by goniometrical observations on the wavelength 6·4 cm. It was shown that the presence of gyrotropic atmosphere or phase temperature effect leads to some polarization of the planet's integral radio-emission (**56**).

A. A. Kaliniak discovered indications of non-stable atmosphere around three Galilean satellites of Jupiter (**57**).

S. K. Vsekhsviatskij (Kiev State University) noticed an unusual activity of Jupiter in 1962–65 and explained it as a display of volcanic activities (**58, 59, 60, 61**).

At the Crimean Astrophysical Observatory considerable intensity fluctuations of the ultra-violet spectral region for equatorial and polar sections of the Jovian disk connected with periods of greater activity in its atmosphere were discovered (**62**).

At the Abastumani Astrophysical Observatory electropolarimetric investigations of the surfaces of Jupiter and Saturn were carried out in the focus of the 40 cm refractor.

Polarization of the planet's integral light is low (from 0·1% to 0·5%) and does not show any evident dependence on the phase angle. Polarization in the central and polar regions is comparatively high and it depends on the phase angle. For all this, for both planets, the southern regions show greater polarization than the northern ones (maximum degree of polarization of the southern region of Jupiter is equal to 9·4%, and of Saturn 11·7%).

Especially high polarization (up to 14–15%) is displayed by the east and west sides of the outer ring of Saturn, and in this respect the west side always dominates over the east side (**63**).

At the Gorky Scientific Research Institute (NIRFI) brightness temperatures of Jupiter and Saturn near 1·25 cm region (ammonia resonance) were estimated. The difference between the brightness temperature and the infra-red temperature of the cloud layer for Jupiter and Saturn makes up 50°K (**64**).

Plasma frequency concerning the origin of decametrical radio-emission of Jupiter, with allowance to the planet's magnetic field, was discussed. The presence of characteristics of decametric emission may be explained if the magnetic field in the planet ionosphere amount to some oersteds and the electron concentration is $3 \cdot 10^5$ electrons cm^{-3}. In particular, this value of concentration is enough to explain direction of emission owing to gyroresonance absorption (**65, 67, 68**).

The movement of ionized gas in Saturn's exosphere under the influence of a force due to its rings, which is perpendicular to the magnetic field, was considered. The character of plasma drain due to the rings, and the magnetic field distortion in case of fully or slightly ionized gas was investigated in detail. The received criteria of considerable distortion of force lines allow (when there are experimental data of orientation of the polarization plane of radio-emission of Saturn) to judge over physical conditions in the planet's exosphere as well as over possibilities of existence of Saturn's radiation belts (**66**).

In September and October 1963 in the U.S.S.R., Jupiter's radio-location on the frequency 700 MHz was carried out. Average reflection factor of Jupiter is equal to 10% (**69**).

On the basis of photometric data analysis M. S. Bobrov (Astronomical Council of the U.S.S.R. Academy of Sciences) showed that Saturn's rings present a system with thickness of many particles. Divergences in observing the phase curve and the theory of 'inter-darkening' are explained by the fact that in the inter-darkening theory dispersion of particle sizes was not taken into account; and taking this fact into account leads to a good agreement between theory and observations. It was shown that neither diffraction of separate particles nor the opposition effect of Hapke and Gerels can explain the observing factors. Problems of ring dynamics were considered (**70**).

M. S. Bobrov gave an account of a number of rare phenomena in the Saturnian system, which will take place in 1966 and worked out a programme for their world-wide patrol observations (**71**).

According to his proposal, IAU Commission 16 circulated proposals dealing with observations of Saturn at the time of the Earth's and Sun's passage across their plane and with observing the non-illuminated side of the rings.

V. I. Moroz identified with CH_4 the strong absorption bands in Saturn's spectrum in the region of 1–2·5μ. In spectra of rings obtained with the 2·6 m reflector at a highly satisfactory signal to noise ratio details typical to reflection from icy particles are very distinctly seen (**3, 51, 72**).

Saturn's brightness temperature measured at FIAN of the U.S.S.R. averaged only over its disk, equals 132 ± 9°K, what is very close to the value received by the measurements in the infra-red region. Allowance for the radio-emission of the ring allows, according to M. S. Bobrov, to adjust the received value with the latest data of measurement in the region of 8–14μ (**73**).

Laboratory and Theoretical Research. Cosmogonical Problems

In the astrophysical laboratory of Pulkovo Observatory a 97 metre tube has been installed and tested; its purpose is investigation of molecular spectra of gas absorption. Changes in equivalent width of oxygen absorption lines (Band A) were studied depending on pressure change in the range from 0·1 to 1 atm for single and triple of passage light through the tube. (Mitrofanova, L. A., Jukova, L. N., Derviz, T. E.) (**74**).

U. V. Alexandrov and V. I. Garaja (Kharkov Astronomical Observatory) made a calculation of polydispersive indicatrices of light dispersion on particles, electrical properties of which differ slightly from the properties of surrounding medium (**75**).

E. G. Janovitskij (Kiev, Golosejevo) received approximate solution of a problem on diffuse reflection and light transmission through planets' atmospheres when dispersion indicatrix is arbitrary (but not very much oblate) (**76**).

V. V. Sobolev (Astronomical Observatory of the Leningrad State University) obtained average optical characteristics of the atmosphere of Venus (dispersion indicatrix, albedo of single dispersion). Meanwhile a modern theory of light dispersion and new observations data about changes in the stellar magnitude of Venus with phase were employed (**77**).

Some works by V. V. Sobolev and I. N. Minin (**78–81**) are devoted to theoretical investigation of light dispersion in the planet's atmosphere with regard to atmosphere layers' curvature. In the paper (**78**) integral and in (**79**) approximate differential equations of the problem are considered. These equations are solved approximately by the method of averaging by angles for homogeneous (**80**) and inhomogeneous (**81**) atmospheres. The obtained results can be applied to problems connected with studying twilight phenomena and planets' luminescence near terminator.

I. N. Minin offered an approximate method of solving the problem of polarized light dispersion in planet's atmosphere taking into account aerosol component and light reflexion from the surface (**82**).

B. Ju. Levin (Institute of Physics of the Earth) made a review of works dealing with the study of the formation process of planets. He also treated the problem of changing the content of metallic iron of primary and secondary origin in planets depending on the distance towards the Sun (**84**).

V. S. Safronov (Institute of Physics of the Earth) made a critical review of modern hypotheses on the origin of the protoplanetary cloud (**85**). The problem of the origin of asteroids and dispersion of their speeds was studied (**86**) as well as the origin of thermal inhomogeneities in the primary Earth connected with the dropping of large bodies into it during its formation (**87**). From the inclinations of the planets' rotation axes it was discovered that the masses of the largest bodies fallen on planets were of the order of a thousandth of their mass (**88**).

E. L. Ruskol (Institute of Physics of the Earth) considered the problem of dissipation of a considerable part of lunar tidal energy in the layer of reduced viscosity in the depth of 100–300 km (**89**).

S. V. Kozlovskaja (Institute of Physics of the Earth) tabulated the most probable mass values and radii of planets and satellites (**90**). From the comparative analysis of the inner structure of the planets' of the terrestrial group, it was found out that the substance of Mars and Venus on the whole is slightly denser than the substance of the Earth (**91**).

S. V. Majeva (Institute of Physics of the Earth) investigated the possibility of heating-up bodies of asteroid sizes by long-lived radioactive elements (**92**). She constructed curves of temperature distribution by depth for equal time moments for Mars and the Moon (**93**).

BIBLIOGRAPHY

Mercury

1. Moroz, V. I. 1963, 'CO_2 in Atmosphere of Mercury', *Astr. Cirk.*, Moscow, no. 270.
2. Moroz, V. I. 1964, 'Infra-Red Spectrum of Mercury (λ 1·0–3·9μ)'. *Astr. Zu.*, **41**, no. 6, 1108.
3. Moroz, V. I. 1965, 'Infra-Red Spectra of Planets, Stars and Nebulae', Thesis, Moscow State University.
4. Kutuza, B. G., Losovskiy, B. J., Salomonovitch, A. E. 1965, 'Measurements of Radio-Emission of Mercury on 8-mm Wave', *Astr. Cirk.*, Moscow, no. 327, 5.

Venus

5. Polojentseva, G. A. 1964, 'About Violet Absorption in Spectrum of Venus', *Izv. glav. astr. Obs. Pulkove*, **23**, no. 175, 75.
6. Kozyrev, N. A. 1966, *Izv. glav. astr. Obs. Pulkove*, **24**, no. 180.
7. Barabashov, N. P. 1965, 'About Photometric Peculiarities of Light Reflection from the Visible Surface of Venus', *Vestnik Karkovsk. gos. Univ.*, no. 4, Ser. Astr., no. 1, 13.
8. Barabashov, N. P., Belkina, I. L. 1965, 'Spectrophotometric Investigations of the Great Dark Spot on Venus', *Vestnik Karkovsk. gos. Univ.*, no. 8, Ser. Astr., no. 2, 3.
9. Barabashov, N. P. 1966, 'Spectrophotometry of the Great Dark Spot on Venus', *Astr. Cirk.*, Moscow, no. 353.
10. Jevsjukov, N. N., Jezersky, V. I. 1965, 'About Optic Properties of the Atmosphere of Venus', *Vestnik Karkovsk. gos. Univ.*, no. 4, Ser. Astr., no. 1, 71.
11. Moroz, V. I. 1964, 'Recent Observations of Infra-Red Spectra of Planets (Venus 1–4μ, Mars 1–4μ and Jupiter 1–6μ)', *Mém. Soc. R. Sci. Liège*, 5ème série, **9**, 520.
12. Moroz, V. I. 1964, 'Infra-Red Spectra and the Problem of Physical Conditions on the Surface of Venus'. Proceedings of the XIII-th International Astronautical Congress, Varna, 1962, Springer Verlag, Wien and New York.
13. Jesipov, V. F., Moroz, V. I. 1963, 'Experiment of Spectrophotometry of Venus and Mars in the Interval of 7000–10 000 Å', *Astr. Cirk.*, Moscow, no. 262, 1.

14. Moroz, V. I. 1963, 'New Observations of Infra-Red Spectrum of Venus', *Astr. Cirk.*, Moscow, no. 262, 6.

15. Moroz, V. I. 1964, 'New Observations of Infra-Red Spectrum of Venus (λ 1·2–3·8μ)', *Astr. Zu.*, **41**, 711.

16. Moroz, V. I. 'Bands CO_2 and Some Optic Properties of Atmosphere of Venus', *Astr. Zu.* (in press).

17. Moroz, V. I. 1964, 'Phase Effect of Intensified Absorption Bands CO_2 in Spectrum of Venus', *Astr. Cirk.*, Moscow, no. 305.

18. Vetuhnovskaja, U. N., Kuzmin, A. D., Kutuza, B. G., Losovskij, B. J., Salomonovitch, A. E. 1963, 'Measurements of Radio-Emission Spectrum of Venus Night Part in centimeter Wave Range'. *Izv. Vyssih Učeb. Zaved. Radiofiz.*, **6**, 1054.

19. Troitskiy, V. S. 1964, 'To the Theory of Radio-Emission of Venus and Mars'. *Izv. Vyssih Učeb. Zaved. Radiofiz.*, **7**, 202.

20. Kuzmin, A. D. 1964, 'About Venus Model with "Cold" Absorbing Atmosphere'. *Izv. Vyssih Učeb. Zaved. Radiofiz.*, **7**, 1021.

21. Basharinov, A. E., Vetuhnovskaja, U. N., Kuzmin, A. D., Kutuza, B. G., Salomonovitch, A. E. 1964, 'Measurements of Brightness Temperature of the Illuminated Part of Venus on 8 mm Wave'. *Astr. Zu.*, **41**, 707.

22. Kuzmin, A. D. 1965, 'To the Theory of Radio-Emission of Venus'. *Izv. Vyssih Učeb. Zaved. Radiofiz.*, **8**, 7.

23. Kuzmin, A. D., Clark, B. G. 'Measurements of Polarization and Distribution of Brightness Temperatures of Venus on 10·6-cm wave'. *Dokl. Akad. Nauk SSSR*, **161**, 551, 1965; *Astr. Zu.*, **42**, 595, 1965; *Astrophys. J.*, **142**, 23, 1965.

24. Kuzmin, A. D. 1965, 'Measurements of Brightness Temperature of the Illuminated Part of Venus on 10·6-cm Wave'. *Astr. Zu.*, **42**, 1281.

25. Basharinov, A. E., Kutuza, B. G. 1965, 'Determination of Parameters of Dielectric Constant of Water Drops in Cloud Formations According to Data of Radio-Astronomical Observations in Medium Wave Range'. Report at the 5th All Union Conference on Radioastronomy. Kharkov, October.

26. Ananov, N. I., Basharinov, A. E., Kirdjashev, K. P., Kutuza, B. G. 1965, 'Fluctuation of Radio-Emission of Cloud Atmosphere in the Millimeter Wave Range'. *Radio-Engineering and Electronic Physics*, **10**, no. 11, 1941.

27. Basharinov, A. E., Kutuza, B. G. 1965, 'About the Origin of Cloud Layer of Venus (according to the results, of radio-astronomical observations)'. Report at the International Symposium on Research of Planets' Surfaces and Atmospheres by Radio-Astronomical Method. Puerto Rico, May.

28. Kotelnikov, V. A., *et al.* 1965, 'Radio-Location Observations of Venus in the Soviet Union in 1964'. Reports *Dokl. Akad. Nauk SSSR*, **163**, 50.

29. Prokofijev, V. K. 1966, 'About Presence of Oxygen in Atmosphere of Venus'. *Izv. Krym. astrofiz. Obs.*, **34**, 243, 1965; **36** (in press).

Mars

30. Morojenko, A. V. 1964, 'Results of Polarimetric Observations of Mars in 1962–63'. Collection 'Physics of the Moon and Planets'. Publishing House 'Naukova Dumka', Kiev, p. 58.

31. Morojenko, A. V. 1960, 'Polarization Properties of Atmosphere and Surface of Mars'. Collection 'Physics of the Moon and Planets', Publishing House 'Naukova Dumka', Kiev.

32. Morojenko, A. V. 1966, 'Investigation of Polarization Properties of Light, Reflected by Different Samples of Earth Bed-Rocks'. Collection 'Physics of the Moon and Planets'. Publishing House 'Naukova Dumka', Kiev.

33. Koval, I. K. 1964, 'Photoelectric Measurements of Brightness Distribution Across Disk of Mars'. *Life Sciences and Space Research*, **7**, 246.

34. Janovitskij, E. G. 1965, 'About Aerosol Component at Atmosphere of Mars'. Collection 'Astrophysical Problems', Publishing House 'Naukova Dumka'.

35. Moroz, V. I. 1963, 'Observation of Infra-Red Spectrum of Mars in the Interval of 1·1–4·1μ'. *Astr. Cirk.*, Moscow, no. 262, 4.

36. Moroz, V. I. 1963, 'Estimation of Carbon Dioxide Gas Content and Total Pressure According to Infra-Red Spectra of Venus and Mars'. *Astr. Cirk.*, Moscow, no. 273.
37. Moroz, V. I. 1964, 'Infra-Red Spectrum of Mars (λ 1·1–4·1μ)'. *Astr. Zu.*, **41**, 350.
38. Vasiltchenko, N. V., Moroz, V. I. 'Spectrum of Mars in the Region of 1·1–2·5μ (opposition of 1965)', *Astr. Zu.* (in press).
39. Kozirev, N. A. 1964, 'Spectral Indications of the Presence of Snow in the Atmosphere of Mars'. Information. *Izv. glav. astr. Obs. Pulkove*, **23**, no. 175, 72.
40. Kutuza, B. G., Losovskij, G. J., Salomonovitch, A. E. 1965, 'Observations of Radio-Emission of Mars on 8-mm Wave'. *Astr. Zu.*, (in press).
41. Kotelnikov, V. A., *et al.* 1963, 'Radio-Location of the Planet of Mars in the Soviet Union'. *Dokl. Akad. Nauk SSSR*, **151**, 811.

Jupiter and Saturn

42. Tejfel, V. G. 1964, 'About Latitudinal Distribution of Methane Absorption Across Disk of Jupiter'. *Astr. Cirk.*, Moscow, no. 296.
43. Tejfel, V. G. 1966, 'Molecular Absorption of Light in Atmosphere of Jupiter and Surface Structure of the Planet Cloud Layer'. Collection 'Research of the Planet of Jupiter', Academy of Sciences of the U.S.S.R.
44. Tejfel, V. G. 1966, 'Spectrophotometry of Methane Absorption Bands on Disk of Jupiter in the Near Infra-Red Region (0·7–1·0μ)'. *Astr. Zu.*, **43**, 154.
45. Tejfel, V. G. 'Comparison of Distribution of Methane Absorption of Disk of Jupiter and Saturn'. Works of the Astrophysical Institute of the Academy of Sciences of the Kazakh SSR (in press).
46. Tejfel, V. G. 'Distribution of Molecular Absorption of Methane across Disk of Jupiter in the Near Infra-Red Region (0·7–1·0μ)'. Works of the Astrophysical Institute of the Academy of Sciences of the Kazakh SSR (in press).
47. Tejfel, V. G. 1964, 'To the Problem of Photometric Properties of the Red Spot on Jupiter', *Astr. Zu.*, **43**, 531.
48. Tejfel, V. G. 'Preliminary Results of Spectrophotometry of the Red Spot on Jupiter'. Works of the Astrophysical Institute of the Academy of Sciences of the Kazakh SSR, (in press).
49. Tejfel, V. G. 1964, 'About Value of Temperature Gradient in beyond Cloud Layer of Atmosphere of Jupiter'. *Astr. Cirk.*, Moscow, no. 298.
50. Pribojeva, N. V. 'Continuous Spectra of Satellites of Jupiter'. Works of the Astrophysical Institute of the Academy of Sciences of the Kazakh SSR (in press).
51. Moroz, V. I. 'Spectra of Jupiter and Saturn in the Region of 1–2·5μ'. *Astr. Zu.* (in press).
52. Moroz, V. I. 1964, 'Infra-Red Spectra of Io and Ganimed. (0·7–2·5μ)'. *Astr. Cirk.*, Moscow, no. 297, 1.
53. Moroz, V. I. 1965, 'Experience of Infra-Red Spectrophotometry of Satellites; the Moon and Galileum Satellites of Jupiter', *Astr. Zu.*, **42**, 1287.
54. Alexandrov, U. V. 1965, 'Results of Surface Photometry of Jupiter'. *Vestnik Karkovsk. gos. Univ.*, no. 8, Ser. Astr., no. 2, 32.
55. Korolkov, D. V., Parijskij, U. N., Timofejeva, G. M. 1964, *Astr. Cirk.*, Moscow, no. 283.
56. Soboleva, N. S., Parijskij, U. N. 1964, *Astr. Zu.*, **41**, 362.
57. Kaliniak, A. A. 1965, 'Data on Galileum Satellites of Jupiter'. *Astr. Zu.*, **42**, 1067.
58. Vsekhsviatskij, S. K. 1965, 'Processes in the System of Jupiter'. *Astr. Cirk.*, Moscow, no. 317.
59. Vsekhsviatskij, S. K. 1965, 'About Nature of Changes on the Surface of Jupiter'. *Astr. Zu.*, **42**, 639.
60. Vsekhsviatskij, S. K. 1965, 'Changes on Jupiter'. Information Bulletin 'Geophysics and Astronomy', Kiev, no. 8, 3.
61. Vsekhsviatskij, S. K. 'Results of Observations of Jupiter in 1962–65'. Works of the Astrophysical Institute of the Academy of Sciences of the Kazakh SSR (in press).
62. Galkin, L. S. 1964, 'About Some Properties of Atmosphere of Jupiter'. *Izv. Krym. astrofiz. Obs.*, **32**, 11; **35** (in press).

63. Bolkvadze, O. R. 'Investigation of Polarization Peculiarities of Jupiter and Saturn.' Works of the Astrophysical Institute of the Academy of Sciences of the Kazakh SSR (in press).
64. Naumov, A. P., Khijniakov, I. P. 1965, 'About Thermal Radiation of Jupiter'. *Astr. Zu.*, **42**, 629.
65. Jelezniakov, V. V. 1965, 'About Generation of Radiation of Jupiter'. *Astr. Zu.*, **42**, 798.
66. Jelezniakov, V. V. 1964, 'About Configuration of Magnetic Field of Saturn'. *Astr. Zu.*, **41**, 995.
67. Zaharov, A. V., Krotikov, V. D., Troitskij, V. S., Tsejtlin, N. M. 1964, 'Results of Measurements of Radio-emission Intensity of a Number of Discrete Sources of the Moon and Jupiter on wave 70, 16 cm'. *Izv. Vyssih Učeb. Zaved. Radiofiz.*, **8**, 553.
68. Krotikov, V. D., Troitskij, V. S., Tsejtlin, N. M. 1964, 'Temperature of Radio-Emission of the Moon and Jupiter on wave 70, 16 cm'. *Astr. Zu.*, **41**, 951.
69. Kotelnikov,V. A., *et al.* 1964, 'Radiolocation of the Planet of Jupiter'. *Dokl. Akad. Nauk SSSR*, **155**, 1937.
70. Bobrov, M. S. 'Modern State of Problem on Structure and Order of Thickness of Saturn's Rings'. Works of the Astrophysical Institute of the Academy of Sciences of the Kazakh SSR (in press).
71. Bobrov, M. S. 'About Observations of Saturn's Rings in the Period of Passing the Earth and Sun through their Plane and in the Period of Visibility of Non-Illuminated Part of Rings'. *Cirk. VAGO* (in press).
72. Gretchushnikov, B. N., Moroz, V. I., Shniriov, G. D. 1964, 'Investigation of Infra-Red Spectrum of Saturn by Fourier Transformation Method'. *Astr. Cirk.*, Moscow, no. 302.
73. Kutuza, V. G., Losovskij, B. L., Salomonovitch, A. E. 1965, 'Radio-Emission of Saturn on 8-mm Wave', *Dokl. Akad. Nauk SSSR*, **161**, 1031.

Laboratory and Theoretical Research, Cosmogonical Problems

74. Mitrofanova, L. A., Jukova, A. N., Derviz, T. E. 1964, 'Installation and Testing of Optic Pipe with Considerable Way of Absorption for Investigation of Molecular Spectra'. *Izv. glav. astr. Obs. Pulkove*, **23**, no. 175, 80.
75. Alexandrov, U. V., Garaja, V. I. 1965, 'Polydispersive Indicatrices of Light Dispersion'. *Vestnik Karkovsk. gos. Univ.*, No. 4, Ser. Astr., no. 1, 91.
76. Janovitskij, E. G. 1964, 'Approximate Solution of Problem on Diffuse Reflection and Light Transmission through Planets' Atmospheres when Dispersion Indicatrix is Arbitrary'. *Vestnik Karkovsk. gos. Univ.* Collection 'Physics of the Moon and Planets', Publishing House 'Naukova Dumka', Kiev, p. 92.
77. Sobolev, V. V. 1964, 'Investigation of Atmosphere of Venus'. *Astr. Zu.*, **41**, 97.
78. Minin, I. N., Sobolev, V. V. 1963, 'To the Theory of Light Dispersion in Planets' Atmospheres'. *Astr. Zu.*, **40**, 496.
79. Sobolev, V. V., Minin, I. N. 1962, 'Dispersion of Light in Spherical Atmosphere', Collection 'Satellites of the Earth', issue 14.
80. Minin, I. N., Sobolev, V. V. 1963, 'Dispersion of Light in Spherical Atmosphere, II', *Kosmichiskie Issledovaniya*, **1**, no. 2, 227.
81. Minin, I. N., Sobolev, V. V. 1964, 'Dispersion of Light in Spherical Atmosphere, III'. *Kosmichiskie Issledovaniya*, **2**, no. 4, 610.
82. Minin, I. N. 1965, 'About Dispersion of Light in Planets' Atmospheres'. *Trudy astr. Obs. Leningrad. Univ.*, **22**, 39.
83. Levin, B. Ju. 1966, 'Accumulation of the Solar Nebula'. *International Geophysical Dictionary*. Pergamon Press.
84. Levin, B. Ju. 1964, 'The Problem of Densities and Composition of Terrestrial Planets in the Light of Modern Ideas on the Origin of Meteorites'. *Icarus*, **3**, 198.
85. Safronov, V. S. 1966, 'The Protoplanetary Cloud and its Evolution'. *Astr. Zu.*, **43**, no. 4.
86. Safronov, V. S. 1966, 'Dispersion of Speeds of Asteroids and the Problem of their Origin'. Collection 'Problems of Motion of Small Bodies of Solar System'. Publishing House of the Academy of Sciences of the Azerbaijan SSR, Baku.

87. Safronov, V. S. 1964, 'Primary Inhomogeneities of the Earth's Mantle'. *Tectonophysics*, 1(3), 217.
88. Safronov, V. S. 1965, 'Sizes of the Largest Bodies Dropped on Planets in the Process of their Formation', *Astr. Zu.*, **42**, 1270.
89. Ruskol, E. L. 1965, 'About Nature of Thermal Inhomogeneities in the Entrails of the Earth'. Information of the Academy of Sciences of the U.S.S.R., Ser. 'Physics of the Earth', No. 4, pp. 1–8.
90. Kozlovskaja, S. V. 1963, 'Masses and Radii of Planets and Satellites'. *Bjull. Inst. teor. Astr.*, **9**, 330.
91. Kozlovskaja, S. V. 1966, 'Models of Inner Formation of the Earth, Venus and Mars'. *Astr. Zu.*, **43**, no. 4.
92. Majeva, S. V. 1963, 'About Possibilities of Heating-up Bodies of Asteroid Sizes with the Help of Long-Lived Radioactive Elements'. Information of the Committee on Physics of Planets, no. 4, p. 40.
93. Majeva, S. V. 'Some Calculations of Thermal History of Mars and the Moon'. *Dokl. Akad. Nauk SSSR*, **159**, 294, 1964; The detailed description may be found in the Information of the Committee on Physics of Planets of the Astronomical Board, issue 5, 1966.

17. COMMISSION DE LA LUNE

PRÉSIDENT: Dr D. H. Menzel, Harvard College Observatory and Smithsonian Astrophysical Observatory, Cambridge, Mass. 02138, U.S.A.

VICE-PRÉSIDENTS: Dr A. Dollfus, Dr K. Kozieł.

SECRÉTAIRE: Dr B. Bell.

COMITÉ D'ORGANISATION: D. Martynov, M. G. J. Minnaert.

MEMBRES: Arthur, Ashbrook, Barabashov, Boneff, Botelheiro, Drofa, Dzapiashvili, Eckert, Ezersky, Fielder, Focas, Gold, Günther, Guth, Habibulin, Hall (R. G.), Heyden, Hirose, Hopmann, Ingrao, Jeffreys, Kopal, Kuiper, Levin, Liddel, Link, Markowitz, Murray, Nefediev, Nicholson, O'Keefe, Pettengill, Potter, Rösch, Sadler (F. M. McBain), Sagan, Sato, Schrutka-Rechtenstamm, Sharonov†, Shoemaker, Sinton, Sytinskaya, Troitsky, Ueta, Urey, Watts, Weimer, Whipple, Whitaker, Yakovkin (A. A.).

Stimulated by the successful missions of several lunar probes, the field of lunar research has developed tremendously since the formation of Commission 17 at the Hamburg Meeting of the IAU. So many papers have appeared that it becomes difficult to organize and impossible in any reasonable time to digest the literature, particularly with the variety of new techniques coming into use. By way of partial compensation we have added to the references identifying numbers (when available) by which an abstract may be found in *Physics Abstracts* (P) and/or one of the NASA Abstract publications (A, N). We also call attention to the NASA continuing bibliography (S). Abstracts for many earlier publications can be found in (U).

We are indebted to Dr E. L. Ruskol for a report on the study of the Moon in the U.S.S.R. from which we insert paragraphs in the appropriate sections below.

A. BOOKS, CONFERENCE PROCEEDINGS

In April 1965, Commission 17 of the IAU joined with NASA to sponsor a symposium on the nature of the lunar surface (**8**). In May 1966, COSPAR sponsored a symposium in Vienna on the Moon and planets (**14**). A number of other books have appeared that deal in whole (**1, 3, 4, 5, 7, 10**) or in part (**2, 9, 11**) with a variety of problems of lunar research. Many of the articles appearing in the proceedings of the conferences (**3, 8, 9**) are referenced individually in the appropriate section below. Kopal (**6**) has published an atlas of lunar photographs from plates taken at the Pic du Midi Observatory.

Scientists of the Central Astronomical Observatory of the Academy of Science of the Ukrainian SSR compiled two books (**12, 13**) on the shape and motion of the Moon, following an all-Union conference on this topic in Kiev in May 1964.

B. LUNAR PROBES

a. Hard Landings

The greatest achievement in lunar research has unquestionably been the successful missions of several Moon probes launched by the United States of America and by the Soviet Union. A preliminary report on the first successful hard-landing probe, Ranger VII on 31 July 1964, was presented at the Hamburg meetings. The data secured by Ranger VII have been published in a series of three photographic atlases (**1**) and a volume of interpretive analysis (**2**). The area selected for impact was a relatively smooth mare, known to have several ray systems and ridges and a relatively high crater density, a region since named Mare Cognitum. 'Craters are the dominant topographic features of the mare surface at all scales down to the smallest features observed in the last photographs (less than 1 m across) . . . Craters are more abundant in the rays than in the areas between the rays. The number of craters increases . . . rapidly with decreasing crater size . . . Most of the craters smaller than 250 or 300 m in diameter . . . have smoothly rounded rims . . . and a large variation in depth-to-diameter ratio . . . Some of these craters are surrounded by bright halos . . . ' (**2**).

The lunar maria can be characterized (**Ha 17**) as 'red' or 'blue' on the basis of a slight shift to the red or blue in their reflected solar spectra. Ranger VII impacted a red mare. Fo Ranger VIII the target of the blue Mare Tranquillitatis was selected, and for Ranger IX the Crater Alphonsus (**5**). The Ranger VIII and IX missions impacted their selected target areas on 20 February 1965 and 24 March 1966, and returned to earth 7137 and 5814 photographs, respectively. These are published in two atlases (**3, 4**), with a separate volume of interpretive analysis (**5**). Most of the chapters of (**5**) as well as other analyses of the Ranger photographs are referenced separately below; see especially refs. (**D 32; Ha 6, 10, 12, 15, 16; Hb 3, 5, 8, 16, 17, 22, 23, 31, 32, 37, 46, 47; Hc 2, 3, 6, 11, 13; I 15, 34, 39, 49, 52, 53**).

Other general discussions of the Ranger missions are presented in (**6, 10, 11, 13**), while (**7, 8, 9, 12, 14, 15**) emphasize the general interpretation of the new high-resolution photographs.

b. Soft Landings

The successful soft landing on the Moon by the Soviet Union's Luna 9 spacecraft on 3 February 1966, enabled earthmen for the first time to see millimeter-sized details of the lunar surface. Luna 9 landed in the western part of Oceanus Procellarum, a region in the immediate neighborhood of the broken ridge that ends southeast of the crater Cavalerius F, so that the photographs cannot with certainty be referred to either a pure mare or a pure mountain surface (**7**).

Preliminary results from Luna 9 have been published in a book (**1**) that contains a description of the apparatus, of the photogrammetric methods used for the construction of topographic charts on the landing place, and an illustrated description of its morphological features. The photographs have a resolution of 1–2 mm for the nearest details. On this scale the surface is very rough indeed, but becomes smoother for the larger scale. This conclusion confirms those reached from earth-based observations. No dust was detected. The bearing strength of the surface is sufficient for the soft landing. However the station later slightly shifted its orientation, probably because of some deformation of the lunar soil.

Gold and Hapke (**5**) interpret the shift of the instrument as evidence that the lunar subsurface may contain many cavities which are in a precarious state of equilibrium. Lipsky infers (**7**) a bearing strength of the surface of several kg/cm^2 from the fact that most of the small rocks have not noticeably depressed the surface under them. He finds 'no fairy castles, no heaps of angular fragments, sand or loose dust or like forms . . . The photographs show that the surface is not the result of fusion of deposited fine dust, for almost all the objects have regular forms, not branched or random ones. Many depressions of various sizes are clearly visible . . . with many rocklike objects' (**7**).

Davies *et al.* (**2**) discuss the signals received from Luna 9 at Jodrell Bank. Fielder *et al.* (**3**) comment on the similarity of the terrain revealed by Luna 9 to the scoriaceous surface of an aa (Hawaii) type lava flow. The resemblance to a lava flow is also noted by Kuiper *et al.* (**6**) who conclude that the surface shown by Luna 9 is a highly vesicular igneous rock, and clean of dust. Gault *et al.* (**4**) infer that the lunar surface consists of non-cohesive to weakly cohesive, poorly sorted fragmental material of unknown source and a depth of at least 20 cm, with most of the fragments having sizes less than 1 cm. Shoemaker *et al.* (**8**) note that the individual fragments could be light and frothy or dense and solid. See also (**I 32, 52**).

On 2 June 1966, the U.S. Surveyor 1 landed softly inside the Flamsteed ghost ring, about 500 miles from the site of Luna 9. By 5 June Surveyor 1 had sent back 2500 pictures, while the final picture count surpassed 10 000. 'Beautifully clear and sharp, the pictures revealed an eerie, rubble-strewn wasteland stretching level in all directions' (**10**).

Analysis and evaluation of the data returned during Surveyor's first five days on the Moon (**9**) indicate that the terrain within 1–2 km of the landing site is 'a gently rolling surface studded with craters, with diameters from a few centimeters to several hundred meters, and littered with fragmental debris ranging in size from less than 1 mm to more than 1 m. The larger craters resemble those seen in the Ranger photographs in shape and distribution . . . The surface is composed of granular material of a very wide size range; coarse blocks of rock and smaller fragments are set in a matrix of fine particles too small to be resolved. This material was disturbed and penetrated by the footpads of the spacecraft to a depth of a few cm . . .' The static bearing capacity of the soil is estimated at about 5 lb/in^2 (1/3 atm). Counts have been made of rocks as a function of size. The exterior of Surveyor 1 has remained clean and essentially dust free. A tiny nitrogen jet that was fired at the surface near one foot of the spacecraft produced no plume of dust visible to the watching camera (**9, 10**). Some material disturbed by the footpads was thrown out to form rays . . . darker than the adjacent surface. Photometric and colorimetric observations were obtained. See also (**I 21, 52**).

On 24 December 1966, the Soviet Union's Luna 13 made a successful soft landing on the Moon, and is sending back pictures of good quality. A probe rod, to test the firmness of the lunar ground, was reported unable to penetrate far into the rocky ground. Luna 13 is reported to carry also a special apparatus for measuring the density of lunar rocks.

c. Far Side of the Moon

In July 1965 the Soviet space probe Zond–3 photographed most of the area of the far side of the Moon not previously recorded by Lunik 3. The new pictures, with a resolution of about 3 km, fully confirm the earlier conclusion that the far side of the Moon has far fewer maria than the near side, and is in general lighter, more mountainous and more densely cratered. Two features not seen on the earthward side are of particular interest: chains of large (10–30 km) craters extending up to 1500 km in length; and numerous large ($\sim$ 500 km) depressions, named thalassoids, in size and shape similar to maria, but different in having crater-scattered floors and lacking the dark color of maria (**1, 4, 5**).

Measurements of the UV spectrum by a spectrophotometer (1900 to 2750Å) and by a spectrograph (2700–3550Å) gave a decrease in albedo from 5% at 3550Å to 3·5% at 3400–3100Å and to 1% at 2850Å. The albedo increases to 2% at 2400Å and decreases again to 1% at 2000Å (**2**). Infrared spectra at 3–4μ were also obtained by Zond–3.

Hartmann (**3**) discusses the origin of lunar basins and thalassoids, the history of huge concentric structures, and the origin of radial and grid systems of lineaments on the lunar surface, with reference to the structures of the far side.

d. Orbiters

On 3 April 1966 the Soviet probe Luna 10 became the first artificial satellite of the Moon, with the orbit inclined 71° 54′ to the lunar equator, with a periselenium distance of 350 km and aposelenium of 1017 km from the surface, and a period of $2^h\,58^m\,15^s$ (**1**). After 460 revolutions, on 30 May 1966, the orbital elements were 72° 2′, 378·7 km, 985·3 km, and $2^h\,58^m\,3^s$. A preliminary analysis of the orbital evolution indicated that gravitational anomalies of the Moon are small (**3**).

Preliminary results of measurements by Luna 10 gave the following magnetic field intensities (in 10^{-5} gauss)

3 April	5 April	8 April	9 April
14	24	18–36	17

The increase in intensity in the period 5–6 April was possibly caused by the passage of the Moon through the magnetic tail of the Earth. Perhaps also as a result of that passage, the registered densities of electrons with energies of the order of 10^4 eV exceeded the interplanetary background by a factor of 70–100. The density of micrometeoroids measured by acoustic counters was nearly 100 times higher than that in the interplanetary space (**2**).

The total γ-radiation intensity of the lunar surface measured by a γ-spectrometer is 1·5–2·0 times that measured over granite rocks. But only 10% of it can originate from the decay of radioactive elements U, Th, K^{40}, and 90% must be attributed to the radioactivity induced by cosmic rays. The abundance of radioactive elements in rocks on maria is similar to that in basalts. On the continents this abundance is lower and corresponds to that in ultra-basic rocks. Local variations in the total intensity of the lunar γ-radiation (induced + natural) do not exceed 40% (**4, 5, 6**).

On 14 August 1966 the United States' Orbiter 1 became an artificial satellite of the Moon in an orbit inclined 12·16° to the equator, a period of $3^h 37^m 45^s$, and periselenium and aposelenium distances of 189 and 1865 km; these figures were changed to $3^h 20^m$, 58 km and 1848 km on 21 August. Orbiter 1 was equipped to monitor meteoroids and radiation, and to photograph the lunar surface, particularly nine possible landing sites for Program Apollo. Orbiter 1 secured over 200 photographs, many of excellent quality (**8, 9**), before it was caused to crash on 29 October.

Tracking data have been obtained to permit detailed study of the Moon's shape and gravitational field. Preliminary indications, from data of Orbiter's first few days, are that the Moon has a 'relatively large pear-shaped component and that the gravitational properties will be of considerable scientific interest' (**10**). See also (**M 13**).

The Moon's third satellite, Luna 11, was placed in an orbit inclined 27° to the lunar equator on 28 August 1966, at a height ranging between 159 and 1200 km. We have no information on its mission. Luna 12 was placed in orbit on 25 October 1966, and has sent back pictures.

On 10 November 1966 Orbiter 2 attained a circumlunar orbit ranging between 1850 and 210 km above the surface, where it served for five days as a passive probe of the lunar gravitational field. On 15 November the low point of the orbit was reduced to 50 km for the primary mission of photographing 13 potential landing sites. Orbiter 2 fell silent on 6 December, after transmitting 95% of its scheduled 211 photographs. To date NASA has released a set of spectacular pictures of the region of the crater Copernicus, seen obliquely from a distance of 240 km and a height of 46 km above the Moon with the crater Fauth in the foreground (**11**).

C. CARTOGRAPHY

Arthur and Kuiper report completion of the scheme of lunar nomenclature for the Moon's visible hemisphere, which was proposed and approved by the IAU at its 1964 Hamburg meeting. This nomenclature is embodied in the catalog, *The System of Lunar Craters* (**1**), which is in four parts and gives positions and dimensions for 17 000 craters, and in a two color map in four sheets (**2**). The compilation of this nomenclature has been a task of some magnitude, and it is hoped that the work will be rewarded by wide acceptance of the new scheme.

Other cartographic work by scientists of the Lunar and Planetary Laboratory include selenodetic measures on three Yerkes photographs by Arthur (**6, 12, 13**), drawings of areas of the northwest (**19**) and southeast (**20**) lunar limbs by Herring. Strom (**32**) has mapped in detail lineaments within 60° of the center of the Moon's face, and delineated four global lineament systems and radial systems associated with four circular maria.

Since the Hamburg meetings, scientists of the U.S. Geological Survey have accelerated their program of geologically mapping the Moon by telescopic means. Thirty-five maps at a scale of 1:1 000 000 are now published (**3**) or available in preliminary form (**A 7a**) out of a planned 44 which will cover most of the earthside hemisphere. Aspects of the geology of selected regions have been discussed (**26, 27, 30, 34, 37, 38, 39**). The geological mapping aims to express the structure, history, and formative processes of near-surface materials, and maps class into units those materials believed to have formed simultaneously in about the same way. The units appear in order of relative age. In addition the scientists have employed photographs taken by Ranger VIII for geologic (**24, 33, Hc 13**) and topographic (**25**) mapping of a region of Mare Tranquillitatis; and by Ranger IX for the floor of Alphonsus (**Hb 8, 13**). See also refs. (**Ha 11, 12, 15, I49**).

The U.S. Air Force, Aeronautical Chart and Information Center has used the Ranger photographs to refine its shaded relief maps of the Moon (**4**). Their previous maps at a scale of 1:1 000 000 were the largest-scale maps previously available for the impact areas of the Rangers. Their new series (**4**) possesses scales of 1:500 000, 1:100 000, 1:10 000, and 1:1000. These maps 'provide in a condensed, but qualitative, form much of the new topographic data acquired by the Ranger Block III missions' (**Ba 5**). References (**15, 17, 23**) discuss various aspects of the ACIC mapping program.

A number of papers discuss various procedures and problems in lunar cartography and selenodesy (**5, 7-11, 14, 18, 21, 22, 31, 35, 36**). Papers (**16, 28, 29**) are concerned with the problem of determining elevations on the Moon. See also (**Ha 5, 8**).

Lisina and Shevchenko developed van Diggelen's photometric method for the study of relief and applied it to maria regions (**40**) and the surroundings of the crater Kepler (**41**). Markov (**42**) found the diameter-depth relation: $\log D = 1{\cdot}0662 \log d + 0{\cdot}6200$, for craters seen on Ranger VII photographs. Mukhamedzhanov and Stanjukovich (**43**) studied the distribution of primary and secondary ejecta of the crater-forming impact which can explain the double rims of certain craters.

Towards an investigation of the geometrical shape of the Moon, Gavrilov (**44**) developed a method of photography of the Moon and of compiling catalogs of selenocentric positions of details of the lunar surface. After analyzing existing catalogs and adding new measures, he constructed a net of basic points. From measures of 16 photographs of the Moon, with the Schrutka-Rechtenstamm and Baldwin Catalogs as reference, Gavrilov *et al.* (**45**) compiled a catalog of space coordinates of 160 basic points. This was extended to a similar catalog of 500 basic points (**46**). Gavrilov (**47**) has compared different systems of position.

Lunar Nomenclature

The enormous amount of new data related to the mapping of lunar craters, obtained from the space programs of the U.S.A. and the U.S.S.R. necessitates a complete review of the problem of nomenclature of various lunar features. Recommendations made by special committees of the National Academies of the U.S.A. and U.S.S.R. furnish a starting point for consideration of the problem by the Sub-committee on Lunar Nomenclature of Commission 17 of the IAU under the chairmanship of Z. Kopal. As the result of correspondence undertaken prior to and meetings held during the General Assembly of the IAU in Prague, recommendations will be made to the General Assembly concerning nomenclature for features of the reverse side of the Moon and also of special features on the visible surface.

A poll of the membership of Commission 17, undertaken by its president, indicates that the majority favors extension of the present system of nomenclature to the reverse side of the Moon, with craters named for deceased scientists, with special attention to astronomers.

D. PHOTOMETRY OF THE MOON

Barabashov and Ezersky studied the microrelief of different lunar regions (**30, 31**) on the basis of V. A. Fedoretz's photometric catalog of 1952. The same authors and Ezerskaya (Fedoretz) (**32**) studied the reflective properties of 31 areas in Mare Nubium and Mare Cognitum using the Ranger VII photographs, and found that their microrelief does not differ considerably from that of the average lunar surface.

Akimov (**33**) deduced a complicated empiric formula for the brightness distribution at different phase angles, and found it in fairly good agreement with his observations. Hämeen-Anttila *et al.* (**8**) studied the shadow effect of surface irregularities, which they found insufficient to explain the observed phase curves, the theoretical curves being more linear and less steep at small phase angles than the observed curves. Hapke (**12**) modified his earlier (**11**) theoretical lunar photometric function, by wrinkling the porous, open surface of the previous model into a series of steep-sided depressions. Oetking (**19**) found that many terrestrial substances, when observed with an instrument of small aperture, show a prominent rise in reflectivity if the direction of observation is within $\pm$ 5° of the direction of the incident light. Hapke (**20**) has commented on this conclusion. Wildey (**29**) made photoelectric measurements in *U*, *B*, and *V* of a number of lunar features extending over a range of morphological types and selenographic coordinates; deviations from the average photometric function have been correlated with stratigraphic class.

Ashby (**1**) determined the average energy reflected by the lunar surface. Rydgren (**21**) used the coronagraph at Anacapri to measure the earthshine on the Moon. Opriatova (**34**) discovered a decrease in albedo towards the ultraviolet in the region 4200–3200 Å.

Coyne (**4**) determined differential colors for 36 areas on the Moon, with respect to a standard area in Mare Serenitatis. He found no dependence of the differential colors on phase angle, a result confirmed by Scott (**22**). Mironova (**38**) studied the spectral distribution of light reflected from 30 small regions of the lunar surface.

Mironova (**17**) and Sergeeva (**23**), from spectrophotometric studies, found signs of luminescence at 4250 Å and 5050 Å in the crater Aristarchus. Petrova (**35**) obtained almost identical spectrophotometric curves for four lunar maria: Tranquillitatis, Crisium, Serenitatis, Sinus Iridum. In the southwest part of Sinus Iridum she found (**36**) a violet emission band (λ_{max} = 4250 Å) similar to that in Aristarchus. Near craters Kepler and Le Monnier a green (λ_{max} = 5350 Å) emission band was observed. She found that some volcanic sublimates containing rock-salt and exposed to ultraviolet illumination showed a luminescence with two maxima, at 4260 Å and 4100 Å, resembling that in Aristarchus. See also (**I 45**).

Sharonov and Sytinskaya (**37, 24, 25, 27**) found that the reflective properties of fresh volcanic crusts studied in field conditions at Kamchatka are similar to those of the Moon, but the roughness of the lunar surface should be much greater. See also (**10, 14**).

Van Diggelen (**5**) studied the radiance of lunar objects at small phase angles on plates taken during the eclipse of 18 November 1956. The photometric and color properties of several lunar eclipses have been observed (**3, 6, 13, 2, 16, 18**). Hansen and Matsushima (**9**) have sought to explain the exceptional darkness and non-reddening of the eclipse of 30 December 1963 by an unusually large extinction in the Earth's upper atmosphere, while Link (**15**) invokes lunar luminescence to explain the difference in brightness of various eclipses. Bell and Wolbach (**39**) call attention to the relation originally found by Danjon between eclipse brightness and color and phase of the sunspot cycle. Matsushima (**40, 41**) finds some correlation between the brightness of the eclipsed Moon and the geomagnetic index K_p.

See Section I and (**J 4, 5**) for additional material on the interpretation of photometric observations.

E. POLARIMETRY

Kokhan (**10**) published a catalog of the polarization and position angles of the polarization plane of light reflected by 35 areas of the lunar surface at phase angles from $-140°$ to $+140°$. Lipsky and Bondarenko (**11**) found that for many lunar objects at phase angles $>30°$ the polarization decreases with increasing wavelength; at $\pm 90°$ it is 10–15% higher in the blue than in the red. On the other hand, at angles $<30°$ the polarization in the red exceeds that in the blue, especially for maria and ray systems.

Avramchuk (**1**) studied the polarization of several lunar areas with color filters and discussed the discrepancies between the data by different authors. Morozhenko (**12**) measured the polarization of light reflected by different terrestrial rocks. Dzapiashvili and Xanfomaliti (**13**) used a new electronic polarovisor to obtain lunar images showing the distribution of position angles of the plane of polarization over the disk. Pospergelis (**14**) designed a new electronic polarimeter for planetary research, which permits the determination of all Stokes' parameters.

Clarke (**2**) has discussed previous data and added new observations on the wavelength dependence of the polarization. Hopfield (**7**) presented a theoretical interpretation for the negative polarization of moonlight at small phase angles. Marin (**8**) made polarization measurements of the Moon and planets in the spectral region from 0·8 to 1 μ.

Gehrels, Coffeen and Owings (**6**) made photoelectric measurements of brightness and polarization on various lunar regions at *UBV* and 0·94 μ wavelengths. Coffeen (**3**) studied five laboratory samples, three of porous dust layers of ground volcanic cinder particles in 'fairy castle' structures, and two of porous but solid lava fragments. He found the wavelength and phase dependence of the lunar polarization most closely matched by the fairy castle structures; the solid fragment was more highly polarized than the Moon and had essentially no wavelength dependence.

Dollfus has presented (**5**) preliminary results of an extensive study of the polarization of 14 selected regions of the lunar surface at Meudon and Pic du Midi. Complete curves of polarization have been obtained at the wavelengths 1·10, 0·95, 0·83, 0·60, and 0·55 μ; measures at 0·40, 0·37, 0·35, and 0·325 μ are in progress. From these data and laboratory study of the polarimetric properties of various substances, Dollfus (**4, 5**) concludes that the Moon's surface is completely covered in all areas by a layer of small uncompacted, highly absorbing dust particles, whose absorptivity has been increased by the action of the solar wind on the lunar surface.

The U.S. Geological Survey has made polarimetric measurements to provide data for the description of lunar geologic units, with emphasis on determination of maximum positive polarization for each unit (**9**); in general the maximum polarization of a geologic unit is inversely related to its albedo, although occasional exceptions occur.

F. THERMAL PROPERTIES AND INFRARED STUDIES

Radiometric and photometric mapping of the Moon at various phases by Shorthill and Saari (**22, 18**) and by Ingrao *et al.* (**5**) has led to the discovery of many thermal anomalies on the surface of the Moon. While detectable at various phases, particularly around the sunset limb, the lunar hot spots are especially striking during total eclipses of the Moon, where they have been extensively studied by Shorthill and Saari (**16, 17, 18, 19, 20, 21, 23**) at the wavelength region 10–12 μ. They report (**18**) that more than 400 hot spots have been identified and cataloged. 'Most of them are craters that appear visually bright at full Moon; some have ray systems, some bright interiors, and some bright rims' (**18**). A substantial percentage were not associated with major ray craters (**16**), with a marked concentration in Mare Tranquillitatis (**21**). Some substantial anomalies are identified with quite small craters (**18**). It is suggested that the thermal anomalies have the properties of bare rock, without an insulating layer of dust. Other scientists (1, 26, 29) suggest that the thermal anomalies may arise from lunar roughness on a centimeter scale rather than from special thermal composition or the presence of local thermal sources.

Markov and Khokhlova (8) measured the emission near $3{\cdot}6\mu$ and 8–13μ during the eclipse of 7 July 1963, and obtained a relatively high albedo at $3{\cdot}6\mu$ of 0·3. They found the maria remained warmer by several degrees than the highlands during the eclipse. They detected (9) also a difference in the heating rate of the east and west limbs of the Moon during this eclipse, an anomaly they attribute to differences in the physical parameters.

Several scientists studied the lunar surface in the near infrared. Moroz (10) measured spectra of selected regions at full Moon and found an increase in albedo with wavelength out to $2{\cdot}2\mu$, which was very similar for a variety of lunar structures—maria, highlands, and bright craters—that differ considerably in absolute albedo. Binder *et al.* (2) found the color of the Moon to be very uniform in the region $1{\cdot}0$–$2{\cdot}2\mu$, with a few exceptions. From observations made on the second flight of Stratoscope II, Wattson and Danielson (28) also find that the lunar albedo is substantially higher in the infrared out to about $2{\cdot}5\mu$ than in the visual. Moroz (10) found that volcanic ash and slag display a variation in albedo with wavelength similar to that of the lunar surface.

Wattson and Hapke (27) and Binder *et al.* (2) investigated the infrared characteristics of terrestrial rocks and simulated lunar surface materials. Terrestrial rock samples were generally much less red than the lunar surface in the $1{\cdot}0$–$2{\cdot}2\mu$ region. However samples of volcanic rock, irradiated with 2·0 keV H^+ ions, corresponding to an equivalent irradiation of the lunar surface for about $2{\cdot}5 \times 10^5$ years, simulated the intensity ratios for the lunar infrared. This experiment supports Hapke's hypothesis that solar irradiation can produce the unusual spectral curve and low albedo of the lunar surface.

Burns and Lyon (3) studied the errors introduced into temperature determinations when one assumes that the emissivity of the lunar surface is independent of wavelength. Their calculations for a variety of substances which might reasonably be expected on the surface of the Moon indicates that the emissivity varies significantly. The deviations from blackbody behavior would give calculated lunar temperatures too low by 3·5 to 5·5%. Murcray (11) found that the observed radiance of the lunar surface in the 8 to $10{\cdot}4\mu$ region was not compatible with gray or blackbody radiation at any temperature. He concluded that the emissivity varied from roughly unity at $8{\cdot}5\mu$ to about 0·9 at 10μ.

In a study of the lunar spectrum in the region 8–11μ, Moroz (10) showed that the absorption band of SiO_2 at 9μ is absent, probably because of the pulverized nature of the surface layer. Bolometric measurements by Chistjakov (32) with a resolution of 45×15 km² did not reveal any thermal anomalies for the crater Alphonsus. Ryadov *et al.* (14) discuss their measurements in the 8–$13{\cdot}5\mu$ region of the integrated thermal radiation from the Moon over various phase angles, and derive an average disk temperature of the full Moon. Observing the Moon in the 16–24μ window, Hunt and Salisbury (4) found anomalies that suggested differences in chemical composition among several features of the lunar surface.

Linsky (6) developed models with temperature-dependent thermal properties to analyze the observations at both the infrared and the radio wavelengths. Troitsky (24) developed a mathematical analysis of the thermal and radio emission data from the surfaces of the Moon and planets. Murray (13) points out some inconsistencies between the radio-derived temperature of 207 °K used in the foregoing (24) model, and infrared data which give a temperature of about 100 °K—as determined independently by Saari (15) and Troitsky (25)—for the antisolar point of the lunar surface.

Measures of the surface temperature during the lunar night were carried out also by Low (7) and by Murray and Wildey (12). The latter used especially sensitive infrared detectors to measure the radiance of the shaded lunar surface, and consider the observed pattern of sunset cooling to be inconsistent with the presence of a thick homogeneous layer of dust. They report a number of thermal anomalies, but no systematic nighttime temperature differences between maria and uplands. The presence of thermal anomalies also suggested the existence of regions completely free from strongly insulating dust.

G. RADIO STUDIES

This section deals with two major fields of investigation: thermal radiation from the Moon at radio (mm and cm) wavelengths, reviewed in (**26, 39, 40**); and the reflection of man-made radar waves by the Moon, reviewed in (**18, 30**). The distance to the Moon has been measured by radar (**11, 24, 42**), and by laser (**29**); reflected light signals were registered by a group of scientists of Lebedev Physical Institute (**45**). Kokurin *et al.* (**46**) discuss methods of determining orbital parameters and the shape of the Moon by optical location.

Thermal. A number of papers (**1, 20, 22, 27, 31, 41, 48**) report observations of thermal radiation at mm wavelengths from the eclipsed Moon. Troitsky (**38**) elaborated a theory for the radioemission of the eclipsed Moon, which he compares with observations of the 1963 and 1964 eclipses at 1·20 mm. He concludes that the average thickness of the porous surface layer is 6 ± 3 meters, and that the density doubles its value at 2–3 cm below the surface, with a decrease by a factor 2 in the absorption coefficient for cm wavelengths.

Observations of thermal radiation from the center of the uneclipsed Moon yielded determinations of the brightness temperature (**27, 35, 47**). Papers (**8, 14, 22, 34**) investigate the relative brightness temperatures of selected regions of the Moon. Gary *et al.* (**14**) have determined contours of 3 mm brightness over the lunar disk at 14 phases of the Moon, and derived phase curves for a number of latitudes. The average phase lag of 22° indicates that the lunar surface has a very low thermal conductivity (**8, 14**). In addition Drake (**8**) reports preliminary evidence for a correlation between 3 mm and infrared thermal anomalies, and stresses that the 3 mm observations require for interpretation at least a two-component model and 1-mm structures all over the surface. In a related study, King *et al.* (**22**) determined relative brightness temperatures and cooling rates of selected lunar regions during the eclipse of 1963 December 30.

Gary *et al.* (**14**) found the brightness temperature of the maria to exceed that of the neighbouring highlands by $2{\cdot}6 \pm 0{\cdot}2°$ K, at 3 mm. Salomonovich (**34, 43**) found that at 8 mm the brightness temperature of the maria exceeded that of the continental regions by $1{\cdot}5 \pm 0{\cdot}5\%$ (averaged over a lunation); for the midnight temperature this excess was 8° K. The corresponding variation in γ is no more than 25%. From a study of the polarization of the lunar radioemission at 6·3 cm and comparison with previous data at 3·2 cm, Gol'nev and Soboleva (**44**) conclude that the surface is rough on the mm scale.

Troitsky and his collaborators at Gorki Radiophysical Institute continued to develop the 'artificial Moon' method for temperature measurements at different radio wavelengths. The temperature gradient with depth was found to be 2–5°/meter, the thermal heat flow from the interior, 1×10^6 cal cm^{-2} s^{-1} (assuming temperature-independent thermal properties of the porous surface layer). The full list of papers published by this group can be found in reviews (**26, 39**).

Radar. In a study of lunar radar echoes by a delay-frequency analysis, Thompson (**37**) derived radar scattering maps of the Moon. He finds that mountainous regions reflect 1·5 to 2 times as much power as maria regions, while certain craters reflect up to 10–20 times the average.

From a review of radar observations Hagfors (**18**) concludes that the lunar surface is covered by a material having an effective dielectric constant of 2·6 and that the surface undulations on the Moon have a mean slope of 11–12° on the scale of about 1 meter. He suggests that the enhanced reflectivity found for young rayed craters is caused by both a higher intrinsic reflectivity and a greater roughness in these features. The results of a study of polarization in radar echoes agree with a model consisting of a tenuous top layer at least 10 cm thick, supported by a denser underlying layer (**17**).

A number of other papers discuss the analysis of radar echoes from the Moon, with emphasis on the laws of geometrical optics in (**32, 33**) and on a distribution of structure sizes in (**2, 3, 23, 4, 5, 19**). Values for the mean slope are derived in (**2, 4, 18, 32**), while (**11, 18, 30, 32**) provide values of the dielectric constant. Katz (**21**) compares the wavelength dependence of the lunar reflectivity with that of various terrestrial surfaces. Taylor (**36**) investigates anomalies in the Faraday rotation of lunar echoes and the possibility of a relationship to the geomagnetic index K_p.

Hagfors (**16**) establishes the relationship between the geometric-optic approach and the autocorrelation approach to the analysis of lunar radar echoes for a Gaussian autocorrelation function when the surface has Gaussian height statistics and introduces deep phase modulation on the incident wave. He finds that, if an appreciable amount of small-scale structure is present, the range of scales responsible for the scattering will include an increasing amount of small-scale structure with increasing angle of incidence. A similar conclusion is reached by Fung and Moore (**12**), who show that large-scale features determine the return at near-normal incidence and small-scale features determine that from near grazing incidence. Evans (**11**) reports measurements at 1130, 68, 23, 10, 3·6 and 0·86 cm. At all six wavelengths it appears that part of the echo arises from a highlight located at the center of the Moon's visible disk. A second component comes almost equally from the remaining parts of the surface. The division of power in the two components changes markedly as the wavelength is reduced. At 68 cm, 80% of the power is returned from the highlight, while at 8·6 mm only 15% can be associated with this component. The angular power spectrum observed for the power from the highlight also changes with wavelength, indicating that the r. m. s. slope of the surface increases as the wavelength is reduced. These observations are interpreted as evidence of a wide range of structure sizes on the Moon. The measurements at decameter wavelengths by Davis *et al.* (**6, 7**) provide additional evidence that at radio frequencies the Moon displays a central highlight that contracts and brightens with increasing wavelength.

H. SURFACE FORMATIONS

a. General

O'Keefe (**13**) has reviewed the present understanding of the structure and origin of the Moon. Fielder (**4**) has reviewed the present state of lunar geology. Kopal (**8**) reviews the definition of lunar coordinates, methods for determining the exact shape of the Moon, the origin of lunar formations, and other problems of lunar topography. McCauley (**11**) discusses the nature of the lunar surface as determined by geologic mapping.

From a theoretical comparison of lunar and terrestrial surfaces on globes of equal size, Brock (**1**) concludes that the coarser tectonic patterns of the Earth and Moon are similar. Dietz (**2**) suggests that there may be structures on Earth that are analogous to both lunar craters and maria. Geyer and Lopik (**5**) discuss application of methods of geophysical analysis to study of the lunar surface. Khodak (**7**) discusses the chief structural elements of both sides of the Moon, including walled plains, craters, maria, valleys and fissures. Krejci-Graf (**9**) discusses primarily rays and crater shapes.

Heacock (**6**) reviews pre-Ranger lunar data and presents a general summary of the Ranger results. Kuiper *et al.* (**10**) discuss particularly the lunar collapse depressions, as well as the structures of maria and craters, and lineaments. Moore (**12**) compares two lunar crater morphologies, based on Ranger IX photographs, and concludes that the lunar surface materials are weakly cohesive to noncohesive. Shoemaker (**15**) discusses the pattern of low ridges and troughs observed on the lunar surface in Ranger photographs, and their relation to the larger lineaments. Urey (**16**) discusses theories on the evolution and thermal history of the Moon with reference to the Ranger photographs.

Whitaker (**17**) summarizes conclusions of the Lunar and Planetary Laboratory on the linear grid, rays, craters, and the absence of dust. He presents evidence for a distinction between 'red' and 'blue' maria, and finds evidence that maria are dust-free lava flows from the sharp boundaries between red and blue mare regions. Van Diggelen (**3**) discusses a statistical method for determining the preferential direction shown by lunar surface formations known as the lunar grid system, and by this method confirms nearly all the systems previously mentioned by Fielder. Rae (**14**) presents a historical review of work on lunar domes, and a catalog of 113 domes, which he finds normally occur only on maria near the borders or on the floors of lava craters. See also (**I 4, 6, 49**).

b. Craters

The numerous papers on lunar craters fall into three major categories: discussion of individual craters, statistics of crater size and distribution, and hypotheses of the origin of craters.

The target of Ranger IX, crater Alphonsus, has been mapped and discussed geologically (**31**). Carr (**8**) suggests that the alignment of small craters in Alphonsus is strong evidence that many craters were formed by mechanisms originating within the Moon. Ronca (**37**) concludes that Alphonsus has been deformed by a strike-slip fault of dextral type. The craters Caramuel (**1, 33**), Dionysius (**42**), and Aristarchus (**D 17, 23**) have also received individual attention. The origin of the central peaks of craters has been discussed in terms of faulting and volcanic extrusion (**9**), and of meteorite impacts (**24**). Pohn (**34**) found many crater slopes to have a steepness of at least 37°.

Hartmann (**19**) finds that the distribution of lunar crater diameters is satisfactorily explained by the impact hypothesis; he attributes a deficiency of small sizes among the oldest craters to a process of erosion. On the other hand, Fielder (**13, 14**) finds that craters cluster more than could be expected on the impact hypothesis and he concludes that a substantial proportion of craters must be of internal origin, an interpretation criticized by Byron (**7**).

Ranger photographs have yielded several studies on the distribution (**5, 32, 46, 47**) and diameter-depth relationship (**3, C 42**) of craters.

Limitations of space prevent us from discussing individually the various papers dealing with aspects of the meteorite hypothesis (**2, 10, 11, 16, 20, 23, 25, 35, 39, 40**), the volcanic hypothesis (**4, 6, 12, 17, 18, 30, 43, 44**) and other problems (**15, 26-29, 36, 38, 41, 45**) of crater formation, except to note Kopal's suggestion (**22**) that many smaller craters may be subsidence phenomena, possibly triggered by Moon-quakes following major impacts, rather than due to secondary impacts. See also (**I 39-41**) and section **J**.

c. Maria

Shoemaker (**11, 12**) discusses the fine structure of Mare Cognitum, defining primary and secondary craters on the basis of morphology and distribution, and commenting upon the rarity of small features of positive relief in the Ranger VII photographs. (However a number of small positive-relief features appear on the Orbiter 2 photographs recently released by NASA.) Other discussions of individual maria include Mare Cognitum (**2, 3**), Mare Tranquillitatis (**13**), and Mare Humorum (**9, 10**); see also (**D 32**). Several papers (**1, 4, 6, 7**) discuss possible origins of the maria, while (**8**) considers the possible origin of rilles, ridges, and domes in relation to maria, and (**5**) compares maria and continental regions with respect to radio emission.

d. Ray Systems

From a study of the faint ray systems around Mare Imbrium by special photographic techniques, Alter (**1**) finds sufficient resemblance between these faint patterns and the patterns of craters and bright ray systems to conclude that they are of the same genus; he considers the existence of the faint patterns argues against any hypothesis of a deep dust layer, and also against the hypothesis that the craterlets observed along bright rays are direct results of secondary impacts.

Also discussed are the bright rays radiating from certain lunar craters (**2**), rille complexes (**3**), ray systems from Ranger VIII photographs (**4**). In an analysis of radial structures surrounding lunar basins, Hartmann (**5**) notes that such systems are more developed around presumably younger basins such as Mare Imbrium and Mare Orientale than around older basins, from which he postulates that conditions for producing radial lineaments were optimal during a relatively short period when the basins were flooded.

I. PHYSICAL PROPERTIES AND CHEMICAL COMPOSITION OF THE LUNAR SURFACE

Ruskol (**47**) reviewed investigations of lunar surface properties carried out in the previous 2–3 years, including results of laboratory simulation of conditions on the lunar surface, and the connection of data on the thermal regime of the Moon's upper layers with the degassing problem of the lunar interior and the history of the lunar atmosphere.

On the basis of photopolarimetric observations (**E 6**), Gehrels (**7**) postulates a model of the lunar subsurface that is smooth and firm with little or no dust but with an optically thin layer of ionized micron-sized particles electrostatically suspended over the surface, and accreted from interplanetary space. Objections have been raised (**28**) concerning the electrostatics of lunar particles and the optical properties of this layer. Gehrels (**8**) replies that most observations can be explained with a tenuous surface texture with interconnected (rather than freely suspended) scattering elements or particles. Sagan (**48**) suggests that the photometric properties of the Moon may arise from infall of zodiacal dust particles rather than by proton irradiation of existing dust. Glaser (**14**) analyzes the implications of the absence of atmospheric pressure for structural, physical, and heat-transfer properties of the lunar surface.

From a review of optical data on the lunar surface and related laboratory studies, Hapke (**27**) finds strong evidence that the surface of the Moon is covered with a layer of fine dust. From examination of Ranger VII pictures (**15**), the distinctive thermal and radar properties of the crater Kepler (**16**), and a review of various data (**17**), Gold finds evidence for a substantial layer of dust on the lunar surface. Other discussions of lunar dust and its properties include (**29, 30, 31, 32, 50, 51, 52**): Urey (**52**) points out that the dust layers postulated by Gold do not necessarily have low physical strength, incapable of supporting a space vehicle. Jaffe (**30–32**) discusses aspects of the strength of a lunar dust surface. Smoluchowski (**51**) suggests that sintering by corpuscular radiation may compact the dust and increase its mechanical strength. O'Keefe (**43**) attributes the softened appearance of the larger craters to a deposit of ash over the lunar surface, or less probably to erosion. Possible mechanisms of erosion on the Moon are treated in (**2, 5, 42**).

In a study of morphological aspects of the lunar crust, Miyamoto considers such problems as the gas content, viscosity, and differentiation of the original magma (**39**), the effects on crater formation of decreasing thickness of a silicic crust (**41**), and suggests that lunar terra is basaltic and maria ultrabasic (**40**). Paper (**46**) considers chemical evolution, and (**M 28**) treats the differentiation of igneous rocks with reference to the Moon.

Laboratory studies have been made of the properties—photometric (**20, 23, 26, 44, 45; D 10, 14, 17, 19, 20, 24, 25, 27**), polarimetric (**E 3, 4, 5**), thermal (**12, 13, 54; F 27**), and miscellaneous (1, 3, 18, 29, 33, 42, 47)—of a variety of substances for comparison with the respective properties of the lunar surface. To match the lunar photometric law, Hapke (23) concludes that a surface must consist of extremely small, rough, nearly opaque objects, arranged into a complex surface of low compaction, such as rock dust, and modified in color and albedo by solar proton bombardment (26).

On the other hand, Halajian (19) finds radiothermal and photometric data to be compatible with a 'homogeneous underdense-cohesive silicate' model, as of rock froth or sintered slag. From thermal data, Glaser *et al.* (13) conclude that the surface is not composed solely of unconsolidated particles, but more likely of sintered materials or highly vesiculated rocks. In a study of 23 samples of terrestrial magmatic rocks, Orlova (44) finds that lunar photometric properties most closely resemble those of volcanic slaggy products. Other Soviet workers (D 14, 24, 25, 27) also find the reflective properties of certain volcanic rocks similar to those of the Moon.

Reference (54) reviews the literature on the thermal conductivity of various substances, while (12) finds that sintered loose perlite has thermal properties similar to the lunar surface. Paper (38) concerns the interpretation of radio and radar data, and (21) considers various properties of the Surveyor 1 landing site.

J. LUNAR VOLCANISM

The last few decades have witnessed a spirited and sometimes acrimonious debate between those who maintain that the lunar craters originated in meteoric impact and those who aver that the craters are of volcanic origin. Data from lunar probes seem about to resolve the question. Although it is still too early to make a definite pronouncement, current evidence suggests that both methods of crater production have been active in producing craters and otherwise molding the lunar surface. Scientists of the U.S. Geological Survey report (**Ha 11**) their mapping to indicate that the lunar surface probably consists of interbedded and interfingering impact and volcanic debris. The observed 'softness' of the high-resolution photographs and the random distribution of crater diameters smaller than 250 meters clearly support the impact theory. On the other hand, the occurrence of large, deep pits with nearly vertical walls, the sharply delineated regions of contrasting color on the surfaces of certain maria, and the scars of field and ghost craters all suggest the action of tectonic forces including extensive lava flows. The observations would appear to rule out the occurrence of extensive transport of dust over the lunar surface. These conclusions are, of course, still tentative. Final decision on the contribution of volcanic processes must await further analysis of existing films and possible exploration of the lunar surface itself by manned or unmanned techniques. In the meanwhile, we should keep an open mind on such questions.

Green (9) reviews the current evidence in support of the hypothesis that volcanic processes have played a major role in the modelling of lunar features. Menzel (24) shows that the Moon, if liquid at any time during its history, would probably have cooled from the outside in. The outgassing processes would have formed a light, foamy lava that would have floated on the heavier magma. Subsequent flows of lava, controlled in part by tides, would account for the appearance of maria and the ghost craters.

Kuiper (11) points out similarities between Liamana Volcano in Hawaii and various lunar features, and suggests that the secondary craters on the slopes of Liamana are probably caused by impact. From a study of the frequency and distribution of crater diameters, Simpson (21, 22) argues that the majority of craters on the Moon and Mars are of volcanic origin. Vonnegut *et al.* (23) suggest that gasses escaping from hot lava may condense to form stable aerosols. O'Keefe (16) interprets a dark stripe seen on certain Ranger VII photographs as evidence for recent lunar volcanism, of intermediate or acid type. On the other hand, Dietz and Holden (**3**) express strong doubt that the Moon has either significant internal convection or currently active volcanos. They argue, from the complete or near absence of water, that explosive volcanism is highly unlikely on the Moon. Section H includes additional references on lunar volcanism.

Recent laboratory studies suggest that the absence of a lunar atmosphere must be given more consideration in theorizing on the origin of lunar formations. Dobar (**4, 5**) finds that a simulated basalt and granite magma, allowed to upwell and solidify in a vacuum, has produced a porous material not found in nature on Earth. The photometric curves of this material resemble closely those from the lunar surface. During the vacuum upwelling, Dobar observed color phenomena similar to those reported around Aristarchus (**K 8**).

In a report on further observations of the luminous glow in the crater Aristarchus, Kozyrev (**10**) finds both C_2 and H_2 emission and suggests the latter may be emitted from a fumarole within the crater area. See also (**K 4**). Middlehurst and Burley (**13, K 3**) made a complete catalog of observed and recorded lunar events—defined as temporary changes in the appearance of a lunar feature—over the past four centuries. Middlehurst (**12**) finds no correlation between the occurrence of such events and sunspot numbers, and suggests the events may arise from tidal cracking or crustal deformations, with a release of hot or cold gasses. In a review of evidence for changes on the Moon, Moore (**15**) doubts the reality of reported structural changes, such as the vanishing of crater Linné, and long-term changes in the reflectivity. But he considers that the transient outbreaks, like those reported in Alphonsus, are probably real. Deák *et al.* (**2**) postulate a volcanic origin for the reddish spots seen (**K 8**) in the vicinity of Aristarchus, and calculate the energy liberated from the lunar interior in such an event. See section K for additional references on transient phenomena.

The origin of tektites found on the surface of the Earth has long been debated. These glassy objects appear to be some form of meteorite. In the absence of any clear-cut alternative theory, several scientists have suggested (**1, 6, 7, 14, 17**) that tektites originated on the Moon, and further, that their study may contribute to understanding the lunar structure (**19, 20**). O'Keefe and Adams (**18**) suggest that tektites originated in a lunar ash flow.

K. LUNAR LUMINESCENCE

The problem of whether the lunar surface produces detectable luminescence has two aspects—the possible existence of transient phenomena induced in some way by flare-associated solar activity; and possible longer period variations related to the sunspot cycle and general level of solar activity. Most of the research deals with the former aspect.

A survey of the historical literature turned up 159 dated and seemingly reliable observations of apparent activity on the Moon over the past four centuries (**3, J 13**). The material was analyzed with respect to solar activity and to tidal action by the Earth (**3**). The observations of color streaks in the Aristarchus area in late 1963 have been reviewed (**8**). A search of the literature turned up 16 additional transient events observed between 1783 and 1963 in the vicinity of Aristarchus (**6**), and negatively correlated with solar activity. A transient increase in brightness around the crater Kepler was observed through a filter centered at 6725Å, while control plates at 5450Å showed no effect (**10, 9**). A subsequent search by similar techniques detected no luminescent phenomena (**14**). However use of a blink comparator technique has revealed additional color events (**20**).

By the 'method of line-depths', a variable luminescence reaching as much as 30% of the continuum was observed near 5450Å, but no luminescence greater than 2% was detected at $H\alpha$ or the Na D-lines (**16**), nor at around 3900Å (**18**). Other observations interpreted as evidence of luminescence at particular wavelengths have been cited above (**D 17, 23, 36**); see also (**I 45**).

The explanation of these transient phenomena remains controversial (**1, 6, 7, 12, 13**; **L 4**). A laboratory study found (**5**) that some materials luminesce strongly in the red when bombarded by 40 keV protons, but this finding was not substantiated by other investigators (**17**). See also (**19**).

A different type of transient phenomenon, the C_2 Swan bands observed by Kozyrev around the crater Alphonsus, has been attributed to explosions of acetylene (**4**).

On the second aspect of the problem, Gehrels *et al.* (**E 6**) report that luminescence was detected in the photometry of various lunar regions, and independently confirmed by the polarimetry—under the assumption that the luminescent light is not appreciably polarized. They found the lunar surface to be 10–20% brighter (and less polarized) in 1956–59 than in 1963 November–1964 January. The effect was fairly constant from day to day, and thought to vary with the solar cycle. This phenomenon may be related to Danjon's (1922) finding of a relation between phase of the sunspot cycle and brightness and color of the eclipsed Moon (**D 15, 39**; **E 6**). Gehrels *et al.* (**E 6**) note that the eclipse of 1965 November 18 was reddish and bright, and that of 1963 December 30 unusually dim. (See also **D 2, 3, 5, 6, 9, 13, 15, 16, 18**.) However opinion differs whether the dark eclipses (**D 9**) or the bright eclipses (**D 15**) require more 'explanation.'

L. LUNAR ATMOSPHERE AND ENVIRONMENT

Safronov and Ruskol (**12**) consider the history of the lunar atmosphere formed by gradual degassing of the interior; they conclude that the Moon never had any appreciable atmosphere and that there has been no liquid water on its surface. Gilvarry (**I 10**) presents evidence in favor of the former presence of water on the Moon. The possibility of detecting a permafrost layer beneath the lunar surface by observations of OH molecules near the Moon has been considered (**I 25, 55**).

Singer (**13**) concludes that the only gas in the vicinity of the Moon comes from interplanetary space, and predicts a peculiar type of 'ionosphere' to exist near the Moon, made up of photoelectrons that have insufficient energy to escape and thus are held back by the Moon's electrostatic field. Hinton and Taeusch (**8**) use a simplified model of the lunar atmosphere to calculate average densities of certain neutral and ionized gases near the Moon's surface as a function of the solar corpuscular flux and taking account of sources of gas in the lunar crust. Michel (**9**) examines the flow of solar wind plasma about the Moon for two limiting cases. Several authors discuss aspects of interaction between the solar wind and the lunar surface (**14**; **I 26, 50, 51**) or lunar field (**1, 3, 4, 10, 11**). See also (**J 8**).

Observations of γ-radiation and magnetic field intensity by Luna 10 have been cited above (**Bd 2, 4, 5, 6**). Luna 9 provides data on the γ-radiation from its vicinity (**15**), much of which is considered a secondary effect of cosmic radiation. Also considered are lunar X-rays (**5, 6, 7**), radioactivity (**I 35**), and expected variations in cosmic ray intensity on the Moon (**2**).

M. INTERNAL STRUCTURE

Levin (**23, 24**) reviews research on such aspects of internal structure and evolution of the Moon as the origin, thermal history, shape, density distribution, and chemical composition of the Moon. Several of these problems are treated also in (**18**). From calculations on the thermal history of the Moon, Levin (**22**) concludes that a period favorable for lava effusions occurred 2·5 to 3·0 × 10^9 years ago, depending on the content of radioactive elements in the Moon. Papers (**18, 27**) also treat the thermal history of the Moon.

Fielder (**8**) considers that Runcorn's theory, in which convection cells keep the Moon's surface buoyed up, provides the only satisfactory explanation for the slight bulge of the Moon in the direction of the Earth. He further suggests (**9**) that this convection may also explain the orientation of the two major systems of pan-lunar lineaments. Some theoretical aspects of convection in the Moon's interior are considered in (**17, 18**). References (**4, 15**) treat radiative heat transfer within the Moon, while (**16, 27**) deal with problems of thermal conduction, (**7**) is a theoretical investigation of the energy balance in the crust, and (**28**) considers processes of igneous rock differentiation as applied to the Moon.

The shape of the Moon has been reviewed (**23, 24**), and recently investigated by means of Orbiter 1 (**13, Bd 10**), and annular eclipses (**3, 5**). Wildey (**29**) reviews other methods. From a comparison of observations and new calculations of the motions of the node and perigee, Eckert (**6**) finds evidence for a large concentration of mass near the surface of the Moon. Koslovskaya (**36**) considers the density distribution along the radius for some models of the Moon. Shimazu (**N 7**) analyzes the viscosity distribution within the Moon. Several papers discuss the shape and internal structure (**11, 13, 20, 21, 24**), the shape and gravity field (**2, 12, 26**), and the distribution of mass in the Moon (**1, 6, 10, 23, 24, 25**).

Levin postulates that the oblate shape of the Moon is a direct consequence of the decrease of surface temperature from the equator to the poles, with a corresponding absence of spherical symmetry in the distribution of internal temperature (**32, 33, A 13**). Safronov (**34, A 13**) considers some isostatically-compensated models of the Moon with a semi-molten core and solid outer parts, the thickness of the latter being greater along the polar axis due to thermal conditions (Levin's suggestion). He showed that such models can account for the observed 'non-equilibrium' oblateness of the Moon. The necessary difference in thickness of the solid layer is in accordance with the calculations on the thermal history of the Moon performed by Levin and Majeva (**22, 35**). Thus, contrary to a widespread opinion, the Soviet scientists find the semi-molten state of the lunar interior, which follows from thermal calculations for the chondritic model of the Moon, does not contradict its non-equilibrium shape.

Zharkov *et al.* (**30**) review recent trends in studies of the interior structure of the Moon and planets by geophysical methods. They consider (**30, 37**) the propagation of seismic waves on the Moon for different temperature conditions of its interior. The existence of a very thick layer of low velocities is suggested. Seismic studies are considered also in (**1, 19**). Zharkov (**30**) considers quantitatively the relationship between the lunar magnetic field and the conditions in the liquid iron core.

N. ORIGIN AND EVOLUTION OF THE MOON

Scientists of the O. J. Schmidt Institute of Physics of the Earth (**5, 6, 11, 12, 13**) made new calculations of the effects of tidal friction in the interior of the Earth and Moon on the evolution of the Moon's orbit. They find that the lunar orbit was formerly smaller, less eccentric, and less inclined to the Earth's equator than now, a result they interpret as favoring the formation of the Moon in the Earth's vicinity, and in contradiction to the Alfvén-Gerstenkorn (**1**) hypothesis on the capture origin of the Moon. The effects of tidal friction are considered also in (**4**), while (**2**) presents the hypothesis of origin by capture in a direct orbit at a distance of about 45 Earth radii when the Earth was rotating in about ten hours. Urey (**10**) reviews, in the light of recent evidence, his 1957 theory that the Moon was one of the primary objects accumulated in primitive gas spheres and is older than the Earth. In (**9**) he discusses evidence for the contamination of the Moon by terrestrial water. Shoemaker (**8**) describes the establishment of a geological time scale for major events on the Moon, by using the technique of stratigraphy.

D. H. MENZEL
President of the Commission

B. BELL
Secretary of the Commission

BIBLIOGRAPHY

Sources of Abstracts

P. *Physics Abstracts*, publ. monthly by the Institution of Electrical Engineers, London.

A. *International Aerospace Abstracts*, publ. semi-monthly by American Institute of Aeronautics and Astronautics, New York, N.Y.

N. *Scientific and Technical Aerospace Reports*, publ. semi-monthly by NASA, Scientific and Technical Information Division, available from U.S. Government Printing Office, Washington, D.C.

S. *Lunar Surface Studies:* a Continuing Bibliography with indexes.
NASA-SP-7003 (01), March 1965;
NASA-SP-7003 (02), April 1966.

U. *Bibliography of Lunar and Planetary Research*, 1960-64. Ed. J. W. Salisbury. AFCRL-66-52, Special Report No. 40. 524 p. (Moon, p. 257-373), January 1966.

A. Books and Conference Proceedings

1. *The Lunar Surface Layer, Materials and Characteristics*. Eds. J. W. Salisbury, P. E. Glaser. New York, Academic Press, 1964, 532 p. (A64-26059)

2. *Physics of the Moon and Planets* (Fizika Luny i Planet). Ed. I. K. Koval. Kiev, Naukova Dumka, 1964, 140 p. (A65-18967)

3. Geological Problems in Lunar Research. *Annals N. Y. Acad. Sci.*, **123**, p. 367-1257, 1965. (A65-34229)

4. *Lunar and Planetary Surface Conditions*. N. A. Weil. New York, Academic Press, 1965, 222 p. (A65-26272)

5. *Lunar Geology*. G. Fielder. London, Lutterworth Press, 1965, 184 p. (A66-27490)

6. *Photographic Atlas of the Moon*. Z. Kopal. New York, Academic Press, 1965, 277 p. (A66-31747)

7. *Astrogeologic Studies – Summary:* Annual Progress Report, 1 July 1964-1 July 1965, 1965, 34 p. (N66-21635)

7a. *Astrogeologic Studies*. Part A: *Lunar and Planetary Investigations*, Annual Progress Report No. 6. 1 July 1964-1 July 1965, 1965, 131 p. (N66-21350)

7b. *Astrogeologic Studies*. Part B: *Crater Investigations*, Annual Progress Report, 1 July 1964-1 July 1965. H. G. Wilshire *et al.*, 1965, 184 p. (N66-21584)

8. *The Nature of the Lunar Surface:* Proceedings of the IAU/NASA 1965 Symposium, Eds. W. N. Hess, D. H. Menzel, J. A. O'Keefe. Baltimore, Johns Hopkins Press, 1966, 320 p. (A66-35437)

9. Proceedings of the Caltech-JPL Lunar and Planetary Conference. Eds. H. Brown, G. J. Stanley, D. O. Muhleman, G. Muench. 15 June 1966, 357 p. (N66-31446) (Summary by J. A. O'Keefe, *Sky and Tel.*, **31**, 10-12, 1966)

10. *La Lune à l'Ère Spatiale*. Ed. J. Rösch. Publ. of the Centre National d'Études Spatiales, Presses Universitaires de France, 1966, 192 p.

11. *Physics of the Moon and Planets*. (Fizika Luny i Planet). Kiev, Naukova Dumka, 1966, 110 p.

12. *Figure and Motion of the Moon* (Figura i drizhenie Luny). Ed. A. A. Yakovkin. Kiev, Naukova Dumka, 1965, 137 p.

13. *Figure and Motion of the Moon* (Figura i drizhenie Luny). Kiev, Naukova Dumka, 1967 (in press).

14. *Moon and Planets*. Ed. A. Dollfus. Amsterdam, North Holland Publ. Co., 1967 (in press).

B. Lunar Probes

a. Hard Landings

1. Ranger VII Photographs of the Moon. Part I: Camera A Series, Sept. 1964, 226 p., NASA-SP-61 (N64-31723); Camera B Series, Dec. 1964, 200 p., NASA-SP-62; Camera P Series, Feb. 1965, 200 p., NASA-SP-63.

2. Ranger VII. Part II: Experimenters' Analyses and Interpretations. R. L. Heacock, G. P. Kuiper, E. M. Shoemaker, H. C. Urey, E. A. Whitaker. *Jet Propulsion Lab. Tech. Rep.* No. 32-700. Feb. 1965, 156 p. (N65-22162)
3. Ranger VIII Photographs of the Moon, Cameras A, B, and P. 1966, 478 p., NASA-SP-111. (N66-25366)
4. Ranger IX Photographs of the Moon, Cameras A, B, and P. 1966, 193 p., NASA-SP-112. (N66-26555)
5. Ranger VIII and IX. Part II: Experimenters' Analyses and Interpretations. R. L. Heacock, G. P. Kuiper, E. M. Shoemaker, H. C. Urey, E. A. Whitaker. *Jet Propulsion Lab. Tech. Rep.* No. 32-800, 1 March 1966, 182 p. (N66-25046)
6. Couderc, P. The Moon photographed by Ranger VII. *L'Astronomie*, **79**, 41. 1965. (P65-24334)
7. Kuiper, G. P. The surface structure of the Moon. p. 99-105 in ref. **A8.** (A66-35441)
8. Kuiper, G. P. Lunar results from Rangers VII to IX. *Sky and Tel.*, **29**, 293, 1965. (A65-23739)
9. Kuiper G. P. Interpretation of the Ranger records. p. 24-29 in ref. **A9.** (N66-31448)
10. Nicks, O. W. The Ranger Lunar Missions. COSPAR, 6th Intern. Space Sci. Symposium, Buenos Aires, May 1965, 54 p. (A65-25918)
11. Schurmeier, H. M., Heacock, R. L., Wolte, A. E. The Ranger missions to the Moon. *Sci. Amer.*, **214**, 52, 1966. (A66-18360)
12. Shoemaker, E. M. The Moon close up. *Nat'l. Geographic*, **126**, No. 5, 690, 1964.
13. Smith, G. M., Vrebalovich, T., Willingham, D. E. Eyes on the Moon. *Astronautics and Aeronautics*, **4**, 74, 82, 1966.
14. Urey, H. C. Observations on the Ranger photographs. p. 3-21 in ref. **A8.** (A66-35438)
15. Urey H. C. Observations on the Ranger VIII and IX pictures. p. 1-23 in ref. **A9.** (N66-31447)

b. Soft Landings

1. The First Panoramic Views of the Lunar Surface (Pervyye Panoramy Lunnoy Poverkhnosti). Moscow, Izd. Nauka, 1966, 132 p. (N66-38486)
2. Davies, J. G., Lovell, B., Pritchard, R. S., Smith, F. G. Observations of the Russian Moon probe Luna 9. *Nature*, **209**, 848, 1966. (A66-23915)
3. Fielder, G., Wilson, L., Guest, J. E. The Moon from Luna 9. *Nature*, **209**, 851, 1966. (A66-23916)
4. Gault, D. E., Quaide, W. L., Oberbeck, V. R., Moore, H. J. Luna 9 photographs – evidence for a fragmental surface layer. *Science*, **153**, 985, 1966. (A66-40017)
5. Gold, T., Hapke, B. W. Luna 9 pictures – implications. *Science*, **153**, 290, 1966. (A66-40522)
6. Kuiper, G. P., Strom, R. G., LePoole, R. S., Whitaker, E. A. Russian Luna 9 pictures – provisional analysis. *Science*, **151**, 1561, 1966. (A66-25137)
7. Lipsky, Y. N. What Luna 9 told us about the Moon. *Sky and Tel.*, **32**, 257, 1966.
8. Shoemaker, E. M., Batson, R. N., Larson, K. B. An appreciation of the Luna 9 pictures. *Astronautics and Aeronautics*, **4**, 40, May 1966.
9. Surveyor I – preliminary results. *Science*, **152**, 1737, 1966. (A66-29657)
10. Pictures from the Moon. *Sky and Tel.*, **32**, 16, 1966.
11. Some Surveyor findings. *Sky and Tel.*, **32**, 63, 1966.

c. Far Side

1. Lipsky, Y. N. Zond-3 photographs of the Moon's far side. *Sky and Tel.*, **30**, 338, 1965. (A66-14023)
2. Acad. Sci. U.S.S.R., Report to COSPAR, Wien, 1966.
3. Hartmann, W. K. Lunar basins, lunar lineaments, and the Moon's far side. *Sky and Tel.*, **32**, 128, 1966. (A66-40544)
4. Kovalevsky, J. The far side of the Moon seen by Zond-3. *L'Astronomie*, **79**, 421, 1965. (A66-21755)
5. Lipsky, Y. N. The far side of the Moon. *Flight International*, **88**, 554, 1965. (A65-34814)

d. Orbiters

1. TASS Reports, *Pravda*, 5 April and 10 April, 1966.
2. Press-conference on Luna–10, *Pravda*, 17 April, 1966.
3. Acad. Sci. U.S.S.R. Report, *Pravda*, 3 June, 1966.
4. Vinogradov, A. P., Surkov, Y. A., Chernov, G. M. *Dokl. Akad. Nauk SSSR*, 1966, in press.
5. Vinogradov, A. P., Surkov, Y. A., Chernov, G. M., Kirnosov, F. F., Nasarkina, G. B. *Geokhimia*, No. 8, 1966.
6. Vinogradov, A. P., Surkov, Y. A. *Pravda*, 14 Sep. 1966.
7. Gors, W., Oberdorfer, H. Luna 10. *Sterne und Weltraum*, **5**, 215, 1966. (A66–40377)
8. Lunar Orbiter surveys the Moon. *Sky and Tel.*, **32**, 192, 1966.
9. Lunar Orbiter photographs of the Moon. *Sky and Tel.*, **32**, 346, 358, 1966.
10. Michael, W. H., Jr., Tolson, R. H., Gapcynski, J. P. Lunar orbiter – tracking data indicate properties of Moon's gravitational field. *Science*, **153**, 1102, 1966. (A66–40302)
11. *New York Times*, 1 Dec. 1966; *Sky and Tel.*, **33**, 22, Jan. 1967.

C. Cartography

1. The System of Lunar Craters.
   Quadrant II. *Communications of the Lunar and Planetary Laboratory*, No. 40, D. W. G. Arthur, A. P. Agnieray, R. A. Horvath, C. A. Wood, C. R. Chapman. March 1964. (N65–18937)
   Quadrant III. *Ibid.*, No. 50. D. W. G. Arthur, A. P. Agnieray, R. H. Pellicori, C. A. Wood, T. Weller. Feb. 1965. (A66–11465)
   Quadrant IV. *Ibid.*, No. 70. D. W. G. Arthur, R. H. Pellicori, C. A. Wood. May 1966. (N66–38712)
2. Lunar Designations and Positions.
   Quadrant II. D. W. G. Arthur, A. P. Agnieray, August 1964;
   Quadrant III. D. W. G. Arthur, A. P. Agnieray, April 1965;
   Quadrant IV. D. W. G. Arthur, R. H. Pellicori, Sep. 1966. Univ. Arizona Press, Tucson.
3. United States Geological Survey Maps of the Moon:
   The Riphaeus Mountains region, Map I–458, R. Eggleton, 1965.
   The Timocharis region, Map I–462, M. H. Carr, 1965.
   The Montes Apenninus region, Map I–463, R. J. Hackman, 1966.
   The Aristarchus region, Map I–465, H. J. Moore, 1965.
   The Pitatus region, Map I–485, N. J. Trask, S. R. Titley, 1966.
   The Mare Serenitatis region, Map I–489, M. H. Carr, 1966.
   The Helvelius region, Map I–491, J. F. McCauley, 1967.
   The Mare Humorum region, Map I–495, S. R. Titley, 1967.
   The Julius Caesar quadrangle, Map I–510, E. C. Morris, D. E. Wilhelms, 1967.
4. Ranger Lunar Charts. United States Air Force, Aeronautical Chart and Information Center, St. Louis. Ranger VII, RLC 1–5, 1964; Ranger VIII, RLC 6–12, 1966; Ranger IX, RLC 13–17, 1966.
5. Arthur, D. W. G. The reduction of measures for position on a single lunar photograph. *Comm. Lunar Planet. Lab.*, **4**, 77, 1965. (A66–29969)
6. Arthur, D. W. G. Selenodetic measures on Yerkes lunar photograph No. 1170. *Ibid.*, **4**, 81, 1965. (A66–29970)
7. Arthur, D. W. G. The computation of selenodetic coordinates using the librations. *Ibid.*, **4**, 89, 1965. (A66–29971)
8. Arthur, D. W. G. Measures on star-trailed lunar photographs and their initial reductions. *Ibid.*, No. 71, Jan. 1966.
9. Arthur, D. W. G. Scale transfer for lunar photographs. *Ibid.*, No. 73, Feb. 1966.
10. Arthur, D. W. G. On the validity of selenodetic positions. *Ibid.*, No. 74, March 1966.
11. Arthur, D. W. G. A method for the determination of the Moon's constants of rotation from measurements on scaled and oriented lunar photographs. *Ibid.*, No. 75, May 1966.

12. Arthur, D. W. G. Selenodetic measures on Yerkes lunar photograph No. 1269. *Ibid.*, No. 76, July 1966.
13. Arthur, D. W. G. Selenodetic measures on Yerkes lunar photograph No. 482. *Ibid.*, No. 77, July 1966.
14. Bird, T. H. Large-scale lunar photogrammetry. *Photogrammetric Engineering*, **32**, 227, 1966. (A66–25601)
15. Bray, T. A., Goudas, C. L. A contour map based on the selenodetic control system of the Aeronautical Chart and Information Center of the U.S. Air Force. *Boeing Math. Note* No. 457, 20 p., March 1966. (N66–31418)
16. Eckhardt, D. H. On the harmonic analysis of absolute lunar elevations. *Icarus*, **5**, 313, 1966. (P66–28583; A66–33019)
17. Eriksen, J. G. Lunar charting. *Air Univ. Rev.*, **16**, 76, 1965. (A65–27043)
18. Goudas, C. L. The selenodetic control system of the U.S. Army Map Service. *Icarus*, **4**, 188, 1965. (A65–26047)
19. Herring, A. K. Preliminary drawings of lunar limb areas. IV. *Comm. Lunar Planet. Lab.*, **3**, 13, 1965. (A66–11496)
20. Herring, A. K. Preliminary drawings of lunar limb areas. V. *Comm. Lunar Planet. Lab.*, **3**, 15, 1965. (A66–11497)
21. Hunt, M. S. Progress in selenodesy. p. 1220–35 in ref. **A3.** (A65–34276)
22. Kopal, Z., Rackham, T. W. Photographic resolution on the lunar surface from ground-based facilities. *Icarus*, **2**, 329, 1963. (A64–14463)
23. Meyer, D. L., Ruffin, B. W. Coordinates of lunar features. *Icarus*, **4**, 513, 1965. (A66–19263; P66–13166)
24. Milton, D. J., Wilhelms, D. E. Geology from a relatively distant Ranger VIII photograph. p. 302–13 in ref. **Ba5.** (N66–25055)
25. Moore, H. J., Lugn, R. V. Experimental topographic map of a small area of the lunar surface from the Ranger VIII photographs. p. 295–302 in ref. **Ba5.** (N66–25054)
26. Morris, E. C., Wilhelms, D. E. Preliminary geologic map of the Julius Caesar quadrangle. ref. **A7,** map supplement.
27. Pohn, H. A. The Serenitatis Bench and the bond formation. p. 9–12, ref. **A7a.**
28. Riesen, A. A partial explanation of the divergence between visual and photographic measurements of relative elevations on the Moon. *Arch. Sci. Switz.*, **17**, 225, 1964. (P65–15825)
29. Rindfleisch, T. Photometric method for lunar topography. *Photogrammetric Engineering*, **32**, 262, 1966. (A66–25603)
30. Rowan, L. C., West, M. Preliminary albedo map of the lunar equatorial belt. p. 101–14 in ref. **A7a.**
31. Samaha, A. H., Kopal, Z. Lunar photography with the 74-inch reflector of the Helwan Observatory at Kottamia. *Icarus*, **5**, 309, 1966. (A66–33018; P66–28585)
32. Strom, R. G. Analysis of lunar lineaments. I–Tectonic maps of the Moon. *Comm. Lunar Planet. Lab.*, **2**, 205, 1964. (A65–23264)
33. Trask, N. J. Preliminary geologic map of a small area in Mare Tranquillitatis. p. 319–26 in ref. **Ba5.** (N66–25057)
34. Trask, N. J. Preliminary report on the geology of the Byrgius region of the Moon. p. 3–8 in ref. **A7a.**
35. Triplet, J.-M. Lunar topography and cartography. *L'Astronomie*, **80**, 183, 1966. (A66–32132).
36. Wildey, R. L. A computer program for the transformation of lunar observations from celestial to selenographic coordinates. *Icarus*, **3**, 136, 1964. (A64–28018)
37. Wilhelms, D. E. Fra Mauro and Cayley formations in the Mare Vaporum and Julius Caesar regions of the Moon. p. 13–27 in ref. **A7a.**
38. Wilhelms, D. E., Masursky, H., Bender, A. B., Ryan, J. D. Preliminary geologic mapping of the easternmost part of the lunar equatorial belt. p. 45–54 in ref. **A7a.**
39. Wilhelms, D. E., Trask, N. J. Compilation of geology in the lunar equatorial belt. p. 29–34 in ref. **A7a.**
40. Lisina, L. R., Shevchenko, V. V. p. 80–91 in ref. **A12.**

41. Lisina, L. R. p. 95–109 in ref. **A11.**
42. Markov, A. V. *Dokl. Akad. Nauk SSSR*, **167**, No. 1, 1966.
43. Mukhamedzhanov, A. K., Stanjukovich, K. P. *Kosmicheskie Issledovania*, **4**, 408, 1966.
44. Gavrilov, I. V. In ref. **A12.**
45. Gavrilov, I. V., Duma, D. P., Kislyuk, V. S. In ref. **A12.**
46. Gavrilov, I. V., Duma, D. P., Kislyuk, V. S. In ref. **A13.**
47. Gavrilov, I. V. In ref. **A13.**

D. Photometry

1. Ashby, N. Energy balance on the lunar surface. *Publ. astr. Soc. Pacif.*, **78**, 254, 1966. (A66–38052)
2. Bouska, J., Mayer, P. Photoelectric observation of the lunar eclipse of 1964 June 24–25. *Bull. astr. Inst. Csl.*, **16**, 252, 1965. (A65–32576)
3. Cohen, H. L., Dyck, H. M., Young, A. Intensity and color gradients in the Earth's umbra of 1964 December 19. *Astr. J.*, **71**, 270, 1966. (P66–28575; A66–32068)
4. Coyne, G. V. Differential colors on the Moon. *Astr. J.*, **70**, 115, 1965. (P65–10162)
5. Diggelen, J. van The radiance of lunar objects near opposition. *Planet. Space Sci.*, **13**, 271, 1965. (P65–27804)
6. Feinstein, A. Photoelectric observations of the Moon during the total lunar eclipse of 1964 June 24–25. *Bull. astr. Inst. Csl.*, **17**, 163, 1966. (A66–39858)
7. Gallouët, L. New measurement of the apparent magnitude of the Sun and of the full Moon. *C. R. Acad. Sci.* (France), **256**, 4593, 1963. (P64–2415)
8. Hämeen-Anttila, K. A., Laakso, P., Lumme, K. The shadow effect in the phase curves of lunar type surfaces. *Ann. Acad. Sci. Fennicae A* VI (Finland), No. 172, 15p. 1965. (P65–27812)
9. Hansen, J. E., Matsushima S. Light illuminance and color in the Earth's shadow. *J. geophys. Res.*, **71**, 1073, 1966. (A66–22570)
10. Hapke, B. W., Van Horn, H. Photometric studies of complex surfaces, with applications to the Moon. *J. geophys. Res.*, **68**, 4545, 1963. (P64–2394)
11. Hapke, B. W. A theoretical photometric function for the lunar surface. *J. geophys. Res.*, **68**, 4571, 1963. (P64–2395)
12. Hapke, B. W. An improved theoretical lunar photometric function. *Astr. J.*, **71**, 333, 1966. (P66–32296)
13. Hardie, R. H. Three-color photometry of the 1960 March lunar eclipse. *Publ. astr. Soc. Pacif.* **76**, 257, 1964. (P65–10164)
14. Lebedeva, I. I. Spectrophotometric comparison of the lunar surface with certain volcanic flows. *Trudy astr. Obs. Leningrad. gos. Univ.*, **21**, 99, 1964. (A65–25695)
15. Link, F. Two remarks concerning the terrestrial shadow on the Moon. *Bull. astr. Inst. Csl.*, **17**, 161, 1966. (A66–39857)
16. Marin, N., Zahn, J. P. The eclipse of the Moon, 1964 June 24–25. *L'Astronomie*, **79**, 14, 1965. (P65–15826)
17. Mironova, M. N. Spectrophotometry of details on the lunar surface. p. 16–29, in ref. **A2.** (A65–18969)
18. Moore, P. The partial lunar eclipse of 1965 June 13–14. *J. Brit. astr. Assoc.*, **75**, 339, 1965. (P66–6060)
19. Oetking, P. Photometric studies of diffusely reflecting surfaces with applications to the brightness of the Moon. *J. geophys. Res.*, **71**, 2505, 1966. (P66–25153)
20. Hapke, B. W. Comments on paper by Philip Oetking, 'Photometric studies of diffusely reflecting surfaces with applications to the brightness of the Moon'. *J. geophys. Res.*, **71**, 2515, 1966.
21. Rydgren, B. Photometric measurements of earthshine on the Moon. *Stockholms Obs. Medd.*, No. 154.
22. Scott, N. W. Colour on the Moon. *Nature*, **204**, 1075, 1964. (P65–10161)
23. Sergeeva, A. N. Spectral investigations of the lunar surface. p. 30–45 in ref. **A2.** (A65–18970)

24. Sharonov, V. V. Some results of photometric and colorimetric comparison of terrestrial volcanic crusts with the lunar surface. p. 740–50 in ref. **A3.** (A65–34248)

25. Sharonov, V. V. A photometric investigation of the presence of outer layers of volcanic origin on the Moon. *Astr. Zu.*, **42,** 136, 1965. (P65–13741; A65–20656)

26. Sytinskaya, N. N. Albedo values for separate features of the lunar surface. *Astr. Zu.*, **40,** 1083, 1963. (P65–4626)

27. Sytinskaya, N. N. Photometric data for and against the presence of widely distributed volcanic activity on the Moon. p. 756–767, in ref. **A3.** (A65–34250)

28. Wildey, R. L. The Moon's photometric function. *Nature*, **200,** 1056, 1963. (P64–23890)

29. Wildey, R. L., Pohn, H. A. Detailed photoelectric photometry of the Moon. *Astr. J.*, **69,** 619, 1964. (A65–26789)

30. Barabashov, N. P., Ezersky, V. I., Prishljak, N. P. *Zirk. Kharkov Obs.*, No. 25, 1962.

31. Barabashov, N. P., Ezersky, V. I. *Vestnik Kharkov Gos. Univ.*, No. 4, issue 1, ser. astr., 22–42, 1965.

32. Barabashov, N. P., Ezersky, V. I., Ezerskaya, V. A. *Ibid.*, issue 2, 12–25, 1965.

33. Akimov, A. A. *Ibid.*, issue 1, 43–61, 1965.

34. Opriatova, V. S. *Ibid.*, 62–64.

35. Petrova, N. N. *Izv. glav. astr. Obs. Pulkove*, **24,** 168, 1965.

36. Petrova, N. N. *Ibid.*, No. 182.

37. Sharonov, V. V. *Izv. Komissii po Fisike Planet*, issue 4, 24, 1963.

38. Mironova, M. N. Spectral reflective characteristics of the lunar surface. p. 5–11 in ref. **A11.**

39. Bell, B., Wolbach, J. G. Lunar eclipses and the forecasting of solar minima. *Icarus*, **4,** 409, 1965. (P66–6121)

40. Matsushima, S. Apparent correlation between the lunar eclipse brightness and the solar wind. *Nature*, **211,** 1027, 1966.

41. Matsushima, S. Variation of lunar eclipse brightness and its association with the geomagnetic planetary index Kp. *Astr. J.*, **71,** 699, 1966.

E. Polarimetry

1. Avramchuk, V. V. Polychromatic polarimetry of some lunar areas. p. 3–15 in ref. **A2.** (A65–18968)

2. Clarke, D. Studies in astronomical polarimetry, III. The wavelength dependence of the polarization of light reflected by the Moon and Mars. *Mon. Not. R. astr. Soc.*, **130,** 83, 1965, (P65–33883)

3. Coffeen, D. L. Wavelength dependence of polarization, IV. Volcanic cinders and particles. *Astr. J.*, **70,** 403, 1965. (P65–27794; A65–32411)

4. Dollfus, A., Geake, J. E. The alteration of the polarimetric properties of the lunar surface under the action of solar protons. *C. R. Acad. Sci.*, **260,** 4921, 1965. (A65–25246)

5. Dollfus, A. The application of polarized light for the study of the surface of the Moon. p. 19–30 in ref. **A8.** (A66–35445)

6. Gehrels, T., Coffeen, T., Owings, D. Wavelength dependence of polarization. III. The lunar surface. *Astr. J.*, **69,** 826, 1964. (P65–10163; A65–30647)

7. Hopfield, J. J. Mechanism of lunar polarization. *Science*, **151,** 1380, 1966. (P66–22071)

8. Marin, M. Photoelectric measurements of polarization with a coudé telescope of 1 m. *Rev. Opt.*, **44,** 115, 1965. (P65–31234)

9. Wilhelms, D. E., Trask, N. J. Polarization properties of some lunar geologic units. p. 63–80 in ref. **A7a.**

10. Kokhan, E. K. *Izv. glav. astr. Obs. Pulkove*, **24,** issue 2, No. 178, 182, 1965.

11. Lipsky, Y. N., Bondarenko, L. N. *Soobšč. gos. astr. Inst. P. K. Sternberga*, 1966 (in press).

12. Morozhenko, A. V. Investigation of polarizable properties of light reflected by various specimens of terrestrial rock. p. 70–82 in ref. **A11.**

13. Dzhapiashvili, V. P., Xanfomaliti, L. V. p. 275 in ref. **A8.**

14. Pospergelis, M. M. *Astr. Zu.*, **42,** 398, 1965.

F. Thermal Properties and Infrared Studies

1. Bastin, J. A. Lunar hot spots. *Nature*, **207**, 1381, 1965. (A66–10884; P66–6071)
2. Binder, A. B., Cruikshank, D. P., Hartmann, W. K. Observations of the Moon and of the terrestrial rocks in the infrared. *Icarus*, **4**, 415, 1965. (P66–6068; A65–35367)
3. Burns, E. A., Lyon, R. J. P. Errors in the measurement of the lunar temperature. *J. geophys. Res.*, **69**, 3771, 1964. (P65–15833; A64–26097)
4. Hunt, G. R., Salisbury, J. W. Lunar surface features: mid-infrared spectral observations. *Science*, **146**, 641, 1964. (P65–4623; A64–28415)
5. Ingrao, H. C., Young, A. T., Linsky, J. L. A critical analysis of lunar temperature measurements in the infrared. p. 185–211 in ref. **A8.** (A66–35447)
6. Linsky, J. L. Models of the lunar surface including temperature-dependent thermal properties. *Harvard Coll. Obs. Sci. Rep.* No. 8, Jan. 1966, 90 p. (N66–21679)
7. Low, F. J. Lunar night-time temperatures measured at 20 microns. *Astrophys. J.*, **142**, 806, 1965. (P65–33929)
8. Markov, M. N., Khokhlova, V. L. Coefficients of emission in the infrared region of the spectrum and differences in the parameter $\gamma = (k\rho c)^{-1/2}$ for the sea and continent regions of the lunar surface. *Dokl. Akad. Nauk SSSR*, **157**, 826, 1964. (A65–17942)
9. Markov, M. N., Khokhlova, V. L. The difference in the rate of heating of the east and west limbs of the Moon after an eclipse. *Astr. Zu.*, **42**, 386, 1965. (P65–27795)
10. Moroz, V. I. Infrared spectrophotometry of the Moon and the Galilean satellites of Jupiter. *Astr. Zu.*, **42**, 1287, 1965. (P66–9182)
11. Murcray, F. H. The spectral dependence of lunar emissivity. *J. geophys. Res.*, **70**, 4959, 1965. (P66–6070)
12. Murray, B. C., Wildey, R. L. Surface temperature variations during the lunar nighttime. *Astrophys. J.*, **139**, 734, 1964. (P64–15929; A64–16828)
13. Murray, B. C. Current problems in the interpretation of lunar physical observations. p. 32–33 in ref. **A9.** (N66–31450)
14. Ryadov, V. Y., Furashov, N. I., Sharonov, G. A. Measurements of the Moon's thermal radiation in the infrared. *Astr. Zu.*, **41**, 112, 1964. (P64–13915)
15. Saari, J. M. The surface temperature of the antisolar point of the Moon. *Icarus*, **3**, 161, 1964. (A64–28020)
16. Saari, J. M., Shorthill, R. W. Thermal anomalies on the totally eclipsed Moon of 1964 December 19. *Nature*, **205**, 964, 1965. (P65–20966)
17. Saari, J. M., Shorthill, R. W. Isotherms in the equatorial region of the totally eclipsed Moon. *Boeing Sci. Res. Doc.* D1–82–0530, Apr. 1966. (N66–31340)
18. Saari, J. M., Shorthill, R. W. Hot spots on the Moon. *Sky and Tel.*, **31**, 327, 1966.
19. Saari, J. M., Shorthill, R. W., Deaton, T. K. Infrared and visible images of the lunar surface during the eclipsed Moon of 1964 December 19. *Icarus*, in press, 1966. *Boeing Sci. Res. Doc.* D1–82–0533. (N66–38578)
20. Shorthill, R. W., Saari, J. M. Recent discovery of hot spots on the lunar surface – a brief report of infrared measurements on the eclipsed Moon. p. 215–28 in ref. **A8.** (A66–35448)
21. Shorthill, R. W., Saari, J. M. Nonuniform cooling of the eclipsed Moon – a listing of thirty prominent anomalies. *Science*, **150**, 210, 1965. (A66–11302)
22. Shorthill, R. W., Saari, J. M. Radiometric and photometric mapping of the Moon through a lunation. p. 722–739 in ref. **A3.** (A65–34247)
23. Shorthill, R. W., Saari, J. M. Recent results of lunar eclipse measurements. (To be publ. in Proc. of 4–6 May 1965 meeting of the Amer. Astronautical Society)
24. Troitsky, V. S. Investigation of the surfaces of the Moon and planets by thermal radiation. p. 34–58 in ref. **A9.** (N66–31451)
25. Troitsky, V. S. Some remarks on new measurements of lunar temperature at the antisolar point. p. 59–60 in ref. **A9.** (N66–31452)
26. Watson, K. Small-scale roughness from lunar infrared emission. p. 55–60 in ref. **A7a.**
27. Wattson, R. B., Hapke, B. W. A comparison of the infrared spectra of the Moon and simulated lunar surface materials. *Astrophys. J.*, **144**, 364, 1966. (P66–22072; A66–28126)

28. Wattson, R. B., Danielson, R. E. The infrared spectrum of the Moon. *Astrophys. J.*, **142**, 16, 1965. (P65–27793)
29. Winter, D. F. On the nonuniform cooling behavior of the eclipsed Moon. *Boeing Sci. Res. Lab. Doc.* D1–82–0531, May 1966, 10 p. (N66–34528)
30. Winter, D. F. Transient radiative heat exchange at the surface of the Moon. *Boeing Sci. Res. Lab. Doc.* D1–82–0559, Aug. 1966, 20 p.
31. Markov, M. N., Khokhlova, V. L. *Izv. Komissii po Fizike Planet*, No. 5, 1966, in press.
32. Chistjakov, Y. N. *Izv. glav. astr. Obs. Pulkove*, **24**, issue 2, No. 178, 175–181, 1965.

G. Radio Studies

1. Baldock, R. V., Bastin, J. A., Clegg, R. E., Gaitskell, J. N., Gear, A. E. Lunar eclipse observations at 1-mm wavelength. *Astrophys. J.*, **141**, 1289, 1965. (P65–18543)
2. Beckmann, P. Radar backscatter from the surface of the Moon. *J. geophys. Res.*, **70**, 2345, 1965. (P65–20964; A65–24794)
3. Beckmann, P., Klemperer, W. K. Interpretation of the angular dependence of back-scattering from the Moon and Venus. *J. Res. nat. Bur. Stand.*, **69D**, 1671, 1965. (A66–20127; P66–13159)
4. Daniels, F. B. Radar determination of lunar slopes: correction for the diffuse component. *J. geophys. Res.*, **68**, 2864, 1963. (P64–2457)
5. Daniels, F. B. Radar reflections from a rough Moon described by a composite correlation function. *J. geophys. Res.*, **68**, 6251, 1963. (P64–13942)
6. Davis, J. R., Rohlfs, D. C. Lunar radio-reflection properties at decameter wavelengths. *J. geophys. Res.*, **69**, 3257, 1964. (P65–15828)
7. Davis, J. R., Rohlfs, D. C., Skaggs, G. A., Joss, J. W. Decameterwave radar studies of the lunar surface. *J. Res. nat. Bur. Stand.*, **69D**, 1661, 1965. (P66–13157; A66–20125)
8. Drake, F. Radio measurements of the Moon. p. 19–30 in ref. **A8.** (A66–35453)
9. Dyce, R. B., Pettengill, G. H. [430 Mc/s] Radar observation of the Moon and the four nearest planets using the Arecibo antenna. *IEEE Northeast Electronics Res. Eng. Meeting*, **17**, 182, 1965. (P66–28581)
10. Evans, J. V., Hagfors, T. On the interpretation of radar reflections from the Moon. *Icarus*, **3**, 151, 1964. (A64–28019)
11. Evans, J. V. Radar studies of the Moon. *J. Res. nat. Bur. Stand.*, **69D**, 1639, 1965. (P66–13156; A66–20124)
12. Fung, A. K., Moore, R. K. Effects of structure size on Moon and Earth radar returns at various angles. *J. geophys. Res.*, **69**, 1075, 1964. (P65–15830)
13. Fung, A. K. Theory of radar scatter from rough surfaces, bistatic and monostatic, with application to lunar radar return. *J. geophys. Res.*, **69**, 1063, 1964. (P65–14109)
14. Gary, B., Stacey, J., Drake, F. D. Radiometric mapping of the Moon at 3 millimeters wavelength. *Astrophys. J. Suppl.*, **12**, 239, 1965. (A66–20389)
15. Giraud, A. A note on the radio reflectivity of the lunar surface. *J. Res. nat. Bur. Stand.*, **69D**, 1679, 1965. (P66–13160; A66–20128)
16. Hagfors, T. Relationship of geometric optics and autocorrelation approaches to the analysis of lunar and planetary radar. *J. geophys. Res.*, **71**, 379, 1966. (P66–19068)
17. Hagfors, T., Brockelman, R. A., Danforth, H. H., Hanson, L. B., Hyde, G. M. Tenuous surface layer on the Moon: evidence derived from radar observations. *Science*, **150**, 1153, 1965. (P66–9181; A66–13895)
18. Hagfors, T. Review of radar observations of the Moon. p. 229–39 in ref. **A8.** (A66–35449)
19. Hayre, H. S. Acoustic simulation of lunar echoes. *J. geophys. Res.*, **70**, 3831, 1965. (P65–30703; A65–32672)
20. Kamenskaya, S. A., *et al.* Observation of radio eclipses of the Moon at millimeter wavelengths. *Izv. Vyssih Učeb. Zaved., Radiofiz.*, **8**, 219, 1965. (P66–28576)
21. Katz, I. Wavelength dependence of the radar reflectivity of the Earth and the Moon. *J. geophys. Res.*, **71**, 361, 1966. (P66–16169)

22. King, H. E., Jacobs, E., Stacey, J. M. A 2·8 arc-min beamwidth antenna – lunar eclipse observations at 3·2 mm. *IEEE Transactions on Antennas and Propagation*, **AP-14**, 82, 1966. (A66–22551)
23. Klemperer, W. K. Angular scattering law for the Moon at 6-meter wavelength. *J. geophys. Res.*, **70**, 3798, 1965. (A65–31280; P66–6069)
24. Kokurin, Y. L., Kurbasov, V. V., Lobanov, V. M., Mozhzherin, V. M., Sukhanovskii, A. N., Chernykh, N. S. Measurement of the distance to the Moon by optical radar. *Zu. eksper. teor. Fiz. Pis'ma*, **3**, 219, 1966. (P66–28578)
25. Krotikov, V. D. Averaging effect of antenna radiation pattern in measurements of the radio emission of the Moon. *Izv. Vyssih Učeb. Zaved., Radiofiz.*, **8**, 453, 1965. (P66–28582; A65–36555)
26. Krotikov, V. D., Troitsky, V. S. Radio emission and nature of the Moon. *Uspekhi Fiz. Nauk*, **81**, 589, 1963. (P64–29471)
27. Low, F. J., Davidson, A. W. Lunar observations at a wavelength of 1 millimeter. *Astrophys. J.*, **142**, 1278, 1965. (P66–13162; A66–15290)
28. Lynn, V. L., Sohigian, M. D., Crocker, E. A. Radar observations of the Moon at a wavelength of 8·6 millimeters. *J. geophys. Res.*, **69**, 781, 1964. (P65–15827)
29. Orszag, A. Moon distance measurement by laser. *J. Res. nat. Bur. Stand.*, **69D**, 1683, 1965. (P66–13161; A66–20129)
30. Pettengill, G. Lunar radar reflections. p. 355–69 in *Solar System Radio Astronomy*, 1965, ed. by J. Aarons. New York: Plenum Press. (A65–27036; P66–25146)
31. Plechkov, V. M. Observations of the lunar eclipse of 1964 June 25, at 1·8 cm. *Astr. Zu.*, **43**, 172, 1966. (P65–19066; A66–23507)
32. Rea, D. G., Hetherington, N., Mifflin, R. The analysis of radar echoes from the Moon. *J. geophys. Res.*, **69**, 5217, 1964. (P65–15829; A65–13509)
33. Rea, D. G., Hetherington, N., Mifflin, R. A note on 'The analysis of radar echoes from the Moon'. *J. geophys. Res.*, **70**, 1565, 1965. (A65–22458; P65–24335)
34. Salomonovich, A. E. Measurements of lunar radio brightness distribution and certain properties of its surface layer. *J. Res. nat. Bur. Stand.*, **69D**, 1618, 1965. (P66–13155)
35. Staelin, D. H., Barrett, A. H., Kusse, B. R. Observations of Venus, the Sun, Moon, and Tau A at 1·18-cm wavelength. *Astr. J.*, **69**, 69, 1964. (P64–13959)
36. Taylor, G. N. Some curious anomalies in the Faraday rotation of lunar echoes. *J. atmos. terr. Phys.*, **27**, 118, 1965. (P65–27802)
37. Thompson, T. W. Lunar mapping by coherent-pulse analysis. *J. Res. nat. Bur. Stand.*, **69D**, 1669, 1965. (P66–13158; A66–20126)
38. Troitsky, V. S. The radio emission of the eclipsed Moon. *Astr. Zu.*, **42**, 1296, 1965. (A66–19459; P66–9179)
39. Troitsky, V. S. Investigation of the surfaces of the Moon and planets by the thermal radiation. *J. Res. nat. Bur. Stand.*, **69D**, 1587, 1965. (P66–13154)
40. Weaver, H. The interpretation of thermal emission from the Moon. p. 295-354 in *Solar System Radio Astronomy*, ed. J. Aarons. New York: Plenum Press, 1965. (A65–27035; P66–16384)
41. Welch, W. J., Thornton, D. D., Winter, S. The variation of lunar radio emission during an eclipse. *J. geophys. Res.*, **70**, 2793, 1965. (P65–27801)
42. Yaplee, B. S., Knowles, S. H., Shapiro, A., Craig, K. J., Brouwer, D. The mean distance to the Moon as determined by radar. *Bull. astr.*, **25**, 81, 1965. (P65–30705)
43. Kislyakov, A. G., Salomonovich, A. E. *Radiofizika*, **6**, No. 3, 431, 1963. (P66–9179; A66–19459)
44. Gol'nev, V. I., Soboleva, N. S. *Izv. glav. astr. Obs. Pulkove*, **23**, No. 172, issue 3; 83–86, 1964.
45. Grasjuk, A. Z., *et al.* *Dokl. Akad. Nauk SSSR*, **154**, 1303, 1964.
46. Kokurin, Y. L., *et al.* *Kosmicheskie–Issledovania*, **4**, issue 3, 412, 1966.
47. Mayer, C. H. Thermal radio radiation from the Moon and planets. *IEEE Trans. Military Electr.*, **Mil-8**, 236, 1964. (P65–7630; A64–28200)
48. Seling, T. V. Observations of total eclipses of the Moon at a wavelength of 1·82 cm. *J. geophys. Res.*, **71**, 3339, 1966. (P66–35301)

H. Lunar Surface Formations

a. General

1. Brock, B. B. Lunar and terrestrial structural models on a common sphere. p. 617–30 in ref. **A3.** (A65–34241)
2. Dietz, R. S. Astroblemes, lunar craters, and maria. p. 895–6 in ref. **A3.** (A65–34257)
3. Diggelen, J. van The linear network of lunar surface features. *Bull. astr. Inst. Netherl.*, **18,** 311, 1966. (A66–38637)
4. Fielder, G. Astrogeology-lunar geology. *Earth-Science Rev.*, **1,** 231, 1966.
5. Geyer, R. A., Van Lopik, J. R. Use of geophysical measurements in lunar surface analysis. p. 1160–74 in ref. **A3.** (A65–34272)
6. Heacock, R. L. Ranger block III. p. 7–34 in ref. **Ba5.** (N66–25047)
7. Khodak, Y. A. Principal structural elements of the Moon and the significance of the geographic-geological approach. p. 641–55 in ref. **A3.** (A65–34243)
8. Kopal, Z. Topography of the Moon. *Space Sci. Rev.*, **4,** 737, 1965. (A66–18265)
9. Krejci-Graf, K. Interpretation of lunar surface features. p. 751–5 in ref. **A3.** (A65–34249)
10. Kuiper, G. P., Strom, R. G., Le Poole, R. S. Interpretation of the Ranger records. p. 35–248 in ref. **Ba5.** (N66–25048)
11. McCauley, J. F. The nature of the lunar surface as determined by systematic geologic mapping, in *The Mantles of the Earth and Terrestrial Planets*, John Wiley and Sons (in press).
12. Moore, H. J. Cohesion of material on the lunar surface. p. 263–270 in ref. **Ba5.** (N66–25050)
13. O'Keefe, J. A. The Moon. p. 631–67 in *Introduction to Space Science*, ed. by W. N. Hess. New York, Gordon and Breach Science Publishers, 1965. (A66–15758)
14. Rae, W. L. Lunar domes. *J. Brit. astr. Assoc.*, **76,** 319, 1966. (A66–41209)
15. Shoemaker, E. M. Lunar patterned ground. p. 275–84 in ref. **Ba5.** (N66–25052)
16. Urey, H. C. Observations on the Ranger VIII and IX pictures. p. 339–62 in ref. **Ba5.** (N66–25060)
17. Whitaker, E. A. The surface of the Moon. p. 79–98 in ref. **A8.** (A66–35440)

b. Craters

1. Allen, D. A. The lunar crater Caramuel. *J. Brit. astr. Assoc.*, **76,** 248, 1966. (A66–35245)
2. Baldwin, R. B. The origin of lunar features. p. 543–546 in ref. **A3.** (A65–34237)
3. Baldwin, R. B. The crater diameter-depth relationship from Ranger VII photographs. *Astr. J.*, **70,** 545, 1965. (P66–6066; A66–12186)
4. Boneff, N. The origin of the lunar craters. p. 524 in ref. **A3.**
5. Brinkmann, Robert T. Lunar crater distribution from the Ranger VII photographs. *J. geophys. Res.*, **71,** 340, 1966. (A66–19417)
6. Bülow, K. von The central problem of selenogeology. *Scientia*, **100,** 5, 1965. (A65–27715)
7. Byron, J. F. Distribution of lunar craters. *Observatory*, **86,** 31, 1966. (A66–24033)
8. Carr, M. H. The structure and texture of the floor of Alphonsus. p. 270–5 in ref. **Ba5.** (N66–25051)
9. Chapman, R. G., Fielder, G. On the central peaks of lunar craters. *Observatory*, **84,** 23, 1964. (P64–23893)
10. Currie, K. L. Analogues of lunar craters on the Canadian shield. p. 915–940 in ref. **A3.** (A65–34259)
11. Danes, Z. F. Rebound processes in large craters. p. 81–99 in ref. **A7a.**
12. Dodd, R. T., Smalley, V. G., Salisbury, J. W., Adler, J. E. M. Crater frequency evidence for volcanism in the lunar highlands. p. 555–61 in ref. **A3.** (A65–34238)
13. Fielder, G. Distribution of craters on the lunar surface. *Mon. Not. R. astr. Soc.*, **129,** 351, 1965. (P65–27803; A65–31182)
14. Fielder, G. Tests for randomness in the distribution of lunar craters. *Mon. Not. R. astr. Soc.*, **132,** 413, 1966. (P66–28584)
15. Ford, B. J. Origin of the lunar craters. *Spaceflight*, **7,** 13, 1965. (A65–15637)

16. Gault, D. E., Quaide, W. L., Oberbeck, V. R. Interpreting Ranger photographs from impact cratering studies. p. 125–40 in ref. **A8.** (A66–35443)

17. Green, J. Interpretation of Ranger VII photographs. p. 999–1002 in ref. **A3.** (A65–34262)

18. Green, J. The Moon's surface. *Int'l. Sci. and Tech.* p. 59–62, 64, 66, 67, 82, 84, Sept. 1966. (A66–40398)

19. Hartmann, W. K. On the distribution of lunar crater diameters. *Comm. Lunar Planet. Lab.*, **2,** 197, 1964. (A65–23263)

20. Hartmann, W. K. Secular changes in meteoritic flux through the history of the solar system. *Icarus*, **4,** 207, 1965. (A65–26048)

21. Hartmann, W. K. Early lunar cratering. *Icarus*, **5,** 406, 1966. (P66–32298; A66–38026)

22. Kopal, Z. The nature of secondary craters photographed by Ranger VII. *Icarus*, **5,** 201, 1966. (P66–25151; A66–29265)

23. Kopal, Z. The nature of secondary craters photographed by Ranger VII. *Boeing Math. Note* No. 430. 42 p., 1965. (N66–14550)

24. Kvíz, Z. On the central peak in lunar craters. *Bull. astr. Inst. Csl.*, **15,** 227, 1964. (A65–15341)

25. Levin, B. J. The modern form of the impact hypothesis of lunar relief formation. *Proc. XIIIth Internat. Astronautical Congress*, Wien: Springer-Verlag, 1964, **I,** p. 11–20. (P65–13740)

26. Marcus, A. A stochastic model of the formation and survival of lunar craters. I – Distribution of diameter of clean craters. *Icarus*, **3,** 460, 1964. (A65–17766)

27. Marcus, A. A stochastic model of the formation and survival of lunar craters. II – Approximate distribution of diameter of all observable craters, *Icarus*, **5,** 165, 1966. (A66–29262)

28. Marcus, A. A stochastic model of the formation and survival of lunar craters. III – Filling and disappearance of craters. *Icarus*, **5,** 178, 1966. (A66–29263)

29. Marcus, A. A stochastic model of the formation and survival of lunar craters. IV – On the nonrandomness of crater centers. *Icarus*, **5,** 190, 1966. (A66–29264)

30. McCall, G. J. H. The caldera analogy in selenology. p. 843–75 in ref. **A3.** (A65–34256)

31. McCauley, J. F. Intermediate-scale geologic map of a part of the floor of Alphonsus. p. 313–19 in ref. **Ba5.** (N66–25056)

32. Miller, B. P. Distribution of small lunar craters based on Ranger VII photographs. *J. geophys. Res.*, **70,** 2265, 1965. (P65–24333; A65–25018)

33. Moore, P. The lunar crater Caramuel. *J. Brit. astr. Assoc.*, **75,** 142, 1965. (P66–6061)

34. Pohn, H. A. New measurements of steep lunar slopes. *Publ. astr. Soc. Pacif.*, **75,** 186, 1963. (P64–7975)

35. Quaide, W. L., Gault, D. E., Schmidt, R. A. Gravitative effects on lunar impact structures. p. 563–72 in ref. **A8.** (A65–34239)

36. Roberts, W. A. Permanent angular displacement and ejecta-induced impulse associated with crater formation. *Icarus*, **4,** 480, 1965. (A66–19260)

37. Ronca, L. B. Structure of the crater Alphonsus. *Nature*, **209,** 182, 1966. (A66–20397)

38. Ronca, L. B., Salisbury, J. W. Lunar history as suggested by the circularity index of lunar craters. *Icarus*, **5,** 130, 1966. (P66–25147)

39. Safronov, V. S. The sizes of the largest bodies which fell on planets during their formation. *Astr. Zu.*, **42,** 1270, 1965. (P66–9183)

40. Shoemaker, E. When the irresistible force meets the immovable object. *Engr. Sci.*, **29,** 11, 1966. (A66–24199)

41. Shteinberg, G. S. Inapplicability of Baldwin's relation for determining the causes of the formation of lunar craters. *Dokl. Akad. Nauk SSSR*, **165,** 55, 1965. (A66–15332)

42. Smalley, V. G. The lunar crater Dionysius. *Icarus*, **4,** 433, 1965. (A65–35371)

43. Smalley, V. G., Ronca, L. B. Structures inside a lunar crater photographed by Ranger VII. *Nature*, **208,** 930, 1965. (P66–16387; A66–17456)

44. Smith, R. L. Terrestrial calderas, associated pyroclastic deposits, and possible lunar applications. p. 241–57 in ref. **A8.** (A66–35450)

45. Tazieff, H. Convective origin of lunar craters. p. 526–7 in ref. **A3.** (A65–34235)

46. Trask, N. J. Size and spatial distribution of craters estimated from the Ranger photographs. p. 252–63 in ref. **Ba5.** (N66–25049)
47. Trask, N. J. Distribution of lunar craters according to morphology from Ranger VIII and IX photographs, *Icarus*, **6,** 270, 1967.
48. Markov, A. V. The relative depths of lunar ring mountains and craters in Mare Nubium. *Dokl. Akad. Nauk SSSR*, **167,** 63, 1966. (A66–27182)

c. Maria

1. Dauvillier, A. Paleovolcanic origins of the lunar seas. p. 516–23 in ref. **A3.** (A65–34234)
2. Jaffe, L. D. Lunar dust depth in Mare Cognitum. *J. geophys. Res.*, **71,** 1095, 1966. (A66–22571)
3. Jaeger, R. M., Schuring, D. J. Spectrum analysis of terrain of Mare Cognitum. *J. geophys. Res.*, **71,** 2023, 1966. (A66–28336)
4. Kopal, Z. On the possible origin of the lunar maria. *Nature*, **210,** 188, 1966. (A66–27900)
5. Losovskii, B. Y., Salomonovich, A. E. The radio emission and differences in the upper layer of 'maria' and 'continental' regions of the Moon. *Astr. Zu.*, **42,** 390, 1965. (P65–27796; A65–26229)
6. Miyamoto, S. Maria surface of the Moon. *J. Internat. Lunar Soc.*, **2,** 132, 1964. (A66–12455)
7. Moore, P. Origin of the lunar maria. *Nature*, **210,** 1347, 1966. (P66–32297)
8. Quaide, W. Rills, ridges, and domes – clues to maria history. *Icarus*, **4,** 374, 1965. (P66–6067; A65–35362)
9. Ronca, L. B. A geological model of Mare Humorum. *Icarus*, **4,** 390, 1965. (A65–35363)
10. Salisbury, J. W., Smalley, V. G., Ronca, L. B. Origin of linear elements on Mare Humorum. *Nature*, **206,** 385, 1965. (A65–24360)
11. Shoemaker, E. M. Preliminary analysis of the fine structure of the lunar surface in Mare Cognitum. p. 23–77 in ref. **A8.** (A66–35439)
12. Shoemaker, E. M. Preliminary analysis of the fine structure of the lunar surface in Mare Cognitum. p. 75–134 in ref. **Ba2.**
13. Shoemaker, E. M. Geology of Mare Tranquillitatis as shown in the high-resolution Ranger VIII photographs. p. 327–32 in ref. **Ba5.** (N66–25058)

d. Ray Systems

1. Alter, D. The faint ray systems. p. 768–75 in ref. **A3.** (A65–34251)
2. Bülow, K. von The riddle of the 'bright rays'. *Sterne und Weltraum*, **5,** 43, 1966. (A66–22770)
3. Carr, M. H. Dark volcanic materials and rille complexes in the north central region of the Moon. p. 35–43 in ref. **A7a.**
4. Rackham, T. W. A note on lunar ray systems. *Icarus*, **4,** 544, 1965. (P66–25152; A66–19265)
5. Hartmann, W. K. Radial structures surrounding lunar basins. II – Orientale and other systems – Conclusions. *Comm. Lunar Planet. Lab.*, **2,** 175, 1964. (A65–23261)

I. Physical Properties and Chemical Composition of the Lunar Surface

1. Azmon, E. Melting temperature of complex silicates. p. 481–94 in ref. **A3.** (A65–34232)
2. Berg, C. A. Lunar erosion and Brownian motion. *Nature*, **204,** 461, 1964. (P65–4627; A65–11379)
3. Cameron, A. G. W., Schneck, P. B. Density of matter assembled from randomly incident particles. *Icarus*, **4,** 396, 1965. (P66–6062; A65–35360)
4. Elston, W. E. Rhyolite ash-flow plateaus, ring-dike complexes, calderas, lopoliths, and Moon craters. p. 817–42 in ref. **A3.** (A65–34255)
5. Fielder, G. Erosion and deposition on the Moon. *Planet. Space Sci.*, **11,** 1335, 1963. (P64–7976)
6. Fielder, G. Strike-slip faulting in the Vaporum region of the Moon. *Quart. J. geol. Soc. Lond.*, **120,** 275, 1964.

7. Gehrels, T. A model of the lunar surface. *Icarus*, **3**, 491, 1964. (A65–17770)

8. Gehrels, T. Some comments on Hapke's comments. *Icarus*, **5**, 160, 1966. (P66–25149)

9. Gehrels, T. Problems of the lunar surface texture in connection with cometary particles. *Mém. Soc. Roy. Liège*, **12**, 481, 1966. (P66–28586)

10. Gilvarry, J. J. Evidence for the pristine presence of a lunar hydrosphere. *Publ. astr. Soc. Pacif.*, **76**, 245, 1964. (P65–10165)

11. Gilvarry, J. J. Observability of indigenous organic matter on the Moon. *Icarus*, **5**, 228, 1966. (P66–28590)

12. Glaser, P. E., Wechsler, A. E. Small-scale structure of the lunar surface. *Icarus*, **4**, 104, 1965. (A65–23490; P65–27797)

13. Glaser, P. E., Wechsler, A. E., Germeles, A. E. Thermal properties of postulated lunar surface materials. p. 656–70 in ref. **A3**. (A65–34244)

14. Glaser, P. E. Implications of high vacuum on the characteristics of the lunar surface. *J. Vacuum Sci. Technol.*, **3**, 181, 1966. (P66–32300)

15. Gold, T. Ranger Moon pictures: implications. *Science*, **145**, 1046, 1964. (A64–24799; P65–7629)

16. Gold, T. Structure of the Moon's surface. p. 345–53 in ref. **A1**. (A64–26073)

17. Gold, T. The Moon's surface. p. 107–21 in ref. **A8**. (A66–35442)

18. Green, J. Selection of rock standards for lunar research. p. 1123–47 in ref. **A3**. (A65–34270)

19. Halajian, J. D. The case for a cohesive lunar surface model. p. 671–710 in ref. **A3**. (A65–34245)

20. Halajian, J. D., Spagnolo, F. A. Photometric measurements of simulated lunar surfaces. *Grumman Res. Dept. Mem.* RM–308, 88 p. Jan. 1966.

21. Halajian, J. D. Mechanical, optical, thermal, and electrical properties of the Surveyor I landing site. *Grumman Aircraft Eng. Report* No. AS 424–4, 32 p., Nov. 1966.

22. Hämeen-Anttila, K. A. A statistical model for the microstructure of the lunar surface. *Ann. Acad. Sci. Fennicae A VI*, 24 p., 1963.

23. Hapke, B. W. Photometric and other laboratory studies relating to the lunar surface. p. 323–44 of ref. **A1**. (A64–26072)

24. Hapke, B. W. Packing properties of fine powders and the depth of the lunar dust layer. *J. geophys. Res.*, **69**, 1147, 1964. (A64–15732; P65–17831)

25. Hapke, B. W., Goldberg, L. Molecular flow through soils and the possibility of lunar permafrost layers. *Trans. Amer. geophys. Un.*, **46**, 139, 1965.

26. Hapke, B. W. Effects of a simulated solar wind on the photometric properties of rocks and powders. p. 711–21 in ref. **A3**. (A65–34246)

27. Hapke, B. W. Optical properties of the Moon's surface. p. 19–30 in ref. **A8**. (A66–35444)

28. Hapke, B. W. Some critical comments on Gehrels' model of the lunar surface. *Icarus*, **5**, 154, 1966. (P66–25148; A66–29260)

29. Jaffe, L. D. Depth of the lunar dust. *J. geophys. Res.*, **70**, 6129, 1965. (A66–16783)

30. Jaffe, L. D. Strength of the lunar dust. *J. geophys. Res.*, **70**, 6139, 1965. (A66–16784)

31. Jaffe, L. D. Bearing strength of 'Fairy Castle' structures. *J. geophys. Res.*, **70**, 6268, 1965. (A66–16788)

32. Jaffe, L. D., Scott, R. F. Lunar surface strength – implications of Luna 9. *Science*, **153**, 407, 1966. (A66–40518)

33. Johnson, R. W. Environmental test criteria for lunar and planetary soils. *Astr. Acta*, **11**, 252, 1965. (A66–28179)

34. Kapp, R. O. A fourth state of matter on the surface of the Moon? *Space/Aeronautics*, **45**, 132, 134, 136, 138, 140, 1966. (A66–20173)

35. Kraner, H. W., Schroeder, G. L., Davidson, G., Carpenter, J. W. Radioactivity of the lunar surface. *Science*, **152**, 1235, 1966. (A66–32942; P66–28588)

36. Levin, B. Y. Nature of the lunar surface layer. *Astr. Zu.*, **40**, 1071, 1963. (P65–4624)

37. Martin, J. J. Scattering from the Moon and other rough surfaces. *J. geophys. Res.*, **71**, 2687, 1966. (P66–25154)

38. Matveyev, Y. G., Suchkin, G. L., Troitsky, V. S. Density variations of lunar matter with depth in the surface layer. *Astr. Zu.*, **42**, 810, 1965. (P65–33927; A66–10274)

39. Miyamoto, S. Morphological aspects of the lunar crust. *Icarus*, **3**, 486, 1964. (A65–17769)
40. Miyamoto, S. Morphological aspects of the lunar crust. II. *Icarus*, **4**, 421, 1965. (A65–35368)
41. Miyamoto, S. Morphological features of the libratory region of the Moon. p. 776–96 in ref. **A3**. (A65–34252)
42. Naughton, J. J., Barnes, I. L., Hammond, D. A. Rock degradation by alkali metals – a possible lunar erosion mechanism. *Science*, **149**, 630, 1965. (A65–33343)
43. O'Keefe, J. A. Lunar ash flows. p. 264–6 in ref. **A8**. (A66–35451)
44. Orlova, N. S. Characteristic light-scattering curves for certain magmatic rocks. *Trudy astr. Obs. Leningrad. gos. Univ.*, **21**, 74, 1964. (A65–25693)
45. Petrova, N. N. Spectral investigations of the lunar surface. *Astr. Zu.*, **43**, 162, 1966. (A66–23506; P66–19065)
46. Ringwood, A. E. Chemical evolution of the terrestrial planets. *Geochim. Cosmochim. Acta*, **30**, 41, 1966. (A66–19075)
47. Ruskol, E. L. Physical properties of the lunar surface. *Kosmicheskie Issledovaniya*, **3**, 395, 1965. (P65–22068; A65–35809)
48. Sagan, C. Photometric properties of Mercury. *Astrophys. J.*, **144**, 1218, 1966.
49. Shoemaker, E. M. Progress in the analysis of the fine structure and geology of the lunar surface from the Ranger VIII and IX photographs. p. 30–1 in ref. **A9**. (N66–31449)
50. Smoluchowski, R. Radiation sintering of lunar dust. *Science*, **150**, 1025, 1965. (P66–13167; A66–13338)
51. Smoluchowski, R. Structure and coherency of the lunar dust layer. *J. geophys. Res.* **71**, 1569, 1966. (A66–24892; P66–22069)
52. Urey, H. C. 'Dust' on the Moon. *Science*, **153**, 1419, 1966. (A66–41309)
53. Walker, E. H. Comments on the photographs obtained by Ranger VII. *Astronautyka*, **8**, 3, 1965. (A65–24419)
54. Wechsler, A. E., Glaser, P. E. Pressure effects on postulated lunar materials. *Icarus*, **4**, 335, 1965. (A65–35359; P66–6063)
55. Werner, M., Gold, T., Harwit, M. On the detection of water on the Moon. Cornell-Sydney University Astronomy Center Report No. 44, 16 p., July 1966.

J. Lunar Volcanism

1. Adams, E. Origin of lunar tektites. *Raumfahrtforschung*, **10**, 105, 1966. (A66–38059)
2. Deák, G., Erdösi, K., Füredi, M., Hédervári, P., Sugár, I. Some energetical calculations regarding the recent volcanic phenomena of the Moon. *Gerlands Beiträge zur Geophysik*, **75**, 122, 1966. (A66–34099)
3. Dietz, R. S., Holden, J. C. Earth and Moon – tectonically contrasting realms. p. 631–40 in ref. **A3**. (A65–34242)
4. Dobar, W. I. Behavior of lava on the lunar surface. p. 495–515 in ref. **A3**. (A65–34233)
5. Dobar, W. I. Simulated basalt and granite magma upwelled in vacuum. *Icarus*, **5**, 399, 1966. (A66–38025)
6. Gilvarry, J. J. The lunar origin of tektites. *Icarus*, **4**, 317, 1965. (A65–29112)
7. Gilvarry, J. J. The lunar origin of tektites. p. 1061–81 in ref. **A3**. (A65–34267)
8. Gold, T. Outgassing processes on the Moon and Venus. p. 249–55 in *The Origin and Evolution of Atmospheres and Oceans*. Eds. Brancazio, P. J., Cameron, A. G. W. New York, John Wiley and Sons, 1964. (A65–33298)
9. Green, J. Tidal and gravity effects intensifying lunar defluidization and volcanism. p. 403–69 in ref. **A3**. (A65–34230)
10. Kozyrev, N. Volcanic phenomena on the Moon. *Nature*, **198**, 979, 1963. (P64–4883)
11. Kuiper, G. P. Volcanic sublimates on Earth and Moon. *Comm. Lunar Planet. Lab.*, **3**, 33, 1965. (A66–11500)
12. Middlehurst, B. M. Transient lunar events: possible causes. *Nature*, **209**, 602, 1966. (P66–25155)
13. Middlehurst, B. M., Burley, J. M. Chronological listing of lunar events. NASA–TM–X–55470. Apr. 1966, 46 p. (N66–24927)

14. Mills, A. A. Tektites – their nature and possible origin. *J. Brit. astr. Assoc.*, **76**, 73, 1966. (A66–27823)
15. Moore, P. An evaluation of the reported lunar changes. p. 797–810 in ref. **A3.** (A65–34253)
16. O'Keefe, J. A. Interpretation of Ranger photographs. *Science*, **146**, 514, 1964. (P65–4622)
17. O'Keefe, J. A. Tektites and impact fragments from the Moon. *Sci. American*, **210**, 50, 1964. (P65–4645)
18. O'Keefe, J. A., Adams, E. W. Tektite structure and lunar ash flows. *J. geophys. Res.*, **70**, 3819, 1965. (A65–32671)
19. O'Keefe, J. A., Lowman, P. D., Jr. Tektites as a guide to the structure of the Moon. *Tectonophysics*, **2**, 319, 1965.
20. O'Keefe, J. A., Adler, I. Lunar structure as deduced from Muong Nong Tektites. p. 92–101 in ref. **A9.** (N66–31456)
21. Simpson, J. F. Evidence for the volcanic origin of lunar and Martian craters. *Earth Planet. Sci. Letters*, **1**, 130, 1966. (A66–33625)
22. Simpson, J. F. Additional evidence for the volcanic origin of lunar and Martian craters. *Earth Planet. Sci. Letters*, **1**, 132, 1966. (A66–33626)
23. Vonnegut, B., McConnell, R. K., Jr., Allen, R. V. Evaporation of lava and its condensation from the vapour phase in terrestrial and lunar volcanism. *Nature*, **209**, 445, 1966. (A66–21288)
24. Menzel, D. H. The surfaces of the Moon, Mars, and Venus. In ref. **A14.**

K. Luminescence and Related Transient Phenomena

1. Anand, S. P. S., Oster, L., Sofia, S. Excitation of lunar luminescence by solar protons. *Nature*, **202**, 1097, 1964. (P65–20965)
2. Brzostkiewicz, S. R. Luminescence of the lunar surface. *Urania*, **36**, 166, 1965. (A65–27709)
3. Burley, J., Middlehurst, B. M. Apparent lunar activity: historical review. *Proc. nat. Acad. Sci.*, **55**, 1007, 1966. (P66–28589)
4. Cohen, A. J. Source of evolution of gas from the lunar crater Alphonsus. *Nature*, **201**, 1015, 1964. (A65–15141)
5. Derham, C. J., Geake, J. E. Luminescence of meteorites. *Nature*, **201**, 62, 1964. (P64–23892)
6. Flamm, E. J., Lingenfelter, R. E. Lunar luminescence. *Nature*, **205**, 1301, 1965. (P66–9180)
7. Geake, J. E. Lunar luminescence. *Nature*, **204**, 866, 1964. (P65–13738; A65–13535)
8. Greenacre, J. C. The 1963 Aristarchus events. p. 811–5, in ref. **A3.** (A65–34254)
9. Kopal, Z., Rackham, T. W. Excitation of lunar luminescence by solar flares. *Nature*, **201**, 239, 1964. (A64–14148; P64–23891)
10. Kopal, Z., Rackham, T. W. Excitation of lunar luminescence by solar activity. *Icarus*, **2**, 481, 1963. (P65–7627)
11. Kopal, Z. Luminescence on the lunar surface. *Sterne und Weltraum*, **5**, 56, 1966. (A66–23943)
12. Kopal, Z. Luminescence of the Moon and solar activity. p. 173–83 in ref. **A8.** (A66–35446)
13. Ney, E. P., Woolf, N. J., Collins, R. J. Mechanisms for lunar luminescence. *J. geophys. Res.*, **71**, 1787, 1966. (P66–22070)
14. Righini, A., Jr., Rigutti, M. Some results of research on lunar luminescence. *Icarus*, **5**, 258, 1966. (P66–28587)
15. Sanduleak, N., Stock, J. Indication of luminescence found in the 1964 December lunar eclipse. *Publ. astr. Soc. Pacif.*, **77**, 237, 1965. (P66–16385; A66–30644)
16. Scarfe, C. D. Observations of lunar luminescence at visual wavelengths. *Mon. Not. R. astr. Soc.*, **130**, 19, 1965. (P65–33928)
17. Schutten, J., Dijk, T. Van Luminescence caused by proton impact with special reference to the lunar surface. *Nature*, **211**, 470, 1966. (P66–32301)
18. Wildey, R. L. Lunar luminescence. *Publ. astr. Soc. Pacif.*, **76**, 112, 1964. (P64–29424)

19. Geake, J. E., Walker, G. The luminescence spectra of meteorites. *Geochim. Cosmochim. Acta*, **30**, 929, 1966. (A66–42363)
20. Moore, P. Color events on the Moon. *Sky and Tel.*, **33**, 27, 1967.

L. Lunar Environment and Atmosphere

1. Aronowitz, L., Milford, S. N. Magnetic shielding of the lunar surface from the solar wind as a function of lunar magnetic moment. *J. geophys. Res.*, **70**, 227, 1965. (A65–15054)
2. Biswas, M. M., Majumdar, D. P. Expected variations of cosmic-ray intensity on the Moon. *Nuovo Cimento*, **36**, 1039, 1965. (P65–18541)
3. Bowen, E. G. Lunar and planetary tails in the solar wind. *J. geophys. Res.*, **69**, 4969, 1964. (P65–15834)
4. Cameron, A. G. W. Particle acceleration in cislunar space. *Nature*, **202**, 785, 1964. (P65–24337)
5. Hayakawa, S. High-energy radiations from the planets and the Moon. p. 298–311 in *Space Exploration and the Solar System*, New York, Academic Press, 1964. (P65–24331; A65–26915)
6. Hayakawa, S. [Estimate of] Lunar X-rays excited by energetic terrestrial electrons, *Rep. Ionosphere, Space Res. Japan*, **19**, 375, 1965. (P66–16386)
7. Haymes, R. C. Juday, R. D. Detectability of lunar X-rays. *Planet. Space Sci.*, **13**. 1249, 1965. (P66–13163; A66–20885)
8. Hinton F. L., Taeusch, D. R. Variation of the lunar atmosphere with strength of the solar wind. *J. geophys. Res.*, **69**, 1341, 1964. (P65–15832)
9. Michel, F. C. Interaction between the solar wind and the lunar atmosphere. *Planet. Space Sci.*, **12**, 1075, 1964. (P65–4628; A65–13344)
10. Osborne, F. J. F., Bachynski, M. P. On a possible lunar magnetospheric configuration, *J. geophys. Res.*, **70**, 4983, 1965. (P66–6064)
11. Palm, A., Silver, S. On observations of lunar magnetic storms. *Planet. Space Sci.*, **12**, 649, 1964. (P64–29423)
12. Safronov, V. S., Rouskol, E. L. The history of the lunar atmosphere and possibility of the presence of ice and organic compounds on the Moon. p. 42–53 in Proc. XIIIth Int. Astronautical Congress, Wien, Springer-Verlag, 1964. (P65–13737)
13. Singer, S. F. Atmosphere near the Moon. *Astronautical Sci.*, **8**, 531, 1963. (P64–13916)
14. Wehner, G. K., Kenknight, C. E., Rosenberg, D. Modification of the lunar surface by the solar-wind bombardment. *Planet. Space Sci.*, **11**, 1257, 1963. (P64–7975)
15. Vernov, S. N., *et al.* Measurement of the intensity of penetrating radiation on the surface of the Moon. *Dokl. Akad. Nauk SSSR*, **169**, 1044, 1966. (A66–41410)

M. Internal Structure

1. Anderson, D. L., Kovach, R. L. The internal structure of the Moon and the terrestrial planets. p. 84–91 in ref. **A9**. (N66–31455)
2. Caputo, M. On the shape, gravity field, and strength of the Moon. *J. geophys. Res.*, **70**, 3993, 1965. (P65–30704)
3. Carson, D., Davidson, M., Goudas, C. L., Kopal, Z., Stoddard, L. G. Lunar profiles determined from annular solar eclipses of 1962 and 1963. *Icarus*, **5**, 334, 1966. (P66–32295)
4. Clegg, P. E., Bastin, J. A., Gear, A. E. Heat transfer in lunar rock. *Mon. Not. R. astr. Soc.*, **133**, 63, 1966. (A66–37260)
5. Davidson, M., Goudas, C. L., Kopal, Z. Lunar profiles determined from annular solar eclipses of 1962 and 1963. *Boeing Document* D1–82–0470, 73 p., 1966.
6. Eckert, W. J. On the motions of the perigee and node and the distribution of mass in the Moon. *Astr. J.*, **70**, 787, 1965. (P66–13165; A66–19352)
7. Enzmann, R. Balance of endogenic to exogenic energy in the crust of the Moon. p. 532–542 in ref. **A3**. (A65–34236)
8. Fielder, G. A case for convection in the Moon. *Nature*, **204**, 171, 1964. (P65–18542; A65–10368)

9. Fielder, G. Convection in the Moon – a boundary condition. *Geophys. J.*, **10**, 437, 1966. (A66–33235)
10. Goudas, C. L. The nonhomogeneity of the Moon. *Icarus*, **4**, 218, 1965. (P65–27800)
11. Goudas, C. L. Note on 'Shape and Internal Structure of the Moon' by Lamar and McGann. *Icarus*, **5**, 99, 1966. (P66–19063; A66–21210)
12. Goudas, C. L. The figure and gravity field of the Moon. p. 27–151 in *Advances in Astronomy and Astrophysics*, **4**, ed. Z. Kopal. New York, Academic Press, 1966. (A66–38822)
13. Goudas, C. L., Kopal, Z., Kopal, Z. The shape of the Moon as deduced from the Orbiter determination of its field. *Boeing Document* D1–82–0570, 1966.
14. Kopal, Z. Stress history of the Moon and of terrestrial planets. *Icarus*, **2**, 376, 1963. (P65–7631; A64–16486)
15. Kopal, Z. Radiative transport of heat in lunar and planetary interiors. *Icarus*, **3**, 8, 1964. (P65–20967)
16. Kopal, Z. Effects of thermal expansion on the moments of inertia of the Moon. *Icarus*, **4**, 166, 1965. (P65–27798)
17. Kopal, Z. Possible effects of convection on lunar moments of inertia. *Icarus*, **4**, 173, 1965. (P65–27799)
18. Kopal, Z. Internal structure of the Moon. *Boeing Sci. Res. Lab. Math. Note*, No. 426, 1965. (N66–15700)
19. Kovach, R. L., Carr, R. E. Free oscillations of the Moon and observations by a long-period seismograph system. *Proc. of the XIIIth Internat. Astronautical Congress*, Wien, Springer-Verlag, **I**, 1–10, 1964. (P65–13736)
20. Lamar, D. L., McGann, J. Shape and internal structure of the Moon. *Icarus*, **5**, 10, 1966. (P66–19062; A66–21203)
21. Lamar, D. L., McGann, J. Reply to 'Note on the Shape and Internal Structure of the Moon' by C. L. Goudas. *Icarus*, **5**, 101, 1966. (P66–19064)
22. Levin, B. J. Thermal history of the Moon and the development of its surface. p. 271–273 in ref. **A8.** (A66–35452)
23. Levin, B. J. The structure of the Moon. *Astr. Zu.*, **43**, 606, 1966. (P66–28579; A66–37038)
24. Levin, B. J. The structure of the Moon. p. 61–76 in ref. **A9.** (N66–31453)
25. MacDonald, G. J. F. On the internal constitution and origin of the Moon. p. 77 in ref. **A9.** (N66–31454)
26. Mikhailov, A. A. Gravity force and configuration of the Moon. *Astr. Zu.*, **42**, 1062, 1965. (A66–12918)
27. Ornatskaia, O. I., Al'ber, I. I. Thermal history of the Moon. *Radiofizika*, **9**, 615, 1966. (A66–37151)
28. Walter, L. S. Lunar differentiation processes. p. 470–80 in ref. **A3.** (A65–34231)
29. Wildey, R. L. Measuring the shape of the Moon. *Sky and Tel.*, **31**, 147, 1966. (A66–22964)
30. Zharkov, V. N., Berikashvili, V. S., Osnach, A. I. Geophysical problems and investigations of the Moon (a review). *Astr. Zu.*, **43**, 622, 1966. (P66–28580)
32. Levin, B. J. *Astr. Zirk.*, No. 285, 1964; *Nature*, **202**, 1201, 1964.
33. Levin, B. J. Proc. Lunar Meeting Roy. Soc., June 1965 (in press).
34. Safronov, V. S. *Icarus* (in press).
35. Majeva, S. V. *Dokl. Akad. Nauk SSSR*, **159**, No. 2, 1964.
36. Koslovskaya, S. V. *Voprosy Kosmogonii*, **8**, 1962.
37. Berikashvili, V. S., Zharkov, V. N., Yanovskaya, T. B. *Izv. Akad. Nauk SSSR*. Ser. fiziki zemli, issue 7, 9–21, 1965.

N. Origin and Evolution of the Moon

1. Alfvén, H. Origin of the Moon. *Science*, **148**, 476, 1965. (P66–6065)
2. Baldwin, R. B. On the origin of the Moon. *J. geophys. Res.*, **71**, 1936, 1966. (P66–22067; A66–26366)

3. Hédervári, P. Instability of small-size planetary cores and the development of the Moon. *Nature*, **202**, 378, 1964. (P64–11889)

4. MacDonald, G. J. F. Earth and Moon: past and future. *Science*, **145**, 881, 1964. (P65–15835)

5. Ruskol, E. L. On the past history of the Earth-Moon system. *Icarus*, **5**, 221, 1966. (P66–28577)

6. Ruskol, E. L. The tidal history and origin of the Earth-Moon system. *Astr. Zu.*, **43**, 829, 1966. (P66–35302)

7. Shimazu, Y. Survival time of lunar surface irregularities and viscosity distribution within the Moon. *Icarus*, **5**, 455, 1966. (A66–38030; P66–32299)

8. Shoemaker, E. M. The geology of the Moon. *Sci. Amer.*, **211**, 38, 1964. (P65–13739)

9. Urey, H. C. Meteorites and the Moon. *Science*, **147**, 1262, 1965. (P66–6072)

10. Urey, H. C. Chemical evidence relative to the origin of the solar system. *Mon. Not. R. astr. Soc.*, **131**, 199, 1966. (P66–13152)

11. Sorokin, N. A. *Astr. Zu.*, **42**, 1070, 1965.

12. Petrova, G. G. Diploma work, Physical Dept., Moscow University, 1965.

13. Ruskol, E. L. *Izv. Akad. Nauk SSSR*, ser. geofiz., issue 2, 216–22, 1963.

19. COMMISSION DE LA ROTATION DE LA TERRE

PRÉSIDENT: Dr B. Guinot, Observatoire de Paris, 61 avenue de l'Observatoire, Paris-14e, France.

VICE-PRÉSIDENT: Dr P. J. Melchior, Observatoire Royal de Belgique, 3, avenue Circulaire, Uccle-Bruxelles 18, Belgique.

COMITÉ D'ORGANISATION: G. Cecchini, E. P. Fedorov, W. Markowitz, H. M. Smith.

MEMBRES: Abraham, Aksentjeva, Atkinson, Billaud, Bonanomi, Brkić, Buchar, Chudovicheva, Danjon †, Demetrescu, Drâmbă, Enslin, Essen, Fichera, Fleckenstein, Gama, Hall (R. G.), Hers, Iijima, Jeffreys, Kalmykov, Koebcke, Kulikov (K. A.), Lagrula, Lederle, Levallois, Mikhailov, Miyadi, Nicolini, Okuda, Opalski, Orlov (B. A.), Orte, Pavlov (N. N.), Popov, Postoiev, Randić, Rice, Romanskaya, Sakharov, Scheepmaker, Sevarlic, Stoyko (A.), Stoyko (N.), Sugawa, Takagi, Tanner, Tardi, Torao, Tsao, Vicente, Witkowski, Young, Yumi.

I. MESURES DE LA LATITUDE ET DU TEMPS UNIVERSEL

(a) *Observations astronomiques en cours*

Nous donnons ci-dessous la liste des stations ayant mesuré l'heure et la latitude durant la période 1963 à 1966.

La station est désignée par le nom de la ville où elle se trouve. On donne les coordonnées approximatives.

La colonne 'instrument' désigne l'instrument *en service à la fin de 1966.* Les abréviations suivantes sont utilisées:

- IP : instrument méridien des passages
- IP Ph : instrument méridien des passages avec enregistrement photoélectrique
- PZT : lunette photographique zénithale.
- LZV : lunette zénithale visuelle des latitudes.

La notation x/y donne l'ouverture x de l'objectif et la longueur focale y, en cm.

Dans la colonne 'grandeur mesurée', φ indique la latitude, T, le temps universel.

Dans la colonne 'renseignements divers', on donne des indications sur la continuité des mesures entreprises et des références d'ouvrages où ont été publiés les résultats détaillés.

Station	Instrument	Grandeur mesurée	Renseignements divers
Alger $+37°, -0^h\ 12^m$	Astrolabe (OPL 8)	φ, T	Depuis 1958, programme inchangé. Publication détaillée en préparation.
Belgrade $+45°, -1^h\ 22^m$	LZV (11/129)	φ	Depuis 1949 Publ.: *Bull. Obs. astr. Belgrade.*
	IP (10/100)	T	Publ.: jusqu'à 1960: *Bull. Obs. astr. Belgrade.*
Besançon $+47°, -0^h\ 24^m$	Astrolabe (OPL 31)	φ, T	De 1959 à 1960, astrolabe prototype. Depuis janvier 1961, astrolabe OPL 31, suivant programme invariable. Publ.: pour 1959–60, *Ann. Obs. Besançon,* **6**, fasc. 1 et 2.

Station	Instrument	Grandeur mesurée	Renseignements divers
Blagovestchensk $+50°, -8^h\ 30^m$	LZV (ZTL 180)	φ	Publ.: 1960–63, Observations with zenith telescopes during the period 1960–63. Moscow, 1964.
Borowa-Gora $+52°, -1^h\ 24^m$	IP Ph (Zeiss 10/100)	T	
Boroviec $+52°, -1^h\ 8^m$	2 IP (Zeiss 10/100) 2 LZV	T φ	Publ.: Astr. Lat. st. of the polish Academy of Sciences, Borowiec. Circular.
Bucarest $+44°, -1^h\ 44^m$	IP Zeiss	T	Depuis 1957. Publ.: Bulletin du Service de l'Heure de l'Obs. Des mesures de latitude ont été effectuées de 1959 à 1961.
Buenos Aires (Inst. géogr.) $-35°, +3^h\ 53^m$	IP	T	Publ.: Bulletin mensuel 'Señales horarias radiotelegraficas'.
Buenos Aires (Obs. nav.) $-35°, +3^h\ 53^m$	2 IP	T	Publ.: Bulletin mensuel de 'Observatorio naval de Buenos Aires, Servicio de Hidrografia Naval' Note: Installation d'un PZT en cours, à la même latitude que Mount Stromlo.
Carloforte $+39°, -0^h\ 33^m$	LZV	φ	Depuis 1900. Station du SIMP.
Dresde $+51°, -0^h\ 55^m$	LZV (Zeiss 10/100)	φ	Publ.: Zirkular, Technische Universität Dresden, Lohrmann-Institut.
Gorki $+56°, -2^h\ 56^m$	LZV	φ	Publ.: Voir Blagovestchensk.
Gaithersbourg $+39°, +5^h\ 9^m$	LZV	φ	Depuis 1900. Station du SIMP
Greenwich (Herstmonceux) $+51°, -0^h\ 1^m$	PZT	φ, T	1. La latitude a été mesurée de 1911 à 1940 avec la lunette zénithale flott. Cookson. 2. L'astrolabe OPL 9 a été utilisé de 1959 à 1963. Publ.: *R. Obs. Bull.*, no. 92. 3. Le PZT fonctionne depuis 1955. Publ.: *R. Obs. Bull.*, tous les 3 mois.
Hambourg (Inst. hydrog.) $+54°, -0^h\ 40^m$	PZT	φ, T	1. Des instruments des passages ont été utilisés de 1955 à 1958. 2. Le PZT est en service depuis 1958. Publ.: 1957–58, Breitbestimmungen (Circ.). 1959–61, Deutsches hydrog. Zeitung. 1961 . . . Time Service Bulletins.
Irkoutsk (Obs.) $+52°, -6^h\ 57^m$	IP Ph	T	Publ.: Jetalonnoe Vremja . . . , Institut de mesures physico-techniques et radio-techniques (Moscou), mensuel.
	LZV (ZTL 180)	φ	Voir Blagovestchensk.

Station	Instrument	Grandeur mesurée	Renseignements divers
Irkoutsk (Mesures) $+52°, -6^h\ 57^m$	2 astrolabes (OPL 22 et 26)	φ, T	Publ.: Voir Blagovestchensk et Irkoutsk (Obs.).
	IP IP Ph	T	
Kasan (Obs. Engelhardt) $+56°, -3^h\ 15^m$	LZV (ZTL 180)	φ	Publ.: voir Blagovestchensk.
Kharkov $+50°, -2^h\ 25^m$	IP Ph	T	Publ.: voir Irkoutsk (Obs.)
Kitab $+39°, -4^h\ 28^m$	LZV	φ	Depuis 1928. Station du SIMP. (a remplacé la station de Tchardjui).
La Plata $-35°, +3^h\ 52^m$	LZV IP	φ T	Depuis 1929. Publ.: Boletin Horario, Observatorio astronomico de la Universidad Nacional.
Le Cap $-34°, -1^h\ 14^m$	Astrolabe (OPL 9)	φ, T	Depuis 1965. Publ.: *Mon. Not. astr. Soc. Sth. Afr.* (vol. XXV, 1966, pour avril 1965–juillet 1966).
Leningrad (Obs. astr.) $+60°, -2^h\ 1^m$	IP Ph	T	Publ.: voir Irkoutsk (Obs.).
Leningrad (Mesures) $+60°, -2^h\ 1^m$	IP IP Ph	T	Publ.: voir Irkoutsk (Obs.).
Milan $+45°, -0^h\ 37^m$	IP Obs. zénithales	φ	Depuis 1961. Publ.: Astronomical Observatory of Milan. Circular no. 18.
	IP (Askania 10/100)	T	Publ.: Circulars. Note: L'installation d'un astrolabe OPL est prévue.
Mizusawa $+39°, -9^h\ 25^m$	LZV	φ	Depuis 1900. Station SIMP.
	LZ flottante	φ	Depuis 1940
	PZT	φ, T	Depuis 1956
	Astrolabe (OPL 34)	φ, T	Depuis 1966
			(pour les instruments depuis 1940, 1956 et 1966) Publ.: (1) IGY and IGC Data on Latitude and Longitude, Part 1, Sc. (Council of Japan) (2) Publ. of the Int. Lat. Obs. of Mizusawa (3) Annual Reports of IPMS (4) Mizusawa Time Service Bull.
Mont Pourpre $+32°, -7^h\ 55^m$	IP	T	

Station	Instrument	Grandeur mesurée	Renseignements divers
Mount Stromlo (Canberra) $-35°$, $-9^h 56^m$	PZT	φ, T	Depuis 1959. Publ.: Temps, Mount Stromlo Bulletins TS1D, jusqu'à 1961, puis Bulletins B. Note: mesures de T.U. faites auparavant avec IP.
Moscou (Obs. astr.) $+56°$, $-2^h 30^m$	IP Ph	T	La latitude a été mesurée également, de 1958 à 1964 avec un instrument ZTL 180 (période de 6 ans).
	PZT	φ, T	Publ.: voir Blagovestchensk et Irkoutsk (Obs.).
Moscou (Mesures) $+56°$, $-2^h 31^m$	IP Ph	T	Publ.: voir Irkoutsk (Obs.).
	Astrolabe (OPL 23)	φ, T	
Neuchâtel $+47°$, $-0^h 28^m$	PZT	φ, T	Depuis 1956. Publ.: Bulletins A et B de l'Obs. de Neuchâtel. Note: Des observations de φ, T ont été faites avec un astrolabe OPL, de 1958 à 1965.
Nikolaiev $+47°$, $-2^h 8^m$	IP Ph	T	Publ.: voir Irkoutsk (Obs.).
Novossibirsk (Mesures) $+55°$, $-5^h 32^m$	2 IP	T	Publ.: voir Irkoutsk (Obs.).
	2 Astrolabes (OPL 25 et 27)	φ, T	
Ottawa $+45°$, $+5^h 3^m$	PZT	φ, T	Depuis 1952. Publ.: Time and Latitude Bull. of the Dominion Obs. (nouvelle réduction, à partir de 1956, disponible sur demande).
Paris $+49°$, $-0^h 9^m$	Astrolabe (OPL 01, puis 35)	φ, T	Depuis 1956. Publ.: Notes et Informations, fasc. 2, 3, 4, 5, 7, 13, 19, 29, jusqu'à 1964. La suite sera publiée dans le Journal des Observateurs, Marseille.
Pecný $+50°$, $-0^h 59^m$	IP Circum zénithol Nušl. Frič	T	Publ.: Bulletins de la Station de l'Heure à Prague. Académie tchécoslovaque des Sciences.
	LZV	φ	
Poltava $+50°$, $-2^h 18^m$	LZV	φ	Publ.: voir Blagovestchensk.
	Astrolabe (OPL)	φ	
Prague $+50°$, $-0^h 58^m$	IP	T	Publ.: Voir ci-dessus pour Pecný.

Station	Instrument	Grandeur mesurée	Renseignements divers
Potsdam +52°, −0h 52m	3 IP Astrolabe (OPL 10)	*T* φ, *T*	Astrolabe, depuis 1957. Publ.: 1957–62, Arbeiten aus dem Geodätischen Institut Potsdam, Nr 1, 4 und 10. Depuis 1962: Astronomische Zeit-und Breitenbestimmungen, Empfangzeiten von Zeitsignalen (Bulletin mensuel).
Poulkovo +60°, −2h 1m	IP Ph 2 LZV (ZTF 135 et ZTL 180)	*T* φ	Publ.: voir Blagovestchensk et Irkoutsk (Obs.).
Quito 0°, +5h 14m	Astrolabe (OPL 13)	φ, *T*	Depuis 1963. Publ.: prévue en 1967, par l'Obs. de Quito.
Richmond +26°, +5h 22m	PZT Astrolabe (OPL)	φ, *T* φ, *T*	Depuis 1949. Publ.: voir Washington. Depuis 1964.
Riga +57°, −1h 36m	IP Ph	*T*	Publ.: voir Irkoutsk (Obs.).
Rio de Janeiro −23°, +2h 53m	IP	*T*	Publ.: Bulletin horaire mensuel.
Saint-Michel +44°, −0h 23m	Astrolabe (OPL 12)	φ, *T*	Depuis 1963. Publ.: prévue en 1967, J. Observateurs, Marseille.
San Fernando +36°, +0h 25m	IP	*T*	Publ.: Bol. del Obs. de San Fernando, Servicio de Hora (mensuel). Note: Installation d'un astrolabe OPL en cours.
Santiago du Chili −33°, −4h 42m	Astrolabe (OPL 3)	φ, *T*	Depuis 1965. Auparavant, depuis 1957, observations de T sur IP. Publ.: Bull. horaires de l'Obs. nat. de Santiago du Chili.
Tachkent +41°, −4h 37m	IP IP Ph }	*T*	Publ.: voir Irkoutsk (Obs.).
Tientsin +39°, −7h 49m	LZV	φ	Note: sur le parallèle des stations du SIMP.
Tokyo +36°, −9h 18m	PZT	φ, *T*	Depuis 1955. Publ.: IGY Data on Long. and Lat. (Science Council of Japan). Post IGY Data on Long. and Lat. (Science Council of Japan). Time Service Bull. Time and Lat. Bull.
Turku +60°, −1h 29m	LZV	φ	Depuis 1960. Publ.: 1960–63. Annales Academiae Scientiarum Fennicae, 1963, 71.

Station	Instrument	Grandeur mesurée	Renseignements divers
Uccle +51°, $-0^h\ 17^m$	2 IP Astrolabe (OPL 7)	T φ, T	Astrolabe, depuis 1962 (1965 pour φ). Publ.: pour T: Bull. horaires de l'Obs. Royal de Belgique.
Ukiah +39°, $+8^h\ 13^m$	LZV	φ	Depuis 1900. Station du SIMP.
Varsovie +52°, $-1^h\ 24^m$	IP LZV	T φ	De 1959 à 1963. Reprise prévue. Depuis 1959. Publ.: Publ. of the chair of geodetic Astronomy of the Warsaw techn. Univ., no. 4. Circulaires de l'Obs. no. 1 à 34.
Washington +39°, $+5^h\ 8^m$	PZT	φ, T	Depuis 1915, pour φ. Publ.: φ, Astr. J., 1110, 1130, 1141, 1144, 1147, 1150, 1154, 1159, de 1939 à 45. T, U.S. Naval Obs. Time Signal Bulletin.
Zikawei +31°, $-8^h\ 6^m$	2 IP Ph Astrolabe (OPL)	T φ, T	

(*b*) *Projets en cours de réalisation*

Plusieurs projets tendent à corriger la mauvaise répartition géographique des instruments et à organiser des chaines d'instruments à latitude égale qui donneront des informations sur la dérive des continents et du pôle.

La lunette zénithale photographique (PZT) d'Ottawa doit être placée à proximité de Calgary (Alberta) sur le parallèle de celle de Herstmonceux. Les programmes seront conjugués. Un nouveau PZT sera installé à Ottawa (nouveau site de l'Observatoire).

De même un nouveau PZT installé à La Plata sera conjugué avec celui de Mount Stromlo.

L'Observatoire de Besançon envisage de transférer son astrolabe à la Terre de Feu ($\varphi = -55°$), par accord avec le service hydrographique de la marine d'Argentine.

Des mesures de temps et latitude ont été effectuées avec un astrolabe OPL, à Sao Paulo ($\varphi = -23°$). Le transfert de l'instrument plus au nord devrait permettre d'avoir de meilleures conditions atmosphériques et il est prévu.

Les mesures de latitude doivent être entreprises à Sopron (Hongrie), avec une lunette zénithale visuelle.

On note, avec la plus grande satisfaction, que la situation des mesures dans l'hémisphère austral s'améliore considérablement, grâce aux efforts faits par plusieurs observatoires d'Amérique du Sud et par la mise en service d'un astrolabe au Cap. Cependant, l'installation de nouvelles stations dans l'hémisphère austral reste souhaitable, afin d'égaliser la répartition en longitude.

II. SERVICE INTERNATIONAL DU MOUVEMENT POLAIRE

(*a*) *Rapport du Directeur* (*Dr S. Yumi*)

The work of the International Polar Motion Service has been carried on smoothly from the beginning of 1962 through this period.

During the past three years, eight stations participated in the service and 38 stations in 21 countries are collaborating in the IPMS at present. They are: Alger, Belgrade, Besançon, Blagovestchensk, Borowiec, Carloforte, Cape, Dresden, Engelhardt, Gaithersburg, Gorky, Greenwich, Hamburg, Irkutsk, Kitab (2 instr.), La Plata, Milan, Mizusawa (3 instr.), Mt Stromlo, Neuchâtel, Ottawa, Paris, Pecný, Poltava (2 instr.), Potsdam, Prague, Pulkovo, Quito, Richmond, San-Fernando, Santiago du Chili (2 instr.), São Paulo, Tokyo, Uccle (2 instr.), Ukiah, U.S.S.R. (Mean Observatory), Warsaw and Washington.

The collected data of all the collaborating stations were arranged and studied during this period so that they could contribute to the determination of the pole's coordinates, the results of which will be published in the next three year period.

Preliminary values of the instantaneous pole referred to the new system 1900–05 of the ILS, calculated using only the results of the five ILS stations on the northern parallel of 39° 8′, and the preliminary summaries of the results of latitude observations made at the collaborating stations were published in the 'Monthly Notes of the IPMS' as a rapid service.

'Annual Report of the IPMS' was prepared to give the full data of the collaborating stations, coordinates of the pole, method of reduction and any other remarks which are considered to be necessary. Three volumes were published for the years 1962, 1963 and 1964 by this time. No time data but only the latitude data were given in them.

Coordinates of the instantaneous pole referred to the new system 1900–05 (Unit: 0″001)

U.T.	*x*	*y*	U.T.	*x*	*y*	U.T.	*x*	*y*
1962·00	− 9	+297	1963·50	+274	+251	1965·00	− 15	+ 49
·05	+ 8	+309	·55	+301	+193	·05	− 73	+ 68
·10	+ 27	+314	·60	+281	+139	·10	−125	+101
·15	+ 47	+312	·65	+237	+ 91	·15	−161	+154
·20	+ 71	+304	·70	+176	+ 46	·20	−183	+223
·25	+ 95	+290	·75	+112	+ 8	·25	−194	+291
·30	+120	+271	·80	+ 48	− 20	·30	−193	+337
·35	+144	+246	·85	− 11	+ 5	·35	−175	+374
·40	+162	+214	·90	− 69	+ 41	·40	−134	+409
·45	+173	+175	·95	−122	+ 78	·45	− 72	+435
·50	+171	+132	1964·00	−171	+120	·50	+ 2	+444
·55	+157	+ 92	·05	−206	+168	·55	+ 74	+434
·60	+128	+ 68	·10	−194	+230	·60	+126	+402
·65	+ 94	+ 60	·15	−169	+294	·65	+166	+354
·70	+ 56	+ 67	·20	−139	+353	·70	+200	+307
·75	+ 17	+ 83	·25	−101	+412	·75	+221	+261
·80	− 19	+104	·30	− 55	+455	·80	+229	+222
·85	− 54	+128	·35	+ 4	+467	·85	+223	+187
·90	− 86	+160	·40	+ 74	+459	·90	+195	+157
·95	−110	+200	·45	+164	+436	·95	+137	+132
1963·00	−121	+248	·50	+214	+394	1966·00	+ 71	+115
·05	−119	+295	·55	+240	+339	·05	+ 30	+103
·10	−105	+329	·60	+241	+275	·10	− 5	+ 98
·15	− 76	+356	·65	+239	+219	·15	− 37	+ 98
·20	− 38	+376	·70	+255	+168	·20	− 64	+105
·25	+ 9	+388	·75	+250	+123	·25	− 88	+120
·30	+ 70	+387	·80	+219	+ 85	·30	−106	+145
·35	+134	+375	·85	+161	+ 60	·35	−117	+178
·40	+191	+349	·90	+ 99	+ 46	·40	(−120)	(+213)
·45	+239	+307	·95	+ 42	+ 43	·45	(−114)	(+252)

A total of 42 255 observations were made at the five ILS stations—Mizusawa, Kitab, Carloforte, Gaithersburg and Ukiah during the period from 6 August 1963 to 5 August 1966. The coordinates of the instantaneous pole, derived from these observations are shown in the following table together with the revised values for the period 1962·00–1963·55. The values for the years 1962–64 are referred to the Annual Report of the IPMS and those for 1965–66 are to the Monthly Notes of the IPMS.

Scientific Council of the Service proposed at its meeting of Berkeley, U.S.A. in 1963 that the star pairs of the current observation programme for the five ILS stations which was adopted at 1955·0 were desirable to be improved and the new ones would be put into practice from 1967·0 on for at least 12 years. New programme and its star pairs were prepared by the Central Bureau and were approved by the majority of the members of the Council and the ILS stations during 1966. There is no change in observation programme—three groups a day—but combinations of stars or of star pairs in a group are changed.

(*b*) *Réunions du conseil scientifique du SIMP*

Ce conseil (B. Guinot, président, E. Fedorov, P. Melchior, W. Markowitz, S. Yumi) s'est réuni en août 1964, août 1965. Il a eu à traiter de la publication des résultats dans les *Monthly Notes of the IPMS* et du renouvellement du programme des stations internationales.

III. SERVICE INTERNATIONAL DES LATITUDES

(*a*) *Période 1941·0–1948·9*

Le Professeur T. Nicolini qui s'est chargé de la réduction des observations faites durant cette période sous la direction de Carnera, annonce que son travail sera achevé avant la fin de l'année 1966. Pour atténuer les incertitudes sur l'inconnue *x*, il a fallu faire appel aux séries discontinues de Kitab et Pino Torinese. Un rapport plus détaillé est donné en annexe au présent rapport, page 390.

(*b*) *Période 1949·0–1961·9*

Le Professeur G. Cecchini a établi le rapport suivant (traduction).

'Comme l'on sait, après la réduction provisoire des observations de latitude accomplies dans les stations boréales internationales de Mizusawa, Kitab, Carloforte, Gaithersburg et Ukiah, au cours des six années 1949–54 et des six années 1955–61 (ces dernières sur un nouveau programme d'observations), on a entrepris l'étude définitive de ces mêmes observations, au nombre de 148 350; étude rendue nécessaire pour les raisons suivantes. Avant tout, les calculs provisoires, bien qu'effectués avec le maximum de soin et d'attention, nécessitent un examen approfondi, afin d'éliminer, chaque fois que possible, l'erreur systématique loin d'être négligeable qui résulte de l'emploi de tours de vis et de coefficients thermiques constants pour chacune des stations: l'important matériel d'observations, sûr, a démontré incontestablement la réalité des remarquables variations du tour de vis, quand, par les approximations successives, on utilise les réductions d'un même couple d'étoiles à la déclinaison moyenne du groupe auquel il appartient, compte tenu du fait que ces réductions devraient donner des résultats identiques pour toutes les stations.

'Cependant la bonne qualité des résultats obenus, pour le but identique s'avère démontrée, particulièrement au cours de la deuxième période, en admettant que les corrections de tour de vis ont été traitées séparément pour les groupes du soir, intermédiaire et du matin.

'En outre, on a dû tenir compte de petits termes négligés dans les réductions provisoires et, particulièrement, du fait que ces mêmes observations de latitude, assignent la valeur 20″.50 à la constante de l'aberration, au lieu de la valeur usuelle 20″.47, ceci en accord avec les résultats d'autres auteurs, à partir d'observations d'un autre type. Ceci, évidemment, constitue un résultat d'une remarquable valeur, encore que la complexité du phénomène de la variation de la latitude terrestre soit telle (spécialement par l'intervention de phénomènes locaux si difficiles à préciser et impossibles à éliminer) que les résultats obtenus puissent bien difficilement être considérés comme entièrement exempts d'erreurs systématiques.

'L'élaboration définitive de tout le matériel observé demandera encore beaucoup de temps; mais, juqu'à présent, on a pu constater que les changements, bien que notables, apportés aux constantes de réduction, n'influeraient pas beaucoup sur la polhodie de toute la période 1949–61, telle qu'elle a été présentée aux Assemblées Générales de l'UAI et de l'UGGI.

'La valeur de la réduction définitive en cours, consistera principalement à permettre l'usage, pour des recherches ultérieures, des résultats obtenus dans un système le plus homogène possible et d'où l'on pourra déduire des conclusions dignes de foi.'

(*c*) *Remerciements*

Nous tenons à remercier les Professeurs T. Nicolini et G. Cecchini, ainsi que la Commission Géodésique Italienne pour la poursuite de cet important travail.

IV. SERVICE INTERNATIONAL RAPIDE DES LATITUDES (SIR)

Madame A. Stoyko et N. Stoyko ont poursuivi le travail du SIR jusqu'en avril 1965. Ensuite, la publication rapide des coordonnées du pôle et des corrections de longitude déduites a été considérée comme faisant partie du travail du Bureau International de l'Heure.

Actuellement (1966), les coordonnées du pôle sont encore déduites des mesures de latitude seulement, par Melle M. Feissel et B. Guinot, à partir des données d'une trentaine de stations. Les calculs et l'édition des circulaires ont été automatisés.

On commence à utiliser l'ensemble des mesures de temps et de latitude pour calculer simultanément les coordonnées du pôle et le temps universel T.U. 1. Voir à ce sujet le rapport du BIH à la Commission 31. Une publication détaillée de ces méthodes est en préparation. On donnera ci-après quelques résultats obtenus.

V. TRAVAUX DE RECHERCHE SUR LA ROTATION DE LA TERRE

(*a*) *Résultats d'ensemble des observations pour la période 1962–66*

La Figure 1 montre la polhodie déduite des observations aux stations du SIMP, la Figure 2, les fluctuations de la durée du jour.

(*b*) *Instruments*

Il n'est pas apparu d'instruments d'un type nouveau, mais l'équipement a fait l'objet d'améliorations techniques. L'emploi de chronographes totalisateurs ou perforateurs, de machines automatiques à mesurer les plaques se généralise.

(*c*) *Erreurs d'observation*

Pour les mesures de la latitude, la qualité des divers instruments reste un objet d'étude. Glagoleva et d'autres auteurs d'U.R.S.S. ont étudié la répartition spectrale des erreurs d'observations. Les instruments récents: PZT, astrolabes et lunettes zénithales modernes (ZTL 180) donnent, dans l'ensemble, des résultats meilleurs que les anciennes zénithales, mais cette amélioration est assez modeste.

Par contre, il y a des différences marquées dans la qualité des mesures de l'heure. Les instruments où le zénith est uniquement défini optiquement par réflexion sur un bain de mercure (astrolabes, PZT) donnent des résultats d'une meilleure précision interne et affectés d'erreurs systématiques plus petites ou plus stables que les instruments des passages classiques.

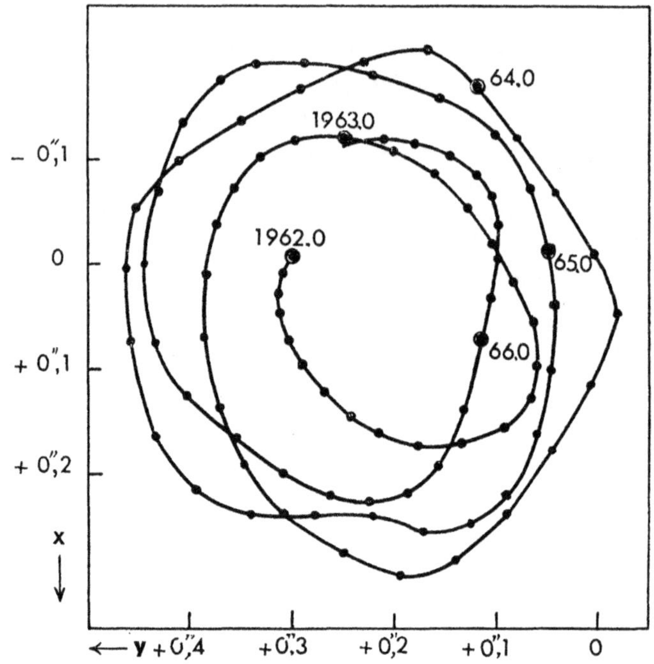

Fig. 1. Polhodie d'après le SIMP.

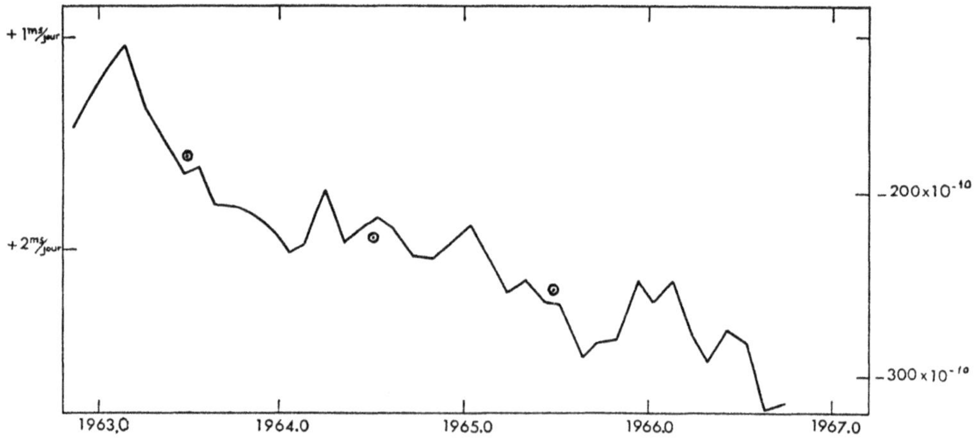

Fig. 2. Variation de la durée du jour de 1963·0 à 1966·7. L'échelle de gauche montre l'excès de la durée du jour sur la durée de 86 400 secondes atomiques (après correction de l'effet saisonnier); l'échelle de droite, le décalage relatif de la fréquence nominale qu'il faudrait adopter pour que le temps atomique suive T.U. 2. Les petits cercles représentent les moyennes annuelles.

Les astrolabes A. Danjon ont fait l'objet d'études quant aux déplacements du plan focal en fonction de la température par Gozhy et Slavinskaja. Scheepmaker a étudié les erreurs personnelles à l'Observatoire de Quito. Les écarts personnels des résultats peuvent atteindre 0",1 en latitude, 10 ms en temps.

Les PZT montrent parfois des erreurs de fermeture anormales en latitude: à Hambourg et à Herstmonceux. En particulier à Herstmonceux, il est nécessaire d'apporter une correction empirique de 0″06 sin *d* (*d* partie décimale de la date julienne) aux latitudes observées: l'origine de cette erreur paraît être instrumentale.

Une limitation à la précision des mesures est dans la turbulence atmosphérique. Les mêmes instruments placés dans des sites différents donnent des résultats d'une valeur parfois très variable, comme l'ont montré le transport du PZT d'Ottawa et des transferts d'astrolabes, suivis par les mêmes observateurs. Les effets de salle ont été étudiés par Goto. Les réfractions à basse altitude par Goto *et al.* L'effet du vent et des anomalies de la réfraction a été étudié par Mischenko. Tuterev a trouvé une corrélation entre la réfraction calculée en tenant compte de l'inclinaison des couches d'air et les résultats de mesures de temps et latitude à Herstmonceux, Poulkovo et Tokyo.

(*d*) *Programmes d'observation, réduction des observations*

La méthode des observations en chaîne restant d'un emploi général pour la réduction des observations à un système homogène, dans les stations indépendantes, a fait l'objet de travaux de Thomas et de Enslin, qui recherchent un programme optimal, compte tenu des conditions météorologiques.

A la méthode classique des paires de Talcott, Kolaczek et Bieniewski ont substitué l'observation de paires de groupes d'étoiles.

Plusieurs observatoires ont dû modifier leurs programmes d'observations, à cause de la précession. Le SIMP a préparé un nouveau programme de 12 ans, à mettre en service en janvier 1967.

Outre les coordinations de programmes mentionnées ci-dessus en 1 (*b*), des observations simultanées des mêmes étoiles avec les lunettes zénithales flottante et visuelle de Mizusawa sont en cours.

Teleki et Sevarlic ont comparé, par des observations simultanées, l'ancien et le nouveau programme du service des latitudes de Belgrade.

Fichera propose une coordination des mesures de l'heure pour les instruments des passages à des latitudes voisines.

Les méthodes de réduction des observations au PZT et à l'astrolabe de Herstmonceux ont été exposées dans deux importants mémoires de Thomas. Takagi a étudié le rôle des erreurs de positionnement du PZT.

La constante d'échelle et le pas de vis des lunettes zénithales ont fait l'objet de nombreux travaux. Le large champ des lunettes zénithales ZTL 180 permet d'observer des paires d'étoiles avec des différences de déclinaison de 1°5; par suite l'échelle et son coefficient thermique sont déterminés avec une grande précision relative et les résultats ne sont pratiquement pas affectés par les erreurs des déclinaisons (Gurstein, Rykhlova et Prodan). N. A. Popov a discuté une nouvelle méthode pour obtenir la constante d'échelle à partir des mêmes observations d'étoiles zénithales.

(*e*) *Révision des catalogues et nouvelles réductions des séries de mesure de temps et latitude*

Poulkovo: Une nouvelle réduction des observations faites avec la lunette ZTF 135, de 1948 à 1960, a été achevée (avec la nouvelle valeur de la constante de l'aberration).

Mount Stromlo: Les résultats du PZT ont été réduits à un système unique, depuis 1959.

Tokyo: Des déclinaisons corrigées sont utilisées depuis 1966 pour le PZT, mais les ascensions droites n'ont pas été changées.

Washington: Toutes les observations de PZT, depuis 1915 sont en cours de révision, afin d'améliorer les positions des étoiles.

Ottawa: Les résultats de la période 1956·0–1961·0 ont été rendus homogènes avec les suivants, par nouvelle mesure des plaques et nouvelle réduction (conclusion à paraître dans le *Dominion Obs. Bull.*).

Rappelons que les programmes d'astrolabe peuvent être conservés invariables et qu'ils sont par conséquent homogènes quand ils sont exprimés dans le système du catalogue de référence (FK4). Cependant certaines stations font usage de corrections de groupe qui sont sujettes à révision.

La grande précision avec laquelle la constante d'échelle des ZTL 180 est mesurée permet d'utiliser ces lunettes pour déterminer les différences de déclinaison des paires d'échelle du SIMP. Des observations, pour ce but, ont été entreprises par quelques observatoires d'U.R.S.S.

Les positions et mouvements propres des étoiles du SIMP (ou SIL) ont été obtenus par Major; l'accord est bon avec les résultats de Mizusawa, Ukiah, Carloforte, de 1906 à 1935.

Les mesures de temps et de latitude sont utilisées pour améliorer le catalogue fondamental (voir Commission 8). Plusieurs catalogues basés sur les mesures d'heure en U.R.S.S. ont été publiés par Afanasyeva *et al.*, Vasiliev, Chelomlitko, Michenko, Yasevich, Blinov; un catalogue de compilation en est déduit par l'Observatoire de Poulkovo (à paraître à la fin de 1966); il doit comprendre 800 étoiles et repose sur plus de 150 000 observations.

(*f*) *Calcul de la polhodie*

La simple considération de l'erreur interne des mesures ne suffit pas à fixer le poids à introduire dans les calculs, comme l'ont montré plusieurs auteurs d'U.R.S.S. Une nouvelle méthode d'estimation des poids a été décrite par Fedorov et Korsun.

Mme A. Stoyko a achevé la détermination de la polhodie, pendant l'AGI et la CIG à partir de 50 instruments. Pour cette période, elle a calculé le terme non polaire *Z*, à partir des 38 meilleures séries d'observations: la demi-amplitude de ce terme (0″038) et sa phase sont en parfait accord avec celles déduites de 60 années d'observations du SIL. *Z* ne dépend pas, en moyenne de la latitude, ni du type d'instrument. Mais quelques stations donnent des valeurs anormales.

B. Guinot et M. Feissel ont déterminé le mouvement du pôle de 1964·0 à 1966·5 à partir de l'ensemble des mesures de temps et de latitude de plus de 60 stations. Dans ce premier travail, les mesures brutes sont utilisées, telles qu'elles sont transmises par les observatoires. L'étude des résidus est en cours. Bien que les résultats de chaque station n'aient pas été lissés, les coordonnées du pôle calculées pour chaque 1/20 d'année varient d'une façon très régulière (Figure 3). La polhodie diffère sensiblement de celle obtenue par le SIMP; son aspect général est plus proche du cercle. La solution essayée avec le terme *Z* des latitudes, donne à *Z* une 1/2 amplitude de 0″02. La phase est différente de celle trouvée à partir des mesures de latitude seules.

(*g*) *Analyses des mesures*

Le problème général de l'analyse des séries temporelles a été traité par Jeffreys dans la dernière édition de son ouvrage 'Theory of Probability'. La méthode d'Orlov, pour séparer les composantes du mouvement du pôle a été généralisée par Iijima qui l'a étendue à toutes les périodicités pratiquement utiles.

Le spectre du mouvement polaire a été étudié par des auteurs d'U.R.S.S. Pour la période 1891·5 à 1962·0, Yashkov, en plus du terme annuel et du terme de Chandler a trouvé des termes à périodes de 1·17, 1·238 et 1·245 an. Une conclusion différente a été atteinte par Fedorov et Yatsin, qui attribuent ces périodes multiples à de considérables variations de phase et d'amplitude de l'oscillation chandlérienne. Le spectre typique de la variation de la latitude montre une décroissance continue de la densité spectrale avec la fréquence et des sommets pour certaines fréquences; ces traits sont communs à toutes les séries d'observations (Glagoleva,

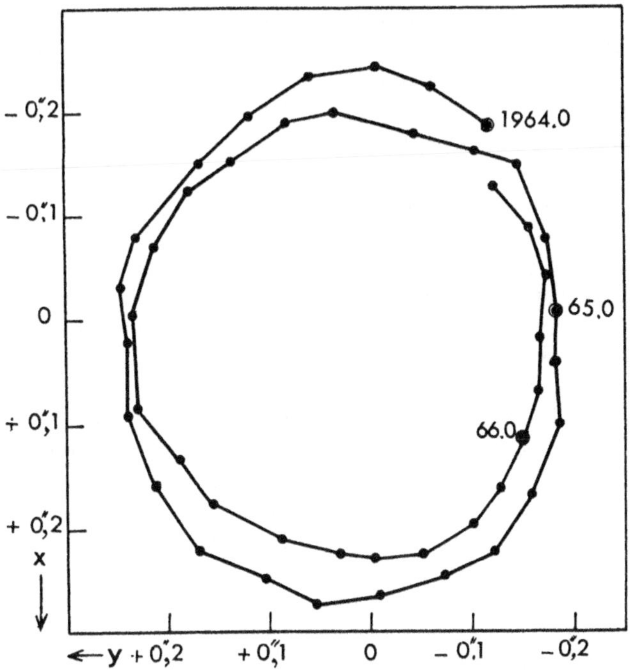

Fig. 3. Polhodie d'après les mesures de temps et de latitude faites en 61 stations. L'origine est arbitraire. Les points sont obtenus sans lissage.

Yatskiv, Panchenko et Korba). Des termes périodiques non polaires sont présents dans les observations de latitude; Major attribue certains d'entre eux à des termes incorrects de la nutation astronomique et de la nutation diurne, l'origine des autres est encore obscure. N. A. Popov a continué ses études de la nutation diurne. Kulagin et Kovbasyuk ont trouvé dans les observations de latitude à Gorki les termes diurne, semi-annuel et semi-mensuel de la nutation, avec des demi-amplitudes de 0",039, 0",039 et 0",015. Nesterov s'est aussi occupé du terme semi-annuel. Rusu a annoncé l'existence d'un terme à période de 0·75 an. Sugawa a étudié les effets lunaires de marée sur les observations faites à Mizusawa, de 1935 à 1961 et Mlle Débarbat, sur celles faites à Paris (temps et latitude) de 1956·5 à 1963·0.

La variabilité de la période de Chandler a été à nouveau mise en évidence par les travaux de Iijima; d'après cet auteur l'ellipse du mouvement annuel du pôle varie considérablement et par suite les positions du pôle d'inertie qu'il en déduit varient d'une année à l'autre; ces variations sont attribuées à des anomalies météorologiques à l'échelle mondiale. Les coordonnées du pôle d'inertie ont été également calculées par Drâmbă et Stănilă.

On n'a pratiquement pas progressé dans la détermination expérimentale des dérives du pôle. Le taux de variation de la latitude moyenne à Paris, de 1956·6 à 1962·7 trouvé par Guinot a été confirmé par Okuda et Sugawa, d'après les observations de Mizusawa et Tientsin; à Paris, cette variation ne s'est pas maintenue par la suite et la latitude moyenne paraît être revenue en 1965 à sa valeur de 1956·6; il serait intéressant de voir si des phénomènes semblables se sont manifestés à Mizusawa et Tientsin.

La vitesse de la rotation de la Terre décroît, depuis le début de 1962 et la durée du jour croît linéairement en fonction du temps de 0·3 ms par an, environ (voir Figure 2); ceci confirme l'hypothèse de Brouwer selon laquelle la Terre ne subit pas de changements brusques de vitesse, mais que de brusques changements d'accélération pourraient survenir. Markowitz situe ces changements d'accélération, depuis juin 1955, en septembre 1957 et en janvier 1962.

A cette accélération d'ensemble se superposent des effets saisonniers, vraisemblablement non reproductibles d'année en année et par suite mal compensés par la formule fixe que le BIH utilise pour le calcul de T.U. 2. Ainsi, Pavlov et Staritsyn attribuent les irrégularités de 1959 et 1961 au printemps précoce sur l'Europe et l'Asie; Iijima *et al.* ont trouvé un bon accord quantitatif entre les irrégularités de la rotation terrestre en 1963 et la situation météorologique anormale.

De nouveaux termes périodiques de la rotation de la terre ont été révélés. Iijima et Okazaki, outre les termes annuel et semi-annuel considèrent un terme à période de 2 ans, de demi-amplitude 9 ms qui pourrait être attribué au changement bi-annuel de la circulation des vents stratosphériques. Belotserkovsky a trouvé des fluctuations à période d'environ 3 mois et dont l'amplitude atteint 4 ms. Mme A. Stoyko a recherché les termes ayant leur origine dans la marée terrestre de périodes mensuelle, semi-annuelle et de 18·6 ans.

Le problème de la dérive des continents continue à partager les opinions. Mme A. Stoyko d'une part, Torao *et al.* d'autre part trouvent que la différence des longitudes augmente entre le Japon et l'Amérique du Nord, qu'elle diminue entre l'Amérique du Nord et l'Europe; l'ordre de grandeur de ces effets est 1 ms par an. Cependant W. Markowitz doute de la validité de tels résultats à cause de la diversité des programmes d'observations astronomiques et il suggère d'établir des programmes plus stricts, permettant d'éliminer les effets des erreurs sur les mouvements propres des étoiles.

Mme A. Stoyko a trouvé que depuis 1900, à un retard aléatoire de la rotation terrestre correspond une augmentation de l'amplitude de la nutation libre.

(*h*) *Travaux théoriques*

Plusieurs auteurs ont continué à rechercher les effets de la constitution fluide de l'intérieur de la Terre: Tarady a discuté le mouvement annuel du pôle, Kakuta a conclu que la rotation irrégulière de la Terre pourrait être partiellement due à des phénomènes hydromagnétiques.

Sekiguchi et Matsumoto ont construit un appareil pour déterminer expérimentalement la relation entre un couple externe et la précession d'un sphéroïde rempli d'un fluide visqueux: ils concluent que le modèle d'une terre solide est convenable pour la théorie de la nutation.

Sekiguchi a développé la théorie de l'excitation et de l'amortissement du mouvement polaire.

Yumi et Nakagawa ont analysé la variation de marée de la gravité; ils ont montré que les anomalies de ce phénomène sont en corrélation avec celles du mouvement polaire. D'autre part, Shimizu a trouvé la période de Chandler dans la variation du niveau de la mer et dans la pression barométrique.

Melchior déduit les amplitudes des nutations astronomiques à courte période de l'observation des marées terrestres: l'existence d'effets dynamiques dûs au noyau liquide de la Terre est, en effet, de nature à altérer les rapports d'amplitude des diverses nutations tels qu'ils sont donnés par la Mécanique Céleste. Les résultats expérimentaux obtenus avec des pendules horizontaux ont confirmé les calculs théoriques sur modèles de Jeffreys-Vicente et de Molodensky.

B. GUINOT
Président de la Commission

BIBLIOGRAPHIE

Afanasyeva, P. M., *et. al.* 1966, *Trudy glav. astr. Obs. Pulkove*, **75**.
Belotserkovsky, D. Y. 1963, Trans. Conf. on *Rotation of the Earth*, Kiev.
Blinov, N. S. 1965, *Bull. Sternberg Inst.*, **140–141**, 3.
Brouwer, D. 1952, *Astr. J.*, **57**, 125.

Chelomlitko, A. P. 1966, *Trudy glav. astr. Obs. Pulkove*, **75**.
Drâmbă, C. 1964, *Studii Cerc. Astr.*, **9**, 1.
Drâmbă, C., Stănilă, G. 1966, *Studii Cerc. Astr.*, **11**, 1.
Enslin, H. 1964, Actes du Congrès international de Chronométrie, **1**, Lausanne.
Enslin, H. 1965, *Jb. deutsch. Ges. Chronom.*, **16**.
Fedorov, E. P., Korsun, A. A. *Var. Lat.*, Kiev, **2** (sous presse).
Fedorov, E. P., Yatskiv, Y. S. 1964, *Astr. Zu.*, Moscou, **41**, 764.
Glagoleva, I. I. 1965, *Var. Lat.*, **1**, 59.
Glagoleva, I. I., Yatskiv, Y. S. 1965, *Lat. Var.*, Kiev, **1**, 70.
Goto, T., *et al.* 1964, *Proc. int. Latit. Obs. Mizusawa*, **4**.
Goto, T., *et al.* 1965, *Proc. int. Latit. Obs. Mizusawa*, **5**.
Goto, T., *et al.* 1966, *Proc. int. Latit. Obs. Mizuaswa*, **6**.
Gozhy, A. V., Slavinskaja, A. A. 1965, *Lat. Var. and Polar Motion*, **N3**, Moscou.
Guinot, B. 1967, *Bull. horaire BIH*, **J13**.
Guinot, B. 1965, *Bull. astr.*, Paris, **24**, 461.
Gurstein, A. A. 1965, *Bull. Sternberg Inst.*, **140–141**, 17.
Gurstein, A. A. 1963, *Astr. Zu.*, Moscou, **40**, 178.
Iijima, S. 1965, *Ann. Tokyo astr. Obs.*, 2nd serie, **9**, no. 4, 145.
Iijima, S., Okazaki, S. 1965, *J. geod. Soc. Japan*, **11**, no. 3–4, 98.
Iijima, S. 1965, *Ann. Tokyo astr. Obs.*, 2nd serie, **9**, no. 4, 155.
Iijima, S., Okazaki, S. *J. geod. Soc. Japan*, **12**, no. 2 (sous presse).
Iijima, S., Matsunami, N., Okazaki, S. 1964, *Ann. Tokyo astr. Obs.*, **8**, no. 4.
Kakuta, C. 1965, *Publ. int. Latit. Obs. Mizusawa*, **5**, no. 1, 17.
Kolaczek, B., Bieniewski. *Publ. Chair geodet. Astr. Warsaw Techn. Univ.* (sous presse).
Kulagin, S. G., Kovbasyuk, L. D. 1964, *Astr. Zu.*, Moscou, **41**, 758.
Major, S. P. 1966, *Problems of Astronomy*, Kiev.
Major, S. P. *Lat. Var.*, Kiev, **2** (sous presse).
Markowitz, W., *et al.* 1964, *Research in Geophysics*, vol. 2, (Massachusetts Inst. of Technology).
Markowitz, W. 1966, Proc. Second int. Symposium on *Recent Crustal Movements*, Helsinki, 241. (*Ann. Acad. Sci. Fennica*, **90**).
Melchior, P. 1965, *Commun. Obs. r. Belgique*, sér. Géophys., **70**, 24.
Melchior, P. 1966, *Geophys. J.R. astr. Soc.*, **12** (sous presse).
Mischenko, M. P., Shirjaev, A. V. 1964, *Trudy astr. Obs. Leningr. gos. Univ.*, **21**.
Nesterov, V. V. 1964, *Bull. Sternberg Inst.*, **134**, 17.
Okuda, T., Sugawa, C. 1964, *J. geod. Soc. Japan*, **10**, 3–4, 207.
Panchenko, N. I., Korba, S. N. *Lat. Var.*, Kiev, **2** (sous presse).
Pavlov, N. N., Staritsyn. 1965, *Lat. Var. and Polar Motion*, **N3**, Moscou.
Popov, N. A. 1965, *Var. Lat.*, Kiev, **1**, 16.
Popov, N. A. 1965, *Commun. Obs. r. Belgique*, sér. Géophys., **69**.
Rusu, L. 1966, *Studii Cerc. Astr.*, Bucarest, **11**, 1, 93.
Rykhlova, L. V., Prodan, Y. I. 1964, *Bull. Sternberg Inst.*, **134**, 33.
Scheepmaker, A. *Bol. astr. Obs. Quito*, Serie A, **1** (à paraître).
Sekiguchi, N., Matsumoto, J. 1965, *Ann. Tokyo astr. Obs.*, **10**, no. 1.
Sekiguchi, N. 1965, *Publ. astr. Soc. Japan*, **17**, no. 2, 136.
Sekiguchi, N. 1965, *Publ. astr. Soc. Japan*, **18**, no. 2 (sous presse).
Shimizu, T. 1963, *Spec. Contr. geoph. Inst. Kyoto Univ.*, **3**, 255.
Stoyko, A. 1966, *Ann. int. geophys. Year*, Vol. Latitude, 2ème partie, 23.
Stoyko, A. 1964, *Commun. Obs. r. Belgique*, **236**, 440.
Stoyko, A. 1966, *Ann. int. geophys. Year*, Longitudes.
Stoyko, A. 1966, *C. r. Acad. Sci.*, Paris, **262**, sér. A et B, no. 16, 1098.
Sugawa, C. 1965, *Publ. int. Latit. Obs. Mizusawa*, **5**, no. 1, 1.
Takagi, S. 1965, *Publ. int. Latit. Obs. Mizusawa*, **5**, no. 1, 43.
Tarady, V. K. 1965, *Astr. Zu.*, **42**.
Taterev, G. S. 1965, *Lat. Var. and Polar Motion*, **N3**, Moscou.
Teleki, G., Ševarlić, B. M. 1964, *Bull. Obs. astr. Belgrade*, **25**, no. 3, 73.
Thomas, D. V. 1964, *R. Obs. Bull.*, **81** (PZT).

Thomas, D. V. 1965, *R. Obs. Bull.*, **92** (astrolabe).
Thomas, D. V. 1967, *R. Obs. Bull.*, **127** (group corrections . . .)
Torao, M., Okazaki, S. 1965, *Tokyo astr. Obs. Bull.*, 2nd ser., **169.**
Torao, M., Okazaki, S. 1964, *Ann. Tokyo astr. Obs.*, **8**, no. 4.
Torao, M., Okazaki, S., Fujii, S. 1964, *J. geod. Soc. Japan*, **10**, no. 3–4, 203.
Vasiliev, V. M., Karetnikova, E. N. 1966, *Trudy glav. astr. Obs. Pulkove*, **75.**
Yasevich, B. V. 1964, *Trudy Taskent astr. Obs.*, Serie II, **10.**
Yashkov, V. Y. 1964, *Astr. Zu.*, Moscou, **41**, 760.
Yatskiv, Y. S. 1965, *Var. Lat.*, Kiev, **1**, 113.
Yatskiv, Y. S. 1965, *Lat. Var.*, Kiev, **1**, 75.
Yumi, S., Nakagawa, I. *J. geod. Soc. Japan*, **12**, no. 2 (sous presse).

ANNEXE. RAPPORT SUR LE MOUVEMENT POLAIRE 1941.0–1948.9 DÉDUIT DES OBSERVATIONS DE TROIS STATIONS

(préparé par T. Nicolini)

Les résultats définitifs du mouvement polaire du SIL dans l'intervalle très critique 1941.0–1948.9 ont été souhaités par l'UAI, l'AIG, et la C.G.Ital. Les calculs provisoires (ainsi que ceux de 1935.0 à 1940.9) avaient été déjà accomplis à l'Observatoire de Capodimonte, et publiés par le regretté Professeur Carnera dans *Contributi Astronomici* de l'Observatoire, nos. 1, 2, 6, vol. **4** (1947, 1948, 1949).

Après sa retraite à Florence, le Professeur Carnera publia le volume IX du SIL (1935.0–1940.9) (1), et se proposa de calculer tout seul la deuxième série définitive 1941.0–1948.9. Il était un grand et infatiguable travailleur, mais deux causes réussirent à plier ses efforts: son âge très avancé, et plus encore les difficultés imposées par le fonctionnement très critique des Stations pendant la guerre mondiale. Il avait réduit sa tâche à la considération de trois Stations, Mizusawa, Gaithersburg, Ukiah, qui avaient observé avec continuité; mais enfin il était convaincu, et il le disait franchement, qu'on ne pouvait espérer rien de sûrement satisfaisant.

Quoiqu'il en soit, il faut donner les résultats possibles, si imparfaits qu'ils soient. Parmi les manuscrits du Professeur Carnera, que le Professeur Dore, Président de la Commission Géodésique Italienne envoya à l'Observatoire de Capodimonte, on n'a trouvé pas trace de calculs utilisables pour le mouvement polaire, mais seulement de longues comparaisons de déclinaisons stellaires (c'était un autre objet d'un intérêt particulier de la part du Professeur Carnera), aussi inachevées. Cependant, heureusement le Professeur Carnera a laissé dans un mémoire (2) et dans le volume IX du SIL les éléments utiles à la poursuite du travail. Ces éléments sont: (1) révolutions définitives des micromètres, données dans le Mémoire avec les coefficients de température; (2) corrections de déclinaison pour chaque couple et pour chaque année, obtenues en poursuivant la table de Carnera (vol. IX du SIL, pp. 18–25) dans l'intervalle 1941.0–1948.9.

On rapporte ici la solution de la polhodie résultant des trois Stations mentionnées. Les demi-révolutions des micromètres adoptées ici (avec quelques ajustements des coefficients de température trouvés par Carnera) sont les suivantes (T = température centigrade), pour tout l'intervalle:

Mizusawa: $R/2 = 19''.97340 - 0''.0003\ (T - 10°)$
(± 0.001)
Gaithersburg: $R/2 = 19''.81260$,, ,,
(± 0.002)
Ukiah: $R/2 = 19''.88007$,, ,,

Disposer de ces éléments, c'est une circonstance heureuse, puisque on peut retenir que les résultats qu'on déduira seront conformes au plan de Carnera, et notre contribution sera restreinte à une complète réélaboration numérique du travail provisoire.

Dans ce rapport, la table numérique qui suit donne les coordonnées X et Y du pôle de rotation pour les époques standard, moyennes (soir + matin)/2.

Polhodie 1941.0–1948.9: Solution par trois Stations, Mizusawa, Gaithersburg et Ukiah, unité 0″001

Fract. d'année	1941		1942		1943		1944	
	X	Y	X	Y	X	Y	X	Y
·060	+103	+ 43	+ 55	+128	+ 89	+207	−161	+273
·143	+ 21	+ 49	+ 66	+ 93	+120	+176	−159	+301
·226	− 8	+107	0	+ 90	+127	+180	− 62	+278
·310	−155	+184	− 45	+137	+ 61	+161	− 17	+289
·393	−139	+221	− 41	+145	+204	+110	+132	+246
·476	− 79	+240	− 55	+148	+165	+ 64	+164	+240
·560	+ 7	+262	+ 55	+ 94	+170	+ 18	+297	+ 61
·643	+ 92	+256	+ 10	+120	+134	+ 34	+200	+ 46
·726	+ 18	+262	+ 59	+134	− 23	+ 52	+ 60	− 12
·810	+ 3	+229	− 22	+149	− 94	+ 90	− 78	− 40
·893	− 1	+185	− 22	+170	−196	+171	−122	− 48
·976	− 14	+163	− 14	+257	−320	+264	−338	+122
	1945		1946		1947		1948	
	X	Y	X	Y	X	Y	X	Y
·060	−380	+227	−203	+ 76	+ 60	− 41	+ 61	+223
·143	−339	+239	−309	+ 73	−129	+ 17	+ 89	+104
·226	−330	+361	−301	+158	−307	+ 66	−119	+152
·310	−162	+365	−280	+283	−368	+215	− 77	+ 56
·393	+ 41	+341	−180	+348	−331	+273	−109	+ 50
·476	+140	+345	−100	+409	−239	+337	− 61	+ 29
·560	+135	+249	− 15	+463	− 98	+341	− 10	+ 57
·643	+ 13	+266	+ 94	+395	− 68	+409	− 25	+137
·726	+191	+ 91	− 57	+475	+ 34	+390	− 19	+169
·810	+194	− 60	+179	+220	+ 81	+358	+ 15	+248
·893	+172	−171	+450	− 44	+135	+273	+ 94	+310
·976	− 88	− 14	+ 33	+ 99	+108	+263	+ 49	+250

La représentation graphique (Figures 1 et 2) montre une allure parfois singulière peu satisfaisante, particulièrement si l'on considère que les résultats séparés des groupes du matin et du soir sont souvent très écartés entre eux. Le défaut est particulièrement apparent dans la coordonnée X, et cela est bien naturellement dû à l'interruption des observations à Carloforte.

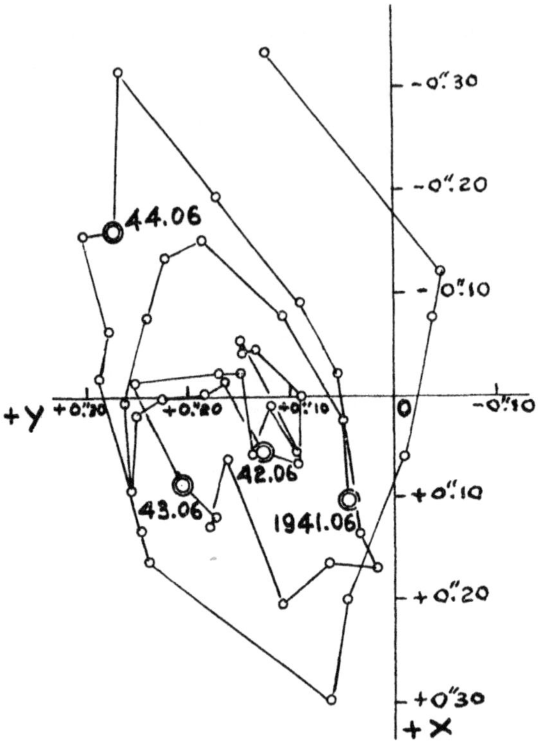

Fig. 1. Polhodie d'après Mizusawa, Gaithersburg et Ukiah, de 1941.0 à 1945.0

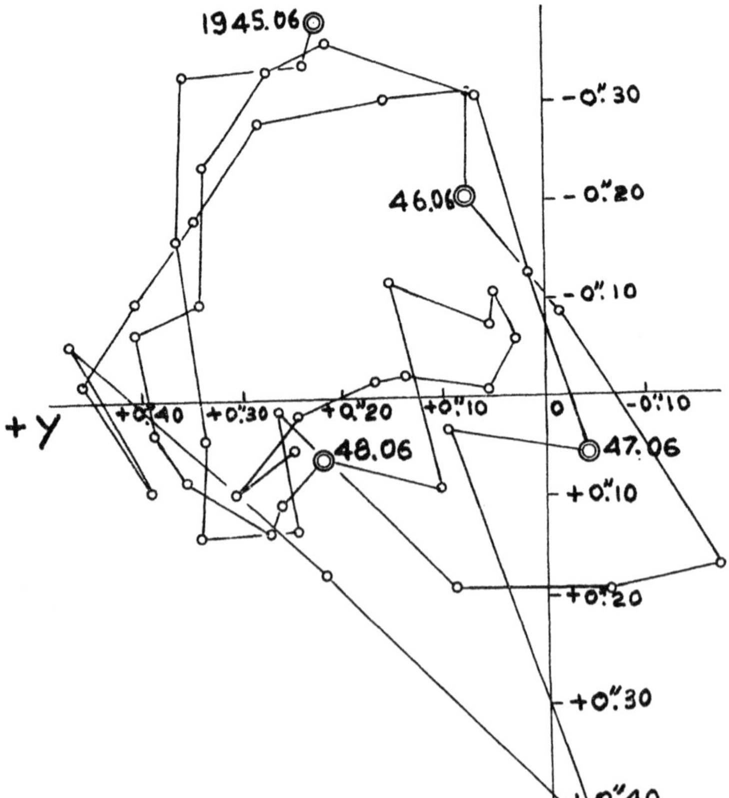

Fig. 2. Polhodie d'après Mizusawa, Gaithersburg et Ukiah, de 1945.0 à 1949.0

A ce point, les doutes sur les résultats obtenus font ressortir l'opportunité d'une nouvelle solution, où les observations lacunaires de Carloforte et Kitab seront prises en considération, contrairement au programme préalablement établi. Au moment de la rédaction de ce rapport, cette nouvelle solution est très avancée.

BIBLIOGRAPHIE

1. 'Risultati del Servizio Internazionale delle Latitudini dal 1935.0 al 1941.0.' Vol. IX S.I.L., F.lli Stianti, Sancasciano 1957.
2. 'Le livelle ed i micrometri dei telescopi zenitali nelle Stazioni Internazionali di Latitudine.' Memorie della Acc. Naz. dei Lincei, serie III, vol. IV, Roma, 1953.

20. COMMISSION DES POSITIONS ET DES MOUVEMENTS DES PETITES PLANETES, DES COMETES ET DES SATELLITES

PRÉSIDENT: Dr P. Herget, Cincinnati Observatory, Observatory Place, Cincinnati, Ohio 45208, U.S.A.

VICE-PRÉSIDENT: Dr S. G. Makover, Institute of Theoretical Astronomy, Mendelejevskaja linia 1, Leningrad 164, U.S.S.R.

COMITÉ D'ORGANISATION: S. Arend, D. Brouwer†, F. K. Edmondson, H. Hirose, H. W. Wood.

MEMBRES: Ahmed, Böhme, Boyer, Bruwer, Candy, Clemence, Cunningham, Evdokimov, Gehrels, Giclas, Gutierrez-Moreno, Heinrich†, Herrick, Hertz, Itzigsohn, Jeffers, Kahrstedt, Kamienski, Kazimirchak-Polonskaya, Kepinski†, König, Kovalevsky, Krésak, Kuiper, Michkovitch, Missana, Orlov (A. A.), Pels†, Popović, Porter, Protitch, Rabe, Rasmusen, Reinmuth, Robertson, Roemer, Schmitt, Schrutka-Rechtenstamm, Schubart, Strobel, Subbotin†, Torroja, Väisälä, Van Biesbroeck, Van Houten, Van Houten-Groeneveld, Wilkins, Yakhontova, Zadunaisky.

INTRODUCTION

The outstanding feature of minor planet research during this triennium has been the completion of orbit improvements on a large scale, both by the Institute of Theoretical Astronomy at Leningrad and by the Cincinnati Observatory. This in turn improves the accuracy and the reliability of the published ephemerides. The increasing impact of the more widespread use of electronic computers is also becoming evident. Much good continues to come from the work of Dr O. Kippes in establishing identifications; and L. Boyer has rejuvenated the program of identities of the late A. Patry. Also considerable recent success along these lines has come from C. M. Bardwell at Cincinnati. The dominant problem continues to be in the area of observations. One reason is that nearly all significant observations now need to be in the range of one or two magnitudes fainter than have formerly been useful. Another is that the efforts to secure a suitable instrument for conducting another survey such as the Yerkes-McDonald survey of 1950–52 have not been successful as yet.

The number of MPC's issued during this three-year period is more than double that of the previous one. From 1961–63 the Cincinnati Observatory computed 154 sets of improved elements of numbered planets. During 1964–66, the Institute of Theoretical Astronomy at Leningrad computed 415 sets (**1, 2, 3, 4, 5**), and the Cincinnati Observatory computed 242 sets (**6**), as well as 129 sets for unnumbered planets (**7**). In passing, we should inscribe a special tribute to the NORC, the Naval Ordnance Research Calculator at Dahlgren, Va. It is estimated that during the 12-year lifetime of this electronic calculator it calculated 10^{10} individual arithmetic operations for the benefit of minor planet research and more than 15 000 extended opposition ephemerides.

Contributions to this draft report have been received from S. Arend, L. Boyer, J. Bruwer, F. K. Edmondson, T. Gehrels, A. Gutierrez-Moreno, S. Herrick, P. Jackson, T. Kiang, J. Kovalevsky, L. Krésak, S. G. Makover, B. G. Marsden, V. V. Michkovitch, B. Milet, L. Oterma, E. Rabe, E. Roemer, G. Schrutka-Rechtenstamm, J. Schubart, C. J. Van Houten, G. A. Wilkins, and H. Wood. In the area of comets, the draft report has been prepared separately by Dr Elizabeth Roemer.

OBSERVATIONS

Observations with the 10-inch Cooke triplet at the Goethe Link Observatory have been at a gradually declining rate (426 plates in 1963, 313 in 1964, and 216 in 1965) but the rate has fallen off even more in recent months, due to the lack of personnel who are interested in this program. The number of accurate positions which have been requisitioned in support of the improvement of elements has increased sharply (more than 300), and such requests may be expected to continue for some time to come. Mrs B. Potter retired in June 1966, and this will adversely affect the program because she has blinked and measured all the plates during the last 17 years.

Somewhat in compensation, an observing program with a 40 cm *f*/4, astrographic camera (100 square degrees) was initiated in 1964 at the Crimean Astrophysical Observatory. In 1965, 92 plates were obtained, with a total of 449 minor planet positions. The plate reductions are computed on the BESM-2 electronic computer by the Institute of Theoretical Astronomy, mainly based on reference stars from the Yale Zone Catalogues. This rate of activity may well be doubled in future years. Observations also continue at Abastumani, Alma-Ata, Tartu, Tashkent and other places (**50** to **69**, incl.). The program of observations of ten selected minor planets for determination of the systematic errors of fundamental catalogues continues successfully. Pulkova, Moscow, Kiev, Tashkent, Nikolajev and several observatories outside the U.S.S.R. all contribute their observations to ITA (**70**, **71**, **72**, **73**, **74**). To date about 5000 observations have been collected and the $(O - C)$'s and normal places have been formed. Pulkova has measured and reduced the plates from the Cape Observatory (**75**, **76**). This program will be carried through three complete revolutions of each planet, and it will be completed by 1969.

Other observatories which have observing programs for minor planets report as follows. At Uccle, observations with the 40 cm double astrograph continue as in former years, at a rate of about 250 accurate positions per year (**77** to **82**, incl.). It is planned to transfer this instrument to a better observing site. The IBM-1620 has been programmed by H. Debehogne and G. Roland to compute accurate plate reductions and checks by Arend's method (**8**), preliminary orbits by the Gauss-Encke method, orbit improvements by variation of geocentric distances, parabolic orbits by the method of Olbers-Banachiewicz, and ephemerides and residuals. At the Republic Observatory, Johannesburg, about 200 accurate positions per year are obtained, and 64 positions were accurately reduced from older plates. The program is aimed at objects brighter than 15·8 magnitude and south of $-20°$. Comets brighter than 14th magnitude are observed in the southern hemisphere (a total of 24 positions). At Santiago, Chile, a Gauthier astrograph was newly mounted in March 1965, and thus far 140 plates have been taken for the planets of the Pulkova program, and (1), (2), (3), (4), and (51). Skalnaté Pleso initiated a program in 1966 with a new 30/150 cm Zeiss Vierlinser. At Turku the 500/1031 mm Anastigmat has been removed because of site deterioration and it is expected to be remounted 27 km to the east. At Tuorla (12 km east of Turku, long. $1^h 29^m 46^s$E, lat. $+60° 24' 57''.5$) an anastigmatic reflector 380/380/1031 mm has been mounted, and test plates reached 16·7–17·0 phot. magn. Another reflector (700/700/1719 mm, with full correcting plate) is not yet finished. A list of 1000 accurate positions from Turku plates during 1938, 39 has been prepared by Mrs Rantaseppa. Miss Roemer's observing program at Flagstaff has been concluded and published (**83**, **84**, **85**). Her new prospects at Tucson include the use of the 61-inch, *f*/13·5 reflector in the Catalina Mts., a 7-inch, *f*/7 Bailey astrograph, and other instruments in the area and at Kitt Peak. At Sydney, Australia, Wood, Robertson, and Sims continue a program of observing all even-numbered minor planets in the southern hemisphere within their magnitude capability, as well as special observing of selected objects for use in fundamental astronomy (**86** to **91**, incl.). It was presumed that the odd-numbered minor planets would be observed at La Plata, but their contributions have greatly diminished in recent years. At Wein, Paul Jackson has undertaken to measure accurately about 500 old plates which are on file, and the reductions will be carried out in cooperation with the Cincinnati Observatory. At Nice, the Zeiss double astrograph has been used by B. Milet since July 1965 to obtain 247 plates with 1071 positions of 200 numbered minor planets, and 38 positions of two comets. He has established a least squares reduction program using Cracovians which permits the determination of the constants to the 6th order.

The Palomar-Leiden survey of faint minor planets has reached the stage where ambiguities of identification have been settled for the vast majority of objects and the elements have been computed. One comet was discovered (**9**). Some plates were re-examined with the help of search ephemerides, and 38 additional objects were found. The remaining objects which afforded only indeterminate orbits will be examined with the view to determining an approximate distance, for photometric purposes. In 1965 six plates were taken by Dr T. Gehrels on the 48-inch Palomar Schmidt, covering two fields in the neighborhood of one of the Trojan libration points. These will be examined at Leiden in the same manner as were the previous plates.

Dr T. Gehrels determined a photoelectric light-curve for the Trojan (624) with the Kitt Peak 84-inch during February 1964. The amplitude of the light-curve is 0·2 magnitude, which is to be compared with the amplitude of 0·8 magnitude found, at the Radcliffe 74-inch, in April 1957. The Trojan apparently has a large obliquity, similar to an effect found for the common asteroids (**10**). During 1965, C. J. Van Houten and I. Van Houten-Groeneveld determined photoelectric light-curves with the 36-inch at Broederstroom, South Africa, for (2), (19), (22), (23), (24), (28), (29), (43), and (54). Also (349) and (354) were observed for the determination of their phase function. Using an additional 1500 magnitude determinations made at Indiana by Mrs D. Owings, Gehrels has prepared an improved list of minor planet absolute magnitudes for approval and adoption, to replace the list adopted in Moscow in 1958. An additional intercomparison of 400 Indiana magnitudes with the Yerkes-McDonald survey of 1950–52 substantiates that the variation of brightness is inversely as the square of the heliocentric distance over the range from 2·2 to 3·1 A.U. (exponent is $2{\cdot}00 \pm 0{\cdot}02$).

ORBITS, EPHEMERIDES

The ITA continues to publish the annual ephemeris volume in the same satsifactory format as in previous years (**11, 12, 13**). All ephemerides are based upon special perturbations unless the elements are designated by A (for absolute or general perturbations) or E (for elliptic orbit or no perturbations). Some planets, marked L, are considered to be lost and no ephemerides are given. In 1967, those not computed at the ITA or the Latvian State University are credited as follows: Cincinnati 230, Tokyo, 60, Shanghai 60, Purple Mt. 5, Herrick 4, J. Schubart 1, P. Naur 1. At ITA the computations are mostly by Cowell's method, including Jupiter and Saturn. All computations on the NORC are by the variation of vectorial constants at a 20-day interval, including all major planets from Venus to Neptune.

Two different methods for the improvement of orbits are used at the ITA (**14**). In the first method only accurate observations are used. This provides improved orbits with a precision comparable to that of the observations. The second simplified method deals with approximate observations, and is convenient for a preliminary improvement of orbits having large $(O - C)$'s, either not observed for many years, or having an insufficient number of accurate observations. Systematic work on the improvement of orbits of such planets is now in progress. Unfortunately the number of planets not observed in the last 10 years has not only not decreased but has actually increased. In the 1967 volume, 170 such planets have been listed. The increase in the number of such planets is probably the result of a general diminuation of observations throughout the world.

At the Cincinnati Observatory there is only one orbit improvement method, which is programmed for the IBM-1620. It allows the operator complete flexibility to reject unsatisfactory observations. Approximate positions are not used if it is possible to have the position remeasured accurately, and sometimes approximate positions are excluded simply because they increase the probable error of the solution. Herget has also programmed the IBM-1410 to compute perturbations at intervals of 4-, 10-, 20-, or 40-days, with any selection of major planets from Mercury to Pluto. The Minor Planet Center has a quick-response capability for computing preliminary orbits on the IBM-1620, using as many observations as may be available (**15**). Orbits for newly discovered objects are also computed at various other places (**16** to **21**, incl.).

At the Latvian State University at Riga, M. A. Dirikis has programmed the BESM-2 for accurate numerical integrations, including perturbations by seven major planets, and also programs for comparison of observations and orbit improvements. Orbits for about 40 planets have been improved and the ephemerides are calculated until 1975. At Heidelberg, J. Schubart has used his program on the Siemens 2002 to compute the orbit of (1221) Amor, using all the observations in five apparitions since 1932. The mass of the Earth had to be introduced as a seventh unknown. It turned out that Amor offers a reliable way to determine the mass of the Earth. The value obtained agrees closely with the result found from radar observations.

This work is one of three examples which have recently required the introduction of a new concept into orbit correction processes, namely the integration of the variational orbit. Schubart found that the successive differential corrections did not converge, simply because the usual method of computing the partial differential coefficients is inadequate. A similar condition was experienced by Cohen and Hubbard when they undertook to improve the orbit of Pluto. The resulting residuals showed a systematic trend with the period of Jupiter, and a satisfactory result was impossible until the variational orbit was integrated. At Cincinnati, Herget has completed a large portion of the work on the orbits of the outer satellites of Jupiter, where the integration of the variational orbit was included as a principal part of the main program. The differential corrections proceeded in a highly satisfactory manner. This problem and other related ones have also been studied by V. A. Izvekov (**22, 23, 24**).

Herrick has provided all the appropriate ephemerides for the close-approach planets, (1566), (1580), (1620), and (1685). Special observations of Icarus were secured at Pretoria in 1966 June and July, and this will strengthen the prediction for the passage in 1968. L. Boyer is engaged in searching for identifications, especially utilizing the circular elements published by the late A. Patry (**25**). Michkovitch reports: (1) The list of the mean periods of quasi-identical oppositions has been completed with the new numbered asteroids; (2) Card catalogues have been completed for the elements Ω, i, a, Φ, ω, ordered according to their ascending values; (3) Studies have been continued of the peculiarities and interesting proximities of quasi-coplanar orbits. Bardwell has recomputed the preliminary orbits of many minor planets with a view to providing reliable ephemerides, both for the past and the immediate future, to aid in identifications and recoveries.

At the Bureau des Longitudes, Paris, Kovalevsky has resumed work on a general theory of Jupiter VIII, aided by a recent series of observations by Miss Roemer. He has also started a general analytical theory of Nereid, for which observations are urgently needed. A. Bec is engaged in a discussion of all the observations of Jupiter's outer satellites VI to XII, which is expected to yield a new determination of the mass of Jupiter. Theoretical work on the motion of the Galilean satellites has been started under S. Ferraz Mello (**26**). A program of photographic observations is being developed, and G. Billaud has made photoelectric observations of eclipses. B. Morando (**27**) has derived a numerical general theory of the motion of Vesta without any secular or mixed terms. A comparison of the theory with observations is to be started soon.

Herget's program for the numerical integration of the orbits of the outer satellites of Jupiter has yielded excellent agreement with observations for Jupiter VIII, Jupiter IX, Jupiter XI, and Jupiter XII, and all the ephemerides have been computed until 2000 A.D. The orbit of Jupiter XII is weakly determined because of the paucity of observations since its discovery in 1951. There is still some problem with the determination of the mass of Jupiter and this must be examined further. J. D. Mulholland has developed a program for deriving a general numerical theory for Jupiter VI, based on rigorous differential equations and successive iterations. Several iterations have been completed, and the results continue to improve, but the agreement with observations is not yet satisfactory.

GENERAL

G. A. Wilkins (**28**) has continued his studies of the orbits of the satellites of Mars. He has obtained orbital elements for Phobos and Deimos by an analysis of all suitable observations from 1877–1928 and has obtained revised estimates of the mass and dynamical flattening of Mars; the values do not differ significantly from those obtained in earlier discussions of such orbital data. He also found that the new orbital elements give a satisfactory fit to observations made in 1941 and 1956 without the postulation of a secular acceleration in the motion of Phobos.

T. Kiang (**29**) has examined the problem of observational selection in the statistics of the orbital elements of asteroids. Reasons are given for believing that the discovery of asteroids, unsystematic as it has been, is nevertheless practically complete down to magnitude 14–15; beyond this we have to reckon with three selection factors (brightness, latitude and seasonal) and their interactions. A technique of 'discovery sequence' is found effective. It is then inferred that selection has caused (A) an overabundance of small inclinations and a spurious second harmonic in the frequency distribution of nodes among asteroids fainter than magnitude 15, and (B) an overabundance of large eccentricities among those fainter than magnitude 16. Inference (A) fully resolves the age-long controversy, initiated by Kleiber in 1886 (**30**) and Bauschinger in 1901 (**31**), over the question whether or not the distribution of nodes agrees with the predictions of the perturbation theory. The well-known discrepancy between observation and theory in the case of the distribution of perihelia is shown to be largely an indirect effect of (B), rather than due to seasonal factors as first conjectured by Brunn in 1906 (**32**); but part of discrepancy remains in the sense that the displacements in the (e,π) plane of the centers of revolution should be some 30% greater than hitherto calculated. There is also a significant, positive correlation between proper eccentricities and inclinations, which suggests the former existence of a resisting medium.

L. Krésak has investigated asymmetries in the asteroid belt, and determined its inner boundary and the elliptic elements of the center. The non-uniformities in the distribution of asteroids in heliocentric latitude and longitude, found by Narin, are explained by observational selection and alignment of the lines of apses to that of Jupiter. This alignment governs the radial asymmetry of the belt in that it produces differences in the asteroid density at the same heliocentric distance but different heliocentric longitude by as much as one order of magnitude. Some implications of this phenomenon are briefly discussed: the region of maximum probability of collisions able to yield Earth-crossing orbits, the seasonal variation of meteor impact on Mars and space probes launched to Mars, and seasonal variation of the discovery conditions of faint asteroids moving in eccentric orbits.

Rabe's work is reported, in part, in Commission 7. In collaboration with Mrs M. P. Francis, he is extending the former work on Eros, including many additional observations up to 1966 and a new integration program for the motion, including the relativity effect on the perihelion.

Herrick has examined and summarized the problem of universal variables for two body motion, differential correction formulas, and the variation of parameters (**33**). He has also discussed two problems associated with the Gaussian constant as it applies to heliocentric and geocentric orbits (**34**, **35**). Annual reviews of minor planet problems have been prepared by N. S. Yakhontova (**36** to **40**, incl.).

Other general works which deal with problems concerning minor planets will be found in the following references (**41** to **49**, incl.).

PROPOSALS

Gehrels proposes the adoption of a new list of absolute magnitudes of the numbered minor planets, to supplant the one adopted in 1958, since the new list will be based on more observations and include recently numbered planets. Presumably the material will be fully prepared by the time of the meeting in Prague.

The question of how to activate effective observing programs reaching to one or two magnitudes fainter than has routinely been achieved in the past will be an important subject of discussion. Chebotarev proposes an annual subsidy for a special program with the 40 cm astrograph of the Crimean Astrophysical Observatory. More details will be available at the time of the meeting.

BIBLIOGRAPHY

1. Khanina, F. B. 1965, *Bull. ITA*, **10**, 1(114), 44.
2. Makover, S. G. 1965, *Bull. ITA*, **10**, 3(116), 181.
3. Khanina, F. B. 1965, *Bull. ITA*, **10**, 6(119), 424.
4. Khanina, F. B. *Bull. ITA*, **10**, 10(123).
5. Baidarova, N. P., Labzina, L. I. *Bull. ITA*, **11**, 1(124).
6. *Minor Planet Circulars*, 2262, 2328, 2427, 2661, 2678.
7. Bardwell, C. *Minor Planet Circulars*, 2486, 2570, 2632.
8. Arend, S. 1933, Réduction de plaques astrophotographiques en coordonnées équatoriales. *Bull. astr. Obs. r. Belgique*, **1**, 9, 199.
9. Van Houten, C. J., Van Houten-Groeneveld, I. 1966, A New Periodic Comet, Observed in 1960. *Bull. astr. Inst. Netherl.*, **18**, 5, 441.
10. Gehrels, T., Owings, D. Photometric Studies of Asteroids. IX. Additional Light-Curves. 1962, *Astrophys. J.*, **135**, 906.
11. Ashkova, N. V. 1966, On the Preparation of the Annual 'Ephemerides of Minor Planets'. Trans. of the Baku Conf. 1962, 17.
12. Shmakova, M. J. Information on the Elements Published in the 'Ephemerides of Minor Planets'. *Bull. ITA*, **11**, 2(125).
13. Shor, V. A. 1964, On the Combined Use of High-Speed Computers and Punched-Card Computing Machines for Calculation and Preparation of Printing 'Ephemerides of Minor Planets'. *Bull. ITA*, **9**, 9(112), 601.
14. Makover, S. G., Beliajev, N. A. 1964, Programme d'intégration numérique des équations du mouvement des petites planètes et de la comparaison avec les observations. *Bull. ITA*, **9**, 8(111), 542.
15. Herget, P. 1965, Computation of Preliminary Orbits. *Astr. J.*, **70**, 1, 1.
16. Abalakin, V. K. 1965, The Elements of the Elliptic Orbit of the New Minor Planet 1963 UA. *Bull. ITA*, **10**, 5(118), 379.
17. Abalakin, V. K. 1964, The Elliptical Orbital Elements of Minor Planets 1961 TA and 1960 OA. *Bull. ITA*, **9**, 9(112), 615.
18. Bohan, N. A. Elements of Circular Orbits of Minor Planets Observed at the Crimean Astrophysical Observatory. *Bull. ITA*, **11**, 2(125).
19. Shmakova, M. 1966, Circular Orbit. Trans. of the Baku Conf. 1962, 48.
20. Debehogne, H. 1965, Orbite elliptique de l'astéroide 1964 OA. *Bull. astr. Obs. r. Belgique*, **6**, 1, 32.
21. Debehogne, H. 1966, Orbite elliptique de l'astéroide 1965 UA. *Bull. astr. Obs. r. Belgique*, **6**, 3, 107.
22. Izvekov, V. A. On the Improvement of Orbits of Minor Planets with Account of Perturbations. *Bull. ITA*, **11**, 1(124).
23. Izvekov, V. A. 1966, Determination of Astronomical Constants from Observations of Minor Planets. Trans. of the Baku Conf. 1962, 44.
24. Izvekov, V. A. 1964, Program for Solution of a System of Equations with Formal Discussion of Data. *Bull. ITA*, **9**, 8(111), 537.

25. Patry, A. 1964, Catalogue d'orbites circulaires. *J. Observateurs*, **47**, 2; 1966, *ibid.* **49**, 1 and 2.
26. Ferraz Mello, S. 1966, *Bull. astr.*, 3, **1**, fasc. 4, in press.
27. Morando, B. 1966, *Bull. astr.*, 3, **1**, fasc. 4, in press.
28. Wilkins, G. A. 1966, The Mantles of the Earth and Terrestrial Planets. Proc. NATO Advanced Study Inst., in press.
29. Kiang, T. 1966, *Icarus*, **5**, 4.
30. Kleiber, J. 1886, *Astr. Nachr.*, **115**, 135.
31. Bauschinger, J. 1901, *Veröff. Königl. Rechen-Inst. Berlin*, 16.
32. Brunn, A. 1906, *Astr. Nachr.*, **172**, 273.
33. Herrick, S. 1965, Universal Variables. *Astr. J.*, **70**, 4, 309.
34. Herrick, S. 1965, Astronomical and 'Astrodynamical' Values of Constants and Ephemeral Data. *Bull. astr.*, **25**, 1, 105 (IAU Symposium no. 21).
35. Herrick, S. 1965, The Fixing of the Gaussian Gravitational Constant and the Corresponding Geocentric Gravitational Constant. *Bull. astr.*, **25**, 1, 95 (IAU Symposium no. 21).
36. Samoilova-Yakhontova, N. S. 1966, Activity of the Institute of Theoretical Astronomy on the Investigation of Minor Planet's Motion. Trans. of the Baku Conf. 1962, 10.
37. Samoilova-Yakhontova, N. S. 1963, Minor Planets (1961). *Bull. ITA*, **9**, 4(107), 219.
38. Samoilova-Yakhontova, N. S. 1965, Minor Planets (1963). *Bull. ITA*, **10**, 3(116), 173.
39. Samoilova-Yakhontova, N. S. Minor Planets (1964). *Bull. ITA*, **10**, 10(123).
40. Samoilova-Yakhontova, N. S. 1964, Minor Planets (1962). *Bull. ITA*, **9**, 6(109), 379.
41. Lazovich, I. P. On the Determination of the Shortest Distance Between Asteroid Orbits with Small Mutual Inclination. *Bull. ITA*, **11**, 1(124).
42. Makover, S. G. 1966, Prospect of Works of the Institute of Theoretical Astronomy on the Investigation of Minor Planet's Motion. Trans. of the Baku Conf. 1962, 10.
43. Orelskaya, V. I. 1966, On the Orientation of Star Catalogues from Observations of Minor Planets. Trans. of the Baku Conf. 1962, 25.
44. Orelskaya, V. I. 1963, On the State of Observations of Selected Minor Planets. *Astr. Circ.*, U.S.S.R., 259, 2.
45. Perlin, F. H. 1964, Tafeln zur Berechnung der Störungen in den Elementen mit Exzentrischer Anomalie als unabhändige Variable. *Bull. ITA*, **9**, 9(112), 607.
46. Sultanov, G. F. 1964, The Disintegration of a Large Primary Body and the Determination of its Orbit. *Proc. Acad. Sci. Azerbaijan SSR*, 3, 115.
47. Sultanov, G. F., Ibrahimov, N. B. 1962, A Catalogue of Constants of Jacobi's Integral for Asteroids. *Trudy Shemaha Astrophys. Obs.*, 2, 23.
48. Tovstik, M. E. Minor Planet (944) Hidalgo. *Bull. ITA*, **11**, 2(125).
49. Debehogne, H. 1966, Les petites planètes. *Ciel et Terre*, **82**, 7 and 8, 229.

Observations

50. Alijeva, N. *Bull. ITA*, **10**, 8(121).
51. Alijeva, N. 1964, *Circ. Tashkent Obs.*, 333.
52. Alijeva, N. *Bull. ITA*, **10**, 8(121).
53. Chernykh, N. S. 1965, *Bull. ITA*, **10**, 3(116), 236.
54. Chernykh, N. S. *Bull. ITA*, **10**, 9(122).
55. Chernykh, L. I., Chernykh, N. S. *Bull. ITA*, **10**, 10(123).
56. Chernykh, L. I., Chernykh, N. S. *Bull. ITA*, **11**, 1(124).
57. Chernykh, L. I., Chernykh, N. S. *Bull. ITA*, **11**, 2(125).
58. Chuadze, A. D. 1964, *Bull. ITA*, **9**, 9(112), 623.
59. Deutsch, A. N. 1966, Trans. of the Baku Conf. 1962, 22.
60. Dzhakusheva, K. G. 1964, *Bull. ITA*, **9**, 9(112), 624.
61. Ishmukhamedov, Kh., Kadyrov, A. 1962, *Circ. Tashkent Obs.*, 322.
62. Khatisov, A. Sh., Beitrishvili, I. R. *Bull. ITA*, **10**, 8(121).
63. Korol, A. K., Nenakhova, E. M., Koval, R. N. 1965, *Bull. ITA*, **10**, 6(119), 450.
64. Napolskij, A. *Bull. ITA*, **10**, 8(121).

65. Raudsaar, H. K. *Bull. ITA*, **10**, 10(123).
66. Raudsaar, H. K. 1964, *Bull. ITA*, **9**, 8(111), 579.
67. Shor, V. A., Feoktistova, S. A. *Bull. ITA*, **10**, 10(123).
68. Sklyarov, Yu., Polyakov, V., Ermilov, G. 1964, *Bull. ITA*, **9**, 8(111), 576.
69. Yurevich, V. A. *Bull. ITA*, **11**, 2(125).
70. Bronnikova, N. M., Kiseleva, T. P., Strugackaja, A. A., Chudovicheva, O. N. 1965, *Bull. ITA*, **10**, 1(114), 81.
71. Bronnikova, N. M., Voronenko, V. I., Strugackaja, A. A., Chudovicheva, O. N. *Bull. ITA*, **10**, 9(122).
72. Ivakina, T. Y., Kalinevich, Th. Th., Fedorova, R. T. 1965, *Bull. ITA*, **10**, 2(115), 164.
73. Perezhogina, T. P. 1964, *Bull. ITA*, **9**, 9(112), 617.
74. Perezhogina, T. P. 1965, *Bull. ITA*, **10**, 5(118), 380.
75. Kiseleva, T. P., Koroleva, L. S., Sokolova, V. A. 1965, *Bull. ITA*, **10**, 1(114), 76.
76. Kiseleva, T. P., Koroleva, L. S., Sokolova, V. A. *Bull. ITA*, **10**, 9(122).
77. Arend, S., Dommanget, J., Roland, G. 1964, *Bull. astr. Obs. r. Belgique*, **5**, 5, 124.
78. Arend, S., Dommanget, J., Debehogne, H., Roland, G. 1965, *Bull. astr. Obs. r. Belgique*, **6**, 1, 2 to 12.
79. Sauval, A. 1965, *Bull. astr. Obs. r. Belgique*, **6**, 1, 13.
80. Arend, S., Debehogne, H., Roland, G. 1966, *Bull. astr. Obs. r. Belgique*, **6**, 3, 78.
81. Arend, S., Roland, G. 1966, *Bull. astr. Obs. r. Belgique*, **6**, 3, 88.
82. Denoyelle, J., Sanders, P., Sauval, A. 1966, *Bull. astr. Obs. r. Belgique*, **6**, 3, 89.
83. Roemer, E. 1965, *Astr. J.*, **70**, 6, 397.
84. Roemer, E., Lloyd, R. E. 1966, *Astr. J.*, **71**, 6, 443.
85. Roemer, E., Thomas, M., Lloyd, R. E. 1966, *Astr. J.*, **71**, 7, 591.
86. Robertson, W. H. 1964, *Sydney Obs. Papers*, 43.
87. Robertson, W. H. 1964, *Sydney Obs. Papers*, 45.
88. Robertson, W. H. 1964, *Sydney Obs. Papers*, 46.
89. Robertson, W. H. 1965, *Sydney Obs. Papers*, 49.
90. Robertson, W. H. 1965, *Sydney Obs. Papers*, 51.
91. Robertson, W. H. 1966, *Sydney Obs. Papers*, 52.

REPORT OF THE WORKING COMMITTEE ON ORBITS AND EPHEMERIDES OF COMETS

CHAIRMAN: E. Roemer.

MEMBERS: Candy, Kępiński†, Kresák, Makover, Porter.

GENERAL

With the widespread availability of electronic computing facilities, a decreasing proportion of the efforts of those engaged in study of the dynamical properties of comets is required to be devoted to routine calculations of perturbations and ephemerides. Almost all current predictions for returns of periodic comets are based on improved starting orbits and allowance is made for perturbations due to most of the major planets. Two or more independent predictions are available for a gratifying number of returning comets. In an increasing number of cases, several returns have been linked, leading to a greater reliability of the predicted positions.

Especially noteworthy have been the recoveries of three 'long-lost' periodic comets, two of them (P/Holmes, P/de Vico-Swift) as a direct consequence of studies by Marsden (**1**). Investigations by Schubart contributed to the recovery of P/de Vico-Swift and were basic to the recovery of P/Tempel-Tuttle, associated with the Leonid meteors (**3**). Adequate searches should be made for several other objects at appropriate times.

Satisfactory progress is being made toward computation of general orbits of adequately observed nearly parabolic comets. These general orbits have in turn been used in several investigations of original and future orbits. A word of caution may be in order, however, in interpretation of general orbits based on only a short observed arc, from which the eccentricity often is only very poorly, or not at all significantly determined.

A Supplement to the Catalogue of Cometary Orbits, 1960, has been published (2) under the sponsorship of the British Astronomical Association, aided by a subvention from the IAU.

ASTROMETRIC OBSERVATIONS

A useful adjunct of transfer of the Telegram Bureau to the Smithsonian Astrophysical Observatory has been the availability of the Baker-Nunn cameras of the SAO satellite tracking stations in obtaining rapid confirmation of newly discovered brighter comets and early approximate positions of useful accuracy. The value of this network has been repeatedly demonstrated.

Important series of observations of brighter comets have been obtained during the triennium at the following observatories:

Abastumani
Alma-Ata
Armagh
Ascot
Ashkhabad
Athens
Barcelona (Fabra)
Bloemfontein (Boyden Sta.)
Bordeaux (Floirac)
Bucharest
Burakan
Canberra (Mt. Stromlo)
Cleveland (Case Inst.)
Mayaki (near Odessa)
Meudon
Nanking
Nice
Prague
Pulkovo
San Juan (Yale-Columbia Sta.)
Saratov
Simeis
Córdoba
Cracow
Dresden
Dushanbe
Flagstaff (Lowell Obs.)
Flagstaff (U.S. Naval Obs. Sta.)
Goloseyevo
Johannesburg (Hartbeespoort Annexe)
Heidelberg-Königstuhl
Kazan
Khabarovsk
Kiev
Leiden
Skalnaté Pleso
Sonneberg
Tartu
Tashkent
Tautenberg
Tucson (Catalina Sta.)
Tokyo (Mitaka, Dodaira, Okayama)
Uccle
Vienna
Washington (U.S. Naval Obs.)

In several cases valuable assistance with plate measurements and reductions is provided by one observatory to another. The Boyden Observatory acknowledges important help from Uccle, while the Cincinnati Observatory regularly assists several observatories with dependence plate reductions.

The amount of time available with major instruments for long-focus astrometric work, especially on fainter comets, minor planets, and satellites, leaves much to be desired. The observed orbital arcs of comets often can be extended by many months, minor planets can be followed longer and at less favorable oppositions, and the faint satellites of major planets can be recorded only with those large telescopes already under great pressure from observing-time demands of other programs. The importance of the longest possible accurately observed arcs to improvement of orbital elements, both of periodic and of nearly parabolic comets, can hardly be overemphasized.

So far as is known, important regular assignments of time for long-focus astrometric observations of the fainter comets are being made currently at only two centers:

Tokyo–Dodaira 36-inch, *f*/5 reflector (Hirose, Tomita, Kozai)
Okayama 74-inch, *f*/4·8 reflector (Hirose, Tomita, Kozai)

Tucson–Catalina Station 61-inch, *f*/13·5 reflector (Van Biesbroeck, Roemer)

The skilful use of the 24-inch *f*/5 reflector at Skalnaté Pleso, an appreciably smaller telescope than those mentioned above, in following comets over the longest possible arc, deserves special mention.

Several important lists of position have been published (**4–10**) in addition to those reported to the IAU Central Bureau.

ORBITS OF SHORT-PERIOD COMETS

With the widespread application of electronic computers to the work of prediction, it has become possible to devote more attention to improvement of starting orbits and to accounting more precisely for perturbations, with resulting greater accuracy of current ephemerides. Orbits of several short-period comets have been improved by linking accurately observations

Table 1

Short-period comets—current predictions

Encke	Makover 1963, *Comet Circ.* (*Kiev*) 4. Kastel 1967, *Handbook, Brit. astr. Assoc.*
Grigg-Skjellerup	Sitarski 1966, *Acta astr.*, **16**, 209.
Tempel 2	Marsden 1967, *Handbook, Brit. astr. Assoc.*
Neujmin 2	Marsden (**1**).
Brorsen	Marsden (**1**).
Tuttle-Giacobini-Kresák	Lea and Milbourn 1967, *Handbook, Brit. astr. Assoc.*; Kresák MS.
Tempel-Swift	Marsden (**1**).
Tempel 1	Marsden (**1**); Schrutka-Rechtenstamm MS.
Forbes	Marsden 1967, *Handbook, Brit. astr. Assoc.*
Perrine-Mrkos	Hirose
Schwassmann-Wachmann 2	Rasmusen ?; Marsden MS.
Biela	Marsden (**1**).
Wirtanen	Julian, Marsden 1967, *Handbook, Brit. astr. Assoc.*
Reinmuth 2	E. Rabe, Ainslie and Christison and Delo 1967, *Handbook, Brit. astr. Assoc.*
Brooks 2	Wilson and Freeman, Buckley 1966, *Handbook, Brit. astr. Assoc.* Vorobjev 1966, *Bull. Engelhardt Obs.*, Kasan; 1966, *Comet Circ.* (*Kiev*) 39.
Harrington 2	Wilson, Marsden 1967, *Handbook, Brit. astr. Assoc.*
Finlay	Porter and Candy 1967, *Handbook, Brit. astr. Assoc.*
Borrelly	Christison and Delo 1966, *Handbook, Brit. astr. Assoc.*
Harrington-Abell	Hasegawa
Arend	Lea and Milbourn 1966, *Handbook, Brit. astr. Assoc.*
Wolf 1	Kamienski and Sitarski 1967, *Acta astr.*, **17**; Harbour, Egerton 1967, *Handbook, Brit. astr. Assoc.*
Tuttle	Abalakin and Belyayev 1966, *Comet Circ.* (*Kiev*) 38; Cundall and Marsden 1966, *Handbook, Brit. astr. Assoc.*; F. Narin 1953, *IAU Circ.*
Neujmin 1	Raudsaar 1966, *Comet Circ.* (*Kiev*) 37.
Halley	Zadunaisky 1966, *Astr. J.* **71**, 20; Brady MS.

at several perihelion passages (**13, 15, 24, 33, 34**). Further studies of P/Halley have been made by Kamieński (**16**), and special studies of P/Brooks 2 by Kastel (**17**), of P/Wolf 1 by Kazimirčak-Polonskaya (**21, 22**), P/Kopff by Kępiński (**23**), of P/Tempel 1 by Schrutka-Rechtenstamm (**31**). Significant basic work on the orbits of certain comets of interest as possible targets of space probes has been done under contract with the U.S. National Aeronautics and Space Administration (**14, 28, 29, 30**), or for the European Space Research Organization (**32**).

Marsden reports that a preliminary study has been made of the past histories of the short-period comets of more than one apparition, using the *n*-body integration program written by J. Schubart and P. Stumpff; it has been established that the comet observed by La Hire in 1678 was not P/de Vico-Swift, as Leverrier supposed. Marsden also proposes to study further the past histories of the short-period comets for data of statistical use in trying to understand their orbital evolution. A careful study of certain comets, especially P/Encke and P/Pons-Winnecke, is to be made, independent of previous or intended work by other investigators, to examine further the question of non-gravitational effects.

Sitarski reports that programs to carry out almost all the orbital computations by digital computer have been prepared. It is intended to investigate non-gravitational anomalies in motions of periodic comets by a linkage of several appearances of the comet. Detailed investigations of the motion of P/Wolf 1, Kopff, and Grigg-Skjellerup are to be undertaken.

An extensive study of the effects of the outer planets on the evolution of the orbits of well-studied short-period comets over an interval of 400 years is being made by Belyayev (**11, 12**) and by Kazimirčak-Polonskaya (**18, 19, 20**).

Kresák (**25, 26**) discusses effects of commensurability in the system of short-period comets, while Makover (**27**) speculates that the hypothesis of capture of comets may not be sufficient to explain the origin of all comets of short period.

ORBITS OF NEARLY PARABOLIC COMETS

An improved orbit of Comet Everhart (1964 b) has been published by Sekanina (**45**). Sitarski (**48**) has studied the circumstances of the close approach of Comet 1759 III to Jupiter.

Definitive orbits have been computed for nearly parabolic comets by Belous (**36, 37**), Galibina and Barteneva (**41**), Grabner (for Comet Peltier, 1952 VI, MS), Hasegawa (**42**), Mamedov (**44**), and Van Biesbroeck (**50**).

Original and future orbits of additional comets have been investigated by Barteneva (**35**), Galibina (**39, 40**) and Sekanina (**46**). Sekanina also has studied the apparently slightly hyperbolic orbits of comets as related to Oort's hypothesis of a cloud of comets at great distance from the Sun (**47**).

Steins and Kronkalne (**49**) examined changes in orbital elements for a complete passage of a comet through the planetary system.

Loss of long-period comets from the solar system has been the subject of investigations by Lyttleton and Hammersley (**43**) and by Brady (**38**). Brady reports that additional nearly parabolic orbits have been integrated subsequent to his published report, to total 100 comets now investigated. It appears that 80% of comets with an osculating eccentricity greater than a critical value of about 1·0003 pick up energy and escape from the solar system.

Marsden reports that it was found from an investigation of the sungrazing comet group that the differences among the orbits of the various members cannot be explained in a simple fashion by planetary perturbations. The family of sungrazers has also been of interest to Harwit (Abstract, Amer. Astr. Soc., Ithaca, N.Y., 1966).

Orbits of long-period comets have been reported under further investigation as follows: 1955 V (Honda) by Hasegawa; 1957 III (Arend-Roland) independently by Hasegawa and by

Candy; 1957 VI (Wirtanen) by Hasegawa; 1962 VIII (Humason) by Marsden; 1963 III (Alcock) by Sekanina; 1963 V (Pereyra) by Sekanina; and 1963 VIII (Kearns-Kwee) by Marsden.

THEORETICAL INVESTIGATIONS

Mrs Samoilova-Jakhontova has reviewed the orbit work on comets conducted at the Institute of Theoretical Astronomy, Leningrad (**51**).

Dynamical problems of the origin of the comet cloud and of the capture of some objects into orbits of shorter period have been the subjects of papers by Harwit (**52**), Kresák (**53**), Laucenieks (**54**), Makover (**55**), Öpik (**56**), Steins (**57**), Vsekhsviatsky (**58**), Whipple (**59**), and Witkowski (**60**).

COMPUTING TECHNIQUES

Mamedov has modified Olbers' method for determination of a parabolic orbit from three observations by use of the electronic computer BESM-2 (**61**) and applied a method for the determination of a nearly parabolic orbit to Comet Borrelly-Daniel, 1909 I (**62**).

Perlin describes a method for the calculation of perturbations of a comet moving in a very eccentric orbit (**64**) and describes auxiliary tables used in calculations by her method (**63**).

Sitarski reports papers by himself (**65, 66**) and by Ziołkowski (**67, 68, 69**) which are concerned with problems associated with computations of disturbed motion by use of the electronic computer.

PROPOSALS

Probably the most serious and immediate problem is that of assuring adequate observation of intrinsically faint comets and of comets remote from the Sun. It is proposed to discuss the steps that may be taken to encourage astrometric observations of comets with large telescopes. Two questions are involved: (1) telescope time for interested observers, and (2) arrangement for measurement and reduction of plates. Because the only suitable source of reference star positions for long-focus observations is the *Astrographic Catalogue*, often unavailable at newer observatories, endorsement of the efforts of Commission 23 toward reprinting of the Catalogue volumes now out of print and toward calculation of plate constants on a reliable standard catalogue system for all zones could prove helpful.

Marsden has called attention to the potential usefulness of a more complete account of all comets ever observed, including those for which orbits could not be calculated. Pre- and post-discovery ephemerides from improved orbits for at least some objects could prove useful, as, for example, in analysis of photometric data. It may be discussed at Prague whether the large effort required by such a compilation would be justified, and if so, what steps could be taken toward its realization.

BIBLIOGRAPHY

(References to *IAU Circulars* and to papers already indexed in the *Astronomischer Jahresbericht* have generally been omitted. See also Reports on the Progress of Astronomy: Comets, 1964, *Quarterly J. R. astr. Soc.*, **5**, 228; 1965, *ibid.*, **6**, 336.)

General

1. Marsden, B. G. 1963, *Astr. J.*, **68**, 795.
2. Marsden, B. G., Porter, J. G., Eds. 1966, *Mem. Brit. astr. Assoc.*, **40**, no. 2, 1966.
3. Schubart, J. 1966, *Science*, **152**, 1236.

Astrometric Observations

4. Bruwer, J. A. 1964, *Republic Obs. Circ.*, **7**, 47; 1965, *ibid.*, **7**, 69.
5. Hirose, H., Tomita, K., Kösai, H. 1965, *Tokyo astr. Bull.*, *Second Ser.*, no. 168, 2061; 1965, *ibid.*, no. 171, 2081.
6. Kresák, L., Antal, M. 1966, *Contr. Skalnaté Pleso Obs.*, **3**, 113.
7. Roemer, E. 1965, *Astr. J.*, **70**, 397.
8. Roemer, E., Lloyd, R. E. 1966, *Astr. J.*, **71**, 443.
9. Roemer, E., Thomas, M., Lloyd, R. E. 1966, *Astr. J.*, **71**, 591.
10. van Genderen, A. M. 1964, *Bull. astr. Inst. Netherlds.*, **17**, 298.

Orbits of Short-Period Comets

11. Belyayev, N. A. 1966, The orbital evolution of Comets Neujmin 2 (1916 II), Comas Solá (1927 III) and Schwassmann-Wachmann 2 (1929 I) for 400 years (1660–2060). Preliminary investigation. *Astr. Zu.*
12. Belyayev, N. A. 1966, The orbital evolution of Daniel's Comet, 1909 IV, for 400 years (1660–2060). Preliminary investigation. *Bull. Inst. teor. Astr.*, **10**.
13. Gaska, S. 1965 (Motion of P/Kopff, 1906 IV, 1939–45). *Bull. Obs. Toruń*, no. 34, 21.
14. Giddis, A. R. 1965, Comet and close-approach asteroid mission study final report, (Science), Vol. 2. *Philco Western Development Laboratories*, WDL-TR 2366.
15. Imnadze, M. P. 1963, Investigations of the movement of the Comet 1947 XIII = 1954 XI Wirtanen. *Trans. Comput. Center Acad. Sci. Georgian SSR*, **3**, 269; 1963, *Publ. Acad. Sci. Georgian SSR*, **30**, 157.
16. Kamieński, M. 1964–65, Perihelion passages of Halley's Comet in ancient times . . . *Poznań Bull. Livraison* XVIII.
17. Kastel, G. R. 1965, On the close approach of Comet Brooks 2 with Jupiter in 1886. *Bull. Inst. teor. Astr.*, **10**, 118; 1966, *Trans. Conf. Baku 1962*, 153.
18. Kazimirčak-Polonskaya, H. I. 1966, Évolution des orbites des comètes à courte période durant 400 ans (1660–2060) et le rôle des planètes extérieures dans leur évolution. *Astr. Zu.*
19. Kazimirčak-Polonskaya, H. I. 1966, Rapprochements des comètes aux grosses planètes. *Trans. Conf. Baku 1962*, 90.
20. Kazimirčak-Polonskaya, H. I. 1967, Quelques problèmes actuels de l'astronomie cométaire du point de vue de la mécanique céleste contemporaine. *Mem. Inst. teor. Astr.*, **12**.
21. Kazimirčak-Polonskaya, H. I. 1967, Précision de la théorie de la comète Wolf 1 pendant la période 1918–25 contenant son rapprochement de Jupiter en 1922. *Mem. Inst. teor. Astr.*, **12**.
22. Kazimirčak-Polonskaya, H. I. 1967, Évolution de l'orbite de la comète Wolf 1 durant 400 ans (1660–2060). Investigations préliminaires. *Mem. Inst. teor. Astr.*, **12**; 1967, Sur la communauté possible de l'origine des comètes Wolf 1 (1884 III) et Barnard (1892 V). *Mem. Inst. teor. Astr.*, **12**.
23. Kępiński, F. 1963, The motion of P/Comet 1906 IV (Kopff) in the years 1958–64, Part 9. *Acta. astr.*, **13**, 87.
24. Khanina, F. B. 1966, The motion of Comet Faye. *Trans. Conf. Baku 1962*, 69.
25. Kresák, L. 1965, The effect of commensurability in the system of short-period comets. *Bull. astr. Inst. Csl.*, **16**, 292.
26. Kresák, L. 1966, On two aspects of evolution of short-period comets. *Mem. in 8° Soc. R. Sci. Liège*, **12**, 459.
27. Makover, S. G. 1966, On the origin of short-period comets. *Bull. Inst. teor. Astr.*, **11**.
28. Narin, F., Pierce, P. M. 1964, Perturbation, sighting and trajectory analysis for periodic comets: 1965–75. Report no. T-7, Astro Sciences Center, IIT Research Institute.
29. Narin, F., Rejzer, B. 1965, Sighting and trajectory analysis for periodic comets: 1975–86. Report no. T-11, Astro Sciences Center, IIT Research Institute.
30. Roberts, D. L. 1964, A compendium of data on some periodic comets. Report no. P-9, Astro Sciences Center, IIT Research Institute.
31. Schrutka-Rechtenstamm, G. 1873, MS. Definitive orbit of P/Tempel 1.

32. Stumpff, P. 1965, Report to ESRO: On the accuracy of the orbits of the periodic comets Brooks 2, Tempel 2, Giacobini-Zinner, and Tuttle-Giacobini-Kresák.

33. Vorobjev, E. A. 1964, (P/Johnson 1949 II, linking apparitions in 1949 and 1956, with perturbations.) *Comet Circ.* (*Kiev*), 6; 1966, (P/Johnson 1963c) *Comet Circ.* (*Kiev*) 39; 1964, *Handbook, Brit. astr. Assoc.*

34. Wiśniewski, W. 1964, P/Wolf-Harrington (Identity of the two comets confirmed by linking observations in 1925 and 1952, with perturbations.) *Acta astr.*, **14**, 129.

Orbits of Nearly Parabolic Comets

35. Barteneva, O. N. 1965, Original and future orbits of the Comet 1950 I (Johnson). *Bull. Inst. teor. Astr.*, **10**, 443.

36. Belous, L. M. 1964, The definitive orbit of Comet Bappu-Bok-Newkirk (1949 IV). *Bull. Inst. teor. Astr.*, **9**, 569.

37. Belous, L. M. 1966, The definitive orbit of Comet Mrkos (1953 II). *Bull. Inst. teor. Astr.*, **10**; 1965, *Comet Circ.* (*Kiev*) 24.

38. Brady, J. L. 1965, Effect of the planetary system on the nearly parabolic comets. *Astr. J.*, **70**, 279.

39. Galibina, I. V. 1964, Original and future orbits of long-period comets. *Bull. Inst. teor. Astr.*, **9**, 465.

40. Galibina, I. V. 1966, Determination of original and future orbits of long-period comets. *Trans. Conf. Baku 1962*, 145.

41. Galibina, I. V., Barteneva, O. N. 1965, Definitive orbit of the Comet 1950 I Johnson. *Bull. Inst. teor. Astr.*, **10**, 192.

42. Hasegawa, I. 1966, (Definitive orbit of Comet Vozárová, 1954 VIII). *Publ. astr. Soc. Japan*, **18**.

43. Lyttleton, R. A., Hammersley, J. M. 1964, The loss of long-period comets from the solar system. *Mon. Not. R. astr. Soc.*, **127**, 257.

44. Mamedov, M. A. 1966, Definitive orbit of Comet 1941 I, Cunningham. *Proc. Acad. Sci. Azerbaijan SSR*, **5**.

45. Sekanina, Z. 1965, Orbit of Comet Everhart (1964 h). *Bull. astr. Inst. Csl.*, **16**, 361.

46. Sekanina, Z. 1966, New original and future comet orbits. *Bull. astr. Inst. Csl.*, **17**, 67.

47. Sekanina, Z. 1966, Hyperbolic comets and Oort's hypothesis of cometary cloud. *Mém. in 8° Soc. R. Sci. Liège*, **12**, 443.

48. Sitarski, G. 1966, Close approach of Parisian comet (1759 III) to Jupiter. *Acta astr.*, **16**, 197.

49. Steins, K., Kronkalne, S. 1964, Changes in orbital elements for a complete comet's passage through the planetary system. *Acta astr. Cracoviae*, **14**, 4, 311.

50. Van Biesbroeck, G. 1964, The orbit of Comet 1947 VI (Wirtanen). *Comm. Lunar and Planetary Laboratory* no. 43.

Theoretical Investigations

51. Samoilova-Jakhontova, N. S. 1966, The work on the motion of comets, conducted at the Institute of Theoretical Astronomy of the Academy of Sciences of the U.S.S.R. *Trans. Conf. Baku 1962*, 54.

52. Harwit, M. 1966, On the transition between long- and short-period comets. *Mém. in 8° Soc. R. Sci. Liège*, **12**, 489.

53. Kresák, L. (Relation between orbits of meteors, comets, and asteroids.) *Smithsonian Contr. Astrophys.*, (in press).

54. Laucenieks, L. K. 1964, On the solution of equations of comets' diffusion. *Sci. Papers of the Latvian Univ.*, **68**, 2, 91.

55. Makover, S. G. 1964, On perturbations of comets by stars. *Bull. Inst. teor. Astr.*, **9**, 525.

56. Öpik, E. J. 1966, The dynamical aspects of the origin of comets. *Mém. in 8° Soc. R. Sci. Liège*, **12**, 523.

57. Steins, K. A. 1964, Evolution of comets' orbit. *Sci. Papers of the Latvian Univ.*, **68**, 2, 39.

58. Vsekhsviatsky, S. K. 1966, Comet cosmology of Lagrange and the problems of the solar system. *Mém. in 8° Soc. Sci. R. Liège*, **12**, 495.

59. Whipple, F. L. 1964, Evidence for a comet belt beyond Neptune. *Proc. National Acad. Sci.*, **51**, 711.

60. Witkowski, J. 1965, Some remarks concerning cometary statistics. *Acta astr.*, **15**, 273

Computing Techniques

61. Mamedov, M. A. 1965, Program for calculation of a parabolic orbit from three observations. *Proc. Acad. Sci. Azerbaijan SSR*, **6**, 57.

62. Mamedov, M. A. 1965, On the computation of a nearly parabolic orbit from three observations of a comet. *Bull. Inst. teor. Astr.*, **10**, 41.

63. Perlin, F. H. 1964, Tafeln zur Berechnung der Störungen in den Elementen mit exzentrischer Anomalie als unabhändige Variable. *Bull. Inst. teor. Astr.*, **9**, 607.

64. Perlin, F. H. 1966, Methoden zur Berechnung der Störungen der Kometen, die sich in sehr exzentrischer Bahnen bewegen. *Trans. Conf. Baku 1962*, 62.

65. Sitarski, G. 1966, A criterion for the automatical change of integration step in Cowell's method. *Acta astr.*, **16**, 201.

66. Sitarski, G. Correction of cometary orbits in rectangular coordinates. *Acta astr.*, (in press).

67. Ziołkowski, K. 1965, Tables of the Jupiter and Saturn coordinates on a punched paper tape. *Computatio*, **3**, 5.

68. Ziołkowski, K. 1965, The integration of the minor planet (or comet) motion equations. *Computatio*, **3**, 89.

69. Ziołkowski, K. 1965, On the integration of equations of celestial mechanics using digital computers. *Computatio*, **3**, 45.

GPSR Compliance

The European Union's (EU) General Product Safety Regulation (GPSR) is a set of rules that requires consumer products to be safe and our obligations to ensure this.

If you have any concerns about our products, you can contact us on ProductSafety@springernature.com

In case Publisher is established outside the EU, the EU authorized representative is:

Springer Nature Customer Service Center GmbH
Europaplatz 3
69115 Heidelberg, Germany

Batch number: 10762209

Printed by Printforce, the Netherlands

TRANSACTIONS
OF THE
INTERNATIONAL ASTRONOMICAL UNION

VOLUME XIIIA – REPORTS

INTERNATIONAL COUNCIL OF SCIENTIFIC UNIONS
INTERNATIONAL ASTRONOMICAL UNION
UNION ASTRONOMIQUE INTERNATIONALE

TRANSACTIONS
OF THE
INTERNATIONAL ASTRONOMICAL UNION

VOLUME XIIIA - REPORTS

REPORTS
ON
ASTRONOMY

Edited by
Luboš PEREK
General Secretary of the Union

PUBLISHED FOR THE INTERNATIONAL COUNCIL OF SCIENTIFIC UNIONS
WITH THE FINANCIAL ASSISTANCE OF UNESCO

SPRINGER-SCIENCE+BUSINESS MEDIA, B.V.
1967

Originally published by D. Reidel Publishing Company in 1967
Softcover reprint of the hardcover 1st edition 1967

Unesco subvention – 1967 – AVS/414/25

ISBN 978-94-010-3478-4 *ISBN 978-94-010-3476-0 (eBook)*
DOI 10.1007/978-94-010-3476-0

PREFACE

The present volume, XIII A, contains the corrected Reports of IAU Commissions, as presented to the XIII General Assembly of the Union, held in Prague, Czechoslovakia, from 22 to 31 August 1967. The volume, entitled "Reports on Astronomy", is an authoritative, detailed, and unique survey of the accomplishments in all branches of astronomy from 1964 to 1967. For technical reasons, the volume also contains the Report of the Executive Committee 1964–1967, which originally was to appear in Volume B.

The page proofs of the present volume, assembled into the Agenda and Draft Reports, had been distributed to the Members of the Union before the General Assembly. Only errors and misprints in the Draft Reports have been corrected in the present volume: additions had to be deferred to volume XIII B in order to speed up publication.

Volume XIII B, subtitled "Proceedings of the Thirteenth General Assembly", will contain: the report of the Inaugural Ceremony and of the sessions of the General Assembly; the resolutions adopted by the General Assembly and by Commissions; the proceedings of Commission meetings, and those parts of the Astronomer's Handbook which underwent changes (i.e. representation of the IAU in international scientific bodies; composition of the Executive Committee; Symposia, Colloquia, and Publications of the Union; list of Commissions and of their membership; list of Members of the Union).

A separate volume, not numbered in the series of IAU Transactions, and entitled "Highlights of Astronomy as presented at the Thirteenth General Assembly of the Union 1967" will contain the Invited Discourses, the proceedings of the Joint Discussions, and those of one or two other meetings.

The proceedings of Symposia No. 33—Physics and Dynamics of Meteors, No. 34—Planetary Nebulae, and No. 35—Development of Solar Active Regions, held immediately after the General Assembly, will be published in the Symposium series of the Union.

The present volume is a collective work of forty Presidents of Commissions. The personal style was preserved in preference to an entirely homogenous presentation. The reports were edited by the former General Secretary, Professor Jean-Claude Pecker, during his term of office. The extremely difficult task of subediting the volume was entrusted to the experience, patience, and competence of Miss Geneviève Drouin. All those who assisted in preparing this volume, as well as the printer, Willmer Brothers Limited, and the publisher, D. Reidel Publishing Company, deserve the gratitude and appreciation of the Union and of the readers as well.

Praha
4 October 1967

L. Perek
General Secretary

CONTENTS

Report of the Executive Committee 1964–1967

Reports of Commissions

RAPPORT DU COMITE EXECUTIF 1964-1967*

ACTIVITÉ DE L'UNION ASTRONOMIQUE INTERNATIONALE

(See English version on page xvii)

Introduction

Ce rapport couvre la période qui s'étend du 1 janvier 1964 au 31 décembre 1966. Par souci de continuité, il a été composé selon des principes identiques à ceux qui ont guidé la rédaction des rapports précédents et notamment celui relatif à la période 1961–63, publié dans le Volume XII B des *Transactions*, et approuvé formellement par la douzième Assemblée Générale.

La période couverte comprend le début de l'année 1964 (jusqu'au 25 août 1964) pendant laquelle le Comité Exécutif actuel n'était pas encore en fonction. Symétriquement, elle exclut le début de l'année 1967, qui pourtant dépend du Comité Exécutif actuel: un rapport complémentaire provisoire, portant sur cette période (1 janvier–21 août 1967) sera oralement présenté devant la treizième Assemblée Générale, et, en cas d'approbation, publié dans le Volume XIII B des *Transactions*.

Pendant la période couverte par le présent rapport, un certain nombre de décisions ont été prises par le Comité Exécutif. Les unes, liées à la douzième Assemblée Générale, ou prises lors de cette Assemblée, ont été incluses dans le Volume XII B des *Transactions*, et ne seront pas à nouveau mentionnées ci-après. Les autres ont été publiées dans le *Bulletin d'Information* de l'UAI et sont à nouveau mentionnées ci-après dans les paragraphes appropriés. Le *Bulletin d'Information* de l'UAI a publié un grand nombre d'informations en provenance des Commissions de l'UAI, des Commissions Inter-Unions, des Services Astronomiques, etc. Ces informations sont essentiellement contenues dans les *Draft Reports* des Commissions appropriées de l'UAI, dans la mesure ou elles restent valides actuellement.

Les problèmes les plus graves rencontrés par l'Union concernent les Publications et les Finances: à chacune de ces deux questions est consacré un appendice au présent rapport. Un troisième appendice est consacré à quelques données relatives aux Organisations Adhérentes et aux Membres de l'Union. De plus, un certain nombre de Commissions Administratives (5, 6, 38, 46) sont considérées comme des comités du Comité Exécutif: leurs rapports complètent le présent rapport, auquel s'ajoutent également le rapport final du Président de l'IUCSTR (Inter-Union Commission on Solar and Terrestrial Relationships), l'UAI ayant joué le rôle d'union-mère de cet organisme inter-union, et le rapport du Président du Groupe de Travail sur les Émulsions Photographiques.

Structure et Administration de l'UAI

L'Union Astronomique Internationale fonctionne depuis la douzième Assemblée Générale dans le cadre des Statuts nouveaux adoptés lors de cette Assemblée.

Administration

Le fait dominant qui a commandé depuis lors la politique du Comité Exécutif est l'augmentation rapide des activités de l'Union: augmentation du nombre de ses Membres, de ses Commissions, des Pays Adhérents, complexité croissante des relations avec les autres organisations.

**Note.* Le texte du Rapport et de ses deux premiers Appendices sont publiés ci-après en deux langues. Seul le texte français fera autorité, en cas de besoin, le texte anglais n'étant qu'une traduction du texte français.

Depuis l'été 1964, afin de faire face à cette situation difficile, une sténo-dactylographe a été adjointe au Secrétaire Adjoint. De plus le Secrétaire Général et le Secrétaire Général Adjoint ont bénéficié des services de plusieurs personnes, à Meudon et à Prague, non salariées par l'UAI, grâce à la compréhension des autorités nationales françaises et tchécoslovaques, ce qui prouve que, même augmenté d'une personne, le personnel administratif de l'UAI est, en nombre, très insuffisant.

Il est apparu au Comité Exécutif qu'il était, pour des raisons financières claires, impossible de faire mieux, dans l'avenir immédiat. C'est pourquoi des solutions ont été recherchées en vue de simplifier l'administration, parfois très lourde, de certains problèmes: c'est en partie pour cette raison que la Liste de Distribution des publications de l'UAI a été considérablement réduite, et que l'édition des Symposiums de l'UAI postérieurs au no. 31 sera confiée, si possible, à une firme privée (voir ci-après page lx).

Réunions du Comité Exécutif

Le Comité Exécutif sortant a tenu sa vingt-cinquième réunion à Hambourg, pendant la douzième Assemblée Générale (24 août-3 septembre 1964), sous la présidence du Professeur V. A. Ambarcumjan. Tous les membres et conseillers étaient présents. Le Comité Exécutif actuel a tenu sa vingt-sixième réunion à Hambourg, le 3 septembre 1964, sous la présidence du Professeur P. Swings, Président. Tous les membres et conseillers étaient présents. Le Comité Exécutif a tenu, sous la présidence du Professeur P. Swings, les deux réunions suivantes: La vingt-septième session a eu lieu à l'Observatoire de Nice, France, du 15 au 19 septembre 1965; tous les membres et conseillers étaient présents sauf les Professeurs G. Haro et V. A. Ambarcumjan. La vingt-huitième session a eu lieu à Prague, Tchécoslovaquie, du 16 an 20 septembre 1966; tous les membres et conseillers étaient présents, sauf les Professeurs Y. Hagihara et V. A. Ambarcumjan, dont l'état de santé avait empêché la participation.

Entre les réunions, les affaires de l'Union ont été conduites par correspondance. Vingt-quatre lettres circulaires ont été envoyées par le Secrétaire Général et le Secrétaire Général Adjoint aux membres du Comité Exécutif.

Membres de l'UAI et Organisations Adhérentes

La Nouvelle-Zélande a adhéré à l'UAI en 1964. Aucune autre modification depuis la onzième Assemblée Générale n'est intervenue, malgré des contacts préliminaires pris avec des représentants qualifiés des pays suivants: Algérie, Cuba, Indonésie, Iran, Niger, Pakistan. La Colombie a demandé son adhésion qui sera proposée à la treizième Assemblée Générale.

L'Appendice III (page lxxi) rappelle la liste des Pays Adhérents et des Organisations Adhérentes correspondantes.

Il est à noter que la création de 'Pays Adhérents Associés' (dont la cotisation aurait été nulle, et qui n'auraient pas eu droit de vote sur les questions ayant des implications financières) a été longuement envisagée. Le Comité Exécutif a créé en son sein en 1965 un groupe de travail chargé d'étudier cette question. Mais après discussion, le Comité Exécutif, en 1966, a décidé de ne demander à l'Assemblée Générale aucune modification des Statuts à cet effet.

Au 1 janvier 1964, le nombre des Membres de l'UAI était de 1276. Depuis cette date, la douzième Assemblée Générale a élu 352 Membres nouveaux, cependant que nous étions informés du décès de 56 Membres; au 31 décembre 1964, le nombre de Membres de l'Union est donc de 1572.

Une liste des Membres du décès desquels nous avons été informés depuis le 1 janvier 1964 figure dans l'Appendice III. Cette liste comprend malheureusement tant d'astronomes distingués qu'il est seulement possible de rendre ici hommage à ceux qui ont contribué à l'administration de l'Union.

Bertil Lindblad, Président de l'Union Astronomique Internationale de 1948 à 1952, a été Directeur de l'Observatoire de Stockholm, à Saltsjöbaden, depuis 1927. Ses travaux dans le domaine de la structure galactique (au sens le plus large) lui ont valu, entre autres distinctions, la médaille d'or de la Royal Astronomical Society, et la Bruce Gold Medal de la Société Astronomique du Pacifique. Il a été Président de l'Académie Royale Suédoise des Sciences. Il laissera le souvenir d'un ami bienveillant et humain, autant que celui d'un astronome distingué (1895–1965).

Robert Methven Petrie, Vice-Président de l'Union Astronomique Internationale de 1958 à 1964, était Directeur du Dominion Astrophysical Observatory, à Victoria, Canada. Ses travaux de spectroscopie stellaire, appliquée notamment aux problèmes des parallaxes spectroscopiques, lui ont valu de nombreuses distinctions. Il était honoré du titre de 'Dominion Astronomer', et était Président de la Royal Astronomical Society of Canada. Ses collègues garderont de lui le souvenir d'un savant éminent, mais aussi d'un ami plein d'humour et de sagesse (1906–1966).

André Danjon, Président de l'Union Astronomique Internationale de 1958 à 1931, était, avant sa retraite, Directeur de l'Observatoire de Paris et Professeur à la Sorbonne depuis 1945. Ses travaux d'optique et de photométrie, puis ses recherches d'astronomie fondamentale, et particulièrement la conception et l'utilisation de l'astrolabe impersonnel à prisme qui porte son nom, lui ont valu de très nombreuses distinctions, notamment la médaille d'or de la Royal Astronomical Society et la médaille d'or du C.N.R.S. Il était Membre de l'Académie des Sciences de Paris, associé étranger de plusieurs institutions académiques, et, en France, il a présidé de nombreux comités. Ses collègues et ses élèves n'oublieront jamais ce qu'ils doivent à son énergie, et en garderont un très ému souvenir (1890–1967).

Commissions de l'UAI

Ce volume des *Agenda and Draft Reports* contient des rapports excellents et détaillés, préparés par les Présidents de chacune des 38 Commissions de l'Union, assistés par leur Vice-Président, leur Comité d'Organisation, les responsables des divers groupes de travail ou comités ayant été créés au sein de ces Commissions par les Commissions elles-mêmes.

Depuis la douzième Assemblée Générale deux nouvelles commissions ont été créées: la Commission 45 (Classifications Spectrales et Indices de Couleur à Plusieurs Bandes) et la Commission 46 (Enseignement de l'Astronomie); cette dernière fonctionne comme comité du Comité Exécutif.

En raison du décès du Professeur R. M. Petrie, le Professeur D. S. Evans a été nommé Président par interim de la Commission 30 (Les Vitesses Radiales), cependant que le Professeur F. L. Whipple était nommé Président par interim de la Commission 6 (Télégrammes Astronomiques) en raison du décès du Dr A. G. Hogg.

Entre les Assemblées Générales, les Commissions ont coopté parmi les Membres de l'UAI un certain nombre de membres. De plus, des Présidents ont souhaité s'adjoindre les services d'astronomes non-membres de l'UAI. En vue de simplifier la procédure, il a été demandé aux Présidents de les considérer pour tous les problèmes pratiques comme des Membres de l'UAI. Des propositions relatives aux nouvelles compositions des Commissions, comprenant ces membres cooptés et les nouveaux Membres élus par l'Assemblée Générale seront faites par les Présidents, les listes publiées dans le Volume XII C des *Transactions* servant de point de départ à ces propositions.

Entre les Assemblées Générales, les Commissions ont pris l'initiative d'un certain nombre de Colloques et de projets financés par l'UAI.

Colloques des Commissions de l'UAI

(tenus entre le 1 janvier 1964 et le 31 décembre 1966).

IAU Colloquium on *The Position of Variable Stars in the Hertzsprung-Russell Diagram*, Bamberg, 10–15 August 1965 (Commissions 27 and 42).

IAU Colloquium on *Interstellar Grains*, Troy, New York, 24–26 August 1965 (Commission 34)

IAU Colloquium on *The Blanketing Effect*, Heidelberg, 17–19 March 1966 (Commission 36)

IAU Colloquium on *Late-Type Stars*, Trieste, 13–16 June 1966 (Commissions 14, 29, 35, 36).

IAU Colloquium on *Atomic Collisions*, Boulder, 11–16 July 1966 (Commission 14).

IAU Colloquium on the *Evolution of Double Stars*, Uccle, 29 August–2 September 1966 (Commission 26).

Entreprises spécifiques des Commissions, approuvées par la douzième Assemblée Générale

La liste de ces entreprises, et des dépenses correspondantes, se trouvent page xxx, dans l'Annexe au présent rapport traitant des Finances de l'Union. Pour la plupart, ces entreprises constituent le prolongement d'entreprises précédentes, et n'appellent aucun commentaire.

Nous noterons cependant que les crédits attribués aux Cartes d'Étoiles Variables de l'Hémisphère Sud, et ceux attribués au Catalogue sur Cartes des Amas et Associations n'ont pas été utilisés. Cette dernière entreprise a dû être retardée, et le Comité Exécutif a donné son accord pour la réalisation, à partir de 1967, de ce projet par les éditions de l'Académie des Sciences de Budapest. La somme attribuée au Dr Bateson pour les Cartes d'Étoiles Variables de l'Hémisphère Sud devra être versée en 1967.

Entreprises spécifiques des Commissions, approuvées, entre les douzième et treizième Assemblées Générales, par le Comité Exécutif

Il s'agit:

(*a*) de la réédition de l'*Atlas of Representative Cometary Spectra*, par l'Institut d'Astrophysique de Liège, demandée par de nombreux chercheurs;

(*b*) de l'édition, par le Dr Sahade, de l'*Information Bulletin for the Southern Hemisphere*, publication très lue dans l'Hémisphère Sud, et particulièrement bien préparée et présentée;

(*c*) de la réédition, sous forme de feuilles mobiles, du livre ancien, et maintenant dépassé, de Rigaux, *Les Observatoires Astronomiques et les Astronomes*, par l'Observatoire Royal de Belgique à Uccle. Cette nouvelle forme de publication permettra l'actualisation permanente de l'ouvrage.

Symposiums

Symposiums de l'UAI

Pendant la période couverte par le présent rapport, les Symposiums suivants ont été organisés par l'UAI:

no.	Sujet	Lieu	Date
23	Astronomical Observations from Space Vehicles	Liège	Août 1964
24	Spectral Classification of Multi-Colour Photometry	Saltsjöbaden	Août 1964
25	The Theory of Orbits in the Solar System and in Stellar Systems	Thessaloniki	Août 1964
26	Abundance Determination in Stellar Spectra	Utrecht	Août 1964
27	The Construction of Large Telescopes	Tucson, Pasadena	Avril 1965

28	Cosmical Gas Dynamics. V. Aerodynamical Phenomena in Stellar Atmospheres	Nice	Septembre 1965
29	Instability Phenomena in Galaxies	Burakan	Mai 1966
30	Determination of Radial Velocities and their Applications	Toronto	Juin 1966
31	Radio Astronomy and the Galactic System	Noordwijk	Septembre 1966

Les Symposiums nos. 23, 24, 25 and 26 ont été organisés peu de temps avant la douzième Assemblée Générale de l'UAI. Le Symposium no. 28 a été organisé en coopération avec l'Union Internationale de Mécanique Théorique et Appliquée (IUTAM), le Symposium no. 31 avec l'Union Radio Scientifique Internationale (URSI).

Le Symposium no. 32, organisé conjointement avec l'Union Internationale de Géodésie et de Géophysique (IUGG), aura lieu à Stresa, Italie, du 21 au 25 mars 1967 et portera sur 'Continental Drift, Secular Motion of the Pole and Rotation of the Earth'.

Les Symposiums nos. 33, 34, 35, portant respectivement sur 'Physics and Dynamics of Meteors', 'Planetary Nebulae', 'The Structure and Development of Solar Active Regions', auront lieu dans la semaine qui suivra la treizième Assemblée Générale de l'UAI à Prague, et se tiendront respectivement à Tatranska Lomnica, Tchécoslovaquie (nos. 33 et 34) et à Budapest, Hongrie (no. 35).

Symposiums auxquels l'UAI a participé

The History of Astronomy (IUPHS-UAI)	Hambourg	Août 1964
The Trajectories of Artificial Celestial Bodies as Determined from Observations (IUTAM-COSPAR-UAI)	Paris	Avril 1965
Planetary Atmospheres and Surfaces as Deduced from Radio Observations (URSI-UAI)	Dorado Beach	Mai 1965
COSPAR Sixth International Space Science Symposium (sessions on Galactic and Extragalactic Phenomena and Space Research) (COSPAR-UAI)	Buenos Ayres	Mai 1966
Solar Terrestrial Physics (URSI, IUGG-IAGA, UAI)	Belgrade	Août 1966

L'UAI était de plus représentée à de nombreux symposiums, et aux assemblées de nombreux organismes de l'ICSU; ces représentations ont été annoncées dans les *Bulletins d'Information* successifs de l'UAI.

Relations avec les autres Organisations

Représentation de l'UAI

1. *Conseil International des Unions Scientifiques.* Le Secrétaire Général (D. H. Sadler) a représenté l'Union à la seconde réunion du Comité Exécutif de l'ICSU à Londres (juin 1964). Le Secrétaire Général a représenté l'Union à la 3ème réunion du Comité Exécutif de l'ICSU à Munich (avril 1965). Le Président (P. Swings) et le Secrétaire Général (J.-C. Pecker) ont représenté l'Union à la onzième Assemblée Générale de l'ICSU à Bombay (janvier 1966). Le Secrétaire Général a représenté l'Union aux 4ème et 5ème réunions du Comité Exécutif de l'ICSU à Bombay (janvier 1966). Le Secrétaire Général Adjoint (L. Perek) a également assisté à la onzième Assemblée Générale de l'ICSU. Le Secrétaire Général a représenté l'Union à la 6ème réunion du Comité Exécutif de l'ICSU à Monaco (septembre 1966).

Le Secrétaire Général est actuellement le représentant désigné de l'Union au Comité Exécutif de l'ICSU.

2. *Organismes de l'ICSU.* L'Union participe aux travaux d'un certain nombre de comités spéciaux et scientifiques, de comités et de commissions interunions fondés sous les auspices de l'ICSU. On trouvera dans le Volume XII C des *Transactions*, p. 13 et sq., des données sur ces organismes.

La liste suivante donne seulement l'état actuel de la représentation de l'UAI au sein de ces organismes.

(*a*) Committee on Space Research (COSPAR): L. Gratton.

(*b*) Special Committee for the International Years of the Quiet Sun (IQSY): G. Righini (Vice-Président); R. Michard (reporter for 'Solar Activity').

(*c*) Comité International de Géophysique (CIG): G. Righini (Vice-Président); R. Michard.

(*d*) Federation of Astronomical and Geophysical Services (FAGS): D. H. Sadler (Vice-President); B. Guinot.

(*e*) Inter-Union Commission on Frequency Allocations for Radioastronomy and Space Science (IUCAF): J. F. Denisse; V. A. Sanamian; F. G. Smith (Chairman); A. Unsöld.

(*f*) Inter-Union Commission on Solar-Terrestrial Physics (IUCSTP): Z. Svestka.

(*g*) Inter-Union Commission on Spectroscopy: B. Edlén; G. Herzberg; J. G. Phillips; M. J. Seaton.

(*h*) Commission Inter-Union sur l'Enseignement des Sciences (CIES-IUCST): E. Schatzman.

(*i*) ICSU Abstracting Board (IAB): J. Kleczek.

(*j*) Committee for Data on Science and Technology (CODATA): Ch. Moore-Sitterly.

3. *Autres organisations.* L'Union est représentée dans:

(*a*) La Fondation Internationale du Pic-du-Midi, par A. Lallemand.

(*b*) Le Comité Consultatif pour la Définition de la Seconde (CCDS), du Bureau International des Poids et Mesures, par W. Markowitz et N. Stoyko.

(*c*) Le Comité Consultatif pour la Définition du Mètre, du Bureau International des Poids et Mesures, par Ch. Moore-Sitterly.

(*d*) Les différents services de FAGS:

Bureau International de l'Heure (BIH): H. M. Smith; N. N. Pavlov.
International Polar Motion Service: (Organizing Committee of IAU Commission 19).
Quarterly Bulletin of Solar Activity: (Organizing Committee of IAU Commission 12).
International Ursigrams and World Days Service (IUWDS): R. Michard.
Solar Particles and Radiations Monitoring Organizations (SPARMO): C. de Jager.

Remarque: Les Commissions Inter-Unions sur les Relations Soleil-Terre (IUCSTR) et sur l'Ionosphère (IUCI) ont achevé leurs travaux. L'Union Astronomique Internationale y était représentée, respectivement par C. W. Allen (Président), R. Giovanelli, E. Mustel, M. Waldmeier pour la première, et par C. W. Allen (Secretary), R. N. Bracewell, V. G. Fesenkov, K. O. Kiepenheuer, pour la seconde.

Relations directes avec l'UNESCO

L'Union a bénéficié de contrats directs avec l'UNESCO, dans des buts variés; ils figurent, ci-après, dans les bilans financiers.

Il s'est agi d'abord d'aider la réunion de deux Symposiums de l'UAI (nos. 23 et 25) sur des sujets d'un intérêt général suffisant pour légitimer l'aide de l'UNESCO.

Ce fut ensuite une aide régulière à la Commission 38, par le moyen d'un contrat forfaitaire, destiné à couvrir une partie des dépenses de la Commission, spécialement celles entraînées par les échanges d'astronomes intéressant des pays en voie de développement.

Enfin, le principe d'une 'Ecole Annuelle de Jeunes Astronomes' a été retenu, à la suite d'une réunion d'un groupe d'experts, travaillant sous l'égide de la Commission 46 de l'UAI. Un crédit prévu à cet effet pour 1966 n'a pu être utilisé en raison du faible délai imparti à la préparation de l'école; un contrat de remplacement a permis l'attribution à un étudiant d'un pays en voie de développement d'une bourse d'études dans un observatoire bien équipé. En 1967, la première école d'été de l'UNESCO-UAI aura lieu à Manchester, sous la direction du Professeur Z. Kopal. Le Secrétaire Général en est le Dr J. Kleczek.

Il s'agit là d'une expérience, qui a été longuement débattue au sein du Comité Exécutif. Celui-ci aurait préféré sans doute que l'UNESCO aidât plutôt l'UAI à assumer ses charges normales, devenues trop lourdes. D'autres collègues pensent qu'une école d'été n'est efficace que si elle joue le rôle de pré-sélection, les meilleurs élèves qu'elle permet de détecter effectuant ensuite de longs séjours dans des observatoires équipés.

Quoiqu'il en soit, les résultats de l'expérience seront examinés avec attention, en vue des actions à venir.

L'UAI est désireuse de maintenir avec l'UNESCO de tels contrats: cette possibilité s'est révélée fructueuse et l'UAI tient à en exprimer à l'UNESCO sa reconnaissance.

Relations avec l'ICSU et ses organismes

L'ICSU est l'organisme dont dépend l'Union le plus directement: il verse à l'UAI la subvention de l'UNESCO (non compris les contrats directs) et coordonne les actions interunions.

Cette aide est précieuse, et nécessaire à l'UAI.

Malheureusement le nombre d'Unions membres de l'ICSU augmente, et la subvention de l'UNESCO à l'ICSU reste sans changement: il est donc difficile que la subvention reçue par l'UAI puisse croître.

De plus la création de nouveaux organismes interunions, en forçant l'UAI à rembourser les frais de ses représentants, qui doivent participer à des réunions assez fréquentes et souvent lointaines, impose une charge à l'Union.

C'est pourquoi, tout en poussant au développement de la coopération interdisciplinaire, l'UAI a toujours eu pour politique de favoriser les efforts de regroupements, et d'oeuvrer pour une grande efficacité. Ainsi a-t-elle joué un rôle important dans la réorganisation des comités intéressés à la physique des relations Soleil-Terre: suppression de l'IUCSTR, de l'IUCI, fin prochaine de CIG et de l'IQSY, création de l'IUCSTP qui devrait avoir avant tout un rôle de coordination: autant de faits récents qui contribueront, sans doute, à l'amélioration de la coopération interdisciplinaire dans un domaine essentiel.

Le Comité Exécutif de l'UAI a aussi souhaité renforcer le rôle des Unions, et par conséquent a toujours manifesté son opposition au développement de comités nationaux indépendants consacrés à des questions interdisciplinaires, les structures nationales devant conserver le caractère interdisciplinaire des structures internationales.

L'UAI a participé, avec de nombreuses Unions (IUTAM, UGGI, URSI) et Comités (COSPAR, IQSY) à l'organisation de symposiums (voir ci-dessus p. xiii).

CONCLUSIONS GÉNÉRALES

La période de trois ans qui vient de s'écouler a été réellement difficile.

Certes, les Membres de l'Union, et parmi eux, principalement les Présidents des Commissions, les Vice-Présidents, les membres des Comités d'Organisation des Commissions, les responsables des symposiums et colloques, les responsables de divers services, les représentants de l'UAI au sein de divers organismes, ont accompli un travail admirable.

Symétriquement, le Bureau Administratif de l'UAI, avec le souci permanent de ne pas excéder les possibilités financières de l'UAI, a fait de son mieux pour les y aider, et pour contribuer à l'essor de l'astronomie mondiale. On a notamment tenté, dans ce but, d'améliorer et de simplifier certaines méthodes administratives, cependant qu'un grand effort était fait pour maintenir les publications de l'UAI à un niveau élevé.

Pourtant tout ne va pas pour le mieux dans le meilleur des mondes: la nouvelle politique des publications de l'UAI, qui trouve son origine dans le désir légitime de diffuser mieux des publications d'une qualité améliorée, est coûteuse. De plus, sa mise en application, qui implique des relations étroites entre le Secrétaire Général, en France, l'imprimeur et l'éditeur, au Royaume-Uni, rencontre des difficultés pratiques, accrues encore par des difficultés internes (d'ailleurs momentanées) à la firme chargée de l'impression, et par le changement, en 1966, du Secrétaire Adjoint de l'UAI. Les publications sont en vérité d'une qualité satisfaisante, mais la production est trop lente, et des difficultés de distribution se présentent. Ces difficultés gênent bien évidemment les relations entre le Bureau Administratif et les Membres de l'UAI. De plus la vente reste insuffisante.

Ces relations, à vrai dire, sont fondées sur une tradition ambiguë. L'UAI est une des rares Unions à être constituée de membres individuels. Ceux-ci ne payent aucune cotisation, aucun droit; mais, de plus en plus nombreux, ils reçoivent gratuitement des publications de plus en plus épaisses, ils utilisent les services d'organismes financés par l'Union, ils bénéficient des contrats scientifiques que l'Union favorise largement.

De plus la contribution des pays est modeste, eu égard aux budgets sans cesse croissants de l'Astronomie dans ces pays, surtout à l'ère du développement rapide de la Radioastronomie et de la Recherche Spatiale.

La contribution annuelle des pays dont la cotisation est la plus élevée est égale à cinq fois le montant du voyage aller et retour entre l'Amérique du Nord et l'Europe; ou au coût de l'impression et de la distribution de deux *Bulletins d'Information*; ou encore à la contribution de l'UAI à l'organisation d'un seul symposium!

Il n'y a que deux solutions possibles.

Ou bien une certaine justice financière est rétablie, et la tradition généreuse de l'Union tempérée: les Membres devraient payer une cotisation; ses publications seraient toutes vendues et non plus partiellement distribuées gratuitement (cette mesure est déjà en cours d'application); les observatoires accepteraient de prendre le relais de l'Union dans le financement des entreprises spécifiques des Commissions.

Ou bien, une certaine libéralité est maintenue, notamment en ce qui concerne l'adhésion des Membres et les entreprises spécifiques: mais les organisations adhérentes devraient accepter une augmentation de leur cotisation (voir ci-dessous page xxv).

Le Comité Exécutif est aujourd'hui convaincu qu'une attitude différente est impossible, et qu'un certain archaïsme trop généreux des budgets de l'UAI n'est plus aujourd'hui de mise. De plus, il est certain que le 'gigantisme' croissant de l'Union devra dans peu de temps faire place à de nouvelles structures. Préparer cette mutation sera le rôle du prochain Comité Exécutif.

REPORT OF THE EXECUTIVE COMMITTEE 1964–1967*

ACTIVITY OF THE INTERNATIONAL ASTRONOMICAL UNION

(voir texte français, page ix)

INTRODUCTION

The present report covers the period from 1 January 1964 to 31 December 1966. For the sake of continuity, it has been prepared in accord with the principles governing previous reports, especially that of the period 1961–1963, formally approved by the twelfth General Assembly, and published in Volume XII B of the *Transactions*.

The period under report includes the first half of the year 1964 (until 25 August 1964) which fell under the responsibility of the former Executive Committee. On the other hand, it does not include the first half of the year 1967, still within the responsibility of the present Executive Committee. An oral complementary report to cover this period (1 January to 21 August 1967) will be presented to the thirteenth General Assembly and, if approved, published in Volume XIII B of the *Transactions*.

The Executive Committee has reached a number of decisions during this period. Those in connection with the twelfth General Assembly, or approved by it, were published in Volume XII B of the *Transactions* and will not be repeated here. Those taken at later dates have been published in the *IAU Information Bulletin*, and will be referred to in the appropriate paragraphs that follow. Reports on the activity of IAU Commissions, Inter-Union Commissions, Astronomical Services, etc., published in the *Information Bulletin*, have been incorporated into the Draft Reports of Commissions, in so far as they still apply.

The most serious problems of the Union are those concerning publication and finance; they are dealt with in two separate appendices.

A third appendix is devoted to factual data concerning Adhering Countries and IAU Members. Four administrative Commissions (5, 6, 38 and 46) are now treated as committees of the Executive Committee; their reports follow immediately after this Report, on page lxxv. The final reports of the President of IUCSTR (Inter-Union Commission on Solar and Terrestrial Relationships), of which the IAU was the parent Union, and of the President of the Working Group on Photographic Materials, are also added to the present Report, on page xcix.

Structure and Administration of the IAU

Since the twelfth General Assembly, the International Astronomical Union has operated in accord with the provisions of the new Statutes, as adopted at that General Assembly.

Administration

The Executive Committee has had to cope with a great intensification in the activity of the Union: the number of its Members increased, as did that of its Commissions and Adhering Organizations; relations with other organizations became more involved.

This difficult situation was met, in the summer of 1964, by adding a shorthand-typist to the existing staff. Moreover, the General Secretary and the Assistant General Secretary had the benefit of the services of several assistants, in both Meudon and Prague, generously

**Note*. The text of the report and of its Appendices I and II are hereafter published in two languages. The French text being the original, from which the English text has been translated, is the only one to be authoritative, in case of need.

provided by the French and Czechoslovak authorities. Evidently, though re-inforced by one additional member, the staff of the Administrative Office of the Union is still largely insufficient.

The Executive Committee realizes that, the financial situation of the Union being as it is, no improvement is possible in the near future. The only remedy seems to be in a simplification of IAU administration, rather cumbersome the last years: partly for this reason, it has been decided to curtail the IAU Distribution List and to arrange for the Symposium volumes to be published by a commercial firm from Symposium no. 32 onwards (see page lxviii below).

Meetings of the Executive Committee

The former Executive Committee held its twenty-fifth meeting in Hamburg during the twelfth General Assembly (from 24 August to 3 September 1964) under the presidency of Professor V. A. Ambarcumjan. All members and councillors were present. The present Executive Committee held its twenty-sixth meeting in Hamburg, on 3 September 1964, under the presidency of Professor P. Swings. All members and councillors were present. The Executive Committee, presided over by Professor P. Swings, met as follows during the period under report: The twenty-seventh meeting was held at the Nice Observatory, France, from 15 to 19 September 1965; all members and councillors were present except Professor G. Haro and Professor V. A. Ambarcumjan. The twenty-eighth meeting was held in Prague, Czechoslovakia, from 16 to 20 September 1966; all members and councillors were present, except Professor Y. Hagihara and Professor V. A. Ambarcumjan, who was prevented from attending by illness.

Between meetings, the business of the Union was conducted by correspondence. Twenty-four circular letters were sent from the General Secretary and the Assistant General Secretary to the members of the Executive Committee.

Members of the IAU and Adhering Countries

New Zealand has adhered to the Union since 1964. No other modification has occurred since the twelfth General Assembly, though preliminary consultations have been held with representatives of the following countries: Algeria, Cuba, Indonesia, Iran, Nigeria and Pakistan. Columbia has applied for adhesion, and its application will be submitted to the thirteenth General Assembly.

The list of Adhering Countries and corresponding Adhering Organizations is given in Appendix III (page lxxi).

It should be noted that much thought was given to the creation of a new category of membership, that of 'Associated Adhering Countries'. Countries adhering in this category would have paid no contribution, but would have had no right to vote on questions with financial implications. The Executive Committee formed a working group from among its members to investigate this problem; but after a lengthy discussion, the Executive Committee decided, in 1966, not to ask the General Assembly to modify the Statutes to this effect.

The number of IAU Members was 1276 on 1 January 1964. Since then, the twelfth General Assembly has elected a further 352 new Members. On the other hand, the General Secretary has been informed of the decease of 56 Members, so that the Union had 1572 Members on 31 December 1964.

The list of the Members whose decease has been notified since 1 January 1964 is given in Appendix III. Unfortunately this list includes so large a number of eminent scientists that it is possible only to pay hereafter separate tribute to the memory of those who have contributed greatly to the administration of the Union:

BERTIL LINDBLAD, President of the International Astronomical Union (1948–1952), was Director of the Stockholm Observatory, Saltsjöbaden, from 1927. His achievements in the domain of the Galactic Structure (in the broadest sense) were rewarded by the award of the Gold Medal of the Royal Astronomical Society, and the Bruce Gold Medal of the Astronomical Society of the Pacific, beside other distinctions. He was President of the Royal Swedish Academy of Sciences. He will be remembered as a good friend, full of understanding and humanity, as well as a distinguished astronomer (1895–1965).

ROBERT METHVEN PETRIE, Vice-President of the International Astronomical Union (1958–1964), was Director of the Dominion Astrophysical Observatory, Victoria, Canada. His work in the domain of stellar spectrography, applied especially to the problem of spectroscopic parallaxes, was rewarded by numerous distinctions. He was honored with the title of 'Dominion Astronomer' and was President of the Royal Astronomical Society of Canada. His colleagues will remember him as an outstanding scholar, but also as a friend full of humour and wisdom (1906–1966).

ANDRÉ DANJON, President of the International Astronomical Union from 1958 to 1961, was, before he retired, Director of the Observatoire de Paris and Professor at the Sorbonne from 1945. His achievements in the domain of photometry, his research into fundamental astronomy and, in particular, his conception and application of the impersonal prismatic astrolabe, which was given his name, brought him many distinctions, including the Gold Medal of the Royal Astronomical Society and the Gold Medal of the C.N.R.S. He was member of the Academie des Sciences de Paris, associate member of numerous foreign academic institutions and president of a number of committees in France. His colleagues and pupils will never forget how much they owe to him, and they will always treasure his memory (1890–1967).

Commissions of the IAU

The volume of *Agenda and Draft Reports* includes as usual the excellent reports prepared by the Presidents of each of the 38 Commissions of the Union, assisted by their Vice-Presidents, Organizing Committees, those responsible for the various working groups or committees formed within the Commissions, and by the Commissions themselves.

At the twelfth General Assembly, two new commissions were established: Commission 45 (Spectral Classification and Multi-Band Colour Indices), and Commission 46 (Teaching of Astronomy); the latter is regarded as a committee of the Executive Committee.

Professor R. M. Petrie has been succeeded by Professor D. S. Evans, nominated as Acting President of Commission 30 (Radial Velocities), and Professor F. L. Whipple has been nominated Acting President of Commission 6 (Astronomical Telegrams) to replace the late Dr A. G. Hogg.

Between the General Assemblies, Commissions have co-opted a number of new members from among the Members of the Union. Moreover, some Presidents wished to obtain assistance from astronomers who are non-members of the Union; they have been asked to regard them as Members for all practical purposes. The Presidents will prepare proposals for the new composition of Commissions including co-opted members and new Members to be elected by the General Assembly. The list published in Volume XII C of the *Transactions* will serve as a starting point for these proposals.

Between the General Assemblies, Commissions organized several Colloquia, and other projects financed by the IAU.

Colloquia of IAU Commissions

(held between 1 January 1964 and 31 December 1966)

IAU Colloquium on *The Position of Variable Stars in the Hertzsprung-Russell Diagram*, Bamberg, 10–15 August 1965 (Commissions 27 and 42).

IAU Colloquium on *Interstellar Grains*, Troy, New York, U.S.A., 24–26 August 1965 (Commission 34).

IAU Colloquium on *The Blanketing Effect*, Heidelberg, 17–19 March 1966 (Commission 36).

IAU Colloquium on *Late-Type Stars*, Trieste, Italy, 13–16 June 1966 (Commissions 14, 29, 35, 36).

IAU Colloquium on *Atomic Collisions*, Boulder, Colorado, U.S.A., 11–16 July 1966 (Commission 14).

IAU Colloquium on the *Evolution of Double Stars*, Uccle, 29 August–2 September 1966 (Commission 26).

Specific Projects of Commissions, approved by the twelfth General Assembly

The list of such projects, with the corresponding expenditure, is given on page xlvi in Appendix I to the present report, dealing with the finances of the Union. These projects are, as a rule, continuations of previous projects and thus need no comment.

It will be seen, however, that the grants allocated to *Southern Variable Stars Charts*, and to the *Catalogue of Star Clusters* have not been used. The latter project has been delayed, and the Executive Committee has now agreed to a proposal that it be published by the Publishing Office of the Academy of Sciences, Budapest, from 1967 onwards. The grant made to Dr Bateson for the *Southern Variable Stars Charts* will have to be paid in 1967.

Specific Projects of Commissions approved, between the twelfth and thirteenth General Assemblies, by the Executive Committee

The projects are as follows:

(*a*) A new edition of the *Atlas of Representative Cometary Spectra* by the Institut d'Astrophysique de Liège, to meet the needs of many scientists.

(*b*) The publication, by Dr Sahade, of the *Information Bulletin for the Southern Hemisphere*. This publication, in great demand in the Southern Hemisphere, is particularly well prepared and presented.

(*c*) A new edition, in the form of a loose-leaf catalogue, of the now outdated catalogue by Rigaux, *Les Observatoires Astronomiques et les Astronomes*; this is a project of the Observatoire Royal de Belgique in Uccle. The revised form of publication will allow the catalogue to be kept up to date.

Symposia

IAU Symposia

During the period covered by the present report, the following symposia were organized by the Union:

no.	Subject	Place	Date
23	Astronomical Observations from Space Vehicles	Liège	August 1964
24	Spectral Classification and Multi-Colour Photometry	Saltsjöbaden	August 1964
25	The Theory of Orbits in the Solar System and in Stellar Systems	Thessaloniki	August 1964
26	Abundance Determination in Stellar Spectra	Utrecht	August 1964
27	The Construction of Large Telescopes	Tucson, Pasadena	April 1965
28	Cosmical Gas Dynamics. V. Aerodynamical Phenomena in Stellar Atmospheres	Nice	September 1965

no.	Subject	Place	Date
29	Instability Phenomena in Galaxies	Bjurakan	May 1966
30	Determinations of Radial Velocities and their Applications	Toronto	June 1966
31	Radio Astronomy and the Galactic System	Noordwijk	September 1966

Symposia nos. 23, 24, 25, and 26 had been held shortly before the twelfth General Assembly of the IAU. Symposium no. 28 was organized in co-operation with the International Union of Theoretical and Applied Mechanics (IUTAM), Symposium no. 31 in co-operation with the International Union of Scientific Radio (URSI).

Symposium no. 32, organized jointly with IUGG, will be held in Stresa, Italy, from 21 to 25 March 1967, and is to deal with 'Continental Drift, Secular Motion of the Pole, and Rotation of the Earth'.

Symposia nos. 33, 34, 35, on 'Physics and Dynamics of Meteors', 'Planetary Nebulae' and 'The Structure and Development of Solar Active Regions' respectively, will be held in Tatranska Lomnica, Czechoslovakia (nos. 33 and 34), and in Budapest, Hungary (no. 35).

Symposia with IAU participation

The History of Astronomy (IUPHS-IAU), Hamburg, August 1964.

The Trajectories of Artificial Celestial Bodies as Determined from Observations (IUTAM-COSPAR-IAU), Paris, April 1965.

Planetary Atmospheres and Surfaces as Deduced from Radio Observations (URSI-IAU), Dorado Beach, Puerto-Rico, U.S.A., May 1965.

COSPAR Sixth International Space Symposium (sessions on Galactic and Extragalactic Phenomena and Space Research) (COSPAR-IAU), Buenos-Aires, May 1966.

Solar-Terrestrial Physics (URSI-IUGG-IAGA-IAU), Belgrade, August 1966.

Moreover, the IAU was represented at numerous Symposia and at Assemblies of many ICSU Organizations; such representations were announced in the *IAU Information Bulletins*.

Relations to Other Organizations

IAU Representation

1. *International Council of Scientific Unions.* The General Secretary (then D. H. Sadler) represented the Union at the second meeting of the ICSU Executive Committee in London (June 1964). The General Secretary represented the Union at the third meeting of the ICSU Executive Committee in Munich, April 1965. The President and the General Secretary represented the Union at the eleventh General Assembly of ICSU in Bombay (January 1966). The General Secretary represented the Union at the fourth and fifth meetings of the ICSU Executive Committee (Bombay, January 1966). The Assistant General Secretary also attended the eleventh General Assembly of ICSU. The General Secretary represented the Union at the sixth meeting of the ICSU Executive Committee, in Monaco, September 1966.

The General Secretary is the present representative of the Union on the Executive Committee of ICSU.

2. *ICSU Organizations.* The Union participates in the activity of a number of Special and Scientific Committees, Inter-Union Commissions and Commissions established and sponsored by ICSU. Particulars of these organizations are given on pages 13 onwards of Volume XII C of the *Transactions*.

The following list gives only the present state of IAU representation in these organizations:

(*a*) Committee on Space Research (COSPAR): L. Gratton.

(*b*) Special Committee for the International Years of the Quiet Sun (IQSY): G. Righini (Vice-President); R. Michard (reporter for 'Solar Activity').

(*c*) Comité International de Géophysique (CIG): G. Righini (Vice-President); R. Michard.

(*d*) Federation of Astronomical and Geophysical Services (FAGS): D. H. Sadler (Vice-President); B. Guinot.

(*e*) Inter-Union Commission on Frequency Allocations for Radioastronomy and Space Science (IUCAF): J.-F. Denisse; V. A. Sanamian; F. G. Smith (Chairman); A. Unsöld.

(*f*) Inter-Union Commission on Solar-Terrestrial Physics (IUCSTP): Z. Svestka.

(*g*) Inter-Union Commission on Spectroscopy: B. Edlén; G. Herzberg; J. G. Phillips; M. J. Seaton.

(*h*) Inter-Union Commission on Science Teaching (CIES-IUCST): E. Schatzman.

(*i*) ICSU Abstracting Board (IAB): J. Kleczek.

(*j*) Committee for Data on Science and Technology (CODATA): Ch. Moore-Sitterly.

3. *Other Organizations.* The Union is represented on the following Organizations:

(*a*) La Fondation Internationale du Pic-du-Midi, by A. Lallemand.

(*b*) Le Comité Consultatif pour la Définition de la Seconde (CCDS) du Bureau International des Poids et Mesures, by W. Markowitz and N. Stoyko.

(*c*) Le Comité Consultatif pour la Définition du Mètre du Bureau International des Poids et Mesures, by Ch. Moore-Sitterly.

(*d*) Various Services of FAGS, that is on the directing boards of:

Bureau International de l'Heure (BIH), by H. M. Smith, N. N. Pavlov.

International Polar Motion Service, by the Organizing Committee of IAU Commission 19.

Quarterly Bulletin of Solar Activity, by the Organizing Committee of IAU Commission 12.

International Ursigrams and World Days Service (IUWDS), by R. Michard.

Solar Particles and Radiations Monitoring Organization (SPARMO), by C. de Jager.

Note: The Inter-Union Commission on Solar and Terrestrial Relationships (IUCSTR), and that on the Ionosphere (IUCI), have terminated their activities. The IAU was represented by C. W. Allen (President), R. Giovanelli, E. Mustel and M. Waldmeier on the former, and by C. W. Allen (Secretary), R. N. Bracewell, V. G. Fesenkov, K. O. Kiepenheuer on the latter.

Direct Relations with UNESCO

The Union had the benefit of various direct contracts with UNESCO, as indicated in the financial report.

The first contract provided for a grant towards two IAU Symposia (nos. 23 and 25), whose subjects were sufficiently broad to justify UNESCO assistance.

The second contract provided for regular financial assistance to Commission 38, partly to cover the Commission's expenditure especially in connection with the exchange of astronomers from or to developing countries.

Finally, an *ad hoc* working group of experts organized by Commission 46 of the IAU, agreed on the organization of an annual 'School of Young Astronomers'. A UNESCO grant provided for this purpose in 1966 could not be fully used owing to the short time available for the preparation of the curriculum. A substitute contract allowed the Union to allocate a scholarship to a student from a developing country towards his studies at a well equipped Observatory. The first UNESCO–IAU Summer School will open in Manchester, in 1967, under the direction of Professor Z. Kopal. The General Secretary of the school is Dr J. Kleczek.

This proposal was studied in great detail by the Executive Committee. The Committee would certainly have preferred that UNESCO, if it had been possible, should have assisted the normal activities of the Union, which have become rather burdensome. Other colleagues believe that a summer school is only justified as a pre-selection agent which would single out the most promising students for a longer stay at well-equipped observatories.

Be it as it may, the results of the experiment will be carefully examined with regard to future policy.

The IAU desires to maintain such contracts with UNESCO, to their mutual benefit. The present possibilities are likely to be fruitful, and the Union wishes to express its thanks and appreciation to UNESCO.

Relations with ICSU and its organisms

The ICSU is the organization on which the Union is most directly dependent: it transmits to the Union the UNESCO subvention (direct contracts not included) and co-ordinates the activity of Inter-Union Organizations.

This help is valuable and necessary to the Union.

Unfortunately the number of Unions within the ICSU increases, while the UNESCO subvention to ICSU remains essentially unchanged: thus the portion of the UNESCO subvention to the IAU is not likely to increase.

The Union is responsible for travel grants for its representatives on the increasing number of Inter-Union bodies whose meetings are very frequent and held at places often very remote.

While continuing to support the development of inter-disciplinary co-operation, the Union has therefore advocated efforts directed towards regrouping and greater efficiency. This was the Union's attitude in the re-organization of the Committees concerned with the physics of solar-terrestrial relationships; leading to the discontinuation of the IUCSTR and IUCI, the imminent termination of the activities of CIG and IQSY, and the formation of IUCSTP, whose principal objective will be the co-ordination of various activities. This recent action of the IAU is certain to contribute to a better inter-disciplinary co-operation in a domain of fundamental importance.

The Executive Committee of the IAU, in its effort to strengthen the position of the Unions, opposed the development of independent national committees to deal with inter-disciplinary problems, as it is desired that national structures should conform to the inter-disciplinary character of international structures.

The IAU participated with numerous Unions (IUTAM, IUGG, URSI) and Committees (COSPAR, IQSY) in the organization of Symposia (see above, page xxi).

GENERAL CONCLUSIONS

The three years covered by the present report were difficult in many respects.

It is certain, on the one hand, that the Members of the Union, and among them especially the Presidents of Commissions, Vice-Presidents, members of Organizing Committees of Commissions, members responsible for the organization of Symposia and Colloquia, and of numerous Special Services, representatives of the Union in different organizations, etc. have achieved admirable results.

The Administrative Office of the Union, always anxious to keep within the narrow limits of the IAU budget, did its utmost to help Members in these efforts and thus to contribute to the promotion of the interests of Astronomy all the world over. In this respect attempts were made to simplify certain administrative procedures, whilst maintaining the high standard of the IAU publications.

On the other hand, however, there is nothing perfect in this best of all worlds: the new IAU publication policy, adopted with the legitimate desire to improve the quality and distribution of IAU publications, is expensive. Moreover, its application, which calls for a narrow contact between the General Secretary in France, and the printers and publishers in the United Kingdom, met with difficulties in practice which were aggravated by a minor temporary crisis in the organization at the printers and by the change in the person of the IAU Assistant Secretary in 1966; the publications have been slow to come out of the press and their distribution has met with difficulties. This, naturally, puts a strain on the relations between Administration and Members. Moreover, the sales have, so far, been unsatisfactory.

In this connection, it should be emphasized that these relations with Members are based on a rather biased tradition. The IAU is one of the few international organizations having individual members. Members of the Union pay no contribution and no fees; yet, steadily increasing in number, they obtain, free of charge, publications ever ampler in volume; and they have the benefit of the numerous services organized and financed by the Union, and of the scientific contacts the Union facilitates.

Moreover, the contributions of Adhering Organizations are modest compared with the sums their countries spend on Astronomy in their own countries, especially in this age of the rapidly-developing Radio-Astronomical and Space Sciences.

The highest single annual national contribution is equal to only five times the return fare between North America and Europe, and corresponds to the cost of printing and distributing two issues of the *Information Bulletin*, which is also equal to the IAU contribution to the costs of a single Symposium!

There are only two possible solutions.

An attempt may be made to introduce more financial justice by relaxing the generous tradition of the Union and having Members pay a contribution; IAU publications being sold to Members and not distributed free of charge (this measure is already being put into operation); observatories assisting the Union by financing part of the Specific Projects of the Commissions.

Or, if a certain liberality is to be maintained, especially as regards membership and Specific Projects, the Adhering Organizations must agree to accept an increase in the contributions that they pay (see below, Appendix I, page xli).

The Executive Committee is convinced that there is no other alternative, and that the traditional generosity of the IAU budgets is obsolete. Moreover, it is certain that the growing 'gigantism' of the Union will soon have to give way to a new structure. It will be the task of the future Executive Committee to work towards this change.

APPENDICE I. RAPPORT FINANCIER

(See English version on page xli)

A. INTRODUCTION

Conformément à la politique financière introduite en 1961, le Comité Exécutif présente aux Membres de l'UAI un seul 'Compte Résumé des Recettes et Dépenses' couvrant la période du 1 janvier 1964 au 31 décembre 1966. Toutes les sommes ont été converties en dollars U.S.; le taux de change adopté, pour suivre les règles de l'Unesco, a été le suivant:

£ sterling: $2·80 $ U.S.: 4·90 francs français

Certaines opérations (virements d'un compte à un autre, remboursements de sommes excédant les besoins) ont été faites au change du jour par les banques: il s'ensuit des erreurs de change.

Le compte résumé a été obtenu à partir des neuf comptes (un pour chaque devise, et un par an) établis à partir des cahiers. Ceux-ci seront disponibles pour examen par le Comité des Finances. Ils ont été examinés et certifiés corrects par les experts comptables ayant travaillé pour l'Union. Le présent rapport se termine (ci-dessous page xxxix) par la déclaration d'expertise pour l'ensemble des comptes des trois années en considération, suivi par des suggestions relatives au budget des années suivantes.

Le rapport ci-après comporte, en commentaire du compte résumé, des notes nécessaires à sa compréhension complète point par point. Les sommes dues à la fin de l'exercice 1966, ou les recettes correspondant à des paiements arriérés pour la période 1961–1964, sont notamment indiquées dans chaque cas. Contrairement à ce qui s'est passé dans la période 1961–1964, ces sommes ne sont pas négligeables, en raison de la modification importante apportée au système des publications de l'Union; nous avons consacré un appendice détaillé à cette question.

Le rapport enfin est conclu par une mise en garde sérieuse, la situation financière étant en effet *particulièrement difficile*, et par des propositions relative au budget de 1967 et des années 1968, 1969, 1970.

B. COMPTE RÉSUMÉ DES RECETTES ET DÉPENSES

Recettes

1. *Cotisations des pays adhérents.* La XIe Assemblée Générale de l'UAI, à Berkeley, a décidé de fixer l'unité de contribution à 600 francs-or pour 1962 et les années suivantes. On rappelle qu'un franc-or est égal à 0,2903225 gramme d'or fin, et à 0,3266 dollar. L'unité de cotisation est donc de $195·96.

A la suite de la modification des statuts adoptée par la XIIe Assemblée Générale, il a été demandé aux Pays Adhérents de choisir une nouvelle catégorie d'adhésion, conformément au nouvel article 4 des Statuts, tel qu'il figure dans le Volume XII C des *Transactions*. Le nombre total d'unités de contribution a donc augmenté nettement, d'une façon qui n'a toutefois pas correspondu entièrement aux espoirs du Comité Exécutif de la période 1961–1964, qui avait suggéré la modification des Statuts, notamment dans ce but.

Il est à noter qu'en 1966 un effort particulier a été fait, avec succès, pour le recouvrement des cotisations dues, et pour un choix meilleur, dans certains cas particuliers, de la catégorie d'adhésion.

Au 31 décembre 1966, *six* pays sont en dette vis-à-vis de l'Union (soit au total 20 unités de contribution dues). *Un* pays en revanche s'est déjà acquitté du montant (6 unités) de la cotisation 1967. Le compte d'un autre pays est créditeur de $140·40.

COMPTE RÉSUMÉ DES RECETTES ET DÉPENSES
(Exprimé en U.S. Dollars)

Pour la période 1964–1966

Recettes

			U.S. Dollars	U.S. Dollars
1. Cotisation des Pays Adhérents	*Nombre d'unités*			
	Dues	*Reçues*		
1963 et antérieures	16	13		
1964	126	119		
1965	152	146		
1966	157	150		
(1967)	0	7		85 341·91
2. Vente des Publications				11 882·78
3. Abonnements aux Publications de l'UAI				4 469·79
4. Agios créditeurs bancaires				
Comptes courants			319·54	
Comptes bancaires			8 513·35	
				8 832·89
5. Subventions de l'UNESCO, accordées par l'intermédiaire de l'ICSU				
Imputation: Symposiums, publications				
1964			14 000·00	
1965			14 000·00	
1966			14 000·00	
				42 000·00
6. Contrats avec l'UNESCO				
Participation à l'organisation du Symposium no. 23			1 154·86	
Participation à l'organisation du Symposium no. 25			4 499·46	
Échange des Astronomes (Commission 38)			2 000·00	
Ecole d'Eté (organisation) (Commission 46)			1 306·00	
Bourse d'étude – 1966 – (Commission 46)			3 000·00	
				11 960·32
7. Divers				
Subvention de l'ICSU pour l'IUCSTR			3 564·40	
Participation de l'URSI à l'organisation du Symposium no. 20			200·00	
Remboursements divers				
– à la Commission 38			3 072·04	
– de la subvention à l'IUCI			180·00	
– du fonds de réserve du Bureau des Télégrammes			2 267·04	
				9 283·48
Total des recettes effectives				$173 771·17
8. Excès des Dépenses sur les Recettes				19 700·84
				$193 472·01
Etat des Comptes Bancaires au 1 janvier 1964				
– en banque (épargne)			62 730·84	
– en banque (comptes courants)			46 798·81	
– en petite caisse			0·37	
				$109 530·02

COMPTE RÉSUMÉ DES RECETTES ET DÉPENSES
(Exprimé en U.S. Dollars)

Pour la période 1964–1966

Dépenses

		U.S. Dollars	U.S. Dollars
1.	Bureau Administratif		30 231·60
2.	Cotisation à l'ICSU		1 490·74
3.	Dépenses de Fonctionnement des Commissions		452·29
4.	Entreprises spécifiques des Commissions approuvées par l'Assemblée Générale		36 514·62
5.	Assemblées Générales – Hambourg, août 1964		
	(*a*) Membres du Comité Exécutif (frais de voyages)	1 714·80	
	(*b*) Présidents des Commissions (frais de voyages)	7 219·41	
	(*c*) jeunes astronomes (frais de voyages)	5 044·71	
	(*d*) divers	1 582·54	
	– Prague, août 1967		
	(*d*) divers (voyage à Prague du Secrétaire Général)	133·26	
			15 694·72
6.	Frais d'impression des Publications éditées par l'UAI		56 712·00
7.	Réunions du Comité Exécutif		
	(*a*) Nice, septembre 1965	3 092·36	
	(*b*) Prague, septembre 1966	2 130·49	
			5 222·85
8.	Organisation des Symposiums et Colloques de l'UAI et Participation à d'autres réunions scientifiques		20 014·30
9.	Publications des Symposiums (lorsque l'éditeur n'est pas l'UAI)		13 088·36
10.	Commissions Interunions – Comités de l'ICSU, etc		7 473·11
11.	Autres frais de représentation		858·00
12.	Entreprises Spécifiques approuvées par le Comité Exécutif		350·00
13.	Programme sous contrat direct avec l'UNESCO		
	(*a*) Commission 38, Echange d'Astronomes	2 000·00	
	(*b*) Commission 46, Organisation des Ecoles d'été	1 306·24	
	(*c*) Commission 46, Bourse d'étude	2 000·00	
			5 306·24
	Total des dépenses effectives		193 408·83
14.	Différence de change dans les virements internes		63·18
		Total	$193 472·01
	Excès des Dépenses sur les Recettes effectives		19 700·84
	Soldes au 31 *décembre* 1966		
	(*a*) existant en banques (épargne)	71 244·19	
	(*b*) existant en banques (comptes courants)	18 518·92	
	(*c*) existant en petite caisse	66·07	
			89 829·18
			$109 530·02

2. *Ventes des publications.* En raison de la complexité des comptes, il est difficile de donner un bilan par ouvrage vendu. Mais nous avons tenté, en un appendice séparé, de présenter un bilan relatif aux opérations (recettes et dépenses) liées aux publications de l'UAI, que l'UAI ait été ou non l'éditeur.

3. *Abonnements aux publications de l'UAI.* Une expérience a été tentée, à partir de l'année 1963: il s'agissait de distribuer les publications de l'UAI contre le versement d'une souscription forfaitaire, à toutes les institutions de la 'liste de distribution'. La difficulté du recouvrement, et la complexité administrative de ces opérations ont conduit à supprimer ces abonnements dès 1966. Voir appendice relatif aux Publications, page lv.

4. *Agios créditeurs bancaires.* Les intérêts sur les comptes d'épargne ont été laissés à ces comptes, et incorporés aux capitaux déposés.

De plus, le compte courant, en francs français, a rapporté des intérêts, comptabilisés séparément.

5. *Subvention de l'UNESCO.* Cette subvention est versée par l'intermédiaire du Conseil International des Unions Scientifiques. Les comptes détaillés relatifs à cette subvention sont adressés chaque année à l'UNESCO. La ventilation des dépenses est laissée à la discrétion de l'UAI pourvu qu'elles restent imputées à des paragraphes budgétaires acceptables par l'UNESCO.

6. *Autres subventions de l'UNESCO.* L'Union a négocié avec l'UNESCO plusieurs contrats séparés, définis par une convention écrite et détaillée. Ces contrats ont été les suivants:

(*a*) Participation à l'organisation du Symposium no. 23, 'Astronomical Observations from Space Vehicles'. (UNESCO/NS/1966/64; 135–5674)
. . . 'The funds shall be used for meeting all or part of the cost of travel of participants'.

Attribution	$1500
Sommes dépensées (figurant au budget)	$1154·86

(le trop perçu a été remboursé à l'UNESCO).

(*b*) Participation à l'organisation du Symposium no. 25, 'The Theory of Orbits in the Solar System and in Stellar Systems'. (UNESCO/NS/1997/64; 135–5675).

Attribution	$4500
Sommes dépensées (figurant au budget)	$4499·46 (Somme réellement encaissée par l'UAI)
Somme remboursée à l'UNESCO à l'expiration du contrat	0

(*c*) Participation aux échanges d'astronomes (Commission 38) (UNESCO/NS/2477/65; ED.SS.0085)

Attribution (1965–1966)	$2000

(Sommes dépensées en totalité. Voir paragraphes 4 et 13 des Dépenses.)

(*d*) Organisation d'une réunion d'experts chargés d'élaborer un programme à long terme d'Ecoles d'Eté Internationales d'Astronomie (Nice, 1965). (Commission 46) (UNESCO/NS/2468/65; EC.SC.0085)

Attribution	$2000
Dépense réelle (figurant au budget)	$1306

(L'UNESCO n'a pas déboursé la différence)

(*e*) Organisation de stages d'études individuels (Commission 46) (UNESCO/NS/3034/66; EC.DOC.5555)

Attribution	$4000
Versé en 1966 (figurant au budget)	$3000
Dépensé en 1966 (figurant au budget)	$2000

Le compte sera clos en 1967, conformément au contrat.

7. *Divers.* Dans ce paragraphe on a groupé:

(*a*) les subventions d'organismes divers.

(*i*) Il s'agit de la subvention de l'ICSU pour le fonctionnement de l'IUCSTR. La somme figurant dans les comptes représente le bilan d'un versement de $3500·00, et d'un reversement de $64·40, conformément au compte détaillé ci-après:

Recettes		*Dépenses*	
Subvention	3500·00	Frais de voyages	1360·63
Reversement par le Professeur Allen	64·40	Publication du Xe Rapport de l'IUCSTR	1194·20
		Frais administratifs et de secrétariat	562·55
	$3564·40		$3117·38

La somme de $64·40, reversée à l'UAI à la fin de l'exercice 1966, sera reversée à l'ICSU en 1967, de même que les sommes non dépensées: le total s'en monte à $447·02, montant du trop perçu à rembourser par l'UAI à l'ICSU.

L'IUCSTR ayant été supprimée en 1966, cette somme ne peut en effet être utilisée dans l'avenir.

(*ii*) Il s'agit aussi de la subvention de l'URSI pour l'organisation du Symposium no. 20. L'URSI a également contribué à l'organisation du Symposium UAI-URSI no. 31, mais cette subvention a été directement versée aux organisateurs. Il en est de même pour la subvention de l'IUTAM au Symposium UAI-IUTAM no. 28.

(*b*) Au moment du déplacement du Bureau des Télégrammes de Copenhague à Cambridge (Mass.), il a été remboursé à l'UAI les sommes non dépensées de roulement, constitué antérieurement à 1961.

La subvention de l'UAI à l'IUCI, commission interunion, supprimée en 1966, a été en partie reversée à l'UAI.

Enfin certains bénéficiaires des bourses de voyage de la Commission 38 ont remboursé une partie des fonds reçus, en raison de modifications intérieures dans leurs projets après le versement des sommes attribuées.

Il est clair que, sur les $9283·48 figurant à ce paragraphe, seulement $3700 sont des recettes effectives; le reste correspond à des reversements et devrait en fait être soustrait des dépenses effectives, et des recettes effectives. Il s'agit d'un trop faible pourcentage (2% environ) de ces sommes pour que nous ayons cru devoir compliquer les comptes en les présentant de cette façon, malgré le caractère plus véridique qu'aurait eu cette présentation.

8. *Excès des dépenses sur les recettes.* Cet excès n'a jamais été aussi important, malgré l'accroissement notable des recettes. Nous y reviendrons ci-après, au paragraphe D, page xxxv.

Dépenses

Les numéros de chaque paragraphe correspondent à ceux du budget adopté par l'Assemblée Générale, à Hambourg, en août 1964.

1. *Dépenses du bureau administratif.* Ces dépenses sont constituées uniquement par les salaires du Secrétaire Adjoint et d'une sténodactylo, par les charges sociales diverses y afférentes, par les frais de poste et télégraphe, ainsi que par l'achat de papeterie.

Bien que très réduites (grâce à l'aide fournie par l'Observatoire Royal de Greenwich, puis par les Observatoires de Nice et de Paris, et par l'Académie des Sciences de Tchécoslovaquie: mobilier, téléphone, personnel, locaux, mis sans contrepartie à la disposition de l'UAI) ces dépenses excèdent très légèrement les prévisions budgétaires, mais de façon non significative.

Entreprises Spécifiques des Commissions

Commission no.	Entreprises		Solde 1964	Attribué 1965–1967	Payé en $ 1964	Payé en $ 1965	Payé en $ 1966	Total $	Solde 1967 $ (*d*)
5	Bibliographie de l'Astronomie			250·00	—	—	—	—	250·00
6	Bureau des Télégrammes			2 000·00	666·66	666·66	666·66	1 999·98	0·02
10	Cartes Synoptiques			2 646·00	882·00	882·00	882·00	2 646·00	—
10	Cartes Héliographiques			980·00	326·66	326·66	326·66	979·98	0·02
16	Centre de Documentation sur les Planètes (Meudon)			2 000·00	—	665·31	665·31	1 330·62	669·38
20	Minor Planet Center			1 500·00	—	—	1 000·00	1 000·00	500·00
23	Carte du Ciel	(*a*)	4 365·91	2 000·00	1 476·54	—	3 467·22	4 943·76	1 422·15
27	English Version of Remarks to the Second Supplement of J.C.V.S.	(*b*)		700·00	—	—	—	—	—
27	English Version of Remarks to the Second Catalogue of Stars suspected of variability	(*b*)		700·00					
27	Southern Variable Stars			1 000·00	—	—	—	—	1 000·00
37	Card Catalogue of Clusters and Associations			1500·00	—	—	—	—	1500·00
38	Exchange of Astronomers	(*c*)	8 006·47	—					
	IAU grant 1964		7 500·00	—	—	—	—	—	—
	Reimbursed grants (1964–1966)		3 072·04	—	—	—	—	—	—
			18 578·51	20 000·00 (+2 000·00 contrat UNESCO)	7 200·58	5 200·67	13 213·03	23 614·28 (+2 000·00 contrat UNESCO)	14 964·23
					Ces sommes comprennent les versements effectués à l'aide du contrat UNESCO de $2 000.				
	Total		22 944·42	35 276·00	10 552·44	6 891·30 (+ 850·00 contrat UNESCO)	19 070·88 (+1 150·00 contrat UNESCO)	36 514·62 (+2 000·00 contrat UNESCO)	20 305·80

2. *Cotisation à l'ICSU.* L'Union verse annuellement à l'ICSU, pour contribuer à ses dépenses administratives, une somme de 2% des cotisations nationales versées à l'UAI l'année précédente.

3. *Dépenses des Commissions.* Cette somme reste minime. Seules les Commissions 42 et 46 en ont fait usage pendant la période 1961–1964; les pièces justificatives sont à la disposition du Comité des Finances.

4. *Entreprises spécifiques des Commissions approuvées par l'Assemblée Générale.* Les détails relatifs aux projets en question figurent dans le rapport du Comité des Finances à la XIIe Assemblée Générale (*Trans. UAI*, XI B, pour 1964, *Trans. UAI*, XII B, 1966, pour 1965 et 1966).

L'état des dépenses pour ces différentes entreprises est donné dans la liste ci-après (voir tableau page xxx).

Cette liste appelle quelques commentaires:

(*a*) La Carte du Ciel fonctionne avec des fonds de roulement, dont un compte détaillé est tenu par le Président de la Commission 23.

(*b*) Les sommes en question ($1400) attribuées par l'UAI ne seront pas en fait dépensées: elles seront dépensées par l'Académie des Sciences de l'U.R.S.S., en contrepartie d'une portion de dettes diverses de l'U.R.S.S. vis-à-vis de l'UAI.

(*c*) Les sommes relatives à l'Echange des Astronomes proviennent, pour $2000, d'un contrat avec l'UNESCO. Nous avons donc retranché cette somme de la colonne 'total dépensé', correspondant à ce paragraphe du budget.

(*d*) Le solde au 1 janvier 1967, représente les sommes dues aux différentes entreprises; il n'est pas certain que toutes ces sommes seront dépensées; il est clair que certaines d'entre elles correspondent à des erreurs de change; d'autres à des entreprises dont la réalisation a dû prendre un certain retard.

5. *Assemblée Générale, Hambourg, août* 1964. En 1964, les dépenses ont été plus faibles qu'en 1961; il y a à cela une raison principale, à savoir la position géographique de Hambourg, plus centrale que celle de Berkeley. De plus la générosité de nos hôtes, comme dans le passé, a permis de limiter les dépenses administratives sur place. Les sommes indiquées en(*d*), Divers, correspondent principalement aux frais de voyage du personnel du secrétariat de l'UAI ayant assisté le Secrétaire Général et le Secrétaire Général Adjoint.

6. *Frais d'impression des publications éditées par l'UAI.* Il s'agit du paragraphe le plus lourd des dépenses, celui où les dépassements budgétaires sont les plus grands. Un chapitre spécial de ce rapport est consacré au problème des publications. On s'y reportera pour plus de détails.

7. *Réunions du Comité Exécutif.* Les dépenses de la réunion de 1964, à Hambourg, sont comprises dans celles de l'Assemblée Générale, paragraphe 5*a*, ci-dessus.

La somme totale dépensée est faible, grâce au fait que les pays hôtes et les institutions de ces pays ont reçu généreusement le Comité Exécutif, mais aussi aux efforts des membres du Comité Exécutif pour combiner la réunion du Comité Exécutif avec quelque autre réunion scientifique, financée par ailleurs.

8. *Organisation des symposiums et colloques.* La majeure partie des dépenses est consacrée aux frais de voyage de certains participants, choisis par le Comité d'Organisation de chaque symposium, et dans une mesure beaucoup plus faible, à des dépenses administratives. Il est souvent arrivé que les sommes attribuées aux différentes réunions par le Comité Exécutif n'aient pu être dépensées dans leur totalité, certaines subventions ayant souvent été reversées en un moment où leur utilisation n'était plus possible. Le tableau ci-dessous donne les détails des sommes dépensées.

Symposium no.	Sujet, Lieu, Date	Paiements effectués
23	Astronomical Observations from Space Vehicles, Liège, août 1964	$ 1 155·05
24	Spectral Classification of Multi-colour Photometry, Saltsjöbaden, août 1964	2 403·89
25	The Theory of Orbits in the Solar System and in Stellar Systems, Thessaloniki, août 1964	5 459·57
26	Abundance Determination in Stellar Spectra, Utrecht, août 1964	300·00
27	The Construction of Large Telescopes, Tucson (Pasadena), avril 1965	00·00
28	Cosmical Gas Dynamics, Aerodynamical Phenomena in Stellar Atmospheres, Nice, septembre 1965	1 450·00
29	Instability Phenomena in Galaxies, Burakan, mai 1966	1 100·01
30	Determination of Radial Velocities, Toronto, juin 1966	1 500·00
31	Radio Astronomy and the Galactic System, Noordwijk, septembre 1966	1 972·25
		$15 340·77

Le Symposium 27 a pu être organisé grâce à la générosité des organisations invitantes, sans qu'aucune subvention de l'UAI ait été nécessaire. Les Symposiums 23 et 25 ont été financés en presque totalité par des contrats spéciaux avec l'UNESCO.

De plus, les sommes suivantes ont été dépensées pour d'autres réunions:

Colloque UAI sur les Étoiles Variables (Bombay 1965)	$ 500·00
Colloque UAI sur les Atmosphères Stellaires (août 1962)	150·00
Colloque UAI sur les Étoiles de Types Avancés (Trieste)	373·53
Colloque UAI sur le 'blanketing effect' (Heidelberg)	500·00
Colloque UAI sur les Étoiles Doubles (Uccle)	500·00
Colloque UAI sur les Collisions Atomiques (Boulder)	500·00
Symposium IUPHS (Histoire de l'Astronomie)	400·00
Symposium IUTAM-COSPAR-UAI (Trajectories of Artificial Celestial Bodies)	1 000·00
Symposium COSPAR-UAI-URSI-UGGI (Solar Terrestrial Physics, Belgrade, 1966)	750·00
	$ 4 673·53

9. *Publication des symposiums, lorsque l'éditeur n'est pas l'UAI.* L'UAI a décidé d'éditer tous les symposiums à partir du Symposium 24. On se reportera à leur sujet à l'appendice consacré aux Publications de l'Union, page lv.

Ce paragraphe concerne donc uniquement les volumes antérieurs:

Symposium no.	Titre	Editeur	Sommes payées
9	Radio Astronomy	Stanford U. Press	$ 80·28
14	The Moon	Academic Press	12·09
19	Site Testing	*Bulletin Astronomique*	1 648·12

Symposium	Titre	Editeur	Sommes payées
20	The Galaxy and the Magellanic Clouds	Australian Academy of Sciences	2 240·00
21	The System of Astronomical Constants	*Bulletin Astronomique*	1 005·83
22	Stellar and Solar Magnetic Fields	North Holland Pub. Co.	3 000·00
23	Astronomical Observations from Space Vehicles	*Annales d'Astrophysique*	5 102.04
			$13 088·36

En ce qui concerne le Symposium 9, il s'agit du rachat par l'UAI du stock, et, pour le Symposium 14, d'un seul volume.

Pour les Symposiums 19, 21, 23, publiés sans frais dans des revues scientifiques, il s'agit de l'achat des exemplaires que l'UAI distribuait gratuitement (participants et auteurs aux symposiums, Liste de Distribution) avant la décision relative à la limitation de cette distribution. Voir pour plus de détails l'appendice consacré aux Publications de l'UAI (page lv). Il n'est pas à espérer de recettes notables issues des ventes de ces volumes, dans les années à venir.

10. *Commissions interunions, comités de l'ICSU, etc.* Les sommes dépensées peuvent l'être soit sous forme d'une contribution forfaitaire au fonctionnement de l'organisme interunion, soit sous forme de remboursement de frais de voyage aux représentants de l'UAI. Le détail des dépenses a été le suivant:

Dépenses diverses de fonctionnement général pour IUCSTR (voyages des représentants de l'UAI, frais administratifs, impression des rapports IUCSTR) (1964, 1965, 1966)	$3 117·38
Frais de représentation (voyage des représentants de l'UAI) à COSPAR (1966)	1 070·00
Frais de représentation à CIG (pour les années antérieures à 1964)	286·14
Cotisations (forfaitaires) à CIES (1964, 1965, 1966)	599·59
à IUCAF (1964, 1965, 1966)	1 200·00
à IUCI (1964, 1965)	200·00
à IQSY (1963, 1964)	1 000·00
	$7 473·11

Plusieurs remarques doivent être faites:

Le cas de l'IUCSTR est spécial, puisque l'UAI y a joué le rôle d'union-mère. La subvention de l'ICSU a été administrée par l'UAI, et les comptes détaillés en sont les suivants, pour les trois années sur lesquelles portent le rapport.

Subvention accordée		Dépenses	
$3000	pour 1964	1964	$1 500·53
500	pour 1966	1965	772·80
remboursement du trop perçu: $64·40	pour 1966	1966	844·05
			$3 117·38

L'IUCI n'a eu qu'une existence éphémère. Sur les $200 versés par l'UAI, $180 nous ont été remboursés: voir Recettes, paragraphe 7.

Des dettes vis-à-vis du CIG (voyages de délégués) correspondant à l'exercice 1963 ont dû être réglées en 1964. Il en est de même pour la cotisation à l'IQSY de 1963, réglée en 1964; nous noterons qu'elle n'a pas été réclamée en 1965 et 1966, et nous la considérons comme due (soit $1500 à régler en 1967, nos engagements vis-à-vis de l'IQSY étant d'un maximum de cinq années).

11. *Autres frais de représentation.* La somme indiquée a permis au Président d'assister à l'Assemblée Générale du Conseil International des Unions Scientifiques à Bombay, en janvier 1966.

Il est à noter que les paragraphes 10 et 11, qui correspondent à ceux du budget, sont facilement confondus; il est proposé dans l'avenir de limiter le paragraphe 10 aux cotisations forfaitaires, et de faire entrer les frais de déplacements des représentants de l'UAI aux réunions des Commissions interunions dans le paragraphe 11.

12. *Entreprises spécifiques approuvées par le Comité Exécutif.* Un petit nombre d'entreprises ont été approuvées par le Comité Exécutif après l'Assemblée Générale de Hambourg.

Commission 15	*Atlas of Representative Cometary Spectra* (Institut d'Astrophysique de Liège) Cette somme n'a pas encoré été utilisée	$500
Commission 29	*Information Bulletin for the Southern Hemisphere* (Dr Sahade) Cette somme correspond à $700 pour 1965, $700 pour 1966; le Comité Exécutif a accepté la demande de Monsieur Sahade de reporter sur 1966–1967 l'utilisation de cette somme. Toutefois les chèques adressés en 1966 à Monsieur Sahade n'ont pas encore été encaissés par lui au 1 décembre 1966. Seule une somme de $350 a été utilisée.	$1400
Commission 5	*Les Observatoires Astronomiques et les Astronomes* (Observatoire Royal de Belgique) Il s'agit de la réédition, sous forme de feuilles mobiles, du livre de Rigaux. Cette somme n'a pas encore été utilisée.	$500

En résumé, $2400 ont été attribués, mais $2050 restent encore à être versés, au cours de l'année 1967. S'il n'en n'était pas ainsi, faute d'un avancement suffisant des projets, l'Assemblée Générale devrait se prononcer sur la reconduction des attributions proposées par le Comité Exécutif.

13. *Programmes sous contrat direct avec l'UNESCO.* L'Unesco a financé, par l'intermédiaire de contrats directs, 5 opérations (voir Recettes, paragraphe 6). Les dépenses relatives aux Symposiums 23 et 25 figurent au paragraphe 8. Le détail des dépenses relatives aux autres contrats figure dans les rapports adressés à l'Unesco et sont à la disposition du Comité des Finances.

14. *Différences de change dans les virements internes.* Cette somme correspond essentiellement à l'opération de change défavorable qui a accompagné, en 1966, le virement d'une importante partie du compte en livres sterling aux deux autres comptes, en francs et en dollars; ce virement a été effectué en un moment où circulaient en Grande-Bretagne des informations faisant état d'une dévaluation possible de la livre sterling.

Nous remarquerons que, par la suite, les factures en livres sterling ont été payées par l'intermédiaire de notre compte français, à un taux de change lui aussi différent du taux de l'Unesco. Si nous avions conservé le compte en livres, le déficit de change serait donc légèrement différent de celui porté dans les comptes. A ces différences de change, ont été incorporés (sans que la séparation soit possible) les changes bancaires des virements internes.

Solde au 31 *décembre* 1966. L'examen des avoirs bancaires fait apparaître un lourd excès des dépenses de $19 700.84 et un accroissement de $8 513.35 des comptes d'épargne.

Le déficit est considérable, malgré une politique d'économie stricte, particulièrement en ce qui concerne le bureau administratif. Il est essentiellement dû (voir ci-après xxxvii) à l'importance des frais d'impression et d'édition des publications de l'UAI.

C. DÉCLARATION D'EXPERTISE RELATIVE AUX COMPTES 1964–1966

Messieurs,

Conformément à la mission que vous avez bien voulu nous confier, nous vous présentons le résultat de notre examen des comptes de l'Union Astronomique Internationale pour la période du 1.1.64 au 31.12.66.

L'examen a porté, notamment, sur les documents suivants:

– comptes des recettes et dépenses pour l'année 1964 vérifiés par Monsieur G. Chambraud, Expert-Judiciaire à Nice (Alpes-Maritimes);
– comptes des recettes et dépenses pour l'année 1965 vérifiés par Monsieur W. E. Copus, F. C. A.-Chartered Accountant, à Mougins (Alpes-Maritimes);
– comptes des recettes et dépenses pour l'année 1966;
– comptes des recettes et dépenses et situation des comptes bancaires pour la période de trois ans: 1964, 1965 et 1966.

Nous avons également examiné les six comptes séparés des dépenses insérées dans le rapport financier du Comité Exécutif.

Les comptes ont été comparés par Messieurs Chambraud (année 1964) et Copus (année 1965) et par nous-mêmes (année 1966) avec les cahiers de comptes, reçus et pièces comptables diverses.

Nous certifions que ces comptes sont en accord complet avec les informations et explications que nous avons demandées pour nous permettre d'effectuer notre examen.

Le compte, ci-inclus, des recettes et dépenses, résumant les comptes des années 1964, 1965 et 1966, est la représentation exacte des dépenses et recettes de l'Union Astronomique Internationale pour la période du premier janvier 1964 au trente et un décembre 1966.

A Paris, le 8 Mars 1967

ROGER BACLE
H. E. C.–Expert-Comptable
Inscrit au Tableau de l'Ordre
de la Région Parisienne

D. CONSIDÉRATIONS RELATIVES AU BUDGET DE 1967

Afin de préparer le budget des années 1968–1970 et un budget additionnel pour 1967 les tableaux ci-dessous rappellent (sans répéter la définition des différents paragraphes) les prévisions budgétaires antérieures et les différences relevées avec les comptes réels.

Remarques

Les prévisions 1967 ont été faites, compte tenu de la situation difficile des finances de l'Union, en astreignant le budget à des économies sévères sur certains chapitres, comme suit.

Révision des prévisions pour la période 1965-1967

Nouvelles prévisions relatives au Budget 1967 (établies en mars 1967)		Nouvelles prévisions relatives au Budget 1965-1967 (établies en mars 1967)			
Recettes	Dépenses	Recettes	$P_2 - P_1$	Dépenses	$P_2 - P_1$
1. 35 000	1. 8 000	1. 93 309	− 691	1. 30 112	+ 112
2. 12 000	2. 500	2. 20 011	−21 989	2. 1 510	− 490
3. 0	3. 0	3. 1 496	− 8 504	3. 277	− 1 722
4. 3 000	4. 10 000	4. 9 146	+ 1 146	4. 35 962	+ 686
5. 14 000	5. 14 000	5. 42 000	0	5. 14 134	+ 134
6. (pour mémoire)	6. 63 600	6. (6 306 + x)	(+ 6 306 + x)	6. 109 153	+79 153
7. (divers)	7. 0	7. (6 083)	(+ 6 083)	7. 5 223	+ 223
	8. 4 500	8. 10 000	+10 000	8. 14 216	− 2 784
	9. 0			9. 10 768	−39 232
	10. 2 000			10. 6 472	− 38
	11. 0			11. 858	− 142
	12. 2 000			12. 2 350	− 2 650
	13. (pour mémoire; égal au §6 des Dépenses, + 1000)			13. (6 306 + x)	(+6 306 + x)
				14. 56	+ 56
64 000	105 600	190 351 + x		237 397 + x	
8. 10 000					
74 000					

P_1 = Prévisions faites en mars 1967. P_2 = Prévisions faites en août 1964.
(Sommes en dollars U.S.)

Le licenciement d'une personne, en janvier 1967 et jusqu'au transfert du Bureau Administratif à Prague, licenciement rendu possible par l'aide apportée par le Personnel de l'Observatoire de Nice, permet de réduire à $8000 les besoins du Bureau Administratif à la charge de l'UAI. Mais les dépenses réelles sont plus élevées; si l'on voulait assurer l'indépendance du Bureau Administratif, il faudrait prévoir $20 000 à 25 000 par an. Si l'on désirait, de plus, améliorer les conditions de travail du Bureau Administratif, il faudrait 2 ou 3 personnes supplémentaires, et un budget de $35 000 à 40 000 par an.

Les soldes des subventions aux entreprises spécifiques se montent à $20 306. Il est suggéré d'économiser sur ces soldes et de ne dépenser que $4500 pour la Commission 38, à quoi s'ajoutera le montant du contrat de l'UNESCO.

Un emprunt fait à l'ICSU, et remboursable d'ici 1972, de $10 000, a de plus été contracté (Recettes, § 8).

Les estimations relatives aux publications (Recettes, §2 et 3; Dépenses, §6 et 9) résultent des calculs décrits ci-après, dans le chapitre consacré à la politique des publications de l'UAI, pp. liii-lx. Ces chiffres peuvent apparaître catastrophiques. Ils sont fort différents de ceux qu'impliquait le Budget établi au mois d'août 1964: les recettes sont insuffisantes et les dépenses excessives. Les ventes n'ont pas correspondu aux espoirs. L'accroissement du volume de littérature imprimée et le coût de la distribution gratuite des *Draft Reports* et des *Transactions* ont pesé largement sur les dépenses.

Les nouvelles estimations pour 1967 permettent de réévaluer les estimations pour la période de trois ans 1965–1967 et de les comparer avec le Budget. Mis à part les subventions de l'UNESCO, non incluses dans le Budget autorisé, nous noterons que les estimations budgétaires de 1964 relatives aux dépenses du Bureau Administratif, de l'Assemblée Générale, du Comité Exécutif, et aux entreprises spécifiques des Commissions, correspondent presque exactement aux estimations actuelles. Des économies importantes ont été réalisées sur l'organisation des Symposiums, grâce à l'aide des organisations nationales, – et sur les entreprises approuvées par le Comité Exécutif. Mais les publications (§6 + 9) montrent un excès de dépenses de $40 000 par rapport aux prévisions de 1964.

Les estimations budgétaires de 1964 relatives aux recettes correspondent presque exactement aux estimations pour les cotisations. Le surplus de $10 000 au §8 correspond au prêt remboursable de l'ICSU. On notera (§2 + 3) un manque à gagner de $30 000 sur les ventes des publications par rapport aux prévisions. Les recettes diverses (§7) correspondent surtout à des reversements de sommes trop perçues au titre des §4 et 8 des dépenses, qui sont donc en fait bénéficiaires d'une économie supplémentaire correspondant à cette somme.

Le tableau ci-dessous montre alors les éléments du calcul du Budget additionnel demandé en 1967.

Calcul du Budget Additionnel 1967

		Recettes	Dépenses	
1.	Montant réel (1962–63)	$96 184	$87 109	
2.	Montant réel 1964	57 420	61 675	
3.	Total réel 1962–64 (1 + 2)	153 604	148 784	
4.	Disponible fin 1964 (Recettes moins dépenses)			$4 820
5.	Budget estimé 1962–64 (estimation février 1964)	143 384	150 219	
6.	Supplément budgétaire autorisé en août 1964 par l'Assemblée Générale (calculé en comparant recettes et dépenses du Budget estimé)			6 500
7.	Budget 1965–1967 approuvé par la XIIe Assemblée Générale	198 000	197 776	
8.	Report de l'avance budgétaire (4 + 6)	11 320		
9.	Budget autorisé 1965–1967 (non compris contrats UNESCO) (8 + 7)	209 320	197 776	
10.	Total subventions UNESCO sous contrat 1965–1966	6 306	5 306	
11.	Montant réel 1965–1966 (y compris contrats UNESCO)	116 351	131 797	
12.	Estimation 1967, sans les contrats UNESCO (estimation mars 1967)	74 000	105 600	
13.	Estimation 1965–1967 (11 + 12) (y compris contrats UNESCO 1965–1966)	190 351	237 397	
14.	Supplément budgétaire demandé à la XIIIe Assemblée Générale (calculé en comparant recettes et dépenses du Budget estimé)			50 000

C'est donc la somme considérable de $50 000 qui sera proposée au Comité des Finances, comme budget additionnel de l'UAI pour 1967. Bien que cette somme soit légitimée, en importante partie, par le fait que des publications payées en 1967 ne produisent de recettes qu'en 1968, il est clair que nous devons proposer une méthode qui permette de corriger cette ponction (aux détriments des sommes figurant dans les comptes d'épargne) dans la période 1968–1970, faute de quoi la situation deviendra réellement catastrophique.

E. CONCLUSIONS ET BUDGET 1968–1970

Il est donc clair qu'en raison principalement de l'accroissement considérable des charges dues aux publications, des sources nouvelles de revenus doivent être trouvées.

Sans entrer ici dans le détail, nous remarquerons que les recettes sont proportionnelles, à peine, au nombre de Membres (proportionnalité acceptable par les Organisations Adhérentes), les dépenses dues à la distribution gratuite, à tous les Membres, de diverses publications (*Draft Reports*, *Transactions*, volumes B et C, *Bulletins d'Information*) sont approximativement proportionnelles au *carré* du nombre de Membres: le volume de la publication augmente, comme le nombre de destinataires, proportionnellement au nombre de Membres.

Le déséquilibre est donc fatal, et s'accroît nécessairement. Il est dans la logique de la croissance de l'Union. On peut sans doute reprocher au Secrétaire Général actuel, responsable des comptes depuis l'été 1964, de n'avoir pas pris conscience assez tôt de cet état de fait et de n'avoir pas suffisamment éclairé le Comité Exécutif à ce sujet. Mais il était inévitable, un jour ou l'autre, de faire face à un tel déséquilibre.

Une première solution est dans une réduction de la générosité de l'Union. Les Membres de l'UAI, sans verser la moindre cotisation, bénéficient d'avantages nombreux, reçoivent des publications gratuites, utilisent les services des Commissions de l'UAI. On pourrait appliquer strictement les Statuts, exiger une rentabilité des entreprises spécifiques des Commissions, réduire l'aide apportée à l'organisation des Assemblées Générales, etc.

Mais il serait souhaitable de réduire le plus possible les effets de l'austérité: priver d'aide les entreprises spécifiques des Commissions, réduire l'aide apportée aux jeunes astronomes, sont des opérations contraires à l'esprit de l'UAI.

Le Secrétaire Général, au nom du Comité Exécutif, proposera donc de tempérer l'austérité grâce à des ressources nouvelles. Conscient de l'effort qu'il demande aux Organisations adhérentes, mais certain que la vie même de l'UAI étant en cause, ces organisations auront à coeur de comprendre la nécessité de cette action, le Comité Exécutif propose l'augmentation de l'unité de contribution jusqu'à sa valeur maximum tolérée par les Statuts (soit de 600 à 900 Francs-or).

Si cette proposition est adoptée par l'Assemblée Générale, alors un budget raisonnable pourra être maintenu, et le budget additionnel de 1967 compensé en très large partie d'ici 1970.

Si cette proposition est rejetée, alors il faudra que l'UAI en mesure les conséquences. La suppression des publications de l'UAI, celle des Membres individuels, peuvent être nécessaires, dès après le Volume XIII des *Transactions* et la XIIIe Assemblée Générale.

Des budgets précis seront proposés, pour chaque éventualité, au Comité des Finances. Comme base de discussion, on peut partir du Budget suivant pour 1968–1970, établi, en ce qui concerne les recettes, à partir du Budget 1967, où le budget des publications correspond à celui du Rapport II ci-dessous sur la politique des publications de l'UAI (page lv), et où des compressions sur les dépenses ont été opérées (notamment au § 4).

	Recettes		Dépenses
1.	\$105 000 + *x*?	1.	\$30 000
2.	73 400	2.	1 500
3.	0	3.	500
4.	8 000	4.	20 000
5.	42 000	5.	14 000
6.	(*x*)	6.	99 000
7.	?	7.	5 000
		8.	20 000
	\$228 400 + *x*?	9.	0
		10.	7 500
		11.	0
		12.	2 000
		13.	(*x*)
		14.	10 000 (Remboursement à l'ICSU)
		15.	50 000 (Remboursement aux comptes épargnes)
			\$259 500

Il est nécessaire d'épargner $30 000, ou de susciter une recette de cet ordre de grandeur.

Solutions possibles:

(*a*) Augmentation de 600 à 900 Francs-or, de la valeur de l'unité de cotisation, sans changement de catégorie d'adhésion des Organisations Adhérentes.

Ressource supplémentaire: $52 000

(*b*) Cotisation des Membres individuels (mesure à introduire) de $10 par an.

Ressource supplémentaire: $48 000 environ

(*c*) Suppression du *Bulletin d'Information*; remplacement par un bulletin ronéotypé.

Économie, environ: $8000

(*d*) Suppression des publications (y compris les *Transactions*, mais à l'exception du *Bulletin d'Information*, des *Draft Reports* 1967, et et du Symposium 32, antérieurs à l'Assemblée Générale de Prague) (voir ci dessous, p. lv.)

Économie, environ: $30 000

D'autres économies devront être trouvées, si les mesures adoptées par l'Assemblée Générale s'avèrent insuffisantes. Il est suggéré à l'Assemblée Générale d'autoriser le Comité Exécutif à étudier et décider ces nouvelles économies, pour un montant de $50 000, si aucune solution n'est définitivement adoptée à l'issue de l'Assemblée Générale.

APPENDIX I. REPORT ON IAU FINANCES

(Voir texte français, page xxv)

A. INTRODUCTION

In accord with the policy adopted in 1961, the Executive Committee presents to the Members of the Union a single 'Summarized Account of Receipts and Expenditure' covering the period from 1 January 1964 to 31 December 1966. All amounts are given in U.S. dollars converted from other currencies at the following UNESCO conversion rates, treated as exact:

£ sterling: $2·80 U.S. $: 4·90 French francs

Certain bank operations, such as transfers from one account to another, reimbursements of excess funds, were effected at current rates of exchange and thus resulted in exchange differences.

The summarized account has been derived from nine (one for each currency and each year) individual accounts. These accounts, which will be available for examination by the Finance Committee, have been audited and certified as correct by the Union's professional accountants. At the end of the present report (page liv) there is printed an auditor's certificate covering the account for the three years under consideration.

The present report includes, appended to the summarized account, explanatory notes to each point concerned; the amounts due at the end of the financial year 1966, or the receipts corresponding to outstanding payments for the period 1961–1964, are given separately for each case. In contrast to 1961–1964, these amounts are not negligible, owing to the important modification in the IAU publication system; this problem is dealt with in more detail in a separate appendix (see page lxiii).

The report ends with a serious warning, the financial situation of the Union being particularly difficult, and with proposals as to the budget for 1967 and the years 1968, 1969, 1970.

B. SUMMARIZED ACCOUNT OF RECEIPTS AND PAYMENTS

(see appended table)

Receipts

1. *Subscriptions of Adhering Countries.* The eleventh General Assembly of the IAU, held in Berkeley, resolved to fix the amount of the unit of contribution at 600 gold francs, the gold franc being equivalent to 0·3266 U.S. dollar. Thus, the unit of contribution is 195·96 U.S. dollars.

Arising from the modifications in the Statutes adopted by the twelfth General Assembly, Adhering Organizations were approached with the request to choose a new category of adherence, in accord with the new article 4 of the Statutes, as published in Volume XII C of the *Transactions*. This appeal led to an increase in the total number of units of contribution which, however, failed fully to live up to the hopes of the Executive Committee in office between 1961 and 1964, the one to propose the modifications adopted.

It should be noted that a special and successful effort was made in 1966 to collect the outstanding contributions and to have the category of adherence raised in certain individual cases.

As on 31 December 1966, six countries still owed contributions to the Union (the units unpaid amounting to 20). On the other hand, one country had already paid its contribution (6 units) for 1967, and the account of another country shows a balance of $140·40 to its credit.

SUMMARIZED ACCOUNT OF RECEIPTS AND PAYMENTS

(in U.S. Dollars)

for the period 1964–1966

Receipts

	U.S. dollars	U.S. dollars
1. Subscriptions from Adhering Countries		

	Number of units due (*outstanding*)	*received*
(1963 and before)	16	13
1964	126	119
1965	152	146
1966	157	150
(1967)	0	7

	U.S. dollars	U.S. dollars
		85 341·91
2. Sales of publications (for details see page lxviii)		11 882·78
3. Subscriptions to IAU publications (for details see page xliv)		4 469·79
4. Interest on savings and other accounts		
Current accounts	319·54	
Saving accounts	8 513·35	
		8 832·89
5. Subvention from UNESCO, through ICSU		
1964	14 000·00	
1965	14 000·00	
1966	14 000·00	
		42 000·00
6. Contracts with UNESCO		
Grant towards the organization of Symposium no. 23	1 154·86	
Grant towards the organization of Symposium no. 25	4 499·46	
Exchange of Astronomers (Commission 38)	2 000·00	
Summer School (Organization) (Commission 46)	1 306·00	
Scholarship 1966 (Commission 46)	3 000·00	
		11 960·32
7. Other receipts		
ICSU grant for IUCSTR	3 564·40	
URSI participation in the organization of Symposium no. 20	200·00	
Various refunds		
– to Commission 38	3 072·04	
– of the IUCI grant	180·00	
– of the Telegram Bureau reserve fund	2 267·04	
		9 283·48
Total of effective receipts		$ 173 771·17
8. Excess of Payments over Receipts		19 700·84
		$ 193 472·01
Bank accounts as on 1 *January* 1964		
in bank (savings)	62 730·84	
in bank (current accounts)	46 798·81	
in petty cash	0·37	
		$ 109 530·02

SUMMARIZED ACCOUNT OF RECEIPTS AND PAYMENTS

(in U.S. Dollars)

for the period 1964–1966

Payments

	U.S. dollars	U.S. dollars
1. Administrative Office		30 231·60
2. Subscription to the ICSU		1 490·74
3. Expenses of Commissions		452·29
4. Specific projects of Commissions approved by the General Assembly (for details see page xlvii)		36 514·62
5. General Assemblies – Hamburg, August 1964		
(*a*) Members of Executive Committee (travel expenses) ..	1 714·80	
(*b*) Presidents of Commissions (travel expenses)	7 219·41	
(*c*) Young astronomers (travel expenses)	5 044·71	
(*d*) Miscellaneous – Prague, August 1967	1 582·54	
(*d*) Other expenses (Travel expenses of General Secretary to Prague)	133·26	
		15 694·72
6. Printing of publications by the IAU (for details see page xlvii)		56 712·00
7. Meetings of Executive Committee		
(*a*) Nice, September 1965	3 092·36	
(*b*) Prague, September 1966	2 130·49	
		5 222·85
8. Organization of IAU Symposia and Colloquia, and participation in other scientific meetings (for details see page xlvii)		20 014·30
9. Publication of Symposia (if the IAU is not the publisher) (for details see page xlviii)		13 088·36
10. Inter-Union Commissions, Committees on the ICSU, etc. (see separate list on page xlix)		7 473·11
11. Other expenses of representation		858·00
12. Specific projects approved by the Executive Committee (for details see page l)		350·00
13. Programmes under direct UNESCO contract		
(*a*) Commission 38, Exchange of Astronomers	2 000·00	
(*b*) Commission 46, Organization of Summer Schools ..	1 306·24	
(*c*) Commission 46, Scholarship	2 000·00	
		5 306·24
Total of effective expenses		$ 193 408·83
14. Exchange losses due to internal transfers		63·18
Total		$ 193 472·01
Excess of payments over receipts, carried down		19 700·84
Balances as on 31 December 1966		
(*a*) in banks (savings)	71 244·19	
(*b*) in banks (current accounts)	18 518·92	
(*c*) in petty cash	66·07	
		89 829·18
		$ 109 530·02

2. *Sales of publications.* Owing to the complexity of the accounts it is difficult to give separate statements for each volume sold. However, an attempt has been made in Appendix II to present a balance-sheet of all operations (receipts and payments) in connection with IAU publications, whether or not the Union was financially responsible.

3. *Subscriptions to IAU publications.* A new procedure was introduced in 1963: it consisted in requesting annual subscriptions from all institutions on the Distribution List. The difficulties connected with collecting these subscriptions and the administrative complexity of the operations led to the termination of this system in 1966. See Appendix II on Publications, page lxiii.

4. *Interest on savings accounts.* The interest on the savings accounts was retained in these accounts and has been incorporated into their respective balances.

In addition, the current account in French francs attracted interest, as shown in a separate account.

5. *Subvention from UNESCO.* This subvention is paid through the International Council of Scientific Unions. Detailed accounts of the expenditure from this subvention are submitted to UNESCO each year. The distribution of the subvention is left to the discretion of the Union, provided the individual allocations are within the categories approved by UNESCO.

6. *Contracts with UNESCO.* The Union negotiated a number of separate contracts with UNESCO, of which the details are as follows:

(*a*) Participation in the organization of Symposium no. 23, 'Astronomical Observations from Space Vehicles' (UNESCO/NS/1966/64; 135–5674)... 'The funds shall be used for meeting all or part of the cost of travel of participants'.

Grant	$1500
Expenditure (as appearing in the accounts)	$1154·86

(the balance was reimbursed to UNESCO).

(*b*) Participation in the organization of Symposium no. 25, 'Theory of Orbits in the Solar System and in Stellar Systems' (UNESCO/NS/1997/64; 135–5675).

Grant	$4500
Expenditure (as appearing in the accounts)	$4499·46 (amount actually received by IAU)
Reimbursed to UNESCO	$ 0

(*c*) Contribution to Exchange of Astronomers (Commission 38) (UNESCO/NS/2477/65; ED.SS. 0085)

Grant (1965–1966)	$2000

(Total grant spent. See headings 4 and 13 of Payments).

(*d*) Organization of meeting of experts to work out long-term programme of International Astronomical Summer Schools (Nice, 1965) (Commission 46) (UNESCO/NS/2468/65; EC.SC. 0085)

Grant	$2000
Expenditure	$1306

(UNESCO did not pay the balance).

(*e*) Organization of individual scholarships (Commission 46) (UNESCO/NC/3034/66; EC.DOC. 5555)

Grant	$4000	
Received in 1966	$3000	as appearing in the accounts
Spent in 1966	$2000	

The account will be closed in 1967, according to contract.

7. *Miscellaneous.* This heading includes:

(*a*) Subventions from various organizations.

(*i*) A subvention from ICSU towards the expenses of IUCSTR. The figure in the accounts is the balance of a receipt of $3500·00 and of a reimbursement of $64·40, as follows:

Receipts		*Payments*	
Grant	$3500·00	Travel expenses	$1360·63
Reimbursement by Professor Allen	64·40	Publication of the Xth IUCSTR Report	1194·20
		Administrative and secretariat expenses	562·55
	$3564·40		$3117·38

The amount of $64·40 reimbursed to the IAU, will be returned to ICSU in the course of 1967, in the same way as other unspent sums: the total balance to refund to ICSU is $447·02. IUCSTR having terminated in 1966, this amount cannot be used in the future.

(*ii*) Subvention from URSI towards the organization of Symposium no. 20.

URSI also contributed to the organization of IAU–URSI Symposium no. 31, but this contribution was paid direct to the organizers. The same applies to the IUTAM contribution to IAU–IUTAM Symposium no. 28.

(*b*) When the Telegram Bureau was transferred from Copenhagen to Cambridge (Massachusetts, U.S.A.), the active balance of the funds attributed to it in 1961 was reimbursed to the Union. The IAU subvention to IUCI, an Inter-Union Commission terminated in 1966, was partly reimbursed to the IAU. Some recipients of travel grants made by Commission 38 later returned part of the funds received, as their personal programmes had changed after receipt of the grants.

It should be emphasized that out of the $9283·48 under this heading only $3700 are actual receipts: the remainder corresponds to reimbursements and should be subtracted from both the actual receipts and payments. This amount represents a very small percentage of the budget (about 2%) and it is preferred not to complicate the accounts by presenting them in this way, though this would add to their accuracy.

8. *Excess of Expenditure over Receipts.* The excess has never yet been so high, in spite of the marked increase in receipts. This problem will be dealt with in more detail in paragraph D, page li.

Payments

The headings have been numbered to correspond to those of the Budget approved by the General Assembly in Hamburg in August 1964.

1. *Expenses of the Administrative Office.* These expenses include only the salaries of the Assistant Secretary and of one typist; social security payments; postage and telegraph charges; and purchase of stationery.

Although much smaller than they might otherwise be, thanks to the material aid of the Royal Greenwich Observatory, the Observatoire de Nice and the Czechoslovak Academy of Sciences (in the form of furniture, telephone, personnel, accommodation), the expenses exceed slightly the budgetary estimates, but not significantly to affect the budget.

2. *Subscription to the ICSU.* The Union makes an annual contribution towards the administrative expenses of ICSU of an amount equal to 2% of the income from subscriptions from Adhering Countries for the previous year.

Specific Projects of Commissions

Commission no.	Projects	Balance 1964	Allocated 1965–1967	Paid in $ 1964	Paid in $ 1965	Paid in $ 1966	Total $	Balance 1967 $ (*d*)
5	Bibliography of Astronomy		250·00	—	—	—	—	250·00
6	IAU Telegram Bureau		2 000·00	666·66	666·66	666·66	1 999·98	0·02
10	Cartes Synoptiques		2 646·00	882·00	882·00	882·00	2 646·00	—
10	Cartes Héliographiques		980·00	326·66	326·66	326·66	979·98	0·02
16	Centre de Documentation sur les Planètes (Meudon)		2 000·00	—	665·31	665·31	1 330·62	669·38
20	Minor Planet Center		1 500·00	—	—	1 000·00	1 000·00	500·00
23	Carte du Ciel (*a*)	4 365·91	2 000·00	1 476·54	—	3 467·22	4 943·76	1 422·15
27	English Version of Remarks to the Second Supplement of J.C.V.S. (*b*)		700·00	—	—	—	—	—
27	English Version of Remarks to the Second Catalogue of Stars suspected of variability (*b*)		700·00					
27	Southern Variable Stars		1 000·00	—	—	—	—	1 000·00
37	Card Catalogue of Clusters and Associations		1 500·00	—	—	—	—	1 500·00
38	Exchange of Astronomers (*c*)	8 006·47	—	—	—	—	—	—
	IAU grant 1964	7 500·00	—	—	—	—	—	—
	Reimbursed grants (1964–1966)	3 072·04	—	—	—	—	—	—
		18 578·51	20 000·00 (+2 000·00 UNESCO contract)	7 200·58	5 200·67	13 213·03	23 614·28 (+2 000·00 UNESCO contract)	14 964·23
				these sums include payments from the $2000 UNESCO contract				
	Total	22 944·42	35 276·00	10 552·44	6 891·30 (+ 850·00 UNESCO contract)	19 070·88 (+1 150·00 UNESCO contract)	36 514·62 (+2 000·00 UNESCO contract)	20 305·80

3. *Expenses of Commissions.* This amount is minimum. Grants were only made to Commissions 42 and 46 over the period 1964–1966. The respective documents will be available to the Finance Committee for examination.

4. *Specific Projects approved by the General Assembly.* The details of these projects are given in the Report of the Finance Committee to the twelfth General Assembly (*Transactions IAU*, XI B, for 1964; *Transactions IAU*, XII B, for 1965 and 1966).

Details of actual payments on the different projects are given in the following list (see table appended).

This list requires the following comments:

(*a*) The publication of the Carte du Ciel is subsidized by funds administrated by the President of Commission 23.

(*b*) The IAU allocations concerned ($1400) will actually not be spent: they are partly compensated by sums due to the Union from the Academy of Sciences of the U.S.S.R.

(*c*) $2000 is a grant attributed by UNESCO contract towards the Exchange of Astronomers. This amount has been subtracted from the 'total of payments' corresponding to this heading of the budget.

(*d*) The balance as on 31 December 1966 is made up of credits in favour of different projects. It is not certain whether all these credits will be spent. Some represent exchange differences, others cover projects not yet realized.

5. *General Assembly, Hamburg, August 1964.* Expenditure was less in 1964 than in 1961 owing to the favourable position of Hamburg, more central than that of Berkeley. Moreover, the generosity of the host country permitted us, as in the past, to limit incidental expenses to a minimum. The items under (*d*), Miscellaneous, correspond, in the main, to travel expenses of the Secretarial staff present at the General Assembly and to those of the Assistant General Secretary.

6. *Printing of publications, published at the expense of the IAU.* This is the weightiest heading of the budget, the one in which over-expenditure is most serious. The problem of publication is dealt with in more detail in a special Appendix to the present report (Appendix II).

7. *Meetings of the Executive Committee.* The expenses arising from the 1964 meeting in Hamburg are included in those for the General Assembly, heading 5, as above. Total expenditure is small owing to the generosity of the host countries and their institutions, and to the efforts of the members of the Executive Committee to combine these with other scientific meetings subsidized from other than IAU sources.

8. *Organization of Symposia and Colloquia.* The bulk of the payments consists of travel grants to participants selected by the Organizing Committee of each Symposium; organizational expenses were much smaller. Allocations were frequently not fully spent: individual grants were often returned to the Union too late to be re-allocated. The following table gives the payments in detail.

Symposium no.	Subject, Place Date	Payments made $
23	*Astronomical Observations from Space Vehicles*, Liège, August 1964	1155·05
24	*Spectral Classification of Multi-Colour Photometry*, Saltsjöbaden, August 1964	2403·89
25	*The Theory of Orbits in the Solar System and in Stellar Systems*, Thessaloniki, August 1964	5459·57
26	*Abundance Determination in Stellar Spectra*, Utrecht, August 1964	300·00

continued on next page

Symposium no.	Subject, Place, Date	Payments made $
27	*The Construction of Large Telescopes*, Tucson (Pasadena), April 1965	0
28	*Cosmical Gas Dynamics, Aerodynamical Phenomena in Stellar Atmospheres*, Nice, September 1965	1450·00
29	*Instability Phenomena in Galaxies*, Burakan, May 1966	1100·01
30	*Determination of Radial Velocities*, Toronto, June 1966	1500·00
31	*Radio Astronomy and the Galactic System*, Noordwijk, September 1966	1972·25
		$15 340·77

Owing to the generosity of the host organizations, the Union incurred no expenses in respect of the organization of Symposium no. 27. Symposia nos. 23 and 25 were financed almost completely from separate UNESCO contracts. In addition, the following amounts were spent on other meetings:

IAU Colloquium on *Variable Stars* (Bamberg, 1965)	$500·00
IAU Colloquium on *Late-Type Stars* (Trieste, 1966)	373·53
IAU Colloquium on *The Blanketing Effect* (Heidelberg, 1966)	500·00
IAU Colloquium on *Double Stars* (Uccle, 1966)	500·00
IUPHS Symposium on *The History of Astronomy* (Hamburg, 1964)	400·00
IUTAM–COSPAR–IAU Symposium on *Trajectories of Artificial Celestial Bodies* (Paris, 1965)	1000·00
COSPAR–IAU–URSI–UGGI Symposium on *Solar Terrestrial Physics* (Belgrade, 1966)	750·00
	$4 673·53

9. *Publication of the proceedings of Symposia not published at the expense of the IAU.* The IAU has decided to publish, at its own expense, all symposia from Symposium no. 24 onwards; for details see Appendix II on the Publications of the Union, page lxiii. Thus, the present paragraph deals only with earlier volumes:

Symposium no.	Title	Published by	Payment in $
9	Radio Astronomy	Stanford U. Press	80·28
14	The Moon	Academic Press	12·09
19	Site Testing	*Bulletin Astronomique*	1648·12
20	The Galaxy and the Magellanic Clouds	Australian Academy of Sciences	2240·00
21	The System of Astronomical Constants	*Bulletin Astronomique*	1005·83
22	Stellar and Solar Magnetic Fields	North Holland Pub. Co.	3000·00
23	Astronomical Observations from Space Vehicles	*Annales d'Astrophysique*	5102·04
			$13 088·36

As to Volume no. 9, the IAU has repurchased the stock; in the case of Volume no. 14 a single copy was concerned.

Symposia nos. 19, 21 and 23 were published free of cost to the Union in scientific journals; the Union purchased the copies it required for free distribution to participants, authors and institutions on the Distribution List, prior to the decision to limit this distribution. For more details see Appendix on Publications, on page lxiii. No appreciable revenue may be expected from future sales of these volumes.

10. *Inter-Union Commissions, ICSU Committees, etc.* The payments were made either as contributions to the expenses of the inter-union bodies concerned, or as refunds of travel expenses to IAU representatives. Expenditure was as follows:

Various contributions to the general expenses of IUCSTR (travel of IAU representatives, organizational expenses, printing of IUCSTR Reports 1964, 1965, 1966)	$3117·38
Representational expenses (travel expenses of IAU representatives) to COSPAR (1966)	1070·00
Representational expenses to CIG (for years preceding 1964)	286·14
Contributions to CIES (1964, 1965, 1966)	599·59
Contributions to IUCAF (1964, 1965, 1966)	1200·00
Contributions to IUCI (1964, 1965)	200·00
Contributions to IQSY (1963, 1964)	1000·00
	$7473·11

The following notes should be added:

IUCSTR is a special case, the IAU having been the parent Union. The ICSU subvention was administrated by the IAU, and the respective accounts are as follows for the three years covered by the present report:

Receipts			Payments	
ICSU grant	$3000	for year 1964	1964	$1500·53
	500	for year 1966	1965	772·80
Refund:	64·40	for year 1966	1966	844·05
				$3117·38

The existence of IUCI was short-lived. From the $200 paid by the IAU, $180 was repaid: see Receipts, paragraph 7.

Our debits to CIG (travel grants to delegates) for the year 1963 had to be paid in 1964. The same applies to the contribution to IQSY for 1963, paid in 1964; it should be noted that this contribution has not been claimed for 1965 and 1966, but it is still as a debit (that is $1500 to be paid in 1967, our engagement to IQSY being for a maximum of five years).

11. *Other representational expenses.* The budget provision permitted the President to participate in the General Assembly of the International Council of Scientific Unions in Bombay, in January 1966.

It should be noted that headings 10 and 11, which correspond to those of the budget, are easily confused: it is proposed henceforth to limit heading 10 to contributions, and to include travel expenses of IAU representatives to meetings of Inter-Union Commissions under heading 11.

12. *Specific projects approved by the Executive Committee.* A small number of projects were authorized by the Executive Committee after the General Assembly in Hamburg, as follows:

Commission 15. *Atlas of Representative Cometary Spectra* (Institut d'Astrophysique de Liège) $500
This sum has not yet been used.

Commission 29. *Information Bulletin of the Southern Hemisphere* (Dr Sahade) $1400
This grant is made up of $700 for 1965 and $700 for 1966; the Executive Committee approved Dr Sahade's proposal to carry over this credit to the years 1966–1967. However, the cheques sent to Dr Sahade had not been passed through the bank as on 31 December 1966. The amount actually paid was $350.

Commission 5. *Les Observatoires Astronomiques et les Astronomes* (Observatoire Royal de Belgique) $500
This is a grant towards a new edition, on cards, of the book by Mr Rigaux. The credit is still unspent.

To sum up, $2400 was granted, but $2050 is still to be paid in the course of 1967. If the actual progress of the projects requires a further deferment of payments, the General Assembly will have to reconsider the allocations made by the Executive Committee.

13. *Programmes under direct UNESCO contracts.* The UNESCO financed five projects by direct contracts (see Receipts, heading 6). Expenditure on Symposia 23 and 25 is included under heading 8. Details of payments in respect of the contracts are given in the reports submitted to UNESCO. They are available to the Finance Committee for examination.

14. *Exchange differences in internal transfers.* The item refers to a currency operation in 1966 when a considerable sum was transferred from the account in pounds sterling to two other accounts, that in French francs and that in U.S. dollars. The transfer was made subsequent to rumours in Great Britain indicating an imminent devaluation of the pound.

It should be noted that invoices in pounds had thus to be paid from our French account at a rate of exchange different from that of UNESCO. If the account in pounds had been maintained, the currency losses would have differed slightly from those actually incurred. Incorporated into the exchange differences are also the bank expenses incurred with the internal transfers (their separation was impossible).

Balance as on 31 December 1966. An examination of the bank balances reveals an excess of $19 700·84 in payments over receipts, and an increase of $8513·35 in the savings accounts.

The deficit is appreciable, in spite of a policy of rigid economies, especially as regards the Administrative Office. It is due, in the main (see below), to the high costs of printing and publishing IAU publications.

C. AUDIT OF THE ACCOUNTS 1964–1966

In compliance with the mission with which you kindly entrusted us, we present the result of our audit of the accounts of the International Astronomical Union for the period from 1 January 1964 to 31 December 1966.

We have examined, in particular, the following documents:

- the accounts of receipts and payments for the year 1964, as audited by Mr G. Chambraud, Chartered Accountant, Nice, Alpes-Maritimes;
- the accounts of receipts and payments for the year 1965, as audited by Mr W. E. Copus, F.C.A., Chartered Accountant, Mougins, Alpes-Maritimes;
- the account of receipts and payments for the year 1966;
- the accounts of receipts and payments and the state of the bank accounts for the period of three years 1964, 1965 and 1966.

We have also examined the six separate accounts of payments included in the financial report of the Executive Committee.

The accounts were compared by Messrs Chambraud (year 1964) and Copus (year 1965) and by ourselves (year 1966) with the account books, receipts and other documents produced.

We certify that the accounts are in full accord with the information and explanations we have required for the purpose of our audit.

The enclosed account of receipts and payments, summarizing the accounts for the years 1964, 1965 and 1966, is an accurate representation of the payments and receipts of the International Astronomical Union for the period from 1 January 1964 to 31 December 1966.

ROGER BACLE

Paris,
8 March 1967

H.E.C.—*Chartered Accountant Inscribed in the Tableau de l'Ordre of the Region of Paris*

(*Note*: The original statement by the Chartered Accountant is written in French, and signed)

D. CONSIDERATIONS AS TO THE BUDGET FOR 1967

In order to prepare the budget for the years 1968–1970, and an additional budget for 1967, the tables below summarize (without repeating the titles of the individual headings) the previous budgetary estimates and the differences as compared with the actual accounts.

Survey of the estimates for the period from 1965 to 1967

New estimates as to the budget for 1967 (as established in March 1967)

	Income		Expenditure
1.	35 000	1.	8 000
2.	12 000	2.	500
3.	0	3.	0
4.	3 000	4.	10 000
5.	14 000	5.	14 000
6.	(pro memoria)	6.	63 600
7.	(miscellaneous)	7.	0
		8.	4 500
		9.	0
		10.	2 000
		11.	0
		12.	2 000
		13.	(pro memoria, equal to heading 6 of income + 1000)
	64 000		105 600
8.	10 000		
	74 000		

New estimates as to the budget for 1965–1967 (as established in March 1967)

	Income	$E_1 - E_2$		Expenditure	$E_1 - E_2$
1.	95 309	− 691	1.	30 112	+ 112
2.	20 011	− 21 989	2.	1 510	− 490
3.	1 496	− 8 504	3.	277	− 1 722
4.	9 146	+ 1 146	4.	35 962	+ 686
5.	42 000	+ 0	5.	14 134	+ 134
6.	(6 306 + x)	(+ 6 306 + x)	6.	109 153	+ 79 153
7.	(6 083)	(+ 6 083)	7.	5 223	+ 223
8.	10 000	+ 10 000	8.	14 216	− 2 784
			9.	10 768	− 39 232
			10.	6 472	− 38
			11.	858	− 142
			12.	2 350	− 2 650
			13.	(6 306 + x)	(+ 6 309 + x)
			14.	56	+ 56
	190 351 + x			237 397 + x	

(All sums in U.S. dollars)

E_1: Estimates made in March 1967
E_2: Estimates made in August 1964

Notes

The estimates for 1967 have been made with due regard to the difficult financial situation of the Union, and impose on the budget rigid economies under certain headings. One employee was discharged in January 1967 and will not be replaced until the Administrative Office moves to Prague. This step could only be taken as a result of the help given by the staff of the Observatoire de Nice. It permitted a reduction of the estimated IAU expenditure on the Administrative Office to $8000. However, this expenditure would rise to about $20 000 to $25 000 a year if the Administrative Office were to be independent of extraneous financial assistance. And if it were desired to improve the working conditions of the Administrative Office, its staff would have to be increased by two or three members, and the budget raised to $35 000–$40 000.

The balance in favour of the Specific Projects is $20 306. It is suggested to economize on this balance by not spending for Commission 38 more than $4500, in addition to the grant from the UNESCO contract.

ICSU has granted the Union a loan of 10 000 U.S. dollars to be repaid by 1972.

The estimates as to publications (Income, headings 2 and 3, Expenditure, headings 6 and 9) result from calculations given in more detail in the Appendix II, on the IAU publication policy, pages lxiii-lxx.

No matter how well justified by the fact that the publications paid for in 1967 will only sell in 1968, these figures may seem disastrous. They differ widely from those estimated in the budget set up in August 1964: receipts are low, expenditure is high. Sales of IAU publications did not live up to expectation. Expenses are enhanced by the increase in the volume of IAU literature and the costs of the free distribution of the *Transactions*.

The new estimates for 1967 permit a re-evaluation of the estimates for the three-year period 1965–1967 and a comparison with the budget. Disregarding the UNESCO grants, not included in the approved budget, the following should be noted:

The budgetary estimates of 1964 as regards the expenses of the Administrative Office, the General Assembly, the Executive Committee and for the Specific Projects of Commissions, correspond almost exactly to the present estimates. Significant economies were achieved in the organization of Symposia (owing to help of national organizations), and on the projects approved by the Executive Committee. On the other hand, publications (headings 6 and 9) show an excess of $40 000 in expenditure over the estimates from 1964.

The budgetary estimates established in 1964 of income correspond almost exactly to the estimates of the contribution by Adhering Organizations. The surplus of $10 000 under heading 8 corresponds to the loan obtained from ICSU. On the other hand, it should be noted (headings 2 and 3) that, in comparison with the estimates, the Union failed to obtain $30 000 through sales of IAU publications. The miscellaneous receipts (heading 7) correspond mainly to reimbursed grants, made under headings 4 and 8 of Expenditure; thus, these headings include additional savings corresponding to this amount.

The details of the additional budget proposed for 1967 are given in the following table:

Additional Budget for 1967

	Income	Expenditure	
1. Actual amount (1962–63)	$96 184	$87 109	
2. Actual amount (1964)	57 420	61 675	
3. Total actual amount 1962–64 (1 + 2)	153 604	148 784	
4. Liquid assets by end 1964 (income minus expenditure)			$4 820
5. Budget estimate 1962–64 (estimate of February 1964)	143 384	150 219	

	Income	Expenditure	
6. Budgetary supplement authorized by General Assembly in August 1964 (calculated from comparison with income and expenditure of budget estimate)			6 500
7. Budget 1965–1967, as approved by the XIIth General Assembly	198 000	197 776	
8. Surplus, from 4 and 6, not used in 1964	11 320		
9. Budget authorized for 1965–1967 (UNESCO contracts excluded) (8 + 7)	209 320	197 776	
10. Total subventions from UNESCO under contract 1965–66	6 306	5 306	
11. Actual amount 1965–66 (UNESCO contracts included)	116 351	131 797	
12. Estimation 1967, without UNESCO contracts (estimation of March 1967)	74 000	105 600	
13. Estimation 1965–67 (11 + 12) (UNESCO contracts 1965–66 included)	190 351	237 397	
14. Budgetary supplement proposed to the XIIIth General Assembly (calculated from comparison of income and expenditure of the budget estimate)			50 000

Thus, the amount of $50 000 will be proposed to the Finance Committee, as an additional IAU budget for the financial year 1967. It need not be emphasized that proposals will have to be made to remedy this situation (to the detriment of the balance on the savings accounts) in the period 1968–1970, if the situation is not to become catastrophic.

E. CONCLUSION AND BUDGET 1968–1970

The estimated expenses, especially high due to the marked increase in the costs of publication, call for new sources of income.

Without entering into more detail it should be noted that the receipts are proportional, or almost so, to the number of Members (such a proportionality should be acceptable to the Adhering Organizations in computing a raise of their annual contribution to the Union) while the expenses increased by the free distribution to all Members of IAU publications (*Draft Reports*, *Transactions* Volumes XII B and C, *Information Bulletins*), are approximately proportional to the square of the number of Members. The amount published increases (like the number of recipients) in proportion to the number of Members.

Thus, an imbalance is inevitable, and it will become even more serious in the future. This, however, is implied in the logic of the Union's growth. One might reproach the present General Secretary, responsible for the accounts since the summer of 1964, with the fact that he did not realize this state of things early enough and thus could not inform the Executive Committee accordingly.

A first solution is to reduce the generosity of the Union. Members of the IAU, without paying any fee, have numerous advantages, in that they receive free of charge IAU publications, and use the services of IAU Commissions. It would be possible to apply the Statutes rigorously, to require that the specific projects of Commissions be profitable, to reduce the contribution to the organization of General Assemblies, etc.

However, it is desirable to minimize the effects of austerity: it is contrary to the spirit of the IAU not to assist the Special Projects of Commissions or to reduce the help given to young astronomers, etc.

Thus, the General Secretary will propose, on behalf of the Executive Committee, to moderate austerity by finding new sources of financial support. Fully aware of the sacrifices asked for from the Adhering Organizations, but certain that the very existence of the IAU is endangered, and that these organizations will not fail to understand the necessity of the measures to be taken, the Executive Committee proposes an increase in the unit of contribution to the maximum value allowed by the Statutes, that is from 600 to 900 gold francs.

If this proposal is approved by the General Assembly, it will be possible to formulate a reasonable budget and to redress the effect of the additional budget for 1967 by 1970.

If this proposal is rejected, it will be necessary for the IAU to face the consequences. It might become necessary to suppress IAU publications, to suppress the individual membership, etc.

Budgets for both alternatives will be proposed to the Finance Committee. Discussions may be based on the following budget proposed for 1968–1970, in which income has been estimated on the same bases as for 1967 and expenditure on publications has been taken from the report on the IAU publication policy (Appendix II); reductions on certain expenses (§ 4 for example) have been taken into account.

	Income		Expenditure	
1.	\$105 000 + x	1.	\$30 000	
2.	73 400	2.	1 500	
3.	0	3.	500	
4.	8 000	4.	20 000	
5.	42 000	5.	14 000	
6.	(x)	6.	99 000	
7.	?	7.	5 000	
		8.	20 000	
	\$228 400 + x	9.	0	
		10.	7,500	
		11.	0	
		12.	2 000	
		13.	(x)	
		14.	10 000	Re-imbursement to ICSU
		15.	50 000	Re-imbursement to savings accounts
			\$259 500	

It is imperative either to save 30 000 U.S. dollars or to raise additional funds of this order of magnitude.

Possible solutions:

(*a*) Increase the unit of contribution by the Adhering Organizations from 600 to 900 gold francs, without change in the category of adherence.
Additional income for three years about: \$52 000.

(*b*) Contribution of \$10 a year by individual Members (to be introduced).
Additional income for three years about: \$48 000.

(*c*) Suppression of the *Information Bulletin* and its replacement by a mimeographed Bulletin.
Economies of about: \$8 000.

(*d*) Suppression of publications (*Transactions* included, with the exception, however, of the *Information Bulletin*, *Draft Reports* 1967, and Symposium no. 32, which are to be prepared previous to the General Assembly in Prague).
Savings of about: \$30 000.

Other possibilities of achieving economy will have to be sought if the measures adopted by the General Assembly prove insufficient. It is accordingly proposed to the General Assembly that it authorize the Executive Committee to study possibilities of finding new ways of saving 50 000 U.S. dollars, and to introduce the necessary measures, if no definitive solution is otherwise adopted by the General Assembly itself.

APPENDICE II. LA POLITIQUE DES PUBLICATIONS DE L'UNION ASTRONOMIQUE INTERNATIONALE

(See English version on page lxiii)

I. TRANSACTIONS, SYMPOSIUMS, BULLETINS D'INFORMATION

Toute publication doit être éditée, imprimée, distribuée, et vendue. Ces quatre fonctions essentielles sont accompagnées de fonctions secondaires, par exemple la publicité liée à la distribution.

Les publications de l'UAI (voir Appendice III.C, page lxxii) sont de trois sortes. On exclura de cette analyse celles que l'UAI subventionne à fonds perdus, sans d'ailleurs s'occuper en rien des problèmes techniques de leur édition. Elles figurent le plus souvent parmi les 'Entreprises Spécifiques des Commissions' (au § 4 ou au § 12 du Budget, Dépenses, voir Appendice I, page xxv), ou encore au § 8 du Budget (Dépenses).

(*a*) *Transactions de l'Union*

(*i*) L'UAI a la responsabilité de l'édition, l'édition scientifique proprement dite étant du ressort des Présidents de Commissions, des responsables des Discussions, etc.

Le manuscrit préparé par l'UAI est imprimé par la firme britannique Willmer Brothers, dont les Membres de l'UAI ont pu apprécier la compétence technique remarquable. Les relations entre l'imprimeur et les auteurs (épreuves, commandes de tirages à part) sont entièrement assurées par les services de l'UAI responsables des Publications, et situés à l'Observatoire de Meudon.

La reliure, les couvertures, la publicité, la vente sont du ressort d'Academic Press, avec qui un contrat définit les relations de l'UAI.

Les *Agenda and Draft Reports*, publiés sans reliure, jouent le rôle d'épreuves pour le Volume A des *Transactions*.

Pour la onzième Assemblée Générale, les *Transactions* ont été publiées en 2 volumes, distribués gratuitement.

Pour la douzième Assemblée Générale, les *Transactions* ont été publiées en 3 volumes; seuls les Volumes B et C ont été distribués gratuitement aux Membres de l'Union, le Volume A ayant été constitué essentiellement par la réimpression des *Draft Reports*, et par les Conférences Solennelles; sans reliure, ces derniers volumes ont été distribués gratuitement aux Membres de l'UAI.

Il faut noter que le Volume B a été distribué gratuitement aux Membres anciens (ceux qui faisaient partie de l'Union lors de la douzième Assemblée Générale) cependant que le Volume C a été distribué gratuitement à tous les Membres de l'UAI, y compris ceux élus lors de la douzième Assemblée Générale.

Il est clair qu'au point de vue financier, les *Transactions* sont une lourde charge pour l'UAI, les ventes ne pouvant compenser que de façon infime les dépenses liées à cette considérable distribution gratuite.

(*ii*) Il est donc envisagé de réduire les charges imposées par la publication des *Transactions*, en supprimant la distribution aux Membres des volumes autre que le Volume B, qui seul, d'après les Statuts, doit leur être distribué gratuitement.

Cependant, le volume total imprimé augmente régulièrement proportionnellement au nombre de Membres,—et même plus vite. De plus le coût de la production a augmenté sensiblement. Il faut donc faire les bilans suivants:

Transactions	Pages	Dépenses	Ventes
Volumes XI A, B (y compris les *Draft Reports*)	1112	$21 930·17 (1961–64)	env. $6000* (1961–66)
Volumes XII A, B, C (y compris les *Draft Reports*)	1690	38 378·52 (1964–66) 4 599·63 (1967)	5 533·47* (1966) +4 000 ? (1967)
Volumes XIII A, B, C (y compris les *Draft Reports*)	2000?	15 000·00 (1967) 50 000·00 (1968–69)	0 (1967) 30 000 ? (1968–70)

* à quoi il faut ajouter environ $2500 en tout, provenant de la contribution forfaitaire versée par les Institutions de la Liste de Distribution (1964–66).

Le bilan, au 31 décembre 1966, dans le cadre de la période 1964–66, se traduit par un excès de dépenses de $28 000 environ.

(*b*) *Symposiums de l'Union*

(*i*) Symposiums jusqu'au No. 23 inclus.

L'édition scientifique est toujours du ressort des responsables de l'organisation du symposium, ou plus précisément de l'un d'entre eux, choisi avec l'accord du Comité Exécutif, comme éditeur scientifique. C'est, depuis 1961, le Secrétaire Général Adjoint qui a la responsabilité, pour l'UAI, de mener à bien les négociations préliminaires à ces désignations.

Jusqu'au Symposium No. 23 inclus, l'édition a été la responsabilité entière de diverses maisons d'éditions privées, ou de journaux scientifiques. Des contrats divers, tous différents, ont lié l'UAI à ces éditeurs.

Sans entrer dans les détails, disons que ce système a conduit à des irrégularités graves dans la date de distribution des symposiums, à des présentations et des formats hétérogènes, parfois même à des volumes n'indiquant aucunement la responsabilité de l'UAI dans la publication. Il s'en est suivi des réclamations nombreuses de la part des bibliothécaires de bibliothèques scientifiques, et une forte limitation des ventes, du fait qu'il ne s'agissait pas d'une véritable collection.

Au point de vue financier, le tableau ci-après montre que cette solution, quoique économique, conduisait néanmoins à une dépense notable pour chaque volume.

Symposiums No.	Contribution de l'UAI
12	$2750 (1961)
13	600 (1961)
14	4300 (1962)
15	100 (1962)
16	2600 (1963)
17	0 (1962)
18	1500 (1963)
19	1650 (1965)
20	2250 (1964)
21	1000 (1965)
22	3000 (1965)
23	5100 (1966)

Les ventes n'ont pratiquement rien rapporté. Dans la plupart des cas, l'UAI s'est bornée à acheter les exemplaires nécessaires aux distributions gratuites, sans plus. Dans d'autres cas, un petit bénéfice a été enregistré. Globalement, la vente des Volumes 12 à 23 (bien qu'il soit difficile de donner des chiffres détaillés) a rapporté à l'Union, pour la période 1961–1966, une somme d'environ $8000.

Il est clair que les abonnements forfaitaires souscrits (§ 3 des Recettes) par les institutions figurant sur la Liste de Distribution de l'UAI ne compensent pas ces dépenses.

Abonnements (pendant la période 1961–1966)

1963	$3271·61
1964, 1965	6367·70
1966	30·11
	$9669·42

Ceci correspond approximativement à 215 institutions cotisant en fait, alors que 264 institutions ont reçu les publications; la difficulté du recouvrement des versements forfaitaires est clairement exprimée par ces chiffres.

De plus, il reste des stocks de certains de ces symposiums, dont la vente peut encore rapporter quelques revenus à l'UAI. En toute objectivité, ces revenus doivent être considérés comme négligeables.

(*ii*) Pour ces diverses raisons, ce système a été abandonné, à partir du Symposium No. 24 inclus. Le Comité Exécutif espérait, d'une part, une régularité plus franche des publications, d'autre part, une vente plus efficace.

A partir du Symposium No. 24, le système adopté est devenu identique à celui adopté pour les *Transactions*. L'éditeur est devenu le Service des Publications de l'UAI, sis à Meudon. L'imprimeur a été la firme Willmer Brothers. La firme Academic Press a eu la responsabilité de la reliure, des couvertures, de la publicité, de la distribution, de la vente.

Certains des espoirs fondés sur ce nouveau système ont été en partie déçus. La lenteur de la publication n'a pas été améliorée (ni d'ailleurs aggravée!): seuls sont sortis des presses à ce jour les Symposiums No. 24, 25, 26.

Les ventes ont été meilleures qu'auparavant; mais le progrès n'est guère sensible: il est vrai qu'un stock important reste à vendre, et que les revenus à en attendre sont sans doute, cette fois, non négligeables, et ont été estimés compte tenu de la publication des volumes au cours de l'été 1966 et des stocks existants.

Enfin la qualité de la publication a été unanimement considérée comme très satisfaisante.

Pour ces symposiums, le détail des comptes est comme suit:

Symposium	Impression	Reliure, distribution	Vente en 1966	Produit des ventes futures
24	$5990·25	$738·55	$2411	env. $1000
25	7800·94	728·34	2434	env. 1000
26	6102·81	716·00	2336	env. 1000

Dépenses:	$22 076·89	Recettes:	$10 400
dont	14 112·92 en 1966	dont	8000 (env.) en 1967
Reste dû env.:	8000 à payer en 1967	et	2400 (env.) en 1968
			(y compris la vente des tirages à part).

Il faut noter que \$2000 environ viennent s'ajouter aux recettes, provenant de la contribution forfaitaire des Institutions de la Liste de Distribution.

Le bilan (arrondi) au 31 décembre 1966 est, en excès de dépenses: \$14 100; mais en 1967 et 1968, il y aura un excès de recettes de \$2400 environ. Le bilan total reste nettement déficitaire.

Les sommes indiquées pour les ventes en 1966 sont des valeurs exactes. Mais ces ventes affecteront seulement les comptes de 1967. Il n'est pas facile de comparer ces comptes aux comptes réels; dans certains cas en effet, les sommes provenant des ventes ont été déduites des factures de la firme Academic Press; les nombres figurant aux recettes (2) et aux dépenses (6) sont donc, en réalité l'un et l'autre sous-estimés, mais le bilan en est correct.

Pour les Symposiums 27 à 31, imprimés en 1966 et 1967, nous avons estimé à \$7000 en moyenne le coût du symposium, et à \$6000 environ les recettes par symposium.

(*iii*) Pour les Symposiums 32 et suivants, une politique différente a été envisagée par le Comité Exécutif. Il a semblé en effet, au vu des résultats de l'expérience commencée avec le Symposium 24, que la faible vente (se traduisant par une lourde insécurité pour l'UAI) était due à deux raisons: la lenteur de la publication, l'excès des distributions gratuites, souvent injustifiées. La suppression de la Liste de Distribution doit aider, en partie seulement, à résoudre ce problème.

La lenteur de la publication des Volumes 24, 25, 26 a été due en partie à un manque de coordination entre l'imprimeur et la maison Academic Press. A la suite de longues discussions, il a été envisagé de passer avec cette firme un contrat valable pour une durée de temps limité, et selon lequel elle prendrait à sa charge la responsabilité complète de l'édition, le service des publications de l'UAI n'intervenant plus que pour assurer les relations des auteurs avec Academic Press. L'UAI recevrait un pourcentage des ventes et ne débourserait en contrepartie que le prix des volumes distribués gratuitement. Les Membres de l'UAI continueraient à pouvoir se procurer à prix réduit les publications de l'Union Astronomique Internationale. Il est prévu, dans des conditions, que la publication de chaque volume de symposium pourrait ne coûter qu'une somme assez faible, de l'ordre de \$1000, d'ailleurs compensée, l'année suivant la publication, par des recettes assez faibles, mais du même ordre de grandeur.

En 1967, ces dépenses seront nulles. En 1968, pour les Symposiums 32 à 35, elles seront de l'ordre de \$4000. Des dépenses du même ordre doivent être prévues pour les années à venir (\$4000 en 1969).

Il est important ici de noter que les publications astronomiques se vendent mal, par rapport, par exemple, aux publications de chimie, ou même de mathématiques. Un éditeur professionnel arrive à une compensation entre différentes sciences; l'UAI, en tant qu'éditeur, est au contraire nécessairement perdante.

Précisons enfin que le présent rapport, et tout particulièrement le présent paragraphe, ne sont que provisoires. Des discussions sont menées actuellement avec Academic Press (et d'autres éditeurs), et un rapport plus actuel sera présenté devant l'Assemblée Générale, au mois d'août 1967.

(*c*) *Bulletin d'Information de l'UAI*

Cette publication, qui compte deux livraisons chaque année, (sans que ce soit une règle!), est éditée et distribuée par le Bureau Administratif; elle est imprimée à Nice par la maison Pierotti depuis 1965, et distribuée gratuitement. Il semble souhaitable de maintenir le système actuel. Le bilan est le suivant depuis 1965 (antérieurement à cette date, les dépenses correspondantes ont été incluses dans le § 1).

	1965	\$2146·89	(No. 13 et 14)
	1966	\$1226·72	(No. 15 et 16)
Prévisions:			
	1967	Paiement du No. 17 publié en 1966, plus deux livraisons environ	\$3000
	1968–1969	4 livraisons environ	\$6000

La somme relative à 1966 est faible, car à partir du No. 15, l'imprimeur n'a plus été chargé de l'expédition qui a été faite par le Bureau Administratif; les frais de poste ont donc été, en 1966, contrairement à ce qui s'est passé en 1965, inclus dans le § 1 des Dépenses.

Les prévisions tiennent compte de l'augmentation du nombre des Membres de l'UAI, mais supposent que le prix des travaux d'impression effectués en Tchécoslovaquie restera le même que celui des travaux effectués en France.

2. LES TIRAGES À PART

Il s'agit essentiellement de la fabrication des tirages à part compensés partiellement par leur vente (un certain nombre de tirages à part étaient édités pour les besoin du Bureau Administratif).

Les sommes dépensées dans ce but sont les suivantes:

		Dépenses	Récupération
Transactions	XII A	£147 10 0	£0 0 0
	XII B	1042 3 2	805 17 7
	XII C	64 18 0	0 0 0
Symposium	24	519 11 6	627 10 3
	25	327 18 3	453 9 3
	26	453 4 9	592 11 9
Struve Requiem		9 5 0	0 0 0
	Total	£2564 10 8	£2479 8 10

Les sommes se compensent à peu près. On notera que les sommes payées par les auteurs sont légèrement, d'après les tarifs utilisés (tarifs moyens), supérieures au prix coûtant.

Nous devons noter que le système de commande des tirages à part a été grandement compliqué par une distinction entre 'offprints' et 'reprints'. Malheureusement, cette distinction, de nature technique, ne permet pas toujours facilement de prévoir si tel article pourra être tiré à part comme 'offprint' ou comme 'reprint'; et les prix en sont fort différents! Le secrétariat de l'UAI a essayé d'établir deux échelles différentes de tarifs; ces tarifs n'ont pu être appliqués de façon stricte; d'où les différences entre les débours et les récupérations.

Par ailleurs, il est souvent clair que des distributions excessives de tirés à part, si elles permettent d'aider les bibliothèques de petits établissements, ou des chercheurs isolés, sont une lourde charge pour l'UAI, en limitant la vente des volumes, et en imposant un travail considérable au Bureau Administratif.

Pour cette raison, les principes suivants seront appliqués, par décision du Comité Exécutif, à partir du Symposium 27 et du Volume XIII des *Transactions*.

(*a*) Un seul type de tirages à part est utilisé.

Il s'agira de documents sans couverture, sans repagination, et sans surcharges imprimées (ce qu'en anglais on désigne par 'offprints' par opposition avec les 'reprints' qui sont repaginés et même recomposés parfois en partie.)

(*b*) Chaque Président de Commission recevra gratuitement 25 exemplaires du Rapport de sa Commission (Volume A) et du compte-rendu des séances (Volume B). Cette quantité devrait suffire, les Membres de l'UAI recevant tous ces textes par ailleurs.

(*c*) Les Présidents des Discussions Communes recevront gratuitement 50 exemplaires de l'ensemble de la Discussion qu'ils auront présidée. Tous les Membres de l'UAI reçoivent ces textes dans le Volume B des *Transactions.*

(*d*) Les auteurs des Conférences Solennelles recevront gratuitement 25 exemplaires de leur Conférence; tous les Membres de l'UAI recevront ces textes en accompagnement du Volume B (ou C dans le cas du Volume XII) des *Transactions.*

(*e*) Les auteurs de communications aux symposiums recevront autant d'exemplaires qu'ils le désireront de leurs communications; les prix leur seront communiqués avec le bulletin de commande qui leur sera adressé au moment de la correction des épreuves. Les discussions ayant eu lieu pendant les symposiums ne feront pas l'objet de tirages à part.

(*f*) Aucune autre des publications de l'UAI (et notamment pas le Volume XII C) ne sera l'objet d'une édition de tirages à part.

3. DISTRIBUTION ET VENTE DES PUBLICATIONS DE L'UAI

La politique traditionnelle de l'UAI était une politique de générosité. Quoi en effet de plus généreux que la distribution gratuite des *Draft Reports* et des *Transactions,* à tous les Membres de l'UAI, – sans aucune contrepartie?

De plus, toutes les publications de l'UAI sont vendues à ses Membres à prix réduits depuis des années.

Le *Bulletin d'Information,* depuis sa création, est aussi distribué gratuitement.

En 1962, il a paru souhaitable de conserver le bénéfice de cette générosité aux institutions dont la liste constitue la Liste de Distribution, mais en la tempérant du paiement forfaitaire d'un abonnement très avantageux aux souscripteurs. Cette Liste de Distribution, en 1966, comprenait 264 institutions; environ 20% de ces institutions ne versaient pas leurs contributions, malgré des rappels fréquents. De plus, il était clair que la liste elle-même n'était pas réellement à jour, ni représentative de l'état mondial des bibliothèques astronomiques. Enfin, le Bureau Administratif devait faire face à une administration très lourde, en raison même des carences de nombreuses institutions de la liste, et de son caractère toujours provisoire.

Enfin, les charges croissantes de l'Union (en conclusion du rapport principal, on trouvera des remarques générales sur ce thème) ont impressionné le Comité Exécutif. Malgré un évident serrement de coeur à l'idée de renoncer, même partiellement à une tradition de générosité, le Comité Exécutif a donc décidé une diminution massive des dépenses liées à la distribution, devenue trop libérale, des publications de l'Union, et ce de trois façons.

(*a*) *Liste de distribution*

Une première mesure est la suppression des avantages des institutions de la Liste de Distribution. Les institutions intéressées, prévenues par circulaire de cette décision regrettable mais nécessaire, se procureront les publications de l'UAI par les moyens commerciaux normaux, à partir du Symposium 27 et du Volume XIII des *Transactions.* Elles continueront à recevoir les *Bulletins d'Information* de l'UAI, mais ne souscriront aucun abonnement.

Cependant 21 institutions situées dans des pays en voie de développement ont été exclues de cette mesure. Une Liste de Distribution réduite a été établie. Les institutions de cette liste reçoivent désormais (sans avoir besoin de souscrire un abonnement), toutes les publications de l'UAI: il a semblé en effet au Comité Exécutif que les ressources des différents pays varient beaucoup d'un pays à l'autre, et que celles de cette liste réduite avaient des difficultés particulières.

Enfin, certaines institutions, appartenant à des pays où des contrôles stricts sont appliqués aux changes, ont des difficultés à souscrire des abonnements, comme à utiliser les moyens commerciaux internationaux normaux, sans pour autant manquer de ressources. Pour ces pays (et d'ailleurs tous les autres), le Comité Exécutif a proposé aux Comités Nationaux d'Astronomie respectifs le paiement d'une souscription forfaitaire (s'ajoutant à la cotisation à l'UAI) pour toutes les institutions de la liste de distribution établie par les comités nationaux intéressés. Ces institutions sont ajoutées à la Liste de Distribution réduite. Elles ne sont actuellement encore qu'au nombre de 7. Mais tous les pays intéressés par ce nouveau système n'ont pas encore pris position à cet égard.

Enfin la Liste de Distribution comprend les membres du Comité Exécutif, et les conseillers (11 personnes), l'UNESCO (10 exemplaires) et certaines institutions internationales comme l'ICSU. La Liste de Distribution contient ainsi environ 100 institutions.

Il est bon de garder en mémoire que le système de publication proposé, à partir du Symposium 32, qui consiste à faire appel aux services d'un éditeur professionnel, aura pour conséquence de contraindre l'UAI à acheter les volumes distribués aux institutions de la Liste de Distribution réduite.

Il a semblé au Comité Exécutif que la nouvelle politique de distribution, nettement plus économique que l'ancienne, d'une administration plus simple, était également plus juste que l'ancienne, dans la mesure où elle favorise (plus qu'auparavant d'ailleurs) les seules institutions ayant réellement des difficultés. Certaines injustices subsistent nécessairement (certaines institutions de pays développés peuvent n'avoir que de très faibles ressources), mais le Comité Exécutif a le sentiment que la nouvelle politique de distribution, d'une impérieuse nécessité, était aussi la moins mauvaise solution possible.

(*b*) *Membres de l'UAI*

L'article 24 du Règlement de l'UAI prévoit que:

'Le volume des *Transactions* de l'Union contenant les comptes rendus de l'Assemblée Générale est fourni gratuitement:

(*a*) à tous ceux qui, à l'époque de la réunion de l'Assemblée Générale, étaient Membres de l'Union, y compris les Membres cooptés au titre de l'article 2 du présent Règlement;

(*b*) aux institutions dont la liste est arrêtée par le Comité Exécutif;

(*c*) aux Organismes ou Comités Nationaux adhérents.'

Ceci veut dire que seul le Volume B des *Transactions*, sous leur forme actuelle, doit être distribué gratuitement aux Membres. Aussi le Comité Exécutif a-t-il dû, à regret, prendre la décision de limiter effectivement cette distribution. Le seul Volume XIII B sera certainement distribué gratuitement; rappelons que déjà le Volume XII A avait été vendu et exclu de la distribution gratuite aux Membres de l'Union Astronomique Internationale. Evidemment, les *Draft Reports* (qui essentiellement constituent les épreuves du Volume A) continueront à être distribués gratuitement; ceci est essentiel à la préparation de l'Assemblée Générale de l'UAI.

(*c*) *Participants aux Symposiums de l'UAI*

Enfin le Comité Exécutif a décidé de ne plus distribuer les volumes des symposiums aux participants, souvent très nombreux, des symposiums de l'UAI, mais seulement à un nombre très réduit d'entre eux, une dizaine environ. Cette décision est évidemment très regrettable: mais là encore, il s'agit d'une nécessité qui a paru impérieuse au Comité Exécutif.

4. PRÉVISIONS FINANCIÈRES RELATIVES AUX PUBLICATIONS

En rassemblant les chiffres ci-dessus, il faut prévoir pour les années 1967 et suivantes, les Dépenses et Recettes suivantes relatives aux publications.

Année		Budget des publications Recettes	Dépenses	Excès de Dépenses
1967	Transactions XII A, B, C	$4 000	$4 600	
	Draft Reports and Transactions XIII A	0	15 000	
	Symposiums 24, 25, 26	8 000	8 000	
	Symposiums 27 à 31	0	33 000	
	Bulletins d'Information	0	3 000	
		12 000	63 600	+$51 600
1968	Transactions XIII A, B, C	26 000	50 000	
	Symposiums 24, 25, 26	2 400	0	
	Symposiums 27 à 31	30 000	2 000	
	Symposiums 32 à 36	0	4 000	
	Bulletins d'Information	0	3 000	+ 600
1969	Transactions XIII A, B, C	2 000	0	
	4 Symposiums	4 000	4 000	
	Bulletins d'Information	0	3 000	+1 000
1970	Transactions XIII A, B, C	2 000	0	
	Draft Reports and Transactions XIV A	3 000	25 000	
	4 Symposiums	4 000	4 000	
	Bulletins d'Information	0	4 000	+24 000
		$73 400	$99 000	+$25 600

Il est clair que ce type d'estimations est hautement imprécis, et qu'elles peuvent être aussi bien trop optimistes que trop pessimistes.

Ces estimations ont servi de base au Budget qui sera proposé au Comité des Finances, lors de la treizième Assemblée Générale, et à l'Assemblée Générale elle-même (voir, ci-dessus pages xxxv et xxxviii).

NOTE AJOUTÉE AUX ÉPREUVES

Une réestimation de certaines des valeurs ci-dessus a été faite par le Secrétaire Général et le Secrétaire Général Adjoint avant la réunion de Prague et a servi de base aux discussions du Comité des Finances. Les nouvelles estimations sont incorporées dans le rapport du Comité des Finances, volume XIII B.

APPENDIX II. PUBLICATION POLICY

(Voir texte français, page lv)

1. TRANSACTIONS, SYMPOSIA, INFORMATION BULLETINS

All IAU publications have to be edited, printed, distributed, and sold. These four principal operations are accompanied by secondary ones, such as publicity.

The IAU publications may be divided into three groups, excluding those subsidized by the Union on a non-profit basis along with the technical problems to which they lead. Payments in respect of the latter are normally included among the 'Specific Projects of Commissions' (headings 4 and 12 of the Budget and Accounts), or in heading 8.

(a) *Transactions of the IAU*

(*i*) The IAU is responsible for the editing of its publications, the scientific editing in the proper sense of the word being the province of the Presidents of Commissions, organizers of Discussions, etc.

The manuscript, sub-edited by the IAU, is printed by the British printers Willmer Brothers, whose outstanding achievements are generally appreciated by Members. The contact between printer and authors (proofs, orders for reprints) is maintained through the IAU Publication Service in Meudon.

Binding, covers, publicity and sales are looked after by Academic Press, with whom the Union has a contract.

The *Agenda and Draft Reports*, published with paper covers, are simultaneously the page proofs of Volume A of the *Transactions.*

The *Transactions* of the eleventh General Assembly were published in two volumes, distributed free of charge to Members of the Union.

The *Transactions* of the twelfth General Assembly were published in three volumes, of which Volumes B and C were distributed free of charge to Members of the Union. Volume A included the re-imposed Draft Reports and the Invited Discourses of which reprints, unbound and without covers, were also distributed to Members free of charge.

It should be noted that Volume B was distributed free only to those Members who belonged to the Union at the time of the twelfth General Assembly, while Volume C was distributed also to new Members, that is those admitted at that General Assembly.

From the economic standpoint, the *Transactions* are obviously a heavy burden on IAU finances, sales making up for only an insignificant part of the large-scale courtesy distribution.

(*ii*) Thus, steps have already been taken to reduce the expenses of the publication of the *Transactions* by discontinuing the free distribution of volumes other than Volume B, the only one required by the Statutes for such distribution.

It should be emphasized that the amount of IAU printed matter increases in proportion to the number of Members, if not faster, and that the rise in production costs is considerable. This gives the following balance:

Transactions	Pages	Cost (in $)	Sales (in $)
Volumes XI A, B (*Draft Reports* included)	1112	21 930·17 (1961–64)	Approx. 6 000* (1961–66)
Volumes XII A, B, C (*Draft Reports* included)	1690	38 378·52 (1964–66) 4 599·63 (1967)	5 533·47*(1966) +4 000 (1967) ?
Volumes XIII A, B, C (*Draft Reports* included)	2000?	15 000·00 (1967) ? 50 000·00 (1968–69) ?	0 30 000 (1968–70) ?

* to which one has to add about $2500, coming from the forfaitary contribution paid by the Institutions on the Distribution List (1964–66).

The balance on 31 December 1966 for the period from 1964 to 1966 shows an excess of expenditures of about $28 000.

(*b*) *IAU Symposia*

(*i*) Symposia up to and including no. 23.

Scientific editing is the responsibility of the organizers of the particular symposium or, more precisely, of one of them, selected with the approval of the Executive Committee as Scientific Editor. Since 1961, the arrangements for the designation of Scientific Editors have been made by the Assistant General Secretary.

Up to and including Symposium no. 23, publication was the exclusive responsibility of different private publishers or scientific journals. The contracts of the Union with such publishers varied widely in scope.

Without going into more details, it should be said that this system led to serious trouble: the distribution dates of symposium volumes were irregular, their presentation and format were non-uniform, IAU prerogatives were not mentioned on them, etc. Complaints by librarians and scientific libraries were numerous, and sales diminished as the symposium volumes formed no uniform series in the proper sense of the word.

The following table shows that, no matter how economical, this system resulted in considerable expenses for each volume.

Symposium no.	IAU contribution
12	$2750 (1961)
13	600 (1961)
14	4300 (1962)
15	100 (1962)
16	2600 (1963)
17	0 (1962)
18	1500 (1963)
19	1650 (1965)
20	2250 (1964)
21	1000 (1965)
22	3000 (1965)
23	5100 (1966)

Sales brought virtually no financial return to the IAU. In most cases they were limited to the purchase made by the Union for free distribution, benefits going to the publisher. In other cases, the records show some very small benefit. Altogether, the income from the sales (although difficult to estimate in detail) brought to the IAU, for the period 1961–66, about $8000.

Subscriptions (heading 3 of Receipts) by institutions on the Distribution List were obviously insufficient to balance expenditure.

Subscriptions from 1961 to 1966

1963	$3271·61
1964, 1965	6367·70
1966	30·11
	$9669·42

This corresponds to about 215 subscribers, while publications were distributed to 264 institutions. The figures are illustrative of the difficulty of collecting outstanding subscriptions.

A number of symposium volumes still in stock may represent some future income to the Union, but probably quite insignificant.

(*ii*) The above considerations called for suppression of the existing system from and including Symposium no. 24. The Executive Committee had hoped for more regularity in production and more efficient sales.

The system adopted for and from Symposium no. 24 onwards is the same as for the *Transactions*. Editing has become an IAU service, and is done at Meudon. The volumes have been printed by Willmer Brothers. Academic Press has been responsible for binding, covers, publicity, distribution and sales.

Many a hope based on this new system was frustrated. Publication was not speeded up (nor slackened, either): so far only Symposia nos. 24, 25 and 26 have come from the press.

Sales were better than previously, but the improvement is not very marked. On the other hand, there still remains a stock of volumes to be sold, and the income may be appreciable. It has been estimated from the sales effected during the summer of 1966.

The quality of the publications has unanimously been considered very satisfactory.

For these Symposia, the following details are given:

Symposium no.	Printing $	Binding, Distribution $	Sales in 1966 $	Future sales $
24	5990·25	738·55	2411	About 1000
25	7800·94	728·34	2434	About 1000
26	6102·81	716·00	2336	About 1000

Payments: 22 076·89
out of which 14 112·92 paid in 1966
and about 8 000·00 to be paid in 1967

Receipts: 10 400
out of which about 8 000 in 1967
and 2 400 in 1968
(sales of reprints included)

It should be noted that about $2000 must be added to the product of sales (1964–65) coming from the forfaitary contribution of the Institutions on the Distribution List.

The balance (rounded off) as on 31 December 1966 shows an excess of $14 100 over revenue. However, it is expected that the years 1967 and 1968 will bring an excess of $2400 over expenditure. Nevertheless, the overall balance shows a large deficit.

The figures given for the sales in 1966 are exact. But they will only be credited to the accounts in 1967.

It is not easy to compare these estimates with the actual accounts, the amounts resulting from sales having, in certain cases, been subtracted from the invoices of Academic Press; thus the figures under heading 2 of Receipts and under heading 8 of Payments are actually underestimates, but the balance is correct.

The estimates for Symposiums volume 27 to 31, printed in 1966 or 1967, are about $7000 for production and $6000 as revenue, as an average for each volume.

(*iii*) As to Symposium volumes from and including no. 32 onwards, the Executive Committee is examining the possibility of introducing a new policy.

Experience with Symposium no. 24 showed that sales are insufficient (adding to the insecurity of the IAU) owing to two facts: the long publication delay and the too liberal courtesy distribution, often unjustified. The abolition of the Distribution List may be a partial solution of the problem.

The long delay in the publication of Symposia nos. 24, 25, and 26 was due, in part, to insufficient coordination between printers and publishers. After lengthy discussions it has been considered to negotiate a contract of limited duration with Academic Press by which it would be fully responsible for publication, so that the Union's only responsibility would consist in arranging contact between authors and publishers. The Union would receive part of the revenue from sales, but would pay for the volumes distributed free. Members would be entitled to purchase IAU publications at reduced prices. It is expected that publication of each Symposium volume would not cost the IAU more than $1000 in such conditions, compensated, the year following publication, by an income of the same order of magnitude, though not very large.

In 1967, this expenditure will be zero. In 1968, for Symposia nos. 32 to 35, it will be of the order of $4000. Payments of the same order have to be expected for the years to come ($4000 in 1969).

In this connection it should be noted that astronomical publications have a thin market in comparison with those of other branches, such as chemistry or even mathematics. A professional publisher may balance his losses in one branch by gains in another, but the Union may only lose as its own publisher.

The present report, and especially the above paragraph, are only tentative. Discussions now under way with Academic Press and other publishers will permit the Executive Committee to prepare a more definitive report before the General Assembly in August 1967.

(*c*) *IAU Information Bulletin*

This publication, which appears twice a year (not, however, compulsorily), is edited and distributed by the Administrative Office. It has been printed by Messrs. Pierotti, Nice, since 1965. It is distributed free of charge. The present system is satisfactory and should be maintained. The balance for 1965 and 1966 is as follows (prior to 1965, the corresponding expenditure was included under heading 1):

1965	$2146·89 (nos. 13 and 14)
1966	1226·72 (nos. 15 and 16)

Estimates:

1967	no. 17 printed in 1966 and about two further editions	$3000
1968–1969	about 4 editions	$6000

The total paid in 1966 is reasonably low, the distribution having been made by the Administrative Office from no. 15 onwards, not by the printer. However, postage for 1966 has been included under heading 1 of Payments, not heading 8 as in 1965.

The estimates have taken into account the anticipated increase in the number of IAU Members, and assume that the costs of printing in Czechoslovakia will be the same as in France.

2. OFFPRINTS AND REPRINTS

The production of offprints and reprints is partly compensated by their sale (a number of offprints is ordered for the use of the Administrative Office).

The balance in this respect is as follows:

	Payments	Receipts
Transactions XII A	£147 10 0	£0 0 0
XII B	1042 3 2	805 17 7
XII C	64 18 0	0 0 0
Symposium no. 24	519 11 6	627 10 3
25	327 18 3	453 9 3
26	453 4 9	592 11 9
Struve Requiem	9 5 0	0 0 0
Total	£2564 10 8	£2479 8 10

Payments and receipts are virtually in balance. It is seen that the prices, as given in the price lists and paid by the authors, are, on the average, slightly above production costs.

The system of ordering and selling prints has been much complicated by the distinction between 'offprints' and 'reprints'. Unfortunately, this distinction, of a purely technical character, does not always indicate whether a certain paper is better printed as an 'offprint' or a 'reprint'; and the price difference is appreciable. The Secretariat undertook to establish two different price lists; these lists, however, could not be applied strictly, and hence the difference between outlay and revenue.

It is certain that excessive distribution of offprints, though welcomed by minor libraries and isolated scientists, is a heavy burden to the Union. Moreover, it reduces the volume of sales and gives much trouble to the Administrative Office.

This prompted the Executive Committee to adopt the following principles for and from Symposium 27 and *Transactions* XIII onwards:

(*a*) Offprints are standardized.

Henceforward only offprints will be used, with no cover, without re-pagination and without added material (in contradistinction to reprints, which are re-paginated and sometimes partly re-composed).

(*b*) Each Commission President will obtain, without charge, 25 copies of the Report of his Commission (Volume A) and of the proceedings of its meetings (Volume B). This number may be sufficient, the Members of the Union receiving all this material, either in advance or in Volume B.

(*c*) The Presidents of Joint Discussions will obtain free of charge 50 copies of the complete Discussion over which they presided. All IAU Members will receive these texts in Volume B of the *Transactions*.

(*d*) The authors of the Invited Discourses will obtain free of charge 25 copies of their discourses. All IAU Members will receive these texts simultaneously with Volume B (or C in the case of *Transactions* XII) of the *Transactions*.

(*e*) The authors of contributions to symposia may obtain any desired number of copies of their contributions: the respective prices will be marked on the order sheet they will receive with the proofs. No offprints will be made of the discussions held during symposia.

(*f*) IAU publications other than those mentioned above, will not be available as offprints (especially not Volume C of *Transactions*).

3. DISTRIBUTION AND SALES OF IAU PUBLICATIONS

The traditional policy of the IAU has been that of generosity. For, actually, what is more generous than to distribute the *Draft Reports* and *Transactions* to all Members without reciprocity?

Moreover, all IAU publications have been sold to Members at reduced prices for years. Ever since it started, the *Information Bulletin* has also been distributed free of charge.

In 1962 it appeared reasonable to maintain the advantage of this generosity as regards the institutions which make up the IAU Distribution List, but to moderate it by introducing a subscription system, very favourable for the subscribers. The Distribution List included 264 subscribers in 1966, of which about 20% failed to pay their fees, in spite of numerous reminders. In addition, it became evident that the list was not up to date and did not represent that state of astronomical libraries the world over. And, eventually, the Administrative Office was faced with difficult administrative problems, for many institutions on the list did not pay, and the list itself was changing continuously.

The ever-increasing responsibilities of the Union (for general notes, see end of the main report) led the Executive Committee to reconsider the position. Although completely reluctant to renounce the tradition of generosity, the Executive Committee resolved to reduce the expenditure incurred through publication in three ways.

(*a*) *Distribution List*

The first step consisted in abolishing the advantages of the institutions on the Distribution List. The institutions concerned, advised of this regrettable but necessary step by circular letters, will have to purchase IAU publications through normal commercial channels, as from Symposium 27 and *Transactions* XIII onwards; they will continue receiving the *IAU Information Bulletin* free of charge.

However, 21 institutions in developing countries were exempt from this measure. A new, curtailed Distribution List was prepared. The institutions of this list will obtain all IAU publications without any subscription fee: the Executive Committee is fully aware that resources vary widely between countries, and that the institutions on the curtailed list have especially great difficulties.

Other institutions, that is those in countries with strict currencies, have difficulties in subscribing to controlled international publications, though they do not lack funds. For such countries (not excluding others), the Executive Committee has proposed that the respective National Committees of Astronomy should contribute to the distribution of IAU publications to their institutions by adhering to the Union in a slightly increased category (*b*). So far only seven such institutions have responded favourably, the other countries having not yet replied and notified their final decision.

The curtailed Distribution List includes also the members and advisors of the Executive Committee (11), UNESCO (10 copies) and certain international organizations, such as ICSU. The curtailed Distribution List thus contains about a hundred addresses.

It should be realized that the publication system proposed as from Symposium 32 onwards (i.e. to use the services of a professional publisher) implies that the Union will have to purchase the volumes distributed to the institutions on the curtailed Distribution List.

The Executive Committee felt that the new distribution policy, much more economical and simpler from the administrative standpoint, is also much fairer than the previous system, as it only favours those institutions facing real difficulties. A certain injustice cannot be avoided (some institutions of developed countries may be very low in funds), but the Executive Committee feels that the new distribution policy, imperative indeed, is also the best solution.

(*b*) *IAU Members*

Article 24 of the By-laws reads as follows:

'The volume of the *Transactions* of the Union containing the Proceedings of the General Assembly is distributed free of cost to:

(*a*) all those who, at the time of the meeting of the General Assembly, were Members of the Union including those co-opted by virtue of article 2 of the present By-laws;

(*b*) institutions on a list approved by the Executive Committee;

(*c*) Adhering Organizations or National Committees.'

Hence it follows that only Volume B of the *Transactions*, in their present form, has to be distributed free to Members. The Executive Committee has unfortunately to agree to a reduction of free distribution as much as possible. Only Volume XIII B will be distributed free of charge; it should be remembered that the first volume distributed through normal commercial channels, and thus exempt from free distribution, was Volume XII A. The *Draft Reports*, which actually are the page proofs of Volume A will continue to be distributed free of charge; the preparation for the General Assembly so requires it.

(*c*) *Participants in IAU Symposia*

The Executive Committee has also agreed to discontinue free distribution of Symposium volumes to participants, often very numerous, in IAU Symposia, and to present courtesy copies only to a limited number (about 10) of them. This decision is very regrettable, but it is a step the Executive Committee felt imperative.

4. FINANCIAL ESTIMATES IN RESPECT OF PUBLICATIONS

An analysis of the figures given above permits the following estimates to be made of receipts from and payments for IAU publications for 1967 and the following years:

Continued overleaf.

		Receipts $	Payments $	Balance $
1967	Transactions XII A, B, C	4 000	4 600	
	Draft Reports and Transactions XIII A	0	15 000	
	Symposia 24, 25, 26	8 000	8 000	
	Symposia 27 to 31	0	33 000	
	Information Bulletins	0	3 000	
		12 000	63 600	+51 600
1968	Transactions XIII A, B, C	26 000	50 000	
	Symposia 24, 25, 26	2 400	0	
	Symposia 27 to 31	30 000	2 000	
	Symposia 32 to 36	0	4 000	
	Information Bulletins	0	3 000	+ 600
1969	Transactions XIII A, B, C	2 000	0	
	4 Symposia	4 000	4 000	
	Information Bulletins	0	3 000	+1 000
1970	Transactions XIII A, B, C	2 000	0	
	Draft Reports and Transactions XIV A	3 000	25 000	
	4 Symposia	4 000	4 000	
	Information Bulletins	0	4 000	+24 000
		$73 400	$99 000	+$25 600

The figures are commented upon in more detail in paragraph (*d*) of the Financial Report, Appendix I. Naturally, such estimates are rather inaccurate and may be either over-optimistic or too pessimistic. They have been taken as basis for the Budget which will be submitted to the Finance Committee during the thirteenth General Assembly and to the General Assembly itself (see above, pages li and liii).

NOTE ADDED IN PROOF

A re-estimation of some of the above values has been made by the General Secretary and the Assistant General Secretary before the meeting in Prague and has been used as a basis for the discussions of the Finance Committee. These new estimations are incorporated in the report of the Finance Committee, volume XIII B.

APPENDIX III

A. LIST OF COUNTRIES ADHERING TO THE UNION

The following is a list of the 43 countries that adhere to the Union in December 1966, giving also the year of adherence, the approximate number of Members, and the Adhering Organizations.

Country	Year	Members	Adhering Organizations
Argentina	1927	14	Comité Nacional de Astronomía, Buenos Aires.
Australia	1939	39	Australian Academy of Sciences, Canberra.
Austria	1955	14	Österreichische Akademie der Wissenschaften, Wien.
Belgium	1920	34	Académie Royale de Belgique, Bruxelles.
Brazil	1961	8	Conselho Nacional de Pesquisas, Rio de Janeiro.
Bulgaria	1957	6	Académie des Sciences Bulgare, Sofia.
Canada	1920	40	Department of Energy, Mines and Resources, Ottawa.
Chile	1947	6	Universidad de Chile, Santiago de Chile.
Czechoslovakia ..	1922	25	Academy of Sciences, Praha.
Denmark	1922	13	Det Kongelige Danske Videnskabernes Selskab, København.
Finland	1948	8	Académie des Sciences et des Lettres, Helsinki.
France	1920	123	Académie des Sciences, Paris.
Germany (GDR)	1951	21	Deutsche Akademie der Wissenschaftenzu Berlin.
Germany (GFR)	1951	100	Rat Westdeutscher Sternwartenzu München.
Greece	1920	11	Academy of Athens, Athens.
Hungary	1947	8	Hungarian Academy of Sciences, Budapest.
India	1946	17	Ministry of Education, Government of India, New Delhi.
Ireland	1947	3	Dublin Institute for Advanced Studies, Dublin.
Israel	1954	2	The Israel Academy of Sciences and Humanities, Jerusalem.
Italy	1921	43	Consiglio Nazionale delle Ricerche, Roma.
Japan	1920	47	Science Council of Japan, Tokyo.
Korea (PDR) ..	1961	3	Academy of Sciences, Phyongyang.
Lebanese (Republic)	1954	1	National Lebanese Committee for Astronomy, Ksara.
Mexico	1921	9	Universidad Nacional de Mexico, Mexico, D.F.
Netherlands ..	1922	47	Koninklijke Nederlandse Academie van Wetenschappen, Amsterdam.
New Zealand ..	1964	2	Royal Society of New Zealand, Wellington.
Norway	1922	10	Det Norske Videnskaps-Akademi, Oslo.
Poland	1922	33	Polska Akademia Nauk, Warszawa.
Portugal	1924	12	Secção Portuguesa das Uniões Internacionais, Astronomica, Geodésica e Geofísica, Lisboa.
Roumania ..	1928	8	Conseil Scientifique de l'Observatoire de Bucarest, Bucarest.
South Africa ..	1938	16	Council for Scientific and Industrial Research, Pretoria.
Spain	1922	14	Comision Nacional de Astronomia, Madrid.
Sweden	1925	26	Kungl. Vetenskaps-Akademien, Stockholm.
Switzerland ..	1923	14	Schweizerische Naturforschende Gesellschaft, Basel.
Taiwan (Rep. of China)	1959	7	Academia Sinica, Nankang, Taipei.
Turkey	1961	10	Türk Astronomi Dernegi, Istanbul.
United Arab Republic	1925	6	Astronomical Center at Helwan, Egyptian Region.
United Kingdom	1920	133	The Royal Society, London.
U.S.A.	1920	396	The National Academy of Sciences, Washington, D.C.

Country	Year	Members	Adhering Organizations
U.S.S.R.	1920	207	Academy of Sciences of the U.S.S.R., Moscow.
Vatican City State	1932	6	Pontificia Accademia delle Science, Città del Vaticano.
Venezuela ..	1953	1	Academia de Ciencias Fisicas, Matematicas y Naturales, Caracas.
Yugoslavia ..	1935	12	Conseil Fédéral pour la Coordination de la Recherche Scientifique, Belgrade.

B. LIST OF DECEASED MEMBERS

The Executive Committee regrets to record the loss by death of the following 58 Members of the Union (at the date of 1967, January 1st).

H. L. Alden	3 February 1964
Sir Edward Appleton	21 April 1965
F. Baldet	8 November 1964
D. Barbier	1 April 1965
A. Bečvař	10 January 1965
D. Brouwer	31 January 1966
L. Carnera	30 July 1962
G. Cassinis	13 January 1964
M. S. Eigenson	15 August 1962
F. Fernandez de la Puente	4 January 1965
E. Finlay-Freundlich	24 July 1964
J. Génard	6 March 1963
J. Gadomski	2 January 1966
A. Gütler	19 March 1964
C. B. Haselgrove	27 May 1964
T. Hanala	10 November 1963
M. B. B. Heath	21 November 1963
W. W. Heinrich	30 May 1965
D. Hendrix	26 December 1961
A. R. Hogg	31 March 1966
I. G. Izsak	21 April 1965
H. S. Jelstrup	6 April 1964
F. Kepinski	8 April 1966
E. A. Kreiken	16 August 1964
D. K. Kulikov	11 June 1964
G. Lemaître	19 June 1966
B. Lindblad	25 June 1965
J. C. Lindsay	26 September 1965
D. B. McLaughlin	8 December 1965

J. E. Mack	15 April 1966
D. D. Maksutov	12 August 1964
E. L. Martin	9 August 1966
W. P. Meggers	19 November 1966
S. K. Mitra	27 February 1967
K. Moder	13 November 1965
J. J. Nassau	11 May 1965
N. Nechvile	5 July 1964
B. A. Orlov	16 September 1963
B. M. Peek	5 May 1965
G. Pels	2 November 1966
J. G. Perel	28 November 1964
R. M. Petrie	8 April 1966
V. F. Proskurin	21 May 1964
J. Racah	28 August 1965
F. Rigaux	21 September 1962
N. Robinson	7 September 1964
A.-M. Rozis-Saulgeot	1965
E. Schoenberg	23 January 1965
F. H. Seares	20 July 1964
V. V. Sharonov	27 November 1964
H. Siedentopf	28 November 1963
E. C. Slipher	7 August 1964
J. Stebbins	16 March 1966
Y. Tsukamoto	20 December 1965
L. Vegard	21 December 1963
J. H. Waddell	1966
A. A. Weiss	30 December 1964
F. Wünschmann	17 January 1966

C. LIST OF IAU PUBLICATIONS

Transactions of the IAU, Volume XII A, pp. 809, price 168s. (Members 104s.)	Editor: J.-C. Pecker Published by Academic Press, Inc. (London) Ltd.
Transactions of the IAU, Volume XII B, pp. 672, price 140s. (Members 100s.)	Edited and published as Volume XII A.
Transactions of the IAU, Volume XII C, pp. 207, price 60s. (Members 40s.)	Edited and published as Volume XII A.

IAU Symposium Volumes

No. 19. *Le Choix des Sites d'Observatoires* (Site Testing), pp. 386, price 40 Frs (Members 24 Frs)	Editors: J. Rösch, G. Courtès, J. Dommanget Reprinted from *Bulletin Astronomique* (Gauthier-Villars, Paris).
No. 20. Joint IAU-URSI Symposium on *The Galaxy and the Magellanic Clouds*, pp. 404, price 105s. (Members 65s.)	Editors: F. J. Kerr and A. W. Rodgers Published by the Australian Academy of Sciences.
No. 21. *Le Système des Constantes Astronomiques* (The System of Astronomical Constants), pp. 330, price 22 Frs (Members 14 Frs)	Editor: J. Kovalevsky Reprinted from *Bulletin Astronomique* (Gauthier-Villars, Paris)
No. 22. *Stellar and Solar Magnetic Fields*, pp. 460, price $15·75 (Members $10·35)	Editor: R. Lüst Published by North-Holland Publishing Company, Amsterdam.
No. 23. Joint IAU-COSPAR Symposium on *Astronomical Observations from Space Vehicles*, pp. 370, price 50 Frs (Members 30 Frs)	Editor: J.-L. Steinberg Reprinted from *Annales d'Astrophysique*.
No. 24. *Spectral Classification and Multicolour Photometry*, pp. 383, price 120s. (Members 80s.)	Editors: K. Lodén, L. O. Lodén, U. Sinnerstad Published by Academic Press, Inc. (London) Ltd.
No. 25. *The Theory of Orbits in the Solar System and in Stellar Systems*, pp. 380, price 120s. (Members 80s.)	Editor: G. Contopoulos Published by Academic Press, Inc. (London) Ltd.
No. 26. *Abundance Determination in Stellar Spectra*, pp. 373, price 120s. (Members 80s.)	Editor: H. Hubenet Published by Academic Press, Inc. (London) Ltd.

IAU Colloquium Volumes

Commission 36: Proceedings of the Colloquium on *The Theory of Stellar Atmospheres*	Editor: C. de Jager Reprinted from *Journal of Quantitative Spectroscopy and Radiative Transfer* (Pergamon Press).
Commission 37: Proceedings of the Colloquium on *Clusters and Stellar Evolutions*	Editors: O. J. Eggen, C. H. Herbig Published as no. 82 of the *Royal Observatory Bulletin*.
Commission 27 and 42: Proceedings of the Colloquium on *The Position of Variable Stars in the Hertzsprung-Russell Diagram*	Published by the Remeis-Sternwarte Bamberg. *Kleine Veröff. Remeis-Sternw.*, **4**, no. 40.

Publications by, or on behalf of, IAU Commissions and by Astronomical Services

Commission 6: *IAU* (Telegram Bureau) *Circulars* nos. 1874–2000	Issued by the IAU Central Bureau for Astronomical Telegrams.

Commission 10: *Heliographic Charts of the Photosphere* (Volume XII, nos. 3 and 4)	Published by the Eidg. Sternwarte in Zürich, under the direction of M. Waldmeier.
Commission 10: *Cartes Synoptiques de la Chromosphère Solaire* (Volume IV, fascicules 3 and 4)	Prepared by Observatoire de Paris at Meudon, under the direction of R. Michard.
Commission 20: *Minor Planet Circulars*	Mimeographed Circulars issued by the Minor Planet Center, Cincinnati.
Commission 23: *Astrographic Catalogues.* Information regarding the different sections (Hyderabad, Perth-Edinburgh, Oxford-Potsdam, Melbourne, Catania, Tacubaya, Uccle-Paris, etc.) may be obtained from the respective Observatories or from the President of Commission 23	P. Sémirot, President of Commission 23, Observatoire de Bordeaux, 33-Floirac, France.
Commission 26: *Circulaires d'Information* no. 34 through no. 41	Prepared by P. Müller, Observatoire de Meudon, France.
Commission 27: *Information Bulletin on Variable Stars,* up to no. 165	Prepared by L. Detre, Konkoly Observatory, Budapest.
The Solar Spectrum 2935 Å *to* 8779 Å	Ch. E. Sitterly, M. G. J. Minnaert, J. Houtgast *National Bureau of Standards Monograph* 61.
Monthly Notes of the IPMS, no. 7 (1964) through no. 10 (1966)	Published by International Polar Motion Service, Mizusawa-shi.
IAU Information Bulletin, nos. 12–17	Prepared by the General Secretary.

5. COMMISSION DES ANALYSES DE TRAVAUX ET DE BIBLIOGRAPHIE

(Comité du Comité Exécutif)

PRÉSIDENT: Dr J. Kleczek, Astronomical Institute, Czechoslovak Academy of Sciences, Observatory Ondřejov, Czechoslovakia.

VICE-PRÉSIDENT: Dr J. B. Sykes, 20 Milton Lane, Steventon, Abingdon, Berkshire, England.

PAS DE COMITÉ D'ORGANISATION.

MEMBRES: Alter, Baldet†, Beer, Belorizky, Bourgeois, Bouska, Collinder, Dermul, Felber, Fricke, Harwood, Heinemann, Henn, Hirose, Luplau-Janssen, Martynov, Ogorodnikov, Pecker (J.-C.), Sticker, Witkowski, Zinner.

FORM OF PAPERS

Not only the content, but also the presentation of a scientific paper is important for the reader. A Style Book for astronomers has been published (*Trans. IAU*, **12 C,** 55 and 101, 1966). It contains:

A. How to prepare manuscripts for printing.

B. Spelling and transliteration.

C. Notations, formulae, units.

D. Abbreviations of titles of scientific periodicals (*Trans. IAU*, **12 C,** 76).

E. How to correct proofs.

For the comments of librarians of astronomical observatories see section C of the Report of the Working Group, which follows. Useful hints for authors may be found also in a booklet published recently by the Royal Society (1965).

It has been remarked that English abstracts in foreign journals are very useful, even if the English is sometimes not very good. Such abstracts should be encouraged. Our Commission recommends also that Russian abstracts be published with astronomical papers.

ECONOMY OF PUBLICATION

In some cases, there is a tendency for the same work to be published at least four times: as research material in a scientific journal; repeated in a slightly different form at one or more symposia, either before or after the previous publication; summarized, or expanded, in one or more progress reports on astronomy; and, finally, incorporated in a textbook. In the opinion of some astronomers, it is the second stage, namely publication in the proceedings of symposia, that is confusing and wasteful. In this connection it would be interesting to extrapolate the gloomy comments of Barnabe Rich from 1613: 'One of the diseases of this age is the multiplicity of books; they doth so overcharge the world that it is not able to digest the abundance of idle matter that is everyday hatched and brought forth into the world.' (Swanson 1966, p. 12).

The average readership of a scientific article is surprisingly low in chemistry and biosciences (Swanson 1966) and this seems also to be true for astronomy. It has been therefore suggested to publish papers only in the form of lengthy (about one page) abstracts, with the full text available on request. The Russian journal *Atomnaya Energiya* regularly publishes some papers in this way. This new possibility of economic publication has of course also its drawbacks.

BIBLIOGRAPHY OF ASTRONOMY 1881–98

All the material concerning this bibliography collected with great patience and diligence by Mr Dermul in Anvers has been transferred to the Observatoire Royal de Belgique. The present state and future plans for publication follow from a quotation by the Director of the Observatory, Professor Velghe (1966): 'L'Observatoire Royal de Belgique envisage la publication de la Bibliographie de l'Astronomie pour la période 1881–98, lorsqu'il sera en possession de toutes les fiches y afférentes. Le Directeur de l'Observatoire Royal fait un pressant appel à tous ceux qui sont engagés à dépouiller la littérature astronomique de leur pays, pour que les fiches lui parviennent avant le ler juin 1967, date à laquelle le fichier sera considéré comme complet. Le projet de publication sera communiqué aux membres de la Commission 5, au cours de la prochaine Assemblée Générale de l'UAI, à Prague.'

ABSTRACTING JOURNALS

Recorded knowledge about the Universe accumulates at an increasing rate while the rate at which it can be read remains constant. In the astronomer's endeavour to keep abreast of the increasing flow of recorded knowledge, the abstracting journals have the most important role. The four abstracting journals for astronomy have continued to abstract world literature on astronomy on a routine basis. They have all become member-services of the ICSU Abstracting Board.

ABSTRACTING BOARD OF THE INTERNATIONAL COUNCIL OF SCIENTIFIC UNIONS (IAB)

The IAU is a member of the IAB. The aim of the IAB is to initiate and further develop the exchange and publication of primary and secondary scientific information. By secondary publications are meant abstracts and indexes drawn up from the already published material.

Since 1964 an annual list of publications issued by ICSU and its members is prepared by IAB. The IAB acts as an advisory body for ICSU and its members on problems of scientific information. It represents the interests of ICSU and its members in other international institutions dealing with scientific information.

There are two categories of members of the IAB:

(1) representatives of international unions and

(2) representatives of the main secondary publications (abstracting and reference services —the so-called member-services). At present there are four member-services for astronomy: *Astronomischer Jahresbericht*, *Bulletin signalétique du Centre National de la Recherche Scientifique*: Section 2, Astronomie, Astrophysique, Physique du Globe, *Physics Abstracts* and *Referativnyi Zhurnal.*

At the July 1966 full board meeting in Paris the IAU was represented by the President of Commission 5. His proposals were all accepted and it was resolved (Resolution 8, cooperation with IAU) that:

(1) The abstracting journal *Astronomischer Jahresbericht* was unanimously elected member-service of the Board for the German language in astronomy.

Bulletin signalétique du CNRS and *Referativnyi Zhurnal* are the two other member-services for this field, for the French and Russian languages respectively.

There is no abstracting journal in the English language covering astronomy as such.

(2) The IAU will keep the IAB informed of collective works and of publications resulting from scientific meetings in the field of astronomy.

The Secretariat will communicate this information to member-services in astronomy with a view to improving the coverage of non-periodical publications in this field by member-services of the Board.

The IAU recommends to member-services for astronomy to envisage the possibility of announcing these publications not only under the names of scientific editors, but also under the names of the individual authors themselves.

(3) With the help of the IAU the Board will try to establish a list of all primary periodicals dealing with astronomy.

The action on hand in the other fields covered by the Board to improve primary publications will be extended to these publications in astronomy.

A Working Group on Documentation in Astronomy was formed at the Hamburg meeting. This Working Group forms a link between IAU Commission 5 and the 'Fédération Internationale des Associations de Bibliothécaires'. On the basis of correspondence with all the astronomical libraries of the world Mme Feuillebois has carefully prepared the hereafter appended information, suggestions and comments for the Draft Report of Commission 5.

J. KLECZEK
President of the Commission

REFERENCES

The Royal Society 1965, *General Notes on the Preparation of Scientific Papers*, 2nd edition, London, 32 pages.

Swanson, D. R. 1966, *Bulletin of the Atomic Scientists*, **22** (2), Feb. 1966, pp. 8–12.

Velghe, A. G. 1966, A note on the astronomical bibliography 1881–98, to be published in the *IAU Information Bulletin*.

APPENDICE. GROUPE DE TRAVAIL SUR LA DOCUMENTATION ASTRONOMIQUE

A. Informations

1. La Fédération Internationale des Associations de Bibliothécaires (FIAB) a créé une sous-section des bibliothèques d'observatoires et de sociétés astronomiques, qui travaillera en étroit rapport avec le groupe de travail, et qui a tenu ses premières réunions pendant la 32ème session de la FIAB à La Haye, les 13 et 15 Septembre 1966.

2. Son statut prévoit, dans le cadre des sessions de la FIAB:

(*a*) une réunion annuelle;

(*b*) une assemblée générale tous les 3 ans (l'année précédant l'Assemblée Générale de l'UAI), pour préparer les travaux du groupe de travail.

3. Le programme des séances tenues à La Haye était le suivant:

(*a*) Présentation de la nouvelle sous-section et établissement des statuts.

(*b*) Communications:

– Emploi de la Flexowriter pour le catalogage en astronomie (Mr Kemp, Edinburgh).

– Etat actuel et problèmes de la bibliographie astronomique (Mme Lavrova, Institut Sternberg, Moscou).

– 'The U.S. Naval Observatory Library at Washington, D.C.' (Mrs L. S. Multhauf, U.S. Naval Observatory).

– Présentation du catalogue par matières et de la liste des 'vedettes-matières' d'astronomie de l'Observatoire de Paris (Mme Feuillebois).

– Classification de la 'Library of Congress' et les vedettes-matières pour l'astronomie. Son emploi dans les bibliothèques américaines (Mr Weigel, University of Michigan).

(*c*) Séance de travail:

– Analyse et critique du 'style book' à paraître dans le volume XII C des *Transactions of the IAU.*

– Publications et distribution des publications d'observatoires.

– Etude des possibilités de limitation des revues commerciales.

– Etablissement d'un Répertoire International des Bibliothèques d'Observatoires et de Sociétés Astronomiques. Un important matériel est déjà réuni. Pour la publication préliminaire de ce répertoire, sur feuillets mobiles, une demande est faite au Conseil Exécutif de la F.I.A.B. d'une subvention de 5000 francs suisses.

B. *Suggestions pour la réunion de la Commission 5 à Prague*

Au cours de l'Assemblée Générale de l'UAI à Prague, le groupe de travail souhaite que soit prévue une séance de travail qui lui permettra de mettre au point le rapport de ses travaux.

C. *Commentaires*

– Le groupe de travail note avec plaisir la publication d'un 'Style Book' dans le volume XII C des *Transactions of the IAU.*

– Elle suggère les commentaires suivants en vue de l'élaboration du volume XIII C: (*N.B.:* Les paragraphes et pages indiqués ci-dessous sont ceux du volume XII C).

1. *Présentation des articles* (*Form of papers*)

– 'Résumés' (paragr. *b*, p. 57; English version, p. 103): Le groupe de travail suggère 'que les articles . . . soient toujours accompagnés d'un sommaire dans la langue employée par l'auteur, ainsi que d'un sommaire plus court, soit en anglais, soit en français (ou au moins dans l'une des quatre langues suivantes: français, anglais, allemand, russe). L'idéal consisterait d'ailleurs, de la part des éditeurs de revues, à insérer dans chaque fascicule des fiches détachables 125 × 75 mm pour chaque article publié (solution déjà adopté par le *Bulletin Astronomique*, 3ème série, 1966.

– 'Bibliographie', paragr. *e*, p. 59; English version, p. 105): Le groupe de travail trouve très souhaitable les recommandations de l'UAI. Toutefois, à propos du paragraphe 'Présentation des références' elle demande à l'UAI:

– d'étudier avec précision la question de la date de la référence: s'agit-il de la date de la soumission du manuscrit ou de celle de l'édition? Cette précision est particulièrement importante lorsqu'il s'agit d'un livre à auteurs multiples ou d'un livre réédité (*iii*, p. 60; English version, p. 106).

– de distinguer (δ, p. 61; English version, p. 107) les mentions 'in preparation' et 'in press', la seconde ne devant être employée que lorsque l'article est effectivement en cours d'impression, c'est-à-dire après correction des épreuves, ce qui permet d'ailleurs d'apporter certaines précisions à cette référence.

– l'indication 'in preparation' ne permettant pas de se reporter à une référence bibliographique utilisable, il serait même préférable de ne la mentionner qu'au cours de l'article et non à la fin.

2. *Translittération*

– Le groupe de travail souhaite que l'Union Astronomique Internationale, conformément à sa vocation d'association internationale, adopte la translittération ISO/R9, 1954.

– Si cette translittération est favorablement accueillie à Prague, elle suggère que le document ISO/R9 soit intégralment reproduit dans le volume XIII C (de préférence au tableau donné p. 69; English version, p. 115).

3. *Abréviations des titres de périodiques scientifiques* (*pp.* 74 *et suiv.*)

Lors de la préparation du volume XII C il n'existait pas de liste internationale d'abréviations des titres de périodiques scientifiques. C'est pourquoi la *World List of Scientific Periodicals* pouvait être recommandée. Mais une liste internationale est actuellement préparé par l'International Standard Organization, tandis que la World List semble en défaveur.

Dans ces conditions, le groupe de travail pense qu'il serait prudent d'attendre la publication de la liste ISO et de surseoir à toute décision jusqu'à cette date.

4. *Revues de références astronomiques* (*Astronomical abstracting journals*)

La création d'un nouvel 'Abstracting Journal' de langue anglaise ne paraît pas souhaitable (Les *Physics Abstracts* analysent d'ailleurs les grandes revues d'astronomie).

Par contre, *l'Astronomischer Jahresbericht*, qui est actuellement la bibliographie la plus complète et la plus utilisée, pourrait être améliorée: (*a*) en abrégeant le délai de publication; (*b*) en translittérant les titres russes au lieu de les traduire en allemand; (*c*) en publiant des 'abstracts' en langue française ou anglaise, ce qui rendrait la consultation beaucoup plus aisée, en particulier pour les jeunes astronomes rarement familiarisés avec la langue allemande; (*d*) en élargissant le domaine des publications analysées: congrès, livres à auteurs multiples; publications d'observatoires (chaque observatoire devrait vérifier que ses publications figurent au paragraphe 4 de *l'Astronomischer Jahresbericht*); (*e*) enfin en étudiant les moyens d'automatiser la bibliographie actuellement existante. Etudier l'automatisation de la bibliographie internationale courante devrait en effet être une préoccupation de la Commission 5 dans un proche avenir. L'idéal serait évidemment de n'avoir qu'une seule bibliographie internationale courante d'astronomie.

5. *Publications*

La question de la distribution des publications d'observatoires, et de la parution de nouvelles revues commerciales, a été longuement discutée à La Haye. Une enquête préalable avait d'ailleurs été faite et nous reproduisons ci-dessous l'essentiel des réponses.

(*a*) *Distribution des publications d'observatoires.* Bien que la multiplicité des publications d'observatoires, généralement constituées de 'reprints', soit un lourd handicap pour les bibliothèques, les bibliothécaires souhaitent en général leur maintien. Ces publications permettent en effet: (*i*) d'avoir une idée d'ensemble du travail effectué dans chaque observatoire; (*ii*) de constituer une monnaie d'échange pour les observatoires qui n'ont pas assez de crédits pour acquérir les revues dont sont extraits les articles reproduits sous forme de 'reprints' (en particulier lorsqu'il s'agit de publications d'Académies ou de revues non astronomiques); (*iii*) de multiplier les possibilités de prêt dans les observatoires où les lecteurs sont très nombreux. Toutefois le groupe de travail suggère que soit réduit dans la mesure du possible le nombre de titres publiés par chaque observatoire. De plus, afin de faciliter la distribution et l'échange des publications d'observatoires, la sous-section des bibliothèques d'observatoires et de sociétés astronomiques de la FIAB prépare pour 1967 une liste mise à jour de toutes les publications non commerciales d'observatoires (y compris les catalogues d'étoiles)*. Cette

*Mme Van de Poll de Lange, bibliothécaire de l'Observatoire de Leyde, a bien voulu se charger de ce travail.

liste précisera, pour chaque publication l'adresse exacte de l'organisme distributeur, ainsi que les numéros disponibles et les numéros épuisés. Cette liste sera présentée à Prague par le groupe de travail.

(*b*) *Etude des possibilités de limitation des revues commerciales.* Le groupe de travail s'inquiète de la parution, au cours de ces dernières années, de nouvelles revues astronomiques prises en charge par des éditeurs commerciaux et qui viennent s'ajouter aux revues déjà existantes. Mais les bibliothécaires sont obligés de constater que les astronomes eux-mêmes souhaitent l'acquisition de ces nouvelles revues. De plus, les revues existantes (comme *l'Astrophysical Journal*) sont de plus en plus volumineuses, et l'on conçoit mal comment elles pourraient s'accroître encore.

Toutefois ce problème doit, semble-t-il, être pris en considération par la Commission 5 de l'UAI: c'est en effet aux astronomes eux-mêmes qu'il appartient de prendre la décision de limiter de telles revues.

G. FEUILLEBOIS

6. COMMISSION DES TELEGRAMMES ASTRONOMIQUES
(Comité du Comité Exécutif)

PRÉSIDENT: Dr A. R. Hogg†.

VICE-PRÉSIDENT: Dr F. L. Whipple, Director of the Smithsonian Observatory, 60 Garden Street, Cambridge, Massachusetts 02138, U.S.A.

DIRECTEUR DU BUREAU: Dr O. Gingerich, Smithsonian Astrophysical Observatory, Cambridge, Massachusetts 02138, U.S.A.

DIRECTEURS ADJOINTS DU BUREAU: Dr R. B. Southworth and Dr B. G. Marsden, Smithsonian Astrophysical Observatory, Cambridge, Massachusetts 02138, U.S.A.

COMITÉ D'ORGANISATION: A. G. P. Velghe, K. A. Thernöe, O. J. Gingerich.

MEMBRES: Buchar, Candy, Cunningham, Hers, Martynov, Roemer.

Le Comité Exécutif de l'Union Astronomique Internationale a appris avec peine le décès du Dr A. R. Hogg, Président de la Commission 6 depuis 1961.

Le Professeur F. L. Whipple a bien voulu accepter de préparer ce rapport et d'assumer le rôle de Président par interim de la Commission. Il a préparé le présent rapport, assisté par le Dr O. Gingerich, Directeur du Bureau des Télégrammes Astronomiques.

INTRODUCTION

At the Twelfth General Assembly of the IAU in Hamburg, arrangements were made to move the Central Bureau from its traditional home in Copenhagen to Cambridge. The gradual increase in the number of subscribers, coupled with the erratic nature of the Bureau's work—periods of intense activity when spectacular comets appear alternating with many weeks of inactivity—made it desirable to locate the Bureau at a larger observatory. In its new headquarters at the Smithsonian Astrophysical Observatory, the Central Bureau can call upon the services of the round-the-clock communication center, the network of astrophysical observing stations (with their Baker-Nunn cameras, which are ideal for confirming bright comets), the computations center, and the publications division.

Dr Gingerich, whose unusually able services brought about the smooth and efficient transfer of the Bureau to its new location, has indicated his intention to resign as the Bureau's Director, since his own interests lie primarily elsewhere. Dr Marsden, who has played an increasingly active role as an Associate Director and who is a specialist in the computation of comet orbits, would appear to be the logical choice for the new Director. Dr Gingerich and Dr Southworth would then serve as Associate Directors. The Commission is truly indebted to Dr Gingerich for his skill and perseverance in effecting the transfer of the Central Bureau.

F. L. WHIPPLE
Acting President of the Commission

REPORT OF THE CENTRAL BUREAU FOR ASTRONOMICAL TELEGRAMS

In its first year in Cambridge, the Central Bureau distributed 64 numbered *Circulars* on 53 occasions. These carried information about 16 supernovae and ten comets discovered during 1965. The most extensive activity resulted from the spectacular sun-grazing comet, Ikeya-Seki, for which 11 *Circulars* were issued in a period of six weeks. In its second year, the Central Bureau issued 39 numbered *Circulars* on 30 occasions. These carried information about five supernovae and six comets discovered during 1966. Since 17 comets are predicted to come to perihelion during 1967, we can anticipate considerable activity in the Bureau.

When the Central Bureau moved to Cambridge, the *Harvard Announcement Cards* were discontinued, and their subscribers began to receive the *Circulars*. This more than doubled the number of subscribers, so that at present over 600 *Circulars* are mailed to 50 countries.

The telegram activity of the Bureau is somewhat harder to measure; 43 'telegram books' were distributed to a large block of domestic subscribers as well as to the AGIWARN system during 1965 and 1966. It is difficult to count the actual number of recipients because the telegrams are relayed to partial lists of subscribers according to various categories. In addition, on special occasions messages were sent to only a few selected observatories.

The international distribution of telegrams has worked out magnificiently through the AGIWARN system of the International Ursigram and World Days Service. In particular, the arrangements made by Dr R. Michard and the effort of Dr P. Simon at Meudon have facilitated the redistribution of most of the European telegrams. Telegrams to South Africa, Japan, and some points of South America have been relayed through the telegram-link connecting the Smithsonian Observatory with its astrophysical observing stations.

The task of collecting advance telegram deposits for the subscribers served from the Meudon Center has proved rather formidable, and several European subscribers are still delinquent in sending in their deposits. Similarly, many observatories have been lax in responding to our expiration notices. Since the *Circulars* are addressed by a computer program, the failure to renew promptly results automatically in an interruption of service.

At the Hamburg Meeting of the IAU, suggestions for an improved telegram code were discussed, and fortunately it was possible to introduce the revisions at the same time that the transfer of the Central Bureau took place. The success of the new code is perhaps most vividly shown by the awkward situation that almost inevitably arises when an occasional astronomer transmits a message in the old form: it is then generally necessary to telegraph back asking for the equinox of the observations. In addition to the explicit inclusion of the equinox in the revised code, the time of observation is given in decimals of day, there is a continuous ordering of data in the accurate position code, and a more precise description of magnitudes is provided. An explanation of the new code appears in the IAU Handbook (*Transactions*, XII C); a list of errata are attached to this report. Corrected reprints of the code are available through the Central Bureau.

During 1965 and 1966, $7328.00 were received for subscriptions to the *Circulars*, which covered only the printing and postage. The continuing operation of the Bureau requires more than the equivalent of one full-time worker, which is supplied by the Smithsonian Astrophysical Observatory. The subvention of $666.00 per year from the IAU pays for very little of the actual operation of the Bureau, but it provides considerable moral support and gives the operation a desirable flexibility that it would not otherwise have. It seems entirely reasonable to ask that the IAU subvention be raised to at least $1000 per year.

OWEN GINGERICH
Director of the Bureau

APPENDIX. ERRATA ET CORRIGENDA FOR IAU TRANSACTIONS XIIC.

p. 33 Strike out category H and for category G read:
G. GAMMA All others, including planetary or lunar phenomena, spectroscopic binaries, etc.

p. 34 The Air Mail rates in the column 'Other' are now $8.50 and $16.50. *Under Middle groups for a position*, the fifth approximate group should read: XNMMP.

p. 35 The fifth line should read:
8. X = X, the character for a nonsignificant digit.

p. 36 The fourth line should read:
K. KKKK = q in A.U.
The last line of the first example should read:
R.A. – $1^{m}.2$, in Dec. + 12′.

p. 37 The first line should read:
3. CANDY COMET CANDY 19503 10208 58335 13625 17656 15092 10640 45059 CANDY.

38. COMMISSION POUR L'ECHANGE DES ASTRONOMES

(Comité du Comité Exécutif)

PRÉSIDENT: Professor M. G. J. Minnaert, Sterrewacht 'Sonnenborgh', Zonnenburg 2, Utrecht, The Netherlands.

VICE-PRÉSIDENT: Professor M. Migeotte, Institut d'Astrophysique, Université de Liège, Cointe-Sclessin, Belgium.

COMITÉ D'ORGANISATION: B. J. Bok, D. Ya. Martynov, T. L. Page.

MEMBRES: Abetti, Babcock (H. D.), Bappu, Bowen, Danjon, Keller, Kienle, Kourganoff, Mikhailov, Miyamoto, Mohr, Righini, Rosseland, Sahade, Stoy, Swings, Thackeray, Witkowski, Wood (F. B.), Wyatt.

MEMBRE CONSULTANT: Fournier d'Albe (E. M.).

In the last report of this Commission, written at the end of 1963, we mentioned with some surprise that the number of travel grants awarded has shown but little variations in the preceding ten years. Since then the situation has drastically changed. In the last three years a total of 52 awards has been made, *twice as much* as in each of the previous periods. It is clear that the wonderful possibilities available have gradually become known and that more and more institutes are taking advantage of them. In 16 other cases the request could not be granted or was sent only for provisional information and did not result in an allocation.

In the period 1964–66 the following grants were made:

125. A British astronomer to Austin.
6. A Czechoslovakian astronomer to Herstmonceux.
7. A Belgian astronomer to Tucson.
8. A British astronomer to Paris.
9. An Italian astronomer to Mt Hamilton.
130. A Japanese astronomer, returning from Heidelberg.
1. A Japanese astronomer to Santa Monica.
2. An English astronomer to Groningen.
3. A Polish astronomer to Sydney.
4. A Belgian astronomer to Mt Stromlo.
5. An Argentine astronomer to Hamburg and Praha.
6. A French astronomer to Mt Palomar.
7. A Czechoslovakian astronomer to Liège.
8. A Japanese astronomer to Pic-du-Midi.
9. A Japanese astronomer to Utrecht.
140. An Indian astronomer to Tucson.
1. A Czechoslovakian astronomer to Victoria.
2. A French astronomer to Cairo.
3. A Dutch astronomer to Nice.
4. An Italian astronomer to Pic-du-Midi.
5. An Italian astronomer to Utrecht.
6. A Belgian astronomer to Edinburgh.
7. An Australian astronomer to McDonald Observatory.
8. An Indian astronomer to Utrecht.
9. An Italian astronomer to Leiden.
150. A Czechoslovakian astronomer to Asiago.

1. A Japanese astronomer to Los Angeles.
2. A Dutch astronomer to Praha.
3. A British astronomer to Mt Hamilton.
4. An Egyptian astronomer to several European observatories.
5. A Finnish astronomer to Meudon.
6. A Greek astronomer to Meudon.
7. An Israeli astronomer, returning from Tucson.
8. An Italian astronomer to Leiden.
9. A French astronomer to Berkeley.
160. A Jugoslavian astronomer to Nijmegen.
1. Two Argentine astronomers to U.S.A. National Radio Observatory.
2. An Indonesian astronomer to Mt Stromlo.
3. Two German astronomers to Berkeley.
4. A Belgian astronomer to Nice.
5. A Greek astronomer to Meudon.
6. A Japanese astronomer to Pasadena.
7. An Indian astronomer to Toronto.
8. An Egyptian astronomer to Herstmonceux.
9. A French astronomer to Tucson.
170. An Indian astronomer, returning from Utrecht.
1. A French astronomer to La Jolla.
2. A Japanese astronomer to Groningen.
3. An Australian astronomer to the U.S.S.R.
4. A British astronomer to Kingston.
5. An Israeli astronomer to Utrecht.
6. An Italian astronomer to the U.S.A.

In surveying this list, we notice that a considerable fraction of the grants, actually 67%, have been allocated for *intercontinental* travel. It seems that gradually more and more opportunities for nearby travels become available, while the IAU is almost the only body from which help on a larger scale can be expected. In the same way we notice that there were no requests concerning travels from one of the U.S.A. states to the other, nor from one of the U.S.S.R. republics to the other. From the total number of grants, 33% concerned travels to the U.S.A.

The purposes for which the subsidies were granted may be summarized as follows:

To get acquainted with the instruments or methods of an institute, specialized in a certain field; or to discuss theoretical problems with the members of such an institute.

To establish contacts between an observer and theorists.

To learn a special technique, in view of an instrument to be erected at the home observatory.

To obtain high dispersion spectra or other observations with big instruments.

To make special observations in the Southern Hemisphere or at a station with an exceptionally clear sky.

To combine a special, self-built apparatus with a big telescope.

To inspect the plate collections.

To make use of a big computer.

To prepare a book, written together with a foreign astronomer. To give final shape to a monograph.

During the three years 1964–66, the grants totalled $22 542·24, including those (see hereafter) payed with the UNESCO subsidy, and taking into account partial reimbursements. Some of the grants listed above where not yet cashed on 31 December, 1966, and are therefore still included in the balance at hand on 31 December, 1966, which is of $14 964·23.

During the years 1965–66 the committee received from UNESCO an important subsidy of 2000 U.S. dollars, to be granted to astronomers of developing countries. (Cf. *IAU Inf. Bull.* no. 16). This subsidy was used to pay the allocations nos. 140, 151, 162 in the above list and the allocation made by the Sub-Commission.

In the course of the year 1966, the Commission has for the first time exceeded to a rather considerable extent its yearly subsidy. This is considered to be a healthy development, since former years have left a certain accumulated balance, which should be gradually used.

As yet, the action of the Commission is still restricted in two respects: (1) only costs of travel are paid, no costs of living at the host observatory; (2) no grants are given for participation in symposia, unless if combined with scientific work of longer duration. Very roughly speaking, the subsidies granted do seldom exceed \$300 $\sqrt{t}$, where t is the duration of the stay, expressed in months. On the whole these restrictions are unavoidable; one might, however, consider the possibility of exceptions in very special cases.

Information concerning the activities of the Commission were given on request to cultural administrations in Belgium and in Peru.

The list of 'Facilities for Visiting Astronomers', prepared for the 1964 meeting as an appendix to the report, has been published only at the end of 1966 in volume XII A of the *Transactions*, p. 82. In order to avoid a similar delay, the next list will be published directly after the Prague meeting in the *IAU Information Bulletin.*

Since the last General Assembly the status of Commission 38 was modified: it became a committee of administrative character, working directly under the direction of the Executive Committee. The Executive Committee will have in future a certain responsibility in determining the general policy of the Commission. The collaboration with the General Secretary proved excellent.

On this same occasion the decision was taken to create a Sub-Commission, intended to help graduate students who want to specialize in astronomy, but who have not yet reached the doctor's degree. (Cf. *IAU Inf. Bull.*, no. 14.)

A report of this Sub-Commission is added below. The coordination of the activity of the Sub-Commission with the work of Commission 46 and its summer-schools will have to be carefully planned.

M. G. J. MINNAERT
President of the Commission

RAPPORT DE LA SOUS-COMMISSION

PRÉSIDENT: Professeur M. Migeotte, Institut d'Astrophysique, Université de Liège, Cointe-Sclessin, Belgique.

MEMBRES: H. Kienle, Mlle E. A. Müller, P. M. Routly.

La Sous-Commission a été créée à l'occasion du congrès de Hambourg 1964, pour faciliter le séjour à l'étranger de *licenciés en sciences*, qui désirent se spécialiser en astronomie et qui n'ont pas la possibilité de faire ces études dans leur pays. On a songé en particulier aux pays en voie de développement.

Deux demandes de subsides ont été reçues par le sous-comité.

(1) Un étudiant du Nigéria, ayant le grade de B.Sc., se proposait d'aller étudier à Berkeley, où il pouvait bénéficier d'un 'research assistantship'. Il a demandé l'intervention de l'UAI pour couvrir ses frais de voyage. Le crédit demandé a été accordé en se

basant sur le curriculum vitae et sur des recommandations très élogieuses. Cependant, par suite de circonstances locales et personnelles, le candidat en question a dû décider d'ajourner d'un an son séjour à l'étranger.

(2) Un étudiant d'Israël a demandé l'intervention du sous-comité pour poursuivre ses études à l'Université de Manitoba (Canada), où il peut obtenir une assistance financière. Il étudiera l'astronomie, en particulier la cosmologie et la relativité générale.—Après avoir examiné le curriculum vitae et vu les recommandations de quatre de ses professeurs, le sous-comité a décidé de mettre à la disposition de ce candidat une somme couvrant les frais de voyage.

Il est clair que la création du sous-comité n'a pas encore suffisamment attiré l'attention, quoiqu'elle ait été annoncée dans le *Bulletin d'Information* no. 14 (juin 1965). D'autre part il est difficile, pour la Sous-Commission, de payer les frais de séjour d'étudiants qui désirent poursuivre leurs études à l'étranger pendant un ou deux ans, vu l'importance des crédits qui devraient être accordés.

M. MIGEOTTE
Président de la Sous-Commission

46. COMMISSION POUR L'ENSEIGNEMENT DE L'ASTRONOMIE

(Comité du Comité Exécutif)

PRÉSIDENT: Professeur E. L. Schatzman, Institut d'Astrophysique, 98bis, Boulevard Arago, Paris 14e, France.

VICE-PRÉSIDENT: Professor Miss E. A. Müller, Observatoire Cantonal, Geneva, Switzerland.

COMITÉ D'ORGANISATION: L. H. Aller, B. J. Bok, E. K. Kharadze, V. Kourganoff, M. G. J. Minnaert, T. L. Page, H. C. Van de Hulst.

MEMBRES: La Commission est en cours d'organisation.

MEMBRES CONSULTANTS: E. P. Levitan, G. Walusinsky.

A la suite de la discussion de Hambourg sur l'Enseignement de l'Astronomie a été créée la Commission 46. La discussion de Hambourg proposait à la Commission des tâches résultant des obligations internationales de l'UAI.

Les problèmes qui se sont posés à la Commission 46 se sont montrés extrêmement difficiles. En effet, la situation de l'Astronomie diffère considérablement d'un pays à l'autre, et on voit jouer l'influence du niveau général de développement, de l'organisation de l'enseignement, des moyens de recherche. Les traditions ont un rôle considérable et, à l'intérieur d'un même pays, la situation peut varier d'une Université à l'autre.

Les problèmes d'un pays en voie de développement ne sont pas ceux des Etats-Unis ou de l'U.R.S.S. Ainsi en est-il par exemple pour l'Inde, où le Président de la Commission a entrepris une enquête spéciale. Sur ces pays, on consultera avec fruit la référence (1).

ENQUÊTE SUR LA SITUATION DE L'ENSEIGNEMENT DE L'ASTRONOMIE

La Commission 46 a fait une enquête sur la situation de l'Enseignement de l'Astronomie au moyen d'un questionnaire paru dans le *Bulletin d'Information* no. 14 de l'UAI. Nous reproduisons ci-dessous le questionnaire:

Questionnaire

1. Name?

2. Institution?

3. Teaching programme
- *(a) Place (observatory, university, institute)?*
- *(b) Subject(s)?*
- *(c) Level (graduate, undergraduate, etc.)?*
- *(d) Schedule (number of hours annually)?*
- *(e) Number of students?*

4. Do you have at your disposal course notes, books, duplicated notes, or notes that could be photographically reproduced?

5. Enrollment:
- *(a) Present (1963–65)?*
  - *M.Sc.: (about) . . .*
  - *Ph.D.: (about) . . .*
- *(b) Expected (1966)?*
  - *M.Sc.: (about) . . .*
  - *Ph.D.: (about) . . .*
- *(c) Desired?*

6. Present schedules at high school or secondary school level?

Les réponses reçues appellent les observations suivantes:

(1) Quarante-deux institutions seulement ont répondu; un grand nombre de centres importants n'ont pas donné de réponse. La liste des institutions ayant répondu est donnée en Annexe I.

(2) Les réponses ont été faites du point de vue du professeur et non pas du point de vue des étudiants. On sait en général le nombre d'heures d'enseignement qui sont données, mais on ne sait pas le nombre d'heures d'enseignement que doivent suivre les étudiants.

Exceptionnellement, le rapport d'Utrecht donne les précisions suivantes:

Undergraduate: 4^h lecture + 3^h practical work, during 2 years.

Graduate: 4^h per week annually; 60^h of observations (solar and stellar) annually; 1^h per week seminar and colloquia; 5 afternoons per week of practical and theoretical work during 3 years.

(3) Quelques institutions disposent de notices détaillées sur l'Enseignement de l'Astronomie (Harvard College Observatory, Faculté des Sciences de Paris), mais il n'est pas possible, la plupart du temps, de savoir exactement quelle est l'organisation de l'Enseignement.

(4) Le project d'échange de notes de cours se heurte à de grandes difficultés:

– la plupart des cours de 3ème cycle ne sont pas rédigés, et les notes de cours ne sont pas disponibles.
– les notes de cours, quand elles existent, sont toujours rédigées dans la langue où est fait l'enseignement, c'est-à-dire en Hollandais, Italien, Polonais, Russe, etc.

(5) Les statistiques sur le recrutement et les perspectives de recrutement ne permettent pas de se faire une idée de la situation mondiale, pour les raisons suivantes:

– les renseignements sont incomplets,
– les différences de niveau sont très grandes d'une institution à l'autre.

Cependant, sur les 42 Institutions ayant répondu au questionnaire, seulement 15 donnent le nombre d'étudiants pour le M.Sc.et le Ph.D. pour les périodes 1963–64, 1964–65 et les perspectives. Les données cumulatives déduites des 15 institutions sont les suivantes:

	1963–64	1964–65	Perspectives
M.Sc.	75	89	124
Ph.D.	148	165	241

Le nombre beaucoup plus élevé de Ph.D. que de M.Sc. provient du fait que dans quelques grandes Universités des Etats-Unis, le degré de 'Master of Science' n'existe pas, la fin des études conduisant à l'obtention d'un doctorat en sciences (par exemple: Harvard, Caltech, Yerkes).

En d'autres termes, dans la plupart des cas, les collègues qui ont répondu au questionnaire l'on fait comme s'ils s'adressaient à un collègue connaissant déjà parfaitement la situation locale. Il est donc extrêmement difficile de rendre compte de l'enquête d'une façon utilisable par les collègues.

L'ASTRONOMIE DANS L'ENSEIGNEMENT SECONDAIRE

La discussion de 1964, à Hambourg, avait montré un très grand intérêt pour les questions relatives à l'Enseignement secondaire. Il sera à nouveau question de ce problème à propos de la CIES (Commission Internationale pour l'Enseignement des Sciences). De même que l'Union Internationale de Mathématiques prend une part considérable dans l'organisation et le renouvellement de l'enseignement mathématique secondaire, de même l'Union Astronomique Internationale a pour devoir de s'occuper des problèmes analogues pour l'astronomie.

Dans beaucoup de pays le Ministère de l'Education Nationale consulte les universités pour l'établissement des programmes pour écoles secondaires. Souvent les comités nationaux d'astronomie ont pris des initiatives pour obtenir des modernisations. De plus, les futurs enseignants des écoles secondaires sont en général formés dans les universités et nous sommes responsables de leur formation scientifique et didactique.

Les problèmes de l'Enseignement Secondaire exigent la collaboration d'astronomes professionnels et d'enseignants, au courant de la pédagogie et au niveau des écoles. Les avis de nos Membres Consultants seront de grande valeur pour une telle collaboration.

D'autre part il est d'un grand intérêt de coopérer pour ces questions avec des représentants d'autres sciences. Nous rejoignons ici le problème de l'Enseignement intégré des sciences dont il a été question à la CIES.

CRÉATION D'UNE ÉCOLE ANNUELLE DE FORMATION

La constitution de la Commission 46 a permis d'entreprendre la réalisation d'un projet auquel s'intéressait particulièrement l'UNESCO: la création d'une Ecole annuelle de formation de jeunes astronomes. Un contrat avec l'UNESCO, pour une somme de $2000, en 1965, a permis la réunion d'une commission d'experts à Nice, les 8 et 9 juillet 1965. Participaient à cette réunion: Mlle Edith Müller, MM. Kharadze, Kourganoff, Liller, Pecker (hôte), Van de Hulst, Woolley, Fournier d'Albe (UNESCO) et Schatzman (président).

A la suite de cette réunion, une résolution a été proposée au Comité Exécutif de l'UAI en vue de la création de ces Ecoles (Annexe II). Après cette résolution une enquête a été faite auprès des diverses institutions astronomiques. On trouvera en Annexe III le rapport résumant les résultats de cette enquête. Une Ecole d'un type un peu différent a été organisée par l'U.R.S.S. à Tashkent en 1966. On trouvera en Annexe IV le rapport sur l'Ecole pour l'observation des satellites artificiels.

Actuellement, le Dr J. Kleczeck a été chargé du secrétariat scientifique de l'Ecole, et les négociations sont en cours avec Manchester pour son organisation en 1967.

Il semble être dans l'esprit de la résolution de Nice que la Commission 46 n'assume plus de responsabilités après la nomination du secrétaire de l'Ecole de jeunes astronomes. Eventuellement, un groupe d'experts désignés par l'UAI pourrait assurer le rôle de conseiller scientifique auprès du Dr Kleczeck.

Le budget 1965–66 de l'UNESCO prévoyait le démarrage des Ecoles d'été dès 1966. Comme cela n'a pas été possible, un contrat a été passé entre l'UNESCO et l'UAI, la Commission 46 étant chargée de l'application de ce contrat, pour permettre l'attribution de bourses à des étudiants de pays en voie de développement; cette solution permettait l'emploi de sommes disponibles, et répondait à un besoin analogue à celui qui avait entraîné la formation des écoles d'été. L'existence de ces bourses a été annoncée par le truchement du *Bulletin d'Information* de l'UAI, mais les demandes d'allocation ont été peu nombreuses; de plus la somme attribuée par l'UNESCO ne permettait d'attribuer qu'une bourse: l'examen des dossiers a permis au Secrétaire Général de l'UAI, après consultation avec le Président de la Commission, d'attribuer cette bourse à Monsieur de la Reza, étudiant bolivien désirant étudier l'astronomie à l'Université de Genève, pendant l'année scolaire 1966–67.

Un contrat avec l'UNESCO, de $8000 par année, est en préparation pour les années 1967 et 1968, pour permettre l'organisation des Ecoles d'été.

COMMISSION INTERNATIONALE POUR L'ENSEIGNEMENT DES SCIENCES (CIES)

La représentation de l'UAI à l'ICSU entraînait immédiatement une demande d'admission de l'UAI à la CIES, émanation de l'ICSU.

La CIES a tenu depuis 1964, trois réunions: Dakar, janvier 1965; Paris, septembre 1965; Paris, septembre 1966, réunions auxquelles l'UAI était représentée par MM. Minnaert et Schatzman (Dakar), par M. Schatzman (Paris, 1965) et par M. Kourganoff (Paris, 1966), représentant le Président de la Commission.

Le colloque et la réunion de Dakar ont fait l'objet d'un compte-rendu publié dans le *Bulletin d'Information* no. 14 de l'UAI et reproduit ici.

Rapport sur la réunion de Dakar

Du lundi 18 janvier au vendredi 22 janvier 1965, a eu lieu à Dakar (Sénégal) le congrès de la CIES. Dans cette Commission collaborent les sections d'enseignement de 10 Unions scientifiques internationales, parmi lesquelles l'Union Astronomique Internationale. La CIES est directement responsable envers le Conseil International des Unions Scientifiques (ICSU).

Le but spécifique de la CIES est l'étude des relations mutuelles entre l'enseignement des mathématiques et des différentes sciences naturelles. Le congrès de Dakar avait encore un but plus étendu, choisi en rapport avec les problèmes locaux: 'L'Enseignement des Sciences et le Progrès Economique'.

Le délégué de l'UAI était le Professeur E. Schatzman, président de la section d'enseignement nouvellement créée; le Professeur M. Minnaert, qui avait déjà collaboré avec la CIES, a pu également assister à ce congrès' Ces deux astonomes étaient les seuls représentants de notre Science. En vue du congrès, des rapports préliminaires avaient été préparés sur l'enseignement de chaque science en particulier. Un tel rapport avait aussi été rédigé pour l'astronomie par M. Minnaert.

Les jours du 14 au 18 étaient consacrés aux séances de groupes d'études, auxquelles prirent part environ 40 personnes. Chaque participant était membre de deux des six groupes. M. Schatzman assista: (1) au groupe C, étudiant les rapports entre les mathématiques et la physique; (2) au groupe E, examinant la question de la formation des techniciens. M. Minnaert prit part: (1) au même groupe C; (2) au groupe D concernant l'enseignement aux niveaux élémentaire et secondaire. Un programme particulièrement détaillé fut élaboré en vue de l'introduction des mathématiques modernes dans l'enseignement secondaire, spécialement pour la section scientifique.

La partie officielle du congrès fut ouverte solennellement en présence du Ministre de l'Education Nationale, de membres de l'Université de Dakar et d'autres personnalités officielles. Environ 75 participants étaient inscrits, parmi lesquels un assez grand nombre de scientifiques et d'enseignants, appartenant à des pays en voie de développement. Pendant le congrès, les rapporteurs des six groupes d'étude firent part au congrès en son entier de leurs conclusions, auxquelles de nombreux compléments et modifications furent présentés par les participants.

Le congrès était bien d'accord sur la nécessité d'enseigner l'astronomie dans toutes les écoles secondaires, pendant un minimum d'une heure par semaine durant une année, soit dans la dernière ou dans l'avant-dernière classe; de cette manière on peut faire usage des connaissances déjà obtenues en physique et en mathématiques. On y ajouta le souhait d'introduire des leçons occasionnelles d'astronomie dans les écoles élémentaires et au commencement des études secondaires.

Lorsque les géologues et les météorologistes proposèrent à leur tour d'introduire des notions de leurs sciences dans l'enseignement secondaire, on considéra la possibilité d'un cours intégré des 'sciences de la Terre' comprenant l'astronomie, la géologie, la géophysique, la météorologie;

chacune de ces sciences étant enseignée par un professeur spécial, mais dans l'ensemble étant strictement coordonné.

Le congrès de Dakar donna également l'occasion de se réunir au comité restreint de la CIES, qui vérifia les comptes et qui décida de proposer à l'ICSU les trois actions suivantes: (1) rédiger les rapports complets des séances du congrès, rapports qui seront présentés à l'UNESCO pour publication; (2) étudier plus en détail les questions concernant la formation des enseignants et des techniciens; (3) préparer un 'Source Book', contenant un grand nombre d'exemples d'applications des mathématiques dans l'enseignement secondaire et dans les deux premières années universitaires pour la Physique, la Mécanique, l'Astronomie et la Cristallographie. Ce livre sera également édité par l'UNESCO.

Une journée spéciale, très intéressante, fut consacrée aux contributions que l'enseignement des sciences peut apporter au développement économique des pays nouveaux.

Questions posées aux réunions de Paris

A la réunion de Paris, 1965, la question principale qui s'est trouvé posée est celle d'un 'Source Book' dans lequel des sujets de problèmes de physique, cristallographie, astronomie, et mécanique seraient proposés et résolus en illustration du cours de mathématiques du niveau 'fin des études secondaires, début des études supérieures'. Le professeur Kristansen, de Copenhague, a été chargé de l'unification de l'ouvrage du point de vue des énoncés et du traitement mathématique de la solution.

En même temps, les représentants de l'UNESCO ont soumis à la CIES un projet d'étude de l'Enseignement intégré des Sciences dans l'Enseignement secondaire.

Ce projet paraît présenter le plus grand intérêt, et il faut signaler à ce sujet le remarquable travail accompli aux Etats-Unis pour donner dans les Ecoles secondaires un enseignement intégré des Sciences de la Terre.

Il paraît difficile que la CIES joue le rôle de rédacteur d'ouvrages d'enseignement secondaire. Il paraît important par contre que la CIES joue le rôle de conseiller scientifique auprès de l'UNESCO. Le problème de l'Enseignement intégré des Sciences dépasse largement par son importance les problèmes de l'Enseignement secondaire. Il se pose comme un problème philosophique; des recherches pédagogiques doivent être entreprises à son sujet, il est extrêmement actuel et fait partie du progrès même des sciences. Il mérite d'être étudié au niveau universitaire, aussi bien du point de vue des débutants que du point de vue de la formation des chercheurs.

L'étude de l'enseignement intégré des sciences pose aussi le conflit entre l'unité de la Science et la spécificité des sciences; d'un côté la profonde unité des processus naturels, de l'autre la spécificité des méthodes associées aux concepts de champ et de force, d'onde, de particule, de processus élémentaire et de processus collectif, qu'il s'agisse des méthodes de mesure, ou des méthodes de traitement théorique de ces concepts.

L'étude de ces questions, à la fois scientifiques et pédagogiques, doit être entreprise avec le but de donner, en 1967, l'enseignement qui permettra aux étudiants de faire des recherches d'avant-garde en 1977.

Un résumé de cette tendance peut être trouvé en particulier dans le programme de travail proposé à la Commission.

TERMES DE RÉFÉRENCE DE LA COMMISSION 46

La Commission 46 n'administre pas les bourses, n'organise pas l'échange des astronomes, n'établit pas la liste des situations disponibles.

La Commission 46 coopère avec la Commission 38 pour élaborer la liste des Institutions, des chercheurs qui y travaillent et des directions de recherches dans lesquelles de jeunes astronomes pourraient être guidés.

La Commission 46 a pour tâche principale l'étude de la pédagogie de l'astronomie en vue de la préparation à la recherche. Elle examine les résultats obtenus au moyen d'une formation physique et mathématique convenable, et le rôle de la formation astronomique au cours des études universitaires.

Elle s'efforcera de promouvoir la publication d'un recueil de questions d'examen, d'exercices de travaux pratiques et de problèmes, afin de faciliter la tâche de formation des astronomes.

Ceci pose le problème de l'enseignement intégré des sciences qui devrait être étudié par la CIES. Toutefois les questions de l'Enseignement secondaire (enseignement moyen, ou secondary level) ne sont pas du ressort de la Commission 46. Celle-ci pourrait tout au plus agir comme conseiller scientifique en ces matières.

On trouvera maintenant quelques propositions.

(1) Le recrutement des astronomes serait facilité par la distribution aux étudiants d'une brochure sur l'astronomie, les matières enseignées, les directions de recherches existantes, les carrières possibles. Cela peut être une affaire locale ou nationale; la Commission 46 lance un pressant appel pour que de telles brochures soient rédigées.

Une telle brochure devrait comporter des conseils aux étudiants sur la meilleure formation.

(2) Le recrutement de bons étudiants et de futurs chercheurs de bonne qualité repose beaucoup plus sur l'organisation d'un enseignement de haut niveau que sur l'existence d'un enseignement élémentaire et de caractère semi-populaire.

(3) L'étude des problèmes pédagogiques peut s'accompagner d'une aide pédagogique aux petites institutions. L'UAI pourrait charger un astronome d'éditer un ouvrage réunissant des questions d'examen, des énoncés d'exercices pratiques et de problèmes, éventuellement accompagnés de leur solution.

A cet égard, il convient de signaler les problèmes posés par Irwin (**2**); la formation des astronomes doit comprendre un entraînement aux techniques d'observations. Cet entraînement ne peut se faire sans instruments modernes. S'il s'agit d'instruments optiques ils doivent aussi opérer dans un ciel clair. On peut signaler à cet égard le succès de l'Observatoire de Genève dans la formation de ses étudiants au moyen de l'instrument de recherche de 1 mètre situé à 360 kilomètres de Genève, à l'Observatoire de Haute-Provence, en France.

E. SCHATZMAN
Président de la Commission

BIBLIOGRAPHIE

1. Salam, A. *Phys. Today*, March 1965.
2. Irwin, J. B. 1966, *Science*, **152**, 1597.
3. McNally, O. 1965, The Education of Astronomers in the United Kingdom, *Q.J.R. astr. Soc.*, **7**, 22.
4. 1965, The Earth Science Curriculum Project Investigating the Earth, The American Society of Physics.

ANNEXE I. LISTE DES INSTITUTIONS AYANT RÉPONDU AU QUESTIONNAIRE

Afrique du Sud: Univ. Obs., Cape Town: Univ. South Africa, Pretoria.
Allemagne (DBR): Univ. Kiel.
Allemagne (DDR): Techn. Univ. Dresden.
Australie: Monash Univ. Clayton; Perth Obs.; Sydney Obs.
Bulgarie: Sofia Univ.

Canada: David Dunlap Obs., Univ. Toronto; Univ. Western Ontario, London.
Espagne: Inst. Marina San Fernando, Cadiz.
Etats-Unis: Boston Univ., Boston, Mass.; Calif. Inst. Techn., Pasadena, Calif.; Yerkes Obs., Univ. Chicago, Illinois; Univ. Colorado, Boulder, Col.; Harvard Univ., Cambridge, Mass.; Indiana Univ., Bloomington, Ind.; University Mich., Ann Arbor, Mich.; Pomona College, Claremont, Calif.; Swarthmore College Sproul Obs., Swarthmore, Penna.
Finland: Astron. Obs., Helsinki.
France: Univ. Aix-Marseille; Univ. Montpellier; Univ. Paris.
Grande-Bretagne: Royal Obs., Edinburgh; Royal Greenwich Obs., Herstmonceux; Univ. London Obs.
Italie: Oss. Astrof. Arcetri, Firenze; Oss. Astron., Trieste.
Japon: Univ. Tohoku, Sendai; Univ. Tokyo; Univ. Kyoto.
Mexique: Univ. Mexico.
Pays-Bas: Kapteyn Astron. Lab., Groningen; Sonnenborgh Obs. Univ., Utrecht.
Pologne: Astron. Obs. Univ. Copernicus, Torun.
Portugal: Astron. Obs. Univ. Lisbonne; Astron. Obs. Univ., Porto.
Suède: Univ. Obs. Stockholm.
Suisse: Univ. Obs. Basel; Obs. Genève.
Turquie: Univ. Ankara.

Received after completion of the report

U.S.A.: Princeton Univ.; Southern Missionary College, Collegedale, Tennessee; UCLA.
U.R.S.S.: Conseil astronomique.

ANNEXE II. RÉUNION DE DÉLÉGUÉS DE LA COMMISSION 46 ET D'EXPERTS

(Nice, 8–9 juillet 1965)

Propositions relatives à la formation des jeunes astronomes

I. La formation de jeunes astronomes exige un entraînement pratique et théorique de longue durée.

Organisée sur le plan international, notamment à l'intention de pays en voie de développement, cette formation pourrait se faire au moyen d'une Ecole de formation pratique d'Astronomie de l'UAI. Cela comporterait une formation essentiellement pratique 'A', reçue pendant 3 mois. Dans cette période, les étudiants seraient astreints à un contrôle strict de leur travail.

Il serait souhaitable que des étudiants reçoivent une formation 'B', en étant envoyés pour une année dans une institution où ils trouveraient une formation pratique et une formation théorique plus spécialisées.

II. L'organisation de cette Ecole Internationale pourrait se faire de la façon suivante:

(1) Trouver, dans chaque institution astronomique, le nombre et le niveau des jeunes astronomes qui pourraient faire l'objet d'un encadrement scientifique sérieux pendant un an.

(2) Choisir, chaque année, le pays et l'institution qui pourraient donner la formation intense de 3 mois. Si cela est à la fois souhaitable et possible, l'institution pourrait inviter des assistants (instructeurs) d'autres institutions.

(3) Adresser aux différents pays et institutions, en vue du recrutement des étudiants, un programme de l'Ecole, indiquant notamment les exigences du niveau (graduate level en mathématiques ou en physique, et au moins une connaissance de base de l'astronomie), les conditions de travail pendant la période d'entraînement intense, les conditions d'admission dans les institutions pour les deux types d'entraînement, une demande séparée devant être faite pour les 2 stages.

III. Le fonctionnement de cette Ecole Internationale demande les conditions suivantes:

(1) Désignation par le Comité Exécutif d'un secrétaire général pour 3 ans, prenant en charge l'organisation des Écoles de formation de l'UAI.

(2) Désignation, dans l'Institution qui prend en charge la formation de 3 mois, d'un directeur.

(3) Préparation d'un budget de l'Ecole. Il conviendra de déterminer la contribution de l'institution recevant l'Ecole de 3 mois et la contribution des institutions recevant les stagiaires d'un an.

(4) Préparation d'un contrat UAI-UNESCO, permettant notamment d'aider au recrutement des étudiants pour le stage de formation intense. On rappelle qu'il existe déjà des ressources appréciables pour les stages de longue durée.

ANNEXE III. RAPPORT SUR L'ENQUÊTE CONCERNANT LA FORMATION DE JEUNES ASTRONOMES

A la date du 19 novembre 1965, j'avais reçu vingt et une réponses à l'enquête commencée en septembre.

Vingt réponses approuvent en termes souvent chaleureux, l'intention de la Commission 46. Toutes les Institutions qui estiment pouvoir contribuer à l'initiative de la Commission 46 indiquent ce qu'elles peuvent faire.

Le classement suivant résume les propositions reçues:

(1) *Ecole de trois mois*

L'Observatoire de Bucarest demande à l'Academie des Sciences de Roumanie d'approuver l'organisation d'une Ecole pratique d'Astrométrie en 1967.

Le Dublin Institute for Advanced Studies (Professeur Wayman) envisage d'accueillir une telle Ecole dans quelques années.

L'Observatoire d'Edinbourg (Professeur Bruck) envisage de participer à l'organisation de ces Ecoles après 1967.

L'Observatoire de Harvard (Professeur Menzel) envisage de tenir une telle Ecole en 1968, moyennant quelques arrangements de durée. La lettre du Professeur Menzel pose clairement le problème du budget de l'Ecole.

Le Joint Institute for Laboratory Astrophysics, Boulder (Professeur Garstang) pourrait tenir une Ecole de Physique solaire après 1966.

Le Department of Astronomy de l'Université de Manchester (Professeur Kopal) pourrait s'engager à tenir une Ecole de trois mois en 1967.

Le Comité National d'Astronomie de Prague (Professeur Sternbeck) envisage de participer à ces Ecoles après 1968.

L'Observatoire du Mont Stromlo (réponse indicative du Professeur Bok) tient chaque année, à plus petite échelle (8 étudiants), une telle Ecole. L'extension paraît possible au cours d'une des prochaines années.

L'Observatoire de Wesleyan (Professeur Page) pourrait prendre une fois cinq étudiants (astrométrie, astrophysique, galaxies).

L'Observatoire de Yerkes (Professeur Morgan) pourrait participer aux Ecoles dans quelques années.

(2) *Entraînement d'un an*

Les Institutions suivantes peuvent contribuer à l'entraînement d'un an:

Royal Greenwich Observatory (Professeur Woolley); jusqu'à trois étudiants par an.

Observatoire de Liège (Professeur Swings): peut accueillir des étudiants à condition que soit résolu le problème financier.

National Radio Observatory (Green Bank, Professeur Heeschen). Un certain nombre de postes de 'Research associate' sont disponibles, pour des étudiants ayant déjà reçu un entraînement. Il est probable que seuls les meilleurs étudiants des Ecoles de l'UAI pourraient bénéficier des bourses du NRAO.

(3) *Autres réponses*

Les institutions suivantes: Berkeley, Johannesburg, Kitt Peak, Royal Society (au nom des Institutions Britanniques), Saltsjöbaden, Torun, pour des raisons variées, ne peuvent prendre part à l'organisation des Ecoles elles-mêmes.

Les réponses laissent penser que le stage de formation de un an serait possible, mais l'absence de précision à ce sujet rend nécessaire une nouvelle enquête.

L'Académie des Sciences de Paris (au nom des Institutions Françaises) fait une réponse évasive; le Professeur Rienäcker, au nom de l'Académie des Sciences de Berlin, annonce une réponse précise avant le 31 décembre 1965 concernant une Ecole en 1967.

En résumé, on peut faire les prévisions suivantes:

1967: Manchester ou Bucarest
1968: Harvard
1969: JILA
1970: Edinburgh ou Mont Stromlo

Il paraît vraisemblable que dès maintenant des plans peuvent être faits pour six Ecoles consécutives.

Il avait été convenu de chercher un secrétaire pour l'Ecole. Après consultation, le Comité Exécutif de l'UAI a désigné le Dr J. Kleczek.

ANNEXE IV. A SUMMER SCHOOL ON SATELLITE TRACKING FOR YOUNG OBSERVERS FROM THE COUNTRIES OF EASTERN EUROPE AND ASIA

(Tashkent, 7–15 September 1965)

In agreement with the plans of multilateral co-operation between the Academies of Sciences of East European and Asian countries on the problem 'Scientific investigations based on satellite tracking data' the Astronomical Council of the U.S.S.R. Academy of Sciences organized in Tashkent jointly with the Tashkent Astronomical Observatory during 7–13 September 1965 a summer school on satellite tracking for young observers. Tashkent was chosen since a good tracking station functions at the Tashkent Astronomical Observatory and because fine weather is typical there for this season, which is very important for training in tracking of satellites. Young astronomers from Bulgaria, Hungary, the DDR, Mongolia, Poland, Romania, the Soviet Union and Czechoslovakia (altogether 40 students) participated at the school.

The program included 14 lectures on general topics related to: methods of satellite tracking with the aid of photographic cameras; reduction methods of data obtained; and possibilities for using the data for scientific purposes. The detailed program was the following.

On 7 September three lectures were delivered on the utilization of precise photographic satellite tracking data for cosmic geodesy and for the study of the figure of the Earth and the Earth atmosphere.

On 8 September the NAPA camera used in the Soviet Union for precise satellite tracking was demonstrated and a review of methods for taking photographs of artificial satellites presented. Besides, the students were informed in a special lecture about possibilities to insure precise timing.

The third and the fourth days (9–10 September) were dedicated to two important items: predictions of satellite passages and methods of the reduction of photographic data. These problems were considered also in two lectures delivered on 13 September.

Then separate groups of students (5–6 persons in a group) were trained daily in handling the cameras, plate reduction, reception of precise time signals and calculation of ephemeris. Each student was given the opportunity to take several photographs of the balloon-satellite Echo-2 and to reduce the obtained plates. The lectures were delivered and the training carried out by the scientific staff of the Astronomical Council of the U.S.S.R. Academy of Sciences of the Institute of Theoretical Astronomy and of the Pulkovo and Tashkent Astronomical Observatories.

As a result, young observers, who as a rule had not been enough experienced in photographic tracking had a good training under the guidance of experienced observers and were given theoretical knowledge necessary for the work in this domain.

On 13 September a closing lecture was given by Professor V. P. Shcheglov, the Director of the Tashkent Observatory on the history of the astronomy of middle Asia. The lecture was followed by a two-day tour to the ancient Observatory of Ulughbok near Samarkand. The students had also the opportunity to see the beautiful architectural monuments of Samarkand.

Sight-seeing tour of the city of Tashkent and a visit to the Institute of Ancient Oriental Manuscripts were also arranged for the participants. Several documentary films about Middle Asia and a new picture about Ulughbok—the ancient founder of the Uzbok astronomy—were shown.

On the closing day all the participants were given special diplomes.

N. P. SLOVOKHOTOVA,
Scientific collaborator
of the Astronomical Council,
U.S.S.R. Academy of Sciences

INTER-UNION COMMISSION ON SOLAR AND TERRESTRIAL RELATIONSHIPS

The Commission held its final meetings in the House of Youth, Belgrade, on 25–26 August 1966, prior to the Inter-Union Symposium on Solar-Terrestrial Physics.

The meetings to be reported are:

(*i*) Business meeting of the Commission.

(*ii*) Open discussion on 'Suitability of Solar Data for Terrestrial Correlations'.

(*iii*) Individual contributions on solar-terrestrial relations.

At the time of the meetings, the membership was as follows:

Members

IAU	IUGG	URSI
C. W. Allen (U.K.), *President*	Mrs N. P. Benkova (U.S.S.R.)	G. M. Allcock (N.Z.)
R. Giovanelli (Australia)	T. Obayashi (Japan)	D. K. Bailey (U.S.A.), *Secretary*
E. R. Mustel (U.S.S.R.)	W. O. Roberts (U.S.A.)	R. Coutrez (Belgium)
M. Waldmeier (Switzerland)	E. J. Vassy (France)	A. H. Shapley (U.S.A.)

Corresponding Members

E. K. Bigg (Australia)
Mrs. H. Dodson-Prince (U.S.A.)
F. Link (Czechoslovakia)
R. Michard (France)
M. Nicolet (Belgium)

There were 155 names on the Roster of Active Workers.

Business Meeting of the Commission (IUCSTR)

After some informal discussion the meeting began at $10^h\ 20^m$, 25 August 1966. It was attended by the following Members or Corresponding Members:

C. W. Allen, *President* (in the Chair), D. K. Bailey, *Secretary*, E. R. Mustel (represented by N. W. Pushkov), T. Obayashi, W. O. Roberts, A. H. Shapley, M. Waldmeier.

The President of the newly formed Inter-Union Commission on Solar-Terrestrial Physics (IUCSTP), Dr H. Friedman, attended by invitation.

It was accepted that since the IUCSTR is to terminate in 1966 by ICSU resolution the present meetings would be the last held by the present Commission.

The report of the previous meeting at Hamburg, 24 August, 1964 (Trans. IAU, **12B**, 651, 1966) was adopted.

The President reported briefly on the use of available funds for travel and subsistence allowances. The funds are from ICSU sources and accounting is in the hands of the IAU Secretariat.

Future organization of solar terrestial relations

The main concern of the meeting was to consider the situation occasioned by various efforts to coordinate the science of solar-terrestrial relations (STR) with the wider science now using the title solar-terrestrial physics (STP). These activities were briefly reviewed and will be recorded. The Working Group on Solar-Terrestrial Physics, mentioned in the report on the Hamburg Meeting (p. 652), had met in September 1964 and recommended the formation of a Symposium Committee on STP to coordinate symposia in this field. This was considered by the IUCSTR through the medium of circulars in November 1964 and February 1965. The IUCSTR approved the Working Groups recommendation but considered it was *not* necessary to form a *new* Symposium Committee. In February 1965 the IUCSTR reported its willingness to modify itself with the view to accepting the responsibility of coordinating STP. This proposal was considered by ICSU at its meeting in Munich, April 1965 and resulted in Resolution 5 (ICSU Executive Committee 3rd Meeting, Summary Record, p. ii, 1965) which stated:

> The Executive Committee resolves that the Inter-Union Commission on Solar and Terrestrial Relationships (IUCSTR) be urged to make appropriate additions to its membership, and modify its terms of reference so as to provide representation for all interested Unions, Commissions of ICSU; and that the Unions and Commissions concerned be likewise urged to coordinate the arranging of symposia through this Commission.

In accordance with this resolution the IUCSTR submitted in July 1965 Draft Proposals for its own modification in order to undertake coordination of STP symposia. Several discussions with interested parties were held and in November 1965 the 'Proposals for an International Solar-Terrestrial Commission (ISTC)' was submitted to ICSU in time for consideration at the ICSU Bombay Meeting in January 1966. However, these proposals were not actually used as basis for the discussions in Bombay. The future of STP research and STP symposia was conjoined with the future of IQSY and led to the Resolution VII (ICSU Eleventh General Assembly, Summary Record, p. 37, 1966) which might be stated in a reduced form as follows:

(1) That the IUCI and IUCSTR be terminated in 1966,

(2) That an Inter-Union Commission on Solar-Terrestrial Physics (IUCSTP) be set up in 1966:

- (*i*) to promote, organize and coordinate international research in solar-controlled disciplines of geophysics,
- (*ii*) to coordinate all ICSU symposia in the field of solar-terrestrial physics, and
- (*iii*) to provide certain advisory services on operational activities arising out of IQSY.

The ICSU resolution was discussed and was considered to be a satisfactory basis for the future promotion of STR. It was noted that no Parent Union had been proposed and satisfaction was expressed that the IUCSTP would be directly responsible to ICSU.

The Commission then discussed what recommendations, arising out of the work of the IUCSTR, should be sent forward to the IUCSTP. The following recommendations were made:

(*a*) That the IUCSTP be asked to accept advisory responsibilities with respect to the *Quarterly Bulletin of Solar Activity* and other relevant Services.

(*b*) That the IUCSTP might at times be expected to promote operational planning, associated working group activities, and timely assessments of particular subjects such as problems arising from overlapping areas of interest in various scientific Unions under ICSU.

(*c*) That the IUCSTP give attention to maintenance of continuity in STP programmes and data.

(*d*) That the Roster of Active Workers maintained by the IUCSTR (containing 155 names) be sent to the IUCSTP Executive. This represents a useful list of scientists actively interested in STR.

(*e*) That, although the Reports **1** to **10** on Solar-Terrestrial Relations constitute an important part of the IUCSTR (and previous JCSTR) output, it is *not* recommended that the IUCSTP necessarily continue such reports.

(*f*) That archive material relating to the IUCSTR (and previous JCSTP) be retained for the present by C. W. Allen at the University of London Observatory. The following spares of the Reports on STR still exist 8 of **2**, 55 of **4**, 15 of **5**, 14 of **6**, 34 of **7**, 29 of **8**, 10 of **9**, and 13 of **10**.

It was recommended that the Report of the Belgrade meetings of IUCSTR should be prepared for presentation to the ICSU Executive Committee meeting in Monte Carlo (October 7–8, 1966), and that the points relevant to the IUCSTP should be sent to its President (Dr H. Friedman) as soon as possible in order that the points may be considered in the planning of future activities.

As a conclusion to the Meeting the Commission noted with satisfaction the creation of the new IUCSTP and commended the President (IUCSTR) for his leadership in the Commission's activities and for his participation in the arrangements for specification of terms of reference for the new Commission which should serve as an excellent basis for future international activities not only in the field of solar-terrestrial physics but also for the continuation and extension of activities of the character that have in the past occupied our Commission.

The activities and responsibilities of the IUCSTR *will close* when the various reports and recommendations have been prepared, circulated and submitted. A statement is to be circulated to the Roster of Active Workers. The President is to report to ICSU in October. He will also arrange with the IAU to close the accounts after the ICSU meeting in October.

Open Discussion on Suitability of Solar Data for Terrestrial Correlations

Meeting held at 2 p.m. 25 August 1966.

Attended by Allen, Bailey, Bernard, Buzchanov, Campbell, Cardús, Davies, Dieminger, Dinulescu, Djurkovič, Ehmert, Fortini, Friedman, Jacchia, Karabin, Knuth, Kundu, Landstreet, Lauter, Lincoln, Lüst, Matsushita, Milogradov, Minnis, Mustel, Obayashi, Parepeanu, Piccardi, Pick, Popovici, Przeuijervic, Pushkov, Ramanathan, Righini, Roberts, Roederer, Sawyer, Senatre, Shapley, Svestka, Tandberg-Hanssen, Valniček, van Zandt, Waldmeier, White, Zhulin.

The President introduced the discussion by saying that the solar data to be discussed related to the extended routine data intended for use rather than exploration or speculation. The purposes of such data could be summarized: (*i*) for studying effects on the Earth and surrounding Earth-space, (*ii*) for coordinating specialized observations and researches, and (*iii*) for studying the Sun itself. The discussion would take this order of priorities.

Points that might be considered in discussing the data were, (*i*) the measurement often was no more than an *indication* of the physical quality required, (*ii*) measurements were frequently subjective and required standardization and comparability, (*iii*) data were often applied through correlations and secondary correlations may require elimination, (*iv*) it was often necessary to segregate a total solar effect into individual activity contributions, (*v*) there is need for numerical indices for descriptively complex phenomena.

The discussion was subdivided into (*i*) generalities, such as the points mentioned above, (*ii*) slowly varying phenomena, active areas, (*iii*) rapidly varying phenomena, flares, and (*iv*) phenomena not associated with active areas including the quiet Sun.

Discussion points

Shapley: Difficulty of dealing with very extensive and almost continuous data. There may be need for more analysis before publishing.

Waldmeier: Filament should be reported in a numerical form suitable for correlation with *M* regions.

Mustel: More whole-disk magnetic field measurements should be undertaken (recommendation A made). Early magnetic storm data should be sorted into recurrent and sporadic.

Lincoln: The CRPL are preparing to publish the Mount Wilson magnetic data.

Pushkov: Introduced the question of solar terrestrial services (recommendation C made).

Valniček: Introduced the question of computer processing of data.

Jacchia: Suggested that steps be taken to segregate the active and quiet Sun components of 10·7 cm radiation (recommendation B made).

Friedman: Referred to the variations in 2 to 10 Å solar emissions and their relations to ionosphere responses. These variations challenge solar observers to detect analogous variations in spectroheliographic material.

Kundu: Suggested 3 cm solar-radio recordings might give indication of similar variations.

Godoli (Submitted): The peculiarities of long-term correlations can be ascribed to the behaviour of solar indices.

IUCSTR Recommendations arising out of Discussion on Solar Data

A. *Solar Disk Magnetic Observations*

In view of the importance of relating the whole picture of the Sun's magnetic field to the magnetic field in the neighbourhood of the Earth and in interplanetary space

It is *recommended:*

That observatories with facilities for high spectroscopic resolution consider the possibility of making magnetic observations of the whole solar disk on a regular basis.

B. *Quiet and Active Radio Observations*

In view of the great use made of solar radio observations at various wavelengths for correlation with solar XUV flux measurements at all stages of the sunspot cycle

It is *recommended:*

That radio observatories offering suitable data be asked to endeavour to express the results quantitatively in such a way that the two components, quiet and active, can be segregated.

C. *International Solar-Terrestial Services*

Considering:

(1) That the collection and publishing of solar-terrestrial data is prepared by many different organizations in several countries,

(2) That this work has received some coordination in the past, but that this coordination needs to be maintained and improved with particular reference to such questions as standards and definitions,

(3) That it is by no means certain that all data thus published are of sufficient use to the scientific and operating communities to justify their publication,

(4) That it is possible that additional observations could usefully be made,

(5) That in the field of meteorology the WMO provides services for all user purposes,

The IUCSTR *recommended:*

(1) That a clear need now exists for a more closely coordinated *international solar-terrestrial services* such that the needs of both the scientific and operational communities requiring solar-terrestrial information will be effectively served.

(2) That the IUCSTP be asked to consider how the need for such services can be met.

Individual Contributions on Solar-Terrestrial Relations

Meeting held at 9.30 a.m. 26 August 1966.

It was an open meeting attended by about 45 persons. The papers submitted were reported very briefly since it may be expected that many of them will be published elsewhere.

T. Obayashi dealt with the identification of *M*-regions from their influence on the diurnal variations of cosmic rays. The phasing of the various disturbances was illustrated on a 'floral' diagram.

S. Matsushita gave an analysis showing the detailed effects on the geomagnetic S_q and L variations that could be associated with solar activity.

M. Pick described a study of the slowly varying component of solar radio emission in relation to the structure of associated centres of solar activity. The distribution in the radio spectrum is found to be related to a factor measuring certain geometric properties of sunspots and their magnetic fields.

E. Tandberg-Hanssen gave results on the magnetic fields of prominences obtained from the High Altitude Observatory magnetograph at Climax.

B. Valniček illustrated the features of an active flare prominence event observed with high spectroscopic resolution in Hα.

G. Piccardi questioned the causality of correlating phenomena. Previously his F-test had been related to flares and he now finds a good relation between the P-tests and proton flares.

P. Bernard discussed a 2-year period in geomagnetic variations but could not ascribe it to a solar activity effect; he find the magnetic field of ionospheric winds quantitatively sufficient to account for it.

Constance Sawyer described the preference for certain heliographic longitudes of 45 proton-flare regions of cycle 19. With 15 famous flares of the last century they define a rotation period of 27·2135 days. The rigid rotation and long lifetime of this pattern suggest a deep-seated magnetic field of large scale.

T. E. van Zandt had studied the variations in the upper atmospheric ratios $O^+/He^+/H^+$ throughout the minimum of the solar cycle. The interpretation of these variations led to the suggestion that the coronal variations lagged a year behind those of the chromosphere. Support for this suggestion was found in the E and F layer critical frequency variations.

C. Popovici, A. Parepeanu and V. Dinulescu considered the major flares during 1958–61 which did *not* give rise to geomagnetic storms. Certain aspects and morphology regularities were detected.

K. Davies studied the sudden ionosphere frequency deviations resulting from flares and showed that the characteristic electron increase occurred mostly above 100 km. Time variation studies suggest that the D and E–F effects originate from separate parts of the solar ionizing spectrum. The relaxation time of the E region is found to be about one minute.

M. L. White described how the mechanism of rotation and drift in the solar photosphere should give rise to a 3-leaf Rossby wave pattern such as occurs in 'dishpan' experiments. Many solar phenomena, including the equatorial drift, can be explained on this basis.

D. H. Menzel submitted (in his absence) a paper on the nature of sunspots and solar activity.

Invited Address

An invited address by Dr E. G. Bowen, Chief of Division of Radiophysics, CSIRO, Sydney, entitled 'Solar and Planetary Interactions' had been planned for the afternoon of 26 August.

Unfortunately this address had to be cancelled as Dr Bowen was unable to make the trip from Australia.

Closure of IUCSTR Meetings

The Belgrade final Meetings of the IUCSTR closed at the end of the Individual Contributions meeting on the morning of 26 August 1966.

REPORT OF THE WORKING GROUP ON PHOTOGRAPHIC MATERIALS

1. The formation of this group dates from a meeting convened during the Hamburg General Assembly, 31 August 1964, by R. H. Stoy, supported by I. S. Bowen and H. Haffner. There was a large and enthusiastic attendance, which passed with acclamation the following resolution:
'In view of the great importance to astronomy of the supply of the most suitable photographic material, this meeting of astronomers from many countries requests the Executive Committee of the International Astronomical Union to set up a working group responsible either to itself or to some nominated Commission to investigate, (*a*) the present supply of photographic materials and the best methods of procuring and handling them, (*b*) the potential consumption of specialized photographic material, with a view to interesting the various manufacturers in its production.'

2. By letter of 15 September 1964, the General Secretary conveyed the approval of the Executive Committee of the IAU to the establishment of the Working Group, with the following membership:

President: R. H. Stoy
Secretary: D. S. Evans
Members: Aly, Mme Andrillat, Bertola, Gollnow, Griffin, Haffner, Mendoza, Mikhailov, Savedoff, Sharpless, Harlan J. Smith, Steinlin, Valniček, Velghe, Wempe.

3. The operations of the Working Group so far undertaken, demonstrate the great extent of the problems involved. Some of these are capable of being solved almost immediately by the action of astronomers themselves. Others require consultation with manufacturers. Yet others, which are concerned with problems of currency allocation, the possibility that some materials may be designated as of strategic importance by manufacturing countries, and so forth, may require discussion at governmental level. It is clear that there is a continued necessity for the work of the Group, and that numbers of problems will require a long time for their evaluation even before solutions can be sought. The investigations undertaken so far have been admirably supported by the manufacturers of photographic materials, and by a considerable number of observatories. The number of the latter is, however, not as great as it should be. It must be emphasized that, in order to get meaningful statistics describing the supply position, it is just as important for observatories not now facing difficulties, to report their situation, as it is for those who have problems in need of solution.

We take this opportunity of thanking those manufacturers and observatories who have collaborated in the work of the Group, and urge those who have not yet sent in reports to do so.

The foregoing remarks make it clear that the present report cannot be comprehensive. Its aim is to draw attention to a few of the problems which have revealed themselves, in the hope that these can be tackled first, leaving others for the continuing work of the Group.

4. A circular letter was sent to all manufacturers known to be active in the production of photographic materials suitable for astronomical purposes. A report incorporating their replies was produced in January 1966 and circulated with a questionnaire to a large proportion of the world's observatories. Copies of these documents and of an analysis of the replies may be had on request from the Cape Observatory.

5. The American Astronomical Society has set up a Committee of its own in the same general field. It should be clear that because of the relatively close proximity of the U.S. and Canadian observatories to the Kodak works at Rochester, N.Y., and because of the large numbers of observatories and astronomical departments of universities in these two countries, the problems encountered are somewhat different from those in the world at large. Nevertheless there are

problems, many of which have been discussed at an expert level in the admirable memoranda prepared for private circulation by Mr William C. Miller of Mount Wilson and Palomar Observatories. It is evident that the U.S. and Canada constitute a natural area for the establishment of a system of cooperative ordering and supply, and the practical problem is one in which the IAU Working Group as such can have very little constructive role, except to offer encouragement.

It is understood that a cooperative ordering scheme is being operated in West Germany and one may be begun in Australia.

A Joint Meeting of the AAS and IAU Groups was held in 1966 July 29, together with representatives of Eastman Kodak, Rochester, at the Mees Observatory, Naples, N.Y. Members of the Working Group who attended were Evans, Sharpless, Smith and Savedoff. The discussion ranged over the whole field of supply and storage. Items from the conference summary which are of particular interest are incorporated in the report referred to in paragraph 4. It seems desirable for the Group to organize discussion leading to definite recommendations on particular topics.

The obvious first choices would seem to be:

(*a*) Standardization of plate sizes for astronomical purposes.

(*b*) Canalization of demand for emulsion varieties to reduce unnecessary diversity.

(*c*) Definition of the next tasks to be undertaken by the Group.

6. *Publications.*

It is understood that Mr William C. Miller may be contemplating the production of a book. At the moment his vast experience and knowledge of photographic emulsions as applied in astronomy is scattered through various published papers and memoranda produced for private circulation. The Working Group considers that a book by Mr Miller would be invaluable, and wishes him success in his labours.

The following is a list of some publications likely to be of interest in the field:

F. Bertola, Sull'aumento di rapidità delle lastre fotographiche per uso astronomico, *Contr. Oss. astrofis. Univ. Padova Asiago*, No. 154, *Mem. Soc. astr. ital.*, **35**, no. 2, 1964.

I. S. Bowen and L. T. Clark, Hypersensitisation and Reciprocity Failure of Photographic Plates, *J. opt. Soc. Am.*, **30**, 508, 1940.

J. D. Fernie, A New Photographic Developer for Astronomical Use, *Publ. astr. Soc. Pacif.*, **74** 238, 1962.

J. D. Fernie and Sandra Holm, Studies in Astronomical Photography. I. On the Use of baked IIa-O Spectroscopic Plates, *J. R. astr. Soc. Can.*, **58**, 13, 1964.

II. On the Use of Diafine Developer, *Ibid.*, **58**, 201, 1964.

A. A. Hoag, Cooled-Emulsion Experiments, *Publ. astr. Soc. Pacif.*, **73**, 301, 1961.

William C. Miller, The Application of Pre-Exposure to Astronomical Photography, *Publ. astr. Soc. Pacif.*, **76**, 328, 1964.

: A Pre-Flasher for Photographic Plates, *Ibid.*, **76**, 433, 1964.

: Memorandum: The Mount-Wilson Palomar Photographic Plate Speed System.

: Preliminary Report on the Reduction of Reciprocity Failure in Astronomical Plates by Controlled Baking, Photographic Laboratory, Mount Wilson and Palomar Observatories, 1956: Appendix, 1966.

N. Richter and W. Hogner, Eine neue Plattensorte für Astroaufnahmen im roten Spektralbereich, *Monatsbericte der Deutschen Akademie der Wissenschaften*, **6**, 809, 1964.

R. Shaw, The Equivalent Quantum Efficiency of Aerial Films, *J. photogr. Sci.*, **13**, 308, 1965.

W. F. Swann, Summary of AAS, IAU, Kodak meeting at Naples, N.Y., 1966.

Report of Meeting on Photographic Plate Testing, Mount Wilson and Palomar Observatories, 1966.

Kodak Plates and Films for Science and Industry.

R. H. STOY
President of the Working Group

21. COMMISSION DE LA LUMINESCENCE DU CIEL

PRÉSIDENT: Dr F. E. Roach, Environmental Science Services Administration, Boulder, Colorado, U.S.A.

VICE-PRÉSIDENT: Professor M. Huruhata, Tokyo Astronomical Observatory, Mitaka-shi, Tokyo, Japan.

COMITÉ D'ORGANISATION: J. W. Chamberlain, J. Dufay, M. F. Ingham, V. I. Krassovsky, F. Link, M. Nicolet.

MEMBRES: Asaad, Barber, Barbier†, Bates, Blackwell, Blamont, Chapman, Cook (A. F.), Dauvillier, Dufay (M.), Elsässer, Elvey, Fesenkov, Fesenkova, Friedman, Galperin, Gartlein†, Gauzit, Grandmontagne, Harang, Haug, Hoffmeister, Hunten, Jarrett, Kaplan (J.), Karyagina, Kastler, Lebedinsky, Megrelishvili, Pearse, Pettit-Knaflich, Richter, Roblev, Seaton, Steiger, Swings, Tousey, Vassy, Wallace.

INTRODUCTION

Included in this draft report are discussions of papers on the night airglow and the twilight glow. Because of the fact that Commission 21 in collaboration with Commission 22 is sponsoring a colloquium in Honolulu on the Zodiacal Light and the Interplanetary Medium from 28 January to 2 February 1967, (one month after this draft report is due) it is planned to postpone definitive discussion of the zodiacal light, gegenschein and the integrated starlight until the time of the general assembly. We are omitting general discussion of the aurora except for some low and mid-latitude phenomena which have both airglow and auroral characteristics. Research reports based on work done in the Soviet Union have been reviewed by N. N. Shefov. His resumé is included as a part of the present report.

The period between the XIIth and the XIIIth General Assemblies includes much of the IQSY (International Year of the Quiet Sun) from 1 January 1964 to 30 December 1965. The level of night airglow research activity during the IQSY was somewhat lower (19 reporting stations) than that during the earlier IGY (28 reporting stations). During the triennium the space age has matured and a beginning has been made on significant observations from orbiting spacecraft. Observations have also continued from rockets as well as balloons.

Recently there has been a polarization of interest in the nightglow by geophysicists, in contrast with the earlier studies in which the astronomical fraternity maintained the initiative. This is a natural evolution as the interest in the nightglow tends to concentrate on the atmospheric reactions and processes responsible for the emissions. The earthbound astronomer maintains his interest because of the fact that the nightglow constitutes a foreground 'haze' which he must pierce. As we proceed to astronomy from spacecraft beyond the nightglow layers we may anticipate that the nightglow will come to be more and more associated with geophysics until such a time as we begin to accumulate data of atmospheric radiations from other planets.

Before beginning the detailed report it seems appropriate to express the personal sorrow and to acknowledge the professional loss in the death of the distinguished former president of Commission 21, Dr Daniel Barbier (1907-65) (*Annales de Géophysique*, **21**, 295, 1965).

GROUP DISCUSSIONS, COLLOQUIA AND SYMPOSIA

During the triennium group discussions have been held, or are planned, as follows:

1965, April 14–16, Boulder, Colorado, U.S.A.

Group discussion participated in by some 70 persons. Subject 'Middle and Low Latitude

Atmospheric Emissions and the Ionosphere'. The deliberations were dedicated to the memory of Dr Daniel Barbier. Proceedings have been prepared by F. E. Roach and K. D. Cole in a non-referencable form which can be obtained by request to F. E. Roach, ESSA, Boulder, Colorado.

1965, August 16–20, Cambridge, Massachusetts, U.S.A.
International Symposium on Aeronomy organized by the International Association of Geomagnetism and Aeronomy (UGGI). Many of the papers have been published in *Ann. Géophys.*, **22**, no. 2, 1966.

1965, September, São Jose dos Campos, Brazil.
Second international symposium on Equatorial Aeronomy. Proceedings and abstracts of the papers have been published (F. de Mendonca, editor) as CNAE-LAFE 32, November, 1965. Many of the full papers have been published in *Ann. Géophys.*, **22**, no. 3, 1966.

1966, August, University of Keele, England.
Advanced Study Institute on Aurora and Airglow. A compilation of proceedings and papers is in press (Billy M. McCormic, editor), Reinhold, 1967.

1967, January 30–February 2, Honolulu, Hawaii, U.S.A.
An international colloquium sponsored by Commissions 21 and 22 of the IAU on The Zodiacal Light and the Interplanetary Medium. It is planned to publish the papers (J. L. Weinberg, editor).

1967, July 17–21, London, England.
Fourth and final IQSY Assembly.

A. AIRGLOW, GENERAL

In preparing this draft report, the procedure was to scan systematically the following journals for 1964, 1965, and 1966: *Annales de Geophysique*, *Astrophysical Journal*, *Journal of Atmospheric and Terrestrial Physics*, *Journal of Geophysical Research*, *Planetary and Space Science*, *Space Science Reviews*. This was supplemented by lists furnished by several colleagues. It is planned to circulate the draft report widely with the hope that the bibliography can be extended to include papers that may have been missed in the first screening.

For convenience the bibliography is arranged to correspond to the text. At the end of each section of the bibliography papers by Soviet colleagues are indicated according to the listing by Shefov.

Of particular interest in the listing of general airglow papers is the comprehensive report by Mme Glaume (1965, reference **A5**). The paper comprises a detailed study of what may be considered the 'classical' nightglow radiation, the 5577Å line of atomic hydrogen. Mme Glaume discussed the temporal variations on the basis of the large library of observational data from Haute-Provence. She also made a comparison of the observations from many stations, using north-south intensity gradients to call attention to evidence that 5577Å maximizes in mid-latitudes, under the influence or control of the Earth's magnetic field.

In addition to the papers which deal with specific problems attention is called to the broader review papers (**A2**, **A6**, **A14**, **A15**).

B. TROPICAL AIRGLOW AND THE IONOSPHERE

The references listed in this section are heavily weighted to the ionosphere, which needs some explanation since this is a report to astronomers. The reason is as follows. In the tropics the nightglow 6300Å and one photometric component of 5577Å are emitted in the F region of the ionosphere by a photochemical process that is directly associated with the equilibrium that maintains the ionospheric electron balance.

$$\begin{cases} O_2 + O^+ \rightarrow O_2^+ + e \\ O_2^+ + e \rightarrow O + O + 6{\cdot}96 \text{ eV} \end{cases}$$

The second of these reactions is capable of exciting the atomic oxygen to the ^1D (1·96 eV) and/or the ^1S (4·16 eV) states from which the 6300Å and/or 5577Å emissions originate. It is possible to derive a relationship between the brightness of the radiations (e.g. 6300) and ionospheric parameters customarily measured on ionograms (reference **B26**). This relationship (known as the 'Barbier formula' after its discoverer) is

$$Q_{6300} = A + B\,(f_0\,F_2)^2 \left[\exp - \left(\frac{h'F - 200}{H}\right)\right]$$

where Q_{6300} is the brightness of the 6300 emission (usually in rayleighs), A is an additive constant corresponding to a background brightness, B is a constant which essentially puts the results in rayleighs, f_0F_2 is the so-called critical frequency, the square of which is proportional to the F-region peak electron density, h'F is the 'virtual height' as measured on an ionogram and H is a scale height, approximately 40 km in the F region.

In the tropics both the nightglow and the ionosphere exhibit dynamic changes. For many years it has been known that the ionosphere has maximum electron densities some 15° from the geomagnetic equator—a phenomenon which is known as the equatorial anomaly. During the triemmium under review a number of papers have been produced on the anomaly (**B1, 6, 7, 8, 9, 10, 11, 16, 17, 18, 19, 20, 22, 23, 24, 28, 30, 32, 33, 35, 36**). The other papers listed discuss the other aspect of the problem, namely the photometric observations of the nightglow which is directly associated both with the above mentioned photochemical reactions and with the morphology of the equatorial ionospheric anomaly.

C. MID-LATITUDE STABLE AURORAL ARCS

The phenomenon which, in the literature has been referred to as 'l'arc auroral stable', M-arc, SAR-arc, has the 'auroral property' of alignment under the control of the geomagnetic field and the airglow property of a low level of excitation. The conference in Boulder (14–16 April 1965) referred to above was devoted, in large part, to a discussion of its properties. It is now apparent that the occurrence of the arcs follows the sunspot cycle as does the visible aurora. Thus the period between the XIIIth and XIVth General Assembly of the IAU should see a revival of occurrence of the arcs. During the triennium under review they have been sparse. Attention is called especially to the papers by K. D. Cole in the attached bibliography.

D. EXCITATION OF 6300Å [OI] BY PHOTOELECTRONS

For decades observers have noted that the 6300Å emission often has a pre-dawn enhancement which is not at all symmetrical with the post-twilight decrease in intensity. The paper by K. D. Cole (**D2**) explains the pre-dawn enhancement as the result of the travel of photoelectrons from the magnetically conjugate region where local dawn starts earlier than at the observing station. The prediction has been confirmed by H. C. Carlson at the Arecibo station in Puerto Rico (**D1**).

E. NIGHTGLOW VARIATIONS, TEMPORAL AND SPATIAL

The papers listed under this category Include discussions of diurnal, seasonal and sunspot cycle variations. Of particular interest are studies of geographical changes and the reader is referred back to the discussion of Mme Glaume's paper (**A5**) under the Airglow, General, discussion.

F. HYDROGEN AIRGLOW

The fluorescent excitation of atomic hydrogen in the outer atmosphere has received significant attention during the triennium. Our Soviet colleagues have been especially active in these studies.

Attention is called to the interesting discussion by Kondo and Kupperian (**F7**) of a suggested interaction between neutral hydrogen and charged particles in the radiation belt based largely on the observational results of Winter (**F8**).

G. AERONOMY AND EXCITATION PROCESSES

The large number of papers cited in this section illustrates the point made earlier that the night airglow center of attention is frequently more geophysical than astronomical. During the triennium the reactions mentioned in section B have been established as of prime importance both in the ionospheric balance and of the emission of the high nightglow. These reactions are discussed in references (**G4, 8, 14, 15, 16, 21, 22, 26, 27**).

An important paper on the rate coefficient of the Chapman reaction ($O + O + O \rightarrow O_2 + O^*$) is that of Young and Black (**G28**; see also **G2**). The question of excitation by energetic particles is actually one of the extension of auroral type mechanisms equatorward of the auroral zone (pertinent references: **G1, 5, 7, 11, 12, 23, 24**).

H. N_2^+ IN THE NIGHTGLOW

The N_2^+ band at 3914Å is characteristic of auroras and of the twilight. Its possible presence in the nightglow is difficult to determine for two reasons: (1) if present at all it is extremely faint and (2) contamination by either auroral or twilight emissions is an observational hazard. Three of the papers cited illustrate the difficulty and the disagreement of observers. Broadfoot and Hunten (**H1**), working with relatively high spectroscopic resolution, report that N_2^+ 3914Å has an intensity *less than* one rayleigh. O'Brien, *et al.* (**H4**) from rocket observations report some five rayleighs. Yano (**H5**) from spectrographic observations in Australia has reported about ten rayleighs.

I. HYDROXYL (OH) NIGHTGLOW

This very intense radiation has been studied with considerable concentration by the Soviet observers (note the eleven references in Shefov's list). It is hoped that this summary may inspire all observers to give more attention to this interesting phenomenon.

J. DOPPLER TEMPERATURES

The determination of temperature based on the measurement of the Doppler width of a spectral line requires high resolution interferometric techniques. Because of the meticulous technology involved in such measurements there have not been extensive measurements on anything like a synoptic basis. The references cited illustrate that good results have been obtained. Of particular interest during the next sunspot maximum will be measurements of 6300Å temperatures in midlatitude auroral arcs (see section C above).

K. INSTRUMENTATION AND TECHNIQUES

Absolute calibration of photometers for low light levels continues as a problem (**K2, 5**). Height measurements by the van Rhjin method are discussed in references **K1** and **K3**. A successful use of triangulation in determining nightglow heights is described in **K6** and **K7**.

L. ROCKET AND SATELLITE OBSERVATIONS

This triennium has witnessed significant observations of the airglow from orbiting spacecraft. An effort has been made to record a definitive list of references (**L6, 7, 8, 9, 10, 11, 12**) in which descriptions occur of the nightglow as seen by astronauts. The references cited are in general available for a moderate fee by application to the Superintendent of Documents, U.S. Government Printing Office, Washington, D.C., 20402. The observations may be summarized by the statement that all of the Mercury and Gemini astronauts have seen the 100-kilometer airglow layer below the spacecraft usually as a structureless band about 2° above the horizon for orbits below 200 kilometers. In the absence of moonlight it is much more conspicuous than the solid earth surface. Two of the astronauts (Schirra and Cooper) have reported sightings at higher elevations than the spacecraft which have been tentatively ascribed to the F-region nightglow. For photographs from a rocket of the 100-kilometer nightglow see reference **L4**.

An interesting observation (reference **L2**) is the isolation of the nightglow 6300Å emission layer together with a 5577Å emission layer in the F region, much fainter than the 5577Å emission near 100 kilometers. This confirms what had been deduced from tropical nightglow studies where sporadic changes in 6300Å nightglow are accompanied by simultaneous changes in 5577Å (see section B).

M. TWILIGHT GLOW

Two excellent review papers (**M1, 2**) are in press. During the period under review some 60 papers on the twilight glow have been published (see the extensive bibliography in **M2**). Attention is called to the significant activity in this field by Soviet investigators.

FRANKLIN E. ROACH
President of the Commission

BIBLIOGRAPHIE

A. Airglow, General

1. Abbott, W. N. 1964, Possible correlations between sporadic-E and low-altitude airglow emission of 5577Å in median latitudes, *J. atmos. terr. Phys.*, **26,** 776.
2. Barbier, D. 1964, Airglow, *Res. Geophys.*, **1,** Chapter 16, Mass. Inst. Tech.
3. Barbier, D. 1965, La lumière du ciel nocturne aux Iles Kerguelen, *Ann. Géophys.*, **21,** 299.
4. Bellew, W., Silverman, S. 1966, General behavior of 6300Å [O I] at Sacramento Peak, New Mexico, *Planet. Space Sci.*, **14,** 407.
5. Glaume, J. 1965, Étude de la raie 5577Å de l'oxygène dans la luminescence atmosphérique nocturne, *Ann. Géophys.*, **21,** 1.
6. Hunten, D. M. 1967, Spectroscopic studies of the twilight airglow, *Space Sci. Rev.*, **6,** 493.
7. Huruhata, M. 1965, Lunar tidal effect on the oxygen green line in the night airglow, *J. geophys. Res.*, **70,** 5979.
8. Huruhata, M. 1964–1965, Japanese contribution to the IQSY.
9. Kulkarni, P. V. 1965, On the height of the 5577Å [O I] airglow layer in Hawaii, *Ann. Géophys.*, **21,** 58.
10. Lagos, P. 1964, The airglow 6300Å [O I] emission the luminosity profile with varying scale height, *J. atmos. terr. Phys.*, **26,** 325.
11. Markham, T. P., Anctil, R. E. 1966, Airborne night airglow measurements in the south atlantic magnetic anomaly, *J. geophys. Res.*, **71,** 997.
12. Murcray, W. B. 1963, Infrared radiation from the atmosphere over the Arctic Ocean, *Science*, **141,** 802.
13. Nguyen-Huu, Doan 1964, Étude spectrophotométrique de la lumière ciel nocturne et crépusculaire dans le visible, *Ann. Géophys.*, **20,** 1.

14. Roach, F. E. 1964, The light of the night sky: astronomical, interplanetary and geophysical, *Space Sci. Rev.*, **3**, 512.
15. Roach, F. E., Smith, L. L. 1967, The worldwide morphology of the atomic oxygen nightglows, in Aurora and Airglow, Proc. of Aurora and Airglow Advanced Study Institute, Ed. Billy M. McCormic, Reinhold, p. 29.
16. Tanabe, H. 1964, Zodiacal light and airglow components at 5300Å, *Publ. astr. Soc. Japan*, **16**, 324.
17. Tanabe, H. 1964, Further evidence for the existence of an airglow radiation near 5300Å independent of 5577Å line, *Publ. astr. Soc. Japan*, **16**, 340.
18. Pertinent papers by Soviet authors: Shefov's list, No. **7, 13, 15, 16, 24, 27, 57, 58.**

B. Tropical airglow and the ionosphere

1. Awajobi, O. A. 1965, On the ionspheric F2-region equatorial anomaly, *J. atmos. terr. Phys.*, **27**, 1309.
2. Barbier, D. 1964, Nouvelles observations de la raie rouge du ciel nocturne en Afrique, *Ann. Géophys.*, **20**, 22.
3. Barbier, D. 1964, L'observation optique de la couche F de l'ionosphère, *Scientia*, Sept.-Oct.
4. Barbier, D. 1964, Observations photométrique d'une perturbation de la haute atmosphere, *Astrophys. norw.*, **9**, 55.
5. Barbier, D. 1965, Deux phenomènes particuliers présentés par la raie rouge de la lumière du ciel nocturne, *Ann. Géophys.*, **21**, 228.
6. Baxter, R. G. 1964, On the theoretical effects of electrodynamic drift in the equatorial F2 layer, *J. atmos. terr. Phys.*, **26**, 711.
7. Baxter, R. G., Kendall, P. C., Windle, D. W. 1965, Power series solution of the F2-layer diffusion equation, *J. atmos. terr. Phys.*, **27**, 1263.
8. Brace, L. H., Spencer, N. W. 1965, Detailed behaviour of the midlatitude ionosphere from the Explorer XVII satellite, *Planet. Space Sci.*, **13**, 647.
9. Bramley, E. N. 1966, Plasma diffusion in the ionosphere, *J. atmos. terr. Phys.*, **28**, 681.
10. Bramley, E. N., Peart, M. 1964, Diffusion and electromagnetic drift in the equatorial F2 region, *J. geophys. Res.*, **69**, 4609 (also in 1965, *J. atmos. terr. Phys.*, **27**, 1201).
11. Burkard, O. M. 1966, Some remarks on papers concerning the geomagnetic anomaly of the F2-layer, *J. atmos. terr. Phys.*, **28**, 1129.
12. Carman, E. H. 1964, Dependence of [O I] 6300Å airglow on latitude as observed from Townsville, Australia, *Nature*, **201**, 595.
13. Carman, E. H. 1965, Seasonal trends in the latitude correlation between ionospheric parameters and [O I] 6300Å airglow observed from Townsville, *J. atmos. terr. Phys.*, **27**, 329.
14. Carman, E. H., Gibson-Wilde, B. C. 1964, Seasonal variation of the [O I] 6300Å airglow at Townsville from isophote sky maps, *J. geophys. Res.*, **69**, 487.
15. Carman, E. H., Kilfoyle, B. P., Gibson-Wilde, B. C. 1964, Pattern of the [O I] 6300Å airglow emitting regions at Townsville, *J. geophys. Res.*, **69**, 4725.
16. Chandra, S. 1964, Plasma diffusion in the ionosphere, I., *J. atmos. terr. Phys.*, **26**, 113 (II. 1965, *J. atmos. terr. Phys.*, **27**, 1159).
17. Chandra, S., Goldberg, R. A. 1964, Geomagnetic control of diffusion in the upper atmosphere, *J. geophys. Res.*, **69**, 3187.
18. Duncan, R. A. 1964, Electrodynamic lifting and the equatorial F region of the ionosphere, *J. geophys. Res.*, **69**, 3741.
19. Gliddon, J. E. C., Kendall, P. C. 1964, Perturbation of a model F2-region by a small arbitrary time-dependent electrodynamic drift, *J. atmos. terr. Phys.*, **26**, 721.
20. Goldberg, R. A., Kendall, P. C., Schmerling, E. R. 1964, Geomagnetic control of the electron density in the F region of the ionosphere, *J. geophys. Res.*, **69**, 417.
21. Hunten, D. M., Godson, W. L. 1967, Upper-atmospheric sodium and stratospheric warnings at high latitudes, *J. Atmos. Sci.*, **24**, 80.

22. Kendall, P. C. 1964, On the equations governing diffusion in the F2 layer, *J. atmos. terr. Phys.*, **26**, 624.
23. Kendall, P. C., Windle, D. W. 1965, A possible explanation of the Appleton anomaly of the F2-layer, I., *J. atmos. terr. Phys.*, **27**, 795 (II. 1965, *J. atmos. terr. Phys.*, **27**, 1163).
24. Mariani, F. 1964, Pitch-angle distribution of the photoelectrons and origin of the geomagnetic anomaly in the F2 layer, *J. geophys. Res.*, **69**, 556.
25. Nagata, T., Ogawa, T. 1964, Synthetic interpretation of the airglow [O I] 6300 Å emission and the ionospheric condition, *Rep. Ionosph. Space Res. Japan*, **18**, 394.
26. Peterson, V. L., Van Zandt, T. E., Norton, R. B. 1966, F-region nightglow emissions of atomic oxygen, I., *J. geophys. Res.*, **71**, 2255.
27. Peterson, V. L., Steiger, W. R. 1966, F-region nightglow emissions of atomic oxygen, II. Analysis of 6300 Å and electron density data, *J. geophys. Res.*, **71**, 2267.
28. Rishbeth, H. 1964, A time-varying model of the ionospheric F2-layer, *J. atmos. terr. Phys.*, **26**, 657.
29. Roach, F. E., Steiger, W. R., Brown, W. E. 1964, A summary of one year of observations of 6300 Å airglow at Haleakala, Tech. Rep. HIG-64-16, Hawaii Institute of Geophys.
30. Sharp, G. W. 1966, Midlatitude trough in the night ionosphere, *J. geophys. Res.*, **71**, 1345.
31. Silverman, S. M. 1965, The equatorial airglow, AFCRL-65-280.
32. Somayajulu, Y. V. 1964, Evidence on the horizontal diffusion of F-region ionization along the magnetic lines of force in equatorial latitudes, *J. geophys. Res.*, **69**, 561.
33. Somayajulu, Y. V. 1964, Some features of the F-region electron density and height variations in the equatorial regions, *J. geophys. Res.*, **69**, 1329.
34. Steiger, W. R., Brown, W. E., Roach, F. E. 1966, The alignment of 6300 Å airglow isophotes in the tropics, *J. geophys. Res.*, **71**, 2846.
35. Swider, W., Jr. 1967, On the decay of the ionospheric airglow 6300 Å line at night, Paper given at meeting of COSPAR, Vienna, 1966, *Space Res.*, **7**, in press.
36. Varnasavang, V., Hargreaves, Sylvia, Kendall, P. C. 1965, The influence of horizontal air motions on F2-region ionization, *J. atmos. terr. Phys.*, **27**, 707.
37. Yoshizaki, W. The effect of ionospheric movements on the emission rate of [O I] 6300 Å airglow, in preparation.
38. Yoshizaki, W. 1965, Binary transport of electrons and ions in the equatorial ionosphere and its effect on the brightness of [O I] 6300 Å airglow at Huancayo, *Rep. Ionosph. Space Res. Japan*, **19**, 299.
39. Pertinent paper by Soviet authors: Shefov's list, no. **23**.

C. Mid-latitude stable auroral arcs

1. Bellew, W., Silverman, S. M. 1965, The dependence of occurrence of midlatitude aurora on the sunspot cycle, *J. geophys. Res.*, **70**, 4985.
2. Carleton, N. P., Roach, J. R. 1965, Spectrographic observations of a mid-latitude red auroral arc, *J. geophys. Res.*, **70**, 1262.
3. Cole, K. D. 1965, On the D_{st} main phase of a magnetic storm and certain associated phenomena, in *Phys. of Geomag. Phenomena*, Ed. S. Matsushita and W. H. Campbell, Academic Press.
4. Cole, K. D. 1965, Stable auroral red arcs, sinks for energy of D_{st} main phase, *J. geophys. Res.*, **70**, 1689
5. Cole, K. D. 1966, Magnetic storms and associated phenomena, *Space Sci. Rev.*, **5**, 699.
6. Cole, K. D. 1966, A magnetic storm experiment for Australia, *The Australian Physicist*, **3**, 27.
7. Cole, K. D. Suggestions for the experimental study of the stable auroral red arc and the ring current, in preparation.
8. Marovich, E. 1967, Fritz Peak observations of stable auroral arcs, *ESSA Tech. Rep.* 16.
9. Marovich, E. On the enhancement of 5577 Å emission associated with sar-arcs, in preparation.

D. *Excitation of* 6300 Å [O I] *by photoelectrons*

1. Carlson, H. C., Jr. 1966, Ionospheric heating by magnetic conjugate-point photoelectrons, *J. geophys. Res.*, **71**, 195 (also paper presented to Aurora-Airglow conference at Keele, England; Reinhold, 1967, p. 643).
2. Cole, K. D. 1965, The predawn enhancement of 6300 Å airglow, *Ann. Géophys.*, **21**, 156.
3. Dalgarno, A., McElroy, M. B. 1965, Ionospheric electron temperatures near dawn, *Planet. Space Sci.*, **13**, 143.
4. Hoegy, W. R., Fournier, J.-P., Fontheim, E. G. 1965, Photoelectron energy distribution in the F-region, *J. geophys. Res.*, **70**, 5464.
5. Tohmatsu, T., Ogawa, T., Tsuruta, H. 1965, Photo-electronic processes in the upper atmosphere, I. Energy spectrum of the primary photo-electrons, *Rep. Ionosph. Space Res. Japan*, **19**, 482.

E. *Nightglow variations, temporal and spatial*

1. Barbier, D. 1965, Variations de l'intensité des principales radiations de la luminescence atmosphérique nocturne avec le cycle solaire, *Ann. Géophys.*, **21**, 265.
2. Bollinger, C. J. 1965, Comments on paper by S. H. Neff entitled 'Observation of the O I 5577 Å nightglow at Christchurch, New Zealand', *J. geophys. Res.*, **70**, 5469.
3. Chiplonkar, M. W., Agashe, V. V., Tillu, A. D. 1965, A spectrophotometric study of the night airglow at Poona, *Indian J. Pure and Applied Phys.*, **3**, 182.
4. Dandekar, B. S., Silverman, S. M. 1964, The effect of solar flares on the [O I] green line of the nightglow, *Planet. Space Sci.*, **12**, 867.
5. Dandekar, B. S. 1965, Study of the [O I] green line at Mount Abu, *J. atmos. terr. Phys.*, **27**, 245.
6. Davis, T. N., Smith, L. L. 1965, Latitudinal and seasonal variations in the night airglow, *J. geophys. Res.*, **70**, 1127.
7. Greenspan, J. A., Stone, C. A. 1965, The longitudinal variation of night airglow intensity in the region of the South Atlantic magnetic anomaly, *J. geophys. Res.*, **69**, 465.
8. Greenspan, J. A. 1966, Synoptic description of the 6300 Å nightglow near 78° west longitude, *J. atmos. terr. Phys.*, **28**, 739.
9. Hernandez, G. J. 1965, Data previous to the International Geophysical Year I: 5577 [O I] line at Sacramento Peak, New Mexico, *AFCRL*–65–319.
10. Hernandez, G. J., Silverman, S. M. 1964, A reexamination of Lord Rayleigh's data on the airglow 5577 Å [O I] emission, *Planet. Space Sci.*, **12**, 97.
11. Neff, S. H. 1965, Observations of the O I 5577 Å nightglow at Christchurch, New Zealand, *J. geophys. Res.*, **70**, 1743.
12. Okuda, M. 1965, The intensity of oxygen red line in night airglow and its dependence on solar activity, *Rep. Ionosph. Space Res. Japan*, **19**, 470.
13. Saito, B., Takahashi, T. 1966, Large amplitude variations of the intensity of oxygen 5577 Å emission in the night airglow and structure of lower thermosphere, *Rep. Ionosph. Space Res. Japan*, **20**, 439.
14. Sandford, B. P. 1964, Aurora and airglow intensity variations with time and magnetic activity at southern high latitudes, *J. atmos. terr. Phys.*, **26**, 749.
15. Silverman, S. M. 1964, A note on the diurnal variation of the 5577 Å [O I] airglow emission, *Planet. Space Sci.*, **12**, 247.
16. Silverman, S. M., Bellew, W. 1966, Magnetic storm enhancement of the 5577 Å [O I] airglow emission intensity, *Ann. Géophys.*, **22**, 458.
17. Smith, L. L., Owen, R. W. 1966, Seasonal variation of nightglow Na I 5890–96 Å, [O I] 5577 Å and [O I] 6300 Å in the tropics, *NBS Tech. Rep.* 329.
18. Pertinent papers by Soviet authors: Shefov's list, No. **17**, **18**, **19**, **20**, **21**, **22**, **25**, **29**, **30**.

F. *Hydrogen airglow*

1. Brandt, J. C. 1964, On diffuse galactic Lyman α in the night sky, *Planet. Space Sci.*, **12**, 650.
2. Donahue, T. M. 1964, Lyman alpha scattering in the Earth's hydrogen geocorona, 3. The extrageocoronal contribution, *J. geophys. Res.*, **69**, 1301.

3. Donahue, T. M. 1964, H alpha excitation in the hydrogen near the Earth. *Planet. Space Sci.*, **12**, 149.
4. Donahue, T. M. 1966, The problem of atomic hydrogen, *Ann. Géophys.*, **22**, 175.
5. Donahue, T. M., Fastie, W. G. 1964, Observation and interpretation of resonance scattering of Lyman α and O I (1300) in the upper atmosphere, *Space Res.*, **4**, 304.
6. Eather, R. H., Sandford, B. P. 1966, The zone of hydrogen emission in the night sky, *Austr. J. Phys.*, **19**, 25.
7. Kondo, Y., Kupperian, J. E. 1967, Interaction of the neutral hydrogen and charged particles in the radiation belt: the consequent Lyman alpha emission, in preparation (to be submitted to *J. geophys. Res.*).
8. Winter, T. C. 1965, The determination of the profile of the north-south hydrogen Lyman alpha emission line, Ph.D. dissertation, UCLA.
9. Pertinent papers by Soviet authors: Shefov's list, no. **31, 32, 33, 34, 35, 36, 37**.

G. Aeronomy and excitation processes

1. Barbier, D. 1964, Possibilité de détection de particules de grande énergie dans la haute atmosphère par des observations de la luminescence nocturne atmosphérique, *Space Res.*, **5**, Proc. 5th int. Space Sci. Sym., Florence, 8–20 May, 1964.
2. Barth, C. A. 1964, Three body reactions, *Ann. Géophys.*, **20**, 182.
3. Barth, C. A. 1966, Nitric Oxide in the upper atmosphere, *Ann. Géophys.*, **22**, 198.
4. Bates, D. R. 1964, Chemical reactions contributing to the nightglow, *Discussions of the Faraday Society*, No. 37, 21.
5. Batey, P. H., Court, G. R., Sayers, J. 1965, Afterglow measurements of the rate coefficients for the reactions $O^+ + O_2 \rightarrow O_2^+ + O$ and $O^+ + N_2 \rightarrow NO^+ + N$, *Planet. Space Sci.*, **13**, 911.
6. Dalgarno, A. 1964, Corpuscular radiation in the upper atmosphere, *Ann. Géophys.*, **20**, 65.
7. DeMore, W. B., Raper, O. F. 1964, Deactivation of $O(^1D)$ in the atmosphere, *Astrophys. J.*, **139**, 1381.
8. Desesquelles, J., Do Cao, G., Dufay, M. 1966, Etude de l'ionization de quelques gas par des ions accélérés H_2^+, *C. r. Acad. Sci.*, Paris, **262**, 1329.
9. Donahue, T. M. 1966, Ionospheric reaction rates in the light of recent measurements in the ionosphere and the laboratory, *Planet. Space Sci.*, **14**, 33.
10. Donahue, T. M. 1967, The sodium airglow, *International Dictionary of Geophysics*, Pergamon.
11. Dufay, M., Druetta, M., Eidelsberg, M. 1965, Sur la luminescence de l'oxygène excitée par choc de particules accélérées sous 500 keV, *C. r. Acad. Sci.*, Paris, **260**, 1123.
12. Dufay, M., Druetta, M., Eidelsburg, M. 1965, Sur l'excitation de l'azote et de l'oxygène par ions accélérés, *C. r. Acad. Sci.*, Paris, **261**, 1635.
13. Dufay, M., Druetta, M., Eidelsberg, M. 1965, Sur la luminescence de l'azote excitée par choc de particules accélérées sous 500 keV, *C. r. Acad. Sci.*, Paris, **260**, 1901.
14. Dufay, M., Desesquelles, J., Druetta, M., Eidelsberg, M. Relatif à l'ensemble des expériences sur l'excitation des spectres de l'oxgène et de l'azote, *Ann. Géophys.*, in press.
15. Fehsenfeld, F. C., Schmeltekopf, A. L., Ferguson, E. E. 1965, Some measured rates for oxygen and nitrogen ion-molecule reactions at atmospheric importance, including $O^+ + N_2 \rightarrow NO^+ + N$, *Planet. Space Sci.*, **13**, 219.
16. Fehsenfeld, F. C., Goldan, P. D., Schmeltekopf, A. L., Ferguson, E. E. 1965, Laboratory measurement of the rate of the reaction $O^+ + O_2 \rightarrow O_2^+ + O$ at thermal energy, *Planet. Space Sci.*, **13**, 579.
17. Fite, W. L. 1964, Charge transfer and ion-atom interchange collisions above thermal energies, *Ann. Géophys.*, **20**, 47.
18. Hunten, D. M., McElroy, M. B. 1966, Quenching of metastable states of atomic and molecular oxygen and nitrogen, *Rev. Geophys.*, **4**, 303.
19. Megill, L. R., Carleton, N. P. 1964, Excitation by local electric fields in the aurora and airglow, *J. geophys. Res.*, **69**, 101.

20. Megill, L. R., Cahn, J. H. 1964, The calculation of electron energy distribution functions in the ionosphere, *J. geophys. Res.*, **69**, 5041.
21. Nathan, K. V. S. K. 1966, Particle heating effects in the upper F-region, *Planet. Space Sci.*, **14**, 717.
22. Nichols, R. W. 1964, Transition probabilities of aeromically important spectra, *Ann. Géophys.*, **20**, 144.
23. Ogawa, T., Tohmatsu, T. 1966, Photoelectronic processes in the upper atmosphere, II. The hydrogen and helium ultraviolet glow as an origin of the night-time ionosphere, *Rep. Ionosph. Space Res. Japan*, **20**, 395.
24. Paulson, J. F. 1964, Low energy charge exchange and ion-molecule reactions, *Ann. Géophys.*, **20**, 75.
25. Poulizac, M.-C., Desesquelles, J., Dufay, M. Sur l'excitation de l'oxyde de carbone par chocs protoniques, *C. r. Acad. Sci.*, Paris, in press.
26. Prag, A. B., Morse, Fred A., McNeal, R. J. 1966, Nightglow excitation and maintenance of the nighttime ionosphere by low-energy protons, *J. geophys. Res.*, **71**, 3141.
27. Schiff, H. I., Megill, L. R. 1964, The influence of metastable oxygen molecules on ozone and airglow, *J. geophys. Res.*, **69**, 5120.
28. Swider, W., Jr. 1965, A study of the nighttime ionosphere and its reaction rates, *J. geophys. Res.*, **70**, 4859.
29. Thomas, L., Norton, R. B. 1966, Possible importance of internal excitation in ion-molecule reactions in the F-region, *J. geophys. Res.*, **71**, 227.
30. Young, R. A., Black, G. 1966, Excitation of the auroral green line in the Earth's nightglow, *Planet. Space Sci.*, **14**, 113.
31. Pertinent papers by Soviet authors: Shefov's list, no. **43, 44, 45, 46.**

H. N_2^+ in the nightglow

1. Broadfoot, L., Hunten, D. M. 1966, N_2^+ emission in twilight, *Planet. Space Sci.*, **14**, 1303.
2. Dalgarno, A., Latimer, I. D., McConkey, J. W. 1965, Corpuscular bombardment and N_2^+ radiation, *Planet. Space Sci.*, **13**, 1008.
3. Dalgarno, A., McElroy, M. B. 1966, Twilight effects of solar ionizing radiation, *Planet. Space Sci.*, **14**, 1321.
4. O'Brien, B. J., Allum, F. R., Goldwire, H. C. 1965, Rocket measurement of midlatitude airglow and particle precipitation, *J. geophys. Res.*, **70**, 161.
5. Yano, Kyo 1966, N_2^+ ions in the nocturnal upper atmosphere at middle latitudes, *Planet. Space Sci.*, **14**, 709.

I. Hydroxyl (OH) Nightglow

1. Huruhata, M. 1965, Correlation between the OH radiation in the night airglow and the height of the 10-millibar surface, *J. geophys. Res.*, **70**, 4927.
2. Nguyen, Huu-Doan 1963, Temperature de rotation des bandes (9–3) et (6–1) de la molecule OH, *C. r. Acad. Sci.*, Paris, **256**, 245.
3. Noxon, J. F. 1964, The latitude dependence of OH rotational temperature in the night airglow, *J. geophys. Res.*, **69**, 4087.
4. Stair, A. T. Jr., Kennealy, J. P., Stewart, S. P. 1965, Observation of the $\Delta v = 1$ sequence of OH produced in the H + O_3 reaction, *Planet. Space Sci.*, **13**, 1005.
5. Wolstencroft, R. D., Brandt, J. C., Rose, L. J. 1966, Anomalous brightness of the post-twilight evening sky in the near infra-red, *Planet. Space Sci.*, **14**, 445.
6. Pertinent papers by Soviet authors: Shefov's list, no. **1, 2, 3, 4, 6, 8, 9, 10, 11, 12, 14.**

J. Doppler temperatures

1. Fymat, A. L., Venketeswaran, S. V. 1965, The Doppler temperatures obtained from the 6300Å line of atomic oxygen in the airglow, *Ann. Géophys.*, **21**, 443.
2. Hernandez, G. J., Turtte, J. P. 1965, Nightglow 5577Å [O I] line kinetic temperatures, *Planet. Space Sci.*, **13**, 901.

3. Hilliard, R. L., Shepherd, G. G. 1966, Upper atmospheric temperatures from Doppler line widths, IV. A detailed study using the [O I] 5577Å auroral and nightglow emission, *Planet. Space Sci.*, **14,** 383.

4. Jarrett, A. H., Hoey, M. J., Paffrath, L. 1964, Observations of the 6300Å O I emission of the nightglow with a pressure scanning Fabry-Perot interferometer, *Planet. Space Sci.*, **12,** 591.

5. Jarrett, A. H., Hoey, M. J. 1965, Temperature determinations of the Earth's upper atmosphere by optical methods, *Proc. R. Soc. London*, Ser. A **288,** 510.

6. Jarrett, A. H., Hoey, M. J. 1966, Observations of the 6300Å O I nightglow emission, *J. atmos. terr. Phys.*, **28,** 175.

K. Instrumentation and techniques

1. Best, G. T. 1965, A new use of the van Rhijn formula for airglow measurements, *Planet· Space Sci.*, **13,** 855.

2. Blacker, H. V., Gadsden, M. 1966, On the calibration of airglow photometers, *Planet. Space Sci.*, **14,** 921.

3. Chiplonkar, M. W., Agashe, V. V., Tillu, A. D. 1966, Heights of the airglow emission layers at Poona, *Indian J. Meteorology Geophys.*, **17,** 95.

4. Dumont, R. 1965, Séparation des composantes atmosphérique, interplanétaire et stellaire du ciel nocturne à 5000Å A. Application à la photométrie de la lumière zodiacale et du gegenschein, *Ann. Astrophys.*, **28,** 265.

5. Kulkarni, P. V., Sanders, C. L. 1964, Use of a radioactivated light source for the absolute calibration of two-colour night airglow photometer, *Planet. Space Sci.*, **12,** 189.

6. Wolff, M. M. 1966, A new attack on height measurement of the nightglow by ground triangulation, *J. geophys. Res.*, **71,** 2743.

7. Wolff, M. M., Garmil, J. Summary: Measurements have been made of the height of the 5577Å nightglow layer using a new method of ground-based triangulation. The theory of the method is discussed and results compared with other methods. Average height obtained was 97 ± 3 kilometers, in preparation.

8. Pertinent papers by Soviet authors: Shefov's list, no. **26, 28.**

L. Rocket and satellite observations

1. Gillett, F. C., Huch, W. F., Ney, E. P., Cooper, G. 1964, Photographic observations of the airglow layer, *J. geophys. Res.*, **69,** 2827.

2. Gulledge, I. S., Packer, D. M., Tilford, S. G. 1966, Rocket measurements of the high altitude [O I] 5577Å and 6300Å night airglow, *Trans. Am. geophys. Un.*, **47,** 74.

3. Hennes, J. P. 1966, Measurement of the ultraviolet nightglow spectrum, *J. geophys. Res.*, **71,** 763.

4. Hennes, J. P., Dunkelman, L. 1966, Photographic observations of nightglow from rockets, *J. geophys. Res.*, **71,** 755.

5. Huruhata, M., Nakamura, T., Tanabe, H. 1966, Rocket observations of emitting height of 6300Å oxygen line in night airglow, *Rep. Ionosph. Space Res. Japan*, **20,** 223.

6. NASA. Results of the first U.S. manned orbital space flight (20 February 1962).
Appendix C, pp. 195–198, Description of the MA–6 Astronomical, Meteorological and Terrestrial Observations by John H. Glenn, Jr.
Appendix D, pp. 199–204, Preliminary report on the results of the MA–6 flight in the field of space science by John A. O'Keefe.

7. NASA SP–6. Results of the second U.S. manned orbital space flight (24 May 1962).
Chapter 4, pp. 35–42, Space Science Report by John A. O'Keefe.
Chapter 7, pp. 69–75, Pilot's flight report by M. Scott Carpenter.

8. NASA SP–12. Results of the third U.S. manned orbital space flight (3 October 1962).
Chapter 4, pp. 37–48, Pilot's flight report by Walter M. Schirra, Jr.

9. NASA SP-45. Mercury project summary including results of the fourth manned orbital flight (15-16 May 1963).
Chapter 19, pp. 327-347, Observations of space phenomena by John A. O'Keefe, L. Dunkelman, S. D. Soules and P. D. Lowman, Jr.
Chapter 20, pp. 349-358, Astronaut's summary flight report by L. Gordon Cooper.
10. NASA. Manned Space Flight Experiments Symposium, Gemini Missions 3 and 4 (18-19 October 1965).
Chapter 1, pp. 1-18, Geo-astronomical observations by L. Dunkelman, J. R. Gill, J. A. McDivitt, F. E. Roach, and E. H. White II.
Gemini 3 –
Gemini 4 – 3-7 June 1965.
11. NASA SP-121. Gemini Midprogram Conference (23-25 February 1966).
(32) Geoastronomical observations, pp. 315-323, by F. E. Roach, L. Dunkelman, J. R. Gill, R. D. Mercer.
(33) Dim light photography, pp. 325-328, by L. Dunkelman, and R. D. Mercer.
12. NASA report dated 1 November 1966.
Interim report, Gemini 9 mission.
Experiment S-1 Zodiacal light photography by E. P. Ney and W. F. Huch, pp. 1-6.
Experiment S-11 Airglow horizon photography by M. J. Koomen and R. T. Seal, Jr.
13. O'Brien, B. J., Allum, F. R., Goldwire, H. C. 1965, Rocket measurements of midlatitude airglow and particle precipitation, *J. geophys. Res.*, **70**, 161.
14. Reed, E. I., Blamont, J. R. 1966, Some results concerning the principal airglow lines as measured from the OGO-II satellite, *NASA* X-613-66-190.
15. Stecher, Th. P. 1965, The spectral energy distribution of the Earth's ultraviolet night airglow, *J. geophys. Res.*, **70**, 2209.

M. Twilight glow

1. Gadsden, M. 1967, Twilight observations (Review paper with 22 references to papers in this field since 1964), *Aurora and Airglow*, Reinhold, p. 109.
2. Hunten, D. M. 1967, Spectroscopic studies of the twilight airglow (Review paper with 60 references to papers in this field and dayglow since 1964), *Space Sci. Rev.*, **6**, 493.
3. Pertinent papers by Soviet authors: Shefov's list, no. **38, 39, 40, 41, 42, 43.**

APPENDIX I. SPECTRAL INVESTIGATIONS OF THE UPPER ATMOSPHERE EMISSIONS IN THE U.S.S.R.

(prepared by N. N. Shefov)

During a period following the IGY and during the IQSY period in the Soviet Union systematic investigations were continued into the properties of the upper atmosphere emissions by means of spectrographic and photoelectric methods. The data obtained for several years at Abastumani have shown that during a period since 1953 the average annual intensity of hydroxyl emission had a maximum in 1958. After a small minimum in 1961-62 the intensity began to increase again (**1**). The study of the seasonal variations of the intensity and rotational temperatures of the OH bands has revealed long ago that there is a maximum in the winter months and a minimum in the summer months. This result has been confirmed by observations lasting many years at many stations (**1-3**). The amplitude of the seasonal variations of the rotational temperature is greater for bands from higher initial vibrational levels.

Statistical processing of the data on OH emission obtained for a number of years has made it possible to reveal periodic variations of mean diurnal intensities and rotational temperatures of OH bands (4, 1), (5, 2), (6, 2) and (7, 3) with periods of 27 days and 29·5 days (the age of the Moon) from observations at Zvenigorod (**2**) and the OH band (9, 3) with a period of 27 days by observations at Yakutsk (**4-6**). 27-day variations in the intensity and rotational temperature of OH bands (4, 1), (5, 2) and (9, 3) take place in antiphase and OH bands (6, 2) and (7, 3) synchronously. According to the data obtained at Yakutsk, along with the 27-day variations in the rotational temperature of the OH band (9, 3), 27-day variations in the atmospheric temperature were also observed at a level of 50 mb and the height of the isobaric surface of 15 mb ($\sim$ 27–28 km).

The variations in the rotational temperature of OH bands (4, 1), (5, 2), (6, 2) and (7, 3) with the Moon's age have their deepest minimum during the full Moon and the least during the new Moon and two maxima in the first and the last quarters. Variations in the intensity of OH bands (6, 2) and (7, 3) are similar. For OH bands (4, 1) and (5, 2) there are almost no intensity variations with the Moon's age.

Studies of the behaviour of the intensity and rotational temperature of OH bands in the presence of noctilucent clouds have shown that the intensity of OH emission, as well as of the atmospheric band (0, 1) O_2 grows during a period of the appearance of noctilucent clouds and sharply decreases the following night (up to 2–3 times with respect of the mean intensity), when noctilucent clouds are already absent. The following night the intensities of OH and O_2 emissions are restored to their mean values. Such a regularity has been observed annually since 1962 (**7–9**). The rotational temperature of OH bands varies little, although it reaches the values of 160–180°K during the period of the emergence of noctilucent clouds. This seems to be accounted for by the fact that the rotational temperature variations reflect changes in the height of rather a wide emitting layer.

The clarification of the character of dirunal variations of OH emission has shown that there are diverse types of variations in intensity and rotational temperature of different bands (**1, 10, 11**). However, most often variations of the same phase of mean values of the intensity and rotational temperatures are observed. Against the background of such variations, fluctuations can take place. Periodic variations of the rotational temperature of OH bands have been revealed with a period of about 1 hour and the modulation depth of 30–40° (**10**). Such variations are observed mainly early at night. During the evening twilight (at the solar depression angle from 7 to 18°) the OH band intensity decreases by 30% on the average. The rotational temperature decreases on the average by 15°. On the average during the morning twilight the intensity and the rotational temperature is lower than in the evening (**12**). From observations in high latitudes no correlation has been observed between variations in the intensity and OH rotational temperature and auroral displays (**11**).

The study of spatial irregularities of OH emission by the data of different stations has shown that co-phaseness and simultaneity of rotational temperature variations have been systematically observed at distances of up to 3000 km (**3, 11**). The patchy structure of the intensity and rotational temperature which was already supposed (**13**) has been revealed from observations at Anastumani (**1**) and Zvenigorod (**10**). A comparison of simultaneous diurnal temperature variations at different points of the sky has provided an opportunity of estimating the velocity of the vertical wind – 1 to 3 m s^{-1} and the horizontal east-west wind – 30 to 200 m s^{-1} (**10**).

Investigations of the dependence between the intensities of OH bands (3, 0), (4, 1), (5, 2), (6, 2), (7, 3), (8, 4), (9, 5) and O_2 (0, 1) revealed earlier have made it possible to find two types of correlations between them (**14**). With the intensity of oxygen emission of less than 0·3–0·5 kilorayleighs the correlation field has the form of a sector (**15**). The intensities of OH bands can reach maximum observed values.

With the oxygen emission intensities of over 0·5 kilorayleighs, the correlation becomes linear rather rapidly up to the maximum observed values of intensities. These data permit one to over the vertical distribution of OH and O_2 emissions.

A great amount of data on spatial and time variations of green and red emissions of atomic oxygen has been obtained in Ashkhabad. Maps of isophotes obtained from observations in

1964 have been published (**16**). The time of the existence of emission patches can vary from a few minutes to a few hours (**17**). The size of patches and their intensities can often pulsate. Periods of 5·5 and 15·5 minutes have been observed. The most typical size of patches is 60–90 km. The minimum size is about 30 km. The distribution of the velocities of the patches has two maxima—about 75 $m\ s^{-1}$ and 150 $m\ s^{-1}$. The velocity of patches increases with the increase of the *K*-index (**18–20**). The greatest velocity of motion is observed in October. A dependence of the intensity of the emission of 5577 Å on lunar time has been revealed. The amplitude of variations is 35 rayleighs and the intensity minimum coincides with the moment of the Moon's lower culmination (**19**). A comparison of the data of observations of red and green emissions has shown the existence of patches of emissions of 6300 Å (situated at a height of 400 km) and of 5577 Å (at a height of 100 km) lying at the same geomagnetic line of force (**17, 21, 22**). In this case small rise is observed in the region of 3914 Å. If the intensity of this additional emission is due to the emission of N_2^+ of 3914 Å, its maximum intensity is not more than 10 rayleighs (**23**).

Investigations of the properties of the continuum of the spectrum of the nightglow have led to the detection of spatial and time variations of both the intensity and spectral distribution of the continuum (**24–27**). Therefore observations of emissions with wide filters at a constant point of the sky do not allow one to determine correctly the intensities of discrete lines (**28**). Sodium emission in the night sky also has a patchy structure. The velocities of the motion of patches are about 70 $m\ s^{-1}$. The motion is mainly eastward and northward (**29**).

As a result of the hydrogen emission observations of many years at Abastumani, data have been obtained on variations of the Hα emission intensity during a solar cycle. In the years of a minimum (1962) its mean yearly intensity was maximum (**30, 31**) and constituted about 20 rayleighs. Such a character of variations gives evidence for the terrestrial origin of Hα emission. Observations of hydrogen emission have made it possible to calculate hydrogen distribution in the atmosphere (**31, 32**) supposing the fluorescence mechanism of solar radiation in the $L\beta$ line. Observations performed by means of the Fabry-Perot interferometer (**33–37**) confirmed the minimum detected earlier at the antisolar point. The width of the Hα line does not exceed 0·3 Å while in emission nebulae the emission line width sometimes exceeds this value. The existence of variations in the Celestial Pole speaks in favour of the emergence of the emission in the Earth's atmosphere. It has been revealed that Hα emission is concentrated in the ecliptic. The intensity on the morning side exceeds several times the Hα intensity on the evening side of the Earth. Radial velocities with the elongation of 30° do not exceed the instrumental limit of 4·5 $km\ s^{-1}$.

Investigations of helium emission of the upper atmosphere have been carried out for several years using spectrographic and photoelectric methods. Variations have been obtained of the intensity of helium emission and the calculated flux of solar ultraviolet radiation (**37, 38**). According to the data of photographic observations at Zvenigorod it turns out that on the average the helium emission intensity for the *K*-index < 4 remains almost constant and only with the *K*-index $\geqslant 4$ a sharp increase in the intensity begins. For the local *K*-index this transition takes place more sharply than for the planetary one. According to observations in high latitudes, the helium emission intensity very often reached several kilorayleighs (**39**). Variations of the helium emission intensity in ordinary twilight have a noticeable evening-morning asymmetry which gave the opportunity to determine the atmospheric temperature, which in the morning turns out to be lower than in the evening (**40**). The calculation has been performed on the basis of the theory of excitation of metastable helium atoms by photoelectrons and fluorescence of these atoms in solar radiation (**41–44**). The evening-morning intensity asymmetry was detected also from observations in high latitudes (**39**).

The estimate of intensities of argon and neon emissions in the upper atmosphere turned out to be very low even in auroras (**45**). However, the existence of argon and neon metastable atoms in the atmosphere can be revealed in ordinary auroras from the properties of variations in the atomic oxygen line of 8446 Å (**46**).

An analysis was made of auroral spectra observed at various stations during several years. The relative and absolute intensities for many auroras have been obtained (**47–49**). The height distribution of intensities of emission of N^+, 5577 Å, 6300 Å, 1 PG N_2 in auroras varies in different auroral forms (**50, 51**). A correlation is observed between height distributions of emissions of 5577 Å, 1 PG N_2^+ and 1 NG O_2^+. However, the relationship between them depends on the form of auroras. The properties and specific features of an aurora of 10–11 February 1958 have been time and again discussed in several papers (**44, 52–55**). The ratios of the intensities of nebular and auroral transitions for oxygen and nitrogen atoms and ions have been studied. The energy released in auroras reach 10–100 erg $cm^{-2}\,s^{-1}$.

On the basis of auroral hydrogen line contours obtained at different zenithal distances a distribution of injected protons by pitch-angles has been calculated as well as the energy spectrum which well agrees with the measured spectrum in the solar wind (**56**).

On determinations of the heights of night sky emissions it was reported in (**57, 58**).

References

Sbornik in the references is for 'Aurorae and Airglow', series 'Results of IGY', Publishing House NAUKA of the U.S.S.R. Academy of Sciences. Letters refer to general bibliography, above, pages 415 to 422.

1. (I) Fishkova, L. M., Marzvaladze, N. M. 1966, Some results of the upper atmosphere hydroxyl emission according to the observations in Abastumani. *Sbornik*, no. 13, 24.

2. (I) Shefov, N. N. 1966, Some properties of the hydroxyl emission. *Sbornik*, no. 13, 21.

3. (I) Sidorov, V. N. Some results of the observations of the hydroxyl emission during IQSY. *Sbornik*, no. 18, in press.

4. (I) Kuzmin, A. I., Krivoshapkin, P. A., Chirkov, N. P., Yarin, V. I. 1963, Investigations of the atmospheric temperature above Yakutsk. Report to the XIIIth IQSY Assemblee, Berkeley.

5. Kuzmin, A. I. 1964, Variations of the high energy cosmic rays. Publishing House, 'NAUKA', Moscow.

6. (I) Yarin, V. I. Connection of the hydroxyl emission with the meteorological conditions above Yakutsk. *Sbornik*, no. 18, in press.

7. (A) Shefov, N. N. 1965, The upper atmosphere emissions and the noctilucent clouds. *Sbornik*, no. 11, 48.

8. (I) Shefov, N. N. 1966, OH emission and noctilucent clouds. Proc. Intern. Conference on Noctilucent Clouds, Tallin.

9. (I) Shefov, N. N. OH variations during the occurrences of the noctilucent clouds. *Sbornik*, no. 18, in press.

10. (I) Taranova, O. G. 1966, Studies of the spatial and temporal properties of the hydroxyl emission. *Sbornik*, no. 13, 10.

11. (I) Fedorova, N. I. 1966, Hydroxyl emission at the high latitudes. *Sbornik*, no. 13, 14.

12. (I) Taranova, O. G., Toroshelidze, T. I. On the measurements of the hydroxyl emission in twilight. *Sbornik*, no. 18, in press.

13. (A) Krassovsky, V. I., Shefov, N. N. 1965, Airglow. *Space Sci. Rev.*, **4**, 176.

14. (I) Taranova, O. G. Connection between the hydroxyl and molecular oxygen emissions. *Sbornik*, no. 18, in press.

15. (A) Shefov, N. N. 1965, Correlations between the upper atmosphere emissions. *Sbornik*, no. 11, 43.

16. (A) Korobeynikova, M. P., Nasirov, G. A., Khamidulina, V. G. 1966, The nightglow emission λ 5577 Å. Tables and isophote maps. Serie 'Results of IQSY', Publishing House VINITI, Moscow.

17. (E) Truttse, Yu. L. 1965, The spatial variations of the oxygen lines intensities. *Sbornik*, no. 11, 52.
18. (E) Nasirov, G. A. 1966, The peculiarities of the spatial pattern structure of the atomic oxygen emission λ5577Å in the nightglow. *Sbornik*, no. 13, 5.
19. (E) Korobeynikova, M. P., Nasirov, G. A., Khamidulina, V. G. Intensity variations and dynamical characteristics of the spatial patches of the emission λ5577Å. *Sbornik*, no. 18, in press.
20. (E) Kalchaev, K. K., Korobeynikova, M. P., Nasirov, G. A., Khamidulina, V. G. Search of the rapid spatial variations of the intensity of the oxygen green line. *Sbornik*, no. 18, in press.
21. (E) Truttse, Yu. L. 1963, Connection between the pattern structures of the green and red emissions of the atomic oxygen in the nightglow. *Astr. Cirk.*, U.S.S.R., no. 258, 1.
22. (E) Truttse, Yu. L. 1964, Connection between green and red oxygen emission patches. *Planet. Space Sci.*, **12**, 725.
23. (B) Krassovsky, V. I., Truttse, Yu. L., Shefov, N. N. 1965, On the mechanism of maintenance of the nocturnal ionosphere. *Space Res.*, **5**, North-Holland Publishing Company, Amsterdam, 43-48.
24. (A) Nasirov, G. A. 1964, On the continuum in the nightglow. *Geomag. Aeronom.*, U.S.S.R., **4**, 1118.
25. (E) Truttse, Yu. L. 1965, The spectral variations of the nightglow continuum. *Sbornik*, no. 11, 6.
26. (K) Migulin, A. V. 1965, On the separation of the nightglow continuum components. *Sbornik*, no. 11, 20.
27. (A) Gindilis, L. M. 1965, The absolute measurements of the nightglow continuum. *Sbornik*, no. 11, 26.
28. (K) Taranova, O. G. 1965, On the statistical treatment of the photoelectric measurements with the filters. *Sbornik*, no. 11, 35.
29. (E) Nasirov, G. A. 1966, Spatial variations of the nightglow in the region λ5893Å. *Sbornik*, no. 13, 8.
30. (E) Fishkova, L. M. 1962, Intensity variations and spatial distribution of the λ6563Å emission. *Abastumansk. astrofiz. Obs. Bjull.*, no. 29, 77.
31. (F) Fishkova, L. M., Marzvaladze, N. M. 1966. On the Hα-emission variations and on the hydrogen distribution in the upper atmosphere and geocorona. *Sbornik*, no. 13, 38.
32. (F) Fishkova, L. M., Marzvaladze, N. M. 1963, On the hydrogen distribution in the upper atmosphere and geocorona. *Astr. Cirk.*, U.S.S.R., no. 253, 1.
33. (F) Shcheglov, P. V. 1963, On the concentration of nightglow Hα-emission to the ecliptic and the radial velocities of this line. *Astr. Cirk.*, U.S.S.R., no. 237, 1.
34. (F) Shcheglov, P. V. 1963, Concentration of nightglow Hα-emission to the ecliptic and the radial velocities of this line. *Nature*, **199**, 990.
35. (F) Shcheglov, P. V. 1963, The flat hydrogen cloud around the Earth. *Priroda*, Moskva, no. 8, 85.
36. (F) Shcheglov, P. V. 1964, Concentration of nightglow Hα-emission to the ecliptic and the radial velocities of this line. *Astr. Zu.*, **41**, 371.
37. (F) Krassovsky, V. I., Shefov, N. N., Vaisberg, O. L. 1966, Atomic hydrogen and helium in the airglow. *Ann. Géophys.*, **22**, 208.
38. (M) Shefov, N. N. 1966, Statistical properties of the helium emission. *Sbornik*, no. 13, 35.
39. (M) Fedorova, N. I. 1966, The twilight helium emission at the high latitudes. *Sbornik*, no. 13, 30.
40. (M) Taranova, O. G. 1966, On the diurnal variations of the helium emission. *Sbornik*, no. 13, 27.
41. (M) Shefov, N. N. 1963, Helium in the upper atmosphere. *Planet. Space Sci.*, **10**, 73.
42. (M) Shefov, N. N. 1963, The behaviour of the helium λ10830Å emission in twilight. *Sbornik*, no. 10, 56.
43. (G) Krassovsky, V. I., Shefov, N. N. 1964, Fast photoelectrons and helium emission in the upper atmosphere. *Planet. Space Sci.*, **12**, 91.

44. (G) Krassovsky, V. I., Shefov, N. N. 1965, On the mechanism of production of metastable orthohelium atoms in the upper atmosphere. *Space Res.*, **5**, North-Holland Publishing Company, Amsterdam, 822-825.
45. (G) Shefov, N. N. 1965, The rare gas emissions in the airglow and aurora. *Sbornik*, no. 11, 65.
46. (G) Galperin, Yu. I., Sobelman, I. I. 1965, On the possible role of the metastable states of argon and neon in the auroral afterglow processes. *Geomagn. Aeronom.*, U.S.S.R., **5**, 729.
47. (N) Yashchenko, I. A. 1965, The relative intensities of the auroral emissions in the visual spectrum. *Sbornik*, no. 11, 70.
48. (N) Shouyskaya, F. K. 1966, Relative populations of the upper vibrational levels in the 2 PG N_2 bands in the aurorae at different latitudes. *Sbornik*, no. 13, 48.
49. (N) Yurchenko, O. T. Absolute intensities of aurorae in Zvenigorod. *Sbornik*, no. 18, in press.
50. (N) Vaisberg, O. L., Shuyskaya, F. K. 1964, Ions N_2^+ density in the high rays in aurora. *Geomagn. Aeronom.*, U.S.S.R., **4**, 785.
51. (N) Fedorova, N. I. Emission height distributions in the low auroral form at the high latitudes. *Sbornik*, no. 18, in press.
52. (N) Krassovsky, V. I. 1964, Polar auroras. *Space Sci. Rev.*, **3**, 232.
53. (N) Krassovsky, V. I. 1965, Problems relating to the power of aurorae. *Space Res.*, **5**, North-Holland Publishing Company, Amsterdam, 71–76.
54. (N) Shouyskaya, F. K. 1966, Observations of the aurora on 11 February 1958 at Roshchino, Leningrad. *Sbornik*, no. 13, 51.
55. (N) Shefov, N. N. Spectral properties of the aurora on 11 February 1958. *Sbornik*, no. 18, in press.
56. (N) Vaisberg, O. L. 1966, Pitch-angle and energy proton distribution in the aurorae. *Sbornik*, no. 13, 42.
57. (A) Tarasova, T. M., Slepova, V. A. 1964, Intensity height distribution of the principal emission of the nightglow. *Geomagn. Aeronom.*, U.S.S.R., **4**, 321.
58. (A) Tarasova, T. M. 1966, Nightglow in the region λ6300Å. *Kosmičeskie Issledovanja*, **4**, no. 2, 242–248.

22. COMMISSION DES MÉTÉORES ET MÉTÉORITES

PRÉSIDENT: Dr P. M. Millman, National Research Council, Ottawa 7, Ontario, Canada.

VICE-PRÉSIDENT: Dr Z. Ceplecha, Astronomical Institute, Czechoslovak Academy of Sciences, Observatory Ondřejov, Czechoslovakia.

COMITÉ D'ORGANISATION: P. B. Babadjanov, C. D. Ellyett, E. L. Fireman, G. Guigay, B. A. Lindblad.

MEMBRES: Abbott, Astapovich, Baldet†, Cook (A. F.), Davies, Fedynsky, Fesenkov, Gauzit, Guth, Halliday, Hawkins, Hemenway, Hey, Hirose, Hoffmeister, Hoppe, Jacchia, Katasev, Kizilirmak, Kostylev, Kresák, Krinov, Levin, Lovell, McCrosky, Nielsen, O'Keefe, Olivier, Öpik, Whipple, Yavnel.

MEMBRES CONSULTANTS: E. Anders (U.S.A.), W. G. Elford (Australia), E. P. Henderson (U.S.A.), M. H. Hey (Great Britain), B. H. Mason (U.S.A.), B. A. McIntosh (Canada), T. N. Nazarova (U.S.S.R.), H. Oleak (Germany, Democratic Republic), K. I. Sztrokay (Hungary), F. Wickman (Sweden).

INTRODUCTION

In the preparation of this report, the following members of Commission 22 assisted by sending me summaries of research carried out, and bibliographies of papers published in the field of meteoritics:—Astapovich, Babadjanov, Ceplecha, Cook, Elford, Ellyett, Halliday, Hirose, Hoppe, Kizilirmak, Kresák, Lindblad, McCrosky, Nielsen, Oleak. I wish to thank all these for their help. I would like also to thank a number of other scientists who have kept me supplied with relevant papers, and pre-prints in advance of publication.

Since the last report of this Commission was completed during the second half of 1964, I have omitted all 1963 references here. For references from 1964 on there will be some omissions in borderline fields such as space, lunar surface studies and geology, and also in the papers of those countries from which I did not receive a complete bibliography.

Research on meteorites, tektites, and the dust collected from polar snows and deep-sea sediments, is covered in the accompanying report for our Committee on Meteorites, by E. L. Fireman. I include here a general treatment of meteor crater research, and the detection of cosmic dust by satellites and spacecraft.

GENERAL

During the last three years there has been considerable activity in the field of meteoritics. New monographs have been published by Hawkins (1964*a*), 'Meteors, Comets, and Meteorites', and by Astapovich (1966*c*), 'Meteors and Meteorites'. Russian editions have been published of 'Meteor Science and Engineering' by D. W. R. McKinley, 'Meteorites' by B. H. Mason, and 'Tektites' by J. A. O'Keefe, while an English edition of 'The Giant Meteorites' by E. L. Krinov has appeared. The third, and completely revised, edition of the Prior-Hey 'Catalogue of Meteorites' became available late in 1966 at the British Museum (Natural History) London. A new catalogue of bright fireballs by A. V. Nielsen (1967) will appear shortly. E. O'Connell (1965) has prepared a very useful catalogue of meteorite craters on the earth's surface, both confirmed and suspected features.

An international symposium on the subject 'Meteor Orbits and Dust', was held at the Smithsonian Astrophysical Observatory, Cambridge, Mass., U.S.A. in August, 1965. The proceedings of this symposium will appear in *Smithsonian Contributions to Astrophysics*, vol. **11**. The Committee on Comets and Meteors in the U.S.S.R. held a session in Kazan in 1965, under the chairmanship of V. V. Fedynsky. The Committee on Meteorites in the U.S.S.R. organized the 11th Conference on Meteorites in Moscow in May 1964, (Yavnel, 1964) and the 12th Conference on Meteorites in Novosibirsk in May 1966. The latter dealt mainly with the problems of the Tunguska event of 1908. A conference on luminous night clouds was held in Tallin, Estonia, in March 1966. The Meteoritical Society held Annual meetings at Tempe, Arizona, October, 1964; Odessa, Texas, October, 1965; and Washington, D.C., November, 1966. The Working Group on Meteorites, organized by UNESCO, held three meetings in Paris, the first in February, 1964, the second in October, 1965, and the third (final meeting) in October, 1966. An international workshop on methods of obtaining winds and densities from radar meteor trails was held at Waltham, Mass., in August, 1966, under the auspices of the Air Force Cambridge Research Laboratories. Meetings of the Organizing Committees of Commission 22 and of the 1967 Symposium on the Physics and Dynamics of Meteors, were held at Cambridge, Mass., August, 1965, and at Smolenice, Czechoslovakia, October, 1966.

Papers in the field of meteoritics now appear in a continually growing list of publications. The following current periodicals are devoted specifically to this subject:—*Meteoritika* and *The Meteoritical Bulletin*, published in Moscow; *Bulletin of the Committee on Comets and Meteors*, published in Dušhanbe; and *Meteoritics*, published by the Meteoritical Society.

One of the trends noticed during the last three years has been a new awareness of the importance of the rapid recovery of meteorites and the study of these objects before the short-lived radio isotopes have dropped below detectable levels. To aid in tracking fireballs bright enough to drop meteorites multiple camera networks have been set up in the mid-western United States (McCrosky and Boeschenstein, 1965) and in Czechoslovakia (Ceplecha and Rajchl, 1965*a*, 1965*b*). The latter is now being extended into the territory of Germany. Tentative plans have been made for a network similar to the above in Ukrainian SSR and in western Canada. The Canadian national committee on meteorites has for some time had an organized system for reporting bright fireballs and recovering newly fallen meteorites (Folinsbee and Bayrock, 1964). The chief subjects dealt with by the UNESCO Working Group on Meteorites have been the facilities for the rapid international exchange of new meteorites, and the legal status of such specimens. The study and identification of terrestrial meteorite craters has been actively pursued.

Manned space flights beyond the immediate environment of the Earth will take place in the near future. As a result of this, interest in the conditions found in interplanetary space has accelerated rapidly, and numerous experiments for the recording of the impacts of space dust have been mounted on satellites and spacecraft. A great deal of work has also been carried out in the field of a quantitative study of the influx of meteorite particles on the Earth, and the distribution of the orbits of these particles in the solar system.

A number of reports appeared summarizing the results of cooperative efforts in various countries. Babadjanov *et al.* (1964) described the IGY meteor observations in the U.S.S.R., and Keay (1965) meteor research in New Zealand. Baldet (1965) wrote a general note on meteor observing. Šimek (1964*a*) has published a convenient table for the calculation of meteor velocities, and the same author (1965*b*) has dealt with the automatic registration of fast random phenomena.

VISUAL AND TELESCOPIC OBSERVATIONS

Visual observations of meteors and fireball reports were collected by the American Meteor Society as in previous years (Olivier, 1964*a*, 1965*a*, 1966). Based on observations received 1958–63 a second catalogue of visual hourly rates was compiled (Olivier 1965*b*). A catalogue of fireball radiants was also published (Olivier, 1964*b*). The collection of group counts of visual meteors, commenced at Ottawa for the IGY, was continued with the cooperation of selected groups and individuals. Brief summaries have been published in Meteor News (Millman, 1964, 1966). Outstanding bright fireballs have been investigated by Oriti (1965), Pokrzywnicki (1965), and Oleak (1966*a*, *b*). Lamar and Romig (1964) discussed anomalous sound effects connected with fireballs. Various observational effects that must be allowed for in the reduction of visual meteor data were studied by Ivanikov (1965*b*), Znojil (1966), and Kvíz (1965*a*, *b*), and the use of the stereographic projection in meteor reductions was described by McIntosh (1965). A new method of making telescopic meteor velocity observations was described by Porubčan (1966), and the spectral energy distribution for telescopic meteors was studied by Ceplecha *et al.* (1965). Observations of telescopic meteors over base lines of various lengths have been reported by Grygar and Kohoutek (1964) and Znojil (1965). In general heights are in the 80 to 90 km range with an average near 85 km. A comparison of telescopic observations with different types of instruments was also made (Grygar and Kohoutek, 1965). Bakharev (1966) has summarized various characteristics of telescopic meteors, in particular with relation to the IQSY observations in the U.S.S.R.

PHOTOGRAPHY

Various techniques for photographing meteors, and the effects of these techniques and weather conditions on the photometry of meteor trails, have been studied by Volynskaya (1964), Kramer (1965*a*), Izraetskaya (1965), and Ivanikov and Khaimov (1965). Results from a number of systematic programs of meteor photography have appeared. Jacchia *et al.* (1967) discussed selected results from 413 precision-reduced meteors photographed with the Super-Schmidt cameras in New Mexico, U.S.A., during the period 1951–54. Shelton (1965) has made a study of the Quadrantid shower, using all photographic data available. The results of the double-station photographic program, commenced at Ondřejov for the IGY, and since continued, were reported by Ceplecha *et al.* (1964) and by Ceplecha and Ježková (1965). Gulmedova *et al.* (1964) reported results of the photography of meteors in the Turkmen SSR. Detailed photometry of 152 photographed meteors was carried out by Ivanikov (1965*a*). Calibration was made through the use of artificial meteors. It was concluded that, for the 30 meteors with bursts, the normal mass loss in a burst is 10–20% of the mass of the meteoroid.

SPECIAL STUDIES

Experiments in the photoelectric recording of meteors brighter than magnitude + 3 were made by Stepanov (1964, 1965). Hicks *et al.* (1967) used an image orthicon system to photograph meteors. The effects of cosmic rays on meteoroids in space were discussed by Tilles (1966). Konstantinov *et al.* (1966) noted a correlation between gamma ray intensity, as recorded by high-altitude aircraft, and the maxima of prominent meteor showers. They suggest the presence of antimatter in some meteoritic material. The relation between meteors and geomagnetic effects is still controversial. Theoretical aspects of this subject have been treated by Ivanov and Medvedev (1964), Marochnik (1964) and Chapman and Ashour (1965). The general conclusion is that it should be possible to detect magnetic effects from large meteors under certain conditions. The possible electric effects of large meteors have also been computed by Ivanov and Medvedev (1965).

METEOR SPECTRA

Active programs of meteor spectroscopy were continued in Canada at the Meanook, Newbrook, and Springhill Meteor Observatories; in Czechoslovakia at the Ondřejov and Skalnaté Pleso Observatories; and in the U.S.S.R. at the Ashkhabad Observatory. Programs on a smaller scale were also carried out at a number of locations in the U.S.A., U.S.S.R., Great Britain, Japan, the Netherlands and India.

Good dispersion spectra of excellent quality, taken at Ondřejov, have been studied intensively by Ceplecha (1964, 1965*a*, *b*, 1966*b*). Rajchl (1964*a*) has discussed the low dispersion spectra from Ondřejov. Excellent Leonid spectra were secured in Japan during the 1965 return (Tomita, 1966), and these are being studied by H. Hirose, T. Nagasawa and K. Tomita. Meteor spectroscopy, carried out in Scotland from 1958 to 1965, is described by Evans and Evans (1966). They include the detailed study of a Lyrid spectrum. Feibelman and Emmerich (1964) secured an Orionid spectrum at Pittsburgh. The accidental recording of a brilliant fireball on an auroral patrol spectrograph in Alaska is described by Romick, *et al.* (1964). Russell (1964) has discussed the spectra of faint Perseids, photographed in California. Nasyrov and Nasyrova (1966) found from 8 Leonid train spectra, photographed at low dispersion, that the peak of luminous intensity was in the orange-red.

Rajchl (1964*b*, 1965) has summarized some of the general characteristics of meteor spectra as related to the physical theory of meteors. Kramer (1966) has studied the spectrum of a meteor that divided into two parts, and Ivanikov (1966) has determined a panchromatic colour index of + 0·2 from the spectrum of a Perseid meteor.

METEOR TRAINS

The study of meteor trains has recently received considerable attention, especially in the U.S.S.R. Savrukhin and his co-workers have published numerous results based on both visual and photographic observations at Ashkhabad (Savrukhin, 1964*a*, *b*, *c*, *d*, *e*, *f*, 1965*a*, *b*, 1966*a*; Savrukhin and Bezuglova, 1964; Savrukhin and Nasyrova, 1964). Heights of trains were measured in the range 80 to 105 km. Small drift velocities were found below 95 km, higher velocities above 95 km. There was good agreement between meteor train drifts and the drift of the ionosphere in the Es region. Initial diameters of trains were found to average 80 metres. A morphological classification of meteor trains was proposed. Astapovich (1966*a*) has found diurnal, annual, and 11-year periodic variations in 150 meteor trains, recorded during 1500 hours of observation with binoculars. Bakharev and Shodiev (1966) have reported on dual-station photography of meteor trains at Dušhanbe, and Shodiev (1965, 1966) on the characteristics of trains of Perseid and Orionid meteors. Gulmedov (1964), Gulmedov and Yushkevich (1964), and Gulmedov and Nasyrova (1966) have studied atmospheric turbulence and diffusion, using both visual and photographic records of meteor trains at Kazan and Ashkhabad. Truttse (1965) has reported the telescopic observation of a train of a meteor with magnitude + 8. Babadjanov and Kramer (1965) have studied meteor wakes through a system of instantaneous photography. They conclude that wakes, from 300 to 1000 metres long, owe their luminosity primarily to particles fragmented from the meteoroid. Afterglow tails have been studied in the laboratory by Fischer and Rüppel (1964).

RADAR RESEARCH

The radar study of meteors is currently active at a number of locations. Delcourt (1967) has described the French equipment, and Neale (1966*a*, *b*) the two Canadian systems. Šimek (1965*a*, 1966*b*), Hajduk (1965*a*, *b*), and Sládek and Šimek (1965) have discussed features of the meteor radar at Ondřejov, Czechoslovakia, and certain methods of record reduction. Various equipment parameters and numerous results from the Harvard College Observatory, Smithsonian Astrophysical Observatory, Radio Meteor Project have appeared in a series of Research Reports (nos. 1–15) prepared under NASA contract and written by a large number of authors. The physical characteristics of 320 faint radio meteors, selected from over 10 000 reduced on the Harvard-Smithsonian meteor project, have been summarized by Verniani, *et al.* (1966). For these objects, with average mass 10^{-4} g, fragmentation was of common occurrence. Average length of trail was 10 km, average duration 0·3 s, and average height

2·5 km higher than the theoretical value. The maximum electron line-density was found to vary as $m_\infty^{0.7} v_\infty^3$, and was larger than its theoretical value by a factor of 2. Median bulk density was 0·8 g cm^{-3}, confirming the fragile structure of these particles. Radio meteor observations at Millstone Hill, on a wavelength of 68 cm, have been described by Evans (1965). General features of radar meteor work have been covered by Keay (1964), Evans (1967), Kashcheev and Lebedinets (1967) and Millman (1967*b*). The relation of meteor ionization to the ionosphere, and a possible connection with sporadic E formation, was treated by Glazov (1964), Rubtsov (1965), and Ovezgeldyyev and Ostanina (1965). Physical problems of the ionization in meteor trails have been attacked by Fialko (1965*a*, *b*), Fialko, *et al.* (1966), Belkovich (1964), Stepanov and Belkovich (1966), Bayrachenko (1964, 1965), Kruchinenko and Terentjeva (1964) and Rice and Forsyth (1964). Greenhow and Watkins (1964) reported on the characteristics of meteor trails observed at a wavelength of 1 metre. Sugar (1964) has written a general survey of radio propagation by means of meteor trail ionization, and the subject has been further discussed theoretically by Třísková (1964, 1965*a*).

The problem of the relation between the meteor echo duration and the corresponding meteoroid mass has claimed the attention of a number of workers. Underdense trails have been studied by Třísková (1965*b*) and Poole (1967*b*), the Geminid shower by Plavcová (1965), and the problem generally by McIntosh (1966). It appears difficult to go from a given echo duration distribution to a unique solution of the mass distribution. In a series of papers Fialko (1965*c*, *d*, *e*, 1966*a*, *b*) and Fialko and Romanyuk (1966), using radar observations from a number of stations in the U.S.S.R., attacked both the problem of the mass distribution and the kinetic energy distribution of meteoroids.

A monumental study of the drift of radio meteor trails was carried out at Kharkov by Kashcheev (1965) and Kashcheev and Tsesevich (1965). Some 500 000 drifts were measured during the IQSY in 1964–65, and the chief periodicity found was one of 12 hours. The initial radius of overdense trails was estimated as 1 metre at 95 km height (Bibarsov, 1966), the drifts of radio trails and visual trains were compared (Bibarsov, 1964), and parameters of turbulent motion were found for the Kharkov radio trails (Delov, 1965). Watkins (1965) discussed the influence of the Earth's magnetic field on radio meteor trails.

PHYSICAL THEORY

A general note on the physics of meteor entry has been published by Romig (1965). The ionizing efficiency of meteors has been discussed by Verniani and Hawkins (1964) who conclude, based on Ottawa radio results, that a meteor of absolute visual magnitude 0, velocity 40 km s^{-1}, and radiant elevation 30°, has an original mass, m_∞, of 0·8 g. Verniani (1965) found that the photographic luminous efficiency coefficient $\tau_{op} = 10^{-19}$, which means that, for a velocity of 40 km s^{-1}, 0·002 of the kinetic energy appears as photographed luminosity. Theoretical luminous efficiencies, based on atomic collisional cross sections, have been given by Derbeneva (1966). Oleak (1966*c*) has published a note on temperature radiation from fireballs.

Dalton (1967), using Öpik's theory, Hawkins and Southworth's 285 sporadic Super-Schmidt meteors, and a five-variate statistical analysis, finds the mass of a meteor of absolute visual magnitude 0, velocity 30 km s^{-1}, as 2·2 g. Verniani (1964) finds densities for faint photographic meteoroids in the range 0·1 to 1·0 g cm^{-3}, with the density for Giacobinid meteoroids possibly as low as 0·001. Markina (1965) has also studied the densities and masses of photographic meteors. Ceplecha (1966*c*) has called attention to the necessity of separating meteors into major groups based on orbital characteristics, before finding mean densities for the corresponding meteoroids. The way in which a meteoroid looses mass has been discussed by Bibarsov (1964) and Sandakova and Sherbaum (1966), and the physical details of fragmentation of a meteoroid by Kruchinenko (1965), Kramer (1965*b*), Vorobjeva and Kramer (1965), Levin and Simonenko (1966), and Lebedinets and Portniagin (1966). The effects of the structure of the meteoroid on the physics of meteor entry have been studied by Jones and Kaiser (1966), and Kolomiets (1966). The determination of atmospheric parameters from meteor theory is the subject of papers by Babadjanov (1965), and Shiryaeva (1965).

METEOR FLUX

In discussing meteor rates, from which we can determine the flux of meteoroids on the Earth, one of the basic problems is the mass distribution. Kalinkov (1964) has studied this on the basis of over 300 000 visual and telescopic meteor observations taken from the literature for the period 1900–61. He found an average value of 2·5 for the ratio between numbers of meteors in successive magnitude intervals from magnitude − 4 to + 9, after eliminating the doubtful area from magnitude + 3 to + 6 where naked-eye observations overlap the telescopic. The magnitude distribution in various meteor streams has been investigated by Kresáková (1966), and for sporadic meteors by Lindblad (1967). The latter finds a ratio over 4 between successive magnitudes in the magnitude range − 3 to + 2. The size distribution of the very much smaller zodiacal particles was discussed by Southworth (1964).

Systematic observations of meteor rates determined by radar have been published for the northern hemisphere by Millman and McIntosh (1964, 1966), Lazarev (1965*a*), and Baldwin and Kaiser (1965). Millman (1965) has listed a simplified radar meteor index, based on the Ottawa meteor patrol. Elford (1967) has compared the published radar rates for the northern and southern hemispheres respectively, and has estimated the incidence of meteors on the Earth. Kresák and Kresáková (1965) have studied the frequency of bright photographic meteors, and the geographic latitude variations in rates have been discussed by Kresák (1964*a*) for meteor showers, and by Halliday (1964) for meteorite falls. The use of the diurnal and seasonal variations in meteor rates to determine meteor velocities has been dealt with by Kohoutek (1964*a*, *b*) and by Kresák and Kresáková (1964).

Pushnoj (1964) has published the details of 1565 telescopic meteors observed in the Crimea during the IGY, over a total observing period of 250 hours. Kresák and Kresáková (1966) have determined the density distribution of telescopic meteors around the Earth's orbit.

Unusually high radar meteor rates in 1963 have been reported for both the northern and southern hemispheres (Ellyett and Keay, 1964*a*; McIntosh and Millman, 1964; McIntosh, 1967). Meteor rates have been further studied in relation to unusual periodicities (Keay *et al.*, 1966), lunar phases (Ellyet and Keay, 1964*b*), and Indian rainfall (Duncan *et al.*, 1965). Non-random aspects of the times of meteorite falls have been investigated by Brown and Goddard (1964), and the past history of the flux of large meteorites by Hartmann (1965*a*, *b*).

METEOR SHOWERS

Special studies have been made of the returns of a number of showers. The most outstanding event for many years past was the 1966 Leonid return. The maximum, near 12 hours U.T. 1966 November 17 was very sharp and well recorded by radar at Ottawa, Ontario, by a visual team at Kitt Peak, Arizona and by arctic stations in the U.S.S.R. It is possible that the shower in 1966 reached a higher peak than that for the 1833 Leonid return. In addition to numerous shower observations published in papers already referenced the following reports on various recognized showers have appeared: Perseids—Abdylov (1965), Poole (1965), Kaiser *et al.* (1966), Kopecký (1966), Kramer and Rudenko (1965); Geminids – Webster *et al.* (1966); Leonids – Martynenko (1965), Nazarenko and Lazarev (1966), Šimek (1966*a*); Lyrids – Bagrov *et al.* (1964), Savrukhin (1966*b*), Šimek (1964*b*); δ Arietids – Terentjeva (1965*a*, *b*); δ Aquarids – Terentjeva (1965*c*); Sextanids – Nilsson (1964*a*); Vela-Puppids – Ellyett and Roth (1964); and minor showers—Terentjeva (1964*a*, 1965*d*), Bakharev (1965). General characteristics of some of the major meteor showers were summarized by Millman (1967*a*).

ORBITS

Detailed studies of the orbits of bright photographic meteors have appeared in a number of papers (Babadjanov, Suslova, Karaselnikova, 1966; Babadjanov, Getman, Karaselnikova, 1966; Babadjanov and Kramer, 1966, 1967; Sandakova, 1964; Kramer and Markina, 1966). The velocities of faint meteors were determined by Hawkins, Lindblad and Southworth (1964). In 45 orbits determined at Odessa, seven were found to be hyperbolic (Vorobjeva and Rudenko, 1965). Kresák (1964*b*) has discussed the relation between orbits and magnitude distributions, Stohl (1967) certain aspects of sporadic meteor activity, and Kvízová (1964) the relation of meteor and comet orbits to the orbit of Mars. Guth (1967) has proposed a cosmic probe to study retrograde meteors, and Kresák (1966) a probe into the orbit of the Sun-grazing comets.

Terentjeva (1964*b*, *c*, 1965*e*, 1966, 1967) has carried out an exhaustive study of the complex of minor shower orbits, their major characteristics, and their relations to the orbits of comets. Kashcheev *et al.* (1965, 1966) have analysed 12 500 orbits determined by radar observations at Kharkov, and have listed 105 meteor streams and associations. Of these 24 were known previously. Pupyshev (1964, 1965) has studied radiant distribution on the basis of radio observations made at Kazan. Radio surveys of meteor streams have also been carried out, for the northern hemisphere by Poole (1967*a*), and for the southern hemisphere by Nilsson (1964*b*).

Lazarev (1965*b*) has proposed an elliptical model of true-radiant distribution in the ecliptic plane, and from this has calculated rates for meteoric particles encountered in space. Ceplecha (1966*a*, 1967) discussed parameters for the classification of the orbits of sporadic meteors. He found four groups of orbits with some correlation between the orbital elements and the physical characteristics of the meteoroids. In determining average values for the physical and dynamical properties of sporadic meteors the existence of these groups should be recognized. Verniani (1967), in commenting on this, takes issue with some of Ceplecha's conclusions. Kresák (1967) has made a detailed study of the relations among various groups of meteor orbits, comet orbits, and asteroid orbits. He used a graphic coordinate plot of semi-major axis, a, against eccentricity, e, to illustrate various empirical relations, evolutionary trends, and observational effects. Five distinct classes of meteor orbits were noted, generically related to either comets or asteroids. Using the orbital elements of meteors Dycus and Bradford (1964) discussed their origin, and pointed out the significance of observational boundary conditions in the statistics of meteor orbits, and the importance of noting orbital perturbations produced in a number of different ways. Öpik (1966), in a comprehensive and thought-provoking paper, has dealt with various problems concerning the orbits of small bodies in the solar system. He suggested a basic cometary origin for the meteorites, but it should be noted that among the comets he included certain peculiar objects usually listed as asteroids with abnormal orbits.

INTERPLANETARY DUST

The general properties of interplanetary space were surveyed in a panel discussion held in Ottawa by the Canadian Aeronautics and Space Institute (Millman *et al.*, 1964). Other general papers have been written by Öpik (1964) and Hawkins (1964*b*). Theoretical studies of meteor flux about an attractive centre in space have been carried out by Hale and Wright (1964), and by Shelton *et al.* (1964). The dynamics of dust particles in space have been considered by Shapiro *et al.* (1967), and by Singer (1967), and additional work in the same general area has appeared by McCracken (1967) and by Hoffmeister (1967). The role of the zodiacal light, as an indicator of the nature and distribution of interplanetary dust, was noted by Weinberg (1967), and Roach (1967). Techniques for detecting micrometeoroids from satellites or rockets have been described by Jennison and McDonnell (1964) and by Wlochowicz (1966). Meteoroid bumpers for spacecraft was the subject of a paper by Rolsten *et al.* (1964).

Interplanetary dust in the stratosphere has been discussed by Newkirk (1967), Rosen and Ney (1967) and Grjebine (1967), and in the upper atmosphere by Newkirk and Eddy (1964), Stepanov and Gusakovskaya (1964), Parthasarathy and Rai (1967), and Soberman (1967). The optical laser is still a controversial technique for recording dust in the upper atmosphere (Fiocco and Colombo, 1964; McCormick *et al.*, 1966; Deirmendjian, 1965) but future developments in this field may be of interest.

Results from the detection of micrometeoroid impacts by satellites and space probes have been given by Nazarova (1964*a*, *b*, 1965, 1967) for the U.S.S.R. vehicles, and by D'Aiutolo (1964), D'Aiutolo *et al.* (1967), Nilsson and Alexander (1967), McCracken *et al.* (1967), Alexander *et al.* (1967), and Secretan (1967) for the U.S.A. vehicles. Some previous results have indicated a high concentration of interplanetary dust near the Earth and a possible concentration around the Moon. In this regard Katasev (1964) has noted probable effects of the Earth's atmosphere. The subject has received a very detailed theoretical treatment by Shapiro, Lautman and Colombo (1966), who conclude that it is quite unlikely that the Earth can maintain dust concentration greater than ten times that of interplanetary space. On the experimental side, recent information concerning the behaviour of acoustic detecting crystals (Nilsson, 1966), suggests that the quantitative accuracy of all impact rates determined by the acoustic method must be viewed with suspicion. All this throws grave doubt on the reality of the Earth's dust cloud. Results obtained with a Pegasus-type satellite, where an impact actually shorts a capacitor, are much more reliable. The quantitative data in this case are in the form of penetration rates for known target thicknesses, and the determination of meteoroid masses is still a problem.

I have made no attempt to reference completely the laboratory study of suspected cosmic dust particles. Representative papers were by Wright and Hodge (1964), Hodge and Wright (1964), Hodge, Wright, Langway (1964), Hemenway *et al.* (1964), Crozier (1966). A number of additional papers appear in the *Smithsonian Contributions to Astrophysics*, vol. **11**, 1967.

METEOR CRATERS

Work on the Tungus impact of 1908 in the U.S.S.R. is still very active and has resulted in a number of recent papers. Astapovich (1965, 1966*b*) has made a new study of the trajectory for this object, and Ivanov (1965) has determined a precise time for the event. The atmospheric effects have been summarized by Vasiljev *et al.* (1965) in two papers, the ionization effects by Zolotov (1965), a detailed account of the resulting forest fire has been given by Kurbatskii (1964), and a model for study of the blast effects built by Zotkin and Tsikulin (1966). Shalimov (1964) and Stanjukovich (1964) have made theoretical calculations on the ablation and melting in a hypothetical core of the Tungus object.

The general problems of meteorite crater recognition have been dealt with by Halliday and Griffin (1964), Beals (1965), Beals and Halliday (1965), Shteinberg (1965), Currie (1964*a*), and McCall (1964). Barringer (1964) has published a list of the known meteorite craters on Earth. Hawkins, Meunier and Rosenthal (1964) have discussed some theoretical calculations bearing on the formation of an impact crater. Other features of crater formation have been noted by Tolansky (1964), Ronca (1966), and Short (1966).

Research on specific craters throughout the world has been published as listed below:— Campo del Cielo in Argentina by Cassidy *et al.* (1965); Wolf Creek in Australia by McCall (1965); Henbury in Australia by Hodge (1965); Lonar in India by Lafond and Dietz (1964); Ries Kessel in Germany by Johnson *et al.* (1964); Faugères-Cabrerolles in France by Beals (1964); Crestone in U.S.A. by Marvin and Marvin (1966). The greatest activity in the field of crater research has been in Canada during recent years. The current Canadian programme has been summarized by Innes (1964), and additional comments on the origin of the Canadian craters have been published by Dence (1965). Research on specific Canadian craters has been published as follows:—Brent by Beck and Logis (1964); Clearwater Lakes by Dence *et al.* (1965); Deep Bay by Innes *et al.* (1964); Lac Couture by Beals *et al.* (1967); New Quebec by Currie (1964*b*); Nicholson Lake and Pilot Lake by Bunch *et al.* (1967); West Hawk Lake by Halliday and Griffin (1966).

The rapid increase of our knowledge concerning the numerous crater features on the Moon and Mars gives clear indication that craters are one of the most important forms of surface feature on small planetary bodies in the solar system. It is to be hoped that intensive study of all terrestrial crater forms will continue.

PETER M. MILLMAN
President of the Commission

REFERENCES

Abdylov, A. 1965, *Izv. Akad. Nauk turkmen. SSR*, no. 3, 113.
Alexander, W. M., Berg, O. E., McCracken, C. W., Secretan, L., Bohn, J. L. 1967, *Smithson. Contr. Astrophys.*, **11**, 227.
Astapovich, I. S. 1965, *Fizika komet meteorov*, Kiev, 105.
Astapovich, I. S. 1966*a*, *Resultaty issledovanij po MGP, issl. meteorov*, no. 1, 7.
Astapovich, I. S. 1966*b*, *Resultaty issledovanij po MGP, issl. meteorov*, no. 1, 152.
Astapovich, I. S. 1966*c*, *Meteors and Meteorites*, Kiev.
Babadjanov, P. B. 1965, *Geomagn. Aeronom.*, **6**, 153.
Babadjanov, P. B., Getman, T. I., Karaselnikova, S. A. 1966, *Bjull. Inst. Astrofiz., Dušanbe-Stalinabad*, no. 43, in press.
Babadjanov, P. B., Kashcheev, B. L., Kramer, E. N., Tsesevich, V. P. 1964, *Geofiz. Bjull.*, no. 14, 83.
Babadjanov, P. B., Kramer, E. N. 1965, *Astr. Zu.*, **42**, 660.
Babadjanov, P. B., Kramer, E. N. 1966, *Astr. Zu.*, **43**, in press.
Babadjanov, P. B., Kramer, E. N. 1967, *Smithson. Contr. Astrophys.*, **11**, 67.
Babadjanov, P. B., Suslova, N. N., Karaselnikova, S. A. 1966, *Bjull. Inst. Astrofiz., Dušanbe-Stalinabad*, nos. 41–42, 3.
Bagrov, A. V., Koval, V. I., Sidorov, A. K. 1964, *Bjull. vses. astr. geod. Obšč.*, no. 35, 34.
Bakharev, A. M. 1965, *Bjull. vses. astr. geod. Obšč.*, no. 37, 29.
Bakharev, A. M. 1966, *Rezultaty issledovanij po MGP, issl. meteorov*, no. 1, 147.
Bakharev, A. M., Shodiev, U. 1966, *Bjull. Inst. Astrofiz., Dušanbe-Stalinabad*, nos. 41–42, 13.
Baldet, F. 1965, *Annuaire du Bureau des Longitudes*, 363.
Baldwin, J. P., Kaiser, T. R. 1965, *Mon. Not. R. astr. Soc.*, **129**, 85.
Barringer, R. W. 1964, *Meteoritics*, **2**, 169.
Bayrachenko, I. V. 1964, *Geomagn. Aeronom.*, **4**, 313.
Bayrachenko, I. V. 1965, *Geomagn. Aeronom.*, **5**, 460.
Beals, C. S. 1964, *Meteoritics*, **2**, 85.
Beals, C. S. 1965, *Ann. New York Acad. Sci.*, **123**, 904.
Beals, C. S., Dence, M. R., Cohen, A. J. 1967, *Publ. Dom. Obs., Ottawa*, **31**, in press.
Beals, C. S., Halliday, I. 1965, *J. R. astr. Soc. Can.*, **59**, 199.
Beck, A. E., Logis, Z. 1964, *Nature*, **201**, 383.
Belkovich, O. I. 1964, *Meteornoe rasprostranenie radiovoln*, Kazan Univ., no. 2, 135.
Bibarsov, R. Sh. 1964, *Bjull. Inst. Astrofiz., Dušanbe-Stalinabad*, no. 37, 34.
Bibarsov, R. Sh. 1966, *Bjull. Inst. Astrofiz., Dušanbe-Stalinabad*, nos. 41–42, 20.
Bibarsov, R. Sh., Isamutdinov, Sh., Chebotarev, R. P. 1965, *Dokl. Akad. Nauk tadz. SSR*, **8**, 18.
Brown, H., Goddard, I. 1964, *Isotopic and Cosmic Chemistry*, p. 365, Amsterdam, North-Holland Pub. Co.
Bunch, T. E., Dence, M. R., Cohen, A. J. 1967, *Am. Mineralogist*, **52**, in press.
Cassidy, W. A., Villar, L. M., Bunch, T. E., Kohman, T. P., Milton, D. J. 1965, *Science*, **149**, 1055.

Chapman, S., Ashour, A. A. 1965, *Smithson. Contr. Astrophys.*, **8**, 181.
Ceplecha, Z. 1964, *Bull. astr. Inst. Csl.*, **15**, 102.
Ceplecha, Z. 1965*a*, *Bull. astr. Inst. Csl.*, **16**, 88.
Ceplecha, Z. 1965*b*, *Meteoritika*, **26**, 66.
Ceplecha, Z. 1966*a*, *Bull. astr. Inst. Csl.*, **17**, 96.
Ceplecha, Z. 1966*b*, *Bull. astr. Inst. Csl.*, **17**, 195.
Ceplecha, Z. 1966*c*, *Bull. astr. Inst. Csl.*, **17**, 35.
Ceplecha, Z. 1967, *Smithson. Contr. Astrophys.*, **11**, in press.
Ceplecha, Z., Grygar, J., Kohoutek, L. 1965, *Bull. astr. Inst. Csl.*, **16**, 123.
Ceplecha, Z., Ježková, M. 1965, *Bull. astr. Inst. Csl.*, **16**, 166.
Ceplecha, Z., Ježková, M., Novák, M., Rajchl, J., Sehnal, L., Davies, J. G. 1964, *Bull. astr. Inst. Csl.*, **15**, 144.
Ceplecha, Z., Rajchl, J. 1965*a*, *Bull astr. Inst. Csl.*, **16**, 15.
Ceplecha, Z., Rajchl, J. 1965*b*, *Meteoritika*, **26**, 119.
Crozier, W. D. 1966, *J. geophys. Res.*, **71**, 603.
Currie, K. L. 1964*a*, *Meteoritics*, **2**, 93.
Currie, K. L. 1964*b*, *Nature*, **201**, 385.
D'Aiutolo, C. T. 1964, *Space Res.*, **4**, 858.
D'Aiutolo, C. T., Kinard, W. H., Naumann, R. J. 1967, *Smithson. Contr. Astrophys.*, **11**, 239.
Dalton, C. C. 1967, *Smithson. Contr. Astrophys.*, **11**, 81.
Deirmendjian, D. 1965, *J. geophys. Res.*, **70**, 743.
Delcourt, J. 1967, *Smithson. Contr. Astrophys.*, **11**, 91.
Delov, I. A. 1965, *Fizika komet meteorov*, Kiev, 113.
Dence, M. R. 1965, *Ann. New York Acad. Sci.*, **123**, 941.
Dence, M. R., Innes, M. J. S., Beals, C. S. 1965, *J. R. astr. Soc. Can.*, **59**, 13.
Derbeneva, A. D. 1966, *Astr. Zu.*, **43**, 454.
Duncan, R. A., Rao, M. S., Lokanadham, B. 1965, *J. atmos. terr. Phys.*, **27**, 1055.
Dycus, R. D., Bradford, D. C. 1964, *Icarus*, **3**, 306.
Elford, W. G. 1967, *Smithson. Contr. Astrophys.*, **11**, 121.
Ellyet, C. D., Keay, C. S. L. 1964*a*, *Science*, **146**, 1458.
Ellyet, C. D., Keay, C. S. L. 1964*b*, *J. geophys. Res.*, **69**, 1397.
Ellyet, C. D., Roth, K. W. 1964, *Austr. J. Phys.*, **17**, 505.
Evans, J. V. 1965, *J. geophys. Res.*, **70**, 5395.
Evans, J. V. 1967, *Smithson. Contr. Astrophys.*, **11**, 133.
Evans, R. L., Evans, T. L. 1966, *J. Br. astr. Ass.*, **76**, 231.
Feibelman, W. A., Emmerich, W. S. 1964, *J. R. astr. Soc. Can.*, **58**, 268.
Fialko, E. I. 1965*a*, *Geomagn. Aeronom.*, **5**, 97.
Fialko, E. I. 1956*b*, *Geomagn. Aeronom.*, **5**, 1118.
Fialko, E. I. 1965*c*, *Astr. Zu.*, **42**, 409.
Fialko, E. I. 1965*d*, *Astr. Zu.*, **42**, 656.
Fialko, E. I. 1965*e*, *Astr. Cirk. Akad. Nauk SSSR*, no. 325, 1.
Fialko, E. I. 1966*a*, *Astr. Zu.*, **43**, in press.
Fialko, E. I. 1966*b*, *Soviet Rad.*, in press.
Fialko, E. I., Moisya, R. I., Kolomiets, G. I., Bairachenko, I. V., Melnik, V., Chaumak, Y. V. 1966, *Meteory*, Kiev. Univ.
Fialko, E. I., Romanyuk, V. F. 1966, *Astr. Cirk. Akad. Nauk SSSR*, in press.
Fiocco, G., Colombo, G. 1964, *J. geophys. Res.*, **69**, 1795.
Fischer, H., Rüppel, W. B. 1964, *Appl. Opt.*, **3**, 769.
Folinsbee, R. E., Bayrock, L. A. 1964, *J. R. astr. Soc. Can.*, **58**, 109.
Glazov, G. N. 1964, *Trudy Tomsk Inst. radioelektroniki elektrotekhniki*, **3**, 11.
Greenhow, J. S., Watkins, C. D. 1964, *J. atmos. terr. Phys.*, **26**, 539.
Grjebine, T. 1967, *Smithson. Contr. Astrophys.*, **11**, 361.
Grygar, J., Kohoutek, L. 1964, *Bull. astr. Inst. Csl.*, **15**, 62.
Grygar, J., Kohoutek, L. 1965, *Bull. astr. Inst. Csl.*, **16**, 273.
Gulmedov, Kh. D. 1964, *Geomagn. Aeronom.*, **4**, 796.

Gulmedov, Kh. D., Nasyrova, L. I. 1966, *Astr. Cirk. Akad. Nauk SSSR*, no. 356, 1.
Gulmedov, Kh. D., Yushkevich, E. S. 1964, *Geomagn. Aeronom*, **4**, 965.
Gulmedova, A. A., Simonenko, A. N., Yushkevich, E. S. 1964, *Izv. Akad. Nauk turkmen. SSR*, no. 3, 131.
Guth, V. 1967, *Smithson. Contr. Astrophys.*, **11**, 119.
Hajduk, A. 1965*a*, *Bull. astr. Inst. Csl.*, **16**, 132.
Hajduk, A. 1965*b*, *Bull. astr. Inst. Csl.*, **16**, 135.
Hale, D. P., Wright, J. J. 1964, *J. geophys. Res.*, **69**, 3709, 3719.
Halliday, I. 1964, *Meteoritics*, **2**, 271.
Halliday, I., Griffin, A. A. 1964, *Meteoritics*, **2**, 79.
Halliday, I., Griffin, A. A. 1966, *J. R. astr. Soc. Can.*, **60**, 59.
Hartmann, W. K. 1965*a*, *Icarus*, **4**, 157.
Hartmann, W. K. 1965*b*, *Icarus*, **4**, 207.
Hawkins, G. S. 1964*a*, *Meteors, Comets, and Meteorites*, McGraw-Hill Co., New York.
Hawkins, G. S. 1964*b*, *A. Rev. Astr. Astrophys.*, **2**, 149.
Hawkins, G. S., Lindblad, B. A., Southworth, R. B. 1964, *Smithson. Contr. Astrophys.*, **8**, 133.
Hawkins, G. S., Meunier, P., Rosenthal, S. 1964, *Geochim. Cosmochim. Acta*, **28**, 1011.
Hemenway, C. L., Soberman, R. K., Witt, G., *et al.* 1964, *Tellus*, **16**, 84.
Hicks, G. T., Barton, G. G., Jr., Dambeck, W. J. 1967, *Smithson. Contr. Astrophys*, **11**, 95.
Hodge, P. W. 1965, *Smithson. Contr. Astrophys.*, **8**, 199.
Hodge, P. W., Wright, F. W. 1964, *J. geophys. Res.*, **69**, 2449.
Hodge, P. W., Wright, F. W., Langway, C. C., Jr. 1964, *J. geophys. Res.*, **69**, 2919.
Hoffmeister, C. 1967, *Smithson. Contr. Astrophys.*, **11**, 229.
Innes, M. J. S. 1964, *Meteoritics*, **2**, 219.
Innes, M. J. S., Pearson, W. J., Geuer, J. W. 1964, *Publ. Dom. Obs., Ottawa*, **31**, 17.
Ivanikov, V. I. 1965*a*, *Bjull. Inst. Astrofiz., Dušanbe-Stalinabad*, nos. 39–40, 3.
Ivanikov, V. I. 1965*b*, *Bjull. Kom. Komet. Meteor.*, no. 10, 13.
Ivanikov, V. I. 1966, *Bjull. Inst. Astrofiz., Dušanbe-Stalinabad*, nos. 41–42, 26.
Ivanikov, V. I., Khaimov, I. M. 1965, *Bjull. Kom. Komet. Meteor.*, no. 10, 35.
Ivanov, K. G. 1965, *Geomagn. Aeronom.*, **5**, 1119.
Ivanov, V. V., Medvedev, J. A. 1964, *Astr. Zu.*, **41**, 1118.
Ivanov, V. V., Medvedev, J. A. 1965, *Geomagn. Aeronom.*, **5**, 284.
Izraetskaya, N. N. 1965, *Bjull. Kom. Komet. Meteor.*, no. 10, 31.
Jacchia, L. G., Verniani, F., Briggs, R. E. 1967, *Smithson. Contr. Astrophys.*, **11**, 1.
Jennison, R. C., McDonnell, J. A. M. 1964, *Planet. Space Sci.*, **12**, 627.
Johnson, G. G., Vand, V., Dachille, F. 1964, *Nature*, **201**, 592.
Jones, J., Kaiser, T. R. 1966, *Mon. Not. R. astr. Soc.*, **133**, 411.
Kaiser, T. R., Poole, L. M. G., Webster, A. R. 1966, *Mon. Not. R. astr. Soc.*, **132**, 225.
Kalinkov, M. 1964, *Astr. Zu.*, **41**, 419.
Kashcheev, B. L. 1965, *Fizika komet meteorov*, Kiev, 81.
Kashcheev, B. L., Lebedinets, V. N. 1967, *Smithson. Contr. Astrophys.*, **11**, 183.
Kashcheev, B. L., Lebedinets, V. N., Lagutin, M. F. 1965, *Dokl. Akad. Nauk SSSR*, **164**, 1256.
Kashcheev, B. L., Lebedinets, V. N., Lagutin, M. F. 1966, *Rezultaty issledovanij po MGP, issl. meteorov*, no. 1, 141.
Kashcheev, B. L., Tsesevich, V. P. 1965, *Issledovanie tsirkulyatsii atmosfery v meteornoi zone*, MGSS istruktsiya, Moscow, Nauka.
Katasev, L. A. 1964, *Space Res.*, **4**, 914.
Keay, C. S. L. 1964, *Astr. J.*, **69**, 484.
Keay, C. S. L. 1965, *Southern Stars* (N.Z.), **21**, 9.
Keay, C. S. L., Ellyett, C. D., Brown, T. A. 1966, *J. geophys. Res.*, **71**, 1409.
Kohoutek, L. 1964*a*, *Bull. astr. Inst. Csl.*, **15**, 20.
Kohoutek, L. 1964*b*, *Bull. astr. Inst. Csl.*, **15**, 256.
Kolomiets, A. R. 1966, *Astr. Cirk. Akad. Nauk SSSR*, nos. 363 and 381.

Konstantinov, B. P., Bredov, M. M., Belyaevskii, A. I., Sokolov, I. A. 1966, *Kosm. Issled.*, **4**, 66.
Kopecký, M. 1966, *Bull. astr. Inst. Csl.*, **17**, 94.
Kramer, E. N. 1965*a*, *Novaya tekhnika v astronomii, Nauka, Moskva-Leningrad*, **57**, 87.
Kramer, E. N. 1965*b*, *Geomagn. Aeronom.*, **5**, 276.
Kramer, E. N. 1966, *Astr. Zu.*, **43**, in press.
Kramer, E. N., Markina, A. K. 1966, *Meteory*, Kiev Univ.
Kramer, E. N., Rudenko, O. A. 1965, *Astr. Zu.*, **42**, 416.
Kresák, L. 1964*a*, *Bull. astr. Inst. Csl.*, **15**, 53.
Kresák, L. 1964*b*, *Bull. astr. Inst. Csl.*, **15**, 190.
Kresák, L. 1966, *Bull. astr. Inst. Csl.*, **17**, 188.
Kresák, L. 1967, *Smithson. Contr. Astrophys.*, **11**, 9.
Kresák, L., Kresáková, M. 1964, *Bull. astr. Inst. Csl.*, **15**, 255.
Kresák, L., Kresáková, M. 1965, *Bull. astr. Inst. Csl.*, **16**, 81.
Kresák, L., Kresáková, M. 1966, *Contr. Skalnaté Pleso*, **3**, 39.
Kresáková, M. 1966, *Contr. Skalnaté Pleso*, **3**, 75.
Kruchinenko, V. G. 1965, *Geomagn. Aeronom.*, **5**, 104.
Kruchinenko, V. G., Terentjeva, A. K. 1964, *Vestnik Kiev Univ., astr.*, no. 6, 98.
Kurbatskii, N. P. 1964, *Meteoritika*, **25**, 168.
Kvíz, Z. 1965*a*, *Bull. astr. Inst. Csl.*, **16**, 151.
Kvíz, Z. 1965*b*, *Bull. astr. Inst. Csl.*, **16**, 263.
Kvíz, Z., Kvízová, J. 1964, *Bull. astr. Inst. Csl.*, **15**, 31.
Kvízová, J. 1964, *Bull. astr. Inst. Csl.*, **15**, 230.
Lafond, E. C., Dietz, R. S. 1964, *Meteoritics*, **2**, 111.
Lamar, D. L., Romig, M. F. 1964, *Meteoritics*, **2**, 127.
Lazarev, R. G. 1965*a*, *Izv. Tomsk Politechn. Inst.*, **131**, 3.
Lazarev, R. G. 1965*b*, *Astr. Zu.*, **42**, 1075.
Lebedinets, V. N., Portniagin, Y. T. 1966, *Bjull. Kom. Komet. Meteor.*, no. 13, 9.
Levin, B. J., Simonenko, A. N. 1966, *Geomagn. Aeronom.*, **6**, 608.
Lindblad, B. A. 1967, *Smithson. Contr. Astrophys.*, **11**, 171.
Manning, L. A. 1964*a*, *Radio Science*, **68D**, 1067.
Manning, L. A. 1964*b*, *Radio Science*, **68D**, 1079.
Markina, A. K. 1965, *Bjull. Kom. Komet. Meteor.*, no. 11, 47.
Marochnik, L. S. 1964, *Geomagn. Aeronom.*, **4**, 193.
Martynenko, V. V. 1965, *Astr. Cirk. Akad. Nauk SSSR*, no. 350, 1.
Marvin, U. B., Marvin, T. C. 1966, *Meteoritics*, **3**, 1.
McCall, G. J. H. 1964, *Nature*, **201**, 251.
McCall, G. J. H. 1965, *Ann. New York Acad. Sci.*, **123**, 970.
McCormick, P. D., Poultney, S. K., Van Wijk, U., Alley, C. O., Bettinger, R. T., Perschy, J. A. 1966, *Nature*, **209**, 798.
McCracken, C. W. 1967, *Smithson. Contr. Astrophys.*, **11**, 213.
McCracken, C. W., Alexander, W. M., Dubin, M. 1967, *Smithson. Contr. Astrophys.*, **11**, 259.
McCrosky, R. E., Boeschenstein, H., Jr. 1965, *J. Soc. Photographic Instrumentation Engineers*, **3**, 127.
McIntosh, B. A. 1965, *J. R. astr. Soc. Can.*, **59**, 155.
McIntosh, B. A. 1966, *Can. J. Phys.*, **44**, 2729.
McIntosh, B. A. 1967, *Smithson. Contr. Astrophys.*, **11**, 201.
McIntosh, B. A., Millman, P. M. 1964, *Science*, **146**, 1457.
Millman, P. M. 1964, *J. R. astr. Soc. Can.*, **58**, 41, 179, 229.
Millman, P. M. 1965, *J. R. astr. Soc. Can.*, **59**, 39, 143.
Millman, P. M. 1966, *J. R. astr. Soc. Can.*, **60**, 151, 242.
Millman, P. M. 1967*a*, *Smithson. Contr. Astrophys.*, **11**, 105.
Millman, P. M. 1967*b*, *Smithson. Contr. Astrophys.*, **11**, 151.
Millman, P. M., Covington, A. E., Carmichael, H., Wlochowicz, R. 1964, *Can. aeronaut. space J.*, **10**, 143.

Millman, P. M., McIntosh, B. A. 1964, *Can. J. Phys.*, **42**, 1730.
Millman, P. M., McIntosh, B. A. 1966, *Can. J. Phys.*, **44**, 1593.
Nasyrova, L. I., Nasyrov, G. A. 1966, *Astr. Cirk. Akad. Nauk SSSR*, no. 370, 1.
Nazarenko, M. K., Lazarev, R. G. 1966, *Astr. Cirk. Akad. Nauk SSSR*, no. 362, 1.
Nazarova, T. N. 1964*a*, *Geofiz. Bjull.*, no. 14, 89.
Nazarova, T. N. 1964*b*, *Space Res.*, **4**, 921.
Nazarova, T. N. 1965, *Issledovanie kosmicheskogo prostranstva*, Nauka, Moskva, 572.
Nazarova, T. N. 1967, *Smithson. Contr. Astrophys.*, **11**, 231.
Neale, M. J. 1966*a*, *Can. J. Phys.*, **44**, 1021.
Neale, M. J. 1966*b*, *Can. J. Phys.*, **44**, 1573.
Newkirk, G., Jr. 1967, *Smithson. Contr. Astrophys.*, **11**, 349.
Newkirk, G., Jr., Eddy, J. A. 1964, *J. Atmospheric Sci.*, **21**, 35.
Nielsen, A. V. 1967, *Catalogue of Paths of Bright Meteors*, Meddelelser fra Ole Rømer-Obs. i Aarhus, in press.
Nilsson, C. S. 1964*a*, *Austr. J. Phys.*, **17**, 158.
Nilsson, C. S. 1964*b*, *Austr. J. Phys.*, **17**, 205.
Nilsson, C. S. 1966, *Science*, **153**, 1242.
Nilsson, C. S., Alexander, W. M. 1967, *Smithson. Contr. Astrophys.*, **11**, 301.
O'Connell, E. 1965, *A Catalog of Meteorite Craters and Related Features with a Guide to the Literature*, P–3087 Rand Corp., Santa Monica, California.
Oleak, V. H. 1966*a*, *Sterne*, **42**, 91.
Oleak, V. H. 1966*b*, *Astr. Nachr.*, **289**, 71.
Oleak, V. H. 1966*c*, *Astr. Nachr.*, **289**, 157.
Olivier, C. P. 1964*a*, *Am. Meteor Soc. Flower Cook Obs. Repr.*, no. 149.
Olivier, C. P. 1964*b*, *Am. Meteor Soc. Flower Cook Obs. Repr.*, no. 146.
Olivier, C. P. 1965*a*, *Am. Meteor Soc. Flower Cook Obs. Repr.*, no. 155.
Olivier, C. P. 1965*b*, *Smithson. Contr. Astrophys.*, **8**, 171.
Olivier, C. P. 1966, *Am. Meteor Soc. Flower Cook Obs. Repr.*, no. 164.
Öpik, E. J. 1964, *Irish astr. J.*, **6**, 260.
Öpik, E. J. 1966, *Adv. Astr. Astrophys.*, **4**, 301.
Oriti, R. A. 1965, *Griffith Observer, Los Angeles*, **29**, 1.
Ovezgeldyyev, O., Ostanina, M. B. 1965, *Geomagn. Aeronom.*, **5**, 356.
Parthasarathy, R., Rai, D. B. 1967, *Smithson. Contr. Astrophys.*, **11**, 365.
Plavcová, Z. 1965, *Bull. astr. Inst. Csl.*, **16**, 127.
Pokrzywnicki, J. 1965, *Bull. Soc. des Amis des Sci. et des Lettres de Poznán*, Ser. B, **18**, 103.
Poole, L. M. G. 1965, *Bull. astr. Inst. Csl.*, **16**, 364.
Poole, L. M. G. 1967*a*, *Smithson. Contr. Astrophys.*, **11**, 157.
Poole, L. M. G. 1967*b*, *Smithson. Contr. Astrophys.*, **11**, 181.
Porubčan, V. 1966, *Bull. astr. Inst. Csl.*, **17**, in press.
Prior, G. T., Hey, M. H. 1966, *Catalogue of Meteorites, 3rd edition*, British Museum, London.
Pupyshev, Y. A. 1964, *Meteornoe rasprostranenie radiovoln*, Kazan Univ., no. 2.
Pupyshev, Y. A. 1965, *Meteornoe rasprostranenie radiovoln*, Kazan Univ., no. 3.
Pushnoj, L. A. 1964, *Bjull. vses. astr. geod. Obšč.*, no. 35, 18.
Rajchl, J. 1964*a*, *Bull. astr. Inst. Csl.*, **15**, 138.
Rajchl, J. 1964*b*, *Bjull. vses. astr. geod. Obšč.*, no. 35, 3.
Rajchl, J. 1965, *Bull. astr. Inst. Csl.*, **16**, 282.
Rice, D. W., Forsyth, P. A. 1964, *Can. J. Phys.*, **42**, 2035.
Roach, F. E. 1967, *Smithson. Contr. Astrophys.*, **11**, 225.
Rolsten, R. F., Hunt, H. H., Wellnitz, J. N. 1964, *Meteoritics*, **2**, 149.
Romick, G. J., Deehr, C. S., Belon, A. E. 1964, *Icarus*, **3**, 164.
Romig, M. F. 1965, *Am. Inst. aeronaut. astronaut. J.*, **3**, 385.
Ronca, L. B. 1966, *Icarus*, **5**, 515.
Rosen, J. M., Ney, E. P. 1967, *Smithson. Contr. Astrophys.*, **11**, 345.
Rubstov, L. N. 1965, *Bjull. Kom. Komet. Meteor.*, no. 10, 38.
Russell, J. A. 1964, *Meteoritics*, **2**, 117.
Sandakova, E. V. 1964, *Vestnik Kiev Univ., astr.*, no. 6, 76.

Sandakova, E. V., Sherbaum, L. M. 1966, *Meteory*, Kiev Univ.
Savrukhin, A. P. 1964*a*, *Izv. Akad. Nauk turkmen. SSR*, no. 1, 127.
Savrukhin, A. P. 1964*b*, *Izv. Akad. Nauk turkmen. SSR*, no. 4, 126.
Savrukhin, A. P. 1964*c*, *Izv. Akad. Nauk turkmen. SSR*, no. 5, 128.
Savrukhin, A. P. 1964*d*, *Izv. Akad. Nauk turkmen. SSR*, no. 6, 122.
Savrukhin, A. P. 1964*e*, *Akad. Nauk SSR, Mezhduvedomstvennyi geofizicheskii institut komitet, Ionosfernye issledovaniia*, no. 13, 158.
Savrukhin, A. P. 1964*f*, *Bjull. vses. astr. geod. Obšč.*, no. 35, 37.
Savrukhin, A. P. 1965*a*, *Bjull. Kom. Komet. Meteor.*, no. 10, 27.
Savrukhin, A. P. 1965*b*, *Izv. Akad. Nauk turkmen. SSR*, no. 1, 127.
Savrukhin, A. P. 1965*c*, *Izv. Akad. Nauk turkmen, SSR*, no. 6, 124.
Savrukhin, A. P. 1966*a*, *Bjull. Kom. Komet. Meteor.*, no. 14, in press.
Savrukhin, A. P. 1966*b*, *Izv. Akad. Nauk turkmen. SSR*, no. 2, 125.
Savrukhin, A. P., Bezuglova, V. 1964, *Izv. Akad. Nauk turkmen. SSR*, no. 2, 129.
Savrukhin, A. P., Nasyrova, L. I. 1964, *Izv. Akad. Nauk turkmen. SSR*, no. 1, 121.
Secretan, L. 1967, *Smithson. Contr. Astrophys.*, **11**, 307.
Shalimov, V. P. 1964, *Meteoritika*, **24**, 70.
Shapiro, I. I., Lautman, D. A., Colombo, G. 1966, *J. geophys. Res.*, **71**, 5695, 5705, 5719, 5733.
Shapiro, I. I., Lautman, D. A., Colombo, G. 1967, *Smithson. Contr. Astrophys.*, **11**, 359.
Shelton, J. W. 1965, *Astr. J.*, **70**, 337.
Shelton, R. D., Stern, H. E., Hale, D. P. 1964, *Space Res.*, **4**, 875.
Shiryaeva, V. P. 1965, *Bjull. Kom. Komet. Meteor.*, no. 12, 38.
Shodiev, U. 1965, *Bjull. vses. astr. geod. Obšč.*, no. 37, 35.
Shodiev, U. 1966, *Bjull. Inst. Astrofiz., Dušanbe-Stalinabad*, nos. 41–42, 31.
Short, N. M. 1966, *Science*, **154**, 382.
Shteinberg, G. S. 1965, *Dokl. Akad. Nauk SSSR*, **165**, 55.
Šimek, M. 1964*a*, *A Table for the Calculation of Meteor Velocities*. Astr. Inst. Csl., Ondřejov.
Šimek, M. 1964*b*, *Bull. astr. Inst. Csl.*, **15**, 202.
Šimek, M. 1965*a*, *Bull. astr. Inst. Csl.*, **16**, 142.
Šimek, M. 1965*b*, *Slaboproudý obzor*, **26**, 449.
Šimek, M. 1966*a*, *Bull. astr. Inst. Csl.*, **17**, 86.
Šimek, M. 1966*b*, *Bull. astr. Inst. Csl.*, **17**, 90.
Singer, S. F. 1967, *Smithson. Contr. Astrophys.*, **11**, 317.
Sládek, J., Šimek, M. 1965, *Bull. astr. Inst. Csl.*, **16**, 285.
Smirnov, V. A. 1964, *Materialy Mezdunarodnogo Geofiziceskogo Goda, Bjull.* no. 6, 16.
Soberman, R. K. 1967, *Smithson. Contr. Astrophys.*, **11**, 323.
Southworth, R. D. 1964, *Ann. New York Acad. Sci.*, **119**, 54.
Stanjukovich, K. P. 1964, *Meteoritika*, **24**, 66.
Stepanov, D. I. 1964, *Itogi nauchnoi konf. Kazanskogo Univ.* 1963, Kazan, 96.
Stapenov, D. I. 1965, *Meteornoe rasprostranenie radiovoln*, Kazan Univ., no. 3.
Stepanov, D. I., Belkovich, O. I. 1966, *Meteornoe rasprostranenie radiovoln*, Kazan Univ., no. 4.
Stepanov, D. I., Gusakovskaya, L. B. 1964, *Meteornoe rasprostranenie radiovoln*, Kazan Univ., no. 2, 156.
Štohl, J. 1967, *Smithson. Contr. Astrophys.*, **11**, 115.
Sugar, G. R. 1964, *Proc. Inst. electr. electron. Eng.*, **52**, 116.
Terentjeva, A. K. 1964*a*, *Astr. Cirk. Akad. Nauk SSSR*, no. 277, 1.
Terentjeva, A. K. 1964*b*, *Informazionnyj Bjull. MGG, Kiev*, no. 6, 11.
Terentjeva, A. K. 1964*c*, *Astr. Cirk. Akad. Nauk SSSR*, no. 307, 4.
Terentjeva, A. K. 1965*a*, *Bjull. Kom. Komet. Meteor.*, no. 11, 33.
Terentjeva, A. K. 1965*b*, *Izv. Akad. Nauk turkmen. SSR*, no. 2, 172.
Terentjeva, A. K. 1965*c*, *Bjull. Kom. Komet. Meteor.*, no. 11, 17.
Terentjeva, A. K. 1965*d*, *Astr. Cirk. Akad. Nauk SSSR*, no. 313, 7.
Terentjeva, A. K. 1965*e*, *Bjull. Kom. Komet. Meteor.*, no. 12, 31.
Terentjeva, A. K. 1966, *Rezultaty issledovanij po MGP, issl. meteorov*, no. 1, 62.

Terentjeva, A. K. 1967, *Smithson. Contr. Astrophys.*, **11**, 109.
Tilles, D. 1966, *Science*, **153**, 981.
Tolansky, S. 1964, *Nature*, **202**, 75.
Tomita, K. 1966, *Astr. Herald, Tokyo*, **59**, 58.
Třísková, L. 1964, *Bull. astr. Inst. Csl.*, **15**, 67.
Třísková, L. 1965*a*, *Bull. astr. Inst. Csl.*, **16**, 146.
Třísková, L. 1965*b*, *Bull. astr. Inst. Csl.*, **16**, 288.
Truttse, Yu. L. 1965, *Izv. Akad. Nauk turkmen. SSR*, no. 6, 124.
Vasiljev, N. V., Kovalevskii, A. F., Zhuravlev, V. K. 1965, *Bjull. vses. astr. geod. Obšč.*, no. 36, 60.
Vasiljev, N. V., Zhuravlev, A. F., Zhuravleva, R. K., Kovalevskii, V. K., Plekhanov, G. F. 1965, *Nochnye svetyashchiesya oblaka i opticheskie anomalii, syvazannye s padeniem Tungusskogo meteorita*, Nauka, Moskva.
Verniani, F. 1964, *Nuovo Cimento, Bologna*, **33**, 1173.
Verniani, F. 1965, *Smithson. Contr. Astrophys.*, **8**, 141.
Verniani, F. 1967, *Smithson. Contr. Astrophys.*, **11**, 61.
Verniani, F., Hawkins, G. S. 1964, *Astrophys. J.*, **140**, 1590.
Verniani, F., Hawkins, G. S., Southworth, R. B. 1966, *J. geophys. Res.*, **71**, 2749.
Volynskaya, L M. 1964, *Bjull. Inst. Astrofiz., Dušanbe-Stalinabad*, no. 37, 26.
Vorobjeva, V. A., Kramer, E. N. 1965, *Geomagn. Aeronom.*, **5**, 269.
Vorobjeva, V. A., Rudenko, O. A. 1965, *Bjull. Kom. Komet. Meteor.*, no. 11, 36.
Watkins, C. D. 1965, *Nature*, **206**, 1027.
Webster, A. R., Kaiser, T. R., Poole, L. M. G. 1966, *Mon. Not. R. astr. Soc.*, **133**, 309.
Weinberg, J. L. 1967, *Smithson. Contr. Astrophys.*, **11**, 203.
Wlochowicz, R. 1966, *Can. J. Phys.*, **44**, 1.
Wright, F. W., Hodge, P. W. 1964, *Ann. New York Acad. Sci.*, **119**, 287.
Yavnel, A. A. 1964, *Priroda, Moskva*, no. 8, 112.
Znojil, V. 1965, *Bull. astr. Inst. Csl.*, **16**, 162.
Znojil, V. 1966, *Bull. astr. Inst. Csl.*, **17**, 287.
Zolotov, A. V. 1965, *Geomagn. Aeronom.*, **5**, 584.
Zotkin, I. T., Tsikulin, M. A. 1966, *Dokl. Akad. Nauk SSSR*, **167**, 59.

REPORT OF THE COMMITTEE ON METEORITES

(Prepared by E. L. Fireman, Chairman of the Committee)

Introduction

A number of important advances in the field of meteoritic research have been made during the past three years. Most important is the evidence that the chondrules in meteorites are the most primitive material extant. The silicate spheroids, called chondrules, which are a common textural feature of meteorites, contain the highest Xe 129/Xe 132 ratios. The excess Xe 129 results from the decay of the extinct radioactive isotope I 129, which has a 17 million year half-life. The K–Ar ages of the chondrules are also older than the age of the bulk meteorite. Xe 129 excesses have been found in most stony meteorites; this excess is the most striking of the xenon anomalies reported. Other xenon anomalies have been attributed to cosmic-ray-produced spallations, neutron captures, or a primordial xenon component that differs isotopically from atmospheric xenon. In the case of the calcium-rich achondrite, Pasamonte, the xenon anomalies provide evidence of another extinct radioactivity, Pu 244, which has an 84 million year half-life. This evidence is not nearly as convincing as that for I 129.

Next in importance to the xenon anomalies are the primordial gas contents of meteorites. Primordial gas was first discovered in 1956 by Gerling and Levsky, who found a large amount of helium in the dark fine-grained phase of Pesyanoe. Seven other meteorites, including Kapoeta and Pantar, with dark and light phases, have been shown to resemble Pesyanoe in primordial gas content. Most stony meteorites do not contain dark and light phases; however, they do contain primordial gas of another type that consists mainly of argon, krypton, and xenon.

The rare-gas work resulted from the application of modern mass-spectrometric techniques to meteorites; the application of another modern technique, the electron-probe analyser, has led to important conclusions regarding the temperature history of meteorites. From the nickel contents at taenite-kamacite borders cooling rates of 1 to 10° C per million years were obtained for the metallic meteorites and also for the metallic phase of stony meteorites. This rate range corresponds to that expected at the centre of an asteroid 50 to 300 km in radius. In the electron-probe analyses of stony meteorites, the homogeneity of the olivine was found to be related to the recrystallization or metamorphism that occurred in the meteorite. The degree of metamorphism led to an improved classification of meteorites that is correlated with their primordial gas content.

Of the controversies that engaged the attention of scientists several years ago, the one over 'organized elements' or life in meteorites has subsided for lack of positive evidence.

The controversy over the origin of tektites has also lessened as more evidence for a terrestrial origin is produced; nevertheless, this subject is still being discussed.

The more important dispute regarding the asteroidal, cometary, or lunar origin of meteorites continues. It is generally believed that meteorites are asteroidal.

The study of meteoritics is now so extensive that even the incomplete bibliographies published by the Air Force Cambridge Research Laboratories (AFCRL–66–5, January, 1966), which cover the three years ending in 1964, contain 524 closely typed pages. Rather than reproduce another bibliography of this kind, we shall arrange the meteorite work alphabetically by countries. The reports received from A. G. Massevitch of the U.S.S.R. and W. G. Elford of Australia are given almost in their entirety. We have attempted to summarize the work carried out in other countries. In those where there is a great deal of activity, e.g., the United States and Germany, we have omitted many references for the sake of brevity. In countries such as Tanzania and Turkey, where little work is done, we have listed any reference we could find.

Australia

A study was made of the abundance of the tin isotopes in meteorites in order to test a theory associated with nucleosynthesis, which suggests that the planetesimals were irradiated with a strong proton flux prior to meteorite parent-body formation (Cleverly, 1965*a*, de Laeter and Jeffery, 1965). As a by-product of isotopic analysis, tin has been redetermined for a number of meteorites with a precision that is probably slightly better than that reported previously.

Attention is at present being directed toward the isotopic abundance of elements of high-ionization potential, and the rare-gas content of newly recovered meteorites. Note also papers by Baker (1962, 1963), Baker *et al.* (1964), Binns (in preparation), Binns *et al.* (1966), Brooks *et al.* (1964), Cleverly (1965*b*), Corbett (1964, in press *a*, in press *b*), de Laeter and Jeffery (in preparation), McCall (1965*a*, *b*, *c*, *d*, *e*, in press *a*, in press *b*), McCall and de Laeter (1965), McCall and Jeffery (1964), McCall and Wiik (1966), and Simmonds (1964).

Austria

A study of the extraordinary meteorite collection in Vienna has been initiated (Dorfler *et al.*, 1965, Kurat and Kurzweil, 1965).

Canada

Important meteorite isotope work was carried out at MacMaster University. Krypton anomalies were discovered in meteorites and interpreted in terms of spallation and neutron capture (Clarke and Thode, 1964*a*, *b*). Sulphur isotopes were measured in various phases of meteorites (Hulston and Thode, 1965, Monster *et al.*, 1965). It was found that the sulphate sulphur was slightly depleted and the troilite sulphur was slightly enriched in S 34. In the bulk meteorite samples of sulphur had the same composition as in the Earth. The recently fallen Peace River meteorite (Folinsbee and Bayrock, 1964) was recovered and distributed for extensive analysis.

Czechoslovakia

A number of papers (Fireman, 1965, Fireman and DeFelice, 1964, Kharitonova, 1965*a*, Rost, 1965) have appeared on further studies of the Příbram meteorite. These include isotopic work, chemical analyses, and microscopic investigations. The European programme to obtain photographic records of meteorite-fireballs is discussed by Ceplecha and Rajchl (1965).

Denmark

The museum collection of meteorites as well as the Greenland meteorites have been catalogued and described (Buchwald, 1965, Buchwald and Munk, 1965).

England

The recently fallen Barwell meteorite was recovered and described (Lancaster-Brown, 1966). Electron-prove analyses (Reed, 1965*a*, *b*) of iron meteorites indicate cooling of the iron bodies at low pressure, i.e., a few kilobars or less; on the other hand, the textures of chondrules (Le Bas, 1966) are taken as evidence for their formation in a high-pressure environment, i.e., approximately 50 kb.

Finland

Pallasites should be classified into two groups: those with rounded olivine grains, like Brenham, into one group, and those with angular olivine grains, like Eagle Station, into the other group (Wahl, 1965).

France

The Bogou and St Severin meteorites were recovered shortly after fall and were examined for cosmic-ray-induced isotopes (Tobailem and Nordemann, 1965). Studies of spherules of possible cosmic origin gave high accretion rates (Grjebine, 1967) of approximately 10^9 tons per year for the Earth.

Germany

Very extensive isotope and mineralogical work is being carried out at Heidelberg and at Mainz. Research on meteorites is also conducted at Cologne and Tübingen. The overall meteorite research in Germany is so extensive that we can give only the most important results. The fact that chondrules have higher Xe 129/Xe 132 ratios than other phases was confirmed (Auer *et al.*, 1965), and the K–Ar ages of chondrules were found to be higher than those in the other phases. The rare-gas contents in a very large number of meteorites (81 chondrites plus other types) were measured at Heidelberg (Kirsten *et al.*, 1963, Zähringer, 1966*b*. The bronzite chondrites were found to have a different distribution of cosmic-ray exposure ages than the hypersthene chondrites. Both kinds of chondrites were found to have the same distribution of radiogenic ages with the exception that the hypersthene chondrites contain a group with K–Ar and U–He ages of $0{\cdot}5 \times 10^9$ years. K–Ar ages for iron meteorites continue to be high, $6{\cdot}3 \pm 0{\cdot}4 \times 10^9$ years (Müller and Zähringer, 1966). The K–Ar and Rb–Sr ages of the Bosumtwi crater glasses were found to be identical to those of the Ivory Coast tektites

(Gentner *et al.*, 1964, Lippolt and Wasserburg, 1966). Metallic spherules were found in the Bosumtwi crater glass, proving that it was an impact crater (El Goresy, 1966). The German investigators therefore strongly believe that tektites are blobs of glass splashed from impact craters. A mass spectrometer and an electron probe were combined into a single instrument for the first time (Zähringer, 1966*a*). Instruments of this type will probably be useful in future studies of the locations of the primordial and other kinds of gas in meteorites. At Mainz, the primordial helium and neon in Pantar were found by fractional etchings (Hintenberger *et al.*, in press) to be located in the outermost layers of the smallest grains in the dark phase, and could be interpreted as solar-wind implantation before the material accreted. The information on helium and neon obtained for a large number of meteorites (Hintenberger *et al.*, 1964) indicated a difference in the cosmic-ray exposure ages of the high-iron and low-iron groups. About 74% of the high-iron chondrites had cosmic-ray exposure ages below 10^7 years, while only 40% of the low-iron group had ages below 10^7 years.

Holland

Research was carried out on the sputtering of meteorites by positive ions (Heymann, 1964).

India

Extensive radioactivity investigations are being made at the Tata Institute in Bombay. Evidence was found for Al 26 in a Pacific sediment (Lal and Venkatavaradan, 1966). This evidence indicates that these sediments contain an extraterrestrial component irradiated by solar protons.

Japan

Mineralogical studies and radioactivity measurements have been made on meteorites (Honda and Arnold, 1964, Mason, 1963, Miyashiro, 1962). It was found (Miyashiro, 1962) that the plagioclase in chondrites belonged either to a high-temperature form or to a transitional state close to it, thus indicating that chondrites were heated at high temperatures and then cooled sufficiently rapidly to preserve high-temperature optics.

Norway

A carbonaceous chondrite of type 2, the Pollen meteorite, was described (Wolff, 1963). The kamacite was found to exist as spherules of diameter 0·01 mm in the olivine grains, and of diameter 0·1 mm in the matrix.

Poland

The meteorites of Poland are described and chemically analysed. The circumstances surrounding the falls and recoveries with maps of the specific locales are also given. A detailed catalogue of the meteorites in the Polish collections has been prepared (Pokrzywnicki, 1964).

Sweden

In collaboration with U.S. investigators, noctilucent cloud particles were collected in Northern Sweden by rockets and were examined with an electron microscope and an electron microprobe (Hemenway *et al.*, 1964*a*, *b*; Witt *et al.*, 1964). Some insight into the nature and composition of the cloud particles was obtained, but the mechanism of the cloud formation remains uncertain. The minerals and the hydrocarbons in the Orguil meteorite were investigated. It is not determined with sufficient certainty whether or not the material is of biogenic indigenous origin (Nagy, 1966).

Switzerland

A large amount of isotope work was done, and some chemical measurements were made in Switzerland. The rare-gas contents of 30 strong meteorites were measured and the results are in good agreement with the Heidelberg measurements. The same correlation of cosmic-ray exposure ages with meteorite classification (Eberhardt *et al.*, in press) was obtained. The Khor Temiki aubrite was found to contain primordial gas with large amounts of helium (Eberhardt *et al.*, 1965*a*) quite similar to Pesyanoe. This primordial gas was found (Eberhardt *et al.*, 1965*b*) to be in the finest grains of the matrix material and located near the surface of the grains. Anomalies were measured in krypton extracted from meteorites, thus confirming the Canadian results (Marti *et al.*, 1966). Xenon anomalies were measured and they were explained as a combination of spallation, neutron capture, and fission. Bromine and chlorine also were measured in meteorites (von Gunten *et al.*, 1965, Wyttenbach *et al.*, 1965).

Tanzania

A list was compiled (Harpum, 1964) of 25 different types of chondrules in Tanzanian chondrites.

Turkey

The Bursa and Cannakkale chondrites are described. Their rare-gas contents were measured and the following results obtained: the He 4 ages of these meteorites are 3·4 and $3 \cdot 5 \times 10^9$ years, the Ar 40 ages are 4·1 and $4 \cdot 4 \times 10^9$ years, and the cosmic-ray exposure ages are 2·0 and 9×10^6 years, for Bursa and Cannakkale, respectively (Heymann, 1966, Kizilirmak, 1965).

U.S.S.R.

Special committees of the U.S.S.R. Academy of Sciences, and other learned societies have published extensively on the subject of meteorites (Aaloe, 1963; Fesenkov and Krinov, 1964, 1965, 1966; Krinov, 1964*a*, 1965*a*, *b*, 1966*a*, *b*, *c*; Plekhanov, 1963; Vasiliev *et al.*, 1965). Translations of English books have also appeared (Mason, 1962, O'Keefe, 1963).

In the summer of 1964 a stony meteorite (achondrite), Pomozdino, weighing 327 g was discovered in the Komi ASSR. The specific feature of this meteorite is the absence of the unified fusion crust. The numerous solidified spatters seen on its crustless portions were formed by the meteorite matter, which fused in passing through the atmosphere. The meteorite was delivered to the Committee on Meteorites of the U.S.S.R. Academy of Sciences. Another stony meteorite (chondrite), Odessa, weighing 1926 g and having an oriented (bun-shaped) form was also delivered to the Committee on Meteorites (Krinov, 1965*c*).

In August and September 1965 the Committee on Meteorites, U.S.S.R. Academy of Sciences, in cooperation with the Commission on Meteorites, Academy of Sciences of Estonia SSR, and the Institute of Geochemistry and Analytical Chemistry, U.S.S.R. Academy of Sciences, conducted an expedition to study the Kaali and Ilumetsa craters. A prospect hole was sunk into the main Kaali crater and the structure of the crater's rim was studied; during these investigations the fragments of the shatter cones caused by the meteorite explosion were found in neighbouring dolomite deposits. A crater 12 m in diameter, the smallest of the group, was fully stripped of the overburden material. The initial form of the crater and the structure of its interior slopes were examined. The dispersed meteoritic matter was extracted from the overburden material. To study the structure of the rims of the Ilumetsa craters, whose meteoritic origin has not yet been definitely established, the researchers made trench-shaped cuttings through the craters' rims. The scientific data gathered by the expedition are now being studied (Krinov, 1966*d*).

The Commission on Meteorites of the Siberian Branch of the U.S.S.R. Academy of Sciences sponsored annual (1964–66) expeditions to the area of the Tunguska fall to collect scientific data. The scientists studied the area of uprooted and flattened forest and observed the character and distribution of the radiation burn, anomalous increment of trees, etc. The data obtained were processed by G. F. Plekhanov, N. V. Vasiliev, and V. G. Fast.

These scientific data, together with material collected earlier, which specified the atmospheric trajectory and orbit of the entering Tunguska substance, indicate that it was a small comet (Ivanov, 1964, Fesenkov, 1966). Experiments carried out with a model and a scaled-down explosion exactly reproduced the actual situation and showed both a blast wave and a ballistic wave effect upon the forest (Zotkin and Tsikulin, 1966).

E. L. Krinov (1964*b*) made a statistical analysis of meteorite falls and finds that occurred in the decade 1953–62.

N. I. Zaslavskaia, I. T. Zotkin, and O. A. Kirova (1964) (the Committee on Meteorites, U.S.S.R. Academy of Sciences) investigated the size distribution of the cosmic globules taken from the Tunguska fallout area and obtained the distribution curve formula.

I. V. Mikheeva and V. D. Kolemensky (1964) (the Leningrad Mining Institute) carried out comparative studies on meteoric and industrial dust by means of X-ray diffraction. For meteoric dust they used the magnetite spherules from the Sikhote-Alin fallout area, and spherules taken from an open-hearth furnace were used for their industrial dust. The data obtained show insignificant differences in the crystal lattice of the magnetite forming the globules.

Kh. A. Viiding (1965) (Tartu University) found meteoric dust (magnetite globules) in the lows of Cambrian sandstones on the territory of Estonia SSR.

V. D. Vilensky (1966) (the Institute of Geochemistry and Analytical Chemistry, U.S.S.R. Academy of Sciences) determined the specific weight of spherical particles collected in the atmosphere.

M. I. D'yakonova and V. Y. Kharitonova carried out systematic chemical analyses of the 15 main components of meteorites. Each new meteorite delivered to the Committee on Meteorites was examined, its chemical composition was determined, and its class and type were determined. They also studied the chemical compositions of both the chondrules and the matrix of meteorites, including the dark and light varieties of some chondrites, and obtained reliable data (D'yakonova, 1964, D'yakonova and Kharitonova, 1964, Kharitonova, 1965*b*).

I. A. Yudin and S. I. Smyshlyaev (the Urals Commission on Meteorites) obtained factual data on the chemical composition and mineragraphic features of opaque minerals of some stony meteorites (Smyshlyaev and Yudin, 1964, Yudin and Smyshlyaev, 1964).

L. G. Kvasha (1965) (the Committee on Meteorites, U.S.S.R. Academy of Sciences) examined the composition of chondrules and obtained results on types of chondrules and on the matrix of meteorites.

A. P. Vinogradov and G. P. Vdovykin (the Institute of Geochemistry and Analytical Chemistry, U.S.S.R. Academy of Sciences) carried out studies on the organic matter of carbonaceous chondrites and arrived at the conclusion that its origin is abiogenic (Vdovykin, 1964*a*, *b*; Vinogradov and Vdovykin, 1964; Vinogradov *et al.*, 1964*a*).

A. A. Yavnel (the Committee on Meteorites, U.S.S.R. Academy of Sciences) studied the distribution of elements in the metal phase of iron meteorites and in chondrites (Yavnel, 1964). He also examined the chemical fractionation in the silicate phase of meteorites (Yavnel, 1966*a*).

Ye. G. Gus'kova (the Leningrad Branch of the Institute of Earth Magnetism, U.S.S.R. Academy of Sciences) studied natural remanent magnetization in iron and stony-iron meteorites (Gus'kova, 1965*a*, *b*).

E. L. Krinov (1965*c*) (the Committee on Meteorites, U.S.S.R. Academy of Sciences) investigated the form, relief, and structure of the fusion crust of some new meteorites. As a result, he found, in particular, that the solidified droplet-globules and little sprays seen spattered out on the crustless meteorite (achondrite), Pomozdino, prove that the crustless portions never had a fusion crust. Hence, the absence of the fusion crust was not due to its peeling off when the meteorite collided with the ground or for any other mechanical reason.

A. P. Vinogradov and I. K. Zadorozhnyi (1964, 1965) measured the cosmogenic, radiogenic, and primary inert gases in stony meteorites.

A. P. Vinogradov and colleagues investigated nuclear reactions in iron meteorites and determined the production rates of radioactive isotopes in chondrites under the influence of cosmic rays (Lavrukhina, 1965*a*, Lavrukhina *et al.*, 1964, Surkov and Nazarkina, 1965, Vinogradov *et al.*, 1964*b*).

G. P. Lovtsus (1964) (the Radium Institute) determined the content and isotopic composition of lead in meteorites.

E. V. Sobotovich (1964) (the Radium Institute) summarized the data on the content of radiogenic and cosmogenic isotopes in meteorites.

A. K. Lavrukhina and colleagues (the Institute of Geochemistry and Analytical Chemistry, U.S.S.R. Academy of Sciences) determined cerium, europium, scandium, barium, and lead contents in the dark and light varieties of the Kunashak and Pervomaiskii Posyoelok chondrites (Lavrukhina *et al.*, 1966).

K. A. Liubarsky (1966*a*, *b*) (the Moscow State University) made a survey of the radiative ages of iron and stony meteorites.

V. G. Fesenkov (1965) (the Committee on Meteorites, U.S.S.R. Academy of Sciences) reviewed the role played by meteorites in the origin of the solar system, and stressed the necessity of taking into account the material composition and structural features of meteorites in connection with the evolution of the solar system.

A. P. Vinogradov (1965) considered the composition of meteorites and their physical and structural features in order to throw light on their origin.

B. J. Levin (1965) (the Institute of Earth Physics, U.S.S.R. Academy of Sciences) reviewed the contemporary information about the conditions of meteorite formations and examined current hypotheses.

I. E. Starik, E. V. Sobotovich, and M. M. Shats (1965) investigated the early history of both terrestrial and cosmic substances according to isotopic dating information.

A. A. Yavnel studied regularities in the composition and structure of meteorites and considered conditions under which the formation of meteorites occurs. He found that chondrites possess no equilibrium conditions in their compounds and concluded therefore that material crystallization in chondrites proceeds comparatively fast (1965, 1966*b*).

A. K. Lavrukhina (1965*b*) studied the effects of nuclear reactions caused by cosmic rays.

G. P. Vdovykin (1964*c*, 1965) investigated the carbonaceous matter of meteorites in connection with their origin.

K. N. Alexeeva (1964) (the Geology Institute, Academy of Sciences of Ukrainian SSR) studied the physical properties of tektites and obtained quantitative characteristics.

G. G. Vorobiev (1964) (the Committee on Meteorites, U.S.S.R. Academy of Sciences) carried out research on the chemical composition of tektites, and determined the beryllium content in tektites and some other glasses. He also, in collaboration with G. Shkrov (Czechoslovakia), investigated the conditions of falling out and transportation of moldavites (Vorobiev and Shkrov, 1965).

United States

Meteorite work in the United States is extensive and varied. The xenon results are most important. The most striking characteristic of xenon in most meteorites is the Xe 129 excess (Reynolds, 1963). This excess differs not only in different meteorites but also in different phases of a single meteorite. The chondrules have the highest Xe 129/Xe 132 ratio (Merrihue, 1966). Since Xe 129 arises from the decay of extinct radioactivity I 129, which has a 17 million year half-life, chondrules were the earliest to form. There is very little primordial gas in chondrules (Merrihue, 1966), which indicates that they formed under conditions that minimized the entrapment of gas. It was recently suggested (Whipple, 1966) that chondrules formed by lightning in the primitive Laplacian-type nebula. Although the Xe 129 is the most striking xenon anomaly, other anomalies exist (Reynolds, 1963). Xenon of the composition measured in carbonaceous chondrites is called primordial.

Deviations from the primordial composition are ascribed to processes such as spallation, neutron capture, fission, and the decay of extinct radioactivities. The calcium-rich achondrite, Pasamonte (Hohenberg *et al.*, in press, Rowe and Bogard, 1966, Rowe and Kuroda, 1965), has fission-type excesses in the Xe 134 and Xe 136 isotopes, which are ascribed to the extinct radioactivity, Pu 244. This Pu 244 decays mainly by alpha emission with a 76 million year half-life, but in 0·3% of the decays undergoes spontaneous fission; Pu 244 can be produced only by rapid successive neutron captures such as might occur in a super-nova explosion. The Pu 244 evidence indicates that formation of the solar system was preceded by a nearby super-nova explosion. It is puzzling (Rowe *et al.*, 1966), however, that the calcium-rich achondrites Lafayette and Nahkla do not contain Xe 134 and Xe 136 excesses indicative of Pu 244, even though they contain Xe 129 excesses. Since Pu 244 should be enriched in calcium-rich achondrites and since the shorter-lived I 129 was present, the longer-lived Pu 244 should also have been present in Lafayette and Nahkla. Excess fission tracks (Fleischer *et al.*, 1966) were observed in minerals from inclusion in the iron meteorites Toluca and Odessa and were attributed to Pu 244; however, these excess fission tracks follow the uranium contents. It is possible that the excess fission is caused by an excess neutron flux (Fireman, 1966) in the past due either to space erosion or to increased solar activity. Toluca and Odessa are large irons where large fluxes of neutrons would be expected. The case for Pu 244 in Pasamonte rests upon ruling out a host of other possibilities that might produce the excess Xe 134 and Xe 136. Because of its importance to theories regarding the formation of the solar system, a great deal of effort is being put into the study of Pu 244.

Rare gases not ascribed to radioactive decay, cosmic-ray or solar-flare interactions, or extinct radioactivities are called primordial. Extensive measurements of primordial gas have been and are being made. This work has been reviewed in a recent article (Pepin and Signer, 1965). Primordial gas has been classified into two types, a solar type with large helium and neon excesses, and a planetary type with argon, krypton, and xenon excesses. The solar type is in the dark phase of meteorites with dark and light phases and in the carbonaceous chondrites. This type of primordial gas increases with the carbon content and decreases with the degree of metamorphism. The primordial gas contents indicate that the individual grains first solidified and were later compacted together to form the parent bodies of meteorites. The individual grains were altered in the highly metamorphosed meteorites but not in the others.

Radioactivity measurements are quite extensive in the United States but have not led to results as interesting as those of the stable rare-gas measurements. To approximately a factor of two, the radioactivity measurements can be explained by the action of a constant cosmic-ray flux on the meteorite. However, factor-of-two discordances exist and may lead to important developments.

Extensive Monte Carlo calculations of changes in the orbits of small bodies with time have been made (Arnold, 1965), and confirm earlier conclusions obtained analytically by Öpik.

A camera network covering a wide area of central United States (Prairie Network) has been set up and should soon give orbits for a considerable number of meteors that result in meteorites (McCrosky and Boeschenstein, 1965).

Almost as extensive as the isotope studies are the electron-probe studies that bear on the temperature and pressure history of meteorites. The rates of cooling of the meteorite parent bodies in the temperature range of 600 to 400°C were obtained from the gradient in the nickel composition at the borders of the two metal phases, taenite and kamacite. The cooling rates were 1 to 10°C per million years (Goldstein and Ogilvie, 1965*a*, *b*; Goldstein and Short, 1966; Goldstein *et al.*, 1965; Wood, 1964, 1965). These rates are consistent with parent bodies of 300-km radius or smaller.

Electron-probe studies have shown that the degree of homogeneity of the olivine and pyroxene is related to other characteristics of metamorphism in meteorites (Dodd and Van Schmus, 1965, Keil and Fredriksson, 1964, Keil *et al.*, 1964). For example, the carbonaceous chondrites, which are primitive materials with no sign of metamorphism and have high primordial gas contents, contain olivine and pyroxene with inhomogeneous chemical compositions. In ordinary chondrites, such as Bruderheim, which show much evidence for recrystallization or metamorphism in their texture and have little or no primordial gas, the olivine has a homogeneous chemical composition throughout. This metamorphic feature has been incorporated into an improved classification scheme (Van Schmus and Wood, in press) that is very useful for systematizing the primordial gas results.

The data obtained from microphones on satellites that led to the conclusion that a dust belt exists around the Earth were found to be uncertain (Nilsson, 1966). The nature and the quantity of cosmic dust arriving at the Earth are uncertain. Merrihue (1964) found an excess He 3 in the gas released at 600° and 1000°C and an excess Ar 36 and Ar 38 in the gas released at 1400°C from a magnetic fraction of a deep-sea sediment. These results were the first positive isotopic evidence for the presence of cosmic dust widely distributed on Earth. The argon anomaly in sea sediment was confirmed by a number of other scientists. There is also an indication for Al 26 in sea sediments (see Lal and Venkatavaradan, 1966). Dust from the polar ice sheets has given isotopes characteristic of an extraterrestrial component (Fireman, 1967, Tilles, 1967), but no Al 26 (Fireman and Langway, 1965). The discrepancy between sea sediment and polar dust Al 26 is probably due to the dissolution of the Al 26-containing material in the ice.

REFERENCES

Aaloe, A. O., ed. 1963, *Meteoritika*, XI.
Alexeeva, K. N. 1964, *Meteoritika*, **24**, 56.
Arnold, J. R. 1965, *Astrophys. J.*, **141**, 1536.
Auer, S., Braun, H. J., Zähringer, J. 1965, *Z. Naturforsch.*, **20a**, 156.
Baker, G. 1962, *Meteoritical Bull.*, no. 24.
Baker, G. 1963, *Meteoritical Bull.*, no. 28.
Baker, G., Gittins, A., Donnelly, T. H. 1964, *Geochim. Cosmochim. Acta*, **28**, 1377.
Binns, R. A. In preparation.
Binns, R. A., Jobbins, E. A., Dines, F. G., Hey, M. H., Reed, S. J. B. 1966, *Mineral. Mag.*, **35**, 881.
Brooks, J. H., Simmonds, N. A. H., Houston, B. R. 1964, *Geol. Survey of Queensland Publ.* no. 320, 7.
Buchwald, V. F. 1965, *Geochim. Cosmochim. Acta*, **29**, 603.
Buchwald, V. F., Munk, S. 1965, Catalogue of Meteorites in the Museum of the Univ. of Copenhagen.
Ceplecha, Z., Rajchl, J. 1965, *Bull. astr. Inst. Csl.*, **16**, 15.
Clarke, W. B., Thode, H. B. 1964*a*, The isotopic composition of krypton in meteorites, in *Isotopic and Cosmic Chemistry*, North-Holland Publ. Co., Amsterdam, 471.

Clarke, W. B., Thode, H. G. 1964*b*, *J. geophys. Res.*, **69**, 3673.
Cleverly, W. H. 1965*a*, *Austr. J. Sci.*, **28**, 126.
Cleverly, W. H. 1965*b*, *J. Roy. Soc. W. Austr.*, **48**, 55.
Corbett, D. W. P. 1964, *Rec. S. Austr. Mus.*, **14**, 695.
Corbett, D. W. P. *Mineral. Mag.*, in press *a*.
Corbett, D. W. P. *Rec. S. Austr. Mus.*, in press *b*.
Dodd, R. T., Jr., Van Schmus, R. 1965, *J. geophys. Res.*, **70**, 3801.
Dorfler, G., Hecht, F., Plochinger, E. 1965, *Tschermaks. Min. Pet. Mitt.*
D'yakonova, M. I. 1964, *Meteoritika*, **25**, 129.
D'yakonova, M. I., Kharitonova, V. Y. 1964, *Meteoritika*, **24**, 37.
Eberhardt, P., Geiss, J., Grögler, N. 1965*a*, *Tschermaks. Min. Pet. Mitt.*, **10**, 535.
Eberhardt, P., Geiss, J., Grögler, N. 1965*b*, *J. geophys. Res.*, **70**, 4375.
Eberhardt, P., Euegster, O., Geiss, J., Marti, K. *Z. Naturforsch.*, in press.
El Goresy, A. 1966, *Earth and planet. Sci. Letters*, **1**, 23.
Fesenkov, V. G. 1965, *Meteoritika*, **26**, 69.
Fesenkov, V. G. 1966, *Astr. Zu.*, **43**, 241.
Fesenkov, V. G., Krinov, E. L., eds. 1964, *Meteoritika*, **25**.
Fesenkov, V. G., Krinov, E. L., eds. 1965, *Meteoritika*, **26**.
Fesenkov, V. G., Krinov, E. L., eds. 1966, *Meteoritika*, **27**.
Fireman, E. L. 1965, *Bull. astr. Inst. Csl.*, **16**, 291.
Fireman, E. L. 1966, *Z. Naturforsch.*, **219**, 1138.
Fireman, E. L. 1967, Proc. Conf. Meteor Orbits and Dust, *Smithson. Contr. Astrophys.*, **11**.
Fireman, E. L., DeFelice, J. C. 1964, *Bull. astr. Inst. Csl.*, **15**, 113.
Fireman, E. L., Langway, C. C., Jr. 1965, *Geochim. Cosmochim. Acta*, **29**, 21.
Fleischer, R. L., Price, P. B., Walker, R. M., Maurette, M., Morgan, G. 1966, *Gen. El. Tech. Inf. Ser.*, no. 66–C–288.
Folinsbee, R. E., Bayrock, L. A. 1964, *J. R. astron. Soc. Can.*, **58**, 109.
Gentner, W., Lippolt, H. J., Müller, O. 1964, *Z. Naturforsch.*, **19a**, 150.
Goldstein, J. I., Hanneman, R. E., Ogilvie, R. E. 1965, *Trans. met. Soc. AIME*, **233**, 812.
Goldstein, J. I., Ogilvie, R. E. 1965*a*, *Geochim. Cosmochim. Acta*, **29**, 893.
Goldstein, J. I., Ogilvie, R. E. 1965*b*, *Trans. met. Soc. AIME*, **233**, 2083.
Goldstein, J. I., Short, J. M. 1966, Goddard Space Flight Center X–641–66–289.
Grjebine, T. 1967, Proc. Symp. Meteor Orbits and Dust, *Smithson. Contr. Astrophys.*, **11**.
von Gunten, H. R., Wyttenbach, A., Scherle, W. 1965, *Geochim. Cosmochim. Acta*, **29**, 475.
Gus'kova, Ye. G. 1965*a*, *Geomagn. Aeronom.*, **5**, 126.
Gus'kova, Ye. G. 1965*b*, *Meteoritika*, **26**, 60.
Harpum, J. R. 1964, Records geol. Survey Tanganyka, 54.
Hemenway, C. L., Fullam, E. F., Skrivanek, R. A., Soberman, R. K., Witt, G. 1964*a*, *Tellus*, **16**, 96.
Hemenway, C. L., Soberman, R. K., Witt, G. 1964*b*, *Tellus*, **16**, 84.
Heymann, A. 1966, *Sci. Rep. Fac. Sci.*, Ege Univ., no. 28, Izmir.
Heymann, D. 1964, *J. geophys. Res.*, **69**, 1941.
Hintenberger, H., König, H., Schultz, L., Wänke, H. 1964, *Z. Naturforsch.*, **199**, 327.
Hintenberger, H., Vilaceh, E., Wänke, H. *Z. Naturforsch.*, in press.
Hohenberg, C. M., Munk, M. N., Reynolds, J. H. *J. geophys. Res.*, in press.
Honda, M., Arnold, J. P. 1964, *Science*, **143**, 203.
Hulston, J. R., Thode, H. G. 1965, *J. geophys. Res.*, **70**, 3475.
Ivanov, K. G. 1964, *Meteoritika*, **24**, 141.
Keil, K., Fredriksson, K. 1964, *J. geophys. Res.*, **69**, 3487.
Keil, K., Mason, B., Wiik, H. B., Fredriksson, K. 1964, *Amer. Mus. Novitates Publ.* no. 2173.
Kharitonova, V. Y. 1965*a*, *Bull. astr. Inst. Csl.*, **16**, 101.
Kharitonova, V. Y. 1965*b*, *Meteoritika*, **26**, 146.
Kirsten, T., Krankowsky, D., Zähringer, J. 1963, *Geochim. Cosmochim. Acta*, **27**, 13.
Kizilirmak, A. 1965, *Meteoritical Bull.* no. 23, Moscow.
Krinov, E. L., ed. 1964*a*, *Meteoritical Bull.*, no. 32.
Krinov, E. L. 1964*b*, *Meteoritika*, **25**, 173.

Krinov, E. L., ed. 1965*a*, *Meteoritical Bull.*, no. 33.
Krinov, E. L., ed. 1965*b*, *Meteoritical Bull.*, no. 34.
Krinov, E. L. 1965*c*, New meteorites in our country, in Earth and Universe, 'Nauka' Publ. House, no. 1.
Krinov, E. L., ed. 1966*a*, *Meteoritical Bull.*, no. 35.
Krinov, E. L., ed. 1966*b*, *Meteoritical Bull.*, no. 36.
Krinov, E. L., ed. 1966*c*, *Meteoritical Bull.*, no. 37.
Krinov, E. L. 1966*d*, New material on meteorite craters, in Earth and Universe, 'Nauka' Publ. House, no. 4.
Kurat, G., Kurzweil, H. 1965, *Ann. Naturkist Wien*, **68.**
Kvasha, L. G. 1965, *Meteoritika*, **26**, 35.
de Laeter, J. R., Jeffery, P. M. 1965, *J. geophys. Res.*, **70**, 2895.
de Laeter, J. R., Jeffery, P. M. In preparation.
Lal, D., Venkatavaradan, V. S. 1966, *Science*, **151**, 1381.
Lancaster-Brown, P. 1966, *Sky Telesc.*, **3**, 7.
Lavrukhina, A. K., Kuznetsova, R. I., Satarova, L. M. 1964, *Geokhimia*, no. 12, 1219.
Lavrukhina, A. K. 1965*a*, Effects of nuclear reactions in stony meteorites, in *Problems of Geochemistry*, 'Nauka' Publ. House, no. 9.
Lavrukhina, A. K. 1965*b*, *Izv. Akad. Nauk SSSR*, geol. ser, **12**, 3.
Lavrukhina, A. K., Kolesov, G. M., Kalicheva, I. S., Akolzina, L. D. 1966, *Geokhimia*, no. 3, 281.
Le Bas, M. J. 1966, *Nature*, **211**, 1355.
Levin, B. J. 1965, *Usp. fiz. Nauk*, **86**, 41.
Lippolt, H. J., Wasserburg, J. G. 1966, *Z. Naturforsch.*, **21a**, 226.
Lovtsus, G. P. 1964, Synopsis of thesis to compete for Candidate's Degree of Chemical Sciences, Leningrad.
Liubarsky, K. A. 1966*a*, *Izv. Akad. Nauk turkmen. SSR*, ser. fiz.-tek., no. 4.
Liubarsky, K. A. 1966*b*, *Izv. Akad. Nauk turkmen. SSR*, ser. fiz.-tek., no. 3.
Marti, K., Eberhardt, P., Geiss, J. 1966, *Z. Naturforsch.*, **21a**, 398.
Mason, B. 1962, *Meteorites*, John Wiley and Sons, New York (transl. from English, ed. by E. L. Krinov, 'Mir' Publ. House, 1965).
Mason, B. 1963, *J. Fac. Sci., Tokyo*, **14**, 445.
McCall, G. J. H. 1965*a*, *Ann. N. Y. Acad. Sci.*, **123**, 970.
McCall, G. J. H. 1965*b*, *Austr. J. Sci.*, **27**, 267.
McCall, G. J. H. 1965*c*, *Meteoritics*, **2**, 315.
McCall, G. J. H. 1965*d*, *Mineral. Mag.*, **35**, 241.
McCall, G. J. H. 1965*e*, *Mineral. Mag.*, **35**, 476.
McCall, G. J. H. *J. R. Soc. W. Austr.*, **49**, in press *a*.
McCall, G. J. H. *J. R. Soc. W. Austr.*, **49**, in press *b*.
McCall, G. J. H., Jeffery, P. M. 1964, *J. R. Soc. W. Austr.*, **47**, 33.
McCall, G. J. H., de Laeter, J. R. 1965, *Spec. Publ. W. Austr. Mus.* no. 3.
McCall, G. J. H., Wiik, H. B. 1966, *J. R. Soc. W. Austr.*, **49**, 13.
McCrosky, R. E., Boeschenstein, H., Jr. 1965, *Smithson. Astrophys. Obs. Spec. Rep.* no. 173.
Merrihue, C. M. 1964, *Ann. N. Y. Acad. Sci.*, **119**, 351.
Merrihue, C. M. 1966, *J. geophys. Res.*, **71**, 263.
Mikheeva, I. V., Kolemensky, V. D. 1964, *Meteoritika*, **25**, 156.
Miyashiro, A. 1962, *Jap. J. Geol. Geog.*, **33**, 235.
Monster, J., Anders, E., Thode, H. C. 1965, *Geochim. Cosmochim. Acta*, **29**, 773.
Müller, O., Zähringer, J. 1966, *Geochim. Cosmochim. Acta*, **30**, 1075.
Nagy, B. 1966, *Geologiska Föreningens i Stockholm Förhandlinger*, **88**, 235.
Nilsson, C. 1966, *Science*, **153**, 1242.
O'Keefe, J. A., ed. 1963, *Tektites*, Univ. Chicago Press, Chicago (transl. from English, ed. by E. L. Krinov, 'Mir' Publ. House, 1966).
Pepin, R. O., Signer, P. 1965, *Science*, **149**, 253.
Plekhanov, G. F., ed. 1963, *Problem of Tunguska Meteorite*, V, Tomsk.
Pokrzywnicki, J. 1964, *Studia geol. Polonica*, **15**.

Reed, S. J. B. 1965*a*, *Geochim. Cosmochim. Acta.*, **29**, 513.
Reed, S. J. B. 1965*b*, *Geochim. Cosmochim. Acta.*, **29**, 535.
Reynolds, J. H. 1963, *J. geophys. Res.*, **68**, 3263.
Rost, R. 1965, *Bull. astr. Inst. Csl.*, **16**, 102.
Rowe, M. W., Bogard, D. D., Kuroda, P. K. 1966, *J. geophys. Res.*, **71**, 4679.
Rowe, M. W., Bogard, D. D. 1966, *J. geophys. Res.*, **71**, 686.
Rowe, M. W., Kuroda, P. K. 1965, *J. geophys. Res.*, **70**, 709.
Simmonds, N. A. H. 1964, *Geol. Survey Queensland Publ.* no. 320, 1.
Smyshlyaev, S. I., Yudin, I. A. 1964, *Meteoritika*, **25**, 121.
Sobotovich, E. V. 1964, *Meteoritika*, **25**, 40.
Starik, I. E., Sobotovich, E. V., Shats, M. M. 1965, in *Problems of Geochemistry*, 'Nauka' Publ. House, 20.
Surkov, Yu. A., Nazarkina, G. B. 1965, *Geokhimia*, no. 8, 918.
Tilles, D. 1967, Proc. Conf. Meteor Orbits and Dust, *Smithson. Contr. Astrophys.*, **11**.
Tobailem, J., Nordemann, D. 1965, *Geochim. Cosmochim. Acta*, **29**, 1317.
Van Schmus, W. R., Wood, J. A. *Geochim. Cosmochim. Acta*, in press.
Vasiliev, N. V., Zhuravlev, V. K., Zhuravleva, R. K., Kovalevsky, A. F., Plekhanov, G. F. 1965, *Nochnye svetyashchiesya oblaka i opticheskie anomalii svyazannye s padeniem Tungusskogo meteorita*, Nauka, Moscow.
Vdovykin, G. P. 1964*a*, *Geokhimia*, no. 7, 678.
Vdovykin, G. P. 1964*b*, *Meteoritika*, **24**, 194.
Vdovykin, G. P. 1964*c*, *Geokhimia*, no. 4, 299.
Vdovykin, G. P. 1965, *Meteoritika*, **26**, 151.
Viiding, Kh. A. 1965, *Meteoritika*, **26**, 132.
Vilensky, V. D. 1966, *Geokhimia*, no. 5, 586.
Vinogradov, A. P., Vdovykin, G. P., Marov, I. V. 1964*a*, *Geokhimia*, no. 5, 395.
Vinogradov, A. P., Lavrukhina, A. K., Revina, L. D. 1964*b*, *Meteoritika*, **24**, 22.
Vinogradov, A. P. 1965, *Geokhimia*, no. 11, 1275.
Vinogradov, A. P., Vdovykin, G. P. 1964, *Geokhimia*, no. 9, 843.
Vinogradov, A. P., Zadorozhnyi, I. K. 1964, *Geokhimia*, no. 7, 587.
Vinogradov, A. P., Zadorozhnyi, I. K. 1965, *Meteoritika*, **26**, 77.
Vorobiev, G. G. 1964, *Meteoritika*, **24**, 51.
Vorobiev, G. G., Shkrov, G. 1965, *Dokl. Akad. Nauk SSSR*, **161**, 63.
Wahl, W. 1965, *Geochim. Cosmochim. Acta*, **29**, 177.
Whipple, F. L. 1966, *Science*, **153**, 54.
Witt, G., Hemenway, C. L., Lange, N., Modin, S., Soberman, R. K. 1964, *Tellus*, **16**, 103.
Wolff, F. C. 1963, *Geochim. Cosmochim. Acta*, **27**, 979.
Wood, J. A. 1964, *Icarus*, **3**, 429.
Wood, J. A. 1965, *Nature*, **208**, 1085.
Wyttenbach, A., von Gunten, H. R., Scherle, W. 1965, *Geochim. Cosmochim. Acta*, **29**, 467.
Yavnel, A. A. 1964, *Meteoritika*, **25**, 75.
Yavnel, A. A. 1965, *Meteoritika*, **26**, 26.
Yavnel, A. A. 1966*a*, *Geokhimia*, no. 3, 291.
Yavnel, A. A. 1966*b*, *Astr. Zu.*, **43**, 647.
Yudin, I. A., Smyshlyaev, S. I. 1964, *Meteoritika*, **25**, 96.
Zähringer, J. 1966*a*, *Earth and Planet. Sci. Letters*, **1**, 20.
Zähringer, J. 1966*b*, *Meteoritika*, **27**, 25.
Zaslavskaia, N. I., Zotkin, I. T., Kirova, O. A. 1964, *Dokl. Akad. Nauk SSSR*, **156**, 47.
Zotkin, I. T., Tsikulin, M. A. 1966, *Dokl. Akad. Nauk SSSR*, **167**, 59.

23. COMMISSION DE LA CARTE DU CIEL

PRÉSIDENT: Professeur P. Sémirot, Directeur de l'Observatoire de l'Université de Bordeaux, 33-Floirac, France.

VICE-PRÉSIDENT: Professor G. J. Järnefelt, Director of the Astronomical Observatory, Helsinki, Finland.

COMITÉ D'ORGANISATION: P. Couderc, A. N. Deutsch, F. Fernandez de la Puente†, H. Kox, W. J. Luyten, H. W. Wood.

MEMBRES: Abhyankar, Bouigue, Danjon†, Eichhorn, Fracastoro, Harris, Landi Dessy, Lourens, Murray, O'Connell, Paloque, Warzée.

C'est avec regret que nous signalons de décès de notre collègue Fernandez de la Puente, Directeur de l'Observatoire de San Fernando. M. Vicente Planelles Ripoll, Capitan de Navio, lui a succédé.

Le Draft Report comprendra deux parties. La première sera un compte rendu d'une réunion tenue à l'Observatoire de Paris. La seconde résumera l'activité des observatoires.

I. COMPTE RENDU DU COLLOQUE DES CONSTANTES DÉFINITIVES DU CATALOGUE PHOTOGRAPHIQUE (C.P.) (HÉMISPHÈRE NORD)

Un colloque destiné à préparer le calcul des constantes définitives pour les zones couvertes par les catalogues AGK2–AGK3 s'est tenu les 28 et 29 janvier 1966 à l'Observatoire de Paris, conformément au voeu de la Commission 23.

M. Hans Kox de l'Observatoire de Hambourg, a présidé les séances de travail. Etaient présents M. Gunther de l'Observatoire de Hambourg, Mme Dampierre, Melles Lanoë, Canavaggia, MM. Bouigue, Couderc, Delhaye, Lacroute, Sémirot, Gallouët, Rapaport.

A. Identification des étoiles AGK2 dans le Catalogue Photographique

M. Dieckvoss a établi, en Algol, un programme permettant de passer des α, δ, de l'AGK2 aux x, y, du Catalogue Photographique en valeurs approchées. Il faut ensuite retrouver les étoiles appartenant à chaque cliché.

Un programme projeté par M. Lacroute diffère de celui de M. Dieckvoss, car on y trouve directement les étoiles par cliché.

Heidelberg possède l'AGK2 sur cartes, mais ce ne sont pas des cartes IBM. Si l'on peut en obtenir copie sur cartes (ou bandes) IBM, le service de la Carte du Ciel de Paris se chargera de calculer les x, y approximatifs des étoiles repères, de les classer par clichés et de les identifier dans le Catalogue Photographique pour tout l'hémisphère Nord.

B. Perforation

Deux sortes de cartes sont prévues:

(1) *les cartes en-têtes de cliché* (*rouges*)

n° du cliché	sur	6	colonnes	de	9 à 14	±	δ et n° de 1 à 180
Date	„	4	„	„	15 18		XX, XX
α	„	9	„	„	19 27		$XX^h\ XX^m\ XX^s$, XXX
δ	„	9	„	„	28 36	±	XX° XX′ XX″, XX

Ces cartes à perforer en priorité, serviront pour le programme d'identification.

(2) *les cartes étoiles (blanches)*

n° AGK2	sur	8	colonnes de	1 à	8	± XX°XXXX$^{bis}_{ter}$
n° du cliché	,,	6	,, ,,	9	14	
date	,,	4	,, ,,	15	18	
n° de l'étoile	,,	5	,, ,,	19	23	
qualité de l'étoile		3	,, ,,	24	26	voir code
m (C.P.)	,,	3	,, ,,	27	29	
x	,,	7	,, ,,	30	36	± XX, XXXX
y	,,	7	,, ,,	37	43	± XX, XXXX
Repérage	,,	1	,, ,,	44		0 pour une étoile quelconque
étoile repérée						1 pour une étoile AGK2

C. Calcul des constantes

M. Dieckvoss doit sortir sur cartes le α, δ de l'AGK2 et de l'AGK3 (rapportés au FK4). Avec les mouvements propres de l'AGK3, on pourra calculer les α, δ à la date de prise du cliché.

On pourra prendre comme centre du cliché celui qui avait été calculé en tant que constante provisoire dans le Catalogue Photographique en le ramenant à l'équinoxe 1950.0.

Résolution

L'Observatoire de Hambourg prend en charge le calcul des constantes pour les zones de déclinaison supérieure à 32°. L'Observatoire de Paris prend en charge le calcul des constantes des zones de déclinaison inférieure à 32°.

(Nous remercions Mademoiselle Lanoë qui nous a communiqué le compte rendu de la séance qui a été reproduit ici dans ses grandes lignes.)

II. ACTIVITÉ DES OBSERVATOIRES

Observatoire Royal de Greenwich

C. A. Murray

Les mouvements propres de plus de 150 variables RR Lyrae ont été déterminés à partir de catalogues et de plaques astrographiques. Pour les étoiles de l'hémisphère Nord, les plaques de la seconde époque ont été prises avec l'astrographe normal de 13 pouces à Herstmonceux en choisissant les centres de plaques de l'ancien catalogue. Pour les zones de Greenwich et d'Oxford, on disposait déjà des plaques originales à Herstmonceux; pour la zone du Vatican, les plaques de la première époque furent aimablement prêtées par le Père D. J. O'Connell; pour les étoiles des autres zones nord, on a pris les coordonnées de la première époque dans les différents volumes du Catalogue Astrographique.

Dans l'hémisphère Sud, les mouvement propres des étoiles des zones de Sydney et Melbourne ont été obtenus par comparaison directe entre les plaques originales et les nouvelles plaques prises à Sydney grâce à l'amabilité de Mr Harley Wood. Pour les autres zones sud, sauf Le Cap, de nouvelles plaques ont été prises au Cap, et les mouvements propres obtenus par comparaison avec les coordonnées publiées.

Tous ces mouvements propres ont été ramenés à un système basé sur un modèle cinématique d'étoiles de référence faibles pour lesquelles on a obtenu par calculs numériques des parallaxes moyennes.

Les résultats, pour plus de 100 étoiles de l'hémisphère Nord, paraîtront bientôt dans un numéro du *R. Obs. Bull.*, et ceux concernant environ 50 étoiles Sud sont en cours de préparation pour la publication. On est en train de prendre maintenant avec le réfracteur de 13 pouces (soit 33 cm) des plaques centrées sur environ 200 variables Nord RR Lyrae; un réseau d'objectif, donnant une réduction de 2·5 magnitudes environ, est utilisé, et on a prévu d'obtenir quatre plaques de chaque variable à l'époque actuelle. Ce programme sera étendu à d'autres classes d'étoiles variables dès que les étoiles RR Lyrae seront terminées.

Un programme de mesure des mouvements propres de plus de 3000 étoiles dans la zone d'Oxford à moins de 5° du Pôle Galactique Nord a été entrepris; cela fait partie d'un projet à exécuter en collaboration avec le Professeur O. J. Eggen.

Des études de mouvements propres d'étoiles dans les basses latitudes galactiques dans la zone d'Oxford sont aussi prévues.

Observatoire de Hambourg

H. Cox, W. Dieckvoss, A. Gunther

Les cartes en-têtes des clichés individuels du Catalogue Astrographique ont été perforées à Bergedorf pour toutes les zones de déclinaison supérieure à + 32°. Les cartes contiennent pour chaque cliché—numéroté selon un système simple agréé à la conférence de Paris en janvier 1966—les points tangentiels, l'époque de la prise du cliché. En outre nous avons ajouté les constantes provisoires imprimées. Les astronomes français feront des listes avec les coordonnées approximatives pour les étoiles de repère tirées de l'AGK2. Les données de l'AGK2 ont été envoyées à Paris par le Rechen-Institut de Heidelberg.

Nous avons utilisé les données d'une première édition de l'AGK3 au nord de + 23° de déclinaison combinées avec les résultats d'une réduction nouvelle des données de l'AGK2 pour l'étude des mouvements propres de quelques amas galactiques (dissertation de M. de Vegt utilisant des plaques du Bergedorfer Schmidt et les clichés du Catalogue Astrographique des zones Rome-Vatican).

Des travaux de Dieckvoss dans un champ autour de α Persei ont montré qu'une méthode d'itérations sera très puissante pour l'amélioration des mouvements propres des étoiles de repère (extraits de l'AGK2 et de l'AGK3). La discussion de 45 clichés de la zone de Catane a montré 4 termes significatifs provenant d'un effet de coma. Les termes de correction publiés par H. Eichhorn pour la zone de Helsingfors donnent une amélioration sensible.

Vers la fin de 1966 nous commencerons à Bergedorf l'étude systématique des zones de déclinaison plus grande que + 32°, selon la résolution adoptée lors de la conférence de Paris, premier pas pour la détermination des constantes définitives.

Observatoire de Catane

M. G. Fracastoro

Avec la collaboration de Mesdames Cristaldi, Campisi et Celebre, M. P. Tempesti détermine les mouvements propres de quelques étoiles du catalogue astrographique de Catane.

Observatoire de Helsinki

G. Järnefelt

Les observations ont été limitées par suite des travaux de réparation exécutés sur l'astrographe. La détermination de mouvements propres sera ultérieurement poursuivie. Voir bibliographie (1) et (2).

Observatoire de Paris

P. Couderc

Conformément aux décisions prises au colloque de 1966, l'Observatoire de Paris a entrepris le calcul des cartes têtes de clichés dans tout l'hémisphère Nord pour la recherche, dans ces clichés du C.P. des étoiles figurant dans l'AGK2 et pouvant servir ultérieurement de repères.

Actuellement, les programmes ont été établis, en vue de leur exécution par l'IBM 7040 de Meudon. Pour Paris, Toulouse, le Vatican et Oxford, l'exécution est en cours.

En outre, Paris fait exécuter des cartes perforées pour toutes les étoiles de sa zone de 17° à 25° par les soins de la maison IBM. Tous les catalogues ont été préparés et les zones 18°, 19°, 20° et 21° ont été intégralement perforées.

Observatoire de Cincinnati

P. Herget

Concerning the Draft Report of Commission 23, we have accomplished to date all of the following:

(1) From the Bordeaux Astrographic Catalog, we have key-punched B.D., α (1900·0), δ (1900·0), x, y for each comparison star.

(2) We have tested these against the stars of the Yale Zone Catalog, and the printed positions against the computed positions. We have detected about 300 typographical errors of all kinds in the Bordeaux catalog: headings, plate constants, α, δ, x, or y.

(3) We have key-punched the 1953 Bordeaux Zone Catalog and tested it against the stars of the Yale Zone Catalog.

(4) We have key-punched the Berlin zones of the AGK1 Catalog from +10° to +20° and it is proofread, but not yet tested.

It is our plan to form a limited 'General Catalog' of these stars on magnetic tape, using the positions of AGK1, AGK2, AGK3, the Yale Zones, the 1953 Bordeaux Zone, and the Astrographic Catalog. By successive approximations, we will compute:

(*a*) the 'General Catalog' positions and proper motions,
(*b*) the positions of the Astrographic Catalog comparison stars from the 'General Catalog',
(*c*) the plate constants of the Astrographic Catalog from the positions of (*b*),
(*d*) the measured positions of the Astrographic Catalog comparison stars from (*c*),
(*e*) Repeat (*a*), (*b*), and (*c*).

This work will be discussed at the Star Catalog Conference at Washington, D.C., in October, and I shall give you a further report of the sessions of Commission 23 in Prague.

Observatoire de Bordeaux

P. Sémirot

Le catalogue des nébuleuses extragalactiques de la zone −5° à −25° de déclinaison choisies pour la détermination absolue des mouvements propres stellaires a été publié (programme de A. N. Deutsch). 521 positions de nébuleuses correspondant à 48 centres de clichés du Catalogue de Tachkent, soit 165 plaques dont 146 ont été prises par Milet, figurent dans ce catalogue. Pour chaque nébuleuse trois clichés ont été pris. Tous les calculs et toutes les mesures ont été exécutés sous la direction de Pourtau (3).

L'observation de certaines petites planètes et des comètes est poursuivie.

Observatoire de Toulouse

R. Bouigue

(1) Pour chaque étoile de repère de la zone de Toulouse, toutes les positions méridiennes connues y compris celles de l'AGK2 ont été perforées sur cartes de manière à déterminer la position 1900·0 (équinoxe 1950·0) et le mouvement propre. Les corrections de catalogues utilisés ont déjà été publiées par M. Paloque.

Les résultats correspondants figurent dans les *Annales de l'Observatoire de Toulouse.*

(1) $\alpha = 0^h$ à 4^h tome XXX, 1964
(2) $\alpha = 4^h$ à 8^h tome XXXI, 1965 (sous presse)

(2) Compte tenu de cette position 1900·0 et du mouvement propre trouvé, la position est calculée pour chaque étoile pour la date d'observation de l'AGK2 et les deux positions sont comparées (1).

Les résultats montrent que l'accord est très acceptable pour un grand nombre d'étoiles.

Actuellement Toulouse achève la tranche 8^h–12^h, au terme de laquelle Bouigue compte faire une étude d'ensemble.

Ce procédé de calcul permettra d'inclure très rapidement les positions de l'AGK3 et de faire de nouvelles comparaisons tant cette fois sur les positions que sur les mouvements propres.

Les tranches 12^h à 24^h ne seraient poursuivies que si cela en valait la peine.

D'autre part Toulouse a perforé et envoyé à P. Couderc les positions provisoires des centres de la carte nous concernant, comme cela avait été décidé lors du colloque de Paris.

Observatoire de San Fernando

V. Planelles

L'Observatoire a obtenu les mouvements propres de 8544 étoiles dont 31 appartiennent à la liste de Plaut. En décembre 1963 les mouvements propres de 10 062 étoiles ont été publiés.

Le programme des nébuleuses extragalactiques a été poursuivi. Trois clichés ont été pris pour 33 centres, deux clichés ont été pris pour 12 centres et 1 cliché a été pris pour 3 centres.

Observatoire de Perth

B. J. Harris

Actuellement le principal travail concerne l'étude des mouvements propres dans la région de l'amas ouvert NGC 6475 (M7) en utilisant les clichés du Catalogue Photographique.

Une étude des données publiées du Catalogue Photographique de Perth a été entreprise pour déterminer les sources des erreurs systématiques et la précision interne que l'on peut atteindre quand il est possible de réviser les constantes.

Les nouveaux bâtiments de l'Observatoire de Perth à Bickley sont presque achevés. Dès la mise en service, les programmes suivants seront entrepris:

– mouvement propres dans les 'cluster areas',
– mouvements propres des variables RR Lyrae (listes de Plaut et de Van Herk),
– observations des nébuleuses extragalactiques du programme de A. N. Deutsch.

Royal Observatory, Observatory, Cape

J. v. B. Lourens

The Cape continues with the Groningen Programme for the determination of proper motions of the 136 Plaut variable stars in the Cape Astrographic Zone $-40°$ to $-52°$. So far proper motions for 73 stars have been determined and modern plates for each of 116 stars have been obtained.

The Cape and Greenwich Observatories are working jointly on proper motion determinations of RR Lyrae variables in the southern hemisphere. 69 plates taken recently of areas outside the C.A.Z. $-40°$ $-52°$ zones have been sent to Greenwich Observatory where their measurement is undertaken. Relative proper motions have been determined for 20 stars at the Cape falling in the C.A.Z. $-40°$ $-52°$ zone.

A few RV Tauri and SRd variables are also required by the Greenwich Observatory for which plates are to be taken at the Cape.

Observatoire de Sydney

H. W. Wood

The last volume of rectangular coordinates of the Sydney Section of the Astrographic Catalogue (Volume 52) was distributed about the time of the Hamburg Meeting. At present we are working on Volume 53 which will contain tabular matter to assist with the use of the Catalogue, a history of the work and discussion of magnitudes and of probable errors.

We are engaged in measuring proper motions of variable stars chiefly from the list of Plaut and of some galactic clusters in our zone. Several of our volumes have been in short supply for some time and we have just re-published Volumes 5 and 6 with corrections and will shortly be doing Volumes 11 and 12.

Observatoire de Leiden

G. van Herk

Van Herk détermine les mouvements propres de 319 étoiles du type RR Lyrae, de 343 étoiles variables à longue période, de 35 étoiles du type T Tauri, de quelques supernovae situées à haute latitude galactique et d'étoiles de l'amas NGC 663.

Actuellement il dispose de 850 plaques prises après 1954 avec le réfracteur ($f = 525$ cm, $\phi = 33$ cm) pour 620 étoiles du programme et il poursuit les observations. Comme étoiles de première époque il utilise les données du Catalogue Photographique.

Van Herk désirerait savoir s'il existe des observatoires possédant des clichés pris avant 1940 avec des échelles de 1 mm pour 35″ à 70″. Dans l'affirmative, ces observatoires accepteraient-ils de prêter leurs clichés ou des reproductions?

Observatoire de Minnesota

M. J. Luyten

Luyten continue ses mesures de mouvements propres des 'étoiles bleues faibles' en utilisant les positions anciennes publiées dans le Catalogue Photographique. Pour la plupart des étoiles, il est possible, avec des intervalles de 60 ans ou plus, d'atteindre des erreurs moyennes égales à $\pm 0\overset{''}{.}003$ par an (**4**).

Observatoire de Strasbourg

P. Lacroute

Lacroute montre que sur un champ de 1200 degrés carrés, grâce aux machines à calculer modernes, en utilisant les recouvrements de clichés et en résolvant l'ensemble en bloc, on peut, soit améliorer beaucoup les constantes des clichés, soit se contenter d'étoiles de repère observées avec moins de précision (**5**).

Lacroute a étudié les problèmes qui se posent dans l'établissement d'un catalogue photographique en vue de rechercher comment opérer pour économiser le travail d'observation tout en obtenant les résultats désirés, au prix de calculs plus lourds (**6**).

Lacroute espère obtenir, grâce aux recouvrements des clichés du Catalogue Photographique et avec l'extrapolation AGK2–AGK3, pour le calcul des coordonnées équatoriales des étoiles repères pour l'instant de la prise des clichés, des constantes aussi bonnes que celles que l'on aurait obtenues, sans recouvrement, avec des clichés de date intermédiaire entre celles de l'AGK2 et l'AGK3.

III. QUESTIONS À DÉBATTRE

(1) La méthode du calcul des constantes étant bien définie, il s'agit d'étudier la forme exacte de la publication des α, δ (bandes magnétiques, cartes, etc.) ou celle des coordonnées standards proposée par Murray.

(2) Calcul des contantes des clichés de l'hémisphère Sud en profitant de l'expérience acquise pour l'hémisphère Nord (proposition de Lacroute).

(3) Machine à mesurer automatique.

(4) Faut-il refaire un catalogue photographique? Avec quels astrographes? Magnitude limite à atteindre?

(5) Questions diverses.

CONCLUSION

Il est enfin permis d'envisager l'utilisation prochaine du Catalogue Photographique, principalement en ce qui concerne la détermination des mouvements propres des étoiles repères. Que tous les observatoires qui ont participé à ce travail de longue haleine soient remerciés, et notre reconnaissance est due particulièrement à l'équipe Kox, Dieckvoss, Gunther et à Lacroute.

P. SÉMIROT

Président de la Commission

BIBLIOGRAPHIE

1. Proper motions of 41 variable stars in the photographic zone of Helsinki Observatory, by Örlander, Lehti, Pipping et Savalius (*Commentationes Physico-Mathematicae*, Societas Scientiarum Fennica, **22**, 2).
2. Proper motions, mean parallaxes and space velocities of RR Lyrae variables, by G. Van Herk (1965, *Bull. astr. Inst. Netherlands*, **18**, 71–105).
3. Catalogue des nébuleuses extragalactiques de la zone $-5°$ à $-25°$ de déclinaison choisies pour la détermination des mouvements propres absolus stellaires (Pourtau, Milet, Observatoire de Bordeaux).
4. A search for faint blue stars, no. XVII, XXI, XXIX, XXXVII (Luyten, W. J., The Observatory, University of Minnesota, Minneapolis).
5. Améliorations des constantes de clichés en utilisant leurs recouvrements (Lacroute, P., 1964, *Ann. Obs. Strasbourg*, **6**, 97–111).
6. Etude sur l'établissement d'un catalogue photographique (Lacroute, P., 1964, *Ann. Obs. Strasbourg*, **6**, 113–127).

24. COMMISSION DES PARALLAXES STELLAIRES ET DES MOUVEMENTS PROPRES

PRÉSIDENT: Professor Dr A. N. Deutsch, Pulkovo Observatory, Leningrad M-140, U.S.S.R.

VICE-PRÉSIDENT: Professor W. J. Luyten, The Observatory, University of Minnesota, Minneapolis 14, Minnesota, U.S.A.

COMITÉ D'ORGANISATION: V. V. Lavdovski, J. V. B. Lourens, W. W. Morgan, C. A. Murray, K. Aa. Strand, P. Van de Kamp, S. Vasilevskis.

MEMBRES: Cecchini, Delhaye, Eichhorn, Franz, Giclas, Gliese, Haas, Harris, Hoffleit, Kohlschüter, König, Lippincott, Meurers, Missana, Moreno, Osvalds, Paloque, Schilt, Smart, Stearns, Vyssotsky, Wagman, Wilkens.

TRIGONOMETRIC PARALLAXES

An essential increase of the parallax determination programme may be stated since the last report, particularly with respect to fainter stars.

Professor K. Aa. Strand reports that since the completion in April 1964 of the 61-inch (155 cm) astrometric reflector of the U.S. Naval Observatory, located at Flagstaff, Arizona, 9000 plates have been obtained for astrometric studies.

In this programme 335 stars are being observed for parallaxes. Most of these stars are in the visual magnitude range 12·0 to 16·5 and were selected from the Luyten and the Giclas proper motion programmes.

A programme of photometry and spectral classification of the parallax stars and their comparison stars have been initiated. On the average, four comparison stars are selected for each parallax star.

Since the installation in April 1966 of a two-coordinate automatic measuring machine, a total of 1500 plates have been measured. An average plate with 3 to 5 comparison stars now takes 5 minutes to measure, including orientation, which is a gain by a factor of 6 in measuring speed as compared to manual measuring techniques.

The measurements also show a gain in accuracy in that the mean error of unit weight in the parallax solutions is reduced by 24% to $\pm$ 0".035 ($\pm$ 2·7 μ).

Electronic computer reduction programmes are used for the computation of parallax factors, plate constants, and the least squares solution for parallax and proper motion. As of 1 October 1966 twenty or more plates have been obtained for 115 parallax series. Twenty-five series have 30 or more plates of the required 50-plate series for a definite parallax determination and these are estimated to be completed within the next year or year and a half.

At the Lick Observatory, the observing programme to measure the parallaxes of 120 faint red dwarfs, white dwarf and U Geminorum stars from plates taken with the automatic camera on the 36-inch (91 cm) refractor, began in 1961 by S. Vasilevskis, is nearing completion. Parallax fields have been photographed in the blue and the yellow with 20-inch Carnegie double astrograph. These will be used not only for the evaluation of magnitude and colour effects but also for investigating the reference frame of the comparison stars in deriving absolute parallaxes (**1**, p. 611).

Professor W. J. Luyten has determined a few preliminary parallaxes with the 48-inch (122 cm) Schmidt telescope, thus indicating that it is at least possible to use Schmidt telescopes for parallax work. Professor Luyten notes that for extremely faint stars—18th magnitude or fainter—there may be no other way at present.

Dr J. V. B. Lourens reports that at the Cape Observatory the present parallax programme consists of two parts:

(*a*) The measurement of plates obtained mainly during the years 1943 to 1951 for the re-determination of parallaxes of nearby stars already observed at the Cape. Since the last report 1567 of these plates have been measured and 83 rejected; about 900 remain to be measured.

(*b*) An observing programme which started at the end of January, 1965. On this programme at present are 102 stars: 57 stars, south of $-20°$ declination, from Gliese's list, 12 stars which have relatively large photometric parallaxes, 18 stars from Parenago's lists and the remainder consisting of stars which are known astrometric binaries, stars included in the programme at the request of other astronomers, e.g. suspected white dwarfs, subdwarfs, etc. and some other stars.

Dr C. A. Murray reports from the Royal Greenwich Observatory that the current programme on the 26-inch (66 cm) refractor has been reduced to about 100 stars, by the exclusion of a large number of stars on the original pre-war Greenwich list for which modern spectroscopic and other evidence suggests that the parallaxes are likely to be small. Two thirds of the present programme consists of stars in Gliese's catalogue for which only spectroscopic parallaxes are available. For about half the stars on the programme, twenty or more plates have been obtained since 1959; measurement and reduction has been started.

First-epoch plates have been obtained with the 26-inch refractor on faint stars with annual proper motions in excess of 1″, which appear from time to time in the lists of Luyten and Giclas; it is hoped to measure parallaxes of these stars from plates to be obtained at the prime focus of the 98-inch (249 cm) Isaac Newton telescope.

At the Leander McCormick Observatory the parallax and proper motion programme is being continued. Emphasis is being placed on the 300 red dwarfs on Vyssotsky's list not previously observed for parallaxes. Trigonometric parallaxes of 36 stars determined with the 26-inch refractor have been printed in *Astr. J.* (**3**). This is the 39th list of parallaxes by the Leander McCormick Observatory, and follows that in *Astr. J.*, **64**, 265, 1959.

At the Allegheny Observatory six definitive and five preliminary parallaxes were determined (from the new parallax plates measured for the determination of white and red dwarf stars) (**1**, p. 581).

At the Sproul Observatory the work in progress includes a comprehensive study of systematic errors in parallaxes with particular attention to the parallax determinations from declination measurements. A high accumulation of photographic plates was maintained; a relation with the continued drought seems to be indicated (**2** p. 780). As before the observation programme includes mainly astrometry of nearby stars.

Professor Van de Kamp reports that the Sproul refractor has been in continuous operation since 1912 and a total of over 66 000 plates have been taken. Between January 1912 and 1 July 1962 a total of about 51 000 plates have been taken; 41 155 were taken over the 25-year period ending 1 July 1962, an average of 1646 annually. Over the four-year period ending 1 July 1962, 15 222 plates were taken, or about 3800 yearly. The number would have been slightly higher except for the dismantling of the telescope on May 9–11 of this year. The telescope has been completely overhauled and modernized. The telescope is now being installed and we should be in full operation again by the first of January 1967.

The astrometric programme is being continued without any substantial changes. We continue to measure parallaxes and mass ratios and search for perturbations. The perturbations of μ Cassiopeiae (**4**) and α Ophiuchi (**5**) both discovered at the Allegheny Observatory have been analysed and published. Studies of several other perturbations are under way.

Recent results for parallax and mass ratio include: ξ Scorpii (**6**); Σ 2173 (**7**); BD + 77° 361 (**8**); White Dwarf CC 398 (**9**); Hu 1176 (**10**); Six nearby stars (**11**); Astrometric Binary CC 1299

(**12**). The parallaxes of two nearby stars, L 1159–16 (parallax 0".210 ± 0·009) and G 175–34 (parallax 0".19) have been measured and published (**13**). Additional information may be found in the report of Commission 26. A study has been made of the effective wavelengths of the Sproul photovisual technique using different emulsions and different filters.

At the Van Vleck Observatory the programme of measuring parallaxes and proper motions is being continued (**15**). The investigations of the method of reduction and of the systematic errors in parallax research are in progress. In an abstract of J. Gibson paper (**14**) a procedure of taking parallax plates of a galactic cluster is proposed. By such observation it may be possible to establish the existence of systematic errors among measures of small parallaxes by different observatories.

At the Pulkovo Observatory parallax work is being started with the 26-inch visual refractor. The first preliminary results of the parallax determination of five stars are printed in *Astr. Zu.* (**16**). The observing programme is now in progress concerning red dwarfs on Vyssotsky's list and some occasional stars of special interest, as for instance the white dwarf LP9–231. In a note on this star W. F. Van Altena (**17**) points out the large discrepancy between the trigonometric parallax determined by Luyten and the photometric parallax. Parallax observations are now under way at the Lick, Yerkes and Pulkovo Observatories.

SPECTROSCOPIC PARALLAXES

As in previous reports we can mention only a few papers concerning the determination of stellar distances by the spectroscopic method.

At the David Dunlap Observatory Dr J. D. Fernie developed a method for determining the distances of R. R. Lyrae variables from just a single observation of V and $B-V$, provided that the period is already known. A feature of the result is that true distance moduli may be obtained without knowledge of the interstellar reddening (**1**, p. 584). H. H. Guetter published a paper on distances of 97 OB stars (**18**). A. Beer determined the photoelectric distances of 461 northern OB stars and the galactic structure from Hγ–luminosities (**19**). R. M. Petrie published 'Spectroscopic absolute magnitudes of early-type stars from hydrogen absorption—a revision' (**20**)).

Quite possibly, I have missed some other papers concerning spectroscopic (and also photometric) parallaxes, but I should like to note that it would be more suitable to report on such investigations connected with spectrophotometric and colourimetric methods in the reports of Commissions 29 or 45. It is desirable that this question be discussed at our meeting in Prague.

PROPER MOTIONS

The following report should be supplemented by reference to the reports of Commissions 8, 23, 26, 27, 34, 37 and perhaps some other commissions since proper motion determinations concern also double stars, variable stars, planetary nebulae, star clusters, etc.

The programme for obtaining absolute proper motions of stars referred to galaxies is being continued.

The Pulkovo programme for photographing the first-epoch plates of selected areas of galaxies and fundamental stars in the declination zone + 90° to − 25° is being completed at the following observatories: Pulkovo, Moscow, Kiev (Golosseyevo), Tashkent, Bucharest, Shanghai (Zô-Sè) and Bordeaux. In all 3500 plates for galaxies and 2500 plates for fundamental stars were obtained. We note with satisfaction the Bordeaux catalogue of the galaxies in the declination zone − 5° to − 25° with equatorial coordinated of 521 galaxies calculated from measured positions on 165 plates taken with the Carte-du-Ciel astrograph (**21**).

At the Taschkent Observatory the determination of proper motions of stars referred to galaxies in 23 selected areas of the Pulkovo programme has been completed and the constants of solar motion and galactic rotation derived. The second-epoch plates have been taken after an interval of 20 years (**22**).

Similar but more extensive work is in progress at the Pulkovo Observatory concerning the reduction of relative to absolute proper motions of stars by means of the galaxies in 89 areas. A communication on this work will be made at the Prague conference.

At the Toulouse Observatory, the photographic observations of galaxies were continued.

In the southern hemisphere the observations of galaxies at the Cerro Calan Observatory (Santiago, Chile) have been renewed after remounting the astrographic refractor at the end of 1964.

Dr B. J. Harris reports that at Bickley—the new site of the Perth Observatory—the observations of the galaxy areas in the declination zone $-25°$ to $-90°$ will be commenced during 1967. Dr Harris hopes that by the time of the General Assembly in Prague the astrographic telescope will be working fully again.

Dr J. Landi Dessy reports that at the Cordoba Observatory the photographic observations of the galaxy areas are in progress and more than 100 plates have been obtained. Seven areas south of $-69°$ declination could not be photographed, as also at Santiago, owing to the mounting of the astrograph.

These seven areas were fully observed at the Cape Observatory.

At Pulkovo, I am preparing a list of fundamental stars in the southern zone $-25°$ to $-90°$ to be proposed for photographic observations in addition to the galaxies.

As to the installation in Chile of the new 700 mm meniscus astrometric telescope (see the previous report of Commission 24 (**23**), it was decided to mount the instrument at Mount El Roble 100 km north of Santiago at an altitude 2200 m. There are certainly great difficulties in building a new observatory on a rocky top where it is necessary to build a new road. Nevertheless the work has advanced well and it appears that the instrument may be installed in 1967.

At the Yale-Columbia Southern Observatory, El Leoncito, Argentina a new 50-cm twin astrograph was mounted in March 1965 (**24**). The chief purpose is to extend to the South Pole the Lick Observatory proper motion programme referring to the faint galaxies.

The determination of absolute proper motions of brighter stars by comparing the two catalogue AGK2 and AGK3 is nearly completed at the Hamburg-Bergedorf Observatory. Being in the FK4 system, they will present the data for comparison with the proper motions derived by galaxies in the northern hemisphere.

The determinations of proper motions of the Carte du Ciel catalogue stars are in progress at the Toulouse Observatory: positions and proper motions of reference stars are determined, see IV Catalogue, revised (**25**).

The large proper motions of about 1000 stars in the astrographic zones $+32°$ and $+33°$ have been determined (**26**). Stars of the Catania astrographic catalogue with large proper motions have been also published (**27**).

The determination of large proper motions of stars fainter than the 8th magnitude (to about the 17th magnitude) with motions $\geqslant 0\overset{''}{.}27$ per year have been continued at the Lowell Observatory (**28**). The number of stars listed in 200 regions is 11 514. The number of different stars is 8151 and newly discovered 4610. Approximately 62 additional regions remain to be studied in order to complete all of the sky from the equator to the north pole.

Professor Luyten reports that the Proper Motion Survey with the 48-inch (122 cm) Schmidt telescope has been vigorously pushed (**29**) and to date more than 100 pairs of plates have been blinked, more than 40 000 proper motion stars found, and motions have been published for more than 15 000 of these, down to the 21st magnitude, photographic. In particular, more than 3000 proper motions have been determined in the region of the Hyades cluster while the

North Polar Cap – north of declination + 75° – has been completely examined. It is hoped that the general catalogue of all proper motions found in this region of 700 square degrees, and containing, probably, some 8000 different stars, will be published before the next General Assembly.

Among the objects found is one star of the fifteenth magnitude with a proper motion of 3"63 annually, the largest motion found in more than 30 years. In addition a large number of new white dwarfs has been found (**30**), bringing the total published at Minnesota to more than 1500. More than 700 faint wide pairs with common proper motion have been found, and data for these are being prepared for publication.

The more than 15 000 motions referred to before have been used to determine the solar motion for extremely faint stars (**31**). Finally, about a dozen stars with $H = m + 5 + 5 \log \mu$ exceeding + 25, indicating absolute magnitudes beyond $M = + 19$ pg, have been found.

J. H. Anderson has discussed the motions of a number of F stars and has shown that, contrary to the spectroscopic claims, these stars do not form a cluster.

At Minnesota, Luyten, Hughes and Smith have determined proper motions for another 800 faint blue stars (**32**), including 300 within the area of the Praesepe cluster, and have shown that quasi-stellars may well become the ideal objects against which to measure absolute proper motions of very faint stars (**33**).

Proper motions have been published for more than 30 U Geminorum stars and ancient novae, from which Kraft and Luyten have made a new determination of the luminosities of the U Geminorum stars (**34**).

Professor Strand reports that as an adjunct to the parallax programme of the U.S. Naval Observatory with the 61-inch (123 cm) astrometric reflector, a proper motion programme is carried out of stars selected from the Luyten and the Giclas proper motion programmes.

A total of 867 stars are in the programme for which first epoch plates have been completed for 225 stars. A completed first epoch consists of two plates of fair quality or better, with at least two exposures on each plate.

In the Special Reprint of the Smithsonian Institute a catalogue of positions and proper motions of 258 997 stars is given. These have been compiled from several catalogues in the FK4 system and reduced to the epoch and equinox of 1950·0 (**35**).

Dr C. A. Murray reports from the Royal Greenwich Observatory that a large programme for determining proper motions of RR Lyrae stars from astrographic material has been carried out. Details have been reported to Commission 23.

In connection with the RR Lyrae programme, Dr Clube has made a study of the relative merits of various methods of determining mean parallaxes of reference stars; the most readily available method is from the observed dispersion of proper motions, and this leads to reliable results provided that care is taken over the rejection of stars with large relative proper motions. Dr Clube has prepared a scheme for the reduction of relative to absolute proper motion, which will appear in *Quart. J. R. astr. Soc.*

Details of work in fields of various star clusters have been reported to Commission 37. Measurement of the Mount Wilson 60-inch (152 cm) plates is continuing.

Details have also been reported to Commission 23 of an investigation of proper motions in the Oxford astrographic zone in the region of the North Galactic Pole.

A repetition of plates on most of the northern Kapteyn Areas, originally obtained at Greenwich before about 1915 is being carried out at Herstmonceux; it is also planned to repeat the Radcliffe proper-motion plates on the 26-inch (66 cm) refractor at Herstmonceux. Details of this work have been reported to Commission 33.

Mr Murray has investigated the proper motions of field stars in the region of M 67. He has constructed a kinematic frame of reference in this field from the observed relative proper motions and two-colour photometry, which agrees well with the fundamental system obtained by correcting GC for known errors.

Determinations of proper motions of variable stars and novae, of planetary nebulae, star clusters and field stars in the surrounding regions have been continued at many observatories.

At the Pulkovo Observatory, members of the galactic clusters NGC 1960, 2099 and 6705 were found on the basis of *UBV* photometry and proper motions (**36**). A study of eight galactic clusters from proper motions and photometric characteristics of separate stars have been carried out by V. V. Lavdovsky (**37**). The proper motions of two planetary nebulae NGC 6853 (Dumbbell Nebula) and NGC 7662 and of 2112 field stars have been determined by O. N. Chudovicheva (**38**). By the same author an investigation of expansion of the Crab Nebula and proper motions of two central stars and 699 field stars have been published (**39**). The proper motion of Nova Lac 1910 and of 1814 field stars have been determined by V. A. Sokolova (**40**).

The revised tables for the reduction of relative to the absolute proper motions have been published using new values of the Oort's constants and Binnendijk's mean parallaxes (**41**).

At the Sternberg Astronomical Institute, N. M. Artjukhina has obtained proper motions of 4350 field stars in the region of the Praesepe cluster (**42**). She has also obtained proper motion of 54 variable stars W UMa type (**43**). The proper motions of three planetry nebulae NGC 6210, 6543 and 6826 and of 298 field stars have been determined by D. K. Karimova (**44**). The proper motions of 65 stars of the globular cluster M 13 and of 48 field stars have been determined by L. P. Panteleeva (**45**). The same author together with Ja. P. Gorelov obtained the absolute proper motion of 90 stars relative to the galaxy NGC 4736 (**46**). E. D. Pavlovskaya determined proper motion of 28 RR Lyrae variable stars (in press).

At the Tashkent Observatory the proper motion of two variable stars SU UMa and AY Lyr and field stars have been obtained by Sh. Primkulov (**47**).

A. Upgren determined proper motions of stars in the galactic cluster M 37 (**48**).

G. Van Herk obtained proper motions, mean parallaxes and space velocity of RR Lyr variables (**49**).

We must cite also:

Murray, C. A., Corbek, P. M., Allchorn, Mary B: Investigation of proper motions in field of the cluster M 67.I: The central region (**50**).

Murray, C. A., Jones, D. H. P., Candy, M. P.: Studies of the globular cluster ω Centauri. III: Proper motions (**51**).

Murray, C. A., Lowne, C. M., Clements, E. D.: Proper motions in the region of the Hyades (**52**).

Wayman, P. A., Symms, Z. S. T., Blukwell, K. C.: Proper motions and radial velocities of Hyades stars (**53**).

Dr T. Shimizu reports about some by-product results from the latitude and the time observations announced by Professor T. Okuda, director of Mizusawa Observatory. Corrections to Boss' catalogue values of proper motions of the observed stars have been derived by analysing the secular change in the yearly mean values of *Z* term. They coincide well with those given in the catalogue of all ILS stars prepared by P. Melchior. Corrections of proper motions in declination to Boss' Catalogue were preliminary determined from the results of seven years latitude observations. Preliminary corrections of proper motions of the Mizusawa PZT stars to the values given in the star list referred to the FK3 system were determined from the results of seven years observations during the period from 1958 to 1964. Details will be published in the coming issue of the *Publ. int. Latit. Obs. Mizusawa* which will appear early in 1967.

A. N. DEUTSCH
President of the Commission

BIBLIOGRAPHY

1. Reports of Observatories, 1964–65. 1965, *Astr. J.*, **70**, 581–665.
2. Further Reports of Observatories. 1965, *Astr. J.*, **70**, 765–783.
3. Osvalds, V. 1966, *Astr. J.*, **71**, 513.
4. Lippincott, S. L., Wyckoff, S. 1964, *Astr. J.*, **69**, 471
5. Lippincott, S. L., Wagman, N. 1966, *Astr. J.*, **71**, 122.
6. Van de Kamp, P., Harrington, B. S. 1964, *Astr. J.*, **69**, 402.
7. Gray, D. F. 1964, *Astr. J.*, **69**, 406.
8. Gray, D. F. 1965, *Astr. J.*, **70**, 304.
9. Gray, D. F. 1965, *Astr. J.*, **70**, 414.
10. Van de Kamp, P., Moore, S. 1966, *Astr. J.*, **71**, 524.
11. Bieger, G. S. 1964, *Astr. J.*, **69**, 804.
12. Bieger, G. S. 1964, *Astr. J.*, **69**, 812.
13. Van de Kamp, P., Yang, C. -Y., Worth, M. D. 1966, *Publ. astr. Soc. Pacif.*, **78**, 167.
14. Gibson, J. 1964, *Astr. J.*, **69**, 138.
15. Osborn, W., Ching-Yew-Yang 1966, *Astr. J.*, **71**, 647.
16. Kanaev, I. I. 1965, *Astr. Zu.*, **42**, 669.
17. Van Altena, W. F. 1966, *Publ. astr. Soc. Pacif.*, **78**, 345.
18. Guetter, H. H. 1964, *Publ. David Dunlap Obs.*, **2**, 405.
19. Beer, A. 1964, *Mon. Not. R. astr. Soc.*, **128**, 261.
20. Petrie, R. M. 1965, *Publ. Dom. astrophys. Obs. Vittoria*, **12**, 317.
21. Milet, B., Pourtau, L., *et al.* 1966, *Publ. Obs. Univ. Bordeaux*, **1**, **2**, **3**.
22. Rahimov, A. G. 1964, *Trudy Tashkent. Astr. Obs.*, ser. II, **10**, 31.
23. Deutsch, A. N. *Trans. IAU.*
24. Brouwer, D. 1965, *Inf. Bull. Southern Hemisph.*, no. 6, 2.
25. Bouigue, R., Dedier, H. 1964, *Ann. Obs. astr. mét. Toulouse*, **30**, 7.
26. Goyal, A. N., Aravamudan, S. 1965, *J. Observateurs*, Marseille, **48**, no. 7–8, 158 and 167.
27. Tempesti, P., *et al.* 1966, *Mem. Soc. astr. ital.*, **37**, 65.
28. Giclas, H. L., Burnham, R., Thomas, N. G. 1964–66, *Lowell Obs. Bull.*, nos. 129, 132 and 136.
29. Luyten, W. 1965, *Proper motion survey*, **6**, **7**, **8**, Minnesota.
30. Luyten, W. 1965, *Publ. astr. Obs. Univ. Minnesota*, **3**, no. 16.
31. Luyten, W. 1966, *Proper motion survey*, **10**, Minnesota,
32. Luyten, W., Smith, J., Hughes, 1966, *A search for faint blue stars*, **37**, **38**.
33. Luyten, W., Smith, J. 1966, *Astrophys. J.*, **145**, 366.
34. Luyten, W., Kraft, R. 1966, *A search for faint blue stars*, **36**.
35. *Spec. Repr. Smithsonian Inst. astrophys. Obs.* 1964, no. 151, 39.
36. Bronnikova, N. M. 1964, *Izv. glav. astr. Obs. Pulkove*, no. 174, 144.
37. Lavdovsky, V. V. 1965, *Ibid.*, no. 176, 138.
38. Chudovicheva, O. N. 1965, *Ibid.*, no. 176, 219.
39. Chudovicheva, O. N. 1966, *Ibid.*, no. 179, 115.
40. Sokolova, V. A. 1966, *Ibid.*, no. 179, 133.
41. Zhukov, L. V. 1966, *Astr. Zu.*, **43**, 1107.
42. Artjukhina, N. M. 1966, *Trudy gos. astr. Inst. Sternberga*, **34**, **35**.
43. Artjukhina, N. M. 1964, *Perem. Zvezdy*, **15**, no. 2.
44. Karimova, D. K. 1965, *Soobshch. gos. astr. Inst. Sternberga*, nos. 139, 142 and 143; 1967, *Ibid.*, no. 149 (in press).
45. Panteleeva, L. P. 1965, *Ibid.*, nos. 140–141.
46. Panteleeva, L. P., Gorelov, Ja. P. 1965, *Ibid.*, nos. 140–141.
47. Primkulov, Sh. 1965, *Cirk. Tashkent. astr. Obs.*, no. 334.
48. Upgren, A. 1965, *Astr. J.*, **70**, 149.
49. Herk van, G. 1965, *Bull. astr. Inst. Netherl.*, **18**, 71.
50. Murray, C. A., *et al.* 1965, *R. Obs. Bull.*, no. 91, E325.
51. Murray, C. A., *et al.* 1965, *Ibid.*, no. 100, E81.
52. Murray, C. A., *et al.* 1966, *Ibid.*, no. 108, E153.
53. Wayman, P. A., *et al.* 1965, *Ibid.*, no. 98, E33.

25. COMMISSION DE PHOTOMETRIE STELLAIRE

PRÉSIDENT: Dr V. B. Nikonov, Crimean Astrophysical Observatory of the Academy of Sciences of the U.S.S.R., P/O Nauchny, Crimea, U.S.S.R.

VICE-PRÉSIDENT: Dr A. W. J. Cousins, Royal Observatory, Observatory, Cape Province, South Africa.

COMITÉ D'ORGANISATION: D. Crawford, R. H. Hardie, G. E. Kron, K. Osawa, E. Rybka, A. G. P. Velghe.

MEMBRES: Aly, Argue, Argyle, Arp, Bahng, Barbier†, Baum, Becker (W.), Behr, Bigay, Bok (B. J.), Bok (P. F.), Borgman, Brück (H. A.), Chis, de Kort, Dessy, Dufay (J.), Eggen, Elvius (T.), Fellgett, Fitch, Gallouët, Gehrels, Golay, Haffner, Hertzsprung, Hiltner, Hoag, Hogg†, Holmberg, Irwin, Jackisch, Javet, Johnson (H. L.), Keller, Kostylev, Kuprevich, McCarthy, McGee (J. D.), Masani, Mayall (M. W.), Mianes, Moreno, Muller (A. B.), Ninger-Kosibowa, Oosterhoff, Payne-Gaposchkin, Pesch, Ramberg, Redman, Schilt, Schmidt (H.), Schmidt-Kaler, Smith (Harlan J.), Smyth, Steinlin, Stock, Stoy, Strohmeier, Wallenqvist, Walraven, Wesselink, Westerlund, Whitford, Zonn.

INTRODUCTION

According to decision of the Hamburg Meeting of Commission 25 its competence is restricted now to photometric and polarimetric instrumentation, observational, extinction and reduction procedures, photometric systems and standards. Extra-terrestrial stellar photometry belongs to its competence too.

The main lines of development of stellar photometry during the last three years were: rapid increase of number and size of photometric telescopes; development of multichannel photoelectric photometers: automation of observational and reduction procedures; further careful studies of the *UBV* photometric system and its extension to the infrared—up to 20 microns (*UBVRIJKLMNQ* system of Johnson); establishment of a common system of the *UBV* photometric standards for northern and southern hemispheres; increasing role of intermediate and narrow band photometric systems.

INSTRUMENTATION AND OBSERVATIONAL METHODS OF STELLAR PHOTOMETRY

Many new photometric telescopes have been designed, constructed and put in operation during the current period.

At the Royal Observatory, Edinburgh, Scotland, a twin photometric telescope has been constructed after a concept by H. Seddon, P. B. Fellgett, W. Matley and W. Jones (**1**). Two 40 cm Ritchey-Chrétien telescopes, one of which can be moved by $\pm$ 5° (with a precision of 6″) orthogonally with respect to the other, are carried by a single mount. Thus, two stars can be compared simultaneously, with consequent virtual cancellation of the effects of variable atmospheric transparency. Photometry is by pulse counting, with a typical limiting exposure of ten minutes for $V = 15^m_\cdot 0$ to a count of 13 binary digits. Measures of adjacent sky brightness are automatically interleaved with those of the stars. Colour filters and focal plane apertures are under remote control, and may be programmed.

A complex of two twin photometric telescopes have been put in operation at the W. Struve Observatory, Tartu, U.S.S.R. Two 48 cm Cassegrain telescopes located in separate domes at a distance of nearly 30 meters are under remote control from the same control panel. This complex permits to carry out simultaneous observations of two stars like the twin telescope of Edinburgh Observatory, just described. The digital technique is used for photoelectric measures (**26, 27, 28**).

At the Pine Bluff Observatory of the University of Wisconsin, U.S.A., a telescope that automatically sets on programmed stars and then makes multicolour photoelectric measures has been installed. The telescope is a 20 cm reflector under the control of a PDP–8 general purpose computer, with punched tape input.

A 150 cm Ritchey-Chrétien telescope with fused quartz optical parts is to be erected on Palomar Mountain for use as a photometric telescope. It is very short owing to a small primary focal length; this and other features permit rapid shifts from one object to another. The telescope will be equipped with a high speed (about $4 \cdot 10^6$ counts per second) pulse counting multichannel photometer.

Ritchey-Chrétien photometric telescopes of 125 cm aperture are being designed for the Crimean and Abastumani Observatories, U.S.S.R. These telescopes would be under control of a general purpose computer. The setting on the stars to be observed and operation of photometric equipment may be programmed. For input of commands and output of measures (observed colours and magnitudes) punched tapes will be used.

The 2·2 meter (87-inch) telescope, now under construction at the University of Arizona, U.S.A., will be erected at Kitt Peak and will be equipped for multichannel photometry.

A 125 cm automatically controlled telescope with a three-channel photometer is now available at the Kitt Peak Observatory, U.S.A. Other photometric telescopes being under operation at this observatory are: two 40 cm reflectors each with single channel photometer, one with automatic digital read out; two 90 cm reflectors both with automatic digital read out and single channel photometers; a 2·1 meter telescope with a three-channel photometer. Now being designed is a 3·75 meter telescope on which construction should begin the summer of 1967.

The one meter telescope 'The Elisabeth Telescope' used for photoelectric work has been put in operation at the Cape Observatory, South Africa, in 1964 (**2, 3**). Increasing artificial illumination of the sky is reducing the effectiveness of this telescope for photometry, and plans are forming to monitor the sky brightness with a small reflector attached to the main instrument for continual monitoring of the sky.

A 90 cm photometric telescope will be erected in 1967 at the Interamerican Observatory at Cerro Tololo, Chile.

The new 75 cm reflector of the Leushner Observatory, University of California, U.S.A., will be equipped for both conventional photometry and for the spectrum scanning. The instrument consists of a combination spectrograph, scanner, photometer and photography unit. The optical system is so designed that the rapid changes can be made from one mode to another. Photoelectric data will be aquired by a pulse counting system which will have digital output in the form of printed tape and punched cards.

Many other smaller photometric telescopes were put in operation during the last three years. A 70 cm Cassegrain telescope—at Simeis Station of the Crimean Observatory, U.S.S.R. It is equipped with single and three channel photometers and integrating polarimeter. Sixty cm Cassegrain telescopes—at Lund Observatory Station, Sweden, Lick Observatory, U.S.A. and at Kenneth Mees Observatory, U.S.A. Two identical 48 cm Cassegrain telescopes have been installed at Abastumani Observatory and at the Southern Station of Leningrad University Observatory (near Burakan Observatory), U.S.S.R. (**4**). A 45 cm reflecting telescope, having coudé and Cassegrain systems, have been erected at the Sverdlovsk University, U.S.S.R. It will be used for multicolour photometry. The 40 cm reflector, built for the University of Washington, has been installed on Rattlesnake Ridge near Hanford on an elevation near to 1 km. The telescope will be equipped with an ordinary photometer for photometry on *UBVR* system. Another photometric telescope of 40 cm diameter was put in operation at the new Smith College Observatory, U.S.A., and at Mount John Station, New Zealand, by Cook and Flower Observatory. A Cassegrain telescope of the same diameter has been constructed at the

Hamburg-Bergedorf Observatory, West Germany; it would be installed in Greece near Korinth at the height of 800 meters. In midsummer 1967 the Remeis Observatory, Bamberg, West Germany, will have a 40 cm telescope of its own at the Boyden Observatory in South Africa for photoelectric measurements in *UBV* system.

A rotatable telescope of Hiltner type, 60 cm in diameter (**5**) has been ordered for the polarimetric observations at the Bingar Station of Mount Stromlo Observatory, Australia.

Some new telescopes suitable for photographic photometry have been installed too. We like to mention here a one meter Schmidt telescope erected at the Baldone Station of the Astrophysical Laboratory of the Latvian Academy of Sciences, U.S.S.R. and a 60 cm Schmidt telescope at Torun Observatory, Poland (**6**).

Very useful reviews of modern light sensitive devices were published by W. A. Baum (**7**) and R. Smith (**8**).

The problem of thermal dependences of spectral responses of photoelectric apparatus is very actual and should be considered with most care (**12, 13, 14, 15, 16**).

A. T. Young at the University of Texas is revising now the photometric equipment of the Department of Astronomy to eliminate sources of inaccuracy arising from temperature dependence of filter transmission and of photomultiplier characteristics. So, photomultipliers will be cooled by thermoelectric means to provide critical temperature control. Solid state amplifiers are being prepared in order to eliminate the effects on amplifier sensitivity that arises from heating from thermal emission type tubes.

Two types of refrigerating systems for photomultipliers have been tested at Castel Gandolfo Observatory. The first, constructed by Corrido Corridi of Roma employs freon as a refrigerating agent. The second, supplied for experimental tests by Siemens Research Laboratory in Erlangen, is based on the Peltier effect. Both systems reach temperatures of the order of -40°C. Such refrigeration lowers the dark current by a factor of 4.10^3. Thermoelectric cooling of an EMI 9558 photomultiplier are being investigated at the University of Michigan too. It is interesting that at the Cape Observatory was found that cooling with the ordinary water ice gives practically adequate results to the dry ice refrigeration.

When using infrared sensitive devices, cooling to much lower temperatures is needed. For this purpose Mrs M. Lunel uses at the Lyon Observatory the following arrangement. Liquid air is evaporated by a heating resistance, and the cold air is circulating around the light sensitive device. The necessary temperature (down to $-$ 150°C) may be stabilized by regulating the intensity of flow of the cold air. The temperature is measured by a thermocouple (copper-constantan). To prevent condensation of water vapour, all the device is evacuated to the pressure of the order of 10^{-5} mm. The described system gives for a PbS cell a gain of order of 50.

Thermal dependence of spectral transmission of glass filters (**16**) and of brightness of radioactive dye light sources (**17**) raise the question of the necessity of temperature control of these components of photoelectric photometers. Such control for a radioactive dye light source is just described (**18**). But the best way, as it was mentioned in our previous report, is the use of light source of high stability not sensitive to temperature changes. A light source of this kind, based on Čerenkov effect, has been constructed and carefully studied at the Geneva Observatory (**19**). A detailed investigation of another type of light sources—electroluminescent lamps—was undertaken by West and Houck at the Washburn Observatory resulting in a monitored light source capable of variates in intensity over a range of 10^5 with negligible variations in spectral energy distribution.

The question on the dependence of photomultiplier's spectral sensitivity from the high voltage applied has been investigated (**20**).

At the Gottingen Observatory Gomez studied the influence of weak magnetic fields on the behaviour of photomultipliers. The most sensitive were the RCA head on type photomultipliers.

It was found that by a special choice of magnetic field it is possible to improve signal to noise ratio.

Any improvements in the construction of interference filters, widely used now in intermediate and narrow-band photoelectric photometry, are very important. Th. Walraven has been working at the Leiden Observatory on the quartz birefringent filters which produces fringes in the spectrum of rectangular shape, instead of a sinusoidal one. The bright fringes are separated by equally wide dark gaps. Consequently the dark gaps are relatively much wider than in the case of sinusoidal profiles. That permits to have in the well known multichannel photometer of Walraven's type smaller dispersion and therefore a smaller size of the instruments or, alternatively, the use of narrower spectral bands. At the Leiden Observatory such filters are now constructed and in particular a photometer is being built, which is intended for measuring simultaneously in the four Strömgren bands as a small portable instrument which can be carried by a 40 cm telescope.

Now, interference and other types of filters for the wavelength region 3000–1700 Å to be used in extra-terrestrial ultraviolet photometry have been developed (**22**).

Looking forward for the needs of photographic intermediate and narrow-band photometry the development of large, stable and highly uniform interference filters is very much needed.

Between the photoelectric photometers constructed during current period we find spectrum scanners suitable for measurements of monochromatic or quasi-monochromatic magnitudes and colours, single and multichannel photometers of different kinds. Different methods for measurement of the photocurrent are being used: DC integrating circuits and pulse counting technique (**23, 24, 25**) are most frequently used at the present time. Digital read out with printing on an ordinary tape or perforated card or tape is very common (**26, 27, 28**).

In their reports many astronomers discussed the problem of the increasing role of multichannel photometry. So, Willstrop emphasizes the importance of simultaneous observations in many colours to eliminate variations in the neutral component of atmospheric absorption and to economize in observing time. He stresses the gain to be derived especially at sites not enjoying the best conditions. Jackisch marks the importance of multicolour photometers for the study of stellar microvariability and for investigations of variables showing very rapid changes in their brightness (flare stars for example).

And now about the new photoelectric photometers.

At the Kodaikanal Astrophysical Observatory two new instruments, a spectrum scanner and a two channel pulse counting photometer are completed and will be in use soon. The scanner employs 1800 lines per mm grating with exit slits that allow the measurements of spectral regions 4 to 80 Å wide. The photometer is designed for use with interference filters on a programme measurement of Hγ line intensities in early-type stars. The filter selectors will, however, allow simultaneous measurements through many other pairs of filters.

A ten-channel spectrometer designed by Willstrop and Beggs is in an advanced stage of development at University of Cambridge Observatory. It is hoped to commence observations soon at the Cape Observatory.

A four-channel photometer designed by Walker has been built in the Dominion Astrophysical Observatory. A small quartz spectrometer employs a grating for dispersion; four windows at the camera focal plane provide selection of the wavelength. The intensity in each wavelength is measured by each of four separate photomultipliers. The multiplier outputs are integrated by condensers and read out with a digital voltmeter.

At the University of Michigan, as communicates Williams, the slot-photometry scanner is being more highly automated and is being fitted with a second channel to achieve some compensation for changes in atmospheric transparency.

At the Crimean Observatory a photoelectric scanner has been put in operation by Dimov and Severny in coudé focus of the Shajn's 2·6 meter telescope. The pulse counting technique is being used.

Zarevsky and Guseva constructed at the Sternberg Astronomical Institute a two-channel photoelectric spectrometer for intermediate and narrow-band photometry. It is especially suitable for line photometry ($H\alpha$, $H\beta$ or $H\gamma$ for example) and was extensively used in 1966 at the Southern Station of Sternberg Institute and at the Crimean Observatory. Pulse counting technique is used.

At Lyon Observatory, Bigay designed a modern two-channel photometer for wide, intermediate and narrow-band photometry. Twelve interference filters may be used in one of the channels and four in the other one. The possibility of rapid changes of field diaphragms is very suitable for photometry of extended objects. Lallemand's photomultipliers are being used and integration of photocurrent is applied. A digital voltmeter gives the possibility to print automatically the results on a punched tape. All reductions are being carried out by an electron computer. This photometer may be used with the 60, 80 and 193 cm telescopes of the Haute-Provence Observatory.

A single-channel photometer has been tested in the coudé focus of the 50 cm telescope of the Crimean Observatory. The integration or pulse counting methods can be used. The results of measurement are registered by a digital voltmeter. All operations with filters and standard light source are being carried out by a remote control system.

The charge integrating photometer, that has been in use for several years with the Steward Observatory of the University of Arizona 90 cm reflector, is being modernized by substitution of a new general purpose data logging system which records on punched paper tape.

A single-channel photometer for six-colour photometry has been modernized at the Observatory of the University of Bordeaux by Mianes. The filters are being changed by remote control and a new, very stable amplifier, using a vibrating condenser, is applied (**30**).

Descriptions of many photoelectric photometers have been published. The photoelectric photometer used for observations in the Golay's multicolour system is described by Rufener (**64**); Argue describes an infrared photometer used for observations of the late type stars at the Kitt Peak Observatory (**212**); Xanfomaliti—a pulse counting photometer of Abastumani Observatory (**31**) and Kubyshkin a photometer with remote control (**32**). Descriptions of photoelectric photometers used at Bosscha (**33**), Pulkovo (**34**), Uppsala (**35**) and Haute-Provence (**36**) observatories have also been published.

Ackermann and Hermann carried out laboratory tests of a photometer designed for the use in 0·8–5·0 microns region (**37**).

A special photoelectric scanner has been constructed at the Lowell Observatory and is being used to carry out exploratory photometric observations of the components of visual double stars too close to be observed by conventional photoelectric technique (**38**). Now a new scanner designed specifically for double star work is being designed. Scans will be produced by a slit of adjustable length and width connected to a motor driven spiral cam. An oscilloscope will be used for monitoring scanner performance during observations. Data will be recorded in machine-readable form for subsequent computer reductions.

At the Arthur J. Dyer Observatory, as communicates Hardie, an interesting experiment in spectral scanning is being performed with an objective grating and a standard photoelectric photometer. The grating gives a dispersion of about 75 Å/mm; slits down to a width of 15 Å can be used, and scanning is done by drifting the telescope in R.A. The method is promising.

The effects of telescopes on the polarization measurements have been discussed by Miller (**39**), and the requirements to telescopes and other polarimetric equipment—by Clarke (**40**). Clarke described also a double-beam polarimeter of new type and discussed the methods of reductions, and the results of measures of the instrumental polarization (41). Another double beam polarimeter has been constructed by Dombrovsky, Shulov and Vasiliev at the Observatory of Leningrad University and by Xanfomaliti at the Abastumani Observatory.

A polarimeter permitting to measure four Stokes's parameters in the wavelength range 0·19–3·0 microns has been constructed at Sternberg Institute by Pospergelis (42).

Polarimeters for multicolour observations were constructed at Harvard (43) and Stockholm (44) observatories; the last one with the sky background compensation.

A polarimeter for the *UBVR* colour system is being installed at the 80 cm telescope of Haute-Provence Observatory (45) and a polarimeter for the near infrared region is being used at one-meter telescope of Meudon Observatory (46).

New polarimeters have been tested and put in operation at Pulkovo (47), Shemaha (48) and Heidelberg (49) observatories.

A polarimeter has been designed by Heara for use on the 40 cm telescope at Pine Bluff and Kitt Peak observatories. It is a one-cell instrument using a Glan-Thomson prism, and having a movable shutter on the diaphragm slide to permit rapid switching between the object being observed and the adjacent sky.

Dollfus described a light-modulator for photoelectric polarimeters (50) and Rudolf at Gottingen Observatory developed special AC apparatus for the polarimetric measurements in red and infrared.

In the field of photographic photometry the work on improvement and automation of iris-photometers has been continued (51, 52, 53, 54, 55, 58). The measurements of stellar magnitudes using a recording microphotometer has been reported (57). The isophotometer of the Astronomical Institute of the University of Amsterdam have been described (58). And an automated instrument counting the number of stars of different magnitudes on a photographic plate has been constructed (59).

Wallenquist has constructed a device for photometric calibration of surface photometric or objective prism measurements by means of Schmidt telescopes. In this device the image from a tube sensitometer, mounted inside the telescope, are projected on the plate during the exposure. This instrument has been tested with the 100/135/300 cm Schmidt telescope of the Uppsala Observatory Kvistaberg Station and it seems to give promising results. At the St Andrews University Observatory, as communicated by Stibbs, a new type of tube sensitometer has been developed making use of an integrating sphere with the interior coated with magnesium oxide. This new sensitometer is expected to be a superior substitute for the use of rhodium plated crossed step wedges.

Some experiments with application of the electronic-image methods to stellar photometry should be mentioned.

Now the electronic camera is in routine operations with the 153 cm (61-inch) astrographic reflector of the Flagstaff Station of the Naval Observatory, communicates Kron. The linear relationship between exposure and density has been verified for electrograms. Hence, the camera should be useful for stellar photometry (experimental use is in progress), and it will certainly be valuable for spectrophotometry, for which it will yield the equivalent of photoelectric scans, but with the advantage of data from all elements of the available spectrum simultaneously. Electronic camera of Kron design is planned to be used with the 75 cm reflector of the Leushner Observatory.

The electronic camera of Lallemand has been applied by Lallemand, Canavaggia and Amiot to the photometry of the faintest stars using Schilt's method of measurements (60).

A photometer is being designed by Livingston and Lynds using an image tube which permits to register single photoelectrons (61).

OBSERVATIONAL, EXTINCTION AND REDUCTION PROCEDURES

The problem of establishing the best observational, extinction and reduction procedures for stellar photoelectric photometry is a very important one.

A very interesting and important summary report of the works done on these topics in English-speaking countries and Japan, has been prepared by Dr R. Hardie. He writes as follows:—

'On the whole, observers appear to be adopting similar procedures in observing standard stars frequently interpressed among programme stars, and in exercising considerable care to minimize residual errors due to extinction, filter-leaks, and calibration. It would appear that greater reliability and wider agreement among observers are being achieved, probably by improved and refined application of principles that are already well recognized rather than by the introduction of any basic new concepts. Some observers are successfully observing programme and standard stars simultaneously through the use of dual photometers (**1**), others are adopting rigorous constant-altitude techniques to improve the reliability of their photometry (**213**). The adaptation of automated data-handling systems to photometric installations in many observatories has promoted more efficient use of observing time, with consequent improvement in securing sufficient control measures of extinction and standard stars.

'Cousins (**62**) has described the procedures used at the Cape in the well known and highly accurate photometry done there by him and his colleagues. In general transfer work, stars are observed there at equal altitudes (around 30° or 40° zenith distance), standard stars being observed after every 2 or 4 programme stars, depending on whether magnitudes or colours respectively are being sought. Thus, accurate extinction coefficients are not required, as approximate or seasonal means are fully adequate for the very minor differential corrections.

'Recognition of the critical parameters required before setting up the observing programme has been stressed by Hardie (**63**) in order to make more efficient use of photometric observing time. By adopting standards either locally (as in variable-star or cluster work) or at equal altitudes, extinction determinations may be dispensed entirely. On the other hand, in general transfer work where these techniques are not feasible, much programme work may be rendered inaccurate without critical and judicious timing of extinction measures. Rapid extinction measures can be used with small investment of time, and experience with this technique reveals that the zero-points are indeed virtually constant and subject to little change. Moreover, appreciable improvement in accuracy of the extinction coefficients, as revealed by the greater constancy and repeatability of the zero-points, results from measures made over as great a range in air-mass as possible, even up to 5 and 6 air masses. Thus Hardie finds that limiting the observations of extinction stars to an air-mass of 2 or 3, as is more customary, is not conducive to the most reliable transfer work.

'In an interesting exchange of views, Fernie (**14**) and Young (**15**) have focussed some attention on temperature effects in photomultipliers. That the effects of these in practical photometry, by its nature a closely controlled comparison process among stars, is of little consequence is clear from their converging statements and from the experience of other photometrists. Hardie comments that he is unable to find among the zero-points any evidence of significant or discernable temperature variations, within his refrigerated photomultiplier housings, to the degree stressed by Young.

'Osawa and his colleagues and Cousins (**62**) report difficulty in transforming to $U-B$, in common with many observers. The causes have been discussed widely and are no doubt due to the critical location of the Balmer discontinuity within the U band, the width of the band, and the arbitrary original definition of the U stellar magnitudes in which the $U-B$ extinction was taken to have no colour dependent term. This term is now well recognized as being complex, and Hardie is re-examining all the original U,B,V stars in an attempt to evaluate how serious its neglect may be.

'Argue (**213**) in a red and infrared photometric survey of G and K stars employed a novel reduction method which, in its initial phases, used all stars to define the systems without special reference to stardard stars. "The reduction was unconventional in that the zero points for the separate nights were brought into mutual consistency by means of every observation (virtually). There were no specially observed 'standard stars' in the early phases of the reduction. Only in the final phases were standards invoked to tie up to the standard system. This procedure made it unnecessary to use up observing time measuring the same standards night after night. It is only necessary to ensure that, at any rate, most stars are measured on more than one night and in such a way that all the nights can be intercompared ultimately, and also that the standards appear somewhere in the final output."

'Cousins (**62**) provides many useful details involved in the measurements and reduction of the photometric data, and the reader is referred to his paper for them.

'A number of observers in private communications have asserted that the reduction of data in strict magnitude form (i.e., U,B,V) is superior to that resulting from one magnitude and several colour-indices (i.e., V, $B-V$, $U-B$). With this assertion Hardie takes strong exception, claiming that the results should be identical, as no variation in arithmetical manipulation can possibly lead to different physical results. Given certain functional relationships involving stellar brightness, temperature, instrumental and physical parameters, equations can be arranged in many convenient forms and necessarily must be equivalent, barring inconsistent definition or determination of the various constants and coefficients.

'Dependence of the extinction coefficients on stellar colour-index is undoubtedly somewhat more complex than is generally represented by a simple linear function of the form $k = k_1 + Ck_2$. However, evidence to date in the $B-V$ colour range is sufficient to establish that this linear relation is substantially adequate, second-order effects being barely, if at all, discernible. On the other hand, it has been well established by several observers that in U or $U-B$ the linear form is entirely inadequate.

'It is emphasized by Hardie that no reduction procedure, however sophisticated, can provide data not inherently contained in observations which are not infinitely precise. Observers have sometimes sought to solve by elaborate simultaneous least-squares solutions for many or all parameters (magnitudes, colours, coefficients, constants, etc.) within a night's or session's observations, without having planned to include critical observations in the programme. For example, transformation constants require a wide range of colour-index simultaneously with no range in air-mass for optimum determination; extinction coefficients (and stable zero-points) require a great range in air-mass, virtually at the same time or within a very short time interval; colour dependent (second-order) extinction terms require very close pairs of wide variety of colour observed repeatedly. If appropriate observations are secured at the telescope during the programme, then the simultaneous reduction for all parameters will indeed provide valuable and reliable data; however, if the needed observations were not secured, the results will be much weakened by the distribution of errors throughout all parameters. Since some parameters (such as transformation constants, and certain second-order extinction coefficients) are inherently more constant and stable than others (such as the principal extinction coefficients) and are thus capable of more precise determination, spurious errors may be associated with them which more properly belong elsewhere. The reduction procedure is merely a convenient mathematical treatment of data of finite accuracy and information content; it cannot take the place of critically and deliberately planned observing procedure.'

Only a few remarks should be added to this general report.

In Europe, where climatic conditions are not so perfect as in western parts of the United States, Chile or South Africa, simultaneous determinations of extinction are necessary to obtain accurate values of extra-atmospheric magnitudes and colours.

A very elaborate method of this kind has been worked out by Ruefener at the Geneva Observatory (**64**). In the U.S.S.R. Nikonov's extinction procedure is generally used. This procedure, permitting to obtain extra-atmospheric magntiudes and colours with minimum of auxiliary observations, becomes extremely simple when applied to quasi-monochromatic systems (spectrophotometry, narrow-band photometry and also intermediate-band one, if the half band width does not exceed approximately 300Å).

But even at such perfect site as Cerre Tololo it was found necessary to use mean values of extinction for every night if the highest accuracy should be attained. A special extinction procedure has been developed by Gutierrez-Moreno, Moreno, Stock, Torres and Wroblewski (**220**). Taking account for the dependence of second-order extinction coefficient from the position of the star on the two-colour diagram they could obtain a really extra-atmospheric system of $U-B$ colours.

The anomalous run of extinction in the ultraviolet region for early-type stars has been also considered by Hardie, Heiser and Weedman (**65**) and Ažusienis and Straižys (**66**).

Some other papers dealing with extinction procedures should be mentioned. Gutièrrez-Moreno and Stock discussed the problem of the accuracy of extinction determinations (**67**). Several methods for determination the extinction have been used by Jerzykiewicz and Serkowski during the course of photometrical study in the B,V system of Neptune, Uranus and F–G stars. They led to essentially the same results (**68**). Przybylski worked out some improvements in the extinction procedure (**69**). Uranova improved Hardie's method (**70**). A paper of Dachs, Haug and Pfleiderer is devoted to extinction measurements by photoelectric star photometry (**71**). A review of extinction procedures was given by Edwin and Berg (**72**).

Special observations of extinction have been carried out at several observatories. A photoelectric scanner was used at the Inter-American Observatory at Cerro Tololo by Gutièrrez-Moreno, Moreno, Stock and Sanduleak to study the wavelength dependence of extinction. The results showed exceptional transparency in ultra-violet (**73**). Głeboçki has determined the wavelength dependence of extinction by photographic spectrophotometry at Torun Observatory (Piwnice) for four different types of weather (**74**). Photographic and photovisual extinction for the same place and the same types of weather were determined photographically by Wegner (**75**). Photoelectric measurements of extinction have been carried out in South Africa by a ESO group; the mean extinction coefficients were obtained for Rockdale Mt; Flatehill and Zeekougat. Kivila and Eelsalu carried out such measurements at W. Struve observatory (Tartu) (**76**). Extinction for some other places were studied too (**77, 78**).

At the Remeis Observatory, Bamberg, Kohler made an investigation of colour extinction with six interference filters. Their bandwidths were between 70 and 100Å and effective wavelengths covered the region from 3530 to 6150Å. It was shown that the observations cannot be interpreted by the well known formula $k_h \sim \lambda^{-\alpha}$ (magnitudes), and the determination of precise extra-atmosphere colour-indices with narrow band filters should be based on special measurements of colour extinction. Studies of extinction carried out at Mt Elbrus (in visual and infrared) showed that even at high altitudes the wavelength dependence of extinction does not follow the λ^{-4} law (**79**).

It is interesting that the eruption of volcano Mt Agunga in Indonesia caused a large increase of extinction in Chile, Australia and South Africa (**86, 69, 87**).

The very important problem of sky background (geophysical, astronomical and interplanetary) has been considered by Roach (**88**). Fluctuations in the sky background as a limiting factor in the photometry of faintest objects were studied by Miller (**89**).

Some instrumental effects in photoelectric photometry were considered: Red-leak corrections in ultraviolet photometry by Shao Cheng-Yuan and Young (**90**) and filter transmission effects on photoelectric measures of rotationally broadened Hβ line strength by Abt and Osmer (**91**).

The problem of accuracy of photographic photometry has been carefully studied by Vanýsek (**92**). The comparison of photographic photometries made in Hamburg and Edinburgh with photoelectric photometry carried out with the new automated telescope of Edinburgh Observatory were published by Lawrence and Reddish (**93**).

The usually adopted method of calibration of the photographic magnitudes is the observation of photoelectric standards in the same fields. But the problem of the transfer of photometric measurements from one field to another is still actual. This problem was studied at the Bonn University Observatory, as it was reported by H. Schmidt. Two open cluster (NGC 744 and 952) with known *UBV* magnitudes are photographed on different halfs of the same plate by a Schmidt telescope. It will be studied whether the magnitudes derived for one cluster from the other agree sufficiently well with the known magnitudes to permit the use of this method for the determination of magnitudes in a star field where no magnitude standards are given. Furthermore, since NGC 957 is close enough to the field to be photographed on the same plate during a single exposure, the field magnitudes thus derived can be compared with those found by the previous method (Karimie).

The problems arising in the photographic photometry of faintest objects have been considered by Eelsalu and Kalv at the W. Struve Observatory, Tartu (**95, 96**).

A photographic method 'Method of star trails' is being used at Tashkent Observatory for detection of rapid variations in star brightness (**97**).

Methods for separating the polarization of stellar radiation from interstellar polarization have been developed by Shakhovskoi at the Crimean Observatory and Vardanian at the Burakan Observatory (**98, 99, 100**).

Turning now to the reduction procedures, it should be mentioned that at several observatories special programmes for reduction of photometric and polarimetric observations have been developed. So, Bertiau worked out at the Castel Gandolfo Observatory programmes for reduction of the *UBV* measures (**101, 102**) and for reduction of iris-microphotometer measurements too, including corrections for varying sky fog (**103**). At Kitt Peak Observatory computer reduction programmes are available for the photoelectric measurements, either with punched cards or with paper tape produced directly at the telescope. A computer programme for reduction of the photographic measurements of magnitudes by means of standard stars has been prepared by Miss Sundman at Stockholm Observatory.

In connection with the problem under consideration Cousins remarks that the hand processing has been found at the Cape Observatory to give more prompt results than machine processing, but machine data processing has been found convenient when promptness is not important, with the final stages of reductions being done by hand to allow individual attention to the observations.

PHOTOMETRIC SYSTEMS

The development of the work on photometric systems were characterized by extension of *UBV* systems to red and infrared, forming a wide band multicolour system of Johnson *UBVRIJKLMNQ*, covering the wavelength range from 3000 Å to 20μ, and by increasing role of other multicolour systems (wide, intermediate and narrow-band ones) devoted, in general, for solution of special problems, and especially the problems of stellar astronomy.

Summarizing the work on wide band systems Dr A. W. J. Cousins writes:

'The *UBV*-system is one most generally used for wide-band photometry by both photoelectric and photographic observers. The *RGU*, *PV* and S'Pg, S'pv (**202**) systems and $(U-B)_c$ (**197, 196**) are still being used, mainly by those who have used them in the past or in continuation of programmes for which they had previously been adopted. More work has been done in the red and infrared; the choice of band depending more on the purpose in view, with less emphasis on standardization than at shorter wavelengths. There has been an increasing tendency to use narrower and more clearly definite bands.

'Several extensive series of *UBV* observations of bright stars have been completed since the last Report (**212, 188, 189, 190, 191, 192, 184, 185, 210, 220**). An important feature of these series is that they include a fair sample of equatorial stars so that there is now, for the first time, a good overlap between the work of observers in the northern and southern hemispheres. These, and earlier series, leave little doubt that V and $B-V$ as observed at the Cape and Cerro Tololo are on essentially the same system as these observed from the northern hemisphere (**200, 184, 185**). There are small local and perhaps seasonal (**15**), differences in zero point between the different series of about the same size as the s.e. of one magnitude or colour index, but there is little correlation between them and it is probable that the zero point errors of the combined series do not reach $0\overset{m}{.}01$ for V and $0\overset{m}{.}005$ for $B-V$ in any part or the equatorial zone, or in the E regions on which the Cape zero points depend. Johnson and associates have shown (**184, 185**) that there are small zero point and colour differences between the Catalina Mountains series and the observations made at Cape, but it is also apparent that there is not complete homogenity between their new observations and earlier measures that are believed to be on the *UBV* system. For most purposes the differences are of no practical importance, but it is now becoming important to discuss whether or not the present definition of the *UBV* system* is adequate.

'The position is less satisfactory for $U-B$ because there is difficulty in reconciling the observations made with different photometers. The transformations from natural to standard system frequently involve both $B-V$ and $U-B$, and the same relation may not be valid for all luminosity classes and different degrees of reddening. There is not much difference between the Cape $U-B$ system and that of the Arizona-Tonantzintla Catalogue (**184, 185**) for main sequence stars earlier than type G and for normal giants of later types, but there are some large discrepancies for high luminosity stars. A natural ultraviolet system depends rather critically on the filters and other components in the photometer and the method of allowing for extinction (**71, 220**). Although no short-wavelength cut-off filter was used for the original U system, it should not be forgotten that the 1P 21 photomultiplier has a glass envelope. The glass must not be too thick, but if the photomultiplier has a quartz window it is important to include glass elsewhere in the optical path. Some observers have pointed out that Schott UG 2 is to be preferred to Corning 9863 glass for the U filter because of the much smaller red leak (**212, 90**).

'The Cape $U-B$ system needs some minor changes (**185**) to make it conform more closely with the system of the Arizona-Tonanzintla Catalogue but after that, and except for the high luminosity stars, there is satisfactory agreement with that catalogue and also with Johnson and Harris, Argue (**212**) and for the later type stars observed at Cerro Tololo (**220**). Stars of the earlier types in this last list are not on the same system. The local zero point errors of these series do not appear to exceed $0\overset{m}{.}01$. There is poor agreement for very red stars between all the lists but the differences appear to be mainly random. Stellar variability and the relatively small amount of ultraviolet light present are certainly contributing factors.

'Young has warned observers using wide-band systems of the danger of seasonal effects due to temperature changes of the filters (**16**) as well as photomultipliers (**13, 15**). Ageing of the aluminium surfaces in reflectors can produce even larger systematic effects, especially in the U band. Ažusienis and Straižys (**104, 105**) have derived corrections to the response curves published by Johnson** to include these and other factors.

*H. L. Johnson, D. L. Harris, 1954, *Astrophys. J.*, **120**, 196.

H. L. Johnson and W. Morgan, 1951, *Astrophys. J.*, **114, 522.
H. L. Johnson, 1955, *Ann Astrophys.*, **18**, 292.

'Van den Bergh finds that $Q = (U-B) - 0{\cdot}72\ (B-V)$ is a good reddening free metallicity index for globular clusters and that the intrinsic colours of globular clusters in M31 and the galaxy are adequately represented by the relation $(B-V)_0 = Q + 1{\cdot}00$. He has also found that the transformation equation from $P-V$ to $B-V$ for globular clusters is $(B-V) = 0{\cdot}07 + 1{\cdot}07\ (P-V)$, which differs significantly from that given by Kron and Mayall* for single stars because of the very different energy distributions and removes an apparent discrepancy between globular cluster observations by Hiltner and those of Kron and Mayall.'

The properties of the *UBV* system and its two-colour diagram have been carefully discussed. At Vilnius Observatory Straižys, Zdanavičius and Ažusienis considered the response curves, relation between 'winter' and 'summer' *UBV* systems, red leak effect, extinction corrections, systematic differences between different catalogues of Johnson and his associates etc. (**104, 105, 106, 107, 108**). The two-colour diagram was considered by Ažusienis, Straižys, Wampler and Nikolov (**109, 110, 111, 112**). The properties of *UBV* system were discussed also by Karyagina and Kharitonov (**113**) and Osawa (**114**). An effect in modification of the system *UBV* by Fraunhofer lines has been studied by Wilday (**115**).

According to opinion of Walraven the connection of *UBV* photometry between Northern and Southern Hemisphere seems rather weak. The residuals of the difference Cape-Leiden in magnitude *V* show variations with right ascension with total amplitude of about $0\overset{m}{.}03$.

The new *UBVRIJKLMNQ* system of Johnson and his associates (**185**)—the extension to longer wavelength of *UBV* system—has been developed. Its response curves are published (**116**), and absolute calibration of this system is carried out (**118**).

A summary of recent infrared astronomical measurements were published by Johnson (**119**). Some very interesting papers dealing with infrared photometry must be mentioned (**120, 121, 122, 123**). In the last paper the first observation in *Q* system ($\lambda_e \sim 20\mu$) is reported.

A theoretical discussion of some problems of wide-band photo-electric photometry has been carried out by Eelsalu (**124**). An investigation of the *RGU* photometric system was published by Straižys (**125**), and the blanketing corrections for this system were considered by Smith and Steinlin (**126**).

Turning to the multicolour systems we like to mention publication of papers presented on the IAU Symposium No. 24: Spectral Classification and Multicolour Photometry (**127**).

The wide-band seven-colour system of Golay has been carefully tested by observations (**128** to **139**).

The intermediate-band systems by Strömgren, Borgman and Walraven are being used now for regular observations.

A four-colour intermediate-band system by Neff (**140**) covering the wavelength region 3330–5500Å has been applied for discrimination between late-type dwarf and giant stars.

At Vilnius Observatory the work on the Straižy's multicolour system has been continued (**141** to **147**). Successful tests of two versions of this seven-colour system (with glass and interference filters) are being carried out at Vilnius and Crimean Observatories.

The problem of optimal choice of colours for multicolour photometry were considered by Steinlin (**148**).

Several narrow-band photometric systems have been developed, studied and used for practical observations.

The narrow-band photometry of hydrogen lines and especially Hβ is widely used, and in particular for determining luminosities of early-type stars (**149** to **156**).

*G. E. Kron, M. W. Mayall, 1960, *Astr. J.*, **65**, 581.

Narrow-band photometry of late-type stars is being developed (**157** to **165**). The work on the infrared narrow band photometry is being carried out also (**166**). Bahng at Dearborn Observatory is working with an infrared narrow band system (*XYZ*), using interference filters centered at 1·21, 1·59 and 2·15μ. The corresponding half-width are 77, 86 and 98 Å. A lead-sulphide cell is used. The *X* and *Z* bands are very similar to *J* and *K* bands of Johnson, but are much narrower—by a factor 5 or so. The *Y* band has no corresponding band in other infrared systems. Bahng's opinion is that this band is very important in stellar photometry, as it coincides very nearly with the H^- opacity window in the stellar atmospheres.

Some other papers dealing with the photometric systems should be mentioned. Photographic system near to *UBV* (**167, 168, 169**) and *RGU* (**170**) have been described. The properties of different photometric systems from the points of view of influence of absorbtion lines (**171**) or measuring of metal content (**172**) have been compared.

Comparison of different photometric systems is considered in (**173, 174, 175, 176**). V. Shevchenko at Tashkent Observatory proposed to use the observations of a regular variable star, having a well determined light curve in a standard photometric system, for determination of a transformation formula connecting instrumental and standard systems (**177**).

The important problem on the relation between heterochromatic and monochromatic magnitudes in three-colour systems have been considered by Straižys (**178**).

The questions of relations between a wide-band and intermediate band system are being discussed by Walraven. He communicates: 'From the observations with the five-channel photometer of the Leiden Southern Station a large number of bright stars were selected, which are also present in the Photometric Catalogue of the Cape Observatory (*R. Obs. Bull.* no. 64).

'The transformation of magnitude *V* of the Leiden *VBLUW* system to the magnitude of the Cape *UBV* system is not linear but reliable.

'Likewise the transformation for colour $B-V$ is possible with reasonable accuracy, although the relation between the wide band *UBV* and intermediate band *VBLUW* system is strongly non-linear and depends also on reddening and luminosities of the stars.

'The transformation of $U-B$ is unreliable due to the width of band *U* of the *UBV* system.'

Walraven remarks also that the effective wavelength of the bands *V* and *B* of the *UBV* photometry, which follows from the transformations of the *VBLUW* system and from the accurately determined effective wavelength of the latter, deviate from the values, which are generally adopted.

In conclusion we like to emphasize that the intermediate and narrow-band systems have a great advantage over the wide-band ones to operate with the quasi-monochromatic magnitudes when it is possible to neglect practically all changes of effective wavelength with spectral type, extinction or interstellar reddening (the half width of response curves does not exceed approximately 300 Å for a quasi-monochromatic system). In these systems reductions for extinction and transformation to a standard system are very simple; the Forbes effect and curvature of reddening lines on two-colour diagrams are absent. As a consequence, a much higher accuracy can be attained.

PHOTOMETRIC STANDARDS

The problem to establish photometric standards of highest accuracy depends not only on the precision of observations and extinction procedure, but, of course, on the internal stability of stellar radiation (stellar microvariability).

Jackisch (Sonneberg Observatory) returned to this problem during his stay at Kottamia Observatory (United Arab Republic). Using a 25 cm reflector he tested photoelectrically 210

A-type stars of III and V luminosity classes to study their stability. The preliminary result is that the variability is being found for 20% of stars studied with amplitudes between $0^m.03$ and $0^m.06$ (**179, 180**).

On the other side the photoelectric observations of Neptune, and Uranus carried out by Jerzikiewicz and Serkowski in 1950–66 years (**68, 221**) were used to study the problem of solar and stellar variability. It was shown that the short period variations of solar brightness do not exceed $0^m.003$. The high stability of solar type stars was stated too. The observations of 15 stars of spectral type F and G in the years 1950–66 indicate that for none of these stars does the standard deviation of the yearly mean magnitude exceed $0^m.008$ and for some stars this deviation is even less than $0^m.004$. The V magnitudes and $B-V$ colours of about 50 F and G stars were determined during the course of this work. Their mean errors do not exceed $0^m.003$, so the third digit beyond the decimal is quite significant.

As communicates Feinstein, it was decided to include in the observing programme of La Plata Observatory nearly 30 bright and very red stars ($B-V > 1{\cdot}6$) to check the possibility of using them as standards if they don't show light variation. The measurements are being carried out in the *UBVRI* system.

Cousins (**181**) and McCullogh (**182**) support the point of view on the general constancy of stellar radiation.

As C. Jaschek reports the reference catalogue of all photoelectric measurements is in progress. At the present it contains about twenty thousand stars and six hundred references. It is hoped that it can be published in 1967.

The practical work on the photometric standards is being summarized by Professor E. Rybka:

'Several catalogues containing the standard magnitudes of bright stars have been published during the past three years.

'In 1964 the third edition of the Catalogue of Bright Stars (**183**) appeared, where magnitudes and colours for nearly 50% of stars have been compiled by Klemola and reduced to a common system, for over 1100 stars the HR magnitudes revised by Rybka have been included. For remaining stars, for which neither photoelectric nor revised by Rybka magnitudes were available, the original HR magnitudes have been retained.

Comprehensive catalogues have been published by the Lunar and Planetary Laboratory of the University of Arizona (**184, 185**). The catalogue (**184**) contains magnitudes V and colour $U-V$, $B-V$, $V-R$, $V-I$ for 1325 naked-eye stars determined in Catalina Mountains and at Tonantzintla observatories. For some southern stars, Cape Observatory values are given. Still more comprehensive photometric lists have been published in (**185**), where results of the Arizona multicolour photometric work has been presented. Of the stars listed in (**185**) 1567 were observed in U, B, V, R and I, 653 in J and K and 268 in L. The southern Cape photometry has been transformed into the *UBV* system and a catalogue of *UBV* photometry on 4780 bright stars over the entire sky was given.

'Photoelectric magnitudes and colours of pairs of 6^m stars of A and K spectral types near the Kapteyn's S.A. 1 – 139 have been determined from observations made at the Crimean Observatory in the years 1959–61. The catalogue (**186**) contains magnitudes and colours of 278 stars both in the natural *ubv* and in the *UBV* systems. It is planned to extend the determination of fainter photometric standards in S.A. at the Abastumani Observatory.

'Astronomers from Cape Observatory continued the *UBV* photometry of bright stars. The Cape observations of "Stars in the Equatorial Zone" were completed in 1964 and have been published (**187, 188, 189, 190, 191**). Publication (**192**) is a continuation of *Royal Observatory Bulletin* no. 64. Observations of HR stars, south of $-10°$ (V, $B-V$ and $(U-B)_c$) are being continued and several lists have been published by Lake and Corben (**193, 194, 195, 196, 197**).'

Dr Cousins remarks: 'The completion of four new series observed elsewhere has at last provided the overlaps that were needed to show that the northern and southern *UBV* systems are in good agreement. These are: "The Arizona-Tonantzintla Catalogue" of Johnson and his colleagues (**185**), Stock and Moreno's observations at Cerro Tololo, Argue's observations of F, G and K type stars at Kitt Peak, and Häggkvist and Oja's observations of bright stars from Uppsala (**210**). These observations have now been incorporated in the photometric catalogue of bright stars being compiled at the Cape Observatory, that includes all stars south of + 10° declination with HR magnitude 5·0 or brighter. The standard errors of the mean magnitudes of nonvariable stars are nearly all less than $\pm$ 0^{m}01. The s.e.'s of $B-V$ and $U-B$ are usually less than $\pm$ 0^{m}005 and $\pm$ 0^{m}015 respectively. The northern and southern V and $B-V$ systems have the same colour equations and mean zero points but the new material has shown the need for some small corrections to the Cape $U-B$. After that there is satisfactory agreement between the Arizona and Cape $U-B$ for main sequence stars earlier than type G (of which there was a fair sample in the Johnson and Harris list) and for class III giants of later type, for which the transformation has previously been uncertain.'

Wallenquist reports on the works on photometry made in Sweden:

'1. Lund Observatory.

Roslund has determined magnitudes and colours photoelectrically for 250 stars of early types in Milky Way field in Scorpius in the *UBV* system (**204**).

2. Stockholm Observatory, Saltsjöbaden.

L. Lodén has reported that a list of photometric standard sequences in the Southern Milky Way is under preparation (covering a.o. the Crux region). For the standard sequences 15–20 stars are selected in each region. The stars have been measured photoelectrically in the *UBV* system at the Boyden Observatory. An extensive programme for photographic photometry in two colours of early and very late stars in Centaurus is in progress.

K. Lodén has determined B and V magnitudes and spectral classes for 450 stars in the S.A. 193 on the basis of plates secured at the Boyden Observatory. The catalogue will be published soon in *Stockholm Observatoriums Annaler*.

Sinnerstad reports that Arcling has determined *UBV* magnitudes photoelectrical for about 30 standard stars for the determination of the atmospheric extinction. Sinnerstad is making photoelectric photometry in the *UBV* system (Strömgren photometer) for trying to find luminosity criteria suitable for Schmidt-spectra.

3. Uppsala Observatory.

T. Elvius reports the following photometric investigations which are carried out at Uppsala under his supervision:

Sjögren: The catalogue (**206**) gives V and $B-V$ for 76 stars in S.A. 8.

T. Elvius is determining *UV*-magnitudes in S.A. 19 on the basis of plates secured at Kvistaberg by means of the large Schmidt telescope.

T. Elvius and Lyngå carried out three-colour photometry of 43 stars in S.A. at southern latitudes (**205**).

T. Elvius and Häggkvist: (**207**) contains V and $B-V$ photoelectrically determined for 75 stars.

Ljunggren determined photoelectrically V and $B-V$ for 298 stars (**208**).

Ljunggren and Oja: (**209**) contains photoelectrically determined V and $B-V$ for stars chiefly situated at low galactic latitudes.

Häggkvist and Oja: (**210**) contains V and $B-V$ for all northern stars brighter than $m = 5{\cdot}25$, according to Catalogue of Bright Stars.

Oja: (**211**) contains spectrophotometric data for 2500 stars situated in the region $80° < l^1 < 100°$ and $|b^1| < 3°$.

Oja: (**211a**) contains *UBR* magnitudes for about 2000 stars within a square degree around the cluster M 103 as well as spectrophotometric data for about 1000 stars.

Häggkvist: (**211b**) contains photoelectric magnitudes and colours for 72 visual binaries.

Eriksson is making a spectral classification of about 3000 stars situated in the neighbourhood of the Southern Galactic Pole. Moreover he has determined photographically *B* and *V* magnitudes for about 10 000 stars (will be published soon).

Linden: photoelectric determinations of *B* and *V* magnitudes for stars in S.A. 18 (60 stars) and S.A. 19 (40 stars).

Ekedahl: photographic determination of m_{pg} and m_{pv} for 273 stars in the S.A. 4 (ready for publication).'

Argue draws attention to his two papers (**212, 213**). The paper (**212**) includes magnitudes and colours for about 130 equatorial stars expressed in the northern *UBV* system, which were observed with the intention of strengthening the tie between the northern and southern system. In (**213**) the magnitudes and colours have been derived from measurements made with an RCA 7102 photomultiplier and the following four filters: 6900 Å interference filter (r); Schott RG8; Schott RG9; 1·00 micron interference filter (i) The r magnitudes and ($r - rg8$) and ($r - rg9$) colours are similar to Kron's R and $(R-I)$ while $(r - i)$ can be transformed to a colour very similar to the Arizona-Tonantzintla $(R - I)$. The catalogue may thus provide a useful selection of sub-standards for later types. The limiting magnitude is about $R = 8^m$.

Argue remarks: 'It might be worth while for the Commission to consider whether the original definition of the *UBV* system in terms of stars mostly situated in the Northern Hemisphere is still adequate? With the rapid increase in activity in the Southern Hemisphere it would be better to redefine it in terms of equatorial stars. The opportunity should also be taken to advocate the use of the UG2 ultraviolet filter in place of the Corning 9863 which requires a considerably larger correction for red leak on late type stars.'

Some other photometric catalogues are to be mentioned: (**214**) contains seven-colour photometry of 342 stars in the Golay's wide-band photometric system.

The catalogue (**215**) includes *BVRI* photometry in Johnson's system for 275 stars mainly between $-25°$ and $-50°$.

The results of *UBV* photometry of 263 stars of B, A0, A1 and A2 spectral types in S.A. 8 ($7^m.8 < m_v < 13^m$ according to BSD) are given in (**216**) and (**217**). And for S.A. 40 in (**218**).

Osawa published the *UBV* photometry for 211 peculiar A stars (**219**).

Van den Bergh informs that at the David Dunlap Observatory McClure is currently carrying out an extensive programme of intermediate band photometry of stars, clusters and galaxies with the 74-inch telescope (λ_{eff} 3800, 4150, 4250, 4500 and 4861 Å); Racine is currently obtaining *UBV* photometry of most of the northern stars which are known to be located in reflection nebulae.

A very precise catalogue for 366 stars have been observed at Cerro Tololo by Gutierrez-Moreno, Moreno, Stock, Torres and Wroblewski in the *UBV* system (**220**).

As it was mentioned above, in the course of photometric observations of the planets Neptune and Uranus, made at the Lowell Observatory for a number of years, 50 stars, mostly of spectral types F and G and situated along the ecliptic, were repeatedly observed in the *B*, *V* system. For the stars observed on about 25 nights the mean errors are about $\pm 0^m.003$ and $\pm 0^m.002$ for V and $B-V$ respectively; for stars observed on about 100 nights these errors are $\pm 0^m.002$ and $\pm 0^m.001$ (**221**). That is, perhaps, the most accurate *B,V* photometry than any other published until now.

As communicates Crawford, at the Kitt Peak Observatory *UBV* measurements have been made for all stars B5 or earlier brighter than $V = 6{\cdot}5$, and of declination above $-10°$.

Some other *UBV* photometries should be mentioned, carried out for specially selected groups of stars (**222** to **223**).

Recent observations of stars in the red and infrared photoelectric systems are published in papers (**234**, **235**, **236**, **237**, **238**, **239**).

Intermediate-band photometric systems: Graham is now carrying out at Kitt Peak Strömgren four-colour photometry for faint high latitude blue stars, as communicates Crawford. Special attention is being paid to stars which have already been studied spectroscopically by Greenstein and others. Crawford and his associates at Kitt Peak are pursuing narrow and intermediate band photometry on the $H\beta$ snd Strömgren four-colour systems for a number of galactic clusters and for the bright field stars. Crawford is preparing a list of standard stars and the discussion of Strömgren four-colour photometry reduction procedures. This work should be in press by the time of the IAU meeting. A catalogue of Strömgren's *ubvy* and $H\beta$ photometry of 749 B5–F0 stars observed by Cameron, will appear in a *Georgetown College Observatory Monograph* in late 1966. This catalogue includes data for approximately 300 Ap stars and 100 Am stars. Seven-colour photometric observations on Strömgren's *ubv* and Crawford's *abc* systems have been continued. Bigay communicates that it is expected to carry out observations in Strömgren's system of stars in S.A. near galactic plane and of Luminous Stars. Some considerations concerning his system were published by Strömgren (**241**).

Borgman applied his seven-colour system for observations of 102 stars belonging to spectral types from A to M, and located nearer than at 20 pcs. Mean errors of colours were found to be from $\pm$ 0^m.008 to $\pm$ 0^m.002 (**242**).

The Walraven's system has been used for observations of high velocity stars by Ponsen and Oosterhoff (**243**, **244**) and by Graham and Lyngå for observations of 454 OB stars in Carina (to be published in a recent Mt Stromlo Memoir).

Hydrogen lines photometry: Crawford and Mancler published a list of standard stars for photoelectric $H\beta$ photometry (**245**). β-indexes have been observed by Crawford for 93 stars in the region of 'Orion belt' (**246**). $H\alpha$ photometry of late-type stars has been carried out by Peat (**247**, **248**).

A summary of all photoelectric observations for stars O9–B3 (III and V luminosity classes) having $H\alpha$ in emission were compiled by Schmidt-Kaler (**249**)

Very important successes were attained in quasi-monochromatic and monochromatic photoelectric photometry (spectrophotometry).

Narrow-band photoelectric colour-indices for 25 bright stars O9–F5 have been observed in 16 bands ($\Delta\lambda \cong 54$Å) in the region 3200–6400Å) by Bahner (**250**). Observations of α Lyrae have been calibrated using a lamp.

Absolute measures of energy distribution in stellar spectra have been carried out by Willstrop (**251**), Oke and Conti (**252**), Kharitonov, Karjagina and Neljubin (**253**, **254**, **255**, **256**, **257**).

Photoelectric spectrophotometry of selected bright southern stars has been carried out by Aller, Falkner and Norton (**258**, **259**), and for F2–K7 stars by Van den Bergh and Sackman (**260**).

Irvin communicates that a programme of photoelectric scanning of all stars south of declination $+20°$ and brighter than visual magnitude 5·0 with spectral type later than A9 has been initiated at Inter-American Observatory at Cerro Tololo and is 45% complete. The scans are from 3100 to 6000Å with a bandpass of 80Å. The region shortward of 4000Å is especially interesting and the Tololo sky is excellent for this purpose.

Standard stars for infrared region have been recommended by D'Agati (**261**).

As Heddle has mentioned (**262**), accurate absolute calibration of photometric data in the visual region are very needed to obtain precise relation between this region and the ultraviolet observed from outside the atmosphere.

At the present time we have no photometric systems for the ultraviolet region with $\lambda < 3000$ Å. But very important photometric observations have been carried out in this region from rockets and satellites (**263, 264, 265, 266, 267, 268, 269**). As an example of a very interesting astronomical result we like to mention here the determination of the law of interstellar extinction for the wavelength down to 2200 Å obtained by Boggess III and Borgman (**270**).

Almar proposed to use a slowly moving satellite with known photometric parameters for calibrating the scale of stellar magnitudes (**271**).

Among several polarimetric measures of stellar radiation we will mention the *UBV* (and for some stars *UBVRI*) observations carried out by Krzeminski and Serkowski at the Lowell Observatory (**272**), the *UBV* Stepien's measures in the polar region (**273**), observations of Kruszewski (**274**) and Serkowski (**275**) at Belgrad Observatory and polarimetric investigations of L. Lodén in the Milky Way (**276**).

Many papers dealing with photographic photometry of different stellar fields, which could not be regarded as sources of photometric standards and belongs to competence of other IAU Commissions were not included in the present report.

President of Commission is very grateful to all members of Commission who kindly sent their reports to him or to members of the Organizing Committee, and he is especially grateful to the members of Organizing Committee for their great work in preparing special parts of the report.

V. B. NIKONOV
President of the Commission

BIBLIOGRAPHY

Instrumentation and observational methods

1. Reddish, V. C. 1966, Twin 16-inch photometric reflectors at Edinburgh. *Sky Telesc*, **32**, 124.
2. Evans, D. S. 1964, The new Cape reflector. *Observatory*, **84**, 222.
3. Stoy, R. H. 1964, The Elizabeth telescope. *Mon. Notes astr. Soc. Sth. Afr.*, **23**, 56.
4. Dombrovsky, V., Gagentorn, V., Gutkevich, S., Poljakova, T., Sveshnikov, M., Shulov, O. A. 1965, A 20-inch reflecting telescope with photometer for photoelectric, colourimetric and polarimetric observations (in Russian). *Uchenye Zapiski Leningrad. Univ.*, no. 328, 83.
5. Hiltner, W. A. 1965, A rotable telescope for polarization studies. *Sky and Telesc.*, **30**, 144.
6. Iwanowska, W. 1963, Telescop Schmidta-Cassegraina w Polsce. *Nauka Polsca*, **11**, no. 2, 95.
7. Baum, W. A. 1964, Photosensitive detectors. *A. Rev. Astr. Astrophys.*, **2**, 165.
8. Smith, R. 1965, Detectors for ultraviolet, visible and infrared radiation. *Appl. Opt.*, **4**, 631.
9. Fain, D. L. 1965, Photomultiplier sensitivity limitations. *J. opt. Soc. Am.*, **55**, 206.
10 Shulov, O. 1965, General parameters of photomultipliers FEU–31 and FEU–64. (In Russian.) *Uchenye Zapiski Leningrad. Univ.*, no. 328, 95.
11. Blackwell, D., Petford, A., McCrea, S. 1965, An image-scrambler for use with a photomultiplier. *Observatory*, **85**, 21.
12. Schmeidler, F. 1965, Temperatureffekte bei Multipliern. *Sterne*, **41**, 211.
13. Young, A. T. 1965, Effects of photomultiplier temperature variations in astronomical photometry. *Mon. Not. R. astr. Soc.*, **129**, 215.

14. Fernie, J. D. 1965, Temperature effects in photomultipliers. *Observatory*, **85**, 169.
15. Young, A. T. 1966, Temperature effects in photomultipliers. *Observatory*, **86**, 71.
16. Young, A. T. 1965, Glass-filter temperature effects in astronomical photometry. *Astr. J.*, **70**, 697.
17. Vitrichenko, E. A. 1966, An investigation of the dependence between the radiation of sources of constant action and temperature (in Russian). *Izv. Krym. astrofiz. Obs.*, **35**, 274.
18. Kutusov, P. S., Mench, K. L. 1966, Use of a luminescent standard of brightness for the sensitivity control of photoelectric photometers (in Russian). *Astr. Zu.*, **43**, 878.
19. Peytremann, E. 1964, Etude et réalisation d'une lampe à effet Čerenkov. *Publ. Obs. Genève*. Ser. A. no. 69, 84.
20. Irvine, W., Picoes C., Charm, J., Lecomte, G. 1964, Effect of high voltage on spectral sensitivity for two photomultipliers. *Astrophys. J.*, **140**, 1629.
21. Korolev, F. A., Klementjeva, A. J., Meshcerjakova, T. F., Shcheglov, P. V. 1963, Multilayer interference filters of high contrast (in Russian). *Astr. Cirk.* no. 272, 1.
22. Ultraviolet interference spectroscopy (Roy. Holloway Coll. London). Space Res. U.K. Rep. 1964–65, 23, 1966.
23. Growes, J. 1966, Amplifier and pulse-counter for fast, low level photometry. *J. scient. Instrum.*, **42**, 430.
24. Artemiev, V. V., Rojnova, I. P. 1963, Photon counting in low level photometry (in Russian). *Izv. Krym. astrofiz. Obs.*, **30**, 297.
25. Dimov, N. A. 1963, Comparison of DC amplification and pulse counting when measuring very weak light fluxes (in Russian). *Izv. Krym. astrofiz. Obs.*, **30**, 308.
26. Veismann, U. 1965, A digitalized printing photoelectric photometer (in Russian), *Novaja Technika v Astronomii*, **2**, 40.
27. Veismann, U., Kübar, T. A photoelectric photometer with the remote control (in Russian). *Publ. Tartu Obs.*, **36** (in press).
28. Veismann, U. Digitalization of photometer's readings (in Russian). *Publ. Tartu Obs.*, **36** (in press).
29. Willstrop, R. V. 1965, A wide-field coma-free all-reflection plane grating spectrometer. *Mon. Not. R. astr. Soc.*, **130**, 233.
30. Mianes, P. 1965, Photométrie en 6 couleurs. Méthode mixte de mesure d'étoiles faibles. *J. Observateurs*, **48**, 103.
31. Xanfomaliti, L. V. 1965, Photometer with pulse counting system 'Volna' PS–5M (in Russian). *Abastumanskaja astrofiz. Obs. Bjull.*, no. 32, 197.
32. Kubyshkin, V. V. 1964, A photoelectric photometer with remote control (in Russian). *Astr. Zu.*, **41**, 738.
33. Nitisastro, S. 1964, The photoelectric instrument at the Bosscha Observatory. *Madjelah inst. Techn. Bandung Proc.*, **2**, 185.
34. Kokhan, E. K. 1966, Photometer AFM–6. Testing observations (in Russian). *Izv. glav. Obs. Pulkove*, **24**, 96.
35. Elvius, T., Eriksson, P., Linden, H., Ljunggren, B. 1964, The new photoelectric photometer of the Uppsala Observatory. Preliminary results for a field in Cassiopeia (S.A. 19). *Ark. Astr.*, **3**, 317.
36. Bigay, J. H. 1963, Mesures photoélectriques en 3 couleurs (*UBV*) d'étoiles O–B et Ao dans les selected areas du plan galactique S.A. 9, 19 et 24. *J. Observateurs*, **46**, 319.
37. Ackermann, C., Hermann, W. 1965, Experimentelle Arbeiten zur lichtelektrische Photometrie im nahen und mittleren Infrarot (0·8 m–5 m). *Mitt. astr. Ges.*, 95.
38. Franz, O. G. 1966, Photometry and astrometry of close double stars by a photoelectric scanning technique. *Lowell Obs. Bull.*, no. 134.
39. Miller, R. H. 1963, The effects of telescopes on astronomical polarization measurements. *Appl. Opt.*, **2**, 61.
40. Clarke, D. 1965, Theoretical considerations in the design of an astronomical polarimeter. *Mon. Not. R. astr. Soc.*, **129**, 71.
41. Clarke, D. 1965, Studies in astronomical polarimeter. II. A double beam polarimeter. *Mon. Not. R. astr. Soc.*, **130**, 75.

42. Pospergelis, M. M. 1965, Electronic polarimeter 'Taimyr' (in Russian). *Astr. Zu.*, **42**, 398.
43. Harvard Observatory Report. 1964, *Astr. J.*, **69**, 660.
44. Loden, L. O. 1964, A photoelectric starlight polarimeter with sky background compensation. *Stockh. Obs. Ann.*, **22**, no. 7.
45. Martel, L., Martel, M-Th. 1964, Polarisation de la lumière des étoiles dans le système *UBVR*, *Ann. Astrophys.*, **27**, 203.
46. Marin, M. 1965, Mesures photoélectriques de polarisation à l'aide d'un télescope coudé de 1 m. *Rev. Opt.*, **44**, 115.
47. Kokhan, E. K. 1964, Testing of the polarimeter of Pulkovo Observatory and of the camera ACU-4 (in Russian). *Izv. glav. astr. Obs. Pulkove*, **23**, 86.
48. Omarov, S., Gadjiev, M. 1966, Polarimetric tests of the 200 mm reflecting telescope of the Shemaha Observatory of the Academy of Sciences of the Azerbajdjan S.S.R. (in Russian). *Soobšč. Shemahinskoj Astrofiz. Obs.*, **5** (in press).
49. Lehnert, Ch., Schmidt, Th. 1965, Ein neues lichtelectrischen Polarisationsphotometer. *Mitt. astr. Ges.*, 81.
50. Dollfus, A. 1963, Un modulateur de lumière polarisée pour polarimètre photoélectrique. *C. r. Acad. Sci. Paris*, **256**, 1920.
51. Seddon, H., Jones, W. 1965, A semi-automatic iris photometer. *Publ. R. Obs. Edinb.*, **5**, 99.
52. Olivier, J. P., Kinman, T. D. 1964, A null indicating meter and other modifications of the Sartorius iris photometer. *Publ. astr. Soc. Pacif.*, **76**, 88.
53. Van Chuan-Zin, Lu dun-Chgen 1965, Improvement in the Sartorius iris photometer and investigation of its errors. *Acta astr. Sinica*, **13**, 57.
54. Urasin, L. A., Urmatsky, A. I. 1965, An iris-type microphotometer (in Russian). *Bjull. astr. Obs. Engelgardta*, no. 38, 101.
55. Urasin, L. A. 1965, An iris photometer on the base of MF-2 microphotometer (in Russian). *Novaja Tekhnika v Astronomii*, **2**, 46.
56. Friedemann, Ch., Schoneich, W. 1963, Uber eine Zusatzeinrichtung zur schleiermessung mit der Irisphotometer. *Astr. Nachr.*, **287**, 187.
57. Wamer, B. 1964, The measurements of stellar magnitudes using a recording microphotometer. *J. Brit. astr. Ass.*, **74**, 99.
58. Koelbloed, D. 1965, The isophotometer of the Astronomical Institute of the University of Amsterdam. *Bull. astr. Inst. Netherl.*, **18**, 62.
59. Kuschbert, D. 1965, Ein Lichtpunktabtastgerat zur automatische Auswertung von Himmelsaufnahmen. *Astr. Nachr.*, **288**, 125.
60. Lallemand, A., Canavaggia, R., Amiot, F. 1966, Détermination des magnitudes d'objets très peu lumineux au moyen de la caméra électronique. *C. r. Acad. Sci. Paris*, **262**, 838.
61. Livingston, W., Lynds, C. 1963, Recent experiments with image-tubes at Kitt Peak National Observatory. *Astr. J.*, **68**, 284.

Observational, extinction and reduction procedures

62. Cousins, A. W. J. 1966, Methods of observation and reduction of photometric observations at the Cape Observatory. MNASSA, **25**, 100.
63. Hardie, R. H. 1966, European Southern Observatory (E.S.O.) colloquium on photometry. Groningen.
64. Rufener, F. 1964, Technique et rèduction des mesures dans un nouveau système de photométrie stellaire. *Publ. Obs. Genève*, no. 66, 413.
65. Hardie, R. H., Heiser, A. M., Weedman, D. W. 1964, Study of anomalous ultraviolet extinction. *Astr. J.*, **69**, 543.
66. Ažusienis, A., Straižys, V. 1964, On the reduction of colour-indices $U-B$ outside the atmosphere (in Russian). *Astr. Circ.*, no. 304, 4.
67. Gutièrrez-Moreno, A., Stock, J. The accuracy of extinction determinations. *Publ. Dep. Astr. Univ. Chile*, no. 1 (in press).
68. Jerzykiewicz, M., Serkowski, K. The Sun as a variable star. III. Photometric observations of Uranus, Neptune, and F and G type stars. *Lowell Obs. Bull.* (in press).

69. Przybylski, A. 1964, The reduction of photometric observations affected by variable extinction (with an application to the effect of the volcanic dust). *Acta astr.*, **14**, 285.

70. Uranova, T. 1966, *Soobšč. gos. astr. Inst. Sternberga* (in Russian), no. 149.

71. Dachs, J., Haug, U., Pfleiderer, J. 1966, Atmospheric extinction measurements by photoelectric star photometry. *J. atmos. terr. Phys.*, **28**, 637.

72. Edwin, A., Berg, R. 1964, Stellar magnitudes and atmospheric extinction. *Astr. J.*, **69**, 557.

73. Gutièrrez-Moreno, A., Moreno, H., Stock, J. 1967, The atmospheric extinction on Cerro Tololo. *Publ. Dep. Astr. Univ. Chile*, no. 1.

74. Głebo̧cki, R. 1965, Spectrophotometric investigations of atmospheric extinction at Pivnice. *Bull. Obs. Torun*, no. 34, 29.

75. Wegner, W. 1965, The atmospheric extinction in photographic and photovisual ranges at the Astronomical Observatory in Pivnice-Torun. *Bull. Obs. Torun*, no. 34, 21.

76. Kivila, A., Eelsalu, H. An attempt to estimate atmospheric extinction for the *BV* photometry at the new Tartu Observatory (in Russian). *Publ. Obs. Tartu*, **36**, (in press).

77. De Vaucouleurs, G. 1965, Atmospheric absorption at McDonald Observatory, 1960–64. *Publ. astr. Soc. Pacif.*, **77**, 5.

78. Sharov, A. S. 1965, Some characteristics of the atmospheric transparency in the region of the high altitude expedition of Sternberg Astronomical Institute (in Russian). *Soobšč. gos. astr. Inst. Sternberga*, no. 139, 3.

79. Kondratyev, K. Y., Andreev, S. D., Badinov, I. Y., Grishechkin, V. S., Popova, L. V. 1965, Atmospheric optics investigations on Mt Elbrus. *Appl. Opt.*, **4**, 1069.

80. Johnson, H. L. 1965, Atmospheric extinction in the infrared. *Comm. Lunar Planet. Lab.*, **3**, 67.

81. Westphal, Y. 1964, New observations of atmospheric emission and absorption in the 9–14 micron region. *Mém. R. Soc. Liège*, **9**, 357.

82. Kondratyev, K. Y., Naylisk, Kh. J. 1963, On the heat emission of the 9·6 μ ozone absorption band in the atmosphere (in Russian). *Problemy Fisiki Atmosfery*, no. 2, 3.

83. Rasool, S. 1964, Intensities of 9·4 μ and 10·4 μ CO_2 Bands. *Mém. R. Soc. Liège*, **9**, 55.

84. Bignell, K., Saiedy, F., Sheppard, P. 1963, On the atmospheric infrared continuum. *J. opt. Soc. Am.*, **53**, 466.

85. Curcio, J., Drummeter, L., Knestrick, G. 1964, An atlas of the absorption spectrum of the lower atmosphere from 5400Å to 8520Å. *Appl. Opt.*, **3**, 1401.

86. Moreno, H., Stock, J. 1964, The atmospheric extinction on Cerro Tolelo during 1963. *Publ. astr. Soc. Pacif.*, **76**, 56.

87. Hill, P. W. 1964, Atmospheric extinction and volcanic ash in the upper atmosphere. *Mon. Notes astr. Soc. Sth Afr.*, **23**, 148.

88. Roach, H. 1964, The light of the night sky: astronomical, interplanetary, and geophysical. *Space Sci. Rev.*, **3**, 512.

89. Miller, R. H. 1963, A limitation on the photometry of faint objects. *Astrophys. J.*, **137**, 1049.

90. Shao, C.-Y., Young, A. T. 1965, Red-leak corrections in ultraviolet photometry. *Astr. J.*, **70**, 726.

91. Abt, H., Osmer, P. 1965, Filter transmission effects on photoelectric measures of rotationally broadened Hβ line strength. *Astrophys. J.*, **141**, 949.

92. Vanýsek, V. 1964, A note on the accuracy of photographic photometry with Schmidt telescope. *Acta Univers. Carolinae, Math. Phys.*, no. 1, 19.

93. Lawrence, L., Reddish, V. 1965, The Cygnus II association. I. Intercomparison of photometries with the Edinburgh and Hamburg Schmidt telescopes. *Publ. R. Obs. Edinb.*, **3**, 279.

94. Bresterhus, E. 1963, Zur Automatisierung photographischer Dreifarbenphotometrie mit Anvendung auf M 37. *Astr. Abhand. Hamburg. Sternw.*, **7**, 15.

95. Eelsalu, H. 1963, On the treatment of stellar-photometrical equations in the case of extremely weak darkenings (in Russian). *Publ. Obs. Tartu*, no. 8, 1.

96. Kalv, P., Eelsalu, H. 1966, An attempt to investigate photometrical properties of faint stellar images on the nonsensibilized photographic plates (in Russian). *Publ. Obs. Turtu*, **35**, 69.

97. Ishcenko, I. M. 1964, Photographic photometry of star trails (in Russian). *Astr. Zu.*, **41**, 138.
98. Shakhovskoi, N. M. 1963, Investigation of the polarization of variable stars. I. (in Russian). *Astr., Zu.*, **40**, 1055.
99. Shakhovskoi, N. M. 1964, An investigation of the polarization of variable stars. II. (in Russian). *Astr. Zu.*, **41**, 1042.
100. Vardanian, R. 1964, The polarization of T and RY Tau. (in Russian). *Soobšč. Bjurak. Obs.*, no. 35, 3.
101. Bertiau, F. C. 1963, A programme for the reduction of three-colour photoelectric observations. *Ric. astr. Spec. Vatic.*, **6**, 443.
102. Bertiau, F. C. 1965, A programme to establish primary standards in independent three-colour stellar photometric systems. *Ric. astr. Spec. Vatic.*, **6**, 611.
103. McCarthy, M., Treanor, P. M. 1964, Stars in the region of Pleiades. *Ric. astr. Spec. Vatic.*, **6**, 535.

Photometric systems

104. Ažusienis, A., Straižys, V. 1966, On the response curves of the *U,B,V,* system (in Russian). *Astr. Circ.* no. 352, 1.
105. Ažusienis, A., Straižys, V. 1966, The corrections of response curves and parameters of the *U,B,V,* system. I. Response curves (in Russian). *Bull. Vilnius Obs.*, no. 16, 3.
106. Ažusienis, A., Straižys, V. 1966, The corrections of response curves and parameters of the *U,B,V,* system. II. Colour-indices (in Russian). *Bull. Vilnius Obs.*, no. 17, 3.
107. Straižys, V., Zdanavičius, K. 1963, The band width effect and the parameters of system *U,B,V,* (in Russian). *Astr. Cirk.*, no. 254, 5.
108. Straižys, V., Zdanavičius, K. 1964, The influence of the band-width effect on the parameters of the *U,B,V,* system (in Russian). *Astr. Zu.*, **41**, 519.
109. Straižys, V. 1963, On the slope of reddening-lines in two-colour diagrams (in Russian). *Astr. Cirk.*, no. 254, 1.
110. Ažusienis, A., Straižys, V. 1964, On the slope of reddening-lines in $U-B$, $B-V$ diagram (in Russian). *Astr. Circ.*, no. 304, 1.
111. Wampler, E. 1964, The curvature of the reddening-line in the *U,B,V,* colour-colour plot. *Astrophys. J.*, **140**, 1615.
112. Nikolov, N. 1964, On the curvature of the reddening-line in the diagram $U-B$, $B-V$ (in Russian). *Astr. Cirk.*, no. 309, 3.
113. Karyagina, Z. V., Kharitonov, A. V. 1965, Some investigations of the *U,B,V,* photometric system (in Russian). *Astr. Zu.*, **42**, 377.
114. Osawa, K. 1963, An experiment concerning the width of the transmission band of the *U* filter. *Contr. Bosscha Obs.*, no. 21, 25.
115. Wilday, R. L. 1965, On line-blanketing normalization of subdwarf photometry. *Astrophys. J.*, **141**, 943.
116. Johnson, H. L. 1965, Interstellar extinction in the galaxy. *Astrophys. J.*, **141**, 923.
117. Johnson, H. L. 1964, The colors, bolometric corrections and effective temperatures of the Bright Stars. *Bol. Obs. Tonantz.*, **3**, 305.
118. Johnson, H. L. 1965, The absolute calibration of the Arizona photometry. *Comm. Lunar Planet. Lab.*, **3**, 73.
119. Johnson, H. L. 1966, Astronomical measurements in the infrared. *A. Rev. Astr. Astrophys.*, **4**, 193.
120. Wilday, R., Murray, B. 1964, Ten micron stellar photometry. First results and future prospects. *Mém. R. Soc. Liège*, **9**, 460.
121. Low, F. J., Johnson, H. L. 1964, Stellar photometry at 10μ. *Astrophys. J.*, **139**, 1130.
122. Johnson, H. L., Mendoza, E., Wisniewski, W. 1965, Observations of 'infrared stars'. *Astrophys. J.*, **142**, 1249.
123. Johnson, H. L., Low, F. J., Steinmetz, D. 1965, Infrared observations of the Neugebauer-Martz-Leighton 'Infrared Star' in Cygnus. *Astrophys. J.*, **142**, 808.
124. Eelsalu, H. Stellar-statistical approach to the wide-band photoelectric photometry (in Russian). *Publ. Obs. Tartu*, **36** (in press).

125. Straižys, V. 1964, An investigation of the *R,G,U* photometric system (in Russian). I. *Astr. Zu.*, **41**, 406; II. *Astr. Zu.*, **41**, 979.
126. Smith, L., Steinlin, U. 1964, Blanketing corrections for the *R,G,U* system. *Z. Astrophys.*, **58**, 253.
127. Spectral Classification and multicolour photometry. *IAU Symposium No.* 24. Eds. K. Lodén, L. Lodén, U. Sinnerstad, p. 383, 1966.
128. Golay, M. 1964, Photométrie en 5 couleurs et propriétés du diagramme $\triangle$, *g* (Note 2). *Publ. Obs. Genève*, no. 66, 19.
129. Golay, M. 1964, Position des étoiles métalliques et des sous-naines dans le diagramme $\triangle$, *g*. *Publ. Obs. Genève*, no. 66, 27.
130. Golay, M. 1964, Etude de quelques étoiles des Pleiades en cinq et sept couleurs. *Publ. Obs. Genève*, no. 68, 29.
131. Golay, M. 1964, Etude de quelques étoiles de Coma Berenices en cinq et sept couleurs. *Publ. Obs. Genève*, no. 68, 64.
132. Golay, M. 1964, Etude de quelques étoiles des Hyades en cinq et sept couleurs. *Publ. Obs. Genève*, no. 70, 179.
133. Golay, M. 1964, Etude de quelques étoiles de Praesepe en cinq et sept couleurs. *Publ. Obs. Genève*, no. 70, 148.
134. Golay, M. 1964, Cas des étoiles doubles dans les diagrammes en plusieurs couleurs. *Publ. Obs. Genève*, no. 68, 1.
135. Golay, M. 1964, Photométrie en cinq et sept couleurs et position des étoiles doubles dans le diagramme $\triangle$, *g*. *Publ. Obs. Genève*, no. 68, 15.
136. Golay, M., Goy, G. 1965, Photométrie en 5 et 7 couleurs d'étoiles proches. *Publ. Obs. Genève*, no. 71, 19.
137. Hauck, B. 1964, Couleurs des divers corps noirs dans le système de l'Observatoire de Genève. *Publ. Obs. Genève*, no. 70, 3.
138. Hauck, B. 1964, Note sur la sous-naine σ Boo dans le système photométrique de l'Observatoire de Genève. *Publ. Obs. Genève*, no. 70, 45.
139. Hauck, B. 1966, Un critère de luminosité dans le système photométrique de l'Observatoire de Genève. *Bull. Soc. Sci. Nat.*, **69**, 181.
140. Neff, J. S. 1966, A sensitive photometric discriminant for late-type dwarf and giant stars. *Astr. J.*, **71**, 203.
141. Straižys, V. 1964, A three-colour system for the determination of interstellar absorption (in Russian). *Astr. Zu.*, **41**, 750.
142. Zdanavičius, K., Straižys, V. 1964, The selection of spectral regions for heterochromatic photometry. II. (in Russian). *Bull. Vilnius Obs,.* no. 11, 1.
143. Straižys, V. 1964, The selection of spectral regions for heterochromatic photometry. III. Observational test of the system *X, Y, Z* (in Russian). *Bull. Vilnius Obs.*, no. 11, 11.
144. Straižys, V., Zdanavičius, K. 1965, The selection of spectral regions for heterochromatic photometry. IV. Two-dimensional classification of red stars (in Russian). *Bull. Vilnius Obs.*, no. 14, 3.
145. Straižys, V. 1966, The selection of spectral regions for heterochromatic photometry. V. The classification of early-type stars and ultraviolet magnitudes (in Russian). *Bull. Vilnius Obs.*, no. 16, 3.
146. Straižys, V., Zdanavičius, K. 1965, Two-dimensional photometric classification of red stars (in Russian). *Astr. Cirk.*, no. 320, 1.
147. Straižys, V. 1963, Optimum wavelengths for determination of interstellar absorption (in Russian). *Astr. Cirk.*, no. 254, 3.
148. Steinlin, U. 1963, Basic remarks about the choice of colors for systems of multicolor photometry. *Contr. Bosscha Obs.*, no. 21, 34.
149. Wood, H. J. 1966, Photoelectric Balmer-line photometry. *Contr. Kitt Peak nat. Obs.*, no. 96.
150. Beer, A. 1966, Calibration problems in hydrogen luminosities. *IAU Symposium.*, no. 24, 322.
151. Fernie, J. D. 1965, An absolute magnitude calibration of the Hβ index for early-type stars. *Astr. J.*, **70**, 575.

152. Abt, H., Golson, J. 1966, On the separation by Balmer-line photometry of high and low luminosity stars having H-alpha in emission. *Astrophys. J.*, **143**, 306.
153. Beer, A. 1964, Photoelectric distances of 461 northern OB stars and galactic structure for Hγ luminosities. *Mon. Not. R. astr. Soc.*, **128**, 261.
154. Andrews, P. J. 1966, Narrow band photoelectric photometry of O and B-type stars. Thesis, Cambridge.
155. Price, M. J. 1966, Narrow-band photoelectric Hα photometry in late-type stars. *Mon. Not. R. astr. Soc.*, **133**, 449.
156. Price, M. J. Narrow-band photometry in the study of stellar populations. Comparison of the Hyades and Coma Berenices open clusters. *Mon. Not. R. astr. Soc.*, **134** (in press).
157. Price, M. J. Narrow-band photometry in the study of stellar populations. The sodium D lines in late-type stars. *Mon. Not. R. astr. Soc.*, **134** (in press).
158. Kovar, R. P. 1965, Population discriminant for late-type dwarfs. *Astr. J.*, **70**, 142.
159. Miner, E. 1966, Narrow-band photometry of F5–K5 stars. *Astrophys. J.*, **144**, 1101.
160. Peat, D. W. 1964, Photometry of the orange-red Ca I triplet in late-type stars. *Mon. Not. R. astr. Soc.*, **128**, 475.
161. Fernie, D. 1965, The λ4226 line as a spectrum-luminosity indicator in late-type stars. *Astr. J.*, **70**, 675.
162. Scarfe, C. D. 1966, Photoelectric observations of Fe I lines in late-type stellar spectra. *Mon. Not. R. astr. Soc.*, **133**, 99.
163. Gyldenkerne, K. 1964, A three-dimensional spectral classification of G and K stars. *Publ. Kbn. Obs.*, no. 178.
164. Gyldenkerne, K. 1964, Photoelectric narrow-band classification of late-type stars. *Publ. Kbn. Obs.*, no. 180.
165. Gyldenkerne, K., Helt, B. 1966, Two-dimensional narrow-band classification of M stars. *IAU Symposium*, no. 24, 162.
166. Barnhart, P., Haynie, W. 1964, A program of stellar narrow-band infrared photometry. *Mém. Soc. R. Sci. Liège*, **9**, 425.
167. Jasevičius, V. 1964, On the three colour system used in Vilnius Observatory and its reduction to the system *U,B,V* (in Russian). *Bull. Vilnius Obs.*, no. 10, 19.
168. Urasin, L. A. 1963, Spectrophotometric and photometric investigations in Galaxy (I) (in Russian). *Bjull. Obs. Engelgardta*, no. 34, 130.
169. Urasin, L. A. 1965, Spectrophotometric and photometric investigations in Galaxy (II) (in Russian). *Bjull. Obs. Engelgardta*, no. 38, 47.
170. Matjagin, V. S. 1965, Three-colour photographic photometry with the Macsutov's meniscous telescope (in Russian). *Trudy astrofiz. Inst. Akad. Nauk Kasakh. SSR*, **5**, 303.
171. Rozis-Saulgeot, A.-M. 1965, Du rôle des raies d'absorption dans les magnitudes photoélectriques: cas du Soleil. *Ann. Astrophys.*, **28**, 562.
172. Wallerstein, G., Helfer, H. Photometric methods for measuring the metal content of K giants. *Astr. J.*, **71**, 350.
173. Ažusienis, A., Juodakas, A. 1964, The relation between magnitudes of Zeiss astrotriplet and the system *B*, *V* (in Russian). *Bull. Vilnius Obs.*, 11, 45.
174. Ažusienis, A. 1965, The correlation between magnitudes m_{pg} and *B* in the case of chromatic aberration (in Russian). *Bull. Vilnius Obs.*, no. 14, 24.
175. Kolesnik, L., Polischuk, E. 1966, Relation of two-colour system obtained at the telescope AZT–2 with *B*, *V* system (in Russian). *Dokl. Ukr. Akad. Nauk*, **9**.
176. Kreiner, J. M. 1964, A comparison of the colour indices obtained with those of Wroclaw system. *Acta astr.*, **14**, 219.
177. Shevchenko, V. Reduction to the *V* system by using colour changes of a variable star (in Russian). *Trudy Tashkent. Obs.*, Ser. II, **2** (in press).
178. Straižys, V. 1964, The relation between heterochromatic and monochromatic magnitudes in three-colour systems (in Russian). *Astr. Zu.*, **41**, 128.

Photometric standards

179. Jackisch, G. 1963, Mikrovariabilität und Zustandsgressen von hellen Sternen. *Veröff. Sternw. Sonneberg*, **5**, 309.

180. Jackisch, G. Lichtelektrische Photometrie von A-Sternen am Kottamia Observatorium der VAR. *Mitt. astr. Ges.*, no. 21 (in press).
181. Cousins, A. W. J. 1965, Micro-variability of Sun and stars. *Irish astr. J.*, **7**, 133.
182. McCullough, J. 1964, A photoelectric search for ultra-short period variable stars. *Astr. J.*, **69**, 251.
183. Hoffleit, D. 1964, *Catalogue of bright stars*. Third revised edition. Yale Univ. Obs.
184. Iriarte, B. Johnson, H. L., Mitchell, R., Wisniewski, W. 1965, Five-color photometry of bright stars. *Sky Telesc.*, **30**, 21.
185. Johnson, H. L., Mitchell, R. I., Iriarte, B., Wisniewski, W. 1966, *UBVRIJKL* photometry of the bright stars. *Comm. Lunar Planet. Lab.*, no. 63.
186. Nekrasova, S. V., Nikonov, V. B., Rybka, E. 1965, Photoelectric magnitudes and colours of photoelectric standards in Selected Areas. II. Catalogue (in Russian). *Izv. Krym. astrofiz. Obs.*, **34**, 69.
187. Cousins, A. W. J. 1963, *UBV* standards in the equatorial zone. *Observatory*, **83**, 254.
188. Cousins, A. W. J. 1963, Photometric data for stars in the equatorial zone (fifth list). *Mon. Notes astr. Soc. Sth. Afr.*, **22**, 130.
189. Cousins, A. W. J. 1964, Photometric data for stars in the equatorial zone (sixth list). *Mon. Notes astr. Soc. Sth. Afr.*, **23**, 10.
190. Cousins, A. W. J. 1964, Photometric data for stars in the equatorial zone (seventh list). *Mon. Notes astr. Soc. Sth. Afr.*, **23**, 175.
191. Cousins, A. W. J. 1965, Photometric data for stars in the equatorial zone (eighth list). *Mon. Notes astr. Soc. Sth. Afr.*, **24**, 120.
192. Cousins, A. W. J., Lake, R., Stoy, R. H. 1966, Photoelectric magnitudes and colours of southern stars. *R. Obs. Bull.*, no. 121.
193. Lake, R. 1963, Photoelectric magnitudes and colours for southern stars. *Mon. Notes astr. Soc. Sth. Afr.*, **22**, 79.
194. Lake, R. 1964, Photoelectric magnitudes and colours for 100 southern stars. *Mon. Notes astr. Soc. Sth. Afr.*, **23**, 14.
195. Lake, R. 1964, Photoelectric magnitudes and colours for 100 southern stars (fifth list). *Mon. Notes astr. Soc. Sth. Afr.*, **23**, 136.
196. Lake, R. 1965, Photoelectric magnitudes and colours for bright southern stars (sixth list). *Mon. Notes astr. Soc. Sth. Afr.*, **24**, 41.
197. Corben, P. M. 1966, Photoelectric magnitudes and colours of bright southern stars. *Mon. Notes astr. Soc. Sth. Afr.*, **25**, 44.
198. Cousins, A. W. J. 1963, Standard magnitudes and magnitude sequences in the southern hemisphere. *Contr. Bosscha Obs.*, no. 21.
199. Cousins, A. W. J. 1966, Fabry photometry of bright southern stars with an appendix giving detailed observations of bright variable stars made at the Cape prior to 1966. Roy. Obs. of Good Hope, pp. 45.
200. Cousins, A. W. J. 1966, On standard magnitudes and $B-V$ colours of bright stars in the southern hemisphere. *Observatory*, **86**, 69.
201. Stoy, R. H. 1965, Photoelectric three-colour magnitudes for southern stars. *Mon. Notes astr. Soc. Sth. Afr.*, **24**, 29.
202. Stoy, R. H. 1966, Cape photographic catalogue for 1950. *Cape Annals*, **21**,
203. Evans, D., Laing, J., Menzies, A., Stoy, R. 1964, Fundamental data for southern stars. *R. Obs. Bull.*, no. 85, 207.
204. Roslund, C. 1964, Investigation of a Milky Way Field in Scorpius. I. Magnitudes and colours of O and B stars. *Ark. Astr.*, **3**, 357.
205. Elvius, T., Lyngå, G. 1965, Three-colour photometry of 43 stars in Kapteyn's Selected Areas at southern galactic latitudes. *Ark. Astr.*, **3**, 467.
206. Sjögren, U. 1964, Photometric and spectrophotometric observations with a discussion of the interstellar absorption in the region of Kapteyn's Selected Area 8. *Ark. Astr.*, **3**, 339.
207. Elvius, T., Häggkvist, J. 1966, Photoelectric measurements of stars in Kapteyn's S.A. of + 60° zone. *Ark. Astr.*, **4**, 49.
208. Ljunggren, B. 1966, A photoelectric and spectrophotometric investigation of a region near the north galactic pole. *Ark. Astr.*, **3**, 535.

209. Ljunggren, B., Oja, T. 1965, Photoelectric measurements of magnitudes and colours for 849 stars. *Ark. Astr.*, **3**, 439.

210. Häggkvist, L., Oja, T. 1966, Photoelectric photometry of bright stars. *Ark. Astr.*, **4**, 137.

211. Oja, T. 1966, A spectrophotometric survey of stars along the Milky Way. (Part II). *Uppsala astr. Obs. Ann.*, **5**, no. 4.

211a. Oja, T. 1966, A photometric study of a region in Cassiopeia. *Uppsala astr. Obs. Medd.*, no. 157.

211b. Häggkvist, L. 1966, Photoelectric magnitudes and colours for visual binaries, *Uppsala astr. Obs. Medd.*, no. 156.

212. Argue, A. N. 1966, *UBV* photometry of 550 F, G and K type stars. *Mon. Not. R. astr. Soc.*, **133**, 475.

213. Argue, A. N. Red and infrared magnitudes and colours for 300 F, G and K type stars. *Mon. Not. R. astr. Soc.*, (in press).

214. Rufener, F., Hauck, B., Goy, G., Peytremann, E., Golay, M. 1964, Catalogue des étoiles mesurées dans le système photométrique de l'Observatoire de Genève. *Publ. Obs. Genève*, no. 66.

215. Iriarte, B. 1965, Fotometria fotoelectrice en *BVRI* para 275 estrellas en su mayoria entre $-25°$ y $-50°$. *Bol. Obs. Tonantzinla*, **4**, 33.

216. Bigay, J. H., Lunel, M. 1965, Photométrie photoélectrique en trois couleurs (*U,B,V*) d'étoiles chaudes dans la 'Selected Area' 8. *C. r. Acad. Sci. Paris*, **260**, 3853.

217. Bigay, J. H., Lunel, M. 1965, Photométrie photoélectrique *UBV* de 263 étoiles B et A de la S.A.8. *J. Observateurs*, **48**, 171.

218. Bigay, J. H. 1966, Photométrie photoélectrique en trois couleurs (*U,B,V*) d'étoiles chaudes dans la SA 40. *C. r. Acad. Sci. Paris*, **262**, 1570.

219. Osawa, K. 1965, Spectral classification and three-color photometry of A-type stars. *Ann. Tokyo astr. Obs.*, **9**, 123.

220. Gutièrrez-Moreno, A., Moreno, H., Stock, J., Torres, C., Wroblewski, H. A system of photometric standards. *Publ. Dep. Astr. Univ. Chile*, no. 1 (in press).

221. Jerzikiewicz, M., Serkowski, K. Fifty standard stars for yellow and blue magnitudes. *Publ. astr. Soc. Pacif.*, (in press).

222. Pfleiderer, J., Dachs, J., Haug, U. 1966, Lichtelektrische UBV-Photometrie von Standardsternen in vier Sternfelder am Äquator. *Z. Astrophys.*, **64**, 116.

223. Harris III, D. L., Upgren, A. R. 1964, Photoelectric magnitudes and colors of stars near the galactic pole. *Astrophys. J.*, **140**, 151.

224. Takayanagi, K., Shimizu, T. 1966, *U,B,V* photoelectric photometry of the Ursa Major group. *Mem. Coll. Sci. Univ. Kyoto*, Ser. A, **31**, no. 2.

225. Hogg, A. R. 1963, A southern photoelectric magnitude sequence. *Publ. astr. Soc. Pacif.*, **75**, 194.

226. Mayer, P. 1964, Photoelectric measurements of OB stars. *Acta Univ. Carolinae Math. Phys.*, no. 1, 25.

227. Hardie, R. H., Heiser, A. M., Tolbert, C. R. 1964, A study of the B stars in the Orion Belt region. *Astrophys. J.*, **140**, 1472.

228. Imagawa, F. 1964, Observational results of three-color photometry for F-type stars. *Dep. Astr. Univ. Kyoto. Repr.*, no. 22, 61.

229. Eggen, O. 1964, Colors, luminosities and motions of the nearer G-type stars. *Astr. J.*, **69**, 570.

230. Przybylski, A., Kennedy, P. 1965, Radial velocities and three-colour photometry of 52 stars with large proper motions. *Mon. Not. R. astr. Soc.*, **129**, 63.

231. Przybylski, A., Kennedy, P. 1965, Radial velocities and three-colour photometry of 166 southern stars. *Mon. Not. R. astr. Soc.*, **131**, 95.

232. Tolbert, C. R. 1964, A *UBV* study of 94 wide visual binaries. *Astrophys. J.*, **139**, 1105.

233. Wildey, R. 1964, The stellar content of χ and h Persei cluster and association. *Astrophys. J. Suppl.*, **8**, 439.

234. Johnson, H. L. 1965, Infrared photometry of M-dwarf stars. *Astrophys. J.*, **141**, 170.

235. Mendoza, E., Johnson, H. L. 1965, Multicolor photometry of carbon stars. *Astrophys. J.*, **141**, 161.
236. Mendoza, E., Eugenio, E., Johnson, H. L. 1965, An investigation of late-type stars. *Bol. Obs. Tonantz.*, **4**, 57.
237. Neugebauer, G., Martz, D., Leighton, R. 1965, Observations of extremely cool stars. *Astrophys. J.*, **142**, 399.
238. Breckinridge, J. B., Kron, G. E. 1964, Red and infrared photometry of double stars. *Publ. astr. Soc. Pacif.*, **76**, 139.
239. Blanco, V. 1964, $B-V$ and $V-I$ colors of giant M stars. *Astr. J.*, **69**, 730.
240. Crawford, D. L., Perry, C. L. 1966, Four-color and Hβ photometry of open clusters. I. The Hyades. *Astr. J.*, **71**, 206.
241. Strömgren, B. 1964, On the chemical composition and kinematics of disk high-velocity stars of the main sequence. *Astrophys. Norv.*, **9**, 333.
242. Borgman, J. 1965, Seven-colour photometry of A, F, G, K and M stars. *Bull. astr. Inst. Netherl.*, **17**, 58.
243. Ponsen, J., Oosterhoff, P. Th. 1966, Five-colour photometry of high-velocity stars and of pulsating variables of different types. *Bull. astr. Inst. Netherl. Suppl.*, no. 1.
244. Oosterhoff, P. Th., Ponsen, J. 1966, Discussion of five-colour observations of stars of high velocity. *Bull. astr. Inst. Netherl.*, **18**, 150.
245. Crawford, D. L., Mancler, J. 1966, Standard stars for photoelectric Hβ photometry. *Astr. J.*, **71**, 114.
246. Crawford, D. L. 1965, Photoelectric Hβ photometry for B stars in the Orion Belt region. *Astr. J.*, **70**, 671.
247. Peat, D. W. 1964, Hα photometry of late-type stars. I, F, G and K stars north of the equator. *Mon. Not. R. astr. Soc.*, **128**, 435.
248. Peat, D. W. 1966, Hα photometry of late-type stars. II. F and G dwarf stars south of the equator. *Mon. Not. R. astr. Soc.*, **131**, 476.
249. Schmidt-Kaler, Th. 1964, Die galaktischen emissions B Sterne. *Veröff. Sternw. Bonn*, no. 70,
250. Bahner, K. 1963, Energy distribution in the spectra of early-type stars. *Astrophys. J.*, **138**, 1314.
251. Willstrop, R. V. 1965, Absolute measures of stellar radiation. II. *Mem. R. astr. Soc.*, **69**, 83.
252. Oke, J. B., Conti, P. S. 1966, Absolute photoelectric spectrophotometry of stars in the Hyades. *Astrophys. J.*, **143**, 134.
253. Kharitonov, A. V. 1964, Energy distribution in the spectra of four stars in absolute units (in Russian). *Izv. Akad. Nauk Kazakh. SSR Fiz. Mat. ser.*, no. 3, 28.
254. Karjagina, Z., Kharitonov, A. V. 1964, Absolute energy distribution in the spectra of 17 stars (in Russian). *Izv. Akad. Nauk Kazakh. SSR Fiz. Mat. ser.*, no. 3, 10.
255. Kharitonov, A. V. 1966, Distribution of energy in spectra of 13 stars in absolute units (fourth list of extra-atmospherical spectrophotometric standards) (in Russian). *Trudy Inst. astrofiz. Akad. Nauk Kazakh. SSR*, **7**, 9.
256. Kharitonov, A. V., Neljubin, N. The method and preliminary results for the energy distribution in the spectrum of α Lyrae standardized by the central part of the solar disk (in Russian). *Trudy Inst. astrofiz. Akad. Nauk Kazakh. SSR*, **8**, (in press).
257. Kharitonov, A. V. Absolute energy distribution in the spectra of 14 bright stars in Pleiades (in Russian). *Astr. Zu.* (in press).
258. Aller, L. H., Faulkner, D. J., Norton, R.H. 1964, Photoelectric spectrophometry of selected southern stars. *Astrophys J.*, **140**, 1609.
259. Aller, L. H., Faulkner, D. J., Norton, R. H. 1966, Photoelectric spectrophotometry of selected southern stars. *Astrophys. J.*, **144**, 1073.
260. Van den Bergh, S., Sackman, I. 1965, Photoelectric spectrophotometry of stars. II. *Astr. J.*, **70**, 353.
261. Walker, R. G., D'Agati, A. 1964, Infrared stellar irradiance. *Appl. Opt.*, **3**, 1289.
262. Heddle, D. W. O. 1964, The importance of absolute photometry. *Ann. Astrophys.*, **27**, 800.

263. Chubb, T. A., Byram, E. 1963, Stellar brightness measurements at 1314 and 1427Å; observation of the O I twilight glow. *Astrophys. J.*, **138**, 617.
264. Bogess, A. 1964, B star colours between 2000 and 3000 angströms. *Ann. Astrophys.*, **27**, 805.
265. Alexander, J. D. H., Bowden, P. J., Heddle, D. W. O. 1964, Southern hemisphere observations of ultraviolet radiation from celestial objects. I. Experimental techniques. *Proc. R. Soc. London*, **A279**, 510.
266. Bless, R. C., Code, A. D., Houck, T. E., McNall, J. F., Taylor, D. J. 1965, Ultra violet observations of stars. *Astr. J.*, **70**, 667.
267. Byram, E. T., Chubb, T. A., Friedman, H. 1964, Measurements of stellar brightness in the far ultraviolet. *Astr. J.*, **69**, 135.
268. Byram, E. T., Chubb, T. A., Werner, M. W. 1965, 1115Å far ultraviolet stellar photometry. *Ann. Astrophys.*, **28**, 594.
269. Stecher, Th. P. 1965, Ultraviolet spectrophotometry of early-type stars. II. *Astr. J.*, **70**, 693.
270. Boggess III, A., Borgman, J. 1964, Interstellar extinction in the middle ultraviolet. *Astrophys. J.*, **140**, 1636.
271. Almar, I. 1964, Bemerkungen über die absolut – Standardisierung der Sternhelligkeiten. *Mitt. und Ergebn. der Satellitbeobachtungsdienst DDR*, no. 8, 49.
272. Krzeminski, W., Serkowski, K. Photometric and polarimetric observations of the nearby strongly reddened open cluster Stock 2. *Astrophys. J.* (in press).
273. Stepień, K. 1965, Polarimetric measurements in the north polar region. *Acta. Astr.*, **15**, 51.
274. Kruszewski, A. 1963, Measurements of polarization in the region of the III Cephei Association. *Acta Astr.*, **13**, 92.
275. Serkowski, K. 1965, Further polarimetric observations of highly polarized stars in two spectral regions. *Acta Astr.*, **15**, 79.
276. Lodén, L. O. 1965, A polarimetric investigation in the Milky Way. *Stockh. Obs. Ann.*, **22**, no. 8.

26. COMMISSION DES ETOILES DOUBLES

PRÉSIDENT: Dr K. Aa. Strand, U.S. Naval Observatory, Washington, D.C. 20390, U.S.A.

VICE-PRÉSIDENT: Dr P. C. Couteau, Observatoire de Nice, Le Mont Gros, 06-Nice, France.

COMITÉ D'ORGANISATION: J. Dommanget, O. J. Eggen, P. G. Kulikovsky, P. Muller, R. M. Petrie†, P. van de Kamp.

MEMBRES: Arend, Baize, Cester, Deutsch (A. N.), Dick, Djurkovic, Finsen, Franz, Güntzel-Lingner, Heintz, Hertzsprung, Hopmann, Jeffers, Johnson (M. C.), Jonckheere, Kuiper, Lippincott, Luplau-Janssen, Luyten, Martin, Mourao, Thé, van Albada, Van Biesbroeck, Van den Bos, Wierzbinski, Wieth-Knudsen, Worley, Zagar.

1. INTRODUCTION

According to the reports which have been received, it appears that the observational efforts in visual double star astronomy have remained at about the same level during the report period as they have been during the preceding period.

Problems in stellar structure and stellar evolution, which receive considerable attention at the present time, are often related to the visual binaries. In these studies, as well as others, it is apparent that there is need for more measures and more observations of astrophysical interest, such as photometric data, spectral classification and radial velocities—especially with instruments of large aperture.

There is still an urgent need for encouragement of young astronomers to take up double-star astronomy and to allow experienced observers in this field increased opportunities to use reflectors of large apertures.

2. DOUBLE STAR CATALOGUES

At the General Assembly in Hamburg, the U.S. Naval Observatory accepted the responsibility for future maintenance of the Index Catalogue of Visual Double Stars and Catalogue of Observations formerly maintained at the Lick Observatory. In accepting the catalogues, the U.S. Naval Observatory has the following obligations:

1. The sole responsibility to maintain the catalogues to date.
2. To furnish the other centers and the Lick Observatory with supplements when new data become available.
3. To furnish individuals with data from the two catalogues at cost.

The other Double Star Centers will be limited for the present to the Observatory at Meudon and to the Royal Greenwich Observatory at Herstmonceux. It may be desirable in the future to also have a center in the southern hemisphere. Lick Observatory will be a depository without the responsibility of the centers.

The obligations of the other centers are:

1. To maintain the catalogue to correspond precisely to the original at the U.S. Naval Observatory.
2. To supply individuals with limited data as requested, at cost.

The transfer of the Catalogues was effected in March 1965, with copies transmitted to the Paris Observatory at Meudon and to the Royal Greenwich Observatory at Herstmonceux.

Duplicates of the data from the catalogue at Washington will be provided at cost, upon request, on punched cards, tabular listings or magnetic tape—depending upon the amount of information required. Requests should be addressed to:

Superintendent
U.S. Naval Observatory
Washington, D.C. 20390, U.S.A.
Attention: Double Star Center

At the Hamburg General Assembly it was further agreed that it was the obligation of the double star observers to communicate to the Center:

1. Any errors found in the Catalogues.

2. Copies of new observational data sent simultaneously to be published. Mr. C. Worley, who is in charge of the Double Star Center, has requested insofar as practicable that observers follow these rules:

(*a*) Send a copy of any manuscript containing new observational data to this office at the same time the manuscript is submitted for printing. Notify Mr Worley upon publication of the data, and give the reference. If possible, send a copy of the publication (or reprint of the paper) to him at your earliest convenience.

(*b*) Since all records are being maintained by the use of electronic computers and associated punch-card equipment, it is necessary to code much material in a form suitable for these machines.

Therefore, when listing a measure, please give the following information:

(1) The right ascension and declination for 1900 as tabulated in Part II of the Index Catalogue of Visual Double Stars, 1961.0 (= IDS). Also, if possible, please give the 2000 position.
(2) the ADS Number;
(3) the component observed, as described by IDS, Part II, under 'MULT';
(4) the date, rounded to two decimals of a year, unless a very rapid binary is being observed, when three decimals should be given;
(5) the position angle as observed, not precessed to a certain epoch;
(6) the separation;
(7) the estimated magnitudes or magnitude difference;
(8) the number of nights included in the mean.

Additional information, such as the close designations of the discoverers (listed on p. xvii of Part I of the Index Catalogue of Visual Double Stars), the discoverers' number, durchmusterung number, etc. are not strictly necessary for maintenance of the catalogue of observations. However, this information is of value in maintaining the Index Catalogue, and hence will be most welcome. The aperture of the telescope, and the type of instrument used for the measure, are also necessary information.

3. COLLOQUIUM 'ON THE EVOLUTION OF DOUBLE STARS'

The response to the invitations to attend the IAU Colloquium at Uccle, Belgium, in September 1966 'On the Evolution of Double Stars' was most gratifying. Of the 65 specialists invited to attend, 39 accepted the invitation, representing 12 countries and 11 IAU Commissions.

In addition to six invited introductory discourses, 25 papers were presented, and five notes by colleagues not present. Time permitted ample discussion of the presented papers.

The proceedings of the colloquium will be published in *Communications de l'Observatoire Royal de Belgique* and the costs will be defrayed by the Belgian government. The volume will be dedicated to Professor G. Van Biesbroeck, in recognition of his devotion to visual observations of double stars of over 60 years.

4. VISUAL OBSERVATIONS

Herstmonceux, 28-inch (71 cm) refractor (Symms): continuation of measurements of pairs with real or suspected motions contained in the list of nearly 500 pairs drawn up when the telescope was re-erected at Herstmonceux in 1957.

Johannesburg, 26½-inch (67 cm) refractor (van den Bos, Knipe, Newburg, Postma): 7775 micrometer measures. Observations have been published in *Republic Observatory Circulars* 123–125. Finsen re-examined 1000 stars in his interferometer survey and discovered two new close pairs. Present status of the survey: 73 new pairs or slightly less than 1% of all stars examined. For 12 pairs it has been possible to compute orbits, three of which have periods as short as 2·65, 2·67 and 3·2 years. The total number of interferometer measures made in the three year period was in excess of 1100, not taking into account many time consuming negative results.

Lamont-Hussey, 27-inch (69 cm) refractor (Holden): continuation of re-measures of selected pairs from the Rossiter's Catalogue with a filar micrometer. It is proposed to complete the re-examination of all hitherto neglected pairs discovered at the observatory, and to commence photographic observations of the wider pairs. Two lists containing 1355 observations were published (1).

Meudon (Muller): The 83-cm refractor has been in regular operation since 1965. Five hundred and eighty measures of close binaries have been made.

Munich, 28-cm refractor (Heintz): observations are continued at a rate of 800 per annum.

Nice, 38-cm refractor (Bertin, Couteau, Morel): 2200 measures of 750 pairs. The program consists of all binaries accessible to the telescope with special emphasis on neglected pairs, such as those discovered by Espin and Jonckheere. The observations are carried out with either a Muller double image micrometer or a filar micrometer. A part of the observations have been published (2).

Since 1963 Couteau has systematically examined, for duplicity, the stars with large proper motion in the Paris Astrographic Zone (+ 17° to + 25° dec.). So far he has examined 4858 stars and found 115 new pairs of which 48 have separations less than 1 inch. The percentage of new binaries remains nearly constant, and amounts to one binary for each 42 stars examined.

Paris, 38-cm refractor (Baize): 1322 measures with filar micrometer of 364 pairs which are considered either in a critical orbital phase, are red dwarf or are neglected pairs. Baize intends to observe over the next years chiefly double stars between + 50° and + 90° declination.

Rio de Janeiro, 46-cm refractor (Mourao): visual observations since 1959. A third series comprising 450 measures of 160 systems has been puslished (3).

Sternberg Institute, 20-cm refractor (Kuročkin, Starikova): measurements with a polarizing micrometer according to design by Muller of 114 systems (4). Measures of additional 76 systems in press.

Tucson, 84-inch (213 cm) Kitt Peak National Observatory reflector and 61-inch (155 cm) Lunar Planetary Laboratory reflector (Van Biesbroeck): 1300 measures of close binaries with special attention to close pairs in critical parts of their orbits. Measures obtained from 1960 to the end of 1965 have been published (5).

Uccle, 45-cm refractor (Arend, Dommanget): no observations were made during the report period, but are expected to be resumed in 1967 after the installation of a new Zeiss objective.

U.S. Naval Observatory, 26-inch (66 cm) and 12-inch (30 cm) at Washington, 61-inch (155 cm) and 40-inch (102 cm) reflectors at Flagstaff (Walker, Worley), 7601 measures, the majority with a digitized micrometer. 4586 measures made with the 26-inch were published by Worley (**6**), and 1013 measures made with the 12-inch were published by Walker (**7**).

5. PHOTOGRAPHIC OBSERVATIONS

The Hertzsprung stereotype plate and card catalogue of relative double star positions resulting from multiple exposure photographic plates, according to the Hertzsprung method, was transferred from Professor Hertzsprung's residence in Tølløse, Denmark to the U.S. Naval Observatory, Washington, D.C. At the same time a total of 2300 plates, consisting primarily of plates taken by Hertzsprung at Potsdam, Johannesburg and Lick, and his collection of measurements on these plates were transferred to the Observatory.

Abastumani, 40-cm refractor (Salukvadze): photographic measures of trapezium type multiple systems. In addition *U*, *B*, *V* photometry were made and spectral types were determined (**8**).

Minnesota, 48-inch (122 cm) Palomar Schmidt telescope (Luyten): nearly 1000 new wide double stars have been discovered during the past three years from blinking of plates of the Palomar Survey. The entire polar cap north of declination + 75° has been completed and 280 double stars discovered, some as faint as the 21st magnitude. Some 20 more pairs, containing one degenerate component, were found and will be observed intensively with a view of obtaining orbital motions and statistical values for the masses. On the basis of what has been discovered so far, it appears that there are some 15–20 000 wide binaries, based upon common proper motions of the components, and brighter than 21st magnitude over the entire sky. At least 400–500 of these will contain one white or degenerate component.

Mount Stromlo, 26-inch (66 cm) refractor (Abraham): some 200 plates have been taken, using the Hertzsprung technique, and a multiple exposure camera designed by Strand. It is intended to measure the plates on a Grubb Parson digitized measuring machine.

Potsdam, 50-cm visual refractor and 20-cm astrograph (Böttger): the program with the first instrument consists of observations of 120 selected binary systems of special interest, using Hertzsprung's technique. Results for 68 systems were published (**9**). Since 1962 139 plates of 41 systems have been taken. The astrographic program consists of photographic observations of 750 (mostly neglected) binaries brighter than tenth magnitude with wide separations larger than 6″.

Pulkovo, 65-cm refractor (Deutsch, Kiselev, Plugin, Kanaev): observations of 75 binary systems with nearly equal magnitudes have been completed. The results for 40 systems have been published (**10**). A program of observations of 61 Cyg, Lalande 21 185, η Cas, and others, is continued for the purpose of analyzing perturbations due to invisible companions.

Sproul, 24-inch (61 cm) refractor (van de Kamp, Lippincott): mass ratios and parallaxes for nearby stars, mostly within 10 parsecs. Recent results include parallax and mass ratio for ξ Sco (**11**), Σ2173 (**12**), BD + 77° 361 (**13**), Hu 1176 (**14**). Studies were made of the photocentric orbits of μ Cas (**15**) and α Oph (**16**). Observations of Barnard's Star have been continued and have so far confirmed the earlier (**17**) orbital analysis.

U.S. Naval Observatory, 26-inch (66 cm) refractor at Washington (Mikesell, *et al.*); 24-inch (61 cm) refractor at Lowell Observatory (Priser, *et al.*): the photographic double star program was continued. Approximately 500 objects are under observation. During the report period 1499 multiple exposure plates were taken at Washington and 955 at the Lowell Observatory. 2454 plates, with a total number of 150 000 exposures, were measured. Automatic data recording at the measuring machine and the use of electronic data processing equipment made it possible to keep plate measurements and reductions up-to-date, avoiding accumulation of a large backlog of unmeasured plates.

Van Vleck Observatory, 20-inch (51 cm) refractor (Gasteyer): relative position of 65 double stars were published from 326 plates (**18**), using the Hertzsprung method and a multiple exposure camera designed by Strand.

Vienna, 32-cm astrograph (Haupt): measures of 250 wide binary systems have been published (**19**). Another series of 120 pairs is near completion.

6. ORBITAL STUDIES

Since the last general assembly Muller has published *Information Circular* Nos. 33–40, reporting a total of 82 new orbits.

Calculations were made by Baize, Couteau, Djurkovic, Finsen, Heintz, Knipe, Mourao, Newburg, Popoplic, Starikova, Symms and van den Bos.

Hopmann published provisional elements of ten binary systems with long periods (**20**).

Miss Zeller studied ξ Cep, ϵ Equ and ϵ Hya (**21**).

Miss Starikova published relative rectilinear motions in 31 double stars (**22**). Similar studies have been published by Arend (**23**), Dommanget (**24**), Dommanget and Nys (**25**), and Nys (**26**).

Fesenko has carried out statistical studies concerning eccentricities of binary orbits (**27**).

Arend, in collaboration with Nys, has worked on computing programs in connection with orbital investigations of double stars (**28**).

Kulikovsky has designed a simple graphical method of determining orbital elements of binaries (**29**).

Dommanget and Nys is preparing a catalogue of ephemerides of radial velocity curves of visual binaries.

Couteau and Mme François have constructed with an IBM calculator an atlas of orbits which contain the apparent ellipse of all known orbits. The part of the orbit which has been effectively observed is specially marked, which shows at a glance the quality of the elements. The atlas also includes an ephemeris for one revolution, the masses and the dynamical parallax. Copies of the atlas can be requested from Observatoire de Nice.

Wempe reports that a card index initially containing some 20 000 single measurements, copied from the Muller catalogue, is continuously being augmented by all accessible publications and now contains almost 35 000 observations. A card index of known orbits is similarly being augmented and now contains 2007 orbit determinations of 559 binaries, 41 of which were published between 1964 and 1966.

7. SPECTROSCOPIC AND PHOTOMETRIC OBSERVATIONS

The Victoria program of determining radial velocities of visual binaries with relative short periods and with large Δm's to ensure only one spectrum in visible is being continued at the Dominion Astrophysical Observatory, Victoria, B.C. Initiated by Miss Underhill, the program has been transferred to the coudé spectrograph of the 48-inch (122 cm) telescope.

The Victoria program of spectroscopic observations of wide visual binaries began with the late Dr Petrie, and is being continued by Batten. A detailed discussion of four systems was published by him (**30**).

Work started by Slettebak (**31**) some years ago on the spectra of visual double stars is being extended by Meisel, who is investigating some 50 visual double star systems, mostly of early type, with preference to those containing Be, O and supergiant components. The spectrograms were taken with the Perkins 72-inch (183 cm) telescope of the Ohio State and Wesleyan Universities at the Lowell Observatory at Flagstaff. In addition to spectral types, luminosity classes and rotational velocities, Meisel has measured Hγ photoelectric indices, Δm's and radial velocities.

Abt and Sanwal at Kitt Peak National Observatory are obtaining spectral types and *U*, *B*, *V* photometry for most of the systems in the Orbit Catalogue (**32**) which do not have such data. In addition, they are obtaining 13·5Å/mm spectra regularly with the 84-inch (213 cm) coudé spectrograph of most of the visual systems having orbits and being brighter than 7·0 magnitudes.

Since 1966 Couteau and Mlle Lacoarret have carried out spectroscopic observations of double stars at the Haute-Provence Observatory with the 193-cm reflector for the purpose of determining relative radial velocities and spectral types.

Franz is using a photoelectric scanner designed by Rakos, in conjunction with the 72-inch (183 cm) Perkins telescope, for obtaining photometric data as well as relative positions of the components of double stars (**33**).

Preliminary results of photographic *B*, *V* photometry of the components of the double cepheid CE Cas in NGC 7790 was obtained by Kholopov and Efremov (**34**).

8. MISCELLANEOUS STUDIES

Kulikovsky has published in the *Information Circular* lists of T-Tauri type stars for testing of duplicity with telescopes of different sizes.

Arend and Dommanget are continuing their study of the evolution of double stars, taking into account mass loss.

Harrington is in the process of developing a theory of the motion in triple systems.

James at Manchester is developing numerical methods for determining the structures of stars distorted by strong gravitational fields and/or rapid rotation. He is extending the work to realistic stellar models with the object of exploring mutual interactions of the components of close binary systems.

Piotrowski derived general formulae for the variation of the elements of a close binary system caused by mass ejection, mass transfer and mass infall (**35**). It was shown that the eccentricity of the orbit has a tendency to decrease if the mass transfer takes place predominantly near periastron.

The problems related to the exchange of matter and period changes in close binaries were reviewed by Kruszewski (**36**).

Further work of the Polish group is reported by Commission 42.

Theoretical work to establish the major properties of tides in close binary systems and to examine the role the tides play in the dynamical and nuclear evolution of the systems has been carried out by Zahn (**37**).

PROPOSALS

The Nice Observatory is proposing to hold a colloquium on the subject of different methods of observing visual binaries. If approved, the colloquium is expected to take place at Nice in the fall of 1969 with an expected attendance of 25 workers in the field.

Couteau proposes the establishment of a working group to select and publish periodically a list of definitive or well-determined orbits. The working group should be composed of a small number of members experienced in observing. The list of orbits could be published in the *Transactions* of the Union with revisions and additions each three years. According to the catalogue of orbits published by Worley (**32**) there are at the present only 43 definitive orbits, so the space requirements for this information would not be excessive.

Muller reports that his orbit catalogue is kept up-to-date with a view of publishing a new edition of the catalogue of ephemerides. He is principally awaiting information as to whether the Commission should designate a committee to select the orbits to be retained for general catalogues of either elements or ephemerides.

Couteau proposes that a census be made of regular double star observers and the instruments which are used primarily for observations of binary stars. He also proposes that a more rapid diffusion of observations for the purpose of preparation of programs should be discussed. Perhaps this could be done by the distribution of copies of manuscripts of observations to regular observers.

Worley would like to hear a discussion of the advisability of further discovery surveys, both of a general and of a special character.

He points out that the Catalogue of Observations includes all known measures of visual binaries south of declination − 20°, but north of this limit it contains, with a few exceptions, only those measures made and published since 1927. He wishes to know if the members of the commission feel that the inclusion of the earlier measures in this catalogue are important enough to justify the large amount of effort that the collection of such a list would entail.

Mourao suggests a discussion of whether three or six color photoelectric observations of binaries should be made.

Holden feels that with the catalogue of observations in machine-readable form, it appears timely to suggest that lists of stars in urgent need of observations should be circulated.

Baize wishes to bring up the problem of 'wholesale' orbit computation with high speed computers leading to inacceptable results.

Arend and Dommanget propose:

1. that the Commission at the Prague meeting call attention to the convention adopted by the Commission at the 1935 Paris meeting in regard to a single system of elements.
2. that the Commission support necessary modifications of the system then adopted for the purpose of avoiding the confusion which originates from the usage of ± in front of the element *i*, as follows:

   i – Inclination at the ascending node. For direct motion, in the first quadrant, for retrograde motion in the second. To be counted from the direction of increasing position angle in the apparent plane, to the direction of motion in true orbital plane.

   ☊ – *Position angle of ascending node for equinox 1900.* When data referring to the third coordinate is not available, the ascending node is always chosen provisionally between 0° and 180°.

   ω – Longitude of periastron. Angle in plane of true orbit counted always in the direction of motion from the node as given, to periastron.
3. that the computers always give both the Campbell and the Thiele-Innes elements, especially *C* and *H*.
4. that to avoid any ambiguity in the position angle of the ascending node, the observed radial velocities with their corresponding residuals should be published as it is required for the observed positions.

In regard to the above proposal, Finsen reports that he would prefer as follows:

i – Inclination is the angle between the line of sight and the positive pole of the true orbit, defined as the normal to the plane of the true orbit from the extremity of which the orbital motion is seen to be direct, i.e., anti-clockwise. It is therefore between 0° and 90° when the motion in the apparent orbit is direct, and between 90° and 180° when it is retrograde, and is always positive.

☊ – *Position angle of the ascending node* or, if it is not possible to identify the ascending node, the node with position angle less than 180°. If the node is the ascending node this is indicated by the unambiguous signs of *C*, *H*, *pL*, *pN*, but it is recommended that it should also be indicated by underlining, thus: $\underline{123^{\circ}\!.4}$ (when written by hand or typed), or by *italics—123°4—*(when printed). If it is not possible to distinguish between the ascending and decending nodes, *C*, *H*, *pL*, *pN* should be given double signs, thus ± or ∓, the upper signs referring to the node as given, i.e., < 180°, and the longitude of periastron associated with it.

If the conventional node (< 180°) is subsequently identified as the ascending node, leave ☊ and ω unchanged, and take the upper signs of *C*, *H*, *pL*, *pN*. If it is identified as the descending node, reverse the quadrants of both ☊ *and* ω, and take the lower signs of *C*, *H*, *pL*, *pN*.

Eggen proposed that the joint working group of this Commission and Commission 42, of which he is chairman, and Cester, Couteau, Eggen, Fredrick and Popper are members, meet at Prague and discuss a course of action.

K. Aa. STRAND
President of the Commission

REFERENCES

1. Holden, F. 1965, *Publ. Obs. Univ. Michigan*, **9**, no. 6.
Holden, F. 1966, *Publ. Obs. Univ. Michigan*, **9**, no. 7.
2. Couteau, P. 1964, *J. Observateurs*, Marseille, **47**, 229.
3. Mourao, R. R. de Freitos 1966, *Publ. Serv. Astr. Obs. Nac. Rio de Janeiro*, no. 20.
4. Kuročkin, N. E., Starikova, G. A. 1964, *Soobšč. gos. astr. Inst. P. K. Sternberga*, no. 130, 22.
Kuročkin, N. E., Starikova, G. A. 1965, *Soobšč. gos. astr. Inst. P. K. Sternberga*, no. 139, 11.
Kuročkin, N. E., Starikova, G. A. 1965, *Soobšč. gos. astr. Inst. P. K. Sternberga*, nos. 142–143, 41.
5. Van Biesbroeck, G. 1966, *Commun. Lunar Planet. Lab.*, 3, 51.
6. Worley, C. 1967, *Publ. U.S. nav. Obs.*, 2nd Ser., **18**, Part VI.
7. Walker, R. L. 1967, *Publ. U.S. nav. Obs.* 2nd Ser., **18**, Part IV.
8. Salukvadze, G. A. 1965, *Bjull. Abastumani Obs.* no. 32, 69.
9. Böttger, G. 1965, *Astr. Nachr.*, **288**, 203.
10. Deutsch, A. N. *et al.* 1965, *Izv. glav. astr. Obs. Pulkove*, **22**, Part 6 (no. 176), 124.
11. van de Kamp, P., Harrington, R. S. 1964, *Astr. J.*, **69**, 402.
12. Gray, D. F. 1964, *Astr. J.*, **69**, 406.
13. Gray, D. F. 1965, *Astr. J.*, **70**, 304.
14. van de Kamp, P., Moore, S. 1966, *Astr. J.*, **71**, 524.
15. Lippincott, S. L., Wyckoff, S. 1964, *Astr. J.*, **69**, 471.
16. Lippincott, S. L., Wagman, N. 1966, *Astr. J.*, **71**, 122.
17. van de Kamp, P. 1963, *Astr. J.*, **68**, 515.
18. Gasteyer, C., Eichhorn, H. 1964, *Astr. J.*, **69**, 762.
19. Haupt, H. 1964, *Mitt. Univ. Sternw. Wien*, **12**, no. 2, 18.
20. Hopmann, J. 1964, *Ann. Univ. Sternw. Wien*, **26**, Part 1.
21. Zeller, G. 1965, *Ann. Univ. Sternw. Wien*, **26**, Part 4.
22. Starikova, G. A. 1965, *Soobšč. gos. astr. Inst. P. K. Sternberga*, nos. 142–143.
Starikova, G. A. 1966, *Soobšč. gos. astr. Inst. P. K. Sternberga*, no. 148.
23. Arend, S. 1966, *Commun. Obs. r. Belgique*, Sér. B., no. 6.
24. Dommanget, J. 1966, *Commun. Obs. r. Belgique*, no. 239.
25. Dommanget, J., Nys, O. 1964, *Ann. Obs. r. Belgique*, 3rd Ser., **9**, Part 6.
26. Nys, O. 1963, *Bull. Astr. Obs. r. Belgique*, **5**, no. 3, 69.
Nys, O. 1964, *Bull. Astr. Obs. r. Belgique*, **5**, no. 5, 146.
27. Fesenko, B. I. 1963, *Vest. Leningrad. gos. Univ.*, no. 1, Part 1, 151.
Fesenko, B. I. 1965, *Vest. Leningrad. gos. Univ.*, no. 1, Part 1, 121.
28. Arend, S. 1966, *Commun. Obs. r. Belgique*, no. 243.
29. Kulikovsky, P. G. 1966, *Soobšč. gos. astr. Inst. P. K. Sternberga*, no. 146.
30. Batten, A. H. 1966, *J. R. astr. Soc. Can.*, **60**, 177.

31. Slettebak, A. 1963, *Astrophys. J.*, **138**, 118.
32. Worley, C. 1963, *Publ. U.S. nav. Obs.*, 2nd Ser., **18**, Part III.
33. Franz, O. G. 1966, *Lowell Obs. Bull.*, no. 134.
34. Kholopov, P. N., Efremov, Ju. I. 1965, *Astr. Circ.*, U.S.S.R., no. 326.
35. Piotrowski, S. L. 1964, *Bull. Acad. pol. Sci. Ser. Sci. math. astr. phys.*, **12**, 323 and 419.
Piotrowski, S. L. 1964, *Acta astr.*, **14**, 251.
36. Kruszewski, A. 1966, *Advances in Astronomy and Astrophysics*, Vol. IV, Ed. Z. Kopal, Academic Press, New York and London, p. 233.
37. Zahn, J. P. 1966, *Ann. Astrophys.*, **29**, 313, 489 and 565.

27. COMMISSION DES ETOILES VARIABLES

PRÉSIDENT: Dr G. H. Herbig, Lick Observatory, University of California, Santa Cruz, California, U.S.A.

VICE-PRÉSIDENT: Dr L. Detre, Director, Konkoly Observatory, Szabadsághegy, Budapest XII, Hungary.

COMITÉ D'ORGANISATION: H. C. Arp, M. W. Feast, P. N. Kholopov, P. Ledoux, Th. Walraven.

MEMBRES: Abbott, Ahnert, Alania, Ashbrook, Balázs-Detre, Banerji, Bateson, Bertaud, Bhatnagar, Canavaggia, Chavira, Chis, Daube-Kurzemniece, de Kock, Deutsch (A. J.), Efremov, Eggen, Eskioğlu, Fitch, Gaposchkin, Geyer, Haro, Harwood, Henize, Hertzsprung, Heyden, Hoffleit, Hoffmeister, Huffer, Huruhata, Ikaunieks, Istchenko, Javet, Jones, Kippenhahn, Kopal, Kopylov, Kordylewski, Kraft, Krogdahl, Kukarkin, Kumsishvili, Lacchini, McNamara, Mannino, Masani, Mavridis, Mayall (M. W.), Miller (W. J.), Nielsen, O'Connell, Odgers, Oosterhoff, Opolski, Oskanjan, Payne-Gaposchkin, Peltier, Piotrowski, Plaut, Plavec, Rakos, Rodgers, Romano, Rosino, Sawyer-Hogg, Schneller, Shapley (H.), Slavenas, Smith (Harlan J.), Strohmeier, Swope, Tempesti, Tsesevich, Vandekerkhove, Van Hoof, Wachmann, Walker (M. F.), Wenzel, Wesselink, Wright (F. W.), Xanthakis, Zinner, Zuckermann, Zwicky.

1. INTRODUCTION

It is obvious that there is no reason to give here an exhaustive listing of all the work in the field of variable stars over the past three years; that service is provided by the regular bibliographic publications such as the *Astronomischer Jahresbericht* or *Referatnyi Zhurnal* (*Astronomiya*). Some 500 references were examined during the preparation of this report, plus many citations of unpublished work furnished by members of the Commission, but I have made no attempt to cover everything. Semi-popular articles, short notes, lists of discoveries of new variables, and routine observations have in most cases not been mentioned unless I have seen something of particular interest in them. Since this must be a subjective judgment, not everyone will agree with my choice in every case. But I hope that no major contributions have been slighted.

This Report is in three parts: the first deals with variable stars in general, and was prepared by the President. Appendix I covers the Spectra of Variable Stars and was written by Dr M. W. Feast, Chairman of that Committee of Commission 27. Appendix II encompasses Variable Stars in Clusters, and was prepared by Dr Helen Sawyer Hogg, Chairman of that Committee. There is inevitably some small overlap between the three parts, and probably to some extent with Reports of other Commissions, but this seems inevitable if one is to present the various topics in proper perspective.

2. ACTIVITIES OF THE COMMISSION

Under the editorship of L. Detre, the very useful *Information Bulletin on Variable Stars* of Commission 27 has continued to appear; through October 1966, some 164 numbers have been distributed from the Konkoly Observatory, generally with a publication time of 2 to 3 weeks. The support of the *Bulletins* has been guaranteed by the Hungarian Academy of Sciences for a further 3-year period. A similar service for longer contributions is provided by the *Mitteilungen über veränderliche Sterne*, edited by W. Wenzel at Sonneberg. The time between receipt of a manuscript and its publication is about two months. Workers in the field of variable stars are invited to make use of either of these series, as appropriate.

The Commission has recently established, through the cooperation of the Royal Astronomical Society, a depository of unpublished photoelectric observations of variable stars. It is located in the R.A.S. Library, Burlington House, London, W.1, England and meets a long-felt need for an archive in which large masses of data, too expensive to publish in detail, can be deposited. Authors are now free to publish only condensations of their observations, accompanied by a statement to the effect that the complete data are on file in the Commission 27 depository, from which a copy can be obtained at cost plus a small charge for postage and handling. This offers an advantage over the system in which the data are supposedly kept on file at the originating observatory, but cannot be found years later. The administrative details of the service will undoubtedly be modified with time and experience so for the present, interested observers are referred to the statement on procedures which appears in *IAU Information Bulletin* No. 18. Some idea of the type of material that is especially appropriate for deposit can be gained by referring to the first three lists of photoelectric data that were placed on file: 3400 observations in three colors of the peculiar rapid variable and short-period eclipsing binary V Sge (**1**); 2400 ultraviolet observations of the rapidly erratic star AE Aqr (**2**), and 990 observations in three colors of the somewhat similar object BD + 14° 341 (**3, 4**).

Since the Berkeley General Assembly in 1961, a working group of the Commission under the Chairmanship of A. J. Wesselink has been studying the problem of nomenclature and identification of variable stars in the Magellanic Clouds, both for those variables already known and those yet to be discovered. It is hoped that the recommendations of this working group can be considered by the full Commission in the near future.

3. MEETINGS AND SYMPOSIA

A large number of the papers presented at the Colloque International sur les Novae, Novoïdes et Supernovae held at Haute-Provence in September 1963 have now appeared in *Ann. Astrophys.*, **27**, 1964. The proceedings of a symposium on Clusters and Stellar Evolution held at Herstmonceux in August 1963, and containing much material on variable stars, have also been published (**5**). At the Hamburg General Assembly in August 1964, Commissions 27, 34, and 40 sponsored a Joint Discussion on 'The Orion Nebula', part of which dealt with problems of variable stars in the Nebula (**6**). Some problems of variable stars in the halo and of hot subluminous variables were discussed at the First Conference on Faint Blue Stars, held in Strasbourg in August 1964 (**7**). In Moscow in November 1964, the Variable Star Commission of the Astronomical Council (Academy of Sciences of U.S.S.R.) held a symposium on 'Variable Stars and Stellar Evolution' (**8**), and another in Odessa in September 1965, on observational problems of variable stars in the context of stellar evolution. A conference in Sverdlovsk in June-July 1966 dealt with variables in clusters and associations, and with eclipsing variables. An IAU Colloquium on 'Variable Stars in the Hertzsprung-Russell Diagram' was held at Bamberg in August 1965 (**9**), the third in the triennial series of Bamberg meetings on variable star topics. The fourth, also under the sponsorship of Commission 27, will convene in Budapest in 1968. The subject of magnetic variables, of interest to this Commission, was discussed at a symposium held in Washington in November 1965, sponsored by the American Astronomical Society and the U.S. National Aeronautics and Space Administration. The proceedings will appear as a special volume (**10**). *IAU Symposium* No. 30 on Radial Velocities, held in Toronto in June 1966, included one session on radial velocities of variable stars, organized on behalf of Commission 27 by J. D. Fernie. The symposium volume is being edited by J. F. Heard and A. H. Batten.

4. CATALOGUES AND BOOKS

The variable star bureau at the Astronomical Council, Academy of Sciences of the U.S.S.R., and Sternberg Institute (Moscow) continued with its indispensable work of systematizing variable star data. Under the authorship of B. V. Kukarkin, P. Kholopov, Yu. N. Efremov and N. Kurochkin, the *Second Catalogue of Suspected Variables* appeared in 1966, containing data on 3907 objects. Compilation of data on the *Second Supplement* to the *General Catalogue of the Variable Stars* (2nd ed., 1958) was completed in 1966 by Kukarkin, Kholopov, M. S. Frolov, *et al.* An index of the increasing magnitude of this task is provided by the number of annual references consulted: 5540 in 1962, 6060 in 1963, 7500 in 1964, 10 530 in 1965. In both instances, English translations of the introduction and remarks will be published with financial assistance by the IAU. All investigators of variable stars are indebted to the Moscow group for its thorough and painstaking work, and to the Astronomical Council for support of the program.

A very useful Atlas of finding charts for RR Lyrae and other types of variable star has been published by Tsesevich and Kazanasmas (**11**). Tsesevich has also written a monograph on RR Lyrae variables (**12**). Much detailed information on variable stars has been presented by Beyer in the new Landolt-Börnstein (**13**). A catalog of variables in double and multiple systems has been compiled by Perova (**14**). A festschrift dedicated to Professor C. Hoffmeister contains many papers on variable stars (**15**). An atlas of the Large Magellanic Cloud has been published by Hodge and Wright (**16**).

The books *Pulsation Theory of Variable Stars* by S. Rosseland, and *The Galactic Novae* by C. Payne-Gaposchkin have been reissued in paperback form by Dover Press, New York.

5. DEVELOPMENTS AND TRENDS IN VARIABLE STAR RESEARCH

Since the beginnings of the subject, the study of intrinsic variable stars has been essentially descriptive. Since there was no real understanding of the causes of variability, there was no real theory. But over the past two decades a new point of view has gradually emerged, namely that the key to intrinsic variability lies in the properties of stellar structure which come about from the predictable evolutionary movement of a star across the Hertzsprung-Russell Diagram. A recent exposition of this point of view is given by Kippenhahn (**9**, p. 7). The concept certainly provides a major simplification in the almost zoological diversity which existed in the classification of variable stars before about 1950. The idea that for example, an individual star in the course of its lifetime may successively become a variable of more than one type, such as at the beginning a T Tau star and much later a long-period variable, is a consequence that would not have been anticipated otherwise. This concept results in the now-familiar HR Diagrams in which strips and domains of specific instabilities are outlined, with the implicit assumption that every star falling therein must be a variable of that particular class. Whether this latter assumption is indeed true is not proven, but one will not be too surprised if it can be violated in special cases, since other parameters than are displayed in the conventional HR Diagram (initial chemical composition, previous history, degree of rotation) may be relevant.

The most convincing demonstration of this point of view is probably provided by recent investigations of the properties of pulsating stars. With large modern computers, the coupled partial differential equations which describe the time-dependent structure and motion of a spherically-symmetric stellar model can be integrated directly. The point of view is that the pulsation of an RR Lyrae or cepheid variable arises not from variation in the rate of energy generation in the core, but by a variable throttling action of layers near the surface upon a constant energy flux. Zhevakin (who reviews the subject as well as his own work in **17**) first pointed out the importance of the ionized helium convection zone for this process. Subsequent investigations by Kippenhahn, Baker, Cox, Iben, Christy, and their associates have extended the subject, which is reviewed thoroughly in the Report of Commission 35. The Munich group has considered simultaneously the evolution of stars of masses 5 to 9 $\odot$ across the upper part of the HR Diagram and the accompanying changes in their pulsation properties (**18**). This makes possible in principle a comparison with the observed behavior of individual cepheids.

Christy (**19**; **9**, p. 77) has carried out detailed calculations of the pulsations of models having a variety of effective temperatures, masses and chemical compositions, with a somewhat more sophisticated consideration of non-linear effects. He has applied this approach to cepheids of intermediate period (**20**), to RR Lyrae stars (**21**), and to W Vir (**22**).

The details of such work do not fall in the domain of Commission 27, but only direct observations of variable stars can provide a basis for improving the theory and insuring that the calculations refer to real stars. It is not always clear from the published papers whether the observational information is adequate for a proper bridge between theory and observation. Dr Christy has very kindly written for inclusion here a brief statement of the present theoretical situation, pointing out especially those quantities which observations must provide.

Theory and Observation of Variable Stars

R. F. Christy

In the past few years there has been very considerable progress in our ability to calculate the behavior of a pulsationally unstable stellar envelope. (See report of Commission 35.) This progress brings the hope that we will be able to make detailed comparison of calculated models with observed variables and thereby determine additional parameters of these stars by observational means.

At present, calculations are being made of various types of variables in a generalized cepheid instability strip. We may hope in the near future to see this progress extended to the red variables. Separate attacks are being made on the β CMa variables. Irregular variables are excluded from this discussion.

Undoubtedly the most important contribution from observation on variable stars is their classification by physical groupings. This work of classification has been going on for a long time and much of our progress is attributable to it; but much more needs to be done. In the case of the variables of spectral type later than A, rotation and magnetic fields do not seem to be essential, so the significant parameters which affect the envelope are the composition (hydrogen, X; helium, Y; and metals, Z), the luminosity, the mass, and the effective temperature. Given these parameters, the structure and stability of the envelope can be calculated. In the early-type variables, rotation or magnetic fields may well be important and we must add these quantities to the list of physical quantities we wish to explore.

There are many different observational classes of variables in the region of $T_{eff} < 10^4$ °K or of spectral type later than A. In the generalized cepheid instability strip there are, in addition to the classical cepheids, the W Vir variables, the RR Lyr variables, the δ Sct variables, the dwarf cepheids, and perhaps also the RV Tau variables. The physical bases for the distinctions between these classes is not always clear and, indeed, the systematics and location of some of these classes is still obscure. It is probable that the log g vs. log T_{eff} diagram is the most suitable one to demonstrate the relationship of these various classes and new information to order them in this diagram is needed.

In addition to the generalized cepheid instability strip, there are several classes of red variables. Among those, the physical distinctions are even more obscure and the delineation of an instability region in some suitable parameter space has hardly begun. There is much yet to be done in defining the various variable types by physical parameters. It is again likely that the log g vs. log T_{eff} diagram would be the most useful one to display these instabilities.

Perhaps the greatest need, from the viewpoint of theory, is for more information on the mean T_{eff} of the various variables of type later than A. The observational problems associated with determining T_{eff} from possible photometric data are different for different categories of variable. The corrections for reddening are very difficult and the blanketing corrections are apt to be important. Nevertheless, the envelope structure is very sensitive to T_{eff} because of the rapid variation of hydrogen opacity and therefore the instability problem is also sensitive to T_{eff}. It has been found that the high T_{eff} boundary of the cepheid instability strip is quite dependent on the helium abundance in the envelope. The accurate location of groups of cepheid-type variables in terms of T_{eff} will therefore determine their helium content. In the case of the various red variables, T_{eff} is very inadequately determined by $B-V$ and a more systematic use of red or infrared colors and a better knowledge of the red and infrared spectrum will be necessary to classify them in T_{eff}.

The accuracy with which T_{eff} and log g are needed is suggested by the approximate relation $\theta_{eff} = 1{\cdot}09 - 0{\cdot}11 \log g - 0{\cdot}20\, Y$ which governs the dependence of the value of T_{eff} at the high temperature boundary of cepheid instability on log g and Y, the helium abundance by mass. The table below shows the content of this expression:

log g	Y	θ_{eff}	T_{eff}
3·0	0·30	0·70	7200°
	0·0	0·76	6640°
2·0	0·30	0·81	6230°
	0·0	0·87	5800°
1·0	0·30	0·92	5480°
	0·0	0·98	5150°

In addition to T_{eff} the envelope structure is primarily determined by the surface gravity g. In some cases, measures of g can be made from the Balmer jump. In other cases, it can be determined from spectral lines. It is likely that the principal distinction between different classes of variables in the generalized cepheid instability region may lie in their differing g (which, of course, reflects their mass to luminosity ratio). The determination of g values, coupled with T_{eff}, may show these differences. Actually the period of the fundamental mode is primarily sensitive to g and only secondarily to the mass. It follows that the period can verify the determination of g by other means or, alternatively, may indicate that the pulsation is not in the fundamental mode.

When it is possible to locate the different variable classes in T_{eff} and g, it may become possible to distinguish additional effects associated with differing compositions or with rotation.

For the early-type variables (β CMa stars and magnetic or spectrum variables), the principal observational problems appear to lie in the acquisition of the basic data on variability rather than in classification of the variables.

In addition to information, discussed above, which could help in determining the basic envelope structure of various instable groups, there is, in many cases, additional detailed information on individual variables that would be of help in evaluating the physical nature of the variability.

In the β CMa variables, it is likely that the stellar surface displays a quite complicated motion that is responsible for the radial velocity and line width variations. It is almost certain that there must be associated variations in the shapes of the line profiles. These line profile shape variations will be, in turn, closely connected with the nature of the surface motions. The surface motions will in turn reflect the type of instability responsible for the variability. It is hoped that some detailed measures of line shape variations can be used to distinguish between different possible explanations of these variables.

In the case of cepheid variables, it is now becoming possible to calculate detailed surface dynamics. Some of these calculations show interesting sub-structure in the light curve and in the velocity curve which is reminiscent of observed bumps in these curves. It is likely that systematic study of this sub-structure (small variations which appear regularly in the light or velocity curves) will soon be interpretable in terms of additional detailed knowledge of the structure of the stellar envelope.

Another concept that has gained force in recent years is that some types of intrinsic variability may be restricted to binary systems, through the tidal effect of one star upon the other. One of the first suggestions of this kind was Meyer's (**23**) in the case of β CMa. It is now clear that many if not all U Gem variables and novae are close binaries. The symbiotic variables also are double, or at least exhibit two dissimilar spectra, which may come from the two nuclei within a common envelope, but how explosive activity is generated in such a system is obscure. Recently the binary hypothesis has been revived by Fitch to explain the long-period modulation in the β CMa variable σ Sco, in the δ Sct stars, and possibly in the RR Lyr stars having secondary periods. The possibility that a tidal instability can be stimulated in a magnetic variable by a close companion has been considered by Ledoux (**24**). The case for tidally-excited pulsation would be strengthened if such variability could be found in spectroscopic binaries of high eccentricity, such as σ Psc and 26 Aql. A possible development of this kind may be Preston's discovery that the δ Sct star δ Del is a double-line binary with a very eccentric orbit (see Sec. 14).

Consideration has also been given to the role played by axial rotation in stellar variability. The apparent preference of β CMa and δ Sct-type variability for slowly-rotating stars is well known, as is the fact that RR Lyr stars always have narrow lines. The relationship between rotation and variability in the Hertzsprung gap has been discussed by Preston (**9**, p. 155) and by Kraft (**25**) from the observational point of view. On theoretical grounds, however, Kippenhahn (**9**, p. 162) doubts that, in that part of the HR Diagram, rotation can affect pulsation to any significant degree.

We should mention a few surprises: quite unexpected observational discoveries made during the past few years in the field of variable stars. Certainly surprising, although the extent of their significance is not yet clear, are the discoveries by Lovell of radio emission from flare stars, by Mendoza of infrared excesses in Orion population stars, by Serkowski of strong polarization in some red variables. Each of these developments is discussed later in this Report.

6. GENERAL PROBLEMS OF STELLAR VARIABILITY

A perennial question is this: if one were able to determine apparent magnitudes to a much higher degree of accuracy, would there then be found a large number of new, intrinsic variable stars? A definite answer to this question may require observations from outside the Earth's atmosphere. Opinions of photoelectric observers have themselves varied. According to Nikonov (**26**), carefully selected standard stars are likely to be constant to at least 0·02 magnitude. Sharov (**27**) has pointed out how small spurious variability can arise as the result of inadequate allowance for atmospheric transparency. Jackisch (**28**) at Sonneberg studied photoelectrically about 156 stars brighter than $m = 4{\cdot}5$, and found that about 22% showed some suspicion of variability, in amount 0·03 to 0·10 magnitude. Some stars now known to be small-range variables were missed (β Cas) but at least one star now known to be a short-period variable was found (κ Boo). Öpik (**29**) has however criticized this work on the basis that, due to the size of the observational error (0·025 magnitude for one night), it is impossible to be certain of the reality of variations of less than about 0·10 magnitude. He recommended that the program be repeated in a better photometric climate. In 1965, Jackisch did make further observations at Kottamia (U.A.R.). Of 210 A-type stars in his program (**30**), 20% proved to be variable with ranges of 0·03–0·06 magnitude, with 2% showing ranges up to 0·15 magnitude.

It is well known that M-type giants are very susceptible to small light variations; a statistical study by Blanco and Plaut (**31**) of old observations by Stebbins and Huffer showed that the percentage of variables increases from 20% at K5 to 100% at M6. The majority have amplitudes less than 0·5 magnitude. A similar study of variability among M giants in Groningen-Palomar Field 2 is being made by Miss Houk and Plaut, and by Miss Houk in VSF 193 in Sagittarius. Small light and radial-velocity variations are common among high-luminosity stars. This was shown also in Jackisch's original investigation, and a new program to observe supergiants in *UBV* has been inaugurated at Sonneberg.

The linear polarization of the light from pulsating variables in general has been discussed by Shakhovskoy (**32**); polarization in magnetic variables and in long-period variables is described in Sections 13 and 11. Since the amount and direction of polarization varies in some stars, Shakhovskoy considers this component to be of stellar origin. The best cases are the RV Tau stars AC Her and R Sct, the M supergiant μ Cep, and the Mira variable RT Cyg. The mechanism suggested for the production of the polarization is Rayleigh scattering by neutral atoms in atmospheric inhomogeneities, and the reorganization of these inhomogeneities by the shock waves at successive maxima is perhaps responsible for the variations. The polarization of μ Cep and other M supergiants has been investigated also by Grigorian (**33**) who first found the effect, and by Vardanian (**34**), who also gives reasons for believing the polarization found in T Tau and RY Tau is intrinsic (**35**).

Theoretical questions of stability, although of course fundamental to the problem of stellar variability, are technically not within the province of Commission 27. Attention is only called here to a review of this subject (**36**) and remarks on the effects of neutrino loss upon stability (**37**), by Ledoux.

7. PHOTOGRAPHIC SURVEYS, SPECIAL FIELDS

A major study by Richter has been completed at Sonneberg on the distribution of variable stars of all types throughout the Galaxy (**38**). Color estimates from Palomar Sky Survey prints make it possible to eliminate erroneously classified stars, and the material collected at Sonneberg in course of the Felderplan is used to allow for incompleteness. The star density and logarithmic density gradients in the solar vicinity for various types of object are derived, as are the total number of each type in the Galaxy, their Population assignments, and the duration of that phase of evolution. Pending the appearance of these results, one of the most authoritative reviews of the distribution of (some kinds of) variable stars are those of Plaut (**39**, **40**). The Sonneberg surveys for new variable stars are being pressed with the new 400/1600 mm astrograph. About 50 areas are under investigation, the regions added most recently being at higher galactic latitudes because the areas at low latitudes are beginning to show signs of exhaustion (**9**, p. 299). The total number of new variables discovered at Sonneberg had reached 10 100 by mid-October 1966.

The very large Palomar-Groningen search for faint variables in four selected regions is continuing under Plaut's direction. The blinking has been finished, and estimates of brightness completed for Fields 1, 2, 4. The results for Field 1 have been published (**41**; see also **9**, p. 297). De Kort is collaborating on Field 4. The Lick survey of variables in high latitudes is discussed in Section 9.

Among the bright stars, a number of short lists of small-amplitude southern variables have been published by the Cape observers Cousins, Lake, Warren and Corben in *Mon. Not. R. astr. Soc. So. Africa.* More recently Cousins has published a list of variables found in his Fabry photometric program at the Cape (**42**). A report by Strohmeier (**9**, p. 302) on the results of the first two years' work at the Bamberg South African Station lists about 50 bright new southern variables. Further discoveries are announced from time to time in the *Inf. Bull. var. Stars.* The photographic program of Weber (at Mainterne) deserves commendation, both for the discovery of new variables and for attention to stars of special interest. Romano (at Asiago) has collected plates of ten high- and five low-latitude fields for the past five years, for the study of variables to $m_{pg} \approx 17{\cdot}5$.

The Leiden work on variables in the southern Milky Way is represented by the work of Kooreman, Charlier, van Houten *et al.* in *Ann. Sterrew. Leiden*, **22**, 1965–66. Oosterhoff and his associates are working on variables in the two Sagittarius 'windows' (**43**). Hoffleit's significant work on faint variables in Sagittarius continues (**44**). This region is important since on account of the richness of the field, one finds here good examples of some types of rare and interesting variables. Wachmann's results on variables in SA 98 and the Cygnus cloud have been published (**45**), and Miller's work on faint variables in Cygnus has also been very fruitful (**46**). Miss Harwood and Mrs Swain continue work on the variables in the Scutum cloud. Investigation now is complete for over half of the new variables found at Nantucket (**47**); the majority are long-period stars.

Outside our Galaxy, the very comprehensive studies by the Gaposchkins and by Hodge and Wright of variable stars in the Magellanic Clouds is described in Section 10. Baade and Swope have published (**48**) their results on variables in three fields in Messier 31, and Tammann and Sandage (**49**) have reported on their study of variables in NGC 2403.

8. VISUAL OBSERVATIONS

The visual programs of the various amateur variable star groups have continued, with a heavy emphasis on observations of long-period variables. The Variable Star Section, R.A.S. of New Zealand, directed by Bateson, is following some 400 southern variables, including all types suitable for visual work. The observations are published in detail in the *Circular* of the Section, as well as occasional discussions of special objects. The Variable Star Section, British Astronomical Association, observes both eruptive and long-period variables; reports appear from time to time in the *J.B.A.A.* The Variable Star Section of the Danish Astronomical Society has continued its activity under the direction of O. Klinting. Its reports appear in *Nord. astr. Tidskr.* The summary reports of the Berliner Arbeitsgemeinschaft für veränderliche Sterne are published in the *Astr. Nachr.*, and give mainly times of maxima and minima of long-period, semiregular, and irregular variables. Mrs Mayall, Director of The American Association of Variable Star Observers, publishes a monthly article in *J. R. astr. Soc. Canada.* Condensed lists of observations are issued as *Quarterly Reports*, but the original observations are available on microfilm by request. The Swedish Astronomical Society, Variable Star Section, publishes its reports in *Gothenburg astr. Not.*

Several members of this Commission are very active visual observers. Beyer publishes much of his work in *Astr. Nachr.* including some observations through blue and red filters. Peltier participates in the AAVSO program, and during this Report period, was responsible for the detection of the flare-up of Nova (GK) Per 1901 in 1966. De Kock (at the Cape) is following some 175 stars, including irregular, semiregular, and Mira types, as well as UV Cet and AE Aqr.

9. RR LYRAE VARIABLES*

The subject of the RR Lyr stars has been reviewed in depth by Preston (**50**) and by Woolley (**51**). The two articles overlap only partially; one emphasizes the astrophysical approach wherein major attention is given to individual stars, while the other represents the statistical and kinematical point of view. A new book on the RR Lyr stars has been written by Tsesevich (**52**).

Systematic searches for RR Lyr stars in high galactic latitudes, to determine the dimensions of the Galaxy at right angles to the plane, are in progress at several institutions. At Sonneberg, Hoffmeister has found many faint RR Lyr stars (e.g. **53**) in the north galactic pole and in the anticenter. The observational work on the Sonneberg variables near the pole is nearing completion (by Löchel, Meinunger, Wenzel, Ziegler, and the Hartha workers). A catalog of these

*Stars with periods shorter than $0\overset{d}{.}25$, except those that are clearly Bailey type *c*'s are discussed under δ Sct stars and dwarf cepheids in Section 14.

stars will appear in *Mitt. veränd. Sterne*, 4. The very extensive Lick program directed by Kinman continues. The methods have been described by Kinman *et al.* (54), and results for a low-latitude field (55) and for three fields near the north galactic pole (56) have been published. An analysis of the latter results in terms of density gradient and force law was carried out. The faintest RR Lyr stars found at Lick near the pole have m_{pg} (mean) = 18·0 approximately. Still fainter variables in the same area, at magnitudes 19 and 20, have been reported by Luyten (57); according to Luyten and Erickson (58) three of these have colors appropriate to RR Lyr stars, but no periods were obtained. Some 51 faint RR Lyr stars were found by Plaut (59) in Field 1 of the Palomar-Groningen survey at $l^{II} = 0°$, $b^{II} = +29°$, and discussed by him in a determination of the density gradient in that direction. Many other new RR Lyr variables have been discovered in the general surveys for variables that are described in Section 7. Mention should be made of the study by Arp (60) of the properties and statistics of RR Lyr stars in the direction of the galactic bulge.

An extensive series of five-color photoelectric observations of pulsating stars (mostly RR Lyr variables) made by Ponsen (61) has been discussed by Oosterhoff and Walraven (62) and by Oosterhoff (63). They have considered particularly the cyclic motion of RR Lyr stars in several kinds of two-color diagram. Further five-color photometry of southern RR Lyr stars by Van Houten and Van Genderen is in process of reduction by Van Herk. A large program of multicolor observations of RR Lyr stars selected from Van Herk's list (64) is being carried out by Johnson, Wisniewski and Fitch. The first observational phase is completed, consisting of *UBV* measures at maximum and minimum for over 100 stars. The observations will appear shortly, and the analysis later. Observations of RR Lyr stars have been published by many observers. Space does not permit them all to be cited here; we mention here only the *UBV* studies of six stars (three in only *BV*) by Jones (65), of five southern variables by Warren (66), and of three southern stars by Breger (67). Miss Capelli (Padova) is observing a number of RR Lyr stars that seem to have ranges either abnormally small or large for their periods. This is part of a thesis study on the statistical properties of the RR Lyr stars. Programs of *UBV* work on RR Lyr variables are underway at Budapest, during which Kanyó (68) has found the very short secondary period of 22ᵈ75 for Z CVn, and Tremko believes that a secondary period of 28ᵈ8 is present (69) in RU Psc. At Edinburgh, an attempt is being made by Smyth to observe a number of brighter variables for which such data is lacking. SW Dra, KN Per, RV UMa have been observed, and W CVn, ST CVn, XZ Dra, CZ Lac, TU Per are on the program. Conventional photographic work has not been entirely neglected: at Toruń, a program for determination of photographic and photovisual light curves of RR Lyr stars is being carried out (70). Robinson (71) has studied the changing period of SZ Hya on Harvard patrol plates. With regard to such effects, Tsesevich pointed out that period variation appears to be a general property of the RR Lyr stars of Population II, while the much less common RR Lyr stars of Population I have constant periods. The same general phenomenon is exhibited in an investigation by Balázs-Detre and Detre (72). They have studied the possibility that the deviations of RR Lyr stars (as well as other types) from a linear ephemeris might be explained by the accumulation of small random errors. They show that this is indeed possible, the important parameter being the standard deviation of the random period noise, and tabulate mean values of this quantity for a number of types of variable star.

The *UBV* colors of RR Lyr stars were studied by Sturch (73). He found that in variables of Bailey types *a* and *b*, $B-V$ and $U-B$ are essentially constant between phases 0·5 and 0·8 (counted from maximum light). The observed $U-B$ is used to correct $B-V$ for line blanketing, which then gives (following allowance for its dependence on period) the interstellar reddening. Frolov (74) had earlier investigated the dependence of $B-V$, after a correction for line strength, upon period.

Preston *et al.* have demonstrated the advantages that come from a combination of time-resolved high-dispersion spectroscopy and simultaneous *UBV* photometry of both singly

(**75**) and doubly-periodic (**76**) RR Lyr stars. The amount of detail that is missed by conventional observations is striking. Their results are discussed in Appendix I.

Fernie determined an empirical period-radius relation for intrinsic variables in general (**77**), and later included allowance for a mass dependence (**78**). It is essentially a substitute for the usual period-density law, and can be expressed as a *P*, *L*, color relationship for several kinds of pulsating stars. Since individual stars follow this relationship not only in the mean but even as they vary, Fernie has used this fact to develop a short method of determining distances and reddenings of RR Lyr stars (**79**) from 'instantaneous' observations. With a similar aim of lessening the amount of observational data required for statistical studies, Breger (**80**) has studied the question whether there is a definite phase in the cycles of all RR Lyr stars at which they have the mean luminosity, or the mean velocity. The answer appears affirmative in both cases, as long as one avoids stars having variable light curves. Demarque and Percy (**81**) have investigated the possibility of explaining Fernie's empirical relationships by studying the adiabatic pulsations of giant star models having a large mass range. Gough *et al.* (**82**) show that the *R*-dependence found by Fernie can be understood in terms of the travel-time of a wave through the atmosphere, but not the mass dependence. The *P*, *L* relation for RR Lyr stars was re-investigated by Frolov (**8**), who found the expression $M_v = -4{\cdot}41 \log P - 0{\cdot}66$. Van den Bergh found (**83**) that the scatter of the *P*, *L* relation in the Magellanic Clouds can be improved if $\langle V \rangle$ is corrected by $-3 \langle B-V \rangle$, which is possible since the constant-*P* and reddening lines are so nearly parallel in a color-magnitude diagram. Wesselink and Shuttleworth (**84**) in their study of 45 variables in the SMC found five variables of periods less than 1^d with $\langle m_{pg} \rangle \approx 18{\cdot}0$. It was concluded that these are members of the cloud with $M_{pg} \approx -0{\cdot}5$. The non-linear pulsation calculations for RR Lyr stars by Christy (**21**) now provide a consistent theoretical foundation with which to compare observational data, although Christy's models are subject to possible future refinement. Christy has already commented very briefly (**9**, p. 111) on the possibility of understanding Fernie's empirical *P*, *R*, *M* relationship in terms of these calculations.

A discussion of the absolute magnitudes of the RR Lyr stars was made by Woolley *et al.* (**85**) on the basis of a statistical comparison of proper motions with radial velocities. A similar discussion was carried out by Van Herk (**64**) who had, however, new proper motions for a large number of stars, from a Mt Wilson program begun by Van Maanen and from a comparison of new Leiden plates with Carte du Ciel positions. The results of the two investigations are therefore independent to some extent. It would be of interest to see if improved allowance for reddening, as is now possible if *UBV* data are available, would sharpen such results. Van Herk is now engaged in the determination of the proper motions of a large number of additional RR Lyr stars, using the Carte du Ciel catalogues or original plates as far as possible. In view of the impending availability of such data, it is very important that radial velocities for many more RR Lyr stars be measured. As Woolley *et al.* point out (**85**), the requirements on velocity precision are not high.

10. CEPHEIDS OF BOTH POPULATIONS

Recent theoretical calculations of stellar pulsation have been reviewed in Section 5, and will be referred to only incidentally here. Mention should be made of the general discussion of evolutionary tracks in the cepheid region by Kippenhahn (**86**) in his review of stellar evolution and variability. The detailed discussion has been published separately for masses of 5·0 and 7·0 ⊙ (**87, 88**) with the results for 9·0 ⊙ yet to appear. Iben has performed similar calculations for masses between 3 and 15 ⊙ (**89**). The correlation of these tracks and the speed of evolution along them with the observed properties of cepheid variables has been investigated by Miss Hofmeister (**90**), Kraft (**25**), Kopylov (**91**), Efremov and Kopylov (**121**), and Iben. Expectation and observation seem to be in acceptable agreement. Reference is made to the use of pulsation theory by Masani *et al.* (**92**) to estimate the mass of a cepheid variable; their procedure is to vary the mass until the fundamental theoretical period matches the observed one.

A very useful catalog of some 10 000 *UBV* observations of over 300 cepheids has been published by Mitchell, Iriarte, Steinmetz, and Johnson (**93**). Approximately 2000 of these observations are new, made by Mitchell *et al.*; the others are published data from a variety of sources, all re-reduced to the *UBV* system. Walraven, Tinbergen, and Walraven (**94**) have observed 24 southern cepheids with periods between 5^d and 79^d in the five-color *UBVWL* system. Breckenridge and Kron (**95**) have published observations in the Stebbins-Whitford six-color system of four southern cepheids, and Nielsen at Aarhus is observing ten bright cepheids photoelectrically in *BV*. Gouseva and Tsarevsky are also carrying out narrow-band photometry of a number of cepheids.

Cepheids that behave in some anomalous manner received attention. Fernie (**96**) pointed out that the LMC cepheid HV 953, of period 48^d, should change in $B-V$ at a rate of $+0{\cdot}26$ magnitude per century if the period change of one day per century reported by Oosterhoff were maintained. However, a subsequent study of HV 953 by Janes (**97**) on Harvard plates shows that the period fluctuates erratically rather than progressively. Fernie, Demers, and Marlborough (**98**) found that the period of the galactic 45^d cepheid SV Vul behaves in the same fashion. Oosterhoff (**99**) investigated the cepheid AP Vel, and found it to be a case like TU Cas and U TrA: a fundamental period of 2 to 3^d beating against a secondary period of 0·70 that length. In the case of AP Vel, the fundamental is $3^d_{\cdot}13$, and the period of the beating $7^d_{\cdot}4$. A photoelectric study of TU Cas was carried out by Vasilianovskaya (**100**). Kapko (**101**) and Huth (**102**) have studied the remarkable 10^d cepheid AP Her, whose period is subject to such large changes. Over the interval of observation, it has varied between $10^d_{\cdot}35$ and $10^d_{\cdot}42$. The period variations of PP Aql, AP Her, and MZ Cyg have been studied by Tsesevich. In this connection, attention is drawn again to the work of Balázs-Detre and Detre (**72** and Section 9) on the representation of such behavior as the result of the accumulation of a series of small random events. The problem has been discussed by Demers (**103**) in the case of three specific cepheids, among them RU Cam, which underwent a remarkable decrease in amplitude in 1965–66. This latter fact was discovered by Demers and Fernie (**104**), and further contributions were made by Wamsteker (**105**), Smak (**106**), and Nikolov (**107**). All the available photoelectric material was discussed by Detre (**108**), who showed that the range seems to vary cyclically with a period of about five years. On this basis, Detre suggests that an increase in the range is to be expected soon; in August 1966 the range was only $\Delta V = 0{\cdot}08$ magnitude as compared to about 1·6 magnitude two years before. RU Cam is being investigated in detail by Fernie.

Fernie has examined the properties of the sixth-magnitude B-type companion of δ Cep (**109**). Largely on the basis of its Hβ strength, he finds it to have an absolute magnitude which is 1·6 magnitude too faint for agreement with the presently-accepted M_v of the cepheid. Fernie therefore believes that the two do not constitute a physical system. In the case of the F-type visual companion of α UMi, Fernie (**110**) assumes the system to be physical and from the further assumption that the companion lies on the zero-age main sequence, finds that $\langle M_v \rangle = -3{\cdot}2$ for the cepheid, which appears reasonable. Fernie uses Miss Roemer's (**111**) data on the long-period spectroscopic orbit to estimate that the invisible spectroscopic companion is probably a main-sequence star of type near F0. Fernie also remarks on the curious fact that although α UMi lies near the center of the cepheid instability strip, it has the smallest known amplitude among cepheids.

As to cepheids in galactic clusters, Landolt has studied S Nor in NGC 6087 and U Sgr in Messier 25 (**112**). Preston (**113**) has obtained the spectral types of many stars in the field of CV Mon, and finds that there are difficulties in assigning the cepheid to cluster membership, if indeed the cluster itself is real. Malik (**114**) has observed the three-day cepheid BY Cas in *UBV*, but the question of its membership in NGC 663 remains unanswered. BQ Ser lies near the cluster IC 4756, but is not a member according to Wenzel, who has finished a discussion of the Sonneberg *UBV* observations. It is a cepheid with variable light curve and/or period, as recognized first by Meinunger (**115**). Smak (**116**) has obtained limited photometric material in *BV* on the cepheids CE*a* and CE*b* Cas, in NGC 7790. Comparison with CF Cas in the same cluster shows that the data on the three variables are internally consistent, and also are in accord with the mean *P*, *L*, color relations for cepheids at large. Efremov and Kholopov (**117**) have also studied CE*a*, *b* Cas, and drew similar conclusions. Westerlund (**118**) has published his investigation of the 41^d cepheid RS Pup and the OB association of which it is apparently a member. This cepheid is further remarkable in the fact that it illuminates a small reflection nebula. Efremov (**119**) has argued in favor of the membership of cepheids in the halos of some clusters and associations, including *h*, χ Per. He also found a dependence of the period of a cepheid upon the age of the cluster to which it belongs (**8, 120**). This is in accord with the production of cepheids from B stars, and provides an estimate of the duration of the cepheid stage. Efremov and Kopylov (**121**) revised the data on the 19 cepheids which possibly are members of 11 clusters and associations. They found that Kopylov's (**122**) calibration of the zero-age main sequence requires that the cepheids be 0·5 magnitude fainter than Kraft's estimate. They confirmed Efremov's earlier estimate of the duration of the cepheid stage, and regard this time (about 10^6 years) as compatible with the time that the star remains in the instability strip according to the tracks calculated by Kippenhahn and his associates. Since the same deviation follows from the relative numbers of B stars and cepheids, Efremov and Kopylov conclude that a majority of the B stars evolve into cepheids. Tsarevsky *et al.* (**123**) believe that as many as 40 cepheids can belong to clusters or their halos.

The Gaposchkins are engaged in a very large program involving all the known variables in both Magellanic Clouds. A discussion of the relationship of light curve to period for cepheids in the SMC has been published (**124**), as well as general accounts of the project (**125**). The complete SMC results will appear soon. Wesselink and Shuttleworth (**84**) have obtained photographic light curves for 39 cepheids in the SMC, and Van Genderen is working on cepheids in the central section of the SMC. Work on the LMC by the Gaposchkins is now underway; over 1000 cepheids alone are on the program. Wright and Hodge have published an Atlas (**16**) of over 2000 variables in the LMC, and are now engaged in an exhaustive study (in *BV*) of 80 variables in a small region north of the bar (**126**). They estimate that only 25% of the variables in the area have been discovered, so that the total number in the LMC to $M = 0$ is about 8000. They have also worked on Population II cepheids in the LMC (**127**). Dickens (**128**) has obtained light curves of ten bright cepheids in the LMC and finds, in accord with earlier work, that the LMC stars are bluer than galactic variables of the same period.

The *P*, *L* and *P*, color laws for W Vir stars in globular clusters have been investigated by Fernie (**129**), who finds that they can be extrapolated to include the RR Lyr stars. The W Vir cepheids occur in metal-poor globular clusters, yet Woolley has pointed out (**130**) that the kinematics of the field W Vir stars are such that they cannot be regarded as the halo analogues of the classical cepheids. Kholopov (**131**) suggested that the W Vir stars may actually be a product of the evolution of classical cepheids. An extensive investigation of the properties of the W Vir stars has been carried out by Demers (**132**). Michalowska-Smak and Smak have published *UBV* observations of eight cepheids considered to be Population II objects (**133**). Kwee is engaged in a photometric study of Population II cepheids. New five-color observations of W Vir have been discussed by Oosterhoff and Walraven (**62**) and narrow-band photometry of a large number of cepheids, including W Vir stars, has been carried out by Williams (**134**) in an effort to measure metal/hydrogen ratios in different parts of the Galaxy. The period variations of W and AL Vir have been investigated by Prikhodko and Chouprina (**135**). On the theoretical side, Christy (**22**) has carried out exploratory calculations on an unstable stellar model of mass 0·88 $\odot$ which exhibits some of the observed properties of W Vir.

The normal colors of cepheids have been reconsidered by Nikolov (**136**) and by Efremov (**137**), who finds from a discussion of the six-color data that in the period range 20 to 30^d the intrinsic colors are 0·15 to 0·20 magnitude redder than obtained by Kraft; Efremov gives the relation: $\langle B-V\rangle_0 = +0{\cdot}21 + 0{\cdot}56 \log P$. Reference has already been made to the work of Fernie (**77, 78**) on the empirical *P*, *R*, mass relationship for pulsating stars. The empirical data on the *P*, *L* relation has been summarized by Efremov (**138**). Both Efremov and Kopylov (**121**) and Genkin (**139**), the latter on the basis of cepheid proper motions, suggested that the zero-point of Kraft's *P*, *L* law requires a correction of + 0·5 magnitude.

We only mention here the use that has been made of the cepheids for investigations of galactic structure. Kraft (**140**) has discussed their distribution in the Galaxy and their lack of concentration to spiral arms. Takase (**141**) has made a kinematical analysis of their motions, and Howard and Kirk (**142**) have determined galactic rotation parameters from the radial velocities.

II. SEMIREGULAR, LONG-PERIOD, AND RED VARIABLES

Smak has given (**143**) a thorough review of the observational problems of the Mira and long-period variables. A major advance in this area was provided by Smak's *UBV* and low-dispersion spectrophotometry of 29 Miras, 12 semiregular M's, and several miscellaneous M stars (**144**). These were accompanied by a rediscussion of radiometric data, and an analysis of the effect of the TiO bands. A later narrow-band photometric study (**145**) was directed toward classification on the basis of TiO strength, with the effect of H emission eliminated; this two-color system was shown to be useful for determination of reddening. Bolometric corrections and effective temperatures for M giants were obtained by Mendoza and Johnson (**146**) from their photometry to 9μ; a new scale of BC's has been established by Smak (**147**) on the basis of those results together with balloon observations from Stratoscope II. Tsesevitch has also studied the problem of BC's for Mira variables. The ubiquity of variability among M giants has again been emphasized by Blanco and Plaut; their results are discussed briefly in Section 6. Maffei (**148**) finds that there is a considerable advantage in searching for very long-period Miras by using the near infrared.

Mrs Mayall has published (**149**) a valuable series of light curves of long-period variables based on visual observations over a long time interval: R Lep, over the years 1854–1962; T Cas, 1889–1964; X Cam, 1904–64; RV Cas, 1906–65; V Hya, 1884–1965; χ Cyg, 1841–1966. Certainly such long series provide one of the best scientific justifications for the emphasis of the amateur associations upon observations of Mira and long-period stars. A good example of the use that can be made of such material is the study by Harrington (**150**) of cycle lengths and heights of maxima for some 165 stars. He finds that if a maximum is earlier than expected, it is likely to be brighter than the average. Miss Houk (**151**) has studied the light curve of the N-type Mira V1280 Sgr, with $P \approx 523^d$. The cycle lengths and the magnitude at maximum vary considerably, but their correlation may be in the opposite sense to that of Harrington. Houk points out that the departures of V1280 Sgr from strict periodicity can be explained formally as the effect of a long secondary period. In fact, if this secondary period were due to the beats between two nearly equal shorter *P*'s, the sense of the correlation between spacings and heights of maxima would be understandable. In some well-studied Miras, the systematic deviations from a linear ephemeris are very large. Huth (**152**) and Schneller (**153**) have discussed the material on W Tau, R Aql and R Hya, in all of which a parabolic term is a clear necessity, although it may be that a limited number of straight-line segments would fit the $O-C$ curve as well. Superimposed upon these large-scale changes are more erratic deviations which may be identified with the accumulation of random errors, as discussed elsewhere (**72**). The dependence of range on period for Mira stars, and its interpretation in terms of age were investigated by Bredichin (**154**).

The variability of M supergiants is of course of a different character than in M giants. Larsson-Leander (155) has reported on a five-year series of *P*, *V* observations of μ Cep, and Magalashvili and Kumsishvili (156) have made photoelectric observations of the supergiants RW Cyg and SU Per. Some tendencies toward cyclic behavior are apparent in these stars, but they are not striking. A periodogram analysis of the older observations of μ Cep by Sharpless *et al.* (157) does, however, reveal a considerable degree of 'hidden' regularity. No further optical information is available on the rapidly variable infrared (9μ) emission reported by Low (158) around μ Cep, α Ori, and α Tau.

Some general properties of the RV Tau stars have been reviewed by Stothers (159), with emphasis upon their galactic distribution. The light curves of two RV Tau stars, EZ Aql and EQ Cas, on the basis of about 18 years of visual observation, have been discussed by Beyer (160) with emphasis on the period changes. Questions of the nature of RV Tau stars, and of Mira and long-period variables with periods near 200^d soon involve the stars of these types in globular clusters. These matters are discussed by Mrs Sawyer Hogg in Appendix II; we only refer here to the useful paper by Stothers (161) on longer-period variables in clusters and the field, and to the remarks by Feast (162) on the same subject.

The kinematics of long-period and Mira stars have been studied by Smak and Preston (163) on the basis of new radial velocities for 270 variables; the discussion was directed toward galactic-dynamic ends so we do not consider it further here. Feast (164) has with new velocities for 53 Me and Se variables in the general direction of the galactic center, discussed the motions in terms of galactocentric orbits; an interesting comparison with the planetary nebulae in the same direction is made. Kreiken (165) has examined the question whether the velocity components in different directions are a function of *P* for stars with $140^d < P < 200^d$. Eskioğlu (166) earlier had obtained a positive correlation between *P* and the tangential component, but Kreiken's results were inconclusive. A major injection of fresh material in this field can be anticipated when the results of Van Herk's current proper motion program for 343 red variables become available.

The significance of the discovery by Serkowski (167) of large amounts (up to 6%) of plane polarization in some Mira stars, variable with phase, is not yet clear. Certainly the observations should be pressed vigorously to many more objects over their complete light cycles. Donn *et al.* (168) have suggested that aligned graphite flakes, condensed in the atmospheres of these stars, could be responsible through the angular dependence of their scattering. One would expect such an effect to be much greater in the carbon variables, but the two carbon stars for which data have been published appear no different in amount than the M stars.

The discovery by Neugebauer *et al.* (169) of a number of visually very faint stars that were of magnitude 0 or 1 at 2μ was most exciting, and for a time it appeared that possibly a new type of stellar object had been discovered. It now appears that that belief was premature. With the exception of one object in Cygnus near $20^h43^m, +40°$ (1900) which may be a heavily-reddened M supergiant, and an object in the Orion Nebula (see Section 15) the other 'infrared stars' appear to be no more than very late-type Mira variables of unusually long period. That such stars can be discovered by conventional infrared photography was demonstrated in the 1930's by Hetzler, and more recently by Haro and Luyten (170) and by Haro and Chavira (171). Reference has already been made to the finding of Maffei (148) that such long-period Miras become very conspicuous in surveys made in the near infrared. The literature on the subject of the 'infrared stars' is already fairly extensive, so that for spectroscopic information we refer to the discussion in Appendix I, and for a fairly complete discussion and bibliography to a review paper by Wing *et al.* (172).

12. β CANIS MAJORIS STARS AND OTHER EARLY-TYPE VARIABLES

Observational activity in the field of β CMa variables has slackened, due at least in part to the loss of the stimulus provided by O. Struve. The photometric program of van Hoof of southern β CMa stars has produced two- and three-color light curves of β CMa itself (**173**), 15 (AO) CMa (**174**), σ Sco (**175**), information on the beat period of τ^1 Lup (**176**). Miss Kupo (**177**) has published a spectrophotometric study of β CMa, and Milone (**178**) has studied the fundamental periods of β CMa as exhibited in the radial velocities. Klock (**179**) has observed KP Per (= HD 21803), and found in addition to the usual interfering fundamental periods near 0ᵈ20, evidence for a still shorter period of 10 to 15 minutes. Rossati (**180**) has observed the same star in two colors but since he regarded the light curve as containing two minima of unequal depth, interpreted it in terms of the double period of 0ᵈ40. Jerzykiewicz (**181**) has observed δ Cet in two colors and compared his results with the older radial velocity observations. From radial velocity observations, van Hoof and Blaauw (**182**) have obtained a provisional period of 0ᵈ153 for 53 Ari.

The origin of the beat phenomena in β CMa stars has generated some discussion. Theoretical studies by Chandrasekhar and Lebovitz (for references see **183**) showed how the effect of rotation could produce two normal modes of nonradial oscillation whose difference might account for the β CMa effect. Mrs Böhm-Vitense (**183**) actually computed the rotational velocities required to reproduce the observations; they were in the range 100–300 km s^{-1}, depending upon details. These velocities v_{rot} were so large as compared to the uniformly small values of $v_{rot} \sin i$ actually observed, that to save the explanation it was necessary to postulate that we observe all the known β CMa stars nearly pole-on (and also that they do not appear to vary appreciably if seen otherwise). However, van Hoof (**184**) pointed out that two of these β CMa stars were also single-line spectroscopic binaries, and if one assumed the orbital and rotational axes to be parallel, then the inclination requirement of the Chandrasekhar-Lebovitz theory led to impossibly high masses for the components, from Kepler's third law. Clement (**185**) has refined the original theory and finds that the required v_{rot}'s become somewhat smaller. It is possible that if all the contributing factors were now to lean in the same direction, the conflict would not be so decisive, but it is certainly not yet resolved. For the clarification of such questions, it is becoming ever clearer that photometric observers have an obligation to settle once and for all the issue: has there been some bias in the search for β CMa variables *against* broad-line stars? That is, are there any cases of rapidly rotating β CMa stars?

Another suggestion as to the origin of the pulsations in β CMa stars has been made by Oleinik and Porfiriev (**186**). They believe that as a result of evolutionary change in the star's structure, meridional circulation will force a drastic equatorial deceleration of the rotation, which in turn will set up radial oscillations of the entire star. This proposal has not yet been explained in detail, although Porfiriev (**187**) has computed the line profiles expected for a star undergoing both pulsation and meridional circulation. Hitotuyanagi and Takeuti (**188**) have obtained a pulsation constant near 0·02 for the β CMa stars from a theoretical mass, L relationship.

Fitch has suggested that all β CMa stars showing beat phenomena do so as the result of the effects of tidal deformation by a faint companion, and not as the result of interference by multiple periods. In the case of σ Sco, a known binary, in which the orbital elements are known, he finds that the periodic phase changes in the 0ᵈ25 β CMa-type cycle do correlate well with the calculated variation of the tide-raising potential.

There is general agreement that the β CMa stars are former B-type dwarfs that have evolved off the main sequence but opinions differ as to their precise location on the theoretical evolutionary tracks. Stothers (**189**) has developed in detail the thesis that they are non-rotators in radial oscillation, evolving across an instability strip in the HR Diagram. Some of the observational issues involved are discussed by van Hoof (**9**, p. 149) and by Odgers (**9**, p. 145). The only safe conclusion at the present time is a familiar one: the information is still inadequate for a decision.

There are, of course other types of intrinsic variability among early-type stars. Feinstein has observed about 70 southern Be stars in *UBV*, and some in *RI* as well. About half of these objects show variations in both *V* and *U*−*B*. Jackisch has observed the Be stars 48 and 53 Per.

13. MAGNETIC AND PECULIAR A-TYPE VARIABLES

The general subject of magnetic variables has been reviewed at length by Ledoux and Renson (**190**), while the theories devised for their explanation have been discussed by Cowling (**191**). Observational aspects, and in particular the location of the magnetic variables in the HR Diagram have been considered by Jarzebowski (**9**, p. 25).

Ledoux (**10**) has suggested a new 'pulsation theory' of magnetic variables, based on the excitation in the presence of rotation of special non-radial modes in an external slightly super-adiabatic layer. This suggestion eliminates the difficulties concerning the excitation mechanism and the order of magnitude of the period (which is of the same order as that of rotation), but the excitation of the low-degree harmonics seems to require the presence of a binary companion in an eccentric orbit. The details of the magnetic variation have not been worked out. Renson has pointed out that the distribution of the periods for about 30 magnetic variables is very much like that of the periods of spectroscopic binaries (with $P < 11^{d}$), which reinforces the binary hypothesis of the magnetic variables. Renson believes that the invisible companions must be highly evolved degenerate stars, originally of higher mass that have transferred much of their material to the present magnetic star. Steinitz however regards the oblique rotator hypothesis as still the most promising explanation, and has reviewed the relevant observational information (**192**).

Renson has given special attention to the reconciliation of the magnetic and photometric observations. Some of his results both published and as yet unpublished are as follows: *HD 10783*: both magnetic field and radial velocity vary with the light period of $4^{d}.1565$, although erratic fluctuations are superimposed. The same period was found also by Van Genderen (**193**). *HR 8861*: The photometric observations of Rakos can be represented by a period of $0^{d}.723$, which is consistent with the period-line width relation (**194**). *HD 8441*: According to Renson the magnetic field varies with the period of the spectroscopic binary, $106^{d}.27$ (**195**, **196**). Steinitz however believes (**197**) that the field period is $2^{d}.9632$. *21 Per*: Renson finds that the magnetic period is $1^{d}.728$, in agreement with the light variation. *HR 7058*: There is some difficulty in fitting the magnetic observations to the light period of $10^{d}.1$. *γ Equ*: a magnetic period of $314^{d} \pm 3^{d}$ has been obtained. *HD 32633*: The magnetic observations can be fitted with a period of $6^{d}.4289$, which is compatible with the light variation. *HR 5597*: both the magnetic and photometric data can be represented with a (slightly variable) period of about $1^{d}.054$. Babcock (**198**) has discussed the case of *53 Cam*, in which the phase difference of light and magnetic field appears to violate a general relationship suggested by Rakos. A new photoelectric light curve and improved period of *ι Cas* have been obtained by Klock (**199**). The rather unusual Ap star HD 221 568 has been found to be variable by Osawa *et al.* (**200**), apparently with a period of about 160 days.

HD 215441 (GL Lac) the magnetic variable of largest known field, has been studied both polarimetrically and photometrically by Polosuhina and Lebedeva (**201**, **202**); they find that there is a linear correlation between the percentage polarization and the magnitude in the $9^{d}.5$ photometric period. At Uppsala, A. Elvius and Engberg are attempting to detect variable polarization in magnetic variables. No strong polarization has been found in any of the stars studied, but in 21 Per polarization changes are indicated that might have a period slightly shorter than the photometric period of Rakos.

14. δ SCUTI VARIABLES AND DWARF CEPHEIDS ($P < 0^d.2$)

An investigation by McNamara (**9**, p. 111) of dwarf cepheids and RR Lyr variables in a narrow-band three-color system has enabled him to determine, under certain assumptions, the luminosities and intrinsic colors. Empirical adjustment of these luminosities makes possible the location of these variables (together with the δ Sct stars, from an earlier investigation) and their constant-P lines on a HR Diagram. A discussion of the relative space densities is also included. One of the distinctive features of the δ Sct stars is that they possess a significant amount of axial rotation, which has long been recognized as an exception among the intrinsic variables. Preston (**9**, p. 155) has developed the idea that pulsation and rapid rotation are incompatible, and specifically that only those stars with sufficiently low rotational velocities are able to pulsate as δ Sct stars. Statistics then available on line widths seemed to support this thesis, but it has recently been discovered that some rapid rotators (such as β Cas (**204**)) are variable. Kraft has considered the similar problem for cepheids (**25**).

Since so many δ Sct variables and dwarf cepheids have been detected accidentally among the bright stars, it is obvious that many more await discovery. A short, rather general program directed to this question by McCullough (**205**) produced no new examples, however. Eggen has suggested a number of possible δ Sct stars on the basis of location in color-luminosity arrays. A considerable number of new short-period variables whose precise classification is not yet certain have been found recently, as the result of specific photoelectric observation of plausible candidates. A list of those announced through November 1966 as certainly variable is as follows:

HR	Name	Period	References
21	β Cas	$0^d.104$	Millis (**203**) (**204**)
1287	44 Tau	·134	Danziger and Dickens (**211**)
1706	14 Aur	·122	(**211**)
2107	1 Mon	·137	(**211**)
3265	—	·13	(**211**)
3888	υ UMa	·13	(**211**)
4715	4 CVn	·1707	Jones and Haslam (**206**), Wachmann (**207**), Smyth (**208**), (**211**)
5005	—	·14	(**211**)
5017	20 CVn	·14	Wehlau *et al.* (**210**), (**211**)
—	HD 116994	·104	Chen (**209**)
5329	κ^2 Boo A	·069	(**203, 28**)
5435	γ Boo	·2903	Magalashvili and Kumsishvili (**212**)

Possibly to be included are HR2539 (59 Aur) and HR3889 (20 Leo) (**211**). As in the case of flare stars and β CMa variables, those concerned with statistical properties of variable stars must be very careful that such lists of discoveries are not biased in some way, as if for example the stars examined were only those having sharp lines, or luminosities above the main sequence. It is an unfortunate fact that by such observational selectivity, an initial bias or prejudice can become hopelessly enshrined in the literature.

Photometric observations of the well-known δ Sct star DQ Cep have been published by Fitch and Wehlau (**213**), and reported by Jenks *et al.* (**214**). The multiple periodicity of δ Del has been studied by Wehlau and Leung (**215**) on the basis of Fourier analysis of new observations on 25 nights. They represented the variation of period $0^d.134$ by superimposing six short-period waves. Jerzykiewicz (**216**) reported that the variation on four nights in 1965 followed a period of $0^{d}.1363$, but that this could be equally well represented as the beat between Eggen's discovery period of $0^{d}.1350$ and a period of 14^{d}, which latter Preston had suspected to be the interval associated with the occasional line-doubling which he had discovered in δ Del. The idea that a tidal resonance effect in a close binary with an eccentric orbit could stimulate a component in the light variation is attractive (see below) but in this case it is not convincing since more recent work by Preston indicates that the period of δ Del is actually about 41 days. In the case of the δ Sct star CC And, Fitch has detected a $10^{d}.5$ modulation of the fundamental period for whose explanation he proposes the tidal effect of a close companion. He makes the same proposal for those β CMa stars with long-period modulations. Such an explanation will be most convincing if a photometric term can be found that is clearly identifiable with an orbital period. No one can argue against an orbital interpretation of a radial velocity variation if *two* spectra are visible. Since this is the case for δ Del, this star will be a crucial test for such proposals.

An analysis has been made by Oosterhoff (**217**) of Ponsen's photoelectric observations of the dwarf cepheid V 703 Sco. The high maxima are separated by an interval of $0^{d}.1152$, but the light curve is subject to major short-period fluctuations which Oosterhoff has now disentangled into primary and beat periods, together with a short period of about $0^{d}.042$ that is also subject to beat phenomena. The dwarf cepheid SZ Lyn ($0^{d}.1205$) has been observed in *UBVR* by Van Genderen (**218**). The probably similar stars YZ Boo ($0^{d}.1041$) and VZ Cnc ($0^{d}.1784$) have been observed by Heiser and Hardie (**219**) and by Wildey (**220**), respectively. The complex photometric behavior of the dwarf cepheid VX Hya has been studied by Fitch (**221**), who compares it to AI Vel. A double harmonic series with slowly changing coefficients fits the variation in V quite accurately; a fundamental period of $0^{d}.2234$ and an overtone of $0^{d}.1727$ was found. Notni (**222**) has studied the rapidly-changing light-curve of AC And, and points out similarities to the dwarf cepheids in its luminosity and color-magnitude variation. The period (sometimes represented as the sum of two variations with periods of $0^{d}.52$ and $0^{d}.71$) is however much longer than the generally accepted upper boundary for that class.

15. ORION POPULATION VARIABLES: T TAURI AND RW AURIGAE STARS, ETC.

Survey work and studies of variable stars occurring in young clusters and associations that have been published in the last three years will be found listed in Appendix II, the Report of the Committee on Variables in Clusters. Non-technical reviews of the present state of the subject have been presented by Kuhi (**223**) and by Hunger (**224**).

Götz (**225**) has made a very important study of the relationship of emission Hα stars to interstellar material, including a thorough examination of the systematics of the known variables in the major associations. One interesting result is that the intensity of Hα emission in individual stars is strongly correlated with the opacity of the dark material upon which they are seen projected. This is certainly connected with the finding by Herbig and Peimbert (**226**) that the number of such stars per unit projected area of the nebulae is also proportional to the opacity. It will be noted that this does not necessarily contradict the earlier belief that such stars prefer the edges of observed areas; obviously the edge regions occupy more square degrees than do the dense centres.

Garzoli and Varsavsky (**227**) have observed, in agreement with an earlier result by Burke, that 21-cm observations of the Taurus dark nebulae indicate no excess or deficiency of neutral H in those regions where emission-line stars are concentrated. Whatever the explanation for this apparent conflict—H_2 formation, very low H I temperature, or some other—it is apparent that direct observation is beginning to provide some useful boundary conditions for theories of star formation.

For such purposes however, the data must be trustworthy. Meinunger has begun a systematic examination of the variations of the rapidly irregular RW Aur stars. His first results (**228**) indicate that a number of T associations must be spurious, since most of their members belong to some other class of variables. Hoffmeister (**229**) has analyzed the light curves of four southern RW Aur variables (T Cha, RU Lup, RY Lup, AK Sco) and finds that quasi-periods of lengths three to five days persist. Hoffmeister proposes that these are due to axial rotation combined with variable, non-uniform distribution of brightness over the stellar surfaces. It is interesting that the rotational velocities implied by Hoffmeister's results are of the same order as the velocities inferred from the line widths observed in a number of other T Tau stars.

Much new data on the variations in color of RW Aur itself have appeared, especially as the result of an intensive campaign carried out in 1962–63 in the U.S.S.R. At Abastumani, spectrophotometric observations were made by Kharadse and Bartaya (**230**), while it was observed in three colors by Mosidze (**231**), and photoelectrically in two colors by Kumsishvili, Magalashvili, and Abuladze (**232**). Observations of RW Aur were also reported by Chugainov and Zaitseva (**233**), by Alksnis and Daube (**234**), by Mirzoian and Kazaryan (**235**) and by Ishchenko and Shevchenko (**236**). An extensive program of three-color photoelectric observations of this and similar stars continue at Sonneberg; the first results for RW Aur were published by Fürtig and Wenzel (**15**, p. 64). A limited series of *UBV* observations of RW Aur were made at Lick by Taylor. Despite all this attention, the star still lacks a firm physical model. The photometric activity proceeds on time scales from a few hours upward, the colors being correlated with the light variation. Efimov (**237**) suspects that there may be a change in the degree of polarization that varies with the $B-V$ color.

Three-color observations of some 26 T Tau stars have been carried out by Smak (**238**), together with a narrow-band substitute for the B color that avoids the strongest emission lines in that spectral region. Even with this refinement, the colors are quite anomalous in the sense that the emission-line stars are too bright in the blue, and of course in the ultraviolet. Spectrophotometric observations of T Tau and RW Aur by Aveni (**239**) indicate that the emission lines contribute about 0·1 magnitude to the B color in those stars. Smak's data indicate that Hβ and Hγ alone can brighten B by 0·3 to 0·4 magnitude in extreme cases. One anticipates that *continuous* emission will add more energy to B and especially to U than do the lines, but no estimates of those amounts have been attempted. Kuhi's (**240**) scanner observations show that the *continuum* ultraviolet excess is well correlated with the intensity of Hα emission; if the proportionality can be determined, perhaps the necessary corrections can be made to the observed colors. Until that is possible, the colors of these objects cannot be placed on theoretical evolutionary tracks with any degree of confidence.

The veiling of the absorption spectra of the T Tauri stars and the excess energy in the near ultraviolet could be reproduced formally by the superposition of a hot continuous spectrum. Gurzadian (**241**) has shown how a blanket of electrons, of appreciable optical thickness, could do just this by, in effect, simply transferring energy from the infrared into the blue and ultraviolet. The sudden appearance of such an electron cloud would then lead to a sudden flare-up in blue-ultraviolet light. But whether this suggestion is compatible with the simultaneous existence of infrared excesses in T Tauri stars is not clear. Kolesnik and Frank-Kamenetsky (**242**, **243**) have indicated how the emission of dense, nonequilibrium plasma clouds ejected into the atmospheres of these stars would have the same effect. Kolesnik (**8**) has applied this idea to a sequence of stars in gravitational contraction. Gershberg (**263**) has applied similar considerations (i.e., that the spectrum is a composite one) to flare stars in some detail.

It is probably too much to expect at the present time that observations can differentiate between mechanisms whose end effects are qualitatively so similar. It is even less hopeful that observations can tell us anything directly about the forces which drive such phenomena. Schatzman (**9**, p. 17) has demonstrated how useful is the concept of surface electromagnetic activity in very young stars, in inducing light variations, mass ejection, nuclear reactions, and in affecting the axial rotation. But the hard facts at our disposal, for testing such proposals, are very few.

The all-too-sparse astrometric information on young stars above the main sequence increased substantially by the astrometric study of NGC 2264 by Vasilevskis, Sanders, and Balz (**244**). This serves to remove a certain number of non-members from the familiar color-magnitude diagram of this important cluster. The motions of 24 emission-Hα stars were measured; two of these stars (unfortunately with motions not of high weight) seem not to be cluster members. A somewhat similar situation exists for several variables in the Orion Nebula, as pointed out by Van Schewick (**245**). In the case of Orion, however, it is not completely certain that two of the three stars in question are truly variable. If a single unquestioned case could be established, it would be of considerable importance. New proper motions for 35 T Tauri stars are being measured by Van Herk. The difficulties with radial velocities make proper motions especially important in studies of the T Tauri stars.

An entirely unexpected development was the discovery by Mendoza (**246**) that many Orion population variables, especially T Tauri stars, have a major excess of radiation in the infrared (0·9 to 5μ) as compared to normal stars of the same spectral types. The most extreme case, R Mon, has been observed also by Low and Smith at 20μ, confirming Mendoza's result that in that case the brightness curve peaks (on a wavelength scale) near 3μ. An obvious explanation is that we are seeing in these stars the contribution from an unresolved source of low temperature and large area. Before Mendoza's discovery, Poveda had suggested (**247**) that such very young stars might be surrounded by very small, dense, dust nebulae. Possibly the infrared excess is due to the thermal emission of such a circumstellar cloud. Indeed this may be supported by the fact that the objects with the largest excesses are just those involved in bright nebulosity (R Mon, T Tau, V 380 Ori). But the effect could also be produced by a 'window' in the continuous absorption coefficient of the stellar photosphere. A decision between these two possibilities could be made, at least in principle, by observation of the degree of veiling of the stellar absorption spectrum in the infrared. It is odd that only in our own visual region do the T Tau stars radiate according to our tastes: toward both shorter and longer wavelengths, they behave in a completely surprising fashion. It is not yet clear whether the discovery by Becklin and Neugebauer (**248**) of an infrared star in the Orion Nebula that is very bright at 2μ yet undetected at 0·8μ, is related to the infrared excess phenomenon.

R Mon has received attention not only on account of its infrared excess, but also because it lies at the apex of Hubble's Variable Nebula NGC 2261 which has itself been observed extensively in recent years. We only refer here to the work of Johnson (**249**) and of Hall (**250**) which concern R Mon to some extent, and to the suggestion by Dibai (**251**) that the formation of such nebulae and their associated stars might be due to the focussing of shock waves generated at an H I–H II boundary.

Finally, the 1936 flareup of FU Ori has been interpreted by Herbig (**252**) in terms of the rapid transit of the forbidden region on the right side of the HR Diagram by a contracting star, which is expected theoretically just prior to its appearance at the top of the vertical branch of its evolutionary track. A refinement of this suggestion has been advanced by Hayashi (**253**), and another interpretation altogether by Poveda (**247**).

16. FLARE AND FLASH VARIABLES

There is reason to suspect that the flare stars are connected with the Orion population variables in the evolutionary sense, and that the so-called flash variables constitute the bridge. (The presently-held distinction between flare and flash stars is explained in **254**). The major observational contributors to this concept have been Haro and his associates, who have given a summary of their recent work (**255**), although full details are still to be published. The entire subject has been reviewed by Haro (**256**). That review also gives an excellent account of the Tonantzintla work. Rosino and his co-workers also continue their program for the discovery of flash variables. In the Orion Trapezium region 14 have been discovered, 13 in the Pleiades, and two in Praesepe. The critical question in these studies is that of the physical membership of these faint variables in the galactic associations and cluster against which they are seen. When the cluster is projected on heavy, relatively nearby obscuration (Orion, NGC 2264) the problem of membership is much simpler than in older clusters (Pleiades, Praesepe), unless proper motions can be used. The best example of the latter is the case of the Hyades, where an astrometric study by van Altena (**257**) demonstrates the possibilities.

The problems of the classical flare stars have been discussed by Andrews (**258**). Photographic observations of UV Cet, YZ CMi, and V 371 Ori were made in 1963–64 by Mosidze and Chuadze (**259**) as part of a photometric campaign to support radio observations of flare stars (see below); they observed no actual flares, but did suspect some small variations. Gershberg and Chugainov (**260**) have observed five flares of AD Leo and six of UV Cet simultaneously with spectrograph and photoelectric photometer. Chugainov (**261**) has followed EV Lac through four flares in BV, and continues observations of flare stars with an interference filter centered on $H\beta$. At Odessa, visual observations of flare stars were carried out by Mandel, Migach, *et al.* In early 1966, Andrews (**262**) found YZ CMi to be extraordinarily active: in 26 days he observed nine flares. There was a suggestion that the time spacings of the flares were not random. Such extended intervals of high activity have been observed in other flare stars, but nothing is known of their systematics. Possibly these intervals of high flare activity bear some analogy to the renewal of flare activity in the Sun with the solar cycle.

Gershberg (**263**), as part of two more general papers on the representation of several kinds of variables as a quasi-normal star plus a volume of hot, optically thin gas, has analyzed quantitatively the light and color curves of several flares. These results have been displayed as movement of the star in the $U-B$, $B-V$ plane, and the decay of the flare as fitted to a recombination law. Mirzoian (**264**) has however questioned this interpretation.

One of the most significant developments in this field is the discovery by Lovell (**265, 266**) that the flares of dMe stars are accompanied by bursts of radio emission; the Sydney observations by Slee *et al.* (**267**) show that the radio flares are detectable over the frequency range from 20 to 1410 MHz. A very critical requirement, the coordination of radio and optical observations, was made possible by arrangements between the Jodrell Bank and Sydney radio observatories on the one hand, and the Crimean, Abastumani, and Odessa observatories in the U.S.S.R., the Belgrade Observatory, the Smithsonian satellite-tracking network, and a group of Australian amateurs on the other. The implication is, of course, that these are surface events like solar flares, although the radio energy radiated in the stellar flares is 10^4 to 10^6 times that of a typical large solar event. Shlysh (**268**) finds that it is possible to explain the radio bursts from flare stars by stimulated emission of 20 MeV electrons at densities of the order of 10^6 cm^{-3}. Vardya (**269**) finds that the alternative explanations of flares in M dwarfs are unacceptable or unlikely, and favors a hydromagnetic mechanism. Lortet-Zuckermann (**270**) has considered the observable consequence of the emission of particles with energies of 10^{-2} to 10 MeV from a flare star. If the star were accidentally immersed in a dense interstellar cloud, then the ejected protons would produce a small H II region of calculable surface brightness around the star, whose $H\alpha$ emission might be detectable under favorable circumstances. The fact that she has been unable to find nebulae around any known flare stars indicates however that the required conditions are somehow not satisfied.

A new flare star (S 9537) has been announced by Hoffmeister (**271**), who has also discovered two others, as yet unpublished. Eggen (**272**) reported a flare of the short-period visual binary BD $-$ 8° 4352 AB. Flare-like activity has been reported in a number of other stars of various spectral types. Andrews (**273**) found that the bright B 8 star HR 1938 (= + 31° 1048) was about three magnitudes brighter than normal on one plate in March 1964; a lesser variation was observed on one night about two weeks later. No further variation has been reported, nor did Mendoza (**274**) find any activity (photoelectrically) on 19 nights in early 1965. The color of the star is normal (**274**) as is the spectrum (Mammano, **275**; Herbig unpublished). Nielsen (**276**) has called attention to an apparently similar event observed spectroscopically in the case of the type A2 binary ζ Boo in 1905. Kuhi (**277**) observed a flare of amplitude about 1·5 magnitude at 3300Å in the type F8 contact binary W UMa. Butler (**278**) observed a $\Delta B = 0{\cdot}1$ magnitude flare in an early G-type star projected on the Small Magellanic Cloud, and a second flare of amplitude $\Delta B = 0{\cdot}8$ magnitude was followed more completely five nights later (**279**). To say the least, this extension of flare-like activity to earlier spectral types than M raises serious problems, if a general explanation is to be found. One is reminded of the lengthening list of stars in which K I emission lines have appeared temporarily (Andrillat and Morguleff, **280**).

17. LATE-TYPE DWARFS, NOT FLARE STARS

Flare stars aside, single M-type dwarfs give the impression that as a class, they are quite constant in light. The fact that this is not true of M giants at the same spectral class indicates that this difference in behavior is a structural characteristic, and has nothing to do with the low temperature or the properties of molecular bands. K- and M-type dwarfs in close binaries are however subject to intrinsic variation. The best documented cases are the eclipsing binary YY Gem (Kron, **281**) and the spectroscopic binary CC Eri (Evans, **282**, **283**). A third example has been discovered by Chugainov (**284**): HD 234677, which subsequently has been found by Kraft and Krzeminski (**285**) to be a double line spectroscopic binary. These variations may be due to non-uniform surface brightness of the components, and apparently change with time. One expects that they will probably be detectable in systems of any orbital inclination, and thus in stars that cannot otherwise be shown to be binaries. An example of a variable M dwarf not known to be binary is Ross 248, found by Kron (**286**) to vary in a manner resembling the secondary variation of YY Gem, and with a period of approximately 120 days.

The three examples of known binaries mentioned above are of types M 0·5 V, K7 V and K6 V. This concentration to early spectral type may reflect nothing more than the fact that these stars are brighter than later M dwarfs, and hence have received more attention. Yet the dM2e spectroscopic binary BD −21° 6267 A (Herbig and Moorhead, **287**; Paczynski, **288**; Chugainov, **289**) has not been found to be variable. The solution of such problems depends upon the detailed examination of more M dwarfs. A very promising beginning has been made by Kraft and Krzeminski (**285**) on the basis of both photoelectric and spectroscopic observations. In any such program, it is recommended that the stars chosen for study be selected from among both emission and non-emission objects, in order that no observational bias arise such as is present in lists of the known flare stars. Although Gabriel (**290**) has found that M dwarfs are vibrationally unstable with respect to the fundamental mode of radial pulsation (in the case of Krüger 60A, type dM4, mass 0·27 ⊙, with a pulsation period of about 10^3 seconds) there is as yet no connection of this result to the kinds of variability actually observed in the M dwarfs. It seems safe to say that for the moment, observers must proceed without theoretical guidance.

18. R CORONAE BOREALIS VARIABLES

R CrB itself has been below maximum light since mid-1962, but with major fluctuations. Miskin observed a rapid oscillation of about 40 minutes in length just before the minimum. The star reached $m_{vis} \approx 14$, and both *UBV* and spectroscopic observations were made at Lick as opportunity permitted. At such deep minima, the lines of [O II] appear in the spectrum, indicating that very low-density envelope is present near the star. Direct photographs obtained with the 120-inch (305 cm) telescope in fairly good seeing, in an attempt to photograph the nebula directly, were not conclusive. Schmidt (**9**, p. 295) has studied a situation in which density fluctuations occur in the outward mass flow from a carbon-rich supergiant. If sporadic density enhancements by a factor of 200 do occur as Schmidt's arguments require, the precipitation of solid carbon particles in them could explain the variations of R CrB-like stars. One problem with the hypothesis of soot formation in R CrB variables is why this characteristic form of light activity is possible in very hot carbon stars like MV Sgr and in stars of intermediate temperature, as R CrB itself, but not in the cool N variables in which one would expect the formation of solids to be easiest.

The light curve of the far southern R CrB star S Aps has been studied by Waters (**291**), who finds indication of a cyclic variation of range about 0·3 magnitude and period 120^d, superimposed upon the R CrB-type activity. Furthermore, there is a tendency for major minima to recur at intervals of about 1300^d. Cyclic activity with length of the order of a few months is known in other R CrB stars, but no suggestion of any periodicity in their minima has been made. In fact, R CrB itself was found by Sterne (**292**) to be an ideally irregular variable star.

The already small number of known R CrB variables has been diminished by one: RZ Vul is believed by Meinunger (**293**) to be a nova-like object with cycle length of about 1500^d. CT Vul is likewise to be withdrawn from the list of possible R CrB stars. In this connection, attention should be drawn to Wenzel's (**294**) photometric investigation of four hydrogen-deficient carbon stars.

The light variation of the hot, hydrogen-deficient carbon star HD 30353, which is also a spectroscopic binary with a period of 360^d, has been studied in *UBV* by Osawa *et al.* (**295**). They find a roughly sinusoidal light curve of amplitude 0·14 magnitude and period 30–40^d, which length is reminiscent of some R CrB stars, as mentioned before. The fact that HD 30353 is a single-line binary (although not a R CrB star) is perhaps relevant to the question of duplicity in H-deficient stars raised by McCrea. However, according to Hill (**296**), there is no evidence for duplicity in a number of other H-deficient objects.

The collection of spectroscopic material on R CrB stars assembled at Lick Observatory over the period 1950–60 is now being studied by J. Moorhead. The spectral type and radial velocity results are supplemented by *UBV* observations of as many stars as can be reached from Mt Hamilton.

19. SYMBIOTIC VARIABLES

The most direct approach to the problems of the symbiotic stars is through their spectra, which aspect is considered in Appendix I. Nevertheless, the stars do vary, in some cases in a rather spectacular fashion. The subject is reviewed by Sahade (**9**, p. 140) from the point of view of the binary hypothesis. Boyarchuk (**297**) has provided a physical model for AG Dra in terms of a K3 III star, an invisible hot star, and a cloud of ionized gas with $T_{\rm eff} = 17\,000°$ and $n_e \geqslant 10^7$ cm^{-3}. Luud and Ilmas provided (**298**) a composite model for AG Peg consisting of a dense planetary nebula with a double nucleus, of types WN6 and M1–3 II–III. Belyakina (**299**) interpreted her three-color photometry of the star, which varied only in the ultraviolet over the period of observation, in terms of this model. A preliminary light curve for AG Dra for the period 1890–1952 was derived by Robinson (**300**) from Harvard plates. The light curve of AS 373 (= MHα 328 − 116) has been extended by Fitzgerald *et al.* (**301**) back to 1920. The star was fainter than $m_{\rm pg} = 15{\cdot}6$ prior to 1948, when it brightened to about 15·0, and then faded slightly before rising to $m_{\rm pg} \approx 12$ in 1964. The object is being followed by Bertaud and at Asiago. Z And became bright in August 1966. The light curve of V407 Cyg has been obtained by Meinunger (**302**), who finds Mira-like behavior with a cycle length of 745^d. Lick spectrograms in 1966 confirm the earlier classification as type Me; perhaps, as Meinunger suggests, the star is similar to RR Tel. It is possible that a study of the object M1–2 studied by O'Dell (**303**) might show it to be a variable of this rather poorly defined (in the photometric sense) family of variables.

20. NOVAE DURING OUTBURST

The distribution of novae in the Galaxy and other systems has been reviewed by Plaut (**304**). During the report period, papers have appeared on the outburst activity of a few older novae, as well as on those which were bright during the triennium. The photometric investigations are listed here; the spectroscopic work is covered by Dr Feast in Appendix I of this Report.

Nova (RS) Oph 1898, 1933, 1958. Six-color photoelectric observations made during the outburst were reported by Svolopoulos (**305**), while the light curve as well as that of *Nova (WZ)Sge 1913, 1946* have been compared by Eskioğlu (**306**) with those of other recurrent novae. The spectrum as well as the reddening of RS Oph have been studied by Dufay and Bloch (**307**).

Nova Cet 1963 was discovered by Strohmeier (**308**). It reached $m = 10$ in September 1963, but no further information on the light curve is available. The star is of some interest because of its high galactic latitude (− 79°).

Nova Her 1963 was observed extensively. *UBV* observations by Chincarini (**309**), by Van Genderen (**310**). and by Almár and Illés-Almár (**311**). Six-color observations of the Nova were made by Breckenridge (**312**). An extensive investigation of the light curve in narrow spectral bands was carried out by Götz (**313**). A light curve has been published by Busch (**314**), and photographic observations by Weber (**315**) and by Bertaud are continuing. A small, possibly interstellar polarization of the Nova was reported by Clarke (**316**).

Nova Pup 1963–1 was discovered by Hoffmeister who also derived the light curve (**317**).

Nova Pup 1963–2 was found by Strohmeier (**318**). It attained magnitude 8 in December 1963. The early photometric history has been investigated by Huth and Hoffmeister (**319**).

Nova Sco 1964 was discovered by Przybylski, who also reported several *UBV* observations (**320**). The only other published data seem to be that on the spectrum, by Wilde (**321**).

S5420 Aur has been described by Hoffmeister (**322**) and Popova (**323**), and classified as either a slow nova like RT Ser with maxima in 1960 and 1964, or as a blue variable of new type. In support of the latter, Popova notes that six maxima have been observed since 1900. Additional observations have been reported by Weber (**324**). The spectrum has been observed by Preston.

Nova (T) Pyx 1890, 1902, 1920, 1944 recurred again in December 1966; the flareup was detected first by Jones.

Friedjung (**325**) has attempted to construct physical models of novae that would represent characteristic features of the light curves and of the spectroscopic evolution. His first paper considers various alternatives for the nature of the ejection, the second deals with the temperature and radius variation, while the third deals with the interpretation of the various absorption systems. Pecker (**326**) has shown how the oscillatory phases of declining novae can be represented by an extension of Grotrian's explanation of the deep minimum of Nova (DQ) Her 1934, through allowance for a time-dependent opacity. Schatzman (**327**) has reviewed in considerable detail the current hypotheses of nova and supernova outbursts, and has presented there some of his own ideas on the subject as well.

21. OLD NOVAE AND HOT SUBLUMINOUS VARIABLES

The discovery that at least some of the quasi-stellar radio sources are variable in light raises the possibility that extraneous objects of this kind may appear in lists of faint variable stars, especially in higher galactic latitudes. Hopefully, such intruders can be recognized once radial velocities and/or radio observations become available.

Except for this future and probably minor complication, the general outline of this subject has not changed over the past three years, so we subdivide it in the same way as in the last Report.

(i) Novae near minimum light

Nova (V528) Aql 1945 was observed photographically by Taffara (**328**) during 1945–63, as it declined to $m = 18$.

Nova (T) CrB 1866, 1946 was observed to undergo a very brief flare by Ianna (**329**). The range was about a magnitude in *U*, and the duration probably less than one hour.

Nova Her 1963 has not yet returned to pre-outburst minimum light. The light is subject to short-period fluctuations; Almár (**330**) also observed a minimum rather like a partial eclipse, of depth 0·1 magnitude. Further observations reported by Chincarini and Howard (**331**) showed no clear evidence of eclipses, although there was some suggestion of repetition after an interval 6^h.

Nova (DI) Lac 1910 has been observed by Mumford (**332**). Erratic fluctuations with a range of 0·26 magnitude were found, but no eclipses.

Nova (V529) Ori 1677, 1740, 1894? has been discussed by Ashbrook (**333**) who concludes that in view of the dubious nature of the evidence, the Nova is non-existent. However, Eggen (**334**) has reported briefly on his photoelectric observations of a magnitude 14 object identified with the Nova which is not variable in *B* and *V*, but fluctuates irregularly in *U*. This conflict should be resolved.

Nova (GK) Per 1901 has been found to 'flicker' with a range of 0·2 to 0·4 magnitude by Mumford (**332**). The star was usually bright in August 1966, according to Peltier (**335**) and Huth (**336**), becoming at least two magnitudes brighter (visually) than usual.

Nova (RR) Pic 1925 was discovered by Van Houten (**337**) to be variable with a period of 0^d145. The light curve is quite unlike those of conventional eclipsing stars, but does resemble those increasingly familiar curves such as VV Pup.

Nova (WZ) Sge 1913, 1946 has been studied in detail by Krzeminski and Kraft (**338**). The system differs from others that have been studied thus far in that the spectrum of what appears to be a o^2 Eri-type white dwarf is present. The fact that primary minimum occurs at elongation, according to the velocity curve, makes necessary a rather complicated model of the system.

Nova (CK) Vul 1670 probably still has not been identified. Observations by Walker (**339**) of a variable discovered by Wachmann in this area show none of the characteristics of old novae.

(ii) Hot variables, and short-period eclipsing systems with additional activity.

BD + 14° 341 is the brightest known member of this class. Unpublished photoelectric observations by Smak have now been analyzed by Williams (**4**). There is evidence for a periodicity of 0^d117 superimposed on more rapid activity. Smak has recently obtained more extensive observations of the star at Haute-Provence, and a thorough analysis of the object is anticipated.

EM Cyg has been found to eclipse; Mumford (**340**) has determined an accurate period (about 0^d291).

A bright, blue irregular variable very near the position of the Scorpius X-ray source was discovered by Osawa and Jugaku (**341**) and independently by Johnson and Stephenson (**342**). The color, irregular variability and spectrum of this object are much like those of other hot subluminous variables, and one does not hesitate in adding this to the class. Why only this particular member should apparently be a strong X-ray emitter is an unanswered question.

V Sge has been discussed in detail by Herbig, Preston, Smak and Paczynski (**1**); it is a double-line eclipsing binary having a period of 0^d514, and showing much non-periodic activity. Romano (**343**) has reexamined his photographic observations of 1956–58, recovered the short-period light curve, detected a number of primary minima, and obtained a slightly improved period. Smak (**344**) has made new photoelectric observations that, if combined with Romano's, suggest that the period may be decreasing slightly. Smak points out the importance of the investigation of times of minima of V Sge on collections of old photographic plates, to determine if this period change is real. It would also be of interest to examine the great number of visual observations of V Sge that have been made by the amateur associations over the past years, to see if significant information on the period length could be obtained.

Ton S 120 (a notation used by Greenstein for entry no. 120 in the list of blue stars found in the south galactic cap by Chavira (**345**)) was observed by Greenstein (**346**) to have a spectrum like that of an old nova or U Gem star. Krzeminski (**347**) found that the star is an eclipsing binary with a period of $0\overset{d}{.}145$.

UX UMa was observed by Bretz through four minima in 1965 in a four-color system.

(*iii*) *U Gem and Z Cam stars*

AE Aqr has been observed photoelectrically and spectroscopically by Walker (**348**), who states that contrary to earlier belief, the spectrum of the secondary star is not visible. Furthermore, the binary period is not $0\overset{d}{.}71$, but near $0\overset{d}{.}4$.

Z Cam is discussed by Mayall (**349**), who shows the very interesting AAVSO light curve for 1945–65.

2·1937 Cet = *HV 8002* was observed by Paczynski (**350**) in *UBV*. The familiar rapid flickering was found, as well as two outbursts. According to Kraft (**351**), the star is a single-line binary of type sdBe and period $0\overset{d}{.}160$.

SS Cyg has been studied statistically by Lortet-Zuckermann (**352**) in order to classify both the major outbursts and the smaller oscillations observed at minimum light. These data were then analyzed from the point of view of the Markov chain theory of successive events (**353**) but it was found that the theory did not represent at all well the succession of outbursts and oscillations actually observed in SS Cyg. A few *UBV* observations of SS Cyg are given by Eggen and Sandage (**354**). Observations of the star in *UBV* were also made by Mirzoian and Bretz during the descent from a maximum in September 1966.

U Gem, an eclipsing binary with period of $0\overset{d}{.}177$, was observed by Mumford (**355**) about a day before the beginning of a major outburst, but a physical interpretation of the changes that he found was not possible. Krzeminski (**356**), from more extensive data, was able to discuss the system in some detail. One interesting feature of his model is that the eruptions occur not on the hotter star but on the larger, cooler component of the system.

EX Hya was observed in *UBV* by Mumford (**357**), and the variations compared by him to those of Nova (WZ) Sge. The star is an eclipsing binary with period of $0\overset{d}{.}0682$; according to Mumford (**340**), the period is quite constant.

Mumford (**358**) has reported the results of extended photoelectric runs on a number of other U Gem or Z Cam stars, all of which were found to 'flicker': AB Dra, RX And, CN Ori, TZ Per, RU Peg, SU UMa, AH Her, and YZ Cnc. The last two stars also have been observed by Miss Bretz (**9**, p. 244); her observations of YZ Cnc continue. Mumford also studied SS Aur, but found no eclipses; this star is a single-lined sdBe binary of period $0\overset{d}{.}180$, according to Kraft (**351**).

A number of new U Gem and Z Cam variables, some of them fairly bright, have been discovered or recognized in the last few years. The brightest are DX And, *11–16* (Weber, **359**); GR 102, *10·9–17·5* (Romano, **360**); GR 91, *10·9*– $<$ *15·0* (Löchel, **361**); and Wr 160, *12·5–17*: (Weber, **362**). Fainter members of the same types have been announced by Hoffmeister (**363**) and by Meinunger (**364**).

The mean absolute magnitudes of the U Gem variables have been redetermined by Kraft and Luyten (**351**) from new proper motions of 25 members, and radial velocities of 11. A mean value of $M_v = +7{\cdot}5$ at minimum light is found, with some evidence for a period-luminosity dependence. An earlier proposal by Kraft on an evolutionary relationship between binaries of the W UMa and U Gem types had been criticized by Popov (**365**) on the basis of dissimilarities in the space distributions and kinematics. Kraft replied to these criticisms (**366**) taking advantage of the newer distance scale, and allowance for the effect of incompleteness upon the space distributions; he believes that his original proposal is substantiated. Kopal, on the other hand, believes that the masses require that these stars were not binary at the time they were on the main sequence (**9**, p. 52).

(*iv*) *Variable nuclei of planetary nebulae*

The outstanding example of this class is FG = 377·1943 Sge. Three-color photoelectric observations by Wenzel (**367**) show that in B and V the star continues to brighten at about 0·05 magnitude per year, but that the star is becoming fainter in U. Lick spectrograms show that this is due to an increase in the Balmer jump. The star is also being observed photometrically by Bertaud. Kohoutek (**368**) has discovered that the nucleus of the (presumed) planetary nebula K1–2 is variable. The secular change in the emission-line spectrum of the bright planetary nubula IC 4997 has been interpreted by Aller and Liller (**369**) as due to the change in electron temperature and density of the gas resulting from expansion. There is some evidence that the radial velocity has also changed (Smak and Preston, **163**), and it might be asked if variability of the exciting star may not also be a contributing factor. The assumption that the hot stars which illuminate planetary nebula are constant, despite the fact that rapid variability is widespread among other subluminous blue stars, should be re-examined.

REFERENCES

1. Herbig, G. H., Preston, G. W., Smak, J., Paczyński, B. 1965, *Astrophys. J.*, **141**, 617.
2. Walker, M. F. Unpublished.
3. Smak, J. Unpublished.
4. Williams, J. O. 1966, *Publ. astr. Soc. Pacif.*, **78**, 279.
5. 1964, *R. Obs. Bull.*, no. 32.
6. 1966, *Trans. IAU*, **12B**, 443.
7. *First Conference on Faint Blue Stars.* 1965, Ed. by W. J. Luyten, Observatory, Univ. of Minnesota.
8. *Problems of Stellar Evolution and Variable Stars.* 1967, Moscow.
9. *Kleine Veröff. Remeis-Sternw.* 1965, **4**, no. 40.
10. *The Magnetic and Related Stars.* 1966, Mono Book Corp., Baltimore, Md.
11. Tsesevich, V., Kazanasmas, M. 1963, *Atlas of Charts for Variables of RR Lyrae Type and Others*, Odessa.
12. Tsesevich, V. 1966, *The RR Lyrae Type Stars*, Kiev.
13. Beyer, M. 1965, *Landölt-Bornstein*, new ser., Group VI, **1**, sec. 6·2. Ed. by H. H. Voigt.
14. Perova, N. B. 1963, *Peremennye Zvezdy Bjull.*, **14**, 357.
15. *Astronomische Abhandlungen* (Professor Dr C. Hoffmeister zum 70 Geburtstage gewidmet) Barth, 1965.
16. Hodge, P. W., Wright, F. W. 1966, *The Large Magellanic Cloud.* Smithsonian Institution, Astrophysical Observatory.
17. Zhevakin, S. A. 1963, *A. Rev. Astr. Astrophys.*, **1**, 367.
18. See **9**, p. 7 for references.
19. Christy, R. F. 1966, *A. Rev. Astr. Astrophys.*, **4**, 353.
20. Christy, R. F. 1966, *Astrophys. J.*, **145**, 340.
21. Christy, R. F. 1966, *Astrophys. J.*, **144**, 108.
22. Christy, R. F. 1966, *Astrophys. J.*, **145**, 337.
23. Meyer, W. F. 1934, *Publ. astr. Soc. Pacif.*, **46**, 202.
24. Ledoux, P. in ref. **10**.

25. Kraft, R. P. 1966, *Astrophys. J.*, **144**, 1008.
26. Nikonov, V. B. 1965, *Trans. IAU*, **12A**, 338.
27. Sharov, A. S. 1963, *Astr. Zu.*, **40**, 754.
28. Jackisch, G. 1963, *Veröff. Sternw. Sonneberg*, **5**, 231.
29. Öpik, E. J. 1964, *Irish Astr. J.*, **6**, 174.
30. Jackisch, G. 1966, *Mitt. astr. Ges.*, **21**.
31. Blanco, V. M., Plaut, L. 1965, *Publ. astr. Soc. Pacif.*, **77**, 466.
32. Shakhovskoy, N. M. 1963, *Astr. Zu.*, **40**, 1055; 1964, **41**, 1042.
33. Grigorian, K. A. 1966, *Soobšč. Bjurak. Obs.*, **37**, 18.
34. Vardanian, R. A. 1966, *Soobšč. Bjurak. Obs.*, **37**, 23.
35. Vardanian, R. A. 1964, *Soobšč. Bjurak. Obs.*, **35**, 3.
36. Ledoux, P. 1965, *Stars and Stellar Systems. VIII. Stellar Structure.* Ed. by L. H. Aller and D. B. McLaughlin, Univ. of Chicago Press, Ch. 10.
37. Ledoux, P. 1964, *Astrophys. norw.*, **9**, 187.
38. Richter, G. 1967, *Veröff. Sternw. Sonneberg*, in press.
39. Plaut, L. 1965, *Stars and Stellar Systems. V. Galactic Structure.* Ed. by A. Blaauw and M. Schmidt, Univ. of Chicago Press, Ch. 13.
40. Plaut, L. 1963, *Bull. astr. Inst. Nethlds.*, **17**, 81.
41. Plaut, L. 1966, *Bull. astr. Inst. Nethlds. Suppl.*, **1**, 105, (no. 33).
42. Cousins, A. W. J. 1966, *R. Obs. Bull.*, no. 122.
43. 1960, *Trans. IAU*, **10**, 409.
44. Hoffleit, D. *Astr. J.*, **68**, 207, 1963; **69**, 301, 1964; **70**, 307, 1965; **71**, 130, 1966.
45. Wachmann, A. A. *Astr. Abh. Hamburg*, **6**, no. 3, 1964; **7**, no. 4, 1964.
46. Miller, W. J. 1965, *Ric. astr. Specola astr. Vatic.*, **7**, 1, 197.
47. Harwood, M. 1962, *Ann. Sterrew. Leiden*, **23**, 387.
48. Baade, W., Swope, H. *Astr. J.*, **68**, 435, 1963; **70**, 212, 1965.
49. Tammann, G. A., Sandage, A. 1966, *Astr. J.*, **71**, 400.
50. Preston, G. W. 1964, *A. Rev. Astr. Astrophys.*, **2**, 23.
51. Woolley, R. 1965, *Quart. J. R. astr. Soc.*, **6**, 125.
52. Tsesevich, V. P. 1966, *The RR Lyrae type Stars.* Kiev.
53. Hoffmeister, C. *Astr. Nachr.*, **287**, 169, 1963; **288**, 49, 1965.
54. Kinman, T. D. *Astrophys. J. Suppl.*, **11**, 199, 1965; Lafler, J., Kinman, T. D. *Astrophys. J. Suppl.*, **11**, 216, 1965.
55. Kinman, T. D., Wirtanen, C. A., Janes, K. A. 1965, *Astrophys. J. Suppl.*, **11**, 223.
56. Kinman, T. D., Wirtanen, C. A., Janes, K. A. 1966, *Astrophys. J. Suppl.*, **13**, 379.
57. Ref. **7**, p. 107.
58. Luyten, W. J., Erickson, R. R. 1965, *Publ. astr. Soc. Pacif.*, **77**, 472
59. Plaut, L. 1966, *Bull. astr. Inst. Nethlds. Suppl.*, **1**, 105.
60. Arp, H. 1965, *Astrophys. J.*, **141**, 43.
61. Ponsen, J. 1966, *Bull. astr. Inst. Nethlds. Suppl.*, **1**, 3.
62. Oosterhoff, P. Th., Walraven, Th. 1966, *Bull. astr. Inst. Nethlds.*, **18**, 387.
63. Oosterhoff, P. Th. 1966, *Vistas in Astr.*, **8**, 179.
64. Van Herk, G. 1965, *Bull. astr. Inst. Nethlds.*, **18**, 71.
65. Jones, D. H. P. 1966, *R. Obs. Bull.*, no. 112.
66. Warren, P. R. 1966, *Mon. Not. R. astr. Soc. So. Africa*, **25**, 103.
67. Breger, M. *Mon. Not. R. astr. Soc. So. Africa*, **23**, 64, 1964; **24**, 27, 1965; **24**, 72, 1965.
68. Kanyo, S. 1966, *Inf. Bull. var. Stars*, no. 146.
69. Tremko, J. 1964, *Mitt. Sternw. Budapest*, no. 55.
70. 1965, *Bull. astr. Obs. N. Copernicus Univ. Toruń*, no. 38.
71. Robinson, L. 1964, *Inf. Bull. var. Stars*, no. 69.
72. Balázs-Detre, J., Detre, L. Ref. **9**, p. 184.
73. Sturch, C. 1966, *Astrophys. J.*, **143**, 774.
74. Frolov, M. S. 1965, *Astr. Cirk.*, no. 314.
75. Preston, G. W., Paczynski, B. 1964, *Astrophys. J.*, **140**, 181.
76. Preston, G. W., Smak, J., Paczynski, B. 1965, *Astrophys. J. Suppl.*, **12**, 99.
77. Fernie, J. D. 1964, *Astrophys. J.*, **140**, 1482.

78. Fernie, J. D. 1965, *Astrophys. J.*, **142**, 1072; ref. **9**, p. 106.
79. Fernie, J. D. 1964, *Astrophys. J.*, **140**, 699.
80. Breger, M. 1964, *Mon. Not. astr. Soc. So. Africa*, **23**, 117.
81. Demarque, P., Percy, J. R. 1965, *Astrophys. J.*, **142**, 1082.
82. Gough, D. O., Ostriker, J. P., Stobie, R. S. 1965, *Astrophys. J.*, **142**, 1649.
83. Van den Bergh, S. 1967, *Stars and Stellar Systems. IX. Galaxies and the Universe*. Ed. by A. R. Sandage, Univ. of Chicago Press, Ch. 15.
84. Wesselink, A. J., Shuttleworth, M. 1965, *Mon. Not. R. astr. Soc.*, **130**, 443.
85. Woolley, R., Harding, G. A., Cassells, A. I., Saunders, J. 1965, *R. Obs. Bull.*, no. 97.
86. Kippenhahn, R. Ref. **9**, p. 7.
87. Kippenhahn, R., Thomas, H-C., Weigert, A. 1965, *Z. Astrophys.*, **61**, 241.
88. Hofmeister, E., Kippenhahn, R., Weigert, A. 1964, *Z. Astrophys.*, **60**, 57.
89. Iben, I. *Astrophys. J.*, **142**, 1447, 1965; **143**, 483, 1966; **143**, 505, 1966.
90. Hofmeister, E. Ref. **9**, p. 224.
91. Kopylov, I. M. 1965, *Izv. Krym. astrofiz. Obs.*, **33**, 286.
92. Masani, A., Nelli, M., Martini, A., Albino, E. 1966, Ref. **9**, p. 255; *Inf. Bull. var. Stars*, no. 163.
93. Mitchell, R. I., Iriarte, B., Steinmetz, D., Johnson, H. L. 1964, *Bol. Tonantzintla*, **3**, 153 (no. 24).
94. Walraven, J. H., Tinbergen, J., Walraven, Th. 1964, *Bull. astr. Inst. Nethlds.*, **17**, 520.
95. Breckenridge, J., Kron, G. E. 1963, *Publ. astr. Soc. Pacif.*, **75**, 285.
96. Fernie, J. D. 1963, *Observatory*, **83**, 259.
97. Janes, K. A. 1964, *Astr. J.*, **69**, 131.
98. Fernie, J. D., Demers, S., Marlborough, J. M. 1965, *Astr. J.*, **70**, 482.
99. Oosterhoff, P. Th. 1964, *Bull. astr. Inst. Nethlds.*, **17**, 448.
100. Vasilianovskaya, O. P. *Bjull. astr. Inst. Tadjik SSR.*, in press.
101. Kapko, Ya. T. 1964, *Lvov Circ.*, no. 41.
102. Huth, H. 1965, *Mitt. veränd. Sterne*, **3**, 25.
103. Demers, S. 1966, *J. R. astr. Soc. Canada*, **60**, 23.
104. Demers, S., Fernie, J. D. 1966, *Astrophys. J.*, **144**, 440.
105. Wamsteker, W. 1966, *Inf. Bull. var. Stars*, no. 128.
106. Smak, J. 1966, *Inf. Bull. var. Stars*, no. 135.
107. Nikolov, N. S. 1966, *Inf. Bull. var. Stars*, no. 152.
108. Detre, L. 1966, *Inf. Bull. var. Stars*, no. 152.
109. Fernie, J. D. 1966, *Astr. J.*, **71**, 119.
110. Fernie, J. D. 1966, *Astr. J.*, **71**, 732.
111. Roemer, E. 1965, *Astrophys. J.*, **141**, 1415.
112. Landolt, A. U. 1964, *Astrophys. J. Suppl.*, **8**, 329, 352 (no. 82).
113. Preston, G. W. 1964, *Publ. astr. Soc. Pacif.*, **76**, 165.
114. Malik, G. M. 1965, *Astr. J.*, **70**, 94.
115. Meinunger, L. 1966, *Mitt. veränd. Sterne*, **3**, 150.
116. Smak, J. 1966, *Acta astr.*, **16**, 11.
117. Efremov, Yu. N., Kholopov, P. N. 1965, *Astr. Zirk.*, no. 326.
118. Westerlund, B. E. 1963, *Mon. Not. R. astr. Soc.*, **127**, 71.
119. Efremov, Yu. N. 1964, *Peremennye Zvezdy Bjull.*, **15**, 242.
120. Efremov, Yu. N. 1964, *Astr. Zirk.*, no. 311.
121. Efremov, Yu. N., Kopylov, I. M. 1967, *Izv. Krym. astrofiz. Obs.*, **36**, 240.
122. Kopylov, I. M. 1964, *Izv. Krym. astrofiz. Obs.*, **32**, 88.
123. Tsarevsky, G. S., Urecke, V., Efremov, Yu. N. 1966, *Astr. Zirk.*, no. 367.
124. Payne-Gaposchkin, C., Gaposchkin, S. 1966, *Vistas in Astronomy*, **8**, 191.
125. Gaposchkin, S. 1966, *Sterne u. Weltraum*, **5**, 90; 1965, *Publ. astr. Soc. Pacif.*, **77**, 127.
126. Hodge, P. W., Wright, F. W. 1966, *Astr. J.*, **71**, 131.
127. Hodge, P. W., Wright, F. W. 1963, *Astr. J.*, **68**, 281.
128. Dickens, R. J. 1966, *Observatory*, **86**, 18.
129. Fernie, J. D. 1964, *Astr. J.*, **69**, 258.
130. Woolley, R. 1966, *Observatory*, **86**, 76.

131. Kholopov, P. N. 1965, *Astr. Zu.*, **42**, 1195.
132. Demers, S. 1966, *Thesis*, Univ. of Toronto.
133. Michalowska-Smak, A., Smak, J. 1965, *Acta. astr.*, **15**, 333.
134. Williams, J. A. 1966, *Astr. J.*, **71**, 615.
135. Prikhodko, A. E., Chouprina, R. I. 1965, *Peremennye Zvezdy Bjull.*, **15**, 597.
136. Nikolov, N. S. 1966, *Astr. Zu.*, **43**, 783.
137. Efremov, Yu. N. 1966, *Peremennye Zvezdy Bjull.*, **16**, 18.
138. Efremov, Yu. N. 1963, *Peremennye Zvezdy Bjull.*, **14**, 444.
139. Genkin, I. L. 1964, *Peremennye Zvezdy Bjull.*, **15**, 120.
140. Kraft, R. P. 1965, *Stars and Stellar Systems. V. Galactic Structure*. Ed. by A. Blaauw and M. Schmidt, Univ. of Chicago Press, Ch. 8.
141. Takase, B. 1963, *Astr. J.*, **68**, 80.
142. Howard, W. E., Kirk, J. G. 1965, *Astr. J.*, **69**, 544.
143. Smak, J. 1966, *A. Rev. Astr. Astrophys.*, **4**, 19.
144. Smak, J. 1964, *Astrophys. J. Suppl.*, **9**, 141.
145. Smak, J. 1966, *Acta astr.*, **16**, 109.
146. Mendoza, E., Johnson, H. L. 1965, *Astrophys. J.*, **141**, 161.
147. Smak, J. 1966, *Acta astr.*, **16**, 1.
148. Maffei, P. 1963, *Mem. Soc. astr. ital.*, **34**, 441, and in press; 1967, *Astrophys. J.*, **147**, 802.
149. Mayall, M. W. *J. R. astr. Soc. Can.*, **57**, 237, 1963; **58**, 239, 283, 1964; **59**, 195, 245, 1965; **60**, 205, 1966.
150. Harrington, J. P. 1965, *Astr. J.*, **70**, 569.
151. Houk, N. 1963, *Astr. J.*, **68**, 253.
152. Huth, H. 1965, *Mitt. veränd. Sterne*, **2**, 186.
153. Schneller, H. 1965, *Mitt. veränd. Sterne*, **3**, 86.
154. Bredichin, M. A. 1963, *Peremennye Zvezdy Bjull.*, **14**, 406.
155. Larsson-Leander, G. 1963, *Ark. astr.*, **3**, 285.
156. Magalashvili, N. L., Kumsishvili, J. J. 1965, *Abastumanskaja astrofiz. Obs. Bjull.*, **32**, 9.
157. Sharpless, S., Riegel, K., Williams, J. O. 1966, *J. R. astr. Soc. Can.*, **60**, 15.
158. Low, F. 1965, *IAU Circ.*, no. 1884–5.
159. Stothers, R. 1964, *Publ. astr. Soc. Pacif.*, **76**, 98.
160. Beyer, M. 1965, *Astr. Nachr.*, **288**, 247.
161. Stothers, R. 1963, *Astr. J.*, **68**, 242.
162. Feast, M. W. 1965, *Observatory*, **85**, 16.
163. Smak, J., Preston, G. W. *Astrophys. J.*, **142**, 943.
164. Feast, M. W. 1966, *Mon. Not. R. astr. Soc.*, **132**, 495.
165. Kreiken, E. A. 1964, *Ann. Astrophys.*, **27**, 1.
166. Eskioğlu, A. N. 1963, *C. r. Acad. Sci. Paris*, **256**, 2118.
167. Serkowski, K. 1966, *Astrophys. J.*, **144**, 857; 1966, *Inf. Bull. var. Stars*, no. 141.
168. Donn, B., Stecher, T. P., Wickramasinghe, N. C., Williams, D. A. 1966, *Astrophys. J.*, **145**, 949.
169. Neugebauer, G., Martz, D. E., Leighton, R. B. 1965, *Astrophys. J.*, **142**, 399.
170. Haro, G., Luyten, W. J. 1962, *Bol. Tonantzintla*, **3**, 113 (no. 22).
171. Haro, G., Chavira, E. 1965, *Astrophys. J.*, **142**, 1249.
172. Wing, R. F., Spinrad, H., Kuhi, L. V. 1967, *Astrophys. J.* **147**, 117.
173. Van Hoof, A. 1963, *Z. Astrophys.*, **57**, 278.
174. Van Hoof, A. 1965, *Z. Astrophys.*, **62**, 174.
175. Van Hoof, A. 1966, *Z. Astrophys.*, **64**, 165.
176. Van Hoof, A. 1964, *Inf. Bull. var. Stars*, no. 63.
177. Kupo, I. D. 1965, *Astr. Zu.*, **42**, 358.
178. Milone, L. A. 1965, *Z. Astrophys.*, **62**, 1.
179. Klock, B. L. 1965, *Astr. J.*, **70**, 476.
180. Rossati, F. 1963, *Mem. Soc. astr. Ital.*, **34**, 347.
181. Jerzykierwicz, M. 1965, *Lowell Obs. Bull.*, **6**, 175 (no. 127).
182. Van Hoof, A., Blaauw, A. 1964, *Bull. astr. Inst. Nethlds.*, **17**, 451.
183. Böhm-Vitense, E. 1963, *Publ. astr. Soc. Pacif.*, **75**, 154.

184. Van Hoof, A. 1964, *Z. Astrophys.*, **60**, 184.
185. Clement, M. *Astrophys. J.*, **141**, 1443, 1965; **144**, 841, 1966.
186. Oleinik, G. T., Porfiriev, V. V. 1963, *Astr. Zu.*, **40**, 774.
187. Porfiriev, V. V. 1963, *Astr. Zu.*, **40**, 579.
188. Hitotuyanagi, Z., Takeuti, M. 1964, *Sci. Rep. Tohoku Univ.*, I, **47**, 159 (= *Sendai astr. Rep.*, no. 88).
189. Stothers, R. 1965, *Astrophys. J.*, **141**, 671.
190. Ledoux, P., Renson, P. 1966, *Ann. Rev. Astr. Astrophys.*, **4**, 293.
191. Cowling, T. G. 1965, *Stars and Stellar Systems. VIII. Stellar Structure.* Ed. by L. H. Aller, D. B. McLaughlin, Chicago Univ. Press, p. 454.
192. Steinitz, R. 1964, *Bull. astr. Inst. Nethlds.*, **17**, 504.
193. Van Genderen, A. M. 1965, *Bull. astr. Inst. Nethlds.*, **18**, 67.
194. Renson, P. 1965, *Bull. Soc. R. Sci. Liège*, **34**, 302.
195. Renson, P. 1966, *Bull. Soc. R. Sci. Liège*, **35**, 244.
196. Renson, P. 1965, *Inf. Bull. var. Stars*, no. 108.
197. Steinitz, R. 1965, *Bull. astr. Inst. Nethlds.*, **18**, 125.
198. Babcock, H. W. 1963, *Publ. astr. Soc. Pacif.*, **75**, 74.
199. Klock, B. L. 1965, *Astr. J.*, **70**, 176.
200. Osawa, K., Nishimura, S., Ichimura, K. 1965, *Publ. astr. Soc. Japan*, **17**, 199.
201. Polosuhina, N. S. 1963, *Astr. Zu.*, **40**, 651.
202. Polosuhina, N. S., Lebedeva, L. 1966, *Astr. Zu.*, **43**, 513.
203. Millis, R. L. 1966, *Inf. Bull. var. Stars*, no. 137.
204. Millis, R. L. 1966, *Publ. astr. Soc. Pacif.*, **78**, 340.
205. McCullough, J. R. 1964, *Astr. J.*, **69**, 251.
206. Jones, D. H. P., Haslam, C. M. 1966, *Observatory*, **86**, 34.
207. Wachmann, A. A. 1966, *Inf. Bull. var. Stars*, no. 138.
208. Smyth, M. J. Unpublished.
209. Chen, F.-Y. 1966, *Inf. Bull. var. Stars*, no. 142.
210. Wehlau, W. H., Chen, S. C. N., Symonds, G. 1966, *Inf. Bull. var. Stars*, no. 143.
211. Danziger, I. J., Dickens, R. J. 1966, *Inf. Bull. var. Stars*, no. 150.
212. Magalashvili, N. L., Kumsishvili, J. J. 1965, *Abastumanskaja astrofiz. Obs. Bjull.*, no. 32, 3.
213. Fitch, W., Wehlau, W. 1965, *Astrophys. J.*, **142**, 1616.
214. Jenks, R. L., Schroeder, L. W., Heiser, A. 1965, *Astr. J.*, **70**, 680.
215. Wehlau, W., Leung, K.-C. 1964, *Astrophys. J.*, **139**, 843.
216. Jerzykiewicz, M. 1966, *Lowell Obs. Bull.*, **6**, 249 (no. 133).
217. Oosterhoff, P. Th. 1966, *Bull. astr. Inst. Nethlds.*, **18**, 140.
218. Van Genderen, A. M. 1964, *Bull. astr. Inst. Nethlds.*, **17**, 243.
219. Heiser, A. M., Hardie, R. H. *Astrophys. J.*, **140**, 694.
220. Wildey, R. L. 1966, *Publ. astr. Soc. Pacif.*, **78**, 132.
221. Fitch, W. S. 1966, *Astrophys. J.*, **143**, 852.
222. Notni, P. 1963, *Astr. Nachr.*, **287**, 65.
223. Kuhi, L. V. 1966, *J. R. astr. Soc. Canada*, **60**, 1.
224. Hunger, K. 1964, *Die Umschau*, **23**, 709.
225. Götz, W. 1965, *Veröff. Sternw. Sonneberg*, **7**, heft 1.
226. Herbig, G. H., Peimbert, M. 1966, *Trans. IAU*, **12B**, 412.
227. Garzoli, S. L., Varsavsky, C. M. 1966, *Astrophys. J.*, **145**, 79.
228. Meinunger, L. 1966, *Mitt. veränd. Sterne*, **3**, 137.
229. Hoffmeister, C. 1965, *Veröff. Sternw. Sonneberg*, **6**, 97; 1966, *Trans. IAU*, **12B**, 276.
230. Kharadse, E. K., Bartaya, R. A. 1964, *Abastumanskaja astrofiz. Obs. Bjull*, no. 30, 3.
231. Mosidze, L. N. 1964, *Abastumanskaja astrofiz. Obs. Bjull.*, no. 30, 21.
232. Kumsishvili, I., Magalashvili, N., Abuladze, O. 1964, *Abastumanskaja astrofiz. Obs. Bjull.*, no. 30, 33.
233. Chugainov, P. F., Zaitseva, G. V. 1962, *Peremennye Zvezdy Bjull.*, **14**, 148.
234. Alksnis, A., Daube, I. 1964, *Peremennye Zvezdy Bjull.*, **15**, 194.
235. Mirzoian, L. V., Kazaryan, E. S. 1965, *Astrofizika*, **1**, 213.

236. Ishchenko, I. M., Shevchenko. V. S. 1966, *Tashkent Trans.*, ser. II, 2.
237. Efimov, Yu. S. *Izv. Krym. astrofiz. Obs.*, **37**, in press.
238. Smak, J. 1964, *Astrophys. J.*, **139**, 1095.
239. Aveni, A. F. 1966, *Astrophys. J.*, **144**, 666.
240. Kuhi, L. V. 1966, *Publ. astr. Soc. Pacif.*, **78**, 430.
241. Gurzadian, G. A. 1966, *Astrofizika*, **2**, 217.
242. Kolesnik, I. G., Frank-Kamenetsky, D. A. 1964, *Astr. Zu.*, **41**, 78.
243. Kolesnik, I. G. 1965, *Astr. Zu.*, **42**, 67.
244. Vasilevskis, S., Sanders, W. L., Balz, A. G. A. 1965, *Astr. J.*, **70**, 797.
245. van Schewick, H. 1966, *Die Sterne*, **42**, 64.
246. Mendoza, E. *Astrophys. J.*, **143**, 1010, 1966; **145**, 660, 1966.
247. Poveda, A. 1965, *Bol. Obs. Tonantzintla*, **4**, 15 (no. 26).
248. Becklin, E. E., Neugebauer, G. 1967, *Astrophys. J.*, **147**, 799.
249. Johnson, H. M. 1966, *Astr. J.*, **71**, 224.
250. Hall, R. G. 1965, *Publ. astr. Soc. Pacif.*, **77**, 158.
251. Dibai, E. A. 1963, *Astr. Zu.*, **40**, 795.
252. Herbig, G. H. 1966, *Vistas in Astr.*, **8**, 109.
253. Hayashi, C. 1966, *Ann. Rev. Astr. Astrophys.*, **4**, 171.
254. Herbig, G. H. 1966, *Trans. IAU*, **12B**, 455.
255. Haro, G., Chavira, E. 1966, *Vistas in Astronomy*, **8**, 89.
256. Haro, G. 1967, *Stars and Stellar Systems. VII. Nebulae and Interstellar Matter.* Ed. by B. Middlehurst, Univ. of Chicago Press, Ch. 4.
257. Van Altena, W. F. 1966, *Astr. J.*, **71**, 482.
258. Andrews, A. D. 1965, *Irish astr. J.*, **7**, 15, 20.
259. Mosidze, L. N., Chuadze, A. D. 1965, *Abastumanskaja astrofiz. Obs. Bjull.*, no. 32, 21.
260. Gershberg, R. E., Chugainov, P. F. *Astr. Zu.*, in press.
261. Chugainov, P. F. 1965, *Izv. Krym. astrofiz. Obs.*, **33**, 215.
262. Andrews, A. D. 1966, *Publ. astr. Soc. Pacif.*, **78**, 324.
263. Gershberg, R. E. *Izv. Krym. astrofiz. Obs.*, **32**, 133, 1964; **33**, 206, 1965.
264. Mirzoian, L. V. 1966, *Astrofizika*, **2**, 121.
265. Lovell, B. 1964, *Observatory*, **84**, 191.
266. Lovell, B., Solomon, L. H. 1966, *Observatory*, **86**, 16.
267. Slee, O. B., Solomon, L. H., Patston, G. E. 1963, *Nature*, **199**, 991.
268. Shlysh, V. I. 1964, *Astr. Zu.*, **41**, 1038.
269. Vardya, M. S. 1966, *Mon. Not. R. astr. Soc.*, **131**, 521.
270. Lortet-Zuckermann, M.-C. Ref. **9**, p. 30; 1965, *C. R. Acad. Sci. Paris*, **260**, 6534.
271. Hoffmeister, C. 1966, *Inf. Bull. var. Stars*, no. 126.
272. Eggen, O. J. 1966, *R. Obs. Bull.*, no. 120, E382.
273. Andrews, A. D. 1964, *Irish astr. J.*, **6**, 212.
274. Mendoza, E. E. 1965, *Astrophys. J.*, **142**, 1270.
275. Mammano, A. 1966, *Inf. Bull. var. Stars*, no. 134.
276. Nielsen, A. V. 1965, *Irish astr. J.*, **7**, 14.
277. Kuhi, L. V. 1964, *Publ. astr. Soc. Pacif.*, **76**, 430.
278. Andrews, A. D. 1966, *Inf. Bull. var. Stars*, no. 156.
279. Andrews, A. D., Butler, C. J., Eksteen, J. P. 1966, *Inf. Bull. var. Stars*, no. 157.
280. Andrillat, Y., Morguleff, N. 1966, *Ann. Astrophys.*, **29**, 17.
281. Kron, G. E. 1952, *Astrophys. J.*, **115**, 301.
282. Evans, D. S. 1959, *Mon. Not. R. astr. Soc.*, **119**, 526.
283. Evans, D. S. 1964, *Mon. Not. R. astr. Soc. So. Africa*, **23**, 68.
284. Chugainov, P. 1966, *Inf. Bull. var. Stars*, no. 122.
285. Krzeminski, W., Kraft, R. P. 1967, *Astr. J.*, **72**, 307.
286. Kron, G. E. 1950, *Astr. J.*, **55**, 69.
287. Herbig, G. H., Moorhead, J. M. 1965, *Astrophys. J.*, **141**, 649.
288. Paczynski, B. Unpublished.
289. Chugainov, P. F. Unpublished.
290. Gabriel, M. 1964, *Ann. Astrophys.*, **27**, 141.

291. Waters, B. H. J. 1966, *R. astr. Soc. New Zeal., Var. Star Sec., Circ.*, no. 119.
292. Sterne, T. E. 1934, *Harv. Bull.*, no. 896, 17.
293. Meinunger, L. 1966, *Mitt. veränd. Sterne*, **3**, 157.
294. Wenzel, W. 1963, *Mitt. veränd. Sterne*, **2**, 34.
295. Osawa, K., Nishimura, S., Nariai, K. 1963, *Publ. astr. Soc. Japan*, **15**, 313.
296. Hill, P. W. 1965, *Observatory*, **85**, 41.
297. Boyarchuk, A. A. 1966, *Astrofizika*, **2**, 101.
298. Luud, L. S., Ilmas, M. E. 1966, *Astrofizika*, **2**, 205.
299. Belyakina, T. S. 1966, *Astrofizika*, **2**, 119.
300. Robinson, L. J. 1963, *Inf. Bull. var. Stars*, no. 73.
301. Fitzgerald, M. P., Houk, N., McCuskey, S. W., Hoffleit, D. 1966, *Astrophys. J.*, **144**, 1135.
302. Meinunger, L. 1966, *Mitt. veränd. Sterne*, **3**, 111.
303. O'Dell, C. R. 1966, *Astrophys. J.*, **145**, 487.
304. Plaut, L. 1965, *Stars and Stellar Systems. V. Galactic Structure.* Ed. by A. Blaauw and M. Schmidt, Univ. of Chicago Press, Ch. 14.
305. Svolopoulos, S. 1966, *Publ. astr. Soc. Pacif.*, **78**, 157.
306. Eskioglu, A. N. 1963, *Ann. Astrophys.*, **26**, 331.
307. Dufay, J., Bloch, M. 1964, *Ann. Astrophys.*, **27**, 462.
308. Strohmeier, W. 1964, *Inf. Bull. var. Stars*, no. 47.
309. Chincarini, G. 1964, *Publ. astr. Soc. Pacif.*, **76**, 289.
310. Van Genderen, A. M. 1964, *Bull. astr. Inst. Nethlds*, **17**, 293.
311. Almár, L., Illés-Almár, E. 1966, *Mitt. Sternw. Budapest*, no. 60.
312. Breckenridge, J. B. 1964, *Publ. astr. Soc. Pacif.*, **76**, 110.
313. Götz, W. 1965, *Mitt. veränd. Sterne*, **3**, 41.
314. Busch, H. 1964, *Mitt. veränd. Sterne*, **2**, 89.
315. Weber, R. 1964–66, *Bull. Station Astrophot. Mainterne*, **1**, pp. 3, 17, 32, 47.
316. Clarke, D. 1964, *Observatory*, **84**, 28.
317. Hoffmeister, C. 1961, *Inf. Bull. var. Stars*, no. 45; 1965, *Astr. Nachr.*, **288**, 147.
318. Strohmeier, W. 1964, *Inf. Bull. var. Stars*, no. 59.
319. Huth, H. Hoffmeister, C. 1964, *Inf. Bull. var. Stars*, no. 60.
320. Przybylski, A. 1964, *IAU Circ.*, nos. 1864, 1876.
321. Wilde, K. 1964, *Publ. astr. Soc. Pacif.*, **77**, 208.
322. Hoffmeister, C. 1965, *Inf. Bull. var. Stars*, no. 93.
323. Popova. M. 1965, *Inf. Bull. var. Stars*, no. 97.
324. Weber, R. 1965, *Bull. Station Astrophot. Mainterne*, **1**, 29.
325. Friedjung, M. *Mon. Not. R. astr. Soc.*, **131**, 447, 1965; **132**, 143, 1966; **132**, 317, 1966.
326. Pecker, J.-C. 1964, *Ann. Astrophys.*, **27**, 287.
327. Schatzman, E. 1965, *Stars and Stellar Systems. VIII. Stellar Structure.* Ed. by L. H. Aller and D. B. McLaughlin, Univ. of Chicago Press, Ch. 6.
328. Taffara, S. 1964, *Mem. Soc. astr. Ital.*, **35**, 125.
329. Ianna, P. A. 1964, *Astrophys, J.*, **139**, 780.
330. Almár, L. 1966, *IAU Circ.*, no. 1953.
331. Chincarini, G., Howard, V. S. 1966, *Inf. Bull. var. Stars*, no. 139.
332. Mumford, G. S. 1966, *Astrophys. J.*, **146**, 411.
333. Ashbrook, J. A. 1963, *Sky and Telescope*, **26**, 80.
334. Eggen, O. J. 1961–62, *Ann. Rep. Director, Mt. W. Palomar Obs.*, p. 16.
335. Peltier, L. 1966, *IAU Circ.*, no. 1968.
336. Huth, H. 1966, *IAU Circ.*, no. 1969.
337. van Houten, C. J. 1966, *Bull. astr. Inst. Nethlds.*, **18**, 439.
338. Krzeminski, W., Kraft, R. P. 1964, *Astrophys. J.*, **140**, 921.
339. Walker, M. F. 1963, *Publ. astr. Soc. Pacif.*, **75**, 458.
340. Mumford, G. S. 1965, *Astr. J.*, **70**, 686
341. Sandage, A. R., *et al.* 1966, *Astrophys. J.*, **146**, 316.
342. Johnson, H. M., Stephenson, C. B. 1966, *Astrophys. J.*, **146**, 602.

343. Romano, G. 1965, *Inf. Bull. var. Stars*, no. 98; 1965, *Contr. Oss. astrofis. Univ. Padova, Asiago*, no. 173.
344. Smak, J. 1966, *Acta Astr.*, **16**, in press.
345. Chavira, E. 1958, *Bol. Obs. Tonantzintla*, no. 17, 15.
346. Greenstein, J. L. 1966, *Astrophys. J.*, **144**, 496.
347. Krzeminski, W. 1966, *Inf. Bull. var. Stars*, no. 160.
348. Walker, M. F. 1965, *Sky and Telescope*, **29**, 23.
349. Mayall, M. W. 1965, *J. R. astr. Soc. Canada*, **59**, 285.
350. Paczynski, B. 1963, *Publ. astr. Soc. Pacific*, **75**, 278.
351. Kraft, R. P., Luyten, W. 1965, *Astrophys. J.*, **142**, 1041.
352. Lortet-Zuckermann, M.-C. 1965, *Ann. Astrophys.*, **27**, 65.
353. Lortet-Zuckermann, M.-C. 1966, *Ann. Astrophys.*, **29**, 205.
354. Eggen, O. J., Sandage, A. R. 1965, *Astrophys. J.*, **141**, 821.
355. Mumford, G. S. 1964, *Astrophys. J.*, **139**, 476.
356. Krzeminski, W. 1965, *Astrophys. J.*, **142**, 1051.
357. Mumford, G. S. 1964, *Publ. astr. Soc. Pacif.*, **76**, 57.
358. Mumford, G. S. 1966, *Astrophys. J.*, **146**, 411.
359. Weber, R. 1963, *Mem. Soc. astr. Ital.*, **34**, 183.
360. Romano, G. 1964, *Mem. Soc. astr. Ital.*, **35**, 101.
361. Löchel, K. 1965, *Mitt. veränd. Sterne*, **3**, 107.
362. Weber, R. 1966, *Inf. Bull. var. Stars*, no. 123; 1966, *Bull. St. Astrophot. Mainterne*, no. 11.
363. Hoffmeister, C. 1965, *Inf. Bull. var. Stars*, no. 80.
364. Meinunger, L. *Mitt. veränd. Sterne*, **2**, 162, 1964; **3**, 110, 1965.
365. Popov, M. V. 1964, *Peremennye Zvezdy Bjull.*, **15**, 115.
366. Kraft, R. P. 1965, *Astrophys. J.*, **142**, 1588.
367. Wenzel, W., Fürtig, W. 1967, *Die Sterne*, **43**, 19.
368. Kohoutek, L. 1964, *Bull. astr. Inst. Czech.*, **15**, 161.
369. Aller, L. H., Liller, W. 1966, *Mon. Not. R. astr. Soc.*, **132**, 337.

APPENDIX I

REPORT OF THE COMMITTEE ON THE SPECTRA OF VARIABLE STARS

(Prepared by M. W. Feast, Chairman)

In the following report purely radial velocity work has generally been omitted as has spectroscopic work on eclipsing binaries (except old novae and U Gem stars). For work published in Russian I have relied heavily on a report prepared by Professor B. V. Kukarkin and Dr Yu. N. Efremov. I am extremely grateful to them for their help. As far as possible titles of papers have been included in the bibliography.

1. *The Mira and Semi-Regular Variables*

Keenan (**1**) has published an important catalogue of the spectra of Me and Se variables. His work covers 253 stars and is based on observations made by many observers over a 50-year period. It is of great value to have this large body of data classified on a uniform system. The data refer almost entirely to Miras north of $-35°$ and it is to be hoped that Keenan will be able to extend this work to southern Miras, possibly using existing material. Keenan shows that considerable caution is necessary in applying normal luminosity criteria to the spectra of Mira variables. This is a problem of very considerable importance especially since the dispersion in absolute magnitude at a given period remains almost entirely a matter of conjecture. Further studies will probably require combined spectroscopic and photometric observations throughout the light cycle for a representative selection of Mira variables. Some of the complications found in the study of Mira variables may well be due to the fact that most observations refer to epochs near light maximum, especially since there is a wide range of amplitudes at any given period. Before spectroscopic luminosity criteria of Mira variables can be adequately calibrated more work will be necessary on the absolute magnitudes of these stars by other methods (statistical parallaxes, membership of Mira variables in clusters and in extragalactic systems (e.g. Magellanic Clouds)).

Of considerable importance is the progress that has been made by Smak in applying multi-color photometry to Mira variables. From combined photometry and spectrophotometry (2) he has been able to establish the effects of both the TiO bands and the emission lines on *U B V* colours. In another paper (3) he showed how the spectral types of long period variables can be derived from narrow-band photometry (centred on TiO, 4955 Å). He also showed that these results may be combined with a further colour to obtain interstellar reddening for these stars. Coupled with this type of work is the problem of the bolometric corrections for the Mira stars (2, 4, 5).

The recent increase in work in infrared stellar spectroscopy both from ground-based telescopes and balloons has provided much new information about the brighter Mira variables. Spinrad (6) has succeeded in detecting the quadrupole rotation-vibration bands of H_2 in some Mira variables. The lines are in the $1\,\mu$ region and he has detected the $S(2)$, $S(3)$ and possibly the $S(5)$ lines of the (2, 0) band. The H_2 lines appear strongest in *o* Cet near minimum light and have also been detected in W Hya and R Aql. The amount of H_2 detected is less than theoretical prediction though additional opacity (such as the pseudo-continuum of H_2O suggested by Auman) may resolve the discrepancy. H_2O in the infrared spectra of Mira variables and other stars has been studied by a number of workers (7–12). The bands become extremely strong in certain of the variables. For instance Woolf *et al.* (12) note that in *o* Cet the bands absorb about 1/4 of the total energy emitted by the star. They point out that these bands are likely to have a marked effect on the structure of the star's atmosphere while no other bands will have any significant effects. Spinrad and Newburn (10) find that the amount of H_2O detected in o Cet is a factor of about 60 less than the theoretical predictions. The amount detected varies with phase in the predicted manner (7). Spinrad *et al.* (7) included Miras of types M, MS and S in their survey. There is an H_2O deficiency in the S stars relative to the M types which they interpret in terms of a reduced *O*/*C* ratio in the S types. Spinrad and Younkin (13) also consider that a change in the *O*/*C* ratio may explain the weakness of the infrared *VO* bands in χ Cyg (MS) relative to *o* Cet and R Leo (M types) in which they are very strong. Keenan (14) has discussed the strength of YO bands in variables of spectral types M and S. He indicates that YO may be a good index of heavy metal abundance and also that the abundances of Y and Zr may not be completely correlated in these stars. Fredrick (15) reports that an unidentified feature at 11 028 Å appears rather abruptly in the spectrum of the Mira variable R Leo as the star approaches minimum light. There is a possibility that the band is due to TiO. For spectroscopic work on Mira variables the computations by Dolan (16) of the abundances of polyatomic molecules will be of value.

Detailed quantitative work on the spectra of Mira variables has, in the past, been limited by difficulties in locating the true continuum level even at very high dispersion. Preston (17) has however shown how the excitation temperature and level of ionization may be estimated from the spectra without an exact knowledge of the continuum level. He finds for *o* Cet near maximum light that $\theta_{exc} = 1{\cdot}98 \pm 0{\cdot}14$ (p.e.) and log $P_e \sim -4$. From this he deduces that H^- is not a major source of opacity in the violet region near maximum light. The study of the physical and chemical state of the atmospheres of the cool carbon variables has been continued by Fujita and his colleagues (18–23) and by Wyller (29). Further progress has also been made with the determination of the C^{12}/C^{13} ratio in these stars (24) (25). The high C^{12}/C^{13} ratio ($\sim 2:1$) in some semiregular carbon variables (e.g. Y CVn and WZ Cas) constitutes a serious problem, possibly indicating spallation processes in the atmospheres of these stars (now or at some time in the past) (**25**). The high lithium abundance of some carbon variables (and also some S stars) (**26, 27, 28**) also raises considerable theoretical problems.

From combined spectroscopic and photometric work Wing, Spinrad and Kuhi (**30**) have been able to reach the important conclusion that the majority of 'infra-red' stars are Mira variables with periods longer than a year. The Mira sequence as usually considered is therefore now extended at the long period end by the infrared stars and at the short-period end by the SRd variables (**31**). Amongst the infrared Mira variables, TX Cam appears to have a normal composition whilst the Taurus object is believed to have a low *O/C* ratio like the S star R And but without, however, any appreciable enrichment of *s*-process elements (**30**).

Several papers (**32–35**) give new spectral classifications for long-period variables. Stephenson (**35**) lists five stars which may belong to the group of stars like the semiregular variable GP Orionis which seem to combine S- and C-type characteristics. Several additional members of this class have been found by Feast as a result of slit spectroscopy of stars in Henize's unpublished catalogue of S-type stars. UY Cen and other members of this group are being studied at coudé dispersion in Pretoria. A number of other spectral classification studies of late-type variables are in progress at the Warner and Swasey Observatory. Terrill and Stephenson are studying spectral changes in 20 S-type and 40 M-type variables using objective prism spectra in the near infrared whilst Miss Houk is obtaining similar data for long period variables in the Groningen-Palomar Field 2 (approx. 120 stars). Miss Houk is also studying magnitude and spectral variability of giant M stars in VSF 193 (Sagittarius) also using objective prism spectra. About 440 stars are included in her survey and the frequency of occurrence of variability as a function of spectral type is under study. Also at Warner and Swasey, Miss Wycoff and Stephenson are obtaining slit spectra of about 80 late-type variables in the near infrared (390Å/mm). Stars to $m_v = 14{\cdot}2$ can be observed by their technique.

Herbig (**36**) searched spectroscopically for physical companions to long-period variables. Several new companions were found but they all appeared to be normal F–G stars. Surprisingly no new hot peculiar companions like o Cet B were found. Stephenson and Blanco (**37**) have shown that the double BD + 15°4134 (A 7 III + Me) is probably optical.

Amongst individual stars observed the following may be mentioned: Spectral changes in R Cam have been discussed by Nassau (**40**). The star resembles an S-type at maximum and an M-type at minimum. Some work on the spectrum of L_2 Pup was published by Janes and Gaposchkin (**39**). Z Cir which has been classified as an eclipsing system or a peculiar R CrB variable on the basis of its light curve was found to have the spectrum of a normal Me variable (**42**).

There are still many unsolved problems connected with the emission lines in long period variables. For instance Preston (cf. **38**) has deduced that the difference $(A-E)$ between the velocities given by the absorption and emission lines is a function of space velocity for Mira variables with $150^d < P < 200^d$. On the other hand the possibility that the strengths of the Balmer emission lines is different in semiregular variables of high and low velocity is raised by the work of Keenan and Young (**41**). Wing (**43**) has shown that some emission lines give evidence of being fluorescent emissions excited by Balmer lines. The large and variable polarization found in the integrated light of Mira variables (e.g. (**44, 45**)) suggests that tests should be made to see whether or not the polarization is confined to the continuum or whether the emission lines also show polarization effects. Herbig (see (**6**)) has suggested that H_2 in Mira variables might be detected by Raman-scattered Mg II 2795·5, 2802·7Å. These lines would be well worth searching for in suitable stars. A number of years ago an unpublished survey by Feast of all unidentified emission lines in the spectra of Mira variables indicated that none of these were likely to be due to Raman scattering by H_2 or by any of several other molecules which were expected to be abundant in these stars.

2. *RV Tauri Stars*

Preston (46) made a spectroscopic study of the RV Tau star U Mon. He finds metallic emission lines during rising light. He I 5876Å emission is also present during rising light whilst Hα emission is found at all phases. The absorption line velocities indicate a velocity gradient in the atmosphere. The slow radial velocity changes have a period of 1560 days. Further work is required to determine definitely whether the long-period variations are due to binary motion or not. Preston finds evidence from the H and K lines for two expanding circumstellar shells. A. Piev (unpublished) has discussed the shock wave interpretation of U Mon. A spectroscopic survey of southern semi-regular variables is being carried out in Pretoria and this has already revealed one or two stars which are probably RV Tau stars.

Spectra and photometry of two irregular variable F-type supergiants are discussed by Smak (125). They are *not* RV Tau stars.

3. *Cepheids*

In two papers Rodgers and Bell (47, 48) have discussed the atmosphere of the southern cepheid β Dor in some detail. The star has approximately solar abundances but the abundance derived from Eu II lines in anomalous and varies with phase (the element is found to be over-abundant at all phases except maximum light). It would be very important to discover the cause of this anomaly especially since Eu is an important element in some discussions of nucleogenesis. In their second paper Rodgers and Bell study the macroturbulence in β Dor as a function of phase and interpret the results in terms of a running wave in the atmosphere.

Some years ago Bahner, Hiltner and Kraft (49) noticed some spectral peculiarities in a few distant cepheids with large light amplitudes. A detailed analysis of one of these stars (TV Cam) has now been published (50). Abt, Osmer and Kraft found that this star has a normal chemical composition and that the marked line weakening is due to exceptionally low atmospheric turbulence. It would be interesting to know whether there is any general correlation between atmospheric turbulence and light amplitude.

The peculiar carbon variable RU Cam which is often grouped with the type II cepheids is being studied at coudé dispersion by Hack. Ahnert (51) has reinvestigated the period-spectrum relationship for cepheids. A number of the stars discussed by him which have anomalous spectra types for their periods would be worth further investigation. Schneider at the Institut d'Astrophysique has classified eight cepheids (in the IAP spectral system) throughout their light cycle. He has been searching for line intensity parameters associated with chemical composition in order to discriminate reddening effects from evolutionary effects.

4. *RR Lyrae Stars, Dwarf Cepheids, δ Scuti Stars*

Preston and Paczynski (52) have made a very detailed study of the atmospheres of singly-periodic RR Lyr variables based on simultaneous spectroscopic and photometric work. This is a continuation of work discussed in the last report. The relation between the hydrogen emission, the Bailey type and the $(U-B)$ excess is discussed. Excitation temperatures are derived from Hγ profiles and equivalent widths. Preston and Paczynski discuss the relationship between the light range and the radial velocity range. RR Lyr stars of all Bailey types seem to define a single relation. The slope of the relation as derived from the metal lines is smaller than that from the hydrogen lines although it exceeds the value derived for classical cepheids. In this connection it may be mentioned that there appears to be considerably more scatter in the light range-velocity range relationship for classical cepheids than was at one time thought (cf. Eggen, Gascoigne and Burr (53)).

Preston, Smak and Paczynski (54) have discussed simultaneous photometry and spectroscopy of RR Lyr itself. They particularly discuss the long period (41^d) variations. In 1962 and 1964 most parameters varied in this long-period cycle. However, in 1963 the long-period variations were not present. A model is advanced based on the observed emission and doubling of the H lines. A critical level of shock wave formation is believed to move up and down in the atmosphere of RR Lyr during the 41-day cycle. Metallic line doubling has also been detected. The effects of Balmer emission, metal line blanketing and effective gravity on the $(U-B)$ colour has been discussed. Preston, Smak and Paczynski also discuss problems of temperature determination in RR Lyr. Mlle Fringant observed spectrophotometrically one total secondary cycle of RR Lyr in 1963. She will investigate the blue and ultraviolet gradients, and the Balmer discontinuity. A detailed study of the RR Lyr variable X Ari has been published by Oke (**55**). This work follows the general lines of his previous studies of RR Lyrae stars and cepheids, and combines photoelectric spectrum scans with accurate radial velocities and Hγ profiles. He finds that the interstellar reddening is somewhat greater than had previously been supposed. This increases the metal abundance as compared with earlier estimates although the star is still metal deficient by a factor of about 100. Oke deduces from his data a mean radius of $4{\cdot}8\ R_{\odot}$, a mean absolute visual magnitude $< M_{\mathrm{v}} >$ of $+\ 0{\cdot}8 \pm 0{\cdot}4$ and a mass of $0{\cdot}4\ M_{\odot}$.

The type *a* RR Lyr variable Su Dra with a well-marked $(U-B)$ excess was at one time thought to be anomalous in that, unlike similar stars, no H emission had been detected during rising light. Preston (**56**) has, however, been able to show that strong H emission is present in this star when adequate dispersion and time-resolution are employed. Preston notes that the effects of Balmer continuum radiation must be taken into account in discussing observations obtained by photoelectric scanning techniques. Spectrophotometry of some RR Lyr stars has been carried out by Alania (**57**).

There is considerable current activity in the photometry and spectroscopy of δ Sct itself, and of suspected δ Sct-type variables. Work at Mt Wilson has been reported by Danziger and Dickens (**58**) and spectroscopic work has been carried out at Herstmonceux (**59**) and Pretoria (**59a**) by Jones. Simultaneous spectroscopy and phtotelectric scans are being made of the dwarf cepheids AI Vel, SX Phe and of ρ Pup by Rodgers, Searle and Bessell.

5. β *Canis Majoris Stars*

Milone (**60**) has discussed radial velocity variations in β CMa itself. He is at present working on a kinematic study of the stars atmosphere using Mt Wilson spectra. Van Hoof (**61**) has discussed theories of the multiperiodicity of β CMa stars in the light of the observed axial rotations. Koupo (**62, 63**) has carried out a spectrophotometric study of β CMa on the basis of 60 slitless spectra. She determined the physical parameters of the atmosphere and established that their variations are related to the period P_2. She interpretes the connection of the light variations with each period of radial velocity variations as an effect due to different oscillation periods at different levels in the stellar atmosphere.

6. *R CrB Variables, Symbiotic Variables and Nova-like Variables*

The R CrB variable MV Sgr has been studied by Herbig (**64**). The object is of great interest since it is much hotter than the general run of R CrB stars. Herbig suggests that its spectrum is much like that to be expected if R CrB itself were to be raised to a temperature of about 20 000°. The spectrum is in fact rather similar to that of the helium star BD + 10°2179 although some emission lines (He I 5875, 6678 and possibly also Hα, [S II] and [O I] are found in the yellow-red region. Herbig points out the importance of investigating whether or not the other known helium-rich stars are variable in light.

A detailed high dispersion analysis of the R CrB star RY Sgr has been carried out by Danziger (**132**). The star is generally similar to R CrB itself. The C/H ratio is found to be 25 : 1 and the C/Fe ratio is 35 times normal.

Dossin has given a summary (**65**) of his spectroscopic investigations of a number of symbiotic stars. Between 3300 Å and 5000 Å more than 850 lines have been measured in the spectrum of Z And including many new lines of Fe II and Cr II. The profiles of hydrogen emission lines indicate that the star is surrounded by a slowly expanding atmosphere. Evidence is found for some effects due to fluorescence. BF Cyg and other stars have been studied by Dossin in the photographic infra-red (6000–9000 Å). New emission lines have been observed and the behaviour of the infra-red O I lines has been particularly discussed. The spectra of the symbiotic stars AG Peg, AG Dra and AX Mon were investigated by Boyarchuk and Pronik (**66, 67, 68**). They found that AG Peg is a double star with components WN6 and M 3 III. The character of the WN6 component coincides with that of the nuclei of planetary nebulae. Both components are immersed in a nebula the mass of which is $10^{-3} M_\odot$. The light and spectrum variations of AG Peg are explained by the flares of the hot component. In AG Dra the cool component has the spectral class K 3 III; the hot component is invisible, its luminosity is probably + 4·5. The parameters of the nebula surrounding the star were determined. Evidence for mass loss by the hot component of AX Mon is found. E. Vanderkerkhove reports that he also has carried out spectrophotometric work on AG Peg. Boyarchuk (**69**) studied the spectra of Z And obtained with a dispersion of 16 Å/mm at the Lick Observatory and estimated a number of parameters of the star's envelope. He suggested that symbiotic stars may be related to the stage of evolution of double stars at which the more massive component becomes a planetary nebula, the less massive component still being in the red giant stage. Changes in the spectrum of Z And since 1959 have been discussed by Bloch (**116**). Mlle Bloch has a number of other symbiotic stars under study. Sahade (**70**) reviewed the data on symbiotic variables and discussed the problem of eruptive variability. Wilde (**99**) reports that AS 210 (an Hα emission object) is possibly a symbiotic object. He finds emission lines of H, He I, He II, [O III] and [Ne III]. Absorption by TiO is not seen although the G band may be present. A spectrogram of 1964 May 10, gives evidence for a flare-up of this star.

There is every hope that a full understanding of the unique object η Car will provide important clues to problems of stellar evolution. Thackeray (**71**) has continued his work on this object which has extended over many years. His most recent paper discusses the spectroscopic complexities of this object. Many new [Fe II] and [Ni II] lines have been identified and an excitation temperature of about 8500° has been derived. Thackeray has identified [N I] 3466 in η Car and has noted various temporal changes in the spectrum. For instance [Ne III] disappeared temporarily in 1965 whilst He I emission weakened. Furthermore the structures of the displaced absorption line varies with time. This work emphasises the need for continued detailed spectroscopic work on this object. Photoelectric spectrophotometry of η Car has been carried out by Rodgers and Searle (**72**). They are able to estimate physical parameters for the emitting volume. Thackeray (**74**) had originally pointed out that O was almost certainly underabundant in η Car. Rodgers and Searle show that in addition, N and S are also underabundant. These workers also discuss in some detail the possible origin of the optical continuum. Photoelectric spectrophotometry of η Car has also been carried out by Aller (**73**). Both Rodgers and Searle, and Aller reach the very interesting conclusion that the gaseous mass associated with η Car is less than one solar mass.

The variable RR Tel which is generally referred to as a slow nova has been studied quantitatively by Friedjung (**75**) on the basis of Thackeray's spectrograms. Half-intensity widths were measured for emission lines and interpreted in terms of a deceleration in the gaseous envelope. Electron temperatures were also discussed.

McCuskey discovered in July 1965 that MHα 328–116 had brightened by about three magnitudes during the preceding three years. The spectrum at maximum has been investigated by Bloch (**76**), FitzGerald *et al.* (**77**) and Rosino (**78**). The object shows affinities to both the slow novae (e.g. RR Tel) and to the symbiotic stars. Many forbidden and permitted emission lines have been detected. The object will undoubtedly be extensively investigated by spectroscopists. Mlle Bloch draws attention to the fact that accurate photoelectric photometry should be carried out on this star to facilitate the discussions of the spectroscopy. She also points out the dearth of photoelectric observations of several other stars she is studying spectroscopically (e.g. 377·1943 Sge which has brightened practically linearly from magnitude 13·5 in 1890 to 10 at present and shows Balmer emission lines (of P Cygni type)). Webster (**124**) has given an account of a number of interesting emission line stars in the southern Milky Way which were originally classified as planetaries. Some of these objects warrant further study. For instance work at Canberra and Pretoria has shown (Henize, 177) to have a rich emission spectrum including [Fe II], [Fe III] and [Fe V]. It is not clear whether this object should be classified with slow novae of the RR Tel type or as a symbiotic object. Studies for possible light variations would be of interest.

Amongst miscellaneous variable stars that have been studied spectroscopically mention should be made of the very luminous irregular variable S Dor in the LMC. Thackeray (**126**) studied spectra at a recent faint phase when [Fe II] was prominent. Spectroscopic resemblances to η Car are noted. Barbon *et al.* (**127**) have obtained a long series of spectra of the Wolf Rayet star HD 50896. Although they find marked radial velocity changes to occur they have been unable to understand them in terms of binary motion. They suggest the possibility that the velocity effects are due to activity in the W-R star's atmosphere. Further work on this star is clearly desirable. Vanderkerkhove is measuring line intensities and gradients in γ Cas whilst Herman (**128**) has discussed variations in the hydrogen profiles in P Cyg.

7. *Flare Stars, T Tauri Stars and Related Variables*

The detailed spectroscopic investigation of mass loss from T Tau stars by Kuhi mentioned in the last report has now been published (**80**). This work has been supplemented by a spectroscopic study of two additional T Tau stars (**81**). The early conclusions have been strengthened by this work. Kuhi has also shown that the Wilson-Bappu relation (on Ca II reversals) does not apply to T Tau stars (**82**). The effect of line emission on $(B-V)$ colours of T Tau stars was studied by Aveni (**83**) who concluded that the effect was sufficient to significantly affect the computed contraction times for some stars.

Simultaneous photoelectric and spectral observations of flare stars was carried out by Gershberg and Chugainov (**84**). Five flares of AD Leo and six of UV Cet were observed. The spectra were obtained with the 2·6-meter Shajn telescope using a grating spectrograph (50 Å/mm) and an image converter. The variations of intensity of hydrogen and helium emission lines and the intensity jumps near the limits of TiO bands were obtained for different phases of flares. Shevchenko (**85, 86**) has investigated the relationship between the amplitudes and spectra of RW Aur stars in certain stellar aggregates. He interprets the results as a relation between the light variations and the relative age of the star during its gravitational contraction. Rachkovskaya (Bartash)((**87, 88**) carried out a spectrophotometric investigation of RW Aur stars of early spectral type. There appeared to be no differences from normal stars. Mirzoyan and Kazaryan investigated the variations in the continuous spectrum of RW Aur in November 1962. An emission at short wavelengths was observed at times of decreasing light. The emission shifted towards the red as it increased its intensity. Shevchenko (**89**) has investigated the distribution of axial velocities of early-type stars in the Orion aggregate. He finds that these stars are intermediate between the Oe and Be stars and the normal OB stars. He has suggested that Be stars could be an early stage of B star evolution and that the Orion Nebula could be formed by the matter thrown off by rapidly rotating Be stars. Bartaya and Kharadze (**90**) found that the ultraviolet continuous emission in RW Aur was always present.

The emission line spectrum (Balmer lines + Ca II, K) of the flare star YZ CMi which has shown some recent activity, has been described by Chincarini and Mammano (**91**). Mammano

(**92**) has confirmed the conclusions reached by observers at Herstmonceux and Edinburgh that BD + 31° 1048 has a normal B8 absorption spectrum despite its recent identification as a flare star by Andrews (**93**).

An important paper on the variable FU Ori has been published by Herbig (**94**). He considers that the most satisfactory explanation of the 1936 flare-up of this star is that it represents a pre-main sequence collapse. Some further discussion of this interpretation is given by Hayashi (**95**). Herbig's investigation includes some detailed spectroscopic work. The low-dispersion spectra of FU Ori resemble a G-type supergiant with abnormally strong hydrogen lines. However, Herbig's work at coudé dispersion shows the spectrum to be quite peculiar. There are two components to the spectra, a high luminosity F type spectrum (F2:p I–II) with a shell spectrum displaced 80 km s^{-1} to the blue. The Li/Ca ratio in FU Ori is found to be about 80 times the solar value (as in the T Tauri stars).

8. *Novae*

A large number of papers dealing with the spectra of novae were presented at the 1963 Haute-Provence Colloquium. These papers have now appeared in print (**95a**) and are not listed separately. Besides this there have been a considerable number of papers published elsewhere—most of these (and many at the Colloquium) refer to the bright Nova Her 1963. Most such papers on this and other novae are relatively brief accounts dealing with the spectral changes over a limited period. No attempt will be made to summarize the individual papers. In the case of Nova Her 1963 in particular, it is to be hoped that someone will undertake the task of bringing together all the data obtained for this star and will make a definitive study of the results. Although strictly the province of Commission 28, it is of interest to note here that the search for Novae in M 31 is being continued by Rosino (**95a**) and that a start has been made in a systematic nova survey in the Magellanic Clouds.

General. Pecker-Wimel discusses (**95a**) the theoretical interpretation of the presence of coronal lines in novae spectra whilst Swings (**95a**) indicates the possibilities in the study of novae spectra that will result from orbiting telescopes. A general discussion of the various absorption systems in the spectra of novae has been given by McLaughlin (**95a**). Andrillat compares the near infrared spectra of Novae Her 1960 and 1963; some features in common with Wolf-Rayet star spectra are noted. Underhill (**95a**) also compares the spectra of Wolf-Rayet stars and novae.

Nova Pup 1963. Short accounts of the spectrum have been published (**97, 98**).

Nova Sco 1964. The spectrum is described by Wilde (**99**).

Nova (RS) Oph 1958. Folkart, Pecker and Pottasch (**95a**) have determined the profiles of Balmer lines. The narrow emission and absorption lines in the nova are also being investigated. Dufay and Bloch (**95a**) discuss the corrections to the colour temperature necessitated by the existence of considerable interstellar reddening. They note the presence of the diffuse interstellar bands in the spectrum. Dufay *et al.* (**95a**) give a very detailed discussion of the evolution of the spectrum. They also discuss the possibility that two lines in the red region are a permitted doublet of Ca x (the doublet is observed in the corona).

Nova Her 1960. The spectral evolution was described by Dufay, Bloch and Chalonge (**95a**).

Nova Gem 1912. McLaughlin (**96**) has given a summary of his work on the early stages of the nova.

Nova Her 1963. Papers in (**95a**) deal with, amongst other things (1) Distribution of light in the continuum; (2) Evolution of the various spectral systems; (3) Infrared spectrophotometry. Some idea of the contents of other papers on this nova (**100–111, 130**) may be obtained from their titles. Boyarchuk and Bartash (**112**) report that the abundance of both H and He is down (relative to Sun) by a factor of about ten compared with N and O. Another spectroscopic investigation covering the period 1963–66 was made by Doroshenko. Other spectroscopic investigations of this nova in the U.S.S.R. are (**113, 114**) (**115**).

Nova (DQ) Her 1934. High-dispersion spectra of this nova taken at Mount Wilson soon after outburst are being studied by Larsson-Leander (**95a**). An analysis of the chemical abundance in the envelope of this nova has been made by Mustel and Baranova (**129**).

9. *Hot Subluminous Variables and Ex-Novae*

Since many variables of this type have been shown to be eclipsing binaries they are now the province of Commission 42. However, some mention should be made of them here.

Krzeminski and Kraft (**116a**) have made a detailed study of the ultra-short period ex-nova WZ Sge based on photoelectric and spectroscopic observations. They find that the W UMa type of eclipse curves and the Hα emission velocities are about 90° out of phase. Maximum velocity of recession occurs near principal minimum. They propose a new model for the system which attributes this effect to a rapidly receding stream ejected from the dark star towards the primary. The properties of the radiation field of the white dwarf primary are derived from the Hγ absorption and are shown to be consistent with the observed geometrical extent of the line emitting region. Krzeminski and Kraft also discuss the geometrical and dynamical parameters of the system.

Van Houten (**117**) has shown that the light of the bright ex-nova RR Pic varies with a period of about 3·5 hours. A series of spectra of this object with short exposures has been obtained in Pretoria to examine for spectral and radial velocity changes in this period. Earlier spectroscopic work (**118**) integrated over a considerable fraction of van Houten's period. Boyarchuk and Mustel (**131**) have investigated Nova Aql 1918 spectroscopically. They find that at present $T = 30\,000°$ and $R = 3 \times 10^{10}$ cm. They have determined the physical characteristics of the envelope.

Kraft and Luyten (**118a**) found the statistical parallaxes of U Gem variables give $M_v = +7{\cdot}5$ whilst spectral classifications of four of the stars give $M_v \sim +5{\cdot}4$ for the red components ($+4^m_{\cdot}8$ for the total light). They interpret this to mean that the stars have a considerable range in absolute magnitudes with the stars of longest orbital period being the brightest. Bretz (**119**) has obtained spectroscopic observations of the U Gem stars YZ Cnc and AH Her. She has not been able so far to find evidence of binary nature for these two stars. Work on SS Cyg was carried out by Mirzoyan and Kalloglyan (**120**) who discuss a non-thermal continuum present near light minimum. Chuvaev has been studying this star spectroscopically with a time resolution of two minutes. Spectroscopic work on Rosino's variable near M 88 (= AL Com) has been carred out by Bertola (**121**) and by Moorhead (**122**). A nearly featureless spectrum is reported with shallow Balmer absorptions. The variable is probably a U Gem star. Efremov and Kholopov (**123**) also reach this conclusion on the grounds of the light variations.

To conclude this section especial mention should be made of the very detailed study of the nova-like object V Sge by Herbig *et al.* (**79**) on the basis of spectroscopic and photometric observations. The object is apparently an eclipsing variable with a period of 1/2 day which undergoes various small- and large-scale fluctuations in light. The spectrum of V Sge resembles a WN5 star but it contains sharp fluorescent O III lines arising in both components of the binary. Most of the emission features are broadened and strengthened during a major outburst. Herbig *et al.* have been able to estimate various physical parameters for the system which they believe to represent an advanced stage in the evolution of a close binary system.

BIBLIOGRAPHY

1. Keenan, P. C. 1966, A Catalogue of Spectra of Mira Variables of Types Me and Se. *Astrophys. J. Suppl.*, **13**, 333.
2. Smak, J. 1964, Photometry and Spectrophotometry of Long Period Variables. *Astrophys. J. Suppl.*, **9**, 141.
3. Smak, J. 1966, On the Spectral Classification and Reddening Determination for Long Period Variables. *Acta astr.*, **16**, 109.
4. Smak, J. 1966, Bolometric Corrections for M type Giants. *Acta astr.*, **16**, 1.
5. Tsessevich, V. P. 1965, On Determination of Temperatures of M Stars. *Astr. Cirk. U.S.S.R.*, no. 339.
6. Spinrad, H. 1966, Observations of Stellar Molecular Hydrogen. *Astrophys. J.*, **145**, 195.
7. Spinrad, H., Pyper, D. M., Newburn, R. L., Younkin, R. L. 1966, Further Studies of the Infrared Spectra of Cool Stars: The Water Deficiency in S Stars and the Variation of Water Abundance with Mira's Phase. *Astrophys. J.*, **143**, 291.
8. Danielson, R. E., Rose, W. K., Schwarzschild, M., Woolf, N. J. 1964, New Infrared Observations with Stratoscope II. *Astr. J.*, **69**, 539 (Abstract).
9. Boyce, P. B., Sinton, W. M. 1964, Infrared Interferometric Spectroscopy of Stars and Planets from 1·9μ to 4·0μ. *Astr. J.*, **69**, 534 (Abstract).
10. Spinrad, H., Newburn, R. L. 1965, A Low Dispersion Spectroscopic Search for Water Vapour in Cool Stars. *Astrophys. J.*, **141**, 965.
11. Moroz, V. I. 1966, Infrared Spectra of Stars (λ 1–2·5μ). *Soviet Astr.*, **10**, 47.
12. Woolf, N. J., Schwarzschild, M., Rose, W. K. 1964, Infrared Spectra of Red Giant Stars. *Astrophys. J.*, **140**, 833.
13. Spinrad, H., Younkin, R. L. 1966, Infrared Bands of Vanadium Oxide in Three Mira Stars. *Publ. astr. Soc. Pacif.*, **78**, 65.
14. Keenan, P. C. 1964, Yttrium Oxide in M and S Type Variable Stars. IAU Symposium no. 24, p. 26.
15. Fredrick, L. W. 1963, An Absorption Feature at 11 028 Å in the Spectrum of R Leonis. *Publ. astr. Soc. Pacif.*, **75**, 413.
16. Dolan, J. P. 1965, Stellar Molecular Abundances. I. Polyatomic Molecules. *Astrophys. J.*, **142**, 1621.
17. Preston, G. W. 1965, A Method for Determining the Excitation Temperatures of Late-Type Stars. *Astrophys. J.*, **142**, 990.
18. Fujita, Y., Tsuji, T. 1964, Some Problems Related to the Chemical Abundance of Cool Stars. IAU Symposium, no. 26, p. 307.
19. Fujita, Y., Yamashita, Y., Kamijo, F., Tsuji, T., Utsumi, K. 1964, Some Problems on the Classification of MS and C Type Stars. IAU Symposium, no. 24, p. 94.
20. Fujita, Y., Tsuji, T. 1965, Spectrophotometry of Y Canum Venaticorum. *Publ. Dom. astrophys. Obs.*, **12**, 339.
21. Fujita, Y., Yamashita, Y., Kamijo, F., Tsuji, T., Utsumi, K. 1965, Comparative Study of the Spectra of Some MS and C Type Stars. *Publ. Dom. astrophys. Obs.*, **12**, 293.
22. Utsumi, K. 1963, Spectral Analysis of Y Canum Venaticorum in the Infrared Region. *Publ. astr. Soc. Japan*, **15**, 482.
23. Fujita, Y. 1964, Spectral Features in the Infrared Region of Some Carbon Stars. III. *Tokyo Contr.*, no. 54.
24. Wyller, A. A. 1966, New C^{13} Indicators in Stellar Spectra. *Astrophys. J.*, **143**, 828.
25. Climenhaga, J. L. 1964, C^{12}/C^{13} Abundance Ratio in Cool Carbon Stars. *Astr. J.*, **69**, 537.
26. Torres-Peimbert, S., Wallerstein, G., Phillips, J. G. 1964, Lithium in Carbon Stars. *Astrophys. J.*, **140**, 1313.
27. Bretz, M. C. 1964, Lithium in S Type Stars. IAU Symposium no. 26, p. 304.
28. Torres-Peimbert, S., Wallerstein, G 1966, Lithium in Carbon Stars. II. *Astrophys. J.*, **146**, 724.
29. Wyller, A. A. 1963, Medium Dispersion Spectra of Two Carbon Stars and One Oxygen Star in the One Micron Region. Liège Symposium, p. 479.

30. Wing, R. F., Spinrad, H., Kuhi, L. V. 1967, Infrared Stars, *Astrophys. J.*, **147**, 117.
31. Feast, M. W. 1965, Long Period Variables in Globular Clusters and in the General Field. *Observatory*, **85**, 16.
32. Nassau, J. J., Stephenson, C. B., Caprioli, G. 1964, Spectral Classifications for New and Unclassified Late-Type Variables, Emission Line Stars, S Stars and Planetary Nebulae. *Astrophys. J.*, **139**, 864.
33. Bidelman, W. P. 1965, Spectral Types of Nineteen Red Variable Stars. *Publ. astr. Soc. Pacif.*, **77**, 392.
34. Warner, B. 1963, Spectral Classifications of Some Southern Late-Type Peculiar Stars. *Mon. Not. R. astr. Soc.*, **126**, 61.
35. Stephenson, C. B. 1965, Spectral Classifications for New or Previously Unclassified Carbon Stars. *Astrophys. J.*, **142**, 712.
36. Herbig, G. H. Physical Companions to Long-Period Variables. Bamberg Colloquium 1965, p. 164. *Kleine Veröff. Remeis-Sternw.*, **4**, no. 40, 1965.
37. Stephenson, C. B., Blanco, V. M. 1965, BD + 15°4134: A Probable Optical Visual Binary including a Long-Period Variable. *Publ. astr. Soc. Pacif.*, **77**, 53.
38. Smak, J. 1966, The Long-Period Variable Stars. *A. Rev. Astr. Astrophys.*, **4**, 19.
39. Janes, K. A., Gaposchkin, S. 1964, The Spectrum of L_2 Puppis. *Astrophys. J.*, **139**, 813.
40. Nassau, J. J. 1964, Spectrum Changes in R Camelopardalis. *Publ. astr. Soc. Pacif.*, **76**, 94.
41. Keenan, P. C., Young, W. 1964, Semi-regular Red Variables with High Velocities and Balmer Emission. *Astr. J.*, **69**, 546.
42. Feast, M. W. 1965, The Spectrum of Z Circini. *Comm.* 27 *IAU Inf. Bull. var. Stars*, no. 87.
43. Wing, R. F. 1964, Excitation of Fluorescent Emission by Balmer Lines in Long Period Variables. *Publ. astr. Soc. Pacif.*, **76**, 296.
44. Serkowski, K. 1966, Intrinsic Polarization in the Mira Variables. *Astrophys. J.*, **144**, 857.
45. Serkowski, K. 1966, Long Period Variables with Large Intrinsic Polarization. *Comm.* 27 *IAU Inf. Bull. var. Stars*, no. 141.
46. Preston, G. W. 1964, Metallic Emission and Other Spectroscopic Peculiarities of U Monocerotis. *Astrophys. J.*, **140**, 173.
47. Rodgers, A. W., Bell, R. A. 1964, The Atmosphere of β Doradus. I. Differential Curves of Growth. *Mon. Not. R. astr. Soc.*, **127**, 471.
48. Bell, R. A., Rodgers, A. W. 1964, The Atmosphere of β Doradus. II. Line Profiles, Half-Widths and Central Depths. *Mon. Not. R. astr. Soc.*, **128**, 365.
49. Bahner, K., Hiltner, W. A., Kraft, R. P. 1962, Colours and Magnitudes for 45 Cepheids of the Northern Milky Way. *Astrophys. J. Suppl.*, **6**, 319.
50. Abt, H. A., Osmer, P. S., Kraft, R. P. 1966, Atmospheric Characteristics of a Cepheid of Unusually Large Light Amplitude: TV Camelopardalis. *Astrophys. J.*, **145**, 479.
51. Ahnert, P. 1965, Period-Spectrum Relation of 253 Cepheids. Bamberg Colloquium 1965, p. 237. *Kleine Veröff. Remeis-Sternw.*, **4**, no. 40.
52. Preston, G. W., Paczynski, B. 1964, Atmospheric Phenomena in the RR Lyrae Stars. I. The Singly Periodic Variables. *Astrophys. J.*, **140**, 181.
53. Eggen, O. J., Gascoigne, S. C. B., Burr, E. J. 1957, Photoelectric Photometry of Galactic Cepheids. *Mon. Not. R. astr. Soc.*, **117**, 406.
54. Preston, G. W., Smak, J., Paczynski, B. 1965, Atmospheric Phenomena in the RR Lyrae Stars. II. The 41-Day Cycle of RR Lyrae. *Astrophys. J. Suppl.*, **12**, 99.
55. Oke, J. B. 1966, A Spectrophotometric Study of X Arietis. *Astrophys. J.*, **145**, 468.
56. Preston, G. W. 1965, Hydrogen Emission in SU Draconis during Rising Light. *Astrophys. J.*, **142**, 1262.
57. Alania, I. F. 1965, Spectral Investigation of DX Del. *Abastumansk. astrofiz. Obs. Bjull.*, **32**, 31.
58. Danziger, I. J., Dickens, R. J. 1966, Observations of Variable F Type Stars with Short Periods. *Comm.* 27 *IAU Inf. Bull. var. Stars*, no. 150.
59. Jones, D. H. P., Haslam, C. M. 1966, A New δ Scuti Variable. *Observatory*, **86**, 34.

59a. Jones, D. H. P. 1966, HR 1653 – A New δ Scuti Variable. *Mon. Notes astr. Soc. Sth. Afr.*, **25**, 97.

60. Milone, L. A. 1965, β Canis Majoris. A Discussion of the Oscillations Affecting its Radial Velocity Variations. *Z. Astrophys.* **62**, 1.

61. van Hoof, A. 1964, On the Axial Rotation and the Multiperiodicity of β Canis Majoris Stars. *Z. Astrophys.*, **60**, 184.

62. Koupo, I. D. 1965, Spectrophotometry of β Canis Majoris. *Astr. Zu.*, **42**, 358.

63. Koupo, I. D. 1964, Notes to the Paper by Van Hoof 'Photoelectric Light Curves of β CMa'. *Variable Stars*, **15**, 222.

64. Herbig, G. H. 1964, MV Sagittarii: A Helium-rich Variable Star. *Astrophys. J.*, **140**, 1317.

65. Dossin, F. V. 1964, New Spectra of Symbiotic Stars. *Astr. J.*, **69**, 137 (Abstract).

66. Boyarchuk, A. A., Esipov, V., Moroz, V. 1966, The Continuous Spectrum of AG Peg. *Astr. Zu.*, **43**, 421.

67. Boyarchuk, A. A. 1966, The Variable Star AG Dra. *Astrofizika*, **2**, 101.

68. Boyarchuk, A. A., Pronik, I. I. 1967, *Izv. Krym. astrofiz. Obs.*, **37**.

69. Boyarchuk, A. A. 1964, The Study of Z Andromedae. *Variable Stars*, **15**, 48.

70. Sahade, J. 1965, Symbiotic Variables. Bamberg Colloquium, 1965, p. 140; *Kleine Veröff. Remeis-Sternw.*, 4, no. 40.

71. Thackeray, A. D. 1966, Coudé Spectra of Eta Carinae and the Strongest Lines of [Fe II] and [Ni II]. *Mon. Not. R. astr. Soc.*, in press.

72. Rodgers, A. W., Searle, L. 1966, Spectrophotometry of the Object η Carinae. *Mon. Not. R. astr. Soc.*, in press.

73. Aller, L. H. 1966, Photoelectric Spectrophotometry of η Carinae. *Proc. nat. Acad. Sci.*, **55**, 671.

74. Thackeray, A. D. 1953, Identification in the Spectra of Eta Carinae and RR Telescopii. *Mon. Not. R. astr. Soc.*, **113**, 211.

75. Friedjung, M. 1966, The Physical Conditions and Velocity Distribution in the Envelope of RR Telescopii. *Mon. Not. R. astr. Soc.*, **133**, 401.

76. Bloch, M. 1966, MHα 328–116 Étoile à Raies d'Émission. *Publ. Haute-Provence*, **8**, no. 18.

77. FitzGerald, M. P., Houk, N., McCuskey, S. W., Hoffleit, D. 1966, A Peculiar Emission Star in Cygnus, MHα 328–116. *Astrophys. J.*, **144**, 1135.

78. Rosino, L. 1965, IAU Circ., no. 1917.

79. Herbig, G. H., Preston, G. W., Smak, J., Paczynski, B. 1965, The Nova-like Variable V Sagittae as a Short-period Eclipsing Binary. *Astrophys. J.*, **141**, 617.

80. Kuhi, L. V. 1964, Mass Loss from T Tauri Stars. *Astrophys. J.*, **140**, 1409.

81. Kuhi, L. V. 1966, Mass Loss from T Tauri Stars. II. *Astrophys. J.*, **143**, 991.

82. Kuhi, L. V. 1965, The Wilson-Bappu Effect in T Tauri Stars. *Publ. astr. Soc. Pacif.*, **77**, 253.

83. Aveni, A. F. 1966, The Effect of Line Emission upon (B – V) Colours of T Tauri Objects. *Astrophys. J.*, **144**, 666.

84. Gershberg, R. E., Chugainov, P. F. *Astr. Zu.*, in press.

85. In *Problems of Stellar Evolution and Variable Stars*. Moscow 1967.

86. Shevchenko, V. S. 1964, The Dependence of Amplitude on Spectral Class for Stars of the RW Auriagae Type and its Possible Evolutional Meaning. *Variable Stars*, **15**, 229.

87. Bartash, T. M. 1965, Spectrophotometric Observation of KX Ori, V 359 Ori and V 372 Ori. *Izv. Krym. astrofiz. Obs.*, **34**, 128.

88. Rachkovskaya, T. M. (Bartash). *Izv. Krym. astrofiz. Obs.*, **36**, 1964 and **37**, 1967.

89. Ishchenko, I. M., Shevchenko, V. S. 1966, *Trudy Tashkent. astr. Obs.*, Ser. II, **11**.

90. Bartaya, R. A., Kharadze, E. E. 1965, On the Investigation of the Spectrum of RW Aur. *Astrofizika*, **1**, 405.

91. Chincarini, G., Mammano, A. 1966, Emission Lines in the Flare Star YZ CMi. *Comm. 27 IAU Inf. Bull. var. Stars*, no. 130.

92. Mammano, A. 1966, Spectrum of the Early-type Flare Star BD + 31° 1048. *Comm. 27 IAU Inf. Bull. var. Stars*, no. 134.

93. Andrews, A. D. 1964, A Suspected Flare Star in Auriga. *Armagh Contr.*, no. 46.
94. Herbig, G. H. 1966, On the Interpretation of FU Orionis. *Vistas in Astronomy*, **8**, 109.
95. Hayashi, C. 1966, Evolution of Proto-Stars. *A. Rev. Astr. Astrophys.*, **4**, 171.
95a. Colloque International sur les Novae, Novoïdes et Supernovae. (Haute-Provence, 1963). *Ann. Astrophys.*, **27**, 1964.
96. McLaughlin, D. B. 1965, An Interpretation of the Complex Early Stages of Nova Geminorum 1912. *Astr. J.*, **70**, 326 (Abstract).
97. Herbig, G. H., Wilde, K. 1964, The Spectrum of Nova Puppis 1963. *Comm. 27 IAU Inf. Bull. var. Stars*, no. 53.
98. Geyer, E. H. 1964, Objective Prism Spectrum of Nova Puppis 1963. *Comm. 27 IAU Inf. Bull. var. Stars*, no. 48.
99. Wilde, K. 1965, The Spectra of Two Emission Line Stars: Nova Scorpii 1964 and AS 210. *Publ. astr. Soc. Pacif.*, **77**, 208.
100. Baschek, B. 1964, The Near Infrared Spectrum of Nova Herculis 1963. *Publ. astr. Soc. Pacif.*, **76**, 22.
101. Seitter, W. 1964, Atlas for Objective Prism Spectra and Information obtainable from Low Dispersion Spectrograms of Nova Herculis 1963. *Astr. J.*, **69**, 149 (Abstract).
102. Friedjung, M., Smith, M. G. 1966, A Spectrophotometric Examination of Herstmonceux Spectra of Nova Herculis 1963. *Mon. Not. R. astr. Soc.*, **132**, 239.
103. Andrillat, Y. 1965, Évolution de Nova Herculis 1963 dans le Proche Infra-rouge de Juillet 1963 à Mai 1965. *Ann. Astrophys.*, **28**, 867.
104. Andrillat, Y. 1965, Variation d'intensité de C IV 5812 et 7726Å dans Nova Herculis 1963. *Ann. Astrophys.*, **28**, 387.
105. Glebocki, R., Smolinski, J., Woszczyk, A. 1964, The Continuous Spectrum of Nova Herculis 1963. *Acta Astr.*, **14**, 301.
106. Smolinski, J., Glebocki, R., Woszczyk, A. 1965, The Emission Line Spectrum of Nova Herculis 1963. *Acta Astr.*, **15**, 15.
107. Ahlmark, L. F., Clemart, K.-A. 1964, A Determination of the Expansion Velocities of Nova Herculis 1963. *Ark. Astr.*, **3**, 333.
108. Heng-rong, Hang, Jing-er, Mo., Ru-liang, Liu 1965, Spectrophotometry of N Her 1963 in its Nebular Stage. *Acta Astr. Sinica*, **13**, 218.
109. Heng-rong, Hang, Bao-an, Yao 1964, Photometric and Spectroscopic Observations of Nova Herculis 1963. *Acta Astr. Sinica*, **12**, 49.
110. Batten, A. H. 1964, Low Dispersion Spectroscopic Observations of Nova Herculis 1963. *J.R. astr. Soc. Can.*, **58**, 253.
111. Chincarini, G. 1964, Lo Spettro della Nova Herculis 1963 nella Zona del Vicino Infrarosso (6700–8700Å). *Contr. Padova Asiago*, no. 152.
112. Bartash, I. M., Boyarchuk, A. A. 1965, Spectrophotometric Observations of N Her 1963. I. The Nebular Stage. *Izv. Krym. astrofiz. Obs.*, **33**, 173.
113. Ivanova, N. L., Kazaryan, M. A., Oganesyan, R. X. 1966, Spectrophotometric Investigation of N Her 1963. *Soobšč. Bjurakansk. Obs.*, **37**, 3.
114. Alania, I. F., Popov, M. V. 1965, Nova Herculis 1963. *Abastumansk. astrofiz. Obs. Bjull.*, **32**, 41.
115. Kakaras, G., Straižys, V. 1964, Spectrophotometry of Nova Herculis 1963. *Vilnius Bull.*, no. 10.
115a. Zdahovičus, K., Kalitis, R. 1964, Three Colour Photoelectric Observations of Nova Herculis 1963. *Vilnius Bull.*, no. 10.
116. Bloch, M. 1964, Changements dans le Spectra de Z And. *Ann. Astrophys.*, **27**, 292.
116a. Krzeminski, W., Kraft, R. P. 1964, Binary Stars amongst Cataclysmic Variables. V. Photoelectric and Spectroscopic Observations of the Ultra-Short Period Binary Nova WZ Sagittae. *Astrophys. J.*, **140**, 921.
117. van Houten, C. J. 1966, A Short Period Light-variation of RR Pictoris. *Bull. astr. Inst. Netherl.*, **18**, 439.
118. Feast, M. W. 1953, The Spectrum of RR Pictoris (Nova 1925) in 1952–53. *Observatory*, **73**, 256.

118a. Kraft, R. P., Luyten, W. J. 1965, Binary Stars amongst Cataclysmic Variables. VI. On the Mean Absolute Magnitude of the U Geminorum Variables. *Astrophys. J.*, **142**, 1041.
119. Bretz, M. 1965, Observations of two Variables of the U Geminorum Type, YZ Cancri and AH Herculis. Bamberg Colloquium 1965, p. 244. *Kleine Veröff. Remeis-Sternw.*, **4**, no. 40.
120. Mirzoyan, L. V., Kalloglyan, N. L. 1965, On Continuous Radiation of SS Cyg. *Astrofizika*, **1**, 385.
121. Bertola, F. 1964, The Variable Star near NGC 4501. *Ann. Astrophys.*, **27**, 298.
122. Moorhead, J. M. 1965, Observations of Rosino's Star near M 88. *Publ. astr. Soc. Pacif.*, **77**, 468.
123. Efremov, Yu. N., Kholopov, P. N. 1966, *Astr. Cirk. USSR*, no. 384.
124. Webster, L. B. 1966, Stellar Planetaries in the Southern Milky Way. *Publ. astr. Soc. Pacif.*, **78**, 136.
125. Smak, J. 1964, CO Aurigae and V 395 Cygni: Two Irregularly Variable F Type Supergiants. *Publ. astr. Soc. Pacif.*, **76**, 40.
126. Thackeray, A. D. 1965, Spectroscopic Variation of S Doradus. *Mon. Not. R. astr. Soc.*, **129**, 169.
127. Barbon, R., Bertola, F., Ciatti, F., Margoni, R. 1965, Spectral Variations of the WR Star HD 50896. *Comm. 27 IAU Inf. Bull. var. Stars*, no. 109.
128. Herman, R. 1964, Variations Spectrales de P Cygni. *Ann. Astrophys.*, **27**, 507.
129. Mustel, E. R., Baranova, L. M. 1966, An Analysis of the Chemical Composition of Nova Envelopes. III. *Astr. Zu.*, **43**, 488.
130. Chalonge, D., Bloch, M. 1964, Evolution du Spectre de Nova Herculis 1963. 89[éme] Congrès des Sociétés Savantes, Lyon, 241.
131. Boyarchuk, A. A., Mustel, E. R. 1965, A Spectroscopic Study of V 603 Aql (N Aql 1918). *Astrofizika*, **1**, 337.
132. Danziger, I. J. 1965, A High Dispersion Spectral Study of RY Sagittarii. *Mon. Not. R. astr. Soc.*, **130**, 199.

APPENDIX II

REPORT OF THE COMMITTEE ON VARIABLE STARS IN CLUSTERS

(prepared by Mrs H. B. Sawyer Hogg, Chairman)

This report will follow the outline adopted for the last IAU report (1). Researches now in progress or published since the preparation of the last report will be considered under the following sections. 1. Variables in globular clusters. 2. Variables in galactic clusters and associations, with a separate subsection by Dr G. H. Herbig on irregular variables in young clusters and associations. 3. Variables in star clusters of external galaxies.

As this report shows, researches on variables in clusters are being carried out at more observatories than formerly participated. However, a large proportion of the contributions still come from Rosino and his colleagues in Italy, and from a group of Soviet astronomers.

1. *Variables in Globular Clusters*

(*a*) *Discovery of new variables and derivation of periods.*

There are now 121 clusters catalogued as globular in and around our galaxy. In these, more than 1700 variables have been published, with about 88 clusters examined; four of these clusters have no variables.

Table 1 lists those clusters, with references, in which data on new variables and their positions, and new or revised periods, have actually been published. In some cases, the variables published are in the field surrounding the cluster and this is indicated by an F. If the manuscript is in the hands of this compiler, the data is included in this published section. Information on clusters under investigation, in which data are not yet available for periods or positions of new variables will be found in Section 1(*j*).

Clusters which are now on record for the first time as searched for variables are NGC 1261, IC 4499, 6352 and 6624 while NGC 5286, formerly recorded as without variables, now has seven. All these clusters are the work of Laborde and Fourcade at Cordoba, as shown in Table 1.

Table 1. Published Data on Individual Variables and Periods

Cluster	New Variables	New Periods	Revised Periods	Investigator	Ref.
NGC					
1261	6			Laborde and Fourcade	2
1904	2			Tsoo	3
3201	3 + 7F			Wilkens	4
5024	3	3		Cuffey	5, 6
	1	1	19	Margoni	7, 8, 9
5139			47	Belserene	10
	8 + 7F			Wilkens	4
			54	Geyer and Szeidl	11, 12, 13
5272			113	Szeidl	14
5286	7			Fourcade and Laborde	15
5466			17	Bartolini, Biolchini, Mannino	16
IC 4499	6			Fourcade and Laborde	17
6121			30	Wilkens	18
6205	2 susp.			Tsoo	3
6341			10	Kheylo	19, 20
6352	2 + 10F			Fourcade and Laborde	15
OHP 1	1 + F			Terzan	21, 22
6402	4	37	3	Sawyer Hogg and Wehlau	23, 24
6522			6	Clube	25
6624	3 + 10F			Laborde and Fourcade	2
6626	1	1		Hoffleit	26
6656	2	2		Hoffleit	27, 28
6712	3	7	4	Sandage, Smith, Norton, Rosino	29, 30, 31, 32
7078			35	Makarova and Akimova	33
Pal 13		4		Ciatti, Rosino, and Sussi	34

Note: Periods classed as 'new' are for those variables which have never had a published period. Some of the 'revised' periods are so different from the original that they are in a sense, new.

The number of periods of variables determined for the first time increases rather slowly, while the redetermination or revision of periods in this interval runs to several hundred.

(*b*) *Clusters of special importance.*

Several globular clusters deserve to be singled out in this report because of the very extensive work which has been carried out on them in the last several years. These include the two richest in variable stars.

(1) *Omega Centauri.* This cluster has been under intensive investigation by the Herstmonceux observers with plates taken at the Cape and Radcliffe observatories in South Africa, for positions, motions, radial velocities and magnitudes. In one of the series of publications on this cluster, Dickens and Saunders (**35**) describe the photometry of 45 RR Lyrae variables in *B* and *V*. They compare the relations between colour, amplitude, period and luminosity with those in

M 3 from the observations of Roberts and Sandage (**36**). Uncertainty in the reddening of ω Cen precludes an exact comparison, but they conclude that apparently the variables in ω Cen are consistently bluer at a given period than those in M 3. If the pulsation theory is correct, the absolute magnitude of the variables in ω Cen is 0·2 magnitude brighter than those of M 3.

Belserene (**10**) studied the phase shifts of 47 RR*ab* variables, see Section 1(*d*).

Wilkens (**4**) has published eight new variables in this cluster, and seven in the nearby field, and continues his work on the periods of 130 variables in the cluster.

Geyer and Szeidl (**12**) reported at the Bamberg Colloquium first results on 45 RR Lyrae stars in this cluster from 38 *B* and 38 *V* plates taken with the ADH telescope of the Boyden Observatory. Their independent investigation proved to have 25 stars in common with that of Dickens and Saunders. In general the agreement between the two results is good. The visual mean intensity by Geyer and Szeidl is brighter, on the average by 0·06 magnitude and the colour bluer by 0·08, though larger deviations occur. Geyer and Szeidl believe that their series of observations double in number those of Saunders and Dickens, and obtained in a shorter period of time, are more homogeneous. They consider that the interstellar absorption in front of this cluster is practically zero. The instability zone of the RR Lyrae stars is sharp, at $B-V = 0{\cdot}15$ and $B-V = 0{\cdot}43$. The paper is in press (**13**).

(2) In *Messier* 3, Szeidl (**14**) has continued work started by Ozsvath, with 214 plates taken at Budapest and 17 supplied by Hamburg. On most plates 117 RR Lyrae stars were measurable. He has investigated 112 variables, of which 36 or 32% show well pronounced light curve changes. These periods have the highest relative abundance in the period interval $0^{d}_{\cdot}47$–$0^{d}_{\cdot}56$. In the period-amplitude relation their greatest amplitudes fit the relation of regular RR Lyrae stars as noted by Preston. In the period-amplitude relation for ordinary RR Lyrae stars outside the centre, a phenomenon found earlier by Belserene, the separation of the RR*ab* stars into a short and long period sequence, is clearly shown.

(3) *NGC* 6712 in the Scutum Cloud is the subject of a consecutive series of three papers by Sandage and Smith (**29**), Sandage, Smith and Norton (**30**), and Rosino (**31, 32**), in which the variables are investigated and placed on the colour-magnitude diagram of the cluster. Three new variables are discovered, which with those found earlier by Sawyer (**37**) and Harwood (**38**) brings the total to 20 (though one may not vary). Of these, at least five have periods over one day. This is proving to be a very interesting cluster. Rosino finds that the period distribution of the RR*a* in this cluster of intermediate metal content is that of Oosterhoff group I, with a mean value $P = 0^{d}_{\cdot}548$.

Recently Feast (**39**) has determined the radial velocity of the Mira-type star CH Scuti, period 190·6 days, and shows it to be a member of the cluster. For it, Feast derives a mean emission velocity of -124 km s^{-1}, an absorption velocity of -115 km s^{-1}, compared with the cluster velocity of -131 ± 22 km s^{-1}.

Worthy of special mention too, is Pal 13, where work on the four variables and the C–M diagram by Ciatti, Rosino, and Sussi (**34**) have shown this to be the faintest ($M_v = -2{\cdot}6$) and least massive ($\mathfrak{M} = 10^3\ \mathfrak{M}_\odot$) globular cluster known.

(*c*) *Long Period Variables in Globular Clusters.*

The radial velocity work of Feast is a major contribution in clearing up the problem of cluster membership of long period variables. Feast (**40**) considers it unlikely that there are any known long period variables in globular clusters with $P \geqslant 220^{d}$. From kinematic reasons he thinks these variables are too young to occur in globular clusters. Var. 6 in M 80 and Var. 4 in M 22 were earlier eliminated from cluster membership by radial velocity determination.

Now Feast (**39**) has studied radial velocities for two variables which were left with doubtful standing. He finds that Var. 2 in NGC 4833 is absolutely ruled out as a cluster member, and he considers Var. 1 in NGC 6397 improbable as a member.

His placing of CH Scuti in NGC 6712 adds another Mira star to the very few which exist in globular clusters. Feast notes that the three variables of 200-day period in 47 Tucanae are of normal luminosity.

(*d*) *Period Changes.*

As the observational material on these clusters accumulates over decades and at observatories with different longitudes, it is becoming easier to detect period changes in the variables in clusters. Such changes appear to be present in most of the clusters thoroughly investigated. In some of the clusters such as M 53, variables with changing periods may outnumber those with constant period.

In ω Cen Belserene (**10**) has studied phase shifts of 47 RR*ab* variables relative to older observations. Of these, 34 have increased in period during the past 70 years, eight have decreased and four have changed in both directions. The median rate of change is + 0·11 day per million years. In M 4 Wilkens (**18**) has determined 30 periods more accurately. Of these, 11 RR*a* have constant period, 12 have variable periods equally divided between increasing and decreasing. The six RR*c* have practically constant periods. In M 53 Margoni (**7, 8, 9**) finds 13 of the variables he has investigated have changing periods.

(*e*) *Photoelectric Photometry and Colours of Variables.*

Some investigations now are not concerned with periods of variables, but rather with getting highly accurate magnitudes at various wavelengths for sample variables in a cluster. Five colour photoelectric photometry has been carried out for some variables in ω Cen and M 4 by Oosterhoff and Walraven (**41**) and Ponsen and Oosterhoff (**42**). Sturch (**43**) has obtained colours on the *U*, *B*, *V* system for several RR Lyrae stars near minimum light in the clusters M3, M 5, M 92 and NGC 6171, with the 84-inch (213 cm) Kitt Peak reflector. The values found for $\delta(U-B)$, a metallic line strength indicator, are well correlated with the population types assigned by previous investigators.

Barbon (**44**) has drawn conclusions about the membership of six variables in M 56 by their position on the colour-magnitude diagram he has determined for the cluster. He finds the cepheid, Var. 1, the semiregulars Vars. 3 and 5, Var. 6, an RV Tauri type to be cluster members, while Var. 10, an RR Lyrae star, and Var. 11 a dwarf cepheid, seem to be excluded from membership.

Demers (**45**) in his study of the population II cepheids has determined new *B*, *V* data for two variables in M 2, Var. 1, a W Vir star, and Var. 11, an RV Tauri star.

(*f*) *Variables in the Surroundings of Globular Clusters.*

NGC 6171 is the centre of an area 10° by 10° in which Kurochkin (**46**) has catalogued 16 variable stars, of which 13 are new variables. He finds some excess of space density of RR Lyrae stars in the vicinity of the cluster.

Clusters set in variable-rich regions near the galactic nucleus have been studied by several investigators. Terzan (**21, 22**) has found over a hundred new variables from plates in the red and infrared at Haute-Provence in the region around the globular cluster Haute-Provence No. 1, the galactic cluster Trumpler 26, and 45 Ophiuchi. Other work includes that in the direction of NGC 6522 by Arp (**47**) and Clube (**25**).

(*g*) *Theoretical Papers and General Relationships.*

Review papers and general articles on variables frequently include much useful information on variables in clusters. Among such works on RR Lyrae variables Preston (**48**) finds a dispersion in M_V of ± 0·3 or 0·4 magnitudes for the RR Lyrae stars in different clusters. He discusses the subgroups of RR Lyrae stars in ω Cen. Three disk clusters, 47 Tuc, NGC 6356 and 6838 have a strong concentration of stars at the red end of the horizontal branch, indicating a small RR Lyrae population. The long period boundary of the RR Lyrae's in M 3 is 0·9, but shorter in ω Cen. There are no criteria to define an RR Lyrae star with a period near 1·0 day. Variables with $P \leqslant 0{\cdot}2$ are a distinct family, dwarf cepheids.

Rosino (**49**) gives a comprehensive discussion of the relation of variables to the metal content of the cluster, the distribution of RR Lyrae periods, and the values of the absolute magnitude of the RR Lyrae stars in various clusters. He finds no correlation of the absolute magnitude of the RR Lyrae variables with the metal class or the age of the cluster.

Kukarkin reports that Frolov (**50**) determined with the help of his version of Wesselink's method the luminosities of RR Lyrae variables in a number of globular clusters.

Woolley (**51**) in his comprehensive paper on the RR Lyrae variables discusses some of those in clusters, especially ω Cen. He notes that the amplitude of variation of Bailey type *a* increases from ½ magnitude to 1½ magnitude as the period decreases. Then a break occurs, and all variables with periods below a certain limit have more or less sinusoidal light curves. Type *c* has a low amplitude of ½ magnitude. The mean absolute magnitude of the two groups is nearly the same.

Christy (**52, 53**) has studied in detail the pulsation in RR Lyrae models. Following Schwarzschild's early suggestion that the RR Lyrae type *c* variables are pulsating in the first overtone, Christy has been able to describe how the transition from the fundamental to the first overtone pulsation occurs. He has also determined luminosities and masses for cluster type variables, discussing the possibility that in a given cluster the RR Lyrae stars may show a range of masses from $0{\cdot}45\mathfrak{M}_{\odot}$ to $0{\cdot}60\mathfrak{M}_{\odot}$.

Kheylo (**54**) investigated the variables in M 92 and M 3.

Thänert (**55**) has compared the mean period of the RR Lyrae stars in a globular cluster with the age of the cluster. Some of the more massive RR Lyrae's in the younger globular clusters have shorter mean periods than the less massive ones of older clusters.

Balazs-Detre and Detre (**56**) in their paper on period changes of variables and evolutionary paths in the HR diagram make special reference to M 3. From Szeidl's work they conclude that the standard deviation of random period fluctuations is strongly dependent on the period of RR*ab* stars, and that a continuous evolution through the RR Lyra gap appears improbable.

RR Lyrae stars in clusters are also considered in the large review papers by Arp (**57**) on globular clusters, and by Plaut (**58**) on variable stars.

Fernie (**59**) in his discussion of the period-radius relation for pulsating stars points out Var. 42 in M 5, with period $25^{\mathrm{d}}_{\cdot}74$ as the only W Virginis star which has both a reliably determined colour and absolute magnitude. He also gives data (**60**) on individual W Vir stars in various clusters.

Stothers (**61**) has given a comprehensive discussion of the RV Tauri stars and their absolute magnitudes in a number of globular clusters. Later, with collaborators, he has continued with an investigation of ultraviolet dwarfs (**62**), concluding that the nova in M 80 was a cluster member, that the U Gem star near M 5 is likely a member of that cluster, and that no W UMa stars exist in globular clusters.

(*h*) *Catalogues of Variable Stars in Globular Clusters.*

The Cordoba Observatory is preparing an atlas of photographs of all the globular clusters south of −29°. Fourcade and Laborde are identifying all the variable stars in and around each cluster. The plates are taken with the 154-cm telescope of Bosque Alegre.

The third catalogue of H. Sawyer Hogg, mentioned in the IAU 1964 report, has been delayed in preparation, but should be published within a year or so.

(*j*) *Reports of Miscellaneous Current Investigations.* (Details still to be published).

NGC 5024 R. Margoni is continuing his investigations of M 53 at Asiago. On a new series of blue and visual (103*a*–D + GG11) plates, period and light curves in two colours have been obtained for most of the variables, and their position on a *C–M* diagram has been derived. A preliminary report was given at the Bamberg Conference (**7**), and another paper is in press.

NGC 5139 H. Wilkens of La Plata is continuing the investigations of 130 variables in this cluster.

NGC 5904 Margoni and Stagni are taking plates at Asiago for the study of the variables in *UBV*.
A long series of plates is being accumulated at Konkoly Observatory for period studies.
Miss C. Coutts is studying a 30-year series of plates taken by H. Sawyer Hogg at the David Dunlap Observatory for determination of period changes.

NGC 6171 A paper by Coutts and Sawyer Hogg on David Dunlap observations of these variables will appear shortly.

NGC 6402 The variables are under investigation on more than 120 blue plates taken with the 122-cm reflector of Asiago from 1962–65, and preliminary periods have been derived for many. Margoni and Stagni are now collecting yellow material for the *V* magnitudes.
Sawyer Hogg and Wehlau, with 40 periods completed in this cluster, (**23, 24**) are continuing their work on the other variables on the more than 250 plates taken over 30 years with the two large Canadian reflectors.
No further observations of the nova found by Sawyer Hogg and Wehlau (**63**) have come to light.

NGC 6760 Taffara has determined periods of the variables in the cluster, and of some field variables.

NGC 6934 Sawyer Hogg plans to have the periods of more than 40 variables in this cluster published within a year.

NGC 6981 Margoni is now examining the accumulated Asiago plates to improve the periods and light curves given by Rosino in 1953.

NGC 7006 With material taken at Mt Wilson by Hubble and Baade in 1931–33, plates obtained at Palomar by Sandage, and at Asiago from 1953–58, Rosino has determined improved periods for 27 RR Lyrae variables, and derived blue and visual light-curves. The paper is in print.

NGC 7078 A long series of plates is being accumulated at Konkoly Observatory for period studies.

NGC 7089 Margoni and Stagni have obtained more than 150 blue and yellow plates to determine light-curves in *B* and *V* and the position of the variables in the colour-magnitude diagram.

NGC 7492 Mrs S. A. Barnes is completing a study of the variables in this cluster at the Goethe Link Observatory.

2. *Variables in Galactic Clusters and Associations*

Part A of this section deals with cepheids, eclipsing stars and miscellaneous types of variables, while Part B by Dr George Herbig is concerned with the irregular variables in young clusters and associations.

Part A

(*a*) *RR Lyrae Stars in Galactic Clusters.*

Probably the most important development in this area of Part A since the Hamburg meeting is that RR Lyrae stars continue to be non-existent in galactic clusters. At the time of the Hamburg meetings great interest was aroused by the announcement by Hoffmeister (**64**) from plates of the Karl Schwarzschild Observatory of four short period variables in the old galactic cluster NGC 188. Two short period variables, the stars SVS 1277 and SVS 1284 were also known in M 67. Now Efremov, Kholopov, Kukarkin and Sharov (**65**) and Kukarkin and Kurochkin (**66**) have showed that these rapid variables in M 67 and NGC 188 are not RR Lyrae stars, but are W UMa stars. This was proved by Kurochkin (**67**) for AH Cnc (SVS 1284) in M 67, and by Kholopov and Sharov (**68**) for four stars in NGC 188. They measured 69 plates of NGC 188 (JD 38389–39270) and obtained the *B*, *V* light curves for these stars, which are typical for W UMa stars.

Some of these stars were discussed at the Variable Star Symposium in Bamberg where Eggen (**69**) reported also that the light curve and colour showed SVS 1284 to be a contact binary, and Mammano reported that a slit spectrum at Asiago revealed it to be type F5–F7.

In NGC 6871 a new RR Lyrae star found by Bakos (**70**) with a period of 0·275 days cannot be a cluster member as its distance is twice that of the cluster.

(*b*) *New Variables in Galactic Clusters.*

Bakos and Rymer (**71**) hunted 19 galactic clusters for variables. Only three new variables (of 11 found) are specifically mentioned. These are a short period eclipsing star in NGC 1893, a flare star in M 67 which subsequently proved (**70**) to be not a cluster member, and the variable mentioned in NGC 6871.

In NGC 7789 a new variable is reported by Starrfield (**72**), a red giant with low amplitude variability. However, Starrfield confirms the earlier findings by Burbidge and Sandage (**73**) that the variable found independently by Weber and Romano (**74**) does not vary.

Smak has commented (**75**) that the number of binaries which should exist in clusters like M 67 and NGC 188 in order to explain the observed number of blue extension stars is much larger than that actually known.

(*c*) *Cepheids in Galactic Clusters.*

In NGC 7790 the Cepheids CE Cas *a* and CE Cas *b* continue to get attention. Following the long series of early observations by Payne-Gaposchkin and Gaposchkin (**76**) in which the cepheids were not separated on Harvard plates, attempts are made to get individual magnitudes. Kukarkin reports that Efremov and Kholopov measured separately the light of each of the components of CE Cas on plates obtained at the Cassegrain focus of the 70–cm reflector in Moscow, with the following results

	CE Cas a	CE Cas b
V mag	10·82	10·90
B–*V*	1·34	1·20

Smak (**77**) notes that a comparison with CF Cas, the third cepheid in this cluster, leads to a conclusion that the existing data on magnitudes and colours of these three cepheids do not show any large discrepancies, both internally, as compared with each other, and in comparison with the mean period-colour-luminosity relations.

Fernie and Marlborough (**78**) have derived absolute magnitudes (M_V) by two independent methods for five classical cepheids whose membership in a galactic cluster is well established.

(d) Variables in the Halos of Galactic Clusters.

Along the lines of previous investigations for globular clusters, the Soviet astronomers have looked for variables in the surroundings of galactic clusters. Efremov (**79, 80**) found that the following cepheids could belong to the halos of open clusters: SZ Tau (NGC 1647), GU Nor (NGC 6067), CG Cas (NGC 7790) Y Sgr (M 25), TY Sct (Tr 35). He found also that modern data confirm Bidelman's suggestion that a number of cepheids from the large surroundings of *h* and χ Per could belong to the cluster, especially SZ Cas. Tsarevsky, Ureche and Efremov (**81**) drew up a list of about 40 cepheids whose relationship to the halos of open clusters needs classification.

The researches of Terzan (**21, 22**) have shown many variables in the region of the galactic centre around the cluster Tr 26.

Part B

Second Supplement to Report on Irregular Variable Stars in Young Clusters and Associations (*Trans. IAU*, **11A**, 275-278, 1962 *and* **12A**, 395-398, 1965)

(Prepared by G. H. Herbig)

This second supplement has the same format as the earlier lists, and the numbering of the bibliography continues on from the last one. It is the result of an examination of the literature up to about November 1966, and thus is an index of the activity in this subject over the past three years. The degree of activity appears to be declining, partly because there has been a

Region	α, δ (1900)	Spectroscopic survey (usually Hα) by:	Slit spectroscopic observations by:	Variable star search or observations by:
IV Cep (C214)	$0^h\,00^m$, + 65°	**134**	**130**	**134**
NGC 225	0 40, + 61	**128**	—	**128**
IC 348, B4	3 38, + 32	—	—	**131**
Tau-Aur dark clouds	4^h – 5^h, + 16° – + 30°	**139, 140**	—	**123, 124, 125 133, 151, 152**
NGC 1579	4 24, + 35	—	—	**136**
IC 405	5 10, + 34	—	—	**136**
(IC 2118)	5 15, – 6	—	**130**	**151**
Orion Nebula	5 30, – 5	—	**142**	**141**
λ Ori, B 35	5 40, + 9	—	**129**	—
IC 2169	6 26, + 10	—	—	**137**
NGC 2245, 2247, IC 446	6 27, + 10	—	—	**137**
NGC 2261	6 33, + 8	—	—	**138, 143**
NGC 2264	6 36, + 10	**126**	**126**	—
IC 2944	11 32, – 62	—	**146**	**145**
ν Sco	16 06, – 19	**147**	—	—
Dark nebula B 44	16 35, – 24	**149**	—	—
Dark nebulae Kh 618, 635, 655*	16 43, – 15	**127**	—	—
Dark nebula B59	17 04, – 27	**148**	—	—
NGC 6383	17 28, – 32	**150**	—	**150**
M 20	17 56, – 23	—	—	**135**
M 16	18 13, – 14	—	—	**132**
M 17	18 15, – 16	—	—	**135**
IC 1396	21 38, + 57	—	—	**144**

* Dark nebulae as numbered in the *Atlas* of J. Khavtassi (Ak. N. Georg. SSR, Tbilisi, 1960).

shift of observational effort away from survey-type work toward more detailed studies of individual stars. The most active survey work for new emission-Hα stars is being carried out by Miss Dolidze (Abastumani) and by The (Lembang), while the photometric work of the Sonneberg group and of Badalian (Burakan) continues to be noteworthy. The searches for faint variables in nebulae by Rosino and his associates (Asiago) continue; an extensive search for flash variables in Orion, the Pleiades, and Praesepe is underway. Similar work is now being pressed vigoriously by Maffei (Roma).

Several reviews of this subject have appeared recently. Special attention should be drawn to a very thorough study by W. Götz: 'Einige Beziehungen der RW-Aurigae-Sterne zur Interstellaren Materie', *Veröff. Sternw. Sonneberg*, **7**, 7, 1965. L. Rosino has reviewed the subject of nebular variables in *Star Evolution* (ed. L. Gratton, Academic Press, 1963), p. 264, and in *Trans. IAU*, **12B**, 450, 1964. A critical examination of the photometric behavior of stars assigned to the RW Aur class, by L. Meinunger in *Mitt. veränd. Sterne*, **3**, 137, 1966, has important implications. Some of these matters are discussed in the preceding Report of Commission 27.

REFERENCES

123. Badalian, G. S. 1963, *Astr. Cirk. URSS*, no. 261, 3.

124. Badalian, G. S. 1964, *Soobšč. Bjurakan Obs.*, no. 36, 71.

125. Badalian, G. S., Erastova, L. K. 1964, *Soobšč. Bjurakan Obs.*, no. 36, 55.

126. Bretz, M. C. 1965, *C. R. Acad. Sci. Paris*, **261**, 2161. = *Publ. Obs. Hte-Provence*, **8**, no. 8.

127. Dolidze, M. V. 1962, *Abastumanskaja astrof. Obs. Bjull.*, no. 30, 71.

128. Dolidze, M. V., Ponomareva, G. A. 1965, *Astr. Zu.*, **42**, 205. Most of the emission-Hα stars listed are probably not associated with the dark material.

129. Herbig, G. H. 1966, *Vistas in Astr.*, **8**, 109.

130. Herbig, G. H. Unpublished.

131. Hoffmeister, C. 1966, *Mitt. veränd. Sterne*, **3**, 172.

132. Impallomeni, A. G., Maffei, P. 1965, *Mem. Soc. astr. Ital.*, **36**, 493.

133. Kirillova, T. S., Pavlovskaya, E. D. 1963, *Peremennye Zvezdy Bjull.*, **14**, 420.

134. MacConnell, D. J. 1967, Thesis, Case Institute.

135. Maffei, P. 1963, *Mem. Soc. astr. Ital.*, **34**, 441.

136. Maffei, P. 1965, *Mem. Soc. astr. Ital.*, **36**, 487.

137. Maffei, P. 1965, *Rapp. inter. Labor. astrof., Roma*, no. 16.

138. Martel, M. T., Rousseau, J. 1963, *Notes et Inform. Obs. Paris*, fasc. XIV, 33.

139. McCarthy, M. F., Treanor, P. J. 1965, *Ric. astr. Spec. Vatic.*, **7**, no. 3. Some of these emission stars are certainly background objects.

140. Metreveli, M. 1966, *Astr. Cirk. URSS*, no. 352, 4.

141. Mirzoian, L. V. 1964, *Peremennye Zvezdy Bjull.*, **15**, 192.

142. Johnson, H. M. 1965, *Astrophys. J.*, **142**, 964.

143. Johnson, H. M. 1966, *Astr. J.*, **71**, 224.

144. Romano, G. 1964, *Mem. Soc. astr. Ital.*, **35**, 97.

145. Thackeray, A. D. 1955, *Mém. Soc. Sci. Liège*, ser. 5, **15**, 437.

146. Thackeray, A. D., Wesselink, A. J. 1965, *Mon. Not. R. astr. Soc.*, **131**, 121.

147. The, Pik-Sin. 1964, *Publ. astr. Soc. Pacif.*, **76**, 293.

148. The, Pik-Sin. 1964, *Contr. Bosscha Obs.*, no. 27.

149. The, Pik-Sin, Lim, Ham-Keng. 1964, *Contr. Bosscha Obs.*, no. 23.

150. The, Pik-Sin. 1965, *Contr. Bosscha Obs.*, no. 32. No emission-Hα stars were found, contrary to evolutionary expectation.

151. Ziener, R. 1964, *Mitt. veränd. Sterne*, **2**, 80.

152. A number of papers report photoelectric and spectrophotometric observations of RW Aur; references are given in the main Commission Report.

3. *Variables in Clusters of External Galaxies*

Once again this discussion is limited by observation to the clusters of the Magellanic Clouds. Gascoigne (**82**) has recently shown that nine globular-like clusters in the Clouds can be divided into three groups. One group of four is comparable with the globular clusters in our Galaxy, and the members contain RR Lyrae stars. A second group of five is in the Small Cloud, as noted by Arp 1962, of intermediate age. There seems so far to be no evidence of variables in these. The third group, numbering three, is all in the Large Cloud and has no recognizable galactic counterparts.

Arp (**83**) has discussed eight cepheids within 5'2 of the center of NGC 1866 which are apparently members. Photographic and photovisual light curves have been obtained for seven, with detailed observational material of Arp and Thackeray to be published later. The periods range from 2·6 days to 3·5, and the apparent magnitude at mean light, $\bar{B}$ = 16·77 to 16·41. From the color-magnitude relation the stars are less than 10^8 years old. According to Arp, these cepheids are a better fit to the period luminosity relation of the Small Cloud than of the Large Cloud. All the cepheids are at essentially the same color index.

Thackeray (**84**) has commented that the discovery of the RR Lyrae stars in NGC 121, 1466 and 2257 has proved that 'old' representatives of Population II similar to those in our galaxy are indeed present in both Clouds.

Wesselink and Shuttleworth (**85**) have discussed the RR Lyrae stars around NGC 121.

REFERENCES

1. Sawyer Hogg, H. B. 1965, *Trans. IAU*, **12A**, 390.
2. Laborde, J. L., Fourcade, C. R. In press.
3. Tsoo, Y-h. Private communication, Purple Mountain Obs.
4. Wilkens, H. 1965, *Mitt. veränd. Sterne, Sonneberg*, **3**, no. 3, 72, 75; *Obs. astr. La Plata Sep.*, no. 61.
5. Cuffey, J. 1965, *Astr. J.*, **70**, 732.
6. Cuffey, J. 1966, *ibid.*, **71**, 514.
7. Margoni, R. 1965, *Kleine Veröff. Remeis-Sternw. Bamberg*, **4**, no. 40, 249.
8. Margoni, R. 1964, *Contr. Oss. astrofis. Univ. Padova Asiago*, no. 150.
9. Margoni, R. 1965, *ibid.*, no. 170.
10. Belserene, E. P. 1964, *Astr. J.*, **69**, no. 7, 475.
11. Geyer, E. H. 1965, *Mitt. astr. Ges.*, 96.
12. Geyer, E. H., Szeidl, B. 1965, *Kleine Veröff. Remeis-Sternw. Bamberg*, **4**, no. 40, 63.
13. Geyer, E. H., Szeidl, B. *Z. Astrophys.* In press.
14. Szeidl, B. 1965, *Mitt. Sternw. Ungar. Akad. Wiss. Budapest-Szabadsghegy*, no. 58.
15. Fourcade, C. R., Laborde, J. R. 1963, *Bol. Inst. Mat. Astr. Fis., Cordoba*, **1**, no. 6.
16. Bartolini, C., Biolchini, R., Mannino, G. 1965, *Publ. Oss. astr. Univer. Bologna*, **9**, no. 4.
17. Fourcade, C. R., Laborde, J. R. 1965, *Mem. Soc. astr. Ital.*, **36**, no. 4, 417.
18. Wilkens, H. 1964, *Obs. Astr. La Plata Sep.*, no. 54.
19. Kheylo, E. S. 1964, *IAU Inf. Bull. var. Stars*, no. 43.
20. Kheylo, E. S. 1965, *ibid.*, no. 104.
21. Terzan, A. 1965, *Publ. Obs. Haute-Provence*, **8**, no. 11.
22. Terzan, A. 1965, *ibid.*, **8**, no. 12.
23. Sawyer Hogg, H. B., Wehlau, A. 1966, *Publ. Dunlap Obs.*, **2**, no. 17.
24. Sawyer Hogg, H. B., Wehlau, A. *Ibid.*, to appear in 1967.
25. Clube, S. V. M. 1965, *R. Obs. Bull.*, no. 95.
26. Hoffleit, D. 1965, *Astr. J.*, **70**, 307.
27. Hoffleit, D. 1964, *Astr. J.*, **69**, 301.
28. Hoffleit, D. 1965, *Sky and Tel.*, **27**, 274.
29. Sandage, A., Smith, L. L. 1966, *Astrophys. J.*, **144**, 886.
30. Sandage, A., Smith, L. L., Norton, R. H. 1966, *Astrophys. J.*, **144**, 894.
31. Rosino, L. 1966, *Astrophys. J.*, **144**, 903.

32. Rosino, L. 1965, *Kleine Veröff. Remeis-Sternw. Bamberg*, **4**, no. 40, 202.
33. Makarova, V. A., Akimova, V. P. 1965, *Peremennye Zvezdy Bjull.*, **15**, 350.
34. Ciatti, F., Rosino, L., Sussi, M. G. 1965, *Kleine Veröff. Remeis-Sternw. Bamberg*, **4**, no. 40, 228.
35. Dickens, R. J., Saunders, J. 1965, *R. Obs. Bull.*, no. 101.
36. Roberts, M., Sandage, A. 1955, *Astr. J.*, **60**, 185.
37. Sawyer, H. B. 1953, *J. R. astr. Soc. Can.*, **47**, 229 = *Dunlap Comm.*, no. 34.
38. Harwood, M. 1962, *Leiden Ann.*, **21**, pt. 8, 387.
39. Feast, M. W. *Observatory*, in press.
40. Feast, M. W. 1965, *Observatory*, **85**, 16; –*Radcliffe Repr.*, no. 40.
41. Oosterhoff, P. Th., Walraven, Th. 1966, *Bull. astr. Inst. Netherlds.*, **18**, 387.
42. Ponsen, J., Oosterhoff, P. Th. 1966, *Bull. astr. Inst. Netherlds., Suppl.*, **1**, no. 1, 3.
43. Sturch, C. *Astrophys. J.* In press.
44. Barbon, R. 1965, *Contr. Oss. astrofis. Univ. Padova Asiago*, no. 175.
45. Demers, S. Doctoral thesis, University of Toronto.
46. Kurochkin, N. E. 1964, *Peremennye Zvedzdy Bjull.*, **15**, 164.
47. Arp, H. 1965, *Astrophys. J.*, **141**, 43.
48. Preston, G. W. 1964, *A. Rev. Astr. Astrophys.*, **2**, 23; *Lick Contr.*, no. 158.
49. Rosino, L. 1965, *Kleine Veröff. Remeis-Sternw. Bamberg*, **4**, no. 40, 98.
50. Frolov, M. S. 1966, *Peremennye Zvezdy Bjull.*, **16**.
51. Woolley, R. v.d. R. 1966, *Quart. J. R. astr. Soc.*, **6**, 125.
52. Christy, R. F. 1966, *Astrophys. J.*, **144**, 108.
53. Christy, R. F. 1966, *A. Rev. Astr. Astrophys.*, **4**, 353.
54. Kheylo, E. S. 1966, Problems of Astrophysics, 124, Kiev.
55. Thänert, W. 1965, *Kleine Veröff. Remeis-Sternw. Bamberg*, **4**, no. 40, 292.
56. Balázs-Detre, J., Detre, L. 1965, *Kleine Veröff. Remeis-Sternw. Bamberg*, **4**, no. 40, 184.
57. Arp, H. 1965, *Stars and Stellar Systems*, **5**, ch. 19, University of Chicago.
58. Plaut, L. *Ibid.*, ch. 13.
59. Fernie, J. D. 1964, *Astrophys. J.*, **140**, 1482.
60. Fernie, J. D. 1964, *Astr. J.*, **69**, 258.
61. Stothers, R. 1964, *Publ. astr. Soc. Pacif.*, **76**, 98.
62. Stothers, R., Chin, C., Chiu, H. In press.
63. Sawyer Hogg, H. B., Wehlau, A. 1964, *J. R. astr. Soc. Can.*, **58**, 163 = *Dunlap Comm.*, no. 97.
64. Hoffmeister, C. 1964, *IAU Inf. Bull. var. Stars*, no. 67.
65. Efremov, Y. N., Kholopov, P. N., Kukarkin, B. V., Sharov, A. S. 1964, *IAU Inf. Bull. var. Stars*, no. 75.
66. Kukarkin, B. V., Kurochkin, N. E. 1966, *Astr. Zu.*, **43**, 83.
67. Kurochkin, N. E. 1965, *IAU Inf. Bull. var. Stars*, no. 79.
68. Kholopov, P. N., Sharov, A. S. 1966, *Astr. Cirk. URSS*, no. 377.
69. Eggen, O. J. 1965, *Kleine Veröff. Remeis-Sternw. Bamberg*, **4**, no. 40, 51.
70. Bakos, G. A. 1965, *Astr. J.*, **70**, 132.
71. Bakos, G. A., Rymer, K. R. 1964, *Astr. J.*, **69**, 531.
72. Starrfield, S. 1965, *Publ. astr. Soc. Pacif.*, **77**, 206 = *Lick. Contr.*, no. 184.
73. Burbidge, E. M., Sandage, A. 1964, *Astrophys. J.*, **69**, 531.
74. Weber, R. 1958, *Astronomie*, **72**, 80.
75. Smak, J. 1966, *Acta Astr.*, **16**, 11.
76. Payne-Gaposchkin, C., Gaposchkin, S. 1963, *Publ. astr. Soc. Pacif.*, **75**, 171.
77. Smak, J. 1966, *Acta Astr.*, **16**, 25.
78. Fernie, J. D., Marlborough, J. M. 1965, *Publ. astr. Soc. Pacif.*, **77**, 218 = *Dunlap Comm.*, no. 111.
79. Efremov, Y. N. 1964, *Astr. Cirk. URSS*, no. 292, 3.
80. Efremov, Y. N. 1964, *Peremennye Zvezdy Bjull.*, **15**, no. 3, 242.
81. Tsarevsky, G. S., Ureche, V., Efremov, Y. N. 1966, *Astr. Cirk. URSS*, no. 367.
82. Gascoigne, S. C. B. 1966, *Mon. Not. R. astr. Soc.*, **134**, 59.
83. Arp, H. 1965, *Kleine Veröff. Remeis-Sternw. Bamberg*, **4**, no. 40, 209.
84. Thackeray, A. D. 1964, *J. R. Astr. Soc. Can.*, **58**, 55.
85. Wesselink, A. J., Shuttleworth, M. 1965, *Mon. Not. R. astr. Soc.*, **130**, 443.

28. COMMISSION DES GALAXIES

PRÉSIDENT: Dr R. L. Minkowski, Radio Astronomy Laboratory, University of California, Berkeley 4, California, U.S.A.

VICE-PRÉSIDENT: Professor G. C. McVittie, University of Illinois Observatory, Urbana, Illinois, U.S.A.

COMITÉ D'ORGANISATION: E. M. Burbidge, E. Holmberg, B. Y. Mills, A. Sandage, A. D. Thackeray.

MEMBRES: Abell, Aly, Ambartsumian, Arp, Baum, Bertola, Bigay, Boischot, Bondi, Bonnor, Burbidge (G. R.), Contopoulos, Dessy, de Vaucouleurs, Edmondson, Eigenson†, Elvius (A. M.), Evans (D. S.), Gascoigne, Gomes, Haro, Heckmann, Heeschen, Heidmann, Hindman, Hoyle, Humason, Johnson (H. M.), Kalinkov, King (I. R.), Kustaanheimo, Laffineur, Layzer, Lemaître†, Lindblad (B.)†, Lindblad (P. O.), McCrea, Madwar, Markarian, Mayall (N. U.), Morgan, Münch (G.), Neyman, Oort, Osterbrock, Page, Pishmish de Recillas, Poveda, Prendergast, Reddish, Rindler, Roberts (M. S.), Robinson, Rosino, Rudnicki, Schmidt (M.), Schücking, Schürer, Sciama, Scott (E. L.), Sěrsic, Shane (C. D.), Shapley (H.), Stibbs, Vandekerkhove, Van den Bergh, Vorontsov-Velyaminov, Westerlund, Whitford, Whitrow, Wild (P.), Wilson (A. G.), Zwicky.

INTRODUCTION

Important progress has been made in almost all parts of the large field covered by the Commission. The most spectacular and dramatic new chapter of extragalactic research has expanded rapidly. Nine quasi-stellar sources only were known at the time of the XIIth General Assembly. The number of known objects of this class now ranges in the hundreds, and the discovery of the related class of radio-quiet quasi-stellar objects has opened a large field for investigations. Detailed observations of the objects do not keep up with the rapid rate of new discoveries, mainly owing to the small number of suitable telescopes, particularly in the southern hemisphere, and the restricted amount of available observing time.

The present report is based mainly on communications from members of the Commission. With the growth of the Commission's domain it was found necessary to distribute the task of compiling the report. Separate reports are given by:

A. Sandage, on: Quasi-stellar Sources and Objects,
Mrs E. M. Burbidge, on: Spectra, Internal Motions and Masses of Galaxies,
the Chairman of the Working Group on Galaxy Photometry, G. de Vaucouleurs,
the Chairman of the Committee on Supernovae, F. Zwicky,
the Chairman of the Committee on the Magellanic Clouds, A. D. Thackeray.

A report on Researches on Galaxies in the U.S.S.R. again has been given by B. Vorontsov-Velyaminov whose continued cooperation deserves the thanks of the Commission.

The response of the members of the Commission to the President's request for contributions to this Report was in general poor and seems to indicate a lack of interest. In particular, almost no contributions were received from those members who worked on theoretical cosmology and relativistic astrophysics. Under these circumstances the President and the Vice-President of the Commission, G. C. McVittie, felt unable to assemble a separate section on these subjects. With regard to these subjects, the Report is incomplete and superficial. Radio observations of galaxies are in the Report of Commission 40 on Radioastronomy.

SYMPOSIA

The First Texas Symposium on Relativistic Astrophysics was held 16 to 18 December, 1963 (1). The contribution by J. A. Wheeler and his co-workers has been published separately in expanded form (**2**). A report on the Second Texas Symposium on Relativistic Astrophysics was held 15 to 19 December, 1964. A brief report has been given by I. Robinson, A. Schild and E. L. Schücking (**3**). A summary report on the Miami Conference on Observational Aspects of Cosmology held 15 to 17 December, 1965 has been given by Th. Page (**4**).

A meeting on Cosmology was held 14 to 16 September, 1964 in Padua as part of the celebration of the 4th centenary of the birth of Galileo Galilei (**5**).

The 13th Solvay Conference in Physics on The Structure and Evolution of Galaxies (**6**) was held in Brussels 21 to 25 September, 1964.

IAU Symposium No. 29 on Instability Phenomena in Galaxies was held in Byurakan, 4 to 13 May, 1966.

CATALOGS

A long felt need has been satisfied by the publication of the Reference Catalog of Bright Galaxies by G. de Vaucouleurs and A. de Vaucouleurs (**7**). I. R. King and A. F. Setteducati (**8**) have computed millimeter coordinates on the Palomar Observatory National Geographic Society Sky Survey charts for the galaxies listed in the Reference Catalog.

Volume V of the Catalog of Galaxies and of Clusters of Galaxies by F. Zwicky and E. Herzog (**9**) has been published. Volume III is in press, and publication of the final Volumes IV and VI is expected in December 1967. C. D. Shane and G. E. Kron are engaged in a photoelectric calibration of Zwicky's photographic magnitudes.

Volume II of the Morphological Catalog of Galaxies by B. Vorontsov-Velyaminov (**10**) has been published. This catalog now covers the sky between the North Pole and declination $-9°$.

A catalog containing luminosity classifications of all known dwarf galaxies with diameters larger than one minute of arc has been prepared by S. van den Bergh. Descriptions and finding charts for 188 OB associations in M31 found on plates taken with the 52-inch (132 cm) Schmidt at the Karl Schwarzschild Observatory have been given by van den Bergh (**11**).

An Atlas of Peculiar Galaxies, mostly from photographs with the 200-inch telescope (508 cm) has been compiled by H. Arp (**12**).

A catalog of 688 emission nebulae in M31 with precisely measured positions by W. Baade has been completed by H. Arp (**13**) who has based a discussion of the spiral structure in M31 on this catalog (**14**).

An Atlas and Catalog of H II regions in galaxies has been compiled by P. W. Hodge (**15**).

M. Pastoriza and J. L. Sérsic have published a list of galaxies with peculiar nuclei in the southern hemisphere (**16**). A catalog of 300 H II regions in NGC 300 has been assembled by J. L. Sĕrsic (**17**).

F. Zwicky has prepared four lists of compact and post-eruptive galaxies which contain 703 objects; these lists are available on request. Additional lists are in preparation.

SPECTROSCOPIC OBSERVATIONS

D. S. Evans and R. S. Malin have continued observations of radial velocities of southern galaxies. Nine galaxies have been observed (**18**) of which only one has a previous velocity. A velocity measure of the globular cluster NGC 6273 is included in the paper. The paper also includes a velocity measure of part of the gaseous ring of 400″ diameter at $5^h43^m.4$, $-67°53'$ in the Large Magellanic Cloud. Observed radial velocities of $+289$ and $+263$ km s^{-1} indicate that this is actually a Cloud structure with a diameter of about 100 pc.

G. Carranza has measured radial velocities of 21 southern galaxies with a spectrograph designed by T. L. Page and attached to the 60-inch (152 cm) reflector at Bosque Allegre (**19**).

The observations are being continued. Spectra of peculiar nuclei of galaxies are being observed by Mrs M. Pastoriza, and some peculiar galaxies are being studied by J. L. Sĕrsic who has completed a study of the interacting galaxy NGC 6438 (**20**).

I. R. King is observing elliptical galaxies for rotation. Some, as observed by R. Minkowski, have pronounced rotation (**21**), while in others no rotation is observable.

V. C. Rubin and W. K. Ford Jr. have obtained low dispersion spectra of faint radio-galaxies, quasi-stellar sources, faint blue objects, and galaxies with telescopes of moderate sizes (< 84 inches, or 213 cm) making use of an RCA cascaded image tube. The spectra extend out to about 8400Å and hence are valuable for recording lines to the red of the normal photographic region (**22, 23, 24**).

G. and A. de Vaucouleurs have new redshifts for about 100 galaxies derived from about 200 spectrograms taken between 1960 and 1966 with the prime focus B spectrograph of the McDonald 82-inch (208 cm) reflector. The list is to be published in 1967. Redshifts have been measured for about 30 galaxies by narrow band (15–20 Å) photoelectric spectrophotometry with the medium dispersion scanner at the Cassegrain focus of the McDonald 82-inch (208 cm) reflector (**25, 26**), and for about 10 galaxies with an experimental double beam photoelectric photometer equipped with a tunable interference filter attached to the McDonald 36-inch (91 cm) reflector (**27**). Both systems offer significant gains in sensitivity and observing time compared to photographic methods.

F. Zwicky has observed spectra of some compact galaxies (**28, 29**). Previously reported results about the great variety of spectra of all types and color combinations exhibited by these objects are claimed to have been confirmed.

GENERAL PROPERTIES AND STRUCTURE

S. van den Bergh (**30**) has discussed the associations OB 184 and OB 185 in Baade's Variable Star Field IV in M31. The same field is rediscussed by R. Racine (**31**) using composite *U*, *B* and *V* plates prepared by combining 48-inch (122 cm) and 52-inch (132 cm) Schmidt plates. Racine finds significant variations of reddening throughout Field IV. The most frequent value of the reddening is found to be $E(B-V) = 0{\cdot}15$ with some values as large as $E(B-V) = 0{\cdot}50$ occuring. It is not known which fraction of this observed reddening is produced in the Galaxy. Photometry of a 17th magnitude RR Lyrae variable projected on M31, which has been found by A. F. J. Moffat, may help to answer this question. Van den Bergh (**32**) has given a color magnitude diagram of the association *OB 78*. Moffat (**33**) has also used 48-inch and 52-inch Schmidt plates and plates obtained with the Kitt Peak 84-inch (213 cm) to study the brightest variables in M31. Moffat's data indicate that the number of cepheids per unit mass with periods larger than 20 days is between one and two orders of magnitude lower in M31 than it is in the Galaxy. There is some indication that the most luminous cepheids occur in those associations which have the smallest diameters.

Aina Elvius and John S. Hall (**34**) have now published the results of photoelectric polarization measurements in integrated light of NGC 5128 and other extragalactic objects, mentioned in *Transactions IAU*, **12A**.

They have later (**35**) made three-color observations of polarization in the light from NGC 1068, 7814 and 3034 (M82). About 5% polarization in ultra-violet light and less at longer wavelengths was observed in the nuclear region of NGC 1068, where M. Walker had already reported strong UV polarization. This polarization is probably due to synchrotron radiation from the nucleus of this Seyfert galaxy. The observed polarization of about 4% in the center of NGC 7814 is probably due to absorption in the dark band of dust clouds hiding the nucleus of this galaxy.

Results of the above mentioned polarimetric investigation on M82 and color data obtained in the same observations were discussed by A. Elvius at the IAU Symposium 29 on Instability Phenomena in Galaxies in Byurakan. The color of the filamentary structures in M82 seems to show considerable ultra-violet excess compared to colors usually found in galaxies. The degree of polarization of light from the outer regions of M82 was observed to be mainly in the range 10 to 25% in the ultra-violet and slightly less at longer wavelengths.

T. Hatanaka, S. Hayakawa, K. Ishida and M. Taketani (**36**) give a general review of the evolution of galaxies. This article is part I containing a summary of observational information and a discussion of the evolution starting with a description of the dynamical behavior of a system of gas clouds. The evolution is followed from a primordial gas of high temperature to the formation of galaxies of various types. Morphological features of clusters of galaxies and of galaxies are shown to be characterized by two parameters, the random velocities of the clouds and the rotational velocity of a system of clouds. The following parts II to VII are reviewed. These are concerned with discussions of the properties of the galactic nucleus, of radio galaxies, and of the structure and evolution of spiral galaxies and the stability of spiral arms.

P. W. Hodge has determined in the Sculptor dwarf galaxy colors and magnitudes of 523 stars (**37**). The resulting color-magnitude diagram is very similar to that of galactic globular clusters. In the slope of the giant branch and in the main period of the RR Lyrae variables, it closely resembles the cluster M3. A distance of 88 kpc is derived.

The distributions of the stars in all six dwarf galaxies of the Local Group has been used by P. W. Hodge (**38**) to compare observed and predicted limiting radii. For the three nearest galaxies, the observed limits agree with those computed, but for the three more distant ones, the observed limits are too small. This was interpreted to be in agreement with the fact that relaxation times computed for these objects are extremely large, 10^{12} to 10^{15} years. The orbital parameters of the galaxies were estimated. For the Fornax galaxy the radial velocity of which is known, parameters for the orbit could be fairly completely determined.

A study of the distribution within galaxies of H II regions detected in an Hα survey of some 91 galaxies (**15**) has been completed by P. W. Hodge, and a discussion of the significance of this to the problem of the location of star formation in galaxies was published (**39**).

P. W. Hodge has examined the three-dimensional shape of irregular galaxies by measuring the axial ratios for 131 irregular galaxies, using the Palomar Sky Survey prints and Lick and Mt Wilson photographs (**40**). The resulting frequency-distribution indicated that most irregular galaxies are flat with true axial ratios in the range 0·2 to 0·4. Irregular galaxies are in this sense less flat than late-type spiral galaxies, according to the general consensus for the latter.

I. R. King has completed photographic photometry of NGC 2300, 4261, 4472, 4621 and 4697. (See report of working group on galaxy photometry.) The radial profiles fit Hubble's law well, except at the very centers. They fit de Vaucouleurs' law less well, and the central deviation is greater. The profiles agree fairly well with the relaxed theoretical models computed by R. W. Michie and P. H. Bodenheimer (**41**) and by I. R. King (**42**). Combination of the photometry with velocity dispersions by R. Minkowski (**43, 44**) gives an average photographic M/L of 50.

In a summary of mass determinations T. L. Page (**45**) shows the statistical agreement between masses of single galaxies and those in pairs. This summary also shows a trend in mass/luminosity ratio to very large values in clusters.

A discussion of the characteristics of elliptical galaxies and the explosive origin of galaxies has been carried out by J. L. Sérsic (**46**).

Extending Brandt's method of analyzing observational velocity curves, B. Takase (**47**) has calculated the mass M, the angular momentum Q and the rotational energy T for about 20 galaxies most of which are Sb and Sc. These values were derived by integrating to the limiting radii which had been homogeneously determined by E. Holmberg. In combination with the corresponding integrated luminosities, the mass to luminosity ratio has also been estimated. The relations such as $Q \propto M^{5/3}$ and $T \propto M^{5/3}$ seem to hold in general from statistics.

J. S. Mathis (**115**) measured helium content of the interstellar gas in two irregular galaxies, NGC 4214 and IC 1569, and also (**116**) in the 30 Dor nebula in the LMC, and found that in all these cases it is about the same as the helium content of the gas in our Galaxy.

RADIO GALAXIES

H. Arp (**48**) has investigated pairs of radio sources which are separated by from 2° to 6° on the sky. In a number of cases peculiar galaxies have been found approximately midway along a line joining the two radio sources. The central peculiar galaxies belong mainly to a certain class in the recently compiled *Atlas of Peculiar Galaxies* (**12**). Among the radio sources so far associated with the peculiar galaxies are at least five known quasars. These quasars are indicated to be not at cosmological distances (that is, redshifts not caused by expansion of the universe) because the central peculiar galaxies are only at distances of 10 to 100 megaparsecs. The absolute magnitudes of these quasars are indicated to be in the range of brightness of normal galaxies and downward. Some of the radio sources which have been found to be associated with peculiar galaxies are galaxies themselves. It is therefore implied that ejection of material took place within or near the parent peculiar galaxies with speeds between 10^2 and 10^4 kilometers per second. After traveling for times of the order of 10^7 to 10^9 years, the luminous matter (galaxies) and radio sources (plasma) have reached their observed separations from the central peculiar galaxy. The large redshifts measured for the quasars would seem to be either (*i*) gravitational, (*ii*) collapse velocities of clouds of material falling toward the center of these compact galaxies, or (*iii*) some as yet unknown cause.

D. S. Heeschen and C. M. Wade (**49**) made a radio survey of galaxies in the Shapley-Ames catalog, at frequencies of 750 and 1400 MHz. The observations are complete to photographic magnitude 11·2. M. L. De Jong (**50, 51**) measured the brightness distribution of a number of galaxies at a frequency of 1415 MHz. He found the radio diameters of the galaxies to range from much less than to about equal to the optical diameters. He also found a somewhat higher than average density of radio sources in the neighborhoods of the galaxies. All of the above observations were made with the NRAO 300-foot (91 m) telescope.

T. A. Matthews (**52**) has studied identified radio-galaxies on photographs with the 48-Schmidt (122 cm) telescope and the 100-inch (254 cm) telescope using fine grain plates. Of about 65 objects studied 10% show absorption features, 7% jets from the nuclear regions, 9% plumes outside the galaxy, 17% structure in the nuclear regions, 9% structure in the envelopes, 47% nonsymmetry in the surface brightness or in the extent of the envelope, and a few show other rarer peculiarities. Matthews suggests that the discussed features are probably caused by the violent explosion which produces the radio source, but no comparable study has been done yet on D galaxies that are not radio sources.

A. Sandage (**53**) has observed redshifts of nine radio galaxies and finds one of them (3C 305) to be a flattened, rotating galaxy with a faint, uniformly textured spiral arm. In this connection it should be emphasized that NGC 1275 was originally interpreted as a tightly wound spiral on the basis of 200-inch photographs (**54**). It has later been classified as ED2 from its appearance on a red 48-inch Schmidt plate (**55**). It is, however, clearly not a spheroidal nebula. The central region surrounding the Seyfert nucleus has a structure which actually is somewhat reminiscent of that found by Sandage in 3C 305, but more irregular, and a spectrum of type A (**56**). The examples of 3C 305 and NGC 1275 show that, as should be expected, the identifications of strong radio sources as spheroidal galaxies on the basis of their appearance on 48-inch Schmidt plates are not entirely reliable. Some galaxies of types E and D with peculiarities may turn out to be spirals or irregulars on more detailed observations.

F. Sato (**57**) has investigated the physical conditions of the radio galaxies NGC 1068, NGC 1275 and Cyg A in terms of the ionization mechanism of hydrogen atoms. Three possible mechanisms were examined, collisions of thermal electrons, thermal radiation from hot stars and ultra-violet synchrotron radiation. The conclusion is that every mechanism meets difficulties. If collisions of thermal electrons play a role in ionization of hydrogen atoms, the kinetic energy of electrons are not high enough to maintain the ionization for the life time of emission line phenomena.

It would seem reasonable to suppose that ultraviolet radiation from hot stars is responsible for the ionization, but this possibility is excluded because of absence of Bowen resonance-fluorescence lines of O III in the case of NGC 1068. If the ionization is produced by synchrotron radiation, relativistic electrons must be provided either by acceleration or by continuous production. Both the acceleration and the production of the relativistic electrons meet difficulties from a point of view of energy balances.

G. de Vaucouleurs has investigated peculiar galaxies showing signs of instability (**58**).

B. E. Westerlund and L. F. Smith (**59**) report identifications and investigations of southern radio sources. An investigation of the Parkes radio source 0521–36 by B. E. Westerlund and N. R. Stokes (**60**) shows that it is a compact (N) galaxy of high optical luminosity and strong radio emission.

D. E. Osterbrock and R. A. R. Parker (**117**) measured the intensities in the spectrum of the nucleus of the Seyfert galaxy NGC 1068, and showed that in the ionized gas $T > 8000°$ and that substantial amounts of neutral gas are mixed into the ionized regions. Frequent collisions between high-velocity clouds probably produce a large part of the observed ionization.

GALAXY COUNTS. CLUSTERS OF GALAXIES

An exhaustive search for dwarf galaxies in the Fornax cluster of galaxies was carried out by P. Hodge, M. Pyper and J. Webb (**61**). Fifty dwarf objects were chosen as probable members of the cluster, and photometry revealed that some are probably sculptor-type dwarfs and others are dwarf irregular galaxies. The spatial distributions of giant and of dwarf galaxies in the cluster are sufficiently different to suggest the presence of equipartition of galaxies of different masses.

The distributions of the stars in all six dwarf galaxies of the Local Group have been used to compare observed and predicted limiting radii (**62**). For the three nearest galaxies, the observed limits agree with those computed, but for the three more distant ones, the observed limits are too small. This was interpreted to be in agreement with the fact that relaxation times computed for these objects are extremely large, 10^{12} to 10^{15} years. The orbital parameters of the galaxies were estimated, and it was found that for the Fornax galaxy, the radial velocity of which is known, parameters for the orbit could be fairly completely determined.

T. Kiang (**63**) has examined the problem of clustering of Abell's rich clusters of galaxies by adapting Neymann and Scott's model of uniform clusters to these objects.

The parameters in the model were determined from the variation of the index of clumpiness K^2 with the cell size z. Random realizations of the model so specified were then effected to produce (*i*) synthetic surface distribution and (*ii*) the variation with separation k of the coefficient of quasi-correlation, $Q(k;z)$ for 2 values of z, and these were then compared with their observed counterparts. Discrepancies of the same sort as found by Neyman and Scott in their examination of the clustering of galaxies show up, namely (*i*) the synthetic surface distribution did not reproduce all the 'clumpiness' of the observed surface distribution and (*ii*) the two synthetic $Q(k)$ curves were close together whereas the observed curves for the same two cell-sizes were far apart. The present result on the clustering of Abell's objects, combined with Neyman and Scott's result on the clustering of galaxies and the phenomenon of sub-clustering (**64**) suggest that galaxies are clustered on all scales. This picture is in no way inconsistent with the apparent absence of clustering among radio sources when the great numerical disparity between radio sources and galaxies is remembered. Further, Kiang believes that while there is clustering on all scales, it does not take the form of physically distinct clusters. Kiang also points out that this picture of indefinite clustering naturally results from the mechanism of continual creation proposed by W. H. McCrea (**65**).

H. Neckel (**66**) finds that the brightness distribution within the galaxies plays a serious role in the determination of the optical thickness of the galactic absorption layer from the latitude variation of galaxy counts and from the mean surface brightness of the galaxies. A value of about $0^m.9$ is found for the photographic optical thickness of the absorption layer. General agreement is achieved between the hitherto different results of (1) Hubble's counts of galaxies, (2) the counts by C. D. Shane, C. A. Wirtanen and U. Steinlin (**67**), and (3) the observations of the mean surface brightness. The value of $0^m.45$ for the absorption at the galactic pole and the color excess $B - V = 0^m.05$ or $0^m.06$, which remains unchanged, lead to a value of about eight for the ratio between total photographic absorption and color excess.

C. D. Shane and G. E. Kron are determining the limiting photographic magnitudes of galaxy counts made with the 20-inch (51 cm) astrograph at the Lick Observatory. While not all of the final corrections have been applied, the average galaxy at the limit of identification is quite close to magnitude 19·0. Shane and Kron have also measured the color of several hundred galaxies in different galactic latitudes. Tentative discussion of only a fraction of the measures reveals a small or even negligible relation between latitude and color.

A new analysis of Hubble's galaxy counts has been carried out by G. de Vaucouleurs and G. Malik (**68**) who found serious bias in earlier analyses; in particular the half thickness of the galactic absorbing layer is $A_B = 0·5$ magnitude (not 0·25 magnitude) in agreement with Shane's analysis of the Lick counts. An improved expression for galactic extinction as a function of latitude and longitude was obtained.

G. de Vaucouleurs (**69**) has derived distance indicators based on magnitudes and diameters for galaxies in small groups or clusters, and distances of 54 nearby groups have been established; these are in good agreement with distances from van den Bergh's luminosity classification.

G. C. Omer, T. L. Page, and A. G. Wilson (**70**) have intercompared three independent counts of the Coma cluster of galaxies. The three different surveys appear to be consistent with each other. The conclusion is reached that the cluster contains about 800 members to photovisual magnitude about 18·8 and has a radius of about 100′. A possible spatial density distribution is derived from the combined data.

A. G. Wilson studied the distribution of rich clusters of galaxies. Regularities in their distances (**71**) and angular separations (**72**) suggest the existence of larger scale structures than any presently recognized. Radio galaxies also appear to share cluster distributions (**73**) in the sense of indicating the existence of large scaled structures. The redshift distributions may be interpreted as indicative of a continuation of a Charlier type hierarchy up to organizations of the order of 500 megaparsecs.

F. Zwicky and his collaborators (**74-78**) have made extensive studies of the distribution of clusters of galaxies which have shown: (*a*) The largest open, medium compact and compact clusters of galaxies at all distances have the same indicative dimensions, (*b*) There is no evidence for any clustering of clusters of galaxies, all irregularities being accounted for by accidental fluctuations as well as the interference of interstellar and intergalactic obscuration and (*c*) The average size of the cluster cells occupied by rich clusters of galaxies, as determined from the distribution of 7000 clusters of galaxies is about 50 million pc, in close agreement with the value determined by Zwicky in 1938 from the distribution of the hundred nearest clusters then known.

Several investigations of individual clusters of galaxies have been carried out by members of Zwicky's group. R. Okroy (**79**) has investigated the Virgo cluster and confirmed a segregation effect: the number of galaxies per square degree increases towards the center for the more luminous galaxies, but decreases for the fainter galaxies. A similar result was obtained by T. Kwast (**80**) for the Hydra I cluster. K. Rudnicki and M. Baranowska (**81**) find the segregation effect in the clusters Zwicky 156–5 and 156–14, but not (**82**) in the cluster Zwicky 97–8 (= 127–2).

COSMOLOGY AND RELATED SUBJECTS

Relevant observational information on the physical conditions in intergalactic space now begins to become available. One of the most important newly discovered phenomena is cosmic background radiation whose significance has been discussed by R. H. Dicke, P. J. E. Peebles, P. G. Roll, D. T. Wilkinson (**83**). This radiation becomes observable at wavelengths below 20 cm. At longer wavelengths galactic and extragalactic emissions dominate the spectrum. The observed background temperatures are:

2·8 (±0·6) °K at 20·7 cm by T. F. Howell and J. R. Shakeshaft (**84**)
3·5 (±1·0) °K at 7·4 cm by A. A. Penzias and R. W. Wilson (**85**)
3·2 (±0·5) °K at 3·2 cm by P. G. Roll and D. T. Wilkinson (**86**)
3·22 (±0·15) °K at 0·26 cm by G. B. Field and J. L. Hitchcock (**87**),

determined from the excitation of interstellar cyanogen. These results suggest that the background radiation has a blackbody spectrum with a temperature close to 3°K. The discordant value

1·7 (±0·4) °K at 1·5 cm by W. J. Welch (**88**)

has quite recently been obtained. Additional observations obviously are needed to establish the nature of the background radiation beyond doubt.

G. J. Whitrow and B. D. Yallop (**89**) have discussed the problem of background radiation in homogeneous isotropic models in which account is taken of possible intergalactic absorption effects due to the finite extent of the galaxies and to intergalactic matter. The possible effect of relaxing the assumption that on the average the galaxies radiate uniformly in time is also considered in a preliminary way.

In objects with a redshift of 2, Lyman-α of hydrogen and the shortward region of the spectrum become accessible to ground based optical observations. If the redshift is cosmological, neutral hydrogen between the source and us will become observable, either as an absorption continuum extending shortward from the redshifted line to its zero-velocity wavelength if the intergalactic medium is uniform (J. C. Gunn and B. A. Peterson, **90**; P. A. G. Scheuer, **91**) or as discrete absorption features if the intergalactic hydrogen is concentrated in clusters of galaxies (J. N. Bahcall and E. E. Salpeter, **92, 93**). Gunn and Peterson find from the lack of strong absorption shortward of Lyman-α that intergalactic hydrogen accounts for less than 10^{-6} or 10^{-7} of the cosmological density to be expected if the acceleration parameter $q_0 = 1/2$. Thus, most of the hydrogen must be ionized if the discordance between the cosmological density and the density of matter in galaxies is to be explained as due to the presence of intergalactic hydrogen. G. B. Field, P. M. Solomon and E. J. Wampler (**94**) show that H_2

molecules also cannot contain a major fraction of the mass. The intergalactic gas then must be highly ionized and at temperature close to the limit of about 10^6 °K that has to be set according to G. B. Field and R. Henry (**95**) if the bremsstrahlung from the intergalactic medium is not to exceed the total flux in the X-ray region. It should be pointed out that a very small value of q_0 cannot be ruled out at present, but it is not suggested by any other evidence.

D. W. Sciama and M. J. Rees (**96, 97, 98, 99**) have made studies of the intergalactic medium in the light of recent observations of the X-ray background and the spectra of quasi-stellar objects with redshifts ~ 2. In the steady state model of the universe an intergalactic hydrogen gas of density $\sim 10^{-29}$ g cm^{-3} is required to have a kinetic temperature $\sim 10^6$ °K, which could plausibly be achieved by cosmic ray heating. If oxygen is present with its normal cosmical abundance, it could be detected in the spectra of sources with a redshift 2·2. However, the steady state model itself is not consistent with recent data on the redshifts of quasi-stellar objects, if these redshifts are cosmological in origin.

Whether the redshifts of quasi-stellar sources and objects is cosmological and whether they are correspondingly large distances is subject to doubt in view of the suggestion by J. Terrell (**100, 101**) that quasi-stellar objects are local, rapidly moving objects. The two different points of view have been discussed by F. Hoyle and G. R. Burbidge (**102**). At this time there is no conclusive argument in favor of one of the competing interpretations.

A. Sandage (**103**) has given an interim report on his recent work on the redshift-apparent magnitude relation for the brightest members of clusters of galaxies and for radiogalaxies. The results are mainly based on photoelectric photometry. The K term needed to correct the photometric data for the effects of redshift has been re-determined. The plot of redshift against the corrected V magnitude for the 1st ranked galaxy in clusters shows a linear relation with surprisingly small dispersion which emphasizes that the absolute magnitude of the brightest galaxy of a cluster is an exceedingly stable statistical parameter. The observed deceleration parameter will be very close to $q_0 = 1$. A correction must be applied, however, to the absolute magnitudes to account for the evolution of the stellar content for the individual galaxies during the light travel time. This correction gives $q_0 = +0{\cdot}2$ based on an evolutionary theory obtained from old star clusters. The uncorrected observations are consistent with a model with positive curvature, while the corrections lead to one with zero or negative curvature.

The redshift-apparent magnitude relation for the radio galaxies fits the relation for the brightest galaxies in clusters closely, although with somewhat larger scatter. The conclusion is that the radio galaxies do not differ significantly in luminosity from the first ranked cluster galaxies and that a galaxy must be supergiant in luminosity and presumably in mass to be a strong radio source.

G. de Vaucouleurs (**103a**) finds that local departures from linearity and isotropy in the redshift-apparent magnitude relation are confirmed by an analysis of improved and enlarged magnitude and velocity data. It follows that the value of the Hubble constant derived from nearby groups or the Virgo cluster is biased and probably too low.

H. Alfvén (**104**) has further studied the properties and evolution of a meta-galaxy containing equal amounts of matter and antimatter. He analyzed the properties of an 'ambiplasma' (containing ionized matter and antimatter) and showed that a magnetized ambiplasma (under the conditions assumed to exist in the universe now) should emit synchrotron radiation but no detectable gamma-radiation. He also suggested mechanisms for separating matter from antimatter. The possible formation of proto-galaxies from an 'ambiplasma' and their subsequent evolution into galaxies was treated by H. Alfvén and A. Elvius (**105**) and reported at the IAU Symposium 29. The annihilation of protons-antiprotons as a possible source of energy for certain astronomical objects was again discussed. A. G. Ekspong, N. K. Yamdagni and

B. Bonnevier (**106**) showed that the radio spectra to be expected from extragalactic objects deriving their energy from the annihilation of nucleon-antinucleons would be similar to radio spectra actually observed for quasi-stellar radio sources and strong radio galaxies. B. Bonnevier (**107**) studied some problems related to the early development of the metagalaxy in the theory of H. Alfvén and O. Klein. His solutions of equations previously derived by Alfvén and Klein indicated that an original contraction of the metagalaxy might be converted to an expansion in agreement with observed redshifts.

R. J. Dickens and S. R. C. Malin (**108**) find that observations of quasi-stellar sources and small-diameter galaxies show no differential aberration with respect to nearby stars. The result contradicts the Ritz theory of light propagation.

P. E. Kuustanheimo (**109, 110, 111**) has developed a new kind of linear gravitational theory in which the gravitational redshift may be non-conservative. This would explain a possible limb effect in the sun, and the route dependence of the redshift could be measured in the laboratory if the redshift experiments of Pound and Rebka based on the Mössbauer effect are repeated using photons not travelling vertically.

D. Sugimoto (**112**) explains phenomena in quasi-stellar sources by the nuclear instability of a star of about 200 $M_{\odot}$ that are formed by the rapid fragmentation of gas of 10^8 $M_{\odot}$ in the nuclei of galaxies.

Zwicky (**113**) has discussed the discovery and some principal characteristics of previously unknown families of compact galaxies. He also has discussed a number of circumstances which will cause an entire compact galaxy to implode or, in some cases at least to liberate and to radiate energy at a much more rapid rate than an ordinary galaxy. He has observed differential redshifts in the spectra of some compact galaxies (**114**) that suggest an explanation in terms of Einstein's gravitational effect and the possibility that large redshifts are only in small part due to the cosmological effect.

R. MINKOWSKI
President of the Commission

REFERENCES

1. *Quasi-Stellar Sources and Gravitational Collapse*, eds. I. Robinson, A. Schild, and E. L. Schücking. (University of Chicago Press, Chicago and London), 1965.
2. Harrison, B. K., Thorne, K. S., Wakano, M., and Wheeler, J. A. 1965, *Gravitation Theory and Gravitational Collapse* (University of Chicago Press, Chicago and London).
3. Robinson, I., Schild, A., Schücking, E. L. 1965, *Physics Today*, **18**, 17.
4. Page, Th. 1966, *Science*, **151**, 1411,
5. *Proceedings of the Meeting on Cosmology* (Barbèra, Firenze), 1966.
6. *The Structure and Evolution of Galaxies* (Interscience Publishers, London, New York, Sydney), 1965.
7. de Vaucouleurs, G., de Vaucouleurs, A. 1964, *Reference Catalog of Bright Galaxies*, (University of Texas Press, Austin).
8. King, I. R., Setteducati, A. F. *Astrophys. J. Suppl.*, (in press).
9. Zwicky, F., Herzog, E. 1966, *Catalog of Galaxies and Clusters of Galaxies*, **Vol. V**, (California Institute of Technology, Pasadena).
10. Vorontsov-Velyaminov, B. A. 1964, *Morphological Catalog of Galaxies*, **Vol. II**, (Moscow).
11. van den Bergh, S. 1964, *Astrophys. J. Suppl.*, **9**, 65.
12. Arp, H. 1966, *Astrophys. J. Suppl.*, **14**, no. 123.
13. Baade, W., Arp, H. 1964, *Astrophys. J.*, **139**, 1027.
14. Arp, H. 1964, *Astrophys. J.*, **139**, 1045.
15. Hodge, P. W. 1966, *An Atlas and Catalog of H* II *Regions in Galaxies*, (University of Washington, Seattle).

16. Pastoriza, M., Sĕrsic, J. L. 1965, *Publ. astr. Soc. Pacif.*, **77**, 287.
17. Sĕrsic, J. L. 1966, *Z. Astrophys.*, **64**, 212.
18. Evans, D. S., Malin, R. S. 1965, *Mon. Notes astr. Soc. Sth. Afr.*, **24**, no. 3, 32.
19. Carranza, G. *Observatory*, (in press).
20. Sĕrsic, J. L. 1966, *Z. Astrophys.*, **64**, 202.
21. King, I. R., Minkowski, R. 1966, *Astrophys. J.*, **143**, 1002.
22. Ford, W. K., Jr., Rubin, V. C. 1965, *Astrophys. J.*, **142**, 1303.
23. Ford, W. K., Jr., Rubin, V. G. 1966, *Astrophys. J.*, **145**, 357.
24. Rubin, V. C., Ford, W. K., Jr. 1966, *Astr. J.*, **71**, 396.
25. de Vaucouleurs, G. 1966, *Astr. J.*, **71**, 183.
26. de Vaucouleurs, G. Galaxy Red Shifts by Photoelectric Spectrophotometry, Proc. IAU Symposium no. 30, Toronto, June 1966, (in press).
27. de Vaucouleurs, G. Radial Velocities of Extended Sources by a Slitless Scanning Technique, Proc. IAU Symposium no. 30, Toronto, June 1966, (in press).
28. Zwicky, F. 1964, *Astrophys. J.*, **140**, 1467.
29. Zwicky, F. 1966, *Astrophys. J.*, **143**, 192.
30. van den Bergh, S. 1964, *Astr. J.*, **69**, 610.
31. Racine, R. (in preparation).
32. van den Bergh, S. 1966, *Astr. J.*, **71**, 219.
33. Moffat, A. F. J. 1966, *M. Sc. Thesis*, Toronto.
34. Elvius, A., Hall, J. S. 1964, *Lowell Obs. Bull.*, **6**, 123.
35. Elvius, A., Hall, J. S. 1965, *Astr. J.*, **70**, 138.
36. Hatanaka, T., Hayakawa, S., Ishida, K., Taketani, M. 1964, *Prog. Theor. Phys.*, Osaka, Suppl. no. 31.
37. Hodge, P. W. 1965, *Astrophys. J.*, **142**, 1390.
38. Hodge, P. W. 1966, *Astrophys. J.*, **144**, 869.
39. Hodge, P. W. 1965, *Science*, **150**, 374.
40. Hodge, P. W. 1965, *Publ. astr. Soc. Pacif.*, **78**, 79.
41. Michie, R. W., Bodenheimer, P. H. 1963, *Mon. Not. R. astr. Soc.*, **126**, 269.
42. King, I. R. 1966, *Astr. J.*, **71**, 64.
43. Minkowski, R. 1962, *Problems of Extragalactic Research*, IAU Symposium no. 15, ed. G. C. McVittie (Macmillan, New York), p. 112.
44. Minkowski, R. unpublished.
45. Page, T. L. 1965, *Smithsonian Astrophys. Obs.*, Special Report, no. 195.
46. Sĕrsic, J. L. *Observatory* (in press).
47. Takase, B. *Publ. astr. Soc. Japan*, (in press).
48. Arp, H. 1966, *Science*, **151**, 1214.
49. Heeschen, D. S., Wade, C. M. 1964, *Astr. J.*, **69**, 277.
50. de Jong, M. L. 1965, *Astrophys. J.*, **142**, 1333.
51. de Jong, M. L. 1966, *Astrophys. J.*, **144**, 553.
52. Matthews, T. A. 1966, *Astr. J.*, **71**, 169.
53. Sandage, A. 1966, *Astrophys. J.*, **145**, 1.
54. Baade, W., Minkowski, R. 1954, *Astrophys. J.*, **119**, 215.
55. Matthews, T. A., Morgan, W. W., Schmidt, M. 1964, *Astrophys. J.*, **140**, 35.
56. Minkowski, R. IAU Symposium 29, *Instability Phenomena in Galaxies*, Byurakan, May, 1966.
57. Sato, F. *Publ. astr. Soc. Japan*, (in press).
58. de Vaucouleurs, G. IAU Symposium no. 29, *Instability Phenomena in Galaxies*, Byurakan, May 1966.
59. Westerlund, B. E., Smith, L. F. 1966, *Austr. J. Phys.*, **19**, 181.
60. Westerlund, B. E., Stokes, N. R. *Astrophys. J.*, **145**, 354.
61. Hodge, P., Pyper, D. M., Webb, C. J. 1965, *Astr. J.*, **70**, 559.
62. Hodge, P. 1966, *Astrophys. J.*, **144**, 869.
63. Kiang, T. 1967, *Mon. Not. R. astr. Soc.*, **135**, 1.
64. van den Bergh, S. 1961, *Publ. astr. Soc. Pacif.*, **73**, 46.
65. McCrea, W. H. 1964, *Mon. Not. R. astr. Soc.*, **128**, 335.

66. Neckel, H. 1965, *Z. Astrophys.*, **62**, 180.
67. Shane, C. D., Wirtanen, C. A., Steinlin, U. 1959, *Astr. J.*, **64**, 197.
68. de Vaucouleurs, G., Malik, G. N. (in preparation).
69. de Vaucouleurs, G. Nearby Groups of Galaxies, Chapter 17 in *Galaxies and the Universe*, eds. A. and M. Sandage (University of Chicago, Chicago), (in press).
70. Omer, G. C., Jr., Page, T. L., Wilson, A. G. 1965, *Astr. J.*, **70**, 440.
71. Wilson, A. G. 1954, *Proc. Nat. Acad. Sci. U.S.A.*, **52**, 847.
72. Wilson, A. G. 1965, *Astr. J.*, **70**, 150.
73. Wilson, A. G. 1966, Proc. 14th Internatl. Astrophys. Symposium, Liège, (in press).
74. Zwicky, F. 1964, *Publ. astr. Soc. Pacif.*, **75**, 373.
75. Zwicky, F., Berger, J. 1965, *Astrophys. J.*, **141**, 34.
76. Zwicky, F., Karpowicz, M. 1965, *Astrophys. J.*, **142**, 625.
77. Zwicky, F., Karpowicz, M. 1966, *Astrophys. J.*, **146**, 43.
78. Zwicky, F., Rudnicki, K. 1966, *Z. Astrophys.*, **64**, 246.
79. Okroy, R. 1966, *Acta astr.*, **16**, 33.
80. Kwast, T. 1966, *Acta astr.*, **16**, 45.
81. Rudnicki, K., Baranowska, M. 1966, *Acta astr.*, **16**, 65.
82. Rudnicki, K., Baranowska, M. 1966, *Acta astr.*, **16**, 55.
83. Dicke, R. H., Peebles, P. J. E., Roll, P. G., Wilkinson, D. T. 1965, *Astrophys. J.*, **142**, 414.
84. Howell, T. F., Shakeshaft, J. R. 1966, *Nature*, **210**, 1318.
85. Penzias, A. A., Wilson, R. W. 1965, *Astrophys. J.*, **142**, 419.
86. Roll, P. G., Wilkinson, D. T. 1966, *Phys. Rev. Lett.*, **16**, 405.
87. Field, G. B., Hitchcock, J. L. 1966, *Astrophys. J.*, **146**, 1.
88. Welch, W. J. 1967, *Astr. J.*, (in press).
89. Whitrow, G. J., Yallop, B. D. 1965, *Mon. Not. R. astr. Soc.*, **130**, 31.
90. Gunn, J. C., Peterson, B. A. 1965, *Astrophys. J.*, **142**, 1633.
91. Scheuer, P. A. G. 1965, *Nature*, **207**, 963.
92. Bahcall, J. N., Salpeter, E. E. 1965, *Astrophys. J.*, **142**, 1677.
93. Bahcall, J. N., Salpeter, E. E. 1966, *Astrophys. J.*, **144**, 847.
94. Field, G. B., Solomon, P. M., Wampler, E. J. 1966, *Astrophys. J.*, **145**, 351.
95. Field, G. B., Henry, R. 1964, *Astrophys. J.*, **140**, 1002.
96. Sciama, D. W., Rees, M. J. 1966, *Astrophys. J.*, **145**, 6.
97. Sciama, D. W., Rees, M. J. 1967, *Astrophys. J.*, (in press).
98. Sciama, D. W. *Cosmological Aspects of High-Energy Astrophysics* in Proc. Internatl. School of Physics 'Enrico Fermi' 12 to 24 July 1965, Course 35, (in press).
99. Sciama, D. W., Rees, M. J. 1966, *Nature*, **211**, 1283.
100. Terrell, J. 1964, *Science*, **145**, 918.
101. Terrell, J. 1966, *Science*, **154**, 1281.
102. Hoyle, F., Burbidge, G. R. 1966, *Astrophys. J.*, **144**, 534.
103. Sandage, A. *Radio Sources and the Expansion of the Universe*, Proc. Conference on Cosmology, Padua, 14–16 September, 1964. (G. Barbèra, Firenze), 104, 1966.
103a. de Vaucouleurs, G. *The Velocity-Distance Relation for Bright Galaxies.* Proc. Conference on Cosmology, Padua, 14-16 September, 1964. (G. Barbèra, Firenze), 104, 1966.
104. Alfvén, H. 1965, *Rev. Mod. Physics*, **37**, 652. (1965, *Astr. Zu.*, **42**, 873).
105. Alfvén, H., Elvius, A. *Proceedings of IAU Symposium no.* 29 (to be publ.).
106. Ekspong, A. G., Yamdagni, N. K., Bonnevier, B. 1966, *Phys. Rev. Lett.*, **16**, 664.
107. Bonnevier, B. 1964, *Arkiv Fysik*, **27**, 305.
108. Dickens, R. J., Malin, S. R. C. 1965, *Observatory*, **85**, 260.
109. Kuustanheimo, P. E. 1965, *Publ. astr. Obs. Helsinki*, no. 1.
110. Kuustanheimo, P. E. 1966, *Publ. astr. Obs. Helsinki*, no. 112.
111. Kuustanheimo, P. E. 1966, *Publ. astr. Obs. Helsinki*, no. 118.
112. Sugimoto, D. 1965, *Progress Theor. Phys.*, **33**, 339.
113. Zwicky, F. *Proc. Conference on Cosmology*, Padua, 14–16 September, 1964. (G. Barbèra, Firenze) 122, 1966.

114. Zwicky, F. 1966, *C. r. Acad. Sci. Paris*, **262**, 1566.
115. Mathis, J. S. 1965, *Publ. astr. Soc. Pacif.*, **77**, 90.
116. Mathis, J. S. 1965, *Publ. astr. Soc. Pacif.*, **77**, 189.
117. Osterbrock, D. E., Parker, R. A. R. 1965, *Astrophys. J.*, **141**, 892.

APPENDIX I. OPTICAL DATA ON QUASI-STELLAR SOURCES

(Prepared by A. Sandage)

Introduction

There has been enormous activity in the study of quasi-stellar radio sources since the 1964 Hamburg meeting. It is now generally accepted that the name 'quasars' be applied to these objects and we shall adhere to this nomenclature in this report. Work on quasars can be conveniently divided into eight sections, touching the subjects of identification, redshifts, absorption lines, photometry, optical variations, interlopers ('radio quiet' quasars, sometimes designated QSG for quasi-stellar galaxies), the controversy of cosmological versus local distances, and theoretical studies of the structure of the sources.

The present summary is a partial literature review of papers appearing between 1963 and 1966. No attempt for completeness was made. It is further confined primarily to optical observations with only a few remarks on interpretation.

Identification

The first quasars were optically identified in 1960. This was achieved when radio positions of high precision became available through the work at the Owens Valley Radio Observatory on a list of high-flux-density sources of small angular diameter, as measured at Jodrell Bank. Direct photographs of several sources in this list were obtained with the 200-inch telescope. Matthews and Sandage (**1**) found that 3C 48, 3C 196, and 3C 286 appeared to coincide with starlike objects at the radio positions (which were accurate to better than $\pm 10''$ arc in both coordinates). These three objects all had peculiar *U*, *B*, *V* colors compared to normal stars. This peculiarity of bright ultraviolet has been one of the main observational methods of optical confirmation of suggested identifications.

The fourth source to be identified was 3C 273 by Schmidt (**2**), who used a precise radio position due to workers at the Owens Valley Radio Astronomy Observatory and to Hazard, Mackey, and Shimmins (**2a**) at Parkes, who obtained the position by a lunar occultation. The 13th magnitude object was bright enough for Schmidt to obtain highly widened spectra of good quality from which he discovered the general property of redshifts (**2**).

The next three sources were found by Ryle and Sandage (**3**) using a two-color photographic method invented by Haro, where an ultraviolet and a blue image of each star in a given field is obtained on the same plate. These plates are searched for extreme ultraviolet objects. Sources 3C 9, 3C 216, and 3C 245 were discovered in this way. Photoelectric photometry confirmed the peculiar colors.

Schmidt and Matthews (**4**) identified 3C 47 and 3C 147 and confirmed from spectrograms that these were quasars.

By the middle of 1964 it was realized that quasars must be numerous. Estimates indicated that about 30% of the 3*C Revised Cambridge Catalogue* were objects of this type. Concentrated efforts of identification were then made at Owens Valley by Matthews and Wyndham, at Mt Wilson and Palomar by Sandage and Véron, at Cambridge by Longair (**5**) and later by Wills and Parker (**6**), and by Bolton and his co-workers at Parkes (**13, 13a, 14**).

Sandage and Wyndham identified nine new sources in late 1964 (7) with 3C 93, 3C 208, 3C 279, 3C 287, MSH 13–011, 3C 345, 3C 380, 3C 446, and CTA 102, where 208 and 287 had previously been suggested by Longair (5). Bolton, Clarke, Sandage, and Véron (8) made the first identifications of fainter radio sources of the Parkes *Catalogue* where they found six sources to be quasars with peculiar colors and with coincident optical and radio positions.

The next identification list was that of Sandage, Véron, and Wyndham (9), where 21 quasars were suggested, 20 of which proved to be correct.

Hazard, Mackey, and Nicholson (**10**), Véron (**11**), and others added source identifications based on highly precise radio and optical positions and the coincidence of blue objects on the Palomar Sky Survey prints. Lynds, Stockton, and Livingston (**12**) identified 3C 351. A long new series of identifications of the Parkes radio catalogue objects was begun by Bolton, Kinman, and their co-workers. A number of identifications were suggested by Bolton and Ekers (**13a**). Bolton and Kinman (**13**) identified 12 new quasars, followed by Bolton *et al.* (**14**) for 15 additional quasars, and by Kinman, Bolton and Sandage (**15**) for an additional 16 sources. Scheuer and Wills (**16**) and Wyndham (**17**) published several identifications made from the *4C Cambridge Radio Catalogue.*

By the end of 1966 over 130 quasars were either positively identified or were highly probable suspects associated with blue optical objects.

Redshifts

The phenomenon of the redshift was discovered by M. Schmidt (**2**) through his interpretation of the spectrum of 3C 273. Greenstein and Matthews (**18**), following the lead of Schmidt, obtained the redshift of 3C 48 in 1963. Confirmation of Schmidt's redshift for 3C 273 was obtained by Oke (**19**) when he found the Hα line at λ 7590Å using a spectrum scanner.

Following these initial discoveries, five independent groups began systematic observations of redshift for the identified sources as they became available. The principal observers in each group were M. Schmidt at Mt Wilson and Palomar, C. R. Lynds at Kitt Peak, E. M. Burbidge at Lick, T. D. Kinman at Lick, and Vera Rubin and W. K. Ford at the Carnegie Institution Department of Terrestrial Magnetism.

Schmidt and Matthews (**4**) obtained z values for 3C 47 and 3C 147. A list of five quasars with very large redshifts (3C 9, 245, 254, 287, and CTA 102) was published by Schmidt (**20**). These objects proved to be crucial because they established the existence of the C III line at λ 1909Å, C IV at λ 1550Å, and finally $L\alpha$ at λ 1216Å for 3C 9. This was the first observation of the Lyman-α line, subsequently repeated many times on other quasars in 1966. The redshift of 3C 9 is $\Delta\lambda/\lambda_0 \equiv z = 2{\cdot}012$ with $L\alpha$ appearing at λ 3666Å.

Oke obtained the redshift of 3C 286 (**21**) by identifying the line, which he had found with a scanner, at λ 3528Å with C III (1909). E. Burbidge (**23**) found the redshift of MSH 14–121, which had been identified by Hazard, Mackey, and Nicholson (**10**). Lynds, Stockton, and Livingston (**12**) obtained redshifts for 3C 245, 279, 286, 334, 345, 351, 380, and 1252 + 11 using, for the first time, an image-intensifier spectrograph. Burbidge and Rosenburg (**22**) found the redshift for 3C 279, confirming the results of Lynds *et al.* (**12**).

Redshifts of six quasars (3C 334, 345, 380, 446, 208, and 0106 + 01) were then published by E. Burbidge (**23**). Schmidt (**24**) contributed a list of 14 redshifts for 3C 249·1, 277·1, MSH 03–19, 1252 + 11, MSH 14–121, CTD 141, 3C 204, 208, 181, 446, 270·1, 432, 454, and 1116 + 12. This was followed by Lynds and Stockton (**25**) for 1116 + 12, and Lynds, Hill, Heere, and Stockton (**26**) for 1217 + 02, 3C 215, 275·1, 261, 4C 39·25, 3C 138, 196, 4C 20·33, 0922 + 14, 0957 + 00, 3C 336, 186, 298, and 280·1.

Ford and Rubin (**27**) used an image-tube spectrograph to obtain redshifts for 1217 + 02, 3C 249·1, and 3C 263. Burbidge and Kinman (**28**) determined redshifts for 14 quasars

(3C 138, 0859–14, 4C 39·25, 1127–14, 3C 261, 1148–00, 4C 31·38, 4C 21·35, 1327–21, MSH 13–011, 1354 + 19, 1454–06, 3C 309·1, and 1510–08). Stockton and Lynds (**29**), and Burbidge, Lynds, and Burbidge (**30**) obtained the redshift of 3C 191 and showed that the spectrum has remarkable absorption lines in addition to the usual emission features.

Schmidt (**31**) published redshifts for 3C 232, 268·4, 288·1, 323·1, 454·3, and 4C 1·4 = PHL 1093. Finally, Lynds (**32**) published a long list of redshifts for 3C 2, 175, 207, 309·1, 454·3, 4C 21·35, 4C 29·68, 4C 37·24, 0056–00, 0122–00, 0159–11, 0350–07, 0403–13, 0736 + 01, 2146–13, 2216–03, PHL 658, PHL 1078, PHL 1093, PHL 1305, and PHL 5200.

A few redshifts are available for radio-quiet quasars (see a following section). Sandage, in a first discussion of the phenomenon (**33**), obtained redshifts for BSO 1, Ton 256, and Ton 730. Kinman (**34**) obtained a redshift for PHL 938, and Hiltner, Cowley, and Schild (**35**) published redshifts for Ton 1542, Ton 1530, together with values for 1217 + 02, 3C 249·1, 277·1, 281, 263, 1252 + 11, 0922 + 14, 0957 + 00, and PHL 1377. Sandage and Luyten (**74**), in their study of a single Haro-Luyten field, found redshifts for PHL 1127, 1194, and 3424, and showed that PHL 1027, 1070, 1072, 1222, 1226, 1237, and 3375 are also radio-quiet quasars, but with only one line visible on their spectra.

This list has 91 individual objects with redshifts, all of which are either radio quasars or radio-quiet quasars. The literature search for this tabulation is complete to December 1966.

Quasars with Absorption Lines

The detection of absorption lines in quasar spectra may prove to be important in unravelling the physics involved in the quasar phenomenon and in studying the intergalactic medium. Theoretical discussions by Bahcall and Salpeter (**36**), among others, suggest some of the possibilities.

Absorption lines have been found in a number of quasars. Some of the lines show little or no displacement relative to the emission lines and appear as central self-absorptions principally in $L\alpha$ and C IV 1550, but also occasionally in other lines. In other objects the lines are highly displaced from the emission components. Among the objects showing absorption are BSO 1 (**33**), PHL 938 (**34**), 3C 270·1 (**24**), 1116 + 12 (**37**), and 3C 191 (**25, 30**), where a large number of displaced lines are present. The most spectacular case is PHL 5200, where Lynds (**38**) has found extremely wide absorption 'bands' extending more than 20 000 km s^{-1} blueward of the C IV (1550), Si IV (1403), and N V (1243) emission lines. Furthermore, three velocity systems appear to exist in this object, as evidenced by three displaced components to many of the emission lines.

Absorption continua have been looked for shortward of $L\alpha$ in an effort to detect intergalactic hydrogen. Gunn and Peterson (**39**) and, independently, Scheuer (**40**) discussed this possibility in regard to Schmidt's spectrum of 3C 9 (**20**), where $L\alpha$ appears at λ 3666Å. There is only a slight intensity jump, if any, below the $L\alpha$ line, a circumstance which allowed these authors to put very low limits on the density of intergalactic neutral hydrogen. A similar study by Field, Solomon, and Wampler (**41**), using scanner observations of 3C 9, failed to detect molecular hydrogen absorption in a band shortward of λ 1108Å. They concluded that the density of H_2 was less than 4×10^{-9} molecules per cm^3, which is 8×10^{-6} of the expected cosmological density.

Photometry

Three-color *UBV* photometry has been obtained for many quasars. The peculiar position in the $U - B$, $B - V$ diagram was one of the original distinguishing features of the objects and helped in their first detection. The photometry has often been published in the identification papers as an aid in the confirmation. Two summary lists (**33, 42**) give the available photometry to January 1966. Additional photometry has been reported in references **13, 14,** and **15**.

Analysis of the *UBV* data by McCrea (**43**), Kardashev and Komberg (**44**), and by Barnes (**45**) showed systematic correlations of the color indices with redshift. These were interpreted by Sandage (**42**) as due to features in the energy distribution being redshifted through the *UBV* filter bands, causing the systematic correlations. Lynds (**46**) and Strittmatter and Burbidge (**47**) extended the analysis to show that these features are undoubtedly the strong emission lines. The existence of these correlations, even with their moderately large dispersion, suggests that the underlying continuum radiation of quasars is statistically similar from object to object. The continuum extends at least to λ 1100 Å, and is still rising in a plot of flux per unit wavelength interval versus wavelength.

Photoelectric spectrophotometry over a wavelength range λλ 3300–9000 Å has been obtained for a small number of quasars by Oke (**48**) and Wampler (**49**). These authors find a relatively smooth continuum energy distribution upon which the strong broad emission lines appear.

Wide-band photometry in the infrared at 2·2 μ and 10 μ has been obtained for 3C 273 by Johnson (**50**) and Low and Johnson (**51**). This work shows that the energy distribution begins to climb toward high values at about 1 μ relative to a black body, fitting nicely onto the radio spectrum at 3 mm, which is now available for 3C 273 and 3C 279 from the work of Epstein (**52**) and Epstein, Oliver, and Schorn (**53**).

Optical and Radio Variations

Soon after the discovery of quasars, optical studies showed that some of the objects vary in optical radiation. Smith and Hoffleit (**54**) and Sharov and Efremov found that 3C 273 has varied by about 1 magnitude during the past 80 years. Source 3C 48 was also shown to vary (**1**) by about 0·4 magnitude during the three-year interval since its discovery. Further study revealed that 3C2, 43, 47, 196, 216, 245, and 454·3 (**55, 56**) were also variable. A beautiful study by Goldsmith and Kinman (**57**) showed that 3C 345 varies by about 1 magnitude with a time scale as short as several weeks. The Burbidges (**58**) reported that the spectral lines, especially Mg II (2800), appear to vary in wavelength on the red wing in this object.

The most spectacular variation was recorded for 3C 446. Sandage found that the object had undergone an outburst of 3·2 magnitude sometime between September 1965 and June 1966. Photoelectric observations (**59**) showed the colors had also changed significantly during the outburst. Analysis of the data by Sandage, Westphal, and Strittmatter (**60**) showed that the color change was very probably due to the decreased influence of the C IV (1550) emission line on the total radiation at the maximum outburst phase. The emission lines apparently remained constant in absolute intensity while the continuum radiation increased. These results were confirmed with scanner observations by Oke (**61**) and Wampler (**62**). The light curve of 3C 446 is highly variable. Kinman, Lamla, and Wirtanen (**63**) followed the object for about 80 days between July and September 1966 and found variations of about 0·5 magnitude in ten days, with a total range of about 2 magnitudes. They also found that the light is 20% polarized, and that the plane of polarization varies from week to week.

Radio variations were first discovered in quasars by Dent (**64**) during his observations of 3C 273, 279, and 345 at 3·75 cm wavelength. Further data by Maltby and Moffet (**65, 66**), Adgie *et al.* (**67**), Moffet (**68**), and others show that many quasars do indeed vary by large factors, and that the percentage variation is highly wavelength-dependent, decreasing toward longer wavelengths. There is, as yet, no indication that the radio and optical variations are coupled.

Radio-Quiet Quasars (QSG)

During a search (**33**) for blue objects associated with radio sources, using Haro's two-color method, a number of blue objects that imitated optical quasars appeared, but did not accurately coincide with the radio direction. These objects are apparently identical in superficial characteristics with the many blue objects found in previous searches of high galactic latitudes by many workers, e.g., Haro and Luyten (**69**). Photoelectric photometry of a special sample of highly ultraviolet objects, found by Sandage and Véron (**70**), showed that many of these objects were above the black-body line of the $U - B$, $B - V$ diagram, contrary to the situation for white dwarfs. Sandage (**33**) concluded that many of the Haro-Luyten objects were quasi-stellar but without radio emission. Zwicky states that he predicted the existence of such objects at the Hamburg meeting of the Union in 1964 and at the Galileo conference (**71**) in Padua in the same year.

Evidence that at least some blue objects were quiet quasars came from redshifts of BSO 1, Ton 256, and Ton 730 (**33**). The conclusions on the statistics, but not on the concept, were criticized by Kinman (**72**) and Lynds and Villere (**73**). Kinman (**34**) found that PHL 938 was radio-quiet and had a redshift of $\Delta\lambda/\lambda_0 = 1{\cdot}93$. Additional radio-quiet quasars were found by Hiltner, Cowley, and Schild (**35**). A study to improve the statistics was made by Sandage and Luyten (**74**) in a single Haro-Luyten field. Ten definite radio-quiet quasars were found, some with large redshifts. The surface density is estimated to be at least 0·3 QSG per square degree to $B = 18$ and perhaps as high as 3 QSG per square degree to $B = 19{\cdot}7$. This estimate to $B = 18$ is about 70 times the surface density of radio quasars to 9 flux units at 178 MHz. It is about ten times smaller than Sandage's first prediction.

Special faint optical surveys for new blue objects have been made by van den Bergh (**75**), Richter (**76**) and his co-workers, and others. It is not known what percentage of these objects will be quasi-stellar.

Special searches for radio emission at faint flux levels have been started by the Bologna group (**78**), by observers at Green Bank, at the University of Illinois, and at Parkes. Bolton surveyed the region near PHL 938 with the Parkes dish, but without success. Ton 256 and Ton 730 were looked for at Arecibo, but again with negative results.

It seems probable that radio-quiet quasars are not a new class of objects but rather are radio quasars near the faint end of the radio-luminosity function. Evidence for this view was presented by Sandage and Luyten (**74**) by showing that a very strong observational selection operates against finding quasars of small absolute radio power when optical identifications are made from existing radio catalogs. The total number of quasars is very large if the estimate of 3 QSG per square degree to $B = 19{\cdot}7$ is correct. There may be more than 100 000 of these objects over the sky to this optical limit.

The Issue of Local vs. Cosmological Distances

There is no definite proof that quasars are at the distances inferred by use of the Hubble law. Neither is there a definite disproof. A crucial experiment has not been performed in this matter.

Two major hypotheses have been discussed and many papers dealing with the subject have been published. The cosmological hypothesis assumes that Hubble's law applies and that the quasars are extremely distant objects partaking in the general expansion of the universe. The second hypothesis assumes the objects are local. If the redshifts indicate Doppler motion, the quasars have either been expelled from our own Galaxy or from a nearby system with velocities ranging from low values ($\Delta\lambda/\lambda_0 = 0{\cdot}153$ for 3C 273) up to 80% the velocity of light (3C 9 and others with $\Delta\lambda/\lambda_0 \simeq 2$). Another version of the local hypothesis asserts that the redshifts are not due to Doppler motion but to some unknown physical cause, unspecified.

The first serious suggestion of the local hypothesis was put forward by Terrell (**79**) on the basis that the linear size of objects must be extremely small if they vary in brightness on a short-time scale. A detailed discussion by Hoyle and Burbidge (**80**) treats the problem from a number of standpoints. Setti and Woltjer (**81**) discuss Terrell's suggestion that quasars are ejected from our Galaxy. These authors show that difficulties exist on Terrell's picture in that the total mass and energy involved in accelerating the objects to their observed velocities is very high. This is especially true if the number of quasars is as large as Sandage and Luyten suggest (**74**).

If the quasars are ejected from other nearby galaxies, and if the wavelength shifts are due to Doppler motion, then a large fraction of the objects should show blueshifts. Faulkner, Gunn, and Peterson (**82**) and Noerdlinger, Jokipii, and Woltjer (**83**) discuss the problem and conclude that in a sample of quasars that is complete to a given optical apparent magnitude there should be far more blueshifted objects than redshifted on the local hypothesis. Because no blueshifts have been found among the 91 objects surveyed, they reject this version of the theory.

A different version by Arp (**84**) attributes quasars *and radio galaxies* to daughter products of explosions of certain peculiar parent galaxies. The excess redshifts of the daughters relative to the parents are speculated to be caused by an unknown phenomenon. The chief difficulty is that radio galaxies are known (Sandage **85**) to follow closely the Hubble law (log $\Delta\ \lambda/\lambda_0$ is proportional to apparent magnitude). These galaxies also show a linear decrease of angular diameter with increasing redshift. If, as Arp suggests, redshifts are not correlated with distance for radio galaxies, then the observed Hubble law and angular diameter relations cannot be explained unless the intrinsic properties of absolute luminosity and linear diameter are themselves regular functions of redshift, which seems quite implausible.

Adherents of the cosmological hypothesis point out that the absolute radio luminosity of quasars, computed with the Hubble law, are the same as the brightest radio galaxies such as Cyg A, 3C 295, and Her A, which are known to be at the Hubble distance. Furthermore, Heeschen (**86**) has shown that the radio surface brightness of quasars forms a continuous sequence with radio galaxies when plotted against absolute radio power computed with the Hubble law. This would not be the case if quasars violated the Hubble law.

The most direct observational evidence bearing on the general question is due to Koehler and Robinson (**87**). The data show that 3C 273 must be either imbedded in or beyond the Virgo cluster. However, this observation is not conclusive because certain versions of the local hypothesis suggest the same thing. What can be said with certainty is that the objects are not extremely local because of the absence of proper motion as measured by Luyten (**88**), Luyten and Smith (**89**), Jefferys (**90**), and Sanders (**91**).

What appears to be needed is a crucial experiment, such as the detection of absorption lines in the spectrum of a quasar due to a foreground cluster of galaxies. Alternately, the discovery of a quasar in a cluster of galaxies would be decisive. This latter test is known to be difficult for $\Delta\ \lambda/\lambda_0 \geqslant 0{\cdot}5$, and is negative in the case of 3C 48 and 3C 273 as shown by Sandage and Miller (**92**).

Controversy is likely to continue until some crucial experiment is devised.

Theoretical Models

Many models have been proposed, and a growing body of literature exists. A summary of the state of the problem in 1964 is given in the symposium volume on *Quasi-Stellar Sources and Gravitational Collapse*, edited by Robinson, Schild, and Schücking and published by the University of Chicago Press.

Early investigations were concerned with the nature of the outer regions producing the line spectrum. Greenstein and Schmidt (**93**), Osterbrock and Parker (**94**), Bahcall (**95**), Burbidge, Burbidge, Hoyle, and Lynds (**96**), Shklovsky (**97**), and others contributed to the subject.

The origin of the continuum radiation, the problem of the small linear diameters, and of the high densities are other parts of the general dilemma. Field (**98**) suggested that the objects are newly-forming galaxies in the state of initial gravitational contraction, and that the energies now seen in the configuration could be accounted for by gravitational energy plus many supernovae. Ne'eman (**99**) suggested that the objects were left over as dense cores from the initial configuration of an inhomogeneously expanding universe.

Other suggestions for the high densities consider the end product of a dynamic evolution of star clusters. Ulam and Walden (**100**), Miller and Parker (**101**), and Gold, Axford, and Ray (**102**) have suggested that evaporation of stars cools the configuration, causing it to condense, evaporate more rapidly, and condense faster until stellar collisions occur, a process which releases the observed energy. Colgate (**103**) considers the energy to be due to supernovae. Alfvén (**104**) and Teller have discussed the possibility of matter and antimatter annihilation.

It is probably fair to state that many theoretical ideas concerning the interpretation of the observational data are in a state of flux.

REFERENCES

1. Matthews, T. A., Sandage, A. R. 1962, *Publ. astr. Soc. Pacif.*, **74**, 406; 1963, *Astrophys. J.*, **138**, 30.
2. Schmidt, M. 1963, *Nature*, **197**, 1040.
2a. Hazard, C., Mackey, M. B., Shimmins, A. J. 1963, *Nature*, **197**, 1037.
3. Ryle, M., Sandage, A. 1964, *Astrophys. J.*, **139**, 419.
4. Schmidt, M., Matthews, T. A. 1964, *Astrophys. J.*, **139**, 781.
5. Longair, M. S. 1965, *Mon. Not. R. astr. Soc.*, **129**, 419.
6. Wills, D., Parker, E. A. 1966, *Mon. Not. R. astr. Soc.*, **131**, 503.
7. Sandage, A., Wyndham, J. D. 1965, *Astrophys. J.*, **141**, 328.
8. Bolton, J. G., Clarke, M. E., Sandage, A. R., Véron, P. 1965, *Astrophys. J.*, **142**, 1289.
9. Sandage, A., Véron, P., Wyndham, J. D. 1965, *Astrophys. J.*, **142**, 1307.
10. Hazard, C., Mackey, M. B., Nicholson, W. 1964, *Nature*, **202**, 227.
11. Véron, P. 1965, *Astrophys. J.*, **141**, 1284.
12. Lynds, C. R., Stockton, A. N., Livingston, W. C. 1965, *Astrophys. J.*, **142**, 1667.
13. Bolton, J. G., Kinman, T. D. 1966, *Astrophys. J.*, **145**, 951.
13a. Bolton, J. G., Ekers, J. 1966, *Austr. J. Phys.*, **19**, 471, 559; **20** (in press).
14. Bolton, J. G., Shimmins, A. J., Ekers, J., Kinman, T. D., Lamla, J., Wirtanen, C. A. 1966, *Astrophys. J.*, **144**, 1229.
15. Kinman, T. D., Bolton, J. G., Clarke, R. W., Sandage, A. 1967, *Astrophys. J.*, **147**, 848.
16. Scheuer, P. A. G., Wills, D. 1966, *Astrophys. J.*, **143**, 274.
17. Wyndham, J. D. 1966, *Owens Valley Obs. Preprint.*
18. Greenstein, J. L., Matthews, T. A. 1963, *Nature*, **197**, 1041.
19. Oke, J. B. 1963, *Nature*, **197**, 1040.
20. Schmidt, M. 1965, *Astrophys. J.*, **141**, 1295.
21. Oke, J. B. 1965, *Astrophys. J.*, **142**, 810.
22. Burbidge, E. M., Rosenburg, F. D. 1965, *Astrophys. J.*, **142**, 1673.
23. Burbidge, E. M. 1965, *Astrophys. J.*, **142**, 1291, 1674; 1966, **143**, 612.
24. Schmidt, M. 1966, *Astrophys. J.*, **144**, 443.
25. Lynds, C. R., Stockton, A. N. 1966, *Astrophys. J.*, **144**, 446.
26. Lynds, C. R., Hill, S. J., Heere, K., Stockton, A. N. 1966, *Astrophys. J.*, **144**, 1244.
27. Ford, W. K., Rubin, V. C. 1966, *Astrophys. J.*, **145**, 357.
28. Burbidge, E. M., Kinman, T. D. 1966, *Astrophys. J.*, **145**, 654.
29. Stockton, A. N., Lynds, C. R. 1966, *Astrophys. J.*, **144**, 451.
30. Burbidge, E. M., Lynds, C. R., Burbidge, G. R. 1966, *Astrophys. J.*, **144**, 447.
31. Schmidt, M. S. 1965–66, *Carnegie Institution of Washington Year Book*, Vol. **65**, Annual Report, Director, Mt Wilson and Palomar Observatories.
32. Lynds, C. R. 1967, *Astrophys. J.*, **147**, 837.
33. Sandage, A. 1965, *Astrophys. J.*, **141**, 1560.

34. Kinman, T. D. 1966, *Astrophys. J.*, **144**, 1232.
35. Hiltner, W. A., Cowley, A. P., Schild, R. E. 1966, *Publ. astr. Soc. Pacif.*, **78**, 464.
36. Bahcall, J. N., Salpeter, E. E. *Astrophys. J.*, **142**, 1677, 1965; **144**, 847, 1966.
37. Bahcall, J. N., Peterson, B. A., Schmidt, M. 1966, *Astrophys. J.*, **145**, 369.
38. Lynds, C. R. 1967, *Astrophys. J.*, **147**, 396.
39. Gunn, J. C., Peterson, B. A. 1965, *Astrophys. J.*, **142**, 1633.
40. Scheuer, P. A. G. 1965, *Nature*, **207**, 963.
41. Field, G. B., Solomon, P. M., Wampler, E. J. 1966, *Astrophys. J.*, **145**, 351.
42. Sandage, A. 1966, *Astrophys. J.*, **146**, 13.
43. McCrea, W. H. 1966, *Publ. astr. Soc. Pacif.*, **78**, 49.
44. Kardashev, N. S., Komberg, B. V. 1966, *Astr. Cirk. SSSR*, no. 357.
45. Barnes, R. 1966, *Astrophys. J.*, **146**, 285.
46. Lynds, C. R. 1966, Paper read at the Bjurakan Conference on Instability Phenomena in Galaxies.
47. Strittmatter, P. A., Burbidge, G. R. 1967, *Astrophys. J.*, **147**, 93.
48. Oke, J. B. 1966, *Astrophys. J.*, **145**, 668.
49. Wampler, E. J. 1967, *Astrophys. J.*, **147**, 1.
50. Johnson, H. L. 1964, *Astrophys. J.*, **139**, 1022.
51. Low, F. J., Johnson, H. L. 1965, *Astrophys. J.*, **141**, 336.
52. Epstein, E. E. 1965, *Astrophys. J.*, **142**, 1282, 1285.
53. Epstein, E. E., Oliver, J. P., Schorn, R. A. 1966, *Astrophys. J.*, **145**, 367.
54. Smith, H. J., Hoffleit, D. 1963, *Nature*, **198**, 650; 1963, *Astr. J.*, **68**, 292; 1964, *Quasi-Stellar Sources and Gravitational Collapse*, chap. 16, eds. Robinson, Schild, and Schücking, University of Chicago Press.
55. Sandage, A. *Astrophys. J.*, **139**, 416, 1964; **144**, 1234, 1966.
56. Sandage, A., Véron, P., Wyndham, J. D. 1965, *Astrophys. J.*, **142**, 1307.
57. Goldsmith, D. W., Kinman, T. D. 1965, *Astrophys. J.*, **142**, 1693.
58. Burbidge, E. M., Burbidge, G. R. 1966, *Astrophys. J.*, **143**, 271.
59. Sandage, A. 1966, *IAU Circ.*, no. 1961.
60. Sandage, A., Westphal, J. A., Strittmatter, P. A. 1966, *Astrophys. J.*, **146**, 322.
61. Oke, J. B. 1967, *Astrophys. J.*, **147**, 901.
62. Wampler, E. J. 1967, *Astrophys. J.*, **147**, 9.
63. Kinman, T. D., Lamla, E., Wirtanen, C. A. 1966, *Astrophys. J.*, **146**, 964.
64. Dent, W. A. 1965, *Science*, **148**, 1458.
65. Maltby, P., Moffet, A. T. 1965, *Science*, **150**, 63.
66. Maltby, P., Moffet, A. T. 1965, *Astrophys. J.*, **142**, 409.
67. Adgie, R. L., Gent, H., Slee, O. B., Frost, A. D., Palmer, H. P., Rowson, B. 1965, *Nature*, **208**, 275.
68. Moffet, A. T. 1966, Paper read at the Bjurakan Conference on Instability Phenomena in Galaxies.
69. Haro, G., Luyten, W. J. 1962, *Bol. Obs. Tonantzintla*, **3**, no. 22, 37.
70. Sandage, A., Véron, P. 1965, *Astrophys. J.*, **142**, 412.
71. Zwicky, F. 1966, *Atti del Convegno sulla Cosmologia*, p. 122, ed. G. Barbèra, Monograf, Firenze.
72. Kinman, T. D. 1965, *Astrophys. J.*, **142**, 1241.
73. Lynds, C. R., Villere, G. 1965, *Astrophys. J.*, **142**, 1296.
74. Sandage, A., Luyten, W. J. 1967, *Astrophys. J.*, **148**, 767.
75. Bergh, S. van den 1966, *Astrophys. J.*, **144**, 866.
76. Richter, N. 1965, *Mitt. Karl-Schwarzschild Obs. Tautenberg*, Nr. 25.
77. Richter, N., Sahakjan, K. 1965, *Mitt. Karl-Schwarzschild Obs. Tautenberg*, Nr. 24.
78. Braccesi, A., Fanti, R., Giovannini, C., Vespignani, G. 1966, *Astrophys. J.*, **143**, 600.
79. Terrell, J. 1964, *Science*, **145**, 918.
80. Hoyle, F., Burbidge, G. R. 1966, *Astrophys. J.*, **144**, 534.
81. Setti, G., Woltjer, L. 1966, *Astrophys. J.*, **144**, 838.
82. Faulkner, J., Gunn, J. E., Peterson, B. A. 1966, *Nature*, **211**, 502.
83. Noerdlinger, P. D., Jokipii, J. R., Woltjer, L. 1966, *Astrophys. J.*, **146**, 523.

84. Arp, H. C. *Science*, **151**, 1214.
85. Sandage, A. 1965–66, *Carnegie Institution of Washington Year Book*, Vol. **65**, Annual Report, Director, Mt Wilson and Palomar Observatories.
86. Heeschen, D. S. 1966, *Astrophys. J.*, **146**, 517.
87. Koehler, J. A., Robinson, B. J. 1966, *Astrophys. J.*, **146**, 488.
88. Luyten, W. J. 1963, *Publ. Obs. Minnesota*, **3**, no. 13.
89. Luyten, W. J., Smith, J. A. 1966, *Astrophys. J.*, **145**, 366.
90. Jefferys, W. H. 1964, *Astr. J.*, **69**, 142.
91. Sanders, W. L. 1966, *Astrophys. J.*, **146**, 609.
92. Sandage, A., Miller, W. C. 1966, *Astrophys. J.*, **144**, 1238.
93. Greenstein, J. L., Schmidt, M. 1964, *Astrophys. J.*, **140**, 1.
94. Osterbrock, D. E., Parker, R. A. R. 1966, *Astrophys. J.*, **143**, 268.
95. Bahcall, J. N. 1966, *Astrophys. J.*, **145**, 684.
96. Burbidge, G. R., Burbidge, E. M., Hoyle, F., Lynds, C. R. 1966, *Nature*, **210**, 774.
97. Shklovsky, I. S. 1965, *Soviet Astr.*, **8**, 638.
98. Field, G. B. 1964, *Astrophys. J.*, **140**, 1434.
99. Ne'eman, Y. 1965, *Astrophys. J.*, **141**, 1303.
100. Ulam, S., Walden, W. 1964, *Nature*, **201**, 1202.
101. Miller, R. H., Parker, E. N. 1964, *Astrophys. J.*, **140**, 50.
102. Gold, T., Axford, W. I., Roy, E. C. 1964, *Quasi-Stellar Sources and Gravitational Collapse*, chap. 8, eds. Robinson, Schild, and Schücking, University of Chicago Press.
103. Colgate, S. 1967, *Astrophys. J.*, **148**, (in press).
104. Alfvén, H. 1965, *Rev. mod. Phys.*, **37**, 652.
105. Teller, E. 1966, Amer. phys. Soc. Meeting.

APPENDIX II. SPECTRA, INTERNAL MOTIONS AND MASSES OF GALAXIES

(Prepared by E. M. Burbidge)

Radial Velocities

The Reference Catalogue of Bright Galaxies, by G. and A. de Vaucouleurs (**1**), contains radial velocities and is thus the most modern compilation of velocities. Many of the entries are new ones, observed by the de Vaucouleurs at McDonald Observatory. Shobbrook (**2**) measured the radial velocities of 19 southern galaxies. Radial velocities of radio galaxies have been measured by Schmidt and Sandage; Schmidt (**3**) published results for 31 radio galaxies and gave also estimates of emission line intensities, which range from zero in two objects with absorption lines only, to the strongest emission lines, comparable with those in Cyg A. Sandage (**4**) published results for nine radio galaxies, for one of which data were also given on the velocity field (see below). Westerlund and Stokes (**5**) studied the optical properties of the radio galaxy PKS 0521–36; it is a highly luminous galaxy with $M_B = 21{\cdot}2$ and has an emission-line spectrum.

Rotation Curves, Masses, and Mass-to-Light Ratios

Density distribution, masses, and mass-to-light ratios were determined by means of rotation curves observed with a long-slit spectrograph at McDonald, Lick, Haute-Provence, and Asiago Observatories. The group consisting of the Burbidges, Prendergast, Rubin, and Crampin obtained results for NGC 3521, 6503, 1792, 613, 4826, 7331, 681, and 972 (**6–13**). Bertola measured the rotation mass, and mass-to-light ratio of NGC 3310 (**14**), and also studied the rotation of the Seyfert and radio galaxy NGC 1068 (**15**). Mme. Duflot measured the rotation of NGC 3034 (M82) (**16**).

A redetermination of the rotation curve of M33 was made by Brandt (**17**) by means of velocities of individual H II regions measured at Mt Wilson. Rotations were detected in the elliptical galaxies NGC 4621, 4697 by means of inclined absorption lines by King and

Minkowski (**18**), thus showing that the inner parts of these ellipticals do possess angular momentum. Velocity dispersions were also measured in these two galaxies. Searle (**19**) measured the rotation of the radio galaxy NGC 1316 (Fornax A) by means of its absorption-line spectrum, and showed that it is rotating about its minor axis, in contradistinction to the situation in NGC 5128 (Centaurus A).

Determination of rotations and total content of neutral atomic hydrogen were made by means of 21-cm measures by a number of workers. Epstein (**20**) discussed all the data available in 1964, on 48 galaxies, of which 3 Sb's, 13 Sc's, and 13 irregulars showed measurable 21-cm emission. He discussed models to explain the observations in a following paper (**21**). Roberts and Höglund and Roberts measured the rotation of NGC 925 (**22**) in a region containing neutral H and lying outside the region covered by the optical measures; the rotation curve was found to continue on out into this region. Roberts measured 21-cm emission in the relatively distant galaxy NGC 5668 (**23**), and made a high-resolution study of M31 (**24**). Rotation curves and total masses were derived for NGC 55, 300, 2403, 3109, 4214, and 4244 by Seielstad and Whiteoak (**25**).

On the theoretical side, Pskovskii (**26**) has collected together the data on masses and has discussed the mass-to-light ratio for galaxies and obtained $L \propto \mathfrak{M}^{1.08}$ for spiral and irregular galaxies. Huang (**27**) used a more indirect argument based on the concept of the luminosity of spirals and irregulars being determined by the neutral H content. Brandt and Scheer (**28**) gave tables for calculating mass distributions, densities, and mass-surface densities from observed rotation curves. Wallace and Copeland (**29**) discussed the velocity field in a model spiral galaxy, and compared it with that of NGC 1084. Kinman (**30**) derived a model for the nucleus of M31 based on the luminosity profile, and showed that $\mathfrak{M}/L$ could be ~50 in the center.

Non-Circular Velocity Fields in Galaxies

A study following up the work of Lynds and Sandage on M82 was made by Burbidge, Burbidge, and Rubin (**31**). It was shown that ejection of gas from the nucleus probably occurred in a cone about the axis of rotation. Non-circular motions were found in the nuclear region of M51 by Burbidge and Burbidge (**32**). NGC 2685, a spindle-shaped galaxy with helical structure about its major axis, was studied by Demoulin (**33**) who found that the gas in the helical windings was rotating about the long axis, as were the stars giving absorption lines in the spindle, while the spindle as a whole was also rotating about its minor axis.

The radio galaxy NGC 1275 was studied by Burbidge and Burbidge (**34**), who mapped out the velocity field of its main component and the component with a velocity displaced by 3000 km s^{-1}, discovered by Minkowski. Sandage (**4**) found that the radio galaxy NGC 305 is rotating and its emission lines reach maximum intensity 2000 pc outside the nucleus, suggesting a cone or jet-like distribution of gas. NGC 305 is an Sa galaxy; Minkowski (**35**) showed that NGC 1275 has Balmer absorption lines suggesting that it also is an Sa type. Münch (**36**) continued observations of gas in the nucleus of M31; the outflow along the minor axis amounts to a flux of 0·1 $\mathfrak{M}_\odot$/year at 100 pc distance from the center. Burbidge and Burbidge (**37**) measured the velocity field in the peculiar double galaxy NGC 4038–9; negative velocities occur predominantly in the dusty or upward-lying regions, suggesting outflow. Walker (**38**) discovered that the violet-displaced component in [O II] λ 3727 in M87 (Virgo A), noted in 1959 by Osterbrock, had in the spring of 1966 changed to a component shifted to the red of the central velocity by an approximately equal amount. He also mapped out four high-velocity turbulent gas clouds in the nucleus of the Seyfert and radio galaxy NGC 1068, and showed that a radial outflow of gas probably occurs in the inner parts, while rotation dominates outside 30″. In the bright Sb galaxy NGC 4736, pronounced 'bumps' were found on the rotation curve. NGC 3432, another galaxy with marked bumps on its rotation curve, was studied by Bertola (**39**).

Two peculiar multiple galaxies were discovered to show very large velocity differences, amounting to 2000–3500 km s^{-1}: NGC 4496, studied by de Vaucouleurs (**40**), and NGC 6438, studied by Sĕrsic (**41**). Both these indicate violent explosive events.

Gas streaming in barred spiral galaxies was studied theoretically by Freeman (**42**), and observations by de Vaucouleurs and himself gave evidence supporting the theoretical work; they indicate that gas flow occurs around the axis of the bar while stars have a retrograde mean motion. The intermediate barred spiral NGC 6181 was found by Burbidge, Burbidge, and Prendergast (**43**) to show a marked velocity peak at one end of the major axis of the bar.

Peculiar Galaxies; Compact Blue Galaxies

Zwicky and Humason (**44**) continued their work on spectra and other characteristics of interconnected galaxies and of galaxies in groups and in clusters. Redshifts and velocity dispersions were measured in a number of systems by Zwicky in a series of papers on compact galaxies and compact parts of galaxies (**45, 46**). In a plot of spectral type against morphological type, Zwicky found galaxies populating the hitherto empty part of the diagram, with compact elliptical-type structures and early spectral types.

Spectroscopic studies of four peculiar galaxies were made by the Burbidges and Mrs Rubin – Mayall's Object and IC 883 (**47**), VV 144 from Vorontsov-Velyaminov's catalogue of interacting galaxies (**48**), and NGC 2188 (**49**). VV 144 has a Seyfert-type spectrum and a long tail or jet.

Partly as a result of searches for quasi-stellar objects, some very compact blue galaxies have been discovered and investigated spectroscopically. Sandage (**50**) studied Ton 730 (a 'blue stellar object' from the Tonanzintla Catalogue). Kinman discovered a similar-looking object with a smaller redshift that has $M_B \approx -14$ and discussed its spectroscopic and optical resemblance to NGC 1569 and NGC 2366 (**72**). Zwicky (**51**) described spectra of several systems. Arp (**52**) discovered a very small condensed double galaxy, with an emission-line spectrum, an absolute magnitude for the two components of only $M_{pg} = -12{\cdot}7$ and $-12{\cdot}8$, and diameters of only 70 pc. In a plot of M_{pg} against diameter, the object lies between globular clusters and small galaxies like NGC 147.

Physical Conditions, Composition, and Distribution of Gas in Galaxies

Courtès and his colleagues have continued their studies of galaxies with very narrow-band filters in the light of Hα. Courtès, Viton, and Véron (**53**) studied NGC 1275, M82, NGC 4258, M51, and NGC 6946. Courtès and Cruvellier (**54**) studied the H II regions in M33. Hα emission extends right into the nucleus, as in NGC 4258. Véron and Sauvayre (**55**) studied NGC 2403, 2903, and 4490. Hodge (**56**) studied the distribution of Hα emission regions in eight irregular galaxies that resemble M82, and Baade and Arp mapped their positions in M31 (**57, 58**).

Burbidge and Burbidge (**59**) made a survey of the existence of and relative line intensities in ionized gas in the nuclei of 85 elliptical, SO, spiral, and irregular galaxies. The intensity ratio Hα/[N II] λ 6583 is $\leqslant 1$ in 100% of the E and 81% of the SO galaxies, and in 55% of the spirals, while this ratio is $\geqslant 3$ in 100% of the irregulars. This ratio is governed by the physical conditions of excitation in the gas, particularly the electron temperatures, but could be affected by the abundance ratio H/N.

A detailed study of the physical conditions and source of ionization in the Seyfert and radio galaxy NGC 1068 was made by Osterbrock and Parker (**60**). The electron temperature and density were derived, and it was shown that the most probable ionizing agent is 20 keV protons which are simply the ionized hydrogen in the moving clouds discovered by Walker, whose velocities are ~2000 km s^{-1}. Pacholczyk and Wisniewski (**61**) measured the infrared radiation in NGC 1068 and found a big excess over that in normal spiral galaxies so that it resembles the quasi-stellar object 3C 273.

Vorontsov-Velyaminov and Dibai (**62**) studied the hydrogen emission-line profiles in the nucleus of the Seyfert galaxy NGC 4151 and found two components, one with a velocity dispersion of 750 km s^{-1} and an electron density $N_e = 2 \times 10^3$ cm^{-3}, and one with 3500 km s^{-1} and $N_e \geqslant 10^7$ cm^{-3}. Dibai (**63**) found that the Seyfert galaxies NGC 1275, 3516, 3227, and 4051 all show two such components, while NGC 1068 had only a low-density component and NGC 7469 has only a high-density one.

Mathis (**64**) studied the He/H abundance ratio in the two gassy irregular galaxies NGC 4214 and IC 1569, which are the most distant objects in which this ratio has been measured. The values were in the same range in which lie determinations in our own Galaxy and in Local Group galaxies. Dickel (**65**) made a spectroscopic study of physical conditions in selected H II regions in the Magellanic Clouds.

Stellar Population of Galaxies

Deutsch (**66**) discussed critically the earlier work of Spinrad on the intensity of the blended Na I D lines as an indicator of the size of the red dwarf stellar population in the nuclei of K-type galaxies. Spinrad and Wood (**67**) measured the MgH intensity in cool stars and galaxies, and showed that dwarf stars must contribute to the integrated light in the nucleus of M31 and, to a lesser extent, probably in M32. Wood (**68**) gave the results of multicolor photoelectric photometry of galaxies, in which narrow-band filter measures of the light in and nearby to spectral features were made in order to build up a synthetic stellar population that would reproduce the measures. Philip and Sanduleak (**69**) used low-dispersion spectra of galaxies for spectral classification. Spinrad (**70**), in a review paper on normal galaxies, discussed the line strength ratios as population indicators, and suggested that strong-line K giants and strong CN giants like μ Leo might make an appreciable contribution to the integrated light of K-nuclei galaxies. On the theoretical side, Gratton (**71**) discussed the mass-to-light ratios of galaxies in relation to the population of evolved or 'dead' stars.

REFERENCES

1. Vaucouleurs, G. de, Vaucouleurs, A. de 1964, *Reference Catalogue of Bright Galaxies*, Texas University Press.
2. Shobbrook, R. R. 1965, *Mon. Not. R. astr. Soc.*, **131**, 293, 365.
3. Schmidt, M. 1965, *Astrophys. J.*, **141**, 1.
4. Sandage, A. R. 1966, *Astrophys. J.*, **145**, 1.
5. Westerlund, B. E., Stokes, N. R. 1966, *Astrophys. J.*, **145**, 354.
6. Burbidge, E. M., Burbidge, G. R., Crampin, D. J., Rubin, V. C., Prendergast, K. H. 1964, *Astrophys. J.*, **139**, 1058.
7. Burbidge, E. M., Burbidge, G. R., Crampin, D. J., Rubin, V. C., Prendergast, K. H. 1964, *Astrophys. J.*, **139**, 539.
8. Rubin, V. C., Burbidge, E. M., Burbidge, G. R., Prendergast, K. H. 1964, *Astrophys. J.*, **140**, 80.
9. Burbidge, E. M., Burbidge, G. R., Prendergast, K. H. 1964, *Astrophys. J.*, **140**, 85.
10. Rubin, V. C., Burbidge, E. M., Burbidge, G. R., Prendergast, K. H. 1965, *Astrophys. J.*, **141**, 759.
11. Rubin, V. C., Burbidge, E. M., Burbidge, G. R., Prendergast, K. H. 1965, *Astrophys. J.*, **141**, 885.
12. Burbidge, E. M., Burbidge, G. R., Prendergast, K. H. 1965, *Astrophys. J.*, **142**, 154.
13. Burbidge, E. M., Burbidge, G. R., Prendergast, K. H. 1965, *Astrophys. J.*, **142**, 649.
14. Bertola, F. 1965, *Contr. Obs. Astrofis. Asiago*, no. 172.

15. Bertola, F. 1965, *Ann. Astrophys.*, **28**, 574.
16. Duflot, R. 1965, *Publ. Obs. Haute-Provence*, **7**, no. 7.
17. Brandt, J. C. 1965, *Mon. Not. R. astr. Soc.*, **129**, 309.
18. King, I. R., Minkowski, R. 1966, *Astrophys. J.*, **143**, 1002.
19. Searle, L. 1965, *Nature*, **207**, 1282.
20. Epstein, E. E. 1964, *Astr. J.*, **69**, 490.
21. Epstein, E. E. 1964, *Astr. J.*, **69**, 521.
22. Höglund, B., Roberts, M. S. 1965, *Astrophys. J.*, **142**, 1366.
23. Roberts, M. S. 1965, *Astrophys. J.*, **142**, 148.
24. Roberts, M. S. 1966, *Astrophys. J.*, **144**, 639.
25. Seielstad, G. A., Whiteoak, J. B. 1965, *Astrophys. J.*, **142**, 616.
26. Pskovskii, J. P. 1965, *Astr. Zu.*, **42**, 323.
27. Huang, S. S. 1966, *Astrophys. J.*, **146**, 25.
28. Brandt, J. C., Scheer, L. S. 1965, *Astr. J.*, **70**, 471.
29. Wallace, J. M., Copeland, J. A. 1964, *Publ. astr. Soc. Pacific*, **76**, 315.
30. Kinman, T. D. 1965, *Astrophys. J.*, **142**, 1376.
31. Burbidge, E. M., Burbidge, G. R., Rubin, V. C. 1964, *Astrophys. J.*, **140**, 942.
32. Burbidge, E. M., Burbidge, G. R. 1964, *Astrophys. J.*, **140**, 1445.
33. Demoulin, M.-H. 1965, *Publ. Obs. Haute-Provence*, **7**, no. 44.
34. Burbidge, E. M., Burbidge, G. R. 1965, *Astrophys. J.*, **142**, 1351.
35. Minkowski, R. 1966, *Proc. IAU Symposium* no. 29, Byurakan.
36. Münch, G. 1966, *Proc. IAU Symposium* no. 29, Byurakan.
37. Burbidge, E. M., Burbidge, G. R. 1966, *Astrophys. J.*, **145**, 661.
38. Walker, M. F. 1966, *Proc. IAU Symposium* no. 29, Byurakan.
39. Bertola, F. 1966, *Proc. IAU Symposium* no. 29, Byurakan.
40. Vaucouleurs, G. de 1966, *Proc. IAU Symposium* no. 29, Byurakan.
41. Sĕrsic, J. L. 1966, *Proc. IAU Symposium* no. 29, Byurakan.
42. Freeman, K. C. *Mon. Not. R. astr. Soc.*, **130**, 183, 1965; **134**, 1, 15, 1966.
43. Burbidge, E. M., Burbidge, G. R., Prendergast, K. H. 1965, *Astrophys. J.*, **142**, 154.
44. Zwicky, F., Humason, M. L. 1964, *Astrophys. J.*, **139**, 269, 1393.
45. Zwicky, F. 1964, *Astrophys. J.*, **140**, 1467.
46. Zwicky, F. 1966, *Astrophys. J.*, **143**, 192.
47. Burbidge, E. M. 1964, *Astrophys. J.*, **140**, 1617.
48. Burbidge, E. M., Burbidge, G. R. 1964, *Astrophys. J.*, **140**, 1307.
49. Rubin, V. C., Burbidge, E. M., Burbidge, G. R. 1964, *Astrophys. J.*, **140**, 1304.
50. Sandage, A. R. 1965, *Astrophys. J.*, **141**, 1560.
51. Zwicky, F. 1965, *Astrophys. J.*, **142**, 1293.
52. Arp, H. C. 1965, *Astrophys. J.*, **142**, 402.
53. Courtès, G., Viton, M., Véron, P. 1964, Proc. Dallas Symposium on *Quasi-Stellar Objects and Gravitational Collapse*, Univ. of Chicago Press, p. 307.
54. Courtès, G., Cruvellier, P. 1965, *Ann. Astrophys.*, **28**, 683.
55. Véron, P., Sauvayre, A. 1965, *Ann. Astrophys.*, **28**, 698.
56. Hodge, P. W. 1966, (in press).
57. Baade, W., Arp, H. C. 1964, *Astrophys. J.*, **139**, 1027.
58. Arp, H. C. 1964, *Astrophys. J.*, **139**, 1045.
59. Burbidge, E. M., Burbidge, G. R. 1965, *Astrophys. J.*, **142**, 634.
60. Osterbrock, D. E., Parker, R. A. R. 1965, *Astrophys. J.*, **141**, 892.
61. Pacholczyk, A. G., Wisniewski, W. Z. 1966, Steward Obs. Preprint no. 4.
62. Vorontsov-Velyaminov, B. A., Dibai, E. A. 1966, *Proc. IAU Symposium* no. 29, Byurakan.
63. Dibai, E. A. 1966, *Proc. IAU Symposium* no. 29, Byurakan.
64. Mathis, J. S. 1965, *Publ. astr. Soc. Pacific*, **77**, 90.
65. Dickel, H. R. 1965, *Astrophys. J.*, **141**, 1306.
66. Deutsch, A. J. 1964, *Astrophys. J.*, **139**, 532.
67. Spinrad, H., Wood, D. B. 1965, *Astrophys. J.*, **141**, 109.
68. Wood, D. B. 1966, *Astrophys. J.*, **145**, 36.

69. Philip, A. G. D., Sanduleak, N. 1966, *Publ. astr. Soc. Pacific*, **78**, 30.
70. Spinrad, H. 1966, *Publ. astr. Soc. Pacific*, **78**, 367.
71. Gratton, L. 1964, *Ann. Astrophys.*, **27**, 114.
72. Kinman, T. D. 1965, *Astrophys. J.*, **144**, 1241.

APPENDIX III. REPORT OF THE WORKING GROUP ON GALAXY PHOTOMETRY

(Prepared by G. de Vaucouleurs, Chairman of the Working Group)

This report covers work on photometry and spectrophotometry of galaxies based on papers received since the Hamburg Meeting and, in particular, on replies to the Circular No. 4 of the working group.

Techniques and Instrumentation

Several new techniques and instruments have been introduced to speed up the tedious methods of photographic photometry.

D. Koelbloed (Amsterdam) has described a scanning isophotometer attachment to a Hilger microphotometer recording by means of electric sparks on 'Teledelthos' paper (**1**).

G. de Vaucouleurs, *et al.* (Texas) have described a double-beam, self-balancing isophotometer operating in either contour following or scanning modes and recording on either the paper chart of a slave x–y plotter or on a photographic plate (**2**).

A scanning 'Isodensitracer' built by Technical Operations Research, for radiography (**3**) could be applied to galaxy photometry.

Approximate isophotal contours have also been produced by composite printing of negative-positive pairs, an old technique recently revived for applications to galaxy photometry (**4, 5**).

A rigorous reduction program ('GALAXY 2') using digital techniques of 'numerical mapping' has been developed at the University of Texas and U.S. National Bureau of Standards to take advantage of the capabilities of large modern computers (**12**). A double-beam, two-coordinate, digitized microdensitometer in construction at the University of Texas will record automatically on punched paper tape the input (x, y and density) for this program. The system is designed to deliver accurate photographic isophotes (corrected for plate errors, with or without smoothing) and integrated magnitudes of galaxies on a mass production basis. Regular operations are expected to begin in 1967.

Detailed Photometry

Detailed surface photometry photographic isophotes (pg-i) or photoelectric scans (pe-s) has been reported as follows:

F. Bertola (Asiago) observed NGC 3432, 4485–90, 4605, 4618–25, 5005, 6503 (pg-i) to supplement spectroscopic rotation data (M/L ratio). The exponential decrease of luminosity in the outer regions of late-type spirals is confirmed.

P. W. Hodge (Seattle, Wash.) has continued his intensive studies of dwarf (E, I), lenticular (SO) and peculiar galaxies, on Mt Stromlo and Mt Wilson plates including NGC 128, 185, 1332 (**6**), the Sculptor dwarf (**7**), and bright and faint galaxies in the Fornax cluster (**8**) (pg-i with pe calibration), work in progress includes Local Group members NGC 6822, IC 1613 and the WLM galaxy (pe-s) and further surface photometry of SO, Irr II and spiral galaxies.

I. R. King (Berkeley) has continued his photographic studies of the brightness distribution in elliptical galaxies. Special attention is paid to maintaining a resolving power of 1″ or less in the central regions – a requirement for proper dynamical interpretation. The work is complete for NGC 2300, 4261, 4472, 4621, and 4697; and the results are being prepared for publication.

Additional material is being reduced for NGC 147, 185, 205, 221, 4365, 4374, 4382, 4406, 4486, 4552, 4589, 4636, 4649, 5846, 6166, 6703, and 7626; less complete material is available for NGC 404, 410, 524, 545–7, 584, 720, 1700, 3379, 4459, 4874, 4889, and 5322. These lists include all E's and round SO's whose internal velocity dispersions are known.

B. E. Markarian (Byurakan Obs., Armenian S.S.R.) reports that with the assistance of E. Y. Hovhannisian and S. N. Arakelian he has carried out detailed two-color photometry (pg-i) of 16 galaxies in the central part of the Virgo cluster (NGC 4371, –74, –88, 4406, –35, –38, –40, –58, –59, –61, –73, –77, –78, –86, 4501) (**9**).

P. Pishmish (Tonantzintla Obs., Mexico) indicates that she has started a four-color study (pg-i with pe calibration in *B*, *V*, *R*, *I*) of six edge-on galaxies (NGC 891, 4244, 4565, 4631 and 5904) and NGC 5128 in infra-red.

G. de Vaucouleurs and collaborators (University of Texas) have continued work on a 'Photometric Atlas of Bright Galaxies' (pg-i and pe-s) which will include about 100 of the brighter northern and southern galaxies. A detailed study of NGC 4486 (pg-i and pe-s) and its compact dwarf companion NGC 4486–B was completed. In collaboration with H. Ables, several Local Group dwarfs including the Sculptor and Fornax systems (pe-s) and IC 10 (**10**) have been observed in the *U*, *B*, *V* system. H. Ables (U.S. Naval Obs., Flagstaff) has started an intensive study of a dozen large, nearby galaxies observed at 21 cm and for which optical data are poor or lacking.

Integral Photometry

As part of work on vol. III, IV and V of the 'Catalogue of Galaxies and Clusters of Galaxies' (to be published in 1966–67) F. Zwicky, E. R. Herzog, C. Kowal and M. Karpowicz have estimated photographic magnitudes of some 10 000 galaxies on 18-inch Schmidt Schraffier films.

G. E. Kron and C. D. Shane (Lick Obs. and U.S. Naval Obs., Flagstaff) are continuing their photoelectric measurements (*P*, *V*) of integrated magnitudes to determine the limiting magnitude and magnitude scale of Zwicky's Catalogue and study galactic extinction. About 400 galaxies have been observed since 1964 with the Lick 120-inch (304 cm) and Flagstaff 40-inch (102 cm) reflectors.

R. R. Shobbrook (Sydney Univ.) measured in 1964–65 with the McDonald 36-inch (91 cm) reflector magnitudes and colors in the *U*, *B*, *V* system for about 30 galaxies in the zone $-20° < \delta < +5°$. These galaxies are in several groups for which he obtained radial velocities with the 82-inch (208 cm) reflector.

G. de Vaucouleurs and collaborators (McDonald Obs.) continued the reduction of *U*, *B*, *V* observations of some 550 galaxies with the 36-inch reflector in progress since 1960 (see 1964 report).

Multicolor Photometry

H. L. Johnson (University of Arizona) has published an eight-color photometry (*UBVRIJKL*) of the nuclear regions of ten bright galaxies extending to 3·4 μ. A stellar mixture including a large abundance of late K and M giants is necessary to explain the infra-red excess (**11**).

W. G. Tifft (University of Arizona) has prepared for publication narrow-band multicolor observations of galaxies made at Lowell Obs. in 1963–64.

S. van den Bergh and R. McClure (D. Dunlap Obs., Toronto) have started a seven-color intermediate band photometry of early type galaxies with a two-channel photometer on the 74-inch reflector. The objective is to obtain information on population content or luminosity from some combinations of color indices.

Spectrophotometry

E. Vandekerkhove (Uccle) has published values of relative gradients of 16 galaxies in the Virgo cluster (**13**) and is continuing work in about 50 objects measured on objective-prism spectra with the Uccle Obs. Schmidt telescope. R. G. Tull and G. de Vaucouleurs (McDonald Obs.) recorded in 1965 the spectra of 48 bright galaxies and 50 standard stars at 15 to 25Å resolution with a photoelectric scanner at the Cassegrain focus of the 82-inch reflector. Energy distribution, equivalent widths (and incidentally radial velocities) are measured and will be analyzed to derive the spectral composition (population content) of galaxies at several stages of the classification sequence.

This report was prepared from information received on or before 8 August, 1966.

REFERENCES

1. Koelbloed, D. 1965, *Bull. astr. Inst. Netherlands*, **18**, 62.
2. Vaucouleurs, G. de, Griboval, P., White, T. 1965, *Publ. astr. Soc. Pacific*, **77**, 115.
3. Miller, C. S., Parsons, F. G., Kofsky, I. L. 1964, *Nature*, **202**, 1196.
4. Högner, W., Richter, N. 1964, *Jena Review*, **187**; 1964, *Astr. Nachr.*, **287**.
5. Hodge, P. W., Brownlee, D. E. 1966, *Publ. astr. Soc. Pacific*, **78**, 125.
6. Hodge, P. W. *et al.* *Astr. J.*, **68**, 691, 1963; *Astrophys. J.*, **140**, 681, 1964; **144**, 875, 1966.
7. Hodge, P. W. *et al.* 1966, *Astr. J.*, **71**, 204.
8. Hodge, P. W., Pyper, D. M., Webb, C. J. 1965, *Astr. J.*, **70**, 559.
9. Markarian, B. E. *Astrophysica*, **1**, 38, 1965; **2**, 53, 1966.
10. Vaucouleurs, G. de, Ables, H. 1965, *Publ. astr. Soc. Pacific*, **77**, 272.
11. Johnson, H. L. 1966, *Astrophys. J.*, **143**, 187.
12. Jones, W. B., Obitts, D., Gallet, R., Vaucouleurs, G. de Astronomical Surface Photometry by Numerical Mapping Techniques, *Publ. Univ. Texas*, Ser. II, **1**, no. 8.
13. Vandekerkhove, E. *Comm. Obs. R. Belgique*, no. 238.

APPENDIX IV. REPORT OF THE COMMITTEE FOR RESEARCH ON SUPERNOVAE

(Period covered from June 1964 to October 1966)

(Prepared by F. Zwicky, Chairman of the Committee)

Fourteen Observatories (eleven and three respectively on the Northern and Southern Hemisphere) have cooperated in the search for supernovae. During the period of 27 months 33 supernovae were found (**5**, **6**, **7**): one each by Arp, Gates, Haro, Hoffmeister, Yakimov, Lovas and Reaves, three by Rosino, four each by Chavira, Rudnicki, and Wild, five by Zwicky, and six by Kowal, all of them observers at the 14 observatories represented on the Committee for Research on Supernovae. This brings the total number of supernovae definitely recorded since 1885 to 184.

A circular letter no. 8 on the supernovae no. 141 (SN–1963s) to no. SN 170 (SN–1965i) was issued by the Chairman of the Committee on 20 July 1965 as well as a general review (**2**).

The brightest supernova found, SN 1964f ($m_p \sim 14{\cdot}0$), in this period appeared in NGC 4303 and is, incidentally, the last of three which have been discovered in this galaxy within 40 years.

Light curves and spectra of various supernovae have been published by a number of investigators (**8**, **9**, **10**, **11**, **12**, **13**). None of the features in the spectra of the supernovae of the type I and IV have as yet been identified.

No emission of radio waves, cosmic rays or X-rays from one of the recent supernovae has as yet been observed.

A study of both the optical and the radio features of some of the remnants of past galactic supernovae by Minkowski (**3**, **4**, **14**) suggests that SN 1006 was an object of type I, SN 1054 cannot be assigned any type as yet, SN 1572 and 1604 are types I, the unobserved SN associated with the radio source Cas A to be classified as type III. η Carinae (according to Thackeray and Zwicky) was possibly type V.

No determination of the frequencies of occurrence of the various types of supernovae has as yet been possible with the exception of the frequency of objects of the type I in large nearby clusters of galaxies. This frequency is of the order of one supernovae per 300 years per average cluster galaxy whose luminosity lies within the first five magnitudes from the brightest galaxy.

REFERENCES

1. Communications présentées au Colloque International sur les *Novae, Novoïdes et Supernovae.* 1965, Monograph of the Centre National de la Recherche Scientifique, Paris.
See especially Zwicky, F., 'Basic Results of the International Search for Supernovae', p. 168.
2. Zwicky, F. 1965, *Supernovae, Stellar Structure*, eds. L. H. Aller and B. M. Middlehurst (University of Chicago Press, Chicago), 367.
3. Minkowski, R. 1964, Supernovae and Supernova Remnants, in *Ann. Rev. Astr. Astrophys.*, **2**, 247.
4. Minkowski, R. *Non-thermal Radiosources, Nebulae and Interstellar Matter*, eds. L. H. Aller and B. M. Middlehurst (University of Chicago Press, Chicago), in press.
5. Zwicky, F. 1964, *Publ. astr. Soc. Pacif.*, **76**, 325.
6. Zwicky, F. 1965, *Publ. astr. Soc. Pacif.*, **77**, 456.
7. Zwicky, F. 1966, *Publ. astr. Soc. Pacif.*, **78**, 471.
8. Aller, L. H., Ross, J. 1965, *Publ. astr. Soc. Pacif.*, **77**, 469.
9. Bertola, F. 1965, *Mem. Soc. astr. ital.*, **36**, 299.
10. Hill, P. W. 1965, *Mon. Not. R. astr. Soc.*, **131**, 155.
11. Zwicky, F. 1964, *Astrophys. J.*, **139**, 514.
12. Zwicky, F., Karpowicz, M. 1964, *Astr. J.*, **69**, 759.
13. Zwicky, F., Karpowicz, M. 1965, *Astr. J.*, **70**, 564.
14. Minkowski, R. 1966, *Astr. J.*, **71**, 371.

APPENDIX V. REPORT OF THE COMMITTEE ON THE MAGELLANIC CLOUDS

(Prepared by A. D. Thackeray, Chairman of the Committee)

1. *General*

Current plans for new equipment in the southern hemisphere, including both large and small telescopes, imply greatly extended research on the Magellanic Clouds in the 1970's. Publication of the Smithsonian Astrophysical Observatory's Atlas of the Large Cloud by Hodge and Wright (**34**) is an important event which will greatly aid both planners and observers. The Atlas contains 81 V charts and 80 B charts at 16″/mm (except for four in the bar at 4″/mm), with limiting magnitudes about 17·0 and 17·5 respectively. The V charts are marked with clusters and emission nebulae, the B charts with variables discovered at the Harvard, Republic and Greenwich Observatories. The introduction, itself a valuable summary of Large Cloud work, contains Tables listing all the variables, clusters etc. with bibliography and other essential data.

Bok (**10**) has written a valuable review article on the two Clouds as of December 1965 which of course presents a much more detailed picture than this report. A colloquium of optical and radio astronomers was held at Mt Stromlo Observatory (**48**) in May 1965. A summary of contributions by Shapley to Magellanic Cloud research has been given by Bok (**12**).

2. *Variables and the P–L Relation*

Accurate two-color photoelectric observations of 14 SMC and 13 LMC cepheids are reported in an important contribution by Gascoigne and Kron (**29**). The SMC *P–L* relation is $< V > = 17{\cdot}85 - 2{\cdot}97 \log P$, which agrees in slope with that found in the LMC by the Herstmonceux group. The colors are systematically bluer than Arp's SMC cepheids. The derived distances are 61 kpc (SMC) and 50 kpc (LMC). Dickens (**15**) reports 2-color photometry of 10 cepheids in Herstmonceux Field 2. The *P–L* relation was discussed at the Herstmonceux Conference on Stellar Evolution (**63**).

Wesselink (**57, 58**) has published results for 45 faint SMC variables, including five with periods shorter than one day, and magnitudes based on measures with objective grating. The predominance of periods less than two days in this rich region is striking. Hodge and Wright (**35**) have discovered 51 new variables in the LMC, while eight new variables in the wing of the SMC are reported by A. D. Andrews (**3**).

The Gaposchkins continue their studies of Harvard variables based on the Arp SMC sequence (**24, 25, 26**). They find that variables with sinusoidal light-curves ($M - m \geqslant 0{\cdot}3$) define a $P - L$ relation lying $0{\cdot}5\ m$ higher than the $P - L$ relation for normal cephoids (**24**). The general relationship of form of light-curve and period is similar to that found in the galaxy. Mrs Gaposchkin (**23**) finds the slope of the $P - L$ relation systematically steeper at the bright end than the faint. S. Gaposchkin (**28**) finds that at any given period the amplitude of a SMC cepheid tends to increase with maximum brightness. He has also studied 33 eclipsing variables in the SMC (**27**); he finds none brighter than $-\ 5{\cdot}0$ (M_{pg}) while relative to the known SMC cepheids they appear to be five times rarer than in the Galaxy.

Gascoigne (**48**) discusses cepheids and stellar birth-rates and stresses the rarity of Cloud cepheids in clusters and associations.

Arp's SMC sequence is being re-observed at Boyden (C. J. Butler) and Radcliffe (P. J. Andrews) Observatories. At Boyden cepheids in two LMC fields and one SMC field are being observed by four-color photometry to 18th magnitude (**4**). Van Genderen (Leiden) is studying SMC cepheids on the basis of plates taken by Arp.

The LMC cepheid HV 953 has been studied over a 60 year span by Janes (**40**) who finds the period fluctuating rather than progressively lengthening.

Arp (**6, 7**) has derived light-curves of cepheids in two colors in NGC 1866, based on Radcliffe plates secured in part by Thackeray. The colors, amplitudes and light curves are essentially independent of period (ranging from 2·64 to 3·53 days).

Short-period variability in some bright LMC supergiants has been looked for in vain by Bok *et al.* (**13**). Spectra of S Dor during a record deep minimum (from which it has not yet recovered) have been described by Thackeray (**52**).

3. *Radio and Gaseous Components*

Detailed results of 21 cm surveys with the 210 ft (64 m) Parkes telescope have been published for the LMC by McGee and Milton (**44, 45, 46, 48**) and for the SMC by Hindman and Balnaves (**31, 32, 33, 48**).

McGee detects 52 large H I complexes in the LMC. He attributes the very complicated velocity pattern to two pairs of spiral arms in planes differing by about 20°. The bridge between SMC and LMC has been resolved with the Parkes equipment into patchy condensations.

Hindman has given further details of the double-peaked 21cm profiles in the SMC. The pattern is attributed to expanding shells; the absence of a non-thermal component associated with supernova shells is to be noted and may perhaps be due to an ageing effect. He has found the first evidence of a rotation curve for the SMC with turn-over points and a rather surprisingly large velocity gradient; the maximum gradient is in p.a. 55°, similar to the major axis of Lindsay's distribution of clusters.

Westerlund and Mathewson (**61**) identify Henize nebulae N49, 63A, 132D in the LMC with non-thermal sources and attribute them to remnants of Type II supernovae.

Broten (**48**) has surveyed the LMC at 11 cm with the Parkes telescope. Over 50 sources are found, with the strongest coinciding with H II regions. Mills (**48**) reports preliminary observations of the Clouds with the EW arm of the Cross radio telescope at 408 MHz.

The H I content of the Clouds is listed for comparison with that of other galaxies by Epstein (**16**).

Dickel (**14**) has determined electron densities and masses of 8 LMC and 1 SMC H II regions from low-resolution scans with results that are matched by many galactic H II complexes (larger and less dense than the Orion nebula). Feast (**19**) in a complementary investigation based on coudé spectra, determines electron densities and masses of 36 LMC and 2 SMC nebulae from measures of the [O II] 3726/3729 ratio combined with Michigan measures of Hα surface brightness; three nebulae are common to Mrs Dickel's list. The general conclusions are similar. Feast's densities run higher, partly at least because his analyzing slit was very much narrower; his total masses run smaller.

Aller and Faulkner (**5**) compare 30 Dor and Eta Car nebulae by photoelectric spectrophotometry; abundances (He/H, O/H), electron temperatures and densities are compared. No differences in abundance ratios greater than 40% can be found. Mathis (**47**) has also studied the inner regions of 30 Dor with results that resemble galactic nebulae fairly closely. Faulkner (**17**) in a further study of 30 Dor compares surface brightness in Balmer lines with radio fluxes. He derives $R = 7$ for the ratio A_v/E_{B-V}.

Feast (**18**) has observed spectra of ten planetaries and finds a rotation curve for the Cloud similar to that from Population I stars. The velocity dispersion is 22 ± 7 km s^{-1}. This program is being continued and Feast has now material on 37 planetaries (both Clouds). Webster (**48**) gives preliminary results of velocities of Cloud planetaries. Bok *et al.* (**11**) have measured radial velocities of nine emission nebulae in the bar of the SMC for comparison with the 21 cm velocities.

A third list of 446 emission-line objects in the LMC discovered on ADH objective prism spectra is given by Andrews and Lindsay (**2**).

4. *Clusters*

Gascoigne (**30**) has completed a major program on color-magnitude diagrams of nine Cloud globular clusters down to $V \sim 19{\cdot}5$. Among these he recognizes three types: (1) NGC 1466, 2257 (both containing known RR Lyr variables) and 1841 have arrays like that of Tifft for NGC 121, and all are similar to galactic halo clusters; (2) Lindsay 1, Kron 3 and NGC 339 (all SMC) have no blue horizontal branch despite blue integrated colours and are regarded as of intermediate age; (3) NGD 2209, 2231 and Hodge 11 (all LMC) differ in having a very weak (or no) giant branch but many faint blue stars; this type has no known galactic counterpart. Gascoigne (**48**) has also measured integrated colors of 60 Cloud globulars and confirms the two-fold grouping in color; there is an almost complete absence of colors $(B - V)$ between 0·30 and 0·60.

Arp (**6, 8**) has completed his study of the *c–m* array of NGC 1866 which contains many red giants ($B - V = +\,0{\cdot}6$ to $+\,1{\cdot}5$) despite its early-type integrated spectrum. He assigns an age a little less than 10^8 years and finds some evidence in favour of evolutionary tracks predicted for stars of mass 5 $M_\odot$. There appears to be a progressive reddening of giant branches of globular clusters in SMC, LMC and the solar neighborhood. Hodge and Sexton

(**36**) have listed 457 new clusters in the LMC. Lindsay (**41**) has commented on the classification of some objects listed as clusters by Lynga and Westerlund. Lindsay (**42**) has also identified most of the 366 NGC objects in the LMC region on ADH plates. He lists 26 (almost entirely in the outer regions) as galaxies.

Renewed attempts at measuring radial velocities of Cloud globulars are being made at the Radcliffe Observatory.

5. *General Field*

(a) c–m arrays and photometry

Woolley and Epps (**64**) have studied colors of stars down to about 16·5 *m* (two magnitudes fainter than an earlier proper-motion study) and conclude that in the relevant LMC field the great majority of stars with $B - V$ between + 0·5 and + 1·5 are foreground. Basinski *et al.* (**9**) have derived a *c–m* array down to $V = 15{\cdot}0$ for a dense region of the SMC. They find numerous stars with $B - V < + 0{\cdot}4$ (approximately equally divided into groups −·4 to 0 and 0 to +·4 and some stars with $B - V > + 1{\cdot}20$). They also attribute stars of intermediate color to the foreground. The same team have continued to study NGC 1929–37 (LMC). According to a preliminary report by Bok (**48**) the core of the region is dominated by blue stars, while further out (after allowance for foreground stars) some yellow-red stars appear which may represent evolved stars expelled from the central core. Hodge (**37**) has published a corrected p–e sequence and *c–m* array for 938 stars in the LMC.

UBV photometry of LMC supersupergiants is being undertaken by P. J. Andrews. Walraven (**55**) is also working on multi-color photometry of Fehrenbach stars. He has observations on his system of 365 and 274 stars in the region of SMC and LMC respectively of which at least 45% are Cloud members.

Westerlund (**48**) divides each of the Clouds into four sub-systems according to age, partly on the basis of colors.

(b) Spectroscopy

Hutchings (**39**) has published measures of Hγ equivalent widths for 149 Cloud supergiants (O to K5). The variation with spectral type shows a maximum near F0. Approximate distance moduli (corrected for absorption) of 18·2 (LMC) and 18·3 (SMC) are derived. Przybylski (**49**) has carried out a curve of growth analysis of the brightest known LMC member, HD 33 579, and finds $\Delta \log N$ (Fe, Cr, Ti/H) relative to the Sun of $- 0{\cdot}2 \pm 0{\cdot}2$. J. A. Williams (**62**) has also studied the metal/H ratio of long period Cloud cepheids by means of four-color intermediate band photometry. He finds no significant differences in the metal indices for either Cloud and the solar neighborhood. L. F. Smith (**50**) has classified 41 Cloud WR stars through narrow band photometry and some spectra. The order of decreasing luminosity appears to be WN7–8–6 (rare)–WC5– WN5. Dessy is obtaining spectra of the brightest Cloud stars at 42 Å/mm for inclusion in an Atlas of Spectra in collaboration with the Jascheks.

(c) Radial Velocities

Fehrenbach (**20**) has recapitulated his group's lists of stars with large and small velocity in the direction of the LMC. Fehrenbach *et al.* (**20**) have published radial velocities of 111 stars in a central region of LMC of which 29 are attributed to the Cloud. Fehrenbach (**21**) reported at IAU Symposium no. 30 on further progress in this program and on the extension of his survey to regions between the Clouds and between LMC and Galaxy where several high velocity stars have been found. Florsch and Carozzi (**22**) have found 33 high-velocity stars between the Clouds of which they regard 21 as probable members.

Slit spectra of all the later type LMC members, $m \leqslant 12$, revealed by Fehrenbach's survey are being obtained at the Radcliffe Observatory. Thackeray (**53**) has published two instances of high-velocity galactic subdwarfs and it is clear that the objective prism velocities are not a completely sure criterion of Cloud membership. Radcliffe slit spectra of SMC stars are also being secured for improved radial velocities.

Wayman (**56**) has proposed a technique for separating foreground stars among the later types through narrow-band photometry.

Gollnow (**48**) has reviewed the general problem of observed radial velocities in the Clouds.

(*d*) *Polarization and Absorption*

Visvanathan (**54, 48**) has published his polarization measures of 30 LMC stars. His discussion suggests that the Cloud is seen nearly face-on with a magnetic field related to Cloud structure. The polarization is correlated with color excesses derived from Radcliffe photometry and spectra. Hutchings (**38**) has published details of his important discovery of 4430 Å absorption in both Clouds. The correlation of strength and color excess is similar to that found for galactic stars.

Wesselink (**59**) has further discussed selection effects due to absorption in the discovery of variables. He is inclined to attribute the displacement of the LMC centre of rotation from the bar to heavy absorption distorting the optical picture.

7. *Summary of Trends and Future Needs*

The comparison of the Clouds with each other and with the Galaxy, particularly as regards cepheids and the *P–L* relation, continues to be of fundamental interest. Extensive statistics of cepheid periods and light-curves may have encouraged early belief in differences in composition, which would serve as a key to evolutionary theory. The differences sought, however, have become elusive. For example, the 'unusual' galactic cepheid TV Cam resembles the SMC cepheids in its light-curve, and was at one time thought to display abundance anomalies in its spectrum. These are now attributed to low turbulent velocity by Abt *et al.* (**1**).

Further study of the *P–L* relation could be planned to examine questions of *tilt* of the two systems, especially in view of Hindman's model of the SMC viewed nearly edge-on. It would be logical if further plans, especially in the choice of fields were coordinated with this problem in mind together with provisions for an adequate degree of overlap.

Uncertainty in the masses of the Clouds, arising largely from uncertainty in their tilts (see Bok (**10**)) should be resolved in order to clarify the problem of tidal interactions between the Clouds and Galaxy. Kerr (**48**) discusses such interactions as possibly associated with the warped H I galactic plane, north-south asymmetry in the H I rotation curve, and the high-velocity thin H I clouds observed at Leiden ($l^{II} = 120$, $b^{II} = +40$).

The complexities of radio patterns revealed by the Parkes telescope have posed many problems whose solution depends in part on the slower approach of optical astronomers. Visvanathan's work on optical polarization deserves to be prosecuted.

Narrow-band photometry could be pushed to fainter limits. Hβ photometry could be used for improved distance moduli; Hutchings' Hγ equivalent widths refer only to the brightest supergiants and may frequently be affected by shell absorption or superposed emission.

The luminosity function in different parts of the Clouds is an important datum in evolutionary theory. It has been discussed by McCuskey (**43**) in the light of former star-counts. There are many suggestions of bursts of local star formation in regions up to 1 kpc across. Even at magnitude 15–16 the contribution of galactic foreground can be very serious and statistical corrections may be misleading. Multicolor photometry at this level would probably offer the safest criterion of membership. For the brightest stars, Fehrenbach's survey supplemented by slit spectra is leading to a fairly complete census.

It is now generally accepted that both Clouds contain Populations I and II. But our knowledge of Population II components is severely limited. Gascoigne's *c–m* arrays of globular clusters underlines the recognized need for radial velocities – an especially difficult problem for the red globulars. Continuation of Dessy's search for long period variables is required and is being attempted at the Radcliffe Observatory.

Perhaps the most neglected field in recent Cloud research has been the search for novae. Several must have been missed since the abandonment of the Harvard patrols in the last decade. A small resumption has been begun by Sterling (**51**) in association with the Radcliffe Observatory. With many telescopes capable of following up a discovery, the importance of immediate blinking of search-plates, with accurate information on position and nearest bright star cannot be overstressed.

BIBLIOGRAPHY

1. Abt, H., Osmer, P. S., Kraft, R. P. 1966, *Astrophys. J.*, **145**, 479.
2. Andrews, A. D., Lindsay, E. M. 1964, *Irish astr. J.*, **6**, 241.
3. Andrews, A. D. 1965, *Irish astr. J.*, **7**, 79.
4. Andrews, A. D. 1966, *Inf. Bull. S. Hemisphere*, no. 8, 23.
5. Aller, L. H., Faulkner, D. J. 1965, *Mon. Not. R. astr. Soc.*, **130**, 393.
6. Arp, H. C. 1965, IAU 3rd Colloquium on Variable Stars. *Kl. Veröff. Remeis Sternw. Bamberg*, **4**, no. 40, 209.
7. Arp, H. C., Thackeray, A. D. *Astrophys. J.* Submitted 1966.
8. Arp, H. C. *Astrophys. J.* Submitted 1966.
9. Basinski, J. M., Bok, B. J., Bok, P. F. 1966, *Mon. Not. R. astr. Soc.*, in press. (Summary also in *Publ. astr. Soc. Pacif.*, **78**, 439, 1966).
10. Bok, B. J. 1966, *Ann. Rev. Astr. Astrophys.*, **4**, 95.
11. Bok, B. J., Gollnow, H., Hindman, J. V., Mowat, M. 1964, *Austr. J. Phys.*, **17**, 404.
12. Bok, B. J. 1965, *Publ. astr. Soc. Pacif.*, **77**, 416.
13. Bok, B. J., Kidd, C., Routcliffe, P. 1966, *Publ. astr. Soc. Pacif.*, **78**, 333.
14. Dickel, H. R. 1965, *Astrophys. J.*, **141**, 1306.
15. Dickens, R. J. 1966, *Observatory*, **86**, 18.
16. Epstein, E. E. 1964, *Astr. J.*, **69**, 503.
17. Faulkner, D. J. 1966, *Mon. Not. R. astr. Soc.*, In press.
18. Feast, M. W. 1964, *Observatory*, **84**, 266.
19. Feast, M. W. 1964, *Mon. Not. R. astr. Soc.*, **128**, 327.
20. Fehrenbach, Ch., Duflot, M., Duflot, A. 1965, *J. Observateurs*, **48**, 185, 199.
21. Fehrenbach, Ch. 1966, *IAU Symposium no. 30 (Toronto)*. In press.
22. Florsch, A., Carozzi, N. 1965, *Comm. ESO*, no. 5.
23. Gaposchkin, C. P. 1965, IAU 3rd Colloquium on Variable Stars, *Kl. Veröff. Remeis Sternw. Bamberg.*, **4**, no. 40, 178.
24. Gaposchkin, C. P., Gaposchkin, S. 1966, *Vistas in Astronomy*, **8**, 191.
25. Gaposchkin, S. 1965, *Publ. astr. Soc. Pacif.*, **77**, 127.
26. Gaposchkin, S. 1966, *Smithson. Contr. Astrophys.*, **9**, 1.
27. Gaposchkin, S. 1965, IAU 3rd Colloquium on Variable Stars, *Kl. Veröff. Remeis Sternw. Bamberg.*, **4**, no. 40, 66.
28. Gaposchkin, S. 1965, IAU 3rd Colloquium on Variable Stars, *Kl. Veröff. Remeis Sternw. Bamberg.*, **4**, no. 40, 76.
29. Gascoigne, S. C. B., Kron, G. E. 1964, *Mon. Not. R. astr. Soc.*, **130**, 333.
30. Gascoigne, S. C. B. 1966, *Mon. Not. R. astr. Soc.* In press.
31. Hindman, J. V. 1964, *Nature*, **202**, 377.
32. Hindman, J. V., Balnaves, K. M. 1966, *Austr. J. Phys. Suppl.* In press.
33. Hindman, J. V. 1966, *Austr. J. Phys.* In press.
34. Hodge, P. W., Wright, F. W. *Smithsonian Astrophysical Observatory*, 1966. See also *Astr. J.*, **64**, 565, 1964.
35. Hodge, P. W., Wright, F. W. 1966, *Astr. J.*, **71**, 131.

36. Hodge, P. W., Sexton, S. 1966, *Astr. J.*, **71**, 363.
37. Hodge, P. W. 1965, *Astrophys. J.*, **141**, 806.
38. Hutchings, J. B. 1966, *Mon. Not. R. astr. Soc.*, **131**, 299.
39. Hutchings, J. B. 1966, *Mon. Not. R. astr. Soc.*, **132**, 433.
40. Janes, K. A. 1964, *Astr. J.*, **69**, 131.
41. Lindsay, E. M. 1964, *Irish astr. J.*, **6**, 233.
42. Lindsay, E. M. 1964, *Irish astr. J.*, **6**, 286.
43. McCuskey, S. W. 1966, *Vistas in Astronomy*, **7**, 141.
44. McGee, R. X. 1964, *Austr. J. Phys.*, **17**, 515.
45. McGee, R. X., Milton, J. A. 1966, *Austr. J. Phys.*, **19**, 343.
46. McGee, R. X., Milton, J. A. 1966, *Austr. J. Phys. Suppl.*, no. 2.
47. Mathis, J. S. 1965, *Publ. astr. Soc. Pacif.*, **77**, 189.
48. Mt Stromlo *Symposium on the Magellanic Clouds.* (Edited Hindman, J. V. & Westerlund, B. E.), 1965.
49. Przybylski, A. 1965, *Nature*, **205**, 163.
50. Smith, L. F. 1966, Private Communication.
51. Sterling, K. L., Wolterbeek, J. 1966, *Inf. Bull. S. Hemisphere*, no. 8, 27.
52. Thackeray, A. D. 1964, *Mon. Not. R. astr. Soc.*, **129**, 169.
53. Thackeray, A. D. 1966, *Observatory*, **86**, 60.
54. Visnanathan, N. 1966, *Mon. Not. R. astr. Soc.*, **132**, 423.
55. Walraven, Th. 1966, Private Communication.
56. Wayman, P. A. 1966, IAU Symposium no. 30 (Toronto), p. 89.
57. Wesselink, A. J., Shuttleworth, M. 1965, *Mon. Not. R. astr. Soc.*, **130**, 443.
58. Wesselink, A. J. 1965, IAU 3rd Colloquium on Variable Stars. *Kl. Veröff. Remeis. Sternw. Bamberg*, **4**, no. 40, 181.
59. Wesselink, A. J. 1966, *Astr. J.*, **71**, 185.
60. Westerlund, B. E. 1964, *Observatory*, **84**, 253.
61. Westerlund, B. E., Mathewson, D. S. *Mon. Not. R. astr. Soc.*, **131**, 371, 1966; *Astr. J.*, **69**, 563, 1964.
62. Williams, J. A. 1966, *Astr. J.*, **71**, no. 3.
63. Woolley, R.v.d.R. 1964, *R. Obs. Bull.*, no. 82, E96.
64. Woolley, R.v.d.R., Epps, E. 1965, *R. Obs. Bull.*, no. 90, E313.

APPENDIX VI. INVESTIGATIONS IN EXTRAGALACTIC ASTRONOMY MADE IN THE U.S.S.R. DURING 1964–66

(Prepared by B. Vorontsov-Velyaminov)

Photometry, Colorimetry, Polarization

Markarian, Oganesian and Arakelian made a detailed photometry and colorimetry of E and SO galaxies (1) and of spiral galaxies (2) in the Virgo cluster. The principal regularities observed in the distribution of brightness and color in the images of galaxies were noted. E and SO galaxies were studied: NGC 4374, 4406, 4435, 4458, 4459, 4461, 4473, 4477, 4478, 4486, and spirals: NCG 4371, 4388, 4402, 4438, 4440, 4501. The main features of their structure were noted. Kalloghlyan (3) measured photographically the surface brightnesses of bars of nine brighter SB galaxies. He found the mean value to be $20^{m}.9$ per square second of arc and concluded that this brightness is constant, which was confirmed by the comparison of the angular dimensions of the bars with the integrated brightness of many SB galaxies. Later Kalloghlyan and Tovmassian (4) made a two-color photometry of 50 SB galaxies and classified their nuclei from the point of view of their structure. Besides, Tovmassian (5) from the observations of 20 SB galaxies discussed the distribution of nuclei depending on their structure. Kalloghlyan (6) performed also the two-color photometry of the interacting galaxy NGC 3656 and described the structural peculiarities of this double system.

Artamonov, Vorontsov-Velyaminov and Dibai (**7**) published the isophotes of the spiral galaxy NGC 2861. Vorontsov-Velyaminov (**8**), using the photographic measurements of stellar magnitudes of 200 galactic nuclei made by Gorbatschev (**9**) and the published photoelectric measurements of galaxies, compiled a catalog, which gives for 173 nuclei their apparent and absolute magnitudes, surface brightnesses, and luminosities per unit volume. They all are similar to the corresponding characteristics of Zwicky's compact galaxies. The luminosity of the nuclei increases with the luminosity of galaxies. Gorbatschev (**10**) studied the distribution of absolute luminosity in the nuclei of NGC 1300, 4051, 4040 and 4548 and, using Spinrad's model of the M31 nucleus, calculated the density distribution and the masses of these galaxies. Vorontsov-Velyaminov's group obtained the isophotes for several peculiar and interacting galaxies (in press).

Gagen-Torn (**11**) from the photographic and photoelectric observations studied the polarization distribution in NGC 3077, which points to the presence of dust in this galaxy. Dombrovsky (**12**) detected photoelectrically polarization in the nuclei of M51 and NGC 1068. Dibai and Schahovskoy (**13**) measured polarization in the nuclei of the Seyfert galaxies, but found it only in the radiogalaxies NGC 1068 and 1275.

Morphology of Galaxies

Vorontsov-Velyaminov described some new morphological types of galaxies, which show their large variety and the repetition of some 'peculiar' forms. Vorontsov-Velyaminov and Davydov (**15**) from the photographs with interference filters found a difference in the distribution of hydrogen and of [O III] in M82. Gurzadian (**16**) explained the presence of the hydrogen absorption lines in the spectrum of M82 not by the presence of numerous hot stars, but by the light absorption, by interstellar hydrogen inside this galaxy and by the presence of relativistic electrons. Markarian (**17**) studied ten irregular galaxies, which he considered to belong to the M82 type. He concluded that the discrepancy between the colour and the spectrum cannot be accounted for by the influence of dust and suggested that the early type of their spectra were due to some unusual emission. Ambartsumian (**18**) analyzing the data on the galaxies of M82 type concluded that they are young and that their luminosity increases in the course of time. The relationship between the physical state of the nucleus and the evolutionary phase of the galaxy was discussed. Markarian (**19**) also investigated other galaxies with anomalous spectral characteristics as regard their type and drew up a list of them. The conclusion was drawn that the peculiarly blue color of their nuclei is due to the additional ultraviolet emission of unusual nature. Vorontsov-Velyaminov (**20**) described eight types of peculiar nuclear regions and established that hot nuclei with traces of spiral structure occur among a third of the galaxies of SAB and SB families. Some of their physical characteristics were determined. Vorontsov-Velyaminov (**21**) showed that the structure of the flat component of many galaxies is in favor of the presence of magnetic-like phenomena: loops, eight-forms, intersections of the arms, etc. However, some data do not confirm the magnetic forces hypothesis. Zasov (**22**) found 13 galaxies where the layer of dark matter is inclined to the equatorial plane and suspected this inclination in 17 galaxies. To explain this he turns to the hydrodynamical interaction of the layer with a current of intergalactic gas outflowing the galaxy.

Ambartsumian, Iskudarian, Shachbazian and Sahakian (**23**) described the powerful associations of 30 Doradus type in the LMC. They call them superassociations. The latter are most frequent in the supergiant Sc and in irregular galaxies. A list of galaxies with superassociations is given. Markarian (**24**) quoted examples of isolated chains of six and more galaxies. The calculation of the probability of an accidental formation leads to the conclusion that the majority of the chains must be real physical systems. Karachenzev (**25**) studied the distribution of galaxies and the structure of the supercluster in Hercules. He (**26**) has also found 165 dwarf galaxies in the Virgo cluster and established their statistical relation to the bright galaxies S and E.

Richter and Sahakjan (**27**) discovered 54 remote blue objects and studied their distribution. Vorontsov-Velyaminov (**27a**) showed that the number of globular clusters relate neither to luminosity nor to mass of the galaxy, though generally there are less globular clusters in small galaxies.

Statistics and Kinematics of Galaxies and of QSS

The Morphological Catalog of galaxies (MCG) provides an exhaustive source for statistical investigations and for the compilation of observational programs. Now with the publication of the II and IV volumes this catalog, compiled by Vorontsov-Velyaminov and Arhipova (**28, 29**), is complete. It gives the coordinates, integrated and surface brightness, dimensions, inclination, description and all other known data for 34 000 galaxies 15^m and brighter to the north of $\delta = -33°$. Among them 1765 interacting, 4700 double and multiple systems are noted. Zvyagina (**30**) from the counts of faint galaxies in small areas around galaxies of 13^m to 14^m finds that statistically more than a half of them have dwarf companions. The number of galaxies with increasing number of companions drops quickly. Denisjuk (**31**) discussed the possibility of finding the orientation of an elliptical galaxy in space with respect to the observer. Roslyakova and Gainullina (**32**) gave a theoretical interpretation to the fact that the symmetry planes of bright E galaxies tend to pass through the center of compact clusters. Agekjan and Sumsina (**33**) found a connection between momenta of apparent and real sphericity of galaxies in 90 clusters. The squares of the sphericity are different, sometimes negative. This points to the preponderance of orientation or to the existence of needle-like galaxies. Veltman (**34**) in his papers, based on the Schuster law, constructed the phase and hydrodynamical models with radially elongated velocity distribution. In the last case the dispersions of velocity are expressed through hypergeometric functions.

Karachenzev concludes that the clusters Virgo (**35**) and Coma (**36**) are not stable. From counts of 54 000 galaxies in the region of the North Galactic Pole he (**37**) finds, using the angular diameters, that the average density of Metagalaxy is $1{\cdot}2 \times 10^{-31}$ g cm^{-3} and makes the inference that the new galaxies are formed in the lapse of time. Pskovsky (**38**) estimated the density and mass of the Coma clusters ($4 \times 10^{14} M_{\odot}$) and, accepting that the clusters occupy 1% of the whole volume, evaluated the density of the Metagalaxy as 7×10^{-30} g cm^{-3}. Karachenzev (**39**) determined the ratio of the virial mass to luminosity for multiple galaxies and clusters and concluded that the non-stability of systems increases with their luminosity. He believed this to be akin to the expansion of the Metagalaxy. Zasov (**40**) studied the correlation of the distance moduli with the red shift of galaxies taking into account the influence of the latter on the apparent luminosity. He has found that the second order term in the relation describing the deceleration of expansion equals $-2{\cdot}2$ (from the luminosity of the 10th galaxy in clusters). Pskovsky (**41**) established the connection of optical limunosity with radio index and obtained the luminosity function of normal galaxies in the photographic and in the radio region. He (**42**) studied the correlation of radio luminosity with the spectral index and linear dimension and obtained the radio-luminosity distribution which proved to be monotonic. Zasov and Ozernoi (**43**) determined the luminosity function of QSS for different models of the Universe in radio and optical frequencies. The continuous transition of the function from radio galaxies to the quasi-stellar radiosources points to their genetic connection.

Spectra and Spectrophotometry of Galaxies and QSS

Dibai and Pronik investigated spectrophotometrically the Seyfert galaxies NGC 1068 (**44**), 1275 (**45**) and others (**46**); Dibai and Vorontsov-Velyaminov (**47**)—NGC 4151; and Dibai,

Esipov and Pronik (48)—NGC 5548. They calculated the physical conditions in the two systems of gas in all these nuclei and the probable excitation mechanism. Dibai and Pronik (49) made also the spectrophotometry of QSS 3C 273 and determined the physical parameters in its gaseous envelope. Dibai and Esipov (50) measured the red shift in the spectrum of 3C 345 and noticed the changes of the profile of Mg II line. They detected Hα emission in the nucleus of NGC 4486, estimating its gas content and obtained (52) the spectra of ten faint blue objects in high galactic latitudes. Object Tonanzintla no. 256 has a spectrum similar to that of the nuclei of the Seyfert galaxies. Pronik (53) obtained the spectrophotometric characteristics of QSS CTA-102 and estimated the mass of gas to be above $10^6 M_\odot$.

Supernovae

Dibai, Esipov and Pronik (54) studied the energy distribution in the spectra of Chavira and Rosino 1965 supernovae. Kukarkin (55) and Ambartsumian (56) noticed the very high frequency of supernova I outbursts in giant spirals of late types and Pskovsky (57) gave the characteristics of an object near radiosource 3C 386 considering it as being a supernova of anno 1230.

Cosmogonic Hypotheses

According to Pickelner's hypothesis (58) the spiral arms appear when the primordial gas in the center is contracting with a subsequent propagation of a condensation wave along the magnetic tube. Their thickness depends on the star density in the disk. This explains the difference in the arm structure in the galaxies of the Sa, Sb and Sc types. The propagation of the condensation wave in the gas cloud which surrounds the galaxy can explain the formation of the tails and of the connecting filaments in the interacting galaxies. Another possibility of their formation was pointed out by Gerschberg (59). It is the twisting of the magnetic field with the relative rotation of galaxies which lie on one and the same tube of force lines. Sb galaxies require the absence of differential rotation and according to Pickelner (60), the explosion in a gaseous proto-galaxy pushes the gas off center creating the homogeneous sphere of old stars. The flow of gases to the ends of the bar, rings and other structures is also explained.

Notice. In this review there are not included: cosmology, radioastronomical studies of galaxies, theoretical speculations concerning QSS, as well as discoveries and photometric observations of supernovae.

REFERENCES

1. Markarian, B. E., Oganesian, E. J., Arakelian, S. N. 1965, *Astrofizika*, **1**, 38.
2. Markarian, B. E., Oganesian, E. J., Arakelian, S. N. 1965, *Astrofizika*, **2**, 55.
3. Kalloghlyan, A. T. 1963, *Soobšč. Bjurakan. Obs.*, **33**, 19.
4. Kalloghlyan, A. T., Tovmassian, H. M. 1964, *Soobšč. Bjurakan. Obs.*, **36**, 31.
5. Tovmassian H. M. 1965, *Astrofizika*, **1**, 197.
6. Kalloghlyan, A. T. 1963, *Soobšč. Bjurakan. Obs.*, **34**, 31.
7. Artamonov, V. P., Vorontsov-Velyaminov, B. A., Dibai, E. A. 1966, *Astr. Zu.*, **43**, 557.
8. Vorontsov-Velyaminov, B. A. 1965, *Astr. Zu.*, **42**, 1168.
9. Gorbatschev, B. I. 1966, *Soobšč. gos. astr. Inst. Sternberga*, no. 148.
10. Gorbatschev, B. I. 1966, *Astr. Zu.*, **42**, 46.
11. Gagen-Torn, V. A. 1966, *Trudy astr. Obs. Leningr. Univ.*, **24**.
12. Dombrovsky, V. A. *Vestnik Leningr. Univ.* (in press).
13. Dibai, E. A., Schahovskoy, N. M. *Astr. Circ. Akad. N. SSSR* (in press).
14. Vorontsov-Velyaminov, B. A. O. Struve's memorial volume. Italy (in press).
15. Vorontsov-Velyaminov, B. A., Davydov, V. D. 1965, *Astr. Circ. Akad. N. SSSR*, no. 316.

16. Gurzadian, G. A. 1963, *Soobšč. Bjurakan. Obs.*, **34**, 36.
17. Markarian, B. E. 1963, *Soobšč. Bjurakan. Obs.*, **34**, 18.
18. Ambartsumian, V. A. 1964, 13ème Conseil de Physique. Solvay, Bruxelles.
19. Markarian, B. E. 1963, *Soobšč. Bjurakan. Obs.*, **34**, 3.
20. Vorontsov-Velyaminov, B. A. IAU Symposium no. 29 (in press).
21. Vorontsov-Velyaminov, B. A. 1964, *Astr. Zu.*, **49**, 814.
22. Zasov, A. V. 1965, *Astr. Circ. Akad. N. SSSR*, no. 334.
23. Ambartsumian, V. A., Iskudarian, S. G., Shachbazian, R. K., Sahakian, K. A. 1963, *Soobšč. Bjurakan. Obs.*, **33**, 3.
24. Markarian, B. E. 1963, *Soobšč. Bjurakan. Obs.*, **33**, 29.
25. Karachenzev, I. D. 1964, *Izv. Akad. Nauk Arm. SSR, ser. fiz.-matem.*, **17**, 109.
26. Karachenzev, I. D. 1965, *Astrofizika*, **1**, 203.
27. Richter, R., Sahakjan, K. 1965, *Mitt. Sternw. Tautenberg.* no. 24.
27a. Vorontsov-Velyaminov, B. A. 1966, *Astr. Zu.*, **43**, 231.
28. Vorontsov-Velyaminov, B. A., Arhipova, V. P. 1964, *Morphological Catalogue of Galaxies.* Vol. II, Moscow.
29. Vorontsov-Velyaminov, B. A., Arhipova, V. P. *Morphological Catalogue of Galaxies.* Vol. IV, Moscow (in press).
30. Zvyagina, E. V. 1966, *Astr. Zu.*, **43**, 34.
31. Denisjuk, E. K. 1966, *Trudy astrofiz. Inst. Akad. N. KazSSR*, **7**, 45.
32. Gainullina, R. H., Roslyakova, T. V. 1965, *Trudy astrofiz. Inst. Akad. N. KazSSR*, **5**, 237.
33. Agekjan, T. A., Sumsina, N. K. (in press).
34. Veltman, U. I. 1964, *Publ. Obs. Tartu*, **34**, 81, 101, 114; 1965, *Soobšč. Obs. Tartu*, no. 18.
35. Karachenzev, I. D. 1965, *Astrofizika*, **1**, 303.
36. Karachenzev, I. D. 1964, *Izv. Akad. Nauk Arm. SSR, Ser. fiz.-matem*, **17**, 104.
37. Karachenzev, I. D. 1964, *Dokl. Akad. N. SSSR*, **158**, 815.
38. Pskovsky, Y. P. 1965, *Astr. Zu.*, **42**, 323.
39. Karachenzev, I. D. 1966, *Astrofizika*, **2**, 1.
40. Zasov, A. V. 1963, *Astr. Zu.*, **40**, 868.
41. Pskovsky, Y. P. 1965, *Vestnik Mosk. gos. Univ.*, no. 1, 3.
42. Pskovsky, Y. P. 1964, *Astr. Zu.*, **41**, 619.
43. Zasov, A. V., Ozernoi, L. M. 1965, *Astr. Circ. Akad. N. SSSR*, no. 334.
44. Dibai, E. A., Pronik, V. I. 1965, *Astrofizika*, **1**, no. 1.
45. Dibai, E. A., Pronik, V. I. *Izv. Krym. astr. Obs.*, **35**, (in press).
46. Dibai, E. A., Pronik, V. I. *Astr. Zu.* (in press).
47. Dibai, E. A., Vorontsov-Velyaminov, B. A. IAU Symposium no. 29 (in press).
48. Dibai, E. A., Esipov, V. F., Pronik, V. I. *Astr. Zu.* (in press).
49. Dibai, E. A., Pronik, V. I. 1964, *Astr. Circ. Akad. N. SSSR*, no. 282, 286.
50. Dibai, E. A., Esipov, V. F. 1965, *Astr. Circ Akad. N. SSSR*, no. 349.
51. Dibai, E. A., Esipov, V. F. *Astr. Zu.* (in press).
52. Dibai, E. A., Esipov, V. F. *Astr. Zu.* (in press).
53. Pronik, V. I. *Astr. Zu.* (in press).
54. Dibai, E. A., Esipov, V. F., Pronik, V. I. *Astr. Zu.* (in press).
55. Kukarkin, B. V. 1965, *Astrofizika*, **1**, 465.
56. Ambartsumian, V. A. 1965, *Astrofizika*, **1**, 473.
57. Pskovsky, Y. P. 1965, *Astr. Zu.*, **42**, 683.
58. Pickelner, S. B. 1965, *Astr. Zu.*, **42**, 515.
59. Gerschberg, R. E. 1965, *Astr. Zu.*, **42**, 330.
60. Pickelner, S. B. 1965, *Astr. Zu.*, **42**, 3.

29. COMMISSION DES SPECTRES STELLAIRES

PRÉSIDENT: Dr J. Sahade, Observatorio Astronómico, La Plata, Argentina.

VICE-PRÉSIDENT: Dr M. W. Feast, Radcliffe Observatory, P.O. Box 373, Pretoria, South Africa.

COMITÉ D'ORGANISATION: Y. Andrillat, A. Boyarchuk, Y. Fujita, J. B. Oke, K. O. Wright.

MEMBRES: Abt, Aller (L. H.), Andrillat (H.), Asaad, Babcock (H. W.), Bappu, Bartaya, Beals, Berger, Bertaud, Bidelman, Bloch, Boggess, Bouigue, Bretz, Brück (H. A.), Burbidge (E. M.), Buscombe, Cayrel (R.), Chalonge, Code, Deutsch (A. J.), Divan, Dobronravin, Dolidze, Dufay (J.), Dunham, Elvius (T.), Evans (D. S.), Fringant, Gollnow, Gorbatsky, Gratton, Greenstein, Griffin (R. F.), Groth, Guthrie, Gyldenkerne, Hack, Haro, Heard, Herbig, Herman, Huang, Hunaerts, Iwanowska, Jaschek (C. O. R.), Jugaku, Keenan, Kienle, King (R. B.), Kopylov, Kraft, Kuprevich, Lamla, Larsson-Leander, McCuskey, McLaughlin†, McNamara, Mannino, Mayall (M. W.), Melnikov, Miczaika, Milligan, Morgan, Münch (L.), Mustel, Nassau†, Nicholls, Nikitin, Oetken, Osawa, Payne-Gaposchkin, Pedoussaut, Platzeck, Przybylski, Richter, Roman, Rosen, Rozis-Saulgeot†, Schalén, Searle, Slettebak, Smak, Stawikowski, Stecher, Stephenson, Strömgren, Swensson, Swings, Taffara, Thackeray, Thompson (G. I.), Underhill, Vandekerkhove, Van t'Veer-Menneret, Voigt, Wehlau, Wellman, Wempe, Westerlund, Willstrop, Wilson (O. C.), Wilson (R.), Wyller.

The writing of the Draft Report of Commission 29 brings to light once again the need of better defining the province of its domain since there is an overlap with areas of other commissions such as 27, 30 and 42, which use stellar spectra for the investigations they are concerned with. In what follows I have tried as far as possible not to include whatever would naturally go into the Draft Reports of other commissions and whatever Dr Feast has already included in his report on the spectra of variable stars for Commission 27. As a result, the report on some items is only fragmentary.

In my task, I have had the help of the Vice-President and of the members of the Organizing Committee who have prepared reports of the work done in the southern hemisphere, Europe (except U.S.S.R.), U.S.S.R., Japan and Canada, respectively. Dr Oke contributed a report on abundances which I have incorporated within the information under this heading. Mme Herman prepared a report of her working group on Be stars from which I have extracted whatever was relevant to our Commission. I am very grateful to all of them for their help. I am also grateful to Drs Oke and Cayrel de Strobel for the reports concerning the Committees they head. They are appended at the end of this report.

The report and the bibliographical references reflect the enormous amount of work in the field of stellar spectra done in the last three years. Commission 29 is very large indeed but it is also most active.

Although there must certainly be better ways of doing it, the report has been divided in two main sections, namely, general spectroscopic studies and abundance determinations. Of course, in trying to fit the report information into the two sections some material that should have perhaps been included in the second one was considered better fitted for the first section.

Last but not least, I would like to express my gratitude to the Vice-President and to the members of the Organizing Committee for their contribution to the bibliography.

I. GENERAL SPECTROSCOPIC STUDIES

(a) *Early-Type Stars*

Miss Underhill (1966*a*) has taken the task of putting together and discussing the observational results of early-type stars and has suggested (Underhill 1965) that the persistence of the Mg II 4481 doublet in O-type stars is due to the fact that the very high ionization potential for Mg III forces most Mg atoms to stay in the first stage of ionization, even at the high temperatures of the O stars.

Underhill and de Groot (1964) have published identification and equivalent widths of the spectral lines of 10 Lacertae in the range λλ 3074–5160 Å; it was possible to identify only about 80% of a total of 2000 lines, and Ne and O were found. An interesting fact found by van Helden (1966) is that in 10 Lacertae the profiles and intensities of O II and Fe III vary in time intervals smaller than ten minutes, while the radial velocities display irregular short period variations, according to Grygar's (1964) measures.

Y. Andrillat and L. Houziaux (1966*a*) have studied 80 B and Be stars in the near infrared, compare, in 20 of them, the homologous lines of the Balmer and Paschen series and determine the intensity variation of the O I triplet at 7772 Å in relation with spectral type and luminosity class (1966*b*).

No differences in the atmospheric parameters for the high latitude O-type stars HD 93 521 and HD 60 848 and for O stars in the galactic plane has been found by Kumaigorodskaya (1967).

P. Smeyers (1966) is interested in determining the profiles of the Mg II λλ 7877 and 7896 Å in a B9 V star in order to compare them with the profiles computed from a model atmosphere.

In regard to the B stars, the late R. M. Petrie (1965*a*) had almost completed his studies of B stars of magnitudes 7 to 9 visible at Victoria and a paper by Petrie and Lee (1966) gives Hγ measures and luminosities for 571 B stars. A revised luminosity-Hγ calibration that agrees well with other luminosity studies was also published by Petrie (1965*b*).

From material on OB stars secured at Pretoria and in Cambridge, Andrews (1965) has found that for class V stars, Hα is about as good an absolute magnitude index as Hβ or Hγ while for supergiants it is more accurate.

Aller and Jugaku (1967) have completed a description of their work on the B stars which has resulted in line intensities and line profiles for λ Orionis, 10 Lacertae, φ Orionis, ξ_1 Canis Majoris, 15 Canis Majoris, HD 36 959, HD 36 960, 22 Orionis and 114 Tauri.

Inspecting the early-type spectra at Victoria, Bidelman has discovered several B-type stars showing variable helium lines. This interesting phenomenon had previously been reported for HR 5378 (Bidelman, 1965) and for the first time in 1957 by Struve, Sahade, Lynds and Huang in Maia (20c Tauri).

At Mt Wilson, Heintze is studying H-line profiles in the sharp line B stars, τ Herculis, ι Herculis, ι Arietis and ζ Draconis to compare them with theoretical line profiles.

(b) *'Blue Stragglers'*

Deutsch is studying the so-called 'blue stragglers' in M 67 and NGC 752; the spectra obtained confirm and extend Wallerstein and Herbig's results that in most of these stars the spectrum lines are appreciable broadened by stellar rotation.

Since the 'blue stragglers' appear to be metamorphs of the late F dwarfs, Deutsch proposes that on the main sequence the solar-type stars suffer a rotational deceleration largely confined to the hydrogen convective zone leaving the radiative core with its angular momentum only slightly diminished even after times of the order of 10^{10} years.

(*c*) *Of and Wolf-Rayet Stars*

Kumaigorodskaya (1964) has investigated the dependences of emission-line intensities with temperature and luminosity for 13 Of stars and has found that these stars can be grouped in two sequences similar as those of the Wolf-Rayet stars. Nitrogen Of stars appear to be hotter than the carbon Of stars.

G. A. Gurzadian and H. S. Tchaurushian (1964) have determined emission line intensities, velocities of expansion of the envelopes and spectrophotometric gradients of the Wolf-Rayet stars HD 165 688, 211 564, 213 049 and R 114, while the energy distribution of the continuous spectrum of V 444 Cygni has been studied by Ivanova *et al.* (1965) at different phases; a decrease of the spectrophotometric temperature at the two minima was found.

Vainu Bappu and Ganesh are concentrating their efforts in the investigation of the Wolf-Rayet stars and are preparing an atlas of tracings of Wolf-Rayet spectra, which will include a wide range of dispersions, from 10 Å/mm to 250 Å/mm. Infrared spectra of HD 192 103 (WC7), HD 162 763 (WC6), HD 68 273 (WC7), HD 192 163 (WN6), HD 50 896 (WN5), HD 93 131 (WN7), HD 92 740 (WN7) and HD 151 932 (WN 7·5) have been taken and line identifications have been completed; this work will be published in the *Kodaikanal Obs. Bull.*

The two Indian astronomers have also determined excitation temperatures of four WC stars and found values which range from 24 000° to 57 000°. On the other hand D. M. Pyper (1966) from their *UBV* colour indices corrected for emission, finds a difference of about 20 000° between the WN and the WC stars, the WN stars being hotter.

At Mt Wilson, Kuhi has studied energy distribution of Wolf-Rayet stars in the range λ 3200 to λ 11 000 with the Cassegrain photo-electric scanner used as a narrow-band photometer and concluded that they are definitely peculiar and different from those of blackbodies; therefore, no unique temperature can be defined. The apparent different sequences upon which the WC and WNs appear to fall can be explained by emission-line contamination. Kuhi's observations confirmed the behaviour in Wolf-Rayet binaries of He II emission phase-wise found previously by Hiltner and also confirmed Sahade's earlier conclusion that ionization decreases outwards in the WR atmospheres; the most highly ionized atoms appear to occur in a geometrically small region and the emitting volume for the lines of lower ionization appear to be very large. Also there seems to be some kind of a reflection effect due to the O star. Moreover, Kuhi (1966) has studied the infrared spectrum, from 8000 to 10 900 Å, of some bright Wolf-Rayet stars.

R. Barbon *et al.* (1965) suggest a strong activity in the envelope of the WR stars while Bertola finds that in HD 50 896 the velocities and the profiles of He I 4100 and N IV 4058 are variables although the radial velocity variations do not suggest binary motion.

Code and Bless (1964) described the infrared spectra, (from 8600 to 10 830 Å) of γ_2 Velorum and developed a model for the WR envelope in which collisional processes by particles ejected into a thermalized envelope play an important role in the ionization equilibrium of the envelope.

A. Underhill (1966*b*) has examined the position of the WR stars in the HR diagram and concluded that it is reasonable to consider them as contracting objects that are still approaching the main sequence, in agreement with a suggestion made for the first time in 1958 by Sahade. In this connection it is interesting to note a discussion by Sahade (1965) who pointed out that the presently accepted value for the density of the WR component of V444 Cygni is not reliable.

Stepien (1964) has discussed the spectroscopic and photometric data in regard to the WR component of V444 Cygni. New WR stars have been reported by The (1964*a*, *b*).

(*d*) *Intermediate and Late-Type Stars*

It is interesting to note that Peat (1964) has been able to establish a way of distinguishing giants from supergiants among the F, G and K stars by plotting the intensity of the Ca II triplet in the red-orange against the intensity of the b-line of Mg II. The same procedure permits the detection of subdwarfs in the case of high-velocity G and K stars.

At the 1965 NASA Symposium, Wright (1965) discussed the low excitation temperatures obtained for A- and F-type stars using the Fe I *f*-values, and at the Trieste Colloquium presented a review paper on the spectra of late-type stars (Wright, 1966).

Clarke and Grainger (1965) have detected differential polarization effects on the profiles of the H β line of γ Ursae Majoris, and at Kodaikanal, Bhatnagar has started a programme of observation of the infrared O I lines in A and F stars at 250 Å/mm.

In regard to the late-type stars, concentrated efforts are being made in Tokyo and in Pasadena. Some of the results obtained in Tokyo by Fujita and his collaborators during the past few years have been recently published (Fujita 1966) and special emphasis is being given now to an identification programme that will proceed towards wavelengths shorter than those covered previously in similar work.

Yamashita has studied the late-type spectra available at Victoria and found several new CH stars. In South Africa, Feast is studying coudé spectra of southern S-type stars.

(*e*) *T Tauri Stars*

Herbig has accumulated a considerable amount of information on the infrared Ca II triplet which are characteristically very strong in T Tauri stars, even though they are very weak in dMe stars that have comparable H and K emission intensities; the effect is probably due to heavy self-absorption in T Tauri stars.

Herbig (1966) has interpreted FU Orionis as a star which has very recently appeared at the top of the vertical branch of its evolutionary track, an interpretation which implies that the high Li abundances observed in T Tauri stars must have its origin either in the original interstellar material or been produced before the star became luminous in the usual sense.

Kuhi has obtained photoelectric spectrum scans of several T Tauri stars from λ 3200 to λ 11 000 and found that many of these stars are excessively red suggesting that the reddening may be of circumstellar origin. The origin of the peculiar excess ultraviolet emission in T Tauri stars is being investigated.

(*f*) *Giant and Supergiant Stars*

Zeinalov and Kopylov (1966) have identified about 600 spectral lines in the spectrum of the A0 Ib star η Leonis and carried out a detailed study of the velocity field in its atmosphere. The microturbulent velocity decreases with optical depth while the macroturbulent velocity goes in the opposite sense.

Variations of the equivalent widths and profiles of Hγ and Hδ and of the equivalent widths of metallic lines in η Leonis were found by Malov and Vitrichenko (1964), who were led to temperature variations of 1000°K, electron pressure variations up to a factor of 2 and radius variations up to about 30%.

Kupo and Mamatkazina (1966) have studied the continuum of HD 21 389, an A0 Ia star.

In a number of K and early M giants of luminosity class III Deutsch has found time variations in the profiles of the central components of Ca II 3933 similar to the ones reported previously by him in many K-type supergiants; the doubly-reversed features, at high spectral resolution, exhibit a surprising degree of fine structure suggesting variable chromospheric activity and sometimes the motion of a plage across the visible hemisphere, as the result of stellar rotation. The observations will be continued in a representative group of giants near K5 III.

Deutsch and Keenan are continuing obtaining spectrograms in the blue and violet (20 Å/mm) that show systematic intensity anomalies in various atomic absorption lines and also in the bands of AlO.

Griffin has prepared – publication is pending – a spectrophotometric atlas of Arcturus covering the wavelength region between 3600 and 8800 Å, and at least comparable in all respects with the Utrecht solar atlas. A catalogue of lines (about 7000) with wavelengths, equivalent widths, central intensities and identifications for $\lambda > 5000$ Å is well advanced. The work of cataloguing and indexing features was undertaken in collaboration with the present Mrs R. E. M. Griffin-Gasson, who has also investigated the atmospheric characteristics of Arcturus. The discovery by Griffin of the H and K emission reversals being variable in time, has been followed up by W. Liller observing H and K emissions in a number of stars with Griffin's photoelectric spectrophotometer. The Atlas of Arcturus has led to the discovery of [O I] and [Fe II] lines in Arcturus (Gasson and Pagel, 1966).

Griffin (1964*a*) found broad features at $\lambda\lambda$ 6318, 6344 and 6362 Å in Arcturus and in eleven other F5 to M0 stars which have been identified as calcium lines broadened by autoionization by Mitchell and Mohler who discovered them independently in the solar spectrum. The line at λ 6362 has been studied by Griffin and Liller with the photoelectric spectrometer just mentioned. Griffin (1964*b*) has also given line identifications of Arcturus in the 8875 to 11 200 Å region.

F. K. Edmonds, Jr. will measure line blanketing in the 1–2·5 μ region, using high-resolution Fourier transform spectroscopic observations obtained by P. Connes of the C.N.R.S. Bellevue Laboratories, in France, and in the $\lambda\lambda$ 3500–8800 Å region, using high-dispersion observations and equivalent width measurements of Griffin and Mrs Griffin-Gasson.

Kubiak (1966) has discussed the influence of TiO bands on magnitudes and colours of M-type giants, by using high-dispersion spectrograms obtained by Preston.

A detailed study of the spectrum of Procyon that comprises equivalent widths of 152 lines in the wavelength range $\lambda\lambda$ 3260–6610 Å and line profiles of Hα, Hβ and Ca II-K, and the measurement of the continuous spectrum and of line blanketing has been published by F. N. Edmonds, Jr. in two papers, the second one in collaboration with F. D. Talbert (Edmonds, 1965; Talbert and Edmonds, 1966).

Edmonds is also working on the photoelectric calibration of photographic equivalent widths of Procyon and Arcturus using the high-dispersion photoelectric scanner attached to the coudé spectrograph at the MacDonald Observatory.

F. Praderie is computing H lines in the atmospheres of AF-type stars.

Sargent (1965*a*) suspects the presence of [N II] at 6583–6548 Å in the spectrum of the G0 Ia star, HR 8752, which would imply the existence of a highly ionized region around the star; the star displays Hα and Hβ emission suggesting superficial mass loss.

Bonsack and Culver have continued their programme to determine, on high dispersion Mt Wilson spectra, wavelengths, equivalent widths and identification of the lines in the spectrum of the normal late G-type giant ϵ Virginis; the measurements and reductions are already complete for 6500 lines in the wavelength range 4000 to 9000 Å.

Smak (1966*a*) has determined a new bolometric corrections scale for M-type giants by using the spectrophotometric data of Stratoscope II and the available six-colour data; and combined his results with recent determinations by Mendoza and Johnson.

(g) *Be and Shell Stars*

A desire to co-ordinate and stimulate work on the Be stars led Mrs Herman to propose in 1961 at Berkeley the formation of a working group to which she is serving as chairman.

A somewhat condensed and free English version of her report follows:

'In spectroscopy or spectrophotometry certain groups are particularly active.

'At La Plata, C. Jaschek, with the collaboration of B. Kucewicz, has continued his search for H α emission in southern B0–A0 stars brighter than 6·5 magnitudes, and a list for the range B0–B5 has been published (Jaschek *et al.*, 1964), while Kucewicz (1965) has summarized the state of the search. C. and M. Jaschek have described several southern Be stars and, in particular, have reported that in 1965 HD 68 980 has displayed a sort of helium envelope with peculiar line profiles (C. and M. Jaschek, 1965; see also M. and C. Jaschek, 1964).

'Miss Underhill has secured a number of spectra of ζ Tauri at Victoria (October and December 1964) and at Kitt Peak (January 1966) and van der Wel, at Utrecht, has made a spectrophotometric study. In 1964 the shell absorption lines are shaded towards the violet while in 1966 the shading was towards the red.

'At David Dunlap, Heard has taken a series of spectrograms of φ Persei at 10 Å/mm.

'L. Houziaux and Mme Y. Andrillat have undertaken the study of the infrared region, specially at Haute-Provence. A report on the measurement of the Paschen lines and those of O_2 at 7771–5 and at 8446, with the aim of determining the envelope electronic temperature and the oxygen abundance, has been published (Andrillat and Houziaux, 1966*b*).

'At present, Houziaux is analysing the spectral variations of some 20 Be stars, between 7 and 8 magnitudes, observed in the last ten years.

'At Crimea, Boyarchuk and Pronik have studied the Hα profiles in HD 217 050, ζ Tauri, φ Persei, ψ Persei, 11 Camelopardalis and c Persei, finding that the wings are determined by radiation damping; they also reached the conclusion that the envelopes of the Be stars are oblated (Boyarchuk and Pronik, 1964). They find that the Be stars are somewhat reddened relative to the normal B stars, a phenomenon probably caused by the envelope; in the case of X Persei and χ Ophiuchi the Balmer decrement for the stars was found steeper than the Balmer decrement for the nebulae, apparently due to the large optical thickness of the Be envelopes for the Balmer lines radiation (Boyarchuk and Pronik, 1965*a*).

'In Italy, M. Hack and her collaborators (Hack *et al.*, 1964; Aydin *et al.*, 1965) have observed ζ Tauri from 1961 through 1964 and found that the envelope contracts and that the velocity of contraction has been decreasing.

'HD 50 138 has been observed at Haute-Provence by Mme Doazan, who found pulsations of the envelope; she has applied Sobolev's method to such an envelope (Doazan, 1965) and has also measured spectra of HD 218 393 taken between 1960 and 1964.

'Miss Lacoarret (1965) has studied the variations of the emission relative to the continuum in HD 168 957, HD 171 406, HD 171 780, HD 175 869, HD 177 648, HD 187 811, HD 191 610 and HD 174 237 and found periods of several years, while the V/R variations in HD 174 237 have a quasi periodicity of several days. She also discusses the characteristics of the envelopes and compares the results obtained by applying Miyamoto-Kogure, Pottasch and Sobolev's methods.

'Miss Delplace (1966) has been concerned also with ζ Tauri and discussed the variation of the Balmer decrement in relation with the variation in radial velocities of the Balmer series. The problem of the periods of the spectral variations is also concerning Peton, who is studying HD 45 910 and HD 4180, and Mme Herman who has observed emission, present again in 1964–65, in HD 142 926; this star appears to have also a small metallic envelope. Lecontel is investigating variations in β Canis Majoris stars in the hope of finding some relationship with the Be stars.

At La Plata, C. Jaschek is undertaking the continuation of the Merrill and Burwell's catalogue and bibliography of Be stars.'

Pik-Sin The (1964*c*, *d*; 1966*a*, *b*; The and Lim, 1964) has continued his search for Hα emission objects and it is rewarding to mention that Bidelman reports that the surveillance of the brighter Be stars carried on at Michigan for so many years by the late D. B. McLaughlin will be continued to some extent. Andrews and Breger (1966) have found that Achernar is a Be star, the brightest known.

At Helwan, R. Goubrous is studying the He I lines in some shell stars.

Mme Herman (1964) has confirmed the existence of double H absorptions in P Cygni, during certain periods and given a qualitative interpretation of the observed profiles in the framework of Sobolev's theory.

Y. Andrillat (1965*b*) has spectrographically observed a potassium eruption in the B9e star 4 Herculis, a phenomenon which is analogous to that observed by Barbier and Morguleff in two dwarfs stars HD 117 043 (dG6) and HD 88 230 (K7V) (Barbier and Morguleff, 1964). Y. Andrillat and N. Morguleff (1966*a*, *b*) have made a comparative study of the three stars known to display the phenomenon; they think that the phenomenon is not of atmospheric origin and plan to study its frequency by means of an objective grating and to continue investigating it in the case of the Sun.

At Abastumani, Alania (1964) has found in the spectrum of MWC 84 that since 1932 there has been an increase of the order of seven times in the ratio of the emission lines at λ 3889 and H β.

Dolidze *et al.* (1965) reports a decrease in the Balmer jump of HD 190073 in 1964–65 and Dolidze, Apriamashvili and Jimsheleishvili have measured the Balmer jump in V425 Cygni.

(*h*) *Ca II Emission*

In Pretoria, Warner has observed over 200 G, K and M stars south of declination $-20°$ for H and K emission; preliminary results show a good correlation between emission intensity and space velocity but there are indications that different stars in a given moving group do not necessarily have the same emission intensity even when they have identical spectral types.

Warner (1966*a*) has detected Ca II emission in Canopus, but this star does not obey the Wilson-Bappu relationship.

Wilson finds that as one proceeds down the main sequence slow rotations set in very abruptly at $B - V$ indices only 0·01 magnitude from the onset of Ca II emission (near F5 V) and that Ca II emission is characteristic of stars that develop chromospheres, chromospheric activity, and 'winds' and that such activity causes a loss of angular momentum with concomitant rotational braking. He is making now a strong effort to study stellar chromospheres by rapid and accurate measurements of the central emission components of H and K.

Wilson is also investigating the frequency distribution of the intensities of the chromospheric H and K emissions in main sequence K- and M-type stars. Since all the available evidence points to a close correlation between chromospheric activity and age for main sequence stars the frequency distribution should approximate the frequency distribution of stellar ages.

Thackeray (1964) has reported the presence of abnormally strong Ca II emission in a young K star that belongs to the Sco-Cen association, and Metrevely (1966) has announced that the K5 star, BD + 28° 637, displayed strong H and Ca II emissions on December 24 1960 while they appear to be absent at other nights.

(*i*) *Chromospheric He* I

Vaugham and Zirin discovered He I λ 10 830 in absorption in λ Andromedae and have looked for the same feature in late-type stars. About 86 stars (G0 to M6) selected from Wilson-Bappu list have been observed and about half of them show He I λ 10 830 implying a hot chromosphere with temperatures exceeding 20 000°K. The strength of the line tends to increase with increasing Ca II H and K emission intensity, and it is violet displaced in virtually all supergiants as well as in some giants of luminosity class III. The He I λ 10 830 feature has been found in at least one F star, namely, α Canis Majoris.

(*j*) *Metallic-Line and Peculiar A Stars*

The field of metallic-line and peculiar A stars has been a most active and a considerable amount of information have been obtained. Review papers have been prepared by Sargent (1964), M. Hack (1965, 1966*a*) and a catalogue of peculiar A stars has been published by Bertaud (1965).

At La Plata, the Jascheks (M. and C. Jaschek, 1965) have analysed the information on the metallic-line and peculiar A stars and suggested that they represent a transient stage in a star's life; furthermore, on kinematical grounds they concluded that the metallic-line stars can not have gone through the giant stage. This conclusion finds support in the finding of an Ap star in I Persei by Kraft and in the presence of an Ap object in the *o* Velorum cluster.

A review paper on the magnetic stars has been produced by Ledoux and Renson (1966).

A preliminary determination of the magnetic field of α^2 Canis Venaticorum has been tried at the Okayama Astrophysical Observatory (Unno and Kato, 1965).

An important result is the discovery in β Coronae Borealis of the presence of a line due to lithium (Garstang, 1964) and of the possibility of variations in its intensity (Wallerstein and Hack, 1964). Hack and Faraggiana (1964) have discussed that discovery, and the results of the investigations of β Coronae Borealis and γ Equulei at Merate.

Adam (1965) has observed the magnetic variable HD 125 248 and found that the reduction in the amplitude of the magnetic field that was first noted by Babcock in 1953 is still continuing; the period, however, has remained remarkably constant over 17 years.

P. Renson (1965) has determined the variations in radial velocity, luminosity and magnetic field and found the variation in magnetic field in function of phase presents an assymetry and resembles the luminosity curve as is true for other magnetic variables.

Bruce Peterson (1966) has investigated the spectrum variables HD 124 224 and 56 Arietis and, from the equivalent widths of the variable lines of Si II and He I, has determined new epochs of maximum strength of the He I lines thus finding that the times of He I maxima coincide with those of minimum temperature determined from $(B - V)$ and from $H\gamma$ profiles as well as with the times of minimum light in V and minimum Si II. The intensity variations in Si II are very nearly in the same phase for lines that differ in excitation potential by 2·7 electron volts. These results conform better to a rigid rotator model than to a magnetic pulsation.

The Ap star HD 221 568 has been studied by Kodaira in the spectral range λλ3800–7000 Å and Kondo (1965) has analysed the continuous spectrum of 63 Tauri finding that $T_{\rm eff} \simeq 7600°$ = 7900° and log $g \simeq 4·0$. Also in Japan, Jugaku and Kikuchi are making a detailed spectrophotometric study of β Orionis on high dispersion Mt Wilson spectrograms that cover the range λλ3100–8800 Å; the spectrum of this star has been investigated by Svolopoulos (1966*b*) on McDonald spectra from 3660 Å to $H\gamma$ line identifications and intensities are given and electron pressure and the amount of hydrogen above the photosphere have been determined.

Osawa (1965) finds that the Ap stars may have ultraviolet excesses that become more pronounced if the colour indices are corrected for line blanketing.

A list of 3000 lines and their equivalent widths for γ Equulei in the range $\lambda\lambda$ 3730–6593 Å has been published by M. Hack and M. S. Matthews (1965) by using material with dispersions of 2·8 to 4·5 Å/mm; only 600 of these lines are unblended.

Bidelman is planning to publish soon line identifications of a number of Ap stars studied on 2 Å/mm coudé Lick plates, beginning with the manganese stars; the analysis of line intensities will be done as a second step.

Bidelman is also studying a number of interesting stars two of which deserve mention at this time, namely, HR4487 (worked out in collaboration with J. Baumert) which appears to have very high abundance of yttrium and zirconium, and HD3473, which appears to be the only example of a Si-Mg star.

Inspecting the spectra of most of the early-type stars investigated at the Dominion Astrophysical Observatory, Victoria, Bidelman has discovered a considerable number of new Ap and Am stars, and E. N. Walker (1966*b*) has recently published a list of 25 new Am stars.

Other interesting results were reported by Bidelman at the 1964 Utrecht Symposium on Abundances.

Svolopoulos (1966*a*) have analysed the spectrum of the manganese star α Andromedae, and given line identification and their intensities; electron pressure and the amount of hydrogen above the photosphere have been determined.

Wehlau (1965) has studied the spectrum of the Ap star χ Serpentis and also that of 73 Draconis (Wehlau 1966), where he found large variations in the K line; a spectrophotometric analysis of 73 Draconis is underway. John Rice is doing similar work on HD 174 650, where large variations in line intensities have been noted. 73 Draconis has been studied also by Galkin (1964) who has determined the variations with phase of the residual intensities of 19 lines, has constructed curves of growth and determined the main atmospheric parameters.

Renson (1966) finds that the magnetic variable HD8441 is a spectroscopic binary with a period of 106·27 days and that the published period of $2^{d}_{\cdot}9632$ for the magnetic variation is due to chance.

Abt (1965) has measured 500 coudé spectra of 55 normal A4-F2 IV, V stars and found no binaries with periods less than 100 days, and 17 objects with periods larger than 100 days; he suggests that all members of close binaries in that temperature range have metallic-line spectra, perhaps because the slow rotation found in close binaries is necessary to produce the unusual metallic-line spectra.

In connection to the question of the rotational velocities of the Ap stars it is interesting to mention that E. N. Walker (1966*a*) in a recent discussion concludes that although some Ap stars may be seen pole on, Ap stars in general cannot be rapidly rotating normal A stars seen pole on.

Baschek and Oke (1965) have shown that after deblanketing corrections are introduced, the metallic-line A stars in the Hyades cluster lie to the left of the main sequence in the colour-magnitude diagram, and Strittmatter and Sargent (1966) find that the same is true for the metallic-line stars in the Coma and Praesepe clusters, which these authors interpret as being due to the slow intrinsic rotation of the metallic-line stars and in agreement with the theoretical result of Roxburgh and Strittmatter that suggest that, at a given colour, more rapidly rotating stars will appear more luminous.

Guthrie (1965*b*) has noted that there seems to be a preference for the manganese stars (it is not so among the silicon stars) to exhibit sharp lines; the difference between the silicon and the manganese stars may be related to the direction of the magnetic field relative to the axis of rotation.

Buscombe (1965*b*) has measured equivalent widths of Hγ and Hδ for a large number of stars in IC2391 including several peculiar A stars.

At Odessa, Komarov (1965) finds that, on the average, the electron densities in the atmospheres of the metallic-line stars do not differ from those of normal main sequence stars of the same hydrogen spectral class.

(*k*) *Subdwarfs*

Warner reports work in progress on several southern F-G subdwarfs. Two galactic subdwarfs in the field of the Large Magellanic Cloud have been discussed by Thackeray (1966*a*).

D. S. Evans is working on a number of problems related to subdwarfs.

(*l*) *Faint Blue Stars*

Greenstein is continuing the low dispersion studies of faint blue stars, in an effort to obtain a more accurate estimate of the percentage of white dwarfs at approximately the 17th magnitude. The mean Einstein gravitational red shift of 53 white dwarfs have been measured by Greenstein and Trimble as $+51$ km s^{-1}, the mean for the DA stars being from $+56$ to $+62$ km s^{-1}. The derived mass of a helium core, about $0{\cdot}87\mathfrak{M}_{\odot}$, is roughly consistent with the photometric radii deduced from luminosities and colours. A further extension of the spectroscopic and colourimetric work by Greenstein and Eggen is resulting in a third list of white dwarfs which will bring the total of completely observed objects to 202.

Greenstein points out that the most interesting future problems will concern with hot white dwarfs with helium and carbon features and the possible effect of line blanketing on the colours of the red white dwarfs; for many of these no lines are visible but the radii deduced from the luminosity and the photoelectric colours are perhaps three times too large, it may be that they are hotter than they seem to be but their colours are perturbed by very greatly broadened lines, as Weidemann has suggested.

A surprising number of the faint blue stars show composite spectra (hot star + G star) and one faint object Ton 120 had broad emission like a U Geminorium star or an old nova.

The spectroscopic distinction between hot subdwarfs from very hot white dwarfs is difficult.

The H and He line widths among subdwarfs vary by large factors. Subdwarfs occur with He I and He II lines nearly as wide as in white dwarfs of type DB and DO and the Stark broadening suggests a range in luminosity by a factor of 1000. It is likely that the majority of the halo blue stars classified as Bp (presumably horizontal-branch stars) are helium-deficient objects.

Greenstein and Oke are continuing their study of horizontal-branch stars in globular clusters and have scanned 17 stars in M92 and several stars in M15. The gravities obtained suggest that these stars may not all have the same mass.

Bell and Rodgers (1964) have discussed the star L97-12; its spectrum appears continuous and the star may be dynamically related to the group of three λ 4670 objects discussed by Bell in 1962.

Greenstein and Eggen's effort in finding new white dwarfs has resulted in the discovery of white dwarfs with small space motion that is of white dwarfs in the younger population indicating that extensive loss of mass can occur as evidenced by Sirius B.

Greenstein and Eggen have studied correlation between colour and spectrum: the cooler white dwarfs that still show H lines lie near $B - V = +0{\cdot}3$ magnitude and $U - B = 0{\cdot}5$ magnitude; the H lines are then extremely sharp, and deceptively weak, but this is not caused by low surface gravity alone but presumably by the depression of the ionization continuum, caused by high pressure.

(*m*) *Halo Population Stars*

Greenstein finds that the most single interesting star of the halo population yet observed is BD + 39°4926, a very metal-poor object with exceptionally strong neutral carbon and oxygen

Under the current interpretation of nucleosynthesis theory BD + 39°4926 may be an object which has formed early and has synthesized its own carbon and oxygen or it may be one with very abnormal, selective, early composition.

(*n*) *Planetary Nebulae*

L. H. Aller has observed the nuclei of NGC40 and BD + 30°3639 with the Lick coudé and measured them with the spectrum scanner; he has also observed, at 16 Å/mm in the photographic region, the central stars of NGC6543, NGC6826, NGC6891 and NGC2392. Fainter nuclei have been observed with the Lick nebular spectrograph at the 3-m reflector: the central stars of NGC6853, NGC6058, IC3568 are essentially absorption-line objects while the nuclei of NGC6751, NGC6905, NGC7026 and IC1747 display interesting emission-line spectra of the Wolf-Rayet type.

F. Bertola (1964) gives the list of emissions in the Wolf-Rayet type nucleus of BAC209, the first one which does not show lines due to C.

C. R. O'Dell (1966) has studied M1-2 which were discovered and identified by Minkowski as a stellar planetary nebula; its spectrum may be explained as arising from a system of a G2 star and a hot companion but we could be dealing, according to O'Dell with a very peculiar single star.

(*o*) *Binary Stars*

Orbital elements in the following Wolf-Rayet objects, γ_2 Velorum (HD68273), HD193928, HD186943, HD211853, HD193576 have been determined by Vainu Bappu and Ganesh (1967*a*, *b*). The period derived for γ_2 Velorum is 78·5 days, which does not agree with the ones suggested by Sahade and by Gaposchkin (around 16 days).

Azimov (1965) has carried out a study of the subgiants components of RS Arietis, S Cancri, RS Canis Venaticorum, SX Cassiopeiae, GC Cassiopeiae, U Cephei, UX Monocerotis, RY Persei and I Sagittae, finding that the electron densities and the spectrophotometric temperatures are higher than those of single subgiants of same spectral types. Spectrophotometric gradients, temperature and physical parameters of the two components of AR Lacertae were also determined (Azimov, 1966).

Thackeray (1965*a*) has studied the spectra of the eclipsing system S Equulei and observed 'satellite' lines during eclipse probably due to gaseous streams; Thackeray's spectrograms are being investigated by Plavec.

Smak (1965) has been able to measure lines of Na I that belong to the secondary component of XZ Sagittarii and made a direct determination of the masses of the two components.

M. Spite-Lebon (1965) has determined the masses of the two components of χ Draconis and Bartoli *et al.* have studied BV382 Cephei.

Sahade (1966*a*) has extended his study of the spectroscopic binary HD698 on the basis of new Mt Wilson material by Jugaku and Sahade and confirms that the lines of the secondary component are absent from the spectrum, reaching the conclusion that the star is another system in which the mass of the fainter component is the larger; emission at H α appears to arise from an expanding envelope around the system. The star (Sahade, 1966*b*) belongs to a group of systems with underluminous massive components (HD47129, AO Cassiopeiae, V453 Scorpii, V448 Cygni, β Lyrae and possibly ϵ Aurigae and W Crucis). One member of this group, V453 Scorpii, has been discussed in detail by Sahade and Frieboes-Conde (1965*b*).

An extensive study of the physical conditions in the atmospheres of HD47129 and AO Cassiopeiae has been undertaken by Galkina (1965, 1967); special attention has been given to the behaviour of the He II 4686 emission. The wavelength dependence of the depth of the eclipse of ϵ Aurigae during the 1955–57 minimum has been investigated by Ivanova; in long wavelengths there is almost no variation with wavelength.

The spectrum of β Lyrae in the visual region has been studied by Sahade and Hernández (Sahade, 1966c); full details will be published later. A new series of spectra of the same region of β Lyrae are being obtained by Batten at Victoria.

Feast (1967) has made a detailed spectroscopic study of the supergiant eclipsing system BL Telescopii, finding that the F-type primary has normal abundances though the *r*-process element europium may be overabundant; the secondary may be either a cool supergiant or a hot subdwarf ionizing a dense H II region. The spectroscopic changes during eclipse are interpreted as centre to limb variations of the primary. BL Telescopii is a 'run away' system and its present state is satisfied by the hypothesis that the secondary became of type II supernova 10^7 years ago. Further work on this system is in progress.

B. F. Peery, Jr. has published his detailed study of VV Cephei where he has found spectroscopic information that was interpreted as evidence of material falling upon the B component. Data on the 1955–57 eclipse of VV Cephei and later data are being studied by Wright (1966*a*).

Odgers and Wright (1965) studied multiple K-lines near the 1963–64 eclipse of ζ Aurigae and Wright is analysing similar data for the 1965 eclipse of 32 Cygni. During the 1963–64 eclipse ζ Aurigae has been also studied by Faraggiana (1965), and by Bardin and Prévot (1964*a*), while Scholz (1965) has measured the equivalent width of the K line in 32 Cygni, and Faraggiana and Hack (1965) have compared the spectra of the K-type component of ζ Aurigae, 31 Cygni and 32 Cygni.

Sahade in collaboration with Hernández, Faÿ and Cohen are carrying out a detailed investigation of δ Librae based on all available spectrograms.

At the David Dunlap Observatory Abhyankar obtained new spectroscopic orbits for S Coronae Borealis and HD 208392. Fletcher (1966) made a new study of the orbit of 47 Andromedae. Heard determined spectroscopic orbits for the W Ursae Majoris star V566 Ophiuchi (1965), for the double-lined B0 IV system HD 217312 in the III Cephei Association (1966*a*), and for HD 131861 (1966*b*). Miss Northcott obtained spectroscopic orbital elements for the double-line binaries HD 15138 and 82780 and also for the double-line eclipsing system HD 200391, for which Bakos has also obtained the photometric elements.

At the Dominion Astrophysical Observatory orbital elements have been determined by Gutmann for ζ^1 Ursae Majoris (1966*a*) and υ Ophiuchi (1966*b*), by Thackeray and Tatum for HD 175544 (1966) and for 31 Cygni by Wright and Huffman. The study of the spectroscopic binaries HD 208947 and HR 8800 by the late R. M. Petrie, the latter with emphasis on the question of the apsidal motion, that were in a well-advanced stage at the time of his passing away, is being completed by Mrs Petrie.

Bidelman *et al.* (1963) have called attention to the fact that HR 4511 is probably a G-type supergiant spectroscopic binary.

A. P. and C. R. Cowley (1966) have reported changes in the spectrum of HR 8164 = Boss 5481 Å which combines features of a sgM and of a B2 V objects.

D. S. Evans has continued his spectroscopic study of the interesting triple system p Velorum.

(*p*) *Miscellaneous Stars*

Webster (1966) has discussed some interesting emission line stars formerly classified as planetary nebulae; at least some of these stars seem likely to be symbiotic objects.

Feast is studying coudé of UY Centauri and several other similar stars.

L. H. Aller and Th. Dunham, Jr. (1966) have described the spectrum of η Carinae in 1961 and, except for a few changes, there is an agreement with Thackeray's 1951 observations.

Extensive work on η Carinae by Thackeray is reported under Commission 27. The same is true for the work done on RR Telescopii.

Objective Prism Work

Objective prism surveys of the southern sky are being continued by Pik-Sin The and his collaborators. Among work in progress there are surveys for M stars in the direction of the south galactic pole, in Carina and towards the galactic centre.

Helium-Weak Stars

Greenstein, Sargent, Zirin and Heintze, Jugaku, Tsuji, Schadee and Boesgaard are devoting time to the problem of the apparently helium-weak stars.

Metal-Weak Stars

H. Bonel is attempting to discover at Michigan metal-weak stars on 10° objective prism plates; suspect ones are later observed with slit spectrographs.

High-Velocity Stars

BD + 37°442, discovered by Fehrenbach and his group, was studied by E. Rebeirot (1966); the star does not show H lines but does show strong ionized helium.

Globular Cluster Stars

Spectra of both the horizontal and the giant branches stars in NGC6397 have been obtained in Pretoria.

Novae

R. Goubrous (Helwan) has studied the behaviour of the H lines in Nova Herculis 1963 and Baschek (1964*a*) its near infrared spectrum; this has been followed by Y. Andrillat (1965*a*) from July 1963 to May 1965.

Batten (1964) has published results from the spectra of Nova Herculis 1963 taken between February and August, and Wright and Bochonko are preparing an atlas of the H line intensities from plates of 6 and 10Å/mm.

Nova Herculis 1963, of which Götz (1965), with the Sonnenberg Schmidt, had secured a spectrum on 2 September 1962, before the outburst, has been the subject of several investigations. At Abastumani Alania and Popov (1965) have obtained a large number of objective prism plates between 25 March and 9 September 1963, deriving an expansion velocity of 1190 km s^{-1}.

B. Folkart *et al.* (1964*a*, *b*) have produced two papers where the results of the measurements of the Balmer line profiles and of the narrow emission and absorption lines in RS Ophiuchi 1958 are given.

Cepheids

Warner (1964) has published a list of wavelengths and identifications in the bright cepheid *l* Carinae which should be generally useful for supergiants of about the same spectral type.

In Canada, Crompton has studied the spectrum of RU Camelopardalis which had been found by Demers and Fernie to have ceased its cepheid variation.

Alania (1965) have found that while the variations in spectral type of the short-period cepheid DX Delphini go from A8 to F8, the spectrophotometric gradient changes a little more that what it would correspond to the range in spectral types.

More information on cepheids will be found in Commission 27 report of the Committee on the spectra of variable stars.

Symbiotic Stars

M. P. Fitzgerald, N. Houk, S. W. McCuskey and D. Hoffleit (1966) give a preliminary line identification list of M H α 382-116 which appears to be similar to a symbiotic variable, spectrum wise.

Long-Period Variables

A review article on the long-period variables including the discussion of some spectroscopic problems has been published by Smak (1966*b*).

Keenan, Deutsch and Garrison are undertaking a programme to observe the intensities of selected atomic lines and molecular bands in a number of the brighter Mira variables over several cycles and compare them with simultaneous photometric data obtained by other observers; the molecular bands under particular study include AlO λ4842 and YO λ6132.

Smak (1964) has discussed the effect of the TiO bands on magnitudes and colours of the Mira type stars as well as their position in the $M_{bol} - T_e$ diagram.

Alania (1966) has completed a study of the continuous spectrum of AC Herculis; at λλ4000–6000 Å the colour temperature varies from 4900° to 6800°.

(*q*) *Miscellaneous studies*

Rachkovskaya (1967) at Crimea has found that the spectra of the Ia-type variable stars AB Aurigae, V 568 Cygni and *o* Herculis do not differ from the spectra of normal stars of same spectral type. Variations with time of the intensities of the emission and absorption components of the H lines were found in the spectrum of AB Aurigae.

Work on the MC Objects

Hutchings (1966) has measured Hγ equivalent widths for many stars in both Clouds and Thackeray (1965*b*) has studied spectroscopic changes in S Doradus.

Infrared Studies

Moroz (1966) has observed infrared spectra (1 to 2·5 μ) of 19 A1–M7 stars. Strong absorptions at 1·5 μ and at 2 μ which appear to be due to wings of stellar H_2O bands, were found in the spectra of the long-period variables *o* Ceti, R Aquilae and X Ophiuchi at light minima. A large number of fine absorption details due to CO were found in the spectra of K2–M5 giants. The energy distribution of red giants has a maximum at 1·6 μ, apparently due to the H- and H_2-ions.

Becklin at Caltech is determining infrared energy distribution out to 4μ for several infrared stars including an infrared object in the Orion Nebula that appears to have a blackbody temperature near 600°K.

McCammon, Neugebauer and Münch have obtained spectra of some of the extremely red objects discovered in the Infrared Sky Survey carried out at Mt Wilson. Among the interesting features found let us mention the sharp discontinuity at $\lambda = 1{\cdot}745\ \mu$ in the N-type stars Y Canis Venaticorum and U Hydrae.

Neugebauer and Münch have observed the spectrum of χ Cygni with the coudé 100-inch and a PbS detecting system.

Rotation

At Crimea, Boyarchuk and Kopylov (1964) have compiled a general catalogue of rotational velocities of 2558 stars, reduced to the Slettebak system. The catalogue is sufficiently complete for O5-G5 stars brighter than 5·5 magnitudes.

E. N. Walker (1965) has studied the variation of axial rotation with spectral type for main sequence stars within the range B9 to A5, and reached the conclusion that there is a preferential rotational velocity for stars of 2 $\mathfrak{M}_\odot$; the material comprises 633 stars.

Slettebak (1966*a*, *b*) has continued his studies on stellar rotation by investigating axial rotation in 56 later Be stars; the frequency distribution of rotational velocities of 42 of those stars suggests that all are rotating with an equatorial velocity near 350 km s^{-1} but have their rotation axes randomly oriented in space. Further, a comparison was made between the largest observed rotational velocities for stars of spectral types O 9·5–F0 and the computed equatorial break up velocities for stars of corresponding spectral types; and the region of minimum difference between the observed and the computed velocities is approximately that occupied by the Be stars.

Slettebak and Wright have estimated rotational velocities in 77 A-type stars near the north galactic pole and are carrying out studies of axial rotation in the Scorpio-Centaurus association and in h and χ Persei.

G. W. Collins II (1965) has extended and refined earlier work on the continuum emission from an early-type rapidly rotating star finding that a grey atmosphere is a moderately good approximation in the visible and a rather poor approximation in the ultraviolet where a variation up to 1·4 magnitudes in the colours, due to the effects of rotation, may be expected.

Collins and Harrington (1966) have computed Hβ profiles for rapidly rotating stars taking into account the effects of gravity darkening, limb darkening and shape distortion. In the range B0 V–B5 V considerable variation in shape with rotational velocity was found for the earlier types, while markedly variations of equivalent widths with rotation was found for the later types; calculations of Hγ profiles are near completion.

Andrews has measured the rotational velocities of B stars in IC2391, IC2602 and in the Scorpio-Centaurus association and was able to show that the faster rotators have the smaller Hα strengths, thus mimicking stars of higher luminosity.

Deutsch has examined the available statistics of rotational line widths at various spectral types along the main sequence and concludes that these statistics are fully consistent with a Boltzmann-type law for the distribution of angular momenta of stars. Deutsch and Kraft have accumulated some line-widths statistics for main sequence F stars.

Kraft and Wrubel have shown empirically that, at a given $B - V$, the slow apparent rotators of the Hyades show an ultraviolet excess relative to the fast rotators for spectral types between A8 V and F5 V, the amount of which is fully predicted by the change in surface brightness and effective temperature as computed by Sweet and Hogg. The implication that appears to result is that mild ultraviolet excesses for field F stars cannot be interpreted directly as evidence of metal deficiency unless the rotation is average to large for that spectral type.

Kraft has made an extensive study of $v \sin i$ in function of M_v for the stars of galactic clusters.

Kraft, C. Anderson and R. Stoeckly have redetermined the rotational velocities for stars in the Pleiades. On the basis of this work and the work of Abt and Hunter the suggestion is made that even after stars arrive on the main sequence, after gravitational contraction, there is still some slowing down of apparent rotation.

Baum, McGee and Kraft have made a study of rotational velocities for Praesepe stars using an electronographic image tube to obtain spectra of stars fainter than $M_v \sim +3$. The rotational velocities of all stars of very young I Persei (α Persei) group down to F5 V were studied by Kraft.

In a rediscussion of the rotational velocities of three T Tauri stars previously studied by Herbig, Kraft finds that angular momentum probably is not conserved for these stars as they contract to the main sequence along their Hayashi tracks. Spectrograms of stars of NGC2264 are obtained to check this point.

Kraft has studied in more detail the rotations of main sequence stars later than F5 V and finds a systematic tendency for higher rotations up to 25 km s^{-1} to occur among the younger dwarfs having K emission than among older dwarfs without K emission. Angular momentum among stars on the main sequence appears to decline with a time scale like that of the decay of stellar chromospheres.

Analysing Slettebak's measures of $v \sin i$, Deutsch can distinguish, near A0, two groups of stars, one the 'Y' (or young population) which comprises about 80% of them, while the remaining 20% constitute the 'O' (old) population. According to Deutsch, the latter, sharp line objects, are on the main sequence for the second time and are metamorphs of originally massive main sequence stars that have lost substantial mass and angular momentum while in the red giant stage. The Ap stars—and similarly the Am stars—are the slowest rotations of the 'Y' population.

Walker and Hodge (1966) have measured equivalent widths and half-widths of the He I lines λ4388 and λ4471 as well as rotational velocities for 425 O9–B5 stars, and Deeming and Walker (1966) have made a statistical study of these data and also of Hγ-measures, and were led to propose a model in which atmospheric turbulence is generated in the equatorial region of a star as a result of rotation.

Turbulence

Conti and Deutsch (1966) have established that the weak-line and strong-line characteristics observed in stars result principally and directly from differences in turbulence and that the weak-line characteristic is very probably just the result of the systematic decay in the microturbulence of a star in a time scale that is short compared to its life time on the main sequence. Deutsch suggests that microturbulence decays for the same reason that chromospheric activity decays. O. C. Wilson has recently shown that this is due to a diminution of the non-radiative energy flux generated in the hydrogen convection zone: this flux excites microturbulence, chromospheric activity and a quasi-steady mass loss similar to the solar wind.

Conti and Deutsch have shown that in the disk population turbulence differences from star to star will change the line strengths in a way that produces ultraviolet excesses or other colour anomalies.

In short, turbulence may be more important than abundance anomalies in producing unusual colours.

In agreement with the above results, in a study of a cepheid of large light amplitude and weak lines, Abt *et al.* (1966) have found that the lines are weak because of low microturbulence rather than low abundances. This arises the question whether microturbulence in stars depends on the location of origin in the Galaxy.

Bonsack and Culver (1966) have determined widths at half-depth and equivalent widths of a selection of weak lines of neutral vanadium in the spectra of 58 stars of types G, K and M that include all luminosity classes, and found that the profiles of the weak lines in K stars are dominated by the motions of large elements of gas and that the velocities of these elements are correlated with stellar luminosity in the same way as are the phenomena at higher levels in the atmosphere which determine the widths of the K line emission and H α absorption.

By comparing the profiles and intensities of the lines with a theoretical model that represents the continuum spectrum, Underhill and de Groot (1964) concluded that the strongest lines in 10 Lacertae are enhanced by microturbulence, evaluated in 15 km s^{-1}, and that the He II intensities are influenced by turbulence and therefore are not good indicators of helium abundance.

Bell and Rodgers (1965) discussed turbulent velocities in the atmosphere of the supergiant δ Canis Majoris from equivalent widths, half widths and central depth measurements of the Fe I and Fe II lines. Microturbulence is of about 12 km s^{-1} for lines of all excitation potentials. Macroturbulence is also independent of excitation potential but probably depends on equivalent width.

Kopylov (1965) has determined the microturbulent velocities in the atmospheres of 68 O6–B5 type stars by the curve-of-growth method and the dependence of the microturbulent velocities with spectral type was analysed in a semiquantitative manner. A detailed analysis of the physical conditions in the atmospheres of 19 O-type stars have been carried out by R. N. Kumaigorodskaya (1965*a*, *b*) who found that the large scale motions exceeds those derived through the curve-of-growth method by a factor of 2 to 3.

Bell and Rodgers have also studied turbulence in the cepheid β Doradus (cf. Commission 27 report of the Committee on the spectra of variable stars), and Boyarchuk and Pronik (1965*b*) have found correlation between the turbulent velocity and the character of the H emission in the spectrum of ζ Tauri.

II. ABUNDANCE DETERMINATIONS

Just before the Hamburg General Assembly, a symposium on 'Abundance Determination in Stellar Spectra' was held in Utrecht. The proceedings of this symposium (IAU Symposium No. 26) have just recently been published as well as the proceedings of a conference on stellar evolution sponsored by the NASA Institute for Space Studies and held in New York in November, 1963, part of which deals with abundances (Stein and Cameron, 1966). A recent survey article by Cayrel and Cayrel de Strobel (1966) deals with many of the current problems.

(*a*) *Early-Type Stars*

Warner (1966*b*) has used his calculated *f*-values of Fe III to determine the iron abundance in γ Pegasi (B2 IV) using the equivalent widths measured by Aller and Jugaku in 1958 and an approximate solar abundance was found.

D. H. McNamara and Mrs L. Sivers are investigating the titanium abundance in the B3 star HD 37 058, the spectrum of which displays very strong Ti II lines at λ3759 and 3761, lines which are absent in other sharp-line stars of comparable spectral type.

L. H. Aller and J. Ross are concerned with A and B stars of peculiar chemical compositions and are reanalysing their vast high-dispersion material by using a better method of reduction of the plates and improved model atmospheres. Preliminary abundance analysis was carried out for υ Herculis, χ Lupi, φ Herculis, ι Coronae Borealis, HR 8349, HR 7664, τ Capricorni, HR 6870 and ι Herculis. In all these objects, except for φ Herculis, for which no Sr, Y or Zr data are available, it is found that Sr, Y and Zr are greatly enhanced in abundance by factors ranging from 10 to 1000. Ti and Mn appear to be enhanced in υ Herculis and the binary star χ Lupi—which has been studied in detail by Robert Wolfe—appears to be a heavy-metal star with prominent Hg, while in φ Herculis, Sc and Cr appear to be enhanced whereas Fe is nearly normal. Reanalysis is essentially completed for ι Coronae Borealis in which both Mn and Fe are substantially enhanced; Ti and Fe are enhanced by factors of 5 and 10 in the phosphorus star HR 8349 and HR 7664. Titanium may be also overabundant in τ Capricorni, whose spectrum is characterized by many weak lines. A remarkable star on the list is HR 6870 = HD 168733 which has strong lines of chlorine; Fe which appears as Fe I, Fe II and Fe III, is enhanced about a hundred times, while titanium and strontium are increased by more than a thousand times over normal abundances. ι Herculis appears to have normal abundances although sulphur may be enhanced.

Conti has studied five extremely sharp-line A stars and found abundance differences that suggest relations with the 'metallic-line' stars of lower temperatures.

Buscombe is studying coudé spectra of southern OB stars with the aim of detecting possible differences in the He/H ratio, and is analysing χ Octantis (B8 V) and κ Eridani (B5 III) for abundances.

Hutchings is determining line intensities and Hγ profiles of several members of NGC 6231 and the results discussed in terms of model atmospheres and in conjunction with photoelectrically measured line intensities at Cambridge.

Henry and Mihalas (1964) have made an extensive study of magnesium in many O9 V to A2 V stars, and find N (Mg)/N (H) = $2 \cdot 2 \times 10^{-4}$ with little scatter. This abundance is an order of magnitude greater than the solar value. In 10 Lacertae, Underhill and de Groot find a magnesium abundance lower by a factor of 10 than that of Henry and Mihalas. Underhill is trying now to determine from Fe III the iron abundance in 10 Lacertae.

An unsuccessful search for deuterium in B, A and F stars was carried out by Peimbert and Wallerstein (1965*b*); in some cases the upper limit is below the terrestrial value.

(*b*) *Helium-Weak Stars*

The important discovery of He^3 in 3 Centauri A by Sargent and Jugaku has been confirmed by Rodgers and Bell (1964), and from the line displacements of the helium lines in several B-type spectra, Gutmann (1966*c*) has concluded that He^3 is almost as abundant in ρ Leonis as in 3 Centauri A. The spectrum of η Leonis has been analysed by J. Hardorp (1963).

Bidelman (1965) has called attention to the variation with time of the strength of the helium lines in HD 125823 (=HR 5378), a member of the Scorpio-Centaurus association; this variation has been confirmed by Thackeray (1966*b*). The Jascheks (1967) have found Ga II in the star which they class with the helium-weak stars, 3 Centauri A, α Sculptoris, 36 Lyncis and α Cancri. The discrepancy between the spectral types of these objects, classified on the basis of the He lines, and their *UBV* colours is being used by the Jascheks at La Plata to search for more members of the group.

B. N. G. Guthrie (1965*a*) has studied some Be stars (HD 37051, α Sculptoris, 20 Tauri and 36 Lyncis) with weak He I lines for their intrinsic colours and strong Sc II, Ti II, Cr II, Fe II and Sr II and also HD 175156 where lines of He I, C II, N II, Mg II, Si II, Si III, S II, Ca II, Ti II and Fe II were identified and reaches the conclusion that the five stars are probably rapidly rotating stars of normal chemical composition viewed pole-on, the line strengths appearing abnormal only because the effective temperature and surface gravity are much higher at the pole than at the equator in rapidly rotating stars.

(*c*) *Hydrogen-Poor Stars*

Several hydrogen-poor, or helium, stars have been subject of investigation. A review article on these stars has been written by M. Hack (1966*b*).

Hill (1964, 1965) has analysed HD 168476, HD 124448 and BD + 10°2139 and the abundance of He, relative to the total mass of the atmosphere, corresponds to that expected for total H exhaustion. The abundance of O is only $\frac{1}{4}$ to $\frac{1}{3}$ normal while C and possibly also Ne and N are overabundant; other light elements appear normal. In HD 168476 the elements of the iron peak appear to be ten times overabundant but the lines of these elements may be formed in an outer cool shell.

HD 96446 has been analysed by Cowley *et al.* (1963) on the basis of high dispersion spectra, and a ratio of H/He $\sim 0 \cdot 38$ was found, line identifications of this star, which include 79 lines attributed to Kr II, have been given by Buscombe (1965*a*).

Warner has observed the four known H-deficient stars with constant brightness and is carrying out an abundance analysis.

(*d*) *Metal-Weak Stars*

M. Spite-Lebon (1965) has made a differential study, relative to the Sun, of χ Draconis A, a metal-weak star and the brighter component of a 280-day period spectroscopic binary, where she found that most of the elements are deficient by a factor of 2 (except calcium, manganese and barium) and that the age of the stars is the same as that of the Sun. Mme Spite-Lebon has determined the masses of the two components.

(e) *Metallic-Line and Peculiar A Stars*

A spectrophotometric investigation in the near infrared of metallic-line stars has led Praderie (1965) to the determination of the abundance of O, N and Mg in those objects. For the stars 63 Tauri, ω Tauri and τ Ursae Majoris the N abundance seems to be somewhat higher than in the Sun, while O appears to be relatively slightly underabundant, in agreement with an earlier result of Sargent and Searle.

In Pasadena, Subhash Chandra has examined the spectra, with particular attention, to the region $\lambda\lambda$3100–3700, of many relatively sharp-line Ap stars and found that strong lines of Be II are present in several Si-λ4200 stars as well as in most Mn stars and that a strong correlation appears to exist between the intensity anomalies in Ca II and Ba II and Mn II and Y II. Less pronounced correlations exist between other pairs of elements and in a number of Ap stars Chandra finds chromium lines (but not iron lines) that lie far below the flat part of the curve of growth in normal stars of the same temperature, and concludes that many abundances are really highly anomalous in Ap atmosphere but that the most conspicuous differences among the Ap stars may result simply from differences in temperature.

Conti (1965a) has studied the abundances in five metallic-line A stars in the Hyades relative to normal F stars in the cluster using semi-empirical model atmospheres. The abundances differences appear to be real: he finds calcium to be deficient, some elements lighter than iron underabundant and some heavier than iron overabundant. Conti (1965*b*) has also made a high dispersion spectroscopic survey of the sharp line early A stars brighter than 5·5 magnitudes and accessible from Mt Wilson and found that nearly half of them have the properties of the metallic-line stars that had been previously identified only in stars as early as A5. Among the stars that Conti finds that are metallic-line objects let us mention Sirius and 68 Tauri. Most of those stars are spectroscopic binaries. Conti's investigation have been extended, in collaboration with S. Strom, also to eight sharp-line early A stars in the Pleiades cluster, several of which appear to be metallic-line objects of early type.

K. Kohl (1964*a*) with the help of non-grey models have determined the abundances of 27 elements in Sirius and the conclusion is also that the star is a metallic-line object; he (Kohl 1964*b*) studied the spectrum of Sirius from 3100 to 8863 Å. Sirius has been also studied by Warner (1966*b*) who rediscussed the abundances of the iron group with the use of a semi-empirical model atmosphere; Warner has concluded that actually Sirius has some of the characteristics of a metallic-line star but the overabundances are less extreme than previously reported by others.

F. Praderie is making a detailed analysis of the metallic-line star ζ Lyrae A.

Searle and Sargent (1964) have compared Si II and Mg II lines in 31 Ap and five normal A- and B-type stars. These lines behave in a very similar fashion, are not sensitive to temperature or electron pressure changes, and therefore should indicate real abundance effects. Among the manganese stars they find silicon and magnesium normal. The stars of the manganese group were found by M. and C. Jaschek and González (1965) to have strong overabundances of Mn, Yt, Ga and Hg; He tending to be weak and Sr and Ti being generally overabundant though varying from star to star. Two stars of the 'λ4200' group (HD 173650 and 192913) were found to be iron-rich and also to have large overabundances of Si, Cr and Hg (M. Jaschek and Z. López García, 1966; A. D. Thackeray, 1966*b*; M. Jaschek, Z. González and C. Jaschek, 1965). The 'λ4200' stars, HD 34452, v Fornacis and 41 Tauri as well as the Cr-Eu star HD 2453, 25354, 42616, 71866 and 135297 are under investigation. Cl II has been found in HD 34452 (M. and C. Jaschek, 1967) and P II in HD 6822 (M. and C. Jaschek, 1964). A review of these investigations shows (1965*b*) that besides general group characteristics there are individual composition anomalies among stars that occupy nearby places in the two colour diagram, implying that there is no close correlation between colours and abundance anomalies.

Mrs Z. López García (1966) studied γ Capricorni on the basis of high-dispersion material from Mt Wilson, and determined atmospheric parameters and found overabundance of Ti, Cr, Fe, Co, Sr, Y, Zr, Ba and La, underabundance of Sc and practically normal abundances of Mg, Ca, Mn and Ni. An investigation of the same star made earlier by Faraggiana (1964) on intermediate dispersion plates, had indicated an overabundance of some metals.

The abundances in the Mn star 53 Tauri has been revised by Warner (1965*a*) by using new *f*-values, and a new determination of the chemical composition of the star has been made by Guthrie (1966), by differential curve-of-growth method with α Lyrae as the comparison star, by using groups of lines of similar mean excitation and ionization potentials and wavelengths. It was found that Mn is overabundant by a factor of about 12, while Mg, Ca, Sc and Fe are deficient and the abundances of Si, Ti, Cr, Sr and Zr are about normal.

The peculiar A star HD89822 (=HR4072) found by Guthrie to be a double-line spectroscopic binary, was analysed by the same investigator from the point of the chemical composition resulting that for the brighter component Si, Mn, Sr and Y are overabundant by factors of about 20, 4, 90, and 200, respectively, while Sc is deficient by a factor of at least 5 and the abundances of Mg, Ti, Cr and Fe are about normal; the relative abundances of Ti, Cr, Fe, Sr and Y are similar in the two components.

Baschek (1964*b*) has determined the Na abundance in peculiar and metallic-line stars.

For the λ4200–Si II stars, silicon is overabundant by factors of 10 to 60 and the ratio of silicon to magnesium is 60 times the normal value, while the C II and He I lines are very weak suggesting abundance deficiencies. An analysis of the manganese star 53 Tauri by Aller and Bidelman (1964) indicates that manganese and gallium are very overabundant, strontium, yttrium and zirconium are somewhat enhanced.

L. H. Auer *et al.* (1966) have carried out a fine abundance analysis of 53 Tauri where they find that all of the metals lighter than titanium have normal abundances, and that He is underabundant by a factor of 5. From titanium onward, the elements show large overabundance factors, except for iron and chromium, both of which are normal; manganese is found to be 120 times overabundant.

Auer (1964) finds, in an analysis of 10 Aquilae, that europium, strontium, magnesium, manganese and cobalt are overabundant, while aluminium, scandium, zinc and barium are underabundant. Searle *et al.* (1966) have studied the iron-peak elements in 18 Ap stars. In manganese stars Mn/Fe is 6 to 60 times normal. The chromium to iron and titanium to iron ratios are abnormal and vary from star to star. In silicon-rich and oxygen-poor stars the iron-peak elements are in normal relative abundance. Wright (1965) has begun analyses of the Ap stars π^1 Bootis, HR4072 and 49 Cancri. The star π^1 Bootis has strong lines of Ga II and Hg II. Yttrium is strong, but ionized lines of the iron group and rare earths are weak.

The Aop star ϵ Ursae Majoris has been analysed by Leushin (1965) at Crimea and physical conditions and relative abundances were determined, Ca deficiency being detected. Log g was found to be 3·5 from the Hβ and Hγ line profiles while the mass-radius relation gives a value of 4·4.

Yamashita (1966) has made an abundance study of the peculiar K-type star 37 Comae Berenices and compared its spectrum with those of the K giants χ and ψ Ursae Majoris; the abundance of Carbon in 37 Comae Berenices was found to be one-seventh the value found for the comparison stars.

In Pretoria Warner has obtained high dispersion spectra of a number of Ap and Am stars and abundances are being derived.

Hyland has in progress in Canberra a large programme of high dispersion spectroscopy and spectral scans of Ap and metallic-line stars in clusters and in the general field.

A search for deuterium in magnetic stars (using the $H\alpha$ transition) by Peimbert and Wallerstein (1965*a*) was negative. They give upper limits for D/H of 7×10^{-4} to 4×10^{-5}.

e.1. Barium Stars

All the stars of this group have been studied by Warner (1965*b*) who finds carbon overabundant by factors 3 to 5, in agreement with previous determinations. The overabundance of heavy elements appears to be well explained by the *s*-process in nucleosynthesis. Danziger (1965*b*) has analysed the Ba II stars HD 116713 and HD 83548 finding somewhat greater overabundances. Among the light elements discussed by Warner and Danziger there is evidence for overabundances of Na, C and N. Warner has also derived absolute magnitudes from the H and K emission widths (Wilson-Bappu effect).

It is of interest to note that the problem raised by the large overabundance of Pr, an *r*-process element, in barium stars seems to have been solved in a theoretical discussion by Reeves (cf. R. Cayrel and G. Cayrel de Strobel, 1966).

In Japan, Nishimura is analysing the star HR 774 with the curve-of-growth method.

e.2. The Holmium Star HD 101065

Przybylski (1966) has continued his work on the strange rare-earth star HD 101065, where most rare-earth elements are very overabundant (by factors of 100 to 4000) although no evidence has been found for ytterbium. The question of lithium in this star has been considered by Warner (1966*c*) and is mentioned in the section on lithium abundances in stellar atmospheres; he is using $H\alpha$ profiles together with model atmosphere calculations to establish the absolute magnitude of the star. Przybylski who is studying HD 9996, a similar star to HD 101065 but not such an extreme case, notes the possibility that HD 101065 should be considered as a metal-poor star.

(*f*) *Late-Type Stars*

F. K. Edmonds, Jr. and S. Strom will do a fine analysis abundance determination for Procyon; Strom will furnish a non-LTE, line blanketed model atmosphere for the star.

G. Cayrel has determined the physical parameters and chemical composition of the atmospheres of seven stars of early K-type, as well as those of a comparison star (ϵ Virginis). Two of the stars are main sequence objects—HD 190404 and HD 219134—which were observed to try to solve the problem posed by O. C. Wilson as to whether the scatter observed on the colour-spectral type diagram for stars later tha G 5 is due to chemical composition differences; the conclusion of Mrs Cayrel's study is that the two stars have the same chemical composition but that microturbulence is much larger in HD 219134 than in HD 190404 and this effect plus the temperature difference found to exist between the two stars, explains their colour difference ($B - V = 0{\cdot}82$ for HD 190404, and $B - V = 1{\cdot}02$ for HD 219134; O. C. Wilson's spectral types for the two stars being almost the same).

The rest of the stars investigated by Mrs Cayrel were giants of luminosity class III and II. A pair of them HD 48781 and HD 94264, classified as 'strong line' and 'weak line' stars, respectively, suggested that the chemical composition as well as other atmospheric parameters were identical in the two stars but microturbulence was smaller in the weak-line star. HD 6437, is metal deficient by a factor of 2 to 3; HD 6833, a typical halo object, is less metal deficient (on the average by a factor of 8) than the three halo K giants previously analysed by Greenstein and his collaborators, but metal deficiencies are not uniform from element to element; HD 35620 has an abnormal spectrum and yielded a slight underabundance of the metals by a factor of 3. In HD 6833 fairly strong interstellar D_1 and D_2 lines were found.

Baschek *et al.* (1967) have carried out a quantitative analysis of the spectrum of the F8 V star β Virginis, which has been described by Namba (1964) from λ3900 to λ6820 Å.

Climenhaga (1965, 1966) has continued his studies of the C^{12}/C^{13} abundance ratio in late-type stars, and, from a study of the red bands, has determined the ratio for 19 Piscium, DS Pegasi and Y Canum Venaticorum.

Wright (1966) has obtained high-dispersion spectra for the G5 stars, 51 Pegasi and 20 Leo Minoris, for an abundance study.

At Canberra, Pagel has secured coudé spectra in the blue and red of a number of solar-type stars for which six-colour and other photometric data indicated that the metal abundance might differ from that of the Sun; these stars are α Centauri A and B, ρ Pavonis, HD 191408, α Phoenicis, ρ Tucanae, 82 Eridani and δ Pavonis); abundance analysis are in progress.

Conti and Greenstein, in co-operation with Spinrad, Vardya and Wallerstein have made an extensive study of O abundance using the [O I] lines at $\lambda\lambda$ 6300 and 6363. The observational data show a steady decrease of [O I] line strength with advancing spectra type and increasing luminosity; it is strongest in high velocity stars, which have a moderate metal deficiency.

Koelbloed has obtained at Mt Wilson spectra of G and K dwarfs, giants and supergiants for chemical abundance analysis.

Greenstein's programme on S-type variables has been supplemented with the observation of high-dispersion spectra of the bright, non variable S star, HR 1105.

Isotope ratios are being investigated by Schadee who has computed the shifts that according to theory should show the molecular bands. He is concerned at present with the search for Si^{28} and Si^{30} in HR 1105 and for Zr isotopes in R Andromedae.

Wyller (1966) has measured wavelengths of lines in the CN bands, particularly in the near infrared, to discover regions where the C^{12}/C^{13} ratio can be measured. In five carbon stars he finds this ratio to vary from 2 (in Y Canis Venaticorum and WZ Cassiopeiae) to 11; the equilibrium ratio for the CNO cycle being 3·4. Wyller suggests that C^{13} in carbon stars is partially synthesized by spallation and processes of neutron irradiation either in the atmosphere of Wolf-Rayet stars or in the carbon stars themselves. Schadee is studying the ultraviolet SiH bands in order to determine if the Si^{28}/Si^{30} ratio can be obtained. The same kind of analysis of the ZrO bands is used to determine how the Zr isotopes can be measured.

Mrs Locanthi has studied ZrO in R Andromedae and in V Cancri.

(g) *Giant and Supergiant Stars*

F. Spite (1964) has dealt with the method to approximately evaluate the metal abundances in some giant stars from low dispersion spectra by measuring the central depths of Hγ and of Ca I 4227.

Przybylski (1965) has made an abundance analysis of HD 33579, a giant in the LMC.

Zeinalov and Kopylov (1966) have made a detailed investigation of the physical conditions of the atmosphere and abundances in the supergiant η Leonis.

At Kiev M. Ja. Orlov (1966) has analysed the spectra of the M-type supergiant μ Cephei (λ 4250–6600 Å) and determined physical conditions in the atmosphere and abundances and discussed the relations between some parameters of the M-type supergiants.

At Crimea, M. E. Boyarchuk (1965) has studied the spectrograms of ρ Cassiopeiae, δ Canis Majoris and γ Cygni and found that in the atmosphere of ρ Cassiopeiae the abundance of the light elements was smaller and that of the heavy elements greater than in the atmospheres of the supergiants δ Canis Majoris and γ Cygni. A. Underhill is attempting to determine from Fe III the iron abundance in γ Pegasi and ρ Cassiopeiae.

(h) *Subdwarfs*

Danziger (1966) has carried out a curve of growth analysis of the bright southern late-type subdwarfs γ Pavonis and ρ^1 Reticuli. The results, which are compatible with *s*-process theory, indicate that the metals lighter than nickel are underabundant by 0·7 (in the logarithm) in γ Pavonis and by 0·4 in ρ^1 Reticuli while the underabundance of carbon is greater (1·3 in γ Pavonis and 0·8 in ρ^1 Reticuli). The relative abundance of Mn to Fe is normal in both stars but the heavy rare-earth metals are more underabundant than the iron group.

(*i*) *High-Velocity Stars*

Koelbloed has studied abundances in two weak-line stars of very high velocity, HD 2665 and HD 6755 on Palomar spectra taken by Greenstein.

HD 116745, discovered by Fehrenbach and Duflot, was found by Sargent (1965*b*) to be metal-deficient; the star must be highly evolved. Metal deficiency attains a factor of 40 in Wilson 10367 (LPM 661) as it results from the study made by Baschek (1965) in the ultraviolet and the blue.

Sargent *et al.* (1964) find a normal chemical composition in 7 Sextantis; earlier the star was supposed to be overabundant in helium and carbon.

At Pretoria, Peat has obtained spectra of six southern high velocity K giants in the 5000–6600 Å region and a detailed analysis is in progress; the low velocity K giant α Indi was observed as a comparison star.

Kodaira (1964) has determined the structure and chemical composition of HD 161817.

(*j*) *Globular Cluster Stars*

Searle and Rodgers (1966) have discussed 86 Å/mm spectra of horizontal branch stars in the globular cluster NGC 6397, finding that the metals are deficient by a factor of 100 while helium is deficient by a much smaller amount. They estimated the masses of these stars as 1·3 $\mathfrak{M}_{\odot}$.

Sargent (1965*b*) has studied Radcliffe spectra of HD 116745, the bright F-type member of ω Centauri, discovered by Fehrenbach and Duflot. The star was found to have the same Hγ profile and hence the same temperature as the normal F5 III star HD 113 537, but to be deficient in Ca and Sr by a factor of between 25 and 60. Because of large number of Ti II lines in its spectrum, Sargent suggests that something more complicated than a simple general metal deficiency will have to be postulated for the star.

Observations of the horizontal-branch stars in globular clusters by Münch and of spectroscopically similar brighter objects in the galactic polar caps by Greenstein do not necessarily support the suggestion from theoretical work that the early He/H ratio was not much lower than it is now in B stars.

There have been several abundance analyses of field stars which are probably analogous to horizontal-branch stars in globular clusters. The early A-type star HD 109995 was analysed by Wallerstein and Hunziker (1964) who found a factor 10 metal deficiency, compared with the Sun. Kodaira (1964) obtained a metal deficiency by a factor 13 in the slightly cooler star HD 161 817.

HD 26 and HD 201 626, studied by Wallerstein and Greenstein (1964), are somewhat to the right of the vertical part of the globular-cluster giant sequence. They find metal deficiencies by factors of 5 and 30 respectively. In both stars the carbon to iron ratio is five times normal and barium, lanthanum, cerium and neodymium are up by a factor 20. Europium is higher than normal by only a factor 2 to 5. The abnormal abundances might be produced by helium burning followed by neutron addition to the iron-peak elements, followed by some mixing to the surface. Searle and Rodgers (1966) have obtained low dispersion spectra of horizontal-branch stars in NGC 6397; the metal-to-hydrogen ratio is down roughly by a factor 100 from the Snu.

(*k*) *Halo Population Stars*

Spectroscopic studies of faint stars in the galactic halo have been pursued by Greenstein (1966). He finds He I and He II lines enhanced by factors of 3 to 5 in sdB and sDO stars. He I lines are weakened by a factor 3 in Bp stars. Berger (1963) and Sargent and Searle (1966) have found groups of hot halo stars in which the He I lines are weak and the hydrogen lines are strong compared with those in normal stars of the same colour and Balmer discontinuity.

W. L. W. Sargent and L. Searle (1966) have compared Feige 11, 36 and 65 stars with τ Scorpii and concluded that at least the three halo stars are helium-deficient, a rough estimate being on those objects where the He/H ratio is lower by a factor of 100 than it is in main-sequence stars of Population I.

B. Pagel (1965*a*, *b*) has derived heavy metals abundance in HD 122563 finding that the ratio Fe/H is about 250 times weaker than in the Sun; relative to the Sun there is a deficiency of barium, cerium and europium.

(*l*) *Miscellaneous Stars*

Two main sequence stars of same spectral type and different age, HR 2217 B and χ Bootis B, were observed by Bardin and Prévot (1964*b*) and found to have identical chemical composition in their atmospheres.

Danziger (1965*a*) has analysed the star RY Sagittarii which is similar to R Coronae Borealis; the abundance ratio of carbon to hydrogen is 25 and the ratio of carbon to iron is 35 times normal while sodium and lithium abundances are high.

l.1. *Cepheids*

In the report of Commission 27 the abundance work on the cepheid β Doradus is discussed.

Kraft and Abt have studied TV Camelopardalis and found that abundance is normal to within a factor of about 2, and that the cause of the weak lines is a low turbulent velocity.

l.2. *Binaries*

A new discussion on β Lyrae from the point of view of the chemical composition is due to Hack and Job (1965).

In regard to the subgiant components of Algol systems it is important to mention an investigation that is being carried out by D. S. Hall who has found, using narrow band photometry technique, that there may be an underabundance of CN in the mentioned components.

Sargent (1966) has analysed both components of ADS 3910, finding that one component is a typical B5 V star and that the companion is deficient in helium by a factor of 100 and in iron and silicon by a factor of 10; the latter star has probably suffered surface activity and may be related to the λ Bootis stars.

(*m*) *Lithium Abundances in Stellar Atmospheres*

Herbig has devoted most of his time to the problem of lithium in F-K stars and its implications. In a first paper (1965) he is concerned with dwarfs in the F5-G8 range; these stars show rather high Li abundances suggesting that convective depletion of Li during contraction cannot have been completed in the mass range concerned and that the variations in Li abundance in F5-G8 dwarfs may be due to a slow process of convective depletion. In a paper on the lithium abundances in F-G and K type subgiants, written in collaboration with R. J. Wolff (1966), it is suggested a mechanism of Li depletion through mass loss which is age-dependent and does not involve continuous Li burning at the base of the convection zone.

A programme that Herbig is carrying out at present is that of Li abundance in the later-type component of composite systems, usually G-K + B-A types.

The problem of the correlation of Li abundance with intensity of H, K emission cores is being dealt with by Herbig by means of an extensive programme of spectrophotometry of the emission cores at 8Å/mm, mainly in F-G dwarfs.

Wyller is concerned with the problem of lithium in carbon stars on account of possible C^{13} N^{14} line contamination in the calcium and lithium lines used for differential comparisons. The results of his study will be presented at the Joint Discussion in Prague on the lithium problem.

Wallerstein *et al.* (1965) have looked at 23 stars in the Hyades and find lithium to calcium ratios from 3 to 100 times that in the Sun. For $B-V \geqslant 0{\cdot}61$ the abundance is correlated with temperature, again suggesting destruction by convection into the interior. Further, Wallerstein (1966*a*) has measured the same ratio in 5 F-type and 6 G-type giants; the F giants show large enhancement of lithium while the G-stars do not. The F component of Capella has 60 times the lithium abundance of the G-star. The change in depth of the convection during the star's evolution is again responsible for the large differences.

In order to test Herbig's theory in regard to correlation between a star's lithium content and age, Wallerstein (1966*b*) has observed visual binaries in which at least one component is of a sufficiently late spectral type (F0 V) that the resonance line of neutral lithium might be present. Six out of nine very young systems have a high lithium content and the upper limits on the lithium content of the other three systems are not in serious disagreement with Herbig's hypothesis. However, several subgiants have a substantial lithium content, in contradiction with Herbig's correlation. As a consequence, Wallerstein suggested a possible mechanism in terms of internal magnetic field that might assist in the preservation of lithium.

Some carbon stars have lithium to calcium ratios ten times larger than in the Sun while others show no lithium (Torres-Peimbert *et al.* 1964).

Rodgers and Bell find the abundance of lithium to be more than ten times the solar value in the 35-day cepheid *l* Carinae. Also an overabundance is found by Danziger and Oke (1966) in VZ Cancri a metal-deficient cluster-type variable. The presence of lithium in this presumably highly evolved star is surprising and suggests that surface reactions may have occurred.

Lithium is one of the few cases where atomic lines show isotope shifts large enough to be measured: the shift between Li^6 and Li^7 is 7 km s^{-1}. Herbig (1964) observed 15 F- and G-type stars and found appreciable Li^6 in a few of them. A substantial extension of this early work on Li isotope ratios in F-K subgiants is in progress, with some improvements in technique.

In two Hyades F-type stars Merchant *et al.* (1965) found no evidence for Li^6. Wallerstein (1965) finds that two metallic-line stars have more Li^6 than similar normal stars and suggests the possibility of spallation reactions. In the sunspot spectrum Schmahl and Schröter (1965) found $Li^6/Li^7 \approx 0{\cdot}05$.

A high dispersion coudé programme in progress at Pretoria, being carried out by Feast, aims at determining the Li abundance and the Li^6/Li^7 ratio in a variety of southern stars. Among them the bright subgiant β Hydri – a member of the ρ Herculis group – with a general metal deficiency of about 2 or 3, according to Rodgers and Bell (1964), a Li/Ca ratio ten times the solar value and a high Li^6/Li^7 ratio (Feast 1967). The Li^6/Li^7 ratio appears to be high for lithium-rich subgiants and low for main sequence stars possibly implying lithium production during evolution off the main sequence.

Warner (1965*c*) has discussed a method of deriving lithium abundances from Li I equivalent widths.

Conti and Danziger have studied further the decrease of lithium with age in field stars and in the Pleiades, and found no difference in the pre-main sequence depletion and that the depletion time scale in G stars seems to be of the order of 10^9 years. The F stars appear to have a considerable range of initial abundance of Li.

The fact that no beryllium has been found in A or F supergiants led Conti to think that evolution might be from right to left in supergiants. He found no beryllium in the G subgiants η Bootis and ζ Herculis, making its presence in δ Eridani even more surprising.

A comparison of the beryllium and lithium abundances in ten F- and G-stars has been made by Merchant (1966) who found a much smaller range of abundances in beryllium than in lithium.

M. Bretz (1966) is interested in the presence of lithium in S stars: no lithium was found in the nine bright stars between types M and S that were studied, except perhaps in HD30959.

JORGE SAHADE
President of the Commission

BIBLIOGRAPHY

Abt, H. A. 1965, *Astrophys. J. Suppl.*, **11**, 429.
Abt, H. A., Osmer, P. S., Kraft, R. P. 1966, *Astrophys. J.*, **145**, 479.
Adam, M. G. 1965, *Observatory*, **85**, 204.
Alania, I. F. 1964, *Astr. Cirk.*, no. 301,
Alania, I. F. 1965, *Abastumansk. astrofiz. Obs. Bjull.*, **32**, 31.
Alania, I. F. 1966, *Abastumansk. astrofiz. Obs. Bjull.*, in press.
Alania, I. F., Popov, M. V. 1965, *Abastumansk. astrofiz. Obs. Bjull.*, **32**, 41.
Aller, L. H., Bidelman, W. P. 1964, *Astrophys. J.*, **139**, 171.
Aller, L. H., Dunham, Jr., Th. 1966, *Astrophys. J.*, **146**, 126.
Aller, L. H., Jugaku, J. 1967, *Publ. Obs. Univ. Michigan*, in press.
Andrews, P. J. 1965, Thesis, Cambridge.
Andrews, P. J., Breger, M. 1966, *Observatory*, **86**, 108.
Andrillat, Y. 1965*a*, *Ann. Astrophys.*, **28**, 867.
Andrillat, Y. 1965*b*, *C. r. Acad. Sci. Paris*, **261**, 221.
Andrillat, Y., Houziaux, L. 1966*a*, *C. r. Acad. Sci. Paris*, **263**, 635.
Andrillat, Y., Houziaux, L. 1966*b*, *J. Observateurs*, in press.
Andrillat, Y., Morguleff, N. 1966*a*, *Ann. Astrophys.*, **29**, 17 (*Publ. Obs. Hte.-Provence*, **8**, no. 17).
Andrillat, Y., Morguleff, N. 1966*b*, *Proc. Trieste Coll. on Late-Type Stars*, in press.
Auer, L. H. 1964, *Astrophys. J.*, **139**, 1148.
Auer, L. H., Mihalas, D. M., Aller, L. H., Ross, J. E. 1966, *Astrophys. J.*, **145**, 153.
Aydin, C., Hack, M., Islik, S. 1965, *Mem. Soc. astr. ital.*, **36**, 331.
Azimov, S. M. 1963, *Izv. glav. astr. Obs. Pulkove*, **23**, 76.
Azimov, S. M. 1965, *Abastumansk. astrofiz. Obs. Bjull.*, **33**, 81.
Azimov, S. M. 1966, *Soobšč. Schemachi astrofiz. Obs.*, no. 5.
Bappu, M. K. Vainu and Ganesh 1967*a*, *Mon. Not. R. astr. Soc.*, in press.
Bappu, M. K. Vainu and Ganesh 1967*b*, *Kodaikanal Obs. Bull.*, in press.
Barbier, D., Morguleff, N. 1964, *C. r. Acad. Sci. Paris*, **258**, 4209 (*Publ. Obs. Hte.-Provence*, **7**, no. 14).
Barbon, R., Bertola, F., Ciatti, F., Margoni, R. 1965, *Comm.* 27 *IAU. Inf. Bull. var. Stars*, no. 109.
Bardin, C., Prévot, L. 1964*a*, *C. r. Acad. Sci. Paris*, **259**, 1007.
Bardin, C., Prévot, L. 1964*b*, *C. r. Acad. Sci. Paris*, **260**, 424.
Baschek, B. 1964*a*, *Publ. astr. Soc. Pacific*, **76**, 22.
Baschek, B. 1964*b*, *Z. Astrophys.*, **60**, 76.
Baschek, B. 1965, *Z. Astrophys.*, **61**, 27.
Baschek, B., Holweger, H., Namba, O., Traving, G. 1967, *Z. Astrophys.*, in press.

Baschek, B., Oke, J. B. 1965, *Astrophys. J.*, **141**, 1404.
Batten, A. H. 1964, *J. R. astr. Soc. Can.*, **58**, 253.
Bell, R. A., Rodgers, A. W. 1964, *Observatory*, **84**, 29.
Bell, R. A., Rodgers, A. W. 1965, *Mon. Not. R. astr. Soc.*, **129**, 127.
Berger, J. 1963, *Publ. astr. Soc. Pacific*, **75**, 393.
Bertaud, C. 1965, *J. Observateurs*, **48**, 211.
Bertola, F. 1964, *Publ. astr. Soc. Pacific*, **76**, 241.
Bidelman, W. P. 1965, *Astr. J.*, **70**, 667.
Bidelman, W. P., Sahade, J., Frieboes-Conde, H. 1963, *Publ. astr. Soc. Pacific*, **75**, 524.
Bonsack, W. R., Culver, R. B. 1966, *Astrophys. J.*, **145**, 767.
Boyarchuk, A. A., Kopylov, I. M. 1964, *Izv. Krym. astrofiz. Obs.*, **31**, 44.
Boyarchuk, A. A., Pronik, I. I. 1964, *Izv. Krym. astrofiz. Obs.*, **31**. 3.
Boyarchuk, A. A., Pronik, I. I. 1965*a*, *Izv. Krym. astrofiz. Obs.*, **33**, 195.
Boyarchuk, A. A., Pronik, I. I. 1965*b*, *Izv. Krym. astrofiz. Obs.*, **34**, 118.
Boyarchuk, M. E. 1965, *Izv. Krym. astrofiz. Obs.*, **34**, 136.
Bretz, M. C. 1966, *Proc. IAU Symposium No.* 26, Ed. H. Hubenet, Academic Press, London, p. 304.
Buscombe, W. 1965*a*, *Mon. Not. R. astr. Soc.*, **129**, 1.
Buscombe, W. 1965*b*, *Mon. Not. R. astr. Soc.*, **129**, 411.
Cayrel, R., Cayrel de Strobel, G. 1966, *A. Rev. Astr. Astrophys.*, **4**, 1.
Clarke, D., Grainger, J. F. 1965, *Nature*, **207**, 1180.
Climenhaga, J. L. 1965, *J. R. astr. Soc. Canada*, **59**, 269.
Climenhaga, J. L. 1966, *Ann. Astrophys.*, in press.
Code, A. D., Bless, R. C. 1964, *Astrophys. J.*, **139**, 787.
Collins II, G. W. 1965, *Astrophys. J.*, **142**, 265.
Collins II, G. W., Harrington, J. P. 1966, *Astrophys. J.*, in press.
Conti, P. S. 1965*a*, *Astrophys. J. Suppl.*, **11**, 47.
Conti, P. S. 1965*b*, *Astrophys. J.*, **147**, 1594.
Conti, P. S., Deutsch, A. J. 1966, *Astrophys. J.*, **145**, 742.
Cowley, A. P., Cowley, C. R. 1966, *Astrophys. J.*, **144**, 824.
Cowley, A. P., Aller, L. H., Dunham, Th. 1963, *Publ. astr. Soc. Pacific*, **75**, 441.
Danziger, I. J. 1965*a*, *Mon. Not. R. astr. Soc.*, **130**, 199.
Danziger, I. J. 1965*b*, *Mon. Not. R. astr. Soc.*, **131**, 51.
Danziger, I. J. 1966, *Astrophys. J.*, **143**, 527.
Danziger, I. J., Oke, J. B. 1966, *Astrophys. J.*, in press.
Deeming, T. J., Walker, G. A. H. 1966, *Z. Astrophys.*, in press.
Delplace, A. M. 1966, *C. r. Acad. Sci. Paris*, **262**, 1562.
Doazan, V. 1965, *Ann. Astrophys.*, **28**, 1.
Dolidze, M. V., Apriamashvili, S. P., Jimsheleishvili, G. N. 1965, *Astr. Cirk.*, no. 340.
Edmonds, F. N. 1965, *Astrophys. J.*, **142**, 278.
Faraggiana, R. 1964, *Mem. Soc. astr. ital.*, **35**, 307.
Faraggiana, R. 1965, *Z. Astrophys.*, **62**, 108.
Faraggiana, R., Hack, M. 1965, *Mem. Soc. astr. ital.*, **36**, 63.
Feast, M. W. 1967, *Mon. Not. R. astr. Soc.*, in press.
Fitzgerald, M. P., Houk, N., McCuskey, S. W., Hoffleit, D. 1966, *Astrophys. J.*, **144**, 1135.
Fletcher, J. M. 1966, *J. R. astr. Soc. Can.*, in press.
Folkart, B., Pecker, J.-C., Pottasch, S. 1964*a*, *Ann. Astrophys.*, **27**, 249.
Folkart, B., Pecker, J.-C., Pottasch, S. 1964*b*, *Ann. Astrophys.*, **27**, 252.
Fujita, Y. 1966, *Vistas in Astronomy*, Ed. A. Beer, Pergamon Press, Oxford, **7**, 71.
Galkin, L. S. 1964, *Izv. Krym. astrofiz. Obs.*, **31**, 23.
Galkina, T. S. 1965, *Izv. Krym. astrofiz. Obs.*, **36**.
Galkina, T. S. 1967, *Izv. Krym. astrofiz. Obs.*, **37**.
Garstang, R. H. 1964, *Observatory*, **84**, 161.
Gasson, R. E. M., Pagel, B. E. J. 1966, *Observatory*, **86**, 196.
Götz, W. 1965, *Sterne*, **41**, 150.

Greenstein, J. L. 1966, *Astrophys. J.*, **144**, 496.
Griffin, R. F. 1964*a*, *Observatory*, **84**, 154.
Griffin, R. F. 1964*b*, *Mon. Not. R. astr. Soc.*, **128**, 187.
Grygar, J. 1964, *Bull. astr. Inst. Netherl.*, **17**, 305.
Gurzadian, G. A., Tchaurushian, H. S. 1964, *Soobšč. Bjurak. Obs.*, **34**, **45**.
Guthrie, B. N. G. 1965*a*, *Publ. R. Obs. Edinburgh*, **3**, 263.
Guthrie, B. N. G. 1965*b*, *Publ. R. Obs. Edinburgh*, **5**, 1.
Guthrie, B. N. G. 1966, *Publ. R. Obs. Edinburgh*, **5**, 181.
Gutmann, F. 1966*a*, *Publ. Dom. astrophys. Obs. Victoria*, **12**, 11.
Gutmann, F. 1966*b*, *Publ. Dom. astrophys. Obs. Victoria*, **12**, 13.
Gutmann, F. 1966*c*, *Publ. astr. Soc. Pacific*, in press.
Hack, M. 1965, *Vistas in Astronomy*, Ed. A. Beer, Pergamon Press, Oxford, **7**, 107. (*Oss. astr. Trieste Publ.*, no. 337.)
Hack, M. 1966*a*, *Proc. NATO Course at Newcastle-upon-Tyne*, Ed. S. K. Runcorn, in press.
Hack, M. 1966*b*, '*A Memorial to Otto Struve*', Ed. M. Hack, Gauthier-Villars, Paris, in press.
Hack, M., Aydin, C., Islik, S. 1964, *Mem. Soc. astr. ital.*, **35**, 331.
Hack, M., Faraggiana, R. 1964, *Nuovo Cim. Suppl.*, **2**, 737.
Hack, M., Job, F. 1965, *Z. Astrophys.*, **62**, 203.
Hack, M., Mathews, M. S. 1965, *Oss. astr. Trieste Publ.*, no. 343.
Hardorp, J. 1963, *Tag. Astr. Gesellsch. Frankfurt a. M.S.I.*, **8–9**, 47.
Heard, J. F. 1965, *J. R. astr. Soc. Can.*, **59**, 258.
Heard, J. F. 1966*a*, *Proc. IAU Symposium* no. 30.
Heard, J. F. 1966*b*, *J. R. astr. Soc. Can.*, **60**, 128.
Helden, R. van 1966, *Bull. astr. Inst. Netherl.*, **18**, 367.
Henry, R. C., Mihalas, D. 1964, *Astrophys. J.*, **140**, 873.
Herbig, G. H. 1964, *Astrophys. J.*, **140**, 702.
Herbig, G. H. 1965, *Astrophys. J.*, **141**, 588.
Herbig, G. H. 1966, *Vistas in Astronomy*, Ed. A. Beer, Pergamon Press, Oxford, **8**, 109.
Herbig, G. H., Woolf, R. J. 1966, *Ann. Astrophys.*, in press.
Herman, R. 1964, *Ann. Astrophys.*, **27**, 507 (*Publ. Obs. Hte-Provence*, **7**, no. 22), 121*e Colloque International C.N.R.S.*, Sept. 3-7, 1963, Ed. C.N.R.S., Paris, p. 157.
Hill, P. W. 1964, *Mon. Not. R. astr. Soc.*, **127**, 113.
Hill, P. W. 1965, *Mon. Not. R. astr. Soc.*, **129**, 137.
Hutchings, J. B. 1966, *Mon. Not. R. astr. Soc.*, **132**, 433.
Ivanova, N. Z., Hovhannessian, R. K., Jepremian, R. A. 1965, *Astrofizika*, **1**, 397.
Jaschek, C., Jaschek, M. 1965, *Publ. astr. Soc. Pacific*, **77**, 376.
Jaschek, C., Jaschek, M. 1967, *Publ. astr. Soc. Pacific*, in press.
Jaschek, C., Jaschek, M., Kucewicz, B. 1964, *Z. Astrophys.*, **59**, 108.
Jaschek, M., González, Z., Jaschek, C. 1965, *Asoc. argent. Astr. Bol.*, no. 8, **41**.
Jaschek, M., Jaschek, C. 1964, *Asoc. argent. Astr. Bol.*, no. 7, 16.
Jaschek, M., Jaschek, C. 1965, *NASA Symp. of Magnetic, Peculiar and Metallic-line A Stars*, in press.
Jaschek, M., Jaschek, C., González, Z. 1965, *Z. Astrophys.*, **62**, 21.
Jaschek, M., López García, Z. 1966, *Z. Astrophys.*, **64**, 217.
Kodaira, K. 1964, *Z. Astrophys.*, **59**, 139.
Kohl, K. 1964*a*, *Z. Astrophys.*, **60**, 115.
Kohl, K. 1964*b*, *Veröff. Inst. theoret. Physik Sternw. Kiel.*
Komarov, N. S. 1965, *Izv. Krym. astrofiz. Obs.*, **33**, 273.
Kondo, M. 1965, *Publ. astr. Soc. Japan*, **17**, 412.
Kopylov, I. M. 1965, *Izv. Krym. astrofiz. Obs.*, **34**, 163.
Kubiak, M. 1966, *Acta astr.*, **16**, in press.
Kucewicz, B. 1965, *Asoc. argent. Astr. Bol.*, no. 10, 63.
Kuhi, L. V. 1966, *Astrophys. J.*, **145**, 715.
Kumaigorodskaya, R. N. 1964, *Izv. Krym. astrofiz. Obs.*, **32**, 108.
Kumaigorodskaya, R. N. 1965*a*, *Izv. Krym. astrofiz. Obs.*, **33**, 242.
Kumaigorodskaya, R. N. 1965*b*, *Izv. Krym. astrofiz. Obs.*, **34**, 177.

Kumaigorodskaya, R. N. 1967, *Izv. Krym. astrofiz. Obs.*, **37**, in press.
Kupo, I. D., Mamatkazina, A. K. 1966, *Trudy Inst. Astrofiz.* (*Alma-Ata*), **8**, in press.
Lacoarret, M. 1965, *Ann. Astrophys.*, **28**, 321.
Ledoux, P., Renson, P. 1966, *A. Rev. Astr. Astrophys.*, **4**, 293.
Leushin, V. V. 1965, *Izv. Krym. astrofiz. Obs.*, **34**, 151.
López García, Z. 1966, Thesis.
Malov, I. F., Vitrichenko, E. A. 1964, *Astr. Zu.*, **41**, 637.
Merchant, A. E. 1966, *Astrophys. J.*, **143**, 336.
Merchant, A. E., Bodenheimer, P., Wallerstein, G. 1965, *Astrophys. J.*, **142**, 790.
Metrevely, M. 1966, *Astr. Cirk.*, no. 352, in press.
Moroz, V. I. 1966, *Astr. Zu.*, **43**, 63.
Namba, O. 1964, *Veröff. Inst. theoret. Physik Sternw. Kiel*,
O'Dell, C. R. 1966, *Astrophys. J.*, **145**, 487.
Odgers, G. J., Wright, K. O. 1965, *J. R. astr. Soc. Can.*, **59**, 115.
Orlov, M. J. 1966, *Astr. Zu.*, **43**, in press.
Osawa, K. 1965, *Ann. Tokyo astr. Obs.*, **9**, 123.
Pagel, B. E. J. 1965*a*, *Nature*, **206**, 282.
Pagel, B. E. J. 1965*b*, *R. Obs. Bull.*, no. 104.
Peat, D. W. 1964, *Mon. Not. R. astr. Soc.*, **128**, 435.
Peimbert, M., Wallerstein, G. 1965*a*, *Astrophys. J.*, **141**, 582.
Peimbert, M., Wallerstein, G. 1965*b*, *Astrophys. J.*, **142**, 1024.
Peterson, B. 1966, *Astrophys. J.*, **145**, 735.
Petrie, R. M. 1965*a*, *Publ. astr. Soc. Pacific*, **77**, 97.
Petrie, R. M. 1965*b*, *Publ. Dom. astrophys. Obs.*, *Victoria*, **12**, 9.
Petrie, R. M., Lee, E. K. 1966, *Publ. Dom. astrophys. Obs.*, *Victoria*, **12**, 16.
Praderie, F. 1965, *C. r. Acad. Sci. Paris*, **261**, 4625.
Przybylski, A. 1965, *Nature*, **205**, 163.
Przybylski, A. 1966, *Nature*, **210**, 20.
Pyper, D. M. 1966, *Astrophys. J.*, **144**, 13.
Rachkovskaya, T. M. 1967, *Izv. Krym. astrofiz. Obs.*, in press.
Renson, P. 1965, *Bull. Soc. r. Sci. Liège*, **34**, 118.
Renson, P. 1966, *Bull. Soc. r. Sci. Liège*, **35**, 244.
Rebeirot, E. 1966, *C. r. Acad. Sci. Paris*, **262**, 1105, *Publ. Obs. Hte-Provence*, **8**, no. 19.
Rodgers, A. N., Bell, R. A. 1964, *Observatory*, **84**, 69.
Sahade, J. 1965, *Observatory*, **85**, 214.
Sahade, J. 1966*a*, '*A Memorial to Otto Sruve*', Ed. M. Hack, Gauthier-Villars, Paris, p. 219.
Sahade, J. 1966*b*, in *Stellar Evolution*, Ed. A. G. W. Cameron and R. F. Stein, Plenum Press, New York, p. 449.
Sahade, J. 1966*c*, *Trans. int. astr. Un.*, **12 B**, 494.
Sahade, J., Frieboes Conde, H. 1965*b*, *Astrophys. J.*, **141**, 652.
Sargent, W. L. W. 1964, *A. Rev. Astr. Astrophys.*, **2**, 297.
Sargent, W. L. W. 1965*a*, *Observatory*, **85**, 33.
Sargent, W. L. W. 1965*b*, *Observatory*, **85**, 116.
Sargent, W. L. W. 1966, *Astrophys. J.*, **144**, 1128.
Sargent, W. L. W., Searle, L. 1966, *Astrophys. J.*, **145**, 652.
Sargent, W. L. W., Searle, L., Wallerstein, G. 1964, *Astrophys. J.*, **139**, 3.
Schmahl, G., Schröter, E. H. 1965, *Z. Astrophys.*, **62**, 143.
Scholz, M. 1965, *Z. Astrophys.*, **61**, 179.
Searle, L. H., Luvigerschansen, W. T., Sargent, W. L. W. 1966, *Astrophys. J.*, **145**, 141.
Searle, L. H., Sargent, W. L. W. 1964, *Astrophys. J.*, **139**, 793.
Searle, L. H., Rodgers, A. W. 1966, *Astrophys. J.*, **143**, 809.
Slettebak, A. 1966*a*, *Astrophys. J.*, **145**, 121.
Slettebak, A. 1966*b*, *Astrophys. J.*, **145**, 126.
Smak, J. 1964, *Astrophys. J. Suppl.*, **9**, 141.
Smak, J. 1965, *Acta astr.*, **15**, 327.
Smak, J. 1966*a*, *Acta astr.*, **16**, 1.

Smak, J. 1966*b*, *A. Rev. Astr. Astrophys.*, **4**, 19.
Smeyers, P. 1966, *Ann. Astrophys.*, **29**, 1.
Spite, F. 1964, *C. r. Acad. Sci. Paris*, **258**, 1410.
Spite-Lebon, M. 1965, *C. r. Acad. Sci. Paris*, **260**, 417.
Stein, R. F., Cameron, A. G. W., Eds. 1966, *Stellar Evolution*, Plenum Press, N. York.
Stepien, K. 1964, *Post. Astr. Polska*, **12**, 85.
Strittmatter, P. A., Sargent, W. L. W. 1966, *Astrophys. J.*, **145**, 130.
Svolopoulos, S. N. 1966*a*, *Ann. Astrophys.*, **29**, 23.
Svolopoulos, S. N. 1966*b*, *Ann. Astrophys.*, **29**, 29.
Talbert, F. D., Edmonds, F. K. 1966, *Astrophys. J.*, **146**, 177.
Thackeray, A. D. 1964, *IAU Symposium no.* 24, p. 44.
Thackeray, A. D. 1965*a*, *Observatory*, **85**, 206.
Thackeray, A. D. 1965*b*, *Mon. Not. R. astr. Soc.*, **129**, 169.
Thackeray, A. D. 1966*a*, *Observatory*, **86**, 60.
Thackeray, A. D. 1966*b*, *Mon. Not. astr. Soc. S. Africa*, **25**, 7.
Thackeray, A. D., Tatum, J. B. 1966, *Publ. Dom. astrophys. Obs., Victoria*, **13**, 2.
The, P.-S. 1964*a*, *Observatory*, **84**, 83.
The, P.-S. 1964*b*, *Observatory*, **84**, 122.
The, P.-S. 1964*c*, *Contr. Bosscha Obs.*, no. 23.
The, P.-S. 1964*d*, *Publ. astr. Soc. Pacific*, **76**, 293.
The, P.-S. 1966*a*, *Contr. Bosscha Obs.*, no. 34.
The, P.-S. 1966*b*, *Contr. Bosscha Obs.*, no. 35.
The, P.-S., Lim, H. K. 1964, *Contr. Bosscha Obs.*, no. 23.
Torres-Peimbert, S., Wallerstein, G., Phillips, J. G. 1964, *Astrophys. J.*, **140**, 1313.
Underhill, A. B. 1965, *Bull. astr. Inst. Netherl.*, **18**, 5.
Underhill, A. B. 1966*a*, *The early-type stars*, Reidel Publishing Co., Dordrecht.
Underhill, A. B. 1966*b*, *Observatory*, **86**, 25.
Underhill, A. B., Groot, M. de 1964, *Rech. astr. Obs. Utrecht*, **17**, 280.
Unno, W., Kato, S. 1965, *Stellar and Solar Magnetic Fields*, Ed. R. Lüst (IAU Symposium no. 22), p. 84.
Walker, E. N. 1965, *Observatory*, **85**, 162.
Walker, E. N. 1966*a*, *Observatory*, **86**, 110.
Walker, E. N. 1966*b*, *Observatory*, **86**, 154.
Walker, G. A. H., Hodge, S. M. 1966, *Publ. Dom. astrophys. Obs., Victoria*, **13**, 2.
Wallerstein, G. 1965, *Astrophys. J.*, **141**, 311.
Wallerstein, G. 1966*a*, *Astrophys. J.*, **143**, 823.
Wallerstein, G. 1966*b*, *Astrophys. J.*, **145**, 759.
Wallerstein, G., Greenstein, J. L. 1964, *Astrophys. J.*, **139**, 1163.
Wallerstein, G., Hack, M. 1964, *Observatory*, **84**, 160.
Wallerstein, G., Herbig, G. H., Conti, P. S. 1965, *Astrophys. J.*, **141**, 610.
Wallerstein, G., Hunziker, W. 1964, *Astrophys. J.*, **140**, 214.
Warner, B. 1964, *Commun. Univ. London Obs.*, **62**.
Warner, B. 1965*a*, *Mon. Not. astr. Soc. South Africa*, **24**, 157.
Warner, B. 1965*b*, *Mon. Not. R. astr. Soc.*, **129**, 263.
Warner, B. 1965*c*, *J. quantit. Spectrosc. radiat. Transfer*, **5**, 693.
Warner, B. 1966*a*, *Observatory*, **86**, 82.
Warner, B. 1966*b*, *Mon. Not. R. astr. Soc.*, **133**, 389.
Warner, B. 1966*c*, *Nature*, **211**, 55.
Webster, L. B. 1966, *Publ. astr. Soc. Pacific*, **78**, 136.
Wehlau, W. H. 1965, *NASA Symp. on Magnetic, Peculiar and Metallic-line A Stars*, in press.
Wehlau, W. H. 1967, *Proc. IAU Symposium no.* 30, Academic Press, London, p. 257.
Wright, K. O. 1965, *NASA Symp. on Magnetic, Peculiar, and Metallic-line A Stars*, in press.
Wright, K. O. 1966, *Ann. Astrophys.*, (Trieste colloquium), in press.
Wyller, A. A. 1966, *Astrophys. J.*, **143**, 828.
Yamashita, Y. 1966, *Publ. Dom. astrophys. Obs., Victoria*, **12**, no. 16.
Zeinalov, S. K., Kopylov, I. M. 1966, *Izv. Krym. astrofiz. Obs.*, **36**, in press.

APPENDIX I: REPORT ON ABSOLUTE SPECTROPHOTOMETRY

(Prepared by J. B. Oke)

Since the last IAU meeting in Hamburg, little fundamental absolute calibration work has been carried out. Material discussed at that time has now appeared in print. Willstrop (1965) has published energy distributions of 215 stars of types O5 to M4 with measures made over successive 50-angström bands from λ4000 to λ6900. He also made comparisons with standard lamps. A sequence of early-type stars near the celestial equator and suitable for absolute spectrophotometric standards has been measured by Oke (1964). All stars have been observed from λ3390 to λ5900 and some out as far as λ10800. It now appears likely that in this list all the absolute spectrophotometric magnitudes below the Balmer jump should be made fainter by approximately 0·06 magnitude on the basis of the best absolute data now available for α Lyrae. Aller, Faulkner, and Norton (1964, 1966) have published absolute fluxes for more than 20 southern stars. They have been tied to the standards of Oke (1964).

The remaining papers on absolute spectrophotometry have been primarily for the purpose of determining fundamental data about stars, i.e. log T_e and log g. These are used to determine accurately the locations of stars in the Hertzsprung-Russell diagram and for obtaining the parameters needed in chemical abundance analyses.

Glushneva (1964) has obtained absolute energy distributions of 14 B 9 to A 7 and seven earlier-type stars from λ3000 to λ4700, using photographic slitless techniques. More recently four additional early-type stars have been studied (Glushneva, 1966). Oke (1965) in a review article discussed some of the observational problems associated with absolute spectrophotometry and gave energy distributions for a number of stars. Baschek and Oke (1965) have measured 11 Am, Ap, and normal A stars from λ3400 to λ10800. Measures have been corrected for absorption lines and a temperature scale is given. Oke and Conti (1966) obtained continuum energy distributions from λ3400 to λ10800 for 24 Hyades stars of spectral types A3 to K0. The relation of $B - V$ to log T_e is given. Whiteoak (1966) obtained absolute fluxes from λ3448 to λ10500 for 38 early-type stars to study the interstellar extinction law. Oke, Greenstein and Gunn (1966) obtained and discussed energy distributions of a number of suspected field horizontal branch stars. Karetnikov and Medvedev (1966) have described a new spectrometer and published energy distributions from λ3800 to λ5800 of α Andromedae, β Arietis and α Cygni. Baschek, Oke and Searle (unpublished) have obtained absolute energy distributions from λ3400 to λ10800 of all known λ Bootis stars.

Aller and Faulkner (1964) obtained spectrophotometric data for the WR star γ Velorum. Kuhi (1966) obtained absolute fluxes from λ3200 to λ11000 of many nothern WR stars. He finds high colour temperatures in the blue and violet and lower ones in the visual and red. Kupo (1965) obtained gradients and Balmer jumps around the cycle of β CMa. Oke (1966) has obtained similar data for the very metal-deficient RR Lyrae star X Arietis.

Rodgers and Searle at Mt Stromlo have scanned η Carinae and discovered a red continuum similar to that of the nucleus of the Seyfert galazy NGC4151 and to the Crab nebula. Comparison of absolute Hα fluxes and radio-thermal emission in the surrounding H II region leads to a value of A_V/E_{B-V} of 6.

In the infrared, Woolf, Schwarzschild and Rose (1964) using Stratoscope II obtained scans from 0·8 to 3μ of 6 K- and M-type stars. These were all compared directly with the A1 star α Canis Majoris. Low and Johnson (1964) have extended their broad-band photometry out to 10μ. They now have sufficient infrared points to represent at least approximately a large part of the spectral energy distribution of cool stars. Wildey and Murray (1964) have also made measurements of 25 stars at 10μ. Johnson (1965) has obtained broad-band colours out to 3·5μ of 43 K- and M-type stars. Mendoza and Johnson (1965) obtained colours out to 10μ for many carbon stars.

BIBLIOGRAPHY

Aller, L. H., Faulkner, D. J. 1964, *Astrophys. J.*, **140**, 167.
Aller, L. H., Faulkner, D. J., Norton, R. H. 1964, *Astrophys. J.*, **140**, 1609.
Aller, L. H., Faulkner, D. J., Norton, R. H. 1966, *Astrophys. J.*, **144**, 1073.
Baschek, B., Oke, J. B. 1965, *Astrophys. J.*, **141**, 1404.
Glushneva, I. N. 1964, *Astr. Zu.*, **41**, 212.
Glushneva, I. N. 1966, *Astr. Zu.*, **43**, 80.
Johnson, H. L. 1965, *Astrophys. J.*, **141**, 170.
Karetnikov, V. G., Medvedev, Y. A. 1966, *Astr. Zu.*, **43**, 873.
Kuhi, L. V. 1966, *Astrophys. J.*, **143**, 753.
Kupo, I. D. 1965, *Astr. Zu.*, **43**, 358.
Low, F. J., Johnson, H. L. 1964, *Astrophys. J.*, **139**, 1130.
Mendoza, E. E., Johnson, H. L. 1965, *Astrophys. J.*, **141**, 161.
Oke, J. B. 1964, *Astrophys. J.*, **140**, 689.
Oke, J. B. 1965, *A. Rev. Astr. Astrophys.*, **3**, 23.
Oke, J. B. 1966, *Astrophys. J.*, **145**, 468.
Oke, J. B., Conti, P. S. 1966, *Astrophys. J.*, **143**, 134.
Oke, J. B., Greenstein, J. L., Gunn, J. 1966, *Stellar Evolution*, Ed. R. F. Stein and A. G. W. Cameron, Plenum Press, New York, p. 399.
Whiteoak, J. B. 1966, *Astrophys. J.*, **144**, 305.
Wildey, R. L., Murray, B. C. 1964, *Astrophys. J.*, **139**, 435.
Willstrop, R. V. 1965, *Mem. R. astr. Soc.*, **69**, 83.
Woolf, N. J., Schwarzschild, M., Rose, W. K. 1964, *Astrophys. J.*, **140**, 833.

After the report by Dr Oke was written the following additional information was received:

Žilevičiute and Straižys (1965) have studied the energy distribution in the continuous spectrum of γ Cassiopeiae during 1964–65.

The energy distribution in the continuous spectrum of RW Aurigae has been studied by Mirzoyan and Kazarian (1965) and by Kharadze and Bartaya (1965). A more or less intense UV-continuous emission is permanently present in the stellar radiation and the energy distribution changes conspicuously with light variation.

At Alma-Ata the work of establishing spectrophotometric standards is in progress. Karjagina and Haritonov (1964) have made absolute measurements of flux at different wavelengths for a number of stars and published results for 34 stars. Haritonov (1964, 1966*a*) has published absolute measurements for 17 field stars and for 14 Pleiades stars (1966*b*).

Haritonov and Neljubin (1966) have made absolute measurements of the fundamental star α Lyrae, calibrating with the integral flux from the centre of the solar disk.

Noskova (1965) has found that the energy distribution of the nuclei of the planetary nebulae NGC40, 3242, 6210, 6543, 6572, 7009, 7662, IC2149, 4593 and BD + 30°3639 is similar to that of a black body.

Glagolevsky (1966) has studied the energy distribution of magnetic and Ap stars and found that on the average the spectrophotometric gradients and the Balmer discontinuities differed from those of normal A stars.

At Alma-Ata, Kozlova and Glagolevsky (1966) have investigated the energy distributions in the continuous spectra of 40 Ap and Am stars within the range $\lambda\lambda$3200–5000 Å and concluded that these stars have the same anomaly in their continuum and that such an anomaly might be connected with the same anomaly in the structure of the atmosphere.

Energy distributions have been measured by Aller in the $\lambda\lambda$3300–5800 range for the following stars: 2 Coronae Borealis, ζ Virginis, HR4072, 15 Cancri, 53 Tauri, η Virginis, HD144206, 112 Herculis, HR8349, τ Capricorni, φ Herculis, ι Herculis, HR8348, π_1 Bootis.

G. Cayrel and A. M. Fringant (1964) have given the results of a spectrophotometric study of two metal deficient stars, HD 221 170 and HD 122 563, which show a difference in blue gradients that suggests a local difference of the absorption coefficient.

BIBLIOGRAPHY

Cayrel, G., Fringant, A. M. 1964, *C. r. Acad. Sci. Paris*, **258**, 3195 (*Publ. Obs. Hte.-Provence*, **7**, no. 34).
Glagolevsky, J. V. 1966, *Astr. Zu.*, **43**, 73.
Haritonov, A. V. 1964, *Izv. Akad. Nauk Kazakh. SSR. Ser. fiz. mat. Nauk Astrofiz.*, **3**, 28.
Haritonov, A. V. 1966*a*, *Trudy astrofiz. Inst. Akad. Nauk Kazakh. SSR*, **7**, 9.
Haritonov, A. V. 1966*b*, *Astr. Cirk*, no. 43, in press.
Haritonov, A. V., Neljubin, N. T. 1966, *Trudy astrofiz. Inst. Akad. Nauk Kazakh. SSR*, **8**, in press.
Karjagina, Z. V., Haritonov, A. V. 1964, *Izv. Akad. Nauk Kazakh. SSR. Ser. fiz. mat.*, **3**, 10.
Kharadze, E. K., Bartaya, R. A. 1965, *Astrofizika*, **1**, 405.
Kozlova, K. I., Glagolevsky, J. V. 1966, *Izv. Inst. Akad. Nauk Kazakh. SSR*, in press.
Mirzoyan, L. V., Kazarian, E. S. 1965, *Astrofizika*, **1**, 213.
Ņoskova, R. I. 1965, *Astr. Zu.*, **42**, 1038.
Žilevičiute, Z., Straižys, V. 1965, *Vilnius Univ. Bull.*, no. 15, p. 49.

APPENDIX 2: REPORT OF THE COMMITTEE ON LINE INTENSITY STANDARDS

(Prepared by G. Cayrel de Strobel, Chairman of the Committee)

During the XIIth IAU General Assembly (Hamburg 1964), K. O. Wright and G. Cayrel de Strobel prepared a list of 20 astronomers who could possibly collaborate in the working group on Line Intensity Standards of Commission 29. These astronomers have been chosen either because of their previous interest in this problem, or because they have high dispersion equipment at their disposal. To these astronomers, G. Cayrel de Strobel sent a circular letter on 31 March 1965 explaining the ever present difficulties of systematic discordances in equivalent widths obtained from different spectrographs.

G. Cayrel de Strobel suggested in her circular the observation of five standard stars with a dispersion equal to or better than 12 Å/mm on II-aO, II-aD, and II-aF plates. These stars are: γ Pegasi B2, 5 IV, 68 Tauri A2 V, 15 Vulpeculae ML, α Canis Minoris F5 IV and ϵ Virginis G9 III.

For calibration, a double calibration has been suggested: one with the local system and one with the Kienle step-filter, which could be sent from one observatory to another.

A list of reasonable unblended lines was enclosed in the circular in order to define not only standard stars, but also preferred standard lines to be used for comparison.

The following remarks are based upon the replies to the circular, made either in writing or in person. Five astronomers did not reply at all.

H. A. Abt (Kitt Peak Observatory) answered only recently that he agreed to collaborate but has not yet begun to observe any of the requested stars.

A. A. Asaad (Helwan Observatory, Cairo) has used the 74-inch coudé spectrograph of the Helwan Observatory to obtain spectra of the five programme stars.

W. Buscombe (Mt Stromlo Observatory) did not answer, but L. Searle promised to observe some of the programme stars with a dispersion of 2·8 and 6·7 Å/mm with the 74-inch coudé spectrograph of Mt Stromlo Observatory. By the way, L. Searle found that the line-intensity programme is too heavy. He believes that three stars would be enough, and he suggested: γ Pegasi B2·5 V, α Canis Minoris F5 IV, and ϵ Virginis G9 III. He thinks also that the number of standard lines for each emulsion should be reduced and called attention to the fact that the circular did not mention the number of plates for each emulsion. He suggested two plates.

R. Cayrel and G. Cayrel de Strobel (Haute-Provence Observatory) sent a graduate student, Miss C. Moreau, to the Haute-Provence Observatory to work on the programme. C. Moreau has observed three stars: 68 Tauri, 15 Vulpeculae and α Canis Minoris. The results obtained are contained in her 'diplôme d'études supérieures' (1966). This work will appear in the *Journal des Observateurs* and an abstract is given here:

68 *Tauri A2 V*

The relation between the equivalent widths of 68 Tauri obtained from spectra taken at Victoria (Wright *et al.*, 1964) and the equivalent widths of 68 Tauri from spectra taken at the Observatory of Haute-Provence (OHP) is given by the formula:

$$W\begin{pmatrix}\text{Victoria}\\ 3{\cdot}4\,\text{Å/mm}\end{pmatrix} = 0{\cdot}89\,W\begin{pmatrix}\text{OHP}\\ 9{\cdot}7\,\text{Å/mm}\end{pmatrix} + 4$$

The comparison between Lick equivalent widths of 68 Tauri taken from Conti *et al.* (1965) and OHP ones gives

$$W\begin{pmatrix}\text{Lick}\\ 10{\cdot}5\,\text{Å/mm}\end{pmatrix} = 1{\cdot}31\,W\begin{pmatrix}\text{OHP}\\ 9{\cdot}7\,\text{Å/mm}\end{pmatrix} - 45$$

15 *Vulpeculae M L*

$$W\begin{pmatrix}\text{Victoria}\\ 7{\cdot}5\,\text{Å/mm at H}\gamma\end{pmatrix} = 0{\cdot}89\,W\begin{pmatrix}\text{OHP}\\ 9{\cdot}7\,\text{Å/mm}\end{pmatrix} + 9$$

$$W\begin{pmatrix}\text{Victoria} & \text{Palomar}\\ 2{\cdot}8\text{ and }3{\cdot}4\,\text{Å/mm}, & 4{\cdot}5\,\text{Å/mm}\end{pmatrix} = 0{\cdot}90\,W\begin{pmatrix}\text{OHP}\\ 9{\cdot}7\,\text{Å/mm}\end{pmatrix} + 2$$

$$W\begin{pmatrix}\text{Mt Wilson}\\ 2{\cdot}8\text{ and }4{\cdot}5\,\text{Å/mm}\end{pmatrix} = 1{\cdot}04\,W\begin{pmatrix}\text{OHP}\\ 9{\cdot}7\,\text{Å/mm}\end{pmatrix} - 18$$

In the two first relations, the equivalent widths of Victoria and Palomar have been taken from K. O. Wright *et al.* (1964) the last one from G. R. Miczaika *et al.* (1956).

α *Canis Minoris F5 IV*

For α Canis Minoris, C. Moreau compared equivalent widths obtained from 3·2 Å/mm OHP spectra with those obtained from 9·7 Å/mm OHP spectra. She obtained:

$$W\begin{pmatrix}\text{OHP}\\ 3{\cdot}2\,\text{Å/mm}\end{pmatrix} = 0{\cdot}95\,W\begin{pmatrix}\text{OHP}\\ 9{\cdot}7\,\text{Å/mm}\end{pmatrix} + 5$$

Then she compared the following sets of equivalent widths

$$W\begin{pmatrix}\text{Victoria}\\ 7{\cdot}5\,\text{Å/mm at H}\gamma\end{pmatrix} = 1{\cdot}17\,W\begin{pmatrix}\text{OHP}\\ 9{\cdot}7\,\text{Å/mm}\end{pmatrix} + 6$$

$$W\begin{pmatrix}\text{from Hiltner and Williams Atlas}\\ 2{\cdot}8\,\text{Å/mm at H}\gamma\end{pmatrix} = 1{\cdot}04\,W\begin{pmatrix}\text{OHP}\\ 9{\cdot}7\,\text{Å/mm}\end{pmatrix} + 46$$

$$W\begin{pmatrix}\text{McDonald}\\ 2{\cdot}8\,\text{Å/mm at H}\gamma\end{pmatrix} = 1{\cdot}18\,W\begin{pmatrix}\text{OHP}\\ 9{\cdot}7\,\text{Å/mm}\end{pmatrix} + 17$$

$$W\begin{pmatrix}\text{Bergedorf}\\ 8\,\text{Å/mm at }H\gamma\end{pmatrix} = 0{\cdot}90\, W\begin{pmatrix}\text{OHP}\\ 9{\cdot}7\,\text{Å/mm}\end{pmatrix} + 46$$

$$W\begin{pmatrix}\text{Asiago}\\ 12\,\text{Å/mm at H}\gamma\end{pmatrix} = 1{\cdot}28\, W\begin{pmatrix}\text{OHP}\\ 9{\cdot}7\,\text{Å/mm}\end{pmatrix} + 22$$

The equivalent widths from Hiltner and Williams, McDonald, Bergedorf and Asiago have been taken from Hiltner and Williams (1946 *Photometric Atlas of Stellar Spectra*), Greenstein (1948), Wellman (1955) and Taffara (1953).

The spectrophotometric calibration has been checked by C. Moreau with the help of Kienle's step-filter. No significant difference was found with the current calibration obtained at the OHP. In the case of α Canis Minoris it is likely that the differences found in equivalent width measurements do not come from photographic effects (Edmonds, 1965; Moreau, 1966) but from the rather subjective way in which the continuum is located and the wings of the lines are drawn. In this connection the following remark can be made: C. Moreau measured a certain number of lines of α Canis Minoris on the Hiltner and Williams atlas. Greenstein independently did the same. The two sets of measurements compared with the OHP measurements are quite different. This result does not involve different instruments but only different personal judgements. By the way, Greenstein and Hiltner (1949) found that the equivalent widths obtained from the Hiltner and Williams atlas should be reduced by about 8%.

G. Cayrel de Strobel has compared equivalent widths from nine spectra of ϵ Virginis taken at Mt Wilson Observatory (Cayrel, G., Cayrel, R., 1963) with those from 11 spectra taken at the Observatory of Haute-Provence (Cayrel de Strobel, 1966). She found:

ϵ Virginis G9 III

$$W\begin{pmatrix}\text{Mt Wilson}\\ 1{\cdot}0\text{ and }2{\cdot}8\,\text{Å/mm}\end{pmatrix} = 0{\cdot}90\, W\begin{pmatrix}\text{OHP}\\ 9{\cdot}7\text{ and }12{\cdot}4\,\text{Å/mm}\end{pmatrix} + 8$$

C. and A. Cowley (McDonald Observatory) have agreed to observe some standard stars with the McDonald 82-inch coudé spectrograph. In the meantime, equivalent widths of α Canis Minoris measured chiefly on 4·8 Å/mm McDonald coudé spectra have been published by F. N. Edmonds (1965). The comparison between McDonald equivalent widths of α Canis Minoris spectra, and the means of equivalent widths found by K. O. Wright for α Canis Minoris from Victoria and Mt Wilson spectra, gives:

$$W\begin{pmatrix}\text{McDonald}\\ 4{\cdot}8\,\text{Å/mm}\end{pmatrix} = 0{\cdot}97\, W\begin{pmatrix}\text{Victoria} & \text{Mt Wilson}\\ 4{\cdot}6\text{ and }2{\cdot}2\,\text{Å/mm}, & 2{\cdot}9\,\text{Å/mm}\end{pmatrix} + 5$$

and the comparison between McDonald and OHP gives

$$W\begin{pmatrix}\text{McDonald}\\ 4{\cdot}8\,\text{Å/mm}\end{pmatrix} = 1{\cdot}06\, W\begin{pmatrix}\text{OHP}\\ 9{\cdot}7\text{ and }12{\cdot}4\,\text{Å/mm}\end{pmatrix} + 11$$

This last result relies unfortunately on a very scarce number of lines (seven lines).

V. Fujita (Tokyo Observatory) wants to collaborate and is eager to receive the step filter to check Tokyo's calibration system.

L. Gratton (Laboratorio de Astrofisica di Frascati-Roma) has not yet any high dispersion equipment at his disposal, but he comments on some of his previous works (Gratton 1953). The equivalent widths of the K2 III p Star α Bootis obtained from McDonald spectra do agree with those of Victoria, but do not agree with those obtained from the Hiltner and Williams atlas. Gratton does not believe in continuous calibration systems but prefers step filter. For high dispersion analysis on G and K stars, he does not trust dispersion less than 10 Å/mm.

J. L. Greenstein (Mt Palomar and Mt Wilson Observatories) transmitted the circular to P. S. Conti, who sent to G. Cayrel de Strobel a few interesting comments but did not say anything about a collaboration on this programme. Conti (1965) found that the equivalent widths of 45 Tauri dF4 from 10 Å/mm Mt Wilson spectra were systematically stronger by 15% than those from 10 Å/mm Lick spectra. Wallerstein and Conti (1964) found also that the equivalent widths of the K0 III giant γ Tauri from 15 Å/mm Mt Wilson spectra were 5% stronger than the Mt Palomar ones (dispersion 13·5 Å/mm) and 25% stronger than the Lick ones (dispersion 16 Å/mm). Helfer and Wallerstein (1964) have compared equivalent widths from 6·0 and 6·7 Å/mm spectra of ϵ Virginis with those obtained by the Cayrels (1963) on 2·8 Å/mm spectra, and they did not find any significant differences.

G. Herbig (Lick Observatory) agreed to collaborate, but on a less heavy programme. He would like to observe one or two stars out of five and to measure a few lines out of the proposed one. Until now only 68 Tauri among the five standard stars has been observed (Conti *et al.*, 1965), and some of its equivalent widths have been measured on Lick 10·5 Å/mm spectra. The comparison between these equivalent widths and those of the Observatory of Haute-Provence has already been quoted.

J. Houtgast (Utrecht Observatory) is always interested in the problem of line intensity standards, but thinks that at Utrecht, he cannot help with it at the present.

B. E. J. Pagel (Herstmonceux Observatory) will collaborate on this programme, but only with blue plates, and will take 10 Å/mm spectra of the standard stars. The dispersion in the red and yellow region is below the requested limit. In his letter Pagel also suggests that spectra be taken of integrated sunlight from sky, Moon or minor planets with the equipment of each of the collaborating observatories. The different set of equivalent widths should then be compared with the excellent photoelectric measurements of MacMath (see for example Müller and Mutschlecner, 1964) which can be converted into values for integrated sunlight by a very simple numerical quadrature. This procedure will tell what is really wrong with everybody's equivalent widths, rather than merely how much they disagree with each other.

J. Sahade (La Plata Observatory) has not yet any high dispersion equipment at his disposal in Argentina but is willing to collaborate securing programme stars spectra whenever he would have observing runs at Mt Wilson. This would be indeed very welcome.

A. D. Thackeray (Radcliffe Observatory) will try to collaborate and take some 6·8 Å/mm spectra in the blue region of a few programme stars, but he does not have facilities for measuring the lines equivalent widths from the plates. G. Cayrel will try to find a student to do this work in Paris.

S. Taffara (Asiago Observatory) found a young astronomer, P. L. Bernacca, who is willing to observe the programme stars and measure the equivalent widths of the requested lines. The calibration system of Asiago has already been checked with Kienle's step filter a few years ago.

K. O. Wright (Victoria Observatory) has done the most accurate and extended work on line intensity standards which has ever been done (Wright *et al.*, 1964). Equivalent widths measured on Victoria, Mt Wilson and Mt Palomar spectra, of twelve main sequence stars B0 to G5 have been compared. The spectral range has been 3900 to 4500 Å, and the number of lines measured passed 500 in the F type stars. The results are already contained in Wright *et al.* (1964) and in the Transactions of the 12th IAU General Assembly. In spite of his already impressive contribution, Wright will collaborate on this programme and get some spectra of the requested stars with a better resolution than the resolution of the spectra he used in his former study. Wright found also that the combination of a II-aD plate from 4800 to 6200 Å with a II-aE plate from 6200 to 6600 Å gives approximately the right density all along the plates whereas the II-aF is a little too weak.

Regarding Pagel's suggestion of using the solar spectrum as a standard, Wright suggests the use of planetary or asteroid spectra. He fears that the spectrum of the daytime sky light may be affected by scattered light.

Some Conclusions on Line Intensity Standards

There is a statement one can make: until now we have not been able to give any easy correlations between the values of equivalent widths obtained with one instrument and the values obtained with another.

K. O. Wright gave us (Taffara, 1962) the correlations between Victoria, Mt Wilson and Mt Palomar, but every time we reduce new material we have to check: (1) if, between the equivalent widths coming from different spectra taken at the same instrument, there are not big systematic differences; (2) if the mean of equivalent widths of the new material has the same value as the mean formerly found for this instrument.

From the paper by Wright *et al.* (1964) we see that the equivalent widths obtained from photoelectric scans are in good agreement with those obtained by photographic means. They tend to be slightly larger than those obtained by Wright and Greenstein at Mt Wilson, but almost coincide with the photographic means of equivalent widths measured on high dispersion spectra.

In the near future it will be necessary to pass slowly to the technique of observing one or two standard stars and of comparing the equivalent widths obtained by photographic means with those obtained from spectrophotoelectric scans. Therefore we suggest that we ask the observatories which already have spectrophotoelectric equipment to scan the standard stars we shall propose. Once we obtain equivalent widths from photoelectric scans of different instruments, we have to see if they agree with each other. This set of equivalent widths will then be compared with the photographic measurements of equivalent widths obtained from high dispersion spectra with different instruments.

The programme has been regarded as too heavy by almost every co-worker. Therefore, it would be perhaps more practical to reduce the programme to one star and the solar spectrum for which excellent photoelectric measurements do already exist.

BIBLIOGRAPHY

Cayrel de Strobel, G. 1966, *Ann. Astrophys.*, **29**, 413.
Cayrel, G., Cayrel, R. 1963, *Astrophys. J.*, **137**, 431.
Conti, P. S. 1965, *Astrophys. J. Suppl.*, **11**, 47.
Conti, P. S., Wallerstein, G., Wing, R. F. 1965, *Astrophys. J.*, **142**, 999.
Edmonds, F. N. 1965, *Astrophys. J.*, **142**, 278.
Gratton, L. 1953, *Me'm. Soc. R. Liège*, **14**, 419.
Greenstein, J. L. 1948, *Astrophys. J.*, **107**, 151.
Greenstein, J. L., Hiltner, W. A. 1949, *Astrophys. J.*, **109**, 265.
Helfer, H. L., Wallerstein, G. 1964, *Astrophys. J. Suppl.*, **9**, 81.
Miczaika, G. R., Franklin, F. A., Deutsch, A. J., Greenstein, J. L. 1956, *Astrophys. J.*, **124**, 134.
Moreau, C. 1966, Diplôme d'Études Supérieures.
Müller, E. A., Mutschlecner, J. P. 1964, *Astrophys. J. Suppl.*, **9**, 1.
Taffara, S. 1953, *Contr. Oss. astrofis. Univ. Padova*, no. 32.
Taffara, S. 1962, *Contr. Oss. astrofis. Univ. Padova*, no. 126.
Wallerstein, G., Conti, P. 1964, *Astrophys. J.*, **140**, 858.
Wellman, P. 1955, *Z. Astrophys.*, **36**, 194.
Wright, K. O., Lee, E. K., Jacobson, I. V., Greenstein, J. L. 1964, *Publ. Dom. Astrophys. Obs., Victoria*, **12**, 173.

30. COMMISSION DES VITESSES RADIALES

PRÉSIDENT: Dr R. M. Petrie†, Dominion Astrophysicist, Dominion Astrophysical Observatory, Victoria, B.C., Canada.

VICE-PRÉSIDENT, PRÉSIDENT PAR INTERIM: Dr D. S. Evans, Royal Observatory, Observatory, Cape Province, South Africa.

SECRÉTAIRE: Dr W. Buscombe, Mount Stromlo Observatory, Canberra, A.C.T., Australia.

COMITÉ D'ORGANISATION: F. K. Edmondson, Ch. Fehrenbach, J. F. Heard, A. D. Thackeray.

MEMBRES: Abt, Batten, Bouigue, Boulon, Duflot, Gollnow, Gratton, Kraft, Northcott, Pagel, Preston, Sahade, Underhill, Wayman.

The Commission suffered a severe loss by the death on 8 April, 1966 of its President, Robert Methven Petrie, a few weeks before his 60th birthday. Well known to astronomers throughout the world, Petrie was, at the time of his death, Dominion Astronomer of Canada and Director of the Dominion Astrophysical Observatory at Victoria, British Columbia. He was a Vice-President of the IAU from 1958 until 1964. He served as a member of Commission 30 from 1950 onwards, being its Vice-President from 1961 to 1964, when he became President. From 1950 to 1955 he was President of the Sub-Commission on Standard Wavelengths.

His astrophysical contributions were numerous and diverse. Within the field of our Commission his influence was of capital importance. His work on standard wavelengths and on methods of measurement of spectra has been of fundamental importance to our knowledge of the velocities of early-type stars. His analyses of the motions of B-stars, and of the equivalent widths of the Balmer lines in their spectra have been corner stones in the edifice of modern ideas concerning galactic structure.

IAU SYMPOSIUM NO. 30

At the time of his death, Petrie was engaged in planning IAU Symposium No. 30, to be held in Toronto from 20 June to 24 June 1966, on the topic, 'Determination of Radial Velocities and their Applications'. His sudden death brought home to the surviving members of the organizing committee, the full realization of their dependence on him. Even so, the ground work which he had laid had been carried out with such foresight and care, that the Symposium took place successfully, very much as he had planned it. The meeting, which was thus transformed into a kind of memorial to our late President, could not have taken place but for the devoted efforts of the organizing committee, especially the Canadian members, and, more particularly, Professors Heard and MacRae of Toronto.

The Symposium took place on the Campus of the University of Toronto, by courtesy of the University authorities, under the presidency of the acting President of the Commission. About 60 astronomers attended.

Since the printed report (1), dedicated to R. M. Petrie, is likely to be published before the Prague General Assembly, no detailed account is needed here. However, because certain contributions were of outstanding interest, while others are not mentioned elsewhere in this report, a partial list of topics discussed is included.

Among new techniques presented, especially notable was the photo-electric method of Griffin. Mrs Rubin and her co-authors represented the growing body of observers using image tubes, particularly for very faint objects. De Vaucouleurs demonstrated velocity measures of galaxies by a scanning technique and by a photoelectric method.

Oscilloscope plate-measuring device were described by Gollnow and by Andrews and their inherent errors discussed by Scarfe.

A paper by the late R. M. Petrie and J. M. Fletcher demonstrated techniques for attaining extremely high accuracy with a coudé spectrograph.

Weaver described his definitive reduction of Trumpler's velocities of stars in clusters. This is likely to become a standard reference for classical photographic methods of radial velocity measurement.

Abt reported the near completion of a catalogue project including a plate-by-plate measurement catalogue of the Mount Wilson material.

The Marseilles-Haute-Provence group, led by Fehrenbach, gave a general conspectus of their objective prism work.

Bertiau described a very comprehensive Fortran programme for the determination of elements of spectroscopic binary orbits.

Dommanget reported the appearance of his catalogue of visual binaries worthy of spectroscopic observation.

There was a considerable interest in the evolutionary significance of the properties of binary stars and several speakers discussed their statistics (e.g. Batten, Blaauw, Bouigue).

In a subsequent brief report to the Executive Committee regret was expressed that there had been no opportunity to welcome to the Symposium colleagues from the countries of eastern Europe and Asia, and it was hoped that in future astronomers from these areas would join in the activities of our Commission. It is therefore very gratifying to receive a report from the U.S.S.R. which is included below.

REPORTS FROM OBSERVATORIES

The next part of the report summarizes, as has become customary, the written reports received from various observatories.

Allegheny Observatory

Beardsley reports: The radial velocity programme continues essentially without change since being reinstituted in 1958. It consists of all stars listed as SB, without orbit in the Wilson Radial Velocity Catalogue, brighter than magnitude 7·0, earlier than G0, and north of $-20°$ declination. A few additional stars fainter than 7·0 or later than G0 have been placed on the programme from time to time. The programme is supplemented by those stars north of $-20°$ listed in the new Bright Star Catalogue as having no radial velocity. A number of Be stars and a group of stars designated as 'Astrometric-Spectroscopic Binaries' are being observed as well.

Instrumentation consists of the Keeler 31-inch (79 cm) reflector and Mellon one-prism spectrograph giving a dispersion of 40 Å/mm at $H\gamma$.

Standard velocity stars are observed at every session in an effort to tie the programme directly to the standard velocity system. Measures are made on a null-projection machine similar to that used at Victoria. A Grant machine will be installed in 1966. With National Science Foundation support the output of radial velocities from Allegheny should be much increased in 1967. To date over 2800 plates have been taken. In the next few years an extensive analysis of the quality of these radial velocities will be undertaken.

It is expected that nearly 3000 radial velocities from the period 1905–17 will be published in 1967.

Cambridge Observatories

Griffin reports that he has developed the Cambridge technique of narrow-band stellar photometry with a slit spectrometer to measure radial velocities photoelectrically (2). A diaphragm containing 234 apertures whose positions correspond to those of strong lines in the blue region ($\lambda\lambda$4370–4830Å) in the spectrum of Arcturus, is matched against the spectra of other late type stars and its position then determines the relative Doppler shift. Using the Cambridge 92 cm-reflector, stars of $m_v = 9$ are measured to about 2 km s^{-1} standard error at the rate of 6 to 8 stars per hour. All stars later than about G5 can be measured using the same diaphragm. Observations of fifth magnitude stars of which the velocities have been determined photographically confirm the freedom of the method from detectable systematic errors, and show a standard deviation of 1 km s^{-1} per observation at this magnitude. Some hundreds of velocities, mainly in the $+15°$ selected areas, have been measured with the new equipment. A series of observations of the spectroscopic binary, 73 Leo, has also been made (**3**).

Mount Stromlo Observatory, Canberra

Observations of B-stars in four selected fields along the Southern Milky Way (at longitudes 17°, 245°, 312° and 328°) which had been started by B. J. Bok, Gollnow and Mowat, are now almost completed. These were made with the grating spectrograph at the Newtonian focus of the 74-inch (188 cm) reflector. Preliminary results for the stars at 245° and 328° (Puppis and Norma) have been given by Mowat at the Mount Stromlo Symposium (**4**) and at IAU Symposium No. 30 (**1**). In both cases the distances derived for these groups of stars from their radial velocities are in very good agreement with those derived from Hβ photometry. The group in Norma appears to belong to the nearer Sagittarius arm rather than to the more distant Norma-Scutum arm.

The observations of selected nebulosities and stars in the Large and Small Magellanic Clouds made with the same instrument (Bok, Gollnow and Mowat) are also practically completed, and some of the results for the Small Cloud have been published in cooperation with Hindman of the Radiophysics Laboratory, Sydney, who made radio measurements in H I (**5**).

The same equipment has been used by Westerlund, Stokes and Lewis to investigate radial velocities of southern galaxies, and of some supernova remnants seen as emission nebulae in the Milky Way and Magellanic Clouds.

The 32-inch (81 cm) camera of the Mount Stromlo coudé spectrograph has been used by Dunham and by Buscombe for studies of velocities of interstellar clouds from measures of the Ca II and Na I absorption lines. Bessell has studied the velocity variations of several dwarf cepheids, mainly with the 10-inch (25 cm) R.E.O.S.C., camera.

The 2-prism Zeiss spectrograph, transferred to the Cassegrain focus of the 50-inch (127 cm) reflector, has been used by Buscombe to complete a study of the cluster, Messier 7, and to obtain velocity measures of B8 stars (in cooperation with Heard) and supergiants (many of which had been recognized photometrically by Walraven). Webster has completed a study of planetary nebulae in the Galaxy and Magellanic Clouds, and Roslund has made observations of B stars in the III Sgr association.

A new grating spectrograph, designed by Meinel, has been brought into use on the 40-inch (102 cm) reflector at Siding Spring for studies of long-period variables by Buscombe, and of galactic B-stars by Lyngå, and of members of the III Pup association by Westerlund.

A more elaborate model of the impersonal setting device with greatly improved electronics has been put into use (Gollnow, Rudge and Thomas, (**1**)).

References (**6**) to (**16**) cover recent publications embodying radial velocity measures which have not previously been noted in these reports. They include velocities for about 550 field stars.

Royal Observatory, Cape of Good Hope

Two more papers in the series of Fundamental Data for Southern Stars, (Nos. 5 and 6) (**17**), (**18**), have been published. Material for a further list is available on file. A considerable number of new sub-dwarfs is included in these lists. Attention has been paid to the rapid handling of data for spectroscopic binaries. Orbital elements for nine stars of which HD 2070, 15064, 123515, 155099, 159656, 183007, 194184, 200334 are single-lined and HD 188164 double-lined, are in press (**19**). During his visit to the University of Texas, Evans collaborated with Andrew T. Young in devising an analogue method for rapid determination of the periods of recurrent phenomena such as velocity variation in spectroscopic binaries (**20**). In addition, computer programmes were written by E. Barker and D. Wells, on the one hand for the iterative determination of orbital elements, on the other, as a development of Lafler and Kinman's method, (**21**) for the determination of periodic times.

The Cape-Radcliffe programmes have been occupied recently to a large extent by further work on a selected list of RR Lyrae stars, for which both photometric and spectroscopic observations are planned. For the list of about 60 stars some 260 spectra have so far been obtained. In recent months D. H. P. Jones, Clube and Menzies have been concerned in this work.

A short list of extragalactic velocities based on spectra obtained with the new Newtonian spectrograph at the Radcliffe Observatory was published by Evans and Malin (**22**). Some further spectra have recently been obtained by D. H. P. Jones.

A note on the cepheid HR 5527 (misprinted as 5227), previously thought to be a double star, HD 130701/2, has been published by Evans (**23**).

Owing to the interruption consequent on his visit to the U.S.A., Evans has not yet completed his new analysis of p Velorum but this should be finished shortly (**1**).

For the same reason the revision of the Mount Wilson Radial Velocity Catalogue has lagged, though Miss Y. Z. R. Thomas has continued work on this. The card catalogue contains some 9500 cards, each referring to a star, about equally divided between stars which have been previously observed, and new stars. It is hoped to report definitive progress on this project at the Prague General Assembly (**1**).

D. H. P. Jones (**24**) has extended the error investigation of Wayman (**25**) to the two Herstmonceux spectrographs and finds that the additional error element to be applied in passing from formal probable error to real standard error has the same value as for the Pretoria cameras. Expressed in terms of microns on the plate it has the same value for all the cameras of all three spectrographs.

D. H. P. Jones and H. C. Lagerweij (**26**) have made a spectroscopic and photometric study of the δ Scuti variable, 1 Monocerotis (HD 40535).

David Dunlap Observatory, Toronto

Since 1964 the Hilger Prism spectrograph at the Cassegrain focus of the 74-inch (188 cm) telescope has been replaced by an all-reflection grating spectrograph which has dispersions of 10 Å/mm and 40 Å/mm. Crampton and others (**1**) have found that the 40 Å/mm grating system is faster than the 66 Å/mm prism system and comparable in accuracy for the determination of radial velocities with the 33 Å/mm prism system.

A Grant oscilloscopic comparator is now in use for measuring spectrograms, and programmes have been devised for the reduction of measurements by electronic computer.

Heard (**27**) has published the radial velocities of 55 Kapteyn Area fundamental stars in high galactic latitude.

Heard and Petrie (**1**) have investigated the radial velocities and other criteria of cluster membership for 77 stars in the field of the α Persei cluster. Spectroscopic binaries were found to be rare in this cluster.

Crampton has completed the observations for a programme of radial velocities of 60 Be stars, the purpose of the investigation being a kinematic study with the help of distances derived from Schmidt-Kaler's absolute magnitude calibration.

Hube has nearly completed a radial velocity programme of 250 B8–B9 stars of the Bright Star Catalogue north of declination $-26°$ and as yet unobserved for radial velocity, and he proposes to extend this study to more southerly B8–B9 stars at the Radcliffe Observatory. He will attempt to correlate his velocity data with distances, age grouping, and rotational velocities.

Dunsink Observatory

Radial velocities of 315 southern stars south of declination $-20°$ of photographic magnitude about 8·5, evenly distributed over the sky, with spectral types in the Henry Draper Catalogue from F5 to Mb have been determined on the system of the General Catalogue of Radial Velocities. The spectra were obtained in 1958–60 at the Radcliffe Observatory, Pretoria as a programme undertaken on behalf of the Royal Greenwich Observatory. Photoelectric measures of U_c, B and V for these stars have been made at the Royal Cape Observatory. A list of the data, including MK spectral types derived visually has been communicated to Evans for inclusion in the revision of the General Radial Velocity Catalogue (**28**). Densitometer tracings of the spectra are now being made at Dunsink for spectral classification purposes.

Royal Greenwich Observatory, Herstmonceux

Woolley and Harding (**29**) have published radial velocities of 96 stars derived from 241 spectra obtained with the coudé spectrographs at Mount Wilson and Palomar, and Radcliffe Observatories. Twelve double stars and nine stars in the Hyades are included. The possibility of a systematic difference between the measured velocities of giants and dwarfs in the same cluster is considered.

Harding has published radial velocities of bright members of the globular cluster ω Centauri (**30**). Woolley and Aly (**31**) have published radial velocities of 18 RR Lyrae stars, for which, in most cases, the velocity curves are obtained.

At Herstmonceux the grating spectrograph with dispersions of 40 and 80Å/mm at the Cassegrain focus of the Yapp 36-inch reflector is being used for the determination of radial velocities of a list of spectroscopic binaries brighter than 7·5 m_{pg}, for which no orbital elements have previously been determined. Radial velocities of 633 stars of HD spectral types B9-A5, brighter than $m_v = 6{\cdot}3$ and north of the equator, have been determined, using the 1 prism spectrograph. Estimates of spectral type and of $|v \sin i|$ have also been made, and the results are in press (D. H. P. Jones, Palmer, Walker and Wallis (**32**)).

A coudé spectrograph with a dispersion of 10Å/mm in the second order in the blue, attached to the 30-inch (76 cm) reflector, has been brought into use. A programme of measurements of radial velocities of IAU standard stars has been undertaken to test the stability of the instrument, and the results, accompanied by a description of the instrument, are in press (Harding, Palmer and Pope (**33**)).

In 1965, D. H. P. Jones observed five stars intensively at 80Å/mm with the Yapp reflector, in a search for short-period velocity variation. Only one star, HD 107904, was found certainly to show variations. It has since been confirmed as a δ Scuti variable by Wachmann (**34**).

By kind permission of the Egyptian Ministry of Scientific Researches, a number of radial velocity programmes is being carried out with the 74-inch (188 cm) reflector at Kottamia, U.A.R., in collaboration with the staff at that Observatory. These include the report on RR Lyrae stars (Woolley and Aly) mentioned above. In addition the following programmes continue:

(*a*) Observations of semi-regular variables brighter than 11·5 m_{pg} not listed in Wilson's Catalogue.

(*b*) Spectroscopic binaries with unknown orbital elements between 7·5 and 11·5 m_{pg}.

(*c*) A0 stars in the north galactic cap ($\alpha = 11^h\ 00^m$ to $14^h\ 30^m$, $\delta = +10°$ to $+50°$, 1900·0), which are in the HD but not in Wilson's Catalogue. Preliminary results from this programme were presented at IAU Symposium No. 30 at Toronto (Woolley, Aly and Asaad (**1**)).

(*d*) K0 stars taken from the HD catalogue which have no known radial velocity, in the areas

$\alpha =$ $3^h\ 20^m$ to $5^h\ 20^m$	:	$\delta = +10°$ to $+20°$ (1900·0)
$6^h\ 00^m$ to $8^h\ 00^m$	:	$+33°$ to $+43°$
$18^h\ 10^m$ to $20^h\ 10^m$	:	$+55°$ to $+65°$
$21^h\ 15^m$ to $00^h\ 15^m$	:	$+29°$ to $+39°$

(*e*) Members of Gliese's catalogue of stars within 20 parsecs, brighter than 11·5 m_{pg} and north of $-30°$, with no known radial velocity.

In August–September 1966, Woolley and Palmer undertook a programme of observations of W Virginis variables, using the 100-inch (254 cm) Newtonian spectrograph at the Mount Wilson and Palomar Observatories.

A line-scanning measuring machine developed in the Observatory workshops has been brought into use.

Marseilles and Haute-Provence Observatories

The delegation attending the Toronto Symposium, headed by Fehrenbach gave a general summary of the work undertaken with the various objective prism telescopes since its inception (**1**). The publication of some of the work has had to wait for the establishment of satisfactory reference star velocities in the areas. Among galactic fields studied, those numbered 5–15 and I–M, together with the special fields designated δ Tau, P Cygni, *h* and χ Persei, and Praesepe have been published. All these, with the exception of the last named, have been observed with the small objective prism (PPO) at Haute-Provence. Results for twenty Selected Areas, observed with the small objective prism, and, in three cases, reobserved with the large objective prism, have been published. These results are included in the Draft Report of Commission 33 (Committee on Selected Areas), where the reference numbers of Selected Areas under study are given. Here also are listed the names of the Parenago areas now being studied with the large objective prism (GPO 3), and that one, ($5^h\ 56^m.0$, $+27°\ 34'$), already published, for which the plates were taken with the small objective prism. References additional to those previously published are Rebeirot (**35**), Fehrenbach *et al.* (**36**).

Mme Rebeirot and Imbert have published a trial list of velocities (33 stars and two standards) obtained with the short focus ($F/1·1$) camera of the Haute-Provence coudé spectrograph (**37**).

La Plata Observatory

Radial velocities have been determined for the following stars:

HD 698 (Spectroscopic binary: Sahade (**38**)); V 453 Scorpii (eclipsing variable: Sahade and Frieboes-Conde (**39**)); 17 Leporis (peculiar star, spectroscopic binary, late-type component: Ringuelet-Kaswalder, Sahade and Wallerstein (**40**)).

Stars in course of measurement are the following:

δ Librae: All the material from Allegheny, Michigan and Mount Wilson is being re-measured by Sahade, Hernández, Faÿ and Cohen.

WY Velorum, symbiotic star, is being studied by Sahade and Hernández, who are also studying the red region of β Lyrae.

The eclipsing variable ϵ Coronae Austrinae is being studied by Sahade, Hernández, Sisteró and Ringuelet-Kaswalder.

PGC 4444 (=HD 159 176), a sixth magnitude spectroscopic binary Oe 5 star, for which an orbit was given by Trumpler, is being reinvestigated by Hernández.

Sahade presented a paper to the Toronto Symposium under the title 'A systematic effect in the radial velocities determined with an oscilloscope-type comparator' (**1**).

Radcliffe Observatory, Pretoria

(1) *B Stars:* A fourth list of radial velocities of southern B stars has been published (**41**). This comprises 73 stars, (including 19 faint companions of visual binaries) selected as members of the Scorpio-Centaurus association. As reported to IAU Symposium No. 30 by Thackeray (**1**), these velocities strongly confirm the concept of group motion, but the radial velocities alone do not determine a convergent with sufficient precision to derive a K-term. The existence or otherwise of a K-term must be decided by determining a convergent from more accurate proper motions.

Feast and Shuttleworth have published extensive analyses of radial velocities of OB stars, Cepheids and clusters (**42, 43**). Velocity dispersions of the various objects, including interstellar gas are compared. A statistical correction for distances of B stars, (which becomes serious beyond 3 kpc), is determined. The ω (R) curve agrees well with the 21 cm curve after this correction is applied. A value of $A = 14{\cdot}3 \pm 0{\cdot}8$ km s^{-1} kpc^{-1} is derived, and $R_0 = 9{\cdot}5 \pm 1{\cdot}0$ kpc. Feast and Thackeray (**44**), point out that the ω (R) curve derived by Münch and Münch can be seriously affected by such statistical corrections to the distances of the stars used.

Thackeray (**45, 46**) has observed radial velocities of eight 9th-magnitude stars for comparison with objective prism velocities by the Fehrenbach group.

A programme is being initiated to observe radial velocities of the remaining 150 early-type southern stars brighter than 7·5 m with unknown radial velocities.

Graham, now at Kitt Peak National Observatory, has discussed nine distant B-type stars in Ara for which radial velocity material was obtained at Pretoria and concludes that the ratio of total to selective absorption is smaller than previously believed (**47**).

(2) *Me variables:* Feast (**48**) has completed an important programme on 51 Me and 2 Se variables towards the galactic centre. The large velocity dispersion and steep velocity gradient with longitude across the galactic centre is closely similar to that exhibited by the planetary nebulae. The velocity dispersion is related to period in the same way as for Me variables in the solar neighbourhood.

Feast (**49, 50**), has disproved the suspected presence of any Me variable in a globular cluster with period appreciably greater than 200 days. This strengthens the belief that there is no essential difference between the Me variables in globular clusters and in the general field. CH Sct (period: 190·6 days), has been shown to belong to the globular cluster NGC6712 despite some opinions to the contrary (**51**).

Feast is currently observing velocities of semi-regular and S-type variables.

(3) *Cepheids:* Feast (**52**) has completed observations of distant (2·5 to 4·5 kpc) cepheids near $l^{II} = 290°$. Velocity curves are derived for 10 cepheids, while observations near maximum light have also been secured for 11 other faint cepheids. The results, compared with velocities of OB stars and H I gas in the same region, confirm the concept of large-scale regional motions which produce humps in the $\omega(R)$ curve. A value of $R_0 = 9{\cdot}8 \pm 1{\cdot}4$ kpc is derived.

T. Lloyd Evans is reobserving some southern cepheids for radial velocity, especially V 636 Sco, whose γ-velocity seems to have changed.

M. Breger has observed the cepheids U Sgr and S Nor at 29 Å/mm throughout their cycles. He also observed the βC Ma variable, α Lup, as well as the variable velocity star, β Cen, which is also suspected to be of this or a closely related class (**53**).

(4) *Clusters:* Thackeray and Wesselink (**54**) have published radial velocities of stars in the so-called cluster IC2944. No clear-cut variation of velocity with magnitude is found. The uncorrected velocity dispersion in this young cluster is 6·4 km s^{-1}. Evidence of expanding gas around the cluster is found.

T. Lloyd Evans is observing radial velocities in NGC6383 and IC2581.

(5) *Binaries:* Orbits of six double-lined spectroscopic binaries of early type have been published: HD140008, 178322 (**55**); δ Pic (**56**); HD147971, 75759 (**57**); HD175544 (**58**). Two of these were Radcliffe discoveries, and two more recent discoveries (HD104631, 161756), are being investigated. Other binaries under investigation include HD101131, 101205, (in IC2944), 93403, 136504, 142096, 155775, and HD161783 (**59, 60**).

Sher observed a number of suspected binaries of relatively small amplitude belonging to the Sco-Cen association.

Feast has studied the remarkable supergiant eclipsing system BL Tel (**61**), a high-velocity star out of the galactic plane.

Thackeray (**62**) observed S Equ during eclipse and found satellite lines during partial phases. Spectroscopic observations of this star have been continued for Dr M. Plavec.

(6) *Miscellaneous:* Thackeray (**63**) has shown that the [Fe II] lines observed during a remarkably deep minimum of S Dor fix the zero-point for radial velocity in the P Cyg contours shown by permitted lines. This radial velocity agrees closely with the mean velocity for the relevant part of the LMC.

Feast has obtained spectra of 10 planetary nebulae in the Large Cloud and obtained evidence of rotation and an indication of velocity dispersion from them (**64**).

Friedjung (**65**) has measured widths of emission lines in RR Tel and interpreted them in terms of his model for continuous ejection of gas from novae.

Slit spectra of stars discovered by the Fehrenbach group as having large radial velocity in the neighbourhood of the Large Magellanic Cloud have proved some of these stars to belong to the galactic foreground, e.g. two sub-dwarfs (**66**).

Daylight spectra of Comet Ikeya-Seki near perihelion showed large velocity shifts, corresponding to the motion towards the Sun for reflection from the nucleus, and to the motion of the comet relative to the Earth for cometary emission of Fe I, Ca II etc. (**67**).

In a recent note Thackeray reports a radial velocity for Proxima Centauri based on emission line measures. The mean value $-15·7 \pm 3·3$ km s^{-1}, is 7 km s^{-1} different from that of α Cen but this is not necessarily significant (**68**).

Thackeray has remeasured some B-type spectra with oscilloscope projection machines and finds that, at least for his measures of diffuse lines, two-way measures are essential to avoid systematic errors. This bears out the experience of Abt (**69**).

Observatoire de Toulouse

La recherche d'étalons pour la mesure des vitesses radiales des étoiles O, B, A signalée dans le précédent rapport a été poursuivie avec le spectrographe à fente, C, de l'Observatoire de Haute-Provence. Le nombre considérable de spectres à mesurer n'a pas encore permis, comme nous l'espérions, de publier des résultats. La méthode consiste en effet à comparer 2 à 2 les 14 spectres pris en quelques minutes sur une étoile brillante de type O, B, ou A. Dans le cas où la vitesse radiale relative est nulle, la mesure absolue de cette vitesse est alors faite sur les 14 spectres.

Des variations inopinées de vitesse radiale ayant été mises en évidence sur plusieurs étoiles, il convient de contrôler ces variations à partir de nouveaux spectres.

Ce n'est que lorsque ce travail d'ensemble sera terminé que nous pourrons en faire connaître les conclusions.

En ce qui concerne les binaires spectroscopiques la mise à jour du catalogue complémentaire a été continuée. Comme les références correspondantes n'ont jamais été publiées dans les 'Draft Reports', une liste complète des contributions de Bouigue, Bouigue et Chapuis, et Pedoussaut (1952–64), est donnée, qui pourra rendre service aux utilisateurs (**70–77**).

U.S.S.R. observatories

Professor E. K. Kharadze reports as follows:

R. Kh. Salman-Zade of the astronomical observatory of Leningrad State University has proposed an improved sheme, suitable for large batches of material, for the calculation of corrections for the diurnal and orbital motions of the observer in the reduction of radial velocities to the Sun (**78**).

R. E. Gershberg and P. V. Shcheglov of the Crimean Astrophysical Observatory, developing a method used by Courtès, have used a Fabry-Pérot étalon, combined with an interference filter centred at Hα, and a contact image tube. They have determined radial velocities and internal motions in six gaseous objects, NGC1976, 6523, 6618, 7000, 7822 and IC1318a. The plates were standardized by means of images from a hydrogen discharge tube (**79**).

L. V. Mirzoyan and E. S. Kazarian have analysed radial velocity data for the O-B stars, long period cepheids and interstellar gas, and have found a negative K-term attributed to non-stationary motions in the Galaxy (**80**).

Sonnenborgh Observatory, Utrecht

Professor Anne B. Underhill reports a study of the O9 V star, 10 Lacertae, using a series of 30Å/mm spectrograms taken in rapid succession. An abstract by Underhill and van Helden has been published (**81**) and a more extended account will appear later (**82**). The spectra, taken at Victoria in November and December 1964 seem to show a rapid and erratic variation of velocity over a range of 10 km s^{-1} although the individual probable error of a single spectrogram is only 2 km s^{-1}. This is accompanied by rapid changes in the strength and shape of weak lines such as those of O II and Fe III. The radial velocities are dependent on measures of the strong lines, chiefly those of H and He I, and changes in them are not so easily observed because of saturation. From the astrophysical standpoint it seems important to repeat these observations with more powerful spectrographic equipment in order to discover what variations of velocity do occur within intervals of less than 10 minutes in the atmosphere of this standard, rather sharp-lined O9 V star.

De Groot, van Helden and Underhill hope to pursue further the radial velocity observation of stars in the direction of IC805 reported in preliminary form at Toronto (**1**).

Dominion Astrophysical Observatory, Victoria B.C.

General: Radial-velocity observations have continued of clusters, spectroscopic binaries and visual binaries. Results on clusters are being reported to Commission 37. Orbital elements have been determined for ξ^1 Ursae Majoris, υ Ophiuchi (**83, 84**); 47 Andromeda (**85**), and HD175544 (**58**). Radial velocities of visual binaries have not yet been published. At the suggestion of Dr Thackeray, a co-operative Victoria-Pretoria programme has been undertaken by Batten and Thackeray to determine radial velocities of 11 bright B8-B9 stars. The aim is to test the apparent correlation between K-term and quality of velocity-determination discussed by Boulon at the Hamburg General Assembly. Observations are complete, but measures have not yet been made. Derivation of galactic-rotation constants from the results of the B star radial-velocity programme was completed by R. M. Petrie before his death, and will be published.

Wavelengths: Selection of wavelengths suitable for velocity measures of G and K spectra was completed by R. M. Petrie shortly before his death, for the highest dispersion (2·1 Å/mm), available at the coudé focus of the 48-inch (122 cm) telescope. It is planned to extend this work to earlier spectral types as soon as possible. J. M. Fletcher is studying the selection of wavelengths, suitable for all spectral types, at the dispersion of 5 and 10Å/mm, also available at the coudé focus. A. H. Batten has been studying the problem of wavelengths in O-type spectra, mainly with reference to the Victoria IM dispersion (30Å/mm at Hγ). A preliminary report has been published (**86**) and the work is continuing.

Accuracy of coudé-spectrographs: Study of the problems reported by Dr Petrie at Hamburg was continued by him until his death, and since then, by Fletcher. Fletcher reported on his progress at the Toronto Symposium (**1**). It appears that the relatively large external errors encountered in coudé spectrographs are due, at least partly, to non-uniform illumination of the collimator arising from failure to achieve exact focus of the stellar image on the slit, or possible from imperfections in the primary mirror of the telescope. It is also important to maintain long-term stability of all parts of the spectrograph. Investigation is continuing to try to estimate the relative importance of these two effects and to ensure that the maximum accuracy of which the spectrograph appears to be capable (m.e. of single plate $\sim \pm 0{\cdot}07$ km s^{-1} for solar-type spectra) is available at all times.

Systematic Errors in Measurement of Double-Line Spectroscopic Binaries: Reports on this work have been presented by Petrie and Andrews (**87**) and by Petrie, Andrews, and Scarfe (**1**). It appears that systematic errors may be made in the measurement of double-line binaries, especially when oscilloscope devices are used to measure the plates. In extreme cases, errors of up to 50% may occur in the determination of the total mass of a binary system.

CONCLUSION

To all appearances the work covered by the field of the Commission is in a healthy state, but in spite of this there is some cause for concern. The determination of radial velocities yields fundamental astronomical data, without which our theoretical understanding of stellar motions and galactic structure would be paralysed. It is essential that this work be continued, and the discussions at Toronto, and the content of the foregoing report, make it clear that those active in the field are well aware of what needs to be done for objects ranging from the brightest stars to the faintest quasars, and of all population and morphological classifications.

Much of the work is still undertaken by classical methods, using photographic plates and slit spectrographs. It is clear that this style of work will remain indispensable for a long time to come. It is a source of great satisfaction to see new methods introduced, such as the photoelectric method of Griffin, the application of image tubes by Lynds, Mrs Rubin and others, and the scanning technique of de Vaucouleurs. The invention of new methods is, however, of little significance unless they are applied on a considerable scale. The example of the work of Fehrenbach's school shows how much meticulous effort needs to be expended on the development of a new technique before it becomes fully productive.

The attendance roll at the Toronto Symposium demonstrated that radial velocity work is being supported by a rather small number of observatories in rather few countries. The analysis presented by Weaver shows, that, certainly when classical methods are employed, it is an essential condition for the production of accurate results that the work be planned on a large scale, and be long continued. Only when a number of observers cooperate, in a long programme, using a spectrograph of which the essential adjustments are left undisturbed, is it usually possible to evaluate with certainty the various systematic and accidental errors for which corrections must be applied. If the equipment is especially designed so as to permit changes from one dispersion to another without modification of the basic adjustments, as is the case for example with the Haute-Provence coudé spectrograph, versatility of equipment can be reconciled with this basic requirement. If the design necessitates an actual removal and re-mounting of the dispersing elements each time a new dispersion is required, then there can be erratic changes of zero-point in the velocity system between one observing run and another, a matter stressed by Edmondson in his contribution to the Toronto Symposium.

These requirements of stability and continuity have meant that successful radial velocity work has usually been confined to those observatories, among which government observatories are the prime example, where long-term routine work can be undertaken. University observatories, catering for the needs of doctoral candidates, have often been less successful in this field. They are more often concerned with versatility in their equipment, and the provision of opportunities for the demonstration of originality on the part of their students within the time-limits for the production of their theses. The production of a considerable number of new and accurate radial velocities is an undoubted contribution to astronomy, but it may not be regarded as a sufficiently original contribution to an academic thesis.

Among astronomers influential in the formulation of astronomical policy there is sometimes a feeling that the two types of requirement are in conflict. It should be no harder to reconcile them in radial velocity work than it is in any other field of astronomy, or indeed, in any other field of science where a number of workers have to share large and expensive equipment. It is, however, highly desirable that some kind of solution be found within the field of work of Commission 30. Unless this is done, the number of observatories participating effectively in radial velocity work is unlikely to increase, and the recruitment of younger astronomers to an interest in the field is likely to suffer. A possible solution in the case of university departments would be for graduate students to pool their requirements in a common programme. They would also pool their observing resources, so that each observer contributed to the programme as a whole, and not merely to that section of it in which he was personally interested. When the individuals came to draw from the pool of results those pertaining to their own programmes, they would obtain measures made under the most favourable conditions all around the sky, and subject to the running statistical control on errors derived from a study of the entire pool.

Commission 30 shares with a number of others the characteristic that it is built round a single basic astronomical operation, of which the fundamental principles have been established for a very long time. The data produced are essential, and there is no question that the basic operations need to be continued for the forseeable future. There is a need to maintain them in institutions which are already active and to introduce them on the necessary scale into others.

DAVID S. EVANS
Acting President of the Commission

REFERENCES

1. Report of *IAU Symposium No.* 30, edited by J. F. Heard and A. H. Batten. Academic Press, London, 1967.
2. Griffin, R. F. A photoelectric radial-velocity spectrometer, *Astrophys. J.* (in press).
3. Griffin, R. F. 1966, The spectroscopic binary 73 Leonis, *Observatory*, **86**, 145.
4. Mowat, M. Optical Determination of Distant Objects, Symposium on *Radio and Optical Studies of the Galaxy*, p. 100, Mount Stromlo Observatory, May 1966.
5. Bok, B. J., Gollnow, H., Hindman, J. V., Mowat, M. 1964, Radial Velocities associated with Selected Emission Nebulae in the S.M.C., *Austr. J. Phys.*, **17**, 404.
6. Buscombe, W., Kennedy P. M. 1962, Two B-Type Spectroscopic Binaries, *Publ. astr. Soc. Pacif.*, **74**, 323.

7. Westerlund, B. E., Danziger, I. J., Graham, J. 1963, Supergiant Stars in the Wing of the Small Magellanic Cloud, *Observatory*, **83**, 74.
8. Buscombe, W. 1963, Radial velocities of Bright Southern Stars, III, *Mon. Not. R. astr. Soc.*, **126**, 29.
9. Kennedy, P. M., Przybylski, A. 1963, Radial velocities of 50 high velocity stars, *Mon. Not. R. astr. Soc.*, **126**, 381.
10. Buscombe, W. 1965, Stellar Spectra in Galactic Cluster IC2391, *Mon. Not. R. astr. Soc.*, **129**, 411.
11. Buscombe, W. 1965, The Scorpio-Centaurus Association (Paper V), *Irish astr. J.*, **7**, 63.
12. Buscombe, W., Kennedy, P. M. 1965, Radial Velocities of 200 Southern B Stars, *Mon. Not. R. astr. Soc.*, **130**, 281.
13. Przybylski, A., Kennedy, P. M. 1965, Radial Velocities and three-colour Photometry of 166 Southern Stars, *Mon. Not. R. astr. Soc.*, **131**, 95.
14. Przybylski, A., Kennedy, P. M. 1965, Radial Velocities and three-colour Photometry of 52 Stars with large Proper Motions, *Mon. Not. R. astr. Soc.*, **129**, 63.
15. Gollnow, H. Radial Velocities in the Magellanic Clouds, Symposium on the *Magellanic Clouds*, p. 12, Mount Stromlo Observatory, May 1965.
16. Buscombe, W., Kennedy, P. M. Optical Insterstellar Lines in the Southern Milky Way. Symposium on *Radio and Optical Studies of the Galaxy*, p. 82, Mount Stromlo Observatory, May 1966.
17. Evans, D. S., Laing, J. D., Menzies, A., Stoy, R. H. 1964, Fundamental Data for Southern Stars (Fifth List). *R. Obs. Bull.* no. 85.
18. Evans, D. S. 1966, Fundamental Data for Southern Stars (Sixth List). *R. Obs. Bull.* no. 110.
19. Evans, D. S. *et al.* *R. Obs. Bull.* (in press).
20. Evans, D. S., Young, A. T. 1966, A Technique for Period Finding. *Observatory*, **86**, 200.
21. Lafler, J., Kinman, T. D. 1965, *Astrophys. J., Suppl.*, **11**, 199.
22. Evans, D. S., Malin, S. R. 1965, Radial Velocities of Southern Galaxies. II. *Mon. Notes astr. Soc. Sth. Afr.*, **24**, 32.
23. Evans, D. S. 1965, A Note on HR5527 = HD130701/2: *Mon. Notes astr. Soc. Sth. Afr.*, **24**, 11.
24. Jones, D. H. P. Personal communication.
25. Wayman, P. A. 1961, Appendix to Fundamental Data for Southern Stars. IV. *R. Obs. Bull.* no. 48.
26. Jones, D. H. P., Lagerweij, H. C. The δ Scuti variable 1 Monocerotis. *Mon. Notes astr. Soc. Sth. Afr.*, (in press).
27. Heard, J. F. 1965, *Publ. Dom. astrophys. Obs.*, **2**, 443.
28. Wayman, P. A. Personal communication.
29. Woolley, R. v. d. R., Harding, G. A. 1965, Radial Velocities of 96 stars derived from 241 spectra obtained with the coudé spectrographs at Mount Wilson and Palomar and Radcliffe Observatories. *R. Obs. Bull.* no. 93.
30. Harding, G. A. 1965, Radial Velocities of bright members of the globular cluster ω Centauri. *R. Obs. Bull.* no. 99.
31. Woolley, R. v. d. R., Aly, K. 1966, Radial Velocity observations of RR Lyrae variables. *R. Obs. Bull.* no. 114.
32. Jones, D. H. P., Palmer, D. R., Walker, E. N., Wallis, R. E. The Radial Velocities, Spectral Types and Projected Rotational Velocities of Six Hundred Bright Northern A Stars. *R. Obs. Bull.* (in press).
33. Harding, G. A., Palmer, D. R., Pope, J. D. *R. Obs. Bull.* (in press).
34. Wachmann, A. A. 1966, IAU *Inf. Bull. Var. Stars*, no. 138.
35. Rébeirot, E. La mesure des vitesses radiales au prisme objectif, XVIII. Classifications spectrales et mesures de vitesses radiales de 112 étoiles situées dans Praesepe. *Publ. Obs. Hte-Provence*, **8**, no. 24, 1966; *J. Observateurs*, **49**, no. 4–5.
36. Fehrenbach, Ch., *et al.* La mesure des vitesses radiales au prisme objectif, XIX. Liste de 893 vitesses radiales determinées au prisme objectif à vision directe. *Publ. Obs. Hte-Provence*, **8**, no. 25, 1966; *J. Observateurs*, **49**, no. 4–5.

37. Rébeirot, E., Imbert, M. 1965, Vitesses radiales obtenues avec la chambre I du spectrographe coudé de l'Observatoire de Haute-Provence. *Publ. Obs. Hte-Provence*, **8**, no. 6; *J. Observateurs*, **48**, 153.
38. Sahade, J. 1966, *Modern Astrophysics*, M. Hack, ed. (Paris: Gauthier-Villars), p. 219.
39. Sahade, J., Frieboes-Conde, H. 1965, *Astrophys. J.*, **141**, 652.
40. Ringuelet-Kaswalder, A., Sahade, J., Wallerstein, G. 1965, *Assoc. argent. Astr. Bol.*, no. 8–9–10, 32.
41. Thackeray, A. D. 1966, Radial Velocities of Southern B Stars, IV. The Sco-Cen Association. *Mem. R. astr. Soc.*, **70**, 73.
42. Feast, M. W., Shuttleworth, M. 1965, Kinematics of B stars, Cepheids, Galactic Clusters and Interstellar Gas in the Galaxy. *Mon. Not. R. astr. Soc.*, **130**, 245.
43. Feast, M. W., Shuttleworth, M. 1966, *Mon. Not. R. astr. Soc.*, **134**, 107.
44. Feast, M. W., Thackeray, A. D. 1965, Rotation of Inner Parts of the Galactic System as determined from OB Stars. *Astrophys. J.*, **142**, 1645.
45. Thackeray, A. D. 1964, Radial Velocities of 5 Stars. *Mon. Notes astr. Soc. Sth. Afr.*, **23**, 81.
46. Thackeray, A. D. 1966, Radial Velocities of 3 Stars. *Mon. Notes astr. Soc. Sth. Afr.*, **25**, 35.
47. Graham, J. A. Observations of some distant B-type stars in Ara. *Bull. astr. Inst. Netherl.* (in press).
48. Feast, M. W. 1966, Kinematics of Me Variables towards the Galactic Centre and a Comparison with the Planetary Nebulae. *Mon. Not. R. astr. Soc.*, **132**, 495.
49. Feast, M. W. 1965, Long Period Variables in Globular Clusters and in the General Field. *Observatory*, **85**, 16.
50. Feast, M. W. 1966, Long Period Variables in the Field of Globular Clusters. *Observatory*, **86**, 120.
51. Feast, M. W. CH Sct as a Member of NGC6712 and the absolute magnitude of the Mira variables. *Observatory* (in press).
52. Feast, M. W. The Radial Velocities of distant Cepheids and the distance to the Galactic Centre. *Mon. Not. R. astr. Soc.* (in press).
53. Breger, M. A Spectroscopic Study of two Southern B-Type Variables. *Mon. Not. R. astr. Soc.* (in press).
54. Thackeray, A. D., Wesselink, A. J. 1965, A Photometric and Spectroscopic Study of Cluster IC2944. *Mon. Not. R. astr. Soc.*, **131**, 121.
55. Thackeray, A. D., Hutchings, J. 1965, Orbits of Two Double-lined Binaries, HD140008 and 178322. *Mon. Not. R. astr. Soc.*, **129**, 191.
56. Thackeray, A. D. 1966, The Spectroscopic Binary, δ Pic. *Mon. Not. R. astr. Soc.*, **131**, 435.
57. Thackeray, A. D. 1966, Orbits of Two-Double-Lined Spectroscopic Binaries. HD147971 and 75759. *Mon. Not. R. astr. Soc.*, **134**, 97.
58. Thackeray, A. D., Tatum, J. B. The B-type Spectroscopic Binary HD175544. *Publ. Dom. astrophys. Obs.* (in press).
59. Thackeray, A. D., Knipe, G. F. G. 1965, The Eclipsing Variable HD161783 (BV 420). *Mon. Notes astr. Soc. Sth. Afr.*, **24**, 109.
60. 1966, *Ibid.*, *Mon. Notes astr. Soc. Sth. Afr.*, **25**, 30.
61. Feast, M. W. The Supergiant Eclipsing System BL Telescopii. *Mon. Not. R. astr. Soc.* (in press).
62. Thackeray, A. D. 1965, Spectra of S Equ during Eclipse. *Observatory*, **85**, 206.
63. Thackeray, A. D. 1965, Spectral Variations of S Dor. *Mon. Not. R. astr. Soc.*, **129**, 169.
64. Feast, M. W. 1964, Planetary Nebulae in the Large Magellanic Cloud. *Observatory*, **84**, 266.
65. Friedjung, M. The Physical Conditions and Velocity Distribution in the Envelope of RR Telescopii. *Mon. Not. R. astr. Soc.* (in press).
66. Thackeray, A. D. 1966, Two New High Velocity Sub-dwarfs. *Observatory*, **86**, 60.
67. Thackeray, A. D., Feast, M. W., Warner, B. 1966, Daytime Spectra of Comet Ikeya-Seki near Perihelion. *Astrophys. J.*, **143**, 276.
68. Thackeray, A. D. Measured Radial Velocities of Proxima Centauri. *Observatory*, (in press).

69. Abt, H. Personal communication.
70. Bouigue, R. 1952, *Ann. Obs. Toulouse*, **21**, 31.
71. Bouigue, R. 1954, *Ibid.*, **22**, 49.
72. Bouigue, R. 1955, *Ibid.*, **23**, 45.
73. Bouigue, R. 1956, *Ibid.*, **24**, 69.
74. Bouigue, R. 1957, *Ibid.*, **25**, 69.
75. Bouigue, R., Chapuis, J. L. 1959, *Ann. Obs. Toulouse*, **27**, 87.
76. Pedoussaut, A. 1963, *Ann. Obs. Toulouse*, **29**, 31.
77. Pedoussaut, A. 1964, *Ibid.*, **30**, 50.
78. Salman-Zade, R. Kh. 1966, *Vestn. Leningrad. gos. Univ.*, p. 66.
79. Gershberg, R. E., Shcheglov, P. V. 1964, *Astr. Zu.*, **41**, 425.
80. Mirzoyan, L. V., Kazarian, E. S. 1965, *Trans. Astrophys. Inst. Kazak Acad. Sci.*, **5**, 224.
81. Underhill, A. B., van Helden, R. C. P. 1966, *Astr. J.*, **71**, 183.
82. Underhill, A. B., van Helden, R. C. P. *Bull. astr. Inst. Netherl.*, **18** (in press).
83. Gutman, F. 1965, *Publ. Dom. astrophys. Obs.*, **12**, 361.
84. Gutman, F. 1966, *Ibid.*, **12**, 391.
85. Fletcher, J. M. 1964, M.Sc. thesis, Toronto, (unpublished).
86. Batten, A. H. 1965, *Publ. astr. Soc. Pacif.*, **77**, 210.
87. Petrie, R. M., Andrews, D. H. 1966, *Astr. J.*, **71**, 175.

31. COMMISSION DE L'HEURE

PRÉSIDENT: Mr H. M. Smith, Royal Greenwich Observatory, Herstmonceux Castle, Hailsham, Sussex, England.

VICE-PRÉSIDENT: Professor Dr N. N. Pavlov, Pulkovo Observatory, Leningrad, U.S.S.R.

COMITÉ D'ORGANISATION: L. Essen, W. Markowitz, N. Stoyko, M. Torao, F. Zagar.

MEMBRES: Abraham, Arbey, Bakulin, Belotserkovski, Billaud, Blaser, Bonanomi, Brkić, Danjon†, Decaux, Delhaye, Demetrescu, Dingle, Drâmbă, Dubois-Chevallier, Enslin, Fernandez de la Puente†, Fuchs, Gama, Gökmen, Gougenheim, Guinot, Hall (R. G.), Hers, Iijima, Koebcke, Lacombe, Lederle, Lorón, Madwar, Melchior, Mikhailov, Miyadi, Opalski, Orte Lledó, Postoiev, Randić, Sandig, Shcheglov, Shirjaev, Sternberk, Stoyko (A.), Tagaki, Tardi, Thomson (M. M.), Tsao, Verbaandert, von der Heide.

CO-ORDINATED TIME SIGNALS

The recommendations of the Xth Plenary Assembly of the International Radio Consultative Committee, Geneva 1963 (Rec. 374) were further modified at the Study Group VII Interim meeting (Monte Carlo 1965) and the XIth Plenary Assembly (Oslo 1966). The effect of the modifications has been to clarify the recommended system U.T.C. (annual offset, plus time adjustments when necessary) and to place responsibility for co-ordination on the Bureau International de l'Heure. In addition, the Oslo recommendation (Doc. VII/1012) makes formal provision for emissions with zero carrier offset but with the rate of the time signals offset, and co-ordinated with those mentioned above; also for experimental emissions in which both the carrier and the rate of the time signals is not offset. In the latter case time adjustments should also be co-ordinated by the BIH.

At the XVth General Assembly of the International Scientific Radio Union (Munich 1966) Commission 1 expressed the opinion that all the methods of operating Standard Frequency services which have been proposed contain defects which will cause increasing difficulties as the use of the services extends; and that these services must inevitably develop towards a system of uniform atomic time and constant frequency.

On the other hand, astronomers concerned with the needs of navigation consider it essential that:

(*a*) the navigational ephemerides continue to use GMT (U.T.) as the time argument.

(*b*) time signals, deviating by not more than $0^s.1$ from U.T.2 continue to be available to navigators, preferably in the present form of U.T.C.

If radio time signals gave only A.T., it would be necessary to communicate, at least every two months, predicted values of U.T. − A.T. by radio broadcast to navigators. This procedure is far less convenient and dependable than the communication of corrections to laboratory-borne physicists. Separate emissions of U.T. and A.T. would provide a possible solution, but might prove impracticable for economic reasons.

Some dozen countries now conform to the U.T.C. co-ordination scheme, and this has led to a drastic reduction in the number of individual corrections to radio time signals published each month by the Bureau International de l'Heure. Moreover it has proved possible to reduce significantly the delay in publishing definitive corrections.

A report prepared by Professor Belotserkovsky (see page 670) describes a parallel scheme, operating in the U.S.S.R., adjacent countries and China, and co-ordinated by the Institut des Mesures Physicotechniques et Radiotechniques (IMPR). It is hoped that the two co-ordinating authorities, BIH and IMPR, may be able to come to some mutual understanding leading to the adoption of a single system in all radio time signal emissions.

CO-OPERATION WITH OTHER INTERNATIONAL UNIONS

The CCIR meetings at Oslo, 1966 were attended by H. M. Smith (representing the IAU) and B. Guinot (representing the BIH). In addition to the resolutions on the U.T.C. system of time signal emissions, co-operation between CCIR, URSI, IUGG and IUPAP was proposed in pursuing studies and experiments relative to the following problems:

(*i*) how to provide both the epoch of U.T. and the international unit of time interval in the same emission;

(*ii*) how the various essential requirements could be met by the emission of a single uniform time scale.

Reference was also made to the continuing problem of interference, both mutual and from other emissions, experienced on standard frequency and time signal services, particularly in the European area. This matter also received attention at the URSI meetings at Munich, 1966 and it is hoped that the various authorities concerned will be able to come to a satisfactory solution.

The CCIR recommended that emitted frequencies should now be maintained constant to $\pm 1 \times 10^{-10}$, which corresponds to a uniformity in the rate of the time signals of about 10 microseconds per day. The tabulated data on standard frequency and time signal emissions was brought up-to-date and the assistance of all authorities responsible is requested in ensuring that all changes in emission schedules are notified.

DEFINITION OF THE SECOND

A further step was taken by the Twelfth General Conference of Weights and Measures (October 1964) towards the definition of the second in terms of an atomic resonance. In recent years, it has been increasingly realized that the second of U.T.2 is unsuitable, owing to its inherent variability, as the basic unit of time interval. The second of E.T., though by definition invariable, is not readily accessible, and is therefore equally unacceptable. Progress in the development of atomic standards of frequency has demonstrated that these may now be used to make available an accurate and convenient unit of time interval. Continuously running atomic standards of frequency may also be used for the establishment of a uniform time scale.

The Twelfth General Conference of Weights and Measures considering the urgency of arriving at an atomic or molecular standard of time interval for the needs of precise metrology and that, in spite of the results obtained in the use of caesium as an atomic frequency standard, the moment has not yet come for the General Conference to adopt a new definition of the second because of new and important progress which may arise from current researches, authorised the International Committee of Weights and Measures to designate atomic or molecular standards of frequency to be employed temporarily. The committee declared that the standard to be so employed is the transition between the hyperfine levels $F = 4$, $M_F = 0$ and $F = 3$, $M_F = 0$ of the fundamental state ${}^2S_{\frac{1}{2}}$ of the atom of caesium 133, not perturbed by external fields, and that the value of 9 192 631 770 hertz is assigned to the frequency of this transition.

It will be noted that the caesium frequency adopted is that determined in 1958, (Markowitz, Hall, Essen and Parry. *Phys. Rev. Lett.*, **1**, 105, 1958), and now in general use. Recent redeterminations which take account of the effects of the adoption of the revised system of astronomical constants have shown that there is no practical need to revise this value (e.g. Nicholson and Sadler, *Nature*, **210**, 187, 1966). At this stage the choice of caesium as an interim standard is favoured on practical grounds: there are many caesium standards in use in many different countries; comparisons between caesium standards of slightly different construction exhibit no significant frequency deviations.

It is not certain that caesium may be the best final choice, as the stabilities now believed to be obtainable under laboratory conditions (month long samples) are quoted by R. F. C. Vessot (Varian Associates) as

Rubidium gas cell	5×10^{-11}
Caesium beam	3×10^{-12}
Hydrogen maser	3×10^{-14}

On the astronomical side it is anticipated that a more accurate relationship between atomic frequencies and the second of E.T. may be possible within the next few years. There are various complications: it has been suggested that relativistic gravitational effects on quantum transitions may produce a periodic term between A.T. and E.T. with an amplitude of 3×10^{-10} and a period of one year, (S. Aoki, *Astr. J.*, **69**, 221, 1964): there are reasons to suppose that there may be an acceleration between A.T. and E.T. On the other hand, the linking of A.T. and E.T., both in the unit of time interval and in epoch, might be a useful simplification and, as the astronomer only requires E.T. to an accuracy of about 0·1 seconds, tolerable mutual departures should be achieved without further adjustment for many decades.

THE EPOCH OF EPHEMERIS TIME

The formal definition of the epoch of Ephemeris Time, as adopted by the General Assembly in 1958 (*Trans. IAU*, **10**, 72, 1960, Resolution No. 2) on the recommendation of Commissions 4 and 31, specifies the measure of E.T. at the instant at which the geometric mean longitude of the Sun was 279° 41′ 48″.04. However, as subsequently pointed out by Professor Danjon and incorporated in the report of the meetings of Commission 4 (*loc. cit.* p. 97), this definition depends on the adopted value of the constant of aberration. The IAU system of Astronomical Constants (*Trans. IAU*, **12**, 593, 1966), which was adopted by the Twelfth General Assembly in Hamburg (*loc. cit.*, p. 95, Resolution No. 4), introduces a change in the adopted value of the constant of aberration from 20″.47 to 20″.496.

It would appear that, to accord strictly with the formal definition, it is necessary to increase the measure of E.T. at any instant by + 0ˢ.63, to apply a correction of − 0″.346 to the mean longitude of the Moon, and to apply smaller (but more complicated) corrections to the ephemerides of the planets.

Pending consideration of the possibility of revising the definition these corrections have not been incorporated into the national and international ephemerides for 1968.

There are several alternative possibilities, one of which is, of course, to adhere strictly to the present definition; these possibilities are being examined by Commission 4 and its considered recommendation will certainly be the subject of discussion with Commission 31 in Prague. If a revised form of definition is recommended, it may be possible and desirable to specify the numerical constants in the definitions of the epochs of *both* Ephemeris Time and Atomic Time in such a way that the two scales can be linked as precisely as their characteristics allow.

TRAVELLING CLOCK EXPERIMENTS

The feasibility of transporting clocks by air in order to make microsecond comparisons between standard clocks in widely separated locations was first demonstrated by Reder and Winkler in 1959.

Since then, a number of quartz and caesium clocks have been used to make precise comparisons between selected pairs of establishments. In addition, the Hewlett Packard Co. have made tours which have included a number of establishments in the U.S.A. and in Europe. In 1964 comparisons were made between U.S.A. and Switzerland. In February–March 1965, a similar experiment included comparisons at 21 establishments in 11 different countries.

The synchronization between California and Japan was also verified by a U.S. Navy experiment employing artificial satellite Relay II. In May–June 1966 a further tour by Hewlett Packard caesium standards embraced 25 establishments in 12 different countries. Comparisons with local standards were made to accuracies of the order of 0·1 microseconds, and the closing error, after one month, was less than one microsecond.

TRAVEL TIME OF RADIO TIME SIGNALS

Direct comparisons between clocks at distant stations, established to microsecond accuracy by means of flying clocks, permit the measurement of travel times of radio time signals and hence the determination of the effective speed of propagation, assuming a great-circle route. In this way trans-Atlantic speeds have been determined at 15 MHz (using WWV) and for VLF radio time signals. Great care must be taken to measure all lags associated with reception and measurement. In general the results for HF emissions (Band 7) confirm the extensive series of tests carried out by the Tokyo Astronomical Observatory. An attempt to measure LF travel times between European stations which had been linked by the 1966 Hewlett Packard flying clock programme proved disappointing, owing to the errors of reception and measurement which are of overriding importance in view of the short radio paths involved. For VLF signals useful supplementary information has been obtained from an extensive study undertaken by the Royal Aircraft Establishment, Farnborough, England, in connection with VLF aircraft navigational aids. For this purpose, phase velocity is of interest, whereas for time service applications it is the group velocity which is important. Results in the 10–14 kHz band are scheduled for publication by Burgess and Rawles in Radio Science. From results so far available, it appears that for a path over the sea, the effective propagation speed at 16 kHz is 298 km/ms.

REPORTS OF OBSERVATORIES AND LABORATORIES

Australia. Canberra

The Mt Stromlo Observatory clock equipment has been rebuilt so that linear phase comparators continuously record the minor differences between co-ordinated systems. New crystals, VLF tracking receivers, steerable antenna and emergency power plant have been installed. A caesium beam frequency standard is on order. Continuous comparisions by landline are made with a rubidium frequency standard at the tracking station D.S.I.F. 42.

Mention should be made of recent progress by independent collaborating organizations. This was discussed at a Symposium on Time and Frequency that was held at the Observatory on 7 and 8 October 1965. The Postmaster-General's Department now maintains continuous time signal transmissions from VNG, Lyndhurst, Victoria. The Weapons Research Establishment, Salisbury, South Australia makes comparisons of VNG, VHP and WWVH, which are sent to the BIH. The National Standards Laboratory, Sydney, is building two hydrogen masers.

Belgium. Uccle

(1) *Instruments:*

Les observations à l'astrolabe Danjon, en service depuis juillet 1962, sont intensifiées depuis janvier 1965. Les enregistrements des observations astronomiques se font sur des chronographes électroniques perforateurs ATEA (de construction belge) qui assurent le millième de seconde. Nous disposons depuis 1960 de 2 oscillateurs à quartz, 1 MHz, Rohde et Schwarz. Les réceptions horaires se font sur oscilloscope au $0^{s}0001$ depuis le 15 février 1965. Un récepteur 'Racal' de 10 kHz à 30 MHz est en service depuis 1965.

(2) *Publications:*

(*i*) L'installation de l'astrolabe Danjon à Uccle, par P. Sanders, *Communication de l'Observatoire Royal de Belgique*, no. 230, 1964.

(*ii*) Déduction du phénomène de précession-nutation à partir de la marée terrestre, par P. Melchior, *Académie Royale de Belgique, Bulletin de la Classe des Sciences*, 5ème Série, **51**, no. 1, 24, 1965.

Brazil. São Paulo

Observations are made with a Danjon astrolabe (OPL 33). The equipment in use includes two Sulzer-Hewlett-Packard quartz clocks, printing chronograph Belin, radio-receiver Collins, integrating chronograph by Ebauches, a Rubidium Frequency Standard (General Technology) and a Dymec VLF receiver. The last three instruments are not yet in regular use. The reductions are performed on an IBM 1620 computer owned by the S.P. University. Observational programme contains 11 groups with 26-28 passages each. The list of stars contains 279 FK4 and 13 FK4 Sup. stars. Regular reporting to the BIH is delayed by acute shortage of observing and auxiliary personnel.

Canada. Ottawa

The former transit and meridian circle rooms have been combined into a thermally controlled time laboratory. A commercially built caesium clock provides a frequency reference to a part in 10^{11}. Its performance is compared daily with the long beam at the National Research Council. Time signal emissions CHU have remained unchanged except for a rubidium vapour atomic standard for direct control of frequency. Voice announcement of time each minute is made in both French and English. Time and Latitude bulletins are published quarterly.

Czechoslovakia. Prague

A new method for the microsecond clock comparison was developed by the Institute of the Radioelectronics and the Astronomical Institute (both of the Czechoslovak Academy of Sciences) and verified in November, 1965 between Prague and Potsdam. Being fully compatible with the ordinary TV transmission the method is based on the use of TV frame-synchronizing pulses. The regular comparisons of the main clocks in Prague and Potsdam are made weekly since January 1966.

France. Besançon

Modification dans l'équipement du service:

On a installé deux nouveaux étalons de fréquence; une horloge Sulzer (2, 5 MHz); un garde-temps Hewlett Packard (5 MHz); un chronographe totalisateur Ebauches; utilisation d'un oscilloscope bi-courbe permettant la mesure des signaux horaires à l'aide d'un signal de référence.

Recherches et expériences:

Mesure des températures à la surface du prisme de l'astrolabe à l'aide de thermistances; étude spéciale du signal HBG émis par l'Observatoire de Neuchâtel; étude de l'évolution du T.U.2 mesuré à Besançon; établissement d'une échelle de temps atomique locale à l'aide du maser à NH_3 du Laboratoire de l'Horloge Atomique de Besançon et tentative d'emploi d'un tube à césium.

Publications:

Etablissement d'une échelle de temps atomique à l'aide du maser M4 du Laboratoire de l'Horloge Atomique, par P. Benevides (*Annales Françaises de Chronométrie*, 3ème trimestre, 1964).

France. Paris

En 1965, l'Observatoire de Paris a été pourvu d'un étalon atomique commercial à jet de césium (Pickard and Burns). Cet étalon sert d'une manière discontinue, pour le contrôle journalier d'une horloge à rubidium. Les phases de 3 stations VLF sont enregistrées en permanence; les résultats, rapportés à l'étalon au césium, sont diffusés auprès des laboratoires intéressés.

Depuis le 1 er avril 1966, les signaux horaires de l'Observatoire de Paris (FTH42, FTK77, FTN87, FTA91) sont émis en temps coordonné dérivé de l'échelle de temps associée à notre étalon à césium. Ils seront réajustés par de petits sauts si nécessaire.

Un nouveau type de chronographe imprimant et totalisateur a été mis au point par Monsieur Parcelier.

Monsieur N. Stoyko et Madame A. Stoyko

Depuis le dernier rapport M. N. Stoyko et Mme A. Stoyko ont déterminé pour le Service International Rapide des Latitudes (SIR) la polhodie interpolée du 17 octobre 1963 au 9 avril 1965 et la polhodie extrapolée du 5 janvier 1964 au 28 juin 1965 (*Circulaires du SIR*, nos. 92–109). De plus, ils ont extrapolé les variations saisonnières de rotation de la Terre pour les années 1964 et 1965 (*Bulletin Horaire*, Série 6, nos. 3 et 8) qui sont nécessaires pour le calcul de T.U.2.

M. et Mme Stoyko ont étudié la variation aléatoire et la variation progressive de la rotation de la Terre en utilisant les étalons atomiques pour la période de 1955 à 1965 inclus. Ils ont trouvé qu'en plus des variations aléatoires à courte durée il existe les variations séculaires et à longue période dans la rotation de la Terre. Ainsi, pour l'étude des influences météorologiques et solaires sur la rotation de la Terre, il faut éliminer préalablement ces variations à longue durée.

M. et Mme Stoyko ont étudié la précision de l'intégration du temps atomique depuis 1955. En comparant les différentes échelles atomiques de temps on constate les variations des différences systématiques de 1 à 3 ms pour un intervalle de 5 dernières années, soit de 0·6 à 2×10^{-11} pour la fréquence. Par conséquent, d'après M. L. Essen, la précision du temps atomique intégré est meilleure que celle avec laquelle on peut se repérer individuellement à la raie de césium.

M. et Mme Stoyko ont étudié les variations séculaires des longitudes dont les résultats ont été présentés au Symposium sur les Mouvements Récents de l'Ecorce Terrestre à Aulanco. Ils ont trouvé un accord avec les résultats de trois Opérations Internationales des Longitudes.

M. N. Stoyko a rédigé le *Bulletin Horaire*, Série 6, les numéros de 3 à 9 en donnant les résultats de l'heure demi-définitive (Observatoire de Paris) jusqu'à juin 1964 inclus. Il a introduit l'utilisation du temps coordonné à l'Observatoire de Paris depuis le 1er novembre 1963.

Il s'occupait jusqu' à octobre 1964 du développement de l'utilisation du temps atomique à l'Observatoire de Paris. Il a tracé un bref historique du Bureau International de l'Heure (40 dernières années de l'activité de cet organisme) en insistant surtout sur le progrès effectué dans la précision de la conservation et de la distribution de l'heure.

Madame Stoyko a comparé l'échelle du temps atomique intégré avec celle du temps des Ephémérides en utilisant les derniers résultats des réductions d'occultations d'étoiles par la Lune pour la période de 1958·5 à 1961·5 qui ont été aimablement communiqués par Mrs Sadler. Ces résultats ont été calculés dans le système d' 'Improved Lunar Ephemeris'. Elle a trouvé pour la fréquence correspondante de césium la valeur de 9 192 631 799 Hz qui est en bon accord avec les résultats publiés par les astronomes japonais (Sh. Iijima, Sh. Aoki et d'autres). Cette valeur diffère presque de 30 Hz de la valeur conventionnelle adoptée (9 192 631 770 Hz).

En utilisant les corrections de la théorie de la Lune de Brown déterminées par M. W. Eckert, Madame Stoyko, aidée par M. N. Stoyko, a trouvé pour la fréquence de l'étalon de césium la valeur de 9 192 631 771 Hz, en utilisant les occultations mondiales de la Lune calculées par l'Observatoire de Greenwich, les observations méridiennes de la Lune à Washington, ainsi que les occultations d'étoiles et les observations méridiennes de la Lune observées au Japon. Cette valeur de fréquence de l'étalon à césium est en bon accord avec la valeur conventionnelle. De plus, Mme Stoyko a calculé les corrections nouvelles du temps des Ephémérides à partir de 1881.

Si le terme qui dépend de sin 2 ($\pi_1 - \Omega$) existe réellement, il doit apparaître dans les valeurs brutes de ΔT de la réduction du temps universel au temps des Ephémérides. En prenant les valeurs brutes annuelles de ΔT depuis 1881, Madame Stoyko a trouvé pour le coefficient de ce terme qui a la période de 3 ans la valeur de $0^s.113$. Cette valeur est en bon accord avec la valeur d'Eckert.

Ainsi, se pose la question du tableau de ΔT qu'on publie dans les grands Annuaires astronomiques (American Ephemeris and Nautical Almanac, Connaissance des Temps, etc.). Il faut qu'on le remplace par un tableau amélioré.

Madame A. Stoyko a tiré des conclusions sur la précision de l'heure définitive en 1962 et 1963. Elle a donné les corrections pour la réduction des heures définitives dans un système uniforme et au pôle moyen de l'époque depuis 1931. De plus, elle a montré le progrès accompli dans la conservation internationale de l'heure (heure définitive) au cours de 42 ans. La précision a augmenté, en moyenne, de 9 fois.

Elle a déterminé la réduction de temps atomique intégré à l'échelle conventionnelle absolue depuis 1961 en utilisant les résultats des étalons atomiques des laboratoires de Boulder, Neuchâtel et Teddington.

Madame Stoyko a fait la discussion des résultats des longitudes pendant l'AGI et la CIG (1957·5–1960·5). Ce mémoire (371 pages) est publié dans les *Annales de l'Année Géophysique Internationale*. Pour cette discussion elle a utilisé le temps atomique intégré et différents systèmes de coordonnées du pôle instantané. Elle a calculé les longitudes en utilisant, en plus des coordonnées du pôle d'après le SIR, les coordonnées du pôle instantané ramenées au pôle moyen de l'époque et les coordonnées du SIL ramenées au pôle nouveau '1900–05'. Toutes les longitudes ont été réduites au catalogue FK4.

En utilisant les résultats de trois Opérations Internationales des Longitudes ramenés tous au catalogue FK4 et au pôle moyen de l'époque, Madame Stoyko a trouvé les variations séculaires des longitudes qui sont d'accord avec les derniers résultats:

(1) Une augmentation de la différence des longitudes entre le Japon et l'Amérique du Nord;

(2) Une diminution de la différence des longitudes entre l'Amérique du Nord et l'Europe.

Madame Stoyko a rédigé le *Bulletin Horaire*, Série G (nos. 23 et 24) et Série H (nos. 1 à 6) qui contiennent les heures définitives pour la période de septembre 1962 jusqu'à décembre 1963.

Germany. Dresden

Coordinated time signals, DIZ (4525 kHz) are controlled by the Geodetic Institute, Potsdam, Department of Astronomy where, in 1964, was installed a new system of three crystal clocks (Normalzeitanlage) by Rohde and Schwarz, Munich. Additional ancilliary equipment and part automation has also been installed.

The astronomical determinations of time have been made at Potsdam. During the period under review 567 observations have been secured with the three available transit instruments and 506 with the Danjon astrolabe. The latter instrument was used in conjunction with an Ebauches automatic printing chronograph presenting mean values.

Forty-three visual observations of occultations were received, reduced and published to assist in the establishment of ephemeris time. Oscillographic measurements of signals received from some 20 transmitting stations are published in the monthly bulletins of the institute. During former years (1957–62) such results were published in a supplement.

A comparison of the standard clocks at Potsdam and Prague has been effected using the television net of directed ultrashort waves. Accuracy attained is of the order 1 μs. Subsequently, similar accuracy was obtained by a simple method of comparison using the normal television programme received at the two stations.

Experiments have been made upon the behaviour of quartz crystals subjected to artificial ageing processes.

Analyses of the time observations have been made to ascertain the effects of systematic influences associated with the type of instrument used, atmospheric effects, etc. Corrections to the places of the stars observed have also been determined.

A series of observations was undertaken in May 1966 for the determination of the longitude differences of the Latitude station Borowiec of the Polish Academy, the Geodetic Institute, Potsdam and the Lohrmann-Institute, Dresden. These observations, which involved the interchange of observers, have been completed.

Germany, Hamburg

Determinations of Universal Time with the PZT were continued. The observed Universal Time was referred to the DHI quartz clocks and to the atomic time system of the Physikalisch-Technische Bundesanstalt, Braunschweig.

The DHI time signals give Co-ordinated Time since the beginning of 1965. The reference frequency is derived from registrations of LF and VLF standard frequency transmissions. Since 1965, a time signal of the English type has been transmitted at 0^h and 12^h U.T. after the Onogo time signal by the station DAN, 2614 kHz.

At the transmitting station of DCF77 a rubidium gas cell is to be installed by the Physikalisch-Technische Bundesanstalt.

Hewlett-Packard flying clock experiments included time comparisons at the DHI. From measurements of WWV HF transmissions there resulted an effective surface speed of 285 km/ms and 283 km/ms, for the March 1965 and June 1966 experiments, respectively. A speed of 291 km/ms was obtained from super-propagation receptions of DAM 16980 kHz transmissions during the northern winter.

Japan, Tokyo

In order to determine the values of ΔT, occultation observations were made by the Hydrographic Office of Japan at the head office in Tokyo and three branch observatories in Japan. The results obtained by the observations with photoelectric registration are as follows,

Epoch	ΔT_0	n	p.e.	limb corr.
1958·5	31·03	28	±0·12	Weimer's
59·5	32·12	36	0·15	,,
60·5	32·04	47	0·15	,,
61·5	32·07	30	0·24	,,
62·5	32·35	25	0·21	,,
63·5	33·03	15	0·14	Watts'
64·5	33·85	43	0·18	,,

The observations were all referred to the NZC catalogue.

Utilizing the various data of ΔT = (E.T. − U.T.2) as compared with those of (A.T. − U.T.2), the value of the caesium frequency was examined in Tokyo. Data of ΔT from 1955 to 1959 (To. Occ., Wld., Occ., To. Mer., Wa. Mer., and Wa. M.C.) and those from 1955 to 1962 (To. Occ. and To. Mer.) gave the values of 9 192 631 799 ± 17 Hz and 9 192 631 794 ± 10 Hz respectively. Moreover, the recent calculation by the use of ΔT from 1955 to 1964 (same series of material) also gave the result of 9 192 631 797 ± 5 Hz (S. Iijima, 1965).

With the introduction of the new astronomical constants it became necessary to apply a correction to the values of ΔT so far obtained (W. Nicholson, 1966). The average rate of $\Delta(\Delta T)$ from 1955 to 1964 amounts to − 29 Hz in a correction to the caesium frequency. Thus the conventional value of 9 192 631 770 Hz was proved appropriate, the error being, by chance, just cancelled by the new astronomical constants.

The time signal JJY on HF standard frequencies have been emitted continuously under the international co-ordination of transmission. Besides these, a LF standard frequency transmission on 40 kHz was commenced in January 1966 with the call sign JAG2AS. These standard frequencies have been monitored by the NH_3 maser type frequency standard operated in the Radio Research Laboratories (R.R.L.). The time signal JAS22 on 16 170 kHz has been emitted exclusively to Europe every day around 12^h00^m U.T., being monitored directly by the Tokyo Astronomical Observatory (T.A.O.).

A clock synchronization experiment was carried out between the U.S.A. and Japan via the artificial satellite Relay II in February 1965. Another synchronization experiment using the flying clock (Hewlett Packard Inc.) was made in parallel in February 1965 and again in June 1966.

Measurements of the apparent velocity of radio time signals have been effected since 1957 between Prague and Tokyo. The available signals are OMA (18 985 kHz) from Prague and JAS22 (16 170 kHz) from Tokyo. From 374 pairs of data so far obtained, a close relationship between the apparent velocity, v_a, and the Wolf sunspot number, r, has been found. (S. Iijima and G. Shibutani, T.A.O., and V. Ptáček, Czechoslovak Academy Astronomical Observatory). The result is given below,

$$v_a = \underset{\pm\,\cdot 1}{284{\cdot}9} - \underset{\pm\,\cdot 002}{0{\cdot}055}\,(r - 100)\ \text{km/ms}$$

The standard deviation of $(O - C)$ for a single datum was ± 2·4 km/ms.

Frequency of GBR (16 kHz) or GBZ (19·6 kHz) has been measured in the T.A.O. with reference to the standard frequency furnished by the R.R.L., since August 1964. The diurnal phase variation of the signal as received was studied. Day to day fluctuation in time of arrival showed a standard deviation of about ± 5 micro-seconds which corresponds to a standard deviation in the derived frequency of about $\pm 6 \times 10^{-11}$. The amplitude and phase relations in the diurnal phase variation have been satisfactorily explained by the distribution of day and night-time zone on the propagation route.

South Africa. Johannesburg

Throughout this period time signals and standard frequencies, as broadcast by station ZUO, were derived from a 100 kHz ring crystal oscillator, supplemented by five GT crystal oscillators. The frequency was regularly adjusted on the basis of daily phase comparisons with the VLF transmissions from GBR and NBA. This was found entirely adequate until the end of 1965, but became unsatisfactory at the beginning of 1966 when GBR transmissions ceased. The diurnal variation of NBA as received at Johannesburg is very nearly two cycles of the present carrier frequency of 24 kHz (as compared to approximately one cycle for GBR) and, as good reception from NBA can be obtained for only part of the day, it is often difficult to determine whether a cycle has been gained or lost during the preceding 24 hours. The stability achieved in 1965 was therefore not maintained in 1966, and time signal deviations have occasionally exceeded 2 milliseconds.

The absolute difference between local and overseas time standards remains in doubt, and while several methods have been proposed for determining this difference, no finality has yet been reached.

To overcome the above difficulties, a caesium beam frequency standard (Hewlett-Packard) was acquired in October 1966, and immediately put into use. It is too early to report on its performance, but results so far have been in every way satisfactory. A more versatile VLF comparison receiver is to be installed shortly. Other changes of a minor nature have been mainly connected with the transfer of time apparatus from the old to the new building.

It would seem desirable that the IAU should once again stress, to the authorities responsible for the transmission of VLF signals, the great value of these signals for frequency and time interval determination throughout the world, and the undesirability of frequent changes in carrier frequencies or transmission periods.

Spain. San Fernando

During the interval 1963–66·5 continued observations of various time signals, made in order to select the most convenient for long distance time synchronization of our crystal clocks, revealed a very good stability for MSF 10 MHz signals during certain intervals in the morning hours. Accuracy obtained was only comparable with that of the LF signals from the HBG new emitter.

To complete synchronization a knowledge of the time of propagation is necessary. To secure this information an experiment in collaboration with the Smithsonian Institution using a portable clock has just been carried out.

The calculations of travel time, based independently on two measurements made at Rugby, resulted in the same value.

MSF 10 MHz Rugby-San Fernando = $0\overset{s}{.}0074$

Assuming that MSF, HBG and WWV were synchronized to each other with an accuracy better than $0\overset{s}{.}0001$, it may be deduced that

HBG	Prangins-San Fernando	= $0\overset{s}{.}0058$
WWV	Greenbelt-San Fernando	= $0\overset{s}{.}0215$

The propagation time of WWV was also determined by the Astrophysical Observing Station which used the same portable clock. Their values agreed with our result based on indirect methods. It is hoped that synchronization between San Fernando and both MSF and HBG will, in future, be maintained to an accuracy of $\pm\ 0\overset{s}{.}0001$.

Switzerland. Neuchâtel

Time determinations have been continued with the PZT. The observations at the Vue-des-Alpes station with the Danjon astrolabe were discontinued in 1965.

Time signal and standard frequency transmission from HBN, based on caesium and thallium beam standards, continued till the end of 1966 when HBN was definitely replaced by HBG.

Since 1 January 1966, a new standard frequency and time signal service has been transmitted by the LF-transmitter HBG on 75 kHz, located at Prangins on the lake of Geneva. The power of 20 kW enables the uninterrupted transmissions to be received throughout Europe, without interference and independent of ionospheric disturbances. Carrier and time signals are based on a rubidium frequency standard having a day to day stability of better than $\pm 10^{-11}$. It is regularly calibrated against caesium frequency standards via VLF and flying clocks, to retain an absolute accuracy of $\pm 2 \times 10^{-11}$.

Taiwan. Professor Tsao

The methods and instrumentation of the time service have continued unchanged.

Turkey. Istanbul

Preliminary observations for time are being made with a Bamberg-Askania transit instrument using an Omega printing chronograph. A second quartz clock by Rhode and Schwarz has been installed.

A Danjon astrolabe will be installed in 1967.

An oscilloscope has been used for the reception of time signals. Continuous frequency comparisons with the 75 kHz transmission of HBG will commence early in 1967.

The time signals transmitted by station TAO3 on 7615 kHz were kept within ± 5 milliseconds of those of FTK77.

U.K. National Physical Laboratory, Teddington

The long caesium beam tube has continued to operate as the standard and the accuracy has been doubled, to one part in 10^{11}, by careful attention to the field conditions. It's precision should enable it to be used with an accuracy of one part in 10^{12}, and a commercial equipment has been purchased so that more extensive experimental work with this object in view can be carried out on the beam standard. Two commercial rubidium gas cells are used as working standards at the NPL and similar standards are being installed to control the MSF standard frequency transmissions. The 60 kHz transmissions have operated 24 hours per day since May 1966. When atomic control is installed in the latter part of 1966 the carrier frequency will be maintained at its nominal value and the offset will be applied to the time signals by means of a continuously operating phase shifter. Experimental work is in progress on a hydrogen maser and a second model is in course of construction.

U.K. Royal Greenwich Observatory

The Greenwich Atomic Time Scale has been published quarterly. Although it has previously been published under other names, e.g. Cs (EB) and Cs (H 12), the G.A. time scale is continuous from June 1955 and has been based on the summations of the rates of quartz clocks calibrated in terms of the caesium atomic frequency standard at the National Physical Laboratory. A commercial caesium beam frequency standard was installed at the Royal Greenwich Observatory in May 1966.

The Royal Greenwich Observatory has continued to participate in the co-ordinated radio transmissions. The GBR 16 kHz transmitter was out of operation for modernization from 1 January 1966 until 30 November 1966. During this period the service of VLF radio time signals was provided by the reserve transmitter Criggion GBZ on a frequency of 19·6 kHz. The MSF standard frequency and time signal service radiated from Rugby on a frequency of 60 kHz was extended from one hour to 24 hours per day from 1 June 1966.

The weekly circulars published jointly by the Royal Greenwich Observatory and the U.S. Naval Observatory give provisional corrections to the time signals in terms of U.T.1., U.T.2 and A.1.

In order to attain co-ordination of radio time signals to 1/10th of a millisecond the Royal Greenwich Observatory has taken part in several experiments using travelling quartz crystal or atomic clocks. VLF phase comparisons have also been used to maintain co-ordination.

U.S. Naval Observatory

The determination of Universal, Ephemeris and Atomic times was continued at Washington, D.C., and Richmond, Florida, utilizing the PZT, Danjon Astrolabe, Dual-Rate Moon-Position camera, and atomic clocks.

The time transmissions of U.S. Naval radio stations and the frequency of 5 VLF stations were controlled to 1×10^{-10}. The maximum accumulation of time difference due to frequency deviations of the VLF transmission is 50 microseconds. The carrier frequency of NSS, Annapolis, Maryland, on 21·4 kHz is derived directly from a caesium beam standard. The East Coast Loran-C chain of the U.S. Coast Guard is controlled in time to 25 microseconds and frequency to 3×10^{-11}.

The carrier frequencies of Omega VLF stations are controlled to 1×10^{-11}. These stations are located in Norway, New York (U.S.A.), Trinidad and Hawaii (U.S.A.). They time-share the frequencies 10·2 and 13·6 kHz. Trinidad also transmits on 12 kHz. Forestport (New York) is scheduled to commence experimental transmissions on 12·4 and 12·5 kHz in June 1967. These frequencies are derived directly from caesium-beam frequency standards.

Relay II satellite was used in clock synchronization experiments made jointly with the Radio Research Laboratories, Tokyo, Japan. An accuracy of 0·1 microsecond was achieved for the satellite ground stations at Mojave and Kashima.

The Naval Observatory has installed a Precise Time Synchronization Service whose mission is to carry time and frequency to remote locations with an accuracy of one microsecond in time and one part of 10^{12} in frequency. Caesium-beam portable clocks are used.

Two PZT's are being constructed in our shop. One is for the Dominion Observatory to be used at Ottawa, Canada. The other will be tested in Washington for about one year and then sent to Punta Indio, Argentina, where it will be operated by the Hydrographic Office of Argentina and the La Plata Observatory.

U.S.S.R.

En U.R.S.S. 12 observatoires et laboratoires spécialisés ont effectué les déterminations astronomiques du temps.

Ce sont essentiellement les instruments photoélectriques de passage et les astrolabes à prisme Danjon qui ont été utilisés. Une attention particulière a été portée sur le perfectionnement de la méthode photoélectrique de l'enregistrement des passages des étoiles.

Les observations de tous les services horaires de l'U.R.S.S., de deux services horaires de la République de Tchécoslovaquie, du service horaire de la République Démocratique Allemande, de deux services horaires de la République Populaire de Chine et de celui de la République Populaire de Pologne (à partir d'août 1965) participant au travail du Service Horaire d'Etat de l'U.R.S.S. au titre de la coopération libre ont été utilisés pour calculer l'heure définitive.

L'heure définitive des radiosignaux a été publiée dans les bulletins mensuels 'L'heure étalon aux moments moyens des émissions des radiosignaux'. Jusqu'à 1966 le calcul de l'heure étalon s'est effectué dans le système T.U.2, à partir de 1966 le calcul se fait dans le système T.U.1. A partir de 1965 les différences entre le temps universel uniforme provisoire T.U.2, déterminé au cours du calcul de l'heure étalon, et le temps atomique T.A.1, basé sur l'Etalon d'Etat de temps et de fréquence de l'U.R.S.S. ont été publiées.

Les transmissions des signaux de l'heure ont été effectuées dans le système du temps atomique uniforme. Les décalages de fréquences introduits pour rapprocher le système du temps atomique au système du temps uniforme provisoire ont été déterminés sur la base des études du changement de la vitesse de la Terre, effectuées régulièrement à l'Institut des Mesures Physicotechniques et Radiotechniques (IMPR).

Pour les années 1964 et 1965 la valeur du décalage de fréquence a été admise égale $+ 200 \times 10^{-10}$ (au lieu de $+ 150 \times 10^{-10}$, recommandée par le BIH). La valeur du décalage de fréquence, admise par les services horaires de l'U.R.S.S., répondait mieux à la vitesse réelle de rotation de la Terre. En 1964 et 1965 les services horaires de l'U.R.S.S. n'ont introduit que trois fois les corrections dans les moments des émissions des signaux soit en tout 200 ms, les services horaires coordonnés ont effectué pour cette période six corrections par 100 ms.

Pour l'année 1966 la valeur du décalage de fréquence ($+ 300 \times 10^{-10}$) d'après les déterminations de I'IMPR a coïncidé avec la valeur recommandée le BIH.

Le lien existant entre les changements de la vitesse de rotation de la Terre, les déformations de l'écorce terrestre et l'activité solaire a été étudié à l'Observatoire de Poulkovo. Les variations saisonnières de la longitude de l'Eurasie relativement à l'Amérique, découvertes par N. Stoyko, sont dues à la circulation saisonnière de l'atmosphère. N. N. Pavlov a découvert que la vitesse de rotation de la Terre varie avec une période égale à celle de l'activité solaire. Ces variations sont dues à l'influence de la circulation atmosphérique. Sur la base des comparaisons des résultats des observations de Poulkovo, de Moscou et de Zi-Ka-Wei ont été découverts des déplacements irréguliers et assez rapides de l'écorce terrestre. Ces déplacements sont également liés aux cycles de courtes périodes de la circulation atmosphérique. La vitesse de ces déplacements est bien plus grande que le terme annuel saisonnier.

Le catalogue de base des ascensions droites de 512 des étoiles de la liste des services horaires de l'U.R.S.S. est achevé. Il est basé sur les trois derniers catalogues du service horaire de Poulkovo, obtenus à l'aide d'instruments photoélectriques de haute qualité et contenant plus de 6000 observations individuelles. La comparaison de ce catalogue avec le FK4, le N30 et les autres catalogues a montré que sa précision est bien supérieure surtout en ce qui concerne les erreurs systématiques.

RAPPORT DU BUREAU INTERNATIONAL DE L'HEURE

Statuts du BIH

Conformément au voeu de FAGS, le BIH a été réorganisé et il a été créé un Comité de Direction comprenant des représentants de l'UAI, de l'UGGI et de l'URSI. Ce comité est ainsi composé en 1966.

H. M. Smith, représentant l'UAI, président
N. N. Pavlov, représentant l'UAI
W. Markowitz, représentant l'UGGI
B. Decaux, représentant l'URSI
B. Guinot, Directeur du Service.

Les buts du service ont été redéfinis. Ils sont:

(*a*) centraliser toutes les déterminations astronomiques ou physiques du temps,

(*b*) discuter ces déterminations et publier la meilleure valeur de l'heure d'émission des signaux horaires dans les échelles de temps astronomiques ou physiques en usage,

(*c*) contribuer à la coordination des émissions de temps en diffusant tous les avis et donnés nécessaires à cet effet,

(*d*) diffuser tous les renseignements utiles aux usagers du temps et fournir sur demande toutes les données sur le temps.

Activités du BIH de 1964 *à* 1966

1. *Temps coordonné*

Le BIH a attaché une importance particulière à la coordination des émissions de signaux horaires. Il a introduit en 1964 la notion du *temps coordonné*, T.U.C., qui est une variante du temps atomique. C'est dans l'échelle T.U.C. que sont émis les *signaux coordonnés*. L'UAI et le CCIR ont chargé le BIH du maintien de l'échelle T.U.C., qui doit être proche de T.U.2. Le BIH a diffusé les avis de sauts et de décalage de l'échelle T.U.C.; il doit, de plus, suggérer les petits ajustements à faire pour ramener dans l'échelle T.U.C. les signaux horaires qui s'en écarteraient.

L'adoption de l'échelle T.U.C. a permis une simplification du Bulletin Horaire qui ne donne plus que de 5 jours en 5 jours les différences T.U.2 – T.U.C., et pour chaque mois l'écart entre T.U.C. et le temps d'émission de chaque signal. Il est remarquable que cet écart a diminué considérablement en 2 ans grâce à des ajustements de nombreux signaux et que maintenant, il est le plus souvent inférieur à 1 ms.

2. *Temps atomique*

L'échelle de temps atomique A.3 a été régulièrement établie. L'intercomparaison des étalons, par l'intermédiaire des fréquences radiodiffusées, est maintenant faite sur ordinateur, ainsi que la formation des échelles de temps associées, en tenant compte de toutes les équations de condition liant les fréquences des étalons et des signaux. L'échelle A.3 a été redéfinie en 1966, comme une échelle moyenne de tous les étalons primaires. Ceci n'a pas affecté sa continuité.

T.U.C. est déduit de l'échelle A.3. On est en train d'étudier le maintien de la coordination des signaux lorsqu'elle a été précédemment établie par une expérience telle que celle des 'horloges volantes' par des liaisons par satellites. Ce maintien est, en principe, assuré par la comparaison des fréquences porteuses des signaux. Si les résultats de cette étude sont satisfaisants, les écarts T.U.C – signaux seront ainsi calculés. Pour l'instant, ces écarts sont encore déterminés en appliquant des corrections de durée de propagation.

Le Bulletin Horaire publie de 5 jours en 5 jours les valeurs de T.U.2 – A.3 et A.3 – T.U.C., ainsi que les écarts avec A.3 des échelles associées à chaque étalon.

3. *Temps universel*

Jusqu'à l'année 1965 inclus, l'heure définitive a été calculée en 3 étapes:

(*a*) calcul par le BIH des coordonnées du pôle et des corrections de longitude, à partir des mesures de latitude,

(*b*) formation, par les stations, des heures demi-définitives, par lissage,

(*c*) calcul par le BIH de l'heure définitive, comme la moyenne des heures demi-définitives

Bien que certaines parties de ces calculs aient été automatisées, il est clair qu'aucun gain de temps sérieux ne peut être espéré sans une modification profonde de ce schéma de calcul.

Nous avons mis au point une méthode pour obtenir en un seul passage en machine, à partir des mesures de latitude et de temps T.U.0, les coordonnées du pôle et la valeur de T.U.1 – T.U.C. Le but que nous nous proposons est d'obtenir l'heure définitive T.U.1 – T.U.C. dans les mêmes délais que ceux que nous utilisons actuellement pour publier les coordonnées du pôle et les corrections de longitude. Pour atteindre ce but, nous avons demandé aux observatoires participants d'envoyer leurs valeurs individuelles de T.U.0 – T.U.C. et de latitude, dans les délais très brefs, en utilisant des formulaires établis conjointement par le SIMP et le BIH. Dans cette nouvelle méthode, les lissages au niveau des stations sont évités autant que possible.

Les coordonnées du pôle et l'heure définitive ont été recalculées par cette méthode, pour 1964 et 1965, à titre d'essai. Voici comme exemple les résultats de ces calculs à partir de 20 stations, dont 13 de latitude et 7 de temps et latitude, avec et sans terme z en latitude. Dans les calculs complets, effectués pour chaque 1/20 d'année, entrent une soixantaine de stations.

Exemple de calcul simultané des coordonnées du pôle et de T.U.2 – T.U.C. à partir des données de 20 stations (20 en latitude, 7 en temps).

Date	avec terme z en latitude				sans terme z en latitude		
	x	y	z	T.U.2 – T.U.C.	x	y	T.U.2 – T.U.C.
	$0''.001$	$0''.001$	$0''.001$	$0^{s}.0001$	$0''.001$	$0''.001$	$0^{s}.0001$
1964, 0	– 200	– 138	+ 003	9039	– 199	– 139	9040
1	– 242	– 24	+ 8	8865	– 238	– 27	8866
2	– 225	+ 127	+ 13	9647	– 219	+ 123	9650
3	– 115	+ 227	– 1	9442	– 115	+ 228	9442
4	+ 18	+ 262	– 20	9214	+ 9	+ 269	9210
5	+ 156	+ 214	– 26	9014	+ 145	+ 223	9010
6	+ 228	+ 93	– 13	8816	+ 222	+ 98	8814
7	+ 237	– 37	+ 13	9577	+ 243	– 41	9579
8	+ 181	– 150	+ 17	9244	+ 188	– 156	9247
9	+ 58	– 208	+ 15	8989	+ 65	– 214	8991
1965, 0	– 62	– 210	+ 18	9726	– 54	– 217	9729
1	– 166	– 143	+ 19	9517	– 158	– 149	9520
2	– 217	– 33	+ 16	0239	– 210	– 39	0241
3	– 197	+ 93	– 6	9898	– 200	+ 95	9897
4	– 143	+ 184	– 28	9612	– 156	+ 193	9607
5	– 40	+ 236	– 21	0253	– 49	+ 243	0249
6	+ 105	+ 196	– 6	9906	+ 102	+ 198	9905
7	+ 185	+ 101	+ 14	9497	+ 192	+ 97	9500
8	+ 226	– 12	+ 18	0085	+ 234	– 18	0088
9	+ 195	– 105	+ 6	9691	+ 197	– 107	9692

Pour l'année 1966, la nouvelle méthode est employée en reconstituant les valeurs de T.U.0 – T.U.C. d'après les heures demi-définitives des stations.

Pour l'année 1967, les valeurs individuelles de T.U.0 – T.U.C. seront reçues directement.

Publication du Bulletin Horaire

Les modifications signalées dans les paragraphes précédents ont permis de réduire le volume du Bulletin Horaire, malgré l'addition d'un tableau récapitulatif des principales émissions de signaux horaires. Ce tableau paraît chaque année et il est tenu à jour par des corrections paraîssant dans chaque Bulletin.

Le Bulletin qui était consacré à l'heure demi-définitive de Paris a été supprimé. Il est provisoirement remplacé par un encart rose qui disparaîtra à son tour quand la publication de l'heure définitive sera suffisamment accélérée.

Personnel

Le personnel participant aux travaux du BIH est, à la fin de 1966:

M. B. Guinot, Directeur
Melle A. M. Desprats, Assistante }
M. E. Eisop, Calculateur } à plein temps
Melle R. Barthalot, Calculatrice }

M. R. Forga, Aide Astronome, Melle M. Feissel, assistante, Mme C. Mulhauser, secrétaire et M. R. Thomas, comptable, ont travaillé à temps partiel ou occasionnellement pour le BIH.

B. GUINOT
Directeur du BIH

33. COMMISSION DE LA STRUCTURE ET DE LA DYNAMIQUE DU SYSTEME GALACTIQUE

PRÉSIDENT: Professor B. J. Bok, Steward Observatory, Tucson, Arizona, U.S.A.

VICE-PRÉSIDENT: Professor Dr G. Contopoulos, Astronomical Department, University of Thessaloniki, Thessaloniki, Greece.

COMITÉ D'ORGANISATION: W. Becker, A. Blaauw, C. O. R. Jaschek, E. K. Kharadze, M. Schmidt.

MEMBRES: Agekian, Ambartsumian, Barbanis, Beer, Bidelman, Bigay, Blanco, Boulon, Crawford, Delhaye, Dzigvashvili, Edmondson, Elsässer, Elvius (A. M.), Elvius (T.), Fehrenbach, Fricke, Gyldenkerne, Haro, Haug, Heckmann, Henize, Hoffleit, Ikaunieks, Irwin, Iwaniszewska-Lubienska, Johnson (H. M.), Kaburaki, Kerr, Kholopov, King (I. R.), Kinman, Kopylov, Kraft, Kukarkin, Kurth, Kuzmin, Lindblad (B.)†, Lindblad (P. O.), Lunel, Luyten, Lynden-Bell, McCarthy, McCuskey, MacRae, Malmquist, Mavridis, Mirzojan, Mohr, Münch (G.), Nahon, Nassau†, Ogorodnikov, Oort, Ovenden, Perek, Petrie (R. M.)†, Pilowski, Priester, Ramberg, Roman, Rudnicki, Schalén, Schilt, Schmidt (H.), Schmidt-Kaler, Shimizu, Stibbs, Strömgren, Vanderlinden, Van Hoof, Velghe, Wayman, Weaver, Westerhout, Westerlund, Woolley.

I. INTRODUCTION

In order to avoid any overlapping with the Reports of related Commissions, we shall strictly limit ourselves in the present Report to matters concerning the Structure and Dynamics of our Galaxy. We shall carefully avoid dealing with topics that are primarily the concern of related Commissions, such as Nos. 25, 27, 29, 30, 34, 37, 40, and 43. Sections I to IV and VI to VIII have been written by Bok; Section V by Becker; and Sections IX and X by Contopoulos. Bok wishes to acknowledge the assistance received from Jane M. Basinski, T. E. Margrave, and G. A. Chapman; Contopoulos, that from B. S. Barbanis and J. Hadjidemetriou. Mr. and Mrs Margrave kindly assumed responsibility for the final editing of the Report. The Report on Selected Areas is, as in the past, appended to the present Report; it was written by Dr T. Elvius, who wishes to draw attention to the increasing importance for galactic research of work on the Kapteyn Selected Areas.

Galactic Structure and Dynamics continues to be a very active and broad field of research. The year 1966 saw the publication of Volume V of *Stars and Stellar Systems* (**1**), which beautifully covers our whole field of research. The President wishes to draw attention to the review of Volume V written by himself (**2**), which was prepared with the work of our Commission in mind. All of us now eagerly await the publication of Volume VII of *Stars and Stellar Systems*.

Several Symposia were held that have advanced research in our areas. The most relevant IAU Symposia were No. 25 on 'The Theory of Orbits in the Solar System and in Stellar Systems' (August 1964, Thessaloniki, Greece), No. 30 on 'Determination of Radial Velocities' (June 1966, Toronto, Canada), and No. 31 (jointly with URSI) on 'Radio Astronomy and the Galactic System' (August 1966, Noordwijk, Holland). IAU Symposia Nos. 24 and 26 are also of special interest to the members of our Commission. A first, general, report on IAU–URSI Symposium No. 31 has been written by Bok (**3**).

During September 1964, the NATO Science Committee organized a two-weeks Summer Course on 'Observational Aspects of Galactic Structure'; many members of our Commission participated, with Blaauw and Mavridis acting as organizers. A volume based upon the lecture notes was prepared by Mavridis and has been made fairly generally available. A second NATO Summer Course was held at Herstmonceux in August–September 1965, on 'The Kinematical and Chemical History of the Galaxy'. Dr Bernard E. Pagel has kindly prepared a summary of the lectures for use by Contopoulos and myself in the preparation of the present Report. Members of Commission 33 will also be interested in the Report of the 'Tenth Herstmonceux Conference' held in April 1966 (**4**); we shall have occasion to refer to it in later Sections.

Two informal Symposia held in recent years concern Commission 33. In August 1964, a 'First Conference on Faint Blue Stars' was held in Strasbourg with about a dozen astronomers attending. The Observatory of the University of Minnesota has published the papers in a separate small volume (1965) with W. J. Luyten as editor. The summaries of the comments made at this Symposium following the presentation of each paper should be read by all members of Commission 33. The second informal 'Pre-Symposium' was one held in May 1966, at Mt Stromlo Observatory on 'Radio and Optical Studies of the Galaxy'. The proceedings have received limited distribution from Mt Stromlo Observatory.

Two other useful reports have been made available for use in the preparation of the present Report. One is a report by Kuzmin, which summarizes advances made over the past three years in the U.S.S.R. in areas of interest to Commission 33. The other report is by Nahon, and it concerns a Colloquium held at Besançon on 'Les Nouvelles Méthodes de la Dynamique Stellaire' (September 1966). The President and Vice-President of Commission 33 have been greatly assisted by these reports. We should also note that Kharadze's Report to the Fourth Full Session of the Commission on Stellar Astronomy of the Astronomical Council, Ac. Sci., U.S.S.R. (1964, Moscow) has now been published (**5**).

Survey articles relating to the work of our Commission have been published in several places. Particularly I wish to draw attention here to Volumes 7 and 8 of *Vistas in Astronomy* (**6**), (**7**). The first volume contains two articles that are especially relevant: one by Iwanowska on 'Statistical Population Indices', the other by McCuskey on 'The Stellar Luminosity Function'. Volume 8 contains the Proceedings of the 'Ejnar Hertzsprung' Symposium, which was held at Flagstaff, Arizona, in June 1964, and which dealt with 'Aspects of Stellar Evolution'. Three more volumes of the *Annual Review of Astronomy and Astrophysics* have appeared; each volume contains several articles of interest to our members.

The Fourteenth International Astrophysical Symposium held at Liège (June 1966) was devoted to problems of special interest to our Commission. The published volume of the Proceedings is awaited with interest.

Brief reference should be made to Joint Discussion A at the Hamburg General Assembly on 'Local Structure and Motions in the Galaxy', which was entirely concerned with problems affecting the work of Commission 33 (**8**). Finally, we should not fail to mention the publication of the New Series of Landolt–Börnstein (edited by Voigt (**9**)) which has several chapters with reference material that should be useful to members of our Commission.

The only major textbook that has been published in the area of our Commission is an English translation of a book by Ogorodnikov (**10**); it makes a very welcome addition to the literature.

BIBLIOGRAPHY

1. Blaauw, A., Schmidt, M. (eds.). 1965, *Galactic Structure: Stars and Stellar Systems*, Volume V, Univ. of Chicago Press.

2. Bok, B. J. 1966, *Sky and Telescope*, **31**, 357.

3. Bok, B. J. 1966, *Sky and Telescope*, **31**, 271 and 341.

4. Report of 'Tenth Herstmonceux Conference'. 1966, *Observatory*, **86**, 180.

5. Kharadze, E. K. 1965, *Abastumani astroph. Obs. Bull.*, no. 33.

6. Beer, A. (ed.). 1966, *Vistas in Astronomy*, Vol. 7, Pergamon Press.
7. Beer, A., Strand, K. Aa. (eds.). *Vistas in Astronomy*, Vol. 8, Pergamon Press.
8. *Trans. IAU*, **12 B**, 389, 1966.
9. Voigt, H. H. (ed.) 1965, *Landolt-Börnstein*, Group VI, Vol. I, Springer-Verlag.
10. Ogorodnikov, K. F. 1965, *Dynamics of Stellar Systems* (Transl. by J. B. Sykes and A. Beer), Pergamon Press.

II. THE OVERALL STRUCTURE OF OUR GALAXY

The major accomplishment of our Commission at the Twelfth General Assembly (Hamburg) of the IAU was agreement among the members present with regard to the adoption of standard values for the basic parameters of our Galaxy:

$R_0 = 10$ kpc.
$\Theta_0 = 250$ km s^{-1},
$A = 15$ km s^{-1} kpc^{-1}, and
$B = -10$ km s^{-1} kpc^{-1}.

So far no major objections to these values have been raised, and there appears to be no need to start discussions anew at the Prague Assembly. The minutes of our Commission Meeting at Hamburg (*Trans. IAU*, **12 B**, p. 314 to 316) adequately summarize the situation. Kuzmin's Report (see Section I) has the following relevant paragraph:

'Theoretical foundations for obtaining a consistent system of galactic parameters on the basis of stellar dynamical equations as well as for constructing empirical models of the Galaxy have been worked out by Kutuzov (**1**) and Einasto (**2**). The latter has constructed a new composite model of the Galaxy. A preliminary consistent system of the local galactic parameters has been published (**3**), where, in particular, a moderate value for the dynamical parameter C has been accepted (cf. Einasto, *Trans. IAU*, **12 B**, p. 436). The system of galactic parameters has been discussed also by Genkin (**4**).'

We note that the 'consistent system of local galactic parameters' by Einasto and Kutuzov (**3**) has the following basic parameters:

$R_0 = 9{\cdot}0 \pm 0{\cdot}4$ kpc,
$\Theta_0 = 235 \pm 10$ km s^{-1},
$A = 15{\cdot}7 \pm 0{\cdot}4$ km s^{-1} kpc^{-1}, and
$B = -10{\cdot}3 \pm 0{\cdot}4$ km s^{-1} kpc^{-1},

in good agreement with the Hamburg values.

The principal difference that remains between the Tartu data and the values proposed by Oort concerns the dynamical value of the mass density in the galactic plane at the position of the Sun, which, according to Oort (**5**) and others, is equal to $0{\cdot}145$ $\odot$/pc^3 and which, according to Einasto and Kutuzov (**3**) and Einasto (**6**), amounts to only $0{\cdot}09$ $\odot$/pc^3. It seems very important that this discrepancy be clarified. The Oort value requires that, in addition to the known stars, atomic hydrogen and cosmic dust, there be a component of unknown origin contributing 30% of the gravitational mass density, which may consist of either gaseous molecules or sub-luminous stars. The lower value of the Tartu observers would imply that almost all of the mass density near the Sun has been accounted for. The discrepancy can be expressed most clearly in terms of Kuzmin's dynamical parameter C, which is defined (**6**) so that $-C^2$ measures the gradient of gravitational acceleration in the z-direction at the Sun, $z = 0$. The Tartu value is about $C = 73$ km s^{-1} kpc^{-1}, whereas the values found by others are generally 25% higher. The only recent value for C that agrees with the Tartu results is that of Stothers and Tech (**7**).

Members of Commission 33 will remember the report by the late Dr R. M. Petrie, presented at the Hamburg Meeting, in which he gave a revised value for A, equal to 15·6 km s^{-1} kpc^{-1}. His widow, Mrs J. K. Petrie, very kindly has written that a more refined study of the earlier material is in progress. She reports the following results obtained by S. C. Morris and herself:

1. From Stellar Velocities, O9–B8, 688 stars:

$$A = 15{\cdot}1 \pm 0{\cdot}4 \text{ km s}^{-1} \text{ kpc}^{-1},$$
$$K = -0{\cdot}6 \pm 1{\cdot}3 \text{ km s}^{-1},$$

2. From Interstellar Velocities, O9–B3, 286 stars:

$$A = 15{\cdot}7 \pm 0{\cdot}5 \text{ km s}^{-1} \text{ kpc}^{-1},$$
$$K = -0{\cdot}6 \pm 0{\cdot}3 \text{ km s}^{-1}.$$

Multi-parameter solutions yield essentially the same results.

W. Fricke reports that he has re-analyzed the McCormick and Cape proper motions, previously studied by Williams and Vyssotsky. On the system of FK4, Fricke finds:

$$A = +12{\cdot}4 \pm 2{\cdot}4 \text{ (m.e.) km s}^{-1} \text{ kpc}^{-1},$$
$$B = -7{\cdot}4 \pm 2{\cdot}4 \text{ (m.e.) km s}^{-1} \text{ kpc}^{-1}.$$

He reports, furthermore, that he has underway an investigation of 500 stars with known proper motions and at distances greater than 500 parsecs from the Sun. These stars, for most of which radial velocities are available, should serve for a study of the distance scale and of star streaming.

In contrast to Fricke's result, we quote a value of B recently obtained by Wayman (**8**), who finds:

$$B = -15 \text{ km s}^{-1} \text{ kpc}^{-1}.$$

This value was obtained from an analysis of high-precision (mean errors less than $\pm$ 0″002 per year) proper motions in the FK4 Catalog for 34 O and B stars at distances greater than 300 parsecs.

A recent result by J. S. Miller of Washburn Observatory deserves to be noted. He has obtained radial velocities for about 30 H II regions for which distances are available. A preliminary analysis shows that deviations from circular motion average less than 10 km s^{-1} and the tentative value for A is 17 km s^{-1} kpc^{-1}; there is a general agreement between the H I and H II velocities.

Attention is drawn to Blaauw's (**9**) useful tabulations and graphs relating to space distributions and motions of various classes of objects in our Galaxy.

Recent work by H. L. Johnson and associates (**10**) has given new arguments in favor of possible revision of the cosmic distance scale. In most earlier work, the assumption was made that the ratio between visual total absorption, A_V, and the color excess, E_{B-V}, was equal to 3·0 or, at most, 3·5. There were some known exceptions to this rule, notably for the Trapezium Cluster in Orion and for a region in Cygnus, but no major variations were suspected elsewhere. The far infra-red measures by Johnson and associates now make it seem likely that greater values of $R = A_V/E_{B-V}$ prevail in many directions close to the galactic plane. Values as great as $R = 6$ are indicated by the infra-red measurements, especially for the sector $100° < l^{II} < 200°$. This article is not the place to enter upon a full discussion of the implications of Johnson's work, but we note that W. Becker (**11**) and others have objected strongly to certain aspects of Johnson's approach. It would be wrong for the members of Commission 33 to ignore the new Johnson results, for, even if they are substantiated only in part, they contribute a major new element in the interpretation of color excess as an indicator of total visual absorption. One consequence of Johnson's work is that within the next decade we may have to revise rather drastically the values for the 'constants' for our Galaxy which were adopted at the Hamburg meeting and which are listed earlier in this Section. It becomes all the more important that, in our analyses relating to the overall structure of our Galaxy, we should, for remote objects affected by the least total absorption, make every effort to obtain the most precise possible color data, referring to the largest available color base. We note that Johnson has repeated the galactic-rotation analysis that he and Svolopoulos (**12**) performed some years ago and we note further that, on the basis of his new distances for open clusters, he finds:

$$A = 20 \text{ km s}^{-1} \text{ kpc}^{-1},$$
$$R_0 = 7{\cdot}8 \text{ kpc}.$$

Continuing attention to the whole problem of the galactic distance scale is obviously a matter of high priority in the years to come.

Most of the current work relating to the central region of our Galaxy belongs properly within the areas covered by Commissions 40 and, to a lesser extent, 34; we shall not trespass unnecessarily. However, we must draw attention here to the far-reaching results that are emerging from the study by radio-astronomical techniques of the radial-velocity fields in the central regions of our Galaxy. The new information comes not only from 21-cm research, but also from studies of OH lines and from high-level recombination transitions in H I that are materially increasing our knowledge in the area. We shall also not report here on the exciting new results obtained for X-ray sources in our Galaxy, but we note that this field may become the concern of our Commission if the suggested identifications with old novae are substantiated.

We should not close this Section without a brief reference to the importance that variable stars, especially RR Lyrae variables, continue to play in the study of the overall properties of our Galaxy. The chapter by Plaut (**13**) illustrates this aspect of variable-star research very well; the matter is also stressed in the series of papers by Kinman and associates (**14**) which are based upon observational work at Lick Observatory.

BIBLIOGRAPHY

1. Kutuzov, S. A. *Tartu astr. Obs. Teated*, no. 9, 1964; also, no. 16, 1965.
2. Einasto, J. E. 1965, *Tartu astr. Obs. Teated*, no. 17.
3. Einasto, J. E., Kutuzov, S. A. 1964, *Tartu astr. Obs. Teated*, nos. 10 and 11.
4. Genkin, I. L. 1965, *Publ. astroph. Inst. Alma-Ata*, **5**, 101.
5. Oort, J. H. 1966, *Trans. IAU*, **12 B**, 315.
6. Einasto, J. E. 1966, *Trans. IAU*, **12 B**, 436.
7. Stothers, R., Tech, J. L. 1964, *Mon. Not. R. astr. Soc.*, **127**, 287.
8. Wayman, P. A. 1966, *Observatory*, **86**, 184.
9. Blaauw, A. 1965, *Galactic Structure: Stars and Stellar Systems*, Vol. V, 438.
10. Johnson, H. L. 1966, *Vistas in Astronomy*, **8**, 133; see also the chapter by H. L. Johnson in the forthcoming Volume VII of *Stars and Stellar Systems, Nebulae and Interstellar Matter* (available in Pre-print form from the Lunar and Planetary Laboratory, Univ. of Arizona, Tucson, Arizona).
11. Becker, W. *Z. Astrophys.*, **64**, 77, 1966; see also *Observatory*, **86**, 180, 1966.
12. Johnson, H. L., Svolopoulos, S. 1961, *Astrophys. J.*, **134**, 868.
13. Plaut, L. 1965, *Galactic Structure: Stars and Stellar Systems*, Vol. V, Chapter 13.
14. Kinman, T. D., Wirtanen, C. A., Janes, K. A. *Astrophys. J. Suppl.*, **11**, no. 100, 1965, and **13**, no. 119, 1966; *Contr. Lick Obs.*, nos. 174 and 202.

III. NEW BASIC MATERIAL

The past three years have witnessed the publication of much useful material for the study of the structure and dynamics of our Galaxy. While detailed and complete references belong in the Reports of other Commissions, we will note here some of the more important information that has come to our attention.

There continues to be much useful activity in the area of spectral classification. C. Jaschek, Condé, and de Sierra have published an extensive 'Catalog of Stellar Spectra Classified in the Morgan-Keenan System' (1). Seven additional parts have appeared of 'Luminous Stars in the Northern Milky Way': Volumes IV, IVa, VI, and VIa by J. J. Nassau, C. B. Stephenson, and D. J. McConnell of the Warner and Swasey Observatory, and Volumes I, III, and V by J. Hardorp, I. Theile, and H. H. Voigt of the Hamburg-Bergedorf Observatory. Stephenson and McConnell are planning to extend the survey over the remaining galactic longitudes $220° < l^{II} < 15°$. The work will be done with the 24–36 inch (61–91 cm) Curtis–Schmidt telescope of the University of Michigan, relocated at the Cerro–Tololo Inter–American Observatory in Chile. Bidelman reports that he expects the Curtis–Schmidt telescope to be in full operation in Chile by April 1967. The University of Michigan astronomers will retain one-third of the time of the instrument; they are planning to undertake a moderately high-dispersion blue survey of the entire southern sky, coupled with special studies for the galactic center and of southern associations. Bidelman reports that the search for Hα emission objects has continued. H. E. Bond of Michigan Observatory has developed a fruitful technique for the discovery of metal-weak stars; four-color measurements are being made for all such stars found on the Schmidt plates with the 10° prism, and radial-velocity measurements are planned for many of the new discoveries. Bidelman (2) has written a comprehensive article on techniques of classification from objective-prism spectra.

Morgan will soon publish a new and revised edition of the *Yerkes Spectral Atlas*. At Steward Observatory, A. B. Meinel is working on an *Atlas of Stellar Spectra* that reaches to 3300 Å in the ultra-violet. We note that some time ago there was held at Kitt Peak National Observatory a very interesting 'Discussion on Spectral Classification' in which some of these problems and projects were summarized (3). Lindoff and Lyngå (4) have published a study of comparisons between spectra classified on the MK, HD, and HDE systems. At the Vatican Observatory, P. J. Treanor, M. F. McCarthy, and F. C. Bertiau have undertaken spectral surveys with the Vatican Schmidt telescope and its 2°5, 4°, and 12° objective prisms.

There is no need to review here in any detail the researches that were reported at IAU Symposium No. 24 (Stockholm, August 1964), since the Symposium Volume has been distributed. We note that there continues to be much activity in the field of photo-electric measurement of brightnesses and colors of stars. Strömgren (5) has written an excellent summary of techniques and results obtained from narrow-band photometry, including Hβ photometry. The work in this area by Strömgren, Crawford, Perry, Graham and associates at Kitt Peak National Observatory is now being extended to stars of southern declinations observed from the Cerro-Tololo Observatory. The techniques of narrow-band photometry are also being applied extensively by Bigay and Mme Lunel at Lyon, by Gyldenkerne in Denmark, and by Golay in Geneva.

Schmidt-Kaler stresses the need for *UBV* photometry to supplement searches by the Warner and Swasey and Hamburg-Bergedorf Observatories for luminous stars. A joint effort is being made by Schmidt-Kaler, Haug, Bigay, and Mme Lunel to obtain *UBV* colors and magnitudes for 1400 stars in the Cleveland-Hamburg lists which lack such data. Cooperation is asked for the observation of the remaining 4800 stars. The European Southern Observatory's new Schmidt telescope will be used for the work on southern spectral classification. C. Jaschek reports that a comprehensive catalog of photo-electric magnitudes and colors is in preparation at La Plata; this catalog is intended to supplement the spectral catalog to which reference was made earlier in this section. At the Stockholm Symposium (IAU No. 24), Bok drew attention to the powerful new search techniques that are available now through combined photographic (*UBV*) and photo-electric work. He concludes that the discovery of OB stars to $V = 15$ and with total absorptions not exceeding $A_v = 5$ magnitudes is now within reach.

Beer is reducing all measures of Balmer line equivalent widths for O7 to B9 to a uniform system, calibrated from cluster data. This material should be useful for the study of local spiral structure.

There has been a comprehensive survey of current work in the field of radial velocities at IAU Symposium No. 30, held at Toronto in June, 1966. The published volume is awaited with interest. Rapid advances in radial-velocity work for faint stars may be expected on the basis of recent developments involving applications of image conversion techniques. The measurement of radial velocities for faint stars of special interest has now become practicable with reasonably short exposure times. Bertiau, Burley, and McCarthy (**6**) have published new galactic coordinates for the 15 106 stars listed in Wilson's *General Catalog of Radial Velocities*. Recently Bertiau has devised a program for selecting and plotting various types and classes of stars according to their galactic coordinates.

For the field of proper motions, we refer to the Report for Commission 24. We note that Fricke has made a study of the distance scale and of galactic-rotation effects based upon proper motions reduced to the FK4 system (**7**). He concludes that mean parallaxes derived from proper motions can be expected to yield reliable results only for stars within 1000 parsecs of the Sun. A rather parallel type of study is reported at the U.S. Naval Observatory, where precision proper motions, based on FK4, are being determined for 1000 stars with known MK classes. The stars are a part of the W^3_{50} Catalog, derived from observations with the 6-inch Transit Circle (**8**). The publication of a new edition of the *Yale Catalogue of Bright Stars* will assist greatly in the furtherance of galactic research (**9**).

There are many special varieties of stars that contribute to our knowledge of galactic structure, but none seems to be quite so useful as are the Wolf-Rayet stars. Their absolute magnitudes are becoming known with increasing precision through the work of Westerlund and Lindsey F. Smith on WR stars in the Magellanic Clouds. The current surveys, coupled with special narrow-band photometry for these stars, provide basic structural information to great distances. Westerlund (**10**) has measured photoelectrically, in a narrow-band four-color system, 35 objects listed by Roberts as WR stars. The as-yet unpublished doctoral thesis of Lindsey F. Smith contains a wealth of new material. One of her tables is a list of absolute magnitudes, color excesses, and estimated distances for 121 galactic WR stars; the results are based mostly on her own system of narrow-band photometry (six interference filters). The search for WR stars is being pushed to the faintest accessible limits in both the northern and southern hemispheres. Because of emission features in the spectra of these stars, their detection to $B = 15$ or $B = 16$ can be achieved with relatively small telescopes .C. B. Stephenson (**11**) has completed an objective-prism survey of WR stars for $10° < l^{II} < 240°$ in which seven new objects were found. None of these are in the range $135° < l^{II} < 225°$; this section of the band of the Milky Way apparently shows a very real gap in the distribution of the WR stars.

BIBLIOGRAPHY

1. Jaschek, C., Condé, H., de Sierra, A. C. 1964, *Publ. Obs. La Plata*, Ser. Astr., **28**, Part II.
2. Bidelman, W. P. 1966, *Vistas in Astronomy*, **8**, 53.
3. Abt, H. A. 1963, *Astrophys. J. Suppl.*, **8**, no. 75.
4. Lindoff, U., Lyngå, G. 1965, *Lund Medd.*, Series II, no. 146.

5. Strömgren, B. 1966, *A. Rev. Astr. Astrophys.*, **4**, 433.
6. Bertiau, F. C., Burley, J. M., McCarthy, M. F. 1965, *Ric. astr. Vatican*, **7**, 21.
7. Fricke, W. *Vistas in Astronomy*, **8**, 205, 1966; *Z. Astrophys.*, **61**, 20, 1965.
8. Adams, A. N., Bestul, S. M., Scott, D. K. 1964, *Publ. U.S. naval Obs.*, Second Series, **19**, Part I.
9. Hofflei D. 1964, *Catalogue of Bright Stars*, Third Revised Edition, Yale Observatory.
10. Westerlund B. E. 1966, *Astrophys. J.*, **145**, 724.
11. Stephenson C. B. 1966, *Astr. J.*, **71**, 477.

IV. CALIBRATION PROBLEMS; LUMINOSITY FUNCTIONS

We noted in our Report for 1964 that problems of the calibration of criteria for absolute magnitudes seemed reasonably well in hand, thanks especially to the work of Blaauw, Voigt, Strömgren and others. The Hyades moving cluster plays an important role in all this work. The recent publication by Hodge and Wallerstein (**1**) of a revised distance for the Hyades produces a correction of 0·39 magnitude to the distance modulus of the Hyades. If accepted, this result would radically change our concepts of the dimensions of our Galaxy and of most basic constants in current use. The zero-age main sequence would have to be revised and the distance from the Sun to the center of the Galaxy might be as great as 12 kpc! Hodge and Wallerstein place their confidence in the published trigonometric parallaxes and point out that small and, in their opinion, permissible systematic errors in proper motions in declination would make the moving-cluster parallax of the Hyades agree with their trigonometric value. This possibility seems supported by results based upon the mass-luminosity relation and by absolute luminosities estimated with the aid of the Wilson-Bappu effect. The required systematic errors in the proper motions seem excessive, but the paper shows once again how urgently we need reliable trigonometric parallaxes for a representative group of fainter members of the Hyades; possibly undetected systematic errors in proper motions may basically affect our work on even the nearest moving clusters.

During the next few years, we should plan to emphasize increasingly studies of nearby moving clusters, with, as the ultimate aim, the obtaining of improved calibrations of absolute magnitudes derived from narrow-band photometry and precise spectral-luminosity classification. The recent announcement (28 December 1966; Am. Astron. Soc. Meeting) by R. F. Garrison of a narrow main-sequence for B2 to F0 members in the inner region of the Scorpius-Centaurus Association is especially significant for future calibration developments.

Stars of various spectral classes have been the subject of extensive studies of their mean absolute magnitudes and dispersions around these mean values. McCuskey and Rubin (**2**) have made an analysis of proper-motion and radial-velocity data for 6th magnitude F0, F2, and F5 stars listed in the Third Edition of the *Yale Catalogue of Bright Stars*. From these data they find $M_v = +2{\cdot}6$ and a dispersion $\sigma = \pm 0{\cdot}9$ magnitude for this group as a whole per unit volume of space. Ljunggren and Oja (**3**) have studied similar problems for the MK and Uppsala giant stars of types G8 to K5; they used trigonometric parallaxes as well as proper motions and radial velocities. They have found indications that the mean absolute magnitudes of these stars are dependent upon their kinematical properties and their distances from the galactic plane. From trigonometric parallaxes they derive for the Uppsala giants, gG8 to gK5, and for the MK giants, G8III to K5III, an absolute magnitude $M_0 = +0{\cdot}9$ (visual), $\sigma = \pm 0{\cdot}6$.

Efforts are continuing to separate stars of given general characteristics into groupings according to chemical composition. A report on some of this work appears as Section V, and there is further information in Section IX. However, we note here that Iwanowska and her colleagues at Torun are developing a scheme according to which stars of certain characteristics are being separated statistically into two main population types: one originating in or near the galactic plane but far from the galactic center, and the other originating in the central regions of our Galaxy. We have already referred in Section I to her general report in *Vistas in Astronomy* (Vol. **7**, p. 133, 1966); more detailed reports are contained in *Torun Observatory Bulletin*, Nos. 32 (1962), 37 (1965), and 41 (1966). Statistical probabilities, called 'statistical population indices', are assigned to each star; these indicate that the object belongs to one of the two classes.

At the end of Section II we referred to the increasing emphasis that is being placed upon the study of RR Lyrae variable stars; therefore, it is important to note that much work on the basic properties of these stars is in progress. The Reports of Commissions 24, 25, 27 and 30 should be consulted for details. Van Herk has published a monumental study of the proper motions, mean parallaxes, and space velocities of RR Lyrae variables (**4**). His derived value for the median photographic absolute magnitude of these stars is $M_{pg} = +0{\cdot}87 \pm 0{\cdot}22$ (m.e.), and the average normal color index is $(B - V)_0 = +0{\cdot}19$. There are prospects of much additional radial-velocity work, especially through the collaborative efforts of the Royal Observatory and of Helwan Observatory (**5**). The results of a careful and extensive investigation of intrinsic *UBV* colors of RR Lyrae variables have been published by Sturch (**6**). Since the final probable error of Sturch's standard $(B-V)$ and $(U-B)$ color indices is only $\pm 0{\cdot}01$ magnitude, these colors should serve admirably for the detection of very small amounts of interstellar reddening and in studies of population differences for various classes of RR Lyrae variables. Photo-electric studies of magnitudes and colors are soon to be published by W. S. Fitch, H. L. Johnson, and W. Z. Wizniewski of the University of Arizona; I. Epstein of Columbia University is also making studies of RR Lyrae variables. Iwaniszewska-Lubienska of Torun Observatory has applied the statistical approach of Iwanowska to the population assignment of RR Lyrae variables and is studying their distribution. R. E. White (Univ. of Illinois, Doctoral Thesis, unpubl.) is attempting a re-analysis of the material used originally by Perek (**7**) and by Plaut and Soudan (**8**). He divides 1500 of the stars listed by them into two groupings: one metal-poor (Preston, $\Delta S \geqq 5$) and the other metal-rich ($\Delta S < 5$). They are considered the Halo–Population II component and the Disk component, respectively. At the position of the Sun, the Disk component outnumbers the Halo component by a factor of 6.

McCuskey (**9**) has written an excellent and comprehensive review of our present knowledge of stellar luminosity functions. It deals exhaustively with the earlier researches in the field, especially with the early work on the General Luminosity Function. The article reviews in considerable detail recent work on the faint end of the General Luminosity Function. The results of the analyses by Luyten, Warner, Upgren, and McCuskey are compared with van Rhijn's Luminosity Function, which is a useful standard for reference purposes. The two diagrams shown in McCuskey's article indicate that the General Luminosity Function, $\varphi(M_v)$, (the number of stars per cubic parsec with absolute visual magnitudes between $M_v - 1/2$ and $M_v + 1/2$), rises steeply until $M_v = +6$, then increases slowly until $M_v = +11$, and from there remains almost constant to $M_v = +16$ (Luyten). If anything, there may be an excess of faint M dwarfs in the range $+11 < M_v < +16$, as indicated by Sanduleak (**10**), who finds very large numbers of M dwarfs in the North Galactic Polar Cap. Useful information is also given regarding variations in the Luminosity Function, both in the galactic plane and for directions at right angles to it. The article concludes with a new determination of the evolutionary 'initial' General Luminosity Function, which agrees closely with an earlier determination by Sandage (**11**). We note that Eelsalu and Jôeveer at Tartu Observatory have investigated the luminosity functions for the A and gK stars (**12, 13**).

This Report is obviously not the place to write at length on the results of proper-motion surveys to very faint limits or on the results for searches of intrinsically very blue and faint stars. Luyten reports the discovery of many such faint blue stars in the galactic polar caps and for selected regions at intermediate galactic latitudes (3000 objects); he reports that the total number of available proper motions for these faint blue stars amounts by now to well over 1000. He concludes that to $B = 16$ the number of white dwarfs is small and does not begin to increase markedly until $B = 18$ or 19. He estimates that his General Proper Motion Survey has yielded to date well over 500 stars with $M_{pg} \leqq +16$. It is highly gratifying to note the rapid advances that are being made in our knowledge of intrinsically faint stars, first, through the proper-motion surveys of Luyten, Giclas and others and, second, through the search for very faint blue stars by Zwicky, Luyten, Haro and others (**14**).

BIBLIOGRAPHY

1. Hodge, P. W., Wallerstein, G. 1966, *Publ. astr. Soc. Pacif.*, **78**, 411.
2. McCuskey, S. W., Rubin, R. H. 1966, *Astr. J.*, **71**, 517.
3. Ljunggren, B., Oja, T. 1965, *Uppsala Obs. Medd.*, no. 150.
4. Van Herk, G. 1965, *Bull. astr. Inst. Netherlands*, **18**, 71.
5. Woolley, R. v. d. R., Aly, K. 1966, *R. Obs. Bull.*, no. 114.
6. Sturch, C. 1966, *Astrophys. J.*, **143**, 774.
7. Perek, L. 1951, *Brno Contr.*, **1**, no. 8.
8. Plaut, L., Soudan, A. 1963, *Bull. astr. Inst. Netherlands*, **17**, 70.
9. McCuskey, S. W. 1966, *Vistas in Astronomy*, **7**, 141.
10. Sanduleak, N. 1964, Doctoral Thesis, Case Inst. of Techn. (unpublished).
11. Sandage, A. R. 1957, *Astrophys. J.*, **125**, 422 and 435.
12. Eelsalu, H., Jôeveer, M. *Tartu Astr. Obs. Publ.*, **34**, 47 and 121, 1964; **35**, 49, 1966.
13. Jôeveer, M., Eelsalu, H., Silvet, H. *Tartu Astr. Obs. Publ.*, **36** (in press).
14. Luyten, W. J. (ed.). 1964, *First Conference on Faint Blue Stars*, Strasbourg, August.

V. SPACE DISTRIBUTION AND CHEMICAL COMPOSITION

(prepared by W. Becker)

Dr W. Becker, one of the members of the Organizing Committee, has prepared the following summary of recent work in an area that is more and more becoming the concern of Commission 33.

'The relation between spatial distribution and chemical composition is of increasing interest. The exact determination of the chemical composition presupposes an intricate photometric analysis of the line spectrum, which has so far only been undertaken for a few objects, mostly in our neighborhood. Therefore the search for relations between distribution and composition is based on indicators for the chemical composition which are calibrated with the help of only very few objects. Such indicators are: the general appearance of a spectrum, an ultraviolet excess, a certain type of light variation, or characteristic types of motion. The results so far are in general limited to analyzing the space distribution of objects of Populations I and II. Primarily, the results concerning the distribution give distances from the galactic center or from the galactic plane.

'General considerations concerning the concept of different populations are given by Blaauw (**1**). Best defined are, for the time being, the relations for globular clusters and RR Lyrae stars because of their easy identification. Sandage (**2**) and G. Burbidge (**3**) have found different relations between age and metal content in globular clusters within or outside the galactic plane. According to Arp (**4**), the globular clusters with higher metal content are closer to the galactic center than the others. According to Böhm-Vitense, Holweger, and Kohl (**5**), the metal content of a cluster depends less upon its age than on its distance from the galactic plane. This is at variance with the results of van den Bergh (**6**), whose study shows that in stars the enrichment of the heavy elements is rather a function of time than of location.

'Plaut (**7**) has given a comprehensive report about the behavior of RR Lyrae stars. According to Geyer (**8**), they belong partly to the Disk Population and partly to the Halo Population, depending on the intensity of their metal lines. Preston (**9**) finds that the separation of the two groups occurs between $P = 0\overset{d}{.}55$ and $0\overset{d}{.}44$, with the RR Lyrae stars of longer period being in the Halo Population. The extension of the galactic halo seems to be considerably larger than was thought until recently; Kinman, Wirtanen, and Janes (**10**) have found increasing values of log N for RR Lyrae stars up to a distance of 25 kpc in the direction of the galactic pole (SA 57).

'Long-period variables seem to be present equally in both populations. According to Plaut (**11**), their mean periods increase with increasing distance from the galactic center. Preston and Wallerstein (**12**) found in two investigations that the halo variables contain about 50 times less metals than do the disk variables.

'The chemical composition of cepheids seems also to be a function of their distribution. Kraft (**13**) believes that cepheids beyond the Perseus Arm have a composition which is different from that of the nearer ones. This could restrict the general validity of the period-luminosity relation.

'Some results have been obtained for field stars. According to Wallerstein (**14**), the central galactic core, as far as it is represented by red giants of Population II, does not reach farther than 1800 pc above the galactic plane. The metal content of these stars seems to range between that of M67 and the halo-type globular clusters. Upgren (**15**) finds that the General Luminosity Function of stars brighter than $M = +15$ is the same to distances of 100 pc from the galactic plane as in the solar neighborhood. Only from about 200 pc outward do halo stars prevail, and from 400 pc outward disk stars disappear entirely. Becker (**16**) uses the possibility of separating disk and halo stars by means of three-color photometry to determine density gradients in the galactic halo. This gradient could be followed in the direction of the anti-center (SA 51) up to a distance of 15 kpc (or 25 kpc from the galactic center).

'General considerations about statistical population indices, which could be useful for studies of spatial distributions, are given by Iwanowska (**17**).'

BIBLIOGRAPHY

1. Blaauw, A. 1960, *Stars and Stellar Systems: Galactic Structure*, Vol. V, 435.
2. Sandage, A. 1962, *Astrophys. J.*, **135**, 349.
3. Burbidge, G. R. 1962, *Ann. Rev. Nuclear Science*, **12**, 507.
4. Arp, H. C. 1965, *Stars and Stellar Systems: Galactic Structure*, Vol. V, 401.
5. Böhm-Vitense, E., Holweger, H., Kohl, K. 1963, *Astrophys. J.*, **138**, 604.
6. van den Bergh, S. 1962, *Astr. J.*, **67**, 486.
7. Plaut, L. 1965, *Stars and Stellar Systems: Galactic Structure*, Vol. V, 267.
8. Geyer, E. 1962, *Kleine Veröff., Remeis-Sternwarte, Bamberg*, no. 34, 121.
9. Preston, G. W. 1964, *A. Rev. Astr. Astrophys.*, **2**, 30.
10. Kinman, T. D., Wirtanen, C. A., Janes, K. A. 1966, *Astrophys. J. Suppl.*, **13**, no. 119.
11. Plaut, L. 1963, *Bull. astr. Inst. Netherlands*, **17**, 75.
12. Preston, G., Wallerstein, G. 1963, *Astrophys. J.*, **138**, 820.
13. Kraft, R. P. 1963, *Stars and Stellar Systems: Basic Astronomical Data*, Vol. III, 421.
14. Wallerstein, G. 1962, *Astr. J.*, **67**, 329.
15. Upgren, A. R. 1963, *Astr. J.*, **68**, 475.
16. Becker, W. 1965, *Z. Astrophys.*, **62**, 54.
17. Iwanowska, W. I. 1966, *Vistas in Astronomy*, **7**, 133.

VI. STELLAR DISTRIBUTION IN LOW GALACTIC LATITUDES

Summaries and General Surveys

McCuskey's chapter on 'The Distribution of the Common Stars in the Galactic Plane' (1) will remain a basic reference for many years to come. It is comprehensive and complete, giving for each separate spectral group the best available picture of the stellar distribution in the galactic plane. McCuskey's analysis shows that, if we did not have distributional data for Extreme Population I, we would really have no inkling that our Galaxy possesses spiral structure. McCuskey has written two further survey papers; the first, an article to which we have already referred, is the chapter on the Luminosity Function (2), and the other is his contribution to Joint Discussion A on the Local Structure of the Galaxy (3). The latter paper deals specifically with the distribution of the A and F stars. In all three summaries it is clear that our knowledge of the stellar distribution for the southern Milky Way still lags behind that for the northern Milky Way; this situation justifies the increasing attention that is being paid to sections at far southern declinations.

It is of interest to note a revived interest in Gould's Belt and related problems. The phenomenon of Gould's Belt, which is inclined by 15° to 18° with respect to the fundamental plane of our Galaxy, is shown most clearly by the brighter OB stars and the Be stars within 400 pc of the Sun, by the nearest extended dark nebulae and the nearer diffuse nebulae, and by the nearby large complexes of H I. To these can be added the I Orionis, the II Persei, and the Scorpius-Centaurus Associations (4). These objects define a compact Local System, possibly a self-contained unit. According to recent work by D. W. Dewhirst, reported during the Tenth Herstmonceux Conference (5), Gould's Belt suggests the existence of a very thin sheet of young objects. The sheet is 700 pc in extent, but only 70 pc thick, and is somewhat like the flat ribbons which are characteristically observed in the outer arms of Sb and Sc spiral galaxies. Some of us have referred to these ribbons emanating from spiral features as 'spurs'. Much attention is being paid to the motions of the stars in this Local System. In 1965, Bonneau studied the motions of the nearby OB stars (6) and found evidence for expansion. Blaauw has confirmed and extended Bonneau's work (7) for the O to B3 stars and he has explored the hypothesis that these stars, or rather the gas from which they were formed, originally occupied a much smaller volume of space than is the case at present. The pre-stellar gas seems to have spread out from a small volume in about 4×10^7 years; this conclusion is supported by the observed velocity gradients exhibited by stars in Gould's Belt. S. V. M. Clube (5) agrees with Blaauw in accepting the hypothesis of a general expansion of the Local System, to which he assigns an age of 3×10^7 years. As a consequence of the large V-component of the stream motion (twice the basic solar motion!), he considers it likely that the absolute magnitudes of B stars in the Scorpius-Centaurus Association may have been underestimated by 0·5 magnitude, and that systematic effects may result in the derived proper motions, precessional constants, and the values of Oort's A and B. Since the Scorpius-Centaurus Association plays, as does the Hyades (see Section IV), an important role in the calibration of absolute magnitudes, these problems are of more than local significance.

We should draw attention once again to the peculiar density distribution of the F0 to F5 stars, which appear to have negative space density gradients in all directions away from the Sun. McCuskey (3) finds an average density at a distance of 500 pc less than one-half that near the Sun. This result has been confirmed in a recent study by McCuskey and Rubin (8). McCuskey reports that the Warner and Swasey Observatory has begun a survey of a belt 12° wide to detect all main-sequence A stars and gK stars to the limit of $V = 13{\cdot}5$ along the central band of the Milky Way. The Warner and Swasey Observatory, in collaboration with Bosscha Observatory, is extending its general spectral-magnitude survey southward to include the following centers:

	l^{II}	b^{II}
LF 11	212°7	+1°2
LF 12	240°8	−0°5
LF 13	281°0	+3°9
LF 14	297°6	+1°4
LF 15	330°3	−2°8

The approach is similar to that followed in the study of luminosity functions for the northern part of the Milky Way (**2**). Pik-Sin The reports that the basic spectral and magnitude plates are already available for four of the five areas. Mavridis (**9, 10**) has summarized his work and that of others on M, C, and S stars for different sections of the Milky Way.

The Basel researches on stellar distribution at low galactic latitudes for rich fields in the Milky Way have thus far produced three papers (**11, 12, 13**). The limiting magnitude is 17. Density gradients have been derived for three fields. A new program has begun to study the density gradients for stars of lower luminosity than OB stars for the direction of a spiral arm (anti-center) and for an inter-arm direction (toward the galactic center).

Much new material has accumulated, and is continuing to accumulate, for the southern Milky Way. Graham and Lyngå (**14**) have published a finding list of 454 OB stars in Carina, and Lyngå (**15**) has published a similar list for the section $l^{II} = 310°$ to 334° (482 stars). Elsässer reports that E. Geyer, G. Klare, and B. Szeidl have in press two major papers with lists of OB stars complete to $m_B = 11{\cdot}5$, in the region $220° < l^{II} < 20°$, $-12°5 < b^{II} < +12°5$ (1660 OB^+ and OB^- stars). Color searches for faint OB stars have been initiated by B. J. Bok (**16**). A comprehensive study of the spatial distribution of OB stars has been completed by Th. Neckel and G. Klare. The resulting distribution, clearly showing three spiral arms, is in general agreement with the model derived by W. Becker from OB star clusters. Much work has been done in recent years by Elsässer and associates at Königstuhl Observatory on problems relating to interstellar absorption, reddening, and polarization, but these researches fall strictly within the domain of Commission 34.

In the Sub-Sections to follow, we shall include the results obtained at observatories in the U.S.S.R. of work in areas contained in the Parenago Plan. We note here that the Observatory of the Ukrainian Academy of Sciences and the Abastumani Observatory have finished their assignments under the plan and that the Crimean Astrophysical Observatory has almost completed its share of the work.

Regional Surveys

(*a*) *The section of the Galactic Center* ($350° < l^{II} < 10°$)

Much work is being done in areas in the general direction of the galactic center on the density gradients for the stars within 1500 pc of the Sun. Reference has already been made to the studies by W. Becker and associates at Basel, and, in passing, we should also refer to the work for Kapteyn Selected Area 158 (**17**); the matter is dealt with in more detail in T. Elvius' Report. The existing results imply that, for this direction, without fail, the star densities decrease with increasing distance from the Sun, as one would expect when proceeding from the position of the Sun, at the inside of a spiral arm, towards an inter-arm region.

McCuskey reports that Peter Wehinger has studied the interstellar reddening and the distribution of M giants for a field centered upon the radio source Sagittarius A at $l^{II} = 0°$. The derived total visual absorptions are as follows:

b^{II}	Distance (kpc)					
	0·5	1·0	2·0	3·0	5·0	7·0
+1°5	2·6	2·6	2·6	2·6	2·6	2·6
−0°3	1·5	1·6	2·4	2·9	3·2	3·2
−1°8	0·9	1·9	1·9	1·9	2·0	2·2

Spectral classification, supplemented by color work, yields the following space densities for the M2 to M10 stars:

Distance (kpc)	0·6	1·0	2·0	3·0	4·0
Space Density (No. per 10^6 pc^3)	35	8·7	8·4	8·2	4·7

While these space densities decrease rapidly with distance from the Sun, they are twice the values derived for Palomar-Groningen Field No. 2 at $l^{II} = 4°$, $b^{II} = +11°$.

(*b*) $10° < l^{II} < 60°$ (*Sct, Aql*)

The Main Astronomical Observatory of the Ukrainian Academy of Sciences has finished work for a large region in Aquila ($l^{II} = 35°$ to $52°$); the area covers about 150 square degrees. The catalogue of stellar spectra and magnitudes and the study of the spatial distribution of stars and obscuring clouds in several areas in the Aquila region have been published by Apriamaschvili and Kuznetsov (**18**), Voroshilov and Poleshchuk (**19**), and Fedorchenko (**20**). Strong obscuration is present at relatively small distances from the Sun.

At Abastumani Observatory, Apriamaschvili (**21**) has classified, in two small areas in Aquila and Scutum ($l^{II} = 29°$ and $37°$), about 1900 stars to 14–15 pg magnitude and, moreover, about 900 faint M stars to the 18th photographic magnitude. The distributions of OB stars and obscuring clouds in both areas to distances of 8–10 kpc have been studied. A strong obscuration at small distances has been detected. OB stars show a remarkable clustering tendency, especially in the southernmost area.

McCuskey reports on an extensive investigation by R. B. Herr for the section $16° < l^{II} < 46°$. The study is based upon spectra of OB stars obtained in the Hamburg-Cleveland survey 'Luminous Stars in the Northern Milky Way' II and IV. It involves luminosity classification and (*U,B*) photometry of these stars. The former was done directly from objective-prism photographs. After being corrected for inter-stellar absorption, the derived space densities for the OB stars in two regions included in the survey are as follows:

	No. Stars per 10^6 pc.	
Distance (kpc)	Region A ($l^{II} = 19°$)	Region B ($l^{II} = 41°$)
0·5	5·4	0·6
1·0	3·5	0·2
1·5	3·1	0·2
2·0	1·8	0·2
2·5	1·1	0·1
3·0	0·7	0·07

McCuskey comments that the OB-star space densities for Region A ($l^{II} = 19°$) run very similar to those for B8 to A0 stars at the same longitude (**22**).

Lisicki (**23**) has studied the absorption and other structural characteristics in a field in Aquila and Sagittarius, and Dolidze (**24**) has completed a survey of M stars for the region near ζ Oph.

(*c*) $60° < l^{II} < 150°$ (*Cyg, Cep, Cas, Per*)

In the U.S.S.R., much attention is being given to the study of the stellar distribution in Kapteyn Selected Area 40. Kolesnik (**25**) has found that the density maximum for the A stars coincides with the obscuring clouds. Kharadze, Apriamaschvili, and Kochlaschvili (**26**) have published a catalog of over 1000 stars near Selected Area 40. Kolesnik (**27**) stresses the urgent need for studies of galactic structure by means of spectroscopic and photometric data for the most distant stars within reach. We note that the Main Observatory of the Ukrainian Academy of Sciences is extending its photometric studies to $V = 14{\cdot}5$–$15{\cdot}0$ in several areas within the section under discussion and the anti-center region ($50° < l^{II} < 180°$). Kharadze reports that Abastumani Observatory, in cooperation with Kiev, is undertaking two-dimensional spectral classification and three-color photometry for eight areas along the galactic equator from Aquila to the anti-center. Two further studies of the Cygnus Region should be noted. Dolidze's (**28**) survey of M stars includes the region of the Cygnus Loop and NGC 6819. M. Lunel (**29**) has made *UBV* measurements of the highly reddened VI Cygni Association.

The program of the Crimean Astrophysical Observatory has included work by Masnauskas (**30**), who has determined the spectra and *B*,*V* magnitudes for about 3600 stars in a 40-square-degree area in Cepheus ($l^{II} = 105°$), and by Brodskaya and Grigorieva (**31**), who have done the same for about 2700 stars in an area of comparable size in Cassiopeia ($l^{II} = 121°$). For the second area, the space distribution of obscuring clouds and OB stars has been studied. The OB stars show a considerable concentration to the Perseus Arm. A similar investigation in a neighboring area containing a stellar association ($l^{II} = 118°$, $b^{II} = +5°$) has been carried out by Raznik (**32**).

B. J. Bok and M. Smith have in progress a photometric study of a number of distant OB stars in a relatively unobscured region in Cepheus; similar work is under way for a region in Monoceros. The search for faint Hα emission objects is continuing. Blanco and Fitzgerald (**33**) have studied the section near ζ Cep, with 26 new stars added to the lists.

There is much progress to report for the section of the Milky Way in Cassiopeia. H. Schmidt is working on a field near $l^{II} = 136°$, $b^{II} = -1{.}^{\circ}5$, where 2000 stars to $V = 15{\cdot}5$ have been measured for a field of 1·6 square degrees. McCuskey and Houk (**34**) have published photographic photometric results for five clusters and 159 OB stars near $l^{II} = 129°$, $b^{II} = -2°$, where they find a high concentration of OB stars between 2 and 3 kpc from the Sun. They conclude that the clusters and the OB stars are probably in the Perseus Arm. Here we also note Ampel's work (**35**), which has yielded spectral-luminosity classes, photographic magnitudes, and colors for 275 OB stars in the Associations III Cassiopeiae, IV Cassiopeiae, and V Cassiopeiae. B stars in the Perseus Arm have been found in considerable numbers by Grigorieva (**36**) at distances of 2 to 3·5 kpc in the section $117° < l^{II} < 124°$. Extensive work on the IV Cephei Association is underway at Warner and Swasey Observatory.

An important study for this section is that published in Oja's doctoral thesis (**37**), which covers the range $112° < l^{II} < 133°$. Spectrophotometric data are now available for 2600 stars with $b^{II} \leqq \pm 3°$. An investigation in depth (**38**) was made for a region near $l^{II} = 128°$, $b^{II} = -2°$. Oja's catalog contains spectrophotometric data (derived from plates taken with the large Bergedorf-Schmidt telescope) and integral photometry (done with the Uppsala double refractor and the Uppsala-Kvistaberg Schmidt telescope) for about 1000 stars brighter than 14·0 magnitude (pg). All types of stars with $M < 0$ have a maximum space density at a distance of 3·2 kpc; this maximum may be identified with the 'spiral arm + I', according to W. Becker's notation, the Perseus Arm.

We note finally that Velghe is continuing his infra-red survey of C, S, and M stars (M2 to M10) to $I = 13{\cdot}5$ for a field in Cygnus and Draco (at $l^{II} = 77°$, $b^{II} = 0°$ to $+22°$). He reports that the *V* and *I* magnitudes are almost ready for publication.

(*d*) *The Anti-Center Region* ($150° < l^{II} < 210°$)

Two recent studies from the U.S.S.R. are to be reported. Kalandadze has published (**39**) a catalog containing 3600 stars for an 80-square-degree region near $l^{II} = 182°$, and Bartkus (**40**) has published one containing 2455 stars for the central part of the Orion Association.

McCuskey has finished a catalog of 3620 stars to $V = 12 \cdot 2$ for an area of 20 square degrees centered upon $l^{II} = 186°$, $b^{II} = +1°$. The catalog has spectral classifications and V magnitudes for all stars and $(B-V)$ values for a representative sample. He expects that a comparison of the spatial stellar distributions for the regions of the center and the anti-center will help clarify the situation regarding the Sun's position relative to the local spiral structure.

In connection with her work on radial velocities of distant OB stars, Rubin (**41**) has begun a search for such stars in the range $139° < l^{II} < 216°$. She uses three-color Palomar-Schmidt plates for eight centers. McCarthy and Treanor (**42**) are studying fields in the anti-center section with the aid of the Vatican Observatory equipment. As part of the Basel program, Fenkart's (**12**) work should again be noted at this point. We should also mention the extensive program on spectral classification, objective-prism radial velocities, and photometry undertaken jointly by the Observatoire de Toulouse (by Bouigue and Mme Lunel) and by the Observatoire de Marseille. Finally, Karlsson (**43**) has published a first paper relating to the study of a region in Monoceros.

(*e*) $210° < l^{II} < 300°$ (*Pup, Vel, Car, Cru*)

Velghe has made an extensive OB-star spectral-color-magnitude survey for the section $257° < l^{II} < 285°$. The results have not yet been published in full. J. Denoyelle has now begun an extensive investigation of colors and magnitudes for 350 OB stars in two low-latitude sections bordering the section studied by Velghe. The work is being done at Boyden Observatory.

Much attention has been paid to the Carina section, $280° < l^{II} < 295°$. Sher (**44**) has published a comprehensive survey for this part of the Milky Way. He shows that the preceding edge of the Carina concentration of stars and gas, located at $l^{II} = 275°$, is real and not an effect caused by local obscuration. The Carina section shows a 'bewildering array of Population I' (Sher), but all clusters with estimated ages less than 10^8 years appear to be at distances in excess of 1500 pc. Recent distance estimates for the η Carinae Nebula (NGC 3372), and some of the clusters almost surely imbedded (Tr. 16) in it, place it at a distance of between 2500 and 3000 pc. Feinstein reports that 2500 pc appears to be the best estimate, which agrees well with his earlier published estimate of 2800 pc (**45**). D. J. Faulkner's value (**46**) is 2500 pc, whereas Graham has given 2800 pc on the basis of $H\beta$ photometry for peripheral stars. These are considerably greater than the earlier values of 1400 to 1800 pc. The gaseous and OB-star concentrations are obviously more distant than the known concentration of late B and early A stars, which occurs within 1000 pc of the Sun. Graham and Lyngå (**47**) have published a survey of faint OB stars in Carina. There was a useful discussion on the distance problems in the Carina section during the Hamburg meeting of Commission 37 (**48**). In his Report to our Commission, Feinstein announces his conclusions, which are, (1) that the main concentration of OB stars is at a distance greater than 1600 pc from the Sun, (2) that the normal value $R = 3$ applies to the highly-luminous stars outside the immediate surroundings of η Carinae, and, (3) that $A_v = 0 \cdot 45$ magnitude kpc^{-1} for the initial 1600 pc.

Westerlund (**49**) has made an infra-red survey of the region of Kapteyn Selected Area 193 at $l^{II} = 293°$. He has compared his results with those obtained earlier for five other fields, and certain patterns emerge. He finds that the M2 and M4 stars form a thin galactic layer with a half-thickness of 250 pc. They tend to concentrate toward spiral arms. The M5 to M6·5 stars show a marked concentration toward the galactic center, but the M7 to M10 stars form a homogeneous thick layer with negligible variations in space density. The carbon stars (Class N) are most likely spiral-arm objects.

(*f*) $300° < l^{II} < 350°$ (*Cen, Cir, Nor, Sco*)

It has become increasingly apparent that the galactic structure in this section is critical for our understanding of the local spiral structure of our Galaxy. Therefore, it is not surprising that much effort is being expended on studies of galactic structure for this section of the Milky Way. We shall leave a discussion of the spiral features for the next Section, but here we will briefly note some other structural features.

Lyngå (**50**) has published a series of five papers dealing with this section. He concentrates his efforts on the sector $311° < l^{II} < 332°$; the concluding paper has a summary of the overall results obtained from this work. He has written another summarizing article (**51**), which serves to complete the picture. Three OB associations have been found in this section, though it is recognized that one of them is very weak and that all three are inconspicuous groupings when compared to the objects generally classified as associations. Lyngå's Catalog for this section contains 482 OB stars. Three young open clusters have been studied in this section, and several new clusters are confirmed or suspected. The H II regions in the section are small and inconspicuous, but it is noted that they are spread fairly evenly in galactic longitude. Lyngå notes that more dramatic and conspicuous features are present for galactic longitudes $l^{II} > 330°$.

For the section $330° < l^{II} < 340°$, the stellar distribution shows the features generally associated with Population I and with spiral structure. B. J. Bok, P. F. Bok, and Graham (**52**) have studied the OB stars within one degree of $l^{II} = 327°$, $b^{II} = 0°$; they used the strength of Hβ as a criterion for absolute magnitude. There appears to be a marked concentration of OB stars at a distance of 2500 pc from the Sun. Graham (**53**) has studied nine distant B stars in Ara at $l^{II} = 337°$ and finds (from radial-velocity data) that these stars are spread over a considerable range of distance. In agreement with Whiteoak's (**54**) earlier results, some of these stars are found at much greater distances than that of the I Ara Association. H. L. Johnson had indicated (**55**) that the ratio of total to selective absorption might be quite large for the Ara B-stars, but Graham points out that his radial-velocity data indicate a large spread in distance and that the more usual ratio $A_V/E_{B-V} = 3{\cdot}0$ seems preferable.

Haug, Pfleiderer, and Dachs (**56**) have studied 105 stars of early spectral class near $l^{II} = 332°$, $b^{II} = -2°$ and 38 stars near $l^{II} = 321°$, $b^{II} = -1{\cdot}5°$. *UBV* magnitudes and colors have been obtained for all stars on the list. Some highly reddened distant stars (to 4 kpc from the Sun) have been detected for the Norma field ($l^{II} = 332°$), but the greatest estimated distance for a star in the Circinus field ($l^{II} = 321°$) is considerably less than 3 kpc. Further work on this section of the Milky Way is in progress. G. Schnur at Heidelberg has reported that he has *UBV* data for 700 stars near $l^{II} = 332°$. Radial-velocity data for ten of the stars studied by Bok, Bok, and Graham (**52**) are now available (**57**). This information confirms the photometric distances.

We should draw attention briefly to the distribution of planetary nebulae. Much new material for the southern sky is accumulating through the work of Henize and Westerlund and also from the color-filter studies of B. Louise Webster. She has noted that the density of planetaries in Norma (near $l^{II} = 330°$) is higher than for the section $270° < l^{II} < 320°$, and also that the planetary nebulae in Norma are found at greater distances from the Sun than in the Carina-Centaurus section. An atlas with photographs of 99 southern planetary nebulae is to be published shortly by Henize and Westerlund.

Roslund (**58**) is continuing his work on spectral classification, magnitudes, and colors for a field in Scorpius. *UBV* magnitudes and colors and Hβ indices are available for 250 early-type stars, and MK classifications are available for 63 of the brighter stars. His results show that these OB stars form a homogeneous groups of young stars. Strong interstellar reddening in the northern part of the region appears to originate from insterstellar dust in the local spiral arm. Bok, Bok, and Graham (**59**) have published *UBV* and Hβ photometry for the I Scorpii Association, for which a distance of 1800 pc is obtained. Graham reports that Hβ photometry for seven open clusters in the southern section of the Scorpius-Centaurus Association is in press. Velghe and P. Sanders have underway a search for and study of new H II regions in Scorpius.

BIBLIOGRAPHY

1. McCuskey, S. W. 1965, *Stars and Stellar Systems: Galactic Structure*, Vol. V, Chapter 1.
2. McCuskey, S. W. 1966, *Vistas in Astronomy*, **7**, 141.
3. McCuskey, S. W. 1966, *Trans. IAU*, **12 B**, 425.
4. Mumford, G. S. 1957, *Sky Telesc.*, **16**, 214.
5. Report of 'Tenth Herstmonceux Conference' 1966, *Observatory*, **86**, 182 to 184.
6. Bonneau, M. 1965, *Ann. Obs. Besançon*, **6**, Fasc. V.
7. Blaauw, A. 1965, *Verslagen Afd. Natuurkunde, Kon. Nederl. Akad. Wetenschappen*, **74**, no. 4.
8. McCuskey, S. W., Rubin, R. H. 1966, *Astr. J.*, **71**, 517.
9. Mavridis, L. N. 1965, Lecture Notes of NATO International Summer Course, *Observational Aspects of Galactic Structure*, Lecture XVI, Athens.
10. Mavridis, L. N. 1966, *Colloquium on Late-Type Stars*, Trieste, 13–17 June.
11. Brodbeck, K. 1963, *Z. Astrophys.*, **58**, 127.
12. Fenkart, R. 1964, *Z. Astrophys.*, **59**, 56.
13. Wick, Ch. 1965, *Z. Astrophys.*, **61**, 110.
14. Graham, J. A., Lyngå, G. 1965, *Memoirs Mt. Stromlo Obs.*, no. 18.
15. Lyngå, G. 1964, *Lund Medd.*, Series II, no. 141.
16. Bok, B. J. 1964, *Astr. J.*, **69**, 533.
17. McCuskey, S. W. 1964, *Astr. J.*, **69**, 104.
18. Apriamaschvili, S. P., Kuznetsov, V. I. 1966, *Abastumani astroph. Obs. Bull.*, **35**, 46.
19. Voroshilov, V. I. 1966, *Problems of Astrophysics* and *Physics of Stars and Diff. Matter*, Naukova Dumka Press, Kiev; Voroshilov, V. I., Poleshchuk, E. P. (in press).
20. Fedorchenko, G. L. *Res. on Physics of Stars and Diff. Matter*, p. 63, 1964, and *Problems of Astrophysics*, Naukova Dumka Press, Kiev, 1966.
21. Apriamaschivili, S. P. 1964, *Abastumani astroph. Obs. Bull.*, **30**, 49; **31**, 27.
22. McCuskey, S. W. 1966, *Trans. IAU*, **12 B**, 425.
23. Lisicki, A. 1965, *Stud. Soc. Sci. Torun*, **3**, no. 6, 1.
24. Dolidze, M. B. 1964, *Abastumani astroph. Obs. Bull.*, **30**, 71.
25. Kolesnik, L. N. 1964, *Res. on Physics of Stars and Diff. Matter*, p. 52, Naukova Dumka Press, Kiev.
26. Kharadze, E. K., Apriamaschvili, S. P., Kochlaschvili 1964, *Abastumani astroph. Obs Bull.*, **31**, 3.
27. Kolesnik, L. N. 1965, *Abastumani astroph. Obs. Bull.*, **33**, 41.
28. Dolidze, M. B. *Abastumani astroph. Obs. Bull.*, **30**, 81, 1964; **32**, 53, 1965.
29. Lunel, M. *J. Observateurs*, (in press).
30. Masnauskas, J. 1964, *Vilnius astr. Obs. Bull.*, no. 12.
31. Brodskaya, E. S., Grigorieva, N. B. *Soobšč. gos. astr. Inst. P. K. Sternberga*, no. 136, 1964; Grigorieva, N. B., *Astr. Zu.*, **41**, 185, 1964; *Izv. Krym. astrofiz. Obs.*, **34**, 238, 1965.
32. Raznik, R. M. 1964, *Izv. Krym. astrofiz. Obs.*, **32**, 165.
33. Blanco, V. M., Fitzgerald, M. P. 1964, *Publ. astr. Soc. Pacific*, **76**, 438.
34. McCuskey, S. W., Houk, N. 1964, *Astr. J.*, **69**, 412.
35. Ampel, R. 1964, *Torun Obs. Bull.*, no. 35, 52.
36. Grigorieva, N. B. 1965, *Izv. Krym. astrofiz. Obs.*, **34**, 238.
37. Oja, T. 1966, *Uppsala Obs. Annals*, **5**, no. 4.
38. Oja, T. 1966, *Uppsala Obs. Medd.*, no. 157.
39. Kalandadze, N. B. 1964, *Abastumani astroph. Obs. Bull.*, **31**, 66.
40. Bartkus, R. *Vilnius astr. Obs. Bull.*, no. 13, 32, 1964; no. 15, 28, 1965.
41. Rubin, V. C. 1965, *Astrophys. J.*, **142**, 934.
42. McCarthy, M. F., Treanor, P. J. *Ric. astr. Vatican*, **6**, 511, 1964; **7**, 177, 1965.
43. Karlsson, B. 1965, *Lund Medd.*, Series II, no. 145.
44. Sher, D. 1965, *Q. J. R. astr. Soc.*, **6**, 299.
45. Feinstein, A. 1963, *Publ. astr. Soc. Pacific*, **75**, 269.
46. Faulkner, D. J. 1963, *Publ. astr. Soc. Pacific.*, **75**, 269.
47. Graham, J. A., Lyngå, G. 1965, *Mt Stromlo Obs. Memoirs*, no. 18.

48. *Trans. IAU*, 1966, **12 B**, 399 and 340.
49. Westerlund, B. E. 1965, *Mon. Not. R. astr. Soc.*, **130**, 45.
50. Lyngå, G. *Lund Medd.*, Series II, nos. 139, 140, and 141, 1964; nos. 142 and 143, 1965.
51. Lyngå, G. Mt Stromlo Observatory Symposium on 'Radio and Optical Studies of the Galaxy', May, 1966, p. 61.
52. Bok, B. J., Bok, P. F., Graham, J. A. 1963, *Publ. astr. Soc. Pacif*, **75**, 514.
53. Graham, J. A. 1966, *Bull. astr. Inst. Netherlands*, **18**, 372.
54. Whiteoak, J. B. 1963, *Mon. Not. R. astr. Soc.*, **125**, 105.
55. Johnson, H. L. 1965, *Astrophys. J.*, **141**, 923.
56. Haug, U., Pfleiderer, J., Dachs, J. 1966, *Z. Astrophys.*, **64**, 140.
57. Mowat, M. Mt Stromlo Observatory Symposium on 'Radio and Optical Studies of the Galaxy', May, 1966, p. 103.
58. Roslund, C. 1966, *Lund Medd.*, Series I, nos. 207, 212, and 213.
59. Bok, B. J., Bok, P. F., Graham, J. A. 1966, *Mon. Not. R. astr. Soc.*, **131**, 247.

VII. THE SPIRAL STRUCTURE OF OUR GALAXY

The picture of the local spiral structure of our galaxy that appears to be generally accepted is that of W. Becker (1), published after our previous Report was completed. It is reproduced, with slight modifications, by Schmidt-Kaler (2). In our previous Report to Commission 33, we listed five basic approaches for the tracing of the spiral arms of our Galaxy. We can add the positioning of dark nebulae as a potential tool for the tracing of spiral arms, especially applicable to the insides of the spiral arms. Observations of faint emission stars, which are easy to detect from objective-prism spectrum photographs, can also prove very helpful; one should, however, bear in mind that Be stars probably have considerable dispersion in absolute magnitude. Analyses of spiral structure should continue to take into account the need for constant reference to the observed spiral features of galaxies other than our own. Becker has suggested that NGC 1232 (Sb) deserves special attention.

Kinematical features should not be overlooked. There is increasing, but conflicting, evidence for differences in velocity between the stars and the gas associated with a given spiral arm. Such effects have been detected by Abt and Bautz (3), who found that the radial velocities of the stars, presumably associated with the Perseus Arm, were 7 to 8 km s^{-1} greater than the corresponding 21-cm H I velocities. They attributed their result as possibly being caused by binary effects in the stars, or atmospheric effects, rather than indicating a difference between stellar and gaseous velocities. In a presentation at the Colloquium held at Besançon on 'Les Nouvelles Méthodes de la Dynamique Stellaire' (September 1966), Dixon states that 'present data indicate that, near the Sun, the interstellar medium lags behind the gravitational circular velocity by 14 km s^{-1}'. A rather different conclusion is drawn by Fletcher (4) in an article that should have been included in the Hamburg Report. When comparing structural features on the basis of observed radial velocities alone, he finds good agreement between optical results for OB stars and radio results from 21-cm profiles. M. Mowat has made a strong plea (5) for more optical radial velocity data in sections of the Milky Way for which the galactic-rotation component of radial velocity varies markedly with distance from the Sun. The comparison between the radio (21 cm) and the optical results becomes then model-independent. However, the spiral pattern that is derived from radial-velocity data is still highly dependent on the law of rotational velocity that is assumed, and ultimately we must refer to the assumed model of rotation for the Galaxy. Large-scale non-circular gaseous motions, which have been observed to run as high as 10 km s^{-1}, can affect the derived pattern very much.

Several important papers on the spiral structure of our Galaxy have come from the U.S.S.R. Kardashev, Lozinskaya, and Sleptsova (6) have re-analyzed all available 21-cm observations and continuum data. They find that the most probable large-scale spiral features of our Galaxy can be represented by a logarithmic spiral with two arms and with an average 'pitch angle' of 84°. However, Kuzmin notes in his Report that these results are contradicted by the work of Kostyakova (**7**), who has analyzed for interstellar absorption the spectrophotometrically-determined color distribution along the band of the Milky Way. While this approach is useful, it is obviously subject to uncertainties of interpretation. No agreement is found with the spiral features derived from 21-cm data. A multi-arm model for the spiral structure has been proposed by Pskovskii (**8**), who derives from optical and radio data a series of logarithmic spirals with an average pitch angle of 72°3. Pavlovskaya and Sharov (**9**) have generally confirmed the work of Pskovskii.

Mrs Pronik (**10**) has drawn attention to unevenness in the velocity picture for the outer spiral arms as shown by the 21-cm profiles. She has found many cases in which the velocities shown by the gas 2°5 north of the galactic equator show different maxima and a different pattern from the gas at the same small angular distance south of the galactic equator.

General calibration problems of absolute magnitudes continue to be of much concern. W. Becker reports that corrected distances for H II regions have been obtained on the basis of Petrie's new calibration of absolute magnitudes of early B-type stars. He mentions that Fenkart has prepared a list of 90 O to B3 stars with $V < 10$, which are obviously responsible for 50 observed H II regions; luminosity estimates and colors are not yet available for these stars, which require urgent attention. The unpublished work of J. S. Miller of Washburn Observatory should fill some of these gaps.

Sharpless is extending his work on M-type supergiants through the application of narrow-band photometric techniques. He expects to identify older spiral features and obtain information on the evolution of the spiral structure of the Galaxy. He is also engaged upon the study of the cosmically very young multiple O-type systems of the Trapezium variety.

Lindsey F. Smith (**11**) has studied the distribution of the Wolf-Rayet stars in the galactic plane. On the basis of newly-established calibrations of absolute magnitude, she has determined fairly precise distances for the known galactic WR stars. These stars appear to fall along the inner edge of the Local Arm and are found in the Sagittarius Arm.

Muhleman and Walker (**12**) have investigated one of the causes for gaps in the H I distribution. They find that the observed gaps in part of the Cygnus Arm can be identified with regions of high ionization, since the concentration of OB stars is large for those regions without H I; these obviously have become H II regions.

In recent years, there has been much discussion about the possible uses of Be stars for studies of spiral structure. Schmidt-Kaler (**13**) and Behr (**14**) have taken opposite points of view on this question. Schmidt-Kaler claims that the Be stars are very well suited for this work, whereas Behr is of the opinion that their dispersions in absolute magnitude are too great for individual stars to serve as spiral tracers; he estimates the average dispersion to be greater than 0·4 magnitude. However, in a later paper (*Mitt. astr. Ges.*, no. 21, 68, 1966), Behr classifies the Be stars as good spiral tracers.

In an as-yet unpublished doctoral thesis (Northwestern University, 1966), J. D. Wray has surveyed the southern Milky Way, $240° < l^{II} < 10°$, for Hα-emission objects, which survey has yielded an additional 870 probable Be stars. The strong concentration to the Carina section has been confirmed and the arrangement of these stars follows the inner spiral arm defined by Morgan, Whitford and Code. The Be stars apparently provide a 0·5 to 1·0 kpc resolution element for the study of intermediate age spiral structure for distances up to 5 kpc from the Sun.

Isserstedt and Schmidt-Kaler (**15**) have analyzed the distribution of reddening as a function of distance from the sun for all directions and in different zones of galactic latitude for $-8°5 < b^{II} < +8°5$. They find two conspicuous 'transparency channels', one centered at $l^{II} = 45°$ and another at $l^{II} = 285°$. They argue in favor of a new interpretation of the surface photometry of the galactic belt by Elsässer and Haug (**16**). The criticisms by Isserstedt and Schmidt-Kaler are, however, rejected in a subsequent paper by Elsässer, Neckel, and Scheffler (**17**); see, however, Behr's comments (*Mitt. astr. Ges.*, no. 21, 76, 1966).

One of the most important investigations published during the past triennium has been a series of papers by Lyngå (**18**). He has studied extensively the sector of the band of the Milky Way, $311° < l^{II} < 330°$, and he has related the results of his work (already reported in part in the preceding Section) to that of Becker and others. In his most recent report (**19**), he has presented his views on the spiral structure for the fourth galactic quadrant. In agreement with Becker (**1**), he concludes that the Sagittarius Arm continues past Norma into Circinus and Centaurus and on into Crux and Carina. This situation is illustrated nicely in his diagram, which shows the location of the spiral tracers in the fourth quadrant; however, one should bear in mind when inspecting this diagram that Lyngå has plotted, with equal emphasis, the weak spiral features in the section $300° < l^{II} < 325°$ and the strong features in Carina, $275° < l^{II} < 300°$, and in Norma and Scorpius, $325° < l^{II} < 350°$. Lyngå mentions a divergence between the optical and the radio picture; the difference in longitude between the optical and the radio spiral arm in Carina amounts to almost 10°. I do not fully agree with this conclusion, for the coincidence between optical and radio features seems reasonably good; Kerr has recently derived a new diagram of spiral structure from 21 cm data, which shows good agreement with the optical result for the Carina section.

The detailed piecing-together of the recognized spiral features of our Galaxy remains one of the principal tasks for the members of Commission 33. The chapters in *Stars and Stellar Systems*, Vol. V, by S. Sharpless (Ch. 7), R. P. Kraft (Ch. 8), F. J. Kerr and G. Westerhout (Ch. 9), and G. Münch (Ch. 10) cover the earlier work in some detail.

While it is obviously important to attempt to establish differences between optical (stars and H II) and radio features (H I, radio continuum), every effort should be made to have the two general patterns agree before we attempt to distinguish differences. The major problem before us is that the pitch angle of the recognized radio spiral arms appears to be near 85°, whereas a preponderance of opinion favors a pitch angle in the range 70°–75° for the optical picture.

We note, first, that it is becoming increasingly apparent that the anlysis of the radio data has by no means been completed. The whole picture may change as it becomes necessary to assume new and different overall dimensions for our Galaxy, and such changes will certainly affect the law of circular galactic rotational velocity. Furthermore, recognition of the existence of large-scale deviations from pure circular motion are leading to fundamental changes in our interpretation of the radio data.

The present situation with regard to the optical picture is that by now we recognize certain spiral features as firmly established. These features are shown most completely in the spiral diagram proposed by W. Becker. However, the pattern may be more confused than is assumed by some. To simplify future optical work, it seems important that we should concentrate our efforts on the least questionable spiral tracers, which are the optically-recognized H II regions and their associated OB stars, and clusters with OB stars. For the present, long-period cepheids and cosmic dust clouds are to be looked upon with a degree of suspicion, since they may not represent tracers for the identical spiral structure which is presumably shown by the radio H I data and by the optical H II features. Interstellar absorption lines should enter the picture much more fully than they do at present. They promise not only to yield useful data on radial velocities of gaseous features associated with the spiral pattern, but they can also provide much information on possible deviations from pure circular motion.

It is well to stress here that we should be inherently suspicious of attempts to draw an overall spiral pattern from the available, limited, optical data. An inspection of the patterns found within 3 to 4 kpc of positions in selected outlying regions of half a dozen of the best-known spiral galaxies shows how easily one might draw the wrong overall spiral diagram on the basis of limited local information. In future optical studies of the spiral structure of our own Galaxy, we should proceed step by step. First, we should follow the pattern set by W. Becker and others, which is to delimit the clearest and cleanest spiral features (spiral knots, or established sections of what appear to be parts of a major arm pattern). However, we should not extrapolate unnecessarily and connect these features into a major, overall, pattern until radio and optical evidence jointly force us to do so. The second step should be to concentrate our efforts on the borders of established concentrations and on improved distances of spiral features for the sections where the situation seems confused. We should attempt to distinguish between spurs and stragglers and true connecting links between sections of a major arm. The third step should be a careful search for faint and remote OB stars and H II regions, and possibly also long-period cepheids, in those parts of the Milky Way for which we are viewing between spiral arms. In all of this work the 21 cm data should be fully considered. This applies especially to the location of true edges of spiral arms. Every effort should be made to distinguish between them and optical effects caused by overlying absorption.

The optical features that seem to be clearly present are the following:

1. The section of the Perseus Arm for $90° < l^{II} < 140°$ at distances from the Sun of 3 kpc and greater;
2. The section of the Local Arm for $60° < l^{II} < 210°$;
3. The section of the Sagittarius Arm for $330° < l^{II} < 30°$ at distances from the Sun of about 2 kpc;
4. The section of the Carina Arm between $280° < l^{II} < 310°$, pointing away from the Sun with established spiral features in the range of distance from the Sun of 1·5 to 6 kpc;
5. The connecting weak link, or weak part of the Carina-Sagittarius Arm, for $310° < l^{II} < 330°$ at distances from the Sun of about 2 kpc.

The recognized gaps in our spiral pattern to distances from the Sun no greater than 4000 pc are the following:

1. $30° < l^{II} < 60°$,
2. $270° < l^{II} < 280°$.

There are several sections of the Milky Way that demand urgent effort on the part of optical astronomers. The principal section of the northern Milky Way that requires prompt attention is at $140° < l^{II} < 180°$, where we must continue to search for evidence for or against an extension of the Perseus Arm; this search may involve detection of optical spiral tracers with distances between 3 and 7 kpc from the Sun. Greatest confusion exists in the section $180° < l^{II} < 270°$. There are some well-established remote features in this section, but it is difficult to judge their importance for the overall pattern. Mowat's (**5**) recent discussion lists one major concentration at great distance; it is the I Puppis Association, at approximately 5 kpc from the Sun ($l^{II} = 245°$). The nearby 'Gum Nebula' is a major local feature in this section. The situation in the next critical section, $280° < l^{II} < 295°$, and its bordering areas, has been well summarized in the recent paper of Sher (**20**). In a way, the most critical section of all is the Norma section, $325° < l^{II} < 335°$. Here we are faced with a choice between two alternatives. The first hypothesis is that at these longitudes our line of sight crosses two spiral arms, first the Sagittarius Arm, passing into Centaurus, second, the Norma-Scutum Arm, an inner spiral arm first suggested by Thackeray (**21**). The alternate hypothesis is that at these longitudes we are viewing tangentially an edge of the Sagittarius Arm pointing away from the Sun. The general consensus among members of our Commission apparently favors the first alternative, but the general appearance in the sky of the Norma section favors the second. Fortunately, the galactic-rotation component of radial velocities varies rapidly with distance from the Sun for the Puppis ($l^{II} = 245°$) and Norma ($l^{II} = 330°$) sections, thus making available an additional parameter for estimating distances. This situation, unfortunately, does not occur in the Carina section (**22**).

In the years to come, it will benefit us to pay increasing attention to the detailed structure of the Local Arm. There are many details here that require close study. The recent studies of Gould's Belt by Dewhirst and by Clube (**23**) show how it outlines a distinct feature of fairly large scale (70 by 700 parsecs) associated with the inner part of the Local Arm. The relative distribution of young stars, gas, and cosmic dust in the Local Arm must be explored in detail if we wish to interpret the difficult evidence presented by the Carina and other sections.

In conclusion, we should stress the increasing need for continued consultation between observers and theoreticians working on problems of the spiral structure of our Galaxy. Section X shows that there are several new approaches to the problems of the dynamical interpretation of spiral structure. In a recent communication, C. C. Lin asked that certain highly specific questions be answered by observers in the field. Some of these questions are as follows:

1. Is the Puppis-Vela Spur an extension of the Perseus Arm?
2. Can additional optical studies clarify the situation for the range $240° < l^{II} < 300°$ and $60° < l^{II} < 90°$?
3. Can one study further the distribution in depth for the range $315° < l^{II} < 340°$, especially for distances in excess of 2 kpc?

The study of the galactic spiral structure is presently passing through a difficult and confused phase, but we now have available the tools which should allow us to resolve most of the observational uncertainties.

BIBLIOGRAPHY

1. Becker, W. 1964, *Z. Astrophys.*, **58**, 202.
2. Schmidt-Kaler, Th. 1966, *Trans. IAU*, **12 B**, 417.
3. Abt, H. A., Bautz, L. P. 1963, *Astrophys. J.*, **138**, 1002.
4. Fletcher, E. S. 1963, *Astr. J.*, **68**, 407.
5. Mowat, M. Mt Stromlo Observatory Symposium on 'Radio and Optical Studies of the Galaxy', May, 1966, p. 100.
6. Kardashev, N. S., Lozinskaya, T. A., Sleptsova, N. F. 1964, *Astr. Zu.*, **41**, 601 (see also *Soviet Astr. A. J.*, **8**, 479, 1965).
7. Kostyakova, E. B. 1965, *Astr. Zu.*, **42**, 537 (see also *Soviet Astr. A. J.*, **9**, 426, 1965).
8. Pskovskii, Y. P. 1965, *Astr. Zu.*, **42**, 1184 (see also *Soviet Astr. A. J.*, **9**, 919, 1966).
9. Pavlovskaya, E. D., Sharov, A. S. 1966, *Astr. Zu.*, **43**, 40 (see also *Soviet Astr. A. J.*, **9**, 30, 1966).
10. Pronik, I I. 1965, *Astr. Zu.*, **42**, 923 (see also *Soviet Astr. A. J.*, **9**, 709, 1966).
11. Smith, L. F. Mt Stromlo Observatory Symposium on 'Radio and Optical Studies of the Galaxy,' May, 1966, p. 96; see also her unpublished Doctoral Thesis, Australian National University, 1966.
12. Muhleman, D. O., Walker, R. G. 1964, *Astr. J.*, **69**, 95.
13. Schmidt-Kaler, Th. *Z. Astrophys.*, **58**, 217, 1964; **63**, 131, 1966.
14. Behr, A. *Z. Astrophys.*, **61**, 182, 1965; **63**, 133, 1966.
15. Isserstedt, J., Schmidt-Kaler, Th. 1964, *Z. Astrophys.*, **59**, 182.
16. Elsässer, H., Haug, U. 1960, *Z. Astrophys.*, **50**, 121.
17. Elsässer, H., Neckel, Th., Scheffler, H. 1965, *Z. Astrophys.*, **63**, 1.
18. Lyngå, G. *Lund Medd.*, Series II, nos. 139, 140, and 141, 1964; and nos. 142 and 143, 1965.
19. Lyngå, G. Mt Stromlo Observatory Symposium on 'Radio and Optical Studies of the Galaxy', May, 1966, p. 61.
20. Sher, D. *Q. J. R. astr. Soc.*, **6**, 299, 1965.

21. Thackeray, A. D. 1956, *Nature*, **178**, 1458.
22. Bok, B. J. 1964, *IAU/URSI Symposium no.* 20, p. 147.
23. Report of 'Tenth Herstmonceux Conference'. 1966, *Observatory*, **86**, 182.

VIII. STELLAR DISTRIBUTION IN HIGH AND INTERMEDIATE GALACTIC LATITUDES

During the past three years, much new observational material has accumulated for the study of density and velocity distributions in the two Galactic Polar Caps, but little or no progress has been made in the theoretical interpretation of new and old data. No clear-cut regularities have been established, and there continues to be much uncertainty about most basic features of the kinematics and dynamics of our Galaxy for directions from the Sun at right angles to the galactic plane. In other words: We cannot report at this stage any real improvement in our knowledge of the force law, $K(z)$, for values of z less than 1000 parsecs, z being the height above or below the galactic plane. Observations of two varieties are required; density distribution functions, $\Delta(z)$, and, preferably for a good sample of the same stars, the corresponding distribution function of velocities perpendicular to the galactic plane, $\varphi(Z)$, where Z is the linear velocity perpendicular to the plane. If $\sigma^2(z)$ represents the square of the dispersion in random velocities Z, at height z, then the basic relation for a 'well-mixed' state of equilibrium perpendicular to the galactic plane is:

$$K(z) = \frac{1}{\Delta(z)} \cdot \frac{\partial}{\partial z} [\Delta(z)\, \sigma^2(z)] \quad , \qquad \text{(A)}$$

which, for the simplest case $\sigma^2(z) = \sigma^2$, of constant value, and for a Gaussian distribution of Z, reduces to:

$$K(z) = \frac{\sigma^2}{\text{Mod}} \cdot \frac{d[\log \Delta(z)]}{dz} \quad , \qquad \text{(B)}$$

where Mod = $\log_{10} e$.

The most extensive recent study has been made in the doctoral thesis of Perry (1), which provides a wealth of new material for the A0 to A2 stars. His basic list of stars consists, first, of all (97) A0 to A2 stars in the North Galactic Polar Cap (NGPC), i.e. within 15° of the North Galactic Pole. To these are added 12 A0 to A2 stars between $m = 9.5$ and 11.0 from the list published for the NGPC by Slettebak and Stock (2). Perry has determined for all 109 stars radial velocities as well as apparent and absolute magnitudes (the latter from narrow-band photometric measurements). He has, in addition, obtained similar material for all A0 to A2 stars (122) with known trigonometric parallaxes greater than 0″.020. Perry's radial velocities increase slowly, but gradually, with distance from the galactic plane; σ^2 equals about 200 for distances $50 < z < 200$ pc, and $\sigma^2 = 700$ at $z = 650$ pc. However, a much smaller value, $\sigma^2 = 44.3$, is found for the sample of the nearby stars, all within 50 pc of the Sun. If the Z velocities for distances greater than $z = 50$ pc are averaged and represented by a single Gaussian distribution function, then their dispersion corrected for accidental errors of measurement comes out to be $\sigma = 11.9$ km s^{-1}. The density function for these A0 to A2 stars shows a strong negative density gradient in the range $0 < z < 350$ pc and a smaller gradient for the range $350 < z < 750$ pc, with the latter the limit of the survey. On the basis of the 'well-mixed' hypothesis, as expressed by equations (*A*) and (*B*), the above results would imply a force function reaching a fairly sharp maximum at $z = 200$ pc. Such a conclusion seems rather improbable, as Perry stresses, but solutions involving several Gaussian components do not seem to alter the situation basically. It is not unlikely that his sample of A stars is not sufficiently ancient to be considered as 'dynamically relaxed', especially since moving groups may contribute more than their share to the sample close to the galactic plane. There is also the possibility of two basically different populations of A stars within 750 pc of the galactic plane.

Stothers and Tech (**3**) have studied the force field perpendicular to the galactic plane and in the solar neighborhood with the aid of all available data on motions and distance moduli for OB stars, excluding recognized members of known moving clusters. They find a surprisingly consistent picture when they derive the force law, $K(z)$, from a plot of Z against z. The force, $K(z)$, increases proportionally to z up to $z = 200$ pc, which is the limit of their survey.

The chapter by Elvius (**4**) has a good summary of the earlier work in the field, especially of the Swedish work, the papers by Oort and Hill, and those by Upgren, to which reference has been made already in earlier Reports. One class of stars to which relatively little attention has been given in recent years are the F stars. Photographic and photo-electric techniques enable us to obtain information on their density distributions. Furthermore, radial-velocity measurements for a representative sample of these F stars can at present be more easily obtained than was the case a few years ago. Work on radial velocities for F stars in the NGPC is well advanced at the Dominion Astrophysical Observatory. Bok and Basinski (**5**) have published B and V magnitudes to $V = 16$ for stars in the range $+0{\cdot}30 < B-V < +0{\cdot}60$ in the SGPC. The space density (per unit volume) drops steadily with increasing z, as is shown in the following tabulation:

z	$\Delta(z)$
100 pc	1·00
200	0·35
500	0·051
1000	0·0123
2000	0·0030

The F stars are probably a rather homogeneous group. A survey of their distribution to $V = 16$ and fainter in the NGPC seems highly desirable. The existing spectral surveys of Upgren (**6**) and of Slettebak and Stock (**2**) reach only to $V = 12$.

The work at the Basel Observatory by W. Becker and associates is progressing well (**7**). Results for the space distribution of Population II stars in the halo are now available for Selected Areas 51 and 57, the latter close to the North Galactic Pole. We refer to T. Elvius' Report on Selected Areas for details, especially regarding the work in progress for intermediate galactic latitudes. It is of interest to note that Terzan at Lyon has completed a three-color photographic survey (at $\lambda = 4400$, 6400, and 8100Å) for Selected Area 57 (**8**).

R. Bouigue at Toulouse reports good progress on the determination of colors, magnitudes, and objective-prism radial velocities for all stars brighter than 10th magnitude in the NGPC.

Blaauw reports that a collaborative project has been undertaken by the Kapteyn Astronomical Laboratory and the Leander McCormick Observatory to investigate the density distribution and kinematic properties of the intermediate-age G and K giants up to distances of several kiloparsecs from the Sun. Ljunggren has completed his studies for the NGPC (**9**). One important conclusion of his work is that there is evidence of irregular absorption and reddening over the NGPC, suggesting a patchy distribution of absorbing material mostly at distances, z, of 200 to 250 pc. We note that Perry's results gave no evidence for marked reddening effects in the NGPC.

We need not deal at length again with the derived value of the mass density near the galactic plane. However, we should draw attention once again to Einasto's spirited defense of the lower value for the density near the galactic plane (**10**). His result is more or less supported by Stothers and Tech (**3**), but is in conflict with most other recent results. However, our rather full description of the difficulties encountered by Perry in his analysis for the A0 to A2 stars shows that the issue is by no means a closed one. Sanduleak's work (**11**) on the faint M stars in the NGPC is relevant at this point. He found over 1200 such stars in an area of 120 square degrees from a survey with the Warner and Swasey Observatory Burrell-Schmidt telescope, a survey complete to $V = 17$. If his M stars were all dwarf M stars, then they may contribute 0·04 $M_\odot$ per cubic parsec to the mass density near the Sun. In his report to our Commission, McCuskey points out that proper-motion surveys do not confirm this large excess of dwarf M stars. If Sanduleak's stars are giant M stars, then they must lie at very great distances from the galactic plane, with a density of one such star per volume of $2{\cdot}5 \times 10^8$ pc^3 at $z = 20\,000$ pc. For the SGPC, on the other hand, Pik-Sin Thé and Blanco find space densities for M dwarfs that are not nearly as great as those of Sanduleak. McCarthy, Bertiau, and Treanor (**12**) have made a complementary survey for the SGPC, using ADH-Schmidt plates, taken with a 3° objective prism, which records the brighter stars of spectral class M2 and later. Their results agree with those of Pik-Sin-Thé and Blanco.

Davis Philip is continuing his comprehensive spectral-color-magnitude program for regions at intermediate galactic latitudes. The complete results have been published (**13**) for a region in Pegasus at $b^{II} = -29°$. His basic material consists of objective-prism plates obtained with the Warner and Swasey Observatory Burrell-Schmidt telescope and of photo-electrically (*UBV*) calibrated photographic magnitudes for these stars. A rapid decrease in space density is found for main-sequence stars of spectral class earlier than G0. At $z = 300$ pc, the space density of these stars is equal to only one-tenth that at $z = 100$ pc, or very much the same result as the one of Basinski and Bok, shown above. Constant density was found for the dwarfs to the limit of the survey. For the late-type giants, the space densities drop off much slower with z than for the early-type stars; at $z = 300$ pc, the average space density is one-half of that at $z = 100$ pc. The study is the first one published for an extensive general survey. The plan is to cover ultimately areas at $b^{II} = +75°, +60°, +45°, +30°, +15°, 0°, -15°, -30°, -45°, -60°, -75°$ for the four galactic longitude belts, $l^{II} = 76°, 290°, 0°$, and 180°. The results will be related to those obtained by radio-astronomical techniques for the distribution of H I (**14**). A remarkable blueing effect found by Davis Philip was traced in part to inconsistencies in spectral classification; the main cause for the effect remains lower metal abundance of the older stars.

J. S. Drilling of Warner and Swasey Observatory is studying the space distribution of stars in an area of 20 square degrees at $l^{II} = 330°$, $b^{II} = -30°$; this is a joint project with Bosscha Observatory. Spectral-photometric data, complete to $V = 10{\cdot}5$ or $11{\cdot}0$, will form the basic material for the analysis to follow.

McCuskey reports preliminary results for the stellar distribution for three of the Groningen-Palomar Fields used in the survey of faint RR Lyrae variables (**15**). Because of the importance of these fields for our future understanding of the overall properties of our Galaxy, we reproduce McCuskey's report in full:

'Absorption characteristics and the space-density distributions of giant M stars and of stars of types B8–A2 and gG8–gK2 have been determined for Fields 1, 2, and 4 of the Groningen-Palomar Survey of Variable Stars. The areas are located at $l^{II} = 0°$, $b^{II} = +28°$; $l^{II} = 3°$, $b^{II} = +11°$; and $l^{II} = 81°$, $b^{II} = +11°$ respectively. Each field covers 5 sq. deg. The principal results of this study, which was made by Dr B. Hidajat, are as follows:

(*a*) Photoelectric photometry on the *UBV* system for selected stars in each region, together with available data for nearby clusters, indicates the following run of interstellar absorption, A_v, magnitudes:

	Distance (pcs)				
Field	200	300	400	500	Max.
1	0·2	0·8	0·8	0·9	0·9
2	0·6	0·7	0·7	0·8	1·4
4	0·1	0·1	0·2	0·3	0·8

The maximum values of A_v refer to distances of the order of 1000–1500 parsecs. Beyond 1500 parsecs the values of A_v were found to be sensibly constant.

(*b*) Space densities, *number of stars per* 10^6 *cubic parsecs*, were derived as functions of distance, as follows:

		Distance (kpc)				
Field		1	2	3	4	5
1	M2–4	3·8	1·4	0·7	0·4	0·3
	M5–10	—	—	—	—	—
2	M2–4	6·9	4·1	2·5	1·4	1·0
	M5–10	0·5	0·5	0·4	0·3	0·2
4	M2–4	2·3	3·0	2·0	1·3	0·6
	M5–10	1·1	0·6	0·3	0·2	0·2

There appears to be an almost complete absence of late M stars in Field 1.

(*c*) The decrease in the numbers of giant M stars with distance from the galactic plane is marked. In Field 4, for example, stars of all classes (M2–M10) diminish from 3·8 stars per 10^6 pc^3 at $z = 0{\cdot}3$ kpc to 0·9 at $z = 1{\cdot}0$ kpc. Similarly, stars of late type (M5–M10) decrease in density from 1·1 at $z = 0{\cdot}2$ kpc to 0·15 at $z = 1{\cdot}0$ kpc.

(*d*) In the galactic meridional plane defined by the Sun, the galactic center, and the galactic poles, the present study, in conjunction with similar studies by other investigators, indicates that the giant stars of the M5–M10 group are distributed symmetrically with respect to the plane of the Galaxy.

(*e*) There is some evidence, based upon the space densities obtained for the B8–A2 and the giant K stars in these three regions, of a correlation between the two groups. Furthermore, some evidence indicates that the A stars are located on the inner edges of the galactic spiral arms as delineated by the interstellar hydrogen. This possibility, along with a similar apparent correlation between the early (M2–M4) giant M star concentrations and the spiral structure, remains to be investigated further.

'The results of this study are being prepared for publication. The study was done under the supervision of Dr V. M. Blanco.'

BIBLIOGRAPHY

1. Perry, C. L. 1965, Doctoral Thesis, Univ. of California (unpublished).
2. Slettebak, A., Stock, J. 1959, *Hamburg Obs. Publ.*, **5**, no. 5.
3. Stothers, R., Tech, J. L. 1964, *Mon. Not. R. astr. Soc.*, **127**, 287.
4. Elvius, T. 1965, *Stars and Stellar Systems: Galactic Structure*, Vol. V, Chapter 3.
5. Bok, B. J., Basinski, J. M. 1964, *Mt Stromlo Obs. Memoirs*, no. 16.
6. Upgren, A. R. 1962, *Astr. J.*, **67**, 37.
7. Becker, W. 1965, *Z. Astrophys.*, **62**, 54.
8. Terzan, A. 1965, *Ann. Astrophys.*, **28**, 935.
9. Ljunggren, B. *Uppsala Medd.*, no. 154, 1966; see also, *Uppsala Medd.*, no. 151, 1965.
10. Einasto, J. 1966, *Trans. IAU*, **12 B**, 436.
11. Sanduleak, N. 1964, Doctoral Thesis, Case Inst. of Techn. (unpublished).
12. McCarthy, M. F., Bertiau, F. C., Treanor, P. J. 1964, *Ric. Astr. Vatican.*, **6**, 571.
13. Davis Philip, A. G. 1966, *Astrophys. J. Suppl.*, **12**, no. 112.
14. Kaftan-Kassim, M. A. 1961, *Astrophys. J.*, **133**, 821.
15. *Trans. IAU*, 1960, **10**, 409.

IX. KINEMATICS

(prepared by G. Contopoulos)

This part of the Report is connected with Commissions 24 (proper motions), 30 (radial velocities) and 34 (kinematics of interstellar matter). Here, however, the kinematics of stars and interstellar matter is considered in connection with the problems of galactic structure and dynamics.

A number of articles in Vol. 5 of '*Stars and Stellar Systems*' are connected with this subject. Delhaye's (**1**) article on the 'Solar Motion and Velocity Distribution of Common Stars' gives a review of the determinations of the solar motion and the dependence of kinematic properties on spectral type and age.

Woolley's (**2**) discussion in 'Motions of the Nearby Stars' includes the galactic orbits of stars, the velocity ellipsoid, the connection between color-magnitude diagrams and orbital characteristics, the motions perpendicular to the galactic plane, and the kinematic properties of special groups of stars.

The 'Moving Groups of Stars' were discussed in detail by Eggen (**3**).

The motions of late-type giants, variable stars, planetary nebulae, high velocity stars, sub-luminous stars, blue stars at high galactic latitudes, globular clusters and interstellar gas in general and hydrogen in particular were considered by Blanco (**4**), Plaut (**5**), Minkowski (**6**), Roman (**7**), Greenstein (**8**), Luyten (**9**), Arp (**10**), Münch (**11**), and Kerr and Westerhout (**12**) respectively.

The relations between stellar kinematics and stellar populations were discussed by Blaauw (**13**).

A number of galactic kinematical problems were discussed by Bok (**14**).

Perek (**15**) published tables for the transformation of space velocities from the equatorial system to the new galactic system.

A. Stars

Solar Motion—Galactic Rotation

New solutions for the solar motion with respect to the local group and other nearby galaxies have been given by Byrnes (**16**). Boulon (**17, 18**) found the influence of the K-term on the error of the determination of the solar motion. Petrie (**19**) determined the solar motion from radial velocities of nearby B stars. A program has been undertaken at Greenwich to find the solar motion with respect to old disk population stars.

Determinations of the constants of galactic motion were made by many authors.

Boulon (**18**) found the values $A = 17{\cdot}0 \pm 1{\cdot}7$, $K = +1{\cdot}2 \pm 3{\cdot}4$ from a study of ten galactic fields.

Howard and Kirk (**20**) discussed the variation of the parameters of galactic rotation with the distance scale, from a study of 144 cepheids and 39 galactic clusters; they adopted the values $R_0 = 11{\cdot}4$ kpc, $A = 13{\cdot}1$, $K = -3{\cdot}3$.

Fricke (**21**) found, on the basis of the FK4 proper motion system, the values $A = 12{\cdot}3 \pm 2{\cdot}4$, $B = -7{\cdot}8 \pm 2{\cdot}4$. Further work, including 500 stars at distances larger than 100 pc, is in progress. A similar analysis by Wayman (**22**) of the proper motions of 34 OB stars, taken from FK4, gave B about equal to -15. An analysis of the radial motions of about 500 B-stars has been done by Petrie (**23**).

A very small value, $A = 7{\cdot}8 \pm 3{\cdot}4$, was found by Feast (**24**) for Me variables; he points out that this value is not basically out of line with $A = 15$, found for OB stars.

The A constant derived from a study of B-stars, cepheids, galactic clusters and interstellar gas by Feast and Shuttleworth (**25**) is $A = 14{\cdot}3 \pm 0{\cdot}8$.

Rubin and Burley (**26**) found the values $A = 14 \cdot 5 \pm 1 \cdot 5$, $B = -9 \cdot 5 \pm 2 \cdot 5$, $K = -2 \pm 1$, from the study of O–B5 stars within 2 kpc from the Sun. The rotation curve for $R > R_0$ looks flatter than usually assumed. Barbier (**27**) found also, from a study of O, B and A stars in Cassiopeia, that, for large R, the velocity curve is above that of the new Schmidt model.

Radial velocities of early-type stars in the Perseus arm were studied by Abt and Bautz (**28**). A discrepancy between the stellar and the 21-cm rotation curve was explained by the fact that the velocities of spectroscopic binaries are systematically too large.

Radial velocities of early-type stars in the direction of the anticenter were determined by Rubin (**29**) and by Christy. Rubin found a rotation curve for $R > R_0$, which agrees with that given by Schmidt.

Bonneau (**30**) has found the relative velocity of the Perseus arm with respect to the Orion arm.

Münch and Münch (**31**) have found that the rotation curve of the inner parts of the galaxy derived from OB stars is below the corresponding 21-cm curve. However Feast, Shuttleworth and Thackeray (**25, 32**) found a better agreement between optical and radio velocity curves after correcting the stellar distances. In addition, they found a variation of the K-term with the distance of B-stars. Further work is being done on OB stars by Bonneau (**30**), Lyngå and Haug, and on cepheids by Feast.

A large K-term in southern B-stars was found by Thackeray (**33**). Mirzoyan and Kazarian (**34**) studied the K-term in the motions of OB-stars and cepheids; they found compatibility with Kerr's expansion.

A critical discussion of earlier determinations of the constant A and the galactic scale was given by Stibbs (**35**).

Massonie (**36**) found a velocity of escape in the Galaxy of the order 340–345 km s^{-1} and a rotational velocity $\Theta_0 = 256$ km s^{-1}.

Velocity Distribution—Velocity Ellipsoid

Eggen (**37**) has studied the kinematics of various groups of stars near the Sun. He has found a remarkable clumpiness in velocity distribution.

Van Wijk, Smith and Daniels (**38**) have computed the velocity distribution of low-velocity stars in the solar neighborhood assuming that the stars were formed with an isotropic, Maxwellian velocity distribution, followed by a dynamical mixing in the radial direction.

Perek (**39**) and Minkowski (**40**) have found a large dispersion of radial velocities of planetary nebulae close to the direction of the galactic center. Minkowski concluded that the motions of the planetaries are not circular. Feast (**24, 41**) explained this phenomenon by assuming that the planetaries cover a wide range of ages. In the case of Me variables he found that the velocity dispersion is remarkably constant over a large range of distances from the center (for a given period). An observed asymmetry of radial velocities indicates an excess of inward motions for $l > 0°$ and of outward motions for $l < 0°$. The velocity ellipsoid of Me variables was also studied.

Schlesinger (**42**) found that the velocity dispersion of metallic-line A-stars is similar to that of the normal A-stars.

The velocity ellipsoid of subdwarfs was investigated by Takayanagi (**43**).

Pavlovskaya (**44**) determined the velocity ellipsoids of stars of various spectral and luminosity classes.

Dahn (**45**) studied the velocity ellipsoid of carbon stars. Rudnicki (**46**) found that carbon stars far from the galactic plane have larger velocity dispersion.

An increase of the velocity dispersion of main sequence stars with height above the galactic plane was found by Murray (47) from a study of proper motions of field stars near M 67. He suggested that most bright red stars and many stars of intermediate colors are distant halo stars. Similar results were found by Cannon for field stars near NGC 752.

The kinematics of Mira variables was studied by Smak and Preston (48). They derived values of A ranging from 15 to 25.

Work on proper motions and radial velocities of RR Lyrae variables has been done by Woolley, Harding, Cassells, Saunders, Aly and Clube (49, 50, 51). This work continues with the aim of finding the orbital elements as functions of the metal abundances, period of the variables, etc. Further work is being done on semi-regular variables.

Van Herk (52) also made an extensive study of the motions of RR Lyrae variables.

A study of the motions of bright A and gK stars near the galactic equator has been made by Eelsalu and Jôeveer (53); in particular they determined the dispersion of velocities in the z-direction.

Shatsova (54) proposed a Planck law to describe the distribution of stellar velocities. A truncated Schwarzschild distribution of velocities was used by King (55), Bouvier (56) and others. Dzigvashvili (57) determined the parameters of the velocity ellipsoid and the rotation velocity according to Chandrasekhar's theory.

Motions of high velocity stars have been determined by Kennedy and Przybylski (58).

Vitrichenko, Gershberg and Metik (59) have studied the motions of 120 high-velocity OB stars; they found that probably all of them originated in groups of hot stars.

Blaauw is investigating the distribution and kinematics of intermediate age G and K giants in collaboration with the Leander McCormick Observatory.

The kinematics of the Gould belt has been studied by Bonneau (30) and by Blaauw (60). Blaauw found from a study of the velocity gradients of the stars in different directions that, about 40 million years ago, the prestellar gas probably occupied a small volume. This work continues, together with Lesh. Clube (61) found that streaming motions associated with the Gould belt may change appreciably the kinematics of nearby low velocity stars.

Mouchet (62) and Martinet (63) have found, from kinematic properties of peculiar A stars, indications that these stars are connected with the Gould Belt.

A few theoretical papers by Nahon (64) and Massonie (65, 66) deal with an application of the least squares method to the analysis of the radial velocities, the distribution of errors and the influence of selection on the determination of mean velocities.

Kinematics, Age and Chemical Composition

General articles on the connection between the kinematic properties of stars, age, and chemical composition were given by Delhaye (1), Cayrel and Kovalevsky (67) and Cayrel (68). Eggen (69, 70) studied the colors, luminosities and motions of more than 1000 nearby A and G stars; many of these stars belong to moving groups. He found that for stars of the same age the chemical composition may vary, but there is a rough correlation of age with metal abundance and with the space motions. The chemical composition and kinematics of disk high-velocity stars of the main sequence were considered by Strömgren (71).

A strong correlation between ultraviolet excess and space motion of high-velocity stars was found by Oosterhoff and Ponsen (72). Similar results were found by Kreiken and Yilmaz (73).

A study of the kinematics of planetary nebulae, carbon stars and RV Tauri stars in connection with population type was carried out by Iwanowska, Kanthak and Boenigk (74, 75, 76).

A correlation between age and velocity dispersion perpendiculary to the galactic plane was found by Dennis (73a).

Kinematic ages of stars were found by Schmidt-Kaler (**77, 78**) from the distribution of different groups in the spiral arms. Connections between the orbital characteristics of the nearby stars and their ages were found by Shimizu and Takahashi (**79**). An explanation of the age effect was based primarily on the variation of the force field.

B. Gas

Galactic Rotation—Non Circular Motions

A detailed discussion of the galactic rotation curve, based on radio observations, has been given by W. W. Shane and Bieger-Smith (**80**). If we assume the irregularities in the rotation curve to be due to lack of hydrogen then an improbable density contrast arises. A more plausible explanation is that there are irregularities in the large-scale motions associated with the spiral structure. Stream motions in the outer edges of the arms were suggested. Such motions were found by Burton (**81**) in the Sagittarius arm. The discrepancy between the northern and southern velocity curves, as given by Kerr (**82**), was verified.

Braes (**83**) tried to find the expansion motion suggested by Kerr, but reached a negative conclusion. The same conclusion was drawn by W. W. Shane (**84**). On the other hand, Locke, Galt and Costain (**85**) found evidence in favor of Kerr's hypothesis, from their study of the distribution of neutral hydrogen in the anticenter direction.

Pismis (**86**) suggested that the minima in the rotation curve are due to the existence of clouds with large radial and z-motions.

Dixon (**87**) found indications that the interstellar medium lags behind the gravitational circular velocity by 14 km s^{-1}, because of non-gravitational forces.

Agekyan, Petrovskaya and Fesenko (**88**) found a similar lag of 32 km s^{-1}. They derived a galactic rotation curve below the Dutch one. Non-circular motions in the outer spiral arms were considered by Pronik (**89**).

Detailed observations of the hydrogen velocity distribution at the anticenter have been done by Lindblad (**90**). Non-circular motions are apparent there.

The problem of the rotation curve has been discussed, along with other problems, by Kerr (**91**).

The smoothing of the 21-cm profiles has been discussed by Wilhelmsson and Winnberg (**92**).

Central Region

The kinematics of the gas near the center of the Galaxy was described by Rougoor (**93**) and by Oort (**94**). There is a 'nuclear disk', inside 800 pc, rotating at about 200 km s^{-1} and there is no evidence that it is expanding. The velocity decreases abruptly within 100 pc from the nucleus. Most of the gas in the nuclear region is expanding with high velocities. The properties of the gas near the center of the Galaxy were considered by Lequeux (**95**) and Kerr (**96**).

Expansion motions have been also found by observations of the OH radiation by Robinson *et al.* (**97**), Goldstein *et al.* (**98**) and Bolton *et al.* (**99**).

High-Velocity Clouds

The evidence for the existence of many high velocity clouds approaching the galactic plane from both hemispheres has been increasing in recent years. Such evidence has been collected by Oort (**100, 101, 102, 103**) and other astronomers in the Netherlands, such as Blaauw, Tolbert, Muller, Raimond, Schwarz, Hulsbosch and W. W. Shane (**104, 105, 106, 107, 108**). Most of these clouds, with velocities above 100 km s^{-1}, come from longitudes between 60° and 200° (i.e. mostly between the direction of the anticenter and the direction of galactic rotation), and latitudes between + 10° and + 80°. The most probable explanation is that they are intergalactic in origin. Other possible explanations, e.g. that they are supernova shells, or remnants of a superexplosion, do not seem very probable.

The flux of the incoming high-velocity hydrogen is of the order of 10^{19} atoms cm^{-2} per million years; this corresponds to an increase of the mass of the disk by $10^5 M_\odot$ per million years. Perhaps the high velocity clouds could replenish the hydrogen of the nuclear part of our Galaxy.

A different explanation for the high velocity clouds at latitudes below 25° was proposed by Habing (**109**), who considers them as belonging to an envelope of the Outer Arm.

Shklovski (**110**), based on a suggestion by Varshalovitch, advanced the hypothesis that these clouds radiate only when they approach the galactic plane, because of a maser effect due to the excitation by Lyman-alpha radiation of hot stars.

Dieter (**111, 112**) found many clouds with negative velocities of a few tens of km s^{-1} near the galactic poles.

Smith (**113**) and Prata (**114**) found clouds with positive radial velocities of the order of 100 km s^{-1}, associated probably with the galactic structure.

A search for OB stars near one of the high-velocity clouds is being made by Svolopoulos.

BIBLIOGRAPHY

1. Delhaye, J. 1965, *Stars and Stellar Systems*, **5**, 61.
2. Woolley, R. v. d. R. 1965, *Stars and Stellar Systems*, **5**, 85.
3. Eggen, O. J. 1965, *Stars and Stellar Systems*, **5**, 111.
4. Blanco, V. M. 1965, *Stars and Stellar Systems*, **5**, 241.
5. Plaut, L. 1965, *Stars and Stellar Systems*, **5**, 267.
6. Minkowski, R. 1965, *Stars and Stellar Systems*, **5**, 321.
7. Roman, N. G. 1965, *Stars and Stellar Systems*, **5**, 345.
8. Greenstein, J. L. 1965, *Stars and Stellar Systems*, **5**, 361.
9. Luyten, W. J. 1965, *Stars and Stellar Systems*, **5**, 393.
10. Arp, H. C. 1965, *Stars and Stellar Systems*, **5**, 401.
11. Münch, G. 1965, *Stars and Stellar Systems*, **5**, 203.
12. Kerr, F. J., Westerhout, G. 1965, *Stars and Stellar Systems*, **5**, 167.
13. Blaauw, A. 1965, *Stars and Stellar Systems*, **5**, 435.
14. Bok, B. J. 1965, NATO Summer Course, 'Observational Aspects of Galactic Structure', Part 29, Athens.
15. Perek, L. 1964, *Publ. astr. Inst. Czech. Ac. Sci.*, no. 48, Praha.
16. Byrnes, D. V. 1966, *Publ. astr. Soc. Pacif.*, **78**, 46.
17. Boulon, J. 1963, *C. r. Acad. Sci.*, Paris, **256**, 5280.
18. Boulon, J. 1963, *J. Observateurs*, **46**, 225.
19. Petrie, R. M. 1964, *Mon. Not. R. astr. Soc.*, **128**, 245.
20. Howard, W. E. III, Kirk, J. G. *Astr. J.*, **69**, 544, 1964; **70**, 434, 1965.
21. Fricke, W. 1966, Besançon Colloquium, 'Nouvelles Méthodes de la Dynamique Stellaire'.
22. Wayman, P. A. 1966, *Quart. J. R. astr. Soc.*, **7**, 138; *Observatory*, **86**, 185, 1966.
23. Petrie, R. M. 1964, *J. R. astr. Soc. Can.*, **58**, 34.
24. Feast, M. W. 1963, *Mon. Not. R. astr. Soc.*, **125**, 367.
25. Feast, M. W., Shuttleworth, M. 1965, *Mon. Not. R. astr. Soc.*, **130**, 243.
26. Rubin, V. C., Burley, J. 1964, *Astr. J.*, **69**, 80.
27. Barbier, M. 1966, *J. Observateurs*, **49**, 269.
28. Abt, H. A., Bautz, L. P. 1963, *Astrophys. J.*, **138**, 1002.
29. Rubin, V. C. 1965, *Astrophys. J.*, **142**, 934.
30. Bonneau, M. 1964, *J. Observateurs*, **47**, 251.
31. Münch, G., Münch, L. 1964, *Astrophys. J.*, **140**, 162.

32. Feast, M. W., Thackeray, A. D. 1965, *Astrophys. J.*, **142**, 1645.
33. Thackeray, A. D. 1966, *Mem. R. astr. Soc.*, **70**, 33.
34. Mirzoyan, L. V., Kazarian, E. S. 1965, *Trudy astrofiz. Inst. Alma-Ata*, **5**, 224.
35. Stibbs, D. W. N. 1964, *Publ. R. Obs. Edinburgh*, **4**, 87.
36. Massonie, J. P. *J. Observateurs*, **48**, 119, 1965; Thesis, *Bull. Inst. Statist. Univ. Paris*, **14**, 1, 1965.
37. Eggen, O. J. 1965, NATO Summer Course 'Kinematic and Chemical History of the Galaxy', Herstmonceux.
38. van Wijk, V., Smith, B. F., Daniels, W. E. 1966, *Astr. J.*, **71**, 401.
39. Perek, L. 1963, *Bull. astr. Inst. Csl.*, **14**, 218.
40. Minkowski, R. 1964, *Publ. astr. Soc. Pacif.*, **76**, 197.
41. Feast, M. W. 1966, *Mon. Not. R. astr. Soc.*, **132**, 495.
42. Schlesinger, B. M. 1966, *Publ. astr. Soc. Pacif.*, **78**, 39.
43. Takayanagi, K. 1965, *Publ. astr. Soc. Japan*, **17**, 71.
44. Pavlovskaya, E. D. 1965, *Trudy astrofiz. Inst. Alma-Ata*, **5**, 69.
45. Dahn, C. C. 1964, *Publ. astr. Soc. Pacif.*, **76**, 403.
46. Rudnicki, K. 1966, *Acta Astr.* **16**, 335.
47. Murray, C. A. 1967, *R. Obs. Bull.* (in press).
48. Smak, J. I., Preston, G. W. 1965, *Astrophys. J.*, **142**, 943.
49. Woolley, R. v. d. R., Harding, G. A., Cassells, A. I., Saunders, J. 1965, *R. Obs. Bull.*, no. 97.
50. Woolley, R. v. d. R., Aly, K. 1966, *R. Obs. Bull.*, no. 114.
51. Clube, S. V. M. 1967, *R. Obs. Bull.* (in press).
52. van Herk, G. 1965, *Bull. astr. Inst. Netherlds*, **18**, 71.
53. Eelsalu, H., Jõeveer, M. *Tartu Astr. Obs. Publ.*, **34**, 47, 1964; **34**, 121, 1964; **35**, 49, 1966.
54. Shatsova, R. B. *Astr. Zu.*, **42**, 160, 581, 1965 = *Sov. Astr.*, **9**, 124, 461, 1966.
55. King, I. 1966, *Astr. J.*, **71**, 64.
56. Bouvier, P. 1966, Besançon Colloquium, 'Nouvelles Méthodes de la Dynamique Stellaire'.
57. Dzigvashvili, R. M. 1965, *Trudy astrofiz. Inst. Alma-Ata*, **5**, 233.
58. Kennedy, P. M., Przybylski, A. 1963, *Mon. Not. R. astr. Soc.*, **126**, 381.
59. Vitrichenko, E. A., Gershberg, R. E., Metik, L. P. 1965, *Izv. Krym. astrofiz. Obs.*, **34**, 193.
60. Blaauw, A. 1965, *Konikl. Nederl. Akad. Wetenschappen, Verslagen Afd. Natuurkunde*, **74**, no. 4.
61. Clube, S. V. M. 1967, *R. Obs. Bull.* (in press).
62. Mouchet, M. T. 1966, *Bull. astr.*, 3e Sér., **1**, 87.
63. Martinet, L. 1966, *Bull. astr.*, 3e Sér., **1**, 95.
64. Nahon, F. 1963, *C. R. Acad. Sci.*, Paris, **257**, 2965, 3305.
65. Massonie, J. P. 1965, *Ann. Sci. Univ. Besançon*, 3e Sér., Fasc. 4.
66. Massonie, J. P. 1966, *C. R. Acad. Sci.*, Paris, **262**, 591, 673.
67. Cayrel, R., Kovalevsky, J. 1964, *Astronomie*, **78**, 437.
68. Cayrel, R. 1965, NATO Summer Course 'Observational Aspects of Galactic Structure', Parts 25, 26, Athens.
69. Eggen, O. J. 1963, *Astr. J.*, **68**, 697.
70. Eggen, O. J. 1964, *Astr. J.*, **69**, 570.
71. Strömgren, B. 1964, *Astrophys. Norv.*, **9**, 333.
72. Oosterhoff, P. T., Ponsen, I. 1966, *Bull. astr. Inst. Netherlds*, **18**, 150.
73. Kreiken, E. A., Yilmaz, N. 1964, *Ann. Astrophys.*, **27**, 672.
73a. Dennis, T. R. 1966, *Astrophys. J.*, **146**, 581.
74. Iwanowska, W., Kanthak, J. 1965, *Bull. Acad. Pol. Sci.*, Ser. Sc. Math. Astr. Phys., **13**, 155.
75. Iwanowska, W., Boenigk, T. *Bull. Acad. Pol. Sci.*, Ser. Sc. Math. Astr. Phys., **13**, 163, 1965; **13**, 169, 1965.
76. Boenigk, T. 1965, *Bull. Acad. Pol. Sci.*, Ser. Sc. Math. Astr. Phys., **13**, 175.

77. Schmidt-Kaler, T. 1964, *Z. Astrophys.*, **58**, 217.
78. Schmidt-Kaler, T. 1965, *Astr. J.*, **70**, 147.
79. Shimizu, T., Takahashi, K. 1964, *Mem. Coll. Sci., Univ. Kyoto*, **A 30**, no. 3, 333.
80. Shane, W. W., Bieger-Smith, G. P. 1966, *Bull. astr. Inst. Netherlds*, **18**, 263.
81. Burton, W. B. 1966, *Bull. astr. Inst. Netherlds*, **18**, 247.
82. Kerr, F. J. 1964, IAU-URSI Symposium no. 20, Canberra.
83. Braes, L. L. E. 1963, *Bull. astr. Inst. Netherlds*, **17**, 132.
84. Shane, W. W. 1965, NATO Summer Course 'Kinematic and Chemical History of the Galaxy', Herstmonceux.
85. Locke, J. L., Galt, J. A., Costain, C. H. 1966, *Astrophys. J.*, **139**, 1066.
86. Pismis, P. 1965, *Bull. Obs. Tonantzintla y Tacubaya*, **4**, 8.
87. Dixon, E. 1966, Besançon Colloquium, 'Nouvelles Méthodes de la Dynamique Stellaire'.
88. Agekyan, T. A., Petrovskaya, I. V., Fesenko, B. I. *Astr. Zu.*, **41**, 1027, 1964 = *Sov. Astr.*, **8**, 823, 1965; *Astrophysika*, **1**, 373, 1965.
89. Pronik, I. I. *Astr. Zu.*, **42**, 923, 1965 = *Sov. Astr.*, **9**, 709, 1966.
90. Lindblad, P. O. 1966, *Bull. astr. Inst. Netherlds, Suppl.*, **1**, 77, 177.
91. Kerr, F. J. 1965, NATO Summer Course 'Observational Aspects of Galactic Structure', Parts 25, 26, Athens.
92. Wilhelmsson, H., Winnberg, A. 1963, *Astr. Notes, Univ. Gothenburg*, no. 8, 20.
93. Rougoor, G. W. 1964, *Bull. astr. Inst. Netherlds*, **17**, 381.
94. Oort, J. H. 1966, IAU-URSI Symposium No. 31, Noordwijk.
95. Lequeux, J. 1966, IAU-URSI Symposium No. 31, Noordwijk.
96. Kerr, F. J. 1965, NATO Summer Course 'Kinematic and Chemical History of the Galaxy', Herstmonceux.
97. Robinson, B. J., Gardner, F. F., van Damme, K. J., Bolton, J. G. 1964, *Nature*, **202**, 989.
98. Goldstein, S. J., Jr., Gundermann, E. J., Penzias, A. A., Lilley, A. E. 1964, *Nature*, **203**, 65.
99. Bolton, J. G., Gardner, F. F., McGee, P. X., Robinson, B. J. 1964, *Nature*, **204**, 30.
100. Oort, J. H. 1966, *Trans. IAU*, **12 B**, 395, 789.
101. Oort, J. H. 1966, *Bull. astr. Inst. Netherlds*, **18**, 421.
102. Oort, J. H. 1966, IAU-URSI Symposium No. 31, Noordwijk.
103. Oort, J. H. 1966, Ventlesen Award Lecture, Columbia University.
104. Blaauw, A. 1966, IAU-URSI Symposium No. 31, Noordwijk.
105. Blaauw, A., Tolbert, C. R. 1966, *Bull. astr. Inst. Netherlds*, **18**, 405.
106. Muller, C. A., Raimond, E., Schwarz, U. I., Tolbert, C. R. 1966, *Bull. astr. Inst. Netherlds, Suppl.* **1**, 215.
107. Hulsbosch, A. N. M., Raimond, E. 1966, *Bull. astr. Inst. Netherlds*, **18**, 413.
108. Shane, W. W. 1966, IAU-URSI Symposium no. 31, Noordwijk.
109. Habing, H. J. 1966, *Bull. astr. Inst. Netherlds*, **18**, 323.
110. Shklovski, I. S. 1966, IAU-URSI Symposium no. 31, Noordwijk.
111. Dieter, N. H. 1964, *Astr. J.*, **69**, 288.
112. Dieter, N. H. 1965, *Astr. J.*, **70**, 552.
113. Smith, G. P. 1963, *Bull. astr. Inst. Netherlds*, **17**, 203.
114. Prata, S. W. 1964, *Bull. astr. Inst. Netherlds*, **17**, 511.

X. DYNAMICS

(prepared by G. Contopoulos)

Oort's (1) chapter on Stellar Dynamics in *Stars and Stellar Systems*, Vol. 5, gives an account of the progress made in many traditional problems of Stellar Dynamics, emphasizing their relations to observation.

At the same time, a number of new developments in Stellar Dynamics should be noted.

(*a*) An approach between Stellar Dynamics and Celestial Mechanics, as indicated by the IAU Symposium no. 25 in Thessaloniki (the Proceedings have been edited by Contopoulos (**2**)). The same trend was evident at the Besançon Colloquium in 1966 (*Bull. astr.*, Paris, 3ème série, **2**, fasc. 1, 1967).

(*b*) The use of experimental methods in Stellar Dynamics, with the help of fast electronic computers;

(*c*) The development of new statistical methods; in particular the use of methods developed in plasma physics;

(*d*) The development of new gravitational theories concerning the formation of galactic structure.

The Gas Dynamical Problems and the Magnetic Fields of our Galaxy have been discussed by Woltjer (**3**).

Three earlier review articles connected with Galactic Dynamics should be mentioned: the first, by King (**4**), refers to the Dynamics of Galaxies in general; the second, by Wentzel (**5**), refers to the Magnetic Fields and Spiral Structure; and the third article, by Contopoulos (**6**), deals with problems of Stellar Dynamics in general.

Other discussions of problems of Stellar Dynamics, and Galactic Dynamics in particular, were given by Oort (**7**), Prendergast (**8**) and Kuzmin (**9**). The connection between Stellar Dynamics and Celestial Mechanics was discussed by Kovalevsky (**10**).

A book on Stellar Dynamics by Ogorodnikov (**11**) appeared in English translation.

Stellar Orbits—Third Integral

Two review articles on the theory of orbits in the Galaxy have been written by Ollongren (**12, 13**).

Contopoulos and Strömgren (**14**) have calculated numerically tables of plane galactic orbits, with the aim of finding the places of formation of stars with ages of a few times 10^8 years. Recent application to a sample of stars indicates that the stars of this age range originated either in the Orion or in the Perseus arm; one can even find the general areas where star formation has been most active.

The perturbations of the orbits, due to the spiral arms, have been considered by Barbanis and by Lin, and they were found to be small in general. An epicyclic theory for orbits in a spiral field has been developed by Woltjer and Barbanis.

Plane orbits in particular models of the Galaxy were studied by Woolley (**15**), Perek (**16**) and Genkin (**17**). Takayanagi (**18**) calculated plane orbits of subdwarfs. He found a concentration of perigalactic distances between 1·5–3·0 kpc; this indicates that the subdwarfs originated in the central bulge.

Orbits in special fields have been considered by Nahon (**19**) and Losco (**20**).

Three-dimensional orbits in particular galactic models were studied theoretically by Perek (**21**) and numerically, with limited accuracy, by Innanen (**22**).

Woolley (**23, 24**) continued his work on the galactic orbits of nearby stars and globular clusters.

The statistics and dynamics of the orbits of globular clusters in the Galaxy have been studied by Henry-Hillaire, and by Ogorodnikov (**25**). Matsunami (**26**) found correlations between various characteristics of the orbits of globular clusters and their physical properties.

Orbits of n gravitating bodies have been calculated numerically by a number of authors such as Sherman and Kinman (**27**), Aarseth (**28**), Wielen (**29**) and Standish and Szebehely, in order to study the evolution of n-body systems.

The problem of the accuracy of the n-body problem calculations and the related problem of the reversibility in stellar dynamical systems was considered by Miller (30). Hayli (31) began n-body problem calculations taking into account also the external galactic field. Lindblad (32) continued his work on the evolution of a system of massive bodies (cloud complexes) in the galactic field. Van Albada (33, 34) and Worall (35) studied the evolution of many small stellar systems; they found that in most cases their final state is only a double star.

Collisionless orbit calculations were made by Aarseth (36) for $n = 1000$. Some insight into the n-body problem can be achieved by one-dimensional calculations, as done by Lecar (37) and by Hohl and Feix (38). One can investigate in this way whether or not there is a tendency to an equilibrium and the nature of this equilibrium, also the breaking up of the system into subsystems, etc.

Of interest are some calculations of the evolution of collapsing collisionless systems by Hénon (39, 40) and Campbell (41), by means of orbit calculations, as well as a few calculations of the acceleration of the rate of collisions at the center of very dense systems by Ulam and Walden (42).

A short discussion of the n-body problem from the numerical point of view has been given by Aarseth and Hoyle (43). An IAU Colloquium on the gravitational n-body problem is scheduled for the summer of 1967 in Paris.

Orbit calculations in connection with the existence of a 'third' integral of motion were made by many authors. Hénon and Heiles (44) have studied the intersections of many orbits by a surface of section in a simple dynamical problem of two degrees of freedom. They found indications for the existence of an isolating or very nearly isolating 'third' integral of motion for small energies, while for large energies the 'third' integral tends to become ergodic. Similar results were reported by Perek and Peterson and by Ollongren at the Thessaloniki 1964 Symposium. Hayli (45) and Aarseth (46) found similar results for high velocity orbits of our Galaxy, e.g. orbits of globular clusters, especially if these orbits pass through the central region of the Galaxy. It seems, however, that a third integral is nearly isolating when the velocities on a meridian plane do not exceed a few hundred km s^{-1}.

Barbanis (47) found empirically a third integral, which persists even when the energy exceeds the energy of escape, and which keeps some stars from escaping to infinity.

A classification of the integrals of motion into isolating, quasi-isolating and ergodic has been given by Contopoulos (48). The third integral in a non-smooth potential has been studied by Contopoulos and Woltjer (49). Other special cases of the third integral have been studied by Andrle (50).

Tables of the coefficients of the third integral in a special case have been prepared by Contopoulos (51). Contopoulos and Moutsoulas (52, 53) have studied the resonance cases and the effects of the small divisors in the third integral. The resonance effects were used by Contopoulos (54) to explain the tube orbits (i.e. elongated orbits near periodic orbits), that were found by Ollongren (12, 13) and others in our Galaxy.

Further work on the integrals of stellar motion has been done by Barbanis (55), Schmeidler (56), Kuzmin (57), Idlis (58), Ogorodnikov (59), Message (60), and Nahon (61). Gustavson (62) found a good agreement between a theoretical form of the third integral and the numerical results of Hénon and Heiles. A comparison of theoretical and experimental results was made also by Roels (63).

Hori and Liu (64) used a third integral in the theory of the velocity ellipsoid.

Recent developments in the theory of the third integral include numerical and theoretical results on the restricted three-body problem by Déprit, Hénon, Bozis and Contopoulos; the theories of satellite orbits around an oblate planet by Vinti, Conley, Kyner, etc., and many papers on non-linear mechanics with applications to various problems of mathematics, physics, chemistry, engineering and astronomy. In particular, the study of adiabatic invariants and experimental results in plasma physics and geophysics, as in the case of mirror machines and the van Allen belts, have much similarity with the astronomical problems mentioned above. No detailed reference will be made to these papers, however.

The theoretical problem of the existence of integral surfaces in non-integrable systems of two degrees of freedom has been solved by Arnol'd and by Moser. Arnol'd (**65**) has written a number of papers on this problem and related problems in recent years. Moser has proved recently that the motions near a periodic orbit or an equilibrium point are quasi-periodic for a set of initial conditions of positive measure. The corresponding expansions of the form of the third integral or the von Zeipel method are convergent. These theorems have been applied to various problems by Moser and Jefferys (**66, 67**). A similar theorem has been proved in a special case by Barrar (**68**).

These theorems, however, are proved only for a very small neighborhood of the periodic orbit or the equilibrium point. In order to study how far the third integral is applicable, numerical methods are necessary. Contopoulos (**69**) found that the breakdown of the third integral for large perturbations is due to a concentration of resonance regions; thus a numerical estimate of the breakdown perturbation can be found.

Models of the Galaxy

A number of mass-models of the Galaxy have been proposed recently.

A simple formula for the force and potential on the plane of symmetry of the Galaxy has been used by Contopoulos and Strömgren (**14**).

Besides the well known new Schmidt (**70**) model, Innanen (**71**) constructed a number of galactic models, consisting of concentric spheroids, and studied their properties. Kuzmin and Kutuzov (**72, 73**) have also considered various models of the Galaxy. Genkin (**74, 75**) has studied models admitting an exact third entegral of motion. Further models have been constructed by Einasto and Kutuzov.

Self-gravitating models of galaxies, satisfying Liouville's and Poisson's equations have been given by Prendergast, Vandervoort (**76**), Aoki (**77**) and Perek (**78**). A self-gravitating model for a flat galaxy with a truncated Schwarzschild distribution of velocities has been given by Ng. Toomre (**106**), Hunter (**108**), and Barbanis and Prendergast have developed methods of finding the potential corresponding to a given density of a flat disk and vice-versa.

Wallace and Copeland (**79**) have studied the velocity field in a particular model of a spiral galaxy. The flux of the angular momentum in a spiral galaxy was considered by Starr and Newell (**80**). The distribution of mass and angular momentum in our Galaxy has been studied by Takase (**81**).

Hodge (**82**) found limits of the mass of our Galaxy between $1{\cdot}5 \times 10^{11} M_{\odot}$ and $6 \times 10^{11} M_{\odot}$.

Data about the galactic field have been derived by King from a study of the boundaries of globular clusters.

Kreiken (**83**) proposed a set of galactic constants R_0, Θ_0, A and B, derived from his polytropic galactic model.

The force perpendicular to the galactic plane has been studied by Stothers and Tech (**84**) through the kinematics of OB stars. Dorchner, Gürtler and Schmidt (**85**) derived, from the magnitude of this force, the density of molecular hydrogen near the Sun; they found that 80% of the hydrogen is in molecular form.

Belton and Brandt (**86**) succeeded in reproducing the rotation curve of our Galaxy by assuming that it consists of a set of spheroidal homoeoids; they concluded that the distribution of the 'missing matter' is like halo population II objects. Various assumptions about the distribution of the 'missing matter' were discussed by Oort (**1, 7**).

Woolley found, through a study of the z-velocities of faint A0 stars, that the total density in the solar neighborhood is smaller than the accepted value; thus the amount of 'missing matter' is less than previously assumed. Similar results were found by Einasto (**87**).

The mass of the galactic nucleus was estimated recently by Lequeux (**88**) to be of the order of $10^8 M_\odot$. Rougoor (**89**) estimated the mass of the neutral atomic hydrogen in the central disk to be $3 \times 10^6 M_\odot$.

Statistical Methods—Stability Problems

Lynden-Bell (**90, 91**) has developed a new statistics, similar to the Fermi-Dirac statistics, to describe the properties of a collapsing galaxy. A numerical example presented by Hénon at the Besançon Colloquium shows partial agreement with this theory.

A statistical theory of the gravitational forces in a stellar field has been given by Camm (**92**).

A Monte-Carlo method to deal with encounters in the n-body problem, was developed by Hénon (**93, 94**). This method appears very promising for obtaining quickly results on the evolution of stellar systems.

Massonie (**95**) began the study of a markovian process that changes progressively the velocity distribution in a stellar system.

Plasma dynamics has initiated the development of new methods in stellar dynamics in recent years.

Prigogine and Severne (**96, 97**) have developed statistical mechanical methods to deal with the gravitational n-body problem. They introduced non-markovian equations to take account of the 'memory' of previous conditions. The n-body system does not tend to an equilibrium state, but its final evolution depends on the initial conditions.

Application of the methods of plasma physics to stellar systems has been made by Lebedev, Maksumov and Marochnik (**98, 99, 100**). In particular they considered the growth of instabilities on a time scale of a few times 10^8 years. On the other hand Antonov has shown that a one-dimensional stellar system is stable.

A general discussion of cooperative phenomena in stellar dynamics has been made by Lynden-Bell (**101**). He discussed, among other subjects, a necessary and sufficient condition, due originally to Antonov (**102**), for the stability of a stellar gas. Further Lynden-Bell continued Antonov's discussion of the entropy in stellar systems.

Gravitational instability plays a role in the formation of galaxies, of large scale structures of galaxies, and of stars. Only work connected with the second problem will be mentioned here.

The influence of rotation on the gravitational instability of a cloud has been considered by Arny (**103**) and by Simon (**104**).

A number of authors have discussed the stability of a disk of stars or gas against axi-symmetric perturbations.

Toomre (**105**) found that a velocity dispersion is needed to stabilize the galactic disk. The minimum velocity dispersion required for stability near the Sun is between 20 and 35 km s^{-1}. Non linear effects and other generalizations were given by Graham (**107**). Hunter (**108**) has studied free modes of oscillation in the plane of infinitesimally thin disks with no velocity dispersion.

The effect of the z-dimension on the stability of a disk of finite thickness was studied by Sweet and McGregor (**109, 110**).

The stability of a rotating disk of gas has been considered by Goldreich and Lynden-Bell (**111**). Lynden-Bell (**112**) has studied the large-scale instabilities of Maclaurin spheroids during collapse.

Owaki (113) has studied the stability of an incompressible galactic nucleus under the influence of the galactic disk.

Ostriker (114) has studied the stability of self-gravitating rings and cylinders.

Non-axi-symmetric instabilities of a generally spiral form are mentioned below.

There is a series of papers in the *Astrophysical Journal* on rotating gaseous masses by Chandrasekhar, Lebovitz, Roberts, Hurley and Limber. These papers, however, are rather loosely connected with our problems.

Dynamics of Spiral Structure

Lin (115, 116) has emphasized the fact that the magnetic field of the Galaxy is quite insufficient to explain the main spiral effects of our Galaxy. On the other hand the winding difficulty of differentially rotating spiral arms is well known. Thus Lin developed a spiral pattern theory, of the type proposed originally by B. Lindblad, assuming that the stars move freely through the spiral pattern, which is a density wave rotating like a rigid body. Lin and Shu (115, 116, 117, 118) have found the response of gas and stars to an imposed rotating spiral field. A possible spiral solution is found when the induced field is equal to the imposed one. This method gives preference to trailing arms.

The effect of a finite thickness of the Galaxy has been considered by Shu.

A similar theory, which includes also non-linear effects and boundary conditions, was developed by Kalnajs (119).

A paper by the late B. Lindblad (120) describes his own views on the circulation of matter in a spiral galaxy.

The angular velocity of the spiral pattern was determined to be about 20 km s^{-1} kpc^{-1} by Strömgren, who compared the present position of the Perseus arm with its position when stars with ages of a few times 10^8 years were formed.

Julian and Toomre (121) have developed a local theory for the formation of trailing spiral structures in differentially rotating stellar disks due to concentrated perturbations. If the perturbation is permanent, e.g. a point mass, the mass of the resulting trailing arms can be larger, by an order of magnitude, than the perturbing mass. If, however, the forcing is temporary, then the resulting spiral response is also temporary.

Rehm (122) found, under various conditions, both trailing and leading spiral patterns.

The problem of a gaseous disk was considered in detail by Goldreich and Lynden-Bell (123, 124) who found trailing (sheared) instabilities. These instabilities are transient, but they are regenerated continuously. Similar views were developed by Toomre.

Reviews of spiral structure theories were given by Lynden-Bell (125) and Prendergast (126).

Spiral instabilities in special cases were considered by Buggish and Kahn (127) and by Helfer (128).

Contopoulos (129) found a new integral of motion in a disk spiral galaxy. Barbanis found numerical evidence from orbit calculations for the existence of such an integral in most cases of actual interest. Such an integral may be useful for the construction of spiral models of the Galaxy.

Some n-body problem calculations are connected with the formation or evolution of spiral structures. Such work has been done by Lindblad (32). Prendergast, as well as Contopoulos and Barbanis, are studying the evolution of a plane galaxy by calculating a very large number of orbits.

A theory of galactic structure based on the accretion of interstellar matter has been developed by Lyttleton and Bondi (130).

Magnetic Fields—Dynamics of Central Region

Lynden-Bell (**131, 132**) has found that a magnetic field can initiate spiral instabilities in a galaxy.

Freeman and Mestel (**133**) indicated the difficulties of constructing a model, in which magnetic forces maintain a differentially rotating spiral pattern.

Pacholczyk (**134**) and Stodolkiewicz (**135**) have studied the stability of the whole Galaxy and, in particular, spiral arms, taking into account both magnetic and gravitational effects.

Similar work for the spiral arms was done by R. Graham (**136**).

A theory of the formation of galactic spiral structure through the contraction of metagalactic gas possessing a magnetic field has been developed by Pikel'ner (**137**).

Fujimoto (**138**) proposed a mechanism to explain the expansion of the 3 kpc arm. He assumed that circular orbits are unstable in this region and an inflow of gas from the halo takes place there. The expansion is produced by a coupling of differential rotation and the magnetic field. Ôki, Fujimoto and Hitotuyanagi (**139**) have developed a theory combining gravitational and magnetic effects in the formation of spiral arms. In addition, Fujimoto (**140**) has considered the formation of dark lanes in a galaxy by non-circular motions of gas.

Setti (**141**) found indications that the gravitational field of a spiral arm stabilizes it when the magnetic field is helical.

Arguments for the existence of a helical field were given by Hornby (**142**). A model of a helical field with inclined windings has been considered by Stepien (**143**).

Greyber (**144**) supported the view that the magnetic field above the galactic plane is opposite in direction to that below.

An attempt to consider the dynamics of a galactic plasma by using an ellipsoidal distribution law of velocities was made by Pratap (**145**).

Sakakibara (**146**) has studied the diffusion of the cosmic rays due to the magnetic field of the Orion arm, which results in an anisotropy of the cosmic ray flux.

Various determinations of the magnetic field in our Galaxy have been reviewed critically by Woltjer (**3, 147**). The arguments for a strong field in the spiral arms, of the order of 3×10^{-5} gauss, arise from observations of synchrotron radiation, and considerations about the Faraday rotation, the alignment of dust grains and the solar wind. Arguments for a weak field of a few times 10^{-6} gauss, are based on the absence of a measurable Zeeman effect and considerations about star formation. The field in the halo seems to be of the order of a few times 10^{-6} gauss.

A recent review of the situation was given by van de Hulst (**148**). Mathewson and Milne have found a spiral pattern by radio studies of the Faraday rotation. Brouw has determined the magnetic field in the galactic spurs. The magnetic field in filamentary structures was studied by Pronik, who confirmed previous conclusions by Shajn.

A number of papers on the magnetic fields of our Galaxy by Jacklyn, Gardner, Whiteoak, Piddington, and Visvanathan were presented at the Mount-Stromlo 'Pre-Symposium' on 'Radio and Optical Studies of the Galaxy', in May 1966.

In some cases the field may be very strong, of the order of 10^{-3} gauss; such indications were found by Davies, de Jager and Verschuur (**149**) by studying the polarization of OH sources.

The magnetic field near the center of the Galaxy was estimated by Downes and Maxwell (**150**) to be between 5×10^{-5} and 5×10^{-4} gauss.

It is not clear how magnetic effects could explain the expansion motions near the center of the Galaxy.

An explanation based on an explosion of the galactic nucleus was proposed by G. Burbidge and Hoyle (**151**), Lequeux (**152**), and Ishida (**153**). Further support of this view was provided at the IAU Symposium no. 31, in Noordwijk.

Pikel'ner (**154**) suggested that the energy needed for the activity of the central region comes from the gravitational contraction of gas clouds.

The phenomena occurring during stellar collisions at the center of the Galaxy have been considered by Spitzer and Saslaw (**155**).

Lequeux (**152**) and Ginzburg and Syrovatskii (**156**) suggested that most of the cosmic rays come from the explosion of the galactic nucleus. This subject has also been studied by Rosen (**157**).

Miscellaneous Problems

Mestel (**158**) has studied the evolution of a rotating spherical cloud into a thin disk. Depending upon the initial density distribution the final result is a spiral or a barred spiral. Lin, Mestel and Shu (**159**) have studied the collapse of a uniform spheroid. Crzedrielski (**160**) has considered the fragmentation of a galaxy during collapse. Milder (**161**) has considered the flattening of a rotating isothermal model.

A virial theorem for a continuous distribution of mass has been given by Camm (**162**).

King (**163**) has given a relation between velocity distributions and spatial gradients in a stellar system. He found the ratio of the axes of the velocity ellipsoid from moment equations.

The problem of the deviation of the vertex has been considered by Genkin (**164**). Kuzmin (**165**) has considered the conditions for the existence of a plane of symmetry in stellar systems.

Kitamura (**166**) has studied the effect of the encounters on a stellar velocity spheroid; in a time of the order of the time of relaxation, the velocity dispersions become appreciably smaller and approach each other.

Lynden-Bell (**167**) has considered the problem of the bending of the galactic plane. He attributed it to a 1° deviation between the galactic axis of symmetry and its angular momentum.

The same effect was attributed to the action of the Magellanic Clouds by Avner (**168**).

Toomre and Hunter have studied this problem from the point of view of the stability of a disk towards bending. Some bending modes are excited by the Magellanic Clouds.

BIBLIOGRAPHY

1. Oort, J. H. 1965, in *Stars and Stellar Systems*, **5**, 455.
2. Contopoulos, G. (ed.) 1966, *The Theory of Orbits in the Solar System and in Stellar Systems*, IAU Symposium No. 25, Academic Press.
3. Woltjer, L. 1965, in *Stars and Stellar Systems*, **5**, 531.
4. King, I. 1963, *A. Rev. Astr. Astrophys.*, **1**, 179.
5. Wentzel, D. G. 1963, *A. Rev. Astr. Astrophys.*, **1**, 195.
6. Contopoulos, G. 1966, *Problems of Stellar Dynamics*, Proc. Space Math. Seminar, **1**, 169, American Mathematical Society.
7. Oort, J. 1965, NATO Summer Course 'Observational Aspects of Galactic Structure', Part 23, Athens.
8. Prendergast, K. 1966, IAU–URSI Symposium no. 31, Noordwijk.
9. Kuzmin, G. G. 1965, *Trudy astrofiz. Inst. Alma-Ata*, **5**, 11, 70.
10. Kovalevsky, J. 1966, Besançon Colloquium 'Nouvelles Méthodes de la Dynamique Stellaire'.
11. Ogorodnikov, K. F. 1965, *Dynamics of Stellar Systems*, Pergamon Press.
12. Ollongren, A. 1965, *A. Rev. Astr. Astrophys.*, **3**, 113.
13. Ollongren, A. 1965, in *Stars and Stellar Systems*, **5**, 501.

14. Contopoulos, G., Strömgren, B. 1965, *Tables of Plane Galactic Orbits*, Inst. Space Studies, New York.
15. Woolley, R. v. d. R. 1964, *Astrophys. Norv.*, **9**, 261.
16. Perek, L. 1964, *Vistas in Astronomy*, **5**, 28.
17. Genkin, I. L. 1966, *Trudy astrofiz. Inst. Alma-Ata*, **7**, 16.
18. Takayanagi, K. 1965, *Publ. astr. Soc. Japan*, **17**, 71.
19. Nahon, F. 1966, Besançon Colloquium 'Nouvelles Méthodes de la Dynamique Stellaire'.
20. Losco, L. 1966, Besançon Colloquium 'Nouvelles Méthodes de la Dynamique Stellaire'.
21. Perek, L. 1966, *Bull. astr. Inst. Csl.*, **17**, 111.
22. Innanen, K. A. 1966, *Z. Astrophys.*, **64**, 445, 457.
23. Woolley, R. v. d. R. 1965, NATO, Summer Course 'Kinematical and Chemical History of the Galaxy', Herstmonceux.
24. Woolley, R. v. d. R. 1967, *R. Obs. Bull.* (in press).
25. Ogorodnikov, K. F. 1965, *Trudy astrofiz. Inst. Alma-Ata*, **5**, 219.
26. Matsunami, N. 1964, *Publ. astr. Soc. Japan*, **16**, 141.
27. Sherman, N. W., Kinman, D. 1964, *Astr. J.*, **69**, 149.
28. Aarseth, S. J. *Mon. Not. R. astr. Soc.*, **126**, 223, 1963; **132**, 35, 1966.
29. Wielen, R. 1966, Besançon Colloquium 'Nouvelles Méthodes de la Dynamique Stellaire'.
30. Miller, R. H. 1964, *Astrophys. J.*, **140**, 250.
31. Hayli, A. 1966, Besançon Colloquium 'Nouvelles Méthodes de la Dynamique Stellaire'.
32. Lindblad, P. O. 1966, Besançon Colloquium 'Nouvelles Méthodes de la Dynamique Stellaire'.
33. van Albada, T. S. 1966, 14th Liège Symposium, *Gravitational Instability and the Formation of Stars and Galactic Structures.*
34. van Albada, T. S. 1966, Besançon Colloquium 'Nouvelles Méthodes de la Dynamique Stellaire'.
35. Worall, G. 1966, 14th Liège Symposium, *Gravitational Instability and the Formation of Stars and Galactic Structures.*
36. Aarseth, S. J. 1966, Besançon Colloquium 'Nouvelles Méthodes de la Dynamique Stellaire'.
37. Lecar, M. 1966, 14th Liège Symposium, *Gravitational Instability and the Formation of Stars and Galactic Structures.*
38. Hohl, F., Feix, M. R. 1966, 14th Liège Symposium, *Gravitational Instability and the Formation of Stars and Galactic Structures.*
39. Hénon, M. 1964, *Ann. Astrophys.*, **27**, 83.
40. Hénon, M. 1966, 14th Liège Symposium, *Gravitational Instability and the Formation of Stars and Galactic Structures.*
41. Campbell, P. M. 1966, *Proc. nat. Acad. Sci.*, **55**, 1.
42. Ulam, S. M., Walden, W. E. 1964, *Nature*, **201**, 1202.
43. Aarseth, S. J., Hoyle, F. 1964, *Astrophys. Norv.*, **9**, 313.
44. Hénon, M., Heiles, C. 1964, *Astr. J.*, **69**, 73.
45. Hayli, A. 1965, *Ann. Astrophys.*, **28**, 49.
46. Aarseth, S. J. 1966, *Nature*, **212**, 57.
47. Barbanis, B. 1966, *Astr. J.*, **71**, 415.
48. Contopoulos, G. 1963, *Astrophys. J.*, **138**, 1297.
49. Contopoulos, G., Woltjer, L. 1964, *Astrophys. J.*, **140**, 1106.
50. Andrle, P. 1966, *Bull. astr. Inst. Csl.*, **17**, 166, 169.
51. Contopoulos, G. 1966, *Astrophys. J. Suppl.*, **13**, 503.
52. Contopoulos, G. 1963, *Astr. J.*, **68**, 763.
53. Contopoulos, G., Moutsoulas, M. *Astr. J.*, **70**, 817, 1966; **71**, 687, 1966.
54. Contopoulos, G. 1965, *Astr. J.*, **70**, 526.
55. Barbanis, B. 1965, *Astr. J.*, **70**, 285.

56. Schmeidler, F. 1963, *Astr. Nachr.*, **287**, 1.
57. Kuzmin, G. G. 1964, *Tartu astr. Obs. Publ.*, **34**, 457.
58. Idlis, G. M. 1965, *Trudy astrofiz. Inst. Alma-Ata*, **5**, 105.
59. Ogorodnikov, K. F. 1965, *Trudy astr. Obs. Leningrad. Univ.*, **22**, 132.
60. Message, J. P. 1966, Besançon Colloquium 'Nouvelles Méthodes de la Dynamique Stellaire'.
61. Nahon, F. 1966, *C. R. Acad. Sci.*, Paris, **262**, 669.
62. Gustavson, F. G. 1966, *Astr. J.*, **71**.
63. Roels, J. 1966, Besançon Colloquium 'Nouvelles Méthodes de la Dynamique Stellaire'.
64. Hori, G., Liu, T. 1963, *Publ. astr. Soc. Japan*, **15**, 100.
65. Arnol'd V. I. *Uspekhi Mat. Nauk*, **18**, no. 5 (113), 13, 1963; **18**, no. 6 (114), 91, 1963; *Sov. Math. Dokl.*, **161**, 331, 1965.
66. Jefferys, W. H. 1966, *Astr. J.*, **71**, 306.
67. Moser, J., Jefferys, W. H. 1966, *Astr. J.*, **71**, 568.
68. Barrar, R. 1966, *Amer. J. Math.*, **88**, 206.
69. Contopoulos, G. 1966, Besançon Colloquium 'Nouvelles Méthodes de la Dynamique Stellaire'.
70. Schmidt, M. 1965, in *Stars and Stellar Systems*, **5**, 513.
71. Innanen, K. A. *Astrophys. J.*, **143**, 150, 153, 1966; *Z. Astrophys.*, **64**, 158, 1966.
72. Kuzmin, G. G. 1966, *Tartu astr. Obs. Publ.*, **35**, 285.
73. Kuzmin, G. G., Kutuzov, S. A. 1966, *Tartu astr. Obs. Publ.*, **35**, 316.
74. Genkin, I. L. 1963, *Soobšč. gos astr. Inst. Sternberg*, no. 129.
75. Genkin, I. L. 1966, *Trudy astrofiz. Inst. Alma-Ata*, **7**, 16.
76. Vandervoort, P. O. 1966, 14th Liège Symposium, *Gravitational Instability and the Formation of Stars and Galactic Structures.*
77. Aoki, S. 1965, *Publ. astr. Soc. Japan*, **17**, 273.
78. Perek, C. 1966, *Bull. astr. Inst. Csl.*, **17**, 333.
79. Wallace, J. M., Copeland, J. A. 1964, *Publ. astr. Soc. Pacif.*, **76**, 315.
80. Starr, V. P., Newell, R. E. 1963, *Publ. astr. Soc. Pacif.*, **75**, 239.
81. Takase, B. 1966, *Publ. astr. Soc. Japan* (in press).
82. Hodge, P. W. 1966, *Proc. astr. Soc. Pacif.*, **78**, 72.
83. Kreiken, E. A. 1964, *Observatory*, **84**, 115.
84. Stothers, R., Tech, J. L. 1964, *Mon. Not. R. astr. Soc.*, **127**, 287.
85. Dorschner, J., Gürtler, J., Schmidt, K. H. 1964, *Astr. Nachr.*, **288**, 149.
86. Belton, M. J. S., Brandt, J. C. 1963, *Ann. Astrophys.*, **26**, 229.
87. Einasto, J. 1964, *Comm. Tartu astr. Obs.*, no. 11, 1.
88. Lequeux, J. 1966, IAU-URSI Symposium no. 31, Noordwijk.
89. Rougoor, G. W. 1966, IAU-URSI Symposium no. 31, Noordwijk.
90. Lynden-Bell, D. 1966, 14th Liège Symposium, *Gravitational Instability and the Formation of Stars and Galactic Structures.*
91. Lynden-Bell, D. 1966, Besançon Colloquium 'Nouvelles Méthodes de la Dynamique Stellaire'.
92. Camm, G. L. 1963, *Mon. Not. R. astr. Soc.*, **126**, 283.
93. Hénon, M. 1966, *C. R. Acad. Sci.*, Paris, **262**, 666.
94. Hénon, M. 1966, Besançon Colloquium 'Nouvelles Méthodes de la Dynamique Stellaire'.
95. Massonie, J. P. 1966, Besançon Colloquium 'Nouvelles Méthodes de la Dynamique Stellaire'.
96. Prigogine, I., Severne, G. 1966, *Physica* (in press).
97. Severne, G. 1966, Besançon Colloquium 'Nouvelles Méthodes de la Dynamique Stellaire'.
98. Marochnik, L. S. *Astr. Zu.*, **41**, 264, 1964 = *Sov. Astr.*, **8**, 202, 1964; *Astr. Zu.*, **43**, 560, 919, 1966.
99. Lebedev, V. I., Maksumov, M. N., Marochnik, L. S. *Astr. Zu.*, **42**, 709, 1965 = *Sov. Astr.*, **9**, 549, 1966.

100. Maskumov, M. N., Marochnik, L. S. *Astr. Zu.*, **42,** 1261, 1965 = *Sov. Astr.*, **9,** 980, 1966. *Dokl. Akad. Nauk SSSR*, **164,** 1019, 1965.
101. Lynden-Bell, D. 1965, Amer. Math. Soc., Summer Seminar on *Relativity Theory and Astrophysics.*
102. Antonov, V. A. 1962, *Vestnik Leningrad. gos. Univ.*, no. 19, 96.
103. Arny, T. T. 1966, 14th Liège Symposium, *Gravitational Instability and the Formation of Stars and Galactic Structures.*
104. Simon, R. 1966, 14th Liège Symposium, *Gravitational Instability and the Formation of Stars and Galactic Structures.*
105. Toomre, A. 1964, *Astrophys. J.*, **139,** 1217.
106. Toomre, A. 1963, *Astrophys. J.*, **138,** 385.
107. Graham, R. 1966, 14th Liège Symposium, *Gravitational Instability and the Formation of Stars and Galactic Structures.*
108. Hunter, C. 1963, *Mon. Not. R. astr. Soc.*, **126,** 299; **129,** 321, 1965.
109. Sweet, P. A., McGregor, D. D. 1964, *Mon. Not. R. astr. Soc.*, **128,** 195.
110. Sweet, P. A., McGregor, D. D. 1964, *Observatory*, **84,** 42.
111. Goldreich, P., Lynden-Bell, D. 1965, *Mon. Not. R. astr. Soc.*, **130,** 97.
112. Lynden-Bell, D. 1964, *Astrophys. J.*, **139,** 1195.
113. Owaki, N. 1966, *Publ. astr. Soc. Japan* (in press).
114. Ostriker, J. 1964, *Astrophys. J.*, **140,** 1056, 1067, 1529.
115. Lin, C. C. 1966, *J. SIAM Appl. Math.*, **14,** 876.
116. Lin, C. C. 1965, Amer. Math. Soc. Seminar on *Relativity and Astrophysics*, Cornell University.
117. Lin, C. C., Shu, F. H. 1964, *Astrophys. J.*, **140,** 646.
118. Lin, C. C., Shu, F. H. 1966, *Proc. nat. Acad. Sci.*, **55,** 229.
119. Kalnajs, A. J. 1965, Thesis, Harvard University.
120. Lindblad, B. 1964, *Astrophys. Norv.*, **9,** 103.
121. Julian, W. H., Toomre, A. 1966, *Astrophys. J.*, **146,** 810.
122. Rehm, R. G. 1965, Thesis, MIT.
123. Goldreich, P., Lynden-Bell, D. 1964, *Observatory*, **84,** 171.
124. Goldreich, P., Lynden-Bell, D. 1965, *Mon. Not. R. astr. Soc.*, **130,** 125.
125. Lynden-Bell, D. 1965, NATO Summer Course 'Kinematical and Chemical History of the Galaxy', Herstmonceux.
126. Prendergast, K. 1966, IAU–URSI Symposium no. 31, Noordwijk.
127. Buggish, H., Kahn, F. D. 1966, 14th Liège Symposium, *Gravitational Instability and the Formation of Stars and Galactic Structures.*
128. Helfer, H. L. 1966, 14th Liège Symposium, *Gravitational Instability and the Formation of Stars and Galactic Structures.*
129. Contopoulos, G. 1966, 14th Liège Symposium, *Gravitational Instability and the Formation of Stars and Galactic Structures.*
130. Lyttleton, R. A., Bondi, H. 1963, *Mon. Not. R. astr. Soc.*, **128,** 207.
131. Lynden-Bell, D. 1966, *Observatory*, **86,** 57.
132. Lynden-Bell, D. 1966, 14th Liège Symposium, *Gravitational Instability and the Formation of Stars and Galactic Structures.*
133. Freeman, K. C., Mestel, L. 1966, *Mon. Not. R. astr. Soc.*, **134,** 37.
134. Pacholczyk, A. G. 1963, *Acta Astr.*, **13,** 1.
135. Stodolkiewicz, J. S. 1963, *Acta Astr.*, **13,** 30.
136. Graham, R. 1966, 14th Liège Symposium, *Gravitational Instability and the Formation of Stars and Galactic Structures.*
137. Pikel'ner, S. B. 1965, *Astr. Zu.*, **42,** 515 = *Sov. Astr.*, **9,** 408, 1965.
138. Fujimoto, M. 1963, *Sci. Rep. Tohoku Univ. (Sendai)*, Ser. 1, **47,** 165.
139. Ôki, T., Fujimoto, M., Hitotuyanagi, Z. 1965, *Suppl. Progr. Theor. Physics*, no. 31, 77.
140. Fujimoto, M. 1965, *Publ. astr. Soc. Japan*, **17,** 231.
141. Setti, G. 1966, *Bull. astr. Inst. Netherlds*, **18,** 51.
142. Hornby, J. M. 1966, *Mon. Not. R. astr. Soc.*, **133,** 213.
143. Stepien, K. 1964, *Acta Astr.*, **14,** 81.

144. Greyber, H. D. 1966, 14th Liège Symposium, *Gravitational Instability and the Formation of Stars and Galactic Structures.*
145. Pratap, R. 1966, 14th Liège Symposium, *Gravitational Instability and the Formation of Stars and Galactic Structures.*
146. Sakakibara, S. 1965, *J. Geomagn. Geoelectricity*, **17**, 99.
147. Woltjer, L. 1964, 13th Solvay Conference, Brussels.
148. van de Hulst, H. C. 1966, IAU-URSI Symposium no. 31, Noordwijk.
149. Davies, R. D., de Jager, G., Verschuur, G. L. 1966, *Nature*, **209**, 974.
150. Downes, D., Maxwell, A. 1966, IAU-URSI Symposium no. 31, Noordwijk.
151. Burbidge, G. R., Hoyle, F. 1963, *Astrophys. J.*, **138**, 57.
152. Lequeux, J. 1963, *Ann. Astrophys.*, **26**, 429.
153. Ishida, K. 1964, *Suppl. Progr. Theor. Physics*, no. 31, 116.
154. Pikel'ner, S. B. 1966, IAU-URSI Symposium no. 31, Noordwijk.
155. Spitzer, L., Saslaw, W. C. 1966, *Astrophys. J.*, **143**, 400.
156. Ginzburg, V. L., Syrovatskii, S. I. 1964, *The Origin of Cosmic Rays*, Pergamon Press.
157. Rosen, S. 1966, 14th Liège Symposium, *Gravitational Instability and the Formation of Stars and Galactic Structures.*
158. Mestel, L. 1963, *Mon. Not. R. astr. Soc.*, **126**, 553.
159. Lin, C. C., Mestel, L., Shu, F. H. 1965, *Astrophys. J.*, **142**, 1431.
160. Grzedzielski, S. 1966, *Mon. Not. R. astr. Soc.*, **134**, 109.
161. Milder M. 1966, *Astrophys. J.*, **145**, 109.
162. Camm, G. L. 1966, Besançon Colloquium 'Nouvelles Méthodes de la Dynamique Stellaire'.
163. King, I. 1965, *Astr. J.*, **70**, 296.
164. Genkin, I. L. 1965, *Trudy astrofiz. Inst. Alma-Ata*, **5**, 23.
165. Kuzmin, G. G. 1964, *Tartu astr. Obs. Publ.*, **34**, 9.
166. Kitamura, S. 1966, *J. astr. Soc. Japan*, **18**, 317.
167. Lynden-Bell, D. 1965, *Mon. Not. R. astr. Soc.*, **129**, 299.
168. Avner, E. S. 1965, *Dissert. Abstr.*, **26**, 27.

B. J. BOK
President of the Commission

APPENDIX. REPORT OF THE COMMITTEE OF 'SELECTED AREAS'

CHAIRMAN: Dr T. Elvius, Astronomiska Observatoriet, Uppsala, Sweden.
MEMBERS: Kharadze, McCuskey, Plaut.

Introductory notes

This committee serves in continuation of the previous Commission 32 and Sub-commission 33c (*Trans. IAU*, **11B**, 333, 1962). For a list of reports prior to those in *Trans. IAU*, **11A**, 389, 1962, and **12A**, 548, 1965, see *Trans. IAU*, **11A**, 397, 1962.

In the fifth volume of the series 'Stars and Stellar Systems' is an appendix by A. Blaauw and T. Elvius (**1**) giving a brief summary of Kapteyn's Plan of Selected Areas with a table of the centres of the areas of his Systematic Plan. In the following these areas are referred to as SA.

At the Hamburg meeting of IAU it was decided that this committee should deal also with plans for co-operative galactic research supplementing the Kapteyn Plan (*Trans. IAU*, **12B**, 317, 1966). In order to avoid waste of space, most reports on such work, e.g. on the results of work in the Parenago Plan (**52**) or on the Cleveland fields, cf. (**53**) are included only in the main report of Commission 33.

Progress of research

1. *Durchmusterung and maps*

H. Vehrenberg has edited an Atlas (**2**), covering the Harvard-Groningen fields of SA 1–206, with pg-magnitudes of star sequences indicated.

2. *Photometry*

Abastumani (*Mt Kanobili*). E. K. Kharadze reports on catalogues which have been published for regions in Parenago's Plan (**52**). E. K. Kharadze, S. P. Apriamashvili and T. A. Kotchlashvili (**3**) give pv-magnitudes for about 1000 stars in Parenago field III (in Cygnus), identical with SA 40. S. P. Apriamashvili (**4**) publishes pg-magnitudes and CI for nearly 1900 stars in Parenago field I (in Aquila and Scutum) which includes SA 110. N. P. Kalandadze's catalogue (**5**) for Parenago field IV (in Taurus), which includes SA 49 and touches SA 50, contains photographic *UBV* data for about 3600 stars. Another group of papers has just been printed: S. P. Apriamashvili and V. I. Kutznetsov give pg- and pv-magnitudes for a region within Parenago field I (in Aquila) at SA 87 (**6**). A catalogue of magnitudes and colours in the *UBV* system for stars in Parenago field III (Taurus) is published in two parts by N. B. Kalandadze (**7**); two selected areas are in this region. A paper by A. D. Chuadze (**8**) gives pg-magnitudes for stars in one of the Parenago fields. Several of the Abastumani papers include spectral data, cf. Section 6.

Basel. Work on three-colour photometry in the *RGU*-system and partly also in the *UBV*-system has continued in SA 51, 54, 57, 71, 82, 94, 107, 133, 141, and 158. It is aimed at a limiting magnitude of 19, but for some fields photoelectric standards are available only to 16 or 17. For SA 51 and 57 the results are ready, cf. (**48**).

Berkeley. King is determining *B*, *V*, and *R* magnitudes in SA 51, 57, and 68, down to the limit of the 120-inch Lick reflector. The faint photometry is based on unpublished *P* and *V* photoelectric scales by Baum.

Cape. Among the 5000 stars included in the catalogue of photoelectric magnitudes and colours of southern stars by A. W. J. Cousins and R. H. Stoy (**9**), there is a considerable number of stars in Kapteyn's SA.

Cracow and Crimea. E. Rybka (Cracow) reports that the catalogue of photoelectric magnitudes and colours for photometric standard stars in Selected Areas 1–139 has now been published (**10**). It is the result of cooperative efforts between S. V. Nekrasova and V. B. Nikonov of the Crimean Astrophysical Observatory and himself. The observations and methods have been previously described (**54**). The catalogue contains magnitudes and colours, both in the natural *ubv* and in the *UBV* systems, for two 6 magnitudes stars in each area, of A and K spectral type respectively. In cooperation with Abastumani Observatory it is planned to extend the scheme to include three A and K type pairs of $V = 7{\cdot}5$, $9{\cdot}0$ and $10{\cdot}5$ in each of SA 1–43. Rybka mentions the desirability of extending his plan also to the remaining southern Selected Areas.

Engelhardt (*Kazan*). L. Urasin reports on the programme mentioned in the previous report. For SA 24–27 the photographic material has been obtained; its relation to the *UBV* system is described in (**11**). For SA 25 a catalogue of magnitudes and colours of 4700 stars has been published (**12**).

Haute-Provence (*Saint Michel*) and *Lyon.* J. H. Bigay reports on the previously announced investigation of O, B, and A0 stars in galactic areas, where photoelectric *UBV* measurements are made. Results have been published for SA 8 (**13, 14**) and previously for SA 9, 19, and 24 (**55**). For SA 40, 49, 64, and 110 the observations are finished and ready for publication. For preliminary reports on SA 40 and 64, cf. (**15, 16**). Observations are going on in SA 74 and 98. For SA 55–58, close to the north galactic pole, a similar programme for about 200 A stars is nearly ready. These investigations are performed in collaboration with Mme M. Lunel.

A programme for narrow-band photometry in nine regions from λ 3600–5000 Å for determination of Strömgren parameters and hydrogen line intensities has been initiated. Photoelectric measurements have been finished for SA 40, 64, 87, and 110, are on way for SA 9, 19, 24, and 49, and are planned for SA 8, 74, and 98.

A. Terzan has published a photometric study of the central condensation of the Galaxy, for which SA 57 was used as standard field (**17**). From plates taken at the Haute-Provence Observatory magnitudes in blue, red, and infrared have been derived for about 300 stars in SA 57, down to the limit 19·3 (pg). Terzan has published a catalogue (**18**) giving blue, red and infrared magnitudes for 2600 stars in the above-mentioned region including SA 157.

Kiev. V. I. Voroshilov reports from the Golossejev Central Astronomical Observatory of the Ukrainian Academy of Sciences on continued investigations into the Parenago fields. Voroshilov has published a catalogue (**19**) of pv-magnitudes for an area centered at $\alpha = 19^h$, $\delta = +11°$ within the Parenago field I, previously treated (**56**). G. L. Fedorchenko has published pg- and pv-magnitudes (**20**) for stars in the region of SA 87 also within Parenago's field I, cf. (**57**). Three-colour photometry will be undertaken for eight regions from Aquila to Cygnus and at the anticentre (including SA 19, 64, and 87) for an investigation in cooperation with Abastumani Observatory, which takes care of the spectral determinations.

Lund. C. Roslund reports on a photometric investigation in cooperation with K. Särg of a part of SA 195, based on plates taken with the Uppsala Schmidt telescope on Mt Stromlo in ultraviolet, blue and yellow light (limiting magnitude 15·2 in V). They take advantage of G. Lyngå's photoelectric sequence Sq III in SA 195 (**21**) derived from Mt Stromlo and Bingar observations.

Mt Palomar (*Pasadena*). H. M. Johnson has taken multicolour plates of SA 132 in connection with a search for the optical identification of the brightest X-ray source in Scorpius.

Mt Stromlo (*Canberra*). B. J. Bok has, in collaboration with Mrs J. Basinsky, completed the photometric work in SA 141, situated at the south galactic pole (**22**). Photographic V and $B-V$ are given for over 1400 stars, and photoelectric V and $B-V$ for 26 standard stars. Observations have been made for SA 161, 169, and 205. B. Westerlund has published an infrared survey of the region of SA 193 (**23**) giving I-magnitudes and $R-I$ colours for about 300 stars.

Radcliffe (*Pretoria*). A. D. Thackeray reports that photoelectric *UBV* photometry is being carried out in SA 162 by P. J. Andrews to $V = 16$.

Stockholm (*Saltsjöbaden*). K. Lodén reports on the programme based on material from the Boyden Observatory (*Trans. IAU*, **11A**, 392, 1962) that for SA 193 photographically determined B and V magnitudes are ready for publication (limiting magnitude: $B = 13·0$; number of stars 447; size of area 1·0 sq. deg.). Photoelectric *UBV* measures for 21 stars serve as standards. For SA 192 the photometric results are nearly completed.

Sternberg (*Moskva*). For a 45 square degrees field centered on $\alpha = 0^h\ 30^m$, and $\delta = +62°$, which includes SA 8, E. S. Brodskaya and N. B. Grigorieva (**22a**) have published a catalogue, with maps, giving, for nearly 2700 stars brighter than magnitude 12, photographically determined V and $B - V$, together with spectral types.

Tartu. In connection with an investigation of eclipsing variables H. Albo has derived a photometric sequence in SA 18 (**24**).

Toulouse. R. Bouigue reports on a spectral and photometric investigation for determination of accurate stellar distances in selected regions. Observations have been performed in $5° \times 5°$ fields at the north galactic pole (with SA 57) and the galactic anticentre direction (including

SA 49). For *UBV* and narrow-band photometry (λ 3525, 3864, 4250, and 4990; band-width 70–90 Å) the same equipment with a Lallemand photomultiplier was used at the Toulouse, Haute-Provence, and Pic du Midi observatories (**25**). The programme is enlarged to include the direction and antidirection of galactic rotation (close to SA 40 and 172 respectively).

Tübingen. U. Haug reports on the photoelectric *UBV* measurements made in South Africa for SA 96, 100, 108, and 112. Results have been published (**26**) for 201 stars in these fields together with 120 stars of the southern *UBV*-system, including the Harvard E-regions; for E 1, 6 and 7 the catalogue gives data for several stars. Some of these stars have also been measured from the northern hemisphere: in SA 100 by Haug at Kitt Peak, Arizona, and in SA 112 by K. Walter at Serra La Nave, Sicily.

Uppsala. For SA 8–19, T. Elvius and L. Häggkvist have published photoelectric measurements of 73 relatively bright stars (**27**). For supplementing previous research (**58**) T. Elvius has derived photographic *U*-magnitudes for the region of SA 19 from plates taken with the Kvistaberg Schmidt telescope. Y. Ekedahl has concentrated his photometric work on SA 4. It should also be mentioned that SA 8 and 19 are covered by the Uppsala Milky Way Survey, Part III, by T. Oja (**28**), who also gives a useful survey of photoelectric data for the investigated galactic strip.

T. Elvius and G. Lyngå have published *UBV* photometry for 43 stars in SA 68, 92, 138, 164, 165, 188, 200, 201, 204, 205, and 206, measured at the Bingar Station of the Mt Stromlo Observatory (**29**). This programme has been continued by Lyngå and C. Roslund at the Siding Spring Observatory. They have also carried on observations with the Uppsala Schmidt on Mt Stromlo for areas at southern latitudes. From plates taken with the same telescope P. I. Eriksson has practically completed photographic determination of blue and yellow magnitudes for nearly 10 000 stars in a 30 sq. deg. field at the south galactic pole, which includes SA 141.

Vatican (*Castel Gandolfo*). M. F. McCarthy and P. J. Treanor have in collaboration with A. R. Upgren (now at van Vleck Observatory) begun a programme for studying stellar densities at intermediate latitudes by spectral and photometric observations within SA 28, 54, 106, and 107, cf. Section 6. Photometric plates have been taken with the 40 cm refractor. Cf. (**59**).

Warner and Swasey (*Cleveland*). S. W. McCuskey has published an investigation of a 25 sq. deg. area centered on SA 158 (**30**). For nearly 1700 stars pg-magnitudes are given and for 275 of them also colours, approximately in the $B-V$ system. The plates were taken at the Bosscha Observatory, Lembang. The investigation published by D. Philip of a region (HLF2) in Pegasus includes the field of SA 90 (**31**). The catalogue gives photographically determined blue magnitudes; for standardization a special *UBV* sequence was observed at the Kitt Peak and McDonald observatories. For further information on the Cleveland fields in low, intermediate and high latitudes reference is made to the main report of Commission 33.

Washington. From the Naval Observatory has been published a catalogue by D. L. Harris III and A. R. Upgren (**32**) which gives photoelectric magnitudes and colours in the *UBV* system for 280 BD-stars in a region at the north galactic pole which is identical with SA 57. The observations were made at the McDonald Observatory. V. M. Blanco reports that J. Priser has at the Flagstaff Station completed a photoelectric survey of 111 stars in even numbered SA of the 0° + 30° and + 60° declination zones (**33**).

Miss N. Roman, now with NASA, reports on her photoelectric work in a number of SA. The *UBV* observations are now reduced for nearly 800 stars in 40 areas. Generally, all proper-motion reference stars are included in the areas where observations have been made.

3. *Variable stars*

Hamburg-Bergedorf. The series of catalogues of variables in the southern part of the Cygnus cloud by A. A. Wachmann (**34**) include several stars in SA 64. Wachmann has also published catalogues of variables in the region of SA 98 in Monoceros (**35**).

Harvard (*Cambridge, Mass.*). Miss M. Olmsted reports that she has obtained photographic and photored light curves for nine cepheids in SA 193 (**35a**). She is engaged in a search for more faint cepheids in the same area, using plates taken with the Baker-Schmidt telescope of the Boyden Observatory.

Haute-Provence (*Saint Michel*) *and Lyon.* Terzan has published identification charts for new variable stars at the central condensation of the galaxy (**36, 37**); the field includes SA 157, cf. (**18**).

Lick (*Mt Hamilton*). T. D. Kinman, C. A. Wirtanen and K. A. Janes have in the course of the RR Lyrae star survey with the 20-inch (51 cm) astrograph published results for three fields near the north galactic pole (**38**); field RR 4 coincides with SA 57 whereas fields RR 2 and 3 are between this and SA 56. For SA 57 a couple of faint stars have been added to the photoelectric sequence by J. Stebbins, A. E. Whitford and H. L. Johnson (**60**). Kinman reports that the extended RR Lyrae programme will include SA 61; a survey of variables of longer period will include SA 27 and 28.

4. *Proper motions and positions*

Bonn. H. van Schewick reports the intention to measure relative proper motions for the northern SA which remain after the publication of data for 18 galactic fields, cf. *Trans. IAU*, **12A,** 550, 1965. SA 72, 95, and 115 have already been measured and relative proper motions have been derived. For SA 40 two more pairs of plates were measured in connection with an investigation of the galactic cluster An. Barkh. I (**39**). Similarly the proper motions of SA 98 have been used for a study of NGC 2301 (**40**). The results of the determination of the solar apex from high-velocity stars, mentioned in the last report, will soon be published. van Schewick has prepared a list of proper motions for variable stars, including several SA objects.

Greenwich (*Herstmonceux*). The possibility of improving the proper motions in SA is considered by C. A. Murray. The old Radcliffe Catalogue (**61**) should be used as 'first epoch' and be combined with plates taken with the 26-inch (66 cm) Greenwich refractor in addition to those obtained with the Radcliffe telescope at London Observatory, cf. *Trans. IAU*, **12A,** 551, 1965. The first area to be investigated is SA 57 at the north galactic pole.

5. *Radial velocities*

David Dunlap (*Richmond Hill*). J. F. Heard has published radial velocities and spectral classes of 55 fundamental stars in the high-latitude areas SA 13–15, 29–35, and 53–60 (**41**).

Haute-Provence (*Saint-Michel*) and *Marseille.* The following information has been obtained from Ch. Fehrenbach, in addition to the report in *Trans. IAU*, **12A,** 551, 1965.

Concerning work with the smaller objective prism (PPO):

(*a*) Results have been published for SA 9, 21, 29, 30, 46, 67, 69, 75, 90, 91, 94, and 113 by Ch. Fehrenbach, M. Duflot, J. Boulon, E. Rebeirot, and C. Lanoë (**42**).

(*b*) Results are close to publication for SA 41 and 43.

(*c*) Measurements are going on for SA 3, 4, 13, 22, 23, 35, 37, 39, 61, 62, 66, 68, 81, 85–88, 98, and 114.

Concerning work with the larger prism (GPO):

(*d*) Results are ready for publication for SA 15 and SA 19, northern part.

(*e*) Measurements are on way for SA 9, 11, 13, 36, 40, 41, 53, 58, 59, 61, 64 centre, 64 north, 81, and 110.

Measurements have been published during the years for altogether 1432 stars, of which 736 belong to types O, B, and A, and 696 to types F, G, K, and M.

Observations have also been made of a certain number of regions within the Parenago fields; results have been published for an area in field IV (**42** p. 185).

Radcliffe (*Pretoria*). A. D. Thackeray reports that slit spectroscopy will be undertaken by T. Lloyd Evans for stars down to V about 12 in SA 141 at the south galactic pole.

6. *Spectral and luminosity classification; absolute magnitudes*

Abastumani (*Mt Kanobili*). Spectral types and luminosity classes are given in the catalogues mentioned above in Section 2, for stars in Parenago fields, including some SA (**3–8**). Investigations are going on for a number of regions in Taurus and Aquila. Two-dimensional spectral classification is planned for a series of areas uniformly distributed in the equatorial plane from Aquila to Cygnus and at the anticentre for which the photometric work will be undertaken by the Golossejevo Observatory at Kiev, cf. Section 2.

David Dunlap (*Richmond Hill*). Spectral classification in the MK system is included in the paper (**41**), cf. Section 5.

Engelhardt (*Kazan*). The previously reported spectrophotometric investigation of SA in the + 45° declination zone is continued, cf. Section 2.

Haute-Provence (*Saint-Michel*) and *Marseille*. In continuation of (**62**) Mme M. Barbier has published two lists of M, S, and C stars in several fields, among them Kapteyn areas in the declination zones + 75°, + 45°, + 30°, + 15° and 0° (**43**). She has also determined absolute magnitudes for 200 O, B, and A stars in SA 8 by spectrophotometric measurements of objective prism plates (**44**).

The catalogues referred to above in Section 5 as (**42**) include spectral and luminosity classification for a considerable number of SA stars. Here should also be mentioned the paper by Mme N. Martin (**45**) on the determination of absolute magnitudes by 'Öhman's method' (**63**, **64**). Her catalogue for the Carina field includes stars in SA 193, whereas SA 18 is just outside the Cepheus field.

Kiev. Spectral classes have now been published by Voroshilov (**46**) for 3500 stars in two of the sub-regions in Aquila within Parenago field I, for which photometry has been reported (**19** and **56**). Cf. also I. I. Pronik's catalogues from the Crimean Astrophysical Observatory (**65**). Spectral classes are also included in Fedorchenko's catalogue for stars in the same Parenago field (**20**), mentioned in Section 2.

Radcliffe (*Pretoria*). MK classification will be included in the programme for SA 141 reported in Section 5. There are also plans for spectroscopy of stars within other southern SA in the R.A. range from 20^h to 2^h.

Sternberg (*Moskva*). Spectral types have been published for a field including SA 8; cf. Section 2.

Stockholm (*Saltsjöbaden*). Spectrophotometrically derived spectral types and luminosity classes are included in the investigation of SA 193 just terminated by K. Lodén, mentioned in Section 2.

Toulouse. Spectrophotometric measurements from objective-prism plates are planned for the fields mentioned in Section 2 for determination of absolute magnitudes from equivalent widths of hydrogen lines.

Uppsala. For the Stockholm and Uppsala SA programmes reported in *Trans. IAU*, **11A**, 394 and 395, 1962, additional plates have been taken with the Uppsala Schmidt telescope on Mt Stromlo by Roslund and Lyngå. Eriksson's spectral measurements for the region of SA 141 are now ready. Oja's catalogue (**28**) mentioned in Section 2 includes data for stars in SA 8 and 19.

Van Vleck (*Middletown, Conn.*) and *Vatican.* A. R. Upgren has in collaboration with McCarthy and Treanor of the Vatican Observatory, cf. Section 2, outlined an observational programme to determine the inclination of plane-parallel layers of stellar density to the galactic plane. The regions chosen are SA 28, 54, 106, and 107. Spectral plates have been taken with the 24-inch (61 cm) Baker-Schmidt telescope of the Dyer Observatory, Nashville, and also, for smaller dispersion, with the Schmidt telescope of the Vatican Observatory.

Warner and Swasey (*Cleveland*). Spectral types and luminosity classes have been published for SA 158 stars by McCuskey (**30**), and for stars in SA 90 by Philip (**31**).

7. *Polarization*

Göttingen. Stars in SA 40 and 97 are included in a polarimetric, photometric and spectroscopic investigation of stars in Cygnus and Orion by I. Appenzeller (**47**).

8. *Investigations based on the material of Selected Areas*

Several of the papers referred to above in the SA report and in the main report of Commission 33 contain investigations of galactic structure based on SA material.

Of special interest for the future is an investigation by W. Becker aiming at derivation of star densities in the galactic halo by means of three-colour photometry. The SA included in the observational programme were mentioned in *Trans. IAU*, **12A**, 548, 1965. A first paper treating SA 51 has been published (**48**), cf. also (**66**).

In (**44**) Mme Barbier investigates the absorption, the stellar distribution and the galactic rotation in the Cassiopeia field SA 8.

Here should also be mentioned the extensive study of stellar distribution near the south galactic pole (**22**) by B. J. Bok and J. Basinski. It is based on photometric measurements in SA 141. The density decrease with z is derived for A and F type stars, thereby supplementing the discussion by T. Elvius (**49**) in his summarizing paper on the distribution of common stars in intermediate and high galactic latitudes, for which data from Selected Areas are fundamental. In this paper Elvius presents some not previously published results derived from the third Stockholm SA catalogue (**67**).

Studies of the distribution of stars and interstellar matter, based on data from the Kapteyn and Parenago fields, have been published by L. N. Kolesnik (**50**) for the Cygnus field SA 40 = Parenago region II), and by Fedorchenko for the Parenago region I in Aquila (**20, 51**).

Desiderata for future work

Several members of the Commission 33 have expressed their view that the Selected Area Committee should once more emphasize the need for more observational data in the SA and similar fields in accordance with the Desiderata of the previous report, *Trans. IAU*, **12A**, 553, 1965.

Especially has been stressed the importance of having, for kinematical studies, more and better proper motions and radial velocities of faint SA stars. The value of such work would be increased if photoelectric data could be obtained for as many as possible of these stars.

It is obviously difficult in many cases to get a satisfactory survey of the existing data. The desire has therefore been expressed that for each individual area summaries should be prepared, facilitating the exploitation of existing material and also the planning for supplementary data.

T. ELVIUS
Chairman of the Committee

BIBLIOGRAPHY

a. New Selected Area data and investigations

1. Blaauw, A., Elvius, T. 1965, *Stars and Stellar Systems*, Gen. eds. G. Kuiper and B. Middlehurst, Vol. **5**, *Galactic Structure*. Eds. A. Blaauw and M. Schmidt, 589, University of Chicago Press, Chicago.
2. Brun, A., Vehrenberg, H. 1965, *Atlas of Harvard-Groningen Selected Areas*, Treugesell-Verlag, Düsseldorf.
3. Kharadze, E. K., Apriamashvili, S. P., Kotchlasvili, T. A. 1964, *Abastumani astrophys. Obs. Bull.*, **31**, 3.
4. Apriamashvili, S.P. 1964, ibid, **31**, 27.
5. Kalandadze, N. B. 1964, ibid., **31**, 66.
6. Apriamashvili, S. P., Kuznetsov, V. I. 1966, ibid., **35**.
7. Kalandadze, N. B. 1966, ibid., **35**, and .
8. Chuadze, A. D. 1966, ibid., **35**.
9. Cousins, A. W. J., Stoy, R. H. 1963, *R. Obs. Bull.*, no. 64.
10. Nekrasova, S. V., Nikonov, V. B., Rybka, E. 1965, *Izv. Krym. astrofiz. Obs.*, **34**, 69.
11. Urasin, L. 1965, *Bull. astr. Obs. Engelhardt* (*Kazan*), **38**, 47.
12. Urasin, L. 1966, *Izv. astr. Obs. Engelhardt* (*Kazan*), no. 35.
13. Bigay, J. H., Lunel, M. 1965, *C.R. Acad. Sci., Paris*, **260**, 3853.
14. Bigay, J. H., Lunel, M. 1965, *J. Observateurs*, **48**, 171 = *Publ. Obs. Hte-Provence*, **8**, no. 7.
15. Bigay, J. H. 1966, *C.R. Acad. Sci. Paris*, **262** Ser. A–B, 1570 = *Publ. Hte-Provence*, **8**, no. 28.
16. Bigay, J. H. 1966, ibid., **263** Ser. A–B, 214 = *Publ. Hte-Provence*, **8**, no. 29.
17. Terzan, A. 1965, *Ann. Astrophys.*, **28**, 935 = *Publ. Obs. Hte-Provence*, **8**, no. 12.
18. Terzan, A. 1966, *J. Observateurs*, **49**, 225 = *Publ. Obs. Hte-Provence*, **8**, no. 12 vis.
19. Voroshilov, V. I. 1966, *Problems of astrophysics*, 136. '*Naukova Dumka*', Kiev.
20. Fedorchenko, G. L. 1964, *Investigations of the physics of stars and diffuse matter*, 63. '*Naukova Dumka*', Kiev.
21. Lyngå, G. 1964, *Acta Univers. Lund*, Sect. II, no. 11 = *Medd. Lunds Obs.*, Ser. II, no. 139.
22. Bok, B.J., Basinsky, J. 1964, *Mem. Mt Stromlo Obs.*, no. 16.

22a. Brodskaya, E. S., Grigorieva, N. B. 1964, *Soobšč. gas. astr. Inst. Sternberga (Moskva)*, no. 136.

23. Westerlund, B. 1965, *Mon. Not. R. astr. Soc.*, **130**, 45 = *Mt Stromlo Obs. Repr.*, no. 116.
24. Albo, H. 1964, *Tartu astr. Obs. Publ.*, **34**, 169.
25. Azzopardi, , Bouigue, R. 1965, *J. Observateurs*.
26. Pfleiderer, J., Dachs, J., Haug, U. 1966, *Z. Astrophys.*, **64**, 116 = *Mitt. astr. Inst. Tübingen*, no. 88.
27. Elvius, T., Häggkvist, L. 1966, *Ark. Astr.*, **4**, no. 3 = *Uppsala astr. Obs. Medd.*, no. 152.
28. Oja, T. 1966, *Nova acta reg. soc. sci. Upsal.*, Ser. IV, **19**, no. 3 = *Uppsala astr. Obs. Ann.*, **2**, no. 4.
29. Elvius, T., Lyngå, G. 1965, *Ark. Astr.*, **3**, no. 32 = *Uppsala astr. Obs. Medd.*, no. 149.
30. McCuskey, S. W. 1964, *Astr. J.*, **69**, 104 = *Warner and Swasey Obs. Repr.*, no. 128.
31. Davis Philip, A. G. 1966, *Astrophys. J. Suppl.*, **12**, 391 = *Warner and Swasey Obs. Repr.*, no. 142.
32. Harris, D. L. III., Upgren, A. R. 1964, *Astrophys. J.*, **140**, 151 = *U.S. nav. Obs. Washington Repr.*, no. 52.
33. Priser, J. B. 1966, *Publ. astr. Soc. Pacific*, **78**, 474, *U.S. Naval Obs. Washington, Repr.*, no. 79.
34. Wachmann, A. A. *Astr. Abh. Hamburg. Sternw.*, **6**, no. 1, 1961; no. 2, 1963; no. 3, 1964; no. 4, 1966.
35. Wachmann, A. A. ibid., **7**, no. 3 and 4, 1964; no. 7, 1966.

35a. Olmsted, M. 1966, *Astr. J.*, **71**, 916.

36. Terzan, A. 1965, *C. R. Acad. Sci. Paris*, **261**, 3974 = *Publ. Obs. Hte-Provence*, **8**, no. 11.
37. Terzan, A. 1966, *J. Observateurs*, **49**, 258 = *Publ. Obs. Hte-Provence*, **8**.

38. Kinman, T. D., Wirtanen, C. A., Janes, K. A. 1966, *Astrophys. J. Suppl.*, **13**, 279 = *Contr. Lick Obs.*, no. 202.
39. van Schewick, H. 1966, *Veröff. Univ. Sternw. Bonn*, no. 74.
40. van Schewick, H. 1966, ibid., no. 76.
41. Heard, J. F. 1965, *Publ. David Dunlap Obs. Univ. Toronto*, **2**, no. 16.
42. Fehrenbach, Ch., Duflot, M., Boulon, J., Rebeirot, E., Lanoë, C. 1966, *J. Observateurs*, **49**, 155, 158, 165, 176, 194, and 205 = *Publ. Obs. Hte-Provence*, **8**, no. 25.
43. Barbier, M. 1966, *J. Observateurs*, **48**, 149, and **49**, 290, = *Publ. Obs. Hte-Provence*, **8**, no. 5 and no. 30, 23.
44. Barbier, M. 1966, *J. Observateurs*, **49**, 269 = *Publ. Obs. Hte-Provence*, **8**, no. 30, 1.
45. Martin, N. 1964, *J. Observateurs*, **47**, 125 = *Publ. Obs. Hte-Provence*, **7**, no. 33.
46. Voroshilov, I. I. 1966, *Physics of stars and interstellar medium*, 92. '*Naukova Dumka*' Kiev.
47. Appenzeller, I. 1966, *Z. Astrophys.*, **64**, 19 = *Veröff. Univ. Sternw. Göttingen*, no. 144.
48. Becker, W. 1965, *Z. Astrophys.*, **62**, 54 = *Mitt. astr-meteor. Anst. Univ. Basel*, no. 40.
49. Elvius, T. 1965, *Stars and Stellar Systems*, **5**, *Galactic Structure*. Eds. A. Blaauw and M. Schmidt, 41, University of Chicago Press, Chicago.
50. Kolesnik, L. N. 1964, *Investigations of the physics of stars and diffuse matter*, 52. '*Naukova Dumka*', Kiev.
51. Fedorchenko, G. L. 1966, *Problems of astrophysics*, 169. '*Naukova Dumka*', Kiev.

b. Other references

52. Parenago, P. P. 1956, *Astr. Zu.*, **33**, 749. See also *IAU Symp.*, no. 7, 61, 1959.
53. McCuskey, S. W. 1965, *Stars and Stellar Systems*, Gen. eds. G. Kuiper and B. Middlehurst, **5**: *Galactic Structure*, Eds. A. Blaauw and M. Schmidt, 1, University of Chicago Press, Chicago.
54. *Trans. IAU*, **12A**, 554, ref. 5, 1965.
55. *Trans. IAU*, **12A**, 554, ref. 11 and 12, 1965.
56. *Trans. IAU*, **12A**, 554, ref. 9a, 1965.
57. *Trans. IAU*, **12A**, 554, ref. 9c, 1965.
58. *Trans. IAU*, **9**, 472, ref. 62, 1955.
59. O'Connell, D. J. K. 1965, *Ann. rep. Vatican Obs.*,
60. *Trans. IAU*, **9**, 470, ref. 16, 1957.
61. *Trans. IAU*, **9**, 471, ref. 36, 1957.
62. *Trans. IAU*, **12A**, 554, ref. 29-31, 1965.
63. Öhman, Y. 1930, *Medd. Astr. Obs. Uppsala*, no. 48.
64. Fehrenbach, Ch. 1958, *Handbuch Physik*, Ed. S. Flügge, **50**, 25. Springer Verl., Berlin.
65. *Trans. IAU*, **12A**, 534, ref. 10, 1965.
66. Becker, W. 1966, *IAU Symp.*, no. 24, 358.
67. *Trans. IAU*, **11A**, 397, ref. 8, 1962.

34. COMMISSION DE LA MATIERE INTERSTELLAIRE ET DES NEBULEUSES PLANETAIRES

PRÉSIDENT: Professor S. B. Pikelner, Sternberg Astronomical Institute, Moscow B-234, U.S.S.R.

VICE-PRÉSIDENT: Professor D. E. Osterbrock, Washburn Observatory, University of Wisconsin, Madison 6, Wisconsin, U.S.A.

MEMBRES: Aller (L. H.), Andrillat (H.), Andrillat (Y.), Behr, Bok (B. J.), Chvojkova, Courtès, Divan, Dunham, Elvius (A. M.), Gehrels, Goldsworthy, Greenberg, Gurzadian, Hall (J. S.), Haro, Henize, Hiltner, Herbig, Johnson (H. M.), Kahn, Kaplan (S. A.), Kerr, Kharadze, Khavtasi, Kipper, Kohoutek, Lambrecht, Liller, Lodén (L.), McCrea, Martel-Chossat, Menon, Minin, Minkowski, Münch (G.), Pottasch, Razmadze (N. A.), Rosino, Rozhkovski, Safronov, Savedoff, Schalén, Schatzman, Schmidt (K. H.), Schmidt-Kaler, Seaton, Serkowski, Sharpless, Shklovsky, Sobolev, Spitzer, Strömgren, Takakubo, Thackeray, Treanor, van de Hulst, van Woerden, Vanýsek, von Hoerner, Vorontsov-Velyaminov, Walker (G. A. H.), Westerhout, Whitford, Zimmermann (H.), Zirin.

INTRODUCTION

New fields of astrophysics, which arised recently, resulted in the enlargement of topics included into the list of problems of Commission 34. As far as the problems connected with quasars, structure of galaxies and their nuclei are concerned the review deals with questions relating to the diffuse medium. The review is partly based on the reports of Commission members, especially on those of Professor Osterbrock and Dr Aller. I am greatly obliged to Professor Greenberg, who kindly wrote a separate review on interstellar grains.

Since 1964 several new books and reviews have appeared:

Structure and Evolution of Galaxies, 1964, *Suppl. Prog. theor. Phys.*, Osaka, no. 31.

Osterbrock, D. E., 1964, *A. Rev. Astr. Astrophys.*, **2**, 95.

Pottasch, S. R., 1965, *Vistas in Astronomy*, **6**, 149.

Kahn, F. D., Dyson, J. E., 1965, *A. Rev. Astr. Astrophys.*, **3**, 47.

Stars and Stellar Systems, V: 1965, *Galactic Structure* Ed. by A. Blaauw and M. Schmidt, Univ. of Chicago Press, Chicago.

Schalén, C., 1965, *Publ. astr. Soc. Pacif.*, **77**, 409.

Kaplan, S. A., Pikelner, S. B., 1965, *Izv. Akad. Nauk SSSR*, fiz. ser., **29**, 1830.

Stars and Stellar Systems, VII: 1966, *Nebulae and Interstellar Matter*, Univ. of Chicago Press, Chicago.

Dieter, N. H., Goss, W. M., 1966, *Rev. mod. Phys.*, **38**, 256.

Kaplan, S. A., Pikelner, S. B., *Interstellar medium*, Harvard, in press.

Symposium on *Radio and Optical Studies of the Galaxy*, 1966, Mount Stromlo Observatory.

IAU Symposium no. 31, 1966, *Radio Astronomy and the Galactic System*, Noordwijk, Sept. (in press).

PLANETARY NEBULAE

During these years spectrophotometric observations of planetary nebulae were continued: absolute intensities of spectral lines and of continuum, in particular in peculiar and stellar-like objects, infrared and radio-observations. The classical theory of nebulae emission was improved. Transition probabilities and collision cross-sections for a number of ions were calculated. Theoretical investigations of the recombination spectra of H, He and of other elements were performed, especially for different l-values and for higher levels, where some anomalies were revealed. The chemical composition, in particular for nebulae of different sub-systems was defined more precisely. However the main progress now takes place in the investigation of the dynamics and evolution of nebulae, of their stratification, expansion, with simultaneous evolution of their central stars. This problem is particularly important, because planetary nebulae are advanced stages of the evolution of some types of stars, and a comparison of the central stars of different-age nebulae allows to find an evolutionary modification in blue dwarfs.

Eighty-six planetaries were identified on Palomar Sky Survey by Abell (**1**). A catalogue of 202 new planetaries on the South Sky was published (**2**), and two nebulae were found (**3**). Perek and Kohoutek have prepared for publication an extensive catalogue with charts, containing more than 1000 planetaries with positions 1950·0, galactic coordinates, sizes, surface brightness, magnitude, classification and information on the central star (**4**). Westerlund and Smith revealed 42 planetaries in LMC (**5**), measured their colours and magnitudes. Spectra of 20 planetaries in other galaxies are being investigated.

The determination of absolute proper motions in the system of FK4 for 16 planetary nebulae, with a mean error of $\pm$ 0″008/year, is reported by Dr Chr. de Vegt.

Phtoelectric and photographic spectrophotometry was carried out for a large number of planetaries by Aller, Bowen, Czyzak, Faulkner, Liller, Kaler, O'Dell, Walker, Kohoutek (**7, 8, 9, 10, 11, 12, 13, 13a**). Application of the Lallemand image converter made it possible to study in detail a visual part of spectra (**14**). Absolute intensities of main lines were measured by photographic method in 65 northern and 90 southern nebulae (in the direction of the Galactic centre) by Vorontsov-Velyaminov, Kostyakova, Dokuchaeva, Arkhipova (**15, 16**).

Infrared line 12·8 μ Ne II was observed in IC 418 by Low (**17**). The flux is 10^{-16} w cm^{-2} in agreement with theoretical estimation $0{\cdot}6 \times 10^{-16}$ w cm^{-2} (**18, 19**). Radio observations of planetary nebulae were carried out by Golnev and Parijsky (**20**), Khromov and others (**21**), Menon and Terzian (**22**).

Models of nebulae based on these observations were constructed (**23, 24**), which are in agreement with optical data.

Transition probabilities for a number of permitted optical and ultraviolet lines of astrophysical interest were calculated by Czyzak, Aller and their associates (**25, 26**). Collision cross-sections for a fine-structure N III, O IV, Ne II were calculated by Osterbrock (**27**), and in (**28**) a general formula for the calculation of critical electron concentration is given. Dalgarno estimated A_{2s1s} for two-photon transition of He I and the distribution of the energy in two-photon emission (**28a**).

Capriotti (**29**) confirmed the conclusion by Osterbrock that $2s$ state is more important for optical depths of Balmer lines than $2p$ state, but the majority of planetaries are optically thin. Pengelly showed that the recombination spectrum of H and He II is not in agreement with observations for upper levels (**30**). It was necessary to take into account collisions transitions $n,l \rightarrow n,l+1$. Such calculations (**31**) including the radiation gave smooth transitions from the lines to the continuum (**32**). Similarly Kaler (**33, 34**) showed that in many planetaries higher members of H, He I, He II series are relatively brighter than in recombination theory. Discrepancies are not connected with parameters of nebulae and their origin is not clear. Clarke made more extensive calculations of stationary populations of sublevels, taking into account all effects. He obtained better agreement with observations than for the pure radiative decrement (**35**).

The temperature of nebulae calculated inclusive the fine structure of ion levels is in agreement with observations (**27**). From the uniform material (**15, 16**), physical parameters of nebulae were calculated (**36**). The importance of interstellar absorption was evaluated in

(37). The temperature and density of IC 4997 were calculated by Gurzadian (38). He also showed that the inner part of ring-shape planetary nebulae is relatively hotter due to high ionization (39). Chemical composition of a number of nebulae (40) and an average composition (41) was calculated by Aller.

The nebula in globular cluster M 15 contains few heavy elements but a normal abundance of He. Apparently He was in primordial matter. It makes the age of globular clusters shorter (42).

The transfer theory for nebulae is developed in (43). In (44) the probability of escape $L\alpha$ quanta is calculated numerically with frequency re-distribution and geometry. Accumulation of $L\alpha$ in many cases greatly exceeds the theoretical value apparently due to H I shell around the nebulae (45). It is supported by the presence of [O I] line which is an indicator of the nebulae optically thick in Lc (46). Mass, T_* and other parameters for such nebulae are calculated in (47, 48). Stratification of nebulae, ionization of He in different layers and for stars of different temperature were evaluated in (49). The space structure of a nebula can be determined from its picture and spectrum (50).

Expansion of nebulae was investigated on radial velocities (51). In several cases it was determined directly from proper motions (52, 53). In many nebulae there is no expansion, apparently due to a dense surrounding medium. Expansion was discussed theoretically in (54). The model of an expanding nebula, i.e. the distribution of the density, temperature and velocity with radial distance and with time was calculated by Mathews (55). He took into account the heating and cooling of the expanding gas and obtained the variation of $H\beta$ profiles and emission measure with time. To receive an agreement with the observations an additional stream of the gas from the central star is needed. This confirms that its activity is continued.

A central star of WN type was first revealed by Bertola (56). Spectrophotometry of nuclei is given in (57). Determination of nebulae distances as well as temperatures and luminosity of their nuclei allows to draw evolutionary tracks from the left end of the horizontal branch to the white dwarfs (58, 59). The evolution of central stars was discussed also in (60). In (61) a convection in cold variable supergiants-embryons of planetaries and the transfer of angular momentum into outer zone were discussed. The importance of rotation for detachment of the envelopes was stressed in (62). Temperatures of the central stars were determined in (63, 64), the theoretical model and the spectrum of the central stars were calculated in (65, 66).

Calculations on the theory of transfer of radiation in resonance lines (67, 68, 69, 70, 71, 72) can be applied to planetary nebulae.

DIFFUSE NEBULAE AND H II REGIONS

Investigations of diffuse nebulae and H II clouds developed mainly in traditional directions. More precise evaluations of the spectrophotometry in optical and infrared lines and in continuous spectrum were carried out, a thermal continuous radio-emission was determined. An investigation of emission radio-lines of highly excited hydrogen is a new step in this field. This allowed to identify distant heavily obscured H II regions and to determine their radial velocities. This also permitted the calculation of temperatures. Some contradictions with the theory stimulated the development of the latter. Questions were elaborated on: dynamics of the nebulae expansion, their interaction with the surrounding gas, the influence of the dust on the structure of ionization and shock fronts, formation of globulae and of elephant trunks, internal motions in nebulae. The structure and origin of cometary nebulae and the fine filamentary structure of reflecting nebulae are also connected with the dynamics of the gas and with magnetic fields. Searches and investigations of emission regions in Magellanic Clouds and in other galaxies were continued.

A catalogue of bright nebulae that appear on the Palomar Sky Survey was prepared by Lynds (**73**). The careful spectrophotometry of Orion Nebula in spectral region 3187–5016Å was made by Kaler, Aller and Bowen (**74**). Relative intensities and identifications of lines up to 1.5×10^{-4} H β are given, O'Dell and Hubbard (**75**) made a photoelectric scanning of lines and continuum of the Orion Nebula. The continuum moving from the Trapezium becomes bluer and brighter relatively to H β. This is due to the light scattered by dust. Reitmeyer (**76**) made the photometry of the Orion Nebula images taken in selected wavelengths. Electron temperature determined from the ratio H β to the Balmer continuum is changed inside the nebula, it is higher where abundance O III is less. Some other results of the paper are not reliable as the influence of interstellar absorption was not taken into account (**76a**). The influence of the density fluctuations on the mass evaluation was discussed by Peimbert (**77**). He obtained a uniform temperature from the comparison O II and O III lines if the abundance of oxygen is constant. Wurm and Perinotto compared the structure of the Orion Nebula in lines and in the continuum (**78**).

Profiles of 10 830Å He I in Orion were obtained by Voughan (**79**). Infrared observations by Moroz showed that, in 0,85–1·7 μ, the main emission of this nebula is produced primarily by the gas, not by the dust (**80**).

The intensity of [O II] and the electron concentration were determined in several bright nebulae (**81**) and the spectrophotometry of the nebulae around η Carinae and 30 Doradus was made (**82**). In the central part of the Lagoon Nebula $T = 9000°$, $N_e = 250$ cm^{-3}, and in two bright condensations—10 000° and 1.6×10^4 cm^{-3} respectively. The mass of the condensations is about 0·7 $M_\odot$ (**83**). Khachikian (**84**) determined the strong polarization of the Omega Nebula (18%) with a constant direction and came to the conclusion that the emission of the Nebula is not connected with the reflection. But the photoelectrical measurements do not give values exceeding 4% (**84a**), and the polarization is undoubtedly of interstellar origin.

Continuous radio-emission of the Rosetta Nebula gave (**85**) its emission measure, and observations of four nebulae on different frequencies showed the thermal origin of their emission (**86**). The thermal character of the spectrum is confirmed for a weak nebula near γ Cyg (**87**). The structure of nebulae in radio-emission and the connection of the structure with stars in Orion are studied in (**88**). It was confirmed that the thermal source Cyg X is a large Strömgren sphere around the association VI Cyg (**89**) which is situated at the distance of 1·5 kps along the arm (**90**). Twenty separate sources on the general phone Cyg X are revealed in (**91**). All of them, except one, have the thermal spectrum. Thirteen galactic radio-sources were investigated at the $\lambda = 6$ cm (**92**), their structure and brightness were determined.

In the central region of the Galaxy there are several sources with thermal spectra (**93, 94**). Their thermal nature is confirmed by the lines of highly excited hydrogen. These lines were predicted theoretically long ago and were revealed recently (**95, 96, 97**). They are an important source of information as they allow to measure the radial velocity of the source and to separate thermal and non-thermal radio-emission. The intensity of these lines as well as the continuum are proportional to the emission measure but their dependence on temperature differs. Therefore their ratio permits to determine T, if the Doppler width is known (**97**). Such measurements for 20 sources gave T from 4000 to 9000°. Earlier, Burbidge and others (**98**) obtained theoretically such low temperatures, taking into account the excitation of the fine-structure of ground states of abundant ions. But they overestimated the collision cross-sections. Osterbrock estimated once again the collision strengths for important transitions and showed that the expected temperatures are between 5000° and 10 000°. Temperatures of low-density nebulae are significantly lower than those of high-density nebulae, such as classical bright planetaries and the centre of the Orion Nebula (**99**). Observations of radio-lines deal with rather dense nebulae. O'Dell (**100**) measured the temperature from [O III] lines in NGC 6618 and NGC 6523 and in three rarefied planetaries and showed that they are in fact significantly lower than in the centre of the dense Orion Nebula. His measurements also gave a higher temperature than the theoretical one (**98**) and that from radio-lines. They agree better with Hjellming's calculations (**101**), who used new models of high temperature stars and obtained T from 5700° to more than 8000° depending on the star temperature.

An explanation of the temperature determined from radio-lines gave Goldberg (**102**). Populations of upper levels are practically the same as the equilibrium ones, especially if collisions are taken into account. However the lower the level, the more is the deviation from the equilibrium due to an increasing role of spontaneous transitions. The ratio of the population of two neighbouring levels deviates from the equilibrium one less than by 10^{-3}, but even such little relative overpopulation of upper levels gives a maser-effect and lines get brighter relative to continuum. It gives an anomalous low calculated temperature.

The widths of the lines of highly excited hydrogen were also in disagreement with theoretical expectations. The accepted theory by Griem, Kolb and Shen (**103**), which takes into account electron impacts besides the statistical stark-effect, suggested a width considerably exceeding the observed value (**104**). Some laboratory experiments with hydrogen plasma also show widths less than the theoretical ones (**105**). These discrepancies were apparently a result of a wrong dependence of broadening by electrons $\lambda_e \sim (n^5 + n'^5)$ in (**103**) and also of an incorrect extrapolation of the results found for the disturbance of one level for the general case of the disturbance of both initial and final levels (see (**106**)). Besides the hydrogen lines, higher transitions of helium were also observed (**107**). Palmer (**107a**) calculated the catalogue of radio-lines of hydrogen and helium.

Temperature of the gas is connected with its movements. High temperature of H II regions in galactic nuclei, determined from [N II]: Hα ratio, is explained in (**98**) by the influence of stellar winds, which is important in the regions with high star density. In (**101**), there was calculated the structure of H II regions which are limited by an ionization front. The absorption of the radiation in the nebula, scattered radiation and cooling due to the excitation of levels and sublevels of iones were taken into account.

The dynamics of the expansion of H II regions was discussed by Axford who took into account the absorption of stellar radiation at the early stage of expansion (**108**) and showed that the stability of weak ionization fronts of D-type increases due to this absorption (**109**). Vandervoort (**110**) calculated the age of the Orion Nebula (14–23 thousand years) from the comparison with the theoretical model. He calculated the early stage of the expansion of H II region (before the radius has reached the Strömgren radius), the propagation of the ionization and shock fronts, and investigated the non-stationary ionization front formed by a star with a varying flux of radiation (**111**). Lasker (**112**) discussed the later stage of expansion of H II regions with the age of about 2 million years. Goldsworthy (**113**) investigated the influence of gravitation on the central part of H II region and found a more precise age of the Orion and Rosetta.

Mathews (**114**) discussed the expansion of the nebula with the massive central star which has a considerable evolution during the expansion of the nebula. He evaluated all main processes of heating and cooling and found a model of physical conditions, velocity field in H II and parameters of a shock wave in H I with time. The maximum temperature is expected immediately behind the ionization front. He believes that the formation of a central hole in Rosetta and in similar objects can be explained by the pressure of 'winds' from the stars which are situated inside or by a condensation of a part of the gas in stars (**115**).

According to radio-observations the low-latitude regions H II are surrounded by H I, the main part of which does not expand. High-latitude regions are not associated with a considerable amount of H I (**116**). Apparently H II regions are immersed into large complexes of neutral hydrogen.

An estimation of the ratio of dust to neutral hydrogen density in Cas and Sge was done in (**116a**).

The presence of dust in the ionization and in shock fronts has an influence on their properties, since the grains absorb radiation and cool the gas. This makes the ionization front slower, especially if the density is low (**117**). The dust increases the compression in shocks (**118**) and has an influence on the dynamics of clouds collisions. The conditions in shocks, on the contrary, are important for the growth and evaporation of grains (**119, 120**). The relative velocity of the dust and gas was also determined. Calculations of the gas resistance under different conditions and with different velocities of grains are given in (**121**).

Gershberg and Sheglov (**122**) revealed with a Fabry-Pérot interferometer fast internal motions $V \approx \pm 50$ km s^{-1} in NGC 6618, $V \approx \pm 25$ km s^{-1} in NGC 6523 and $V \approx \pm 10$ km s^{-1} in NGC 7000 (see also (**123**)). Sheglov revealed weak components with V up to ± 100 km s^{-1} in Omega and Orion (**124**). The origin of these movements is not clear yet, they may be connected with outbursts of supernovae in nebulae (**122**), but the necessary outburst frequency is too high. A similar method was developed by Flynn and Ring (**125**) for the Orion Nebula. The general character of internal motions was evaluated here (**126**).

The presence of dust in diffuse nebulae was examined on the basis of H β and continuum ratio (**127**). The light reflected by dust is present in all three nebulae studied, so that the disintegration of dust in H II regions and its repelling by radiation pressure are rather slow.

The statistics of reflecting nebulae on Palomar Sky Survey was carried out by Dörschner and Gürtler (**128, 129**). They compared the dimensions of the nebulae with stellar spectra and with dimensions of H II regions around B–stars. Previous calculations of Strömgren's radii were overestimated, new ones were calculated. Cometary-like nebulae adjoin to the reflective ones. Hall showed that the polarizations of the variable nebula R Mon is chiefly radial, it increased with distance from the star and reaches as much as 34%. The colour along the axis first turns blue and then red (**130**). The spectra of cometary nebulae were investigated by Dibai and Esipov (**131, 132**). Usually it resembles the spectrum of the star but S 129 turned out to be a common gas nebula. The morphological features of this nebula may be explained by the focusing of the converged shock (**133**). Khachikjan and Parsamian showed that the polarization of NGC 2261 does not change for six years (**134**). They carried out the colourimetry of an anonimous nebula and came to the conclusion that its emission is not reflected light (**135**). There was compiled a list of 23 nebulae on Palomar Sky Survey (**136**). Vardanian (**137**) showed that the tails of cometary nebulae are parallel to the mean direction of the stellar polarization, i.e. their directions may be connected with the magnetic fields of the spiral arms. The fine filamentary structure of reflecting nebulae is apparently a sequence of the interchange instability, when the cloud without the field penetrates into the magnetic field of spiral arms (**138**). After the cloud disintegrates into the filaments, the field penetrates into them and fixes ions and especially charged grains.

H II zone around the Sun was calculated on the basis of the emission spectrum of the Sun in the ultraviolet and X-region (**139**). Due to a weak absorption for the quanta of high energy the ionization decreases continuously. The size of the zone is several hundred of astronomical units. A similar value is found in (**140**) where the electron concentration is inferred from $L\alpha$ observations. The cut-off of the spectrum of cosmic radio noise near 4–5 MHz confirmed the presence of the absorbing zone H II with the emission measure from 8 to 165 in different l (**141, 142**).

If it is not a cloud surrounding the Sun, then $< N_e >$ in the galactic disk should be about 0·1 cm^{-3}. However, the uniform distribution is quite improbable in this case. They are not common H II clouds, but more probable quasi-regular small-scale fluctuations of the gas between clouds.

H II regions in the Magellanic Clouds were studied by Feast (**143**), Dickel (**144**), Mathis (**145**). Temperatures and densities, non-homogeneities, and masses were determined. The decrement of 30 Doradus gave the extinction. The abundance of He is the same as in galactic nebulae (**145**). Baade and Arp (**146**) measured the positions of emission nebulae in M 31 which were used for drawing the pieces of arms (**147**). The catalogue of nebulae in M 33 (**148**) and in three other galaxies (**149**) is based on pictures in Hα. Sěrsic compiled a catalogue of 85 H II regions in NGC 300 (**150**).

DISTRIBUTION AND PHYSICAL CONDITIONS IN INTERSTELLAR GAS. MOLECULES

The general distribution of gas in the galactic disk had been investigated before. Therefore now only separate features and details were investigated, more attention was paid to deviations from the average picture. Relations of hydrogen and associations were studied, statistical parameters of clouds were determined (on hydrogen and Ca II lines). A great progress was achieved in the investigation of clouds on intermediate and high latitudes. Some peculiarities found here have not been definitely explained yet.

The temperature of the interstellar medium evaluated from radio observations depends on the kind of lines (emission or absorption) observed and on the distribution of the temperature in dense and rarefied clouds. The temperature itself largely depends on the presence of H_2 molecules. There are many papers with calculations of the abundance of these molecules, but there are no definite results yet. It may be concluded that the abundance of H_2 is relatively low in common clouds and high in dense globulae, but the total abundance in the disk hardly exceeds the abundance of atomic hydrogen. Methods of direct observations of H_2 ultraviolet lines are discussed. Among other molecules, especially interesting is OH, whose lines show extraordinary properties, which are qualitatively explained now as the result of stimulated emission (maser-effect) and sometimes as that of the presence of the magnetic field.

The rotation curve of the Galaxy was redetermined from new observations by Shane and Bieger-Smith (**151**). They showed that this curve is systematically different for the Milky Way in the Northern and Southern Hemispheres. The analysis also indicates the presence of systematic streaming motions with velocities about 10 km s^{-1} and breadths of the order of 100 pcs. This gives difficulties in the determination of the distribution of gas. Pronik analysed the published data of radio observations and showed that the rotation curves are not the same in regions with positive and negative latitudes (**152**). Streams along the Sagittarius arm were found by Burton (**153**). Lindblad has observed and analysed profiles 21 cm in the anticentral region, represented there in terms of gaussian components (**154**) and studied the structure and motions of the gas in this region (**155**). Asymmetry of the velocity field on both sides of the galactic centre was confirmed by Kerr and Hindman (**156**).

Using data on the cross-sections of the Galaxy in (r, z)-planes through the Sun, published earlier by Westerhout, van Woerden re-determined the general distribution of neutral hydrogen in the Galaxy (**157**). The thickness of the hydrogen layer increases markedly with increasing distance from the galactic centre, the amount of the neutral hydrogen in the Galaxy may be 5 to 7×10^9 solar masses.

From the analysis of optical and radio data Pskovsky (**158**), and Pavlovskaya and Sharov (**159**) came to the conclusion that our Galaxy is probably a multi-arm rather than usual spiral.

Gosachinsky (**160**) showed that far away from the centre in $l = 10$ to $40°$ hydrogen has the structure of sheets inclined to the Galactic plane. The cause of deviation of the disk from the plane was discussed by Elwert and Hablick (**161**). They think that the gravitational attraction of the Magellanic Clouds may have a resonance effect on the gas layer, which later on leads to the observed deviation.

Smith (162) and Prata (163) found hydrogen concentrations with peculiarly high velocities. The nature of these objects is unclear, they may be similar to the gas complexes discussed in (164) which possibly are connected with the associations in Perseus. The distribution of H I in the region of the galactic anticentre and in the vicinity of super-nova remnant IC 443 (165), in the region of Taurus (166) was studied. A reverse correlation of absorption and of H I concentration was revealed in the latter paper so that in heavy clouds H_2 consists of 85% of the whole mass. Habing (167) revealed at the intermediate latitudes gas with velocities corresponding to those of the outer parts of the Galaxy. This gas may extend several kiloparsecs from the galactic plane; it may be a distorted part of a far arm.

A connection of gas with stellar associations was studied for the associations Mon I and especially for Mon II. Raimond (168) found there condensations of H I with total mass of about 20 000 $M_{\odot}$. Such condensations had been mentioned before, for instance gas was found in the cluster NGC 2244 which is a part of Mon II (169). An expanding shell was also revealed around this association (170). But Bystrova, Gosachinsky and Ryzhkov (171) believe that there is neither a concentration of gas towards the association nor an expanding shell. They interpret the maximum of hydrogen emission as a bridge between two arms. Investigations of H I in other associations are carried out mainly in the Netherlands (172). Goldstein (173) showed that the total mass of H I in the globular cluster M 13 is less than 140 $M_{\odot}$.

The distribution of H II is determined from thermal radio emission (cf. Draft Report of Commission 40), as well as from the absorption of low-frequency cosmic radio noise (142). The ring of H II in the inner part of the Galaxy (3·5–4·5 kpcs) may also be revealed on the deficiency of H I (174). The motion of a star with Strömgren sphere through the gas forms a shock (175).

Cloud structure and kinematics of the gas observed on intermediate latitudes, i.e. near the limits of the disk, were investigated by Takakubo (176) with the decomposition of the profiles into gaussian components (177). There was revealed a high correlation of the low-velocity peak of H I and the main components of Ca II line, in spite of the difference in directions. The correlation of the secondary peak of H I profile and the high-velocity Ca II component is poor, indicating that high-velocity clouds are small in size. Clark (178) determined densities, masses and other parameters of the clouds from the absorption components in spectra of five bright radio sources. He found on the average 4·1 clouds per kpc in the galactic plane. There is no systematic deviation from pure rotation—the line in the spectrum of Sgr A is symmetrical about standard frequency. The layer of the clouds is very thin and dense, H I clouds were not observed in absorption on high latitudes. The different width of absorption and emission lines is due to the fact that absorption components are formed in cool clouds, while emission components—also in the hot rarefied gas between clouds; this gas is transparent due to its high temperature.

A preliminary report on observations of optical lines for the comparison with radio observations is performed by van Woerden (179). The velocity dispersion of Ca II interstellar lines (6·9 km s^{-1}) is less than that of young stars (180). The dependence of the intensities of Ca II lines on the distance from the stars was studied by Wilkens (181). Buscombe and Kennedy (182) summarized all published radial velocities of Ca II lines on the southern Sky, together with new observations and compared them with new distances of OB stars. Na I lines were also studied.

Livingston and Lynds (183), and Hobbs (184) measured interstellar lines in the spectra of bright stars in particular α Cyg, with very high wavelength resolution, and showed that there are much more components than usually are observed with lower resolution. A superfine structure was also observed in some clouds. It means that internal turbulent velocities are less than 0,64 km s^{-1} (184). Smith (185) showed that when analysing profiles one should take into account the expansion of the clouds along the interstellar magnetic fields. The profile may be in such case with two peaks and will be interpreted as two clouds.

Resonance interstellar absorption lines of O I, S II, Al III, C II in the region 1260–1720Å were observed with rockets by Morton and Spitzer (**186**) in the spectra of δ and π Sco.

A comprehensive investigation of H I at high latitudes ($b > +40°$) was carried out mainly by Dutch radioastronomers. A systematic survey resulted in the discovery of several new clouds approaching with high velocity (**187**). Blaauw and Tolbert (**188**) discussed the intermediate-velocity features up to $\pm$ 70 km s^{-1}. They suggested that an inflow of matter from $l \approx 120°$, $b \approx +15°$ penetrates close to the Sun. The total mass involved in disturbances is evaluated about 2600 $M_\odot$. The phenomenon is much more pronounced at the high negative velocities and is restricted to a smaller area on the sky. The high-velocity clouds were studied by Hulsbosch and Raimond (**189**). They determined for two clouds sizes of $-3\overset{\circ}{.}5$ and $1\overset{\circ}{.}3$ and masses of $-2500\, r^2\, M_\odot$ and $150\, r^2\, M_\odot$, where r is the distance to the clouds in kpcs. An investigation of high-velocity clouds near N and S galactic poles was carried out by Dieter (**190**). Two peaks with velocities -20 and -50 km s^{-1} are observed near the North pole. Masses and densities of the clouds are less than common ones. The picture in the southern hemisphere is similar, but there are no such definite streams as in the northern one. Some separate clouds with velocities up to -90 km s^{-1} were observed. The width of the lines corresponds to $T \geqslant 150°$, the velocity dispersion is about 2·5 km s^{-1}.

Oort considered several possible interpretations of these phenomena (**191**): motions in the galactic corona, connected with explosions in the disk or with giant associations, the inflow of metagalactic gas into the Galaxy, and so on. However, the flux of the flowing matter should be too high in this case, and we expect a considerable influence of it on the dynamics of the gas in the galactic disk. Moreover, too high density of metagalatic matter is necessary for this explanation. A possible exchange of the gas between different parts of the disk and outbursts of the gas from the galactic layer were discussed. It is difficult to explain the formation of condensed clouds with a high density and fast internal motions. The possibility cannot be excluded that the clouds are near-by and are supernova remnants. But a comparison with the data on Ca II lines speaks in favour of larger distances. Shklovsky supposed that there is a maser-effect in clouds, which increases the black-body radiation on the line frequency.

The inversion of H I atoms population may be created under some conditions with $L\alpha$ line from H II regions near the limits of the disk (**192**). The possibility of such inversion was studied by Varshalovich (**193**).

The distribution of the mass of clouds $\rho(M) \sim M^{-\beta}$ was calculated by Sheffler (**194**) from the distribution of the extinction. He found $\beta \approx 2$ for $M < 5 \times 10^3\, M_\odot$ and 1 to 1·5 for $M > 5 \times 10^3\, M_\odot$. The mass of a common cloud is 20–30 $M_\odot$ and the diameter is about 3 pcs. A general statistical model of clouds and of their transformation on the basis of former ideas by Oort and Spitzer was constructed by Field and Saslaw (**195**). They considered collisions of clouds, their confluence due to dissipative processes, gradual growth of the mass, formation of massive complexes. The appearance of young stars gives H II regions and the expanding shell of H I divided into little clouds run away from the complex. The calculations of the kinetics of this processes gave the spectrum of mass of clouds as $M^{-3/2}$ in agreement with (**194**). Inelastic collisions of clouds were considered also in (**196**) and (**197**). The propagation of the shock wave, heating and cooling of the gas up to the standard external pressure, a possibility of gravitational instability in the compressed region were discussed in the latter paper. Shock waves with emission, were discussed by Kaplan and Klimishin (**198**). They take into account the non-stationary scattering of light and heating of the gas before the front. Numerical calculations of the structure of the waves with the emission in a partly ionized gas were done by Kaplan and Podstrigach (**199**). The structural function of the turbulence was inferred from the analysis of interstellar lines by Kaplan and Klimishin (**200**).

Pikelner discussed general dynamics of clouds in the magnetic field of spiral arms (**201**). Common clouds are supported by the field. They have oscillations across the magnetic lines and the collisions in these directions are elastic. They move rather freely along the lines and the collisions are inelastic. The complexes grow, come down to the galactic plane and collect gas until the beginning of stars formation and flying away of the gas.

The temperature in H I regions used to be determined from saturated emission lines which give the harmonic mean $T \approx 125°$. Shuter and Verschuur (**202**) separated absorption lines 21 cm in the spectrum of radio sources into gaussian components, which they treat as thermal ones. The width of the components gives temperatures of clouds from 25 to 120° and N_H. The calculated harmonic mean temperature for all clouds and for 20% invisible clouds appears to be less than 70°. This value differs from the emission radio temperature, because the main contribution to the absorption gives cold clouds and the contribution to the emission gives, besides cold clouds, also hot clouds and the gas between clouds.

Different harmonic temperatures for emission and absorption lines with some given distribution of H I are obtained also by Marx (**203**). An influence of the cloud structure on the measurements of the temperature and density with 21 cm line was discussed by Rohlfs (**204**). The heating of the rarefied gas between the clouds in the absence of ultraviolet radiation may be performed by superthermal particles from 10 to 100 MeV (**205**).

The radiation field in different parts of the Galaxy was calculated by Zimmermann (**206**) and Kaplan (**207**). The density of the energy is higher in the halo where interstellar absorption is ineffective (**207**). Cooling of interstellar gas was reconsidered by Dalgarno and Rudge (**208**). They take into account excitations of C^+ and Si^+ in collisions with hydrogen atoms followed by spin exchange. The process appears from 3 to 5 times more effective than electronic collisions.

The temperature depends markedly on the presence of H_2 molecules, which cool the gas effectively if $T > 60°$. Cooling is a result of shock collisions 0–2 and spontaneous transition 2–0. Transitions 2–1 and 1–0 have insignificant probability (**209**). Transitions 2–0 give the line 28 μ, which can be principally observed. Its intensity in the galactic plane should be comparable with $B_\nu(50°)$. The possibility of observations of ultraviolet Lyman line H_2 have been discussed by Spitzer, Dressler and Upson (**210**). The curve of growth oscillator strength and equivalent width were calculated when different concentrations of H_2 were present.

Direct observations of H_2 have not been performed yet. The relative deficiency of H I in heavy absorbing clouds speaks in favour of the presence of molecules. For instance, there is relatively few H I in the cloud obscuring 'Omega' (**211**). The other method of evaluation is from the gravitational potential in the disk. Such estimates were done, for instance in (**212**), on the base of the comparison of velocities and Z-distribution of H I clouds and cepheids. The estimates gave for H_2 abundance of 0·8 to H. However, the evolutional effect for cepheids (their age is comparable with the time of motions from the galactic plane) and magnetic forces for clouds were not taken into account.

The origin of H_2 is usually connected with recombinations on the surface of an interstellar dust grain. Atoms are bound with the grain surface by van der Waals forces, i.e. by physical absorption. However in (**212a**) it was shown that due to zero-point energy, physical absorption of H atoms by grains with $T > 8°K$ is negligible. The possibility of chemical absorption is not studied yet.

In this connection the considerably less effective mechanism by Stecher and Williams (**213**) should be discussed. This mechanism is an outflow of H_2 from the atmospheres of cold giants and supergiants with stellar winds.

Very strong interstellar CN lines were revealed in the group of stars situated in the emission nebula (**214**). Propagation of the ionization front in the cloud with a high concentration of dust resulted in the evaporation of grains and dissociation of molecules of the stream. In general, CN lines are observed in complexes abundant with dust. One of the levels of molecule CN is very close to the ground state (the difference of the energy corresponds to 0·254 cm). Shklovsky (**215**), and Field and Hitchcock (**216**) pointed out independently that the population of this level is very sensitive to the density of cosmic radiation. The fine structure of CN absorption-line shows that the population of the second level corresponds to the temperature of exciting radiation about 3° (**215, 217**). This confirms the presence of black-body radiation in the Universe, as all other sources of radiation in interstellar space are much weaker.

OH molecule radio lines of which $\lambda = 18$ cm were revealed in 1963 is especially interesting. Observations of these absorption lines in spectrum of Sag A showed some peculiarities (**218, 219**). The lines are very strong, wide and displaced relatively to the position, expected from the velocities of hydrogen features. Relative intensities of quadruplets do not correspond to the theoretical values. First, there were attempts to explain this by the effect of saturation (**220**), but they were not a success. The distribution of OH in the vicinity of the galactic centrum is irregular, there are many clouds with different velocities, which are not in accordance with the velocities from 21 cm line (**221**).

It was an unexpected discovery that OH is observed in the emission (**222**) near large H II regions. The lines were very bright, the excitation temperature was more than 9°, i.e. much higher than that far from H II. One component of the quadruplet consisted of several very strong peaks with the width of about 0·4 km s^{-1} that corresponds to an unexpectedly low temperature. Some other components of quadruplet were not observed at all. Similar peculiarities were observed in the Orion Nebula and in others. It was revealed further that the emission in some components is strongly lineary polarized and the value and direction are different for the clouds with different velocities (**223**). All these properties were confirmed and investigated in detail in many papers (see Draft Report of Commission 40). More than 50% of all investigated H II regions show anomal OH emission. The size of the sources is usually less than 20″ (Burke), several details show almost 100% circular polarization (Barrett).

The interpretation of anormal intensity and of narrows of some components is connected with maser-effect. The excitation of upper levels may be due to ultraviolet emission of the surrounding H II regions (**224**), with their $L\alpha$ radiation (**225a**). According to (**225**) the maser-effect is connected with the infrared radiation of the hot pre-stellar bodies which contain OH molecules inside. Effects of polarization are connected with magnetic field. In different components maser-effect can be different and probably it is effective mainly along the magnetic field, giving circular, but not elliptic polarization (Ginzburg). The details of this process are not clear yet.

There were some attempts to observe the ultraviolet absorption line of OH 3080Å in the spectrum of O6 star (**226**). The line was absent and the upper limit of abundence of OH was 9 times less than may be expected from the radio data. Apparently OH is concentrated in small regions which are not projected on the star. Abundance of OH in the interstellar gas is determined mainly by exchange reactions with H_2 and other molecules at temperature of about 1000° connected with shock waves (**227**).

SUPERNOVAE REMNANTS, MAGNETIC FIELD OF THE GALAXY, DYNAMICS AND ORIGIN OF SPIRAL ARMS

Hogg (**237**) showed that the radio source IC 443 has a shell-like structure, coinciding with a region of optic brightness. He measured also radio radiation from a great number of other possible remnants of supernovae. Davies (**238**) investigated a part of a spur, visible from England, and showed that its outer edge is sharp and the spur itself is similar to Cygnus loop by its structure and brightness. Brosche (**239**) investigated the expanding of a shell of Crab Nebula by proper motions. He confirmed that the motion is accelerated. Heiles (**240**) investigated a shell expanding in interstellar gas. He showed that Shklovskii's automodel solution is true only as long as the temperature behind the front does not fall below 10^7 °K. After this stage the emission of ions O VIII and others becomes essential. However in (**240**) the isothermicity of plasma behind the front is proved incorrectly. In the layers, near to the front, Coulomb interactions cannot establish the equi-partition. The electron temperature should be determined rather by different instabilities. Sheglov (**241**), discovered in NGC 6990 radiation of Fe X with intensity of 3–5 relays, proved experimentally the high electron temperature behind the front in Cygnus Loop.

The influence of the magnetic field of the Galaxy on the expanding shells of supernovae was considered by scientists of the Princeton group (**242, 243**). They considered a case of a strong and weak field—the last one is closer to reality. The field is compressed strongly only near a shell-piston, and in the main part of interstellar gas, compressed by a wave, it is strengthened slightly. The instability of Rayleigh-Taylor takes place near the boundary, it explains the filamentary structure. The Crab Nebula is the most interesting supernova remnant. Here, radio emission is concentrated on the centre and it has not a shell structure. Shklovsky showed that, on the whole, relativistic electrons, but not protons, are generated in it, so such a nebulae cannot be a source of usual cosmic rays (**244**). Both the origin of relativistic particles in remnants and the formation of the magnetic field in it are left to be a problem, because a magnetic flux, for example that of the Crab Nebula, could not be taken from a star or interstellar medium. Kardashov (**245**) supposed that in the period of a stellar collapse a rotating instability occurs, a shell, connected with a star by a magnetic field, is separated, and the fast rotation of the central nucleus winds the lines of force. The rotation is maintained by compression and takes place till now. Observations show that for 10 years the optical polarization of Crab Nebula has half decreased and that the nebula radiation has a component with an elliptic polarization (**247**). A checking of these wonderful results is very desirable.

Westerlund and Mathewson (**248**) discovered in LMC three nebulae—remnants of supernovae of type II. Two of them and many associations are situated in the ring H I by a diameter of about 1 kpc—apparently it is a remnant of an explosion of a massive supernova of the first generation. The third remnant is in the other ring, containing associations and strong non-thermal source. The mass of a large shell is about $3 \times 10^7 M_{\odot}$, therefore, the original mass is about $10^5 M_{\odot}$, the age is about 3 million years with medium density of 1 cm^{-3}. Apparently, a series of superexplosions results in conditions for the formation of associations and individual stars.

The interstellar polarization gives one of the methods of the investigation of the magnetic field. Martel (**249**) investigated a dependence of the polarization on a wavelength. Studying a correlation with the galactic longitude, Lodén (**250**) investigated in great detail the polarization of accidentally selected stars in 19 regions of the sky. Serkowski (**251, 252**) determined a dependence of the polarization on longitude and investigated in detail a region including a cluster and association of Cyg VI. Dombrovski (**253**) found a region of homogeneous polarization in a district of the cluster NGC 1502. Pronik (**254**) compared the polarization with both the velocities of Ca II and the shape of dark nebulae. The field turned to be very irregular, often directed differently above and under the galactic plane, and in some parts it was not parallel to the galactic plane and not directed along the arms. Visvanathan (**255**) investigated 30 stars in LMC. The maximal polarization is $0^m.068$; $(p/A_v)_{max} = 0.07$ on the average all over a cloud. The polarization plane does not change in large regions. On the average, lines of force are parallel to the plane and to the structure of the nebula, forming a very flat disk. Zwicky (**256**) found in M 82 three types of polarization—common, along a plane, radial around some luminous spots and directed along filaments and light band, thrown out by an explosion.

The polarization of non-thermal radiation gives information on a magnetic field of the Galaxy. Mathewson and Milne (**257**) found that regions of high polarization of radiophone coincide with that of weak Faraday rotation for distant radio sources, it is naturally explained by low depolarization on these places. They found that the direction of polarization, as a rule, lies in a band with width of 60° relative to a big circle, passing through galactic poles and the equator point with $l = 160°$. It has been confirmed in (**258**) and indicates the presence of a sufficiently homogeneous field in the arm.

The rotation of a polarization plane of extragalactic sources takes place, as it has been shown before, in the Galaxy, and the field is parallel to the galactic plane (**259**). However, the field structure is compound and does not correspond to a simple picture of a tube of parallel lines of force. However, the authors did not divide north and south latitudes, where a rotation has a different sign. Indeed, Morris and Berge (**260**) showed that the rotation is carried out by lines of force, which are parallel in a middle of the arm, but have an opposite direction in the northern and southern hemispheres. Moreover, RM is 1·5 to 2 times as much in the S hemisphere than in the N. Gardner and Davies (**261**) made these results more precise and found that there is a third narrow region along the equator where the rotation is especially strong and has one sign.

Bologna and others (**262**) investigated the depolarization of extragalactic sources. The polarization increases towards the equator and, therefore, it is determined by the Galaxy. The depolarization grows with the increase of an angular dimension of the source. These phenomena can be in principle explained if fluctuations which are less than 0·3 pcs and with $N_e\tilde{H} \approx 3 \times 10^6$ are superimposed on the general field. Maltby arrives at smaller dimensions (0·1 pcs) and expresses his doubt that the depolarization is in the Galaxy (**263**). Such a doubt is expressed in (**259**), too.

Measurements of Zeeman's effect in 21 cm line (**264, 265, 266**) gave $B \leqslant (5\text{–}7 \times 10^{-6}$ G in the cloud, but this result depended on the adopted direction of the complete vector $\boldsymbol{B}$.

The obtained data of observations do not yet allow to construct a picture of the field, but they exclude some simple models proposed before. Stępień (**267**) and Hornby (**268**) consider that observations are presented best by lines of force which form (inside the arm) a spiral, stretched by the differential rotation, thus the plane of spires form with an axis of the arm an acute angle. Woltjer (**269**) considered the field as consisting of weak quasi-homogeneous (B later 5×10^{-6} G) (Symposium no. 31), on which the strong chaotic field are superimposed. The lines of force of this field are parallel to the axis of the arm, but changing the sign at a distance of about 0·1 pcs. This little scale is necessary for the explanation of a regular distribution of the rotation measure with the position on the sky. It is supposed that the field has no dynamic meaning on the gas. Using this picture, Wentzel (**270**) applied Petchek's mechanism to the interstellar gas—the compression of gas between the regions of the opposite fields and ejection of compact clouds from there. However Pikelner (**201**) points out a number of dynamic difficulties in the picture of the strong irregular field. Besides that, the explanation of the rotation of the polarization plane in the presence of small irregularities needs improbably high electronic concentration, if gas is neutral in the main. As a whole, the problem should be additionally investigated—both by observation and theoretically.

The problem of the origin and existence of spiral arms is of great interest. The gravitation theory is widely used. Small disturbances in the gas of gravitating stars are considered (**271**). The density fluctuations propagate as a heavy sound. The field of these fluctuations attracts stars with low velocity dispersion and the gas fluctuations can have a spiral form under certain conditions.

By the gas compression the field in it takes a prolonged orientation. The arms are not wound, because perturbations move relative to stars. The investigation of perturbations in a rotating disk, its stability, axis-symmetric oscillations was carried out in many papers (**272,**

273, 274, 275, 276). The gas flows into the arm on the one side, on the back one it flows out. True, it is not proved yet whether this process could take place in the non-linear approximation, when the gas density is comparable to or exceeds the stellar density of the background. The decay and formation of arms and indirect influence of the field on the relaxation of star velocities in linear approximation are considered in (**277, 278**). Pikelner (**138**) considered another model, where the field has a definite dynamic role. It explains the thickness of arms in different galaxies and at different distances from the centres, allows to understand the opposite direction of lines of force and explain a number of other peculiarities. However, this theory, as others, which consider the arm as a conservative formation, cannot explain a persistence of the spiral shape. Probably, there should be a synthesis of both points of view, where the field plays an essential dynamic role, but the gas flows through the arm. Richter and Wallis (**279**) supposed that, in the Galaxy, the gas forms rings, not spread out by the differential rotation, and arms are formed by crossing these rings with fronts of shock waves, originated from explosions in the centre.

Now the formation of the field of arms is usually connected with a condensation of metagalactic field together with the gas when forming of the Galaxy. Piddington (**280**) showed that the strength of the magnetic field in the disk depends considerably on the relative orientation of the primary field and rotation axis.

Barred spirals are also considered within the limits of the gravitating theory (**281, 282, 283**), the angular momentum of individual stars is not constant in this scheme because of the attraction of a bar, and streams in the outer directions appear. However, the stability of the system is not proved and, in general, the whole scheme needs artificial suppositions and special initial conditions. Pikelner (**284**) supposed that in SB, especially in SBc, there is an almost homogeneous sphere of old stars, in the field of which the rotation is rigid body. Such a sphere is formed as the result of the explosion which has taken place in the Galaxy at the very early stage of its formation. Streams along the bar, condensations on its ends, the transition of the bar into spiral arms, etc. are explained within the limits of this scheme.

The accretion of intergalactic gas, flowing into the Galaxy and winding can form in principle a spiral arm (**285**). However, this hardly explains the symmetry and regular structure of the arms.

The formation of bridges between interacting galaxies and tails is discussed in (**284, 286, 287**). Zasov (**288**) supposed that intergalactic gas flows to bridges, penetrates into them and then falls to galaxies. The statistics of distortions of gaseous layers of galaxies (it is often in small close groups) are given in (**288**).

THE INTERACTION OF THE GAS WITH COSMIC RAYS. INTERGALACTIC GAS, GAS IN QUASI-STELLAR SOURCES

The interaction of cosmic rays with interstellar medium becomes apparent not only in nuclear collisions and in collisions with moving magnetizing clouds, but also in the interaction with plasma waves which should always be present in the gas with microscopic motions. It was investigated in detail for different conditions by Tsitovitch (see a review (**289**)), a possibility of the acceleration of relativistic particles was determined. Ginzburg (**289a**) showed that the interaction with plasma waves explained the isotropy of cosmic rays. Parker (**290**) supposed that the formation of the galactic corona can be connected with the instability of the gas with the field, retaining cosmic rays.

The existence of the metagalactic gas is inferred sometimes from cosmological considerations, according to the value of critical density, necessary for the noticeable slow-down of the metagalaxy expansion. However, such a slow-down has only been proved indefinitely so far, that is why direct observations of the gas is very important. Kowal (**291**) has discovered dark spots on the background of three galaxies, but they are smaller than galaxies and in two cases they are connected with the galactic structure, that is why they cannot be considered as metagalactic. Gunn and Peterson (**292**) investigated a weak depression in the spectrum of the distant quasars 3C 9 beginning from $L\alpha$. It corresponds to $n_H \approx 6 \times 10^{-11}$ cm^{-3} at this distance, it is considerably smaller than the critical density even taking into consideration the ionization when $T = 2 \times 10^6$. However, such an estimate depends strongly on the extrapolation of the continuous spectrum, and the latest data indicate that absorption is somewhat smaller. Bahcall and Salpeter (**293**) indicated that if gas is concentrated in clusters, lines but not depressions should be in the spectrum of quasars. The absence of H_2 1108 Å line means that the abundance of molecular hydrogen is small and, apparently, the intergalactic gas is ionized completely. The absence of H_2 is shown in (**294**), too. The low concentration of atomic hydrogen follows also from measurements of 21 cm line from Cyg A and Vir A– $\tau < 0{\cdot}001$ and 0·02 correspondingly. In the second case it means that hydrogen does not absorb also in this galaxy itself – the spin temperature is high because of the strong radiation and the gas is transparent (**295**).

The ionization of hydrogen can be connected only with high kinetic temperature. The temperature conditions of metagalactic hydrogen was discussed by Ginzburg and Ozernoy (**296**). They took into account the emission of hydrogen and cooling due to the general expanding of gas, not being a part of clusters of galaxies. Heating is produced by cosmic and sub-cosmic rays, the existence of which is probable in the rarefied gas. Explosions of radiogalaxies give the directed streams of relativistic particles, which generate plasma waves, heating the gas. This mechanism is especially effective in clusters. In the result of it the temperature in a cluster depends on the history of the given gas, and between the clusters the temperature, apparently, is about 10^5 or somewhat higher. A similar calculation, taking into account helium, but without the general expanding, was carried out by Gould and Ramsay (**297**). The upper limit of the temperature of the intergalactic gas when $n \approx 10^{-5}$ cm^{-3} is determined by the absence of the noticeable background of X-rays. Different mechanisms of production of X- and γ rays in the Galaxy and Metagalaxy are discussed by Ginzburg and Syrovatski (**298**), Tucker and Gold (**299**) (see in detail Commission 44). Modern measurements of X-rays give the upper limit of $T < 3 \times 10^6$ (**300**). The absence of the background of X-radiation shows also that the quantity of relativistic electrons is considerably smaller in the Metagalaxy than in the Galaxy, otherwise there should be a strong inverse Compton-effect for photons of the relict black-body radiation (**301**). This fact and the existence of black-body radiation itself, exclude steady state cosmology.

Black-body radiation was discovered at centimeter waves in 1965 (**302, 303**). Its spectrum and the energy density correspond to $T = 3$°K. It is a remnant of the black-body radiation of high temperature at the initial period of the Universe expansion (**304**). From the theory of the hot Universe it follows, that about 10% of He atoms should be in the primordial gas. The possibility of observations of He^3 in metagalactic and interstellar gas is discussed by Sunyayev (**305**).

Emission lines of different ions are observed in quasars (**306, 307, 308, 309**). The analysis shows that it is a shell with a standard chemical composition and possible filamentary structure, surrounding the nucleus of quasars and expanding with the velocity of about 1000 km s^{-1}. Both the conditions of ionization and excitation and the gas temperature are studied, but here there are yet many indefinite parameters (**310, 211, 312, 313**), in particular, conditions of excitation. The ultra-violet part of the spectrum of synchroton radiation is calculated on the basis of the optic part by Greenstein (**314**). The absorption lines of expanding shells are observed in some quasars (**315, 316**).

Dibai, Pronik, Esipov carried out the spectrophotometry of nuclei of Seyfert galaxies with strong emission lines, calculated physical conditions in gas, discussed the probable mechanisms of excitation (317, 318, 319). Besides that, they analysed spectra of quasi-stellar sources (320, 321), estimated the mass of the gas.

GRAVITATIONAL CONDENSATION, FORMATION OF STARS AND GALAXIES

The general theoretical consideration of the gravitational instability of the non-uniform medium was carried out by Ozernoy (322) and Simon (323). The influence of the magnetic field, when the compression occurs along lines of force or independently, is estimated in (324). In the first case the critical mass is two times less. The thermal condition during a protostar compression, escape of energy in H_2 lines, the role of H_2 dissociation are considered by Gould (325). When $n < 100\ \text{cm}^{-3}$ and $T \approx 100°$ compression turns into free fall, in spite of the possible heating by fast particles and turbulence (325). When $n > 100\ \text{cm}^{-3}$ superelastic collisions decrease cooling and compression becomes slower, turning into oscillations (326). McNally (327) took into consideration that the centre of a cloud (from which a cluster forms) is more dense. It cools quicker, that is why a compression occurs there also quicker and a dense nucleus forms. In protoclusters with $M < 10^4 M_\odot$ the central temperature is only about 100°, cooling occurs slower, and no nucleus forms.

The thermal instability is possible on scales smaller than the critical Jeans' length. Field (328) takes into account the sound velocity, heat conductivity, magnetic field, rotation, external gravitational field; and Hunter, the gravitation of fluctuations themselves (329). Such an instability can give condensations in nebulae, the solar corona and partly facilitate the formation of galaxies. But it is effective only in special conditions, that is why its role in the formation of stars and galaxies is hardly considerable. Condensations in nebulae are also formes not due to thermal instability (330).

The compression of a cloud, when there is an external pressure, can be not quasistatic, but with the formation of a cumulative shock wave, forming a region of high pressure in the centre of a cloud (331). It facilitates the stellar formation in the centres of globules, submerged into nebulae. The formation of pre-stellar bodies in a cloud, compressed by a shock wave, is calculated also by Kossatsky (332).

Massive clouds fragments during compression. This process was considered on a simple model when $p = 0$ for different types of disturbances (333). The statistics of fragments according to masses allow to explain the function of stellar masses (besides massive stars) in clusters (334). The integration of fragments is taken into consideration in (335).

Grzedzielski (336) examined the fragmentation of protogalaxy with large dispersion of internal velocities. Besides the condensation of globules, the appearance of young stars in old clusters can be connected with the accretion of gas by stars (337). Earlier, this idea was expressed by Lebedinsky. As the result of condensation of globules or accretion non-stationary red dwarfs and Herbig-Haro objects are formed with small mass. Schatzman and Magnan (338) supposed that the excitation of these objects is produced by protons of high energy and Lortet-Zuckermann estimated the emission of the interstellar matter in the neighbourhood of flare stars (339).

The dense stellar nuclei of galaxies lose stars and compress until the collisions of stars would be essential (340, 341). The gas appears, it compresses in a thin disk, and stars form again (342). This may result in a marked luminosity increase.

A number of review papers, covering the investigations of Japanese scientists in this field (carried out in the main till 1964) is included in the collection 'The structure and evolution of galaxies'. The evolution of galaxies, radiogalaxy, the formation of spiral arms and dynamic model of barred galaxies, explosions of galactic nuclei, the stability of spiral arms are included in this review.

Lynden-Bell (**343**) considered the large-scale instabilities, appearing during the collapse of the rotating self-gravitating spheroid without internal pressure. The formation of the bar distributed along the diameter is possible in this process. From the other point of view the formation of bar and spiral arms can also be connected with the fact that the gravitating condensation compresses the field at first in the central, more dense, nucleus, and magnetic forces give the initial disturbance and stimulate the condensation of the gas situated along the tube of lines of force (**138, 284**).

The discovery of the black-body relict radiation allows to make a considerable progress in the understanding of the early stage of the formation of galactic clusters.

The conception of Jeans' instability raises a question on the initial equilibrium state, which could not be the state one. Peebles showed that in the hot expanding Universe the gas was connected at first with the radiation by way of a friction. Condensations could not compress because the radiation pressure prevented it (**344**). Osernoy (**345**) considered in detail the development of these instabilities, their temperature condition, the radiation output in the period of gas-recombination, the influence of different mechanisms of heating and cooling of the gas at different stages of the compression and fragmentation. Zeldovich, Novikov and Doroshkevich (**346, 347**) investigating the same problem, have shown that during the gas-recombinations the density is still so large that fragments with $M \approx 10^5 M_\odot$ are formed, i.e. still not clusters of galaxies. The energy resulting from the condensation of the first fragments, which may be quasars, heats the gas and prevents the condensation of other fragments. This energy creates conditions when the large masses of the galactic clusters type can be condensed.

S. B. PIKELNER
President of the Commission

BIBLIOGRAPHY

1. Abell, G. O. 1964, *Publ. astr. Soc. Pacif.*, **76**, 382.
2. Henize, K. G. 1964, *Publ. astr. Soc. Pacif.*, **76**, 385.
3. Apriamashwily, S. P. 1965, *Astr. Circ.*, SSSR, no. 318.
4. Perek, L., Kohoutek, L. In press.
5. Westerlund, B. E., Smith, L. F. 1964, *Mon. Not. R. astr. Soc.*, **127**, 449.
6. Andrillat, Y., Andrillat, H. Unpublished.
7. Aller, L. H., Kaler, J. B. 1964, *Astrophys. J.*, **139**, 1074.
8. Aller, L. H., Kaler, J. B. 1964, *Astrophys. J.*, **140**, 936.
9. Czyzak, S. J., Aller, L. H., Kaler, J. B., Faulkner, D. J. 1966, *Astrophys. J.*, **143**, 327.
10. Aller, L. H., Kaler, J. B., Bowen, I. S. 1966, *Astrophys. J.*, **144**, 291.
11. Liller, W., Aller, L. H. 1966, *Mon. Not. R. astr. Soc.*, **132**, 337.
12. O'Dell, C. R. 1964, *Publ. astr. Soc. Pacif.*, **76**, 308.
13. Aller, L. H., Kaler, J. B. 1964, *Astrophys. J.*, **140**, 621.
14. Aller, L. H., Walker, M. F. 1965, *Astrophys. J.*, **141**, 1318.
15. Vorontsov-Velyaminov, B. A., Kostyakova, E. B., Dokuchaeva, O. D., Arkhipova, V. P. 1964, *Astr. Zu.*, **41**, 255; 1965, **42**, 464, 730; 1967, **44**, in press.
16. Vorontsov-Velyaminov, B. A., Kostyakova, E. B., Dokuchaeva, O. D., Arkhipova, V. P. 1964, *Astr. Circ.*, SSSR, no. 305; 1965, no. 348.
17. Gould, R. J. 1966, *Astrophys. J.*, **143**, 603.
18. Gould, R. J. 1963, *Astrophys. J.*, **138**, 1308.
19. Gould, R. J. 1964, *Ann. Astrophys.*, **27**, 815.
20. Golnev, V. Ya., Parijsky, Yu. N. 1964, *Astr. Zu.*, **41**, 846.
21. Khromov, G. S., Indisov, O. S., Matveyenko, L. I., Turevsky, V. M., Sholomitsky, G. B. 1965, *Astr. Zu.*, **42**, 1120.
22. Menon, T. K., Terzian, Y. 1965, *Astrophys. J.*, **141**, 745; Terzian, Y., 1966, *Astrophys. J.*, **144**, 657.
23. Khromov, G. S. 1964, *Astr. Zu.*, **41**, 71; 1965, **42**, 918.

24. Osterbrock, D. E. 1965, *Astrophys. J.*, **141**, 1285.
25. Chapman, R. D., Clarke, W. N., Aller, L. H. 1966, *Astrophys. J.*, **144**, 376.
26. Czyzak, S. J., Krueger, T. K. 1963, *Mon. Not. R. astr. Soc.*, **126**, 177.
27. Osterbrock, D. E. 1965, *Astrophys. J.*, **142**, 1423.
28. Sampson, D. H. 1966, *Astrophys. J.*, **144**, 96.
28a. Dalgarno, A. 1966, *Mon. Not. R. astr. Soc.*, **131**, 311.
29. Capriotti, E. R. 1964, *Astrophys. J.*, **139**, 225; **140**, 632.
30. Pengelly, R. M. 1964, *Mon. Not. R. astr. Soc.*, **127**, 145.
31. Pengelly, R. M., Seaton, M. J. 1964, *Mon. Not. R. astr. Soc.*, **127**, 165.
32. Seaton, M. J. 1964, *Mon. Not. R. astr. Soc.*, **127**, 177.
33. Kaler, J. B. 1964, *Publ. astr. Soc. Pacif.*, **76**, 231.
34. Kaler, J. B. 1966, *Astrophys. J.*, **143**, 722.
35. Clarke, W. H. 1967, *Mon. Not. R. astr. Soc.*, in press.
36. Kostyakova, E. B. 1967, *Astr. Zu.*, **44**, in press.
37. Arkhipova, V. P. 1967, *Astr. Zu.*, **44**, in press.
38. Gurzadian, G. A. 1965, *Astrofizika*, **1**, 225.
39. Gurzadian, G. A. 1965, *Astrofizika*, **1**, 91.
40. Aller, L. H. 1964, *Publ. astr. Soc. Pacif.*, **76**, 279.
41. Aller, L. H. 1964, *Astrophys. norw.*, **9**, 293.
42. O'Dell, C. R., Peimbert, M., Kinman, T. D. 1964, *Astrophys. J.*, **140**, 119.
43. Wilson, S. J., Sen, K. K. 1965, *Ann. Astrophys.*, **28**, 855.
44. Capriotty, E. R. 1965, *Astrophys. J.*, **142**, 1101.
45. O'Dell, C. R. 1965, *Astrophys. J.*, **142**, 1093.
46. Khromov, G. S. 1965, *Astr. Zu.*, **42**, 543.
47. Herman, R. J., Seaton, M. J. 1966, *Mon. Not. R. astr. Soc.*, **132**, 15.
48. Seaton, M. J. 1966, *Mon. Not. R. Soc.*, **132**, 113.
49. Hummer, D. G., Seaton, M. J. 1964, *Mon. Not. R. astr. Soc.*, **127**, 217.
50. Vorontsov-Velyaminov, B. A. 1966, *Astr. Zu.*, **43**, 691.
51. Osterbrock, D. E., Miller, J. S., Weedman, D. W. *Astrophys. J.*, in press.
52. Liller, W. 1965, *Publ. astr. Soc. Pacif.*, **77**, 25.
53. Liller, M. H., Welther, B. L., Liller, W. 1966, *Astrophys. J.*, **144**, 280.
54. Fowler, R. G. 1963, *Publ. astr. Soc. Pacif.*, **75**, 36.
55. Mathews, W. G. 1966, *Astrophys. J.*, **143**, 173.
56. Bertola, F. 1964, *Publ. astr. Soc. Pacif.*, **76**, 241.
57. Noskova, R. I. 1965, *Astr. Zu.*, **42**, 1038.
58. Abell, G. O. 1966, *Astrophys. J.*, **144**, 259.
59. Harman, R. J., Seaton, M. J. 1964, *Astrophys. J.*, **140**, 824.
60. Osterbrock, D. E. 1966, *in Stellar Evolution*, Ed. R. F. Stein, A. G. W. Cameron, New York, p. 381.
61. Deeming, T. J. 1965, *Publ. astr. Soc. Pacif.*, **77**, 443.
62. Limber, D. N. 1964, *Astrophys. J.*, **140**, 386.
63. Bogorodskii, O. F., Turchaninova, S. V. 1964, *Vest. Kiev. Univ.* (Ser. Astr.), no. 6.
64. Turchaninova, S. V. 1964, *Vest. Kiev Univ.* (Ser. Astr.).
65. Böhm, K. H., Deinzer, W. 1965, *Z. Astrophys.*, **61**, 1.
66. Böhm, K. H., Deinzer, W. 1966, *Z. Astrophys.*, **63**, 177.
67. Ivanov, V. V. 1965, *Uchen. Zap. Leningr. Univ.*, no. 328, 44.
68. Ivanov, V. V., Nagirner, D. I. 1965, *Astrofizika*, **1**, 143.
69. Ivanov, V. V., Shcherbakov, V. T. 1965, *Astrofizika*, **1**, 22, 31.
70. Nagirner, D. I. 1964, *Astr. Zu.*, **41**, 669.
71. Sobolev, V. V. 1966, *Astrofizika*, **2**, 135.
72. Vasiljev, A. P., Cogan, V. I. 1966, *Dokl. Akad. Nauk SSSR*, in press.
72a. Van de Hulst, H. C., Terhoeve, F. G. 1966, *Bull. astr. Inst. Netherl.*, **18**, 377.
73. Lynds, B. T. 1966, *Astrophys. J. Suppl.*, **12**, no. 105.
74. Kaler, J. B., Aller, L. H., Bowen, I. S. 1965, *Astrophys. J.*, **141**, 912.
75. O'Dell, C. R., Hubbard, W. B. 1965, *Astrophys. J.*, **142**, 591.
76. Reitmeyer, W. L. 1965, *Astrophys. J.*, **141**, 1331.

76a. Boyce, P. B. 1966, *Astrophys. J.*, **144**, 833.
77. Peimbert, M. 1966, *Astrophys. J.*, **145**, 75.
78. Wurm, K., Perinotto, M. 1965, *Z. Astrophys.*, **62**, 30.
79. Voughan, A. H. 1964, *Astr. J.*, **69**, 151.
80. Moroz, V. I. 1963, *Astr. Zu.*, **40**, 788.
81. Davies, L. B., Ring, J., Selby, M. J. 1964, *Mon. Not. R. astr. Soc.*, **128**, 399.
82. Faulkner, D. J., Aller, L. H. 1965, *Mon. Not. R. astr. Soc.*, **130**, 121.
83. Grachev, N. I., Dibai, E. A. 1966, *Astr. Zu.*, **43**, 60.
84. Khachikian, E. E. 1964, *Soobšč. Bjurak. Obs.*, **35**, 25.
84a. Schakhowskoi, N. M., Sheglov, P. V. 1967, *Astr. Circ.*, SSSR, in press.
85. Bottinelli, L., Gouguenheim, L. 1964, *Ann. Astrophys.*, **27**, 885.
86. Terzian, Y. 1965, *Astrophys. J.*, **142**, 135.
87. Feibelman, W. A. 1966, *Astrophys. J.*, **144**, 1222.
88. Thompson, A. R., Krishnan, T. 1965, *Astrophys. J.*, **141**, 19.
89. Véron, P. 1965, *Ann. Astrophys.*, **28**, 391.
90. Dickel, H. R., Yang, K. S., Dickel, J. R. 1966, *Astrophys. J.*, **143**, 218.
91. Downes, D., Rinehart, R. 1966, *Astrophys. J.*, **144**, 937.
92. Mezger, P. G., Henderson, A. P. *Astrophys. J.*, in press.
93. Hollinger, J. P. 1965, *Astrophys. J.*, **142**, 609.
94. Downes, D., Maxwell, A. 1966, *Astrophys. J.*
95. Dravskih, A. F., *et al.* 1965, *Dokl. Akad. Nauk SSSR*, **163**, 332.
96. Sorochenko, R. L., Borozdich, E. V. 1965, *Dokl. Akad. Nauk SSSR*, **163**, 603.
97. Höglund, B., Mezger, P. G. 1965, *Science*, **150**, 339; 1966, *Astrophys. J.*
98. Burbidge, G. R., Gould, R. J., Pottasch, S. R. 1963, *Astrophys. J.*, **138**, 945.
99. Osterbrock, D. E. 1965, *Astrophys. J.*, **142**, 1423.
100. O'Dell, C. R. 1966, *Astrophys. J.*, **143**, 168.
101. Hjellming, R. M. 1966, *Astrophys. J.*, **143**, 420.
102. Goldberg, L. 1966, *Astrophys. J.*, **144**, 1225.
103. Griem, H. R., Kolb, A. C., Shen, K. I. 1959, *Phys. Rev.*, **116**, 4; Griem, H. R., 1960, *Astrophys. J.*, **132**, 883; 1967, *Astrophys. J.*, **147**, 1092.
104. Palmer, P., Zuckerman, B. 1966, *Nature*, **209**, 1118.
105. Pfennig, H., Treftz, E., Vidal, C. R. 1966, *J. quantit. Spectrosc. radiat. Transfer.*, **6**, 557.
106. Minaeva, L. A., Sobelman, I. I., Sorochenko, R. L. 1967, *Astr. Zu.*, **44**, no. 4.
107. Lilley, A. E., Palmer, P., Penfield, H., Zuckerman, B. 1966, *Harvard Coll. Obs. Announc. Cards*, 121.
107a. Palmer, P. 1966, *Harvard Coll. Obs. Announc. Cards*, 123.
108. Axford, W. I. 1964, *Astrophys. J.*, **139**, 761.
109. Axford, W. I. 1964, *Astrophys. J.*, **140**, 112; *Astr. J.*, **69**, 133.
110. Vandervoort, P. O. 1964, *Astrophys. J.*, **139**, 869, 889.
111. Vandervoort, P. O. 1965, *Astrophys. J.*, **142**, 507, 521.
112. Lasker, B. M. 1966, *Astrophys. J.*, **143**, 700.
113. Goldsworthy, F. A. 1966, *J. Inst. Math. Appl.*, **2**, part 4.
114. Mathews, W. G. 1965, *Astrophys. J.*, **142**, 1120.
115. Mathews, W. G. 1966, *Astrophys. J.*, **144**, 206.
116. Riegel, K. W. 1966, Dissertation, Univ. Maryland, U.S.A.
116a. Ampel, R., Iwaniszewska, C. 1962, *Bull. astr. Obs. Univ. N. Copernicus*, no. 27.
117. Kovar, N. S. 1964, *Astr. J.*, **69**, 548.
118. Gidalevich, E. Y. 1965, *Astr. Zu.*, **42**, 932; 1966, **43**, 553, 1018.
119. Gidalevich, E. Y. 1965, *Astrofizika*, **1**, 475; 1966, **2**, 286.
120. Gidalevich, E. Y. 1966, *Izv. astrofiz. Inst. Akad. Nauk Kazah. SSR*, **7**, 3, 12.
121. Baines, M. J., Williams, I. P., Asebiome, A. S. 1965, *Mon. Not. R. astr. Soc.*, **130**, 63.
122. Gershberg, R. E., Sheglov, P. V. 1963, *Astr. Zu.*, **41**, 425.
123. Sheglov, P. V. 1963, *Astr. Circ.*, SSSR, no. 266.
124. Sheglov, P. V. 1967, *Astr. Zu.*, **45**, in press.
125. Flynn, F. H., Ring, J. 1965, *Mon. Not. R. astr. Soc.*, **130**, 1.

126. Flynn, E. H. 1965, *Mon. Not. R. astr. Soc.*, **130**, 9.
127. O'Dell, C. R., Hubbard, W. B., Peimbert, M. 1966, *Astrophys. J.*, **143**, 743.
128. Dorschner, J., Gürtler, J. 1966, *Astr. Nachr.*, **289**, 57.
129. Dorschner, J. *Astr. Nachr.*, in press.
130. Hall, R. C. 1964, *Astrophys. J.*, **139**, 759.
131. Dibai, A. E., Esipov, V. F. 1965, *Astr. Zu.*, **42**, 281.
132. Dibai, E. A. 1966, *Astr. Zu.*, **43**, 903.
133. Dibai, E. A. 1966, *Astr. Zu.*, **43**, 52.
134. Khachikjan, E. E., Parsamian, E. S. 1964, *Soobšč. Bjurak. Obs.*, **35**, 71.
135. Khachikjan, E. E., Parsamian, E. S. 1965, *Astrofizika*, **1**, 417.
136. Parsamian, E. S. 1965, *Izv. Akad. Nauk Armjan. SSR*, **18**, no. 2.
137. Vardanian, R. A. 1965, *Astrofizika*, **1**.
138. Pikelner, S. B. 1965, *Astr. Zu.*, **42**, 515.
139. Williams, R. E. 1965, *Astr. J.*, **142**, 314.
140. Lenchek, A. M. 1964, *Ann. Astrophys.*, **27**, 219.
141. Alexander, J. K., Stone, R. G. 1965, *Astrophys. J.*, **142**, 1327.
142. Ellis, G. R. A., Hamilton, P. A. 1966, *Astrophys. J.*, **143**, 227; **146**, 78.
143. Feast, M. W. 1964, *Mon. Not. R. astr. Soc.*, **127**, 195; **128**, 327.
144. Dickel, H. R. 1965, *Astrophys. J.*, **141**, 1306.
145. Mathis, J. S. 1965, *Publ. astr. Soc. Pacif.*, **77**, 189.
146. Baade, W., Arp, H. 1964, *Astrophys. J.*, **139**, 1027.
147. Arp, H. 1964, *Astrophys. J.*, **139**, 1045.
148. Courtès, G., Gruvellier, P. 1965, *Ann. Astrophys.*, **28**, 683.
149. Véron, P., Souvayre, A. 1965, *Ann. Astrophys.*, **28**, 698.
150. Sěrsic, J. L. 1966, *Z. Astrophys.*, **64**, 212.
151. Shane, W. W., Bieger-Smith, G. P. 1966, *Bull. astr. Inst. Netherl.*, **18**, 263.
152. Pronik, I. I. 1965, *Astr. Zu.*, **42**, 923.
153. Burton, W. B. 1966, *Bull. astr. Inst. Netherl.*, **18**, 247.
154. Lindblad, P. O. 1966, *Bull. astr. Inst. Netherl. Suppl.*, **1**, 77, 177.
155. Lindblad, P. O. 1966, *Bull. astr. Inst. Netherl.*, in press.
156. Kerr, F. J., Hindman, J. V. 1966, in Symposium on *Radio and Optical Studies of the Galaxy*, Mt Stromlo, Australia.
157. Van Woerden, H. 1967, *Bull. astr. Inst. Netherl.*, in press.
158. Pskovsky, Y. P. 1965, *Astr. Zu.*, **42**, 1184.
159. Pavlovskaya, E. D., Sharov, A. S. 1966, *Astr. Zu.*, **43**, 40.
160. Gosachinsky, I. V. 1966, *Astr. Zu.*, **43**, 284.
161. Elwert, G., Hablick, D. 1965, *Z. Astrophys.*, **61**, 273.
162. Smith, G. P. 1963, *Bull. astr. Inst. Netherl.*, **17**, 203.
163. Prata, S. W. 1964, *Bull. astr. Inst. Netherl.*, **17**, 511.
164. Van Woerden, H., Hack, M., Blaauw, A. 1967, *Bull. astr. Inst. Netherl.*, in press.
165. Locke, J. L., Galt, J. A., Costain, C. H. 1964, *Astrophys. J.*, **139**, 1066, 1071.
166. Garzoli, S. L., Varsavsky, C. M. 1966, *Astrophys. J.*, **145**, 79.
167. Habing, H. J. 1966, *Bull. astr. Inst. Netherl.*, **18**, 323.
168. Raimond, E. 1966, *Bull. astr. Inst. Netherl.*, **18**, 191; Ibid. *Suppl.*, **1**, 33.
169. Davies, R. D., Tovmassian, H. M. 1963, *Mon. Not. R. astr. Soc.*, **127**, 45, 61.
170. Girnstein, H. G., Rohlfs, K. 1964, *Z. Astrophys.*, **59**, 83.
171. Bystrova, N. V., Gosachinsky, I. V., Ryzhkov, N. F. 1966, *Izv. glav. astr. Obs. Pulkove*, **24**, 130.
172. Kaper, H. G., Smits, D. W., Schwarz, U. J., Takakubo, K., Van Woerden, H. *Bull. astr. Inst. Netherl.*, in press.
173. Goldstein, S. J. 1964, *Astrophys. J.*, **140**, 802.
174. Rasiwala, M. 1965, *C. r. Acad. Sci.*, Paris, **261**, 3291, 3755.
175. Ariskin, V. I. 1965, *Astr. Zu.*, **42**, 939.
176. Takakubo, K. 1967, *Bull. astr. Inst. Netherl.*, **19**, 125.
177. Takakubo, K., Van Woerden, H. 1966, *Bull. astr. Inst. Netherl.*, **18**, 488.
178. Clark, B. G. 1965, *Astrophys. J.*, **142**, 1398.

179. Van Woerden, H. 1964, *A. Rep. Mt Wilson and Palomar Obs.*, Carnegie Inst. Year Book, **63**, 23.

180. Feast, M. W., Shuttleworth, M. 1965, *Mon. Not. R. astr. Soc.*, **130**, 245.

181. Wilkens, H. 1965, La Plata, Ser. Circ. no. 17.

182. Buscombe, W., Kennedy, P. M. 1966, *Mt Stromlo Mimeogram* no. 9; *Publ. astr. Soc. Pacif.*, **78**.

183. Livingston, W. C., Lynds, C. R. 1964, *Astrophys. J.*, **140**, 818.

184. Hobbs, L. M. 1965, *Astrophys. J.*, **142**, 160.

185. Smith, A. 1964, *Mon. Not. R. astr. Soc.*, **127**, 347.

186. Morton, D. C., Spitzer, L. 1966, *Astrophys. J.*, **144**, 1.

187. Muller, C. A., Raimond, E., Schwarz, U. J., Tolbert, C. R. 1966, *Bull. astr. Inst. Netherl. Suppl.*, **1**, 213.

188. Blaauw, A., Tolbert, C. R. 1966, *Bull. astr. Inst. Netherl.*, **18**, 405.

189. Hulsbosch, A. N. M., Raimond, E. 1966, *Bull. astr. Inst. Netherl.*, **18**, 413.

190. Dieter, N. H. 1965, *Astr. J.*, **70**, 552.

191. Oort, J. H. 1966, *Bull. astr. Inst. Netherl.*, **18**, 421.

192. Shklovsky, I. S. 1967, *Astr. Zu.*, **44**.

193. Varshalovich, D. A. 1965, *Astr. Zu.*, **42**, 557; 1966, *Zu. eksp. teor. Fiz. Lett.*, **4**, 180; *Zu. eksp. teor. Fiz.*, no. 1.

194. Scheffler, H. 1966, *Z. Astrophys.*, in press.

195. Field, G. B., Saslaw, W. C. 1965, *Astrophys. J.*, **142**, 568.

196. Kogure, T. 1965, *Publ. astr. Soc. Japan*, **17**, 385.

197. Zimmerman, H. 1967, *Astr. Nachr.*; 1966, Proc. of 14. Int. Astr. Symp., Liege.

198. Kaplan, S. A., Klimishin, I. A. 1964, *Astr. Zu.*, **41**, 657.

199. Kaplan, S. A., Podstrigach, T. S. 1965, *Astr. Zu.*, **42**, 552.

200. Kaplan, S. A., Klimishin, I. A. 1964, *Astr. Zu.*, **41**, 274.

201. Pikelner, S. B. 1967, *Astr. Zu.*, **44**, in press.

202. Shuter, W. L. H., Verschuur, G. L. 1964, *Mon. Not. R. astr. Soc.*, **127**, 387.

203. Marx, S. 1965, *Astr. Nachr.*, **288**, 155.

204. Rohlfs, K. 1966, *Z. Astrophys.*, **63**, 207.

205. Takayanagi, K. 1964, *Joint Inst. Lab. Astrophys. Publ.*, no. 19.

206. Zimmermann, H. 1964, *Astr. Nachr.*, **288**, 95.

207. Kaplan, S. A. 1965, in *Dynamika Zvezdnykh Sistem i Fyzika Gazovykh Tumannostej*, Alma-Ata, U.S.S.R.

208. Dalgarno, A., Rudge, M. R. H. 1964, *Astrophys. J.*, **140**, 800.

209. Raich, J. C., Good, R. H., Jr. 1964, *Astrophys. J.*, **139**, 1004.

210. Spitzer, L., Dressler, K., Upson, W. L. 1964, *Publ. astr. Soc. Pacif.*, **76**, 387.

211. Gosachinsky, I. V. 1965, *Astr. Zu.*, **42**, 929.

212. Dorschner, J., Gürtler, J., Schmidt, K. H. 1965, *Astr. Nachr.*, **288**, 149.

212a. Knaap, H. F. P., van den Meijdenberg, C. J. N., Beenakker, J. J. M., Van de Hulst, H. C. 1966, *Bull. astr. Inst. Netherl.*, **18**, 256.

213. Stecher, T. P., Williams, D. A. 1966, *Publ. astr. Soc. Pacif.*, **78**, 76.

214. Münch, G. 1964, *Astrophys. J.*, **140**, 107.

215. Shklovsky, I. S. 1966, *Astr. Cirk.*, SSSR, no. 364.

216. Field, G. B., Hitchcock, J. L. 1966, *Phys. Rev. Lett.*, **16**, 817.

217. Field, G. B., Herbig, G., Hitchcock, J. L. 1966, *Astrophys. J.*

218. Robinson, B. J., Gardner, F. F., Van Damme, K. J., Bolton, J. G. 1964, *Nature*, **202**, 989.

219. Goldstein, S. J., Gundermann, E. J., Penzias, A. A., Lilley, A. E. 1964, *Nature*, **203**, 65.

220. McGee, R. X., Robinson, B. J., Gardner, F. F., Bolton, J. G. 1965, *Nature*, **208**, 1193.

221. Bolton, J. G., Gardner, F. F., McGee, R. X., Robinson, B. J. 1964, *Nature*, **204**, 30.

222. Weaver, H. F., Williams, D. R. W., Dieter, N. H., Lum, W. T. 1965, *Nature*, **208**, 29.

223. Weinreb, S., Meeks, M. I., Carter, J. C., Barrett, A. H., Rogers, A. G. E. 1965, *Nature*, **208**, 440.

224. Perkins, F., Gold, T., Salpeter, E. E. 1966, *Astrophys. J.*, **145**, 361.

225. Shklovsky, I. S. 1966, *Astr. Circ.*, USSR, no. 372.
225a. Cook, A. H. 1966, *Nature*, **210**, 611.
226. Goss, M. M., Spinrad, H. 1966, *Astrophys. J.*, **143**, 989.
227. Carroll, T. O., Salpeter, E. E. 1966, *Astrophys. J.*, **143**, 609.
228. Nguen-Hun-Doan 1965, *C. r. Acad. Sci.*, Paris, **260**, 2715.
229. Stoeckly, R., Dressler, K. 1964, *Astrophys. J.*, **139**, 240.
230. Hanasz, J. 1965, *Acta Astr.*, **15**, 31.
231. Romashin, G. S. 1965, *Trudy astr. Obs. Leningr. Univ.*, **22**, 104; 1966, **23**.
232. Hutchings, J. B. 1966, *Mon. Not. R. astr. Soc.*, **131**, 299.
233. Malville, J. M. 1964, *Astrophys. J.*, **139**, 198.
234. Herzberg, G. 1966, in IAU Symposium no. 31, *Radio Astronomy and the Galactic System*, Noordwijk, Netherlands, in press.
235. Roberts, M. S. 1966, *Astrophys. J.*, **144**, 639.
236. Gottesman, S. T., Davies, R. D., Reddish, V. C. 1966, *Mon. Not. R. astr. Soc.*, **133**, 359.
236a. McGee, R. X., Milton, J. A. 1966, *Austr. J. Phys. Astrophys. Suppl.*, no. 2.
237. Hogg, D. E. 1966, *Astrophys. J.*, **144**, 819.
238. Davies, R. D. 1964, *Mon. Not. R. astr. Soc.*, **128**, 173.
239. Brosche, P. 1966, *Z. Astrophys.*, **64**, 1.
240. Heiles, C. 1964, *Astrophys. J.*, **140**, 470.
241. Sheglov, P. V. 1967, *Astr. Zu.*, **44**, in press.
242. Bernstein, I. B., Kulsrud, R. M. 1965, *Astrophys. J.*, **142**, 479.
243. Kurlsrud, R. M., Bernstein, I. B., Kruskal, M., Fanucci, J., Ness, N. 1965, *Astrophys. J.*, **142**, 491.
244. Shklovsky, I. S. 1966, *Astr. Zu.*, **43**, 10.
245. Kardashov, N. S. 1964, *Astr. Zu.*, **41**, 807.
246. Khachikian, E. E., Bërngen, F. 1966, *Astr. Nachr.*
247. Michelkin, E. G., Jacusheva, K. G. 1966, *Dokl. Akad. Nauk SSSR*, **168**, 304; *Astr. Zu.*, **43**, 543.
248. Westerlund, B. E., Mathewson, D. S. 1966, *Mon. Not. R. astr. Soc.*, **131**, 371.
249. Martel, L., Martel, M.-Th. 1964, *Ann. Astrophys.*, **27**, 203.
250. Lodén, L. O. 1965, *Stockholm Obs. Ann.*, **22**, no. 8.
251. Serkowski, K. 1965, *Acta Astr.*, **15**, 78.
252. Serkowski, K. 1965, *Astrophys. J.*, **141**, 1340.
253. Dombrovski, V. A., Gagen-Torn, V. A. 1964, *Trudy astr. Obs. Leningr. Univ.*, **20**, 75.
254. Pronik, I. I. 1966, *Astr. Zu.*, **43**, 291.
255. Visvanathan, N. 1966, *Mon. Not. R. astr. Soc.*, **132**, 423.
256. Zwicky, F. 1964, *Astrophys. J.*, **139**, 1394.
257. Mathewson, D. S., Milne, D. E. 1965, *Austr. J. Phys.*, **18**, 635.
258. Mathewson, D. S., Broten, N. W., Cole, D. J. 1966, *Austr. J. Phys.*, **19**, 93.
259. Seielstad, G. A., Morris, D., Radhakrishnan, V. 1964, *Astrophys. J.*, **140**, 53.
260. Morris, D., Berge, G. L. 1964, *Astrophys. J.*, **139**, 1388.
261. Gardner, F. F., Davies, R. D. 1965, *Austr. J. Phys.*, **19**, 129.
262. Bologna, J. M., McClain, E. F., Rose, W. K., Sloanaker, R. M. 1965, *Astrophys. J.*, **142**, 106.
263. Maltby, P. 1965, *Astrophys. J.*, **142**, 621.
264. Davies, R. D., Shuter, W. L. H., Slater, C. H., Wild, P. A. T. 1963, *Mon. Not. R. astr. Soc.*, **126**, 353.
265. Davies, R. D., Shuter, W. L. H. 1963, *Mon. Not. R. astr. Soc.*, **126**, 369.
266. Werschuur, G. L. 1965, Thesis, Univ. of Manchester.
267. Stępień, K. 1964, *Acta Astr.*, **14**, 81.
268. Hornby, J. M. 1966, *Mon. Not. R. astr. Soc.*, **133**, 213.
269. Woltjer, L. 1965, In *Stars and Stellar Systems*. Vol. V, Univ. of Chicago Press, Chicago.
270. Wentzel, D. G. 1966, *Astrophys. J.*, **145**, 595.

271. Lin, C. C., Shu, F. H. 1964, *Astrophys. J.*, **140**, 646; 1966, *Proc. nat. Acad. Sci. Am.*, **55**, 229.
272. Toomre, A. 1964, *Astrophys. J.*, **139**, 1217.
273. Faller, A. J. 1964, *Astr. J.*, **69**, 138.
274. Lindblad, B. 1963, *Stockholm Obs. Ann.*, **22**, 5.
275. Julian, W. H., Toomre, A. 1966, *Astrophys. J.*
276. Hunter, C. 1965, *Mon. Not. R. astr. Soc.*, **129**, 321.
277. Goldreich, P., Lynden-Bell, D. 1965, *Mon. Not. R. astr. Soc.*, **130**, 97, 125.
278. Lynden-Bell, D. 1966, *Observatory*, **86**, 57.
279. Richter, G., Wallis, G. *Mitt. Heinrich-Hertz-Inst. Berlin-Adlershof*, in press.
280. Piddington, J. H. 1964, *Mon. Not. R. astr. Soc.*, **128**, 345.
281. Danby, J. M. A. 1965, *Astr. J.*, **70**, 501.
282. Freeman, K. C. 1965, *Mon. Not. R. astr. Soc.*, **130**, 183; 1966, **133**, 47; **134**, 1, 15.
282a. Freeman, K. C., Mestel, L. 1966, *Mon. Not. R. astr. Soc.*, **134**, 37.
283. Prendergast, K. H. 1964, *Astr. J.*, **69**, 147.
284. Pikelner, S. B. 1965, *Astr. Zu.*, **42**, 3.
285. Lyttlton, R. A., Bondi, H. 1964, *Mon. Not. R. astr. Soc.*, **128**, 207.
286. Vorontsov-Velyaminov, B. A. 1964, *Astr. Zu.*, **41**, 814.
287. Gershberg, R. E. 1965, *Astr. Zu.*, **42**, 330.
288. Zasov, A. V. 1965, *Astr. Zu.*, **42**, 959; 1967, **44**.
289. Tsitovich, V. N. 1966, *Usp. fiz. Nauk*, **89**, 89.
289a. Ginzburg, V. L. 1965, *Astr. Zu.*, **42**, 1129.
290. Parker, E. N. 1965, *Astrophys. J.*, **142**, 584.
291. Kowal, Ch. T. 1964, *Astr. J.*, **69**, 757.
292. Gunn, J. E., Peterson, B. A. 1965, *Astrophys. J.*, **142**, 1633.
293. Bahcall, J. N., Salpeter, E. E. 1965, *Astrophys. J.*, **142**, 1677.
294. Field, G. B., Solomon, P. M., Wampler, E. J. 1966, *Astrophys. J.*, **145**, 351.
295. Davies, R. D., Jennison, R. C. 1964, *Mon. Not. R. astr. Soc.*, **128**, 123, 132.
296. Ginzburg, V. L., Ozernoy, L. M. 1965, *Astr. Zu.*, **42**, 943.
297. Gould, R. J., Ramsay, W. 1966, *Astrophys. J.*, **144**, 587.
298. Ginzburg, V. L., Syrovatski, S. I. 1964, *Usp. fiz. Nauk*, **84**, 201; *Zu. eksp. teor. Fiz.*, **46**, 1865.
299. Tucker, W. H., Gould, R. J. 1966, *Astrophys. J.*, **144**, 244.
300. Field, G. B., Henry, R. C. 1964, *Astrophys. J.*, **140**, 1002.
301. Fazio, G. G., Stecker, F. W., Wright, J. P. 1966, *Astrophys. J.*, **144**, 611.
302. Penzias, A. A., Wilson, R. W. 1965, *Astrophys. J.*, **142**, 419.
303. Roll, P. G., Wilkinson, D. T. 1966, *Phys. Rev. Lett.*, **16**, 405.
304. Dicke, R. H., Peebles, J. E., Roll, P. G., Wilkinson, D. T. 1965, *Astrophys. J.*, **142**, 414.
305. Sunyayev, R. A. 1966, *Astr. Zu.*, **43**, 1242.
306. Greenstein, J. L., Schmidt, M. 1964, *Astrophys. J.*, **140**, 1.
307. Schmidt, M., Matthews, T. A. 1964, *Astrophys. J.*, **139**, 781.
308. Schmidt, M. 1965, *Astrophys. J.*, **141**, 1295.
309. Osterbrock, D. E., Parker, S. A. R. 1966, *Astrophys. J.*, **143**, 268.
310. Lynds, C. R., Sandage, A. R. 1963, *Astrophys. J.*, **137**, 1005.
311. Shklovsky, J. S. 1964, *Astr. Zu.*, **41**, 801.
312. Oke, J. B. 1965, *Astrophys. J.*, **141**, 6.
313. Lynds, C. R., Hill, S. J., Heere, K., Stockton, A. N. 1966, *Astrophys. J.*, **144**, 1244.
314. Greenstein, J. L. 1964, *Astrophys. J.*, **140**, 666.
315. Bahcall, J. H., Peterson, B. A., Schmidt, M. 1966, *Astrophys. J.*, **145**, 369.
316. Stockton, A. N., Lynds, G. R. 1966, *Astrophys. J.*
317. Dibai, E. A., Pronik, V. I. 1965, *Astrofizika*, **1**, 78.
318. Dibai, E. A., Pronik, V. I. 1967, *Astr. Zu.*, **44**, in press.
319. Dibai, E. A., Esipov, V. F., Pronik, V. I. 1967, *Astr. Zu.*, **44**, in press.
320. Pronik, V. I. 1966, *Astr. Zu.*, **43**, 1207.
321. Dibai, E. A., Esipov, V. F. 1967, *Astr. Zu.*, **44**, no. 1.
322. Ozernoy, L. M. 1964, *Astr. Zu.*, **41**, 494.

323. Simon, R. 1964, *Ann. Astrophys.*, **27**, 191; 1965, **28**, 40, 625.
324. Strittmatter, P. A. 1966, *Mon. Not. R. astr. Soc.*, **131**, 491; **132**, 359.
325. Gould, R. J. 1964, *Astrophys. J.*, **140**, 638.
326. Simoda, M., Kikuchi, S., Unno, W. 1966, *Publ. astr. Soc. Japan*, **18**, 31.
327. McNally, D. 1964, *Astrophys. J.*, **140**, 1088.
328. Field, G. B. 1965, *Astrophys. J.*, **142**, 531.
329. Hunter, J. H. 1966, *Mon. Not. R. astr. Soc.*, **133**, 239.
330. Sofia, S. 1966, *Astrophys. J.*, **145**, 84.
331. Dibai, E. A., Kaplan, S. A. 1964, *Astr. Zu.*, **41**, 652.
332. Kossatsky, K. 1967, *Acta Astr.*
333. Arny, T., Weymann, R. 1964, *Astr. J.*, **69**, 530.
334. Auluck, F. C., Kothari, D. S. 1965, *Z. Astrophys.*, **63**, 9.
335. Nakano, T. 1966, *Progress theor. Phys.*, Osaka, **36**, no. 3.
336. Grzedzielski, S. 1966, *Mon. Not. R. astr. Soc.*, **134**, 109.
337. Dorschner, J., Gürtler, J., Schmidt, K. H. 1965, *Astr. Nachr.*, **288**, 159.
338. Schatzman, E., Magnan, C. 1965, *C. r. Acad. Sci.*, Paris, **260**, 6289; **261**, 3294.
339. Lortet-Zuckerman, M. C. 1965, *C. r. Acad. Sci.*, Paris, **260**, 6534.
340. Zeldovich, V. B., Podurets, M. V. 1965, *Astr. Zu.*, **42**, 963.
341. Muller, R. H., Parker, E. N. 1964, *Astrophys. J.*, **140**, 50.
342. Spitzer, L., Saslaw, W. C. 1966, *Astrophys. J.*, **143**, 400.
343. Lynden-Bell, D. 1964, *Astrophys. J.*, **139**, 1195.
344. Peebles, P. F. 1965, *Astrophys. J.*, **142**, 1317.
345. Osernoy, L. M. 1966, in *Variable Stars and Stellar Evolution*, Moscow.
346. Zeldovich, V. B., Novikov, I. D. 1966, IAU Symposium no. 29, *Instability Phenomena in Galaxies*, Bjurakan Obs.
347. Doroshkevich, A. G., Zeldovich, V. B., Novikov, I. D. 1967, *Astr. Zu.*, **44**, in press.

INTERSTELLAR GRAINS AND RELATED TOPICS

(prepared by J. Mayo Greenberg)

Diffuse Nebulae

A. Elvius and J. S. Hall have made observation of the color and polarization of the reflection nebulae NGC 2068, NGC 7023 and the Merope nebula in three spectral regions (**1, 2**). The polarization was found always to increase with wavelength and its direction was usually radial with respect to the illuminating star except in the region south of Merope where the polarization tended to follow the filamentary structure. The colors of the nebulae NGC 2068 and NGC 7023 were bluer than the (reddened) light of the illuminating stars. For Merope the color of the nebula was found to increase with distance from Merope. O'Dell (**3**) and Roark and Greenberg (**4, 5**) have reported on observations on Merope and have confirmed that the nebula is bluer than the star and that it becomes redder with increasing distance from the star. A preliminary analysis by O'Dell (**3**) indicated that the nebular particles are not small metallic or dielectric ones. Greenberg and Roark (**4, 5**) using a more detailed analysis found that the nebular particles are probably intermediate sized dielectric grain, perhaps slightly larger than the average of those which give rise to interstellar extinction. Greenberg and Roark found that the density of grains in Merope is at least ten times greater than in a standard interstellar cloud and one thousand times greater than in the average interstellar medium.

Dorschner and Gürtler (**6**) have made a statistical analysis of 196 reflection nebulae on the charts of the Palomar Sky Survey. They consider a possible cosmogonic relation between stars and dust clouds.

Minin (**7**) considered spherically symmetric cloud models and by comparing theoretical with observational results on polarization and brightness for several nebula concluded that if the scattering particles are dielectric they are smaller than $a = 0{\cdot}07$ micron.

Vanysek (9) made photoelectric measurements of NGC 7023 including the ultraviolet region. By comparing his observations and those of others with a theoretical spherical cloud model he found that the scattering particles are slightly larger than those commonly used to account for the interstellar extinction. The observations show the blue color of the nebula returning to the star color (becoming redder) at increasing distances from the star.

R. C. Hall (10) has also found this color variation in Hubble's Variable Nebula NGC 2261 and has shown from polarization measurements of the wavelength dependence that the light of NGC 2261 is due to a reflection process rather than to synchrotron emission.

A number of investigations of dust in H II region have been made. O'Dell and Hubbard (11) have observed the strength and energy distribution of the continuous radiation of the Orion nebula and corrected for the expected atomic component. The effective gas-to-dust ratio appeared to decrease sharply in the outer regions. Possible interpretation based on effects of radiation pressure on and evaporation of grains have been discussed. Further analyses of the physical interactions of the dust in H II regions have been made (12, 13, 14). The possibility of accounting for the anomalous reddening law in the Trapezium have been examined. It has been shown by Greenberg (15) that no minor modification of normal size distribution of dielectric grain can account for this anomaly. It has been shown that the grains must be about ten times larger than the average in interstellar space.

Gürtler (16) has estimated the value of the albedo and its wavelength dependence for the contribution to the nebular continuum in the Orion nebula by the scattering off dust particles.

Dorschner (17) has compared sizes of reflection nebulae around B stars with the expected radii of H II regions and finds a discrepancy with the radii of the Strömgren spheres.

Ahmed, Lawrence and Reddish (18) examined the nebulosity in the Pleiades and found evidence for wavelength independent absorbing matter.

Boyce (19) showed that ignoring effect of interstellar reddening in the Orion nebula can lead to anomalous results.

Extinction and the Distribution of Interstellar Grains

Some very interesting results have been obtained on extinction in the infrared (20, 21, 22, 23, 24). The review articles by Johnson (20, 21) cover a wide range of observations. The most peculiar extinction curves appear to be for φ Persei and for α Leonis the latter showing gray extinction from the far infrared out to the near ultraviolet. For many stars there is a sharp rise in the extinction up to the M color and what may be referred to as a knee in the extinction curve in the infrared. Comparison of extinction made by variable-extinction methods, cluster diameters methods and the color-differences method was made. Johnson (22) and Johnson and Mendoza (23) found essential agreement between the color-difference method determination of $R = A_V/E_{B-V} > 3$ for M type supergiants and the cluster diameter value of $R = 5{\cdot}6$ in Perseus. Similar agreement has been found in Cepheus. In any case a wide variety of values of R have been obtained with a value of $R = 3$ appearing to be an approximate lower bound. Further discussion of this question of the variability of extinction laws in the giving rise to values of R from 3 (in most regions) to 7 (in the Orion nebula region) has been given by Borgman (24).

Observations of interstellar reddening in Cygnus and Perseus have been made by Nandy (25, 26, 27). The results are based on spectrophotometric measurements of O and early B type stars using spectra taken with the Schmidt telescope of the Royal Observatory in conjunction with an objective prism. In both regions the mean reddening curve can be represented by two straight lines which intersect near λ 4300 Å but the ratio of the slope of the ultraviolet part is smaller by 30% in Perseus than that in Cygnus. Schalén (28) discussed the method of using spectrophotometric measurements of objective prism plates for making detailed studies of the wavelength dependence of interstellar absorption.

Underhill and Walker (29) compared the measured spectral intensity distribution for O stars with those from model atmosphere calculations and found that the mean extinction curve showed a slope discontinuity at λ 4445 Å.

Whiteoak (30) used a photoelectric spectrum scanner to find wavelength dependence of extinction at 30 wavelengths in the spectral region 3448 Å $<\lambda \leqq$ 10 500 Å. Stars in Perseus, Monoceros, Cygnus and Cepheus were investigated. Perseus and Cepheus are similar. Stars inside and outside the Orion nebula give the usual difference in reddening. Monoceros lies between Orion and the others. No correlation of variation with galactic structure was found.

Guthrie and Nandy (31) combine their extinction results with those of Mendoza (32) to arrive at a mean extinction law in the Pleiades, similar to that in Cygnus. They found a value of $R = 3{\cdot}3$ as against the value $R = 4{\cdot}2$ given by Mendoza. In any case the value of R is greater than the average as might be expected in the regions of hot stars.

Wampler (33) discussed the curvature (or lack of curvature) of the reddening line in the *UBV* color-color plot.

Becker (34) showed that in studying the color magnitude diagrams of clusters and associations the value of $R = A/E_{(B-V)}$ is increased by including nonphysical members. Becker showed that this spuriously large value of R is produced by the distance effect in the reddening of field stars and that statistical methods may be used to distinguish nonphysical members which affect the scatter in R.

Krzeminski and Serkowski (35) applied the variable extinction method to members of the cluster Stock 2 and obtained the ratio of total to selective extinction $R = A_V/E_{(B-V)} = 3{\cdot}4 \pm 0{\cdot}1$.

Lodén (36) has estimated the value of $R = 3{\cdot}9 = A_V/E_{(B-V)}$ in the region of the association Pup I.

Neckel (37, 38) using the galaxy count method found a value of $R = A_V/E_{(B-V)} = 4$ and indicated the possibility that there may be an appreciable amount of neutral absorbing material at high galactic latitudes whose relative abundance perhaps increases with increasing distance from the galactic plane.

Glushneva (39) showed by spectrophotometry of the pair ϵ Per and ρ Per out to $\lambda = 3000$ Å that the absorption follows a λ^{-1} law and does not begin to bend over as indicated by Stebbins and Whitford.

Boggess and Borgman (40) extended the extinction measurements to the wavelength bands centered at 2600 and 2200 Å. The extinction continues to rise more or less linearly out to the farthest measurement.

Stecher (41, 42) observed extinction in the far ultraviolet out to $\lambda = 1200$ Å and found not only the generally tendency to increasing extinction with decreasing wavelength but also a possible detailed hump-like feature at $\lambda^{-1} = 4{\cdot}4\ \mu^{-1}$.

Oja (43, 44) has made spectrophotometric studies of over 2500 stars in the area $80° < l^{\mathrm{I}} < 100°\ |\ b^{\mathrm{I}}\ | < 3°$. The absorbing clouds appear to concentrate in the spiral arms ('0' and '+ I') i.e., concentrated within distances less than 1 kpc and some near 3 kpc respectively. Diameters of some clouds near the Sun have been estimated to have 40 pc diameter. Stars brighter than $m_{\mathrm{pg}} = 14$ in a one-degree field in Cassiopeia are under investigation (44).

Ekedahl (45) found a dust cloud with a fairly sharp border in Kapteyn Selected Area 4. Absorption is 0·4 magnitude in V at 400 pc with no significant absorption outside the cloud. T. Elvius has studied the distribution of stars and interstellar dust clouds in a Milky Way field around the cluster NGC 7654.

Ljunggren (46, 47) found no continuous absorbing layer toward the North Galactic Pole. The photoelectric and spectrophotometric investigation showed interstellar reddening to set in at a distance of about 200–250 pc with average absorption in the layer of about 0·4 magnitude in *V*. Sjögren (**48**) has investigated absorption in Sel. Area 8 ($l^{II} = 125°$, $b^{II} = -2°$) and has discussed available polarization data in this area. Absorbing matter near the South Galactic Pole is being investigated by P. I. Erikson.

Scheffler (**49**) derived frequency distributions for color excesses of stars with $|z| < 75$ pc and six distance intervals from data on 4700 stars. For distance less than 1·3 kpc, he obtained statistically = 5 clouds/kpc. From 1·3 to 4 kpc the average absorption (in $l^{II} = 160° \ldots 360° \ldots 20°$) was only about one fifth that for $r < 1{\cdot}3$ kpc.

From data of about 4700 stars and galactic clusters Neckel (**50**) has derived a spatial distribution of absorbing matter. It was shown that for $150° \leqslant l^{II} \leqslant 330°$ the absorption is generally smaller than in the remaining Milky Way. Within the galactic plane the dust is concentrated in clouds ranging from 100 to 1000 pc in diameter. Neckel derives an absorption layer of about $\beta = 40$ pc. Elsässer, Neckel and Scheffler (**51**) have reinvestigated insterstellar reddening and its relationship to spiral structure and disagree with conclusions of Isserstedt and Schmidt-Kaler (**52**).

Straizys (**53**) has found that narrow band magnitudes at 4000 Å and 4500 Å are indispensable in two-index diagrams for a single valued determination of interstellar absorption for the whole range of spectral types.

Grigor'eva (**54**) showed that absorption in the brighter portion of the region $117° \leqslant l^{II} < 124°$ varies between $1^m\!.7$ for 1700 pc to $2^m\!.4$ for 4500 pc and in the darker portion from $2^m\!.6$ to $3^m\!.2$. The study was based on 2816 O–B stars investigated by Brodskoya and Grigor'eva.

Herzog (**55**) has found in the Palomar Sky Survey plates three cases for direct measurement of absorption in intergalactic bridges.

Gosachinskii (**56**) showed that the dense cloud obscuring much of the Omega Nebula (HGC 6618) is associated with a reduced density of neutral hydrogen atoms as inferred from observations of the 21 cm line in absorption. Schmidt (**57**) has studied the correlation between dust and gas in Andromeda.

The problem of correlation of gas and dust is complicated by the possible existence of undetectable molecules which may form on interstellar grains. The H_2 molecule and its formation on grains has therefore come under investigation. K. H. Schmidt (**58**) has found the probability of formation of H_2 molecules on dielectric grains to be sufficiently large that the indicated percentage of interstellar molecular hydrogen is as much as 80%. On the other hand Knaap, van den Meijdenberg, Beenakker and van de Hulst (**59, 60**) showed that at grain temperatures above 8°K the chance of simultaneous presence of two adsorbed H atoms on the grain is too small for efficient recombination. Stecher and Williams (**61**) have applied the theory of chemical reaction between gas atoms and dust atoms. Rates of formation and destruction of molecules on graphite flake grains are discussed for varying conditions in the interstellar medium.

Polarization and Galactic Magnetic Field Structure

Behr (**62**) found that the color dependence of polarization is not uniform in all parts of the sky. A weak dependence upon galactic longitude may be indicated however a better correlation can be found if the angle between the line of sight and the lines of the magnetic field (assuming Davis-Greenstein mechanism) is regarded.

Serkowski (**63**) found that the ratio of polarization in the yellow to that in the blue is generally greater than one. The smallest ratio of yellow to blue polarization was observed for the association VI Cygni and for the Star 55 Cygni both situated in the region of anomalous reddening in Cygnus. An attempt to associate both of these effects with grain size or character is natural. Correlation between polarization and reddening was found for cluster M 25 and for the association VI Cygni. For the stars observed the degree of polarization ellipticity was less than 0·05%. Serkowski, Chojnacki and Rucinski (64) found a correlation between the ratio of yellow to blue polarization with the ratio of total of selective extinction. Higher values of P_V/P_B are associated with increasing values of $A_V/E_{(B-V)}$. Krzeminski and Serkowski (35) find no correlation of polarization with reddening in the cluster Stock 2. The microscale 0·3 pc was found for the fluctuations in the polarizing medium.

Gehrels and Meltzer (65) made measures of the wavelength dependence of polarization of eight stars. Coyne and Gehrels (66) observed the wavelength dependence of polarization of 18 stars for seven wavelengths between $1{\cdot}05\mu \leqslant \lambda^{-1} \leqslant 3{\cdot}04\ \mu^{-1}$. The results conclusively indicate a lack of uniformity in the wavelength dependence of polarization, and a correlation with variability of the reddening curve. An overall average of this data has been made by Greenberg (15) and shown to be almost identical with that of Visvanathan (67) and furthermore both give a maximum at $\lambda^{-1} \simeq 2\ \mu^{-1}$. Coyne and Gehrels noted that variability of polarization position angle with wavelength. Visvanathan found no general distinction between polarization in H I and H II regions.

Gehrels and Silvester (68) found ten stars showing an appreciable wavelength dependence of position angles. These stars are mostly at distances greater than 0·6 kpc.

Lodén (69), by means of the polarization parameters of the stars of highest polarization in 19 sample fields, has studied correlation between polarization parameters and galactic coordinates. A first impression was that the degree of polarization statistically depends more upon the distance to the stars than upon any other parameter. The polarization has a flat maximum within the visual region.

Visvanathan (70) has observed polarization for 30 stars in the large Magellanic Cloud. The maximum value of P/A_V was found to be 0·07 and the average of P/A_V was 0·05 indicating a high degree of ordering and also that the cloud is seen nearly face on. Polarization alignment was uniform over distances of the order of kiloparsecs and seemed to be related to the cloud structure.

Pronik (71) analyzed polarization of starlight, the motions of interstellar Ca II clouds and Milky Way photographs and showed that these all indicate an irregular galactic magnetic field in the solar neighborhood.

Appenzeller (72) has made polarization measures for 320 stars in Cygnus and Orion. A system of 20 polarimetric standard stars was established. Using polarization measures Appenzeller (73) found the magnetic field in Cygnus to be rather uniform and parallel to the galactic plane and to $l = 45°$. In Orion aggregate about 140 pc below the plane the magnetic field is almost perpendicular to the plane.

Serkowski (74) found that the Mira variables have intrinsic polarization whose values are always greater in ultraviolet than in the blue and yellow spectral regions. For several stars the position angle is different in each spectral region, changing gradually from yellow to ultraviolet.

Theories of Interstellar Grains and Interpretations of Extinction and Polarization

The physical, chemical and optical properties of interstellar grains have received active attention and a wide range of detailed calculations have been made.

The problem of the source of nuclei upon which grains may grow has been considered by Kamijo (75) who calculated the rate of growth and ejection of small icy (mainly SiO_2) particles formed in the circumstellar space of M giant stars.

Friedemann and Schmidt (76) have reinvestigated the work of Hoyle and Wickramasinghe on the evolution of graphite particles in the atmospheres of carbon stars. They estimate that particles of some 10^{-6} cm are or are not able to escape in a time equal to the mean period of variation depending on whether the accommodation coefficient for carbon atoms is greater or less than 10^{-3}. Donn, Wickramasinghe, Hudson, and Stecher (77) have re-examined the Hoyle-Wickramasinghe mechanism for condensation of graphite in N stars. The formation of graphite in amorphous polycrystalline, needle, and plate shapes was considered. It is of interest to note in this connection that Mendoza and Johnson (78) assume that carbon stars are not affected by interstellar extinction because of their position in the galaxy with respect to the B and A stars observed by Eggen.

The growth and destruction of ice mantles on interstellar graphite grains has been calculated by Wickramasinghe (79, 80) who showed that the process of sputtering by the gas atoms in a heated cloud is to be considered as the principal destruction mechanism.

Light scattering and extinction by graphite spheres using Mie theory has been calculated for a wide variety of sizes and for various postulated optical properties of graphite (80, 81, 82, 83, 84, 15, 85). Nandy and Wickramasinghe (86) have shown that calculated extinction by graphite spheres (84) with radii $0{\cdot}05 - 0{\cdot}07\mu$ yield good fits with the observed reddening in Cygnus and Perseus. Lenham and Treherne (87) have experimentally obtained somewhat different optical constants for graphite than have been used in most of the scattering calculations.

In view of the unlikelihood that pure graphite grains could exist indefinitely in space, model calculations of grain consisting of graphite cores with dielectric mantles have been made. Various observed extinction curves have been matched in varying degrees of precision using spherical graphite core-ice mantle grains by Greenberg (8), Wickramasinghe, Dharmawardhana and Wyld (88). Wickramasinghe (79), Schalén (89, 90). Schalén (90) and Greenberg (15) have also considered metallic iron cores. Stecher and Donn (91) have indicated a possible connection between detailed structure of the extinction in the ultraviolet at $\lambda^{-1} = 4{\cdot}4\mu^{-1}$ with graphite optical properties.

The polarization of small graphite flakes has been shown by Greenberg (92) to follow a λ^{-1} law rather than have a flat maximum in the visual. Subsequent approximate calculations on larger oblate spheroidal graphite grains by Wickramasinghe, Donn, Stecher and Williams (93), by Wickramasinghe (94) and by Wickramasinghe and Krishna Swamy (95) on dielectric coated graphite platelets indicate a decrease of polarization toward shorter wavelengths. In the former paper (93), it is assumed that the small graphite interstellar particles tend to be spherical and that mainly the larger ones are formed as platelets. Consequently the extinction is due mainly to the small particles and the polarization to the larger ones which now require a high degree of orientation and a field of 10^{-5} gauss. The coated graphite grains (94, 95) have been shown by Greenberg (96) to require fields considerably in excess of $B = 10^{-5}$ gauss assuming that the approximate extinction values in (95) are correct.

The estimates of magnetic field strengths required to orient the interstellar grains have been improved on in an interesting extension of the Davis-Greenstein theory by Spitzer and Jones (97).

Wickramasinghe, Ray and Wyld (98) considered the effects of radius independent destruction mechanism on ice grains and showed that dielectric spherical grains could not account for the observed high ultraviolet extinction.

Danielson, Woolf and Gaustad (99) in looking for a strong ice absorption band in the infrared at $3{\cdot}1\mu$ have suggested an upper limit of 1/4 of the interstellar extinction being due to H_2O ice.

Field, Partridge and Sobel (100) have considered the effects of absorption spectra of solid methane, ammonia and water ice on the ultraviolet extinction by various sized spherical ice grains.

Aivazyan (101) has developed an approximation for computing extinction by spherical interstellar particles having a small absorptivity and real values of m in the range $1{\cdot}15 < m < 1{\cdot}50$.

Pronik (102) has considered a range of causes of variation in the reddening law and has found that the more dense dust clouds consist, on the average, of larger particles.

The development by Lind and Greenberg (103) of a procedure for calculating the scattering by arbitrarily oriented long cylindrical particles of any given sizes and optical properties has been applied to calculation of the polarization and extinction by elongated dielectric grains. The microwave analog method has been shown by Lind, Wang and Greenberg (104) and by Greenberg, Lind, Wang and Libelo (105) to be capable of producing almost complete and very accurate information on extinction by arbitrarily oriented particles. A wide range of microwave experimental data has been accumulated for both elongated and for flat particles (15). Greenberg and Shah (106, 107) have used perfect Davis-Greenstein orientation of cylindrical dielectric particles to calculate simultaneous reddening and wavelength dependence of polarization. More realistically oriented cylindrical particles have been considered by Greenberg (96) to determine both the wavelength dependence of polarization and extinction as a function of the strength of the orienting magnetic field. It was shown that a field of 10^{-5} gauss is required to produce a reasonable ratio of polarization to extinction. Effects of magnetic field tangling (partial disordering of field lines) were also considered. Extensive computation of polarization and extinction by varying size distribution, optical properties, and degrees of orientation of cylindrical particles have been performed (15) and good agreement has been simultaneously found for the extinction as observed out to the ultraviolet and for the average of the wavelength dependence of polarization observed by Gehrel and Visvanathan. Earlier calculations (8, 108) on two and three parameter size distribution of spherical dielectric particles produced too small extinction in the ultraviolet.

Stein (109) has examined the possibility of detecting infrared radiation at wavelengths greater than 10 microns and from a study of its spectral distribution obtain information about the grain composition and temperature.

Irvine (110) has calculated radiation pressures, and symmetry factors for spherical particles. Radiation pressure calculation have been applied by O'Dell (12) and Greenberg and Lind (111) to dust in H II regions. The motion of small bodies in tenuous gases has been carefully studied by Baines, Williams and Asebiomo (112) and approximate expressions for drag resistance over a range of translation speeds were found which should be suitable for application to astronomical problems. Resistance tables have been obtained by Dorschner (113).

Thompson (114) has performed experiments on the chemical effects of far ultraviolet radiation on molecules containing the cosmically abundant atoms H, O, C and N.

Gidalevich (115) has examined the dust-gas interaction and the radiation pressure effects on dust in the propagation of shock and ionization front in a H I medium.

Gehrels (116) has postulated scattering by loose molecules on grain surfaces to account for the wavelength dependence of polarization.

Cernuschi, Marsicano and Kimel (117) have proposed the condensation of interstellar metallic grains in expanding supernova clouds. Magnetic fields are assumed to saturate these small spherical particles which then combine to form elongated grains which are shown to account for the accepted values of interstellar polarization.

Ireland, Nandy, Reddish and Wickramasinghe (118) attribute the wavelength variation of polarization position angle to a time variation in the orienting magnetic field and to the fact that smaller particles have shorter orientation times than the larger particles. The plane of polarization is observed to rotate with wavelength with the opposite sense on opposite sides of $l^{II} = 145°$.

Donn, Stecher, Wickramasinghe and Williams (119) attribute the variable intrinsic polarization in Mira variables to small graphite flakes.

Diffuse Interstellar Lines

The observational data on diffuse lines has continued to accumulate and the information has become somewhat better defined. Although several new and interesting proposals as to their origin have been suggested the theoretical situation remains a mystery.

Stoeckly and Dressler (**120**) made photoelectric intensity measurements of λ 4430 for 59 stars and suggested that a correlation between weakened λ 4430 absorption relative to extinction and high negative velocity Ca II velocities implied that λ 4430 originates in very small grain which are evaporated near hot stars.

Wilson (**121**) has made some remarks on the interstellar absorption bands.

Treanor (**122**) pointed out the complementarity in the use of low and high dispersion methods of studying λ 4430.

A fairly symmetric λ 4430 profile has been reconfirmed by Wampler (**123, 124**) who noted possible differences from star to star. Wampler found no evidence for the extended wings and also no correlation with polarization. A possible detection of λ 4430 where there is apparently negligible color excess indicated that the agency responsible for λ 4430 may not be bound physically to the extinction producing grains.

On the other hand a high correlation—almost 1:1—between λ 4430 strength and color excess was found for 147 intermediate and late type B stars in Cygnus, Cepheus and Cassiopeia by Kristenson and Rudkjöbing (**125**). This work has been extended (**126**) to 506 O, B, and A stars.

Walker (**127**) has suggested that the apparent extended wings of the λ 4430 diffuse line are due to the slope discontinuity in the extinction curve at approximately this wavelength.

Herbig (**128**) found no evidence for wings or fine structure in the profile of λ 4430 in the star HD 183143. The wavelength of the band minimum was found to be near 4428·0 Å. Herbig (**129**) pointed out the strong correlation of a variety of 26 diffuse lines or bands including λ 4430 and suggested these are all attributable to the same origin. He also noted that they all fall in the narrow energy range 1·9 to 2·8 eV with 2·8 eV corresponding to the slope discontinuity in the extinction curve.

Unsöld (**130**) proposed that very small metallic particles within grains would act like plasmas and that they would give rise to λ 4430. A further discussion of this suggestion was made by van de Hulst (**131**).

A novel proposal for the source of the diffuse bands has been made by F. M. Johnson (**132**) who suggested either electron-vibrational absorption spectra of free radicals or inverse Raman absorption spectra, the latter leading to consideration of such molecules as cyclopentane, cyclopropane and others.

BIBLIOGRAPHY

Books and review articles relating to report on interstellar grains

Dieter, N. H., Goss, W. M. 1966, Recent Work on the Interstellar Medium, *Rev. mod. Phys.*, **38**, 256.

Greenberg, J. M. Interstellar Grains, in *Stars and Stellar Systems*, Vol. VII, Univ. of Chicago Press (in press).

Greenberg, J. M., Roark, T. P. (Editors). Proceedings of the IAU Colloquium on *Interstellar Grains*, NASA Publication (in press).

Johnson, H. L. Interstellar Extinction, in *Stars and Stellar Systems*, Vol. VII, Univ. of Chicago Press (in press).

Johnson, H. M. Diffuse Nebulae, in *Stars and Stellar Systems*, Vol. VII, Univ. of Chicago Press (in press).

Lambrecht, H. 1965, Interstellarer Raum, Landolt-Börnstein, Springer-Verlag, Berlin-Göttingen-Heidelberg.

Lambrecht, H., Schmidt, K. H. *Das interstellare Medium*, Akademische Verlagsgesellschaft, Leipsig (in press).
Lynds, B. T. 1965, Catalogue of Bright Nebulae. *Astrophys. J. Suppl.*, **12**, no. 105.
Pratt, N. M., Reddish, V. C. (Editors). 1964, Astronomical Discussions on Problems of the Interstellar Medium and Galactic Structure, *Proc. R. Obs. Edinb.*, **4**.
Wickramasinghe, N. C. *Interstellar Grains*, Chapman and Hall Co. (in press).

Other references

1. Elvius, A., Hall, J. S. 1965, *Astr. J.*, **70**, 138.
2. Elvius, A., Hall, J. S. *Lowell Obs. Bull.*, **6**, 257, 1966; IAU Coll. on *Interstellar Grains* Troy, N.Y., 1965 (in press).
3. O'Dell, C. R. 1965, *Astrophys. J.*, **142**, 604.
4. Roark, T. P., Greenberg, J. M. 1965, IAU Coll. on *Interstellar Grains*, Troy, N.Y. (in press).
5. Greenberg, J. M., Roark, T. P. *Astrophys. J.* (In press).
6. Dorschner, J., Gürtler, J. 1966, *Astr. Nachr.*, **289**, 57.
7. Minin, I. N. 1964, *Astr. Zu.*, **41**, 622.
8. Greenberg, J. M. Proc. IAU Symposium no. 24, 1964, Academic Press, 1966.
9. Vanysek, V. 1965, IAU Coll. on *Interstellar Grains*, Troy, N.Y. (in press).
10. Hall, R. C. 1965, *Publ. astr. Soc. Pacif.*, **77**, 158.
11. O'Dell, C. R., Hubbard, W. B. 1965, *Astrophys. J.*, **142**, 591.
12. O'Dell, C. R. 1965, IAU Coll. on *Interstellar Grains*, Troy, N.Y. (in press).
13. O'Dell, C. R., Hubbard, W. B., Peimbert, M. 1966, *Astrophys. J.*, **143**, 743.
14. Krishna Swamy, K. S. 1965, *Publ. astr. Soc. Pacif.*, **77**, 164.
15. Greenberg, J. M. Interstellar Grains, Chapter VI, Vol. VII, *Stars and Stellar Systems* (in press). Univ. of Chicago Press.
16. Gürtler, J. 1964, *Astr. Nachr.*, **288**, 105.
17. Dorschner, J. *Astr. Nachr.* (in press).
18. Ahmed, F., Lawrence, L. C., Reddish, V. C. 1965, *Publ. R. Soc. Edinb.*, **3**, 187.
19. Boyce, P. B. 1966, *Astrophys. J.*, **144**, 831.
20. Johnson, H. L. Interstellar Extinction, Vol. VII of *Stars and Stellar Systems* (in press). Univ. of Chicago Press.
21. Johnson, H. L. 1965, In *Vistas in Astronomy*, Pergamon Press.
22. Johnson, H. L. 1965, *Astrophys. J.*, **141**, 923.
23. Johnson, H. L., Mendoza, V., Mendoza, E. E. 1966, *Ann. Astrophys.*, **29**, 525.
24. Borgman, J. 1965, *Observatory*, **85**, no. 944.
25. Nandy, K. 1964, *Publ. R. Soc. Edinb.*, **4**, 57.
26. Nandy, K. 1965, IAU Coll. on *Interstellar Grains*, Troy, N.Y. (in press).
27. Nandy, K. 1965, *Publ. R. Obs. Edinb.*, **5**, 13.
28. Schalén, C. 1964, *Publ. R. Obs. Edinb.*, **4**, 47.
29. Underhill, A. B., Walker, G. A. H. 1966, *Mon. Not. R. astr. Soc.*, **131**, 475.
30. Whiteoak, J. B. 1966, *Astrophys. J.*, **144**, 305.
31. Guthrie, B. N. G., Nandy, K. 1966, *Observatory*, **86**, no. 951.
32. Mendoza, V. E. E. 1965, *Bol. Obs. Tonantzintla Tacubaya*, **4**, 3.
33. Wampler, G. J. 1964, *Astrophys. J.*, **140**, 1615.
34. Becker, W. 1966, *Z. Astrophys.*, **64**, 77.
35. Krzeminski, W., Serkowski, K. *Astrophys. J.* (in press).
36. Lodén, L. O. 1966, *Ark. Astr.*, **4**, no. 5.
37. Neckel, H. 1965, IAU Coll. on *Interstellar Grains*, Troy, N.Y. (in press).
38. Neckel, H. 1965, *Z. Astrophys.*, **62**, 180.
39. Glushneva, I. N. 1964, *Astr. Zu.*, **41**, 181.
40. Boggess, A., Borgman, J. 1964, *Astrophys. J.*, **140**, 1636.
41. Stecher, T. P. 1965, *Astrophys. J.*, **142**, 1683.
42. Stecher, T. P. 1965, IAU Coll. on *Interstellar Grains*, Troy, N.Y. (in press).
43. Oja, T. *Uppsala astr. Obs. Ann.*, **5**, no. 4 (in press).

44. Oja, T. *Ark. Astr.*, **4** (in press).
45. Ekedahl, Y. *Ark. Astr.* (in press).
46. Ljunggren, B. 1965, *Ark. Astr.*, **3**, no. 37.
47. Ljunggren, B. 1966, Univ. Uppsala, Diss. in Science, no. 66.
48. Sjögren, U. 1964, *Ark. Astr.*, **3**, no. 27.
49. Scheffler, H. 1966, *Z. Astrophys.*, **63**, 267.
50. Neckel, T. 1966, *Z. Astrophys.*, **63**, 221.
51. Elsässer, H., Neckel, T., Scheffler, H. 1965, *Z. Astrophys.*, **63**, 1.
52. Isserstedt, J., Schmidt-Kaler, T. 1964, *Z. Astrophys.*, **59**, 182.
53. Straizys, V. 1964, *Astr. Zu.*, **41**, 750.
54. Grigor'eva, N. B. 1964, *Astr. Zu.*, **41**, 185.
55. Herzog, E. R. 1965, *Publ. astr. Soc. Pacif.*, **77**, 119.
56. Gosachinskii, I. V. 1965, *Astr. Zu.*, **42**, 929.
57. Schmidt, K. H. 1964, *Astr. Nachr.*, **288**, 19.
58. Schmidt, K. H. 1965, IAU Coll. on *Interstellar Grains*, Troy, N.Y. (in press).
59. Knaap, H. F. P., van den Meijdenberg, Beenakker, J. J. M., van de Hulst, H. C. 1965, IAU Coll. on *Interstellar Grains*, Troy, N.Y. (in press).
60. Knaap, H. F. P., van den Meijdenberg, Beenakker, J. J. M., van de Hulst, H. C. 1966, *Bull. astr. Inst. Netherl.*, **18**, 256.
61. Stecher, T. P., Williams, D. A. 1966, *Astrophys. J.*, **146**, 88.
62. Behr, A. 1965, IAU Coll. on *Interstellar Grains*, Troy, N.Y. (in press).
63. Serkowski, K. 1965, *Astrophys. J.*, **141**, 1340.
64. Serkowski, K., Chojnacki, W., Rucinski, S. 1965, IAU Coll. on *Insterstellar Grains*, Troy, N.Y. (in press).
65. Gehrels, T., Meltzer, A. S. 1966, *Astr. J.*, **71**, 111.
66. Coyne, S. J., Coyne, G. V., Gehrels, T. 1966, *Astr. J.*, **71**, 355.
67. Visvanathan, N. 1965, Mt. Stromlo, Thesis.
68. Gehrels, T., Silvester, A. B. 1965, *Astr. J.*, **70**, 579.
69. Lodén, L. O. 1965, *Stockholm Obs. Ann.*, **22**, no. 8.
70. Visvanathan, N. 1966, *Mon. Not. R. astr. Soc.*, **132**, 423.
71. Pronik, I. I. 1965, *Astr. Zu.*, **43**, 291.
72. Appenzeller, I. 1966, *Z. Astrophys.*, **64**, 269.
73. Appenzeller, I. 1966, *Z. Astrophys.*, **64**, 296.
74. Serkowski, K. 1966, *Astrophys. J.*, **144**, 857.
75. Kamijo, F. 1963, *Publ. astr. Soc. Japan*, **15**, 440.
76. Friedemann, Ch., Schmidt, K. H. *Astr. Nachr.* (in press).
77. Donn. B., Wickramasinghe, N. C., Hudson, J. P., Stecher, T. P. *Astrophys. J.* (in press).
78. Mendoza, V., Mendoza, E. E., Johnson, H. L. 1965, *Astrophys. J.*, **141**, 161.
79. Wickramasinghe, N. C. 1965, IAU Coll. on *Interstellar Grains*, Troy, N.Y. (in press).
80. Wickramasinghe, N. C. 1965, *Mon. Not. R. astr. Soc.*, **131**, 177.
81. Wichramasinghe, N. C., Guillaume, C. 1965, *Nature*, **207**, 366.
82. Wichramasinghe, N. C. 1966, *Mon. Not. R. astr. Soc.*, **131**, 263.
83. Krishna Swamy, K. S., Wickramasinghe, N. C. 1966, *Mon. Not. R. astr. Soc.*, **132**, 193.
84. Grevesse-Guillaume, C., 1966, *Mon. Nat. R. astr. Soc.*, **132**, 471.
85. Friedemann, C., Schmidt, K. H. 1966, *Mitt. Univ.-Sternwarte Jena*, no. 72.
86. Nandy, K., Wickramasinghe, N. C. 1965, *Publ. R. Obs. Edinb.*, **5**, no. 3.
87. Lenham, A. P., Treherne, D. M. 1966, *Observatory*, **86**, no. 950.
88. Wickramasinghe, N. C., Dharmawardhana, M. W. C., Wyld, C. 1966, *Mon. Not. R. astr. Soc.*, **134**, 25.
89. Schalén, C. 1965, *Publ. astr. Soc. Pacif.*, **77**, 409.
90. Schalén, C. 1965, *Ark. Astr.*, **4**, 1.
91. Stecher, T. P., Donn, B. 1965, *Astrophys. J.*, **142**, 1681.
92. Greenberg, J. M. 1966, *Astrophys. J.*, **145**, 57.
93. Wickramasinghe, N. C., Donn, B., Stecher, T. P., Williams, D. A. 1966, *Nature*, **212**, 167.

94. Wickramasinghe, N. C. *Interstellar Grains*, Chapman and Hall Co. (in press).
95. Wickramasinghe, N. C., Krishna Swamy, K. S. *Astrophys. J.* (in press).
96. Greenberg, J. M. 1966, Proc. IAU Symposium no. 31.
97. Spitzer, Jr., L., Jones, R. V. 1965, IAU Coll. on *Interstellar Grains*, Troy, N.Y. (in press).
98. Wickramasinghe, N. C., Ray, W. D., Wyld, C. 1966, *Mon. Not. R. astr. Soc.*, **132**, 137.
99. Danielson, R. E., Woolf, N. J., Gaustad, J. E. 1965, *Astrophys. J.*, **141**, 116.
100. Field, G. B., Partridge, R. B., Sobel, H. 1965, IAU Coll. on *Interstellar Grains*, Troy, N.Y. (in press).
101. Aivazyan, G. M. 1965, *Astr. Zu.*, **42**, 986.
102. Pronik, I. I. 1964, *Astr. Zu.*, **41**, 963.
103. Lind, A. C., Greenberg, J. M. 1966, *J. Appl. Phys.*, **37**, 3195.
104. Lind, A. C., Wang, R. T., Greenberg, J. M. 1965, *Appl. Opt.*, **4**, 1555.
105. Greenberg, J. M., Lind, A. C., Wang, R. T. 1965, Libelo, Proc. Interdisciplinary Conf. on *Electromagnetic Scattering*, II, Gordon and Breach (in press).
106. Greenberg, J. M., Shah, G. 1965, IAU Coll. on *Interstellar Grains*, Troy, N.Y. (in press).
107. Greenberg, J. M., Shah, G. 1966, *Astrophys. J.*, **145**, 63.
108. Greenberg, J. M. 1964, As reported in H. C. van de Hulst, *Publ. R. Obs. Edinb.*, **4**, 13.
109. Stein, W. 1966, *Astrophys. J.*, **144**, 318.
110. Irvine, W. M. 1965, *J. opt. Soc. Am.*, **55**, 16.
111. Greenberg, J. M., Lind, A. C. 1965, IAU Coll. on *Interstellar Grains*, Troy, N.Y. (in press).
112. Baines, M. J., Williams, I. P., Asebiomo, A. S. 1965, *Mon. Not. R. astr. Soc.*, **130**, 63.
113. Dorschner, J. *Astr. Nachr.* (in press).
114. Thompson, W. 1965, IAU Coll. on *Interstellar Grains*, Troy, N.Y. (in press).
115. Gidalevich, E. Ya. 1966, *Astr. Zu.*, **42**, 932.
116. Gehrels, T. 1966, *Astr. J.*, **71**, 62.
117. Cernuschi, F., Marsicano, F. R., Kimel, I. 1965, *Ann. Astrophys.*, **28**, 860.
118. Ireland, J. G., Nandy, K., Reddish, V. C. *Nature* (in press).
119. Donn, B., Stecher, T. P., Wichramasinghe, N. C., Williams, D. A. 1966, *Astrophys. J.*, **145**, 949.
120. Stoeckly, R., Dressler, K. 1964, *Astrophys. J.*, **139**, 240.
121. Wilson, R. 1964, *Publ. R. Obs. Edinb.*, **4**, 67.
122. Treanor, S. J., P. J. 1964, *Publ. R. Obs. Edinb.*, **4**, 81.
123. Wampler, E. J. 1965, IAU Coll. on *Interstellar Grains*, Troy, N.Y. (in press).
124. Wampler, E. J. 1966, *Astrophys. J.*, **144**, 921.
125. Kristenson, H., Rudkjöbing, M. 1965, *J. Observateurs*, **48**, 107.
126. Baerentzen, J., Gammelgaard, P., Hilberg, T., Jorgensen, K. F., Kristenson, H., Nissen, P. E., Rudkjöbing, M. (In press).
127. Walker, G. A. H. 1966, *Observatory*, **86**, no. 952.
128. Herbig, G. H. *Astrophys. J.* (in press).
129. Herbig, G. H. Proc. IAU Symposium no. 31. (in press).
130. Unsöld, A. 1964, *Publ. R. Obs. Edinb.*, **4**, 35.
131. van de Hulst, H. C. 1964, *Publ. R. Obs. Edinb.*, **4**, 13.
132. Johnson, F. M. 1965, IAU Coll. on *Interstellar Grains*, Troy, N.Y. (in press).

NOTE

Addenda to this Report of Commission 34 will appear in Volume XIII B of the *Transactions of the IAU*.

35. COMMISSION DE LA CONSTITUTION DES ETOILES

PRÉSIDENT: Professor P. Ledoux, Institut d'Astrophysique, Université de Liège, Cointe-Sclessin (Liège) Belgium.

VICE-PRÉSIDENT: Professor C. Hayaschi, Department of Nuclear Science, Faculty of Science, Kyoto University, Kyoto, Japan.

MEMBRES: Biermann, Bondi, Brownlee, Burbidge (G. R.), Cameron, Chandrasekhar, Christy, Cowling, Cox (A. N.), Demarque, Dingens, Epstein, Ezer, Ferraro, Fowler, Gamow, Gjellestad, Henyey, Hitotuyanagi, Hoyle, Kaplan (S. A.), Kippenhahn, Krook, Kumar, Kushwaha, Lebedinsky, McCrea, Marx, Masani, Massewitch, Mestel, Reeves (H.), Reiz, Rosseland, Roxburgh, Ruben, Salpeter (E. E.), Savedoff, Schatzman, Schwarzschild, Strömgren, Suda, Sweet, Tayler, Temesvary, Tuominen, Van der Borght, Weigert, Wrubel, Zhevakin.

As will be apparent from the Report, notable progress has been accomplished in many directions during the period under review especially in the field of stellar evolution, where some of the most interesting phases are being tackled leading in some cases to the discovery of new forms of instability. Interaction between dynamical instability towards non-radial oscillations and convective instability has been clarified and delicate forms of instabilities in presence of superadiabatic gradients have been explored. Interest in stellar rotation has increased and has brought sometimes unexpected results as the resulting increase of the critical mass of white dwarfs. Neutrino astrophysics raises fascinating questions. Non-linear methods have definitely entered the field of variable star theories. The attack of the supernovae phenomenon takes by and by a quantitative aspect. The fundamentals of magnetohydrodynamics as applied to stellar structure are emerging. A proper theory of stellar convection may still be a long way off but progress is undoubtedly on the way. The discovery of quasi-stellar-sources and the attempts to find an interpretation in terms of very large masses as well as the interest in stars of very high densities have opened a whole new field in the theory of stellar structure where the general relativity terms become important. No doubt, the recent discoveries of cold infrared objects and of concentrated X-ray sources will create new spectacular developments in the activities of our Commission and may already require new sections in the next of its reports to the IAU.

I. NUCLEAR ENERGY GENERATION, NEUTRINO EMISSION AND NUCLEOSYNTHESIS

In this section as in many others of this report, due to the growth of the subject or of its complexities, it seems impossible to try to account for many of the relevant papers beyond repeating essentially the titles, often quite explicit, and which will be found, in such cases, in the bibliography.

Comprehensive review articles attempting to bring up to date previous compilations and presenting a critical survey of the field of nuclear energy generation and low energy nucleosynthesis in stellar interiors have been prepared by Reeves (**1**).

Experimental results of interest for nuclear astrophysics have been reported by Pearson and Spear (**2**), Larson and Spear (**3**), Domingo (**4**) and McNally (**5**). New experimental cross-sections as well as the modern theory of nuclear beta-decay have been used to discuss the termination of the proton-proton chain in stellar interiors (**6, 7, 8**), the results in reference (**8**) implying an increase of the Be^7 (p, γ) B^8 reaction leading to a doubling of the possible solar neutrinos flux. Different points concerning the nuclear reaction rates in stellar conditions are examined in papers (**9–13**), the last one, in particular, presenting methods for finding accurate nuclear reaction cross-sections averaged over resonances for astrophysical purposes. The evaluation of these rates has been extended by Wolf (14) and by Van Horn (15) to situations arising in the interior of a white dwarf when the Coulomb energies between neighboring nuclei are large compared to their kinetic energies. Bahcall (16) has studied the rates of capture of free electrons by nuclei in stars and especially their dependence on temperature and density, considering different applications. More recently (17), he has discussed the physical basis for exchange and overlap corrections in these electron capture processes.

As far as nucleosynthesis in stars is concerned, Bashkin (18) has prepared a critical survey of the litterature up to 1962–63. A general article on the origin of the elements was also written by Fowler (19). Caughlan (20) has studied in detail the abundances of the nuclei involved in the CNO bi-cycle as equilibrium is approached and the implications for the interpretation of the ratios of C^{13} to C^{12} and of nitrogen to oxygen observed in typical carbon stars. The effects of combined hydrogen and helium burning in the core of a population II red-giant star have also been discussed (21, 22) in an attempt to interpret the anomalous abundances of carbon, barium and the rare-earths in these stars. Wielding (23) reports work in progress on alpha capture reactions at high temperatures expected to exist in supernovae, using recent nuclear data. The problem of nucleosynthesis of elements heavier than the iron group by neutron capture on both slow and fast time scales has been rediscussed by Seeger and Fowler (24) using new cross-section data. On this basis, they find that the solar-system *r*-process material must have been submitted to extreme conditions which could have been realized in an object with a mass of the order of $10^6\ M_{\odot}$. On the other hand, Reeves (25) has reviewed the different neutron-emitting processes in ordinary stars and has evaluated the flux of neutrons per metal atom made available at different stages of evolution through these processes which in general do not overlap. The capture of positrons by *s*-elements during the carbon and oxygen burning stages has been considered by Reeves and Stewart (26) as a possible source of the *p*-elements. Shaw, Clayton and Michel (27) have developed the theory of photon induced β-decay in stellar interiors which may also play an interesting role in that respect.

Calculations of nuclear abundances at statistical equilibrium for 333 sets of values of the temperature, density and ratio of total number of protons to total number of neutrons are reported by Clifford and Tayler (28) and detailed results are given for 96 of the cases. The importance of the equilibrium process for the iron group elements has also been discussed by Fowler (29) and Tsuruta and Cameron (30) have investigated the shift of the Fe equilibrium peak to higher mass number at high densities. The approach to nuclear statistical equilibrium has been studied by Truran, Cameron and Gilbert (31) by solving the nuclear rate equations under conditions of advanced stellar evolution and supernova shock waves.

The formation of the light elements D, Li, Be, B in the solar system by spallation reactions in metric-sized planetesimals as suggested first by Fowler, Greenstein and Hoyle has been further discussed by Mitler (32) and by Burnett and Fowler (33). Recently the problem of the formation of these elements and its meaning for the solar system and stars, in particular T Tauri stars, has also been the object of much interest and research in the group around Reeves and Schatzman in Orsay and Paris (34).

A new attempt to explain the anomalous abundances in peculiar stars of type A and B is due to Fowler, E. M. Burbidge, G. R. Burbidge and Hoyle (35). Considering that these peculiar abundances can only be explained by the combined effect of violent surface spallation (underabundance of light elements especially helium) and a process of rapid neutron capture (high abundance of heavy elements in the rare-earth region) followed by mixing, they have suggested that these stars have returned to the neighborhood of the main sequence after a long and complicated evolution. Some of their conclusions have been criticized by Truran and Cameron (36). Brancazio is reported as re-examining the feasibility of surface nuclear reactions to explain the extreme abundance anomalies in magnetic Ap stars. The problem of surface nuclear reactions has also been reviewed in general by Audouze (**37**) who finds interesting correlations between abundances and the energy of the processes involved.

The general problem of dating the time of nucleosynthesis in our Galaxy by a comparison of the cosmoradiogenic abundances of daughter species with the abundances of their radioactive parents has been discussed by Clayton (**38**).

Let us note also that the problem of the origin of helium has reached at present an extremely interesting stage. The difficulties encountered up to now by all schemes attempting to explain the formation in stars of the bulk of the galactic helium (**39**) have led Hoyle and Tayler (**40**) to revert to a cosmological origin of this element. The subsequent discovery of the background radiation at 3°K, in providing evidence for the conditions prevailing very early in the expansion of the universe have considerably reinforced the plausibility of such an origin. But, on the basis of an ordinary isotropic relativistic expansion, we are now faced with a very large pregalactic helium abundance, perhaps too large according especially to some recent observational evidence, and cosmologists are trying to devise anisotropic models capable of reducing it (**41**). On the other hand, it seems unavoidable that some helium be produced in the galaxy itself and Boury and Ledoux (**42**) are investigating the possible effect, in that respect, of large mass stars evolving homogeneously and shedding their excess of mass above the critical mass corresponding to vibrational instability, the latter decreasing rapidly as hydrogen is transformed into helium.

The role of neutrinos in astrophysics and cosmology has been the object of intense activity and quite a few general summaries (**43–47**) have appeared since 1964 discussing both the nuclear and astronomical aspects of the problem.

Bahcall has pursued his investigations on the emission of solar neutrinos and their possible detection (**48, 49**) and the problem has been discussed also by Pochoda and Reeves (**50**) and Kuzmin (**51**).

Bahcall and Frautschi (**52**) have also considered the possibility of neutrino emission by strong radio-sources and observational tests are discussed.

Neutrino opacity and scattering in a variety of stellar conditions have been discussed with a view to applications to the detection of solar neutrinos, the escape of neutrinos from stars, neutrino scattering in cosmology and the energy deposition in supernovae explosions (**53, 54, 55, 56**). Sakashita and Nishida (**57**) have computed the energy loss rate by neutrino emission from excited nuclei due to the interaction of the type (n, n) (ν, ν) which was suggested by Pontecorvo, showing that it is at most comparable with other neutrino processes.

Inman and Ruderman (**58**) have computed the neutrino pair enission from coherent electron excitation (transverse plasmons) for a range of temperatures and densities relevant to stellar evolution. Correction to the longitudinal and transverse neutrino emissivities of Adams, Ruderman and Woo in a stellar plasma are given by Zaidi (**59**).

The URCA process has been rediscussed, taking into account the pair production at high temperature, by Pinaev (**60**) and Wataghin (**61**) and applications have been considered by Masani *et al.* (**62**).

BIBLIOGRAPHY

1. Reeves, H. 1965, Stellar Energy Sources, in *Stars and Stellar Systems*, vol. VIII, Ed. L. H. Aller and D. B. McLaughlin, Univ. of Chicago Press, Chicago, p. 113; 1966, Nuclear Energy Generation in Stars and some aspects of Nucleosynthesis, *Stellar Evolution*, Ed. R. F. Stein and A. G. W. Cameron, Plenum Press, New York, p. 83.

2. Pearson, J. D., Spear, R. H. 1964, A Study of the Gamma Radiation produced in the Alpha-Particle Bombardment of O^{16}, *Nucl. Phys.*, **54**, 434.

3. Larson, J. D., Spear, R. H. 1964, Gamma Radiation from the Alpha-Particle Bombardment of C^{12}, *Nucl. Phys.*, **56**, 497.

4. Domingo, J. J. 1965, The effect of the 3·10 MeV State of Fe^{17} on the O^{16} (p, γ) Reaction, *Nucl. Phys.*, **61**, 39.
5. McNally, J. H. 1966, Reactions leading to Ar^{37} and K^{37} Energy Levels, *Nucl. Phys.*, (in press).
6. Parker, P. D., Bahcall, J. N., Fowler, W. A. 1964, *Astrophys. J.*, **139**, 602.
7. Bahcall, J. N., Wolf, R. A. 1964, *Astrophys. J.*, **139**, 909.
8. Parker, P. D. 1966, *Astrophys. J.*, **145**, 960.
9. Paquette, G., Reeves, H. 1964, The reaction 3 $He^4 \rightarrow C^{12}$, *Astrophys. J.*, **140**, 1319.
10. Vogt, E., Michaud, G., Reeves, H. 1965, Effect of the Optical Model on Barrier Penetration, *Phys. Lett.*, **19**, 570.
11. Bahcall, J. N. 1966, Non-Resonant Nuclear Reactions at Stellar Temperature, *Astrophys. J.*, **143**, 259.
12. Tombrello, T. A. 1965, The Capture of Protons by Be^7, *Nucl. Phys.*, **71**, 459.
13. Truran, J. W., Hansen, C. J., Cameron, A. G. W., Gilbert, A. 1966, Thermonuclear Reactions in Medium and Heavy Nuclei, *Can. J. Phys.*, **44**, 151.
14. Wolf, R. A. 1965, Rates of Nuclear Reactions in Solid-like-Stars, *Phys. Rev.*, **137**, B 1634.
15. Van Horn, H. M. 1966, *Astr. J.*, **71**, 388.
16. Bahcall, J. N. 1964, *Astrophys. J.*, **139**, 318.
17. Bahcall, J. N. 1965, *Nucl. Phys.*, **71**, 267.
18. Bashkin, S. 1965, The Origin of the Chemical Elements, in *Stars and Stellar Systems*, Vol. VIII, Eds. L. H. Aller and D. B. McLaughlin, Univ. of Chicago Press, Chicago, p. 1.
19. Fowler, W. A. 1964, *Proc. nat. Acad. Sci.*, **52**, 524.
20. Caughlan, G. R. 1965, *Astrophys. J.*, **141**, 688.
21. Caughlan, G. R., Fowler, W. A. 1964, *Astrophys. J.*, **139**, 1180.
22. Caughlan, G. R., Fowler, W. A., Talbot, R. J. 1964, *Astrophys. J.*, **140**, 380.
23. Wielding, T. 1966, in preparation.
24. Seeger, P. A., Fowler, W. A. 1965, *Astrophys. J. Suppl.*, **11**, no. 97, 121; 1966, *Astrophys. J.*, **144**, 822.
25. Reeves, H. 1966, *Astrophys. J.*, **146**, 447.
26. Reeves, H., Stewart, P. 1965, *Astrophys. J.*, **141**, 1432.
27. Shaw, P. B., Clayton, D. D., Michel, F. C. 1965, *Phys. Rev.*, **140**, 1433.
28. Clifford, F. E., Tayler, R. J. 1966, *Mem. R. astr. Soc.*, **69**, 21.
29. Fowler, W. A. 1964, The Iron Group Elements and the Equilibrium Process in Nucleosynthesis, IAU Symposium no. 26 on *Abundance Determination in Stellar Spectra*, Utrecht, Holland.
30. Tsuruta, S., Cameron, A. G. W. 1966, *Can. J. Phys.*, **43**, 2056.
31. Truran, J. W., Cameron, A. G. W., Gilbert, A. 1966, *Can. J. Phys.*, **44**, 563.
32. Mitler, H. E. 1964, *Phys. Rev.*, **136**, B 298.
33. Burnett, D. S., Fowler, W. A. 1965, *Geochim. Cosmochim. Acta*, **29**, 1200; cf. also Burnett, D. S., Lippolt, H. J., Wasserburg, G. J. 1966, The Relative Isotropic Abundance of K^{40} in Terrestrial and Meteoritic Samples, *J. geophys. Res.*, **71**, 1429.
34. Bernas, R., Gradsztajn, E., Reeves, H., Schatzman, E. 1966-67, On the Nucleosynthesis of Lithium, Beryllium and Boron (to be published); Audouze, J., Gradsztajn, E., Reeves, H. 1966-67, The Energetics of Light Elements Formation and the Role of Neutron Secondary reactions (to be published); Audouze, J., Reeves, H. 1966-67, The Spallation of Light Nuclei (oriented towards an estimation of unmeasured cross-sections of astrophysical importance, to be published).
    Schatzman, E. 1966, On the Diffusion in the Convective Zone of the Elements formed by Spallation at the Stellar Surface (to be published).
    Many of these investigations have been circulated as Orsay preprints.
35. Fowler, A. W., Burbidge, E. M., Burbidge, G. R., Hoyle, F. 1965, *Astrophys. J.*, **142**, 423.
36. Truran, J. W., Cameron, A. G. W. 1965, Proc. Symp. on *Magnetic Stars and other Peculiar and Metallic-Line A Stars*, Goddard Space Flight Center, Greenbelt (to be published).
37. Audouze, J. (thèse de 3ème cycle).

38. Clayton, D. D. 1964, *Astrophys. J.*, **139**, 637; 1964, *Science*, **143**, 1281.
39. cf. for instance: Truran, J. W., Hansen, C. J., Cameron, A. G. W. 1965, *Can. J. Phys.*, **43**, 1616.
40. Hoyle, F., Tayler, R. J. 1964, *Nature*, **203**, 1108.
41. Hawking, S. A., Tayler, R. J. 1966, *Nature*, **209**, 1278.
Thorne, K. S. 1966, Caltech Orange Reprint.
42. Boury, A., Ledoux, P. 1967 (in preparation).
43. Fowler, W. A., Hoyle, F. 1964, Neutrino Processes and Pair Formation in Massive Stars and Supernovae, *Astrophys. J.*, *Suppl.* **9**, no. 91, 201.
44. Fowler, W. A., Vogl, J. L. 1964, Nuclear and Neutrino Processes in Stars and Supernovae, *Lectures in Theoretical Physics*, vol. VI, The University of Colorado Press, Boulder, Colorado.
45. Bahcall, J. N. 1964, Observational Neutrino Astronomy: *a* ν– Review, presented at the Second Texas Symp. on Relativistic Astrophysics, Austin; cf. also *Science*, **147**, 115, 1965.
46. Chiu, H. Y. 1966, Neutrinos in Astrophysics, *Stellar Evolution*, Eds. R. F. Stein and A. G. W. Cameron, Plenum Press, p. 175; 1966, Neutrinos in Astrophysics and Cosmology, *A. Rev. nucl. Sci.* (in press); 1965, *Neutrino Astrophysics*, Documents on Modern Physics, Gordon and Breach, New York.
47. Masani, A., Gallino, R., Silvestro, G. 1965, L'Astronomia del neutrino, *Nuovo Cim. Suppl.*, **3**, (also: *Contr. Oss. astr. Milano-Merate*, Nuova Ser., no. 245).
48. (*a*) Bahcall, J. N. 1964, Solar Neutrinos. I. Theoretical, *Phys. Rev. Lett.*, **12**, 300; (*b*) 1964, Solar Neutrinos Cross-sections and Nuclear Beta Decay, *Phys. Rev.*, **135**, B 137; 1964, Neutrino Spectroscopy of the Solar Interior, *Phys. Lett.*, **13**, 332.
49. Bahcall, J. N., Barnes, C. A. 1964, Importance of a Study of the Decay of Ca^{37} to the Problem of Detecting Solar Neutrinos, *Phys. Lett.*, **12**, 48.
50. Pochoda, P., Reeves, H. 1964, *Planet. Space Sci.*, **12**, 119.
51. Kuzmin, V. A. 1965, Neutrino production in the $He^3(p, e^+\nu)$ He^4 reaction in Solar interiors, *Phys. Rev. Lett.*, **17**, 27.
52. Bahcall, J. N., Frautschi, S. C. 1964, *Phys. Rev.*, **135**, B 788.
53. Euwema, R. N. 1964, Neutrino Absorption in Dense Matter, *Phys. Rev.*, **133**, B 1046.
54. Bahcall, J. N. 1964, Neutrino Opacity, I. Neutrino-Lepton Scattering, *Phys. Rev.*, **136**, B 1164; 1964, Neutrino Opacity. II. Neutrino-Nucleon Interactions, *Phys. Rev.*, **136**, B 1547. cf. also Colgate, S. A., White, R. H. 1966, The Hydrodynamic Behavior of Supernova Explosions, Sect. VII a, b, c, *Astrophys. J.*, **143**, 626.
55. Hebb, J. 1966, Radiative Corrections to the Neutrino-Electron Interaction, *Astrophys. J.*, (in press).
56. Chin, C.-W., Chiu, H.-Y., Stothers, R. 1965, Astrophysical Evidence for the $(e\nu_e)$ $(e\ \nu_e)$ interaction, Internat. Conf. on Weak Interaction, October 25–27, 1965, Argone Nat. Lab.; cf. also, 4th Eastern U.S. Theor. Phys. Conf., State Univ. of New York, Nov. 1965.
57. Sakashita, S., Nishida, M. 1964, *Progr. theor. Phys.*, **31**, 727; 1965, *ibid.* **34**, 188.
58. Inman, C. L., Ruderman, M. A. 1964, *Astrophys. J.*, **140**, 1025.
59. Zaidi, M. H. 1965, *Nuovo Cim.*, **40**, 502.
60. Pinaev, V. S. 1964, *Soviet Phys. JETP*, **18**, 377.
61. Wataghin, G. 1965, *Nuovo Cim.*, **36**, 296.
62. Masani, A., Gallino, R., Silvestro, G. 1965, Neutrino Emission in the late Stages of Massive Stars Evolution, *Nuovo Cim.*, **38**, 1142.

II. RELATIVISTIC STELLAR MODELS (SUPERDENSE OR SUPERMASSIVE STARS)

In the last three or four years, much effort has been devoted to what is now called Relativistic Astrophysics. In the field, the 'Texas Symposia' are becoming institutional (1). From our viewpoint, three main subfields can be distinguished, although some overlapping is unavoidable: study of the very dense configurations, attempts to interpret quasi-stellar objects in terms of huge supermassive stars and the study of gravitational collapse. Very masterful reviews, some of them giving more emphasis on the superdense (neutron) stars, have been published by

Zel'dovitch and Novikov (**2, 3**), Harrison, Thorne, Wakano and Wheeler (**4**), Wheeler (**5**) and Thorne (**6, 7**). They contain very extensive bibliography. We will give here a good deal of typical titles but without any claim to exhaustivity. We rather refer the reader back to the reviews above.

Let us first turn towards usual stellar evolution; it is currently thought that the final term in a star's life lies in the white dwarf stage, the neutron stage (both stable below some mass) or gravitational collapse, the fate of the star depending on its mass when it reaches the end point of thermonuclear evolution. The neutron star or the hyperon star have been given a great deal of attention, both from the point of view of the structure and of the properties of matter: various equations of state and forms of the interaction potential have been used. Their stability, their oscillations and their energy emissivity in the form of photons (surface X-rays), of neutrinos and of gravitational waves have also been the subject of much research. Some of the works involve rotation and non-zero temperature (for all these points, *cf.* references **2** to **41**).

The onset of dynamical instability due to General Relativity effects in supermassive 'stars' is still being contemplated as possibly significant for the quasi-stellar objects and the strong radio-sources. The state of affairs by 1964 was described by Fowler on various occasions (**43, 44, 45**). Since that time, more investigations on the structure of supermassive stars and relativistic configurations have been carried out, some of them on very general grounds; for instance, Buchdahl (**46**) has improved the general inequalities he had derived in 1959 (**47**) for regular fluid spheres.

Some simplified models more easily amenable to mathematical treatment: relativistic polytropic and 'adiabatic' structures have been studied by Gratton and Giannone (**49**), Van der Borght (**50**) and Tooper (**51, 52, 53**) who also checked the validity of the post Newtonian approximation in the binding energy calculation (**54**) as compared with its 'standard model' (**53**). Tooper considers in detail the conditions of dynamical instability for his various models. Some of his work (**52**) also applies to the superdense stars considered above. As far as the theory of stellar interiors is concerned, more physical considerations have been developed by Hämeen-Anttila and Anttila (**55, 56, 57, 58**) and are also described briefly in a very intuitive manner in Thorne's reviews (**6, 7**). Thorne has studied locally the convection criterion in General Relativity establishing in all generality the validity of Schwarzschild criterion (**59**) already recovered by Chandrasekhar (**60**) in the post-Newtonian approximation, as a result of his general analysis of radial and non-radial oscillations of spherical gaseous masses. Sato (**61**) has studied non rotating massive stars and gives conditions for the onset of the instability. Let us also recall the work of Bardeen *et al.* (**32**) giving methods for studying the radial pulsations of General Relativistic Models, and a talk by Bardeen at the Second Texas symposium (**62**). The pulsations and evolution of supermassive stars have also been studied by Osaka who finds the expected vibrational instability (**63**). As far as quasistellar objects are concerned, an oscillatory explanation of the variations of 3C 273 has been attempted by Anand (**64**) and a model based on relaxation oscillations in supermassive stars was proposed by Fowler (**65**) but was criticized by Zel'dovitch and Novikov (**2**). Finally, extra factors have been taken into account, rotation by Chandrasekhar (**66**), Roxburgh (**67**), Dubney and Roxburgh (**68**), and magnetoturbulent energy by Anand (**69**) and Bardeen and Anand (**70**). Those factors can significantly delay the setting in of dynamical instability and allow the extension of stable models with energy generation by thermonuclear reactions to much larger masses. However, as has been pointed out on various occasions, these models would then become strongly vibrationally unstable even well before the nuclear energy generation accounts for the total luminosity of the star (**63, 71**).

As has been indicated, stars more massive than a certain critical mass (corresponding to about twice the baryon content of the Sun) cannot settle down into an equilibrium configuration when they arrive at the end-point of thermonuclear evolution if they are prevented to get rid of the excess mass. If this is possible, as in the case studied by Colgate and White (**72**) in their work on supernovae, they end as neutron stars. If matter ejection does not take place, the body should undergo a limitless collapse. On the other hand, supermassive stars, albeit tenuous and containing considerable nuclear energy storage, are also dynamically unstable in the absence of rotation or magnetic fields. All this has led to much current interest in gravitational collapse. Again, the reviews referred to at the beginning of this subsection are of prime interest. A very brief but clear description of the collapse is given at the end of Wheeler's paper (**5**). An extensive treatment in terms of Kruskal metric (**73**) is found in Thorne's Les Houches lectures (**7**). The book (**4**) by Harrison *et al.* does not limit itself to astronomical considerations (e.g. the quantum probability of a mass however small to undergo gravitational collapse by tunnel effect is taken in consideration). Let us also mention the anticollapse theory, which is not the time reversal of the contraction, briefly reviewed by Zel'dovitch and Novikov (**2**).

Dynamical calculations of General-Relativistic collapse, ignoring heat transfer by neutrinos, radiation etc., have been made by May and White (**74**) who do not find very different critical masses, compared with the static case (**5**). They use equations obtained also by Misner and Sharp (**75**). Misner (**76**) and Podurets, Zel'dovitch and Gusseinov (**77, 78, 79**) have studied collapse with escaping neutrinos. Numerical calculations have been carried out by Imshennick, Nadezhin, Gurevitch and Gusseinov (**80, 81**). Collapse with radiation has been treated analytically by Vardya (**82**). Nariai and Tomita have studied several problems appearing in the General Relativistic treatment of collapsing supermassive stars. Nariai (**83**) has first found the formulation in which Einstein's equations have no singularity at the stellar surface. Using this, Nariai and Tomita (**84, 85**) have proposed a way to treat the collapse system free from the Schwarzschild singularity. They have also shown (**86**) that the total energy loss by neutrino emission during the collapse is smaller than one tenth of the mass energy if the stellar mass is inferior to $10^6 M_\odot$. The behavior of the central region of a collapsing star with spherical symmetry and a finite pressure gradient has also been studied, touching upon the case of non-adiabatic collapse (**87**). Possible relevance of the C-field has been investigated (**88**). McVittie (**89**) has also treated some aspects of the question.

Rotation (**90**), magnetohydrodynamics (**91**), scalar interactions in superdense matter (**92**) have also been taken into consideration. Bonnor (**93**) finds that charged interior solutions may be relevant to the problem of the collapse. Harrison (**94**) discusses the meaningfulness of the study in an external system of the ultimate fate of a gravitationally collapsing body in the frame of different models of the universe.

Synge (**95**) has discussed the asymptotic behavior of photons escaping from gravitationally intense stars as $2\,GM/Rc^2$ tends towards unity. In this connexion, the problem of invisible ('hidden') masses in the universe has been brought up again. This has been discussed by Hoyle, Fowler, Burbidge and Burbidge (**96**), and Gratton (**97**). Zel'dovitch and Gusseinov (**98**) suggest the possibility of detecting collapsed stars in spectroscopic binaries.

BIBLIOGRAPHY

1. Robinson, I., Schild, A., Schücking, E. I. Eds., 1965, *Quasi Stellar Sources and Gravitational Collapse* including the Proc. of the first Texas Symp. on Relativistic Astrophysics, Dallas, 1963, University of Chicago Press, Chicago.
2. Zel'dovitch, Ya. B., Novikov, I. D. 1964, Relativistic Astrophysics I. *Usp. Fiz. Nauk*, **84**, (translated in *Soviet Phys. Usp.*, **7**, 763, 1965).
3. Zel'dovitch, Ya. B., Novikov, I. D. 1965, Relativistic Astrophysics II. *Usp. Fiz. Nauk*, **86**, 447 (translated in *Soviet Phys. Usp.*, **8**, 522, 1966).
4. Harrison, B. K., Thorne, K. S., Wakano, M., Wheeler, J. A. 1965, *Gravitation Theory and Gravitational Collapse*, Univ. of Chicago Press, Chicago.

5. Wheeler, J. A. 1966, Superdense Stars, in *A. Rev. Astr. Astrophys.*, Ed. L. Goldberg, **4**, 393.
6. Thorne, K. S. 1965, *The General Relativistic Theory of Stellar Structure and Dynamics in High Energy Astrophysics*, Course 35, Proc. Intern. School Phys. Enrico Fermi, Varenna, Ed. L. Gratton.
7. Thorne, K. S. 1966, *The General Relativistic Theory of Stellar Structure and Dynamics in High Energy Astrophysics*, Proc. Les Houches 1966 Summer School, Ed. C. de Witt, Blackie and Sons Ltd, Glasgow and London.
8. Chiu, H. Y. 1964, Supernovae, Neutrinos and Neutron Stars, *Ann. Phys.*, **26**, 364.
9. Chiu, H. Y., Salpeter, E. E. 1964, Surface X-Ray Emission from Neutron Stars, *Phys. Rev. Lett.*, **12**, 413.
10. Gratton, L., Szamosi, G. 1964, Remarks on hyperdense matter and the final state of matter, *Ann. Astrophys.*, **27**, 533.
11. Just, K. 1964, Baryon Conservation in Superdense Stars, *Phys. Lett.*, **13**, 219.
12. Kaplan, S. A. 1964, Problems of Cosmogony (*Voprosy Kosmogonii*), **10**, 58.
13. Bahcall, J. N., Wolf, R. A. 1965, Neutron Stars. I. Properties at Absolute Zero Temperature, *Phys. Rev.*, **140**, B 1445.
14. Bahcall, J. N., Wolf, R. A. 1965, Neutron Stars. II. Neutrino Cooling and Observability, *Phys. Rev.*, **140**, B 1452.
15. Bahcall, J. N., Wolf, R. A. 1965, Neutron Stars, *Phys. Rev. Lett.*, **14**, 343.
16. Bahcall, J. N., Wolf, R. A. 1965, An Observational Test of Theories of Neutrino Star Cooling, *Astrophys. J.*, **142**, 1254.
17. Cameron, A. G. W. 1965, Cosmic-Ray Production by Vibrating Neutron Stars, *Nature*, **206**, 1342.
18. Cameron, A. G. W. 1965, Possible Magnetospheric Phenomena Associated with Neutron Stars, *Nature*, **205**, 787.
19. Chin, C. W. 1965, Gravitational Radiation from a Spinning Ellipsoid of Uniform Density *Phys. Rev.*, **139**, B 761.
20. Ellis, D. G. 1965, Neutrino Emission by the URCA Process in Neutron Stars, *Phys. Rev.*, **139**, B 754.
21. Finzi, A. 1965, Cooling of a Neutron Star by the URCA Process, *Phys. Rev.*, **137**, B 472.
22. Harrison, B. K. 1965, Asymptotic Behavior of cold Superdense Stars, *Phys. Rev.*, **137**, B 1644.
23. Harrison, B. K. 1965, Equation of State of Matter at supernuclear Density, *Astrophys. J.*, **142**. 1643.
24. Inman, C. L. 1965, Baryon Stars, *Astrophys. J.*, **141**, 187.
25. Kaplan, S. A., Petrukhin, N. S. 1965, *Astr. Zu.*, **42**, 74.
26. Kuchowicz, B. 1965, Neutrinos in Superdense Matter: Massless Neutrino in a perfect fluid Sphere, *Acta Astr. Polon.*, **15**, 297.
27. Michel, F. C. 1965, Gravitational Effects of Luminosity, *Phys. Rev.*, **140**, B 514.
28. Pacini, F. 1965, Equilibrium for Neutron Stars, *Mem. Soc. astr. ital.*, **36**, 323.
29. Thorne, K. S. 1965, Gravitational Collapse and the Death of a Star, *Science*, **150**, 1671.
30. Tsuruta, S., Cameron, A. G. W. 1965, Cooling of Neutron Stars, *Nature*, **207**, 364.
31. Tsuruta, S., Wright, J. P., Cameron, A. G. W. 1965, Oscillation Periods of Neutron Stars, *Nature*, **206**, 1137; cf. also: Misner, C. W., Zapolsky, H. S. High Density Behavior and Dynamical Stability of Neutron Stars Models, *Phys. Rev. Lett.*, 1964, **12**, 635; *ibid.*, **13**, 122; 1964, Critical Mass Limit and Stability of some Relativistic Stellar Models, *Bull. Amer. phys. Soc.*, **9**, 422.
32. Bardeen, J. M., Thorne, K. S., Meltzer, D. W. 1966, A Catalogue of Methods for studying the normal Modes of Radial Pulsation of General Relativistic Stellar Models, *Astrophys. J.*, **145**, 505.
33. Finzi, A. 1966, Vibrational Energy of Neutron Stars and the Exponential Light Curve of Type I Supernovae 1966, *Phys. Rev. Lett.*, **15**, 599.
34. Hansen, C. J. 1966, Neutrino Energy Losses from Vibrating Neutron Stars, *Nature*, **211**, 1069.
35. Meltzer, W. D., Thorne, K. S. 1966, Normal Modes of Radial Pulsation of a Star at the End Point of Thermonuclear Evolution, *Astrophys. J.*, **146**, 514.

36. Ostriker, J. P., Bodenheimer, P., Lynden-Bell, D. 1966, Equilibrium Models of Differentially Rotating Zero-Temperature Stars, *Phys. Rev. Lett.*, **17**, 816.
37. Pacini, F. 1966, Size of a Superdense Star and Interactions between elementary Particles, *Nuovo Cim.*, **37**, 767.
38. Surdin, M. 1966, X-Ray emission from very dense Stars, *Nature*, **209**, 62.
39. Tsuruta, S., Cameron, A. G. W. 1966, The Rotation of Neutron Stars, *Nature*, **211**, 356.
40. Tsuruta, S., Cameron, A. G. W. 1966, Cooling and Detectability of Neutron Stars, *Can. J. Phys.*, **44**, 1863.
41. Tsuruta, S., Cameron, A. G. W. 1966, Some Effects of Nuclear Forces on Neutron Star Models, *Can. J. Phys.*, **44**, 1895.
42. Wolf, R. A. 1966, Some Effects of the Strong Interactions on the Properties of Neutron Star Matter, *Astrophys. J.*, **146**, 834.
43. Fowler, W. A. 1964, Massive Stars, Relativistic Polytropes and Gravitational Collapse, *Rev. mod. Phys.*, **36**, 545; *ibid.*, 1104.
44. Fowler, W. A. 1965, Massive Stars and Strong Radio-sources, read 24 April 1964 at the Am. Philos. Soc. (Proc. **109**, n°3).
45. Fowler, W. A. 1966, The General Relativistic Instability of Massive Stars, in Joint Discussion of Commission 28 and 40 on Radio Galaxies, *Trans. IAU*, **12B**, Academic Press, New York, p. 581; 1965, Proc. Thirteenth Conf. Phys. Brussels 1964, Interscience Pub., New York, p. 67.
46. Buchdahl, H. A. 1966, General Relativistic Fluid Spheres, *Phys. Rev.*, **116**, 1027.
47. Buchdahl, H. A. 1959, General Relativistic Fluid Spheres. II. General Inequalities for Regular Spheres, *Astrophys. J.*, **146**, 275.
48. Buchdahl, H. A. 1964, A relativistic fluid sphere resembling the Emden polytrope of Index 5, *Astrophys. J.*, **140**, 1512.
49. Gratton, L., Giannone, P. 1965, Very large masses in adiabatic Equilibrium, *Mem. Soc. astr. ital.*, **36**, no. 4.
50. Van der Borght, R. 1966, General Relativistic Fluid Spheres with Variable Polytropic Index, *J. austr. math. Soc.*, **6**, 139.
51. Tooper, R. F. 1964, General Relativistic Polytropic Fluid Spheres, *Astrophys. J.*, **140**, 434.
52. Tooper, R. F. 1965, Adiabatic Fluid Spheres in General Relativity, *Astrophys. J.*, **142**, 1541.
53. Tooper, R. F. 1966, The Standard model for Massive Stars, *Astrophys. J.*, **143**, 465.
54. Tooper, R. F. 1966, On the Binding Energy of Massive Stars in the Post Newtonian Approximation of General Relativity, *Astrophys. J.*, **146**, 282.
55. Hämeen-Anttila, K. A., Anttila, E. 1966, The Internal Constitution of Superstars. I. Theory, *Ann. Acad. Sci. Fennicae*, VI, *Physica* no. 191 (*Publ. Aarne Karjalainen Obs.* no. 5).
56. Hämeen-Anttila, K. A. 1966, The Relativistic Rosseland Mean Opacity, *Ann. Acad. Sci. Fennicae*, VI, *Physica* no. 192 (*Publ. Aarne Karjalainen Obs.* no. 6).
57. Hämeen-Anttila, K. A., Anttila, E. 1966, The Internal Constitution of Superstars. II. Convective Stars (*Publ. Aarne Karjalainen Obs.* no. 7, in press).
58. Hämeen-Anttila, K. A. 1966, Relativistic Theory of Supermassive Stars (*Publ. Aarne Karjalainen Obs.*, no. 8, in press).
59. Thorne, K. S. 1966, Validity in General Relativity of the Schwarzschild Criterion for Convection, *Astrophys. J.*, **144**, 201.
60. Chandrasekhar, S. 1965, The Stability of Gaseous Mass for Radial and Non-Radial Oscillations in the Post-Newtonian Approximation of General Relativity, *Astrophys. J.*, **142**, 1519; cf. also: *ibid.*, **142**, 1488; *ibid.*, 1964, **140**, 417.
61. Sato, H. 1966, General Relativistic Instability of Supermassive Stars, *Progr. theor. Phys.*, **35**, 241.
62. Bardeen, J. M. 1964, General Relativistic Hydrodynamics near Equilibrium, Second Texas Symp. Relativistic Astrophysics, Austin.
63. Osaki, Y. 1966, Pulsation and Evolution of Supermassive Stars, *Publ. astr. Soc. Jap.*, **18**, (in press); 1966, Note on the Solution of the Pulsation Equation of Stars in the Post-Newtonian Approximation of General Relativity (in press); cf. also Navez, J. 1966, Oscillations de la sphère homogène relativiste, *Bull. Soc. Roy. Sci. Liège*, **35**, 711.
64. Anand, S. P. S. 1964, Oscillatory Model for Radio Source 3C 273, *Nature*, **201**, 1016.

65. Fowler, W. A. 1964, A Talk presented at the Belfer Science forum of Yeshiva University, New York; 1966, The Stability of Supermassive Stars, *Astrophys. J.*, **144**, 180.

66. Chandrasekhar, S. 1965, The Post Newtonian Effects of General Relativity on the Equilibrium of Rotating Bodies, I. The McLaurin Spheroids and the Virial Theorem, *Astrophys. J.*, **142**, 1513.

67. Roxburgh, I. W. 1965, Effects of Rotation on the Stability of very massive Stars, *Nature*, **207**, 363.

68. Durney, B., Roxburgh, I. W. 1965, Stability of Rotating Massive Stars in General Relativity, *Nature*, **208**, 1305.

69. Anand, S. P. S. 1965, Magnetoturbulent Energy and Stability of Supermassive Stars, *Nature*, **207**, 1345; cf. also Pacholczyk, A. G. 1965, *Astrophys. J.*, **142**, 805; Layzer, D. 1965, *Astrophys. J.*, **141**, 837.

70. Bardeen, J. A., Anand, S. P. S. 1966, Upper Limit on the Binding Energy of a Magnetoturbulent Supermassive Star, *Astrophys. J.*, **144**, 953.

71. Debaise, R. 1966, Thèse de licence, Univ. de Liège.

72. Colgate, S. A., White, R. H. 1966, The Hydrodynamic Behaviour of Supernovae Explosions, *Astrophys. J.*, **143**, 626.

73. Kruskal, M. D. 1960, Maximal Extension of Schwarzschild Metric, *Phys. Rev.*, **119**, 1743.

74. May, M. M., White, R. H. 1966, Hydrodynamical Calculations of General Relativistic Collapse, *Phys. Rev.*, **141**, 1232.

75. Misner, C. W., Sharp, D. H. 1964, Relativistic Equations for Adiabatic Spherically Symmetric Gravitational Collapse, *Phys. Rev.*, **136**, B 571.

76. Misner, C. W. 1965, Relativistic Equations for Spherical Gravitational Collapse with Escaping Neutrinos, *Phys. Rev.*, **137**, B 1360.

77. Podurets, M. A. 1964, Asymptotic Behavior of the Optical Luminosity of a Star in Gravitational Collapse in terms of General Relativity Theory, *Astr. Zu.*, **41**, 1090, (*Soviet Astr.*, **8**, 868, 1965).

78. Podurets, M. A. 1967, *Astr. Zu.* (to be published).

79. Zel'dovitch, Ya. B., Gusseinov, O. K. 1965, *Dokl. Akad. Nauk SSSR*, **162**, 791.

80. Imshennik, V. S., Nadezhin, D. K. 1964, Gas Dynamical Model of a Type II Supernova Outburst, *Astr. Zu.*, **41**, 829 (*Soviet Astr.*, **8**, 664, 1965).

81. Gurevich, V. U., Gusseinov, O. 1965, *JETP Lett.*, **2**, 113.

82. Vardya, P. C. 1966, An analytical Solution for Gravitational Collapse with Radiation, *Astrophys. J.*, **144**, 943.

83. Nariai, H. 1965, On the Boundary Conditions in General Relativity, *Progr. theor. Phys.*, **34**, 173.

84. Nariai, H., Tomita, K. 1965, On the Problem of Gravitational Collapse, *Progr. theor. Phys.*, **34**, 155.

85. Nariai, H., Tomita, K. 1965, A simple and well-adjusted Exterior Metric for a collapsing or anticollapsing Star, *Progr. theor. Phys.*, **34**, 1046.

86. Nariai, H. Tomita, K. 1965, On the Applicability of a Dustlike Model to a Collapsing or Anticollapsing Star at High Temperature, *Progr. theor. Phys.*, **35**, 777.

87. Nariai, H., Tomita, K. 1966, Behavior of the Central Region in a Collapsing Star with Spherical Symmetry and Pressure Gradient, *Progr. theor. Phys.*, **35**, 786.

88. Nariai, H., Tomita, K. 1965, On the Gravito-C-Field in an Empty Region with Spherical Symmetry, *Progr. theor. Phys.*, **33**, 322.

89. McVittie, G. C. 1966, An example of Gravitational Collapse in General Relativity, *Astrophys. J.*, **143**, 682.

90. Wagoner, R. W. 1965, Rotation and Gravitational Collapse, *Phys. Rev.*, **138**, B 1583.

91. Cocke, W. J. 1966, Gravitational Collapse and Magnetohydrodynamics, *Phys. Rev.*, **145**, 1000.

92. Pacini, F. 1966, Gravitational Collapse and Scalar Interactions in Superdense Matter, *Ann. Astrophys.*, **29**, 193.

93. Bonnor, W. B. 1964, Gravitational Collapse, *Nature*, **204**, 868.

94. Harrison, E. 1964, Gravitational Collapse, *Nature*, **204**, 1179.

95. Synge, J. L. 1966, The Escape of Photons from Gravitationally Intense Stars, *Mon. Not. R. astr. Soc.*, **131**, 463.
96. Hoyle, F., Fowler, W. A., Burbidge, G. R., Burbidge, E. M. 1964, On Relativistic Astrophysics, *Astrophys. J.*, **139**, 909.
97. Gratton, L. 1964, The Number of Dead Stars and the Mass-to-Light Ratio in the Universe, *Ann. Astrophys.*, **28**, 115.
98. Zel'dovitch, Ya. B., Gusseinov, O. Kh. 1965, *JETP Lett.*, **1**, 11 (English Translation: **1**, 109, 1965).

III. OPACITY, EQUATION OF STATE, CONDUCTIVITY, ETC.

The important work of A. N. Cox *et al.* on stellar absorption coefficients and opacities has now been published (**1**, **2**, **3**) and contains a great deal of useful information not only on opacities but also on ionization, conductivity, equation of state, etc. A review of the general problem has also been published by Mayer (**4**).

Chiu (**5**) has calculated the Rosseland mean opacity due to Compton scattering in the semi-relativistic, semi-degenerate region. We may also report evaluation of thermal conductivity of a fully ionized gas by Sundaresan and Ta-You Wu (**6**) and by Wyller (**7**) in the case of degenerate stellar cores.

Rouse (**8**) has extended his investigations on ionization equilibrium and equations of state to high densities introducing in Saha equation a factor depending on the density and representing the probability that an ion can exist in a mean atomic volume related to a density-dependent normalization of the isolated atom wave-function. A paper by Stewart and Pyatt (**9**) on the lowering of the ionization potentials in plasma of finite density provides, apart from interesting result, both formal and numerical, a critical survey and significant references for a problem which is particularly important for the very low mass stars [cf. Gabriel (**10**)]. A treatment of the same problem by Harris (**11**) seems to suffer from counting twice the electrostatic corrections.

A general review of the physics of a degenerate gas by Guess (**12**) covers practically all aspects relevant to this section. Kaminishi (**13**) has studied the electron degeneracy at extremely high temperatures including the effect of pair creation and shown that the critical stage of degeneracy is given by $\rho/\mu_e \propto T^2$ instead of $\rho/\mu_e \propto T^3$. In the field of very high density, let us note also the work by Suda (**14**) who has studied the nuclear size correction to the relativistic Thomas-Fermi equation of state.

Kaminishi (**15**) has considered a gas at very high temperature composed of iron nuclei, α-particles, neutrons, electron-pairs and radiation. In particular, he has computed the adiabatic exponent which becomes smaller than 4/3 over a wide range of the dissociation of iron into helium suggesting the possibility of dynamical instability. However, in stellar models, according to Colgate and White (**16**), it does not mean that the corresponding region is always large enough to destabilize the star.

In recent years, the sensitivity of the internal structure to surface boundary condition especially for cold stars has been emphasized on different occasions and, in some cases, opacities, adiabats, equation of state in the very external layers are badly needed. Auman (**17**) has calculated the opacity in the infrared due to water vapor and found it to be important for $T < 3360°$K and dominant for $T < 2520°$K. Vardya (**18**) has also done a considerable amount of valuable work in this field. Much other work about opacities in the external layers has been done concerning the effect of various negative ions, free-free transitions of electrons, etc., but a review of these investigations will appear in reports of other Commissions.

BIBLIOGRAPHY

1. Cox, A. N. 1965, Stellar Absorption Coefficients and Opacities, in *Stars and Stellar Systems*, **8**, Eds. L. H. Aller and D. B. McLaughlin, Univ. of Chicago Press, Chicago, p. 195; 1966, Radiative Absorption and Opacity Calculations, in *Stellar Evolution*, Eds. R. F. Stein and A. G. W. Cameron, Plenum Press, New-York, p. 123; 1964, *J. quantit. Spectrosc. radiat. Transfer*, **4**, 737.
2. Cox, A. N., Stewart, J. H., Eilers, D. D. 1965, Effects of bound-bound Absorption on stellar Opacities, *Astrophys. J. Suppl.*, **11**, 1.
3. Cox, A. N., Stewart, J. H. 1965, Radiative and Conductive Opacities for eleven Astrophysical Mixtures, *Astrophys. J. Suppl.*, **11**, 22.
4. Mayer, H. L. 1964, *J. quantit. Spectrosc. radiat. Transfer*, **4**, 585.
5. Chiu, C. W. 1965, *Astrophys. J.*, **142**, 1481.
6. Sundaresan, M. K., Ta-You Wu 1964, *Can. J. Phys.*, **41**, 794.
7. Wyller, A. A. 1964, On thermal Conductivity in degenerate stellar Cores, *Astrophys. Norv.*, **9**, 79 (volume dedicated to S. Rosseland on the occasion of his 70th birthday).
8. Rouse, C. A. 1964, *Astrophys. J.*, **139**, 339.
9. Stewart, J. C., Pyatt, K. D. Jr. 1966, *Astrophys. J.*, **144**, 1203.
10. Gabriel, M. 1966, Structure et stabilité vibrationnelle des étoiles de faible masse, Thesis, Liège.
11. Harris, G. M. 1964, *Phys. Rev.*, **133**, A 427.
12. Guess, A. W. 1966, The Relativistic Degenerate Gas, in *Adv. Astr. Astrophys*, **4**, Ed. Z. Kopal, New York, p. 153; cf. also Kippenhahn, R., Thomas, H. C. 1964, *Z. Astrophys.*, **60**, 19.
13. Kaminishi, K. 1965, *Progr. theor. Phys.*, **34**, 1048.
14. Suda, K. 1965, *Publ. astr. Soc. Japan*, **17**, 427.
15. Kaminishi, K. 1965, *Progr. theor. Phys.*, **34**, 424.
16. Colgate, S. A., White, R. H. 1966, The hydrodynamic Behavior of Supernovae Explosions, Sects. V and VI, *Astrophys. J.*, **143**, 626.
17. Auman, J. R. 1966, *Astrophys. J. Suppl.* (in press).
18. Vardya, M. S. 1964, Atmospheric Rosseland mean Opacities, *Astrophys. J. Suppl.*, **8**, 277; 1965, Pressure Dissociation and molecular Hydrogen, *Mon. Not. R. astr. Soc.*, **129**, 345; 1965, A Method for Estimating H_2O Absorption, *J. quantit. Spectrosc. radia. Transfer*, **5**, 867; 1966, Pressure Dissociation and High Density Adiabats, *Mon. Not. R. astr. Soc.*, **133**, 145; 1966, March with Depth of molecular Abundances in the outer Layers of K and M Stars, *Mon. Not. R. astr. Soc.* (in press); 1966, Sources of Opacity in late-type Stars, a review in Proc. of the IAU Symposium on Late-Type Stars (in press).

IV. CONVECTION AND SURFACE BOUNDARY CONDITIONS

The meaning of Schwarzschild's criterion and its justification have been discussed by Lebovitz (1) in the general frame of the variational and energy methods applied to linear non-radial perturbations and it has been shown that it is both a sufficient and necessary condition in the adiabatic approximation.

This approach through the linear modes (cf. also sect. VII) has also been used by Saslaw and Schwarzschild (2) to show that the penetration of the convective currents in the stable region surrounding a convective core is negligible. This result is also confirmed by physical arguments about the entropy variation of the moving elements. In this respect, one may keep in mind that, in other contexts this kind of argument does not always seem to work (3).

The usual mixing-length theory of turbulent convection which has been submitted to a thorough analysis on different occasions by Spiegel (4) still seems to remain the main tool for the building up of convection zones in stars including sometimes various refinements of the type introduced first by Vitense and Böhm.

The latter are significant mainly in the external layers and there it might still be possible to improve the description of the structure of a convective zone, as well as of the penetration in the stable surroundings, by coupling of the linear modes analysis with some of the most significant non-linear effects (**5, 6, 7**). In this context, an interesting question is whether the usual homogeneous turbulent convection picture remains valid or whether it has to be superseded by one in which organized large scale temporary circulations prevail instantaneously at least in certain regions (**8**).

Surface boundary conditions for late-type stellar models and applications to the post-main-sequence evolution of stars with masses of 2·5 and 5 $M_{\odot}$ have been discussed by Henyey, Vardya and Bodenheimer (**9**). Tanaka (**10**) has computed boundary conditions for pure helium stars ($1 \leqslant M/M_{\odot} \leqslant 15$, electron scattering, radiation pressure included) and has determined the narrow allowed region in the HR diagram using the same procedure as Hayashi and Hôshi. Auman and Bodenheimer (**11**) in studying the early pre-main-sequence evolution including the effects of the opacity due to water vapor have found, for masses between 0·5 and 8·0 $M_{\odot}$, effective temperatures as low as 2600°K for reasonable values of the parameters in the mixing-length theory of convection.

The formation of zones with variable mean molecular weight $\bar{\mu}$ and their interaction with convection or 'semi-convection' is a delicate problem which takes many aspects in the course of stellar evolution. Baglin (**12**) has discussed the conditions for the setting in of convective instability in presence of a gradient of chemical composition due to gravitational stratification. In cases where the external convection zone extends down to a region of increasing $\bar{\mu}$, Unno (**13**) finds that penetrative convection will wash out the local gradient of $\bar{\mu}$, the edge of the convection zone being determined again by Schwarzschild's criterion. On the other hand, the difficulties connected with the transition between a convective core ($\bar{\mu}_i$) and a radiative envelope ($\bar{\mu}_e < \bar{\mu}_i$) when scattering is the main source of opacity (large masses) have been, as in the past, mainly solved by the introduction of 'semi-convective' zones (cf. however (**14**) where another type of solution is adopted) and Stothers (**15**), in particular, has given detailed attention to the problem.

However, in thermohaline convection, Veronis (**16**) finds that, while the presence of a stabilizing salt gradient deletes considerably the setting in of convective instability towards *infinitesimal* perturbations, overstability and *finite* amplitude convective instability occur practically as soon as the usual criterion of stability is violated. Similar unpublished results have been obtained by Kato and it would certainly be worthwhile to discuss carefully the implications of these results for the stellar case. According to Kato and Stern, the reverse situation with a stabilizing temperature gradient and destablizing $\bar{\mu}$-gradient, while stable for small adiabatic perturbation may also develop instabilities. A similar result was found also by Zahn (**17**) who points out that it may be important in presence of meridional circulations where destabilizing $\bar{\mu}$-gradients can arise.

Let us note also that direct numerical integrations (**18**) have been used recently to tackle the problem of convection in various simplified situations. Although we are still a long way of being able to handle the case of turbulent convection in stars, progress in computing techniques and computing machines may open a new interesting approach which should at least allow us to 'experiment' efficiently with the fundamental parameters of the problem.

The influence of rotation on the onset of convective instability through the *g*-modes of linear non-radial oscillations of polytropes has been discussed by M. J. Clement (**19**) by means of a variational principle. The effects of overshooting from a convective zone in the presence of rotation have been considered by Roxburgh (**20**). Demarque and Roeder (**21**) have attempted to connect the rotation of stars with their chromospheric properties and the existence of an outer convection zone. Smith (**22**) has studied the behavior of the circulation currents in fast rotating stellar models close to the surface and he suggests that the non-local nature of radiative transfer might be the most efficient factor in limiting the velocity to a finite value there.

For the interaction between convection and magnetic fields, cf. Sect. VII.

BIBLIOGRAPHY

1. Lebovitz, N. R. 1965, *Astrophys. J.*, **142**, 229; *ibid.*, **142**, 1257; 1966, *Astrophys. J.* (in press).
2. Saslaw, W. S., Schwarzschild, M. 1965, *Astrophys. J.*, **142**, 1468.
3. Cf. Moore, D. W. 1965, Proc. 2nd IAU-IUTAM Symp. on *Aerodynamical Phenomena in Stellar Atmospheres*, Nice (in press).
4. Spiegel, E. A. 1966, Energy Transport By Turbulent Convection, *Stellar Evolution*, Ed. R. F. Stein and A. G. W. Cameron, Plenum Press, New York, p. 143; Proc. 2nd IAU-IUTAM Symp. on *Aerodynamical Phenomena in Stellar Atmospheres*, Nice (in press); cf. also: Cowley, C. R. 1966, A possible Method for the Estimation of the Mixing-Length in turbulent Convection, *Astr. J.*, **71**, 382, where the mixing-length is found to be a function of the root-square of the Rayleigh number;
   Elässser, K. 1966, *Z. Astrophys.*, **63**, 65, where the Reynolds stresses tensor for non-isotropic turbulence is derived using methods of statistical mechanics and clarifying the assumptions necessary to get Wasiutynski's form of the tensor.
5. Böhm, K. H. 1963, *Astrophys. J.*, **137**, 881; *ibid.*, **138**, 297; a general review by the same author is presented in Proc. 2nd IAU-IUTAM Symp. on *Aerodynamical Phenomena in Stellar Atmospheres*, Nice 1965 (in press).
6. Spiegel, E. A. 1964, *Astrophys. J.*, **139**, 959; 1965, *Astrophys. J.*, **141**, 1068.
7. Kohl, K. 1966, *Z. Astrophys.*, **64**, 472.
8. Walter, K. 1965, *Ann. Astrophys.*, **28**, 648.
9. Henyey, L., Vardya, M. S., Bodenheimer, P. 1965, *Astrophys. J.*, **142**, 841.
10. Tanaka, Y. 1967 (in preparation).
11. Auman, J. R., Bodenheimer, P. 1967 (in preparation).
12. Baglin, A. 1964, *Ann. Astrophys.*, **27**, 389.
13. Unno, W. 1965, *Z. Astrophys.*, **61**, 268.
14. Iben, I. Jr. 1966, *Astrophys. J.*, **143**, 516.
15. Stothers, R. 1963, *Astrophys. J.*, **138**, 1074; 1965, *ibid.*, **141**, 671; 1966, *ibid.*, **143**, 91.
16. Veronis, G. 1965, Sears Foundation: *J. Mar. Res.*, **23**, 1.
17. Zahn, J. P. 1967 (in preparation).
18. Deardorff, J. W. 1963, *J. atm. Sci.*, **20**, 235;
   Fromm, J. E. 1964, *Methods in Computational Physics*, 3, Academic Press, New York;
   Veronis, G. 1966, Courant Inst. Math. Sci., report NYO 1480 61, Mathematics.
19. Clement, M. J. 1965, *Astrophys. J.*, **142**,, 243.
20. Roxburgh, I. W. 1965, *Mon. Not. R. astr. Soc.*, **130**, 223.
21. Demarque, P., Roeder, R. C. 1967, *Astrophys. J.* (in press).
22. Smith, R. C. 1966, *Z. Astrophys.*, **63**, 166.

V. STELLAR STRUCTURE AND EVOLUTION

Apart from a general treatise by Menzel, Bhatnagar and Sen (**1**) devoted to the physics of stellar interiors, we may mention also detailed review articles on stellar models for main sequence stars and sub-dwarfs (**2**), on stellar evolution and age determination (**3**), on the physics and constitution of white dwarfs (**4**), on the theories of novae and supernovae (**5**), on mass loss from stars (**6**) and on the evolution of protostars (**7**). Many other general articles of interest will be found in the proceedings of the International School 'Enrico Fermi', XVIII course (**8**) as well as in the proceedings of a NASA meeting (**9**).

Henyey and his group of associates have pursued the development of their method to compute evolutionary sequences (**10**, **11**) and comments or improvements have also been published by various authors (**12**, **13**, **14**, **15**).

Meggitt (16) has given the seven first terms of a series development around the center of a convective core. Tayler (17) has extended the criterion of Naur and Osterbrock for the occurrence of convective cores. Sack (18) has presented a method for the adjustment of radiative and convective zones in the external layers. Faulkner *et al.* (19) have discussed the surface boundary conditions in stars with an external convection zone.

Pre-main sequence contraction

Hayashi's theory of convective gravitational contraction has been applied to a large domain of stellar masses and has been generally confirmed. However, due to the difficulty of describing with precision the phases of dynamical instability (cf. sect. VI) preceding the sequence of contracting quasi-static models, the initial luminosity of these models remains in doubt. A first attack (20) shows that it must be quite high which seems also very likely on the basis of work by Bodenheimer (21) and von Sengbusch (22, 23).

Cameron and Ezer (24) who have considered again the problem of the contraction of small mass stars confirm that the latter cannot be stopped by nuclear energy generation if $M \lesssim 0{\cdot}1\ M_{\odot}$. After going through a stage of sub-luminous red dwarfs (25), these stars end as black dwarfs detectable as planetary companions of stars of small mass (26). For masses $0{\cdot}1 < M/M_{\odot} < 0{\cdot}3$, main sequence stars are wholly convective.

The gravitational contraction of stars of intermediary masses ($0{\cdot}4 \leqslant M/M_{\odot} \leqslant 10$) has been studied by many authors (27–34). Iben (28) has carefully followed the initial evolution of the abundances of the C-N-O group of elements and Bodenheimer *et al.* (29) have specially discussed the influence of these abundances on the end of the contraction phase. Bodenheimer has followed the combustion of Li (30) and of D and Be (31). Ezer and Cameron (32, 33) have more particularly concentrated on the contraction of solar mass stars and find that the final model properties and the contracting time vary appreciably especially with the treatment adopted for the external layers (convection, opacity).

The contraction of massive stars ($10 \leqslant M/M_{\odot} \leqslant 100$) is little affected by Hayashi's theory as they spend, at most, a few hundred years in the entirely convective stage (34). Tanaka and Sakashita (35) have found that the convective core appears already during the contraction because of the high relative radiation pressure if $M > 40\ M_{\odot}$. According to Meggitt (36), the contraction time increases again when $(M/M_{\odot}) > 32$.

Penston (37) has shown that the observations of NGC6530 and NGC2364 are compatible with Hayashi's theory.

McCrea (38) has attempted to explain the stars above the turn off points in some clusters in terms of mass exchange between the components of a close binary while Williams (39) and Dorschner *et al.* (40) attempt to account for their existence in terms of accretion.

Schatzman (41, 42) has studied the influence of a rapid rotation coupled with magnetic activity on stellar contraction.

Penston (43) has suggested an interpretation of the Neugebauer-Martz-Leighton objects as stars in formation dissipating their excess angular momentum.

Main-sequence

Main sequence models are computed essentially in view of testing the influence of various parameters. According to Morris and Demarque (44) models with $X = 0{\cdot}63$ and $Z = 0{\cdot}03$ are in good agreement with the observations. Mrs Massevitch and her associates have shown (45, 46, 47) that the homogeneous models on the main sequence are quite sensitive to the initial chemical composition. The results have also been compared with observational data (48) and the possible sources of disagreement, discussed (49, 50). For masses $M \leqslant 2{\cdot}5\ M_{\odot}$, the external convective zone is not deep enough, according to Tutukov (51), to allow enough depletion of Li^7 to account for Herbig's observations (52).

We may also report studies of the effects on models around the mid-main-sequence of different factors such as opacity and isotopic abundances (**53, 54**), of the generation of energy outside the convective core (**55**), of the pressure of radiation (**56**), of the bound-bound transitions (**57**), of the central convection on energy generation (**58**). Boury (**59**) has shown that a variable scattering opacity modifies appreciably the models of massive stars.

A comparison (**60, 61**) between models computed following Henyey's method (relaxation) and ordinary models shows satisfactory agreement although the first method seems to be more sensitive to the number of mass shells adopted.

Detailed models for $M = 1{\cdot}48\ M_\odot$ (**61**), $0{\cdot}6\ M_\odot$ (**53**), $0{\cdot}27\ M_\odot$ (Kr 60 A) and $0{\cdot}16\ M_\odot$ (Kr 60 B) (**63**) have also been constructed.

Post main-sequence evolution

Stothers (**64, 65, 66**) has followed the evolution of a $30\ M_\odot$ star of Population I up to the point where He is exhausted at the centre. He has also discussed (**67**) the development of semi-convective zones in stars with masses between 45 and $1000\ M_\odot$ as they burn their hydrogen. Evolutionary sequences covering about the same interval have also been constructed by Kotok (**68–72**) for $M = 15{\cdot}6$, 20 and $30\ M_\odot$. The relative contributions of the different energy sources during these phases have been compared (**73**). Van der Borght (**74**) has studied the phase of hydrogen burning in heavy mass stars (40, 60, 70, $120\ M_\odot$) composed initially of pure hydrogen.

Iben (**75–78**) has followed in great detail the evolution of Population I stars with masses 3, 5, 9 and $15\ M_\odot$ up to the exhaustion of He at the centre taking into account all the possible nuclear reactions. He has also pushed the evolutions of 1, 1·25 and $1{\cdot}5\ M_\odot$ stars up to the red giant stage (**79**).

Evolution through the carbon flash has been studied by Kippenhahn *et al.* (**80, 81**) neglecting the neutrino emission. In taking the latter into account, Weigert (**82**) has encountered 'thermal' instabilities when He burns in a shell. Hofmeister (**83, 84**) has followed the evolution of 5, 7 and $9\ M_\odot$ stars up to and through the He burning phases for $X = 0{\cdot}602$, $Z = 0{\cdot}044$ and $X = 0{\cdot}739$, $Z = 0{\cdot}021$. She finds that the loops of the evolutionary paths in the HR diagram during He-burning are very sensitive to the initial chemical composition. According to Dluzhnevskaya (**85**) and Varskowsky (**86**), the comparison between the number of stars observed and the number predicted by theory reveals a serious lack of agreement in the Hertzsprung-gap. The situation improves if the evolutionary loops are removed but agreement could also be obtained by using a variable initial luminosity function. The models so constructed have been used (**87, 88, 89**) to deduce some properties of the β Cephei stars ($X \simeq 0{\cdot}7$, $M \simeq 4{\cdot}5\ M_\odot$; life-time $\simeq 1{\cdot}5\ 10^6$ years). It was also found (**90, 91**), in comparing ordinary models and those given by Henyey's method that a divergence sets in after H-burning in the centre affecting the shape of the evolutionary loops in the HR diagram.

Hydrogen burning has also been the object of a variety of other papers (**92–98**). Reiz *et al.* (**98**) find that, for stars of spectral type later than about A7, age determinations could be significant only after the location of the zero-age main sequence has been fixed taking variations in initial chemical composition into account.

Solar models and solar evolution have been studied (**99, 100**) especially to determine the neutrinic luminosity (**101, 102, 103**) of the Sun in view of its possible detection by appropriate observations (**104**).

The helium flash has been studied in detail by Härm and Schwarzschild (**105**) and by Thomas (**106, 107**) for a $1{\cdot}3\ M_\odot$ star of population II. Thomas shows that, if energy losses by neutrinos are taken into account, He-burning starts in a shell around the centre. In general, it appears that the flash is much less violent than anticipated and does not lead to a mixing of the envelope into regions where nuclear reactions can take place. Sugimoto (108) finds that if conduction is taken into account in the core, the flash leads to the appparition of a convection zone which, under certain conditions, could lead to a real explosion of the core. According to Eggleton (109), the same conduction could, for certain chemical compositions, start the flash outside the core.

A considerable amount of work (110–118) has also been devoted to the interpretation of the horizontal branch of the globular clusters, Faulkner (113) and Faulkner and Iben (114) finding, in particular, that it is facilitated by adopting a much larger initial abundance of He than is customary for population II stars.

The effects of mass loss on stellar evolution have continued to be studied by different authors (119–124) and the possible effects of a variable G on the evolution of the Sun (125, 126) have been considered.

A beginning has been made on the evolution of pure helium stars (127–130) and some models for pure carbon stars have also been built (131, 132).

The very important problem of the evolution of close binary systems taking into account the exchange of mass between the components has received serious attention by Kippenhahn and Weigert (133), by Paczynski (134, 135) and Zialkowski (136, 137) and by Efrenov (138). In view of tackling the evolution of these systems after the start of helium burning, Giannone (139) has constructed main-sequences for models comprising a helium core and an envelope with high hydrogen content for various values of the ratio of the mass of the core to the total mass.

Last stages of stellar evolution

The idea that planetary nebulae constitute an intermediary stage in the evolution towards the white dwarf stage has received much attention from a theoretical as well as from an observational point of view and has been particularly developed by Savedoff (140). Vila (141) and Chin, Chiu and Stothers (142) have followed the evolution of hot stars of 0·7 to 2 $M_{\odot}$ with high initial luminosity after complete exhaustion of the nuclear fuel. In the HR diagram, their models do indeed evolve from the region of planetary nebulae to the white dwarf region. They also find that neutrino losses accelerate considerably the contraction and lead to better agreement with observations. Rose (143) finds that models of small mass stars with a helium envelope in which nuclear reactions take place evolve like planetary nebulae. According to Kahoutek (144) the latter could be produced by novae while Deeming (145) considers that they could issue from infrared supergiants. Let us recall however that some opposition to these views has been expressed by Woolf (146).

Mestel and Ruderman (147) have evaluated the energy content of white-dwarfs and their rate of cooling.

The possible influence of mass loss on the formation of white dwarfs has been considered by Auer and Woolf (148) and by Williams (149). The evolution of stars with masses in excess of the Chandrasekhar limit, during the phases preceding a supernova explosion has been studied by Chiu (150) and by Rakavy and Shaviv (151).

Takarada *et al.* (152) have evaluated the relation between central densities and temperatures in polytropic or isothermal configurations to estimate the general features of advanced stages of evolution.

BIBLIOGRAPHY

1. Menzel, D., Bhatnagar, P. L., Sen, H. K. 1963, *Stellar Interior*, vol. VI, in the series The International Astrophysics Series, Chapman and Hall Ltd., London.
2. Strömgren, B. 1965, Stellar Models for Main-Sequence Stars and Subdwarfs, in *Stellar Structure*, vol. VIII, in the series *Stars* and *Stellar Systems*, Ed. Aller, L. H. and McLaughin, D. B.; The University of Chicago Press, Chicago and London, chap. 4, p. 269.

3. Sears, R. L., Brownlee, B. B. 1965, Stellar Evolution and Age Determination, in *Stellar Structure*, vol. VIII, in the series *Stars and Stellar Systems*, Ed. Aller, L. H. and McLaughin, D. B.; The University of Chicago Press, Chicago and London, chap. 11, p. 575.
4. Mestel, L. 1965, The Theory of White Dwarfs, in *Stellar Structure* vol. VIII, in the series *Stars and Stellar Systems*, Eds. Aller, L. H. and McLaughin, D. B., The University of Chicago Press, Chicago and London, chap. 5, p. 297.
5. Schatzman, E. 1965, Theory of Novae and Supernovae, in *Stellar Structure*, vol. VIII, in the series *Stars and Stellar Systems*, Eds. Aller, L. H. and McLaughin, D. B., The University of Chicago Press, Chicago and London, chap. 6, p. 327.
6. Weymann, R. 1963, Mass Loss from Stars, *A. Rev. Astr. Astrophys.*, **1**, Ed. Goldberg, L., Rev. Inc., Palo Alto, Calif., p. 97.
7. Hayashi, C. 1966, Evolution of Protostars, *A. Rev. astr. Astrophys.*, **4**. Ed. Goldberg, L., Rev. Inc., Palo Alto, Calif. p. 171.
8. Gratton, L. 1963, *Star Evolution*, Academic Press, New York and London.
9. Stein, R. F., Cameron, A. G. W. 1966, *Stellar Evolution*, Plenum Press, New York.
10. Henyey, L. G., Forbes, J. E., Gould, M. L. 1964, *Astrophys. J.*, **139**, 306.
11. Henyey, L. G., Vardya, M. S., Bodenheimer, P. 1965, *Astrophys. J.*, **142**, 841.
12. Larson, R. B., Demarque, P. 1964, *Astrophys. J.*, **140**, 524.
13. Paczynski, B. 1966, *Postepy Astr.*, **14**, 3.
14. Virgopia, M. 1966, *Mem. Soc. astr. ital.*, **37**, 3.
15. Hofmeister, E., Kippenhahn, R. Weigert, A., Methods of Stellar Evolution Calculations, vol VII, in the series *Methods of Computational Physics*, Eds. Alder, B., Fernbach, S., Rotenberg, M., Academic Press, New York and London (in press).
16. Meggitt, S. M. A. 1966, *Aust. J. Phys.*, **19**, 606.
17. Tayler, R. J. 1966, *Mon. Not. R. astr. Soc.* (to be published).
18. Sack, M. 1966, *Ann. Astrophys.*, **29**, 633.
19. Faulkner, J., Griffiths, M., Hoyle, F. 1966, *Mon. Not. R. astr. Soc.*, **129**, 363.
20. Hayashi, C., Nakano, T. 1965, Thermal and Dynamical Properties of a Protostar and its Contraction to the Stage of Quasi-Static Equilibrium, *Progr. theor. Phys.*, **34**, 754.
21. Bodenheimer, P. 1966, The influence of initial conditions on pre-main-sequence calculations, *Astrophys. J.*, **144**, 709.
22. Sengbusch K. von. 1966, *Mitt. astr. Ges.*, **21**.
23. Sengbusch K. von, Temesvary, S. 1966, Schematic pre-main-sequence Evolution, in *Stellar Evolution*, Eds. Stein, R. F., Cameron, A. G. W., Plenum Press, New York, p. 209.
24. Ezer, D., Cameron, A. G. W. 1966, The Early contracting Phases of Stars, in the Proc. IAU Symp. on Late-Type Stars, Trieste (in press).
25. Kumar, S. S. 1964, *Observatory*, **84**, 18.
26. Kumar, S. S. 1964, *Z. Astrophys.*, **58**, 248.
27. Hayashi, M. 1965, *Publ. astr. Soc. Japan*, **17**, 2.
28. Iben, I. Jr. 1965, *Astrophys. J.*, **141**, 993.
29. Bodenheimer, P., Forbes, J. E., Gould, M. L., Henyey, L. G. 1965, *Astrophys. J.*, **141**, 1019.
30. Bodenheimer, P. 1965, *Astrophys. J.*, **142**, 451.
31. Bodenheimer, P. 1966, *Astrophys. J.*, **144**, 103.
32. Ezer, D., Cameron, A. G. W. 1963, *Icarus*, **1**, 422.
33. Ezer, D., Cameron, A. G. W. 1965, *Can. J. Phys.*, **43**, 1497.
34. Ezer, D., Cameron, A. G. W. 1966, The contraction phase of Solar Evolution, in *Stellar Evolution*, Eds. Stein, R. F., Cameron, A. G. W., Plenum Press, New York, p. 203.
35. Tanaka, Y., Sakashita, S. 1964, *Publ. astr. Soc. Pacific*, **16**, 173.
36. Meggitt, S. M. A. 1965, *Austr. J. Phys.*, **18**, 297.
37. Penston, M. V. 1964, *Observatory*, **84**, 141.
38. McCrea, W. H. 1964, *Mon. Not. R. astr. Soc.*, **128**, 147.
39. Williams, I. P. 1964, *Ann. Astrophys.*, **27**, 198.
40. Dorschner, J., Gürtha, J., Schmidt, K. H. 1965, *Astr. Nachr.*, **288**, 159.

41. Schatzman, E. 1963, The early Stages of Stellar Evolution, in *Star Evolution*, Ed. Gratton, L., Academic Press, New York and London, p. 177.
42. Schatzman, E. 1966, Role of Magnetic Activity during Stellar Formation, in *Stellar Evolution*, Eds. Stein, R. F., Cameron, A. G. W., Plenum Press, New York, p. 215.
43. Penston, M. V. 1966, *Observatory*, **86**, 121.
44. Morris, S. C., Demarque, P. 1966, *Z. Astrophys.*, **64**, 238.
45. Massevitch, A. G., Ruben, G. V., Lomnev, S. P., Popova, E. I. 1965, *Nauchnye Informatsii Astr. Sov. Acad. Nauk SSSR* (Scientific Informations of the Astr. Council USSR Acad. Sci.), **1**, 2.
46. Massevitch, A. G., Ruben, G. V., 1966, *Nauchnye Informatsii Astr. Sov. Acad. Nauk. SSSR*, **3**, 37.
47. Popova, E. I. 1967, Sbornik 'Peremennye Zvezdy i Zvezdnaya Evolutsiia' (Variable Stars and Stellar Evolution) 'Nauka', Moscow (in press).
48. Massevitch, A. G., Popova, E. I. 1967, *Nauchnye Informatsii Astr. Sov. Acad. Nauk SSSR*, **6** (in press).
49. Ruben, G., Massevitch, A. G. 1966, *Astr. Zu.*, **43**, 499.
50. Ruben, G. 1966, *Mitt. astr. Ges.*, **19**.
51. Tutukov, A. V. 1966, *Nauchnye Informatsii Astr. Sov. Acad. Nauk SSSR*, **4**.
52. Weymann, R., Sears, R. L. 1965, *Astrophys. J.*, **142**, 174.
53. Fritze, K. 1966, *Monatsber. Deutschen Acad. Wiss. Berlin*, **8**, 7.
54. Ziolkovsky, G. 1966, *Acta Astr. Pol. Acad. Nauk* (in press).
55. Bahng, J. 1965, *Astrophys. J.*, **140**, 1041.
56. Auman, J. R., Bahng, J. 1965, *Astrophys. J.*, **142**, 170.
57. Imshennick, V. S., Nadezhin, D. K. 1966, *Astr. Zu.*, **43**.
58. Gabriel, M., Ledoux, P., Denis, C. 1965, *Ann. Astrophys.*, **28**, 643.
59. Boury, A. 1964, *Astrophys. J.*, **140**, 1322.
60. Imshennick, V. C., Kotok, E. V., Nadezhin, D. K. 1965, *Nauchnye Informatsii Astr. Sov. Accd. Nauk SSSR*, **1**, 48.
61. Frantsman, Yu. L. 1966, *Nauchnye Informatsii Astr. Sov. Acad. Nauk SSSR*, **3**, 52.
62. Bennick, A. H., Motz, L. 1965, *Astrophys. J.*, **141**, 195.
63. Gabriel, M. 1966, Internal Structure and Vibrational Stability of wholly convective Red Dwarfs, Coll. on Late-Type Stars, Trieste, p. 347.
64. Stothers, R. 1963, *Astrophys. J.*, **138**, 1074.
65. Stothers, R. 1964, *Astrophys. J.*, **140**, 510.
66. Stothers, R. 1966, *Astrophys. J.*, **143**, 91.
67. Stothers, R. 1966, *Astrophys. J.*, **144**, 954.
68. Kotok, E. V., Nadezhin, D. K. 1967, *Nauchnye Informatsii Astr. Sov. Acad. Nauk SSSR*, **6**.
69. Kotok, E. V. 1965, *Astr. Zu*, **42**, 1221.
70. Kotok, E. V. 1966, *Nauchnye Informatsii Astr. Sov. Acad. Nauk SSSR*, **3**, 1.
71. Kotok, E. V. 1966, *Astr. Zu.*, **43**, 316.
72. Kotok, E. V. 1966, *Astr. Zu.*, **43**, 1025.
73. Nadezhin, D. K. 1966, *Nauchnye Informatsii Astr. Sov. Acad. Nauk SSSR*, **4**.
74. Van der Borght, R. 1964, *Austr. J. Phys.*, **17**, 165.
75. Iben, I. Jr. 1964, *Astrophys. J.*, **140**, 1631.
76. Iben, I. Jr. 1966, *Astrophys. J.*, **143**, 483.
77. Iben, I. Jr. 1966, *Astrophys. J.*, **143**, 505.
78. Iben, I. Jr. 1966, *Astrophys. J.*, **143**, 516.
79. Iben, I. Jr. 1967, *Astrophys. J.*, **147**, 624.
80. Kippenhahn, R., Thomas, M. C., Weigert, A. 1965, *Z. Astrophys.*, **61**, 241.
81. Kippenhahn, R., Thomas, M. C., Weigert, A. 1966, *Z. Astrophys.*, (in press).
82. Weigert, A. 1965, *Mitt. astr. Ges.*, **19**, 61.
83. Hofmeister, E. 1965, *Mitt. astr. Ges.*, **19**, 90.
84. Hofmeister, E. 1965, Proc. IAU Coll. Bamberg, p. 224. *Kl. Veröff. Remeis-Sternw. Bamberg*, **4**, no. 40.
85. Dluzhnevskaya, O. B. 1966, *Nauchnye Informatsii Astr. Sov. Acad. Nauk SSSR*, **3**, 77.

86. Varskovsky, Y. I. 1966, *Nauchnye Informatsii Astr. Sov. Acad. Nauk SSSR*, **4**.
87. Percy, J. R., Demarque, P. *Astrophys. J.* (in press).
88. Kopylov, I. M. 1966, *Izv. Krym. astrofiz. Obs.*, **36**.
89. Efremov, Yu. N., Kopylov, I. M. 1966, *Izv. Krym. astrofiz. Obs.*, **36**.
90. Ziolkovsky, G. 1966, *Acta Astr. Pol. Acad. Nauk* (in press).
91. Frantsman, Yu. L. 1967, *Nauchnye Informatsii Astr. Sov. Acad. Nauk SSSR*, **6**.
92. Auman, J. R. 1965, *Astrophys. J.*, **142**, 462.
93. Morris, S. C., Demarque, P. 1965, *Astr. J.*, **70**, 145.
94. Hallgren, E. L., Demarque, P., *Astrophys. J.* (in press).
95. Hallgren, E. L. 1966, *Astrophys. J.* (in press).
96. Paczynski, B. 1966, *Postepy Astr.*, **14**, 3.
97. Massevitch, A. G., Ruben, G. V., Lomnev, C. P., Frantsman, Yu. L. 1966, *Nauchnye Informatsii Astr. Sov. Acad. Nauk SSSR*, **4**.
98. Reiz, A., Petersen, J. O., Hejlesen. (To be published.)
99. Demarque, P., Perey, S. R. 1964, *Astrophys. J.*, **140**, 524.
100. Chiu, H. Y. 1965, Proc. First Conf. on Planetology and Space Mission Planning.
101. Sears, R. L. 1964, *Astrophys. J.*, **140**, 477.
102. Sears, R. L. 1966, Solar Models and Neutrino Fluxes, in *Stellar Evolution*, Eds. Stein R. F., Cameron, A. G. W., Plenum Press, New York, p. 245.
103. Pochoda, P., Reeves, M. 1964, *Planet. Space Sci.*, **12**, 119.
104. Bahcall, J. H., Davis, R. 1966, in *Stellar Evolution*, Eds. Stein, R. F., Cameron, A. G. W., Plenum Press, New York, p. 241.
105. Härm, R., Schwarzschild, M. 1966, *Astrophys. J.*, **145**, 496.
106. Thomas, H. C. 1965, *Mitt. astr. Ges.*, **19**, 93.
107. Thomas, H. C. 1966, *Mitt. astr. Ges.*, **21**.
108. Sugimoto, D. 1964, *Progr. theor. Phys.*, **32**, 703.
109. Eggleton, P. P. 1966, *Mon. Not. R. astr. Soc.*, **132**, 479.
110. Osaki, Y. 1963, *Publ. astr. Soc. Japan*, **15**, 498.
111. Suda, K., Virgopia, M. 1966, *Astrophys. J.*, **143**, 75.
112. Suda, K., Virgopia, M. 1966, *Astrophys. J.*, **143**, 87.
113. Faulkner, J. 1966, *Astrophys. J.*, **144**, 978.
114. Faulkner, J., Iben, I. Jr. 1966, *Astrophys. J.*, **144**, 995.
115. Petersen, J. O. 1966, *Publ. og mindre Medd. Københavns Obs.*, no. 186.
116. Jørgensen, M. E., Petersen, J. O. 1967 (to be published).
117. Hayashi, C., Hoshi, R., Sugimoto, D. 1965, *Progr. theor. Phys.*, **34**, 885.
118. Sugimoto, D., Yamanoto, Y. 1966, *Progr. theor. Phys.*, **36**, 17.
119. Stothers, R. 1966, On the Theory and Evolution of completely mixed Stars with Mass Loss, *Mon. Not. R. astr. Soc.*, **131**, 253.
120. Tanaka, Y. 1966, *Publ. astr. Soc. Japan*, **18**, 47.
121. Tanaka, Y. 1966, *Progr. theor. Phys.*, **36**.
122. Hartwick, F. D. A. 1966, *Astrophys. J.* (in press).
123. Dluzhnevskaya, O. B. 1965, *Nauchnye Informatsii Astr. Sov. Acad. Nauk SSSR*, **2**, 1.
124. Dluzhnevskaya, O. B. 1966, *Astr. Zu.*, **43**, 1226.
125. Ezer, D., Cameron, A. G. W. 1966, *Can. J. Phys.*, **44**, 593.
126. Roeder, R. C., Demarque, P. 1966, *Astrophys. J.*, **144**, 1016.
127. Divine, M. 1965, *Astrophys. J.*, **142**, 725.
128. Giannone, P., Giannuzzi, M. A. 1965, *Nuovo Cim.*, ser. X, **36**, 1267.
129. Giannone, P. 1966, *Z. Astrophys.*, **64**, 229.
130. Cimino, M., Giannone, P., Giannuzzi, M. A., Masani, A., Virgopia, M. 1964, *Nuovo Cim.*, ser. X, **33**, 663.
131. Deinzer, W., Salpeter, E. E. 1965, *Astrophys. J.*, **142**, 813.
132. Noels-Grötsch, A., Boury, A., Gabriel, M. 1967, *Ann. Astrophys*, **30**, 13.
133. Kippenhahn, R., Weigert, A. 1966, *Mitt. astr. Ges.*, **21**.
134. Paczynski, B. 1966, IAU Coll., *The Evolution of Double Stars*, Uccle (to be published).
135. Paczynski, B. 1966, *Acta Astr.* (2 papers in press).
136. Paczynski, B., Ziolkowski, J. 1966, *Acta Astr.* (in press).

137. Ziolkowski, J. 1966, *Acta Astr.* (in press).
138. Efremov, Yu. M. 1966, *Astr. Zu.* (in press).
139. Giannone, P. 1967, *Z. Astrophys* (in press).
140. Savedoff, M. P. 1966, *Astrophys. J.*, **145**, 648.
141. Vila, S. C. 1966, *Astrophys. J.*, **146**, 437.
142. Chin, C. W., Chiu, H. Y., Stothers, R. 1966, *Ann. Astrophys.*, **39** (in press).
143. Rose, W. K. 1966, *Astrophys. J.*, **144**, 1001.
144. Kohoutek, L. 1964, *Bull. astr. Inst. Czech.*, **15**, 81.
145. Deeming, T. J. 1965, *Publ. astr. Soc. Pacific*, **77**, 443.
146. Woolf, M. J. 1966, *Astrophys. J.*, **145**, 649.
147. Mestel, L., Ruderman, M. A. 1967, *Mon. Not. R. astr. Soc.* (in press).
148. Auer, L. H., Woolf, M. J. 1965, *Astrophys. J.*, **142**, 982.
149. Williams, J. P. 1965, *Ann. Astrophys.*, **28**, 672.
150 Chiu, Y. U. 1966, Pre-Supernova Evolution, in *Stellar Evolution*, Eds. Stein, R. F., Cameron, A. G. W., Plenum Press, New York, p. 179.
151. Rakavy, G., Shaviv, G. 1966, *Astrophys. J.* (in press).
152. Takarada, K., Sato, H., Hayashi, C. 1966, *Progr. theor. Phys.*, **36**, 504.

VI. STELLAR STABILITY AND STELLAR PULSATIONS; SHOCK WAVES AND NON-STATIONARY PROCESSES IN STARS

Interesting developments have occurred in the field of stellar stability during the period under review and we shall try to situate them with respect to the classical subdivisions of the subject: dynamical stability, vibrational (or pulsational) stability and secular (or thermal) stability, reviewed for instance in (1).

Note. Although probably difficult to enforce, a convention might be useful here. If the time dependence is taken to be of the form exp $(i\,\sigma\,t)$, then:

(*a*) *dynamical stability* or *instability* might be reserved to characterize the behavior of purely *conservative* systems in *absolute equilibrium* (adiabatic perturbations, real eigenvalue of the form $\lambda = \sigma_a^2$).

(*b*) *Vibrational stability* or *instability*, to characterize the *damping* or *amplification*, under the influence of *non-conservative factors*, of the oscillations found under (*a*) when the system is dynamically stable (non-adiabatic perturbations, complex eigenvalues of the form $\lambda = a + ib$, $\sigma = \pm\,(\sigma_a' + i\,\sigma')$, $\sigma'(+$ or $-)$ due to the contributions of the non-conservative terms.

(*c*) *Secular stability* or *instability* to characterize the amplification or damping of slow motions (for which the inertial terms are negligible) under the influence of the non-conservative effects (non-adiabatic slow perturbations, σ of the form $i\,\alpha$, α : + or $-$); *thermal stability* or *instability*, to characterize the more complicated situation where non-conservative thermal factors (for instance in some cases neutrino losses) can have very strong effects, leading to motions on a time-scale sometimes comparable to that of dynamical motions (inertial terms not necessarily negligible).

(*d*) In the case of conservative systems in relative equilibrium (for instance, in presence of rotation) it happens that some eigenvalues become complex, due to a failing of the dynamical stability of the corresponding absolute equilibrium, the extra-relative force (for instance, due to rotation) providing the restoring force responsible for the real part of λ. In such cases (stability of rotating fluid masses for instance), the word *overstability* has been used in recent years and it could perhaps be kept to characterize this kind of situations although it was coined originally, and not too happily, by Eddington to denote what is called here *vibrational instability*. A somewhat closer case of instability of relative equilibrium may also occur in more complex circumstances as in the Rayleigh problem for convection in presence of rotation or magnetic field and has generally been called there too, overstability.

Dynamical Stability and adiabatic oscillations

In the case of purely *radial* oscillations of 'classical' stars, the essential result (namely that dynamical instability occurs only if $\Gamma < 4/3$ in an appreciable part of the star), has been known for some time but work has continued on interesting applications at the very beginning of stellar evolution (**2**), when ionization of H, He and dissociation of H_2 reduces Γ below 4/3 or towards the very end (**3**) when nuclear equilibrium tends to be established between Fe, α-particles and neutrons or later between He, protons, neutrons and electrons pairs lowering again Γ below 4/3. Equilibrium between electron pairs and radiation may also lead to low values of Γ and recently Rakavy and Shaviv (**8**) have found that this can render a large mass star dynamically unstable before nuclear equilibrium gets established.

The newly discovered dynamical instability due to General Relativity corrections in large mass or very dense stars has been the object of further investigations reviewed mostly in Section II. Let us add Chandrasekhar's discussion (**4**) of this factor in ordinary white dwarfs and the investigations of Baglin (**5**) who has further taken into account the effects of finite temperature and of equilibrium between electrons and nuclei and gives critical masses and radii beyond which dynamical instability occurs for different values of the mean atomic weight.

The overtone radial adiabatic oscillations of massive stars have been studied numerically by Van der Borght (**6**) and the results compared with the analytical asymptotic behavior. Bhatia and Kushwaha (**7**) have studied the radial adiabatic oscillations of a generalized Roche model.

From a methodological point of view, the energy method seems to have become more popular (**8, 9**) following a very clear unpublished presentation by Dyson. In (**8**) a beginning is made in the discussion of the third order terms.

Non-radial oscillations have been the object of concentrated numerical attacks by Smeyers (**10**) and by Van der Borght and Wan Fook Sun (**11**) taking into account the full 4th order problem when the perturbation of the gravitational potential is not neglected.

The application (**12**) of the variational principle derived by Chandrasekhar for non-radial oscillations has led directly or indirectly to some clarifications. As far as the interaction between dynamical instability towards these oscillations and Schwarzschild's criterion for convective stability is concerned we have already referred, in Section IV, to Lebovitz's proof of the existence of unstable *g*-modes as soon as Schwarzschild's criterion is violated in a region however small. Ledoux and Smeyers (**13**) have discussed the complete spectrum in such cases and shown that the *g*-spectrum splits then into two spectra both converging to zero, one positive (stable) corresponding to oscillations with large amplitudes and nodes in the convectively stable region and rapidly decreasing amplitudes in the convectively unstable region and one negative (unstable) with amplitudes whose behavior is exactly the opposite. The latter are actually the modes studied by Schwarzschild and Saslaw (cf. Section IV) in relation with penetrative convection. Detailed examples by Smeyers (**10**, *b*) are in the course of publication. Furthermore, from the work of Chandrasekhar and Lebovitz, it became clear (cf. Robe, **14**) that the so-called *f*-modes constitute really the extension to compressible configurations of the unique Kelvin spectrum for the incompressible sphere (one mode only per value of the degree l of the spherical harmonics). For $l = 1$, this *f*-spectrum vanishes in compressible as well as in incompressible configurations while in compressible masses the *p*- and *g*-spectra subsist for this value of l (**10**). Robe and Brandt (**15**) have applied the variational method using a five parameter trial function to polytropes of indexes $n = 1, 2, 3, 4$ and find that the method does not seem to converge for the *f*-mode when $n > 3$. This is probably due to the peculiar behavior, in highly concentrated models, of the eigensolutions for the *f*- and first *p*-modes which acquire 'spurious' nodes as shown in detail by Robe (**16**).

As an illustration of the possible effects of a finite rotation on non-radial oscillations, Cretin and Tassoul (**17**) have studied the case of a homogeneous compressible cylinder rotating uniformly around its axis with angular velocity Ω. They find that the azimuthal degeneracy is lifted by rotation and that, for Ω smaller than some critical value, the g-modes (which in absence of rotation are unstable in this case) have complex frequencies, the real part increasing with Ω. Of course the imaginary part corresponds to the energy available due to the strong convectively unstable stratification in the model. One may expect this result to remain valid for the unstable g-modes of any configuration comprising a region in which Schwarzschild criterion for convective stability is violated and Ledoux (**18**) has suggested that this might provide an explanation of both the maintenance and the order of magnitude of the periods of the oscillations observed in magnetic variable stars. Robe (**19**) is extending the work of Cretin and Tassoul and of Ostriker (**20**) to general polytropic cylinder with and without rotation and tries to disentangle, in these cases, the gravitational instabilities proper from the convectively unstable gravity-modes (g-modes). Zahn (**21**) is tackling again the more realistic problem of the oscillations of a rotating star especially in view of possible resonance in close binaries.

A review of the principle and of some applications of the higher order virial method has been prepared by Chandrasekhar (**22**) and new applications have been the subject of different papers (**23**) (**24**) which, although somewhat formal, provide interesting examples of detailed analysis of complex situation in presence of uniform or differential rotation and tidal effects.

Let us note also an extension of Chandrasekhar's variational principle to the 6th order problem of the oscillations of an elastic sphere by Ottelet (**25**) and others to the oscillations of a gaseous mass in presence of a rotation by Clement (**26**) or of a magnetic field by Kovetz (**27**).

Vibrational Stability and non-adiabatic oscillations

Boury (**28**) has shown that the critical mass associated with the vibrational instability of helium stars is about 8 to 10 $M_\odot$ as compared to some 60 $M_\odot$ for stars on the main sequence burning hydrogen. Gabriel and Bovie-Thirion (**29**) have discussed the importance of the vibrational instability which manifests itself in a small mass star going through the phase of deuterium burning as a function of the initial abundance of the latter. Perdang (**30**) has discussed in great detail the behavior and the effects of the 3α reactions in pulsating stars. Gabriel (**31**) has confirmed, on the basis of fairly realistic models, the significant vibrational instability of small mass stars ($M \leqslant 0{\cdot}27\, M_\odot$) on or close to the main sequence due to both the energy generation and the low values of the Γ's in the external ionisation zones of H and He. Noels, Boury and Gabriel (**32**) have shown that quasi-static models of pure carbon stars are strongly vibrationally unstable in presence of neutrino emission however small their masses.

Epstein (**33**) has attacked numerically the general linear problem of the non-adiabatic radial oscillations treated as a complex differential system of 4th order with complex eigenvalues.

Unno (**34**) is reporting detailed work on the rôle of a convective envelope upon the vibrational stability of a star, based on a generalization of Vitense mixing-length theory of convection to the time-dependent case. In a similar line, Gough (**35**) finds that energy transport by convection in the external layers can have, depending on the circumstances, either a stabilizing or a destabilizing influence and he suggests that the latter may be related to the variability of the long period variables.

Kopal (**36**) has studied some aspects of the interactions of matter with radiation in the fundamental mode of radial oscillations and finds that they can be significant only in models with very high central condensation.

An attempt at evaluating the effect of gravitational contraction keeping the terms in the velocity of contraction as well as the time variations of ρ and p was made by Meurice (**37**) for a star close enough to the main sequence to be wholly radiative but the thermonuclear energy generation being still negligible. A rough evaluation of the very complicated expression of the coefficient of vibrational instability failed to reveal any significant effect. A somewhat similar problem has been considered by Kato and Unno (38) both for radial and non-radial oscillations around a quasi-equilibrium state in which the nuclear energy generated is not exactly balanced by the heat flux. They find that the region of decreasing entropy (as the star evolves) works in general to excite pulsations.

Let us turn now to investigations directly oriented towards the problem of the excitation of the pulsations of variable stars. Alyoshin (39) concluded on the basis of an approximate model of a cepheid that the ionization zones of H and He I could not lead alone to the excitation of pulsations. Takeuti (40) has written down a boundary condition at the photospheric level of the pulsating radiative atmosphere of a cepheid. Unno (41) has discussed the linear radial pulsation of such an atmosphere to obtain boundary conditions for the pulsation of the interior of a cepheid model. He finds that, very generally, the first overtone as well as the fundamental mode are excited. The paper contains also a discussion of the propagation of running waves in a corona and remarks on the breaking down of the linear theory in the ionization zones as well as suggestions for an approximate non-linear treatment of the external zones. Simon (42) has also analyzed the effects, both from the points of view of dissipation and of phase-shifts, of progressive waves on the non-adiabatic oscillations of an atmosphere surrounded by a hot corona discussing especially the case of resonance between a proper frequency of the atmosphere and that of the exciting mechanism. In an extension of their previous work, containing many improvements, Baker and Kippenhahn (43) have discussed very fully the vibrational stability of actual models along their evolutionary tracks across the cepheid strip confirming the instability due essentially to the second helium ionization although the first helium and hydrogen ionization contribute to it especially for the first overtone. Hofmeister (44) has amplified their investigation of the problem producing a theoretical period-luminosity relation in good agreement with observations and discussing the period changes and the period-frequency relation as they follow from the theoretical evolutionary sequence.

In an interesting paper, dealing with a simple one-zone model, Baker (45) has isolated the essential physics at the basis of the pulsational instability in cepheids and Unno and Kamijo (46) have generalized somewhat his discussion. Baker has also investigated the pulsational properties (47) of outer layer models of RR-Lyrae stars.

The vibrational stability towards non-radial oscillations of large mass stars has been studied by Wan (48) who shows that the coefficient of vibrational stability increases with the mass for the *f*- and *g*-modes. It decreases for the *p*-modes when the mass increases but so slowly that it is not likely that any significant critical mass should exist.

Applications to variable stars

As far as the *excitation* of linear oscillations is concerned, the problem has already been dealt with at the end of the subsection on vibrational stability.

There have been however some other applications of the linear theory mainly concerning the periods given by the eigen-value problem corresponding to the adiabatic approximation. The essential point is to obtain a reasonable model for the internal structure of the variable considered either by direct evolutionary computations or by approximate evolutionary considerations. If the luminosity and radius can be determined more or less directly from the observations, then one can try to adjust the mass and to some extent the model, its chemical composition, etc. in such a way that the theoretical period, corresponding in most cases to the fundamental mode, be just equal to the observed period. Investigations of this type have been reported for η Aquiliae, RT Aurigae, X Cygni, Mira Ceti and some RR Lyrae stars by Masani *et al.* (49) who seem however to have used an unduly complicated approach intended, in principle, to yield also the effects of the non-adiabatic terms. Jörgensen and Petersen (50) have applied the method to derive indications on the masses of RR Lyrae stars of type *a*, *b* and *c* in different clusters.

The same authors (50) have also discussed the secular changes in the periods of β Cephei stars on the basis of actual evolutionary sequences. The smallness of the observed changes leads them to conclude that the β Cephei phase must occur essentially in the slow main-sequence stage a conclusion in agreement with the values of the periods themselves. They also note that the ratio of the periods of the first overtone to that of the fundamental mode does not agree with those derived by van Hoof from the observations. Stothers (51) has also discussed this problem and using Van Hoof's data finds that β Cephei stars of lower mass should lie closer to the initial main sequence. For the same stars Hitotuyanagi and Takeuti (52) find that the pulsation constant should be 0·021.

The interpretation of Fernie's period-radius relation for variable stars has been discussed by Demarque and Percy (53) and by Gough, Ostriker and Stobie (54).

Van der Borght and Murphy (55) have studied the non-linear adiabatic oscillations of a 10 $M_{\odot}$ star up to the third order terms in $(\delta r/r)$ including up to the 5th overtone as spatial coefficients in the development of $(\delta r/r)$. They obtain appreciable skewness of the variations comparable to that observed in cepheids.

Alyoshin (56) has determined the amplitudes of the auto-oscillations of the model whose vibrational instability was discussed in (39) and has discussed the growth of their asymmetry from the interior to the surface as well as the variation of the phase-lag. In another paper (57) taking into account the effects of convective transfer, he finds that, in order to get self-excited oscillations and the right-phase shift, the ratio of the mixing-length to the scale height should be smaller than some upper limit of the order of unity.

The direct computational attack of the full non-linear non-adiabatic problem of stellar pulsation has been reviewed by Christy with emphazis on computational methods and RR Lyrae stars (58) and on the cepheid instability strip (59). In these review articles, as well as in the important paper on RR Lyrae models (60), the author has attempted to clarify as far as possible the physical nature of the effects at the origin of the different characteristics of the computed self-oscillations. A comparison of his own work with the somewhat different approaches due to Zhevakin and Cox and Whitney, reveals that the criteria to which they lead for the location of the Cepheid pulsational strip in the HR diagram are of remarkably similar physical content and require essentially that the depth of He II ionization be such that the heat content of the overlying layers is about 1/4 of the heat radiated in a period. Christy points out also the weaknesses of the present calculations, as for instance the lack of an adequate treatment of time-varying convection in the external instability-layers associated with the ionization of H and He and the limitations of the technique by which they are carried out which may be responsible for the difficulties in relating the finite self-excited modes found (sometimes the fundamental, sometimes the first overtone) to the linear instability which in some cases extends to many overtones. In some ways, this is somewhat reminiscent of the weakly non-linear mechanical systems whose linear modes have uncommensurable periods and in which limit cycles, if they exist, are close to one or more of these linear modes (61). Another aim of Christy's non-linear pulsation computations has always been to derive some of the unknown parameters like total mass or the abundance of He for a given variable star by adjusting these parameters so that the theoretical pulsation fits as well as possible with the observed one and recent extensions in these directions have been made to cepheids (62) and RV Tauri stars (63). Let us note also a new attempt (64) to include a more detailed treatment of radiative transfer and of the development of ionization and shock fronts in the external layers of RR Lyrae models.

A review of some aspects of the non-linear problem insisting more on the growth from linear instability to finite amplitude for a stellar envelope with the position of and the luminosity at the bottom fixed can also be found in a paper by J. P. Cox (65) and a detailed investigation has now been published (66). It is remarked there that the rather small limiting finite amplitude suggests that the limiting mechanism is determined primarily by a saturation effect of the driving mechanisms (varying opacity and low value of κ in the He^+ ionization zone). This paper contains also a discussion of the theoretical location of the cepheid strip in the HR diagram. A note by Christy (**67**) offers some comments and comparisons with his own work.

The thermally excited non-linear oscillator studied by Moore and Spiegel (**68**) is physically rather directly related to the kind of vibrational instability arising for acoustic modes (or non-radial p-modes) in a super-adiabatic region. This may however be directly relevant for some variable stars and, in any case, as noted by the authors, this oscillator provides a good illustration of the variety of behavior, including irregular variability, which can be generated by a single mechanism. In further work, with Baker (**69**), they have extended their non-linear analysis to a model more akin to the radial pulsation of cepheids.

We may perhaps add here references to work on the interpretation of the activity of flare-stars (**70, 71**) and to a review of observational data and theoretical inferences concerning magnetic variable stars (**72**).

Secular and Thermal Stability

These questions which up to recent times had received relatively little detailed attention have suddenly come to the fore with the numerical discovery of secularly unstable phases in the course of the evolution of models with non-degenerate helium burning shells. On the occasion of the first case discovered, Schwarzschild and Härm (**73**) developed a linear analysis of the situation of the type suggested by Ledoux (**74**), but including the terms due to entropy variations, which illustrates the origin and characteristics of the responsible incipient secular instability. Since then, this type of linear analysis has been repeated by Rose (**75**) for similar models and Gabriel (**76**) has found that models with an isothermal core followed by a shell burning zone become secularly unstable, when the core reaches the Schönberg-Chandrasekhar limit, towards a displacement corresponding to a contraction of the core and an expansion of the envelope. Quite generally, at least the linear secular instabilities in question appears for perturbations more general (often with nodes inside the star) than the homologous modification which for so long was the only one considered in the discussion of secular stability. The elucidation of some of the mathematical properties of the problem would be very welcome.

Kippenhahn (**77**) and Weigert (**78**) on their side, encountered the same kind of problem in following the late evolution of a 5 $M_\odot$ star but found numerically that the thermal runaway gave rise to some kind of 'relaxation oscillations' of long periods ($\sim$ 3000 years). Recently Schwarzschild and Härm (**79**) found that, in their case also, the secular instability discovered earlier leads indeed to long period 'relaxation oscillations' which, after about a dozen relaxation cycles (at which time the helium burning shell is still entirely non-degenerate), causes mixing of hydrogen into the hot interior. Rose (**75**/*b*) was led to similar conclusions in his study of models of helium shell-burning stars and points out that the origin of planetary nebulae and the occurrence of novae may be associated with this instability. It has subsequently been shown (**80**) that thermal instability can lead to vibrational instability, a circumstance that provides further physical basis for the above suggestion concerning planetary nebulae and novae. Rakavy and Shaviv (**9**) have also encountered significant thermal instabilities in presence of neutrino emission.

It is obvious that a new and important field of research has been opened here for both the linear and non-linear theories of stellar stability and that a careful choice of descriptive terms will help avoid confusion.

Shock waves, Novae, Supernovae and other sources of extreme energy release

It is quite beyond the present reviewer's ability to cover the abundant literature on shocks in all the various circumstances that have been considered. Many of these refer to conditions not specially typical of stellar interiors and we shall have to ignore them even if something could be learned from them for the problems of interest to this Commission.

Much of the typical work has been associated with the discussion of novae or supernovae explosions and a general review of the relevant theories is due to Schatzman (**81**). As to the effects of nuclear reactions started by the collapse of the initial configuration, the problem has been reviewed by Fowler and Hoyle (**82**) and it has been discussed again by Masani *et al.* (**83**) who took also into account radiation, pair production, relativity effects and neutrino emission. Colgate and White (**84**) came to the conclusion that thermonuclear reactions had a negligible effect in circumstances likely to occur in supernovae and that the reversal of the collapsing motion in the envelope and the formation there of strong relativistic shock waves was due to the deposition of energy by neutrinos. Masani *et al.* (**85**) have extended their computations to the case of very high densities such as might occur in neutron stars. Kalenichenko and Porfiriev (**86**) have followed numerically the outburst of a massive hydrogen star caused by an instantaneous release of a large amount of energy at the centre and compared the luminosity curve with observational data for supernovae type II.

Finzi and Wolf (**87**) have found two types of massive (1·4 and 1·2 $M_{\odot}$) white dwarfs models which, by implosion at the end of a slow contraction (due to inverse β processes), could account for type I supernovae and their frequency.

Let us also note some papers (**88, 89**) treating some consequences of the current ideas on supernovae and refer to a general discussion of the subject (**90**).

Schatzman (**91**) has considered a mechanism in which resonance could lead to ordinary novae explosions and Kahoutek (**92**) has also contributed to the problem of the origin of novae and planetary nebulae. Ishizuka, Hashimoto and Ono (**93**) have generalized the quasi-stationary method of Ono *et al.* to the case of oblique shock propagation and discussed in detail the behavior of a shock in polytropic gases. Simon (**94**) has discussed the stability of shocks by the normal-modes method. Let us also mention a few investigations (**95–98**) directed towards applications to the external layers of variable stars or which might be of interest in this respect.

Before leaving this section, let us add a word about some other aspects of oscillations typical of the very external layers but which might be significant for the problems of internal structure either because they might remain significant at greater depth or because they might play a rôle in the determination of surface boundary conditions when the latter become critical for the internal structure. In the first group, I want to mention especially one result of Spiegel (**99**) who finds that, in a superadiabatic region, acoustic modes can be excited directly by conduction. By analogy, one might expect that non-radial p-modes in the interior could become vibrationally unstable due to the same mechanism. In the second group, which will certainly be reviewed in some other Commission, let us simply cite by title, a few papers (**100–104**) devoted to the excitation of waves and oscillations in a stable atmosphere by the convection and turbulence in an underlying convection zone.

BIBLIOGRAPHY

1. (*a*) Ledoux, P. 1965, Stellar Stability, in *Stars and Stellar Systems*, **8**, Eds. L. H. Aller and D. B. McLaughlin, Univ. of Chicago Press, Chicago, Chap. X., p. 499; (*b*) 1964, Some general Comments on the Problem of Stellar Stability, *Astrophys. Norv.*, **9**, 187 (volume dedicated to Professor S. Rosseland on the occasion of his 70th birthday).

2. Cf. for instance: Hayashi, C., Nakano, T. 1965, *Progr. theor. Phys.*, **34**, 754.

3. Kaminishi, K. 1965, *Progr. theor. Phys.*, **34**, 424; Colgate, S. A., White, R. H. 1966, *Astrophys. J.*, **143**, 626; Barbon, R., Dallaporta, M., Perinotto, M., Sussi, M. G., *Mem. Soc. astr. ital.*, **36**, 127, 1966. For the effects of nuclear excitation on the Γ's, cf.: Perdang, J. 1966, *Bull. Soc. R. Sci. Liège*, **35**, 369.

4. Chandrasekhar, S. 1964, *Phys. Rev. Lett.*, **12**, 114; Chandrasekhar, S., Tooper, R. F., 1964, *Astrophys. J.*, **139**, 1396.

5. Baglin, A. 1964, *C. r. Acad. Sci. Paris*, **258,** 5801; 1965, *ibid.*, **260,** 2429; 1966, *Ann. Astrophys.*, **29,** 103.
6. Van der Borght, R. 1964, *Bull. Acad. R. Belgique*, Cl. Sci., 5ème sér., **50,** 959.
7. Bhatia, K., Kushwaha, R. S. 1964, *Appl. Sci. Res. A. Netherl.*, **12,** 454.
8. Stothers, R., Frogel, J. A. 1966, The Linear Star Model and Dynamical Instability, *Astrophys. J.* (in press).
9. Rakavy, G., Shaviv, G., 1966, Instabilities in Highly Evolved Stellar Models, *Astrophys. J.* (in press).
10. Smeyers, P. 1966, Thesis, Université de Liège. This work contains detailed results for two values of $\beta_c = 0{\cdot}4$ and $0{\cdot}8$ or $(M/M_{\odot})\bar{\mu}^2 = 9{\cdot}85$ and $77{\cdot}7$ and for f, p and g-modes up to the 4th order (exceptionally g^{10}, g^{14}, g^{18}) for spherical harmonics of degree $l = 1$ to 8. The behavior of the solutions close to the singularities at the centre and the surface are also studied in great detail. In this last respect cf. also:
Lebovitz, N. R. 1965, *Astrophys. J.*, **142,** 229.
Partial results are being published separately:
(*a*) Smeyers, P. 1966, Oscillations linéaires et adiabatiques d'une sphère homogène, *Bull. Acad. R. Belgique*, Cl. Sci., 5ème Sér., **52,** 1126; (*b*) 1966, Oscillations non radiales et convection dans les étoiles, *Ann. Astrophys.*, **29,** 539.
(*c*) 1967, Oscillations non radiales adiabatiques d'étoiles massives, *Bull. Soc. R. Sci. Liège* (in preparation).
11. Van der Borght, R., Wan, F. S. 1965, *Bull. Acad. R. Belgique*, Cl. Sci., 5ème sér., **51,** 978, in which the results are given for $(M/M_{\odot})\bar{\mu}^2$ = 10, 20, 30, 40, 50, 60, 70, 80; 1966, *Austr. J. Phys.*, **19,** 467 which contains an analysis of Cowling's perturbation method for correcting the results of the second order problem of the effects of the perturbation of the gravitational potential.
12 Chandrasekhar, S., Lebovitz, N. R. 1964, *Astrophys. J.*, **140,** 1517.
13. Ledoux, P., Smeyers, P. 1966, *C. r. Acad. Sci. Paris*, **262,** 841.
14. Robe, H. 1965, *Bull. Acad. R. Belgique*, Cl. Sci., 5ème sér., **51,** 598.
15. Robe, H., Brandt, L. 1966, *Ann. Astrophys.*, **29,** 571.
16. Robe, H. 1966 (in preparation).
17. Cretin, M., Tassoul, J. L. 1965, *Ann. Astrophys.*, **28,** 982.
18. Ledoux, P. 1965, Oscillation theories of Magnetic Variable Stars, in Proc. Symp. on *Magnetic Stars and other Peculiar and Metallic-Line Stars*, Goddard Space Flight Center, Greenbelt, Maryland (to be published).
19. Robe, H. 1966 (in preparation).
20. Ostriker, J. 1964, *Astrophys. J.*, **140,** 1529.
21. Zahn, J. P. 1966, *C. r. Acad. Sci. Paris*, **263,** 1077.
22. Chandrasekhar, S. 1964, The Higher Order Virial Equations and their Applications to the Equilibrium and Stability of rotating Configurations, in *Lectures in Theoretical Physics*, vol. 6. The University of Colorado Press, Boulder.
23 Chandrasekhar, S. 1964, The Equilibrium and the Stability of the Darwin Ellipsoids, *Astrophys. J.*, **140,** 599; 1965, The Equilibrium and the Stability of the Dedekind Ellipsoids, *ibid.*, **141,** 1043; 1965, The Equilibrium and the Stability of the Riemann Ellipsoids, I., *ibid.*, **142,** 890; 1966, II, *ibid.*, **145,** 842.
24. Lebovitz, N. R. 1966, On Riemann's Criterion for Liquid Ellipsoids, *Astrophys. J.*, **145,** 872.
25. Ottelet, I. 1966, *Astrophys. J.*, **143,** 253.
26. Clement, M. J. 1965, *Astrophys. J.*, **141,** 210.
27. Kovetz, K. 1966; *Astrophys. J.*, **146,** 462.
28. Boury, A. 1965, *Ann. Astrophys.*, **28,** 353.
29. Gabriel, M., Bovie-Thirion, C. 1965, *Bull. Acad. R. Belgique*, Cl. Sci., 5ème sér., **51,** 1330.
30. Perdang, J. 1965, *Bull. Soc. R. Sci. Liège*, **34,** 736.
31. Gabriel, M. 1966, Thesis, Université de Liège; cf. also:
1966, Stellar Models for wholly convective red Dwarfs and their vibrational Stability, Colloquium on Late-Type Stars, Trieste, 13–17 June 1966 (in press).

32. Noels-Grötsch, A., Boury, A., Gabriel, M. 1967, *Ann. Astrophys.* (in press).
33. Epstein, I. 1967 (in preparation).
34. Unno, W. 1967 (in preparation).
35. Gough, D. O. 1967 (in preparation).
36. Kopal, Z. 1965, *Z. Astrophys.*, **61**, 156.
37. Meurice, P. 1964, Mémoire de Licence, Université de Liège.
38. Kato, S., Unno, W. 1966 (preprint).
39. Alyoshin, V. I. 1964, *Astr. Zu.*, **41**, 1056.
40. Takeuti, M. 1964, *Publ. astr. Soc. Japan*, **16**, 64.
41. Unno, W. 1965, *Publ. astr. Soc. Japan*, **17**, 205.
42. Simon, R. 1964, *Astrophys. Norv.*, **9**, 113 (volume dedicated to Professor S. Rosseland on his 70th birthday).
43. Baker, N., Kippenhahn, R. 1965, *Astrophys. J.*, **142**, 868.
44. Hofmeister, E. 1965, Thesis (München); 1965, *Mitt. astr. Ges.* no. 19, 90; 1965, Proc. IAU Coll. Bamberg, p. 224. *Kl. Veröff. Remeis-Sternw. Bamberg*, **4**, no. 40.
45. Baker, N. 1966, Simplified Models for Cepheid Instability, in *Stellar Evolution*, Eds. R. F. Stein and A. G. W. Cameron, Plenum Press, New York, p. 373.
46. Unno, W., Kamijo, F. 1966; *Publ. astr. Soc. Japan*, **18**, 23.
47. Baker, N. 1965, Proc. IAU Coll. Bamberg, p. 121. *Kl. Veröff. Remeis-Sternw. Bamberg*, **4**, no. 40.
48. Wan Fook Sun 1966, Non-radial oscillations and vibrational stability of massive stars initially composed of pure hydrogen, Thesis Australian National Univ., Canberra; *Bull. Acad. R. Belgique*, Cl. Sci., 5ème sér. (in press).
49. Masani, A., Martini, A., Nelli, M., Albino, E. 1965, IAU Coll. on The Position of Variable Stars in the H–R diagram, Bamberg 1965, *Kl. Veröff. Remeis-Sternwarte Bamberg*, **4**, no. 40; 1966, *Inf. Bull. var. Stars*, Com. 27 IAU, no. 162.
50. Jörgensen and Petersen 1966 (University Observatory, Copenhagen, in preparation).
51. Stothers, R. 1965, *Astrophys. J.*, **141**, 671.
52. Hitotuyanagi, Z., Takeuti, M. 1963, *Sci. Rep. Tôhoku Univ.*, Ser. I., **47**, 159.
53. Demarque, P., Percy, J. R. 1965, *Astrophys. J.*, **142**, 1082.
54. Gough, D. O., Ostriker, J. P., Stobie, R. S. 1965, *Astrophys. J.*, **142**, 1649.
55. Van der Borght, R., Murphy, J. O. 1966, *Mon. Not. R. astr. Soc.*, **131**, 225.
56. Alyoshin, V. I. 1964, *Astr. Zu.*, **41**, 201.
57. Alyoshin, V. I. 1966, *Astr. Zu.* (in press).
58. Christy, R. F. 1964, *Rev. mod. Phys.*, **36**, 555; also, *Computational Methods in Stellar Pulsation*, in the series Method of Computational Physics, eds. Alder, B., Fernback, S., Rotenberg, M., Academic Press, New York and London (to be published).
59. Christy, R. F. 1966, *A. Rev. Astr. Astrophys.*, **4**, 353; cf. also, 1966, *Stellar Evolution*, Eds. R. F. Stein and A. G. W. Cameron, Plenum Press, New York, p. 359; 1966, The Cepheid Instability Strip, Comm. 122nd Meeting AAS, Cornell Univ., July 25–28, 1966.
60. Christy, R. F. 1966, *Astrophys. J.*, **144**, 108.
61. Cf. for instance: Hale, J. K. 1963, *Oscillations in nonlinear Systems*, McGraw-Hill series in Advanced Mathematics with Applications, McGraw-Hill, New York, Chap. 10, p. 89; Cesari, I. 1963, *Asymptotic Behavior and Stability Problems in ordinary differential Equations*, Springer-Verlag, ch. 3, §8·5.
62. Christy, R. F. 1966, *Astrophys. J.*, **145**, 337.
63. Christy, R. F. 1966, *Astrophys. J.*, **145**, 340.
64. Castor, J. I. 1966, Atmospheric Dynamics in a Model RR Lyrae Star, Thesis (Caltech).
65. Cox, J. P. 1966, *Stellar Evolution*, Eds, R. F. Stein and A. G. W. Cameron, Plenum Press, New York, p. 347.
66. Cox, J. P., Cox, A. N., Olsen, K. H., King, D. S., Eilers, D. D. 1966, *Astrophys. J.*, **144**, 1038; King, D. S., Cox, J. P., Eilers, D. D. 1966, *Astrophys. J.*, **144**, 1069; also King, D. S. 1966, *Astr. J.*, **71**, 3901.
67. Christy, R. F. 1966, *Astrophys. J.*, **144**, 1212.
68. Moore, D. W., Spiegel, E. A. 1966, *Astrophys. J.*, **143**, 871.

69. Baker, N. H., Moore, D. W., Spiegel, E. A. 1966, Nonlinear Oscillations in the One-Zone Model for Stellar Pulsation, comm. 122nd Meeting AAS, Cornell University, 25–28 July 1966.
70. Vardya, M. S. 1966, On M Dwarf Flare Stars, *Mon. Not. R. astr. Soc.*, **131**, 521.
71. Poveda, A. 1964, A theoretical Domain for Flare Stars in the H–R diagram, *Nature*, **202**, 4314.
72. Ledoux, P., Renson, P. 1966, *A. Rev. Astr. Astrophys.*, **4**, 293.
73. Schwarzschild, M., Härm, R. 1965, *Astrophys. J.*, **142**, 855.
74. Cf. ref. (1), § 6 and also, 1960, *Bull. Acad. R. Belgique*, Cl. Sci., 5ème sér., **46**, 429.
75. Rose, W. K. 1966*a*, *Astrophys. J.*, **144**, 1001; 1966*b*, *Astrophys. J.*, **146**, 838.
76. Gabriel, M. 1967 (in preparation).
77. Kippenhahn, R. 1965, *Mitt. astr. Ges.*, 1965, p. 53.
78. Weigert, A. 1965, *Mitt. astr. Ges.*, 1965, p. 61.
79. Schwarzschild, M., Härm, R. 1966 (in preparation).
80. Rose, W. K. 1966 (in preparation).
81. Schatzman, E. 1965, Theory of Novae and Supernovae, in *Stars and Stellar Systems*, **8**, Eds. L. H. Aller and D. B. McLaughlin, The Univ. Chicago Press, Chicago, Chap. 6, p. 327; also, 1963, White Dwarfs and Type I Supernovae, in *Star Evolution*, Ed. L. Gratton, Academic Press, p. 389.
82. Fowler, W. A., Hoyle, F. 1964, *Astrophys. J. Suppl.*, **9**, 252;
Fowler, W. A. 1964, *Rev. mod. Phys.*, **36**, 545.
83. Masani, A., Borla, V., Ferrari, A., Martini, A. 1966, Shock Waves at high Temperatures, Grup. Ital. Fis. Cosm., Rapp. Interno, no. 26.
84. Colgate, S. A., White, R. H. 1966, *Astrophys. J.*, **143**, 626.
85. Masani, A., Borla, V., Ferrari, A., Martini, A. 1966, Shock Waves at high Densities, Grup. Ital. Fis. Cosm., Rapp. Interno, no. 29.
86. Kalenichenko, V. V., Porfiriev, V. V. 1966, *Astr. Zu.*, **43**, 740.
87. Finzi, A., Wolf, R. A. 1966, Orange preprint, Caltech.
88. Colgate, S. A., Cameron, A. G. W. 1963, Enhancement of Light Output from a Supernova, *Nature*, **200**, 870.
89. Stothers, R. 1963, Neutrino Emission, Mass Loss and the Frequency of Supernovae, *Astrophys. J.*, **138**, 1085.
90. Colloque International sur les novae, novoides et supernovae, St Michel de Haute-Provence, 3–7 September 1963, in Coll. Intern. CNRS, Presses Univ., Paris.
91. Schatzman, E. 1964, *Ann. Astrophys.*, **27**, 590.
92. Kahoutek, L. 1964, *Bull. astr. Inst. Csl.*, **15**, 81.
93. Ishizuka, T., Hashimoto, Y., Ono, Y. 1964, *Progr. theor. Phys.*, **32**, 207.
94. Simon, R. 1964, *J. Méc.*, **9**, 113.
95. Bird, G. A. 1963, The Behavior of Shock Waves in a gravitational Atmosphere, *Astrophys. J.*, **139**, 675.
96. Bhatnagar, M. S., Kushwaha, R. S. 1964, Remarks on the theory of decaying shocks in the atmosphere of β Cephei Stars, *Proc. nat. Acad. Sci. India*, Sect. A, **34**, 181.
97. Skalafuris, 1965, The Structure of a Shock Front in atomic Hydrogen, II, The Region of Internal Relaxation, *Astrophys. J.*, **142**, 351.
98. Caloi, V., Conforto, G. 1966, Teoria non lineare delle oscillazioni nell'atmosfera di una variabile, *Mem. Soc. astr. ital.*, **37**, no. 3 (also *Contr. Lab. astrofis. Frascati Roma*).
99. Spiegel, E. A. 1964, *Astrophys. J.*, **139**, 959; cf. also ref. (**68**) above.
100. Moore, D. W., Spiegel, E. A. 1964, The Generation and Propagation of Waves in a compressible Atmosphere, *Astrophys. J.*, **139**, 48.
101. Kato, S. 1966, On the atmospheric oscillations excited by turbulence, *Astrophys. J.*, **143**, 372.
102. Souffrin, P. 1966, Hydrodynamique d'une atmosphère perturbée par une zone convective turbulente sous-jacente, *Ann. Astrophys.*, **29**, 55.
103. Townsend, A. A. 1966, Internal Waves produced by a convective Layer, *J. Fluid Mech.*, **24**, 307.
104. Bird, G. A. 1965, The Equilibrium State of a Shock heated Atmosphere, *Astrophys. J.*, **141**, 1455.

VII. EFFECTS OF ROTATION, MAGNETIC FIELDS, AND EXTERNAL GRAVITATIONAL FIELDS

The effects of rotation on various types of stellar models have been studied extensively but although the information gathered is valuable, the impression persists that one is still some way from a completely satisfactory physical solution.

Roxburgh (1) has built models in which the angular velocity field adjusts itself so as to drive no meridional circulation. Although there is a typical difference in the resulting angular velocity distribution for upper and lower main sequence stars it is not sufficient to explain the observed difference between the two classes of stars.

Polytropes in fast uniform rotation have been discussed by Duheert (2) ,James (3), Roberts, Hurley and Limber (4) and Stoeckly (5) has constructed, by direct numerical methods, polytropic models in fast, non-uniform rotation. Monoghan and Roxburgh (6) have developed an approximation method based on the use of a first order perturbation technique in the interior and the neglect of the self-gravitation in the outer envelope (Roche model). This method has been extended by Roxburgh, Griffith and Sweet (7) to the construction of more realistic models in fast uniform rotation.

Models for early main-sequence stars in non rigid rotation with an increase in angular velocity towards the surface have also been obtained by Kapylov (8) while Porfirjev (9) has investigated a Roche model with barotropic rotation. Rotational oscillations connected with meridian circulations have been discussed (10) in view of the interpretation of the β CMa stars. Radial velocity curves for eclipsing variables have been computed taking into account an equatorial deceleration (11, 12). A relativistic generalization of the Roche model for applications to neutron stars has also been developed (9).

The question of meridian circulations or the Von Zeipel-Eddington problem has been surveyed by Mestel (13) with special reference to the problem of self-consistent circulation patterns, the effect on stellar evolution and possible observable surface effects. Steenbeck and Krause (14) have studied the differential rotation with equatorial acceleration associated, in a perfect fluid sphere, with a weak meridian circulation flowing on the surface towards the equator and descending there. Mestel (15) has confirmed Öpik's result that, in a uniformly rotating radiative envelope, the meridian circulation pattern breaks up into two zones, the velocity becoming large for low surface densities. Smith (16) suggests that the velocity singularity arising at the surface when the density tends to zero can be removed by taking into account the non local nature of radiative transfer in the external layers. The problem of the effect of rotation in the external layers has also been rediscussed by Osaki (17) using the appropriate equations of radiative transfer. He finds that two steady states are possible depending on whether the viscosity is negligible (redistribution of angular velocity so that the divergence of the radiative flux and the meridional circulation vanish) or not (the non-vanishing divergence of the radiation flux drives meridional circulations which maintain differential rotation).

Anand (18) has also studied polytropic models in uniform rotation but including applications to rotating white dwarfs in which case he finds that, for maximum rotation compatible with stable equilibrium, Chandrasekhar's limiting mass M_3 is increased by 3·5% in agreement with an earlier result of Roxburgh (19) who, furthermore, discussed the possibility for a star of any mass to attain the white dwarf stage through mass loss by rotational instability. This interesting result was already apparent in James' investigation (3) although the critical mass itself was not derived by this author. The discussion of the structure of fast rotating white dwarfs has been refined by Monoghan (20). Differential rotation in white dwarfs has been taken into account by Ostriker, Bodenheimer and Lynden-Bell (21) who find that, in this case, equilibrium models exist with masses considerably greater than M_3. This important result suggests that, for a star above M_3, there is an alternative to mass loss or rapid collapse in the final stages of stellar evolution associated with the redistribution of angular momentum. Papers on the self-consistent field method used numerically (**22**) and on a number of physically realistic white-dwarf models (**23**) are in preparation.

A number of investigations were devoted to the evaluation of the effects of rotation on the observable surface characteristic ($L-T_{\text{eff}}$, $M_{\text{bol}}-B-V$) and on the position of the model in the HR diagram (**24–26**). In the case of an age determination of the Pleiades on the basis of isochrones derived from the evolutionary paths of non-rotating models, it was found necessary to make appropriate corrections for effects of rotation (**27**).

We may perhaps add a few references to semi-empirical papers presenting or discussing collections of data of direct interest for the theory. A discussion, on a large statistical basis, of interrelations between the distribution of rotational velocities on one side and evolutionary paths, early-type stars variability and 'peculiarities' on the other is due to Paczynski (**28**). A study of the distribution of rotational velocities with mass by Kopylov (**29**) shows a maximum for stars of 5–9 $M_{\odot}$, but, on the main sequence, even this maximum is smaller than the critical velocity leading to rotational instability. In the same paper a comparison between observed velocities and predicted ones along early evolution assuming conservation of angular momentum reveals good agreement except for stars B9–A3 (region of the magnetic and peculiar A-stars) for which the observed velocities along the evolutionary paths become rapidly smaller than the predicted ones. Sletteback (**30**) situates the minimum difference between calculated minimum equatorial breakup velocities and observed velocities at the Be stars.

The problem of the interaction between stellar rotation and magnetic fields is more or less unescapable and the generation of a toroidal magnetic field in an initially non-magnetic rotating star by the battery effect of the electron partial pressure has been discussed in detail by Roxburgh (**31**) and the discussion has been extended by Roxburgh and Strittmatter (**32**) to the structure of the rotation and magnetic field inside non-uniformly rotating early main sequence stars where the limit on the growth of the thermally generated toroidal magnetic field comes from the Hall effect rather than the ohmic field. Carr (**33**) finds that balancing, on the average, the extra-force introduced by general relativity in a non-uniformly rotating star by a magnetic force, yields a reasonable value of the magnetic field.

Equilbrium models of polytropes with magnetic fields of various geometries have been discussed by Monoghan (**34**) and Roxburgh (**35**).

In his lectures on the theories of stellar magnetism (**36**), Mestel has reviewed the general problem of the origin of stellar magnetic field and has discussed especially various questions: rôle of 'weak' and 'strong' primeval fields during the Hayashi convective contraction phase including magnetic braking; the appearance or non-appearance of a field above the stellar surface; the reduction in the Cowling decay-time of a strong field; the oblique rotator and the magnetic binary-star models of magnetic variables. Cowling (**37**) has written a very clear account of the fundamentals of the subject and a general review is also given in the article by Ledoux and Renson (**38**).

The study of the interaction between magnetic field and convection has also been pursued. Parker (**39**) and Weiss (**40**) have shown how in a zone which is the seat of laminar convection, a 'weak' field is steadily pushed towards the edge or concentrated in localized flux ropes within the zone. Gough and Tayler (**41**) find that simple local criteria (applicable in absence of intrinsic hydromagnetic instabilities) suggest a high degree of suppression of convection in a sunspot model. This work is being extended by Gough to polytropic atmospheres. This last problem has also been considered by Gusseinov (**42**) who shows, on the basis of convection criteria for the increase of vortexes of different scales, that small ones are depressed by the magnetic field in high atmospheric layers while, in some cases, the magnetic field adds to the convective instability. The influence of radiation has also been considered (**43**). Rüdler (**44**) finds that only completely disordered turbulence intensifies the decay of the mean field. In other cases, especially when one of the two types of helical motion is preferred, for instance as a consequence of the Coriolis forces in a rotating body (planet or star), dynamo effects can arise (**45, 46, 47**) yielding different types of solutions, some periodic, depending on the distribution of rotation inside the configuration.

As far as the theory of magnetic variable stars is concerned, one may record a suggestion by Steinitz (**48**) appealing to the over-stability due to rotation. Although the case discussed by Steinitz is rather far removed from any likely physical model, fundamentally the same over-stability could act in more realistic circumstances and the advantages of an oscillacion theory based on this mechanism have been discussed by Ledoux (**49**). In such a theory, the selection of appropriate modes of oscillation would be greatly helped by the presence of a companion as in the excentric double-star hypothesis suggested and developed by Renson (**50**). This last theory presents intrinsically many advantages and the possibility to explain, on its own, field-reversals if the angular momentum and magnetic axes make a large angle although, in this case, some of the ordinary physical objections to the oblique rotator would probably persist. A discussion of the comparative merits of variable stars theories can be found at the end of the review article by Ledoux and Renson (**38**).

We have already covered some of the work on double-stars and their evolution in Section V. Roxburgh, (**51**) in investigating the pre-main-sequence evolution of a rotating non-magnetic star finds that rotational instability is likely to occur for $M > 0{\cdot}8\ M_{\odot}$ when a radiative core develops and that this could lead to the formation of close binaries for $M < 4\ M_{\odot}$ and wide pairs for $M > 4\ M_{\odot}$. Apsidal motion in close binary system is discussed by Peraiah (**52**) who shows that if the components rotate faster than they revolve, the derived central densities could be higher than usually found. Zahn (**53**) has discussed the principal properties of tides in close binaries in absence of synchronism between rotation and orbital motion and explains three well-known effects on this basis: circularity of the orbits of binaries when at least one of the components possess an external convection zone, the presence of a superficial turbulence zone on star built on the Cowling model, the peculiar evolution of some systems when the secondary is a sub-giant. Kopal (**54**) has surveyed the general problem in a review paper.

BIBLIOGRAPHY

1. Roxburgh, I. W. 1964, *Mon. Not. R. astr. Soc.*, **128,** 157 and 237.
2. Duhurt, M. 1964, Thèse de 3ème cycle, Paris.
3. James, R. A. 1964, *Astrophys. J.*, **140,** 552.
4. Hurley, M., Roberts, P. H. 1964, *Astrophys. J.*, **140,** 583; 1966, *Astrophys J. Suppl.*, **11,** 95; Limber, D. N., Roberts, P. H. 1965, *Astrophys J.*, **141,** 1439.
5. Stoeckly, R. 1965, *Astrophys. J.*, **142,** 208.
6. Monoghan, J. J., Roxburgh, I. W. 1965, *Mon. Not. R. astr. Soc.*, **131,** 13.
7. Roxburgh, I. W., Griffith, J. S., Sweet, P. A. 1965, *Z. Astrophys.*, **61,** 203.
8. Kopylov, I. M. 1967, *Astr. Zu* (in press).
9. Porfirjev, V. V. 1966, Sbornik 'Voprony Astrophysiki' (Problems of Astrophysics), Publ. Naukova Dumka, Kiev.
10. Porfirjev, V. V. 1966, Sbornik, Physics of Stars and of Interstellar Medium, Publ. Naukova Dumka, Kiev; 1963, *Astr. Zu.*, **40,** 579.
11. Oliinik, T. T., Porfirjev, V. V. 1963, *Astr. Zu.*, **40,** 774.
12. Kalenichenko, V. V., Porfirjev, V. V. 1965, *Astr. Zu.*, **42,** 560.
13. Mestel, L. 1966, Meridian Circulation in Stars, in *Stars and Stellar Systems*, **8,** Eds. L. H. Aller and D. B. McLaughlin, Univ. Chicago Press, Chicago, Chap. 5, p. 465.
14. Steenbeck, M., Krause, F. 1965, *Mber. dtsch. Akad. Wiss. Berlin*, **7.**
15. Mestel, L. 1966, *Z. Astrophys.*, **63,** 196.
16. Smith, R. C. 1966, *Z. Astrophys.*, **63,** 161.
17. Osaki, Y. 1966, *Publ. astr. Soc. Japan*, **18,** 7.
18. Anand, S. P. S. 1966, *Astrophys. J.* (in press).
19. Roxburgh, I. W. 1965, *Z. Astrophys.*, **62,** 134.

20. Monoghan, J. J. 1966, *Mon. Not. R. astr. Soc.*, **132**, 305.
21. Ostriker, J. P., Bodenheimer, P., Lynden-Bell, D. 1966, *Phys. Rev. Lett.*, **17**, 816.
22. Ostriker, J. P., Mark, Bodenheimer, P. 1967 (in preparation).
23. Ostriker, J. P., Bodenheimer, P. 1967 (in preparation).
24. Roxburgh, I. W., Strittmatter, P. A. 1965, Rotational Spread of the main sequence, *Z. Astrophys.*, **63**, 15.
25. Roxburgh, I. W., Strittmatter, P. A. 1966, On Stellar Rotation, V. The Structure of rotating Stars, *Mon. Not. R. astr. Soc.*, **133**, 345.
26. Rubin, R. 1966, Effects of stellar Rotation on apparent bolometric Luminosity, *Mon. Not. R. astr. Soc.*, **133**, 339.
27. Hazlehurst, J., Thomas, H. C. 1966, *Mitt. astr. Ges.*, no. 21.
28. Paczynski, B. 1966, *Act. Astr. Polska Acad. Nauk*, **16.**
29. Kopylov, I. M. 1967, Sbornik 'Variable Stars and Stellar Evolution', Nauka, Moscow.
30. Sletteback, A. 1966, *Astrophys. J.*, **145**, 126.
31. Roxburgh, I. W. 1966, *Mon. Not. R. astr. Soc.*, **132**, 20.
32. Roxburgh, I. W., Strittmatter, P. A. 1966, *Mon. Not. R. astr. Soc.*, **133**, 1.
33. Car, W. J. 1965, *Phys. Rev.* **140**, B237.
34. Monoghan, J. J. 1965, *Mon. Not. R. astr. Soc.*, **131**, 105; 1966, *ibid.*, **132**, 1; *ibid.*, **134**, 275.
35. Roxburgh, I. W. 1966, *Mon. Not. R. astr. Soc.*, **132**, 347.
36. Mestel, L. Theories of Stellar Magnetism, Lectures at Enrico Fermi School on *Plasma Astrophysics*, Varenna 1966 (to be published).
37. Cowling, T. G. 1965, Magnetic Stars, in *Stars and Stellar Systems*, Eds. L. H. Aller and D. B. McLaughlin, Univ. Chicago Press, Chicago, chap. 8, 425.
38. Ledoux, P., Renson, P. 1966, Magnetic Stars, *A. Rev. Astr. Astrophys.*, **4**, 293.
39. Parker, E. N. *Stellar and Solar Magnetic Fields*, IAU Symp. no. 22, Ed. Lüst, R., North Holland Publ. Co., p. 232.
40. Weiss, N. O. 1964, *Phil. Trans.*, **A256**, 99; *Mon. Not. R. astr. Soc.*, **128**, 225; 1966, *Proc. R. Soc. London*, **293**, 310.
41. Gough, D. O., Tayler, R. J. 1966, *Mon. Not. R. astr. Soc.*, **133**, 85.
42. Gusseinov, O. Kh. 1966, *Astr. Zu.*, **43**, 772.
43. Zeldovitch, Ya. B., Gusseinov, O. Kh. 1966, *Astr. Zu.*, **43**, 313.
44. Rädler, K. H. 1966, Dissertation, Univ. Jena.
45. Steenbeck, M., Krause, F., Rädler K. H. 1966, *Z. Naturf.*, **21a**, 369.
46. Steenbeck, M., Krause, F. 1966, *Z. Naturf.*, **21a**, 1285.
47. Steenbeck, M., Krause, F. 1965, *Mber. dtsch. Akad. Wiss. Berlin*, **7**, 335.
48. Steinitz, R. 1964, Studies on Magnetic Stars. Thesis, Univ. Leiden; 1965, *Stellar and Solar Magnetic Fields*, IAU Symp. no. 22, Ed. R. Lüst, North Holland Publ. Co., p. 117.
49. Ledoux, P. 1965, Oscillation Theories of Magnetic Variable Stars, in Proc. Symp. on *Magnetic Stars and the Peculiar and Metallic-Line Stars*, Goddard Space Flight Center, Greenbelt, Maryland (to be published).
50. Renson, P. 1963, *Bull. Soc. R. Sci. Liège*, **32**, 664.
51. Roxburgh, I. W. 1966, *Astrophys. J.*, **143**, 111.
52. Peraiah, A. 1966, *Z. Astrophys.*, **64**, 27.
53. Zahn, J. P. 1966, *Ann. Astrophys.*, **29**, 313, 489 and 565.
54. Kopal, Z. 1965, Internal Structure of the Stars and apsidal Motions, in *Adv. Astr. Astrophys.* **3**, Ed. Z. Kopal, Acad. Press, New York and London, p. 89.

VIII. GRAVITATIONAL INSTABILITY — ORIGIN OF STARS

The activity in this field has continued to increase probably because of various favorable factors: better and more detailed observational data on the interstellar matter and better knowledge of the physical processes taking place in it as well as the possibility of new significant observations and tests in the near future, the discovery of very cold infrared objects, new information on the structure of galaxies, a better grasp of factors arising in magnetohydrodynamics and their possible interactions with rotation and finally the access to powerful computers and the resulting enticement to tackle more realistic problems including non-linear ones.

As a result, new factors like thermal instabilities have entered the field and others like magnetic fields and rotation have been the object of new and more precise discussions. One may also note a shift of the attention and interest towards the larger scale problem of the formation of the characteristic structures of galaxies. In this sense, many investigations are strictly outside the competence of this Commission and we shall only survey roughly part of this work when the methods used may be of some interest for the problem of the formation of stars themselves.

The general problem of gravitational instability and of star formation has been reviewed by Spitzer (1), Layzer (2) and Mestel (3). Each author emphazises different aspects but all voice a certain uneasiness with the pioneering treatment of Jeans. The difficulties arise in defining properly the initial state of the system whose stability should be studied. As a finite configuration which has relaxed completely on its own and in all directions to equilibrium (mechanical and thermal) cannot exhibit gravitational instability at least for infinitesimal perturbations, a certain measure of lack of equilibrium (motions, thermal inequalities, extra gravitational field due, for instance, to pre-existing stars, etc.) should be present and this certainly increases the difficulties. Spitzer insists perhaps more on the physics of the problem. Layzer's article contains, apart from a general critical review, a clear account of his own views which place 'gravitational clustering' at the origin of all cosmical condensations. It is further assumed that interactions among pre-stars in a pre-cluster can inhibit their gravitational contraction for periods comparable to the age of a galaxy so that systems of all levels of the hierarchy are actually born almost simultaneously. In Mestel's article, the emphasis is laid rather on anisotropic factors like centrifugal force and magnetic field, especially the latter, and a number of interesting problems are raised in this connection.

Another general review by Woltjer (4) is directed more at the problem of the formation of large scale structures in a galaxy and especially spiral arms.

In an attempt to elucidate some aspects of Jeans's criterion, Simon has followed the development in time of a local perturbation in Jeans medium (5) showing that, whatever the initial scale, it always leads to instability after a time which is the longer the smaller the initial dimensions. In the case of the isothermal stratified nebulae, he has established the existence of a critical wavelength of maximum instability.

Lynden-Bell (7) and Lin, Mestel and Shu (8) have followed the collapse of a uniform non-rotating pressure-free spheroid showing that any initial excentricity is rapidly amplified an oblate spheroid becoming a disk and a probate spheroid a spindle. C. Hunter (9) has studied the formation, inside a large spherical cloud contracting as a whole, of gravitational sub-condensations and finds that rotation has a negligible influence on the process.

The effects of cooling on the contraction of interstellar cloud with an initial polytropic density distribution have been discussed by McNally (10) and Gould (11) finds that a molecular hydrogen protostar ($M > 1\ M_\odot$) would collapse due to the 28μ radiation. The general problem of the thermal stability of a dilute gas in mechanical and thermal equilibrium has been studied exhaustively by Field (12) who finds that, under a wide range of conditions, thermal instabilities exist capable of giving rise, although in times rather long, to astronomically significant condensations of higher density and lower temperature than the surroundings. In a following paper with Saslaw (13), using Oort's model, they find, for the rate of star formation ($\propto \rho^2$) and for the cloudmass spectrum, results in reasonable agreement with observations. The problem of the interaction of this thermal instability with gravitational instability and possibly magnetic fields has been elaborated further by J. H. Hunter (14) who shows that thermal instability at pressure equilibrium followed by gravitational collapse can lead to the formation of significant condensations of the order of $100\ M_\odot$ in 10^9 to 10^{10} years. Thermal effects have also been taken into account by Simoda, Kikuchi and Unno (15) in studying the homologous contraction of an interstellar cloud represented initially by a negative polytrope ($n = -3{\cdot}32$) which ultimately reaches a stage of free-fall collapse except if the density is too high (> 100 cm^{-3}) in which case oscillations occur. In a paper in preparation however, Kikuchi (16) finds that homologous contraction is not a good approximation and studies the effects of non-homology.

The effects of rotation, in a more or less classical approach, have been discussed in a series of papers (17–20). Some of the results here, as in the investigations of Toomre (21), of Lin and Shu (22) and of Goldreich and Lynden-Bell (23), might be more significant for the formation of large scale structures rather than stars.

Mestel finds that only in special cases, does the spin of sub-condensation in a non-uniformly rotating medium have the same sign as the hydrodynamic vorticity.

The general influence of magnetic fields is taken into account in (24) and (25). Some of the conclusions of the linear analysis are discussed by Strittmatter (26) who shows that, even if the linear criteria is not modified by the presence of a magnetic field, contraction in this case will not go on indefinitely. In another paper (27), the same author discusses the gravitational contraction of a cloud in presence of a frozen-in magnetic field taking into account the possibility of anisotropic flow along the field lines. He finds that a change by a factor 4 in the magnetic flux will make all the difference between complete gravitational collapse and the establishment of an equilibrium state at almost the initial lateral dimensions. Considering an anisotropic plasma, Gliddon (28) has shown that, in that case, hose instability as well as gravitational instability may occur, the criterion for the latter depending now on the magnetic field.

The magnetic field of a contracting gas cloud has been studied by Mestel (29) who finds that, if the flux-freezing constraint is strictly enforced, the non-homologous contraction of a gas to form a cloud yields a distorted field that exerts strong pinching forces just beyond the cloud radius. Relaxation of the flux-freezing allow the field lines to snap, yielding a magnetically isolated cloud. A simple illustration model for this snapping has been worked out (30) while, in another investigation (31). the relative importance in this context of the Sweet mechanism, ambipolar diffusion and the Petschek mechanism are being discussed. Of course once magnetic detachment has occurred, transport of angular momentum from the cloud ceases and the possibility of magnetic braking vanishes. However the latter will persist as long as the cloud field is undetached and its effects are being studied in two cases: gravitational collapse (32) and slow contraction due to a sufficiently strong centrifugal force (33).

Mestel (34) has also considered the effect of magnetic braking on the evolution of a protostar in the Hayashi phase assuming a strong stellar wind. He shows that, in the case of a double star system with a strong coupling between orbital and spin motions, this effect could lead to the formation of a close binary. The question of magnetic coupling between the stars is also considered.

The question of the ultimate fate of fragments formed by sub-condensation in a contracting cloud has been the object of some controversy, the linear contraction time-scales being of the same order for the fragments and the cloud as a whole. Layzer (35) has advocated other factors favoring the coalescence of the fragments by collisions. Hunter (36) has shown that the effects of non-linear terms invalidate some of Layzer's arguments and Nakano (37) assuming that the initial large cloud is endowed with some angular momentum finds that, despite some coalescence and disintegration of fragments, the theory can lead to mass-functions which are not incompatible with observations in galactic clusters.

Arny (38) has introduced an interesting approach covering the essential non-linear and thermal effects for studying fragmentation and the growth of sub-condensations in large clouds which may themselves be in internal motion and especially in contraction. He has also considered the same problem as that tackled by Nakano and arrives at more detailed and somewhat different conclusions (**39**).

Finally, let us mention that the possibility of star formation from superdense matter has been considered by Imshennik, Nadezhin, Pinaev and others (**40**).

BIBLIOGRAPHY

1. Spitzer, L. 1967, Dynamics of Interstellar Matter and the Formation of Stars, in *Stars and Stellar Systems*, **7** (to be published); 1965, Physical Processes in Stars Formation, in *The Structure and Evolution of Galaxies*, Proc. 13th Conf. Phys. Brussels, 1964, Interscience Publ., London-New York-Sydney, p. 46.

2. Layzer, D. 1964, The Formation of Stars and Galaxies: Unified Hypotheses, in *A. Rev. Astr. Astrophys.*, Ed. L. Goldberg, **2**, p. 341.

3. Mestel, L. 1965, Problems of Star Formation, I and II, *Quart. J.R. astr. Soc.*, **6**; 161 and 265.

4. Woltjer, L. 1965, Dynamics of Gas and Magnetic Fields; Spiral Structure, in *Stars and Stellar Systems*, **5**, Ed. Blaauw and Schmidt, The Univ Chicago Press, Chicago, chap. 23, p. 531.

5. Simon, R. 1963, *Ann. Astrophys.*, **26**, 456; 1964, *ibid.* **27**, 191.

6. Simon, R. 1965, *Ann. Astrophys.*, **28**, 40.

7. Lynden-Bell, D. 1964, *Astrophys. J.*, **139**, 1195.

8. Lin, C. C., Mestel, L., Shu, F. H. 1965, *Astrophys. J.*, **142**, 1431.

9. Hunter, C. 1963, *Astrophys. J.*, **139**, 570.

10. McNally, D. 1964, *Astrophys. J.*, **140**, 1088.

11. Gould, R. J. 1964, *Astrophys. J.*, **140**, 638.

12. Field, G. B. 1965, *Astrophys. J.*, **142**, 531.

13. Field, G. B., Saslaw, W. S. 1965, *Astrophys. J.*, **142**, 568.

14. Hunter, J. H. 1966, *Mon. Not. R. astr. Soc.*, **133**, 239.

15. Simoda, M., Kikuchi, S., Unno, W. 1966, *Publ. astr. Soc. Japan*, **18**, 31.

16. Kikuchi, S. 1966 (in preparation).

17. Tassoul, J. L. 1964, Sur la stabilité gravitationnelle d'un plasma cylindrique animé d'un mouvement de rotation uniforme, II, *Ann. Astrophys.*, **27**, 150.

18. Simon, R. 1965, Sur l'instabilité gravitationnelle d'une couche fluide incompressible en rotation uniforme, *Bull. Acad. R. Belgique*, Cl. Sci., 5ème sér., **51**, 170.

19. Simon, R. 1965, Gravitational Instability in a Plane gaseous Medium in non-uniform Rotation, *Ann. Astrophys.*, **28**, 626; 1966, Gravitational Instability in the Isothermal Stratified Nebula in uniform Rotation (in preparation).

20. Simon, R., Kollasch, C. 1966, Instabilité gravitationnelle d'un cylindre non-homogène compressible (in preparation).

21. Toomre, A. 1964, *Astrophys. J.*, **139**, 1217.

22. Lin, C. C., Shu, F. H. 1964, *Astrophys. J.*, **140**, 646; 1966, *Proc. nat. Acad. Sci.*, **55**, 229.

23. Goldreich, P., Lynden-Bell, D. 1965, *Mon. Not. R. astr. Soc.*, **130**, 97, and 125.

24. Pacholczyk, A. G., Stodolkiewics, J. S. 1963, On the Gravitational Instability of some magnetohydrodynamical Systems of Astrophysical Interest, III, *Act .Astr. Pol.*, **13**, 30.

25. Uberoi, C., Raghavachar, M. R. 1964, *J. ind. Inst. Sci.*, **46**, 91.

26. Strittmatter, P. A. 1966, *Mon. Not. R. astr. Soc.*, **131**, 491.

27. Strittmatter, P. A. 1966, *Mon. Not. R. astr. Soc.*, **132**, 359.

28. Gliddon, J. E. C. 1966, *Astrophys. J.*, **145**, 583.

29. Mestel, L. 1966, *Mon. Not. R. astr. Soc.*, **133**, 265.

30. Mestel, L., Strittmatter, P. A. 1966 (in preparation).

31. Mestel, L., Strittmatter, P. A. 1966 (in preparation).

32. Mestel, L., Strittmatter, P. A. 1966 (in preparation).

33. Mestel, L., Gillis, J. 1966 (in preparation).

34. Mestel, L. 1967, Proc. 14th Internat. Astrophys. Liège Symp., *Mem. Soc. R. Sci. Liège*, (in press). The 3rd part of this volume is devoted to the formation of stars and further contributions to the problem under review will be found there.
35. Layzer, D. 1963, *Astrophys. J.*, **137**, 351.
36. Hunter, C. 1964, *Astrophys. J.*, **139**, 570.
37. Nakano, T. 1966, *Progr. theor. Phys.*, **36**, 515.
38. Arny, T. T. 1966, *Astrophys. J.*, **145**, 572; 1967, *Ann. Astrophys.* (in press).
39. Arny, T. T. 1966, *Astr. J.*, **71**, 377.
40. Imshennik, V. I., Nadezhin, D. K. 1966, *Astr. Zu.*, (in press); Imshennik, V. I., Nadezhin, D. K., Pinaev, V. S. 1966, *Astr. Zu.* **43**, 1215; Eminzade, T. A., Gusseinov, O. Kh. 1964, *Izv. Akad. Nauk Azerb. SSR, Ser. Fiz.-Tekn.-Matem. N.*, no. 4, 115.

In closing this report, it is a pleasure to thank all the members of Commission 35 who have kindly communicated information on their work. I am especially grateful to Mrs Massevitch and Dr Hayashi who have prepared very complete surveys of the relevant literature and of work in progress in their respective countries. Finally, I want to acknowledge the efficient help of Drs A. Boury and M. Gabriel in preparing this report. I could not hope to have covered *all* contributions of significant interest for the Commission and emphasis may not always have been distributed rightly but, omissions or bias certainly were involuntary and I hope that I may count on the forbearance of all.

P. LEDOUX
Président de la Commission

36. COMMISSION DE LA THEORIE DES ATMOSPHERES STELLAIRES

PRÉSIDENT: Professor Dr K. H. Böhm, Lehrstuhl für Theoretische Astrophysik, Universität Heidelberg, Mönchhofstrasse 12–14, 69 Heidelberg, Germany.

VICE-PRÉSIDENT: Professor Dr A. B. Underhill, Astronomical Observatory 'Sonnenborgh', Zonnenburg 2, Utrecht, The Netherlands.

COMITÉ D'ORGANISATION: C. de Jager, V. V. Sobolev, R. N. Thomas, H. Van Regemorter, M. H. Wrubel.

MEMBRES: Aller (L. H.), Athay, Barbier†, Biermann, Böhm-Vitense, Busbridge, Carson, Cayrel (G.), Cayrel (R.), de Feiter, Edmonds, Elste, Gingerich, Godoli, Gökdoğan, Greenstein, Hack, Hardorp, Henyey, Hitotuyanagi, Hotinli, Houtgast, Houziaux, Huang, Hunger, Ivanov, Jefferies, Johnson (H. R.), Kopylov, Kushwaha, Menzel, Mirzojan, Miyamoto, Müller (E. A.), Münch (G.), Mustel, Neven, Page, Pagel, Pecker (J.-C.), Pecker-Wimel, Petrie (J. K. McD.), Phillips, Pottasch, Rudkjöbing, Saito (S.), Schatzman, Seaton, Sitnik, Stibbs, Strömgen, Swihart, Traving, Ueno, Unno, Unsöld, Van t'Veer (Fr.), Vardya, Weidemann, Wright (K.O.), Wyller, Zirker, Zwaan.

1. INTRODUCTION

Commission activity between the assemblies

A conference on the theory of line formation ('Second Harvard-Smithsonian Conference on Stellar Atmospheres') was held in Cambridge, Mass., on 20–22 January 1965. The meeting was organized by the Smithsonian Astrophysical Observatory and the Harvard College Observatory. A fairly large number of papers were presented. There was a strong emphasis on non-LTE line formation problems for 'few level' atoms.

An IAU Colloquium on the 'Blanketing Effect' (organized by this Commission) was held at the Heidelberg University on 17–19 March 1966. Four review papers and 12 shorter communications were presented. The main emphasis of the talks and discussion centered around:

(1) The applicability of the present form of the line broadening theory to stellar atmospheres, especially to low density regions. (There was no doubt about the applicability under usual laboratory conditions.)

(2) The possibilities of developing a non-LTE line blanketing theory.

(3) Statistical methods of treating a very large number of lines in LTE blanketing effect computations.

Other symposia related to the theory of stellar atmospheres

They included:

An IAU Symposium (no. 26) on 'Abundance Determinations in Stellar Spectra', held in Utrecht, 10–14 August 1964. This meeting was sponsored by the IAU Commissions 29 and 36. There were sessions on methods of abundance analysis, in which problems of the theory of stellar atmospheres were discussed.

The 5th Symposium on Cosmical Gas Dynamics, held in Nice, September 1965 (IAU Symposium no. 28, sponsored in part by IUTAM) included a very detailed discussion of the hydrodynamics of the solar atmosphere (including convection, convective overshoot, and oscillatory motion) and of atmospheres of pulsating stars.

It has been attempted to list the books and the proceedings of symposia which have appeared during the last three years and which are related to the work of this Commission, in section I of the bibliography.

II. BASIC THEORY OF STELLAR ATMOSPHERES

(a) *Theory of Radiative Equilibrium*

Reviews of radiative transfer problems have been given by D. G. Hummer (**37**), in Sampson's book (**1**), and in a very comprehensive way by J. C. Pecker (**62, 61**).

The problem of constructing good non-grey radiative equilibrium model atmospheres (eventually including non-LTE effects) still plays a fundamental role. The Avrett-Krook method has continued to be used successfully in many applications. A new and very elegant temperature correction procedure has been suggested by L. B. Lucy (**48**). The most attractive feature of this method is its theoretical simplicity. K. H. Böhm and W. Deinzer (**16**) have worked out a Fortran programme of this method and have made the first numerical applications. They found that Lucy's method is a very effective temperature correction procedure, unless the scattering contribution to the source function becomes very strong in most parts of the atmosphere. In the meantime L. G. Henyey (**32**) has developed a new temperature correction method, which looks very promising, but which has not yet been tested numerically (K. H. Böhm is preparing such a test). Henyey's method is based on the exact conversion of the monochromatic error ΔF_ν into the monochromatic Planck function correction ΔB_ν.

Another temperature correction procedure based on a certain type of linearization of the equation of transfer has been suggested by P. Feautrier (**28**). A simple modification of the Avrett-Krook method has been suggested by T. L. Swihart (**70**).

A combination of an improved flux iteration method and a revised Λ-iteration scheme has been worked out by T. Tsuji (**74**). It has been successfully tested in the case of a picket-fence model. S. Matsushima and Y. Terashita (**84**) have studied the influence of errors in the absorption coefficient on non-grey models. A useful numerical comparison of non-grey models found by slightly different methods has been given by Gingerich, Mihalas, Matsushima and Strom (**31**).

K. H. Böhm (**15**) found that under certain conditions a temperature inversion can occur in the uppermost layers (where the electron scattering contribution varies rapidly) of very hot non-grey model atmospheres.

Since E. Hopf's early work it has been clear that it must be possible to find an exact solution for at least one standard non-grey problem, namely the LTE picket-fence model. This aim has been achieved recently by C. E. Siewert and P. F. Zweifel (**65**) and by J. C. Stewart (**67**). See also J. C. Stewart (**68**).

A considerable amount of research effort is going into the field of non-LTE radiative transfer. R. Wildt (**77, 78**) is continuing his basic studies of grey non-LTE atmospheres. His approach is based completely on the thermodynamics of the radiation field. He showed that the unattainability of detailed balancing follows from thermodynamics.

An interesting approach to the computation of non-LTE atmospheric structure has been suggested by W. Kalkofen (cf. **44**). The method permits the numerical computation of the temperature stratification and the radiation field in not too high layers in stellar atmospheres. It is based on the (reasonable) assumption that in deeper layers, in which the important continua are formed, the conditions in the strong lines are already very close to detailed balancing. If that is true the line transfer problem is decoupled from the continuum transfer. A very useful discussion of this approach has been given by D. Mihalas (**50**). Numerical applications have been made (see Mihalas and Stone (**49**) and other sections of this report).

A general survey of the present state of the non-LTE approach has been given in R. N. Thomas' book (**72**). He naturally emphasizes line transfer problems (see section II. *f* of this report), but he also includes the interaction between lines and continua. See also Warner (**76**).

A very useful report on the general theory of the emission coefficient has been given by D. G. Hummer (**36**).

A coupled line-continuum transfer problem for a three-level atom has been treated by W. Kalkofen and Avrett (**3**), Y. Cuny (**20**), and by Johnson and Klinglesmith (**6**). A seven-level problem has been solved by T. Kogure (**46**) in connection with the problem of the Balmer decrement in Be stars. A possible interpretation of the oscillatory phase in novae based on non-LTE effects has been suggested by J.-C. Pecker (**63**). The non-LTE formation of the Lyman-continuum in an isothermal atmosphere has been studied by R. D. Dietz and L. L. House (**86**). Non-LTE problems in Wolf-Rayet stars have been considered by Rublev (**64**).

I. Kolesnik and D. A. Frank-Kamenetsky (**47**) have studied the non-LTE formation of recombination continua in non-stable stars. A related problem has been treated by L. M. Biberman, V. S. Vorobyev and A. N. Lagar'kov (**14**). Deviations from LTE in a pure hydrogen plasma are treated in (**34**).

The problem of finding the temperature stratification of an atmosphere in the presence of line blanketing has found renewed and strong interest during the last years. Computational programmes incorporating LTE blanketing by many lines have achieved a rather high degree of perfection, largely due to the work of S. E. Strom and his collaborators (cf. S. E. Strom and R. L. Kurucz (**85**)). O. Gingerich (**89**) has studied, under which conditions a schematic treatment is sufficient. There is general agreement that the LTE treatment overestimates the blanketing effect. The question is: How much? R. N. Thomas (**73**), R. Cayrel (**2**) and H. Frisch (**88**) argue that the error is probably very large. The problem could be definitely settled only, if we would understand better the formation of subordinate lines in a stellar atmosphere. A blanketing problem has also been treated by T. L. Swihart (**69**).

Considerable progress has been made in the mathematical theory of radiative transfer. V. A. Ambarzumian has developed a method for applying principles of invariance to non-linear transfer problems (**4, 5**). His ideas have been also applied by Nikoghossian to a multi-level line transfer problem (**56**) and by N. B. Yengibarian (**82**). Bellman, Kalaba, Kagiwada and Ueno (**7, 8, 9, 10, 11**) have continued to develop the invariant imbedding technique and applied it to a number of problems (e.g. finite slab, one-dimensional medium with a moving boundary). A different procedure of solving transfer problems in a slab geometry has been applied by R. Bellman, H. Kagiwada and R. Kalaba (**12**). They have (together with S. Ueno) computed graphs of the X- and Y-functions for a wide range of slab thicknesses and albedos (**13**). Extensive tables of Chandrasekhar's X- and Y-functions have also been computed by J. L. Carlstedt and T. W. Mullikin (**17**). Finite atmospheres with coherent isotropic scattering have been studied by V. V. Sobolev (**66**) for the limiting case of large optical thickness. He finds that the resolvent of the integral equation determining $S(\tau)$ can be expressed in terms of quantities characteristic of the semi-infinite atmosphere.

J. I. F. King, R. V. Sillars and R. H. Harrison (**45**) have published tables of Hopf's q-function with 8-digit accuracy.

The problem of light scattering in a finite atmosphere has been investigated by V. V. Ivanov (**40, 41, 42**) and by Ivanov and Leonov (**43**). He finds an exact solution for the resolvent of the integral equation for the source function in a finite atmosphere. Tables of Ambarzumian's functions for anisotropic scattering have been computed by I. N. Minin, A. G. Piliposian and N. A. Shidlovskaja (**52**). A small-angle approximation for the source function in certain scattering problems has been discussed by S. Ueno (**75**).

A detailed study of the grey transfer problem with spherical symmetry is due to R. J. Chapman (**18**).

Other problems in the mathematical theory of transfer have been treated by Collins and Code (**19**), Dave and Walker (**21, 22**), P. Feautrier (**27, 29**), Horak and Janousek (**35**), Hummer

(38), Irvine (39), Mullikin (87), Nagirner (53, 54, 55), Obridko (57), Pahor and Kuscer (58), Pavlov (59, 60), Wilson and Sen (79, 80) and by Yanovitsky (81).

The problem of the conversion of the observed centre-to-limb variation into $S(\tau)$ has been rediscussed (using new mathematical methods) by Delache (23, 24, 25) and by David (83).

Delache has also given a qualitative discussion of radiative transfer in an atmosphere with a stellar wind (26).

Time-dependent radiative transfer has been investigated by I. N. Minin (51) (in connection with the close binary problem) and by T. L. Swihart (71). A study of wave propagation in a random medium by U. Frisch (30) may permit application to radiative transfer problems.

(b) Continuous absorption coefficients and some other basic data

There have been further improvements of the theory of the H^--absorption, especially for the free-free transitions. We mention the work by T. L. John (10, 11, 12), by T. Ohmura (32) and by N. A. Doughty and P. A. Fraser (8). The bound-free transitions of H^- have been reinvestigated by Doughty, Fraser and McEachran (7). M. Weinberg and R. S. Berry (36) have drawn attention to the possible importance of a forbidden bound-free continuum of H^- included by perturbing neighbouring particles in a gas of fairly high density. V. Myerscough's, (34) preliminary results indicate that this forbidden continuum will be of importance in the deeper atmospheric layers of a white dwarf like van Maanen 2 and probably also in the upper transition layer of the solar hydrogen convection zone.

There has been a very fast growing interest in the opacity sources in late-type stellar atmospheres. J. R. Auman (1) has succeeded in deriving absorption coefficients due to H_2O for the application to cool stellar atmospheres. A method for estimating the H_2O absorption has been given by M. S. Vardya (23). The problem of the possible importance of negative ions other than H^- and H_2 in late-type stars is of considerable interest. It has been studied by Vardya in (26) and in combination with the general problem of molecular abundances in late-type stars in (27, 28). The Cl^--ion shows a rather large abundance in very cool atmospheres (27, 24). The absorption coefficient of Cl^- has been computed by R. Kandel (13). V. G. Buslavsky (5) has studied opacity sources in late-type stars in general. The presence of a transmission window at $1{\cdot}65\mu$ in late-type stars has been discussed by Vardya (20).

In a semi-emperical investigation M. S. Vardya and K. H. Böhm (22) have looked for undiscovered sources of opacity in an M2 main sequence atmosphere. They find that the question could be answered only in a conclusive way, if we could determine the effective temperature with an error not essentially larger than $\pm$ 100°K.

K. Kodaira (29) finds that the depression at λ 2085 in the solar spectrum can be attributed to the ionization edge of Al I $3p^2P^0$.

It has been found that free-free absorption of He^- can be very important in hydrogen deficient stellar atmospheres. The He^--absorption coefficient has been computed by M. McDowell, M. Williamson and V. Myerscough (15) and by W. B. Sommerville (19).

Work on the calculation of absorption coefficients using the quantum defect method has progressed considerably. The papers of G. Peach (16, 33) seem to be especially useful for astrophysical applications. O. Bély (2) has developed a generalized quantum defect method for hydrogen-like ions. F. Praderie (17) has studied the absorption cross-section for neutral carbon, D. Schlüter (35) has developed a somewhat different method and applied it to Ne I, Ar I, Kr I and Xe I.

New tables of atmospheric opacities have been given by M. S. Vardya (21), A. Cox and J. Stewart (6) and by G. Bode (3).

Monochromatic atmospheric absorption coefficients have been computed by G. Bode (**3**) and by J. Baerentzen and E. Mikkelsen (**31**).

O. Gingerich (**30**) has given a summary of very useful interpolation formulae for absorption coefficient computations. A description of the simplified treatment of line absorption in blanketing computations is due to Vardya (**25**).

The problem of autoionization and dielectronic recombination in the outer layers of a star has continued to receive considerable attention. We mention the work of A. Burgess (cf. **4**), L. Goldberg, A. K. Dupree and J. W. Allen (**9**) and W. van Rensbergen (**18**).

The excitation of H-atoms by protons has been studied by McCarrol and Salin (**14**).

(c) *Hydrodynamic Phenomena in Stellar Atmospheres*

The proceedings of the Joint Discussion on this subject have been published in *Transactions IAU*, **12B**, 531. Hydrodynamic problems of stellar atmospheres have been also discussed at the Nice symposium (see section I of this report).

Convection, Turbulence:

E. A. Spiegel (**43, 44**) summarized the present possibilities for a physical theory of turbulent convection in the Boussinesq approximation.

A general survey of the available data on models of the solar hydrogen convection zone has been given by K. H. Böhm (**4**). He has also discussed convection overshoot at the upper (**4**) and lower (**4, 6**) boundaries of the convection zone. The latter problem has been recently investigated by K. Kohl (**55**). The physical mechanism of convective penetration has been discussed by D. W. Moore (**31**). The possible importance of convective overshoot for the atmospheric Lithium depletion has been studied by R. Weymann and R. L. Sears (**50**) and by Böhm (**6**). The general problem of convective dilution of elements has been studied by E. Schatzman (**39**). Problems of the outer convection zone in connection with the specification of the outer boundary conditions for internal structure calculations have been discussed by L. H. Henyey, M. S. Vardya and P. Bodenheimer (**13**) and by J. Faulkner, K. Griffiths and F. Hoyle (**10**). A semiempirical discussion of stellar convection zones has been given by O. C. Wilson (**51**).

D. Mihalas (**29**) has treated the transition region between a *nongrey* radiative equilibrium atmosphere and a convection zone described by the mixing length theory. A related problem has been studied by N. Sack (**37**). K. Böhm (**5**) has suggested that M-dwarf atmospheres may show an observable deviation from a radiative equilibrium stratification due to the fairly large convective energy transport in these atmospheres. Swihart and Margrave (**45**) are studying inhomogeneities in the solar atmosphere.

Very interesting work on the deformation of magnetic fields by convection has been done by N. O. Weiss (**48, 49**).

K. Walter (**47**) has proposed a very unconventional theory of the outer solar convection zone.

Polytropic atmospheres are still a very useful tool for studies of convection in compressible atmospheres: E. A. Spiegel (**42**) has investigated convective instability in such atmospheres using a perturbation treatment with respect to layer thickness. As an alternative he has explored a variational approach and the WKB approximation. The onset of convection in a polytropic atmosphere with a uniform magnetic field has been studied by S. A. Kaplan and N. S. Petrukhin (**17**). They have derived a simple instability criterion for this case. A generalization of the Schwarzschild criterion for an atmosphere with a magnetic field has been discussed by D. O. Gough and R. J. Taylor (**12**). A. Baglin (**1**) has generalized the Schwarzschild criterion for the case in which a separation of elements occurs in a strong gravitational field. S. Yamaguchi (**52**) used a variational technique to study the convective instability in a layer with depth dependent superadiabatic gradient. Limitations of the applicability of the Schwarzschild criterion have been discussed by G. F. Sitnik (40).

W. Deinzer (7) has used the suppression of convection by a magnet field in a heuristic way to construct a magnetohydrostatic model of a sunspot.

Convective overstability in a geometrically thin medium with a superadiabatic gradient has been investigated by S. Kato (18).

A simple model permitting a physical explanation of overstability has been developed by D. W. Moore and E. A. Spiegel (32).

J. Baerentzen (2) has given new formulae for the computation of the adiabatic gradient and the specific heat.

Turbulence in stellar atmosphere has been discussed by Hollandsky (15), Hollandsky and Chub (16), I. M. Kopylov (22), P. E. Nissen (33) and E. C. Olson (34, 35). O. P. Hollandsky (14) has discussed the possible influence of the radiation pressure on the sound velocities in the atmospheres of hot supergiants.

Oscillations

Considerable attention has been paid to the problem of the generation of the photospheric oscillation or waves by the granulation. M. J. Lighthill (25) has given a very comprehensive and illuminating survey of this problem. S. Kato (19) has investigated atmospheric oscillations excited by turbulence near the critical frequency.

He has also studied (20, 21) the response of an unbounded atmosphere to a point disturbance. F. Meyer and H. U. Schmidt (28) have made extensive computations of the response of the solar atmosphere to a granular disturbance. P. Souffrin (41) has studied the variation of hydrodynamic waves with distance from the hydrogen convection zone. He has generalized diagnostic diagrams for an isothermal atmosphere to include radiative dissipation. He finds that gravity waves should be unimportant in the photosphere because of their strong radiative damping. A standing-mode gravity wave in the chromosphere has been studied by Y. Uchida (46).

Empirical data on photospheric oscillations have been gathered, reduced and interpreted in a very careful way by F. N. Edmonds (8, 9), by P. Mein (27) and by J. B. Zirker (53). A very interesting new observational study of the velocity fields in the photosphere has been made by E. Frazier (56). He finds a definite splitting of the maximum of the oscillation power spectrum into two peaks. He was also able to detect convective overshoot into the uppermost layers of the photosphere.

Other problems

Heating of an atmosphere by shock waves has been investigated by G. A. Bird (3), by M. Kuperus (57) and by M. Saito (38). The propagation and coupling between the three magnetohydrodynamic modes in a stratified medium has been studied by U. Frisch (11).

S. V. Rublev (36) has investigated the dynamics of Wolf-Rayet atmospheres. An important empirical study of the mass loss of T-Tauri stars is due to L. Kuhi (23).

Other hydrodynamic problems have been treated by Lemaire (24), McLean (26), and Grover and Hardy (54).

(*d*) *Thermodynamic Problems of Stellar Atmospheres*

In this section we shall discuss very briefly only those papers which have not yet been considered in connection with either the computation of absorption coefficients or with the convection problem.

B. Schlender and G. Traving (3) have shown that the approximate computation of partition functions can be greatly simplified if one uses a Chebyshev approximation. Applications of this method can be found in the paper by G. Traving, B. Baschek and H. Holweger (5). The problem of pressure dissociation and its influence on molecular partition functions has been investigated by M. S. Vardya (7, 9). He has also (8) computed some basic thermodynamic quantities for a mixture of H, He and 16 other elements (with solar photospheric abundance ratios) allowing for different stages of ionization and the formation of H_2, H_2^+ and H^-.

Molecular abundances in stellar atmospheres have been calculated by Tsuji (6), by Wyller and Morris (10) and by J. F. Dolan (1). Wyller and Morris have studied 15 diatomic and 16 triatomic types of molecules in the atmospheres of carbon stars. Dolan and Vardya (Ref. 27, Section (*b*)) has also paid special attention to polyatomic molecules.

Nonequilibrium ionization in an atmosphere due to the ejection of 'plasma bunches' has been considered by I. G. Kolesnik (2). Studying the importance of inelastic collisions, S. Souffrin (4) has reached the conclusion that in normal stellar chromospheres there will be no observable deviation from the Maxwellian distribution of particles.

(*e*) *Extended Envelopes*

Most of the work done in this field belongs either to the subject of other commissions or has been treated in other sections of this report. In this section we have just listed a few references, which otherwise could not be easily fitted into our classification scheme. These papers contain discussions of (*a*) the influence of the solar wind on abundance determinations in the solar corona (J. C. Brandt, 1), (*b*) problems of expanding stellar envelopes (Gorbatzky, 2) (*c*) a study of shells of Be stars (Lacoarret, 4, (*d*) the extended chromosphere of 31 Cygni (Magnan, 7) and (*e*) a few problems of solar and stellar chromospheres and coronas (Hughes, 3; Livshitz, 5; Livshitz and Nikolsky, 6; and Athay, 8).

(*f*) *General Theory of Line Profiles and Line Formation*

Problems of the physical theory of the line absorptions coefficient mostly belong to the field of Commission 14. I shall mention only a few references to work being done by Commission members or to papers related to this work.

A very useful survey of the theory of line broadening and its astronomical applications has been given by H. van Regemorter (47). The same author (together with S. Bréchot, 46) has derived a generalization of the impact broadening theories of Lindholm and of Baranger, taking into account (among other effects) inelastic collisions. H. Pfennig (38) and H. Pfennig, E. Trefftz and C. R. Vidal (39) have drawn attention to the fact that until recently the Kolb-Griem Stark broadening theory had not been tested in the laboratory in the density range which is of interest in the theory of most stellar atmospheres. Recent experiments of Ferguson, Schlüter and Vidal in the range $N_e \approx 10^{13}\,\mathrm{cm}^{-3}$ give (according to Pfennig, Trefftz and Vidal, 39) results which are much closer to the quasistatic theory than to the Kolb-Griem theory. The theoretical reason for this result is not yet understood. The Stark broadening functions for the hydrogen lines have been recomputed by F. N. Edmonds, H. Schlüter and D. C. Wells (16). They take into account the effects of correlation and shielding. The results are represented graphically for the Lyman-Balmer-Paschen- and Brackett-lines up to $n = 18$. The problem of determining the electron density from line merging due to Stark effect has been reinvestigated by L. N. Kurochka (29) and C. R. Vidal (49). Kurochka (30) has also considered the possible influence of the Doppler effect on line merging. The Influence of Stark broadening on abundance determinations has been studied by D. Mugglestone and B. J. O'Mara (32). New Computations of Voigt functions are due to C. Chiarella and A. Reichel (12).

A general survey of stellar atmospheres and their line spectra has been given by K. O. Wright (51). A series of papers on the measurements of lifetimes of excited levels has been introduced by E. Schatzman (41).

The following work has been done on LTE line formation (mainly in connection with the problem of atmospheric abundance analysis): Baschek, Holweger and Traving (**9**) have published their Algol programme for the quantitative analysis of stellar spectra. This programme has been translated into Fortran by Peytremann (**55**). 'Differential curves of growth' have been discussed by Baschek and Traving (**10**). They have published tables and graphs of the derivatives of Unsöld's curve of growth. Pagel has applied the differential curve of growth method (**37**). F. N. Edmonds (**17**) has compared curves of growth and contribution curves for the same line, but for different fine analysis methods. Mugglestone and O'Mara (**31**) have considered the influence of saturation effects on the abundances derived from faint and medium-strong lines. They especially discuss the determination of the oxygen and notrogen abundance in the solar photosphere.

A very useful and interesting theoretical study of the broadening of lines by the presence of a turbulent velocity field has been made by G. Traving (**44**). He has investigated in quantitative detail the influence of the scale of the hydrodynamic motion on the curve of growth. 'Micro-turbulence' and 'macroturbulence' in the usual sense appear as limiting cases in his theory. Y. Yamashita (**52**) has studied the curve of growth for lines formed by pure absorption in an atmosphere where $S(\tau)$ has a linear + exponential depth dependence and where (κ_ν/κ) can be described by a step function. Unno and Yamashita (**45**) have studied curves of growth for line profiles, which have been truncated at an artificial continuous level. Tables for the calculation of the formation depth are given by Balli, Kiral and J. C. Pecker (**8**).

A considerable amount of work has been done on noncoherent scattering in resonance lines: D. I. Nagirner and V. V. Ivanov (**34**) have considered the non LTE line formation with complete redistribution for a two-level atom in an infinite medium. D. I. Nagirner (**33**) has found an explicit expression for the resolvent of the integral equation for the source function for coherent and for completely noncoherent scattering. Ivanov and Nagirner (**24**) have shown that the radiation emerging from an atmosphere in a resonance line can be expressed in terms of a generalized H-function. A large number of solutions of the two-level line transfer problem with complete redistribution has been given by Avrett and Hummer (**7**) and by Avrett (**6**). Hummer and Rybicki (**21**) have studied the non-LTE line formation for a two-level atom with spatial variation of the Doppler width (see also Hummer, **53**). The problem of *general* non-coherent scattering (i.e. without *complete* redistribution in frequency) has been reviewed by D. G. Hummer (**20**).

S. Dumont (**14**) has developed the discrete ordinate method for computing the integrals in the source function for complete non-coherent scattering.

K. D. Abhyankar (**2, 3**) has studied thoroughly the Schuster problem for an expanding atmosphere.

The general problem of non-LTE line formation has been extensively discussed in R. N. Thomas' book (**42**) and in his article (**43**). J. T. Jefferies (**25**) has surveyed the multi-level problem in the theory of line formation. A review of the computational problems in the theory of non-LTE line formation has been given by Hummer and Rybicki (**22**). The problem of the thermalization length has been studied by D. G. Hummer and J. C. Stewart (**23**). A new method for solving the transfer problem in the non-LTE case has been suggested by E. Böhm-Vitense (**11**).

Under which conditions are the source functions for the lines of a given multiplet equal at a given depth? This important question has been discussed by J. Waddell (**50**), R. G. Athay (**3**) and most recently by E. H. Avrett (54). In a series of papers R. G. Athay (**4, 5**) has developed a new and simplified theory for calculating the equivalent width of lines formed under non-LTE conditions. The method seems to be especially suitable for lines formed in a chromosphere. The possibility that doubly excited levels may play an important role in the formation of the H- and K-emission cores has been discussed by L. Goldberg (**19**).

H. R. Johnson (**26, 27**) has studied the non-LTE formation of the NaD lines in the solar photosphere. S. Dumont (**15**) has considered two- and three-level approximations for Ca II and Mg II. Y. Cuny (**13**) finds in a study of the non-LTE formation of the Balmer lines in the Sun that the residual intensities are almost independent of the chromospheric temperature. In an empirical investigation, Gökdoğan and J.-C. Pecker (**18**) find systematic differences between the curves of growth for different multiplets, even if these multiplets have the same excitation potential.

The important problem of radiative transfer in lines in the presence of a magnetic field has been studied by V. N. Obridko (**36, 35**) and D. N. Rachkowski (**40**).

III. MODEL ATMOSPHERES AND THEORY OF SPECTRA FOR INDIVIDUAL STARS OR GROUPS OF STARS, METHODS OF ABUNDANCE DETERMINATIONS

(*a*) *'Normal' Stars (except Sun) and General Literature on Methods of Abundance Determination*

A general survey of methods and results of abundance determinations has been given by A. Unsöld (**72**). The method of model atmospheres has been discussed by Baschek, Kegel, Bode, Kodaira, Kohl and Traving (**7**). A critical summary of abundance determinations for early-type stars has been given by G. Traving (**78**), the corresponding discussion for stars of type F0-G5 is due to G. Wallerstein (79). B. E. J. Pagel (**57**) gave a critical summary of abundance determination for spectral type G8 and later. The problem of determining basic atmospheric parameters for a chemical analysis has been discussed by C. R. and A. P. Cowley (**49**). A critical investigation of the theoretical considerations influencing the determination of a model atmosphere and the prediction of line profiles has been given by A. B. Underhill in (**69**) and to a certain extent in (**70**). The possibilities of a spectral analysis using hydrogen lines only have been studied by E. Böhm-Vitense (**14**). D. Fischel (**21**) has investigated the influence of the chemical composition on the structure and the metal index of a stellar atmosphere. Tables for the limb darkening for a grid of stellar atmospheres have been published by O. Gingerich (**29**).

Rotating atmospheres and their radiation have been discussed by G. W. Collins (**18**), Y. Osaki (**56**) and J. Hardorp and P. Strittmatter (**32**). A thorough discussion of the comparison of computed fluxes of blanketing models with observations is due to A. Underhill (**70**). E. Peytremann (**58**) is trying to interpret the Geneva seven-colour photometry. He uses model atmospheres and takes the blanketing effect into account. The possible importance of the blanketing effect for the UV spectrum of γ Pegasi has been discussed by Böhm-Vitense, Pyper and Wallerstein (**13**).

Problems of the empirical determination of stellar atmospheres have been studied in great detail by F. van't Veer (**3**) and has been applied by him to the derivation of sunspot models.

Very hot stellar atmospheres

Model atmospheres for neutron stars ($T_{\rm eff} \sim 5 \times 10^6$ °K) have been investigated by Orszag (**55**) and D. C. Morton (**52**). Mrs K. B. Gebbie (**80**) and K. Böhm and W. Deinzer (**11, 12** and unpublished results) have computed model atmospheres for central stars of planetary nebulae. Böhm's and Deinzer's non-grey model atmospheres cover the O'Dell- and the Harman-Seaton sequence. The effective temperature ranges from 38 000°K to $1{\cdot}8 \times 10^5$ °K. A study of hot stars of low luminosity is due to V. V. Sobolev (**60**).

Normal early type stellar atmospheres:

A new fine analysis of τ Scorpii has been made by M. Scholz (**59**). He has determined a new non-grey model atmosphere ($T_{\rm eff} = 32\,800$°K, $\log g = 4{\cdot}2$) using Lucy's method. (See section II.*a* of this report.) D. Mihalas and his collaborators have made very extensive studies of early type stars. They have computed a set of model atmospheres, hydrogen line profiles and helium line strengths (Mihalas, **45**), determined the temperature stratification for a B1 star taking into account line blanketing (Mihalas and Morton, **50**), determined abundances of He (Mihalas, **46**) and of Mg (Henry and Mihalas, **33**). Furthermore Mihalas (**48, 51**) has applied W. Kalkofen's (see section II.*a*) method in order to study the non-LTE formation of the hydrogen continua in early-type stars. Adams and Morton (**1**) have computed the temperature stratification for a B4 star including the effects of line blanketing. Morton (**53, 54**) has also investigated the theoretical line profiles in the ultraviolet spectra of early-type stars.

A. B. Underhill (**71**) and Guillaume, van Rensbergen and Underhill (**31**) have determined the theoretical near ultraviolet spectra of early B stars. They find that broadening effects due to microturbulence are more important in determining the line spectrum than changes in the $p - T$ relation of the model.

A theoretical model of a B 1·5 star with line blanketing has been computed by C. Guillaume (**81**).

Strom and Avrett (**62, 64**) have studied the temperature stratification for a number of early-type stars. Kalkofen and Strom (**38, 63**) used Kalkofen's theory (see above) to compute the influence of non-LTE effects on the stellar continuum fluxes for spectral types B5–A5. They find Balmer discontinuities which are smaller than in the LTE case.

I. M. Kopylov (**42**) has worked out an improved version of Unsöld's method of determining electron densities in hot stars. He finds corrections to Unsöld's values ranging from $-0{\cdot}45$ (O stars) to $-1{\cdot}5$ (F–stars) in $\log n_e$.

M.-O. Baylac (**9**) has investigated the continuous spectrum of B0–F0 stars in the photographic infrared. She finds that the measured Pachen discontinuities are always larger than the calculated ones.

A and F Stars

K. O. Wright (**76**) has studied the excitation temperatures of A– and F–stars. The temperature stratification for A stars, including effects of line blanketing, has been calculated by D. Mihalas (**47**).

Conti, Wallerstein and Wing (**20**) have derived abundances for the main-sequence stars of type A to K in the Hyades. Baschek and Oke (**8**) have determined effective temperatures and surface gravities for normal A, for Ap and Am stars. K. Kohl (**40**) made a fine analysis of Sirius and essentially confirmed the fact that his star has peculiar abundances. These are the subject of a new study by Cl. Chevalier (**17**). L. Houziaux (**35, 36**) has carried out a theoretical investigation of the Lyman lines in A (and B) stars. He has also studied the profiles of the infrared oxygen triplet at λ 7772 in spectrum of Vega (**37**). F. Edmonds (**22, 23**) and Edmonds and Talbert (**24**) have made a very thorough study of Procyon. Edmonds computed a set of eight non-grey models for this purpose. He finds an effective temperature of 6780 °K. Baschek, Holweger, Namba and Traving (**6**) have made a detailed analysis of the spectrum and the atmosphere of β Virginis. They find an overabundance of a factor of about two (with respect to the Sun) for all elements from C to La.

Spectral Type G and later

The influence of the surface gravity on the UBV colours of G stars has been discussed by Baschek (**5**); Matsushima and Terashita (**43, 44**) have computed non-grey model atmospheres for solar type stars. The formation of the forbidden [O I] lines in late-type stellar atmospheres has been discussed by R. E. M. Gasson and B. E. J. Pagel (**28**) and by P. S. Conti *et al.* (**61**). Gasson (**27**) in her thesis has made a detailed analysis of the spectrum of Arcturus.

G. Cayrel (**16**) has studied differentially the physical parameters and chemical composition of seven early-type K stars (five of these are giants) and the comparison star ϵ Virginis. Model atmospheres for M stars have been derived by J. R. Auman (**4**), T. Tsuji (**68,** see also **67**), by M. S. Vardya and R. Kandel (**73**), and by V. G. Buslavsky (**15**). K. S. K. Swamy (**65**) has computed and discussed the profiles of strong lines in K dwarfs. Problems of the curve of growth analysis in cool stars have been studied by Fujita and Tsuji (**26**) and by Yamashita (**77**). L. W. Fredrich, H. R. Johnson and T. Fay (**25**) discussed the infrared spectrum of cool stars. They find that most observed features in the region 8500–12 000 Å can be attributed to CN.

White Dwarfs

Model atmospheres for white dwarfs of type DA have been calculated by Terashita and Matsushima (**66**). Model atmospheres consisting either only of helium or of hydrogen have been derived by A. K. Kolesov (**41**) for the temperature range 12 000–20 000°K. He finds that convection is very important for helium white dwarfs. A detailed study of the blanketing in white dwarfs has been made by V. Weidemann (**75**).

Other problems

Studies of atmospheric Li abundances have been of very great interest during the last three years. We mention only the two fundamental papers by Herbig (**34**) and by Wallerstein, Herbig and Conti (**74**). The production of Li, Be, B by spallation processes in stellar atmospheres has been discussed by Bernas, Gradstajn, Reeves and Schatzman (**10**). In this connection we should also mention the work by J. Audouze (**2**).

(*b*) *Sun*

In this section we quote only some papers which have some relation to the general theory of stellar atmospheres. Other investigations on the solar atmosphere are the subject of the report of Commission 12.

A new and very elaborate semiempirical investigation of the stratification of the solar atmosphere has been made by H. Holweger (**10**). He concludes that the local temperature at $\bar{\tau} \sim 10^{-6}$ is about 3900°K. According to Holweger all non-LTE effects in the solar atmosphere are too small to influence abundance determinations in a perceptible way. The latter question has also been studied in detail by Müller and Mutschlecner (**19**). B. E. J. Pagel (**21**) has discussed the maximal values of departures from LTE compatible with the empirical data on excitation temperatures. A new study of the solar spectrum for the purpose of new abundance determinations has been started by L. H. Aller (**1, 17, 31**) and his collaborators. He is paying special attention to the less abundant elements lke Sb, Ba, Be, Cd, Nb and recently yttrium (see **31**). He considers (1) an extrapolation of the Pierce-Waddell model to higher layers and (2) Pagel's model joined to the one determined from centre-to-limb variation at 9000 Å and 16 000 Å by Pierce and Waddell. C. R. Cowley (**5**), and Müller and Mutschlecner (**19a**) has studied the nickel abundance; King, Lawrence and Link (**15**) have determined the gallium, silver and indium abundances in the solar atmosphere. E. A. Müller (**20**) has made new abundance determinations for a few elements. She finds a new value for Li: $\log (N_{\mathrm{Li}}/N_{\mathrm{H}}) = 11{\cdot}2$. Schmahl and Schröter (**25**) derive $-$ 11·4 for the same ratio from sunspot spectra. Goldberg, Kopp and Dupree (**9**) have rediscussed the important problem of the Fe abundance determination in the photosphere. Excitation temperatures have been determined by Teplitskaya (**26**).

Baschek, Holweger and Müller (**2**) are studying the centre-to-limb variation of weak and medium-strong O I and a few C I lines. C. de Jager and L. Neven (**12**) have made centre-to-limb observations of 50 infrared lines and have discussed (**13**) the centre-to-limb variation of the infrared carbon multiplet. The second paper is the first application of a new method for the systematic analysis of Fraunhofer lines belonging to one multiplet. A theoretical study of blends of metallic lines has been undertaken by Fujita (6). An investigation of the limb intensity of Hα is due to P. R. Wilson (29). The averaging process for photospheric lines which show fluctuations has been discussed by the same author (28). Simplified blanketing models of the photosphere have been computed by Gingerich (8).

J. Houtgast (11) has derived absolute intensities in the range 3000–4000 Å of the solar spectrum at about 200 wavelengths.

Non-LTE line formation processes in the solar atmosphere have been studied by several authors:

P. Chamaraux (3) and H. R. Johnson (14) have made a non-LTE computation of the NaD lines. They find large departures from LTE outside the temperature minimum. According to these investigations the NaD lines are collisionally dominated. A related study is due to B. J. O'Mara (17), who has also computed Ca line profiles. A theoretical study of the Ca II and Mg II lines has been made by S. Dumont (4).

R. E. M. Gasson (7) has confirmed the multiple character of the damping branches of the solar curve of growth.

W. Unno (27) has worked out a generalization of Goldberg's method (for the determination of the depth dependence of turbulence) which includes the effects of macroturbulence. A critical discussion of the inference of velocity fields from line asymmetries is due to Kulander and Jefferies (16).

Zwaan (30) has made a very detailed semi-empirical study of sunspot models. Miyamoto (18) has studied the thermal instability of chromospheric material.

Basic problems of line formation in the solar corona have been investigated by Pottasch (22, 23) and by C. Pecker-Wimel and R. N. Thomas (24).

(c) *Atmospheres of Typical Population II Objects and of Peculiar Stars*

Baschek (3) has studied the atmospheric structure and the abundances of the high velocity star Wilson 10 367. Kodaira (16) has investigated the atmosphere of HD 161 817. Kodaira finds a logarithm of the underabundance of metals of − 1·11 (with respect to the Sun). For Wilson 10 367, Baschek gets a corresponding value of − 1·6. Pagel and Powell (19) find that the cool subdwarf HD 25 329 is metal deficient by a factor 20. They make allowance for the complex character of the curve of growth. R. Cayrel (9) is in the process of re-examining the location of Subdwarfs in the (M_{bol} − log T_{eff})-plane. A detailed analysis of the metal deficient star χ Dra has been made by M. Spite (22). She finds metal deficiencies of about a factor 2. Two population II K giants, HD 6497 and HD 6833 have been studied by G. Cayrel (unpublished).

A considerable amount of work is being done on peculiar A stars: Auer, Mihalas, Aller and Ross (2) have made a fine analysis of the manganese star 53 Tauri. Mihalas and Henshaw (17) have made an abundance analysis of C II, Mg II, O I, Si II and Si III in Ap stars. L. H. Aller (1) is carrying out an extensive programme of abundance determinations in magnetic stars using model atmosphers. Searle, Lungershausen and Sargent (21) have determined abundances of iron-peak elements in peculiar A stars.

E. Böhm-Vitense (using the oblique rotator hypothesis) has constructed magnetic fields for peculiar A stars which permit an explanation of the observed field variations (5) and also of the velocity and intensity variation of lines (4, 5). She also gives an explanation of the crossover effect, using the same form of the magnetic field (7).

Atmospheres and chemical abundances for metallic line stars have been investigated by P. S. Conti (11) and F. Praderie (20). A theoretical study, in which he tries to explain seven-colour photometric results (Golay system) on metalic line stars, has been done by B. Hauck (14). An analysis of two Ba II stars has been carried out by I. J. Danziger (12). A fine analysis of three Centauri A has been made by J. Hardorp (**13**). He confirms the large He^3 abundance determined earlier by Sargent and Jugaku and finds $T_{eff} = 19\,400°K$, $\log g = 3{\cdot}87$.

Model atmospheres for helium stars have been computed by D. Klinglesmith (**18**), E. Böhm-Vitense (**8**), K. Hunger and D. van Blerkom (**15**). Klinglesmith's computations cover the range $16\,000°K \leq T_{eff} \leq 30\,000°K$; $2{\cdot}5 \leq \log g \leq 4{\cdot}0$. Böhm-Vitense has studied the range $7300°K \leq T_{eff} \leq 12\,900°K$; $2{\cdot}0 \leq \log g \leq 4{\cdot}5$. Hunger and von Blerkom have computed a non-grey model for $T_{eff} = 18\,000°K$; $\log g = 3{\cdot}5$ (corresponding roughly to Klemola's star).

Climenhaga (**10**) and Wyller (**23**) have studied carbon isotopic abundances in carbon stars. A. Wyller (**24**) finds that the lithium and calcium lines used in the differential abundance analysis are possibly strongly contaminated by $C^{13}N^{14}$ lines in carbon stars.

K.-H. BÖHM
President of the Commission

BIBLIOGRAPHY

I. Books, Proceedings of Colloquia and Symposia

1. Proc. First Harvard-Smithsonian Conference on Stellar Atmospheres (Model Stellar Atmospheres), Cambridge (Mass.) 1964.
2. Proc. Second Harvard-Smithsonian Conference on Stellar Atmospheres: The Formation of Spectrum Lines, ed. E. H. Avrett, O. Gingerich, C. A. Whitney, Cambridge (Mass.) 1965.
3. Proc. IAU Colloquium: *The Blanketing Effect*, ed. K.-H. Böhm, *J. quantit. Spectrosc. radiat. Transfer*, **6**, no. 5, 1966.
4. Thomas, R. N. 1965, *Non-Equilibrium Thermodynamics in the Presence of a Radiation Field*, University of Colorado Press.
5. *The Solar Spectrum*, ed. C. de Jager. D. Reidel Publ. Co., Dordrecht-Holland, 1965.
6. Proc. IAU Symposium no. 26: *Abundance Determinations in Stellar Spectra*, ed. H. Hubenet, Academic Press, 1966.
7. Sobolev, V. V., Gorbatzky, V. G., Ivanov, V. V. (ed.) 1966, *Theory of Stellar Spectra*, 'Nauka' Publ. Co., Moscow.
8. Sampson, D. H. 1965, *Radiative Contributions to Energy and Momentum Transport in a Gas*, Interscience Publ.
9. Underhill, A. B. 1966, *The Early-Type Stars*, Reidel Publ. Co., Dordrecht.
10. Proc. IAU Symposium No. 22: *Stellar and Solar Magnetic Fields*, ed. R. Lüst, North Holland Pub. Co., Amsterdam, 1965.
11. Review articles:
Pecker, J.-C. Model Atmospheres, in *A. Rev. Astr. Astrophys*, **3**, 135.
Regemorter, H. van Spectral Line Broadening, *A. Rev. Astr. Astrophys.*, **3**, 71.
Cayrel, R., Cayrel, G. 1966, Abundance Determination from Stellar Spectra, *A. Rev. Astr. Astrophys.*, **4**, 1.

II. *Theory of Stellar Atmospheres*

(*a*) *Theory of radiative equilibrium (including blanketing effect, but generally excluding other problems of line formation).*

1. Sampson, D. H. 1965, *Radiative Contributions to Energy and Momentum Transport in a Gas.* Interscience Publishers.
2. Cayrel, R. 1966, The Line Blanketing Problem in the Non-LTE Approach. *J. quantit. Spectrosc, radiat. Transfer*, **6**, 621.
3. Kalkofen, W., Avrett, H. 1965, Solution of the Line and Continuum Transfer Problem for a 3-Level Atom. Proc. 2nd Harvard-Smiths. Conf. on Stellar Atmospheres, p. 249.
4. Ambarzumian, V. A. 1964, On a Non-Linear Problem of Light Scattering in a Turbid Medium. *Dokl. Akad. Nauk Armjam. SSR*, **38**, 225.
5. Ambarzumian, V. A. 1964, On one Case of Diminishing of Optical Thickness of a Medium under the Action of Radiation. *Dokl. Akad. Nauk Armjan. SSR*, **39**, 159.

6. Johnson, H. R., Klinglesmith, D. A. 1965, Proc. 2nd Harvard-Smithson. Conf. on Stellar Atmospheres, 221.
7. Bellman, R., Kalaba, R., Ueno, S. 1964, Invariant Imbedding and Time-Dependent Scattering of Light in a One-Dimensional Medium. *J. Math. Analysis and Applications*, **9**, 424.
8. Bellman, R., Kalaba, R., Ueno, S. 1964, Invariant Imbedding and Scattering of Light in a One-Dimensional Medium with a Moving Boundary. The Rand Corp., Mem. RM-3960-ARPA.
9. Bellman, R., Kalaba, R., Ueno, S. 1964, Invariant Imbedding and a Resolvent of the Photon-Diffusion Equation. The Rand Corp., Mem. RM-3937-ARPA.
10. Bellman, R., Kalaba, R., Kagiwada, H., Ueno, S. The Invariant Imbedding Equation for the Dissipation Function of a Homogeneous Finite Slab. The Rand Corp., Mem. RM-4651-NASA. *Q. J. Appl. Math.* (in press).
11. Bellman, R., Kagiwada, H., Kalaba, R. 1966, Invariant Imbedding and a Reformulation of the Internal Intensity Problem in Radiative Transfer. *Mon. Not. R. astr. Soc.*, **132**, 183.
12. Bellman, R., Kagiwada, H., Kalaba, R. 1966, Numerical Results for the Auxiliary Equation of Radiative Transfer. *J. quantit. Spectrosc. rad. Transfer*, **6**, 291.
13. Bellman, R., Kagiwada, H., Kalaba, R., Ueno, S. 1966, Numerical Results for Chandrasekhar's *X*- and *Y*-Functions of Radiative Transfer. *J. quantit. Spectrosc. rad. Transfer*, **6**, 479.
14. Biberman, L. M., Vorobyev, V. S., Lagar'kov, A. N. 1965, Transfer of Radiation in an Ionization Continuum. *Optica i Spectroscopia*, **19**, 326.
15. Böhm, K.-H. 1965, Temperaturinversion in einer Atmosphare im nichtgrauen Strahlungsgleichgewicht. *Z. Astrophys.*, **62**, 167.
16. Böhm, K.-H., Deinzer, W. 1965, Atmosphären von Zentralsternen planetarischer Nebel, I. *Z. Astrophys.*, **61**, 1.
17. Carlstedt, J. L., Mullikin, T. W. 1966, Chandrasekhar's *X*- and *Y*-Functions. *Astrophys. J.*, **12**, 449.
18. Chapman, R. J. 1966, Radiative Transfer in Extended Stellar Atmospheres. *Astrophys. J.*, **143**, 61.
19. Collins, G. W. II, Code, A. D. 1965, Some Numerical Methods for the Solution of the Equation of Transfer. *Astrophys. J.*, **142**, 1576.
20. Cuny, Y. 1964, Solution of the Transfer Problem. Proc. Second Harvard Smithson. Conf. *C. r. Acad. Sci. Paris*, **258**, 3292.
21. Dave, J. V., Walker, W. H. 1966, Convergence of the Iterative Solution of the Auxiliary Equations for Rayleigh Scattering. *Astrophys. J.*, **144**, 798.
22. Dave, J. V. 1964, Meaning of Successive Iteration of the Auxiliary Equation in the Theory of Radiative Transfer. *Astrophys. J.*, **140**, 1292.
23. Delache, Ph. Comment on the Inversion of the Darkening Law. *Astrophys. J.*, (submitted f. publication).
24. Delache, Ph. 1966, Une Généralisation de l'Approximation de Barbier-Eddington. *Ann. Astrophys.*, **29**.
25. Delache, Ph. 1965, *C. r. Acad. Sci. Paris*, **261**, 643-646.
26. Delache, Ph. Symp. on Interdisciplinary Asp. of Rad. Transfer, Philadelphia, Feb. 1966.
27. Feautrier, P. 1964, A Procedure for Computing the Mean Intensity and the Flux. Proc. First Havard-Smithson. Conf., Cambridge, p. 80.
28. Feautrier, P. 1964, A Method for Improving the $T(x)$-Law. Proc. I. Harvard-Smithson. Conf. on Stellar Atmospheres, Cambridge, p. 108.
29. Feautrier, P. 1964, Sur la Résolution Numérique de l'Equation de Transfert. *C. r. Acad. Sci. Paris*, **258**, 3189.
30. Frisch, U. 1967 (unpublished).
31. Gingerich, O., Mihalas, D., Matsushima, S., Strom, S. 1965, An Archetype Non-Grey Stellar Atmosphere. *Astrophys. J.*, **141**, 316.
32. Henyey, L. G. Radiative Transfer, I: The Flux-Temperature Relation and Non-Grey Stellar Atmospheres. Submitted to *Astrophys. J.*

34. Hoang, Binh-Dy, McCarroll, R. 1966, Méthode Simple de Calcul des Ecarts à l'Equilibre Thermodynamique Local. *C. r. Acad. Sci. Paris*, **262**, 476

35. Horak, H. G., Janousek, A. L. 1965, The *H*- and Related-Functions for $p(\cos\,\theta) = 1 + {}_1P^1 (\cos\,\theta) + {}_2P_2 (\cos\,\theta)$. *Astrophys. J. Suppl.*, **11**, 101.

36. Hummer, D. G. 1965, The Emission Coefficient. Proc. 2nd Harvard-Smithson. Conf. on Stellar Atmospheres, 13.

37. Hummer, D. G. 1966, Introduction to Radiative Transfer Problems in Stellar Atmospheres. Presented at the Symp. on Inter-Disciplinary Aspects of Rad. Energy Transfer, Philadelphia. February.

38. Hummer, D. G. 1964, The Mean Number of Scatterings by a Resonance-Line Photon. *Astrophys. J.*, **140**, 276.

39. Irvine, W. M. 1966, The Distribution of Photon Optical Paths in a Scattering Atmosphere. *Astrophys. J.*, **144**, 1140.

40. Ivanov, V. V. 1965, Scattering of Light in an Atmosphere of Finite Optical Thickness. *Sov. Astr. A. J.*, **8**, 874.

41. Ivanov, V. V. 1964, Light Scattering in a Plane Layer. *Astr. Zu.*, **41**, 44.

42. Ivanov, V. V. 1964, On the Problem of Light Scattering in the Atmosphere of Finite Optical Thickness. *Astr. Zu.*, **41**, 1097.

43. Ivanov, V. V., Leonov, V. V. 1965, Light Scattering in an Optically Thick Atmosphere with a Non-Spherical Indicatrix. *Izv. Akad. Nauk, Ser. Phys. Atm.*, **1**, 803.

44. Kalkofen, W. 1966, Deviations from LTE in Stellar Atmospheres. *J. quantit. Spectrosc. radiat. Transfer*, **6**, 633.

45. King, J. I. F., Sillars, R. V., Harrison, R. H. 1965, Hopf *q*-Function Evaluated to Eight-Digit Accuracy. *Astrophys. J.*, **142**, 1655.

46. Kogure, T. 1967, The Radiation Field and the Theoretical Balmer Decrement of Be-Stars *Publ. astr. Soc. Japan* (in preparation).

47. Kolesnik, I. G., Frank-Kemenetsky, D. A. 1964, Analysis of the Energy Distribution near the Balmer Limit in Spectra of Non-Stable Stars. *Astr. Zu.*, **41**, 178.

48. Lucy, L. B. 1964, A Temperature Correction Procedure. Proc. First Harvard-Smithson. Conf. on Stellar Atmospheres, p. 93.

49. Mihalas, D., Stone, M. Effects of Departures from LTE on Hydrogen and Helium Continua in Early-Type Stars. In preparation.

50. Mihalas, D. 1966, LTE Model Atmospheres with Line Blanketing: Review and Future Prospects. *J. quantit. Spectrosc. radiat. Transfer*, **6**, 581.

51. Minin, I. N. 1964, Non-Stationary Radiation Field of a Semi-infinite Medium. *Dokl. Akad. Nauk SSSR*, **154**, 1059.

52. Minin, I. N., Piliposian, A. G., Shidlovskaja, N. A. 1964, Tables of Ambartzumian Functions for Anisotropic Scattering. *Trudy astr. Obs. Leningrad. gos. Univ.*, **20**, 12.

53. Nagirner, D. I. 1964, On the Solution of Integral Equations of the Theory of Light Scattering. *Astr. Zu.*, **41**, 669.

54. Nagirner, D. I., Ivanov, V. V. 1966, Transfer of Resonance Radiation in an Infinite Medium, I. *Astrofizika*, **2**, 5.

55. Nagirner, D. I., Ivanov, V. V. 1966, Transfer of Resonance Radiation in an Infinite Medium, II. *Astrofizika*, **2**.

56. Nikoghossian, A. G. 1965, Polychromatic Diffuse Reflection of Light from an Infinitely Deep One-Dimensional Medium. *Astrofizika*, **1**, 285.

57. Obridko, V. N. 1965, The Matrix of Radiative Scattering in a Magnetic Field. *Astr. Zu.*, **42**, 102.

58. Pavhor, S., Kuščer, I. 1966, On the Non-Uniqueness of Solutions of Chandrasekhar's *S*-Equation in Radiative Transfer. *Astrophys. J.*, **143**, 888.

59. Pavlov, V. E. 1964, Atmospheric Scatterings Law in the Optical and Ultraviolet Regions of the Spectrum. *Astr. Zu.*, **41**, 546.

60. Pavlov, V. E. 1964, Atmospheric Indicatrix of Scattering in the Region of Small and Large Scattering Angles. *Astr. Zu.*, **41**, 122.

61. Pecker, J.-C. La Structure de la Photosphère. In: *The Solar Spectrum*, ed. by C. de Jager, p. 29.

62. Pecker, J.-C. 1965, Model Atmospheres. *A. Rev. Astr. Astrophys.*, **3,** 135.
63. Pecker, J.-C. 1964, Sur une Interprétation possible de la Phase Oscillatoire des Novae. *Ann. Astrophys.*, **27,** 287.
64. Rublev, S. V. 1965, Characteristics of the He III Zone in the Atmospheres of the WR Stars. *Astr. Zu.*, **42,** 718.
65. Siewert, C. E., Zweifel, P. F. 1966, An Exact Solution of Equations of Radiative Transfer for Local Thermodynamic Equilibrium in the Non-Grey Case. Picket Fence Approximation. *Ann. Phys.*, **36,** 61.
66. Sobolev, V. V. 1964, Diffusion of Radiation in a Plane Layer of Large Optical Thickness. *Dokl. Akad. Nauk SSSR*, **155,** 316.
67. Stewart, J. C. 1966, (unpublished).
68. Stewart, J. C. 1964, Non-Grey Radiative Transfer. *Rad. Transfer*, **4,** 723.
69. Swihart, T. L. 1966, Structure of the Solar Photosphere, I. Blanketing and Convection. *Astrophys, J.*, **143,** 358.
70. Swihart, T. L. 1965, A Modification of the Krook-Avrett Temperature Correction Procedure. *Astrophys. J.*, **141,** 821.
71. Swihart, T. L. 1964, Investigation of the Time-Dependent Heating of a Grey Atmosphere. *Astrophys. J.*, **139,** 772.
72. Thomas, R. N. 1965, Some Aspects of Non-Equilibrium Thermodynamics in the Presence of a Radiation Field. Univ. of Colorado Press.
73. Thomas, R. N. 1965, Comment on the Difference between Non-LTE and a Pure Absorption Model for the Line Blanketing Effect. *Astrophys. J.*, **141,** 333.
74. Tsuji, T. 1965, Non-grey Model Atmospheres in Radiative Equilibrium. *Publ. astr. Soc. Japan*, **17,** 152.
75. Ueno, S. 1967, Small-Angle Approximation of the Source Function in the Equation of Radiative Transfer. *J. atmos. Sci.*, to be published.
76. Warner, B. 1964, Departures from Local Thermodynamic Equilibrium in the Solar Atmosphere. *Mon. Not. R. astr. Soc.*, **128,** 63.
77. Wildt, R. 1964, Thermodynamics of the Grey Atmosphere, I: Reversible Adiabatic Process. *Astrophys. J.*, **140,** 1343.
78. Wildt, R. 1966, Thermodynamics of the Grey Atmosphere, II: Unattainability of Detailed Balancing. *Astrophys. J.*, **143,** 363.
79. Wilson, S. J., Sen, K. K. 1964, Modified Spherical Harmonic Method for Solving the Equation of Transfer with Anisotropic Scattering. *Ann. Astrophys.*, **27,** 46.
80. Wilson, S. J., Sen, K. K. 1964, Modified Spherical Harmonic Method and Spherical Geometry, I. *Ann. Astrophys.*, **27,** 654.
81. Yanovitsky, E. G. 1964, The Diffusion of Radiation in a Plane Layer with Isotropic Scattering. *Astr. Zu.*, **41,** 898.
82. Yengibarian, N. B. 1965, On a Non-Linear Problem of Radiative Transfer. *Astrofizika*, **1,** 297.
83. David, K. H. 1965, Näherungen von $S(\tau)$, $J(\tau)$, und $\pi F(\tau)$, dargestellt durch die Intensitäten auf der Sonnenscheibe. *Z. Astrophys.*, **61,** 237.
84. Matsushima, S., Terashita, Y. 1964, Effect of Errors in the Absorption Coefficient on the Non-Grey Model Stellar Atmospheres. *Astrophys. J.*, **140,** 285.
85. Strom, S. E., Kurucz, R. L. 1966, A Statistical Procedure for Computing Line-Blanketed Model Stellar Atmospheres with Applications to the F5 IV Star Procyon. *J. quantit. Spectrosc. radiat. Transfer*, **6,** 591.
86. Dietz, R. D., House, L. L. 1965, The formation of the Lyman continuum: isothermal atmosphere, *Astrophys. J.*, **141,** 1393.
87. Mullikin, T. W. 1966, The complete Rayleigh-scattered field within a homogeneous plane-parallel atmosphere. *Astrophys. J.*, **145,** 886.
88. Frisch, H. 1966, The Effect of Line Absorption on the Temperature at the Surface of the Solar Photosphere. *J. quantit. Spectrosc. radiat. Transfer*, **6,** 629.
89. Gingerich, O. 1966, Blanketing Approximations for Solar Models. *J. quantit. Spectrosc. radiat. Transfer*, **6,** 609.

(b) Continuous absorption coefficients and some other basic data

1. Auman, J. R. *Astrophys, J. Suppl.* no. 127, (in press).
2. Bély, O. 1965, Extension de la théorie généralisée du défaut quantique au cas des ions hydrogénoïdes I, II. *Ann. Astrophys.*, **28**, 567 and 843.
3. Bode, G. 1965, Die kontinuierliche Absorption von Sternatmosphären. Sonderdruck, Kiel.
4. Burgess, A. 1965, Dielectronic Recombination in the Corona. *Ann. Astrophys.*, **28**, 774.
5. Buslavsky, V. G. 1964, On the Sources of Opacity in the Atmospheres of Late-Type Stars. *Vestnik Leningrad. gos. Univ.*, **13**, 114.
6. Cox, A. N., Stewart, J. N. 1965, Radiative and Conductive Opacities for Eleven Astrophysical Mixtures. *Astrophys. J. Suppl.*, **11**, 94.
7. Doughty, N. A., Fraser, P. A., McEachran, R. P. 1966, The Bound-Free Absorption Coefficient of the Negative Hydrogen Ion. *Mon. Not. R. astr. Soc.*, **132**, 255.
8. Doughty, N. A., Fraser, P. A. 1966, The Free-Free Absorption Coefficient of the Negative Hydrogen Ion. *Mon. Not. R. astr. Soc.*, **132**, 267.
9. Goldberg, L., Dupree, A. K., Allen, J. W. 1965, Collisional Excitation of Autoionizing Levels. *Ann. Astrophys.*, **28**, 585.
10. John, T. L. 1964, The Free-Free Transitions of the Negative Hydrogen Ion in the Exchange Approximation. *Mon. Not. R. astr. Soc.*, **128**, 93.
11. John, T. L. 1966, The Free-Free Transitions of an Electron in the Field of a Hydrogen Atom at Infra-red Frequencies. *Mon. Not. R. astr. Soc.*, **131**, 315.
12. John, T. L. 1966, The Limiting Behaviour of the Absorption Cross Sections of the Negative Hydrogen Ion. *Mon. Not. R. astr. Soc.*, **133**, 447.
13. Kandel, R. 1966, L'absorption continue de l'ion négatif du chlore. *C. r. Acad. Sci. Paris*, **262**, 373.
14. McCarrol, R., Salin, A. 1966, L'excitation de l'hydrogène par protons. *Ann. Astrophys.*, **29**, 283.
15. McDowell, M. R. C., Williamson, J. H., Myerscough, V. P. 1966, Continuous He^- Absorption. *Astrophys. J.*, **144**, 827.
16. Peach, G. 1965, A General Formula for the Calculation of Absorption Cross-Sections for Free-Free Transitions in the Field of Positive Ions. *Mon. Not. R. astr. Soc.*, **130**, 361.
17. Praderie, F. 1964, Calcul de quelques sections de photoionisation du carbone neutre. *Ann. Astrophys.*, **27**, 129.
18. Rensbergen, W. van 1967, The Influence of Doubly Excited Levels on the Ionization Formula for the Solar Corona. *Solar Physics*, **1** (in press).
19. Sommerville, W. B. 1965, The Continuous Absorption Coefficient of He^-. *Astrophys. J.*, **141**, 811.
20. Vardya, M. S. 1964, Opacity at λ 1·65μ in Late-Type Stars. Proc. 1st Harvard-Smithsonian Conf., p. 50.
21. Vardya, M. S. 1964, Atmospheric Rosseland Mean Absorption Coefficients. *Astrophys. J.*, **8**, 277.
22. Vardya, M. S., Böhm, K.-H. 1965, Pressure Broadening and Opacity in the M2 V star HD 95735. *Mon. Not. R. astr. Soc.*, **131**, 89.
23. Vardya, M. S. 1965, A Method for Estimating H_2O Absorption. *J. quantit. Spectrosc. radiat. Transfer*, **5**, 867.
24. Vardya, M. S. 1966, Negative Chlorine Ion and M-Spectral Type Stars. *Mon. Not. R. astr. Soc.*, **132**, 475.
25. Vardya, M. S. 1966, Simplified Description of the Line Absorption Coefficient in Blanketing Effect Computation. *J. quantit. Spectrosc. radiat. Transfer*, **6**, 539.
26. Vardya, M. S. 1967, Role of Negative Ions in Late-Type Stars. *Mem. R. astr. Soc.* (in press).
27. Vardya, M. S. 1966, March with Depth of Molecular Abundances in the Outer Layers of K and M Stars. *Mon. Not. R. astr. Soc.*, **134**, 347.
28. Vardya, M. S. 1967, Sources of Opacity in Late-Type Stars, a Review. Proc. of the IAU Symp. on Late-Type Stars (in press).

29. Kodaira, K. 1965, Quantitative Analyse des Intensitätssprunges im Sonnenspektrum bei λ 2085 Å. *Z. Astrophys.*, **60**, 240.
30. Gingerich, O. 1964, Review of Opacity Calculations. Proc. 1st Harvard-Smiths. Conf. on Stellar Atmosph., p. 17.
31. Baerentzen, J., Mikkelsen, E. 1967, Unpublished.
32. Ohmura, T. 1964, Evaluation of free-free absorption coefficient of the negative ion. *Astrophys, J.*, **140**, 282.
33. Peach, G. 1966, Total Continuous Absorption Coefficients for Complex Atoms. Preprint.
34. Myerscough, V. P. 1966, Unpublished.
35. Schlüter, D. 1965, Die Berechnung der Übergangswahrscheinlichkeiten von Seriengrenzkontinua mit Anwendung auf die schweren Edelgase. *Z. Astrophys.*, **61**, 67.
36. Weinberg, M., Berry, R. S. 1966, Forbidden Continuum: Free-Bound Transitions in Hydrogen Plasmas. *Phys. Rev.*, **144**, 75.

(c) Hydrodynamic phenomena in stellar atmospheres

1. Baglin, A. 1964, Condition d'existence d'une zone convective dans une atmosphère soumise au triage gravitationnel. *Ann. Astrophys.*, **27**, 389.
2. Baerentzen, J. 1965, The Adiabatic Temperature Gradient and the Specific Heat c_p in the Outer Hydrogen-Helium Convection Zones. *Z. Astrophys.*, **62**, 221.
3. Bird, G. A. 1965, The Equilibrium State of a Shock-heated Atmosphere. *Astrophys. J.*, **141**, 1455.
4. Böhm, K.-H. 1967, The Outer Solar Convection Zone, in Proc. IAU Symp. no. 28 'Aerodynamics in Stellar Atmospheres' (in press).
5. Böhm, K.-H. 1964, Significance of Uncertainties in the Physical Theory of Radiative Transfer. Proc. 1st Harvard-Smiths. Conf. on Stellar Atmosph., p. 155.
6. Böhm, K.-H. 1966, Die Lithiumhäufigkeit und die Struktur der äusseren Konvektionszone in sonnenähnlichen Sternen. *Z. Naturforschung*, **21a**, 1107.
7. Deinzer, W. 1965, On the Magneto-Hydrostatic Theory of Sunspots. *Astrophys. J.*, **141**, 548.
8. Edmonds, F. N., Jr. 1966, A Coherence Analysis of Fraunhofer-line Fine Structure and Continuum Brightness Fluctuations near the Centre of the Solar Disk, II. *Astrophys. J.*, **144**, 733.
9. Edmonds, F. N., Jr., McCullough, J. R. 1966, The Evidence for an Oscillatory Component in Solar Granulation Brightness Fluctuations. *Astrophys. J.*, **144**, 754.
10. Faulkner, J., Griffiths, K., Hoyle, F. 1965, On the Surface Boundary Conditions for Stars. *Mon. Not. R. astr. Soc.*, **129**, 363.
11. Frisch, U. 1964, Couplage d'ondes magnétodynamiques en milieu stratifié. *Ann. Astrophys.*, **27**, 224.
12. Gough, D. O., Taylor, R. J. 1966, The Influence of a Magnetic Field on Schwarzschild's Criterion for Convective Instability in an Ideally Conducting Fluid. *Mon. Not. R. astr. Soc.*, **133**, 85.
13. Henyey, L. G. Vardya, M. S., Bodenheimer, P. 1965, The Calculation of Model Envelopes. *Astrophys. J.*, **142**, 841.
14. Hollandsky, O. P. 1964, Sound Velocities in the Atmospheres of Hot Supergiants. *Publ. Krym. astrofiz. Obs.*, **32**, 127.
15. Hollandsky, O. P. 1965, On Supersonic Turbulence in the Atmospheres of Hot Supergiants. *Publ. Krym. astrofiz. Obs.*, **33**, 266.
16. Hollandsky, O. P., Chub, A. T. 1966, Investigation of the Turbulent Spectrum in Application to Stellar Atmospheres. *Publ. Krym. astrofiz. Obs.*, **35**, 34.
17. Kaplan, S. A., Petrukhin, N. S. 1965, On the Theory of Convection in a Polytropic Atmosphere with a Uniform Magnetic Field. *Astr. Zu.*, **42**, 74.
18. Kato, S. 1966, The Effect of Compressibility upon Oscillatory Convection. *Publ. astr. Soc. Japan*, **18**.
19. Kato, S. 1966, On the Atmospheric Oscillations Excited by Turbulence. *Astrophys. J.*, **143**, 372.

20. Kato, S. 1966, The Response of an Unbounded Atmosphere to a Point Disturbance, I. Time-Harmonic Disturbances. *Astrophys. J.*, **143**, 893.

21. Kato, S. 1966, The Response of an Unbounded Atmosphere to Point Disturbances, II. Impulsive Disturbances. *Astrophys. J.*, **144**, 326.

22. Kopylov, I. M. 1964, Turbulent Motions in Stellar Atmospheres, in *Problems of Magnetohydrodynamics and Cosmic Gas Dynamics*, pp. 74–106, Fizmatgiz Publ. Co., Moscow.

23. Kuhi, L. V. 1964, Mass loss from T Tauri Stars. *Astrophys. J.*, **140**, 1409.

24. Lemaire, J. 1966, Evaporative and Hydrodynamical Atmospheric Models. *Ann. Astrophys.*, **29**, 197.

25. Lighthill, M. I. 1967, Waves in the Sun's Atmosphere in Proc. IAU Symposium no. 28, *Aerodynamic Phenomena in Stellar Atmospheres* (in press).

26. McLean, D. 1964, Le champ magnétique dans un fluide de conductivité infinie placé dans un champ gravitationnel uniforme. *Ann. Astrophys.*, **27**, 195.

27. Mein, P. 1966, Champ macroscopique des vitesses dans l'atmosphère solaire, d'après les mesures de déplacement des raies de Fraunhofer. *Ann. Astrophys.*, **29**, 153.

28. Meyer, F., Schmidt, H. U. 1966, Response of Solar Atmosphere to granular disturbance. *Trans. IAU*, **12B**, 559. See also *Z. Astrophys.*, **65**, 274.

29. Mihalas, D. 1965, Constant Flux Convective Models Atmospheres. *Astrophys. J.*, **141**, 564.

30. Miyamoto, S. 1965, Thermal Stability of the Chromosphere and Spicule. *Publ. astr. Soc. Japan*, **17**, 127.

31. Moore, D. W. 1967, Penetrative Convection, in Proc. IAU Symposium no. 28, *Aerodynamic Phenomena in Stellar Atmospheres* (in press).

32. Moore, D. W., Spiegel, E. A. 1966, A Thermally Excited Non-linear Oscillator. *Astrophys. J.*, **143**, 871.

33. Nissen, P. E. 1965, Turbulence in the Solar Photosphere as derived from the Study of Centre-to-Limb Variations of Equivalent Widths of Ti I Lines. *Ann. Astrophys.*, **28**, 556.

34. Olson, E. C. 1966, The Depth Dependence of Solar Turbulence. *Astrophys. J.*, **143**, 912.

35. Olson, E. C. 1966, The Effect of Acoustic Waves and Convection on Solar Line Asymmetries. *Astrophys. J.*, **143**, 904.

36. Rublev, S. V. 1964, On the Dynamic State of the Atmospheres of WR Stars. *Astr. Zu.*, **41**, 1063.

37. Sack, N. 1967, Studies in detailed numerical techniques for the fitting of a radiative zone to a convection zone. *Ann. Astrophys.* (in press).

38. Saito, M. 1965, On Shock Propagation and the Chromospheric Structure of Late-Type Stars. *Publ. astr. Soc. Japan*, **17**, 107.

39. Schatzman, E. 1967, Convective Transport of Elements Formed at the Surface of the Stars and their Dilution in the Convection Zone. Unpublished.

40. Sitnik, G. F. 1963, Stability Criterion for Hydrostatic Equilibrium. *Astr. Zu.*, **40**, 413.

41. Souffrin, P. 1966, Hydrodynamique d'une atmosphère perturbée par une zone convective sous-jacente. *Ann. Astrophys.*, **29**, 55.

42. Spiegel, E. A. 1965, Convective Instability in a Compressible Atmosphere, I. *Astrophys J.*, **141**, 1068.

43. Spiegel, E. A. 1966, Energy Transport by Turbulent Convection. In *Stellar Evolution* (eds. Cameron, A. G. W., Stein, R. F.) Plenum Press, New York.

44. Spiegel, E. A. 1967, The Theory of Turbulent Convection. Proc. IAU Symposium, no. 28: *Aerodynamic Phenomena in Stellar Atmospheres* (in press).

45. Swihart, T. L., Margrave, T. E. 1967, Inhomogeneities in the solar photosphere. Unpublished.

46. Uchida, Y. 1965, Standing Mode of Compressional-body Gravity Wave in the Solar Chromosphere. *Astrophys. J.*, **142**, 335.

47. Walter, K. 1965, Geordnete Strömungen in instabilen Schichten grosser Ausdehnung mit Anwendung auf stellare Konvektionszonen. *Ann. Astrophys.*, **28**, 648.

48. Weiss, N. O. 1964, Magnetic flux tubes and convection in the Sun. *Mon. Not. R. astr. Soc.*, **128**, 225.

49. Weiss, N. O. 1966, The Expulsion of Magnetic Flux by Eddies. *Proc. R. Soc. London*, Ser. A, **293**, 310.

50. Weymann, R., Sears, R. L. 1965, The Depth of the Convective Envelope on the Lower Main Sequence and the Depletion of Lithium. *Astrophys. J.*, **142**, 174.
51. Wilson, O. C. 1966, Stellar Convection Zones, Chromospheres, and Rotation. *Astrophys. J.*, **144**, 695.
52. Yamaguchi, S. 1967, Convective Instability in a Layer of Varying Superadiabatic Gradient. *Publ. astr. Soc. Japan.* (in press).
53. Zirker, J. B. 1964, On the Periods of Large-Amplitude Photospheric Oscillations. *Ann. Astrophys.*, **27**, 429.
54. Grover, R., Hardy, J. W. 1966, The propagation of shocks in exponentially decreasing atmospheres. *Astrophys. J.*, **143**, 48.
55. Kohl, K. 1966, Das Eindringen von Strömungen in die stabile Schicht unterhalb der solaren Wasserstoffkonvektionszone. *Z. Astrophys.*, **64**, 472.
56. Frazier, E. N. 1966, An Observational Study of the Hydrodynamics of the Lower Solar Atmosphere. Thesis, Berkeley.
57. Kuperus, M. 1965, The Transfer of Mechanical Energy in the Sun and the Heating of the Corona. Thesis, Utrecht.

(d) Thermodynamic problems of stellar atmospheres

1. Dolan, J. F. 1965, Stellar molecular abundances, I. Polyatomic molecules. *Astrophys. J.*, **142**, 1621.
2. Kolesnik, I. G. 1965, Enhancement of ionization equilibrium in a stellar atmosphere. *Astr. Zu.*, **42**, 63.
3. Schlender, B., Traving, G. 1965, Tschebycheff-Approximation von Zustandssummen. *Z. Astrophys.*, **61**, 92.
4. Souffrin, S. 1965, Influence des collisions inélastique sur la fonction de distribution des électrons dans des conditions astrophysiques. *Ann. Astrophys.*, **28**, 399.
5. Traving, G., Baschek, B., Holweger, H. 1966, Tabellen für die Berechnung von Zustandssummen. *Abh. Hamburger Sternwarte*, **8**, Nr. 1.
6. Tsuji, T. 1964, Molecular Abundance in Stellar Atmospheres. *Ann. Toyko astr. Obs.*, **9**, 1.
7. Vardya, M. S. 1964, Pressure Dependent Partition Functions and Equilibrium Constants for Molecular Hydrogen. Proc. 1st Harard Smithsonian Conf. on Stellar Atmospheres, p. 5.
8. Vardya, M. S. 1965, Thermodynamics of a Solar Composition Gaseous Mixture. *Mon. Not. R. astr. Soc.*, **129**, 205.
9. Vardya, M. S. 1965, Pressure Dissociation and Molecular Hydrogen. *Mon. Not. R. astr. Soc.*, **129**, 345.
10. Wyller, A., Morris, S. 1967 (submitted to *Ann. Astrophys.*).

(e) Extended envelopes

1. Brandt, J. C. 1966, Chemical composition of the photosphere and corona as influenced by the solar wind. *Astrophys. J.*, **143**, 265.
2. Gorbatzky, V. G. 1965, On the Influence of Electron Collisions of Balmer Line Intensities in Spectra of Moving Stellar Envelopes. *Astrofizika*, **1**, 129.
3. Hughes, C. J. 1965, Profiles of Lyman-Alpha scattered by free electrons in the solar corona. *Astrophys. J.*, **142**, 321.
4. Locoarret, M. 1965, Enveloppes d'étoiles Be: variations spectrales et caractéristiques physiques. *Ann. Astrophys.*, **28**, 321.
5. Livshitz, M. A. 1963, An Estimate of the Electron Densities in Stellar Chromospheres. *Astr. Zu.*, **40**, 1065.
6. Livshitz, M. A., Nikolsky, G. M. 1964, The $n_e - T$ relation in the Transition Region between the Chromosphere and Corona. *Astr. Zu.*, **41**, 75.
7. Magnan, Chr. 1965, Les Couches Extérieures de la Composante de Type K de 31 Cygni. *Ann. Astrophys.*, **28**, 512.
8. Athay, R. G. 1966, Radiative Energy Loss from the Solar Chromosphere and Corona. *Astrophys. J.*, **146**, 223.

(*f*) *General theory of line profiles and line formation*

1. Abhyankar, K. D. 1964, On the Schuster Problem for a Moving Atmosphere. I, II. *Astrophys. J.*, **140**, 1353, 1368.
2. Abhyankar, K. D. 1965, On the Schuster Problem for a Moving Atmosphere, III. *Astrophys. J.*, **141**, 1056.
3. Athay, R. G. 1964, Source-Function Equality in Multiplets. *Astrophys. J.*, **140**, 1579.
4. Athay, R. G. 1965, Theoretical Line Intensities. I. Strong Emission Lines. II. Excitation of Chromospheric He II and Hydrogen. III. Solar UV Lines and Continua of H, He I, and He II and the Chromospheric Model. *Astrophys. J.*, **142**, 724, 732, 755.
5. Athay, R. G. 1966, Theoretical Line Intensities. IV. Source Functions and Equivalent Widths. *Astrophys. J.*, **144**, 1159.
6. Avrett, E. H. 1965, Solutions of the Two-Level Line Transfer Problem with Complete Redistribution. Proc. 2nd Harvard-Smithson. Conf. on Stellar Atmospheres, p. 101.
7. Avrett, E. H., Hummer, D. G. 1965, Non-coherent Scattering, II: Line Formation with a Frequency Independent Source Function. *Mon. Not. R. astr. Soc.*, **130**, 295.
8. Balli, E., Kiral, A., Pecker, J.-C. 1965, Abaque pour le calcul des profondeurs de formation. *Publ. Istanbul. Univ. Obs.*, no. 83.
9. Baschek, B., Holweger, H., Traving, G. 1966, Ein Algol-Programm für die Analyse von Sternspektren. *Abh. Hamburger Sternwarte*, **8**, no. 1.
10. Baschek, B., Traving, G. 1964, Differentielle Wachstumskurven. *Z. Astrophys.*, **59**, 276.
11. Böhm-Vitense, E. 1965, A Differential Equation Solution of the Non-LTE Line Transfer Problem. Proc. 2nd Harvard-Smithson. Conf. on Stellar Atmospheres, p. 165.
12. Chiarella, C., Reichel, A. 1966, On the Evaluation of Voigt Spectral Line Functions. *Mon. Not. R. astr. Soc.*, **134**, 83.
13. Cuny, Y. 1967 (in preparation).
14. Dumont, S. 1966, *C. r. Acad. Sci. Paris*, **262**, 740.
15. Dumont, S. 1966, *C. r. Acad. Sci. Paris*, **263**, 85
16. Edmonds, F. N., Schlüter, H., Wells, D. C. 1967, Hydrogen-Line Stark Broadening Functions. *Mem. R. astr. Soc.* (in press).
17. Edmonds, F. N. 1967 (work in progress).
18. Gökdoğan, N., Pecker, J.-C. 1964, Note sur l'utilisation des courbes de croissance. *Ann. Astrophys.*, **27**, 417.
19. Goldberg, L. 1964, The Origin of the Emission Reversals in the Fraunhofer H- and K-Lines. *Astrophys. J.*, **140**, 384.
20. Hummer, D. G. 1965, General Noncoherent Scattering. Proc. 2nd Harvard Smithson. Conf. on Stellar Atmospheres, p. 143.
21. Hummer, D. G., Rybicki, G. B. 1966, Non LTE Line Formation with Spatial Variations of the Doppler Width. *J. quantit. Spectrosc. radiat. Transfer*, **6**, 661.
22. Hummer, D. G., Rybicki, G. B. 1967, Computational Methods for Non-LTE Line-Transfer Problems. In *Method in Computational Physics*, **7**, Academic Press (in press).
23. Hummer, D. G., Stewart, J. C. 1967, Thermalization Lengths and the Homogeneous Transfer Equation in Line Formation. Submitted to *Astrophys. J.*
24. Ivanov, V. V., Nagirner, D. I. 1965, H-functions in the Theory of Transfer of Resonance Radiation. *Astrofizika*, **1**, 143.
25. Jefferies, J. T. 1965, Multilevel Problems. Proc. 2nd Harvard-Smithson. Conf. on Stellar Atmospheres, p. 177.
26. Johnson, H. R. 1964, A Theoretical Study of the Excitation of Sodium in the Solar Photosphere. *Ann. Astrophys.*, **27**, 695.
27. Johnson, H. R. 1965, Sodium Equilibrium and the NaD Lines. Proc. 2nd Harvard-Smithson. Conf. on Stellar Atmospheres, Spec. Report, **174**, 333.
28. Johnson, H. R., Klinglesmith, D. 1965, The Coupled Line Transfer Problem for Hydrogen. Proc. 2nd Harvard Smithson. Conf. on Stellar Atmospheres, p. 221.
29. Kurochka, L. N. 1965, Electron Density Determination based on Emission Balmer Lines. *Astrofizika*, **1**, 369.
30. Kurochka, L. N. 1966, Doppler-Effect and the Last Resolved Balmer Line. *Astrofizika*, **2**, 131.

31. Mugglestone, D., O'Mara, B. I. 1965, Saturation Effects in Stellar Atmospheric Absorption Lines. *Mon. Not. R. astr. Soc.*, **129**, 41.
32. Mugglestone, D., O'Mara, B. I. 1966, The Influence of Stark Broadening on Abundance Determinations. *Mon. Not. R. astr. Soc.*, **132**, 87.
33. Nagirner, D. I. 1964, On the Solution of Integral Equations of the Theory of Light Scattering. *Astr. Zu.*, **41**, 669.
34. Nagirner, D. I., Ivanov, V. V. 1966, Transfer of Resonance Radiation in an Infinite Medium, I, II. *Astrofizika*, **2**, 5 and **2**, no. 2.
35. Obridko, V. N. 1965, The Matrix of Radiative Scattering in a Magnetic Field. *Astr. Zu.*, **42**, 102.
36. Obridko, V. N. 1965, Formation of Solar Absorption Lines with a Split upper Level in the Presence of a Magnetic Field. *Astr. Zu.*, **42**, 502.
37. Pagel, B. E. J. 1964, *R. Obs. Bull.*, no. 87.
38. Pfennig, H. 1966, Comparison of Quasistatic Line Profiles with Measurements of Balmer and Paschen Hydrogen Lines and Lines of the Diffuse Series of Neutral Helium. *J. quantit. Spectrosc. radiat. Transfer*, **6**, 549.
39. Pfennig, H., Trefftz, E., Vidal, C. R. 1966, A Critical Note on the up to now Available Stark Broadening Theories of Hydrogen. *J. quantit. Spectrosc. radiat. Transfer*, **6**, 557.
40. Rachkowski, D. N. 1963, Theory of Formation of Absorption Lines in the Presence of a Magnetic Field. *Publ. Krym. astrofiz. Obs.*, **30**, 267.
41. Schatzman, E. 1965, Mesures absolues des durées de vie d'états excités. But et méthodes. *Ann. Astrophys.*, **28**, 111.
42. Thomas, R. N. 1965, Some Aspects of Non-Equilibrium Thermodynamics in the Presence of a Radiation Field. Univ. of Colorado Press.
43. Thomas, R. N. 1965, Computation of the Line Source Function: A Review of the Physical Problem. Proc. 2nd Harvard-Smithson. Conf., p. 71.
44. Traving, G. 1964, Über die Bildung von Fraunhoferlinien in turbulenten Sternatmosphären. *Z. Astrophys.*, **60**, 167.
45. Unno, W., Yamashita, Y. 1966, On Curves of Growth for Line Profiles Truncated at Artificial Continuous Levels. In *The Solar Spectrum* (ed. by C. de Jager), p. 102.
46. Van Regemorter, H., Bréchot, S. 1964, L'élargissement des raies spectrales par chocs, I et II. *Ann. Astrophys.*, **27**, 432, 739.
47. Van Regemorter, H. 1965, Spectral Line Broadening. *Ann. Rev. Astr. Astrophys.*, **3**, 71.
49. Vidal, C. R. Determination of Electron Densities from Line Merging. *J. quantit. spectrosc. radiat. Transfer*, **6**, 575, 1966.
50. Waddell, J. III 1964, Reply to Athay's paper: 'Source Function Equality in Multiplets'. *Astrophys. J.*, **140**, 1586.
51. Wright, K. O. 1966, Stellar Atmospheres and their Spectra. *J. R. Astr. Soc. Can.*, **60**, 97.
52. Yamashita, Y. 1964, A Theoretical Study on the Curve of Growth, II. *Publ. astr. Soc. Japan*, **16**, 77.
53. Hummer, D. G. 1964, The Mean Number of Scattering by a Resonance-Line Photon. *Astrophys. J.*, **140**, 276.
54. Averett, E. H. 1966, Source Function Equality in Multiplets. *Astrophys. J.*, **144**, 59.
55. Peytremann, E. 1966, Obs. de Genève, Unpublished.

III. *Model Atmospheres and Theory of Spectra for Individual Stars or Groups of Stars, Methods of Abundance Determination*

(*a*) *'Normal' stars (except Sun) and general literature on methods of abundance determination*

1. Adams, T., Morton, D. C. In preparation.
2. Audouze, J. 1967, unpublished.
3. van't Veer, F. Modèles empiriques d'atmosphères stellaires. I. Introduction et première application; II. Étude statistique, modèle d'une tache solaire; III. Comparaison des différents modèles pour les taches solaires. 1966, *Ann. Astrophys.*, **29**, 7, 119 and 223.
4. Auman, J. R. 1966, Model Stellar Atmosphere with $T_{\text{eff}} = 2500°K$. Paper presented at IAU Sym. on Late-Type Stars, Trieste.

5. Baschek, B. 1966, On the Influence of Surface Gravity on UBV Colours of G-Type Stars. *J. quantit. Spectrosc. radiat. Transfer*, **6**, 701.
6. Baschek, B., Holweger, H., Namba, O., Traving, G. 1967, Quantitative Analyse des Spektrums des F8-Sterns Beta Virginis. *Z. Astrophys.*, **65**, 418.
7. Baschek, B., Kegel, W., Bode, G., Kodaira, K., Kohl, K., Traving, G. 1964, Quantitative Interpretation of Stellar Spectra using Model Atmospheres. Proc. 1st Harvard-Smithson. Conf. on Stellar Atmospheres, p. 254.
8. Baschek, B., Oke, J. B. 1965, Effective Temperatures and Gravities of Ap, Am, and Normal A-Type Stars. *Astrophys. J.*, **141**, 1404.
9. Baylac, M.-O. 1966, In preparation.
10. Bernas, R., Gradstajn, E., Reeves, H., Schatzman, E. 1967, On the Nucleosynthesis of Lithium, Beryllium and Boron. *Ann. Phys.*, New York (to be published).
11. Böhm, K.-H., Deinzer, W. 1965, Atmosphären von Zentralsternen planetarischer Nebel, I. *Z. Astrophys.*, **61**, 1.
12. Böhm, K.-H., Deinzer, W. 1966, Modellatmosphären für die O'Dell-und die Harman-Seaton-Sequenz. *Z. Astrophys.*, **63**, 177.
13. Böhm-Vitense, E., Pyper, D. M., Wallerstein, G. 1964, Spectrum of γ Pegasi between 3300 and 3030Å. *Astrophys. J.*, **140**, 807.
14. Böhm-Vitense, E. 1966, Spectral Analysis by means of the Hydrogen Lines. Struve Memorial Volume, Ed. G. Herbig (in press).
15. Buslavsky, V. G. 1964, Spectral Energy Distribution of Late-Type Stars. *Astr. Zu.*, **41**, 631.
16. Cayrel, G. 1966, Analyse détaillée des atmosphères de sept étoiles K. *Ann. Astrophys.*, **29**, 413.
17. Chevalier, Cl. Critical Analysis of the Peculiar Abundances in Sirius A. To be published in *Ann. Astrophys.*
18. Collins, G. W. II 1965, Continuum Emission from a Rotating Non-grey Stellar Atmosphere. *Astrophys. J.*, **142**, 265.
19. Collins, G. W. II, Harrington, J. P. 1966, Theoretical H-Beta Line Profiles and Related Parameters for Rotating B-Stars. *Astrophys. J.*, **146**, 152.
20. Conti, P. S., Wallerstein, G., Wing, R. F. 1965, The Composition of Main-Sequence Stars of Types A–K in the Hyades Cluster. *Astrophys. J.*, **142**, 999.
21. Fischel, D. 1964, On the Influence of the Chemical Composition on the Structure and Metal Index of Stellar Atmospheres. *Astrophys. J.*, **140**, 221.
22. Edmonds, F. N. 1964, A Set of Non-Grey Model Atmospheres for Procyon. *Astrophys. J.*, **140**, 902.
23. Edmonds, F. N. 1966, The Determination of the Ultraviolet Continuum of Procyon from Balmer Line Profiles. Proc. IAU Symposium no. 26, *Abundance Determination in Stellar Spectra*, p. 159.
24. Edmonds, F. N., Talbert, F. D. 1966, A Spectroscopic Analysis of Procyon, II. The Continuous Spectrum and Line Blanketing. *Astrophys. J.*, **146**, 1.
25. Fredrich, L. W., Johnson, H. R., Fay, T. 1966, Infrared Spectroscopy with a Mica Window Tube. *Adv. Electronics Electron Phys.*, **22**; *Astr. J.*, **71**, 384.
26. Fujita, Y., Tsuji, T. 1966, Some Problems Related to the Chemical Abundance in Cool Stars. Proc. IAU Symposium, no. 26 (Academic Press).
27. Gasson, R. E. M. 1966, Thesis, Cambridge Univ.
28. Gasson, R. E. M., Pagel, B. E. J. 1966, *Observatory*, **86**, 196.
29. Gingerich, O. 1966, Limb Darkening for a Grid of Stellar Atmospheres. *Astrophys. J.*, **144**, 1213.
30. Gökdoğan, N., Pecker, J.-C. 1964, Note sur l'utilisation des courbes de croissance. *Ann. Astrophys.*, **27**, 417.
31. Guillaume, C., van Rensbergen, W , Underhill, A. B. 1965, The Theoretical Near Ultraviolet Spectrum of a B2 Star. *Bull. astr. Inst. Netherl.*, **18**, 106.
32. Hardorp, J., Strittmatter, P. 1967 (in preparation).
33. Henry, R. C., Mihalas, D. 1964, The Abundance of Magnesium in the Atmospheres of O and B Stars. *Astrophys. J.*, **140**, 873.
34. Herbig, G. H. 1965. Lithium Abundances in F5–G8 Dwarfs. *Astrophys. J.*, **141**, 588.

35. Houziaux, L. 1964, Sur la possibilité d'observer les raies de Lyman dans les étoiles du type voisin de Ao V. *C. r. Acad. Sci. Paris*, **258**, 3635.
36. Houziaux, L. 1965, Les raises de Lyman dans le spectre des étoiles B de la séquence principale. *C. r. Acad. Sci. Paris*, **260**, 420.
37. Houziaux, L. 1966, Profil du triplet de O I à 7772 Å dans le spectre de Véga. *C. r. Acad. Sci. Paris*, **263**, 118.
38. Kalkofen, W., Strom, S. E. 1966, The Effects of Deviations from LTE and Line Blanketing on Stellar Atmospheres in the Range B5 to A5. *J. quantit. Spectrosc. radiat. Transfer*, **6**, 653.
40. Kohl, K. 1964, Die Atmosphäre des Sirius. *Z. Astrophys.*, **60**, 115.
41. Kolesov, A. K. 1964, Continuous Spectra of White Dwarfs. *Astr. Zu.*, **41**, 240.
42. Kopylov, I. M. 1966, Electron Densities in the Atmospheres of Hot Stars, II. *Publ. Krym. astrofiz. Obs.*, **35**, 11.
43. Matsushima, S. 1967, Non-Grey Model Atmospheres for Solar Type Stars, I. The Role of the Hydrogen Molecules as a Source of Continuous Opacity. Submitted to *Ann. Astrophys.*
44. Matsushima, S., Terashita, Y. 1967, Non-Grey Model Atmospheres of Solar Type Stars, II. The Source of Opacity in the Ultraviolet Continuum and its Effect on the Radiation Field in other Spectral Regions. Submitted to the *Ann. Astrophys.*
45. Mihalas, D. 1965, Model Atmospheres and Line Profiles for Early-Type Stars. *Astrophys. J. Suppl.*, **9**, 321.
46. Mihalas, D. 1965, A Spectroscopic Estimate of the Atmospheric Properties of O and B Stars. *Astrophys. J.*, **140**, 885.
47. Mihalas, D. 1966, Balmer-Line-blanketed Model Atmospheres for A-Type Stars. *Astrophys. J. Suppl.*, **13**, 1.
48. Mihalas, D. 1967 (submitted to *Astrophys. J.*).
49. Cowley, C. R., Cowley, A. P. 1966, Selection of Data for Abundance Determinations. Proc. IAU Symposium no. 26, p. 43 (Academic Press).
50. Mihalas, D., Morton, D. C. 1965, A Model for a B1 V Star with Line Blanketing. *Astrophys. J.*, **142**, 253.
51. Mihalas, D. 1967, Statistical-Equilibrium Model Atmospheres for Early-Type Stars, I. Hydrogen Continua. Submitted to the *Astrophys. J.*
52. Morton, D. C. 1964, The X-ray Emission of Neutron Stars. *Ann. Astrophys.*, **27**, 813.
53. Morton, D. C. 1964, Theoretical Line Profiles in the Ultraviolet Spectra of Early-Type Stars. *Ann. Astrophys.*, **27**, 797.
54. Morton, D. C. 1965, Theoretical Line Profiles in the Ultraviolet Spectra of Early-Type Stars. *Astrophys. J.*, **141**, 73.
55. Orszag, S. A. 1965, Neutron-Star Atmospheres. *Astrophys. J.*, **142**, 473.
56. Osaki, Y. 1966, On the Atmosphere of a Rotating Star. *Publ. astr. Soc. Japan*, **18**, 7.
57. Pagel, B. E. J. 1966, Critical Summary of Abundance Determinations in Stars of Type G and later. Proc. IAU Symposium no. 26, *Abundance Determination in Stellar Spectra*, p. 272 (Academic Press).
58. Peytremann, E. 1967 (unpublished).
59. Scholz, M. 1967, Eine Analyse des Spektrums des Bo V-Stern τ Scorpii. *Z. Astrophys.*, **65**, 1.
60. Sobolev, V. V. 1965, On the Spectra of Hot Stars of Low Luminosity. *Trudy astr. Obs. Leningrad. gos. Univ.*, **22**, 3.
61. Conti, P. S., Greenstein, J. L., Spinrad, H., Wallerstein, G., Vardya, M. S. 1967, *Astrophys. J.*, **148**, 105.
62. Strom, S. E., Avrett, E. H. 1966, The Temperature Structure of Early-Type Model Stellar Atmosphere, II. A Grid of Stellar Models. *Astrophys. J. Suppl.*, **12**, 103.
63. Strom, S. E., Kalkofen, W. 1966, The Effect of Departures from LTE on the Stellar Continuum Fluxes in the Spectral-Type Range B5–Ao. *Astrophys. J.*, **144**, 76.
64. Strom, S. E., Avrett, E. H. 1964, The Temperature Structure of Early-Type Model Stellar Atmospheres, I. Detailed Analysis of a 10 000°K Model. *Astrophys. J.*, **140**, 1381.
65. Swamy, K. S. K. 1966, Profiles of Strong Lines in K-Dwarfs. *Astrophys. J.*, **145**, 174.

66. Terashita, Y., Matsushima, S. 1966, The Structure of White Dwarf Atmospheres. I. Model Atmospheres and Hydrogen-Line Profiles for DA-Type Stars. *Astrophys. J. Suppl.*, **13**, 461.
67. Tsuji, T. 1966, Some Problems on the Atmospheric Structures of Late-Type Stars. *Proc. Japan Acad.*, **42**, 258.
68. Tsuji, T. 1966, The Model Atmospheres of M-Type Stars. *Proc. Japan Acad.*, **42**, 7.
69. Underhill, A. B. 1966, A Critique of Methods based on Complex Representations of Stellar Atmospheres. Proc. IAU Symposium no. 26, p. 118.
70. Underhill, A. B. 1966, Computation of Emergent Flux and Intensities for Blanketing Effect Models and Comparison with Observation. *J. quantit. Spectrosc. radiat. Transfer*, **6**, 675.
71. Underhill, A. B. The Near Ultraviolet Spectrum from a B Star. Submitted to the *J. R. astr. Soc. Can.*
72. Unsöld, A. The Chemical Composition of the Stars. *Angew. Chemie* (International Edition), **3**, 481.
73. Vardya, M. S., Kandel, R. On Model Atmospheres for Late-Type Stars. 1967, *Ann. Astrophys.*, **30**, 111.
74. Wallerstein, G., Herbig, G. H., Conti, P. S. 1965, Observations of the Lithium Content of Main Sequence Stars in the Hyades. *Astrophys. J.*, **141**, 610.
75. Weidemann, V. 1966, Blanketing in High Gravity Objects. *J. quantit. Spectrosc. radiat. Transfer*, **6**, 691.
76. Wright, K. O. 1967, Excitation Temperatures of A and F-Stars. In *The Magnetic and Related Stars*, Mono Book Corporation, Baltimore (in press).
77. Yamashita, Y. 1965, Curve of Growth Analysis of M-Type Stars. *Publ. astr. Soc. Japan*, **17**, 25.
78. Traving, G. 1966, A Critical Summary of the Results from Early-Type Stars. Proc. IAU Symposium no. 26, p. 213.
79. Wallerstein, G. 1966, Chemical Analysis of Stars of Type F0–G5. Proc. IAU Symposium no. 26, p. 238.
80. Gebbie, K. B. 1966, *Mon. Not. R. astr. Soc.* (in press).
81. Guillaume, C. 1965, Theoretical Model of a B 1·5 Star with Line Blanketing. *Bull. astr. Inst. Netherl.*, **18**, 175.

(*b*) *Sun*

1. Ross, J., Aller, L. H. 1967, *Astr. J.* (in press.)
2. Baschek, B., Holweger, H., Müller, E. A. 1967 (in preparation).
3. Chamaraux, P. 1966, *Ann. Astrophys.* (in press).
4. Dumont, S. 1966, Profils théoriques des raies de résonances des ions Ca II et Mg II. *C. r. Acad. Sci. Paris*, **263**, 85.
5. Cowley, C. R. 1966, The Abundance of Nickel in the Solar Atmosphere. *Astrophys. J.*, **143**, 352.
6. Fujita, C. 1964, Theoretical Study on Blends of Pairs of Metallic Lines in the Solar Spectrum. *Publ. astr. Soc. Japan*, **16**, 301.
7. Gasson, R. E. M. 1967 (unpublished).
8. Gingerich, O. 1966, Blanketing Approximations for Solar Models. *J. quantit. Spectrosc. radiat. Transfer*, **6**, 609.
9. Goldberg, L., Kopp, R. A., Dupree, K. 1964, The Abundance of Iron in the Solar Photosphere. *Astrophys. J.*, **140**, 707.
10. Holweger, H. 1967, Ein empirisches Modell der Sonnenatmosphäre mit lokalem thermodynamischen Gleichgewicht. *Z. Astrophys.*, **65**, 365.
11. Houtgast, J. 1965, Absolute Intensities in high dispersion solar spectra between λ 3000 and λ 4000. *K. ned. Akad. Wet. Proc. Sect. Sci.*, Ser. B., **68**, no. 5.
12. de Jager, C., Neven, L. Submitted to *Bull. astr. Inst. Netherl. Suppl.*
13. de Jager, C., Neven, L. 1967, to be published in *Solar Physics*, **1**.
14. Johnson, H. R. 1964, A Theoretical Study of the Excitation of Sodium in the Solar Photosphere. *Ann. Astrophys.*, **27**, 659.

15. King, R. B., Lawrence, G. M., Link, J. K. 1965, Solar Abundances of Gallium, Silver and Indium. *Astrophys, J.*, **142**, 386.
16. Kulander, J. L., Jefferies, J. T. 1966, Inference of Velocities from Line Asymetries. *Astrophys. J.*, **146**, 194.
17. O'Mara, B. J. 1967, Line Profiles of Sodium and Calcium in the Sun. *Astr. J.*, in press.
18. Miyamoto, S. Thermal Stability of the Chromosphere and Spicule. *Publ. astr. Soc. Japan*, **17**, 127.
19. Müller, E. A., Mutschlecner, J. P. 1964, Effects of Deviations from LTE on Solar Abundances. *Astrophys. J. Suppl.*, **9**, 85.
19a. Müller, E., Mutschlecner, J. P. 1967 (in press).
20. Müller, E. A. 1967, in J. Xanthakis, *Solar Physics*, Wiley and Sons.
21. Pagel, B. E. J. 1965, Deductions as to the Accuracy of LTE from Excitation Temperature Measurements, and a Comment on the Abundance of 'Trace Elements'. Proc. 2nd Harvard-Smithson. Conf. on Stellar Atmospheres, p. 425.
22. Pottasch, S. R. 1964, On the Chemical Composition of the Solar Corona. *Mon. Not. R. astr. Soc.*, **128**, 73.
23. Pottasch, S. R. 1965, On the Determination of the Solar Chemical Composition from a Study of the Ultraviolet Resonance Lines. *Ann. Astrophys.*, **28**, 148.
24. Pecker-Wimel, C., Thomas, R. N. 1965, Possible Implications of the Observed Line Profile of (λ 6374) Fe x. *Astrophys. J.*, **141**, 834.
25. Schmahl, G., Schröter, E. H. 1965, Die Li-Häufigkeit und das Isotopenverhältnis Li^6/Li^7 in Sonnenflecken. *Z. Astrophys.*, **62**, 143.
26. Teplitskaya, R. B. 1964, Determination of the Excitation Temperature in the Solar Atmosphere. *Astr. Zu.*, **41**, 907.
27. Unno, W. 1966, Influence of Macroturbulence on Chromospheric Doppler Velocities from the Multiplet Method. In *The Solar Spectrum* (ed. C. de Jager), p. 120.
28. Wilson, P. R. 1965, On Photospheric Mean Values. *Astrophys. J.*, **142**, 1195.
29. Wilson, P. R. 1966, The Interpretation of Limb Intensity Profiles. *Astrophys. J.*, **146**, 241.
30. Zwaan, C. 1965, Sunspot Models. *Rech. astr. Obs. Utrecht*, **17**, part 4.
31. Czyzak, S. J., Krueger, T. K., Ross, J., Aller, L. H. 1967 (unpublished).

(*c*) *Atmospheres of typical population II objects and of peculiar stars*

1. Aller, L. H., Ross, J. 1966, Proc. AAS–NASA Symp. *Magn. and other Peculiar and Metallic Lune A Stars.*
2. Auer, L. H., Mihalas, D., Aller, L. H., Ross, J. E. 1966, The Manganese Star 53 Tauri. II. Fine Analysis. *Astrophys. J.*, **145**, 153.
3. Baschek, B. 1965, Quantitative Analyse des Spektrums des Schnelläufers Wilson 10 367. *Z. Astrophys.*, **61**, 27.
4. Böhm-Vitense, E. 1966, Zur Interpretation der Geschwindigkeits- und Intensitätsvariationen der Linien von ^{2}CVn. *Z. Astrophys.*, **64**, 326.
5. Böhm-Vitense, E. Line Variations of Stellar Magnetic Fields and the Oblique Rotator Theory. Proceedings of Newcastle Nato Advanced Study Institute on 'Planetary and Stellar Magnetism' (in press).
6. Böhm-Vitense, E. On the Spectra of the Magnetic A Stars. Struve Memorial Volume, Editor H. Hack (in press).
7. Böhm-Vitense, E. 1966, The Crossover Effect in α ^{2}CVn (in preparation).
8. Böhm-Vitense, E. 1966, On the Atmospheres of Helium Stars (in preparation).
9. Cayrel, R. 1966 (unpublished).
10. Climenhaga, I. L. 1967, The C^{12}/C^{13} Abundance Ratio in the Red Carbon Stars 19 Piscium, DS Pegasi and Y Canum Venaticorum. Proc. IAU Symp. on Late-Type Stars (in press).
11. Conti, P. S. The Atmospheres of the Metallic-Line Stars in the Hyades. *Astrophys. J. Suppl.*, **11**, 95, 1965.
12. Danziger, I. J. 1965, A High Dispersion Spectral Analysis of the Ba II Stars HD 11 6713 and HD 83 548. *Mon. Not. R. astr. Soc.*, **131**, 51.

13. Hardorp, J. 1966, Fine Analysis of the B4p Star 3 Centauri A. *Z. Astrophys.*, **63**, 137.
14. Hauck, B. 1967, Thesis, Observatoire de Genève.
15. Hunger, K., van Blerkom, D. A Pure Helium Model Atmosphere. Submitted to *Z. Astrophys.*
16. Kodaira, K. 1964, Atmosphärenstruktur und chemische Zusammensetzung des Schnelläufers HD 161 817. *Z. Astrophys.*, **59**, 139.
17. Mihalas, I., Henshaw, J. L. 1966, Studies of Peculiar A Stars III Model-Atmosphere Line-Strength Calculations with Application to the Ap Stars. *Astrophys. J.*, **144**, 25.
18. Klinglesmith, D. Thesis, Indiana University.
19. Pagel, B. E. J., Powell, A. L. T. 1966, *R. Obs. Bull.* (in press).
20. Praderie, F. 1965, Sur le spectre infrarouge de quelques étoiles à raies métalliques. *C. r. Acad. Sci. Paris*, **261**, 4625.
21. Searle, L., Lungershausen, W. T., Sargent, W. L. W. 1966, Studies of Peculiar A Stars, IV. The Relative Abundances of Four Iron-peak Elements. *Astrophys. J.*, **145**, 141.
22. Spite, M. 1967, To be published in *Ann. Astrophys.*
23. Wyller, A. 1966, New C^{13} Indicators in Stellar Spectra. *Astrophys. J.*, **143**, 828.
24. Wyller, A. 1966 (unpublished).

37. COMMISSION DES AMAS STELLAIRES ET DES ASSOCIATIONS

PRÉSIDENT: Dr M. F. Walker, Lick Observatory, Santa Cruz, California, U.S.A.

VICE-PRÉSIDENT: Dr G. Alter, Bet Yizhaq, near Natanya, Israel.

COMITÉ D'ORGANISATION: W. Becker, O. J. Eggen, A. R. Hogg†, P. N. Kholopov, L. Rosino, H. H. Swope.

MEMBRES: Balázs, Barkhatova, Blaauw, Cuffey, Feast, Gascoigne, Golay, Haffner, Hénon, Hoag, Hopmann, Johnson (H. L.), King (I. R.), Larsson-Leander, Lavdovski, Lohmann, Mavridis, Menon, Meurers, Pishmish de Recillas, Reddish, Ruprecht, Sandage, Sawyer-Hogg, Serkowski, Shapley (H.), Smyth, Steinlin, Takase, The, Vanderlinden, Van Schewick, von Hoerner, Wallenquist, Weaver, Westerlund.

INTRODUCTION

The data contained in this report have, as in preceding years, been taken from two sources: (1) Information received from astronomers active in the field of Commission 37 in response to a circular letter mailed 1 July 1966. (2) A search of the literature. The report is divided into three main sections, dealing with associations, open clusters, and globular clusters. The section on associations has been prepared by Dr J. Ruprecht, and that on globular clusters by Professor L. Rosino. Both sections are included here as received. As in previous reports, the data for the sections on associations and open star clusters have been condensed in so far as possible into two tables. Clusters and associations belonging to the Magellanic clouds or other extragalactic systems are omitted from this report since they are reported in Commission 28.

CATALOGS

G. Alter, H. Sawyer Hogg and J. Ruprecht (**1, 2, 3**) have published three additional supplements (Numbers 7–9) to the *Catalog of Star Clusters and Associations*. At the XIIth General Assembly, it was decided by Commission 37 that the publication of a new edition of this catalog should be undertaken. I am happy to announce that negotiations for the publication of the work have now been completed. The authors of the new edition will be G. Alter, B. Balázs, J. Ruprecht, and V. Vanýsek and the Catalog will be published by the Publishing House of the Hungarian Academy of Sciences. It is expected that the manuscript will be sent to press in the summer or fall of 1967. Owing to the appearance of the new edition of the Catalog, no additional Supplements to the first edition will be issued.

The most useful *Atlas of Open Clusters South of* $-45°$ *Declination*, mentioned in the last report, has now been published by A. R. Hogg and V. O. Hunt. The *Atlas* consists of photographic prints of the fields of 133 clusters from plates taken with the Mt Stromlo 74-inch reflector. The limiting magnitude is about m_{pg} = 16 to 17, and the print scale is 15″/mm. In addition, a *Catalog of Open Clusters South of* $-45°$ *Declination* has been published by A. R. Hogg (**4**) giving the coordinates, apparent diameters, total number of stars, and estimated number of member stars for each of the clusters in the *Atlas*.

After many years in preparation, Trumpler's extensive spectroscopic and photometric observations of open clusters have now been prepared for publication by H. F. Weaver. According to Weaver; 'the manuscript, including explanatory text, is now complete and in form to go to the printer, probably soon after 1 January 1967.'

Ruprecht (**5**) has classified all of the known open star clusters on the Trumpler system.

Sher (6) has published descriptions of all clusters (40 in number) which have been listed as occurring near eta Carina.

Barkhatova and Syrovoi have completed the compilation of the atlas, mentioned in the preceding report, of color-magnitude diagrams of clusters and associations. This atlas has been compiled from observations published up to and including 1962.

ASSOCIATIONS

(Prepared by J. Ruprecht)

Recent results and work in progress are summarized in Table 1 which is self-explanatory.

A. H. Batten (Dominion Astrophysical Observatory, Victoria) has continued observations on Cyg OB 3 (NGC 6871), Cyg OB 1 (NGC 6913) and Cyg OB 4. This work is being done in collaboration with C. H. Costain, of the Dominion Radio Astrophysical Observatory, who

Table 1. Associations

Name	Observer and reference (see end of table)	Photometry, method and limiting magnitude			Other data obtained, remark
Vul OB 4	Bartaya, Kharadze (1)				list of 1240 stars
Cyg OB 1	Batten				s.t., r.v.
Cyg OB 2	Crawford	*uvby*, Hβ			
Cyg OB 3	Batten				s.t., r.v. material collected
	de Vegt				expansion age = $1 \cdot 6 \times 10^6$ y
Cyg OB 4	Batten				s.t., r.v. indicate not physica group
Cyg OB 7	Walker, G. A. H.				s.t., equivalent widths of H
Cep OB 3	de Vegt (2)				expansion age = $1 \cdot 6 \times 10^6$ y
Cas OB 2	Reddish, Lawrence, Pratt (3)	*UBV*	pg+pe	21	CE, structure, rotation
Cas OB 4	Walker (4)				s.t., r.v., equivalent widths o Hγ
	MacConnell (5)	*UBV*	pg	13·0	44 probable members; s.t.
Cas III, Cas OB 4, Cas OB 7	Bartaya, Kharadze (6)				list of 376 O–B8 stars
Mon I	Schmidt-Kaler	*uvby*, Hβ	pe		43 stars
Ori OB 1	Lodén, Morgan (7)	*UBV*			35 stars in two clusterings
	Crawford (8)	Hβ	pe	9	
Pup OB 1	Lodén (9)	*UBV*			s.t., total/selective absorption
	Lodén (10)	*UBV*			MK classification, more tha one assoc.?
Cen OB 2	Thackeray, Wesselink (11)	*UBV*		15	spectroscopy of brightest sta
Ara OB 1	Graham (12)	*VBLUW*	pe	11	r.v. and Walraven
	Brück, Smyth (13)	*UBR*	pg+pe	15·7	22 OB stars
	Brück, Smyth	*UBV*	pg	17	plates taken
Sco OB 1	Bok, Bok, Graham (14)				photometry for 9 stars
	Braes, Hovenier (15)				possible supernova remnant
	Andrews	*UBV*	pg	18	s.t. of selected stars to V=1
			pe	17	
Sco OB 2	Thackeray (16)				r.v. (confirmed concept of communal motion)
	Graham (17)	Hβ	pe	5	33 stars in southern section
Sco OB 4	Lyngå (18, 19)	*UBV*	pe	11·9	part in Norma

Abbreviations and References to Table 1

abs.	= absorption	pg	= photographic
ass.	= association	r.v.	= radial velocities
CE	= color excess	s.t.	= spectral types
pe	= photoelectric		

1. *Bull. Abastumani astrophys. Obs.*, no. 34, 109.
2. 1966, *Z. Astrophys.*, **64**, 268.
3. 1965, *R. Obs. Bull.*, **5**, no. 8, 111.
4. 1965, *Astrophys. J.*, **141**, 660.
5. *Astr. J.*, in press.
6. *Bull. Abastumani astrophys. Obs.*, no. 34, 105.
7. 1966, *Vistas in Astronomy*, **8**, 83.
8. *Astr. J.*, in press.
9. 1966, *Ark. Astr.*, **4**, 65.
10. 1966, *Astrophys. J.*, **141**, 668.
11. 1965, *Mon. Not. R. astr. Soc.*, **131**, 121.
12. 1966, *Bull. astr. Inst. Netherl.*, **18**, 372.
13. 1966, *R. Obs. Bull.*, **5**, no. 10, 195.
14. 1966, *Mon. Not. R. astr. Soc.*, **131**, 247
15. 1966, *Bull. astr. Inst. Netherl.*, **18**, 294.
16. 1966, *Mem. Soc. astr. Ital.*, **70**, 33, and *IAU Symposium No.* 30.
17. *Mon. Not. R. astr. Soc.*, in press.
18. 1964, *Acta Univ. Lund. II*, no. 11 = *Lund Obs. Medd. II*, no. 139.
19. 1964, *Acta Univ. Lund. II*, no 13 = *Lund Obs. Medd. II*, no. 140.

plans 21 cm observations of these regions. Neither the radial velocities, nor the spectroscopic absolute magnitudes give any reason for supposing that the four stars supposed to form Cyg OB 4 are in any way related. The observations on Cyg OB 3 are not yet reduced.

Special consideration has been devoted to the association Cyg OB 2 at the Royal Observatory Edinburgh. L. C. Lawrence and V. C. Reddish (**7**) have shown that photographic *UBV* photometry of this association and the nearby cluster NGC 6910 present systematic differences from the published photoelectric magnitudes used for calibration purposes. No source of error could be found in the photographic photometry large enough to account for the discrepancies. The coming into operation of the twin 16-inch (41 cm) photoelectric telescope at Edinburgh enabled the photoelectric magnitudes to be measured; the results showed that the discrepancies had been due to errors in the published photoelectric magnitudes.

The dependence of interstellar reddening on luminosity among O and B stars has been investigated by V. C. Reddish (**8**). The increase of interstellar reddening of stars with increasing stellar luminosity has been found in all clusters and associations containing stars brighter than $M_v = -9$ (ages $< 10^5$ yr). The effect is found most strongly in Cygnus OB 2 association. It has been concluded that very young early-type stars are surrounded by obscuring clouds having diameter $\leqslant 1$ pc and masses probably around 30 $\mathfrak{M}_{\odot}$.

Research on the association Cyg OB 2 has been continued by V. C. Reddish, L. C. Lawrence and N. M. Pratt (**9**). Star counts and *UBV* photometry show that the association contains 3000 stars brighter than $m_r = 20$, 900 brighter than $m_{pg} = 21$ and 400 brighter than $B = 16{\cdot}5$, of which at least 300 are OB stars. The distance is 2·1 kpc, and the shape is elliptical, 29 pc by 17 pc. The total mass is between $1{\cdot}6 \times 10^4$ and $6{\cdot}3 \times 10^4$ $\mathfrak{M}_{\odot}$, with 60% of the mass in obscuring clouds surrounding the stars. The association may be rotating about an axis in the Galactic plane with an equatorial speed of 2 to 4 km s^{-1}.

D. MacConnell (University of Michigan) found the distance (825 pc) and probable age

(6×10^5 to 2×10^6 yr) of the association Cep OB 4. Some of its members have been found probably to belong to the T Tauri class. One faint flare star has been discovered.

R. Ampel (**10**) has examined the distribution of 275 OB stars within the associations Cas III, Cas OB 4, Cas OB 5, using objective prism spectra, and photographic and photovisual photometry calibrated on photoelectric standards.

An investigation of P. Maffei (Rome) on the variables of the Orion nebula with plates obtained at the Hamburg Observatory in the years 1920–21 is very far advanced. The main purpose of this work, which is being done in collaboration with A. Blagikh, is to compare the behavior of the variables of this association at different times.

From Kitt Peak D. L. Crawford reports that the Hβ-photometry, combined with the *UBV* photometry of Hardie *et al.* for stars in the Orion Belt region, indicate that these stars are approximately all the same age but not at the same distance.

A. D. Thackeray and A. J. Wesselink (**11**) derived distance 2·0 kpc of the association Cen OB 2 (IC 2944). The O-type stars show no large Trumpler red shift. There is evidence of expanding gas around the association.

An extensive investigation of the proper motions and photometric properties of the association Sco OB 1 and the associated cluster NGC 6231 by L. Braes is now approaching its completion and will soon be published in the *Bull. astr. Inst. Netherlands*. It is based on early and modern Cape Observatory Photographs. It results in separation of association and field stars and in a determination of the systematic motion of different parts of the association.

C. Roslund (**12**) reports that the possible association Sco OB 4, found by Pik-Sin The (**13**), extends outside the field surveyed by The, at least to the south. The areal distribution of the O and early B type stars indicates that the nucleus of the association is to be sought in the nebular region of NGC 6334. Physical relationship between the young stars and the nebulosity was demonstrated.

Research by T. S. van Albada and A. Blaauw on the binary properties in the nearest O association reported earlier, has been continued. The observations of spectroscopic binaries for determination of periods and amplitudes have been extended to the Scorpio-Centaurus cluster. Van Albada is now also engaged in theoretical work on the dynamics of small young groups in which binary formation may occur as a consequence of multiple encounters.

G. A. H. Walker, assisted by S. M. Hodge, has attempted to extend the investigation of the possible differences between the width of Hγ-line-spectral type relations for different clusters. They have measured Hγ equivalent widths and estimated spectral types, both from low dispersion spectra in the associations Cas OB 6 (around IC 1805), Aur OB 2 (NGC 1893), Ser OB 1 (NGC 6611), Vul OB 1 (NGC 6823), Cyg OB 2, Cyg OB 1 (NGC 6910) and the components of HD 5005.

L. L. E. Braes and J. W. Hovenier (**14**) have concluded that there may be a close relationship between observed X-ray sources and supernova remnants in OB associations nearer than 2 kpc.

A review of properties of the O associations within 1000 pc from the Sun has been published by A. Blaauw (**15**). Space distribution, stellar content, and internal motions of the associations are discussed. Special attention is given to the occurrence of subgroups of different ages and their properties.

An investigation of the internal motions of several of the nearest O associations, based on the most recently determined meridian circle positions is now being carried out by A. Blaauw in collaboration with several associates.

E. L. Chentson (Sverdlovsk, U.S.S.R.) conducted spectrophotometric studies of supergiant stars ranging from class B8 to K2 which are possible members of the stellar associations Ori OB 1, Perseus OB 1 and the cluster NGC 457.

Byurakan Astrophysical Observatory has made some progress in the study of stellar associations (L. V. Mirzoyan, E. S. Kazaryan). To check the previously obtained 'hyperbolic' law of distribution of the OB stars around the nucleus of a synthetic association, a distribution of 98 probable OB star members of the Per OB 1 association around their double nucleus (h and χ Per) was deduced (**16**). The dependence of the age of OB stars on distance from the nuclei has been studied and the period of the semiage of these stars (5×10^6 to 2×10^7 yr) has been estimated. The observed distribution of the OB stars around the nuclei is satisfactorily represented by exponential laws of aging of stars.

U. Haug (**17**) in collaboration with J. Pfleiderer and J. Dachs has investigated the distribution of 105 early spectral type stars in Norma ($l = 332°$, $b = -2°$) and 38 similar stars in Circinus near $l = 321°$, $b = -1{\cdot}5°$. According to a preliminary statistical discussion, some highly reddened stars, situated near the Hα-emission regions Mt Stromlo nos. 101, 102 and 106, may have high luminosities and distances of about 4 kpc.

J. A. Graham (**18**) studied three groups of OB type stars which appear to be physically associated with Wolf-Rayet type stars. Using distances based on Hβ and U, B, V, photometry for the OB stars, he found a mean absolute magnitude of $-6{\cdot}4$ for 7 Wolf-Rayet stars of the type found in young associations.

L. O. Lodén has paid particular attention during the examination of the material from the Stockholm Observatory Milky Way Survey to stellar associations. A special program is devoted to the detection of associations of intermediate type stars and very open clusters, which can only be detected by means of spectroscopic material.

Mrs H. Antalová (Skalnaté Pleso, Czechoslovakia) has investigated the distribution of the O–B2 stars in the direction of the galactic center by means of photographic *UBV*-photometry. Two groups seem to be shown in the distances of 1·2 to 1·4 kpc and 3·0 kpc respectively.

The survey of faint Hα-emission stars in southern Milky Way fields containing bright and/or dark nebulosities is being continued by Pik-Sin The and the results for the following regions have already been published: Ophiuchus (**19, 20**), Scorpius (**21**). These groups of Hα-emission stars, which are seen projected against bright and/or dark nebulosities, are probably members of T-associations.

P. Maffei (**22**) has discovered 13 variable stars, most of them of RW Aur type, on 204 plates (blue and infrared) obtained at the Observatories of Asiago and Loiano, in the field of NGC 2169. This field (where some emission stars were already known) contains therefore a T association which, perhaps, is distinct from Mon T_1. Some other new T Tau variables have been found in the fields of NGC 1579, IC 405 and HGC 6611 (**23, 24**).

OPEN CLUSTERS

So far as possible, information relating to recent or current observations of open clusters has been collected in Table 2; the discussion in the text is limited to those investigations that are not adequately described by the line-entries in the table. Omitted from the table are: clusters 'to be observed' for which no material has so far been obtained, radio observations, observations of nebulae associated with clusters, and general discussion papers or catalogs of clusters. While an attempt has been made to make the list of published investigations reasonably complete, no effort has been made to include here *every* published paper, since we have available the *Catalog of Clusters* and its Supplements. In general, all references to papers published in 1964, 1965 and 1966 and received by the closing date of the literature search (15 September 1966) are included. Some references to investigations published in 1963 are included except for those already published in the last report, and a few 1962 publications not quoted in the last report are also included, although no effort was made to obtain completeness for this year. References not available in the Lick Observatory library have been omitted except in those cases where they were referred to by astronomers who responded to the circular letter. It is sometimes difficult to determine, in the case of clusters which are the nuclei of associations, whether the observations should more properly be listed in Table 1 or Table 2. Since the two tables were prepared independently, some duplication of material will be found.

Table 2. Open Clusters

NGC	Observer and references (see end of table)	Photometry, method and limiting magnitude			Other data obtained, remarl
188	Greenstein, Keenan (1)			14·7	s.t., r.v. red giants
	Hoffmeister (2)				4 var. stars
	Efremov, *et al.* (3)				var. stars = W UMa type
	Barkhatova, Zinina (4)	*UBV*		17	$c-m$, $L(M)$, structure, var. stars
	Sharov (5)	*B*		18	structure
	Sharov (6)	*BV*	pg	15	stars in region
	Golay	$UBB_1B_2VV_1G$	pe	10	obs. in prog.
225	Dolidze, Ponomareva (7)				45 emHα stars in region
457	Lavdovsky (8)				p.m., mem., c.l. mot.
	Morris				s.t., r.v., EW Hγ
	de Vegt				abs. p.m., sp. v.
	Grigoryan				pol.
559	Lindoff	*UBV*	pg + pe	14·5	plates taken
581	McCuskey, Houk (9)	*UBV*	pg	12·5	
	Oja (10)			15·5	p.m.
	Oja	*UBR*			2000 stars w/in 1 sq. deg., s. 1000 stars
	Lavdovsky (8)				p.m., mem., cl. mot.
Tr 1	McCuskey, Houk (9)	*UBV*	pg	12·5	
	Oja (10)			15·5	p.m.
	Lavdovsky (8)				p.m., mem., cl. mot.
654	McCuskey, Houk (9)	*UBV*	pg	12·5	
659	McCuskey, Houk (9)	*UBV*	pg	12·5	
663	McCuskey, Houk (9)	*UBV*	pg	12·5	
	Morris				s.t., r.v., EW Hγ
752	Kholopov, Artyukhina (11)	*B*		16	$L(M)$
	Barkhatova (12)				cl. rot., age
	Gunn, Kraft (13)				chem. comp. 2 stars
	Wallerstein, Conti (14)				chem. comp. 1 star
	Wallerstein, *et al.* (15)				r.v. 1 star
	Cannon	*BV*			p.m. 238 stars
	McNamara	*uvby*, Hβ	pe		
	Golay	$UBB_1B_2VV_1G$	pe	10	
Stock 2	Krzeminski, Serkowski	*UBV*		13	pol.
869 = *h* Per	Pratt (16)	*UBV*	pg	12·5	pol. in *B*, *V*
h + χ Per	Serkowski (17)		pe	11·7	pol. in *B*, *V*
	Wildey (18)	*UBV*	pg + pe	18	$c-m$, age
	Schild (19)			12	s.t., HR diag.
	Crawford	*uvby*, Hβ			obs. in prog.
	Golay	$UBB_1B_2VV_1G$	pe	10	obs. in prog.

GC	Observer and references (see end of table)	Photometry, method and limiting magnitude			Other data obtained, remarks
C 1805	Vasilevskis, *et al.* (**20**)			14·2	p.m., mem.
	Underhill				r.v.
	Golay	$UBB_1B_2VV_1G$	pe	10	
	de Vegt				abs. p.m., sp.v.
	Walker, G. A. H.				s.t., EW Hγ
57	Larsson-Leander	BV	pg+pe	16	s.t.
39	Mathews (**21**)	V	pg	13	
C 1848	de Vegt				p.m. plates taken
C 348	Golay	$UBB_1B_2VV_1G$	pe	10	obs. in prog.
Per	Guthrie (**22**)				Hγ line int.
	Crawford	*uvby* Hβ			obs. in prog.
	Heard, Petrie				s.t., r.v. 77 stars
leiades	McCarthy, Treanor (**23**)			17·3	s.t. M stars
	Guthrie, Nandy (**24**)				tot./sel. absorption
	Abt, *et al.* (**25**)				r.v. 46 stars for s. bin.
	Crawford	*uvby*, Hβ	pe		material collected
	Golay (**26**)	$UBB_1B_2VV_1G$	pe	10	
502	de Vegt				abs. p.m., sp. v.
513	Barkhatova, Zakharova	UBV		16·5	$c-m$, $L(M)$, structure
528	Mathews (**21**)	V	pg	13	
	Barkhatova, Kuzmina	BV		16·5	$c-m$, $L(M)$, structure
	Larsson-Leander	BV	pg+pe	16	s.t.
545	Barkhatova, Kuzmina, Shashkina	UBV		16·5	$c-m$, $L(M)$, structure
	Golay	$UBB_1B_2VV_1G$	pe	10	
yades	Helfer, Wallerstein (**27**)				chem. comp. 4 giants
	Conti, *et al.* (**28**)				Li/Ca 23 stars
	Conti (**29**)				chem. comp. metallic line stars
	Wallerstein, Conti (**14**)				chem. comp. 1 star
	Wallerstein, *et al.* (**15**)				r.v. 1 star
	Kraft (**30, 31**)				rot., effect on colors
	Crawford (**32**)	*ubvy*, Hβ	pe		132 stars
	Woolley, Harding (**33**)				r.v. 9 stars
	Morgan, Hiltner (**34**)				s.t. 79 stars
	Luyten (**35, 36**)			21	p.m.
	Giclas, *et al.* (**37**)			17	p.m.
	Wayman, *et al.* (**38**)				p.m. 231 stars
	Murray, *et al.* (**39**)			15·5	p.m. 69 stars
	Golay (**40**)	$UBB_1B_2VV_1G$	pe	10	
47	Golay	$UBB_1B_2VV_1G$	pe	10	
62	Golay	$UBB_1B_2VV_1G$	pe	10	
64	Börngen (**41**)	U	pg	14·5	
	Fang	UGR	pg	16	
48	Grigoryan				pol.
93	Bakos, Rymer (**42**)				var. star $P=75$ min.
	Serkowski (**17**)		pe	11·7	pol. in B, V
	Walker, G. A. H.				s.t., EW Hγ
12	Mathews (**21**)	V	pg	13	
76	Johnson (**43**)				s.t., r.v. 45 stars
	Walker, M. F.	UBV	pe	14	
	Walker, M. F. (**44, 45, 46**)			16·5	s.t., r.v. UV excess stars
sel 4	Svolopoulos (**47**)	UGR	pg	19	

NGC	Observer and references (see end of table)	Photometry, method and limiting magnitude			Other data obtained, remarl
2099	West (**48**)	*UBV*		17·7	
	Upgren (**49**)				p.m. 600 stars
	Jeffries (**50**)				p.m. 242 stars
	Artyukhina, Kholopov (**51**)			17·0	structure
King 8	Svolopoulos (**47**)	*UGR*	pg	18	
2129	Wick (**52**)	*UGR*	pg	17	*L*(*M*)
2168	Becker	*UGR*			
	Douglass	*BV*	pg		s.t., r.v., *L*(*M*)
	Golay	$UBB_1B_2VV_1G$	pe	10	
	Morris				s.t., r.v., EW Hγ
2169	de Vegt				p.m. plates taken
	Golay	$UBB_1B_2VV_1G$	pe	10	
2244	de Vegt				p.m. plates taken
	Golay	$UBB_1B_2VV_1G$	pe	10	
	Morgan, *et al.* (**53**)				s.t. OB stars
	Walker, M. F.	*UBV*	pg + pe	16	material collected
	Serkowski (**17**)		pe	11·7	pol. in *B*, *V*
An (van den Bergh)	Preston (**54**)			15	s.r. 19 stars
2264	Bretz (**55**)				em. Hα stars
	Golay	$UBB_1B_2VV_1G$	pe	10	
	Graham	Hβ	pe	9	9 stars
	McNamara	*uvby*, Hβ			mem. of A–F stars
	Morgan, *et al.* (**53**)				s.t. OB stars
	Vasilevskis, *et al.* (**56**)			16	p.m. 245 stars. mem.
	Walker, M. F. (**44, 45, 46**)			16·5	s.t., r.v. UV excess stars
2281	Alcaino (**57**)	*UBV*	pg + pe		121 stars
	Wallerstein, Conti (**14**)				chem. comp. 2 stars
	Wallerstein, *et al.* (**15**)				r.v. 2 stars
Cr 121	Feinstein	*UBV*	pe	9·5	also bright stars in vicinity
2401	Basel Observatory	*UGR*	pg	18	
2412	Basel Observatory	*UGR*	pg	18	
2421	MacConnell				3 em. Hα stars
2422	Serkowski (**17**)		pe	11·6	pol. in *B*, *V*
2420	West (**48**)	*UBV*		19	
Mel 71	Basel Observatory	*UGR*	pg	18	
Haffner 12	Basel Observatory	*UGR*	pg	18	
2437	Riddle	*UBV*	pg + pe	16·5	s.t.
2451	Williams	*UBV*		12	82 stars. 'Cl.' = nucleus of 2° dia.
	Feinstein (**58**)	*UBV*	pe	9·5	s.t. 8 stars
2467	Lodén (**59**)	*UBV*	pe	13·5	not a physical cl.
2533	Lindoff	*UBV*	pg + pe	14·5	material collected
2548	Wallerstein, Conti (**14**)				chem. comp. 1 star
	Wallerstein, *et al.* (**15**)				r.v. 1 star
2567	Lindoff	*UBV*	pg + pe	14·5	material collected
2571	Lindoff	*UBV*	pg + pe	14·5	material collected
Praesepe	Wallerstein, Conti (**14**)				chem. comp. 2 stars
	Crawford	*uvby*, Hβ			material collected
	Golay (**40**)	$UBB_1B_2VV_1G$	pe	10	
IC 2391	Buscombe (**60**)				s.t., r.v., abs. mag., EW H Hδ
	Graham	Hβ	pe	7	11 stars

NGC	Observer and references (see end of table)	Photometry, method and limiting magnitude			Other data obtained, remarks
2682	Greenstein, Keenan (1)				s.t.
	Eggen, Sandage (61)	*UBV*	pe	16·6	$c-m$, line blanketing
	Pesch				s.t., r.v. 8 stars
	Golay	$UBB_1B_2VV_1G$	pe	10	obs. in prog.
	Tammann	*UGR*		15	
	Bakos (62)				var. star not mem.
	Kholopov (63)	*BV*	pg	13	all stars w/in $r=16'$
	Kholopov, Artyukhina (64)	*B*	pg	17	$L(M)$
	Murray, *et al.* (65)			16·5	p.m. in $27'\times27'$ area
	Murray				p.m. in $1°\times1°$ area
Tr 12	Lindoff	*UBV*	pg	15	material collected
IC 2581	Evans, Lloyd	*UBV*	pg+pe	15	r.v. 2 stars. Obs. in prog.
3293	Andrews	*UBV*	pg+pe	17	pe to 16. Obs. in prog.
Cr 232	Feinstein	*UBV*	pe	11	connected to Tr 16
Tr 16	Feinstein (66)	*UBVRI*	pe	11	extension of work in (66)
Cr 234	Feinstein	*UBV*	pe	11	connected to Tr 16
Tr 17	Sher (67)	*UBV*	pg+pe	15	real cl.?
3496	Sher (67)	*UBV*	pg+pe	15	real cl.?
An(Feinstein)	Feinstein (68)	*UBV*	pe	10	around He star HD 96446
3572	Schmidt, Santanilla (69)	*UBV*	pg	13	$c-m$ 50 stars
Tr 18	Schmidt, Santanilla (69)	*UBV*	pg	13	$c-m$ 84 stars
	Graham	Hβ	pe	10	6 stars
3590	Schmidt, Santanilla (69)	*UBV*	pg	13	$c-m$ 36 stars
3603	Sher (67)	*UBV*	pg+pe	15	
Mel 105	Sher (67)	*UBV*	pg+pe	15	
3766	Sher (67)	*UBV*	pg+pe	15	
IC 2944	Thackeray, Wesselink (70)	*UBV*	pe	15	61 stars, s.t., r.v., 24 stars
4052	Engver (71)	*UBV*	pg+pe	14·5	
4103	Wesselink	*BV*			300 stars
Coma	Crawford (72)	*uvby*, Hβ	pe		37 stars
Berenices	Golay (26)	$UBB_1B_2VV_1G$	pe	10	
	Mendoza (73)	*UBVRI*	pe	11	s.t. 54 stars
An (Upgren)	Upgren, Rubin (74)	*UBV*	pe	10	s.t., r.v. 7 stars
	Anderson (75)				p.m. cl. not real
4755	Wesselink	*BV*			obs. in prog.
	Graham	Hβ	pe	8	5 stars
5316	Rahim (76)	*UBV*	pg+pe	15·6	
5606	Lyngå (77)	*UBV*	pg+pe	13·1	
5749	Lyngå (77)	*UBV*	pg+pe	12·1	
5822	Symth	*UBR*	pg+pe	15	plates taken
5823	Smyth	*UBR*	pg+pe	15	plates taken
Pismis 20	Lyngå (77)	*UBV*	pg	14·6	
Lyngå 3	Lyngå (77)	*UBV*	pg	15·2	
6031	Lindoff	*UBV*	pg	16	
Lyngå 7	Lyngå (77)	*UBV*	pg+pe	16·5	
6067	Engver (71)	*UBV*	pg+pe	14	
	Wesselink	*BV*			background stars
6087	Landolt (78)	*BV*	pg	14·3	c–m, $L(M)$
	Graham	Hβ	pe	10	18 stars
6124	The (79)	*BV*	pg+pe	14	
6134	Smyth	*UBR*	pg+pe	15	plates taken
6167	Brück, Smyth (80)	*UBR*	pg+pe	15·7	940 stars

NGC	Observer and references (see end of table)	Photometry, method and limiting magnitude			Other data obtained, remarks
Wolf 630 mov. cl.	Eggen (81)	*UBV*	pe		mem.
6231	Feinstein	*UBVRI*	pe	10	material collected
	Andrews	*UBV*	pg + pe	17	s.t. to $V = 12$
Tr 24	Feinstein	*UBV*	pe	9	material collected
6242	Andrews	*UBV*	pg + pe	17	obs. in prog.
6268	Andrews	*UBV*	pg + pe	17	obs. in prog.
H 10	Lyngå (77)	*UBV*	pg + pe	9·5	
6383	The (82)	*BV*	pg	14	
	Evans	*UBV*	pg + pe	16	pg to 19, s.t. to 10
	Graham	Hβ	pe	11	
Tr 28	The, Stokes	*UBV*	pg + pe	16	obs. in prog.
6404	The, Stokes	*UBV*	pg + pe	16	obs. in prog.
6405	Talbert (83)	*UBV*	pe	14	26 stars. age
6416	The, Stokes	*UBV*	pg + pe	16	obs. in prog.
6425	The, Stokes	*UBV*	pg + pe	16	obs. in prog.
Basel 5	Svolopoulos (84)	*UGR*	pg	15	
IC 4665	Abt, Snowden (85)			11·6	s.t., r.v., 35 stars
	Alcaino (86)	*UBV*	pe	11·5	37 stars *c–m*
	McCarthy, O'Sullivan	*BV*	pg		obj. prism s.t.
	Crawford	*uvby* Hβ			material collected
	Golay	$UBB_1B_2VV_1G$	pe	10	
6451	Svolopoulos (84)	*UGR*	pg	15	
Tr 31	Svolopoulos (84)	*UGR*	pg	15	
6514	Maffei (87)				var. stars
6520	Svolopoulos (84)	*UGR*	pg	15	
6530	Hiltner, *et al.* (88)			10·8	s.t. 25 stars
Cr 469	Grubissich (89)	*UGR*	pg	17	
6603	Grubissich	*UGR*	pg	18	
6611	Crawford	*uvby* Hβ			obs. in prog.
	Walker, G. A. H.				s.t., EW Hγ
	Grigoryan				pol.
6618	Maffei (87)				var. stars
Tr 33	Grubissich (89)	*UGR*	pg	18	
6633	Wallerstein, Conti (14)				chem. comp. 1 star
	Wallerstein, *et al.* (15)				r.v. 1 star
	Golay	$UBB_1B_2VV_1G$	pe	10	
	Grigoryan				pol.
	Mathews (21)	*V*	pg	13	
IC 4725	Graham	Hβ	pe	10	16 stars
	Grubissich	*UGR*	pg	16	
	Landolt (90)	*BV*	pg	14	1401 stars, *c–m*, *L*(*M*), age
	Serkowski (17)		pe	11·7	pol. in *B*, V
	Wallerstein, Conti (14)				chem. comp. 2 stars
IC 4756	Alcaino (86)	*UBV*	pe	11·0	100 stars *c–m*
	Hildajat	*BVI*	pg		obs. in prog.
6683	Yilmaz (91)	*UGR*	pg	15	
Tr 35	Yilmaz (92)	*UBV*	pg	15	
6694	Grubissich	*UGR*	pg	16	
Basel 1	Grubissich (93)	*UGR*	pg	15·5	
	Fenkart (94)	*UGR*	pg	17·5	
6704	Grubissich (93)	*UGR*	pg	16	
6709	Mathews (21)	*V*	pg	13	

NGC	Observer and references (see end of table)	Photometry, method and limiting magnitude			Other data obtained, remarks
δ Lyrae cl.	Bronkalla (**95**)	*UBV*	pg + pe	12·9	260 stars. Cl. not real
6755	Mathews (**21**)	*V*	pg	13	
	Svolopoulos (**96**)	*UGR*	pg	17·5	
6756	Svolopoulos (**96**)	*UGR*	pg	18	
6771	Batten				s.t., r.v., EW Hγ
6791	Kinman (**97**)	*BV*	pg + pe	20	s.t., 16 stars. Old cl.
6811	Chincarini, Walker, M. F.	*UBV*	pe	16	
6819	Dolidze, Lypareva (**98**)				14·6 red stars in area
	Barkhatova, *et al.* (**99**)	*UBV*	pe	16·5	120 stars. Similar to NGC 7789
	Lindoff	*UBV*	pg + pe	14·5	plates taken
6823	Mathews (**21**)	*V*	pg	13	
	Serkowski (**17**)		pe	11·7	pol. in *B*, *V*
	Walker, G. A. H.				s.t., EW Hγ
6834	Basel Observatory	*UBV*	pg	16	
6871	de Vegt				p.m. plates taken
	Batten (**100**)				s.t., r.v., EW Hγ 11 stars
	Golay	$UBB_1B_2VV_1G$	pe	10	
	Grigoryan	*B*			pol.
	Mathews (**21**)	*V*	pg	13	
	Serkowski (**17**)		pe	11·7	pol. in *B*, *V*
6882/5	Mathews (**21**)	*V*	pg	13	
	Elst (**101**)	*I*	pg		plates taken
	Grigoryan				pol.
6883	Mathews (**21**)	*V*	pg	13	
6910	Crawford	*uvby* Hβ			obs. in prog.
	Grigoryan				pol.
	Lawrence, Reddish (**102**)	*UBV*	pg + pe	15	
	Mathews (**21**)	*V*	pg	13	
	Walker, G. A. H.				s.t., EW Hγ
6913	Crawford	*uvby* Hβ			obs. in progress
	Batten				s.t., r.v., EW Hγ
	Grigoryan				pol.
6939	Artyukhina, Kholopov (**103**)	*B*		16·6	star counts
	Barkhatova, Steinberg (**104**)	*UB*		16·5	structure
6940	Larsson-Leander (**105**)	*BV*	pg + pe	15	*c–m*, mag. errors
7062	Fenkart (**106**)	*UBV*	pg	16	*c–m*
	Grigoryan				pol.
7067	Becker (**107**)	*UBV*	pg	16	
7086	Sobhy	*UBV*	pg	16	
7092	Evans, Meadows (**108**)				r.v. 16 stars, sp. mot.
7142	Sharov (**109**)	*B*		17·2	structure
7160	de Vegt				p.m. plates taken
	Golay	$UBB_1B_2VV_1G$	pe	10	
7209	Lavdovsky (**8**)				p.m., mem., cl. mot.
IC 1434	Basel Observatory	*UBV*	pg	16	
	Larsson-Leander	*BV*	pg + pe	16	s.t.
7235	Becker (**110**)	*UBV*	pg	16	
7243	Artyukhina, Kholopov (**111**)	*B*		16	structure, 18 var. stars
	Mathews (**21**)	*V*	pg	13	
7245	Basel Observatory	*UBV*	pg	16	
7261	de Vegt				p.m. plates taken

NGC	Observer and references (see end of table)	Photometry, method and limiting magnitude			Other data obtained, remark
7296	Basel Observatory	*UBV*	pg	16	
7380	de Vegt				p.m. plates taken
	Mathews (**21**)	*V*	pg	13	
	Morris, Walker, G. A. H.				s.t., r.v., EW Hγ
King 19	Grubissich (**112**)	*UBV*	pg	16	
7510	Mathews (**21**)	*V*	pg	13	
Mark 50	Grubissich (**112**)	*UBV*	pg	16	
7654	Bond				2 Be stars
	Elvius	*UV*	pg		stars around cl.
H 21	Barkhatova, Zhelvanova (**113**)	*VY*		16·8	mult. cl. with NGC 7788, 779
7788	Becker (**114**)	*UBV*	pg	16	
	Barkhatova, Zhelvanova (**113**)	*VY*		16·8	mult. cl. with H 21, NGC 779
	Mathews (**21**)	*V*	pg	13	
7789	Starrfield (**115**)	*BV*	pg + pe	12·8	var. star search
	Woolley				p.m. 2000 stars
7790	Barkhatova, Zhelvanova (**113**)	*VY*		16·8	mult. cl. with H 21, NGC 778

Abbreviations to Table 2

abs. = absolute
cl. = cluster
$c-m$ = color-magnitude diagram
em. = emission
EW = equivalent width (of)
HR diag. = Hertzsprung-Russell diagram
int. = intensity
L(*M*) = luminosity function
mem. = members, membership
mot. = motion
mult. = multiple
obs. = observations
obj. = objective
P = period
pe = photoelectric
pg = photographic
p.m. = proper motion
pol. = polarization
prog. = progress
rot. = rotation
r.v. = radial velocity
s. bin. = spectroscopic binary (ies)
sel. = selective
sp. v. = space velocity
s.t. = spectral type
tot. = total
var. = variable
w/in = within

References for Table 2

1. Greenstein, J. L., Keenan, P. C. 1964, *Astrophys. J.*, **140**, 673.
2. Hoffmeister, C. 1964, *Infor. Bull. var. Stars, Comm. 27 IAU*, no. 67.
3. Efremov, Y. N., Kholopov, P. N., Kukarkin, B. V., Sharov, A. S. 1964, *ibid.*, no. 75.
4. Barkhatova, K. A., Zinina, M. B. 1965, *Astr. Circ.*, no. 314.
5. Sharov, A. S. 1965, *Astr. Zu.*, **42**, 1045.
6. Sharov, A. S. 1965, *Soobsch. astr. Inst. Sternberg.*, no. 142, 20.
7. Dolidze, M. V., Ponomareva, G. A. 1965, *Astr. Zu.*, **42**, 205.
8. Lavdovsky, V. V. 1964, *Izv. Pulkova*, **23**, no. 171.
9. McCuskey, S. W., Houk, N. 1964, *Astr. J.*, **69**, 412.
10. Oja, T. 1966, *Ark. Astr.*, **4**, no. 2 = *Medd. Uppsala Obs.*, no. 153.
11. Kholopov, P. N., Artyukhina, N. M. 1964, *Astr. Zu.*, **41**, 743.
12. Barkhatova, K. A. 1963, *Astr. Papers Ural Univ.*, no. 1.
13. Gunn, J. E. Kraft, R. P. 1963, *Astrophys. J.*, **137**, 307.
14. Wallerstein, G., Conti, P. 1963, *Astrophys. J.*, **140**, 858.

15. Wallerstein, G., Westbrooke, W., Hannibal, D. 1963, *Publ. astr. Soc. Pacif.*, **75**, 522.
16. Pratt, N. 1965, in *Report of the Astronomer Royal for Scotland*, p 2.
17. Serkowski, K. 1965, *Astrophys. J.*, **141**, 1340.
18. Wildey, R. L. 1964, *Astrophys. J. Suppl.*, **8**, 439.
19. Schild, R. E. 1965, *Astrophys. J.*, **141**, 979.
20. Vasilevskis, S., Sanders, W. L., van Altena, W. 1965, *Astr. J.*, **70**, 806.
21. Mathews, R. T. 1963, *Publ. Goodsell Obs.*, no. 14.
22. Guthrie, B. N. G. 1964, *Quart. J. R. astr. Soc.*, **5**, 360.
23. McCarthy, M. F., Treanor, P. J. 1964, *Ric. astr. Spec. Vatic.*, **6**, 535.
24. Guthrie, B. N. G., Nandy, K. 1966, *Observatory*, **86**, 78.
25. Abt, H. A., Barnes, R. C., Biggs, E. S., Osmer, P. S. 1965, *Astr. J.*, **70**, 317.
26. Golay, M. 1964, *Publ. Obs. Genève*, no. 68.
27. Helfer, H. L., Wallerstein, G. 1964, *Astrophys. J. Suppl.*, **9**, 81.
28. Conti, P. S., Wallerstein, G., Herbig, G. H. 1965, *Astrophys. J.*, **141**, 610.
29. Conti, P. 1965, *Astrophys. J. Suppl.*, **11**, 47.
30. Kraft, R. P. 1965, *Astrophys. J.*, **142**, 681.
31. Kraft, R. P., Wrubel, M. H. 1965, *Astrophys. J.*, **142**, 703.
32. Crawford, D. L. 1966, *Astr. J.*, **71**, 206.
33. Woolley, R., Harding, G. A. 1965, *R. Obs. Bull.*, no. 93, 380.
34. Morgan, W. W., Hiltner, W. A. 1965, *Astrophys. J.*, **141**, 177.
35. Luyten, W. J. 1965, *Proper Motion Survey*, Univ. of Michigan, **V**.
36. Luyten, W. J. 1966, *ibid.*, IX.
37. Giclas, H. L., Burnham, R., Thomas, N. G. 1962, *Lowell Obs. Bull.*, **6**, 257.
38. Wayman, P. A., Symms, L. S. T., Blackwell, K. C. 1965, *R. Obs. Bull.*, no. 98.
39. Murray, C. A., Lowne, C. M., Clements, E. A. 1965, *R. Obs. Bull.*, no. 108.
40. Golay, M. 1964, *Publ. Obs. Genève*, no. 70.
41. Börngen, F. 1964, *Astr. Nachr.*, **288**, 39.
42. Bakos, G. A., Rymer, K. R. 1964, *Astr. J.*, **69**, 531.
43. Johnson, H. M. 1965, *Astrophys. J.*, **142**, 964.
44. Walker, M. F. 1962, *Adv. in Electronics*, **16**, 341.
45. Walker, M. F. *ibid.*, in press.
46. Walker, M. F. 1963, *Astr. J.*, **68**, 298.
47. Svolopoulos, S. N. 1965, *Z. Astrophys.*, **61**, 97.
48. West, F. R. 1964, *Astr. J.*, **69**, 562.
49. Upgren, A. R. 1965, *Astr. J.*, **70**, 149.
50. Jeffries, W. H. 1962, *Astr. J.*, **67**, 532.
51. Artyukhina, N. M., Kholopov, P. N. 1963, *Astr. Zu.*, **40**, 1101.
52. Wick, C. 1965, *Z. Astrophys.*, **61**, 110.
53. Morgan, W. W., Hiltner, W. A., Neff, J. S., Garrison, R., Osterbrock, D. E. 1965, *Astrophys. J.*, **142**, 974.
54. Preston, G. W. 1964, *Publ. astr. Soc. Pacif.*, **76**, 165.
55. Bretz, M. C. 1965, *C. r. Acad. Sci. Paris*, **261**, 2161.
56. Vasilevskis, S., Sanders, W. L., Balz, A. G. A. 1965, *Astr. J.*, **70**, 797.
57. Alcaino, G. 1964, *Astr. J.*, **69**, 529.
58. Feinstein, A. 1966, *Publ. astr. Soc. Pacif.*, **78**, 301.
59. Lodén, L. O. 1966, *Ark. Astr.*, **4**, 65 = 1966, *Medd. Stockholm Obs.*, no. 152.
60. Buscombe, W. 1965, *Mon. Not. R. astr. Soc.*, **129**, 411.
61. Eggen, O. J., Sandage, A. R. 1964, *Astrophys. J.*, **140**, 130.
62. Bakos, G. A. 1965, *Astr. J.*, **70**, 132.
63. Kholopov, P. N. 1965, *Astr. Zu.*, **42**, 148.
64. Kholopov, P. N., Artyukhina, N. M. 1964, *Astr. Zu.*, **41**, 968.
65. Murray, C. A., Corben, P. M., Allchorn, M. R. 1965, *R. Obs. Bull.*, no. 91.
66. Feinstein, A. 1963, *Publ. astr. Soc. Pacif.*, **75**, 492.
67. Sher, D. 1965, *Mon. Not. R. astr. Soc.*, **129**, 237.
68. Feinstein, A. 1964, *Observatory*, **84**, 111.
69. Schmidt, H., Santanilla, G. D. 1964, *Veröff. Univ. Sternw. Bonn*, no. 71.

70. Thackeray, A. D., Wesselink, A. J. 1965, *Mon. Not. R. astr. Soc.*, **131**, 121.
71. Engver, N. 1966, *Ark. Astr.*, **4**, 53.
72. Crawford, D. L. 1963, *Publ. astr. Soc. Pacif.*, **75**, 410.
73. Mendoza, E. E. 1963, *Bol. Obs. Tonantzintla Tacubaya*, **3**, 137.
74. Upgren, A. R., Rubin, V. C. 1965, *Publ. astr. Soc. Pacif.*, **77**, 355.
75. Anderson, J. H. 1966, *Publ. astr. Soc. Pacif.*, **78**, 256.
76. Rahim, M. A. 1966, *Astr. Nachr.*, **289**, 41.
77. Lyngå, G. 1964, *Acta Univ. Lund II*, no. 11, no. 13.
78. Landolt, A. U. 1964, *Astrophys. J. Suppl.*, **8**, 329.
79. The, P-S. 1965, *Contr. Bosscha Obs.*, no. 33.
80. Brück, H. A., Smyth, M. J. 1966, *Publ. R. Obs. Edinb.*, **5**, 195.
81. Eggen, O. J. 1965, *Observatory*, **85**, 191.
82. The, P-S. 1965, *Contr. Bosscha Obs.*, no. 32.
83. Talbert, F. D. 1965, *Publ. astr. Soc. Pacif.*, **77**, 19.
84. Svolopoulos, S. N. 1966, *Z. Astrophys.*, **64**, 67.
85. Abt, H. A., Snowden, M. S. 1964, *Astrophys. J.*, **139**, 1139.
86. Alcaino, P. 1965, *Lowell Obs. Bull.*, **6**, 167.
87. Maffei, P. 1963, *Mem. Soc. astr. Ital.*, **34**, 441.
88. Hiltner, W. A., Morgan, W. W., Neff, J. S. 1965, *Astrophys. J.*, **141**, 183.
89. Grubissich, C. 1964, *Z. Astrophys.*, **58**, 276.
90. Landolt, A. U. 1963, *Astrophys. J. Suppl.*, **8**, 352.
91. Yilmaz, F. 1966, *Z. Astrophys.*, **64**, 61.
92. Yilmaz, F. 1966, *ibid.*, **64**, 54.
93. Grubissich, C. 1965, *Z. Astrophys.*, **60**, 249.
94. Fenkart, R. P. 1965, *Z. Astrophys.*, **62**, 90.
95. Bronkalla, W. 1963, *Astr. Nachr.*, **287**, 249.
96. Svolopoulos, S. N. 1965, *Z. Astrophys.*, **61**, 105.
97. Kinman, T. D. 1965, *Astrophys. J.*, **142**, 655.
98. Dolidze, M. V., Lypareva, N. 1965, *Bull. Abastumani astrophys. Obs.*, **32**, 53.
99. Barkhatova, K. A., Shashkina, L. P., Vasilevsky, A. S. 1963, *Astr. Papers Ural Univ.*, no. 1.
100. Batten, A. J. 1965, *Astr. J.*, **70**, 318.
101. Elst, E. 1964, *Mitt. astr. Gesell.*, no. 17, 133.
102. Lawrence, L. C., Reddish, V. C. 1965, *Publ. Edinburgh Obs.*, **3**, 275.
103. Artyukhina, N. M., Kholopov, P. N. 1965, *Soobsch. astr. Inst. Sternberg*, no. 142, 3.
104. Barkhatova, K. A., Steinberg, M. K. 1963, *Astr. Papers Ural Univ.*, no. 1.
105. Larsson-Leander, G. 1964, *Astrophys. J.*, **140**, 144.
106. Fenkart, R. P. 1963, *Mem. Soc. astr. Ital.*, **36**, 291 = 1965, *Contr. astroph. Obs. Asiago*, no. 181.
107. Becker, W. 1963, *Mem. Soc. astr. Ital.*, **36**, 285 = 1965, *Contr. astroph. Obs. Asiago*, no. 180.
108. Evans, T. L., Meadows, A. J. 1964, *Observatory*, **84**, 148.
109. Sharov, A. S. 1964, *Astr. Zu.*, **41**, 974.
110. Becker, W. 1963, *Mem. Soc. astr. Ital.*, **36**, 271 = 1965, *Contr. astroph. Obs. Asiago*, no. 178.
111. Artyukhina, N. M., Kholopov, P. N. 1965, *Astr. Zu.*, **42**, 1050.
112. Grubissich, C. 1965, *Z. Astrophys.*, **60**, 256.
113. Barkhatova, K. A., Zhelranova, E. I. 1963, *Astr. Papers Ural Univ.*, no. 1.
114. Becker, W. 1963, *Mem. Soc. astr. Ital.*, **36**, 277 = 1965, *Contr. astroph. Obs. Asiago*, no. 179.
115. Starrfield, S. 1965, *Publ. astr. Soc. Pacif.*, **77**, 206.

New Clusters

Twenty-two new open clusters have been found by Hogg (**25**) on plates taken with the 74-inch (188 cm) reflector of the fields of clusters in Trumpler's list for declination $< -45°$.

Lyngå (26) has surveyed the region from Centaurus to Norma, finding 14 new clusters.

Czernik (27) has found 45 new clusters, and King (28) five new clusters, both from studies of the plates of the Palomar Sky Survey.

Lodén reports that in the course of the Stockholm Observatory survey of the Southern Milky Way, a number of new clusters and associations have been found. Information on these objects (coordinates, size, number of stars, photographic and photovisual magnitudes to $m_{pg} \sim 14$ and rough spectral types of the brightest stars) will be supplied by him on request.

PHOTOMETRIC WORK

Important photometric investigations of old clusters have been made by Eggen and Sandage (29) and by Kinman (30).

In an extensive study of NGC 2682, Eggen and Sandage find that the discontinuity in the evolving portion of the main-sequence, previously suspected in NGC 752 and NGC 2477, is apparently real and may be related to a rapid evolution of the star between the depletion of hydrogen in the core and the onset of hydrogen burning in the shell. In addition, proper motion and spectroscopic data confirm the membership of the 'blue stragglers'-stars on or near the main-sequence above the turn-off point. Those could be either stars younger than the majority of members or objects in which complete mixing occurs. The data have also been used to derive reddening and blanketing corrections.

Kinman has observed NGC 6791 which proves to be similar in age to NGC 188. Novel features of this cluster include: (1) A sequence of blue subgiants whose membership is established from radial velocity measures, and (2) an incipient blueward extension of the red-giant branch, perhaps due to stars in a post red-giant stage.

Gray (31) has computed integrated magnitudes and colors of 67 open clusters from published photometry of their individual stars.

Schmidt-Kaler has extended Gray's work by calculating integrated magnitudes and colors of clusters having very large diameters. Large diameter clusters lie on a rather well defined intrinsic line in the color-color diagram. In the color-magnitude diagram, they are bounded by a well defined upper envelope: $M_v = -9{\cdot}4 + 4{\cdot}5\ (B-V)_0$. Application of these results to the photoelectric photometry of the 'globular' clusters in M 31 and M 33 shows that ten clusters in M 31 and eleven in M 33 are young open clusters, and yield results for the reddening of these objects of $E_{B-V} = 0{\cdot}12 \pm 0{\cdot}04$ and $0{\cdot}08 \pm 0{\cdot}03$ for M 31 and M 33, respectively. The absorption-corrected distance moduli obtained by this method are: $m - M = 24{\cdot}6 \pm 0{\cdot}3$ for M 31 and $24{\cdot}5 \pm 0{\cdot}4$ for M 33.

Schmidt-Kaler has also observed integrated magnitudes and colors (*UBV*, *ubvy* Hβ) of 41 open clusters photoelectrically at Kitt Peak using different diaphragms. The measures appear to give rather accurate reddening estimates and estimates of metal content. Also observed were 25 dBe stars in open clusters, to locate their position in the color magnitude diagram. These stars are found to occupy a strip about 1·5 magnitudes above the zero-age main-sequence.

Schmidt-Kaler and Vogt have completed the iris-photometer and coordinate measurement of all stars (about 4000) within 1° around h and χ Per down to 17·5 magnitude.

Smak (32) has observed T Tauri stars and faint stars in NGC 2264 and NGC 6530 using a special blue filter to eliminate the effect of line emission. He finds that for the 30 stars measured in each cluster down to $V = 15$, the B observations are not appreciably affected by line emission, unlike the 26 T Tau stars in the Taurus cloud also observed with this system. However, it has been pointed out by Aveni (33) that the special filter used by Smak does not eliminate all T Tau emission lines and is thus only partly effective in distinguishing line-emission effects.

Becker (34) has shown that too large a value of the ratio of total to selective absorption may be obtained because of: (1) intrinsic scatter in the color-magnitude diagrams of clusters and associations and, (2) the admixtures of non-member stars. It is concluded that the large ratio of total to selective absorption found by Johnson in III Cep and I Ara may result from the inclusion of non-members.

During the present report period, there has been a marked increase in the amount of work done on the polarization of the light of cluster stars, as indicated in Table 2. Serkowski (35) finds that a correlation between reddening and polarization is found only for IC 4725 and the association VI Cyg, while Grigoryan concludes that stellar (as opposed to interstellar) polarization exists for the stars of NGC 457.

Spectrographic Work

McCarthy and Treanor (36) have published a catalog of 125 red stars which they have observed in the vicinity of the Pleiades using the Vatican Schmidt. Of these, 27 stars fainter than $B = 16$ are considered to be probable members of the cluster. Image tube spectra of three of the latter group have been obtained with the Perkins-Lowell 72-inch (183 cm) reflector (37) and each of them shows the characteristic features of an M dwarf. For six stars of the Vatican catalog, flares have been observed by Haro at Tonantzintla and Rosino at Asiago.

Radial velocity observations of the OB stars in the direction of IC 1805 are being studied by Underhill. Preliminary results indicate that these stars are spread along a string and do not form a compact cluster near IC 1805.

M. F. Walker has continued spectrographic observations of UV excess stars in the Orion nebula and NGC 2264. Particular attention was given to SU Ori, which is one of the more active variables of the 'inverse P Cyg' or 'YY Ori' type. It was found that a correlation exists between the brightness of the star and the visibility of the redward-displaced absorption lines, these lines being visible at maximum and invisible at minimum light.

Roxburgh, *et al.* (38) have described a method whereby it is possible to discriminate between line broadening due to the degree of intrinsic rotation and to the aspect effect for stars in clusters having the same chemical composition, age, and distance. The method has been applied to stars in Praesepe, and the zero-rotation main-sequence derived.

Weaver and Ebert (39) have used the existing spectral types and photoelectric photometry of clusters to recalibrate the absolute magnitudes of the MK spectral types of stars in the range from O5 to A0.

Greenstein (40) has studied spectra of the bright red giants in NGC 188 and NGC 2682. The spectra are normal except that the color excess indicated by the spectral types and colors of the giants is greater than that of stars near the turnoff point of the main sequence.

Variable Stars

Since the subject of variable stars in clusters is covered in the report of Commission 27, a detailed account will not be given here. However, several of the investigations of variables in open clusters are referred to in Table 2. In addition, Efremov (41) has studied the Cepheids known to be related to clusters and has found a weak correlation between the periods of the Cepheids and the ages of their clusters, shorter periods corresponding to greater age. Starikova (42) also finds an indication of a correlation between the periods of Cepheids in clusters and the absolute integrated magnitude of the cluster.

Proper Motions

Guetter and Upgren report that first epoch plates are being taken of a number of open clusters with the 61-inch (155 cm) Astrometric Reflector of the U.S. Naval Observatory. After second epoch plates have been obtained, all plates will be measured on the automatic measuring engine for proper motion. The limiting measurable magnitude is about 18. Three-color photometry as well as radial velocities of the brighter stars will be obtained for some of the clusters in the interval between the first and second epoch plates. For some clusters, an extended region surrounding the cluster is being covered by more than one field in order to determine whether the cluster possesses a halo of faint stars of small mass. The existence of such a halo has been suggested by several investigators as a result of equipartition of energy among the member stars and from the disruption of the clusters by passing stars and clouds and the tidal force of the Galaxy. The small measuring errors achievable with automatic measurement of the plates will make it possible to obtain reliable separation of members from field stars with an interval between epochs of about 10 to 15 years, in most cases. For a few nearby clusters, an interval of about 30 years should make it possible to observe differential motions between different types of stars in the clusters.

Lavdovsky (**43**), continuing the work described in the last report, has used the proper motions from his catalog (**44**) to study the membership of stars in several clusters. In particular, he finds that NGC 6882 and NGC 6885 are two clusters at different distances and that NGC 1907 and NGC 1912 form a single system.

Vasilevskis, Saunders and Balz (**45**) have determined relative proper motions of 245 stars in the extremely young cluster NGC 2264. One hundred and forty-five stars were found to probability of have a membership greater than 0·5. The results confirm the conclusion by M. Walker (**46**) that stars later than A0 lie above the main-sequence; the five yellow giants in the cluster are found to be field stars, however. This result is extremely important since it provides a second case in which the membership of the faint stars lying above the main-sequence of a young cluster has been definitively established, the first being the study of the Orion nebula cluster by Strand (**47**). The observations of NGC 2264 have the advantage of extending to a fainter limiting absolute magnitude than do those in Orion and are thus capable of demonstrating the membership both of very red stars lying far to the right of the main-sequence in the color-magnitude diagram, and a number of relatively bluish stars which lie below the main-sequence. Proper motions in a second very young cluster, IC 1805, have also been measured by Vasilevskis, Sanders, and van Altena (**48**). In this case, determination of the color-magnitude diagram of the members is difficult owing to variable reddening. However, there is some indication that the age of the cluster may be less than that of NGC 2264.

As described in the last report, second epoch astrometric plates were obtained at Mt Wilson by Woolley and Murray of fields photographed by van Maanen with the 60-inch (152 cm) reflector. Reduction of this material has continued, and the proper-motions referred to in Table 2 for the clusters: Hyades, NGC 752, NGC 2168, NGC 2682, and NGC 7789, by Allchorn, Blackwell, Cannon, Clements, Corben, Lowne, Murray, Symms, Wayman, and Woolley, are derived from this material.

At Hamburg, de Vegt is investigating absolute proper-motions and space-velocities of a number of clusters; these are enumerated in Table 2. The material used in all of these studies is the following: First epoch, newly reduced data from the Carte du Ciel plates (Vatican zone). Second epoch, plates taken in 1965 with the Hamburg Schmidt telescope. It has been found that by using only the inner region of the Schmidt plates, and with a difference in epochs of 50 to 60 years, the cluster motion may be determined with a mean error of $\pm$ 0″.002 per year.

Luminosity Function

The structure of NGC 188 has been studied by Sharov (**49, 50**) and by Barkhatova and Zinina (**51**) as indicated in Table 2. These authors find that the cluster, like NGC 752, NGC 2099 and NGC 2682 discussed in the last report, consists of a nucleus and a wide corona. The luminosity function of NGC 188 has also been determined and is found to have two maxima, at $M_B = +3$ and $M_B = +2$. The 'blue-stragglers' are found to be members of the cluster.

Similar studies of the structure of three clusters have been carried out by Kholopov and Artyukhina. The references are given in Table 2. The results of this study may be summarized in the following table:

NGC	Radius of Nucleus	Radius of Corona	Number of Stars
752	25′ – 35′	114′	610 to B = 15·8
6939	16′	85′	3400 to B = 16·8
7243	20′	75′	130 to B = 12·9 (counts to B = 16·1)

In addition, the same authors have studied the luminosity function of NGC 2682 and its variation with distance from the center of the cluster.

The bright end of the luminosity function in 52 galactic clusters has been studied by Gray (**52**) who finds a change in slope with age of the cluster which can be explained in terms of the evolution of the stars. A similar investigation has been made by Martinet (**53, 54, 55**) who, starting from the observed luminosity functions in open clusters, has derived the form of the initial luminosity function, taking account both of the ejection of stars of small mass from the cluster and the evolution of the stars of large mass during the life of the cluster.

Dynamical Models

Bouvier (**56, 57**) has studied the steady state of isolated unrelaxed spherical systems along lines similar to those followed independently by Veltmann (**58**) Classification of the various possible anisotropies in the velocity distribution was made and compared with the generalized Schuster model. Bouvier (**59**) has also investigated the effect of a finite velocity of escape on the Schwarzschild ellipsoidal law for the velocity distribution in a system having no mean motions or undergoing differential rotation.

King (**60, 61**) has calculated self-gravitating cluster models whose cutoff velocity is determined by a finite tidal limit. Additional models have been calculated with a mixture of nine stellar groups of different mass. Models with isotropic velocity distributions fit the extensive star counts that are now available. The only discrepancy is an excess of brightness at the centers of the most concentrated clusters. This cannot be accounted for by anisotropy of the velocity distribution, which has very little effect in the cluster center. There must instead be an excess of mass at the cluster center. The observations can be satisfied by models that attribute the extra mass either to gas or to binaries concentrated at the center.

Prata is modifying King's models to include anisotropy in the velocity distributions.

Radio Observations

Howard and Westerhout (**62**) are conducting a program to determine the neutral hydrogen content of open clusters using the NARO 300-feet (91 m) radio telescope. Fifty clusters with diameters $\leqslant$ 20′ have been observed and H I has been detected in about two-thirds of these. H I is found in both young and old clusters, but few young clusters do not have H I.

Schwartz (**63**) has observed 13 open clusters and their immediate surroundings with the Bonn 25-meter radio telescope for interstellar neutral hydrogen. The expected profile has been calculated from the profiles at positions of equal distance from the cluster center and from parabolic fits for diagonals through the cluster center. H I is present in all clusters studied that have an age of 25×10^6 years or less, while no cluster older than 150×10^6 years had detectable amounts of H I. The optical velocities of the clusters are within 5–10 km s^{-1} of the means velocities of the H I clouds associated with the clusters. In some cases (e.g. NGC 129) there are indications of an expansion with a velocity of about 7 km s^{-1}. In general, the neutral hydrogen gas extends over a region which is about three times the optical diameter of the cluster.

Space Distribution

Clube has examined the radial velocities of B stars in Eggen's catalog and proper motions of A stars in the *Cape Catalog of Proper Motions of* 20 843 *Stars*. This has revealed a common streaming motion for members of the Gould Belt, equal to that of the Pleiades group. Gould's Belt is regarded as being a narrow projection from the lower edge of the Orion spiral arm, expanding towards $l_{II} = 0°$, $b_{II} = +20°$. The kinematic age of Gould's Belt is about $3{\cdot}7 \times 10^7$ years.

The galactic structure in the region from Centaurus to Norma has been investigated by Lyngå (**64**) on the basis of cluster data, while Alter (**65**) has identified seven clusters with a spiral feature extending from Carina to Scutum.

Evolution

A complete discussion of stellar evolution is given in the report of Commission 35, so that only selected references will be mentioned here.

The problem of the spread in ages of stars in clusters has continued to attract considerable attention. Williams (**66**) has constructed a theoretical color-magnitude diagram for a cluster having a spread in the ages of formation of its stars, and finds that this spread in age assumption does not explain the existence of the 'blue straggler' stars. The problem of the unusually long main-sequence in objects like the Pleiades and h and χ Per has been considered by Reddish (**67**) who concludes that in such objects the bright stars may have been formed by accretion from the fainter ones. Observational evidence on this point may be provided by the work of Golay (**68**) who has observed the Pleiades on his seven-color system and finds an indication that the cluster may contain two groups of stars having different ages.

The problem of the pre-main-sequence stages of stellar evolution also continues to be of great importance. Penston (**69**) has reviewed the comparison of the various theories of pre-main-sequence evolution with the observations of NGC 2264 and NGC 6530 and concludes that the evidence favors the Hayashi theory rather than Henyey-type contraction or the theory of Faulkner, Griffiths, and Hoyle. Iben and Talbot (**70**) have calculated star formation rates in young clusters where most of the stars are still in the gravitational contraction stage. They find that the hypothesis of a spread in age or of continual creation does not conflict with the short time-scale associated with the Hayashi tracks. A difficulty in all of the theories remains the existence, particularly in NGC 2264, of faint member stars near and even below the main-sequence. While stars near the main-sequence might be interpreted as being older than the other members of the cluster, no explanation of the stars below the main-sequence exists. However, spectrographic observations by M. F. Walker (**71**, and unpublished) show that many of the faint stars in NGC 2264 and the Orion nebula, including those below the main-sequence, possesses a blue continuum. Thus, for these stars, at least, the observed $B-V$ colors will give too high an effective temperature. Correction for this effect will probably move these stars from below the main-sequence to a position far enough to the right to accord with a model involving a reasonable spread in ages of the member stars.

Dluzhnevskaya (**72**) has studied the distribution functions of the colors of stars in clusters with ages between 0 and 10^8 years. Two cases were considered: (1) evolution with constant mass, and (2) evolution with mass loss. The second case was found to give better agreement between theory and observations, but it is recognized that one weakness of the investigation is the assumption of a time-independent initial luminosity function.

Wildey has continued his investigation of h and χ Per with particular emphasis on the apparent disagreement between the observed ratio of the numbers of early to late supergiants and the much smaller ratio resulting from the rapid evolution due to the neutrino pair production and photoneutrino processes predicted by the conserved-vector current theory of weak-coupling interactions. Statistical studies of the distributions of the early and late supergiants near h and χ Per indicate that the stars are indeed part of the same spatial system, while color photographs indicate that incompleteness is not a significant source of error in the observed early to late ratio.

Kholopov (**73, 74**) regards open and globular clusters as links in a single evolutionary chain, the differences between them being determined chiefly by their mass and age. The general character of the evolution of stars does not depend on their chemical composition. There is no single-valued relationship between the metal content and the age of stars. The relationship between the metal content of the stars in globular clusters and the maximum distance of the latter from the center of the Galaxy indicates that the stars were born from a diffuse medium in the proto-galaxy in which differentiation of elements according to weight had already occurred.

GLOBULAR CLUSTERS

(Prepared by L. Rosino)

Arp (**75**) has published in Vol. V of *Stars and Stellar Systems* a chapter on the intrinsic and kinematical properties of the globular clusters of our own Galaxy. Table 1 in his article gives for 120 clusters the best data obtainable on absorption, diameter, distance modulus, integral magnitudes, colors, spectral type and radial velocity.

Integral Photometry

Van den Bergh (**76**) has determined *UBV* magnitudes for 49 globular clusters at Cerro Tololo inter American Observatory. He has found that the observed colors may be used to define a reddening-independent metallicity index $Q = (U-B) - 0{\cdot}72\ (B-V)$, which can easily be correlated with other indicators of metal content. For 14 metal-rich clusters the radial velocity dispersion has been found to be 75 km s^{-1} compared to 162 km s^{-1} for 56 metal poor clusters. The intrinsic colors of globular clusters in M 31 and the Galaxy are well represented by the relation $(B-V) = Q + 1{\cdot}00$. For 63 clusters with $\csc|b| < 10$ van den Bergh has obtained $E(B-V) = 0{\cdot}060 \csc|b| - 0{\cdot}009$. Finally there is some indication that the frequency function of Q values in M 31 differs from that in the Galaxy in the sense that M 31 contains fewer metal-poor clusters than does the Galaxy. The data, however, are insufficient to establish this point with certainty.

P. McClure (**77**) has used the 74-inch (188 cm) reflector of the D. Dunlap Observatory to show that the intermediate-band color index c ($\lambda\,3800/\lambda\,4150$) may be used to measure the metallicity of globular clusters.

Researches on the colors of globular clusters have also been made by Koehler (**78**) at Mt Stromlo Observatory who has found correlations of the G-band index with metallic content and intrinsic colors. Schmidt-Kaler (**79**) has observed at Kitt Peak 36 globular clusters north of − 30° in Stromgren's system *uvby* using various diaphragms. For 15 of these Hβ has been measured photoelectrically. The observations appear to give a measure of the metal content and reddening and an unambiguous spectral-type-equivalent of the clusters. The intrinsic colors of 20 globular clusters have been determined by comparison with distant early-type stars or galaxies. The globular clusters occupy a triangular region in the *UBV* two-color diagram partially overlapping with that of intermediate-age and old open clusters. The median absolute magnitude of RR Lyrae in seven globular clusters of known reddening appears to be $\bar{M}_v = +0{\cdot}35$. Menzies at Mt Stromlo has also observed *UBV* colours of globular clusters and hopes to publish these shortly. Integrated *UBV* magnitudes of the Fornax dwarf galaxy have been reported by Hodge (**80**). The absolute magnitude is $M_v = -8{\cdot}0$ (in our Galaxy $M_v \sim -8{\cdot}2$).

Light distribution, Luminosity function

Mlle Rousseau (**81**) in Haute-Provence has measured with a series of diaphragms of different size *UBV* magnitudes of 22 globular clusters. I. R. King (**82**) has reported on the photoelectric observations of nine globular clusters (NGC 5024, 5272, 5904, 6205, 6218, 6254, 6341, 7078, 7089) made at Mt Wilson through five concentric circular apertures of various diameters and also with the same small-sized diaphragm at different points of the cluster. Profiles giving the surface brightness as a function of the distance from the center have been derived. Additional photoelectric measurements have been made by J. L. Hitchcock in three clusters. King and his associates have also completed their star-counting program in globular clusters and the results should be in print by the time this report appears. The program includes 164 counts in 54 clusters.

The luminosity function for M 13 has been obtained down to $V \sim 19$ by means of star counts on plates taken with the 74″ Okayama reflector by M. Simoda and H. Kimura (**83**). Particular attention has been paid to the counting method. Counts have been made within a ring having an outer radius of 8′ and an inner radius of 2′ to 4′. Significant differences have been found between the luminosity function of M 13 and that of M 3 as determined by Sandage (**84**), the luminosity function in M 13 having a steeper gradient than in M 3. By combining the observed luminosity function with the population II stellar models by Faulkner and Iben (**85**) some interesting results are obtained on the lifetime for horizontal branch stars.

Photoelectric surface photometry in the *PVI* system has been made by Kron (**86**) at Mt Stromlo on the rich globular cluster 47 Tucanae. The observations confirm that the radial surface brightness gradient follows de Vaucouleurs law (**87**) as already shown by Gascoigne and Burr (**88**). Kron, however, finds that the central brightness of 47 Tuc is nearly as high as is that of the more condensed elliptical galaxies. The radial color gradient is small as previously found by Kron and Mayall (**89**) for 48 clusters.

The space distribution of bright stars in a cluster (M 92) has been determined by star counts by Blagikh (**90**) who has compared his results with those obtained by Kholopov (**91**) for M 3. Equidensity curves of M 13 have been derived by Kadla from plates taken at Tautenburg and Flagstaff; the author has pointed out the dependence of these curves by the axial ratio *b*/*a*.

Color-magnitude diagram

Much work has been done in this important field which is strictly connected with the problems of stellar evolution. Sandage and Katem (**92**) have determined *B* and *V* magnitudes for 293 stars in the moderately strong-lined globular cluster NGC 6171 (Morgan's class V). A similar research has been made by Sandage and Smith (**93**) on 304 stars of the cluster NGC 6712, which also belongs to Morgan's class V. The principal differences presented by the $c-m$ diagrams of these clusters in comparison with halo clusters are: (*a*) strong concentration of stars on the red side of the RR Lyrae gap; (*b*) little steepness of the giant branch which, at $(B-V)^{\circ} = +1{\cdot}4$, rises only 2 magnitudes above the horizontal branch, while in halo clusters it rises 3 magnitudes; (*c*) Intrinsic redward displacement of the giant and subgiant sequences in comparison with clusters having lower metal abundance.

Arp (**94**), in a photometric research on the properties of the Galactic nucleus, has given the color-magnitude diagram of NGC 6522. The absence of stars on the blue side of the RR Lyrae gap identifies the cluster as a metal-rich system such as NGC 6356 and 47 Tuc. After correction for reddening, a distance of 7·5 kpc is found.

Sandage and Walker (**95**) have determined *UBV* magnitudes for stars brighter than $V \sim 17$ in M 92. The horizontal branch redward of the variable star gap is absent, as is characteristic of low metal-abundance clusters. A new and unexpected feature in the color distribution for stars in the subgiant and asymptotic-giant branch is the fact that stars of equal $B-V$ in the two branches differ by as much as 0·25 in $U-B$. An unknown magnitude source of continuous opacity or abnormal line strengths for stars in one branch compared with the other could explain the difference. Similar results have been found also in an analysis of three-color magnitudes for M 3 (**96**).

Extensive photometric studies on individual stars in globular clusters have been made by Cuffey (**97, 98**) on M 53 (*B*, *V*, $c-m$ diagram; color of bright stars in nucleus) and in ω Centauri by Woolley (**99**), Dickens and Saunders (**100**). Geyer (**101**) has also done three-color photometry (*UBV*) for 400 stars in ω Centauri from 10·9 to 15·7 *V* on plates taken at the Boyden Observatory. The color-magnitude diagram has the characteristic features of a halo cluster, while the two-color diagram shows an effect similar to that found by Sandage and Walker in M 92.

Color-magnitude diagrams have been published for NGC 6522 (**102**), and for the loose cluster Pal. 13 (**103**). Work is in progress for M 15 (Sandage and Katem), 47 Tuc (**104**), NGC 4372 and NGC 7099 (P. J. Andrews, Radcliffe Observatory) (**105**). A good deal of work is currently in progress at the Mt Stromlo Observatory. (**106**), as shown in the following table:

Cluster	*Observer*	*Photometry*	*Method*	*Limiting V Magnitude*
NGC 362	Menzies	*UBV*	pe + pg	16
3201	Menzies	*UBV*	pe + pg	16
5139	Brooke	*VRI*	pe	14
6266	Gascoigne	*VB*	pe + pg	17·5
6397	Gascoigne	*UBV*	pe	16
6397	Gascoigne and ass.	*VRI*	pe	14
6723	Menzies	*UBV*	pe + pg	19
104	Brooke	*VRI*	pe	14
104	Menzies	*UBV*	pe	16

Spectra

An extensive study of the spectra of horizontal-branch stars in the nearest globular cluster, NGC 6937 ($m - M = 12{\cdot}0 \pm 0{\cdot}3$) has been made by Searle and Rodgers (**107**) at Mt Stromlo. The stars have been shown to be metal-deficient by a factor of about 100 and evidence has been presented that they are also helium-deficient. Masses of 1·3 $\mathfrak{M}_{\odot}$ have been found by discussing the hydrogen line strength. Using spectral data by Helfer, Wallerstein and Greenstein (**108**), Pagel (**109**) has redetermined abundances in the red giant Barnard III–13 in M 92. Results confirm that the Fe/H ratio is about 1/200 of the solar value and indicate unchanged solar abundances relative to Fe for Mg, Ti, V, Cr, Ni, Sr, Y and Zr. Barium seems over-deficient relative to iron by a factor of about 3.

Other spectroscopic investigations of the brightest stars in the clusters NGC 4372 and 7099 are in progress at the Radcliffe Observatory.

Proper motions, Radial velocities, Dynamics

A series of dynamical researches on ω Centauri have been made in the last three years. Harding (**110**) on spectra taken at the Radcliffe Observatory. has determined 74 radial velocities of 40 individual stars, of which 31 proved to be cluster members. Proper motions of individual stars have been published in *R. O. Annal* no. 2 and discussed by Murray, Jones and Candy (**111**). Through a statistical procedure, cluster members were separated from field stars. A paper of Woolley and Dickens (in press) will finally discuss the structure and dynamics of the cluster. A comparison of stellar distribution with radius will be made with the theoretical models.

Proper motions in NGC 6522 (**112**) have been determined by Clube and are in progress for 1200 stars of 47 Tuc.

Theoretical studies on the dynamics of globular clusters have been published by Matsumani (**113**) who has shown correlations between kinematical and physical characteristics (energy, angular momentum, integrated spectral type, diameter and concentration class) and by Hénon (**114, 115**) who has computed theoretical models for isolated clusters. Hénon is presently working on a new approach to cluster dynamics, using the Monte Carlo method (**116**). The approach achieves a substantial reduction of computing time and will probably permit the treatment of the full dynamical problem without restrictive hypotheses.

Variable stars

The work on variable stars in globular clusters will be extensively reported by H. Sawyer Hogg in Commission 27; only a short and incomplete account is given here.

New variable stars in or near globular clusters have been found by Wilkens (**117**) (ω Cen, NGC 3201), Kurochkin (**118**) (NGC 6171), Fourcade and Laborde (**119, 120, 121**) (NGC 5286, NGC 6352, NGC 1261, NGC 6624, IC 4499). An extensive study of the RR Lyrae variables in M 3 has been made by Szeidl (**122**) who has discussed light curves and periods for 112 stars of this type. Thirty-six of them show well pronounced changes. Wilkens (**123**) has examined the variations of periods of RR Lyrae stars in M 4. An investigation on the period changes of 150 variables in ω Cen, also by Wilkens, is just coming to its end. Studies on the variations of periods and light curves have also been made by Makarova and Akimova (**124**) (M 15) and by Margoni (M 53) (**125**). Photometry of RR Lyrae stars in clusters and period determinations have been made: in NGC 6712 by Sandage, Smith and Norton (**126**) and by Rosino (**127**); in ω Cen by Dickens and Saunders (**100**); in M 53 by Margoni (**128**); in M 14 by Sawyer Hogg and Welhau (**129**); in NGC 7006 by Rosino and Ciatti (**130**).

The long-period variables in globular clusters have been studied by Feast (**131, 132, 133**). By a critical examination of the existing data and by radial velocity determinations made at the Radcliffe Observatory he concludes that there are probably no known Me (Mira Type) variables with periods greater than about 220 days in globular clusters.

W. Lohmann (**134**) has made a general study of the period-luminosity relation for RR Lyrae stars in globular clusters, using the published material. For some clusters he has not found any definite linear relation between m and log P. For other clusters he has found relations which, however, are different from cluster to cluster and not so clearly defined as for the Galactic RR Lyrae stars.

Miscellaneous

McCarthy and Treanor (135) have studied the faint cluster Abell 2 on direct plates taken in infrared with the Vatican Schmidt. The interstellar absorption on the cluster is likely to amount to 2 magnitudes. They conclude that the object may be a peculiar intergalactic cluster or an old open cluster located at the edge of the Galaxy.

Roeder (136) has examined the effect of helium content on the theoretical deviation curves. Since the helium content is known only with little accuracy, the distance modulus of globular clusters derived by comparison of the theoretical curves with the observations may be affected by an uncertainty of 0·2 magnitude, in addition to the errors of observation. Kholopov (137) has advanced some alternatives to the hypotheses of Schwarzschild and Sandage on the evolution of star clusters. One of the alternatives is that there may not be a unique initial luminosity function for all clusters.

Vogt (138) has investigated the interstellar hydrogen content in M 3, M 5 and M 13 by measures of the 21-cm line radiation. He has concluded that upper limits for the mass of neutral atomic hydrogen in these clusters are between 100 and 180 $\mathfrak{M}_{\odot}$. A similar result has been found by Goldstein (139) in M 13 with the Harvard 60 ft (9 m) reflector and maser radiometer.

Vorontsov-Velyaminov (140) has examined from a statistical point of view the globular clusters which surround many elliptical galaxies. He does not find any correlation between their number and the luminosity or mass of the parent galaxy.

Proposals

H. Wilkens (La Plata Observatory) proposes that equatorial and galactic coordinates of globular clusters be given in future for the Equinox 2000·0. He makes some interesting remarks on the subject and suggests that a Table for the rapid conversion of α and δ (2000) into l_{II} and b_{II} be calculated.

MISCELLANEOUS

The application of electronography to stellar photometry is being studied by Lallemand, *et al.* (141) and by Kron and M. F. Walker. As a consequence of the linear relation, in electronography, between plate density and image intensity, Lallemand found a linear relationship between image density measured with a Schild-type photometer and the intensity of each star over a range of 2·5 magnitudes, beginning at the plate limit. Kron and Walker have found that by making a densitometer trace of each star image, it is possible to obtain a linear relation between intensity and area under the image profile over a range of at least 5 magnitudes when the electronic image is recorded on Ilford G5 plates developed 5 minutes in D–19. A larger linear range may exist, since linearity holds on electronographic plates to densities of at least 3. The importance of such a technique in cluster photometry is obvious. Using an electronographic-type image-tube, it is only necessary to set up a few bright standards in the cluster; magnitudes of all fainter stars down to the plate limit may then be determined from the electronographic plates. In addition, the fine grain (track size) of the electronographic plates should make possible both higher accuracy of measurement and a fainter detection limit than are possible photographically, while the high speed of the electronographic system will make it possible for a Cassegrain reflector of moderate aperture to reach faint stars heretofore only observable with the largest reflectors.

ACKNOWLEDGEMENTS

It is a pleasure to thank all of those who have contributed to the preparation of this report by replying to the circular letter. Special thanks are due to Dr J. Ruprecht and Professor L. Rosino, members of the Organizing Committee, for preparing the sections on associations and globular clusters, respectively, and to Mrs Sethanne Hayes who conducted the literature search for the preparation of Table 2.

M. F. WALKER
President of the Commission

REFERENCES

1. Alter, G., Sawyer Hogg, H., Ruprecht, J. 1965, Suppl. 7, *Appendix to Bull. astr. Inst. Czech.*, **16**, no. 1.
2. Alter, G., Sawyer Hogg, H., Ruprecht, J. 1966, Suppl. 8, *ibid.*, **17**, no. 1.
3. Alter, G., Sawyer Hogg, H., Ruprecht, J. Suppl. 9, *ibid.*, in press.
4. Hogg, A. R. 1965, *Mem. Mt. Stromlo Obs.*, **4**, no. 2.
5. Ruprecht, J. 1966, *Bull. astr. Inst. Czech.*, **17**, 33 (no. 1).
6. Sher, D. 1964, *Observatory*, **84**, 263.
7. Lawrence, L. C., Reddish, V. C. 1965, *R. Obs. Bull.*, **3**, no. 9.
8. Reddish, V. C. 1966, *Mon. Not. R. astr. Soc.*, in press.
9. Reddish, V. C., Lawrence, L. C., Pratt, N. M. *R. Obs. Bull.*, **5**, no. 8.
10. Ampel, R. 1964, *Acta Astr.*, **14**, 52.
11. Thackeray, A. D., Wesselink, A. J. 1965, *Mon. Not. R. astr. Soc.*, **131**, 121.
12. Roslund, C. 1966, *Ark. Astr.*, **4**, 101 = *Lund Obs. Medd.*, I, no. 213.
13. The, Pik-Sin 1961, *Contr. Bosscha Obs.*, no. 12.
14. Braes, L. L. E., Hovenier, J. W. 1966, *Nature*, **209**, 360.
15. Blaauw, A. 1964, *A. Rev. Astr. Astrophys.*, **2**, 217.
16. Mirzoyan, L. V. 1964, *Soobsch. Byurakan Obs.*, **35**, 75.
17. Haug, U., Pfleiderer, J., Dachs, J. 1966, *Z. Astrophys.*, **64**, 140.
18. Graham, J. A. 1965, *Observatory*, **85**, 196.
19. The, Pik-Sin 1964, *Contr. Bosscha Obs.*, no. 23.
20. The, Pik-Sin 1964, *Contr. Bosscha Obs.*, no. 27.
21. The, Pik-Sin 1964, *Publ. astr. Soc. Pacif.*, **76**, 293.
22. Maffei, P. *Mem. Soc. astr. Ital.*, in press.
23. Maffei, P. 1965, *Mem. Soc. astr. Ital.*, **36**, no. 4.
24. Impallomeni, A., Maffei, P. 1965, *Mem. Soc. astr. Ital.*, **36**, no. 4.
25. Hogg, A. R. 1965, *Publ. astr. Soc. Pacif.*, **77**, 440.
26. Lyngå, G. 1964, *Acta Univ. Lund.*, II, no. 13 = *Medd. Lund. Obs.*, II, no. 140.
27. Czernik, M. 1966, *Acta Astr.*, **16**, 93.
28. King, I. 1966, *Publ. astr. Soc. Pacif.*, **78**, 81.
29. Eggen, O. J., Sandage, A. R. 1964, *Astrophys. J.*, **140**, 130.
30. Kinman, T. D. 1965, *Astrophys. J.*, **142**, 655.
31. Gray, B. 1965, *Astr. J.*, **70**, 362.
32. Smak, J. 1964, *Astrophys. J.*, **139**, 1095.
33. Aveni, A. F. 1966, *Astrophys. J.*, **144**, 666.
34. Becker, W. 1966, *Z. Astrophys.*, **64**, 77.
35. Serkowski, K. 1965, *Astrophys. J.*, **141**, 1340.
36. McCarthy, M. F., Treanor, P. J. 1964, *Ric. astr. Spec. Vatic.*, **6**, 535.
37. McCarthy, M. F., Treanor, P. J., Ford, W. K. 1966, *Colloquium on Late-Type Stars*, Trieste, June, in press.
38. Roxburgh, I. W., Sargent, W. L. W., Strittmatter, P. A. 1966, *Observatory*, **86**, 118.
39. Weaver, H. F., Ebert, A. 1964, *Publ. astr. Soc. Pacif.*, **76**, 6.
40. Greenstein, J. L. 1964, *Astrophys. J.*, **140**, 544.
41. Efremov, U. N. 1964, *Variable Stars*, **15**, 242.

42. Starikova, G. A. 1965, *Astr. Zu.*, **42**, 1057.
43. Lavdovsky, V. V. 1965, *Izv. Pulkova*, **22**, 138.
44. Lavdovsky, V. V. 1963, *Trud. Pulkova*, II, 73.
45. Vasilevskis, S., Sanders, W. L., Balz, A. G. A. 1965, *Astr. J.*, **70**, 797.
46. Walker, M. F. 1956, *Astrophys. J. Suppl.*, **2**, 365.
47. Strand, K. Aa. 1958, *Astrophys. J.*, **128**, 14.
48. Vasilevskis, S., Sanders, W. L., van Altena, W. F. 1965, *Astr. J.*, **70**, 806.
49. Sharov, A. S. 1965, *Astr. Zu.*, **42**, 1045.
50. Sharov, A. S. 1965, *Soobsch. astr. Inst. Sternberg.*, no. 142, 20.
51. Barkhatova, K. A., Zinina, M. B. 1965, *Astr. Circ.*, no. 314.
52. Gray, D. F. 1963, *Astr. J.*, **68**, 572.
53. Martinet, L. 1964, *Publ. Obs. Genève*, no. 70.
54. Martinet, L. 1965, *ibid.*, no. 71.
55. Martinet, L. 1966, *ibid.*, no. 72.
56. Bouvier, P. 1964, *IAU Symposium no.* 25, p. 57.
57. Bouvier, P. 1965, *Publ. Obs. Genève*, no. 71, 13.
58. Veltmann, V. I. 1964, *Publ. Tartu Astr. Obs.*, **34**, 99.
59. Bouvier, P. *Bull. astr.*, Paris, in press.
60. King, I. 1965, *Astr. J.*, **70**, 376.
61. King, I. 1966, *ibid.*, **71**, 64.
62. Howard, W. E., Westerhout, G. 1965, *Astr. J.*, **70**, 679.
63. Schwartz, R. *Forschungsberichte des Landes Nordheim–Westfalen*, in press.
64. Lyngå, G. 1965, *Acta Univ. Lund*, II, no. 3 = *Medd. Obs. Lund*, II, no. 142.
65. Alter, G. 1965, *Z. Astrophys.*, **60**, 288.
66. Williams, I. P. 1964, *Mon. Not. R. astr. Soc.*, **128**, 389.
67. Reddish, V. C. 1964, *Observatory*, **84**, 73.
68. Golay, M. 1964, *Publ. Obs. Genève*, no. 68.
69. Penston, M. V. 1964, *Observatory*, **84**, 141.
70. Iben, I., Talbot, R. J. 1966, *Astrophys. J.*, **144**, 968.
71. Walker, M. F. 1962, *Adv. in Electronics*, **16**, 34.
72. Dluzhnevskaya, O. B. 1965, *Astrofizika*, **1**, 235.
73. Kholopov, P. N. 1965, *Astr. Zu.*, **42**, 148.
74. Kholopov, P. N. 1965, *ibid.*, **42**, 369.
75. Arp, H. 1965, *Star and Stellar Systems*, Vol. V., Chapter 19 (Edited by Kuiper and Middlehurst).
76. van den Bergh, S. *Comm. D. Dunlap Obs.*, in press.
77. Private Communication.
78. Koehler, J. A. 1965, *Observatory*, **85**, 197.
79. Private Communication.
80. Hodge, P. W. 1965, *Astrophys. J.*, **141**, 308.
81. Rousseau, J. 1964, *Ann. Astrophys.*, **27**, 681.
82. King, I. R. 1966, *Astr. J.*, **71**, 276.
83. Simoda, M., Kimura, H. 1966, *Goddard Space Flight Center, NASA, Preprint.*
84. Sandage, A. 1954, *Astr. J.*, **59**, 162.
85. Faulkner, J., Iben, I. 1966, *Astrophys. J.*, **144**, 995.
86. Kron, G. E. 1966, *Publ. astr. Soc. Pacif.*, **78**, 461.
87. de Vaucouleurs, G. 1953, *Mon. Not. R. astr. Soc.*, **113**, 134.
88. Gascoigne, S. C. B., Burr, E. J. 1956, *Mon. Not. R. astr. Soc.*, **116**, 570.
89. Kron, G. E., Mayall, N. U. 1960, *Astr. J.*, **65**, 581.
90. Blagikh, A. P. 1966, *Soviet Astr.* (Eng. trans.), **10**, 108.
91. Kholopov., P. N. 1955, *Astr. Circ.*, **32**, 4.
92. Sandage, A., Katem, B. 1964, *Astrophys. J.*, **139**, 1088.
93. Sandage, A., Smith, L. L. 1966, *Astrophys. J.*, **144**, 886.
94. Arp, H. 1965, *Astrophys. J.*, **141**, 1.
95. Sandage, A., Walker, M. 1966, *Astrophys. J.*, **143**, 313.
96. *Annual Report Mt. Wilson and Pal.*, 1964–65, p. 23.

97. Cuffey, J. 1965, *Astr. J.*, **70**, 732.
98. Cuffey, J. 1966, *Astr. J.*, **71**, 215.
99. Woolley, R. 1966, *Greenwich Annals*, **2**.
100. Dickens, R. J., Saunders, J. 1965, *R. Obs. Bull.*, no. 101.
101. Geyer, E. H. Private Communication.
102. Clube, S. V. M. 1965, *R. Obs. Bull.*, no. 95.
103. Ciatti, F., Rosino, L., Sussi, M. G. 1965, *3rd Colloquium on Variable Stars*, Bamberg.
104. Private Communication.
105. Andrews, P. J. Private Communication.
106. Gascoigne, S. C. B. Private Communication.
107. Searle, L., Rodgers, A. W. 1966, *Astrophys. J.*, **143**, 809.
108. Helfer, H. L., Wallerstein, G., Greenstein, J. L. 1959, *Astrophys. J.*, **129**, 700.
109. Pagel, B. E. J. Private Communication.
110. Harding, G. A. 1966, *R. Obs. Bull.*, no. 99.
111. Murray, C. A., Jones, D. H. P., Candy, M. P. 1965, *R. Obs. Bull.*, no. 100.
112. Clube, S. V. M. 1966, *R. Obs. Bull.*, no. 111.
113. Matsumani, N. 1964, *Publ. astr. Soc. Jap.*, **16**, 141.
114. Hénon, M. 1964, *Ann. Astrophys.*, **27**, 83.
115. Hénon, M. 1965, *Ann. Astrophys.*, **28**, 62.
116. Hénon, M. 1966, *C. r. Acad. Sci. Paris.*, **262**, 666.
117. Wilkens, H. 1965, *Mitt. veränd. Sterne.*, Nr. 3.
118. Kurochkin, N. E. 1964, *Var. Stars*, **15**, 164.
119. Fourcade, C. R., Laborde, J. R. 1965, *Mem. Soc. astr. Ital.*, **36**, 417.
120. Fourcade, C. R., Laborde, J. R. 1964, *Bol. Inst. Mat. Astr. Fis., Córdoba*, **1**, 6.
121. Fourcade, C. R., Laborde, J. R. 1966, *Mem. Soc. astr. Ital.*, **37**, 251.
122. Szeidl, B. 1965, *Mitt. Budapest*, Nr. 58.
123. Wilkens, H. 1964, *Mitt. veränd. Sterne.*, Nr. 5.
124. Makarova, V. A., Akimova, V. P. 1965, *Var. Stars*, **15**, 350.
125. Margoni, R. In press.
126. Sandage, A., Smith, L. L., Norton, R. H. 1966, *Astrophys. J.*, **144**, 894.
127. Rosino, L. 1966, *Astrophys. J.*, **144**, 903.
128. Margoni, R. 1965, *3rd Colloquium on Variable Stars*, Bamberg.
129. Sawyer Hogg, H., Welhau, A. 1966, *Publ. D. Dunlap Obs.*, II, 17.
130. Rosino, L., Ciatti, F. In press.
131. Feast, M. W. 1965, *Observatory*, **85**, 16.
132. Feast, M. W. 1966, *Observatory*, **86**, 120.
133. Feast, M. W. Private communication.
134. Lohmann, W. 1965, *Astr. Nachr.*, **289**, 1/2.
135. McCarthy, M. F., Treanor, P. J. 1964, *Ric. astr. Spec. Vatic.*, **6**, 511.
136. Roeder, R. C. 1965, *Astr. J.*, **70**, 793.
137. Kholopov, P. N. 1965, *Soviet Astr.* (Eng. trans.), **9**, 290.
138. Vogt, N. 1966, *Forschungsberichte des Landes Nordheim–Westfalen.*
139. Goldstein, S. J. 1964, *Astrophys. J.*, **140**, 802.
140. Vorontsov-Velyaminov, B. A. 1966, *Soviet Astr.* (Eng. trans)., **10**, 1.
141. Lallemand, A., Canavaggia, R., Amiot, F. 1966, *C. r. Acad. Sci. Paris*, **262**, 838.

40. COMMISSION DE LA RADIOASTRONOMIE

PRÉSIDENT: Professor M. Ryle, Cavendish Laboratory, Free School Lane, Cambridge, England.

VICE-PRÉSIDENT: Dr J. P. Wild, Radiophysics Laboratory, C.S.I.R.O., University Grounds, Chippendale, N.S.W., Australia.

COMITÉ d'ORGANISATION: A. Boischot, F. T. Haddock, F. J. Kerr, C. A. Muller, Y. N. Parijsky.

MEMBRES: Akabane, Atanasijević, Avignon, Baldwin, Baltá, Barrett, Blum, Bolton, Bracewell, Brown (R. Hanbury), Burbidge (G. R.), Cardus, Carr, Christiansen, Chvojkova, Cohen, Costain, Coutrez, Daene, Davies (J. G.), Davies (R. D.), de Groot, Denisse, Dewhirst, Dodson-Prince, Drake, Edmondson, Elwert, Eriksen, Field, Findlay, Firor, Fokker, Friedman, Galt, Ginzburg, Gold, Gorgolewski, Grahl, Hachenberg, Hack, Hagen, Harang, Harrower, Heeschen, Heidmann, Hewish, Hey, Hill, Howard (W. E.), Ikhsanova, Kahn, Kakinuma, Kardashev, Kazès, Khaikin, Korolkov, Kraus, Kundu, Laffineur, Lequeux, Le Squeren, Letfus, Lilley, Little (A. G.), Locke, Lovell, McClain, McGee (R. X.), McKinley, MacRae, Macris, McVittie, Mannino, Maxwell, Mayer, Menon, Mills, Minkowski, Moriyama, Oort, Osterbrock, Pick-Gutmann, Priester, Rabben, Righini, Roberts (J. A.), Rydbeck, Sanamian, Scheuer, Schmidt (M.), Scott (P. F.), Seeger, Shakeshaft, Shklovsky, Simon (P.), Slee, Smerd, Smith (A. G.), Smith (F. G.), Smith (Harlan J.), Soboleva, Stahr-Carpenter, Steinberg, Swarup, Takakura, Thompson (A. R.), Tlamicha, Townes, Troitsky, Tuominen, Van Bueren, Van de Hulst, Van Woerden, Varsavsky, Vitkevitch, Wade, Walsh, Weaver, Westerhout, Yen, Zhelezniakov.

A. INTRODUCTION

Radio observations are now relevant in so many branches of astronomy that I have followed the example of Dr J. F. Denisse and asked certain members of the Commission to prepare reviews covering each of the main fields. J. P. Wild has written on solar radio astronomy and the interplanetary medium, J. E. Baldwin and F. J. Kerr on the galactic continuum and spectral line radiation respectively, J. R. Shakeshaft on extragalactic radio observations and cosmology, and C. A. Muller on instrumentation.

Radio studies of the planets are included in the report to Commission 16, and those of the Moon in the report to Commission 17. By agreement with Dr R. L. Minkowski, his report to Commission 28 covers *optical* work on extragalactic radio sources and also, as an exception, both optical and radio studies of the Magellanic Clouds.

Such is the rate of growth of the subjects covered by our Commission, that there is not space enough for a critical survey of the field. In the following sections, therefore, we have tried to draw attention to all the relevant literature even though this has meant only a very brief discussion of the results achieved. It would be helpful if the Commission were to consider, during the General Assembly, the form which the Report might most usefully take in the future.

The striking progress during the past three years has been possible because of the existence of clear frequency bands; a number of recent events have emphasized the fact that our work will only continue if we are able to maintain effective protection against interfering transmissions and other disturbing influences. The organization set up to allocate frequencies for radio astronomy and for the registration of frequencies in use by astronomers has been described by Smith-Rose (1964) and Smith and Smith-Rose (1966), but its success depends on the active participation of individual astronomers, both through their national authorities and internationally.

Attention must also be paid to other forms of activity which may give rise to man-made radiation capable of masking the natural radio signals. The radiation caused by high energy electrons released above the atmosphere by a high-altitude nuclear explosion has already caused concern.

The continuing expansion of radio astronomy is also demonstrated by a number of reports discussing future national and international programmes (e.g. Gold and Messel, 1964; Whitford *et al.* 1964; Report of Fleck Committee, 1965).

I should like to draw attention to the obituary (Lovell, 1964) of Joe Pawsey—President of Commission 40 for the 1955 and 1958 General Assemblies, who was not only a great pioneer in our subject but a great leader of men.

Finally I must express my sincere appreciation to the contributors to this report, and especially to J. R. Shakeshaft who has done much of the editorial work.

References

Fleck, Lord *et al.* 1965, *Radio Astronomy*, H.M. Stationery Office, London.

Gold, T., Messel, H. 1964, *Nature,* **204,** 18.

Lovell, A. C. B. 1964, *Biog. Mem., Roy. Soc.*, **10,** 229.

Smith-Rose, R. L. 1964, *Nature,* **203,** 7.

Smith, F. G., Smith-Rose, R. L. 1966, *IUCAF Document* no. 87, *URSI Information Bulletin* no. 155, p. 49.

Whitford, A. E. *et al.* 1964, *Ground based Astronomy, a ten-year programme,* Nat. Acad. Sci., Nat. Res. Council, Washington, D.C.

B. SOLAR RADIO EMISSION

(prepared by J. P. Wild)

1. *Introduction*

This report covers papers published during the three years between late 1963 and late 1966, a period of low solar activity.

The period has been one of intensive studies of the interplanetary medium by satellite techniques, which are themselves outside the field of this report. But three new techniques of radio astronomy are now being developed to extend these studies closer to the Sun and to the region in which the corona proper merges with the interplanetary medium where the solar wind blows, *viz.*: the use of scintillations of radio sources of small angular diameter to study the irregularities and hence, with the aid of plausible assumptions, the electron density itself (**1**); the extension of spectrographic observations of type III solar bursts to frequencies below 1 MHz by means of satellite-borne equipment (**2, 3**); and the use of solar radar to detect the solar wind (**4**) and study irregularities (**5, 6, 7**).

Special mention may also be made of the emergence of a profound dilemma regarding radio observations and the coronal magnetic field. One finds results obtained by different methods leading to grossly differing values, which roughly sort themselves into two groups with values (at heights of half to one solar radius) of (*a*) $\lesssim 1$ gauss, (*b*) $\gtrsim 10$ gauss. Group (*a*) is supported by the interpretation of type II bursts as disturbances of high enough magnetic Mach number to maintain themselves as shock waves (**8, 9, 10**), by the supposition that the newly discussed 'chains' of type I bursts correspond to movements at the Alfvén velocity (**11**), and by simple *a priori* arguments on the likely coronal effects of sunspot fields. Group (*b*) is supported by gyro-synchrotron theories of type IV and other bursts, by Takakura's theory of type I bursts (**12**), and by the Sturrock-Tidman formula for band-splitting in type II bursts (**13**). The resolution of this question is vital to both coronal studies and the understanding of the physical processes of bursts.

2. *Reviews*

Comprehensive text-books have been published on solar radio emission by Zhelezniakov (**14**) and Kundu (**15**), the former dealing equally with observational and theoretical aspects, the latter with greater emphasis on the observed phenomena which are described in great detail. Review papers have appeared on solar radio astronomy in general (**16, 17**), on the quiet Sun (**18**), on radio investigations of the outer corona and interplanetary medium (**19**), on the slowly-varying component (**18, 20, 21, 22**), on the burst component (**23, 24, 25, 26, 27**) and its generation (**28, 29, 30**), on the study of the corona from burst observations (**31**), on the solar flare phenomenon at radio frequencies (**32**), on narrow-band spectral studies of bursts (**33, 34**), on noise storms (**35**) and type IV bursts (**36**), on microwave bursts (**37**), on the relation of bursts with X-rays (**38, 39, 40**), on eclipse observations (**41**) and on solar radar (**42**).

3. *The Quiet Sun*

Determinations of the Sun's brightness temperature have been made at wavelengths of 1·3 mm (**43**), 1·8 mm (**44**), 3·2 mm (**45**), 4 mm (**46**), 1·2 cm (**47, 48**), 1·8 cm (**49**), 2 cm (**48**). The basic thermal component has been investigated at centimetre and decimetre wavelengths (**50, 51**) and the coronal component calculated at 9 cm (**52**). The ellipticity of the Sun has been investigated at 3·2 cm (**53**) and the two-dimensional brightness distribution at various wavelengths (see *Slowly-varying component*). A polarization technique has been used to estimate the Sun's general magnetic field (**54**).

Quiet-sun theory has been developed and applied to make further evaluations of the coronal temperature, yielding the markedly different values of 2×10^6 °K (**55**) and $6{\cdot}4 \times 10^5$ °K (**56**). A temperature model of the lower chromosphere has been derived from observations at millimetre wavelengths (**57**).

Numerous results were published relating to the study of the chromosphere and corona (including plage regions) by eclipse observations: 1954 June 30 (**58**); 1961 February 15 (**59**); 1962, July 31 (annular eclipse) (**60, 61**); 1962 February 5 (**62**); 1963 July 20 (**63, 64, 65, 66, 67** and **68**); 1965 May 30 (**69**); and conclusions based on more than one eclipse (**70, 71**).

Further investigations have been made of the outer corona (interplanetary medium or 'super corona') based on observations of its occultation of the Crab Nebula (**72, 73, 74, 75, 76, 77, 78, 79, 80, 81, 82, 83, 84, 85**) and other sources out to 80 solar radii (**86**). A new method has been developed for studying the interplanetary medium by the observation of radio sources of very small angular size (**1**).

4. *Active centres and the slowly-varying component*

A detailed theory of the origins of the slowly-varying component in terms of the bremsstrahlung and synchroton radiating mechanisms has been given (**87**).

High resolution maps have been published and discussed at wavelengths of 3 mm (**45**) and 8 mm (**88, 89**), and other observations of active regions at 8 cm (**90**), 32 cm (**91**), 1·3 m (**92**) and 1·5 m (**93**). Spectra of individual centres have been obtained from observations of seven wavelengths between 2 and 21 cm (**94**).

Active centres responsible for radio emission at 3 cm wavelength have been found to possess distinctive optical properties—bipolar spots with penumbra large compared with their separation (**95**).

A bipolar structure has been found in certain polarized active centres at 3 cm (**96**). Further papers have been published on the general characteristics and interpretation of the active regions (**97, 98**); on the statistical variation of the slowly-varying component with sunspot number (**99, 100**); on an apparent east-west symmetry in the emissions (**101, 102**); and on the application of ray-tracing to investigate the effects of coronal scattering (**103**).

Statistical analyses have been made of the variation of the Sun's microwave radiation with the 11-year solar cycle (**104, 105, 106, 107, 108, 109**), the 27-day period of rotation (**110, 111, 112**) and other periodicities (**113, 114**). The intensity (**115, 116**), spectral slope (**117**) and position shifts (**118**) of the slowly-varying component have been found to be criteria for the occurrence of proton and type IV events; also the intensity for the occurrence of SID's (**119**). The inexactness of the correlation between the intensity and the Sun's total radiation in the extreme ultra-violet has been explored (**120**).

5. *Microwave bursts*

Extensive data on microwave bursts obtained at the Heinrich Hertz Institute have been published (**121**), and an investigation carried out of the microwave spectrum (**122**). Investigations have been made into the relation of microwave bursts with X-rays (**123, 124**) and its interpretation (**125, 126, 127, 128**), and various aspects of flares (**129, 130, 131, 132**), including type II and type IV bursts and SWF's (**131, 132**).

The wave-interaction theory for interpreting the polarization of microwave bursts has been re-examined (**133**). A theory has been given by which filaments of flares might give rise to accelerated particles and radio emission at multiples of the gyro frequency (**134**).

6. *Noise storms and type I bursts*

The elusive origin of type I bursts has been discussed in terms of a theory involving plasma radiation from electrons accelerated by two Alfvén wave packets (**12**), of the relative roles of incoherent synchrotron radiation and plasma radiation (**135**), and of instabilities arising in the interaction between coronal electron streamers and MHD shock waves (**136**). A new line of evidence on the origin of the bursts, relating to the tendency of the bursts to appear in long 'chains' with slow frequency drift, has been explored (**11, 137**). Some general conclusions drawn from 169 MHz observations with the Nançay interferometer (**138**) include the result that the majority of storms are initiated by flares. Observations have been reported on the polarization (**139, 140**), and narrow-band spectrum (**34, 141**) of the bursts, on their directivity (**142**) (which is found to be broader than that of the background continuum), on their relation with active regions responsible for proton flares (**143**), and on weak bursts with the aid of special equipment (**34, 144**). A single-frequency classification of storms has been given (**145, 146**).

7. *Type II bursts*

Theories have been given to interpret some of the main characteristics of type II bursts: the nature of the causal shock wave (**9, 13, 135, 147**), the intensities of fundamental and harmonic and the phenomenon of band-splitting (**9, 13, 135**), and inferences on the coronal magnetic field (**8, 13, 148**).

Other theoretical work, relevant to the same topic but not specifically related to type II bursts, has been published on the emissivity of a plasma at fundamental and harmonic frequencies (**149**), on the interaction of plasma and electromagnetic waves (**150, 151**), and on the growth and ultimate limitation of unstable electron plasma oscillations (**152**).

Comprehensive observations on the nature and velocity of the type II sources have been reported (**8**), and it has been shown that certain features (in particular 'herring-bone' structure) is sometimes highly polarized (**153**). Conclusions have been drawn on the dimensions of the exciting agency (**154**) and inferences made on harmonic structure and band-splitting from observations at a low (29 MHz) single frequency (**155**). The association between flares and type II bursts has been investigated and found to increase at the minimum of the solar cycle (**156**).

8. *Type III and type V bursts*

Spectral observations have been used to investigate the speed and acceleration of the sources of type III and type V bursts (**157**) and the general characteristics and interpretation of type V bursts (**158**).

A detailed investigation of the properties of type III and V bursts and their relation with microwave bursts has been made using single-frequency data in the metre and decametre range extending over a five-year period (**159, 160**). Low-frequency observations have been described in the 20–30 MHz range (**161**), at a fixed frequency of 26 MHz (**162**), at five fixed frequencies between 6 and 35 MHz (**163**), and at frequencies between 0·6 and 10 MHz obtained with the Alouette satellites (**2, 3**). The occurrence of type III bursts has been found to be associated with filaments (**164**) and to be correlated with the width of the Hα line of the flare (**163**). Further derivations from type III observations have been made on the height and scatter of the burst sources (**165**), other properties including the electron temperature of the corona (**166**), and the variation of electron density above a coronal streamer (**167**).

The majority of type III bursts at 74 and 200 MHz have been found to be partially elliptically polarized, and estimates of the coronal magnetic field have been deduced by comparing the Faraday rotation at the two frequencies (**168**). The bursts have been found to be more strongly polarized in their sharp short-lived features than in their diffuse background features, suggesting a two-component structure (**169**). A study of polarization with respect to the position of sources on the disk has suggested an origin in the differential absorption of two wave modes at the harmonics of the gyro frequency (**170**). Polarization data at low frequencies (23·5 and 39·6 MHz) have been tabulated and discussed (**171**).

Further contributions have been made to the theory of type III bursts in terms of electron streams (**172, 173**), and the escape of an initially isotropic burst of electrons in the vicinity of magnetically neutral planes has been shown to give rise to an outward-travelling pulse of electrons consistent with that required to explain a type III burst (**174**).

The origin of reverse-drift pairs connected with type III bursts has been discussed (**135**). A new type of burst consisting of two closely-spaced narrow-band components (distinct from drifting pairs) has been reported (**175**).

A well-documented case (**176**) was reported of 18 MHz type III bursts being observed at night-time. They were observed during daylight hours at Boulder and Hawaii, at 7 a.m. at Manila and midnight at Rome. The result was published in 1966, on April 1.

9. *Type IV bursts*

Two detailed observational studies of type IV bursts have been described: one gives data on the position, movement, angular size and polarization of 24 type IV sources at metre wavelengths (15–210 MHz) and establishes a clear distinction between 'stationary' and 'moving' phenomena (**177**); the other gives details of the complex polarization behaviour of 18 events (**178**) and discusses their origin in terms of alternative theories of gyro and plasma radiation (**179**). Two further examples of the comparatively rare 'moving' type IV bursts have been described (**180, 162**). An unusual form of fine-structure in type IV bursts has been noted; it consists of 'spikes' (superimposed on the continuum) which are preceded by a 'dip', interpreted as an absorption effect (**181**).

A model of a type IV burst has been proposed (**182**) to account for an E–W asymmetry in time of onset (**188**). The synchrotron theory of type IV bursts has been discussed and the distinction between decimetre-wave and metre-wave ('storm continuum') components emphasized (**183**). The influence of the medium on synchrotron radiation from relativistic electrons has been examined (**184**).

A careful investigation has been made of the optical characteristics of active centres and flares that produce type IV bursts: specific types of configuration have been found to be most likely to produce the bursts (**186, 187**). A study of type IV-associated flares has been made with data covering the last 11-year cycle (**185**). Further optical relationships have been deduced from onset times of the different burst components (**188, 189**); the relation of the bursts with the 'flare nimbus' phenomenon has been discussed (**190**) The detailed relationships have been explored of type IV bursts with geomagnetic storms (**191, 192**) and with PCA events (**191**), and of type IV-producing active centres with recurrent geomagnetic-storm centres (**193**). Further lists of type IV events have been compiled (**194, 195, 196**).

The dynamic spectrum of several large outbursts (not only of type IV) have been described in detail: 12 events in the 2–3 metre band and their connection with flares, recorded in Crimea (**197**); one event in the metre-decimetre range, and its connection with X-rays and protons, recorded at Fort Davis (**198**); and two events in the range 5–2000 MHz recorded at Dapto (**199**). The tendency for outstanding solar events, including 200 MHz outbursts, to avoid the peak of the solar cycle has been noted (**200**).

10. *Radar Observations*

The 38·26 MHz radar observations at El Campo taken between 1961 and 1964 have been reported in detail (**4, 5, 201**). They indicate systematic outward motions, consistent with solar wind theory, of 16 km s^{-1} in the corona at a calculated height of 0·5 $R_{\odot}$ above the photosphere, and random mass motions of about 40 km s^{-1} at quiet times, increasing by a factor of 2 to 3 at other times. A coronal model can be found that gives consistency between the radar measurements and probe measurements of the solar wind by Mariner II (**6**). Allowance for the effects of scattering at coronal irregularities has been considered theoretically (**7**).

11. *Instrumental developments*

Descriptions have been given of radiometers operating at wavelengths of 8 mm (**202**) and 3 cm (**203, 204**); spectrographs operating in the frequency range 200–2000 MHz (parametric up-converter) (**199, 205**), 160–320 MHz (60-channel) (**206**), 35–250 MHz (simultaneous sweep over 5 bands), 7·6–41 MHz (interferometric) (**207**), and 200 kHz–6 MHz (satellite-borne) (**208**); interferometers at 225 MHz (**209**) and 136 MHz (**210**); high-resolution multi-element instruments at frequencies of 9·4 GHz (**211**), 5 GHz (**212**), 2·7 GHz (**213, 214**), 408 MHz (**215**) and 240 MHz (**216**); a radioheliograph giving high-resolution 80 MHz pictures at intervals of 1 second (**217**); and a fast narrow-band polarimeter at 74 MHz (**218**). Systems using conical scanning have been described for locating microwave centres in the Sun and measuring polarization (**219, 220, 221, 222**). The design of radiometers for precise measurements of the solar radio flux has been discussed (**223**).

The long series of Canadian observations of 10 cm solar flux has now been extended to three Canadian observatories to extend the hours of operation (**214**). The east-west aperture of the El Campo solar radar antenna has been doubled (**224**).

BIBLIOGRAPHY

1. Hewish, A., Scott, P. F., Wills, D. 1964, *Nature*, **203**, 1214; Hewish, A., Dennison, P. A., Pilkington, J. D. H. 1966, *Nature*, **209**, 1188.
2. Hartz, T. R. 1964, *Ann. Astrophys.*, **27**, 831.
3. Nelms, G. L., Barrington, R. E., Belrose, J. S., Hartz, T. R., McDiarmid, I. B., Brace, L. H. 1966, *Can. J. Phys.*, **44**, 1419.
4. James, J. C. 1964, *IEEE Trans. Ant. Propag.*, AP-12, 876.
5. Chisholm, J. H., James, J. C. 1964, *Trans. Amer. Geophys. Union*, **45**, 352.

6. Brandt, J. C. 1964, *Science*, **146**, 1671.
7. Merkulenko, V. E. 1964, *Geomagn. Aeronom.*, **4**, 825.
8. Weiss, A. A. 1965, *Austr. J. Phys.*, **18**, 167.
9. Zaitsev, V. V. 1965, *Astr. Zu.*, **42**, 740 = 1966, *Soviet Astr.*, **9**, 572.
10. Fomichev, V. V., Chertok, I. M. 1965, *Astr. Zu.*, **42**, 1256.
11. Wild, J. P., Tlamicha, A. 1964, *Nature*, **203**, 1128; 1965, *Bull astr. Inst. Csl.*, **16**, 73.
12. Takakura, T. 1963, *Publ. astr. Soc. Japan*, **15**, 462.
13. Tidman, D. A. 1965, *Planet. Space Sci.*, **13**, 781.
14. Zhelezniakov, V. V. 1964, *Radio emission from the Sun and Planets*, Izd. 'Nauka', Moscow, pp. 560.
15. Kundu, M. R. 1965, *Solar Radio Astronomy*, pp. 660, Interscience Publishers, New York.
16. Smerd, S. F. 1964, *Research in Geophysics*. Vol. 1, *Sun, upper atmosphere and space*. Ed. Odishaw, H., MIT Press, Cambridge, Mass., pp 65–97.
17. Boischot, A., Denisse, J. F. 1964, *Adv. Electronics Electron Phys.*, **20**, 147.
18. Christiansen, W. N. 1966, *Progress in Radio Science* 1960–1963, Vol. 5, *Radio Astronomy*. Ed. Herbays, E. *et al.*, Elsevier Publ. Co., Amsterdam, pp. 45–63.
19. Hewish, A. 1965, *Solar System Radio Astronomy*. Ed. Aarons, J., Plenum Press, New York, pp. 255–266.
20. Hachenberg, O. 1965, *Solar System Radio Astronomy*. Ed. Aarons, J., Plenum Press, New York, pp. 95–108.
21. Swarup, G. 1964, AAS/NASA Symposium on the *Physics of Solar Flares*. Ed. Hess, W. N., NASA, Washington, D.C., pp. 179–198.
22. Pick, M. 1965, *Solar System Radio Astronomy*. Ed. Aarons, J., Plenum Press, New York, pp. 81–93.
23. Maxwell, A. 1965, *The Solar Spectrum*. Ed. de Jager, C., D. Reidel Publ. Co., Dordrecht-Holland, pp. 342–397.
24. Maxwell, A., Defouw, R. J., Cummings, P. 1964, *Planet. Space Sci.*, **12**, 435.
25. Wild, J. P. 1964, AAS/NASA Symposium on the *Physics of Solar Flares*. Ed. Hess, W. N., NASA, Washington, D.C., pp. 161–177.
26. Warwick, J. W. 1965, *Solar System Radio Astronomy*. Ed. Aarons, J., Plenum Press, New York, pp. 131–170.
27. Hatanaka, T. 1966, *Progress in Radio Science* 1960–1963. Vol. 5, *Radio Astronomy*. Eds. Herbays, E. *et al.*, Elsevier Publ. Co., Amsterdam, pp. 70–78.
28. Smerd, S. F. 1965, *The Solar Spectrum*. Ed. de Jager, C., D. Reidel Publ. Co., Dordrecht-Holland, pp. 398–407.
29. Smerd, S. F. 1964, AAS/NASA Symposium on the *Physics of Solar Flares*. Ed. Hess, W. N., NASA, Washington, D.C., pp. 343–355.
30. Schatzman, E. 1965, *The Solar Spectrum*. Ed. de Jager, C., D. Reidel Publ. Co., Dordrecht-Holland, pp. 313–341.
31. Takakura, T. 1966, *Space Sci. Rev.*, **5**, 80.
32. McLean, D. J. 1965, *Solar System Radio Astronomy*. Ed. Aarons, J., Plenum Press, New York, pp. 117–130.
33. Elgaroy, O. 1965, *Solar System Radio Astronomy*. Ed. Aarons, J., Plenum Press, New York, pp. 201–223.
34. de Groot, T. 1966, *Rech. Astr. Obs. Utrecht*, **18**, 1.
35. Fokker, A. D. 1965, *Solar System Radio Astronomy*. Ed. Aarons, J., Plenum Press, New York, pp. 171–199.
36. Kundu, M. R. 1965, *The Solar Spectrum*. Ed. de Jager, C., D. Reidel Publ. Co., Dordrecht-Holland, pp. 408–412.
37. Hachenberg, O. 1965, *Solar System Radio Astronomy*. Ed. Aarons, J., Plenum Press, New York, pp. 241–254.
38. de Jager, C. 1965, *Introduction to Solar Terrestrial Relations*. Ed. Ortner, J. *et al.*, D. Reidel Publ. Co., Dordrecht-Holland, pp. 86–95.
39. Kundu, M. R. 1963, *Space Sci. Rev.*, **2**, 438.
40. Kundu, M. R. 1964, AAS/NASA Symposium on the *Physics of Solar Flares*. Ed. Hess, W. N., NASA, Washington, D.C., pp. 335–342.

41. Castelli, J. P., Aarons, J. 1965, *Solar System Radio Astronomy*. Ed. Aarons, J., Plenum Press, New York, pp. 49–79.
42. Eshleman, V. R. 1965, *Solar System Radio Astronomy*. Ed. Aarons, J., Plenum Press, New York, pp. 267–293.
43. Fedoseev, L. I. 1963, *Izv. Vyssih Učeb. Zaved. Radiofiz.*, **6**, 655.
44. Naumov, A. I. 1963, *Izv. Vyssih Učeb. Zaved. Radiofiz.*, **6**, 848.
45. Simon, M. 1965, *Astrophys. J.*, **141**, 1513.
46. Kislyakov, A. G., Plechkov, V. M. 1964, *Izv. Vyssih Učeb. Zaved. Radiofiz.*, **7**, 46.
47. Staelin, D. H., Barrett, A. H., Kusse, B. R. 1964, *Astr. J.*, **69**, 69.
48. Wulfsberg, K. N., Short, J. A. 1965, *U.S. Air Force Cambridge Res. Labs. Phys. Sci. Res. Papers* no. 77, AFCRL-65-75. pp. 7.
49. Tsuchiya, A., Nagane, K. 1965, *Publ. astr. Soc. Japan*, **17**, 86.
50. das Gupta, M. K., Basu, D. 1965, *Nature*, **208**, 739.
51. Olmr, J., Tlamicha, A. 1965, *Bull. astr. Inst. Csl.*, **16**, 250.
52. Barkov, V. I. 1965, *Astr. Zu.*, **42**, 749.
53. Borovik, V. N., Veisig, G. S. 1963, *Soln. Dann. Bjull.*, no. 10, 73.
54. Belov, I. F. 1965, *Izv. Vyssih Učeb. Zaved. Radiofiz.*, **8**, 400.
55. Oster, L., Sofia, S. 1965, *Astrophys. J.*, **141**, 1139.
56. Fokker, A. D. 1966, *Bull. astr. Inst. Netherl.*, **18**, 359.
57. Zheleznyakov, V. V. 1964, *Astr. Zu.*, **41**, 1021 = 1965, *Soviet Astr.*, **8**, 819.
58. Tiainen, P. O. 1964, *Ann. Acad. Sci. Fennicae*, Ser. A, VI, Phys., no. 149, pp. 8.
59. Drago, F. G., Noci, G., Piattelli, M. 1964, *Ann. Astrophys.*, **27**, 708.
60. Peterova, N. G., Molchanov, A. P., Nagnibeda, V. G. 1963, *Soln. Dann. Bjull.*, No. 8, 68.
61. Veisig, G. S., Molchanov, A. P. 1963, *Soln. Dann. Bjull.*, no. 11, 58.
62. Covington, A. E. 1963, *J. R. astr. Soc. Can.*, **57**, 253.
63. McNarry, L. R., Bhonsle, R. V., Viner, M. R. 1964, *J. R. astr. Soc. Can.*, **58**, **221**.
64. Kennedy, W. A. G., Gagnon, H. P. A., Covington, A. E. 1963, *Bull. Radio Elect. Engng. Div. NRC Canada*, **13**, 26.
65. Medd, W. J., Higgs, L. A. 1963, *Bull. Radio Elect. Engng. Div. NRC Canada*, **13**, 29.
66. Tolbert, C. W., Krause, L. C., Straiton, A. W. 1964, *Astrophys. J.*, **140**, 306.
67. Miner, R. J. 1963, U.S. Air Force Cambridge Res. Labs. Office Aerospace Res. Upper Atmos. Phys. Lab. Project 4603. Res. Rep. AFCRL-63-796. pp. 16.
68. Asper, H. K., Bracewell, R. N., Deuter, J., Krishnan, T., Picken, J., Rose, S. 1964. *J. geophys. Res.*, **69**, 1805.
69. Apushkinskii, G. P., Molchanov, A. P. 1965, *Soln. Dann. Bjull.*, No. 7, 61.
70. Castelli, J. P., Cohen, H. W., Straka, R. M., Aarons, J. 1963, *Icarus*, **2**, 317.
71. Molchanov, A. P. 1964, *Dokl. Akad. Nauk SSSR*, **158**, 302.
72. Hewish, A., Wyndham, J. D. 1963, *Mon. Not. R. astr. Soc.*, **126**, 470.
73. Ryzhkov, N. F., Pariiskii, Yu. N., Egorova, T. M., Gosachinskii, I. V., Bystrova, N. V. 1963, *Soln. Dann. Bjull.*, No. 7, 67.
74. Basu, S., Castelli, J. P. 1963, *IEEE Trans. Ant. Propag.*, AP-11, 614.
75. Clark, B. G., Wyndham, J. D. 1963, *Nature*, **200**, 766.
76. Hughes, M. P., Downes, D., Murray, C. 1964, *Astrophys. J.*, **139**, 404.
77. Erickson, W. C. 1964, *Astrophys. J.*, **139**, 1290.
78. Le Squeren-Malinge, A. M. 1964, *Ann. Astrophys.*, **27**, 183.
79. Vitkevich, V. V. 1964, *Dokl. Akad. Nauk SSSR*, **156**, 1065.
80. Gol'nev, V. Ya., Pariiskii, Yu. N., Soboleva, N. S. 1964, *Izv. glav. astr. Obs. Pulkove*, **23**, 22.
81. Vitkevich, V. V. 1964, *Astr. Zu.*, **41**, 684 = 1965, *Soviet Astr.*, **8**, 545.
82. Babii, V. I., Vitkevich, V. V., Vlasov, V. L., Gorelova, M. V., Sukhovei, A. G. 1965, *Astr. Zu.*, **42**, 107 = 1965, *Soviet Astr.*, **9**, 81.
83. Lotova, N. A. 1965, *Izv. Vyssih Učeb. Zaved. Radiofiz.*, **8**, 441.
84. Castelli, J. P. 1965, *Nature*, **205**, 1301.
85. Vitkevich, V. V. 1966, *Radiotekh. Elektron.*, **11**, 623.
86. Slee, O. B. 1966, *Planet. Space Sci.*, **14**, 255.
87. Zheleznyakov, V. V. 1963, *Astr. Zu.*, **40**, 829 = 1964, *Soviet Astr.*, **7**, 630.

88. Khangil'din, U. V. 1964, *Astr. Zu.*, **41**, 302 = 1964, *Soviet Astr.*, **8**, 234.
89. Khangil'din, U. V. 1965, *Trudy Fiz. Inst. P.N. Lebedeva Akad. Nauk SSSR*, **28**, 179.
90. Urbarz, H. W. 1966, *Nature*, **210**, 891.
91. Sholomitskii, G. B., Kuril'chik, V. N., Matveenko, L. I., Khromov, G. S. 1964, *Astr. Zu.*, **41**, 823.
92. Landini, M., Piatelli, M. 1964, *Atti Accad. naz. Lincei Rc.*, Ser. 8, **36**, 352.
93. Mannino, G., Impallomeni, A. G. 1963, *Mem. Soc. astr. ital.*, **34**, 1.
94. Achmedov, Sh., Borovik, V. N., Korzhavin, A. N., Nagnibeda, V. G., Peterova, N. G., Spitkovskii, V. M. 1966, *Soln. Dann. Bjull.*, No. 2, 62.
95. Avignon, Y., Martres, M. J., Pick, M. 1966, *Ann. Astrophys.*, **29**, 33; 1965, *C. r. Acad. Sci. Paris*, Group 3, **261**, 1465.
96. Tanaka, H., Steinberg, J. L. 1964, *Ann. Astrophys.*, **27**, 29.
97. Grebinskii, A. S., Molchanov, A. P. 1964, *Geomagn. Aeronom.*, **4**, 635.
98. Abbasov, A. R. 1965, *Vest. Leningrad. gos. Univ.*, Ser. Mat., Mekh. i Astr. No. 4, 140.
99. Soboleva, N. S. 1965, *Izv. glav. astr. Obs. Pulkove*, **24**, 73.
100. Borovik, V. N., Korzhavin, A. N., Peterova, N. G. 1965, *Soln. Dann. Bjull.*, no. 10, 67.
101. Fokker, A. D. 1963, *Bull. astr. Inst. Netherl.*, **17**, 214.
102. Fokker, A. D. 1963, *Bull. astr. Inst. Netherl.*, **17**, 219.
103. Fokker, A. D. 1965, *Bull. astr. Inst. Netherl.*, **18**, 111.
104. Tanaka, H. 1964, *Proc. Res. Inst. Atmos. Nagoya Univ.*, **11**, 41.
105. Kruger, A., Kruger, W., Wallis, G. 1964, *Z. Astrophys.*, **59**, 37.
106. das Gupta, M. K., Basu, D. 1964, *Nature*, **203**, 626.
107. Altschuler, M. D., Sastry, Ch. V. 1965, *Nature*, **206**, 1035.
108. Covington, A. E., Locke, J. L. 1965, *J. R. astr. Soc. Can.*, **59**, 179.
109. Gnevyshev, M. N. 1965, *Astr. Zu.*, **42**, 488 = 1965, *Soviet Astr.*, **9**, 387.
110. Nicolet, M. 1963, *J. geophys. Res.*, **68**, 6121.
111. Sastry, Ch. V. 1966, *J. atmos. terr. Phys.*, **28**, 117.
112. Tanaka, H., Kakinuma, T. 1966, *Rep. Ionosph. Space Res. Japan*, **20**, 22.
113. das Gupta, M. K., Basu, D. 1964, *J. atmos. terr. Phys.*, **26**, 135.
114. Belmont, A. D., Dartt, D. G., Ulstad, M. S. 1966, *J. atmos. Sci.*, **23**, 314.
115. Caroubalos, C. A. 1965, *Solar System Radio Astronomy*. Ed. Aarons, J., Plenum Press, New York, pp. 109–116.
116. Caroubalos, C. A. 1964, AAS/NASA Symposium on the *Physics of Solar Flares*. Ed. Hess, W. N., NASA, Washington, D.C., pp. 203–205.
117. Tanaka, H., Kakinuma, T. 1964, *Rep. Ionosph. Space Res. Japan*, **18**, 32.
118. Eliseev, G. F., Moiseev, I. G. 1965, *Izv. Krym. astrofiz. Obs.*, **34**, 3.
119. Mitra, A. P., Subrahmanyam, C. V., Karabin, M. 1964, *J. atmos. terr. Phys.*, **26**, 1138.
120. Anderson, A. D. 1964, *J. atmos. Sci.*, **21**, 1.
121. Hachenberg, O., Furstenberg, F., Helms, B., Kruger, A. 1963, *Solar outbursts in the IGY*, 2 Vols., Heinrich-Hertz-Inst., Berlin.
122. Hachenberg, O., Kruger, A. 1964, *Z. Astrophys.*, **59**, 261.
123. Kawabata, K. 1966, *Rep. Ionosph. Space Res. Japan*, **20**, 118.
124. Eryushev, N. N., Eliseeva, L. A. 1965, *Izv. Krym. astrofiz. Obs.*, **34**, 42.
125. Korchak, A. A. 1965, *Geomagn. Aeronom.*, **5**, 32.
126. Korchak, A. A. 1965, *Kosm. Issled.*, **3**, 751.
127. Korchak, A. A. 1965, *Geomagn. Aeronom.*, **5**, 601; Korchak, A. A. 1965, *Izv. Akad. Nauk SSR*, Ser. Fiz., **29**, 1813; Korchak, A. A., Ponomarenko, Yu. B. 1966, *Geomagn. Aeronom.*, **6**, 3.
128. Elwert, G. 1964, AAS/NASA Symposium on the *Physics of Solar Flares*. Ed. Hess, W. N., NASA, Washington, D.C., pp. 364–370.
129. Haurwitz, M. W. 1964, *Astrophys. J.*, **140**, 1236.
130. McKenna, S. M. P. 1965, *Mon. Not. R. astr. Soc.*, **129**, 437.
131. Harvey, G. A. 1964, *Astrophys. J.*, **139**, 16.
132. Harvey, G. A. 1965, *J. geophys. Res.*, **70**, 2961.
133. Zheleznyakov, V. V., Zlotnik E. Ya. 1964, *Astr. Zu.*, **40**, 485.

134. Moore, R. L., Johnston, J. R. 1964, AAS/NASA Symposium on the *Physics of Solar Flares*. Ed. Hess, W. N., NASA, Washington, D.C., pp. 371–376.
135. Zheleznyakov, V. V. 1965, *Astr. Zu.*, **42,** 244 = 1965, *Soviet Astr.*, **9,** 191.
136. Trakhtengerts, V. Yu. 1966, *Astr. Zu.*, **43,** 356.
137. Hanasz, J. 1966, *Austr. J. Phys.*, **19,** 635.
138. Le Squeren, A. M. 1964, AAS/NASA Symposium on the *Physics of Solar Flares*. Ed. Hess, W. N., NASA, Washington, D.C., pp. 199–202.
139. McNarry, L. R. 1964, *Bull. Radio Elect. Engng. Div. NRC Canada,* **14,** 30.
140. Tsuchiya, A. 1963, *Publ. astr. Soc. Japan,* **15,** 368.
141. Tsuchiya, A. 1962, *Contr. Dep. astr. Univ. Tokyo,* **13,** 92.
142. Yurovskaya, L. I. 1964, *Izv. Krym. astrofiz. Obs.*, **32,** 76.
143. Levitskii, L. S. 1965, *Izv. Krym. astrofiz. Obs.*, **34,** 16.
144. Durasova, M. S., Lavrinov, G. A., Shumkina, V. M., Yudin, O. I. 1964, *Geomagn. Aeronom.*, **4,** 938.
145. Tlamicha, A., Krivsky, L., Olmr, J. 1964, *Bull. astr. Inst. Csl.*, **15,** 49.
146. Tlamicha, A., Olmr, J. 1964, *Bull. astr. Inst. Csl.*, **15,** 133.
147. Pikel'ner, S. B., Gintsburg, M. A. 1963, *Astr. Zu.*, **40,** 842 = 1964, *Soviet Astr.*, **7,** 639.
148. Takakura, T. 1964, AAS/NASA Symposium on the *Physics of Solar Flares*. Ed. Hess, W. N., NASA, Washington, D.C., pp. 383–385.
149. Sturrock, P. A., Ball, R. H., Baldwin, D. E. 1965, *Physics Fluids,* **8,** 1509.
150. Terashima, Y., Yajima, N. 1963, *Progress theor. Phys.*, Osaka, **30,** 443.
151. Hughes, W. F., Young, F. J. 1964, *J. Fluid Mech.*, **19,** 11.
152. Drummond, W. E., Pines, D. 1964, *Ann Phys.*, New York, **28,** 478.
153. Stewart, R. T. 1966, *Austr. J. Phys.*, **19,** 209.
154. Moiseev, I. G. 1963, *Izv. Krym. astrofiz. Obs.*, **30,** 141.
155. Krishnamurthi, M., Sastry, G. S., Rao, T. S. 1965, *Austr. J. Phys.*, **18,** 473.
156. Drago, F. G., Tagliaferri, G. L. 1966, *Z. Astrophys.*, **63,** 202.
157. Stewart, R. T. 1965, *Austr. J. Phys.*, **18,** 67.
158. Weiss, A. A., Stewart, R. T. 1965, *Austr. J. Phys.*, **18,** 143.
159. Daene, H., Kruger, A. 1966, *Astr. Nachr.*, **289,** 105.
160. Daene, H., Kruger, A. 1966, *Astr. Nachr.*, **289,** 117.
161. Riihimaa, J. J. 1963, *Ann. Acad. Sci. Fennicae*, Ser. A, VI, no. 131, p. 1.
162. Malitson, H. H., Erickson, W. C. 1966, *Astrophys. J.*, **144,** 337.
163. Artem'eva, G. M., Benediktov, E. A., Rapoport, V. O. 1965, *Astr. Zu..*, **42,** 1011 = 1966 *Soviet Astr.*, **9,** 780.
164. Tlamicha, A., Takakura, T. 1963, *Nature,* **200,** 999; Takakura, T., Tlamicha, A. 1964, AAS/NASA Symposium on the *Physics of Solar Flares*. Ed. Hess, W. N., NASA, Washington, D.C., pp. 207–208.
165. Morimoto, M. 1964, *Publ. astr. Soc. Japan,* **16,** 163.
166. Elgaroy, O., Rodberg, H. 1963, *Astrophys. norw.*, **8,** 271.
167. Morimoto, M. 1963, *Ann. Tokyo astr. Obs.*, **8,** 125.
168. Bhonsle, R. V., McNarry, L. R. 1964, *Astrophys. J.*, **139,** 1312.
169. Rao, U. V. G. 1965, *Austr. J. Phys.*, **18,** 283.
170. Enome, S. 1964, *Publ. astr. Soc. Japan,* **16,** 135.
171. Daene, H., Voigt, W. 1965, *Publ. astrophys. Obs. Potsdam,* **31,** 1.
172. Sturrock, P. A. 1964, AAS/NASA Symposium on the *Physics of Solar Flares*. Ed. Hess, W. N., NAS.A, Washington, D.C., pp. 357–363.
173. Baldwin, D. E 1964, *Phys. Lett.*, **12,** 202.
174. Weiss, A. A., Wild, J. P. 1964, *Austr. J. Phys.*, **17,** 282.
175. Ellis, G. R. A., McCulloch, P. M. 1966, *Nature,* **211,** 1070.
176. Fortini, T., Garcia, C., Gay, M., Glover, F. 1966, *J. geophys. Res.*, **71,** 1938.
177. Weiss, A. A. 1963, *Austr. J. Phys.*, **16,** 526.
178. Kai, K. 1965, *Publ. astr. Soc. Japan,* **17,** 294.
179. Kai, K. 1965, *Publ. astr. Soc. Japan,* **17,** 309.
180. Philip, K. W. 1964, *Astrophys. J.*, **139,** 723.
181. Aller, H. D., Jensen, C. J., Malville, J. M. 1966, *Nature,* **209,** 1014.

182. Sakurai, L. 1965, *J. geophys. Res.*, **70**, 3235.
183. Böhme, A. 1964, *Beitr. Plasmaphys.*, no. 3, 57.
184. Zhelezniakov, V. V., Trakhtengerts, V. Yu. 1965, *Astr. Zu.*, **42**, 1005 = 1966, *Soviet Astr.*, **9**, 775.
185. Krivsky, L., Kruger, A. 1966, *Bull. astr. Inst. Csl.*, **17**, 243.
186. Avignon, Y., Martres, M. J., Pick, M. 1964, *Ann. Astrophys.*, **27**, 23.
187. Avignon, Y., Caroubalos, C., Martres-Tropé, M. J., Pick, M. 1964, AAS/NASA Symposium on the *Physics of Solar Flares*. Ed. Hess, W. N., NASA, Washington, D.C., pp. 25–26.
188. Sakurai, K. 1964, *Rep. Ionosph. Space Res. Japan*, **18**, 366.
189. Akin'yan, S. T., Dolginova, Yu. N. 1965, *Results of Researches on the Program of the International Geophysical Year*. Izd. 'Nauka', Moscow, pp. 183–198.
190. Reid, J. H. 1963, *Dunsink Obs. Publ.*, **1**, 91.
191. Bell, B. 1963, *Smithson. Contrib. Astrophys.*, **8**, 119.
192. Böhme, A. 1963, *Monatsber. Deutschen Akad. Wiss. Berlin*, **6**, 51.
193. Lapointe, S. M. 1964, *Ann. Géophys.*, **20**, 454.
194. Fritzová-Svestková, L., Hrebík, F. 1964, *Bull. astr. Inst. Csl.*, **15**, 222.
195. Kai, K. 1965, *Ann. Tokyo astr. Obs.*, **9**, 195.
196. Svetska, A., Olmr, J. 1966, *Bull. astr. Inst. Csl.*, **17**, 4.
197. Yurovskaya, L. I. 1965, *Izv. Krym. astrofiz. Obs.*, **34**, 9.
198. Maxwell, A. 1963, *Planet. Space Sci.*, **11**, 897.
199. Suzuki, S., Attwood, C. F., Sheridan, K. V. 1964, *Observatory*, **84**, 55.
200. das Gupta, M. K., Basu, D. 1965, *J. atmos. terr. Phys.*, **27**, 1029.
201. Chisholm, J. H., James, J. C. 1964, *Astrophys. J.*, **140**, 377.
202. Efanov, V. A., Moiseev, I. G. 1965, *Izv. Krym. astrofiz. Obs.*, **34**, 53.
203. Koch, G. 1963, *Jena Rev.*, **8**, 259.
204. 1963, *Alta Frequenza*, **32**, 863.
205. Suzuki, S., Attwood, C. F., Sheridan, K. V. 1966, *IEEE Trans. Ant. Propag.* AP-14, 91.
206. van Nieuwkoop, J. 1964, *Nature*, **201**, 380.
207. Lee, R. H., Warwick, J. W. 1964, *Radio Science*, **68D**, 807.
208. Donegan, R., Chomet, M. 1966, *IEEE Int. Convention Record*, **14**, 121.
209. Bertini, S., Piattelli, M. 1962, *Ricerca Sco. Suppl.*, **1**, 96.
210. Smith, J. R. 1965, *J. Br. astr. Ass.*, **75**, 87.
211. Tanaka, H., Kakinuma, T. 1965, *Proc. Res. Inst. Atmos. Nagoya Univ.*, **12**, 27.
212. Tanaka, H. 1966, *Proc. Res. Inst. Atmos. Nagoya Univ.*, **13**, 49.
213. Covington, A. E., Legg, T. H. 1965, *Bull. Radio Elect. Engng. Div. NRC Canada*, **15**, 6.
214. Covington, A. E., Legg, T. H., Bell, M. B. 1966, *Bull. Radio Elect. Engng. Div. NRC Canada*, **16**, 9.
215. Clavelier, B. 1966, *C. r. Acad. Sci. Paris*, Ser. B, **262**, 225.
216. Daene, H., Paschke, J., Voigt, W. 1965, *Monatsber. Deutschen Akad. Wiss. Berlin*, **7**, 120.
217. Wild, J. P. 1965, *Proc. R. Soc. London*, Ser. A, **286**, 499.
218. Bhonsle, R. V., McNarry, L. R. 1964, *Can. J. Phys.*, **42**, 292.
219. Belov, I. F. 1964, *Geomagn. Aeronom.*, **4**, 1106.
220. Molchanov, A. P. 1964, *Izv. glav. astr. Obs. Pulkove*, **23**, 215.
221. Molchanov, A. P., Vyatkina, V. M. 1964, *Izv. glav. astr. Obs. Pulkove*, **23**, 162.
222. Borovik, V. N. 1963, *Soln. Dann. Bjull.*, no. 12, 60.
223. Yurovskii, Yu. F. 1965, *Izv. Krym. astrofiz. Obs.*, **34**, 60.
224. Devane, M. E. 1965, Mass. Inst. Technol. Lincoln Lab. Group Rep. 1965-12, pp. 20.

C. CONTINUUM RADIATION FROM THE GALAXY

(prepared by J. E. Baldwin)

1. *Introduction*

Since 1963 the main lines of development in studies of the continuum radiation from the Galaxy have been the extension of surveys to both lower and higher frequencies than previously employed, accurate measurements of the galactic radio spectrum including the measurements at low frequencies from artificial satellites, attempts to relate this spectrum to observations of the flux of cosmic ray electrons at the Earth, observations of the polarization of the galactic continuum, attempts to determine the strength and form of the galactic magnetic field from polarization studies and other evidence, investigations of the nature of the North Galactic Spur, studies of H II regions, planetary nebulae, supernova remnants and the radio source near the galactic nucleus.

Many topics of current interest were discussed in the Symposium on Radio and Optical Studies of the Galaxy (Mt Stromlo Observatory 1966) and at the IAU Symposium No. 31 on Radio Astronomy and the Galactic System at Noordwijk in August 1966 (to be published). A number of papers relevant to cosmic ray electrons in the Galaxy were presented at the Ninth International Conference on Cosmic Rays, London 1965.

2. *Background Surveys*

The state of knowledge at the beginning of this period has been summarized by Mills (1964) and Pawsey (1965) and a compilation of the earlier surveys was made by Altenhoff *et al.* (1963). Continuum surveys covering large areas of sky have been completed in the southern hemisphere at 4·7 MHz by Ellis *et al.* (1963) and Ellis and Hamilton (1966*a*), at 30 MHz by Mathewson *et al.* (1965), at 144 MHz by Mackrell (1963), at 408 MHz by Komesaroff (1966) and at 960 MHz by Nicolson (1965). In the northern hemisphere similar extensive surveys have been made at 38 MHz by Kenderdine (1963*a*) and Williams *et al.* (1966), at 327 MHz by Braccesi and Vespignani (1964), at 400 MHz by Seeger *et al.* (1965) and at 610 MHz by Moran (1965). Smaller regions of the northern sky have been surveyed with somewhat higher resolution at 178 MHz by Crowther and Clarke (1966), at 610 MHz by MacLeod *et al.* (1965), at 600 and 1415 MHz by Dixon *et al.* (1965) and Scheer and Kraus (1966) and at 3200 MHz by Raghava Rao *et al.* (1965). In the southern sky Beard (1966*b*) made observations at 2650 MHz whilst more restricted observations were made by Klemperer and Ochs (1965) at 50 MHz and by Malumian (1963) at 960 MHz.

Further analysis of the thermal component of the radiation at low galactic latitudes has been discussed by Braude and Vaisberg (1964*a*), Ellis and Hamilton (1964) and (1966*b*), Razin and Khizhnyakov (1965) and Ariskin (1965). The broad features of the distribution of ionized hydrogen throughout the galactic disk seem now to be well established. By contrast, the distribution of non-thermal radiation in the galactic halo is not at all well established (Large 1966, Turner and Burke 1965) but observational discrepancies may soon be resolved. Observations of nearby spiral galaxies (Mills and Glanfield 1965) have been invoked to support the view that the Galaxy may have a radio halo.

3. *Spectrum of the Continuum radiation*

Considerable efforts have been made since 1963 to determine the galactic radio spectrum with some accuracy and, for observations with fairly low resolving powers, brightness temperatures accurate to better than 5% have now been determined at several frequencies. For observations with high resolving power it is doubtful whether accuracies of brightness temperatures of 10% have yet been realized at any frequency.

At the lowest frequencies, below about 10 MHz, much use has been made of observations with single dipoles mounted on artificial satellites. Observations have been reported in the frequency range 0·2–10 MHz by Hartz (1964), Harvey (1965), Hugill and Smith (1965), Huguenin *et al.* (1964), Huguenin and Papagiannis (1965), Lund (1965), Slysh (1965), Alexander and Stone (1965), Benediktov *et al.* (1965) and a discussion given of the results expected for various models of the Galaxy by Smith (1965). Lenchek (1964) discusses the effects of the solar H II region on the low frequency spectrum. These observations, and also ground based observations at low frequencies (Ellis, G. R. A 1964, 1965, Parthasarathy and Lerfald 1965) show a considerable scatter and the details of the low frequency spectrum have still to be established. Dipole observations at higher frequencies (10–100 MHz) now yield satisfactory agreement between those in the northern hemisphere (Andrew 1966) and those in the southern hemisphere (Wielebinski and Yates 1965, Yates and Wielebinski 1965, 1966). Measurements with scaled antennas of higher gain have been made over the frequency range 13–404 MHz by Turtle *et al.* (1962), Purton (1966) and Bridle (1967) and with comparable resolving power by Braude and Vaisberg (1964*b*). Recent observations by Penzias and Wilson (1966) at 4080 MHz offer the prospect of extending the galactic spectrum to much higher frequencies. The preliminary values suggest that the spectral variation of brightness temperature $T \propto \nu^{-2.4}$ at frequencies below 100 MHz steepens to a variation of perhaps $\nu^{-2.9}$ at the highest frequencies. An interpretation of the spectrum in relation to the galactic magnetic field has been given by Turtle (1963).

4. *Polarization of the continuum radiation and the galactic magnetic field*

Following the detection of polarization of the galactic background radiation, extensive surveys have been completed in both the northern and southern hemispheres. In the north surveys have been made at 408 MHz by Berkhuijsen and Brouw (1963), at 610 MHz by Berkhuijsen *et al.* (1964), at 408 MHz by Wielebinski and Shakeshaft (1964) and at 1407 MHz by Bingham (1966) and in the south at 408 MHz by Mathewson and Milne (1965) and at 408 and 620 MHz by Mathewson *et al.* (1966). The concentration on a great circle of areas of high polarization has been noted by Mathewson and Milne (1964). Faraday rotation in the interstellar medium leads to depolarization of the galactic continuum as discussed by Razin and Khroulyov (1965*a*, 1965*b*). Reviews of the present state of galactic polarization studies are given by Gardner (1966) and Gardner and Whiteoak (1966).

The detection of Zeeman splitting in absorption features in the 21 cm line of hydrogen was claimed by Davies *et al.* (1963*a*, 1963*b*, 1963*c*) but observations by Weinreb (1962) did not support this claim and Verschuur (Noordwijk Symposium) has now withdrawn the original results. Another attempt to measure the strength of the field by Faraday rotation in the 21 cm absorption spectrum of the Crab Nebula was unsuccessful (Morris *et al.* 1963). From the first studies of the rotation measures of extragalactic sources, Morris and Berge (1964) deduced the structure of the local magnetic field and this has been extended by Gardner and Davies (1966) and Whiteoak (1966) while Maltby (1965) has discussed the variation of source polarization with galactic latitude. The structure of the field in the local spiral arm has also been discussed by Mathewson (1966) and a detailed model of its form deduced from an analysis by Hornby (1966).

The strength of the magnetic field has again been the subject of speculation (Sironi 1965, Felten 1966). More general discussion of the physics of the magnetic field in the Galaxy have been given by Woltjer (1965*a*, 1965*b*).

5. *The galactic spurs*

Continuum surveys specifically aimed at elucidating the nature of the North Galactic Spur (Haslam *et al.* 1964, Large *et al.* 1966, Quigley and Large 1966, Davies 1964) have lent support to the belief that this object may be a nearby supernova remnant. Attempts to observe optical emission lines from the region of the spur (Davies *et al.* 1963*d*, Lozinskaya and Shcheglov 1965) were unsuccessful but there appears to be some correlation of the spur with neutral hydrogen (Lozinskaya 1964, Meaburn 1965*a*). Quigley and Haslam (1965) found that the spurs could be represented by arcs of small circle on the sky. Meaburn (1965*b*) found evidence of optical emission lines close to the Cetus arc. Ellis, F. E. (1965) has made first epoch observations to detect proper motion of the spur. The possible association of the North Galactic Spur with the Sco X-ray source was discussed by Shklovsky (1964) and by Quigley and Haslam (1964). Quite different interpretations of the spurs as a feature of the local spiral arm (Rougoor 1966) or as the result of instabilities in the local magnetic field (Parker 1965) have been put forward and their true nature has not yet been determined with certainty.

6. *Galactic radio sources*

Work in this field has expanded very considerably covering H II regions, planetary nebulae, supernova remanants, the galactic nucleus and flare stars.

Catalogues of sources have been prepared from surveys (Wilson 1963, Bennett 1964) and observations of numbers of sources made over the wavelength range 0·8–21 cm (Barrett *et. al.* 1965, Golnev *et al.* 1965, Nguyen-Quang-Rieu 1963, Baars *et al.* 1965*a*, 1965*b*).

Observations of several H II regions are discussed by Menon (1964), Higgs and Thibodeau (1965), Terzian (1965), and Le Marne (1966*a*). Studies have also been made of individual objects—the Rosette nebula (Botinelli and Gouguenheim 1964), W49 (Akabane and Kerr 1965), NGC 3603 (Sher 1965), the Carina nebula (Beard 1966*a*), W51 and W43 (Bystrova *et al.* 1963*a*, 1963*b*), W3, W4 and W51 (Bourgois *et al.* 1965) and Orion and M17 (Kuzmin and Salomonovich 1963). Considerable attention has been paid to the Cygnus X region with studies by Pike and Drake (1964), Yang and West (1964), Higgs *et al.* (1964), Veron (1965), Higgs (1966), Dickel *et al.* (1966), Downes and Rinehart (1966) and Wendker (1966). Observations of the Cygnus Y region have been completed by Ko (1963) and Yang and Dickel (1965).

Detection of the thermal radiation from planetary nebulae has been described by Golnev and Parijsky (1964), Khromov (1964, 1965), Khromov *et al.* (1965), Menon and Terzian (1965), Davies *et al.* (1965), Ehman (1965), Kaftan-Kassim (1965, 1966), Osterbrock (1965), Slee and Orchiston (1965), Le Marne (1966*b*) and Terzian (1966).

Supernova remnants continue to play an important role observationally as objects of study and theoretically as sources of cosmic rays. Reviews of the subject include those by Minkowski (1964, 1965*a*), Hoyle (1965), Shklovsky (1966*a*) and Boischot and Lequeux (1964).

The Crab nebula is the most comprehensively observed remnant. Its known spectrum has now been extended to short centimetric and millimetric wavelengths (Lazarevskii *et al.* 1963, Troitsky *et al.* 1964, Staelin *et al.* 1964, Tolbert and Straiton 1965, Tolbert 1965) and has been discussed by Branson (1965) and Shklovsky (1966*b*). Gordon (1965, 1966) has discussed the expected secular variation of its flux density. Lunar occultations have frequently been employed to study its angular structure (Andrew *et al.* 1964, Matveenko *et al.* 1965, Matveenko and Khromov 1965, Matveenko 1966*a*, Matveenko and Artuh 1966, Lastochkin *et al.* 1965, Krishnan and Zisk 1966, Gotwols *et al.* 1966, Taylor 1966, Davies *et al.* 1966 and Artuh *et al.* 1966). Evidence from interferometric measures (Hewish and Okoye 1964) and from interplanetary scintillation (Hewish and Okoye 1965, Malitson *et al.* 1966) have revealed the existence at long wavelengths of a very small source of high brightness in the Crab nebula. Kronberg (1966) has described a possibly similar source at 1422 MHz. More details of the polarization of the Crab nebula have been obtained at centimetric wavelengths (Golnev *et al.* 1964, Gardner 1965, Matveenko 1966*b*). Further discussions of the nature of the Crab nebula have been presented by Johnson (1963), Gould (1965) and Shklovsky (1965).

Penzias and Wilson (1965) supply a 4080 MHz point on the spectrum of Cas A and a number of other measurements are included in papers on radio source spectra referred to in the report on extragalactic radio astronomy. Its spectrum is discussed by Malumian (1966), secular variations in its flux density by Lastochkin and Stankevich (1964), Sanamian and Aslanian (1965) and Sanamian (1966) and measurements of the distribution of brightness over it by Ryle *et al.* (1965) and Zisk (1966*a*, 1966*b*). A further suggestion concerning its history has been put forward by Minkowski (1966*a*). The distance of Tycho Brahe's supernova has been measured (Menon and Williams 1966) and subsequently modified (Menon, Noordwijk Symposium 1966) and an occultation of Kepler's supernova has been observed (Talen 1965).

Among the older supernova remnants which have received attention are the Cygnus Loop (Kenderdine 1963*b*, Poveda 1965, Ramana 1966), IC 443 (Howard and Dickel 1963, Hogg 1964, Higgs 1965), W44 (Scheuer 1963, Dagkesamansky and Korchak 1963, Bystrova *et al.* 1963*c*, 1965, Hollinger and Hobbs 1966, Bondar *et al.* 1965), HB21 (Crowther 1965), HB3 and HB9 (Bazelyan *et al.* 1966). In addition to studies of well known remnants, searches for previously unknown remnants have been vigorously pursued. Several have been suggested by Davies (1963) and Hogg (1966) and a number of individual identifications have been made – W22 (Heidmann and Heidmann 1964, Heidmann 1965), W28 (Courtès *et al.* 1964), a source in Auriga (Dickel *et al.* 1965), OA 184 (Dickel and Yang 1965), Vela X (Milne 1966) and the supernova of A.D. 1006 (Goldstein 1965, Goldstein and Yoke 1965, Gardner and Milne 1965, Minkowski 1965*b*, 1966*b*). Mathewson *et al.* (1963) have discussed a supernova remnant in the Large Magellanic Cloud. Ancient records of novae and supernovae have been reviewed by Xi Ze-Zong and Po Shu-jen (1966). A low upper limit has been set to the flux density of Nova Herculis by Haddock *et al.* (1963).

Work on the region of the nucleus of the Galaxy has been reviewed by Burke (1965). Recent observations of this region have been mainly at short wavelengths (Ryzhkov *et al.* 1964, Maxwell and Downes 1964, Downes *et al.* 1965, Hollinger 1965, Sinclair 1966, Kerr and Sinclair 1966) but some observations have also been made at longer wavelengths (Malumian 1965, Little 1966).

Studies of flare stars have now shown good correlation between the optical and radio events. (Lovell *et al.* 1963, Slee *et al.* 1963, Lovell and Chugainov 1964, Lovell *et al.* 1964, Lovell 1964, Migach 1965, Lovell and Solomon 1966). The nature of the radio emission from flare stars has been discussed by Slysh (1964).

A search for radio emission from the star α Orionis was unsuccessful (Kellerman and Pauliny-Toth 1966).

7. *Cosmic rays and the Galaxy*

The relation of the galactic radio emission to the cosmic rays and their origin remains one of the central problems of galactic research. Considerable progress has been made through the detection and measurement of cosmic ray electrons and positrons, by the spectral measurements discussed in section 3 and through theoretical analyses of the production and behaviour of cosmic ray electrons in the Galaxy.

Synchrotron radiation and the origin of cosmic rays have recently been reviewed by Ginzburg and Syrovatsky (1965*a*, 1965*b*) and Ginzburg (1965).

Measurements of the cosmic ray electron energy spectrum up to energies of 100 BeV have been achieved (Agrinier *et al.* 1965, Shong *et al.* 1964, L'Heureux and Meyer 1965, Daniel and Stephens 1965, Meyer 1965, Bland *et al.* 1966) giving reasonably good agreement with the observed radio spectrum for a magnetic field of about 10^{-5} gauss. Further evidence of the sidereal anisotropy of cosmic rays has been presented (Jacklyn 1966).

Several authors have analysed the generation of cosmic ray electrons (Lequeux 1963, Tsytovich 1963, 1964, Ginzburg and Syrovatsky 1964, Pollack and Fazio 1965, Abraham *et al.* 1966, Spergel and Scanlon 1966, Kurilchik 1966) particular attention being paid to their origin as secondaries, which is now shown to be unlikely. The physical processes acting on the electrons to modify their spectrum have also been discussed (Getmantsev and Tokarev 1964, Ohtrami and Kogure 1964, Souffrin and Chesnay 1965, Ramaty and Lingenfelter 1966*a*,

1966*b*, Daniel and Stephens 1966). The motion in the galactic magnetic field of the cosmic ray particles generated in supernovae may produce observational effects such as the galactic spurs (Oda and Hasegawa 1963). Another possibility is that the interaction of the cosmic ray particles with the field and the inter-stellar gas may produce spur shaped protuberances in the field (Parker 1965, 1966*a*, 1966*b*).

A new radio astronomical technique for the detection of extensive air showers has been developed (Smith *et al.* 1965, Jelley 1965). Radio pulses at a wavelength of 6·8 m have been observed coinciding with the arrival of showers of energy 10^{16} – 10^{17} eV.

BIBLIOGRAPHY

Abraham, P. B., Brunstein, K. A., Cline, T. L. 1966, *Phys. Rev.*, **150**, 1088.

Agrinier, B., Koechlin, Y., Parlier, B., Vasseur, J., Bland, C. J., Boella, G., Degli Antoni, G., Dilworth, C., Scarsi, L., Sironi, G. 1965, Proc. Int. Conf. Cosmic Rays, p. 331.

Akabane, K., Kerr, F. J. 1965, *Austr. J. Phys.*, **18**, 91.

Alexander, J. K., Stone, R. G. 1965, *Astrophys. J.*, **142**, 1327.

Altenhoff, W., Haslam, C. G. T., Large, M. I., Moran, M., Wendker, H. 1963, *Mitt. astr. Inst. Univ. Münster*, no. 9, 1.

Andrew, B. H. 1966, *Mon. Not. R. astr. Soc.*, **132**, 79.

Andrew, B. H., Branson, N. J. B. A., Wills, D. 1964, *Nature*, **203**, 171.

Ariskin, V. I. 1965, *Astr. Zu.*, **42**, 939.

Artuh, V. S., Vitkevich, V. V., Vlasov, V. I., Kafarov, G. A., Matveenko, L. I. 1966, *Astr. Zu.*, **43**, 13.

Baars, J. W. M., Mezger, P. G., Wendker, H. 1965*a*, *Z. Astrophys.*, **61**, 134.

Baars, J. W. M., Mezger, P. G., Wendker, H. 1965*b*, *Astrophys. J.*, **142**, 122.

Barrett, A. H., Kutuza, B. G., Matveenko, L. I., Salomonovich, A. E. 1965, *Astr. Zu.*, **42**, 527.

Bazelyan, L. L., Braude, S. Ya., Men, A. V. 1966, *Astr. Zu.*, **43**, 740.

Beard, M. 1966*a*, Sympos. on Radio and Opt. Studies of the Galaxy, Mt. Stromlo, p. 58.

Beard, M. 1966*b*, *Austr. J. Phys.*, **19**, 141.

Benediktov, E. A., Getmantsev, G. G., Sazonov, Yu. A., Tarasov, A. F. 1965, *Kosm. Issled.*, **3**, 617.

Bennett, A. S. 1964, *Mon. Not. R. astr. Soc.*, **127**, 3.

Berkhuijsen, E. M., Brouw, W. N. 1963, *Bull. astr. Inst. Netherl.*, **17**, 185.

Berkhuijsen, E. M., Brouw, W. N., Muller, C. A., Tinbergen, J. 1964, *Bull. astr. Inst. Netherl.*, **17**, 465.

Bingham, R. G. 1966, *Mon. Not. R. astr. Soc.*, **134**, 327.

Bland, C. J., Boella, G., Antoni, G. D., Dilworth, C., Scarsi, L., Sironi, G., Agrinier, B., Koechlin, Y., Parlier, B., Vasseur, J. 1966, *Phys. Rev. Lett.*, **17**, 813.

Boischot, A., Lequeux, J. 1964, *Ann. Astrophys.*, **27**, 514.

Bondar', L. N., Krotikov, V. D., Stankevich, K. N., Tseitlin, N. M. 1965, *Izv. Vyssih. Učeb. Zaved. Radiofiz.*, **8**, 437.

Botinelli, L., Gougenheim, L. 1964, *Ann. Astrophys.*, **27**, 685.

Bourgois, G., Eskioglu, A. N., Kazès, I., Nguyen-Quang-Rieu 1965, *Notes Inf. Publ. Obs. Paris*, Fasc. 27.

Bracessi, A., Vespignani, G. 1964, *Nuovo Cim.*, **31**, 310.

Branson, N. J. B. A. 1965, *Observatory*, **85**, 250.

Braude, S. Ya., Vaisberg, V. V. 1964*a*, *Izv. Vyssih. Učeb. Zaved. Radiofiz.*, **7**, 193.

Braude, S. Ya., Vaisberg, V. V. 1964*b*, *Izv. Vyssih. Učeb. Zaved. Radiofiz.*, **7**, 1032.

Bridle, A. H. 1967, *Mon. Not. R. astr. Soc.*, in press.

Burke, B. F. 1965, *A. Rev. Astr. Astrophys.*, **3**, 275.

Bystrova, N., Gosatchinsky, I., Egorova, T., Ryzhkov, N. 1963*a*, *Astr. Tsirk.*, no. 245.

Bystrova, N., Gosatchinsky, I., Egorova, T., Ryzhkov, N. 1963*b*, *Astr. Tsirk.*, no. 245.

Bystrova, N. V., Gosatchinsky, I. V., Egorova, T. M., Ryzhkov, N. F. 1963*c*, *Astr. Tsirk.*, no. 269.

Bystrova, N. V., Gosatchinsky, I. V., Egorova, T. M., Ryzhkov, N. F. 1965, *Izv. glav. astr. Obs. Pulkove*, **24**, 202.
Courtès, G., Véron, P., Viton, M. 1964, *Ann. Astrophys.*, **27**, 330.
Crowther, J. H. 1965, *Observatory*, **85**, 110.
Crowther, J. H., Clarke, R. W. 1966, *Mon. Not. R. astr. Soc.*, **132**, 405.
Dagkesamansky, R. D., Korchak, A. A. 1963, *Astr. Zu.*, **40**, 582.
Daniel, R. R., Stephens, S. A. 1965, *Phys. Rev. Lett.*, **15**, 769.
Daniel, R. R., Stephens, S. A. 1966, *Phys. Rev. Lett.*, **17**, 935.
Davies, R. D. 1963, *Observatory*, **83**, 172.
Davies, R. D. 1964, *Mon. Not. R. astr. Soc.*, **128**, 173.
Davies, R. D., Shuter, W. L. H., Slater, C. H., Verschuur, G. L., Wild, P. A. T. 1963*a*, *Mon. Not. R. astr. Soc.*, **126**, 343.
Davies, R. D., Shuter, W. L. H., Slater, C. H., Wild, P. A. T. 1963*b*, *Mon. Not. R. astr. Soc.*, **126**, 353.
Davies, R. D., Shuter, W. L. H. 1963*c*, *Mon. Not. R. astr. Soc.*, **126**, 369.
Davies, R. D., Brown, R. H., Meaburn, J. E. 1963*d*, *Observatory*, **83**, 179.
Davies, R. D., Gardner, F. F., Hazard, C., Mackey, M. B. 1966, *Austr. J. Phys.*, **19**, 409.
Davies, J. G., Ferriday, R. J., Haslam, C. G. T., Moran, M., Thomasson, P. 1965, *Nature*, **206**, 809.
Dickel, J. R., Yang, K. S. 1965, *Astrophys. J.*, **142**, 1642.
Dickel, J. R., McGuire, J. P., Yang, K. S. 1965, *Astrophys. J.*, **142**, 798.
Dickel, H. R., Yang, H. S., Dickel, J. R. 1966, *Astrophys. J.*, **143**, 218.
Dixon, R. S., Meng, S. Y., Kraus, J. D. 1965, *Nature*, **205**, 755.
Downes, D., Maxwell, A., Meeks, M. L. 1965, *Nature*, **208**, 1189.
Downes, D., Rinehart, R. 1966, *Astrophys. J.*, **144**, 937.
Ehman, J. R. 1965, *Astr. J.*, **70**, 674.
Ellis, F. E. 1965, *Astr. J.*, **70**, 674.
Ellis, G. R. A. 1964, *Nature*, **204**, 171.
Ellis, G. R. A. 1965, *Mon. Not. R. astr. Soc.*, **130**, 429.
Ellis, G. R. A., Green, R. J., Hamilton, P. A. 1963, *Austr. J. Phys.*, **16**, 545.
Ellis, G. R. A., Hamilton, P. A. 1964, *Nature*, **204**, 272.
Ellis, G. R. A., Hamilton, P. A. 1966*a*, *Astrophys. J.*, **143**, 227.
Ellis, G. R. A., Hamilton, P. A. 1966*b*, *Astrophys. J.*, **146**, 78.
Felten, J. E. 1966, *Astrophys. J.*, **145**, 589.
Gardner, F. F. 1965, *Austr. J. Phys.*, **18**, 385.
Gardner, F. F. 1966, Sympos. on Radio and Opt. Studies of the Galaxy, Mt Stromlo, p. 19.
Gardner, F. F., Davies, R. D. 1966, *Austr. J. Phys.*, **19**, 129.
Gardner, F. F., Milne, D. K. 1965, *Astr. J.*, **70**, 754.
Gardner, F. F., Whiteoak, J. B. 1966, *A. Rev. Astr. Astrophys.*, **4**, 245.
Getmantsev, G. G., Tokarev, Yu. V. 1964, *Geomagn. Aeronom.*, **4**, 243.
Ginzburg, V. L. 1965, *Astr. Zu.*, **42**, 1129.
Ginzburg, V. L. Syrovatsky, S. I. 1964, *Astr. Zu.*, **41**, 430.
Ginzburg, V. L. Syrovatsky, S. I. 1965*a*, *A. Rev. Astr. Astrophys.*, **3**, 297.
Ginzburg, V. L., Syrovatsky, S. I. 1965*b*, Proc. Int. Conf. Cosmic Rays, p. 53.
Goldstein, B. R. 1965, *Astr. J.*, **70**, 105.
Goldstein, B. R., Yoke, H. P. 1965, *Astr. J.*, **70**, 748.
Golnev, V. Y., Parijsky, Y. N. 1964, *Astr. Zu.*, **41**, 846.
Golnev, V. Y., Parijsky, Y. N., Soboleva, N. S. 1964, *Izv. glav. astr. Obs. Pulkove*, **23**, 17.
Golnev, V. Y., Lipovka, N. M., Parijsky, Y. N. 1965, *Astr. Zu.*, **42**, 902.
Gordon, I. M. 1965, *Astr. Tsirk.*, no. 321.
Gordon, I. M. 1966, *Astr. Zu.*, **43**, 754.
Gotwols, B. L., Erickson, W. C., Fremouw, E., Owren, L. 1966, *Publ. astr. Soc. Pacific*, **78**, 199.
Gould, R. J. 1965, *Phys. Rev. Lett.*, **15**, 577.
Haddock, F. T., Howard, W. E., Malville, J. M., Seling, T. V. 1963, *Publ. astr. Soc. Pacific*, **75**, 456.

Hartz, T. R. 1964, *Nature*, **203**, 173.
Harvey, C. C. 1965, *Ann. Astrophys.*, **28**, 248.
Haslam, C. G. T., Large, M. I., Quigley, M. J. S. 1964, *Mon. Not. R. astr. Soc.*, **127**, 273.
Heidmann, N. 1965, *Ann. Astrophys.*, **28**, 521.
Heidmann, N., Heidmann, J. 1964, *Astronomie*, **78**, 416.
Hewish, A., Okoye, S. E. 1964, *Nature*, **203**, 171.
Hewish, A., Okoye, S. E. 1965, *Nature*, **207**, 59.
Higgs, L. A. 1965, *J. R. astr. Soc. Can.*, **59**, 56.
Higgs, L. A. 1966, *Mon. Not. R. astr. Soc.*, **132**, 67.
Higgs, L. A., Broten, N. W., Medd, W. J., Raghavarao, R. 1964, *Mon. Not. R. astr. Soc.*, **127**, 367.
Higgs, L. A., Thibodeau, E. G. 1965, *Observatory*, **85**, 46.
Hogg, D. E. 1964, *Astrophys. J.*, **140**, 992.
Hogg, D. E. 1966, *Astrophys. J.*, **144**, 819.
Hollinger, J. P. 1965, *Astrophys. J.*, **142**, 609.
Hollinger, J. P., Hobbs, R. W. 1966, *Science*, **153**, 1633.
Hornby, J. M. 1966, *Mon. Not. R. astr. Soc.*, **133**, 213.
Howard, W. E., Dickel, H. R. 1963, *Publ. astr. Soc. Pacific*, **75**, 149.
Hoyle, F. 1965, in *The Structure and Evolution of Galaxies*, Interscience Publishers, p. 95.
Hugill, J., Smith, F. G. 1965, *Mon. Not. R. astr. Soc.*, **131**, 137.
Huguenin, G. R., Lilley, A. E., McDonough, W. H., Papagiannis, M. D. 1964, *Planet. Space Sci.*, **12**, 1157.
Huguenin, G. R., Papagiannis, M. D. 1965, *Ann. Astrophys.*, **28**, 239.
Jacklyn, R. M. 1966, *Nature*, **211**, 690.
Jelley, J. V. 1965, Proc. Int. Conf. on Cosmic Rays, p. 698.
Johnson, H. M. 1963, *Publ. nat. Radio Astr. Obs.*, **1**, 261.
Kaftan-Kassim, M. A. 1965, *Astr. J.*, **70**, 680.
Kaftan-Kassim, M. A. 1966, *Astrophys. J.*, **145**, 658.
Kellerman, K. I., Pauliny-Toth, I. I. K. 1966, *Astrophys. J.*, **145**, 953.
Kenderdine, S. 1963*a*, *Mon. Not. R. astr. Soc.*, **126**, 41.
Kenderdine, S. 1963*b*, *Mon. Not. R. astr. Soc.*, **126**, 55.
Kerr, F. J., Sinclair, M. W. 1966, *Nature*, **212**, 166.
Khromov, G. S. 1964, *Astr. Zu.*, **41**, 71.
Khromov, G. S. 1965, *Astr. Zu.*, **42**, 918.
Khromov, G. S., Indisov, O. S., Matveenko, L. I., Turevsky, V. M., Sholomitsky, G. B. 1965, *Astr. Zu.*, **42**, 1120.
Klemperer, W. K., Ochs, G. R. 1965, *Astr. J.*, **70**, 141.
Ko, H. C. 1963, *Nature*, **200**, 1193.
Komesaroff, M. M. 1966, *Aust. J. Phys.*, **19**, 75.
Krishnan, T., Zisk, S. H. 1966, *Astr. J.*, **71**, 167.
Kronberg, P. P. 1966, *Nature*, **212**, 1557.
Kurilchik, V. A. 1966, *Astr. Zu.*, **43**, 1317.
Kuzmin, A. D., Salomonovich, A. E. 1963, *Astr. Tsirk.*, no. 260.
Large, M. I. 1966, Sympos. on Radio and Opt Studies of the Galaxy, Mt Stromlo, p. 36.
Large, M. I., Quigley, M. J. S., Haslam, C. G. T. 1966, *Mon. Not. R. astr. Soc.*, **131**, 335.
Lastochkin, V. P., Stankevich, K. S. 1964, *Astr. Zu.*, **41**, 769.
Lastochkin, V. P., Lukin, E. B., Stankevich, K. S., Tseitlin, N. M. 1965, *Astr. Zu.*, **42**, 705.
Lazarevskii, V. S., Stankevich, K. S., Troitskii, V. S. 1963, *Astr. Zu.*, **40**, 12.
Le Marne, A. E. 1966*a*, Sympos. on Radio and Opt. Studies of the Galaxy, Mt Stromlo, p. 66.
Le Marne, A. E. 1966*b*, *Observatory*, **86**, 148.
Lenchek, A. M. 1964, *Ann. Astrophys.*, **27**, 219.
Lequeux, J. 1963, *Ann. Astrophys.*, **26**, 429.
L'Heureux, J., Meyer, P. 1965, *Phys. Rev. Lett.*, **15**, 93.
Little, A. G. 1966, Sympos. on Radio and Opt. Studies of the Galaxy, Mt Stromlo, p. 113.
Lovell, A. C. B. 1964, *Observatory*, **84**, 191.
Lovell, A. C. B., Chugainov, P. F. 1964, *Nature*, **203**, 1213.

Lovell, A. C. B., Whipple, F. L., Solomon, L. H. 1963, *Nature*, **198**, 228.
Lovell, A. C. B., Whipple, F. L., Solomon, L. H. 1964, *Nature*, **201**, 1013.
Lovell, A. C. B., Solomon, L. H. 1966, *Observatory*, **86**, 16.
Lozinskaya, T. A. 1964, *Astr. Tsirk.*, no. 299.
Lozinskaya, T. A., Shcheglov, P. V. 1965, *Astr. Tsirk.*, no. 327.
Lund, D. S. 1965, *Astr. J.*, **70**, 683.
Mackrell, T. F. 1963, *New Zealand J. Geol. Geophys.*, **6**, 728.
MacLeod, J. M., Swenson, G. W., Yang, K. S., Dickel, J. R. 1965, *Astr. J.*, **70**, 756.
Malitson, H. H., Stone, R. G., Erickson, W. C. 1966, *Astr. J.*, **71**, 391.
Maltby, P. 1965, *Astrophys. J.*, **142**, 621.
Malumian, V. H. 1963, *Soobšč. Bjurak. Obs.*, **33**, 95.
Malumian, V. H. 1965, *Astrofizica*, **1**, 347.
Malumian, V. H. 1966, *Astrofizica*, **2**, 128.
Mathewson, D. S. 1966, *Astr. J.*, **71**, 170.
Mathewson, D. S., Milne, D. K. 1964, *Nature*, **203**, 1273.
Mathewson, D. S., Milne, D. K. 1965, *Austr. J. Phys.*, **18**, 635.
Mathewson, D. S., Healey, J. R., Westerlund, B. E. 1963, *Nature*, **199**, 681.
Mathewson, D. S., Broten, N. W., Cole, D. J. 1965, *Austr. J. Phys.*, **18**, 665.
Mathewson, D. S., Broten, N. W., Cole, D. J. 1966, *Austr. J. Phys.*, **19**, 93.
Matveenko, L. I. 1966*a*, *Astr. Tsirk.*, no. 358.
Matveenko, L. I. 1966*b*, *Astr. Tsirk.*, no. 360.
Matveenko, L. I., Khromov, G. S. 1965, *Astr. Tsirk.*, no. 343.
Matveenko, L. I., Artuh, V. S. 1966, *Astr. Zu.*, **43**, 275.
Matveenko, L. I., Martirosyan, R. M., Sorochenko, R. L. 1965, *Astr. Zu.*, **42**, 316.
Maxwell, A., Downes, D. 1964, *Nature*, **204**, 665.
Meaburn, J. 1965*a*, *Nature*, **207**, 179.
Meaburn, J. 1965*b*, *Nature*, **208**, 575.
Menon, T. K. 1964, *IEEE Trans. Mil. Electron.*, MIL-8, 247.
Menon, T. K., Terzian, Y. 1965, *Astrophys. J.*, **141**, 745.
Menon, T. K., Williams, D. R. W. 1966, *Astr. J.*, **71**, 392.
Meyer, P. 1965, Proc. Int. Conf. Cosmic Rays, p. 61.
Migach, Yu. E. 1965, *Perem. Zvezdy*, **15**, 400.
Mills, B. Y. 1964, *A. Rev. Astr. Astrophys.*, **2**, 185.
Mills, B. Y., Glanfield, J. R. 1965, *Nature*, **208**, 10.
Milne, D. K. 1966, Sympos. on Radio and Opt. Studies of the Galaxy, Mt Stromlo, p. 74.
Minkowski, R. 1964, *A. Rev. Astr. Astrophys.*, **2**, 247.
Minkowski, R. 1965*a*, in *The Structure and Evolution of Galaxies*, Interscience Publishers, p. 89.
Minkowski, R. 1965*b*, *Astr. J.*, **70**, 755.
Minkowski, R. 1966*a*, *Nature*, **209**, 1339.
Minkowski, R. 1966*b*, *Astr. J.*, **71**, 371.
Moran, M. 1965, *Mon. Not. R. astr. Soc.*, **129**, 447.
Morris, D., Berge, G. L. 1964, *Astrophys. J.*, **139**, 1388.
Morris, D., Clark, B. G., Wilson, R. W. 1963, *Astrophys. J.*, **138**, 889.
Nguyen-Quang-Rieu 1963, *Notes Inf. Publ. Obs. Paris*, Fasc. 11.
Nicolson, G. D. 1965, *Publ. astr. Soc. Pacific*, **77**, 260.
Oda, M., Hasegawa, H. 1963, Proc. Int. Conf. Cosmic Rays, 3, p. 370.
Ohtani, H., Kogure, T. 1964, *Publ. astr. Soc. Japan*, **16**, 206.
Osterbrock, D. E. 1965, *Astrophys. J.*, **141**, 1285, 1965.
Parker, E. N. 1965, *Astrophys. J.*, **142**, 584.
Parker, E. N. 1966*a*, *Astrophys. J.*, **144**, 916.
Parker, E. N. 1966*b*, *Astrophys. J.*, **145**, 811.
Parthasarathy, R., Lerfald, C. M. 1965, *Mon. Not. R. astr. Soc.*, **129**, 395.
Pawsey, J. L. 1965, in *Galactic Structure*, Eds. Blaauw, A., Schmidt, M., Vol. V of *Stars and Stellar Systems*, Univ. of Chicago Press, Chicago, p. 219.
Penzias, A. A., Wilson, R. W. 1965, *Astrophys. J.*, **142**, 1149.

Penzias, A. A., Wilson, R. W. 1966, *Astrophys. J.*, **146**, 666.
Pike, E. M., Drake, F. D. 1964, *Astrophys. J.*, **139**, 545.
Pollack, J. B., Fazio, G. G. 1965, *Astrophys. J.*, **141**, 730.
Poveda, A. 1965, *Bol. Obs. Tonantzintla Tacubaya*, **4**, 49.
Purton, C. R. 1966, *Mon. Not. R. astr. Soc.*, **133**, 463.
Quigley, M. J. S., Haslam, C. G. T. 1964, *Nature*, **203**, 1272.
Quigley, M. J. S., Haslam, C. G. T. 1965, *Nature*, **208**, 741.
Quigley, M. J. S., Large, M. I. 1966, *Mon. Not. R. astr. Soc.*, **134**, 239.
Raghava Rao, Medd, W. J., Higgs, L. A., Broten, N. W. 1965, *Mon. Not. R. astr. Soc.*, **129**, 159.
Ramana, K. V. V. 1966, *Astr. J.*, **71**, 395.
Ramaty, R., Lingenfelter, R. E. 1966*a*, *Phys. Rev. Lett.*, **17**, 1230.
Ramaty, R., Lingenfelter, R. E. 1966*b*, *J. geophys. Res.*, **71**, 3687.
Razin, V. A., Khizhnyakov, I. P. 1965, *Izv. Vyssih Učeb. Zaved. Radiofiz.*, **8**, 822.
Razin, V. A., Khroulyov, V. V. 1965*a*, *Izv. Vyssih Učeb. Zaved. Radiofiz.*, **8**, 857.
Razin, V. A., Khroulyov, V. V. 1965*b*, *Izv. Vyssih Učeb. Zaved. Radiofiz.*, **8**, 1063.
Rougoor, G. W. 1966, *Astrophys. J.*, **144**, 852.
Ryle, M., Elsmore, B., Neville, A. C. 1965, *Nature*, **205**, 1259.
Ryzhkov, N. F., Egorova, T. M., Gosatchinsky, I. V., Bystrova, N. V. 1964, *Izv. glav. astr. Obs. Pulkove*, **23**, no. 3.
Sanamian, V. A. 1966, *Astrofizica*, **2**, 124.
Sanamian, V. A., Aslanian, A. M. 1965, *Astrofizica*, **1**, 247.
Scheer, D. J., Kraus, J. D. 1966, *Astr. J.*, **71**, 179.
Scheuer, P. A. G. 1963, *Observatory*, **83**, 56.
Seeger, Ch. L., Westerhout, G., Conway, R. G., Hoekema, T. 1965, *Bull. astr. Inst. Netherl.*, **18**, 11.
Sher, D. 1965, *J. R. astr. Soc. Can.*, **59**, 67.
Shklovsky, I. S. 1964, *Astr. Tsirk.*, no. 298.
Shklovsky, I. S. 1965, *Dokl. Akad. Nauk SSSR*, **160**, 54.
Shklovsky, I. S. 1966*a*, *Astr. Zu.*, **43**, 747.
Shklovsky, I. S. 1966*b*, *Astr. Zu.*, **43**, 10.
Shong, J. A. de, Hildebrand, R. H., Meyer, P. 1964, *Phys. Rev. Lett.*, **12**, 3.
Sinclair, M. W. 1966, Sympos. on Radio and Opt. Studies of the Galaxy, Mt Stromlo, p. 106.
Sironi, G. 1965, *Nuovo Cim.*, **39**, 372.
Slee, O. B., Solomon, L. H., Patston, G. E. 1963, *Nature*, **199**, 991.
Slee, O. B., Orchiston, D. W. 1965, *Austr. J. Phys.*, **18**, 187.
Slysh, V. I. 1964, *Astr. Zu.*, **41**, 1038.
Slysh, V. I. 1965, *Kosm. Issled.*, **3**, 760.
Smith, F. G. 1965, *Mon. Not. R. astr. Soc.*, **131**, 145.
Smith, F. G., Porter, N. A., Jelley, J. V. 1965, Proc. Int. Conf. on Cosmic Rays, p. 701.
Souffrin, S., Chesnay, N. 1965, *C. r. Acad. Sci. Paris*, **261**, 3541.
Spergel, M. S., Scanlon, J. H. 1966, *Nuovo Cim.*, **42A**, 228.
Staelin, D. H., Barrett, A. H., Kusse, B. R. 1964, *Astr. J.*, **69**, 69.
Talen, J. L. 1965, *Astr. J.*, **70**, 332.
Taylor, J. H. 1966, *Astrophys. J.*, **146**, 646.
Terzian, Y. 1965, *Astrophys. J.*, **142**, 135.
Terzian, Y. 1966, *Astrophys. J.*, **144**, 657.
Tolbert, C. W. 1965, *Nature*, **206**, 1304.
Tolbert, C. W., Straiton, A. W. 1965, *Astr. J.*, **70**, 177.
Troitsky, V. S., Tseitlin, N. M., Porfiryev, V. A. 1964, *Astr. Zu.*, **41**, 446.
Tsytovich, V. N. 1963, *Astr. Zu.*, **40**, 612.
Tsytovich, V. N. 1964, *Astr. Zu.*, **41**, 7.
Turner, K. C., Burke, B. F. 1965, *Astr. J.*, **70**, 332.
Turtle, A. J. 1963, *Mon. Not. R. astr. Soc.*, **126**, 405.
Turtle, A. J., Kenderdine, S., Pugh, J. F., Pauliny-Toth, I. I. K. 1962, *Mon. Not. R. astr. Soc.*, **124**, 297.

Véron, P. 1965, *Ann. Astrophys.*, **28**, 391.
Weinreb, S. 1962, *Astrophys. J.*, **136**, 1149.
Wendker, H. 1966, *Mitt. astr. Inst. Univ. Münster*, no. 10.
Whiteoak, J. B. 1966, Sympos. on Radio and Opt. Studies of the Galaxy, Mt Stromlo, p. 21.
Wielebinski, R., Shakeshaft, J. R. 1964, *Mon. Not. R. astr. Soc.*, **128**, 19.
Wielebinski, R., Yates, K. W. 1965, *Nature*, **205**, 581.
Williams, P. J. S., Kenderdine, S., Baldwin, J. E. 1966, *Mem. R. astr. Soc.*, **70**, 53.
Wilson, R. W. 1963, *Astr. J.*, **68**, 181.
Woltjer, L. 1965*a*, in *The Structure and Evolution of Galaxies*, Interscience Publishers, p. 30.
Woltjer, L. 1965*b*, in *Galactic Structure*, eds. Blaauw, A. and Schmidt, M. Vol. V of *Stars and Stellar Systems*, Univ. of Chicago Press, Chicago, p. 531.
Xi Ze-zong, Po Shu-jen 1966, *Science*, **154**, 597.
Yang, K. S., West, L. A. 1964, *Astr. J.*, **69**, 246.
Yang, K. S., Dickel, J. R. 1965, *Astr. J.* **70**, 300.
Yates, K. W., Wielebinski, R. 1965, *Nature*, **208**, 64.
Yates, K. W., Wielebinski, R. 1966, *Austr. J. Phys.*, **19**, 389.
Zisk, S. H. 1966*a*, *Astr. J.*, **71**, 189.
Zisk, S. H. 1966*b*, *Science*, **153**, 1107.

D. RADIO LINE EMISSION AND ABSORPTION IN THE GALAXY

(prepared by F. J. Kerr)

Several general reviews have appeared during the three-year period. The distribution of interstellar hydrogen in the Galaxy has been surveyed by Westerhout (1964) and Kerr and Westerhout (1965), and the structure and evolution of the Galaxy by Oort (1965) and Bok (1964). Radio line emission and absorption by the various constituents of the interstellar gas have been reviewed by Kerr (1966), and recent work on the interstellar medium in general by Dieter and Goss (1966). The whole subject was extensively discussed at IAU Symposium no. 31, held at Noordwijk, August–September 1966. The proceedings of this symposium will be published.

1. *21-cm Hydrogen Line*

Hydrogen line studies are in the transition stage where a large amount of new observational data is now available from high-resolution surveys, but detailed study of the material has not yet proceeded very far. Much of the work published in this three-year period has been concerned with earlier low-resolution work.

The concluding papers describing the results of a low-resolution survey of the southern sky have been published by McGee *et al.* (1964, 1966). P. O. Lindblad (1966*a*) has presented the results of the 1953–55 Kootwijk survey in a new form, as isophote maps of optical depth in velocity-latitude planes. Makarova (1964) has measured profiles along the galactic equator at 10° intervals in longitude.

Two new high-resolution surveys have been carried out. The Maryland-Green Bank survey has been published in part (Westerhout 1966), covering 60% of the sky in the strip $l^{II} = 11°$ to $235°$, $b^{II} = -1°$ to $+1°$, in the form of contour maps. The Parkes survey (Kerr 1966*b*) covers the longitude range 185°–63° over a varying range of latitude; these results will be published in 1967.

Several authors have discussed the problems of the large-scale distribution and motion of the hydrogen, but a new coherent picture of the whole system has not yet appeared. The most recent review of this part of the subject has been given by P. O. Lindblad (1966*b*). The variation of the rotation curve with galactocentric azimuth and the large-scale deviations from circular motion require the development of more complex models than have been used so far.

Shane and Bieger-Smith (1966) have made a new determination of the galactic rotation curve for northern longitudes. They discuss possible relationships between the irregularities in the curve and the spiral structure. The earlier 21-cm data on rotation have been reinterpreted by Agekyan *et al.* (1964) and Zhu Ci-Sheng (1964). Münch and Münch (1964) and Feast and Thackeray (1965) have compared rotation curves derived from optical and radio data. The latter obtained good agreement after applying statistical distance corrections to the stellar measurements. In a rediscussion of Muller and Westerhout's catalog, Pronik (1965) finds evidence for departures from circular motion in the outer spiral arms.

Comparisons have been made between the spiral structure patterns derived from optical and radio results by Becker (1964), Kostyakova (1965), and Behr (1965). The agreement is generally poor. Multi-arm models have been developed from earlier 21-cm observations by Kardashev *et al.* (1964), Pskovskii (1965), and Pavlovskaya and Sharov (1966). Mass models for the Galaxy have been derived by Schmidt (1965) and Innanen (1966), mainly on the basis of 21-cm data.

New observations on the shape of the outer part of the galactic hydrogen layer have been made by Lozinskaya and Kardashev (1964). The gravitational effect of the Megellanic Clouds has been considered by Avner (1964) and Elwert and Hablick (1965), who show that a systematic distortion of the observed order of magnitude could be produced in this way under certain conditions. Lynden-Bell (1965) has discussed another possibility, postulating a 1° deviation between the Galaxy's symmetry axis and that of its angular momentum. Zasov (1965) has searched the Palomar atlas for galaxies with distorted disks, and found that they are most frequently associated with close groups or small clusters.

Detailed studies of the galactic centre region have been described by Rougoor (1964) and Kerr (1966*a*), and structural models proposed. Kerr has suggested that the high-velocity hydrogen may be in a bar, coming out from the nuclear disk at an angle to the galactic plane. Shane has found material away from the plane at forbidden velocities, which suggests ejection of gas from the nucleus (Oort 1966*c*). Ariskin (1965) and Komesaroff (1966) have discussed possible relationships between the '3-kpc arm' of hydrogen and structural features found in the continuum.

Burton (1966) has made a detailed study of the longitude range 43°–56°, and has identified several structural features, including a stream of hydrogen apparently related to the Sagittarius arm which has a systematically higher velocity than the arm itself. Hill and Kerr (1964) have given additional evidence that a spiral arm is seen tangentially in the Norma region. Several studies of special regions have been reported, namely the anticentre (Locke, Galt and Costain 1964*a*), the continuum spur (Lozinskaya 1964), the Perseus region (Rickard 1965), and the region $l^{II} = 10°–40°$, Gosachinskii (1966). Dieter (1964, 1965) has made a detailed study of the north and south galactic polar caps. Muhleman and Walker (1964) have described a relationship between OB stars and neutral hydrogen in the Cygnus arm, while Balázs (1965) found a concentration of stars that coincides with two H I complexes found by Blaauw near h and χ Persei. McCarthy and Treanor (1965), on the other hand, found no stellar evidence for spiral structure related to Lindblad's narrow hydrogen arm out of the plane near the anticentre.

The Dutch group have made a detailed study of hydrogen clouds at high and moderate velocities at high latitudes. The observational results have been reviewed by Blaauw (1966), and the results are given in detail by Muller *et al.* (1966), Blaauw and Tolbert (1966), and Hulsbosch and Raimond (1966). The data for two individual clouds of particular interest are described by Smith (1963) and Prata (1964). Many of the clouds are found over a broad region centred at about $l^{II} = 120°$, $b^{II} = +40°$, and most move with negative radial velocities, but some clouds are found in other regions, and a few have positive velocities. Oort (1964, 1966*a*, 1966*b*, 1966*d*) has discussed a number of possible interpretations, and considers that the clouds are composed of gas falling in from outside the Galaxy, and decelerated by the galactic interstellar gas. Meaburn (1965) has looked for a relationship between the high-velocity clouds and several loops of continuum emission. Habing (1966) has studied a continuous feature of high-velocity gas at intermediate latitudes, which is probably in the outer part of the Galaxy.

Several studies have been made of the two associations in Monoceros, including the cluster NGC 2264 (Rohlfs 1963; Girnstein and Rohlfs 1964; Bystrova, Gosachinskii and Ryzhkov 1964; Viennot 1965; Raimond 1966*a*, *b*). The most detailed investigation was that of Raimond, who examined the interrelationships between H I, H II, stars and dust in the region. More general surveys of hydrogen in clusters have been carried out by Davies and Tovmassian (1963), Howard and Westerhout (1965), and Tovmassian (1966). Hydrogen was found in most of the clusters examined, including both young and old clusters. Riegel (1966) made 21-cm line observations of 27 small-diameter H II regions, and detected neutral hydrogen associated with four of them. Locke, Galt, and Costain (1964*b*) found an H I cloud in the direction of the nebulosity IC443, which may explain the apparent retardation of the expanding supernova shell. Neutral hydrogen has again been looked for in globular clusters without success. Goldstein (1964*a*) has set an upper limit of about 150 $M_{\odot}$ on the H I content of M13, and Robinson (1966) a limit of 50 $M_{\odot}$ for 47 Tuc and ω Cen.

Evidence on the small-scale cloud structure and local kinematical properties has recently been reviewed by van Woerden (1966). The results of a large survey at Dwingeloo, in which profiles were analyzed into Gaussian components, have been discussed by van Woerden (1963), Takakubo (1963*a*) and Terauti (1963). Cloud models were set up and tested against the observations with varying degrees of success. Another study of the density fluctuations has been discussed by Rohlfs (1964), Grahl (1965), and Rohlfs (1966). Heiles (1966) has surveyed a moderately large region to examine the cloud structure. He identified a number of individual clouds and 'cloudlets', but concluded that 90% of the hydrogen in his region was in a smooth distribution. A narrow emission line has been found which extends over a region of large angular size (Goldstein 1964*b*; Goldstein and Welch 1966). This may come from a discrete H I cloud with a small velocity dispersion.

The temperature of the hydrogen, and the possibility of temperature variations, have been discussed by Marx (1965), and the dispersion of cloud velocities by Wilhelmsson and Winnberg (1963) and Gershberg (1964). A cooling process for the gas involving spin-flip collisions between atomic hydrogen and positive ions of carbon and sulphur has been considered by Dalgarno and Rudge (1964).

Several comparisons have been made between gas and dust for small regions. Pronik (1963) reported a variation in the gas-to-dust ratio from 70 to 150 over the region of the Omega nebula, while Gosachinskii (1965) considered that the dust cloud obscuring much of the nebula is associated with a reduced H I density. Similar comparisons have been made for the direction of the Crab nebula by Brodskaya (1963) and for a region in Cassiopeia by Grigoreva (1964, 1965). In the latter case, dust and atomic hydrogen concentrations were found to coincide. On the other hand, Garzoli and Varsavsky (1966) reported an inverse correlation for a region in Taurus, from which they concluded that there is a significant abundance of molecular hydrogen in these dust clouds.

Howard, Wentzel and McGee (1963) compared the radial velocities of optical and radio interstellar lines, using the data available at that time, and found a fair correlation between the hydrogen and calcium velocities. The theoretical and observed ratios of neutral and ionized number densities were found to agree for sodium but not for calcium. Takakubo (1963*b*) compared a high-dispersion spectrum of the K line in ϵ Ori with a hydrogen profile, and obtained partial agreement. He was able to show from the line widths that the internal motions in the clouds must be mainly nonthermal.

Many hydrogen absorption observations have been made during the survey period, but the results of recent work with large telescopes on a great range of sources have not yet been published. Clark (1965) has found absorption effects in 15 out of 21 sources. He developed a model in which the absorption was produced by a population of cool dense clouds ($T_s < 100°K$) immersed in a general medium at a high temperature ($T_s > 1000°K$). Shuter and Verschuur (1964) studied the three strongest sources with a high frequency resolution, and also found evidence for cool absorbing clouds.

Absorption studies have been made of a number of individual sources, mainly for distance determination. These are the Omega nebula (Bystrova *et al.* 1963*a*, 1964*a*), Tycho's supernova (Menon and Williams 1966), W49 (Akabane and Kerr 1965), several other Westerhout sources examined by Bystrova *et al.* (1963*b*, 1964*b*, 1965), and 3C 273 (Robinson *et al.* 1963; Williams 1965). Absorption in the galactic centre region has been studied by Egorova (1963, 1964), Ryzhkov *et al.* (1964), and Kerr (1966*a*). In this region there are several continuum components, and the absorption pattern is quite complicated. One theoretical study has been published (Smith 1964), on the absorption profiles to be expected for an expanding H I cloud.

No further observations have been published of the Zeeman effect in interstellar hydrogen. The conclusion from work carried out so far is that the magnetic field inside absorbing hydrogen clouds is less than 5×10^{-6} gauss.

A number of precise measurements have been reported for the frequency of the hyperfine transition of ground-state atomic hydrogen. The most recent determinations are those by Beehler *et al.* (1966), Johnson and McGunigal (1966), Peters *et al.* (1965), and Ramsey (1965). A mean of their results gives a free-space value of:

$$1420\ 405\ 751{\cdot}786 \pm 0{\cdot}01 \text{ Hz},$$

corrected to zero magnetic field and zero absolute temperature.

2. *Hydroxyl Lines*

The outstanding development in radio line studies in the last three years has been the detection of the group of four Λ-doubling lines near 18 cm produced by interstellar OH radicals. A recent review has been given by Robinson (1966). The lines have been observed both in absorption and emission, and many unexpected results have been reported. The first discovery was made by Weinreb, Barrett, Meeks, and Henry (1963), who found the 1665 and 1667 MHz lines in absorption in the spectrum of Cassiopeia A. Other early absorption observations were made by Bolton *et al.* (1964*a*), Dieter and Ewen (1964), Weaver and Williams (1964), and Barrett, Meeks, and Weinreb (1964); the values reported for line strengths and shapes were approximately as expected.

Detailed studies of OH absorption effects in and around the direction of Sagittarius A revealed the existence of intense concentrations of OH in the galactic centre region, (Robinson *et al.* 1964; Goldstein *et al.* 1964; Bolton *et al.* 1964*b*; Bystrova *et al.* 1964). The strength of the absorption in this region led to the detection of the other two lines of the group at 1612 and 1720 MHz (Gardner *et al.* 1964). The intensity ratios between the four lines were found to be generally lower than expected theoretically, and to vary considerably from point to point over the region. The distribution of OH in the central region is quite different from that of the hydrogen. Some reasons for the observed systematic increase in the relative abundance of OH towards the galactic nucleus have been suggested by Carroll and Salpeter (1966).

OH emission was not at first detected, but in observations with a narrow-band receiver Weaver *et al.* (1965) discovered some very narrow emission lines in a number of sources Further observations by Weinreb *et al.* (1965), Zuckerman *et al.* (1965), McGee *et al.* (1965)

and Weaver *et al.* (1966) showed the great complexity of the emission phenomena. Narrow lines are emitted in very localized portions of H II regions, often near their peripheries. The intensity ratios of the four lines of the group are highly variable from source to source, and even from one emission component to another in the same source, clearly indicating non-equilibrium populations.

Linear polarization of many of the emission components has been reported by Weinreb *et al.* (1965), Zuckerman *et al.* (1965), Davies *et al.* (1966), and Williams *et al.* (1966). Circular polarization has also been found in some sources by Davies *et al.* (1966), and Barrett and Rogers (1966*a*). The polarization is often very strong, with great differences in degree and direction for neighbouring emission components. A simple Zeeman interpretation is not able to account for the observed phenomena. Possible interpretations have been proposed by Meeks *et al.* (1966), Verschuur (1966) and Cook (1966*b*).

Interferometric studies by Rogers *et al.* (1966) and Cudaback *et al.* (1966) have shown that some OH emission features have an angular size well under one minute of arc; consequently their brightness temperatures must be very high. Dieter *et al.* (1966) have reported secular variations in individual emission components in two sources over a period of a few weeks. These results have not yet been confirmed by other workers.

The basic theory of OH absorption and the early observations was reviewed by Barrett (1964). Attempts have been made to explain the anomalous emission phenomena in terms of a maser action in which population inversion is produced by ultraviolet or infrared radiation (Cook 1966*a*; Shklovskii 1966; Perkins, Gold, and Salpeter 1966). Symonds (1965) discussed the anomalous intensities in terms of the abnormal formation of OH molecules through a process involving negative oxygen ions.

Precise laboratory frequency measurements have been reported for the four lines by Radford (1964), while Powell and Lide (1965) have made an improved measurement of the OH dipole moment. Goss and Weaver (1966) have calculated improved values for the Einstein A coefficients.

The possibility of detecting lines from other isotopic species of OH has been discussed by Barrett and Rogers (1964), and the radio detection of interstellar $O^{18}H^{1}$ has recently been announced (Barrett and Rogers 1966*b*).

3. *Hydrogen and Helium Recombination Lines*

Kardashev predicted in 1959 that high-excitation lines of hydrogen should be detectable in H II regions. The first observation of such a line was made by Dravskikh *et al.* (1964*a*, 1964*b*, 1965), who detected an emission line at 5763 MHz, corresponding to the transition $n = 105$ to $n = 104$, in the Omega nebula. On more recent usage, this is called the H 104 α line.

The most extensive work has been done on the H 109 α line at 5009 MHz, which has been detected in emission from sixteen sources, for which distances, Doppler widths and electron temperatures have been determined (Höglund and Mezger 1965*a*, *b*; Mezger 1966; Mezger and Höglund 1966). Other lines that have been reported are H 90 α (Sorochenko and Borodzich 1965), H 156 α and H 158 α (Lilley *et al.* 1966*a*), and H 166 α (Palmer and Zuckerman 1966). In addition, three lines from neutral helium, He 156 α, He 158α, and He 159 α, have been detected by Lilley *et al.* (1966*b*).

The derivation of electron temperatures and other parameters from observations of these lines has been discussed by Sorochenko (1965) and Goldberg (1966). A table of frequencies for the hydrogen series has been computed by Pulley and Shuter (1966).

The detection of the recombination lines has provided a new method for the study of ionized regions, and also a new approach to the large-scale kinematics of the Galaxy.

4. *Other Lines*

Encouraged by the unexpected strength of the OH lines, several workers have attempted to detect the CH line near 10 cm wavelength, but no detailed accounts have yet been published. One major difficulty has been the uncertainty of the line frequency, for which new determinations have been given by Douglas and Elliott (1965) and Goss (1966). The electric dipole moment of the ground electronic state of CH has been measured by Phelps and Dalby (1966).

Malville (1964) has discussed the possibility of detecting two lines near 5 and 6 cm from metastable hydrogen molecules. General reviews of other line possibilities have been published by Robinson and McGee (1965) and Kerr (1966).

BIBLIOGRAPHY

Reviews

Bok, B. J. 1964, *Galactic and Extragalactic Structure*, pp. 75, Tata Inst. Fundamental Res., Bombay.
Dieter, N. H., Goss, W. M. 1966, *Rev. mod Phys.*, **38**, 256–297.
Kerr, F. J., Westerhout, G. 1965, Distribution of Interstellar Hydrogen, in *Galactic Structure*, Eds. Blaauw, A., Schmidt, M., Univ. of Chicago Press, Chicago, Chap. 9, pp. 167–202.
Kerr, F. J. 1966, Chap. 12 in *Nebulae and Interstellar Matter*, Vol. 7 of *Stars and Stellar Systems*, Ed. Kuiper, G. P., Middlehurst, B. M., Univ. of Chicago Press, Chicago (in press).
Oort, J. H. 1965, *Trans. IAU*, **12A**, 789.
Westerhout, G. 1964, *IEEE Trans. Ant Propag*, AP-12, 954–963.

21-*cm Hydrogen Line*

Agekyan, T. A., Petrovksaya, I. V., Fesenko, B. I. 1964, *Astr. Zu.*, **41**, 1027 = 1965, *Soviet Astr.*, **8**, 823.
Akabane, K., Kerr, F. J. 1965, *Austr. J. Phys.*, **18**, 91.
Ariskin, V. I. 1965, *Astr. Zu.*, **42**, 939 = 1966, *Soviet Astr.*, **9**, 722.
Avner, C. 1964, Ph.D. thesis, Univ. Illinois.
Balázs, B. 1965, *Z. Astrophys.*, **62**, 6.
Becker, W. 1964, *Z. Astrophys.*, **58**, 202.
Beehler, R. and 11 others. 1966, *Proc. Inst. electr. electron. Eng.*, **54**, 301.
Behr, A. 1965, *Z. Astrophys.*, **62**, 157.
Blaauw, A., Tolbert, C. R. 1966, *Bull. astr. Inst. Netherl.*, **18**, 405.
Blaauw, A. 1966, IAU Symp. no. 31 (in press).
Brodskaya, E. S. 1963, *Izv. Krym. astrofiz. Obs.*, **30**, 126.
Burton, W. B. 1966, *Bull. astr. Inst. Netherl.*, **18**, 247.
Bystrova, N. V., Gosachinskii, I. V., Egorova, T. M., Ryzhkov, N. F. 1963*a*, *Astr. Tsirk.*, no. 244, 1.
Bystrova, N. V., Gosachinskii, I. V., Egorova, T. M., Ryzhkov, N. F. 1963*b*, *Astr. Tsirk.*, no. 269, 1.
Bystrova, N. V., Gosachinskii, I. V., Egorova, T. M., Ryzhkov, N. F. 1964*a*, *Izv. glav. astr. Obs. Pulkove*, **23**, 111.
Bystrova, N. V., Gosachinskii, I. V., Egorova, T. M., Ryzhkov, N. F. 1964*b*, *Izv. glav. astr., Obs. Pulkove*, **23**, 116.
Bystrova, N. V., Gosachinskii, I. V., Ryzhkov, N. F. 1964, *Astr. Tsirk.*, no. 307, 1.
Bystrova, N. V., Gosachinskii, I. V., Egorova, T. M., Ryzhkov, N. F. 1965, *Izv. glav. astr. Obs. Pulkove*, **24**, 202.
Clark, B. G. 1965, *Astrophys. J.*, **142**, 1398.
Dalgarno, A., Rudge, M. R. H. 1964, *Astrophys. J.*, **140**, 800.
Davies, R. D., Tovmassian, H. M. 1963, *Mon. Not. R. astr. Soc.*, **127**, 45.

Dieter, N. H. 1964, *Astr. J.*, **69**, 288.
Dieter, N. H. 1965, *Astr. J.*, **70**, 552.
Egorova, T. M. 1963, *Astr. Zu.*, **40**, 382 = *Soviet Astr.*, **7**, 290.
Egorova, T. M. 1964, *Izv. glav. astr. Obs. Pulkove*, **24**, 77.
Elwert, G., Hablick, D. 1965, *Z. Astrophys.*, **61**, 273.
Feast, M. W., Thackeray, A. D. 1965, *Astrophys. J.*, **142**, 1645.
Garzoli, S. L., Varsavsky, C. M. 1966, *Astrophys. J.*, **145**, 79.
Gershberg, R. E. 1964, *Izv. Krym. astrofiz. Obs.*, **31**, 100.
Girnstein, H. G., Rohlfs, K. 1964, *Z. Astrophys.*, **59**, 83.
Goldstein, S. J. 1964*a*, *Astrophys. J.*, **140**, 802.
Goldstein, S. J. 1964*b*, *Proc. Inst. electr. electron. Eng.*, **52**, 1046.
Goldstein, S. J., Welch, B. J. 1966, *Astr. J.*, **71**, 297.
Gosachinskii, I. V. 1965, *Astr. Zu.*, **42**, 929 = 1966, *Soviet Astr.*, **9**, 714.
Gosachinskii, I. V. 1966, *Astr. Zu.*, **43**, 284.
Grahl, B. H. 1965, *Kleinheubacher Berichte*, **10**, 205.
Grigoreva, N. B. 1964, *Astr. Zu.*, **41**, 185 = *Soviet Astr.*, **8**, 139.
Grigoreva, N. B. 1965, *Izv. Krym. astrofiz. Obs.* **34**, 238.
Habing, H. J. 1966, *Bull. astr. Inst. Netherl.*, **18**, 323.
Heiles, C. 1966, Ph.D. thesis, Princeton Univ.
Hill, G., Kerr, F. J. 1964, *Publ. astr. Soc. Pacific*, **76**, 354.
Howard, W. E., Wentzel, D. G., McGee, R. X. 1963, *Astrophys. J.*, **138**, 988.
Howard, W. E., Westerhout, G. 1965, *Astr. J.*, **70**, 679.
Hulsbosch, A. N. M., Raimond, E. 1966, *Bull. astr. Inst. Netherl.*, **18**, 413.
Innanen, K. A. 1966, *Astrophys. J.*, **143**, 153.
Johnson, E. E., McGunigal, T. E. 1966, *U.S. NASA Tech. Note* TN D–3292.
Kardashev, N. S., Lozinskaya, T. A., Sleptsova, N. F. 1964, *Astr. Zu.*, **41**, 601 = 1965, *Soviet Astr.*, **8**, 479.
Kerr, F. J. 1966*a*, IAU Symp. No. 31 (in press).
Kerr, F. J. 1966*b*, IAU Symp. No. 31 (in press).
Komesaroff, M. M. 1966, *Austr. J. Phys.*, **19**, 75.
Kostyakova, E. B. 1965, *Astr. Zu.*, **42**, 537 = 1966, *Soviet Astr.*, **9**, 426.
Lindblad, P. O. 1966*a*, *Bull. astr. Inst. Netherl. Suppl.*, **1**, 77.
Lindblad, P. O. 1966*b*, IAU Symp. No. 31 (in press).
Locke, J. L., Galt, J. A., Costain, C. H. 1964*a*, *Astrophys. J.*, **139**, 1066.
Locke, J. L., Galt, J. A., Costain, C. H. 1964*b*, *Astrophys. J.*, **139**, 1071.
Lozinskaya, T. A., Kardashev, N. S. 1964, *Soobšč. gos. astr. Inst. P.K. Sternberga*, no. 131, 37.
Lozinskaya, T. A. 1964, *Astr. Tsirk*, no. 299, 1.
Lynden-Bell, D. 1965, *Mon. Not. R. astr. Soc.*, **129**, 299.
Makarova, S. P. 1964, *Astr. Zu.*, **41**, 608 = 1965, *Soviet Astr.*, **8**, 485.
Marx, S. 1965, *Astr. Nachr*, **288**, 155.
McCarthy, M. F., Treanor, P. J. 1965, *Ric. astr. Specola astr. Vatic.*, **7**, 177.
McGee, R. X., Milton, J. A. 1964, *Austr. J. Phys.*, **17**, 128.
McGee, R. X., Milton, J. A., Wolfe, W. 1966, *Austr. J. Phys. Suppl.* no. 1, 1–11.
Meaburn, J. 1965, *Nature*, **207**, 179.
Menon, T. K., Williams, D. R. W. 1966, *Astr. J.*, **71**, 392.
Muhleman, D. O., Walker, R. G. 1964, *Astr. J.*, **69**, 95.
Muller, C. A., Raimond, E., Schwarz, U. J., Tolbert, C. R. 1966, *Bull. astr. Inst. Netherl. Suppl.*, **1**, 213.
Münch, G., Münch, L. 1964, *Astrophys. J.*, **140**, 162.
Oort, J. H. 1964, *Verslag Afd. Nat., Kon. Nederl. Akad. Wet.*, Amsterdam, **73**, 94.
Oort, J. H. 1966*a*, *Trans. IAU*, **12 B**, 395.
Oort, J. H. 1966*b*, *Bull. astr. Inst. Netherl.*, **18**, 421.
Oort, J. H. 1966*c*, IAU Symp. No. 29 (in press).
Oort, J. H. 1966*d*, IAU Symp. No. 31 (in press).
Pavlovskaya, E. D., Sharov, A. S. 1966, *Astr. Zu.*, **43**, 40 = *Soviet Astr.*, **10**, 30.
Peters, H. E., Holloway, J., Bagley, A. S., Cutler, L. S. 1965, *Appl. Phys. Lett.*, **7**, 34.

Prata, S. W. 1964, *Bull. astr. Inst. Netherl.*, **17**, 511.
Pronik, I. I. 1963, *Izv. Krym. astrofiz. Obs.*, **30**, 118.
Pronik, I. I. 1965, *Astr. Zu.*, **42**, 923 = 1966, *Soviet Astr.*, **9**, 709.
Pskovskii, Y. P. 1965, *Astr. Zu.*, **42**, 1184 = 1966, *Soviet Astr.*, **9**, 919.
Raimond, E. 1966*a*, *Bull. astr. Inst. Netherl.*, **18**, 191.
Raimond, E. 1966*b*, *Bull. astr. Inst. Netherl. Suppl.*, **1**, 33.
Ramsey, N. F. 1965, *Metrologia*, **1**, 7.
Rickard, J. J. 1965, *Astr. J.*, **70**, 688.
Riegel, K. W. 1966, Ph.D. thesis, University of Maryland.
Robinson, B. J., van Damme, K. J., Koehler, J. A. 1963, *Nature*, **199**, 990.
Robinson, B. J. 1966, *Astrophys. J.* (in press).
Rohlfs, K. 1963, *Mitt. astr. Ges.*, p. 122.
Rohlfs, K. 1964, *Z. Astrophys.*, **59**, 102.
Rohlfs, K. 1966, *Z. Astrophys.*, **63**, 207.
Rougoor, G. W. 1964, *Bull. astr. Inst. Netherl.*, **17**, 381.
Ryzhkov, N. F., Egorova, T. M., Gosachinskii, I. V., Bystrova, N. V. 1964, *Izv. glav. astr. Obs. Pulkove*, **23**, 3.
Schmidt, M. 1965, Rotation Parameters and Distribution of Mass in the Galaxy, in *Galactic Structure*, Eds. Blaauw, A., Schmidt, M., Univ. of Chicago Press, Chicago, Chap. 22, pp. 513–530.
Shane, W. W., Bieger-Smith, G. P. 1966, *Bull. astr. Inst. Netherl.*, **18**, 263.
Shuter, W. L. H., Verschuur, G. L. 1964, *Mon. Not. R. astr. Soc.*, **127**, 387.
Smith, A. 1964, *Mon. Not. R. astr. Soc.*, **127**, 347.
Smith, G. P. 1963, *Bull. astr. Inst. Netherl.*, **17**, 203.
Takakubo, K. 1963*a*, *Sci. Rep. Tohoku Univ.*, First Ser., **47**, 65.
Takakubo, K. 1963*b*, *Sci. Rep. Tohoku Univ.*, First Ser., **47**, 108.
Terauti, R. 1963, *Sci. Rep. Tohoku Univ.*, First Ser., **47**, 114.
Tovmassian, H. M. 1966, IAU Symp. No. 31 (in press).
van Woerden, H. 1963, *Astr. J.*, **68**, 296.
van Woerden, H. 1966, *Trans. IAU*, **12 B**, 391.
Viennot, L. 1965, *Ann. Astrophys.*, **28**, 1008.
Westerhout, G. 1966, *The Maryland-Green Bank galactic 21-cm line survey*, first edition. Univ. Maryland.
Wilhelmsson, H., Winnberg, A. 1963, *Astr. Notes*, no. 8, 20–31.
Williams, D. R. W. 1965, Galactic Interstellar Absorption Lines in the Spectrum of 3C273 in *Quasi-Stellar Sources and Gravitational Collapse*, Ed. Robinson, I., Schild, A., Schücking, E. L., Univ. of Chicago Press, Chicago, Chap. 14, pp. 213–217.
Zasov, A. V. 1965, *Astr. Zu.*, **42**, 959.
Zhu Ci-Sheng 1964, *Acta astr. Sin.*, **12**, 113.

Hydroxyl Lines

Barrett, A. H., Meeks, M. L., Weinreb, S. 1964, *Nature*, **202**, 475.
Barrett, A. H., Rogers, A. E. E. 1964, *Nature*, **204**, 62.
Barrett, A. H. 1964, *IEEE Trans. Ant. Propag.*, AP–12, 822.
Barrett, A. H., Rogers, A. E. E. 1966*a*, *Nature*, **210**, 188.
Barrett, A. H., Rogers, A. E. E. 1966*b*, IAU Symp. No. 31 (in press).
Bolton, J. G., van Damme, K. J., Gardner, F. F., Robinson, B. J. 1964*a*, *Nature*, **201**, 279.
Bolton, J. G., Gardner, F. F., McGee, R. X., Robinson, B. J. 1964*b*, *Nature*, **204**, 30.
Bystrova, N. V., Egorova, T. M., Ryzhkov, N. F., Barrett, A. H. 1964, *Astr. Tsirk.*, no. 310, 5.
Carroll, T. O., Salpeter, E. E. 1966, *Astrophys. J.*, **143**, 609.
Cook, A. H. 1966*a*, *Nature*, **210**, 611.
Cook, A. H. 1966*b*, *Nature*, **211**, 503.
Cudaback, D. D., Read, R. B., Rougoor, G. W. 1966, *Phys. Rev. Lett.*, **17**, 452.

Davies, R. D., de Jager, G., Verschuur, G. L. 1966, *Nature*, **209**, 974.
Dieter, N. H., Ewen, H. I. 1964, *Nature*, **201**, 279.
Dieter, N. H., Weaver, H., Williams, D. R. W. 1966, *Astr. J.*, **71**, 160.
Gardner, F. F., Robinson, B. J., Bolton, F. G., van Damme, K. J. 1964, *Phys. Rev. Lett.*, **13**, 3.
Goldstein, S. J., Gundermann, E. J., Penzias, A. A., Lilley, A. E. 1964, *Nature*, **203**, 65.
Goss, W. M., Weaver, H. 1966, *Astr. J.*, **71**, 162.
McGee, R. X., Robinson, B. J., Gardner, F. F., Bolton, J. G. 1965, *Nature*, **208**, 1193.
Meeks, M. L., Ball, J. A., Carter, J. C., Ingalls, R. P. 1966, *Science*, **153**, 978.
Perkins, F., Gold, T., Salpeter, E. E. 1966, *Astrophys. J.*, **145**, 361.
Powell, F. X., Lide, D. R. 1965, *J. chem. Phys.*, **42**, 4201.
Radford, H. E. 1964, *Phys. Rev. Lett.*, **13**, 534.
Robinson, B. J., Gardner, F. F., van Damme, K. J., Bolton, J. G. 1964, *Nature*, **202**, 989.
Robinson, B. J. 1966, IAU Symp. No. 31 (in press).
Rogers, A. E. E., Moran, J. M., Crowther, P. P., Burke, B. F., Meeks, M. L., Ball, J. A., Hyde, G. M. 1966, *Phys. Rev. Lett.*, **17**, 450.
Shklovskii, I. S. 1966, *Astr. Tsirk.*, no. 372.
Symonds, J. L. 1965, *Nature*, **208**, 1195.
Verschuur, G. L. 1966, *Observatory*, **86**, 55.
Weaver, H. F., Williams, D. R. W. 1964, *Nature*, **201**, 380.
Weaver, H. F., Williams, D. R. W., Dieter, N. H., Lum, W. T. 1965, *Nature*, **208**, 29.
Weaver, H. F., Williams, D. R. W., Dieter, N. H. 1966, *Astr. J.*, **71**, 184.
Weinreb, S., Barrett, A. H., Meeks, M. L., Henry, J. C. 1963, *Nature*, **200**, 829.
Weinreb, S., Meeks, M. L., Carter, J. C., Barrett, A. H., Rogers, A. E. E. 1965, *Nature*, **208**, 440.
Williams, D. R. W., Dieter, N. H., Weaver, H. 1966, *Astr. J.*, **71**, 186.
Zuckerman, B., Lilley, A. E., Penfield, H. 1965, *Nature*, **208**, 441.

Hydrogen and Helium Recombination Lines

Dravskikh, Z. V., Dravskikh, A. F. 1964*a*, *Astr. Tsirk.*, no. 282, 2.
Dravskikh, Z. V., Dravskikh, A. F., Kolbasov, V. A. 1964*b*, *Astr. Tsirk*, no. 305, 2.
Dravskikh, A. F., Dravskikh, Z. V., Kolbasov, V. A., Misezhnikov, G. S., Nikulin, D. E., Shteinshleiger, V. B. 1965, *Dokl. Akad. Nauk SSSR*, **163**, 332.
Goldberg, L. 1966, *Astrophys. J.*, **144**, 1225.
Höglund, B., Mezger, P. G. 1965*a*, *Science*, **150**, 339, 347.
Höglund, B., Mezger, P. G. 1965*b*, *Astr. J.*, **70**, 678.
Lilley, A. E., Menzel, D. H., Penfield, H., Zuckerman, B. 1966*a*, *Nature*, **209**, 468.
Lilley, A. E., Palmer, P., Penfield, H., Zuckerman, B. 1966*b*, *Nature*, **211**, 174.
Mezger, P. G. 1966, *Astr. J.*, **71**, 171.
Mezger, P. G., Höglund, B. 1966, *Astrophys. J.* (in press).
Palmer, P., Zuckerman, B. 1966, *Nature*, **209**, 1118.
Pulley, H. C., Shuter, W. L. H. 1966, *Can. J. Phys.*, **44**, 1663.
Sorochenko, R. L. 1965, *Trudy fiz. Inst. Akad. Nauk SSSR*, **28**, 90.
Sorochenko, R. L., Borodzich, E. V. 1965, *Dokl. Akad. Nauk SSSR*, **163**, 603.

Other Lines

Douglas, A. E., Elliot, G. A. 1965, *Can. J. Phys.*, **43**, 496.
Goss, W. M. 1966, *Astrophys. J.*, **145**, 707.
Kerr, F. J. 1966, Chap. 12 in *Nebulae and Interstellar Matter*, Vol. 7 of *Stars and Stellar Systems*, Eds. Kuiper, G. P., Middlehurst, B. M., Univ. of Chicago Press, Chicago, (in press).
Malville, J. M. 1964, *Astrophys. J.*, **139**, 198.
Phelps, D. H., Dalby, F. W. 1966, *Phys. Rev. Lett.*, **16**, 3.
Robinson, B. J., McGee, R. X. 1965, *Inf. Bull. Southern Hemisphere*, no. 6, 37.

E. EXTRAGALACTIC RADIO ASTRONOMY: COSMOLOGY

(Prepared by J. R. Shakeshaft)

1. *General Remarks*

The past three years have seen considerable progress in studies of extra-galactic radio sources, with the completion of large-scale surveys in both hemispheres, with improvements in positional accuracies and, consequently, the number of optical identifications, and with advances in knowledge of the spectra, brightness distributions and polarization of sources. Our understanding of these phenomena, however, remains rudimentary and it is not yet clear at the time of writing, for instance, what relationship, if any, exists between quasi-stellar sources (QSS) and radio galaxies, nor even whether the QSS are local objects or at cosmological distances.

The situation as of December 1963 is described at length in the proceedings of the first Texas Symposium (Robinson *et al.* 1965). Other reviews include those by Priester and Rosenberg (1964), Bolton (1965), Ryle (1966), Schmidt (1966), Smith, H. J. (1966), and Burbidge, E. M. (1967). The proceedings of the NASA Symposium held in New York in December 1962 have been edited by Maran and Cameron (1964).

2. *Surveys, accurate positions and identifications*

All the information available in May 1964 on 1292 discrete sources was collated by Howard and Maran (1965), but since then Pauliny-Toth, Wade and Heeschen (1966) have used the N.R.A.O. 92 m antenna to check the 3C survey, the 4C survey has been completed in two parts (Pilkington and Scott 1965, Gower *et al.* 1967) with a total of 4843 sources in the range $-05° < \delta < +80°$ and having flux densities $> 2 \times 10^{-26}$ w m^{-2} Hz^{-1} at 178 MHz, and the Parkes catalogue has been published in four parts (Bolton *et al.* 1964, Price and Milne 1965, Day *et al.* 1966, Shimmins *et al.* 1966*a*) with a total of 1673 sources in the range $-90° < \delta < +20°$.

Further surveys have been carried out by Heeschen and Wade (1964) to determine the radio emission from bright galaxies, by Braccesi *et al.* (1965) at 408 MHz for the range $-30° < \delta < -20°$, by Davis *et al.* (1965) at 400 MHz, by Kellermann and Read (1965) at 1421 MHz which is, so far, the highest frequency at which a substantial area of sky has been surveyed in detail, by MacLeod *et al.* (1965) at 610·5 MHz, by Dixon *et al.* (1965) and Nash (1965) at 600 and 1415 MHz, by Kraus *et al.* (1966) at 1415 MHz, by Williams, P. J. S., *et al.* (1966) of 1067 sources at 38 MHz and by Kenderdine *et al.* (1966) with the aperture synthesis telescope at Cambridge. The latter is the 5C·1 survey at 408 MHz which lists 106 sources in an area of about 12 square degrees, the weakest having flux densities $S \approx 25 \times 10^{-29}$ w m^{-2} Hz^{-1} and positional accuracies $\approx 5''$ arc.

In order to identify sources (Minkowski 1964) with galaxies fainter than 16^m–17^m better positional accuracies are required than are, in general, provided by the large-scale surveys. Special observations have therefore been made at the Owens Valley Radio Observatory (Fomalont *et al.* 1964, Wyndham and Read 1965), at Pulkovo (Bystrova *et al.* 1964, Gol'nev and Parijskij 1965, Gol'nev *et al.* 1965), at the Mullard Radio Astronomy Observatory (Clarke, M. E., 1964, Wills and Parker 1966) and at Parkes (Shimmins *et al.* 1966*b*). These have reduced the errors to values typically $10''$–$15''$. For smaller numbers of sources positions have been determined to within a few seconds of arc by the method of lunar occultation (references later) and by interferometric techniques such as those at the Royal Radar Establishment, Malvern (Adgie 1964, Adgie and Gent 1966), at the N.R.A.O., Green Bank (Wade *et al.* 1965, Clark, B. G. and Hogg 1966) and at Cambridge (Parker *et al.* 1966). These measurements rely upon calibration of the equipment by means of accurately determined optical positions of sources (e.g. Griffin 1963, Véron 1966*a*).

Identification work following from these and other surveys has been described by Shobbrook (1963), Hazard *et al.* (1964) Ryle and Sandage (1964), de Vaucouleurs (1965), Longair (1965), Shakeshaft and Longair (1965), Shobbrook and Hunstead (1966) and Wyndham (1965, 1966); the latter paper gives finding-charts and descriptions of objects in the neighbourhood of most of the sources in the revised 3C survey. Many identifications have been suggested from study of the Parkes catalogues by Bolton *et al.* (1965*a*, 1965*b*), Ekers and Bolton (1965), Bolton and Ekers (1966*a*, 1966*b*, 1966*c*, 1966*d*), Bolton and Kinman (1966), Bolton *et al.* (1966), Clarke, M. E. *et al.* (1966), Westerlund and Smith (1966), both with galaxies and also blue stellar objects. Comparison of lists of such objects with the 4C catalogue has also revealed significant coincidences (Aizu 1966, Scheuer and Wills 1966, Wills 1966*a*, Caswell and Wills 1967), although for the great majority of them the radio emission is either extremely weak or undetectable (Dickel and McVittie 1965, Braccesi *et al.* 1966*a*, 1966*b*, Kellermann and Pauliny-Toth 1966*a*, Kinman 1966, Shapiro and Weinreb 1966).

Many of the identifications recently proposed remain to be confirmed, or otherwise, by improved radio positions and optical studies; this work is described in the report to Commission 28.

3. *Spectra: flux density variations*

Because of their uniformity the spectral results by Conway *et al.* (1963) (see also Long *et al.* 1963, Conway 1964*a*, Conway and Moran 1964) have been widely used, although further measurements (Long and Shuter 1963, Razin and Fedorov 1963, Kellermann 1964*a*, 1964*b*, Zaharov *et al.* 1964, Boischot *et al.* 1965, Clarke, M. E., 1965, Conway *et al.* 1965, Vetuhnovskaja and Kuz'min 1965, Artjuh *et al.* 1966) are now available. Kellermann (1964*c*) has reviewed some of these new data and their interpretation.

Howard *et al.* (1965) have also derived a uniform flux density system based on many earlier surveys but present studies (e.g. Long *et al.* 1966) make use of the more reliable surveys recently completed (see section (2)).

In virtually all of these papers, the flux densities quoted have been measured by comparison with those of intense sources. To calibrate them, absolute flux density measurements (Findlay 1966) are required. These are not easy and for low frequencies the errors are still 10–20%. At high frequencies, however, the use of horn antennas (Lastochkin *et al.* 1963*a*, 1963*b*, Findlay *et al.* 1965, Medd and Ramana 1965*a*, 1965*b*, Allen and Barrett 1966, Wilson and Penzias 1966) has enabled the errors to be reduced to a few percent only.

The range of frequencies over which flux densities have been measured has been extended; at low frequencies (< 38 MHz) by the group at Kharkov (Bazeljan *et al.* 1964, Bazeljan *et al.* 1965*a*, 1965*b*, Braude 1966*a*), by Erickson and Cronyn (1965) at 26 MHz, Andrew (1967) at 13·1 MHz and by Guidice (1966). At high frequencies (> 3 GHz) observations have been made by Broten *et al.* 1965, Dent and Haddock (1965*a*, 1966), Gol'nev and Parijskij (1965), Williams, D. R. W. *et al.* (1965), Kellermann (1966*a*), Kostenko and Matveenko (1966), Maxwell and Rinehart (1966) and Yokoi *et al.* (1966). From these and other measurements it has become clear (e.g. Howard *et al.* 1964) that the spectra of a number of sources cannot be represented by the usual power law. In the microwave range, for example, the fluxes of 3C 84 (NGC 1275) and 3C 279 increase with increasing frequency (Dent and Haddock, 1965*b*). This phenomenon is associated with time variations of flux (see later) and small angular size (Barber *et al.* 1966).

At low frequencies it is found that some sources exhibit a relative reduction of flux (Williams, P. J. S. 1963) which is correlated with high brightness temperatures. This result has been interpreted in terms of synchroton self-absorption (Slyš 1963) though there are other possibilities (Gajlitis and Cytovič 1963, Braude 1965*a*, Scheuer 1965*a*, Hornby and Williams 1966). In other cases there is a relative increase at low frequencies which can be attributed to components in the source with different spectral indices. Two examples are 3C 273 (Bailey *et al.* 1964, Conway *et al.* 1964, Šolomickij *et al.* 1965*a*, Hazard *et al.* 1966*a*, von Hoerner 1966) and 3C 84 (Baars *et al.* 1965, Roger *et al.* 1965, Slepcova *et al.* 1965, Braude 1966*b*). Šolomickij *et al.* (1964*a*, 1964*b*, 1965*b*) have found some other sources with apparently peculiar spectra but confirmation of this work is required.

The spectra of some of the stronger sources have been studied in detail (Braude 1963, 1965*b*, Hrulev 1963, Baars *et al.* 1964, Lastočkin 1964, Slyš 1964, Stankevič 1964, Bazeljan *et al.* 1965*c*, 1966, Braude *et al.* 1966, Clark, T. A., 1966). A particularly interesting one is 1934–63, the radio spectrum of which is broadly peaked near 1400 MHz and falls off rapidly above and below this frequency (Bolton *et al.* 1963, Kellermann 1966*b*). Šklovskij (1965*a*, 1965*b*) points out that there may be a detectable secular variation of its spectrum. This has not yet been found but the reality of irregular flux variations in certain other sources is now well established (after searches by Mayer *et al.* 1965), following the work of Dent (1965) on the quasi-stellar objects 3C 273, 279 and 345. It appears (Maltby and Moffet 1965*a*, Bartlett 1966, Kellermann and Pauliny-Toth 1966*b*, Moffet 1966*a*, Pauliny-Toth and Kellermann 1966, Yang and Dickel 1966) that the magnitude of the variations in these sources tends to increase with frequency above about 1 GHz; at centimetric and millimetric wavelengths there can be very rapid changes (Epstein 1965*a*, 1965*b*, Low 1965, Epstein *et al.* 1966). Another variable source (Dent 1966) is 3C 84, particularly significant because of its association with the Seyfert galaxy NGC 1275 rather than a quasi-stellar object.

The first QSS reported as variable was CTA–102, (Šolomickij 1965*a*, 1965*b*) from observations at 920 MHz. His later results (1966*a*) indicate a very regular fluctuation of intensity of a circularly-polarized component with a period 102 days. No other observers (Boischot 1965, Caswell and Wills 1965, Maltby and Moffet 1965*b*, Bologna *et al.* 1966, Nicholson, G. D. 1966) have detected any variation, though none has used precisely the same frequency or polarization.

The evolutionary changes to be expected in source spectra are the subject of papers by Gajlitis and Cytovič 1964, Ohtani and Kogure (1964), Lequeux (1965), Souffrin and Chesnay (1965), de la Beaujardière (1966) and Kellermann (1966*c*, 1966*d*).

Studies of the radio spectra of 'normal' galaxies have been made by Hazard (1963), Kuril'čik (1964, 1965*a*, 1966*a*), Kenderdine and Baldwin (1965) and Kuril'čik and Sycko (1965).

4. *Brightness distribution*

Work on the structure of extragalactic sources has been reviewed by Moffet (1964*a*, 1966*b*), Hazard (1965) and Palmer (1965).

Information on the brightness distribution in relatively extended sources can be obtained directly by means of antennas with narrow beams (Boischot *et al.* 1963, Little, A. G. 1963, Balklav 1964, Ko 1964, Labrum *et al.* 1964, Little, A. G. *et al.* 1964, Milne and Scheuer 1964, Parijskij and Prozorov 1964, Parijskij and Timofeeva 1964, Boischot and Kazès 1965, Broten *et al.* 1965, Scheuer 1965*b*, Thompson and Krishnan 1965, Little, A. G. and Bracewell 1966), but for higher resolution it is necessary to employ interferometric techniques or the method of lunar occultation (for zodiacal sources, Nicholson, W. 1965). A discussion of errors in measuring diameters has been given by von Hoerner (1965). The principal developments in interferometry have been the use of extended base-lines (Adgie *et al.* 1965, Anderson *et al.* 1965, Barber *et al.* 1966, Palmer 1966), revealing that some sources, particularly QSS, have structure subtending 0".1 arc or less; and the production of a pencil-beam as small as 23" by means of 'aperture synthesis' (Ryle *et al.* 1965*a*, 1965*b*, Macdonald *et al.* 1966, Shakeshaft 1966), which has shown, *inter alia*, that the QSS 3C 47 has a double structure indistinguishable from that characteristic of most radio galaxies, and that some radio galaxies, e.g. 3C 465, have multiple components. Other interferometric work included that by Maltby *et al.* (1963) who discuss 24 identified sources in detail, by Moffet (1964*b*, 1964*c*), Tanaka *et al.* (1965), Zisk (1965), Clark, B. G. and Hogg (1966), Hogg (1966) and Wade (1966) who has found fine structure in each of the two radio components of Cyg A.

The technique for analysing observations of a lunar occultation devised by Scheuer (1963) and developed by von Hoerner (1964*a*) has been applied to observations of 3C 273 by Bailey *et al.* (1964), Conway *et al.* (1964), Hughes (1965), Scheuer (1965*a*, 1965*c*), Hazard *et al.* (1966*b*) and von Hoerner (1966). Occultation observations of other sources include those of CTA 21 (Clarke, R. W. and Batchelor 1965, Taylor 1966), 3C 444 (de Jong 1966*a*, Taylor 1966), MSH 19–27 (Brown 1966*a*), MSH 21–23 (Brown 1966*b*), 0938 + 18 and 1101 + 11 (Cohen and Kundu 1966), NGC 6523 (Larionov and Šolomickij 1966, Šolomickij 1966*b*), 3C 17 and 3C 132 (Taylor 1966). The prediction of occultations of radio sources for 17 radio observatories has been continued by the Nautical Almanac Office of the Royal Greenwich Observatory; details are given in the Explanatory Supplement.

Information on small-scale structure (< 1″ arc) in sources can come from two other types of observation: (*a*) that of interplanetary scintillation due to the solar wind (Hewish *et al.* 1964, Cohen 1965, Hogg and Menon 1966, Sinigaglia 1966, Vitkevič *et al.* 1966). The manner in which scintillations vary with the angular distance between a source and the Sun is dependent upon the angular size and structure of the source (Cohen *et al.* 1966, Little, L. T. and Hewish 1966); (*b*) low-frequency spectral cut-offs. If it is assumed that such a cut-off is due to synchroton self-absorption then there is a relationship between the brightness temperature, red-shift, magnetic field and cut-off frequency for the source (e.g. Hornby and Williams 1966). Knowledge of angular sizes can then put upper limits on the field strengths in sources (Williams, P. J. S., 1966*a*) or, alternatively, assumptions about field strengths give upper limits to angular sizes. Williams, P. J. S. (1966*b*) has shown in this way that the ratio of separation to component sizes in the double source 3C 343/343·1 (previously studied by Moffet 1965) is at least 2000:1. Gardner and Bolton (1965) have described another widely-spaced double source MSH 13–33.

The dependence of source structure upon wavelength has been considered by Moffet and Palmer (1965) and by Jennison (1965).

5. *Polarization*

The whole subject of polarization, for the galactic continuum radiation as well as for sources, has been reviewed recently by Gardner and Whiteoak (1966).

Further observations of the linearly polarized radiation from sources have been made by Seielstad and Wilson (1963), Seielstad *et al.* (1963), Conway (1964*b*), Haddock and Hobbs (1964), Hollinger *et al.* (1964), Mayer (1964), Mayer *et al.* (1964), Morris and Berge (1964), Morris *et al.* (1964*a*), Radhakrishnan (1964), Roberts (1964), Rose (1964), Bologna *et al.* (1965), Gol'nev and Soboleva (1965), Boland *et al.* (1966), Gardner and Davies (1966) and Maltby and Seielstad (1966). These are nearly all measurements at various centimetric wavelengths of the percentage polarization and position angle of the electric vector for the integrated radiation from sources. It will clearly be vital to a full understanding of sources to know the distribution of polarization across them as a function of wavelength (as already determined for Centaurus A by the group at Parkes, e.g. Cooper *et al.* 1965). Examples of such measurements are those for Cygnus A with the fan-beam of the Pulkovo antenna (Soboleva and Timofeeva 1963, Soboleva 1966) for Pictor A (Broten *et al.* 1965), for Fornax A (Gardner and Price 1967), and for the source MSH 13–33 (IC 4296) (Gardner and Davies 1964) in which the intrinsic position angle of the electric vector from each of the three widely separated components is closely parallel to the axis of the source. Interferometric studies are, however, necessary in most cases in order to obtain higher resolution. These have so far been carried out only at the Owens Valley Radio Observatory (Morris *et al.* 1964*b*, 1964*c*, Seielstad 1966) and at Parkes (referred to by Gardner and Whiteoak 1966).

It is now clear that nearly all of the Faraday rotation observed for sources occurs in the Galaxy and can give information on magnetic field structure therein (Ginzburg and Pisareva 1963; see Section (C) for other references), but the decrease of polarization with increasing wavelength must be largely due to Faraday rotation occurring within the source; this has been considered by Pacholczyk (1963), Burn and Sciama (1964), Morris *et al.* (1964*d*), Slyš (1965) and Burn (1966) among others. The relation between polarization and other source properties has been discussed by Bolton *et al.* (1964), Moffet (1964*b*), Morris and Berge (1964), Seielstad *et al.* (1964), Maltby (1966) and Davies (1966*a*). The most convincing correlation found is that between intrinsic position angle and that of the major axis of elongated sources. The orientation difference is related to the brightness temperature of the source.

Mayer *et al.* (1964) have put low upper limits on possible circularly polarized radiation from a number of sources at λ 3·15 cm.

6. *'Normal' galaxies*

It appears (e.g. Matthews *et al.* 1964, Longair and Scott 1965, Matthews 1965) that a more or less clear distinction may be drawn between the 'normal' galaxies which have been detected and 'radio galaxies', with enhanced emission. The former are spirals or irregulars and have radio luminosities less than $\approx 10^{23}$ w Hz^{-1} $ster^{-1}$ at 178 MHz while the latter are nearly all D or E galaxies (though see Sandage 1966*a*) and have radio luminosities greater than this value. Possible intermediate cases have, however, been found by Tovmasjan (1966*a*, 1966*b*), and Heeschen and Wade (1964) believe that there is a smooth transition.

Heeschen and Wade (1964) have observed 515 galaxies in the Shapley-Ames catalogue north of $\delta = -19° 23'$ and found detectable continuum emission from many of the spirals. Observations have also been made by de Jong (1965, 1966*b*) and by Mills and Glanfield (1965). Individual 'normal' galaxies studied include

M 31 (NGC 224): Kraus (1963, 1964*a*, 1964*b*), Leibacher (1964), MacLeod (1964), Dickel *et al.* (1965), Kraus and Dixon (1965), Kraus *et al.* (1966), Makarenko (1966),
M 33 (NGC 598): Venugopal (1963),
M 83 (NGC 5236): Kuril'čik (1965*b*),

but the situation with regard to the presence or absence of radio 'haloes' in such objects remains controversial.

More detailed work on the distribution of neutral hydrogen in galaxies other than our own is now under way (Epstein 1964*a*, 1964*b*, Seielstad and Whiteoak 1965). The subject has been reviewed by Roberts (1966*a*, 1967). Apart from the Magellanic Clouds, reviewed in the report to Commission 28, M 31 is the object most studied (Argyle 1965, Brundage and Kraus 1966, Davies 1966*b*, Gottesman *et al.* 1966, Roberts 1966*b*) but NGC 55 (Seielstad and Whiteoak 1965, Robinson and van Damme 1966), NGC 925 (Hoglund and Roberts 1965), NGC 3109 (Seielstad and Whiteoak 1965, Botinelli *et al.* 1966, van Damme 1966), M 33 (Davies 1966*b*, Meng and Kraus 1966), NGC 5668 (Roberts 1965) and galaxies in the Virgo cluster (Robinson and Koehler 1965) have also been observed, among others (Davies *et al.* 1963).

It is often possible to derive a rotation curve for each galaxy (though see Elvius 1964) but it is found in addition that (*a*) the hydrogen usually extends well beyond the optically visible region, (*b*) in nine out of the ten objects studied with sufficient resolution the neutral hydrogen is concentrated to a ring of radius 5–10 kpc in the galactic disk. The relationship between the hydrogen and spiral arms as indicated by OB stars etc. is not yet clear.

Roberts (1966c) has been able to put upper limits of 10^{-3} atoms cm^{-3} on the neutral hydrogen density in the halo regions of NGC 4244 and NGC 7640.

7. *The nature of radio sources*

An understanding of the nature of radio sources requires the solution of two main problems, i.e. the origin of the very large energies involved, and the mode of evolution of the sources.

It is now widely accepted that 'violent events' of some kind (Burbidge, G. R. *et al.* 1963) must be responsible for the energy generation but, until the relationship between QSS and radio galaxies is clarified, we cannot be sure that the same type of event gives rise to each, though this is suggested by the fact that the energies are comparable. These violent events may give rise to cosmic rays (Aizu *et al.* 1965*a*, Burbidge, E. M. and Burbidge, G. R. 1965*a*).

Theories of the origin of radio sources have been reviewed by Burbidge, E. M. and Burbidge, G. R. (1964, 1965*b*), Burbidge, G. R. (1964) and Burbidge, G. R. and Burbidge, E. M. (1965). The following possibilities have been discussed: gravitational collapse, either of gas (Fowler 1964, 1965*a*, 1965*b*, Ginzburg 1964, 1965, Ginzburg and Ozernoy 1964, Miller and Parker 1964, Hoyle and Fowler 1965, Herzenstein and Aivazyan 1966, Wagoner 1966) or due to stellar collisions (Hoyle 1964, Ulam and Walden 1964, Woltjer 1964*a*, Gold, T., *et al.* 1965, van den Bergh 1965*a*); delayed expansion from a condensed state (Novikov 1964, Ne'eman 1965, Zel'dovič and Novikov 1966*a*, 1966*b*); collective star formation (Cameron 1964, 1965, Field 1964*a*, 1964*b*, Sugimoto 1965); magnetic field models (Kardašev 1964, Piddington 1964, 1966, Layzer 1965, Sturrock 1965, 1966); matter-antimatter annihilation (Gray and Browne 1965, Ekspong *et al.* 1966); sudden creation (Stothers 1965, 1966) and a model involving quarks (Blatt and Gutfreund 1966). None of them has gained any general acceptance.

The properties and nature of QSS have been reviewed by Schmidt (1964, 1966*a*) and by Burbidge, E. M. and Burbidge, G. R. (1966). A crucial issue is whether QSS are at the cosmological distances suggested by the large red-shifts found in many cases, or whether they could be comparatively local (within $\approx$ 10 Mpc). This question has been reviewed recently by Hoyle and Burbidge (1966*a*), by Schmidt (1966*a*) who favours the cosmological hypothesis and by Terrell (1966) who does not. Evidence and discussion on the matter includes that put forward by Robinson *et al.* (1963*a*), Terrell (1964), Dent (1965), Field (1965), Šklovskij (1965*c*, 1965*d*, 1966*a*), Williams, D. R. W. (1965), Barnothy, J. M. (1966*a*, 1966*b*, 1966*c*), Barnothy, M. F. and Barnothy, J. M. (1966), Faulkner *et al.* (1966), Hillier (1966), Hoyle and Burbidge (1966*b*, 1966*c*, 1966*d*), Hoyle *et al.* (1966), Hunter *et al.* (1966), Koehler and Robinson (1966), Lozinskaja (1966), Noerdlinger (1966), Odgers and Stewart (1966), Rees (1966), Rees and Sciama (1966*a*), Sandage and Miller (1966), Setti and Woltjer (1966), Strittmatter *et al.* (1966), Woltjer (1966) and Zapolsky (1966) in addition to that on flux variations etc. mentioned elsewhere.

Evidence not already cited which must be taken into account in considerations of the nature of extragalactic sources is that of their association with clusters of galaxies (Matthews *et al.* 1964, Pilkington 1964, Heidmann 1965, Ko 1965, Rogstad *et al.* 1965, Šolomickij and Kokin 1965, van den Bergh 1965*b*, Fomalont and Rogstad 1966, Wills 1966*b*), and with certain peculiar galaxies (Arp 1966, Holden 1966*a*, Lynden-Bell *et al.* 1966). An association is definitely established in the former case, and not so definitely in the latter, but in neither case is the significance entirely clear. A statistical investigation (Holden 1966*b*) of the 4C catalogue reveals no significant clustering on any scale from $0\overset{\circ}{.}5$ to 60°.

The luminosity distribution and luminosity function of sources have been discussed by Pskovskij (1964, 1965) Tovmasjan (1964), Longair and Scott (1965), Zasov and Ozernoy (1965) and Longair (1966*a*) and seem now to be fairly well established. They are closely related to the lifetimes of sources (Schmidt 1966*b*). Other statistical correlations between the physical properties of sources have been reviewed by Lequeux (1966) and discussed by Aizu *et al.* (1964*a*, 1965*b*), Zasov (1964) and Heeschen (1966), but at present this approach seems less promising than studies of individual sources in as much detail as possible, e.g. those by Heidmann (1963*a*, 1963*b*), by Matthews and Sandage (1963) of 3C 48, 3C 196 and 3C 286; by Greenstein and Schmidt (1964) of 3C 48 and 3C 273; by Matthews (1964, 1966) and Matthews *et al.* (1964) of the characteristics of galaxies identified with sources; by van der Laan (1963*a*), and by Cooper *et al.* (1965) of Centaurus A.

Although the origin of the energy remains a mystery, it may be possible to devise models of sources which will account for their observed properties. For radio galaxies, such attempts have been made by Kardašev 1963, van der Laan (1963*b*, 1964, 1966*a*), Šolomickij (1964), Gold, T. (1965), Kogure (1965, 1966), Pacholczyk (1965) and Kuril'čik (1966*b*) (see also Šklovskij 1963, 1966*b*, Woltjer 1964*b*, Sato 1966, Syrovatskij 1966).

More authors have been brave (or rash) enough to voice opinions on the nature of QSS, including Aizu *et al.* (1964*b*), Anand (1964), Ginzburg *et al.* (1964, 1965), Hoffmann (1964, 1966), Šklovskij (1964*a*, 1964*b*, 1964*c*, 1966*c*), Ahmad (1965), Kalinkov and Sadowsky (1965), Menon (1965), Ozernoy (1965, 1966), Rees and Sciama (1965*a*, 1965*b*), Williams, I. P. (1965), Zel'dovič (1965), Zel'dovič and Novikov (1964), Boccaletti *et al.* (1966), Falla (1966), Gordon (1966), Heidmann (1966), Ozernoy and Čertoprud (1966), Pfleiderer and Grewing (1966) and van der Laan (1966*b*).

Although there are strong reasons for believing that the synchroton mechanism is of prime importance in the radio radiation of QSS, other mechanisms may have a role (Gold, L. and Moffat 1964, Surdin 1965, Ginzburg and Ozernoy 1966*a*, 1966*b*, Gold, L. 1966, McCray 1966).

8. *Cosmology*

The implications of radio astronomical investigations in cosmology have been reviewed recently by Minkowski (1966) and in more detail by Davidson and Narlikar (1966) and Scheuer (1967). As Scheuer points out there are three distinct ways in which radio observations may be used to test cosmological models:

(*a*) *By determining limits for the mean density of the Universe.*

Attempts to detect emission or absorption by intergalactic neutral hydrogen have been made by Goldstein (1963), Davies and Jennison (1964) and Koehler and Robinson (1966) with no clearly positive results. The interpretation is discussed by Davies (1964) and Koehler (1966). Robinson *et al.* (1963*b*) and Koehler and Robinson (1966) have, however, found absorption due to hydrogen in the Virgo cluster of galaxies, although Goldstein (1966) could not detect this in emission, setting an upper limit of $5{\cdot}6 \times 10^{12}$ solar masses.

It may well be that the intergalactic medium is highly ionized (Ginzburg and Ozernoy 1965, Gunn and Peterson 1965, Scheuer 1965*d*, Gould and Ramsay 1966, Rees and Sciama 1966*b*). The free-free emission to be expected from it in this case has been considered by Field and Henry (1964), Kaufman (1965) and Weymann (1966) but has not been detected. Other possible effects are Thomson scattering (Kardašev and Šolomickij 1965), free-free absorption (Sciama 1964*a*) and dispersion (Haddock and Sciama 1965) in the intergalactic medium.

(*b*) *By measurement of residual radiation from the highly condensed stage of an evolving cosmology.*

An unexpected and comparatively intense microwave component has been found with a brightness temperature $\approx 3°K$ and a spectrum close to that of a black-body (Penzias and Wilkinson 1965, Howell and Shakeshaft 1966, Roll and Wilkinson 1966). A further point on the spectrum may come from observations of interstellar CN at λ 2·6 mm (Field and Hitchcock 1966*a*, 1966*b*, Thaddeus and Clauser 1966). The most plausible explanation (Dicke *et al.* 1965) is that this is residual radiation from a primordial fireball. The significance is discussed by Felten (1965), Kaufman (1965), Peebles (1965, 1966*a*, 1966*b*), Layzer (1966), O'Connell (1966), Peebles and Dicke (1966), and Sciama (1966) among others. Weymann (1966) points out that the intergalactic medium may cause modifications to the black-body spectrum.

(*c*) *By studying the development of the radio source population.*

There seems now to be substantial agreement between independent observers about the log N–log S relationship for sources. The largest range of flux densities is covered by the Cambridge observations (Gower 1966) but the slope of approximately − 1·8 for bright sources has also been found by Bolton *et al.* (1964) and Price and Milne (1965) in the Southern hemisphere. In addition Braccesi *et al.* (1966*c*) have presented some preliminary results. The interpretation of the number-counts has been discussed by Clarke, R. W. *et al.* (1963), Scott (1963), Davidson and Davies (1964*a*, 1964*b*), Scheuer and Shakeshaft (1964), Sciama (1964*b*, 1964*c*), Davidson (1966), McVittie and Schusterman (1966), Sciama and Saslaw (1966) and, in most detail, by Longair (1966*a*) who has used a digital computer to derive distributions of sources having luminosities and red-shifts chosen to reproduce the observed source counts, luminosity distribution and integrated sky brightness (see also Zasov 1966). It is clear that geometrical effects do not explain the number-counts, and source evolution is important, probably of the QSS (Bolton 1966, Longair 1966*a*, 1966*b*, Roeder and Mitchell 1966, Rowan-Robinson 1966, Sciama and Rees 1966, Véron 1966*b*, 1966*c*) which, together with the 3°K black-body radiation, is a powerful argument against the original form of the steady-state theory (Hoyle 1965, Sciama and Rees 1966) if the cosmological nature of QSS is confirmed.

Other evidence on cosmology from study of sources could come from variations of radio spectra (Ko 1966) or angular diameter (von Hoerner 1964*b*) at different flux levels, but the interpretation of such observations will be even more difficult (Scheuer 1967) than that of the number-counts. In addition Sandage (1964) is making use of identified radio galaxies to determine the *m*–*z* relationship.

BIBLIOGRAPHY

Adgie, R. L. 1964, *Nature*, **204**, 1028.
Adgie, R. L., Gent, H. 1966, *Nature*, **209**, 549.
Adgie, R. L., Gent, H., Slee, O. B., Frost, A. D., Palmer, H. P., Rowson, B. 1965, *Nature*, **208**, 275.
Ahmad, S. M. W. 1965, *Nuovo Cim.*, **37**, 1761.
Aizu, K. 1966, *Publ. astr. Soc. Japan*, **18**, 219.
Aizu, K., Fujimoto, Y., Hasegawa, H., Kawabata, K., Taketani, M. 1964*a*, *Progress theor. Phys.*, Osaka, Suppl. No. 31, 35.
Aizu, K., Fujimoto, Y., Hasegawa, H., Hayakawa, S., Taketani, M. 1964*b*, *Progress theor. Phy.*, Osaka, **32**, 971.
Aizu, K., Taketani, M., Fujimoto, Y., Hasegawa, H., Kawabata, K. 1965*a*, Proc. Int. Conf. Cosmic Rays, p. 129.
Aizu, K., Fujimoto, Y., Hasegawa, H., Taketani, M. 1965*b*, *Progress theor. Phys.*, Osaka, **33**, 996.
Allen, R. J., Barrett, A. H. 1966, *M.I.T. Res. Lab. Electron.*, *Quart. Prog. Rep.* No. 82, 17.
Anand, S. P. S. 1964, *Nature*, **201**, 1016.
Anderson, B., Donaldson, W., Palmer, H. P., Rowson, B. 1965, *Nature*, **205**, 375.
Andrew, B. H. 1967, *Astrophys. J.*, **147**, 423.
Argyle, E. 1965, *Astrophys. J.*, **141**, 750.
Arp, H. 1966, *Science*, **151**, 1214.

Artjuh, V. S., Vitkevič, V. V., Dagkesamanskij, R. D. 1966, Fiz. Inst. Akad. Nauk SSSR, report.
Baars, J. W. M., Mezger, P. G., Wendker, H. 1964, *Astr. J.*, **69**, 531.
Baars, J. W. M., Mezger, P. G., Wendker, H. 1965, *Nature*, **205**, 488.
Bailey, J. A., Branson, N. J. B. A., Elsmore, B., Scheuer, P. A. G. 1964, *Nature*, **201**, 755.
Balklav, A. E. 1964, *Izv. Akad. Nauk Latv. SSR*, Ser. Fiz. tekh. Nauk, No. 4, 47.
Barber, D., Donaldson, W., Miley, G. K., Smith, H. 1966, *Nature*, **209**, 753.
Barnothy, J. M. 1966*a*, *Observatory*, **86**, 115.
Barnothy, J. M. 1966*b*, *Astr. Nachr.*, **289**, 127.
Barnothy, J. M. 1966*c*, *Astr. J.*, **71**, 154.
Barnothy, M. F., Barnothy, J. M. 1966, *Astr. J.*, **71**, 155.
Bartlett, J. F. 1966, *Astr. J.*, **71**, 155.
Bazeljan, L. L., Braude, S. Ja., Vajsberg, V. V., Krymkin, V. V., Men', A. V., Sodin, L. G. 1964, *Dopov. Akad. Nauk. Ukr. RSR*, no. 11, 1464.
Bazeljan, L. L., Braude, S. Ja., Vajsberg, V. V., Krymkin, V. V., Men', A. V., Sodin, L. G. 1965*a*, *Astr. Zu.*, **42**, 618.
Bazeljan, L. L., Braude, S. Ja., Krymkin, V. V., Men', A. V., Sodin, L. G. 1965*b*, *Dopov. Akad. Nauk. Ukr. RSR*, no. 5, 580.
Bazeljan, L. L., Braude, S. Ja., Men', A. V. 1965*c*, *Astr. Cirk.*, no. 328.
Bazeljan, L. L., Braude, S. Ja., Men', A. V. 1966, *Astr. Cirk.*, no. 357.
Blatt, J. M., Gutfreund, H. 1966, *Phys. Lett.*, **23**, 94.
Boccaletti, D., De Sabbata, V., Gualdi, C. 1966, *Nuovo Cim.*, **42 B**, 175.
Boischot, A. 1965, *C. r. Acad. Sci. Paris*, **261**, 647.
Boischot, A., Kazès, I. 1965, *Ann. Astrophys.*, **28**, 75.
Boischot, A., Ginat, M., Kazès, I. 1963, *Ann. Astrophys.*, **26**, 85.
Boischot, A., Ginat, M., Kazès, I. 1965, *Ann. Astrophys.*, **28**, 84.
Boland, J. W., Hollinger, J. P., Mayer, C. H., McCullough, T. P. 1966, *Astrophys. J.*, **144**, 437.
Bologna, J. M., McClain, E. F., Rose, W. K., Sloanaker, R. M. 1965, *Astrophys. J.*, **142**, 106.
Bologna, J. M., McClain, E. F., Sloanaker, R. M. 1966, *Science*, **153**, 294.
Bolton, J. G. 1965, Extragalactic Radio Sources, in *The Structure and Evolution of Galaxies*, Interscience Publishers, p. 107.
Bolton, J. G. 1966, *Nature*, **211**, 917.
Bolton, J. G., Ekers, J. 1966*a*, *Austr. J. Phys.*, **19**, 275.
Bolton, J. G., Ekers, J. 1966*b*, *Austr. J. Phys.*, **19**, 471.
Bolton, J. G., Ekers, J. 1966*c*, *Austr. J. Phys.*, **19**, 559.
Bolton, J. G., Ekers, J. 1966*d*, *Austr. J. Phys.*, **19**, 713.
Bolton, J. G., Kinman, T. D. 1966, *Astrophys. J.*, **145**, 951.
Bolton, J. G., Gardner, F. F., Mackey, M. B. 1963, *Nature*, **199**, 682.
Bolton, J. G., Gardner, F. F., Mackey, M. B. 1964, *Austr. J. Phys.*, **17**, 340.
Bolton, J. G., Clarke, M. E., Sandage, A., Véron, P. 1965*a*, *Astrophys. J.*, **142**, 1289.
Bolton, J. G., Clarke, M. E., Ekers, R. D. 1965*b*, *Austr. J. Phys.*, **18**, 627.
Bolton, J. G., Shimmins, A. J., Ekers, J., Kinman, T. D., Lamla, E., Wirtanen, C. A. 1966, *Astrophys. J.*, **144**, 1229.
Botinelli, L., Gouguenheim, L., Heidmann, J., Heidmann, N., Weliachew, L. 1966, *C. r. Acad. Sci. Paris*, **263B**, 223.
Braccesi, A., Ceccarelli, M., Fanti, R., Gelato, G., Giovannini, C., Harris, D., Rosatelli, C., Sinigaglia, G., Volders, L. 1965, *Nuovo Cim.*, **40b**, 267.
Braccesi, A., Fanti, R., Giovannini, C., Vespignani, C. 1966*a*, *Astrophys. J.*, **143**, 600.
Braccesi, A., Ceccarelli, M., Fanti, R., Giovannini, G., Ficarra, A. 1966*b*, *Astrophys. J.*, **144**, 821.
Braccesi, A., Cecarelli, M., Fanti, R., Giovannini, C. 1966*c*, *Nuovo Cim.*, **16B**, 284.
Braude, S. Ja. 1963, *Dopov. Akad. Nauk. Ukr. RSR*, **2**, 193.
Braude, S. Ja. 1965*a*, *Astr. Zu.*, **42**, 1150.
Braude, S. Ja. 1965*b*, *Izv. Vyssih Učeb. Zaved. Radiofiz.*, **8**, 229.
Braude, S. Ja. 1966*a*, IAU Symp. No. 29 (in press).

Braude, S. Ja. 1966*b*, *Nature*, **210**, 80.
Braude, S. Ja., Men', A. V., Rjabov, B. P. 1966, *Astr. Cirk.*, no. 387.
Broten, N. W., Cooper, B. F. C., Gardner, F. F., Minnett, H. C., Price, R. M., Tonking, F. G., Yabsley, D. E. 1965, *Austr. J. Phys.*, **18**, 85.
Brown, L. W. 1966*a*, *Nature*, **210**, 1102.
Brown, L. W. 1966*b*, *Astrophys. J.*, **146**, 605.
Brundage, W. D., Kraus, J. D. 1966, *Science*, **153**, 411.
Burbidge, E. M. 1967, Quasi-Stellar Objects, in *A. Rev. Astr. Astrophys.*, **5** (in press).
Burbidge, E. M., Burbidge, G. R. 1964, *IEEE Trans. Mil. Electron.*, MIL–8, 165.
Burbidge, E. M., Burbidge, G. R. 1965*a*, Proc. Int. Conf. Cosmic Rays, p. 92.
Burbidge, E. M., Burbidge, G. R. 1965*b*, Theories of the Origin of Radio Sources, in *Quasi-Stellar Sources and Gravitational Collapse*, Ed. Robinson, I., *et al.*, Univ. of Chicago Press, Chicago, p. 291.
Burbidge, E. M., Burbidge, G. R. 1966, IAU Symp. No. 29 (in press).
Burbidge, G. R. 1964, Mechanisms of Energy Generation in Extragalactic Radio Sources, in *Physics of Nonthermal Radio Sources*, Eds. Maran, S. P., Cameron, A. G. W., Goddard Inst. Space Studies, NASA SP–46, p. 123.
Burbidge, G. R., Burbidge, E. M. 1965, Theories of the Origin of Radio Sources, in *The Structure and Evolution of Galaxies*, Interscience Publishers, p. 137.
Burbidge, G. R., Burbidge, E. M., Sandage, A. R. 1963, *Rev. mod. Phys.*, **35** 947.
Burn, B. J. 1966, *Mon. Not. R. astr. Soc.*, **133**, 67.
Burn, B. J., Sciama, D. W. 1964, On the Interpretation of Faraday Rotation Effects Associated with Radio Sources, in *Physics of Nonthermal Radio Sources*, Eds. Maran, S. P., Cameron, A. G. W., Goddard Inst. Space Studies, NASA SP–46, p. 139.
Bystrova, N. V., Gosačinskij, I. V., Egorova, T. M., Ryžkov, N. F. 1964, *Izv. glav. astr. Obs. Pulkove*, no. 177, 73.
Cameron, A. G. W. 1964, Collective Star Formation as a Contributor to Extragalactic Radio, Sources, in *Physics of Nonthermal Radio Sources*, Eds. Maran, S. P., Cameron, A. G. W., Goddard Inst. Space Studies, NASA SP–46, p. 133.
Cameron, A. G. W. 1965, *Nature*, **207**, 1140.
Caswell, J. L., Wills, D. 1965, *Nature*, **206**, 1241.
Caswell, J. L., Wills, D. 1967, *Mon. Not. R. astr. Soc.*, in press.
Clark, B. G., Hogg, D. E. 1966, *Astrophys. J.*, **145**, 21.
Clark, T. A. 1966, *Astr. J.*, **71**, 158.
Clarke, M. E. 1964, *Mon. Not. R. astr. Soc.*, **127**, 405.
Clarke, M. E. 1965, *Observatory*, **85**, 67.
Clarke, M. E., Bolton, J. G., Shimmins, A. J. 1966, *Austr. J. Phys.*, **19**, 375.
Clarke, R. W., Batchelor, R. A. 1965, *Nature*, **207**, 511.
Clarke, R. W., Scott, P. F., Smith, F. G. 1963, *Mon. Not. R. astr. Soc.*, **125**, 195.
Cohen, M. H. 1965, *Nature*, **208**, 277.
Cohen, M. H., Kundu, M. R. 1966, *Astrophys. J.*, **143**, 621.
Cohen, M. H., Gundermann, E. J., Hardebeck, H. E., Harris, D. E., Salpeter, E. E., Sharp, L. E. 1966, *Science*, **153**, 745.
Conway, R. G. 1964*a*, A Survey of the Spectra of Radio Sources, in *Physics of Nonthermal Radio Sources*, Eds. Maran, S. P., Cameron, A. G. W., Goddard Inst. Space Studies, NASA SP–46, p. 1.
Conway, R. G. 1964*b*, Recent Polarization Work at Jodrell Bank, in *Physics of Nonthermal Radio Sources*, Eds. Maran, S. P., Cameron, A. G. W., Goddard Inst. Space Studies, NASA SP–46, p. 69.
Conway, R. G., Moran, M. 1964, *Mon. Not. R. astr. Soc.*, **127**, 377.
Conway, R. G., Kellermann, K. I., Long, R. J. 1963, *Mon. Not. R. astr. Soc.*, **125**, 261.
Conway, R. G., Haslam, C. G. T., Kronberg, P. P., Slater, C. H. 1964, *Nature*, **201**, 756.
Conway, R. G., Daintree, E. J., Long, R. J. 1965, *Mon. Not. R. astr. Soc.*, **131**, 159.
Cooper, B. F. C., Price, R. M., Cole, D. J. 1965, *Austr. J. Phys.*, **18**, 589.
Davidson, W. 1966, *Mon. Not. R. astr. Soc.*, **132**, 389.
Davidson, W., Davies, M. 1964*a*, *Mon. Not. R. astr. Soc.*, **127**, 241.

Davidson, W., Davies, M. 1964*b*, *Mon. Not. R. astr. Soc.*, **128**, 363.
Davidson, W. J., Narlikar, J. V. 1966, *Rep. Prog. Phys.*, **29**, 539.
Davies, R. D. 1964, *Mon. Not. R. astr. Soc.*, **128**, 133.
Davies, R. D. 1966*a*, IAU Symp. No. 29 (in press).
Davies, R. D. 1966*b*, IAU Symp. No. 29 (in press).
Davies, R. D., Jennison, R. C. 1964, *Mon. Not. R. astr. Soc.*, **128**, 123.
Davies, R. D., Gottesman, S. T., Reddish, V. C., Verschuur, G. L. 1963, *Observatory*, **83**, 245.
Davis, M. M., Gelato-Volders, L., Westerhout, G. 1965, *Bull. astr. Inst. Netherl.*, **18**, 42.
Day, G. A., Shimmins, A. J., Ekers, R. D., Cole, D. J. 1966, *Austr. J. Phys.*, **19**, 35.
de Jong, M. L. 1965, *Astrophys. J.*, **142**, 1333.
de Jong, M. L. 1966*a*, *Astr. J.*, **71**, 373.
de Jong, M. L. 1966*b*, *Astrophys. J.*, **144**, 553.
de la Beaujardière, O. 1966, *Ann. Astrophys.*, **29**, 345.
Dent, W. A. 1965, *Science*, **148**, 1458.
Dent, W. A. 1966, *Astrophys. J.*, **144**, 843.
Dent, W. A., Haddock, F. T. 1965*a*, Radio Spectra of a Variety of Non-thermal Radio Sources, in *Quasi-Stellar Sources and Gravitational Collapse*, Eds. Robinson, I. *et al.*, Univ. of Chicago Press, Chicago, p. 381.
Dent, W. A., Haddock, F. T. 1965*b*, *Nature*, **205**, 487.
Dent, W. A., Haddock, F. T. 1966, *Astrophys. J.*, **144**, 568.
de Vaucouleurs, A. 1965, *Observatory*, **85**, 75.
Dicke, R. H., Peebles, P. J. E., Roll, P. G., Wilkinson, D. T. 1965, *Astrophys. J.*, **142**, 414.
Dickel, J. R., McVittie, G. C. 1965, *Astrophys. J.*, **142**, 1302.
Dickel, J. R., MacLeod, J. M., Swenson, G. W. 1965, *Science*, **150**, 883.
Dixon, R. S., Meng, S. Y., Kraus, J. D. 1965, *Nature*, **205**, 755.
Ekers, R. D., Bolton, J. G. 1965, *Austr. J. Phys.*, **18**, 669.
Ekspong, A. G., Yamdagni, N. K., Bonnevier, B. 1966, *Phys. Rev. Lett.*, **16**, 664.
Elvius, A. 1964, *Nature*, **201**, 171.
Epstein, E. E. 1964*a*, *Astr. J.*, **69**, 490.
Epstein, E. E. 1964*b*, *Astr. J.*, **69**, 521.
Epstein, E. E. 1965*a*, *Astrophys. J.*, **142**, 1282.
Epstein, E. E. 1965*b*, *Astrophys. J.*, **142**, 1285.
Epstein, E. E., Oliver, J. P., Schorn, R. A. 1966, *Astrophys. J.*, **145**, 367.
Erickson, W. C., Cronyn, W. M. 1965, *Astrophys. J.*, **142**, 1156.
Falla, D. F. 1966, *Nature*, **211**, 165.
Faulkner, J., Gunn, J. E., Peterson, B. A. 1966, *Nature*, **211**, 502.
Felten, J. E. 1965, *Phys. Rev. Lett.*, **15**, 1003.
Field, G. B. 1964*a*, *Nature*, **202**, 786.
Field, G. B. 1964*b*, *Astrophys. J.*, **140**, 1434.
Field, G. B. 1965, *Science*, **150**, 78.
Field, G. B., Henry, R. C. 1964, *Astrophys. J.*, **140**, 1002.
Field, G. B., Hitchcock, J. L. 1966*a*, *Phys. Rev. Lett.*, **16**, 817.
Field, G. B., Hitchcock, J. L. 1966*b*, *Astrophys. J.*, **146**, 1.
Findlay, J. W. 1966, Absolute Intensity Calibrations in Radio Astronomy, in *A. Rev. Astr. Astrophys.*, **4**, 77.
Findlay, J. W., Hvatum, H., Waltman, W. B. 1965, *Astrophys. J.*, **141**, 873.
Fomalont, E. B., Rogstad, D. H. 1966, *Astrophys. J.*, **146**, 528.
Fomalont, E. B., Matthews, T. A., Morris, D., Wyndham, J. 1964, *Astr. J.*, **69**, 772.
Fowler, W. A. 1964, *Trans. IAU*, **12 B**, 581.
Fowler, W. A. 1965*a*, Massive Stars, Relativistic Polytropes, and Gravitational Radiation, in *Quasi-Stellar Sources and Gravitational Collapse*, Eds. Robinson, I. *et al.*, Univ. of Chicago Press, Chicago, p. 51.
Fowler, W. A. 1965*b*, *Proc. Amer. phil. Soc.*, **109**, 181.
Gajlitis, A. K., Cytovič, V. N. 1963, *Izv. Vyssih Učeb. Zaved. Radiofiz.*, **6**, 1103.
Gajlitis, A. K., Cytovič, V. N. 1964, *Astr. Zu.*, **41**, 452.

Gardner, F. F., Bolton, J. G. 1965, *Austr. J. Phys.*, **18**, 671.
Gardner, F. F., Davies, R. D. 1964, *Nature*, **201**, 144.
Gardner, F. F., Davies, R. D. 1966, *Austr. J. Phys.*, **19**, 441.
Gardner, F. F., Price, R. M. 1967, *Austr. J. Phys.* (in press).
Gardner, F. F., Whiteoak, J. B. 1966, *A. Rev. Astr. Astrophys.*, **4**, 245.
Ginzburg, V. L. 1964, *Dokl. Akad. Nauk. SSSR*, **156**, 43.
Ginzburg, V. L. 1965, On the Magnetic Fields of Collapsing Masses and on the Nature of Superstars, in *Quasi-Stellar Sources and Gravitational Collapse*, Eds. Robinson, I. *et al.*, Univ. of Chicago Press, Chicago, p. 283.
Ginzburg, V. L., Ozernoy, L. M. 1964, *Zu. eksp. teor. Fiz.*, **47**, 1030.
Ginzburg, V. L., Ozernoy, L. M. 1965, *Astr. Zu.*, **42**, 943.
Ginzburg, V. L., Ozernoy, L. M. 1966*a*, *Izv. Vyssih Učeb. Zaved. Radiofiz.*, **9**, 221.
Ginzburg, V. L., Ozernoy, L. M. 1966*b*, *Astrophys. J.*, **144**, 599.
Ginzburg, V. L., Pisareva, V. V. 1963, *Izv. Vyssih Učeb. Zaved. Radiofiz.*, **6**, 877.
Ginzburg, V. L., Ozernoy, L. M., Syrovatskij, S. I. 1964, *Dokl. Akad. Nauk. SSSR*, **154**, 557.
Ginzburg, V. L., Ozernoy, L. M., Syrovatskij, S. I. 1965, On the Mechanism of Radiation of the Galaxy 3C 273 B, in *Quasi-Stellar Sources and Gravitational Collapse*, Eds. Robinson, I. *et al.*, Univ. of Chicago Press, Chicago, p. 277.
Gold, L. 1966, *Nature*, **210**, 927.
Gold, L., Moffatt, J. W. 1964, *Phys. Lett.*, **9**, 138.
Gold, T. 1965, Proc. Int. Conf. Cosmic Rays, p. 132.
Gold, T., Axford, W. I., Ray, E. C. 1965, The Stability of Multistar Systems, in *Quasi-Stellar Sources and Gravitational Collapse*, Eds. Robinson, I. *et al.*, Univ. of Chicago Press, Chicago, p. 93.
Goldstein, S. J. 1963, *Astrophys. J.*, **138**, 978.
Goldstein, S. J. 1966, *Science*, **151**, 71.
Gol'nev, V. Ja., Parijskij, Ju. N. 1965, *Astr. Zu.*, **42**, 305.
Gol'nev, V. Ja., Soboleva, N. S. 1965, *Astr. Zu.*, **42**, 694.
Gol'nev, V. Ja., Novožilova, G. G., Parijskij, Ju. N. 1965, *Izv. Vyssih Učeb. Zaved. Radiofiz.*, **8**, 183.
Gordon, I. M. 1966, *Astr. Cirk.*, no. 374.
Gottesman, S. T., Davies, R. D., Reddish, V. C. 1966, *Mon. Not. R. astr. Soc.*, **133**, 359.
Gould, R. J., Ramsay, W. 1966, *Astrophys. J.*, **144**, 587.
Gower, J. F. R. 1966, *Mon. Not. R. astr. Soc.*, **133**, 151.
Gower, J. F. R., Scott, P. F., Wills, D. 1967, *Mem. R. astr. Soc.* (in press).
Gray, G. K., Browne, P. F. 1965, *Nature*, **206**, 175.
Greenstein, J. L., Schmidt, M. 1964, *Astrophys. J.*, **140**, 1.
Griffin, R. F. 1963, *Astr. J.*, **68**, 421.
Guidice, D. A. 1966, *Nature*, **211**, 57.
Gunn, J. E., Peterson, B. A. 1965, *Astrophys. J.*, **142**, 1633.
Haddock, F. T., Hobbs, R. W. 1964, *Astr. J.*, **69**, 140.
Haddock, F. T., Sciama, D. W. 1965, *Phys. Rev. Lett.*, **14**, 1007.
Hazard, C. 1963, *Mon. Not. R. astr. Soc.*, **126**, 489.
Hazard, C. 1965, The Structure of the Extra-galactic Radio Sources, in *Quasi-Stellar Sources and Gravitational Collapse*, Eds. Robinson, I. *et al.*, Univ. of Chicago Press, Chicago, p. 135.
Hazard, C., Mackey, M. B., Nicholson, W. 1964, *Nature*, **202**, 227.
Hazard, C., Gulkis, S., Bray, A. D. 1966*a*, *Nature*, **212**, 461.
Hazard, C., Gulkis, S., Bray, A. D. 1966*b*, *Nature*, **210**, 888.
Heeschen, D. S. 1966, *Astrophys. J.*, **146**, 517.
Heeschen, D. S., Wade, C. M. 1964, *Astr. J.*, **69**, 277.
Heidmann, J. 1963*a*, *Publ. Obs. Paris*, **14**, 49.
Heidmann, J. 1963*b*, *Publ. Obs. Paris*, **14**, 55.
Heidmann, J. 1965, *Ann. Astrophys.*, **28**, 380.
Heidmann, J. 1966, *C. r. Acad. Sci. Paris*, **262B**, 1112.

Herzenstein, M. E., Aivazyan, Yu. M. 1966, *Phys. Lett.*, **22**, 62.
Hewish, A., Scott, P. F., Wills, D. 1964, *Nature*, **203**, 1214.
Hillier, R. R. 1966, *Nature*, **212**, 1334.
Hoffmann, B. 1964, *Science*, **144**, 319.
Hoffmann, B. 1966, *Science*, **152**, 671.
Hogg, D. E. 1966, *Nature*, **212**, 1336.
Hogg, D. E., Menon, T. K. 1966, *Astr. J.*, **71**, 388.
Höglund, B., Roberts, M. S. 1965, *Astrophys. J.*, **142**, 1366.
Holden, D. J. 1966*a*, *Observatory*, **86**, 229.
Holden, D. J. 1966*b*, *Mon. Not. R. astr. Soc.*, **133**, 225.
Hollinger, J. P., Mayer, C. H., Mennella, R. A. 1964, *Astrophys. J.*, **140**, 656.
Hornby, J. M., Williams, P. J. S. 1966, *Mon. Not. R. astr. Soc.*, **131**, 237.
Howard, W. E., Maran, S. P. 1965, *Astrophys. J. Suppl.*, **10**, no. 93, 1.
Howard, W. E., Dennis, T. R., Maran, S. P., Aller, H. D. 1964, *Nature*, **202**, 862.
Howard, W. E., Dennis, T. R., Maran, S. P., Aller, H. D. 1965, *Astrophys. J. Suppl.*, **10**, no. 93, 331.
Howell, T. F., Shakeshaft, J. R. 1966, *Nature*, **210**, 1318.
Hoyle, F. 1964, *Nature*, **201**, 804.
Hoyle, F. 1965, *Nature*, **208**, 111.
Hoyle, F., Burbidge, G. R. 1966*a*, *Astrophys. J.*, **144**, 534.
Hoyle, F., Burbidge, G. R. 1966*b*, *Nature*, **210**, 1346.
Hoyle, F., Burbidge, G. R. 1966*c*, *Nature*, **212**, 1223.
Hoyle, F., Burbidge, G. R. 1966*d*, *Nature*, **212**, 1334.
Hoyle, F., Fowler, W. A. 1965, Report on the Properties of Massive Objects, in *Quasi-Stellar Sources and Gravitational Collapse*, Eds. Robinson, I. *et al.*, Univ. of Chicago Press, Chicago, p. 17.
Hoyle, F., Burbidge, G. R., Sargent, W. L. W. 1966, *Nature*, **209**, 751.
Hrulev, V. V. 1963, *Izv. Vyssih Učeb. Zaved. Radiofiz.*, **6**, 398.
Hughes, M. P. 1965, *Nature*, **207**, 178.
Hunter, J. H., Sofia, S., Fletcher, E. 1966, *Nature*, **210**, 346.
Jennison, R. C. 1965, *Nature*, **207**, 740.
Kalinkov, M., Sadowsky, L. 1965, *C. r. Acad. Sci. Paris*, **260**, 4917.
Kardašev, N. S. 1963, *Astr. Zu.*, **40**, 965.
Kardašev, N. S. 1964, *Astr. Zu.*, **41**, 807.
Kardašev, N. S., Šolomickij, G. B. 1965, *Astr. Cirk.*, no. 336.
Kaufman, M. 1965, *Nature*, **207**, 736.
Kellermann, K. I. 1964*a*, *Astr. J.*, **69**, 205.
Kellermann, K. I. 1964*b*, *Publ. Owens Valley rad. Obs.*, **1**, No. 1.
Kellermann, K. I. 1964*c*, *Astrophys. J.*, **140**, 969.
Kellermann, K. I. 1966*a*, *Austr. J. Phys.*, **19**, 577.
Kellermann, K. I. 1966*b*, *Austr. J. Phys.*, **19**, 195.
Kellermann, K. I. 1966*c*, IAU Symp. No. 29 (in press).
Kellermann, K. I. 1966*d*, *Astrophys. J.*, **146**, 621.
Kellermann, K. I., Pauliny-Toth, I. I. K. 1966*a*, *Nature*, **212**, 781.
Kellermann, K. I., Pauliny-Toth, I. I. K. 1966*b*, IAU Symp. No. 29 (in press).
Kellermann, K. I., Read, R. B. 1965, *Publ. Owens Valley rad. Obs.*, **1**, no. 2.
Kenderdine, S., Baldwin, J. E. 1965, *Observatory*, **85**, 24.
Kenderdine, S., Ryle, M., Pooley, G. G. 1966, *Mon. Not. R. astr. Soc.*, **134**, 189.
Kinman, T. D. 1966, *Astrophys. J.*, **144**, 1232.
Ko, H. C. 1964, *Observatory*, **84**, 30.
Ko, H. C. 1965, *Astr. J.*, **70**, 681.
Ko, H. C. 1966, *Astrophys. J.*, **145**, 936.
Koehler, J. A. 1966, *Astrophys. J.*, **146**, 504.
Koehler, J. A., Robinson, B. J. 1966, *Astrophys. J.*, **146**, 488.
Kogure, T. 1965, *Publ. astr. Soc. Japan*, **17**, 252.
Kogure, T. 1966, *Publ. astr. Soc. Japan*, **18**, 243.

Kostenko, V. I., Matveenko, L. I. 1966, *Astr. Zu.*, **43**, 280.
Kraus, J. D. 1963, *Nature*, **198**, 844.
Kraus, J. D. 1964*a*, *Nature*, **202**, 269.
Kraus, J. D. 1964*b*, *Nature*, **202**, 1202.
Kraus, J. D., Dixon, R. S. 1965, *Nature*, **207**, 587.
Kraus, J. D., Dixon, R. S., Fisher, R. O. 1966, *Astrophys. J.*, **144**, 559.
Kuril'čik, V. N. 1964, *Astr. Cirk.*, no. 308.
Kuril'čik, V. N. 1965*a*, *Astr. Zu.*, **42**, 1138.
Kuril'čik, V. N. 1965*b*, *Astr. Cirk.*, no. 329.
Kuril'čik, V. N. 1966*a*, *Astr. Zu.*, **43**, 3.
Kuril'čik, V. N. 1966*b*, *Astr. Zu.*, **43**, 732.
Kuril'čik, V. N., Sycko, G. A. 1965, *Astr. Zu.*, **42**, 531.
Labrum, N. R., Krishnan, T., Payten, W. J., Harting, E. 1964, *Austr. J. Phys.*, **17**, 323.
Larionov, M. G., Šolomickij, G. B. 1966, *Astr. Cirk.*, no. 358.
Lastočkin, V. P., Porfir'ev, V. A., Stankevič, K. S., Troickij, V. S., Holodilov, N. N., Cejtlin, N. M. 1963*a*, *Izv. Vyssih Učeb. Zaved. Radiofiz.*, **6**, 629.
Lastočkin, V. P., Plankin, E. S., Stankevič, K. S. 1963*b*, *Izv. Vyssih Učeb. Zaved. Radiofiz.*, **6**, 631.
Lastočkin, V. P., Sorin, Ju. M., Stankevič, K. S. 1964, *Astr. Zu.*, **41**, 770.
Layzer, D. 1965, *Astrophys. J.*, **141**, 837.
Layzer, D. 1966, *Nature*, **211**, 576.
Leibacher, J. 1964, *Astr. J.*, **69**, 374.
Lequeux, J. 1965, *Ann. Astrophys.*, **28**, 360.
Lequeux, J. 1966, *Ann. Astrophys.*, **29**, 533.
Little, A. G. 1963, *Astrophys. J.*, **137**, 164.
Little, A. G., Bracewell, R. N. 1966, *Austr. J. Phys.*, **19**, 421.
Little, A. G., Cudaback, D. D., Bracewell, R. N. 1964, *Proc. nat. Acad. Sci. Am.*, **52**, 690.
Little, L. T., Hewish, A. 1966, *Mon. Not. R. astr. Soc.*, **134**, 221.
Long, R. J., Shuter, W. L. H. 1963, *Observatory*, **83**, 239.
Long, R. J., Haseler, J. B., Elsmore, B. 1963, *Mon. Not. R. astr. Soc.*, **125**, 313.
Long, R. J., Smith, M. A., Stewart, P., Williams, P. J. S. 1966, *Mon. Not. R. astr. Soc.*, **134**, 371.
Longair, M. S. 1965, *Mon. Not. R. astr. Soc.*, **129**, 419.
Longair, M. S. 1966*a*, *Mon. Not. R. astr. Soc.*, **133**, 421.
Longair, M. S. 1966*b*, *Nature*, **211**, 949.
Longair, M. S., Scott, P. F. 1965, *Mon. Not. R. astr. Soc.*, **130**, 379.
Low, F. J. 1965, *Astrophys. J.*, **142**, 1287.
Lozinskaja, T. A. 1966, *Astr. Cirk.*, no. 368.
Lynden-Bell, D., Cannon, R. D., Penston, M. V., Rothman, V. C. A. 1966, *Nature*, **211**, 838.
McCray, R. 1966, *Science*, **154**, 1320.
Macdonald, G. H., Neville, A. C., Ryle, M. 1966, *Nature*, **211**, 1241.
MacLeod, J. M. 1964, *Science*, **145**, 389.
MacLeod, J. M., Swenson, G. W., Yang, K. S., Dickel, J. R. 1965, *Astr. J.*, **70**, 756.
McVittie, G. C., Schusterman, L. 1966, *Astr. J.*, **71**, 137.
Makarenko, E. N. 1966, *Astr. Cirk.*, no. 374.
Maltby, P. 1966, *Astrophys. J.*, **144**, 219.
Maltby, P., Moffet, A. T. 1965*a*, *Science*, **150**, 63.
Maltby, P., Moffet, A. T. 1965*b*, *Astrophys. J.*, **142**, 409.
Maltby, P., Seielstad, G. A. 1966, *Astrophys. J.*, **144**, 216.
Maltby, P., Matthews, T. A., Moffet, A. T. 1963, *Astrophys. J.*, **137**, 153.
Maran, S. P., Cameron, A. G. W. (Eds.). 1964, *Physics of Nonthermal Radio Sources*, Goddard Inst. Space Studies, NASA SP–46.
Matthews, T. A. 1964, Radio Properties of Identified Extragalactic Sources, in *Physics of Nonthermal Radio Sources*, Eds. Maran, S. P., Cameron, A. G. W., Goddard Inst. Space Studies, NASA SP–46, p. 79.
Matthews, T. A. 1965, *Astr. J.*, **70**, 144.

Matthews, T. A. 1966, *Astr. J.*, **71**, 169.
Matthews, T. A., Sandage, A. R. 1963, *Astrophys. J.*, **138**, 30.
Matthews, T. A., Morgan, W. W., Schmidt, M. 1964, *Astrophys. J.*, **140**, 35.
Maxwell, A., Rinehart, R. 1966, *Astr. J.*, **71**, 927.
Mayer, C. H. 1964, N.R.L. Polarization Results, in *Physics of Nonthermal Radio Sources*, Eds. Maran, S. P., Cameron, A. G. W., Goddard Inst. Space Studies, NASA SP–46, p. 61.
Mayer, C. H., Hollinger, J. P., Allen, P. J. 1963, *Astrophys. J.*, **137**, 1309.
Mayer, C. H., McCullough, T. P., Sloanaker, R. M. 1964, *Astrophys. J.*, **139**, 248.
Mayer, C. H., McCullough, T. P., Sloanaker, R. M., Haddock, F. T. 1965, *Astrophys. J.*, **141**, 867.
Medd, W. J., Ramana, K. V. V. 1965*a*, *Astr. J.*, **70**, 327.
Medd, W. J., Ramana, K. V. V. 1965*b*, *Astrophys. J.*, **142**, 383.
Meng, S. Y., Kraus, J. D. 1966, *Astr. J.*, **71**, 170.
Menon, T. K. 1965, *Nature*, **206**, 810.
Miller, R. H., Parker, E. N. 1964, *Astrophys. J.*, **140**, 50.
Mills, B. Y., Glanfield, J. R. 1965, *Nature*, **208**, 10.
Milne, D. K., Scheuer, P. A. G. 1964, *Austr. J. Phys.*, **17**, 106.
Minkowski, R. 1964, The Identification Problem, in *Physics of Nonthermal Radio Sources*, Eds. Maran, S. P., Cameron, A. G. W., Goddard Inst. Space Studies, NASA SP–46, p. 87.
Minkowski, R. 1966, Radio Observations and Cosmology, in Proc. Int. Conf. Cosmology (Padua 1964), Ed. Barbera, G., Bologna Monograf, Bologna.
Moffet, A. T. 1964*a*, Brightness Distributions in Nonthermal Radio Sources, in *Physics of Nonthermal Radio Sources*, Eds. Maran, S. P., Cameron, A. G. W., Goddard Inst. Space Studies, NASA SP–46, p. 19.
Moffet, A. T. 1964*b*, *Trans. IAU*, **12 B**, 569.
Moffet, A. T. 1964*c*, *Science*, **146**, 764.
Moffet, A. T. 1965, *Astrophys. J.*, **141**, 1580.
Moffet, A. T. 1966*a*, IAU Symp. No. 29 (in press).
Moffet, A. T. 1966*b*, *A. Rev. Astr. Astrophys.*, **4**, 145.
Moffet, A. T., Palmer, H. P. 1965, *Observatory*, **85**, 45.
Morris, D., Berge, G. L. 1964, *Astr. J.*, **69**, 641.
Morris, D., Radhakrishnan, V., Seielstad, G. A. 1964*a*, *Astrophys. J.*, **139**, 758.
Morris, D., Radhakrishnan, V., Seielstad, G. A. 1964*b*, *Astrophys. J.*, **139**, 551.
Morris, D., Radhakrishnan, V., Seielstad, G. A. 1964*c*, *Astrophys. J.*, **139**, 560.
Morris, D., Radhakrishnan, V., Seielstad, G. A. 1964*d*, *Astr. J.*, **69**, 145.
Nash, R. T. 1965, *Astr. J.*, **70**, 846.
Ne'eman, Y. 1965, *Astrophys. J.*, **141**, 1303.
Nicholson, G. D. 1966, *Nature*, **210**, 611.
Nicholson, W. 1965, *H.M. Nautical Almanac Office, Tech. Note* no. 3.
Noerdlinger, P. D. 1966, *Astrophys. J.*, **143**, 1004.
Novikov, I. D. 1964, *Astr. Zu.*, **41**, 1075.
O'Connell, R. F. 1966, *Phys. Rev. Lett.*, **17**, 1232.
Odgers, G. J., Stewart, R. W. 1966, *Nature*, **211**, 1284.
Ohtani, H., Kogure, T. 1964, *Publ. astr. Soc. Japan*, **16**, 206.
Ozernoy, L. M. 1965, *Dokl. Akad. Nauk SSSR*, **163**, 54.
Ozernoy, L. M. 1966, *Astr. Zu.*, **43**, 300.
Ozernoy, L. M., Čertoprud, V. E. 1966, *Astr. Cirk.*, no. 388.
Pacholczyk, A. G. 1963, *Nature*, **200**, 765.
Pacholczyk, A. G. 1965, *Astrophys. J.*, **142**, 1141.
Palmer, H. P. 1965, *Contemp. Phys.*, **6**, 401.
Palmer, H. P. 1966, IAU Symp. No. 29 (in press).
Parijskij, Ju. N., Prozorov, V. A. 1964, *Izv. glav. astr. Obs. Pulkove*, no. 172, 9.
Parijskij, Ju. N., Timofeeva, G. M. 1964, *Astr. Zu.*, **41**, 3.
Parker, E. A., Elsmore, B., Shakeshaft, J. R. 1966, *Nature*, **210**, 22.
Pauliny-Toth, I. I. K., Kellermann, K. I. 1966, *Astrophys. J.*, **146**, 634.
Pauliny-Toth, I. I. K., Wade, C. S., Heeschen, D. S. 1966, *Astrophys. J. Suppl.*, **13**, no. 116, 65.

Peebles, P. J. E. 1965, *Astrophys. J.*, **142**, 1317.
Peebles, P. J. E. 1966*a*, *Phys. Rev. Lett.*, **16**, 410.
Peebles, P. J. E. 1966*b*, *Astrophys. J.*, **146**, 542.
Peebles, P. J. E., Dicke, R. H. 1966, *Nature*, **211**, 574.
Penzias, A. A., Wilson, R. W. 1965, *Astrophys. J.*, **142**, 419.
Pfleiderer, J., Grewing, M. 1966, *Science*, **154**, 1452.
Piddington, J. H. 1964, *Mon. Not. R. astr. Soc.*, **128**, 345.
Piddington, J. H. 1966, *Mon. Not. R. astr. Soc.*, **133**, 163.
Pilkington, J. D. H. 1964, *Mon. Not. R. astr. Soc.*, **128**, 103.
Pilkington, J. D. H., Scott, P. F. 1965, *Mem. R. astr. Soc.*, **69**, 183.
Price, R. M., Milne, D. K. 1965, *Austr. J. Phys.*, **18**, 329.
Priester, W., Rosenberg, J. 1964, *Extragalactic Radio Sources*, NASA, Goddard Instit. for Space Studies, New York, pp. 114. (Also NASA Tech. Note TN D–2888, July 1965).
Pskovskij, Ju. P. 1964, *Astr. Zu.*, **41**, 619.
Pskovskij, Ju. P. 1965, *Vest. Mosk. gos. Univ.* (Ser. 3, Astr. Fiz.), no. 1, 3.
Radhakrishnan, V. 1964, Polarization Studies of Discrete Radio Sources, in *Physics of Nonthermal Radio Sources*, Ed. Maran, S. P., Cameron, A. G. W., Goddard Inst. Space Studies, NASA SP–46, p. 53.
Razin, V. A., Fedorov, V. T. 1963, *Izv. Vyssih Učeb. Zaved. Radiofiz.*, **6**, 1052.
Rees, M. J. 1966, *Nature*, **211**, 468.
Rees, M. J., Sciama, D. W. 1965*a*, *Nature*, **207**, 738.
Rees, M. J., Sciama, D. W. 1965*b*, *Nature*, **208**, 371.
Rees, M. J., Sciama, D. W. 1966*a*, *Nature*, **211**, 805.
Rees, M. J., Sciama, D. W. 1966*b*, *Astrophys. J.*, **145**, 6.
Roberts, J. A. 1964, Linear Polarization Effects in Radio Sources, in *Physics of Nonthermal Radio Sources*, Eds. Maran, S. P., Cameron, A. G. W., Goddard Inst. Space Studies, NASA SP–46, p. 37.
Roberts, M. S. 1965, *Astrophys. J.*, **142**, 148.
Roberts, M. S. 1966*a*, IAU Symp. No. 31 (in press).
Roberts, M. S. 1966*b*, *Astrophys. J.*, **144**, 639.
Roberts, M. S. 1966*c*, *Phys. Rev. Lett.*, **17**, 1203.
Roberts, M. S. 1967, Radio Observations of Neutral Hydrogen in Galaxies, in *Galaxies and the Universe*, Eds. Sandage, A. and M., Vol. 9, of *Stars and Stellar Systems*, Univ. of Chicago Press, Chicago, Chap. 11 (in press).
Robinson, B. J., Koehler, J. A. 1965, *Nature*, **208**, 993.
Robinson, B. J., van Damme, K. J. 1966, *Austr. J. Phys.*, **19**, 111.
Robinson, B. J., van Damme, K. J., Koehler, J. A. 1963*a*, *Nature*, **199**, 990.
Robinson, B. J., van Damme, K. J., Koehler, J. A. 1963*b*, *Nature*, **199**, 1176.
Robinson, I., Schild, A., Schücking, E. L. (Eds.). 1965, *Quasi-stellar Sources and Gravitational Collapse*, Univ. of Chicago Press, Chicago.
Roeder, R. C., Mitchell, G. F. 1966, *Nature*, **212**, 165.
Roger, R. S., Costain, C. H., Purton, C. R. 1965, *Nature*, **207**, 62.
Rogstad, D. H., Rougoor, G. W., Whiteoak, J. B. 1965, *Astrophys. J.*, **142**, 1665.
Roll, P. G., Wilkinson, D. T. 1966, *Phys. Rev. Lett.*, **16**, 405.
Rose, W. K. 1964, Linear Polarization of Discrete Radio Sources by Use of a 9·4 cm. Maser, in *Physics of Nonthermal Radio Sources*, Eds. Maran, S. P., Cameron, A. G. W., Goddard Inst. Space Studies, NASA SP–46, p. 75.
Rowan-Robinson, M. 1966, *Nature*, **212**, 1556.
Ryle, M. 1966, Developments in Radio Source Work since 1960, in *Progress in Radio Science 1960–63*, Vol. V, *Radio Astronomy*, Ed. Herbays, E., Elsevier Publ. Co., Amsterdam, p. 17.
Ryle, M., Sandage, A. R. 1964, *Astrophys. J.*, **139**, 419.
Ryle, M., Elsmore, B., Neville, A. C. 1965*a*, *Nature*, **205**, 1259.
Ryle, M., Elsmore, B., Neville, A. C. 1965*b*, *Nature*, **207**, 1024.
Sandage, A. 1966*a*, *Astrophys. J.*, **145**, 1.
Sandage, A. 1966*b*, Radio Sources and the Expansion of the Universe, in Proc. Int. Conf. Cosmology (Padua 1964), Ed. Barbera, G., Bologna Monograf, Bologna.

Sandage, A., Miller, W. C. 1966, *Astrophys. J.*, **144**, 1238.
Sato, F. 1966, *Publ. astr. Soc. Japan*, **18**, 229.
Scheuer, P. A. G. 1963, *Austr. J. Phys.*, **15**, 333.
Scheuer, P. A. G. 1965*a*, Radio Structure of 3C 273 and Spectra of Radio Sources, in *Quasi-Stellar Sources and Gravitational Collapse*, Eds. Robinson, I. *et al.*, Univ. of Chicago Press, Chicago, p. 373.
Scheuer, P. A. G. 1965*b*, *Austr. J. Phys.*, **18**, 77.
Scheuer, P. A. G. 1965*c*, *Mon. Not. R. astr. Soc.*, **129**, 199.
Scheuer, P. A. G. 1965*d*, *Nature*, **207**, 963.
Scheuer, P. A. G. 1967, Radio Astronomy and Cosmology, in *Galaxies and the Universe*, Eds. Sandage, A. and M., Vol. 9 of *Stars and Stellar Systems*, Univ. of Chicago Press, Chicago, Chap. 9 (in press).
Scheuer, P. A. G., Shakeshaft, J. R. 1964, *Mon. Not. R. astr. Soc.*, **128**, 361.
Scheuer, P. A. G., Wills, D. 1966, *Astrophys. J.*, **143**, 274.
Schmidt, M. 1964, *Trans. IAU*, **12 B**, 571.
Schmidt, M. 1966*a*, IAU Symp. No. 29 (in press).
Schmidt, M. 1966*b*, *Astrophys. J.*, **146**, 7.
Sciama, D. W. 1964*a*, *Nature*, **204**, 767.
Sciama, D. W. 1964*b*, *Observatory*, **84**, 261.
Sciama, D. W. 1964*c*, *Mon. Not. R. astr. Soc.*, **128**, 49.
Sciama, D. W. 1966, *Nature*, **211**, 277.
Sciama, D. W., Rees, M. J. 1966, *Nature*, **211**, 1283.
Sciama, D. W., Saslaw, W. C. 1966, *Nature*, **210**, 348.
Scott, P. F. 1963, *Mon. Not. R. astr. Soc.*, **127**, 37.
Seielstad, G. A. 1966, *Astr. J.*, **71**, 180.
Seielstad, G. A., Whiteoak, J. B. 1965, *Astrophys. J.*, **142**, 616.
Seielstad, G. A., Wilson, R. W. 1963, *Nature*, **198**, 274.
Seielstad, G. A., Morris, D., Radhakrishnan, V. 1963, *Astrophys. J.*, **138**, 602.
Seielstad, G. A., Morris, D., Radhakrishnan, V. 1964, *Astrophys. J.*, **140**, 53.
Setti, G., Woltjer, L. 1966, *Astrophys. J.*, **144**, 838.
Shakeshaft, J. R. 1966, IAU Symp. No. 29 (in press).
Shakeshaft, J. R., Longair, M. S. 1965, *Observatory*, **85**, 30.
Shapiro, I. I., Weinreb, S. 1966, *Astrophys. J.*, **143**, 598.
Shimmins, A. J., Day, G. A., Ekers, R. D., Cole, D. J. 1966*a*, *Austr. J. Phys.*, **19**, 837.
Shimmins, A. J., Clarke, M. E., Ekers, R. D. 1966*b*, *Austr. J. Phys.*, **19**, 649.
Shobbrook, R. R. 1963, *Observatory*, **83**, 36.
Shobbrook, R. R., Hunstead, R. W. 1966, *Observatory*, **86**, 204.
Sinigaglia, G. 1966, *Nature*, **212**, 601.
Slepcova, N. F., Šolomickij, G. B., Kuril'čik, V. N. 1965, *Astr. Cirk.*, no. 335.
Slyš, V. I. 1963, *Nature*, **199**, 682.
Slyš, V. I. 1964, *Astr. Cirk.*, no. 300.
Slyš, V. I. 1965, *Astr. Zu.*, **42**, 689.
Smith, H. J. 1966, *Appl. Opt.*, **5**, 1701.
Soboleva, N. S. 1966, *Astr. Zu.*, **43**, 266.
Soboleva, N. S., Timofeeva, G. M. 1963, *Dokl. Akad. Nauk SSSR*, **153**, 555.
Souffrin, S., Chesnay, N. 1965, *C. r. Acad. Sci. Paris*, **261**, 3541.
Stankevič, K. C. 1964, *Izv. Vyssih Učeb. Zaved. Radiofiz.*, **7**, 202.
Stothers, R. 1965, *Nature*, **206**, 82.
Stothers, R. 1966, *Mon. Not. R. astr. Soc.*, **132**, 217.
Strittmatter, P., Faulkner, J., Walmsley, M. 1966, *Nature*, **212**, 1441.
Sturrock, P. A. 1965, *Nature*, **205**, 861.
Sturrock, P. A. 1966, *Nature*, **211**, 697.
Sugimoto, D. 1965, *Progress theor. Phys.*, Osaka, **33**, 339.
Surdin, M. 1965, *Planet. Space Sci.*, **13**, 861.
Syrovatskij, S. I. 1966, *Astr. Zu.*, **43**, 340.
Šklovskij, I. S. 1963, *Astr. Zu.*, **40**, 972.

Šklovskij, I. S. 1964a, *Astr. Zu.*, **41**, 176.
Šklovskij, I. S. 1964b, *Astr. Zu.*, **41**, 801.
Šklovskij, I. S. 1964c, *Nature*, **201**, 588.
Šklovskij, I. S. 1965a, *Nature*, **206**, 176.
Šklovskij, I. S. 1965b, *Astr. Zu.*, **42**, 30.
Šklovskij, I. S. 1965c, *Astr. Cirk.*, no. 332.
Šklovskij, I. S. 1965d, *Astr. Zu.*, **42**, 893.
Šklovskij, I. S. 1966a, *Astr. Cirk.*, no. 372.
Šklovskij, I. S. 1966b, *Astr. Cirk.*, no. 369.
Šklovskij, I. S. 1966c, IAU Symp. No. 29 (in press).
Šolomickij, G. B. 1964, *Astr. Cirk.*, no. 284.
Šolomickij, G. B. 1965a, *Inf. Bull. Variable Stars*, no. 83.
Šolomickij, G. B. 1965b, *Astr. Zu.*, **42**, 673.
Šolomickij, G. B. 1966a, *Astr. Cirk.*, no. 359.
Šolomickij, G. B. 1966b, *Astr. Cirk.*, no. 358.
Šolomickij, G. B., Kokin, Y. F. 1965, *Astr. Zu.*, **42**, 674.
Šolomickij, G. B., Kuril'čik, V. N., Matveenko, L. I., Hromov, G. S. 1964a, *Astr. Zu.*, **41**, 823.
Šolomickij, G. B., Kuril'čik, V. N. 1964b, *Astr. Cirk.*, no. 283.
Šolomickij, G. B., Slepcova, N. F., Matveenko, L. I. 1965a, *Astr. Zu.*, **42**, 1135.
Šolomickij, G. B., Eršov, N. V., Saharov, A. E., Larionov, M. G. 1965b, *Astr. Cirk.*, no. 335.
Tanaka, H., Kakinuma, T., Yamashita, T., Uchida, H., Inaba, H., Tsuru, H., Takahashi, K., Yoshikawa, S., Hozumi, H., Hayashi, H. 1965, *Proc. Res. Inst. Atmos. Nagoya Univ.*, **12**, 35.
Taylor, J. H. 1966, *Astrophys. J.*, **146**, 646.
Terrell, J. 1964, *Science*, **145**, 918.
Terrell, J. 1966, *Science*, **154**, 1281.
Thaddeus, P., Clauser, J. F. 1966, *Phys. Rev. Lett.*, **16**, 819.
Thompson, A. R., Krishnan, T. 1965, *Astrophys. J.*, **141**, 19.
Tovmasjan, G. M. 1964, *Izv. Akad. Nauk Arm. SSR*, Ser. fiz. mat., **17**, 137.
Tovmasjan, G. M. 1966a, IAU Symp. No. 29 (in press).
Tovmasjan, G. M. 1966b, *Austr. J. Phys.*, **19**, 565.
Ulam, S. M., Walden, W. E. 1964, *Nature*, **201**, 1202.
van Damme, K. J. 1966, *Austr. J. Phys.*, **19**, 687.
van den Bergh, S. 1965a, *Astr. J.*, **70**, 124.
van den Bergh, S. 1965b, *Astrophys. J.*, **141**, 1579.
van der Laan, H. 1963a, *Mon. Not. R. astr. Soc.*, **126**, 519.
van der Laan, H. 1963b, *Mon. Not. R. astr. Soc.*, **126**, 535.
van der Laan, H. 1964, A Model of Radio Galaxies, in *Physics of Nonthermal Radio Sources*, Eds. Maran, S. P., Cameron, A. G. W., Goddard Inst. Space Studies, NASA SP–46, p. 147.
van der Laan, H. 1966a, *Z. Astrophys.*, **64**, 16.
van der Laan, H. 1966b, *Nature*, **211**, 1131.
Venugopal, V. R. 1963, *Publ. astr. Soc. Pacific*, **75**, 404.
Véron, P. 1966a, *Astrophys. J.*, **144**, 861.
Véron, P. 1966b, *Nature*, **211**, 724.
Véron, P. 1966c, *Ann. Astrophys.*, **29**, 231.
Vetuhnovskaja, Ju. N., Kuz'min, A. D. 1965, *Trudy fiz. Inst. Akad Nauk SSSR*, **28**, 155.
Vitkevič, V. V., Antonova, T. D., Vlasov, V. I. 1966, *Dokl. Akad. Nauk SSSR*, **168**, 55.
von Hoerner, S. 1964a, *Astrophys. J.*, **140**, 65.
von Hoerner, S. 1964b, *IEEE Trans. Mil. Electron.*, MIL–8, 282.
von Hoerner, S. 1965, *Astrophys. J.*, **142**, 1265.
von Hoerner, S. 1966, *Astrophys. J.*, **144**, 483.
Wade, C. M. 1966, *Phys. Rev. Lett.*, **17**, 1061.
Wade, C. M., Clark, B. G., Hogg, D. E. 1965, *Astrophys. J.*, **142**, 406.
Wagoner, R. 1966, *Phys. Rev. Lett.*, **16**, 249.
Westerlund, B. E., Smith, L. F. 1966, *Austr. J. Phys.*, **19**, 181.

Weymann, R. 1966, *Astrophys. J.*, **145**, 560.
Williams, D. R. W. 1965, Galactic Interstellar Absorption Lines in the Spectrum of 3C 273, in *Quasi-Stellar Sources and Gravitational Collapse*, Eds. Robinson, I. *et al.*, Univ. of Chicago Press, Chicago, p. 213.
Williams, D. R. W., Welch, W. J., Thornton, D. D. 1965, *Publ. astr. Soc. Pacific*, **77**, 178.
Williams, I. P. 1965, *Phys. Lett.*, **14**, 19.
Williams, P. J. S. 1963, *Nature*, **200**, 56.
Williams, P. J. S. 1966*a*, *Nature*, **210**, 285.
Williams, P. J. S. 1966*b*, *Observatory*, **86**, 67.
Williams, P. J. S., Kenderdine, S., Baldwin, J. E. 1966, *Mem. R. astr. Soc.*, **70**, 53.
Wills, D. 1966*a*, *Observatory*, **86**, 245.
Wills, D. 1966*b*, *Observatory*, **86**, 140.
Wills, D., Parker, E. A. 1966, *Mon. Not. R. astr. Soc.*, **131**, 503.
Wilson, R. W., Penzias, A. A. 1966, *Astrophys. J.*, **146**, 286.
Woltjer, L. 1964*a*, *Nature*, **201**, 803.
Woltjer, L. 1964*b*, Dynamics of Magnetic Fields in Radio Sources, in *Physics of Nonthermal Radio Sources*, Eds. Maran, S. P., Cameron, A. G. W., Goddard Inst. Space Studies, NASA SP–46, p. 155.
Woltjer, L. 1966, *Astrophys. J.*, **146**, 597.
Wyndham, J. D. 1965, *Astr. J.*, **70**, 384.
Wyndham, J. D. 1966, *Astrophys. J.*, **144**, 459.
Wyndham, J. D., Read, R. B. 1965, *Astr. J.*, **70**, 120.
Yang, K. S., Dickel, J. R. 1966, *Astr. J.*, **71**, 188.
Yokoi, H., Satoh, T., Yamada, M. 1966, *Publ. astr. Soc. Japan*, **18**, 271.
Zaharov, A. V., Kromikov, V. D., Troickij, V. S., Cejtlin, N. M. 1964, *Izv. Vyssih Učeb. Zaved. Radiofiz.*, **7**, 553.
Zapolsky, H. S. 1966, *Science*, **153**, 635.
Zasov, A. V. 1964, *Astr. Cirk.*, no. 291.
Zasov, A. V. 1966, *Astr. Zu.*, **43**, 7.
Zasov, A. V., Ozernoy, L. M. 1965, *Astr. Cirk.*, no. 334.
Zel'dovič, Ja. B. 1965, *Astr. Zu.*, **42**, 283.
Zel'dovič, Ja. B., Novikov, I. D. 1964, *Dokl. Akad. Nauk SSSR*, **158**, 811.
Zel'dovič, Ja. B., Novikov, I. D. 1966*a*, IAU Symp. No. 29 (in press).
Zel'dovič, Ja. B., Novikov, I. D. 1966*b*, *Astr. Zu.*, **43**, 758.
Zisk, S. H. 1965, *Astr. J.*, **70**, 334.

F. RADIOASTRONOMICAL INSTRUMENTS

(prepared by C. A. Muller)

In the last three years a number of new radiotelescopes have come into use. General trends in the design of new radiotelescopes and in receiver techniques are higher operating frequencies and the increased use of digital techniques, including the use of on-line computers for the control of telescopes, receivers and data processing equipment. In this report we will summarize the major developments in the fields of radiotelescopes and receivers.

1. *Filled-aperture telescopes*

Several new large reflector telescopes have come into use, whose design reflects the general tendency towards increased surface accuracies and higher operating frequencies. Recently completed instruments in the U.S.A. include a 43 metre reflector (**1**) (Greenbank), a 36 meter reflector (Haystack) (**2**) and a 25 meter telescope (N.R.L.) (**3**), all having good efficiencies at centimeter wavelengths. In Canada the 46 meter reflector (**4**) at Algonquin Park, also operating at cm wavelengths has been completed, while in the United Kingdom the new Jodrell Bank Mk II (**5**) telescope with an elliptically shaped reflector, 24 by 36 meter has come into regular use.* Millimeter wavelength instruments include two 5 meter reflectors (**6, 7**) used at 3 millimeters, and an 11 meter reflector (**8**) under construction, all in the U.S.A.

* See Addendum on page 908.

The fixed-tiltable reflector telescopes of Ohio (**9**) and of Nançay (France) (**10**) are now completed, while the Pulkovo telescope in the U.S.S.R. has been reconstructed for use at centimeter wavelengths (**11**).

New plans include a 90 meter telescope in Western Germany (**12**).

2. *Unfilled-aperture telescopes*

The clear distinctions between the various types of unfilled-aperture telescopes gradually seem to vanish, as several new instruments can be used in different modes of operation, corresponding to previously separate instruments. Examples are the compound interferometers in Japan, which also can be used as crosses (see below), the 10 cm multielement interferometer under construction at Algonquin (**4**), and the proposed eight-element telescope for Owens Valley (U.S.A.) (**13**), that includes the possible use as a total power array, as a cross, as a compound interferometer and as an Earth rotation synthesis instrument. Earth rotation synthesis has been shown to be a powerful approach to pencil-beam unfilled-aperture telescopes and is incorporated now in several new designs.

The Cambridge (U.K.) One-Mile telescope (**14**) consisting of three 18 meter reflectors, is now in full operation on 408 and 1407 MHz with beamwidths of 80″ and 23″ respectively. A similar instrument using 12 telescopes 25 meters in diameter with the same beamwidth at 21 cm is under construction near Westerbork in the Netherlands. The original Christiansen grating cross is being converted into a pair of compound interferometers, which system will give a 40″ beam at 21 cm with the use of Earth rotation synthesis (**15**). A compound interferometer that should give a 4′ beam at 200 MHz is under construction near Peking (China) (**16**). At Greenbank a two-element interferometer, using 25 meter reflectors and operating at 10 cm (**17**), has been operated as a prototype study for a very large array telescope. This interferometer is now being extended to a three element system. At Owens Valley the reflectors of the two-element interferometer have been resurfaced for better efficiency at 10 cm (**18**). A 40 meter reflector is being added to the system.

The large cross telescopes near Kkarkov and Serpukhov (**19, 20**) in the U.S.S.R., near Medicina (Italy) (**21**) and near Canberra (Australia) (**22**) are now nearing completion and in partial operation. Low frequency cross arrays operating on 10 and 22 MHz have been built near Penticton (Canada) (**23**). A 500 meters long reflector similar to one arm of a cross is under construction in India, which instrument will be used primarily for lunar occultation studies, operating at 240 MHz. Descriptions have been published for the Clark Lake decametric array (**24**) and a radar astronomy array operating in the range 20–55 MHz (**25**), both in the U.S.A.

The solar ring telescope (**26**) near Culgoora (Australia), consisting of 96 reflectors 13 meters in diameter arranged in a ring 3 kilometers in diameter, is expected to be completed during 1967.

New telescopes for solar observations in Japan include the enlarged compound interferometer at Toyokawa (**27, 28**), which has a 21″ fan beam on 3 cm, but which also can be operated as a T-shaped interferometer with a 1·5′ pencil beam. A similar instrument is under construction for a wavelength of 8 centimeter, while at Mitaka an eight-reflector grating interferometer with a 4′ beam at 17 GHz is being built. Plans exist for a large grating interferometer at 160 MHz at Nobeyama with either a 1·5′ fanbeam or a 4′ by 10′ pencil beam (**29**).

A long-baseline interferometer has been operated in the United Kingdom between Jodrell Bank and Malvern at 21 and 11 cm wavelength with satisfactory phase stability (**30**). Very-long-baseline experiments will be undertaken in Canada and in the U.S.A., using completely separate receivers with atomic frequency standards as local oscillators and intermediate-frequency videotape-recording and correlation between the recorded signals with a large computer. The Canadian project will use a 3000 kilometer baseline between Algonquin and Penticton and will operate at 448 MHz (**31**).

Equipment for absolute flux density measurements using standard horn antennas is in use at Greenbank (**32**), in Canada and in Japan, the latter being used for solar flux measurements at 1000, 2000 and 3750 MHz (**33**). A new method for gain calibrations of large reflectors (**34**) using an artificial satellite, has been used for calibration of a 20 meter reflector in Japan.

3. *Receivers*

The use of low-noise receivers with masers, parametric amplifiers and tunnel diode amplifiers has further increased. A survey of the use of these amplifiers has been made recently by Robinson (**35**). Descriptions of receivers using maser amplifiers (**36, 37, 38, 39, 40, 41**) or parametric amplifiers (**42, 43, 44**) have been given by several authors. In Australia a solar spectrograph using an up-converter (**45**) as a first stage is in regular use. A new multichannel high sensitivity spectrograph is under construction in the Netherlands (**46**).

Important advances have been made in the techniques for accurate antenna temperature measurements as has been shown by the recent observations of a weak isotropic component in the cosmic background radiation in the U.S.A. (**47, 48**) and the United Kingdom (**49**).

A number of multichannel receivers is in use for H- (**50**) and OH-line (**51**) observations. Some H-line receivers use autocorrelation techniques for determining the spectral distribution either in analog (**52**) or in digital form (**53**). A 20-channel receiver (**54**) for the Nançay telescope uses correlator channels.

In Italy several articles have been published on the design of the receiver system (**55, 56, 57, 58**) for the large Bologna cross. In Canada studies have been made on several types of phase comparator systems (**59**) for use in large arrays. In the U.S.A. a discussion of the problems of polarization distribution measurements (**60**) has been published.

A description has been given of the receiver system operating between 0·7 and 3·5 MHz (**61**), used in the satellite Ariel II.

BIBLIOGRAPHY

1. Small, M. M. 1965, *Sky Telesc.*, **30,** 267.
2. Weiss, H. G. 1965, *IEEE Spectrum,* **2,** no. 10, 44.
3. McClain, E. F. 1966, *Sky Telesc.*, **32,** 4.
4. 1965, *Phys. Today,* **18,** no. 11, 84.
5. Lovell, A. C. B. 1964, *Nature,* **203,** 11.
6. King, H. E., Jacobs, E., Stacey, J. M. 1966, *IEEE Trans. Ant. Propag.*, AP–14, 82.
7. Tolbert, C. W., Straiton, A. W., Krause, L. C. 1965, *IEEE Trans. Ant. Propag.*, AP–13, 225.
8. Heeschen, D. S. 1965, *Astr. J.*, **70,** 778.
9. Kraus, J. D. 1963, *Sky Telesc.*, **26,** 12.
10. Boischot, A., Ginat, M., Parise, M. 1964, *Notes Inf. Publ. Obs. Paris,* **21,** no. 3.
11. Parijskij, Ju. N. 1966, private communication.
12. Hackenberg, O. 1966, Report of German National Committee to URSI General Assembly, Munich.
13. 1966, Proposal for Construction of a Multi-element Interferometer at the Owens Valley Radio Observatory, Calif. Inst. Tech., Pasadena.
14. Ryle, M. 1962, *Nature,* **194,** 517.
15. Christiansen, W. N., Wellington, K. J. 1966, *Nature,* **209,** 1173.

16. Christiansen, W. N. 1966, URSI General Assembly, Munich.
17. Wade, C. M., Clark, B. G., Hogg, D. E. 1965, *Astrophys. J.*, **142**, 406.
18. Moffet, A. T. 1964, *Nerem Record*, **6**, 96.
19. Vitkevič, V. V. 1961, *Vest. Akad. Nauk SSSR*, **5**, 23.
20. Kalachov, P. 1963, *Proc. Instn. Radio Eng. Austr.*, **24**, 237.
21. Braccesi, A., Cecarelli, M. 1961, *Nuovo Cim.*, Ser. X, **23**, 208.
22. Mills, B. Y., Aitchison, R. E., Little, A. G., McAdam, W. B. 1963, *Proc. Instn. Radio Eng. Austr.*, **24**, 156.
23. Costain, C. H. 1966, URSI General Assembly, Munich.
24. Erickson, W. C. 1965, *IEEE Trans. Ant. Propag.*, AP–13, 422.
25. Howard, H. T. 1965, *IEEE Trans. Ant. Propag.*, AP–13, 365.
26. Wild, J. P. 1961, *Proc. R. Soc. London*, Ser. A, **262**, 84.
27. Tanaka, H., Kakinuma, T. 1965, *Proc. Res. Inst. Atmos. Nagoya Univ.*, **12**, 27.
28. Tanaka, H. 1966, *Proc. Res. Inst. Atmos. Nagoya Univ.*, **13**, 49.
29. Tanaka, H. 1966, Report of Japanese National Committee to URSI General Assembly, Munich.
30. Frost, D. A., Palmer, H. P. 1966, *Sky Telesc.*, **32**, 21.
31. Yen, J. L., Chisholm, R. M. 1966, URSI General Assembly, Munich.
32. Findlay, J. W., Hvatum, H., Waltman, W. B. 1965, *Astrophys. J.*, **141**, 873.
33. Tanaka, H., Kakinuma, T. 1966, *Proc. Res. Inst. Atmos. Nagoya Univ.*, **13**, 41.
34. Yokoi, H., Satoh, T., Yamada, M. 1966, *Publ. astr. Soc. Japan*, **18**, 271.
35. Robinson, B. J. 1966, URSI General Assembly, Munich. To be published.
36. Martirosian, R. M., Prokhorov, A. M., Sorochenko, R. L. 1965, *Dokl. Akad. Nauk SSSR*, **156**, 1326.
37. Martirosian, R. M., Prokhorov, A. M., Sorochenko, R. L. 1965, *Izv. Vyssih Učeb. Zaved. Radiofiz.*, **8**, 699.
38. Karlov, N. V., Martirosian, R. M., Sorochenko, R. L. 1965, *Radiotekh. Elektron.*, **10**, 40.
39. Shteinshleiger, V. B., Afanas'ev, O. A., Misezhnikov, G. S., Rozenberg, Ya. I. 1964, *Pribory Tekh. Eksper.*, no. 5, 136.
40. Dravskikh, A. F., Dravskikh, Z. V., Kolbasov, V. A., Misezhnikov, G. S., Nikulin, D. E., Shteinshleiger, V. B. 1965, *Dokl. Akad. Nauk SSSR*, **163**, 332.
41. Tanaka, H., *et al.* 1965, *Proc. Res. Inst. Atmos. Nagoya Univ.*, **12**, 35.
42. Gardner, F. F., Milne, D. K. 1963, *Proc. Instn. Radio Eng. Austr.*, **24**, 127.
43. Cooper, B. F. C., Cousins, T. E., Gruner, L. 1964, *Proc. Instn. Radio Eng. Austr.*, **25**, 221.
44. Hughes, M. P., Moley, E., Parenti, D. R., Whelehan, J. J. 1965, *IEEE Trans. Ant. Propag.*, AP–13, 432.
45. Suzuki, S., Attwood, C. F., Sheridan, K. V. 1966, *IEEE Trans. Ant. Propag.*, AP–14, 91.
46. van Nieuwkoop, J. 1964, *Nature*, **201**, 380.
47. Penzias, A. A., Wilson, R. W. 1965, *Astrophys. J.*, **142**, 419.
48. Roll, P. G., Wilkinson, D. T. 1966, *Phys. Rev. Lett.*, **16**, 405.
49. Howell, T. F., Shakeshaft, J. R. 1966, *Nature*, **210**, 1318.
50. Muller, C. A., Raimond, E., Schwarz, U. J., Tolbert, C. R. 1966, *Bull. astr. Inst. Netherl. Suppl.*, **1**, 213.
51. Dieter, N. H., Weaver, H., Williams, D. R. W. 1966, *Sky Telesc.*, **31**, 132.
52. Argyle, E. 1965, *Astrophys. J.*, **141**, 750.
53. Weinreb, S. 1963, *M.I.T. Res. Lab. Electron. Tech. Rep.* no. 412.
54. Blum, E. J., Delannoy, J., Le Roux, E., Weliachew, L. 1966, *C. r. Acad. Sci. Paris*, **262**, 1640.
55. Gelato, G., Rosatelli, C., Sinigaglia, G. 1963, *Alta Frequenza*, **32**, 628.
56. Gelato, G., Rosatelli, C., Sinigaglia, G. 1964, *Alta Frequenza*, **33**, 346.
57. Gelato, G., Rosatelli, C., Sinigaglia, G. 1964, *Alta Frequenza*, **33**, 675.
58. Sinigaglia, G. 1966, *Electron. Communicator*, **1**, 3.
59. Legg, T. H. 1965, *IEEE Trans. Ant. Propag.*, AP–13, 428.
60. Morris, D., Radhakrishnan, V., Seielstad, G. A. 1964, *Astrophys. J.*, **139**, 551.
61. Hugill, J., Smith, F. G. 1965, *Mon. Not. R. astr. Soc.*, **131**, 137.

G. GENERAL BIBLIOGRAPHY

Two further volumes, for 1962 and 1963, of the bibliographical review edited by Martha Stahr Carpenter have been published by Cornell University. Enquiries should now be made to Bibliography of Natural Radio Emission from Astronomical Sources, Hollister Hall, Cornell University, Ithaca, New York 14850.

Various books and symposium proceedings are referred to in the specialized bibliographies, some others to which attention should be drawn include:—

Ginzburg, V. L., Syrovatskii, S. I. 1964, *The Origin of Cosmic Rays*, transl. Massey, H. S. H., ed. ter Haar, D. Pergamon Press.
Hack, M. 1964, *Esplorazione Radioastronomiche*, Editione Boringhieri, Torino.
Herbays, E. (ed.). 1966, *Radio Astronomy* (*Progress in Radio Science* 1960–63, Vol. 5), Elsevier Publ. Co., Amsterdam–London–New York.
Jennison, R. C. 1966, *Introduction to Radio Astronomy*, Newnes, London.
Kaplan, S. A. 1966, *Elementary Radio Astronomy*, (in Russian), Moscow.
Kraus, J. D. 1966, *Radio Astronomy*, McGraw-Hill Book Company, New York.
Smith, F. G. 1966, *Radio Astronomy*, Penguin Books, 3rd edn.

M. RYLE
President of the Commission

ADDENDUM

On page 904, two lines from bottom, before: "Millimeter wavelength instruments . . . ", add: 'A 30-meter reflector in the Argentine has been brought into operation in 1966 at 10 and 20 centimeters. (Varsavsky, private communication).'

41. COMMISSION DE L'HISTOIRE DE L'ASTRONOMIE

PRÉSIDENT: Professor Dr E. Rybka, Director, Astronomical Observatory of Cracow University, Cracow, Poland.

VICE-PRÉSIDENT: Professor Dr B. Sticker, Direktor des Instituts für Geschichte der Naturwissenschaften, Hartungstrasse 5, Hamburg 13, West Germany.

COMITÉ D'ORGANISATION: H. Dingle, A. V. Douglas, O. J. Gingerich, V. L. Tchenakal, K. Yabuuti.

MEMBRES: Abetti, Ashbrook, Baehr, Beer, Birkenmajer, Collinder, Dick, Dijksterhuis†, Erpylew, Evans (D. S.), Felber, Ferrari d'Occhieppo, Filliozat, Fleckenstein, Hellman, Hirose, Kamieński, King (H. C.), Kulikowsky, Link, Michel, Michkowitch, Nielsen, Norlind, Nørlund, Omer, Ovenden, Pelseneer, Perel†, Petri, Pogo, Samaha, Slavenas, Whitrow, Zagar, Zinner.

MEMBRES CONSULTANTS: Burke-Gaffney, Daumas, Drake, Forbes, Freiesleben, Hartner, Horsky, Hoskin, Kennedy, Maddison, Millas-Vallicrosa, Musto, Needham, Pedersen, Price, Przypkowski, Ronan, Ronchi, Russel, Santillana, Thorndike†, Waerden van der, Woolf, Wright.

I. INTRODUCTION

After the XIIth General Assembly of IAU in Hamburg (1964) a new category of consulting members has been constituted. Commission 41 has 24 consulting members. We suffered a loss by the death of the active member of our Commission, Ju. G. Perel (U.S.S.R.), who died 28 November 1964.

The investigations on the history of astronomy in the years 1964–66 concerned all branches of this division of science, viz. the history of astronomy in different countries, the historical development of the astronomical problems, history of calendar, history of observatories and societies, history of astronomical instruments and auxilliary apparatus and so on. The results were presented at congresses and symposia and were published in papers and books. Very many papers and books appeared on the occasion of 400th anniversary of Galilei.

The present draft report contains information, besides those from members, on the results of international conferences, which were held independently of IAU.

II. GENERAL HISTORY OF ASTRONOMY

According to Resolution 3 of the Commission, adopted at the Meeting of 26 August 1964 in Hamburg, a tentative scheme of the four-volume international monograph has been drawn up by the President. The scheme was discussed at the enlarged meeting of the Organizing Committee on 28 August 1965 during the International Congress of the History of Science in Kraków. A detailed plan has been drawn up for Volumes I and II of the proposed monograph, namely, Vol. I—Antiquity and the Middle-Ages till the Renaissance; Vol. II—Astronomy of the Renaissance epoch and the following centuries till Newton. Volume III is planned generally generally for the astronomy of the XVIIIth and XIXth centuries, and Volume IV—for the astronomy of the XXth century (up to the Second World War).

Remarks expressed by the Vice-President Dr B. Sticker, who could not attend the Congress of the History of Science in Poland in August 1965, as well as those sent by J. Veselovsky from Leningrad, and the remarks expressed by the participants of the meeting of 28 August 1965 in Kraców, have been taken into account in the plan published in the Information Circular of Commission 41 No. 10 (October 1965). In 1966 the President received by mail further remarks, which have been published in the Information Circular of the Commission 41 No. 12 (October 1966).

A further discussion on the plan of the international monograph on the history of astronomy is planned during the XIIIth General Assembly of IAU in Prauge (August 1967). It would be very desirable that the national committees of the history of astronomy in countries, where they exist, expressed their opinions on the proposed elaboration of the general history of astronomy.

III. CONGRESSES AND SYMPOSIA

A. Just before the XIIth General Assembly of IAU, an international symposium on the history of astronomy was held in Hamburg (22–24 August 1964), organized by the International Union for the History and Philosophy of Sciences (IUHPS), in cooperation with Commission 41 of the IAU. Two groups of problems were discussed at the symposium: (1) the relations between the instrumental and scientific progress of astronomy, (2) the practical and fundamental problems of the astronomical historiography. More than 50 participants took part from 16 countries. The exhibition 'Documenta Astronomica' was opened on 23 August 1964 in the Hamburg Museum for ethnology and prehistory. Twenty-five contributions were presented together on the meetings of the Symposium. Their titles were published in the Information Circular No. 7 of Commission 41 (December 1964) and additional information was given in the Information Circular No. 8 (February 1965). The contributions will appear in Vol. 9 of *Vistas in Astronomy* (Pergamon Press, London) in 1967.

B. The XIth International Congress for History of Science was held in Poland (Warszawa—Kraków) from 24 to 29 August 1965. The history of astronomy formed a part of section 3 of the Congress as a separate sub-section. E. Rybka was the president of the sub-section. Twenty-eight contributions from different branches of the history of astronomy were delivered at the meetings (Warszawa–Kraków) of the sub-section of the history of astronomy. The titles of the contributions were printed in the Information Circular No. 11 of Commission 41 (December 1965). Besides the meetings of the sub-section of the history of astronomy a symposium was held in Toruń (26–27 August 1965) on the 'Traditional and Innovatory Elements in the Cosmology of Nicolaus Copernicus'. The symposium assembled about 150 participants from 36 countries.

C. The symposium 'The classical antiquity and the beginning of the modern science' was organized during the International Congress 'Antiquitas Graeco–Romana ac tempora nostra' (12–16 April 1966, Brno, Czechoslovkia). Seven contributions on the history of astronomy were delivered on the symposium. Their titles were published in the Information Circular No. 13 of Commission 41 (November 1966).

IV. BIBLIOGRAPHY

Dr P. G. Kulikowsky (Moscow), former President of Commission 41, continued the preparation and publication of the bibliography of books and papers on the history of astronomy, according to Resolution No. 2 of Commission 41 (Hamburg 1964). Three issues of the bibliography were published in 1965 and 1966: (1) the bibliography for the year 1964, (2) the bibliography for the first half of the year 1965, (3) the bibliography for the second half of the year 1965. The bibliography for the year 1966 is in preparation and is intended to be published at the beginning of 1967. In this work, Dr P. G. Kulikovsky has been aided by Mrs N. Lavrova.

Serious obstacles have been encountered by Dr P. G. Kulikovsky in his work on the bibliography. In spite of his frequent request to all members of the Commission, he did not receive necessary information on new publications concerning the history of astronomy, which made it impossible to compile a complete bibliography. Dr P. G. Kulikovsky could not edit two issues of the bibliography for the year 1966. He intends to edit this bibliography in one issue at the beginning of 1967. The problem of the compilation of the bibliography on the history of astronomy will be discussed at a meeting of Commission 41 in Prague (August 1967). *Note*: The Bibliography edited by Dr Kulikovsky will be denoted hereafter by: *Bib. Kul.*

V. INFORMATION CIRCULAR

The President continues the publication of the Information Circular, which was commenced by the former President P. G. Kulikovsky. The circular is printed in Kraców under the auspices of the Committee of the History of Sciences and Techniques of the Polish Academy of Sciences. From December 1964 till the end of 1966 seven issues of the circulars (Nos. 7 to 13) have been distributed to members of the Commission. The circulars contained news on the history of astronomy, information on congresses and symposia, anniversaries, correspondence, a discussion on the general history of astronomy, personal notes, etc.

VI. REPORTS FROM MEMBERS

The President requested all members and consultants to send him information on their activity in the history of astronomy. Twenty-eight answers have been received, which are summarized in this section. The reports have been arranged according to countries, in alphabetical order. The complete bibliography has not been compiled because it has been done in the *Bib. Kul.* The following reports have been based on letters and on reprints received by the President. In the selected bibliography there are quoted generally any papers omitted in the *Bib. Kul.* 1967–65.

1. *Austria*

K. Ferrari d'Occhieppo has undertaken a reinvestigation of the 'Star of Magi' by going back to the Greek and even the old Syrian texts of the gospels compared with all information available from cuneiform tablets and other sources of that age. Three papers have been published on this problem (*Bib. Kul.* 1965 II, p. 5), (**1, 2**). Besides, K. Ferrari d'Occhieppo has written articles on Austrian astronomers and representatives of allied sciences for the 'Oesterr. Biographisches Lexikon' edited by the Oesterr. Akademie d.Wissenschaften and for the 'Neue deutsche Biographie' now edited by the Bayrische Akademie d.Wissenschaften, Munich. His activity concerns also the studies on the life and activities of the Austrian astronomer Franz Ritter von Schaub (1817–71).

2. *Belgium*

H. Michel (Bruxelles) is working on the problems concerning the history of astronomical instruments. Three books, which appeared in 1965–66, concerned the problems of the scientific instruments in art and history, the measures of time and sundials. Two books are intended to be published still in 1966 (**3, 4, 5, 6, 7**).

3. *Canada*

Many articles on the history of astronomy were printed in the *Journal of the Royal Astronomical Society of Canada*. Their authors were: A. D. Thackerey, A. V. Douglas, R. M. Petrie, P. v. de Kamp, I. C. McLennan, J. Low and J. J. Ruiz. The titles have been cited in *Bib. Kul.* for the years 1964–65 and 1966. Besides, M. W. Burke-Gaffney (Halifax) published the Biography of Boutet de Saint Martin (**8**). A. V. Douglas (Kingston) wrote on the scientific writings (**9**). She informs that the Royal Astronomical Society of Canada is undertaking to compile the history of astronomy in Canada.

4. *Czechoslovakia*

K. Fischer (Prague) published articles on astrological illustrations (**10**) and on mathematico-physical instruments in Bratislava (**11**). Z. Horsky (Prague) published a paper on the famous Prague clock (Orloj) (**12**), and on the astronomical sextant of Bürgi (**13**).

F. Link (Prague) in the second part of his paper 'Observations et Catalogue des aurores boréales en Occident 1601–1700' gives the contents of observations of more than 200 aurorae (**14**). The third part comprising the XVIIIth century is in preparation.

5. *Denmark*

Axel V. Nielsen (Aarhus) continues his studies on Ole Roemer. The paper 'Ole Roemer and his Meridian Circle' is intended to appear in *Vistas in Astronomy*.

6. *Germany*

(*a*) *Democratic Republic*

K. Biermann (Berlin) wrote on C. F. Gauss, T. Clausen and F. W. Bessel (*Bib. Kul.* 1965 I, p. 5).

J. Dick (Berlin) published three papers on the history of Berlin Observatory (*Bib. Kul.* 1964, p. 1 and 4) and (**15**).

J. Felber (Berlin) published a comprehensive volume containing W. Foerster's correspondence on the reform of Easter (*Bib. Kul.* 1965 II, p. 8).

(*b*) *Federal Republic*

U. Baehr (Heidelberg) published a biography of F. K. Ginzel (**16**). He took part in the publication concerning the Aztec text of Chimalpahin (**17**) edited by G. Zimmermann. He also wrote on the reform of calendar (**18**).

Ch. Schmidt-Schönbeck wrote the book on the history of physics and astronomy in Kiel University (see below, § VIII, ref. **10**).

B. Sticker (Hamburg) worked on general problems from the history of natural science. The titles are given in the Supplementary Bibliography (**19, 20, 21, 22, 23, 24, 25, 26**). There are in press: Volume **9**, *Vistas in Astronomy* containing the contributions from Hamburg Symposium on the History of Astronomy (1964) and a paper on Galilei (Firenze–Symposium 1964). He prepares the texts for the history of the World-conception.

B. Sticker informs that Encyclopaedia Medii Aevi (Editor A. Bruckner, Basel) is in preparation. There are planned the history of mathematics, of exact sciences and of natural philosophy under the guidance of J. O. Fleckenstein.

E. Zinner (Bamberg) published the astronomical bibliography in Germany from the Renaissance epoch (*Bib. Kul.*, 1964, p. 3), and a continuation of the list of old manuscripts in many European libraries (**27**).

7. *Italy*

G. Abetti (Arcetri) writes:

'An extensive work on the "Astronomy" of Dante Alighieri's Divina Commedia has been published in various numbers of the Italian Review *Physis* for 1965 by Professor Ideale Capasso of Genova. It is the history and explanations of the astronomical observations made by Dante in his journey through Inferno, Purgatorio and Paradiso."

He wrote four articles on the history of astronomy (**28, 29, 30, 31**).

V. Ronchi (Firenze) studied many problems from the history of Lionardo da Vinci, Della Porta, Kepler, Galilei, Torricelli, Boscovich, Amici. His studies concerned the lenses and telescopes. He wrote a comprehensive paper on the history of the telescope (**32**) and three papers from the history of optical research (**33**, **34**, **35**). Besides, V. Ronchi wrote a contribution concerning Newton (**36**).

See also *Bib. Kul.*, 1965 II, p. 19–20.

F. Zagar (Milano) edited the first volume of the correspondence of G. V. Schiaparelli on Mars (**37**). Vol. II is in press.

8. *Japan*

H. Hirose (Mitaka, Tokyo) discussed the ancient determinations of winter solstice with the use of a gnomon (**38**, **39**). His paper (**40**) contains the comparison of ancient lunar eclipse observations made in Yedo-Period with Oppolzer's Canon der Finsternisse and discusses the accuracy of time keeping. The paper (**41**) deals with the comparison of the text of Nigi-Ryakusetu (brief lecture on Earth and Heaven) prepared by K. Kobayashi before 1715, with that of Pedro Gomez's 'De Sphaera'. The author concludes that the text of Nigi-Ryakusetsu may be considered as an ancient translation of the Latin original of 'De Sphaera' into Japanese. Gomez's text is now available in printed form as a supplement to the Japanese translation by S. Obara S.J. (Kirishitan Kenkyu, No. 10, 1965, Tokyo). Hirose's paper (**42**) contains a brief discussion on the European influence on astronomical and geodetic instruments made in old Japan together with brief descriptions of some old instruments still preserved in Japan. The paper (**43**) deals with the discussion of the meridian observations of the Sun made at the old observatory of Tokugawa Shogunate during 1798–1843, before the introduction of modern or European technique.

In (**44**) H. Hirose discusses the background of *Kwansei* Calendar Reform of Japan in connection with the importation of European astronomy into Japan. Relations between court astronomers and people of 'new theory' are also investigated in the paper. The main part of the small book (**45**) is devoted to the Japanese translation of chapters 1 to 10 of Book I of 'De Revolutionibus' and to the diffusion of the geocentric theory into ancient Japanese people.

K. Yabuuti (Kyoto) has sent the following general information on the progress of the research on the history of astronomy led in Japan in 1964–1966. 'Although there are not so many scholars who are working actively on the researches of history of astronomy they always contact each other and are studying mostly in the field of Japanese and Chinese astronomy. Between 1964 and 1966 three exhibitions concerning the history of astronomy were held. The first one (1964) has been an exhibition of telescopes, which memorialized the 400th anniversary of Galileo Galilei. The second one (1965) has been an exhibition for the memory of the famous astronomer H. Sibukawe on the occasion of the 250th anniversary of his death. Both exhibitions were opened at the Tokyo National Science Museum. The third one (1966) was opened at the Osaka Municipal Art Museum for the 150-th anniversary of the death of the Japanese astronomer S. Hazama.

'On the request of the International Union of the History of Sciences a survey of scientific instruments of historical interest in Japan has been carried on by Japanese Committee on the History of Sciences and K. Yabuuti is taking charge of the part of astronomical instruments.'

Several monographs and catalogues (**46**, **47**, **48**, **49**, **50**, **51**) have been published in Japanese. K. Yabuuti wrote two articles (**52**, **53**). Besides the articles of H. Hirose and K. Yabuuti those of I. Imai (**54**), S. Nakayama (Tokyo) (**55**, **56**), K. Saito and S. Sinozawa (Tokyo) (**57**), T. Watanabe (Tokyo) (**58**, **59**) and K. Yamada (Kyoto) (**60**) should be mentioned.

9. *Poland*

J. Dobrzycki (Warszawa) published a paper on the theory of precession in the mediaeval astronomy (*Bib. Kul.* 1965 II. p. 4) and worked on Copernicus.

M. Kamieński (Warszawa) published a paper on Halley's Comet (*Bib. Kul.* 1965 II p. 7).

T. Przypkowski (Jędrzejów) published a paper on the Gnomonics of Hevelius (**61**). A second paper (**62**) is in press. He prepared for the 'History of Astronomy in Poland' its section since the middle of XVIth century till the middle of XVIIIth century.

E. Rybka (Kraków) published his contribution held at the Hamburg Symposium 1964 (**63**).

10. *South Africa*

D. S. Evans (Cape), assisted by B. M. Evans, T. J. Deeming and S. Goldfarb, is editing for publication the diaries of Sir John Herschel at the Cape of Good Hope (1833–38).

11. *Sweden*

P. Collinder (Uppsala) has elaborated a bibliography of astronomical literature printed in Sweden in 1881–98 with historical notes. He continued his research on the accuracy of astronomical observation in Antiquity and published the paper on Dicaearchus (*Bib. Kul.* 1965 II p. 3).

12. *Switzerland*

J. O. Fleckenstein-Gallo (Basel) published a book on science and politics (**64**) and papers on papyrology (**65**) and the Galilean phases of Venus (**66**).

B. L. van der Waerden (Zurich) published a book on the beginning of astronomy (*Bib. Kul.* 1965 II p. 3).

13. *United Kingdom*

E. G. Forbes (Edinburgh) worked on the history of Nautical Almanac (**67**), on Bradley's observations (**68**) and on history of chronometers (**69**). A paper on Tobias Mayer's Lunar Tables is in press.

M. A. Hoskin (Cambridge) has been working in two main fields:

(*a*) A study of Thomas Wright of Durham, who is often falsely credited with the view that the Sun belongs to a system of stars which is disk-shaped. M. A. Hoskin plans to publish a large quantity of supplementary material, recently rediscovered, prepared by Wright for a second edition of his 'An Original Theory'.

(*b*) A study of the history of ideas of nebulae and galaxies between 1822 and 1925.

He published three papers (**70, 71,** and **72**).

H. C. King (London) published two books which deal with aspects of the history of astronomy. The book 'Exploration of the Universe' (*Bib. Kul.* 1965 II p. 2) is a survey of the development of knowledge of the astronomical universe from earliest times to the present. The book 'The world of the Moon' contains sections on early ideas and a history of selenography (**73**).

J. Needham (Cambridge) continues his investigations on the history of astronomy in China and Korea. He published three papers (**74, 75, 76**) with his collaborators. Two articles are in press (**77, 78**).

G. J. Whitrow (London) published a paper on Galilei (*Bib. Kul.* 1965 I p. 7) and wrote a paper 'Kant and the Extragalactic Nebulae' (a lecture at the meeting of the Royal Astronomical Society, September 1960). He is engaged in preparing a new edition of W. Hastie's translation into English of Kant's 'Allgemeine Naturgeschichte und Theorie des Himmels'.

14. *U.S.A.*

O. Gingerich (Cambridge, Mass.) has continued searching for applications of high-speed computers to the history of astronomy. This has resulted in various tables of planetary or lunar positions, such as the extensive computation of Lunar positions, and the extensive computation of Lunar Visibilities in Ancient Babylon. He has recomputed all the planetary tables from Kepler's 'Rudolphine Tables' and his conclusions concerning the high accuracy of this work were reported at the XIth Congress for History of Science (Poland, August 1965). He has currently undertaken detailed computer investigations of the 'Alphonsine Tables' and the 'Prutenic Tables'.

In addition he has begun a translation and commentary of Kepler's 'Astronomia Nova'. Also he is preparing a reprint edition with coloured illustrations of Tycho Brahe's 'Astronomiae Instauratae Mechanica'. He published three papers (**79, 80, 81**). Two papers 'Applications of high-speed computers to the history of astronomy' and 'A study of Kepler's Rudolphine tables' are in press.

C. D. Hellman (New York) is at work on a Tycho Brahe source book and encyclopaedia articles on Tycho Brahe and Renaissance astronomy. She published three contributions at symposia or congresses (**82, 83, 84**) and three encyclopaedia articles. She wrote seven reviews of books.

D. M. Musto (New Haven) continued his work in the general area of the rise of American Astronomy. He is making a detailed study of John Quincy Adams who as President of United States and as a private citizen was a great promotor of astronomy in the United States.

D. J. de Solla Price (New Haven) has written with A. Aaboe the paper on the Qualitative Measurements in Antiquity (**85**).

H. Woolf (Baltimore, Maryland) has been working on the history of astrophysics (**86, 87, 88**).

15. *U.S.S.R.*

Extensive investigations have been carried out on the history of astronomy in U.S.S.R. The list compiled by P. G. Kulikovsky (Moscow) for the years 1965–66, sent to the President, contains 64 titles of books and papers. The activity has been concentrated on the following branches of the history of astronomy.

1. *General works* (A. I. Eremeeva, B. V. Kukarkin, P. G. Kulikovsky, N. I. Nevskaja, K. F. Ogorodnikov).
2. *Antique Astronomy* (V. M. Brabitch, B. E. Tumanian, I. N. Veselovsky).
3. *Astronomy in XVIth–XVIIIth centuries* (N. I. Nevskaja, Yu. G. Perel).
4. *Astronomy in particular countries* (A. G. Abramian, T. N. Klado, Yu. H. Kopelevich, V. K. Kuzakov, P. K. Prüller, D. O. Svyatsky).
5. *Asiatic Astronomy* (L. S. Baranovskaja, U. Iliasov, T. M. Kary-Niazov).
6. *History of the astronomical problems* (O. D. Dokuchaeva, A. E. Medunin, Z. S. Parshina, V. N. Pipunyrov, R. B. Shatzova, F. A. Shibanov, I. P. Spitzberg, N. I. Suvorov, B. M. Tcherniagin).
7. *History of observatories* (A. A. Mikhailov, P. Miursepp, I. Rabinowitch, G. A. Zhelnin).
8. *History of instruments* (G. G. Georgobiani, P. G. Kulikovsky, P. Miursepp, I. M. Tabinovitch, V. P. Shtcheglov, V. L. Tchenakal, B. E. Tumanian).
9. *Biography and Personal Notes* (V. K. Abalakin, A. F. Bogorodsky, Yu. N. Efremov, T. N. Klado, S. N. Korytnikov, P. G. Kulikovsky, B. Yu. Levin, O. A. Melnikov, N. N. Michelson, P. V. Miursepp, Yu. G. Perel, L. N. Radlova, Z. K. Sokolovskaja-Novokshanova, V. L. Tchenakal, N. A. Tchernega, I. A. Tiulina, G. A. Zhelnin).

V. L. Tchenakal (Leningrad) reports that he has been working on M. V. Lomonosov and Russian astronomy in XVIIIth century, on astronomical instruments and sundials of U.S.S.R. peoples and other problems.

The titles of the books and papers with bibliographical details have been given in the Bibliography edited by P. G. Kulikovsky (*Bib. Kul.*) in Moscow for the years 1964–66.

The edition of the 'Istorikoastronomitcheskije issledovania' (Historic-astronomical investigations) have been continued (editor P. G. Kulikovsky). In 1966, Vol. IX of this publication has been issued. It contains a comprehensive paper of D. O. Svyatsky 'Essays on the History of Astronomy in ancient Russia' and many other articles, as:

B. V. Kukarkin 'First steps in the development of astronomy', P. K. Prüller 'Estonian people's astronomy', W. Petri (DBR) 'Astronomical contents of the first book of the Kalachakra tantra', Yu. G. Perel 'On the problem about the time of the recognition of Copernicus in Russia', O. D. Dokuchaeva 'Contradiction in the estimation of first photographs of celestial bodies', A. F. Bogorodsky, N. A. Tchernega 'Robert Filipp Fogel (1859–1929)', F. A. Shibanov 'From the history of practical astronomy, geodesy and cartography in Russia', K. A. Zvonerev 'V. U. Kavraysky', V. Z. Tchenakal 'New data on the system of telescopes by Short', V. P. Shtcheglov 'Reflector by James Short No 1341', V. M. Brabitch 'On the imprint of Zodiacal circle on the Roman contorniates', V. L. Tchenakal 'Jan Baptist Sniadecki and Johann Elert Bode', Letters by M. O. Nyrén to D. J. Dubiago.

16. *Yugoslavia*

V. V. Michkovitch (Beograd) continues his work on the development of astronomy in Serbia. He published a paper on the unification of the calendar (**89**).

VII. GENERAL REMARKS

M. W. Burke-Gaffney (Halifax, Canada) writes:

'During recent years I have noticed in History of Science Societies (and Academies) an increase in the number of papers on the Philosophy of the History of Science. I am one of those who hold that we should not philosophize until we have the facts. I would wish to see Commission 41 continue adhering to facts, as it has done in the past. I am of opinion that in this way we shall not be doing just the same work as, say, the History of Astronomy branch of the International Union of History of Sciences and that the need of Commission 41 may become more apparent, as others branch out into speculations.'

B. Sticker (Hamburg) remarks that the Resolution no. 7 (concerning instruments of historical interest) should be reminded at the General Assembly 1967. Perhaps all older observatories (for example those founded before 1900) should be requested, with the aid of IAU, to give attention to the resolution. It is desirable that the most important old instruments be registered. The observatories ought to be requested for information on what is being done for the preservation of and care on the old instruments of historical interest.'

D. F. Musto (New Haven, U.S.A.) writes:

'I would suggest the study of comparative development of astronomy in various nations. This would be more possible than with other sciences because the building of observatories, their records and support are more obtainable than in such disciplines as chemistry, medicine, etc. Also, the History of Astronomy is the longest permitting comparisons with the development in advance.'

E. Forbes (Edinburgh) puts his attention on the microfilming of documents preserved in the Royal Greenwich Observatory. He writes:

'A problem which in my opinion merits international persuasion, if not cooperation, is that of microfilming for the purpose of general distribution the valuable and unique documents presently preserved in the Record Room of the Royal Greenwich Observatory, Herstmonceux Castle. A complete list of these has recently been issued, for information and research only, by the Public Record Office in London. The papers contain original manuscript observations etc. by the Astronomers Royal, John Flamsteed, Edmond Halley, James Bradley, Nevil Maskelyne, John Pond and George Airy together with a great deal of personal and private correspondence—Airy's in particular being remarkably comprehensive, covering a wide range in astronomy and general science during a period of great industrial and economic development.

'In addition to the above-mentioned data, there is a wealth of information on 18th century astronomical developments contained in the Board of Longitude papers at the Royal Greenwich Observatory.'

VIII. ANY BOOKS ON THE HISTORY OF ASTRONOMY

1. Abetti, G. 1964, *Storia dell' astronomia*. Firenze, Vallecchi.
2. Bernal, J. 1965, *Science in history*. 3rd edition, London, Watts.
3. 1963–65, *A general history of the Sciences*. Vol. 1 to 4 (translated from French). London, Thames and Hudson.
4. Crombie, A. C. 1964, *Von Augustinus bis Galilei. Die Emanzipation der Naturwissenschaft*. Aus der Engl., Köln.
5. King, H. C. 1964, *Exploration of the Universe*. London, Secker and Warburg.
6. Koyré, Alexandre 1964, Mélanges publiés à l'occasion de son soixante-dixième anniversaire. Vol. I: *L'aventure de la science*, Vol. II: *L'aventure de 'esprit*. Paris, Hermann.
7. Ley, W. 1965, *Die Himmelskunde. Eine Geschichte der Astronomie von Babylon bis zum Raumzeitalter*. Übers. aus dem Engl., Düsseldorf–Wien, Econ–Verl.
8. Peuckert, W. E. 1965, *L'astrologie. Son histoire, ses doctrines*. Trad. de l'allemand. Paris.
9. Reichen, Ch. A. 1964, *Histoire de l'astronomie*. Paris, Levallois–Perret (Seine).
10. Schmidt-Schönbeck, Ch. 1965, *300 Jahre Physik und Astronomie an der Kieler Universität*. Verlag Ferdinand Hirt, Kiel.

IX. CONCLUDING REMARKS

This report does not comprise all problems on the history of astronomy from the years 1964–66, but also those in which the members of our Commission have been engaged and results of their investigations communicated to the President.

Concerning the further development of the organization of works on the history of astronomy, it is very desirable to organize national committees for the research of the history of astronomy in particular countries. The preservation and care on astronomical documents and the instruments of historical interest seems an urgent problem. Therefore the organization of museums, where the instruments may be preserved and investigated, is very desirable.

There is a need for a further collaboration with the International Union for the History and Philosophy of Sciences and other organizations which are interested in the History of Astronomy.

It would be desirable to organize under the auspices of IAU the edition of the General History of Astronomy.

X. PROPOSED AGENDA

1. Adoption of Draft Report and discussion on the policy of Commission 41.
2. Discussion on the need of Annual Bibliography.
3. Discussion on the need of Information Circular.
4. Discussion on the scheme of the General History of Astronomy.

5. A discussion on the instruments of historical interest.
6. Microfilming of documents on the history of astronomy.
7. Communications on the development of studies on the history of astronomy.
8. Proposed resolutions.
9. Membership of the Commission.

A SUPPLEMENTARY BIBLIOGRAPHY

Austria

1. Ferrari d'Occhieppo, K. Der Messias-Stern unter neuer astronomischen und archäologischen Gesichtspunkten. *Vierteljahrsschrift Wiener Kath. Akad.*, **15**, 3–19.
2. Ferrari d'Occhieppo, K. 1966, *Die Rolle des Mars bei der grossen Konjunktion im Jahre 7 v. Ch.*

Belgium

3. Michel, H. 1965, *Les instruments des Sciences dans l'art et l'histoire.* A de Visscher, Ed Bruxelles.
4. Michel, H. 1965, *Messen über Zeit und Raum.* Chr. Belser Verl., Stuttgart.
5. Michel, H. 1966, *Scientific instruments.* Barrie and Rockliff, London.
6. Michel, H. 1966, *Strumenti scientifici.* Bramante Edit., Milano.
7. Michel, H. 1966, *Les cadrans solaires de Max Elskamp.* Musée de la Vie Wallonne, Liège.

Canada

8. Burke-Gaffney, M. W. 1966, *Boutet de Saint-Martin. Dictionary of Canadian Biography,* **1**, 119, Toronto, University of Toronto Press.
9. Douglas, A. V. 1966, Litterary History of Canada. Chapter 24. *Scientific Writings,* pp. 445–456. University of Toronto Press.

Czechoslovakia

10. Fischer, K. 1964, Some unpublished astrological illustrations from Central and Eastern Europe. *J. Warburg Courtauld Inst.*, **27**, 311–312.
11. Fischer, K. 1965, *Matematicko-fyzikálne pristroje v zbierkach meského muzea w Bratislave.* Bratislava, 179–200.
12. Horský, Z., Prochazka, Emanuel 1964, Prazsky Orloj. Sborník pro dějiny přírodnich věd a techniky. *Acta historiae rerum naturalium nec non technicarum,* **9**, 83–146.
13. Horský, Z. 1965, Sextant astronomique de Bürgi. *Acta historiae rerum naturalium nec non technicarum. Special issue,* **1**, 125–129.
14. Link, F. 1965, Observations et Catalogue des aurores boréales apparues en Occident de 1601–1700, seconde partie. *Geofysikální Sborník,* no. 212. Praha.

Germany

(*a*) *Democratic Republic*

15. Dick, J. 1965, Wer kennt Jean Philippe Baratier? *Sterne,* **41**, 233–238.

(*b*) *Federal Republic*

16. Baehr, U. 1964, *Friedrich Karl Ginzel. Neue Deutsche Biographie.*
17. Baehr, U., *et al.* 1965, Die Relationen Chimalpahins zur Geschichte Mexico's. Universität Hamburg, *Abhandlungen aus dem Gebiet der Auslandskunde,* **69**.
18. Baehr, U. 1966, Kalenderreform, *Sterne und Weltraum,* **10**, 963.
19. Sticker, B. 1964, Die Stellung der Geschichte der Naturwissenschaften im Rahmen unserer heutigen Universitäten. *Philosophia Naturalis,* **8**, 109–116.

20. Sticker, B. 1964, Herschel's Cosmology. *History of Science*, **3**, 91–101.
21. Sticker, B. 1964, *Documenta Astronomica*, **23**, 8–25, 9; 1964, Katalog, F. Steiner Verlag, Wiesbaden.
22. Sticker, B. Ansätze zur historischer Denkweise in der Naturforschung an der Wende vom 18 zum 19 Jahrhundert. II. Die Konzeption der Entwicklung von Sternen und Sternsystemen durch Wilhelm Herschel. *Sudhoffs Archiv.*, **47**, 26–35.
23. Sticker, B. 1965, 'Artificial' and 'natural' classifications of celestial bodies in the work of William Herschel. Ithaca, 1962, 729–731.
24. Sticker, B. 1965, Die Erweiterung des Erfahrungsbereichs des Menschen durch die Naturwissenschaften im 19. Jahrhundert. *Technik-Geschichte*, **32**, 131–161.
25. Sticker, B. 1965, Über das Verhältnis zwischen der Geschichte der Mathematik und der Geschichte der Naturwissenschaft. Philosophischer Eros im Wandel der Zeit. 125–141. München.
26. Sticker, B. 1966, Die Smithsonian Institution. Bild der Wissenschaft, 454–461.
27. Zinner, E. 1964, Die Bibliographie der Astronomie. Aus alten Handschriften. II. Naturforschende Gesellschaft Bamberg, 39, Bericht.

Italy

28. Abetti, G. 1965, *La data della visione dantesca. Luci ed Immagini.* Firenze.
29. Abetti, G. 1965, *Galileo Galilei.* Celebrazione del IV centenario della nascita. Accademia Nazionale dei Lincei, Roma.
30. Abetti, G. 1965, Paolo dal Pozzo Toscanelli astronomer and the problem of longitudes. *Atti Fondazione G. Ronchi*, Sett. Ott., Firenze.
31. Abetti, G. 1966, The sunspots from Galileo to George Hale. Atti del Convegno sulle macchie solari. Barbéra ed. Firenze.
32. Ronchi, V. 1966, Storia del cannocchiale, Miscellanea Galileiana Pontificiae Academiae Scientiarum Scripta Varia, no. 27, 727–848.
33. Ronchi, V. 1965, Un aspect peu connu de l'activité de Léonard de Vinci dans le domaine de l'optique. *Atti della Fondazione Giorgio Ronchi*, **20**, 394–405.
34. Ronchi, V. 1966, New history of the optical microscope. *Indian J. Hist. Sci.*, **1**, 46–59.
35. Ronchi, V. 1966, Rapporti fra scienza e tecnica nell'antichità e nel medio evo. *Responsabilità del Sapere*, **77**, **78**, 3–21.
36. Ronchi, V. 1965, Il contributo di Isacco Newton alla formulazione della legge dell' attrazione universale. *Cultura e scuola*, no. 16, 249–257.
37. Zagar, F. 1964, Corrispondenza su Marte di G. V. Schiaparelli, Vol. I: 1877–89. Pisa.

Japan

38. Hirose, H. 1964, On the determination of the date of winter solstice of Shou-shih-li of Yüan and Zyōky–Rekiō, one of Japanese national calendars. Private Publication, no. 4, (in Japanese).
39. Hirose, H. 1964, On the determination of the data of winter solstice of Hōryaku-Reki and its correction volume issued in Meiwa years. Private Publication, no. 5, (in Japanese).
40. Hirose, H. 1964, On Lunar Eclipse Observations made during Yedo-Period. Private Publication, no. 5, (in Japanese).
41. Hirose, H. 1966, On the Foundation of Nagasaki-School of Astronomy. Part I. On 'Nigi-Ryakusetsu'. *Rep. Soc. histor. Res. Dutch School Material Japan*, no. 157.
42. Hirose, H. 1965, On astronomical and geodetic instruments used in Yedo-Period. 'Biblia' of Tenri Library, no. 32, 130, (in Japanese).
43. Hirose, H., Manabe, R. 1965, On the meridian observations of the Sun made at the Observatory of Tokugawa Shogunate. *Tokyo astr. Obs. Rep.*, **13**, 660, (in Japanese).
44. Hirose, H. 1964, General aspect of calendar problem after Hōryaku calendar reform and problematic points concerning Kwansei calendar reform. *Rep. Soc. histor. Res. Dutch School Material Japan*, no. 157, (in Japanese).
45. Hirose, H. 1965, *Copernicus*, Tokyo.

46. Murayama, S. 1964, *On the progress of telescope in Japan.*
47. Watanabe, T. 1965, *Forerunner of astronomy in Japan, Sibukawa Harumi and progress of astronomy in Edo Period* (1603–1867) *of Japan.*
48. Hazama, H. 1966, *Catalogue of the Exhibition of Hazama Sigetomi with his chronological table.*
49. Yabuuti, K., Soda, H. 1964, *Scientific instruments in Japan.*
50. Simamura, H., ed. 1965, *Nippon Kagaku-gizyutusi Taikei*, **14**, (in Japanese).
51. *Cntalogue of terrestrial and celestial globes* (ed. by the Tenri Central Library), 1964, (in Japanese).
52. Yabuuti, K. 1965, On the celestial globes in the Tenri Central Library. *Biblia*, **32**, 102–111, (in Japanese).
53. Yabuuti, K. 1966, Astrology of western origin in ancient Japan. *Scientia* (Italy), **101**, 353–358.
54. Imai, I. 1965, On the celestial globes of W. Blaeu in the Tenri Central Library. *Biblia*, **32**, 113–129, (in Japanese).
55. Nakayama, S. 1965, Japanese activity in the history of astronomy during the early half of 1960's. *Jap. Studies in the History of Science*, no. **4**, 1–19.
56. Nakayama, S. 1965, Characteristics of Chinese calendrical science. *Jap. Studies in the History of Science*, no. 4, 124–131.
57. Saito, K., Sinozawa, S. 1966, On the records of solar eclipse observations by Japanese astronomers and geophysicists since Meiji (1868–). *Tokyo astr. Obs. Rep.*, **14**, 1–78, (in Japanese).
58. Watanabe, T. 1965, On the observation of solar and lunar eclipses in Edo Period. *J. Tokyo Mercantile Marine College*, **16**, 31–51, (in Japanese).
59. Watanabe, T. 1965, On the eclipses and calendar reform in Edo Period. *J. Tokyo Mercantile Marine College*, **14**, 53–66, (in Japanese).
60. Yamada, K. 1965, On the early development of astrophysics. *J. History of Science*, no. 73, 9–17, (in Japanese).

Poland

61. Przypkowski, T. 1964, Gnomonics of John Hevelius. Ithaca Proceedings of the Xth Intern. Congress of History of Science.
62. Przypkowski, T. 1967, The art of sundials in Poland from the thirteenth to the nineteenth century. *Vistas in Astronomy*, **9**, Pergamon Press.
63. Rybka, E. 1967, The influence of the Cracow intellectual climate at the end of the fifteenth century upon the origin of the heliocentric system. *Vistas in Astronomy*, **9**. Pergamon Press.

Switzerland

64. Fleckenstein, J. O. 1965, Wissenschaft und Politik von Galilei bis Einstein. München.
65. Fleckenstein, J. O. 1967, Papyrologie und astronomiegeschichtliche Quellen. *Vistas in Astronomy*, **9**, Pergamon Press.
66. Fleckenstein, J. O. 1966, Les phases galiléennes de la planète Vénus, Giornate Galileane, Firenze.

United Kingdom

67. Forbes, E. G. 1965, The Foundation and Early Development of the Nautical Almanac. *J. Inst. Navig.*, **18**, 391–401.
68. Forbes, E. G. 1965, Dr Bradley's Astronomical Observations. *Quart. J. R. astr. Soc.*, **6**, 321–328.
69. Forbes, E. G. 1966, The Origin and Development of the Marine Chronometer. *Ann. Sci.*, **22**, 1–25.
70. Hoskin, M. A., Jones, Ch. 1965, Problems in late Renaissance Astronomy. Le Soleil à la Renaissance. Bruxelles-Paris, 19–31.

71. Hoskin, M. A. 1966, Stellar Distances: Galileo's Method and its Subsequent History. *Indian J. Hist. Sci.*, **1**, 22–29.
72. Hoskin, M. A. 1966, Introduction to facsimile of Clavis Coelestis by Thomas Wright of Durham. London.
73. King, H. C. 1965, The world of the Moon. Barrie and Rockliff, London.
74. Needham, J. 1965, Time and Eastern Man. Royal Anthropological Institute. Occasional Paper no. 21.
75. Needham, J. 1964, Science and Society in China and the West. *Science Progress*, **52**, 50–65.
76. Beer, A., Ho Ping-Yü, Lu Gwei-Djen, Needham, J., Pulleyblank, E. G., Thompson, G. I. 1964, An 8th Century Meridian Line: I-Hsing's Chain of Gnomons and the Pre-history of the Metric System. *Vistas in Astronomy*, **4**, Pergamon Press.
77. Needham, J., Lu, G. D. The Optick Artists of Chiangsu. Proceedings of the Royal Microscopical Society (in press).
78. Needham, J., Lu, G. D. A Mid-Eighteenth Century Korean Astronomical Screen from the Royal Palace at Seoul. *Physics* (in press).

U.S.A.

79. Gingerich, O. 1964, The computer versus Kepler. *Amer. Scient.*, **52**, 218–226.
80. Gingerich, O. 1965, Lunar visibilities in ancient Babylon. *Isis*, **56**, 69.
81. Gingerich, O. 1965, Eleven-digit regular sexagesimals and their reciprocals. *Trans. Amer. Phil. Soc.*, **55**, part 8.
81. Gingerich, O. (*a*) Applications of high-speed computers to the history of astronomy. *Vistas in Astronomy*, **9** (in press). (*b*) A study of Kepler's Rudolphine tables. Organum (in press).
82. Hellman, C. D. Galileo's sources for and use of data concerning the Nova of 1572 and the Comet of 1577 (in press).
83. Hellman, C. D. The role of Measurement in the Downfall of a System: Some Examples from 16th Century Comet and Nova Observations. *Vistas in Astronomy*, **9** (in press).
84. Hellman, C. D. The Supernova of 1572 and the Comet of 1577 and the Alteration and Transmission of Ideas about Comets. Acts of XIth Intern. Congress of the History of Sciences (in press).
85. de Solla Price, D. J., Aaboe, A. 1964, Qualitative Measurement in Antiquity. L'aventure de la science, Mélanges Alexandre Koyré, 1, Histoire de la Pensée, **12**.
86. Woolf, H. 1964, The Beginnings of Astronomical Spectroscopy. L'aventure de la science, A. Koyré, ed. 619.
87. Woolf, H. 1965, Quelques aspects de l'astrophysique au début du XIXème siècle. Conférence au Palais de la Découverte, 6. XI. Paris.
88. Woolf, H. 1966, Introduction to History of Physical Astronomy, by Robert Grant. The Sources of Science, no. 38.

U.S.S.R.

Sixty-four papers listed in the Bibliography edited by P. G. Kulikovsky, Moscow, 1964, 1965 and 1966.

Yugoslavia

89. Michkovitch, V. V. 1966, Sur l'unification du calendrier civil. Glas de l'Académie Serbe des Sciences et des Arts, 263, Classe des Sciences math. et natur., no. 28, 93–197, (in Serbian, résumé français).

E. RYBKA
President of the Commission

42. COMMISSION DES ETOILES DOUBLES PHOTOMETRIQUES

PRÉSIDENT: Dr J. E. Merrill, Morrison Observatory, Central Methodist College, Fayette, Missouri, U.S.A.

VICE-PRÉSIDENT: Professor Dr V. P. Tsesevich, Director, Astronomical Observatory, Odessa, U.S.S.R.

COMITÉ D'ORGANISATION: G. Larsson-Leander, D. J. K. O'Connell, M. Plavec, H. Schneller†, F. B. Wood, K. O. Wright.

MEMBRES: Bappu, Batten, Beer, Binnendijk, Blitzstein, Broglia, Cester, Chis, Cillié, Cousins, Dadaev, de Kort, Ferrari d'Occhieppo, Fracastoro, Fredrick, Fresa, Gaposchkin (S.), Gyldenkerne, Hardie, Heard, Herczeg, Hogg (A. R.)†, Huang, Huffer, Huruhata, Irwin, Kitamura, Koch, Kopal, Krat, Kron (K. G.), Kwee, Lavrov, Linnell, McLaughlin†, Magalashvili, Martynov, Nelson, Ovenden, Pearce, Petrie†, Piotrowski, Plaut, Popovici, Popper, Rakos, Sahade, Schmidt (H.), Serkowski, Shapley (M. B.), Shulberg, Sinvhal, Strohmeier, Szafraniec, van Hoof, Walter, Wellmann, Wesselink, Whitney (B. S.), Widorn.

JOINT WORKING GROUP WITH COMMISSION 26: O. J. Eggen, *Chairman;* Cester, Couteau, Fredrick, Popper.

JOINT WORKING GROUP WITH COMMISSION 30: A. H. Batten, *Chairman;* Abt, Buscombe, Heard, Martynov, Northcott, Popper, Sahade, Thackeray.

INTRODUCTION

For this report, account has been taken of literature and information received up to the middle of September 1966, and of some somewhat later material. Special thanks are due the members of the Organizing Committee who found time in their crowded schedules to compose various sections of the text and compile the relevant tables of references included therein. The work has been very considerably expedited by use of the five issues (numbers 7, 8, 9, 10, 11) of the *Bibliography and Program Notes for Eclipsing Binaries* covering the interval, with the categorizing numbers there assigned to each item listed. We are indebted to the following contributors to the bibliography for their efforts:

Dr Fracastoro, Italy
Dr Hardie, North America
Dr Kitamura, Japan
Mrs Kron, P.A.S.P.
Dr Kwee, Western Europe
Dr Plavec, Eastern Europe
Dr Schneller†, Germany
Dr Sahade, Southern Hemisphere
Dr Shulberg, U.S.S.R.
Dr Sinvhal, India and Indonesia

In accordance with Commission desires, the form of this report is similar to that of previous ones. The total number of references surveyed, which is a rough gauge of the activity in the field, is over fourteen hundred, as compared with about eight hundred for the 1964 issue. Readers should bear in mind that the various sections here set out are by no means mutually exclusive; a given published article will often contain material pertinent to two, or even three, of our sub-fields.

OBSERVATIONAL TECHNIQUES AND NEW PHOTOMETRIC DATA

Fletcher (1) used precise photoelectric measurements of equivalent widths to investigate the variation with phase of the equivalent width of the He I λ4771 line in Algol, and used the observations for a study of the Algol system.

Przybylski (2) prepared a computer program for the reduction of photometric observations affected by variable extinction. He assumed a linear change with time for the extinction coefficient, and in this way obtained a considerable increase in the precision of the results. The method, which can be applied generally, was developed to deal with the greatly increased, and varying, extinction caused by the eruption of Mt Agung volcano in March 1963. Przybylski found that the volcanic dust in the atmosphere did not appreciably affect the extinction coefficient of $B - V$ and $U - B$, in other words that the dust acts as a neutral filter.

Bertiau (3) developed a program for the establishment of independent photometric three-color systems. The program, written for the IBM 1620 computer, derives a system of colors and magnitudes for a limited number of standard stars observed during several nights. The observations of all the nights are treated in a single least squares solution. The magnitudes and colors of the standards enter directly as unknown quantities in the fundamental equations. The normal equations permit a unique solution for the colors and magnitudes of the standards. The program was prepared originally to establish standards for O'Connell's observations of ζ Aurigae made with interference filters (4), but it can be applied to any three-color photometric system.

At Flower and Cook Observatory Blitzstein, Mrs Savage, Jurkevich and A. J. Harris have developed computer programs of various sorts for the reduction of the raw observations to light curve data. Blitzstein has completed the Pierce photometer so that it punches cards at the telescope; it is now also suitable for dual operation, simultaneously observing the variable and the comparison star. The completely digitized card-punching photometer, including a digitized clock, constructed by Petty, has been installed at the Mount John Observatory. Also at Mount John, a mounting has been built for four patrol cameras loaned by Strohmeier; a joint photographic program with the Bamberg Observatory is to be initiated.

Shao and Young (5) have studied the Red Leak corrections in ultraviolet photometry, and Young (6) has made further studies of temperature effects in photoelectric photometry generally.

Table 1a is a continuation of the Table on page 691 of *Trans. IAU*, **12A**, and lists both published photoelectric observations and those we have been informed to be in progress. It reflects the steadily increasing activity in the field, containing 217 binaries with 348 references, as compared with 139 stars and 203 references in the previous list. We are especially pleased to note the evidence it contains of the activity in eclipsing binaries at Kodaikanal Observatory, important to us particularly because of the 'longitude gap' which has existed. Table 1b lists visual and photographic estimates of magnitude. Note also that additional series of observations are implied by publication of new light elements as listed in Table 3.

Table 1a. New Photometric Observations

Star	References
RT And	Shao, C. Y. (Harvard) (3 colors).
	Dean, C., Nelson, B. (San Diego) (*UBV*).
	Miner, E. D., *Astrophys. J.*, **144,** 1101, 1966.
	Pohl, E., Kizilirmak, I., *Astr. Nachr.*, **288,** 69, 1965.
AB And	Kalchaev, K., Truttse, Yu., *Perem. Zvezdy*, **15,** 487, 1965.
	Purgathofer, A., Widorn, T., *Mitt. Univ. Sternw. Wien*, **12,** no. 6, 1964.
	Purgathofer, A., Widorn, T., *Sber. öst. Akad. Wiss.*, Wien, Math.-nat. Kl., Abt. 2, 171, 217, 1964.
AD And	Rucinski, S., *Acta. astr.* (in press).
AN And	Schneller, H. (Potsdam) Beginning obs.
BX And	Todoran, I., *Bukarest Studii*, **71,** 1965.
o And	Jackisch, G., *Veröff. Sternw. Sonneberg*, **5,** 286, 1963.
Y Aql	Breinhorst, R., (Bonn Univ. Obs.) 3 color pe obs.

Star	References
OO Aql	Bookmyer, B. B., *Astr. J.*, **68**, 741, 1963.
	Pohl, E., Kizilirmak, I., *Astr. Nachr.*, **288**, 69, 1965.
KO Aql	Cristaldi, S., *Mem. Soc. astr. ital.*, **37**, 205, 1966.
	Filatov, G. S., *Bull. Dushanbe* **35**, 32, 1963.
V337 Aql	Catalano, F., Rodono, M. (Catania) (*V*).
V889 Aql	Semeniuk, I., (*BV*).
RS Ari	Miner, S. D., *Astrophys. J.*, **144**, 1101, 1966.
RX Ari	McCluskey, G. E., *Astr. J.*, **68**, 741, 1963.
SS Aur	Mumford, G. S., *Astr. J.*, **69**, 270, 1964.
SX Aur	Fliegel, H. F., *Astr. J.*, **68**, 740, 1963.
AR Aur	Johansen, K., (Brorfelde Obs.) Peaks 3500, 4050, 4730, 5520 Å and Hβ.
	O'Connell, D. J. K., (Spec. Vatic.) *UBV*.
BF Aur	Mannino, G., (Bologna-Loiano) (*UBV*).
β Aur	Johansen, K., (Brorfelde Obs.) Peaks 3500, 4050, 4730, 5520Å and at Hβ.
ε Aur	Chen, K. Y., *Astr. J.*, **68**, 740, 1963 (infrared).
	Low, F. J., Mitchell, R. I., *Astrophys. J.*, **141**, 327, 1965.
	Mitchell, R. I., *Astrophys. J.*, **140**, 1607, 1964 (9 colors).
ζ Aur	Bappu, M. K. V., Doss, A. T., Viswanadham, P., *Observatory*, **85**, 85, 1965.
	Chen, K. Y., *Astr. J.*, **68**, 740, 1963 (infrared).
	Fracastoro, M. G., Catalano, S., *Mem. Soc. astr. ital.*, **36**, 99, 1965; interference filters, 4 colors.
	Hayashi, M., (Kyoto).
	Jackisch, G., *Veröff. Sternw. Sonneberg*, **5**, 264, 1963.
	Kondo, Y., Harris, A. J., *Astr. J.*, **69**, 409, 1964.
	Nelson, B., (San Diego) (*UBV*).
	O'Connell, D. J. K., *Ric. astr. Spec. Vatic.*, **6**, 523, 1964 (*UBV* and interference filters).
	Schneller, H., *Mitt. veränd. Sterne*, **2**, 178, 1965. Obs. continuing.
	Shao, C. Y., *Astr. J.*, **69**, 858, 1964.
	van Genderen, A. M., *Bull. astr. Inst. Netherl.*, **17**, 446, 1964.
	Zhang, H.-z., Liu, X-f., *Acta astr. Sinica*, **13**, 102, 1965.
η Aur	Fracastoro, M. G., Catalano, S., *Mem. Soc. astr. ital.*, **36**, 105, 1965. Variation confirmed.
	Jackisch, G., *Veröff. Sternw. Sonneberg*, **5**, 264, 1963, finds no var.
θ Aur	Jackisch, G., *Veröff. Sternw. Sonneberg*, **5**, 267, 1963, finds no var.
XY Boo	Wood, R., *Observatory*, **85**, 258, 1965.
ZZ Boo	(Flower and Cook) 2 colors.
AC Boo	Binnendijk, L., *Astr. J.*, **70**, 201, 1965.
	Mauder, H., *Z. Astrophys.*, **60**, 222, 1964.
44 i Boo	Catalano, S., Saitta, T., *Mem. Soc. astr. ital.*, **35**, 43, 1964.
	Chen, K. Y., Rekenthaler, D. A. (Florida) (*BV*).
	Purgathofer, A., Widorn, T., *Mitt. Univ. Sternw. Wien*, **12**, no. 6, 1964.
	Purgathofer, A., Widorn, T., *Sber. öst. Akad. Wiss.*, Wien, Math.-nat. Kl. Abt. 2, 171, 217, 1964.
	Schneller, H., (Potsdam).
Z Cam	Mumford, G. S., *Astr. J.*, **69**, 270, 1964.
SV Cam	Nelson, B., *Astr. J.*, **69**, 271, 1964; *Publ. astr. Soc. Pacif.*, **75**, 18, 1965.
TU Cam	Margoni, R., (Asiago).
	West, R., *Publ. astr. Soc. Pacif.*, **76**, 49, 1964; *Medd. Københavns Obs.*, no. 181, 1964.
S Cnc	(Flower and Cook).
TX Cnc	Binnendijk, L., *Astr. J.*, **68**, 740, 1963.
	Chou, K. C., *Astr. J.*, **68**, 676, 1963.
	Kitamura, M., (Mt Dodaira).

Star	References
TZ Cnc	Koch, R. H., Lampert, G., *Publ. astr. Soc. Pacif.*, **76**, 254, 1964.
WY Cnc	Chambliss, C. R., *Astr. J.*, **70**, 741, 1965.
R CMa	Kitamura, M., Sato, H., (Mt Dodaira) *UBV* and interference filters.
	Knipe, G. F. G., *Johannesburg Circ.*, no. 122, 21.
29 CMa	Doss, A. T. (Kodaikanal) 2 colors.
XZ CMi	Chou, K. C., *Astr. J.*, **68**, 676, 1963.
	Wilson, R. E., *Astr. J.*, **68**, 740, 1963; **71**, 32, 1966.
RS CVn	Chisari, D., Lacona, G., *Mem. Soc. astr. ital.*, **36**, 463, 1965.
	(Flower and Cook).
	Nelson, B., (San Diego) (*UBV*).
	Catalano, S., Rodono, M., (Catania) (*V*).
TV Car	Devinney, E. J., *Astr. J.*, **68**, 741, 1963.
GV Car	Landolt, A. U., (Cerro Tololo) (*UBV*).
RZ Cas	Burke, E. W., Jr, Henderson, T. C., Boy, W. R., (King College).
	Burke, E. W., Jr, Rolland, W. W., *Astr. J.*, **71**, 38, 1966.
	Plavec, M., *Bull. astr. Inst. Csl.*, **15**, 23, 1964.
	Shao, C. Y., (Harvard) 3 colors.
SX Cas	(Flower and Cook).
	Miner, S. D., *Astrophys. J.*, **144**, 1101, 1966.
	Nelson, B., (San Diego) (*UBV*).
	Shao, C. Y., *Astr. J.*, **70**, 147, 1965.
	Widorn, T., (Univ.-Sternwarte Wien).
TV Cas	Nelson, B., Sievers, J., (San Diego) (*UBV*).
BM Cas	Shao, C. Y., (Harvard) 3 colors.
BY Cas	Malik, G. M., *Astr. J.*, **70**, 94, 1965.
CW Cas	Broglia, P., *Mem. Soc. astr. ital.*, **35**, 23, 1964.
	Hogg, A. R., *Astr. J.*, **71**, 242, 1966.
DO Cas	Winkler, L., *Astr. J.*, **71**, 40, 1966.
RR Cen	Knipe, G. F. G., *Astrophys. J.*, **142**, 1068, 1965.
SV Cen	Irwin, J. B., *Astr. J.*, (in press).
BF Cen	Landolt, A. U., (Cerro Tololo) (*UBV*).
U Cep	Eggen, O. J., *Astr. J.*, **68**, 493, 1963.
	Long, R., Nelson, B., (San Diego) (*UBV*).
	Miner, E. D., *Astrophys. J.*, **144**, 1101, 1966.
VV Cep	Peery, B. F., *Astrophys J.*, **144**, 672, 1966.
VW Cep	Gyldenkerne, K., Jaeger, J. E., Johansen, K. T., *Publ. Københavns Obs.*, 177, 1963.
	Kwee, K. K., *Bull. astr. Inst. Netherl., Suppl.*, **1**, 245, 1966; 265, 1966.
	Rossati, F., *Mem. Soc. astr. ital.*, **36**, 109, 1965.
XX Cep	Angione, R., Nelson, B., (San Diego) (*UBV*).
CQ Cep	Hiltner, W. A., Schild, R. E., *Astrophys. J.*, **143**, 1008, 1966.
	Pyper, D. M., *Astrophys. J.*, **144**, 13, 1966.
CX Cep	Hiltner, W. A., Schild, R. E., *Astrophys. J.*, **143**, 1008, 1966.
	Pyper, D. M., *Astrophys. J.*, **144**, 13, 1966.
EG Cep	Cochran, G., Nelson, B., (San Diego) (*UBV*).
EI Cep	Abrami, A., *Mem. Soc. astr. ital.*, **37**, 369, 1966.
EK Cep	Ebbighausen, E. G., *Astr. J.*, **71**, 642, 1966.
9 Cha	Lagerweij, H. C., Cousins, A. W. J., (Cape Obs.) (*UBV*).
RZ Com	Binnendijk, L., *Astr. J.*, **69**, 154, 1964.
	Chou, K. C., *Astr. J.*, **68**, 676, 1963.
RW CrA	van Houten, C. J., van Houten-Groeneveld, J., *Inf. Bull. So. Hemisphere*, **8**, 25, 1966 (Leiden So. Station of Leiden Obs.).
ε CrA	Churms, J., (Cape Obs.).
	Cousins, A. W. J., *Mon. Notes astr. Soc. Sth. Afr.*, **23**, 24, 1964.

Star	References
ε CrA	Hernandez, C., (La Plata).
	(Flower and Cook) 2 colors.
U CrB	Batten, A. H., *Contr. Dom. astrophys. Obs.*, **85**, 1964.
	Catalano, S., Cristaldi, S., Lacona, G., (Catania).
RT CrB	Miner, E. D., *Astrophys. J.*, **144**, 1101, 1966.
VY Cru	van Houten, C. J., van Houten-Groeneveld, J., *Inf. Bull. So. Hemisphere*, **8**, 25, 1966 (Leiden So. Station of Leiden Obs.).
ZZ Cru	van Houten, C. J., van Houten-Groeneveld, J., *Inf. Bull. So. Hemisphere*, **8**, 25, 1966 (Leiden So. Station of Leiden Obs.).
DK Cyg	Binnendijk, L., *Astr. J.*, **69**, 157, 1964.
EM Cyg	Lowell Obs., *Astr. J.*, **68**, 645, 1963.
	Mumford, G. S., *Astr. J.*, **69**, 270, 1964; **71**, 243, 1966.
GO Cyg	Mannino, G., *Mem. Soc. astr. ital.*, **34**, 191, 1963.
	Purgathofer, A., Widorn, T., *Mitt. Univ. Sternw. Wien*, **12**, no. 6, 1964.
KR Cyg	Vetešník, M., *Bull. astr. Inst. Csl.*, **16**, 326, 1965.
KU Cyg	Popper, D. M., *Astrophys. J.*, **139**, 143, 1964.
MY Cyg	Filatov, G. S., *Bull. Dushanbe*, **35**, 32, 1963.
V380 Cyg	Semeniuk, I., (*V*).
V382 Cyg	Landolt, A. U., *Astr. J.*, **69**, 549, 1964; **71**, 242, 1966; *Astrophys. J.*, **140**, 1494, 1964.
V401 Cyg	Purgathofer, A., *Z. Astrophys.*, **59**, 29, 1964.
V444 Cyg	Demers, S., Fernie, J. D., *Publ. astr. Soc. Pacif.*, **76**, 350, 1964.
	Gusseinzade, A. A., *Perem. Zvezdy*, **15**, 555, 1965.
	Hiltner, W. A., Mook, D. E., *Astrophys. J.*, **143**, 1008, 1966.
	Pyper, D. M., *Astrophys. J.*, **144**, 13, 1966.
	Tcherepashchuk, A. M., *Perem. Zvezdy*, **15**, 549, 1965.
V477 Cyg	Chisari, D., Saitta, T., *Oss. astrofis. Catania Pubbl.*, **54**, 1963.
	O'Connell, D. J. K., (Spec. Vatic.) *UBV*.
	Rodono, M., (Catania) (*V*).
	Semeniuk, I., (*V*).
V478 Cyg	Schneller, H., (Potsdam) l.c. completed.
V548 Cyg	Rodono, M., (Catania) (*V*).
V695 Cyg	Herczeg, T., Schmidt, H., *Z. Astrophys.*, **55**, 162, 1962.
V828 Cyg	Wenzel, W., *Mitt. veränd. Sterne*, **3**, 71, 1965.
V836 Cyg	Harris, A. J., (Flower and Cook).
σ Cyg	Jackisch, G., *Veröff. Sternw. Sonneberg*, **5**, 286, 1963.
31 Cyg	Jackisch, G., *Veröff. Sternw. Sonneberg*, **5**, 283, 1963.
	O'Connell, D. J. K., *Ric. astr. Spec. Vatic.*, **6**, 499, 1964.
32 Cyg	(Flower and Cook).
	Herczeg, T., Schmidt, H., *Z. Astrophys.*, **57**, 254, 1963.
	Jackisch, G., *Veröff. Sternw. Sonneberg*, **5**, 283, 1963.
TW Dra	Miner, E. D., *Astrophys. J.*, **144**, 1101, 1966.
TZ Dra	Plavec, M., *Bull. astr. Inst. Csl.*, **15**, 23, 1964.
WW Dra	Koch, R. H., *Astr. J.*, **69**, 269, 1964.
	Miner, E. D., *Astrophys. J.*, **144**, 1101, 1966.
AI Dra	McCluskey, G. E., *Astr. J.*, **68**, 741, 1963.
S Equ	Catalano, S., Rodono, M., (Catania) (*B*, *V*).
	Plavec, M., *Bull. astr., Inst. Csl.*, **15**, 23, 1964.
RU Eri	Chou, K. C., *Astr. J.*, **68**, 676, 1963.
UX Eri	Binnendijk, L., (Flower and Cook) (*BV*).
YY Eri	Bhattacharyya, J. C., *Kodaikanal Obs. Bull.* (in press).
AS Eri	Filatov, G. S., *Bull. Dushanbe*, **35**, 32, 1963.
U Gem	Bornhurst, L. C., *J. R. astr. Soc. Can.*, **60**, 32, 1966.
	Krzeminski, W., *Astr. J.*, **68**, 639, 1963.

Star	References
U Gem	Krzeminski, W., *Astrophys. J.*, **142**, 1051, 1965.
	Lowell Obs., *Astr. J.*, **68**, 645, 1963.
	Mumford, G. S., *Astr. J.*, **69**, 270, 1964; *Astrophys. J.*, **139**, 476, 1964; *Kitt Peak nat. Obs. Contr.* no. 48.
	Paczyński, B., *Acta. astr.*, **15**, 305, 1965.
RW Gem	Smak, J., *Acta astr.*, **14**, 101, 1964.
AL Gem	Koch, R. H., *Astr. J.*, **68**, 785, 1963.
RU Gru	van Houten, C., van Houten-Groeneveld, J., *Inf. Bull. So. Hemisphere*, **8**, 25, 1966 (Leiden So. Station of Leiden Obs.).
Z Her	(Flower and Cook).
RX Her	Nelson, B., (San Diego) (*UBV*).
UX Her	Gordon, K. C., Kron, G. E., *Astr. J.*, **70**, 100, 1965; *Lick. Obs. Bull.* no. 588.
AK Her	Bertiau, F. C., *Ric. astr. Spec. Vatic.*, **6**, 487, 1963.
	Purgathofer, A., Widorn, T., *Mitt. Univ. Sternw. Wien*, **12**, no. 6, 1964.
	Purgathofer, A., Widorn, T., *Sber. öst. Akad. Wiss.*, Wien, Math.-nat. Kl., Abt. 2, 171, 217, 1964
DI Her	(Flower and Cook).
	Paczyński, B., (*V*).
	Semeniuk, I., (*BV*).
	Shao, C. Y. (Harvard) (*UBV*).
DQ Her	Bornhurst, L. C., *J. R. astr. Soc. Can.*, **60**, 32, 1966.
	Walker, M. E., *Astr. J.*, **69**, 677, 1964.
HS Her	(Flower and Cook).
NQ Her	Rossati, F., *Mem. Soc. astr. ital.*, **35**, 81, 1964.
	Schneller, H., *Mitt. veränd. Sterne*, **3**, 4, 1965 (not variable).
u Her	Catalano, S., (Catania).
Nova Her 1963	Chincarini, G., Howard, V. S., *I.B.V.S.*, **139**, 1966.
	Mumford, G. S., *Sky Telesc.*, **31**, 214, 338, 1966.
VZ Hya	Popper, D. M., *Astrophys. J.*, **141**, 126, 1965.
EX Hya	Mumford, G. S., *Publ. astr. Soc. Pacif.*, **76**, 57, 1964; *Kitt Peak nat. Obs. Contr.*, no. 50; *Astr. J.*, **71**, 243, 1966.
SS Lac	Tashpulatov, N., *Perem. Zvezdy*, **15**, 390, 1965.
SV Lac	Paczyński, B. (*UBV*).
SW Lac	Bookmyer, B. B., *Astr. J.*, **70**, 133, 415, 1965.
	Purgathofer, A., Widorn, T., *Mitt. Univ. Sternw. Wien*, **12**, no. 6, 1964.
	Purgathofer, A., Widorn, T., *Sber. öst. Akad. Wiss.*, Wien, Math-nat. Kl., Abt. 2, 171, 217, 1964.
	(U.S. Naval Obs.) *Astr. J.*, **68**, 676, 1963.
AR Lac	Karle, J. H., *Publ. astr. Soc., Pacif.*, **74**, 244, 1962.
	Miner, E. D., *Astrophys. J.*, **144**, 1101, 1966.
	Nelson, B., (San Diego) (*UBV*).
CO Lac	Semeniuk, I., (*BV*).
	Serkowski, K., (*V*).
EM Lac	Broglia, P., (Merate) (*BV*).
UV Leo	Broglia, P., *J. Observateurs*, **48**, 124, 1965.
	McCluskey, G. E., *Astr. J.*, **68**, 741, 1963.
	Popper, D. M., *Astrophys. J.*, **141**, 126, 1965.
	Purgathofer, A., Widorn, T., *Mitt. Univ. Sternw. Wien*, **12**, no. 6, 1964.
	Purgathofer, A., Widorn, T., *Sber. öst. Akad. Wiss.*, Wien, Math-nat. Kl., Abt. 2, 171, 217, 1964.
AC Leo	Chou, K. C., *Astr. J.*, **68**, 676, 1963.
	Schneller, H., (Potsdam) (not variable).
RS Lep	Klepczynski, W. J., Wood, F. B., *Astr. J.*, **69**, 92, 1964.
17 Lep	Widing, K. G., *Astrophys. J.*, **143**, 121, 1966.

Star	References
δ Lib	Koch, R. H., *Astr. J.*, **70**, 130, 1962.
RR Lyn	Linnell, A. P., *Astr. J.*, **68**, 620, 1963; **69**, 270, 1964; **70**, 582, 1965; **71**, 458, 1966.
SW Lyn	Gleim, J., *Astr. J.*, **68**, 741, 1963.
FL Lyr	Cristaldi, S., *Mem. Soc. astr. ital.*, **36**, 77, 1965.
	Miner, E. D., *Astrophys. J.*, **144**, 1101, 1966.
NY Lyr	Makarenko, E. N., *Perem. Zvezdy*, **15**, 446, 1965.
β Lyr	Appenzeller, I., *Astrophys. J.*, **141**, 1390, 1965.
	Belton, M. J. S., Woolf, N. J., *Astrophys. J.*, **141**, 145, 1965.
	Chia-Shang, Chang, *Acta astr. Sinica*, **7**, 203.
	Gyldenkerne, K., Jaeger, J. E., *Publ. Københavns Obs.*, 177, 1963.
	Herczeg, T., *Veröff. astr. Inst. Univ. Bonn*, **69**, 1964.
	Serkowski, K., *Astrophys. J.*, **142**, 793, 1965.
ζ Lyr	Jackisch, G., *Veröff. Sternw. Sonneberg*, **5**, 281, 1963.
31 Men	Lagerweij, H. C., Cousins, A. W. J., (Cape Obs.) (*UBV*).
RU Mon	Martynov, D., *Astr. Zu.*, **42**, 1209, 1965.
	Martynov, D., *I.B.V.S.*, **102**, 1965.
RW Mon	Smak, J., *Astr. J.*, **69**, 677, 1964.
IM Mon	Sanyal, A., Sinvhal, S. D., *Observatory*, **84**, 211, 1964.
	Sanyal, A., Mahra, H. S., Sanwal, N. B., *Bull. astr. Inst. Csl.*, **16**, 209, 1965.
U Oph	Koch, R. H., *Astr. J.*, **69**, 650, 1964; **70**, 582, 1965.
	(Flower and Cook).
WZ Oph	Popper, D. M., *Astrophys. J.*, **141**, 126, 1965.
V502 Oph	Chou, K. C., *Astr. J.*, **68**, 676, 1963.
	Magalashvili, N. L., Kumsishvili, J. I., *Bull. Abastumani*, **30**, 39, 1964.
	Wilson, R. E., *Astr. J.*, **68**, 740, 1963.
V566 Oph	(Flower and Cook).
	Purgathofer, A., Widorn, T., *Mitt. Univ. Sternw. Wien*, **12**, no. 6, 1964.
	Purgathofer, A., Widorn, T., *Sber. öst. Akad. Wiss*, Wien, Math-nat. Kl., Abt. 2, 171, 217, 1964.
	Schnell, A., Widorn, T., *Mitt. Univ. Sternw. Wien*, **12**, 125, 1965.
V839 Oph	Wilson, R. E., O'Toole, W., *Publ. astr. Soc. Pacif.*, **77**, 58, 1964.
VV Ori	Marasso, R., Nelson, B., (San Diego) (*UBV*).
	Sivaraman, K. R., (Kodaikanal) 4740 Å, 1/2 intensity: 47 Å.
CN Ori	Mumford, G. S., *Astr. J.*, **69**, 270, 1964.
FT Ori	Cristaldi, S., *Mem. Soc. astr. ital.*, **37**, 205, 1966.
U Peg	Purgathofer, A., Widorn, T., *Mitt. Univ. Sternw. Wien*, **12**, no. 6, 1964.
	Purgathofer, A., Widorn, T., *Sber. öst. Akad. Wiss*, Wien, Math-nat. Kl., Abt. 2, 171, 217, 1964.
AW Peg	Fresa, A., (Capodimonte) (*B*).
DI Peg	Chou, K. C., *Astr. J.*, **68**, 676, 1963.
EE Peg	Bakos, G. A., *Publ. Dunlap Obs.*, **2**, 431, 1965.
	Catalano, S., Rodono, M., (Catania), (*B*, *V*).
	Fresa, A., (Capodimonte).
ι Peg	Koch, R. H., *Astr. J.*, **69**, 269, 1964.
RW Per	Miner, E. D., *Astrophys. J.*, **144**, 1101, 1966.
RY Per	Linnell, A. P., *Astr. J.*, **68**, 620, 1963; **69**, 270, 1964; **70**, 582, 1965.
	Miner, E. D., *Astrophys. J.*, **144**, 1101, 1966.
TZ Per	Mumford, G. S., *Astr. J.*, **69**, 270, 1964.
AB Per	Miner, E. D., *Astrophys. J.*, **144**, 1101, 1966.
AG Per	Semeniuk, I., (*V*).
DM Per	Catalano, F., Catalano, S., Rodono, M., (Catania) (*BV*).

Star	References
GK Per	Mumford, G. S., *Astr. J.*, **69**, 270, 1964.
KP Per	Rossati, F., *Mem. Soc. astr. ital.*, **34**, 349, 1963; *Torino Contr.* no. 37, 1963.
β Per	Chen, K.-Y., Reuning, E. G., *Astr. J.*, **71**, 283, 1966.
	Cristaldi, S., Fracastoro, M. G., Sobieski, S., (Catania and Flower and Cook).
	Herczeg, T., *Veröff. astr. Inst. Univ. Bonn*, **69**, 1964.
ζ Per	Jackisch, G., *Veröff. Sternw. Sonneberg*, **5**, 290, 1963.
o Per	Jackisch, G., *Veröff. Sternw. Sonneberg*, **5**, 290, 1963.
RR Pic	van Houten, C. J., *Bull. astr. Inst. Netherl.*, **18**, 439, 1966.
δ Pic	Cousins, A. W. J., *Mon. Notes astr. Soc. Sth. Afr.*, **23**, 67, 1964; *Mon. Not. R. astr. Soc.*, **131**, 443, 1966.
UU Psc	Schneller, H., (Potsdam) Beginning obs.
VV Pup	Walker, M. E., *Astr. J.*, **69**, 677, 1964; *Lick Obs. Contr.* no. 179; *Mitt. Budapest Obs.*, **57**, 1965.
AU Pup	Chambliss, C. R., (Flower and Cook).
U Sge	Miner, E. D., *Astrophys. J.*, **144**, 1101, 1966.
V Sge	Herbig, G., Preston, G., Smak, J., Paczyński, B., *Astrophys. J.*, **141**, 617, 1965.
WZ Sge	Krzeminski, W., Kraft, R. P., *Astrophys. J.*, **140**, 921, 1964.
Nova W2 Sge	(Lowell), *Astr. J.*, **68**, 645, 1963.
V701 Sco	Leung, K.-c., (Flower and Cook).
ER Sct	O'Connell, D. J. K., (Spec. Vatic.) (*UBV*).
W Ser	Leung, K.-c., (Flower and Cook).
CV Ser	Demers, S., Fernie, J. D., *Publ. astr. Soc. Pacif.*, **76**, 350, 1964.
	Pyper, D. M., *Astrophys. J.*, **144**, 13, 1966.
X Tri	Miner, E. D., *Astrophys. J.*, **144**, 1101, 1966.
	Plavec, M., *Bull. astr. Inst. Csl.*, **15**, 23, 1964.
RS Tri	Plavec, M., *Bull. astr. Inst. Csl.*, **15**, 23, 1964.
W UMa	Binnendijk, L., *Astr. J.*, **71**, 340, 1966.
	Blitzstein, W., Chou, K. C., *Astr. J.*, **69**, 365, 1964.
	Cester, B., (Trieste) (*UBV*).
	Kuhi, L., *Publ. astr. Soc. Pacif.*, **76**, 430, 1964.
SU UMa	Mumford, G. S., *Astr. J.*, **69**, 270, 1964.
UV UMa	Broglia, P., (Merate) (*BV*).
UX UMa	Krzeminski, W., Walker, M. F., *Astrophys. J.*, **138**, 166, 1963.
TY UMa	Broglia, P., (Merate) (*BV*).
VV UMa	Wilson, R. E., *Astr. J.*, **68**, 740, 1963; **70**, 368, 1965; *Georgetown Obs. Repr.*, Ser. II, 22.
W UMi	Devinney, E. J., *Astr. J.*, **68**, 741, 1963.
S Vel	Sisteró, R., (Cordoba) pe obs. at Cerro Tololo Interamerican Obs.
AL Vel	Cousins, A. W. J., (Cape Obs.) (*UBV*).
	Wood, F. B., Richardson, R. R., *Astr. J.*, **69**, 297, 1964.
AO Vel	van Houten, C. J., van Houten-Groeneveld, J., *Inf. Bull. So. Hemisphere*, **8**, 25, 1966 (Leiden So. Station of Leiden Obs.).
γ Vel	Ganesh, K. S., (Kodaikanal).
AG Vir	Fliegal, H. F., *Astr. J.*, **68**, 740, 1963.
	Purgathofer, A., Widorn, T., *Mitt. Univ. Sternw. Wien*, **12**, no. 6, 1964.
	Purgathofer, A., Widorn, T., *Sber. öst. Akad. Wiss.*, Wien, Math.-nat. Kl., Abt. 2, 171, 217, 1964.
AH Vir	Purgathofer, A., Widorn, T., *Mitt. Univ. Sternw. Wien*, **12**, no. 6, 1964.
	Sievers, J., Nelson, B., (San Diego) (*UBV*).
	Truttse, Yu. L., Kalchaev, K. K., *Perem. Zvezdy*, **15**, 314, 1965.
BH Vir	Koch, R. H., *Astr. J.*, **69**, 650, 1964; **70**, 582, 1965; **71**, 241, 1966.

Star	References
Z Vul	Broglia, P., *J. Observateurs*, **47**, 99, 1964.
	Paczyński, B., *J. Observateurs*, **47**, 101, 1964.
RR Vul	Filatov, G. S., *Bull. Dushambe*, **35**, 32, 1963.
BP Vul	Sturch, G., Moorehead, J., *Astr. J.*, **70**, 612, 1965.
DR Vul	Krzeminski, W., (*V*).
	O'Connell, D. J. K., (Spec. Vatic.) (*UBV*).
	Semeniuk, I., (*BV*).
BD +70° 1183	Mannino, G., (Loiano) (*UBV*).
BD +69° 389	Harris, A. J., (Flower and Cook).
=BV312	Mannino, G., (Loiano) (*BV*).
	Schneller, H., (Potsdam) lc completed.
BD +57° 1546	
=BV277	Meinunger, L., *Mitt. veränd. Sterne*, **3**, 16, 1965.
BD +54° 49	
=BV328	Kondo, Y., *Veröff. Remeis-Sternw. Bamberg*, **5**, 18, 1964. Not eclipsing.
BD +45° 1570	
=SVS1396	Lange, G. A., *Astr. Circ.* (U.S.S.R.), no. 246, 1963.
BD +34° 1051	
=HD35652	Mayer, P., *I.B.V.S.*, **49**, 1964; *Publ. astr. Soc. Pacif.*, **77**, 436, 1965.
BD +33° 4252	Kondo, Y., *Astr. J.*, **71**, 54, 1966.
=BV342	Mannino, G., (Loiano) (*UBV*).
	Rodono, M., (Catania).
	Schneller, H., (Potsdam) lc completed.
BD +30° 2163	Kalish, M. S., *Publ. astr. Soc. Pacif.*, **77**, 36, 1965.
	Kondo, Y., (Flower and Cook).
	Paczyński, B., *Astr. J.*, **69**, 124, 1964.
HD1372	van Houten, C. J., van Houten-Groeneveld, J., *Inf. Bull. So. Hemisphere*, **8**, 25, 1966 (Leiden S. Station of Leiden Obs.).
HD128220	Guinan, E., (Flower and Cook).
	Schneller, H., (Potsdam) No ecl. min.
HD161783	van Houten, C. J., van Houten-Groeneveld, J., *Inf. Bull. So. Hemisphere*, **8**, 25, 1966 (Leiden S. Station of Leiden Obs.).
HD187418	van Houten, C. J., van Houten-Groeneveld, J., *Inf. Bull. So. Hemisphere*, **8**, 25, 1966 (Leiden S. Station of Leiden Obs.).
HD200391	Bakos, G. A., Northcott, R., (David Dunlap).
HR5687	Smith, L. F., *Publ. astr. Soc., Pacif.*, **78**, 168, 1966.
HR7484	
=HD185912	Snowden, M. S., *I.B.V.S.*, **127**, 1966.
HR8164	(Loiano).
ADS9537 A,B.	Batten, A. H., Hardie, R. H., *Astr. J.*, **70**, 666, 1965.
BV169	(Flower and Cook).
BV267	Kondo, Y., *Astr. J.*, **71**, 46, 1966.
	Mannino, G., (Loiano) (*BV*).
BV332	Harris, A. J., (Flower and Cook).
BV357	Chambliss, C. R., (Flower and Cook).
BV382	Bartolini, C., Mammano, A., Mannino, G., Margoni, R., an early type contact system.
	Gleim, J., (Flower and Cook) (*UBV*).
BV419	Chambliss, C. R., (Flower and Cook) 300 obs., per 15 hrs.
BV430	Chambliss, C. R., Rogers, M. J. (Flower and Cook) 1200 obs.
BV513	Leung, K.-c., (Flower and Cook) 2 colors.
BV516	Chambliss, C. R., (Flower and Cook) 500 obs.
BV544	Leung, K.-c., (Flower and Cook) 500 obs.

Table 1b. New brightness determinations other than photoelectric

Star	References
S 5043 Aps	Hoffmeister, C., *Veröff. Sternw. Sonneberg*, **6**, 3, 1965.
TT Aur	Karetnikov, V. G., *Perem. Zvezdy*, **14**, 38, 1962.
	Obůrka, O., *Bull. astr. Inst. Csl.*, **15**, 26, 1964.
RZ Cas	Karetnikov, V. G., *Perem. Zvezdy*, **14**, 38, 1962.
AZ Cas	Weber, R., *Bull. Mainterne*, **3**, 1966.
V 364 Cas	Borsdiko, V. I., *Bull. Tadjik Obs.*, **38**, 43, 1965.
AI Cep	Albo, Ch., *Publ. Tartu Obs.*, **34**, 169, 1964.
CQ Cep	Ishchenko, I. M., *Trudy Tashkent Obs.*, **9**, 1, 1963.
DK Cep	Albo, Ch., *Publ. Tartu Obs.*, **34**, 169, 1964.
DL Cep	Albo, Ch., *Publ. Tartu Obs.*, **34**, 169, 1964.
Z Cir	van Hoof, A., *I.B.V.S.*, **41**, 1964 (not eclipsing).
Y Crv	Meinunger, L., *Mitt. veränd. Sterne*, **2**, 47, 1964.
RZ Cyg	Ignatieva, L. V., *Perem. Zvezdy*, **14**, 119, 1962.
UZ Cyg	Ishchenko, I. M., *Trudy Tashkent Obs.*, **9**, 1, 1963.
DK Cyg	Ishchenko, I. M., *Trudy Tashkent Obs.*, **9**, 1, 1963.
DL Cyg	Ishchenko, I. M., *Trudy Tashkent Obs.*, **9**, 1, 1963.
MR Cyg	Ishchenko, I. M., *Trudy Tashkent Obs.*, **9**, 1, 1963.
V 753 Cyg	Wawriniak, P. F., *Circ. Lwow Obs.*, no. 39–40.
P 5535 Cyg = 173·1935 = CVS 5365	Meinunger, L., *Mitt. veränd. Sterne*, **7**, 14, 1965,
XY Dra	Meinunger, L., *Mitt. veränd. Sterne*, **3**, 15, 1965.
SU For	Omay, G., *Astr. J.*, **70**, 484, 1965.
RT Lac	Ishchenko, I. M., *Trudy Tashkent Obs.*, **9**, 1, 1963.
AR Lac	Ishchenko, I. M., *Trudy Tashkent Obs.*, **9**, 1, 1963.
AW Lac	Ishchenko, I. M., *Trudy Tashkent Obs.*, **9**, 1, 1963.
CM Lac	Ishchenko, I. M., *Trudy Tashkent Obs.*, **9**, 1, 1963.
LZ Per	Ahnert, P., *Mitt. veränd. Sterne*, **3**, 7, 1965.
V Sge	Romano, G., *I.B.V.S.*, **98**, 1965.
SY Vul	Meinunger, L., *Mitt. veränd. Sterne*, **3**, 23, 1965.
BP Vul	Huth, H., *Mitt. veränd. Sterne*, **3**, 36, 1965.
BD +57° 1546 Dra = BV 277	Meinunger, L., *Mitt. veränd. Sterne*, **3**, 16, 1965.
CoD −42° 9876 = S 5001	Strohmeier, W., *Mitt. veränd. Sterne*, **3**, 2, 1965.
BV 36	Meinunger, L., *Mitt. veränd. Sterne*, **3**, 21, 1965.
BV 159, 343, 344	Nikulina, T. C., *Bull. Dushanbe*, **35**, 45, 1963.
СПЗ 1284	Kuročkin, N. E., *I.B.V.S.*, **79**, 1965.
СПЗ 1394	Nikolov, N. S., *Perem. Zvezdy*, **14**, no. 4, 336, 1963.
СПЗ 1399	Popov, M. V., Kazmina, E. I., *Perem. Zvezdy*, **15**, no. 1, 109, 1064.
GR 109 = SVS 548	Romano, G., *Coelum*, **33**, no. 3–4, 1965.
GR 116	Romano, G., *I.B.V.S.*, **106**, 1965.
WR 153	Weber, R.
Zi 1591 = CVS 4506	Meinunger, L., *Mitt. veränd. Sterne*, **3**, 18, 1965.

References dealing with several stars each

Cousins, A. W. J., Fabry Photometry of Bright Southern Stars of Cape of Good Hope. Royal Obs. 1966.

Gaposchkin, S., *Kl. Veröff. Remeis-Sternw. Bamberg*, Bd IV, Nr. 40, 1965.
Miller, W. J., Wachmann, A. A., *Ric. astr. Spec. Vatic.*, **6**, 413, 1963; **6**, 497, 1963.
Miller, W. J., *Ric. astr. Spec. Vatic.*, **6**, 591, 1964; **7**, 197, 1965; **7**, 217, 1966.
Plaut, L., *Bull. astr. Inst. Netherl., Suppl.*, **1**, no. 3, 1966.
Szafraniec, R., *Acta astr., Suppl.*, Part 6, 1963. Cracow vis. observations Sct-Vul.

NEW SPECTROGRAPHIC DATA

In Table 2 are listed references to spectroscopic data made available since the previous report. Of the stars showing atmospheric eclipses, there are reports of the 1962 eclipse of 32 Cygni and the 1963–64 eclipse of ζ Aurigae; Peery has published a new orbit and discussion for VV Cephei, and Wright is working up the Victoria results for VV Cephei and 31 Cygni and for the 1965 eclipse of 32 Cygni. Popper continues to improve the orbits for binaries for which accurate masses and dimensions can be determined. The Asiago and Toronto observers report more orbits for Bamberg Variables, and Götz and Wenzel have given spectral types for faint variables. Gyldenkerne has called attention to the possibility of fruitful results from the fall 1966 eclipse of ζ Aurigae.

Among more general papers, Karetnikov, Lavrov and Martynov have published volumes 3 and 4 of their *Bibliography of Spectroscopic Binary Orbits*, Perova has published data on the components in systems with known or suspected variables, MacDonald has listed spectra of 179 eclipsing binary stars, and Koch, Olson and Yoss have discussed the luminosities, spectra and rotation of 19 eclipsing systems. Popper has surveyed the spectra of 38 eclipsing binaries at Mt Stromlo, and also has found (private communication) that the following northern stars have double lines: PW Her, BP Vul, EE Reg, HD 90242 (BV 701), and BD + 47°781 (BV 307). Oke has published observed energy distributions for early-type stars and Aller, Faulkner and Morton have given energy distributions for southern stars. Several papers on Wolf Rayet stars, including eclipsing stars, have appeared: classifications by Hiltner and Schick, energy distributions by Kuhi and colour indices by Popper. Azimov has discussed the sub-dwarf components in close binaries, and Kraft has related novae to U Geminorum stars.

Table 2. Spectroscopic Data on Eclipsing Binary Stars

Star	References	Remarks
V822 Aql	Fitzgerald, P., *Publ. Dom. astrophys. Obs.*, **2**, 417, 1964	Orbit
AE Aqr	Walker, M. F., *Sky Telesc.*, **29**, 23, 1965	
ε Aur	Morris, S. C., *Astr. J.*, **70**, 685, 1965	Orbit, dimensions
ζ Aur 1963–64 eclipse	Bardin, C., Prévot, L., *Publ. Hte-Provence*, **7**, no. 32, 1964	
	Faraggiana, R., *Z. Astrophys.*, **62**, 99, 1965	
	Faraggiana, R., Hack, M., *Z. Astrophys.*, **64**, 48, 1966	Comp. 31, 32 Cyg
	Hardop, J., Herczeg, T., Scholz, W., *Z. Astrophys.*, **64**, 97, 1966	
	Kitamura, M.	Okayama observations
	Mammano, A., *et al.*	Asiago observations
	Odgers, G. J., Wright, K. O., *J.R. astr. Soc. Can.*, **59**, 115, 1965	
TU Cam	Margoni. R.. (Asiago). Herczeg, T., (Hamburg)	Current program

Star	References	Remarks
R CMa	Kitamura, M., Sato, H.,	Okayama observations in progress
SX Cas	Fredrick, L. W., (Leander McCormick)	Infra red ob.
YZ Cas	Dombrovsky, V. A.,	
	Publ. Leningrad Obs., **21**, 19, 1964	Limb darkening
	Perry, C. L., Stone, S. N.,	
	Publ. astr. Soc. Pacif., **78**, 5, 1966	Orbit
AO Cas	Hovhannisian, R. H.,	
	Mitt. Burakan, **35**, 37, 1964	O.D.
U Cep	Andrews, D. H., Batten, A. H., Plavec, M.,	
	Publ. astr. Soc. Pacif., **78**, 68, 1966	No Hα emission
	Batten, A. H.,	Victoria observations
VV Cep	Peery, B. F., Jr.,	
	Astrophys. J., **144**, 672, 1966	Orbit, discussion
	Wright, K. O.,	
	Trieste Colloq.	D.A.O. observations
EK Cep	Ebbighausen, E.G.,	
	Astr. J., **71**, 730, 1966	Orbit, discussion
9 Cha= HD 39780	Jones, D. H. P., (Radcliffe)	Double-lined spectra
ε CrA	Hernandez, C., (LaPlata)	Current observations
S CrB	Abhyankar, A. D., (Dunlap Plates)	New spec. orbit
U CrB	Batten, A. H., (D.A.O.)	Victoria observations
α CrB	Ebbighausen, E. G., (Kitt Peak)	Orb els.
Y Cyg	Petrie, R. M., (D.A.O.)	Victoria observations
KU Cyg	Popper, D. M.,	
	Astrophys. J., **141**, 314, 1965	Revision: Dimensions
MY Cyg	Popper, D. M.,	
	Astr. J., **69**, 677, 1964	2 comp. will give mass
57 Cyg	Ovenden, M. W.,	
	Mon. Not. R. astr. Soc., **126**, 77, 1963	Reflection effect
32 Cyg	Faraggiana, R., Gokgoz, A., Hack, M., Kendis, I.,	
1962 ecl.	*Mem. Soc. astr. ital.*, **36**, 63, 1965	
	Scholz, M.,	
	Z. Astrophys., **61**, 179, 1965	
	Wright, K. O.,	
	J. R. astr. Soc. Can., **60**, 89, 1966	
S Dor	Thackeray, A. D.,	Emission (max.) compared with min.
	Mon. Not. R. astr. Soc., **129**, 169, 1965	
AI Dra	Ebbighausen, E. G., (Kitt Peak)	Orb. els.
CC Eri	Evans, D. S.,	
	Mon. Notes Astr. Soc. Sth. Afr., **23**, 67, 1964	
DQ Her	Larsson-Leander, G.,	High dispersion 1935 Mt Wilson spectra
	Ann. Astrophys., **27**, 587, 1964	
VZ Hya	Popper, D. M.,	
	Astrophys. J., **141**, 126, 1965	Orbit, masses
UV Leo	Popper, D. M.,	
	Astrophys. J., **141**, 126, 1965	Orbit, masses
17 Lep	Widing, K.,	Main sequence A+, Sub-giant M
	Astrophys. J., **143**, 121, 1966	
	Sahade, J., (LaPlata)	Current program
δ Lib	McLaughlin, D. B.,	
	Astrophys. J., **138**, 1309, 1963	Rv differences

Star	References	Remarks
β Lyr	Batten, A. H., (D.A.O.)	
	Hack, M.,	
	Z. Astrophys., **62**, 203, 1065	Composition
	Herczeg, T.	Put at disposal of Dr Svolopoulos (Athens)
	Koch, R. H.,	Scanner; 3rd light source
	Astr. J., **70**, 324, 1965	
31 Men = HD 75747	Jones, D. H. P., (Radcliffe)	Current program
WZ Oph	Popper, D. M.,	
	Astrophys. J., **141**, 126, 1965	Orbit, masses
V566 Oph	Heard, J. F.,	
	J. R. astr. Soc. Can., **59**, 258, 1965	Orbit for W UMa system
VV Ori	Sivaraman, K. R., (Kodaikanal)	Current program
DN Ori	Smak, J.	
	Publ. astr. Soc. Pacif., **76**, 210, 1964; *Acta. astr.*, **15**, 327, 1965	Masses: subgiant 10^m overluminous secondary
EE Peg	Bakos, G. A.,	
	Publ. David Dunlap Obs., **2**, 431, 1965	Orbit
IZ Per = BV224 = HD9234	Yavuz, I., (Hamburg)	Orbit
β Per	Fletcher, E. S.,	
	Astr. J., **69**, 357, 1964	Helium cloud model
δ Pic	Thackeray, A. D.	
	Mon. Not. R. astr. Soc., **131**, 435, 1966	Orbit, mass ratio
V Sge	Herbig, G., Preston, G. W., Smak, J., Paczyński, B.,	
	Astrophys, J., **141**, 617, 1965	Nova-like; masses
WZ Sge	Krzeminski, W., Kraft, R. P.,	Stream; W. dwarf pair
	Astrophys. J., **140**, 921, 1964	
XZ Sge	Smak, J.,	
	Acta. astr., **15**, 327, 1965	Inverse *M–L* relation
V453 Sco	Sahade, J., Frieboes-Condé, H.,	
	Astrophys. J., **141**, 652, 1965	Orbit, model
RW Tau	Makarenko, E. N.,	
	Perem. Zvezdy, **14**, 214, 1962	Orbit
BL Tel	Feast, M. W.,	
	Mon. Notes astr. Soc. Sth. Afr., **25**, 5, 1966; *Mon. Not. R. astr. Soc.*, (in press)	Supergiant; semi-detached dark (?) secondary
TX UMa	Grewing, M., Herczeg, T.,	
	Z. Astrophys., **65**, 1966	Els: apsidal motion
S Vel	Sisteró, R., (Cordoba)	Current program
AL Vel	Wesselink, A. J.,	
	Mon. Not. R. astr. Soc., **127**, 105, 1964	Model: reflection effect
γ Vel	Ganesh, K. S., (Kodaikanal)	Current program, spec. orb.
BH Vir	Abt, H. A.,	
	Publ. astr. Soc. Pacif., **77**, 367, 1965	Orbit
B1985	Cowley, A. P.,	
	Astrophys. J., **142**, 299, 1965	Orbit: atmos. eclipse (?)
B5481 = HR8164	Mammano, A.,	Asiago orbit
GC22276	Makarenko, E. N.,	
	Perem. Zvezdy, **14**, 214, 1962	Revised orbit

Star	References	Remarks
BD +46° 985	Margoni, R., *et al.*, *I.B.V.S.*, **131**, 1966	Asiago orbit
BD +54° 2193	Snowden, M. S., *I.B.V.S.*, **127**, 1966 (Kitt Peak)	Vel. curve
HD 35652	Fletcher, J. M. (D.A.O.)	New spec els.
	Mayer, P. (Prague)	β Lyr binary
HD 200391	Bakos, G. A., Northcott, R., (David Dunlap)	New spec. els.
HD 205372	Batten, A. H.,	Discontinued at D.A.O.
ADS 9537	Batten, A. H., Hardie, R. H., *Astr. J.*, **70**, 666, 1965	2 contact binaries
BV 241, 342, 374	Fitzgerald, P. *Publ. David Dunlap Obs.*, **2**, 417, 1964	Orbits
BV 267	Margoni, R., Mammano, A., Biolchini, R., Bartolini, C. *I.B.V.S.*, **131**, 1966	Prelim. spec. els.
BV 312, 374, 412	Margoni, R., Mammano, A., Stagni, R.,	Asiago Orbits
BV 382	Bartolini, G., Mammano, A., Mannino, G., Margoni, R. *Asiago Contr.*, 168, 1965	Orbits, disc.

General papers and papers treating several stars

Aller, L. H., Faulkner, D. J., Morton, R. H., *Astrophys. J.*, **144**, 1073, 1966.
Artiukhina, N. M., *Perem. Zvezdy*, **14**, 225, 1962.
Azimov, S. M., *Nach. Ak. Wiss. Azerbaijan*, no. 4, 109, 1964.
Götz, W., Wenzel, W., *Mitt. veränd. Sterne*, **2**, 85, 1964; *Mitt. veränd. Sterne*, **3**, 133, 1966.
Hiltner, W. A., Schick, R. E., *Astrophys. J.*, **143**, 770, 1966.
Karetnikov, V. C., Lavrov, N. V., Martynov, D. Ya., *Bibliography of Spectroscopic Binary Orbits*, Vols. 3 and 4.
Koch, R. H., Olson, E. C., Yoss, K. M., *Astrophys. J.*, **141**, 955, 1965.
Kuhi, L. V., *Astrophys. J.*, **143**, 753, 1966.
Kraft, R. P., *Astrophys. J.*, **139**, 457, 1964.
MacDonald, D. D., *Publ. Leander McCormick Obs.*, **12**, 51, 1964.
Oke, J. B., *Astrophys. J.*, **140**, 689, 1964.
Perova, N. B., *Perem. Zvezdy*, **14**, 357, 1963.
Popper, D. M., *Inf. Bull. So. Hemisphere*, no. 7, Oct. 1965; also, *Astr. J.*, **71**, 175, 1966.
Pyper, D. M., *Astrophys. J.*, **144**, 13, 1966.

TIMES OF MINIMA AND PERIOD VARIATIONS

Table 3 contains the impressive number of over 400 eclipsing binaries whose periods were determined, revised, or discussed in the past triennium. It is to be understood that the entries in this Table represent in fact three rather different categories of observations.

In the first group, we have stars for which the period of eclipses is determined for the first time, usually from estimates on patrol photographic plates. This kind of work is directly connected with the discoveries of new eclipsing binaries. Here the largest number of new systems has again been supplied by the three German observatories: Bamberg (Strohmeier and collaborators), Sonneberg (Hoffmeister, Ahnert, Huth, Meinunger), and Hamburg (Wachmann). The first two recently shifted their surveys to the southern hemisphere, where the harvest will be particularly rich.

In the second category, we have observations of times of minima of eclipsing binaries known for some time. Here the purpose is not only to improve on the first and of necessity preliminary values of the periods, but also to detect whatever variations of period may occur. This field is now dominated by three large amateur groups. One of them, mostly American amateurs, is coordinated by Williams, the reductions are made by Robinson and Ashbrook, and the results are published by Robinson in the *Information Bulletin on Variable Stars* (*IBVS*) in Budapest. The results of German observers have been published by Dueball and Lehmann, and by Pohl and Kizilirmak, in the *Astronomische Nachrichten.* A third group works in Czechoslovakia under the direction of Obůrka, and publishes in the *Bulletin of the Institutes of Czechoslovakia.*

Thanks to these visual observers, many systems are being almost continually surveyed and larger period changes cannot escape detection. However, for some purposes, the precision of photoelectric work is required. Naturally an exact determination of period is a necessary condition for the derivation of photometric and spectrographic elements, but in order to study the period changes we need repeated observations. In the past triennium, systematic work in this field was reported by Fitch (Steward Observatory, Arizona), Purgathofer and Widorn (Vienna), and Plavec (Ondřejov). It is hoped that more observatories will join this program, which is now being coordinated by a special committee under the chairmanship of Plavec consisting of Koch, Kordylewski, Schneller, and Szafraniec. The systems so far chosen for concentrated effort on times of minima are given in the following list:

Stars for Observations of Times of Minima

Star	Max.	A_1	D	P	Reason for inclusion
RT And	$9^m.0$	$1^m.1$	2^h9	$0^d.63$	Short minima. Period constant?
WW Aur	5·7	0·7	6·4	2·53	Check on constancy of period
i Boo	6·6	0·4	W	0·27	Bright W UMa system
RS CVn	8·0	1·3	12·8	4·80	Periodic changes?
R CMa	6·2	0·6	4	1·14	Representative of subclass
RZ Cas	6·4	1·4	4·8	1·20	Easily observable
TV Cas	7·3	1·1	7	1·81	Period constant?
U Cep	6·7	3·1	10	2·49	Large variations
VW Cep	7·8	0·4	2	0·28	Bright W UMa system
U CrB	7·6	1·4	10·6	3·45	Large variations
Y Cyg	7·0	0·6	7	3·00	Apsidal motion
V477 Cyg	8·3	0·9	3·6	2·35	Apsidal motion indicated
W Del	9·4	3·3	12·3	4·81	Apsidal motion? Large changes
TW Dra	7·8	1·7	10	2·81	Large variations
TX Her	8·3	0·7	4·6	2·06	MS system with variable period
AR Lac	7·3	0·9	8·5	1·98	Sudden change of period
β Lyr	3·4	0·9	β	12·93	Secular increase of period?
U Oph	5·8	0·7	7	1·68	MS system with variable period
V451 Oph	7·8	0·6	9·5	2·20	Apsidal motion indicated
β Per	2·2	1·3	9·8	2·87	Longest series of observations
U Sge	6·4	2·7	13	3·38	Suitable for complex studies
RW Tau	8·0	4·3	9	2·77	Gaseous ring
X Tri	8·9	3·1	5	0·97	Accurate timing of minima
TX UMa	6·8	2·1	12	3·06	Periodic changes? Apsidal motion
VV UMa	10·1	0·9	3·4	0·69	Short-period Algol system

The problem of the causes of the period variations remains to a large extent still open. Some astronomers are of the opinion that at least some of the observed variations are independent of the physical nature of the components of the system in question. In particular, the Cracow astronomers K. Kordylewski (7), Z. Kordylewski (8), and Szafraniec (9) expressed the opinion that the variation of period is correlated with the spherical or galactic coordinates. Again, certainly some period changes are due to apsidal motion (see next section), and some to the presence of a third body in the system; the question is, how much can be explained in these ways. Illés-Almár and Almár (10) suggested that the presence of a third body might be a good explanation in more cases than believed, and in particular suggested that W Del, Z Dra, TX Her, and RV Lyr may be systems of this kind. Schneller (11) opposes this view in each of the above-mentioned cases; he argues that the variations of period are in most cases not periodic but rather cyclic. The same opinion was expressed by Plavec (12, 13).

It appears from the work by Schneller and by Plavec that strongest variations of period occur in systems with unstable components. F. B. Wood's idea that these variations are connected with mass transfer or mass loss appears to be much more strongly supported by observational evidence now than it was in 1950. Considerable progress has been achieved recently in the theoretical foundation of this hypothesis. S.-S. Huang (14) studied modes of mass ejection and their influence on period variations. A very detailed analysis of the problem of period changes due to mass exchange and mass loss was done by Piotrowski (15) and also by Kruszewski (16). The whole theory was reviewed by Kruszewski (17).

The problem is now how to connect the observed variations with the theory. The formulae derived by the authors just mentioned offer a wide variety of results according to the initial conditions. If we assume that the mass loss is due to ejections from the surface of the unstable component, we have the question of whether the areas of ejection are distributed over the surface at random. If so, we shall need a statistical theory; in this connection, a recent paper by Balasz-Detre and Detre (18) is very interesting; they point out the importance of the theory of random walk for the studies of period changes of variable stars in general. On the other hand, many aspects of the problem of period changes of eclipsing binaries appear now to be more directly connected with their dynamics and evolution, and further progress depends much on our understanding why and how the mass transfer goes on.

For Table 3 we have adopted a new form of referencing which we hope will simplify somewhat a rather complicated situation. The first part of the Table, arranged by star, lists the authors and if necessary their (sometimes several) papers a, b, etc.; these papers are then cited by author and letter in the 'Bibliography to Table 3' immediately following that Table.

Table 3. Times of minima, light elements, and discussions of periods of eclipsing binary systems

Star	Authors
RT And	Ahnert (j), Dueball and Lehmann, Obůrka (a, b, c, d), Pohl and Kizilirmak, Robinson, (b, d, e, f), *Sky and Telescope* (a, b)
TW And	Obůrka (a)
XZ And	Dueball and Lehmann, Obůrka (c, d), Pohl and Kizilirmak, Robinson (d, e, f), Sakharov, *Sky and Telescope* (b), Szafraniec (a)
AB And	Dueball and Lehmann, Kalchaev and Truttse, Pohl and Kizilirmak, Purgathofer and Widorn
BO And	Meinunger (C)
BX And	Obůrka (c), Robinson (e)
RY Aqr	Obůrka (d)
CX Aqr	Robinson (d, e), Sakharov, *Sky and Telescope*
V389 Ara	Hoffmeister
77 G Ara (BV420)	Thackeray and Knipe
RX Ari	McCluskey (a)
FK Aql	Szafraniec (a)

Star	Author
GU Aql	Gessner (a)
KO Aql	Cristaldi (b), Filatov, Pohl and Kizilirmak, Robinson (b)
KP Aql	Dueball and Lehmann, Pohl and Kizilirmak
OO Aql	Dueball and Lehmann, Fitch, Obůrka (c), Pohl and Kizilirmak, Robinson (b, d, e), *Sky and Telescope* (a, b, f)
QS Aql	Knipe (b)
V337 Aql	Catalano and Rodono
V342 Aql	Baldwin and Robinson, Robinson (b)
V343 Aql	Huth (c)
V346 Aql	Cristaldi and Walter, Dueball and Lehmann, Huth (c), Obůrka (c), Pohl and Kizilirmak, Robinson (b, d, e)
V348 Aql	Meinunger (x)
V406 Aql	Gessner (a)
V408 Aql	Gessner (a)
V413 Aql	Gessner (a)
V418 Aql	Gessner (a)
V557 Aql	Meinunger (y)
V589 Aql	Meinunger (y)
V640 Aql	Meinunger (z)
V688 Aql	Huth (c)
V869 Aql	Gessner (a)
RY Aur	Kordylewski, K. (a)
SX Aur	Dueball and Lehmann
TT Aur	Dueball and Lehmann, Obůrka (a, b, c, d)
WW Aur	Dueball and Lehmann, Fitch, Robinson (b, f), *Sky and Telescope* (b, f), Kristenson
AR Aur	Robinson (f)
EQ Aur	Meinunger (d)
ζ Aur	Hardorp, Herczeg and Scholz, Schneller (c)
θ Aur	Jackisch
TY Boo	Szafraniec
TZ Boo	Dueball and Lehmann
XY Boo	Wood, R.
ZZ Boo	Obůrka (c), Pohl and Kizilirmak
AC Boo	Binnendijk, Maunder
i Boo	Catalano and Saitta, Fitch, Karetnikov (c), Pohl and Kizilirmak, Purgathofer and Widorn, Schneller (d, f)
Y Cam	Obůrka (d), Robinson (d, f)
RZ Cam	Pohl and Kizilirmak
SV Cam	Dueball and Lehmann, Obůrka (a, c, d), Pohl and Kizilirmak, Robinson (b, d, e, f), *Sky and Telescope* (b, f)
TU Cam	West
AL Cam	Quester, Braune
S Cnc	Kordylewski, K. (a)
TW Cnc	Model
TX Cnc	Kitamura
UU Cnc	Beyer
WY Cnc	Chambliss, Dueball and Lehmann
RS CVn	Dueball and Lehmann, Rodono (unpubl.)
R CMa	Dueball and Lehmann, Fitch, Knipe (a), Obůrka (b, c), Robinson (b)
RX CMa	Kordylewski, K. (a)
SW CMa	Ziegler
EG CMa	Kordylewski, K. (a)
XZ CMi	Dueball and Lehmann, Wilson (b)

Star	Author
YY CMi	Dueball and Lehmann
AK CMi	Kordylewski, K. (a), Obůrka (d)
RZ Cas	Burke and Rolland, Chang-chun, Chupilko, Dueball and Lehmann, Karetnikov (a), Kiperman (a, c), Obůrka (a, b, c, d), Plavec, Robinson (a, b, c, d, e, f), *Sky and Telescope* (a, b), Vikhristjuk, Kristenson
TV Cas	Chupilko, Dueball and Lehmann, Karetnikov (b), Kiperman (c), Obůrka (b, c, d), Pohl and Kizilirmak, Robinson (e, f), *Sky and Telescope* (b), Kristenson
TW Cas	Dueball and Lehmann, Obůrka (c, d), Pohl and Kizilirmak
YZ Cas	Dueball and Lehmann
AB Cas	Dueball and Lehmann, Pohl and Kizilirmak, Robinson (e, f)
BM Cas	Beyer
BY Cas	Malik
CW Cas	Broglia (a)
DO Cas	Obůrka (b, c)
EK Cas	Beyer
GR Cas	Gessner (b)
IV Cas	Dueball and Lehmann
RR Cen	Knipe (c)
SV Cen	Irwin (1966, in press)
U Cep	Dueball and Lehmann, Lukatskaya and Lirovchenko, Obůrka (b, c, d), Pankow, Robinson (b, f), Szafraniec
SU Cep	Dueball and Lehmann, Obůrka (c)
VW Cep	Chupilko, Dueball and Lehmann, Kiperman (c), Kwee, Obůrka (c)
WY Cep	Kordylewski, K. (a), Obůrka (c, d)
WZ Cep	Obůrka (c, d)
XX Cep	Dueball and Lehmann, Kordylewski, K. (a), Obůrka (b, c), Pohl and Kizilirmak, Robinson (b, e, f)
XY Cep	Obůrka (d)
XZ Cep	Obůrka (a, c)
ZZ Cep	Obůrka (a, b), Dueball and Lehmann, Robinson (f)
CW Cep	Kiperman (c), Vikhristjuk
EG Cep	Dueball and Lehmann (=200 Dra), Obůrka (c), Pohl and Kizilirmak
RW Cet	Pohl and Kizilirmak
RZ Com	Dueball and Lehmann, Fitch
ε CrA	Cousins (a)
U CrB	Fitch (a)
RW CrB	Fitch (a)
Y Crv	Meinunger (c)
Y Cyg	Ahnert (d), Chupilko, Dueball and Lehmann, Obůrka (a, c), Schneller (a)
RR Cyg	Pohl and Kizilirmak
RZ Cyg	Ignatieva
SW Cyg	Todoran (a), Schneller (a)
WZ Cyg	Obůrka (d)
ZZ Cyg	Todoran (b)
BR Cyg	Dueball and Lehmann, Fitch, Obůrka (b, c), Pohl and Kizilirmak
CG Cyg	Obůrka (c, d)
DK Cyg	Obůrka (c)
EE Cyg	Wachmann
EM Cyg	Mumford (b)
FR Cyg	Wachmann
HL Cyg	Wachmann
GO Cyg	Fitch, Obůrka (c), Purgathofer and Widorn
KR Cyg	Fitch

Star	Author
MR Cyg	Lavrov (a, c)
MY Cyg	Filatov, Obůrka (c), Pohl and Kizilirmak
V388 Cyg	Obůrka (c)
V401 Cyg	Purgathofer, Wachmann
V445 Cyg	Vam Vjet Chin
V466 Cyg	Pohl and Kizilirmak
V477 Cyg	Ahnert (d), Pohl and Kizilirmak
V548 Cyg	Dueball and Lehmann, Pohl and Kizilirmak
V753 Cyg	Wawriniak
31 Cyg	O'Connell, Wright
W Del	Huth (c), Schneller (b)
RR Del	Huth (c)
TY Del	Dueball and Lehmann, Huth (c), Pohl and Kizilirmak, Robinson (e)
DM Del	Obůrka (c)
BN Del	Meinunger (z)
BO Del	Meinunger (A)
EX Del	Karamish
Z Dra	Obůrka (c, d), Pohl and Kizilirmak, Robinson (b, d, f), Schneller (b), Todoran (a)
RZ Dra	Obůrka (b, c, d)
TW Dra	Kordylewski, K. (a), Obůrka (b, c, d), Robinson (d), Schneller (g)
TZ Dra	Dueball and Lehmann, Plavec, Pohl and Kizilirmak
UZ Dra	Dueball, Dueball and Lehmann, Obůrka (c), Pohl and Kizilirmak
WW Dra	Koch
XY Dra	Meinunger (k)
AI Dra	Dueball and Lehmann, Pohl and Kizilirmak, Robinson (b, d, e), *Sky and Telescope* (e)
S Equ	Dueball and Lehmann, Obůrka (b), Plavec
W Equ	Huth (c, d)
WX Eri	Dueball and Lehmann
YY Eri	Bhattacharyya
AS Eri	Filatov
BC Eri	Fitch (b)
BZ Eri	Meinunger (e)
SU For	Omay
U Gem	Bornhurst, Krzeminski, Mayall
SX Gem	Obůrka (c)
TX Gem	Dueball and Lehmann
YY Gem	Dueball and Lehmann
AL Gem	Dueball and Lehmann
BD Gem	Ahnert (a)
DP Gem	Meinunger (s)
DV Gem	Hildebrandt
EU Gem	Meinunger (t)
FT Gem	Meinunger (s)
Z Her	Fitch (a), Robinson (b)
RX Her	Dueball and Lehmann, Fitch, Obůrka (c), Pohl and Kizilirmak, Vikhristjuk
SZ Her	Obůrka (c, d), Pohl and Kizilirmak, Robinson (b, d, e), *Sky and Telescope* (a, b, f)
TX Her	Dueball and Lehmann, Fitch, Obůrka (c, d), Pohl and Kizilirmak, Schneller (b)
UX Her	Dueball and Lehmann, Fitch, Obůrka (a, b, c, d)

Star	Author
AK Her	Bertiau, Dueball and Lehmann, Fitch, Obůrka (a, c, d), Pohl and Kizilirmak, Purgathofer and Widorn
BC Her	Fridel
CC Her	Obůrka (c)
DQ Her	Bornhurst
GL Her	Kordylewski, Z.
V342 Her	Kapko
V388 Her	Pohl and Kizilirmak
u Her	Fitch (a), *Sky and Telescope* (d)
RX Hya	Kordylewski, K. (a)
TY Hya	Kordylewski, K. (a)
EU Hya	Kordylewski, K. (a)
EX Hya	Mumford (a)
SS Lac	Tashpulatov
SW Lac	Bookmyer, Dueball and Lehmann, Fitch, Lang and Vetešník, Obůrka (a, b, d), Pohl and Kizilirmak, Purgathofer and Widorn, Robinson (b, d, e, f)
VX Lac	Obůrka (c, d), Szafraniec
AR Lac	Ahnert (e), Dueball and Lehmann, Obůrka (b, c)
BP Lac	Gessner (c)
BS Lac	Gessner (c)
BW Lac	Gessner (c)
CM Lac	Ahnert (e), Dueball and Lehmann, Obůrka (a, b), Pohl and Kizilirmak, Vikhristjuk
Y Leo	Chis and Pal, Dueball and Lehmann, Obůrka (c), *Sky and Telescope* (f), Robinson (f)
UV Leo	Dueball and Lehmann, Obůrka (b, c), Pohl and Kizilirmak, Purgathofer and Widorn, Robinson (d), McCluskey (b)
UX Leo	Kordylewski, K. (b)
XY Leo	Dueball and Lehmann
XZ Leo	Dueball and Lehmann
AM Leo	Dueball and Lehmann, Pohl and Kizilirmak
T LMi	Kordylewski, K. (a)
SS Lib	Fitch, Robinson (b)
δ Lib	Dueball and Lehmann, Fitch, Robinson (b), *Sky and Telescope* (f)
RR Lyn	Dueball and Lehmann, Linnell
RV Lyr	Schneller (b)
TZ Lyr	Obůrka (b, c, d)
UZ Lyr	Dueball and Lehmann, Fitch, Obůrka (b, c, d), Pohl and Kizilirmak
AK Lyr	Wachmann
BV Lyr	Wachmann
EP Lyr	Wachmann
FL Lyr	Dueball and Lehmann, Cristaldi (a), Pohl and Kizilirmak, Robinson (b, d, e)
NV Lyr	Wachmann
NY Lyr	Makarenko
β Lyr	Herczeg (a), Pohl and Kizilirmak, Robinson (e), *Sky and Telescope* (c)
RU Mon	Martynov (a, b)
RW Mon	Ahnert (a)
UY Mon	Ahnert (a)
AT Mon	Meinunger (a)
HK Mon	Meinunger (u)
HY Mon	Meinunger (b)
IM Mon	Sanyal and Mahra and Sanwal

Star	Author
IS Mon	Olüniak
U Oph	Dueball and Lehmann, Fitch, Obůrka (c), Pohl and Kizilirmak, Robinson (b, e)
RV Oph	Robinson (b), *Sky and Telescope* (b)
V415 Oph	Meinunger (v)
V451 Oph	Fitch (a), Kiperman (c), Obůrka (c), Vikhristjuk
V502 Oph	Fitch (a), Obůrka (c)
V508 Oph	Fitch (a), Karetnikov (a)
V566 Oph	Kiperman (b), Obůrka (b, c), Purgathofer and Widorn, Robinson (c), Vikhristjuk
V575 Oph	Meinunger (v)
V839 Oph	Obůrka (c), Wilson and O'Toole
V916 Oph	Meinunger (w)
Z Ori	Pochkareva
CQ Ori	Ahnert (c), Kordylewski, K. (a)
FF Ori	Ahnert (b), Kordylewski, K. (b)
FI Ori	Dueball and Lehmann, Kordylewski (b)
FT Ori	Cristaldi (b)
V392 Ori	Meinunger (D)
U Peg	Dueball and Lehmann, Obůrka (a, b, c, d), Pohl and Kizilirmak, Purgathofer and Widorn.
TY Peg	Obůrka (b), Huth (f)
UX Peg	Huth (f), Obůrka (a, d), Pohl and Kizilirmak
AQ Peg	Huth (f), Szafraniec
AT Peg	Ahnert (f), Cristaldi and Walter, Obůrka (a, c), Pohl and Kizilirmak
AW Peg	Huth (f)
BB Peg	Huth (f), Strelkova
BG Peg	Huth (f), Szafraniec
DI Peg	Dueball and Lehmann, Huth (f), Pohl and Kizilirmak
DK Peg	Dueball and Lehmann
EE Peg	Bakos, Dueball and Lehmann, Vikhristjuk
EH Peg	Huth (e)
Z Per	Obůrka (a)
RT Per	Obůrka (b), Robinson (b, e, f), Todoran (a)
RV Per	Kordylewski, K. (a)
RW Per	Kordylewski, K. (a)
RY Per	Kordylewski, K. (a)
ST Per	Ahnert (i), Dueball and Lehmann, Pohl and Kizilirmak, Robinson (f)
XZ Per	Pohl and Kizilirmak
BO Per	Löchel
BP Per	Löchel
DD Per	Obůrka (a)
EN Per	Meinunger (o)
HV Per	Meinunger (k)
LZ Per	Ahnert (i)
β Per	Dueball and Lehmann, Herczeg (a), Pohl and Kizilirmak, Robinson (e, f), *Sky and Telescope* (b)
δ Pic	Cousins (b)
SX Psc	Kordylewski, K. (a)
XZ Pup	Kordylewski, K. (a)
WX Sgr	Fitch (a)
AI Sgr	Kordylewski, K. (a)
V505 Sgr	Cristaldi and Walter, Robinson (b, d), *Sky and Telescope* (b), Vikhristjuk

Star	Author
U Sge	Dueball and Lehmann, Fitch, Obůrka (c), Pohl and Kizilirmak, Robinson (b)
V Sge	Romano (a, b, c)
SY Sge	Huth (c)
UZ Sge	Huth (c)
RS Sct	Obůrka (c, d)
VZ Sct	Kordylewski, K. (a)
V Ser	Fitch (a)
RW Tau	Dueball and Lehmann, Obůrka (c), Robinson (e, f)
CT Tau	Ahnert (h), Beyer
EN Tau	Meinunger (p)
EO Tau	Meinunger (r)
X Tri	Dueball and Lehmann, Obůrka, Plavec, Pohl and Kizilirmak, Robinson (e, f), Sakharov, *Sky and Telescope* (b)
RS Tri	Plavec
RW Tri	Mandel
W UMa	Blitzstein and Chou, Dueball and Lehmann, Fitch, Joshi, Obůrka (b, c, d), Kristenson
RW UMa	Romano (a)
TX UMa	Dueball and Lehmann, Herczeg (b), Obůrka (a, b, c)
TY UMa	Pohl and Kizilirmak
UX UMa	Dueball and Lehmann, Mandel, Obůrka (c, d), Pohl and Kizilirmak
VV UMa	Wilson (a)
XY UMa	Kordylewski, K. (a)
ZZ UMa	Strohmeier *et al.* (b)
AA UMa	Strohmeier and Ott
W UMi	Dueball and Lehmann
RT UMi	Götz and Wenzel
W Vir	Prikhodko, Tchuprima
AG Vir	Purgathofer and Widorn
AH Vir	Dueball and Lehmann, Fitch, Purgathofer and Widorn, Truttse and Kalchaev
BD Vir	Kordylewski, K. (a)
BH Vir	Dueball and Lehmann
Z Vul	Broglia (b), Ahnert (g), Dueball and Lehmann, Obůrka (b, c, d), Pohl and Kizilirmak
RR Vul	Huth (c)
RS Vul	Fitch (a), Vikhristjuk
SY Vul	Meinunger (u)
XZ Vul	Lavrova, Wachmann
AT Vul	Lavrov (b), Lavrova, Huth (c)
AW Vul	Huth (c)
AX Vul	Huth (c)
AY Vul	Huth (c)
BE Vul	Huth (c)
BO Vul	Huth (c), Obůrka (c)
BP Vul	Huth (a, b)
BS Vul	Huth (c)
BV Vul	Huth (c)
DR Vul	Obůrka (b), Schneller (a)
BV7 Cep	Meinunger and Wenzel
BV36 UMa	Meinunger (m)
BV159	Nikulina
BV267	Kondo, Margoni *et al.*

Star	Author
BV 277 Dra	Meinunger (l)
BV 305	Obůrka (c)
BV 312 Tau	Dueball and Lehmann, Robinson (b), *Sky and Telescope* (b)
BV 332	Obůrka (c)
BV 342	Kondo
BV 343	Nikulina
BV 344	Nikulina
BV 374	Strohmeier (d)
BV 375	Strohmeier (d)
BV 376	Strohmeier (d)
BV 382 Cep	Ruiz and Stokes, Bartolini *et al.*
BV 384	Obůrka (d)
BV 412	Schneller (f)
BV 544 Oph	Robinson (e)
BV 623	Strohmeier (c, e)
CVS 134 Cas	Reiche (a)
CVS 160 Cas	Meinunger (e)
CVS 235 Per	Meinunger (i)
CVS 449 Ori	Meinunger (g)
CVS 891 Gem	Meinunger (f)
P 2289 Aqr	Meinunger (A)
S 3948 Gem	Tschäpe
S 3965 Ori	Meinunger (A)
S 4382 Aql	Gessner (a)
S 4417 Aql	Gessner (a)
S 4420 Aql	Gessner (a)
S 4969 Cen	Hoffmeister
S 4997 Cir	Hoffmeister
S 5001 Lup	Hoffmeister
S 5351 TrA	Hoffmeister
S 5375 Ind	Hoffmeister
S 5454 Cam	Meinunger (B)
S 7830 Cyg	Gessner (d)
S 7831 Dra	Gessner (d)
S 7840 Dra	Gessner (d)
S 7847 Cyg	Gessner (d)
S 7849 Dra	Gessner (d)
S 7870 Cyg	Gessner (e)
S 7881 Cyg	Gessner (e)
S 7882 Cyg	Gessner (e)
S 7885 Cyg	Gessner (e)
S 7886 Cyg	Gessner (e)
S 7889 Cyg	Gessner (e)
S 7895 Cyg	Gessner (e)
S 7896 Cyg	Gessner (e)
S 7898 Cyg	Gessner (e)
SVS 645 Del	Kordylewski, Z.
16·1929	Strohmeier (a)
84·1930 Aql	Gessner (a)
107·1930 Aql	Gessner (a)
13·1936 Per	Löchel
35·1936 Aur	Kuhn
166·1940 Cyg	Reiche (b)
76·1942 Lac	Gessner (c)

Star	Author
77·1942 Lac	Gessner (c)
222·1943 Ori	Ziener
173·1945 Cyg	Meinunger (j)
CPD −67° 3312 = HD 156545	Deuricks, Briers
HD 161 783	van Houten
HD 187 418	van Houten
HBV 337, 338, 340, 341, 344, 347, 349, 350, 352, 354, 355, 356, 357, 360, 361, 362, 363, 365, 368, 369, 371, 374, 375	Wachmann (a)
HBV 376, 377, 378, 379, 380, 381, 382, 383, 384, 385, 386, 387, 388, 389, 390, 391, 392, 393, 394, 395, 396, 397, 398, 399, 400, 401	Wachmann (b)

BIBLIOGRAPHY TO TABLE 3

Ahnert, P., *Mitt. veränd. Sterne*, (a) **2**, 61; (b) **2**, 63; (c) **2**, 64; (d) **2**, 170; (e) **2**, 171; (f) **2**, 173; (g) **2**, 175; (h) **2**, 181; (i) **3**, 7; (j) **3**, 131.
Bakos, G. A., *Publ. David Dunlap Obs.*, **2**, 431, 1965.
Baldwin, M. E., Robinson, L. J., *I.B.V.S.*, **92**, 1965.
Bartolini, C., Mammano, A., Mannino, G., Margoni, R., *Kl. Veröff. Remeis-Sternw. Bamberg*, Bd IV, no. 40, 1965.
Bhattacharyya, J. C., *Kodaikanal Obs. Bull.* (in press).
Bertiau, F. C., *Ric. astr. spec. Vatic.*, **6**, 21, 1963.
Beyer, M., *Astr. Nachr.*, **288**, 87, 1965.
Binnendijk, L., *Astr. J.*, **70**, 201, 1965.
Blitzstein, W., Chou, K. C., *Astr. J.*, **69**, 365, 1964.
Bochkareva, L. Y., *Perem. Zvezdy*, **15**, 324, 1965.
Bookmyer, B. B., *Astr. J.*, **70**, 415, 1965.
Bornhurst, L. C., *J. R. astr. Soc. Can.*, **60**, 32, 1966.
Broglia, P., (a) *Mem. Soc. astr. ital.*, **35**, 23, 1964; (b) *J. Observateurs*, **47**, 99, 1964.
Burke, E. W., Rolland, W. W., *Astr. J.*, **71**, 38, 1966.
Chang-chun Huang, *Acta astr. Sinica*, **7**, 15.
Catalano, S., Saitta, T., *Mem. Soc. astr. ital.*, **35**, 43, 1964.
Chambliss, C. R., *Astr. J.*, **70**, 741, 1965.
Chis, G., Pal, A., *Studii Cerc. Astr. Seism.*, Bucarest, **9**, 14, 1964.
Chupilko, V. A., *Perem. Zvezdy*, **14**, 503, 1965.
Cousins, A. W. J., (a) *Mon. Notes astr. Soc. Sth. Afr.*, **23**, 24, 1964; (b) *Mon. Notes astr. Soc. Sth. Afr.*, **23**, 67, 1964.
Cristaldi, S., (a) *Mem. Soc. astr. ital.*, **36**, 77, 1965; (b) *Mem. Soc. astr. ital.*, **37**, 205, 1965.
Cristaldi, S., Walter, K., *Mem. Soc. astr. ital.*, **36**, 197, 1965.
Deuricks, R., Briers, R., *I.B.V.S.*, **136**, 1966.
Dueball, J., *Sterne*, **40**, 108, 1964.
Dueball, J., Lehmann, P. B., *Astr. Nachr.*, **288**, 167, 1965.
Filatov, G. S., *Bull. Dushanbe*, **35**, 32, 1963.
Fitch, W. S., (a) *Astr. J.*, **69**, 316, 1964; (b) *I.B.V.S.*, **132**, 1966.
Fridel, J. W., *Lwow Circ.*, no. 39–40.

Gessner, H., (a) *Mitt. veränd. Sterne*, **2**, 25, 1963; (b) *Mitt. veränd. Sterne*, **2**, 37, 1963; (c) *Mitt, veränd. Sterne*, **2**, 43, 1964; (d) *Mitt. veränd. Sterne*, **2**, 95, 1964; (e) *Mitt. veränd. Sterne*, **2**. 183, 1965.
Götz, W., Wenzel, W., *Mitt. veränd. Sterne*, **2**, 85, 1964.
Hardop, J., Herczeg, T., Scholz, M., *Z. Astrophys.*, **64**, 97, 1966.
Herczeg, T., (a) *Veröff. Univ. Sternw. Bonn*, **69**, 1964; (b) *Z. Astrophys.*, (in press).
Hildebrandt, G., *Mitt. veränd. Sterne*, **2**, 169, 1965.
Hoffmeister, C., *Veröff. Sternw. Sonneberg*, **6**, 3, 1965.
Huth, H., (a) *I.B.V.S.*, **96**, 1965; (b) *Mitt. veränd. Sterne*, **3**, 36, 1965; (c) *Mitt. veränd. Sterne*, **3**, 120, 1965; (d) *Mitt. veränd. Sterne*, **3**, 125, 1965; (e) *Mitt. veränd. Sterne*, **3**, 127, 1965; (f) *Mitt. veränd. Sterne*, **3**, 169, 1965.
Ignatieva, L. V., *Perem. Zvezdy*, **14**, 119, 1964.
Irwin, J. B. (in press).
Jackisch, G., *Veröff. Sternw. Sonneberg*, **5**, 264, 1963.
Joshi, S. C., *Bull. astr. Inst. Csl.*, **17**, 31, 1966.
Kalchaev, K., Truttse, Ju. L., *Perem. Zvezdy*, **15**, 487, 1966.
Kalish, M. S., *Publ. astr. Soc. Pacif.*, **77**, 36, 1965.
Kapko, J. T., *Lwow Circ.*, no. 39–40.
Karamish, V. F., *Astr. Circ.* (U.S.S.R.), 315, 1965.
Karetnikov, V. G., (a) *Perem. Zvezdy*, **14**, 348, 1964; (b) *Perem. Zvezdy*, **14**, 513, 1964; (c) *Perem. Zvezdy*, **15**, 501, 1966.
Kiperman, M. E., (a) *Perem. Zvezdy*, **14**, 345, 1964; (b) *Perem. Zvezdy*, **14**, 429, 1964; (c) *Perem. Zvezdy*, **15**, 218, 1965.
Knipe, G. F. G., (a) *Johannesburg Circ.*, **122**, 21; (b) *Mon. Notes astr. Soc. Sth. Afr.*, **24**, 92, 1965; (c) *Astrophys. J.*, **142**, 1068, 1965.
Koch, R. H., *Astr. J.*, **69**, 269, 1964.
Kondo, Y., *Astr. J.*, **71**, 54, 1966.
Kordylewski, K., (a) *I.B.V.S.*, **46**, 1964; *Astr. Abh. Leipzig*, p. 90, 1965.
Kordylewski, Z., *Acta. astr.*, **14**, 223, 1964.
Kristenson, H., *Bull. astr. Inst. Csl.*, **17**, 123, 1966.
Krzeminski, W., *Astrophys. J.*, **142**, 1051, 1965.
Kuhn, L., *Mitt. veränd. Sterne*, **2**, 83, 1964.
Kwee, K. K., *Bull. astr. Inst. Netherl.*, **18**, 448, 1966.
Lang, K., Vetešnik, M., *Bull. astr. Inst. Csl.*, **17**, 21, 1965.
Lavrov, M. I., (a) *Trudy Kazan Univ.*, **85**, 1963; (b) *Trudy Kazan Univ.*, **86**, 1963; (c) *Bjull. astr. Obs. Engel'gardta*, **38**, 3, 1966.
Lavrova, N. V., *Bjull. astr. Obs. Engel'gardta*, **38**, 32, 1966.
Linnell, A. P., *Astr. J.*, **71**, 458, 1966.
Löchel, K., *Mitt. veränd. Sterne*, **3**, 69, 1965.
Lukatskaya, F. J., Lirovchenko, L. N., *Perem. Zvezdy*, **15**, 527, 1966.
McCluskey, G. E., Jr. (a) *Astr. J.* **71**, 527, 1966; (b) *Astr. J.*, **71**, 536, 1966.
Makarenko, E. N., *Perem. Zvezdy*, **15**, 446, 1966.
Malik, G. M., *Astr. J.*, **70**, 94, 1965.
Mandel, O. E., *Perem. Zvezdy*, **15**, 474, 1966.
Martynov, D. Ya., (a) *I.B.V.S.*, **102**, 1965; (b) *Astr. Zu.* (U.S.S.R.), **42**, 1209, 1965.
Mauder, H., *Z. Astrophys.*, **60**, 222, 1964.
Mayall, M. W., *J. R. astr. Soc. Can.*, **58**, 143, 1964.
Mayer, P., *Publ. astr., Soc. Pacif.*, **77**, 436, 1965.
Meinunger, L., *Mitt. veränd. Sterne*, (a) **2**, 44; (b) **2**, 45; (c) **2**, 47; (d) **2**, 153; (e) **2**, 155; (f) **2**, 157; (g) **2**, 160; (h) **2**, 161; (i) **2**, 163; (j) **3**, 14; (k) **3**, 15; (l) **3**, 16; (m) 3, 21; (n) **3**, 23; (o) **3**, 139; (p) **3**, 141; (r) **3**, 142; (s) **3**, 143; (t) **3**, 144; (u) **3**, 146; (v) **3**, 149; (w) **3**, 150; (x) **3**, 152; (y) **3**, 153; (z) **3**, 154; (A) **3**, 155; (B) **3**, 197; (C) **3**, 198; (D) **3**, 199; (E) **3**, 200.
Meinunger, L., Wenzel W., *Mitt. veränd. Sterne*, **2**, 164, 1965.
Model, A., *Mitt. veränd. Sterne*, **2**, 6, 1964.
Mumford, G. S., (a) *Astr. J.*, **70**, 686, 1965; (b) *Astr. J.*, **70**, 865, 1965.
Nikulina, T. G., *Bull. Dushanbe*, **35**, 45, 1963.

Obůrka, O., (a) *Bull. astr. Inst. Csl.*, **15**, 26, 1964; (b) *Bull. astr. Inst. Csl.*, **15**, 250, 1964; (c) *Bull. Inst. Csl.*, **16**, 212, 1965; (d) *Contr. Public Obs. and Planetarium Brno.*, no. 2, 1966.
O'Connell, D. J. K., *Ric. astr. Spec. Vatic.*, **6**, no. 23, 1964.
Olüniak, G. T., *Lwow Circ.*, no. 39–40.
Omay, C. G., *I.B.V.S.*, **65**, 1964.
Paczynski, B., *Astr. J.*, **69**, 124, 1964.
Pankow, M., *Acta. astr.*, **15**, 73, 1965.
Plavec, M., *Bull. astr. Inst. Csl.*, **15**, 23, 1965.
Pochkareva, Z., *Perem. Zvezdy*, **15**, 437, 1966.
Pohl, E., Kizilirmak, I., *Astr. Nachr.*, **288**, 65, 1965.
Prikhodko, A. E., Tchuprima, R. J., *Perem. Zvezdy*, **15**, 597, 1966.
Purgathofer, A., *Z. Astrophys.*, **59**, 29, 1964.
Purgathofer, A., Widorn, T., *Mitt. Univ. Sternw. Wien*, **12**, 6, 1964.
Quester, W., Braune, W., *I.B.V.S.*, **116**, 1965.
Reiche, J., (a) *Mitt. veränd. Sterne*, **2**, 165, 1965; (b) *Mitt. veränd. Sterne*, **2**, 168, 1965.
Robinson, L. J., (a) *I.B.V.S.*, **84**, 1965; (b) *I.B.V.S.*, **111**, 1965; (c) *I.B.V.S.*, **112**, 1965; (d) *I.B.V.S.*, **114**, 1965; (e) *I.B.V.S.*, **119**, 1965; (f) *I.B.V.S.*, **129**, 1966.
Romano, G., (a) *I.B.V.S.*, **98**, 1965; (b) *Contr. Asiago*, 173; (c) *Mem. Soc. astr. ital.*, **35**, 7, 1965.
Ruiz, J. J., Stokes, A. J., *Sky Telesc.*, **29**, 393, 1965.
Sakharov, V. S., *Perem. Zvezdy*, **15**, 97, 1964.
Sanyal, A., Mahra, H. S., Sanwal, N. B., *Bull. astr. Inst. Csl.*, **16**, 209, 1965.
Schneller, H., (a) *Astr. Nachr.*, **287**, 183, 1963; (b) *Astr. Nachr.*, **288**, 1, 1964; (c) *Mitt. veränd. Sterne*, **2**, 178, 1965; (d) *I.B.V.S.*, **57**, 1964; (e) *I.B.V.S.*, **61**, 1964; (f) *Astr. Nachr.*, **288**, 183, 1965; (g) *Mitt. veränd. Sterne*, **3**, 179, 1966.
Sky Telesc., (a) **27**, 58, 1964; (b) **29**, 254, 1965; (c) **31**, 54, 1966; (d) **31**, 246, 1966; (e) **31**, 314, 1966; (f) **28**, 244, 1964.
Strelkova, E. P., *Perem. Zvezdy*, **14**, 501, 1966.
Strohmeier, W., (a) *I.B.V.S.*, **33**, 1963; (b) *Veröff. Remeis-Sternw. Bamberg*, **5**, 15, 1962; (c) *Sky Telesc.*, **29**, 345, 1965; (d) *Astr. Abh. Leipzig*, p. 119, 1965; (e) *I.B.V.S.*, **89**, 1966.
Strohmeier, W., Ott, H., *I.B.V.S.*, **38**, 1964.
Szafraniec, R., *Acta. astr.*, **16**, no. 2, 1966.
Tashpulatov, N., *Perem. Zvezdy*, **15**, 390, 1966.
Thackeray, A. D., Knipe, G. F. G., *Mon. Notes astr. Soc. Sth. Afr.*, **24**, 109, 1965 and **25**, 30, 1966.
Todoran, I., (a) *Studia Univ. Babes-Bolyai Cluj*, ser. math-phys., **2**, 51, 1965; (b) *Studii Cerc. Astr. Seism.*, Bucarest, **10**, 83, 1965.
Truttse, Yu. L., Kalchaev, K. K., *Perem. Zvezdy*, **15**, 314, 1965.
Tschäpe, R., *Mitt. veränd. Sterne*, **2**, 78, 1964.
Vam Vjet Chin, *Perem. Zvezdy*, **14**, 262, 1964.
van Houten, C. J., van Houten-Groeneveld, J., *Inf. Bull. So. Hemisphere*, **8**, 25, 1966.
Vikhristjuk, S. S., *Perem. Zvezdy*, **15**, 218, 1965.
Wachmann, A. A., (a) *Hamburg Abh.*, **6**, 3, 1964; (b) *Hamburg Abh.*, **7**, 3, 1964.
Wawriniak, P. F., *Lwow Circ.*, no. 39–40.
West, R., *Publ. Medd. Københavens Obs.*, 181, 1964.
Wilson, R. E., (a) *Astr. J.*, **70**, 368, 1965; (b) *Astr. J.*, **71**, 32, 1966.
Wilson, R. E., O'Toole, W., *Publ. astr. Soc. Pacif.*, **77**, 58, 1965.
Wood, R., *Observatory*, **85**, 258, 1965.
Ziegler, E., *Mitt. veränd. Sterne*, **2**, 185, 1965.
Ziener, R., *Mitt. veränd. Sterne*, **2**, 94, 1964.

Data for numerous new eclipsing binaries are given in:

Deurick, R., Briers, R., *I.B.V.S.*, **136**, 1966.
Gaposchkin, S., *Kl. Veröff. Remeis-Sternw. Bamberg*, Bd IV, no. 40, 1965.
Knigge, R., *I.B.V.S.*, **117**, 1965
Mayer, P., *I.B.V.S.*, **49**, 1964 and *Publ. astr. Soc. Pacif.*, **77**, 436, 1965.
Kuročkin, N. E., *I.B.V.S.*, **79**, 1965.

Nikolov, N. S., *Perem. Zvezdy*, **14**, 336, 1964.
Popov, M. V., Kazmina, E. I., *Perem. Zvezdy*, **15**, 109, 1965.
Romano, G., *Coelum*, **33**, 3, 1965; *I.B.V.S.*, **106**, 1965.
Schöffel, E., *I.B.V.S.*, **118**, 1965.
Schöffel, E., Köhler, U., *I.B.V.S.*, **77**, 1965; **91**, 1965; **100**, 1965.
Smith, L. F., *Publ. astr. Soc. Pacif.*, **78**, 1966.
Snowden, M. S., *I.B.V.S.*, **127**, 1966.
Strohmeier, W., *I.B.V.S.*, **47**, 1964; **51**, 1964; **54**, 1964; **55**, 1964; **89**, 1965.
Strohmeier, W., Fischer, H., Ott, H., *I.B.V.S.*, **120**, 1966.
Strohmeier, W., Busch, H., Häusler, K., *I.B.V.S.*, **144**, 1966.
Strohmeier, W., Knigge, R., Ott, H., *I.B.V.S.*, **62**, 1964; **66**, 1964; **70**, 1964; **74**, 1964; **81**, 1965; **86**, 1965; **107**, 1965; **115**, 1965; **140**, 1966.
Weber, R., *I.B.V.S.*, **64**, 1964.
Wolff, S. C., Wallerstein, G., Sandage, A. R., *Publ. astr. Soc. Pacif.*, **77**, 372, 1965.

APSIDAL MOTION

Martynov (**19**) has found that the apsidal motion of RU Mon is faster than had previously been supposed, being 45″ per period of revolution, so that $U/P = 29\,000$. Martynov derives an effective polytropic index of 3·1. Several systems have been examined by Semeniuk (**20**); he derived new values of the apsidal motion period for V 380 Cyg, CO Lac, AG Per and DR Vul. Two further stars, V 889 Aql and DI Her, show slow but detectable apsidal motion. The case of V 477 Cyg is more complicated, because the apsidal motion is apparently distorted by period changes of some other kind. O'Connell is investigating V 477 Cyg, V 526 Sgr, and DR Vul for apsidal motion.

Among the new eclipsing binaries, two systems appear promising for apsidal motion because of large eccentricity: FT Ori (**21**) and AL Cam (**22**).

It would be very important for understanding the internal constitution of the subgiants to detect with certainty apsidal motion in some of the well-known Algol-like systems. However, recent work by Herczeg on Algol and TX UMa, led him to the conclusion that the case is still rather uncertain and that only observations of the secondary minima can decide it.

The problem of apsidal motion, in both its theoretical and its observational aspects, was surveyed by Kopal (**23**). An important chapter in this paper deals with comparison of the results with the theory of stellar constitution.

Table 4. Investigations on apsidal motion

Star	References
V889 Aql	Semeniuk, I., *Acta. astra.* (in press).
Y Cyg	Schneller, H., *Astr. Nachr.*, **287**, 183, 1963.
MR Cyg	Lavrov, M. I., *Trudy Kazan Univ.*, **85**, 1963.
V380 Cyg	Semeniuk, I., *Acta. astr.* (in press).
V477 Cyg	Ahnert, P., *Mitt. veränd. Sterne*, Sonneberg, **2**, 63.
	Semeniuk, I., *Acta. astr.* (in press).
CO Lac	Semeniuk, I., *Acta. astr.* (in press).
DI Her	Semeniuk, I., *Acta astr.* (in press).
AG Per	Semeniuk, I., *Acta. astr.* (in press).
RU Mon	Martynov, D. Ya., *I.B.V.S.*, **102**, 1965; *Astr. Zu.*, **42**, 1209.
β Per	Herczeg, T., *Bonn Veröff.*, 69, 1964.
HV7498 Sgr	Swope, H. H., *Inf. Bull. So. Hemisphere*, **8**, 13, 1966.
TX UMa	Herczeg, T., *Z. Astrophys.* (in press).
DR Vul	Semeniuk, I., *Acta. astr.* (in press).
	Schneller, H., *Astr. Nachr.*, **287**, 183, 1963.

CO-ORDINATED OBSERVING PROGRAMS

Results of earlier co-ordinated programs have appeared in the recent literature. Table C lists those publications concerning Epsilon and Zeta Aurigae and 31 and 32 Cygni whether or not these were made as part of the collaborative program. Dr Gyldenkerne, the co-ordinator, reports that observations on all these stars is continuing at Brorfelde with some participants from the 1963–64 campaign joining in.

For newer work, the committee on co-ordinated programs decided at the 1964 meetings to concentrate on only two systems. Dr Koch agreed to serve as co-ordinator for SX Cas. He reports:

'The cooperative program for observing SX Cas attracted considerable initial interest from photoelectrical observers and little response from spectroscopists. From the observations presently in hand several *U*, *B*, *V* light curves can be compiled with suitable phase coverage. The completeness of Hβ light curves is unknown at this time. Spectrographic results may be limited to the valuable series obtained by Joy several years ago. These have been measured but not yet discussed in detail.'

For W Serpentis, Dr Sahade served as co-ordinator and selected the period 17 June–3 July, 1966 as the interval for concentrated work. A number of observers agreed to participate, but the degree of successful coverage is not yet known. Four reports have been received at this writing; they were from Flagstaff, Cerro Tololo, Abastumani and Mount John (New Zealand).

An effort not planned at the meeting was caused by the announcement in *IAU circular* 1954, by Dr George Wallerstein, that spectroscopic data indicated that HD 128220 might be an eclipsing binary with a period of 870 days. Eclipse was predicted for about 6 June, 1966, with an uncertainty of at least ten days. One component of the system is reported as a type O subdwarf. Because of this and because the next opportunity to observe an eclipse will be five years later, it was decided to try to organize a collaborative effort. Dr E. G. Ebbighausen agreed to act as co-ordinator and a number of observers agreed to participate. Most reports are not yet available at the time of writing, but Schneller writes that his observations do not confirm an eclipse minimum.

At the Commission's request, Dr Batten issued an appeal for interested amateurs to assist in a photometric patrol on U Coronae Borealis. The response was disappointing; Dr Batten's own spectrographic work on the system continues as planned, and it would be distressing indeed if a photometric disturbance were missed.

Table C

Star	References
ϵ Aur	Huang, S., *Astrophys. J.*, **141**, 976, 1965.
	Low, F. S., Mitchell, R. I., *Astrophys. J.*, **141**, 327, 1965.
	Mitchell, R. I., *Astrophys. J.*, **140**, 1607, 1964.
ζ Aur	Bappu, M. K. V., Doss, A. T., Viswanadham, P., *Observatory*, **85**, 85, 1965.
	Bardin, C., Prévot, L., *Publ. Obs. Hte-Provence*, **7**, no. 32, 1964.
	Faraggiana, R., *Z. Astrophys.*, **62**, 99, 1965.
	Faraggiana, R., Hack, M., *Z. Astrophys.*, **64**, 48, 1966.
	Fracastoro, M. G., Catalano, S., *Mem. Soc. astr. ital.*, **36**, 4, 1965.
	Hardop, J., Herczeg, T., Scholz, M., *Z. Astrophys.*, **64**, 97, 1966.
	Jackisch, G., *Veröff. Sternw. Sonneberg*, **5**, 264, 1963.
	Kondo, Y., Harris, A. J., *Astr. J.*, **69**, 409, 1964.
	O'Connell, D. J. K., *Ric. astr. Spec. Vatic.*, **6**, no. 25, 1964.
	Odgers, G., Wright, K. O., *J. R. astr. Soc. Can.*, **59**, 115, 1965.
	Schneller, H., *Mitt. veränd. Sterne*, **2**, 178, 1965.

Star	References
ζ Aur	Shao, C. Y., *Astr. J.*, **69**, 858, 1965.
	Van Genderen, A. M., *Bull. astr. Inst. Netherl.*, **17**, 446, 1964.
	Zhang, C-y., Liu, X-f., *Acta astr. Sinica*, **13**, 102, 1965.
31 Cyg	Faraggiana, R., Hack, M., *Mem. Soc. astr. ital.*, **34**, 4, 1963.
	Hack, M., *Mem. Soc. astr. ital.*, **33**, 189, 1962.
	Jackisch, G., *Veröff. Sternw. Sonneberg*, **5**, 283, 1963.
	O'Connell, D. J. K., *Ric. astr. Spec. Vatic.*, **6**, no. 23, 1964.
	Saito, M., *Publ. astr. Soc.* Japan, **17**, 107, 1965.
32 Cyg	Faraggiana, R., Gokgoz, A., Hack, M., Kendis, I., *Mem. Soc. astr. ital.*, **36**, 63, 1965.
	Kwee, K. K., van Genderen, A. M., *Bull. astr. Inst. Netherl.*, **17**, 53, 1963.
	Jackisch, G., *Veröff. Sternw. Sonneberg*, **5**, 283, 1963.
	Scholz, M., *Z. Astrophys.*, **61**, 179, 1965.
	Wright, K. O., *J. R. astr. Soc. Can.*, **60**, 89, 1966.

DETERMINATION OF PHOTOMETRIC ORBITS

In Table 5 are listed those systems for which photometric orbits have been determined since the previous report, in addition to the systems treated in papers cited in the running references for this section.

One case of 'conventional' solution deserves note. Chen and Reuning (**24**), having observed Algol at 0·60 μ (orange) and in the infrared region from 0·87 to 2·5 μ, deduced relative dimensions in agreement with earlier values. They also discuss the ellipticity, limb-darkening and surface luminosity for the infrared case. These results together with those of the several good earlier investigations, now give a photometric description of this eclipsing binary over the entire spectrum from the ultraviolet to the infrared.

The problem of determination of eclipsing binary elements has been investigated by several persons and some new methods and approaches developed.

Grygar (**25**) considered non-linear distribution of brightness on the disks of O9–A3 stars using Hunger's, Underhill's, Mihalas's and Strom's models of stellar atmospheres. He computed coefficients of limb-darkening for 314 cases of 41 models, and compared the theoretical results with the observed quantities for 9 stars. He found good agreement for A0 stars but the representation for B0 stars not as good. Two explanations for the discrepancy seem possible: either that the B0 stars have cold envelopes or that present theory does not account exactly for the scattering of light by free electrons.

Davis (**26**) computed extensive tables of the function $p\ (x,\ k,\ \alpha^{oc})$ for cosine darkening and x = 0·1, 0·2, . . ., 0·9, 1·0, and similar calculations were made by Linnell (**27**). Jurkevich (**28**) developed explicit expressions for $d\alpha^{x}/dk$ and $d\alpha^{x}/dp$ and computed tables for them. Harris (**29**) developed computer programs for generalized linear least squares rectification of light curves using nine Fourier coefficients.

Shulberg and Tabachnik (**30**) developed a method for determination of elements by electronic computers, and applied it to W Del and YZ Cas. Tabachnik has extended the process to the determination of darkening coefficient and has shown that the calculation gives a unique solution.

Minin (**31**) investigated theoretically the reflection effect in close binary systems. Sobieski (**32**) studied the transport of energy through both gray and non-gray stellar atmospheres for determination of the reflection coefficient, and has applied his method to several systems.

Shulberg (**33**), continuing his study of cases of extended atmosphere, applied his method to a photographic light curve of UX Mon. He found that the G component of the system passes through the extended atmosphere of the A component. Tcherepashchuk (**34**) has developed a new method of solution for light curves of systems which have one component with extended atmosphere. He shows that the solution of a certain integral equation gives enough information for the determination of the geometric parameters of the system and also the darkening function and the absorbing properties of an extended atmosphere. The method yielded satisfactory results on *B*, *V* light curves of V444 Cyg.

Kitamura (**35**) published his method of solution based on Fourier transformation of the light curve. He introduced an incomplete Fourier transformation, and computed tables of the special functions involved.

Kříž (**36**) considered a model for semi-detached systems in which one component is spherical and the other may be approximated by a tri-axial ellipsoid. He applied his method to RW Tau. Horak (**37**) extended the method to include solutions for systems having $i \neq 90°$, and investigated UX Her.

Wilson (**38**) offered a quick method for determination of the parameter *k* from light-curves which show two sufficiently deep minima. Tsesevich has extended Schneller's method for treating β Lyr and W UMa type stars, and computed convenient tables for direct calculations; the paper awaits publication.

Table 5. New solutions of photometric curves

Star	References
AD And	Rucinski, S., (in press in *Acta. astr.*).
BX And	Todoran, J., *Publ. Obs. Astr. Cluj*, no. 44, 1965 (pg).
Y Aql	Breinhorst, R., (Bonn Univ. Obs.).
V346 Aql	Cristaldi, S., Walter, K., *Astr. Nachr.*, **287**, 207, 1963.
RX Ari	McCluskey, Jr, G. E., *Astr. J.*, **71**, 527, 1966.
IU Aur = HD35 652	Mayer, P., *Publ. astr. Soc. Pacif.*, **77**, 436, 1965 (*UBV*).
AC Boo	Binnendijk, L., *Astr. J.*, **70**, 201, 1965 (yellow, blue).
	Maunder, H., *Z. Astrophys.*, **60**, 222, 1964 (*B*, *V*).
44i Boo	Catalano, S., Saitta, T., *Mem. Soc. astr. ital.*, **35**, 43, 1964 (red).
S Cnc	Irwin, J. B., *Astrophys. J.*, **138**, 1104, 1963 (differential coefficient of darkening).
TX Cnc	Kitamura, M., (unpublished).
WY Cnc	Chambliss, C. R., *Astr. J.*, **70**, 741, 1965 (two colors).
CW Cas	Broglia, P., *Mem. Soc. astr. ital.*, **35**, 23, 1964; *Milano-Merate Contr.* no. 226, 1964 (pe obs.).
DO Cas	Winkler, L., *Astr. J.*, **71**, 40, 1966.
RR Cen	Knipe, G. F. G., *Astrophys. J.*, **142**, 1068, 1965.
RZ Com	Binnendijk, L., *Astr. J.*, **69**, 154, 1964.
DK Cyg	Binnendijk, L., *Astr. J.*, **69**, 157, 1964.
KR Cyg	Vetešnik, M., *Bull. astr. Inst. Csl.*, **16**, 326, 1965.
V382 Cyg	Landolt, A. U., *Astrophys. J.*, **140**, 1494, 1964 (*UBV*).
V444 Cyg	Tcherepashchuk, A. M., *Astr. Zu.*, **43**, 517, 1966.
V386 Cyg	Harris, A. J. (unpublished).
EK Cep	Ebbighausen, E. G., *Astr. J.*, **71**, 642, 1966.
CC Eri	Evans, D. S., *Mon. Notes astr. Soc. Sth. Afr.*, **23**, 67, 1964.
YY Eri	Binnendijk, L., *Astr. J.*, **70**, 209, 1965 (yellow, blue).
U Gem	Krzeminski, W., *Astrophys. J.*, **142**, 1051, 1065.
	Mumford, G. S., *Astrophys. J.*, **139**, 476, 1964 = *Kitt Peak nat. Obs. Contr.*, **48**.
UX Her	Horak, T., *Bull. astr. Inst. Csl.*, **17**, 27, 1966.
	Gordon, K. C., Kron, G. E., *Astr. J.*, **70**, 100, 1965 (two colors).
DI Her	Khivrenko, A. P., *Perem. Zvezdy*, **15**, 84, 1964.
EX Hya	Mumford, G. S., *Astrophys. J.*, **139**, 476, 1964 = *Kitt Peak nat. Obs. Contr.*, **50**.

Star	References
SW Lac	Bookmyer, B. B., *Astr. J.*, **70**, 415, 1965 (two colors).
CO Lac	Semeniuk, J., (unpublished).
Y Leo	Chis, G., Pal, A., *Studii Cerc. Astr. Seism.*, **9**, 14, 1964.
UV Leo	McCluskey, Jr, G. E., *Astr. J.*, **71**, 536, 1966.
	Popper, D. M., *Astrophys. J.*, **141**, 126, 1965.
RS Lep	Klepczynski, W. J., Wood, F. B., *Astr. J.*, **69**, 92, 1964.
RR Lyn	Linnell, A. P., *Astr. J.*, **71**, 458, 1966.
FL Lyr	Cristaldi, S., *Mem. Soc. astr. ital.*, **36**, 77, 1965.
RU Mon	Martynov, D. J., *Astr. Zu.*, **42**, 1209, 1965 (phe).
WZ Oph	Popper, D. M., *Astrophys. J.*, **141**, 126, 1965 (yellow).
V566 Oph	Binnendijk, L., *Astr. J.*, **70**, 209, 1965 (yellow, blue).
Z Ori	Bochkareva, L. G., *Perem. Zvezdy*, **15**, 437, 1965 (pgr).
DI Peg	Rucinski, S., (Warsaw Obs.).
EE Peg	Bakos, G. A., *Publ. David Dunlap Obs.*, **2**, 431, 1965.
RT Scl	Bookmyer, B. B., (in press *Astr. J.*).
V499 Sco	Bookmyer, B. B., (in press *Astr. J.*).
RW Tau	Kříž, S., *Bull. astr. Inst. Csl.*, **16**, 118, 1965.
V Tri	Ureche, V., *Publ. Astr. Obs. Cluj*, no. 40, 1964 (pgr).
W UMa	Binnendijk, L., *Astr. J.*, **71**, 340, 1966 (two colors).
VV UMa	Wilson, R. E., *Astr. J.*, **70**, 368, 1965.
AL Vel	Wesselink, A. J., *Mon. Not. R. astr. Soc.*, **127**, 107, 1964.
BV267	Kondo, Y., *Astr. J.*, **71**, 46, 1966.
BV332	Harris, A. J. (unpublished).
BV342	Kondo, Y., *Astr. J.*, **71**, 54, 1966.
BV412	Harris, A. J. (unpublished).

ABSOLUTE DIMENSIONS OF ECLIPSING SYSTEMS

In Table 6 we list most of the determinations of absolute dimensions published since the previous report; some also occur in the papers cited in this section. To the continuing study of normal and of specially selected systems is now added some little emphasis on U Gem stars and old novae.

Krzeminski found in 1961 that U Gem is an eclipsing variable. The star has been further studied recently by Krzeminski (**39**), by Mumford (**40**), and by Paczyński (**41**). It appears well established that U Gem is a system which consists of a flare star and a white dwarf with a gaseous ring.

The following old novae have been investigated: WZ Sge by Krzeminski and Kraft (**42**), and EX Hya by Mumford (**43**). In an investigation of the physical parameters for novae Paczynski (**44**) has given a critical discussion of their masses.

Herbig, Preston, Smak and Paczynski (**45**) discovered that the irregular variable star V Sge is an eclipsing binary. From a detailed study of it they find the system parameters significantly variable.

Popper (**46**) continues his important study of special eclipsing systems with new determinations of absolute dimensions for WZ Oph, UV Leo and VZ Hya. New photoelectric observations of VZ Hya are needed.

Using 268 Michigan spectrograms Peery (**47**) has computed the absolute orbit of VV Cep. His discussion of the structure of the atmosphere and envelope of the M giant has shown that the density decreases with height more slowly than in the case of 31 Cyg. From analysis of his interference-wedge light curves of V444 Cyg at $\lambda\lambda$ 4686, 4795, 4440 and 4260Å, Tcherepashchuk (**48**) has constructed a model of the system and observed the variations of intensity of lines caused by physical changes in the system. Smak (**49**) finds that both components of RW Gem have colors which are abnormal for their radii and masses.

Bookmyer's careful study (**50**) of the light curve of SW Lac and of its variations opens up by implication a field of considerable potential value in the investigation of W UMa type systems. She finds that these variations are consequences of irregular changes in the system which are traceable in the system parameters computed at selected epochs. Her proposal that there be organized a program of simultaneous detailed photometric and spectroscopic observations of SW Lac will be discussed at the Prague sessions.

Finally, Binnendijk (**51**) finds that the increase of period observed in W UMa is a 'consequence' of the flare observed by Kuhi.

Table 6. New determinations of absolute dimensions of eclipsing binaries

Star	References
KO Aql	Roxburgh, J. W., *Astr. J.*, **70**, 690, 1965.
QS Aql	Chincarini, G., Margoni, R., Taffara, S., (unpublished).
V603 Aql	Kraft, R. P., *Astrophys. J.*, **139**, 457, 1964.
V822 Aql	Fitzgerald, P., *Publ. David Dunlap Obs.*, **2**, 417, 1964.
T Aur	Kraft, R. P., *Astrophys. J.*, **139**, 457, 1964.
	Walker, M., *Astrophys. J.*, **138**, 313, 1963.
TT Aur	Lavrov, M. I., *Publ. Kazan Univ.* for 1963, p. 188, 1964.
VV Cep	Peery, B. F., *Astrophys. J.*, **144**, 672, 1966.
EK Cep	Ebbighausen, E. G., *Astr. J.*, **71**, 642, 1966.
RZ Com	Binnendijk, L., *Astr. J.*, **69**, 154, 1964.
T CrB	Kraft, R. P., *Astrophys. J.*, **139**, 457, 1964.
MR Cyg	Lavrov, M. I., *Bull. Engelhardt Obs.* (Kazan), no. 38, 3–26.
V382 Cyg	Landolt, A. U., *Astrophys. J.*, **140**, 1494, 1964.
WW Dra	Kříž, S., *Bull astr. Inst. Csl.*, **16**, 306, 1965.
YY Eri	Binnendijk, L., *Astr. J.*, **70**, 209, 1965.
U Gem	Paczyński, B., *Acta astr.*, **15**, 305, 1965.
RW Gem	Smak, J., *Acta astr.*, **14**, 101, 1964.
UX Her	Gordon, K. C., Kron, G. E., *Astr. J.*, **70**, 100, 1965.
DQ Her	Kraft, R. P., *Astrophys. J.*, **139**, 457, 1964.
VZ Hya	Popper, D. M., *Astrophys. J.*, **141**, 126, 1965.
SW Lac	Bookmyer, B. B., *Astr. J.*, **70**, 415, 1965.
Y Leo	Chis, G., Pal, A., *Studii Cerc. Astr. Seism.*, **9**, no. 1, 14, 1964.
UV Leo	Popper, D. M., *Astrophys. J.*, **141**, 126, 1965.
FL Lyr	Cristaldi, S., *Mem. Soc. astr. ital.*, **36**, 77, 1965.
RU Mon	Martynov, D. J., *Astr. Zu.*, **42**, 1209, 1965.
WZ Oph	Popper, D. M., *Astrophys. J.*, **141**, 126, 1965.
V566 Oph	Heard, J. F., *J. R. astr. Soc. Can.*, **59**, 258, 1965.
DN Ori	Smak, J., *Astr. J.*, **69**, 677, 1964; *Publ astr. Soc. Pacif.*, **76**, 210, 1964; *Acta astr.*, **15**, 327, 1965.
EE Peg	Bakos, G. A., *Publ. David Dunlap Obs.*, **2**, 431, 1965.
GK Per	Kraft, R. P., *Astrophys. J.*, **139**, 457, 1964.
IZ Per	Yavuz, I., (unpublished).
δ Pic	Thackeray, A. D., *Mon. Not. R. astr. Soc.*, **131**, 435, 1966.
V Sge	Herbig, G., Preston, G. W., Smak, J., Paczyński, B., *Astrophys. J.*, **141**, 617, 1965.
WZ Sge	Kraft, R. P., *Astrophys. J.*, **139**, 457, 1964.
	Krzeminski, W., Kraft, R. P., *Astrophys. J.*, **140**, 921, 1965.
XZ Sgr	Smak, J., *Acta astr.*, **15**, 327, 1965.
V1017 Sgr	Kraft, R. P., *Astrophys. J.*, **139**, 457, 1964.
V453 Sco	Sahade, J., *Astrophys. J.*, **141**, 652, 1965.
AL Vel	Wesselink, A. J., *Mon. Not. R. astr. Soc.*, **127**, 107, 1964.

Star	References
BH Vir	Abt, H. A., *Publ. astr. Soc. Pacif.*, **77**, 367, 1965.
BV 241	Fitzgerald, P., *Publ. David Dunlap Obs.*, **2**, no. 14, 1964.
BV 342	Kondo, Y., *Astr. J.*, **71**, 54, 1966.
BV 344	Yavuz, I., (unpublished).
BV 374	Fitzgerald, P., *Publ. David Dunlap Obs.*, **2**, no. 14, 1964.
BV 382 Cep = GK Cep	Bartolini, C., Mammano, A., Mannino, G., Margoni, R., *Contr. Asiago*, 168, 1965; *I.B.V.S.*, **85**, 1965.

ASTROPHYSICAL AND DYNAMICAL INVESTIGATIONS

The fission theory of the origin of binary stars was revived by Roxburgh (**52**). He investigated the pre-main-sequence evolution of rotating non-magnetic stars, formed from interstellar clouds. After the end of the Hayashi track, each element of the growing radiative core conserves its angular momentum. For masses $> 0{\cdot}8\ M_{\odot}$ rotational instability in the core is likely to occur, and this is probably leading to a splitting of the star into two components. Assuming conservation of angular momentum on fission, it is shown that stars with mass $< 4\ M_{\odot}$ can form a contact binary (of the W UMa type), whereas more massive stars will produce separate binaries.

This new version of the fission theory is largely based on recent studies of rapidly rotating masses of gas. Polytropic models in solid-body rotation were treated by Hurley and Roberts (**53**), as a sequel to previous work by Roberts. James (**54**) studied the polytropic case as well as the white-dwarf case. For the study of rapidly rotating polytropes, Monaghan and Roxburgh (**55**) developed an approximation technique, which is well suited for investigating rotating stellar models. This technique was used by Roxburgh, Griffith and Sweet (**56**) to derive models for rapidly rotating main-sequence stars, and by Roxburgh (**57**) for models of rotating white dwarfs. Polytropic models with fast, non-uniform rotation were studied by Stoeckly (**58**); he found that instability may develop and that non-uniform rotation could be a factor in double-star formation. Chandrasekhar has completed (**59**) the extensive and important series of investigations he initiated in 1960 on the ellipsoidal figures of equilibrium of homogeneous masses. Lebowitz (**60**) has specified the circumstances under which Riemann's energy criterion for stability breaks down.

At the 1966 Liège symposium on 'Gravitational instability and the formation of stars and galactic structures' three papers were given which deal with the origin of binary stars. Roxburgh (**61**) discussed the influence of rotation and magnetic fields on the formation of single and multiple stars. Mestel (**62**) outlined a mechanism by which a close binary might be formed from a pair of proto-stars. Such a system is during the Hayashi phase subject to very efficient tidal friction, assuring close synchronism of orbital and rotatory motion. A small amount of mass loss can yield a large loss of angular momentum, if the emitted mass is magnetically coupled to the proto-stars. Provided the magnetic braking more than offsets the increase of spin due to contraction, the loss of rotatory angular momentum is made up at the expense of orbital angular momentum, and the two stars will steadily approach each other. Van Albada (**63**) described numerical integrations of the N-body problem which show that a random initial structure alters into a hierarchial one with a close binary in the center. The escape process is strongly connected with the formation of binaries, due to the tendency to redistribute the binding energy among fewer and fewer stars. It was suggested that this phenomenon may have been operating after the fragmentation of a dense cloud into proto-stars. In this way it may also account for the formation of wide binaries.

A general and rather detailed review article on the evolution of close binaries was written by Hack (**64**). At the Herstmonceux 1962 colloquium on 'Clusters and stellar evolution' McCrea (**65**) and others discussed the status of the stars forming the blue extension of the main sequence beyond the turn-off point in certain old clusters. The consensus was that they probably represent close binaries, the evolution of which have been halted because of considerable mass transfer, a hypothesis further elaborated by McCrea (**66**). According to McCrea the components that have suffered great mass loss would show high helium abundance. Hill (**67**) found that several helium stars, with no detectable hydrogen, show no signs of duplicity, while some helium-rich stars are in fact binaries. Smak (**68**) made a prediction of the number of semi-detached systems to be expected in the blue extensions and reached satisfactory agreement with the observed numbers. Only a few of these binary stars should, however, be eclipsing variables. Williams (**69**) advanced the theory that the blue-extension stars are single objects, which have accreted material ejected from super-giants in the clusters. Coudé spectrograms of blue 'stragglers' in the old clusters M 67 and NGC 752, taken recently by Deutsch (**70**), indicate that the lines are highly broadened by rotation. Deutsch supposes that the stars represent metamorphs of late F-dwarfs, i.e. F-dwarfs that in the red-giant phase have shed a substantial part of their masses, including the slowly rotating outer parts, and returned to the neighborhood of the main sequence as chemically homogeneous objects. That would imply that at least some late-type stars have rapidly rotating interiors. Jaschek and Jaschek (**71**) studied the evolution of close binary systems using the limiting equipotential surface. It was shown that if the limiting sizes for the components are taken into account, it is understandable why the proportion of spectroscopic binaries is decreasing towards the later spectral types. Paczyński and Ziolkowski report that they are engaged in stellar model computations, in an attempt to follow the evolution of close binary components. We note that Zeldovich and Guseynov (**72**) pointed out the exciting possibility that invisible massive components of some binaries may be collapsed stars, which manifest themselves only by gravitational fields.

Eclipsing binaries of the W UMa type have recently been discovered in the old clusters M 67 and M 188 (**73, 74**). Kuročkin and Kukarkin (**75**) discussed the implication of this for some problems of stellar evolution, and concluded, inter alia, that the W UMa-type systems must be very stable.

Zahn (**76**) considered the influence of tides on the angular momentum of close binaries, his conclusion being that due to this mechanism alone, rotation and orbital motion cannot remain synchronized. Dziembowski is at present working on the problem of tidal friction. Peraiah (**77**) obtained a new expression for the angular velocity of the components of close binaries, which shows that synchronism may not hold in most cases. For the W UMa systems Smak (**78**) found that the rotational momentum is more than 100 times smaller than the orbital momentum. In an investigation of the mass-luminosity relationship for components of W UMa stars Osaki (**79**) obtained equality between luminosity ratio and mass ratio. To explain this mass-luminosity behaviour the internal structure was discussed. The magneto-gravitational stability of close binary systems was studied by Santarelli (**80**), using the method of small oscillations. An expression was derived, from which follows that an assumed degree of asynchronism between rotation and orbital motion is related to a minimum value of the magnetic field, and vice versa. Under certain assumptions the component in contact with the limiting equipotential surface was found unstable for perturbations having wave-lengths exceeding a critical value. Numerical examples are given for ER Ori and RZ Cas, a contact and a semi-detached system, respectively. Equations for the internal constitution of components in close binaries were deduced by Dziembowski (**81**). The importance of the binaries in the study of perturbations of spherical equilibrium stars was stressed by Schwarzschild (**82**) in his concluding remarks to the Joint Discussion C at the Hamburg General Assembly.

The theory of apsidal motion and its relation to stellar structure was given a general and detailed treatment by Kopal (**83**). Mean $\bar{k}$ of the constants of internal structure of both components in a large number of systems were derived, using a statistical relation between rotational and angular velocities. From apsidal motion data for RU Mon Martynov (**84**) obtained an effective polytropic index of 3·1 for both components. Peraiah (**85**) deduced expressions for the apsidal motion, assuming that the components rotate faster than is demanded by synchronism. Effective polytropic indices, computed for a number of binaries, give larger central condensation than in the case of synchronism.

Problems of mass transfer have during the last few years been treated by a number of authors. Reference to the paper by Plavec, Sehnal and Mikulas (**86**) on trajectories of particles ejected from L_1 in RW Tau was given already in the previous report of this Commission. Plavec and Kříž (**87**) have made supplementary calculations for RW Tau, as well as calculations for models resembling U Sge and U Cep. The pattern of trajectories is the same, although the mass ratio varies between 0·2 and 0·5. If ejections take place over the whole surface of the secondary, the following phenomena are predicted: (1) a variable extended atmosphere around the secondary, (2) a broad and dense stream flowing to the primary from the adjacent face of the secondary, (3) under favorable circumstances a transient or semi-permanent ring, formed by particles ejected at fairly high speeds into certain directions, (4) a certain amount of particles escaping from the system, if the ejection velocities are high enough.

Particle ejection due to non-synchronism between rotation and orbital motion was investigated by Kruszewski (**88**). The ejecting star was supposed to fill the limiting equipotential surface, ejection taking place at L_1. Numerical calculations of trajectories were made for various values of the non-synchronism parameter and for a variety of mass ratios. In addition to a particle stream ending on the smaller component, there is also a stream returning to the ejecting component. This latter stream always exists in the case of non-synchronism. It is suggested that the impact area, which is asymmetrically situated with respect to the line connecting the centers of the two stars, should be locally heated. Koch (**89**) postulated the asymmetrical shape of the secondary minimum of many close binaries to be evidence that the heated impact area is eclipsed in a number of systems. From the aspect of the asymmetry it was possible to discern the sense of rotation of the spectroscopically invisible component, and sometimes to set limits to its angular velocity.

Huang (**90**) studied several problems in connection with the gaseous flow in close binary systems, and derived a set of differential equations, which take into account the collision of particles but not the pressure. These equations thus form an intermediate between the particle and hydrodynamic treatment of mass motion. The solution of the equations gave results similar to those obtained previously by Prendergast from the hydrodynamic equations.

At the Joint Discussion C in Hamburg 1964 mass transfer was discussed by Plavec (**91**) and others. The restricted problem of three bodies, basic to most of the investigations quoted above, was extended by Lebrun and Robe (**92**) to the case when the finite masses are ellipsoids revolving in circular orbits.

Limber (**93**) studied surface forms and properties relating to mass loss in the general case of non-synchronous rotation. Roche models were used, and the deviation from uniform rotation was neglected in a first approximation of the surface forms. The nature and magnitude of the deviations from this approximation were then considered.

The effects of isotropic mass loss upon the orbital elements were treated by Chiara (**94**), Hadjidemetriou (**95**), Martin (**96**), and Omarov (**97**), assuming various laws for the mass loss. Huang (**98**) studied the transfer of angular momentum between ejected particles and the binary system. Recently, Huang (**99**) advanced the idea that observed axial rotation in excess of synchronism is only a superficial phenomenon, caused by impact of particles carrying a large amount of angular momentum. He also states that an unpublished study shows that no equilibrium surface exists for a component rotating non-synchronously. Kruszewski (**100**) studied period changes caused by the exchange of matter according to his computations of trajectories. The ejection of material, its fall on the other component and perturbations during flight were considered. In close connection with this work, Piotrowski (**101**) gave a detailed discussion of the effects of mass transfer with special attention to changes in period and eccentricity. An exhaustive review article by Kruszewski (**102**) on exchange of matter and period changes has recently appeared. Observed period changes of Y Leo were discussed by Chiş and Pál (**103**) and interpreted in terms of mass ejection from the secondary component. Similar interpretations were made by Schneller for i Boo (**104**), TW Dra (**105**) and, in as yet unpublished papers, for β Lyr and V 401 Cyg.

Porfiryev and Kalenichenko (**106**) made a theoretical study of radial velocity curves for various rotation laws and axial orientations. Shakhovski (**107**) discussed results of polarization measurements for 17 eclipsing binaries, including 5 stars with variable polarization. Mechanisms responsible for the polarization are photospheric scattering, the reflection effect, and, most important, scattering in a gaseous envelope. Orbital orientations and total masses of the envelopes were derived for RY Per and β Lyr. A new effect, a pronounced ultra-violet excess, was found by Popper and Walker (**108**) in some eclipsing variables.

A review paper on the problem of β Lyr, with the inclusion of new results, was given by Sahade (**109**). Reuning and Springer (**110**) gave a distorted-polytrope representation of the components. Woolf (**111**) found that the primary is slightly less distorted than the Roche model. He supports Huang's suggestion that the secondary is a star imbedded in a disk, and states that at least half the mass of the secondary is in the disk. The masses for primary and secondary component are found as 2·05 and 11·74 $M_{\odot}$, respectively. A model was worked out also by Sahade and Hernandez (**112**).

For ϵ Aur Huang (**113**) proposed a new model analogous to that of β Lyr; the secondary is supposed to be a rotating gaseous disk, which appears opaque when viewed edge-on. Roxburgh (**114**) found that KO Aql has an undersize sub-giant secondary in pre-main-sequence contraction. Smak (**115**) reported evidence of low surface gravity for both components of RW Gem. Various problems of the Algol systems were reviewed by Fracastoro (**116**).

The ultra-short period binaries from the class of cataclysmic and related variables have been the subject of a large number of theoretical investigations. Kraft (**117**) published a comprehensive review on our knowledge of this topic about 1962, and gave a shorter account in Hamburg at the Joint Discussion C (**118**). In this latter paper Kraft especially considered period changes due to mass ejection from the red component. Mumford (**119**) wrote a series of popular papers on these stars, dealing also with the theoretical side. Paczyński (**120**) developed a theory for the outbursts, as caused by mass transfer from the red to the blue component. Kraft (**121**), in a paper on observations of 10 old novae, discussed models, the period-luminosity relation for cataclysmic variables, causes of outbursts, and population membership. Kraft and Luyten (**122**), from data on proper motions and radial velocities, determined the mean absolute magnitude of the U Gem stars as 7·5. From this and from kinematical and other data for W UMa stars Kraft (**123**) concluded that the U Gem stars are probably descendants of W UMa systems. Gorbatzky (**124**) investigated the effect on stellar spectra of the disk-like envelopes, using SS Cyg as an example. Krzeminski (**125**) gave an interpretation of U Gem, and a detailed model for the nova-like variable V Sge was derived by Herbig, Preston, Smak and Paczyński (**126**). Another nova-like variable, 17 Lep, was treated by Widing (**127**).

Turning to new developments in the field of element computation, we note that Wilson (**128**) outlined a quick solution method in the case of complete eclipses. Tcherepaschuk (**129**) described a method of solution of light curves when one of the components has an extended atmosphere. Kitamura (**130**) published an exhaustive account of the element determination by means of Fourier transforms of the light curves, together with examples of practical analysis. Tables of the characteristic function of the eclipse and the related delta-functions, to be used in this method, have also been computed by Kitamura (**131**). Tabachnik and

Schulberg (132) published a computer program for the case of total and annular eclipses of spherical components with circular orbits.

Members of the Flower and Cook group have developed computer programs for solutions under the Russell Model, with each program representing a single step of the sequence and so re-runnable at need. The aim is not to substitute the computer for the astronomer, but to enable calculations of all kinds to be made rapidly and then presented to the astronomer for his judgment. These stage-wise programs have not been published in detail, as this seems inefficient, but are available on request.

Linnell (133) wrote a group of computer programs for ${}^{0}\alpha$, ${}^{10}\alpha^{oc}$, ${}^{10}\alpha^{tr}$, ${}^{10}\alpha^{ann}$, ${}^{x}\alpha^{tr}$, and ${}^{x}\alpha^{oc}$. Each program calculates a value of that function for arbitrary permissible values of the parameters x, p, k. Programs were also made for the inverse functions ${}^{x}\varphi^{oc}$, ${}^{x}\varphi^{tr}$, ${}^{x}\chi^{oc}$, and ${}^{x}\chi^{tr}$. Samples of calculated tables have been published (134). Jurkevich (135) developed explicit expressions for the derivatives of ${}^{x}\alpha$ in terms of k and p, and tabulated these quantities. Tables for calculating the light equation due to the orbital motion of the Earth, were published by Schneller (136).

A new method for determining the limb-darkening from light-curves was developed by Wilson (137). Hosokawa (138) considered the coefficient of limb-darkening for bright main-sequence stars. Non-linear limb-darkening laws were computed by Grygar (139) for 41 model atmospheres (range O5–A3) constructed by Hunger, Mihales, Saito, Strom and Underhill. Gingerich (140) computed limb-darkening tables for a series of his model atmospheres (range $T_{\rm eff}$ = 10 000°K to 7000°K). Kříž (141) made a new approach to the limb-darkening of RW Tau.

Minin (142) treated the theory of the reflection effect. Sobieski (143) obtained detailed numerical solutions of the radiative transfer problem for selected systems of close binaries in order to study the wave-length dependence of the phase law and to compute values of the net reflection coefficient A_1.

Ebbighausen and Gange (144) made a spectrographic investigation of the motion of Algol AB in the triple system. The mass ratio m_{AB}/m_C was found to be 3·57 ± 0·16. On the theoretical side Brazhnikova and Babinchuk (145) have investigated aspects of the recurrent problem of photometric spectroscopic anomalies due to 'periastron effects'.

Finally, still another paper in the lengthening series of studies of the *UBV* photometric system, has been published by Karyagina and Kharitonov (146).

STATISTICAL AND THEORETICAL INVESTIGATIONS

A number of interesting papers have appeared relating to the statistics of binary systems. Abt (147) has discussed the proportion of binaries among normal A-type stars and, with others, (148) has found 5 binaries among 47 Pleiades studied. The Jascheks (149) related the lower proportion of binaries in late-type stars to their evolution, while Sahade (150) has called attention to several systems (HD 698, HD 47129, AO Cas, V 453 Sco, V 448 Cyg and β Lyr) all having underluminous, massive components and therefore perhaps a significant group evolutionwise. Szafraniec (151) has studied the dependence of period on galactic latitude and Cester (152) has found that the mass-luminosity relation for binary stars requires different segments for bright and for faint stars. Plavec (153) has suggested a two-dimensional system of classification for close binary systems based on Kopal's criterion of size relative to the Roche surface and Sahade's criterion of their position on the HR diagram.

Peraiah (154) has found that the velocity of rotation of binary components is frequently greater than suggested by the period. Porfiryev (155) has given formulae for line profiles computed for various velocities in systems with various orientations. Koch (156) has noted that the sense of rotation in a binary system can be determined from the asymmetry of the light curve during eclipse and Mumford (**157**) has discussed the orientation of binary orbits. Huang (**158**) has shown that the origin of the angular momentum in binary systems is not from galactic rotation, but must have arisen from random motions in the pre-stellar medium.

There have been several papers discussing the relation between W UMa stars, U Gem stars and main sequence stars. Kuročkin (**159**) finds that the ages of W UMa stars are the same as main sequence stars; Artiukhina (**160**) finds that their solar motions and absolute magnitudes agree; and Smak (**161**) has compared W UMa stars with other stars of the same mass. Kraft (**162**) finds that the kinematics of W UMa stars and U Gem stars are nearly the same, but Popov (**163**) finds that their space distribution in the z-component is different. Kraft and Luyten (**164**) found a statistical parallax for U Gem stars which gave a mean absolute magnitude $M_v = 7{\cdot}6$ for stars with a period of $5^h\,32^m$.

The *Sixth Catalogue of the Orbital Elements of Spectroscopic Binaries* is being prepared for press at the time of writing and should be distributed soon after the Prague Assembly. The Catalogue will appear in the Publications of the Dominion Astrophysical Observatory. It will contain orbital elements for nearly 740 spectroscopic binary systems—an increase of 50% over the Fifth Catalogue of Moore and Neubauer—as well as references to about 60 other systems for which orbital elements are not yet well enough known to be included in the Catalogue.

Schneller writes that work is going forward well on the third edition of his 'Geschichte und Literatur der veränderlichen Sterne' and that the first volume is expected to appear early in 1967. F. B. Wood states that the Flower and Cook Card Catalog on Eclipsing Binaries is expanding at a rapid rate, and reminds all workers in the field that its thousands of entries are at their disposal.

The Third Colloquium on Variable Stars (**165**) held at Bamberg in 1965, on 'The Position of Variable Stars in the Hertzsprung-Russell Diagram' resulted in review papers by Binnendijk, Eggen, Kopal, Plavec and Grygar, Mayall, and Herczeg concerned mainly with summarizing in a new and rather productive fashion material on close binaries already published in conventional form.

Finally, Batten (**166**) has summarized the state of our statistics on all double stars in a way that both shows the high degree of observational selection present and suggests some courses of action for closing the gaps between the number of systems known, the number studied, and the aspects of study needing emphasis.

BIBLIOGRAPHY

1. Fletcher, E. S. 1964, *Astr. J.*, **69**, 357.
2. Przybylski, A. 1964, *Acta astr.*, **14**, 285.
3. Bertiau, F. C. 1964, *Ric. astr. Spec. Vatic.*, **6**, 611.
4. O'Connell, D. J. K. 1964, *Ric. astr. Spec. Vatic.*, **6**, 523.
5. Shao, C. Y., Young, A. T. 1965, *Astr., J.*, **70**, 692, 727.
6. Young, A. T. 1965, *Astr. J.*, **70**, 687; 1965, *Mon. Not. R. astr. Soc.*, **129**, 215; 1966, *Observatory*, **86**, 71.
7. Kordylewski, K. *Mitt. Budapest.*
8. Kordylewski, Z. 1965, *Acta astr.*, **15**, 133, 1965.
9. Szafraniec, R. 1966, *Acta astra.*, **16**, 3.
10. Illés-Almár, E., Almár, I. 1963, *Acta astr.*, **13**, 72.; 1963, *Acta astr.*, **13**, 75.
11. Schneller, H. 1964, *Astr. Nachr.*, **288**, 1.
12. Plavec, M. 1964, *Trans. IAU*, **12B**, 508.
13. Plavec, M. 1966, *ASP Leaflets* no. 440 and 441.
14. Huang S.-S. 1963, *Astrophys. J.*, **138**, 471.
15. Piotrowski, S. L. 1964, *Acta Astr.*, **14**, 251.
16. Kruszewski, A. 1964, *Acta Astr.*, **14**, 241.

17. Kruszewski, A. 1966, *Adv. Astr. Astrophys.*, **4**, 233.
18. Balász-Detre, J., Detre, L. 1965, *Kl. Veröff. Bamberg*, IV/40, 184.
19. Martynov, D. Ya. 1965, *Astr. Zu.*, **42**, 1209.
20. Semeniuk, I. *Acta astr.* (in press).
21. Cristaldi, S. 1965, *Mem. Soc. astr. ital.*, **37**, 205.
22. Quester, W., Braune, W. 1965, *I.B.V.S.*, **116**.
23. Kopal, Z. 1965, *Adv. Astr. Astrophys.*, **3**, 89.
24. Chen, K-y., Reuning, E. G. 1966, *Astr. J.*, **71**, 283.
25. Grygar, J. 1965, *Bull. astr. Inst. Csl.*, **16**, 195.
26. Davis, W. D. 1964, *Astrophys. J.*, **68**, 277.
27. Linnell, A. P. 1965; *Astrophys. J. Suppl.*, **11**, 185; 1965, **12**, 288; 1966, **13**, 413, 428, 445.
28. Jurkevich, J. 1964, *Astr. J.*, **69**, 787.
29. Harris, A. J. 1964, *Astr. J.*, **69**, 787.
30. Shulberg, A. M., Tabachnik, V. M. 1965, *Astr. Zu.*, **42**, 590.
31. Minin, I. N. 1965, *Astrophysica*, **1**, 275.
32. Sobieski, S. 1965, *Astrophys. J. Suppl.*, **12**, 263, 276.
33. Shulberg, A. M. 1965, *Perem. Zvezdy*, **14**, no. 3, 186.
34. Tcherepashchuk, A. M. 1966, *Astr. Zu.*, **43**, 517.
35. Kitamura, M. 1965, *Adv. Astr. Astrophys.*, **3**, 27.
36. Kříž, S. 1965, *Bull. astr. Inst. Csl.*, **16**, 118.
37. Horak, T. 1966, *Bull. astr. Inst. Csl.*, **17**, 27.
38. Wilson, R. E. 1966, *Publ. astr. Soc. Pacif.*, **78**, 223.
39. Krzeminski, W. 1965, *Astrophys. J.*, **142**, 1051.
40. Mumford, G. S. 1964, *Astrophys. J.*, **139**, 2 = *Kitt Peak nat. Obs. Contr.*, no. 48.
41. Paczyński, B. 1965, *Acta astr.*, **15**, 305.
42. Krzemiński, W., Kraft, R. P. 1965, *Astrophys. J.*, **140**, 921.
43. Mumford, G. S. 1964, *Publ. astr. Soc. Pacif.*, **76**, 57.
44. Paczyński, B. 1965, *Acta astr.*, **15**, 197.
45. Herbig, G., Preston, G. W., Smak, J., Paczyński, B. 1965, *Astrophys. J.*, **141**, 617.
46. Popper, D. M. 1965, *Astrophys. J.*, **141**, 126.
47. Peery, B., Jr. 1966, *Astrophys. J.*, **144**, 672.
48. Tcherepashchuk, A. M. 1966, *Astr. Circ.*, no. 384.
49. Smak, J. 1964, *Acta astr.*, **14**, 101.
50. Bookmyer, B. B. 1965, *Astr. J.*, **70**, 415.
51. Binnendijk, L. 1966, *Astr. J.*, **71**, 340.
52. Roxburgh, I. W. 1966, *Nature*, **208**, 65; 1966, *Astrophys. J.*, **143**, 111.
53. Hurley, M., Roberts, P. H. 1964, *Astrophys. J.*, **140**, 583; 1965, *Astrophys. J. Suppl.*, **11**, 95.
54. James, R. A. 1964, *Astrophys. J.*, **140**, 552.
55. Monaghan, J., Roxburgh, I. W. 1965, *Mon. Not. R. astr. Soc.*, **131**, 13.
56. Roxburgh, I. W., Griffith, J. S., Sweet, P. A. 1965, *Z. Astrophys.*, **61**, 203.
57. Roxburgh, I. W. 1965, *Z. Astrophys.*, **62**, 134.
58. Stoeckly, R. 1965, *Astrophys. J.*, **142**, 208.
59. Chandrasekhar, S. 1964, *Astrophys. J.*, **140**, 599; 1965, **142**, 890; 1966, **145**, 842.
60. Lebowitz, N. R. 1966, *Astrophys. J.*, **145**, 878.
61. Roxburgh, I. W. 1966, Liège symposium.
62. Mestel, L. 1966, Liège symposium.
63. van Albada, T. S. 1966, Liège symposium.
64. Hack, M. 1963, *Star Evolution* (ed. Gratton), p. 452, Academic Press.
65. McCrea, W. M. *et al.* 1964, *R. Obs. Bull.*, no. 82.
66. McCrea, W. M. 1964, *Mon. Not. R. astr. Soc.*, **128**, 147.
67. Hill, P. W. 1965, *Observatory*, **85**, 41.
68. Smak, J. 1966, *Acta astr.*, **16**, 25.
69. Williams, I. P. 1964, *Ann. Astrophys.*, **27**, 198; 1965, **28**, 46.
70. Deutsch, A. J. 1966, Am. astr. Soc., 121st meeting.
71. Jaschek, C., Jaschek, M. 1965, *Z. Astrophys.*, **62**, 15.

72. Zeldovich, Ya. B., Guseynov, O. H. 1966, *Astrophys. J.*, **144**, 840.
73. Efremov, Y. N., Kholopov, P. N., Kukarkin, B. V., Sharov, A. S. 1964, *I.B.V.S.*, **75**.
74. Kuročkin, N. E. 1965, *I.B.V.S.*, **79**.
75. Kuročkin, N. E., Kukarkin, B. V. 1966, *Astr. Zu.*, **43**, 83 = 1966, *Soviet Astr.*, **10**, 64.
76. Zahn, J.-P. 1965, *C.R. Acad. Sci.*, Paris, **260**, 413, (Group 3) = *Contr. Inst. astrophys. Paris*, Sér. A, no. 319.
77. Peraiah, A. 1965, *Z. Astrophys.*, **62**, 48.
78. Smak, J. 1964, *Acta astr.*, **14**, 97.
79. Osaki, Y. 1965, *Publ. astr. Soc. Japan*, **17**, 97.
80. Santarelli, A. 1965, *Mem. Soc. astr. ital.*, **36**, 151.
81. Dziembowski, W. 1963, *Acta astr.*, **13**, 157.
82. Schwarzschild, M. 1966, *Trans. IAU*, **12B**, 527, Academic Press.
83. Kopal, Z. 1965, *Adv. Astr. Astrophys.*, **3**, 89.
84. Martynov, D. J. 1965, *Astr. Zu.*, **42**, 1209 = 1966, *Soviet Astr.*, **9**, 939.
85. Peraiah, A. 1966, *Z. Astrophys.*, **64**, 27.
86. Plavec, M., Sehnal, L., Mikulás, J. 1964, *Bull. astr. Inst. Csl.*, **15**, 171.
87. Plavec, M., Křiž, S. 1965, *Bull. astr. Inst. Csl.*, **16**, 297.
88. Kruzewski, A. 1964, *Acta astr.*, **14**, 231.
89. Koch, R. H. 1965, *Astr. J.*, **70**, 681.
90. Huang, S.-S. 1965, *Astrophys. J.*, **141**, 201.
91. Plavec, M. 1966, *Trans. IAU*, **12B**, 508, Academic Press.
92. Lebrun, C., Robe, H. 1964, *Bull. Acad. R. Belg.*, 5e Sér., **50**, 315 = Inst. Astrophys. Liège, Coll, 8°, no. 479.
93. Limber, D. N. 1963, *Astrophys. J.*, **138**, 1112.
94. Chiara, L. 1963, *Mem. Soc. astr. ital.*, **34**, 299.
95. Hadjidemetriou, J. D. 1963, *Icarus*, **2**, 440; 1966, **5**, 34; 1966, *Z. Astrophys.*, **63**, 116.
96. Martin, E. L. 1962, *Mem. Soc. astr. ital.*, **33**, 383 = 1964, *Ciel et Terre*, **80**, **1**.
97. Omarov, T. B. 1964, *Astr. Zu.*, **41**, 170 = 1964, *Soviet Astr.*, **8**, 127.
98. Huang, S.-S. 1963, *Astrophys. J.*, **138**, 481.
99. Huang, S.-S. 1966, *Astr. J.*, **71**, 165; 1966, *Sky Telesc.*, **31**, 215.
100. Kruszewski, A. 1964, *Bull. Acad. pol.*, Sér. math. astr. phys., **12**, 317; 1964, *Acta astr.*, **14**, 241.
101. Piotrowski, S. L. 1964, *Bull. Acad. pol.*, Sér. math. astr. phys., **12**, 323; 1965, **12**, 419; 1964, *Acta astr.*, **14**, 251.
102. Kruszewski, A. 1966, *Adv. Astr. Astrophys.*, **4**, 233.
103. Chi,, G., Pál, A. 1964, *Studii Cerc. Astr.*, *Seism.*, Bucarest, **9**, 15.
104. Schneller, H. 1965, *Astr. Nachr.*, **288**, 183.
105. Schneller, H. 1966, *Mitt. veränd. Sterne*, **3**, 179.
106. Porfiryev, V. V., Kalenichenko, V. V. 1964, *Astr. Zu.*, **41**, 858 = 1965, *Soviet Astr.*, **8**, 687.
107. Shakhovski, N. M. 1964, *Astr. Zu.*, **41**, 1042 = 1965, *Soviet Astr.*, **8**, 833.
108. Popper, D. M., Walker, M. F. 1963, *Astr. J.*, **68**, 543.
109. Sahade, J. 1966, *Trans. IAU*, **12B**, 494, Academic Press.
110. Reuning, E. G., Springer, J. F. 1964, *Astr. J.*, **69**, 148.
111. Woolf, N. J. 1965, *Astrophys. J.*, **141**, 155.
112. Sahade, J., Hernandez, C. *Bol. Asoc. astr. Argent.*, no. 8.
113. Huang, S.-S. 1965, *Astrophys. J.*, **141**, 976.
114. Roxburgh, I. W. 1965, *Astr. J.*, **70**, 690.
115. Smak, J. 1964, *Acta astr.*, **14**, 101.
116. Fracastoro, M. G. 1966, *Trans. IAU*, **12B**, 501, Academic Press.
117. Kraft, R. P. 1963, *Adv. Astr. Astrophys.*, **2**, 43.
118. Kraft, R. P. 1966, *Trans. IAU*, **12B**, 519, Academic Press.
119. Mumford, G. S. 1962, *Sky Telesc.*, **23**, 71, 135; 1963, **26**, 190.
120. Paczyński, B. 1965, *Acta astr.*, **15**, 89, 197.
121. Kraft, R. P. 1964, *Astrophys. J.*, **139**, 457.
122. Kraft, R. P., Luyten, W. J. 1965, *Astrophys. J.*, **142**, 1041.

123. Kraft, R. P. 1965, *Astrophys. J.*, **142**, 1588.
124. Gorbatzky, V. G. 1964, *Astr. Zu.*, **41**, 849 = 1965, *Soviet Astr.*, **8**, 680.
125. Krzeminski, W. 1965, *Astrophys. J.*, **142**, 1051.
126. Herbig, G. H., Preston, G. W., Smak, J., Paczyński, B. 1965, *Astrophys. J.*, **141**, 617.
127. Widing, K. G. 1966, *Astrophys. J.*, **143**, 121.
128. Wilson, R. E. 1966, *Publ. astr. Soc. Pacif.*, **78**, 223.
129. Tcherepashchuk, A. M. 1966, *Astr. Zu.*, **43**, 517.
130. Kitamura, M. 1965, *Adv. Astr. Astrophys.*, **3**, 27.
131. Kitamura, M. Tokyo astr. Obs. (unpublished).
132. Tabachnik, V. M., Shulberg, A. M. 1965, *Astr. Zu.*, **42**, 590 = 1965, *Soviet Astr.*, **9**, 467.
133. Linnell, A. P. 1964, *Astr. J.*, **69**, 143.
134. Linnell, A. P. 1965, *Astrophys. J., Suppl.* **11**, 185; 1965, **12**, 288; 1966, **13**, 413, 428, 445.
135. Jurkevich, I. Flower and Cook Obs. (unpublished).
136. Schneller, H. 1965, *Astr. Nachr.*, **288**, 189.
137. Wilson, R. E. Georgetown Univ. *Bull. Am. phys. Soc.*, Series II, **11**, no. 4, 519.
138. Hosokowa, Y. (unpublished).
139. Grygar, J. 1965, *Bull. astr. Inst. Csl.*, **16**, 195.
140. Gingerich, O. 1966, *Astrophys. J.*, **144**, 1213.
141. Kříž, S. 1965, *Bull. astr. Inst. Csl.*, **16**, 118.
142. Minin, I. N. 1965, *Astrofizika*, **1**, 173 = 1965, *Astrophysics*, **1**, 104.
143. Sobieski, S. 1966, *Astrophys. J. Suppl.*, **12**, 263, 276.
144. Ebbighausen, E. G., Gange, J. J. 1963, *Publ. Dom. astrophys. Obs.*, **12**, 151.
145. Brazhnikova, E. F., Babinchuk, S. V. 1965, *Soviet Astr.*, **9**, 521 = 1965, *Astr. Zu.*, **42**, 678.
146. Karyagina, Z. V., Kharitonov, A. V. 1965, *Soviet Astr.*, **9**, 297 = 1965, *Astr. Zu.*, **42**, 377.
147. Abt, H. A. *Astrophys. J. Suppl.*, **11**, no. 102.
148. Abt, H. A., Barnes, R. C., Biggs, C. S., Osmer, P. S. 1965, *Astrophys. J.*, **142**, 1604.
149. Jaschek, C., Jaschek, M., 1965, *Z. Astrophys.*, **62**, 15.
150. Sahade, J. 1966, *Stellar Evolution*, (Cameron and Stein, ed.), p. 449.
151. Szafraniec, R. 1966, *Acta astr.*, **16**, 3.
152. Cester, B. 1965, *Astrophys. J.*, **62**, 191.
153. Plavec, M. 1964, *Bull. astr. Inst. Csl.*, **15**, 156.
154. Peraiah, A. 1965, *Z. Astrophys.*, **62**, 48.
155. Porfiryev, V. V., Kalenichenko, V. V. 1965, *Sov. Astr.*, **8**, 687.
156. Koch, R. H. 1965, *Astr. J.*, **70**, 681.
157. Mumford, G. S. 1966, *Sky Telesc.*, **31**, 213.
158. Huang, S. S., Wade, C. W. 1966, *Astrophys. J.*, **143**, 146.
159. Kuročkin, N. E., Kukarkin, B. V. 1966, *Astr. Zu.*, **42**, 83.
160. Artiukhina, N. M. 1964, *Perem. Zvezdy*, **15**, 127.
161. Smak, J. 1962, *Acta. astr.*, **14**, 97.
162. Kraft, R. P. 1965, *Astrophys. J.*, **142**, 1588.
163. Popov, M. V. 1964, *Perem. Zvezdy*, **15**, 115.
164. Kraft, R. P., Luyten, W. J. 1965, *Astrophys. J.*, **142**, 1041.
165. 1965, *Kleine veröff. Remeis-Sternw. Bamberg*, IV, no. 40.
166. Batten, A. H. 1967, *A. Rev. Astr. Astrophys.*, **5**, (in press).

JOHN E. MERRILL
President of the Commission

43. COMMISSION DE LA MAGNETO-HYDRODYNAMIQUE ET DE LA PHYSIQUE DES GAZ IONISES

PRÉSIDENT: Professor T. G. Cowling, Department of Mathematics, The University, Leeds 2, England.

VICE-PRÉSIDENT: Professor L. Davis, Jr., Physics Department, California Institute of Technology, Pasadena, California, U.S.A.

COMITÉ D'ORGANISATION: H. Alfvén, L. Biermann, A. B. Severny.

MEMBRES: Bel, Bhatnagar, Bumba, Chandrasekhar, Csada, Edmonds, Ferraro, Ginzburg, Jensen, Kushwaha, Ledoux, Lehnert, Lüst, Piddington, Pikel'ner, Prendergast, Schlüter, Shklovsky, Simon (R.), Smerd, Spitzer, Sweet, Tayler, Trehan, Van Bueren, Wentzel.

INTRODUCTION

This report deals only with those aspects of magnetohydrodynamics and plasma physics that are specifically of interest to astronomy. The emphasis is on theory; however, a number of observational papers are mentioned, both as background and because of interesting suggestions which they contain. Purely mathematical or physical theory is normally included only when its astronomical relevance is clear.

The topics discussed include solar and stellar magnetism, interplanetary and galactic fields, and the origin and propagation of cosmic rays. Problems referring to the magnetosphere are in general excluded as falling within geophysics; however, the tail of the magnetosphere and the sheath between the magnetospheric boundary and its associated bow shock are included, being the seat of processes which may have parallels in interstellar space. In spite of the significance of cosmic rays in X-ray sources, radiation is in general considered only in so far as it assists the diagnosis of charged-particle properties.

During the period covered by the report (1964–66), the proceedings of a number of conferences of interest to Commission 43 have been published. The most important of these were the 1964 conference on the Solar Wind (**1**) and the 1965 conference on Cosmic Rays (**2**).

THE SUN

Field structure

Much recent observational work has been concerned with the microstructure of solar fields. Severny (**3**) has written two detailed reports on the subject; Bumba (**4**) has given a general discussion of the relation between field and velocity structures. Severny (**5**) found that, whereas at low resolution the polar field may appear coherent, at high resolution it consists of elements differing in size, field strength and polarity. Stenflo (**6**) confirmed this result, finding that the mean field measured increases with decreasing slit area. Severny (**7**) has also studied the fine structure in height, and the corresponding shear of the field.

Leighton (**8**) developed Babcock's theory of the solar cycle, assuming the dispersal of magnetic regions to be the combined effect of non-uniform rotation and a random-walk process due to supergranulation currents. Sheeley (**9**) interpreted results for bipolar spots along these lines, pointing out that the fact that the numbers of polar faculae lag 90° in phase behind the sunspot numbers is consistent with the polar fields being fields of following spots carried to the poles in a few years by a random walk (see also Bumba and Howard (**10**)). Similarly Hyder (**11**) suggested that the 'rush to the poles' by prominence filaments synchronizes with the reversal of polar fields; Godoli and Mazzuconi (**12**) found that polar surges have maximum frequency at solar minimum, when the polar field attains its maximum.

The connection between supergranulation and field strengths has been stressed in a number of papers. Bumba and Howard (13) found that active regions and spots tend to develop in the spaces between supergranules. Simon and Leighton (14) found that the boundary between supergranulation cells, where the motion is downward, is associated with enhanced magnetic fields. Theoretical discussions of the concentration of fields at the cell boundaries were given by Weiss (15) and Clark (16).

Filaments, prominences and corona

Chvojková (17) discussed a filamentary model consisting of a magnetic arch filled with charged particles supported at a high level by diamagnetic repulsion, as in a mirror device. If the supporting force is sufficiently strong, a slight variation of conditions may lead to an eruption. An inverse process, in which a slight disturbance leads to a 'dumping' of trapped particles, was suggested by Elliott (18) as a possible explanation of solar flares.

Interpretations of arched prominences as bundles of lines of force were also suggested by Ivanchuk (19), Doherty and Menzel (20), and Jefferies and Orrall (21). Of these, Jefferies and Orrall interpreted prominences as loops along which flare plasma is returned to the photosphere; Doherty and Menzel discuss primarily the cooling of filaments with a longitudinal magnetic field. Howard and Harvey (22) found observationally that, whereas dark fibrils are perpendicular to the isogauss lines of the horizontal field, filaments are parallel to these. This suggests a filamentary model closer to the earlier models of Kippenhahn and Schlüter or of Dungey; a development of Dungey's model was discussed by McLean (23). Kink instability was invoked by Hirayama (24) to explain prominence blow-offs. Magnetic 'hills' in the chromosphere above active regions, with force-free fields, were suggested by Iospha *et al.* (25).

The magnetic field in the corona has mainly been studied through the polar rays, regarded as indicators of field directions. Kopecký and Suda (26) analysed the variation of direction of the rays with latitude and height; Saito (27) and Stoddard and Carson (28) interpreted the field as that of a bar magnet of variable length, the length being greatest near sunspot minimum. The connection of the coronal field with that at the surface was studied by Harvey (29) and by Pikelner (30), but in general terms only; its connection with that of the solar wind has likewise been studied in general terms only, e.g. by Mustel (31), van Heuvel (32) and Mogilevsky (33).

The effects of a magnetic field in limiting motions and wave dissipation in the chromosphere have been considered respectively by Kleczek (34) and by Livshitz *et al.* (35). Chromospheric and coronal heating by magnetosonic waves were considered by Pikelner and Livshitz (36), Lüst and Scholer (37) and Parker (38). Meyer (39) interpreted certain fast disturbances observed by Moreton as chromospheric effects of magnetosonic waves propagated in a channel between corona and chromosphere.

Sunspots; solar cycle

Model calculations of the field due to sunspots have been made by Schmidt (40) and Oster (41).

A number of papers on the solar cycle have already been mentioned in the section on Field Structure. Chvojková (42) developed Babcock's theory of the cycle, in Kopecký's modified form, in discussing the generation of toroidal fields at sunspot latitudes; see also Tuominen (43). Godoli applied the Babcock-Kopecký theory to the law of sunspot drift with latitude (44). Kopecký (45) attempted to explain a possible 80-year cycle on the basis of a theory similar to that for the 11-year cycle. Iroshnikov (46) proposed an oscillatory instability in the presence of differential rotation and a magnetic field to explain the 11-year cycle. Kozhevnikov (47) suggested that spot fields, before their convection to the surface, are concentrated near the bottom of the convection zone, not distributed throughout that zone. Romanchuk (48) attempted to explain solar fields by the interplay of fields in the corona and the photosphere.

Many sunspot observations have been concerned with motions near the spots. Moreton and Gopasyuk (49) found that in the September 1963 spot group the relative motion of spots was in the direction of the transverse field. Gopasyuk (50) found that plasma inflow occurs at photospheric levels when the spot field is increasing. Schröter (51) inferred from observations that the Evershed effect is due to the upward motion of bright material, and the downward motion of dark, in fine structures. Livshitz and Pikelner (52) discussed possible diamagnetic ejection from sunspot regions. Bumba, Kopecký and Kuklin (53) considered the life history of a sunspot, emphasizing the fine structure and the relation to supergranulation and pores. Kopecký and Kuklin (54) applied formulae for the conductivity of an ionized gas to the time of decay of a sunspot; it should be noted, however, that the results are very sensitive to the level at which decay is supposed to occur.

Deinzer (55) gave a magnetohydrostatic theory of sunspot equilibrium, similar in spirit to the earlier theory of Chitre; he used a mixing length whose ratio to the scale height is smaller in the spot than outside, because of magnetic inhibition of convection. Gough and Tayler (56) obtained sufficient criteria for the stabilization of thermal instability by a magnetic field; they applied this to sunspots, confirming the large reduction in heat convection. Danielson's Stratoscope observations (57), however, suggest that small-scale convection is present even in a spot umbra.

Solar flares

Many observers have stressed the intimate connection between solar flares and filaments. Smith and Ramsey (58) pointed out the changes in filamentary structure just before and at the commencement of a flare, and found that flares could set filaments oscillating. Kleczek (59) found that loop prominence systems develop from flares, apparently as a result of the propagation of flare particles along nearly stationary field-lines. Possibly a similar transmission along field-lines may be responsible for the observation by Moreton and Severny (60) that flares first brighten at two points on opposite sides of the line where the vertical field vanishes. Haurwitz (61) suggested that a loop-like field, as revealed by dark filaments, was essential for flare acceleration of electrons to high energies.

Theories of flares are manifold. They mostly rely on magnetic instabilities – exceptions are the theory of Hunter (62), which supposes shock waves to be generated in the chromosphere by a thermal instability, and that of Jacobsen and Carlquist (63), which relies on a circuit interruption more appropriate to laboratory discharges than to solar plasma. Sakurai (64) assumed a collapse of a magnetic field due to changes in the pressures exerted on a magnetic tube by the surrounding plasma; his discussion is, however, mainly descriptive. Kawabata (65) thought in terms of the collapse of a coronal condensation after the ejection of a cloud which becomes magnetically isolated as a consequence of suprathermal turbulence near a neutral point. Wentzel (66) interpreted a flare as arising from the collapse of an unstable chromospheric region towards a surface of minimum but non-zero field.

The main difficulty of magnetic flare theories is to find a sufficiently rapid mechanism for converting magnetic energy into heat. Flows near a neutral line leading to the breaking of field-lines have been considered by Syrovatsky (67). Two other theories deserve a special mention. The first is that of Petschek (68), which seeks to provide the necessary dissipation through standing hydromagnetic shock waves. Shock waves almost certainly do accompany flares; however, Sweet and Green (69) give arguments suggesting that Petschek's flare picture involves internal inconsistencies, and needs at least to be seriously modified. The second is that of Sturrock and Coppi (70), in which gravitational instability of gas supported by a sheared magnetic field sparks off a more violent instability of plasma broken into fine filaments. They claim that this mechanism can give the observed time-scale of rise to flare maximum, but so far it has not been studied in full detail.

Flares emitting solar cosmic rays were found by Warwick (**71**) to be those in which initially there is a steep magnetic gradient. Wentzel (**66**), on the other hand, suggested that the distinguishing feature of such flares might be their occurrence at a high level. The mechanism of generation of the cosmic rays is possibly Fermi acceleration between converging shocks—cf. Sakurai (**72**)—but discussions of the acceleration process do not in general go far into detail.

INTERPLANETARY PLASMA; THE SOLAR WIND

General properties

The Proceedings of the 1964 Solar Wind conference (**1**) gives a good general introduction to present knowledge about interplanetary plasma.

Recent work has confirmed that the solar wind blows all the time, but with substantial variations in velocity, density, and 'temperature'. Electron-densities quoted are 3 to 10 per cm^3, with peaks up to 25 per cm^3 (**73, 74, 75**). Velocities are found to fluctuate between 300 and 800 km s^{-1} ((**74, 75**); see also (**181**)); their directions may deviate 5 to 10° from the radial (**76**), mainly in the sense indicating partial co-rotation with the Sun. The temperature of thermal motions parallel to the magnetic field may be five times that of the thermal motions perpendicular to it (**76**). The magnetic field is, on an average, about 5 γ in magnitude, directed at about 45° to the solar direction (**77, 78**); it also appears to possess a persistent component perpendicular to the ecliptic, which is very difficult to comprehend (**79**). (These results all come from satellites in the Earth's neighbourhood, i.e. not far from the ecliptic.) A recent account of the normal form of the solar wind is that of Coleman *et al.* (**80**).

Although the direction of the magnetic field is, apart from random fluctuations, reasonably constant and along the expected spiral, it reverses in sign from time to time, corresponding to the polarity of the region with which it is linked. Wilcox and Ness (**81**) found evidence of a persistent magnetic structure, co-rotating with the Sun, in which the field was directed alternately towards and away from the Sun, each for some days on end; cf. also Coleman *et al.* (**82**), Ness (**83**). Ness and Wilcox (**79**) and Bumba and Howard (**183**) have also been able to relate the field with that at the centre of the Sun's disk 4–5 days earlier; this is the field of a large magnetic region, not the local field of sunspots. In this connection, see also Mustel (**31**), whose conclusions are somewhat different.

Co-rotation of the field pattern does not by itself imply rotational motions; the familiar garden-hose effect leads to a co-rotating pattern even if the motion is wholly radial. However, the plasma motion near the Sun must have a share of the solar rotation—cf. the model by Stern (**84**). The results of Wolfe *et al.* (**76**) and of the Vela experimenters suggest a corotational velocity of order 15 km s^{-1} near the Earth.

Theory of the solar wind

Parker (**85**) has considerably developed his theory of the solar wind, applying it also to stellar winds. He has usually considered a hydrodynamic spherical expansion of a high-temperature corona up to supersonic velocities, the temperature being maintained at a high value by heat conduction. Magnetic fields and rotation were usually not considered. However, in one paper (**86**) he considered expansion along magnetic filaments rooted in local areas of the solar surface. The filaments were conceived as expanding, with increasing distance from the Sun, to fill the whole of space, and to become tangled together beyond about 1 A.U. by Helmholtz instability.

Spherical expansion was also discussed by a number of other authors (**87**), usually with neglect of viscosity. Whang, Liu and Chang (**88**) found that viscosity leads to a much reduced asymptotic velocity of outflow. Meyer and Schmidt (**89**) showed that a radial magnetic field greatly reduces the effects of viscosity. In any case, the effects of viscosity and thermal conduction, as well as the anisotropy due to a magnetic field, are so great beyond (say) 10 $R_\odot$ that a normal hydrodynamic treatment becomes suspect. However, Brandt and Cassinelli (**90**), treating the wind as an exospheric problem, obtained results differing not too widely from those of the hydrodynamic treatment. Kalinin and Mogilevsky (**182**) have developed the hypothesis that M-region disturbances arise from chains of independent elements ('plasmoids') generated in active regions of the Sun, and travelling radially through the quiet solar wind, each carrying its own force-free field with it.

A corotating magnetic field considerably complicates the mathematical problem. In the first of two papers (**91**), Pneuman gave a picture of the field in meridian planes at solar minimum as roughly dipolar (at least at high latitudes) out to several solar radii, then becoming radial where the flow is dominant. In his second paper he suggested that there is corotation of the outflowing material out to about 3 $R_\odot$. King and Carovillano (**92**) considered the interaction between the magnetic field and the angular momentum of the outflowing gas; Brandt (**93**) considered the corresponding braking of the solar rotation, finding a braking time of 7×10^9 years if the material corotates partially to 1 A.U. In a more detailed discussion, Weber and Davis (**94**) considered a two-dimensional model representing the flow and field in the equatorial plane, paying special attention to the points where the Alfvén velocity is reached. They estimated that the degree of corotation near the Sun, though with less than the solar angular velocity, still gave a braking time of 5×10^9 years.

Shock waves

Interplanetary shock waves can arise when a faster solar stream overtakes a slower (especially after a flare) or when the solar wind impinges on a barrier. Such shock waves are collisionless; the plasma particles can be regarded as forming a normal gas only insofar as they are tied together by the magnetic field.

Sturrock and Spreiter (**95**) have given an explanation of a negative impulse following immediately after the initial positive impulse at the sudden commencement of a geomagnetic storm, in terms of a pair of shock waves due to ejected flare plasma. Colburn and Sonett (**96**) suggested similar shock pairs at the edge of an M-region stream. Simon and Axford (**97**) obtained a similarity solution of the hydromagnetic equations which involved such shock pairs.

The bow shock at the Earth's magnetosphere fluctuates with the strength of the solar stream. Its reality is best shown by the sudden increase in field-strength, accompanied by a transition to a disordered field, as the shock front is penetrated by the plasma (**77**, **98**). A hydromagnetic discussion of the shock front has been given by Lees (**99**); see also Obayashi (**100**), Sozou (**101**) and Spreiter, Summers and Alksne (**102**). A number of possible instabilities have been suggested as contributing to the highly disturbed nature of the region inside the shock front (**103**).

'Spikes' of energetic electrons (energies > 30 keV) are observed near the bow shock (**104**). These are believed to be generated locally by a Fermi mechanism between converging shock fronts, e.g. when trapped between the bow shock and an approaching wave ((**105**), and (**2**), p. 105). Similar energetic electrons are also found in the tail of the magnetosphere (**106**); these may be those accelerated in the bow region, carried down into the tail, but may also be accelerated in the neutral sheet of the tail ((**2**), p.147). Apparent variations of electronic energies behind the bow shock are sometimes to be interpreted in terms of diffusion (**107**) or diamagnetic effects (**108**).

Interaction with the Moon, planets and comets

The tail of the magnetosphere extends downstream from the Earth for a distance of at least 50–60 Earth radii ((**109**), and (**1**) p.231). The Earth's magnetosphere can only in a restricted sense be regarded as the prototype for magnetospheres of the Moon, planets and comets, which often possess no appreciable magnetic fields of their own.

A lunar magnetosphere was inferred by Ness *et al.* (**77**, **110**) from observations of an enhanced irregular magnetic field, apparently in the Moon's wake. Aronowitz and Milford (**111**) worked out how the magnetic shielding of the Moon's surface from the solar wind would vary with the lunar magnetic moment. However, there is no evidence of any appreciable lunar field, and the solar wind is likely to strip away any appreciable atmosphere (**112**). As Gold pointed out ((**1**), p. 381) the lunar magnetosphere is likely to be formed from lines of force of the solar wind, hung up on the Moon because of its finite conductivity.

Venus and Mars do not appear to possess magnetic fields, and no appreciable magnetospheric effects have been observed for either. The radio emissions from Jupiter strongly suggest that it has a magnetic field. No direct observation of its magnetosphere is at present possible; but the correlation of variations in its radio emission with the position of Io suggests some influence of Io on the plasma round Jupiter, related in a highly directional manner to the magnetic field ((**113**), pp. 134, 137 and 141).

Cometary tails provide our strongest evidence of a solar wind at high solar latitudes. An appreciable part of the interaction between the solar wind and cometary tails is hydromagnetic ((**1**), pp. 355, 365, and 373). However, emission of plasma by the comet may also lead to a (collisionless) bow shock in the front of the comet, and a magnetosphere with some features in common with that of the Earth (Biermann *et al.* (**114**), Axford (**115**), Marochnik (**116**)).

COSMIC RAYS

Cosmic rays and the solar wind

General accounts of the interaction of cosmic rays with the solar wind have been given by Anderson ((**1**), p. 53) and Quenby ((**2**), p. 3).

In addition to flare particles, protons of energy 1 – 100 MeV appear to be emitted continuously by the Sun, and to be guided in co-rotating magnetic fields (Fan *et al.* (**117**)). They are strongly collimated along the magnetic field, presumably because of decreasing pitch angles as the magnetic field decreases. They come from a mean direction midway between those of the Sun and of the magnetic field. Rapid intensity-oscillations in them suggest waves in the solar wind (see also Bryant *et al.* (**118**)).

Sakurai (**119**) divides solar protons into those propagating direct from the Sun and those trapped in plasma clouds. A better model for energetic particles is to regard them as propagating along the lines of force, but subject to magnetic scattering. The scattering may be due to shock waves (**120**), or to magnetic irregularities of dimensions about 10^6 km (**121**). Scattering leads to diffusion which, because of the magnetic constraints, is strongly anisotropic; many of the characteristics of solar cosmic ray events can be accounted for adequately by models which take this anisotropy into account (**122**). The diffusion coefficient can be related to the (observable) magnetic field variations (**123**). The different diffusive properties of particles of different mass may lead to differences in composition of solar cosmic rays; but in general the composition of solar cosmic rays is similar to that of normal solar gas ((**1**), p. 215, and (**124**)).

Galactic cosmic rays—for the moment this is taken to refer to all cosmic rays coming from outside the solar system—are affected by scattering (**125**) and by the corotation of the interplanetary magnetic field (**126**). The latter is calculated as producing a diurnal variation of amplitude 0·6%, in good agreement with observation ((**2**), pp. 26, 213 and 231). While within the solar stream cavity, cosmic rays may undergo appreciable losses in energy through Fermi deceleration between diverging wave fronts (Parker (127)). This may contribute to explaining the observed cosmic-ray modulation in the solar cycle; the intensity at the Earth is 5% less than that in interstellar space at solar minimum, and may be 20% less at solar maximum ((2), pp. 97, 261 and 292).

Cosmic rays give information as to the size of the cavity occupied by the solar wind in interstellar space. To produce the observed scattering of solar cosmic rays, a radius of only 1·5–2 A.U. is required (128). However, to produce the long-term modulation, a radius above 20 A.U. may be needed ((2), pp. 3, 173 and 180); the precise value is uncertain, depending on factors like the anisotropy of cosmic ray diffusion, or the friction between cosmic rays and the solar wind.

Origin of cosmic rays

As the results for the solar wind indicate, particle acceleration to high energies is a normal process in the universe. Among the mechanisms studied in the period under review are the following. Hayakawa *et al.* (129) surveyed the usual electromagnetic mechanisms (Fermi, betatron); Wentzel (120, 130) considered Fermi acceleration between shocks; Tsytovich (131) studied acceleration by plasma waves, including waves in supernovae; Colgate ((2), p. 112) invoked supernova blast waves; Sturrock (132) considered plasma waves, accelerating particles riding their crests; Finzi (133) invoked neutron stars; and Burbridge and Hoyle (134) considered explosions in the nuclei of galaxies. Supernova explosions appear to be the most commonly invoked, but even as regards these the mechanism of acceleration requires detailed study.

The seat of origin of non-local cosmic rays may be galactic (Ginzburg and Syrovatsky: (135) and (2), p. 53) or metagalactic (E. M. and G. R. Burbridge, (2), p. 92). One criterion used by Ginzburg and Syrovatsky for distinguishing between the two is the unobservability of any appreciable background flux of radiation from intergalactic space; however, this is not altogether decisive, as the radiation would come from cosmic ray electrons, not heavy particles, and equipartition of magnetic energy with the other forms of energy present is assumed. Another criterion is the ease with which cosmic rays can enter or leave the Galaxy. Ginzburg (136) suggested that where cosmic rays exit from a strong field into a weak, instabilities tangle the field and lead to an effective closure; thus cosmic rays from any galaxy tend to be pent in that galaxy. However, this is not the only possibility; Parker ((137) and (2), p. 126) suggested that cosmic rays will inflate the galactic field into a large-size halo from which they can ultimately escape.

Because of the intimate connection between cosmic rays and X-rays, X-ray observations may ultimately help to decide the location of cosmic ray sources. The possible mechanisms of generation of X-rays have been considered by Gould and Burbidge (138) and Hayakawa *et al.* (139); other papers on radiation due to high-energy particles are given in (140) and (2), pp. 68 and 132.

STELLAR AND INSTELLAR FIELDS

Magnetic stars

Little work on magnetic stars can be reported here. Ledoux and Renson (141) have reviewed the properties of such stars, with special attention to magnetic variables. Ledoux (142) has suggested an explanation of magnetic variation in terms of the control of unstable *g*-modes of oscillation by rotation. Steinitz (143) has found possible periods of magnetic variation for two stars regarded by Babcock as irregular; but Renson (144) has pointed out the danger of spurious periods due to statistical fluctuations. Fowler *et al.* (145), giving up the idea of neutron reactions due to surface electromagnetic acceleration, now suggest the anomalous abundances in magnetic stars to be due to internal nuclear reactions. Smith (**184**) has considered the decay of toroidal magnetic fields, obtaining decay times up to $3{\cdot}79 \times 10^9$ years as maximum.

The generation of stellar fields has been considered by Mestel (**146**); of the three mechanisms discussed (fossil, battery and dynamo) he regards the fossil theory as offering the best hope of explaining strongly-magnetic stars. The battery theory (Biermann's currents due to non-uniform rotation) has been considered by Roxburgh and Strittmatter (**147**), who derived toroidal fields of order 3×10^6 gauss in rapidly rotating stars in favourable circumstances. The fossil theory was considered by Pneuman and Mitchell (**148**), who obtained a reasonable field for the Sun by assuming that the field was not frozen into the contracting gas cloud until a temperature of 1500°K was reached. Steenbeck *et al.* (**149**) have discussed a dynamo mechanism based on the interaction of a turbulent fluid motion with a field with non-vanishing mean value. In both this theory and one by Csada (**150**), based on eigenfunction analysis, it is not clear how far the results may be affected by a 'rounding-off' approximation.

Galactic and intergalactic plasma

A review of observations of magnetic fields in the Galaxy has been given by Davies ((**2**), p. 55); see also papers listed under (**151**). The field as seen from the Sun is mainly along the local spiral arm, though with superposed irregularities; fields at high positive galactic latitudes may be oppositely directed to those at other latitudes. The magnitude of the field in clouds is probably between 5×10^{-6} and 10^{-5} gauss.

The opposition of a magnetic field to gravitational collapse is of interest at a number of levels. Mestel (**152**) considered the contraction of a prestellar cloud, assuming strict flux freezing; inside the cloud his results were in general agreement with those obtained from the virial theorem, but outside he found a layer of near discontinuity ready for field reconnection. Strittmatter (**153**) considered contraction of a mass of gas when its magnetic field is too strong to permit radial contraction; he found that, though initial contraction along the lines of force was possible, this led to little improvement in the ability to form stars. He also investigated how far the Jeans criterion for gravitational instability should be regarded as affected by a magnetic field.

On a larger scale, Setti (**154**) argued that the gravitational field of stars will stabilize a helical magnetic field in a spiral arm. Parker (**155**) pointed out that a magnetic field confined to the galactic plane by the weight of plasma would be subject to Rayleigh-Taylor instability; this would concentrate gas into pockets (clouds) suspended in the field. Gershberg (**156**) considered the possibility that intergalactic 'bridges' might be explained by pinching due to the twisting of lines of force.

The interplay of magnetic fields, rotation and gravitational contraction is often invoked in connection with the formation of galaxies and quasi-stellar sources. Piddington (**157**) explained the difference between normal and radio galaxies in terms of the angle θ between the initial directions of the field and the axis of rotation. Featureless flat disks are produced if θ is small; if it is nearly a right angle, the field is wound into a spiral form, from which ultimately it buckles explosively to produce a radio galaxy. In a quasar there are alterations of field-winding and explosive acceleration. Pikelner (**158**), who developed a similar theory, pointed out the possibility of field reversals in a single spiral arm for intermediate values of θ. He stressed the possible effects of instabilities at surfaces of field reversal. Layzer (**159**), in a less developed theory of similar type, emphasized the role of magnetoturbulence in tangling the field lines, which afterwards straighten explosively, leading to collapse. Bardeen and Anand (**160**) criticized Layzer's theory on the ground that the mass that can be temporarily stabilized against collapse by magnetoturbulence is much reduced when one takes account of supersonic dissipation. Sturrock (161) sought the source of energy of a quasar in a flare-like snapping of the field-lines connecting a contracting mass to the surrounding plasma.

Ginzburg and Ozernoy, in a number of papers (162), examined the possibility that quasars are due to contraction which approaches the relativistic limit. The contraction reinforces the magnetic field, though in the later stages the field may with draw towards the Schwarzschild surface. The collapse was supposed to be halted by a regular circulatory motion in the magnetic field; this leads to a configuration called a magnetoid, which was identified as a quasar. Shock waves and ejected matter were regarded as responsible for most of the radiation. An earlier paper by Kardashev (163) also considered contraction towards the relativistic limit.

Other papers on explosive phenomena are as follows. Kulsrud *et al.* (164) considered the explosion of a supernova into the interstellar field, piling up lines of force before it which, through Rayleigh-Taylor instability, produce strong filamentary fields. Pikelner (165) discussed the formation of arms in a barred spiral after an explosion which compresses the lines of force. Chvojkova (166) suggested that in a planetary nebula there are intense plasma jets from the poles of the central star, which are guided towards the equatorial plane along lines of force; the streams may build magnetic arches or unwinding spirals. Wentzel (167) applied Petschek's mechanism for relinking magnetic field-lines to the formation, acceleration and detachment of interstellar clouds.

GENERAL

Alfvén and Elvius (168) have further developed the suggestion that the universe is composed of 'ambiplasma', consisting of equal amounts of matter and antimatter. These are supposed to be segregated in regions separated by 'Leidenfrost' layers. The present expansion of the universe is ascribed to radiation pressure due to particle annihilation in an initial denser universe. Chiu (169) suggested, however, that the large initial annihilation of matter would lead to a present background radiation-density greater than is observed.

Kaplan and Tsytovich prepared for me a detailed report on recent Soviet work on high-frequency processes in plasma. For the reasons noted in the Introduction to this report, no details (other than bibliographical) of this work can be given. It refers to the generation of plasma waves by streams of fast charged particles (170), turbulence (171) or radiation (172). Parker (173) has also considered possible waves in a gas composed of a superposed cosmic ray plasma and a normal one, and Tidman (174) has discussed the damping of plasma motions through interaction with cosmic rays.

Certain other papers whose interest is mainly mathematical are similarly briefly mentioned. These are on wave propagation (175), stability of a fluid layer (176), a polytropic atmosphere (177) or a fluid cylinder (178), transport processes in a plasma (179), and fire-hose instability (180).

BIBLIOGRAPHY

1. *The Solar Wind* (Mackin, R. J. and Neugebauer, M., editors): Pergamon Press, 1966.
2. *Proc. 9th Int. Conf. Cosmic Rays*: London, The Physical Society, 1965.
3. Severny, A. B. 1964, *Space Sci. Rev.*, **3**, 451.
   Severny, A. B. 1966, *Usp. fiz. Nauk*, **88**, 3.
4. Bumba, V. 1966, *39th Enrico Fermi School of Physics* (*Plasma Astrophysics*) Varenna.
5. Severny, A. B. 1965, *Astr. Zu.*, **42**, 217.
   Severny, A. B. 1966, *Izv. Krym. astrofiz. Obs.*, **35**, 97, and **37** (in press).
6. Stenflo, J. O. 1966, *Ark. Astr.*, **4**, 173.
7. Severny, A. B. 1966, *Astr. Zu.*, **43**, 465.
8. Leighton, R. B. 1964, *Astrophys. J.*, **140**, 1547.
9. Sheeley, N. R. 1964, *Astrophys. J.*, **140**, 731.
   Sheeley, N. R. 1966, *Astrophys. J.*, **144**, 723.

10. Bumba, V., Howard, R. 1965, *Astrophys. J.*, **141**, 1502.
11. Hyder, C. L. 1965, *Astrophys. J.*, **141**, 272.
12. Godoli, G., Mazzuconi, F. 1967, *Astrophys. J.*, **147**, 1131.
13. Bumba, V., Howard, R. 1965, *Astrophys. J.*, **141**, 1492, and **142**, 796.
14. Simon, G. W., Leighton, R. B. 1964, *Astrophys. J.*, **140**, 1120.
15. Weiss, N. O. 1964, *Mon. Not. R. astr. Soc.*, **128**, 225; *Phil. Trans. R. Soc.* Lond. A, **256**, 99.
16. Clark, A. 1964, *Phys. Fluids*, **7**, 1299.
Clark, A. 1965, *Phys. Fluids*, **8**, 644.
17. Chvojková, E. 1965, *Bull. astr. Inst. Csl.*, **16**, 49, 63 and 70.
18. Elliott, H. 1964, *Planet. Space Sci.*, **12**, 657.
19. Ivanchuk, V. I. 1964, *Geomagn. Aeronom.*, **4**, 25.
20. Doherty, L. R., Menzel, D. H. 1965, *Astrophys. J.*, **141**, 251.
21. Jefferies, J. T., Orrall, F. Q. 1965, *Astrophys. J.*, **141**, 505 and 519.
22. Howard, R., Harvey, J. W. 1964, *Astrophys. J.*, **139**, 1328.
23. McLean, D. 1964, *Ann. Astrophys.*, **27**, 195.
24. Hirayama, T. 1966, *Publ. astr. Soc. Japan* (in press).
25. Iospha, B. A., Mogilevsky, E. I., Obridko, W. N. 1964, *Space Res.*, **4**, 789.
26. Kopecký, M., Suda, J. 1966, *Bull. astr. Inst. Csl.*, **17**, 144.
27. Saito, K. 1965, *Publ. astr. Soc. Japan*, **17**, 1.
28. Stoddard, L. G., Carson, D. G. 1966, *Astrophys. J.*, **145**, 796.
29. Harvey, J. W. 1965, *Astrophys. J.*, **141**, 832.
30. Pikelner, S. B. 1965, *Vistas in Astron.*, **6**, 131.
31. Mustel, E. R. 1965, *Astr. Zu.*, **42**, 473.
32. van Heuvel, E. P. J. 1966, *Bull. astr. Inst. Netherl.*, **18**, 161.
33. Mogilevsky, E. I. 1964, *Geomag. Aeronom.*, **4**, 166.
34. Kleczek, J. 1964, *Astrophys. J.*, **140**, 742.
35. Livshitz, M. A., Obridko, V. N., Pikelner, S. B. 1966, *Astr. Zu.*, **43**, 1135.
36. Livshitz, M. A. 1964, *Astr. Zu.*, **41**, 274.
Pikelner, S. B., Livshitz, M. A. 1964, *Astr. Zu.*, **41**, 1006.
37. Lüst, R., Scholer, M. 1966, *Z. Naturf.*, **21a**, 1908.
38. Parker, E. N. 1964, *Astrophys. J.*, **140**, 1170.
39. Meyer, F. 1966, *Mitt. astr. Ges.*, **21**, 90.
40. Schmidt, H. U. 1965, *Mitt. astr. Ges.*, **20**, 89.
41. Oster, L. 1964, *Ann. Astrophys.*, **27**, 14.
42. Chvojková, E. 1965, *Bull. astr. Inst. Csl.*, **16**, 57.
43. Tuominen, J. 1964, *Z. Astrophys.*, **60**, 231.
44. Godoli, G. 1965, *Atti Accad. naz. Lincei Rc.*, **37**, 155 and 197.
45. Kopecký, M. 1964, *Bull. astr. Instr. Inst. Csl.*, **15**, 178.
46. Iroshnikov, R. 1965, *Astr. Zu.*, **42**, 494.
47. Kozhevnikov, N. I. 1965, *Astr. Zu.*, **42**, 78.
48. Romanchuk, P. R. 1965, *Astr. Zu.*, **42**, 775.
49. Moreton, G. E., Gopasyuk, S. I. 1967, *Proc. astr. Soc. Australia*, **1**, 8.
50. Gopasiuk, S. I. 1967, *Izv. Krym. astrofiz. Obs.*, **37**, 29.
51. Schröter, E. H. 1965, *Z. Astrophys.*, **62**, 228 and 256.
52. Livshitz, M. A., Pikelner, S. B. 1964, *Astr. Zu.*, **41**, 464.
53. Bumba, V., Kopecký, M., Kuklin, G. V. 1966, *Bull astr. Inst. Csl.*, **17**, 57.
54. Kopecký, M., Kuklin, G. V. 1966, *Bull astr. Inst. Csl.*, **17**, 57.
55. Deinzer, W. 1965, *Astrophys. J.*, **141**, 548.
56. Gough, D. O., Tayler, R. J. 1966, *Mon. Not. R. astr. Soc.*, **133**, 85.
57. Danielson, R. E. 1964, *Astrophys. J.*, **139**, 45.
58. Smith, S. F., Ramsey, H. E. 1964, *Z. Astrophys.*, **60**, 1.
Ramsey, H. E., Smith, S. F. 1966, *Astr. J.*, **71**, 197.
59. Kleczek, J. 1964, *Astrophys. J.*, **140**, 746.
60. Moreton, G. E., Severny, A. B. 1967, *Solar Phys.*, **2** (in press).
61. Haurwitz, M. W. 1964, *Astrophys. J.*, **140**, 1236.

62. Hunter, J. H. 1966, *Icarus*, **5**, 321.
63. Jacobsen, C., Carlquist, P. 1964, *Icarus*, **3**, 270.
64. Sakurai, K. 1965, *Rep. Ionosph. Space Res. Japan*, **19**, 408.
65. Kawabata, K.-A. 1966, *Rep. Ionosph. Space Res. Japan*, **20**, 107.
66. Wentzel, D. G. 1964, *Astrophys. J.*, **140**, 1563.
67. Syrovatsky, S. I. 1966, *Astr. Zu.*, **43**, 340; *Zu. eksp. teor. Fiz.*, **50**, 1133.
68. Petschek, H. E. 1964, *Symposium on Solar Flares* (NASA, Washington), p. 425.
69. Sweet, P. A., Green, R. M. 1967, *Astrophys. J.* **147**, 1153.
70. Sturrock, P. A., Coppi, B. 1966, *Astrophys. J.*, **143**, 3.
71. Warwick, C. C. 1966, *Astrophys. J.*, **145**, 215.
72. Sakurai, K. 1965, *Publ. astr. Soc. Japan*, **17**, 403.
Sakurai, K. 1966, *Publ. astr. Soc. Japan*, **18**, 77.
73. Staff of Stanford Radar Astronomy Centre. 1966, *J. geophys. Res.*, **71**, 3325.
Serbu, G. P., Maier, E. J. R. 1966, *J. geophys. Res.*, **71**, 3755.
74. Lazarus, A. J., Bridge, H. S., Davis, J. 1966, *J. geophys. Res.*, **71**, 3787.
Neugebauer, M., Snyder, C. W. 1966, *J. geophys. Res.*, **71**, 4469.
75. Wolfe, J. H., Silva, R. W., Myers, M. A. 1966, *J. geophys. Res.*, **71**, 1319.
76. Wolfe, J. H., Silva, R. W., McKibbin, D. D., Mason, R. H. 1966, *J. geophys. Res.*, **71**, 3329.
77. Ness, N. F., Scearce, C. S., Seek, J. B. 1964, *J. geophys. Res.*, **69**, 3534.
78. Greenstadt, E. W. 1966, *Astrophys. J.*, **145**, 270.
79. Ness, N. F., Wilcox, J. M. 1965, *Astrophys. J.*, **143**, 23.
80. Coleman, P. J., Smith, E. J., Davis, L., Jones, D. E. 1966, *Space Res.*, **6**, 907.
81. Wilcox, J. M., Ness, N. F. 1965, *J. geophys. Res.*, **70**, 5793.
82. Coleman, P. J., Davis, L., Smith, E. J., Jones, D. E. 1966, *J. geophys. Res.*, **71**, 2831.
83. Ness, N. F. 1966, *J. geophys. Res.*, **71**, 3319.
84. Stern, D. 1964, *Planet. Space Sci.*, **12**, 961.
85. Parker, E. N. 1964, *Astrophys. J.*, **139**, 72, 93 and 951.
Parker, E. N. 1964, *Planet. Space Sci.*, **12**, 451.
Parker, E. N. 1964, *Space Sci. Rev.*, **4**, 666.
Parker, E. N. 1965, *Astrophys. J.*, **141**, 1463.
Parker, E. N. 1966, *Astrophys. J.*, **143**, 32.
86. Parker, E. N. 1964, *Astrophys. J.*, **139**, 690.
87. Dahlberg, A. 1964, *Astrophys. J.*, **140**, 268.
Mustel, E. R. 1964, *Space Res.*, **4**, 77.
Brandt, J. C., Michie, R. W., Cassinelli, J. P. 1965, *Icarus*, **3**, 270.
Carovillano, R. L., King, J. H. 1965, *Astrophys. J.*, **141**, 526.
Scarf, F. L., Noble, L. M. 1965, *Astrophys. J.*, **141**, 1479.
Whang, Y. C., Chang, C. C. 1965, *J. geophys. Res.*, **70**, 4175.
88. Whang, Y. C., Liu, C. K., Chang, C. C. 1966, *Astrophys. J.*, **145**, 255.
89. Meyer, F., Schmidt, H. U. 1966, *Mitt. astr. Ges.*, **21**, 97.
90. Brandt, J. C., Cassinelli, J. P. 1966, *Icarus*, **5**, 47.
91. Pneuman, G. W. 1966, *Astrophys. J.*, **145**, 242 and 800.
92. King, J. H., Carovillano, R. L. 1966, *Astrophys. J.*, **146**, 372.
93. Brandt, J. C. 1966, *Astrophys. J.*, **144**, 1221.
94. Weber, E. J., Davis, L. 1967, *Astrophys. J.*, **148**, 217.
95. Sturrock, P. A., Spreiter, J. R. 1965, *J. geophys. Res.*, **70**, 5345.
96. Colburn, D. S., Sonett, C. P. 1966, *Space Sci. Rev.*, **5**, 439.
97. Simon, M., Axford, W. I. 1966, *Planet. Space Sci.*, **14**, 901.
98. Ness, N. F., Scearce, C. S., Cantarano, S. 1966, *J. geophys. Res.*, **71**, 3305.
99. Lees, L. 1964, *A.I.A.A. Jour.*, **2**, 1576.
100. Obayashi, T. 1964, *J. geophys. Res.*, **69**, 861.
101. Sozou, C. 1965, *J. geophys. Res.*, **70**, 4165.
102. Spreiter, J. R., Summers, A. L., Alksne, A. A. 1966, *Planet. Space Sci.*, **14**, 223.

103. Bernstein, W., Fredricks, R. W., Scarf, F. L. 1964, *J. geophys. Res.*, **69**, 1201.
Scarf, F. L., Bernstein, W., Fredricks, R. W. 1965, *J. geophys. Res.*, **70**, 9 and 21.
Fredricks, R. W., Scarf, F. L. 1965, *J. geophys. Res.*, **70**, 4765.
Cordey, J. G. 1965, *J. geophys. Res.*, **70**, 1278.
104. Frank, L. A., van Allen, J. A. 1964, *J. geophys. Res.*, **69**, 4923.
Anderson, K. A., Harris, H. K., Paoli, R. J. 1965, *J. geophys. Res.*, **70**, 1039.
Fan, C. Y., Gloeckler, G., Simpson, J. A. 1966, *J. geophys. Res.*, **71**, 1837.
105. Jokipii, J. R. 1966, *Astrophys. J.*, **143**, 961, and **145**, 616.
Chen, C. S., Chang, C. C. 1966, *J. geophys. Res.*, **71**, 241.
106. Anderson, K. A. 1965, *J. geophys. Res.*, **70**, 4741.
107. Jokipii, J. R. 1966, *J. geophys. Res.*, **71**, 3173.
108. Anderson, K. A., Ness, N. F. 1966, *J. geophys. Res.*, **71**, 3705
109. Ness, N. F. 1965, *J. geophys. Res.*, **70**, 2989.
Ness, N. F., Behannon, K. W., Cantarano, S. C., Scearce, C. S. 1966, *Goddard Space Flight Centre, Rep.* X-612-66-529.
Dolginov, S. S., Yeroshenko, E. G., Zhuzgov, L. N., Pushkov, N. F. 1966, *Dokl. Akad. Nauk SSSR*, **170**, 574.
Gringauz, K. I., Bezrukikh, V. V., Khoklov, M. Z., Zastenker, G. N., Remisov, A. P., Musatov, L. S., 1966, *Dokl. Akad. Nauk SSSR*, **170**, 570.
110. Ness, N. F. 1965, *J. geophys. Res.*, **70**, 517.
111. Aronowitz, L., Milford, S. N. 1965, *J. geophys. Res.*, **70**, 227.
112. Hinton, F. L., Taeusch, D. R. 1964, *J. geophys. Res.*, **69**, 1341.
113. *Proc. Caltech –JPL Lunar and Planetary Conf.*, Sept. 1965.
114. Biermann, L., Brosowski, B., Schmidt, H. U. 1967, *Solar Physics*, **1**, 254.
115. Axford, W. I. 1964 *Planet. Space Sci.*, **12**, 719.
116. Marochnik, L. S. 1964, *Astr. Zu.*, **41**, 733.
117. Fan, C. Y., Lamport, J. E., Simpson, J. A., Smith, D. R. 1966, *J. geophys. Res.* **71**, 3289.
118. Bryant, D. A., Cline, T. L., Desai, U. D., McDonald, F. B. 1965, *Astrophys. J.*, **141**, 478.
119. Sakurai, K. 1965, *Planet. Space Sci.*, **13**, 745.
120. Wentzel, D. G. 1964, *Astrophys. J.*, **140**, 1013.
121. Bartley, W. C., Bukata, R. P., McCracken, K. G., Rao, U. R. 1966, *J. geophys. Res.*, **71**, 3297.
122. Fibich, M., Abraham, P. R. 1965, *J. geophys. Res.*, **70**, 2475.
Axford, W. I. 1965, *Planet Space Sci.*, **13**, 1301.
123. Gloeckler, G., Jokipii, J. R. 1966, *Phys. Rev. Lett.*, **17**, 203.
124. Biswas, S., Fichtel, C. E. 1964, *Astrophys. J.*, **139**, 941.
125. Gloeckler, G. 1965, *J. geophys. Res.* **70**, 5333.
Axford, W. I. 1965, *Space Res.*, **4**, 612.
126. Parker, E. N. 1964, *Planet. Space Sci.*, **12**, 735.
Axford, W. I. 1965, *Planet. Space Sci.*, **13**, 115.
127. Parker, E. N. 1965, *Planet. Space Sci.*, **13**, 9.
Parker, E. N. 1966, *Planet. Space Sci.*, **14**, 371.
128. Brandt, J. C. 1964, *Space Res.*, **4**, 828.
129. Hayakawa, S., Nishimura, J., Obayashi, H., Sato, M. 1964, *Progress. theor. Phys., Osaka, Suppl.* no. 30, 86.
130. Wentzel, D. G. 1965, *J. geophys. Res.*, **70**, 2716.
131. Tsytovich, V. N. 1964, *Astr. Zu.*, **40**, 471 and **41**, 7.
Tsytovich, V. N. 1965, *Astr. Zu.*, **42**, 33.
Tsytovich, V. N. 1966, *Usp. fiz. Nauk.*, **89**, 89.
132. Sturrock, P. A. 1966, *Phys. Rev.*, **141**, 186.
133. Finzi, A. 1964, *Astrophys. J.*, **139**, 1398.
134. Burbidge, G. R., Hoyle, F. 1964, *Proc. Phys. Soc. London*, **84**, 141.
135. Ginzburg, V. L., Syrovatsky, S. I. 1967, *IAU Symposium* No. 31, p. 411.
136. Ginzburg, V. L. 1965, *Astr. Zu.*, **42**, 1129.
137. Parker, E. N. 1965, *Astrophys. J.*, **142**, 584.

138. Gould, R. J., Burbidge, G. R. 1965, *Ann. Astrophys.*, **28**, 171.
139. Hayakawa, S., Matsuoka, M., Sugimoto, D. 1966, *Space Sci. Rev.*, **5**, 109.
140. Pollack, J. B., Fazio, G. G. 1965, *Astrophys. J.*, **141**, 730.
Zheleznyakov, V. V. 1965, *Astr. Zu.*, **42**, 96.
Zheleznyakov, V. V. 1966, *Astr. Zu.*, **43**.
Shklovsky, I. S. 1965, *Astr. Zu.*, **42**, 683.
Felten, J. E. 1966, *Astrophys. J.*, **144**, 241 and **145**, 589.
Kaplan, S. A. 1966, *Astr. Zu.*, **43**,
141. Ledoux, P., Renson, P. 1966, *A.. Rev. Astr. Astrophys.*, **4**, 293.
142. Ledoux, P. In *The Magnetic and Related Stars* (Rep. of the 1965 AAS-NASA Symposium: Mono Book Corpn., Baltimore), (in press).
143. Steinitz, R. 1964, *Bull. astr. Inst. Netherl.*, **17**, 504.
Steinitz, R. 1965, *Bull. astr. Inst. Netherl.*, **18**, 125.
144. Renson, P. 1966, *Bull. Soc. r. Sci. Liège*, **35**, 244.
145. Fowler, W. A., Burbidge, E. M., Burbidge, G. R., Hoyle, F. 1965, *Astrophys. J.*, **142**, 423.
146. Mestel, L. 1966, *39th Enrico Fermi School of Physics* (Plasma Astrophysics), Varenna.
147. Roxburgh, I. W. 1966, *Mon. Not. R. astr. Soc.*, **132**, 201.
Roxburgh, I. W., Strittmatter, P. A. 1966, *Mon. Not. R. astr. Soc.*, **133**, 1.
148. Pneuman, G. W., Mitchell, T. P. 1965, *Icarus*, **4**, 494.
149. Steenbeck, M., Krause, F., Rädler, K.-H. 1966, *Z. Naturf.*, **21a**, 369.
Steenbeck, M., Krause, F. 1966, *Z. Naturf.*, **21a**, 1285.
150. Csada, I. K. 1966, *Bull. astr. Inst. Csl.*, **17**, 131.
151. Morris, D., Berge, G. L. 1964, *Astrophys. J.*, **139**, 1388.
Berkhuisen, E. M., Brouw, W. N., Muller, C. A., Tinbergen, J. 1964, *Bull. astr. Inst. Netherl.*, **17**, 465.
Pronik, I. I. 1966, *Astr. Zu.*, **43**, 291.
Mathewson, D. S., Broten, N. W., Cole, D. J. 1966, *Austr. J. Phys.*, **19**, 93.
Gardner, F. F., Davies, D. R. 1966, *Austr. J. Phys.*, **19**, 129.
152. Mestel, L. 1966, *Mon. Not. R. astr. Soc.*, **133**, 265.
153. Strittmatter, P. A. 1966, *Mon. Not. R. astr. Soc.*, **131**, 491 and **132**, 359.
154. Setti, G. 1965, *Bull. astr. Inst. Netherl.*, **18**, 51.
155. Parker, E. N. 1966, *Astrophys. J.*, **145**, 811.
156. Gershberg, R. E. 1965, *Astr. Zu.*, **42**, 330.
157. Piddington, J. H. 1964, *Mon. Not. R. astr. Soc.*, **128**, 345.
Piddington, J. H. 1966, *Mon. Not. R. astr. Soc.*, **133**, 163.
158. Pikelner, S. B. 1965, *Astr. Zu.*, **42**, 515.
159. Layzer, D. 1965, *Astrophys. J.*, **141**, 837.
160. Bardeen, J. M., Anand, S. P. S. 1966, *Astrophys. J.*, **144**, 953.
161. Sturrock, P. A. 1966, *Nature*, **211**, 697.
162. Ginzburg, V. L. 1964, *Dokl. Akad. Nauk SSSR*, **156**, 43.
Ginzburg, V. L. 1965, *Quasi-stellar sources and gravitational collapse* (Chicago), p. 283.
Ginzburg, V. L., Ozernoy, L. M. 1964, *Zu. eksp. teor. Fiz.*, **47**, 1030.
Ginzburg, V. L., Ozernoy, L. M. 1966, *Astrophys, J.*, **144**, 599.
Ozernoy, L. M. 1966, *Astr. Zu.*, **42**, 300. In Trans. Symp. *Variable Stars and Stellar Evolution* (Moscow, Nauka).
163. Kardashev, N. S. 1964, *Astr. Zu.*, **41**, 643.
164. Kulsrud, R. D., Bernstein, I. B., Kruskal, M., Fanucci, J., Ness, N. 1965, *Astrophys. J.*, **142**, 491.
165. Pikelner, S. B. 1965, *Astr. Zu.*, **42**, 3.
166. Chvojková, E. 1944, *Bull. astr. Inst. Csl.*, **15**, 241.
167. Wentzel, D. G. 1966, *Astrophys. J.*, **145**, 595.
168. Alfvén, H. 1965, *Rev. mod. Phys.*, **37**, 652.
Elvius, A., Alfvén, H. 1966, *Proc. IAU Symposium No.* 29.
169. Chiu, H.-Y. 1966, *Phys. Rev. Lett*, **17**, 712.

170. Achiezer, A. I., Achiezer, I. A., Polovin, R. V., Sitenko, A. G., Stepanov, K. N. 1964, *Collective oscillations in plasma* (Atomisdat, Moscow).
Galeev, A. A., Karpman, V. I., Sagdeev, R. Z. 1965, *Nucl. Fusion*, **5**, 20.
Kadomtsev, B. B. 1964, *Voprosy th. Plazmy*, **4**, 188.
Tsytovich, V. N. 1966, *Astr. Zu.*, **43**, 528.
171. Vedenov, A. A. 1965, *Theory of turbulent plasma* (Nauka, Moscow).
Tsytovich, V. N. 1964, *Astr. Zu.*, **41**, 993.
Tsytovich, V. N. 1965, *Astr. Zu.*, **42**, 33.
172. Tsytovich, V. N. 1965, *Zu. eksp. teor. Fiz.*, **35**, 773.
Liperovsky, V. A., Tsytovich, V. N. 1966, *Izv. Vyssih Učeb. Zaved. Radiofizika*, **9**, 469.
173. Parker, E. N. 1965, *Astrophys. J.*, **142**, 1086.
Parker, E. N. 1966, *Astrophys. J.*, **144**, 916.
174. Tidman, D. A. 1966, *Astrophys. J.*, **144**, 615.
175. Uberoi, C. 1964, *Ind. J. pure appl. Math.*, **2**, 133.
Uberoi, C., Devanathan, C. 1964, *Proc. Ind. Acad. Sci.*, **60**, 42.
Frisch, U. 1964, *Ann. Astrophys.*, **27**, 224.
Getling, A. V. 1965, *Astr. Zu.*, **42**, 689.
Tandon, J. N., Bajaj, N. K. 1966, *Mon. Not. R. astr. Soc.*, **132**, 285.
176. Chakraborty, B. B. 1964, *Ind. J. Phys.*, **38**, 490.
Uberoi, C. 1964, *J. Ind. Inst. Sci.*, **46**, no. 1.
Uberoi, C., Raghavachar, M. R. 1964, *J. Ind. Inst. Sci.*, **46**, no. 4.
177. Kaplan, S. A., Petrukhin, N. S. 1965, *Astr. Zu.*, **42**, 74.
178. Tassoul, J. L. 1964, *Ann. Astrophys.*, **27**, 150.
179. Devanathan, C., Uberoi, C., Bhatnagar, P. L. 1965, *J. Ind. Inst. Sci.*, **47**, no. 3.
Devanathan, C., Raghavacnar, M. R., Babu, R. 1966, *J. Ind. Inst. Sci.*, **48**, nos. 2 and 3.
180. Gliddon, J. E. 1966, *Astrophys. J.*, **145**, 583.
Lerche, I. 1966, *Astrophys. J.*, **145**, 811.
181. Bezrukikh, V. V., Gringauz, K. I., Musatov, L. S., Rybchinsky, R. E., Kokhlov, M. Z. 1965, *Dokl. Akad. Nauk SSSR*, **163**, 873.
Gringauz, K. I., Bezrukikh, V. V., Musatov, L. S., 1966, *Belgrade Solar-Terrestrial Physics Symposium*.
182. Kalinin, Y. D., Mogilevsky, E. I. 1965, *Investigations of Outer Space*, (Nauka, Moscow).
183. Bumba, V., Howard, R. 1966, *Astrophys. J.*, **143**, 592.
184. Smith, T. S. 1964, *Astrophys. J.*, **139**, 767.

T. G. COWLING
President of the Commission

44. COMMISSION DES OBSERVATIONS ASTRONOMIQUES AU-DEHORS DE L'ATMOSPHERE TERRESTRE

PRÉSIDENT: Professor L. Goldberg, Harvard College Observatory, Cambridge, Massachusetts 02138, U.S.A.

VICE-PRÉSIDENT: Mrs. Dr A. G. Massevitch, Astronomical Council, U.S.S.R. Academy of Sciences, Moscow V-312, U.S.S.R.

COMITÉ D'ORGANISATION: A. Dollfus, R. Lüst, F. G. Smith, P. Swings, H. C. Van de Hulst.

MEMBRES: Blamont, Boneff, Bosman-Crespin, Boyd, Butler, Chamberlain, Chubb, Code, Coutrez, Danielson, Deutsch (A. J.), Fazio, Friedman, Ginzburg, Gold, Gratton, Hack, Haddock, Hartz, Heddle, Herzberg, Hinteregger, Houziaux, Ivanov-Kholodny, Kraushaar, Kuiper, Kupperian, Lines, Link, Lovell, McGee (J. D.), Mandel'Shtam, Mehltretter, Meinel, Migeotte, Mikhailov, Monfils, Müller (E. A.), Newkirk, Nicolet, Nicolsky, Öhman, Parkinson, Pecker (J.-C.), Pounds, Priester, Reeves (E.M.), Righini, Rogerson, Roman, Rosen, Schwarzschild, Smith (Henry J.), Tousey, Walsh, Whipple, Wilson (A. G.), Wilson (R.), Zhongolovich, Zwicky.

Introduction

Many important advances in the techniques of space astronomy have occurred since the last General Assembly. The most widely publicized developments have been those connected with the exploration of the Moon and Mars, where the exploits of such spacecraft as Luna, Zond, Surveyor, Lunar Orbiter and Mariner in revealing the details of the lunar and Martian landscapes have seemed little short of miraculous. The photographs obtained during these flights have already been extensively disseminated and the results of more detailed analysis are best left to Commissions 16 and 17. Although the lunar and planetary developments have been of the highest scientific importance, the space program has also had a number of other successes of equal scientific significance, although characterized by a much more modest scale of engineering achievement. Many types of satellites and space probes have provided detailed measurements of the interplanetary plasma and its associated magnetic fields. The discovery of a sizable number of extra-solar X-ray sources has already established a new branch of astronomy, which has attracted numerous experimenters and captured the attention of many theoretical physicists. Far ultraviolet stellar spectra have been photographed for the first time, both from sounding rockets and with cameras operated by astronauts in the Gemini spacecraft. Solar X-ray astronomy has also advanced very rapidly. X-ray photographs of the Sun are approaching in resolution those made in visible light. Solar X-ray spectra now reveal emission lines down to a wavelength of 6 Å and seem to defy all limitations imposed by 'reasonable' coronal temperatures.

Unlike lunar and planetary astronomy, which has profited greatly from advances in large-rocket propulsion, solar and stellar astronomy continue to derive most of their benefits from small rockets and satellites. The usefulness of sounding rockets will be even greater when triaxial pointing controls, pioneered in the U.K. with the Skylark rocket, become more generally available. On the whole, the large orbiting observatories have not yet produced the results expected of them both because of technical difficulties and also a certain amount of bad luck.

In recognition of the expanding volume of results and publication in space astronomy, this year's Draft Report has been sub-divided into three separate reports prepared by L. Goldberg, L. Houziaux and R. Lüst, and dealing respectively with solar astronomy, galactic astronomy and solar corpuscular radiation. I am grateful to Drs Houziaux and Lüst for their valuable contributions.

Appendices to this Report, dealing with Soviet investigations, are added at the end.

L. GOLDBERG
President of the Commission

SOLAR ASTRONOMY

(by Leo Goldberg)

1. *Introduction*

Anyone wishing to become acquainted with the literature of space astronomy will find many useful review articles and volumes at his disposal. Most of the papers presented at IAU Symposium No. 23, held at Liège in August 1964 have been published in the *Annales d'Astrophysique* and collected into a single volume edited by J. Steinberg (**1**). The proceedings of the COSPAR symposia held in Florence, Italy in 1964, in Buenos Aires, Argentina in 1965 and in Vienna, Austria in 1966 should also be consulted. A good review of the problems of solar physics and of the relevance of space techniques to their solution is contained in the Proceedings of the Summer Conference held in June 1965 at Woods Hole, Massachusetts (**2**). Other parts of the same volume contain similar reports published in behalf of optical stellar astronomy, radio astronomy, and X-ray astronomy. The end product of each report is a series of recommendations for the long range development of large facilities to be deployed in space by 1980. Several reviews deal with the XUV solar spectrum and its interpretation (**3–6**) while others are more concerned with techniques (**7–10**).

Every effort has been made, in the preparation of this report, to include all results of solar observations from space vehicles that were published or otherwise brought to my attention by preprints or correspondence prior to 1 December 1966. The Bibliography will be found to be more complete in sections dealing with techniques and results than in those concerned with interpretation, which is more fully treated in the reports of other Commissions.

2. *Spacecraft, Instrumentation, Future Experimemts*

2.1 *NASA projects*

Plans for future experiments in space astronomy in the U.S.A. and in Europe were described in detail at the meeting of Commission 44 during the 12th General Assembly and published in Volume XIIB of the *Transactions of the IAU*. The following up-to-date revision of the NASA program has been supplied through the courtesy of Drs H. J. Smith and N. G. Roman. Satellite vehicles now being employed for the purposes of astronomy include the Orbiting Astronomical Observatory (OAO), Orbiting Solar Observatory (OSO), Orbiting Geophysical Observatory (OGO), Radio Astronomy Explorer (RAE), Apollo, Solar Explorer and Apollo Telescope Mount (ATM). The Advanced Orbiting Solar Observatory (AOSO) program was cancelled in 1965 and many of the experiments transferred to other spacecraft, such as the ATM, a manned solar observatory designed to provide angular resolution of better than five seconds of arc. Radio Astronomy Explorer is an orbiting radio telescope composed of two 750 foot (23 m) V antennas intended both for mapping galactic noise between one and seven MHz and for measuring radio bursts from the Sun and Jupiter in the range 300 kHz to 7 MHz. The first ATM mission will carry five instruments as follows:

1. White light coronagraph	High Altitude Observatory
2. Ultraviolet coronal spectrographs	Naval Research Laboratory
3. Spectrographic X-ray telescope	American Science and Engineering Co.
4. Ultraviolet scanning spectrometers	Harvard College Observatory
5. High resolution X-ray telescope	Goddard Space Flight Center

Four launches of OSO spacecraft, designated D, E I, F and G, have been approved, with experiments as shown in the following table:

OSO–D

Pointed section	
Spectroheliometer, 3–75 Å	Giacconi and Paolini, American Science and Engineering Co.
Crystal Spectrometer 0–8 Å.	Chubb, Friedman, Kreplin, Meekins, Naval Research Laboratory.
Spectrometer-Spectroheliometer 300 Å–1300 Å.	Goldberg, Reeves, Parkinson, Harvard College Observatory.
Wheel section	
Solar X-ray Photon Counter, 1–70 Å.	Boyd, University College, London; Pounds, University of Leicester.
Location Soft X-ray Sources.	Gursky, Giacconi, American Science and Engineering Co.
Lyman-α in Earth's Corona.	Mange, Chubb, Friedman, Naval Research Laboratory.
Solar X-ray Monitoring.	Kreplin, Naval Research Laboratory.
High-energy Protons and Electrons.	Waggoner, Univ. of California.
Solar Monitoring He II 304 Å.	Boyd, Stewardson, University College London.

OSO–E I

Pointed section	
Solar Spectroscopic Monitoring, 250 Å–1300 Å.	Hinteregger, Hall, Air Force Cambridge Research Laboratory.
Solar Spectroscopic Monitoring 1–400 Å.	Neupert, White, Goddard Space Flight Center.
Wheel section	
Earth Albedo 3200 Å–7800 Å.	Ames Research Center.
X- and Gamma-rays from Cosmic and Solar Sources, 7–190 KeV.	Peterson, Univ. of California.
Arrival Direction Cosmic Gamma Rays > 100 MeV.	Kraushaar, Clark, Garmire, Mass. Inst. of Technology.
Low Energy Solar X-rays, 8–14 Å.	Teske, Univ. of Michigan.
Cosmic and Solar Charged Particles, Gamma Rays.	Kaplon, Deney, Denrik, Univ. of Rochester.

OSO–F

Pointed section	
Location Regions on Sun of Soft X-rays 3–18 Å.	Boyd, Willmore, University College, London; Pounds, University of Leicester.
XUV Spectroheliograph, 284–1216 Å.	Purcell, Detwiler, Tousey, Naval Research Laboratory.
Solar Spectroscopic Monitoring 1–400 Å.	Neupert, White, Goddard Space Flight Center.

Wheel section

Profile Solar and Geocoronal Lyman-α.	Blamont, Centre National Recherche Scientifique.
Solar X-ray Monitoring 0·5–60 Å.	Chubb, Kreplin, Friedman, Naval Research Laboratory.
Low Energy Solar and Cosmic Gamma Rays.	Frost, Horstman, Roth, Goddard Space Flight Center.
Polarization Zodiacal Light.	Ney, Univ. of Minnesota.
Solar XUV Monitoring, 280–1030 Å.	Rense, Parker, Univ. of Colorado.

OSO–G

Pointed section

Offset Pointing Spectrometer-Spectroheliometer, 300–1300 Å.	Goldberg, Reeves, Parkinson, Huber, Harvard College Observatory.
Location and Spectra Solar X-rays 1–60 Å.	Kreplin, Chubb, Meekins, Friedman, Naval Research Laboratory.

Wheel section

Solar XUV Monitoring in Lines of He, O and N.	Boyd, Woodgate, University College London.
Solar X-ray Line Monitoring, 16–40 Å.	Argo, Bergy, Evans, Henke, Los Alamos Scientific Laboratory.
Brightness, Polarization Zodiacal Light.	Carroll, Aller, Rutgers University.
Solar X-rays and Gamma-rays, 2–200 KeV.	Brini, University of Bologna.
High Energy Neutrons 20–130 MeV.	Leavitt, University of New Mexico.

2.2 *ESRO projects*

Dr B. Bolin has communicated an account of ESRO spacecraft and experiments, to be launched during the next few years. Those concerned with solar astronomy are as follows:

ESRO I: polar orbit of 90° inclination; apogee 1500 km; perigee 275 km; total weight 78 kg; scheduled for launching September-November 1967. Experiments include the measurement of auroral particles and solar protons in five experiments by three different groups, namely, (1) Kiruna Geophysical Observatory, Dr. Hultqvist; (2) Science Research Council, Slough, Dr Dalziel; (3) Ionosphere Laboratory, Lyngby. Mr Peterson is coordinator of all five experiments.

ESRO II: near polar orbit of 98° inclination; spin stabilized; apogee 1100 km; perigee 350 km; scientific payload 20 kg; scheduled for launching, March 1967; intended for studies in the field of cosmic rays and solar astronomy, of cosmic rays in interplanetary space, modulation mechanism of primary cosmic radiation; electron component, solar corpuscles and electromagnetic radiation and trapped protons of the inner Van Allen belt. Solar experiments include (1) Study of the variations of X-rays below 20 Å from the quiet Sun, with five proportional counters, by Boyd, University College, London and Stewardson, University of Leicester; (2) Study of the variations of X-rays in the region 44–70 Å, with two proportional counters, by de Jager, Utrecht Observatory; (3) Measurement of flux and energy spectrum of solar protons (35 MeV–1 GeV) with telescope comprising solid state detectors, by Labeyrie, Centre d'Etudes Nucléaires de Saclay.

TD–1: the first ESRO medium size stabilized (Sun pointing) satellite; near polar circular orbit of 98° inclination; altitude 560 km; scientific payload 80 kg; scheduled for launching in early 1970. Two solar experiments are (1) Monitoring of solar X-rays, 40–300 KeV with scintillation counter, by de Jager, Utrecht Observatory and (2) X-ray spectrometry 3–30 KeV, with proportional counter, by Labeyrie, Saclay. A second satellite in the series, TD–2, is a solar maximum satellite, to be launched

TD–1: by an improved Thor Delta during the latter part of 1969 into an orbit of as yet undetermined inclination, with apogee 1200 km and perigee 350 km. The satellite will carry a number of solar experiments: (1) High resolution spectrometry of solar X-ray emission with a crystal spectrometer, by Stewardson, University of Leicester; (2) Monitoring of the profile of the solar Lyman–α line, with a hydrogen or deuterium cell and photomultipliers, by Blamont, C.N.R.S.; (3) Monitoring of solar particles (13–160 MeV protons) with a solid-state detector telescope, by de Jager, Utrecht Observatory; (4) Monitoring of solar particles (0·6–28 MeV protons) with a scintillator employing solid state detectors, by Lüst, Max Planck Institute, Munich; (5) Extreme ultraviolet solar spectrography, with a spectroheliograph, by Boyd, University College, London.

HEOS–A: inclination of orbit 33°; apogee 240 000 km; perigee 200 km; spin stabilized, scheduled for launching in July 1968, will carry eight experiments for the measurement of particles and fields in the interplanetary medium.

2.3 *Other proposed experiments*

2.3.1 *Utrecht Observatory.* According to C. de Jager, the new Space Research Laboratory of Utrecht Observatory participated or is participating in the following projects: (1) Rocket observations of soft solar X-rays (44–60 Å), in the French national program, and with ESRO (launchings in 1964, 1965, 1966). Precise measurements of the soft X-ray flux were obtained during the years of solar minimum activity. (2) A satellite experiment in the same wavelength band has been prepared for launching in the ESRO II satellite in April 1967. (3) Balloon launchings for the detection of hard solar X-ray bursts between 20 and 600 KeV take place regularly. (4) An extreme ultraviolet and X-ray solar heliograph will be flown early in 1967 in a sun-pointing NASA rocket at White Sands. The heliograph is based on the Fresnel-type zone plate principle and is designed to obtain solar images in Si IX 51 Å; Fe VIII–X 180 Å; He II 304 Å and He I 584 Å. (5) A high-resolution solar spectrophotometer for the region 44–64 Å is in preparation for flight in an ESRO sun-pointing rocket in November 1966.

2.3.2 *University of Tokyo.* Dr Z. Suemoto reports that an institute of space and aeronautical science was established in 1964 in the University of Tokyo to promote the further development of space and aeronautical science in Japan. It is expected that the facilities of the new institute will be utilized by scientists in relevant fields of research throughout Japan. The new institute has a launching site at Uchinoura Kagoshima in Kyushu, which began operation late in 1963. Typical sounding rockets now in use in Japan are the K–9M, which carries a payload of 60 kg to a maximum height of 350 km and the L–3 with a payload of 110 kg and a maximum height of 1100 km. The so-called MU rockets, which are capable of launching satellites, are under development and satellite projects are being seriously considered. To date only some preliminary experiments on the height variation of the solar Lyman–α flux have been performed (11).

2.3.3 *Stockholm Observatory.* Dr Y. Öhman reports that he is designing a spectrograph for recording the spectrum of various parts of the solar surface from an unstabilized rocket. The instrument uses a large number of small lenses placed on a spherical surface. The focus of all these lenses is situated in the center of curvature. Here a very small (about 0·5 mm) reflecting sphere is used, functioning as a slit for a spectrograph. By using a continuously moving film, tracings are obtained when the various lenses happen to project the solar image on the small sphere. The instrument is mainly intended for the study of the Mg II lines at 2800 Å in solar flares. Dr K. Fredga is developing an improved instrument for recording solar images in the Mg II line at 2802·7 Å for flight in the near future. J. O. Stenflo has started the design of a spectrograph, which he hopes to use in space research for studying solar magnetic fields in small structure elements of the solar surface.

2.3.4 *Meudon Observatory.* Dr A. Dollfus reports as follows on his plans to photograph the solar corona from high altitude balloons:

'En ce qui concerne plus particulièrement les activités de mon laboratoire, je prépare une série de vols en ballons, depuis le Centre de Lancement de Aire-sur-Adour (Landes) pour la photographie des grands jets de la couronne solaire. L'instrument est un petit coronographe précédé d'une poutrelle portant un cache à 4 mètres en avant de l'objectif et destiné à porter ombre sur celui-ci, afin de réduire la lumière diffusée à un facteur 10^{-8}; le contour de ce cache est formé de 360 dents de 0·45 mm de profondeur afin de diminuer la brillance de la lumière réfractée par ce contour. La nacelle stabilisée et le dispositif de pointage automatique ont été construits par la Société Française 'Compagnie des Compteurs'. Le ballon de 50 000 m^3 en feuille plastique de polyéthylène de 40μ d'épaisseur atteint l'altitude de 30 000 mètres. Le premier vol est prévu pour le début de l'année 1967.'

2.3.5 *University College, London.* At the University College, London, R. L. F. Boyd and his collaborators are planning to make accurate absolute measurements of the flux of solar Lyman α at intervals in the solar cycle. An experiment to study the center to limb variation in Lyman α was to be flown during the annular solar eclipse of 15 May 1966. The University College group in collaboration with Leicester are preparing satellite-borne instruments containing arrays of proportional counters to study the wavelength bands 1–20Å and 44–70Å on board OSO–D and the band 1–20Å on ESRO II.

2.3.6 *Imperial College.* Dr R. Speer, Imperial College, U.K., has called attention to a unique opportunity to observe XUV radiation at the solar limb with very high spatial resolution during the solar eclipse of March 1970, which passes over the Wallops Island Launching Facility of NASA. He comments that:

'The azimuth, geographic coordinates and speed of the umbra permit trailing stabilised rocket trajectories capable of yielding chromospheric limb scan rates in the range 0·2 to 0·4 seconds of arc per second of time. It is proposed that, at minimum, three stabilised Aerobee 350 rockets be launched, carrying modified STRAP control systems. These launchings will occur at equal intervals during a three-minute period centred around maximum eclipse. It is suggested at this stage that a carefully chosen selection of emissions be made from ions of H I, He I, He II and the Li-like iso-electronic series, together with a portion of the continuum at 1400Å–1600Å associated with the chromospheric/photospheric temperature reversal. A measurement of the Balmer α line of helium II at 1640Å, if combined with observations of the other helium II lines outside of eclipse would provide the first direct measurement of helium abundance in the chromosphere.'

2.3.7 *Geneva Observatory.* Professor M. Golay is preparing to launch a rocket equipped to carry out photoelectric photometry of the Sun in several regions of the UV spectrum.

3. *Solar Ultraviolet, X-rays and Gamma Rays*

Most observations of the Sun's XUV radiation during the past two to three years have been made in the short wavelength region below 500Å. Important gains have been made in spectroscopic resolution, in the mapping of the ultraviolet solar spectrum above the limb, in the spatial resolution of ultraviolet and X-ray images and in X-ray spectroscopy. Most of the results have been obtained by the use of sounding rockets, but small satellites have played an important role in the monitoring of X-rays.

3.1 *Solar Spectra*

The high resolution échelle spectrograph first flown in 1961 by the Naval Research Laboratory was reflown on 19 November 1964 and the solar spectrum extended from 2200Å to 2100Å (3). No measurements from either of these spectra have been reported but there are plans to publish a list of lines with identifications and an atlas showing intensity profiles. With the aid of a new tri-axial stabilization system developed for the Skylark rocket, the Culham Laboratory flew a spectrograph on 9 April 1965 with the entrance slit placed about ten arc sec outside the solar limb (**12, 13**). Photographs of the ultraviolet spectrum between 950 Å and 2950 Å show about 300 emission lines, including intersystem transitions in C III, N IV, O V and new forbidden transitions in Fe XI and Fe XII (**14**). New measurements of absolute fluxes in the solar XUV spectrum were made by photoelectric scanning from 310 Å to 55 Å (**15, 16, 17**) and by photography in the spectral intervals 250 Å to 370 Å and 33 Å to 70 Å (**18**). Absolute fluxes in the spectral region from 1750 Å to 1 Å have been tabulated for relatively quiet solar conditions (**19**).

The first recordings of solar X-ray spectra with a Bragg crystal spectrometer were carried out in 1963 to a short-wavelength limit of 13 Å (**20**). In the U.S.S.R., photographs and spectra of the Sun in short-wave ultraviolet and X-radiation were obtained with apparatus on board two geophysical rockets, which attained heights of nearly 500 km on 20 September and 1 October 1965 (**21**). During the 1 October flight, photographs and spectra of the Sun were obtained for the first time during the beginning stages of an X-ray flare. During the flare, the intensity of the Fe XVII line at $\lambda = 15{\cdot}0$ Å increased by two orders of magnitude as compared with its intensity in the absence of solar flares. A review of the variability of the solar X-ray spectrum below 15 Å has been published (**22**), based on U.K. observations with sounding rockets and the Ariel satellite during the period 1959–63. H. Friedman reports the observation by NRL, on 4 October 1966, of solar X-ray emission lines identified with Na XI, Mg XI, Al XII, Si XIII and Si XIV in the spectral region 6 Å to 8 Å.

3.2 *Ultraviolet images*

The first monochromatic photographs of the Sun in the Mg II line at 2802·7 Å were obtained with a Cassegrain-Maksutov telescope and a Solč-type birefringent filter on board an Aerobee rocket on 12 April and 2 December 1965 (**23**). A photoionization detector behind a pinhole placed at the focus of a Cassegrain telescope of two seconds of arc resolution recorded the fine structure of the Sun in Lyman α during an Aerobee flight on 20 October 1965 (**24**). Rapid progress is being made in the improvement of techniques for solar imaging at extreme ultraviolet and soft X-ray wavelengths. The Culham Laboratory in the U.K. has developed a compact extreme ultraviolet spectroheliograph in which a pinhole camera is combined with a plane diffraction grating used at grazing incidence (**13**). The instrument was flown in a stabilized Skylark rocket on 9 April 1965 and recorded monochromatic images at 304 Å (He II), 171 Å (Fe IX), and strongly limb brightened emission in the wavelength band between 60 Å and 150 Å. High quality monochromatic images have recently been obtained by the Naval Research Laboratory with a normal incidence concave grating spectroheliograph in such lines as Fe XV 284 Å, Fe XVI 335 Å 361 Å, He II 304 Å, and many others, the most recent flight having taken place on 28 April 1966 (**25**). The emission in the strongest lines was observed to extend five arc minutes or more beyond the limb. Accordingly, a photographic extreme ultraviolet heliograph, consisting of an off-axis paraboloidal mirror, with filters transmitting a band from 171 Å to about 400 Å, was flown on 27 July 1966 and revealed emission above active regions extending at least to thirty arc minutes above the limb (**26**).

3.3 *X-ray images*

The distribution of X-ray emitting sources over the solar disk has been studied in four different ways: (1) from the widths of X-ray emission lines observed with a Bragg crystal spectrometer (**20**), (2) by means of slit scans of the solar disk in broad X-ray wavelength bands (**20**), (3) by pinhole photography (**27–30**) and (4) by photography with a grazing incidence imaging telescope (**31, 32**). The newest observations (**32**) suggest that X-ray images may soon be recorded with spatial resolution of one second of arc or better.

3.4 *Monitoring of Solar X-ray Emission*

The Solrad satellites of the Naval Research Laboratory have been monitoring the total flux from the Sun in the three wavelength ranges 2–8 Å, 8–20 Å and 44–60 Å. Results have been reported from three such satellites, 1963–21–C (**33**) 1964–01–D (**34**, **35**, **36**), and 1965–16–D (**37**). Observations made by the Injun I satellite during the second half of 1961 at wavelengths shorter than 14 Å have also been reported (**38**). The first orbiting solar observatory monitored solar X-rays at 2–8 Å and at 20–100 keV (**39**). The 2–8 Å X-ray flux was found to be highly variable, comprising a slowly varying component that correlated well with plage activity and additional variations in time periods ranging from one second to several hours (**40**). A number of high energy bursts accompanying solar flares were observed in the 20–100 keV range (**41**). Observations of solar X-rays in wavelength bands peaked at about 2 Å and about 4 Å were made from a Vela satellite during the period 18 October to 1 November 1963 (**42**). An important set of measurements of the spectral energy distribution of solar flare X-rays was carried out with a low-resolution proportional counter spectrometer for the region 4–14 Å on board the U.K. satellite Ariel I between 26 April and 1 November 1962 (**43**, **44**). Soft X-radiation from the Sun in two spectral intervals, 2–10 Å and 8–18 Å, were carred out during flights of the cosmic stations Electron II and Electron IV in the U.S.S.R. (**45**). Measurements with the Electron II station extended from 30 January to 16 March 1964. The X-ray flux was well correlated with the area of active spots, with the flux of radio radiation at 10·7 cm wavelength and with the flux of radio radiation from discrete sources at 6·6 cm wavelength. Many X-ray flares were observed both with and without the accompaniment of chromospheric flares. Two cases were observed in which X-ray flares were associated with an increase in the flux of heavy nuclei with $Z \geqslant 15$ (**46**). No indication of a statistically significant flux of solar gamma rays in the energy range up to 10 MeV could be detected from balloon measurements of the quiet Sun (**47**, **48**).

4. *Visible and Infrared Radiation*

Observations of visible and infrared solar radiation and of the zodiacal light have been carried out from aircraft, balloons, rockets and satellites. The solar spectrum from 140–1000 cm^{-1} was recorded with a Michelson interference spectrometer flown in a balloon on 10 August 1965 (**49**). Jet aircraft were employed at the eclipse of 20 July 1963 for the observation both of the coronal spectrum λ3700–λ4900 (**50**) and of the solar corona at large distances from the Sun (**51**). Coordinated observations of the corona from the solar limb to the zodiacal light were also carried out during the same eclipse from the ground, from a balloon at 100 000 feet (30 km) altitude and from a jet aircraft (**52**). White light coronagraphs were also flown in stratospheric balloons on 5 March 1964 (**53**), 3 June 1965 and 1 July 1965 (**54**), and from Aerobee rockets on 28 June 1963 (**55**) and in 1966 (**56**). The balloon observations in 1965 were closely coordinated with ground observations including those during the total solar eclipse of 30 May 1965. During this eclipse, profiles of the emission line λ5303 were obtained with a Fabry-Perot interferometer (**57**) from an aircraft instrumented by the Los Alamos Scientific Laboratory. On 14 June 1966, the outer solar corona was photographed with the Surveyor I television camera on the surface of the Moon after the solar disk had set behind the western horizon (**58**). The corona was traced out to an estimated 3 to 4 solar radii from the center of the disk and a very prominent coronal streamer, corroborated by ground-based measurements, was also apparent.

A Japanese K–9M rocket was flown on 26 July 1965 with equipment to measure the zodiacal light photoelectrically at three wavelengths, 4300 Å, 5300 Å and 6000 Å (**59**). The measured region was approximately 40° to 15° from the Sun, which covers the well-known gap between the observation of the zodiacal light and of the corona. The zodiacal light was also observed at 5000 Å and 4200 Å from a balloon flown by H. Tanabe and M. Huruhata of the Tokyo

Observatory at an altitude of 25 km on 27 September 1966. Photoelectric measurements of the zodiacal light were obtained from apparatus in the wheel section of OSO II, beginning in February 1965 (**60**).

5. *Laboratory and Theoretical Spectra, Identifications*

The very large number of unidentified lines in the ultraviolet solar spectrum continues to challenge astrophysicists and laboratory spectroscopists. It has been suggested that many of the unidentified solar emission lines below $\lambda 1500$ may be attributed to the attachment of quarks to ions of C, N and O (**61, 62**). Considerable progress has been made in the identification of lines in the region 25–250 Å (**63–70**). All of the strong lines in the solar spectrum between 167 Å and 220 Å have now been classified and identified by means of a number of ingenious laboratory techniques, (**71–74**). At the Culham Laboratory the wave numbers of lines are traced back along an isoelectronic sequence to well-classified ions. Another way of differentiating between atomic spectra and high stages of ionization is to photograph the spectra separately with two sources of considerably different excitation (**75–78**). Nearly all of the strong lines from 167 Å to 188 Å in the solar spectrum arise from the ions Fe VIII–Fe XII. Intense lines between 188 Å and 220 Å are shown to be due to unclassified lines of Fe XII–Fe XIV. The majority of the emission lines in the solar spectrum in the range 60 Å–170 Å are due to Fe and Ni, mostly the former. Other solar emission lines in the region from 310 Å to 55 Å have been identified on the basis of term differences in the tables of atomic energy levels (**79**).

The resonance lines of Fe XIV have not been observed in the laboratory and therefore their solar identification presents a special problem. One aid to identification is the requirement that two lines having a common upper state and terminating in the ground term must be separated by an amount corresponding to the wave number of the forbidden line 5303 Å (**3**). The intensity variations of extreme ultraviolet lines observed from the OSO–1 satellite also provide added clues to the identifications (**80–82**). It has been shown that the very low electron density in the solar corona relative to laboratory sources causes anomalies in the relative intensities within certain multiplets, notably the P–D multiplet of Fe XIV at 210–220 Å (**73**).

Shock-excited absorption spectra of solar-abundant atoms and molecules are being produced in the Shock Tube Laboratory of the Harvard College Observatory to assist in the identification of the solar spectrum between 3000 Å and 1500 Å. The identification of the fourth positive system of CO in the solar spectrum by this technique (**83**) has now been confirmed in detail at the Naval Research Laboratory (**84**).

6. *Problems of Interpretation*

6.1 *Opacity near* 1600 Å

The recognition that bound-free absorption by Si I and Mg I (**85**) and band absorption by the fourth positive system of CO (**83**) are the major sources of opacity in the spectral region 1550 Å–1800 Å has provided a fresh basis for the interpretation of solar radiation from the region of the temperature minimum at the interface photosphere-chromosphere. New discussion of possible models of the region of the temperature minimum has been carried out in conjunction with careful laboratory measurements of the relevant cross-sections (**86**). Improved measurements of energy distribution and limb darkening are required for a definitive model. Various methods of correcting for instrumental broadening of the solar ultraviolet limb darkening profile have been investigated (**87**). Theoretical profiles for the solar autoionizing lines of Al I at $\lambda 1930$ show that at the limb the profiles are extremely sensitive to the choice of model and may possibly exhibit emission reversals (**88**). The profiles of Lyman α scattered by free electrons in the solar corona have been calculated for a number of points above the solar limb (**89**).

6.2 *Emission line intensities*

A new method for computing the total intensities of emission lines formed out of thermodynamic equilibrium (**90**) has been applied to compute the intensities of the solar ultraviolet lines and continua of H, He I and He II (**91, 92**). Another investigation deals with the formation of the Lyman continuum in an isothermal atmosphere (**93**). Both electron temperature and density may be inferred from the observed total intensities of Lyman α and β (**94**). Extreme ultraviolet line intensities inferred theoretically from radio emission measures are at least ten times stronger than those observed (**95**). The discrepancy can be eliminated by invoking an inhomogeneous model. The method of analysis originally developed by Pottasch for the derivation of chemical abundances in the solar corona from far ultraviolet line intensities has been rediscussed in two separate investigations in which the ratios Fe to Si (**96**) and Fe to H (**97**) are found to be approximately ten times greater than in the photosphere. Other investigators obtain relative abundances of heavy elements (**98**) and abundances relative to H (**99**) which are consistent with those found in the photosphere. The abundances of C, Mg, S and Al relative to Si also seem to be consistent with photospheric abundances according to line intensities observed in the spectral regions 40–62 Å and 256–356 Å (**100**).

6.3 *X-ray emission*

The interpretation of the X-ray emission from the quiet Sun and from flares has received a great deal of attention. Solar X-radiation has been classified as quasi-thermal (emitted from quiet Sun, active regions and flares) or nonthermal (emitted from solar flares only) with a further sub-division for flare associated bursts (**101**). Detailed calculations (**102, 103**) support the assumption of a thermal origin of X-ray radiation from the quiet Sun and from active regions. Above 15 Å, the main contribution to the spectrum is from lines of C, N and O. At wavelengths shorter than 15 Å, the chief contribution is from recombination of highly ionized species of the same and other elements (**102, 104**). In the absence of solar flares, the radiative flux consists of a quasi-constant component generated in the undisturbed region of the corona on which is superposed a slowly varying component generated in the hotter, denser, active regions of the corona. It can be shown that when the emission measure and the electron temperature of active regions in the corona are derived from radio spectroheliograms of the Sun at a wavelength of 10·7 cm, the calculated and experimental values of the X-ray flux are in good agreement (**103, 105**). A number of detailed theoretical investigations concerned with possible mechanisms for the origin of X-ray flares have been carried out in the U.S.S.R. Compton scattering of thermal photons by relativistic electrons has been examined (**106–108**) as well as the thermal synchrotron radiation of accelerated electrons (**109**). The inverse Compton effect has been criticized on the grounds that any quantity of relativistic electrons producing a measureable amount of inverse Compton photons will always produce a greater flux of high energy Bremsstrahlung photons than has been observed (**110**). It has been suggested that electrons with energy in the range 10–30 keV, which are produced during solar flares, may ionize atoms through K-shell ionization and that the resulting X-rays would contribute to the total X-ray emission of flares (**111**). The expectations with respect to gamma-ray emission have been calculated for a simple model of a solar flare (**112**). The detection of such radiation from a flare seems feasible.

6.4 *Ionization equilibrium*

The importance of dielectronic recombination as a major factor in the theory of ionization equilibrium of the solar corona is now firmly established (**113**). A convenient general formula has been provided for the calculation of the relevant rates which may surpass the corresponding radiative rates by 1 to 3 orders of magnitude (**114**). The collisional excitation of autoionizing levels and subsequent ionization may in individual cases increase the rate of ionization by about a factor of 2 (**115**).

BIBLIOGRAPHY

1. Steinberg, J. L., Ed. IAU Symposium No. 23, *Astronomical Observations from Space Vehicles*. Reprinted from *Ann. Astrophys.*, 1964, **27**, no. 6 and 1965, **28**, nos. 1, 2, 3, 4.
2. *Space Research, Directions for the Future*. Part 2, *Solar Astronomy*. National Academy of Sciences – National Research Council, Wash., D.C., Jan. 1966.
3. Tousey, R., Austin, W. E., Purcell, J. D., Widing, K. G. 1965, The extreme ultraviolet emission from the Sun between the Lyman-alpha lines of H I and C VI, *Ann. Astrophys.*, **28**, 755 (also in 1964, IAU Symposium no. 23, Liège).
4. Allen, C. W. 1965, The interpretation of the XUV solar spectrum, *Space Sci. Rev.*, **4**, 91.
5. Pottasch, S. R. 1964, On the interpretation of the solar ultraviolet emission line spectrum, *Space Sci. Rev.*, **3**, 816.
6. Mandel'Shtam, S. L. 1965, X-ray emission of the Sun, *Space Sci. Rev.*, **4**, 587.
7. Rense, W. A. 1966, Techniques for rocket solar UV and for UV spectroscopy, *Space Sci. Rev.*, **5**, 234.
8. Boyd, R. L. F. 1965, Techniques for the measurement of extra-terrestrial soft X-radiation, *Space Sci. Rev.*, **4**, 35.
9. Smith, R. A. 1965, Detectors for ultraviolet, visible, and infrared radiation, *Appl. Opt.*, **4**, 631.
10. Hinteregger, H. E. 1965, Absolute intensity measurements in the extreme ultraviolet spectrum of solar radiation, *Space Sci. Rev.*, **4**, 461.
11. Higashino, I., Masuoka, T., Hisada, H., Sasanuma, M., Oshio, T. 1966, An observation of solar XUV radiation by ionization chambers, *Mem. of the Faculty of Engineering*, Osaka City Univ., **8**.
12. Black, W. S., Booker, D., Burton, W. M., Jones, B. B., Shenton, D. B., Wilson, R. 1965, Solar spectroscopy in the extreme ultraviolet using stabilized Skylark rockets, *Nature*, **206**, 654.
13. Burton, W. M., Wilson, R. 1965, Observations of the Sun in the extreme ultraviolet from a stabilized Skylark rocket, *Nature*, **207**, 61.
14. Burton, W. M., Ridgely, A., Wilson, R. 1966, The ultraviolet emission spectrum of the solar chromosphere and corona. Preprint CLM-P109, Culham Laboratory. Submitted to *Mon. Not. R. astr. Soc.*
15. Hinteregger, H. E., Hall, L. A., Schweizer, W. 1964, Solar XUV-spectrum from 310Å–55Å, *Astrophys. J.*, **140**, 319.
16. Hall, L. A., Schweizer, W., Heroux, L., Hinteregger, H. E. 1965, Solar XUV spectrum of March, 1964, *Astrophys. J.*, **142**, 13.
17. Hall, L. A., Schweizer, W., Hinteregger, H. E. 1965, Long-term variation of solar extreme ultraviolet fluxes, *J. geophys. Res.*, **70**, 2241.
18. Austin, W. E., Purcell, J. D., Tousey, R., Widing, K. G. 1966, Coronal emission line intensities in the extreme ultraviolet, *Astrophys. J.*, **145**, 373.
19. Hinteregger, H. E., Hall, L. A.. Schmidtke, G. 1965, Solar XUV radiation and neutral particle distribution in July 1963 thermosphere, *Space Res.*, **5**, 1175.
20. Blake, R. L., Chubb, T. A., Friedman, H., Unzicker, A. E. 1965, Spectral and photometric measurements of solar X-ray emission below 60Å, *Astrophys. J.*, **142**, 1.
21. Zhitnik, I. A., Krutov, V. V., Malyavkin, L. P., Mandel'Shtam, S. L., Cheremukhin, G. S. (in press).
22. Culhane, J. L., Willmore, A. P., Pounds, K. A., Sanford, P. W. 1964, Variability of the solar X-ray spectrum below 15Å, *Space Res.*, **4**, 741.
23. Fredga, K. 1966, Monochromatic pictures of the Sun in the Mg II line at 2802.7 Å, *Astrophys. J.*, **144**, 854.
24. Sloan, W. A. 1966, Rocket measurements of the solar surface in the light of Lyman alpha, 1216Å, *Astr. J.*, **71**, 399.
25. Purcell, J. D., Tousey, R., Widing, K. G. Some new extreme ultraviolet spectroheliograms, paper presented at AAS Special Meeting on Solar Astronomy, Boulder, Colorado, 3–5 Oct. 1966.

26. Purcell, J. D., Tousey, R. XUV heliograms, paper presented at AAS Special Meeting on Solar Astronomy, Boulder, Colorado, 3–5 Oct. 1966.
27. Zhitnik, I. A., Krutov, V. V., Malyaukin, L. P., Mandel'Shtam, S. L. 1964, Solar images in the far ultraviolet part of the spectrum, *Cosmic Res.*, **2**, 801, translated from *Kosm. Issled.*, **2**, 920, 1964.
28. Russell, P. C. 1965, Soft X-ray image of the Sun, *Nature*, **205**, 684.
29. Russell, P. C. 1965, Further soft X-ray images of the Sun, *Nature*, **208**, 281.
30. Russell, P. C., Pounds, K. A. 1966, Improved resolution X-ray photographs of the Sun, *Nature*, **209**, 490.
31. Giacconi, R., Reidy, W. P., Zehnpfennig, T., Lindsay, J. C., Muney, W. S. 1965, Solar X-ray images obtained using grazing incidence optics, *Astrophys. J.*, **142**, 1274.
32. Underwood, J. H., Muney, W. S. X-ray photographs of the solar corona, paper presented at AAS Special Meeting on Solar Astronomy, Boulder, Colorado, 3–5 Oct. 1966.
33. Thomas, L. 1964, Observations of solar X-ray emission in the band 44–60Å from U.S. Naval Research Laboratory satellite 1963–21–C, *Nature*, **203**, 962.
34. Kreplin, R. W. 1965, NRL solar radiation monitoring satellite: description of instrumentation and preliminary results, *Space Res.*, **5**, 951.
35. Thomas, L., Venables, F. H., Williams, K. M. 1965, Measurements of solar X-ray fluxes by the U.S. Naval Research Laboratory satellite 1964–01–D, *Planet. Space Sci.*, **13**, 807.
36. Landini, M., Piattelli, M., Righini, G., Russo, D., Tagliaferri, G. L. 1964, Some preliminary results from the solar monitoring satellite NRL 1964 1 D obtained at the Arcetri station, *Ann. Astrophys.*, **27**, 765 (also in 1964, IAU Symposium No. 23, Liège).
37. Friedman, H. Monitoring of X-rays by Solrad-8 (1965–16–D), Report of Seventh COSPAR Space Science Symposium, Vienna, Austria 1966.
38. Van Allen, J. A., Frank, L. A., Maehlum, B., Acton, L. W. 1965, Solar X-ray observations by Injun 1, *J. geophys. Res.*, **70**, 1639.
39. Lindsay, J. C. 1965, Comments on X-rays emitted by the Sun, *Ann. Astrophys.*, **28**, 586 (also in 1964, IAU Symposium No. 23, Liège).
40. White, W. A. 1964, *Space Res.*, **4**, 771.
41. Frost, K. J. 1965, Comments on high energy X-ray bursts observed by OSO–1, *Space Res.*, **5**, 513.
42. Conner, J. P., Evans, W. D., Montgomery, M. D., Singer, S., Stogsdill, E. E. 1965, Solar flare X-ray emission measurements, *Space Res.*, **5**, 546. Singer, S., Conner, J. P., Evans, W. D., Montgomery, M. D., Stogsdill, E. E. 1965, and Plasma observations at 10^5 km, *Space Res.*, **5**, 546.
43. Bowen, P. J., Norman, K., Pounds, K. A., Sanford, P. W., Willmore, A. P. 1964, Measurements of the solar spectrum in the wavelength band 4 to 14Å, *Proc. R. Soc. London Ser. A*, **281**, 538.
44. Pounds, K. A. 1965, Recent solar X-ray studies in the United Kingdom, *Ann. Astrophys.*, **28**, 132 (also in 1964, IAU Symposium No. 23, Liège).
45. Tindo, I. P. 1965, *Investigation of Cosmic Space*, Nauka, Moscow.
46. Kurnosova, L. V., Mandel'Shtam, S. L., Razorenov, L. A., Tindo, I. P., Fradkin, M. I. 1966, *Kosm. Issled.*, **4**, 170.
47. Frost, K. J., Rothe, E. D., Peterson, L. E. 1966, A search for the quiet-time solar gamma rays from balloon altitudes, *J. geophys. Res.*, **71**, 4079.
48. Peterson, L. E., Schwartz, D. A., Pelling, R. M., McKenzie, D. 1966, The upper limit solar gamma-ray spectrum to 10 MeV, *J. geophys. Res.*, **71**, 5778.
49. Beer, R. 1966. Decrement of the solar continuum in the far infra-red, *Nature*, **209**, 1226.
50. Deutsch, A. J., Righini, G. 1964, An airborne observation of the coronal spectrum at the eclipse of 20 July, 1963, *Astrophys. J.*, **140**, 313.
51. Blackwell, D. E., Petford, A. D. 1966, Observations of the 1963 July 20 solar eclipse. I. Spectroscopic separation of the F and K components of the solar corona at large distances from the Sun, *Mon. Not. R. astr. Soc.*, **131**, 383.
52. Gillett, F. C., Stein, W. A., Ney, E. P. 1964, Observations of the solar corona from the limb of the Sun to the zodiacal light, July 20, 1963, *Astrophys. J.*, **140**, 292.

53. Newkirk, G., Jr., Bohlin, J. D. 1965, Coronascope II: Observation of the white light corona from a stratospheric balloon, *Ann. Astrophys.*, **28,** 234 (also in 1964, IAU Symposium No. 23, Liège).
54. Bohlin, J. D., Hansen, R. T., Newkirk, G. A. The three dimensional structure of a coronal streamer and the inner solar wind, paper presented at AAS Special Meeting on Solar Astronomy, Boulder, Colorado, 1–3 Oct. 1966.
55. Tousey, R. 1965, Observations of the white light corona by rocket, *Ann. Astrophys.*, **28,** 600 (also in 1964, IAU Symposium No. 23, Liège).
56. Koomen, M. J., Purcell, J. D., Tousey, R. The white light corona from $R = 4$ to 11·5 R. Paper presented at AAS Special Meeting on Solar Astronomy, Boulder, Colorado, 1–3 Oct. 1966.
57. Liebenberg, D. H. Coronal emission line profiles of λ5303 obtained at the 30 May 1965 total solar eclipse, paper presented at 123rd meeting of the Amer. Astr. Soc., Los Angeles, 27–30 Dec. 1966.
58. Gunn, J. E., Norton, R. H. Surveyor I observations of the solar corona, paper presented at AAS Special Meeting on Solar Astronomy, Boulder, Colorado, 1–3 Oct. 1966.
59. Tanabe, H., Huruhata, M. In preparation.
60. Ney, E. P., Huch, W. F. 1966, Observations of the sky, Earth, and airglow from OSOB, *Astr. J.*, **71,** 393 (also in AAS Meeting, Hampton, Virginia, March 1966).
61. Sinanoğlu, O., Skutnik, B., Tousey, R. 1966, Search for quarks in the far ultraviolet solar spectrum, *Phys. Rev. Lett.*, **17,** 785.
62. Bennett, W. R., Jr. 1966, Detection of quarks in the ultraviolet solar spectrum. *Phys. Rev. Lett.*, **17,** 1196.
63. Alexander, E., Feldman, U., Fraenkel, B. S. 1964, A method of differentiating between atomic spectra of high degrees of ionization, *J. quantit. Spectrosc. radiat. Transfer*, **4,** 501,
64. Feldman, U., Fraenkel, B. S. 1966, Unclassified $3p^6\ 3d$–$3p^5\ 3d\ 4S$ transition lines of Fe VIII, Mn VII, and Cr VI, *Astrophys. J.*, **145,** 959.
65. Fawcett, B. C., Gabriel, A. H., Irons, F. E., Peacock, N. J., Saunders, P. A. H. 1966, Extreme ultra-violet spectra from laser-produced plasmas, *Proc. Phys. Soc.*, **88,** 1051.
66. Deutschman, W. A., House, L. L. 1966, New inner-shell resonance lines in highly ionized sulfur and chlorine, *Astrophys. J.*, **144,** 435.
67. House, L. L., Deutschman, W. A., Sawyer, G. A. 1964, Comparison of solar and laboratory iron spectra in the vacuum ultraviolet, *Astrophys. J.*, **140,** 814.
68. Elton, R. C., Kolb, A. C., Austin, W. E., Tousey, R., Widing, K. 1964, Origin of certain solar emission lines between 170 and 220Å, *Astrophys. J.*, **140,** 390.
69. House, L. L. 1964, The laboratory production of solar iron lines in the vacuum ultra-violet, *Ann. Astrophys.*, **27,** 763 (also in 1964, IAU Symposium No. 23, Liège).
70. Mandel'Shtam, S. L., Fedoseyev, S. P., Kononov, E. Ya., Lebedév, S. V. 1965, Laboratory reproduction of the short-wavelength portion of the solar spectrum, *Optics Spectrosc.*, **18,** 522.
71. Gabriel, A. H., Fawcett, B. C., Jordan, C. 1965, Classification of iron lines in the spectrum of the Sun and Zeta in the range 167Å to 220Å, *Nature*, **206,** 390.
72. Fawcett, B. C., Gabriel, A. H. 1965, New spectra of the iron transition elements of astrophysical interest, *Astrophys. J.*, **141,** 343.
73. Gabriel, A. H., Fawcett, B. C., Jordan, C. 1966, Classification of iron VIII to XII and XIV lines in the solar XUV spectrum and their isoelectronic sequences from argon to nickel, *Proc. phys. Soc.*, **87,** 825.
74. Fawcett, B. C., Gabriel, A. H. 1966, Spectra from $3p^n - 3p^{n-1}\ 3d$ transitions of iron period elements in Cl I and S I isoelectronic sequences, *Proc. phys. Soc.*, **88,** 262.
75. Feldman, U., Fraenkel, B. S., Hoory, S. 1965, Identifications of solar ultraviolet lines resulting from a study of the Ar I and K I isoelectronic sequences, *Astrophys. J.*, **142,** 719.
76. Alexander, E., Feldman, U., Fraenkel, B. S., Hoory, S. 1965, $3p^6 - 3p^5\ 3d$ transitions of Fe IX and Ni XI in the solar spectrum, *Nature*, **206,** 175.
77. Cowan, R. D., Peacock, N. J. 1965, Identification of some intense iron lines in the solar spectrum near 170Å, *Astrophys. J.*, **142,** 390.

78. Gabriel, A. H., Fawcett, B. C. 1965, Identification of the solar spectrum in the region 60–170 Å, *Nature*, **208**, 808.
79. Zirin, H. 1964, Identification of extreme ultraviolet solar emission lines, *Astrophys. J.*, **140**, 1332.
80. Neupert, W. M. 1965, Intensity variations in the solar extreme ultraviolet spectrum observed by OSO–1, *Ann. Astrophys.*, **28**, 446 (also in 1964, IAU Symposium No. 23, Liège).
81. Neupert, W. M., Smith, E. V. P. 1964, Possible identification of permitted Fe XIV lines in the extreme ultraviolet solar spectrum, *Astr. J.*, **69**, 554 (also in 1964, paper presented at 116th meeting AAS, Flagstaff).
82. Stockhausen, R. 1965, The UV lines of Fe XIV in the solar corona, *Astrophys. J.*, **141**, 277.
83. Goldberg, L., Parkinson, W. H., Reeves, E. M. 1965, Carbon monoxide in the ultraviolet solar spectrum, *Astrophys. J.*, **141**, 1293.
84. Porter, J. R., Tilford, S. G., Widing, K. G. Carbon monoxide in the solar ultraviolet spectrum (in press).
85. Rich, J. C., Gingerich, O. 1966, Metallic continuous absorption coefficients in the solar ultraviolet, *Astr. J.*, **71**, 161.
86. Rich, J. C. Silicon and carbon monoxide absorption in the solar ultraviolet spectrum, Ph.D. Thesis, Harvard University, Nov. 1966.
87. Lacis, A. A., Matsushima, S. 1965, Method of correction for instrumental broadening to determine the solar ultraviolet limb-darkening, *Astr. J.*, **70**, 142 (also in 1964, paper presented at 117th meeting AAS, Montreal).
88. Withbroe, G. L. 1966, Theoretical profiles for the solar auto-ionization lines of Al I, *Astrophys. J.*, **146**, 294.
89. Hughes, C. J. 1965, Profiles of Lyman-alpha scattered by free electrons in the solar corona, *Astrophys. J.*, **142**, 321.
90. Athay, R. G. 1965, Theoretical line intensities: I. Strong enission lines; II. Excitation of chromospheric He II and hydrogen, *Astrophys. J.*, **142**, 724.
91. Athay, R. G. 1965, Theoretical line intensities: III. Solar UV lines and continua of H, He I, and He II and the chromospheric model, *Astrophys. J.*, **142**, 755.
92. Athay, R. G. 1966, Theoretical line intensities: IV. Source functions and equivalent widths, *Astrophys. J.*, **144**, 1159.
93. Dietz, R. D., House, L. L. 1965, The formation of the Lyman continuum: isothermal atmosphere, *Astrophys. J.*, **141**, 1393.
94. Hearn, A. G. 1966, An interpretation of the total intensities of the Lyman α and β lines of hydrogen emitted by the Sun, U.K. Atomic Energy Authority, Research Group, Culham.
95. Suemoto, Z., Moriyama, F. 1964, Spotty appearance of the solar disk as inferred from the comparison between EUV and radio intensities, *Ann. Astrophys.*, **27**, 775 (also in 1964, IAU Symposium No. 23, Liège).
96. Jordan, C. 1966, The relative abundance of silicon, iron, and nickel in the solar corona, *Mon. Not. R. astr. Soc.*, **132**, 463.
97. Jordan, C. 1966, The abundance of iron in the solar corona, *Mon. Not. R. astr. Soc.*, **132**, 515.
98. Athay, R. G. 1966, Theoretical line intensities: V. Solar UV emission lines of heavy elements, *Astrophys. J.*, **145**, 784.
99. Dupree, A. K., Goldberg, L. 1967, Solar abundance determination from ultraviolet emission lines, *Solar Physics*, **1**, 229.
100. Widing, K. G., Porter, J. R. 1965, Silicon in the solar corona, *Ann. Astrophys.*, **28**, 779 (also in 1964, IAU Symposium No. 23, Liège).
101. de Jager, C. 1965, Solar X-radiation, *Ann. Astrophys.*, **28**, 125 (also in 1964, IAU Symposium No. 23, Liège).
102. Mandel'Shtam, S. L. 1965, On X-ray radiation of the quiet Sun, *Ann. Astrophys.*, **28**, 614 (also in 1964, IAU Symposium No. 23, Liège).
103. Mandel'Shtam, S. L. 1965, *Space Sci. Rev.*, **4**, 587.

104. Fetisov, E. P. 1964, Solar coronal radiation shortward of 10 Å, *Soviet Astr.*, **8**, 231, translated from *Astr. Zu.*, **41**, 299, 1964.

105. Mandel'Shtam, S. L., Prokudina, V. S., Tindo, I. P., Fetisov, E. P. 1965, X-rays from the quiet Sun, *Cosmic Res.*, **3**, 601, translated from the Russian *Kosm. Issled.*, **3**, 737, 1965.

106. Zheleznyakov, V. V. 1965, The mechanism of γ-ray emission of solar flares, *Astr. Zu.*, **42**, 96.

107. Shklovsky, I. S. 1964, The inverse Compton effect as a possible cause of the X-ray radiation of solar flares, *Nature*, **202**, 276.

108. Shklovsky, I. S. 1965, Nature of solar X-ray emission, *Soviet Astr.*, **8**, 538, translated from *Astr. Zu.*, **41**, 676, 1964.

109. Korchat, A. A. 1965, The solar flares of 28 September 1961, and 20 March 1958, *Cosmic Res.*, **3**, 751, 1965.

110. Acton, L. W. 1964, Inverse Compton effect as a possible cause of the X-ray radiation of solar flates, *Nature*, **204**, 64.

111. Acton, L. W. 1965, Contribution of characteristic X-rays to the radiation of solar flares, *Nature*, **207**, 737.

112. Fazio, G. G., Dolan, J. F. 1965, The gamma-ray spectrum of the Sun, *Rev. Geophys.*, **3**, 319.

113. Burgess, A. 1965, Dielectronic recombination in the corona, *Ann. Astrophys.*, **28**, 774 (also in 1964, IAU Symposium No. 23, Liège).

114. Burgess, A. Dielectronic recombination, 1965 Proc. Second Harvard-Smithsonian Conference on Stellar Atmospheres, p. 47.

115. Goldberg, L., Dupree, A. K., Allen, J. W. 1965, Collisional excitation of autoionizing levels, *Ann. Astrophys*, **28**, 589.

ASTRONOMIE GALACTIQUE

(préparé par L. Houziaux)

I. *Véhicules Spatiaux, Instruments et Projets*

I.1. *Observations astronomiques sur orbite (OAO, U.S.A.)*

Aucun exemplaire de cette série de plates-formes hautement stabilisées n'a pu encore être mis en oeuvre avec succès. Le premier OAO emportait à son bord un ensemble de télescopes destinés à la photométrie des étoiles et nébuleuses entre 1000 et 4200 Å. Le 'Célescope' de la Smithsonian Institution avait été remplacé par:

(*a*) un détecteur de rayons gamma de haute énergie (préparé par le M.I.T.),

(*b*) un détecteur de rayons X, de la Lockheed Missiles and Space C°,

(*c*) un détecteur de photons ayant des énergies comprises entre 2000 et 18 000 eV, mis au point par le Goddard Space Flight Center. Malheureusement, quelques heures après la mise sur orbite, le 8 avril 1966, une température excessive a endommagé les batteries du véhicule et la mission d'observation n'a pu être exécutée.

Une nouvelle tentative de lancement doit avoir lieu dans le courant de 1968. La charge utile sera composée des deux expériences prévues primitivement pour l'OAO I (Célescope du Smithsonian Astrophysical Observatory et photométrie ultraviolette des étoiles de l'Université de Wisconsin).

Deux autres véhicules OAO sont actuellement prévus. Le premier emportera un télescope Cassegrain de 36 pouces (91 cm) destiné à la spectrométrie à haute résolution d'étoiles, de nébuleuses et de galaxies.

Cette expérience est préparée par le Goddard Space Flight Center. L'Observatoire de l'Université de Princeton (L. Spitzer et J. B. Rogerson) (1) se propose de placer sur orbite, en 1970, un télescope de 40 pouces (102 cm) équipé d'un spectrographe d'un pouvoir résolvent de 10^4 et permettant la spectrophotométrie stellaire dans le domaine de 1000Å à 3000Å.

La Nasa étudie actuellement la possibilité et l'intérêt d'effectuer des observations astronomiques à partir de satellites habités. Ce programme débuterait par une version de l' 'Apollo Telescope Mount' (Programme d'Application Apollo) permettant l'observation des étoiles. En particulier, l'Observatoire de Princeton projette la réalisation d'un télescope dont le pouvoir de résolution serait limité par la diffraction, et stabilisé à 0"01. Il pourrait être utilisé soit par un observateur humain, soit de façon atomique.

I.2. *Projets d'expériences par satellites dans le cadre de l'Organisation Européenne de Recherches Spatiales* (2)

L'Organisation Européenne de Recherches Spatiales se propose de mettre sur orbite, d'ici à 1970, cinq satellites. Les deux premiers (dénommés ESRO I et ESRO II) doivent être lancés par des fusées américaines 'Scout' sur des orbites polaires. Les trois autres satellites (dénommés HEOSA, ESTER et ESSOR) seront placés sur leur orbite par des fusées Thor-Delta améliorées.

En ce qui concerne l'astronomie galactique, nous noterons les expériences suivantes:

ESRO II: (apogée 1100 km, périgée 350 km, poids: 73 kg, date prévue pour le lancement: avril 1967).

'Mesure du flux et de la distribution d'énergie des électrons produits par le rayonnement cosmique primaire dans le domaine d'énergie des GeV'. Cette expérience est préparée par le Département de Physique de l'Université de Leeds (P. L. Marsden).

ESTER: (orbite circulaire quasi-polaire à 560 km, stabilisation par pointage vers le Soleil et par gyroscopes, poids 400 kg, date prévue pour le lancement: fin 1970).

(*a*) 'Spectrométrie de charge des particules cosmiques primaires'. Centre d'Etudes Nucléaires de Saclay, France (J. Labeyrie).

(*b*) 'Spectrométrie des photons X d'énergie comprise entre 3 et 30 keV provenant de l'espace'. Centre d'Etudes Nucléaires de Saclay, France (J. Labeyrie).

(*c*) 'Spectrographie stellaire dans l'ultraviolet', Observatoire d'Utrecht (Pays-Bas) (C. de Jager).

(*d*) 'Levé du Ciel dans l'ultraviolet entre 1250 et 3000 Å et dans l'infrarouge entre 1 et 3 microns'. Observatoire Royal d'Edimbourg et Institut d'Astrophysique de l'Université de Liège.

L'expérience (*c*) doit permettre l'obtention des spectres d'environ 200 étoiles avec une résolution de 1 Å, dans trois régions spectrales de 100 Å de large, situées entre 2000 et 3000 Å. Quant à l'expérience (*d*) (3) son but essentiel est de permettre de dresser une carte du Ciel dans quatre bandes de longueurs d'onde (de 100 Å ou 200 Å de large) situées entre 1250 Å et 3000 Å, ainsi que de procéder au recensement des sources célestes émettant une partie appréciable de leur rayonnement dans l'infrarouge. Le télescope de 25 cm de diamètre qui pointe dans une direction opposée à la Terre, est fixe par rapport au satellite et le balayage du Ciel est obtenu par le mouvement même du satellite, dont l'orbite doit précesser de 1° par jour.

L'Organisation Européenne prévoit également le lancement vers 1972 d'un grand satellite astronomique (800 kg) destiné à la spectrographie à haute résolution des étoiles entre 950 et 3000 Å.

I.3. *Autres projets*

Plusieurs séries de fusées sondes doivent être lancées par l'Organisation Européenne de Recherches Spatiales en 1967 et 1968. Parmi les expériences intéressant l'astronomie galactique, nous noterons:

(*a*) trois expériences de photométrie et spectrophotométrie à larges bandes stellaires dans l'ultraviolet, placées sur des fusées 'Skylark' non stabilisées (Observatoire Royal d'Edimbourg, J. W. Campbell et G. C. Sudbury).

(*b*) deux expériences de photographie du Ciel dans l'UV à l'aide d'un télescope de Schmidt, à placer sur des fusées stabilisées (Observatoire Royal d'Edimbourg, H. E. Butler).

(*c*) une expérience de spectrographie stellaire dans l'ultraviolet dans la région de 1200 à 3000 Å, avec une résolution de 1 Å, nécessitant l'utilisation d'une fusée Skylark pointant avec une précision de 5′ d'arc (U.K. A.E.E., Laboratoire de Culham, R. Wilson). D'autres expériences de spectroscopie stellaire dans l'ultraviolet sont aussi projetées par le Max Planck Institut für Extraterrestrische Physik, München, et l'Institut d'Astrophysique de l'Université de Liège.

A Joddrell Bank, un groupe de chercheurs prépare pour le satellite UK–3, une version améliorée de l'instrument destiné à un levé du ciel sur 1 ou 2 MHz, qui a déjà volé à bord d' 'Ariel 2'. Cette nouvelle expérience a pour but de mesurer l'émission radio d'origine galactique à des fréquences trop basses pour être observées du sol. Le pouvoir de résolution doit être suffisant pour arriver à séparer du flux isotrope les composantes provenant du plan et du halo galactiques.

Aux Etats-Unis plusieurs groupes étudient la possibilité de mettre sur orbite une antenne de grandes dimensions, pour étudier le ciel dans la bande de 1 MHz. Cette antenne pourrait atteindre un diamètre de plusieurs milliers de pieds de diamètre. Le groupe de radio astronomie de l'Université de Michigan a mis à l'étude la réalisation d'un 'kilometer-wave orbiting telescope' (KWOT). Il s'agirait d'un treillis de 10 km de diamètre supporté par quatre satellites auxiliaires. Ce treillis serait attaché en son centre à un observatoire principal. Le système tournerait sur lui-même avec une période de deux heures et l'axe de rotation pourrait etre incliné au moyen de fusées auxiliaires, de sorte que le ciel tout entier pourrait être étudié. La résolution angulaire à 1 MHz serait suffisante pour mesurer plusieurs douzaines de radiosources galactiques et extragalactiques. On pourrait aussi établir la carte du bruit radio d'origine cosmique à différentes fréquences entre 0·1 et 10 MHz.

II. *Résultats*

II.1. *Rayonnement ultraviolet d'origine galactique*

A. Observations. Une fusée Skylark, lancée de la base de Woomera, le 14 juillet 1965, a permis au groupe de recherches spatiales de l' 'University College' de Londres d'obtenir les flux émis par des étoiles chaudes dans trois régions spectrales entre 1450 et 2800 Å. Les trois photomètres étaient placés au foyer d'un télescope de 35 cm aligné selon l'axe du véhicule. L'attitude de la fusée était déduite de signaux fournis par un senseur lunaire, des senseurs d'horizon et des magnétomètres. La brillance du ciel entre 1650 et 3000 Å a été mesurée à Woomera le 25 mars 1966 par le groupe de l'Observatoire Royal d'Edimbourg.

Le groupe de l'Observatoire de Kitt Peak a lancé du champ de tir de White Sands (N.M.) un spectromètre stellaire. L'instrument était destiné à obtenir des spectres à moyenne résolution d'étoiles de la constellation d'Orion dans le domaine spectral 1050 à 2000 Å. Le spectromètre photoélectrique était attaché à un télescope de 30 cm de diamètre.

Au début de juin 1965, une équipe de l'Observatoire de Princeton, dirigée par D. C. Morton, a obtenu les premiers spectres photographiques d'étoiles chaudes montrant des raies d'absorption dans l'ultraviolet. L'instrument, lancé de la base de White Sands (N.M.) se composait de deux réseaux-objectifs montés sur une plate-forme stabilisée. Le système de pointage, fourni par la Space General C8 a permis de garder le spectrographe dirigé sur l'étoile avec une précision de $\pm$ 16″, ce qui correspond à une résolution de 1 Å.

Le 2 juin 1965, les spectres de δ et π Scorpii ont été obtenus aux longueurs d'onde supérieures à 1260 Å, à l'aide d'une chambre photographique de Schmidt, ouverte à *f*/2, pourvue d'une lame correctrice en fluorure de calcium (7). 29 raies d'absorption ont pu être mesurées dans ces deux étoiles B de la séquence principale. Les raies de Si IV à 1394 et 1403Å, ainsi que la raie de C IV à 1549Å ont été identifiées comme étant d'origine stellaire. Par contre, les raies de O I à 1302Å, C II à 1335Å, Si II à 1527Å et Al II à 1672Å sont probablement dues à l'absorption par le milieu interstellaire.

Un spectrographe similaire lancé le 13 octobre 1965 et pourvu d'une lame correctrice en fluorure de lithium, a permis de photographier les spectres de δ, ϵ, ζ, η, ι et κ Orionis à des longueurs d'onde supérieures à 1200Å (**8**). Les images d'ordre zéro d'autres étoiles du champ ont servi à calibrer les spectres en longueurs d'onde, avec une précision avoisinant l'angström. Des raies d'émission observées à 1403 Å et à 1549 Å dans le spectre de ϵ et ζ Orionis ont été identifiées comme appartenant respectivement aux ions Si IV et C IV. Des raies d'absorption intenses des deux composantes du doublet de Si IV et du doublet non résolu de C IV observés dans ϵ, ζ et ι Orionis sont fortement déplacées vers le violet et indiquent des vitesses de 1800, 1900 et 3800 km s^{-1} respectivement. Le doublet de Si IV dans δ Orionis révèle une vitesse d'approche de 1400 km s^{-1}. Ces raies de résonance, qui apparaissent en absorption, doivent trouver leur origine dans des couches gazeuses s'échappant des étoiles à grande vitesse, indiquant par là une perte de masse pour ces géantes et supergéantes. Une raie d'absorption, large de 9Å, présente à 1216Å dans δ et ζ Ori, est attribuée par Morton à $L\alpha$ interstellaire.

Un nouveau vol a eu lieu le 24 mai 1966. Le spectrographe, ouvert à $f/2$, contenait cette fois uniquement des éléments d'optique par réflexion et a permis l'obtention du spectre de ζ Ophiuchi. Une erreur de pointage dans le système de contrôle d'attitude a déplacé la région spectrale et l'a amenée entre 2250 et 3670Å, où aucune raie n'était suffisamment intense pour être mesurée avec quelque précision. Lors du vol du 20 septembre 1966, de nouveaux spectres ont été photographiés à l'aide d'un instrument similaire aux précédents, mais où le réseau et les deux miroirs avaient été traités au fluorure de lithium, afin de donner un pouvoir réflecteur maximum aux courtes longueurs d'onde. Morton a obtenu ainsi les spectres de γ, δ, ϵ, ζ, η, ι et σ Orionis à partir de 1100Å. En plus des raies observées précédemment, on peut y découvrir de nombreuses raies nouvelles en absorption (notamment C III à 1175Å), grâce à l'amélioration de la résolution et l'élargissement du domaine spectral exploré.

Une nacelle de ballon partiellement stabilisée a été lancée à diverses reprises par le groupe de recherches de l'Observatoire de Genève (**11**), sous la direction de M. Golay. Ces lancements, effectués à partir de la base française d'Aire-sur-Adour, à l'aide de ballons de 50 000 m^3 ont montré qu'il était possible d'effectuer avec cette technique des observations de photométrie ultraviolette jusqu'à 2204Å. Un vol effectué en automne 1966 a permis l'obtention de spectres à faible dispersion grâce au petit télescope de Schmidt, auquel était adjoint un prisme-objectif. K. Hallam du Goddard Space Flight Center a préparé un télescope pour la spectroscopie ultraviolette par satellite. L'instrument a été mis sur orbite au début de février 1965, dans la partie inférieure du satellite OSO B2; les résultats sont actuellement à l'étude.

Le laboratoire de Physique Appliquée de la Johns Hopkins University a obtenu, de son côté, des résultats intéressants (**6**) à partir d'un télescope monté sur satellite et équipé d'un photomètre à bande passante de 204Å centrée sur 1376Å. Un indice de couleur ultraviolet ($m_{1376} - V$) a pu être calculé pour 96 étoiles. Ces données ont permis de déduire une échelle de température pour les étoiles chaudes, échelle qui indique une diminution de 2000 à 3000° par rapport aux températures actuellement admises. L'extinction interstellaire, à 1376Å, tirée des mêmes observations se monte à $10{\cdot}7 \pm 6$ magnitudes, dans une échelle ou $A_V/E(B - V) = 3{\cdot}1$ et $E(B - V) = 1$.

Lors du vol de l'engin Gemini 10, le 19 juillet 1966, les astronautes Young et Collins ont photographié le spectre ultraviolet de la partie sud de la Voie Lactée. Cette observation a été faite dans le cadre du programme du Centre de Recherche Astronomique Lindheimer, Northwestern University, Ill. (K. G. Henize et L. R. Wackerling) (**9**). La caméra Maurer, de

73 mm de distance focale était munie d'un réseau par réflexion et d'une lentille transmettant l'ultraviolet jusqu'à 2200 Å. Les spectres ont été pris du dehors de la cabine Gemini. Pendant que le pilote, à l'aide de fusées auxiliaires, annulait les oscillations du véhicule, le second astronaute pointait la région du ciel choisie et la photographiait. Par suite d'un déplacement du réseau, la région photographiée est celle de γ Velorum et non celle de β Crucis comme il avait été prévu; la résolution spectrale a été fortement réduite. Les spectres de 54 étoiles ont néanmoins été obtenus.

Pendant la mission de Gemini 11, les observations ont été continuées avec un matériel plus perfectionné. Les astronautes Conrad et Gordon ont pu disposer d'un réseau objectif (dispersion 180 Å/mm) et d'un prisme-objectif permettant une dispersion de 1500 Å/mm à 2500 Å. K. G. Henize rapporte que les spectres de 99 étoiles ont été obtenus à l'aide du réseau-objectif dans les régions de λ Scorpii, Canopus et ϵ Orionis. La magnitude visuelle 5·0 peut être atteinte pour les étoiles B des premiers types. Le doublet de Mg II à 2799 Å est très intense dans le spectre de Canopus, de même que les raies ultimes de Fe II vers 2400 Å. Le spectre de Sirius montre aussi les raies de Mg II, ainsi que la discontinuité de Balmer. Les spectres obtenus au prisme-objectif montrent clairement la discontinuité de Balmer pour les étoiles B. Les multiplets dûs aux métaux apparaissent dans les étoiles de type F entre 2400 et 2800 Å, ainsi que dans 48 Librae. μ Leporis, étoile à manganèse, montre une discontinuité importante de l'hélium, tandis que l'étoile de Wolf-Rayet HD 156385 révèle vers 2300 Å une émission due à C III (2297 Å).

E. R. Mustel rapporte que des observations de champs stellaires dans les régions 3700–5500 Å et 2200–3000 Å ont eu lieu à partir du satellite 'Cosmos 51', à une altitude de 300 à 400 km. La brillance du ciel s'est révélée être de deux à trois fois plus forte que prévue. Elle serait d'environ 190 étoiles de magnitude 10 par degré carré (**10**).

Au Goddard Space Flight Center, A. Boggess (**12**) a obtenu de nouvelles mesures photométriques d'étoiles B à 2100 et 1300 Å. A 2100 Å, des étoiles de même type spectral présentent des variations d'intensité importantes, ce qui n'est pas le cas à 1300 Å.

A. Boggess a pu aussi photographier à l'aide d'un réseau-objectif des spectres d'étoiles dans la région d'Orion entre 4000 et 1300 Å. La résolution est de l'ordre de 8 à 10 Å. T. Stecher (**13**) a continué ses recherches photométriques sur les étoiles chaudes dans l'ultraviolet, en vue d'en déduire une échelle de température.

Byram, Chubb et Werner (**14**) ont mesuré la brillance d'une soixantaine d'étoiles chaudes dans la bande 1050 à 1180 Å; ils trouvent un déficit en ultraviolet stellaire par rapport aux modèles d'atmosphère.

T. Gehrels rapporte qu'un prototype de polarimètre, composé de deux télescopes Cassegrain de 7·5 cm de diamètre, équipés de prismes de Wollaston, a été essayé sur des ballons à haute altitude. La brillance du ciel a été mesurée à 2200 et 2820 Å, et la polarisation de la lumière de la Lune a été évaluée à 2900 Å. Lors d'un vol à 36·5 km d'altitude du Polariscope (réflecteur de 71 cm de diamètre), Gehrels a pu mesurer à 2200 et 2820 Å la polarisation de la lumière de ζ Ophiuchi par le milieu interstellaire.

B. *Interpretations et prédictions du spectre ultraviolet*. Les problèmes généraux ont été examinés par Seaton (**15**). Boggess et Borgman (**16**) ont déterminé l'extinction de la matière interstellaire à partir de l'observation de 6 étoiles O9·5 – B1, à 2600 et 2200 Å. Les deux points obtenus ne semblent pas s'accorder avec la courbe théorique de Van de Hulst, l'extinction continuant à augmenter lorsque la longueur d'onde décroît. Spitzer, Dressler et Upson (**17**) ont calculé les largeurs équivalentes des raies interstellaires de H_2 qui pourraient apparaître en absorption dans le spectre d'étoiles O et B. Brandt (**18**) suggère de mesurer l'abondance de l'hydrogène moléculaire interstellaire par l'étude de la raie Raman de H_2 à 1280 Å excitée par $L\alpha$.

Byram, Chubb et Friedmann (**19**) n'ont pu confirmer, après un second vol, la présence d'émission nébulaire autour de Spica. Cette émission, si elle existe, a une intensité inférieure à 10^{-5} erg cm^{-2} s^{-1} $ster^{-1}$ entre 1225 et 1345 Å. Signalons encore différentes études théoriques sur les propriétés absorbantes des grains de graphite (Wickramasinghe, **20, 21**), de l'ion He^- (Somerville, **22**) et de la quasi-molécule d'hydrogène (Solomon, **23**). La présence de cette quasi-molécule ne peut expliquer l'absorption trouvée vers 2400 Å dans le spectre des étoiles B, pas plus que le complexe He H^+ (Werner, **24**). L'influence de l'effet de serre dû aux raies d'absorption intenses dans l'ultraviolet sur la structure atmosphérique des étoiles B a été étudiée par Morton (**25**), Mihalas et Morton (**26**) ainsi que par Avrett et Strom (**27**), et Guillaume (**28**). Les profils des raies spectrales ultraviolettes ont été calculés par Morton (**29**), Guillaume, Van Rensbergen, Underhill (**30**) et Houziaux (**31**). Elst (**32**) a prédit l'absorption de rayonnement par les raies spectrales dans le domaine 1900 à 3000 Å pour une étoile B. En comparant les résultats d'observations photométriques d'étoiles B, Heddle (**33**) montre l'importance d'effectuer des mesures absolues du flux. E. M. Burbidge (**34**) a discuté l'aspect du spectre ultraviolet des galaxies et la possibilité d'observer $L\alpha$ dans leur spectre. Nicol'ski (**35**) a calculé des modèles pour des régions entre la photosphère et l'atmosphère extérieure des étoiles. Houziaux (**36**) a publié des prédictions de flux de rayonnement stellaire pour divers types spectraux et différentes magnitudes visuelles dans l'ultraviolet, en tenant compte de l'effet de l'absorption interstellaire. Houziaux et Honnay (**37**) ont étudié l'influence de l'erreur de pointage et du manque de stabilité d'un télescope à bord d'un satellite sur la précision des mesures photométriques ultraviolettes, dans le cas de l'amas de Praesepe. Agapov et ses associés ont examiné la question des observations d'étoiles à partir d'un télescope placé sur la Lune (**38**).

II.2. *Rayons X et γ d'origine galactique*

L'étude des rayons X et γ d'origine galactique, tant du point de vue de l'observation que du point de vue de la théorie, a fait de grands progrès depuis la dernière assemblée de l'UAI.

Lors du symposium de Liège, en 1964, Bowyer, Byram, Chubb et Friedmann (**39**) ont donné les résultats de mesures obtenues à l'aide de compteurs de Geiger montés sur des fusées non stabilisées. Dix sources ont été identifiées, dont la nébuleuse du Crabe et la source intense Sco X–1. Fisher, Clark, Meyerott et Smith (**40**) font part du repérage d'une source de rayons X mous vers $\alpha = 23^h 40^m$ et $\delta = +78°$. Fisher et Meyerott (**41**) ont aussi mesuré le rayonnement X dans la bande 2–8 Å; ces résultats ont été contestés par Bowyer (**42**). Fisher, Johnson, Jordan et Meyerott (**43**) ont découvert huit sources à faible latitude galactique, mais aucune source n'a été détectée dans la direction de la supernova de Kepler ou du centre galactique. De nouvelles observations ont été faites par Byram, Chubb et Friedmann (**44**) en 1965. L'émission de rayons provenant de restes de supernovae a été examinée par Morrison et Sartri (**45**) et par Peterson, Jacobson et Pelling (**46**) (nébuleuse du Crabe). La distribution des émetteurs de rayonnement X dans la Galaxie a été étudiée par Johnson (**47**), qui suggère l'existence de deux types de sources: des sources formant un anneau autour du centre galactique, et des restes de supernovae. Un balayage du ciel a été effectué dans le domaine d'énergie 20–200 keV par Brini et ses associés (**48**). Un article récent de Giacconi et Gursky résume la situation (**49**). Hayakawa (**50**) a pu déterminer l'allure du spectre pour le rayonnement X galactique et pour les sources Sco X–1, Tau X–1 et les sources situées dans le Cygne. La source céleste la plus intense, appelée Sco X–1 a fait l'objet des travaux de plusieurs groupes (**51, 52, 53**). Découverte en 1962, sa position a été précisée récemment à moins d'une minute d'arc près (Gursky *et al.* **54**). Son identification a été faite par Oda en juin 1966, à l'Observatoire de Toyko, et confirmée par Sandage à l'Observatoire du Palomar. Il s'agit d'un objet de 13ème magnitude, très bleu, dont le spectre, qui contient des raies d'émission d'hydrogène et d'hélium ressemble à celui d'une vieille nova. Cette étoile présente des variations de brillance pouvant aller jusqu'à une magnitude en 24 heures. Sa variabilité a été confirmée par l'examen des plaques du Harvard College Observatory. Sur la base de considérations relatives à l'indice de couleur de cet objet, Matsuoka, Oda et Ogawara déduisent qu'il serait situé à 200 pc environ. L'histoire de la découverte et de l'identification de Sco X-1 est rapportée par Gursky (**55**). Sco X-1 a été identifiée indépendamment par H. M. Johnson et B. Stevenson.

La nébuleuse du Crabe a été étudiée à partir de ballons dans le domaine d'énergie supérieure à 15 keV par Clark (**56**). Les techniques expérimentales pour l'étude du rayonnement X d'origine extra-terrestre ont été revues par Boyd (**57**) et Mayer (**58**).

En ce qui concerne les rayons γ, Helmken rapporte que des observations ont été faites à partir de fusées dans le domaine 500 MeV à 5 GeV. Pour le rayonnement d'énergie supérieure à 10^{11} eV, il compte observer la distribution de la radiation Čerenkov dans le ciel nocturne à l'aide d'un collecteur de 34 pouces (86 cm) en construction au Mont Hopkins (Arizona). Ce système, ouvert à $f/0{\cdot}7$ et composé de 252 miroirs hexagonaux, peut être dirigé à un dixième de degré près vers n'importe quel point du ciel La sensibilité du système va jusqu'à 10^{-10} à 10^{-11} photons $cm^{-2}\,s^{-1}$ pour les énergies supérieures à 10^{11} eV.

C. Fichtel et D. Kniffen ont recherché les sources ponctuelles de rayons γ, en utilisant des émulsions nucléaires orientées à partir de ballons. Le domaine d'énergie étudié va de 10 à $2{\cdot}10^3$ MeV, l'efficacité étant de l'ordre de 10% entre 10 et 50 MeV. Cobb, Duthrie et Stewart (**59**) ont estimé le flux émis dans les rayons γ par la nébuleuse du Crabe.

Des résultats d'observations de rayons γ sont donnés par Kraushaar (**60**), Long, Porter, Weeks, Fruin et Jelley (**61**), Ögelman, Delvaille, Greisen (**62**) (à l'aide de ballons), et Kraushaar, Clark, Garmire, Helmken, Higbie et Agogino (**63**). Balasubrahmanyan, Hogge, Ludwig et McDonald ont étudié le rayonnement cosmique de faible énergie (20 MeV à 1 BeV) en utilisant une combinaison de compteurs à scintillation et de compteurs Čerenkov à partir des satellites OGO et IMP et de ballons. Le spectre décroît rapidement en intensité dans la région inférieure à 100 MeV par nucléon pour les protons et les particules alpha.

Il n'y a pour l'instant aucune explication satisfaisante de l'émission de rayons X, surtout pour les sources intenses comme Sco X–1. Les différentes hypothèses (étoiles à neutrons, couronnes stellaires, rayonnement synchroton) ont été envisagées et étudiées par de nombreux auteurs (références **64** à **93**).

II.3. *Rayonnement infrarouge*

Les principaux résultats ont été obtenus par l'équipe de Princeton, à l'aide du Stratoscope II, un télescope de 36 pouces (91 cm) monté à bord d'un ballon. N. J. Woolf (**94, 95, 96**) a exposé au symposium de Liège les résultats du second vol de cet instrument, où ont pu être observés les spectres de Mira Ceti, Aldébaran, μ Geminorum, ρ Persei, R. Leonis, μ Cephei, Bételgeuse. Les bandes les plus remarquables sont celles de H_2O, à 1·4, 1·9 et 2·7 microns, qui absorbent 1/4 à 1/3 du continuum local. L'intensité de ces bandes augmente fortement avec la luminosité intrinsèque de l'étoile. Danielson, Gaustad et Woolf ont cherché à identifier la bande à 3·1 μ due à la glace interstellaire, mais la présence de cette bande n'a pu être établie avec certitude. Danielson a étudié le problème de l'opacité due à la molécule d'eau pour les étoiles de basse température (1700 à 3400 °K). Cette opacité est très importante et surpasse toutes les autres sources connues à des températures inférieures à 25000 °K.

Gould (**97**) a étudié théoriquement la présence et l'intensité de raies d'émission infrarouges provenant des régions H II, des enveloppes stellaires, des régions H I interstellaires, des proto-étoiles pendant la phase de contraction, des nébuleuses planétaires et de l'hydrogène moléculaire interstellaire. Stein (**98**) s'est intéressé à l'émission infrarouge par des grains interstellaires pour les longueurs d'onde supérieures à 10 μ.

II.4. *Rayonnement de basse fréquence*

Lors du symposium de Liège, F. G. Smith (**99**) a passé en revue les observations aux ondes de basse fréquence obtenues à bord de véhicules spatiaux, et les a confrontées aux résultats obtenus à partir du sol. Au-dessous de 5 MHz, le rayonnement est principalement d'origine extragalactique. Au-dessous de 2 MHz, on observe une absorption qui paraît isotrope et due à une région H II centrée sur le Soleil. Huguenin et Papagiannis (**100, 101**) ont observé le rayonnement hertzien dans l'espace aux longueurs d'onde supérieures à 30 m. Aux altitudes comprises entre 3000 et 11 000 km, le flux a une valeur voisine de celle qui est prévue sur 2·2 MHz, alors que sur 0·7 MHz, le niveau reçu dépasse la valeur prévue de 15 db. Il est possible que cette dernière observation puisse s'expliquer par le rayonnement cyclotron des ceintures de Van Allen et de la ceinture artificielle. Hartz (**102**) a rapporté les observations faites à partir du satellite canadien Alouette II entre 1·5 et 10 MHz. La région du ciel la plus brillante sur 2·3 MHz est centrée sur le pôle galactique sud. La région la moins brillante à 2·3 MHz a pour coordonnées 9^h00^m et $+ 75°$. La courbe représentant la température de brillance en fonction de la fréquence présente une pente de − 1·3 à 1·5 MHz, − 1·7 à 2·3 MHz et − 2·2 à 5 M Hz.

Alexander et Stone (**103**) ont mesuré à l'aide de fusées le bruit cosmique à 1·91, 2·85, 3·60 et 4·70 MHz.

Smith (**104**) a analysé les résultats obtenus à l'aide d'instrumentation placée à bord du satellite Ariel II, qui a relevé la brillance du ciel à 1–2 MHz. La brillance mesurée est compatible avec la contribution attendue de l'espace extragalactique. Mais, aux fréquences inférieures à 2 MHz, sa diminution donne à penser qu'elle est absorbée par l'hydrogène ionisé de notre Galaxie.

Weyman et Chapman (**105**) ont étudié la possibilité de détecter les couronnes stellaires aux radiofréquences.

BIBLIOGRAPHIE

I. Vehicules Spatiaux, Instruments et Projets

1. Hemstreet, H. S., Loening, W. F. 1966, Princeton Advanced Satellite Study Final Report, Perkin-Elmer Corp., Norwalk, Conn., 584p.

2. European Space Research Organisation 1966, General Report 1964–65, Paris, 36 rue la Pérouse, p. 143.

3. Houziaux, L., Ed. 1965, Second Report on the S1/S2/S68 Scanning Experiment for ESRO stabilised Satellite TD.1, Liège, Institut d'Astrophysique.

4. Hallam, K., Mangus, J. 1965, An Ultraviolet Spectrophotometer for Satellite Astronomy, in *Jap. J. Appl. Phys.*, **4,** suppl. 1.

II. Résultats

II.1. Rayonnement Ultraviolet d'Origine Galactique

5. United Kingdom Report on Space Research, 1965–66, 1966, British National Committee on Space Research, The R. Soc., pp. 30–33.

6. United States Space Science Program 1966, Report to Cospar, Ninth Meeting Vienna, May 1966, Nat. Acad. Sci., Nat. Res. Council, pp. 1–3.

7. Morton, D. C., Spitzer, L. 1966, Line Spectra of Delta and Pi Scorpii in the far ultraviolet, *Astrophys. J.*, **144,** 1.

8. Morton, D. C. 1967, *Astrophys. J.*, **147,** 1017.

9. Henize, K. G., Wackerling, L. R. 1966, Stellar Ultraviolet Spectra from Gemini 11, *Sky and Telescope,* **32,** 204.

10. Dinov, N. A., Severny, A. B. Communication présentée au Symposium International Lunar Laboratory, Athènes, septembre 1965.

11. Report on Swiss Space Research, présenté au 9ème Cospar Vienne, 1966.

12. Boggess, A. III. 1964, B Star Colours between 2000 and 3000 angströms, (IAU Symp. No. 23, Liège, 1964) *Ann. Astrophys.*, **27,** 805.

13. Stecher, T. P. 1964, Ultraviolet Spectrophotometry of Early-Type Stars, *Astr. J.*, **69,** 558.

14. Byram, E. T., Chubb, T. A., Werner, M. W. 1965, 1115 Å for Ultraviolet Stellar Photometry, *Ann. Astrophys.*, **28**, 594.
15. Seaton, M. J. 1965, Far Ultraviolet Radiation: Stars, Nebulae and Interstellar Matter, *Ann. Astrophys.*, **28**, 151.
16. Boggess, A. III., Borgman, J. 1964, Interstellar Extinction in the Middle Ultraviolet, *Astrophys. J.*, **140**, 1636.
17. Spitzer, L., Jr., Dressler, K., Upson, W. L. 1964, Theoretical Equivalent Widths of the Interstellar H_2 Lines, *Publ. astr. Soc. Pacific*, **76**, 387.
18. Brandt, J. C. 1965, On Interstellar Raman Line H_2 Excited by Lyman-α, *Publ. astr. Soc. Pacific*, **77**, 212.
19. Byram, E. T., Chubb, T. A., Friedman, H. 1964, On the Absence of a Nebular Glow Around α Virginis in the Far Ultraviolet, *Astrophys. J.*, **139**, 1135.
20. Wickramasinghe, N. C. Oct. 7, 1965, Graphite Grains and Graphite Core-Ice Mantle Grains, Orange Aid Preprint Series in Nuclear Astrophysics (submitted to *Transactions of IAU*: Colloquium of Interstellar Grains).
21. Wickramasinghe, N. C. 1966, On the Optics of Small Graphite Spheres I, *Mon. Not. R. astr. Soc.*, **131**, 263.
22. Somerville, W. B. 1965, The Continuous Absorption Coefficient of He^-, *Astrophys. J.*, **141**, 811.
23. Solomon, P. M. 1964, The Absorption Coefficient of Quasi-Molecular Hydrogen, *Astrophys. J.*, **139**, 999.
24. Werner, M. W. 1964, He H^+ and Flux Deficiences in the Region $\lambda\lambda$ 1600–2400 of the Spectra of Early-Type Stars, *Astrophys. J.*, **139**, 1014.
25. Morton, D. C. 1964, The Effect of Line-Blanketing on Ultraviolet Stellar Radiation, *Astrophys. J.*, **139**, 1383.
26. Mihalas, D. M., Morton, D. C. 1965, A Model for a B1 V Star with Line Blanketing, *Astrophys. J.*, **142**, 253.
27. Avrett, E. H., Strom, S. E. 1964, Comparison Between Model Atmospheres and Spectra from Early-Type Stars, *Ann. Astrophys.*, **27**, 781.
28. Guillaume, C. 1966, Theoretical Model of a B 1·5 star with Line Blanketing, *Bull. astr. Inst. Netherl.*, **18**, 175.
29. Morton, D. C. 1965, Theoretical Line Profiles in the Ultraviolet Spectra of Early-Type Stars, *Astrophys. J.*, **141**, 73.
30. Guillaume, C., Van Rensbergen, W., Underhill, A. B. 1965, The Theoretical Near Ultraviolet Spectrum of a B2 Star, *Bull. astr. Inst. Netherl.*, **18**, 106.
31. Houziaux, L. 1965, The Lyman Lines in the Spectrum of Main Sequence Early-Type Stars, *Acad. r. Belg. Bull. Cl. Sci.*, **60**, 699.
32. Elst, E. W. 1966, The Predicted Line Blocking in the Spectral Region 1900 to 3000 Å of a B Star, *Bull. astr. Inst. Netherl.* (à paraître).
33. Heddle, D. W. O. 1964, The Importance of Absolute Photometry, *Ann. Astrophys.*, **27**, 800.
34. Burbidge, E. M. 1965, Ultraviolet Emission from Galaxies, *Ann. Astrophys.*, **28**, 164.
35. Nikol'ski, G. M. 1964, Structure and Far Ultraviolet Radiation of the Upper Stellar Atmosphere. II. Distribution of Electron Concentration and Temperature, *Soviet Astr.*, **8**, 193.
36. Houziaux, L. 1966, Revised Estimates of the Amounts of Radiation to be Expected from Stars in the Ultraviolet, *ESRO Scient. Memo* no. 3, Paris, 1-14.
37. Houziaux, L., Honnay, S. 1966, Photométrie ultraviolette des amas galactiques à partir de satellites artificiels, *Bull. Acad. r. Belg. Cl. Sci.*, 947–968.
38. Agapov, E. S., Anisimov, V. P., Brainfel'd, S. G., Mozherin, V. M., Prokofiev, V. V., Simenink, S. M. Sept. 1965, Travail présenté au 16ème congrès international d'astronautique à Athènes.

II.2. Rayons X et γ d'Origine Galactique

39. Bowyer, S., Byram, E. T., Chubb, T. A., Friedman, H. 1965, Observational Results of X-ray Astronomy, *Ann. Astrophys.*, **28**, 791.

40. Fisher, P. C., Clark, G. W., Meyerott, A. J., Smith, K. L., 1964, Observational Results on X-rays, *Ann. Astrophys.*, **27**, 809.

41. Fisher, P. C., Meyerott, A. J. 1964, Stellar X-ray Emission, *Astrophys. J.*, **139**, 123.

42. Bowyer, S. 1964, An Alternate Interpretation of the paper 'Stellar X-ray Emission' by Fisher and Meyerott, *Astrophys. J.*, **140**, 820.

43. Fisher, P. C., Johnson, H. M., Jordan, W. C., Meyerott, A. J., Acton, L. W. 1966, Observations of Cosmic X-rays, *Astrophys. J.*, **143**, 203.

44. Byram, E. T., Chubb, T. A., Friedman, H. 1966, Cosmic X-ray Sources, Galactic and Extragalactic, *Science*, **152**, 66.

45. Morrison, P., Sartri, L. 1965, X-ray Emission from Remnants of Type-I Supernovas, *Phys. Rev. Lett.*, **14**, 771.

46. Peterson, L. E., Jacobson, A. S., Pelling, R. M. 1966, Spectrum of Crab Nebula X-rays to 120 keV, *Phys. Rev. Lett.*, **16**, 142.

47. Johnson, H. M. 1966, The Distribution of X-ray Sources in the Galaxy, *Astrophys. J.*, **143**, 261.

48. Brini, D., Cattani, D., Ciriggi, U., Fuligni, F., Galli, M., Gandolfi, A., Moretti, C. 1965, Results of an X-ray Scanning of the Sky in the 20–200 keV Energy Range, *Ann. Astrophys.*, **28**, 1034.

49. Giacconi, R., Gursky, H. 1965, Observations of X-ray Sources Outside the Solar System, *Space Sci. Rev.*, **4**, 151.

50. Hayakawa, S., Matsuoka, M., Yamashita, K. 1966, Observations of Energy Spectra of Galactic X-rays, *Rep. Ion. and Space Res. in Japan*, **20**, no. 4.

51. Bowyer, S., Byram, E. T., Chubb, T. A., Friedman, H. 1964, X-ray Emission from the Direction of Scorpius, *Astr. J.*, **69**, 135.

52. Chodil, G., Jopson, R. C., Mark, H., Steward, F. D., Swift, C. D., X-ray Spectra from Scorpius (Sco–XR–1) and the Sun observed above the Atmosphere, *Phys. Rev. Lett.*, **15**, 605.

53. Manley, O. P. 1966, X-ray Emission from Sco X–1, *Astrophys. J.*, **144**, 1253.

54. Gursky, H., Giacconi, R., Gorenstein, P., Waters, J. R., Oda, M., Bradt, H., Garmire, G., Sreekantan, B. V. 1966, A Measurement of the Location of the X-ray Source Sco X–1, *Astrophys. J.*, **146**, 310.

55. Gursky, H. 1966, The Identification of the X-ray Source in Scorpius, *Sky Telesc.*, **32**, 252.

56. Clark, G. W. 1965, Balloon Observation of the X-ray Spectrum of the Crab Nebula above 15 keV, *Phys. Rev. Lett.*, **14**, 91.

57. Boyd, R. L. F. 1965, Techniques for the Measurement of Extraterrestrial soft X-Radiation, *Space Sci. Rev.*, **4**, 35.

58. Mayer, U. 1964, Optical Focussing with Soft X-rays and in the Extreme Ultraviolet, *Space Sci. Rev.*, **3**, 781.

59. Cobb, R., Duthrie, J. G., Stewart, J. 1965, Spark-chamber Experiment on Cosmic Gamma Rays, *Phys. Rev. Lett.*, **15**, 507.

60. Kraushaar, W. L. 1965, Cosmic Gamma Rays Experimental, *Ann. Astrophys.*, **28**, 202.

61. Long, C. D., Porter, N. A., Weeks, T. C., Fruin, J. F., Jelley, J. V. 1965, Astronomical Observations from Space Vehicles: Upper Limits to the Flux of High Energy γ-rays from Selected Quasi-Stellar and other Sources using the Čerenkov Technique, *Ann. Astrophys.*, **28**, 263.

62. Ögelman, H. B., Delvaille, J. P., Greisen, K. I. Search for Discrete Sources of Cosmic Gamma Rays above 1 GeV, *Phys. Rev. Lett.*, **16**, 491.

63. Kraushaar, W., Clark, G. W., Garmire, G., Helmken, H., Higbie, P., Agogino, M. 1965, Explorer XI Experiments on Cosmic Gamma Rays, *Astrophys. J.*, **141**, 845.

64. Bahcall, J. N., Wolf, R. A. 1965, An Observational Test of Theories of Neutron-Star Cooling, *Astrophys. J.*, **142**, 1254.

65. Burbidge, G. R., Gould, R. J., Tucker, W. H. 1965, Galactic X-ray Sources, *Phys. Rev. Lett.*, **14**.

66. Clayton, D. D., Craddock, W. L. 1965, Radioactivity in Supernova Remnants, *Astrophys. J.*, **142**, 189.

67. Dolan, J. F. 1965, The Expected Polarization of X-rays from Celestial Sources, *Astr. J.*, **70**, 137.
68. Finzi, A. 1965, Cooling of a Neutron Star by the 'Urca' Process, *Phys. Rev.*, **137**, B472–B476.
69. Garmire, G., Kraushaar, W. L. 1965, High Energy Cosmic Gamma Rays, *Space Sci. Rev.*, **4**, 123.
70. Ginzburg, V. L., Syrovatskii, S. I. 1964, Nature of Discrete X-ray Sources, *Soviet Phys.*, **9**, 831.
71. Ginzburg, V. L., Syrovatskii, S. I. 1964, Gamma-rays and Cyclotron Radiation X-rays of Galactic and Metagalactic Origin, *JETP*, **18**, 245.
72. Gould, R. J. 1965, Origin of Cosmic X-rays, *Phys. Rev. Lett.*, **15**, 511.
73. Ginzburg, V. L., Syrovatskii, S. I. 1964, Some Problems of Gamma and X-ray Astronomy, *Soviet Phys. Usp.*, **7**, 696.
74. Gould, R. J., Burbidge, G. R. 1965, High Energy Cosmic Photons and Neutrons, *Ann. Astrophys.*, **28**, 171.
75. Gould, R. J. 1965, High-Energy Photons from the Compton-Synchroton Process in the Crab Nebula, *Phys. Rev. Lett.*, **15**, 577.
76. Gould, R. J., Sciama, D. W. 1964, Cosmic X-ray and Infrared Rays as Tools for Exploring the Large Scale Structure of the Universe, *Astrophys. J.*, **140**, 1634.
77. Gurzadyan, G. A. 1965, Possibility of X-ray Emission by Cosmic Radio Sources, *Societ Phys. Dokl.*, **9**, 841.
78. Hovenier, J. W. 1966, X-ray Spectra Produced by Thermal Bremsstrahlung, *Bull. astr. Inst. Netherl.*, **18**, 185.
79. Hoyle, F. 1965, Origin of Cosmic X-rays, *Phys. Rev. Lett.*, **15**, 131.
80. Jelley, J. V. 1966, High-Energy γ-ray Absorption in Space by a 3·5°K Microwave Field, *Phys. Rev. Lett.*, **16**, 479.
81. Manley, O. P. 1966, Early Radiation Field and High-Energy Electrons in Supernovae, *Nature*, **209**, 901.
82. Morton, D. C. 1964, Neutron Stars as X-ray Sources, *Astrophys. J.*, **140**, 460.
83. Nikol'skii, G. M. 1964, Soft X-radiation from Stars, *Soviet Phys. Dokl.*, **8**, 646.
84. Orzag, S. A. 1965, Neutron-Star Atmospheres, *Astrophys. J.*, **142**, 473.
85. Overbeck, J. W. 1965, Small-Angle Scattering of Celestial X-rays by Interstellar Grains, *Astrophys. J.*, **141**, 864.
86. Pacini, F. 1966, High-Energy Astrophysics and a Possible Sub-nuclear Energy Source, *Nature*, **209**, 389.
87. Shklovsky, I. S. 1964, On the Nature of the X-Ray Sources, *Astr. Zu.*, **42**, 287.
88. Surdin, M. 1966, X-ray Emission from very Dense Stars, *Nature*, **209**, 62.
89. Vlasov, N. A. 1965, Optical Search for Antimatter in the Universe, *Soviet Astr.*, **8**, 715.
90. Woltjer, L. 1964, X-rays and Type I Supernova Remnants, *Astrophys. J.*, **140**, 1309.
91. Lüst, R. 1966, Theoretical Aspects of Celestial Gamma-rays, ESRO SM–60, Paris.
92. Biermann, L. 1966, Theoretical Aspects of Extraterrestrial X-rays of non Solar Origin, ESRO SM–58, Paris.
93. Lequeux, J. 1966, Rayons cosmiques et rayonnement synchroton, ESRO SM–54, Paris.

II.3. Rayonnement Infrarouge

94. Woolf, N. J. 1965, Infrared Spectra of Stars, Planets and the Moon from Stratoscope II, *Ann. Astrophys.*, **28**, 218.
95. Woolf, N. J., Schwarzschild, M., Rose, W. K. 1964, Infrared Spectra of Red-Giant Stars, *Astrophys. J.*, **140**, 833.
96. Danielson, R. E., Woolf, N. J., Gaustad, J. E. 1965, A Search for Interstellar Ice Absorption in the Infrared Spectrum of Mu Cephei, *Astrophys. J.*, **141**, 116.
97. Gould, R. J. 1964, Infrared Emission Lines from Celestial Sources, *Ann. Astrophys.*, **27**, 815.
98. Stein, W. 1966, Infrared Radiation from Interstellar Grains, *Astrophys. J.*, **144**, 318.

II.4. Rayonnement de Basse Fréquence

99. Smith, F. G. 1964, Low Frequency Radio Astronomical Observations from Rockets and Satellites, *Ann. Astrophys.*, **27**, 819.

100. Huguenin, G. R., Papagiannis, M. D. 1965, Spaceborne Observations of Radio Noise from 0·7 to 7·0 MHz and their Dependence on the Terrestrial Environment, *Ann. Astrophys.*, **28**, 239.

101. Huguenin, G. R., Lilley, A. E., Donough, W. H. Papagiannis, M. D. 1964, Measurements of Radio Noise at 0·700 Mc and 2·200 Mc from a high-altitude Rocket Probe, *Planet. Space Sci.*, **12**, 1157.

102. Hartz, T. R. 1964, Observations of the Galactic Radio Emission between 1·5 and 10 MHz from the Alouette Satellite, *Ann. Astrophys.*, **27**, 823.

103. Alexander, J. K., Stone, R. G. 1965, Rocket Measurements of Cosmic-Noise Intensities Below 5 Mc/s, *Astrophys. J.*, **142**, 1327.

104. Smith, F. G. 1965, Cosmic Radio Noise as Measured in the Satellite Ariel II, *Mon. Not. R. astr. Soc.*, **131**, 145.

105. Weyman, R., Chapman, G. 1965, On the Possibility of Detecting Stellar Coronas at Radio Frequencies. *Astrophys. J.*, **142**, 1268.

SOLAR CORPUSCULAR RADIATION

(prepared by R. Lüst)

1. *Introduction*

During the last three years our knowledge about the solar corpuscular radiation—now called the solar wind—has increased considerably due to many measurements carried out from satellites and space probes. Table 1 gives a list of the satellites and space probes which had instruments on board for measuring the interplanetary plasma and magnetic fields.

Table 1

Satellites and space probes equipped for measuring interplanetary plasma and magnetic fields

Vehicle	Launch date	Apogee	Plasma instrument	Magnetometer
Pioneer 1	Oct. 11, 58	18·6 R_e	—	Search coil
Lunik 1	Jan. 2, 59		Plasma trap	Triaxial fluxgate
Lunik 2	Sept. 12, 59		Plasma trap	Triaxial fluxgate
Lunik 3	Oct. 4, 59		Plasma trap	Triaxial fluxgate
Pioneer 5	March 11, 60		—	Search coil
Venus 1	Feb. 12, 61		Plasma trap	?
Explorer 10	March 25, 61	46·6 R_e	Plasma cup	Rubidium
Mariner 2	Aug. 26, 62		Electr. analyzer	Fluxgates
Explorer 14	Oct. 3, 62	16·4 R_e	Electr. analyzer	Triaxial fluxg.
Vela 1, 2	Oct. 17, 63	17 to 20 R_e	Electr. analyzer	Search coil
IMPI(=Ex.18)	Nov. 27, 63	31 R_e	Plasma cup and electr. analyzer	Rubidium fluxg.
Vela 3, 4	July 17, 64	17 to 20 R_e	Electr. analyzer	Search coil
OGO 1	Sept. 5, 64	24·4 R_e	Electr. analyzer	Search coil
IMP 2	Oct. 4, 64	16 R_e	Electr. analyzer	Rubidium fluxg.
Mariner 4	Nov. 28, 64		Plasma cup	Helium
IMP 3	May 29, 65	41·7 R_e	Electr. analyzer	Rubidium fluxg.
Vela 5, 6	July 20, 65	17 to 20 R	Electr. analyzer	Search coil
Venus 3	Nov. 16, 65		Plasma trap	?
Pioneer 6	Dec. 16, 65		Plasma cup and electr. analyzer	Fluxgate

In this report only the results obtained from the measurements by space craft will be mentioned, whereas the numerous other sources of information about the solar corpuscular radiation (e.g. observations of comets, of the refraction and scintillation of radio sources, of the propagation and modulation of energetic particles) will not be discussed here. This material would go beyond the scope of Commission 44. It is anticipated that some of these results and also the theoretical interpretation of all observations will be included in the reports of the other relevant commissions.

Although some results will be mentioned which have already been obtained before the dateline of the last draft report, this summary will concentrate on the data given in publications between 1964 and 1966. Furthermore we shall review only the observations obtained in the interplanetary space. Nothing will be said about the most interesting phenomena related to the interaction of the solar wind with the magnetosphere.

2. *Observations of the particle flux*

2.1 *General remarks*

The solar corpuscular radiation has been investigated by means of space vehicles with plasma probes for measuring primarily the ion component, although some of these are able to detect also electrons. The expected mass motions should be more easily detectable from ion measurements, since the thermal motion of the ions is much smaller than that of the electrons assuming equal temperatures for both components.

Two different kinds of plasma probes have flown in space vehicles. One is a plasma cup which is used essentially as an integral analyzer for the energy with a wide angle of acceptance for the arriving particles. Such instruments have been developed by the group at MIT (Binnetti, Bridge *et al.* (1)) and flown first in Explorer X and later in several other satellites and space probes. Similar devices called ion traps have been used in Russian spacecraft (Gringauz *et al.* (2–4)).

The other type is an electrostatic analyzer used as a differential analyzer for the energies with a narrow angle of acceptance. Such an instrument was first used by the Jet Propulsion Laboratory group on board of Mariner 2 (Neugebauer *et al.* (5)) and later in other spacecraft by the Ames Research Center group (Wolfe *et al.* (6)) and by the Los Alamos group (Coon (7)).

Until now the time resolution of most of the measuring instruments was rather low due to the instrumental, power or telemetry limitations. For instance for the plasma probe of MIT on board of IMP–1 a complete energy scan required 2·8 minutes and was repeated every 5·5 minutes. For the electrostatic analyzer of Ames Research Center on board of IMP–1 a complete process of data acquisition and readout for a complete cycle energy and angular scan required 5 minutes and 28 seconds. For the later spacecraft these times have been shortened, e.g. during real time tracking of the Vela-2 satellites the time required to acquire information for a complete energy spectrum and direction was 128 seconds. But because successive portions of the data are acquired at one-second intervals, a change in the character of the radiation can sometimes be sensed with one second time resolution.

Until recently the instruments were also very limited as far as measurements of the angular distribution of the distribution function are concerned. Thus the wide aperture ion traps on the Luniks, Explorer X, IMP–1 etc. obviously could not investigate details of the particle distribution, and the narrow aperture electrostatic analyzer in Mariner 2 was constrained to look only at the Sun. Similarly the analyzer on IMP–1 yielded only very crude angular distributions (i.e. three sectors in spacecraft longitude, essentially no resolution in spacecraft latitude).

The analyzers on OGO–1, Vela 2A, 2B, 3A, 3B, had a much better longitudinal resolution, but again the observed current represents an integral over the solar plasma direction of arrival with respect to the spacecraft latitude. The first experiment capable of obtaining the complete angular distribution of the solar wind is the Ames Research Center experiment on board of Pioneer 6.

These facts about the instruments so far used should be taken into account in the following discussion of the results.

The first measurements of the solar wind were carried out by Gringauz *et al.* (**2–4**) with Lunik 2 and 3 and with the first Russian Venus rocket and by Bonnetti, Bridge and co-workers with Explorer X. Gringauz measured a flux of about 10^8 particles $cm^{-2} s^{-1}$, which was confirmed by Bonnetti, Bridge and co-workers. The measurements of the latter group indicated a definite wind that came approximately from the Sun with a velocity of about 300 km s^{-1}.

At present we have measurements available from Mariner 2 (20 August–30 December 1962) (Neugebauer and Snyder (**8**)), IMP–1 (27 November 1964–middle of February 1965) (Bridge *et al.* (**9**), Wolfe *et al.* (**6**)), IMP–2 (4 October–6 October 1964) (Wolfe *et al.* **10**)), Mariner 4 (January–March 1965) (Bridge *et al.* (**9**)), Vela 2, and 3 (1964, 1965) (Coon (**7**), Hundhausen *et al.* (**11**), Strong *et al.* (**12**)), OGO–1 (Wolfe *et al.* (**10**)), Pioneer 6 (Wolfe *et al.* (**14**), Bridge *et al.* (**9**)) which cover longer periods. The principal results from these observations during the stated periods are the following.

2.2 *Velocity*

Mariner 2: The daily average velocity was 504 km s^{-1}, with the lowest value between 306 to 318 km s^{-1} and the highest value between 815 and 842 km s^{-1}. Several streams of hot, high velocity plasma were observed to recur at 27-day intervals. The quiet, between-stream, solar wind velocity was in the range of 320 to 340 km s^{-1}. The velocity was independent of the distance from the Sun between 0·7 and 1·0 A.U. (Neugebauer and Snyder (**8**)). The relationship between the plasma velocity v and the density sum of the planetary magnetic activity index ΣK_p can be expressed as follows: $v = (8{\cdot}44 \pm 0{\cdot}74)\, \Sigma K_p + (330 \pm 17)$ [km s^{-1}] (Snyder *et al.* (**15**)).

IMP–1: Average velocity 378 km s^{-1}, ranging from less than 300 to less than 570 km s^{-1} during relatively quiet periods. During disturbed times the velocity ranged from 560 to 750 km s^{-1}. The average direction of the ion flux coincided usually with the apparent direction from the Sun. The fluctuations in the arrived angle were about ± 15 degrees (Wolfe *et al.* (**6**)).

IMP–2: Velocity during geomagnetically disturbed periods 712 ± 13 km s^{-1} (Wolfe *et al.* (**10**)).

Mariner 4: Velocity between 275 km s^{-1} and 600 km s^{-1}. At the time of passing the planet Mars (15 July 1965) the velocity was 330 km s^{-1}.

Vela 2, *Vela* 3: Quiet solar wind velocity (derived from the minimum) 320–330 km s^{-1}. Usually the velocity is enhanced and variable, maximum velocity is 720 km s^{-1}, the peaks of the velocity generally recur at intervals of 27 days, persisting through several solar rotation periods. Direction varies over a range of about 15° (Coon (**7**)). At present some 12 000 individual measurements of the angle of arrival have been made from data obtained between July 1964, and July 1965 by the Vela 2 satellites. The average flow for these cases is from 2°6 east of the Sun (after correction for aberration due to the Earth's orbital motion) (Strong (**16**)).

Venus 3: Velocities in the range of 300 km s^{-1} to 415 km s^{-1} (Gringauz *et al* (**13**)).

Pioneer 6 (16 *December* 1965): Data not yet available. Flow direction has significant deviations from the nearly radial direction in the ecliptic plane and fluctuations as high as ± 5° are observed. Northerly as well as southerly flow with respect to the ecliptic plane has been observed while the flow direction during quiet times remained steady over periods of many hours. During periods apparently associated with solar and terrestrial disturbances, the flow direction toward much greater variability (Wolfe *et al.* (**14**)).

2.3 *Density*

Mariner 2: Near 1 A.U. the average density was approximately 5 protons cm^{-3}. The density was usually highest at the leading or western edge of each stream, with a maximum value of 80 protons cm^{-3}. Otherwise the density varied inversely with the plasma velocity. The lowest values were between 0·081 and 0·094 protons cm^{-3}, the highest between 70 and 88 protons cm^{-3}. The density changed approximately as the inverse square of the distance (Neugebauer and Snyder (**8**)).

IMP–1: The density rarely exceeded 1 to 3 ions cm^{-3}, except in association with geomagnetic storms (Wolfe *et al.* (**6**)).

IMP–2: Density during geomagnetically disturbed period was $2{\cdot}46 \pm 0{\cdot}5$ protons cm^{-3} (Wolfe *et al.* (**10**)).

Vela 2, *Vela* 3: Typical ion densities were to 3 to 5 cm^{-3} (Coon (**7**)).

Venus 3: The ion density was within the limits of 2·4 cm^{-3} and 55 cm^{-3} (Gringauz *et al* (**13**)).

Pioneer 6: Data on the ion density not yet available. The electron density had been determined by a radio propagation experiment from Pioneer 6 and simultaneously by ionosphere measurements using signals from two Beacon satellites. The average interplanetary electron number density near the earth orbit was between 8 and 9 cm^{-3} in the time interval from 20 February to 9 April 1966.

2.4 *Temperature*

In most cases it is not easy to determine the temperature from the measured distribution function which is not really a Maxwellian distribution. Furthermore the later more refined measurements show that the temperature is not equal in different directions. Most of the early observations of the random motions were made from differential energy spectra reflecting the particle flux as a function of energy in some fixed directions. An alternate method of temperature determination involves the measurement of the particle flux in different directions. The width of the resulting 'angular distribution' is related to the ratio of the mean random velocity to the bulk velocity and hence to a 'temperature' in the direction normal to the bulk velocity.

Mariner 2: The high-velocity streams were hotter than the surrounding plasma. The average daily-average value of the proton temperatures T_p was between 1·51 and $1{\cdot}85 \times 10^5$ °K (on 17 December 1962), the minimum value was 3×10^4 °K for the end of the 'between-stream' periods. There was no obvious radial dependence (Neugebauer and Snyder (**8**)).

The energy density of proton random motions was never more than a small fraction of the energy density of bulk motion (in the average between 0·013 and 0·017, the maximum value was between 0·014 and 0·115, the actual minimum could not be determined). The spectra had usually high-energy tails which became more pronounced at the higher plasma velocities. It is not clear if these high energy tails are connected with the anisotropic distributions observed by Vela 3 and Pioneer 6.

IMP–2: The proton temperature was $2{\cdot}0 \pm 0{\cdot}5 \times 10^5$ °K. Ratio of thermal energy density to kinetic energy density was about 7×10^{-3} (Wolfe *et al.* (**10**)).

Vela 2, *Vela* 3: The proton temperatures in a transverse direction ranged from 8×10^5°K. down to 6×10^3°K. The low temperatures (typically near 2×10^4°K) were observed when the solar wind velocity was high. During magnetically quiet periods, that is when K_p was low, the solar wind velocity was low (~ 325 km s^{-1}) and very steady, and the temperature was at the low end of its range, lying close to 10^4°K. At the onset of a magnetically disturbed period within the magnetosphere (when K_p becomes high), the solar wind velocity rose as high as 700 km s^{-1} and the temperature above 10^5°K (Coon (**7**), Strong *et al.* (**12**)).

The temperature in different directions could be investigated in more detail by Vela 3. The ion temperature was higher along the field ($T_{\parallel}$) than transverse to this direction ($T_{\perp}$) The most probable value of T_{max}/T_{av} found in the Vela 3 observations is 1·4; this corresponds to $T_{max}/T_{min} = 2$ for a symmetric distribution.

Pioneer 6: The ratio $K = T_{\parallel}/T_{\perp}$ is investigated in detail. On 26 December 1965, K was found to be 5 and the angle between the flow velocity and the magnetic field about 60°. This high value of K is not an isolated case. In the data thus far analyzed, the results imply that at times $T_{\parallel}$ may exceed $T_{\perp}$ by as much as one order of magnitude. $K = 5$ is a common situation. These anisotropies are higher than those measured by Vela 3. This difference may, however, not be real but caused by different techniques used to measure 'temperatures' (Wolfe *et al.* (**14**)).

2.5 *Composition*

Mariner 2: The ratio of α-particles to protons n_{α}/n_{p} was observed to be 0·046 ± 0·038. A model in which the protons and α-particles have equal thermal velocities gives a better fit to the observed spectra than does an equal temperature model (Neugebauer and Snyder (**8**)).

IMP–2: $n_{\alpha}/n_{p} \approx 0{\cdot}02$ (Wolfe *et al.* (**10**)).

Vela 2, *Vela* 3: The observations showed that the α-particle proportion is variable, from less than 1% up to ~ 20%, and the estimated average was 3% to 6%. From Vela 3 observations it was found that the ratio of the number densities of the alphas and protons varied from 0·00 to 0·15 with a mean value of 0·042 (Coon (**7**), Hundhausen *et al.* (**11**)).

Pioneer 6: Here it was possible to detect a third ionic species, He^{+}. The following data for the composition were given for one date of observation (26 December 1965): H^{+} ~ 91·3%, He^{++} ~ 8·6% and He^{+} ~ 0·1%. In the data analyzed so far the He^{++} and He^{+} contents show a large variance being at times almost unobservable (Wolfe *et al.* (**14**)).

3. *Observations of the magnetic fields*

3.1 *General remarks*

The magnetic fields in the interplanetary space have been measured during the last years in great detail and accuracy. The time resolution of the measurements is in most cases higher than that of the plasma measurements. A number of different instruments have been used, like fluxgate magnetometers, search-coils, proton precession magnetometers, helium magnetometers and rubidian vapor magnetometers.

3.2 *Average field and its direction*

Even before the solar wind properties were determined, the interplanetary magnetic field was measured by Coleman *et al.* (**17, 18**) with Pioneer 5. More precise and long-term measurements have been made by Mariner 2, IMP–1, Mariner 4, IMP–3, and Pioneer 6 (Coleman *et al.* (**19, 20**), Ness (**21, 22, 23**)). Because the interplanetary field has an energy much smaller (~ 1%) than the solar wind energy densities, it is carried along by the solar wind. Field values were usually between 2 and 7 gammas and averaged near 5 gammas. The field pointed roughly away from the Sun or towards the Sun with the theoretically expected angle associated with the Archimedian spiral angle.

3.3 *Origin of the field*

Ness and Wilcox (**24**) have investigated the origin of the interplanetary magnetic field. The direction of the field shows a strong recurrence tendency with an interval of 27 days which corresponds to the rotational period of the solar equatorial region as seen from the Earth and suggests a solar origin of the interplanetary field. A cross correlation function constructed for the IMP–1 data yields the result (Ness and Wilcox (**25**)) that the time of effects for a coherent peak is 4·5 days. This leads to an average plasma velocity of approximately 385 km s^{-1} reported by the MIT plasma experiment in IMP–1 for the same time interval.

The solar latitude of the regions that constitute the source of the interplanetary magnetic field has been investigated in terms of the solar rotation. As seen from the Earth, regions near the solar equator rotate with a period of about 27 days, whereas regions at high latitude have a slower rotation velocity, with a period of about 30 days. This study suggests that the latitude of the solar source was within 10 or 15 degrees of the center of the visible disk or of the equator.

3.4 *Sector structure and filaments*

The interplanetary magnetic field exhibited a well defined 27-day periodicity. This periodicity was interpreted as evidence for a connection between the field at 1 A.U. and the large-scale patterns in the solar magnetic field that were observed between August and November 1962 during the flight of Mariner 2 (Coleman *et al.* (**26**)). Two reversals of the polarity occurred during each of the solar rotations. These reversals occurred at the time of a local minimum of the plasma velocity and were followed within 2·5 days by a local maximum of the velocity.

The sector structure has been observed best between November 1963 and February 1964 during the flight of IMP–1 (Ness and Wilcox (**25**)). A correlation was established between the field pattern indicated by the IMP–1 data and patterns in the photospheric field. The interplanetary field was highly organized during this time on a 27-day basis into four segments within each of which the direction of the field was most of the time either positive or negative. Three sectors had a duration of 7.7 days, and one was 3·8 days long. There was a coherent variation of the interplanetary magnetic field strength with the planetary magnetic index K_p within each sector region and also a similar variation of the flux and the density throughout the sector structure.

During the flight of Mariner 4 (Coleman *et al.* (**26**)), one year later, the variation in the recorded field over 27-day periods showed much less of the well defined periodicity that characterized the earlier observations. It is suggested that the pattern becomes less stable after the minimum in the cycle of the solar activity, which occurred during the period from May through November 1964, and may well become still less regular as the Sun becomes more active.

Wilcox and Ness (**27**) have suggested that this sector structure may be an important aspect for the development of M-region type storms which have yet to be identified with some defined features on the surface of the Sun. The results of Mariner 2, IMP–1 and Mariner 4 support a model in which the high velocity streams in the solar wind originate in regions throughout which the magnetic field has a single polarity.

In this connection the earlier work by Babcock and Simpson (**28**) should be remembered, namely the relation of the 27-day recurring cosmic ray modulation with 'unipolar magnetic regions' on the Sun and M-region type geomagnetic storms.

In connection with the observed sector structure during the time of IMP–1 it is of interest that a recurring stream of protons of a few MeV energy is almost entirely contained within one sector (Fan *et al.* (**29**), and Bryant *et al.* (**30, 31, 32**)).

The collimation of solar cosmic rays with energies between 7·5 and 90 MeV has been increased during the flight of Pioneer 6 (McCracken *et al.* (**33**), and Bartley *et al.* (**34**)). For a certain period of a flare event (29 December 1965), the azimuth of the magnetic field vector and the azimuth of the cosmic ray anisotropy were highly correlated. Despite major changes in the interplanetary field direction, the cosmic ray anisotrophy remained well aligned with the field.

Also the measurements by Fan *et al.* (**35**) showed the collimation of cosmic ray particles by the magnetic fields. This group was able to prove that these particles were protons (not electrons) and that the largest effects extend down to 1 MeV energy.

From these and other measurements (Ness (**36**)) it is concluded that within the interplanetary medium numerous separate bundles of magnetic flux exist that are adjacent to and in equilibrium with each other. These separate bundles are convected outwards by the solar wind but still retain, in general, an Archimedian spiral configuration in the average, although the bundles are intertwined with each other.

Simultaneous observations of proton intensity-time distributions from Pioneer 6 and IMP–3 at Earth confirm that spatial features in the magnetic field with dimensions as small as 4×10^5 km persist for time intervals of at least two hours in which the magnetic field is in co-rotation with the Sun (Fan *et al.* (**35**)).

3.5 *Fluctuations*

Besides the measurements of the strength of the magnetic field and its direction it is important to show its variations in time and space. We know that the field fluctuations, and the measurements of the power spectra will give us indications about the kind of waves that are propagating and also about the instabilities which might be present influencing the particle distribution function. Also correlations between different locations in space are indicative for the propagation mechanism in the interplanetary space. Furthermore we expect not only small amplitude waves, but also those with large amplitudes and finally shock waves.

The problem of discontinuities in the solar wind has been in particular investigated by Colburn and Sonett (**37**). We refer to his recent excellent review paper, where many details can be found about the observations as well as about the theoretical aspects. Coleman (**38–40**) has studied the variations in the interplanetary magnetic field during the time of flight of Mariner 2. He obtained power spectra of the fluctuations for the total field as well as for the three components of the field. The power densities in the spectra are steeply decreasing functions of the frequency f with a dependence between f^{-1} and f^{-2} between 100 and 500 cycles per day. Coleman established also the existence of coherent variations in the field and in the radial component of the plasma velocity. The results suggested that the variations in the interplanetary field are predominantly changes in field orientation rather than in the strength of the field. This indicates that the oscillations are probably transverse waves.

A power spectrum of the fluctuating field has also been measured during the flight of OGO–1 and of Pioneer 6. It should be mentioned in this connection that from the IMP–1 data (Wolfe (**6**)) it is apparent that rapid (period less than 5 minutes) small-amplitude fluctuations were always present in the velocity of the solar wind even during geomagnetically quiet periods. These fluctuations increased in amplitude, however, during geomagnetic storms.

The usefulness of simultaneous measurements of the interplanetary magnetic field has been demonstrated by Ness (**36**), who presented preliminary results obtained from Pioneer 6 and from IMP–3 at two separate points of observation during the same time at distances from 350 000 to 700 000 km from the Earth. The general similarity but time delayed feature of a reversal of the magnetic field when observed by the two satellites on opposite sides of the Earth and separated by $1{\cdot}3 \times 10^6$ km can be interpreted only in terms of the corotation of the interplanetary magnetic field structure. Radial propagation of the field reversal would lead to a time delay substantially less than five minutes, compared to the observed time delay of 57·5 minutes. In this connection also the observed large scale oscillations of proton intensity period of 10^3 s) should be mentioned (Fan *et al.* (**35**)).

Also Smith *et al.* (**41**) have carried out simultaneous measurements of the interplanetary magnetic field from OGO–1 and Mariner 4. Data were available during the intervals from 28 November to 11 December 1964 and from 8 March to 4 May 1965. During the first interval the separation between the two spacecraft increased from about 1/3 to 5 million km and during the second interval from 46 to 112 million km.

4. *Summary*

Summarizing all these measurements carried out so far the following can be said about the properties of the solar wind: The quiet solar wind velocity is of the order of 300–350 km s^{-1}

near the orbit of Earth with a direction almost away from the Sun (taking into account the appropriate correction for aberration due to the orbital motion of the Earth). Its density is of the order of 5 protons cm^{-3} and the temperature probably closer to 10^4 °K than to 10^5 °K. The temperature parallel to the magnetic field is higher than that transverse to the field, very often by a factor of five α-particles (He^{++}) amounting to about 5% in numbers of the protons and perhaps 0·1% of He^+.

The wind is highly variable in its velocity (up to about 850 km s^{-1}), in its direction (± 15°), in its temperature (up to 9×10^5 °K), and in its composition (up to 20% α-particles). Often there exist streams of hot, high velocity plasma which recur at 27-day intervals. So far no measurements are available concerning the period of maximum solar activity.

The averaged magnetic field has a spiral structure with a strength of about 5 gammas. Very often it shows 27-day periodicity and a certain sector structure.

BIBLIOGRAPHY

Review Articles, Proceedings of Symposia and Conferences, Books

Dressler, A. J. 1966, Solar wind and interplanetary magnetic field, submitted to *Rev. Geophysics*.

Haerendel, G. 1965, *Interplanetary space*, Landolt-Börnstein, **1**, 233.

Hess, W. N., Mead, G. D. 1966, Particles and field research, A Bibliography, NASA SP–7026.

Lüst, R. 1966, The properties of the interplanetary medium, presented at the Inter-Union Symposium on *Solar-Terrestrial Physics*, Belgrade.

Mackin, Jr., R. J., Neugebauer, M. 1966, The solar wind, Proceedings of a Conference at Caltech, sponsored by J. P. Lab., Pasadena, Pergamon Press.

Ness, N. F. 1966, Measurements of the magnetic fields in interplanetary space and the magnetosphere, Proc. of the Ninth International Conference on Cosmic Rays, London, **1**, 14.

Parker, E. N. 1963, *Interplanetary dynamical processes*, Interscience Div. of J. Wiley and Sons, N.Y.

Proceedings of the Ninth International Conference on *Cosmic Rays* 1966, The Inst. of Physics and Phys. Society, London, **1**, **2**.

References

1. Bonnetti, A., Bridge, H. S., Lazarus, A. J., Rossi, B., Scherb, F. 1963, Explorer 10 plasma measurements, *J. geophys. Res.*, **68**, 4017.
2. Gringauz, K. I., Bezrukikh, V. V., Ozerov, V. D., Rybchinskii, R. E. 1960, A study of the interplanetary ionized gas, high-energy electrons, and corpuscular radiation from the Sun by means of the three electrode traps for charged particles on the second Soviet cosmic rocket, *Dokl. Acad. Nauk SSSR*, **131**, 1301.
3. Gringauz, K. I. 1961, Some results of experiments in interplanetary space by means of charged particle traps on Soviet space probes, *Space Res.*, **2**, 539.
4. Gringauz, K. I., Kurt, V. G., Morog, V. I., Sklovskij, I. C. 1961, Sbornik, *Iskusstvennye Sputniki Zemli*, Izd. An. SSSR, **6**, 108.
5. Neugebauer, M., Snyder, C. W. 1962, Solar plasma experiment: Preliminary Mariner 2 observations, *Science*, **138**, 1095.
6. Wolfe, J. H., Silva, R. W., Myers, M. A. 1966, Observations of the solar wind during the flight of IMP–1, *J. geophys. Res.*, **71**, 1319.
7. Coon, J. H. 1966, Vela satellite measurements of particles in the solar wind and the distant geomagnetosphere, *Space Res.*, **6**.
8. Neugebauer, M., Snyder, C. W. 1966, Mariner 2 observations of the solar wind, I. Average properties, to be published in *J. geophys. Res.*
9. Bridge, H., Egidi, A., Lazarus, A., Lyon, E., Jacobson, L. 1965, Preliminary results of plasma measurements on IMP–A, *Space Res.*, **5**, 969.

10. Wolfe, J. H., Silva, R. W., Myers, M. A. 1966, Preliminary results from the Ames Research Center plasma probes observations of the solar wind geomagnetic field; interaction region on IMP-2 and OGO-1. NASA Space Sci. Div., Ames Res. Center, Calif.

11. Hundhausen, A. J., Ashbridge, J. R., Bame, S. J., Gilbert, H. E., Strong, I. B. 1966, Vela 3 satellite observations of solar wind ions: A preliminary report, Univ. of Calif., Los Alamos Sci. Lab., Los Alamos.

12. Strong, I. B., Ashbridge, J. R., Bame, S. J., Heckmann, H. H., Hundhausen, A. J. 1966, Measurements of proton temperatures in the solar wind, *Phys. Rev. Lett.*, **16**, 631.

13. Gringauz, K. I., Bezrukikh, V. V., Musatov, L. S. 1966, Observations of the solar wind by Venus 3 probe, presented at the Interunion Symposium on *Solar Terrestrial Physics*, Belgrade.

14. Wolfe, H. H., *et al.* 1966, The compositional, anisotropic and non-radial flow characteristics of the solar wind, *J. geophys. Res.*, **71**, 3329.

15. Snyder, C. W., Neugebauer, M., Rao, U. R. 1963, The solar wind velocity and its correlation with cosmic ray variations and with solar and geomagnetic activity, *J. geophys. Res.*, **68**, 6361.

16. Strong, I. B., Ashbridge, J. R., Bame, S. J., Heckmann, H. H., Hundhausen, A. J. 1966, Measurements of proton temperatures in the solar wind, *Phys. Rev. Lett.*, **16**, 631.

17. Coleman, P. J., Jr., Sonett, C. P., Judge, D. L., Smith, E. J. 1960, Some preliminary results of the Pioneer 5 magnetometer experiment, *J. geophys. Res.*, **65**, 1865.

18. Coleman, P. J., Jr., Davis, L., Jr., Sonett, C. P. 1960, Steady component of the interplanetary magnetic field: Pioneer 5, *Phys. Rev. Lett.*, **5**, 43.

19. Coleman, P. J., Jr., Davis, L., Jr., Smith, E. J., Sonett, C. P. 1962, The mission of Mariner 2: Preliminary observations of interplanetary magnetic fields, *Science*, **138**, 1099.

20. Coleman, P. J., Jr., Smith, E. J., Davis, L., Jr., Jones, D. E. 1966, Measurements of fields in the vicinity of the magnetosphere and in interplanetary space: Preliminary results from Mariner 4, *Space Res.*, **6**.

21. Ness, N. F., Scearce, C. S., Seek, J. B. 1964, Initial results of the IMP-1 magnetic field experiment, *J. geophys. Res.*, **69**, 3531.

22. Ness, N. F., Scearce, C. S., Seek, J. B., Wilcox, J. M. 1965, A summary of results from the IMP-1 magnetic field experiment, presented at Buenos Aires, COSPAR Conference.

23. Ness, N. F., Scearce, C. S., Cantarano, S. 1966, Preliminary results from the Pioneer 6 magnetic field experiment, *J. geophys. Res.*, **71**, 3305.

24. Ness, N. F., Wilcox, J. M. 1964, Solar origin of the interplanetary magnetic field, *Phys. Rev. Lett.*, **13**, 461.

25. Ness, N. F., Wilcox, J. M. 1966, Extension of the photospheric magnetic field into interplanetary space, *Astrophys. J.*, **143**, 23.

26. Coleman, P. J., Jr., Davis, L., Jr., Smith, E. J., Jones, D. E. 1966, Variations in the polarity distribution of the interplanetary magnetic field, *J. geophys. Res.*, **71**, 2831.

27. Wilcox, J. M., Ness, N. F. 1965, Quasi stationary corotating structure in the interplanetary medium *J. geophys. Res.*, **70**, 5793.

28. Babcock, H. W., Babcock, H. D., Simpson, J. A. 1955, Association of a 'unipolar' magnetic region on the Sun with changes of primary cosmic-ray intensity, *Phys. Rev.*, **98**, 4402.

29. Fan, C. Y., Gloeckler, G., Simpson, J. A. 1966, Protons and helium nuclei within interplanetary regions which correlate with the Sun, Proc. of the Int. Conf. on *Cosmic Rays*, **1**, 109.

30. Bryant, D. A., Cline, T. L., Desai, U. D., McDonald, F. B. 1962, Explorer 12 observations of solar cosmic rays and energetic storm particles after the solar flare of September 28, 1961, *J. geophys. Res.*, **67**, 4983.

31. Bryant, D. A., Cline, T. L., Desai, U. D., McDonald, F. B. 1963, New evidence for long-lived solar streams in interplanetary space, *Phys. Rev. Lett.*, **11**, 144.

32. Bryant, D. A., Cline, T. L., Desai, U. D., McDonald, F. B. 1965, Studies of solar protons with Explorers 12 and 14, *Astrophys. J.*, **141**, 478.

33. McCracken, K. G., Ness, N. F. 1966, The collimation of cosmic rays by the interplanetary magnetic field, *J. geophys. Res.*, **71**, 3315.
34. Bartley, W. C., Bukata, R. P., McCracken, K. G., Rao, U. R. 1966, Anisotropic radiation fluxes of solar origin, *J. geophys. Res.*, **71**, 3297.
35. Fan, C. Y., Lamport, J. E., Simpson, J. A., Smith, D. R. 1966, Anisotropy and fluctuations of solar proton fluxes of energies 0·6–100 MeV measured on the Pioneer 6 space-probe, *J. geophys. Res.*, **71**, 3289.
36. Ness, N. F. 1966, Simultaneous measurements of the interplanetary magnetic field, *J. geophys. Res.*, **71**, 3319.
37. Colburn, P. S., Sonett, C. P. 1966, Discontinuities in the solar wind, *Space Sci. Rev.*, **5**, 439.
38. Coleman, P. J., Jr. 1966, Variations in the interplanetary magnetic field: Mariner 2, Part 1, Observed properties. Inst. of Geophys. and Planet. Phys., Univ. of Calif., no. 501.
39. Coleman, P. J., Jr. 1966, Variations in the interplanetary magnetic field: Mariner 2, Part 1. Observed properties. Inst. of Geophys. and Planet. Phys., Univ. of Calif., no. 504.
40. Coleman, P. J., Jr. 1966, Variation in the interplanetary magnetic field: Mariner 2, Part 3. Some effects upon solar cosmic rays, Inst. of Geophys. and Planet. Phys., Univ. of Calif., no. 514.
41. Smith, E. J., Holzer, R. E., Coleman, P. J., Jr., Davis, L., Jr., Siscoe, G. S. 1966, A comparison of simultaneous interplanetary magnetic field measurements: OGO–1 and Mariner 4, Inter Union Symposium on *Solar Terrestrial Physics*, Belgrade.

APPENDIX I. INVESTIGATIONS OF INTERPLANETARY PLASMA IN THE U.S.S.R. IN 1965 AND 1966

(prepared by K. Gringauz)

In 1965 and 1966, theoretical and experimental studies of problems connected with interplanetary plasma were continued in the U.S.S.R.

E. R. Mustel has analysed the role of active regions of the Sun in production of quasi-stationary plasma flows in interplanetary space (1). Two hypotheses were considered. According to one of these hypotheses, quasi-stationary flows of solar corpuscles appear in the most active regions, and, according to the other, active regions only deflect fluxes generated in undisturbed regions of the Sun (hypothesis of the 'escape cone'). In (1) various observational results were used (data derived from direct measurement of fluxes of solar plasma and interplanetary magnetic fields as well as measurements of cosmic rays of solar origin, of radiowave absorption in auroral zones) and a conclusion was made that all data fully confirm the first hypothesis (of active regions).

Studies of statistical very rich results of magnetic measurements performed mainly by Mariner–2 and IMP–1 space probes, gave grounds to Yu. D. Kalinin and E. I. Mogilevsky for extension of a hypothesis on a 'plasmoid' macroscopic structure of corpuscular fluxes generated in active regions of the Sun (2). According to the above mentioned hypothesis, this flux consists of a chain of independent elements, each of which moves radially and possesses its own quasi-force-free magnetic field (satisfying the condition (rot $\mathbf{H}\cdot\mathbf{H}$) = 0).

Each of these elements plasmoids (M-elements, according to the authors' terminology) retains the angular momentum of the Sun's rotation. Thus the whole chain is placed along Archimedes' spiral, the slope of whose tangent is determined by radial velocity. The dimensions of each M-element increase with time. The chain of the M-element is propagating in Parker's quiet solar wind.

The above hypothesis seems to be of importance for elucidating the mechanisms of the appearance of disturbances in the Earth's magnetosphere.

Interesting data relating to interplanetary plasma are obtained during radioastronomical observations.

T. D. Antonova, V. V. Vitkevich and V. I. Vlasov (**3**) revealed the scintillation of source 3C 48 at a wavelength of 3·5 m with a period of 3 seconds determined by inhomogeneities of electron concentration in interplanetary plasma.

V. V. Vitkevich and V. I. Vlasov (**4**) reported the first results of the experiment on determining the velocity of motion of inhomogeneities in interplanetary plasma derived from data of synchronous observations of fluctuations in intensities of radioemission of sources 3C-144, 3C-43 and C3-147 at three points situated at apexes of an equilateral triangle with side of about 220 km.

For the observational period from July 1966 to August 1966 the greatest velocities were 350 km s^{-1}. At velocities of 220–350 km s^{-1}, the directions of velocities are close to the direction away from the Sun. The dispersion of directions was $\pm$ 30°. According to the authors, the characteristic size of inhomogeneities was about 400 km with dispersion by a factor of 1·5 to 2. The greater portion of inhomogeneities moves together with a common flux of solar plasma. On the average a direct correlation is observed between the dimensions of the inhomogeneities and their velocities.

The results of measuring the solar plasma fluxes (solar wind) were processed. The measurements were performed from Zond-2 by V. V. Bezrukikh, K. I. Gringauz *et al.* (**5**). Zond-2 carried both traps which recorded the integral flux of the solar wind ions and an ion trap of a modulation type oriented on the Sun which made it possible to obtain energy spectra of ions with energies of up to 3·6 keV. These experiments showed that the absence of geomagnetic disturbances corresponds to very small fluxes of solar wind ions ($\sim 2 \times 10^7$ cm^{-2} s^{-1}), while during geomagnetic storms ($K_p = 5$) fluxes of solar plasma ions near the Earth's orbit amount to 10^9 cm^{-2} s^{-1}. K. I. Gringauz, V. V. Bezrukikh and L. C. Musatov processed also the results of measurements of solar wind energy spectra performed by means of a sun-oriented ion modulation-type trap from Venus–3 at the close of 1965 and in early 1966 (**6**). These measurements have made it possible to draw the following basic conclusions:

In all cases, when the instrument for studying the solar wind was switched on, ion fluxes were registered. The total flux of ions with energies of about 3600 eV for a period from 16 November 1965 to 7 January 1966 was within the limits of $1{\cdot}5 \times 10^8$ cm^{-2} s^{-1} to 2×10^8 cm^{-2} s^{-1}. The velocities of the main (largest) components of the spectrum for this period did not exceed 510 km s^{-1}. Often the fluxes close by their magnitudes were recorded in different energy ranges (the width of each range was about 450 eV). In these cases the determination of some directed velocity characterizing the solar wind becomes difficult. It is possible that some spectra of this type correspond to the penetration of the fast solar plasma flux into one which moves slower. Although large ion fluxes ($N = 10^9$ cm^{-2} s^{-1}) most often correspond to high K_p-indexes, apparently one is unable to establish a correlation between some structural parameter of the undisturbed solar wind and the intensity of geomagnetic disturbances. It is possible, that geo-effectiveness of solar wind fluxes is greatly affected by the orientation of the vector of the interplanetary magnetic field in the vicinity of the Earth's magnetosphere.

Measurements of the solar plasma fluxes in the neighbourhood of the Moon performed by means of charged traps from Luna–10, the first artificial moon satellite, in April and May 1966, by Gringauz, Bezrukikh *et al.* (**7, 8**), have shown that near the Moon, there is a disturbed region in which plasma fluxes move in the direction differing from the direction of solar wind in undisturbed interplanetary space. Some signs were also revealed that, each month, the Moon in a period close to full moon, approximately for four days, is situated in the tail of the Earth's magnetosphere where undisturbed wind does not penetrate.

References

1. Mustel, E. R. 1965, *Astr. Zu.*, **42**, 473.
2. Kalinin, Yu. D., Mogilevsky, E. I. 1965, *Issled Kosmich. Prostr.* (Investigations of Outer Space), 'Nauka' Publishing House.
3. Antonova, T. D., Vitkevich, V. V., Vlasov, V. I. 1966, Preprint, Lebedev Physical Institute, Academy of Sciences of the U.S.S.R.,
4. Vitkevich, V. V., Vlasov, V. I. 1966, Paper presented at the Belgrade Symposium on *Solar Terrestrial Physics.* (Preprint, Ledebev Physical Institute, Academy of Sciences of the U.S.S.R.
5. Bezrukikh, V. V., Gringauz, K. I., Musatov, L. S., Rybchinsky, R. E., Khokhlov, M. Z. 1965, *Dokl. Akad. Nauk SSSR*, **163**, 873.
6. Gringauz, K. I., Bezrukikh, V. V., Musatov, L. S. 1966, Paper presented at the Symposium on *Solar Terrestrial Physics*, Belgrade.
7. Gringauz, K. I., Bezrukikh, V. V., Remizov, A. P., Khokhlov, M. Z., Musatov, L. S. 1966, *Dokl. Akad. Nauk SSSR*, **170**, 570.
8. Gringauz, K. I., Bezrukikh, V. V., Khokhlov, M. Z., Zastenker, G. N., Remizov, A. P., Musatov, L. S. 1966, *Dokl. Akad. Nauk SSSR*, **170**, 1306.

APPENDIX II. INVESTIGATIONS OF NEUTRAL HYDROGEN IN THE REGION OF THE EARTH AND IN INTERPLANETARY SPACE

(prepared by V. G. Kurt)

During 1964–66 in the Soviet Union, investigations of scattered solar $L\alpha$ radiation were made with devices mounted on cosmic rockets.

V. G. Kurt and V. V. Katiushina carried out three experiments on the geophysical rockets. $L\alpha$ radiation and radiation in lines of the triplet O I λ1300Å were measured at the height up to 500 km (6 June 1963; 25 October 1963; 24 December 1963). During these experiments, the Sun was at the height of $+ 5°$, $+ 3°$ and $- 17°$ above the horizon. The following data were obtained: on the distribution of hydrogen in the Earth's atmosphere, on its full contents and on the effect of the dependence of optical thickness on the solar zenith distance in the place of observations. With similar devices V. G. Kurt carried out observations of scattered $L\alpha$ radiation in the upper atmosphere of the Earth and in the interplanetary space. The observations were carried out from the board of automatic interplanetary station 'Zond–1' launched on 2 April 1964. These measurements showed that thickness of neutral hydrogen up to 6·5 terrestrial radii is equal to 20 cm^{-3} with $R = 42\,000$ km. The intensity of scattered radiation in interplanetary space is $1{\cdot}5 \times 10^{-4}$ erg cm^{-2} s^{-1} $ster^{-1}$ that corresponds to the concentration of interplanetary hydrogen $\sim 10^{-2}$ cm^{-3} with the temperature $\sim 10^5$ °K. More thorough investigations were carried out from the board of 'Venera–2' and 'Venera–3' 12 and 16 November 1965). The concentration of $L\alpha$ radiation to the plane of the ecliptic was not discovered. From these measurements, the concentration of hydrogen was determined up to 20 $R_\oplus$, equal on this distance to about 3 cm^{-3}. The upper limit of the intensity in the interval 1225–1340Å, equal to about 10^{-24} w m^{-2} Hz^{-1} $ster^{-1}$ (3×10^{-7} erg cm^{-2} s^{-1} $ster^{-1}$) was determined.

V. V. Katiushina solved the problem of the scattering of the radiation in the lines O I λ1300Å in the Earth atmosphere, taking into account the absorption by the molecular oxygen in the coherent approximation.

S. A. Kaplan and V. G. Kurt constructed an approximated theory of $L\alpha$ radiation transfer in the spherical atmosphere taking into account the Earth's shadow. They also explained the character of transfer of the radiation in lines O I λ1300Å in the flat atmosphere without taking into consideration redistribution according to frequencies.

T. A. Germogenova solved the problem of the diffusion of $L\alpha$ quanta in the zone H I taking into consideration zone H II in the assumption of coherency using this calculation.

V. G. Kurt and T. A. Germogenova obtained the lower limit of the dimension of the zone H II near the Sun, as equal to $10^3 - 10^4$ A.U. They showed that $L\alpha$ radiation observed in the solar system cannot be completely explained by terms of the scattering on cold galactic hydrogen. The whole effect is explained by hot ($T \sim 10^5 - 10^6$ °K) interplanetary hydrogen with the concentration 10^{-2} to 10^{-3} cm^{-3}.

References

1. Babichenko, S. I., Karpinsky, I. P. *et al.* 1965, *Cosmicheskie issledovaniia*, **3**, 237.
2. Katiushina, V. V., Kurt, V. G. 1965, *Cosmicheskie issledovaniia*, **3**, 243.
3. Kaplan, S. A., Kurt, V. G. 1965, *Cosmicheskie issledovaniia*, **3**, 256.
4. Kaplan, S.A., Kurt, V. G. 1965, *Cosmicheskie issledovaniia*, **3**, 251.
5. Kurt, V. G. 1966, *Cosmicheskie issledovaniia*, **4**, 111.
6. Kurt, V. G. 1966, *Space Res.*, **6**, Buenos-Aires.
7. Katiushina, V. V. 1966, *Cosmicheskie issledovaniia*, **4**, 619.
8. Kurt, V. G. 1967, *Space Res.*, **7**, Vienna.
9. Kurt, V. G., Germogenova, T. A. 1967, *Astr. Zu.*, **44**, 2.
10. Germogenova, T. A. 1966, *Astrofisica*, **2**, 3, 251.
11. Kurt, V. G. 1965, *Issledovaniia cosmicheskogo prostranstva*, Nauka, Moscow, 576.
12. Kaplan, S. A., Katiushina, V. V., Kurt, V. G. 1964, *Space Res.*, **5**, Warszawa.
13. Katiushina, V. V. 1965, *Cosmicheskie issledovaniia*, **3**, 248.

APPENDIX III. SKY LUMINOSITY IN THE ULTRAVIOLET REGION OF THE SPECTRUM

(prepared by A. B. Severny)

Sky luminosity in the ultraviolet region of the spectrum ($\lambda = 2800$ Å) was measured with the astrophotometer (AF–3) supplied with two photomultipliers: 57 for the measurings in the ultraviolet region and 15—for the measurings in the visible region. The device was mounted on the satellite 'Cosmos–51' launched on 10 December 1964. The device operated during three months on the heights about 300 km.

The following results were obtained:

The lowest sky luminosity in the ultraviolet region is approximately equal to the luminosity of 190 stars of 10^m per square degree. This luminosity exceeds 2 to 3 times the expected luminosity of these stars derived from their energy distribution along the spectrum. Considerable fluctuations were observed of the ratio of the luminosity in the visible region to the luminosity in the ultraviolet as well as a discrepancy with the value of this ratio obtained from theory.

APPENDIX IV. DETERMINATION OF SPATIAL DENSITY OF METEORIC DUST IN SPACE

(prepared by T. N. Nazarova)

During 1963–66, measurements were conducted to continue the work aimed for determining spatial density of meteoritic dust in space. Data were obtained along the flight trajectories of the probes in the direction away from the Earth's orbit towards the Sun and away from the Sun and in the vicinity of the Moon.

From the experiments performed during the above-mentioned period and earlier, a conclusion may be drawn that non-uniformity in the distribution of meteoritic matter in space is rather the rule than the exception.

Apparently meteoric matter is united in space into more or less dense formations with linear dimensions varying within wide limits.

During the experiment with Luna 10 satellite, aggregations of particles were recorded in the vicinity of the Moon.

Note—Addenda to this Report of Commission 44 will appear in Volume XIII B of the *Transactions of the IAU*.

45. COMMISSION DES CLASSIFICATIONS SPECTRALES ET INDICES DE COULEUR A PLUSIEURS BANDES

PRÉSIDENT: Dr W. P. Bidelman, The Observatory, University of Michigan, Ann Arbor, Michigan, U.S.A.

VICE-PRÉSIDENT: Professor Ch. Fehrenbach, Directeur de l'Observatoire de Marseille, 2 Place Le Verrier, 13-Marseille-IV, France.

COMITÉ D'ORGANISATION: W. W. Morgan, R. O. Redman, B. Strömgren, Th. Walraven, B. Westerlund.

MEMBRES: Bappu, Barbier†, Bartaya, Becker (W.), Beer, Blanco, Borgman, Boulon, Buscombe, Canavaggia, Chalonge, Crawford, Dessy, Divan, Duflot, Elvius (T.), Evans (D. S.), Feast, Golay, Griffin, Gyldenkerne, Hack, Hallam, Henize, Herbig, Hoag, Iwanowska, Jaschek (C. O. R.), Johnson (H. L.), Keenan, Kharadze, Kopylov, Kron (G. E.), Lindblad (B.)†, Lodén, McCarthy, McCuskey, McNamara, Malmquist, Meinel, Nassau†, Oosterhoff, Osawa, Petrie (R. M.)†, Preston, Reddish, Roman, Schmidt-Kaler, Seitter, Sharpless, Sinnerstad, Sinvhal, Slettebak, Steinlin, Stephenson, Thackeray, Thé, Treanor, Van den Bergh, Velghe, Wilson (O. C.), Yoss.

It is our sad duty to record at the outset of this first report of Commission 45 the loss by death of four of the Commission's most distinguished members: Daniel Barbier, Bertil Lindblad, Jason J. Nassau, and Robert M. Petrie. And, though not Commission members, it is appropriate to add to this list the names of Dean B. McLaughlin and Joel Stebbins. Each of these six men was, in his own way, a real pioneer in the fields of interest to this Commission. Their work will long live in the memory of their successors. Our deepest sympathy goes to their families and colleagues.

INTRODUCTION

The large amount of activity in the fields of spectral classification and multicolor photometry during the past three years certainly attests to the wisdom of the Union in setting up Commission 45. New and most unusual types of objects have been discovered (quasi-stellar sources, X-ray sources, infrared stars) that must be observed and whose spectra must be classified, and new spectral regions have opened up for study in a most dramatic fashion. After all, who would have thought a few years ago that the Lyman-α line would ever be observed from the surface of the Earth, or that one would become accustomed to casually plotting the energy curve of an object using a wavelength scale ranging over a factor of 10^7! At the same time previous classification methods have been critically examined and improved and applied to an ever-growing variety of astronomical problems.

THE BIBLIOGRAPHY

It is hoped that the accompanying bibliography, which is a continuation of two previous efforts along the same line (*Trans. int. astr. Un.*, **11A**, p. 335, 1962, and **12A**, p. 479, 1965), will prove useful. Inadequacies are the fault of the compiler and of those who did not bring papers published in obscure journals to his attention. The arrangement and choice of material included are, of course, the responsibility of the compiler alone.

Prospective users are warned, specifically, of the following limitations of the bibliography:

1. Numerous publications reporting colloquia or symposia are noted simply by a single title. Individual contributions in these volumes are *not* noted separately under any heading.

For example, papers given in the recent IAU symposia at Canberra, Utrecht, and Saltsjöbaden are not listed separately. To have done so would have been to increase substantially an already over-long list of references.

2. The bibliography does not contain references to papers giving a classification or photometry for a single object. This policy has no doubt resulted in the omission of a number of significant papers.

3. References are not, in general, given to papers dealing with three-color wide-band photometry. It can be argued that this exclusion is unwise. However, it appears that most of these papers deal with clusters, variable stars, or galactic research, and it is hoped that adequate references to them will be found in other Commission reports. Details concerning all types of photometric systems will, of course, be found in the report of the Commission on Stellar Photometry.

In the construction of the bibliography it has proved useful to attempt to distinguish between papers that actually list classifications and those that provide fundamental spectroscopic data that will, no doubt, be used eventually for the classification of previously unstudied objects. Many of the latter papers contain new and exciting results. The large number of recent papers dealing with the problem of the abundances of lithium and beryllium in stellar spectra, the first steps in the investigation of the ultraviolet spectra of stars, and the spectacular new studies of the astronomical infrared which have led to the discovery of H_2 and H_2O in the spectra of the cool stars all belong to this category.

Section 6, which deals with calibration problems, has been broadened to include the calibration of stellar temperatures and atmospheric chemical compositions as well as of luminosities alone.

UNPUBLISHED WORK

One of the disadvantages of listing the unpublished work known to the President is that of completely leaving out a very large proportion of the work that is, in fact, going on or planned by members of the Commission. The reader is strongly urged, thus, to consult, in addition to the bibliography, the reports of other Commissions, the annual reports of the various observatories, and also J. Sahade's *Information Bulletin for the Southern Hemisphere*, if he wishes to be reasonably well informed of the current state of research in any specific field.

COMMENTS ON THE PAST THREE-YEARS' WORK

It is clearly impossible to summarize adequately the papers published during the past three years that involve work of interest to Commission 45. A few comments may, however, be appropriate:

Though there was much activity in objective-prism work, new techniques (**7, 18, 84**) were the exception rather than the rule. Stephenson has given (**522**) an interesting summary of the astrophysical utilization of objective-prism spectra. Spectral data on surprisingly faint stars were shown to be obtainable with a 24-inch (61-cm) telescope (**82**), and the possibilities of moderate-dispersion objective-prism equipment for accurate spectral classification were emphasized (**40**). Several investigators (**3, 12**) pointed out the advantages of fairly high dispersion in surveys for bright-line objects, while the usefulness of the visual region for discovery of S-type stars was noted at the Warner and Swasey Observatory (**79**).

New phenomena were noted, new classification systems proposed, and substantial improvements in classification accuracy made (**118, 148, 149, 171, 172, 173, 178, 188**). As is evident from the bibliography, much of the effort of large telescopes was spent on the quasi-stellar sources, and there is now enough material available to justify a rough attempt at spectral classification. A fairly large number of informative papers have dealt with the properties of the faint blue stars and related objects (**106, 125, 136, 137, 139, 146, 147, 161, 185**). An interesting development of the problem of the metallic-line stars was the demonstration (**119**) that the weakness of λ4246 of Sc II can conveniently be used to detect marginal cases. And the infrared stars have been shown, at least in some cases (**216, 298**), to be not quite so unusual as they first appeared to be.

The current situation in narrow- and intermediate-band photoelectric work has been well summarized by Strömgren (**261**). New techniques have been invented (**220, 239, 240, 244, 245, 250, 259, 263, 266**), while previous ones have been applied to many normal and peculiar (**228, 260, 267**) objects. Extensive investigations involving Hα, the sodium D lines, and other spectral features have continued at Cambridge.

A very large amount of highly accurate multicolor photometry has been carried out in the past three years by Johnson and his collaborators. The infrared photometry of normal stars (**290, 310**), M dwarfs (**292**), galaxies (**294**), carbon stars (**311**), infrared stars (**298, 302, 308**), and the truly spectacular results found for several T Tauri objects (**309**) represent a very substantial increase in our knowledge. Yet the limitations of the multi-color, as compared with the spectroscopic, approach are clearly evident in Figure 3 of Johnson's paper (**298**), which shows very similar energy curves, as measured through his filters over a large wavelength range, for a highly reddened late B-type supergiant and a much less reddened M2 supergiant!

Spectrophotometric methods have yielded a number of interesting new results, especially involving the equivalent widths of Hγ in Magellanic Cloud stars (**352**) and the discovery that the width of Hα (**355**) and of other lines as well (**394**) can be used as a luminosity indicator in the later-type stars. While much use has been made of the equivalent widths of the various hydrogen lines as an index of luminosity, the surprising results of Lacoarret (**359**) and of Walker (**375**) on the variability of the profiles of these lines in the B stars leads one to ask whether this method is really as reliable as it is often claimed to be.

It has become increasingly evident that even among stars that are not noticeably peculiar photometric data are often unfortunately a rather poor substitute for a spectrogram. That this is true in the later spectral types has been recognized for some time. However it has been generally assumed that Strömgren's m_1 value is a good index of the metal/hydrogen ratio in F- and G-type stellar atmospheres. This assumption has recently been called into question by Conti and Deutsch (**457**), who have pointed out that the value of the turbulent velocity is also of major importance in this connection. Though the situation is not entirely clear at this time, it is evident that the correct interpretation of even quite sophisticated photometric data is a very difficult problem, and one not to be attacked lightly.

A controversy affecting stellar luminosities has recently arisen through Hodge and Wallerstein's claim (**461**) that the distance of the Hyades is some 18% larger than has been previously believed. The effects of this change on galactic and extragalactic distance scales are, of course, very substantial. It may be hoped that this matter will be settled soon. The effect of stellar rotation on the position of stars in the H–R diagram has also come in for very considerable discussion (**404, 456, 462, 464, 467, 470, 474**). It will probably prove desirable to set up standards of differing rotational velocity for the most accurate classification of early-type stars.

CATALOGUES OF SPECTROSCOPIC DATA

Several very valuable compilations of spectral classifications and photometric data have been published during the past three years. It is already evident that these are proving of great value to workers in our field.

1. Miss Hoffleit's new edition of the Yale *Catalogue of Bright Stars* (**503**) now contains either MK spectral types or at least spectroscopic indications of luminosity (*c*,*g*,*d*) for a large proportion of its 9110 stars.

2. The exhaustive *Catalogue of Stellar Spectra Classified in the Morgan-Keenan System* (**506**) compiled by Jaschek, Conde, and de Sierra is an essential aid to the stellar astronomer. A second edition including new data is eventually planned.

3. The large compilation of stellar rotational velocities by Boyarchuk and Kopylov (**485**), the list of photoelectric Hβ values of 1217 bright stars published by Crawford and his collaborators (**232**), and Keenan's catalogue giving spectral types from individual spectrograms of 253 Mira variables (**156**), are all most useful pieces of work.

4. Two extensive and very useful catalogues of photoelectric multicolor data have been published by the Arizona group:

(*a*) Five-color photometry (*UBVRI*) of the 1325 brightest stars north of $\delta = -50°$ (**288**). This is largely complete to $m_v = 5{\cdot}0$.

(*b*) Multicolor data (five colors or more) for 1567 stars of the *Bright Star Catalogue* (**304**) Six hundred and fifty-three of these have been measured in *J* and *K* ($1{\cdot}25$ and $2{\cdot}2\mu$), and 268 in *L* ($3{\cdot}4\mu$). Also *UBV* data are given for nearly 5000 bright stars over the entire sky on the basis of a new reduction of available Cape reflector *U* values.

Those who have labored long and diligently to prepare these data catalogues merit our heartiest congratulations and thanks.

Two additional catalogues of spectroscopic data are also in the planning stage:

1. (Jaschek): A new catalogue and bibliography of Be stars, continuing the work of Merrill and Burwell, is being prepared. It is hoped that this will be ready in 1967.

2. (Stephenson): A comprehensive catalogue of the carbon stars is being compiled at the Warner and Swasey Observatory. Spectral plates are being used to check ambiguities in published data. While it was originally planned to publish this catalogue in the near future, it has been decided to delay publication until more data for stars of the Southern Hemisphere become available.

Both Dr Jaschek and Dr Stephenson will appreciate receiving unpublished data for inclusion in their catalogues, and will be happy to make information available upon request.

Despite the numerous stellar data catalogues now available or in preparation, it is unfortunately true that there is still much labor involved in obtaining relevant information concerning miscellaneous spectroscopic work, duplicity, variability, proper-motion and radial-velocity data, etc., for specific objects. This situation is made even more complicated than it needs to be by lack of agreement on a common equinox for tabulated positions. The President has a strong preference for the use of 1900 coordinates in data catalogues, and feels that to use any other equinox is to invite chaos; positional catalogues are, of course, a different matter.

The total number of individual stars in the Milky Way system for which spectroscopic information of some sort is now available appears to be of the order of at least 800 000, and if we add in the remaining unclassified stars for which various types of significant data exist we arrive at a figure of the order of approximately one million for the number of objects that should be contained in a master stellar data catalogue. It can be argued that not all of the information that might be contained in such a catalogue is especially useful (!), but certainly data have a surprising way of becoming useful just when we least expect it.

There already exists a fairly comprehensive spectroscopic information card catalogue at Michigan for those normal stars for which two-dimensional spectral classifications exist and for a large variety of spectroscopically peculiar objects. However, this contains a mere 25 000 or so stars, a far cry from a million. And even this has proved troublesome to keep up to date.

Yet the President feels that the time may be ripe for at least investigating the possibility of establishing a general stellar information center that would eventually become a comprehensive storehouse of information—giving data and the sources thereof—on the various inhabitants of the Galaxy. Perhaps the Commission on Galaxies would like to start thinking about a similar data center for galaxies. Presumably by the time either center becomes organized the question of which center should include the quasi-stellar sources will be settled for good!

SPECTRAL ATLASES

At the moment a number of spectral atlases are being prepared:

1. (Fehrenbach): An atlas of stellar spectra obtained with the small objective prism will appear shortly in the *Journal des Observateurs*; prints from this atlas will be available separately at cost.

2. (Landi Dessy and Jaschek): The Córdoba spectral atlas at 42Å/mm is approaching completion and it is hoped to be able to exhibit it at the Prague meeting. As far as possible the MK criteria have been retained; this has been feasible in general in the earlier spectral classes.

3. (Meinel and Christy): A new atlas containing spectra of about 300 stars, equally divided between normal and peculiar objects, is being prepared. The plate material covers a large spectral range.

4. (Morgan, Keenan, and Hiltner): Work is continuing on the revision of the MK classification system in connection with the preparation of the new edition of the Yerkes Spectral Atlas. New criteria have been introduced that permit a smoother and more precise classification for the O and B stars. A series of standard coeval spectral sequences will be incorporated in the Atlas.

Members of the Commission await the appearance of these atlases with great anticipation. The need for these has been the subject of repeated concern at recent IAU meetings.

W. P. BIDELMAN
President of the Commission

BIBLIOGRAPHY (1964–66)

1a. *Classifications Employing Visual Inspection of Objective-Prism or Slitless Spectrograms: Emission-Line Objects*

1. Andrews, A. D., Lindsay, E. M. 1964, Third list of emission-line objects in the Large Magellanic Cloud. *Irish astr. J.*, **6**, 241.
2. Apriamashvili, S. 1965, Three new planetary nebulae. *Astr. Cirk.* no. 318, 1.
3. Bidelman, W. P. 1964, Two new emission-line stars. *Astrophys. J.*, **139**, 405.
4. Bidelman, W. P. 1965, Observations of emission-line stars. *Publ. astr. Soc. Pacif.*, **77**, 388.
5. Blanco, V. M., FitzGerald, M. P. 1964, Faint Hα-emission objects near ζ Cephei. *Publ. astr. Soc. Pacif.*, **76**, 438.
6. Bond, H. E., Bidelman, W. P. 1966, On two nonexistent Wolf-Rayet stars. *Publ. astr. Soc. Pacif.*, **78**, 261.
7. Bretz, M. C. 1965, Application du réducteur focal aux étoiles à émission dans la région de NGC 2264. *Publ. Obs. Hte-Provence*, **8**, no. 8.
8. Dolidze, M. V., Ponomareva, G. A. 1965, New emission stars in Cassiopeia. *Soviet Astr.*, **9**, 157.
9. Fehrenbach, Ch., Duflot, M. 1964, Objets à émission du Grand Nuage de Magellan sur des clichés de prisme-objectif. *J. Observateurs*, Marseille, **47**, 85.
10. Frantsman, Yu. L. 1962, Determination of the coordinates of planetary nebulae from photographs taken with an objective prism. *Soviet Astr.*, **6**, 198.

11. Kohoutek, L. 1965, Hamburg Schmidt-camera survey of faint planetary nebulae. *Bull. astr. Inst. Csl.*, **16**, 221.
12. McCarthy, M. F., Treanor, P. J. 1965, Structural detail in a region near the anticentre. *Ric. astr. Specola astr. Vatic.*, **7**, 177.
13. Peimbert, M., Bátiz, G. 1960, Nuevas nebulosas planetaries. *Bol. Obs. Tonantzintla Tacubaya* no. 19, 19.
14. Peimbert, M., Costero, R. 1961, Nuevas nebulosas planetarias. *Bol. Obs. Tonantzintla Tacubaya*, **3**, 33 (no. 21).
15. Perek, L. 1963, Planetary nebulae in the central region of the galaxy. *Bull. astr. Inst. Csl.*, **14**, 201.
16. Schmidt-Kaler, Th. 1964, Die dBe-Sterne, eine natürliche Spektralgruppe geringer Leuchtkraft-Streuung. *Z. Astrophys.*, **58**, 241.
17. Sher, D. 1965, Distances of five open clusters near Eta Carinae. *Mon. Not. R. astr. Soc.*, **129**, 237.
18. Stephenson, C. B. 1966, Search for new northern Wolf-Rayet stars. *Astr. J.*, **71**, 477.
19. The, P.-S. 1964, Faint stars with Hα emission near ν Scorpii. *Publ. astr. Soc. Pacif.*, **76**, 293.
20. The, P.-S. 1964, Thirty-six new planetary nebulae. *Contr. Bosscha Obs.* no. 26.
21. The, P.-S. 1964, Note on a group of faint Hα-emission stars associated with a dark nebulosity in Ophiuchus. *Contr. Bosscha Obs.* no. 27.
22. The, P.-S. 1964, Eleven new planetary nebulae. *Contr. Bosscha Obs.* no. 28.
23. The, P.-S. 1964, Identification charts for planetary nebulae. *Contr. Bosscha Obs.* no. 29.
24. The, P.-S. 1966, New H-alpha emission stars in the region of the OB association Sgr III. *Contr. Bosscha Obs.* no. 34.
25. The, P.-S. 1966, A catalogue of new H-alpha emission stars in the Eta Carinae Nebula region. *Contr. Bosscha Obs.* no. 35.
26. The, P.-S., Lim, H.-K. 1964, New Hα emission stars in a region to the east of the T-association near Rho Ophiuchi. *Contr. Bosscha Obs.* no. 23.
27. Vandervort, G. L. 1964, Nine new planetary nebulae. *Contr. Bosscha Obs.* no. 30.
28. Voigt, H. H. 1965, *Luminous stars in the northern Milky Way. I. Supplement.* Hamburg-Bergedorf: Hamburger Sternwarte and Warner and Swasey Obs.
29. Westerlund, B. E., Smith, L. F. 1964, Planetary nebulae in the Large Magellanic Cloud. *Mon. Not. R. astr. Soc.*, **127**, 449.
30. Westerlund, B. E., Smith, L. F. 1964, Wolf-Rayet stars in the Large Magellanic Cloud. *Mon. Not. R. astr. Soc.*, **128**, 311.
31. Wray, J. D. 1966, A new catalogue of Hα emission objects in the southern hemisphere. *Astr. J.*, **71**, 403.

Unpublished Work

(Bidelman): Numerous new Hα-emission stars, many fairly bright, have been noted on 10° Curtis Schmidt plates and are awaiting publication. Three stars in the faint unstudied galactic cluster NGC 2421 were detected as emission objects by D. MacConnell.

(The): Approximately 90 bright-line stars, m_v = 12 to 16, have been found by B. Hidajat in a region of about 60 square degrees in Ara.

1b. *Classifications Employing Visual Inspection of Objective-Prism or Slitless Spectrograms: General Work*

32. Apriamashvili, S. P. 1964, Photographic magnitudes, color indices, spectra, and luminosities of faint stars in Aquila and Scutum. *Abastumansk. astrofiz. Obs. Gora Kanobili Bjull.* no. 31, 27.
33. Apriamashvili, S. 1965, Three new variable stars. *Astr. Cirk.* no. 317, 1.
34. Ažusienis, A., Bartkevičius, A. 1964, Spectral types and magnitudes of comparison stars for RS, TZ, WY, BY Andromedae and SVS 1452. *Astr. Obs. Biul., Vilnius*, **11**, 31.
35. Ažusienis, A., Bartkevičius, A. 1966, Photometry and spectrophotometry of RS and TZ Andromedae. *Astr. Obs. Biul., Vilnius*, **17**, 32.

36. Balázs, B. 1965, Luminous stars in a region south of h and chi Persei. *Z. Astrophys.*, **62**, 6.
37. Barbier, M. Liste et classification d'étoiles M, S et C nouvelles. *J. Observateurs*, Marseille, **48**, 149, 1965 and **49**, 290, 1966.
38. Bartkus, R. 1964, Catalogue of 2455 stars in the middle part of the Orion association. *Astr. Obs. Biul., Vilnius*, **13**, 29.
39. Bertiau, F. C., McCarthy, M. F., Treanor, P. J. 1965, M stars in the region of the south galactic pole. *Astr. J.*, **70**, 133.
40. Bidelman, W. P. 1966, Accurate spectral classification by objective-prism techniques. *Vistas in Astr.*, A. Beer and K. Aa. Strand, eds., **8**, 53.
41. Bidelman, W. P., Victor, R. C. 1966, Twenty-three stars with peculiar spectra. *Publ. astr. Soc. Pacif.*, **78**, 550.
42. Bouigue, R., Boulon, J., Pedoussaut, A. 1963, Contribution aux recherches de photométrie photoélectrique dans la Galaxie. *Publ. Obs. Hte-Provence*, **6**, no. 38.
43. Boulon, J. 1963, Étude photométrique et cinématique de dix champs galactiques. II. *J. Observateurs*, Marseille, **46**, 225.
44. Brodskaya, E. S., Grigoreva, N. B. 1964, Spectra, visual magnitudes and colour indices of 2816 stars in Cassiopeia. *Soobšč. gos. astr. Inst. P. K. Sternberga* no. 136.
45. Dolidze, M. V. 1964, M stars in the region of ζ Ophiuchi. *Abastumansk. astrofiz. Obs. Gora Kanobili Bjull.* no. 30, 71.
46. Dolidze, M. V. 1964, The Cygnus loop region in red and infrared light. *Abastumansk. astrofiz. Obs. Gora Kanobili Bjull.* no. 30, 81.
47. Dolidze, M. V., Jimsheleishvili, G. N. 1966, New possible star clusters. *Astr. Cirk.* no. 382, 7.
48. Dolidze, M. V., Jimsheleishvili, G. N. 1966, Spectra of two cool stars. *Astr. Cirk.* no. 387, 7.
49. Dolidze, M. V., Lipaeva, N. A. 1965, Red stars in the area of NGC 6819. *Abastumansk. astrofiz. Obs. Gora Kanobili Bjull.* no. 32, 53.
50. Dombrovskii, V. A., Hagen-Torn, V. A. 1964, Space distribution of stars, dust and interstellar polarization in the region of the open cluster NGC 1502. *Trudy astr. Obs. Leningr. gos. Univ.*, **20**, 75.
51. Fedorchenko, G. L. 1962, Spectral classes of 1717 stars in a region centered at $\alpha = 19^h$, $\delta = +15°$. *Izv. glav. astr. Obs. Kiev*, **4**, no. 2.
52. Fehrenbach, Ch., Duflot, M. 1962, Deux étoiles à grande vitesse découvertes dans le ciel austral. *Commun. europ. sth. Obs.* no. 2.
53. Fehrenbach, Ch., Duflot, M., Boulon, J., Rebeirot, E., Lanoë, C. 1966, Liste de 893 vitesses radiales déterminées au prisme objectif à vision directe. *J. Observateurs*, Marseille, **49**, 155.
54. Fehrenbach, Ch., Duflot, M., Duflot, A. 1965, Vitesses radiales dans la direction du Grand Nuage de Magellan. *J. Observateurs*, Marseille, **48**, 185 and 199.
55. Florsch, A., Carozzi, N. 1965, Étoiles à grande vitesse entre les Nuages de Magellan. *Commun. europ. sth. Obs.* no. 5.
56. Götz, W., Wenzel, W. 1964, Spektralamplituden von 5 RR-Lyrae-Sternen. *Mitt. veränderl. Sterne*, **2**, 87.
57. Götz, W., Wenzel, W. 1964–66, Spektraltypen von Veränderlichen. *Mitt. veränderl. Sterne*, **2**, 49 and 85; **3**, 1 and 133.
58. Graham, J. A., Lyngå, G. 1965, A survey of faint OB stars in Carina. *Mem. Mt Stromlo Obs.* no. 18.
59. Hardorp, J., Theile, I., Voigt, H. H. 1964–65, *Luminous stars in the northern Milky Way. III* and *V*. Hamburg-Bergedorf: Hamburger Sternwarte and Warner and Swasey Obs.
60. Haro, G., Chavira, E. 1966, Flare stars in stellar aggregates of different ages. *Vistas in Astr.*, A. Beer and K. Aa. Strand, eds., **8**, 89.
61. Haug, U., Pfleiderer, J., Dachs, J. 1966, Sterne frühen Spektraltyps in Norma und Circinus. *Z. Astrophys.*, **64**, 140.
62. Hoffleit, D. 1965, Twenty-four new variable stars in Sagittarius. *Astr. J.*, **70**, 307.

63. Jasevičius, V. 1964, Three colour photometry and spectral classification of stars in the region of NGC 133, NGC 146 and K–14. *Astr. Obs. Biul., Vilnius*, **13**, 1.

64. Kalandadze, N. B. 1964, *U, B, V* magnitudes, color indices, spectra and luminosities of stars in Taurus. *Abastumansk. astrofiz. Obs. Gora Kanobili Bjull.* no. 31, 66.

65. Kalandadze, N. B. 1966, Notes on three stars. *Astr. Cirk.* no. 355, 3.

66. Kharadze, E. K., Apriamashvili, S. P., Kochlashvili, T. A. 1964, Photovisual magnitudes, spectra, and luminosities of stars in Cygnus. *Abastumansk. astrofiz. Obs. Gora Kanobili Bjull.* no. 31, 3.

67. Klare, G., Schaifers, K. 1965, Survey for red dwarfs in the solar neighborhood. *Observatory*, **85**, 202.

68. Lodén, L. O., Sundman, A. 1966, List of early (O–B) and late (M) type stars in a region in Centaurus. *Stockholm Obs. Ann.*, **22**, no. 10.

69. Lyngå, G. 1964, Studies of the Milky Way from Centaurus to Norma. III. OB stars. *Medd. Lunds astr. Obs.*, Ser. 2, no. 141.

70. MacDonald, D. D. 1964, Spectra of 179 eclipsing binaries. *Publ. Leander McCormick Obs.*, **12**, 51.

71. Maffei, P. 1966, Variabili nebulari nella zona di IC 2169. *Mem. Soc. astr. ital.*, **37**, 459.

72. Maffei, P. 1966, Osservazioni blue infrarosse di variabili a lungo periodo in Monoceros. *Mem. Soc. astr. ital.*, **37**, 475.

73. Mašnauskas, J. 1964, Stellar spectra, magnitudes and colour indices in a region centered at $\alpha = 22^h32^m$, $\delta = +58°00'$ (1950). *Astr. Obs. Biul., Vilnius*, **12**, 1.

74. McCarthy, M. F., Bertiau, F. C., Treanor, P. J. 1964, M stars in the region of the south galactic pole. *Ric. astr. Specola astr. Vatic.*, **6**, 571.

75. McCarthy, M. F., Treanor, P. J. 1964, M stars in the region of the Pleiades. *Ric. astr. Specola astr. Vatic.*, **6**, 535.

76. McCuskey, S. W. 1964, Stellar space density near Selected Area 158. *Astr. J.*, **69**, 104.

77. McCuskey, S. W., Houk, N. 1964, Galactic structure in Cassiopeia. *Astr. J.*, **69**, 412.

78. Nassau, J. J., Stephenson, C. B. 1963, *Luminous stars in the northern Milky Way. IV.* Hamburg-Bergedorf: Hamburger Sternwarte and Warner and Swasey Obs.

79. Nassau, J. J., Stephenson, C. B., Caprioli, G. 1964, Spectral classification for new or unclassified late-type variables, emission-line stars, S stars, and planetary nebulae. *Astrophys. J.*, **139**, 864.

80. Nassau, J. J., Stephenson, C. B., MacConnell, D. J. 1965, *Luminous stars in the northern Milky Way. VI.* Hamburg-Bergedorf: Hamburger Sternwarte and Warner and Swasey Obs.

81. Nassau, J. J., Velghe, A. G. 1964, Objective-prism spectra in the near infrared. *Astrophys. J.*, **139**, 190.

82. Pesch, P. 1965, Objective-prism spectra of some stars of low luminosity. *Astrophys. J.*, **141**, 1582.

83. Philip, A. G. D. 1966, The stellar space distribution in Pegasus at $b^{II} = -29°$. *Astrophys. J. Suppl. Ser.*, **12**, 391 (no. 112).

84. Philip, A. G. D., Sanduleak, N. 1966, Low-dispersion spectra of stars and galaxies. *Publ. astr. Soc. Pacif.*, **78**, 30.

85. Plaut, L. 1966, Variable stars in a field centered at $l = 0°$, $b = +29°$. *Bull. astr. Inst. Netherl. Suppl. Ser.*, **1**, 105 (no. 3).

86. Raznik, R. M. 1963, Spectra, magnitudes and color indices of stars in an area centered at $\alpha = 0^h0^m$, $\delta = +66{\overset{\circ}{.}}5$. *Izv. Krym. astrofiz. Obs.*, **30**, 131.

87. Raznik, R. M. 1964, The distribution of dust and hot stars in the vicinity of the O-association centered at $\alpha = 23^h59^m$, $\delta = +67{\overset{\circ}{.}}1$. *Izv. Krym. astrofiz. Obs.*, **32**, 165.

88. Rebeirot, E. 1966, Classifications spectrales et mesure de vitesses radiales de 112 étoiles situées dans Praesepe. *J. Observateurs*, Marseille, **49**, 139.

89. Rebeirot, E., Imbert, M. 1965, Vitesses radiales obtenues avec la chambre I du spectrographe coudé de l'Observatoire de Haute-Provence. *J. Observateurs*, Marseille, **48**, 153.

90. Salukvadze, G. N. 1965, Relative positions, magnitudes, and spectral classes of Trapezium type multiple systems. *Abastumansk. astrofiz. Obs. Gora Kanobili Bjull.* no. 32, 69.

91. Seitter, W. 1964, Atlas for objective-prism spectra and information obtainable from low-dispersion spectrograms of Nova Herculis 1963. *Astr. J.*, **69,** 149.

92. Stephenson, C. B. 1965, Spectral classifications for new or previously unclassified carbon stars. *Astrophys. J.*, **142,** 712.

93. Stephenson, C. B., Nassau, J. J., MacConnell, D. J. 1965, *Luminous stars in the northern Milky Way. VIa.* Identification charts. Cleveland: Warner and Swasey Obs.

94. Straižys, V. 1964, Catalogue of F–M stars in the southern part of the Orion association. *Astr. Obs. Biul., Vilnius,* **9,** 4.

95. The, P.-S. 1965, The colour-magnitude diagram of the galactic cluster NGC 6383. *Contr. Bosscha Obs.* no. 32.

96. The, P.-S. 1965, A photometric study of the galactic cluster NGC 6124. *Contr. Bosscha Obs.* no. 33.

97. Treanor, P. J. 1966, Spectra of a fast-moving binary. *Observatory,* **86,** 152.

98. Wachmann, A. A. 1963–64, Die Veränderlichen im Südteil der Cygnuswolke. *Astr. Abh. Hamburg Sternw. Bergedorf,* **6,** no. 2 and 3.

99. Wenzel, W. 1961, Two variables of Beta Lyrae type with long periods. *Inf. Bull. var. stars* no. 1.

100. Westerlund, B. E. 1963, An OB association in the region of RS Puppis. *Mon. Not. R. astr. Soc.,* **127,** 71.

101. Westerlund, B. E. 1965, An infrared survey of the region of SA 193. *Mon. Not. R. astr. Soc.,* **130,** 45.

Unpublished Work

(Albers): One hundred and nineteen of the Hetzler 'infrared' stars have been classified on plates obtained some years ago at the Warner and Swasey Observatory. In agreement with the results of C. F. Rust (*Astrophys. J.*, **88,** 525, 1938), the majority of these stars are found to be normal late M-type stars; a few are carbon or S-type objects.

(Bartaya): Additional classification work has been carried out at Abastumani and at the Sternberg Institute as follows:

Author	Region	No. of Stars
Kalandadze, N. B.	$\alpha = 4^h26^m$, $\delta = +19^\circ$ $\alpha = 4^h40^m$, $\delta = +29^\circ$	3400
Metreveli, M. D.	$\alpha = 4^h30^m$, $\delta = +25\overset{\circ}{.}5$	3000
Apriamashvili, S. P.	$\alpha = 19^h00^m$, $\delta = +5^\circ$	4000
Apriamashvili, S. P., and Kuznetsov, V. I.	$\alpha = 19^h20^m$, $\delta = +17^\circ$	500
Uranova, T. A.	$\alpha = 19^h15^m$, $\delta = +11^\circ$ $\alpha = 18^h21^m$, $\alpha = +22\overset{\circ}{.}5$	1500
Bartaya, R. A., and Kharadze, E. K.	Vul I Cas III, Cas IV, Cas VII	1300 400
Kostiakova, E. B.	Orion	7000
Voroshilov, V. I.	Aquila	——

Also M. A. Shiukashvili has examined criteria for quantitative two-dimensional classification of objective-prism spectra of F0–G5 stars.

(Bidelman): F- and G-type stars of low metal abundance can be readily detected on high-quality plates taken with the 10°-prism combination of the Curtis Schmidt. A program of discovery and further study of such objects is now being carried out by H. Bond. Another student, J. Schmitt, is engaged in an extensive investigation of K-type stars possessing abnormally strong CN bands, and is detecting many such stars on available objective-prism plates.

The visual and infrared regions of Curtis Schmidt spectra are the subjects of detailed examination by D. MacConnell and H. Albers. Much profitable work remains to be done in these spectral regions.

(Iwanowska): Observations for the Toruń spectral sky survey are being continued. The Milky Way belt accessible from Toruń within latitudes ± 6° has been covered with photometrically calibrated exposures. A number of plates have also been taken at higher latitudes.

Spectrophotometric gradients for about 60 F, G, and K stars in the blue-yellow-red region have been determined by R. Głebocki. He is attempting to distinguish stellar population types from these observations.

(Lodén): The pre-classification of stars, by visual inspection of their objective-prism spectra, for the selection of interesting objects involved in the Stockholm Southern Milky Way survey, is being carried out in collaboration with A. Sundman. A special study of the possibility of performing accurate spectral classification by visual inspection of microphotometric records of objective-prism spectra is also under way. Spectral classes have been derived in this manner by K. Lodén for 447 stars in S.A. 193 from plates taken at the Boyden Observatory.

(McCarthy): M. F. McCarthy and P. J. Treanor, in cooperation with A. Upgren of the U.S. Naval Observatory, have begun a study to determine the degree of inclination of layers of equal star density at intermediate galactic latitudes. Plates centered on Selected Areas 28, 54, 106, and 107 have been obtained at the Dyer and Vatican Observatories.

M and emission-line stars are under study in the Cygnus region at $l = 88°$, $b = 0°$, which was searched earlier for carbon stars. Also, the search for concentrations of Population I objects in regions where high-velocity hydrogen gas clouds have been detected is continuing.

(McCuskey): P. Pesch has shown that the infrared Ca II triplet can be detected in emission on widened spectra of bright T Tauri stars taken with the 4° prism of the Burrell Schmidt. Its possible detection in very faint stars is a matter for further study.

As part of a continuing study of dM stars in the Hyades, Pesch has investigated the use of the sodium D lines in low-dispersion spectra as a luminosity discriminant. Due to a nearby TiO band, the D lines cannot be seen in dwarf M stars later than M1 on plates taken with the 4° prism; on dense plates taken with the 6°-prism combination, however, they are visible in Barnard's star, M5 V.

Infrared spectra of some of the stars observed photometrically in **(302)** are also being investigated by Pesch. Studies of the available spectra appear to indicate that the Haro-Chavira objects are predominantly M-type stars rather than carbon stars.

(Stephenson): Plates covering the blue and visual spectral regions have been used by C. B. Stephenson and C. L. Terrill for the classification of 34 additional unusual objects; 26 of these are new S or M–S stars.

The vicinities of the stronger X-ray sources are being searched from the 8th to the 15th magnitude for peculiar spectra, at dispersions of 300 to 1100 Å/mm.

C. L. Terrill is working up the infrared plates of about 50 late-type variable stars accumulated with the Burrell Schmidt over the past ten years. New observations to fill phase gaps in the earlier material are also being made.

(The): Surveys of M stars are in progress in the direction of the south galactic pole, near the open cluster M6, and in four regions in and around the Eta Carinae Nebula. Almost all of the plates necessary for extending McCuskey's work on the stellar luminosity function to the southern hemisphere have been obtained.

B. Hidajat has classified about 950 stars later than M5 as well as about 300 stars of classes B8–A2 and G8 III–K2 III in an area of 25 square degrees at $l = 359°$, $b = -10\overset{\circ}{.}6$.

(Velghe): An infrared survey of cool stars at $l = 77°$ and $0° < b < +22°$, based on plates taken at the Warner and Swasey Observatory, is in progress. Also a study of criteria for the classification of M, S, and C stars in the visual and infrared at various dispersions is continuing.

2. *Classifications Employing Visual Inspection of Slit Spectrograms*

102. Abt, H. A., Snowden, M. S. 1964, The galactic cluster IC 4665. *Astrophys. J.*, **139**, 1139.

103. Andrillat, Y. 1964, Comparaison spectrale des Novae Herculis 1960 et 1963 et caractères Wolf-Rayet de leurs spectres dans le proche infrarouge. *Ann. Astrophys.*, **27**, 486.

104. Appenzeller, I. 1966, Polarimetrische, photometrische und spektroskopische Beobachtungen von Sternen im Cygnus und Orion. *Z. Astrophys.*, **64**, 269.

105. Bennett, N. W. W., Evans, D. S., Laing, J. D. 1963, The southern spectroscopic binaries HD 16589, 24202, 153890 and 197649. *R. Obs. Bull.* no. 78.

106. Berger, J. 1963, Radial velocities and spectra of faint blue stars at high galactic latitudes. *Publ. astr. Soc. Pacif.*, **75**, 393.

107. Bertola, F. 1963, Three faint Wolf-Rayet stars. *Mem. Soc. astr. ital.*, **34**, 257.

108. Bertola, F. 1964, The supernovae in NGC 1073 and NGC 1058. *Ann. Astrophys.*, **27**, 319.

109. Bertola, F. 1965, Observations of four supernovae. *Mem. Soc. astr. ital.*, **36**, 299.

110. Bidelman, W. P. 1965, Spectral types of nineteen red variable stars. *Publ. astr. Soc. Pacif.*, **77**, 392.

111. Burbidge, E. M. 1965, Redshifts of the quasi-stellar radio sources 3C 334, 3C 345, 3C 380, and a discussion of the possible redshift of 3C 446. *Astrophys. J.*, **142**, 1674.

112. Burbidge, E. M. 1966, Two quasi-stellar radio sources with large redshifts. *Astrophys. J.*, **143**, 612.

113. Burbidge, E. M., Burbidge, G. R. 1965, Ionized gas in the nuclei of elliptical, So, spiral, and irregular galaxies. *Astrophys. J.*, **142**, 634.

114. Burbidge, E. M., Kinman, T. D. 1966, Redshifts of fourteen quasi-stellar radio sources. *Astrophys. J.*, **145**, 654.

115. Buscombe, W. 1965, The Scorpio-Centaurus association. *Irish astr. J.*, **7**, 63.

116. Buscombe, W. 1965, Stellar spectra in galactic cluster IC 2391. *Mon. Not. R. astr. Soc.*, **129**, 411.

117. Buscombe, W., Kennedy, P. M. 1965, Radial velocities of 200 southern B stars. *Mon. Not. R. astr. Soc.*, **130**, 281.

118. Chandra, S. 1966, Abundance anomaly correlations in peculiar A-type stars, and a new scheme for their classification. *Astr. J.*, **71**, 848.

119. Conti, P. S. 1965, The early A stars. I. Rotation and metallicism. *Astrophys. J.*, **142**, 1594.

120. Copeland, J. A., Heard, J. F. 1963, Spectroscopic studies of 60 Be stars over a period of 24 years. *Publ. David Dunlap Obs.*, **2**, 315.

121. Cousins, A. W. J., Lake, R., Stoy, R. H. 1966, Photoelectric magnitudes and colours of southern stars. II. *R. Obs. Bull.* no. 121.

122. Cowley, A. P., Cowley, C. R. 1964, Thirty-seven new peculiar and metallic-line A stars. *Publ. astr. Soc. Pacif.*, **76**, 119.

123. Cowley, A. P., Cowley, C. R. 1965, Slit spectra of some peculiar and metallic-line A stars. *Publ. astr. Soc. Pacif.*, **77**, 184.

124. Dibai, E. A., Esipov, V. F., Pronik, V. I. 1966, Spectra of the Chavira and Rosino supernovae. *Astr. Zu.*, **43**, 907.

125. Eggen, O. J., Greenstein, J. L. 1965, Spectra, colors, luminosities, and motions of the white dwarfs. *Astrophys. J.*, **141**, 83.

126. Eggen, O. J., Greenstein, J. L. 1965, Observations of proper-motion stars. II. *Astrophys. J.*, **142**, 925.

127. Evans, D. S. 1966, Fundamental data for southern stars (sixth list). *R. Obs. Bull.* no. 110.

128. Evans, D. S., Laing, J. D., Menzies, A., Stoy, R. H. 1964, Fundamental data for southern stars (fifth list). *R. Obs. Bull.* no. 85.

129. Feast, M. W. 1966, The kinematics of Me variables towards the galactic centre and a comparison with the planetary nebulae. *Mon. Not. R. astr. Soc.*, **132**, 495.

130. Feast, M. W. 1966, Long period variables in the field of globular clusters. *Observatory*, **86**, 120.
131. Feast, M. W., Thackeray, A. D. 1963, Radial velocities of southern B stars determined at the Radcliffe Observatory. *Mem. R. astr. Soc.*, **68**, 173.
132. Feinstein, A. 1966, The open cluster NGC 2451. *Publ. astr. Soc. Pacif.*, **78**, 301.
133. FitzGerald, M. P. 1964, Spectrographic orbits of the eclipsing systems V822 Aquilae, BV 241, BV 342, BV 374. *Publ. David Dunlap Obs.*, **2**, 415.
134. Ford, W. K., Rubin, V. C. 1965, Low-dispersion image-tube spectra in the red: 3C 33, 3C 48, Ton 256, and an infrared star. *Astrophys. J.*, **142**, 1303.
135. Ford, W. K., Rubin, V. C. 1966, Quasi-stellar objects with small redshifts: 1217 + 02, 3C 249·1, and 3C 263. *Astrophys. J.*, **145**, 357.
136. Greenstein, J. L. 1966, The nature of the faint blue stars. *Astrophys. J.*, **144**, 496.
137. Greenstein, J. L., Eggen, O. J. 1966, The calibration of the Hertzsprung-Russell diagram for sub-luminous stars. *Vistas in Astr.*, A. Beer and K. Aa. Strand, eds., **8**, 63.
138. Greenstein, J. L., Keenan, P. C. 1964, A spectroscopic study of NGC 188. *Astrophys. J.*, **140**, 673.
139. Greenstein, J. L., Minkowski, R. 1964, The central stars of planetary nebulae of low surface brightness. *Astrophys. J.*, **140**, 1601.
140. Greenstein, J. L., Münch, G. 1966, The weakness of helium lines in globular cluster and halo B stars. *Astrophys. J.*, **146**, 618.
141. Greenstein, J. L., Schmidt, M. 1964, The quasi-stellar radio sources 3C 48 and 3C 273. *Astrophys. J.*, **140**, 1.
142. Gurzadian, G. A., Chavushian, O. S. 1964, Emission line intensities for four faint Wolf-Rayet stars. *Soobšč. Bjurak. Obs.*, no. 36, 45.
143. Heard, J. F. 1965, The radial velocities and spectral classes of 55 Kapteyn Area fundamental stars in high galactic latitudes. *Publ. David Dunlap Obs.*, **2**, 441.
144. Herbig, G. H. 1965, Physical companions to long-period variables. *Kleine Veröff. Remeis-Sternw.* no. 40, 164.
145. Herbig, G. H. 1966, On the interpretation of FU Orionis. *Vistas in Astr.*, A. Beer and K. Aa. Strand, eds., **8**, 109.
146. Hill, P. W., Hill, S. R. 1966, Faint blue stars in the far southern hemisphere. *Mon. Not. R. astr. Soc.*, **133**, 205.
147. Hiltner, W. A., Cowley, A. P., Schild, R. E. 1966, Observations of quasi-stellar and Haro-Luyten blue objects. *Publ. astr. Soc. Pacif.*, **78**, 464.
148. Hiltner, W. A., Morgan, W. W., Neff, J. S. 1965, The HR diagram of NGC 6530. *Astrophys. J.*, **141**, 183.
149. Hiltner, W. A., Schild, R. E. 1966, Spectral classification of Wolf-Rayet stars. *Astrophys. J.*, **143**, 770.
150. Hiltner, W. A., Schild, R. E., Jackson, S. 1964, Spectra of six faint Wolf-Rayet stars. *Astrophys. J.*, **139**, 763.
151. Honeycutt, R. K., McCuskey, S. W. 1966, Identification of Am and Ap stars from low-dispersion objective-prism spectra. *Publ. astr. Soc. Pacif.*, **78**, 289.
152. Jaschek, C., Jaschek, M. 1965, Spectra of some southern Be stars. *Publ. astr. Soc. Pacif.*, **77**, 376.
153. Jaschek, C., Jaschek, M., Kucewicz, B. 1964, A survey of southern Be stars. *Z. Astrophys.*, **59**, 108.
154. Jaschek, M., Jaschek, C. 1964, Spectroscopic Notes: I. HD 133738 and HD 6882. *Asoc. argent. Astr. Bol.* no. 7, 16.
155. Johnson, H. M. 1965, The spectra and radial velocities of stars in the Orion Nebula cluster. *Astrophys. J.*, **142**, 964.
156. Keenan, P. C. 1966, A catalogue of spectra of Mira variables of types Me and Se. *Astrophys. J. Suppl. Ser.*, **13**, 333 (no. 118).
157. Keenan, P. C., Landi Dessy, J. 1966, Variations in spectral type of the Mira variables R Cen and T Cen. *Astrophys. J.*, **146**, 583.
158. Keenan, P., Young, W. 1964, Semiregular red variables with high velocities and Balmer emission. *Astr. J.*, **69**, 546.

159. Kennedy, P. M., Przybylski, A. 1963, Radial velocities of 50 high velocity stars. *Mon. Not. R. astr. Soc.*, **126,** 381.
160. Kinman, T. D. 1965, The star cluster NGC 6791. *Astrophys. J.*, **142,** 655.
161. Kinman, T. D. 1965, The nature of the fainter Haro-Luyten objects. *Astrophys. J.*, **142,** 1241.
162. Kraft, R. P. 1964, Binary stars among cataclysmic variables. III. Ten old novae. *Astrophys. J.*, **139,** 457.
163. Kuhi, L. V. 1963, Spectral classification of the late component of stars with composite spectra. *Publ. astr. Soc. Pacif.*, **75,** 448.
164. Kuhi, L. V. 1964, Mass loss from T Tauri stars. *Astrophys. J.*, **140,** 1409.
165. Landi Dessy, J., Keenan, P. C. 1966, Spectral types on the MK system for forty-three bright southern stars K2–M6. *Astrophys. J.*, **146,** 587.
166. Lodén, L. O. 1966, A study of NGC 2467 and the association Pup I. *Ark. Astr.*, **4,** 65.
167. Lynds, C. R., Hill, S. J., Heere, K., Stockton, A. N. 1966, New spectroscopic observations of fourteen quasi-stellar sources. *Astrophys. J.*, **144,** 1244.
168. Lynds, C. R., Stockton, A. N., Livingston, W. C. 1965, New spectroscopic observations of quasi-stellar sources. *Astrophys. J.*, **142,** 1667.
169. Mendoza, E. E. 1963, Estudio espectroscópico y fotométrico del cúmulo estelar en Coma Berenices. *Bol. Obs. Tonantzintla Tacubaya,* 3, 137 (no. 23).
170. Mertz, L., Coleman, I. 1966, Infrared spectrum of the Taurus red object. *Astrophys. J.*, **143,** 1000.
171. Morgan, W. W., Hiltner, W. A. 1965, The HR diagram of the Hyades. *Astrophys. J.*, **141,** 177.
172. Morgan, W. W., Hiltner, W. A., Neff, J. S., Garrison, R., Osterbrock, D. E. 1965, The H–R diagrams of NGC 2244 and NGC 2264. *Astrophys. J.*, **142,** 974.
173. Morgan, W. W., Lodén, K. 1966, Some characteristics of the Orion association. *Vistas in Astr.*, A. Beer and K. Aa. Strand, eds., **8,** 83.
174. Moroz, V. I. 1966, Infrared spectra of stars (λ 1–2·5 μ). *Soviet Astr.*, **10,** 47.
175. Münch, G. 1966, Absorption lines from intermediate-latitude interstellar complexes. *Publ. astr. Soc. Pacif.*, **78,** 305.
176. Münch, G., Münch, L. 1964, Radial velocities of distant OB stars. *Astrophys. J.*, **140,** 162.
177. Münch, G., Scargle, J. D. 1965, The spectra of two extremely red objects. *Astrophys. J.*, **142,** 401.
178. Osawa, K. 1965, Spectral classification and three-color photometry of A-type peculiar stars. *Ann. Tokyo astr. Obs.*, **9,** 123.
179. Philip, A. G. D. 1966, High galactic latitude A stars. *Astr. J.*, **71,** 395.
180. Preston, G. W. 1964, Spectral types of stars in the field of CV Monocerotis. *Publ. astr. Soc. Pacif.*, **76,** 165.
181. Przybylski, A., Kennedy, P. M. 1965, Radial velocities and three-colour photometry of 52 stars with large proper motions. *Mon. Not. R. astr. Soc.*, **129,** 63.
182. Przybylski, A., Kennedy, P. M. 1965, Radial velocities and three-colour photometry of 166 southern stars. *Mon. Not. R. astr. Soc.*, **131,** 95.
183. Pyper, D. M. 1966, The effective temperatures of Wolf-Rayet stars as derived from their *UBV* color indices corrected for emission. *Astrophys. J.*, **144,** 13.
184. Rubin, V. C., Ford, W. K. 1966, Image tube spectra of quasi-stellar objects. *Astr. J.*, **71,** 396.
185. Sandage, A. 1965, The existence of a major new constituent of the universe: the quasi-stellar galaxies. *Astrophys. J.*, **141,** 1560.
186. Sandage, A. 1966, Redshifts of nine radio galaxies including the abnormal system 3C 305. *Astrophys. J.*, **145,** 1.
187. Schild, R. E. 1965, Spectral classification in h and χ Persei. *Astrophys. J.*, **142,** 979.
188. Schild, R. E. 1966, Be stars in the region of h and χ Persei. *Astrophys. J.*, **146,** 142.
189. Schlesinger, B. M. 1966, Motions of normal and metallic-line A stars. *Publ. astr. Soc. Pacif.*, **78,** 39.

190. Schmidt, M. 1964, Optical spectra of radio galaxies. In *Physics of Nonthermal Radio Sources*, S. P. Maran and A. G. W. Cameron, eds. Washington: Goddard Institute for Space Studies, p. 111.

191. Schmidt, M. 1965, Optical spectra and redshifts of 31 radio galaxies. *Astrophys. J.*, **141**, 1.

192. Schmidt, M. 1965, Large redshifts of five quasi-stellar sources. *Astrophys. J.*, **141**, 1295.

193. Schmidt, M. 1966, Redshifts of fourteen quasi-stellar radio sources. *Astrophys. J.*, **144**, 443.

194. Schmidt, M., Matthews, T. A. 1964, Redshifts of the quasi-stellar radio sources 3C 47 and 3C 147. *Astrophys. J.*, **139**, 781.

195. Searle, L., Rodgers, A. W. 1966, The horizontal-branch stars of the globular cluster NGC 6397. *Astrophys. J.*, **143**, 809.

196. Shobbrook, R. R. 1966, Spectra and radial velocities of nineteen southern galaxies. *Mon. Not. R. astr. Soc.*, **131**, 293.

197. Slettebak, A. 1966, Axial rotation in the later B-type emission-line stars. *Astrophys. J.*, **145**, 121.

198. Spinrad, H., Wood, D. B. 1965, Magnesium hydride in cool stars and galaxies. *Astrophys. J.*, **141**, 109.

199. Stephenson, C. B. 1965, Spectroscopic observations of 'ce' stars in the *Luminous Stars in the Northern Milky Way* catalogues. *Astrophys. J.*, **142**, 309.

200. Thackeray, A. D. 1962, New members of the Small Magellanic Cloud. *Mon. Notes astr. Soc. Sth. Afr.*, **21**, 74.

201. Thackeray, A. D. 1964, Radial velocities of five stars. *Mon. Notes astr. Soc. Sth. Afr.*, **23**, 81.

202. Thackeray, A. D. 1966, Orbits of two double-lined spectroscopic binaries, HD 147971 and 75759. *Mon. Not. R. astr. Soc.*, **134**, 97.

203. Thackeray, A. D. 1966, Radial velocities of southern B stars determined at the Radcliffe Observatory. IV. The Sco-Cen association. *Mem. R. astr. Soc.*, **70**, 33.

204. Thackeray, A. D. 1966, Radial velocities of three stars. *Mon. Notes astr. Soc. Sth. Afr.*, **25**, 35.

205. Thackeray, A. D. 1966, Two new high-velocity subdwarfs. *Observatory*, **86**, 60.

206. Thackeray, A. D., Wesselink, A. J. 1965, A photometric and spectroscopic study of the cluster IC 2944. *Mon. Not. R. astr. Soc.*, **131**, 121.

207. Upgren, A. R., Rubin, V. C. 1965, An old open cluster near the north galactic pole. *Publ. astr. Soc. Pacif.*, **77**, 355.

208. Walker, M. F. 1966, Ultraviolet excess in T Tauri stars. In *Stellar Evolution*, R. F. Stein and A. G. W. Cameron, eds. New York: Plenum Press, p. 405.

209. Walker, N. 1966, A list of newly identified metallic line A stars. *Observatory*, **86**, 154.

210. Webster, L. B. 1966, Stellar planetary nebulae in the southern Milky Way. *Publ. astr. Soc. Pacif.*, **78**, 136.

211. Westerlund, B. E. 1963, Three-colour photometry of early-type stars near the galactic poles. *Mon. Not. R. astr. Soc.*, **127**, 83.

212. Wilde, K. 1965, The spectra of two emission-line stars: Nova Scorpii 1964 and AS 210. *Publ. astr. Soc. Pacif.*, **77**, 208.

213. Wilson, O. C. 1964, The distribution of intensities of bright H and K in dK stars and the rate of star production in the Galaxy. *Publ. astr. Soc. Pacif.*, **76**, 28.

214. Wilson, O. C. 1966, Stellar convection zones, chromospheres, and rotation. *Astrophys. J.*, **144**, 695.

215. Wilson, O. C., Skumanich, A. 1964, Dependence of chromospheric activity upon age in main-sequence stars: additional evidence. *Astrophys. J.*, **140**, 1401.

216. Wing, R. F., Spinrad, H., Kuhi, L. V. 1966, Spectrophotometric studies of three extremely red stars. *Astr. J.*, **71**, 187.

217. Wolff, S. C., Wallerstein, G., Sandage, A. R. 1965, Low-dispersion spectroscopic observations of proper motion stars with ultraviolet excesses. *Publ. astr. Soc. Pacif.*, **77**, 370.

218. Zwicky, F. 1964, Basic results of the international search for supernovae. *Ann. Astrophys.*, **27**, 300.

219. Zwicky, F. Compact galaxies and compact parts of galaxies. *Astrophys. J.*, **140**, 1467, 1964, and **143**, 192, 1966.

Unpublished Work

(Bidelman): J. Robinson has obtained slit spectra in the visual region of more than 100 of the brighter early-type stars in a 60° by 60° region at the anticenter. Several new bright-Hα stars were found. Several other spectral classification programs are under way with the 37-inch reflector.

A comprehensive re-discussion of the problem of the spectral classification of the carbon stars, with special reference to possible luminosity effects, is currently being made by C. P. Gordon. Her investigation is based on spectra in the visual region of approximately 115 stars at dispersions of 88 to 103 Å/mm as well as spectra in the blue of some 30 additional objects.

(Blanco): At the Flagstaff Station of the U.S. Naval Observatory J. Priser, H. Guetter, and V. Blanco have undertaken the classification of all the parallax and comparison stars in the observatory's astrometric program. Present plans call for the classification of 322 parallax stars and about four comparison stars in each case. The stars range from m_v = 12·0 to 16·5.

J. Christy has assigned MK classes to 24 OB stars in the anticenter direction. Approximately 260 stars selected from the Appendix to IAU Symposium No. 1 (*Co-ordination of Galactic Research*, 1955) are being classified by H. Guetter. For classification work the U.S. Naval Observatory now has a new Cassegrain spectrograph similar in design to the Mt Wilson 60-inch Cassegrain one, and a reducing camera with objective gratings designed by A. Meinel.

(Heard): Fifty-seven stars in the field of the α Persei cluster have been classified on the MK system. This work will appear in a joint paper with the late R. M. Petrie.

(Jaschek): A survey of southern B5–A0 stars brighter than magnitude 6·5, an extension of (**153**), is in progress. Of a total of 1100 stars, 550 have already been observed at Hα at a dispersion of 180 Å/mm. Nine emission objects were found during the survey, four being previously known.

Spectral classification of southern HR stars at 90 Å/mm is continuing.

(Keenan): Classification criteria that can be used for late-type stars on plates ranging from 10 to 100 Å/mm are being re-examined. Spectrograms at 10 Å/mm permit the use of a number of atomic-line ratios to determine whether the MK subtypes previously established from the intensity of TiO bands are consistent for stars of high and low velocities.

(McCarthy): Image-tube spectra of three of the faint Pleiades stars listed in the Vatican catalogue show dwarf characteristics.

(McCuskey): In collaboration with N. Houk 115 OB stars brighter than m_{pg} = 11·5 in the LF5 region in Cassiopeia are being observed at 130 Å/mm.

An emission-like feature near λ6400 noted in low-dispersion objective-prism spectra of M stars is being investigated by R. K. Honeycutt with the slit spectrograph at a dispersion of 390 Å/mm. This feature is probably not a true emission but instead appears to be associated with the incipient appearance of the red ZrO band at λ6474.

(Meinel): New ultraviolet spectral criteria for the classification of middle-type stars are being established by D. C. Barry, who is utilizing primarily plates taken with the new Steward Observatory Cassegrain spectrograph.

(Morgan): Attention has recently been drawn to the importance of the concept of 'coeval sequences' in stellar associations and clusters. Even if investigation is restricted to 'young' star clusters, the spectra of O and B stars sometimes differ greatly from cluster to cluster; this effect is sometimes encountered between differing parts of the same stellar association. For the most precise spectral classification, it now becomes necessary to refer newly-classified spectra to some one of a series of coeval sequences. A short discussion of this subject is included in M. Hack's *Memorial Volume for Otto Struve* (in press).

A paper by R. F. Garrison entitled 'Some Characteristics of the B and A Stars in the Upper Scorpius Complex' is also awaiting publication.

(Schmidt-Kaler): Some 30 dwarf Be stars as well as 12 B- and A-type foreground stars in and around M31 are being classified on David Dunlap Observatory spectrograms.

(Sharpless): The possibility of using 300Å/mm infrared spectra of carbon stars for purposes of spectral classification is being investigated. It is hoped to shed light on the question of the apparent clusterings of these objects. Infrared classification of symbiotic stars is also planned.

(Slettebak): Spectral types of 77 A-type stars near the north galactic pole have been determined by A. Slettebak and R. Wright.

(Stephenson): N. B. Sanwal has completed the classification of nearly 200 visual binaries of small angular separation expected to include an off-main-sequence component. The purpose of the program was the estimation of mass as a function of position in the H–R diagram. Comparison of the results with theoretical evolutionary tracks is consistent with, but does not demand, the conclusion that evolution up to the red-giant stage takes place at constant mass.

In another program involving binary stars, visual binaries containing B-type stars are being observed. Some 40 or 50 such systems will ultimately be studied.

The brighter B to F supergiants of the *Luminous Stars* catalogues are being observed, with emphasis on those reported as having Hα in emission; the latter are observed both in the blue and at Hα. Curiously enough, most of these have not shown bright Hα on our slit spectrograms.

S. Wyckoff is observing approximately 80 late-type variables in the infrared at 390Å/mm using a grating blazed for this spectral region. Satisfactory spectra of stars as faint as $m_v = 14$ can be obtained with the 36-inch reflector in three-hour exposures.

(van den Bergh): *UBV* photometry and MK spectral classifications are currently being obtained by R. Racine for approximately 100 stars situated in reflection nebulae.

(Westerlund): A new system of classification for Wolf-Rayet stars has been devised by L. F. Smith, now at the University of California, Los Angeles. For WC stars her system agrees, in principle, with that of Hiltner and Schild (**149**). For WN stars it is substantially different; the differences lie in the attitudes adopted towards the spectra of binary stars. In Miss Smith's system, it is considered that the spectrum of a Wolf-Rayet star in a binary system is modified only in the strengths of the emission lines with respect to the continuum. The spectrum may still be classified according to the relative strengths of the emission lines, and it is given a classification explicitly indicating its binary nature.

(Yamashita): Seventy-five carbon stars for which Victoria spectrograms exist have been classified in the C system. Intensities of the various molecular bands have been measured. Three new CH stars and three possible members of this group were found. Plots of the temperatures obtained by Mendoza and Johnson versus the C subtypes seem to indicate that there are two groups in the carbon stars. No other characteristic difference, however, has been found spectroscopically between these two groups.

A total of 269 bright M-type stars have been classified in the MK system. The relation between the assigned luminosity classes and Mt Wilson spectroscopic absolute magnitudes shows considerable scatter.

3a. *Classifications Employing Spectrophotometric Measurements: Narrow- and Intermediate-Band Photoelectric Work*

220. Abt, H. A., Golson, J. C. 1966, On the separation by Balmer-line photometry of high- and low-luminosity stars having H-alpha in emission. *Astrophys. J.*, **143**, 306.

221. Abt, H. A., Osmer, P. S. 1965, Filter transmission effects on photoelectric measures of rotationally broadened Hβ line strengths. *Astrophys. J.*, **141**, 949.

222. Barbier, D., Morguleff, N. 1964, Spectrophotométrie photoélectrique à bande passante étroite des étoiles des premiers types spectraux. *Publ. Obs. Hte-Provence*, **7**, no. 15.

223. Beer, A. 1964, Photoelectric distances of 461 northern OB-stars and galactic structure from Hγ-luminosities. *Mon. Not. R. astr. Soc.*, **128**, 261.

224. Bigay, J.-H. 1964, Mesures photoélectriques simultanées dans deux domaines spectraux voisins; application à la détermination des magnitudes absolues des étoiles chaudes. *C. r. Acad. Sci. Paris*, **258**, 5804.

225. Bok, B. J., Bok, P. F., Graham, J. A. 1963, The distance of a group of early-type stars in Norma. *Publ. astr. Soc. Pacif.*, **75**, 514.
226. Bok, B. J., Bok, P. F., Graham, J. A. 1966, A photometric study of the I Scorpii association. *Mon. Not. R. astr. Soc.*, **131**, 247.
227. Borgman, J., Blaauw, A. 1964, Luminosities and photometric distances of early-type stars. *Bull. astr. Inst. Netherl.*, **17**, 358.
228. Cameron, R. C. 1966, An investigation of the magnetic and related stars through photoelectric narrow-band photometry. *Georgetown Obs. Monogr.* no. 21.
229. Crawford, D. L. 1966, Photometry of B stars. In *Stellar Evolution*, R. F. Stein and A. G. W. Cameron, eds. New York: Plenum Press, p. 395.
230. Crawford, D. L., Barnes, J. 1966, Photoelectric *uvby* and Hβ photometry for the Praesepe open cluster. *Publ. astr. Soc. Pacif.*, **78**, 442.
231. Crawford, D. L., Barnes, J. V. 1966, Photoelectric Hβ photometry for B stars in the Orion Belt region. *Astr. J.*, **71**, 610.
232. Crawford, D. L., Barnes, J. V., Faure, B. Q., Golson, J. C., Perry, C. L. 1966, Photoelectric Hβ photometry for 1217 stars brighter than $V = 6{\cdot}5$ mag. *Astr. J.*, **71**, 709.
233. Crawford, D. L., Mander, J. V. 1963, Photometry of the Coma cluster. *Publ. astr. Soc. Pacif.*, **75**, 410.
234. Crawford, D. L., Mander, J. 1966, Standard stars for photoelectric Hβ photometry. *Astr. J.*, **71**, 114.
235. Crawford, D. L., Perry, C. L. 1966, Four-color and Hβ photometry of open clusters. I. The Hyades. *Astr. J.*, **71**, 206.
236. Crawford, D. L., Strömgren, B. 1966, Comparison of the Hyades, Coma, and Pleiades clusters based on photoelectric *u,v,b,y* and Hβ photometry. *Vistas in Astr.*, A. Beer and K. Aa. Strand, eds., **8**, 149.
237. Fernie, J. D. 1966, The λ4226 line as a spectrum-luminosity indicator in late-type stars. *Mon. Not. R. astr. Soc.*, **132**, 485.
238. Graham, J. A. 1966, Intermediate-band photometry of high-latitude blue stars. *Publ. astr. Soc. Pacif.*, **78**, 433.
239. Gyldenkerne, K. 1964, A three-dimensional spectral classification of G and K stars. *Publ. mind. Medd. Kbh. Obs.* no. 178.
240. Gyldenkerne, K. 1964, Photoelectric narrow-band classification of late-type stars. *Publ. mind. Medd. Kbh. Obs.* no. 180.
241. Hoag, A. A. 1966, Open cluster distance moduli. *Vistas in Astr.*, A. Beer and K. Aa. Strand, eds., **8**, 139.
242. Hoag, A. A., Applequist, N. L. 1965, Distance moduli of open clusters. *Astrophys. J. Suppl. Ser.*, **12**, 215 (no. 107).
243. Koehler, J. A. 1965, The intrinsic $(B-V)$ colours of globular clusters. *Observatory*, **85**, 197.
244. Kovar, R. 1964, Infrared photometry of M stars. *Astr. J.*, **69**, 548.
245. Kovar, R. P. 1965, Population discriminant for late-type dwarfs. *Astr. J.*, **70**, 142.
246. Lasker, B. M. 1966, Preliminary report on intermediate-bandwidth photometry of E and So galaxies. *Astr. J.*, **71**, 862.
247. Liller, W. 1966, The central depth of the λ6362 auto-ionization feature of neutral calcium in stellar spectra. *Astrophys. J.*, **146**, 719.
248. McNamara, D. H. 1965, The position of the dwarf cepheids in the HR diagram. *Kleine Veröff. Remeis-Sternw.* no. 40, 111.
249. Miner, E. D. 1966, Narrow-band photometry of F5–K5 stars. *Astrophys. J.*, **144**, 1101.
250. Neff, J. S. 1966, A sensitive photometric luminosity discriminant for late-type dwarf and giant stars. *Astr. J.*, **71**, 202.
251. Peat, D. W. 1964, Hα photometry of late-type stars. I. F, G and K type stars north of the equator. *Mon. Not. R. astr. Soc.*, **128**, 435.
252. Peat, D. W. 1964, Photometry of the orange-red Ca I triplet in late-type stars. *Mon. Not. R. astr. Soc.*, **128**, 475.
253. Peat, D. W. 1966, Hα photometry of late-type stars. II. F and G type dwarf stars south of the equator. *Mon. Not. R. astr. Soc.*, **131**, 467.

254. Philip, A. G. D. 1966, High galactic latitude A stars. *Astr. J.*, **71**, 867.
255. Price, M. J. 1966, Narrow band photoelectric Hα photometry in late-type stars. *Mon. Not. R. astr. Soc.*, **133**, 449.
256. Price, M. J. 1966, Narrow band photometry in the study of stellar populations. I. The sodium D lines in late-type stars. *Mon. Not. R. astr. Soc.*, **134**, 135.
257. Price, M. J. 1966, Narrow band photometry in the study of stellar populations. II. Comparison of the Hyades and Coma Berenices open clusters. *Mon. Not. R. astr. Soc.*, **134**, 171.
258. Scarfe, C. D. 1966, Photoelectric observations of Fe I lines in late-type stellar spectra. *Mon. Not. R. astr. Soc.*, **133**, 99.
259. Smak, J. 1966, On the spectral classification and reddening determination for long-period variables. *Acta astr.*, **16**, 109.
260. Strömgren, B. 1964, On the chemical composition and kinematics of disk high-velocity stars of the main sequence. *Astrophys. norw.*, **9**, 333.
261. Strömgren, B. 1966, Spectral classification through photoelectric narrow-band photometry. *A. Rev. Astr. Astrophys.*, **4**, 433.
262. van den Bergh, S. 1965, A comparison of the line-blanketing indices Δ and m_1. *Publ. astr. Soc. Pacif.*, **77**, 221.
263. Westerlund, B. E. 1966, Multicolor photometry of northern Wolf-Rayet stars. *Astrophys. J.*, **145**, 724.
264. Williams, J. A. 1966, Metal-to-hydrogen ratios in the Magellanic Clouds. *Astr. J.*, **71**, 186.
265. Williams, J. A. 1966, Metal-to-hydrogen ratios in the Galaxy as indicated by narrow-band photometry of cepheids. *Astr. J.*, **71**, 615.
266. Wood, D. B. 1966, Multicolor photoelectric photometry of galaxies. *Astrophys. J.*, **145**, 36.
267. Wood, H. J. Balmer line strengths in peculiar A stars. *Astr. J.*, **69**, 564, 1964, and **70**, 151, 1965.
268. Wood, H. J. 1965, Photoelectric Balmer-line photometry. *Publ. Leander McCormick Obs.*, **15**, 5.

Unpublished Work

(Blanco): Hγ photometry and MK classifications for about 100 early-type stars in the I Lac and α Per associations are being carried out at the U.S. Naval Observatory by H. Guetter.

(Gyldenkerne): Narrow-band photometry has been continued for late-type stars. Most of the data has been obtained by Copenhagen and Brorfelde staff members working at the Lowell Observatory in Flagstaff.

For the G and K stars P.U.B.S. Jacobsen made tests of the transmission bands of the *k*, *n*, *m*, *g* classification indices. Observations in an improved system have been carried out for bright stars and calibration objects. In collaboration with P. Dickow and P. Rasmussen the new system has been investigated preliminarily along the lines followed in (**239**). The improved third parameter C′ correlates linearly with [Fe/H] determined by Wallerstein and Helfer in a range from the Hyades value to about 1·0 smaller. The scatter in the relation does not exceed the measurement uncertainty. The photometric accuracy of C′ corresponds to a mean error of ± 0·03 in [Fe/H]. The photometric accuracy of the absolute magnitudes is estimated to be ± $0^m.3$ (m.e.).

B. Helt has observed about a hundred bright M0–M6·5 giants and supergiants with the classification indices described at the Stockholm Symposium in order to investigate this two-dimensional classification method thoroughly. The color-index differences m_2(4370, 4800, 5350) (spectral-type index) and m_4(4510, 4970, 5500) (mainly luminosity-dependent) are being used in extended observations of bright stars and selected variable stars. Photoelectric scans have been obtained to study the possibility of an improvement of the classification sensitivity of these indices.

(McNamara): Dwarf cepheids, W Ursae Majoris variables, and RR Lyrae stars are currently being observed in the Strömgren four-color system. Narrow-band filters have been employed to measure the β index and $(U-B)_0$ color indices of the β Cephei variables. These data have been used to define the period-luminosity relationship of these stars. Globular clusters are also being studied by narrow-band methods: the strength of the G band and Hγ are used to obtain information on spectral type, and a metal-index indicator is also measured.

(Redman): P. J. Andrews has completed the measurement and discussion of Hα intensities in about 1000 OB stars, 300 of which were measured at the Radcliffe Observatory. For class V stars the Hα strength is about as accurate an absolute-magnitude criterion as is Hβ or Hγ, and for supergiants (where the line is partly in emission) it is more accurate. Stellar age appears to have but little effect on the hydrogen line-absolute magnitude relation, but stellar rotation introduces a considerable scatter.

Intensities of three Si III lines and of λ4471 of He I have been measured both photoelectrically and photographically by J. B. Hutchings in various O and B stars, especially those of high luminosity. He has also measured Hγ profiles. Model atmospheres are being discussed using these measurements.

The λ6305 Sc I line has been measured by R. F. Griffin in 300 G and K stars using a 1·5 Å bandwidth. The line shows a very steep increase in strength with diminishing temperature.

It is our experience that satisfactory discussion of narrow-band measurements is hampered by lack of calibration data, especially masses and absolute magnitudes in class III stars, and by lack of detailed theoretical models of late-type atmospheres in general.

(Schmidt-Kaler): Strömgren four-color photometry of 36 globular and 41 open clusters is being carried out utilizing different apertures, and also Hβ photometry for 15 globular and 41 open clusters.

(Sinnerstad): Recently I have been working on intermediate and narrow-band photometry of early-type stars. The instrument used is a two-channel photometer attached to the 1-m reflector of the Stockholm Observatory. Measurements have been carried out in Strömgren's u, v, b, y system and in two other systems defined by the following filter characteristics:

I		II	
λ max	half-width	λ max	half-width
3530	70	3540	70
3860	100	3740	90
4240	80	4030	30
5010	70	4510	70

The first system is identical with that used in the ζ Aurigae program of 1963–64. In the second system, which aims at deriving luminosity criteria directly applicable to Schmidt spectra, the filters 3540, 4030 and 4510 yield a measure of the magnitude of the Balmer discontinuity, while the filters 3740, 4030 and 4510 yield a measure of its effective location.

Most of my effort has been devoted to determinations of line intensities by means of simultaneous measurements and the narrow-band techniques. Up to May 1966 Hβ was measured in 131 stars of spectral types O6–F8.

An extensive observational program for measuring the line intensities of Hγ, Hβ and He I λ4471 with the same techniques is now going on in collaboration with J. Arkling. On the program are B0–B5 stars brighter than $7^m\!.5$ with an extension to fainter magnitudes in some associations such as I Ori and I Gem. The linear relation between the preliminarily measured intensities of λ4471 and the equivalent widths as determined on high-dispersion spectrograms indicates a scatter of ± 0·11 E.A. The Mg II λ4481 line influencing the λ4471 measurements contributes to this scatter. As the Mg II line, however, is rather weak in main sequence B0–B5 stars and mainly varies with spectral type, no serious effects are likely to be introduced in the determinations of helium abundance.

(Strömgren): An interesting application of narrow-band methods to the study of the A-type stars has recently been made by R. C. Henry, who has photoelectrically determined K-line strengths in numerous normal and metallic-line stars. A number of stars classified as Am were found to have normal K-line strengths.

(van den Bergh): A seven-color system with intermediate-width bands centered at λλ3500, 3800, 4150, 4250, 4500, and 4861 wide and narrow is currently being applied by R. McClure to the study of stars, globular clusters, and galaxies at David Dunlap Observatory.

(Westerlund): The following three photometric systems have been used to assist in the solution of particular astrophysical problems. It is assumed that a pre-classification has been carried out either by objective-prism techniques, by two-color photography, or even by *U*, *B*, *V* photoelectric photometry. The aim is to obtain accurate and detailed information for a large number of faint objects for which moderate-dispersion slit spectra can not be obtained without using an excessive amount of observing time:

The photometric systems are:

I. (*a*). A six-color narrow-band system, in which the effective wavelengths of four colors agree well with those in Strömgren's four-color system. The peak transmission wavelengths and the half-widths of the filters are: *y*, 5480Å, 100; *b*, 4700, 100; *v*, 4090, 100; u_1, 3780, 30; *u*, 3640, 100; u_2, 3550, 100.

(*b*). The six-color system with a red (*R*) and an infrared (*I*) wide-band filter added: *R*, 7000Å; *I*, 8800Å, approximately.

(*c*). *y*, *b*, *v*, u_2 from I(*a*) with narrow-band filters added for measuring Hβ and selected He I and He II lines. For certain programs narrow-band observations of the interstellar K line and of λ4430 are added.

II. A five-color narrow-band system for studies of Wolf-Rayet stars. The filters are chosen so that emission features are avoided as far as possible. The colors are: *r*, 5995, 100; *v*, 5180, 130; *b*, 4270, 60; *u*, 3640, 100; *u'*, 3500, 80.

III. A seven-color narrow-band system for studies of planetary nebulae. The filters are: lines, N_1; Hβ; Hγ; λ3727; continuum, 5300, 140; 4200, 70; 3500, 90.

Details can be found in a contribution to the February 1966 Groningen ESO colloquium.

(Williams): I have devised an intermediate-band system of five bandpasses between λ4400 and λ7500 from which three indices are formed. The first indicates the strength of weak lines; the second the strength of medium-strong lines; and the third a color temperature for the star. The system was constructed from an analysis of the distribution of weak-line, medium-strong-line, and strong-line strength with wavelength in the Sun. The system should be applicable to F- and G-dwarf stars. Observations are now being obtained of stars that have been analyzed spectroscopically in order to verify the action of the indices and to calibrate them. It is planned to observe a large number of F- and G-dwarf stars to obtain the distribution of M/H ratios and microturbulent velocities in these stars.

3*b*. *Classifications Employing Spectrophotometric Measurements: Multicolor Work*

269. Barnhart, P. E., Mitchell, W. E. 1966, Infrared astronomical photometry. *Contr. Perkins Obs.*, Ser. 2, no. 16.

270. Becker, W. 1965, Versuch einer empirischen Bestimmung der Sterndichte im galaktischen Halo mit Hilfe der Dreifarben-Photometrie. *Z. Astrophys.*, **62**, 54.

271. Blanco, V. M. 1964, $B-V$ and $V-I$ colors of giant M stars. *Astr. J.*, **69**, 730.

272. Boggess, A. 1964, B star colours between 2000 and 3000 angströms. *Ann. Astrophys.*, **27**, 805.

273. Boggess, A., Borgman, J. 1964, Interstellar extinction in the middle ultraviolet. *Astrophys. J.*, **140**, 1636.

274. Breckinridge, J. B., Kron, G. E. 1963, Color excesses of four southern hemisphere cepheids. *Publ. astr. Soc. Pacif.*, **75**, 285.

275. Breckinridge, J. B., Kron, G. E. 1963, A six-color criterion for blanketing. *Publ. astr. Soc. Pacif.*, **75**, 408.

276. Canavaggia, R., Mianes, P. 1964, Les couleurs intrinsèques des supergéantes comparées à celles des Céphéides. *C. r. Acad. Sci. Paris*, **258**, 833.

277. Canavaggia, R., Mianes, P. 1964, Photométrie en six couleurs. Un défaut de rayonnement dans le visible est mis en évidence chez les étoiles K par comparaison avec les Céphéides et avec Groombridge 1830. *C. r. Acad. Sci. Paris*, **258**, 5808.

278. Cayrel de Strobel, G. 1966, Analyse détaillée des atmosphères de sept étoiles K. *Ann. Astrophys.*, **29**, 413.

279. Conti, P. S. 1965, The atmospheres of the metallic-line stars in the Hyades. *Astrophys. J. Suppl. Ser.*, **11**, 47 (no. 95).

280. Feinstein, A. 1966, Photoelectric observations of southern late-type stars. *Inf. Bull. south. Hemisphere* no. 8, 30.

281. Golay, M. 1964, Photométrie en 5 couleurs et propriétés du diagramme Δ, *g*. Positions des étoiles métalliques et des sous-naines dans le diagramme Δ, *g*. *Publ. Obs. Genève* no. 66.

282. Golay, M. 1964, Cas des étoiles doubles dans les diagrammes en plusieurs couleurs. Photométrie en 5 et 7 couleurs et position des étoiles doubles dan le diagramme Δ, *g*. Étude de quelques étoiles des Pléiades en 5 et 7 couleurs. Étude de quelques étoiles de Coma Berenices en 5 et 7 couleurs. *Publ. Obs. Genève* no. 68.

283. Golay, M. 1964, Étude de quelques étoiles de Praesepe en 5 et 7 couleurs. Étude de quelques étoiles des Hyades en 5 et 7 couleurs. *Publ. Obs. Genève* no. 70.

284. Golay, M., Goy, M. 1965, Photométrie en 5 et 7 couleurs d'étoiles proches. *Publ. Obs. Genève* no. 71.

285. Graham, J. A. 1966, Observations of some distant B-type stars in Ara. *Bull. astr. Inst. Netherl.*, **18**, 372.

286. Hauck, B. 1966, Un critère de luminosité dans le système photométrique de l'Observatoire de Genève. *Publ. Obs. Genève* no. 72.

287. Iriarte, B. 1965, Fotometría fotoeléctrica en *BVRI* para 275 estrellas comprendidas en su mayoría entre $-25°$ y $-50°$. *Bol. Obs. Tonantzintla Tacubaya*, **4**, 33 (no. 26).

288. Iriarte, B., Johnson, H. L., Mitchell, R. I., Wiśniewski, W. Z. 1965, Five-color photometry of bright stars. *Sky Telesc.*, **30**, 21.

289. Johnson, H. L. 1962, Infrared stellar photometry. *Astrophys. J.*, **135**, 69.

290. Johnson, H. L. 1964, The colors, bolometric corrections and effective temperatures of the bright stars. *Bol. Obs. Tonantzintla Tacubaya*, **3**, 305 (no. 25).

291. Johnson, H. L. 1965, The absolute calibration of the Arizona photometry. *Commun. lunar planet. Lab.*, **3**, 73 (no. 53).

292. Johnson, H. L. 1965, Infrared photometry of M-dwarf stars. *Astrophys. J.*, **141**, 170.

293. Johnson, H. L. 1965, Interstellar extinction in the Galaxy. *Astrophys. J.*, **141**, 923.

294. Johnson, H. L. 1966, Infrared photometry of galaxies. *Astrophys. J.*, **143**, 187.

295. Johnson, H. L. 1966, The bolometric corrections and effective temperatures of two giant stars in the globular cluster M3. *Astrophys. J.*, **146**, 613.

296. Johnson, H. L. 1966, The law of interstellar extinction. *Vistas in Astr.*, A. Beer and K. Aa. Strand, eds., **8**, 133.

297. Johnson, H. L. 1966, Astronomical measurements in the infrared. *A. Rev. Astr. Astrophys.*, **4**, 193.

298. Johnson, H. L. 1966, Infrared stars. *Sky Telesc.*, **32**, 73.

299. Johnson, H. L., Borgman, J. 1963, The law of interstellar extinction. *Bull. astr. Inst. Netherl.*, **17**, 115.

300. Johnson, H. L., Meinel, A. B. 1963, Infrared color index of M31 at 2·2 microns. *Astrophys. J.*, **138**, 1317.

301. Johnson, H. L., Mendoza, E. E. 1966, The law of interstellar extinction in Perseus. *Ann. Astrophys.*, **29**, 525.

302. Johnson, H. L., Mendoza, E. E., Wiśniewski, W. Z. 1965, Observations of 'infrared stars'. *Astrophys. J.*, **142**, 1249.

303. Johnson, H. L., Mitchell, R. I. 1963, Stellar photometry at 5 microns. *Astrophys. J.*, **138**, 302.

304. Johnson, H. L., Mitchell, R. I., Iriarte, B., Wiśniewski, W. Z. 1966, *UBVRIJKL* photometry of the bright stars. *Commun. lunar planet. Lab.*, **4**, 99 (no. 63).

305. Kron, G. E. 1963, The color of the Sun. *Publ. astr. Soc. Pacif.*, **75**, 288.

306. Kruszewski, A. 1966, An experimental four-colour photometry. *Acta astr.*, **16**, 285.

307. Low, F. J., Johnson, H. L. 1964, Stellar photometry at 10 μ. *Astrophys J.*, **139**, 1130.

308. Mendoza, E. E. 1965, Multicolor photometry of TX Camelopardalis – a Hetzler object. *Bol. Obs. Tonantzintla Tacubaya,* **4,** 51 (no. 27).

309. Mendoza, E. E. 1966, Infrared photometry of T Tauri stars and related objects. *Astrophys. J.,* **143,** 1010.

310. Mendoza, E. E., Johnson, H. L. 1965, An investigation of late-type stars. *Bol. Obs. Tonantzintla Tacubaya,* **4,** 57 (no. 27).

311. Mendoza, E. E., Johnson, H. L. 1965, Multicolor photometry of carbon stars. *Astrophys. J.,* **141,** 161.

312. Mianes, P. 1963, Photométrie en six couleurs de 34 céphéides. *Notes Inf. Publ. Obs. Paris* no. 12.

313. Mianes, P. 1963, Excès de couleur et excès d'ultra-violet des céphéides – céphéides d'amas galactiques. *Notes Inf. Publ. Obs. Paris* no. 14, 63.

314. Mianes, P. 1965, Photométrie en 6 couleurs. Méthode mixte de mesure d'étoiles faibles. *J. Observateurs,* Marseille, **48,** 103.

315. Mitchell, R. I. 1964, Nine-color photometry of ϵ Aur, 0·35–9·5 μ. *Astrophys. J.,* **140,** 1607.

316. Morgan, W. W., Neff, J. S. 1964, Stellar population of the nuclear region of the Andromeda Nebula. *Astr. J.,* **69,** 145.

317. Oosterhoff, P. Th. 1966, Five-color photometry of high velocity stars and pulsating variables. *Vistas in Astr.,* A. Beer and K. Aa. Strand, eds., **8,** 179.

318. Oosterhoff, P. Th., Ponsen, J. 1966, Discussion of five-colour observations of stars of high velocity. *Bull. astr. Inst. Netherl.,* **18,** 150.

319. Oosterhoff, P. Th., Walraven, Th. 1966, Discussion of photo-electric five-colour observations of different types of pulsating variables. *Bull. astr. Inst. Netherl.,* **18,** 387.

320. Ponsen, J., Oosterhoff, P. Th. 1966, Five-colour photometry of high-velocity stars and of pulsating variables of different types. *Bull. astr. Inst. Netherl. Suppl. Ser.,* **1,** 3 (no. 1).

321. Rousseau, J. 1965, Application de la photométrie en six couleurs à l'étude de la composition chimique des naines G et K voisines du Soleil. *Publ. Obs. Hte-Provence,* **8,** no. 13.

322. Rousseau, J. 1966, Photométrie en six couleurs. Effet de microturbulence chez les étoiles K0 et K1 des Hyades et chez les étoiles K3 de O. C. Wilson. *C. r. Acad. Sci. Paris,* **262,** 1101 and 1170.

323. Rufener, F., Hauck, B., Goy, G., Peytremann, E., Golay, M. 1964, Catalogue des étoiles mesurées dans le système photométrique de l'Observatoire de Genève. *Publ. Obs. Genève* no. 66.

324. Smak, J. 1964, On the colors of T Tauri stars and related objects. *Astrophys. J.,* **139,** 1095.

325. Smak, J. 1966, Bolometric corrections for M-type giants. *Acta astr.,* **16,** 1.

326. Stebbins, J., Kron, G. E. 1964, Six-color photometry of stars. XI. Black-body color temperatures of 25 stars. *Astrophys. J.,* **139,** 424.

327. Straižys, V. The selection of spectral regions for heterochromatic photometry. *Astr. Obs. Biul., Vilnius,* **6,** 1, 1963; **11,** 11, 1964; **15,** 3, 1965.

328. Straižys, V. 1964, The relation between heterochromatic and monochromatic magnitudes in three-color systems. *Soviet Astr.,* **8,** 96.

329. Straižys, V. 1964, A study of the *R, G, U* photometric system. I. Dependence of the main parameters of the system on spectral class and interstellar absorption. *Soviet Astr.,* **8,** 317.

330. Straižys, V. 1965, Three-color system for the determination of interstellar absorption. *Soviet Astr.,* **8,** 596.

331. Straižys, V., Zdanavičius, K. 1964, The influence of bandwidth on the *U, B, V* parameters. *Soviet Astr.,* **8,** 413.

332. Straižys, V., Zdanavičius, K. 1965, The selection of spectral regions for heterochromatic photometry. *Astr. Obs. Biul., Vilnius,* **14,** 3.

333. Straižys, V., Zdanavičius, K. 1965, Two-dimensional photometric classification of red stars. *Astr. Cirk.* no. 320, 1.

334. van den Bergh, S. 1964, Five-color photometry of variable star field IV in M31. *Astr. J.*, **69**, 610.
335. Walraven, J. H., Tinbergen, J., Walraven, Th. 1964, Five-colour observations of 24 classical cepheids. *Bull. astr. Inst. Netherl.*, **17**, 520.
336. Wildey, R. L., Murray, B. C. 1964, 10-μ photometry of 25 stars from B8 to M7. *Astrophys. J.*, **139**, 435.
337. Zdanavičius, K., Straižys, V. 1964, The selection of spectral regions for heterochromatic photometry. *Astr. Obs. Biul., Vilnius*, **11**, 1.

Unpublished Work

(Bahng): Infrared photometry of late-type stars is being carried out in three bands with interference filters and a lead-sulfide detector. The effective wavelengths of the bands are 1·21, 1·59, and 2·15 μ. Preliminary results are:

The energy distribution of G, K, and M stars in the infrared shows a considerable deviation from that of a black body. The deviation is most apparent at the 1·59 μ band, in a sense that there is an excess continuum emission in this band relative to the other two bands. This is interpreted as due to the H^- opacity minimum in the atmospheres of these stars. There is some indication that the three-color observations in the infrared may be used to establish a luminosity criterion; but more observations, particularly of dwarf G, K, and M stars, are needed to be certain of this.

The infrared colors of carbon stars show a poor correlation with the Keenan-Morgan temperature class for stars of C4 or later.

(Canavaggia): Additional results of the work of Mlle J. Rousseau will soon be forthcoming.

(Elvius): Multi-band photoelectric measurements of relatively bright stars are being made by L. Häggkvist with the refractor of the Uppsala Observatory.

3*c*. *Classifications Employing Spectrophotometric Measurements: Miscellaneous Methods*

338. Ampel, R. 1964, Cassiopeia associations: Cas III, Cas IV and Cas V. *Acta astr.*, **14**, 52.
339. Azimov, S. M. 1963, Physical properties of subgiant members of eclipsing binaries. *Izv. glav. astr. Obs. Pulkove*, **23**, pt. 2, 76.
340. Barbier, M. 1966, Absorption, structure et rotation de la Galaxie dans une région de Cassiopée. *J. Observateurs*, Marseille, **49**, 269.
341. Bardin, C., Prévot, L. 1965, Comparaison des spectres d'étoiles de même type et d'âges différents. *Publ. Obs. Hte-Provence*, **7**, no. 39.
342. Bartash, T. M. 1965, Spectrophotometric observations of KX Ori, V359 Ori and V372 Ori. *Izv. Krym. astrofiz. Obs.*, **34**, 128.
343. Batten, A. H. 1965, The bright stars in the region of NGC 6871. *Astr. J.*, **70**, 318.
344. Batten, A. H. 1966, Four interesting visual binary systems. *J. R. astr. Soc. Can.*, **60**, 177.
345. Batten, A. H., Petrie, R. M. 1966, Spectrographic observations of visual binaries. *Trans. int. astr. Un.*, **12B**, 476.
346. Buscombe, W., Dickens, C. R. 1964, Luminosity classification of G-type stars. *Mon. Not. R. astr. Soc.*, **128**, 499.
347. Cayrel, G., Fringant, A.-M. 1964, Étude spectrophotométrique de deux étoiles très déficientes en métaux. *Contr. Inst. Astrophys. Paris*, A, no. 308.
348. Deutsch, A. J. 1964, D-line intensities in the spectra of galaxies. *Astrophys. J.*, **139**, 532.
349. Elvius, T., Lyngå, G. 1965, Three-colour photometry of 43 stars in Kapteyn's Selected Areas at southern galactic latitudes. *Ark. Astr.*, **3**, 467.
350. Guetter, H. H. 1964, Distances of 97 OB stars. *Publ. David Dunlap Obs.*, **2**, 403.
351. Gurzadian, G. A., Oganesian, R. Kh. 1964, Spectrophotometric measurements of hypothetical former nuclei of planetary nebulae. *Soobšč. Bjurak. Obs.*, **35**, 43.
352. Hutchings, J. B. 1966, Equivalent widths of Hγ in stellar spectra of the Magellanic Clouds. *Mon. Not. R. astr. Soc.*, **132**, 433.

353. Karlsson, B. 1966, Investigations of a region in Monoceros. I. Determination of monochromatic magnitudes and hydrogen line depths of B-, A- and F-type stars. *Medd. Lunds astr. Obs.*, Ser. 2, no. 145.
354. Koch, R. H., Olson, E. C., Yoss, K. M. 1965, A spectrographic investigation of some bright eclipsing binaries. *Astrophys. J.*, **141**, 955.
355. Kraft, R. P., Preston, G. W., Wolff, S. C. 1964, The width of Hα as a discriminant of luminosity in the spectra of late-type stars. *Astrophys. J.*, **140**, 235.
356. Kuhi, L. V. 1965, The Wilson-Bappu effect in T Tauri stars. *Publ. astr. Soc. Pacif.*, **77**, 253.
357. Kumaigorodskaya, R. N. 1964, Some properties of the emission spectra of Of stars. *Izv. Krym. astrofiz. Obs.*, **32**, 108.
358. Kumaigorodskaya, R. N. 1965, An investigation of the physical conditions in the atmospheres of O stars. *Izv. Krym. astrofiz. Obs.*, **33**, 242.
359. Lacoarret, M. 1965, Enveloppes d'étoiles Be: variations spectrales et caractéristiques physiques. *Ann. Astrophys.*, **28**, 321.
360. Ljunggren, B. 1966, A photoelectric and spectrophotometric investigation of a region near the North Galactic Pole. *Ark. Astr.*, **3**, 535.
361. Lohmann, W. 1963, Der Einfluss der interstellaren Absorption auf die Farbhelligkeiten kugelförmiger Sternhaufen. *Z. Astrophys.*, **58**, 23.
362. Martin, N. 1964, Détermination de magnitudes absolues par la méthode d'Öhman.—Application à l'étude de la structure galactique dans deux régions du ciel. *J. Observateurs*, Marseille, **47**, 125.
363. Oja, T. 1966, A spectrophotometric survey of stars along the Milky Way. *Uppsala astr. Obs. Ann.*, **5**, no. 4.
364. Oja, T. 1966, A photometric study of a region in Cassiopeia. *Ark. Astr.*, **4**, 199.
365. Roslund, C. 1966, Investigations of a Milky Way field in Scorpius. II. Hydrogen line intensities and spectral types of O and B stars. *Ark. Astr.*, **4**, 73.
366. Sargent, W. L. W. 1966, The spectra and compositions of the components of ADS 3910. *Astrophys. J.*, **144**, 1128.
367. Sargent, W. L. W., Searle, L. 1966, Spectroscopic evidence on the helium abundance of stars in the galactic halo. *Astrophys. J.*, **145**, 652.
368. Schneider, M. 1966, Application de méthodes numériques à la classification de céphéides. *Ann. Astrophys.*, **29**, 297.
369. Sjögren, U. 1964, Photoelectric and spectrophotometric observations with a discussion of the interstellar absorption in the region of Kapteyn's Selected Area 8. *Ark. Astr.*, **3**, 339.
370. Smak, J. 1964, Photometry and spectrophotometry of long-period variables. *Astrophys. J. Suppl. Ser.*, **9**, 141 (no. 89).
371. Spite, F. 1964, Évaluation approximative de l'abondance en métaux de quelques étoiles géantes à partir de spectres à petite dispersion. *Publ. Obs. Hte-Provence*, **7**, no. 6.
372. van den Bergh, S., Sackmann, I. J. 1965, Photoelectric spectrophotometry of stars. II. *Astr. J.*, **70**, 353.
373. Van't Veer, F. 1966, L'automatisation de la classification spectrale tri-dimensionnelle de Paris. *Ann. Astrophys.*, **29**, 293.
374. Walker, G. A. H. 1964, The ratio of total to selective absorption for the Orion Nebula. *Astrophys. J.*, **140**, 1613.
375. Walker, G. A. H. 1965, The Cepheus IV association. *Astrophys. J.*, **141**, 660.
376. Walker, G. A. H., Hodge, S. M. 1966, Equivalent widths and half-widths of the λ4388 and λ4471 He I lines, rotational velocities, and λ4430 central depths for 450 O to B5 stars. *Publ. Dom. astrophys. Obs.*, Victoria, **12**, 401.

Unpublished Work

(Chalonge): Work in progress includes the classification of blue halo stars brighter than $m_v = 11$, companions of O stars in binary systems and clusters (Mlle M. L. Burnichon), faint stars of h and χ Persei (Mlle L. Divan), little-reddened stars of the Orion association (Mlle M. L. Burnichon) and stars of the cluster NGC 2264.

(Elvius): A large number of spectral plates of various Selected Areas have been obtained with the Schmidt telescope of the Uppsala Southern Station; for details see the Report of Commission 33. A field in S. A. 19, including the cluster NGC 7654, has also been investigated.

The region around the radio source Centaurus A is being studied by A. Elvius on the basis of spectral plates obtained by C. Roslund.

(Henize): W. Hughes has measured equivalent widths of Hγ and Hδ for 20 B-type members of the I Lacerta association from 4°-prism plates taken with the Michigan Curtis Schmidt. These results were compared with values obtained from 17 Å/mm coudé spectrograms. It was concluded that (1) the equivalent widths for strong lines can be measured from the objective-prism plates to somewhat better than ± 15 %, and (2) measuring the equivalent widths gives somewhat better results than simply determining the central depths, if the prism plates are taken in good seeing.

(Schmidt-Kaler): Spectral scans from λ3500 to λ4500 of about 50 supergiants of classes F to K and of five cepheids, obtained with a resolution of 20 Å with the David Dunlap Observatory 74-inch telescope, are now being utilized for quantitative spectral classification.

4a. Data Relevant to Spectral Classification: Continuous Spectra

377. Aller, L. H., Faulkner, D. J., Norton, R. H. Photoelectric spectrophotometry of selected southern stars. *Astrophys. J.*, **140**, 1609, 1964, and **144**, 1073, 1966.

378. Bahner, K. 1963, Energy distribution in the spectra of early-type stars. *Astrophys J.*, **138**, 1314.

379. Baschek, B., Oke, J. B. 1965, Effective temperatures and gravities of Ap, Am, and normal A-type stars. *Astrophys. J.*, **141**, 1404.

380. Glagolevskii, Yu. V. 1966, Some results of observations of the continuous spectra of magnetic and peculiar stars. *Soviet Astr.*, **10**, 56.

381. Glushneva, I. N. Ultraviolet spectrophotometry of some hot stars. *Soviet Astr.*, **8**, 163, 1964, and **10**, 61, 1966.

382. Kharitonov, A. V. 1963, Extra-atmospheric spectrophotometric standards. Energy distribution in the spectra of selected stars in cgs units. *Soviet Astr.*, **7**, 258.

383. Kondo, M. 1965, Continuous spectrum of 63 Tauri. *Publ. astr. Soc. Japan*, **17**, 412.

384. Kuhi, L. V. 1966, Wolf-Rayet stars. I. The continuous energy distribution. *Astrophys. J.*, **143**, 753.

385. Kuhi, L. V. 1966, Ultraviolet continuous emission in T Tauri stars. *Publ. astr. Soc. Pacif.*, **78**, 430.

386. Oke, J. B. 1964, Photoelectric spectrophotometry of stars suitable for standards. *Astrophys. J.*, **140**, 689.

387. Oke, J. B. 1965, Absolute spectral energy distribution in stars. *A. Rev. Astr. Astrophys.*, **3**, 23.

388. Oke, J. B. 1966, Photoelectric spectrophotometry of quasi-stellar sources. *Astrophys. J.*, **145**, 668.

389. Oke, J. B., Conti, P. S. 1966, Absolute photoelectric spectrophotometry of stars in the Hyades. *Astrophys. J.*, **143**, 134.

390. Whiteoak, J. B. 1966, The wavelength dependence of interstellar extinction. *Astrophys. J.*, **144**, 305.

391. Willstrop, R. V. 1965, Absolute measures of stellar radiation. II. *Mem. R. astr. Soc.*, **69**, 83.

Unpublished Work

(van den Bergh): A catalogue of absolute energy distributions of 154 stars is being prepared for publication in collaboration with G. L. Hagen.

4b. Data Relevant to Spectral Classification: Line Spectra

392. Anderson, C. M., Stoeckly, R., Kraft, R. P. 1966, A redetermination of rotational velocities in the Pleiades. *Astrophys. J.*, **143**, 299.

393. Andrillat, Y., Houziaux, L. 1966, Intensité du triplet de O I à λ7772 Å dans le spectre de quelques étoiles des premiers types. *C. r. Acad. Sci., Paris*, **263**, 635.

394. Bonsack, W. K., Culver, R. B. 1966, Line widths and turbulence in K-type stars. *Astrophys. J.*, **145**, 767.

395. Boyce, P. B., Sinton, W. M. 1965, Infrared spectroscopy with an interferometer. *Sky Telesc.*, **29**, 78.

396. Conti, P. S., Danziger, I. J. 1966, The abundances of lithium and beryllium in some F dwarfs and K giants. *Astrophys. J.*, **146**, 383.

397. Danielson, R. E., Woolf, N. J., Gaustad, J. E. 1965, A search for interstellar ice absorption in the infrared spectrum of Mu Cephei. *Astrophys. J.*, **141**, 116.

398. Danziger, I. J., Conti, P. S. 1966, Lithium observations in the Pleiades and in F stars in the field. *Astrophys. J.*, **146**, 392.

399. Fay, T. D., Fredrick, L. W., Johnson, H. R. 1966, Line identifications in the infrared spectra of carbon stars. *Astr. J.*, **71**, 384.

400. Fujita, Y. Spectral features in the infrared region of some carbon stars. *Contr. Dep. Astr. Univ. Tokyo* no. 36, 1963, and no. 54, 1964.

401. Fujita, Y., Utsumi, K. 1963, Spectral features in the infrared region of some carbon stars. *Contr. Dep. Astr. Univ. Tokyo* no. 45.

402. Fujita, Y., Yamashita, Y., Kamijo, F., Tsuji, T., Utsumi, K. 1965, Comparative study of the spectra of some M-, S-, and C-type stars. *Publ. Dom. astrophys. Obs.*, Victoria, **12**, 293.

403. Griffin, R. F. 1964, Very broad features in the spectra of late-type stars. *Observatory* **84**, 154.

404. Guthrie, B. N. G. 1965, Recognition of rapidly rotating pole-on B-type stars. *Publ. R. Obs. Edinburgh*, **3**, 261.

405. Herbig, G. H. 1964, Apparent lithium isotope ratios in F5–G8 dwarfs. *Astrophys. J.*, **140**, 702.

406. Herbig, G. H. 1965, Lithium abundances in F5–G8 dwarfs. *Astrophys. J.*, **141**, 588.

407. Jaschek, M., García, Z. L. 1966, Spectroscopic studies of peculiar A-type stars. II. Two iron-rich stars of the Si-λ4200 group. *Z. Astrophys.*, **64**, 217.

408. Jaschek, M., Jaschek, C., González, Z. 1965, Spectroscopic studies of peculiar A-type stars. I. The manganese group. *Z. Astrophys.*, **62**, 21.

409. Kraft, R. P. 1965, Comparison of rotational velocities in the Hyades and Coma clusters. *Astrophys. J.*, **142**, 681.

410. Kraft, R. P., Wilson, O. C. 1965, The strength of the Li I line in the spectra of main-sequence F- and G-type stars. *Astrophys. J.*, **141**, 828.

411. Kuhi, L. V. 1966, Wolf-Rayet stars. II. The infrared spectrum. *Astrophys. J.*, **145**, 715.

412. McCarthy, M. F., Rubin, V. C. 1963, Classification of G-type stars in the near ultraviolet region. *Ric astr. Specola astr. Vatic.*, **6**, 431.

413. Merchant, A. E. 1966, Beryllium in F- and G-type dwarfs. *Astrophys. J.*, **143**, 336.

414. Merchant, A. E. 1966, Abundance of lithium in early M-type stars. *Astr. J.*, **71**, 170.

415. Morton, D. C., Spitzer, L. 1966, Line spectra of Delta and Pi Scorpii in the far-ultraviolet. *Astrophys. J.*, **144**, 1.

416. Parsons, S. B. 1964, Near-infrared classification of A and F stars. *Astrophys. J.*, **140**, 853.

417. Popov, V. S. 1966, Spectrophotometry of the line spectrum of some β Cephei stars. *Izv. glav. astr. Obs. Pulkove*, **24**, pt. 4, 116.

418. Searle, L., Lungershausen, W. T., Sargent, W. L. W. 1966, Studies of the peculiar A stars. IV. The relative abundances of four iron-peak elements. *Astrophys. J.*, **145**, 141.

419. Searle, L., Sargent, W. L. W. 1964, Studies of the peculiar A stars. II. The silicon-abundance anomaly. *Astrophys. J.*, **139**, 793.

420. Sinton, W. M. 1966, Intensities of stellar bands from interferometric spectra. *Astr. J.*, **71**, 398.

421. Spinrad, H. 1966, Observations of stellar molecular hydrogen. *Astrophys. J.*, **145**, 195.

422. Spinrad, H., Newburn, R. L. 1965, A low-dispersion spectroscopic search for water vapor in cool stars. *Astrophys. J.*, **141**, 965.

423. Spinrad, H., Pyper, D. M., Newburn, R. L., Younkin, R. L. 1966, The water deficiency in S stars and the variation of water abundance with Mira's phase. *Astrophys. J.*, **143**, 291.

424. Spinrad, H., Younkin, R. L. 1966, Infrared bands of vanadium oxide in three Mira stars. *Publ. astr. Soc. Pacif.*, **78,** 65.
425. Torres-Peimbert, S., Wallerstein, G. 1966, Lithium in carbon stars. II. *Astrophys. J.*, **146,** 724.
426. Torres-Peimbert, S., Wallerstein, G., Phillips, J. G. 1964, Lithium in carbon stars. *Astrophys. J.*, **140,** 1313.
427. Wallerstein, G. 1966, Lithium in giant stars of types F0 to G5. *Astrophys. J.*, **143,** 823.
428. Wallerstein, G. 1966, Observations of the lithium content of stars in visual binary systems. *Astrophys. J.*, **145,** 759.
429. Wallerstein, G., Herbig, G. H., Conti, P. S. 1965, Observations of the lithium content of main-sequence stars in the Hyades. *Astrophys. J.*, **141,** 610.
430. Warner, B., Garstang, R. H. 1963, Equivalent width measurements 5880-6867Å in late-type stars. *Commun. Univ. London Obs.* no. 60.
431. Wilson, O. C. 1964, Chromospheric activity and lithium. *Publ. astr. Soc. Pacif.*, **76,** 238.
432. Wilson, O. C., Baum, W. A., Ford, W. K., Purgathofer, A. 1965, A preliminary investigation of lithium in main sequence visual binaries. *Publ. astr. Soc. Pacif.*, **77,** 359.
433. Woolf, N. J. 1965, Infra-red spectra of stars, planets and the Moon from Stratoscope II. *Ann. Astrophys.*, **28,** 218.
434. Woolf, N. J., Schwarzschild, M., Rose, W. K. 1964, Infrared spectra of red-giant stars. *Astrophys. J.*, **140,** 833.
435. Wright, K. O., Jacobson, T. V. 1966, Measurements of line-ratios used for luminosity classification in stellar spectra of types F to K. *Publ. Dom. astrophys. Obs.*, Victoria, **12,** 373.
436. Wright, K. O., Lee, E. K., Jacobson, T. V., Greenstein, J. L. 1964, Line intensities in the spectra of representative stars of spectral types B to G. *Publ. Dom. astrophys. Obs.*, Victoria, **12,** 173.
437. Wyller, A. A. 1966, New C^{13} indicators in stellar spectra. *Astrophys. J.*, **143,** 828.
438. Yamashita, Y. 1965, Curve of growth analyses of M-type stars. II. Twelve M-type stars in the red and infrared regions. *Publ. astr. Soc. Japan*, **17,** 27.
439. Zirin, H., Vaughan, A. H. 1966, Helium 10 830Å absorption in late-type stars. *Astr. J.*, **71,** 188.

Unpublished Work

(Slettebak): Studies are presently under way on the axial rotation of A-type stars in the north galactic pole region. Rotational velocities are also being determined for members of the Scorpio-Centaurus association and the h and χ Persei clusters.

5. *Papers Dealing with the Relation of Colors to Spectra*

440. Alexander, J. B., Stansfield, R. 1966, The $B-V$ of the Sun. *R. Obs. Bull.* no. 119.
441. Arp, H. 1965, Properties of the galactic nucleus in the direction of NGC 6522. *Astrophys. J.*, **141,** 43.
442. Aveni, A. F. 1966, The effect of line emission upon the $B-V$ colors of T Tauri objects. *Astrophys. J.*, **144,** 666.
443. Eggen, O. J., Sandage, A. R. 1964, New photoelectric observations of stars in the old galactic cluster M 67. *Astrophys. J.*, **140,** 130.
444. Lynds, C. R., Villere, G. 1965, On the interpretation of the integral count-apparent magnitude relation for Haro-Luyten objects. *Astrophys. J.*, **142,** 1296.
445. Marlborough, J. M. 1964, Frequency of ultraviolet excesses among late-type dwarfs in the solar neighborhood. *Astr. J.*, **69,** 215.
446. Nevsky, M. Y., Rublev, S. V. 1963, On the influence of bright bands on stellar magnitudes and colour indexes of WR stars. *Astr. Cirk.* no. 241, 1.
447. Philip, A. G. D. 1965, An anomalous bluing effect in the $B-V$ colors of distant A stars. *Astr. J.*, **70,** 687.
448. Rozis-Saulgeot, A.-M. 1965, Du rôle des raies d'absorption dans les magnitudes photoélectriques: cas du Soleil. *Ann. Astrophys.*, **28,** 562.

449. Sandage, A. 1964, Results of a pilot program to discover new subdwarfs in the solar neighborhood. *Astrophys. J.*, **139**, 442.

450. Sandage, A., Smith, L. L. 1966, The color-magnitude diagram of the metal-rich globular cluster NGC 6712. *Astrophys. J.*, **144**, 886.

451. Sargent, W. L. W., Searle, L. 1966, The interpretation of B star spectra having weak helium lines. *Observatory*, **86**, 27.

452. Smith, L. L., Steinlin, U. W. 1964, Blanketing corrections for the *RGU* system. *Z. Astrophys.*, **58**, 253.

453. Underhill, A. B. Interstellar extinction. *Observatory*, **84**, 35, 1964. See also *Observatory*, **85**, 39, 1965.

454. Wildey, R. L. 1965, On line-blanketing normalization of subdwarf photometry. *Astrophys. J.*, **141**, 943.

6. *Papers Dealing with Calibration Problems*

455. Behr, A. Emissions–B–Sterne und galaktische Struktur. *Z. Astrophys.*, **61**, 182, 1965, and **63**, 133, 1966.

456. Collins, G. W., Harrington, J. P. 1966, Theoretical H-beta line profiles and related parameters for rotating B stars. *Astrophys. J.*, **146**, 152.

457. Conti, P. S., Deutsch, A. J. 1966, Color anomalies and metal deficiencies in solar-type disk-population stars. *Astrophys. J.*, **145**, 742.

458. Fernie, J. D. 1965, An absolute magnitude calibration of the Hβ index for early-type stars. *Astr. J.*, **70**, 575.

459. Fischel, D. 1964, On the influence of the chemical composition on the structure and metal index of stellar atmospheres. *Astrophys. J.*, **140**, 221.

460. Graham, J. A. 1965, The absolute magnitude of some galactic Wolf-Rayet stars. *Observatory*, **85**, 196.

461. Hodge, P. W., Wallerstein, G. 1966, A suggested revision in the distance to the Hyades and its implications. *Publ. astr. Soc. Pacif.*, **78**, 411.

462. Ireland, J. G. 1966, The effect of rotation on stellar luminosity. *Publ. R. Obs. Edinburgh*, **5**, 63.

463. Kelsall, T., Strömgren, B. 1966, Calibration of the Hertzsprung-Russell diagram in terms of age and mass for main-sequence B and A stars. *Vistas in Astr.*, A. Beer and K. Aa. Strand, eds., **8**, 159.

464. Kraft, R. P., Wrubel, M. H. 1965, The effect of rotation on the colors and magnitudes of A- and F-type stars in the Hyades. *Astrophys. J.*, **142**, 703.

465. Ljunggren, B., Oja, T. 1966, Absolute magnitudes of late-type giant stars. *Ark. Astr.*, **3**, 501.

466. McCuskey, S. W., Rubin, R. H. 1966, Mean absolute magnitude for stars of spectral class F. *Astr. J.*, **71**, 517.

467. McNamara, D. H. 1963, Dependence of the hydrogen line index β on rotational velocity. *Publ. astr. Soc. Pacif.*, **75**, 454.

468. Petrie, R. M. 1965, Spectroscopic absolute magnitudes of early-type stars from hydrogen absorption – a revision. *Publ. Dom. astrophys. Obs.*, Victoria, **12**, 317.

469. Roslund, C. 1966, Investigations of a Milky Way field in Scorpius. III. Galactic structure from *UBV* and Hβ photometries of O and B stars. *Ark. Astr.*, **4**, 101.

470. Roxburgh, I. W., Strittmatter, P. A. 1965, Rotational spread of the main sequence. *Z. Astrophys.*, **63**, 15.

471. Rublev, S. V. 1963, Spectrophotometric temperatures, absolute magnitudes, and intrinsic color indices of Wolf-Rayet stars. *Soviet Astr.*, **7**, 75.

472. Schmidt-Kaler, Th. Emissions–B–Sterne und galaktische Struktur. *Z. Astrophys.*, **58**, 217, 1964, and **63**, 131, 1966.

473. Stienon, F. M., Strom, S. E. 1964, Observational and predicted contours of Hγ. *Astr. J.*, **69**, 559.

474. Strittmatter, P. A., Sargent, W. L. W. 1966, Stellar rotation and the position of the metallic-line stars in the color-magnitude diagram. *Astrophys. J.*, **145**, 130.

475. Strömgren, B. 1964, Comparison of observed and theoretically calculated intensities in the continuous spectra of main-sequence B stars. *Rev. mod. Phys.*, **36**, 532.

476. Strömgren, B. 1966, Age determination for main-sequence B, A, and F stars. In *Stellar Evolution*, R. F. Stein and A. G. W. Cameron, eds. New York: Plenum Press, p. 391.

477. Wallerstein, G. 1964, Deduction of metal abundances by Borgman's photometric system. *Publ. astr. Soc. Pacif.*, **76**, 175.

478. Wallerstein, G., Helfer, H. L. 1966, Photometric methods for measuring the metal content of K giants. *Astr. J.*, **71**, 350.

479. Weaver, H., Ebert, A. 1964, A recalibration of the absolute magnitudes of early-type stars classified on the MK system. *Publ. astr. Soc. Pacif.*, **76**, 6.

Unpublished Work

(Schmidt-Kaler): An analysis of the OB stars classified in the Hamburg-Warner and Swasey Observatory *Luminous Stars* volumes indicates the following:

The stars classified OB^+ or OB with continuous emission at the Balmer limit (ce) but no Hα emission represent a natural group with $\overline{M}_v = -5{\cdot}3$, $\sigma_M = \pm 0{\cdot}6$, $\overline{(B-V)}_0 = -0{\cdot}28$, $\sigma_{B-V} = \pm 0{\cdot}04$, comprising MK types O5–9 and O9·5–B1 Iab–IV. It is a group of rapid rotators *without* large shells, in contrast to the dBe stars. A calibration of the remaining groups is given in the following table:

Group		M_v	$(B-V)_0$
OB^- ce	no Hα em.	−3·9	−0·28
OB^+ no ce	,,	−6·2	− ·24
OB ,, ,,	,,	−5·0	− ·26
OB^- ,, ,,	,,	−3·9	− ·27

7. *General Papers, Catalogues, Symposia*

480. Abt, H. A. 1966, The metallic-line stars. *Vistas in Astr.*, A. Beer and K. Aa. Strand, eds., **8**, 75.

481. Bertaud, Ch. 1965, Catalogue et bibliographie des étoiles A à spectre particulier. Deuxième supplément. *J. Observateurs*, Marseille, **48**, 211.

482. Blaauw, A. 1965, The concept of stellar populations. In *Galactic Structure*, A. Blaauw and M. Schmidt, eds. Chicago: University of Chicago Press, p. 435.

483. Böhm, K.-H., ed. 1966, Proceedings of the IAU colloquium on the blanketing effect. *J. quantit. Spectrosc. radiat. Transfer*, **6**, 534.

484. Bosman-Crespin, D., ed. 1964, *Les spectres infrarouges des astres*. Univ. Liège Inst. Astrophys. Coll. 8° no. 472.

485. Boyarchuk, A. A., Kopylov, I. M. 1964, A general catalogue of rotational velocities of 2558 stars. *Izv. Krym. astrofiz. Obs.*, **31**, 44.

486. Burbidge, G., Hoyle, F. 1966, The problem of the quasi-stellar objects. *Scient. Am.*, **215**, no. 6.

487. Buscombe, W. 1964, Photometric and spectroscopic data for supergiants of later types. *Mt Stromlo Obs. Mimeog.* no. 8.

488. Cayrel, R., Cayrel de Strobel, G. 1966, Abundance determinations from stellar spectra. *A. Rev. Astr. Astrophys.*, **4**, 1.

489. Coutrez, R. 1963, *La classification stellaire*. Univ. Libre Bruxelles Inst. Astr. et Astrophys. A, no. 5.

490. Deeming, T. J. 1964, Stellar spectral classification. I. Application of component analysis. *Mon. Not. R. astr. Soc.*, **127**, 493. See also *Observatory*, **84**, 47, 1964.

491. Deutsch, A. J. 1966, Blue stragglers in old open clusters: implications of their spectra. *Astr. J.*, **71**, 383.

492. Eggen, O. J. 1964, A catalogue of high-velocity stars. *R. Obs. Bull.* no. 84.

493. Eggen, O. J. 1965, Some observational aspects of stellar evolution. *A. Rev. Astr. Astrophys.*, **3**, 235.

494. Eggen, O. J., Herbig, G. H., eds. 1964, Clusters and stellar evolution: a colloquium. *R. Obs. Bull.* no. 82.

495. Fehrenbach, Ch. 1966, Objective prisms and measurement of radial velocities. *Adv. Astr. Astrophys.*, **4**, 1.

496. Fujita, Y. 1966, Spectroscopic study of late-type stars. *Vistas in Astr.*, A. Beer, ed., **7**, 71.

497. Gingerich, O. 1964, Laboratory exercises in astronomy – spectral classification. *Sky Telesc.*, **28**, 80.

498. Green, L. C. 1966, The peculiar A stars. *Sky Telesc.*, **31**, 84.

499. Greenstein, J. L. 1965, Subluminous stars. In *Galactic Structure*, A. Blaauw and M. Schmidt, eds. Chicago: University of Chicago Press, p. 361.

500. Guthrie, B. N. G., Harris, J., Nandy, K., Reddish, V. C. 1965, A stepwedge calibrator for use in a telescope. *Observatory*, **85**, 209.

501. Hack, M. 1965, Light elements in stellar atmospheres. *Sky Telesc.*, **29**, 347, and **30**, 9.

502. Hack, M. 1966, Magnetic stars and metallic line stars. *Vistas in Astr.*, A. Beer, ed., **7**, 107.

503. Hoffleit, D. 1964, *Catalogue of Bright Stars*. New Haven, Conn.: Yale University Obs.

504. Hubenet, H., ed. 1966, *Abundance Determinations in Stellar Spectra*. IAU Symposium no. 26. London: Academic Press.

505. Iwanowska, W. 1966, Statistical population indices. *Vistas in Astr.*, A. Beer, ed., **7**, 133.

506. Jaschek, C., Conde, H., de Sierra, A. C. 1964, Catalogue of stellar spectra classified in the Morgan-Keenan system. *Obs. astr. Univ. nac. La Plata, Ser. astr.*, **28**, no. 2.

507. Kerr, F. J., Rodgers, A. W., eds. 1964, *The Galaxy and the Magellanic Clouds*. IAU Symposium no. 20. Canberra: Australian Academy of Science.

508. Lindoff, U., Lyngå, G. 1966, Comparisons between MK, HD and HDE spectral classes. *Medd. Lunds astr. Obs.*, Ser. 2, no. 146.

509. Lindsay, E. M. 1964, Spectral classification and structural type of galaxies. *Irish astr. J.*, **6**, 256.

510. Lodén, K., Lodén, L. O., Sinnerstad, U., eds. 1966, *Spectral Classification and Multicolor Photometry*. IAU Symposium no. 24. London: Academic Press.

511. Lüst, R., ed. 1965, *Stellar and Solar Magnetic Fields*. IAU Symposium no. 22. Amsterdam: North-Holland Publ. Co.

512. Luyten, W. J., ed. 1965, *First conference on faint blue stars*. The Observatory: University of Minnesota.

513. Nielsen, A. V. 1963, Contributions to the history of the Hertzsprung-Russell diagram. *Medd. Ole Rømer Obs.* no. 30.

514. Petrie, R. M. 1965, Spectroscopic observations of B-type stars. *Publ. astr. Soc. Pacif.*, **77**, 97.

515. Roman, N. G. 1965, High-velocity stars. In *Galactic Structure*, A. Blaauw and M. Schmidt, eds. Chicago: University of Chicago Press, p. 345.

516. Sargent, W. L. W. 1964, The atmospheres of the magnetic and metallic-line stars. *A. Rev. Astr. Astrophys.*, **2**, 297.

517. Sargent, W. L. W. 1965, A possible relationship between the peculiar A stars and the λ Boötis stars. *Astrophys. J.*, **142**, 787.

518. Scarfe, C. D. 1966, Experiments with principal component analysis of narrow-band photoelectric data. *Observatory*, **86**, 13.

519. Schmidt-Kaler, T. 1964, Die galaktischen Emissions–B–Sterne. *Veröff. astr. Inst. Univ. Bonn* no. 70.

520. Skumanich, A. 1965, Population differences among bright G dwarfs and Ca II emission reversals. *Astr. J.*, **70**, 692.

521. Spinrad, H. 1966, Normal galaxies in the post-Baade era. *Publ. astr. Soc. Pacif.*, **78**, 367.

522. Stephenson, C. B. 1966, Astrophysical investigations utilizing objective prisms. *Vistas in Astr.*, A. Beer, ed., **7**, 59.

523. Thackeray, A. D. 1963, Recent developments in studies of the Magellanic Clouds. *Adv. Astr. Astrophys.*, **2**, 263.

524. Thackeray, A. D. 1965, Spectral classification of southern OB stars. *Observatory*, **85**, 201.

525. The, P.-S., ed. 1963, Proceedings of a symposium on standards for stellar photometry and spectral classification. *Contr. Bosscha Obs.* no. 21.

526. Underhill, A. B. 1966, Suggestions regarding the effective temperatures and bolometric corrections for B-type stars. *Vistas in Astr.*, A. Beer and K. Aa. Strand, eds., **8**, 41.

527. Underhill, A. B. 1966, *The Early-Type Stars*. Dordrecht: D. Reidel Publ. Co.

528. Warner, B. 1965, The barium stars. *Mon. Not. R. astr. Soc.*, **129**, 263.

529. Zwicky, F. 1965, Supernovae. In *Stellar Structure*, L. H. Aller and D. B. McLaughlin, eds. Chicago: University of Chicago Press, p. 367.

GPSR Compliance

The European Union's (EU) General Product Safety Regulation (GPSR) is a set of rules that requires consumer products to be safe and our obligations to ensure this.

If you have any concerns about our products, you can contact us on ProductSafety@springernature.com

In case Publisher is established outside the EU, the EU authorized representative is:

Springer Nature Customer Service Center GmbH
Europaplatz 3
69115 Heidelberg, Germany

Batch number: 10762209

Printed by Printforce, the Netherlands